The Organic Chemistry of Sugars

The Organic Chemistry of Sugars

Edited by
Daniel E. Levy
Péter Fügedi

Boca Raton London New York

A CRC title, part of the Taylor & Francis imprint, a member of the Taylor & Francis Group, the academic division of T&F Informa plc.

Published in 2006 by
CRC Press
Taylor & Francis Group
6000 Broken Sound Parkway NW, Suite 300
Boca Raton, FL 33487-2742

Printed in the United States of America on acid-free paper
10 9 8 7 6 5 4 3 2 1

International Standard Book Number-10: 0-8247-5355-0 (Hardcover)
International Standard Book Number-13: 978-0-8247-5355-9 (Hardcover)
Library of Congress Card Number 2005049282

Library of Congress Cataloging-in-Publication Data

The organic chemistry of sugars / edited by Daniel E. Levy & Péter Fügedi.
p. cm.
Includes bibliographical references and index.
ISBN 0-8247-5355-0
1. Carbohydrates. 2. Glycosides. 3. Oligosaccharides. I. Levy, D.E. (Daniel E.) II. Fügedi, Péter, Ph. D.

QD321.074 2006
547'.78--dc22 2005049282

Taylor & Francis Group
is the Academic Division of T&F Informa plc.

Dedications

This book is dedicated to those who devoted their careers to the advancement of the organic chemistry of sugars

and to

Jennifer, Aaron, Joshua and Dahlia

Enikő and Péter

for their love, understanding and support during the preparation of this work,

and to

the memory of Ákos.

Foreword

From a historical perspective, no single class of organic compounds has shared the same impact on the evolution of stereochemistry as sugar molecules. Compared with the remarkable synthesis of the first natural product, urea, by Friedrich Wöhler in 1828, the total synthesis of glucose by Emil Fischer in 1895 was a hallmark event in the annals of organic synthesis. As biological activity began to be associated with more complex natural products such as alkaloids, steroids and various metabolites by the middle of the twentieth century, interest in sugars as small molecule polyols shifted to the study of polysaccharides and their degradation products.

By the mid-1960s, synthetic carbohydrate chemistry was confined to a small subgroup of organic chemists, who studied methods of interconversion and functional group manipulation in conjunction with the structure elucidation of antibiotics containing sugars. Soon, most naturally occurring sugars, including deoxy, aminodeoxy and branched ones, had been synthesized. As a result, sugar molecules had become ideal substrates to test out new bond-forming methods, particularly because of their conformational properties, and the propensity of spatially predisposed hydroxyl groups. Sugars became a playground to validate concepts related to anchimeric assistance in conjunction with the synthesis of aminodeoxy component sugars in various natural products.

An altogether different view of sugars and their potential as chiral building blocks was introduced in the mid-1970s. This was to have an important impact on the thought process relating to organic synthesis in general. This marked the beginning a new era of rapprochement, integrating sugar chemistry in mainstream organic chemistry. Not only were the sugar components of complex natural products readily made by synthesis, but the entire framework of the "non-sugar," and admittedly the more challenging part, could also be made from sugar building blocks or "chirons."

By the 1980s, the advent of reagent methodology and asymmetric synthesis once again shifted the paradigm of thinking in considering complex natural product assembly from smaller components. Today, it is more practical, in many cases, to consider other innovative approaches to total synthesis without necessarily relying on sugars as chiral, nonracemic starting materials. In fact, *de novo* syntheses of even rare sugars is now possible by relying on efficient catalytic asymmetric processes. In a different context, the unique chemical and physicochemical properties of sugars have propelled them into new and exciting areas of application in molecular biology, drug design, materials, and other fields of direct impact on our quality of life.

A renaissance period for sugars is in full swing with the creation of new subdisciplines that bridge chemistry and biology. New areas relating to glycochemistry and glycobiology have emerged in conjunction with the important interface with proteins, nucleic acids, and other biological macromolecules. The history of sugar chemistry has come full circle since the grandeur of the Emil Fisher era, and the exciting, purely chemical activities of the latter part of the twentieth century. Sugar chemistry has emerged as a pivotal link between molecular recognition and biological events in conjunction with vital life processes.

The preceding preamble to a sugar chemistry panorama was necessary for me to introduce this timely monograph to the readers. In *The Organic Chemistry of Sugars*, authors/editors Daniel Levy and Péter Fügedi have captured the beauty of this panorama in a collection of 16 authoritative chapters covering the essence of almost every aspect of synthetic sugar chemistry.

By focusing on the "organic chemistry" aspect of sugars, the monograph takes the form of a text book in certain chapters, providing excellent coverage of traditional and contemporary methods to manipulate, use, and exploit sugar molecules. With the availability of this monograph, the knowledge base of modern carbohydrate chemistry will be considerably richer for the practitioners of this time-honored and venerable branch of organic chemistry.

Stephen Hanessian
April, 2005

Preface

During my early studies, I observed a natural reluctance of organic chemistry students to embrace carbohydrate chemistry. Understandably, this component of organic chemistry is intimidating because of the presence of multiple and adjacent stereogenic centers and the high degree of polarity these compounds possess. In fact, carbohydrate chemistry was all but glossed over in my sophomore organic chemistry class and, in later courses, there was no effort to address this topic in greater detail. Graduate school did not even have courses designed to fill this void.

Outside of my coursework, I was fortunate to have found mentors interested in the synthesis, manipulation and incorporation of heterocycles and sugars into more complex molecules. It was through my laboratory experience that I began to appreciate the beauty of sugars and the ease with which they could be manipulated. Consequently, I found myself being drawn into industry, and incorporating my interests into the design of biologically useful mimics of sugars. I found opportunities to try to dispel the perception that sugars/carbohydrates belong in a class outside of mainstream organic chemistry. It is my hope that this book will finally accomplish that goal.

In order to address the above objective, this book is designed to first introduce the reader to traditional carbohydrate chemistry and the modern developments we have seen in this area. Next, the reader's attention is drawn away from the carbohydrate nature of sugars towards how sugars can be manipulated similarly to small organic molecules. Sugars are presented as tools where their natural chirality and multiple stereogenic centers are used to the advantage of asymmetric syntheses and the total syntheses of simple and complex molecules. Finally, discussion turns to advanced topics including discussions of combinatorial chemistry, glycoproteins, and glycomimetics.

Part I, comprising five chapters, begins with a historical perspective of carbohydrate chemistry. The following four chapters introduce the reader to mainstream carbohydrate chemistry beginning with the discovery, significance and nomenclature of carbohydrates. Following a discussion on protecting group strategies, this section concludes with chapters on glycosylation techniques and oligosaccharide synthesis.

Part II, consisting of four chapters, considers the conversion of sugars and carbohydrates to molecules that have lost some of the features that define carbohydrates. In Chapter 6, the reader is introduced to strategies enabling the substitution of sugar hydroxyl groups to new groups of synthetic or biological interest. Chapter 7 continues this approach through the special case of substituting the glycosidic oxygen with carbon. Chapter 8 extends the treatment of *C*-glycosides to a discussion of cyclitols and carbasugars where the endocyclic oxygen is replaced with carbon. Finally, Chapter 9 elaborates on the carbasugar discussions by expanding into other types of endocyclic heteroatom substitutions.

Comprising four chapters, Part III moves from the topic of transforming sugars to the actual uses of sugars in mainstream organic chemistry. Chapter 10 reviews the extensive use of these readily available asymmetric molecules as chiral auxiliaries and ligands for use in chiral catalysis. Chapter 11 discusses the exploitation of these molecules as convenient starting materials for the synthesis of complex targets bearing multiple stereogenic centers. Chapter 12 utilizes principles set forth in previous chapters to describe approaches towards the syntheses of notable carbohydrate containing natural products. Finally, Chapter 13 presents approaches towards the asymmetric synthesis of monosaccharides and related molecules.

In Part IV, additional topics are presented that focus on new and emerging technologies. In Chapter 14, approaches to combinatorial carbohydrate chemistry are considered, while Chapter 15 focuses on the biological importance and chemical synthesis of glycopeptides. Finally, Chapter 16 presents the philosophy and chemistry behind the medicinally interesting concept of glycomimetics.

It is my hope that, through this work, the perception of a distinction between sugar chemistry and organic chemistry will be eliminated, and that organic, medicinal and carbohydrate chemists will begin to embrace the organic chemistry of sugars as a broadly useful tool presenting solutions to many complex synthetic challenges.

Daniel E. Levy

About the Editors

Daniel E. Levy first became interested in carbohydrates at the University of California at Berkeley where he studied the preparation of 4-amino-4-deoxy sugars from amino acids under the direction of Professor Henry Rapoport. Later, Dr. Levy pursued his Ph.D. at the Massachusetts Institute of Technology, under the direction of Professor Satoru Masamune, where he studied sugar modifications of amphotericin B and compiled his thesis on the total synthesis of calyculin A beginning with gulose analogs. Upon completion of his Ph.D. in 1992, Dr. Levy joined Glycomed where he pursued the design and synthesis of novel glycomimetics, based on pharmacophores identified from the sialyl Lewis^x tetrasaccharide and GDP-L-Fucose, for the treatment of cancer and inflammatory disorders. He later moved to COR Therapeutics where he pursued carbocyclic AMP analogs as inhibitors of type V adenylyl cyclase. Additional areas of research include the design of matrix metalloproteinase inhibitors and ADP receptor antagonists. During his tenure at Glycomed, Dr. Levy co-authored a book entitled "The Chemistry of *C*-Glycosides" (1995, Elsevier Sciences) and collaborated with Dr. Péter Fügedi in the development and presentation of short courses entitled "Modern Synthetic Carbohydrate Chemistry" and "The Organic Chemistry of Sugars" through the American Chemical Society Continuing Education Department. Dr. Levy is currently pursuing the design of novel kinase inhibitors at Scios, Inc.

Péter Fügedi received his chemistry diploma in 1975 from the L. Kossuth University in Debrecen, Hungary. Following his undergraduate work, he earned his Ph.D. in 1978 from the Institute of Biochemistry of the same university. Through 1989, Dr. Fügedi continued research at the Institute of Biochemistry. Concurrently, he pursued additional research activities in the laboratories of Professors Pierre Sinaÿ and Per J. Garegg. In 1989, Dr. Fügedi joined Glycomed, Inc. in Alameda, CA. On returning to Hungary in 1999, he joined the Chemical Research Center of the Hungarian Academy of Sciences in Budapest where he is currently leading the Department of Carbohydrate Chemistry.

During his career, Dr. Fügedi has introduced new methodologies for the protection of carbohydrates, developed new reagents, pioneered glycosylation methods and synthesized biologically active oligosaccharides and glycomimetics. His current research interests are oligosaccharide synthesis, glycosaminoglycan oligosaccharides, orthogonal protection strategies and the study of enzyme inhibitors. Among his publications, Dr. Fügedi co-authored "Handbook of Oligosaccharides, Vols. I–III" (CRC Press, 1991) and has written many book chapters.

Contributors

Prabhat Arya
Chemical Biology Program
Steacie Institute for Molecular Sciences
National Research Council of Canada
100 Sussex Drive, Ottawa
Ontario K1A 0R6, Canada

Yves Chapleur
Groupe SUCRES
UMR CNRS - Université Henri Poincaré
Nancy 1, BP 239
F-54506 Vandoeuvre, France

Françoise Chrétien
Groupe SUCRES
UMR CNRS - Université Henri Poincaré
Nancy 1, BP 239
F-54506 Vandoeuvre, France

Beat Ernst
Institute of Molecular Pharmacy
Pharmacenter of the University of Basel
Klingelbergstrasse 50
CH-4056 Basel, Switzerland

Robert J. Ferrier
Industrial Research Ltd.
PO Box 31-310
Lower Hutt, New Zealand

Péter Fügedi
Chemical Research Center
Hungarian Academy of Sciences
P.O. Box 17
H-1525 Budapest, Hungary

Bartlomiej Furman
Institute of Organic Chemistry
Polish Academy of Sciences
PL-01-224
Warsaw, Poland

Peter Greimel
Glycogroup
Institut für Organische Chemie
Technische Universität Graz
Stremayrgasse 16
A-8010 Graz, Austria

Stephen Hanessian
Université de Montréal
Department of Chemistry
C.P. 6128, Succursale Centre-Ville
Montreal, Quebec
H3C 3J7, Canada

Jan Kihlberg
Umeå University
Department of Chemistry
Organic Chemistry
SE-901 87 Umeå, Sweden

Hartmuth C. Kolb
Department of Molecular and Medicinal Pharmacology
UCLA
6140 Bristol Parkway
Culver City, CA 90230, USA

Horst Kunz
Institut für Organische Chemie
Universität Mainz
Duesbergweg 10-14
D-55128, Mainz, Germany

János Kuszmann
IVAX Drug Research Institute
P.O.B. 82
H-1325 Budapest, Hungary

Daniel E. Levy
Scios, Inc.
Department of Medicinal Chemistry
6500 Paseo Padre Parkway
Fremont, CA 94555, USA

Mickael Mogemark
Umeå University
Department of Chemistry
Organic Chemistry
SE-901 87 Umeå, Sweden

Stefan Oscarson
Department of Organic Chemistry
Arrhenius Laboratory
Stockholm University
S-106 91 Stockholm, Sweden

Norbert Pleuss
Institut für Organische Chemie
Universität Mainz
Duesbergweg 10-14
D-55128, Mainz, Germany

Bugga VNBS Sarma
Senior Scientist, R&D
SaiDruSyn Laboratories Ltd.
ICICI Knowledge Park, Turkapally
Hyderabad, India

Oliver Schwardt
Institute of Molecular Pharmacy
Pharmacenter of the University of Basel
Klingelbergstrasse 50
CH-4056 Basel, Switzerland

Pierre Sinaÿ
École Normale Supérieure
Département de Chimie
24, rue Lhomond
75231 Paris Cedex 05, France

Matthieu Sollogoub
École Normale Supérieure
Département de Chimie
24, rue Lhomond
75231 Paris Cedex 05, France

Josef Spreitz
Glycogroup
Institut für Organische Chemie
Technische Universität Graz
Stremayrgasse 16
A-8010 Graz, Austria

Friedrich K. (Fitz) Sprenger
Glycogroup
Institut für Organische Chemie
Technische Universität Graz
Stremayrgasse 16
A-8010 Graz, Austria

Arnold E. Stütz
Glycogroup
Institut für Organische Chemie
Technische Universität Graz
Stremayrgasse 16
A-8010 Graz, Austria

Kazunobu Toshima
Department of Applied Chemistry
Faculty of Science and Technology
Keio University
3-14-1 Hiyoshi, Kohoku-ku
Yokohama 223-8522, Japan

Pierre Vogel
Laboratoire de Glycochimie et
de Synthèse Asymétrique
Ecole Polytechnique Fédérale
de Lausanne, BCH
CH-1015 Lausanne-Dorigny, Switzerland

Tanja M. Wrodnigg
Glycogroup
Institut für Organische Chemie
Technische Universität Graz
Stremayrgasse 16
A-8010 Graz, Austria

Gernot Zech
Institut für Organische Chemie
Universität Mainz
Duesbergweg 10-14
D-55128, Mainz, Germany

Table of Contents

Part II
From Sugars to Sugar-Like Structures to Non-Sugars

Part I

A Discussion of Carbohydrate Chemistry

1 An Historical Overview

Robert J. Ferrier

CONTENTS

1.1 INTRODUCTION

Figure 1.1 shows this writer's perspective of the developmental phases of the subject of this book. Generally, each phase melds gradually with its neighbors, but an exception is the beginning of the *Fischer era*, which commenced precisely in 1891 with Emil Fischer's assignments of the relative configurations of the monosaccharides [1]. No other step has been of such fundamental importance to carbohydrate chemistry — and also to organic chemistry in general — because it established the validity of the van't Hoff–Le Bel postulate (1874) of the tetrahedral carbon atom. Given the complexities and practical difficulties of the chemistry involved in these discoveries, and the dearth of applicable techniques, it is quite extraordinary that Fischer and his students also made major contributions to the chemistry of the purine group of nitrogen heterocyclic compounds, to amino acids and proteins, and to fats and tannins. In 1902, Emil Fischer was awarded the second Nobel Prize in Chemistry for his work on sugars and purines; the first went to van't Hoff the previous year.

Frieder Lichtenthaler (Darmstadt), a modern authority on Fischer, has written extensively on his work [1–4]. In addition, Horst Kunz (Mainz) has commemorated the 150th anniversary of his birth and the centenary of the award of his Nobel Prize in a biographical essay, which provides additional insight into the man and his science [5].

The evolutionist Stephen Jay Gould has hypothesized in an essay on the subject of the origins of baseball that humankind is more comfortable with the idea that the important components of life, and life itself, arose from creationary origins, rather than by evolutionary development [6]. If this is so, we will be content that the effective creation of the organic chemistry of sugars occurred in 1891, with Fischer's assignments. However, two points should be noted here. Firstly, important results of some preliminary work were available to Fischer, and second, he identified only the relative configuration of each sugar; the determination of the absolute configurations took a further half century [7]. During the course of the succeeding phases, it seems that progress has been made largely by evolutionary means and at a rapidly increasing pace. One should not overlook, however, that much of the evolutionary process has been stimulated greatly by specific *creations*, notably those referred to in Section 1.5 and Section 1.6.

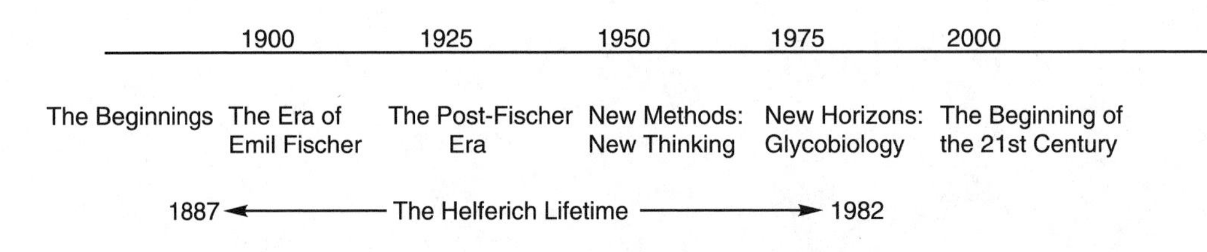

FIGURE 1.1 Phases of the development of the organic chemistry of sugars.

1.2 THE BEGINNINGS

Sugars have been known to humankind since prehistoric times, with Stone Age rock paintings recording the harvesting of honey (a mixture mainly of glucose, fructose and the disaccharide sucrose), and ancient Egyptian hieroglyphics depicting various features of its processing. Likewise, the use of honey in India is reported as far back as records go, and in biblical references, in Old Testament times, Palestine was a land flowing with milk and honey.

The cultivation of sugar cane, and the use of its sucrose component for sweetening purposes, seem to have spread from northeastern India, where sugar canes were established by about AD300, to China and westward to Egypt and beyond. Sugar refineries using sugar cane became commonplace in the developing world, and by the end of the 18th century, sugar beet had been established in Europe as a source crop, with the growth of cane confined to tropical or semitropical regions.

During the developmental stages of the sugar industry, chemistry was in its infancy and progress was made by pragmatic empirical methods, which became an art form that has been followed rather faithfully ever since. Certainly, such an attitude would not have helped the introduction of chemical science. Toward the end of the 19th century, some key sugar manufacturing countries became interested in rationalizing international trade, and initiated a conference in 1863 to which France, Belgium, Holland, and Britain sent delegates. Successive meetings were held in 1864, 1871, 1873, and 1875. A major issue was the means to be used for the evaluation of the refinery products, and at the 1873 conference, the use of the polarimeter was first advocated for this purpose. Prices were to be determined by application of the measured optical activities of samples and adjusted according to other analytical data, for example, ash content. However, the British representatives did not agree to the use of analytical data and were "particularly suspicious of the use of the polarimeter as being open to fraud, and as putting too much power in the hands of the chemist" [8]. Indeed, this comment reflects the lack of faith that early industrialists had in the role of chemistry in the production of one of the most valuable and purest mass-produced organic compounds. Fortunately, however, polarimetry was available very early to organic chemists, and its use proved crucial to the elucidation of the structures of the sugars without which progress in the development of an understanding of their organic chemistry would not have been possible.

By about 1870, glucose and galactose were recognized as similar but distinct sugars, the former having been isolated from raisins in the 18th century and named *dextrose*. Fischer, however, referred to it by its now accepted name. In addition, the ketose fructose and the disaccharides lactose, maltose, and sucrose were known. Of critical importance to the Fischer work were the characterizations of glucose and galactose as derivatives of *n*-hexanal, and of fructose as one of hexan-2-one, as established by Heinrich Kiliani just as Fischer was beginning to tackle the detailed structural problems. The straight-chain nature of these sugars was established by the conversion of glucose and galactose separately to cyanohydrins, by treatment with hydrogen cyanide, and the hydrolyses of these products to aldonic acids followed by reduction with hydrogen iodide and red phosphorus to *n*-heptanoic acid. Similar treatment of fructose gave 2-methylhexanoic acid and, consequently, it was identified as a 2-ketohexose (hex-2-ulose). Considering the probable low efficiencies of these processes, and the difficulty of characterizing the deoxygenated products, this was a considerable feat in itself. The belief of Bernhard Tollens (1893) that sugars existed in cyclic hemiacetal forms was a further matter of great relevance to Fischer's work.

It was in Würzburg in 1884 that Emil Fischer and his students turned their attention to the prodigiously difficult task of bringing together the incoherent knowledge of the chemistry of the sugar family and to elucidating the specific structures of all the members. This had to be done without an accepted understanding of the stereochemistry of the carbon atom, with few developed applicable chemical reactions and almost no characterized reference compounds. Furthermore, no techniques other than crystallization were available for the separation of mixed products and for their purification, and crystallization had to be applied to a series of compounds with notoriously poor crystallizing properties. However, Fischer and his coworkers had one key physical technique available to them — polarimetry — and, most significantly, the new reagent phenylhydrazine that Fischer had discovered earlier, which was to prove invaluable. Alas, it also proved to be toxic, and exposure to it over the years caused Fischer major health problems.

1.3 THE ERA OF EMIL FISCHER

Emil Fischer (Figure 1.2) was born in 1852 near Bonn and died in 1919. His publications span the period from 1875, when he reported phenylhydrazine for the first time, until 1921.

Because the Fischer proof of the structures of the monosaccharides clearly comes into the *classical* part of organic chemical history, having been described time and again in organic and bio-chemical texts (e.g., Ref. 9) and in detail on the centenary of the proof [1], the full argument is not repeated here. Instead, emphasis has been placed on the limited nature of the chemical reactions available for use and the problems with their application at that time in chemical history. In this way, attention is drawn to the immense difficulty of the structural assignment problem and the brilliance of the Fischer solution.

It became evident soon after the beginning of the project in 1884 that the van't Hoff–Le Bel theory predicting the tetrahedral nature of the saturated carbon atom was critical to the solution of the problem, and that acceptable conventions for the description and representation of acyclic

FIGURE 1.2 Emil Fischer. (Reproduced with the kind permission of the Edgar Fahs Smith Collection at the University of Pennsylvania Library.)

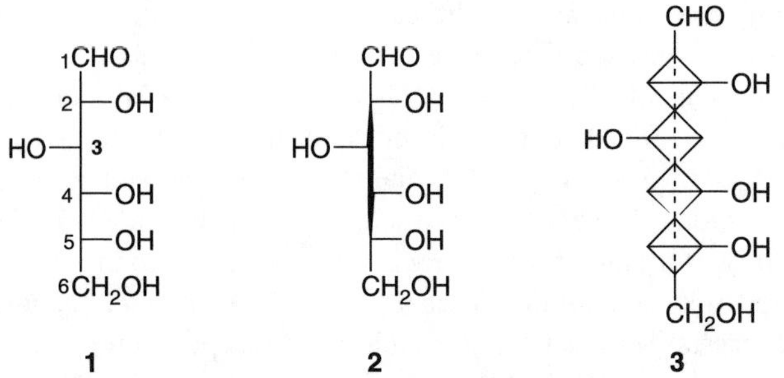

FIGURE 1.3 Representations of acyclic D-glucose.

compounds containing several chiral centers had to be developed. Initially, van't Hoff used a system rather like the Cahn–Ingold–Prelog method to describe the stereochemistry at chiral centers in acyclic molecules. Each was given a + or − sign by application of stated conventions, and natural glucose, which has four such centers in the acyclic form, was thereby accorded the − + + + configuration. After adopting this procedure in a benchmark paper in 1891, Fischer immediately concluded that it was too difficult to apply, and thus prone to error, so instead turned to his own *Fischer projections*. By use of this approach, D-glucose (Figure 1.3) was eventually depicted as **1**, a simplified form of **2** and **3**, the last two indicating the convex arc of carbon atoms, and the first implying both this and projection onto the plane of the paper.

It is of historical significance that Fischer arbitrarily chose to draw the structure of natural glucose with the hydroxyl group at the highest numbered chiral center (C-5) on the right-hand side in the projection. This designation resulted in Fischer assigning it to the D-series according to the Rosanoff convention of 1906, which is still in use. X-ray crystallographic methods eventually showed the Fischer selection to be correct [7]. Although the Rosanoff device halves the overall naming problem for the sugars, the choice of D and L was unfortunate because it causes confusion with d and l, and many compounds belonging to the D set are levorotatory (l) and vice versa.

On heating with phenylhydrazine, glucose and the ketose fructose were soon found to give the same 1,2-bishydrazone **4** (Figure 1.4), or *phenylosazone* by conventional carbohydrate nomenclature. This observation indicated that both sugars possessed the same configurations at C-3–C-5 and, in this way, the interrelating of the structures of different monosaccharides began. Shortly afterward, D-mannose was discovered as the product of selective nitric acid oxidation of D-mannitol, a known plant product. This alditol, fortunately, is one of only two D-hexitols to give the same hexose on selective oxidation at either of its primary positions. The aldohexose, soon to become available from plant sources, also gave D-glucosazone **4** on treatment with phenylhydrazine, thus identifying it as the C-2 epimer of natural glucose. Consistent with this, these

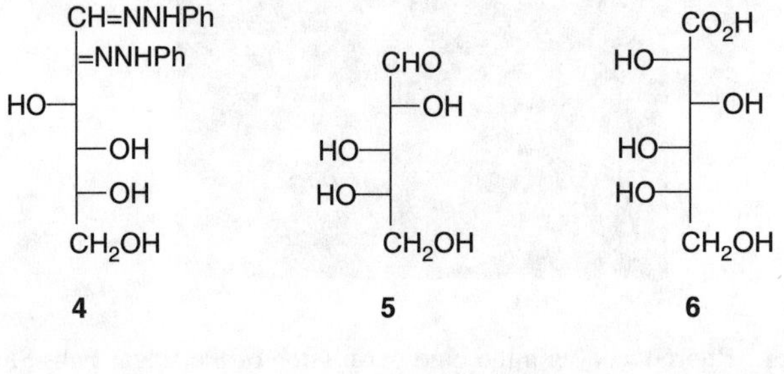

FIGURE 1.4 D-Glucose phenylosazone (**4**), L-arabinose (**5**), L-gluconic acid (**6**).

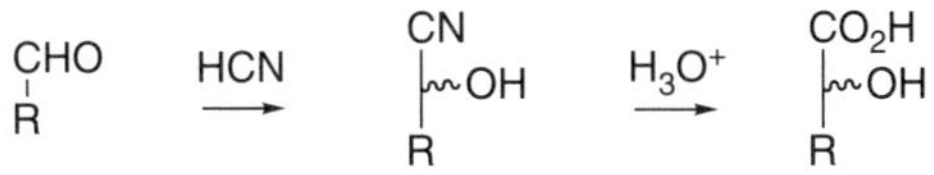

CHO HCN CN OH H_3O^+ CO_2H OH R R R

SCHEME 1.1 Kiliani ascent from an aldose to a higher aldonic acid.

aldoses gave different monophenylhydrazones on careful treatment with phenylhydrazine at reduced temperatures. Fischer acknowledged that this affirmed the van't Hoff–Le Bel tetrahedral carbon proposition and confirmed the prediction that with four chiral centers, the acyclic aldohexoses would have $2^4 = 16$ stereoisomeric forms (eight D and eight L, according to Rosanoff). It seems remarkable that mannose could be obtained by this oxidation method given that it would have been difficult to follow its production and to isolate the product — especially as it would be subject to further oxidation to the aldonic acid.

Fischer applied the Kiliani procedure (Scheme 1.1) to arabinose **5**, an aldopentose obtained from sugar beet, and this produced the enantiomer **6** of the hexonic acid derivable from natural glucose (Figure 1.4). This observation led to the conclusion that the D-enantiomers of arabinose, glucose, mannose, and fructose all share the same configurations at their three highest numbered chiral centers. Moreover, the arabinose obtained from beet was the L-isomer.

At this point, the relationships of some of the known sugars had been established, but the solutions to the complete configurational problems remained some way off. Polarimetry was to play a key role. As implied above, it was used to show that the hexonic acid **6**, obtained from natural arabinose by the Kiliani method, was the enantiomer (with identical physical properties but equal and opposite optical activities) of that derived by oxidation of natural glucose. Now, polarimetry gave information vital to the establishment of the structures of the aldopentoses since the 1,5-dicarboxylic acids obtainable from them by nitric acid oxidation have revealing geometric characteristics. While **7** (Figure 1.5), derived from D-arabinose (and also from D-lyxose), is optically active, those from D-ribose and D-xylose (**8** and **9**) are *meso* compounds and optically inactive. The configurational possibilities for arabinose and hence glucose, mannose, and fructose were therefore narrowed significantly. The final step could be taken when two of the 1,6-dicarboxylic acids obtained from the D-aldohexoses were recognized as enantiomers. The acids were therefore **10** and **11** and could be derived only from glucose and gulose. Determining which was which followed from the known relationship between D-glucose and D-arabinose. Perhaps the nitric acid oxidation, by which these dicarboxylic acids were made, was one of the more straightforward of the reactions used in Fischer's laboratory. However, even this reaction must have been far from trivial given the problems inherent in its application to precious starting materials, difficulties with following the reactions, and the isolation and purification of the products.

Notwithstanding the various intellectual and practical challenges involved, the correct relative configurations were assigned to the family of aldoses (and hence ketoses), and a major milestone

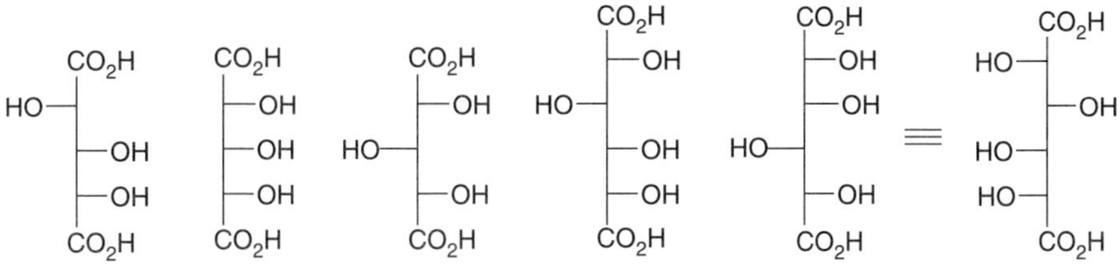

CO_2H HO OH OH CO_2H 7

CO_2H OH OH OH CO_2H 8

CO_2H OH HO OH CO_2H 9

CO_2H OH HO OH OH CO_2H 10

CO_2H OH OH HO OH CO_2H ≡ CO_2H HO OH HO HO CO_2H 11

FIGURE 1.5 D-Arabinaric acid (**7**), D-ribaric acid (**8**), D-xylaric acid (**9**), D-glucaric acid (**10**), L-glucaric acid (**11**).

was reached in natural product chemistry. Fischer has gone down in history as a major figure and, in relation to carbohydrate chemistry, he is unquestionably "the father figure." The importance of his work was recognized immediately and, in 1892, the year following the publication of his solution to the relative configuration problem, he was appointed to the senior chair of chemistry in Germany in the University of Berlin at the age of 40. Ten years later, Fischer was awarded the Nobel Prize.

In Berlin, Fischer directed further work on the organic chemistry of carbohydrates, which included studies on the synthesis of sugars from simple noncarbohydrates, on amino sugars, glycals, glycosides, oligosaccharides and polysaccharides, nucleosides, and several *O*-linked sugar derivatives [10]. Furthermore, through his investigations on amino acids and peptides (and also glycerides), he established these topics as formal branches of organic chemistry. He had seemingly mastered, to the greatest extent possible at the time, almost all the fields of accessible macromolecular natural products; the nucleic acids, however, had to wait. Yet this is not all. His work with simple chiral natural products led him to take key early steps in biochemistry — in particular, into investigations of enzyme action and hence his *lock and key* analogy for the relationship between enzymes and their substrates [3].

1.4 THE POST-FISCHER ERA

During the course of his career, Emil Fischer trained and hosted over 300 Ph.D. students and postdoctoral visitors from around the world in his laboratories, thereby establishing unprecedented influence not only within German natural product chemistry but on the subject internationally. Four of Fischer's coworkers shared the distinction of Nobel Laureates in chemistry (Otto Diels, the Diels–Alder reaction; Hans Fischer, phorphyrin synthesis; Fritz Pregl, microanalytical methods; Adolf Windaus, vitamin D), and two were honored with the Medicine and Physiology Prize (Karl Landsteiner, blood group compounds; Otto Warburg, respiratory enzymes). Among these colleagues, several had their own progeny who were similarly recognized.

Carbohydrate chemistry was further developed in Germany by direct Fischer descendants who went on to contribute to this diverse discipline by means of their own university research groups. A study of the topics pursued by three of the most prominent Fischer prodigies provides an impression of how the subject developed and expanded at that time.

Karl Freudenberg (1886–1983) pursued his Ph.D. (1910) under Fischer and later went on to investigate cyclic acetals and tosyl esters of sugars. From the latter, he introduced deoxyhalogeno and aminodeoxy sugar derivatives [11]. Some of his work involved natural products and he and his students discovered the first naturally occurring branched-chain sugar hamamelose (2-*C*-hydroxymethyl-D-ribose, **12**) in witch hazel tannins (Figure 1.6). Additionally, they worked on the synthesis of natural disaccharides and made several contributions to the understanding of the chemistry of the polysaccharides cellulose and amylose, the latter being the helical component of starch. Notably, the cyclodextrins (then known as Schardinger dextrins), which have received much attention in recent times, were also examined.

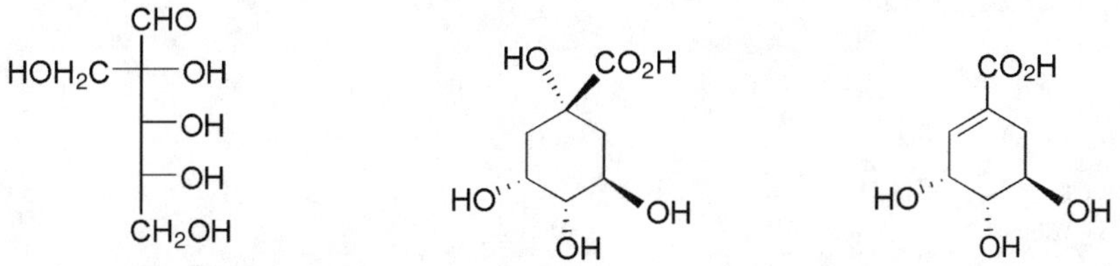

FIGURE 1.6 Hamamelose (**12**), quinic acid (**13**), shikimic acid (**14**).

Burckhardt Helferich (1887–1982) [12,13], who also pursued his Ph.D. under Fischer (1911), examined the fundamental issue of the cyclic/acyclic character of hydroxyaldehydes, and concluded that the common free sugars would favor pyranoid, rather than furanoid, structures. This matter later caused some disagreement between Hudson and Haworth (see below). Helferich was also involved in the early syntheses of purine nucleosides and, for many years, in aspects of glycoside and oligosaccharide syntheses, and he developed the use of trityl ether and mesyl ester derivatives. Additionally, work begun by Fischer on sugars with C–C double bonds (notably, glycals, which have 1,2-endocyclic double bonds, and hexopyranoid derivatives with 5,6-exocyclic unsaturation) was advanced. Helferich's research on the glycosidase group of enzymes also draws attention to early advances made in the biochemical aspects of carbohydrate chemistry.

A notable feature of the lives of Freudenberg and Helferich is their length (97 and 95 years, respectively). They were infants when Fischer started work on the structure of sugars and they lived throughout all but the most recent phases of the history of carbohydrate chemistry. Figure 1.1 indicates this with regard to the latter and, in the Postscript, Helferich's photograph illustrates an observation on the sociology of the subject. These and other chemists passed on the great tradition of carbohydrate chemistry in Germany, and some key figures there today can be identified as Fischer chemical "grandchildren" or "great-grandchildren."

Another chemist with the closest possible links to the great man should be noted at this point — his son, Hermann Fischer (1888–1960) [14,15]. Although he did not complete his Ph.D. work under his father's guidance, Hermann worked with him as a postdoctoral associate just before World War I. After the War, he returned to his father's laboratories before beginning an independent career after the latter's death (1919). He must have inherited the courage to tackle difficult chemical challenges because he started work on the C_3 and C_4 sugar phosphates. Additionally, he pursued studies of the naturally occurring cyclohexane-based quinic acid **13** and shikimic acid **14**, both of which are formed biosynthetically from the above-mentioned phosphates.

Hermann Fischer had an entirely different academic career compared with Freudenberg and Helferich, who remained in their homeland. On the advice of his father, Hermann did some undergraduate work at Cambridge University and then, in the 1930s, took up university positions in Basel and Toronto before joining the new Biochemistry Department at the University of California, Berkeley in 1948. There he continued work at the organic–biochemistry interface, contributing to topics in biosynthesis and developing methods for the synthesis of aminodeoxy sugars. Notably, he was the first to accomplish the conversion of hexose derivatives to inositols. In Berkeley, the name Fischer is honored through the Emil and Hermann Fischer Library, which contains a 4000-volume collection established by the elder Fischer and donated to and maintained by his son.

While Hermann Fischer's career illustrates the impact his father had on carbohydrate chemistry outside of Germany, he was one of many who helped develop the subject internationally having been exposed to the German chemical environment during the first half of the 20th century. Hermann Fischer recognized the American Claude Hudson (1881–1952) as a true successor to his father as a leader of the subject, even though the link between the two was apparently not direct. Having completed his M.Sc. at Princeton University in 1902, Hudson was inclined toward physical chemistry and chose the study of the mutarotation of lactose for his research topic. This is no doubt the reason why, when visiting Germany following his graduation, he chose the van't Hoff laboratory in Berlin for a short stay. On his return to the U.S., he applied van't Hoff's ideas on optical superposition (the additivity of the effects of chiral centers on optical activity) to many sugars and their cyclic derivatives. What emerged was Hudson's *Isorotation Rule* [16] which, although only empirical and not fully reliable, served as a means of anomeric configurational assignment until other physical methods (notably, ^1H NMR spectroscopy and x-ray diffraction analysis) became available several decades later.

For most of his career, Hudson served in the laboratories of the U.S. Government (Bureau of Chemistry, National Bureau of Standards and National Institutes of Health), but this did not detract from his continuing fundamental work with sugars. Although much of it (including some enzyme research) was physical in character, he and his coworkers made a great deal of progress in organic chemistry. They took considerable interest in the ketoses, including heptuloses that were found in plants and others that were made by the novel bacterial oxidation of alditols. Additional studies were carried out on various rearrangements and cleavages of sugars under alkaline conditions, the use of periodate oxidations and higher sugars. Of historical significance is the opinion apparently held in the Hudson laboratory (and no doubt elsewhere) that osazones (of such importance to Fischer in his structural assignments) were overrated as crystalline derivatives of sugars, but the phenylosotriazoles, obtained from them by oxidation, were much more suitable.

Following the German pattern, carbohydrate chemistry spread and became firmly established in North America through the impact of a key figure — in this case Hudson, and his many coworkers. Two of these colleagues are particularly worthy of note: Horace Isbell (1898–1992), who worked with Hudson at the National Bureau of Standards, and Melville Wolfrom (1900–1969), who also trained with Hudson and succeeded him as a major figure in carbohydrate chemistry in the U.S. during his tenure as Professor of Chemistry (The Ohio State University from 1940 until his death).

Elements of the history of carbohydrate chemistry in Britain have interesting parallels to that in the U.S.A. In the former case, the key player was Norman Haworth (1883–1950, later Sir Norman), a close contemporary of Hudson and the first British organic chemist to be awarded the Nobel Prize for Chemistry (for his work on ascorbic acid, 1937). Like Hudson, he studied in Germany in the early 20th century. Also like Hudson, he did not go to the Fischer laboratory, but instead chose to work with Otto Wallach in Göttingen on terpenes, which was his area of primary interest in 1910. At that time, it would have been almost impossible for him to have been unaware of Fischer's work and of the developments in structural carbohydrate chemistry. On his appointment to St. Andrews University in 1912, he found work in this field proceeding there under James Irvine (1877–1952, later Sir James), who also had studied in Germany (Johannes Wislicenus, Leipzig), and was adopting Purdie's methylation reaction for the structural analysis of sugars. (At the time, Thomas Purdie was Head of the Department of Chemistry at St. Andrews).

Major features of the Haworth work at St. Andrews, and later in Birmingham on his appointment there as professor, were the application of methylation analysis to the structures of disaccharides and the discovery (as a parallel to that of Helferich) that pyranoid rings are generally favored relative to the five-membered forms of sugars. A point of historical interest (and a cause of a certain amount of controversy at the time) relates to the early inconsistencies in the assignment of ring size to some compounds following either application of the Hudson Isorotation Rule or Haworth's methylation analysis, the latter ultimately proving to be the more reliable. Vitamin C was examined and synthesized in benchmark developments in organic chemistry. As with the case of Hudson, Haworth influenced chemists from all parts of the world. His impact on carbohydrate chemistry in Britain was enormous and, like Hudson and Fischer, his progeny went on to expand knowledge through the direction of large research groups. In this regard, Sir Edmund Hirst (1898–1975) and Maurice Stacey (1907–1994) were notable, with their teams in Edinburgh and Birmingham, respectively, contributing vastly to structural plant and animal polysaccharide work, as well as to small molecule carbohydrate research.

While the above discussion illustrates the main international research lines that led back directly to Germany (and to Fischer in particular), there were others leading from the farthest parts of the world. Stephen Angyal (Australia), who studied for his Ph.D. (1937) under Géza Zemplén (Hungary), a postdoctoral student of Fischer (1907–1909), is one such example. In the middle of the 20th century many important contributions were also made in countries such as Canada, France, Hungary, Sweden and, of course, Germany by people who had strong historical scientific links to the beginnings in Germany. The former U.S.S.R. had barely been involved, but this was to change thanks to Nicolai Kochetkov's extensive contributions over the latter half of the century. It appears he may have been exceptional in not having historical connections to Fischer.

1.5 NEW METHODS: NEW THINKING

By the 1950s, great progress had been made with the development of the basic chemistry of the sugars, the principles of which had been applied to the study of oligosaccharides and some of the simpler polysaccharides. The groundwork had been laid by the careful and precise application of the methods inherited from the German originators. *Advances in Carbohydrate Chemistry* (a title that had *and Biochemistry* added for Volume 24 in 1969) had been launched in 1945 to provide comprehensive literature reviews and compendia of data covering the important components of the subject. The first volume of 11 chapters commenced with "The Fischer Cyanohydrin Synthesis and the Configurations of Higher-Carbon Sugars and Alcohols" by Claude Hudson. Additional chapters covered features of basic sugar chemistry, natural products, biochemistry and higher polysaccharides. Among the treatments of monosaccharide topics, chapters on "Thio- and Seleno-Sugars" and "Carbohydrate Orthoesters" were included. With respect to natural products, other reports focused on "The Carbohydrate Components of the Cardiac Glycosides" and "The Chemistry of the Nucleic Acids." On the subject of biochemistry, there was a report on "Metabolism of the Sugar Alcohols and Their Derivatives." Finally, concerning higher polysaccharides, topics including aspects of the chemistry of starch, cellulose and plant polyuronides were covered. All the authors were American, presumably to assist with a smooth launch of the new series, but this trend was short-lived with more than half the chapters in Volume 2 coming from non-American contributors. While the publication soon became truly international, the highly professional American influence on it has remained strong.

Advances helped enormously with the assessment of aspects of a rather compact field; one almost entirely the province of specialists. In the main, people contributed in specific areas and seldom ventured outside — especially to noncarbohydrate organic chemistry. To a large extent, this was because of their training and experience, with the majority of investigators having been taught in carbohydrate research environments by carbohydrate chemists. These mentors, in turn, probably came from the same tradition of carbohydrate chemistry, the lineage often being traceable to Germany in the early 1900s. While this may seem surprising, given Fischer's breadth of interest, it was probably inevitable that *big schools* grew to address the enormous challenges of the field. Organic chemists from other backgrounds seldom joined because they were deterred by the water-soluble and poor crystallizing properties of many carbohydrate compounds. Perhaps, however, the real distinguishing feature of the sugars is their high density of chiral centers and the unique, and in some ways unfortunate, methods used in their nomenclature. For example, D has to be distinguished from d and (+). Additionally, the designation α, applied at the anomeric center, can change to β with stereochemical inversion at a distant chiral carbon atom. Such special features and tricks of the trade are not designed to attract newcomers (or students). Whatever the cause of the perceived separate nature of carbohydrate chemistry, it has been largely overcome (see Section 1.6), but a search of recent issues of the *Journal of Natural Products* is still unlikely to locate papers on carbohydrate compounds. Exceptions are found, however, when sugars are attached, as is often the case, to species such as terpenoids. It is interesting to note that one finds no concrete indication that the journal excludes research in the carbohydrate area.

In the early 1950s, practitioners in the carbohydrate field were largely still rather classical in their thinking, and few considered or investigated such matters as the effects of the *conformations* of sugar derivatives (despite Haworth's introduction of the word to chemistry in 1929), or intramolecular effects such as hydrogen bonding or the influences of neighboring groups on reactions. An exception, however, was Horace Isbell (mentioned in Section 1.4) who, because of his

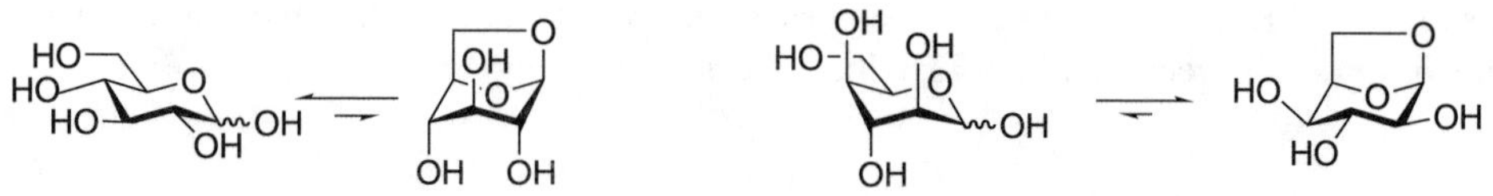

FIGURE 1.7 Equilibria between D-glucopyranose (**15**) and D-idose (**17**), and their respective 1,6-anhydro derivatives (**16** and **18**).

affiliation, published mainly in the *Journal of Research of the National Bureau of Standards* and, consequently, may not have received due recognition for his work. He thought in three dimensions and in terms of reaction mechanisms, and appears to have been appreciably ahead of his contemporaries in these regards.

However, changes were coming because new methodologies were introduced in the 1950s, which increased, more than could have been imagined, the ease of separating mixtures and structurally analyzing their components. Together, chromatographic methods and NMR spectroscopy were to become immeasurably powerful. The former allowed precise quantitative work and facilitated the fractionation of mixtures and the isolation of pure components. Coupled with other physical methods, including mass spectrometry (newly introduced as a tool in organic chemistry), the latter provided structural analytical power previously undreamed of.

By this time, conformational analysis had been introduced largely by Sir Derek Barton through his work with steroids, and was applied rigorously to the sugars by Stephen Angyal, who developed methods for calculating the energy differences between the isomeric forms and the conformations of sugars in water. Thus, he quantified the steric and stereoelectronic factors contributing to the energies of sugar isomeric and conformational states [17]. With this knowledge, it became possible, for example, to account for the observation that, under acidic conditions, D-glucose **15** and its 1,6-anhydropyranose derivative **16** (Figure 1.7) give an equilibrated mixture containing 0.2% of the anhydride, whereas the analogous equilibrium for D-idose **17** consists predominantly of the analogous anhydride **18** (86%).

The newly introduced chromatographic and NMR techniques made it possible to follow in detail the progress of complex carbohydrate reactions — some of which had interested Fischer. For example, gas and paper chromatographic methods, coupled with the use of ^{14}C-labeled sugars, allowed the identification and quantification of the products formed during the course of reactions of sugars with simple alcohols under acid conditions (Fischer glycosidation). They permitted confirmation of Fischer's conclusion that the furanosides are produced initially under kinetic control, and isomerize relatively slowly to the six-membered glycosides under thermodynamic control. In addition, the new methods allowed the identification of the sugar dialkyl acetals, which are formed in small proportions among the early products. Thus, Fischer's belief that these products preceded the furanosides (M.L. Wolfrom, personal communication, 1968) was disproved. The acyclic compounds are co-kinetic products with the furanosides, and the overall reactions have the form indicated in Scheme 1.2 [18,19].

During this part of the development of sugar chemistry, it emerged that reactions proceeding by way of carbocations could be very complex and, in consequence, sometimes synthetically useful. For example, the ion **19**, derived by the treatment of penta-*O*-acetyl-β-D-glucose with antimony

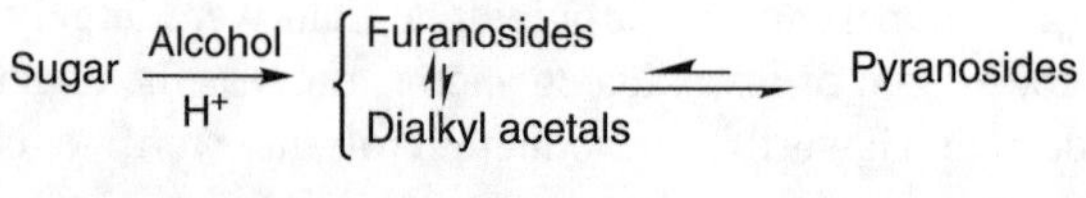

SCHEME 1.2 The path of the Fischer glycosidation of sugars.

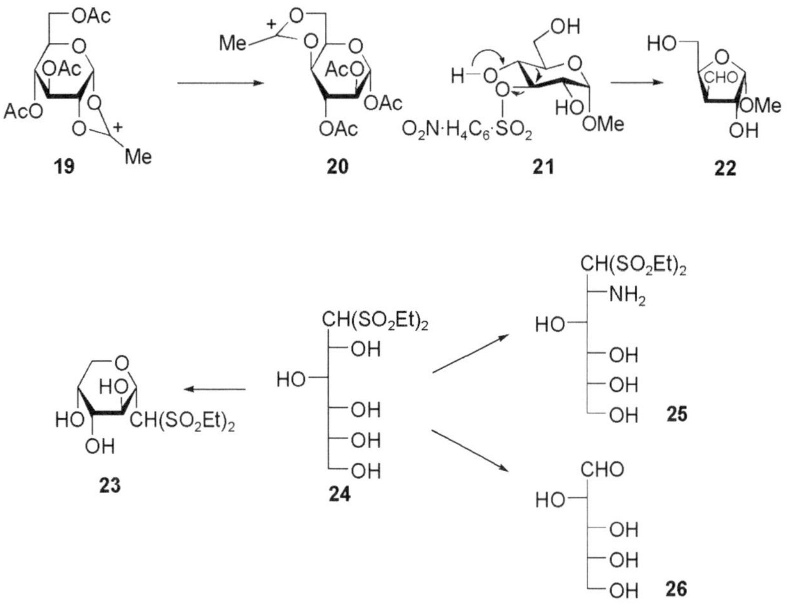

FIGURE 1.8 Some rearrangements of sugar derivatives by way of ionic processes.

pentachloride at low temperatures, yields a succession of isomeric bicyclic acetoxonium ions, including **20** (Figure 1.8). Each of these ions is formed following participation by a neighboring acetoxy group and with a configurational inversion. The D-*ido* species **20** crystallizes to give a 73% yield of its hexachloroantimonate salt, thus opening a useful novel route to the rare D-idose [20]. Sulfonate ester **21** does not undergo direct displacement of the ester group on heating with sodium acetate in water. Instead, a ring contraction to **22** occurs (43% yield) [21]. Sugars were therefore becoming more amenable to controlled structural modification, and this was also the case with their reactions that proceed by way of carbanionic intermediates. For example, on treatment with strong ammonia, bis-sulfone **24** undergoes β-elimination of water from the C1–C2 positions followed by addition of the base to give compound **25**. Alternatively, on treatment with dilute ammonia, **24** is deprotonated at the hydroxyl group of C2, the C1–C2 bond is cleaved and D-arabinose **26** is produced. This way provides a new method of descending the aldose series. Under acidic conditions, compound **24** gives 2,6-anhydride **23** (also a *C-glycoside*), probably by way of the 1,2-ene [22,23].

A plethora of modified sugars were made at this stage of the development of the subject: extended chain, branched chain, *C*-glycosidic, carbocyclic and unsaturated compounds. Additionally, many products having amino-, thio-, and halogeno-substituents replacing one or more hydroxyl groups of sugars were produced. These developments were largely in response to the finding that many sugars with modified structures are metabolites produced by microorganisms. In a high proportion of cases, these types of natural products were biologically active (particularly having antibiotic properties), and the subsequent drive to produce them and their analogs synthetically was extensive. In this field, Japanese chemists made important discoveries of such antibiotics as the kanamycins, kasugamycin, formycin, bleomycin and the anthracyclines. The extent of the contribution of the leader Hamao Umezawa (1914–1986) [24] can be gauged by the fact that he and his coworkers published 1200 papers in the field. Beyond this, the nucleosides remained an appreciable synthetic challenge because of their potential bioactivity either as derivatives or as components of oligonucleotides. Methods were developed to make innumerable analogs of the natural compounds with various modifications to the sugar or the base components, the anti-HIV agent AZT **27** (Figure 1.10) being the outstanding example of a pharmaceutical agent discovered using this strategy.

Given the advances made during this period by chemists around the world, it is extraordinary that one person stood out so clearly as the intellectual leader in the field. Raymond Lemieux (1920–2000) [25], the most loyal of Canadians and a former postdoctoral associate of Wolfrom (Ohio), studied at McGill University (Montreal) where he received his Ph.D. (1946) under Clifford Purves, a chemical descendant of Hudson. Lemieux had a profound and intuitive appreciation of the chemical nature of the electrons, atoms, bonds, functional groups and stereochemistry of sugars, which he applied to the development of their chemistry and latterly to the understanding of the details of their interactions with proteins in fundamental biological processes. His deep awareness of mechanistic chemistry allowed him to attain an early Holy Grail of sugar chemistry in 1953 — the rational synthesis of sucrose (isolated as its octaacetate with 5.5% yield) [26]. Nearly 50 years later, a method that afforded an 80% yield was developed, indicating how far the methodologies of sugar chemistry had progressed in that interval [27].

Glycoside synthesis was a Lemieux interest for many years. He was particularly impressed by the need for a method of making 1,2-*cis*-related products (e.g., α-glucosides and β-mannosides), and he solved the problem by introducing the concept of halide ion catalysis using glycosyl halides with nonparticipating groups at C2 [28]. During his considerations of chemistry at the anomeric center, he recognized the anomeric, the reverse anomeric and the *exo*-anomeric effects, which have become vitally important to an appreciation of mechanistic and stereochemical aspects of the organic chemistry of the sugars [25].

With his close interest in stereochemistry and appreciation of physical methods, it was fortunate that Lemieux could be involved from the outset in the application of NMR spectroscopy in the field. Ottawa, where he was on the faculty of the university, also housed the National Research Council Laboratories with the spectroscopists Harold Bernstein, William Schneider, Rudolf Kullnig (a Lemieux Ph.D. student) and a very early 40 MHz instrument (Figure 1.9). The first carbohydrate spectra were recorded there. The paper with these collaborators, reporting 40 MHz studies of pyranose sugar acetates, showed the significance of chemical shifts for determining the nature of

FIGURE 1.9 Raymond Lemieux (left) and Rudolf Kullnig, a graduate student "strongly oriented toward physics" and a coauthor of the famous 1958 paper, with the 40 MHz Varian NMR spectrometer on which the first spectra of sugar derivatives were recorded. It "had to be operated with the skill and patience of a brain surgeon" [25]. (Photograph taken about the time the work was done.) (Reproduced with the kind permission of Mrs. Jeanne Lemieux).

constituent groups and of $^1H,^1H$ couplings for assigning configurations and conformations [29]. No other paper has had more impact on the organic chemistry of sugars since Fischer's seminal publications on the configurations of the aldoses.

Coupled with his achievements in carrying out highly regio and stereo selective additions to glycals (1,2-unsaturated pyranose derivatives) to give glycosides of 2-amino-2-deoxy sugars, NMR spectroscopy enabled Lemieux to lead the subject into the next phase of its development.

1.6 NEW HORIZONS: GLYCOBIOLOGY

By the end of the 20th century, sugar derivatives had become much more manipulatable and diverse. This was largely a result of the gain of control of many of their ionic reactions, particularly those proceeding by way of carbocationic intermediates and involving molecular rearrangements. In addition, by the end of the century, new procedures based on controlled free radical means of generating carbon–carbon bonds, had become available to organic chemists [30,31], and were highly applicable to sugar derivatives. With the new methodologies, the elaboration of carbon skeletons was permitted via the introduction of new and multiple substituents joined by C–C bonds. As an example, the glycoside **28**, readily made from tri-*O*-acetyl-D-glucal and 2-bromoethanol combined in the presence of a Lewis acid catalyst, was converted to **29** (Figure 1.10). This transformation was accomplished utilizing tri-*n*-butyltin hydride and a free radical initiator and produced a 64% yield. When repeated in the presence of methyl acrylate ($CH_2{=}CHCO_2Me$) as a radical trap, this reaction gave the di-branched product **30** in 53% yield [32]. Commonly, by-products are formed in reactions of this type and arise generally from the involvement of competitor intermediate radicals. For example, during the formation of **30**, the ethyl glycoside analog of the starting material **28** is produced by quenching the first radical formed with hydride following bromine radical abstraction. While this lack of specificity is a weakness of such processes, it often can be tolerated given that the important C–C bond formations are constrained to occur regio- and stereo-specifically. Application of this type of technology has permitted the elaboration of the sugar-derived **31** to the tetracyclic **32** and hence the cedranoid sesquiterpene (–)-α-pipitzol **33** [33].

Radical intermediates are also involved in such relatively new carbohydrate processes as photochemical transformations [34] and the direct radical substitution of certain sugar ring

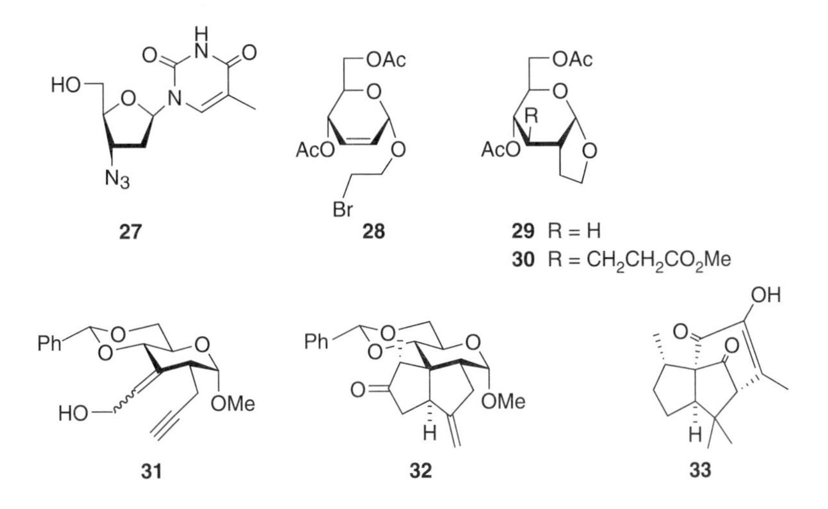

FIGURE 1.10 The nucleoside analog AZT (**27**), the specific branched-chain compounds **29** and **30**, and the sesquiterpene (–)-α-pipitzol (**33**, made from **31** and **32**) available from sugar starting materials.

hydrogen atoms by bromine atoms [35]. Both of these methodologies open new possibilities for synthesis. Likewise, carbenes have played their part in more modern sugar organic chemistry, for example, with glycosylidene species (carbenes at the anomeric position, available by photolysis of corresponding *spiro*-diazirines) being inserted into the O–H bonds of alcohols to open new routes to *O*-glycosides [36].

This period also saw the overdue use in the field of many organometallic reagents and the consequently increased synthetic power available to the chemist [37]. The importance of metathesis processes, particularly as carried out with Grubbs' catalysts, is particularly worthy of note.

Although the field's growth in the last three decades of the 20th century was generally in line with developments in organic chemistry, it became subject to two major influences, both of which were driven by considerations outside carbohydrate chemistry. First, chemists both within and outside the field came to recognize that the functionality and stereochemistry of the common monosaccharides could be manipulated to make them valuable and inexpensive starting materials for the synthesis of a myriad of enantiomerically pure, noncarbohydrate compounds ranging from the almost trivial to the extremely complex [38]. A semiquantitative indication of the rate of development in this area is given by the observation that the Royal Society of Chemistry's Specialist Periodical Reports on carbohydrate chemistry, which aims to record all relevant publications in a given year, shows that in 1973 nine sugar to nonsugar transformations were described, while approximately 100 such reports appeared in the final year of the century. Chemists brought up in the carbohydrate tradition and, very importantly many from other backgrounds were active in this area, and the latter brought appreciable invigoration to carbohydrate chemistry. Unfortunately, there has been a negative consequence of the mixing of the cultures because many systems, in addition to those accepted in the field for decades, are now used for representation of sugar structures. While this may be acceptable in principle, improper use of various methods is not uncommon in the literature, and that is highly undesirable. In particular, some authors do not pay sufficient attention to the representation of the absolute configuration of compounds, and continue to use incorrect or ambiguous procedures.

Not surprisingly, Lemieux was involved and pointed the way with his synthesis of the optically active (*R*)-1-deuterioethanol **34** [39]. However, in this area, (+)-furanomycin **35** (Figure 1.11), a bacterial metabolite with antibiotic activity, is more representative of the type of compound now available from sugars; in this case, L-xylose [40]. Other examples of compounds made from sugars are 7-deoxypancratistatin **36** (made from D-gulonic acid), which has antineoplastic and antiviral activities [41], and the highly antifungal soraphen $A_{1\alpha}$ **37** [42]. In each of the structures **35** through **37**, the carbon atoms of the sugars used as starting materials are indicated by their carbohydrate number. In the soraphen case, D-glucose and D-mannose were employed to provide the northwest and east sections, respectively.

The second and much more powerful force driving sugar organic chemistry ahead was provided by the discovery that the carbohydrates play vital roles in the control of many key biological

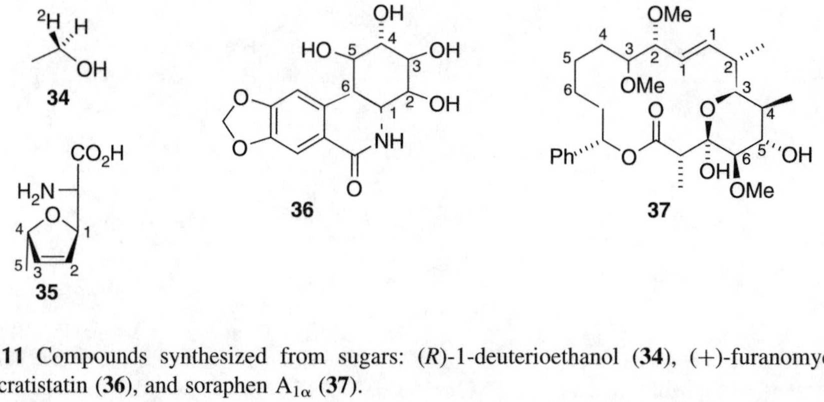

FIGURE 1.11 Compounds synthesized from sugars: (*R*)-1-deuterioethanol (**34**), (+)-furanomycin (**35**), 7-deoxypancratistatin (**36**), and soraphen $A_{1\alpha}$ (**37**).

processes by acting as reciprocating compounds with proteins in molecular recognition events. *Glycoscience* and *glycobiology* had been born [43]. A striking example of its importance accounts for how fish eggs lying in open water are fertilized only by milt of the same species. That is, multiple highly specific binding forces hold together the involved proteins and oligosaccharide-containing compounds on the egg and sperm surfaces in a specific manner like that used in Velcro.

Lemieux's contribution to the use of sugars for making other types of chiral compounds was significant, but much less important than his leadership of the subject into the study of the involvement of sugars in molecular recognition events. His brilliant work in this area seemed to represent a logical culmination of his previous discoveries, among which were the development of synthetic methods for amino-sugars, glycosides and oligosaccharides, contributions to conformational analysis and recognition of stereoelectronic effects, and the applications of NMR spectroscopy [25]. He now used advanced NMR methods and developed calculations (hard-sphere *exo*-anomeric effect [HSEA]) for determining the preferred solution conformations of specifically synthesized oligosaccharides, and immunochemical methods for preparing monoclonal antibodies and quantifying their binding to the antigens and also to specifically monodeoxygenated derivatives of them. For example, the human blood group B determinant, trisaccharide glycoside α-L-Fuc-(1→2)-[α-D-Gal-(1→3)]-β-D-Gal-OMe **38** (Figure 1.12) and several monodeoxy derivatives were examined as their complexes with a monoclonal antibody, and a detailed picture of the sugar–protein binding resulted [44]. A most important generalization that emerged from such work was that associative interactions between nonpolar regions of both carbohydrates and their complementary proteins are involved, and in consequence, the interactions between the polar groups are strengthened. Another typical piece of Lemieux's insight focuses on the role water plays in the complexation of sugars with other compounds [45].

Most binding of this kind involves oligosaccharide constituents of glycoproteins or other glycoconjugates on cell surfaces, and proteinaceous lectins on the surfaces of the binding partners. For example, pathogenic bacteria and biological toxins are bound and ingested by macrophages and, at the onset of the inflammatory reaction, *selectins* on endothelial cell surfaces bind leukocytes of the blood. These leukocytes carry the Lewis^{x} tetrasaccharide antigen α-NeuAc-(2→3)-β-D-Gal-(1→4)-[α-L-Fuc-(1→3)]-D-GlcNAc, sLe^{x}, **39**, and this results in the leukocytes migrating to neighboring tissues [43]. Associations of these kinds can be involved in the initiation steps of many diseases such as stomach cancer, influenza and cholera, and biological processes like cancer metastasis. An understanding of the initiating binding processes at the molecular level provides new strategies for the development of protections from and cures for diseases. Clearly, molecular competitors for either the carbohydrate or protein components of the binding pairs offer

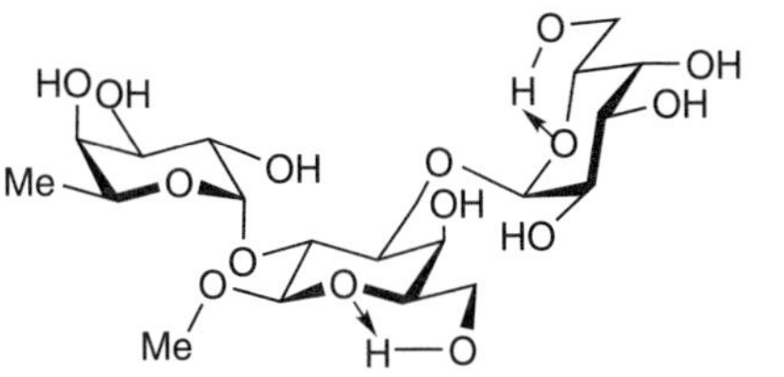

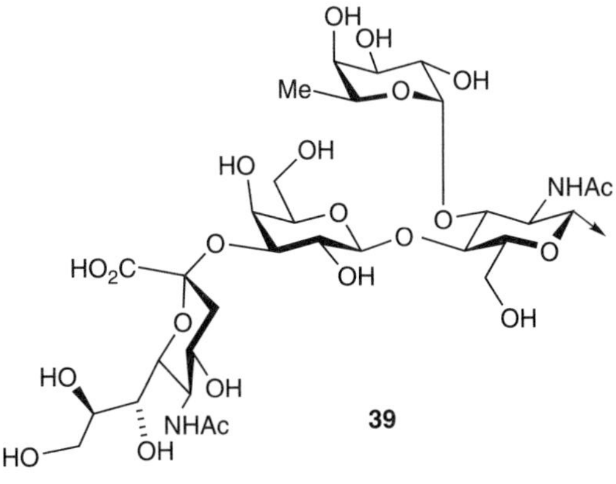

FIGURE 1.12 The human blood group B determinant (**38**) and the Lewis^{x} tetrasaccharide antigen (sLe^{x}, **39**).

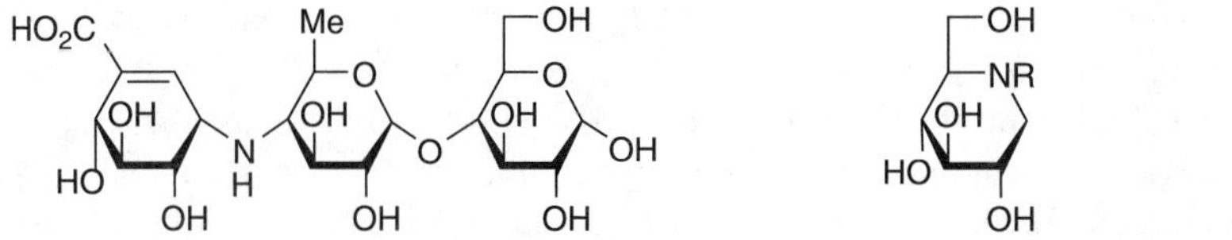

FIGURE 1.13 Acarbose (**40**) and deoxynojirimycin (**41**).

possibilities as inhibitors, and the need for such compounds has led to the production of an extensive range of carbohydrate mimetics [43].

A somewhat different approach to new pharmaceuticals depends upon the use of specific enzyme inhibitors. One example is the α-glycosidase inhibitor acarbose **40** (Figure 1.13), isolated from a microbiological fermentation broth and used for the treatment of diabetes [46]. It binds to the active site of the enzyme that cleaves oligosaccharides derived from starch, and since it is not a substrate, it acts as an inhibitor. However, it is outdone in this regard by a further Bayer AG bacterial product deoxynojirimycin **41**, which had been synthesized in the Hans Paulsen group in Hamburg (previously directed by Kurt Heyns, a Fischer chemical "grandson") 10 years earlier [47]. Paulsen had conducted extensive synthetic studies in the field of sugar analogs with a nitrogen atom in place of oxygen as the ring hetero-atom, and thus provided a fine example of the confluence of academic research in synthesis and commercially driven biological investigations. Bayer chemists studied hundreds of analogs of deoxynojirimycin, resulting in the compound **42** being marketed for the treatment of noninsulin-dependent diabetes [48].

Special interest has been taken in the inhibition of enzymes involved in cleaving the glycosidic bonds of neuraminic acid (a complex C_9-carbon sugar acid), because this process is critical to the spread of infection after the influenza virus binds to the host cell. Following the analysis of the structure of complexes of the enzyme and substrate (bound in a distorted conformation), unsaturated compounds, for example **43** (Figure 1.14), were made as mimics of the latter in the reaction transition state and found to be potent neuraminidase inhibitors and anti-influenza compounds [49].

One of many other types of carbohydrate/noncarbohydrate interactions that lead to bioactivity involves the aminoglycoside class of antibiotics, of which neamine **44** is a simple example. They bind to a specific part of the decoding region of prokaryotic ribosomal RNA to induce their activity, and this has led to the synthesis of analogs designed to have stronger binding [50].

Glycobiology has provided the greatest ever inducement to organic chemists and biochemists to find means of preparing compounds designed to interfere with defined carbohydrate-dependent processes and, hopefully, to have specific biological activities required by both preventative and curative medical science.

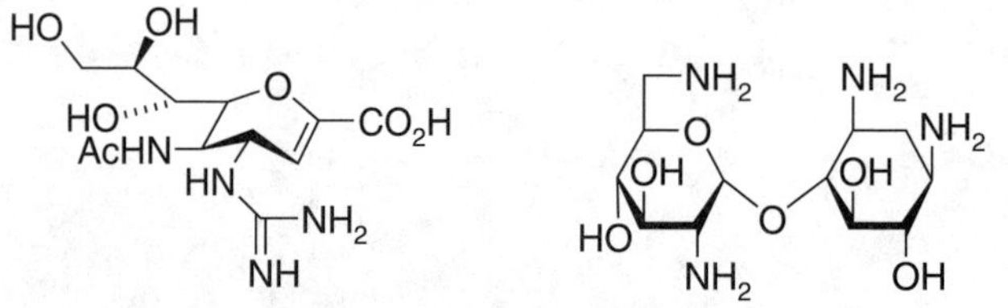

FIGURE 1.14 Relenza™ (**43**), an anti-flu medicine and Neamine (**44**), an antibiotic.

1.7 THE BEGINNING OF THE 21ST CENTURY

Undoubtedly, sugar derivatives will have a future in aspects of nanotechnology as indicated, for example, by the first applications of carbohydrate microarrays [51], and by the complexes made by threading carbon nanotubes and conjugated polymers through cyclodextrins rings [52]. However, it is much clearer that the tasks handed over from the 20th century and the future *raison d'être* for sugar organic chemistry lie in providing for medical and biological sciences. Further developments in methodology are required for this to occur because the traditional synthesis techniques are cumbersome given that the simplest unicellular organisms can put together specific compounds of vast structural complexity with high efficiency. Some progress of the required type has been made, and a thematic issue of *Chemical Reviews*, published in 2000, took as its full title *Carbohydrate Chemistry — A Formidable Scientific Frontier Becomes Friendlier* [53].

Oligosaccharide synthesis is of immense importance to glycobiology, and the biomimetic enzymatic processes already developed will no doubt be improved and conceivably coupled with selective sequential gene expression to afford means of preparing complex hetero-oligosaccharides to order [54]. In the meantime, the overdue production of oligosaccharides by automated chemical synthesis on polymer supports has begun, [55]. and can hopefully be predicted to become very important, as such methods for making oligopeptides and oligonucleotides have been for decades. They could incorporate newer approaches to the establishment of glycosidic bonds such as the use of intramolecular reactions [56] and the selection of the most compatible donor and acceptor pairs from computer programs [54]. These procedures also may be applicable in more extensive "one-pot" homogeneous procedures. A short essay, "Sweet Synthesis", deals with state of developments in the art of oligosaccharide synthesis at the beginning of the millennium [57].

Danishefsky's work on carbohydrate-based anticancer vaccines is a fine example of the combined use of chemical and immunological methods in a new approach to addressing aspects of the cancer problem, and may well point the way for future work. In this approach, synthetic oligosaccharide antigens are coupled with suitable protein or ceramide carriers to give glycoprotein or glycolipid (e.g., **45**) (Figure 1.15) conjugates for use as vaccines. In the illustrated example, the tetrasaccharide is based on the Lewis^{y} blood group determinant, which elicits antibodies against colon and liver cancers and is involved in metastases of prostate cancer and in the development of ovarian tumors [58].

As the understanding of specific enzymatic processes develops, both in the sense of their mechanisms and of their significance in human and pathogen biochemistry, the need for highly efficient and specific inhibitors will increase. An example of what is likely to be developed in the

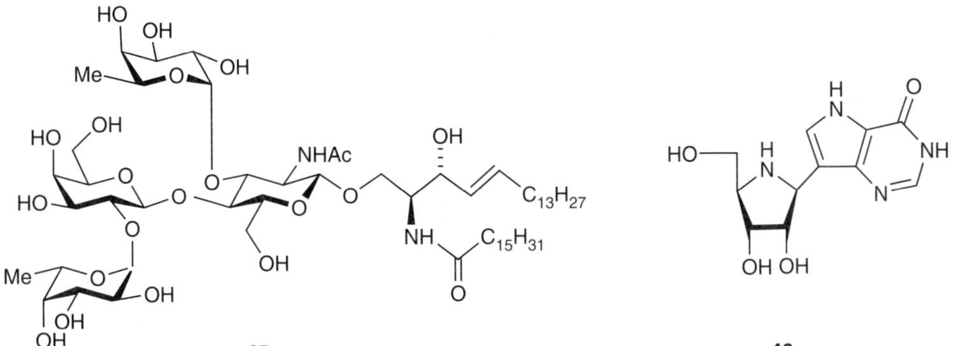

FIGURE 1.15 A synthetic Lewis^{y} blood group determinant/ceramide antigen (**45**) and the synthetic specific enzyme inhibitor immucillin H (**46**).

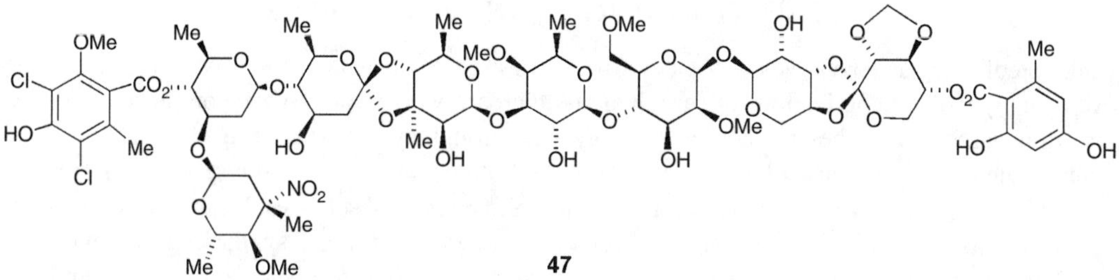

FIGURE 1.16 Everninomycin 13,384-1 (**47**).

future is immucillin H **46** (now Fodosine™, the hydrochloride), a specifically designed picomolar inhibitor of the enzyme purine nucleoside phosphorylase, which catalyzes the phosphorolysis of deoxyguanosine. Deficiency of the enzyme leads to accumulation of the nucleoside and consequent T-cell apoptosis, which is desirable for control of T-cell cancers, autoimmune diseases and transplant rejection. The nitrogen-in-the-ring *C*-nucleoside, which was rationally designed as a transition state inhibitor and made by rational synthesis, results in selective apoptosis of rapidly dividing human T-cells [59]. It is in clinical trials for use against T-cell cancers.

It appears that the challenges facing the organic synthesis chemist will become more severe in the area of antibiotic chemistry with the emergence of increased resistance to such natural products as the glycopeptide, vancomycin. Although this antibiotic has remained effective against evolving bacteria, new drug-resistant compounds are required, and everninomycin 13,384-1 (Ziracin, **47**; Figure 1.16), another natural product, has proved of potential value. While its total synthesis can be acclaimed as a major feat [60], its extremely complex oligosaccharide-based structure would make its production on a commercial scale a vast challenge. Whether formal organic synthesis will be able to develop to the point of overcoming such problems, or whether compounds of this level of complexity will become available in bulk by application of other methods, remain questions for the new millennium.

1.8 POSTSCRIPT

Over the years, activity in carbohydrate chemistry increased to such a point that a series of international conferences was begun in 1960; the 22nd conference was held in 2004. Twelve countries have hosted the meetings and participant numbers have increased from fewer than 300 at the beginning to over 1000. At the largest meetings, the number of countries represented was over 40. To a large extent, this reflects the increasing importance of the field and, notably, the advent of glycoscience and glycobiology. Additionally, European and several national series of meetings are held regularly.

Carbohydrate Research, the first primary journal dedicated to carbohydrate chemistry, was initiated in 1965, to be followed by *The Journal of Carbohydrate Chemistry* in 1982 and *Carbohydrate Letters* in 1996. The importance of the fast growing field of glycobiology is reflected by the commencement of the publication of two journals devoted to this area; *Glycoconjugate Journal* in 1983 and *Glycobiology* in 1990. The annual *Advances in Carbohydrate Chemistry* (which had *and Biochemistry* added to the title in 1969) continues to provide comprehensive reviews. It was first published in 1945 and was initiated with Melville Wolfrom and Ward Pigman as editors, the former remaining in the role until his death in 1969. He therefore made the remarkable contribution to the subject of editing the first 24 volumes. Since then, the great majority of the editing responsibility has been carried admirably by the late Robert Tipson and by Derek Horton.

Each year since 1967, The Royal Society of Chemistry has published Specialist Periodical Reports on carbohydrate chemistry, which aims to cover, with some commentary, every paper published in the field within the previous year. While early volumes were about 300 pages in length, Volume 15 consisted of 1100 pages and appeared in two parts, with the macromolecules book representing over 70% of the total. Clearly, the rate of growth in carbohydrate research has outpaced the ability of this type of survey to cope, and from then on, only the monosaccharide/oligosaccharide sections continued. They categorize material mainly according to classes of compounds (e.g., glycosides, branched-chain sugars and nucleosides), but other sections, such as applications of physical methods and the use of sugars in chiral synthesis, also appeared. These volumes therefore act as a unique search tool, allowing easy access to work done in each area in any year. They also permit an assessment of general trends within any of the areas over 34 years. Alas, the burgeoning of relevant science, and the load required to review it in this way, overtook the stamina of the reviewers, and Volume 34, which covers developments published in 2000, is the last of the series. The mass of information now continuously appearing demands that automated means should be used, not just for data collection and presentation, but also for its searching.

It is the writer's opinion that excellent relationships, communication and co-operation in the field have significantly helped the development of the subject. This has been greatly assisted by the opportunities provided by the meetings noted above and also, perhaps, by a general loosening of the social formality that, in times gone by, may have inhibited scientific interactions — especially those involving younger scientists.

The photograph of Burckhardt Helferich (Figure 1.17), who worked with Emil Fischer (Section 1.4), is provided to illustrate two points: firstly, that the history of the subject is not much longer than the lifetime of one of its early contributors (Figure 1.1), and secondly, that relative informality has commonly characterized personal interactions in recent decades. How warm and encouraging he looks in contrast to the traditional early chemists as suggested by their forbidding photographs. Helferich gives the impression of being the type of person who would have greatly enjoyed the international meetings and the friendships they engender, and who would have been encouraging to students in the way that most modern scientists seem to be. An obituary suggests that this impression may be correct [12]. However, he is reported to have been "not much of a traveler on the scientific circuit" [13].

FIGURE 1.17 Burckhardt Helferich. (Reproduced with the kind permission of Professor Frieder Lichtenthaler).

ACKNOWLEDGMENTS

Frieder Lichtenthaler, Stephen Angyal and Richard Furneaux are thanked for their help. The author accepts responsibility for the accuracy of the above and for opinions expressed.

REFERENCES

1. Lichtenthaler, F W, Emil Fischer's establishment of the configuration of the sugars: a centennial tribute, *Angew. Chem. Int. Ed.*, 31, 1541–1556, 1992.
2. Lichtenthaler, F W, Emil Fischer, his personality, his achievements, and his scientific progeny, *Eur. J. Org. Chem.*, 24, 4095–4122, 2002.
3. Lichtenthaler, F W, 100 Years "Schlüssel-Schloss-Prinzip": what made Emil Fischer use this analogy? *Angew. Chem. Int. Ed.*, 33, 2364–2374, 1994.
4. Jaenicke, L, Lichtenthaler, F W, A Kaiser Wilhelm Institute for Cologne! Emil Fischer, Konrad Adenauer, and the Meirowsky Endowment, *Angew. Chem. Int. Ed.*, 42, 722–726, 2003.
5. Kunz, H, Emil Fischer — unequalled classicist, master of organic chemistry research, and inspired trailblazer of biological chemistry, *Angew. Chem. Int. Ed.*, 41, 4439–4451, 2002.
6. Gould, S J, *Bully for Brontosaurus*, Vintage, London, p. 57, 2001.
7. Bijvoet, J M, Peerdemann, A F, van Bommel, A J, Determination of the absolute configuration of optically active compounds by means of X-rays, *Nature*, 168, 271–272, 1951.
8. Deerr, N, *The History of Sugar*, Chapman & Hall, London, 1949, p. 506.
9. Morrison, R T, Boyd, R N, *Organic Chemistry*, 5th ed., Allyn and Bacon, Boston, pp. 1291–1298, 1987.
10. Freudenberg, K, Emil Fischer and his contribution to carbohydrate chemistry, *Adv. Carbohydr. Chem.*, 21, 1–38, 1966.
11. Lichtenthaler, F W, Karl Freudenberg, *Carbohydr. Res.*, 164, 3–9, 1987.
12. Lichtenthaler, F W, Burckhardt Helferich, *Carbohydr. Res.*, 164, 9–13, 1987.
13. Stetter, H, Burckhardt Helferich, *Adv. Carbohydr. Chem. Biochem.*, 45, 1–6, 1987.
14. Lichtenthaler, F W, Fischer, H O L, *Carbohydr. Res.*, 164, 14–22, 1987.
15. Sowden, J C, Hermann Otto Laurenz Fischer, *Adv. Carbohydr. Chem.*, 17, 1–14, 1962.
16. Hudson, C S, The significance of certain numerical relations in the sugar group, *J. Am. Chem. Soc.*, 31, 66–86, 1909.
17. Angyal, S J, The composition and conformation of sugars in solution, *Angew. Chem. Int. Ed. Engl.*, 8, 157–166, 1969.
18. Heard, D D, Barker, R, An investigation of the role of dimethyl acetals in the formation of methyl glycosides, *J. Org. Chem.*, 33, 740–746, 1968.
19. Ferrier, R J, Hatton, L R, Aspects of the alcoholysis of D-xylose and D-glucose; the role of the acyclic acetals, *Carbohydr. Res.*, 6, 75–86, 1968.
20. Paulsen, H, Cyclic acyloxonium ions in carbohydrate chemistry, *Adv. Carbohydr. Chem. Biochem.*, 26, 127–195, 1971.
21. Austin, P W, Buchanan, J G, Saunders, R M, Rearrangement in the solvolysis of some carbohydrate nitrobenzene-*p*-sulfonates, *J. Chem. Soc. C*, 372–377, 1967.
22. MacDonald, D L, Fischer, H O L, The degradation of sugars by means of their disulfones, *J. Am. Chem. Soc.*, 74, 2087–2090, 1952.
23. Barker, R, MacDonald, D L, The sulfones derived from the dithioacetals of certain hexoses, *J. Am. Chem. Soc.*, 82, 2297–2301, 1960.
24. Tsuchiya, T, Maeda, K, Horton, D, Hamao Umezawa, *Adv. Carbohydr. Chem. Biochem.*, 48, 1–20, 1990.
25. Lemieux, R U, *Explorations with Sugars; How Sweet I Was*, American Chemical Society, Washington, D.C., 1990.
26. Lemieux, R U, Huber, G, A chemical synthesis of sucrose, *J. Am. Chem. Soc.*, 75, 4118–4119, 1953.
27. Oscarson, S, Sehgelmeble, F W, A novel β-directing fructofuranosyl donor concept. Stereospecific synthesis of sucrose, *J. Am. Chem. Soc.*, 122, 8869–8872, 2000.

28. Lemieux, R U, Hendriks, K B, Stick, R V, James, K, Halide ion catalyzed glycosidation reactions. Syntheses of α-linked disaccharides, *J. Am. Chem. Soc.*, 97, 4056–4062, 1975.
29. Lemieux, R U, Kullnig, R K, Bernstein, H J, Schneider, W G, Configurational effects on the proton magnetic resonance spectra of six-membered ring compounds, *J. Am. Chem. Soc.*, 80, 6098–6105, 1958.
30. Giese, B, *Radicals in Organic Synthesis: Formation of C–C Bonds*, Pergamon Press, Oxford, 1986.
31. Curran, D P, The design and application of free radical chain reactions in organic synthesis, *Synthesis*, 417–439; 489–513, 1988.
32. Ferrier, R J, Petersen, P M, The application of unsaturated carbohydrates to the synthesis of doubly and triply branched derivatives, *Tetrahedron*, 46, 1–11, 1990.
33. Pak, H, Canalda, I I, Fraser-Reid, B, Carbohydrates to carbocycles: a synthesis of (−)-α-pipitzol, *J. Org. Chem.*, 55, 3009–3011, 1990.
34. Binkley, R W, Photochemical reactions of carbohydrates, *Adv. Carbohydr. Chem. Biochem.*, 38, 105–193, 1981.
35. Somsák, L, Ferrier, R J, Radical-mediated brominations at ring positions of carbohydrates, *Adv. Carbohydr. Chem. Biochem.*, 49, 37–92, 1991.
36. Vasella, A, Witzig, C, α-D-Selectivity in the glycosidation by carbenes derived from 2-acetamido-hexoses, *Helv. Chim. Acta*, 78, 1971–1982, 1995.
37. Hyldtoft, L, Madsen, R, Carbohydrate carbocyclization by a novel zinc-mediated domino reaction and ring-closing olefin metathesis, *J. Am. Chem. Soc.*, 122, 8444–8452, 2000.
38. Hale, K J, Monosaccharides: use in the asymmetric synthesis of natural products, In *Rodd's Chemistry of Carbon Compounds*, Suppl. 2, Vol. 1E, 1F, 1G, Sainsbury, M, Ed., Elsevier, Amsterdam, pp. 315–437, 1993.
39. Lemieux, R U, Howard, J, The absolute configuration of *dextro*-1-deuterioethanol, *Can. J. Chem.*, 41, 308–316, 1963.
40. Zhang, J, Clive, D L J, Synthesis of (+)-furanomycin: use of radical cyclization, *J. Org. Chem.*, 64, 1754–1757, 1999.
41. Keck, G E, Wager, T T, McHardy, S F, A second generation radical-based synthesis of (+)-7-deoxypancratistatin, *J. Org. Chem.*, 63, 9164–9165, 1998.
42. Abel, S, Faber, D, Hüter, O, Giese, B, Total synthesis of soraphen $A_{1\alpha}$, *Synthesis*, 188–197, 1999.
43. Sears, P, Wong, C-H, Carbohydrate mimetics: a new strategy for tackling the problem of carbohydrate-mediated biological recognition, *Angew. Chem. Int. Ed.*, 38, 2301–2324, 1999.
44. Lemieux, R U, Venot, A P, Spohr, U, Bird, P, Mandal, G, Morishima, N, Hindsgaul, O, Bundle, D R, The binding of the B human blood group determinant by hybridoma monoclonal antibodies, *Can. J. Chem.*, 63, 2664–2668, 1985.
45. Lemieux, R U, How water provides the impetus for molecular recognition in aqueous solution, *Acc. Chem. Res.*, 29, 373–380, 1996.
46. Suami, T, Ogawa, S, Chemistry of carba-sugars (pseudo-sugars) and their derivatives, *Adv. Carbohydr. Chem. Biochem.*, 48, 21–90, 1990.
47. Stütz, A E, Ed., *Iminosugars as Glycosidase Inhibitors; Nojirimycin and Beyond*, Wiley, Weinheim, 1999.
48. Stütz, A E, Ed., *Iminosugars as Glycosidase Inhibitors; Nojirimycin and Beyond*, Wiley, Weinheim, p. 169, 1999.
49. Sears, P, Wong, C-H, Carbohydrate mimetics: a new strategy for tackling the problem of carbohydrate-mediated biological recognition, *Angew. Chem. Int. Ed. Engl.*, 38, 2317, 1999.
50. Sears, P, Wong, C-H, Carbohydrate mimetics: a new strategy for tackling the problem of carbohydrate-mediated biological recognition, *Angew. Chem. Int. Ed. Engl.*, 38, 2316, 1999.
51. Freemantle, M, Insulated molecular wires, *Chem. Eng. News*, 17, 18 November, 2002.
52. Cacialli, F, Wilson, J S, Michels, J J, Daniel, C, Silva, C, Friend, R H, Severin, N, Samori, P, Rabe, J P, O'Connell, M J, Taylor, P N, Anderson, H L, Cyclodextrin-threaded conjugated polyrotaxanes as insulated molecular wires with reduced interstrand interactions, *Nat. Mater*, 1, 160–164, 2002.
53. Bashkin, J K, guest editor, Frontiers in Carbohydrate Research, *Chem. Rev.*, 100, 4265–4711, 2000.
54. Koeller, K M, Wong, C-H, Synthesis of complex carbohydrates and glycoconjugates: enzyme-based and programmable one-pot strategies, *Chem. Rev.*, 100, 4465–4493, 2000.

55. Plante, O J, Palmacci, E R, Seeberger, P H, Automated solid-phase synthesis of oligosaccharides, *Science*, 291, 1523–1527, 2001.
56. Jung, K-H, Müller, M, Schmidt, R R, Intramolecular *O*-glycoside bond formation, *Chem. Rev.*, 100, 4423–4442, 2000.
57. Houlton, S, Sweet synthesis, *Chem. Br.*, 46–49, 2002, April.
58. Danishefsky, S J, Allen, J R, From the laboratory to the clinic: a retrospective on fully synthetic carbohydrate-based anticancer vaccines, *Angew. Chem. Int. Ed.*, 39, 836–863, 2000.
59. Schramm, V L, Development of transition state analogues of purine nucleoside phosphorylase as anti-T cell agents, *Biochim. Biophys. Acta*, 1587, 107–117, 2002.
60. Nicolaou, K C, Rodríguez, R M, Mitchell, H J, Suzuki, H, Fylaktakidou, K C, Baudoin, O, van Delft, F L, Conley, S R, Jin, Z, Total synthesis of everninomycin 13,384-1, *Chem. Eur. J.*, 6, 3095–3185, 2000.

2 Introduction to Carbohydrates

János Kuszmann

CONTENTS

2.1 DEFINITIONS AND CONVENTIONS

The generic term *carbohydrates* covers a fairly well-defined group of organic substances — namely, aliphatic polyhydroxy aldehydes and ketones — as well as compounds obtained from them by reduction of the carbonyl groups (alditols) or oxidation of one or more terminal groups to form carboxylic acids. Derivatives are formed by replacing one or more hydroxyl groups with hydrogen (deoxy derivatives) or with amino, thiol or similar heteroatomic groups. Due to the presence of multiple functional groups, carbohydrates contain several chiral centers and thus form many stereoisomers. They can be divided into two groups: the *monosaccharides*, which cannot be split into smaller subunits by treatment with aqueous acids, and the complex saccharides, which are formed by two or more monosaccharides via an acetal (glycosidic) linkage and can be split into the former by hydrolysis. The complex saccharides comprise *oligosaccharides* and *polysaccharides*, depending on the number of subunits attached to each other. With the increase in the number of subunits, there is a gradual change in chemical and physical properties. Polysaccharides are typical macromolecules; their behavior contrasts sharply with that of the monosaccharides.

Originally, the English term *carbohydrate* was used for monosaccharides, the molecular formula of which corresponded to $C_n(H_2O)_n$. Because this formula can be regarded as the hydrate of a carbon atom, this class of compounds was called *hydrate de carbone* in French and *Kohlenhydrate* in German.

The word *sugar* is also used for carbohydrates in general, but the general public understands this to mean *table sugar*, which is actually a disaccharide named sucrose (saccharose), containing two subunits: an aldehydo sugar (glucose) and a keto sugar (fructose).

In the early 19th century, individual sugars were often named for their sources, for example, grape sugar for glucose, fruit sugar for fructose and cane sugar for sucrose. Before we knew the actual structures of these compounds, their chemical names were often deduced from their physical properties. Accordingly, in 1866 Friedrich August Kekulé proposed the name *dextrose* for glucose because this compound was dextrorotatory (its solution turned the plane of the polarized light to the right). For the same reason, fructose was named *laevulose* because its solution turned the same light to the left. A consistent systematization of carbohydrates can be attributed to Emil Fischer, who began studying carbohydrates in 1880. By adopting the concepts of stereochemistry, developed in 1874 by Jacobus Henricus van't Hoff and Joseph Achille Le Bel, within 10 years Fischer had assigned the relative configuration of most known sugars, thereby establishing their structural relationships.

2.2 ACYCLIC DERIVATIVES

Carbohydrates are synthesized in plants from CO_2 and H_2O via a multistep biochemical process, the stoichiometry of which can be expressed by the following general equation:

$$CO_2 + H_2O \rightarrow CH_2O + O_2$$

Accordingly, the simplest way of "constructing" the family of sugars is to start from the general formula of the carbohydrates $C_n(H_2O)_n$, gradually increasing the number of n. This way, the first member would be CH_2O representing the well-known formaldehyde. Actually, in 1861 Alexander Butlerov showed that, under basic conditions, formaldehyde undergoes an acyloin type polymerization, resulting in a syrupy mixture of polyhydroxy aldehydes and ketones, with a sweet taste. A general scheme of this process allows us to build up the skeleton of both the polyhydroxy aldehydes and the polyhydroxy ketones (Figure 2.1).

In the first step, two formaldehyde molecules combine to yield glycoladehyde, to which the next formaldehyde can be attached either at the terminal hydroxymethyl group, resulting in an *aldose*, or at the carbonyl group, forming a *ketose*. In the first case, the central carbon atom is already a stereogenic center; accordingly, the molecule is a chiral one. The generic names of these open chain compounds depend on the number of carbon atoms incorporated in the skeleton. The aldehydes are assigned the suffix, *-ose* and the ketones are assigned the suffix *-ulose*. Thus, aldoses are classified as trioses, tetroses, pentoses, hexoses, heptoses, and so on, whereas ketoses are classified as tetruloses, pentuloses, hexuloses, heptuloses, and so on.

When the carbonyl group is reduced to an alcohol, the polyhydroxy alcohols obtained are assigned the suffix *-itol*. Monocarboxylic acids, formally derived from aldoses by replacement of the aldehyde group with a carboxy group, are termed *aldonic acids*. Monocarboxylic acids, formally derived from aldoses by replacing the terminal CH_2OH group with a carboxy groups, are termed *uronic acids*. Dicarboxylic acids, in which both terminal groups are replaced by carboxy groups, are called *aldaric acids* (Figure 2.2).

2.2.1 RULES OF THE FISCHER PROJECTION

Introduction of a chiral center will result in stereoisomers; the number of stereoisomers is 2^n where n corresponds to the number of the chiral atoms. Before their real structure was known,

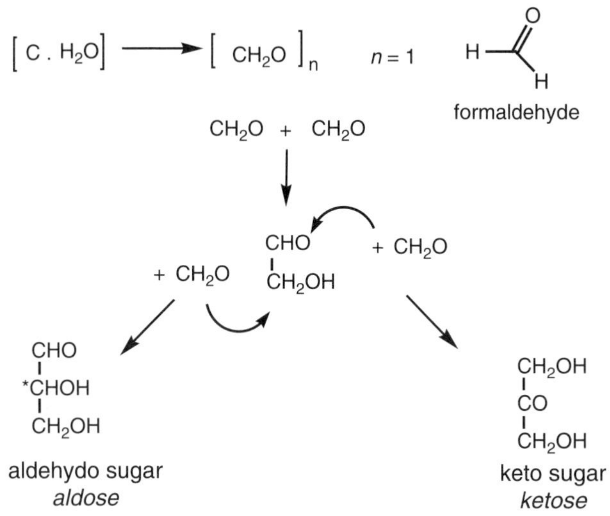

FIGURE 2.1 Formation of polyhydroxy-aldehydes and ketones from formaldehyde.

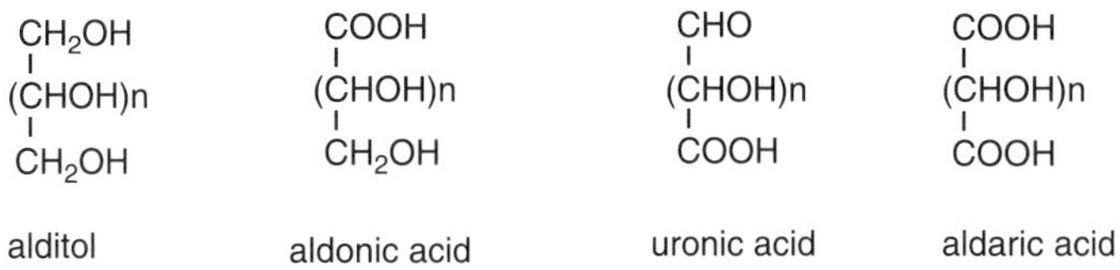

FIGURE 2.2 Carbohydrate derivatives with different oxidation states.

all isolated pentoses and hexoses were given trivial names and characterized by their optical rotation, assigned the prefix (*d*) or (+) when dextrarotatory, and (*l*) or (−) when levorotatory. Later, after it was realized that there is no strict correlation between the sign of rotation and the structure of the more complex carbohydrates, Fischer created a system for drawing sugar structures. He took glyceraldehyde as the point of reference and established the following rules (Figure 2.3):

1. The tetrahedron incorporating the stereogenic carbon atom should be placed on the paper with its edge vertical at the back and the edge in front horizontal (Figure 2.3a).
2. Move the carbon into the plane of the paper, such that the two vertical bonds will point below the plane of the paper (dotted lines) and the two horizontal bonds will point upward (full lines) (Figure 2.3b).
3. Put the carbon atoms of the chain on the vertical bonds such that the aldehyde group is on the upper end and the hydroxymethyl group is on the lower end. Numbering of these carbon atoms starts at the carbonyl group (Figure 2.3c).
4. The stereoisomer, in which the hydroxyl group of the stereogenic center points to the right, is assigned the prefix "D" (*dexter*, after the Latin for right) and the other one (the mirror image) is assigned the prefix "L" (*laevus*, after the Latin for left). According to newer regulations, these prefixes should be written in small capital letters; that is, as "D" and "L" (Figure 2.3d).
5. If more than one stereogenic center is present in the molecule, then the reference atom should have the highest number of stereogenic centers; that is, it should be the most distant from the carbonyl group.

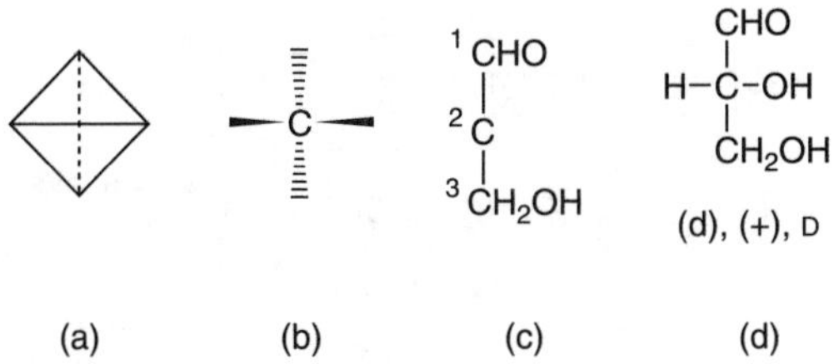

FIGURE 2.3 Basic principles of the Fischer convention.

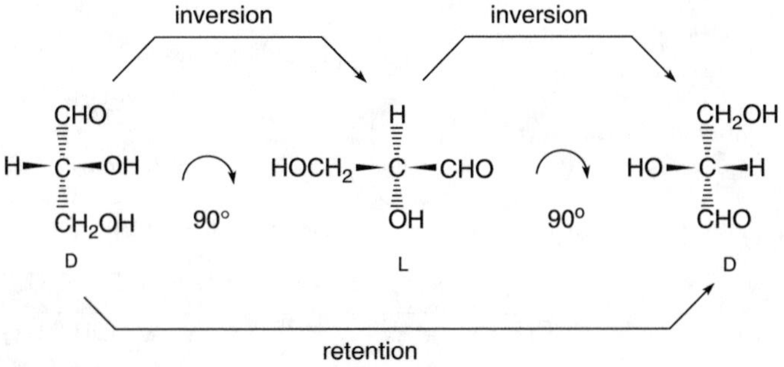

FIGURE 2.4 Transformations in the Fischer projection.

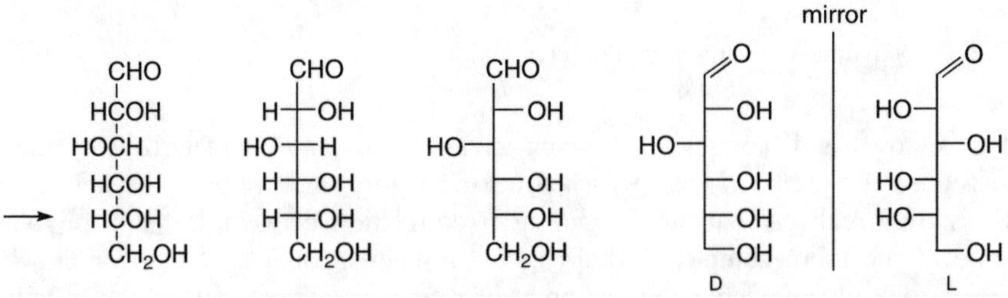

FIGURE 2.5 Different drawings of D- and L-glucose in the Fischer projection; → = stereogenic reference center.

A consequence of the Fischer projection rules is that such formulae must not be rotated 90°, because this would mean an inversion of the configuration of the stereogenic center, but they can be rotated 180° without altering the configuration (double inversion = retention) (see Figure 2.4).

For the sake of simplicity, the original Fischer structures are often drawn in abbreviated form, neglecting the chemical bonds or even the carbon atoms and the attached hydrogen atoms. Accordingly, D-glucose is sometimes drawn as shown in Figure 2.5.

2.2.2 TRIVIAL AND SYSTEMATIC NAMES

By establishing the stereochemical relationship between monosaccharides, Fischer was able to construct a "family tree" of carbohydrates. Figure 2.6 shows the tree for the D-series, together with their trivial names. Sometimes (especially in oligo- and polysaccharides), rather than full names abbreviated names consisting of the first three letters of their trivial names are used for pentoses and hexoses. The only exception is "D-Glc," used for D-glucose, because the abbreviation "Glu" had already been introduced by peptide chemists to denote glutamic acid. Figure 2.7 shows the most common C-2 carbonyl ketoses along with their trivial names.

The trivial names of the aldoses can be used as prefixes to determine the stereochemistry for other chiral compounds. However, in this case, the *-ose* suffix is omitted and the prefix is written in italics (Figure 2.8). In this respect, it should be mentioned that the chiral centers can be separated by one or more nonchiral centers. If the number of chiral centers exceeds four (e.g., in heptose, octose, nonose, etc. derivatives), a multiple configurational prefix is added to the stem name (Figure 2.9). In this case, the first four chiral centers are selected, followed by the remaining ones. However, the name of the compound starts with the prefix of the chiral center with the highest location. In the case of ketoses, if the carbonyl group is not at C-2, then its location should also be given in the name.

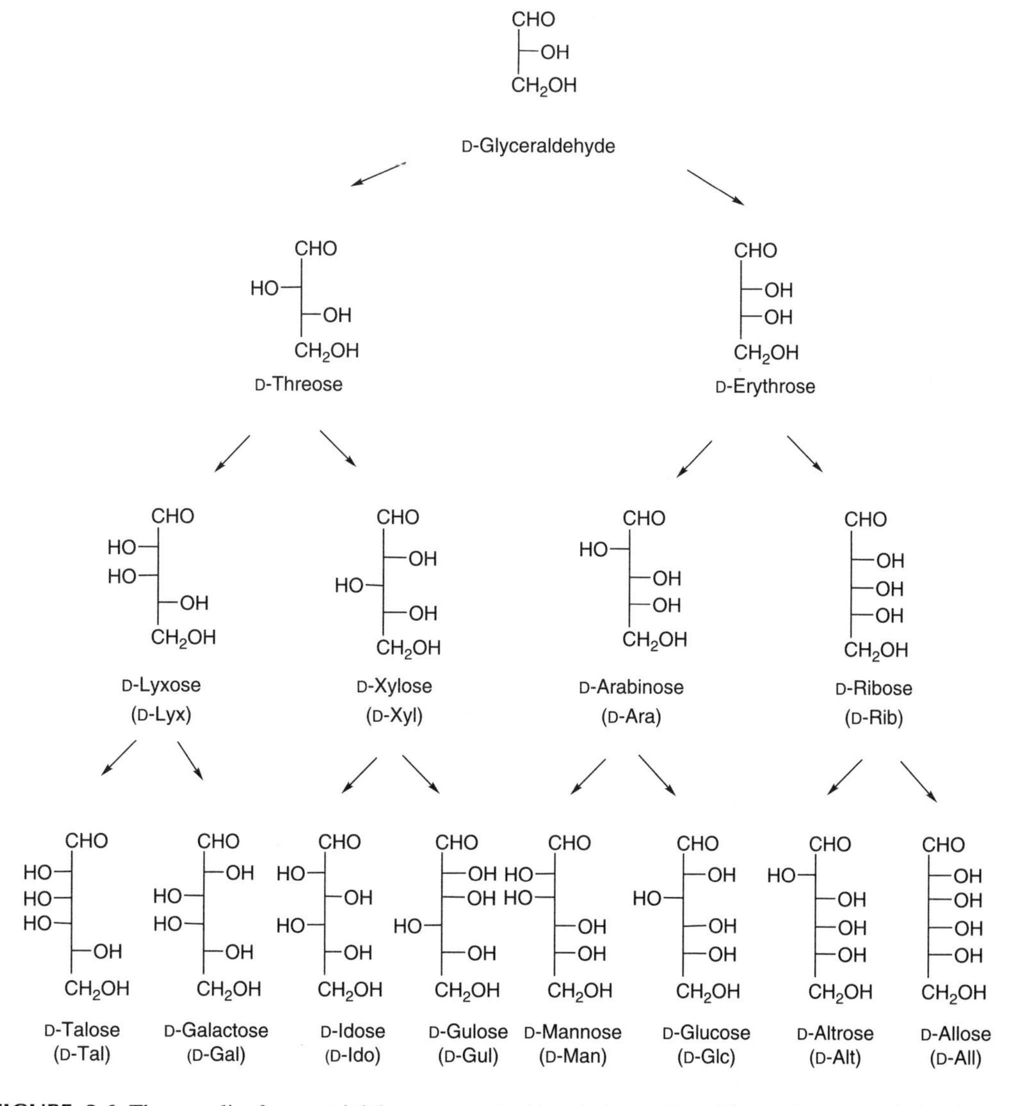

FIGURE 2.6 The acyclic forms, trivial names and abbreviations of D-aldoses drawn as their Fischer projections.

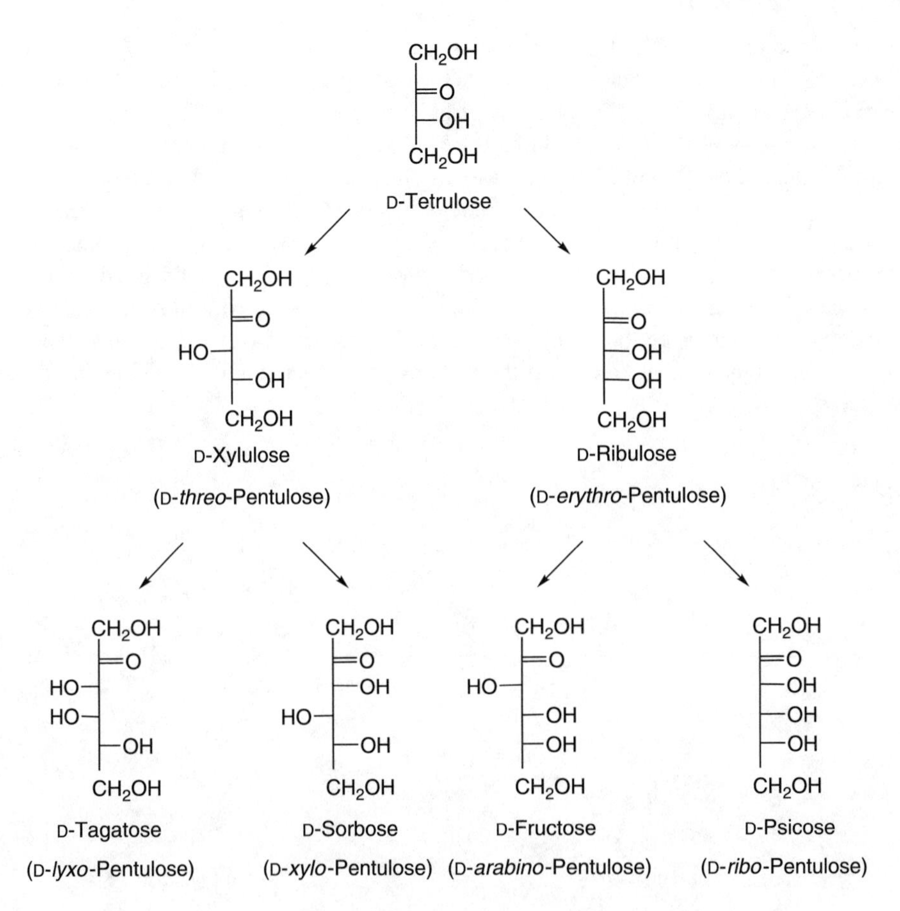

FIGURE 2.7 Trivial and systematic names of ketoses drawn as their Fischer projections.

2-Amino-3-bromo-D-*erythro*-butanoic acid 2,4-Dibromo-L-*threo*-pentanoic acid

FIGURE 2.8 The general use of carbohydrate configurational prefixes.

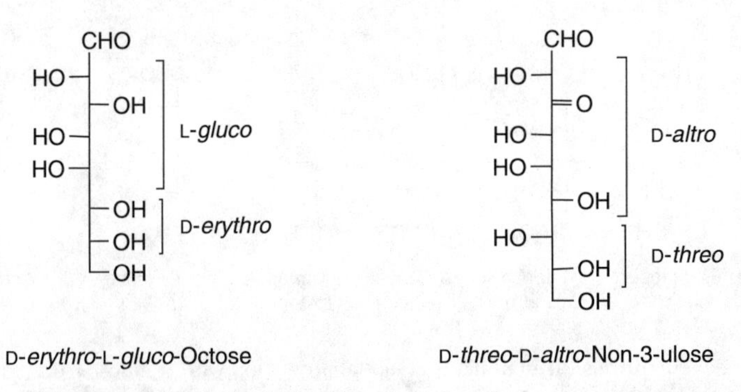

FIGURE 2.9 Systematic names of carbohydrates with more than four stereogenic centers.

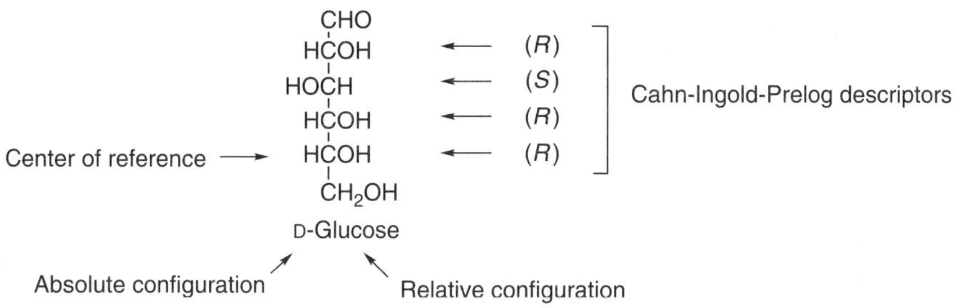

FIGURE 2.10 The absolute and relative configuration of carbohydrates.

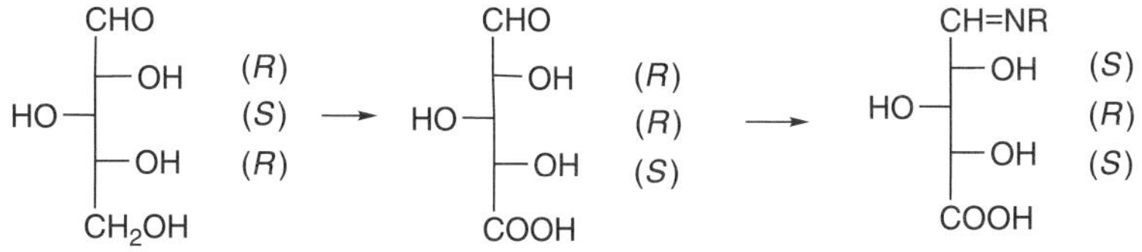

FIGURE 2.11 Change of chirality on derivatization according to the Cahn–Ingold–Prelog nomenclature.

2.2.3 Absolute and Relative Configuration

In the trivial name of any monosaccharide, the prefix D refers to the absolute configuration of the stereogenic center with the highest location (the center of reference), whereas the trivial name itself defines the stereochemistry of all other stereogenic centers in relation to the aforementioned one (Figure 2.10). Both the configuration of stereogenic centers and the names of carbohydrates can be generated using the Cahn–Ingold–Prelog nomenclature. For example, D-glucose would be (2*R*,3*S*,4*R*,5*R*)-pentahydroxyhexanal; however, the *RS*-system is generally used only for chiral substituents (e.g., a benzylidene group) attached to the carbohydrate moiety in their derivatives. A drawback of the *RS*-system is that the stereogenic centers of analogs of identical configuration, which differ only in the substitution of a nonchiral (e.g., terminal) carbon atom, can have different prefixes (Figure 2.11).

2.2.4 Depiction of the Conformation of Open Chain Carbohydrates

Fischer projections are suitable for fixing the configuration of the open chain monosacharides, but the extended form of the vertical carbon chain involves the "bending" of the C–C–C bonds to an angle of 180° when, in reality, this angle is approximately 120°. Therefore, to depict the true stereochemistry (the conformation) of sugars, the structures should be drawn in a different way. Figure 2.12 shows the conformation using the example of D-glyceraldehyde (which has only a single conformation). When the tetrahedron of the central stereogenic carbon atom of the molecule (Fischer Rule a) is turned around the vertical edge to the left such that the central carbon atom is in the plane of the paper, the C–C–C skeleton will be in the same plane, while the hydrogen will be under this plane and the hydroxyl group will be above it. In this view, the bonds of these two groups will overlap. Therefore, it is advisable to depict only one of them, usually that of the more important group. In this case, the OH group is shown, and its stereochemistry is indicated by using a wedge for its valence. A consequence of this projection is that one can turn that formula to any degree as long as the tetrahedron is not turned out of the plane (see the left "circle" in Figure 2.12). Alternatively, the central stereogenic carbon atom of the molecule can be turned around the vertical edge to the right, but, in this case, the linkage to the OH group will point under the plane of the paper and

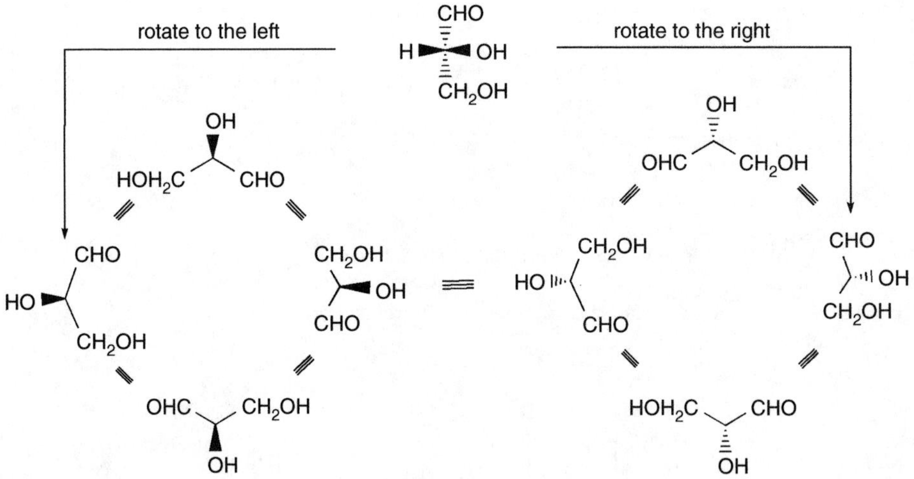

FIGURE 2.12 The conformation of D-glyceraldehyde.

should thus be indicated by a dashed line. This formula can also be turned in the plane of the paper to any degree (right "circle" in Figure 2.12). One should keep in mind that all these projections represent the same molecule, and the valence of the OH group (full or dashed line) depends on the side at which the stereogenic carbon was turned.

When the same process is applied to a molecule with two stereogenic centers (i.e., D-erythrose), the molecule can be turned to its left side, resulting in a "sickle" conformation in which both OH groups are above the plane (solid lines). Alternatively, rotation to its right side places both OH groups under the plane of the paper (dashed lines; Figure 2.13, upper row). On the other hand, it is possible to turn these two stereogenic centers in opposite directions resulting in a "zigzag" conformation. In this case, the two hydroxyl groups will be placed on the opposite sides of the plane of the paper (Figure 2.13, lower row). All four depictions represent the same compound, but in solution, the molecule adopts the "zigzag" conformation. This is especially true when the two HO groups have bulky substituents. Figure 2.14 shows analogous transformations of D-threose.

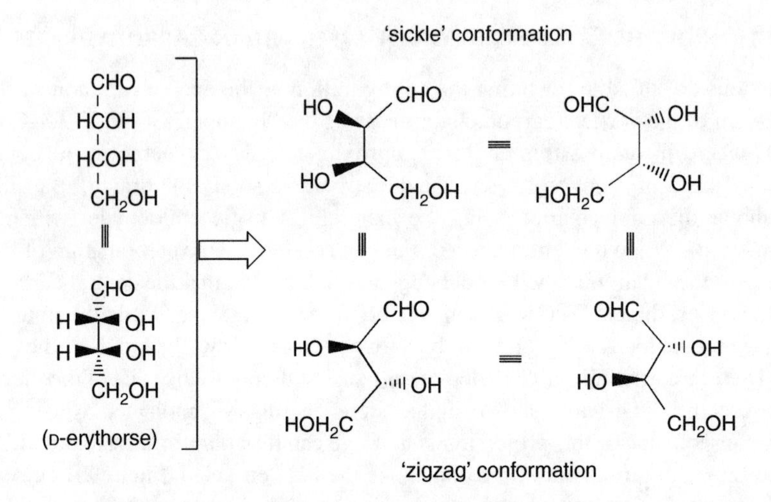

FIGURE 2.13 The Fischer projection and the conformations of D-erythrose.

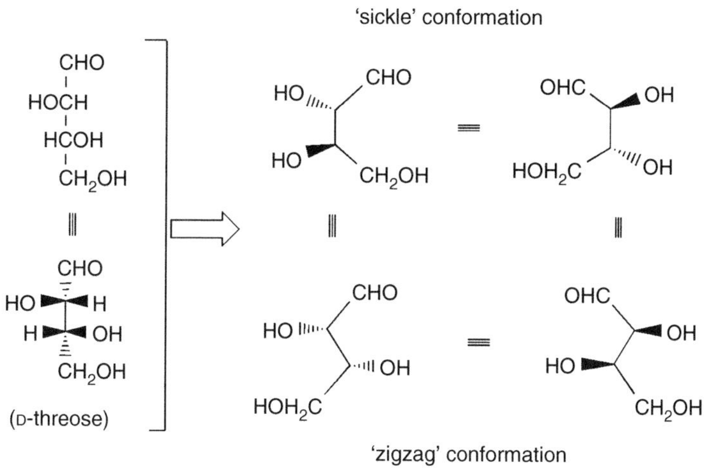

FIGURE 2.14 The Fischer projection and the conformations of D-threose.

2.2.5 The Newman Projection

Although the conformations of the open chain carbohydrates can be described correctly by the projection mentioned above, it is not easy to judge from these conformations the real arrangement (i.e., the dihedral angle of two selected vicinal substituents). This difficulty can be overcome by applying the *Newman projection,* shown in Figure 2.15, using the example of D-erythrose. When depicted in the zigzag conformation, the plane of the carbon atoms is drawn in perspective, all bonds of the stereogenic centers can be seen (a). The next step is to place a nontransparent disk across the C2–C3 bond, which will hide the central part of the linkages of the C-3 atom (b). When the molecule is turned such that the C2–C3 bond becomes vertical to the plane of the paper, the stereogenic centers will overlap. While all three connecting linkages of the C-3 atom can be seen, those of C-2 will be partly covered by the disk (c). Consequently, it is easy to distinguish between the substituents attached to these stereogenic centers. From drawing *c* it can be seen that in the zigzag conformation, the dihedral angle of the two adjacent OH groups is 180°, while in the Fischer projection, they seem to be in a *cis*-relation. Of course, as rotation around the C2–C3 bond is free, the molecule can adopt any conformation. Consequently, the dihedral angle of these two groups can take any value. This is depicted in Figure 2.16 for two vicinal substituents (X and Z), where X, which belongs to C-2, is kept in place, and Z, attached to C-3, is rotated. All arrangements in which these two substituents are on the same side of a plane dissecting horizontally the C2–C3 axis are in a *syn* relation, whereas those on the opposite side occupy an *anti* position. When they are in the same horizontal plane (i.e., their dihedral angle is 0 or 180°), this arrangement is called *periplanar*, and any other arrangement is called *clinal*. Accordingly, the six positions shown in Figure 2.16 are called *synclinal* (two of them), *synperiplanar*, *anticlinal* (two of them) and *antiperiplanar*.

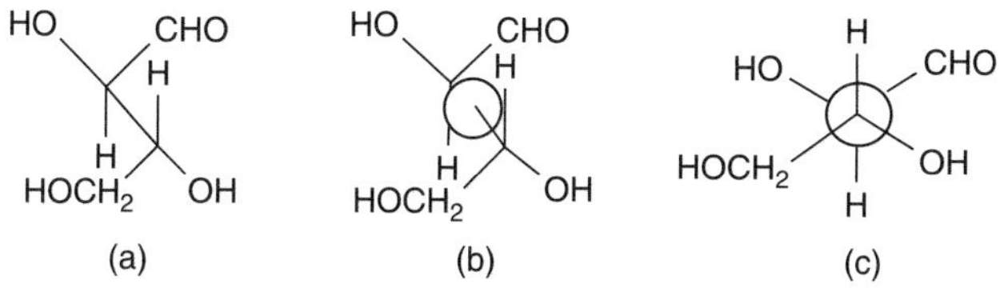

FIGURE 2.15 Construction of the Newman projection of D-erythrose.

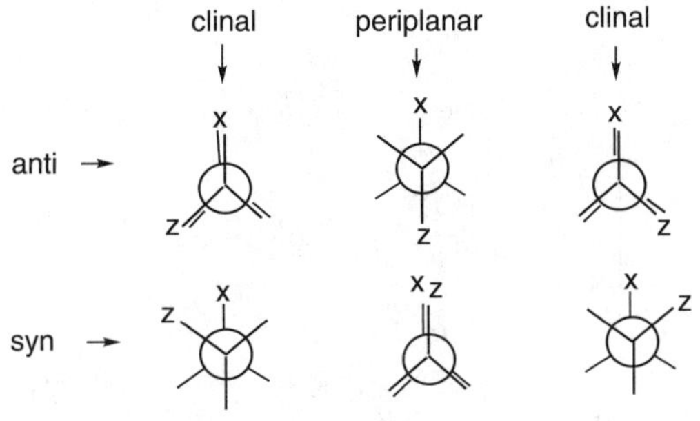

FIGURE 2.16 Conformational arrangements of two vicinal substituents.

2.3 CYCLIC DERIVATIVES

It is well known that carbonyl groups react readily with water and alcohols to form hydrates and hemiacetals, respectively (Figure 2.17). As in carbohydrates, a carbonyl and a hydroxyl group are both present. The "alcohol addition" takes place as an intramolecular reaction, and hemiacetals are formed via ring closure. In the general formula in Figure 2.17, *n* is most often 2 or 3. Consequently, a five-membered *furan* (named for the heterocyclic compound tetrahydrofuran) or a six-membered *pyran* structure (named for tetrahydropyran) results. When $n = 4$, a seven-membered ring called *septan* is formed, but such structures are quite rare among carbohydrates.

2.3.1 Rules of the Fischer Projection

Extending the Fischer projection to cyclic forms of sugars, a long, multiple-bended bond has to be drawn between the hydroxyl group involved in the ring and the former carbonyl group (Figure 2.18). As the bond is a continuation of the carbon skeleton, it must enter the carbonyl group from above. Consequently, the hydrogen in aldoses, and the terminal CH_2OH group in ketoses should be moved to one side, and the newly formed hydroxyl group should be moved to the other side of the carbon atom of the original carbonyl group. The cyclization results in the formation of a new chiral center designated the *anomeric* carbon. The introduction of this new stereogenic center results in two diastereomers. The prefix α is assigned to the diastereomer in which the newly formed hydroxyl group is cis-related to the hydroxyl group of the most distant stereogenic center (which determines the configuration) in the Fischer projection. The other diastereomer (or *anomer*) in which these two hydroxyl groups are related trans, is assigned the prefix β. The same holds for

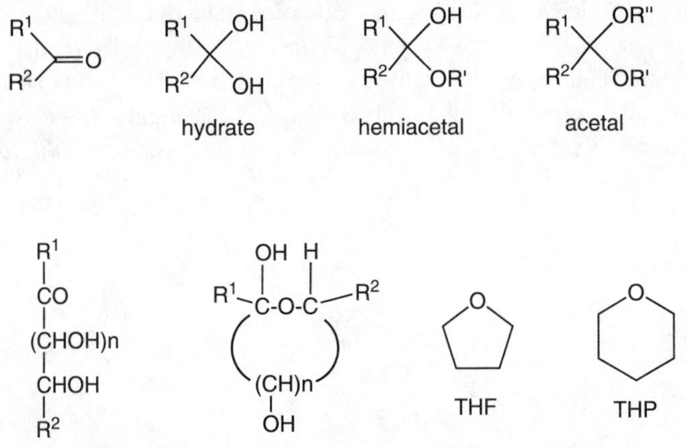

FIGURE 2.17 Cyclic hemiacetal structures; THF = tetrahydrofuran; THP = tetrahydropyran.

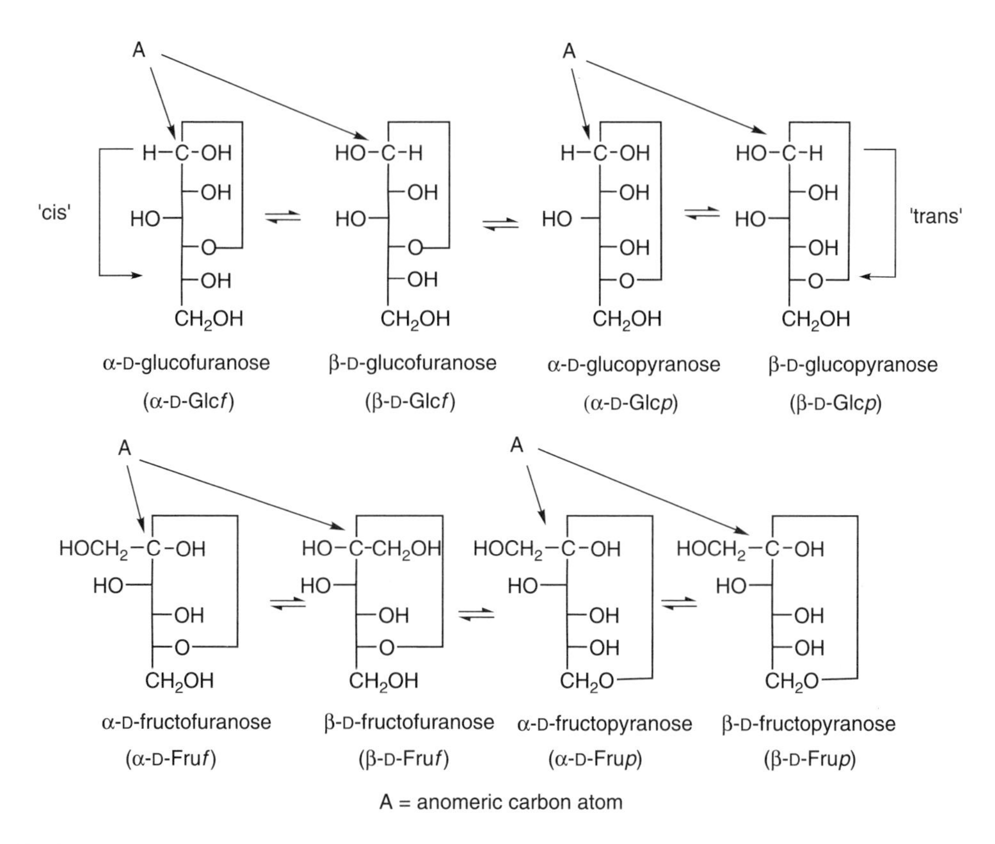

FIGURE 2.18 The cyclic hemiacetal structures of D-glucose and D-fructose.

the L-isomers. Consequently, the mirror image (called *enantiomer* or *antipode*) of any α-D-stereoisomer is the corresponding α-L-isomer (and not the β-L). When using the abbreviated names of the carbohydrates, the furanose and pyranose forms can be indicated by adding an italic *f* or *p* respectively, to the three-letter symbol of the name.

2.3.2 Mutarotation

When a crystalline optically active compound is dissolved in a solvent, and this solution shows a time-dependent change of the optical rotation approaching a certain equilibrium value, this phenomenon is called *mutarotation*. Mutarotation is typical for crystalline free sugars, a freshly prepared solution of which (in water or other solvents) illustrates this phenomenon. In crystalline state, each compound is a discrete stereoisomer in which both the ring size and the configuration at the anomeric center are fixed. In solution, however, the hemiacetal ring opens and an equilibration between anomers, and eventually between furanose and pyranose forms, takes place. This is accompanied by a change in optical rotation, because each component of this equilibrium has its own specific rotation. The final value of the optical rotation of the solution will therefore be a weighted average sum of the individual specific rotations. In the case of glucose, the α-D-glucopyranose (which can be obtained on crystallization from acetic acid) has a specific rotation of +112°, while the crystalline β-D-glucopyranose (which can be obtained on crystallization from aqueous ethanol) has a specific rotation of +19°. When either of them is dissolved in water, at 20°C an equilibrium value of +52.7° is reached after about 3 h. Because in this equilibrium only the two pyranose anomers are appreciably present, their ratio can be calculated from the equilibrium rotation and corresponds to α:β = 38:62. Concentrations of the individual isomers of various sugars in the equilibrium depend on the structure of the individual monosaccharide as shown in

TABLE 2.1
Distribution of the Different Isomers of Some Monosaccharides in Aqueous Solution at Equilibrium

	Pyranose		Furanose		
Carbohydrate	**α-Anomer (%)**	**β-Anomer (%)**	**α-Anomer (%)**	**β-Anomer (%)**	**Acyclic (%)**
Glucose	38.0	62.0	0.5	0.5	0.002
Mannose	65.6	34.5	0.6	0.3	0.005
Galactose	30.0	64.0	2.5	3.5	0.02
Rhamnose	65.5	34.5	0.6	0.3	0.005
Fructose	2.5	65.0	6.5	25.0	0.8
Xylose	36.5	63.0	0.3	0.3	0.002
Ribose	31.5	58.5	6.4	13.5	0.05

Table 2.1. It should be noted that in the D-series the optical rotation of the α-anomers is always a more positive value than that of the corresponding β-anomer (Hudson's rule).

2.3.3 The Haworth Projection

In order to avoid unrealistic bended linkages in the Fischer projection, in 1929 Norman Haworth suggested a representation in which the ring is oriented almost perpendicular to the plane of the paper, but viewed from slightly above. As a consequence, the edge closer to the viewer is drawn below the more distant edge — often with a heavy line. The ring oxygen should be placed at the upper right-hand corner for pyranoses (Figure 2.19) and at the top of the five-membered ring for furanoses (Figure 2.20). The carbon atoms of the ring should be numbered clockwise. To transform a Fischer projection into a Haworth representation, the original Fischer formula (Figure 2.19a) has to be modified the following way:

1. In the pyranose form the oxygen at C-5 must be exchanged with the terminal hydroxymethyl group.
2. Since, according to the Fischer convention, this exchange would mean an inversion of configuration, in order to restore the original configuration, a second substitution (double inversion = retention) has to be carried out at the same carbon atom (i.e., the hydrogen and the hydroxymethyl group also have to be interchanged (Figure 2.19b). In Haworth formulae of D-pyranoses, the anomeric hydroxyl and the terminal hydroxymethyl groups are called *trans* in the α-anomers and *cis* in the β-anomers (Figure 2.19c). For simplicity, the hydrogens are usually not shown (Figure 2.19d). The same rules apply to the furanose structures (Figure 2.20).

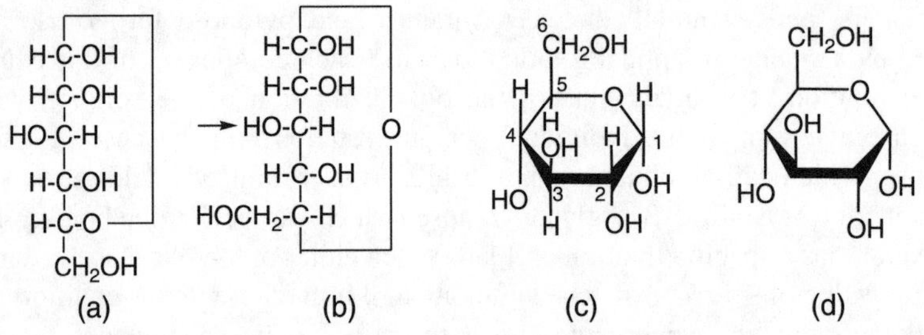

FIGURE 2.19 The Fischer (a), modified Fischer (b), Haworth (c) and simplified Haworth (d) representations of α-D-glucopyranose.

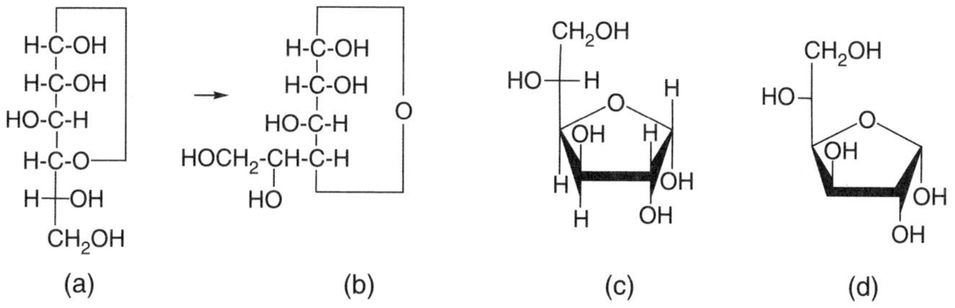

FIGURE 2.20 The Fischer (a), modified Fischer (b), Haworth (c) and simplified Haworth (d) representations of α-D-glucofuranose.

2.3.4 The Mills Projection

John Mills applied the method described above with open chain derivatives (see Section 2.2.4) to cyclic structures, putting the atoms of the hemiacetal ring into the plane of the paper and the attached substituents above (thickened bonds) or below (dashed bonds) this plane. For simplicity, only substituents other than hydrogen are depicted (Figure 2.21). In the case of the α-D-glucopyranose (Figure 2.21a), all six atoms of the ring are shown as in the plane of the paper, and the hydroxymethyl group is attached to C-5 by a wedge. According to Miller's, original suggestion the five-membered ring of the corresponding furanose should be drawn the same way (Figure 2.21b), but in this structure, C-4 carries a further chiral center (C-5) as substituent, the stereochemistry of which should also be specified. As the chirality of any atom should be indicated by one stereo bond only, the stereochemistry of C-4 has to be given by depicting its stereochemistry inside the ring (thickened bond toward the ring oxygen). Nevertheless there are still two possible depictions of the molecule (Figure 2.21c and Figure 2.21d), differing only in the conformation of the side chain.

The structures of all pentoses and hexoses in their furanose and pyranose forms are shown in Figure 2.22.

2.3.5 The Reeves Projection

Both the Haworth and Mills representations imply a planar ring, but, in fact, monosaccharides assume nonplanar conformations. Richard Reeves first applied the nonplanar ring conformations of cyclohexane to depict the stereochemistry of pyranoses but kept Haworth's guidelines (i.e., the stereoview of the ring and the position of the ring oxygen). Originally, the two stable chair conformations (Figure 2.23a) and (Figure 2.23b) were taken into consideration, and they were named *C*1 and 1*C*. To avoid confusion, the positions of the C-1 and C-4 atoms in these structures were taken as reference points, and marked as $^4C_1$ and $_4C^1$. According to the current convention,

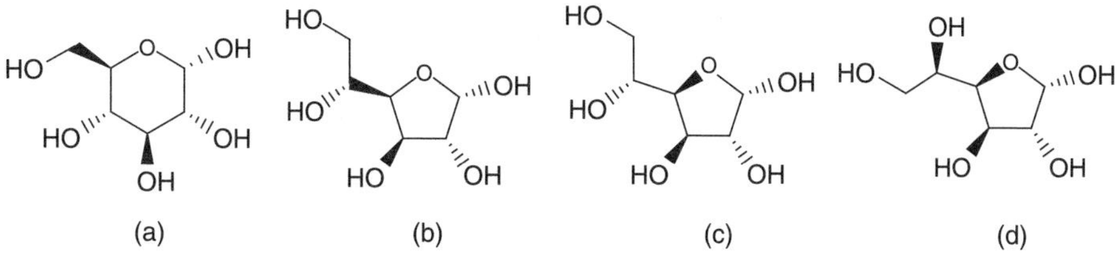

FIGURE 2.21 Mills projections of α-D-glucopyranose (a) and α-D-glucofuranose (b, c and d).

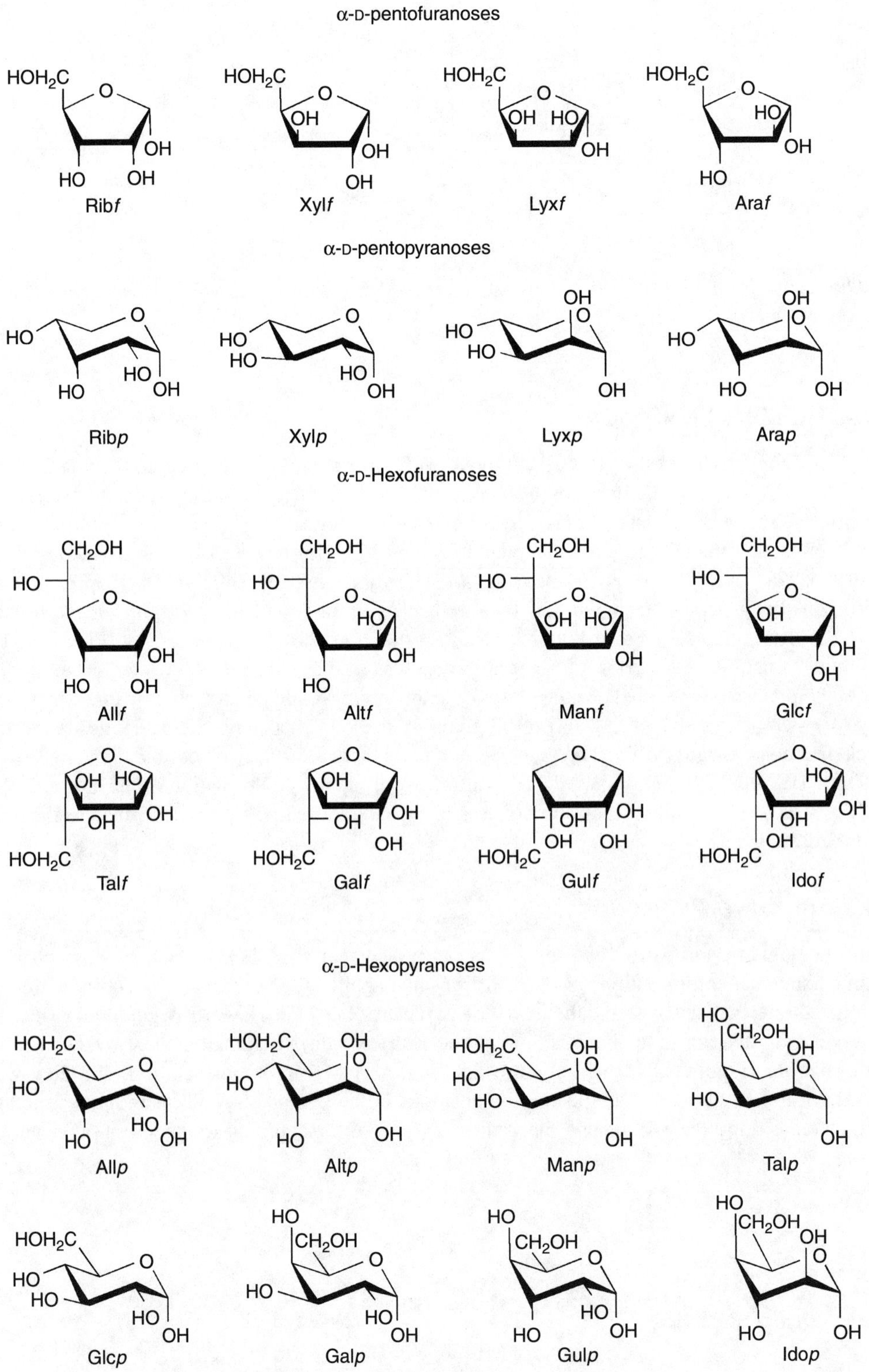

α-D-pentofuranoses

Rib*f* Xyl*f* Lyx*f* Ara*f*

α-D-pentopyranoses

Rib*p* Xyl*p* Lyx*p* Ara*p*

α-D-Hexofuranoses

All*f* Alt*f* Man*f* Glc*f*

Tal*f* Gal*f* Gul*f* Ido*f*

α-D-Hexopyranoses

All*p* Alt*p* Man*p* Tal*p*

Glc*p* Gal*p* Gul*p* Ido*p*

FIGURE 2.22 Furanose and pyranose structures of all α-D-pentoses and α-D-hexoses.

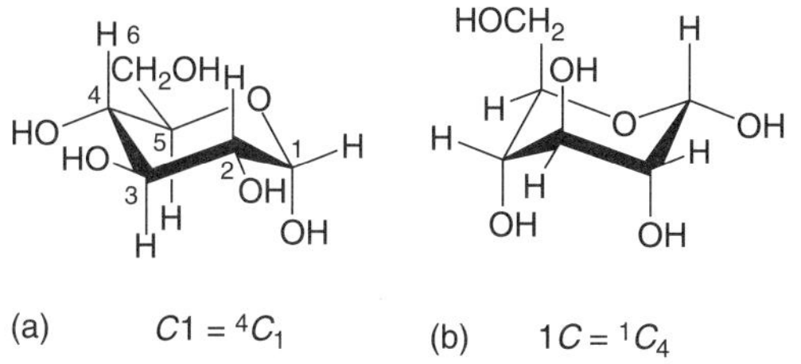

(a) $C1 = {}^4C_1$ (b) $1C = {}^1C_4$

FIGURE 2.23 The two chair conformations of α-D-glucopyranose.

the number of the superscript has to be mentioned first; consequently, the symbols to be used are ${}^4C_1$ and ${}^1C_4$ (Figure 2.23).

2.3.6 Conformations of the Six-Membered Rings

It is obvious that although the two chair conformations differ in energy, this energy barrier is usually not high enough to prevent conformational mobility. Although the molecule will usually adopt one conformer, depending on the stereochemistry of the individual monosaccharide, the other conformer might be present in an equilibrium. Furthermore, interconversion of the two chair conformations involves the rotation of the different ring atoms around the connecting bonds, and this stepwise process results in several conformations that differ in terms of the position of the ring atoms regarding the plane of reference. The main conformations were given individual names and descriptors, such as *half-chair* (*H*), *boat* (*B*), *envelope* (*E*), and *skew* (*S*). These variants are distinguished by the locants of those ring atoms that lie outside a reference plane. The locants of ring atoms that lie outside the reference plane from which numbering appears clockwise (i.e., "above" the plane) are written as superscripts that precede the descriptor; those locants that lie on the other side (i.e., "below" the plane) are written as subscripts that follow the descriptor. In the chair conformation (*C*), the reference plane is defined by two parallel ring sides, chosen such that the lowest-numbered carbon atom in the ring (usually C-1) is exoplanar. In the boat conformation (*B*), the plane is defined by the two parallel sides of the boat. Consequently, both out-of-plane atoms are either above or under this plane. In the skew conformation (*S*), three adjacent atoms and the remaining nonadjacent one define the reference plane, and one of the two extraplanar atoms should be the lowest-numbered one. The reference plane of the half-chair conformation (*H*) is defined by four adjacent coplanar atoms with the remaining two atoms on opposite sides of this plane. In the envelope conformation (*E*), the reference plane is defined by five adjacent coplanar atoms (Figure 2.24). All of these conformations can be arranged into a complex map representing a continuous transformation from one into the other (Figure 2.25).

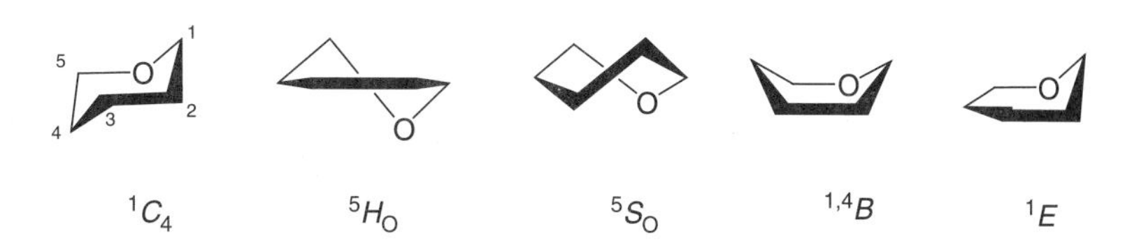

FIGURE 2.24 Variants of the six-membered ring conformations: chair (*C*), half-chair (*H*), skew (*S*), boat (*B*), and envelope (*E*).

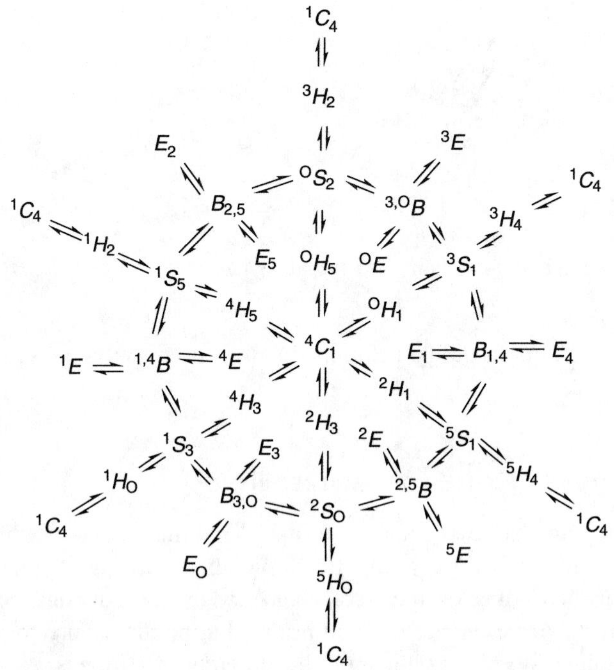

$^{1}C_4$
$^{3}H_2$
E_2 ^{3}E
$^{O}S_2$ $^{1}C_4$
$^{1}C_4$ $B_{2,5}$ $^{3,O}B$ $^{3}H_4$
$^{1}H_2$ E_5 $^{O}H_5$ ^{O}E $^{3}S_1$
$^{1}S_5$
$^{4}H_5$ $^{O}H_1$
$^{4}C_1$
^{1}E $^{1,4}B$ ^{4}E E_1 $B_{1,4}$ E_4
$^{2}H_1$
$^{4}H_3$
$^{2}H_3$ ^{2}E $^{5}S_1$
$^{1}S_3$ E_3
$^{1}H_O$ $^{5}H_4$
$B_{3,O}$ $^{2}S_O$ $^{2,5}B$
$^{1}C_4$ $^{1}C_4$
E_O ^{5}E
$^{5}H_O$
$^{1}C_4$

FIGURE 2.25 Conformational map of pyranoid ring interconversions.

2.3.7 Conformations of the Five-Membered Rings

The five-membered furanose ring can adopt two main conformations: the envelope (*E*), in which the reference plane is formed by four adjacent coplanar atoms with only one atom outside it (above or below the plane); and the twist (*T*), where the reference plane is formed by three adjacent coplanar atoms with the remaining two atoms placed on the opposite sides of it (Figure 2.26). Accordingly, there are ten different envelope conformations (^{1}E, E_1, ^{2}E, E_2, ^{3}E, E_3, ^{4}E, E_4, ^{O}E, and E_O) and ten different twist conformations ($^{O}T_1$, $^{1}T_O$, $^{1}T_2$, $^{2}T_1$, $^{2}T_3$, $^{3}T_2$, $^{3}T_4$, $^{4}T_3$, $^{4}T_O$, and $^{O}T_4$), which can be interconverted by pseudorotation of the furanoid ring (Figure 2.27).

2.3.8 Conformations of the Seven-Membered Rings

The seven-membered septanoid ring can adopt four conformations called chair (*C*), boat (*B*), twist-chair (*TC*) and skew (*S*). Representatives of each type, in which the ring oxygen occupies the upper right-hand corner, are depicted in Figure 2.28. In the chair and boat conformations, the reference plane is formed by four adjacent coplanar atoms, which are separated by one and two atoms, respectively. These separating atoms are placed on the opposite sides of the plane in the chair

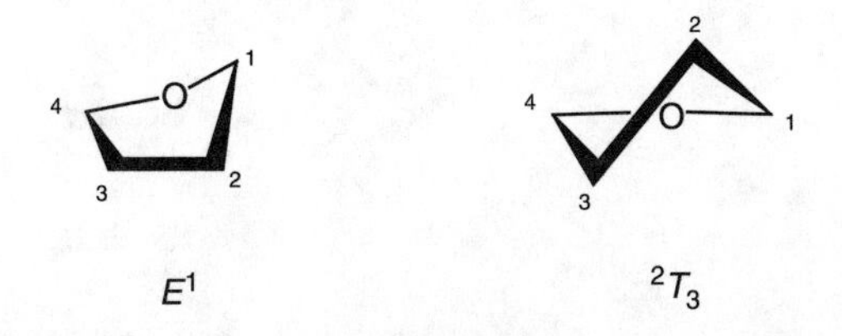

FIGURE 2.26 Variants of the five-membered ring conformations: envelope (*E*) and twist (*T*).

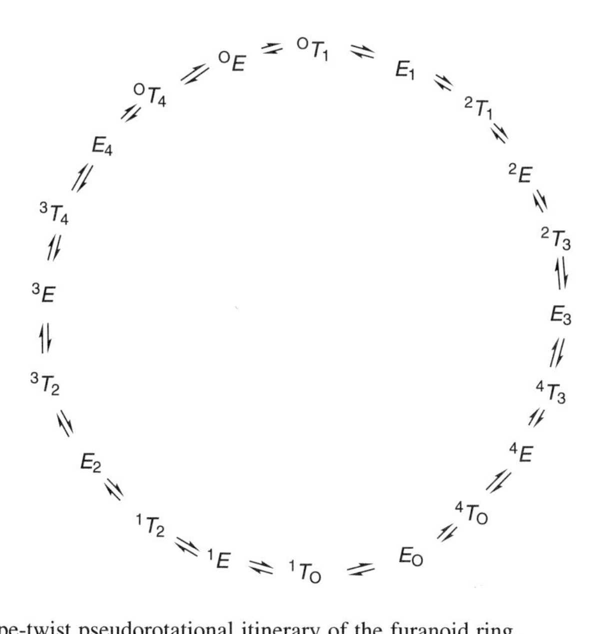

FIGURE 2.27 Envelope-twist pseudorotational itinerary of the furanoid ring.

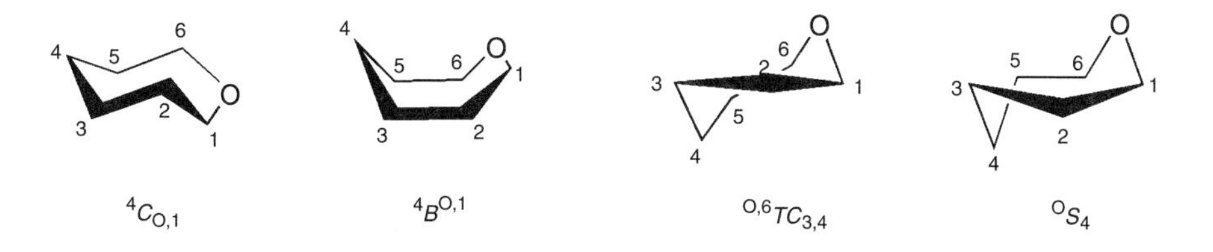

FIGURE 2.28 Four variants of the seven-membered ring conformations: chair (*C*), boat (*B*), twist-chair (*TC*) and skew (*S*).

conformation and on the same side in the boat conformation. In the twist-chair conformation, the reference plane is chosen so that three adjacent atoms and the midpoint of the bond connecting the opposite two atoms are in the same plane, while each of the neighboring pairs of the remaining four atoms is placed on the opposite side of this plane. In the skew conformation, three adjacent atoms and the opposite two atoms form the reference plane, while the remaining atoms are placed on the opposite side of it. Because the seven-membered ring is very flexible, the different conformations are interconvertible via pseudorotation. The chair/twist-chair pseudorotational itinerary of the septanoid ring is shown in Figure 2.29. We know from experience that none of the aldohexoses exists to a large degree in aqueous solution as a septanose. However, when the hydroxyl groups at C-4 and C-5 are blocked, for example, as in 2,3,4,5-tetra-*O*-methyl-D-glucose, then septanose rings may be formed.

2.3.9 Conformations of Fused Rings

Of the theoretically possible bicyclic systems arising via combinations of five- and six-membered rings, we discuss only three of the most common in carbohydrate chemistry here. These are the 5 + 5, 5 + 6 and 6 + 6 variations, in which the two fused rings have only two adjacent common atoms, representing the [3,3,0]octane, the [4,3,0]nonane and the [4,4,0]decane systems. Theoretically, all three can form *cis* or *trans* fused systems, but because of the high steric strain of a *trans* fused 5 + 5 bicyclic system, this combination can be disregarded (Figure 2.30).

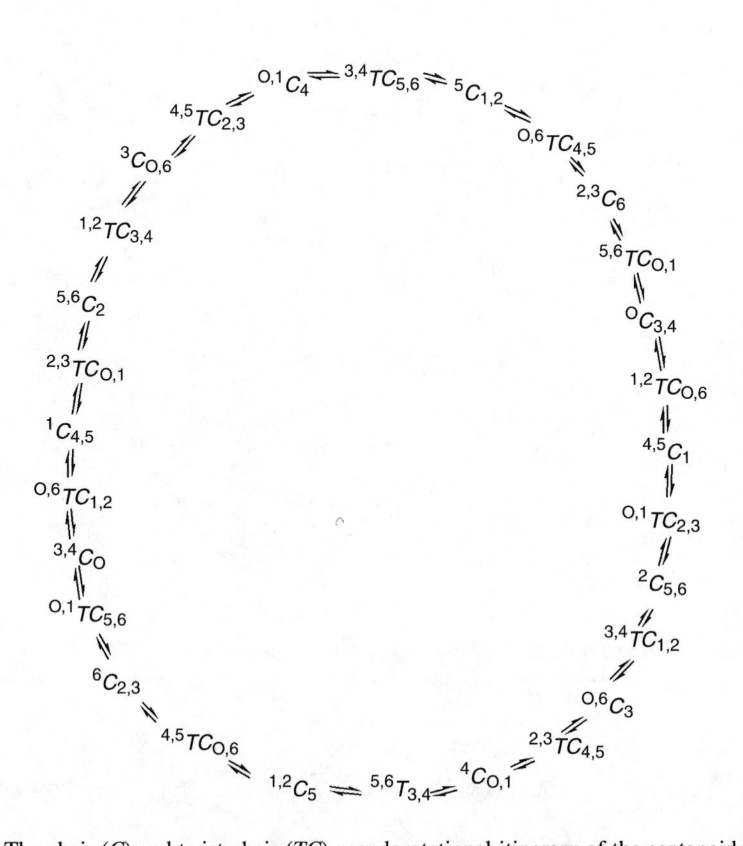

FIGURE 2.29 The chair (*C*) and twist-chair (*TC*) pseudorotational itinerary of the septanoid ring.

When two five-membered rings are *cis* fused, for example, as in 1,4:3,6-dianhydro-hexitols, the common bridge atoms restrict the mobility of the individual atoms, and therefore only the combination of the two envelope conformations has to be considered (Figure 2.31).

In a combination of a five- and a six-membered ring, both the *cis* and the *trans* anellations are viable. However, a precondition of the latter is that the six-membered ring must be able to adopt a conformation in which the connecting bonds are diequatorially oriented (Figure 2.32). As a consequence, the pseudorotation of the six-membered ring becomes restricted, which is energetically unfavored. Therefore, if there are three adjacent hydroxyl groups on a pyranose

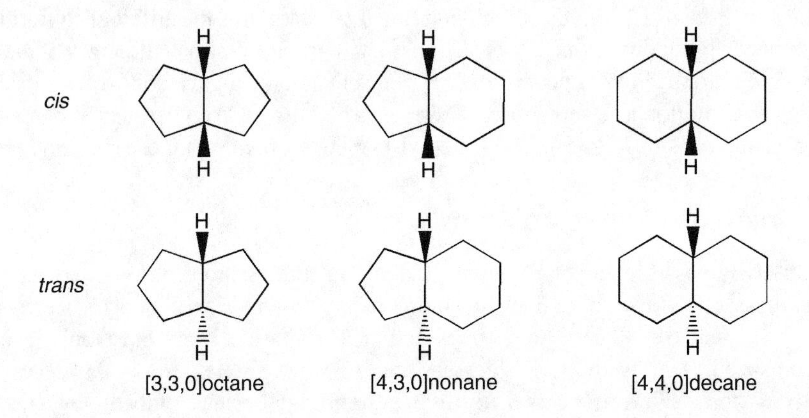

FIGURE 2.30 *Cis* and *trans* fused bicyclic systems.

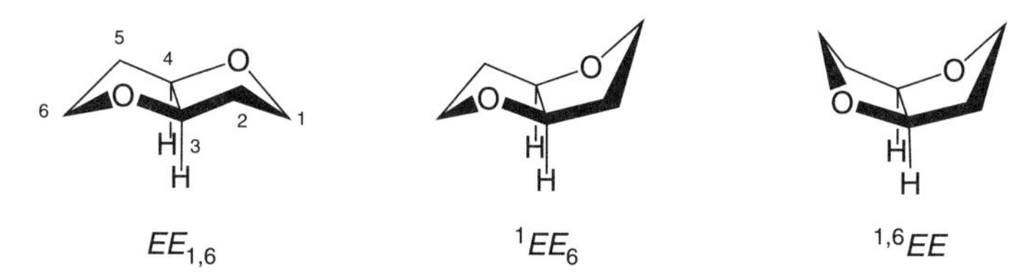

FIGURE 2.31 Conformations of two *cis*-fused five-membered rings.

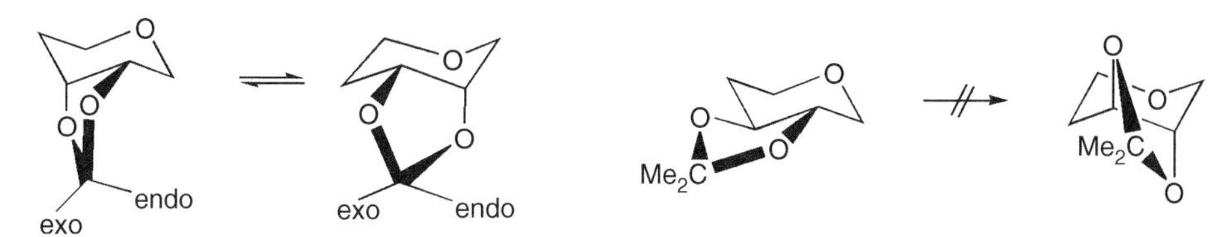

FIGURE 2.32 *Cis* and *trans* anellated five- and six-membered rings.

ring, then five-membered acetals are usually formed by connecting the *cis* related ones. Note that, in this case, the environment of the substituents of the acetal carbon differs significantly since one substituent points under the plane of the six-membered ring (it is called *endo*), whereas the other points away from it (*exo*).

For the combination of two six-membered rings, similar restrictions on the five- and six-membered anellation exist. In the *trans* fused system, the six-membered ring must be able to adopt a conformation in which the connecting bonds are diequatorially oriented. The most common carbohydrate derivatives belonging to these types of compounds are the cyclic *O*-benzylidene acetals, in which the bulky phenyl substituent prefers the equatorial arrangement, and thereby restricts the conformational mobility (Figure 2.33).

2.3.10 Steric Factors

The benefit of the Reeves projection of the pyranose ring is that it approaches the real shape of the molecule which enables us to determine the dihedral angles and the relative distances of the different substituents. This is especially useful for interpreting the NMR spectra, which depends greatly on these steric factors, especially the coupling constants of the protons attached to ring carbons. In accordance with the Karplus equation, the general rule $J_{a,a} > J_{a,e} \sim J_{e,e}$ is valid for the coupling constants of vicinal protons because their dihedral angles are 180 or 60°, with the respective coupling constants of 7 to 10 Hz and 1 to 4 Hz, respectively. When the chair conformation of a given pyranose isomer is transfomed into its counterpart ($^{1}C_4 \leftrightarrow {}^{4}C_1$), the axially oriented substituents will be moved to an equatorial position and vice versa (Figure 2.34). That means that the relative steric arrangement of any vicinal substituent will also be changed; that is, the *trans* related substituents will undergo an axial,axial ↔ equatorial,equatorial interchange, while the *cis* related substituents will undergo an

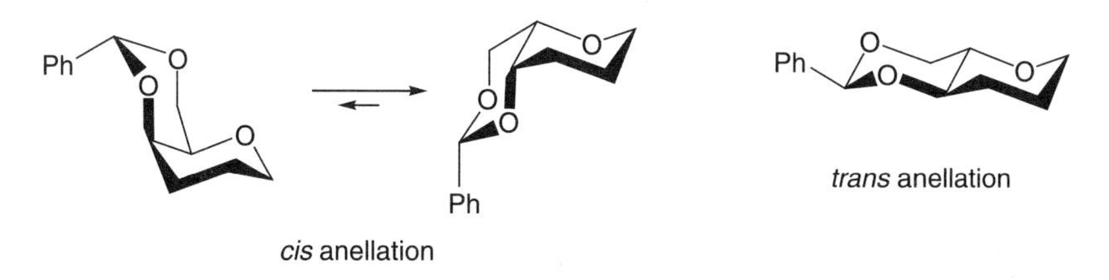

FIGURE 2.33 *Cis* and *trans* anellated 4,6-*O*-benzylidene-pyranoses.

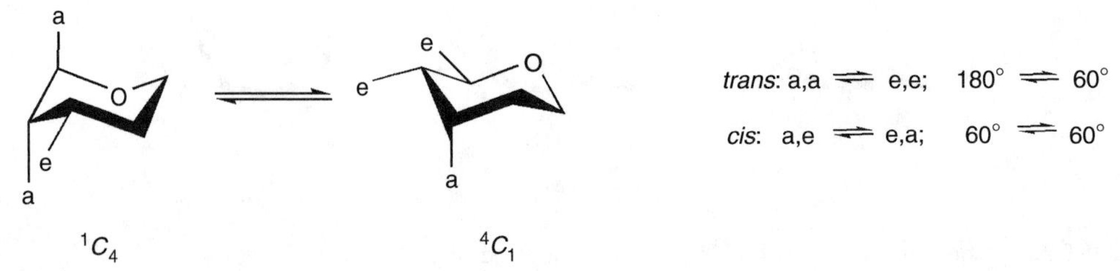

trans: a,a ⇌ e,e; 180° ⇌ 60°

cis: a,e ⇌ e,a; 60° ⇌ 60°

FIGURE 2.34 Change of dihedral angles upon changing the ring conformation.

axial,equatorial ↔ equatorial,axial one. However, the former interconversion means a change in the torsional angle from 180 to 60°, while in the latter case the torsional angle remains practically unchanged (60°). Due to this change in the dihedral angles, which is reflected in the value of their coupling constants, and with the help of NMR data, it is relatively easy to establish the dominant conformer of any hexopyranose (e.g., the $^4C_1$ and $^1C_4$ conformations of D-glucose and D-idose in Figure 2.35), or the relative proportion of these two conformers in equilibrium.

Another important fact is that the conformation adopted by any pyranoid derivative depends on both steric and electronic interactions. In this section, only the former will be discussed. Two main steric factors influence the stability of any conformation: (1) the destabilizing effect of an axial hydroxyl, especially hydroxymethyl group, and (2) the so-called 1,3-diaxial interaction called the Hassel–Ottar effect. Because the terminal hydroxymethyl group is quite a bulky substituent, it tends to occupy an equatorial position in most hexopyranoses. Consequently, all D-isomers exist predominantly as the $^4C_1$ conformers (see D-glucose in Figure 2.35). On the other hand, the presence of two axial groups in 1,3-position on the same side of the pyranose ring will exert an even greater destabilizing effect. However, these two effects can compete with each other (e.g., in D-idopyranose; see Figure 2.35) where, in the $^4C_1$ conformation, the hydroxyl groups at C-2 and C-4 are in an energetically unfavored 1,3-diaxial arrangement, whereas the $^1C_4$ conformation is destabilized by the axially oriented hydroxymethyl group. As a result, this molecule exists in a conformational equilibrium.

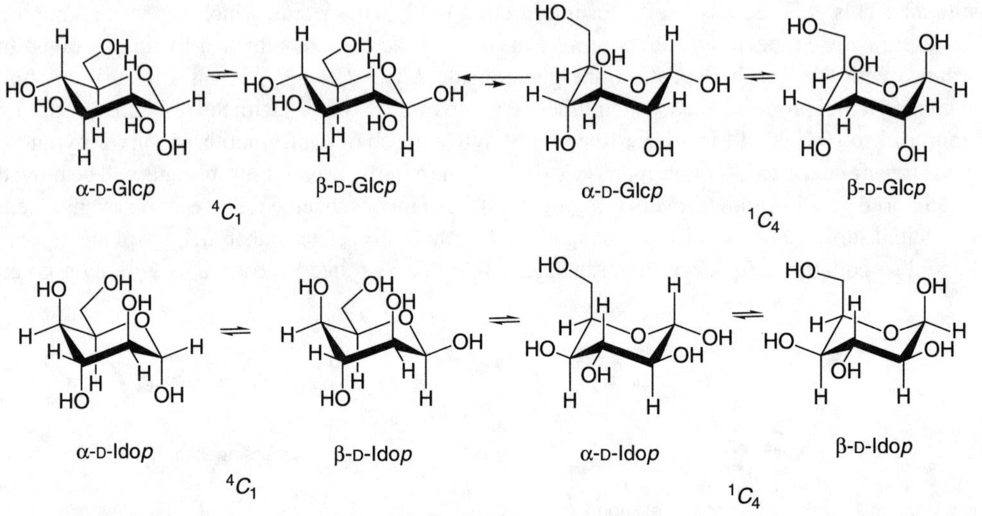

FIGURE 2.35 The two chair conformations of α- and β-D-glucopyranose and α- and β-D-idopyranose.

2.3.11 THE ANOMERIC AND *EXO*-ANOMERIC EFFECTS

As mentioned in the previous section, the axial orientation of a hydroxyl group is unfavored. Consequently, D-glucopyranose should be expected to form the β-anomer, in which all substituents are equatorially oriented in the $^4C_1$ conformation. However, in aqueous solution, the α- and β-anomers are present in a ratio of approximately 1:2. Thus, there must be another operating effect that competes successfully with the aforementioned destabilizing one. Raymond Lemieux called this the *anomeric effect*. It is generally valid for all molecules with two (or more) heteroatoms linked to a tetrahedral center. The unusual conformational behavior of this class of compounds containing the C–X–C–Y moiety, where X = N, O or S, and Y = Br, Cl, F, N, O or S, is called as the generalized anomeric effect. Pyranoses and their derivatives also belong to this class of compounds. The magnitude of the anomeric effect is fairly closely related to the polarity of the –C–Y bond, which increases in proportion to the electronegativity of the Y substituent. This is clearly demonstrated by the equilibrium concentration of the α,β-anomers of some glucopyranose derivatives (Figure 2.36).

There are two alternative explanations for the origin of the anomeric effect: (1) the dipole–dipole interaction, and (2) the stereoelectronic effect. In both cases, the two nonbonding electron pairs on the endocyclic sp^3 hybridized oxygen atom play an important role. They form a dipole that points in the exocyclic direction (Figure 2.37). The polarized bond between the anomeric carbon and the exocyclic heteroatom attached to it form another dipole. The two dipoles are almost parallel and point in the same direction in the $^4C_1$ conformation of the β-D-glucopyranose anomers. Interaction of these two dipoles is energetically unfavored. In the corresponding α-D-anomers, these two dipoles point away from each other. Therefore, their dipole–dipole interaction is small. According to the stereoelectronic interpretation, the nonbonding electrons of the endocyclic oxygen atom (the *p*-orbital of which is axially oriented) are *synperiplanar* to the antibonding orbital of the anomeric substituent when it is in an α-position. As a result, the two orbitals can mix, establishing a $n \rightarrow \sigma^*$ interaction, which results in a shortening of the ring–O–C1 bond and a lengthening of the C1–X bond (Figure 2.38). It should be mentioned that these effects occur only when the anomeric

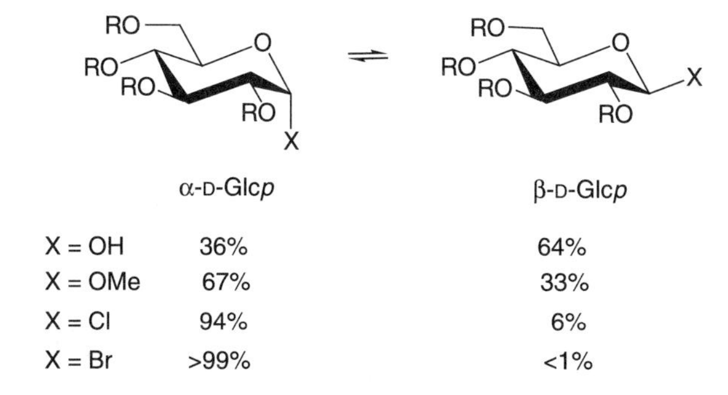

FIGURE 2.36 The α,β-anomeric ratio of some D-glucopyranose derivatives.

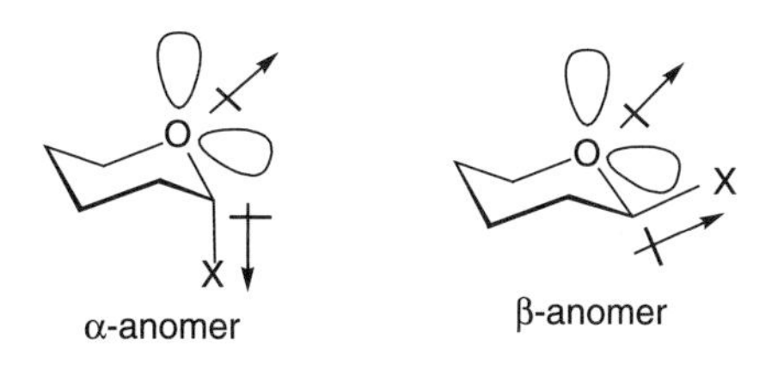

FIGURE 2.37 The dipole moments in the $^4C_1$ conformation of the α- and β-anomers.

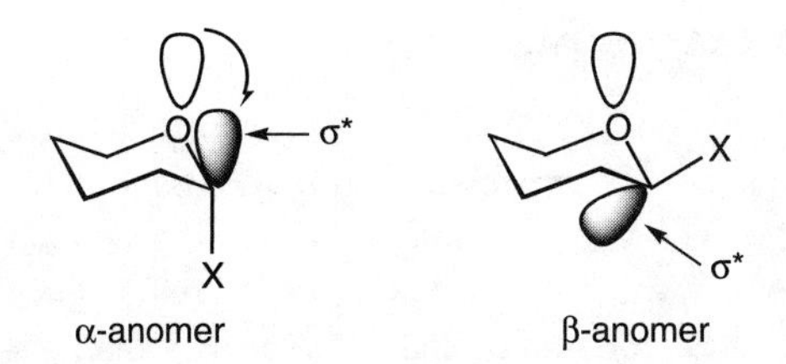

FIGURE 2.38 Orientation of the *p* and σ* orbitals in the $^4C_1$ conformation of α- and β-anomers.

C–X bond is axially oriented; that is, they apply only to the $^4C_1$ conformation of the α-D-anomers and to the $^1C_4$ conformation of the β-D-anomers.

The *exo*-anomeric effect of glycosides can be similarly explained. Owing to the rotation around the C1–O(exo) bond, the nonbonding electrons of the anomeric oxygen atom can occupy different positions (Figure 2.39 and Figure 2.40). From the Newman projection of the three rotamers, in which the O–R bond of the glycoside bisects the bonds of the substituents [C2 and O(endo)] of the anomeric carbon atom C1, it is evident that in rotamer "a," neither of these nonbonding orbitals can overlap with the σ* orbital of the O(endo)–C1 bond. However, in two rotamers ("b" and "c"), such an overlap is possible. The difference between them is the position of the R-substituent of the anomeric oxygen atom, which, in both anomers, occupies an antiperiplanar position to the

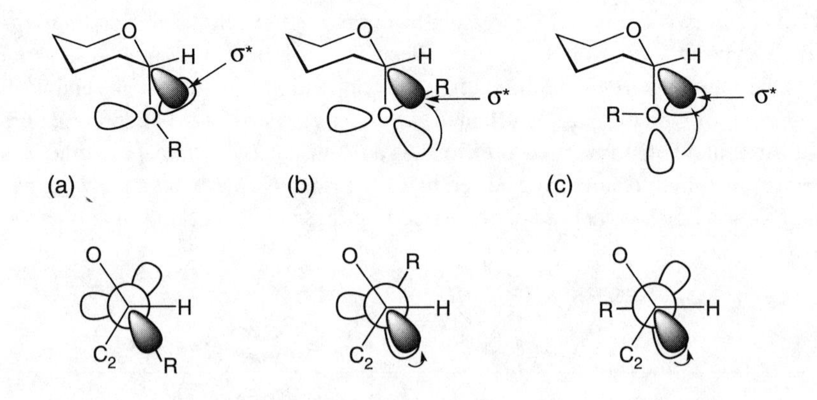

FIGURE 2.39 Three C1 → O_{exo} rotamers of the α-D-anomer and their Newman projections.

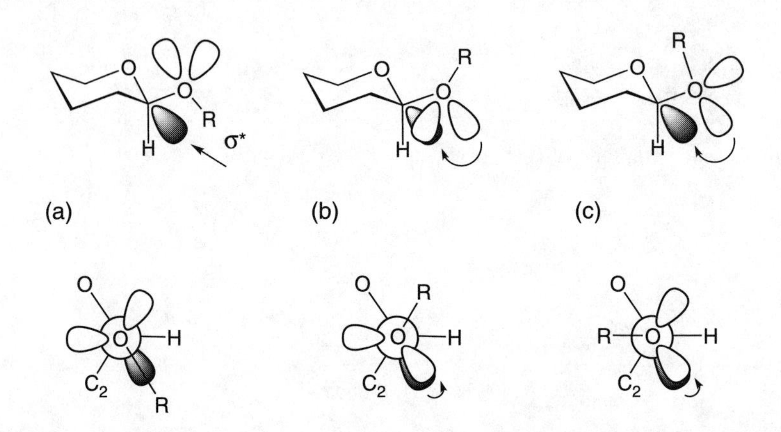

FIGURE 2.40 Three C1 ← O_{exo} rotamers of the β-D-anomer and their Newman projections.

C1–C2 bond in "b," whereas in "c," it is antiperiplanar to the C1–H bond. The latter arrangement is energetically unfavored as the relatively bulky R substituent is placed directly under the plane of the pyranose ring in the α-D-anomers, while in the β-D-anomers it will be placed over it. Therefore, in both anomers, the "b" rotamer is the preferred one.

2.4 DEFINITION AND NOMENCLATURE OF DI- AND OLIGOSACCHARIDES

2.4.1 DISACCHARIDES

A disaccharide is a compound that is formed from two monosaccharide units by elimination of one molecule of water. At least one carbon atom involved in the acetal bridge (called glycosidic linkage)

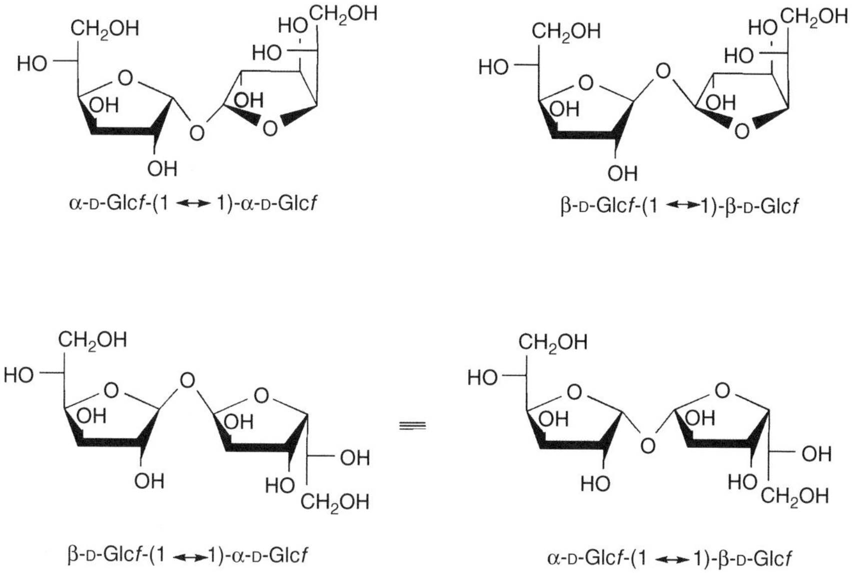

α-D-Glc*f*-(1 ↔ 1)-α-D-Glc*f*

β-D-Glc*f*-(1 ↔ 1)-β-D-Glc*f*

β-D-Glc*f*-(1 ↔ 1)-α-D-Glc*f* ≡ α-D-Glc*f*-(1 ↔ 1)-β-D-Glc*f*

FIGURE 2.41 The three possible nonreducing disaccharides of D-glucofuranose.

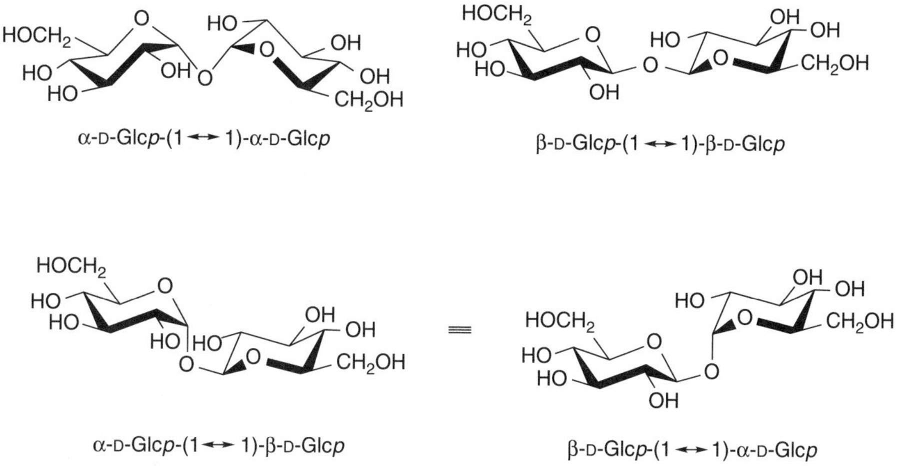

α-D-Glc*p*-(1 ↔ 1)-α-D-Glc*p*

β-D-Glc*p*-(1 ↔ 1)-β-D-Glc*p*

α-D-Glc*p*-(1 ↔ 1)-β-D-Glc*p* ≡ β-D-Glc*p*-(1 ↔ 1)-α-D-Glc*p*

FIGURE 2.42 The three possible nonreducing disaccharides of D-glucopyranose.

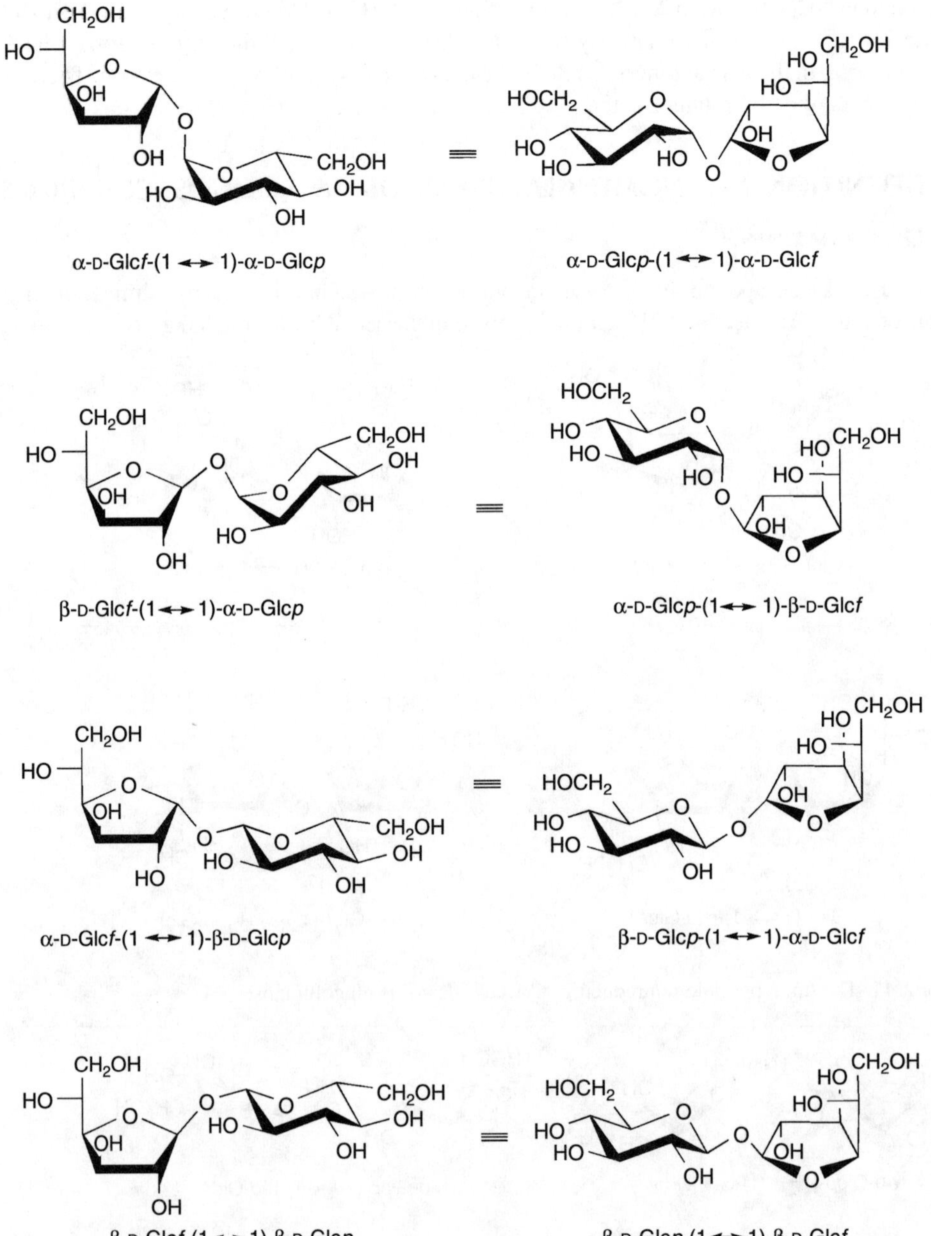

α-D-Glc*f*-(1 ↔ 1)-α-D-Glc*p* ≡ α-D-Glc*p*-(1 ↔ 1)-α-D-Glc*f*

β-D-Glc*f*-(1 ↔ 1)-α-D-Glc*p* ≡ α-D-Glc*p*-(1 ↔ 1)-β-D-Glc*f*

α-D-Glc*f*-(1 ↔ 1)-β-D-Glc*p* ≡ β-D-Glc*p*-(1 ↔ 1)-α-D-Glc*f*

β-D-Glc*f*-(1 ↔ 1)-β-D-Glc*p* ≡ β-D-Glc*p*-(1 ↔ 1)-β-D-Glc*f*

FIGURE 2.43 The four possible isomers of nonreducing disaccharides of D-glucofuranose and D-glucopyranose.

must be an anomeric one. When two anomeric carbon atoms are connected by an acetal bridge, nonreducing disaccharides are formed, while connecting between an anomeric and a nonanomeric carbon atom leads to reducing disaccharides.

The monosaccharide units can be present in both their furanose and pyranose forms, and the anomeric configuration can be both α and β in the case of nonreducing disaccharides. When both monosaccharides have the same configuration (e.g., D-glucose), there are several variations. When both monosaccharide units form furanose rings, three isomers — two symmetric (α-D-glucofuranosyl α-D-glucofuranoside and β-D-glucofuranosyl β-D-glucofuranoside), and one asymmetric (α-D-glucofuranosyl β-D-glucofuranoside) — can be formed, where the α-D-glucofuranosyl β-D-glucofuranoside is identical to the β-D-glucofuranosyl α-D-glucofuranoside isomer (Figure 2.41). To avoid these cumbersome names, the abbreviated versions of the monosaccharides can be used, where the locants of the glycosidic linkage are separated by a doubleheaded arrow and should be given in brackets between the two abbreviated names. Hence, the three isomers mentioned above can be abbreviated as α-D-Glc*f*-(1↔1)-α-D-Glc*f*; β-D-Glc*f*-(1↔1)-β-D-Glc*f*; and β-D-Glc*f*-(1↔1)-α-D-Glc*f*, which is identical to the α-D-Glc*f*-(1↔1)-β-D-Glc*f* isomer. The same holds for the three possible disaccharide isomers of the corresponding D-glucopyranosides (Figure 2.42). When the two glucose units differ in their ring size, four different disaccharides can be formed: α-D-Glc*f*-(1↔1)-α-D-Glc*p*, which is identical to α-D-Glc*p*-(1↔1)-α-D-Glc*f*; β-D-Glc*f*-(1↔1)-α-D-Glc*p*, which is identical to α-D-Glc*p*-(1↔1)-β-D-Glc*f*, α-D-Glc*f*-(1↔1)-β-D-Glc*p*, which is identical to β-D-Glc*p*-(1↔1)-α-D-Glc*f*; and β-D-Glc*f*-(1↔1)-β-D-Glc*p*, which is identical to β-D-Glc*p*-(1↔1)-β-D-Glc*f* (Figure 2.43). When two different monosaccharide units form a nonreducing disaccharide, all formulas depicted in Figure 2.41 through Figure 2.43 will represent different isomers. Thus, theoretically, $4 + 4 + 8 = 16$ compounds can be formed.

In the reducing disaccharides, the glycosidic linkage is formed between the anomeric carbon of one unit and a nonanomeric hydroxyl group of the other monosaccharide. As illustrated above, the glycosyl unit can be present as a furanoside or a pyranoside in both anomeric arrangements (α or β). That means that the combination of these isomers with any given hydroxyl group of the other molecule will result in four isomers. As in a hexose, there are five

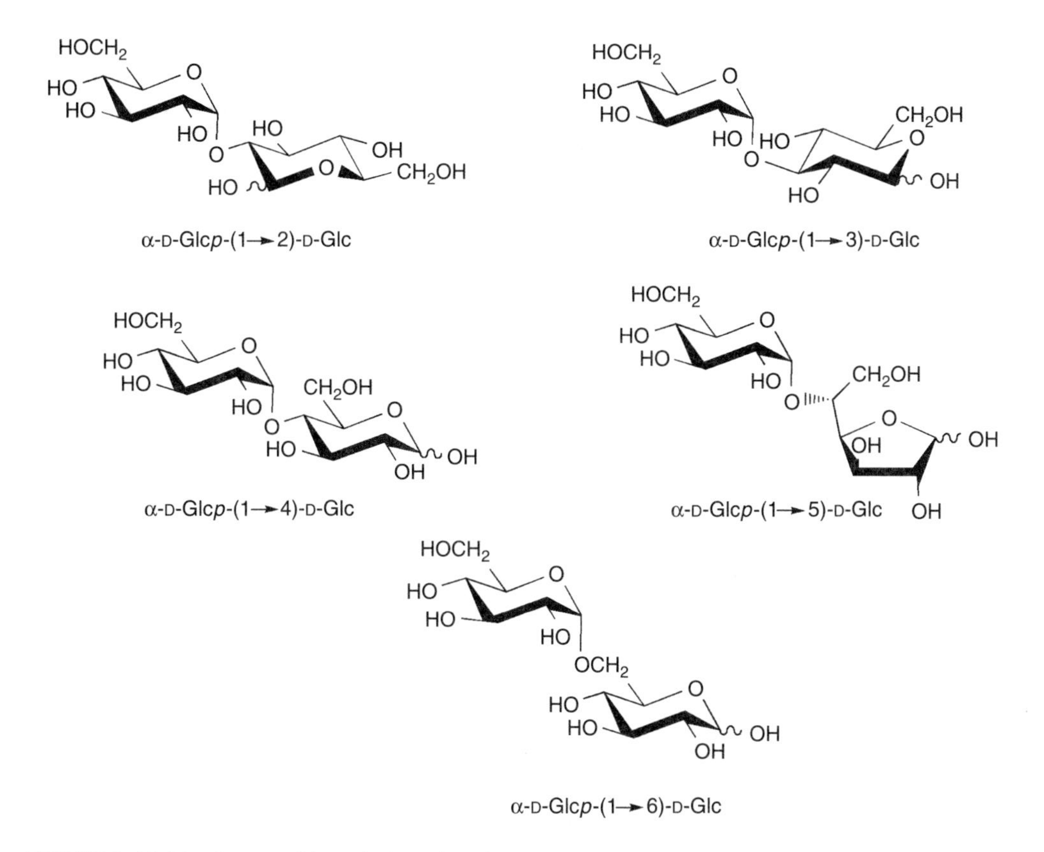

FIGURE 2.44 The five possible reducing disaccharides of α-D-Glc*p*-D-Glc.

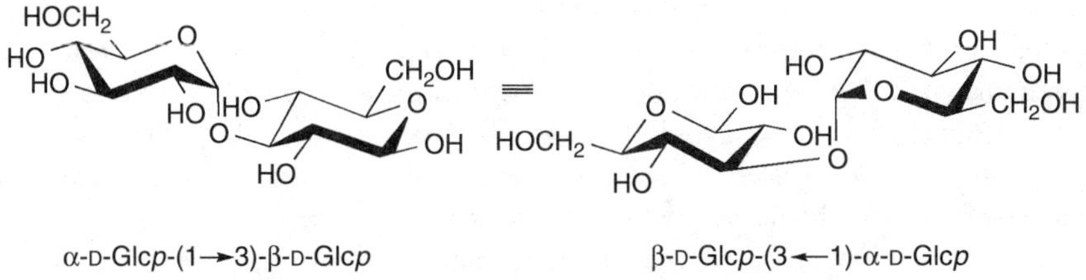

α-D-Glcp-(1→3)-β-D-Glcp

β-D-Glcp-(3←1)-α-D-Glcp

FIGURE 2.45 Two possible depictions of the reducing disaccharide α-D-Glc*p*-(1 → 3)-β-D-Glc*p*.

different hydroxyl groups available for glycosylation (apart from the anomeric one, which, upon glycosylation, will result in a nonreducing disaccharide). Thus, theoretically, $4 \times 5 = 20$ different isomers can be formed if both monosaccharides have the same configuration (e.g., D-glucose). As an example, only the five possible α-D-Glc*p* → D-Glc isomers are depicted in Figure 2.44. When the two monosaccharide units differ in their configuration, the number of possible isomers increases to 40. Because even the reducing end of the disaccharide can form α- and β-anomers as well as furanose and pyranose structures, the total amount of the theoretically possible disaccharide isomers will be 160. In the abbreviated names of these glycosides, a singleheaded arrow must be drawn between the locants of the glycosidic linkage, pointing in the direction of the nonanomeric hydroxy group. This is usually on the right-hand side, but sometimes it is necessary to depict them in reverse order (Figure 2.45) to provide a better view (especially in the case of branched oligosaccharides). If the exact structure of the reducing unit is known, then it should be given in the name.

It should be mentioned that many of the disaccharides have well-established trivial names. Some of them are listed below, together with their abbreviations:

Cellobiose	β-D-Glucopyranosyl-(1 → 4)-D-glucose	β-D-Glc*p*-(1 → 4)-D-Glc
Maltose	α-D-Glucopyranosyl-(1 → 4)-D-glucose	α-D-Glc*p*-(1 → 4)-D-Glc
Gentiobiose	β-D-Glucopyranosyl-(1 → 6)-D-glucose	β-D-Glc*p*-(1 → 6)-D-Glc
Isomaltose	α-D-Glucopyranosyl-(1 → 6)-D-glucose	α-D-Glc*p*-(1 → 6)-D-Glc
Lactose	β-D-Galactopyranosyl-(1 → 4)-D-glucose	β-D-Gal*p*-(1 → 4)-D-Glc
Melibiose	α-D-Galactopyranosyl-(1 → 6)-D-glucose	α-D-Gal*p*-(1 → 6)-D-Glc
Trehalose	α-D-Glucopyranosyl-(1 ↔ 1)-α-D-glucopyranose	α-D-Glc*p*-(1 ↔ 1)-α-D-Glc*p*
Sucrose	β-D-Fructofuranosyl-(2 ↔ 1)-α-D-glucopyranose	β-D-Fru*f*-(2 ↔ 1)-α-D-Glc*p*

2.4.2 Oligosaccharides

When three or more monosaccharides are connected to one another via a glycosidic linkage, oligosaccharides are formed. As already mentioned for disaccharides, this acetal linkage can connect either the two anomeric carbon atoms of the terminal unit when nonreducing oligosaccharides are formed, or an anomeric and a nonanomeric carbon atom when reducing oligosaccharides are formed. Depending on the number of units, they are called trisaccharides, tetrasaccharides, pentasaccharides, and so on. There is no strict borderline between higher oligosaccharides and the polymers called polysaccharides; however, the term "oligosaccharide" is commonly used to refer to a defined structure.

Mono- and oligosaccharides frequently occur in nature linked to noncarbohydrates at their reducing end. The moiety at the reducing end is called the *aglycon*, and the remaining portion is the glycone part. Numbering of oligosaccharide carbon atoms use simple arabic numbers (1 to *n*) for the unit at the reducing end, and primed and double-primed numbers, respectively, for the

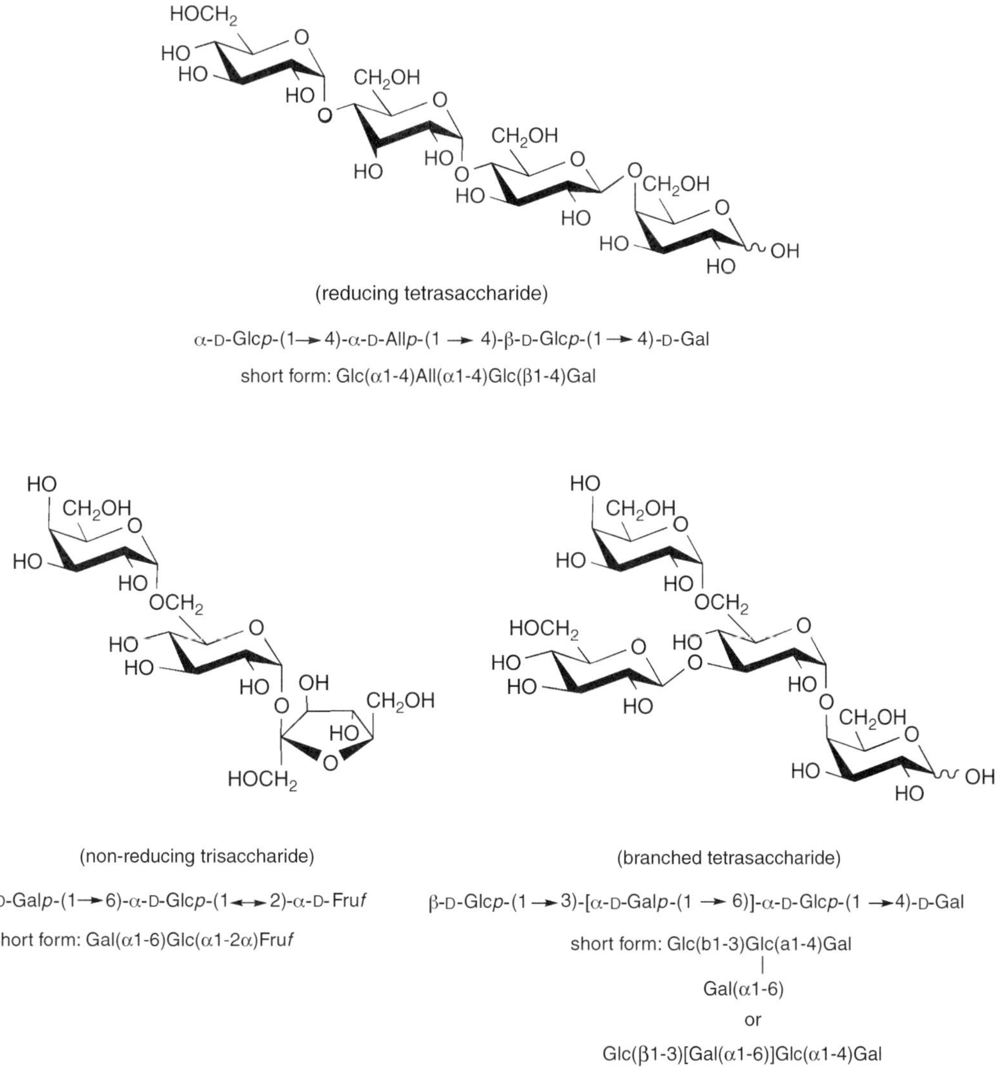

(reducing tetrasaccharide)

α-D-Glc*p*-(1→4)-α-D-All*p*-(1 → 4)-β-D-Glc*p*-(1→4)-D-Gal

short form: Glc(α1-4)All(α1-4)Glc(β1-4)Gal

(non-reducing trisaccharide)

α-D-Gal*p*-(1→6)-α-D-Glc*p*-(1↔2)-α-D-Fru*f*

short form: Gal(α1-6)Glc(α1-2α)Fru*f*

(branched tetrasaccharide)

β-D-Glc*p*-(1→3)-[α-D-Gal*p*-(1 → 6)]-α-D-Glc*p*-(1 →4)-D-Gal

short form: Glc(b1-3)Glc(a1-4)Gal
|
Gal(α1-6)

or

Glc(β1-3)[Gal(α1-6)]Glc(α1-4)Gal

FIGURE 2.46 Reducing, nonreducing and branched oligosaccharides.

glycone constituents. Because oligosaccharides can form both linear and branched molecules, in the latter case the primed numbers should be given to that unit with lower locants at the branch point. For ease of understanding, it is advisable to indicate the glycosidic linkages in the name by giving the numbers of the connecting atoms in brackets. Moreover, in branched oligosaccharides, the branches should be enclosed in square brackets. If two chains are of equal length, then the one with lower locants at the branch point is regarded as the parent one (Figure 2.46).

For longer sequences, it is desirable to use more condensed forms, omitting the configurational signal as well as the ring size. It is widely understood that the configuration is D (with the exception of fucose and iduronic acid, which are usually L), and that the rings are in pyranose form unless otherwise specified. The anomeric descriptor is written in parentheses with locants. Branches can be indicated on the same line by using square brackets, or they can be written on a separate line under (or over) the name of the branching unit.

FURTHER READING

Lichtenthaler, F W, Emil Fischer's proof of the configuration of sugars, *Angew. Chem. Int. Ed. Engl.*, 31, 1541–1556, 1992.

Lindhorst, T K, *Essentials of Carbohydrate Chemistry and Biochemistry*, Wiley-VCH, Weinheim, 2000, pp. 1–217.

McNaught, A D, Nomenclature of carbohydrates, *Carbohydr. Res.*, 297, 1–92, 1997.

Stoddart, J F, *Stereochemistry of Carbohydrates*, Wiley-Interscience, New York, 1971, pp. 1–249.

Tvaroska, I, Bleha, T, Anomeric and Exo-Anomeric Effects in Carbohydrate Chemistry, *Adv. Carbohydr. Chem. Biochem.* 47, 45–123, 1989.

3 Protective Group Strategies

Stefan Oscarson

CONTENTS

3.1 INTRODUCTION

Protecting groups and protection strategies are important components of all total syntheses of organic molecules. This is particularly true in carbohydrate chemistry and oligosaccharide synthesis because of the large number of functional groups present. Most of them are of the same sort — hydroxyl groups — which necessitates (sometimes quite laborious) regioselective protecting strategies. Several books have been written on protecting groups in organic synthesis [1–4], and new groups and methodologies are continually being developed [5,6]. A large part of this literature deals with protecting group manipulations applied to carbohydrate chemistry. In addition to the problems associated with protecting the various functional groups so as to expose the functionality to be reacted (most often glycosylation of a hydroxyl group), several other issues have to be considered. Protecting groups not only protect; they also confer other effects to the molecule. For example, they can increase or decrease the reactivity and they can also participate in reactions, thus affecting the stereochemical outcomes. Important examples of such effects are that acylated glycosyl donors are less reactive than their alkylated counterparts. Furthermore, the use of 2-*O*-participating groups in donors has been developed to ensure 1,2-*trans* selectivity in glycosylation reactions. These factors should be taken into account when planning the protective group strategy. However, due to the large number of protecting groups present in a fully protected sugar, it is almost impossible to predict all effects introduced by the protecting group pattern, and often unexpected reactivity is found requiring the study of alternative strategies.

Despite the above complexities, this chapter aims to impart general synthetic strategies for most sugars and oligosaccharide structures through the use of some basic, well-proven protecting groups, coupled with general strategies towards regioselectivity. The discussion begins with an outline of frequently used protecting groups in carbohydrate chemistry, briefly surveying conditions for their introduction, stability and removal. The following sections summarize standard methodologies for the regioselective introduction of hydroxyl protecting groups. Examples of the application of these methods on common monosaccharides (glucose, galactose, mannose) and disaccharides (lactose, sucrose) are then given. Finally, different protecting group strategies employed in published syntheses of a complex nonasaccharide, trimeric Lewis^{x}, are compared.

3.2 PROTECTING GROUPS

The protecting groups used in carbohydrate chemistry are the same as in any other area of organic chemistry. The conditions for their introduction and removal and their stability and orthogonality are also the same (information which can be found in books on protecting groups) [1–4]. The difference in carbohydrate chemistry is the vast number of protecting groups needed and the continuous necessity for regioselective protection. By far the most important protecting groups in carbohydrate chemistry are those used for the protection of hydroxyl groups, including the anomeric hemiacetal. Amino-protecting groups (for amino-deoxy sugars) and carboxyl protecting groups (for uronic and ulosonic acids) are also of interest.

3.2.1 Hydroxyl Protecting Groups

The large numbers of hydroxyl groups present in carbohydrates, combined with the necessity to protect specific groups regioselectively, thus allowing the generation of oligosaccharides, makes it absolutely necessary to have a know-how of the possibilities and strategies. In oligosaccharide synthesis beyond the level of disaccharides, temporary and permanent (persisting) protecting groups must frequently be distinguished. Permanent protecting groups will remain through all synthetic steps until the liberation of final target oligosaccharide. The removal of temporary protecting groups some time during the synthetic sequence will reveal a free hydroxyl group, most often to produce a glycosyl acceptor ready for glycosylation (Scheme 3.1).

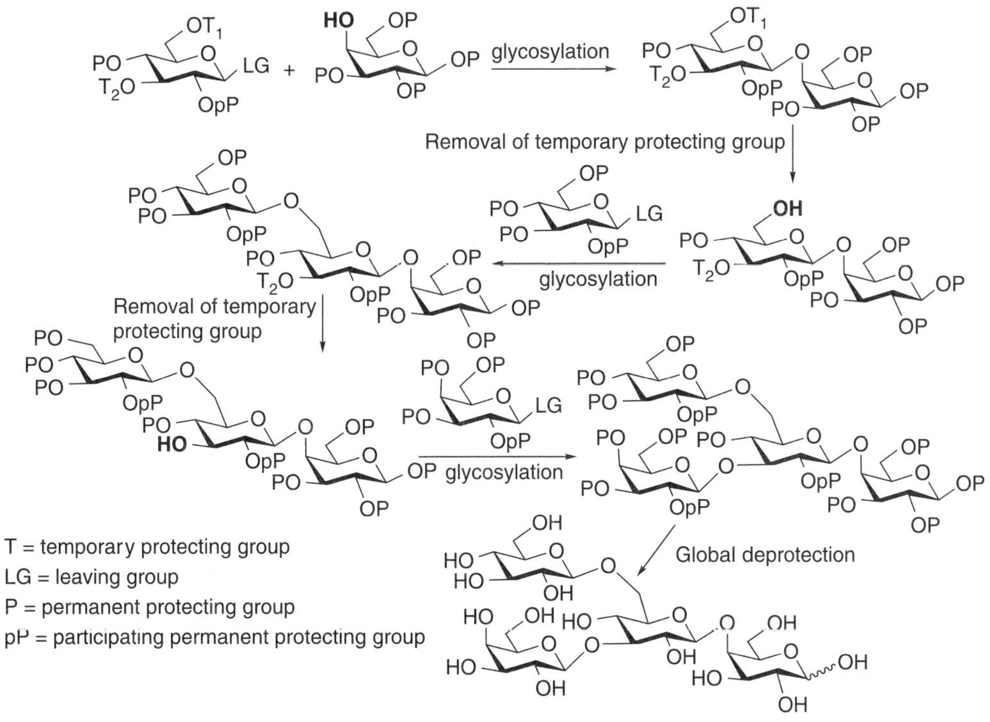

T = temporary protecting group
LG = leaving group
P = permanent protecting group
pP = participating permanent protecting group

SCHEME 3.1 Schematic synthesis of a branched tetrasaccharide.

The two different types of protecting groups require different qualities. It must be possible to introduce and remove permanent groups in bulk with regiocontrol and high efficiency. Obviously, they should also be stable to glycosylation conditions and conditions used for the removal (and introduction) of temporary groups. Temporary protecting groups should be stable during all reactions from their introduction until their own removal, which should be performed under conditions that do not affect anything else in the molecule. In branched structures or in structures with substituents (phosphates, sulfates, acetates etc.), it is necessary to employ several different temporary groups, which then should be orthogonal to each other (i.e., able to be removed selectively in the presence of each other). Preferably only one type of permanent protecting group should be used, to allow only one final deprotection step, an aspiration seldom possible to fulfill.

Below is a list of the most commonly used hydroxyl protecting groups including conditions for their introduction and removal. For a discussion of their regioselective introduction, see Section 3.3. A rough distinction between permanent and temporary protecting groups is also made.

3.2.1.1 Permanent Protecting Groups

Very few of the large numbers of protecting groups available fulfill the criteria for permanent protecting groups. More or less only acetates, benzoates, benzyl ethers and some acetals have the stability and at the same time the efficient introduction/deprotection properties needed to make them suitable for this purpose. Thus, these protecting groups are very much the foundation for all protecting group strategies in oligosaccharide synthesis. Efficient poly-protection of oligo- and polysaccharides are frequently performed using methyl and trimethylsilyl (TMS) ethers, especially for gas chromatographic purposes. However, neither of these groups can be used in synthetic schemes, since they are either too stable (methyl ethers) or too unstable (TMS ethers).

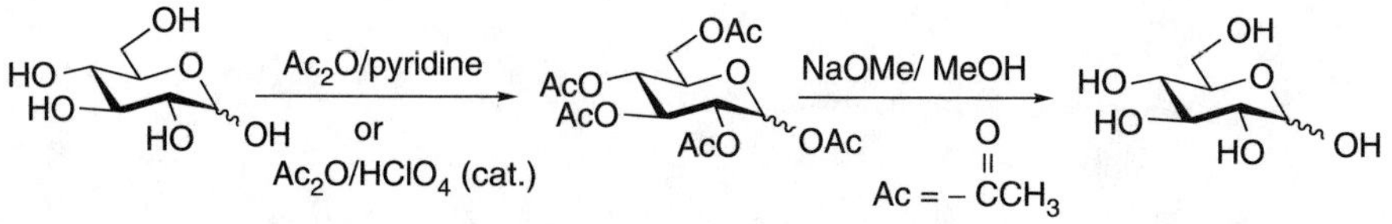

SCHEME 3.2 Acetylation and deacetylation of D-glucose.

3.2.1.1.1 Esters (Acetates and Benzoates)

Since both acetates and benzoates are acyl groups, their introduction and removal are performed under similar conditions. Standard conditions for esterification, which can be performed on reducing sugars, are the acyl chloride (especially for benzoates) or the anhydride (especially for acetates) in pyridine (Scheme 3.2). Addition of other bases or a cosolvent are common variations. The acetylation can also be performed under acidic conditions in an efficient and fast reaction. These esters' acid stability is quite high, but they are only moderately base stable; acetates are the more labile. Additionally, esters (especially acetates) have a tendency to migrate, both under acidic and basic (fastest) conditions [7]. This is a concern in partially protected derivatives, and results in a mixture with the most stable compound preponderant. Thus, in *cis*-hydroxyls, there is normally a preferred migration from the axial position to the equatorial one and in 4,6-diols the migration goes from O-4 to O-6 preferentially.

Standard conditions for the removal of ester protecting groups include a catalytic amount of sodium methoxide in methanol (Zemplén conditions), sometimes with a cosolvent. Additionally, other basic conditions such as ammonia in MeOH may be useful.

Acetates and benzoates are excellent protecting groups. They can be introduced and removed in high yields in large numbers under mild conditions. Drawbacks as permanent protecting groups include their relatively limited base stability and their tendency to migrate.

3.2.1.1.2 Ethers (Benzyl Groups)

Benzylations are usually performed under strongly basic conditions. Standard conditions are benzyl bromide and sodium hydride in a nonprotic polar solvent (Scheme 3.3). Less basic alternatives, allowing for the presence of ester groups, are the use of Ag_2O as base or benzyl trichloroacetimidate or benzyl triflate as reagents. These latter reagents can be activated under acidic or neutral conditions. While providing useful alternatives, these methods are usually less efficient, and can be used for the introduction of only one or, at most, a few benzyl groups.

Benzyl ethers are highly stable, generally succumbing only to strongly acidic (especially acetolysis) conditions. Benzyl ethers are normally deprotected through catalytic hydrogenolysis employing various Pd-catalysts. Problems might be encountered, especially if the molecule contains functionalities that can poison the catalyst (e.g., thio and amino groups). An alternative is a Birch-type reduction (Na/liquid ammonia). These conditions quickly remove benzyl ethers and are surprisingly mild towards other functionalities such as phosphates and even the hemiacetals of reducing sugars [8].

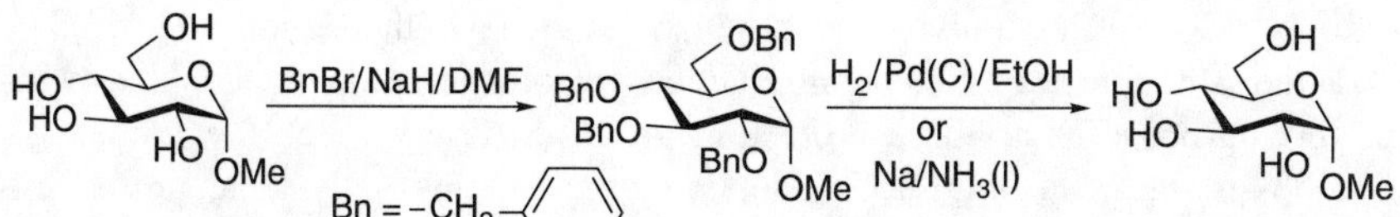

SCHEME 3.3 Benzylation and debenzylation of methyl α-D-glucopyranoside.

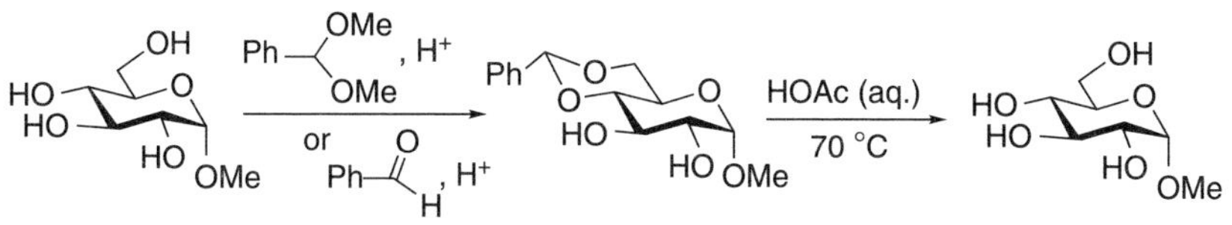

SCHEME 3.4 Benzylidenation and debenzylidenation of methyl α-D-glucopyranoside.

Thus, benzyl ethers are also excellent protecting groups. They are stable towards almost any conditions and are easily removed under essentially neutral conditions. The one drawback is the rather harsh conditions usually employed for their introduction.

3.2.1.1.3 Cyclic Acetals

Of the possible acetals, benzylidene and isopropylidene have been utilized most frequently in carbohydrate chemistry for simultaneous protection of two hydroxyls. They are introduced by standard acetalization conditions; that is, either the aldehyde or the dimethoxy acetal is employed as reagent together with some acid catalyst (Scheme 3.4) [9]. With isopropylidene acetals, the methoxy propenyl ether has been utilized to provide the kinetic product, and formation of benzylidene acetals has been achieved under basic conditions using α,α-dibromotoluene in refluxing pyridine [10].

Acetal cleavage is achieved by acid hydrolysis; standard conditions are acetic acid (70% aqueous) at elevated temperature or trifluoroacetic acid (TFA, 90% aqueous) at 0°C.

The advantage of acetals as protecting groups is their easy regioselective introduction in combination with the number of possible subsequent modifications (of benzylidene acetals) yielding various protecting group patterns (Section 3.3).

3.2.1.2 Temporary Protecting Groups

As opposed to the small selection qualifying for permanent protecting groups, there is a plethora of possible temporary protecting groups, and new ones are continuously being developed. However, many of these have unsuitable or similar stability profiles and features, or are difficult to use (noncommercial). Thus, the number of really efficient, orthogonal temporary protecting groups is not high.

3.2.1.2.1 Esters

Acetates can be used as temporary protecting groups in the presence of benzoates. Acetates are more labile than benzoates, and can be removed selectively both under acidic conditions (HCl/MeOH) [11] or basic conditions ($Mg(OMe)_2$/MeOH) [12]. Obviously, benzoates can also be used as temporary protecting groups (e.g., with benzyls as permanent protecting groups), but there is often a demand for ester groups removable in the presence of both acetates and benzoates. Chloroacetate and levulinate (4-oxopentanoate) esters are examples of such ester groups commonly used in oligosaccharide synthesis (Figure 3.1). Because of the base-lability of these latter esters, caution has to be taken when introduced. Often, a cosolvent is used with pyridine.

Chloroacetates are traditionally removed by treatment with thiourea or hydrazine acetate. The supposed carcinogenicity of these reagents makes substitutes attractive. Alternatives are various

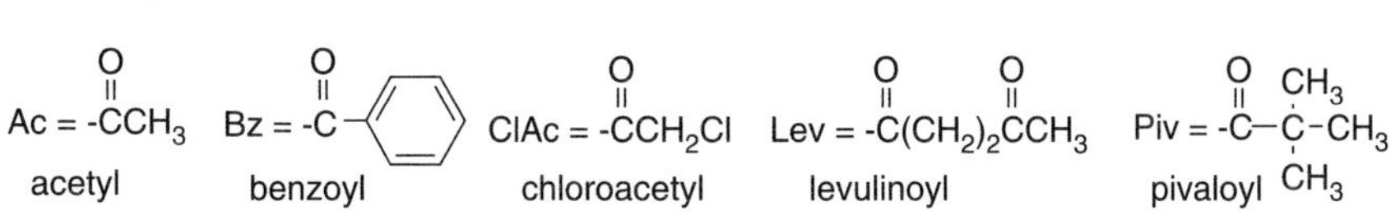

FIGURE 3.1 Ester protecting groups in common use.

weak bases such as wet pyridine or DABCO [13]. Levulinoyl esters are usually removed by mild hydrazine treatment.

Although not formally a temporary protecting group, pivaloate esters are regularly used in oligosaccharide synthesis, the advantage being that given their steric bulk, they are easily introduced selectively in primary positions. Furthermore, as 2-*O*-participating group in donors, they prevent the formation of orthoester side products (see Chapter 4) [14].

3.2.1.2.2 Ethers

The most commonly used ether protecting groups aside from benzyl ethers are *p*-methoxybenzyl, allyl, trityl and silyl ethers (Figure 3.2), which can all be removed in the presence of the standard permanent protecting groups. In their introduction, there is a clear distinction between the *p*-methoxybenzyl and allyl ethers (usually formed under strongly basic conditions) and the trityl and silyl ethers (usually formed under weakly basic conditions — typically with trityl or silyl chloride in pyridine). Thus, there are no problems associated with tritylation or silylation of ester-containing derivatives, whereas allylation or *p*-methoxybenzylation of such derivates often requires special conditions (compare benzylations above).

Removal of *p*-methoxybenzyl ethers is accomplished through hydrolysis under oxidative or acidic conditions. Treatment with DDQ or CAN in wet dichloromethane or acetonitrile is a standard condition. Sometimes, the acid lability of these ethers is a problem during glycosylations.

Another substituted benzyl ether in occasional use is the *p*-chlorobenzyl ether, mainly because of the high crystallinity and added stability of these derivatives. Recently, a methodology to remove *p*-chloro- and bromobenzyl ethers in the presence of both benzyl and *p*-methoxybenzyl groups was reported [15]. It employed a two-stage deprotection scheme where the *p*-halobenzyl group was first converted to an arylamine by Pd-catalyzed amination. The arylamine was then selectively cleaved under very mild acidic conditions.

Cleavage of allyl ethers is another example of a two-stage deprotection. Initial rearrangement of the allyl ether using strong base or metal-complex catalysts (e.g., Wilkinson's catalyst) is followed by Lewis acid-catalyzed hydrolysis of the obtained vinyl ether in a one-pot reaction. Recently, SmI_2 was reported as an efficient deallylation reagent following a different (radical) mechanism [16].

There are numerous types of silyl ethers, but *tert*-butyldimethylsilyl- or -diphenylsilyl groups are the most frequently used — mainly due to their suitable stability for oligosaccharide synthesis (i.e., not too acid sensitive). A problem encountered with silyl ethers is that, like esters, they might

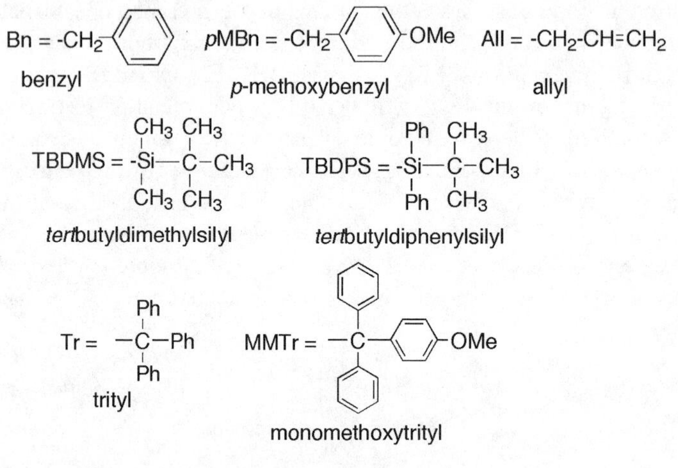

Bn = $-CH_2-C_6H_5$ benzyl
pMBn = $-CH_2-C_6H_4-OMe$ *p*-methoxybenzyl
All = $-CH_2-CH=CH_2$ allyl
TBDMS = $-Si(CH_3)_2-C(CH_3)_3$ *tert*butyldimethylsilyl
TBDPS = $-Si(Ph)_2-C(CH_3)_3$ *tert*butyldiphenylsilyl
Tr = $-C(Ph)_3$ trityl
MMTr = $-C(C_6H_5)_2-C_6H_4-OMe$ monomethoxytrityl

FIGURE 3.2 Commonly used ether protecting groups.

migrate to neighboring free hydroxyl groups under basic conditions [17]. Silyl ethers are removed by treatment with fluoride ion. Basic, neutral or acidic conditions are possible and are dependent on the type of fluoride reagent used. This difference is of special importance in ester-containing derivatives where acyl migration might be a severe problem — especially under basic conditions. Sometimes, silyl ethers are sufficiently stable and easy to introduce and remove, thus making them useful as bulk permanent protecting groups. However, it is often difficult to find the right balance between stability and ease of introduction to qualify them as such protecting groups. For example, per-silylation of glucose with *tert*-butyldimethylsilyl groups is possible but due to the size of the protecting group, a conformational change in the pyranose ring is necessary to accommodate all bulky groups [18].

Cleavage of trityl ethers is accomplished by acid hydrolysis. Use of the mono- or di-methoxy analogs increase the acid lability of the trityl ethers, but none of the trityl ethers is stable enough to withstand normal glycosylation conditions, and they are only used as intermediates to construct building blocks to be used in subsequent oligosaccharide syntheses.

3.2.1.2.3 Acetals

Among the acetals protecting two hydroxyl groups, an important alternative to benzylidene acetals is the *p*-methoxybenzylidene acetal (Figure 3.3). This is more acid labile and can be removed in the presence of the parent benzylidene acetals. Other options are acrolein acetals and the newly introduced 2-naphthyl- and 9-anthracylmethylene acetals [19,20]. All of these acetals exhibit the same regioselectivity (for vicinal *cis*-diols) as benzylidene acetals. In contrast, the newly developed dispiroketal (dispoke), cyclohexane-1,2-diacetal (CDA) and butane-1,2-diacetal (BDA) groups show completely different regioselectivity and provide simultaneous protection of vicinal *trans*-diols [21], thus making them most important additions to protecting group strategies. Cyclohexylidene acetals are occasionally used, most often as an alternative to benzylidene acetals, but with a slightly different conformational impact on the saccharide derivative [22,23].

Finally, silyl acetals (Figure 3.3) are in use as viable alternatives to the acetals discussed thus far. A representative of this group is the 1,3-(1,1,3,3-tetraisopropyl-disiloxanylidene) acetal (TIPDS). The dichloride and a base are used for formation and, as for silyl ethers, fluoride ion is used for removal.

Mono-hydroxyl protecting acetals, mainly methoxymethyl (MOM), benzyloxymethyl (BOM) and tetrahydropyranyl (THP) groups, have only been used sporadically in oligosaccharide chemistry, and their use will probably decrease due to the expected carcinogenicity of the used reagents (MOM-Cl and BOM-Cl) or the resulting diastereomeric mixtures (THP-acetals).

3.2.1.2.4 Miscellaneous Protecting Groups

Other protecting groups used are orthoesters and carbonates. Both are of interest mainly as cyclic derivatives protecting two hydroxyl groups. Orthoesters are formed by using trialkylorthoesters

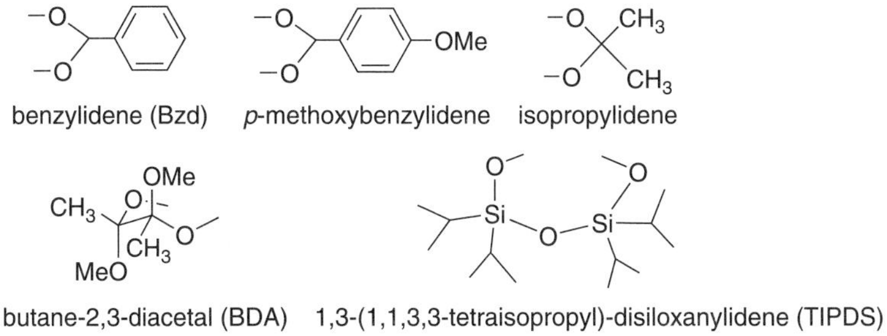

benzylidene (Bzd) *p*-methoxybenzylidene isopropylidene

butane-2,3-diacetal (BDA) 1,3-(1,1,3,3-tetraisopropyl)-disiloxanylidene (TIPDS)

FIGURE 3.3 Acetal protecting groups in common use.

and an acid catalyst (see Scheme 3.19). 1,2-Orthoesters are formed from the corresponding 1-bromo-2-*O*-ester compound in the presence of an alcohol under basic conditions. Orthoesters are too acid labile to endure glycosylation conditions and, like trityl ethers, are mainly used as intermediates in the formation of different synthetic building blocks. The orthoesters are usually not completely removed but are opened up to give a hydroxyl-ester derivative (see Section 3.3.1.6).

Cyclic carbonates are introduced through the use of phosgene or a phosgene equivalent and a base (pyridine). They are most stable to acid hydrolysis and are removed under rather strongly basic conditions. Interestingly, they can protect both *cis*- and *trans*-vicinal diols.

Recently, the use of Fmoc carbonates has been reported [24], mainly in the context of solid-phase synthesis approaches. This is relevant because the removal of the Fmoc group can easily be monitored and quantified. Fmoc groups are typically removed by treatment with triethylamine or piperidine (in CH_2Cl_2).

3.2.2 Anomeric (Hemiacetal) Protecting Groups

Many of the groups used for protection of the anomeric hemiacetal group are the same as those used for nonanomeric hydroxyl groups (i.e., acetyl and benzoyl esters, benzyl, allyl and silyl ethers, and isopropylidene and benzylidene acetals). Conditions for their introduction are also the same; the one exception relates to the alkyl ethers, which are normally formed under various glycosylation reactions (e.g., Fischer conditions, using the alcohol as aglycon). Their removal is performed as discussed above. Anomeric acyl groups are more base labile than nonanomeric esters and can be removed selectively under weakly basic conditions (e.g., piperidine in THF; see Section 3.3.2). Surprisingly, anomeric allyl groups were found to be inert to SmI_2-promoted deallylation [16].

More specific anomeric protecting groups are the 2-trimethylsilylethyl (TMSE) [25] and *p*-methoxyphenyl glycosides [26], both introduced by glycosylation reactions on protected derivatives (Scheme 3.5). The former is cleaved by treatment with BF_3-etherate conditions in which TBDMS ethers survive. *p*-Methoxyphenyl glycosides are cleaved by CAN-oxidation similarly to *p*-methoxybenzyl ethers.

1,6-Anhydro bridges can be considered as protecting groups for the anomeric center (and the 6-hydroxyl group). 1,6-Anhydro derivatives are formed by an internal displacement reaction, where the leaving group can be in either the 1- or the 6-position [27]. Examples are base treatment of phenyl β-glycosides or 6-*O*-tosyl-derivatives with a free hemiacetal group, and Lewis acid treatment of methyl glycosides at elevated temperatures (Scheme 3.6) [28]. The 1,6-anhydro-bridge is stable to most reaction conditions, but very susceptible to mild acetolysis conditions to give the 1,6-diacetate [27].

Glycals (1,2-unsaturated derivatives) might also be viewed as anomeric protecting groups. They are produced from the corresponding acetobromosugars by zinc-mediated elimination reactions. Glycals are moderately stable and can be transformed into 2-acetamidosugars (Section 3.2.3) [29] or via stereoselective epoxidation into β-D-1,2-*trans*-glycosides (Section 3.4.2.1) [30].

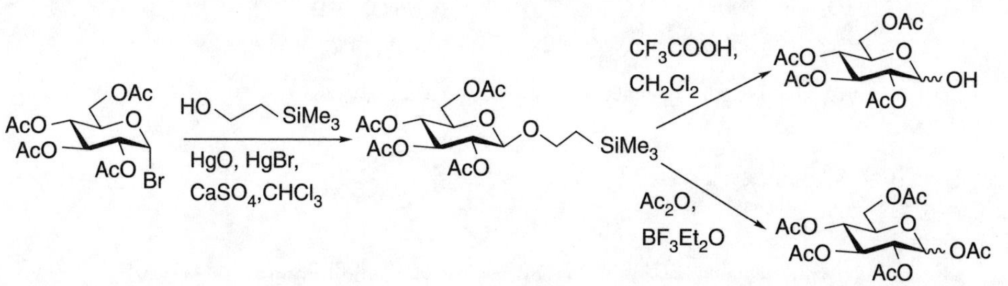

SCHEME 3.5 Formation and cleavage of a TMSE-glycoside.

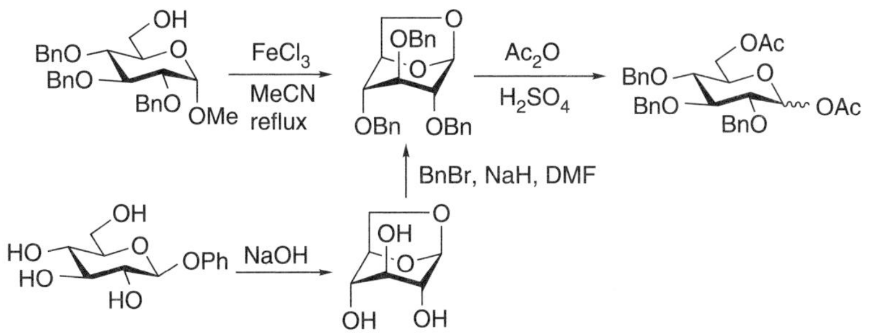

SCHEME 3.6 Formation and opening of a 1,6-anhydro derivative.

Traditional glycosyl donors, such as glycosyl halides, are quite unstable, so the purpose of an anomeric protecting group is often to allow protecting group manipulations elsewhere in the molecule prior to its removal and transformation of the derivative into a glycosyl donor. One of the major improvements in contemporary oligosaccharide synthesis has been the development of stable anomeric protecting groups that can be directly activated and used as donors in glycosylation reactions without the need for deprotection/transformation steps. This shortens the synthesis; however, perhaps even more importantly, this also allows for much greater variation in the protecting group pattern in the building block donors. This is possible since the activation conditions are milder than most of the required transformation conditions, most of which are strongly acidic (typically HBr/HOAc). Milder conditions (slightly basic) are employed in the transformation of the hemiacetal to a trichloroacetamidate donor [31], which is an alternative to direct activation, even in the presence of acid-labile protecting groups.

Alkyl and aryl thioglycosides and pentenyl glycosides, easily formed from the corresponding peracetylated or acetobromo sugars [32,33], are stable under almost all reaction conditions used in protecting group manipulations, but are easily activated by various chemoselective promoters and function then as glycosyl donors. If required, they can then yield the free hemiacetal by using water as the aglycon.

3.2.3 Amino Protecting Groups

The number of amino groups in carbohydrates is much fewer than the number of hydroxyl groups, so a less complicated strategy for their protection and deprotection is usually needed. Of course, this is only part of the strategy and must be harmonized to the protection group strategies of the other functional groups. Selective protection of amino groups in the presence of hydroxyl groups is trivial, because of their higher nucleophilicity. This specificity is further enhanced due to the higher base stability of amides as compared to esters.

The most abundant amino-deoxy-sugars in nature are the 2-acetamido-2-deoxy derivatives with *gluco*, *galacto* or *manno* configurations. Since both α- and β-anomers are found, an important feature of the protecting group used for the 2-amino-function is to allow for the stereoselective formation of the correct anomer during glycosylation reactions (i.e., participating groups such as amides for 1,2-*trans*-linkages and nonparticipating groups for 1,2-*cis*-linkages). However, due to the nucleophilic free electron pair of amines, these groups are "participating" [34]. Presently, there is only one generally useful amine precursor that is nonparticipating — namely, the azido (N_3) group.

As mentioned, most of the natural amino groups are in the form of acetamides, so the optimal strategy would be to use these directly in the various building blocks needed for oligosaccharide syntheses, thus minimizing protecting group manipulations. However, the 2-acetamido group is a problem in glycosyl donors, since it participates but often then forms the rather stable oxazoline

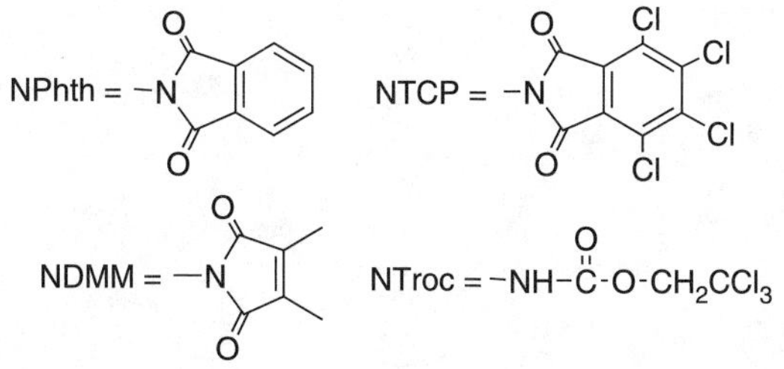

FIGURE 3.4 Amino protecting groups in common use.

instead of the desired glycoside (see Chapter 4). Thus, a number of 2-amino protecting groups have been designed with the ability to participate but not form oxazolines (Figure 3.4) [35].

The first of such participating amino protecting groups introduced, and still much in use, was the phthalimido group. It is incorporated in a two-step procedure beginning with acylation by phthalic anhydride. Subsequent ring closure and simultaneous acylation of the sugar hydroxyl groups is achieved through treatment with acetic anhydride in pyridine (Scheme 3.7).

The phthalimido group is stable to most reaction conditions used in oligosaccharide synthesis. However, problems are sometimes encountered during the removal of the phthaloyl group, the cleavage of which requires strongly basic conditions. Typical reagents include hydrazine, hydroxylamine and alkyldiamines. These are generally used in large excesses and at elevated temperatures, which sometimes results in low yields of the deprotected derivatives. Accordingly, modified phtalimides and other types of imides have been developed and utilized as amino protecting groups. Tetrachlorophtalimides (TCP) [36] and dimethylmaleoides (DMM) [37] are two such examples. They are formed in the same way as phthalimides, but their cleavage is possible using milder conditions. Hence, they can be removed in the presence of phthalimides. Another frequently used (participating) amino protecting group in oligosaccharide synthesis is the trichloroethylcarbamate (Troc) group.

As mentioned, the only nonparticipating amino protecting group is the azido group. This is introduced by an azidonitration reaction of the corresponding glycal. Although this reaction is completely regioselective, it suffers in stereoselectivity (Scheme 3.8). Thus, an alternative strategy can be found in the displacement reaction between a 2-*O*-leaving group (usually a triflate) and

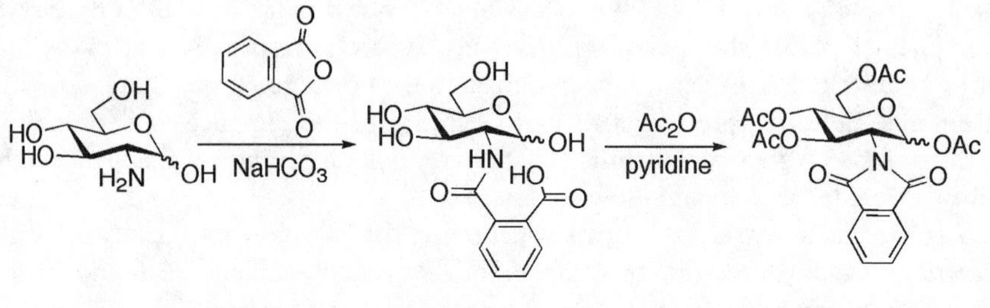

SCHEME 3.7 Formation of 1,3,4,6-tetra-*O*-acetyl-2-deoxy-2-phthalimido-D-glucopyranose.

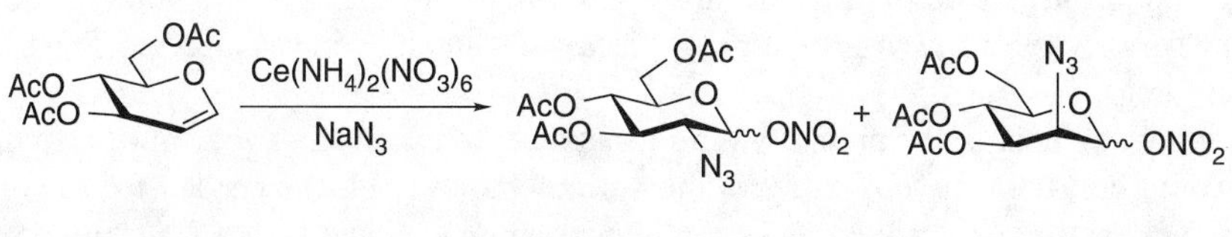

SCHEME 3.8 Azidonitration of 3,4,6-tri-*O*-acetyl-D-glucal.

an azide anion. The yields in the latter reaction are high, except in the case of α-*manno*-precursors. The azido group is stable to a wide range of reaction conditions, but is also readily reduced to an amino group by a variety of reagents (e.g., $H_2/Pd/C$, $NaBH_4/NiCl_2$, H_2S, Ph_3P/H_2O).

3.2.4 Carboxyl Protecting Groups

Carbohydrate carboxyl groups are present in natural oligosaccharides as uronic and ketoaldonic acids. These are found in nature as components of heparin (glucuronic and iduronic acid), bacterial lipopolysaccharides (Kdo and sialic acid) and human glycan structures (sialic acid; Figure 3.5).

As with amino groups, the protection of carboxyl groups is less strategically complicated due to their lower abundance. Almost exclusively, ester groups have been utilized as protection for the carboxyl group in oligosaccharide synthesis, and the variation in the esters used is not extensive. By far the most utilized is the methyl ester, which is essentially the standard carboxyl protecting group. Alternatives are benzyl and *tert*-butyl esters, both possible to cleave in the presence of other esters — a significant strategy when preparing ester-containing target structures. Additionally, internal lactones have been utilized (albeit infrequently) as protecting groups, thus simultaneously blocking the carboxylic acid and one of the hydroxyl groups in the molecule [38]. The carboxyl esters are formed from either the reaction between a carboxylate and an alkyl halide or the reaction between a carboxylic acid and an alcohol catalyzed either by an acid or a condensation reagent (typically DCC). By using an excess of the alcohol, the esterification can be performed in the presence of unprotected carbohydrate hydroxyl groups. A most efficient synthesis of methyl glucuronate is the basic methanolysis of the commercial 3,6-lactone (Scheme 3.9). Both methyl and *tert*-butyl esters are stable under Zemplén conditions (MeO^-/MeOH/solvent), whereas benzyl esters are transesterified to form the methyl ester. Methyl and benzyl esters are removed by saponification using sodium or lithium hydroxide. An alternative method for the cleavage of benzyl ethers is hydrogenolysis, and *tert*-butyl esters are removed under acidic conditions; typically, TFA in CH_2Cl_2.

From a strategic point of view, the problem with synthesis of carboxylic acid-containing oligosaccharides is mainly the different (often less reactive) and less investigated qualities of carboxylic acid-containing donors and acceptors as compared to neutral sugars. Considering protecting group strategies, the difficulties are not in the protection or deprotection of the carboxyl function (ester formation and cleavage are rather simple reactions), but rather in the compatibility

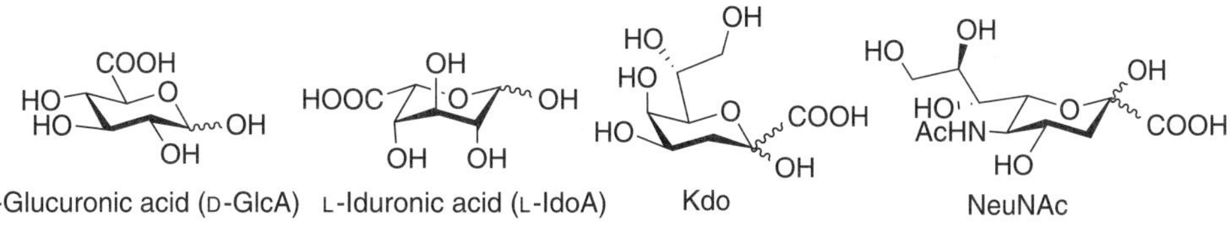

D-Glucuronic acid (D-GlcA) L-Iduronic acid (L-IdoA) Kdo NeuNAc

FIGURE 3.5 Some carboxyl-containing monosaccharides.

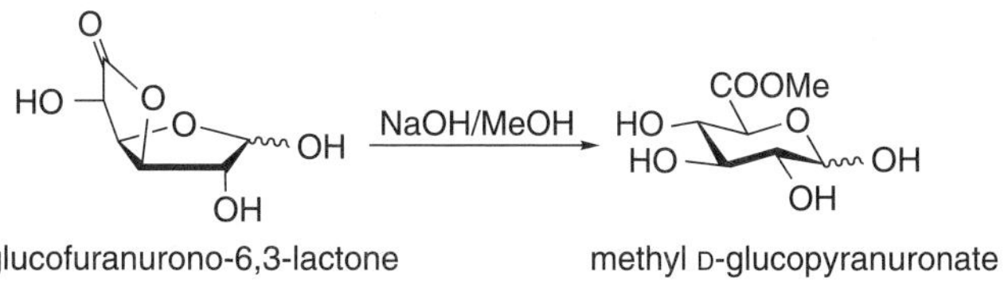

D-glucofuranurono-6,3-lactone → (NaOH/MeOH) → methyl D-glucopyranuronate

SCHEME 3.9 Formation of the methyl ester of D-glucuronic acid.

between the carboxyl and the hydroxyl protecting groups and their introduction and removal. Owing to the risk of ester cleavage and β-eliminations, strong basic conditions have to be avoided (alkylations are often difficult and low-yielding).

3.3 SELECTIVE PROTECTION METHODOLOGIES (REGIOSELECTIVE PROTECTION OF HYDROXYL GROUPS)

3.3.1 Selective Protection

In spite of recent developments, efficient regioselective glycosylations are only rarely possible. Therefore, one major part of oligosaccharide synthesis is the regioselective protection of saccharides. Several techniques have been developed over the years [39], and the most common are discussed below.

3.3.1.1 Utilizing the Different Reactivity of OH-Groups

To simplify, the reactivity order of hydroxyl groups in saccharides is primary OH > equatorial OH > axial OH, mainly due to steric factors [7]. Thus, the use of a bulky reagent (e.g., triphenylmethyl or *tert*-butyldiphenylsilyl chloride) and various basic conditions give a high yield of the 6-*O*-monoprotected hexose derivative. This is especially useful if the anomeric position is already protected (Scheme 3.10). Only under quite forced conditions is disubstitution possible. Additionally, regioselective acylation of the primary position is possible using less reactive reagents (e.g., acylimidazoles or mild conditions). Finally, regioselective acylations catalyzed by enzymes (lipases) are possible with the most common sites of reaction at primary positions [40].

Direct one-step protection (acylations, etherifications) of an unprotected saccharide is sometimes possible to produce a derivative with only one hydroxyl free. However, since the reactivity difference between the secondary hydroxyl groups is not that large, usually a complex mixture of differently protected compounds is obtained. Needless to say, given the need for extensive chromatographic separations and subsequent characterizations, this approach is rarely used in protecting group strategies.

Acetals, however, are easily regioselectively introduced (Schemes 3.11 and 3.12) [9]. Benzylidene acetals are preferentially formed as six-membered dioxane-type acetals (i.e., hexopyranosides form 4,6-*O*-benzylidene derivatives), whereas isopropylidene acetals are more stable as five-membered dioxolane acetals formed on vicinal *cis*-diols. However, the selectivity is not complete and methyl mannoside, for example, produces mixtures of the mono- and di-acetal compounds. The use of methyl propenyl ether yields mainly the kinetic product (the 4,6-*O*-isopropylidene acetal). Complementary to these traditional acetals is a rather recently developed acetal — the BDA-acetal [21]. In contrast to other acetals, this exclusively forms acetals between

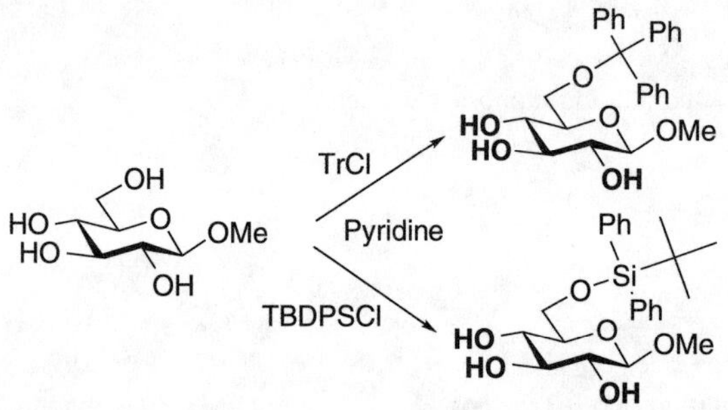

HO HO OH O OMe OH
TrCl Pyridine TBDPSCl
Ph Ph O Ph HO HO O OMe OH
Ph Si Ph O HO HO O OMe OH

SCHEME 3.10 Examples of selective protection of primary hydroxyl groups.

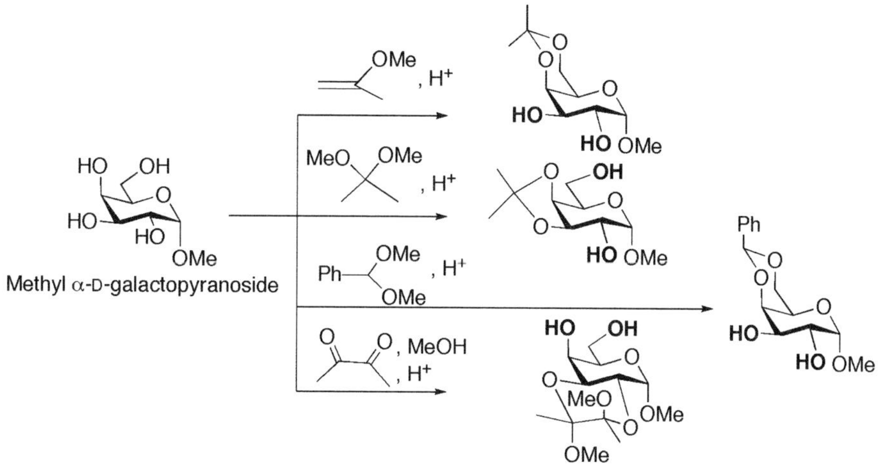

SCHEME 3.11 Acetal formation on methyl α-D-galactopyranoside.

vicinal *trans*-diols. This complementary regioselectivity, combined with the fact that they directly protect two hydroxyl groups and that benzylidene acetals can be transformed in various regioselective ways (Section 3.3.1.5), makes acetals most important in regioselective protecting manipulations.

3.3.1.2 Stannyl Activation

By reacting the hydroxyl groups of saccharides with tin oxide reagents, stannylene ethers and acetals are formed, which enhances the nucleophilicity of the oxygens in a regioselective way and makes the consecutive regioselective acylation or alkylation of saccharides possible [41]. Either a bis(trialkyltin) oxide or a dialkyltin oxide reagent is used. The former yields trialkylstannyl ethers, whereas the latter produces dialkylstannylene acetals. However, the exact structures of these are quite complex and not completely understood. By far the most used alkyl group in both these types of reagents is the *n*-butyl group. The activation can be performed in various solvents, the most common being methanol or toluene (benzene). In the latter case, a Dean-Stark trap is often used to remove the water formed in the reaction. The standard conditions are reflux conditions or microwave heating. Subsequent treatment of the saccharide tin complex with various electrophiles

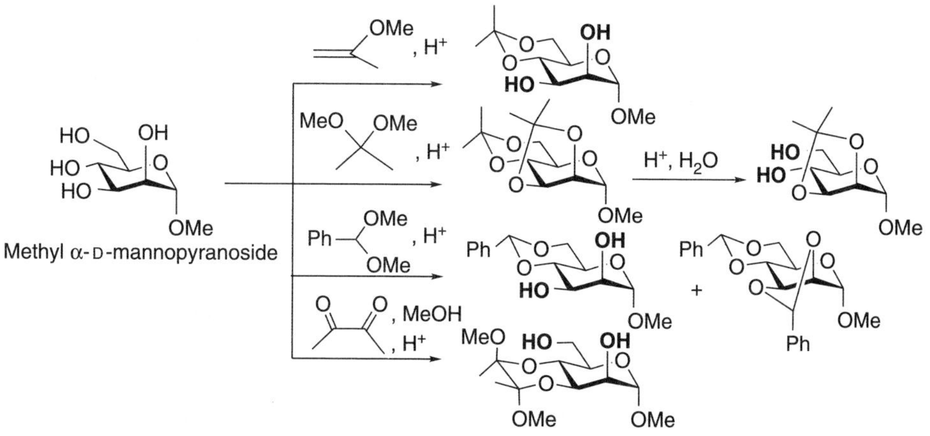

SCHEME 3.12 Acetal formation on methyl α-D-mannopyranoside.

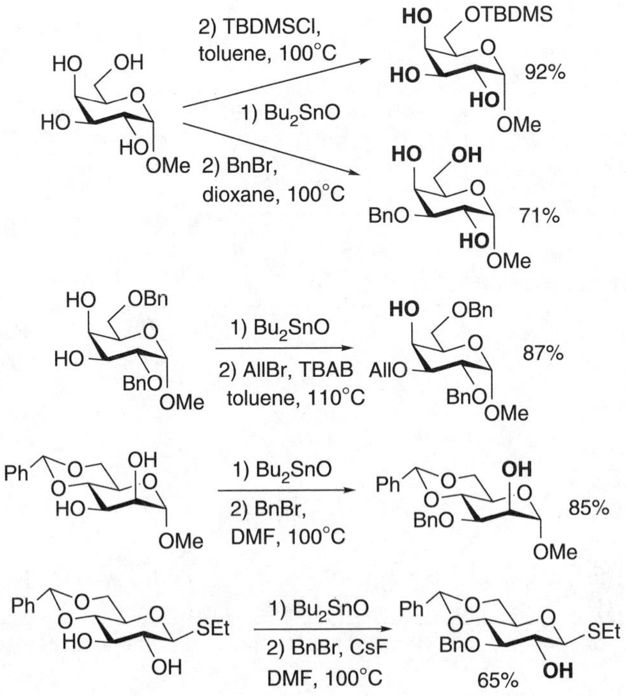

SCHEME 3.13 Examples of stannyl activated regioselective protection.

(acyl chlorides, silyl chlorides, alkyl halides) yields the corresponding esters or silyl or alkyl ethers. In this step, various solvents have been employed, usually DMF or toluene/benzene. The conditions for ester and silyl ether formation are mild (room temperature/a few hours), whereas alkylation requires rather forced conditions (reflux/several days). The addition of nucleophiles (e.g., bromide, iodide or fluoride ion) enhances the rate of the reaction.

The regioselectivity associated with stannyl activation is much the same irrespective of which type of alkyltin derivative is used in the activation step. The primary hydroxyl group and the equatorial hydroxyl group in a vicinal *cis*-dioxygen configuration are activated (Scheme 3.13). If both these motifs are present and compete, as with methyl β-D-galactopyranoside, then the selectivity depends on the electrophile and additives. Bulky groups prefer the primary position. These simple general rules are almost always correct, but the degree of selectivity is also dependent on other structural features such as anomeric and other protecting groups, and additives [42]. Thus, the selectivity is often lowered for thioglycosides and a 4,6-*O*-benzylidene or a 4,6-di-*O*-benzyl protection can affect the 2,3-selectivity.

3.3.1.3 Phase-Transfer Alkylations and Acylations

In this technique, a two-phase system (H_2O/CH_2Cl_2) is employed. The aqueous phase contains a base (NaOH, 5%) and the organic phase an electrophile, usually an alkyl halide. The derivative to be protected, usually a diol, is partitioned between the two phases. In the water phase, one of the hydroxyl groups of the reactant is deprotonated by the base. A phase-transfer reagent (most often a tetrabutylammonium salt) then transfers the oxyanion to the organic phase where it is alkylated. Owing to the higher lipophilicity of this selectively protected derivative, it is much less redistributed to the water phase and, accordingly, disubstitution is usually prohibited. However, disubstitution can be obtained by the use of higher base concentration (50%) or increase of the volume of the water phase. Typical reaction conditions involve refluxing for several days.

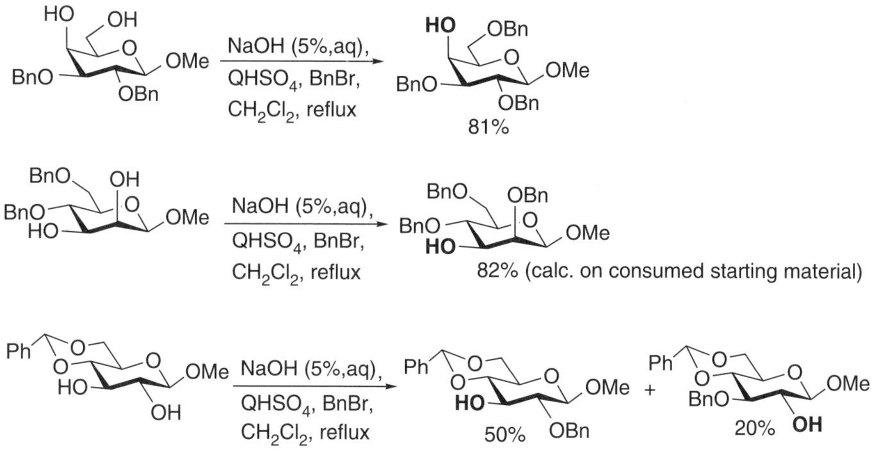

SCHEME 3.14 Examples of phase-transfer benzylations.

Regarding regioselectivity, primary hydroxyl groups are preferentially protected. Furthermore, with respect to 2,3-diols, the 2-hydroxyl group, being more acidic because of its proximity to the ring oxygen, is usually reacted (Scheme 3.14). The selectivity varies among the various sugars and, for 2,3-diols, mannose derivatives often give the best results. This is interesting because in mannose, the 2-hydroxyl group is axial. In this context, the use of phase-transfer methodologies is complementary to those involving tin activation.

Owing to the strongly basic conditions employed, esters are generally not compatible with this technique. However, base-stable esters, such as tosylates, are efficiently formed and with high regioselectivity (Scheme 3.15) [43]. Compared with the alkylations (compare tin activation), this reaction is much faster and is complete after a few hours at room temperature.

3.3.1.4 Cu(II) Activation

Another methodology applied to the monosubstitution of diols is the use of copper complexation of dianions. The dianion is first formed by reaction of a diol with two equivalents of NaH. The copper complex is then formed by addition of a copper salt. Reaction of the copper complex with various electrophiles (alkyl halides, acyl chlorides) then gives the selectively protected products. As with the phase-transfer technique, very little disubstitution is observed. However, as illustrated in Scheme 3.16, the regioselectivity is reversed (i.e., 4,6-diols give mainly 4-substitution and 2,3-diols give mainly 3-substitution). Using this technique, both alkylations (benzylation, allylation) and acylations (acetylation, benzoylation, pivaloylation) have been carried out. As usual, the degree of selectivity depends on reaction conditions and structural factors [44].

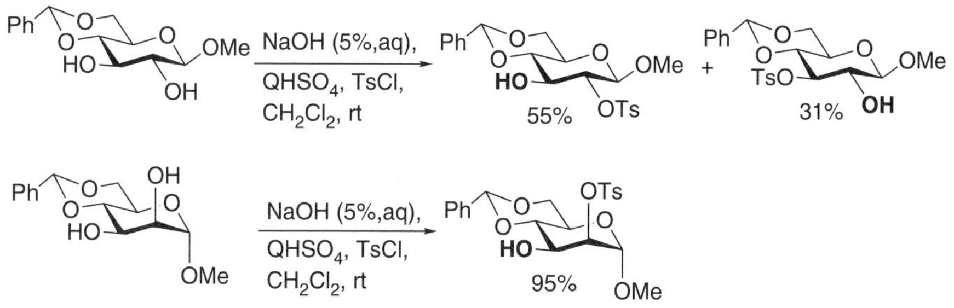

SCHEME 3.15 Examples of phase-transfer tosylations.

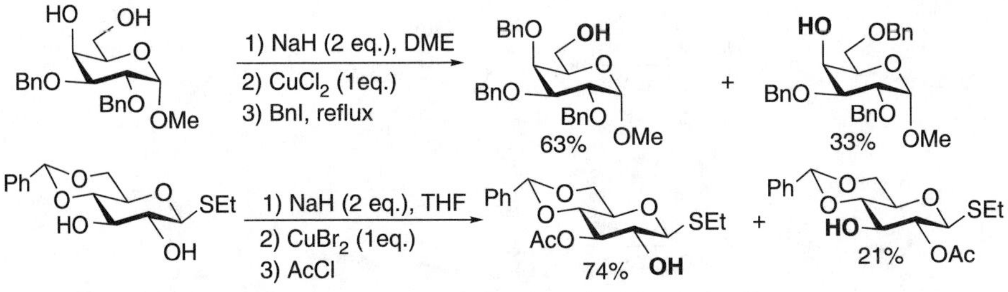

1) NaH (2 eq.), DME
2) $CuCl_2$ (1eq.)
3) BnI, reflux
63% 33%

1) NaH (2 eq.), THF
2) $CuBr_2$ (1eq.)
3) AcCl
74% 21%

SCHEME 3.16 Examples of regioselective protection from copper complexes.

3.3.1.5 Reductive Opening of Acetals

As mentioned above, acetals are among the most important protecting groups in protective group strategies because of their easy regioselective introduction. Their importance was further enhanced when methods became available to open up the formed acetal to yield alkyl ethers and a free hydroxyl group (Scheme 3.17) [45,46]. The reagents used are a hydride reagent in combination with a Lewis acid (or a proton acid). First, combinations of $LiAlH_4/AlCl_3$ were employed. From 4,6-*O*-benzylidene acetals, this yields the 4-*O*-benzyl derivative with high selectivity, especially from precursors with bulky substituents in the 3-position. In dioxolane benzylidene acetals of *cis*-diols (e.g., 2,3-*manno*- or 3,4-*galacto*), the selectivity depends on the configuration of the acetal with *exo*-phenyl derivative giving the equatorial benzyl ether and *endo*-phenyl the axial one with absolute selectivity. (Rule of thumb: *eXo* gives aXial hydroXyl.) A drawback of this methodology is that it is not compatible with various other functionalities (e.g., ester protecting groups). Later, $NaCNBH_3/HCl$ mixtures were introduced. For 4,6-*O*-benzylidene acetals, this gave the opposite selectivity (i.e., the 6-*O*-benzyl ether), whereas the selectivity was the same for dioxolane acetals. This reagent is compatible with esters and also with allyl groups, which allowed the regioselective opening of acrolein acetals with the same regioselectivity. A slight modification also made opening of *p*-methoxybenzylidene acetals possible. By changing the Lewis acid and solvent, either selectivity for 4,6-acetals could be obtained. By changing only the solvent, the same flexibility was shown for benzylidene acetals and the $Me_3NBH_3/AlCl_3$ reagent. THF as solvent gave the 6-*O*-benzyl ether in a slow and mild reaction, whereas toluene (or diethyl ether/CH_2Cl_2 mixtures) gave the 4-*O*-benzyl ether in a fast reaction accompanied by some acetal hydrolysis. Subsequently, there has been a continuous development of new reagents, all of which are variations of the same general theme (Lewis acid/hydride reagent), optimizing yield and selectivity for specific derivatives.

Benzylidene acetals can also be opened under oxidative conditions, typically NBS in CCl_4, to give benzoyl ester protected halogen derivatives, and thereby providing an entry into deoxy carbohydrate compounds (Scheme 3.18) [45]. For 4,6-*O*-benzylidene derivates, the regioselectivity is high for the 4-*O*-benzoyl-6-bromo-6-deoxy derivative. Preprotection of the 2- and 3-hydroxyl groups usually increases the yield in the oxidative cleavage reaction.

3.3.1.6 Orthoester Opening

As mentioned earlier, orthoesters are acid labile and, accordingly, not stable under glycosylation conditions. However, they are important intermediates in protecting group schemes to create building blocks to be used in subsequent glycosylations. One of the major advantages of orthoesters is that they can be regioselectively opened by mild acid hydrolysis to yield the corresponding ester derivative exposing a hydroxyl for further reactions (Scheme 3.19) [47]. Of special significance is that the selectivity is opposite to many other methods, since opening of an orthoester protecting a vicinal *cis*-diol gives the ester on the axial hydroxyl group. In addition, chloroacetates can be regioselectively introduced using this methodology [48].

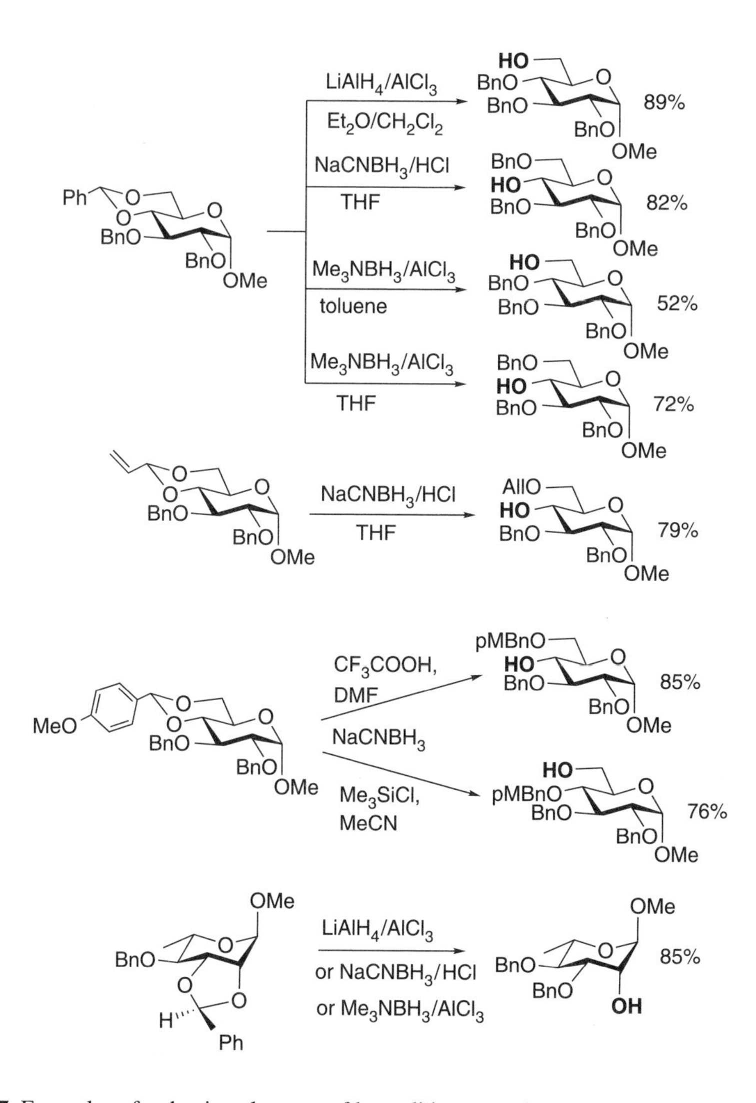

SCHEME 3.17 Examples of reductive cleavage of benzylidene acetals.

NBS, CCl_4
$CaCO_3$
α 15%; β quantitative

NBS, CCl_4
$CaCO_3$
65%

SCHEME 3.18 Examples of oxidative cleavage of benzylidene acetals.

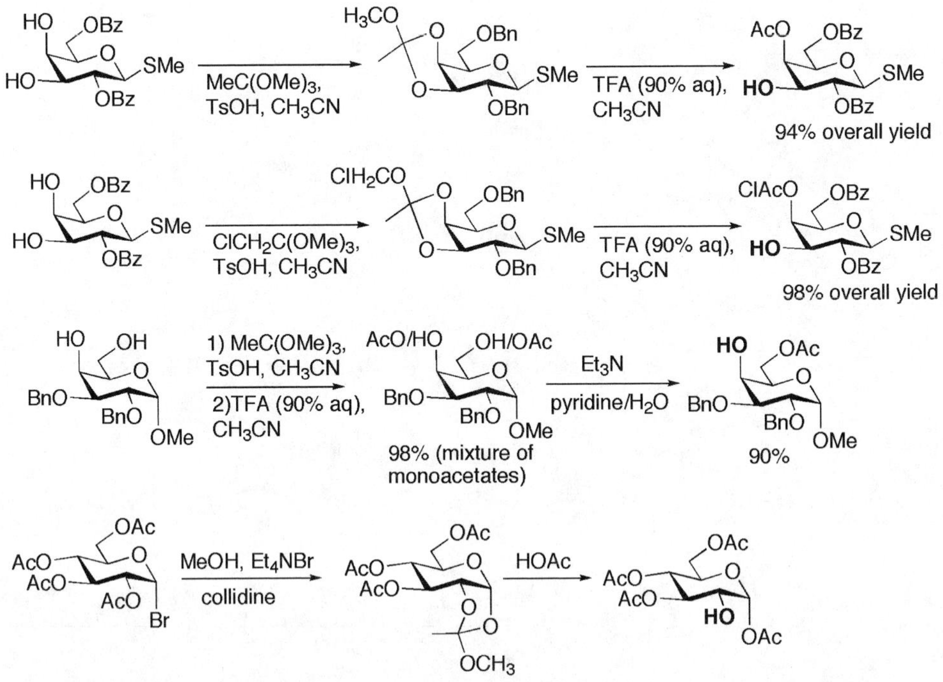

SCHEME 3.19 Examples of orthoester formation and openings.

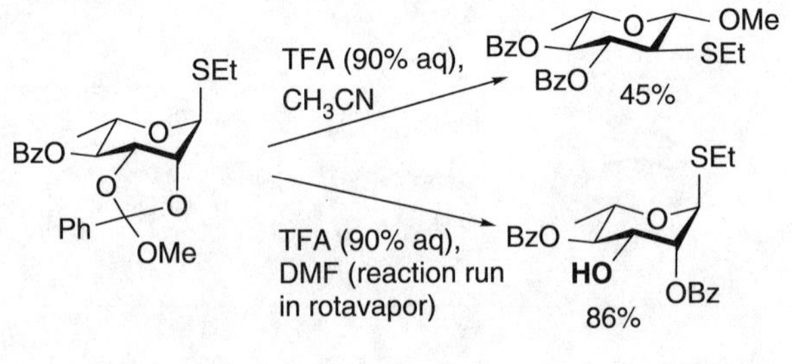

SCHEME 3.20 Examples of orthoester openings in thioglycosides.

Care has to be taken to avoid acyl migration to the uncovered equatorial hydroxyl group. However, since the migration is rather slow under acidic conditions, this can normally be avoided. Another solution is to use orthobenzoates instead of orthoacetates, since benzoates migrate more slowly than acetates. In 4,6-*O*-orthoesters, the opening gives a mixture of the 4-*O*- and the 6-*O*-ester, which makes it less useful for synthetic purposes. However, utilizing successive acyl migration, a good selectivity for the 6-*O*-acetate can be obtained [49]. When 2,3-*O*-orthoesters of thioglycosides are opened, care must be taken to avoid participation of the sulfur (Scheme 3.20). Continuous removal of the methanol formed in combination with DMF as solvent favors the formation of the selectively protected thioglycoside [50,51].

3.3.2 Selective Deprotection

The methods to achieve regioselectively protected derivatives by selective deprotection are less common. However, a few standard methods utilize this approach [52]. The rate difference in

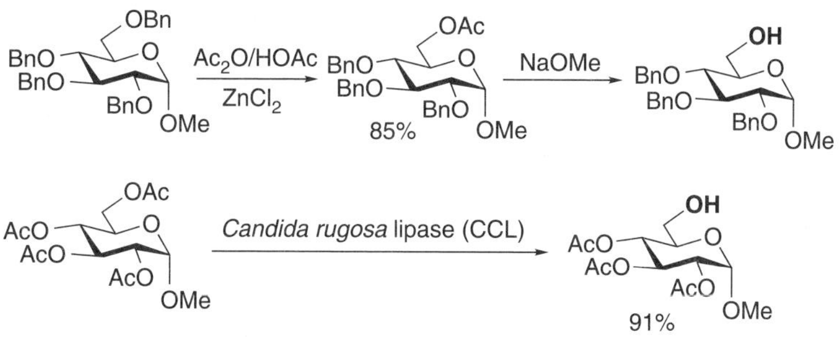

SCHEME 3.21 Examples of regioselective removal of primary protecting groups.

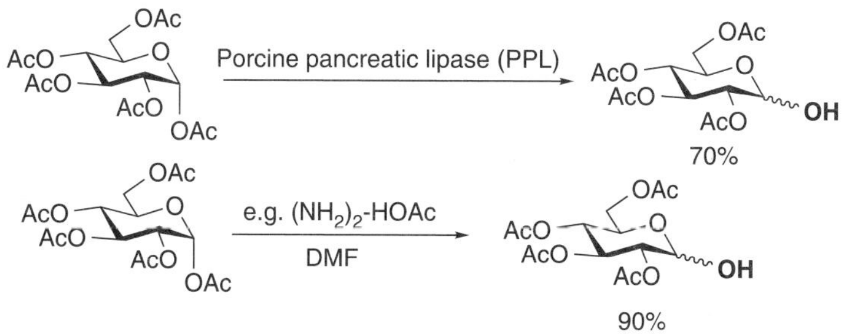

SCHEME 3.22 Examples of regioselective removal of anomeric protecting groups.

acetolysis of primary (as compared with secondary) benzyl ethers is high enough to allow selective removal (Scheme 3.21). The obtained 6-*O*-acetate can then be removed to expose the 6-hydroxyl group.

Selective deacylations of primary ester groups in the presence of secondary are also possible using enzymes (i.e., various lipases) [40]. Other lipases show selectivity for the anomeric position (Scheme 3.22). Also, as mentioned above, anomeric esters are more labile than other esters and can be removed selectively by mild base treatment. Furthermore, anomeric silyl ethers can be removed selectively on treatment with mild acid.

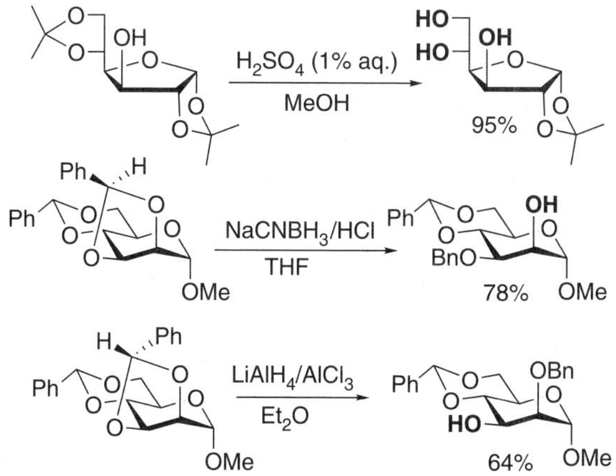

SCHEME 3.23 Examples of regioselective removal and opening of acetal protecting groups.

Dioxane isopropylidene acetals are less stable than dioxolane ones. Hence, mild acid hydrolysis of 2,3;4,6-di-*O*-isopropylidene-mannopyranosides gives the corresponding 2,3-monoacetal derivative in good yield (Scheme 3.12). An even larger difference in stability is found between the two dioxolane isopropylidene acetals in 1,2;5,6-di-*O*-isopropylidene-α-D-glucofuranose (due to larger stability of the bis-fused ring system), where the 5,6-acetal can be removed almost exclusively (Scheme 3.23). Selective hydrolysis of 2,3;4,6-di-*O*-benzylidene-mannopyranosides is not feasible. However, selective reductive acetal opening of the 2,3-acetal is possible, yielding 4,6-*O*-benzylidene-2- (from the *endo*-phenyl) or -3- (from the *exo*-phenyl) -*O*-benzyl derivates [45].

3.4 SELECTIVE PROTECTION STRATEGIES

3.4.1 MONOSACCHARIDES

When planning a protecting group strategy for an oligosaccharide synthesis, several factors must be considered, such as substituents and functional groups in the target structures, anomeric configuration in the monosaccharide units and the order in which to introduce these. Irrespective of the choice of a block or linear synthesis, most protecting group manipulations will be performed on monosaccharide derivatives (if commercial, di- and trisaccharides can be utilized), and, ideally, only simple deprotection reactions have to be performed at the oligosaccharide level. Below are examples of strategies concerning how to protect regioselectively the three most common monosaccharides (mannose, galactose and glucose) to obtain a derivative with one hydroxyl group free and using the groups and methodologies discussed here. In some cases, several suggestions are given, allowing for the flexibility needed in the continuing oligosaccharide synthesis (i.e., 2-nonparticipating groups in donors for 1,2-*cis*-linkages and participating groups for 1,2-*trans*-linkages, ether protecting groups to increase reactivity and acyl protecting groups to decrease reactivity, and compatibility and orthogonality between permanent and temporary protecting groups).

Some protection strategies are the same for all of the starting monosaccharides. This is especially true for pathways to free primary 6-OH-groups. Other strategies vary depending on the precursor. Since the reactivity differences between secondary hydroxyls are not that large, one-step selective protection is rarely used. Instead, multistep sequences, most often based on regioselective acetalization and subsequent regioselective monoprotection of obtained diols, are the standard methodologies.

6-OH free: The selective protection of the primary position of all these three monosaccharides can be performed in a number of ways (see above). Standard methods include tritylation or silylation followed by acetylation, benzoylation or benzylation and detritylation or desilylation. The reaction sequence can be performed on a large scale, and the only problem is the care needed to avoid the possibility of 4- to 6-acetyl migration during 6-*O*-deprotection. In addition, 4,6-*O*-benzylidenation, followed by 2,3-diprotection and reductive benzylidene opening using a proper reagent, gives the 6-OH derivative — this time with the possibility of having orthogonal protecting groups in the 4- and in the 2,3-positions.

4-OH free: Here, for all the three monosaccharides, a derivative with the 4-OH free can be obtained from the 4,6-*O*-benzylidene derivative through regioselective reductive opening. Formation of the 4,6-acetal with consecutive 2,3-protection (acetylation, benzoylation, benzylation) and reductive opening using various reagents yields the desired compounds. In mannosides, the selective formation of the 4,6-*O*-acetal is not trivial, but can be accomplished with a 50 to 70% yield.

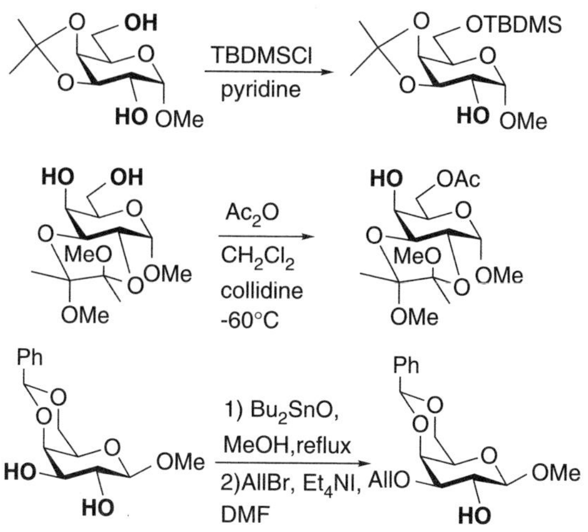

SCHEME 3.24 Examples of regioselective protection of D-galactose.

3.4.1.1 Galactosides

For galactose, the 4-hydroxyl is axial and least reactive, at least towards acylation. Benzoylation at low temperature with limited amounts of benzoyl chloride gives a good yield of the 2,3,6-tri-*O*-benzoylated derivative. Other strategies are two-step syntheses based on a regioselective acetalization step followed by a regioselective protection of a diol. Hence, the 4,6-*O*-benzylidene, the 3,4-*O*-isopropylidene and the 2,3-BDA-acetals (Scheme 3.11) are all excellent precursors for subsequent reactions. Protections of the primary positions in the two latter compounds yield derivatives with a free 2-OH and a free 4-OH, respectively (Scheme 3.24). The regioselective protection of the 2,3-diol in the 4,6-*O*-benzylidene derivative is slightly more complex, and various reagents give quite different results. However, using tin activation, good yields of 3-*O*-protection and thus derivatives with a free 2-OH are obtained.

Other possibilities are four-step sequences comprising full protection of the acetal diols, followed by removal of the acetal and, finally, regioselective protection of the obtained diol (Scheme 3.25). For example, 2,6-protection (acetylation, benzoylation or benzylation) of a 3,4-*O*-isopropylidene derivative and successive acetal cleavage yields the 3,4-diol. Subsequent 3-*O*-protection via tin-activated silylation, alkylation or acylation, or 4-*O*-protection via orthoester or benzylidene (*endo*) formation-opening sequences is then possible.

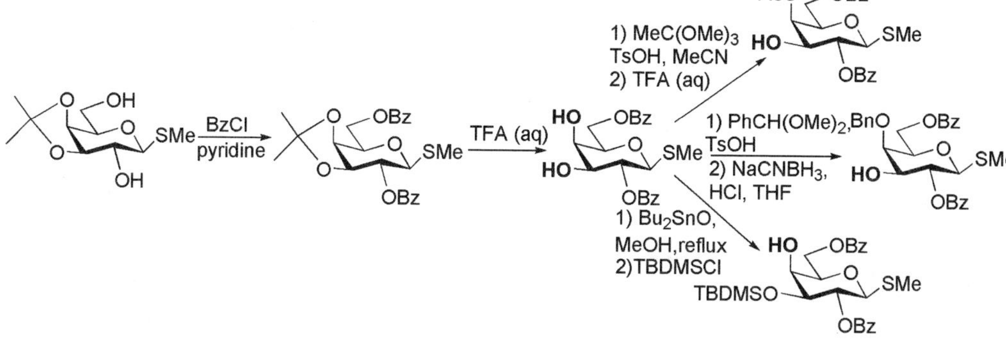

SCHEME 3.25 Further examples of regioselective protection of D-galactose.

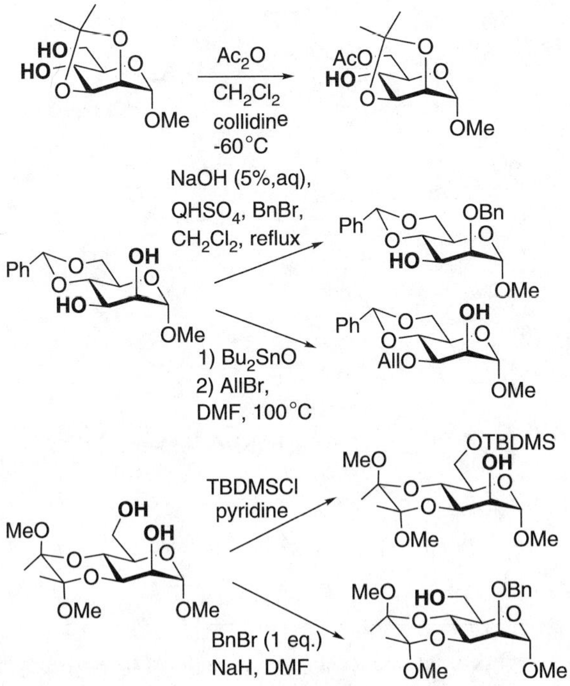

SCHEME 3.26 Examples of regioselective protection of D-mannose.

3.4.1.2 Mannosides

As with galactose, isopropylidene and BDA acetals can be used but with opposite regioselectivity (Scheme 3.26). The BDA-acetal will form between the 3- and the 4-*trans* hydroxyl groups. Isopropylidenation gives the 2,3;4,6-di-*O*-acetal derivative, mild hydrolysis of which produces a fair yield of 2,3-*O*-isopropylidene derivative. Protection of the primary position then gives the 2-OH or the 4-OH compounds, respectively. Surprisingly, benzylation (but not *p*-methoxybenzylation) of the BDA-acetal derivative was found to give predominantly the 2-*O*-benzyl derivative and, accordingly, an alternative path to a 6-OH compound. The 4,6-*O*-benzylidene derivative can be either 3-*O*-protected (tin activation) or 2-*O*-protected (phase-transfer, orthoester opening). Similar acceptors can be obtained by selective opening of the 2,3;4,6-di-*O*-benzylidene derivative (Scheme 3.23).

For the BDA-acetal, 2,6-protection and acetal removal yields the 3,4-diol. This diol can then be 3-*O*-protected through tin-activated silylation, alkylation or acylation (Scheme 3.27). Other ways to diols are tin activation and subsequent alkylation yielding the 3,6-protected 2,4-diol derivative, which can be transformed into a 3,6-diol. Opening of 2,3;4,6-di-*O*-benzylidene derivatives yields various diols depending on the reagents used and the stereochemistry of the 2,3-*O*-acetal (e.g., the 2,4-diol).

3.4.1.3 Glucosides

Protection of glucose to obtain one free hydroxyl group, especially 2-OH or 3-OH, is the most difficult case because all hydroxyl groups are equatorial. Derivatives with free 6- and 4-OH groups can be obtained according to the general procedure, and a 2,3-diol is easily obtained as the 4,6-benzylidene derivative. However, the reactivity between these two hydroxyl groups differs depending on the reaction conditions used, and must frequently be optimized for each

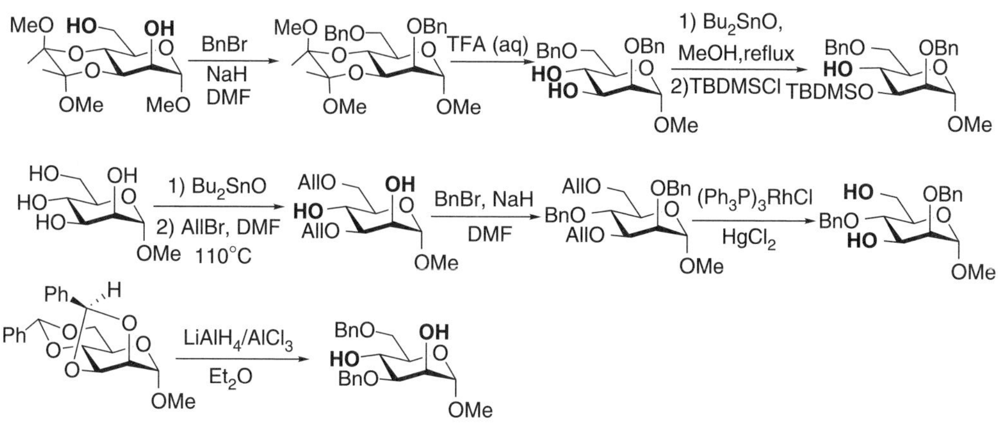

SCHEME 3.27 Further examples of regioselective protection of D-mannose.

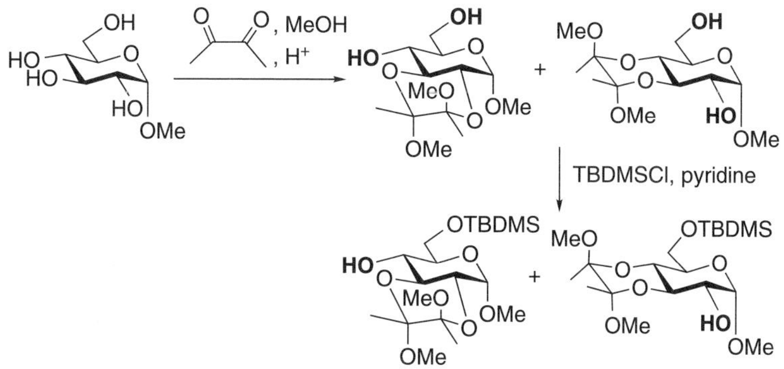

SCHEME 3.28 Examples of regioselective protection of D-glucose.

reaction and precursor. Standard methodologies are those mentioned above (i.e., phase-transfer benzylations, copper activation and tin activation). Introduction of a BDA-acetal gives a 1:1 mixture of the 2,6- and 4,6-diols, which can be further transformed into 2-OH and 4-OH acceptors, respectively (Scheme 3.28).

3.4.2 Disaccharides

Because selective protections of disaccharides (containing more hydroxyl groups) are generally more difficult than monosaccharides, there must be either very safe and high-yielding methods or a cheap starting material available to make this approach worthwhile. Otherwise, the advantage of gaining one glycosidation step is quickly counteracted by the more complex protecting group strategy needed at the disaccharide level. Two very cheap starting materials are lactose and sucrose. Lactose is perhaps the most common structural motif in natural carbohydrate structures and glycoconjugates (e.g., human glycolipids and glycoproteins), and several strategies to protect it regioselectively for continued oligosaccharide syntheses have been developed. Sucrose derivatives found in nature, on the other hand, are mainly various fatty acid esters. Here, the selective protection must be performed at the disaccharide level because the glycosidic linkage in sucrose are most difficult to make chemically. Below are examples of the selective protection of lactose and sucrose. In most cases, the methodologies discussed above are adequate but, occasionally, unique solutions have to be developed.

3.4.2.1 Lactose

The obvious targets in protective group strategies for lactose and lactosides are the primary hydroxyl groups and the axial 4′-hydroxyl group. Because the linkage to the glucose unit occupies the 4-position, it is not difficult to differentiate between the two primary positions (e.g., by the introduction of a 4′,6′-*O*-benzylidene acetal). Protection of residual hydroxyl groups, followed by benzylidene opening either way, then yield the 6′-OH and the 4′-OH derivatives (Scheme 3.29). Selective protection (silylation, tritylation) of the benzylidene compound gives orthogonal 6-OH-protection and, hence, a route to compounds with a free 6-OH group [53].

Selective 6′-*O*-silylation is also possible using tin activation, offering an alternative way to 6′-OH derivatives [54]. Thermodynamic isopropylidenation gives, as discussed, the 3′,4′-acetal, which then can be transformed into 3′-OH derivatives or 4′-OH derivatives, as described for galactose derivatives above (Scheme 3.30) [55]. Often, the 3′,4′-diol is used directly as an acceptor in 3′-*O*-regioselective glycosidations.

Routes to 2′-OH and 3-OH acceptors, however, are not directly apparent. Two ways to 2′-OH lactose derivatives are polyisopropylidenation [53] or consecutive use of lipases [56]. The first strategy exploited the finding that isopropylidenation of lactose under forced conditions yields the

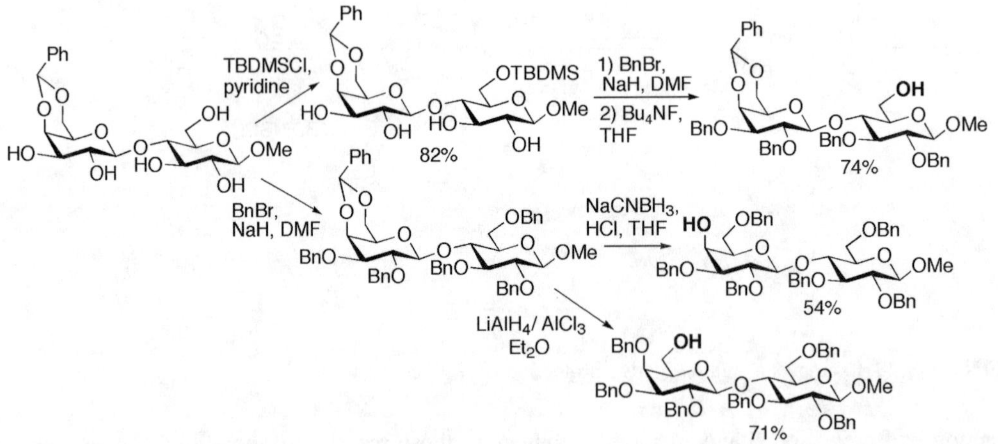

TBDMSCl, pyridine; 82%
1) BnBr, NaH, DMF 2) Bu_4NF, THF; 74%
BnBr, NaH, DMF
$NaCNBH_3$, HCl, THF; 54%
$LiAlH_4$/ $AlCl_3$ Et_2O; 71%

SCHEME 3.29 Strategies to 6-, 4′- or 6′-OH methyl lactoside derivatives from a 4′,6′-*O*-benzylidene precursor.

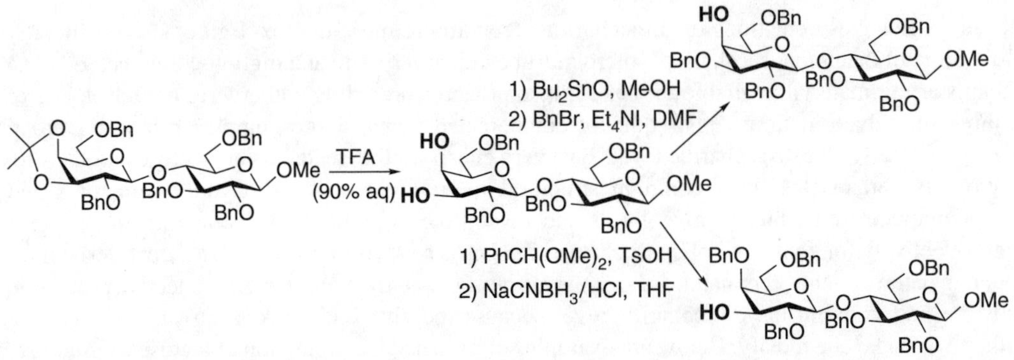

TFA (90% aq)
1) Bu_2SnO, MeOH 2) BnBr, Et_4NI, DMF
1) $PhCH(OMe)_2$, TsOH 2) $NaCNBH_3$/HCl, THF

SCHEME 3.30 Strategies to 4′- and 3′-OH lactoside derivatives from a 3′,4′-*O*-isopropylidene precursor.

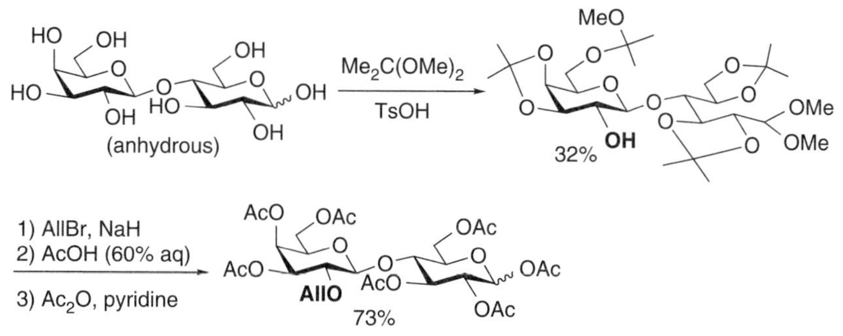

SCHEME 3.31 Strategy to an orthogonally 2′-*O*-protected lactose derivative.

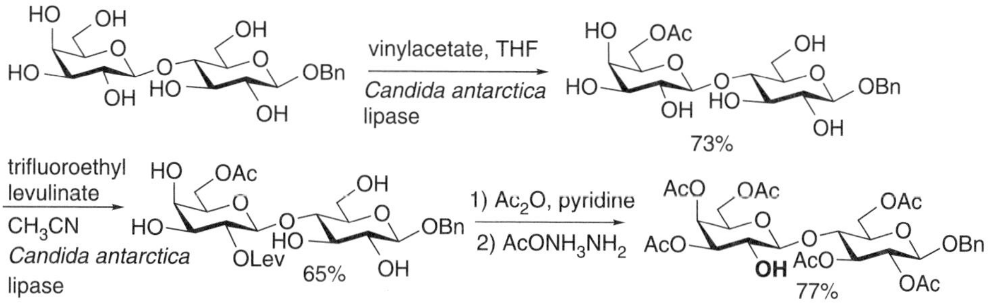

SCHEME 3.32 Strategy to a 2′-OH lactoside compound.

2,3;5,6;3′,4′;6′-tetra-*O*-isopropylidene-1,1-dimethyl acetal with a free 2′-hydroxyl group in a one-step reaction in an attractive yield (Scheme 3.31). Allylation, acetal hydrolysis and peracetylation then gave a 2′-orthogonally protected derivative, which could be further processed to acceptors or donors.

In the latter approach, lipase-assisted acetylation of benzyl lactoside was found to be selective for the 6′-OH group (Scheme 3.32). Subsequent lipase acylation (this time using a levulinate as reagent) gave high regioselectivity for the 2′-hydroxyl group. Consecutive peracetylation and removal of the levulinoyl ester (hydrazine acetate treatment) then gave the 2′-OH derivative in high overall yield.

It has also been found that the 3-OH group is the least reactive of the hydroxyl groups in lactose under many reaction conditions (Scheme 3.33). Accordingly, direct benzoylation of lactose yielded the 3-OH derivative in 24% crystalline yield [7], tin activated benzoylation of TMSE β-lactoside gave 85% of the 3-OH lactoside [57] and phase-transfer benzylation of benzyl lactoside produced 26% of the corresponding 3-OH compound [39].

Pathways to selective formation of 2-OH derivatives can utilize protected glycal derivatives of lactose (lactal). Epoxidation and subsequent epoxide opening with alcohols occurs regioselectively to give 2-OH-derivatives. If the epoxidation is not stereospecific, then both *manno*- and *gluco*-configurations will be obtained. However, using dimethyldioxirane, an excellent selectivity for the β-lactoside configuration was obtained (Scheme 3.34) [58].

3.4.2.2 Sucrose

Sucrose is probably the cheapest starting material in the world, available in vast quantities from sugar canes and beets. Nonetheless, its regioselective protection is a real challenge.

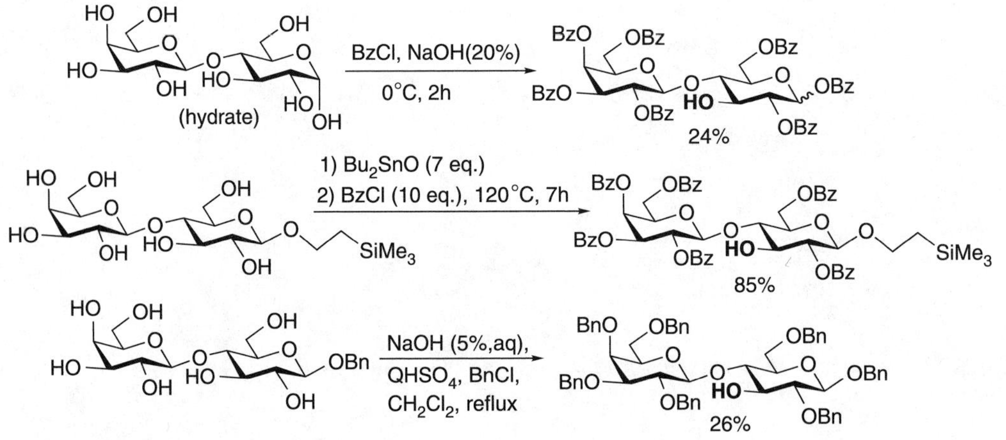

SCHEME 3.33 Strategies towards 3-OH lactose derivatives.

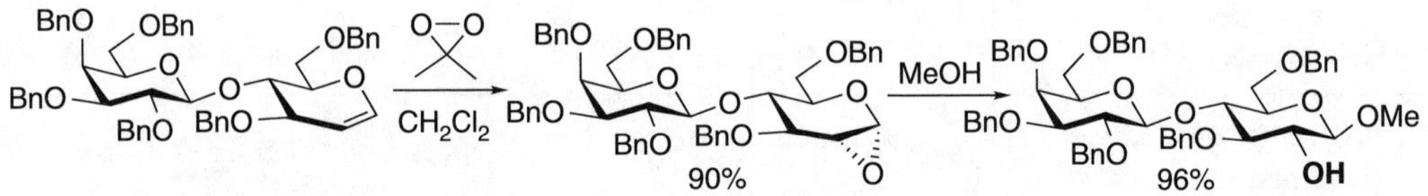

SCHEME 3.34 Strategy to a 2-OH lactoside derivative.

Of the eight hydroxyl groups, the only obvious targets for regioselective protection/deprotection are the primary hydroxyl groups, among which different reactivity is observed with the 1′-hydroxyl being the least reactive. Thus, tritylation of sucrose using different amounts of reagent yields the 6- and 6′-mono-, 6,6′-di- or the 1′,6′,6-trisubstituted derivative [59]. After acetylation followed by detritylation under conditions which promote 4→6 acetyl migration, the latter can give 2,3,6,3′,4′-penta-*O*-acetylsucrose, with the 4,1′ and 6′ hydroxyl groups free (Scheme 3.35). Chlorination and deacetylation of this intermediate give sucralose,

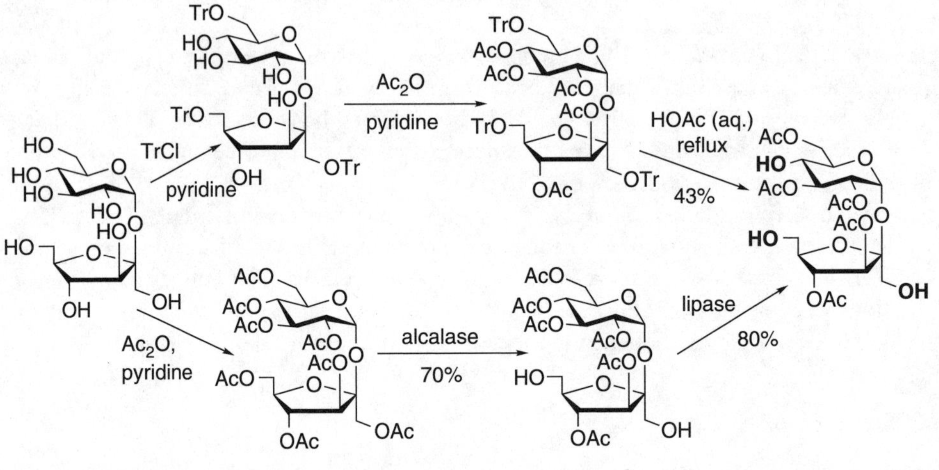

SCHEME 3.35 Strategies towards 2,3,6,3′,4′-penta-*O*-acetyl-sucrose.

a commercial sweetener many times sweeter than sucrose [60]. An alternative approach to this intermediate (and sucralose) is to make a regioselective deacetylation of peracetylated sucrose using enzymes in a two-step synthesis. The second reaction is performed under conditions that also facilitate the 4→6 acetyl migration [61].

Fatty acid esters of sucrose are common natural products and flavor components in many plants. To synthesize derivatives where different fatty acids esterify the various hydroxyl groups (especially in the glucose part), a sophisticated protecting group pattern had to be introduced. Such a pattern would preferably possess orthogonal protecting groups at all glucose hydroxyl groups. When reduced in practice, only protecting groups and methods described herein were used, but the regioselectivity in several of the steps had to be investigated [62,63]. The outcome of isopropylidenation of sucrose is not obvious. Methyl glucopyranosides form the 4,6-acetal, whereas methyl fructofuranosides are essentially inert. Accordingly, sucrose forms the 4,6-acetal. However, under more forced conditions, the 4,6;2,1′-di-*O*-isopropylidene derivative, containing a surprisingly stable eight-membered ring acetal, is formed in good yield (50%; Scheme 3.36). Acetylation yields a crystalline compound, which is easily isolated. In addition, it was found that the 3-*O*-acetyl group is much more base stable than the others and mild deacetylation thus yielded the 3′,4′,6′-triol in high yield, which allowed the introduction of permanent benzyl protecting groups at these positions. Because an acetate was present in the precursor, Ag_2O was used as base in the alkylation reaction to avoid concomitant deacetylation.

The acetate was then changed into an allyl group. To introduce a benzyl group also in the 1′-position, selective hydrolysis of the 1′,2-acetal was attempted without success. Hence, both isopropylidene acetals were removed. Subsequent *p*-methoxybenzylidene acetal formation was regioselective for the 4,6-positions producing the 1′,2-diol. Phase-transfer benzylation of this diol gave a good selectivity for the secondary 2-OH group. Hence, a regioselective phase-transfer

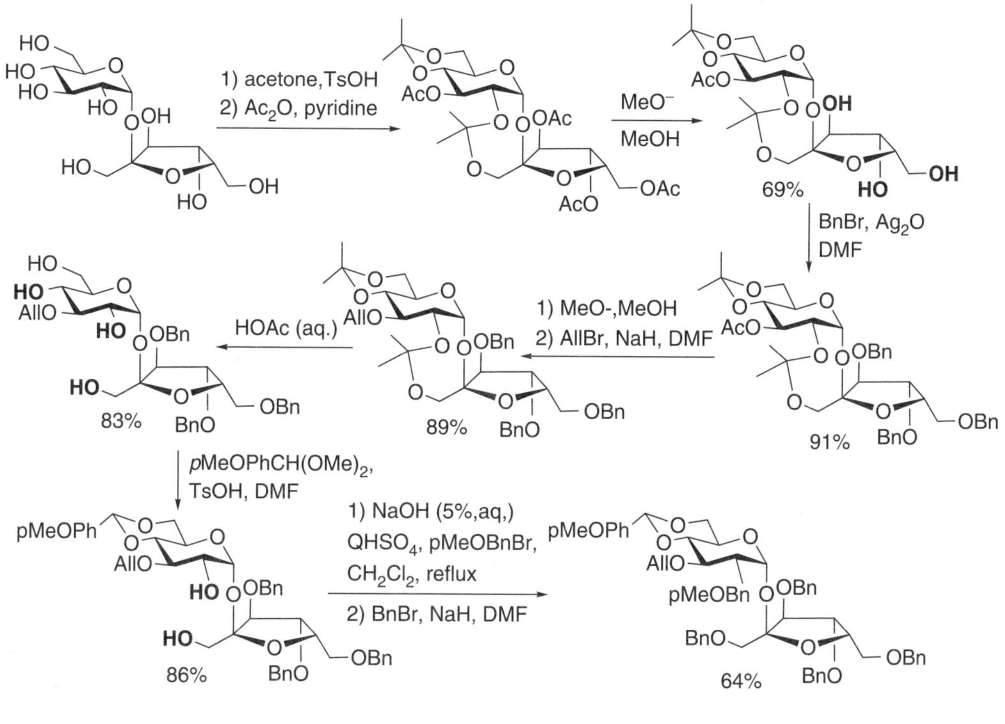

SCHEME 3.36 Regioselective protection of sucrose.

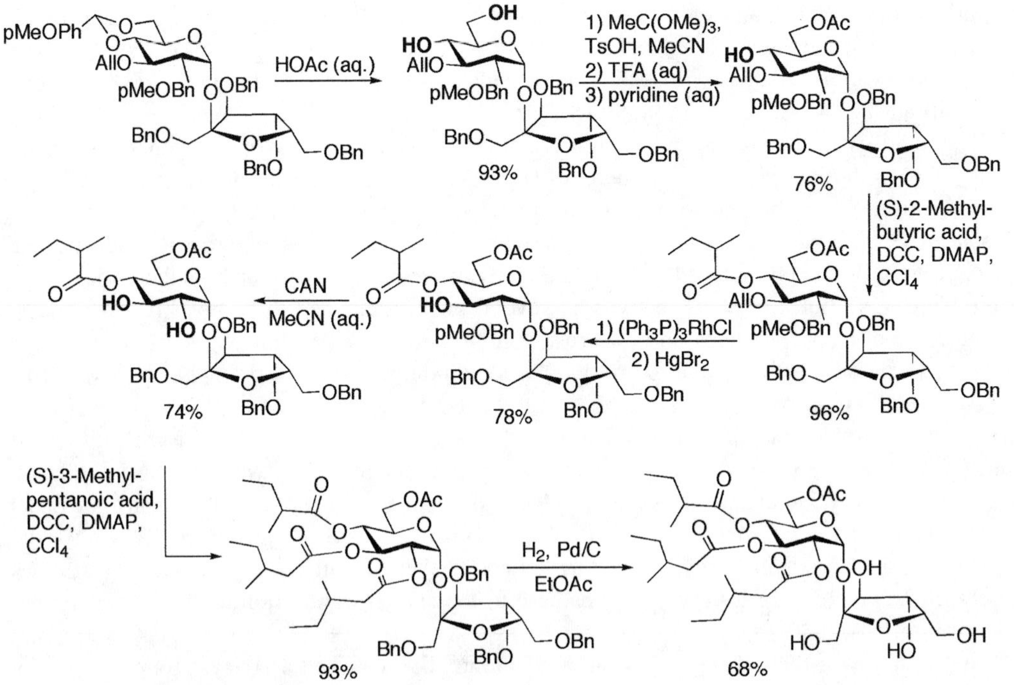

SCHEME 3.37 Synthesis of partially esterified sucrose derivatives.

p-methoxybenzylation was first performed and was followed by a benzylation to get to the desired intermediate, with permanent benzyl protecting groups on all fructose hydroxyl groups and temporary orthogonal groups on all glucose hydroxyl groups. From this intermediate, any derivative with a different fatty acid ester substitution pattern in the glucose part can be produced. An example is shown in Scheme 3.37 [63].

3.4.3 Oligosaccharides

3.4.3.1 Lewis^x

Given that the Lewis^x structure (Figure 3.6) was found to have several interesting biological functions (such as being tumor-associated and involved in inflammation processes), the syntheses of various derivatives of this structure became of interest. During the 1990s, several syntheses of the trimeric structure were published [64–68]. All approaches to this complex target structure generally only used protecting groups and methodologies described in this chapter. However, as expected, the strategies varied. All approaches started from monosaccharide building blocks, although it should be possible to start from a disaccharide lactosamine precursor. Apparently, the advantage of simpler regioselective protection of monosaccharides was more important than gaining one glycosylation step. Three of the syntheses utilized a disaccharide building block corresponding to the lactosamine moiety, whereas the other two used a trisaccharide building block comprising the complete Lewis^x structure. The former syntheses then required two temporary protecting groups, one at the 3-position of GlcNAc for later introduction of the fucose unit and one at the 3-position of Gal for couplings between the blocks. Many approaches acknowledge that, in glycosylations of galactose 3,4-diol or 2,3,4-triol acceptors, position 3 is by far the most reactive. A short discussion and comparison of three of

β-D-Gal*p*-(1→4)-β-D-GlcNAc*p*
↑
α-L-Fuc*p*
Lewis^{x} trisaccharide

β-D-Gal*p*-(1→4)-β-D-GlcNAc*p*-(1→3)-β-D-Gal*p*-(1→4)-β-D-GlcNAc*p*-(1→3)-β-D-Gal*p*-(1→4)-β-D-GlcNAc*p*
↑ ↑ ↑
α-L-Fuc*p* α-L-Fuc*p* α-L-Fuc*p*
Lewis^{x} trimer nonasaccharide

FIGURE 3.6 Lewis^{x} trisaccharide and trisaccharide trimer.

the approaches is presented below. The discussion is mainly focused on the protecting group strategies employed.

The first published synthesis [64] introduced the complete protecting group pattern at the monosaccharide level using standard methodology (Scheme 3.38). The Gal moiety was protected at the anomeric position as a methyl thioglycoside. Benzylidenation (4,6-protection), regioselective monochloroacetylation (temporary 3-*O*-protection) and subsequent acetylation (2-*O*-participating protection) then gave the desired protection pattern. Moreover, the anomeric position of the $GlcNH_2$ was protected as a thioglycoside and the amine as a phtalimide. Benzylidenation (6-*O*-protection, temporary 4-*O*-protection), *p*-methoxybenzylation (temporary 3-*O*-protection) and reductive benzylidene opening gave a 4-OH acceptor. Transformation of the galactose thioglycoside to the corresponding bromide donor and coupling then yield the key disaccharide precursor. Coupling to a spacer and removal of the temporary chloroacetyl protecting group gives a disaccharide acceptor, the coupling of which (to the same donor) yields a tetrasaccharide. Reiteration of the procedure yields a hexasaccharide with the intrinsic possibility to continue this type of elongation. Instead, the other temporary protecting group, the *p*-methoxybenzyl group, is removed to give a triol acceptor (87% yield), which is fucosylated to give the target nonasaccharide after deprotection. All the yields in the protecting group manipulations are high to excellent, the lowest being the regioselective chloroacetylation (59%).

Another synthesis using disaccharide building blocks was performed using minimal protecting group manipulations [68], made possible by employing an enzyme in the fucosylation step (avoiding the necessity for the regioselective introduction of a temporary protecting group in the $GlcNH_2$ 3-position) and by taking advantage of regioselective glycosylation and deacetylation reactions (Scheme 3.39). The $GlcNH_2$ moiety carried a *tert*-butyldimethylsilyl anomeric and a Troc amino protecting group. Regioselective 6-*O*-silylation (TBDPS) gave the 3,4-diol, which was directly used as an acceptor to produce a lactosamine structure in 50% yield. Deacetylation, regioselective 6′-*O*-silylation and acetylation then gave the key intermediate, from which the anomeric silyl group is removed selectively, and the resulting hemiacetal transformed into a trichloroacetamidate donor. This is coupled with allyl alcohol to yield the allyl glycoside, the mild selective deacetylation of which yields a 2′,3′,4′-triol. After acetylation, regioselective glycosylation yields the desired tetrasaccharide (67%). Once more, mild deacetylation was regioselective, exposing only four hydroxyl groups. Again, regioselective glycosylation was performed to give the desired hexasaccharide. Complete deprotection and enzymatic fucosylation then produced the target structure. Although conceptually elegant, using regioselective reactions late in the synthesis can be ineffective. Indeed, in this case, the yield in the hexasaccharide formation was only 35%, and the enzymatic introduction of the fucosyl moieties was performed in a 41% yield. In addition, the yield in the deprotection steps was disappointing (18%).

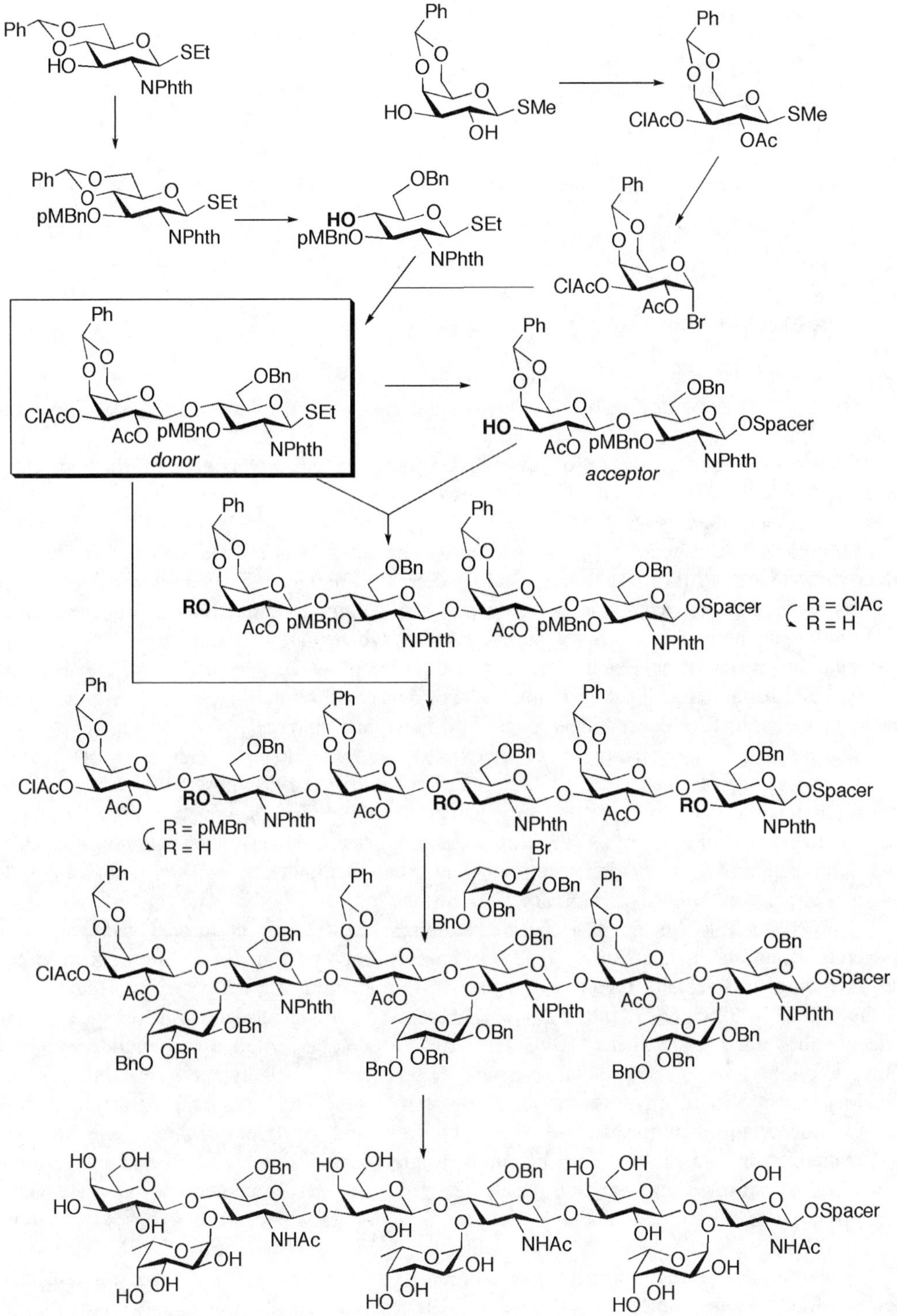

SCHEME 3.38 Synthesis of trimeric Lewis^{x} (Norberg et al.).

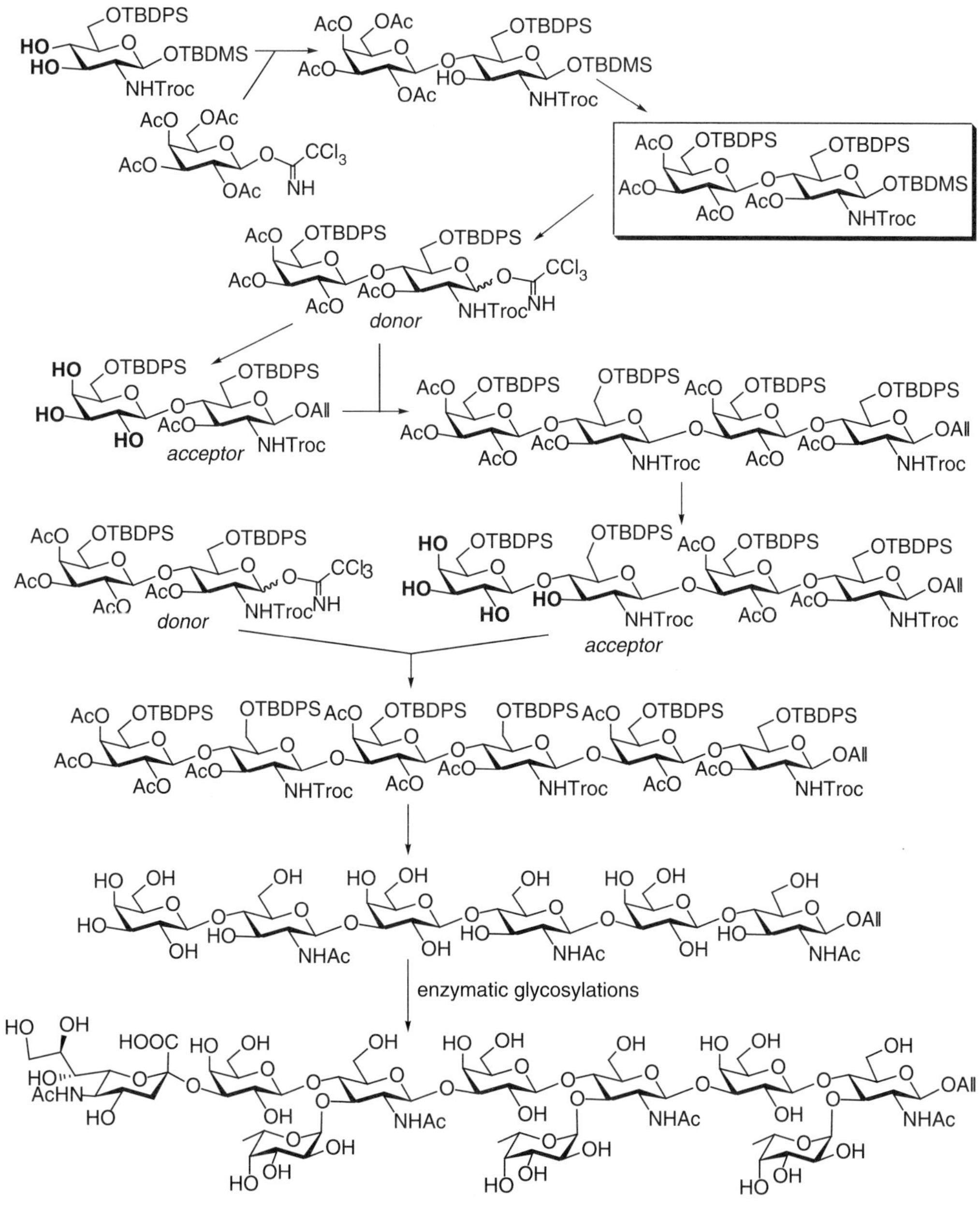

SCHEME 3.39 Synthesis of trimeric Lewis^{x} (Wong et al.).

In one of the trisaccharide building block approaches (Scheme 3.40) [66], the $GlcNH_2$ moiety was protected with a TBDMS group at the anomeric position, a benzylidene acetal (6-*O*-protection, temporary 4-*O*-protection) and with a 2-azido group. The regioselective protection of the galactose moiety was performed at the trisaccharide level. Fucosylation of the $GlcN_3$ acceptor was followed by reductive opening of the benzylidene to yield a new acceptor, which was galactosylated to give the trisaccharide key intermediate. This was transformed to a donor by the removal of the anomeric silyl group and subsequent transformation into the trichloroacetimidate. Alternate conversion to an acceptor was accomplished by deacetylation

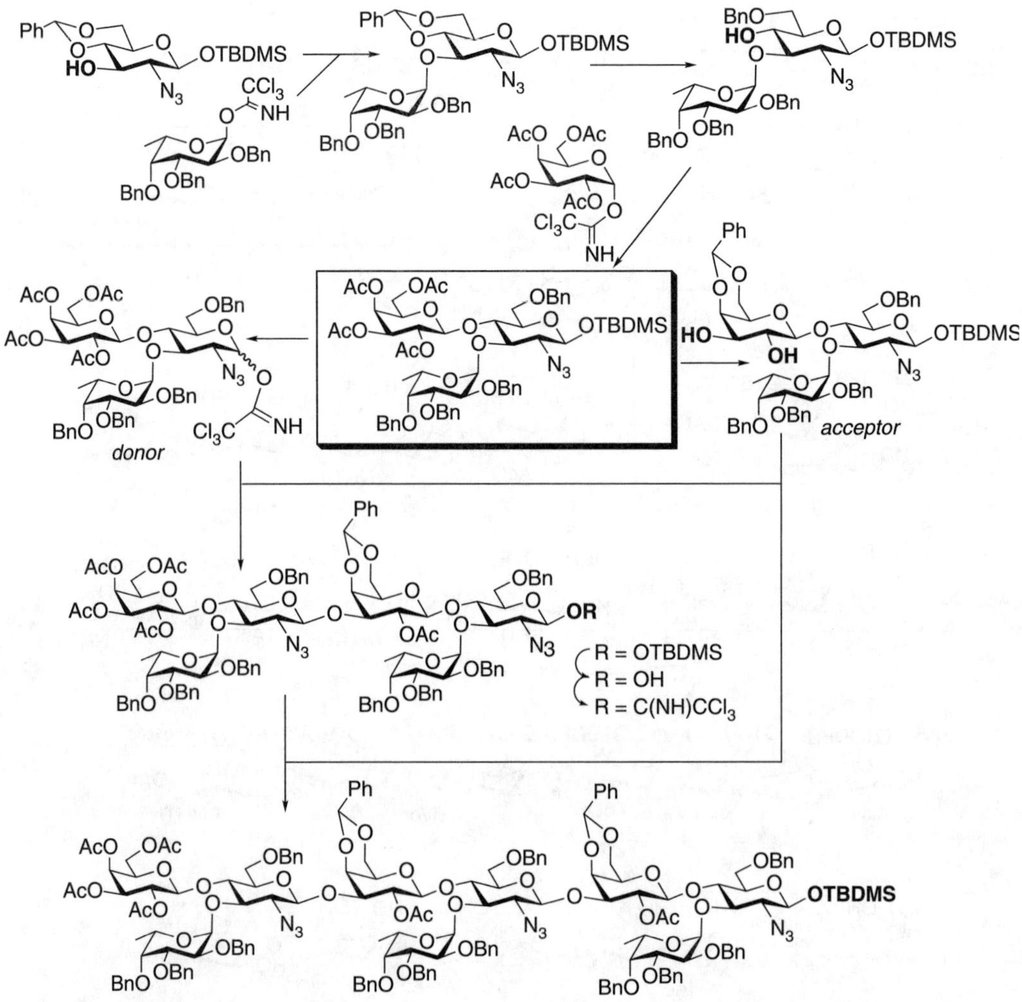

SCHEME 3.40 Synthesis of trimeric Lewis^x (Schmidt et al.).

and subsequent benzylidenation. Coupling of these blocks gave a hexasaccharide, which was similarly changed into a donor and coupled to the above trisaccharide acceptor to yield the trimeric Lewis^x nonasaccharide. Although a 2-nonparticipating group (N_3) was used, high β-selectivity was obtained. Again, transformation into a donor, coupling to an acceptor and deprotection gave the target structure in a most efficient synthesis. Further, in this synthesis, the yields in the protection/deprotection steps are very high, with the lowest being the reductive opening of the benzylidene acetal (67%).

3.5 SUMMARY AND CONCLUSIONS

The examples in this chapter confirm the earlier statement that, with a good knowledge of a rather limited amount of standard protecting groups, coupled with selected regioselective methodologies for their introduction, efficient protective group strategies (even for complex oligosaccharide structures) can be designed and accomplished. However, a lot of hard work, experimentation and optimization is usually involved.

REFERENCES

1. Green, T, Wuts, P G M, *Protective Groups in Organic Synthesis*, 3rd ed., Wiley, New York, 1999.
2. Hanson, J R, *Protecting Groups in Organic Synthesis*, Sheffield Academic Press, New York, 1999.
3. Robertson, J, *Protecting Group Chemistry*, Oxford University Press, New York, 2000.
4. Kocienski, P, *Protecting Groups*, Georg Thieme Verlag, Stuttgart, 2000, corrected edition.
5. Jarowicki, K, Kocienski, P, Protecting groups, *Contemp. Org. Synth.*, 2, 315–336, 1995; see also 3, 397–431, 1996; 4, 454, 1997; *J. Chem. Soc. Perkin. Trans.* 1, 4005–4037, 1998; 1589–1616, 1999; 2495–2527, 2000; 2109–2135, 2001.
6. Clarke, P A, Martin, W H C, Synthetic methods part (v) protecting groups, *Annu. Rep. Prog. Chem. Sect. B*, 99, 84–103, 2003.
7. Haines, A, Relative reactivity of hydroxyl groups in carbohydrates, *Adv. Carbohydr. Chem. Biochem.*, 33, 11–109, 1976.
8. Iseloh, U, Dudkin, V, Wang, Z G, Danishefsky, S, Reducing oligosaccharides via glycal assembly: on the remarkable stability of anomeric hydroxyl groups to global deprotection with sodium in liquid ammonia, *Tetrahedron Lett.*, 43, 7027–7030, 2002.
9. deBelder, A N, Cyclic acetals of the aldoses and aldosides, *Adv. Carbohydr. Chem. Biochem.*, 20, 219–302, 1965; see also 34, 179–241, 1977.
10. Garegg, P J, Swahn, C G, Benzylidenation of diols with α,α-dihalotoluenes in pyridine, *Methods Carbohydr. Chem.*, 8, 317–319, 1980.
11. Byramova, N E, Ovchinnikov, M V, Backinowsky, L V, Kochetkov, N K, Selective removal of *O*-acetyl groups in the presence of *O*-benzoyl groups by acid-catalyzed methanolysis, *Carbohydr. Res.*, 124, C8–C11, 1983.
12. Josephson, S, Bundle, D R, Artificial carbohydrate antigens: the synthesis of the tetrasaccharide repeating unit of *Shigella flexneri O*-antigen, *Can. J. Chem.*, 57, 3073–3079, 1979.
13. Lefeber, D J, Kamerling, J P, Vliegenthart, J F G, The use of diazabicyclo[2.2.2]octane as a novel highly selective dechloroacetylation reagent, *Org. Lett.*, 2, 701–703, 2000.
14. Vlahov, J, Snatzke, G, An improved synthesis of β-glucosiduronic acid derivatives, *Liebigs. Ann. Chem.*, 570–574, 1983.
15. Obadiah, J P, Buchwald, S L, Seeberger, P H, Halobenzyl ethers as protecting groups for organic synthesis, *J. Am. Chem. Soc.*, 122, 7148–7149, 2000.
16. Dahlén, A, Sundgren, A, Lahmann, M, Oscarson, S, Hilmersson, G, SmI_2/water/amine mediates cleavage of allyl ether protected alcohols — application in carbohydrate synthesis and mechanistic considerations, *Org. Lett.*, 5, 4058–4088, 2003.
17. Jones, S S, Reese, C B, Migration of *tert*-butyldimethylsilyl protecting groups, *J. Chem. Soc. Perkin. Trans.* 1, 2762–2764, 1979.
18. Broddefalk, J, Bergquist, K E, Kihlberg, J, Use of acid-labile protective groups for carbohydrate moieties in synthesis of glycopeptides related to type II collagen, *Tetrahedron*, 54, 12047–12070, 1998.
19. Borbas, A, Szabo, Z B, Szilagyi, L, Benyei, A, Liptak, A, Dioxane-type (2-naphthyl)methylene acetals of glycosides and their hydrogenolytic transformation into 6-*O*- and 4-*O*-(2-naphthyl)methyl (NAP) ethers, *Tetrahedron*, 58, 5723–5732, 2002.
20. Ellervik, U, 9-Anthraldehyde acetals as protecting groups, *Tetrahedron Lett.*, 44, 2279–2281, 2003.
21. Ley, S V, Baeschlin, D K, Dixon, D J, Foster, A C, Ince, S J, Priepke, H W M, Reynolds, D J, 1, 2-Diacetals: a new opportunity for organic synthesis, *Chem. Rev.*, 101, 53–80, 2001.
22. Garegg, P J, Iversen, T, Johansson, R, Synthesis of disaccharides containing β-D-mannopyranosyl groups, *Acta Chem. Scand. Ser. B*, 34, 505–508, 1980.
23. Ito, Y, Ohnishi, Y, Ogawa, T, Nakahara, Y, Highly optimized β-mannosylation via 4-methoxybenzyl-assisted intramolecular aglycon delivery, *Synlett*, 10, 1102–1104, 1998.
24. Wu, X, Grathwohl, M, Schmidt, R R, Efficient solid-phase synthesis of a complex, branched *N*-glycan hexasaccharide: use of a novel linker and temporary-protecting-group pattern, *Angew. Chem. Int. Ed.*, 41, 4489–4493, 2002.
25. Jansson, K, Ahlfors, S, Frejd, T, Kihlberg, J, Magnusson, G, Dahmen, J, Noori, G, Stenvall, K, 2-(Trimethylsilyl)ethyl glycosides. 3. Synthesis, anomeric deblocking, and transformation into 1,2-*trans* 1-*O*-acyl sugars, *J. Org. Chem.*, 53, 5629–5647, 1988.

26. Zhang, Z, Magnusson, G, Conversion of *p*-methoxyphenyl glycosides into the corresponding glycosyl chlorides and bromides, and into thiophenyl glycosides, *Carbohydr. Res.*, 295, 41–55, 1996.
27. Cerny, M, Stanek, J, Jr., 1,6-anhydro derivatives of aldohexoses, *Adv. Carbohydr. Chem. Biochem.*, 34, 23–177, 1977.
28. Åberg, P, Ernst, B, Facile preparation of 1,6-anhydrohexoses using solvent effects and a catalytic amount of a Lewis acid, *Acta Chem. Scand.*, 48, 228–233, 1994.
29. Lemieux, R U, Ratcliffe, R M, The azidonitration of tri-*O*-acetyl-D-galactal, *Can. J. Chem.*, 57, 1244–1251, 1979.
30. Seeberger, P H, Bilodeau, M T, Danishefsky, S J, Synthesis of biologically important oligosaccharides and other glycoconjugates by the glycal assembly method, *Aldrichim. Acta*, 30, 75–92, 1997.
31. Schmidt, R R, Kinzy, W, Anomeric-oxygen activation for glycoside synthesis: the trichloroacetimidate method, *Adv. Carbohydr. Chem. Biochem.*, 50, 21–177, 1977.
32. Garegg, P J, Thioglycosides as glycosyl donors in oligosaccharide synthesis, *Adv. Carbohydr. Chem. Biochem.*, 52, 179–205, 1997.
33. Fraser-Reid, B, Udodong, U E, Wu, Z, Ottosson, H, Merritt, J R, Rao, C S, Roberts, C, Madsen, R, *n*-Pentenyl glycosides in organic chemistry: a contemporary example of serendipity, *Synlett*, 12, 927–942, 1992.
34. Jiao, H, Hindsgaul, O, The 2-*N,N*-dibenzylamino group as a participating group in the synthesis of β-glycosides, *Angew. Chem. Int. Ed.*, 38, 346–348, 1999.
35. Banoub, J, Boullanger, P, Lafont, D, Synthesis of oligosaccharides of 2-amino-2-deoxy sugars, *Chem. Rev.*, 92, 1167–1195, 1992.
36. Debenham, J S, Madsen, R T, Roberts, C L, Fraser-Reid, B, Two new orthogonal amine-protecting groups that can be cleaved under mild or neutral conditions, *J. Am. Chem. Soc.*, 117, 3302–3303, 1995.
37. Aly, M R E, Castro-Palomino, J C, Ibrahim, E S I, El-Ashry, E S H, Schmidt, R R, The dimethylmaleoyl group as amino protective group. Application to the synthesis of glucosamine-containing oligosaccharides, *Eur. J. Org. Chem.*, 11, 2305–2316, 1998.
38. Kornilov, A V, Sherman, A A, Kononov, O, Shashkov, A S, Nifant'ev, N E, Synthesis of 3-*O*-sulfoglucuronyl lacto-*N*-neotetraose 2-aminoethyl glycoside and biotinylated neoglycoconjugates thereof, *Carbohydr. Res.*, 329, 717–730, 2000.
39. Stanek, J Jr., Preparation of selectively alkylated saccharides as synthetic intermediates, *Top. Curr. Chem.*, 154, 209–256, 1990.
40. Kadereit, D, Waldmann, H, Enzymatic protecting group techniques, *Chem. Rev.*, 101, 3367–3396, 2001.
41. Grindley, T B, Applications of tin-containing intermediates to carbohydrate chemistry, *Adv. Carbohydr. Chem. Biochem.*, 53, 17–142, 1998.
42. Kaji, E, Shibayama, K, In, K, Regioselectivity shift from β-(1→6) to β-(1→3)-glycosylation of non-protected methyl β-galactopyranosides using the stannylene activation method, *Tetrahedron Lett.*, 44, 4881–4885, 2003.
43. Garegg, P J, Iversen, T, Oscarson, S, Monotosylation of diols using phase-transfer catalysis, *Carbohydr. Res.*, 53, C5–C7, 1976.
44. Osborn, H M I, Brome, V A, Harwood, L M, Suthers, W G, Regioselective C-3-*O*-acylation and *O*-methylation of 4,6-*O*-benzylidene-β-D-gluco- and galactopyranosides displaying a range of anomeric substituents, *Carbohydr. Res.*, 332, 157–166, 2001.
45. Gelas, J, The reactivity of cyclic acetals of aldoses and aldosides, *Adv. Carbohydr. Chem. Biochem.*, 39, 71–156, 1981.
46. Garegg, P J, Regioselective cleavage of *O*-benzylidene acetals to benzyl ethers, In *Preparative Carbohydrate Chemistry*, Hanessian, S, Ed., Marcel Dekker, New York, pp. 53–68, 1997.
47. Lemieux, R U, Driguez, H, Chemical synthesis of 2-*O*-(α-L-fucopyranosyl)-3-*O*-(α-D-galactopyranosyl)-D-galactose. Terminal structure of the blood-group B antigenic determinant, *J. Am. Chem. Soc.*, 97, 4069–4075, 1975.
48. Oscarson, S, Tedebark, U, Synthesis and acidic opening of chlorinated carbohydrate orthoacetates, *J. Carbohydr. Chem.*, 15, 507–513, 1996.

49. Oscarson, S, Szönyi, M, Acidic opening of 4,6-*O*-orthoesters of pyranosides, *J. Carbohydr. Chem.*, 8, 663–668, 1989.
50. Pozsgay, V, A simple method for avoiding alkylthio group migration during the synthesis of thioglycoside 2,3-orthoesters. An improved synthesis of partially acylated 1-thio-α-L-rhamnopyranosides, *Carbohydr. Res.*, 235, 295–302, 1992.
51. Auzanneau, F I, Bundle, D R, Incidence and avoidance of stereospecific 1,2 ethylthio group migration during the synthesis of ethyl 1-thio-α-L-rhamnopyranoside 2,3-orthoester, *Carbohydr. Res.*, 212, 13–24, 1991.
52. Haines, A, The selective removal of protecting groups in carbohydrate chemistry, *Adv. Carbohydr. Chem. Biochem.*, 39, 13–70, 1981.
53. Kitov, P I, Bundle, D R, Synthesis and structure-activity relationships of di- and trisaccharide inhibitors for *Shiga*-like toxin type 1, *J. Chem. Soc. Perkin. Trans.*, 1, 838–853, 2001.
54. Glen, A, Leigh, D A, Martin, R P, Smart, J P, Truscello, A M, The regioselective *tert*-butyldimethylsilylation of the 6′-hydroxyl group of lactose derivatives via their dibutylstannylene acetals, *Carbohydr. Res.*, 248, 365–369, 1993.
55. Dahmen, J, Gnosspelius, G, Larsson, A, Lave, T, Noori, G, Paalsson, K, Frejd, T, Magnusson, G, Synthesis of di-, and tri-, and tetrasaccharides corresponding to receptor structures recognized by *Streptococcus pneumoniae*, *Carbohydr. Res.*, 138, 17–28, 1985.
56. Rencurosi, A, Poletti, L, Panza, L, Lay, L, Improvement on lipase catalysed regioselective *O*-acylation of lactose: a convenient route to 2′-*O*-fucosyllactose, *J. Carbohydr. Chem.*, 20, 761–765, 2001.
57. Zhang, Z, Wong, C H, Regioselective benzoylation of sugars mediated by excessive Bu_2SnO: observation of temperature promoted migration, *Tetrahedron*, 58, 6513–6519, 2002.
58. Upreti, M, Ruhela, D, Vishwakarma, R A, Synthesis of the tetrasaccharide cap domain of the antigenic lipophosphoglycan of *Leishmania donovani* parasite, *Tetrahedron*, 56, 6577–6584, 2000.
59. Khan, R, The chemistry of sucrose, *Adv. Carbohydr. Chem. Biochem.*, 33, 235–294, 1976.
60. Khan, R, Chemical and enzymic transformations of sucrose, *Int. Sugar J.*, 96, 12–17, 1994.
61. Ong, G T, Wu, S H, Wang, K T, Preparation of 2,3,6,3′,4′-penta-*O*-acetyl sucrose, the precursor of sucralose, by enzymic methods, *Bioorg. Med. Chem. Lett.*, 2, 161–164, 1992.
62. Garegg, P J, Oscarson, S, Ritzén, H, Partially esterified sucrose derivatives: synthesis of 6-*O*-acetyl-2, 3,4-tri-*O*-[(*S*)-3-methylpentanoyl]sucrose, a naturally occurring flavour precursor of tobacco, *Carbohydr. Res.*, 181, 89–96, 1988.
63. Oscarson, S, Ritzén, H, Synthesis of 6-*O*-acetyl-2,3,4-tri-*O*-[(*S*)-2-methylbutyryl]sucrose and the three regioisomers of 6-*O*-acetyl-2,3,4-*O*-[(*S*)-2-methylbutyryl]-di-*O*-[(*S*)-3-methylpentanoyl]-sucrose, *Carbohydr. Res.*, 284, 271–277, 1996.
64. Nilsson, M, Norberg, T, Synthesis of carbohydrate derivatives corresponding to a tumor-associated glycolipid: a trimeric Lewis^x nonasaccharide and a trimeric *N*-acetyl lactosamine hexasaccharide, *J. Carbohydr. Chem.*, 8, 613–627, 1989.
65. Nicolaou, K C, Caulfield, T J, Kataoka, H, Stylianides, N A, Total synthesis of the tumor-associated Le^x family of sphingolipids, *J. Am. Chem. Soc.*, 112, 3693–3695, 1990.
66. Toepfer, A, Kinzy, W, Schmidt, R R, Efficient synthesis of the Lewis antigen X (Le^x) family, *Liebigs. Ann. Chem.*, 449–464, 1994.
67. Bröder, W, Kunz, H, Glycosyl azides as building blocks in convergent syntheses of oligomeric lactosamine and Lewis^x saccharides, *Bioorg. Med. Chem.*, 5, 1–19, 1997.
68. Koeller, K M, Wong, C H, Chemoenzymatic synthesis of sialyl-trimeric-Lewis^x, *Chem. Eur. J.*, 6, 1243–1251, 2000.

4 Glycosylation Methods

Péter Fügedi

CONTENTS

4.1 INTRODUCTION

In the overwhelming majority of carbohydrates found in nature, the carbohydrates do not occur in a free form, but the monosaccharides are linked to each other or to other types of compounds (aglycones) by glycosidic bonds. Compounds possessing *O*- (**1**), *S*- (**2**), or *N*-glycosidic linkages (**3**) (Figure 4.1) occur in nature. Among them, *O*-glycosidic compounds are the most abundant and important. As such, this chapter focuses on the preparation of *O*-glycosides.

Glycosylation reactions play a central role in carbohydrate chemistry for several reasons. The rapidly growing interest in the biological roles of glycoconjugates (glycoproteins, glycolipids and proteoglycans) [1,2] requires homogeneous materials to study. However, these are available only with great difficulty from this complex and diverse class of highly heterogeneous compounds. Chemical syntheses directed towards these complex structures is therefore a major tool in the progress of glycobiology, and an essential part of the chemical synthesis is the assembly of oligomeric or polymeric structures by linking the monomers with glycosidic bonds.

As the anomeric center is the most reactive among the several functional groups of a monosaccharide, almost all carbohydrate syntheses require some temporary protection of the anomeric center for chemical transformations at other positions. Frequently, an early step in multistep carbohydrate synthesis is the formation of a simple glycoside for temporary protection of the anomeric center. Fixing the anomeric configuration in the form of a glycoside is also useful in model compounds to avoid the complexity resulting from the equilibration of free sugars into different ring sizes and anomeric configurations.

Most importantly, simple, efficient and stereoselective creation of glycosidic linkages is a major challenge to organic and bioorganic chemistry in its own right. In contrast to peptide syntheses, strict control of stereochemistry is required in glycosylation reactions. As the stereochemical outcome is delicately influenced by a large number of factors, and minor changes in the structures of the glycosyl donors and acceptors might be detrimental, understanding these influences and developing more

Abbreviations: Ac: acetyl; Bn: benzyl; BSP: 1-benzenesulfinyl piperidine; BTIB: bis(trifluoroacetoxy)iodobenzene; DAST: (diethylamino)sulfur trifluoride; DDQ: 2,3-dichloro-5,6-dicyano-*p*-benzoquinone; DMDO: dimethyldioxirane; DMTSF: dimethyl(methylthio)sulfonium tetrafluoroborate; DMTST: dimethyl(methylthio)sulfonium triflate; DTBMP: 2,6-di-*tert*-butyl-4-methylpyridine; DTBP: 2,6-di-*tert*-butylpyridine; DTBPI: 2,6-di-*tert*-butylpyridinium iodide; FDCPT: 1-fluoro-2,6-dichloropyridinium triflate; FTMPT: 1-fluoro-2,4,6-trimethylpyridinium triflate; IDCP: iodonium dicollidine perchlorate; IDCT: idonium dicollidine triflate; LPTS: 2,6-lutidinium *p*-toluenesulfonate; LTMP: lithium tetramethylpiperidide; Me: methyl; MPBT: *S*-(4-methoxyphenyl) benzenethiosulfinate; NBS: *N*-bromosuccinimide; NIS: *N*-iodosuccinimide; NISac: *N*-iodosaccharin; PPTS: pyridinium *p*-toluenesulfonate; TBPA: tris(4-bromophenyl)ammoniumyl hexachloroantimonate; Tf: trifluoromethanesulfonyl; TMTSB: methyl-bis(methylthio)sulfonium hexachloroantimonate; TMU: tetramethylurea; Tr: trityl; TTBP: 2,4,6-tri-*tert*-butylpyrimidine.

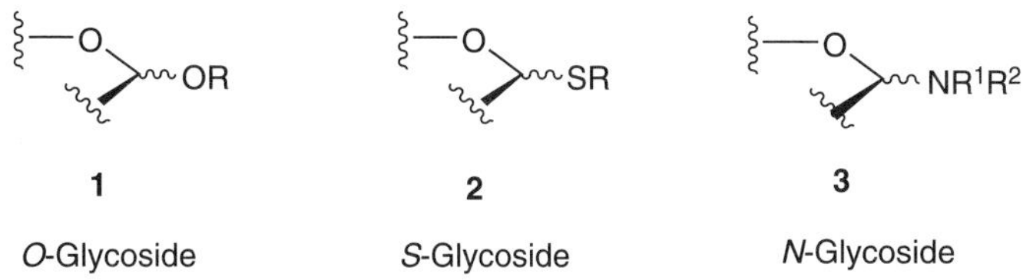

1 O-Glycoside
2 S-Glycoside
3 N-Glycoside

FIGURE 4.1 *O*-, *S*- and *N*-glycosides.

4 + ROH → 1

SCHEME 4.1 Glycoside synthesis by nucleophilic substitution at the anomeric carbon.

efficient and generally applicable glycosylation reactions has long been – and still remains – a major challenge not only for carbohydrate chemistry, but for mainstream organic chemistry as well.

Since the first historical glycoside syntheses by Michael [3] and Emil Fischer [4,5], followed by the seminal work of Koenigs and Knorr [6], a very large number of glycosylation methods have been developed, and the field has been extensively reviewed elsewhere [7–40]. This chapter focuses on recent advances in the progress of glycoside syntheses, including significant earlier developments in the field.

The overwhelming majority of glycosylations are performed by nucleophilic substitution reactions at the anomeric carbon (Scheme 4.1). These reactions are discussed first by classifying them according to the nature of the leaving group. This is followed by other, mechanistically different, methods of glycosylation. A detailed discussion of glycosylation methods will be preceded by general principles of sterocontrol in glycosylations.

4.2 STEREOCHEMICAL ASPECTS OF GLYCOSIDE BOND FORMATION

Stereoselective formation of the glycosidic bond at the chiral anomeric carbon atom is a crucial issue in glycosylation reactions. Problems of stereocontrol arise as a result of the difficulties in accomplishing clean S_N2 reactions with *O*-nucleophiles at the anomeric center. The ready participation of the ring oxygen in nucleophilic displacements gives a considerable S_N1 component to these reactions rendering stereocontrol difficult.

In relation to the substituent at the neighboring carbon, the glycosidic linkage can either be 1,2-*trans* as in β-D-gluco- (**5**), β-D-galacto- (**6**), α-D-manno- (**7**) and α-L-rhamnopyranosides (**8**), or it is 1,2-*cis* as in α-D-gluco- (**9**), α-D-galacto- (**10**), β-D-manno- (**11**) and β-L-rhamnopyranosides (**12**) (Figure 4.2).

The stereoselective synthesis of 1,2-*trans* glycosides can be achieved by neighboring group participation. As shown in Scheme 4.2, promoter-assisted departure of the leaving group from the glycosyl donor **13** results in the formation of the carbocation **14**, which stabilizes by mesomeric release of electrons from the adjacent ring oxygen to give the oxocarbenium ion **15**. Reaction of the oxocarbenium ion with the hydroxyl component is expected to occur both from the bottom (path *a*) and upper face (path *b*) of the molecule, affording the *cis*- (**17**) or *trans*-glycoside (**18**), respectively. In the presence of neighboring group active substituents (such as acyloxy groups) at the adjacent carbon, however, the acyloxonium ion **16** is formed readily. Reaction of the hydroxyl nucleophile with the acyloxonium ion can take place from the upper face of the ring (path *c*), and therefore results in the formation of *trans*-glycoside **18** with regeneration of the acyloxy group.

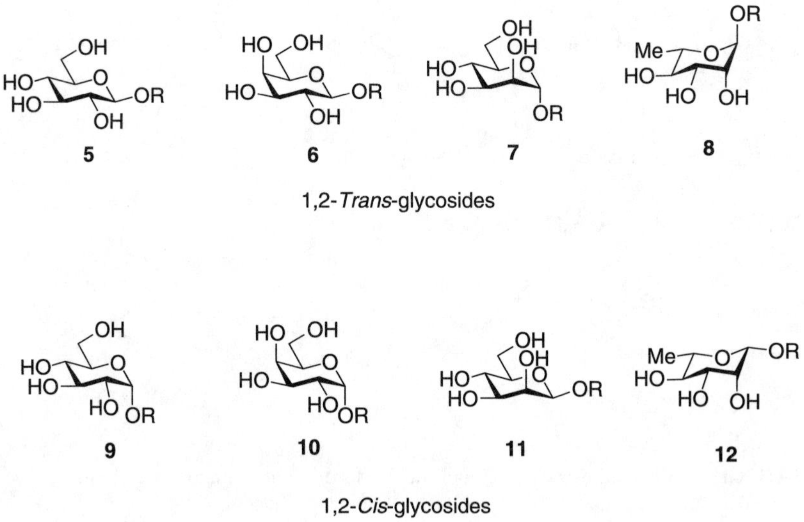

1,2-*Trans*-glycosides

1,2-*Cis*-glycosides

FIGURE 4.2 Examples of 1,2-*trans*-and 1,2-*cis*-glycosides.

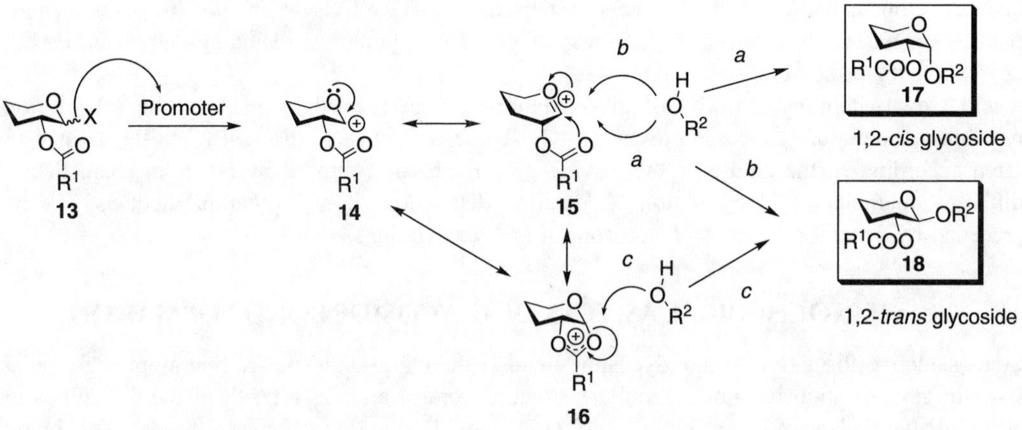

SCHEME 4.2 Synthesis of 1,2-*trans* glycosides by neighboring group participation.

Glycosylations with neighboring group participation provide excellent stereoselectivities for *trans*-glycoside synthesis with various types of glycosyl donors. The formation of *cis*-glycosides, observed in certain cases (particularly with unreactive substrates), can be explained by the reaction taking place with the more reactive **15** via path *a*. The most commonly used substituents to achieve neighboring group participation are the *O*-acetyl, *O*-benzoyl, and *O*-pivaloyl substituents for hydroxyls, and the *N*-phthaloyl and *N*-trichloroethoxycarbonyl derivatives for amino groups.

In compounds without neighboring substituents, such as 2-deoxy sugars or *N*-acetylneuraminic acid, stereocontrolling auxiliaries (such as selenophenyl, thiophenyl or halogen substituents) are sometimes intentionally introduced to achieve stereocontrol in glycosylation reactions [41–43]. In glycosylations using these derivatives (**19**), the auxiliaries control the stereoselectivity by neighboring group participation (via three-membered intermediates, **20**) affording *trans*-glycosides **21**, from which the auxiliaries are reductively removed (→**22**) at a later stage of the synthesis (Scheme 4.3).

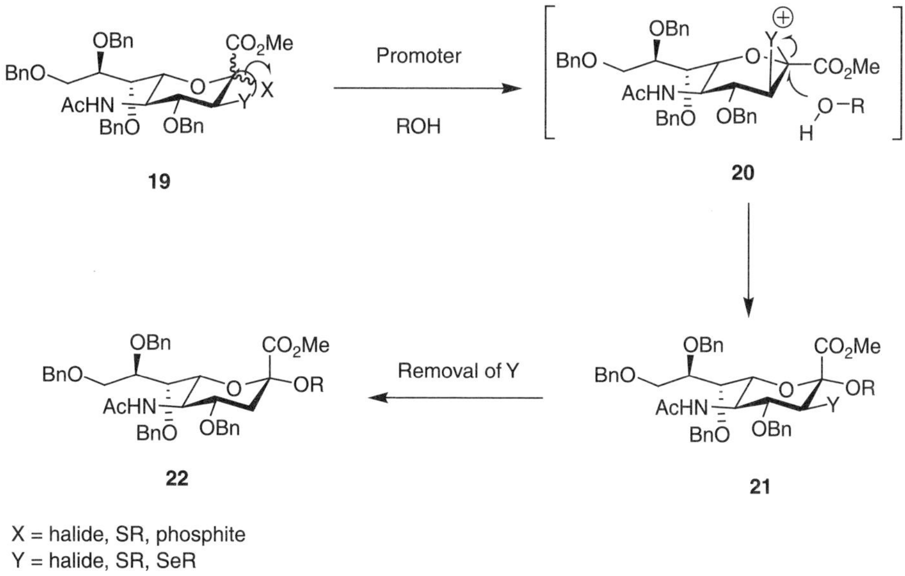

X = halide, SR, phosphite
Y = halide, SR, SeR

SCHEME 4.3 Glycoside synthesis by stereocontroling auxiliaries.

A prerequisite for the synthesis of 1,2-*cis* glycosides is the use of non-neighboring group active substituents. Ether-type protecting groups have a low tendency to participate and therefore are commonly used for this purpose. Benzyl ethers and substituted benzyl ethers (such as 4-methoxybenzyl) are the most frequently used derivatives for hydroxyl groups, whereas for amino sugars, the azido function serves as an excellent non-participating masked form of the amino group.

However, using only a non-participating protecting group is insufficient to guarantee stereoselective *cis*-glycoside formation. As mentioned above, given the tendency of the ring oxygen to participate, and also the fairly weak nucleophilicity of the alcoholic hydroxyl groups, the glycosylation reactions often follow S_N1 mechanisms. Early attempts focusing on changes in the non-neighboring active protecting groups and in the catalyst system resulted in only partial successes [10]. From both a practical and a theorectical perspective, a major breakthrough in *cis*-glycoside synthesis was the introduction of the *in situ* anomerization concept by Lemieux and coworkers [44]. In this halide ion-catalyzed glycosylation starting from the more stable α-glycosyl halide **23**, an equilibrium is established by the addition of halogenide ions (Bu_4NBr) with the more reactive β-glycosyl halide **28** through the intermediates **24** → **25** → **26** → **27**, as shown in Scheme 4.4. The energy barrier of the nucleophilic substitution of the β-halide **28** with the alcohol leading to the *cis*-glycoside **31** (via **27** → **26** → **29** → **30**) is lower than that of the corresponding reaction of the α-halide **23** to the *trans*-glycoside **34** (via **24** → **32** → **33**). The net stereochemical result is retention with the formation of a *cis*-glycoside.

This principle was generalized to other glycosylation reactions by Paulsen [11]. When using the more powerful catalysts of Koenigs–Knorr reactions, it is important that the difference in the energy barriers leading to the *cis*- and *trans*-glycosides is maintained. If the catalysis is too strong, then the difference in the rates of formation of **31** and **34** will decrease, resulting in a loss of stereoselectivity. Therefore, to achieve stereoselective glycosylation, the catalyst should be selected in accord with the reaction partners [11]. The reactivities of glycosyl donors and glycosyl acceptors are strongly influenced by their protecting groups. For this reason, optimization of catalysts and reaction conditions are frequently required for a particular glycosylation reaction.

The reaction conditions also significantly influence the stereoselectivity of glycosylations. Solvents are particularly important in this respect. To achieve stereoselectivity, reactions should

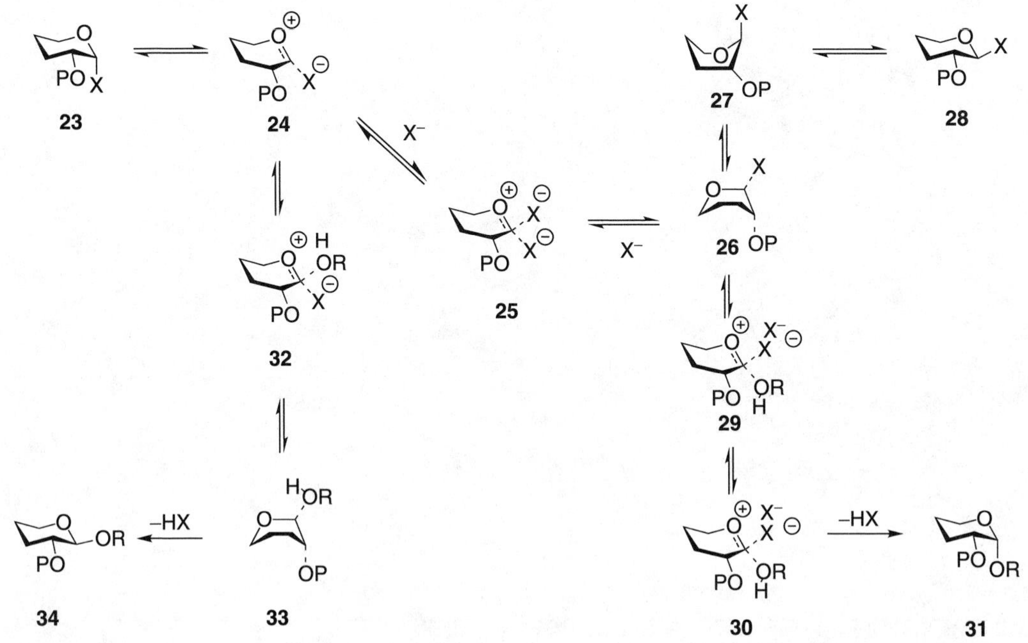

SCHEME 4.4 Halide ion-catalyzed glycosylation.

proceed through the tight ion pairs shown in Scheme 4.4. Therefore, solvents of low polarity, such as dichloromethane, 1,2-dichloroethane and toluene, are used most commonly. However, solvents can also participate in the glycosylation reaction, thereby altering stereoselectivity. Ether-type solvents have a tendency to shift stereoselectivities towards the formation of *cis*-glycosides. In this respect, diethyl ether and dioxane are frequently used. Their effect on the stereoselectivity is rationalized by the preferential formation of the equatorial oxonium ion type intermediate **36** from the oxocarbenium ion **35**. Intermediate **36** then reacts to form the axial glycoside **37** (Scheme 4.5). Alternatively, when using nitrile-type solvents, such as acetonitrile at low

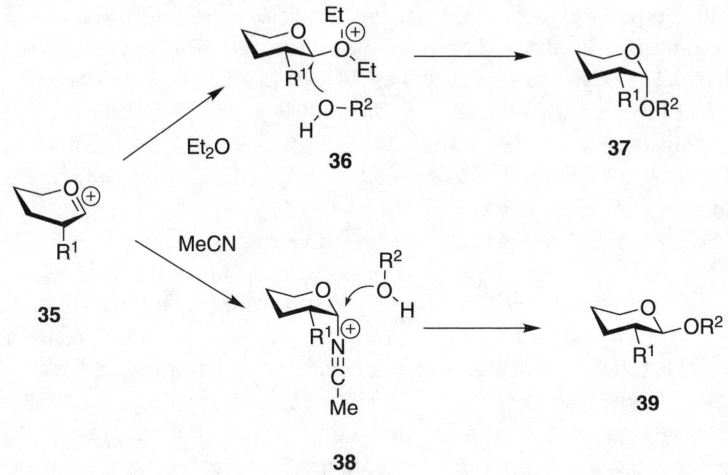

SCHEME 4.5 Solvent participation in glycosylations.

temperature, shifts in the selectivity towards the formation of equatorial glycosides **39** via the preferential formation of the nitrilium ion **38** is observed.

The structures of the glycosyl donors and acceptors and the choice of the promoter and solvent are the major factors influencing the stereoselectivity of glycosylations. Additionally, a number of other parameters including temperature, pressure, concentration and even the sequence of addition of the reactants also exert effects. A detailed discussion of the effects of all these parameters are beyond the scope of this chapter and can be found in various reviews [11,38,39].

The importance of the choice of the glycosylation catalysts and reaction conditions is illustrated by an example in Scheme 4.6. Reaction of the glycosyl donor **40** with the same glycosyl acceptor **41** performed under different conditions produced excellent yields of either the α-linked disaccharide **42α** or the β-linked isomer **42β** in a highly stereoselective manner [45].

A different approach to control the stereoselectivity of glycosylations is through the use of heterogeneous catalysis. Catalysts such as silver silicate were developed for this purpose [46]. Reactions of glycosyl halides on the surface of silver silicate are thought to proceed by a concerted mechanism providing, for example, β-D-mannopyranosides **44** from α-D-mannosyl bromide **43** (Scheme 4.7).

A number of indirect routes for the preparation of *cis*-glycosides are also available; their common principle is illustrated in Scheme 4.8. In these approaches, advantage is taken of the ready formation of *trans*-glycosides **46** by neighboring group participation from glycosyl donors **45**, which possess selectively removable protecting groups at O-2. After selective removal of this protecting group, the C-2 configuration of the resulting **47** is inverted by nucleophilic displacement or oxidation–reduction type reactions, thus creating the *cis*-glycoside **48**.

Although tremendous progress has been made in the synthesis of *cis*-glycosides, no successful general method for their preparation has emerged. This does not mean that these compounds are not accessible by chemical synthesis. However, their syntheses often require substantial know-how and systematic research.

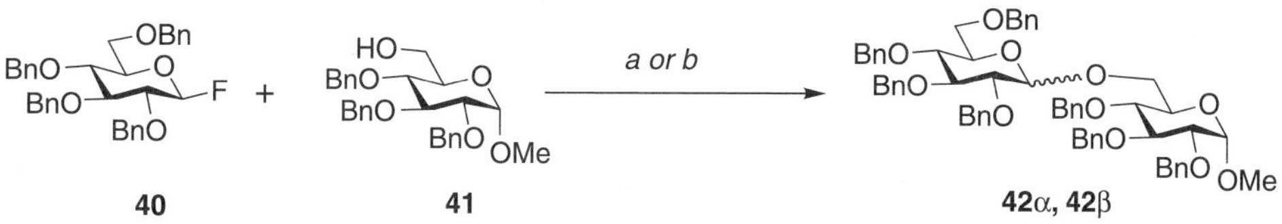

40 + **41** → (*a or b*) **42α, 42β**

a) $HClO_4$, Et_2O, RT — *a)* 98%, α : β = 92 : 8

b) $HB(C_6F_5)_4$, BTF-*t*BuCN, 0°C — *b)* 99%, α : β = 7 : 93

SCHEME 4.6 The effect of the promoter and the reaction conditions on the stereoselectivity of glycosylations.

43 → (Ag-silicate) → **44**

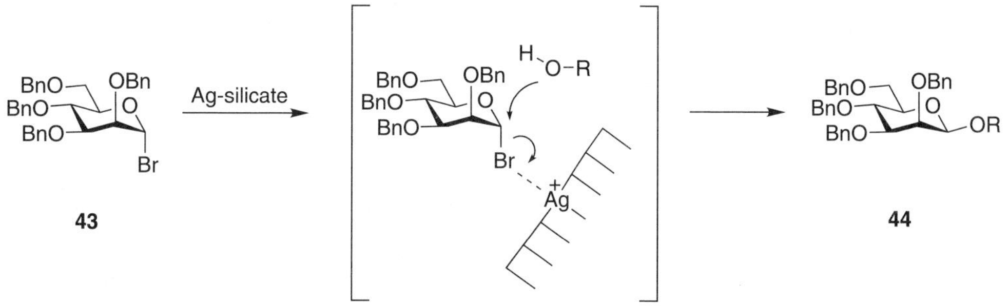

SCHEME 4.7 Synthesis of *cis*-glycosides by heterogeneous catalysis.

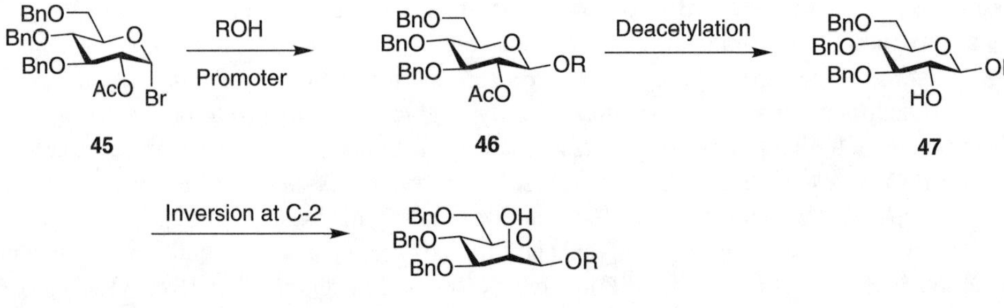

SCHEME 4.8 Synthesis of β-D-mannopyranosides by configurational inversion at C-2.

4.3 GLYCOSYLATIONS BY NUCLEOPHILIC SUBSTITUTIONS AT THE ANOMERIC CARBON

A wide variety of anomeric substituents are used as leaving groups in glycosylation reactions taking place by nucleophilic substitution at the anomeric carbon. In this section, this type of glycosylation with the most frequently used classes of glycosyl donors (such as glycosyl halides, thioglycosides, glycosyl imidates, glycosyl esters, *O*-glycosides, 1,2-anhydro sugars, etc.) will be discussed. The different types of glycosyl donors are grouped according to the heteroatom attached to the anomeric center (Figure 4.3).

4.3.1 Synthesis of Glycosides from Glycosyl Halides

4.3.1.1 Glycosyl Bromides and Chlorides

Glycosyl bromides and chlorides were the first (and for a long time, practically the only) type of glycosyl donors used for the synthesis of complex glycosides. This archetype of glycosyl donors combines high reactivity with reasonable stability. Because of the anomeric effect, glycosyl halides with axial halogens are more stable than the equatorial isomers. Their stability and reactivity is also greatly influenced by the protecting groups on the sugar ring. For example, electron-withdrawing

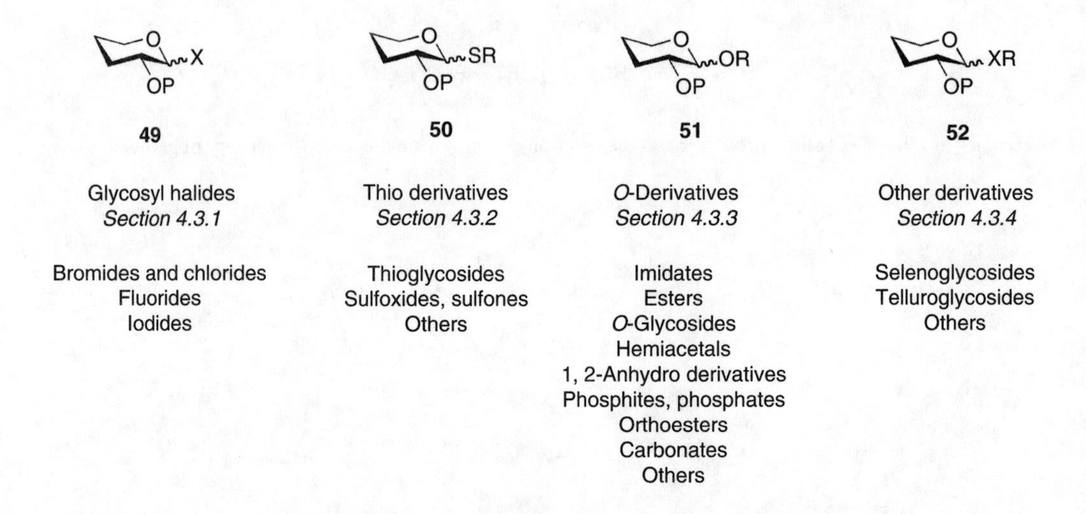

FIGURE 4.3 Types of glycosyl donors.

(acyl) groups decrease, whereas electron-donating (ether-type) protecting groups increase their reactivity.

Glycosylations with glycosyl bromides and chlorides in the presence of heavy metal salts, known as the Koenigs–Knorr reaction [6,9,10], is one of the oldest glycosylation methods. Over the years, a wide variety of promoters to activate glycosyl halides in glycosylation reactions have been developed (Table 4.1). The heavy metal salts frequently used as promoters include Ag_2CO_3 and Ag_2O [6,9,10], silver salts of hydroxy carboxylic acids, such as silver salicylate [9], silver imidazolate [47], $Hg(CN)_2$ and $HgBr_2$ [48–50], HgO and $HgBr_2$ [51], HgI_2 [52] and $CdCO_3$ [53]. $AgClO_4$ [54–57] and AgOTf [58,59] are the most efficient of the heavy metal promoters, and the latter is the most frequently used. Heterogeneous catalysts such as silver silicate [46], silver zeolite [60] and silver silica–alumina [61] were developed particularly for the synthesis of the otherwise hard to synthesize β-mannosidic and β-rhamnosidic linkages [46,60,62–65].

Besides heavy metal salts, Lewis acids such as $SnCl_4$ [66], $BF_3{\cdot}Et_2O$ [66], $ZnCl_2$ [67], $ZnCl_2$–silver imidazolate [47] $ZnCl_2$–TrCl [67], $Sn(OTf)_2$ [68,69] or, more recently, $Cu(OTf)_2$ [70] and $InCl_3$ [71] have also been introduced as promoters. Activation of glycosyl halides with Tf_2O [72]

TABLE 4.1
Activations of Glycosyl Bromides and Chlorides

X (Br or Cl) —ROH / Promoter→ OR

Promoter	Ref.
Ag_2O, Ag_2CO_3	[6,9,10]
Ag salicylate	[9]
Ag imidazolate	[47]
$AgClO_4$	[54–57]
AgOTf	[58,59]
$Hg(CN)_2$, $HgBr_2$	[48–50]
HgO, $HgBr_2$	[51]
HgI_2	[52]
$CdCO_3$	[53]
Ag silicate	[46]
Ag zeolite	[60]
Ag silica alumina	[61]
$SnCl_4$	[66]
$BF_3{\cdot}OEt_2$	[66]
$ZnCl_2$	[67]
$ZnCl_2$–Ag imidazolate	[47]
$ZnCl_2$–TrCl	[67]
$Sn(OTf)_2$	[68,69]
$Cu(OTf)_2$	[70]
$InCl_3$	[71]
Tf_2O	[72]
TfOH	[45,73]
I_2–DDQ	[74]
IBr	[75]
Et_4NBr, Bu_4NBr	[44]
R_4NX, aq. base	[89–94]
$-e$	[96]
Δ	[97–100]

and even the protic acid TfOH [45,73] has also been reported, although moderate yields were obtained with the latter.

Originally used in catalytic amounts in combination with silver oxide and silver carbonate in Koenigs–Knorr reactions, iodine was shown to promote glycosylations either alone or in combination with DDQ [74]. Iodine bromide was reported to be a more efficient promoter [75].

The use of acid scavengers and drying agents in the reaction mixtures in combination with the promoters is advantageous, and often necessary. Metal carbonates and oxides (Ag_2O, Ag_2CO_3, HgO, $CdCO_3$) not only serve as promoters, but they also neutralize the acid released in the reaction. Organic bases, such as *s*-collidine, lutidine, tetramethylurea (TMU), the sterically hindered 2,6-di-*tert*-butyl-4-methylpyridine (DTBMP), 2,6-di-*tert*-butylpyridine (DTBP) and the recently introduced 2,4,6-tri-*tert*-butylpyrimidine (TTBP) [76] serve the same purpose. Drying agents such as Drierite or activated molecular sieves are used to assure strictly anhydrous conditions.

It should be noted that glycosylation reactions of glycosyl halides with participating acyl groups at O-2 in the presence of excess base lead to orthoesters instead of glycosides. Therefore, the amount of base in glycosylation reactions should be limited. However, orthoesters can also serve as glycosyl donors [77,78] because they can be rearranged to glycosides by Lewis acids (Section 4.3.3.7) [77–79]. For the regioselective glycosylation of polyhydroxy systems, the regioselective formation of orthoesters under mild conditions can be utilized [80,81]. The orthoesters are then rearranged to glycosides, providing regioselective glycosylation by a two-step process.

A different type of activation — conversion of perbenzylated α-glycosyl bromides via their highly reactive β-anomers into glycosides — was introduced by Lemieux [44]. In these cases, promoters such as Bu_4NBr or Et_4NBr in combination with Hünig's base are used. These halide ion-catalyzed glycosylations give 1,2-*cis*-glycosides in excellent stereoselectivity. Unfortunately, the scope of the method is limited to reactive glycosyl halides and acceptors. Nevertheless, its use permitted elegant syntheses of biologically important oligosaccharides [82–84]. Advantage is taken of the excellent stereoselectivity of the halide ion-catalyzed glycosylations by other methods. In these cases, alternate glycosyl donors are used, from which the glycosyl halides are generated *in situ*. Thus, glycosylations with thioglycoside donors promoted by $CuBr_2$–Bu_4NBr [85] or DMTST–Bu_4NBr [86] (see Section 4.3.2.1) as well as glycosylations with lactols assisted by Ph_3P–CBr_4 [87,88] proceed through glycosyl halide intermediates.

For the glycosylation of phenols with glycosyl halides, phase-transfer conditions using aqueous sodium or potassium hydroxide and quaternary ammonium salts have also been reported [89–94]. By using phase-transfer catalysis, a method for solid-phase synthesis of aryl glycosides was developed [95].

Moreover, electrochemical [96], thermal [97–100] and high-pressure glycosylations [101,102] have also been described.

Glycosyl halides, for a long time practically the only type of glycosyl donors, have now lost their predominant role in glycosylations. This is partly because of attempts to avoid the use of the toxic, expensive and often light- and moisture-sensitive heavy metal salts used for their activation. Additionally, glycosyl halides are moderately stable. As such, hardly any chemical transformation can be performed on their non-anomeric regions without damaging the highly reactive halide at the anomeric center. Therefore, for glycosyl halides that possess an elaborate protecting group pattern, the halide is introduced immediately before the glycosylation step. However, purification of the halides from these reactions is often problematic. As for the parent compounds (e.g., thioglycosides and *n*-pentenyl glycosides [NPGs]) of these transformations, effective glycosylation methods have been developed more recently. These methods are more often used directly, thereby avoiding preparation of glycosyl halides. As simple glycosyl halides such as the peracetylated and perbenzylated ones are relatively easily accessible, they still retain importance and are used as glycosyl donors.

4.3.1.2 Glycosyl Fluorides

Glycosyl fluorides are too unreactive to be activated under standard Koenigs–Knorr conditions, and, for this reason, their use as glycosyl donors was explored relatively late, after the development of glycosyl bromides. They were introduced in 1981 [103], and glycosyl fluorides are currently among the most frequently used donors.

Compared with glycosyl bromides and chlorides, glycosyl fluorides have increased stability. These compounds have long shelf lives and survive chromatography, which compares very favorably with glycosyl bromides. Several methods for the preparation of glycosyl fluorides are known [104,105]. They are commonly prepared by the reaction of a protected sugar with a free anomeric hydroxyl group with diethylaminosulfur trifluoride (DAST). Treatment of the benzylated D-glucose derivative **49** with DAST in THF at low temperature produces the fluoride **50** in excellent yield and high β selectivity [106] (Scheme 4.9).

Most importantly, glycosyl fluorides can be activated under specific conditions that most protecting groups survive. The first useful activation of glycosyl fluorides for glycosylation was introduced by Mukaiyama [103] using a combination of $SnCl_2$ with $AgClO_4$. In the original catalyst system, $AgClO_4$ is frequently replaced by AgOTf [107,108], and the use of the $SnCl_2$–$TrClO_4$ [109] promoter system has also been proposed. The different promoter systems for the activation of glycosyl fluorides are summarized in Table 4.2.

Noyori and coworkers introduced the use of SiF_4 [110] and TMSOTf [110], whereas $BF_3{\cdot}Et_2O$ was introduced independently by Nicolaou [111], Kunz [112,113] and Vozny [114]. Additionally, TiF_4 (with or without $AgClO_4$) [115,116] and SnF_4 [115] were also found to be effective promoters.

The use of metallocene complexes of the type Cp_2MCl_2 ($M = Zr$, Hf, Ti) in combination with $AgClO_4$, AgOTf or $AgBF_4$ was introduced by Suzuki and coworkers [117–120], and these reagents later found wide application. Other promoters developed include Me_2GaCl and Me_2GaOTf [121], the Bu_2SnCl_2–$AgClO_4$ system [122] and Bu_3SnOTf [43,123]. The use of the metal triflates $Sn(OTf)_2$ [43,123] and, more recently, that of $Cu(OTf)_2$ [70] has also been reported.

Additionally, Wessel et al. introduced Tf_2O as a powerful reagent for glycosylations with glycosyl fluoride donors [72,124,125]. The sequence TMSOTf $<$ $SnCl_2$–AgOTf $<$ TiF_4 $<$ Tf_2O was suggested for the relative reactivity of different promoters. Alternatively, $LiClO_4$ was proposed as a very mild catalyst for fucosylations under neutral conditions [126].

A variety of rare earth salts including $La(ClO_4)_3$ and $Yb(OTf)_3$ used either alone [127–129] or in combination with Lewis acids such as $ZnCl_2$ or $Sn(OTf)_2$ [129,130] were also found to be effective catalysts. A different type of activation using heterogeneous catalysis with sulfated zirconia (SO_4/ZrO_2), Nafion-H or montmorillonite K-10 was developed by Toshima [131,132].

Catalytic amounts of the protic acids TfOH, $HClO_4$ and $HB(C_6F_5)_4$ were recently used by Mukaiyama and coworkers [45,73,133–135]. In addition, $TrB(C_6F_5)_4$ [134,136] and related carbocationic tetrakis(pentafluorophenyl)borates [137], as well as the $SnCl_2$–$AgB(C_6F_5)_4$ [138] or $SnCl_4$–$AgB(C_6F_5)_4$ systems [139], were also found to activate glycosyl fluorides for glycosylations.

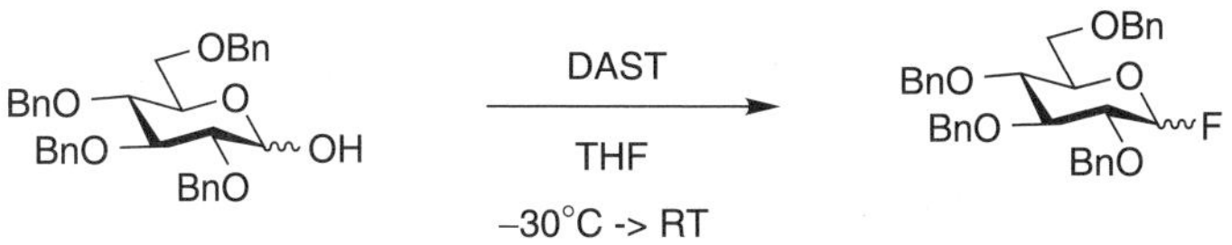

OBn
BnO
BnO
O
OH
BnO
49
DAST
THF
−30°C -> RT
OBn
BnO
BnO
O
F
BnO
50
99%
α : β = 1 : 7.7

SCHEME 4.9 Synthesis of glycosyl fluorides.

TABLE 4.2
Activations of Glycosyl Fluorides

ROX, Promoter

Promoter	X	Ref.
$SnCl_2$–$AgClO_4$	H	[103]
$SnCl_2$–AgOTf	H	[107,108]
$SnCl_2$–$TrClO_4$	H	[109]
SiF_4	TMS	[110]
TMSOTf	TMS	[110]
$BF_3{\cdot}Et_2O$	H, TMS	[111–114]
TiF_4	H, TMS	[115,116]
TiF_4–$AgClO_4$	H	[116]
SnF_4	H, TMS	[115]
Cp_2MCl_2–$AgClO_4$, $AgBF_4$, AgOTf	H	[117–120]
Me_2GaCl, Me_2GaOTf	H	[121]
Bu_2SnCl_2–$AgClO_4$	H	[122]
Bu_3SnOTf	H	[43,123]
$Sn(OTf)_2$	H	[43,123]
$Cu(OTf)_2$	H	[70]
Tf_2O	H	[72,124,125]
$LiClO_4$	H	[126]
$La(ClO_4)_3$	H, TMS	[127–129]
$Yb(OTf)_3$	H, TMS	[127–129]
$Yb(OTf)_3$–$ZnCl_2$	H	[128,129]
$La(ClO_4)_3$–$Sn(OTf)_2$	H	[130]
SO_4/ZrO_2	H	[131,132]
Nafion–H	H	[132]
Montmorillonite K-10	H	[132]
TfOH	H	[45,73,133–135]
$HClO_4$	H	[45,73,133–135]
$HB(C_6F_5)_4$	H	[45,73,133–135]
$TrB(C_6F_5)_4$	H	[134,136]
$SnCl_2$–$AgB(C_6F_5)_4$	H	[138]
$SnCl_4$–$AgB(C_6F_5)_4$	H	[139]

Glycosyl fluorides are also useful for the preparation of aryl glycosides [140–142]. $BF_3{\cdot}Et_2O$ in combination with DTBMP gave good yields and stereoselectivity not only for glycosylation of phenols, but also for carboxylic acids [143].

Glycosyl fluorides have found use in the synthesis of a wide range of complex natural products. In Nicolaou's synthesis of avermectin B_{1a} [144] (Scheme 4.10), the glycosyl fluoride **52** was prepared from the thioglycoside **51** by reaction with NBS and DAST. Glycosylation of **53** with **52** created the disaccharide thioglycoside **54** (α anomer exclusively), which was activated as the fluoride **55**. Glycosylation of the aglycone was accomplished in a stereoselective manner, and the resulting **56** was deprotected to give avermectin B_{1a} **57**.

1,2-Seleno migration of selenoglycosides leading to glycosyl fluorides was reported by Nicolaou [145]. Thus, the selenoglycoside **58** gave the fluoride **59** on treatment with DAST (Scheme 4.11). The 2-deoxy-2-seleno-glycosyl fluorides reacted with alcohols such as **60** in the presence of $SnCl_2$ as a promoter (without the need for $AgClO_4$) to afford 1,2-*trans*-glycosides **61**.

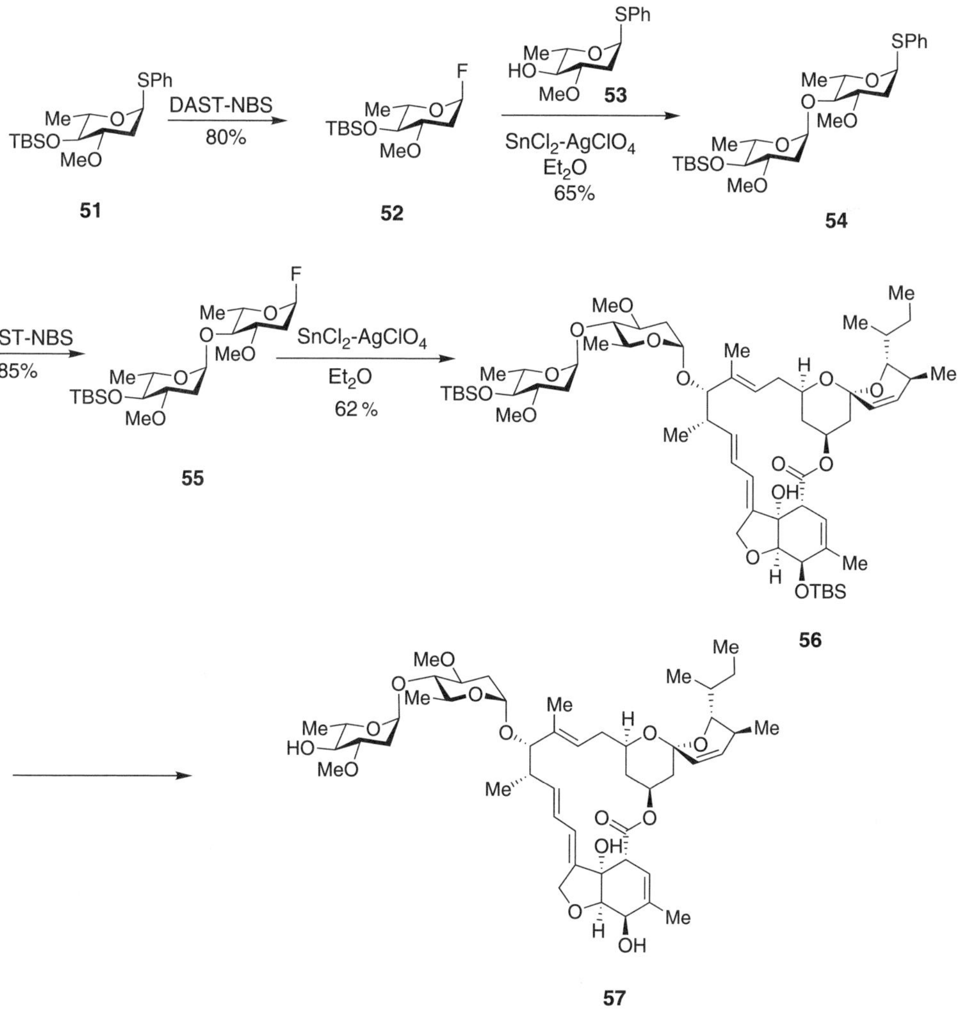

SCHEME 4.10 Synthesis of avermectin B_{1a}.

The selenophenyl substituent, which served as stereodirecting auxiliary (Section 4.2), was then removed to give the 2′-deoxy disaccharide **62**.

The above strategy was extensively used in the total synthesis of everninomicin 13,384-1 [146–149] (Schemes 4.12 and 4.13). In the synthesis of the $A_1B(A)C$ fragment, the 2-thiophenyl fluoride **63** was coupled with C unit acceptor **64** to give the disaccharide **65**. This was converted to the 2′-deoxy derivative **66**, and, after protecting group manipulations, the resulting **67** was acylated with A_1 unit fluoride **68** to give **69**. The selenophenyl glycoside was introduced at this stage, and **70** was glycosylated with the A unit fluoride **71**. The selenoglycoside **72** was transformed to the glycosyl fluoride **73** for final coupling. Glycosylation of the $DEFGHA_2$ fragment **74** with **73** promoted by $SnCl_2$ gave **75** in a 70% yield, which was then converted to everninomycin 13,384-1 **76**.

Although emphasis is placed on the donor properties of glycosyl fluorides in this section, it should be mentioned that, in contrast to glycosyl bromides, glycosyl fluorides could also serve as glycosyl acceptors in oligosaccharide syntheses. They can be glycosylated by a variety of methods including thioglycoside and trichloroacetimidate donors. Moreover, relatively unreactive

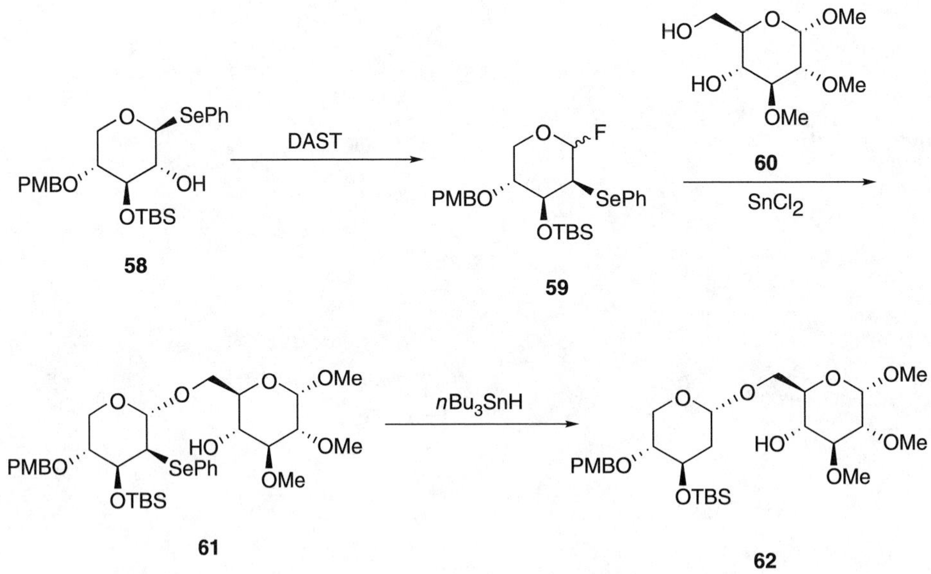

SCHEME 4.11 Use of glycosyl fluorides obtained by 1,2-seleno migration.

(*disarmed*) glycosyl fluorides can be glycosylated with those that are reactive (*armed*) [150] (for the *armed–disarmed* concept, see Chapter 5).

4.3.1.3 Glycosyl Iodides

Glycosyl iodides have been known for a long time [151]. However, because of their instability, they have been used only occasionally as glycosyl donors. Since the introduction of the halide ion-catalyzed glycosylations [44], a number of glycosylations actually proceeded through *in situ* generated glycosyl iodide intermediates by activating glycosyl bromides with tetraalkylammonium iodides. Recently, improved methods for the preparation of glycosyl iodides have been developed [152–156], and a number of glycosyl iodides were prepared and isolated. Although benzylated glycosyl iodides are too unstable and should be used directly, acyl protected derivatives are considerably less labile, and by using strongly electron-withdrawing protecting groups, stable, crystalline glycosyl iodides can be prepared [156].

In most glycosylation reactions performed so far with glycosyl iodides, benzylated derivatives were used. Either an organic base was applied as a promoter [157], or the reactions were essentially a modified version of the halide ion-catalyzed glycosylations [158–163]. These reactions gave *cis*-glycosides in high stereoselectivity. Synthesis of a *trans*-glycoside is shown in Scheme 4.14. As illustrated, reaction of the stable D-glucuronosyl iodide **77** with 3-*O*-pivaloyl-morphine **78** was promoted by iodine and afforded the glycoside **79** in a 55% yield [156]. After removal of the protecting groups, morphine-6-glucuronide **80** was obtained.

4.3.2 Synthesis of Glycosides from Anomeric Thio Derivatives

Several classes of anomeric thio derivatives are used as glycosyl donors; an overview of the major classes is given in the following section.

4.3.2.1 Thioglycosides

Thioglycosides are readily prepared using a variety of methods, but mostly by nucleophilic substitutions at the anomeric center. They are prepared most commonly from anomeric 1,2-*trans*-acetates

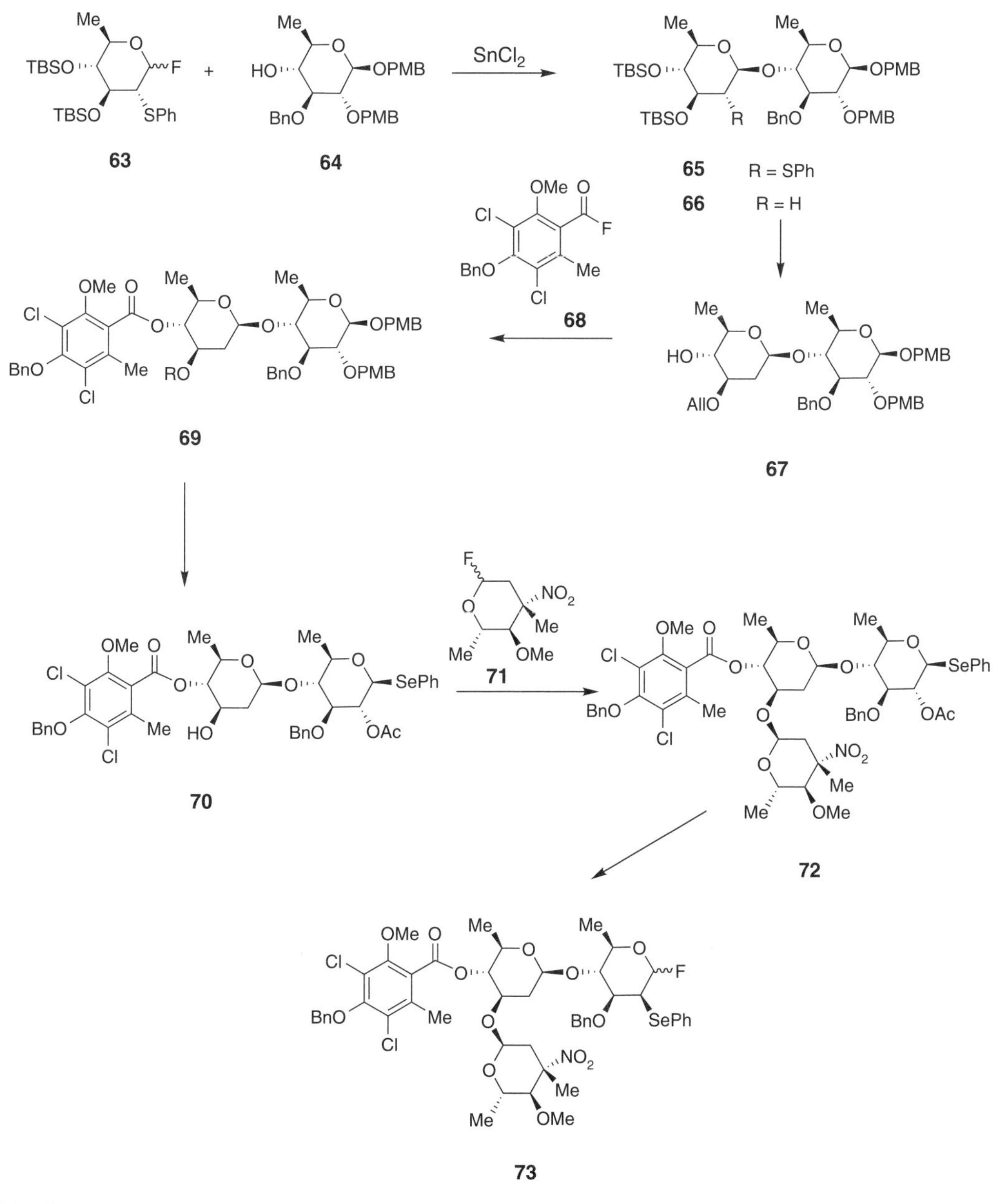

SCHEME 4.12 Synthesis of the $A_1B(A)C$ fragment of everninomicin 13,384–1.

by reaction with thiols or with trimethylsilyl ethers of thiols in the presence of Lewis acids such as $BF_3 \cdot Et_2O$, TMSOTf, $SnCl_4$, and so on. They can also be obtained from the reaction of other common glycosyl donors (glycosyl halides, trichloroacetimidates, etc.) with thiols, from the opening of the oxirane ring of 1,2-anhydro sugars, or by alkylations of 1-thiosugars. Thioglycosides show remarkable stability; not only do they have long shelf lives, but they also tolerate very diverse chemical manipulations leaving the thioglycoside function intact. Importantly, most of the common carbohydrate protecting group manipulations can be performed on thioglycosides, a feature making the preparation of highly functionalized thioglycoside donors possible. Moreover, they are inert under several glycosylation conditions, so thioglycosides can serve as glycosyl acceptors in the assembly of oligosaccharide blocks. These oligosaccharide blocks can be directly used as glycosyl

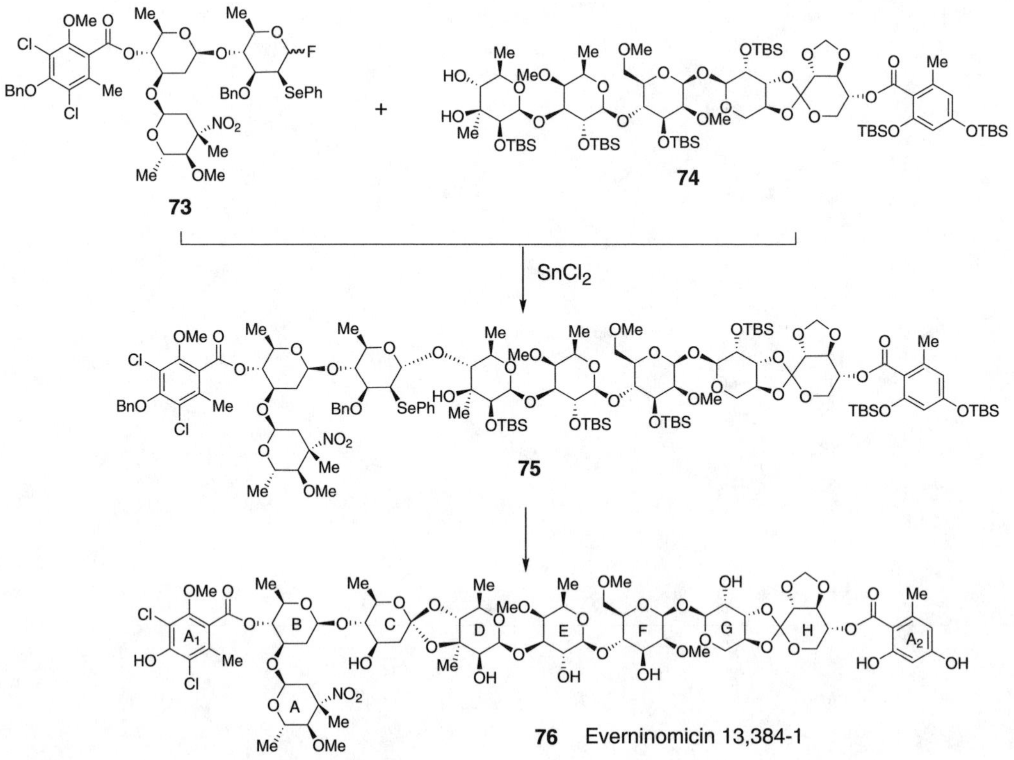

SCHEME 4.13 Synthesis of everninomicin 13,383–1.

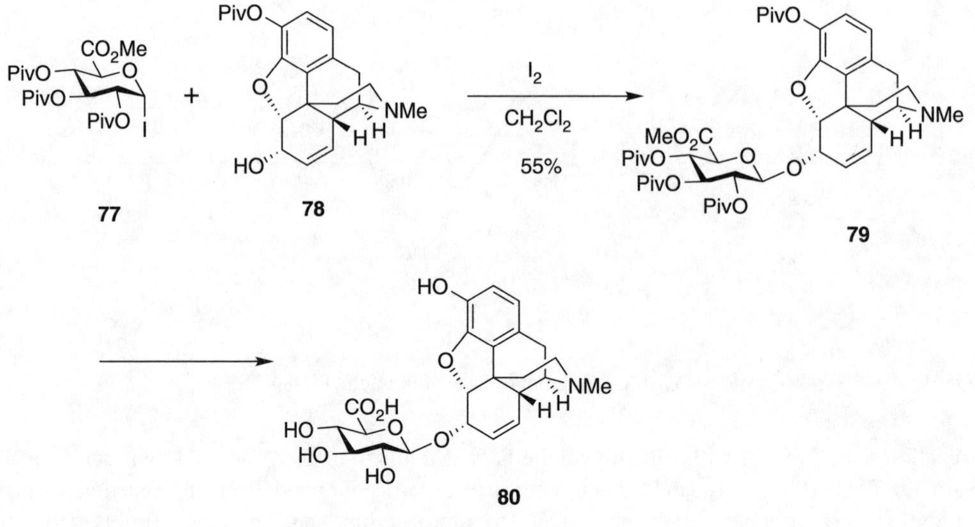

SCHEME 4.14 Synthesis of morphine-6-glucuronide using glycosyl iodide.

donors in the next step. In addition, thioglycosides can also directly be converted into other types of glycosyl donors, making them a very versatile class of compounds.

Despite their stability, thioglycosides **2** can be activated with thiophilic reagents, typically by soft electrophilic reagents, under mild conditions (Scheme 4.15). The sulfonium ion **81** formed in the reaction is a better leaving group, so loss of the sulfonium ion with the assistance of the ring

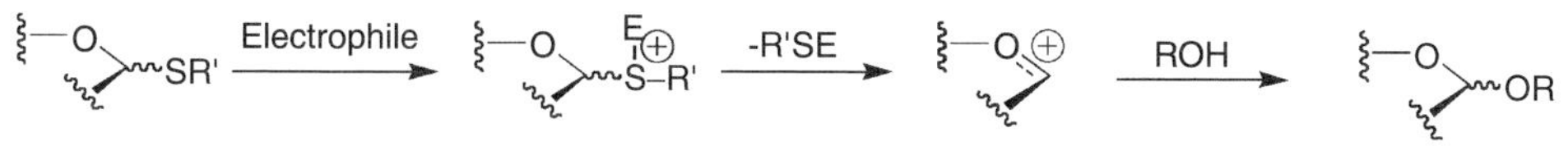

SCHEME 4.15 Activation of thioglycosides for glycosylations by electrophilic reagents.

oxygen or a neighboring group leads to the common intermediates of glycosylation reactions **82**, which will react with the *O*-nucleophile to afford the *O*-glycosides **1**.

The common promoters of thioglycoside activation are summarized in Table 4.3.

Early attempts for the activation of thioglycosides used heavy metal salts. In the first successful attempt, $HgSO_4$ was employed [164]. Other mercury salts, such as $Hg(OAc)_2$ [164], $HgCl_2$ [164, 165], $Hg(OBz)_2$ [166] and PhHgOTf [167], were later tried. These activations gave moderate yields and were not powerful enough to be of general use. To facilitate these reactions, the concept of remote activation using heterocyclic thioglycosides was introduced [168] and pyridyl, benzthiazolyl, pyrimidinyl and phenyltetrazolyl thioglycosides were employed with promoters such as $Hg(NO_3)_2$ [168], $Cu(OTf)_2$ [169], AgOTf [170–172] and $Pb(ClO_4)_2$ [170,173]. Heavy metal salt activation of thioglycosides was employed in the synthesis of such complex natural products as erythromycin [170], digitoxin [165] and avermectin oligosaccharides [173]. Activation with $AgPF_6$ for the preparation of 2-deoxy glycosides was recently reported [174].

Despite the successes in natural products' syntheses, the above activations were limited in scope. Starting from the mid-1980s, however, a series of highly powerful activation methods of thioglycosides was developed. The first method of this kind, based on the alkylation of the thioglycoside sulfur by MeOTf, was introduced by Lönn [175–177]. This method proved to be widely applicable and remains extensively used. With the much milder alkylating agent MeI, alkylation can be performed on 2-pyridyl thioglycosides [178–182]. This type of thioglycoside is more reactive than simple alkyl or aryl derivatives, and their activation with MeI affords 1,2-*cis*-glycosides with good stereoselectivities.

Another concept for the activation of thioglycosides is their reaction with soft sulfur electrophiles. The first reagents of this type were dimethyl(methylthio)sulfonium triflate (DMTST, **83**) and the corresponding tetrafluoroborate (DMTSF, **84**) (Figure 4.4) [86,183–185], which are powerful methylsulfenylating agents (E = SMe) (Scheme 4.15), and are still extensively used.

Notable applications of DMTST-promoted glycosylations are the stereoselective synthesis of α-neuraminic acid linkages [186,187] and the first high-yielding synthesis of sucrose (Scheme 4.16) [188]. In this synthesis, the D-glucose derivative **86** was glycosylated with the fructofuranose thioglycoside **85**, which has the β-directing silyl group, using DMTST to give the disaccharide **87** in an 80% yield. After deprotection, sucrose **88** was obtained in quantitative yield.

Other alkylsulfenylating agents such as MeSBr and MeSOTf [189,190], as well as PhSOTf [191–193] also found use. Activation by alkylsulfenylation is also achieved with the PhthNSEt–$TrB(C_6F_5)_4$ system [194]. In a reverse type of situation, sulfenyl derivatives of the hydroxy compounds are reacted with thioglycosides under Lewis acid catalysis [195]. More recently, other powerful sulfur reagents such as *S*-(4-methoxyphenyl) benzenethiosulfinate (MPBT) **89** (Figure 4.4) [196], 1-benzenesulfinyl piperidine (BSP) **90** [197] and diphenyl sulfoxide [198], all in combination with triflic anhydride, were introduced.

Related to the activation of thioglycosides by sulfenylation are selenylation type reactions (E = SeR) (Scheme 4.15). Arylselenylating type promoters used with thioglycosides include benzeneselenenyl triflate [199,200] or the *N*-phenylselenylphtalimide-trimethylsilyltriflate [201] and *N*-phenylselenylphtalimide-magnesium perchlorate [202] systems.

Halonium ions are another type of electrophile used for thioglycoside activation. The use of NBS, originally introduced by Hanessian for the synthesis of simple glycosides from a 2-pyridinyl

TABLE 4.3
Activations of Thioglycosides

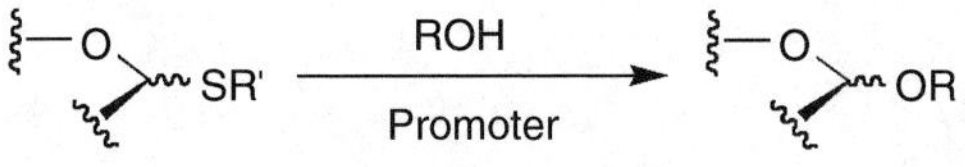

SR′ —(ROH, Promoter)→ OR

Promoter	R′	Ref.
$HgSO_4$		[164]
$Hg(OAc)_2$		[164]
$HgCl_2$		[164,165]
$Hg(OBz)_2$		[166]
PhHgOTf		[167]
$Hg(NO_3)_2$	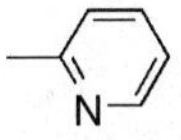 N	[168]
$Cu(OTf)_2$	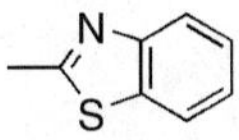 N, S	[169]
AgOTf	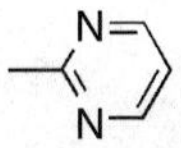 N, N	[170]
	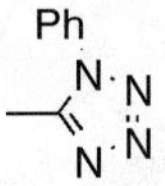 Ph, N, N, N, N	[171, 172]
	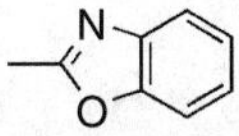 N, O	[246,247]
$Pb(ClO_4)_2$	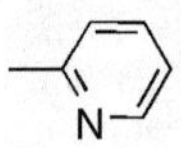 N	[170, 173]
$AgPF_6$		[174]
MeOTf		[175–177]
MeI	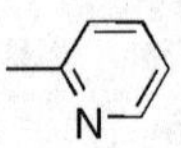 N	[178–182]
DMTST		[86,183–185]
DMTSF		[183]
MeSBr		[189, 190]
MeSOTf		[189, 190]
PhSOTf		[191–193]
$PhthNSEt–TrB(C_6F_5)_4$		[194]
$MPBT–Tf_2O$		[196]
$BSP–Tf_2O$		[197]
$Ph_2SO–Tf_2O$		[198]
PhSeOTf		[199,200]
PhSeNPhth–TMSOTf		[201]
$PhSeNPhth–Mg(ClO_4)_2$		[202,221]
NBS	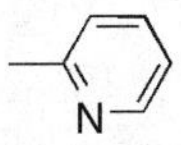 N	[168,203]
NBS–TfOH		[204]
$NBS–Ph_2IOTf\ (Bu_4NOTf)$		[205]
$NBS–LiClO_4\ (LiNO_3)$		[205]

Continued

TABLE 4.3
Continued

Promoter	R′	Ref.
NBS–TMSOTf ($Sn(OTf)_2$)		[206]
IDCP		[208–210]
IDCT		[211]
NIS–TfOH		[212–214]
NIS–$Yb(OTf)_3$		[215]
NIS–$LiClO_4$		[216]
NIS–$TrB(C_6F_5)_4$		[207]
NISac		[217]
PhIO–Tf_2O		[218]
BTIB		[219]
HTIB		[219]
PhIO–$Sn(OTf)_2$		[220]
PhIO–$SnCl_2$–$AgClO_4$		[220]
PhIO–TMSCl–$AgClO_4$		[206,221]
PhIO–$SnCl_4$–$AgClO_4$		[221]
I_2		[222]
IBr		[75,223,224]
ICl–AgOTf		[225]
IBr–AgOTf		[226]
I_2–DDQ–$TrB(C_6F_5)_4$		[227]
SO_2Cl_2–TfOH		[228,229]
Selectfluor-$BF_3{\cdot}Et_2O$		[230]
FTMPT		[231]
FDCPT		[231]
$CuBr_2$–Bu_4NBr		[85]
$CuBr_2$–Bu_4NBr–AgOTf		[85]
DMTST–Bu_4NBr		[86]
Br_2–AgOTf		[235]
$NOBF_4$		[236,237]
TBPA		[238,239]
$TrB(C_6F_5)_4$–$NaIO_4$		[240]
-*e*		[241–243]

thioglycoside [168], as a bromonium ion source was extended by Nicolaou [203] for the synthesis of more complex structures using phenyl thioglycosides. NBS is frequently used together with different additives to increase yields and stereoselectivity. A wide range of these additives have been tried, including triflic acid [204], Ph_2IOTf, Bu_4NOTf, Bu_4NClO_4, $LiNO_3$, $LiClO_4$, silicagel [205], TMSOTf, $Sn(OTf)_2$ [206], and $TrB(C_6F_5)_4$ [207].

Several iodine compounds are used as promoters. Iodonium dicollidine perchlorate (IDCP) was introduced first as an iodonium source [208–210]. The corresponding triflate (IDCT) can also be used [211]. A very efficient promoter system, *N*-iodosuccinimide in the presence of a catalytic amount of triflic acid, was introduced independently by van Boom [212] and Fraser-Reid [213,214]. Glycosylation promoted by NIS–TfOH are frequently used as they proceed at low temperatures within a short time, and are capable of activating a wide variety of glycosyl donors with various acceptors. Triflic acid is frequently replaced by different Lewis acids such as TMSOTf, TESOTf, AgOTf, $BF_3{\cdot}Et_2O$, lanthanide triflates [215], $LiNO_3$ or $LiClO_4$ [216] or $TrB(C_6F_5)_4$ [207]. Another iodinating agent *N*-iodosaccharin (NISac) (**91**) (Figure 4.4) was recently proposed as a

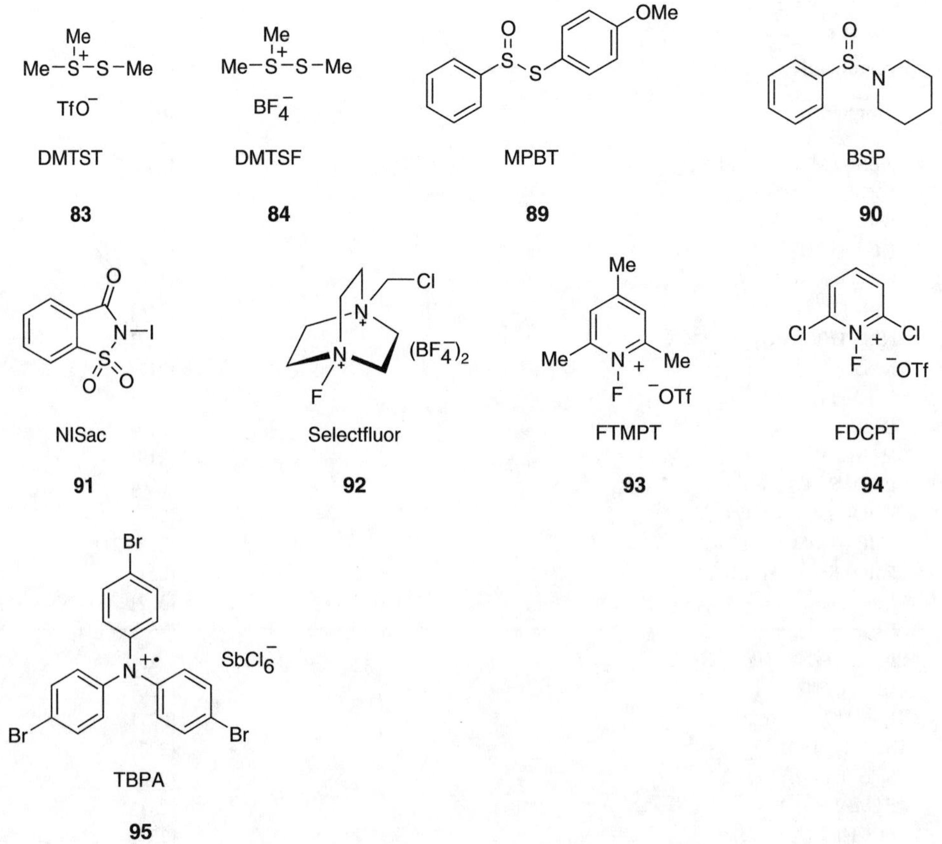

DMTST **83**, DMTSF **84**, MPBT **89**, BSP **90**, NISac **91**, Selectfluor **92**, FTMPT **93**, FDCPT **94**, TBPA **95**

FIGURE 4.4 Structures of promoters for thioglycosides.

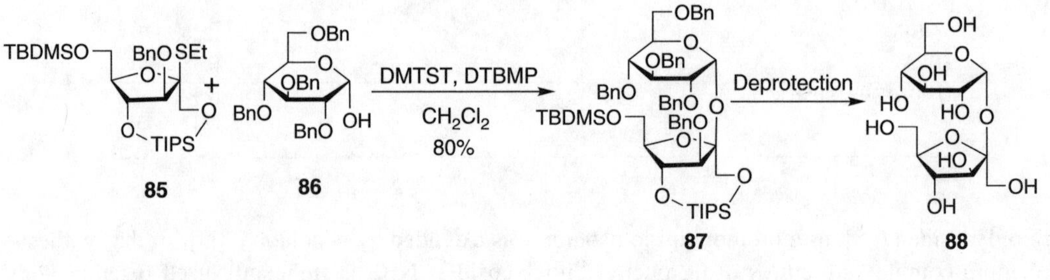

DMTST, DTBMP, CH_2Cl_2, 80%; Deprotection

85 + **86** → **87** → **88**

SCHEME 4.16 Synthesis of sucrose.

replacement for NIS to reduce occasional by-products caused by the nucleophilicity of succinimide formed in NIS-promoted reactions [217].

Hypervalent iodine reagents also found application for thioglycoside activation. Iodosobenzene was used in combination with triflic anhydride [218], whereas the acyl containing bis(trifluoro-acetoxy)iodobenzene (BTIB) and hydroxy(tosyloxy)iodobenzene (HTIB) [219] could be used by itself. Just as in the NIS–TfOH system, in the PhIO–TfOH system, triflic acid is often replaced with various coactivators, such as TMSOTf, $Sn(OTf)_2$, $SnCl_2$–$AgClO_4$, Cp_2ZrCl_2–$AgClO_4$ [220], $TMSClO_4$ [206,221], and $SnCl_4$–$AgClO_4$ [221].

Iodine promotes glycosylation reactions of perbenzylated thioglycosides [222]. The interhalogen compounds of iodine, IBr and ICl are more active, capable of promoting glycosylations with acylated thioglycosides [75,223,224]. As the reaction of these interhalogen compounds gives glycosyl halides as primary products, their combined use with AgOTf was a logical extension. The ICl–AgOTf combination was reported as particularly efficient and practical [225], and for optimization of related reagents for sialylation reactions, IBr–AgOTf gave the best results [226]. The ternary promoter system I_2–DDQ–$TrB(C_6F_5)_4$ has also been described [227].

Other halogenating systems including sulfuryl chloride-triflic acid [228,229], selectfluor **92** (Figure 4.4) with $BF_3 \cdot Et_2O$ [230] and the 1-fluoropyridinium triflates FTMPT **93** and FDCPT **94** [231] have also been reported.

The conversion of thioglycosides into glycosyl halides by halogens was known before thioglycosides could be used directly as glycosyl donors [232,233]. Several methods take advantage of the *in situ* generation of glycosyl halides from thioglycosides, and the halides then react further under Koenigs–Knorr or Lemieux conditions. An octasaccharide was synthesized by using bromine and collidine in 1972 [234]. Ogawa and coworkers [85] reported the use of $CuBr_2$–Bu_4NBr in combination with typical Koenigs–Knorr catalysts such as AgOTf, $HgBr_2$ and silver zeolite. Application of $CuBr_2$–Bu_4NBr alone [85], just as the use of DMTST–Bu_4NBr system [86], can be considered as the thioglycoside version of the halide ion-catalyzed glycosylations [44]. Activation of thioglycosides by Br_2–AgOTf and Br_2–$Hg(CN)_2$ also takes advantage of the ready formation of glycosyl bromides from thioglycosides [235].

Other methods of activation include the use of nitrosyl tetrafluoroborate [236,237], the one electron transfer reagent tris(4-bromophenyl)ammoniumyl hexachloroantimonate (TBPA, **95**) (Figure 4.4) [238,239] or the $TrB(C_6F_5)_4$–$NaIO_4$ system [240] for activation. Electrochemical glycosylations using thioglycosides [241–243] and high-pressure glycosylations [244] are also reported.

A unique feature of thioglycosides is that, in contrast to glycosyl halides for example, their reactivity can be tuned not only by the protecting groups of the sugar but also by the choice of the aglycone. Thioglycosides with different heterocyclic aglycones, such as benzothiazol-2-yl [169,244], pyridin-2-yl [168,170,173,178], pyrimidin-2-yl [168,245], imidazol-2-yl [168], 1′-phenyl-1*H*-tetrazolyl [171,172] and, most recently, benzoxazol-2-yl [246,247] derivatives, have been used for glycosylations. These heterocyclic thioglycosides can be activated by the remote activation concept very often by reagents, which do not necessarily activate common alkyl and aryl thioglycosides. (This is indicated in Table 4.3, which shows the structure of the "special" aglycone of the thioglycoside.) Thus, pyrimidinyl thioglycosides, for example, were activated by TMSOTf, which does not activate alkyl/aryl thioglycosides, to give glycosides in good yields [245]. The recently introduced benzoxazolyl thioglycosides (SBox glycosides) could be activated with such reagents as AgOTf and $ZrCl_4$–Ag_2CO_3 [246,247]. As shown in Scheme 4.17, the thioglycoside acceptor **97** could be glycosylated with SBox thioglycoside **96** to produce the disaccharide **98** in excellent yield as the promoter does not activate the ethylthioglycoside function of the acceptor [247].

The reactivity of thioglycosides can also be decreased, if desired, by making the thioglycoside aglycone more bulky, or more electron withdrawing. Another way of tuning the reactivity of

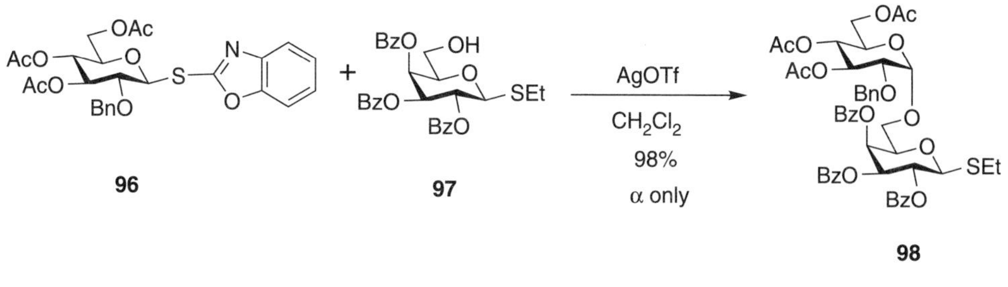

OAc, AcO, AcO, O, BnO, S, N, O, **96** + BzO, OH, O, BzO, BzO, SEt, **97** → AgOTf, CH_2Cl_2, 98%, α only → OAc, AcO, AcO, O, BnO, BzO, O, O, BzO, BzO, SEt, **98**

SCHEME 4.17 Glycosylation with SBox thioglycoside.

thioglycosides is by the selection of protecting groups. In these ways, thioglycosides could be efficiently applied in armed–disarmed type glycosylations.

The use of thioglycosides in glycosylations has been reviewed [13,22].

4.3.2.2 Glycosyl Sulfoxides and Sulfones

Glycosyl sulfoxides and sulfones are readily prepared by the oxidation of thioglycosides. Sulfoxides are generally formed as diastereomeric mixtures, and the mixture is used without separation of the diastereomers in glycosylation reactions.

Glycosyl sulfoxides as glycosyl donors were introduced by Kahne [248]. When activated by triflic anhydride or catalytic amounts of triflic acid, glycosyl sulfoxides react with unreactive substrates giving glycosides in good yields. A notable application of this reagent is the first one-pot synthesis of an oligosaccharide, the cyclamycin trisaccharide (Scheme 4.18) [249]. Treatment of a mixture of the sulfoxides **99**, **100** and thioglycoside **101** with TfOH resulted in the formation of the trisaccharide **103** isolated in a 25% yield. The reaction is based on the reactivity difference of the two sulfoxide donors, where the *p*-methoxyphenyl sulfoxide **100** reacts faster than **99**. Thus, the disaccharide **102** forms first, and is subsequently glycosylated by **99**.

The reaction is reported to proceed through the glycosyl α-triflate, which forms the basis of a very useful method for the synthesis of β-mannopyranosides [192].

Various promoters for the activation of glycosyl sulfoxides have been introduced and are summarized in Table 4.4.

Triflic anhydride in the presence of a base such as DTBMP is the most frequently used promoter with sulfoxides. It has also been used with lanthanide triflates as coactivators [215]. Promoters other than triflic anhydride for the activation of sulfoxides for glycosylation include TfOH [249,250], TMSOTf in the presence of triethylphosphite [251], the heteropolyacid $H_3PW_{12}O_{40}$ [252], the zirconocene reagent Cp_2ZrCl_2–$AgClO_4$ [253] and iodine and iodine monobromide [254,255].

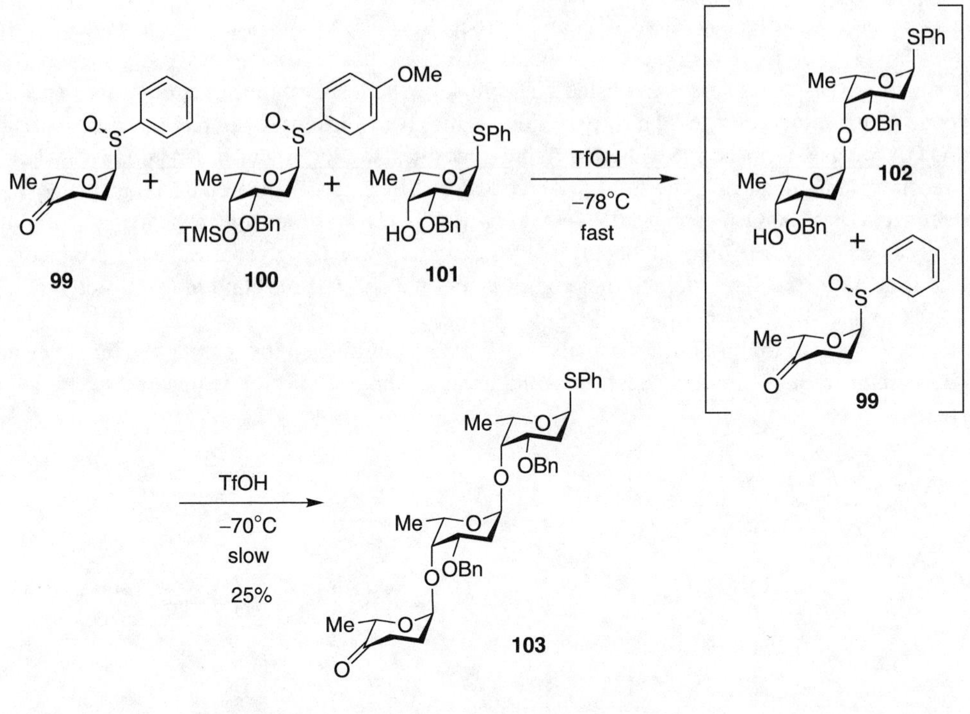

SCHEME 4.18 One-pot synthesis of ciclamycin trisaccharide.

TABLE 4.4
Activations of Glycosyl Sulfoxides

(Scheme: glycosyl sulfoxide (S(O)R') + ROH, Promoter → glycoside (OR))

Promoter	Ref.
Tf_2O	[248]
TfOH	[249,250]
$Tf_2O-Yb(OTf)_3$	[215]
TMSOTf	[251]
$H_3PW_{12}O_{40}$	[252]
$Cp_2ZrCl_2-AgClO_4$	[253]
I_2	[254,255]
IBr	[255]
Nafion-H	[256]
SO_4/ZrO_2	[256]

Activation of glycosyl sulfoxides by heterogeneous acid catalysis using Nafion-H or sulfated zirconia has also been described [256].

Compared with glycosyl sulfoxides, glycosyl sulfones are far less reactive. Nevertheless, activation of sulfones of some 2-deoxy sugars has been successfully used in glycosylations using $MgBr_2{\cdot}Et_2O$ [257,258]. The recently introduced pyridyl sulfones reacted under $Sm(OTf)_3$ promotion to give oligosaccharides [259].

4.3.2.3 Other Anomeric Thio Derivatives

Several types of compounds with anomeric C–S bonds, together with other functionalities, can also be used as glycosyl donors. Their activation methods are summarized in Table 4.5.

Glycosyl dithiocarbonates (xanthates) **104** (Table 4.5) were introduced by Sinaÿ and coworkers [260]. DMTST- and $Cu(OTf)_2$-promoted reactions of xanthates have been used for the synthesis of *cis*-glycosides of 2-amino-2-deoxy sugars [261,262]. Xanthates can be effectively activated with different alkylsulfenylating agents besides DMTST [260]. In fact, MeSOTf [263] and PhSOTf [191] were also recommended. Reaction of xanthates with alkylsulfenylating agents forms the basis of effective methods for α-sialylations [191,260,263]. Additional promoters used with glycosyl xanthates are AgOTf [264,265], NIS–TfOH [266], ICl–AgOTf [225,226], IBr–AgOTf [226], and I_2–AgOTf [226].

Dithiocarbamates of the type **105** (Table 4.5) were shown to be effective glycosyl donors not only with typical thioglycoside promoters (such as MeOTf and DMTST), but also with Lewis acids such as AgOTf, $SnCl_4$ and $FeCl_3$ [267]. Similarly, glycosylation reactions with the corresponding diethyl dithiocarbamates **106** were promoted with NIS–TfOH, IDCP [268,269], AgOTf [264,265, 268], TMSOTf [268], and $BF_3{\cdot}Et_2O$ [268,269].

Glycosyl thiocyanates **107** have been used to glycosylate tritylated acceptors using catalytic amounts of $TrClO_4$ as promoter [270,271]. With thiocyanates that have a nonparticipating protecting group at O-2, these glycosylations showed remarkably high stereoselectivities, giving 1,2-*cis*-glycosides.

Other types of thio compounds have also been proposed as glycosyl donors. Glycosyl disulfides **108** are activated by typical thioglycoside promoters such as NIS–TESOTf [272]. Glycosyl sulfimides **109**, which can be considered as the nitrogen analogs of sulfoxides, were shown to behave as glycosyl donors in $Cu(OTf)_2$-promoted reactions [273].

TABLE 4.5
Activations of Anomeric Thio Derivatives

(glycosyl–X) + ROH —Promoter→ (glycosyl–OR)

X	Promoter	Ref.
—S–C(=S)–OEt **104**	DMTST	[260–262]
	$Cu(OTf)_2$	[261,262]
	MeSOTf	[263]
	PhSOTf	[191]
	AgOTf	[264,265]
	NIS–TfOH	[266]
	ICl–AgOTf	[225,226]
	IBr–AgOTf	[226]
	I_2–AgOTf	[226]
—S–C(=S)–N(piperidino) **105**	MeOTf	[267]
	DMTST	[267]
	AgOTf	[267]
	$SnCl_4$	[267]
	$FeCl_3$	[267]
—S–C(=S)–NEt_2 **106**	NIS–TfOH	[268,269]
	IDCP	[268,269]
	AgOTf	[264,265,268]
	TMSOTf	[268]
	$BF_3 \cdot Et_2O$	[268,269]
—S–C≡N **107**	$TrClO_4$	[270,271]
—S–S–R' **108**	NIS–TESOTf	[272]
—S^+(–NTs^-)–Et **109**	$Cu(OTf)_2$	[273]
—S–P(=S)–$(OR)_2$ **110**	AgF	[274]
	NIS	[275]
	IDCP	[275]
	AgOTf	[265]
	$AgClO_4$	[276]
	DMTST	[277]
—S–P(=NPh)–$(NMe_2)_2$ **111**	LPTS–Bu_4NI	[278,279]

Glycosyl thio compounds with phosphorus in the leaving group have also been studied. Dithioposphates **110** were activated by AgF [274], NIS and IDCP [275], AgOTf [265], $AgClO_4$ [276] or DMTST [277]. Glycosylation reactions of phosphorodiamidimidothioates **111** were promoted by 2,6-lutidinium *p*-toluenesulfonate (LPTS) and Bu_4NI [278,279] to give *cis*-glycosides.

4.3.3 Synthesis of Glycosides from Anomeric *O*-Derivatives

The free hydroxyl group of lactols can also be replaced to form glycosides. Additionally, it is frequently substituted first, in order to increase its leaving group character. The major classes of anomeric *O*-derivatives that are used as glycosyl donors are described below.

4.3.3.1 Glycosyl Imidates

4.3.3.1.1 Glycosyl Acetimidates

Glycosyl acetimidates **114** (Scheme 4.19) were introduced by Sinaÿ and coworkers [280]. They are prepared by the reaction of perbenzylated glycosyl chlorides, such as **112**, with

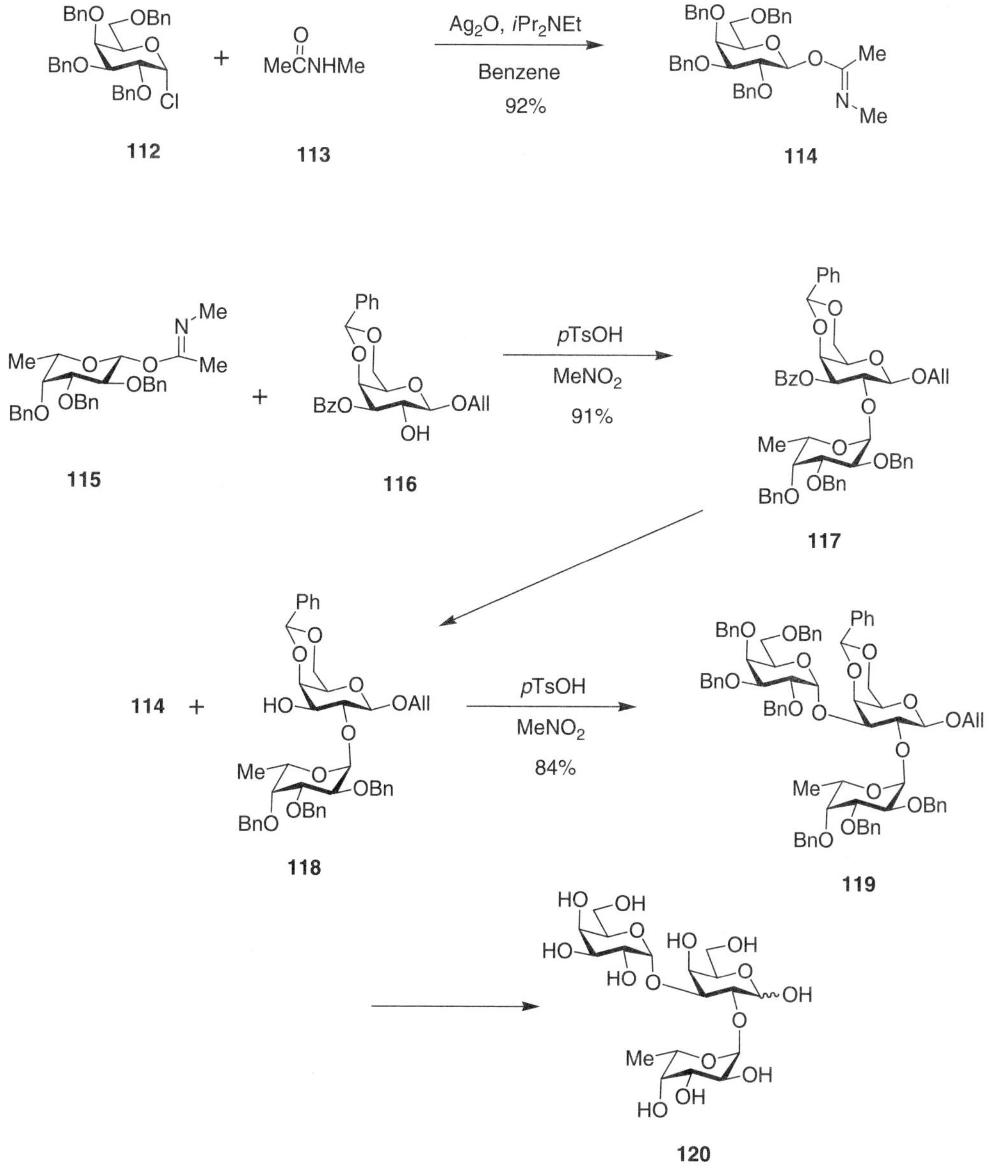

SCHEME 4.19 Glycosylation with glycosyl acetimidates.

N-methylacetamide **113**. The acetimidates react in the presence of *p*TsOH as a promoter to give 1,2-*cis*-glycosides in excellent stereoselectivity [280,281]. The synthesis of the blood group B determinant, oligosaccharide **120** [282], using Sinaÿ's imidate procedure is illustrated in Scheme 4.19.

4.3.3.1.2 Glycosyl Trichloroacetimidates

The original imidate procedure was further developed by Schmidt [283] with the introduction of trichloroacetimidates. The electron-deficient nitrile, trichloroacetonitrile, readily adds to the free hydroxyl of lactols under basic conditions. In the presence of a weak base, such as potassium carbonate, the β-imidate **122** can be isolated as the kinetic product, whereas the use of strong bases, such as sodium hydride or DBU, results in the formation of the thermodynamically more stable α-trichloroacetimidates **123** [284,285] (Scheme 4.20).

Glycosyl trichloroacetimidates are relatively stable under basic or neutral conditions, but react readily under acidic conditions. With various acidic nucleophiles, such as carboxylic acids and phosphoric acid derivatives, the corresponding glycosyl esters are formed without any additional catalyst [12,283,286,287]. Reaction with non-acidic *O*-nucleophiles proceeds in the presence of catalytic amounts of Brønsted or Lewis acids. Originally, *p*TsOH and $BF_3 \cdot Et_2O$ were used [283], while the latter and TMSOTf [288,289] are currently the most frequently employed promoters. Glycosylations with these promoters take place at low temperatures under mild conditions. Neighboring group participation results in the formation of 1,2-*trans*-glycosides, whereas with nonparticipating substituents, the use of $BF_3 \cdot Et_2O$ favors the formation of the reaction product with inversion. Moreover, the use of TMSOTf in ether preferentially affords *cis*-glycosides [12]. Thus, $BF_3 \cdot Et_2O$-promoted reaction of the α-imidate **123** with **41** afforded the disaccharide **124** with high β-selectivity [283], whereas reaction of the β-imidate **125**, assisted by TMSOTf, resulted in the formation of the α-glycoside **126** as the major product (Scheme 4.21).

Besides the activators mentioned thus far, several additional promoters have been introduced and are summarized in Table 4.6. These include protic and Lewis acids (such as TfOH [290,291]), pyridinium *p*-toluenesulfonate (PPTS) [292,293], and $ZnBr_2$ [294]. A different type of activation, based on reversible reaction of the promoter with the glycosyl acceptor, was developed by Schmidt [295] and based on this principle, chloral was introduced as a promoter. Activation under essentially neutral conditions can also be achieved using $LiClO_4$ [126,158].

Various metal triflates such as AgOTf [296,297], $Sn(OTf)_2$ [298,299], $Sm(OTf)_3$ [300], $Yb(OTf)_3$ [301], and $Cu(OTf)_2$ [70] have also been recommended and used. Anhydrous acids such

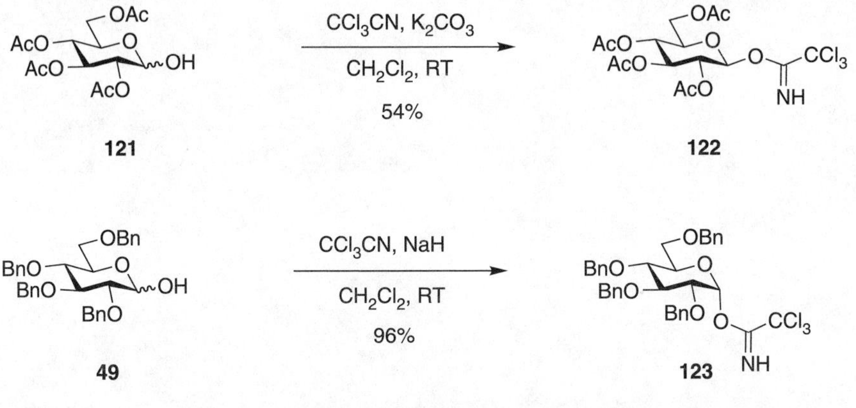

121 → 122: CCl_3CN, K_2CO_3 / CH_2Cl_2, RT / 54%

49 → 123: CCl_3CN, NaH / CH_2Cl_2, RT / 96%

SCHEME 4.20 Synthesis of glycosyl trichloroacetimidates.

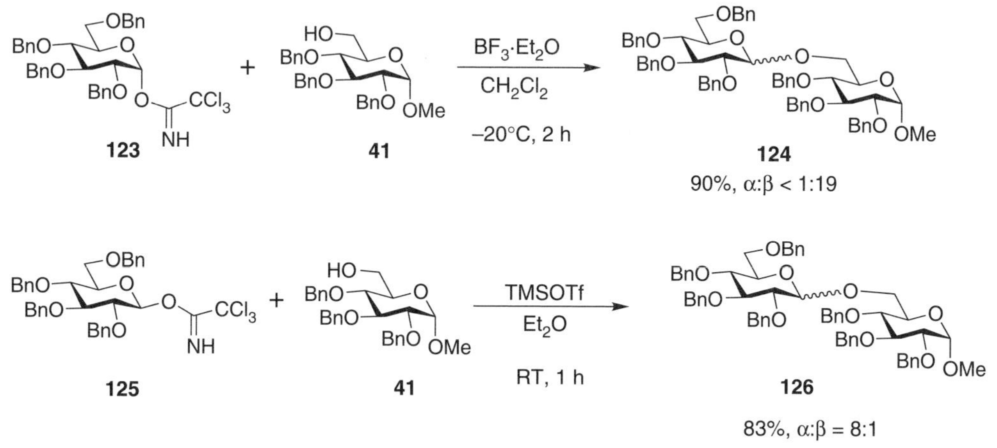

SCHEME 4.21 Stereoselective glycosylations with glycosyl trichloroacetimidates.

as $HClO_4$ and $HB(C_6F_5)_4$ [45,135], as well as the iodine-triethylsilane system (a source of anhydrous HI) [302], were also found to activate trichloroacetimidates for glycosylations. Iodine alone has also been described as capable of activating trichloroacetimidates [254]. Furthermore, molecular sieves, which are commonly used as drying agents in glycosylation reactions can serve as

TABLE 4.6
Activations of Glycosyl Trichloroacetimidates

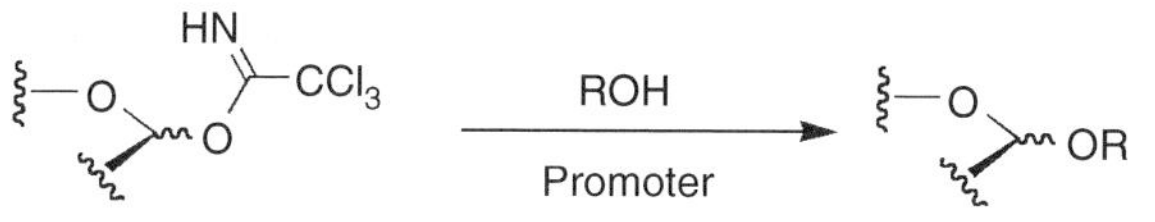

Promoter	Ref.
*p*TsOH	[283]
BF_3·Et_2O	[283]
TMSOTf	[288,289]
TfOH	[290,291]
PPTS	[292,293]
$ZnBr_2$	[294]
CCl_3CHO	[295]
$LiClO_4$	[126,158]
AgOTf	[296,297]
$Sn(OTf)_2$	[298,299]
$Sm(OTf)_3$	[300]
$Yb(OTf)_3$	[301]
$Cu(OTf)_2$	[70]
$HClO_4$	[45,135]
$HB(C_6F_5)_4$	[45,135]
I_2–Et_3SiH	[302]
I_2	[254]
MS AW 300	[303]
Acylsulfonamides	[304]

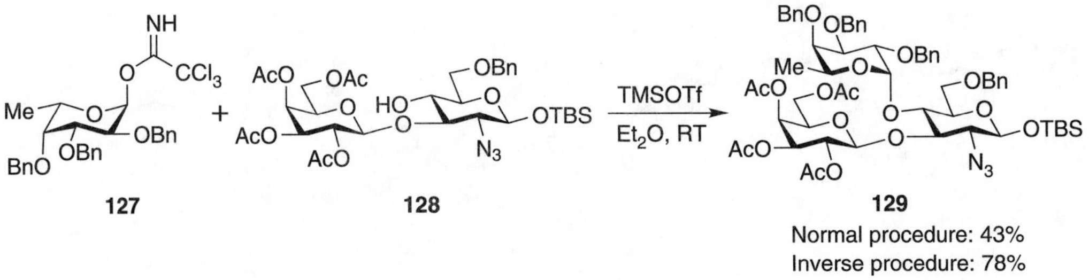

127 128 129

TMSOTf

Et_2O, RT

Normal procedure: 43%
Inverse procedure: 78%

SCHEME 4.22 The normal and the inverse procedure.

promoters. Acid-washed molecular sieves (MS AW 300) have recently been shown to activate trichloroacetimidates without any additional promoters [303]. Finally, the use of acyl sulfonamides as catalysts has been reported [304].

Besides the introduction of new promoters, another technical development related to glycosyl trichloroacetimidates as glycosyl donors was the introduction of the *inverse procedure* [305]. In the original procedure, the promoter is added to a mixture of the glycosyl donor and acceptor. It was assumed that highly reactive glycosyl trichloroacetimidates might partially decompose prior to being exposed to the glycosyl acceptor. In support of this hypothesis, adding the glycosyl donor to a mixture of the acceptor and promoter resulted in increased yields. Thus, glycosylation of the disaccharide acceptor **128** with the reactive fucose imidate **127** gave the trisaccharide **129** in a 43% yield by the normal procedure, whereas **129** was obtained in a 78% yield by the inverse procedure (Scheme 4.22).

Glycosyl trichloroacetimidates are among the most frequently used glycosyl donors today, and they have been successfully used in the syntheses of a large number of complex natural products, as, for example, glycosphingolipids [306,307], amphotericin B [292,293], everninomicin 13,384-1 **80** [146–149], calicheamicin [308,309], eleutherobin [310,311], vancomycin [312–314], and woodrosin [315,316]. As a simple example of these complex syntheses, the solid-phase synthesis of macrophylloside D heptaacetate [317] is shown in Scheme 4.23. As illustrated, glycosylation of the resin-bound carboxylic acid **133** with the imidate **131**, having a temporary protecting group at O-6, gave the ester **134**. After desilylation, the alcohol was glycosylated with **132** to give **135**, which was cleaved from the resin by selenoxide elimination affording **136** in 18% overall yield.

The trichloroacetimidate procedure has been reviewed at various stages of its development [12,318–320].

4.3.3.1.3 *Glycosyl Trifluoroacetimidates*

Glycosyl trifluoroacetimidates have recently been introduced as glycosyl donors [321]. The compounds are prepared by the reaction of the hemiacetal hydroxyl group with *N*-substituted trifluoroacetimidoyl halides **138**, providing anomeric mixtures of the imidates **139** (Scheme 4.24).

The conversion of trifluoroacetimidates **139** into glycosides **1** has been promoted by the same reagents used for trichloroacetimidates. Thus, TMSOTf [321–324], $Yb(OTf)_3$ [325,326], I_2–Et_3SiH [302], and MS AW 300 [303] were used for this purpose.

The utility of glycosyl trifluoroacetimidates was demonstrated in the preparation of saponins. The synthesis of dioscin **146** [322] is shown in Scheme 4.25. As illustrated, diosgenin

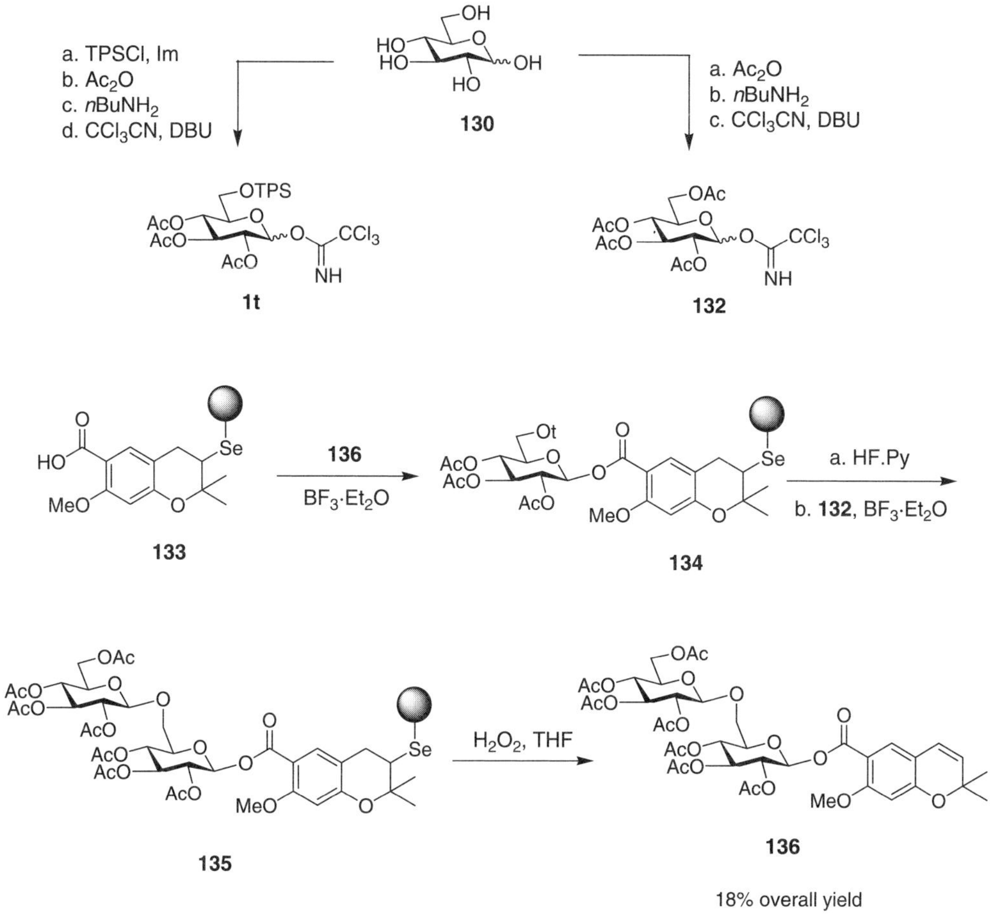

SCHEME 4.23 Solid-phase synthesis of macrophylloside D heptaacetate.

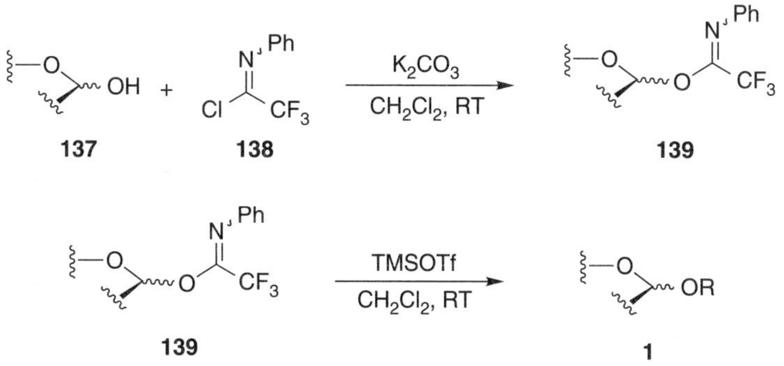

SCHEME 4.24 Synthesis of and glycosylation with glycosyl trifluoroacetimidates.

141 was glycosylated with the benzoylated trifluoroacetimidate **140** to produce the glucoside **142** in excellent yield. Protecting group exchange afforded the 3,6-di-*O*-pivaloate **143**, which was bis-rhamnosylated with the imidate **144** to give **145**. After deprotection, dioscin **146** was obtained.

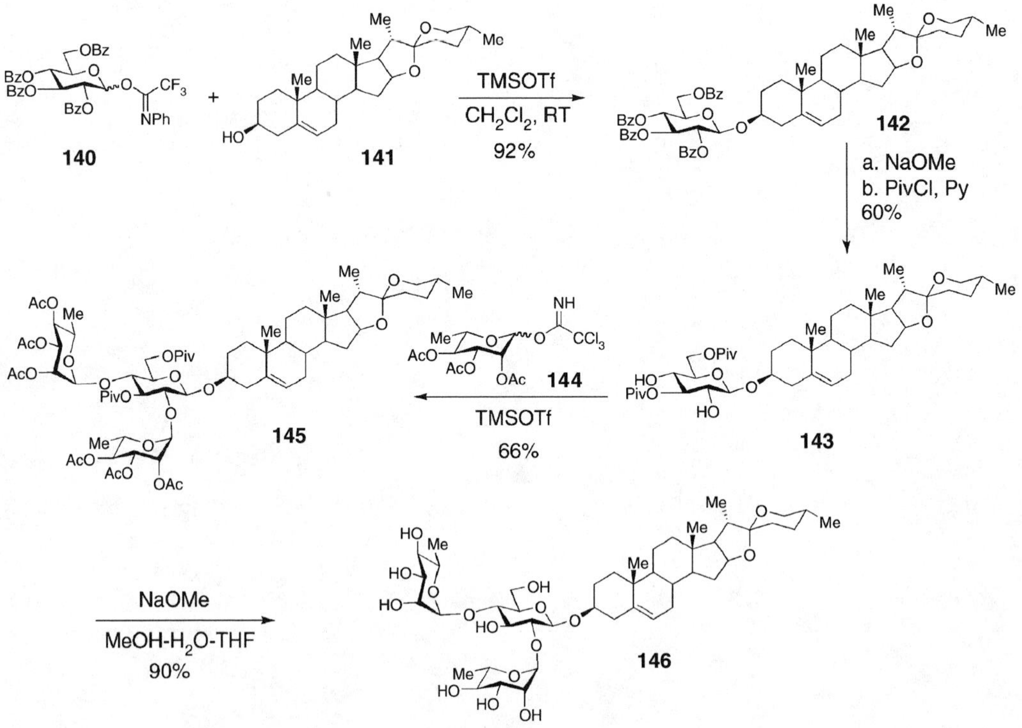

SCHEME 4.25 Synthesis of dioscin.

4.3.3.2 Glycosyl Esters

Glycosyl carboxylates are readily available. In fact, the peracetylated derivatives are particularly easy to prepare on a large scale, and this class of compounds has long drawn attention as potential glycosyl donors. However, the acyloxy group is not as good as leaving group as the trichloroacetimidoyl, for example, so their activation generally requires harsher conditions.

Helferich introduced peracetylated sugars for the glycosylation of phenols using *p*TsOH or $ZnCl_2$ as promoters [327]. Several other Lewis acids were also employed successfully for glycosylations with glycosyl carboxylates (Table 4.7). These include $SnCl_4$ [328,329], $FeCl_3$ [330–333], $BF_3{\cdot}Et_2O$ [334,335], TMSOTf [79,336], $TrClO_4$ [337], $Sn(OTf)_2$ [338], $Cu(OTf)_2$ [70,338], and the $Zn(OTf)_2$–TMSCl combination [339]. Among the above promoters, TMSOTf is quite effective and is widely used. For example, the lactosamine derivative **149** was obtained by glycosylation of the *N*-acetyl-glucosamine derivative **148** with penta-*O*-acetyl-β-D-galactopyranose **147** (Scheme 4.26) [336].

The use of two- and three-component promoter systems, such as $SnCl_4$–$Sn(OTf)_2$–$LiClO_4$ [340], $SnCl_4$–$AgClO_4$ [341,342], $GaCl_3$–$AgClO_4$ [342], $SiCl_4$–$AgClO_4$ [343,344], Ph_2SnS–$AgClO_4$ [344], $AgClO_4$–$LiClO_4$ [344], and Lawesson's reagent (**150**) (Figure 4.5), in combination with $AgClO_4$ [344], for the glycosylations of silylated acceptors allows the use of catalytic amounts of Lewis acids instead of the stoichiometric or higher amounts required by the previous procedures. Furthermore, the stereoselectivity of glycosylations was also influenced by the choice of the proper promoter system. Specifically, formation of α-glucosides was reported with the $SnCl_4$–$AgClO_4$ system in ether, whereas the β-glucosides were obtained using $SiCl_4$–$AgClO_4$ as promoter in acetonitrile [343]. Similarly, the addition of Bu_4NBr to promoters like TMSOTf and

TABLE 4.7
Activations of Glycosyl Esters

Promoter	Acyl	X	Ref.
$ZnCl_2$	Ac	H	[327]
*p*TsOH	Ac	H	[327]
$SnCl_4$	Ac	H	[328,329]
$FeCl_3$	Ac	H	[330–333]
$BF_3{\cdot}Et_2O$	Ac	H	[334,335]
TMSOTf	Ac, Bz	H, $SnBu_3$	[79,336,349]
$TrClO_4$	$COCH_2Br$	H	[337]
$Sn(OTf)_2$	COPyr	H	[338]
$Cu(OTf)_2$	COPyr, Ac	H	[70,338]
$Zn(OTf)_2$–TMSCl	Ac, *p*NBz	H	[339]
$SnCl_4$–$Sn(OTf)_2$–$LiClO_4$	Ac, $COCH_2Br$	TMS	[340]
$SnCl_4$–$AgClO_4$	Ac	TMS	[341,342]
$GaCl_3$–$AgClO_4$	Ac	TMS	[342]
$SiCl_4$–$AgClO_4$	(Methoxyethoxy)acetyl	TMS	[343,344]
Ph_2SnS–$AgClO_4$	$COCH_2I$	H, TMS	[344]
$AgClO_4$–$LiClO_4$	$COCH_2I$	TMS	[344]
150–$AgClO_4$	$COCH_2I$	H, TMS	[344]
TMSOTf–Bu_4NBr	Ac	H	[345]
$BF_3{\cdot}Et_2O$–Bu_4NBr	Ac	H	[345]
$Yb(OTf)_3$	Ac	Borate	[346]
$HClO_4$	Ac	H	[45,135,347]
Perfluorooctanesulfonic acid	Ac	H	[347]
TfOH	Ac	H	[73,134]
$HB(C_6F_5)_4$	Ac	H	[45,135]
Montmorillonite K-10	Ac	H	[348]

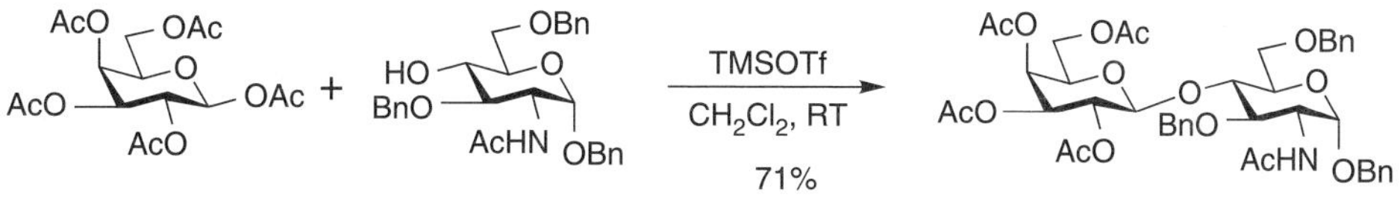

SCHEME 4.26 Synthesis of *N*-acetyllactosamine using glycosyl ester.

$BF_3{\cdot}Et_2O$ resulted in a shift of stereoselectivity towards β-glycoside formation in the synthesis of sordaricin [345].

Boric acid esters of phenols have increased reactivity compared with the parent compounds. As such, triaryloxyboranes were glycosylated with glycosyl acetates using catalytic amounts of $Yb(OTf)_3$ [346]. Strong protic acids, such as $HClO_4$ [45,135,347], perfluorooctanesulfonic acid

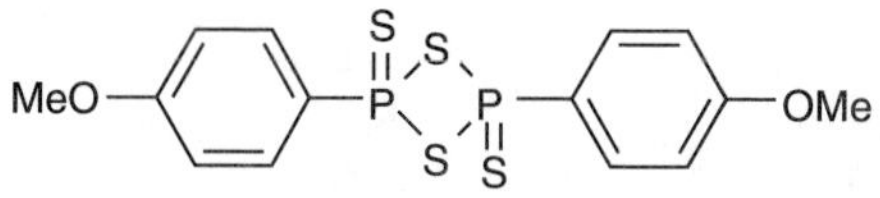

FIGURE 4.5 Lawesson's reagent.

[347] as well as TfOH [73,134] and $HB(C_6F_5)_4$ [45,135], also activate glycosyl acetates for glycosylations.

The use of heterogeneous catalysis with montmorillonite K-10 [348] was reported for the preparation of some simple glycosides.

Among the various possible acyl leaving groups, glycosyl acetates are the most common. The use of other acyl groups was found to be more advantageous in some cases — for example, bromoacetyl [337], iodoacetyl [344] or (2′-methoxyethoxy)acetyl [343]. Other common acyl groups such as benzoyl [349], *p*-nitrobenzoyl [339,350] and trichloroacetyl [351] have also been used. A special case is the use of pyridine-2-carboxylate esters based on the remote activation concept [338] (Section 4.3.2.1). Similarly, the use of pent-4-enoyl esters [352] is based on the specific activation of *n*-pentenyl glycosides (see Section 4.3.3.3.1).

Because the acyloxy group is not as good a leaving group as the trichloroacetimidoyl, for example, these reactions proceed under strongly acidic conditions. Thus, the protecting groups both in the aglycone and in the glycosyl donor molecule should be acid stable. Under the strongly acidic conditions, anomerization of the β-glycosides to the thermodynamically more stable axial glycosides might occur. Furthermore, 1,2-*trans*-acetates react significantly faster than the corresponding 1,2-*cis* derivatives, so their use is more advantageous. As 2-deoxy sugars are generally more reactive glycosyl donors than common hexoses, an important application of this reaction is in the synthesis of 2-deoxy-glycosides, which proceeds relatively easily. Because 2-deoxy sugars are important constituents of different groups of antibiotics, several examples of these transformations can be found in the antiobiotics field [14,26]. An example from the area of anthracycline antibiotics is shown in Scheme 4.27 [350].

As illustrated, glycosylation of the daunomycinone derivative **152** with the L-daunosamine *p*-nitrobenzoate **151** afforded the glycoside **153** as only the α-isomer in a 99% yield. This was deprotected giving 4-demethoxydaunorubicin **154**.

4.3.3.3 *O*-Glycosides

The aglycone of certain *O*-glycosides can also function as a leaving group providing opportunities for transglycosylation reactions.

4.3.3.3.1 n-Pentenyl Glycosides

This class of glycosyl donors, introduced by Fraser-Reid [353], provides excellent stability under a variety of protecting group manipulations. *n*-Pentenyl glycosides **155** are activated by halogenation of the double bond (→**156**), which results in cyclization (→**157**) and the release of the aglycone in the form of **158**, thus providing active glycosylating species **159** (Scheme 4.28).

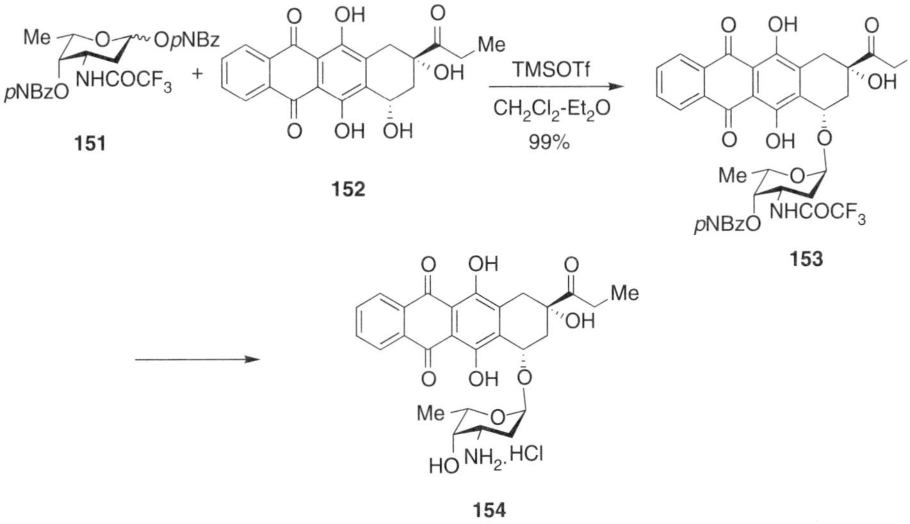

SCHEME 4.27 Synthesis of 4-demethoxydaunorubicin.

These glycosylation reactions are promoted by NBS, iodonium dicollidine perchlorate (IDCP) or the corresponding triflate (IDCT) [353]. Iodine has also been reported to promote the reaction [254]. NIS, in combination with catalytic amounts of a protic acid such as TfOH [213], or Lewis acids such as TESOTf [17], $Sn(OTf)_2$ [354] or $BF_3 \cdot Et_2O$ [355], was introduced as a more active catalyst. Recently, glycosylation with NIS under microwave irradiation has been developed [356].

A major difference in the reactivity of *n*-pentenyl glycosides possessing benzyl protecting groups compared with acylated glycosides was observed. Specifically, the acylated derivatives react at much slower rates. This observation resulted in the development of the *armed–disarmed* concept, in which a disarmed (acylated) *n*-pentenyl glycoside with a free hydroxyl group could be glycosylated with an armed (benzylated) *n*-pentenyl glycoside without self-condensation of the previous one [357,358]. As shown in Scheme 4.29, the armed pentenyl glycoside **161** was coupled with the disarmed pentenyl glycoside **162** using the moderately powerful IDCP promoter affording the disaccharide **163** in a 63% yield [359]. Disarmed pentenyl glycosides can also be activated

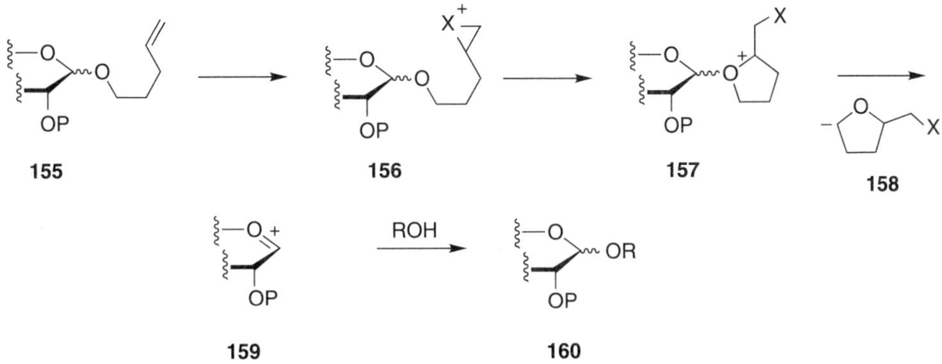

SCHEME 4.28 Glycosylations with *n*-pentenyl glycosides.

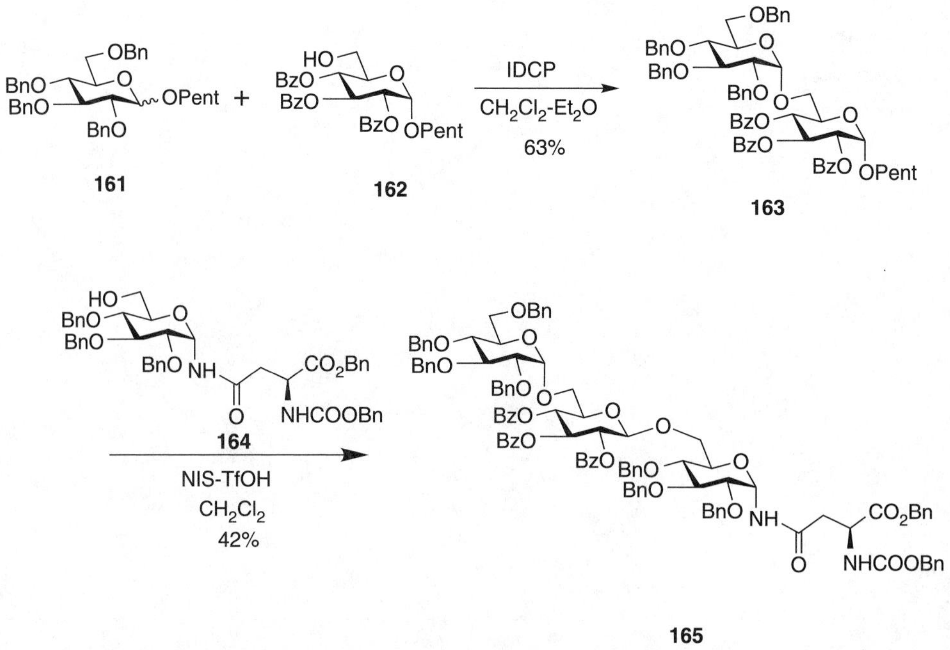

SCHEME 4.29 Oligosaccharide synthesis by armed-disarmed glycosylations.

using the more powerful NIS–TfOH catalyst system. As such, the disarmed disaccharide **163** was coupled directly with the aspartyl derivative **164** to afford the trisaccharide **165**.

The reactivity differences caused by the nature of protecting groups or torsional effects [17] are not specific to *n*-pentenyl glycosides, but turned out to be of a fairly general nature. Armed–disarmed type glycosylations could be performed on other types of glycosyl donors, including thioglycosides and glycosyl fluorides.

The *n*-pentenyl glycosylation was used in the synthesis of biologically important molecules [360], and this method has been thoroughly reviewed elsewhere [17,361].

4.3.3.3.2 Enol Ether-Type Glycosides

The use of an enol ether-type glycoside as a glycosyl donor was described by Schmidt and coworkers [362] through the use of enol ethers formed in the addition of a sugar hemiacetal to ethyl phenylpropiolate. Sinaÿ introduced isopropenyl glycosides [363] obtainable from glycosyl acetates by Tebbe's reagent. These compounds react in the presence of TMSOTf or $BF_3{\cdot}Et_2O$ to give glycosides (Scheme 4.30).

A variety of other electrophiles, such as NIS–TfOH, Tf_2O, AgOTf, and DMTST, have also been shown to be effective activators of isopropenyl glycosides [364,365].

Boons introduced substituted allyl glycosides **169** (Scheme 4.31), which are converted to the enol ethers **170** prior to glycosylation [366,367]. An allyl ether-type glycoside **171** can be glycosylated with the enol ether-type **170**, providing a coupling product **172** suitable for further glycosylation after conversion to the enol ether. The allyl ether-type glycoside can be considered as a latent form of a glycosyl donor, which can be efficiently isomerized to the active vinyl ether form. Thus, a useful latent–active pair for multiple glycosylation reaction sequences is produced. These glycosides were used with TMSOTf or NIS–TMSOTf promoters [366–368].

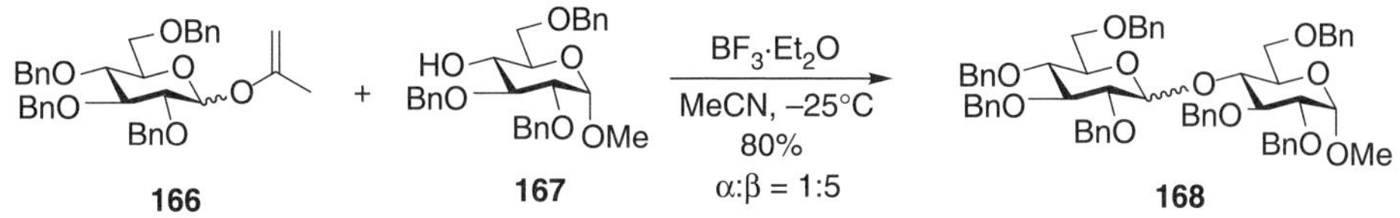

BnO, BnO, OBn, O, O, BnO **166** + HO, BnO, OBn, O, BnO, OMe **167** BF_3·Et_2O MeCN, −25°C 80% α:β = 1:5 → **168**

SCHEME 4.30 Glycosylation using isopropenyl glycoside.

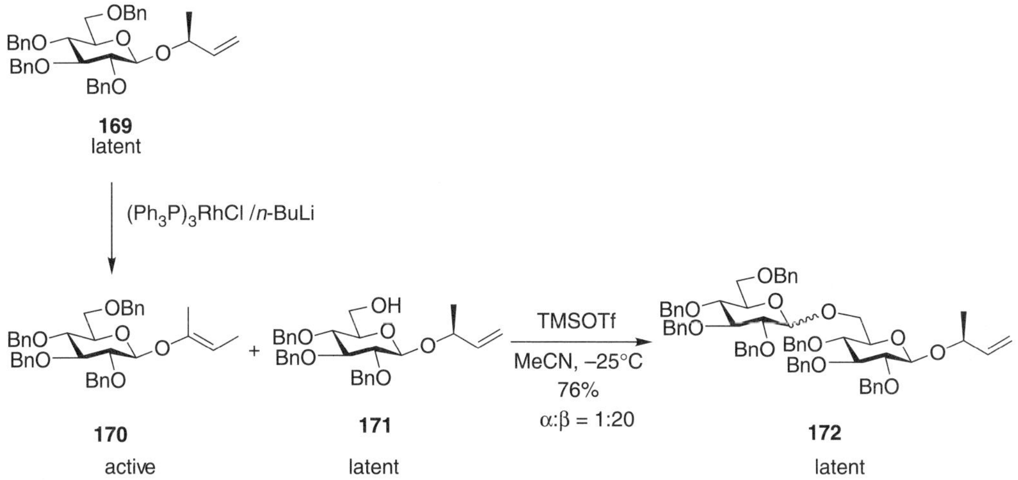

169 latent

$(Ph_3P)_3RhCl$ /*n*-BuLi

170 active + **171** latent → TMSOTf, MeCN, −25°C, 76%, α:β = 1:20 → **172** latent

SCHEME 4.31 Vinyl glycosides in latent-active glycosylation.

4.3.3.3.3 Heteroaryl Glycosides

Glycosides of some heterocycles have also been investigated as glycosyl donors. For example, perbenzylated 2-pyridyl glycosides were activated by electrophiles, such as MeOTf and Et_3O·BF_4, to give mixtures of *cis*- and *trans*-glycosides [369]. Schmidt and coworkers studied the related 2-pyrimidinyl [362], and other heteroaryl glycosides [370]. These donors were activated by TMSOTf.

Based on the concept of remote activation, first applied to pyridine thioglycosides [168], Hanessian introduced 3-methoxy-2-pyridyl (MOP) glycosides [31]. These glycosides react in MeOTf-, $Cu(OTf)_2$-, or $Yb(OTf)_3$-promoted reactions to give glycosides [361,371–373]. Interestingly, unprotected MOP glycosides **173** could also be used as donors. In fact, using an excess of glycosyl acceptor **174**, glycoside **175** was obtained in a reasonable yield [31,361] (Scheme 4.32).

4.3.3.3.4 Dinitrosalicylate Glycosides

A new class of *O*-glycoside glycosyl donors — methyl dinitrosalicylate (DISAL) glycosides — has recently been introduced [374]. Because of the strongly electron-withdrawing groups in the aglycone, these glycosides show increased reactivity compared with common glycosides. DISAL glycosides can be activated not only with strong Lewis acid promoters, such as $FeCl_3$, TMSOTf or BF_3·Et_2O, but they also react in the presence of the mild $LiClO_4$ promoter [374]. Moreover, perbenzylated DISAL glycosides react under neutral or mildly basic conditions [37,375,376].

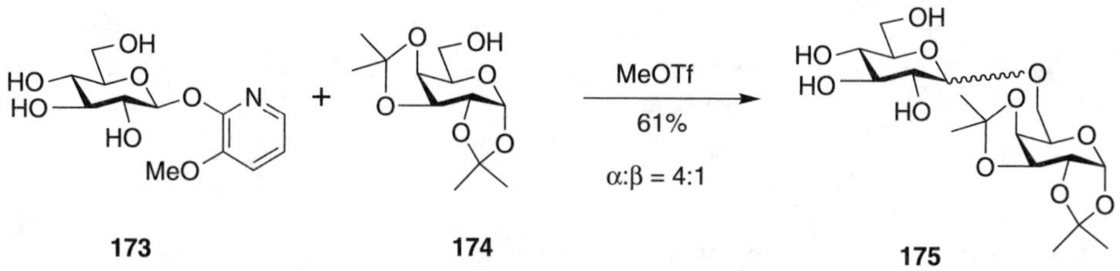

SCHEME 4.32 Glycosylation with MOP glycoside.

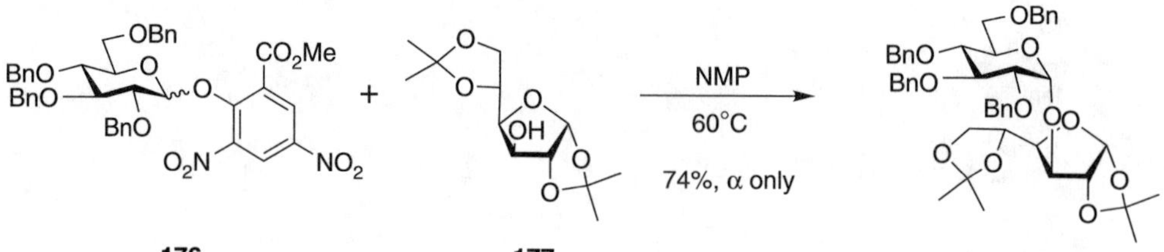

SCHEME 4.33 Glycosylation with DISAL glycoside.

Thus, reaction of the perbenzylated DISAL glycoside **176** with diisopropylidene-α-D-glucofuranose **177** in *N*-methyl-pyrrolidone at 60°C, without any added promoter, afforded the corresponding α-linked disaccharide **178** [376] (Scheme 4.33).

4.3.3.3.5 2′-Carboxybenzyl Glycosides

2′-Carboxybenzyl (CB) glycosides **180** are prepared in the form of their benzyl esters **179** (Scheme 4.34). The free carboxylic acids react in the presence of Tf_2O and DTBMP as a base to give glycosides **182** in good yields with the formation of the lactone **183** as a by-product [377–379]. The method was highly effective for the synthesis of difficultly accessible β-mannopyranosides. For example, in the glycosylation shown in Scheme 4.34, only the β-glycoside was formed, and **182** was obtained in a 91% yield. The carboxybenzyl glycoside benzyl esters **179** and the carboxylic acid **180** constitute another pair of latent–active glycosyl donors.

4.3.3.3.6 Other O-Glycosides

Glycosylation reactions using *O*-glycosides that do not fall in the above categories have also been described. Thus, 3,4-dimethoxybenzyl glycosides of 2-deoxy sugars reacted in DDQ-promoted reactions to give glycosides [380]. Bromobutyl glycosides, both protected and unprotected were also recently used as glycosyl donors [381]. 6-Nitro-2-benzothiazolyl glycosides, which can be considered the cyclic analogs of thioformimidates (see Section 4.3.3.8), were recently introduced [382]. Furthermore, electrochemical glycosylations of aryl glycosides have been described [383].

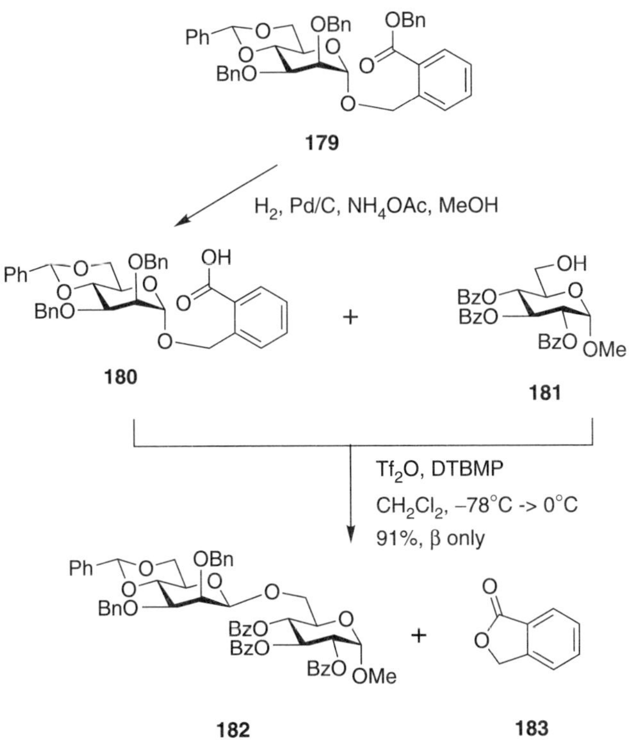

SCHEME 4.34 Glycosylation with CB glycoside.

4.3.3.4 Hemiacetals

4.3.3.4.1 Glycosylation of Unprotected Sugars — Fischer Glycosylation

The direct conversion of the hemiacetal hydroxyl group to a glycoside is one of the earliest glycosylation methods. Concerning the reaction of free sugars with alcohols in the presence of acid catalysts, the well-known Fischer glycosylation [4,5] constitutes a very useful method for the synthesis of glycosides of such simple alcohols as methyl, benzyl or allyl alcohol. Although the reaction leads to a mixture of the possible furanosides and pyranosides, in equilibrium, the ratio of the thermodynamically more stable axial glycosides of pyranosides is sufficiently high to be of use. On the other hand, furanosides, being the kinetic products of the reaction, are often accessible under kinetic control in reasonable yields [7,8,31]. A large number of modifications of the original reaction have been reported [384–386]. The use of some Lewis acids ($FeCl_3$, $BF_3{\cdot}Et_2O$) instead of protic acids has permitted the preparation of several furanosides and some pyranosides in good selectivity [387–390].

To obtain glycosides of predefined ring size, protected derivatives are used in most methods. Activation of the anomeric hydroxyl is generally achieved by its *in situ* conversion into some activated derivative, such as a glycosyl halide or a sulfonate. In contrast, for example, to Koenigs–Knorr reactions, these procedures are performed in a one-pot manner. There is no need to isolate the activated intermediate. In the following section, the different glycosylation methods with hemiacetal glycosyl donors are organized according to the (presumed) intermediates used for their activation. In all of these methods, care has to be taken to avoid self-condensation and formation of nonreducing oligosaccharides because of competition of the hemiacetal as a nucleophile with the intended glycosyl acceptor. This is commonly achieved by using an excess of the glycosyl acceptor.

TABLE 4.8
Activations of Hemiacetals

Promoter	Additive	X	Ref.
$MsOH-CoBr_2$	Bu_4NBr, Et_4NClO_4	H	[391,392]
$TMSBr-CoBr_2$	Bu_4NBr	H	[393–395]
Tf_2O-*s*-collidine	Bu_4NBr	H	[396,397]
PPh_3-CBr_4		H	[88]
Tf_2O		H	[398–401]
Ms_2O		H	[397]
$NsCl-AgOTf-Et_3N$	$AcNMe_2$	H	[402–407]
TsCl–aq. NaOH	$BnEt_3NCl$	H	[408]
PPh_3–DEAD or DIAD	$HgCl_2$, $HgBr_2$	H	[409–412]
n-Bu_3PO-Tf_2O		H	[415]
Ph_2SO-Tf_2O	2-Chloropyridine, TTBP	H	[416,417]
Me_2S-Tf_2O	TTBP	H	[419]
TMSOTf	Pyridine	H	[420,421]
PPTS		H	[422]
191–Tf_2O	$i Pr_2NEt$	H, TMS	[423]
$Ph_2SnS-Tf_2O$	$LiClO_4$	H, TMS	[424]
$Ph_2SnS-AgClO_4$		H, TMS	[344]
150–$AgClO_4$		H, TMS	[344]
$MeOCH_2CO_2H-Yb(OTf)_3$		H	[425]
$TMSCl-Zn(OTf)_2$		H	[426]
$TMS_2O-Sn(OTf)_2$		H	[427]
$TMSCl-Sn(OTf)_2$		H	[428]
$Cu(OTf)_2$		H	[70]
$TrClO_4$		H	[429]
$TrB(C_6F_5)_4$	$LiClO_4$, $LiNTf_2$	H	[429,430]
DCCI–CuCl		H	[432]
Phosphine/Cu(II) complex		H	[433]
Rhodium(III)-triphos		H	[434]

The various promoter systems for the activation of hemiacetals in glycoside synthesis are summarized in Table 4.8.

4.3.3.4.2 Activation via Glycosyl Halides

Several methods of glycosylations with hemiacetals **137** as starting materials proceed via glycosyl halide intermediates **184** (Scheme 4.35).

Koto and coworkers developed different dehydrative glycosylation procedures based on this principle. The promoter system $MsOH-CoBr_2$ [391] with the additives Bu_4NBr or Et_4NClO_4 provided 1,2-*cis*-glycosides through the intermediacy of glycosyl bromides [392]. Similarly, the ternary system $TMSBr-CoBr_2-Bu_4NBr$ affected the same kind of transformations [393–395]. Perlin and coworkers [396,397] activated hemiacetals with triflic anhydride and *s*-collidine in the presence of Bu_4NBr to convert the glycosyl triflate intermediate into glycosyl bromide for use in halide ion-catalyzed glycosylations. The use of carbon tetrabromide with PPh_3 has recently been

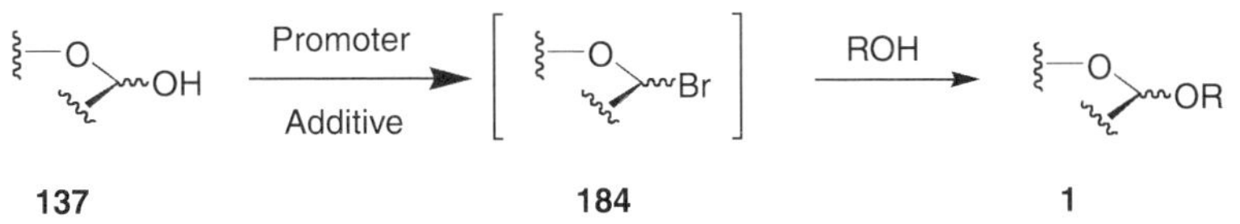

Promoter / Additive → [184] → ROH → 1
137 184 1

SCHEME 4.35 Activation of hemiacetals via glycosyl halides.

reported as a convenient way of activating hemiacetals through glycosyl bromides in halide ion-catalyzed reactions [88].

4.3.3.4.3 Activation via Glycosyl Sulfonates

An alternative to the above methods is the activation of hemiacetals via glycosyl sulfonates **185** (Scheme 4.36).

The activation of hemiacetals to glycosyl halides via glycosyl sulfonates was mentioned in the preceding section [396,397]. Activation with triflic anhydride without halide ions also affects glycoside formation. Although the Tf_2O-*s*-collidine system without halide ions was found to be unsuitable for glycoside synthesis [396,397], Tf_2O without any added base promotes these reactions. In the absence of added nucleophiles, nonreducing disaccharides were formed in high yields [398,399], whereas crosscoupled products were obtained in the presence of hydroxylic components [400,401]. Methanesulfonic anhydride also promotes glycosylations [397].

Koto et al. [402–407] introduced the *p*-nitrobenzenesulfonyl chloride AgOTf, Et_3N promoter system with the use of the additive $AcNMe_2$ for the synthesis of *cis*-glycosides.

In a different approach, phase-transfer catalyzed glycosylation of simple alcohols using tetra-*O*-benzyl-D-glucopyranose and *p*-toluenesulfonyl chloride in the presence of $BnEt_3NCl$ has been also reported [408].

4.3.3.4.4 Activation via Oxophosphonium Intermediates

Glycosylations proceeding through oxophosphonium intermediates **186** (Scheme 4.37) frequently rely on the Mitsunobu reaction or some modified version thereof.

As the Mitsunobu reaction requires weakly acidic nucleophiles, and since phenols have the necessary acidity, aryl glycosides [409–411] have been prepared this way. However, alcohols are

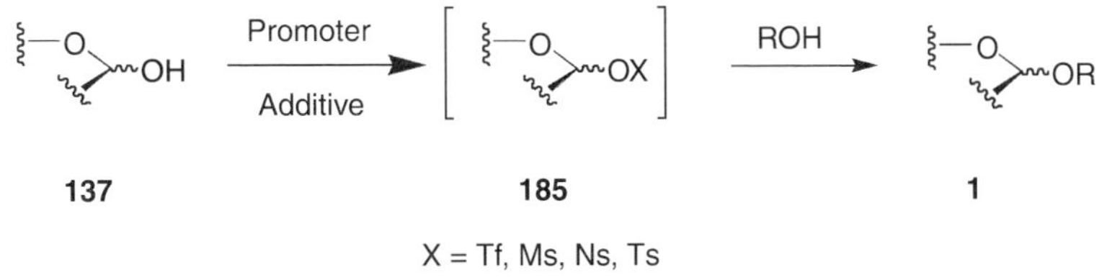

Promoter / Additive → [185] → ROH → 1
137 185 1
X = Tf, Ms, Ns, Ts

SCHEME 4.36 Activation of hemiacetals via glycosyl sulfonates.

Promoter / Additive → [186, $OPR'_3^{\oplus}$] → ROH → 1
137 186 1

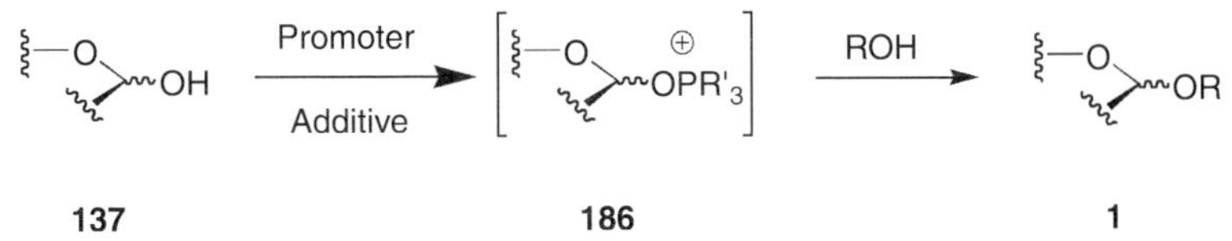

SCHEME 4.37 Activation of hemiacetals via oxophosphonium intermediates.

not acidic enough to be useful acceptors. This problem was circumvented with the addition of mercuric halides to the Mitsunobu reaction mixture [412]. *N*-Hydroxyphthalimide is also a good nucleophilic partner in this reaction, and was used in the preparation of some simple hydroxylamines [413], as well as in the synthesis of complex molecules such as calicheamycin γ_{1a} [414].

Mukaiyama reported an alternative method for the generation of oxophosphonium type intermediates using the reaction of tributylphosphine oxide and triflic anhydride, and applied this reaction to the synthesis of ribofuranosides [415].

4.3.3.4.5 Activation via Oxosulfonium Intermediates

Gin and coworkers [416] reported activation of hemiacetals via oxosulfonium salts **187** (Scheme 4.38).

Reaction of the hemiacetal with the salt generated from treating diphenyl sulfoxide with triflic anhydride provides the oxosulfonium ion intermediate **187**, which, in turn, reacts with nucleophiles providing glycosides [416,417]. This glycosylation method is quite effective and provides a variety of glycosides in good yields using a slight excess of acceptors. It is noteworthy that the reaction could be used with monosaccharide acceptors possessing both a free secondary and a free anomeric hydroxyl group **189** to give coupling on the secondary hydroxyls **190**, thereby providing a way for iterative dehydrative glycosylations (Scheme 4.39) [418].

In addition to sulfoxides, sulfides activated by triflic anhydride also provide oxosulfonium ions. In these reactions, the triflic anhydride serves as the oxidant. Thus, hemiacetals were activated by dimethyl sulfide and triflic anhydride providing glycosides in good yields [419].

4.3.3.4.6 Activation with Lewis Acids

Catalysis with protic acids has traditionally been used in Fischer glycosylations. As with the recent variations of the above procedure, Lewis acid catalysis is also valuable for dehydrative glycosylations with protected derivatives (Scheme 4.40). Trimethylsilyl triflate has been shown to promote reactions of glucuronic acid hemiacetals [420], and pyridine and TMSOTf was used for the synthesis of *cis*-glucosides [421]. Pyridinium *p*-toluenesulfonate (PPTS) was used for the synthesis of glycosides of the biologically important Kdo [422].

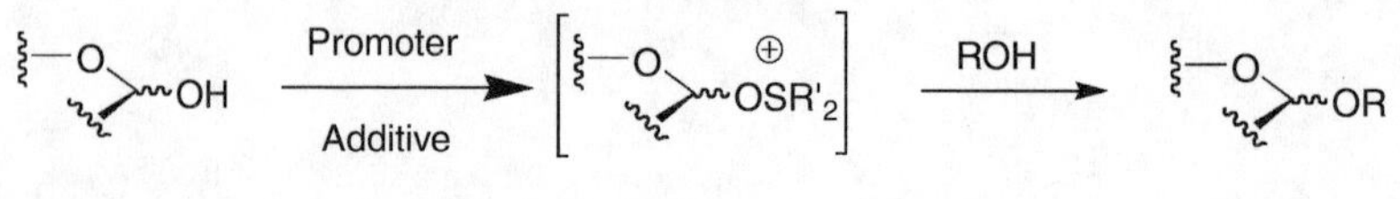

SCHEME 4.38 Activation of hemiacetals via oxosulphonium intermediates.

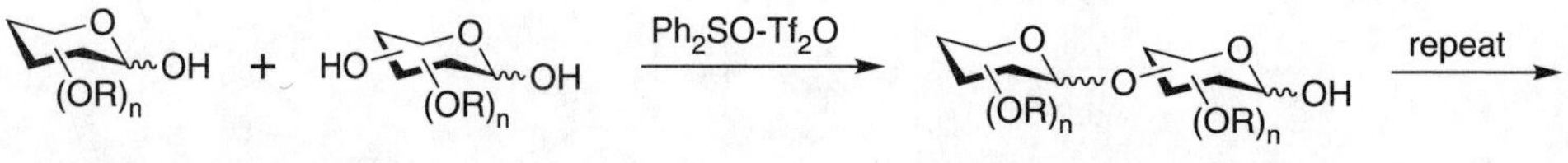

SCHEME 4.39 Iterative dehydrative glycosylation.

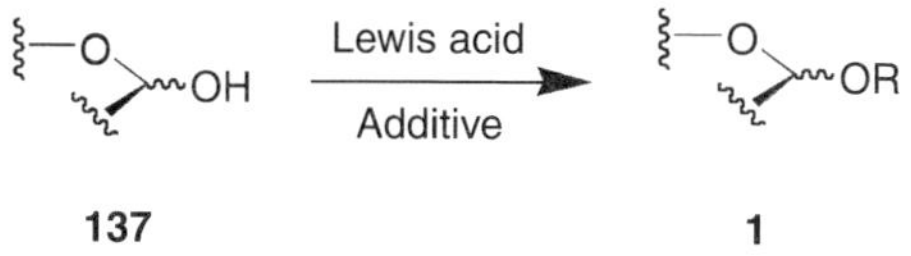

SCHEME 4.40 Lewis acid-promoted glycosylation of hemiacetals.

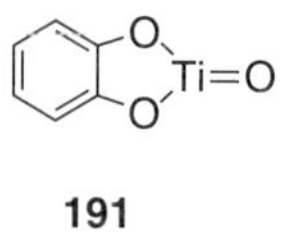

FIGURE 4.6 Oxotitanium reagent.

Mukaiyama proposed an oxotitanium reagent **191** (Figure 4.6) in the presence of triflic anhydride [423] and diphenyltin sulfide with triflic anhydride [424] for the synthesis of β-ribofuranosides. The addition of $LiClO_4$ to the latter promoter system resulted in the formation of α-ribofuranosides [424]. Mukaiyama and coworkers [344] also reported the use of diphenyltin sulfide with silver salts such as $AgClO_4$, or Lawesson's reagent **150** in combination with silver salts.

Inanaga and coworkers [425] introduced the combination of $Yb(OTf)_3$ and methoxyacetic acid. Silylating agents in combination with Lewis acids have also been proposed. Thus, $Zn(OTf)_2$ and TMSCl were introduced by Susaki [426]. Mukaiyama [427] used TMS_2O with a variety of Lewis acids including $Sn(OTf)_2$. Moreover, $Sn(OTf)_2$ was also used in combination with TMSCl [428]. Another metal triflate, $Cu(OTf)_2$, has been shown to be effective without silylating agents [70]. Trityl salts, such as $TrClO_4$ and $TrB(C_6F_5)_4$, were introduced by Mukaiyama [429,430] for the synthesis of ribofuranosides. The stereoselectivity of the reaction could be influenced by the addition of various lithium salts where $LiNTf_2$ was found to be an effective additive for the synthesis of α-ribofuranosides [430].

A variety of Lewis acids, including TMSOTf, $BF_3{\cdot}Et_2O$, $SnCl_4$, and $TiCl_4$, were found to be applicable to the glycosylation of the reactive ketose derivative **192** (Scheme 4.41) [431]. In these glycosylations, proceeding through the tertiary oxocarbenium ion **193**, only the *cis*-glycoside **194** was formed.

4.3.3.4.7 Other Methods of Activation

Hemiacetals can be activated for glycosylations by reaction with carbodiimides via glycosyl isourea intermediates and different aryl glycosides were prepared this way [432].

In another approach for the activation of hemiacetals, a phosphine/Cu(II) complex [433] and a rhodium catalyst have been proposed recently [434]. A detailed discussion on the use of hemiacetals in glycosylation reactions can be found in a recent review [435].

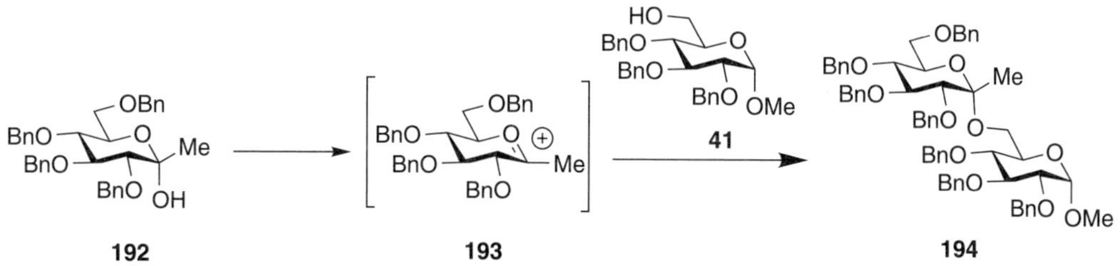

SCHEME 4.41 Stereoselective glycosylation of a ketose hemiacetal.

4.3.3.5 1,2-Anhydro Derivatives

Compounds possessing a reactive oxirane ring at the anomeric center, for example, 1,2-anhydro sugars such as Brigl's anhydride **195** (Scheme 4.42), have been known for a long time. Opening of the epoxide by alcohols or phenols could be performed by heating, and the reaction results in the formation of glycosides. This chemistry was used, among others, in the first synthesis of sucrose **88** by Lemieux [436,437].

Acid or Lewis-acid catalyzed self-condensation reactions of 1,2-anhydro sugars were studied for the synthesis of polysaccharides [438]. Polysaccharides of relatively low molecular weights (< 10,000) were obtained, and the anomeric configurations could be varied depending on the promoter and temperature [439].

1,2-Anhydro sugars recently came into the limelight because of work by Danishefsky and coworkers, who developed a highly efficient procedure for their preparation via oxidation of glycals **198** with dimethyldioxirane (DMDO) (Scheme 4.43) [440]. Reactions of the epoxides, promoted by $ZnCl_2$ in tetrahydrofuran or dichloromethane, produced 1,2-*trans*-glycosides [440]. Glycosylation of glycals by this method gives a reiterative glycosylation strategy for the synthesis of oligosaccharides (Scheme 4.43). As shown in Scheme 4.43, glycosylation of the glycal **200** with the epoxide **199** creates the disaccharide glycal **201**. Subsequent conversion into the epoxide **203** was followed by further glycosylation to afford **204**. This technology was also applied

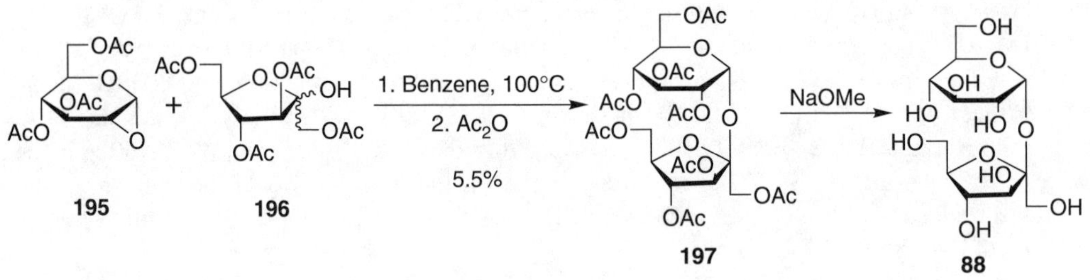

SCHEME 4.42 Lemieux's sucrose synthesis.

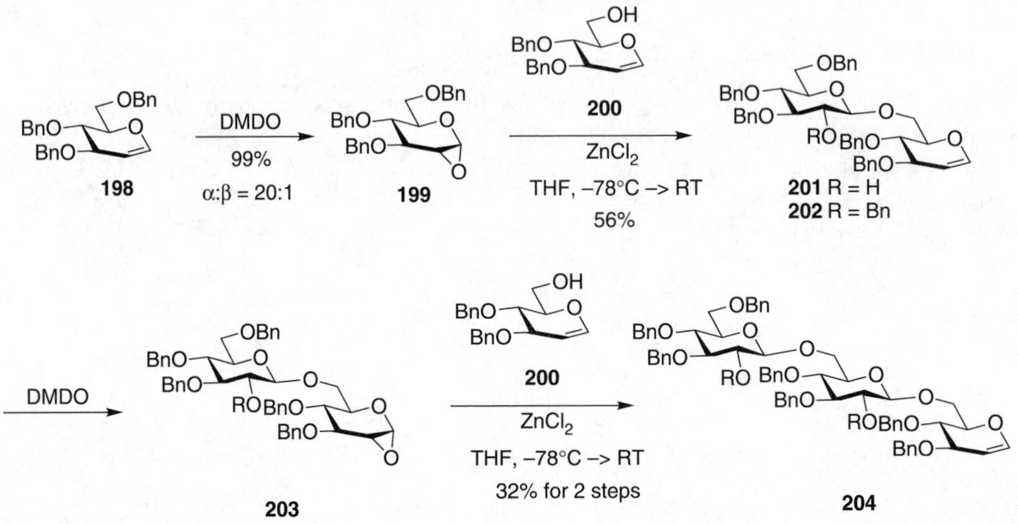

SCHEME 4.43 Iterative glycosylation with glycals.

to solid-phase synthesis [441]. Unfortunately, through experience, it was established that the glycosylation reaction is not always stereospecific [21,442].

Glycosylation reactions with 1,2-anhydro sugars are most frequently promoted by $ZnCl_2$. Additionally, $Zn(OTf)_2$ has been used as a promoter [443–445], whereas $AgBF_4$-promoted reactions with tributylstannylated alcohols produced *cis*-glycosides [446]. Glycosylations of phenols could be performed under alkaline conditions using K_2CO_3 and 18-crown-6 in acetone [447].

A potentially useful feature of the glycosides formed in glycosylations with 1,2-anhydro sugars is that they possess a free 2-OH ready for further transformations, for example, into 2-deoxy glycosides [448], or to configurational inversion as in the synthesis of β-D-mannosides from β-D-glucosides [449]. 1,2-Anhydro sugars are very versatile because they can also be readily converted into other types of glycosyl donors, such as thioglycosides, glycosyl fluorides and NPGs [450].

4.3.3.6 Glycosyl Phosphites, Phosphates and Other Phosphorus Compounds

A variety of glycosyl derivatives with both trivalent and pentavalent phosphorus atoms in the leaving group is known and was investigated as glycosyl donors.

4.3.3.6.1 Glycosyl Phosphites

Glycosyl phosphites were introduced by Schmidt [451–453], Wong [454–457] and Watanabe [458,459]. A variety of dialkyl and cyclic phosphites has proved suitable for the reaction [459,460]. In fact, diethyl and dibenzyl phosphites are the most frequently used. Glycosyl phosphites react with alcohols at low temperatures in the presence of catalytic amounts of Lewis acid promoters such as TMSOTf, $BF_3{\cdot}Et_2O$ and $Sn(OTf)_2$, thus creating glycosides. Phosphites are also activated by other promoters, which are summarized in Table 4.9.

TABLE 4.9
Activations of Glycosyl Phosphites

OP(OR')$_2$ —(ROX, Promoter)→ OR

Promoter	Ref.
TMSOTf	[451,452,454–456]
$Sn(OTf)_2$	[452]
$BF_3{\cdot}Et_2O$	[453]
$ZnCl_2$	[459]
$ZnCl_2$–$AgClO_4$	[458,459,461]
$BiCl_3$	[458,459]
NIS–TfOH	[458,459]
MeOTf	[458,459]
TfOH	[457,462]
$HNTf_2$	[463]
$Ba(ClO_4)_2$	[464]
$LiClO_4$	[464]
DTBPI–Bu_4NI	[159]
I_2	[254]
Montmorillonite K-10	[466–468]

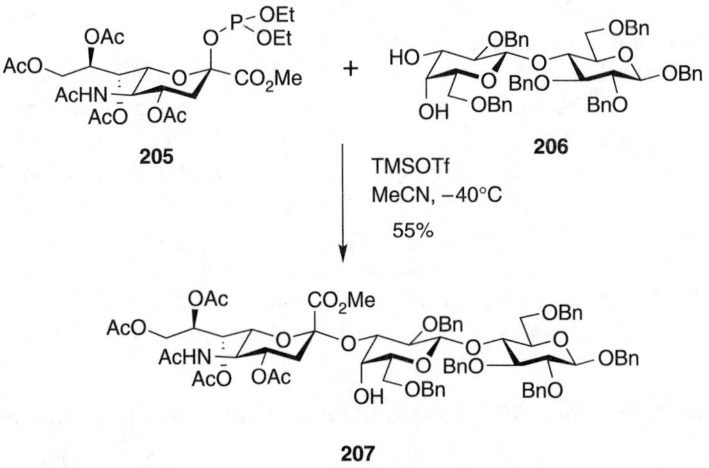

SCHEME 4.44 Synthesis of 3-*O*-neuraminyllactose using glycosyl phosphite.

The typical activators are Lewis acids such as $ZnCl_2$ [459], $ZnCl_2$–$AgClO_4$ [458,459,461], and $BiCl_3$ [458,459]. Other promoters include the NIS–TfOH system [458,459], MeOTf [458,459], as well as strong protic acids such as TfOH [457,462] and $HNTf_2$ [463]. The perchlorates $Ba(ClO_4)_2$ and $LiClO_4$ [464] and the binary promoter system of 2,6-di-*tert*-butylpyridinium iodide (DTBPI) with Bu_4NI [159] are very mild promoters for glycosylations with glycosyl phosphites. Reactions with the latter promoter furnished *cis*-glycosides, whereas conditions for highly selective synthesis of *trans*-glycosides were also developed [465]. Additional methods of activation include the use of iodine [254] and activation by heterogeneous catalysis using montmorillonite K-10 [466–468].

Besides the common organic solvents used for glycosylations, syntheses of glycosides from glycosyl phosphites in ionic liquids were recently reported [463]. Glycosyl phosphites are frequently used for sialylations [451,454] and a typical example is shown in Scheme 4.44. As illustrated, reaction of the β-neuraminyl phosphite **205** with the lactose diol **206** afforded the 3-*O*-neuraminyllactose derivative **207** in a 55% yield [460].

4.3.3.6.2 Glycosyl Phosphates

Glycosyl phosphates as glycosyl donors were introduced by Hashimoto [469]. Their glycosylation reaction is promoted most frequently by TMSOTf [469]. Less common is the use of TDSOTf [277], $BF_3{\cdot}Et_2O$ [277], ZnI_2 [277], TfOH [277] or $LiClO_4$ [470] (Table 4.10).

Most frequently, diphenyl and di-*n*-butyl phosphates are used. As such, Singh and coworkers [471–474] introduced the cyclic propane-1,3-diyl phosphates. Additionally, Seeberger and coworkers [277,475–482] have recently applied glycosyl phosphates for the synthesis of various oligosaccharides both in solution and in solid-phase syntheses.

Glycosyl phosphates were applied in the total syntheses of natural products such as bleomycin A_2 [483] (Scheme 4.45). As illustrated, the reaction of the D-mannosyl phosphate donor **208** with the L-gulose acceptor **209** gave the disaccharide **210** in a 93% yield. After debenzylation and reacetylation, the disaccharide was activated as the phosphate **211**. Coupling of **211** with the L-histidine derivative **212** afforded the glycoside **213** with good stereoselectivity despite the presence of the nonparticipating glycosyl group at O-2. After peptide coupling and deprotection steps, bleomycin A_2 **214** was obtained. Additional applications of glycosyl phosphates have been covered in recent review articles [35,36].

TABLE 4.10
Activations of Glycosyl Phosphates

Promoter	Ref.
TMSOTf	[469]
TDSOTf	[277]
$BF_3{\cdot}Et_2O$	[277]
ZnI_2	[277]
TfOH	[277]
$LiClO_4$	[470]

4.3.3.6.3 Other Phosphorus Compounds

Besides glycosyl phosphites and phosphates, a variety of other phosphorus containing leaving groups have been designed and proposed. These are summarized in Table 4.11.

Among the compounds possessing trivalent phosphorus, *N*,*N*-diisopropyl phosphoramidites **215** of 2-deoxy sugars were reported to give mainly α-glycosides in TMSOTf-promoted reactions [484]. A different type of compound, a perbenzylated glycosyl diphenylphosphinite **216**, was recently reported to give *cis*-glycosides in high stereoselectivity in MeI-promoted reactions [163].

Among glycosyl donors possessing pentavalent phosphorus, glycosyl dimethylphosphinothioates **217** were introduced by Inazu and coworkers, and were used with $AgClO_4$ [485–487], I_2–$TrClO_4$ [486–489] or $TrClO_4$ [490] activators.

P,*P*-Diphenyl-*N*-tosyl-phosphinimidates **218** were used with $BF_3{\cdot}Et_2O$ and TMSOTf promoters [491–493], and *N*,*N*,*N'*,*N'*-tetramethylphosphorodiamidates **219** were also activated by the same catalysts [279,493,494]. These latter compounds are more stable than the previous types, and can also serve as glycosyl acceptors under certain conditions. For example, they could be glycosylated with glycosyl phosphites, diphenylphosphinimidates and phosphorodiamidimidothioates [279]. Moreover, they could also be used in armed–disarmed type glycosylations [279,495].

Glycosyl *N*-phenyl diethyl phosphorimidates **220** were reported to afford *trans*-glycosides with TMSOTf in propionitrile at low temperature, whereas their reaction with LPTS and Bu_4NI produced *cis*-glycosides [496].

Additional types of phosphorus donors include the glycosyl diphenylphosphinates **221**, which were used in TMSOTf-promoted reactions [471,474], whereas reaction of difluorophosphates **222** was assisted by $AgClO_4$ [497].

Phosphorus-containing leaving groups with glycosyl–sulfur linkages such as dithiophosphates and phosphorodiamidimidothioates were discussed in Section 4.3.2.3.

4.3.3.7 Orthoesters and Related Derivatives

Glycosyl orthoesters of the type **223** (Table 4.12) have been introduced for the syntheses of 1,2-*trans*-glycosides [77,78]. Orthoesters can be converted into the corresponding glycosides by

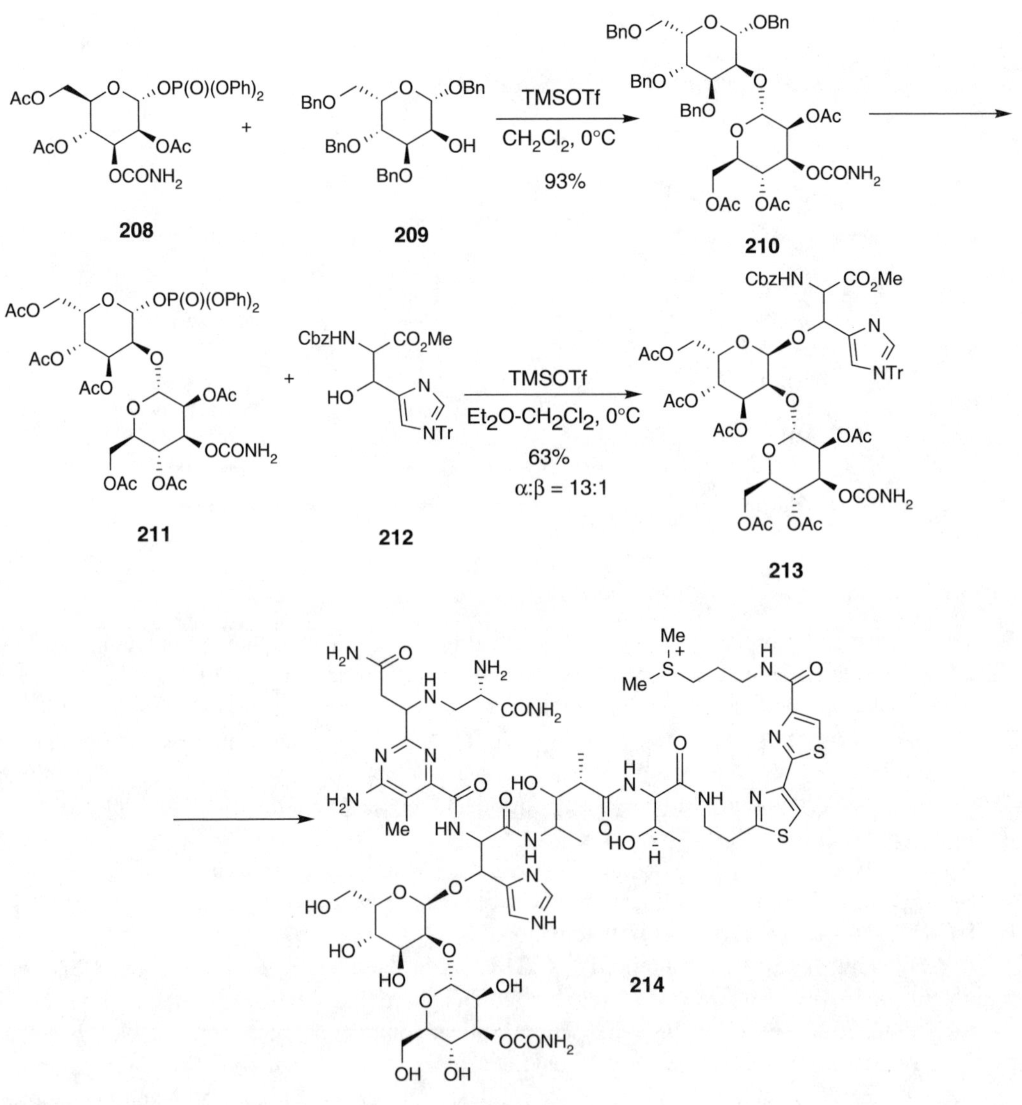

SCHEME 4.45 Synthesis of bleomycin A_2 using glycosyl phosphates.

Lewis acids such as $HgBr_2$ [77] and 2,6-dimethylpyridinium perchlorate [78]. These promoters were later largely replaced by the highly effective TMSOTf [79].

Modifications of the original orthoester glycosylation methods were developed to eliminate their disadvantages. Among them, the use of 1,2-cyanoethylidene derivatives **224** is particularly noteworthy. Cyanoethylidene derivatives react with tritylated alcohols in the presence of catalytic amounts of $TrBF_4$, $TrClO_4$ [498,499], TrOTf [500], or AgOTf [501], to give 1,2-*trans* glycosides. To improve the stereoselectivity in cases when it was unsatisfactory, the effect of substituents in the acyl group and the use of TrOTf [500], and high-pressure conditions [502] were found to be advantageous. An important application of this glycosylation procedure is the synthesis of regular polysaccharides with oligosaccharide repeating units. Thus, a series of bacterial antigenic polysaccharides has been synthesized this way [503,504].

The synthesis of the *Streptococcus pneumoniae* type 14 polysaccharide is shown in Scheme 4.46 [505]. As illustrated, condensation of the lactosaminyl bromide **225** with the lactose

TABLE 4.11
Activations of Phosphorus Compounds

Glycosyl–X + ROH → (Promoter) → Glycosyl–OR

X	Promoter	Ref.
$-O-P(OEt)-NiPr_2$ **215**	TMSOTf	[484]
$-O-P-Ph_2$ **216**	MeI	[163]
$-O-P(=S)-Me_2$ **217**	$AgClO_4$	[485–487]
	$I_2-TrClO_4$	[486–489]
	$TrClO_4$	[490]
$-O-P(=NTs)-Ph_2$ **218**	$BF_3 \cdot Et_2O$	[491–493]
	TMSOTf	[492]
$-O-P(=O)-(NMe_2)_2$ **219**	$BF_3 \cdot Et_2O$	[493,494]
	TMSOTf	[279,494,495]
$-O-P(=NPh)-(OEt)_2$ **220**	TMSOTf	[496]
	$LPTS-Bu_4NI$	[496]
$-O-P(=O)-Ph_2$ **221**	TMSOTf	[471,474]
$-O-P(=O)-F_2$ **222**	$AgClO_4$	[497]

TABLE 4.12
Activations of Orthoesters

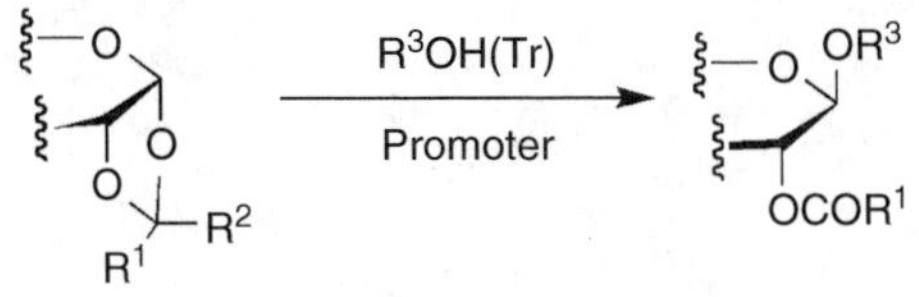

$R^3OH(Tr)$ / Promoter

R^1		R^2	Promoter	Ref.
Me,Ph		*O*-alkyl, *O*-glycosyl	$HgBr_2$	[77]
	223		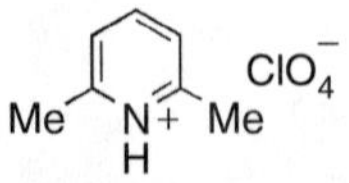 2,6-dimethylpyridinium perchlorate (ClO_4^-)	[78]
			TMSOTf	[79]
Me		CN	$TrBF_4$	[498,499]
	224		$TrClO_4$	[498,499]
			TrOTf	[500]
			AgOTf	[501]
Me		SEt, SC_6H_4Me	$TrClO_4$	[506,507]
	231		NIS–TfOH	[508]
			I_2	[254]
Me,Ph		$O(CH_2)_3CH{=}CH_2$	NIS–TfOH	[509]
	232		NIS–TBSOTf	[510]
			NIS–$BF_3{\cdot}Et_2O$	[511]
			NIS–$Yb(OTf)_3$	[511]
			NBS–TESOTf	[512]
			IDCT	[512]
			NIS-microwave	[356]
t-Bu		ON=C(Me)Ph	$BF_3{\cdot}Et_2O$	[513]
	233			

cyanoethylidene derivative **226** under Helferich conditions gives the tetrasaccharide **227**. The acetyl group was selectively removed and was replaced with a trityl group. Subsequent polycondensation of the resulting **228** gave **229**, which was deprotected to obtain the regular heteropolysaccharide **230** with a tetrasaccharide repeating unit.

Related to the above cyanoethylidene derivatives, thioorthoesters **231** also behave similarly, and can be activated with catalytic amounts of $TrClO_4$ to glycosylate trityl ethers [506,507]. By using the thioglycoside promoters NIS–TfOH [508] or I_2 [254], glycosyl acceptors with free hydroxyls were glycosylated.

n-Pentenyl orthoesters [509] of the type **232** have the additional feature that they can be activated by the same type of promoters used for *n*-pentenyl glycosides. Namely, *N*-iodosuccinimide (NIS) with different Lewis acids, such as TBSOTf [510], $BF_3{\cdot}Et_2O$ or $Yb(OTf)_3$ [511]. The use of NBS–TESOTf, IDCT [512] or NIS under microwave irradiation [356] has been reported also.

An additional type of orthoester was reported by Kunz [513]. Specifically, the oximate orthoester **233** reacted in $BF_3{\cdot}Et_2O$-promoted reactions to give *trans*-glycosides.

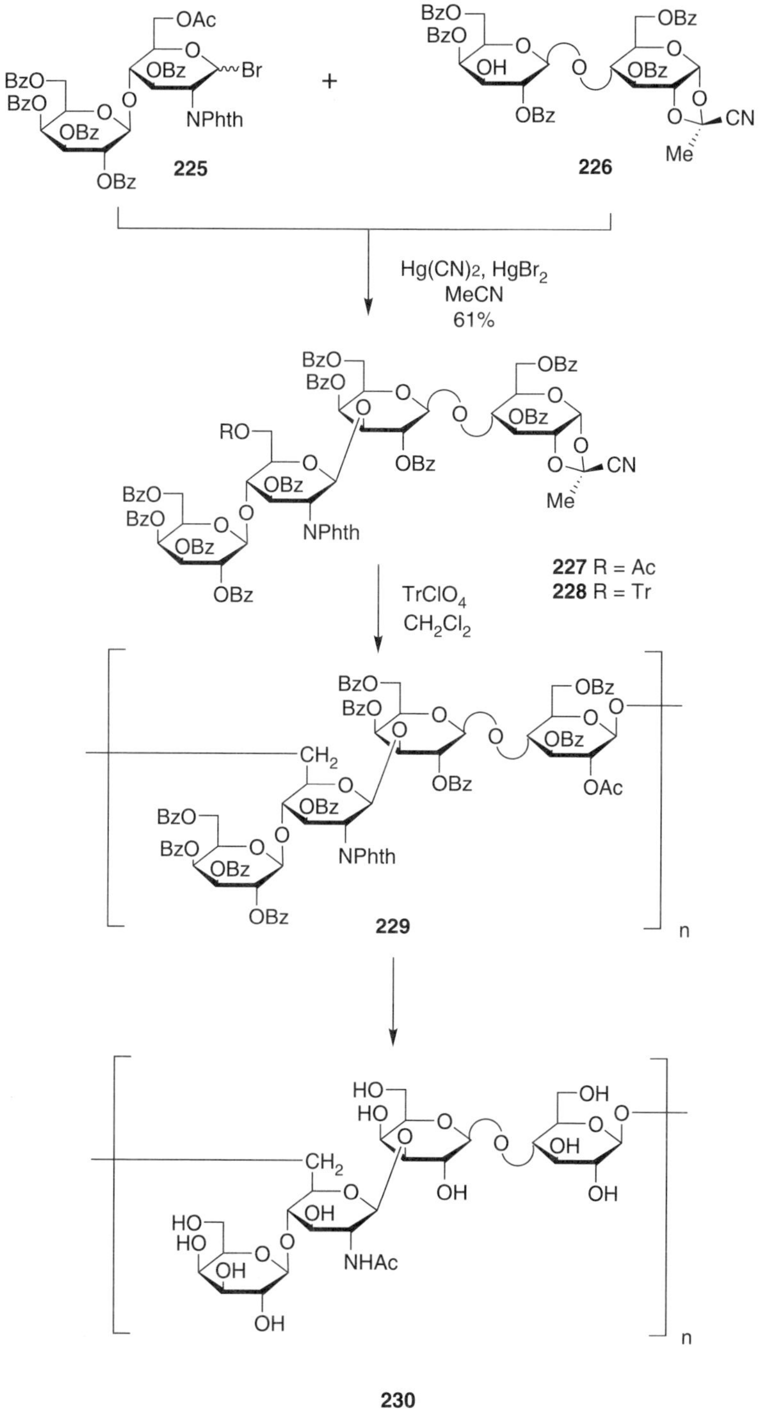

SCHEME 4.46 Synthesis of *Streptococcus pneumoniae* type 14 polysaccharide using cyanoethylidene derivative.

Once a type of prolifically used glycosyl donors, orthoesters are now utilized less frequently. Their formation, however, is occasionally encountered in different glycosylation reactions, raising the possibility that they are intermediates in glycoside syntheses with neighboring group participation. In cases when unwanted orthoesters are isolated from glycosylation reactions instead of the desired glycosides, the orthoesters can be further transformed into the target compounds by the methods mentioned in this section.

4.3.3.8 Carbonates and Related Derivatives

Among the different glycosyl carbonates and related derivatives summarized in Table 4.13, anomeric aryloxycarbonyl derivatives **234** were used in thermal glycosylation reactions to give

TABLE 4.13
Activations of Glycosyl Carbonates

#	X	Y	R′	Promoter	Ref.
234	O	O	Ph	Δ	[514,515]
				$TrB(C_6F_5)_4$	[516–519]
				TfOH	[45,73,134]
235	O	O	Glycosyl	TMSOTf	[520]
				$SnCl_4$–$AgClO_4$	[521]
				Cp_2HfCl_2–$AgClO_4$	[521]
236	O	O	Isopropenyl	TMSOTf	[363]
237	O	imidazol-1-yl		$ZnBr_2$	[522]
238	S	imidazol-1-yl		$AgClO_4$	[523,524]
246	O	S	2-pyridyl	AgOTf	[372,525–527]
247	S	S	Me	$BF_3{\cdot}Et_2O$	[528]
248	O	NH	Ph, alkyl	$BF_3{\cdot}Et_2O$	[529]
249	O	NH	$-CH_2CH{=}CH_2$	DMTST	[530]
				TMTSB	[530,531]
				IDCP	[530,531]
250	O	*N*-alkyl	Ts	TMSOTf	[532]
251	$NPhCF_3(p)$	S	$BnCF_3(p)$	TfOH	[533,534]
				$HB(C_6F_5)_4$	[533,534]
				TMSOTf	[534]
				$TrB(C_6F_5)_4$	[534]

glycosides in moderate yields [514]. In the absence of added nucleophiles, the phenyl glycoside was obtained from the 1-*O*-phenoxycarbonyl derivative, whereas crosscoupling products were isolated in the presence of different phenols both from the phenoxycarbonyl and the methoxycarbonyl derivatives [515].

It was later shown that the glycosylation with phenyl carbonates are efficiently promoted by $TrB(C_6F_5)_4$ [516–519] or by TfOH [45,73,134]. Mixed carbonates **235** in which the carbonate group bridges monosaccharide units were used with trialkylsilyl triflates [520] and $SnCl_4$–$AgClO_4$ or Cp_2HfCl_2–$AgClO_4$ [521] promoters.

Isopropenyl carbonates **236** were introduced by Sinaÿ [363] and were used with TMSOTf as a promoter. Reactions of imidazole carbonate derivatives **237** were assisted by $ZnBr_2$ [522], whereas for imidazole thiocarbonate derivatives **238**, $AgClO_4$ was used as the promoter [523, 524]. These methods were used in the total synthesis of avermectin B1a **245** by Ley [523,524] (Scheme 4.47).

Hanessian introduced 2-pyridylthiocarbonates **246**, the so-called TOPCAT donors, and used them with AgOTf as a promoter [372,525–527].

Glycosyl *O*-xanthates **247**, activated with $BF_3{\cdot}Et_2O$, have been reported by Pougny [528]. Similarly, glycosyl carbamates **248** also provided glycosides in $BF_3{\cdot}Et_2O$-promoted reactions [529].

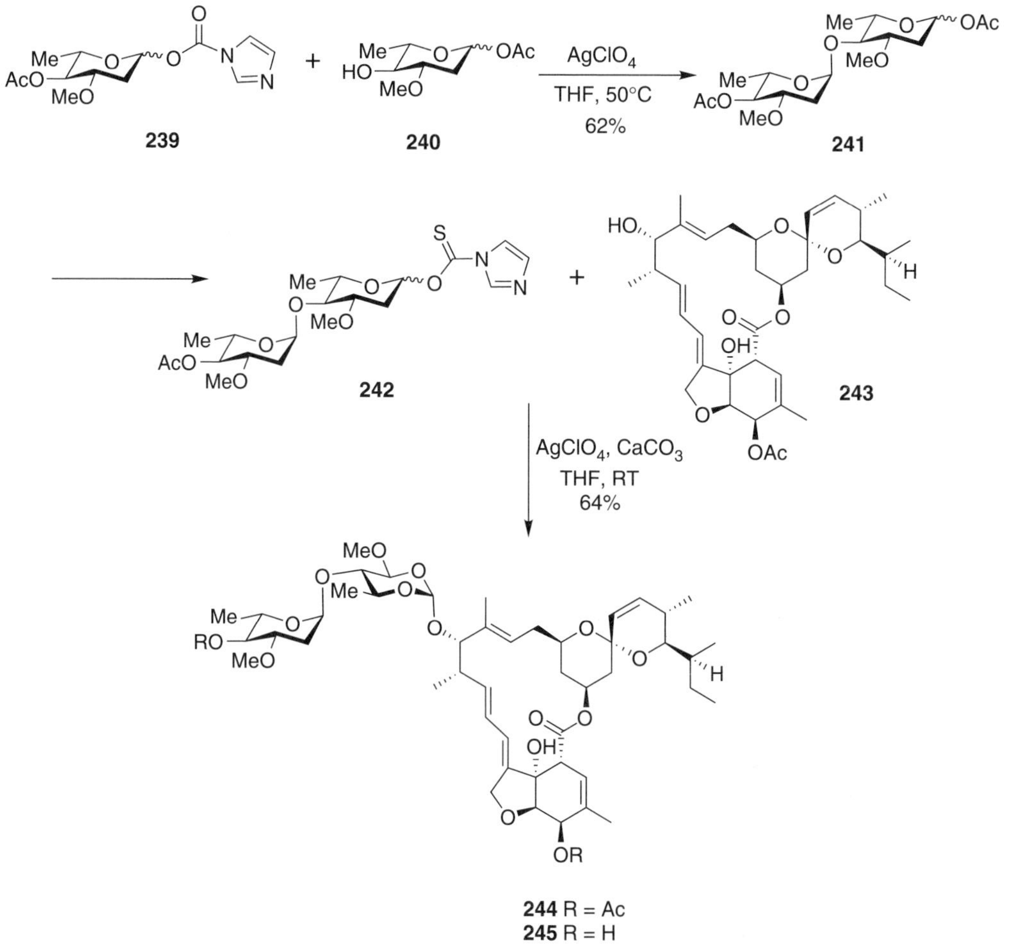

SCHEME 4.47 Synthesis of avermectin B1a using glycosyl carbonate donors.

N-Allylcarbamates **249** could be efficiently activated with different electrophiles, including sulfenylating agents as DMTST, methyl-bis(methylthio)sulfonium hexachloroantimonate (TMTSB) or the iodonium ion source IDCP to furnish glycosides [530,531]. The activation in these reactions is based on the addition of the electrophile to the double bond, followed by a cyclization analogous to that in the activation of *n*-pentenyl glycosides. Glycosyl sulfonylcarbamates **250** have been recently introduced [532], and were used with TMSOTf in glycosylation reactions. Interestingly, the reactivity of this type of glycosyl donor can be tuned with the variation of the alkyl substituent on the nitrogen.

A new class of glycosyl donors, thioformimidates **251**, was introduced by Mukaiyama [533]. These derivatives could be used for glycosylations with catalytic amounts of strong protic acids such as TfOH and $HB(C_6F_5)_4$, and with a variety of Lewis acids including TMSOTf and $TrB(C_6F_5)_4$ [534]. Mostly TfOH was used in later works. In contrast to trichloroacetimidates, disarmed thioformimidates could also serve as glycosyl acceptors in glycosylations with armed thioformimidates [534–536], and conditions were developed for the stereoselective synthesis of both *trans*- and *cis*-glycosides from the same glycosyl donor [535,536].

4.3.3.9 Silyl Ethers

Analogous to the hemiacetals, anomeric silyl ethers can also serve as glycosyl donors. Glycosyl silyl ethers were first used as glycosyl donors to glycosylate acetals of carbonyl compounds [537–540]. Thus, the iridoid glycoside **255** was obtained by glycosylation of the acetyl derivative **253** with the 1-*O*-trimethylsilyl-β-D-glucopyranose derivative **252**, followed by deacetylation of **254** (Scheme 4.48) [539].

For the glycosylation of hydroxyl compounds, the glycosyl acceptor is also used in a silylated form. Thus, aryl glycosides were obtained by reacting 1-*O*-trimethylsilylglucose derivatives with trimethylsilylated phenols using TMSOTf as a promoter [541]. Reaction of glycosyl trimethylsilyl ethers with *tert*-butyldimethylsilyl ethers of carbohydrates afforded oligosaccharides [542]. However, in the absence of a glycosyl acceptor, self-condensation of the glycosyl silyl ether leads to nonreducing trehalose type oligosaccharides [543].

Of the possible silyl ethers, trimethylsilyl and *tert*-butyldimethylsilyl ethers are used most frequently. The use of the more stable *tert*-butyldimethylsilyl ethers is preferred in several cases [544–547]. The various methods for the activation of glycosyl silyl ethers in glycosylation reactions are summarized in Table 4.14.

The most commonly used promoter is TMSOTf. Other activators include $BF_3{\cdot}Et_2O$ [548] or TMSOTf in combination with Ph_2SnS alone [549] or with Ph_2SnS and $LiClO_4$ [549]. The ternary system $SiCl_4$ and $AgClO_4$ in combination with Ph_2SnS [549] was also described.

Glycosylations with silyl ethers of different 2-deoxy sugars proved to be a valuable procedure in the syntheses of antibiotics [544,545,547,550].

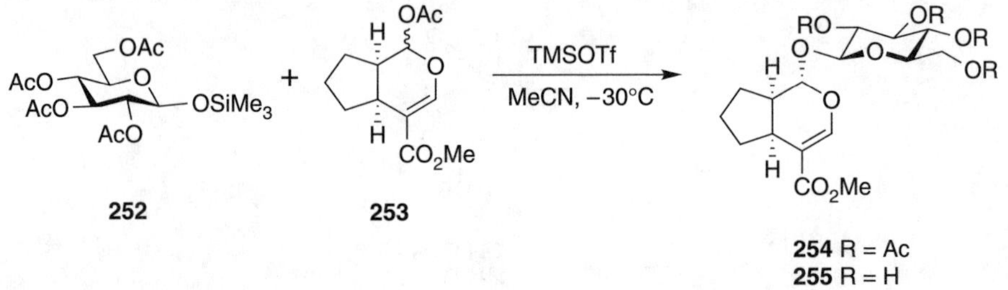

252 **253** **254** R = Ac **255** R = H

SCHEME 4.48 Synthesis of iridoid glycoside using silyl ether.

TABLE 4.14
Activations of Glycosyl Silyl Ethers

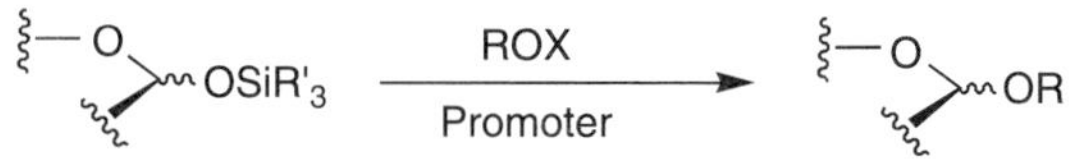

Promoter	X	Ref.
	TMS	[541]
TMSOTf	TBDPS	[542]
	H	[544–546]
$BF_3 \cdot Et_2O$	H	[548]
Ph_2SnS–TMSOTf	TMS	[549]
Ph_2SnS–TMSOTf–$LiClO_4$	TMS	[549]
Ph_2SnS–$SiCl_4$–$AgClO_4$	TMS	[549]

4.3.3.10 Oxazolines

Although most classes of glycosyl donors can be applied to different types of sugars, the specific structural features of a certain carbohydrate might affect the applicability of the different activation methods. Thus, carbohydrates with 2-acylamino-2-deoxy functions, which include such biologically significant monosaccharides as *N*-acetyl-D-glucosamine and *N*-acetyl-D-galactosamine, behave differently in glycosylation reactions to their parent compounds possessing hydroxyl groups in position 2.

Because the *N*-acyl group is a good participating group, departure of the anomeric leaving group from a glycosyl donor **256** of these sugars leads to the ion **257** through neighboring group participation (Scheme 4.49). However, this ion, which corresponds to the acyloxonium ion **16** (Scheme 4.2) of the corresponding hydroxy sugars can lose a proton and become stabilized as the oxazoline **258**. Oxazolines are stable compounds and do not react to form glycosides as readily as their oxygen analogs. Therefore, the use of those *N*-acyl groups (including the most commonly occurring *N*-acetyl) that readily form oxazolines is generally avoided and other *N*-protected derivatives, such as *N*-phthaloyl, *N*-trichloroethoxycarbonyl and *N*-allyloxycarbonyl are used.

On the other hand, oxazolines can be activated to form *trans*-glycosides **259** if required. Early methods used activation by *p*TsOH at elevated temperatures [551,552]. Other promoters were later

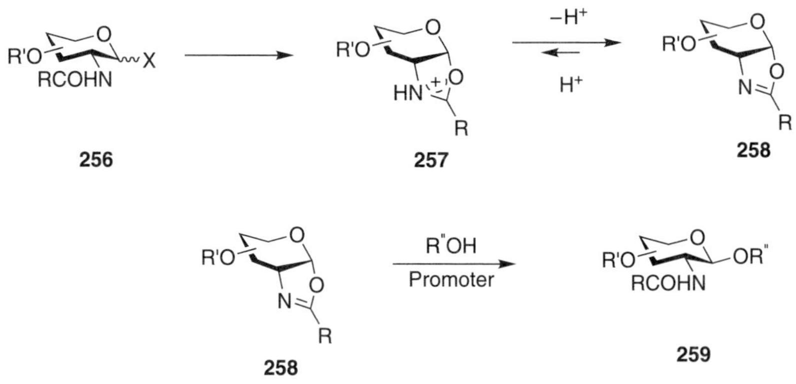

SCHEME 4.49 Formation of and glycosylations with oxazolines.

introduced and include $FeCl_3$ [331,332] and TMSOTf [79]. Milder activators such as pyridinium *p*-toluenesulfonate (PPTS) [553] — the related pyridinium triflate (with microwave irradiation) [554] — along with $CuCl_2$ and $CuBr_2$ [555] were recently described.

The main advantage of the use of oxazolines as donors is the direct access to the *trans*-glycosides of amino sugars in their natural *N*-acetyl form. This is undermined, however, by the harsh conditions required, and use of the method is limited to molecules possessing low acid sensitivity and high nucleophilicity.

4.3.3.11 Other Glycosyl Donors with Glycosyl–Oxygen Bonds

Rounding out the selection of glycosyl donors mentioned to this point, some additional examples deserve mention. Beginning with anhydrosugars, there are alternatives to the 1,2-anhydro analogs already discussed. These include 1,6-anhydro sugars, which can also serve as glycosyl donors. The ring-opening polymerizations of these compounds for the preparation of polysaccharides has been studied in detail [438,503].

Glycosyl sulfonates as glycosyl donors have been investigated by Schuerch [556,557]. These studies led to a synthesis of β-D-mannopyranosides [558–560]. Though glycosyl sulfonates are not prepared separately these days, several glycosylation methods have been reported to proceed through sulfonates as intermediates [34,192,193,248,377] (see also Section 4.3.3.4.3). The use of the related glycosyl sulfates as glycosyl donors has also been described [561].

4.3.4 Synthesis of Glycosides from Donors with Other Heteroatoms at the Anomeric Center

4.3.4.1 Selenoglycosides and Telluroglycosides

The successful use of thioglycosides in glycosylation reactions prompted the introduction of seleno- [562,563] and telluroglycosides [564]. Selenoglycosides were originally used for glycosylations with the AgOTf–K_2CO_3 [562,563] promoter system and their methods of activation are summarized in Table 4.15.

TABLE 4.15
Activations of Selenoglycosides

(Scheme: glycosyl–SePh + ROH —Promoter→ glycosyl–OR)

Promoter	Ref.
AgOTf–K_2CO_3	[562,563]
IDCP	[565]
IDCT	[566]
NIS	[566,567]
NIS–TfOH	[565]
MeOTf	[568]
$h\nu$	[569]
TBPA	[570]
I_2	[254,571]

Not unexpectedly, the promoters used for the activation of thioglycosides, including IDCP [565], IDCT [566], NIS [566,567], the NIS–TfOH system [565], and MeOTf [568], also promote the reactions of selenoglycosides. Activation of selenoglycosides by photo-induced electron transfer [569] and the single electron transfer reagent TBPA **99** [570] were also reported. Selenoglycosides can also be activated by I_2 [254,571].

Selenoglycosides can be selectively activated in the presence of thioglycosides. However, they can be glycosylated by glycosyl halides or trichloroacetimidates. These kinds of differences could be used in block syntheses of complex oligosaccharides [568,572].

Telluroglycosides are activated by the same type of promoters, including NBS, NIS and NIS–TMSOTf [564], and electrochemical activation has also been reported [573]. Telluroglycosides are even more reactive than selenoglycosides. In fact, they can be selectively activated in the presence of the latter [574].

4.3.4.2 Glycosyl Donors with Nitrogen at the Anomeric Center

Compared with other heteroatoms, only a few types of glycosyl donors have glycosyl–nitrogen linkages. Among these, glycosyl sulfonamides with a 2-iodo substituent (such as **260**) are prepared by the IDCP-promoted addition of benzenesulfonamide to glycals **198** (Scheme 4.50). Treatment of the sulfonamide with lithium tetramethylpiperidide (LTMP) results in the formation of the unstable aziridine **261**, which reacts with alcohols in the presence of AgOTf to form *trans*-glycosides of the 2-amino sugar **262**. This latter step is analogous to glycosylations using 1,2-anhydro sugars (Section 4.3.3.3.5).

In this sequence of reactions, sulfonamidoglycosylation of glycals was introduced by Danishefsky [575]. Analogous reactions using glycosyl azides have been previously reported [576].

N-Glycosyl-1,2,3-triazoles, obtainable from glycosyl azides, were used for glycoside synthesis in TMSOTf-promoted reactions [577]. Thus, the reaction of **263** with carminomycinone **264** produced the glycoside **265** in excellent yield [577] (Scheme 4.51).

The use of *N*-glycosyl amides as glycosyl donors has been recently reported [578]. These amides **266** are activated by the Appel reagent to form the intermediate imidoyl bromide **267**. The addition of alcohols and AgOTf to the mixture resulted in the formation of glycosides **268** (Scheme 4.52).

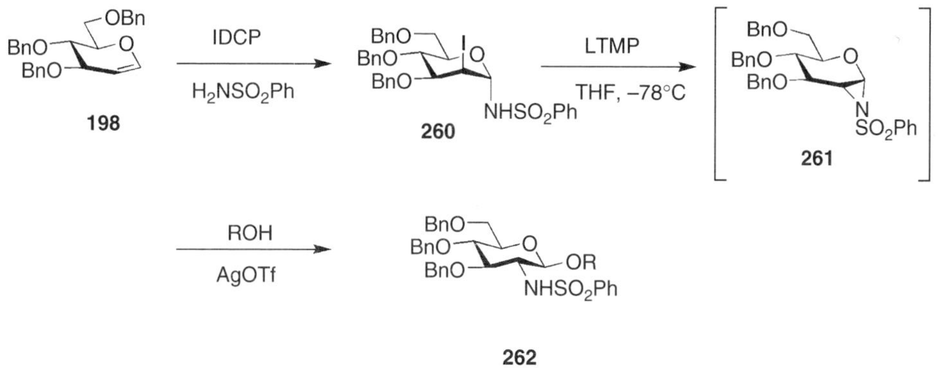

SCHEME 4.50 Sulfonamidoglycosylation of glycals.

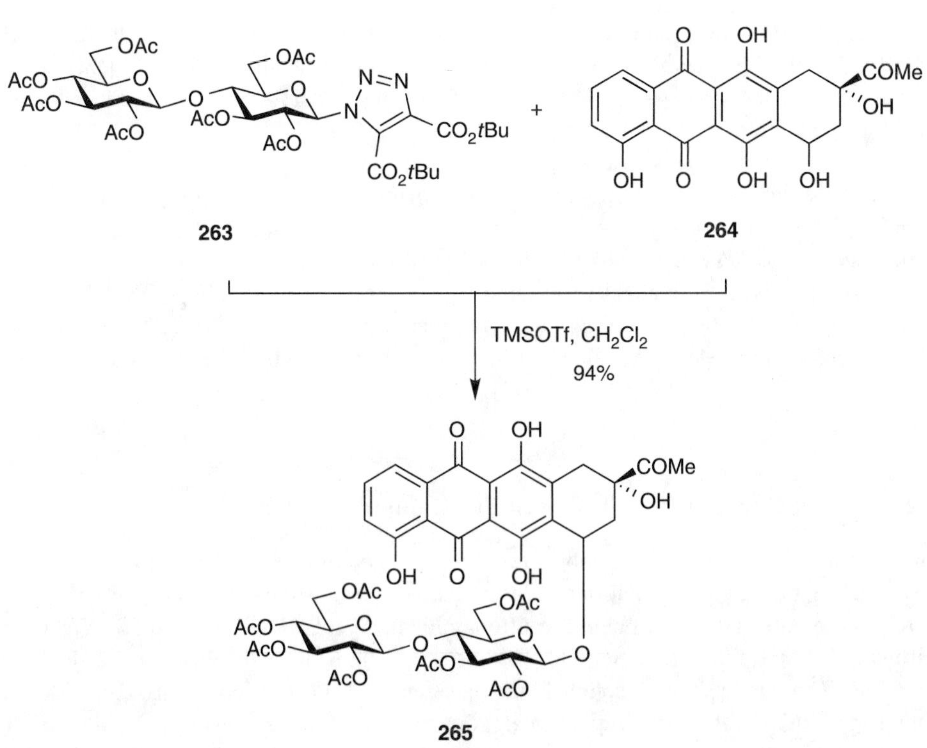

SCHEME 4.51 Glycosylation with *N*-glycosyl triazole.

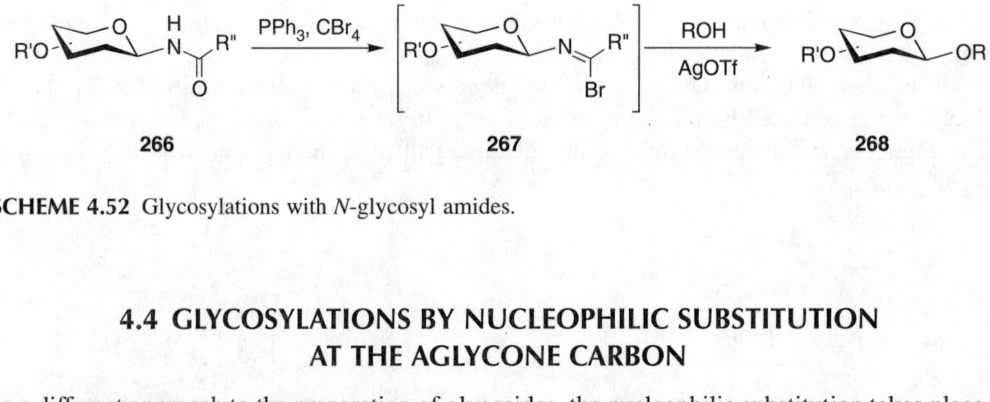

SCHEME 4.52 Glycosylations with *N*-glycosyl amides.

4.4 GLYCOSYLATIONS BY NUCLEOPHILIC SUBSTITUTION AT THE AGLYCONE CARBON

In a different approach to the preparation of glycosides, the nucleophilic substitution takes place at the carbon atom of the aglycone rather than at the anomeric carbon (Scheme 4.53). In contrast to the methods discussed in Section 4.3, there is no scission of the glycosyl–oxygen bond in these reactions. Instead, the R–X bond is cleaved.

These glycosylation are commonly referred to as *anomeric* O-*alkylations* [12,318] and proceed under alkaline conditions, whereas most glycosylations discussed so far take place under acidic conditions. The reaction requires the activation of the aglycone (generally in the form of a triflate), and the glycosyl unit is activated in the form of alkoxide to increase the nucleophilicity of the anomeric hydroxyl. Although there is retention of the anomeric oxygen in the reaction, the anomeric configuration is not necessarily retained because the α- (**269**) and β-alkoxides (**271**) are in equilibrium through the acyclic intermediate **270** (Scheme 4.54).

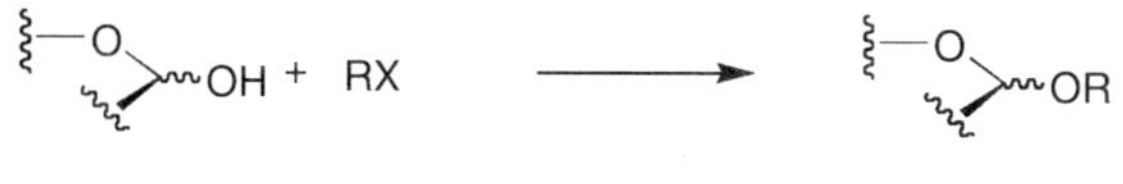

SCHEME 4.53 Glycoside synthesis by nucleophilic substitution at the aglycone carbon.

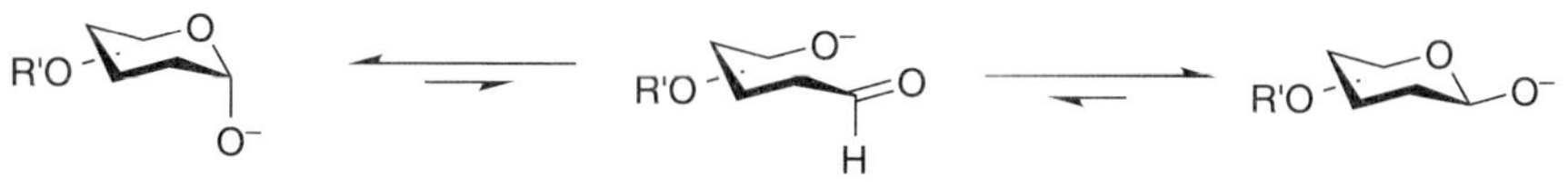

SCHEME 4.54 Equilibrium of anomeric alkoxides.

Generally, the products derived from the open-chain form are insignificant in the reaction mixture and stereocontrol between the anomeric forms can be achieved. It was observed that the anomeric oxygen in the equatorial alkoxide **276** has enhanced nucleophilicity compared with the axial [284]. Often referred to as the *kinetic anomeric effect*, this phenomenon can govern the diastereoselective formations of the equatorial glycosides [579]. The anomeric effect, however, favors the formation of the thermodynamically more stable axial glycosides and chelation control can also play a role in determining anomeric stereoselectivity.

An example of highly selective formation of both anomers is given in Scheme 4.55 [580]. As illustrated, treatment of the D-ribose derivative **272** with KO*t*Bu, followed by the addition of the triflate **274**, gave the β-linked disaccharide **275**. On the other hand, NaH-assisted reaction of the hemiacetal **276** with a bulky protecting group and the same triflate afforded the α-linked compound

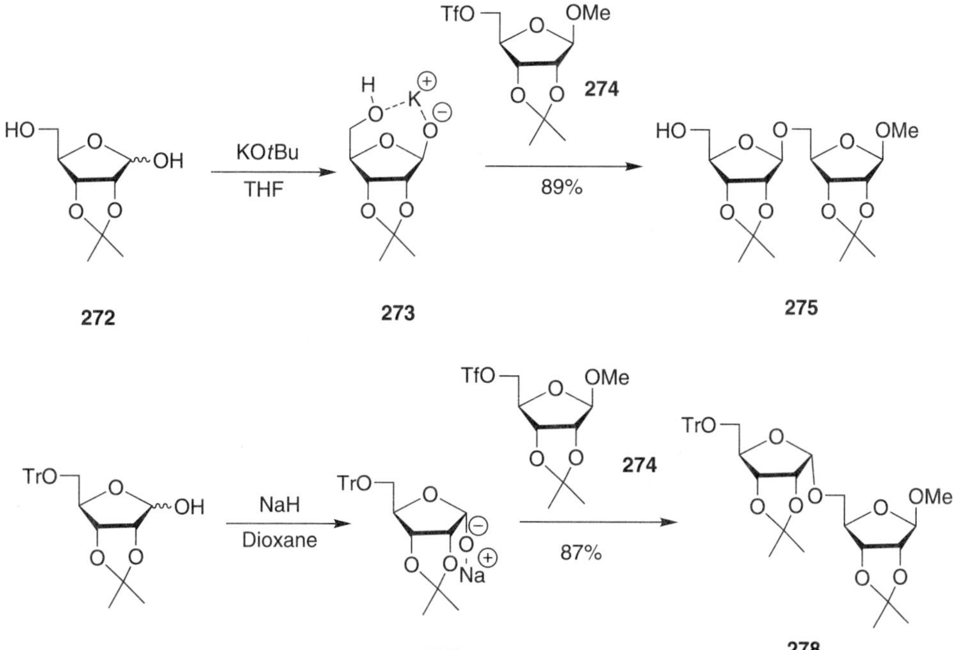

SCHEME 4.55 Synthesis of α- and β-D-ribofuranosides by anomeric *O*-alkylation.

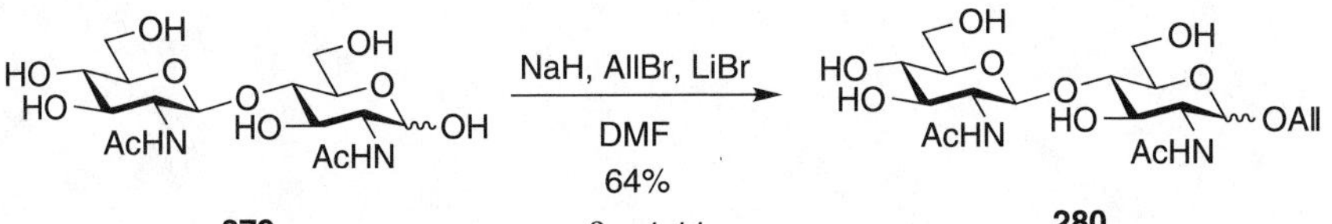

279 NaH, AllBr, LiBr, DMF, 64%, $\alpha:\beta = 1:11$ → 280

SCHEME 4.56 Regioselective glycoside synthesis by anomeric *O*-alkylation.

278. The difference in stereoselectivity was explained by the formation of differing intramolecular metal complexes (**273** and **277**).

Most glycosylations by anomeric *O*-alkylations were performed on triflates of primary alcohols. Nevertheless, reactions with secondary alkyl triflates, including carbohydrates, were also described [581]. These reactions afford inverted configurations in the aglycone.

A noteworthy feature of these glycosylation reactions is that the alkylations take place regioselectively at the anomeric oxygen on compounds possessing multiple free hydroxyl groups [582]. This can even be applied to unprotected sugars [583]. For example, allylation of the disaccharide chitobiose **279** with NaH in the presence of LiBr afforded the glycoside **280** in a 64% yield in a highly regio- and stereoselective manner [584] (Scheme 4.56).

4.5 SYNTHESIS OF GLYCOSIDES BY ADDITION REACTIONS

Besides nucleophilic substitutions, additions to the double bond of 1,2-unsaturated sugars (glycals) constitute an attractive route for the synthesis of glycosides. Glycals react in these reactions either by a 1,2-addition or by an allylic rearrangement. The following discusses these two reactions in turn.

Most synthetic methods utilizing 1,2-additions to glycals result in the formation of 2-deoxy-2-halogeno sugars or related derivatives (e.g., a 2-selenophenyl derivative). These are then converted to the 2-deoxy sugars by reductive methods. The various glycosylation methods leading to 2-deoxy glycosides are summarized in Table 4.16.

Addition of iodine in the presence of silver salts to the double bond of 3,4,6-tri-*O*-acetyl-D-glucal was studied by Lemieux [585], and the use of IDCP was introduced [586,587]. Other halogenating agents that proved to be of practical use were later added. Thus, NBS was proposed by Tatsuta [588] and NIS by Thiem [589,590]. These reactions give the *trans*-addition product preferentially, for example, the 2-deoxy-2-iodo-α-D-mannopyranoside **283** was obtained from glucal **281** and the 2,3-anhydro derivative **282** [589] (Scheme 4.57).

In later developments, the use of tin alkoxides of the acceptors to increase their nucleophilicity was introduced [591]. Danishefsky [592] expanded the use of IDCP-promoted reactions to oligosaccharide syntheses, showing that glycals are also amenable to armed–disarmed type glycosylations.

The 2-halogeno-2-deoxy glycosides formed in these reactions are readily converted into 2-deoxy sugars, and a number of biologically important 2-deoxy oligosaccharides were synthesized this way [14,26]. However, these glycosides are also good precursors to fully functionalized sugars and synthetic strategies for this were developed [593]. Another application of halogen additions — the sulfonamidoglycosylation of glycals — was discussed in Section 4.3.4.2. Alkoxymercurations are another means of preparing 2-deoxy derivatives from glycals [594,595].

Activation of glycals by selenylation was introduced by Sinaÿ and coworkers [41] by the sequential addition of PhSeCl and the glycosyl acceptor. A two-step version of this activation with the isolation of the intermediate glycosyl acetate [596] has also been reported.

Activation by alkylsulfenylation was reported by Ogawa [597], who used alkylsulfenylated acceptor derivatives in the presence of TMSOTf. In a two-step sequence, PhSCl was added to the

TABLE 4.16
Activations of Glycals for 1,2-Additions

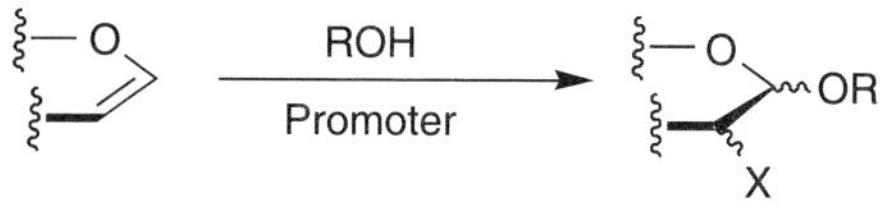

Promoter	X	Ref.
IDCP	I	[586,587,592]
NBS	Br	[588]
NIS	I	[589,590]
$Hg(OAc)_2$	HgOAc	[594,595]
PhSeCl	SePh	[41,596]
ROSPh–TMSOTf	SPh	[597]
$(PhS)_2S^+Ph\ SbCl_6^-$	SPh	[599,600]
CSA	H	[602–604]
TsOH	H	[605]
AG50 WX2	H	[606]
$Ph_3P{\cdot}HBr$	H	[607]
BCl_3	H	[608]
BBr_3	H	[608]
CAN	H	[609]
287, Tf_2O, R′CONHTMS	NHCOR′	[610,611]
Ar_2SO, Tf_2O, MeOH, $ZnCl_2$	OH	[612,613]
$PhI(OCOR')_2$, $BF_3{\cdot}Et_2O$, TfOH	OCOR′	[614]

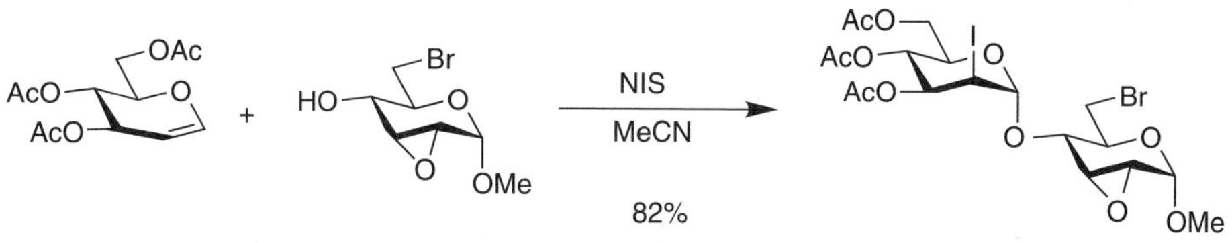

SCHEME 4.57 NIS-promoted glycosylation of glycals.

double bond and glycosylations were performed with the trichloroacetimidates of the 2-deoxy-2-phenylthio derivatives [598]. A different type of reagent, the phenylbis(phenylthio)sulfonium salt **285** (Scheme 4.58) was introduced by Franck [599,600]. Reactions with this promoter also led to 2-phenylthio derivatives **286**.

The addition of hydrogen to C-2 (instead of the heteroatoms discussed so far) leads directly to 2-deoxyglycosides. Various sulfonic acids have been used as activators for this purpose, including MsOH [601], CSA [602–604], *p*TsOH [605], and dehydrated AG50 WX2 resin [606]. In addition to these, triphenylphosphine hydrobromide [607], BCl_3 or BBr_3 [608], and ceric ammonium nitrate (CAN) [609] are effective for this purpose.

The direct synthesis of fully functionalized sugars possessing amino or hydroxyl functions at C-2 has also been studied. Thus, activation of glycals by thianthrene-5-oxide **287** and Tf_2O, followed by treatment with an amide nucleophile and a glycosyl acceptor, led directly to various 2-acylamido glycosides **288** [610,611] (Scheme 4.59).

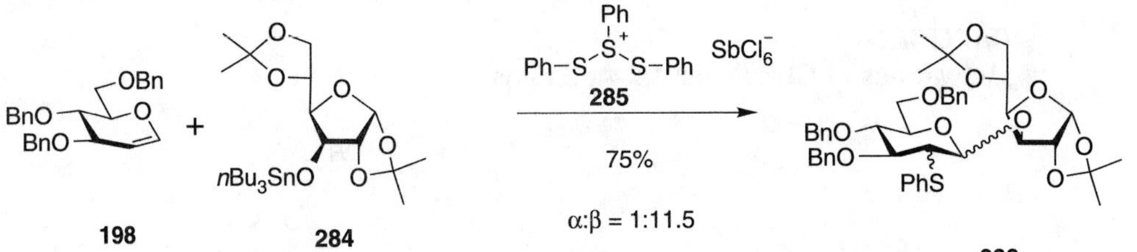

SCHEME 4.58 Synthesis of 2-deoxy-2-phenylthio glycoside using bis(arylthio)sulfonium salt promoter.

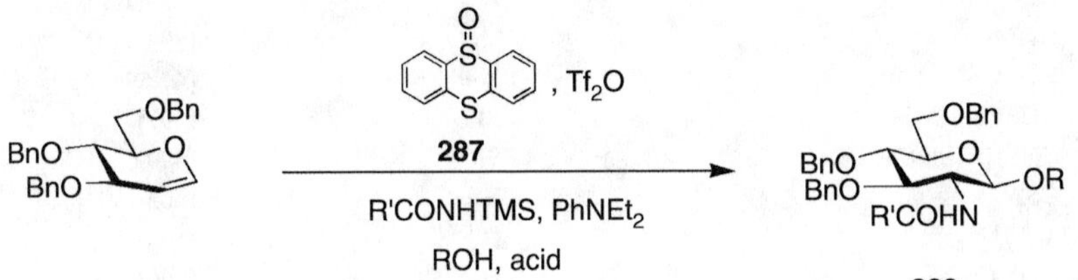

SCHEME 4.59 Acetamidoglycosylation with glycals.

Analogously, conditions were developed for the direct synthesis of 2-hydroxy glycosides from glycals using aryl sulfoxides and Tf_2O activation [612,613]. The preparation of 2-acyloxy glycosides using phenyl iodonium reagents was also studied [614].

A different type of glycosylation reaction of glycals takes place with allyl rearrangement and leads to 2,3-unsaturated glycosides [615]. These glycosylations, often referred to as the Ferrier rearrangement, are summarized in Table 4.17.

TABLE 4.17
Activations of Glycals for Allylic Rearrangement

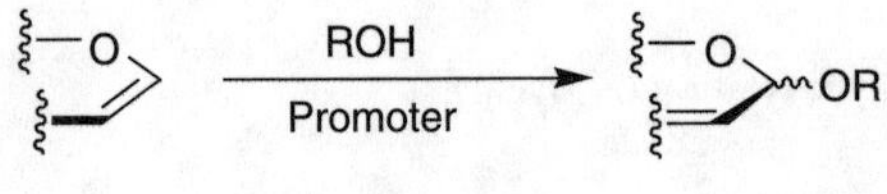

Promoter	Ref.
$BF_3 \cdot Et_2O$	[615–617]
$SnCl_4$	[618,619]
IDCP	[620,621]
DDQ	[622]
Montmorillonite K-10	[623]
I_2	[624]
$FeCl_3$	[625]
$InCl_3$	[626,627]
$Sc(OTf)_3$	[628]
I_2–MeCOSH	[155]
$Yb(OTf)_3$	[629]
TMSOTf	[630]

Boron trifluoride etherate [615–617] and $SnCl_4$ [618,619] are the catalysts that are traditionally used in this reaction. Recently described promoters include IDCP [620,621], DDQ [622], montmorillonite K-10 [623], I_2 [624], $FeCl_3$ [625], $InCl_3$ [626,627], $Sc(OTf)_3$ [628], the I_2–thiolacetic acid system [155], $Yb(OTf)_3$ [629], and TMSOTf [630]. The 2,3-unsaturated glycosides produced in these reactions can easily be converted into the 2,3-dideoxy derivatives or to a variety of other products by functionalization of the double bond.

4.6 OTHER GLYCOSYLATION METHODS

Several different approaches that do not fall into the above categories have also been developed and used to varying degrees. Selected examples of this very diverse area are shown below to illustrate that the formation of glycosidic bonds can be approached by various, less conventional routes.

Vasella and coworkers [631,632] introduced glycosylation reactions proceeding via glycosyl carbenes. The carbenes were generated from glycosyl diazirines, and the regioselectivity of their reactions with acceptors that have multiple hydroxyl groups was strongly influenced by the hydrogen bonding properties of the acceptor hydroxyls. For example, reaction of the carbene **290**, formed from the diazirine **289**, with methyl orsellinate **291** gave the 4-*O*-glycoside **292** due to the hydrogen bonding of the 2-OH to the carbonyl group [633] (Scheme 4.60).

In other approaches, the glycosidic bond is synthesized in some masked form. Thus, in Barett's reductive glycosylation [634] (Scheme 4.61), an ester bond is created first at the anomeric center using the uronic acid derivative **293**. Thionation of the carbonyl group of **294** was followed by reductive desulfurization to give the disaccharide **296**.

The glycosidic bond is generated from an orthoester function in another approach [635] (Scheme 4.62). The spiro orthoester **299** obtained from the lactone **297** and the diol **298** was subjected to reductive ring opening with $LiAlH_4$–$AlCl_3$ and the disaccharide **300** with a β-mannosidic linkage was obtained in a highly regio- and stereoselective manner.

Cyclizations are also used to create glycosidic linkages. In the example given in Scheme 4.63, the *E* (**303**) and *Z* (**304**) enol ethers were obtained from the reaction of the aldehyde **301** with the phosphonate **302** [636]. Oxymercuration–demercuration of **303** and **304** gave the β-linked (**305**) and α-linked (**306**) KDO disaccharides, respectively.

Finally, it should be mentioned that enzymatic syntheses of glycosides and oligosaccharides developed rapidly, having provided an additional tool for the preparation of these compounds. As further discussion of this subject is beyond the scope of this chapter, the reader is referred to overviews given elsewhere [637–640] as starting points for further reading.

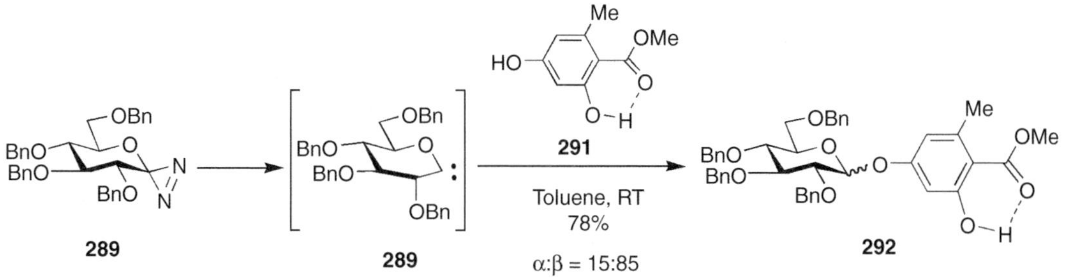

SCHEME 4.60 Glycosylation with glycosyl carbene.

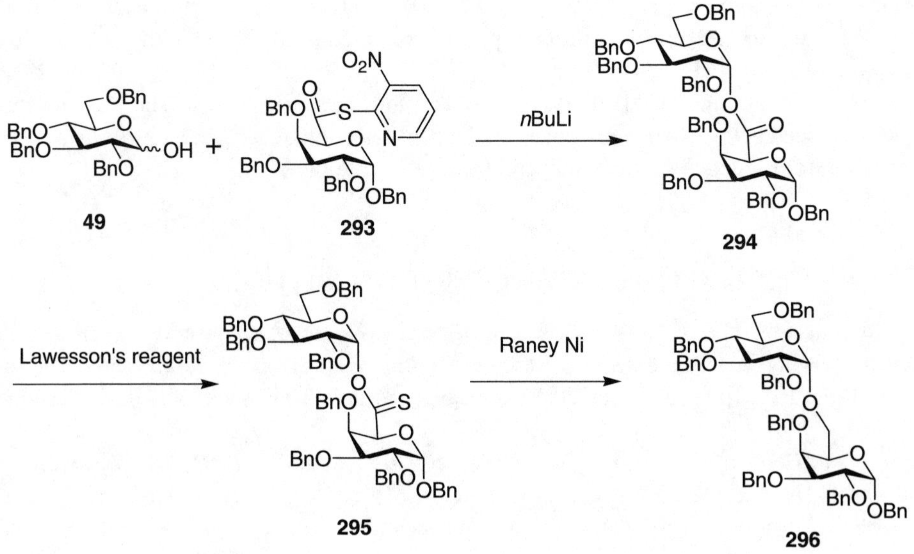

SCHEME 4.61 Reductive glycosylation.

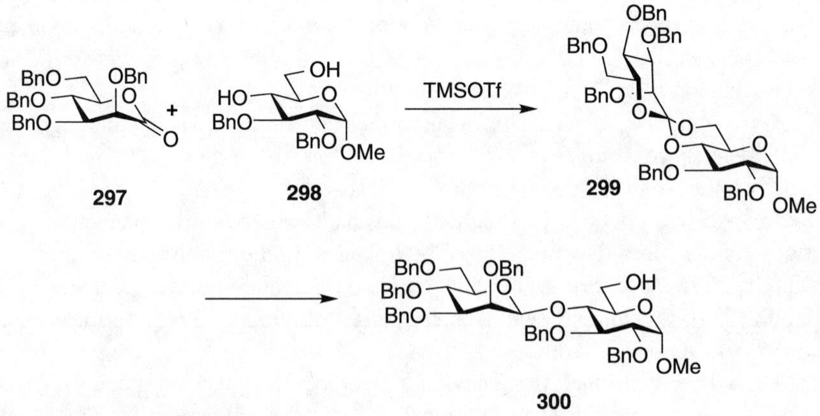

SCHEME 4.62 Glycoside synthesis by reductive ring opening.

4.7 SUMMARY AND OUTLOOK

As a consequence of the interest in glycosides and oligosaccharides, a large number of glycosylation methods have been developed over the years. Existing methods allow the efficient preparation of almost any glycosidic linkages, although, in some cases, syntheses might require experimentation and optimization. Nevertheless, syntheses of highly complex molecules have been successfully accomplished. The earlier belief about the field being in the hands of specialists does not seem to be justified now; nevertheless, a good know-how and sophistication is definitely an advantage concerning the fine details of different targets and approaches.

The *holy grail* of glycosylations — the generally applicable, stereoselective and technically simple glycosylation method — has yet to be found. Indeed, the search will undoubtedly continue in the future. Nevertheless, even if a single general method is developed, because of the highly

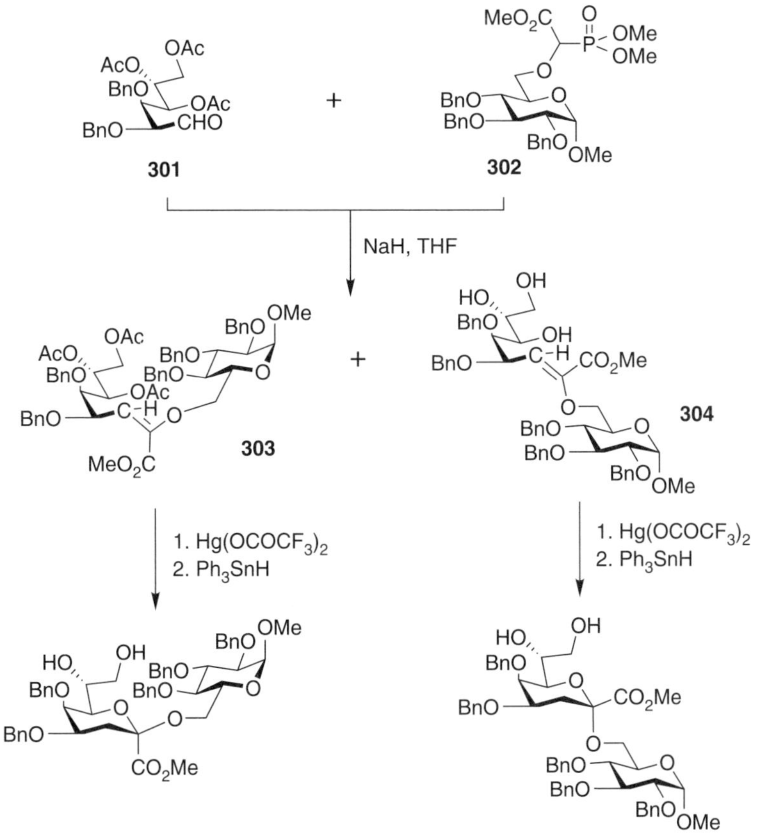

SCHEME 4.63 Stereoselective synthesis of KDO glycosides by cyclization.

diverse structures of carbohydrates and aglycones, the use of certain methods in some special areas might still give better results.

From a theoretical perspective, in developing improved glycosylations, a better understanding of the delicate details of the nucleophilic substitution paradigm might contribute significantly for future advances. From a practical perspective, introductions of new types of glycosyl donors and promoters, and the development of improved reaction conditions for their use, is expected to continue. Additionally, the introduction of new methods, where the underlying chemistry is different from the present ones, can contribute to the progress in this field.

REFERENCES

1. Dwek, R A, Glycobiology: toward understanding the function of sugars, *Chem. Rev.*, 96, 683–720, 1996.
2. Varki, A, Biological roles of oligosaccharides: all of the theories are correct, *Glycobiology*, 3, 97–130, 1993.
3. Michael, A, On the synthesis of helicin and phenolglucoside, *Am. Chem. J.*, 1, 305, 1879.
4. Fischer, E, Ueber die Glucoside der Alkohole, *Ber. Dtsch. Chem. Ges.*, 26, 2400–2412, 1893.
5. Fischer, E, Ueber die Verbindungen der Zucker mit den Alkoholen und Ketonen, *Ber. Dtsch. Chem. Ges.*, 28, 1145–1167, 1895.

6. Koenigs, W, Knorr, E, Ueber einige Derivate des Traubenzuckers und der Galactose, *Ber. Dtsch. Chem. Ges.*, 34, 957–981, 1901.
7. Bochkov, A F, Zaikov, G E, *Chemistry of the O-glycosidic bond: formation and cleavage*, Pergamon Press, Oxford, 1979.
8. Ferrier, R J, Newer observations on the synthesis of *O*-glycosides, *Fortsch. Chem. Forsch.*, 14, 389–429, 1970.
9. Wulff, G, Röhle, G, Results and problems of *O*-glycoside synthesis, *Angew. Chem. Int. Ed. Engl.*, 13, 157–170, 1974.
10. Igarashi, K, The Koenigs–Knorr reaction, *Adv. Carbohydr. Chem. Biochem.*, 34, 243–283, 1977.
11. Paulsen, H, Advances in selective chemical syntheses of complex oligosaccharides, *Angew. Chem. Int. Ed. Engl.*, 21, 155–173, 1982.
12. Schmidt, R R, New methods for the synthesis of glycosides and oligosaccharides — Are there alternatives to the Koenigs–Knorr method? *Angew. Chem. Int. Ed. Engl.*, 25, 212–235, 1986.
13. Fügedi, P, Garegg, P J, Lönn, H, Norberg, T, Thioglycosides as glycosylating agents in oligosaccharide synthesis, *Glycoconjugate J*, 4, 97–108, 1987.
14. Thiem, J, Klaffke, W, Syntheses of deoxy oligosaccharides, In *Carbohydrate Chemistry*, *Topics in Current Chemistry*, Vol. 154, Thiem, J, Ed., Springer, Berlin, pp. 285–332, 1990.
15. Okamoto, K, Goto, T, Glycosidation of sialic acid, *Tetrahedron*, 46, 5835–5857, 1990.
16. Banoub, J, Boullanger, P, Lafont, D, Synthesis of oligosaccharides of 2-amino-2-deoxy sugars, *Chem. Rev.*, 92, 1167–1195, 1992.
17. Fraser-Reid, B, Udodong, U E, Wu, Z, Ottosson, H, Merritt, J R, Rao, C S, Roberts, C, Madsen, R, *n*-Pentenyl glycosides in organic chemistry: a contemporary example of serendipity, *Synlett*, 927–942, 1992.
18. Toshima, K, Tatsuta, K, Recent progress in *O*-glycosylation methods and its application to natural products synthesis, *Chem. Rev.*, 93, 1503–1531, 1993.
19. Barresi, F, Hindsgaul, O, Chemically synthesized oligosaccharides, 1994. A searchable table of glycosidic linkages, *J. Carbohydr. Chem.*, 14, 1043–1087, 1995.
20. Whitfield, D M, Douglas, S P, Glycosylation reactions — present status, future directions, *Glycoconjugate J.*, 13, 5–17, 1996.
21. Danishefsky, S J, Bilodeau, M T, Glycals in organic synthesis: the evolution of comprehensive strategies for the assembly of oligosaccharides and glycoconjugates of biological consequence, *Angew. Chem. Int. Ed. Engl.*, 35, 1380–1419, 1996.
22. Garegg, P J, Thioglycosides as glycosyl donors in oligosaccharide synthesis, *Adv. Carbohydr. Chem. Biochem.*, 52, 179–205, 1997.
23. Fraser-Reid, B, Madsen, R, Campbell, A S, Roberts, C S, Merritt, J R, Chemical synthesis of oligosaccharides, In *Bioorganic Chemistry: Carbohydrates*, Hecht, S M, Ed., Oxford University Press, Oxford, pp. 89–133, 1999.
24. Nicolaou, K C, Bockovich, N J, Chemical synthesis of complex carbohydrates, In *Bioorganic Chemistry: Carbohydrates*, Hecht, S M, Ed., Oxford University Press, Oxford, pp. 134–173, 1999.
25. Osborn, H M I, Khan, T H, Recent developments in polymer supported syntheses of oligosaccharides and glycopeptides, *Tetrahedron*, 55, 1807–1850, 1999.
26. Marzabadi, C H, Franck, R W, The synthesis of 2-deoxyglycosides: 1988–1999, *Tetrahedron*, 56, 8385–8417, 2000.
27. Davis, B G, Recent developments in oligosaccharide synthesis, *J. Chem. Soc. Perkin. Trans. 1*, 2137–2160, 2000.
28. Gridley, J J, Osborn, H M I, Recent advances in the construction of β-D-mannose and β-D-mannosamine linkages, *J. Chem. Soc. Perkin. Trans. 1*, 1471–1491, 2000.
29. Jung, K-H, Müller, M, Schmidt, R R, Intramolecular *O*-glycoside bond formation, *Chem. Rev.*, 100, 4423–4442, 2000.
30. Seeberger, P H, Haase, W-C, Solid-phase oligosaccharide synthesis and combinatorial carbohydrate libraries, *Chem. Rev.*, 100, 4349–4393, 2000.
31. Hanessian, S, Lou, B, Stereocontrolled glycosyl transfer reactions with unprotected glycosyl donors, *Chem. Rev.*, 100, 4443–4463, 2000.
32. Boons, G-J, Demchenko, A V, Recent advances in *O*-sialylation, *Chem. Rev.*, 100, 4539–4565, 2000.
33. Seeberger, P H, Solid phase oligosaccharide synthesis, *J. Carbohydr. Chem.*, 21, 613–643, 2002.

34. Crich, D, Chemistry of glycosyl triflates: synthesis of β-mannopyranosides, *J. Carbohydr. Chem.*, 21, 667–690, 2002.
35. Palmacci, E R, Plante, O J, Seeberger, P H, Oligosaccharide synthesis in solution and on solid support with glycosyl phosphates, *Eur. J. Org. Chem.*, 595–606, 2002.
36. Vankayalapati, H, Jiang, S, Singh, G, Glycosylation based on glycosyl phosphates as glycosyl donors, *Synlett*, 16–25, 2002.
37. Jensen, K J, *O*-Glycosylations under neutral or basic conditions, *J. Chem. Soc. Perkin. Trans. 1*, 2219–2233, 2002.
38. Demchenko, A V, 1,2-*cis* *O*-Glycosylation: methods, strategies, principles, *Current Org. Chem.*, 7, 35–79, 2003.
39. Demchenko, A V, Stereoselective chemical 1,2-*cis* *O*-glycosylation: from "Sugar Ray" to modern techniques of the 21st century, *Synlett*, 1225–1240, 2003.
40. Seeberger, P H, Solid-phase oligosaccharide synthesis, In *Carbohydrate-Based Drug Discovery*, Wong, C H, Ed., Wiley-VCH Verlag, Weinheim, Germany, pp. 103–127, 2003.
41. Jaurand, G, Beau, J-M, Sinaÿ, P, Glycosyloxyselenation-deselenation of glycals: a new approach to 2′-deoxy-disaccharides, *J. Chem. Soc. Chem. Commun.*, 572–573, 1981.
42. Nicolaou, K C, Ladduwahetty, T, Randall, J L, Chucholowski, A, Stereospecific 1,2-migrations in carbohydrates. Stereocontrolled synthesis of α- and β-2-deoxyglycosides, *J. Am. Chem. Soc.*, 108, 2466–2467, 1986.
43. Ito, Y, Ogawa, T, An efficient approach to stereoselective glycosylation of *N*-acetylneuraminic acid: use of phenylselenyl group as a stereocontrolling auxiliary, *Tetrahedron Lett.*, 28, 6221–6224, 1987.
44. Lemieux, R U, Hendriks, K B, Stick, R V, James, K, Halide ion catalyzed glycosidation reactions. Syntheses of α-linked disaccharides, *J. Am. Chem. Soc.*, 97, 4056–4062, 1975.
45. Jona, H, Mandai, H, Chavasiri, W, Takeuchi, K, Mukaiyama, T, Protic acid catalyzed stereoselective glycosylation using glycosyl fluorides, *Bull. Chem. Soc. Jpn.*, 75, 291–309, 2002.
46. Paulsen, H, Lockhoff, O, Bausteine von Oligosacchariden XXX. Neue effektive β-Glycosidsynthese für Mannose–Glycoside. Synthesen von Mannose–haltigen Oligosacchariden, *Chem. Ber.*, 114, 3102–3114, 1981.
47. Garegg, P J, Johansson, R, Samuelsson, B, Silver imidazolate-assisted glycosidations. Part 6. Synthesis of 1,2-*trans*-linked disaccharides, *Acta Chem. Scand. B*, 36, 249–250, 1982.
48. Helferich, B, Wedemeyer, K-F, Zur Darstellung von Glucosiden aus Acetobromoglucose, *Liebigs Ann. Chem.*, 563, 139–145, 1949.
49. Helferich, B, Jung, K-H, Zur Darstellung von Phenol-α-Glykosiden, *Liebigs Ann. Chem.*, 589, 77–81, 1954.
50. Helferich, B, Berger, A, Über die Synthese von Glucuroniden, *Chem. Ber.*, 90, 2492–2498, 1957.
51. Schroeder, L R, Green, J W, Koenigs–Knorr syntheses with mercuric salts, *J. Chem. Soc. C*, 530–531, 1966.
52. Bock, K, Meldal, M, Mercury iodide as a catalyst in oligosaccharide synthesis, *Acta Chem. Scand. B*, 37, 775–783, 1983.
53. Conrow, R B, Bernstein, S, Steroid Conjugates. VI. An improved Koenigs–Knorr synthesis of aryl glucuronides using cadmium carbonate, a new and effective catalyst, *J. Org. Chem.*, 36, 863–870, 1971.
54. Bredereck, H, Wagner, A, Faber, G, Ott, H, Rauther, J, Eine vereinfachte Oligosaccharid-Synthese, *Chem. Ber.*, 92, 1135–1139, 1959.
55. Bredereck, H, Wagner, A, Kuhn, H, Ott, H, Oligosaccharidsynthesen II. Synthesen von Di- und Trisacchariden des Gentiobiosetyps, *Chem. Ber.*, 93, 1201–1206, 1960.
56. Bredereck, H, Wagner, A, Geissel, D, Gross, P, Hutten, U, Ott, H, Oligosaccharidsynthesen III. Synthesen α- und β-konfigurierter Disaccharide, *Chem. Ber.*, 95, 3056–3063, 1962.
57. Bredereck, H, Wagner, A, Geissel, D, Ott, H, Oligosaccharidsynthesen IV. Synthesen α- und β-konfigurierter Disaccharide, *Chem. Ber.*, 95, 3064–3069, 1962.
58. Lemieux, R U, Takeda, T, Chung, B Y, Synthesis of 2-amino-2-deoxy-β-D-glucopyranosides. Properties and use of 2-deoxy-2-phtalimidoglycosyl halides, *ACS. Symp. Ser.*, 39, 90–115, 1976.
59. Hanessian, S, Banoub, J, Chemistry of the glycosidic linkage. An efficient synthesis of 1,2-*trans*-disaccharides, *Carbohydr. Res.*, 53, C13–C16, 1977.

60. Garegg, P J, Ossowski, P, Silver zeolite as promoter in glycoside synthesis. The synthesis of β-D-mannopyranosides, *Acta Chem. Scand. B*, 37, 249–250, 1983.
61. van Boeckel, C A A, Beetz, T, A note on the use of porous silver silicates as promoter in carbohydrate coupling reactions, *Recl. Trav. Chim. Pays-Bas*, 106, 596–598, 1987.
62. Paulsen, H, Kutschker, W, Lockhoff, O, Bausteine von Oligosacchariden, XXXIII. Synthese von β-Glycosidisch verknüpften L-Rhamnose-haltigen Disacchariden, *Chem. Ber.*, 114, 3233–3241, 1981.
63. Paulsen, H, Kutschker, W, Bausteine von Oligosacchariden, XLV. Synthese einer verzweigten Tetrasaccharid-Einheit der *O*-Spezifischen Kette des Lipopolysaccharides aus *Shigella flexneri* Serotyp 6, *Liebigs Ann. Chem.*, 557–569, 1983.
64. Paulsen, H, Lebuhn, R, Bausteine von Oligosacchariden, XLVII. Synthese von Tri- und Tetrasaccharid-Sequenzen von *N*-Glycoproteinen mit β-D-Mannosidischer Verknüpfung, *Liebigs Ann. Chem.*, 1047–1072, 1983.
65. Paulsen, H, Heume, M, Nürnberger, H, Synthese der verzweigten Nonasaccharid-Sequenz der bisected Struktur von *N*-Glycoproteinen, *Carbohydr. Res.*, 200, 127–166, 1990.
66. Ogawa, T, Matsui, M, Approach to synthesis of glycosides: enhancement of nucleophilicity of hydroxyl-groups by trialkylstannylation, *Carbohydr. Res.*, 51, C13–C18, 1976.
67. Higashi, K, Nakayama, K, Soga, T, Shioya, E, Uoto, K, Kusama, T, Novel stereoselective glycosidation by the combined use of a trityl halide and Lewis acid, *Chem. Pharm. Bull.*, 38, 3280–3282, 1990.
68. Lubineau, A, Malleron, A, Stannous triflate mediated glycosidations. A stereoselective synthesis of β-D-glucosides, *Tetrahedron Lett.*, 26, 1713–1716, 1985.
69. Lubineau, A, Le Gallic, J, Malleron, A, Stannous triflate mediated glycosidations. A stereoselective synthesis of 2-amino 2-deoxy-β-D-glucopyranosides directly with the natural *N*-acetyl protecting group, *Tetrahedron Lett.*, 28, 5041–5044, 1987.
70. Yamada, H, Hayashi, T, A substrate-unspecified glycosylation reaction promoted by copper(II) trifluoromethanesulfonate in benzotrifluoride, *Carbohydr. Res.*, 337, 581–585, 2002.
71. Mukherjee, D, Ray, P K, Chowdhury, U S, Synthesis of glycosides via indium(III) chloride mediated activation of a glycosyl halide in neutral condition, *Tetrahedron*, 57, 7701–7704, 2001.
72. Dobarro-Rodriguez, A, Trumtel, M, Wessel, H P, Triflic anhydride: an alternative promoter in glycosidations, *J. Carbohydr. Chem.*, 11, 255–263, 1992.
73. Mukaiyama, T, Jona, H, Takeuchi, K, Trifluoromethanesulfonic acid (TfOH)-catalyzed stereoselective glycosylation using glycosyl fluoride, *Chem. Lett.*, 696–697, 2000.
74. Kartha, K P R, Aloui, M, Field, R A, Iodine: a versatile reagent in carbohydrate chemistry. III. Efficient activation of glycosyl halides in combination with DDQ, *Tetrahedron Lett.*, 37, 8807–8810, 1996.
75. Kartha, K P R, Field, R A, Glycosylation chemistry promoted by iodine monobromide: efficient synthesis of glycosyl bromides from thioglycosides, and *O*-glycosides from disarmed thioglycosides and glycosyl bromides, *Tetrahedron Lett.*, 38, 8233–8236, 1997.
76. Crich, D, Smith, M, Yao, Q, Picione, J, 2,4,6-Tri-*tert*-butylpyrimidine (TTBP): a cost effective, readily available alternative to the hindered base 2,6-di-*tert*-butylpyridine and its 4-substituted derivatives in glycosylation and other reactions, *Synthesis*, 323–326, 2001.
77. Kochetkov, N K, Khorlin, A J, Bochkov, A F, A new method of glycosylation, *Tetrahedron*, 23, 693–707, 1967.
78. Kochetkov, N K, Bochkov, A F, Sokolovsakaya, T A, Snyatkova, V J, Modifications of the orthoester method of glycosylation, *Carbohydr. Res.*, 16, 17–27, 1971.
79. Ogawa, T, Beppu, K, Nakabayashi, S, Trimethylsilyl trifluoromethanesulfonate as an effective catalyst for glycoside synthesis, *Carbohydr. Res.*, 93, C6–C9, 1981.
80. Ogawa, T, Katano, K, Matsui, M, Regio- and stereo-controlled synthesis of core oligosaccharides of glycopeptides, *Carbohydr. Res.*, 64, C3–C9, 1978.
81. Zhu, Y, Kong, F, Regio- and stereoselective synthesis of 1→6 linked manno-, gluco-, and galactopyranose di-, tri-, and tetrasaccharides via orthoester intermediates, *J. Carbohydr. Chem.*, 19, 837–848, 2000.
82. Lemieux, R U, Driguez, H, Chemical synthesis of 2-acetamido-2-deoxy-4-*O*-(α-L-fucopyranosyl)-3-*O*-(β-D-galactopyranosyl)-D-glucose. The Lewis a blood-group antigenic determinant, *J. Am. Chem. Soc.*, 97, 4063–4069, 1975.

83. Lemieux, R U, Driguez, H, The chemical synthesis of 2-*O*-(α-L-fucopyranosyl)-3-*O*-(α-D-galactopyranosyl)-D-galactose. The terminal structure of the blood-group B antigenic determinant, *J. Am. Chem. Soc.*, 97, 4069–4075, 1975.
84. Lemieux, R U, Bundle, D R, Baker, D A, Properties of a synthetic antigen related to human blood-group Lewis a, *J. Am. Chem. Soc.*, 97, 4076–4083, 1975.
85. Sato, S, Mori, M, Ito, Y, Ogawa, T, An efficient approach to *O*-glycosides through $CuBr_2$-Bu_4NBr mediated activation of glycosides, *Carbohydr. Res.*, 155, C6–C10, 1986.
86. Andersson, F, Fügedi, P, Garegg, P J, Nashed, M, Synthesis of 1,2-*cis*-linked glycosides using dimethyl(methylthio)sulfonium triflate as promoter and thioglycosides as glycosyl donors, *Tetrahedron Lett.*, 27, 3919–3922, 1986.
87. Shingu, Y, Nishida, Y, Dohi, H, Matsuda, K, Kobayashi, K, Convenient access to halide ion-catalyzed α-glycosylation free from noxious fumes at the donor synthesis, *J. Carbohydr. Chem.*, 21, 605–611, 2002.
88. Shingu, Y, Nishida, Y, Dohi, H, Kobayashi, K, An easy access to halide ion-catalytic α-glycosylation using carbon tetrabromide and triphenylphosphine as multifunctional reagents, *Org. Biomol. Chem.*, 1, 2518–2521, 2003.
89. Brewster, K, Harrison, J M, Inch, T D, Snthesis of aryl β-D-glucopyranosides and aryl β-D-glucopyranosiduronic acids, *Tetrahedron Lett.*, 52, 5051–5054, 1979.
90. Dess, D, Kleine, H P, Weinberg, D V, Kaufman, R J, Sidhu, R S, Phase-transfer catalyzed synthesis of acetylated aryl β-D-glucopyranosides and aryl β-D-galactopyranosides, *Synthesis*, 883–885, 1981.
91. Kleine, H P, Weinberg, D V, Kaufman, R J, Sidhu, R S, Phase-transfer-catalyzed synthesis of 2,3,4,6-tetra-*O*-acetyl-β-D-galactopyranosides, *Carbohydr. Res.*, 142, 333–337, 1985.
92. Loganathan, D, Trivedi, G K, Phase-transfer-catalyzed D-glucosylation: synthesis of benzoylated aryl β-D-glucopyranosides and β-D-glucopyranosyl-substituted cinnamates, *Carbohydr. Res.*, 162, 117–125, 1987.
93. Lewis, P, Kaltia, S, Wähälä, K, The phase transfer catalysed synthesis of isoflavone-*O*-glucosides, *J. Chem. Soc. Perkin. Trans. 1*, 2481–2484, 1998.
94. Carrière, D, Meunier, S J, Tropper, F D, Cao, S, Roy, R, Phase transfer catalysis toward the synthesis of *O*-, *S*-, *Se*- and *C*-Glycosides, *J. Mol. Catal. A: Chemical*, 154, 9–22, 2000.
95. Zenkoh, T, Tanaka, H, Setoi, H, Takahashi, T, Solid-phase synthesis of aryl *O*-glycoside using aqueous base and phase transfer catalyst, *Synlett*, 867–870, 2002.
96. Hamann, C H, Polligkeit, H, Wolf, P, Smiatacz, Z, An electrochemical synthesis of methyl α-isomaltoside, *Carbohydr. Res.*, 265, 1–7, 1994.
97. Nishizawa, M, Kan, Y, Yamada, H, A simple metal free 2′-discriminated glucosidation procedure, *Tetrahedron Lett.*, 29, 4597–4598, 1988.
98. Nishizawa, M, Kan, Y, Shimomoto, W, Yamada, H, α-Selective thermal glycosidation of rhamnosyl and mannosyl chlorides, *Tetrahedron Lett.*, 31, 2431–2434, 1990.
99. Nishizawa, M, Imagawa, H, Kubo, K, Kan, Y, Yamada, H, Improved synthesis of α-cycloawaodorin, *Synlett*, 447–448, 1992.
100. Nishizawa, M, Shimomoto, W, Momii, F, Yamada, H, Stereoselective thermal glycosylation of 2-deoxy-2-acetoamino-3,4,6-tri-*O*-acetyl-α-D-glucopyranosyl chloride, *Tetrahedron Lett.*, 33, 1907–1908, 1992.
101. Dauben, W G, Köhler, P, High-pressure glycosylations of unreactive alcohols and the formation of *N*-glycosyl collidinium salts, *Carbohydr. Res.*, 203, 47–56, 1990.
102. Sasaki, M, Gama, Y, Yasumoto, M, Ishigami, Y, Glycosylation reaction under high-pressure, *Tetrahedron Lett.*, 31, 6549–6552, 1990.
103. Mukaiyama, T, Murai, Y, Shoda, S-I, An efficient method for glycosylation of hydroxy compounds using glucopyranosyl fluoride, *Chem. Lett.*, 431–432, 1981.
104. Shimizu, M, Togo, H, Yokoyama, M, Chemistry of glycosyl fluorides, *Synthesis*, 799–822, 1998.
105. Toshima, K, Glycosyl halides and anomeric esters as donors, In *Glycoscience — Chemistry and Chemical Biology*, Fraser-Reid, B, Tatsuta, K, Thiem, J, Eds., Springer Verlag, Berlin, Heidelberg, pp. 584–625, 2001.
106. Posner, G H, Haines, S R, A convenient, one-step, high-yield replacement of an anomeric hydroxyl group by a fluorine atom using DAST. Preparation of glycosyl fluorides, *Tetrahedron Lett.*, 26, 5–8, 1985.

107. Ogawa, T, Takahashi, Y, Glucan synthesis. 5. Total synthesis of α-cyclodextrin, *Carbohydr. Res.*, 138, C5–C9, 1985.
108. Takahashi, Y, Ogawa, T, Glucan synthesis. 6. Total synthesis of cyclomaltohexaose, *Carbohydr. Res.*, 164, 277–296, 1987.
109. Mukaiyama, T, Hashimoto, Y, Shoda, S, Stereoselective synthesis of 1,2-*cis*-glycofuranosides using glycofuranosyl fluorides, *Chem. Lett.*, 935–938, 1983.
110. Hashimoto, S, Hayashi, M, Noyori, R, Glycosylation using glucopyranosyl fluorides and silicon-based catalysts. Solvent dependency of the stereoselection, *Tetrahedron Lett.*, 25, 1379–1382, 1984.
111. Nicolaou, K C, Chucholowski, A, Dolle, R E, Randall, J L, Reactions of glycosyl fluorides. Synthesis of *O*-, *S*-, and *N*-glycosides, *J. Chem. Soc. Chem. Commun.*, 1155–1156, 1984.
112. Kunz, H, Sager, W, Stereoselective glycosylation of alcohols and silyl ethers using glycosyl fluorides and boron-trifluoride etherate, *Helv. Chim. Acta*, 68, 283–287, 1985.
113. Kunz, H, Waldmann, H, Directed stereoselective synthesis of α- and β-*N*-acetyl neuraminic acid-galactose disaccharides using 2-chloro and 2-fluoro derivatives of neuraminic acid allyl ester, *J. Chem. Soc. Chem. Commun.*, 638–640, 1985.
114. Vozny, Y V, Galoyan, A A, Chizhov, O S, A novel method for *O*-glycoside bond formation — reaction of glycosylfluorides with trimethylsilyl ethers, *Bioorg. Khim.*, 11, 276–278, 1985.
115. Kreuzer, M, Thiem, J, Aufbau von Oligosacchariden mit Glycosylfluoriden unter Lewissäure-Katalyse, *Carbohydr. Res.*, 149, 347–361, 1986.
116. Jünnemann, J, Lundt, I, Thiem, J, 2-Deoxyglycosyl fluorides in oligosaccharide synthesis, *Liebigs Ann. Chem.*, 759–764, 1991.
117. Matsumoto, T, Maeta, H, Suzuki, K, Tsuchihashi, G, New glycosidation reaction 1. Combinational use of Cp_2ZrCl_2–$AgClO_4$ for activation of glycosyl fluorides and application to highly β-selective glycosidation of D-mycinose, *Tetrahedron Lett.*, 29, 3567–3570, 1988.
118. Suzuki, K, Maeta, H, Matsumoto, T, Tsuchihashi, G, New glycosidation reaction 2. Preparation of 1-fluoro-D-desosamine derivative and its efficient glycosidation by the use of Cp_2HfCl_2–$AgClO_4$ as the activator, *Tetrahedron Lett.*, 29, 3571–3574, 1988.
119. Matsumoto, T, Maeta, H, Suzuki, K, Tsuchihashi, G, First total synthesis of mycinamicin IV and VII. Successful application of new glycosidation reaction, *Tetrahedron Lett.*, 29, 3575–3578, 1988.
120. Suzuki, K, Maeta, H, Suzuki, T, Matsumoto, T, Cp_2ZrCl_2–$AgBF_4$ in benzene: a new reagent system for rapid and highly selective α-mannoside synthesis from tetra-*O*-benzyl-D-mannosyl fluoride, *Tetrahedron Lett.*, 30, 6879–6882, 1989.
121. Kobayashi, S, Koide, K, Ohno, M, Gallium reagents in organic synthesis: dimethylgallium chloride and triflate as activators in glycosidation using glycopyranosyl fluorides, *Tetrahedron Lett.*, 31, 2435–2438, 1990.
122. Maeta, H, Matsumoto, T, Suzuki, K, Dibutyltin diperchlorate for activation of glycosyl fluoride, *Carbohydr. Res.*, 249, 49–56, 1993.
123. Nakahara, Y, Ogawa, T, Synthesis of (1→4)-linked galacturonic acid trisaccharides, a proposed plant wound-hormone and a stereoisomer, *Carbohydr. Res.*, 200, 363–375, 1990.
124. Wessel, H P, Comparison of catalysts in α-glucosylation reactions and identification of triflic anhydride as a new reactive promoter, *Tetrahedron Lett.*, 31, 6863–6866, 1990.
125. Wessel, H P, Ruiz, N, α-Glucosylation reactions with 2,3,4,6-tetra-*O*-benzyl-β-D-glucopyranosyl fluoride and triflic anhydride as promoter, *J. Carbohydr. Chem.*, 10, 901–910, 1991.
126. Böhm, G, Waldmann, H, Synthesis of glycosides of fucose under neutral conditions in solutions of $LiClO_4$ in organic-solvents, *Tetrahedron Lett.*, 36, 3843–3846, 1995.
127. Kim, W-S, Hosono, S, Sasai, H, Shibasaki, M, Rare-earth perchlorate catalyzed glycosidation of glycosyl fluorides with trimethylsilyl ethers, *Tetrahedron Lett.*, 36, 4443–4446, 1995.
128. Hosono, S, Kim, W-S, Sasai, H, Shibasaki, M, A new glycosylation procedure utilizing rare-earth salts and glycosyl fluorides, with or without the requirement of Lewis acids, *J. Org. Chem.*, 60, 4–5, 1995.
129. Kim, W-S, Hosono, S, Sasai, H, Shibasaki, M, Rare earth salts promoted glycosidation of glycosyl fluorides, *Heterocycles*, 42, 795–809, 1996.
130. Kim, W-S, Sasai, H, Shibasaki, M, β-Selective glycosylation with α-mannosyl fluorides using tin(II) triflate and lanthanum perchlorate, *Tetrahedron Lett.*, 37, 7797–7800, 1996.

131. Toshima, K, Kasumi, K, Matsumura, S, Novel stereocontrolled glycosidations of 2-deoxyglucopyranosyl fluoride using a heterogeneous solid acid, sulfated zirconia (SO_4/ZrO_2), *Synlett*, 813–815, 1999.
132. Toshima, K, Kasumi, K, Matsumura, S, Novel stereocontrolled glycosidations using a solid acid, SO_4/ZrO_2, for direct syntheses of α- and β-mannopyranosides, *Synlett*, 643–645, 1998.
133. Jona, H, Takeuchi, K, Mukiyama, T, 1,2-*cis* Selective glycosylation with glycosyl fluoride by using a catalytic amount of trifluoromethanesulfonic acid (TfOH) in the coexistence of molecular sieve 5 Å (MS5 Å), *Chem. Lett.*, 1278–1279, 2000.
134. Mukaiyama, T, Takeuchi, K, Jona, H, Maeshima, H, Saitoh, T, A catalytic and stereoselective glycosylation with β-glycosyl fluorides, *Helv. Chim. Acta*, 83, 1901–1918, 2000.
135. Jona, H, Mandai, H, Mukaiyama, T, A catalytic and stereoselective glycosylation with glucopyranosyl fluoride by using various protic acids, *Chem. Lett.*, 426–427, 2001.
136. Takeuchi, K, Mukaiyama, T, Trityl tetrakis(pentafluorophenyl)borate catalyzed stereoselective glycosylation using glycopyranosyl fluoride as a glycosyl donor, *Chem. Lett.*, 555–556, 1998.
137. Yanagisawa, M, Mukaiyama, T, Catalytic and stereoselective glycosylation with glycosyl fluoride using active carbocationic species paired with tetrakis(pentafluorophenyl)borate or trifluoromethanesulfonate, *Chem. Lett.*, 224–225, 2001.
138. Mukaiyama, T, Maeshima, H, Jona, H, Catalytic and stereoselective glycosylation with disarmed glycosyl fluoride by using a combination of stannous(II) chloride ($SnCl_2$) and silver tetrakis(pentafluorophenyl)borate [$AgB(C_6F_5)_4$] as a catalyst, *Chem. Lett.*, 388–389, 2001.
139. Jona, H, Maeshima, H, Mukaiyama, T, Catalytic and stereoselective glycosylation with disarmed glycosyl fluoride having phthaloyl or dichlorophthaloyl protected amino function using 1:2 combination of stannic chloride and silver tetrakis(pentafluorophenyl)borate, *Chem. Lett.*, 726–727, 2001.
140. Vozny, Ya V, Kalicheva, I S, Galoyan, A A, Sugar fluorides as glycosylating agents. 2. Synthesis of aromatic glycosides using boron-trifluoride etherate, *Bioorg. Khim.*, 10, 1256–1259, 1984.
141. Matsumoto, T, Katsuki, M, Suzuki, K, Rapid *O*-glycosidation of phenols with glycosyl fluoride by using the combinational activator, Cp_2HfCl_2-$AgClO_4$, *Chem. Lett.*, 437–440, 1989.
142. Yamaguchi, M, Hiroguchi, A, Fukuda, A, Minami, T, Novel synthesis of aryl 2,3,4,6-tetra-*O*-acetyl-D-glucopyranosides, *J. Chem. Soc. Perkin. Trans. 1*, 1079–1082, 1990.
143. Oyama, K, Kondo, T, Highly efficient β-glucosylation of the acidic hydroxyl groups, phenol and carboxylic acid, with an peracetylated glucosyl fluoride using a combination of $BF_3{\cdot}Et_2O$ and DTBMP as a promoter, *Synlett*, 1627–1629, 1999.
144. Nicolaou, K C, Dolle, R E, Papahatjis, D P, Practical synthesis of oligosaccharides. Partial synthesis of avermectin B1a, *J. Am. Chem. Soc.*, 106, 4189–4192, 1984.
145. Nicolaou, K C, Mitchell, H J, Fylaktakidou, K C, Suzuki, H, Rodriguez, R M, 1,2-Seleno migrations in carbohydrate chemistry: solution and solid-phase synthesis of 2-deoxy glycosides, orthoesters, and allyl orthoesters, *Angew. Chem. Int. Ed.*, 39, 1089–1093, 2000.
146. Nicolaou, K C, Rodriguez, R M, Mitchell, H J, Suzuki, H, Fylaktakidou, K C, Baudoin, O, van Delft, F L, Total synthesis of everninomicin 13,384-1 — part 1: retrosynthetic analysis and synthesis of the $A_1B(A)C$ fragment, *Chem. Eur. J.*, 6, 3095–3115, 2000.
147. Nicolaou, K C, Mitchell, H J, Fylaktakidou, K C, Rodriguez, R M, Suzuki, H, Total synthesis of everninomicin 13,384-1 — part 2: synthesis of the $FGHA_2$ fragment, *Chem. Eur. J.*, 6, 3116–3148, 2000.
148. Nicolaou, K C, Mitchell, H J, Rodriguez, R M, Fylaktakidou, K C, Suzuki, H, Conley, S R, Total synthesis of everninomicin 13,384-1 — part 3: synthesis of the DE fragment and completion of the total synthesis, *Chem. Eur. J.*, 6, 3149–3165, 2000.
149. Nicolaou, K C, Fylaktakidou, K C, Mitchell, H J, van Delft, F L, Rodriguez, R M, Conley, S R, Jin, Z, Total synthesis of everninomicin 13,384-1 — part 4: explorations of methodology; stereocontrolled synthesis of 1,1′-disaccharides, 1,2-seleno migrations in carbohydrates, and solution- and solid-phase synthesis of 2-deoxy glycosides and orthoesters, *Chem. Eur. J.*, 6, 3166–3185, 2000.
150. Barrena, M I, Echarri, R, Castillon, S, Synthesis of disaccharides by selective metallocene promoted activation of glycosyl fluorides, *Synlett*, 675–676, 1996.
151. Fischer, E, Fischer, H, Über einige Derivate des Milchzuckers und der Maltose und über zwei neue Glucoside, *Ber. Dtsch. Chem. Ges.*, 43, 2521–2536, 1910.

152. Thiem, J, Meyer, B, Synthesen mit Iod- und Bromtrimethylsilan in der Saccharidchemie, *Chem. Ber.*, 113, 3075–3085, 1980.
153. Gervay, J, Nguyen, T N, Hadd, M J, Mechanistic studies on the stereoselective formation of glycosyl iodides: first characterization of β-D-glycosyl iodides, *Carbohydr. Res.*, 300, 119–125, 1997.
154. Caputo, R, Kunz, H, Mastroianni, D, Palumbo, G, Pedatella, S, Solla, F, Mild synthesis of protected α-D-glycosyl iodides, *Eur. J. Org. Chem.*, 3147–3150, 1999.
155. Chervin, S M, Abada, P, Koreeda, M, Convenient, in situ generation of anhydrous hydrogen iodide for the preparation of α-glycosyl iodides and vicinal iodohydrins and for the catalysis of Ferrier glycosylation, *Org. Lett.*, 2, 369–372, 2000.
156. Bickley, J, Cottrell, J A, Ferguson, J R, Field, R A, Harding, J R, Hughes, D L, Kartha, K P. R, Law, J L, Scheinmann, F, Stachulski, A V, Preparation, X-ray structure and reactivity of a stable glycosyl iodide, *Chem. Commun.*, 1266–1267, 2003.
157. de la Fuente, J M, Penades, S, Synthesis of Le^x-neoglycoconjugate to study carbohydrate–carbohydrate associations and its intramolecular interaction, *Tetrahedron: Asymmetry*, 13, 1879–1888, 2002.
158. Schmid, U, Waldmann, H, *O*-Glycoside synthesis with glycosyl iodides under neutral conditions in 1 M $LiClO_4$ in CH_2Cl_2, *Liebigs Ann. Chem.*, 2573–2577, 1997.
159. Tanaka, H, Sakamoto, H, Sano, A, Nakamura, S, Nakajima, M, Hashimoto, S, An extremely mild and stereocontrolled construction of 1,2-*cis*-α-glycosidic linkages via benzyl-protected glycopyranosyl diethyl phosphites, *Chem. Commun.*, 1259–1260, 1999.
160. Hadd, M J, Gervay, J, Glycosyl iodides are highly efficient donors under neutral conditions, *Carbohydr. Res.*, 320, 61–69, 1999.
161. Lam, S N, Gervay-Hague, J, Solution-phase hexasaccharide synthesis using glucosyl iodides, *Org. Lett.*, 4, 2039–2042, 2002.
162. Lam, S N, Gervay-Hague, J, Solution- and solid-phase oligosaccharide synthesis using glucosyl iodides: a comparative study, *Carbohydr. Res.*, 337, 1953–1965, 2002.
163. Mukaiyama, T, Kobashi, Y, Shintou, T, A new method for α-selective glycosylation using a donor, glycosyl methyldiphenylphosphonium iodide, without any assistance of acid promoters, *Chem. Lett.*, 32, 900–901, 2003.
164. Ferrier, R J, Hay, R W, Vethaviyasar, N, A potentially versatile synthesis of glycosides, *Carbohydr. Res.*, 27, 55–61, 1973.
165. Tsai, T Y R, Jin, H, Wiesner, K, A stereoselective synthesis of digitoxin. On cardioactive steroids. 13., *Can. J. Chem.*, 62, 1403–1405, 1984.
166. van Cleve, J W, Reinvestigation of the preparation of cholesteryl 2,3,4,6-tetra-*O*-benzyl-α-D-glucopyranoside, *Carbohydr. Res.*, 70, 161–164, 1979.
167. Garegg, P J, Henrichson, C, Norberg, T, A reinvestigation of glycosidation reactions using 1-thioglycosides as glycosyl donors and thiophilic cations as promoters, *Carbohydr. Res.*, 116, 162–165, 1983.
168. Hanessian, S, Bacquet, C, Lehong, N, Chemistry of the glycosidic linkage. Exceptionally fast and efficient formation of glycosides by remote activation, *Carbohydr. Res.*, 80, C17–C22, 1980.
169. Mukaiyama, T, Nakatsuka, T, Shoda, S, Efficient glucosylation of alcohol using 1-thioglucoside derivative, *Chem. Lett.*, 487–490, 1979.
170. Woodward, R B, Logusch, E, Nambiar, K P, Sakan, K, Ward, D E, Au-Yeung, B-W, Balaram, P, Browne, L J, Card, P J, Chen, C H, Chenevert, R B, Fliri, A, Frobel, K, Gais, H-J, Garratt, D G, Hayakawa, K, Heggie, W, Hesson, D P, Hoppe, D, Hoppe, I, Hyatt, J A, Ikeda, D, Jacobi, P A, Kim, K S, Kobuke, Y, Kojima, K, Krowicki, K, Lee, V J, Leutert, T, Malchenko, S, Martens, J, Matthews, R S, Ong, B S, Press, J B, Babu, T V R, Rousseau, G, Sauter, H M, Suzuki, M, Tatsuta, K, Tolbert, L M, Truesdale, E A, Uchida, I, Ueda, Y, Uyehara, T, Vasella, A T, Vladuchick, W C, Wade, P A, Williams, R M, Wong, H N C, Asymmetric total synthesis of erythromycin. 3. Total synthesis of erythromycin, *J. Am. Chem. Soc.*, 103, 3215–3217, 1981.
171. Tsuboyama, K, Takeda, K, Torii, K, Ebihara, M, Shimizu, J, Suzuki, A, Sato, N, Furuhata, K, Ogura, H, A convenient synthesis of *S*-glycosyl donors of D-glucose and *O*-glycosylations involving the new reagent, *Chem. Pharm. Bull.*, 38, 636–638, 1990.

172. Takeda, K, Tsuboyama, K, Torii, K, Furuhata, K, Sato, N, Ogura, H, A convenient synthesis of *S*-glycosyl donors of sialic acid and their use for *O*-glycosylation, *Carbohydr. Res.*, 203, 57–63, 1990.
173. Wuts, P G M, Bigelow, S S, Total synthesis of oleandrose and the avermectin disaccharide, benzyl α-L-oleandrosyl-α-L-4-acetoxyoleandroside, *J. Org. Chem.*, 48, 3489–3493, 1983.
174. Lear, M J, Yoshimura, F, Hirama, M, A direct and efficient α-selective glycosylation protocol for the kedarcidin sugar, L-mycarose: $AgPF_6$ as a remarkable activator of 2-deoxythioglycosides, *Angew. Chem. Int. Ed.*, 40, 946–949, 2001.
175. Lönn, H, Synthesis of a tri-saccharide and a hepta-saccharide which contain α-L-fucopyranosyl groups and are part of the complex type of carbohydrate moiety of glycoproteins, *Carbohydr. Res.*, 139, 105–113, 1985.
176. Lönn, H, Synthesis of a tetra-saccharide and a nona-saccharide which contain α-L-fucopyranosyl groups and are part of the complex type of carbohydrate moiety of glycoproteins, *Carbohydr. Res.*, 139, 115–121, 1985.
177. Lönn, H, Glycosylation using a thioglycoside and methyl trifluoromethanesulfonate — a new and efficient method for *cis* and *trans* glycoside formation, *J. Carbohydr. Chem.*, 6, 301–306, 1987.
178. Reddy, G V, Kulkarni, V R, Mereyala, H B, A mild general method for the synthesis of α-linked disaccharides, *Tetrahedron Lett.*, 30, 4283–4286, 1989.
179. Mereyala, H B, Reddy, G V, Stereoselective synthesis of α-linked saccharides by use of per-*O*-benzylated 2-pyridyl-1-thio hexopyranosides as glycosyl donors and methyl iodide as an activator, *Tetrahedron*, 47, 6435–6448, 1991.
180. Ravi, D, Kulkarni, V R, Mereyala, H B, A mild general-method for the synthesis of α-2-deoxy-disaccharides: synthesis of L-oleandrosyl-L-oleandrose from D-glucose, *Tetrahedron Lett.*, 30, 4287–4290, 1989.
181. Mereyala, H B, Reddy, G V, Directed, iterative, stereoselective synthesis of oligosaccharides by use of suitably 2-*O*-substituted 2-pyridyl 1-thioglycopyranosides on activation by methyl iodide, *Tetrahedron*, 47, 9721–9726, 1991.
182. Mereyala, H B, Kulkarni, V R, Ravi, D, Sharma, G V M, Rao, B V, Reddy, G B, Stereoselective synthesis of α-linked 2-deoxysaccharides and furanosaccharides by use of 2-deoxy-2-pyridyl-1-thiopyranosides and 2-deoxy-2-pyridyl-1-thiofuranosides as donors and methyl iodide as an activator, *Tetrahedron*, 48, 545–562, 1992.
183. Fügedi, P, Garegg, P J, A novel promoter for the efficient construction of 1,2-*trans* linkages in glycoside synthesis, using thioglycosides as glycosyl donors, *Carbohydr. Res.*, 149, C9–C12, 1986.
184. Andersson, F, Birberg, W, Fügedi, P, Garegg, P J, Nashed, M, Pilotti, Å, Dimethyl(methylthio)-sulfonium triflate as a promoter for creating glycosidic linkages in oligosaccharide synthesis, *ACS. Symp. Ser.*, 386, 117–130, 1989.
185. Fügedi, P, Dimethyl(methylthio)sulfonium trifluoromethanesulfonate, In *e-EROS, Electronic Encyclopedia of Reagents for Organic Synthesis*, Paquette, L A, Ed., Wiley Interscience, New York, 2002, http://www.mrw.interscience.wiley.com/eros/eros_articles_fs.html.
186. Murase, T, Ishida, H, Kiso, M, Hasegawa, A, A facile regio- and stereo-selective synthesis of α-glycosides of *N*-acetylneuraminic acid, *Carbohydr. Res.*, 184, C1–C4, 1988.
187. Hasegawa, A, Ogawa, M, Kojima, Y, Kiso, M, Synthetic studies on sialoglycoconjugates 36: α-selective glycoside synthesis of *N*-acetylneuraminic acid with the secondary hydroxyl group in D-glucopyranose, 2-acetamido-2-deoxy-D-glucopyranose and D-galactopyranose derivatives, *J. Carbohydr. Chem.*, 11, 333–341, 1992.
188. Oscarson, S, Sehgelmeble, F W, A novel β-directing fructofuranosyl donor concept. Stereospecific synthesis of sucrose, *J. Am. Chem. Soc.*, 122, 8869–8872, 2000.
189. Dasgupta, F, Garegg, P J, Alkyl sulfenyl triflate as activator in the thioglycoside-mediated formation of β-glycosidic linkages, *Carbohydr. Res.*, 177, C13–C17, 1988.
190. Dasgupta, F, Garegg, P J, Use of the methylsulfenyl cation as an activator for glycosylation reactions with alkyl (aryl) 1-thioglycopyranosides: synthesis of methyl *O*-(2-acetamido-2-deoxy-β-D-glucopyranosyl)-(1→6)-*O*-α-D-glucopyranosyl-(1→2)-α-D-glucopyranoside, a derivative of the core trisaccharide of *E. coli* K12, *Carbohydr. Res.*, 202, 225–238, 1990.

191. Martichonok, V, Whitesides, G M, Stereoselective α-sialylation with sialyl xanthate and phenylsulfenyl triflate as a promotor, *J. Org. Chem.*, 61, 1702–1706, 1996.
192. Crich, D, Sun, S, Direct chemical synthesis of β-mannopyranosides and other glycosides via glycosyl triflates, *Tetrahedron*, 54, 8321–8348, 1998.
193. Crich, D, Sun, S, Direct formation of β-mannopyranosides and other hindered glycosides from thioglycosides, *J. Am. Chem. Soc.*, 120, 435–436, 1998.
194. Jona, H, Takeuchi, K, Saitoh, T, Mukaiyama, T, Effective activation of armed thioglycoside with a new combination of trityl tetrakis(pentafluorophenyl)borate [$TrB(C_6F_5)_4$] and *N*-(ethylthio)phthalimide (PhthNSEt), *Chem. Lett.*, 1178–1179, 2000.
195. Ito, Y, Ogawa, T, Sulfenate esters as glycosyl acceptors: a novel approach to *O*-glycosides from thioglycosides and sulfenate esters, *Tetrahedron Lett.*, 28, 4701–4704, 1987.
196. Crich, D, Smith, M, *S*-(4-Methoxyphenyl) benzenethiosulfinate (MPBT)/trifluoromethanesulfonic anhydride: a convenient system for the generation of glycosyl triflates from thioglycosides, *Org. Lett.*, 2, 4067–4069, 2000.
197. Crich, D, Smith, M, 1-Benzenesulfinyl piperidine/trifluoromethanesulfonic anhydride: a potent combination of shelf-stable reagents for the low-temperature conversion of thioglycosides to glycosyl triflates and for the formation of diverse glycosidic linkages, *J. Am. Chem. Soc.*, 123, 9015–9020, 2001.
198. Codée, J D C, Litjens, R E J N, den Heeten, R, Overkleeft, H S, van Boom, J H, van der Marel, G A, Ph_2SO/Tf_2O: a powerful promotor system in chemoselective glycosylations using thioglycosides, *Org. Lett.*, 5, 1519–1522, 2003.
199. Ito, Y, Ogawa, T, Benzeneselenenyl triflate as a promoter of thioglycosides: a new method for *O*-glycosylation using thioglycosides, *Tetrahedron Lett.*, 29, 1061–1064, 1988.
200. Ito, Y, Ogawa, T, Numata, M, Sugimoto, M, Benzeneselenenyl triflate as an activator of thiglycosides for glycosylation reactions, *Carbohydr. Res.*, 202, 165–175, 1990.
201. Shimizu, H, Ito, Y, Ogawa, T, PhSeNPhth–TMSOTf as a promotor of thioglycoside, *Synlett*, 535–536, 1994.
202. Fukase, K, Nakai, Y, Kanoh, T, Kusumoto, S, Mild but efficient methods for stereoselective glycosylation with thioglycosides: activation by [*N*-phenylselenophthalimide-$Mg(ClO_4)_2$] and [PhIO-$Mg(ClO_4)_2$], *Synlett*, 84–86, 1998.
203. Nicolaou, K C, Seitz, S P, Papahatjis, D P, A mild and general-method for the synthesis of *O*-glycosides, *J. Am. Chem. Soc.*, 105, 2430–2434, 1983.
204. Sasaki, M, Tachibana, K, Nakanishi, H, An efficient and stereocontrolled synthesis of the nephritogenoside core structure, *Tetrahedron Lett.*, 32, 6873–6876, 1991.
205. Fukase, K, Hasuoka, A, Kinoshita, I, Aoki, Y, Kusumoto, S, A stereoselective glycosidation using thioglycosides, activation by combination of *N*-bromosuccinimide and strong acid salts, *Tetrahedron*, 51, 4923–4932, 1995.
206. Egusa, K, Fukase, K, Nakai, Y, Kusumoto, S, Stereoselective glycosylation and oligosaccharide synthesis on solid support using a 4-azido-3-chlorobenzyl group for temporary protection, *Synlett*, 27–32, 2000.
207. Takeuchi, K, Tamura, T, Mukaiyama, T, Stereoselective glycosylation of thioglycosides promoted by respective combinations of *N*-iodo- or *N*-bromosuccinimide and trityl tetrakis(pentafluorophenyl)borate. Application to one-pot sequential synthesis of trisaccharide, *Tetrahedron Lett.*, 124–125, 2000.
208. Veeneman, G H, van Boom, J H, An efficient thioglycoside-mediated formation of α-glycosidic linkages promoted by iodonium dicollidine perchlorate, *Tetrahedron Lett.*, 31, 275–278, 1990.
209. Zuurmond, H M, van der Laan, S C, van der Marel, G A, van Boom, J H, Iodonium ion-assisted glycosylation of alkyl (aryl) 1-thio-glycosides: regulation of stereoselectivity and reactivity, *Carbohydr. Res.*, 215, C1–C3, 1991.
210. Smid, P, de Ruiter, G A, van der Marel, G A, Rombouts, F M, van Boom, J H, Iodonium-ion assisted stereospecific glycosylation: synthesis of oligosaccharides containing α(1→4)-linked L-fucopyranosyl units, *J. Carbohydr. Chem.*, 10, 833–849, 1991.
211. Veeneman, G H, van Leeuwen, S H, Zuurmond, H, van Boom, J H, Synthesis of carbohydrate-antigenic structures of *Mycobacterium tuberculosis* using iodonium ion promoted glycosidation approach, *J. Carbohydr. Chem.*, 9, 783–796, 1990.

212. Veeneman, G H, van Leeuwen, S H, van Boom, J H, Iodonium ion promoted reactions at the anomeric centre. II. An efficient thioglycoside mediated approach toward the formation of 1,2-*trans* linked glycosides and glycosidic esters, *Tetrahedron Lett.*, 31, 1331–1334, 1990.
213. Konradsson, P, Mootoo, D R, McDevitt, R E, Fraser-Reid, B, Iodonium ion generated *in situ* from *N*-iodosuccinimide and trifluoromethanesulphonic acid promotes direct linkage of disarmed pent-4-enyl glycosides, *J. Chem. Soc. Chem. Commun.*, 270–272, 1990.
214. Konradsson, P, Udodong, U E, Fraser-Reid, B, Iodonium promoted reactions of disarmed thioglycosides, *Tetrahedron Lett.*, 31, 4313–4316, 1990.
215. Chung, S-K, Park, K-H, A novel approach to the stereoselective synthesis of β-D-mannopyranosides, *Tetrahedron Lett.*, 42, 4005–4007, 2001.
216. Imai, H, Oishi, T, Kikuchi, T, Hirama, M, Concise synthesis of 3-*O*-(2-*O*-α-D-glucopyranosyl-6-*O*-acyl-α-D-glucopyranosyl)-1,2-di-*O*-acyl-*sn*-glycerols, *Tetrahedron*, 56, 8451–8459, 2000.
217. Aloui, M, Fairbanks, A J, *N*-Iodosaccharin: a potent new activator of thiophenylglycosides, *Synlett*, 797–799, 2001.
218. Fukase, K, Hasuoka, A, Kinoshita, I, Kusumoto, S, Iodosobenzene-triflic anhydride as an efficient promoter for glycosidation reaction using thioglycosides as donors, *Tetrahedron Lett.*, 33, 7165–7168, 1992.
219. Sun, L, Zhao, K, Stabilization of glycosyl sulfonium ions for stereoselective *O*-glycosylation, *Tetrahedron Lett.*, 35, 7147–7150, 1994.
220. Fukase, K, Kinoshita, I, Kanoh, T, Nakai, Y, Hasuoka, A, Kusumoto, S, A novel method for stereoselective glycosidation with thioglycosides: promotion by hypervalent iodine reagents prepared from PhIO and various acids, *Tetrahedron*, 52, 3897–3904, 1996.
221. Fukase, K, Nakai, Y, Egusa, K, Porco, J A, Kusumoto, S, A novel oxidatively removable linker and its application to α-selective solid-phase oligosaccharide synthesis on a macroporous polystyrene support, *Synlett*, 1074–1078, 1999.
222. Kartha, K P R, Aloui, M, Field, R A, Iodine: a versatile reagent in carbohydrate chemistry II. Efficient chemospecific activation of thiomethylglycosides, *Tetrahedron Lett.*, 37, 5175–5178, 1996.
223. Cura, P, Aloui, M, Kartha, K P R, Field, R A, Iodine and its interhalogen compounds: versatile reagents in carbohydrate chemistry XII. Tuning promoter reactivity for thioglycoside activation, *Synlett*, 1279–1280, 2000.
224. Kartha, K P R, Cura, P, Aloui, M, Readman, S K, Rutherford, T J, Field, R A, Observations on the activation of methyl thioglycosides by iodine and its interhalogen compounds, *Tetrahedron: Asymmetry*, 11, 581–593, 2000.
225. Ercegovic, T, Meijer, A, Magnusson, G, Ellervik, U, Iodine monochloride/silver trifluoromethanesulfonate (ICl/AgOTf) as a convenient promoter system for *O*-glycoside synthesis, *Org. Lett.*, 3, 913–916, 2001.
226. Meijer, A, Ellervik, U, Study of interhalogens/silver trifluoromethanesulfonate as promoter systems for high-yielding sialylations, *J. Org. Chem.*, 67, 7407–7412, 2002.
227. Takeuchi, K, Tamura, T, Jona, H, Mukaiyama, T, A novel activating agents of disarmed thioglycosides, combination of trityl tetrakis(pentafluorophenyl)borate, iodine and 2,3-dichloro-5,6-dicyano-*p*-benzoquinone (DDQ), *Chem. Lett.*, 692–693, 2000.
228. Lönn, H, Sulfurylchloride trifluoromethanesulfonic acid — a novel promoter system for glycoside synthesis using thioglycosides as glycosyl donors, *Glycoconjugate J.*, 4, 117–118, 1987.
229. Kallin, E, Lönn, H, Norberg, T, Glycosidations with thioglycosides activated by sulfuryl chloride trifluoromethanesulfonic acid — synthesis of a human-blood group-B trisaccharide glycoside, *Glycoconjugate J.*, 5, 3–8, 1988.
230. Burkart, M D, Zhang, Z, Hung, S-C, Wong, C-H, A new method for the synthesis of fluoro-carbohydrates and glycosides using selectfluor, *J. Am. Chem. Soc.*, 119, 11743–11746, 1997.
231. Tsukamoto, H, Kondo, Y, 1-Fluoropyridinium triflates: versatile reagents for transformation of thioglycoside into *O*-glycoside, glycosyl azide and sulfoxide, *Tetrahedron Lett.*, 44, 5247–5249, 2003.
232. Weygand, F, Ziemann, H, Bestmann, H J, Eine neue Methode zur Darstellung von Acetobromzuckern, *Liebigs Ann. Chem.*, 91, 2534–2537, 1958.

233. Weygand, F, Ziemann, H, Glykosylbromide aus Äthylthioglykosiden, II, *Liebigs Ann. Chem.*, 657, 179–198, 1962.
234. Koto, S, Uchida, T, Zen, S, Syntheses of isomaltose, isomaltotetraose, and isomaltooctaose, *Chem. Lett.*, 1049–1052, 1972.
235. Kihlberg, J O, Leigh, D A, Bundle, D R, The in situ activation of thioglycosides with bromine: an improved glycosylation method, *J. Org. Chem.*, 55, 2860–2863, 1990.
236. Pozsgay, V, Jennings, H J, A new method for the synthesis of *O*-glycosides from *S*-glycosides, *J. Org. Chem.*, 52, 4635–4637, 1987.
237. Pozsgay, V, Jennings, H J, Synthetic oligosaccharides related to group B streptococcal polysaccharides. 3. Synthesis of oligosaccharides corresponding to the common polysaccharide antigen of group B streptococci, *J. Org. Chem.*, 53, 4042–4052, 1988.
238. Marra, A, Mallet, J-M, Amatore, C, Sinaÿ, P, Glycosylation using a one-electron-transfer homogeneous reagent: a novel and efficient synthesis of β-linked disaccharides, *Synlett*, 572–574, 1990.
239. Zhang, Y-M, Mallet, J-M, Sinaÿ, P, Glycosylation using a one-electron-transfer, homogeneous reagent. Application to an efficient synthesis of the trimannosyl core of *N*-glycosylproteins, *Carbohydr. Res.*, 236, 73–88, 1992.
240. Uchiro, H, Mukaiyama, T, An efficient method for catalytic and stereoselective glycosylation with thioglycosides promoted by trityl tetrakis(pentafluorophenyl)borate and sodium periodate, *Chem. Lett.*, 121–122, 1997.
241. Amatore, C, Jutand, A, Mallet, J-M, Meyer, G, Sinaÿ, P, Electrochemical glycosylation using phenyl *S*-glycosides, *J. Chem. Soc. Chem. Commun.*, 718–719, 1990.
242. Balavoine, G, Gref, A, Fischer, J, Lubineau, A, Anodic glycosylation from aryl thioglycosides, *Tetrahedron Lett.*, 31, 5761–5764, 1990.
243. Mallet, J-M, Meyer, G, Yvelin, F, Jutand, A, Amatore, C, Sinaÿ, P, Electrosynthesis of disaccharides from phenyl or ethyl 1-thioglycosides, *Carbohydr. Res.*, 244, 237–246, 1993.
244. Gama, Y, Yasumoto, M, A highly stereoselective synthesis of α-glucosides from 1-thioglucoside derivative under high pressure, *Chem. Lett.*, 319–322, 1993.
245. Chen, Q, Kong, F, Stereoselective glycosylation using fully benzylated pyrimidin-2-yl 1-thio-β-D-glycopyranosides, *Carbohydr. Res.*, 272, 149–158, 1995.
246. Demchenko, A V, Kamat, M N, De Meo, C, S-Benzoxazolyl (SBox) glycosides in oligosaccharide synthesis: novel glycosylation approach to the synthesis of β-D-glucosides, β-D-galactosides, and α-D-mannosides, *Synlett*, 1287–1290, 2003.
247. Demchenko, A V, Malysheva, N N, De Meo, C, S-Benzoxazolyl (SBox) glycosides as novel, versatile glycosyl donors for stereoselective 1,2-*cis* glycosylation, *Org. Lett.*, 5, 455–458, 2003.
248. Kahne, D, Walker, S, Cheng, Y, van Engen, D, Glycosylation of unreactive substrates, *J. Am. Chem. Soc.*, 111, 6881–6882, 1989.
249. Raghavan, S, Kahne, D, A one-step synthesis of the ciclamycin trisaccharide, *J. Am. Chem. Soc.*, 115, 1580–1581, 1993.
250. Alonso, I, Khiar, N, Martin-Lomas, M, A new promoter system for the sulfoxide glycosylation reaction, *Tetrahedron Lett.*, 37, 1477–1480, 1996.
251. Sliedregt, L A J M, van der Marel, G A, van Boom, J H, Trimethylsilyl triflate mediated chemoselective condensation of arylsulfenyl glycosides, *Tetrahedron Lett.*, 35, 4015–4018, 1994.
252. Nagai, H, Matsumura, S, Toshima, K, A novel promoter, heteropoly acid, mediated chemo- and stereoselective sulfoxide glycosidation reactions, *Tetrahedron Lett.*, 41, 10233–10237, 2000.
253. Wipf, P, Reeves, J T, Glycosylation via $Cp_2ZrCl_2/AgClO_4$-mediated activation of anomeric sulfoxides, *J. Org. Chem.*, 66, 7910–7914, 2001.
254. Kartha, K P R, Karkkainen, T S, Marsh, S J, Field, R A, Iodine and its interhalogen compounds: versatile reagents in carbohydrate chemistry XIII. General activation of armed glycosyl donors, *Synlett*, 260–262, 2001.
255. Marsh, S J, Kartha, K P R, Field, R A, Iodine: a versatile reagent in carbohydrate chemistry. Part XV. Observations on iodine-promoted β-mannosylation, *Synlett*, 1376–1378, 2003.
256. Nagai, H, Kawahara, K, Matsumura, S, Toshima, K, Novel stereocontrolled α- and β-glycosidations of mannopyranosyl sulfoxides using environmentally benign heterogeneous solid acids, *Tetrahedron Lett.*, 42, 4159–4162, 2001.

257. Brown, D S, Ley, S V, Vile, S, Preparation of cyclic ether acetals from 2-benzenesulfonyl derivatives: a new mild glycosidation procedure, *Tetrahedron Lett.*, 29, 4873–4876, 1988.
258. Brown, D S, Ley, S V, Vile, S, Thompson, M, Use of 2-phenylsulfonyl cyclic ethers in the preparation of tetrahydropyran and tetrahydrofuran acetals and in some glycosidation reactions, *Tetrahedron*, 47, 1329–1342, 1991.
259. Chang, G X, Lowary, T L, A glycosylation protocol based on activation of glycosyl 2-pyridyl sulfones with samarium triflate, *Org. Lett.*, 2, 1505–1508, 2000.
260. Marra, A, Sinaÿ, P, A novel stereoselective synthesis of *N*-acetyl-α-neuraminosyl-galactose disaccharide derivatives, using anomeric *S*-glycosyl xanthates, *Carbohydr. Res.*, 195, 303–308, 1990.
261. Marra, A, Gauffeny, F, Sinaÿ, P, A novel class of glycosyl donors: anomeric *S*-xanthates of 2-azido-2-deoxy-D-galactopyranose derivatives, *Tetrahedron*, 47, 5149–5160, 1991.
262. Sinaÿ, P, Recent advances in glycosylation reactions, *Pure Appl. Chem.*, 63, 519–528, 1991.
263. Lönn, H, Stenvall, K, Exceptionally high yield in glycosylation with sialic acid. Synthesis of a GM_3 glycoside, *Tetrahedron Lett.*, 33, 115–116, 1992.
264. Bogusiak, J, Szeja, W, Synthesis of glycofuranosides from *S*-glycofuranosyl dithiocarbonates (xanthates) and dithiocarbamates, *Carbohydr. Res.*, 295, 235–243, 1996.
265. Bogusiak, J, Szeja, W, Studies on the synthesis of 1,2-*cis* pentofuranosides from *S*-glycofuranosyl dithiocarbamates, dithiocarbonates and phosphorodithioates, *Carbohydr. Res.*, 330, 141–144, 2001.
266. Sliedregt, L A J M, van Rosenberg, S M W, Autar, R, Valentijn, A R P M, van der Marel, G A, van Boom, J H, Piperi, C, van der Merwe, P A, Kuiper, J, van Berkel, T J C, Biessen, E A L, Design and synthesis of a multivalent homing device for targeting to murine CD22, *Bioorg. Med. Chem.*, 9, 85–97, 2001.
267. Fügedi, P, Garegg, P J, Oscarson, S, Rosen, G, Silwanis, B A, Glycosyl 1-piperidinecarbodithioates in the synthesis of glycosides, *Carbohydr. Res.*, 211, 157–162, 1991.
268. Pastuch, G, Wandzik, I, Szeja, W, (5-Nitro-2-pyridyl) 1-thio-β-D-glucopyranoside as a stable and reactive acceptor, *Tetrahedron Lett.*, 41, 9923–9926, 2000.
269. Bogusiak, J, Szeja, W, Block synthesis of oligosaccharides. Part 1: preparation of furanosyl-1-thiopyranosides, *Tetrahedron Lett.*, 42, 2221–2223, 2001.
270. Kochetkov, N K, Klimov, E M, Malysheva, N N, Novel highly stereospecific method of 1,2-*cis*-glycosylation. Synthesis of α-D-glucosyl-D-glucoses, *Tetrahedron Lett.*, 30, 5459–5462, 1989.
271. Kochetkov, N K, Klimov, E M, Malysheva, N M, Demchenko, A V, A new stereospecific method for 1,2-*cis*-glycosylation, *Carbohydr. Res.*, 212, 77–91, 1991.
272. Davis, B G, Ward, S J, Rendle, P M, Glycosyldisulfides: a new class of solution and solid phase glycosyl donors, *Chem. Commun.*, 189–190, 2001.
273. Chery, F, Cassel, S, Wessel, H P, Rollin, P, Synthesis of anomeric sulfinimides and their use as a new family of glycosyl donors, *Eur. J. Org. Chem.*, 171–180, 2002.
274. Bielawska, H, Michalska, M, 2-Deoxyglycosyl phosphorodithioates. A novel type of glycosyl donor. Efficient synthesis of 2′-deoxydisaccharides, *J. Carbohydr. Chem.*, 10, 107–112, 1991.
275. Laupichler, L, Sajus, H, Thiem, J, Convenient iodonium-promoted stereoselective synthesis of 2-deoxy-α-glycosides by use of *S*-(2-deoxyglycosyl) phosphorodithioates as donors, *Synthesis*, 1133–1136, 1992.
276. Bielawska, H, Michalska, M, Highly stereoselective synthesis of 2-deoxy-α-glycosides and α-disaccharides, *Tetrahedron Lett.*, 39, 9761–9764, 1998.
277. Plante, O J, Palmacci, E R, Andrade, R B, Seeberger, P H, Oligosaccharide synthesis with glycosyl phosphate and dithiophosphate triesters as glycosylating agents, *J. Am. Chem. Soc.*, 123, 9545–9554, 2001.
278. Hashimoto, S, Honda, T, Ikegami, S, An extremely mild and general method for the stereocontrolled construction of 1,2-*cis*-glycosidic linkages via *S*-glycopyranosyl phosphorodiamidimidothioates, *Tetrahedron Lett.*, 31, 4769–4772, 1990.
279. Hashimoto, S, Sakamoto, H, Honda, T, Ikegami, S, Oligosaccharide synthesis based on glycosyl donors and acceptors carrying phosphorus-containing leaving groups, *Tetrahedron Lett.*, 38, 5181–5184, 1997.
280. Pougny, J-R, Sinaÿ, P, Reaction d'imidates de glucopyranosyle avec l'acetonitrile. Application synthetiques, *Tetrahedron Lett.*, 4073–4076, 1976.

281. Pougny, J-R, Jacquinet, J-C, Nassr, M, Duchet, D, Milat, M-L, Sinaÿ, P, Novel synthesis of 1,2-*cis*-disaccharides, *J. Am. Chem. Soc.*, 99, 6762–6763, 1977.
282. Jacquinet, J-C, Sinaÿ, P, Synthese des substances de groupe sanguin. 9. Une synthese du 2-*O*-(α-L-fucopyranosyl)-3-*O*-(α-D-galactopyranosyl)-D-galactose, le determinant antigenique du groupe sanguin B, *Tetrahedron*, 35, 365–371, 1979.
283. Schmidt, R R, Michel, J, Facile synthesis of α- and β-*O*-glycosyl imidates; preparation of glycosides and disaccharides, *Angew. Chem. Int. Ed. Engl.*, 19, 731–732, 1980.
284. Schmidt, R R, Michel, J, Glycosylimidates. 11. Direct *O*-glycosyl trichloroacetimidate formation. Nucleophilicity of the anomeric oxygen atom, *Tetrahedron Lett.*, 25, 821–824, 1984.
285. Schmidt, R R, Michel, J, Roos, M, Glycosyl imidates. 12. Direct synthesis of *O*-α- and *O*-β-glycosyl imidates, *Liebigs Ann. Chem.*, 12, 1343–1357, 1984.
286. Schmidt, R R, Michel, J, Glycosylimidates. 17. *O*-(α-D-Glucopyranosyl)trichloroacetimidate as a glucosyl donor, *J. Carbohydr. Chem.*, 4, 141–169, 1985.
287. Schmidt, R R, Stumpp, M, Glycosylimidate, 10. Glycosylphosphate aus Glycosyl(trichloroacetimidaten), *Liebigs Ann. Chem.*, 680–691, 1984.
288. Schmidt, R R, Grundler, G, Glycosylimidates, 6. α-Linked disaccharides from *O*-(β-D-glycopyranosyl) trichloroacetimidates using trimethylsilyl trifluoromethanesulfonate as catalyst, *Angew. Chem. Int. Ed. Engl.*, 21, 781–782, 1982.
289. Grundler, G, Schmidt, R R, Glycosyl imidates 13. Application of the trichloroacetimidate procedure to 2-azidoglucose and 2-azidogalactose derivatives, *Liebigs Ann. Chem.*, 2, 1826–1847, 1984.
290. Fügedi, P, Synthesis of 4-*O*-(α-L-rhamnopyranosyl)-D-glucopyranuronic acid, *J. Carbohydr. Chem.*, 6, 377–398, 1987.
291. Fügedi, P, Nánási, P, Szejtli, J, Synthesis of 6-*O*-α-D-glucopyranosylcyclomaltoheptaose, *Carbohydr. Res.*, 175, 173–181, 1988.
292. Nicolaou, K C, Daines, R A, Chakraborty, T K, Ogawa, Y, Total synthesis of amphotericin B, *J. Am. Chem. Soc.*, 109, 2821–2822, 1987.
293. Nicolaou, K C, Daines, R A, Ogawa, Y, Chakraborty, T K, Total synthesis of amphotericin B. 3. The final stages, *J. Am. Chem. Soc.*, 110, 4696–4705, 1988.
294. Urban, F J, Moore, B S, Breitenbach, R, Synthesis of tigogenyl β-*O*-cellobioside heptaacetate and glycoside tetraacetate via Schmidt's trichloroacetimidate method; some new observations, *Tetrahedron Lett.*, 31, 4421–4424, 1990.
295. Schmidt, R R, Gaden, H, Jatzke, H, New catalysts for the glycosyl transfer with *O*-glycosyl trichloroacetimidates, *Tetrahedron Lett.*, 31, 327–330, 1990.
296. Douglas, S P, Whitfield, D M, Krepinsky, J J, Silver trifluoromethanesulfonate(triflate) activation of trichloroacetimidates in glycosylation reactions, *J. Carbohydr. Chem.*, 12, 131–136, 1993.
297. Wei, G, Gu, G, Du, Y, Silver triflate. A mild alternative catalyst for glycosylation conditions using trichloroacetimidates as glycosyl donors, *J. Carbohydr. Chem.*, 22, 385–393, 2003.
298. Castro-Palomino, J C, Schmidt, R R, Glycosylimidates, 72. *N*-Tetrachlorophthaloyl-protected trichloroacetimidate of glucosamine as glycosyl donor in oligosaccharide synthesis, *Tetrahedron Lett.*, 36, 5343–5346, 1995.
299. Bartek, J, Müller, R, Kosma, P, Synthesis of a neoglycoprotein containing the Lewis X analogous trisaccharide β-D-Gal*p*NAc-(1→4)[α-L-Fucp-(1→3)]-β-D-Glc*p*NAc, *Carbohydr. Res.*, 308, 259–273, 1998.
300. Adinolfi, M, Barone, G, Guariniello, L, Iadonisi, A, Efficient activation of armed glycosyl trichloroacetimidates with $Sm(OTf)_3$ in the stereoselective glycosidation of saccharidic acceptors, *Tetrahedron Lett.*, 41, 9005–9008, 2000.
301. Adinolfi, M, Barone, G, Iadonisi, A, Mangoni, L, Schiattarella, M, Activation of disarmed 2-*O*-alkoxycarbonylated glycosyl trichloroacetimidates with lantanide triflates: an efficient approach for the synthesis of 1,2-*trans* glycosides, *Tetrahedron Lett.*, 42, 5967–5969, 2001.
302. Adinolfi, M, Barone, G, Iadonisi, A, Schiattarella, M, Iodine/triethylsilane as a convenient promoter system for the activation of disarmed glycosyl trichloro- and *N*-(phenyl)trifluoroacetimidates, *Synlett*, 269–270, 2002.
303. Adinolfi, M, Barone, G, Iadonisi, A, Schiattarella, M, Activation of glycoyl trihaloacetimidates with acid-washed molecular sieves in the glycosidation reaction, *Org. Lett.*, 5, 987–989, 2003.

304. Griswold, K S, Horstmann, T E, Miller, S J, Acyl sulfonamide catalysts for glycosylation reactions with trichloroacetimidate donors, *Synlett*, 1923–1926, 2003.
305. Schmidt, R R, Toepfer, A, Glycosylimidates, 50. Glycosylation with highly reactive glycosyl donors: efficiency of the inverse procedure, *Tetrahedron Lett.*, 32, 3353–3356, 1991.
306. Toepfer, A, Schmidt, R R, An efficient synthesis of the Lewis X (Le^x) antigen family, *Tetrahedron Lett.*, 33, 5161–5164, 1992.
307. Mayer, T G, Kratzer, B, Schmidt, R R, Synthesis of a GPI anchor of yeast (*Saccharomyces cerevisiae*), *Angew. Chem. Int. Ed. Engl.*, 33, 2177–2181, 1994.
308. Halcomb, R L, Boyer, S H, Danishefsky, S J, Synthesis of the calicheamicin aryltetrasaccharide domain bearing a reducing terminus — coupling of fully synthetic aglycone and carbohydrate domains by the Schmidt reaction, *Angew. Chem. Int. Ed. Engl.*, 31, 338–340, 1992.
309. Nicolaou, K C, Schreiner, E P, Iwabuchi, Y, Suzuki, T, Total synthesis of calicheamicin dynemicin hybrid molecules, *Angew. Chem. Int. Ed. Engl.*, 31, 340–342, 1992.
310. Nicolaou, K C, van Delft, F, Ohshima, T, Vourloumis, D, Xu, J, Hosokawa, S, Pfefferkorn, J, Kim, S, Li, T, Total synthesis of eleutherobin, *Angew. Chem. Int. Ed. Engl.*, 36, 2520–2524, 1997.
311. Nicolaou, K C, Ohshima, T, Hosokawa, S, van Delft, F L, Vourloumis, D, Xu, J Y, Pfefferkorn, J, Kim, S, Total synthesis of eleutherobin and eleuthosides A and B, *J. Am. Chem. Soc.*, 120, 8674–8680, 1998.
312. Nicolaou, K C, Mitchell, H J, van Delft, F L, Rübsam, F, Rodriguez, R M, Expeditious routes to evernitrose and vancosamine derivatives and synthesis of a model vancomycin aryl glycoside, *Angew. Chem. Int. Ed.*, 37, 1871–1874, 1998.
313. Nicolaou, K C, Mitchell, H J, Jain, N F, Winssinger, N, Hughes, R, Bando, T, Total synthesis of vancomycin, *Angew. Chem. Int. Ed.*, 38, 240–244, 1999.
314. Nicolaou, K C, Mitchell, H J, Jain, N F, Bando, T, Hughes, R, Winssinger, N, Natarajan, S, Koumbis, A E, Total synthesis of vancomycin — part 4: attachment of the sugar moieties and completion of the synthesis, *Chem. Eur. J.*, 5, 2648–2667, 1999.
315. Fürstner, A, Jeanjean, F, Razon, P, Wirtz, C, Mynott, R, Total synthesis of woodrosin I — part 1: preparation of the building blocks and evaluation of the glycosylation strategy, *Chem. Eur. J.*, 9, 307–319, 2003.
316. Fürstner, A, Jeanjean, F, Razon, P, Wirtz, C, Mynott, R, Total synthesis of woodrosin I — part 2: final stages involving RCM and an orthoester rearrangement, *Chem. Eur. J.*, 9, 320–326, 2003.
317. Nicolaou, K C, Pfefferkorn, J A, Cao, G-Q, Selenium-based solid-phase synthesis of benzopyrans I: applications to combinatorial synthesis of natural products, *Angew. Chem. Int. Ed.*, 39, 734–739, 2000.
318. Schmidt, R R, Recent developments in the synthesis of glycoconjugates, *Pure Appl. Chem.*, 61, 1257–1270, 1989.
319. Schmidt, R R, Kinzy, W, Anomeric-oxygen activation for glycoside synthesis — the trichloroacetimidate method, *Adv. Carbohydr. Chem. Biochem.*, 50, 21–123, 1994.
320. Schmidt, R R, Jung, K-H, Oligosaccharide synthesis with trichloroacetimidates, In *Preparative Carbohydrate Chemistry*, Hanessian, S, Ed., Marcel Dekker, New York, pp. 283–312, 1997.
321. Yu, B, Tao, H, Glycosyl trifluoroacetimidates. Part 1: preparation and application as new glycosyl donors, *Tetrahedron Lett.*, 42, 2405–2407, 2001.
322. Yu, B, Tao, H, Glycosyl trifluoroacetimidates. 2. Synthesis of dioscin and xiebai saponin I, *J. Org. Chem.*, 67, 9099–9102, 2002.
323. Sun, J, Han, X, Yu, B, Synthesis of a typical *N*-acetylglucosamine-containing saponin, oleanolic acid 3-yl α-L-arabinopyranosyl-(1→2)-α-L-arabinopyranosyl-(1→6)-2-acetamido-2-deoxy-β-D-glucopyranoside, *Carbohydr. Res.*, 338, 827–833, 2003.
324. Zhang, Z, Yu, B, Total synthesis of the antiallergic naphtho-α-pyrone tetraglucoside, cassiaside C_2, isolated from cassia seeds, *J. Org. Chem.*, 68, 6309–6313, 2003.
325. Adinolfi, M, Barone, G, Iadonisi, A, Schiattarella, M, Efficient activation of glycosyl *N*-(phenyl)trifluoroacetimidate donors with ytterbium(III) triflate in the glycosylation reaction, *Tetrahedron Lett.*, 43, 5573–5577, 2002.
326. Adinolfi, M, Iadonisi, A, Schiattarella, M, An approach to the highly stereocontrolled synthesis of α-glycosides. Compatible use of the very acid labile dimethoxytrityl protecting group with $Yb(OTf)_3$-promoted glycosidation, *Tetrahedron Lett.*, 44, 6479–6482, 2003.

327. Helferich, B, Schmitz-Hillebrecht, E, Eine neue Methode zur Synthese von Glykosiden der Phenole, *Chem. Ber.*, 66, 378–383, 1933.
328. Lemieux, R U, Shyluk, W P, A new synthesis of β-glucopyranosides, *Can. J. Chem.*, 31, 528–535, 1953.
329. Hanessian, S, Banoub, J, Chemistry of glycosidic linkage. *O*-Glycosylations catalyzed by stannic chloride, in the D-ribofuranose and D-glucopyranose series, *Carbohydr. Res.*, 59, 261–267, 1977.
330. Zemplén, G, Synthesen in der Kohlenhydrat–Gruppe mit Hilfe von sublimiertem Eisenchlorid, I: Darstellung der Bioside der α-Reihe, *Chem. Ber.*, 62, 985–990, 1929.
331. Kiso, M, Anderson, L, The ferric chloride-catalyzed glycosylation of alcohols by 2-acylamido-2-deoxy-β-D-glucopyranose 1-acetates, *Carbohydr. Res.*, 72, C12–C14, 1979.
332. Kiso, M, Anderson, L, Synthesis of disaccharides by the ferric chloride-catalyzed coupling of 2-acylamido-2-deoxy-β-D-glucopyranose 1-acetates to protected sugar acceptors, *Carbohydr. Res.*, 72, C15–C17, 1979.
333. Lerner, L M, Ferric chloride-molecular sieve-catalyzed formation of a nonreducing disaccharide derivative, *Carbohydr. Res.*, 207, 138–141, 1990.
334. Magnusson, G, Noori, G, Dahmén, J, Frejd, T, Lave, T, BF_3-Etherate induced formation of 2,2,2-trichloroethyl glycopyranosides. Selective visualization of carbohydrate derivatives on TLC plates, *Acta Chem. Scand. B*, 35, 213–216, 1981.
335. Dahmen, J, Frejd, T, Magnusson, G, Noori, G, Boron trifluoride etherate-induced glycosidation: formation of alkyl glycosides and thioglycosides of 2-deoxy-2-phthalimidoglycopyranoses, *Carbohydr. Res.*, 114, 328–330, 1983.
336. Paulsen, H, Paal, M, Lewissäure-Katalysierte Synthesen von Di- und Trisaccharid-Sequenzen der *O*- und *N*-Glycoproteine. Anwendung von Trimethylsilyltrifluoromethanesulfonat, *Carbohydr. Res.*, 135, 53–69, 1984.
337. Mukaiyama, T, Kobayashi, S, Shoda, S-I, A facile synthesis of α-glucosides and α-ribosides from the corresponding 1-*O*-acyl sugars and alcohols in the presence of trityl perchlorate, *Chem. Lett.*, 907–910, 1984.
338. Koide, K, Ohno, M, Kobayashi, S, New glycosylation reaction based on a remote activation concept — glycosyl 2-pyridinecarboxylate as a novel glycosyl donor, *Tetrahedron Lett.*, 32, 7065–7068, 1991.
339. Susaki, H, Higashi, K, Novel α-mannoside synthesis promoted by the combination of trimethylsilyl chloride and zinc triflate, *Chem. Pharm. Bull.*, 41, 201–204, 1993.
340. Mukaiyama, T, Shimpuku, T, Takashima, T, Kobayashi, S, Stereoselective 1,2-*cis* glycosylation reaction of 1-*O*-acetylribose with silylated nucleophiles by the promotion of a new catalyst system, *Chem. Lett.*, 145–148, 1989.
341. Mukaiyama, T, Takashima, T, Katsurada, M, Aizawa, H, A highly stereoselective synthesis of α-glucosides from 1-*O*-acetyl glucose by use of tin(IV) chloride–silver perchlorate catalyst system, *Chem. Lett.*, 533–536, 1991.
342. Mukaiyama, T, Katsurada, M, Takashima, T, The evaluation of catalysts generated from Lewis acids and silver perchlorate in a highly stereoselective glycosylation of 1-*O*-acetyl-D-glucose, *Chem. Lett.*, 985–988, 1991.
343. Matsubara, K, Sasaki, T, Mukaiyama, T, Catalytic stereoselective synthesis of α- or β-glucosides using a single glycosyl donor, 1-*O*-2′-(2′-methoxyethoxy)acetyl-glucopyranose derivative, *Chem. Lett.*, 1373–1376, 1993.
344. Shimomura, N, Mukaiyama, T, Stereoselective syntheses of α-D- and β-D-ribofuranosides catalyzed by the combined use of silver salts and their partners, *Bull. Chem. Soc. Jpn.*, 67, 2532–2541, 1994.
345. Coterón, J M, Chiara, J L, Fernández-Mayoralas, A, Fiandor, J M, Valle, N, Stereocontrolled glycosylation of sordaricin in the presence of ammonium salts, *Tetrahedron Lett.*, 41, 4373–4377, 2000.
346. Yamanoi, T, Yamazaki, I, The catalytic synthesis of aryl *O*-glycosides using triaryloxyboranes, *Tetrahedron Lett.*, 42, 4009–4011, 2001.
347. Yokoyama, Y, Hanamoto, T, Jin, X L, Jin, Y Z, Inanaga, J, Perchloric acid in 1,4-dioxane and perfluoro-octanesulfonic acid as practical catalysts for the stereoselective glycosylation of 1-*O*-acetyl-glycosides, *Heterocycles*, 52, 1203–1206, 2000.

348. Florent, J-C, Monneret, C, Stereocontrolled route to 3-amino-2,3,6-trideoxy-hexopyranoses. K-10 Montmorillonite as a glycosidation reagent for acosaminide synthesis, *J. Chem. Soc. Chem. Commun.*, 1171–1172, 1987.
349. Kovác, P, Efficient chemical synthesis of methyl β-glycosides of β-(1→6)-linked D-galacto-oligosaccharides by a stepwise and a blockwise approach, *Carbohydr. Res.*, 153, 237–251, 1986.
350. Kimura, Y, Suzuki, M, Matsumoto, T, Abe, R, Terashima, S, Trimethylsilyl trifluoromethanesulfonate (trimethylsilyl triflate) as an excellent glycosidation reagent for anthracycline synthesis. Simple and efficient synthesis of optically pure 4-demethoxydaunorubicin, *Chem. Lett.*, 501–504, 1984.
351. Lu, Y-P, Li, H, Cai, M-S, Li, Z-J, Synthesis of a divalent glycoside of an α-galactosyl disaccharide epitope involved in the hyperacute rejection of xenotransplantation, *Carbohydr. Res.*, 334, 289–294, 2001.
352. Lopez, J C, Fraser-Reid, B, *n*-Pentenyl esters versus *n*-pentenyl glycosides. Synthesis and reactivity in glycosidation reactions, *J. Chem. Soc. Chem. Commun.*, 159–161, 1991.
353. Fraser-Reid, B, Konradsson, P, Mootoo, D R, Udodong, U, Direct elaboration of pent-4-enyl glycosides into disaccharides, *J. Chem. Soc. Chem. Commun.*, 823–825, 1988.
354. Pathak, A K, Pathak, V, Seitz, L, Maddry, J A, Gurcha, S S, Besra, G S, Suling, W J, Reynolds, R C, Studies on (β,1 → 5) and (β,1 → 6) linked octyl Gal_f disaccharides as substrates for mycobacterial galactosyltransferase activity, *Bioorg. Med. Chem.*, 9, 3129–3143, 2001.
355. Fraser-Reid, B, Lopez, J C, Radhakrishnan, K V, Mach, M, Schlueter, U, Gomez, A M, Uriel, C, Unexpected role of O-2 protecting groups of glycosyl donors in mediating regioselective glycosidation, *J. Am. Chem. Soc.*, 124, 3198–3199, 2002.
356. Mathew, F, Jayaprakash, K N, Fraser-Reid, B, Mathew, J, Scicinski, J, Microwave-assisted saccharide coupling with *n*-pentenyl glycosyl donors, *Tetrahedron Lett.*, 44, 9051–9054, 2003.
357. Mootoo, D R, Konradsson, P, Udodong, U, Fraser-Reid, B, Armed and disarmed *n*-pentenyl glycosides in saccharide couplings leading to oligosaccharides, *J. Am. Chem. Soc.*, 110, 5583–5584, 1988.
358. Fraser-Reid, B, Wu, Z, Udodong, U E, Ottoson, H, Armed/disarmed effects in glycosyl donors: rationalization and sidetracking, *J. Org. Chem.*, 55, 6068–6070, 1990.
359. Ratcliffe, A J, Konradsson, P, Fraser-Reid, B, *n*-Pentenyl glycosides as efficient synthons for promoter-mediated assembly of *N*-α-linked glycoproteins, *J. Am. Chem. Soc.*, 112, 5665–5667, 1990.
360. Mootoo, D R, Konradsson, P, Fraser-Reid, B, *n*-Pentenyl glycosides facilitate a stereoselective synthesis of the pentasaccharide core of the protein membrane anchor found in *Trypanosoma brucei*, *J. Am. Chem. Soc.*, 111, 8540–8542, 1989.
361. Fraser-Reid, B, Madsen, R, Oligosaccharide synthesis by *n*-pentenyl glycosides, In *Preparative Carbohydrate Chemistry*, Hanessian, S, Ed., Marcel Dekker, New York, pp. 339–356, 1997.
362. Vankar, Y D, Vankar, P S, Behrendt, M, Schmidt, R R, Glycosylimidates. 51. Synthesis of β-*O*-glycosides using enol ether and imidate derived leaving groups. Emphasis on the use of nitriles as a solvent, *Tetrahedron*, 47, 9985–9992, 1991.
363. Marra, A, Esnault, J, Veyrieres, A, Sinaÿ, P, Isopropenyl glycosides and congeners as novel classes of glycosyl donors: theme and variations, *J. Am. Chem. Soc.*, 114, 6354–6360, 1992.
364. Chenault, H K, Castro, A, Glycosyl transfer by isopropenyl glycosides: trisaccharide synthesis in one pot by selective coupling of isopropenyl and *n*-pentenyl glycopyranosides, *Tetrahedron Lett.*, 35, 9145–9148, 1994.
365. Chenault, H K, Castro, A, Chafin, L F, Yang, J, The chemistry of isopropenyl glycopyranosides. Transglycosylations and other reactions, *J. Org. Chem.*, 61, 5024–5031, 1996.
366. Boons, G-J, Isles, S, Vinyl glycosides in oligosaccharide synthesis (part 1): a new latent-active glycosylation strategy, *Tetrahedron Lett.*, 35, 3593–3596, 1994.
367. Boons, G-J, Isles, S, Vinyl glycosides in oligosaccharide synthesis. 2. The use of allyl and vinyl glycosides in oligosaccharide synthesis, *J. Org. Chem.*, 61, 4262–4271, 1996.
368. Bai, Y, Boons, G-J, Burton, A, Johnson, M, Haller, M, Vinyl glycosides in oligosaccharide synthesis (part 6): 3-buten-2-yl 2-azido-2-deoxy glycosides and 3-buten-2-yl 2-phthalimido-2-deoxy glycosides as novel glycosyl donors, *J. Carbohydr. Chem.*, 19, 939–958, 2000.

369. Nikolaev, A V, Kochetkov, N K, Use of (2-pyridyl)-2,3,4,6-tetra-*O*-benzyl-β-D-glucopyranoside in the synthesis of 1,2-*cis*-bound disaccharides, *Izv. Akad. Nauk. SSSR., Ser. Khim.*, 2556–2565, 1986.
370. Huchel, U, Schmidt, C, Schmidt, R R, Synthesis of hetaryl glycosides and their glycosyl donor properties, *Eur. J. Org. Chem.*, 1353–1360, 1998.
371. Hanessian, S, Saavedra, O M, Mascitti, V, Marterer, W, Oehrlein, R, Mak, C-P, Practical syntheses of B disaccharide and linear B type 2 trisaccharide — non-primate epitope markers recognized by human anti-α-Gal antibodies causing hyperacute rejection of xenotransplants, *Tetrahedron*, 57, 3267–3280, 2001.
372. Hanessian, S, Huynh, H K, Reddy, G V, Duthaler, R O, Katopodis, A, Streiff, M B, Kinzy, W, Oehrlein, R, Synthesis of Gal determinant epitopes, their glycomimetic variants, and trimeric clusters — relevance to tumor associated antigens and to discordant xenografts, *Tetrahedron*, 57, 3281–3290, 2001.
373. Lou, B, Huynh, H K, Hanessian, S, Oligosaccharide synthesis by remote activation: *O*-protected 3-methoxy-2-pyridyloxy (MOP) glycosyl donors, In *Preparative Carbohydrate Chemistry*, Hanessian, S, Ed., Marcel Dekker, New York, pp. 413–430, 1997.
374. Petersen, L, Jensen, K J, DISAL glycosyl donors for efficient glycosylations under acidic conditions: application to solid-phase oligosaccharide synthesis, *J. Chem. Soc. Perkin. Trans. 1*, 2175–2182, 2001.
375. Laursen, J B, Petersen, L, Jensen, K J, Intramolecular glycosylation under neutral conditions for synthesis of 1,4-linked disaccharides, *Org. Lett.*, 3, 687–690, 2001.
376. Petersen, L, Jensen, K J, A new, efficient glycosylation method for oligosaccharide synthesis under neutral conditions: preparation and use of new DISAL donors, *J. Org. Chem.*, 66, 6268–6275, 2001.
377. Kim, K S, Kim, J H, Lee, Y J, Lee, Y J, Park, J, 2-(Hydroxycarbonyl)benzyl glycosides: a novel type of glycosyl donors for highly efficient β-mannopyranosylation and oligosaccharide synthesis by latent-active glycosylation, *J. Am. Chem. Soc.*, 123, 8477–8481, 2001.
378. Kim, K S, Park, J, Lee, Y J, Seo, Y S, Dual stereoselectivity of 1-(2′-carboxy)benzyl 2-deoxyglycosides as glycosyl donors in the direct construction of 2-deoxyglycosyl linkages, *Angew. Chem. Int. Ed.*, 42, 459–462, 2003.
379. Kim, K S, Kang, S S, Seo, Y S, Kim, H J, Lee, Y J, Jeong, K-S, Glycosylation with 2′-carboxybenzyl glycosides as glycosyl donors: scope and application to the synthesis of a tetrasaccharide, *Synlett*, 1311–1314, 2003.
380. Inanaga, J, Yokoyama, Y, Hanamoto, T, Utility of 3,4-dimethoxybenzyl (DMPM) glycosides. A new glycosylation triggered by 2,3-dichloro-5,6-dicyano-*p*-benzoquinone (DDQ) oxidation, *Chem. Lett.*, 85–88, 1993.
381. Davis, B G, Wood, S D, Maughan, M A T, Towards an unprotected self-activating glycosyl donor system: bromobutyl glycosides, *Can. J. Chem.*, 80, 555–558, 2002.
382. Mukaiyama, T, Hashihayata, T, Mandai, H, Glycosyl 6-nitro-2-benzothiazoate. A highly efficient donor for β-stereoselective glycosylation, *Chem. Lett.*, 32, 340–341, 2003.
383. Noyori, R, Kurimoto, I, Electrochemical glycosylation method, *J. Org. Chem.*, 51, 4320–4322, 1986.
384. Wessel, H P, Use of trifluoromethanesulfonic acid in Fischer glycosylations, *J. Carbohydr. Chem.*, 7, 263–269, 1988.
385. Defaye, J, Gadelle, A, Pedersen, C, Carbohydrate reactivity in hydrogen-fluoride. 7. Hydrogen fluoride-catalyzed formation of glycosides. Preparation of methyl 2-acetamido-2-deoxy-β-D-gluco- and β-D-galacto-pyranosides, and of β-(1→6)-linked 2-acetamido-2-deoxy-D-gluco- and D-galacto-pyranosyl oligosaccharides, *Carbohydr. Res.*, 186, 177–188, 1989.
386. Konradsson, P, Roberts, C, Fraser-Reid, B, Conditions for modified Fischer glycosidations with *n*-pentenol and other alcohols, *Recl. Trav. Chim. Pays-Bas*, 110, 23–24, 1991.
387. Lubineau, A, Fischer, J-C, High-yielding one-step conversion of D-glucose and D-galactose to the corresponding α and β methyl-D-glucofuranosides and galactofuranosides, *Synth. Commun.*, 21, 815–818, 1991.
388. Ferrières, V, Bertho, J-N, Plusquellec, D, A new synthesis of *O*-glycosides from totally *O*-unprotected glycosyl donors, *Tetrahedron Lett.*, 36, 2749–2752, 1995.
389. Bertho, J-N, Ferrières, V, Plusquellec, D, A new synthesis of D-glycosiduronates from unprotected D-uronic acids, *J. Chem. Soc. Chem. Commun.*, 1391–1393, 1995.

390. Ferrières, V, Bertho, J-N, Plusquellec, D, A convenient synthesis of alkyl D-glycofuranosiduronic acids and alkyl D-glycofuranosides from unprotected carbohydrates, *Carbohydr. Res.*, 311, 25–35, 1998.
391. Koto, S, Morishima, N, Zen, S, Direct glucosidation of tetra-*O*-benzyl-α-D-glucopyranose by system of methanesulfonic acid and cobalt(II) bromide, *Chem. Lett.*, 1109–1110, 1976.
392. Koto, S, Morishima, N, Zen, S, A stereoselective one-stage α-glucosylation with 2,3,4,6-tetra-*O*-benzyl-α-D-glucopyranose and a mixture of methanesulfonic acid, cobalt(II) bromide, and tetraethylammonium perchlorate, *Bull. Chem. Soc. Jpn.*, 55, 1543–1547, 1982.
393. Morishima, N, Koto, S, Kusuhara, C, Zen, S, One-stage α-glucosylation using tetra-*O*-benzyl-α-D-glucose and mixture of trimethylsilyl bromide, cobalt(II) bromide, tetrabutylammonium bromide, and molecular sieve, *Chem. Lett.*, 427–428, 1981.
394. Koto, S, Morishima, N, Kusuhara, C, Sekido, S, Yoshida, T, Zen, S, Stereoselective α-glucosylation with tetra-*O*-benzyl-α-D-glucose and a mixture of trimethylsilyl bromide, cobalt(II) bromide, tetrabutylammonium bromide, and a molecular sieve. A synthesis of 3,6-di-*O*-(α-D-glucopyranosyl)-D-glucose, *Bull. Chem. Soc. Jpn.*, 55, 2995–2999, 1982.
395. Hirooka, M, Mori, Y, Sasaki, A, Koto, S, Shinoda, Y, Morinaga, A, Synthesis of β-D-ribofuranosyl-(1→3)-α-L-rhamnopyranosyl-(1→3)-L-rhamnopyranose by in situ activating glycosylation using 1-OH sugar derivative and $Me_3SiBr-CoBr_2-Bu_4NBr$-molecular sieves 4A system, *Bull. Chem. Soc. Jpn.*, 74, 1679–1694, 2001.
396. Leroux, J, Perlin, A S, New synthesis of glycosides. Reactions of trifluoromethanesulfonic anhydride at anomeric center, *Carbohydr. Res.*, 47, C8–C10, 1976.
397. Leroux, J, Perlin, A S, Synthesis of glycosyl halides and glycosides via 1-*O*-sulfonyl derivatives, *Carbohydr. Res.*, 67, 163–178, 1978.
398. Pavia, A A, Rocheville, J-M, Ung, S N, Nouvelle méthode de synthèse stéréosélective de glycosides. Synthèse des α,α-tréhalose, analogues *galacto*, *manno* et autres α-D-glycosides, *Carbohydr. Res.*, 79, 79–89, 1980.
399. Pavia, A, Ung-Chhun, S N, Mécanisme de la réaction de glycosylation par réversion acide en présence d'anhydride trifluorométhane sulfonique agent de condensation, *Can. J. Chem.*, 59, 482–489, 1981.
400. Lacombe, J M, Pavia, A A, Rocheville, J M, Un nouvel agent de glycosylation: l'anhydride trifluorométhanesulfonique. Synthèse des α et β *O*-glycosyl-L-sérine, -L-thréonine et L-hydroxyproline, *Can. J. Chem.*, 59, 473–481, 1981.
401. Lacombe, J M, Pavia, A A, Chemical synthesis of some monogalactosyl and digalactosyl *O*-glycopeptides, *J. Org. Chem.*, 48, 2557–2563, 1983.
402. Morishima, N, Koto, S, Zen, S, Dehydrative α-glucosylation using a mixture of *p*-nitrobenzenesulfonyl chloride, silver trifluoromethanesulfonate, *N,N*-dimethylacetamide, and triethylamine, *Chem. Lett.*, 1039–1040, 1982.
403. Koto, S, Morishima, N, Owa, M, Zen, S, A stereoselective α-glucosylation by use of a mixture of 4-nitrobenzenesulfonyl chloride, silver trifluoromethanesulfonate, *N,N*-dimethylacetamide, and triethylamine, *Carbohydr. Res.*, 130, 73–83, 1984.
404. Koto, S, Kusunoki, A, Hirooka, M, In situ activating glycosylation of 6-deoxysugars: synthesis of *O*-α-D-fucosyl-(1→4)-*O*-α-D-fucosyl-(1→4)-*O*-α-D-quinovosyl-(1→4)-D-quinovose, *Bull. Chem. Soc. Jpn.*, 73, 967–976, 2000.
405. Koto, S, Hirooka, M, Yoshida, T, Takenaka, K, Asai, C, Nagamitsu, T, Sakuma, H, Sakurai, M, Masuzawa, S, Komiya, M, Sato, T, Zen, S, Yago, K, Tomonaga, F, Syntheses of penta-*O*-benzyl-*myo*-inositols, *O*-β-L-arabinosyl-(1→2)-*sn*-*myo*-inositol, *O*-α-D-galactosyl-(1→3)-*sn*-*myo*-inositol, and *O*-α-D-galactosyl-(1→6)-*O*-α-D-galactosyl-(1→3)-*sn*-*myo*-inositol, *Bull. Chem. Soc. Jpn.*, 73, 2521–2529, 2000.
406. Hirooka, M, Terayama, M, Mitani, E, Koto, S, Miura, A, Chiba, K, Takabatake, A, Tashiro, T, Synthesis of methyl *O*-α-D-mannosyl-(1→4)-[(3-*O*-methyl-α-D-mannosyl)-(1→4)-]$_n$3-*O*-methyl-α-D-mannosides ($n = 0$, 1, and 2) via dehydrative glycosylation, *Bull. Chem. Soc. Jpn.*, 75, 1301–1309, 2002.
407. Koto, S, Hirooka, M, Yago, K, Komiya, M, Shimizu, T, Kato, K, Takehara, T, Ikefuji, A, Iwasa, A, Hagino, S, Sekiya, M, Nakase, Y, Zen, S, Tomonaga, F, Shimada, S, Benzyl derivatives of *N*-2,4-dinitrophenyl-D-glucosamine and their use for oligosaccharide synthesis, *Bull. Chem. Soc. Jpn.*, 73, 173–183, 2000.

408. Szeja, W, A convenient synthesis of α-D-glucopyranosides, *Synthesis*, 223–224, 1988.
409. Grynkiewicz, G, Novel synthesis of aryl glycosides, *Carbohydr. Res.*, 53, C11–C12, 1977.
410. Kometani, T, Kondo, H, Fujimori, Y, Boron trifluoride-catalyzed rearrangement of 2-aryloxytetrahydropyrans: a new entry to *C*-arylglycosidation, *Synthesis*, 1005–1007, 1988.
411. Roush, W R, Lin, X-F, A highly stereoselective synthesis of aryl 2-deoxy-β-glycosides via the Mitsunobu reaction, *J. Org. Chem.*, 56, 5740–5742, 1991.
412. Szarek, W A, Jarrell, H C, Jones, J K N, Synthesis of glycosides: reactions of anomeric hydroxyl group with nitrogen–phosphorus betaines, *Carbohydr. Res.*, 57, C13–C16, 1977.
413. Grochowski, E, Jurczak, J, A new class of monosaccharide derivatives: *O*-phthalimidohexoses, *Carbohydr. Res.*, 50, C15–C16, 1976.
414. Nicolaou, K C, Groneberg, R D, Novel strategy for the construction of the oligosaccharide fragment of calicheamicin $\gamma_{1\alpha}$. Synthesis of the ABC skeleton, *J. Am. Chem. Soc.*, 112, 4085–4086, 1990.
415. Mukaiyama, T, Suda, S, Diphosphonium salts as effective reagents for stereoselective synthesis of 1, 2-*cis*-ribofuranosides, *Chem. Lett.*, 1143–1146, 1990.
416. Garcia, B A, Poole, J L, Gin, D Y, Direct glycosylations with 1-hydroxy glycosyl donors using trifluoromethanesulfonic anhydride and diphenyl sulfoxide, *J. Am. Chem. Soc.*, 119, 7597–7598, 1997.
417. Garcia, B A, Gin, D Y, Dehydrative glycosylation with activated diphenyl sulfonium reagents. Scope, mode of C(1)-hemiacetal activation, and detection of reactive glycosyl intermediates, *J. Am. Chem. Soc.*, 122, 4269–4279, 2000.
418. Nguyen, H M, Poole, J L, Gin, D Y, Chemoselective iterative dehydrative glycosylation, *Angew. Chem. Int. Ed.*, 40, 414–417, 2001.
419. Nguyen, H M, Chen, Y, Duron, S G, Gin, D Y, Sulfide-mediated dehydrative glycosylation, *J. Am. Chem. Soc.*, 123, 8766–8772, 2001.
420. Fischer, B, Nudelman, A, Ruse, M, Herzig, J, Gottlieb, H E, Keinan, E, A novel method for stereoselective glucuronidation, *J. Org. Chem.*, 49, 4988–4993, 1984.
421. Koto, S, Yago, K, Zen, S, Tomonaga, F, Shimada, S, α-D-glucosylation by 6-*O*-acetyl-2,3,4-tri-*O*-benzyl-D-glucopyranose using trimethylsilyl triflate and pyridine. Synthesis of α-maltosyl and α-isomaltosyl α-D-glucosides, *Bull. Chem. Soc. Jpn.*, 59, 411–414, 1986.
422. Horito, S, Tada, M, Hashimoto, H, A facile synthesis of 3-deoxy-α-D-*manno*-2-octulopyranosides using pyridinium *p*-toluenesulfonate, *Chem. Lett.*, 117–120, 1991.
423. Suda, S, Mukaiyama, T, Stereoselective synthesis of 1,2-*trans*-ribofuranosides from 1-hydroxy sugars by the use of [1,2-benzenediolato(2-)-*O,O′*]oxotitanium and trifluoromethanesulfonic anhydride, *Chem. Lett.*, 431–434, 1991.
424. Mukaiyama, T, Koki, M, Suda, S, An efficient method for the stereoselective synthesis of 1,2-*cis*- and 1,2-*trans*-ribofuranosides from 1-hydroxy ribofuranose by the use of diphenyltin sulfide and trifluoromethanesulfonic anhydride, *Chem. Lett.*, 981–984, 1991.
425. Inanaga, J, Yokoyama, Y, Hanamoto, T, Catalytic *O*- and *S*-glycosylation of 1-hydroxy sugars, *J. Chem. Soc. Chem. Commun.*, 1090–1091, 1993.
426. Susaki, H, Glycosidation with 1-hydroxy sugars as glycosyl donors promoted by trimethylsilyl chloride and zinc triflate, *Chem. Pharm. Bull.*, 42, 1917–1918, 1994.
427. Mukaiyama, T, Matsubara, K, Hora, M, An efficient glycosylation reaction of 1-hydroxy sugars with various nucleophiles using a catalytic amount of activator and hexamethyldisiloxane, *Synthesis*, 1368–1373, 1994.
428. Uchiro, H, Miyazaki, K, Mukaiyama, T, Catalytic stereoselective synthesis of α-D-galactopyranosides from 2,3,4,6-tetra-*O*-benzyl-D-galactopyranose and several alcoholic nucleophiles, *Chem. Lett.*, 403–404, 1997.
429. Uchiro, H, Mukaiyama, T, Trityl salt catalyzed stereoselective glycosylation of alcohols with 1-hydroxyribofuranose, *Chem. Lett.*, 79–80, 1996.
430. Uchiro, H, Mukaiyama, T, A significant effect of a lithium salt in the stereocontrolled synthesis of α-D-ribofuranosides, *Chem. Lett.*, 271–272, 1996.
431. Li, X, Ohtake, H, Takahashi, H, Ikegami, S, A facile synthesis of 1′-*C*-alkyl-α-disaccharides from 1-*C*-alkyl-hexopyranoses and methyl 1-*C*-methylhexopyranosides, *Tetrahedron*, 57, 4297–4309, 2001.

432. Tsutsumi, H, Ishido, Y, Partial protection of carbohydrate-derivatives. 5. Synthesis of phenyl D-glucopyranosides; nucleophilic-substitution of *O*-(2,3,4,6-tetra-*O*-benzyl-D-glucopyranosyl)-pseudoureas by phenols, *Carbohydr. Res.*, 88, 61–75, 1981.
433. Suzuki, T, Watanabe, S, Yamada, T, Hiroi, K, Dehydrative glycosylation of tri-O-benzylated 1-hydroxyribofuranose catalyzed by a copper(II) complex, *Tetrahedron Lett.*, 44, 2561–2563, 2003.
434. Wagner, B, Heneghan, M, Schnabel, G, Ernst, B, Catalytic glycosylation with rhodium(III)-triphos catalysts, *Synlett*, 1303–1306, 2003.
435. Gin, D, Dehydrative glycosylation with 1-hydroxy donors, *J. Carbohydr. Chem.*, 21, 645–665, 2002.
436. Lemieux, R U, Huber, G, A chemical synthesis of sucrose, *J. Am. Chem. Soc.*, 75, 4118, 1953.
437. Lemieux, R U, Huber, G, A chemical synthesis of sucrose. A conformational analysis of the reactions of 1,2-anhydro-α-D-glucopyranose triacetate, *J. Am. Chem. Soc.*, 78, 4117–4119, 1956.
438. Schuerch, C, Synthesis and polymerization of anhydro sugars, *Adv. Carbohydr. Chem. Biochem.*, 39, 157–212, 1981.
439. Trumbo, D L, Schuerch, C, Steric control in the polymerization of 1,2-anhydro-3,4,6-tri-*O*-benzyl-β-D-mannopyranose, *Carbohydr. Res.*, 135, 195–202, 1985.
440. Halcomb, R L, Danishefsky, S J, On the direct epoxidation of glycals: application of a reiterative strategy for the synthesis of β-linked oligosaccharides, *J. Am. Chem. Soc.*, 111, 6661–6666, 1989.
441. Danishefsky, S J, McClure, K F, Randolph, J T, Ruggeri, R B, A strategy for the solid-phase synthesis of oligosaccharides, *Science*, 260, 1307–1309, 1993.
442. Timmers, C M, van der Marel, G A, van Boom, J H, A note on the solid-phase synthesis of oligosaccharides by the *Danishefski* approach, *Recl. Trav. Chim. Pays-Bas*, 112, 609–610, 1993.
443. Liu, K K C, Danishefsky, S J, A striking example of the interfacing of glycal chemistry with enzymatically mediated sialylation — a concise synthesis of GM_3, *J. Am. Chem. Soc.*, 115, 4933–4934, 1993.
444. Randolph, J T, Danishefsky, S J, Application of the glycal assembly strategy to the synthesis of a branched oligosaccharide: the first synthesis of a complex saponin, *J. Am. Chem. Soc.*, 115, 8473–8474, 1993.
445. Randolph, J T, Danishefsky, S J, First synthesis of a digitalis saponin. Demonstration of the scope and limitations of a convergent scheme for branched oligosaccharide synthesis by the logic of glycal assembly, *J. Am. Chem. Soc.*, 117, 5693–5700, 1995.
446. Liu, K K C, Danishefsky, S J, A direct route from 1α,2α-anhydroglucose derivatives to α-glucosides, *J. Org. Chem.*, 59, 1895–1897, 1994.
447. Dushin, R G, Danishefsky, S J, Stereospecific synthesis of aryl β-glucosides: an application to the synthesis of a prototype corresponding to the aryloxy carbohydrate domain of vancomycin, *J. Am. Chem. Soc.*, 114, 3471–3475, 1992.
448. Gervay, J, Danishefsky, S, A stereospecific route to 2-deoxy-β-glycosides, *J. Org. Chem.*, 56, 5448–5451, 1991.
449. Liu, K K C, Danishefsky, S J, Route from glycals to mannose β-glycosides, *J. Org. Chem.*, 59, 1892–1894, 1994.
450. Gordon, D M, Danishefsky, S J, Displacement reactions of a 1,2-anhydro-α-D-hexopyranose: installation of useful functionality at the anomeric carbon, *Carbohydr. Res.*, 206, 361–366, 1990.
451. Martin, T J, Schmidt, R R, Efficient sialylation with phosphite as leaving group, *Tetrahedron Lett.*, 33, 6123–6126, 1992.
452. Müller, T, Schneider, R, Schmidt, R R, Utility of glycosyl phosphites as glycosyl donors — fructofuranosyl and 2-deoxyhexopyranosyl phosphites in glycoside bond formation, *Tetrahedron Lett.*, 35, 4763–4766, 1994.
453. Müller, T, Hummel, G, Schmidt, R R, Glycosyl phosphites as glycosyl donors — a comparative study, *Liebigs Ann. Chem.*, 325–329, 1994.
454. Kondo, H, Ichikawa, Y, Wong, C-H, β-Sialyl phosphite and phosphoramidite: synthesis and application to the chemoenzymatic synthesis of CMP-sialic acid and sialyl oligosaccharides, *J. Am. Chem. Soc.*, 114, 8748–8750, 1992.
455. Sim, M M, Kondo, H, Wong, C-H, Synthesis and use of glycosyl phosphites: an effective route to glycosyl phosphates, sugar nucleotides, and glycosides, *J. Am. Chem. Soc.*, 115, 2260–2267, 1993.

456. Kondo, H, Aoki, S, Ichikawa, Y, Halcomb, R L, Ritzen, H, Wong, C-H, Glycosyl phosphites as glycosylation reagents: scope and mechanism, *J. Org. Chem.*, 59, 864–877, 1994.
457. Lin, C-C, Shimazaki, M, Heck, M-P, Aoki, S, Wang, R, Kimura, T, Ritzen, H, Takayama, S, Wu, S-H, Weitz-Schmidt, G, Wong, C-H, Synthesis of sialyl Lewis X mimetics and related structures using the glycosyl phosphite methodology and evaluation of E-selectin inhibition, *J. Am. Chem. Soc.*, 118, 6826–6840, 1996.
458. Watanabe, Y, Nakamoto, C, Ozaki, S, Glycosylation based on phosphite chemistry, *Synlett*, 115–116, 1993.
459. Watanabe, Y, Nakamoto, C, Yamamoto, T, Ozaki, S, Glycosylation using glycosyl phosphite as a glycosyl donor, *Tetrahedron*, 50, 6523–6536, 1994.
460. Martin, T J, Brescello, R, Toepfer, A, Schmidt, R R, Synthesis of phosphites and phosphates of neuraminic acid and their glycosyl donor properties — convenient synthesis of GM_3, *Glycoconjugate J.*, 10, 16–25, 1993.
461. Corey, E J, Wu, Y-J, Total synthesis of (±)-paeoniflorigenin and paeoniflorin, *J. Am. Chem. Soc.*, 115, 8871–8872, 1993.
462. Cheng, Y-P, Chen, H-T, Lin, C-C, A convenient and highly stereoselective approach for α-galactosylation performed by galactopyranosyl dibenzyl phosphite with remote participating groups, *Tetrahedron Lett.*, 43, 7721–7723, 2002.
463. Sasaki, K, Nagai, H, Matsumura, S, Toshima, K, A novel greener glycosidation using an acid-ionic liquid containing a protic acid, *Tetrahedron Lett.*, 44, 5605–5608, 2003.
464. Schene, H, Waldmann, H, Activation of glycosyl phosphites under neutral conditions in solutions of metal perchlorates in organic solvents, *Eur. J. Org. Chem.*, 1227–1230, 1998.
465. Hashimoto, S, Umeo, K, Sano, A, Watanabe, N, Nakajima, M, Ikegami, S, An extremely mild and stereocontrolled construction of 1,2-*trans*-β-glycosidic linkages capitalizing on benzyl-protected glycopyranosyl diethyl phosphites as glycosyl donors, *Tetrahedron Lett.*, 36, 2251–2254, 1995.
466. Nagai, H, Matsumura, S, Toshima, K, Environmentally benign and stereoselective formation of β-*O*-glycosidic linkages using benzyl-protected glucopyranosyl phosphite and montmorillonite K-10, *Tetrahedron Lett.*, 43, 847–850, 2002.
467. Nagai, H, Matsumura, S, Toshima, K, Environmentally benign and stereoselective construction of 2-deoxy and 2,6-dideoxy-β-glycosidic linkages employing 2-deoxy and 2,6-dideoxyglycosyl phosphites and montmorillonite K-10, *Chem. Lett.*, 1100–1101, 2002.
468. Nagai, H, Matsumura, S, Toshima, K, Environmentally benign and stereoselective formation of β-mannosidic linkages utilizing 2,3-di-*O*-benzyl-4,6-*O*-benzylidene-protected mannopyranosyl phosphite and montmorillonite K-10, *Carbohydr. Res.*, 338, 1531–1534, 2003.
469. Hashimoto, S, Honda, T, Ikegami, S, A rapid and efficient synthesis of 1,2-*trans*-β-linked glycosides via benzyl or benzoyl-protected glycopyranosyl phosphates, *J. Chem. Soc. Chem. Commun.*, 685–687, 1989.
470. Böhm, G, Waldmann, H, *O*-Glycoside synthesis under neutral conditions in concentrated solutions of $LiClO_4$ in organic solvents employing benzyl-protected glycosyl donors, *Liebigs Ann. Chem.*, 613–619, 1996.
471. Hariprasad, V, Singh, G, Tranoy, I, Stereoselective *O*-glycosylation reactions employing diphenylphosphinate and propane-1,3-diyl phosphate as anomeric leaving groups, *Chem. Commun.*, 2129–2130, 1998.
472. Singh, G, Vankayalapati, H, A new glycosylation strategy for the synthesis of mannopyranosides, *Tetrahedron: Asymmetry*, 11, 125–138, 2000.
473. Vankayalapati, H, Singh, G, Stereoselective synthesis of α-L-Fucp-(1,2)- and -(1,3)-β-D-Galp(1)-4-methylumbelliferone using glycosyl donor substituted by propane-1,3-diyl phosphate as leaving group, *J. Chem. Soc. Perkin. Trans. 1*, 2187–2193, 2000.
474. Vankayalapati, H, Singh, G, Tranoy, I, Stereoselective *O*-glycosylation reactions using glycosyl donors with diphenylphosphinate and propane-1,3-diyl phosphate leaving groups, *Tetrahedron: Asymmetry*, 12, 1373–1381, 2001.
475. Plante, O J, Andrade, R B, Seeberger, P H, Synthesis and use of glycosyl phosphates as glycosyl donors, *Org. Lett.*, 1, 211–214, 1999.
476. Plante, O J, Palmacci, E R, Seeberger, P H, Formation of β-glucosamine and β-mannose linkages using glycosyl phosphates, *Org. Lett.*, 2, 3841–3843, 2000.

477. Hewitt, M C, Seeberger, P H, Automated solid-phase synthesis of a branched *Leishmania* cap tetrasaccharide, *Org. Lett.*, 3, 3699–3702, 2001.
478. Hewitt, M C, Seeberger, P H, Solution and solid-support synthesis of a potential leishmaniasis carbohydrate vaccine, *J. Org. Chem.*, 66, 4233–4243, 2001.
479. Love, K R, Andrade, R B, Seeberger, P H, Linear synthesis of a protected H-type II pentasaccharide using glycosyl phosphate building blocks, *J. Org. Chem.*, 66, 8165–8176, 2001.
480. Plante, O J, Palmacci, E R, Seeberger, P H, Automated solid-phase synthesis of oligosaccharides, *Science*, 291, 1523–1527, 2001.
481. Melean, L G, Love, K R, Seeberger, P H, Toward the automated solid-phase synthesis of oligoglucosamines: systematic evaluation of glycosyl phosphate and glycosyl trichloroacetimidate building blocks, *Carbohydr. Res.*, 337, 1893–1916, 2002.
482. Hunt, D K, Seeberger, P H, Linker influence on the stereochemical outcome of glycosylations utilizing solid support-bound glycosyl phosphates, *Org. Lett.*, 4, 2751–2754, 2002.
483. Boger, D L, Honda, T, Total synthesis of bleomycin A_2 and related agents. 4. Synthesis of the disaccharide subunit: 2-*O*-(3-*O*-carbamoyl-α-D-mannopyranosyl)-L-gulopyranose and completion of the total synthesis of bleomycin A_2, *J. Am. Chem. Soc.*, 116, 5647–5656, 1994.
484. Li, H, Chen, M, Zhao, K, Stereoselective linkages of 2-deoxyglycosides with hindered acceptors, *Tetrahedron Lett.*, 38, 6143–6144, 1997.
485. Inazu, T, Hosokawa, H, Satoh, Y, Glucosylation using glucopyranosyl dimethylphosphinothioate, *Chem. Lett.*, 297–300, 1985.
486. Yamanoi, T, Nakamura, K, Sada, S, Goto, M, Furusawa, Y, Takano, M, Fujioka, A, Yanagihara, K, Satoh, Y, Hosokawa, H, Inazu, T, New synthetic methods and reagents for complex carbohydrates. 7. Syntheses and glycosylation reactions of glycopyranosyl dimethylphosphinothioate series having a nonparticipating group at the C-2 position, *Bull. Chem. Soc. Jpn.*, 66, 2617–2622, 1993.
487. Yamanoi, T, Nakamura, K, Takeyama, H, Yanagihara, K, Inazu, T, New synthetic methods and reagents for complex carbohydrates. 8. Stereoselective α- and β-mannopyranoside formation from glycosyl dimethylphosphinothioates with the C-2 axial benzyloxyl group, *Bull. Chem. Soc. Jpn.*, 67, 1359–1366, 1994.
488. Inazu, T, Yamanoi, T, A novel glycosylation reaction of 2-amino-2-deoxy-D-glucopyranose using dimethylphosphinothioate, *Chem. Lett.*, 69–72, 1989.
489. Yamanoi, T, Nakamura, K, Takeyama, H, Yanagihara, K, Inazu, T, 1,2-*Cis*-β-mannopyranoside formation by the dimethylphosphinothioate method, *Chem. Lett.*, 343–346, 1993.
490. Yamanoi, T, Inazu, T, A convenient 2-deoxy-α-D-glucopyranosylation reaction using dimethylphosphinothioate method, *Chem. Lett.*, 849–852, 1990.
491. Hashimoto, S, Honda, T, Ikegami, S, A mild and rapid 1,2-*trans*-glycosidation method via benzoyl-protected glycopyranosyl *P,P*-diphenyl-*N*-(*p*-toluenesulfonyl)phosphinimidates, *Heterocycles*, 30, 775–778, 1990.
492. Hashimoto, S-I, Honda, T, Ikegami, S, An efficient construction of 1,2-*trans*-β-glycosidic linkages via benzyl-protected glycopyranosyl *P,P*-diphenyl-*N*-(*p*-toluenesulfonyl)-phosphinimidates, *Chem. Pharm. Bull.*, 38, 2323–2325, 1990.
493. Hashimoto, S, Honda, T, Ikegami, S, A new and general glycosidation method for podophyllum lignan glycosides, *Tetrahedron Lett.*, 32, 1653–1654, 1991.
494. Hashimoto, S, Yanagiya, Y, Honda, T, Harada, H, Ikegami, S, An efficient construction of 1,2-*trans*-β-glycosidic linkages capitalizing on glycopyranosyl *N,N,N′,N′*-tetramethylphosphoroamidates as shelf-stable glycosyl donors, *Tetrahedron Lett.*, 33, 3523–3526, 1992.
495. Hashimoto, S, Sakamoto, H, Honda, T, Abe, H, Nakamura, S, Ikegami, S, Armed-disarmed glycosidation strategy based on glycosyl donors and acceptors carrying phosphoroamidate as a leaving group: a convergent synthesis of globotriaosylceramide, *Tetrahedron Lett.*, 38, 8969–8972, 1997.
496. Pan, S, Li, H, Hong, F, Yu, B, Zhao, K, Glycosyl donors with phosphorimidate leaving groups for either α- or β-glycosidation, *Tetrahedron Lett.*, 38, 6139–6142, 1997.
497. Neda, I, Sakhaii, P, Wassmann, A, Niemeyer, U, Günther, E, Engel, J, A practical synthesis of benzyl α- and allyl β-D-glucopyranosides regioselectively substituted with $(CH_2)_3OH$ groups: stereocontrolled β-galactosidation by cation π-interaction, *Synthesis*, 1625–1632, 1999.
498. Bochkov, A F, Kochetkov, N K, New approach to synthesis of oligosaccharides, *Carbohydr. Res.*, 39, 355–357, 1975.

499. Betaneli, V I, Ovchinnikov, M V, Backinowsky, L V, Kochetkov, N K, Glycosylation by 1,2-*O*-cyanoethylidene derivatives of carbohydrates, *Carbohydr. Res.*, 76, 252–256, 1979.
500. Kochetkov, N K, Betaneli, V I, Kryazhevskikh, I A, Ott, A Ya, Glycosylation by sugar 1,2-*O*-(1-cyanobenzylidene) derivatives: influence of glycosyl-donor structure and promoter, *Carbohydr. Res.*, 244, 85–97, 1993.
501. Kochetkov, N K, Nepogod'ev, S A, Backinowsky, L V, Synthesis of cyclo-[(1→6)-β-D-galactofurano]-oligosaccharides, *Tetrahedron*, 46, 139–150, 1990.
502. Kochetkov, N K, Zhulin, V M, Klimov, E M, Malysheva, N N, Makarova, Z G, Ott, A Ya, The effect of high-pressure on the stereospecificity of the glycosylation reaction, *Carbohydr. Res.*, 164, 241–254, 1987.
503. Kochetkov, N K, Synthesis of polysaccharides with a regular structure, *Tetrahedron*, 43, 2389–2436, 1987.
504. Kochetkov, N K, Microbial polysaccharides: new approaches, *Chem. Soc. Rev.*, 19, 29–54, 1990.
505. Kochetkov, N K, Nifant'ev, N E, Backinowsky, L V, Synthesis of the capsular polysaccharide of *Streptococcus pneumoniae* type 14, *Tetrahedron*, 43, 3109–3121, 1987.
506. Kochetkov, N K, Backinowsky, L V, Tsvetkov, Yu E, Sugar thio orthoesters as glycosylating agents, *Tetrahedron Lett.*, 3681–3684, 1977.
507. Backinowsky, L V, Tsvetkov, Y E, Balan, N F, Byramova, N E, Kochetkov, N K, Synthesis of 1,2-*trans*-disaccharides via sugar thio-orthoesters, *Carbohydr. Res.*, 85, 209–221, 1980.
508. Zuurmond, H M, van der Marel, G A, van Boom, J H, Iodonium ion-promoted glycosidation of sugar 1,2-thio-orthoesters, *Recl. Trav. Chim. Pays-Bas*, 110, 301–302, 1991.
509. Roberts, C, Madsen, R, Fraser-Reid, B, Studies related to synthesis of glycophosphatidylinositol membrane-bound protein anchors. 5. *n*-Pentenyl ortho esters for mannan components, *J. Am. Chem. Soc.*, 117, 1546–1553, 1995.
510. Anilkumar, G, Nair, L G, Fraser-Reid, B, Targeted glycosyl donor delivery for site-selective glycosylation, *Org. Lett.*, 2, 2587–2589, 2000.
511. Jayaprakash, K N, Radhakrishnan, K V, Fraser-Reid, B, Ytterbium(III) trifluoromethanesulfonate for specific activation of *n*-pentenyl orthoesters in the presence of acid sensitive functionalities, *Tetrahedron Lett.*, 43, 6953–6955, 2002.
512. Mach, M, Schlueter, U, Mathew, F, Fraser-Reid, B, Hazen, K C, Comparing *n*-pentenyl orthoesters and *n*-pentenyl glycosides as alternative glycosyl donors, *Tetrahedron*, 58, 7345–7354, 2002.
513. Kunz, H, Pfrengle, W, Effective 1,2-*trans*-glycosylation of complex alcohols and phenols using the oximate orthoester of *O*-pivaloyl glucopyranose, *J. Chem. Soc. Chem. Commun.*, 713–714, 1986.
514. Inaba, S, Yamada, M, Yoshino, T, Ishido, Y, Synthetic studies by the use of carbonates. IV. A new method for the synthesis of glycosyl compounds by the use of 1-*O*-aryloxycarbonyl sugar derivatives, *J. Am. Chem. Soc.*, 95, 2062–2063, 1973.
515. Ishido, Y, Inaba, S, Matsuno, A, Yoshino, T, Umezawa, H, Synthetic studies with carbonates. Part 11. D-Glucopyranose 1-carbonate derivatives; pyrolytic behavior and use for glucosylation of phenols, *J. Chem. Soc. Perkin. Trans. 1*, 1382–1390, 1977.
516. Mukaiyama, T, Miyazaki, K, Uchiro, H, Highly stereoselective synthesis of 1,2-*trans*-glycosides using *p*-chlorobenzylated glycosyl carbonate as glycosyl donor, *Chem. Lett.*, 635–636, 1998.
517. Takeuchi, K, Tamura, T, Mukaiyama, T, The trityl tetrakis(pentafluorophenyl)borate catalyzed stereoselective glycosylation using new glycosyldonor, 3,4,6-tri-*O*-benzyl-2-*O*-*p*-toluoyl-β-D-glucopyranosyl phenylcarbonate, *Chem. Lett.*, 122–123, 2000.
518. Mukaiyama, T, Ikegai, K, Jona, H, Hashiyata, T, Takeuchi, K, A convergent total synthesis of the mucin related F1α antigen by one-pot sequential stereoselective glycosylation, *Chem. Lett.*, 840–841, 2001.
519. Hashihayata, T, Ikegai, K, Takeuchi, K, Jona, H, Mukaiyama, T, Convergent total syntheses of oligosaccharides by one-pot sequential stereoselective glycosylations, *Bull. Chem. Soc. Jpn.*, 76, 1829–1848, 2003.
520. Iimori, T, Shibazaki, T, Ikegami, S, A novel intramolecular decarboxylative glycosylation via mixed carbonate, *Tetrahedron Lett.*, 37, 2267–2270, 1996.
521. Iimori, T, Azumaya, I, Shibazaki, T, Ikegami, S, An α-selective glycosylation via decarboxylation of mixed carbonate catalyzed by the combination of Lewis acid and silver perchlorate, *Heterocycles*, 46, 221–224, 1997.

522. Ford, M J, Ley, S V, A simple one-pot, glycosidation procedure via (1-imidazolylcarbonyl) glycosides and zinc bromide, *Synlett*, 255–256, 1990.
523. Ford, M J, Knight, J G, Ley, S V, Vile, S, Total synthesis of avermectin B1a. Synthesis of the carbohydrate bis-oleandrose fragment and coupling to the avermectin B1a aglycone, *Synlett*, 331–332, 1990.
524. Ley, S V, Armstrong, A, Diez-Martin, D, Ford, M J, Grice, P, Knight, J G, Kolb, H C, Madin, A, Marby, C A, Mukherjee, S, Shaw, A N, Slawin, A M Z, Vile, S, White, A D, Williams, D J, Woods, M, Total synthesis of the anthelmintic macrolide avermectin B1a, *J. Chem. Soc. Perkin. Trans. 1*, 667–692, 1991.
525. Lou, B, Huynh, H K, Hanessian, S, Oligosaccharide synthesis by remote activation: *O*-protected glycosyl 2-thiopyridylcarbonate donors, In *Preparative Carbohydrate Chemistry*, Hanessian, S, Ed., Marcel Dekker, New York, pp. 431–448, 1997.
526. Lou, B, Eckhardt, E, Hanessian, S, Oligosaccharide synthesis by selective anomeric activation with MOP- and TOPCAT-leaving group, In *Preparative Carbohydrate Chemistry*, Hanessian, S, Ed., Marcel Dekker, New York, pp. 449–466, 1997.
527. Hanessian, S, Mascitti, V, Rogel, O, Synthesis of a potent antagonist of E-selectin, *J. Org. Chem.*, 67, 3346–3354, 2002.
528. Pougny, J-R, Anomeric xanthates – a new activation of the anomeric center for rapid glycosylation, *J. Carbohydr. Chem.*, 5, 529–535, 1986.
529. Prata, C, Mora, N, Lacombe, J-M, Maurizis, J C, Pucci, B, Stereoselective synthesis of glycosyl carbamates as new surfactants and glycosyl donors, *Tetrahedron Lett.*, 38, 8859–8862, 1997.
530. Kunz, H, Zimmer, J, Glycoside synthesis via electrophile-induced activation of *N*-allyl carbamates, *Tetrahedron Lett.*, 34, 2907–2910, 1993.
531. Herzner, H, Eberling, J, Schultz, M, Zimmer, J, Kunz, H, Oligosaccharide synthesis via electrophile-induced activation of glycosyl-N-allylcarbamates, *J. Carbohydr. Chem.*, 17, 759–776, 1998.
532. Hinklin, R J, Kiessling, L L, Glycosyl sulfonylcarbamates: new glycosyl donors with tunable reactivity, *J. Am. Chem. Soc.*, 123, 3379–3380, 2001.
533. Mukaiyama, T, Chiba, H, Funasaka, S, Catalytic and stereoselective glycosylation with a novel and efficient disarmed glycosyl donor: glycosyl *p*-trifluoromethylbenzylthio-*p*-trifluoromethylphenyl formimidate, *Chem. Lett.*, 392–393, 2002.
534. Chiba, H, Funasaka, S, Mukaiyama, T, Catalytic and stereoselective glycosylation with glucosyl thioformimidates, *Bull. Chem. Soc. Jpn.*, 76, 1629–1644, 2003.
535. Chiba, H, Funasaka, S, Kiyota, K, Mukaiyama, T, Catalytic and chemoselective glycosylation between armed and disarmed glycosyl *p*-trifluoromethylbenzylthio-*p*-trifluoromethylphenyl formimidates, *Chem. Lett.*, 746–747, 2002.
536. Chiba, H, Mukaiyama, T, Highly α-stereoselective one-pot sequential glycosylation using glucosyl thioformimidate derivatives, *Chem. Lett.*, 32, 172–173, 2003.
537. Tietze, L-F, Fischer, R, Stereoselective synthesis of β-glucosides with 1,1′-diacetal structure, *Angew. Chem. Int. Ed. Engl.*, 20, 969–970, 1981.
538. Tietze, L-F, Fischer, R, Glycoside syntheses. 2. Stereoselective synthesis of α-glucosides with a 1,1′-diacetal structure, *Tetrahedron Lett.*, 22, 3239–3242, 1981.
539. Tietze, L-F, Fischer, R, Iridoids.19. Stereoselective synthesis of iridoid glycosides, *Angew. Chem. Int. Ed. Engl.*, 22, 888, 1983.
540. Tietze, L F, Fischer, R, Guder, H J, Goerlach, A, Neumann, M, Krach, T, Glucosidation. 6. Synthesis of acetal-α-glucosides. A stereoselective entry into a new class of compounds, *Carbohydr. Res.*, 164, 177–194, 1987.
541. Tietze, L-F, Fischer, R, Guder, H-J, Einfache und stereoselektive synthese von α- und β-phenylglykosiden, *Tetrahedron Lett.*, 23, 4661–4664, 1982.
542. Nashed, E M, Glaudemans, C P J, Coupling reactions of *O*-(trimethylsilyl) glycosides and 6-*O*-(tert-butyldiphenylsilyl)-protected galactosides in the presence of trimethylsilyl triflate. A new method of forming β-(1 → 6)-oligosaccharidic linkages, *J. Org. Chem.*, 54, 6116–6118, 1989.
543. Yoshimura, J, Hara, K, Sato, T, Hashimoto, H, The highly stereoselective synthesis of perbenzylated α,α-trehalose and its D-*galacto* and D-*manno* analogs, *Chem. Lett.*, 319–320, 1983.
544. Kolar, C, Kneissl, G, Semisynthetic rhodomycins: novel glycosylation methods for the synthesis of anthracycline oligosaccharides, *Angew. Chem. Int. Ed. Engl.*, 29, 809–811, 1990.

545. Kolar, C, Kneissl, G, Knödler, U, Dehmel, K, Semisynthetic ε-(iso)rhodomycins: a new glycosylation variant and modification reactions, *Carbohydr. Res.*, 209, 89–100, 1990.
546. Priebe, W, Grynkiewicz, G, Neamati, N, 2-Deoxy-1-*O*-silylated-β-hexopyranoses. Useful glycosyl donors and synthetic intermediates, *Tetrahedron Lett.*, 32, 2079–2082, 1991.
547. Chen, G-W, Kirschning, A, First preparation of spacer-linked cyclic neooligoaminodeoxysaccharides, *Chem. Eur. J.*, 8, 2717–2729, 2002.
548. Qiu, D-X, Wang, Y-F, Cai, M-S, Studies on glycosides. 6. Facile synthesis of alkyl α- and β-D-glucopyranosides, *Synth. Commun.*, 19, 3453–3456, 1989.
549. Mukaiyama, T, Matsubara, K, Stereoselective glycosylation reaction starting from 1-*O*-trimethylsilyl sugars by using diphenyltin sulfide and a catalytic amount of active acidic species, *Chem. Lett.*, 1041–1044, 1992.
550. Kolar, C, Dehmel, K, Moldenhauer, H, Gerken, M, Semisynthetic β-rhodomycins: a new approach to the synthesis of 4-*O*-methyl-β-rhodomycins, *J. Carbohydr. Chem.*, 9, 873–890, 1990.
551. Zurabyan, S E, Volosyuk, T P, Khorlin, A J, Oxazoline synthesis of 1,2-*trans*-2-acetamido-2-deoxyglycosides, *Carbohydr. Res.*, 9, 215–220, 1969.
552. Zurabyan, S E, Antonenko, T S, Khorlin, A Ya, Oxazoline synthesis of 1,2-*trans*-2-acetamido-2-deoxyglycosides. Glycosylation of secondary hydroxyl groups in partially protected saccharides, *Carbohydr. Res.*, 15, 21–27, 1970.
553. Yohino, T, Sato, K, Wanme, F, Takai, I, Ishido, Y, Efficient catalysis by pyridinium sulfonate in glycosylation involving an oxazoline intermediate derived from per-*O*-acetyl-*N*-acetyllactosamine and *N,N'*-diacetylchitobiose, *Glycoconjugate J.*, 9, 287–291, 1992.
554. Mohan, H, Gemma, E, Ruda, K, Oscarson, S, Efficient synthesis of spacer-linked dimers of *N*-acetyllactosamine using microwave-assisted pyridinium triflate-promoted glycosylations with oxazoline donors, *Synlett*, 1255–1256, 2003.
555. Wittmann, V, Lennartz, D, Copper(II)-mediated activation of sugar oxazolines: mild and efficient synthesis of β-glycosides of *N*-acetylglucosamine, *Eur. J. Org. Chem.*, 1363–1367, 2002.
556. Eby, R, Schuerch, C, The use of 1-*O*-tosyl-D-glucopyranose derivatives in α-D-glucoside synthesis, *Carbohydr. Res.*, 34, 79–90, 1974.
557. Lucas, T J, Schuerch, C, Methanolysis as a model reaction for oligosaccharide synthesis of some 6-substituted 2,3,4-tri-*O*-benzyl-D-galactopyranosyl derivatives, *Carbohydr. Res.*, 39, 39–45, 1975.
558. Srivastava, V K, Schuerch, C, Synthesis of β-D-mannopyranosides by glycosidation at C-1, *Carbohydr. Res.*, 79, C13–C16, 1980.
559. El Ashry, E S H, Schuerch, C, Synthesis of standardized building-blocks as a β-D-mannosyl donors with a temporary protection to be 3,6-di-*O*-glycosyl acceptors, for constructing the inner core of glycoproteins and artificial antigens, *Bull. Chem. Soc. Jpn.*, 59, 1581–1586, 1986.
560. Awad, L F, El Ashry, E S H, Schuerch, C, A synthesis of methyl 3-*O*-(β-D-mannopyranosyl)-α-D-mannopyranoside from sulfonate intermediates, *Bull. Chem. Soc. Jpn.*, 59, 1587–1592, 1986.
561. Cipolla, L, Lay, L, Nicotra, F, Panza, L, Russo, G, Glycosyl sulfates as glycosyl donors, *Tetrahedron Lett.*, 35, 8669–8670, 1994.
562. Mehta, S, Pinto, B M, Phenylselenoglycosides as novel, versatile glycosyl donors. Selective activation over thioglycosides, *Tetrahedron Lett.*, 32, 4435–4438, 1991.
563. Mehta, S, Pinto, B M, Novel glycosidation methodology. The use of phenyl selenoglycosides as glycosyl donors and acceptors in oligosaccharide synthesis, *J. Org. Chem.*, 58, 3269–3276, 1993.
564. Yamago, S, Kokubo, K, Murakami, H, Mino, Y, Hara, O, Yoshida, J, Glycosylation with telluroglycosides. Stereoselective construction of α- and β-anomers, *Tetrahedron Lett.*, 39, 7905–7908, 1998.
565. Zuurmond, H M, van der Klein, P A M, van der Meer, P H, van der Marel, G A, van Boom, J H, Iodonium ion-mediated glycosidations of phenyl selenoglycosides, *Recl. Trav. Chim. Pays-Bas*, 111, 365–366, 1992.
566. Heskamp, B M, Veeneman, G H, van der Marel, G A, van Boeckel, C A A, van Boom, J H, Design and synthesis of a trisubstrate analog for α(1→3)fucosyltransferase: a potential inhibitor, *Tetrahedron*, 51, 8397–8406, 1995.
567. Bols, M, Synthesis of kojitriose using silicon-tethered glycosidation, *Acta Chem. Scand.*, 50, 931–937, 1996.

568. Baeschlin, D K, Chaperon, A, Green, L G, Hahn, M G, Ince, S J, Ley, S V, 1,2-Diacetals in synthesis: total synthesis of a glycosylphosphatidylinositol anchor of *Trypanosoma brucei*, *Chem. Eur. J.*, 6, 172–186, 2000.
569. Furuta, T, Takeuchi, K, Iwamura, M, Activation of selenoglycosides by photoinduced electron transfer, *Chem. Commun.*, 157–158, 1996.
570. Mehta, S, Pinto, B M, Tris(4-bromophenyl)aminium hexachloroantimonate-mediated glycosylations of selenoglycosides and thioglycosides. Evidence for single electron transfer?, *Carbohydr. Res.*, 310, 43–51, 1998.
571. MacCoss, R N, Brennan, P E, Ley, S V, Synthesis of carbohydrate derivatives using solid-phase work-up and scavenging techniques, *Org. Biomol. Chem.*, 1, 2029–2031, 2003.
572. Grice, P, Ley, S V, Pietruszka, J, Osborn, H M I, Priepke, H W M, Warriner, S L, A new strategy for oligosaccharide assembly exploiting cyclohexane-1,2-diacetal methodology: an efficient synthesis of a high mannose type nonasaccharide, *Chem. Eur. J.*, 3, 431–440, 1997.
573. Yamago, S, Kokubo, K, Hara, O, Masuda, S, Yoshida, J, Electrochemistry of chalcogenoglycosides. Rational design of iterative glycosylation based on reactivity control of glycosyl donors and acceptors by oxidation potentials, *J. Org. Chem.*, 67, 8584–8592, 2002.
574. Stick, R V, Tilbrook, D M G, Williams, S J, The selective activation of telluro- over seleno-β-D-glucopyranosides as glycosyl donors: a reactivity scale for various telluro, seleno and thio sugars, *Aust. J. Chem.*, 50, 237–240, 1997.
575. Griffith, D A, Danishefsky, S J, Sulfonamidoglycosylation of glycals — a route to oligosaccharides with 2-aminohexose subunits, *J. Am. Chem. Soc.*, 112, 5811–5819, 1990.
576. Lafont, D, Guilloux, P, Descotes, G, A new synthesis of 1,2-*trans*-2-acetamido-2-deoxyglycopyranosides via 1,2-*trans*-2-deoxy-2-iodoglycosyl azides, *Carbohydr. Res.*, 193, 61–73, 1989.
577. Petö, C, Batta, G, Györgydeák, Z, Sztaricskai, F, Glycoside synthesis with anomeric 1-*N*-glycobiosyl-1,2,3-triazoles, *J. Carbohydr. Chem.*, 15, 465–483, 1996.
578. Pleuss, N, Kunz, H, *N*-Glycosyl amides: removal of the anomeric protecting group and conversion into glycosyl donors, *Angew. Chem. Int. Ed.*, 42, 3174–3176, 2003.
579. Klotz, W, Schmidt, R R, Anomeric *O*-alkylation of *O*-acetyl-protected sugars, *J. Carbohydr. Chem.*, 13, 1093–1101, 1994.
580. Schmidt, R R, Reichrath, M, Facile, highly selective synthesis of α- and β-disaccharides from 1-*O*-metalated D-ribofuranoses, *Angew. Chem. Int. Ed. Engl.*, 18, 466–467, 1979.
581. Tsvetkov, Y E, Klotz, W, Schmidt, R R, Anomeric *O*-alkylation. 9. Disaccharide synthesis via anomeric *O*-alkylation, *Liebigs Ann. Chem.*, 371–375, 1992.
582. Schmidt, R R, Klotz, W, Glycoside bond formation via anomeric *O*-alkylation: how many protective groups are required?, *Synlett*, 168–170, 1991.
583. Klotz, W, Schmidt, R R, Anomeric *O*-alkylation, 11. Anomeric *O*-alkylation of *O*-unprotected hexoses and pentoses — convenient synthesis of decyl, benzyl, and allyl glycosides, *Liebigs Ann. Chem.*, 683–690, 1993.
584. Vauzeilles, B, Dausse, B, Palmier, S, Beau, J-M, A one-step β-selective glycosylation of *N*-acetyl glucosamine and recombinant chitooligosaccharides, *Tetrahedron Lett.*, 42, 7567–7570, 2001.
585. Lemieux, R U, Levine, S, The products of the Prevost reaction on D-glucal triacetate, *Can. J. Chem.*, 40, 1926–1932, 1962.
586. Lemieux, R U, Levine, S, Synthesis of alkyl 2-deoxy-α-D-glucopyranosides and their 2-deuterio derivatives, *Can. J. Chem.*, 42, 1473–1480, 1964.
587. Lemieux, R U, Morgan, A R, The synthesis of β-D-glucopyranosyl 2-deoxy-α-D-*arabino*-hexopyranoside, *Can. J. Chem.*, 43, 2190–2198, 1965.
588. Tatsuta, K, Fujimoto, K, Kinoshita, M, Umezawa, S, Novel synthesis of 2-deoxy-α-glycosides, *Carbohydr. Res.*, 54, 85–104, 1977.
589. Thiem, J, Karl, H, Schwentner, J, Synthese α-verknüpfte 2′-deoxy-2′-iododisaccharide, *Synthesis*, 696–698, 1978.
590. Thiem, J, Karl, H, Synthesen von methyl-3-*O*-(α-D-olivosyl)-α-D-olivoside, *Tetrahedron Lett.*, 4999–5002, 1978.
591. Thiem, J, Klaffke, W, Facile stereospecific synthesis of deoxyfucosyl disaccharide units of anthracyclines, *J. Org. Chem.*, 54, 2006–2009, 1989.

592. Friesen, R W, Danishefsky, S J, On the controlled oxidative coupling of glycals: a new strategy for the rapid assembly of oligosaccharides, *J. Am. Chem. Soc.*, 111, 6656–6660, 1989.
593. Friesen, R W, Danishefsky, S J, On the use of the haloetherification method to synthesize fully functionalized disaccharides, *Tetrahedron*, 46, 103–112, 1990.
594. Takiura, K, Honda, S, Stereochemistry of the methoxymercuration of glycal acetates, *Carbohydr. Res.*, 21, 379–391, 1972.
595. Takiura, K, Honda, S, Hydroxy- and acetoxy-mercuration of D-glucal triacetate, *Carbohydr. Res.*, 23, 369–377, 1972.
596. Perez, M, Beau, J-M, Selenium-mediated glycosidations: a selective synthesis of β-2-deoxyglycosides, *Tetrahedron Lett.*, 30, 75–78, 1989.
597. Ito, Y, Ogawa, T, Novel approaches to glycoside synthesis. Sulfenate esters as glycosyl acceptors: a novel approach to the synthesis of 2-deoxyglycosides, *Tetrahedron Lett.*, 28, 2723–2726, 1987.
598. Preuss, R, Schmidt, R R, Glycosylimidates. 38. A convenient synthesis of 2-deoxy-β-D-glucopyranosides, *Synthesis*, 694–697, 1988.
599. Ramesh, S, Kaila, N, Grewal, G, Franck, R W, Aureolic acid antibiotics: a simple method for 2-deoxy-β-glycosidation, *J. Org. Chem.*, 55, 5–7, 1990.
600. Grewal, G, Kaila, N, Franck, R W, Arylbis(arylthio)sulfonium salts as reagents for the synthesis of 2-deoxy-β-glycosides, *J. Org. Chem.*, 57, 2084–2092, 1992.
601. Ciment, D M, Ferrier, R J, Unsaturated carbohydrates. Part IV. Allylic rearrangement reactions of 3,4,6-tri-*O*-acetyl-D-galactal, *J. Chem. Soc. C*, 441–445, 1966.
602. Wakamatsu, T, Nakamura, H, Naka, E, Ban, Y, Synthetic studies on antibiotic macrodiolide: synthesis of the A-segment of elaiophylin, *Tetrahedron Lett.*, 27, 3895–3898, 1986.
603. Tatsuta, K, Kobayashi, Y, Gunji, H, Masuda, H, Synthesis of oleandomycin through the intact aglycone, oleandolide, *Tetrahedron Lett.*, 29, 3975–3978, 1988.
604. Toshima, K, Tatsuta, K M, Total synthesis of elaiophylin (azalomycin-B), *Bull. Chem. Soc. Jpn.*, 61, 2369–2381, 1988.
605. Tu, C J, Lednicer, D, Total synthesis of 4-demethoxy-13-dihydro-8-nordaunomycin, *J. Org. Chem.*, 52, 5624–5627, 1987.
606. Sabesan, S, Neira, S, Synthesis of 2-deoxy sugars from glycals, *J. Org. Chem.*, 56, 5468–5472, 1991.
607. Bolitt, V, Mioskowski, C, Lee, S-G, Falck, J R, Direct preparation of 2-deoxy-D-glucopyranosides from glucals without Ferrier rearrangement, *J. Org. Chem.*, 55, 5812–5813, 1990.
608. Toshima, K, Nagai, H, Ushiki, Y, Matsumura, S, Novel glycosidations of glycals using BCl_3 or BBr_3 as a promoter for catalytic and stereoselective syntheses of 2-deoxy-α-glycosides, *Synlett*, 1007–1009, 1998.
609. Pachamuthu, K, Vankar, Y D, Ceric ammonium nitrate-catalyzed tetrahydropyranylation of alcohols and synthesis of 2-deoxy-*O*-glycosides, *J. Org. Chem.*, 66, 7511–7513, 2001.
610. Di Bussolo, V, Liu, J, Huffman, L G, Gin, D Y, Acetamidoglycosylation with glycal donors: a one-pot glycosidic coupling with direct installation of the natural C(2)-*N*-acetylamino functionality, *Angew. Chem. Int. Ed.*, 39, 204–207, 2000.
611. Liu, J, Gin, D Y, C2-Amidoglycosylation. Scope and mechanism of nitrogen transfer, *J. Am. Chem. Soc.*, 124, 9789–9797, 2002.
612. Kim, J-Y, Di Bussolo, V, Gin, D Y, Stereoselective synthesis of 2-hydroxy-α-mannopyranosides from glucal donors, *Org. Lett.*, 3, 303–306, 2001.
613. Honda, E, Gin, D Y, C-2 Hydroxyglycosylation with glycal donors. Probing the mechanism of sulfonium-mediated oxygen transfer to glycal enol ethers, *J. Am. Chem. Soc.*, 124, 7343–7352, 2002.
614. Shi, L, Kim, Y-J, Gin, D Y, C2-Acyloxyglycosylation with glycal donors, *J. Am. Chem. Soc.*, 123, 6939–6940, 2001.
615. Ferrier, R J, Prasad, N, The application of unsaturated carbohydrates to glycoside syntheses: 6-*O*-α-D-mannopyranosyl-, 6-*O*-α-D-altropyranosyl-, and 6-*O*-(3,6-anhydro-α-D-glucopyranosyl)-D-galactose, *J. Chem. Soc. Chem. Commun.*, 476–477, 1968.
616. Descotes, G, Martin, J-C, Sur l'isomérisation du 1,5-anhydro-3,4,6-tri-*O*-benzyl-1,2-didésoxy-D-*arabino*-hex-1-énitol en présence d'acides de Lewis, *Carbohydr. Res.*, 56, 168–172, 1977.
617. Klaffke, W, Pudlo, P, Springer, D, Thiem, J, Artificial deoxy glycosides of anthracyclines, *Liebigs Ann. Chem.*, 509–512, 1991.

618. Grynkiewicz, G, Priebe, W, Zamojski, A, Synthesis of alkyl 4,6-di-*O*-acetyl-2,3-dideoxy-α-D-*threo*-hex-2-enopyranosides from 3,4,6-tri-*O*-acetyl-1,5-anhydro-2-deoxy-D-*lyxo*-hex-1-enitol (3,4,6-tri-*O*-acetyl-D-galactal), *Carbohydr. Res.*, 68, 33–41, 1979.
619. Bhate, P, Horton, D, Priebe, W, Allylic rearrangement of 6-deoxyglycals having practical utility, *Carbohydr. Res.*, 144, 331–337, 1985.
620. Lopez, J C, Fraser-Reid, B, *n*-Pentenyl esters facilitate an oxidative alternative to the Ferrier rearrangement. An expeditious route to sucrose, *J. Chem. Soc. Chem. Commun.*, 94–96, 1992.
621. Lopez, J C, Gomez, A M, Valverde, S, Fraser-Reid, B, Ferrier rearrangement under nonacidic conditions based on iodonium-induced rearrangements of allylic *n*-pentenyl esters, *n*-pentenyl glycosides, and phenyl thioglycosides, *J. Org. Chem.*, 60, 3851–3858, 1995.
622. Toshima, K, Ishizuka, T, Matsuo, G, Nakata, M, Kinoshita, M, Glycosidation of glycals by 2,3-dichloro-5,6-dicyano-*p*-benzoquinone (DDQ) as a catalytic promoter, *J. Chem. Soc. Chem. Commun.*, 704–706, 1993.
623. Toshima, K, Ishizuka, T, Matsuo, G, Nakata, M, Practical glycosidation method of glycals using montmorillonite K-10 as an environmentally acceptable and inexpensive industrial catalyst, *Synlett*, 306–308, 1995.
624. Koreeda, M, Houston, T A, Shull, B K, Klemke, E, Tuinman, R J, Iodine-catalyzed Ferrier reaction. 1. A mild and highly versatile glycosylation of hydroxyl and phenolic groups, *Synlett*, 90–92, 1995.
625. Masson, C, Soto, J, Bessodes, M, Ferric chloride: a new and very efficient catalyst for the Ferrier glycosylation reaction, *Synlett*, 1281–1282, 2000.
626. Babu, B S, Balasubramanian, K K, Indium trichloride catalyzed glycosidation. An expeditious synthesis of 2,3-unsaturated glycopyranosides, *Tetrahedron Lett.*, 41, 1271–1274, 2000.
627. Das, S K, Reddy, K A, Roy, J, Microwave-induced, $InCl_3$-catalyzed Ferrier rearrangement of acetylglycals: synthesis of 2,3-unsaturated glycopyranosides, *Synlett*, 1607–1610, 2003.
628. Yadav, J S, Reddy, B V S, Murthy, C V S R, Kumar, G M, Scandium triflate catalyzed Ferrier rearrangement: an efficient synthesis of 2,3-unsaturated glycopyranosides, *Synlett*, 1450–1451, 2000.
629. Takhi, M, Abdel-Rahman, A A H, Schmidt, R R, Highly stereoselective synthesis of pseudoglycals via $Yb(OTf)_3$ catalyzed Ferrier glycosylation, *Synlett*, 427–429, 2001.
630. Abdel-Rahman, A A H, Winterfeld, G A, Takhi, M, Schmidt, R R, Trichloroacetimidate as a leaving group in the Ferrier rearrangement: highly stereoselective synthesis of pseudogalactal glycosides, *Eur. J. Org. Chem.*, 713–717, 2002.
631. Briner, K, Vasella, A, Clycosylidene carbenes. A new approach to glycoside synthesis. Part 1, *Helv. Chim. Acta*, 71, 1371–1382, 1989.
632. Vasella, A, Glycosylidene carbenes, In *Bioorganic Chemistry: Carbohydrates*, Hecht, S M, Ed., Oxford University Press, Oxford, pp. 56–88, 1999.
633. Briner, K, Vasella, A, Glycosylidene Carbenes. 2. Synthesis of *O*-aryl glycosides, *Helv. Chim. Acta*, 73, 1764–1778, 1990.
634. Barrett, A G M, Lee, A C, Redox glycosidation: stereoselective syntheses of 1→6 linked disaccharides via thionoester intermediates, *J. Org. Chem.*, 57, 2818–2824, 1992.
635. Ohtake, H, Ichiba, N, Ikegami, S, A highly stereoselective construction of β-glycosyl linkages by reductive cleavage of cyclic sugar ortho esters, *J. Org. Chem.*, 65, 8171–8179, 2000.
636. Paquet, F, Sinaÿ, P, New stereocontrolled approach to 3-deoxy-D-*manno*-2-octulosonic acid containing disaccharides, *J. Am. Chem. Soc.*, 106, 8313–8315, 1984.
637. Wong, C-H, Halcomb, R L, Ichikawa, Y, Kajimoto, T, Enzymes in organic synthesis: application to the problems of carbohydrate recognition, *Angew. Chem. Int. Ed. Engl.*, 34, 412–432, 1995.
638. Wong, C-H, Halcomb, R L, Ichikawa, Y, Kajimoto, T, Enzymes in organic synthesis: application to the problems of carbohydrate recognition (part 2), *Angew. Chem. Int. Ed. Engl.*, 34, 521–546, 1995.
639. David, S, Augé, C, Gautheron, C, Enzymic methods in preparative carbohydrate chemistry, *Adv. Carbohydr. Chem. Biochem.*, 49, 175–327, 1991.
640. Koeller, K M, Wong, C-H, Synthesis of complex carbohydrates and glycoconjugates: enzyme-based and programmable one-pot strategies, *Chem. Rev.*, 100, 4465–4493, 2000.

5 Oligosaccharide Synthesis

Péter Fügedi

CONTENTS

5.1 INTRODUCTION

It is now well established that glycoconjugates play essential roles in a variety of biological processes, including fertilization, embryogenesis, cell proliferation and cell adhesion [1,2]. Changes in carbohydrate structures are also associated with disease states. The remarkable changes in cell surface carbohydrates that occur during tumor progression are of notable importance. The biological activities that carbohydrates play in these processes are typically linked not to monosaccharide units, but to oligosaccharide structures of glycoconjugates. Just as in the fields of nucleic acids and peptides, the synthesis of oligosaccharides is an essential tool in studying the biochemistry and biology of vital processes.

Compared with other natural oligomers, such as peptides and oligonucleotides, oligosaccharides have much greater structural diversity. Aside from the number of monomers, the differences in anomeric configuration, ring size and, especially, the possibly different attachment points of the monomers (including the occurrence of branched structures) dramatically increase the diversity of oligosaccharides. Whereas only six tripeptides can be formed from three different amino acids,

Abbreviations: Ac: acetyl; AIBN: azobisisobutyronitrile; All: allyl; Bn: benzyl; Bz: benzoyl; ClAc: chloroacetyl; DAST: diethylaminosulfur trifluoride; DBU: 1,8-diazabicyclo[5.4.0]-undec-7-ene; DDQ: 2,3-dichloro-5,6-dicyano-*p*-benzoquinone; DMDO: dimethyldioxirane; DMTST: dimethyl(methylthio)sulfonium trifluoromethanesulfonate; Fmoc: 9-fluorenyl-methoxycarbonyl; HDTC: hydrazine dithiocarboxylate; IDCP: iodonium di-collidine perchlorate; Lev: levulinoyl; MBz: 4-methylbenzoyl; Me: methyl; MEK: methyl ethyl ketone; MP: 4-methoxyphenyl; NBS: *N*-bromosuccinimide; NIS: *N*-iodosuccinimide; Pent: *n*-pentenyl; Pfp: pentafluorophenyl; Ph: phenyl; Phth: phthaloyl; Piv: pivaloyl; PMB: 4-methoxybenzyl; TBAF: tetrabutylammonium fluoride; TBDMS: *tert*-butyldimethylsilyl; TBDPS: *tert*-butyldiphenylsilyl; TCA: trichloroacetyl; TES: triethylsilyl; Tf: trifluoromethanesulfonyl; TMS: trimethylsilyl; Tol: 4-methylphenyl; Tr: trityl; Troc: 2,2,2-trichloroethoxycarbonyl; Ts: tosyl.

1056 trisaccharides can possibly be formed from three different monosaccharides, taking into account the pyranose forms only [3]. The differences are even more pronounced in the biologically important size range; the number of structurally possible hexasaccharides in mammalian systems was calculated to be in the order of 10^{12} [4].

As a result of the much higher complexity of structures, it is not surprising that the synthesis of oligosaccharides is more demanding than that of peptides or oligonucleotides of the same size. Whereas in the synthesis of oligosaccharides the ring sizes, linkage positions and anomeric stereoselectivity should be strictly controlled, there are no such requirements in, for example, peptide synthesis. As a consequence, although syntheses of peptides and oligonucleotides are easily performed today by automated techniques, no such routine and generally applicable synthetic methods are available for oligosaccharides. Nevertheless, a very large number of oligosaccharides have been synthesized, and syntheses of large and highly complex structures, some even exceeding 20-mer [5], have been published.

Given the complexity of the field and the vast amount of approaches used, this chapter does not attempt to cover all aspects of oligosaccharide synthesis. Instead, the basic concept of oligosaccharide synthesis is demonstrated on the preparation of a simple disaccharide. Furthermore, strategies and selected techniques for the synthesis of higher and structurally more complex oligosaccharides are discussed.

5.2 GENERAL CONCEPT OF OLIGOSACCHARIDE SYNTHESIS

As mentioned in the introduction, the sequence, the ring sizes of the monosaccharides, the linkage positions and the anomeric stereochemistry should be controlled in the synthesis of oligosaccharides. The sequence, which involves the directionality of the monosaccharide units, is assured by differentiating the coupling components as glycosyl donor and glycosyl acceptor, respectively. The donor, corresponding to the nonreducing end part, will be attached through its anomeric hydroxyl to some of the alcoholic hydroxyl groups of the acceptor, which represents the reducing end part. (It should be noted that in the synthesis of nonreducing oligosaccharides, the acceptor also reacts at its anomeric hydroxyl group.) The proper ring sizes are guaranteed by using protected derivatives of the required ring size. Regiochemical control can also be assured in this way — using a partially protected acceptor with a single hydroxyl group. The control of anomeric stereochemistry can be achieved by careful choice of the various glycosylation methods, protecting groups and reaction conditions.

This concept is illustrated by one of the early oligosaccharide syntheses, the synthesis of the disaccharide gentiobiose (β-D-Glc*p*-(1 → 6)-D-Glc) by Helferich [6] (Scheme 5.1). The synthesis consists of the following parts: preparation of the glycosyl donor, preparation of the partially protected acceptor, coupling of the glycosyl donor with the acceptor and deprotection of the resulting coupling product.

As illustrated, the glycosyl donor, acetobromoglucose (**2**), is readily obtained from D-glucose (**1**) by acetylation, followed by replacement of the anomeric acetoxy group with bromine. The glycosyl acceptor (**3**), having only the primary hydroxyl free, was obtained from **1** by regioselective tritylation at O-6, followed by acetylation and detritylation. Coupling of the acceptor (**3**) with the donor (**2**) under Koenigs–Knorr conditions afforded the protected disaccharide (**4**) from which the free disaccharide (**5**) is obtainable by deacetylation. In the final product, the β anomeric configuration of the interglycosidic linkage was provided by neighboring group participation in the glycosylation step. The ring sizes and the linkage position in the coupling product were assured by the use of appropriately protected derivatives, and also by the stability of the protecting groups in the glycosylation step. However, it should be noted that neither the ring size nor the anomeric configuration of the reducing-end unit in the unprotected derivative (**5**) is fixed, as the free hemiacetal might undergo ring-opening and anomerization, resulting in different isomeric forms.

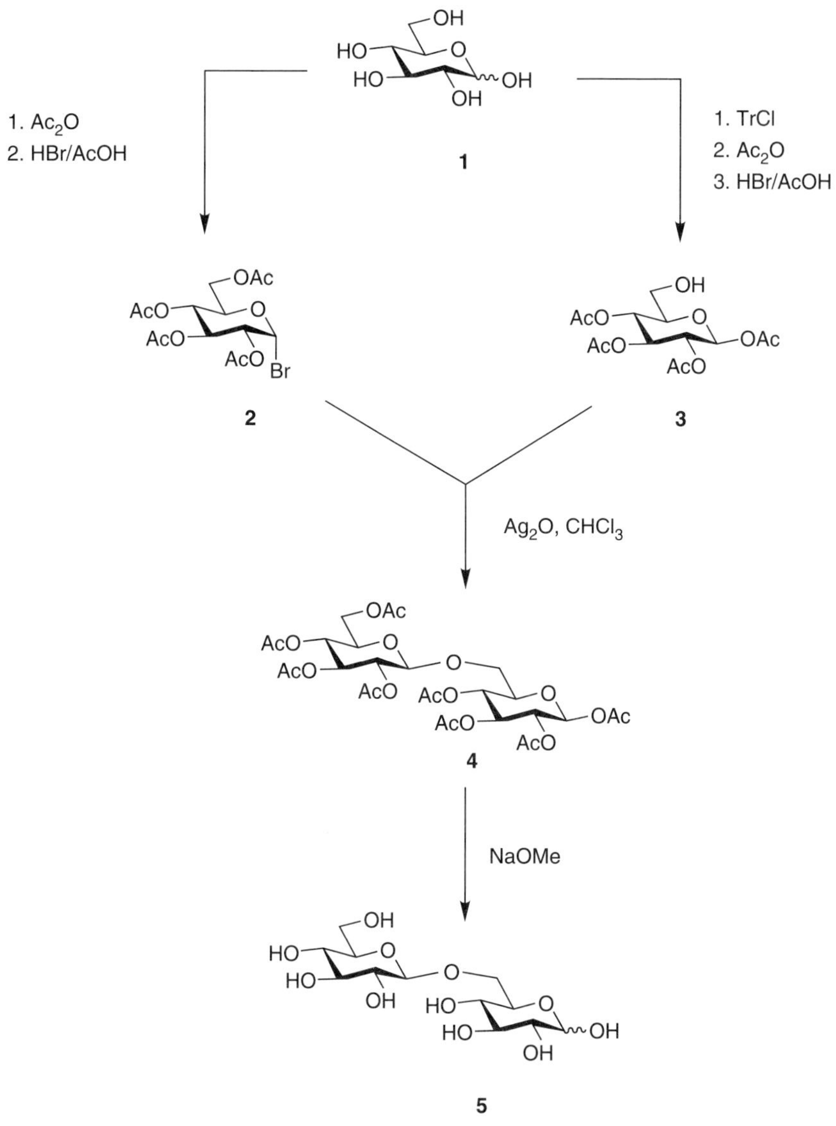

SCHEME 5.1 Synthesis of gentiobiose.

To avoid this problem, oligosaccharides are often synthesized in the form of some of their glycosides and not as free sugars.

5.3 STEPWISE AND BLOCK SYNTHESES OF OLIGOSACCHARIDES

For the synthesis of higher oligosaccharides, two major strategies can be considered. These include the stepwise synthesis and the block synthesis. These two strategies are illustrated for the synthesis of a linear *ABCD* tetrasaccharide in Scheme 5.2 and Scheme 5.3.

In the stepwise synthesis (Scheme 5.2), the reducing end unit (**7**) is first glycosylated by the donor (**6**) corresponding to the *C* unit, and the monosaccharide donors (**10** and **13**) of the further units are then added one by one to the growing oligosaccharide chain. Each coupling step is followed by the removal of a temporary protecting group to liberate the acceptor hydroxyl group for

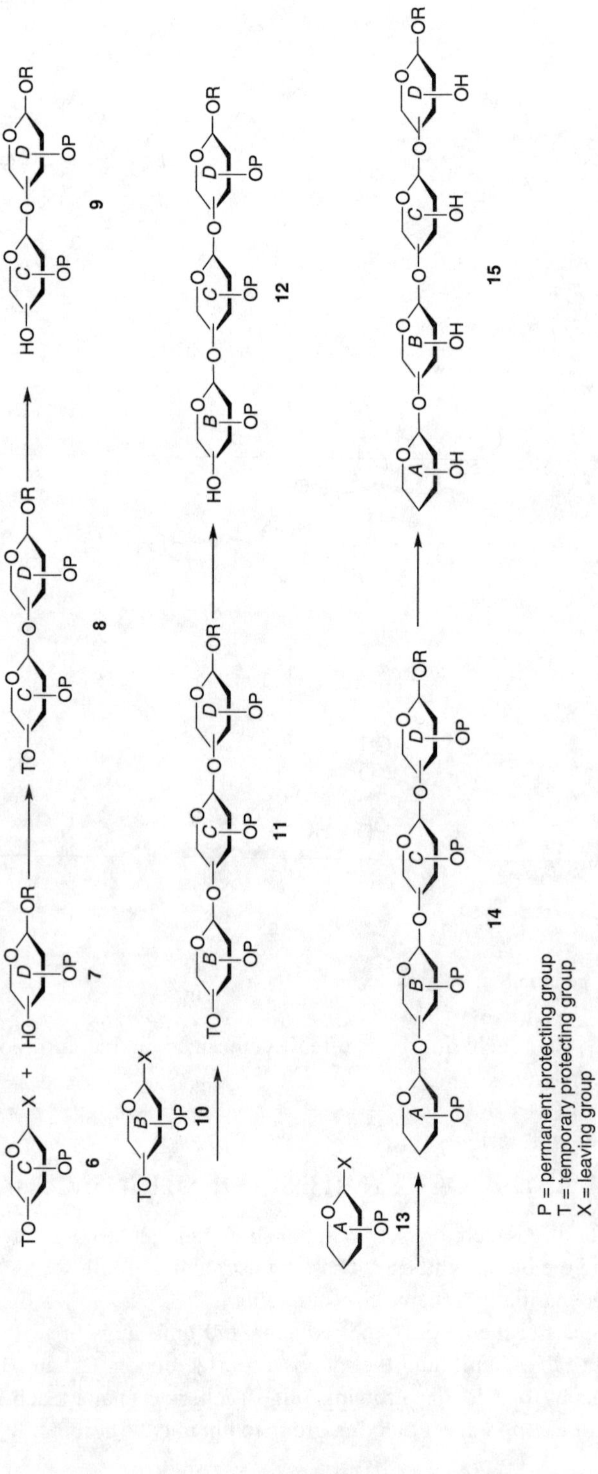

P = permanent protecting group
T = temporary protecting group
X = leaving group

SCHEME 5.2 Stepwise synthesis of oligosaccharides.

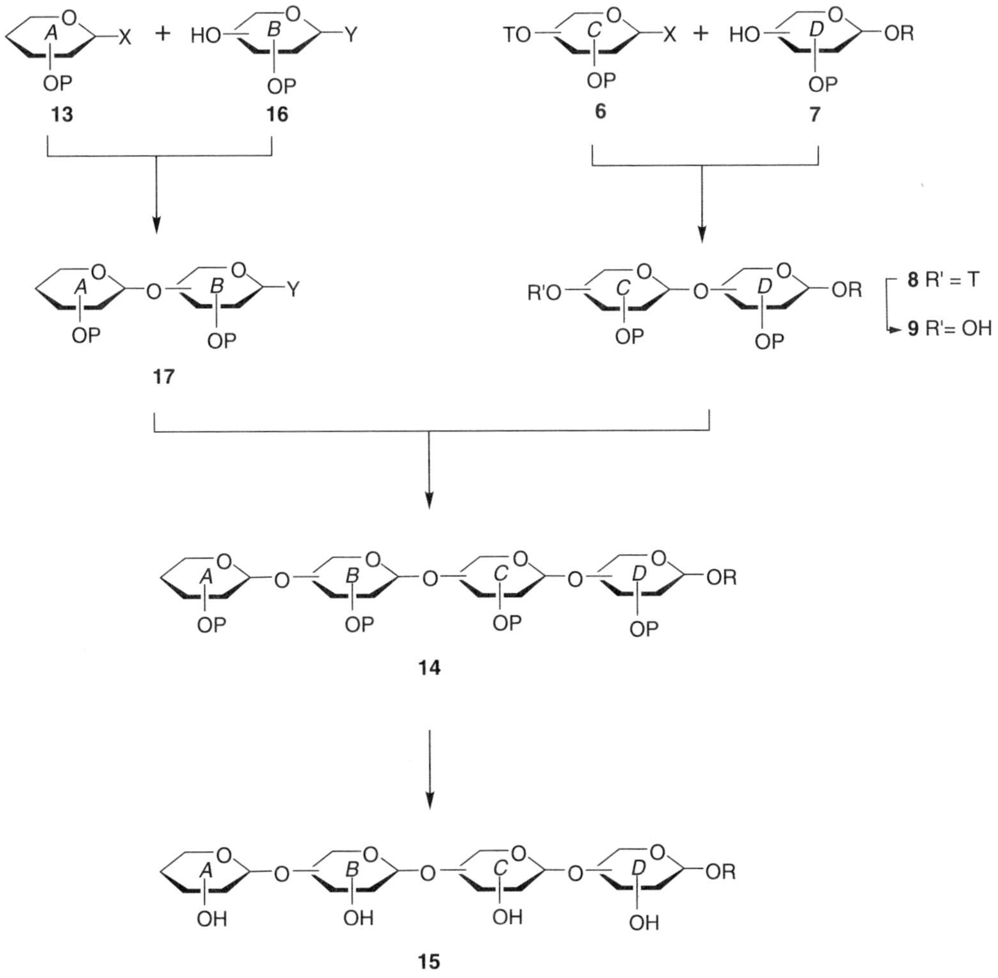

SCHEME 5.3 Block synthesis of oligosaccharides.

the next coupling. Removal of the permanent protecting groups from the protected *ABCD* tetrasaccharide (**14**) affords the free tetrasaccharide (**15**) as its R glycoside.

In the block synthesis, the target oligosaccharide is assembled from a few smaller oligosaccharides. As shown in Scheme 5.3, for the synthesis of an *ABCD* oligosaccharide the *AB* (**17**) and *CD* (**8**) disaccharide blocks are synthesized in parallel and coupled after a single deprotection step (**8** → **9**) to afford the protected tetrasaccharide (**14**).

Nicolaou's synthesis [7] of the sialyl-Lewis^x tetrasaccharide is discussed in detail as an example of the stepwise route (Scheme 5.4).

As illustrated, condensation of the D-galactosyl fluoride (**18**) with the *N*-acetyl-D-glucosamine acceptor (**19**) using $SnCl_2$–$AgClO_4$ as a promoter afforded the lactosamine derivative (**20**). After removal of the temporary allyl protecting group, the resulting **21** was glycosylated with the L-fucopyranosyl fluoride (**22**) to give the trisaccharide (**23**). Deacetylation afforded the triol (**24**), which was coupled with the sialic acid derivative (**25**) possessing a stereodirecting thiophenyl auxiliary. The α-(1→3)-linked tetrasaccharide (**26**) was obtained in a highly regio- and stereoselective manner. Reductive desulfurization resulted in the formation of the δ-lactone (**27**). Alkaline hydrolysis of the lactone, followed by catalytic hydrogenolysis of the benzyl groups, gave the sialyl-Lewis^x tetrasaccharide (**28**).

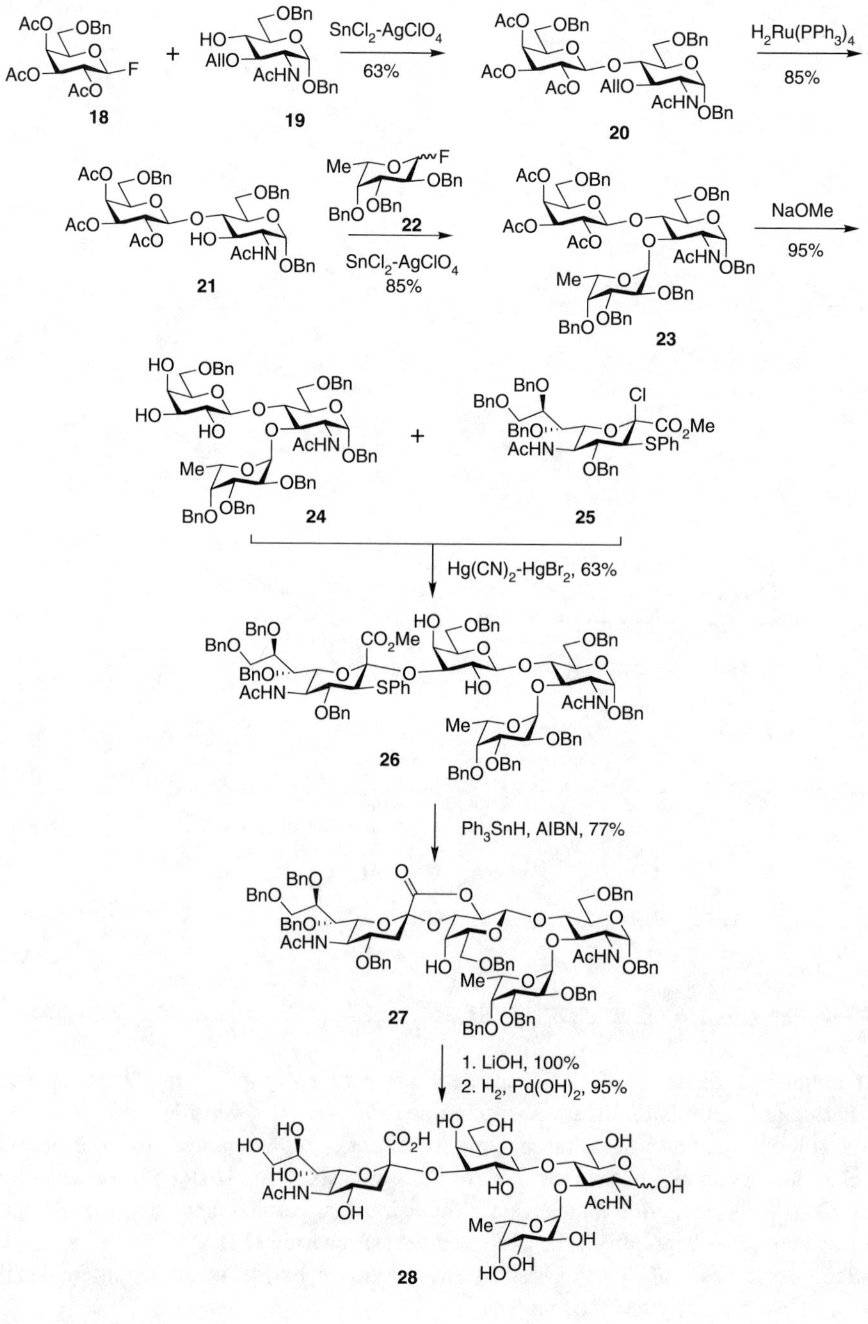

SCHEME 5.4 Synthesis of sialyl-Lewis^x.

The block synthesis strategy is demonstrated here by the synthesis of the dimeric-Lewis^x octasaccharide (**29**) described by Schmidt and coworkers [8,9]. As shown in the retrosynthetic analysis (Scheme 5.5), the octasaccharide (**29**) should be accessible from the tri- (**30**) and disaccharide (**31**) units, represented by the blocks **32** and **33**. The trisaccharide block can be disconnected from the L-fucosyl (**34**) and D-galactosyl (**35**) imidates and the 2-azido-2-deoxy-D-glucose derivative (**36**).

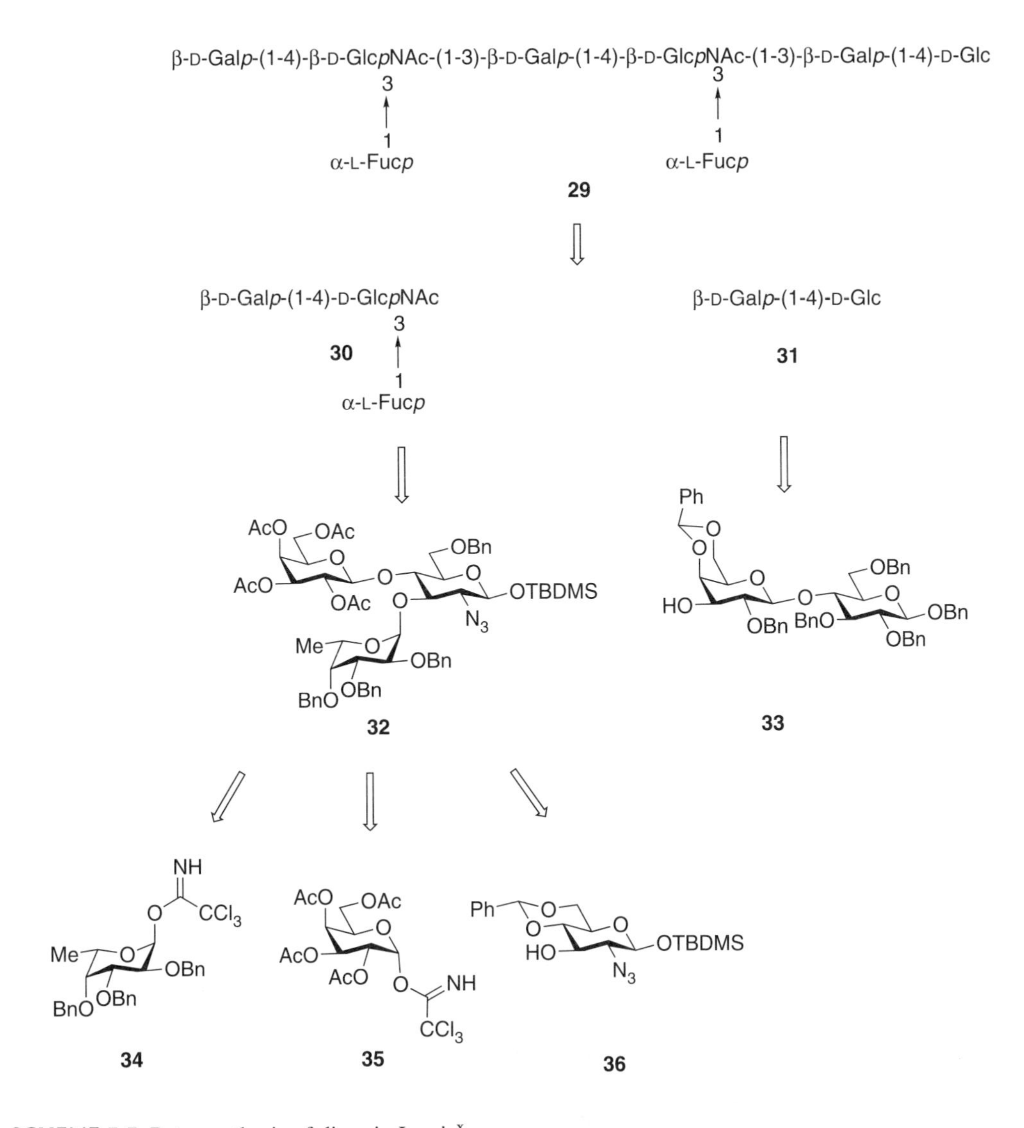

SCHEME 5.5 Retrosynthesis of dimeric-Lewis^x.

For the synthesis of the block, **36** was glycosylated with **34** using the "inverse procedure" [10] affording a high yield of disaccharide (**37**) (Scheme 5.6). Reductive ring-opening of the benzylidene acetal gave the 4-OH derivative (**38**), which was coupled with the imidate (**35**) to yield the trisaccharide block (**32**).

Both the glycosyl donor (**39**) and the glycosyl acceptor (**40**) were obtained from this trisaccharide block (Scheme 5.7). Removal of the *tert*-butyldimethylsilyl group gave the hemiacetal, which was treated with trichloroacetonitrile to give the imidate (**39**). Zemplén deacetylation and subsequent benzylidenation afforded the diol (**40**), which served as the glycosyl acceptor in the next step. Low-temperature coupling of **39** with **40** in acetonitrile, followed by acetylation, gave the β-(1→3)-linked hexasaccharide (**41**). The regio- and stereoselective formation of **41** is due to the higher reactivity of O-3 of the galactose unit and to nitrilium ion intermediates as a result of solvent participation. The hexasaccharide (**41**) was activated for further glycosylation by converting it to the trichloroacetimidate (**43**) via the hemiacetal (**42**). Reaction of the imidate (**43**) with the lactose acceptor (**33**) gave the protected octasaccharide (**44**), which was transformed into the free octasaccharide (**29**) via the peracetylated derivative.

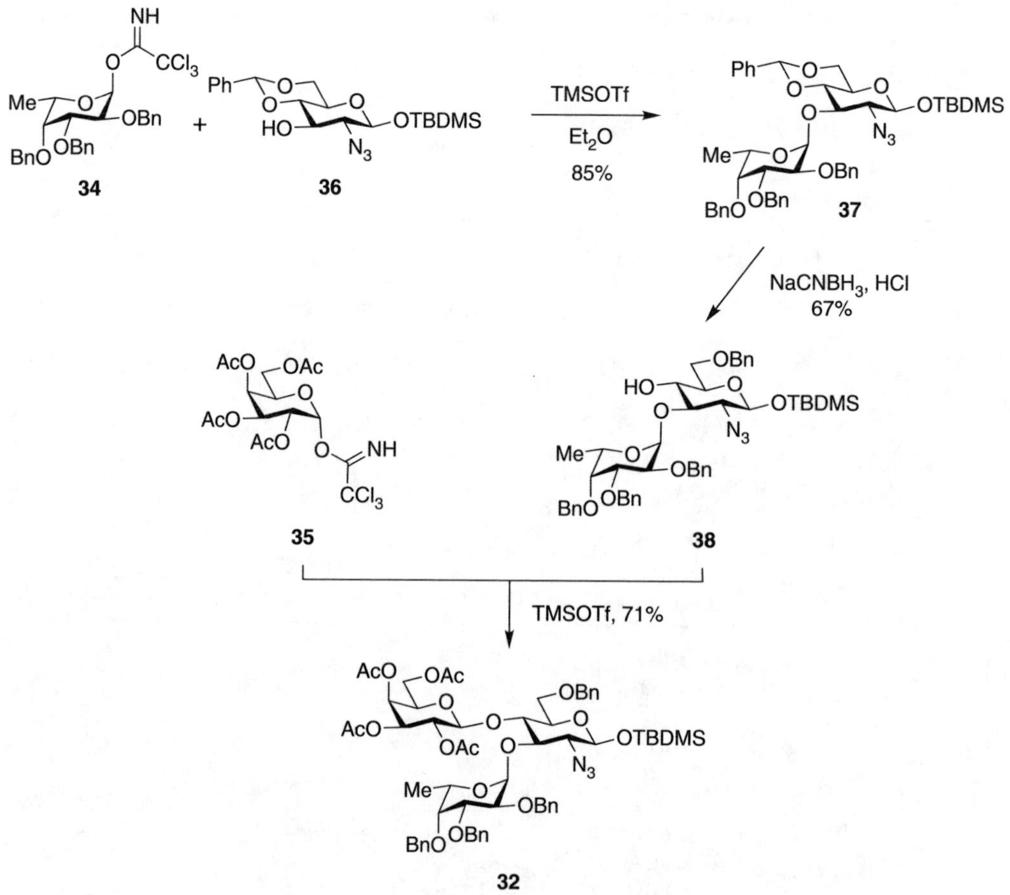

SCHEME 5.6 Synthesis of the trisaccharide building block of dimeric-Lewis^{x}.

Using the same or similar strategies, higher homologs and a series of related structures have also been reported [8,9,11–13].

In comparing stepwise and block syntheses, the stepwise strategy is primarily suitable for smaller oligosaccharides. For larger structures, the linearity of the synthetic sequence makes this method lengthy and less efficient. Another problem associated with the stepwise synthesis is that the sequence of steps is dictated by the structure of the target compounds, and critical steps often must be performed at late stages of the syntheses where tedious chromatographic separations can be problematic. In contrast, block syntheses reduce the overall number of steps and the convergent nature of this strategy makes it more time efficient. Also, block syntheses afford greater flexibility in synthetic design. Specifically, critical steps can be performed at early stages on smaller molecules, thereby facilitating chromatographic separations. Also, for the assembly of the oligosaccharides from the blocks, those glycosylations can be chosen which are easy to perform and commonly produce high yields.

Stepwise syntheses of oligosaccharides were frequently used in the Koenigs–Knorr era of glycosylations. Block syntheses were often difficult since the preparation of glycosyl bromides requires relatively harsh conditions and this type of glycosyl donor suffers from instability; therefore, stepwise synthesis was favored. However, with the advent of new stable glycosyl donors, the way for block syntheses was opened and oligosaccharide synthesis is no longer restricted to the stepwise strategy.

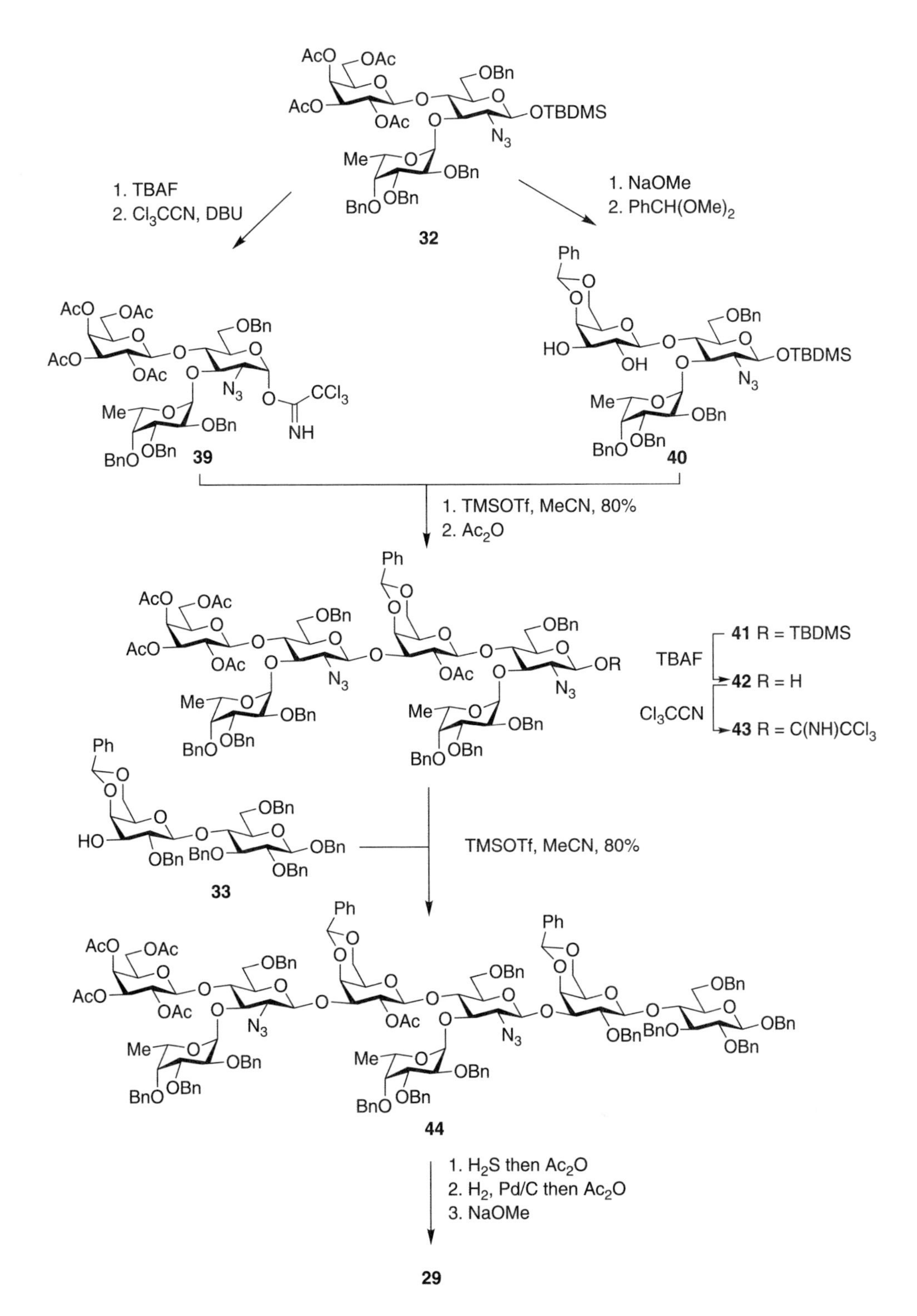

SCHEME 5.7 Synthesis of dimeric-Lewis^{x}.

5.4 GLYCOSYLATION STRATEGIES IN BLOCK SYNTHESES

The reducing end unit of a block and its anomeric substituent (Y in Scheme 5.3) has to fulfill opposing requirements. It should be stable enough to serve as a glycosyl acceptor in the synthesis of the block and, at the same time, it should be reactive enough to serve as a glycosyl donor in the assembly of the oligosaccharide from the blocks. Several solutions to this problem have been developed and these will be discussed in the following sections.

5.4.1 REACTIVATION BY EXCHANGE OF THE ANOMERIC SUBSTITUENT

One solution to this problem is to exchange the anomeric substituent after the synthesis of the block. The reducing end is protected by a temporary protecting group (T in Scheme 5.8) during the synthesis of the block. This protecting group is then removed from the block (**46**), which is reactivated by converting it into a glycosyl donor (**47**) with a leaving group (X) at the reducing end. Glycosylation is then performed with **47** to give the protected oligosaccharide (**49**).

This strategy was followed in the example shown in Scheme 5.5, Scheme 5.6 and Scheme 5.7 using the *tert*-butyldimethylsilyl ether as the temporary protecting group. Silyl ethers, allyl, *p*-methoxyphenyl and 2-(trimethylsilyl)ethyl groups are all commonly used for temporary protection of the anomeric position.

A drawback of this strategy is that the conversion of the block into a glycosyl donor requires several steps, which are especially undesirable in the case of larger fragments. Fortunately, other strategies have been developed which eliminate the need for separate activation steps.

5.4.2 SEQUENTIAL GLYCOSYLATIONS WITH DIFFERENT TYPES OF GLYCOSYL DONORS

In this strategy, a set of different types of glycosyl donors is used, one of which can be glycosylated with the other. As Scheme 5.9 shows, the type 2 donor (**50**), which has an anomeric substituent that withstands glycosylation with a type 1 donor (**13**), is initially used as an acceptor to provide the block **51**. This block then can be used directly as a donor, without any replacement of the anomeric substituent, to provide the oligosaccharide (**49**).

An example of this strategy is the synthesis of the heptasaccharide (**52**) which has phytoelicitor activity [14]. As shown in the retrosynthetic analysis (Scheme 5.10), the heptasaccharide (**52**) is disconnected into the trisaccharide thioglycoside (**53**) and the simple D-glucose derivative (**54**). The trisaccharide block (**53**) is available from the glycosyl bromides (**55**, **56**) and the thioglycoside (**57**).

Silver triflate-promoted glycosylation of **57** with the bromide (**55**) afforded an 81% yield of disaccharide (**58**) (Scheme 5.11). The 6-OH derivative (**59**) was prepared by reductive opening of

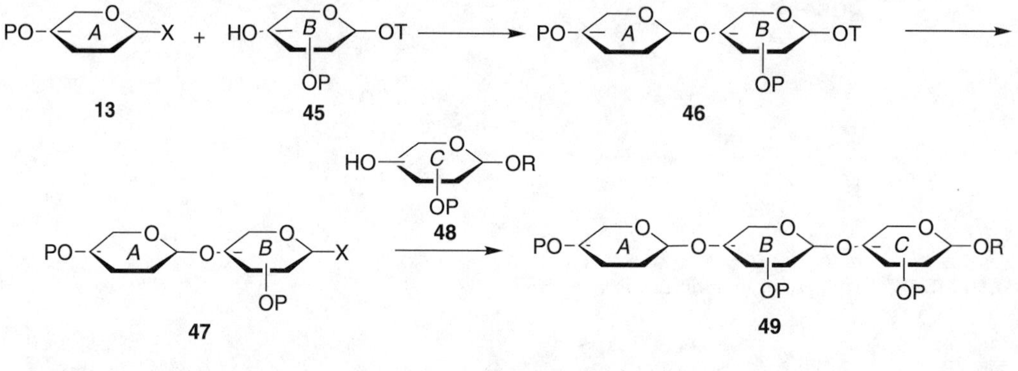

PO–A–X + HO–B(OP)–OT → PO–A–O–B(OP)–OT →
13 45 46

PO–A–O–B(OP)–X + HO–C(OP)–OR (48) → PO–A–O–B(OP)–O–C(OP)–OR
47 49

SCHEME 5.8 Block synthesis by reactivation via exchange of the anomeric substituent.

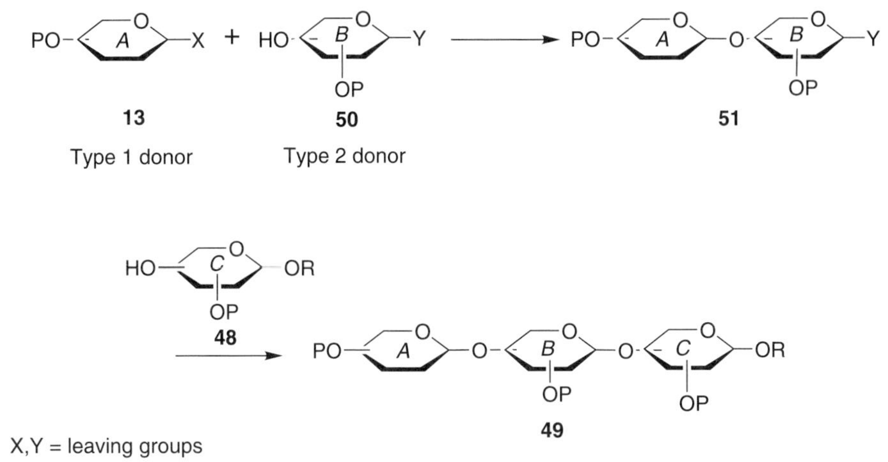

PO–A–X + HO–B(OP)–Y → PO–A–O–B(OP)–Y

13 Type 1 donor; 50 Type 2 donor; 51

HO–C(OP)–OR **48** → PO–A–O–B(OP)–O–C(OP)–OR **49**

X,Y = leaving groups

SCHEME 5.9 Block synthesis by sequential glycosylations with different types of glycosyl donors.

the benzylidene ring, then was glycosylated with the glycosyl bromide (**56**) having a temporary chloroacetyl protecting group giving the trisaccharide block (**53**). This was used directly in the next step, in methyl triflate-promoted coupling with the glycosyl acceptor (**54**), which afforded an 83% yield of tetrasaccharide (**60**). Removal of the temporary chloroacetyl group provided the tetrasaccharide acceptor (**61**), which was glycosylated with the common trisaccharide donor (**53**) resulting in the protected heptasaccharide (**62**). Conventional deprotection then afforded the free oligosaccharide (**52**).

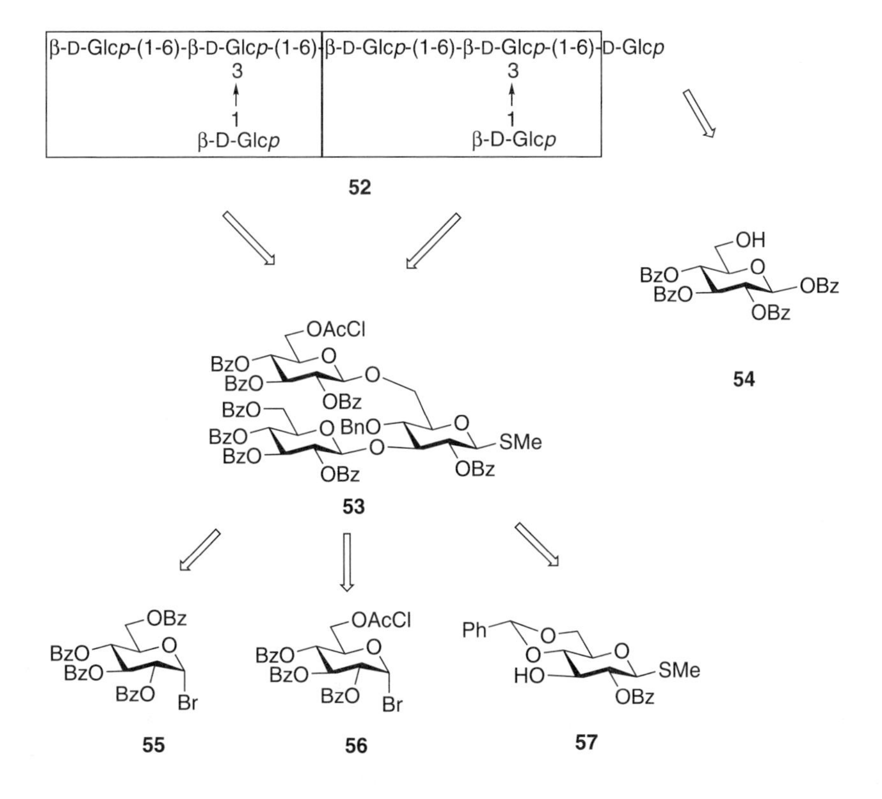

SCHEME 5.10 Retrosynthesis of the phytoelicitor active heptasaccharide.

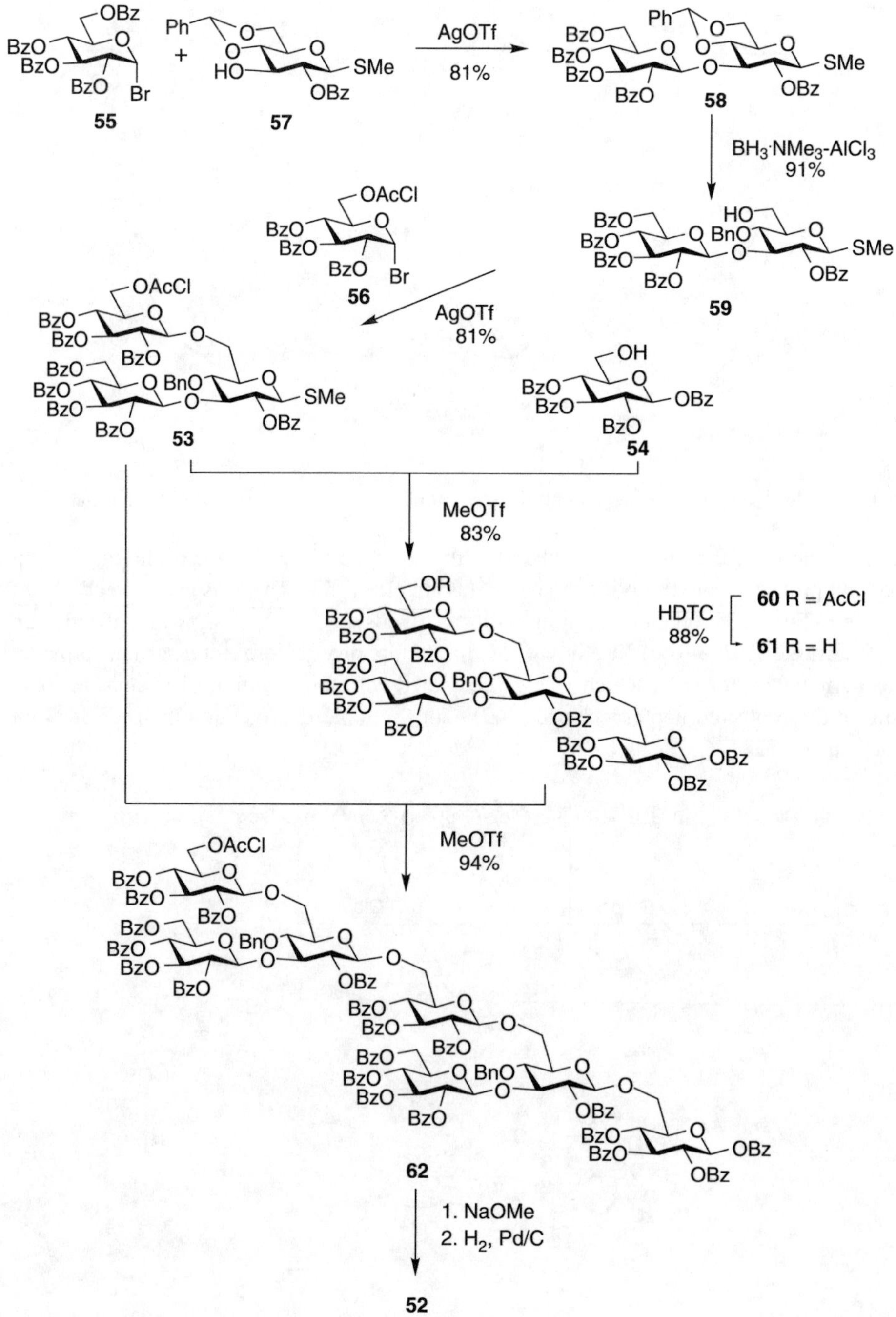

SCHEME 5.11 Synthesis of the phytoelicitor active heptasaccharide.

Higher homologs and other structurally related oligosaccharides have also been synthesized using this same strategy [14–16].

As certain types of glycosyl donors (such as glycosyl bromides, trichloroacetimidates and phosphites) are too reactive to withstand glycosylations, they cannot be used as acceptors. Consequently, they are used as type 1 donors in this strategy. However, other types of common glycosyl donors can be used as glycosyl acceptors. In fact, thioglycosides and *n*-pentenyl glycosides are particularly valuable type 2 donors. The strategy is not limited to the two levels of reactivity or the two types of glycosyl donors shown in the above examples as, with a careful match of donor types and promoters, multiple reactivity levels can be found and used in oligosaccharide syntheses.

5.4.3 Two-Stage Activation

In this strategy introduced by Nicolaou and coworkers [17], two types of anomeric substituents are used. One type (**50**) having Y at the anomeric center serves as a glycosyl acceptor. The other one (**13**), possessing X, is used as a glycosyl donor (Scheme 5.12). After glycosylating **50** with **13**, the anomeric substituent (Y) in the coupling product (**51**) is converted back into X (→ **47**), so the resulting block can be used as a donor in further glycosylations. Using acceptor **63** of the same type as before makes the process suitable for further reiteration.

Nicolaou used glycosyl fluorides as donors and thioglycosides as acceptors, and the procedure took advantage of the ready conversion of thioglycosides into glycosyl fluorides with NBS and DAST [17]. The two-stage activation procedure is illustrated by the synthesis of a phytotoxic oligosaccharide (**78**), a member of rhynchosporosides [18] (Scheme 5.13).

Activation of the cellobiose thioglycoside (**65**) with NBS and DAST gave the fluoride (**66**), which was coupled with the thioglycoside acceptor (**67**) to provide the trisaccharide (**68**). This was again activated as the fluoride (**69**) and was coupled with the common acceptor (**67**) to give the tetrasaccharide (**70**). The 1,2-propanediol glycoside (**75**) was obtained by glycosylating **74** with the thioglycoside-derived fluoride (**73**). The coupling product (**75**) was deacetylated. The resulting **76** was glycosylated with the tetrasaccharide fluoride (**71**) obtained from the thioglycoside (**70**) to afford **77**. The pentasaccharide glycoside (**78**) was obtained after deprotection.

The two-stage activation strategy has also been employed for the synthesis of Lewis^{x} [19–21] and globotriaosylceramide-related structures [22,23].

A different version of two-stage glycosylations was introduced by Danishefsky [24]. His strategy converted glycals to 1,2-anhydrosugars via epoxidation with dimethyldioxirane (DMDO). The 1,2-anhydro derivatives act as glycosyl donors in $ZnCl_2$-promoted glycosylations. By using glycals as glycosyl acceptors in these glycosylations, a reiterative process was developed.

An additional version of the two-stage activation used glycosyl sulfoxides to glycosylate thioglycoside acceptors [25–28]. The thioglycosides are then readily activated as sulfoxides using oxidation with *m*-chloroperbenzoic acid.

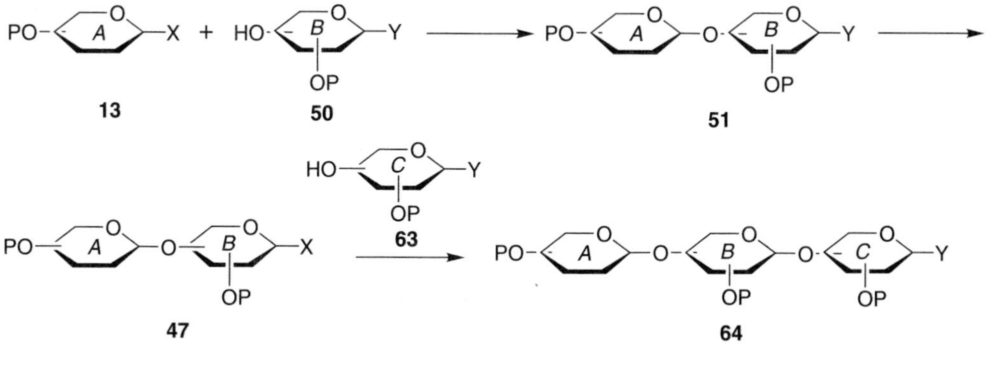

PO–A–X + HO–B(OP)–Y → PO–A–O–B(OP)–Y →
13 50 51
HO–C(OP)–Y 63
PO–A–O–B(OP)–X → PO–A–O–B(OP)–O–C(OP)–Y
47 64

SCHEME 5.12 Two-stage activation.

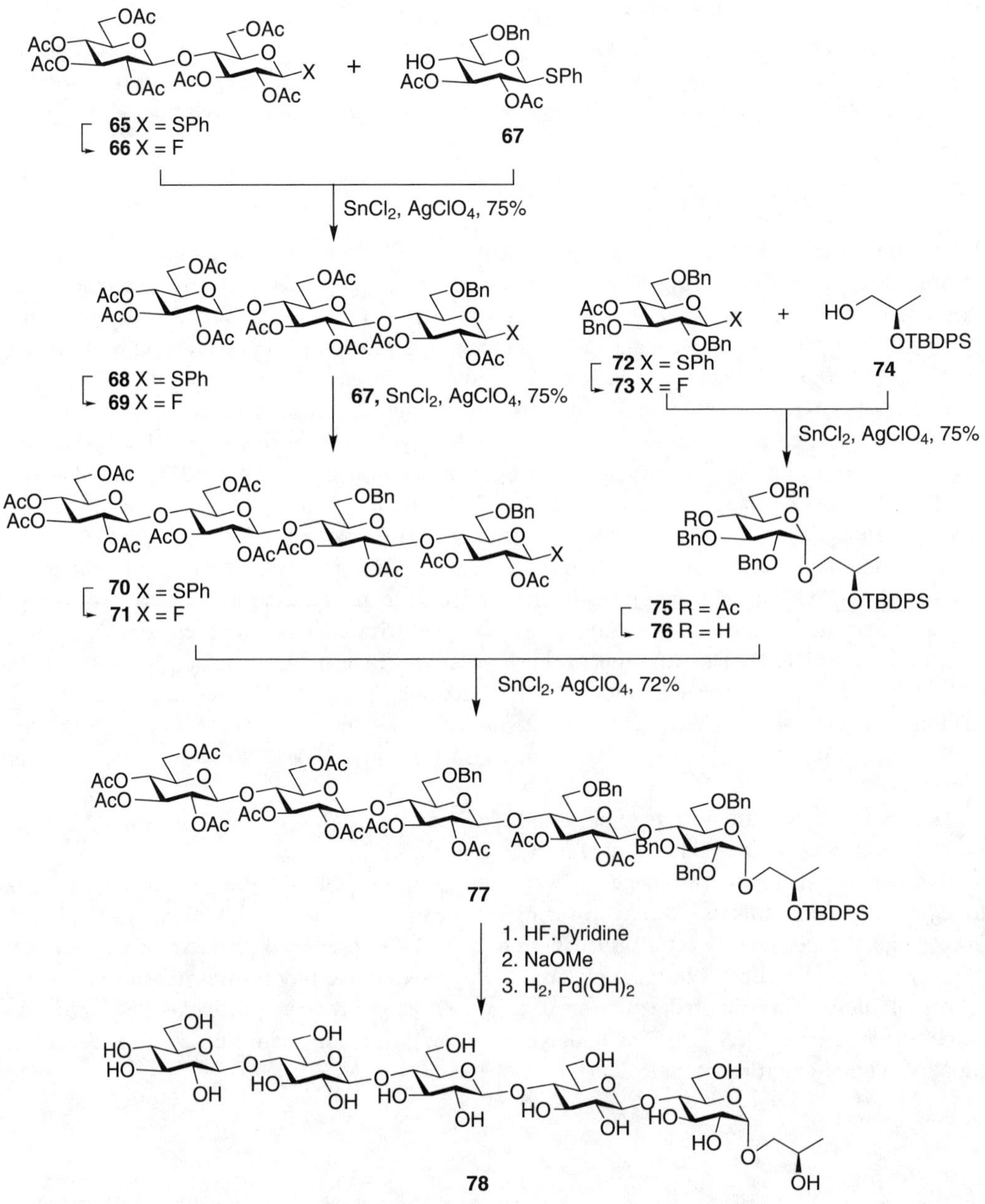

SCHEME 5.13 Synthesis of a rhynchosporoside oligosaccharide.

5.4.4 Orthogonal Glycosylations

In the orthogonal glycosylation strategy introduced by Ogawa and coworkers [29], two types of compounds and two sets of glycosylation conditions are used. Both types of compounds function as glycosyl donors and also as glycosyl acceptors. Type 1 compound (**13**) is activated as a donor under the first set of conditions to glycosylate type 2 compound (**50**) (Scheme 5.14). Under a different set of conditions, the type 2 compound is activated as a donor to glycosylate a type 1 acceptor (**79**). Alternating use of the two types of compounds under the two different conditions of activation allows a reiterative process.

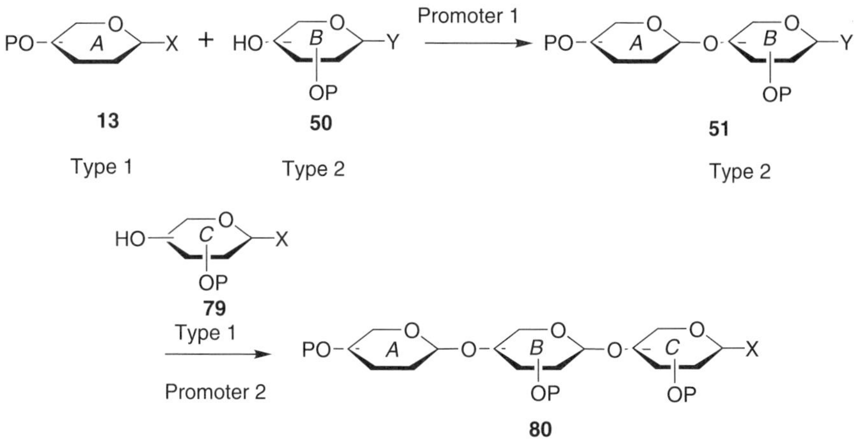

SCHEME 5.14 Orthogonal glycosylation.

The orthogonal glycosylation strategy was based on the finding that glycosyl fluorides can be glycosylated with thioglycosides and *vice versa*. As shown in Scheme 5.15, the fluoride (**82**) was glycosylated with the thioglycoside (**81**) using *N*-iodosuccinimide and silver triflate as promoters [29]. The resulting glycosyl fluoride (**83**) was directly used to glycosylate the thioglycoside (**84**) in a hafnocene-promoted reaction, giving the trisaccharide thioglycoside (**85**). Coupling of **85** with the

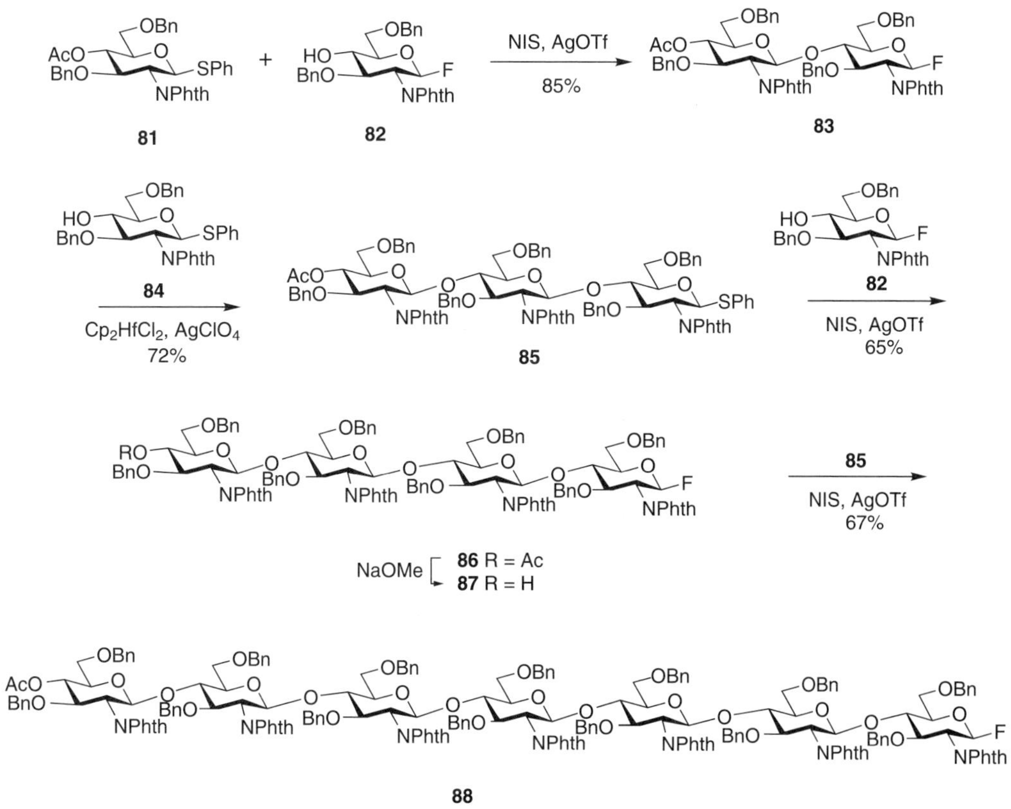

SCHEME 5.15 Synthesis of D-glucosamine oligomers by orthogonal glycosylation.

common fluoride acceptor (**82**) afforded the tetrasaccharide (**86**). After deacetylation, the tetrasaccharide fluoride (**87**) was employed as the acceptor. Final glycosylation with the trisaccharide thioglycoside (**85**) afforded the protected glucosamine heptamer (**88**).

The orthogonal glycosylation strategy was also employed in the synthesis of the blood group B determinant [30] and was applied to polymer-supported synthetic strategies [31].

5.4.5 Armed–Disarmed Glycosylations

In the armed–disarmed glycosylation strategy introduced by Fraser-Reid and coworkers [32–34], only one type of anomeric group is used. The strategy is based on the observation that the substituents of the sugar ring, particularly those at C-2, greatly influence the reactivity in glycosylation reactions. In general, electron withdrawing substituents, such as acyl groups, deactivate (disarm) the glycosylating capability, whereas electron donating substituents (ethers) activate (arm) the anomeric center. The reactivity differences can be sufficient to perform chemoselective glycosylations.

Thus, as shown in Scheme 5.16, an armed donor (**13**) can be coupled with a disarmed acceptor (**89**) without selfcondensation of the latter. The resulting disarmed compound (**90**) can be used as a glycosyl donor in further coupling, either by arming it by exchanging the protecting groups or by using a more powerful promoter capable of activating the disarmed compound.

As illustrated in Scheme 5.17, coupling of the benzylated (armed) pentenyl glycoside (**92**) with the acetylated (disarmed) one (**93**) using the mild IDCP promoter afforded the disaccharide (**94**) as an anomeric mixture [32]. The disarmed disaccharide (**94**) can be further used as a donor; its reaction with **95** by activation with the powerful NIS–TfOH promoter system [33,34] afforded the trisaccharide (**96**).

The armed–disarmed concept was originally described for *n*-pentenyl glycosides. [32,33] However, it was soon found that the phenomenon is quite general. Subsequently, armed–disarmed glycosylations with various types of glycosyl donors, including thioglycosides, [35,36] glycals [37], glycosyl fluorides [38], selenoglycosides [39], hemiacetals [40] glycosyl phosphorodiamidimidothioates, [41,42] and glycosyl thioformimidates [43–45], have been reported.

Besides the electron donating/withdrawing nature of the substituents, torsional effects also influence the reactivity of glycosyl donors and cyclic acetals have been reported to have a disarming effect [46]. An example of tuning, the reactivity by cyclic acetal formation [47] is given in Scheme 5.18. As shown, the armed selenoglycoside (**97**) reacted with **98**, which is disarmed due to its cyclohexane-1,2-diacetal, to give the disarmed selenoglycoside (**99**) — a building block in the synthesis of a high mannose type nonasaccharide.

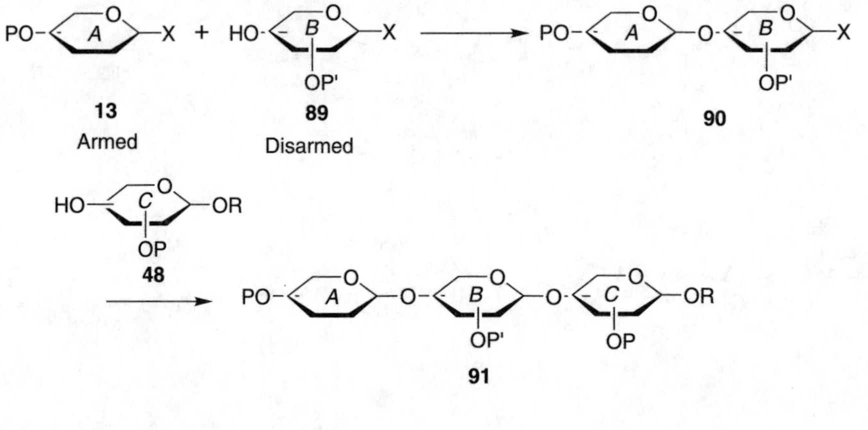

PO–A–X + HO–B(OP')–X → PO–A–O–B(OP')–X

13 Armed **89** Disarmed **90**

HO–C(OP)–OR **48** → PO–A–O–B(OP')–O–C(OP)–OR **91**

SCHEME 5.16 Armed–disarmed glycosylation.

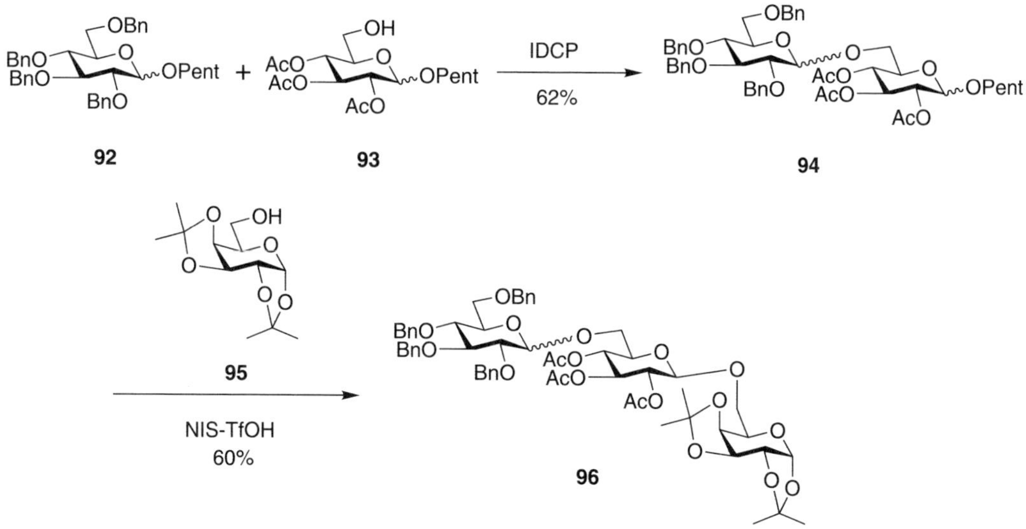

SCHEME 5.17 Armed–disarmed glycosylation based on the electron withdrawing/donating properties of substituents.

Thioglycosides offer another way to modulate reactivity in armed–disarmed glycosylations via changes in their aglycone component. Boons and coworkers [48,49] have reported disarming thioglycosides by increasing the steric bulk of the aglycone. Thus, glycosylation of the dicyclohexylmethyl thioglycoside (**101**) with the ethyl thioglycoside (**100**) using IDCP as the promoter afforded the disaccharide (**102**) [49] (Scheme 5.19). The dicyclohexylmethyl thioglycoside (**102**) was activated using NIS-TfOH to react with the benzoylated thioglycoside (**103**), affording the trisaccharide thioglycoside (**104**). Interestingly, the yields and stereoselectivities of the couplings varied depending on the anomeric configuration of the dicyclohexylmethyl thioglycosides (**101** and **102**).

The armed–disarmed glycosylation strategy relies on the reactivity differences of the coupling partners, which was qualitatively judged from their substitution pattern in the above examples. Quantitative reactivity data of a series of thioglycosides were recently measured and a database of relative reactivities was established [50,51]. Using the reactivity data, a computer program was developed which guides the selection of building blocks in armed–disarmed glycosylations.

5.4.6 Active–Latent Glycosylations

In active–latent glycosylations, the glycosyl donor capability of an acceptor is turned on by a slight chemical modification in its aglycone. Glycosylation of the latent compound (**106**) by the active

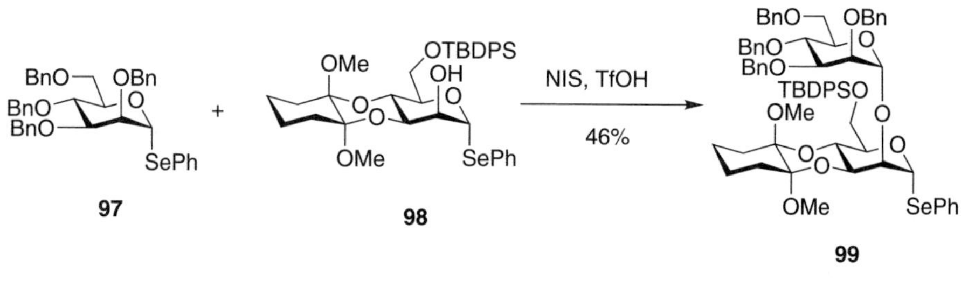

SCHEME 5.18 Armed–disarmed glycosylation based on torsional effects.

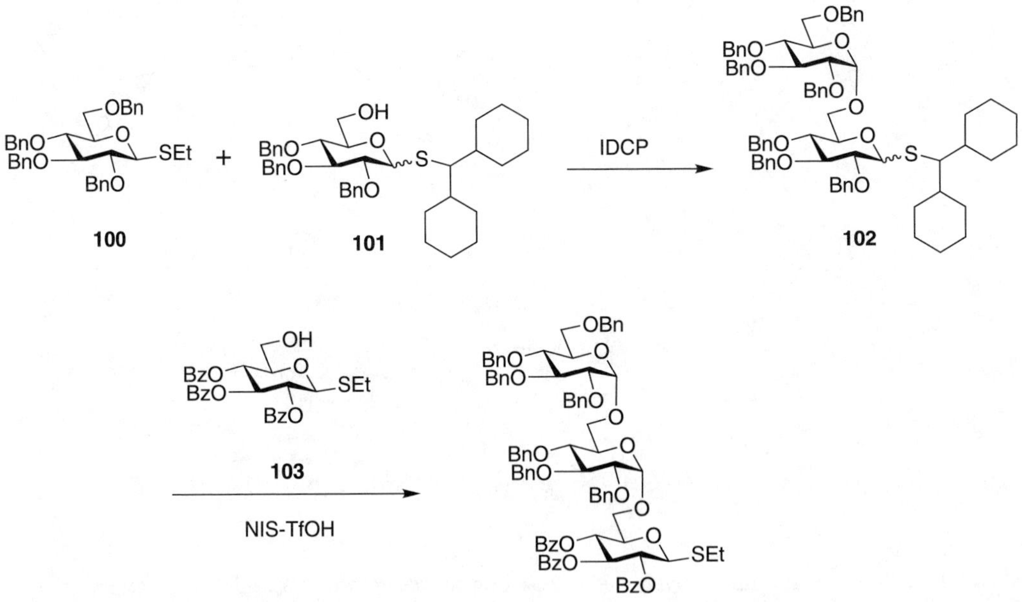

SCHEME 5.19 Armed–disarmed glycosylation based on the bulkiness of the leaving group.

donor (**105**) affords the latent disaccharide (**107**) (Scheme 5.20). Conversion of its R^2 aglycone into R^1 transforms **107** into the active form (**108**), which can be used as a glycosyl donor in further chain elongations.

p-Nitrophenyl thioglycosides are strongly deactivated as glycosyl donors due to their electron withdrawing nitro substituent. The nitro group can readily be transformed into the electron donating *N*-acetyl group, and it is possible to condense the latent *p*-nitrophenyl with the active *p*-acetamidophenyl thioglycoside [52,53]. The process is illustrated by the synthesis of the Lewis^x pentasaccharide [54] (Scheme 5.21).

As shown, the *p*-nitrophenyl thioglycoside (**109**) was glycosylated with the thiophenyl galactoside (**110**) to give the β-(1→4) linked disaccharide (**111**). Fucosylation of the free O-3 with

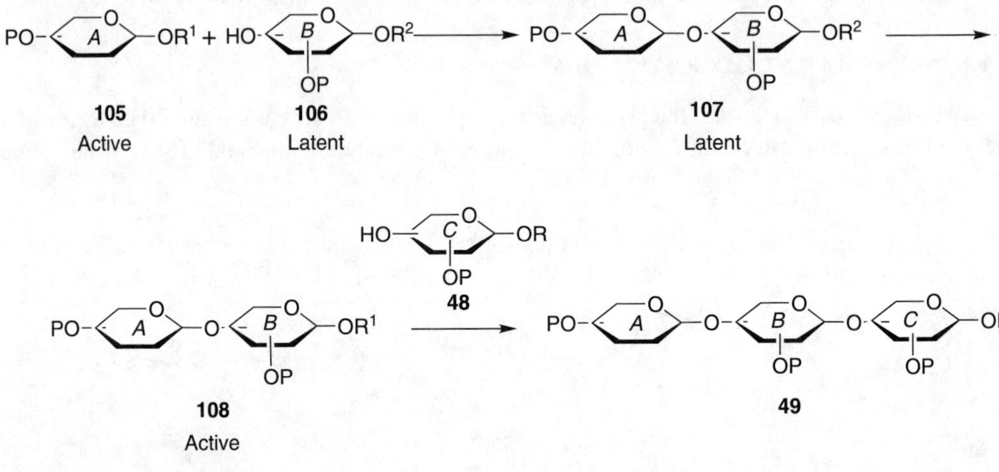

SCHEME 5.20 Active–latent glycosylation.

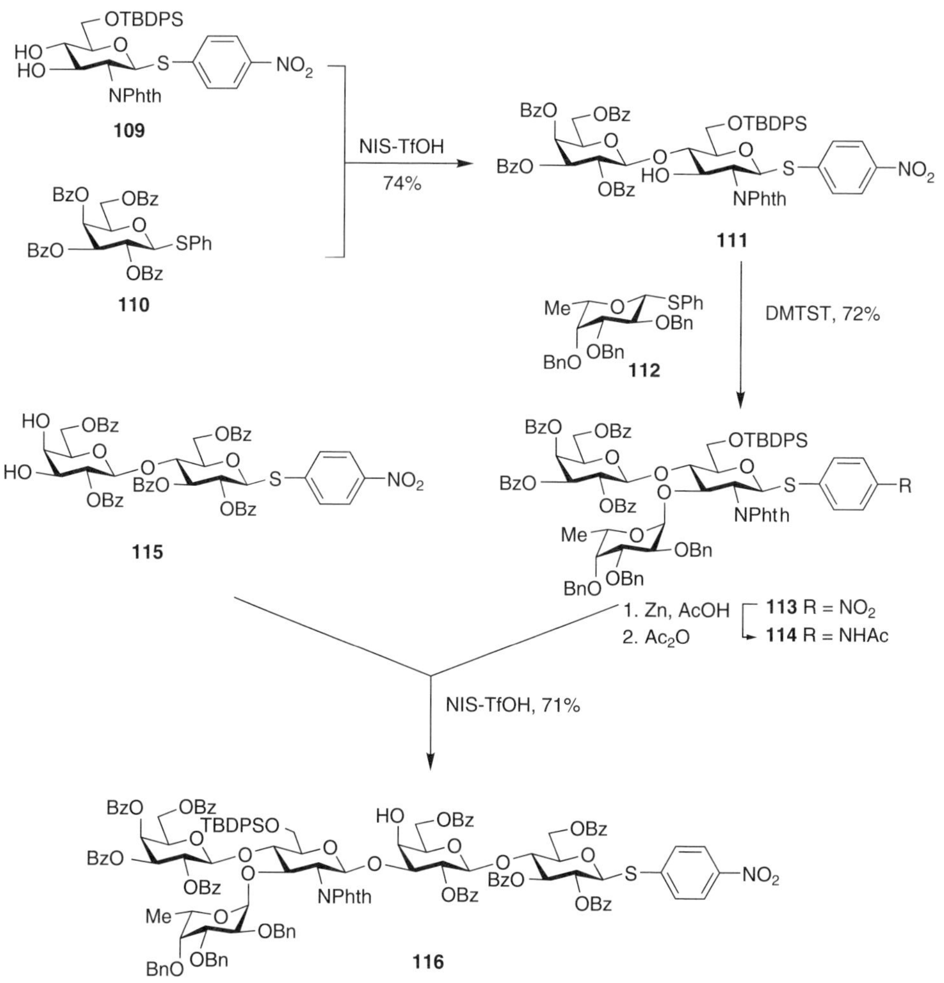

SCHEME 5.21 Synthesis of the Lewis[x] pentasaccharide by active–latent glycosylation.

the thioglycoside donor (**112**) using DMTST as promoter afforded the trisaccharide (**113**). The latent **113** was converted into the active thioglycoside (**114**), which was used as a glycosyl donor in the glycosylation of the *p*-nitrophenyl thiolactoside (**115**). The resulting latent pentasaccharide (**116**) is amenable for further glycosylation after transformation into the active form.

An active–latent glycosylation method has also been developed in connection with pentenyl glycosides. The "sidetracked" [55] 4,5-dibromopentanyl glycosides, such as **121** (Scheme 5.22), are readily obtained by bromination of pentenyl glycosides and can be transformed back into their parent compounds by reductive debromination. The use of 4,5-dibromopentanyl and pentenyl glycosides as a latent–active pair is illustrated by the synthesis of the tetrasaccharide linkage region joining glycosaminoglycans to their core proteins [56] (Scheme 5.22).

As illustrated, glycosylation of the pentenyl galactoside (**118**) with the glucuronic acid imidate (**117**) gave the disaccharide (**119**). The reducing-end disaccharide block was prepared by coupling the latent 4,5-dibromopentanyl xyloside (**121**) with the active pentenyl glycoside (**120**). After removal of the temporary chloroacetyl group from **122**, the resulting latent **123** was glycosylated with the active pentenyl glycoside (**119**). For further coupling, the latent tetrasaccharide (**124**) was

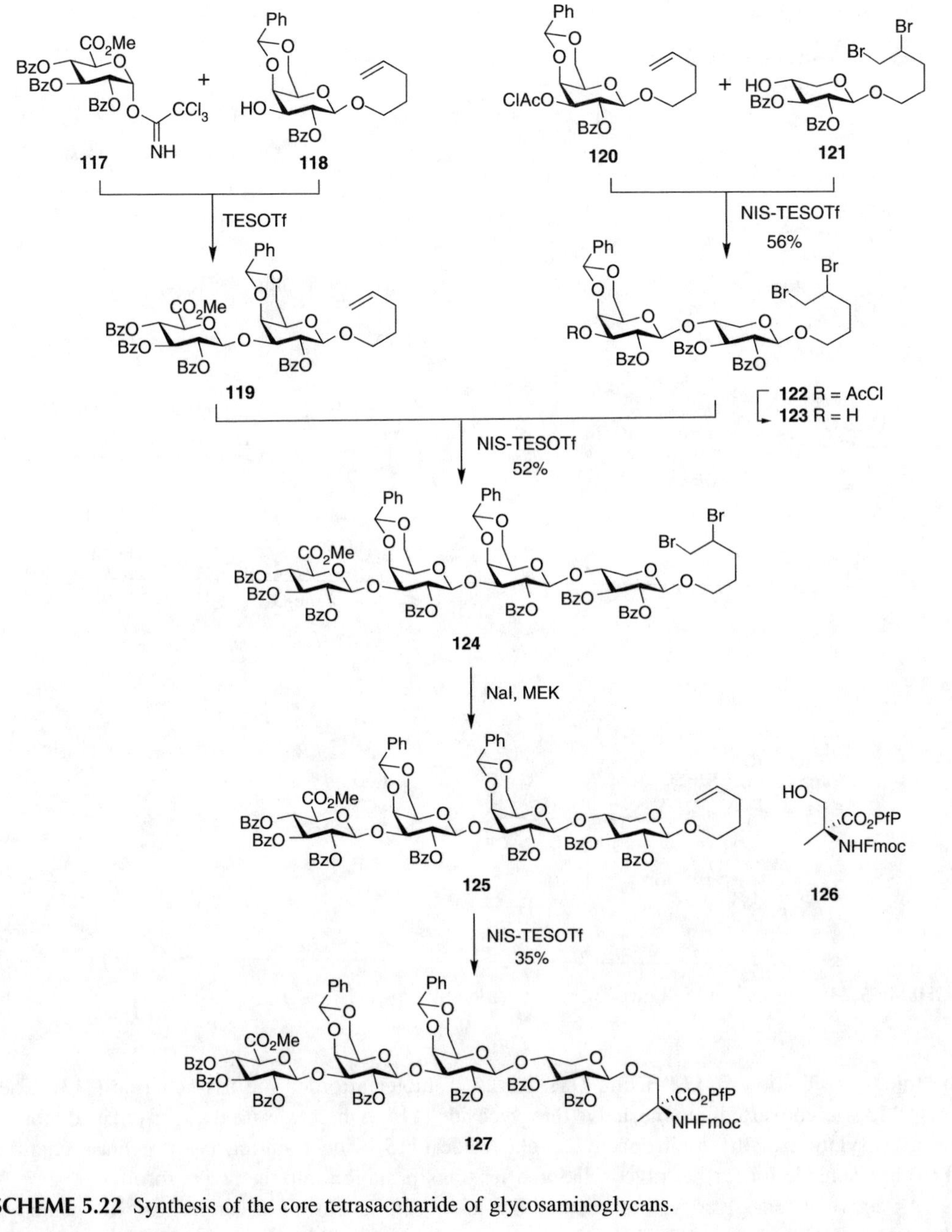

SCHEME 5.22 Synthesis of the core tetrasaccharide of glycosaminoglycans.

converted into the active block (**125**). Glycosylation of the serine derivative (**126**) with **125** then afforded the glycosylated amino acid (**127**).

Boons and coworkers [57–60] introduced alkenyl glycosides for latent–active glycosylations. The latent 3-buten-2-yl glycoside (**128**) can be isomerized to the active 2-buten-2-yl derivative (**129**) using Wilkinson catalyst (Scheme 5.23). Coupling of the active **129** with the latent **130** derived from **128** by deacetylation gave the β-linked disaccharide (**131**) [61]. The latent disaccharide was transformed into the active donor (**133**) by isomerization and to the glycosyl acceptor (**132**) by deacetylation. Coupling of the active **133** with the latent **132** afforded the tetrasaccharide **134** in an α:β ratio of 1:8.

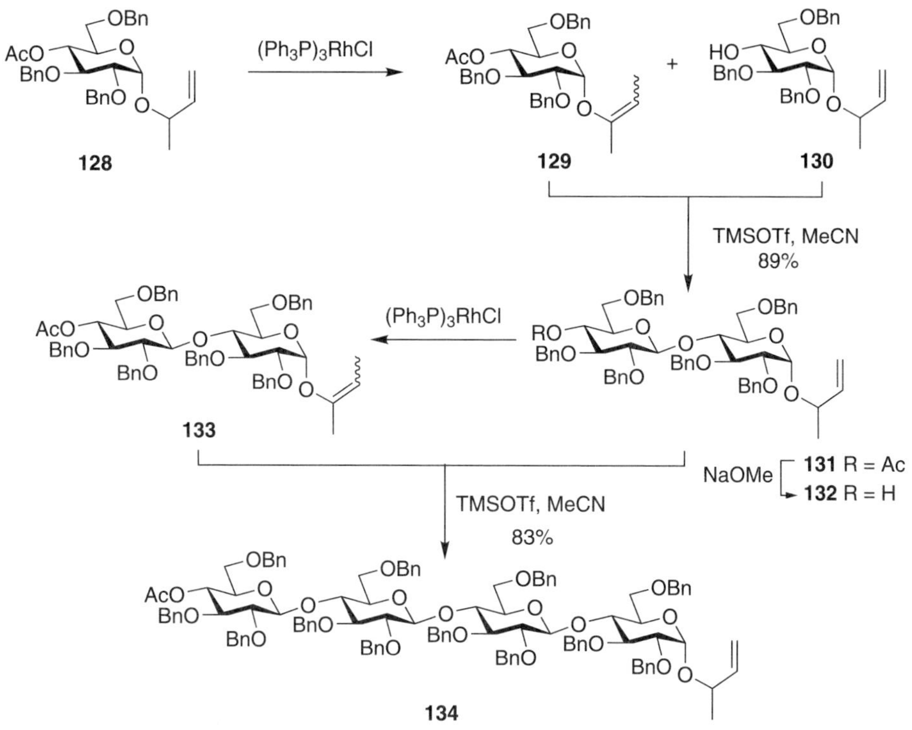

SCHEME 5.23 Active–latent glycosylation using alkenyl glycosides.

A new type of latent–active glycosyl donors, 2′-carboxybenxyl glycosides and their benzyl esters, have been recently introduced [62–64]. The active carboxylic acids are readily available from the latent benzyl esters by catalytic hydrogenation in the presence of ammonium acetate. In an example [62] shown in Scheme 5.24 the latent mannoside (**136**) was glycosylated with the active donor (**135**) to provide exclusively the β-linked disaccharide (**137**) in excellent yield. The latent

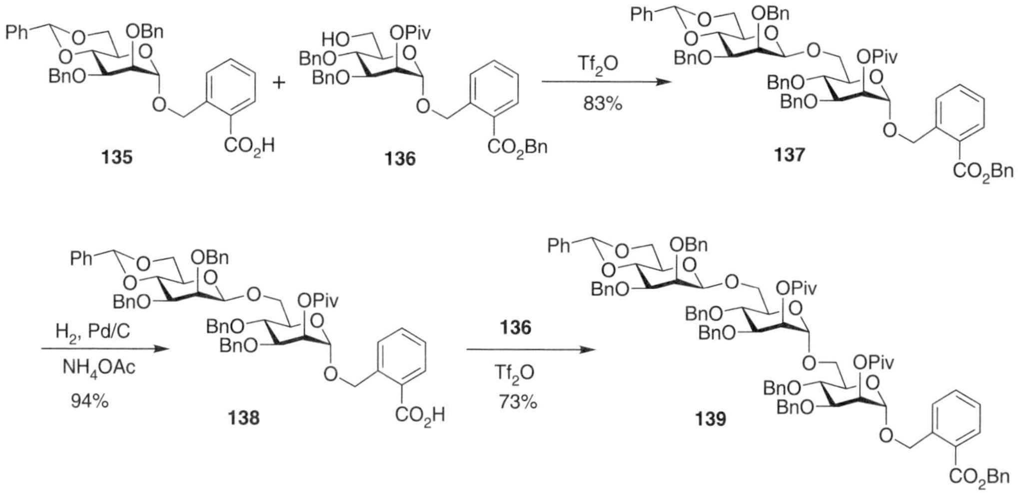

SCHEME 5.24 Active–latent glycosylation using 2′-carboxybenzyl glycosides.

disaccharide was converted into the active carboxylic acid (**138**) by selective hydrogenolysis of the ester benzyl group. Glycosylation of the common latent acceptor (**136**) with the active disaccharide (**138**) then gave the α-linked trisaccharide (**139**) as a result of neighboring group participation by the pivaloyl group.

5.5 METHODS AND TECHNIQUES IN OLIGOSACCHARIDE SYNTHESIS

Besides general synthetic strategies, some new methods and techniques are also worthy of discussion. These nontraditional methods have already contributed, and can be expected to contribute in the future, to the progress of oligosaccharide synthesis.

5.5.1 INTRAMOLECULAR AGLYCONE DELIVERY

The intramolecular aglycone delivery method was developed to improve stereocontrol and yields in difficult glycosylations. It was introduced by Hindsgaul and coworkers [65] for the preparation of β-D-mannopyranosides. The strategy (Scheme 5.25) involves the initial covalent attachment of the glycosyl acceptor to a group on O-2 of the glycosyl donor. Activation of the anomeric leaving group in the tethered derivative (**140**) by an electrophile is then expected to proceed intramolecularly via a concerted mechanism. Thus, the mechanistic intermediate **141** results in formation of **142**. On quenching with water, **142** releases O-2 and yields the glycoside **143**.

This concept was implemented using isopropylidene ketal type tethering [65]. As shown in Scheme 5.26, the 2-*O*-acetyl thioglycoside (**144**) was transformed into the isopropenyl ether (**145**) using Tebbe's reagent. Acid-catalyzed addition of the glycosyl acceptor (**146**) afforded the tethered derivative (**147**). Glycosylation by activation of the thioglycoside with NIS afforded the β-mannopyranoside (**148**) with excellent stereoselectivity.

The method worked well on simple molecules; however, its extension to higher oligosaccharides and complex structures had limited success [66].

In a different version of the same concept, Stork [67,68] and Bols [69–72] introduced silylene type tethering with glycosyl sulfoxide [67,68] or thioglycoside [69–72] donors. The silicon-tethered acceptor approach was also extended to the synthesis of α-D-gluco- [70,72] and α-D-galactopyranosyl linkages in oligosaccharides [70,72].

To avoid the problems associated with the mixed ketal formation in Hindsgaul's procedure, Ito and Ogawa introduced *p*-methoxybenzylidene acetal type tethering [73]. This linker was originally employed with glycosyl fluoride donors [73], but significantly improved yields were later obtained with thioglycosides [74]. The method could be successfully used in the synthesis of complex oligosaccharides, such as the synthesis of the core pentasaccharide unit of *N*-glycoproteins [75,76] (Scheme 5.27).

As illustrated, the trisaccharide donor (**149**), with a *p*-methoxybenzyl ether function at O-2, and the chitobiose acceptor (**150**) were tethered using DDQ. The resulting mixed acetal (**151**) was

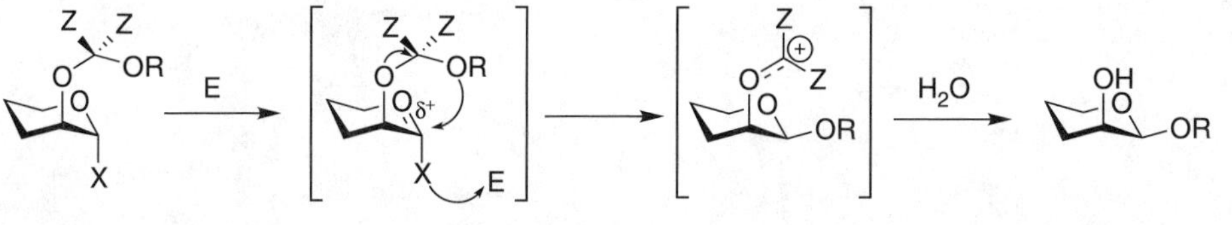

SCHEME 5.25 Intramolecular aglycone delivery.

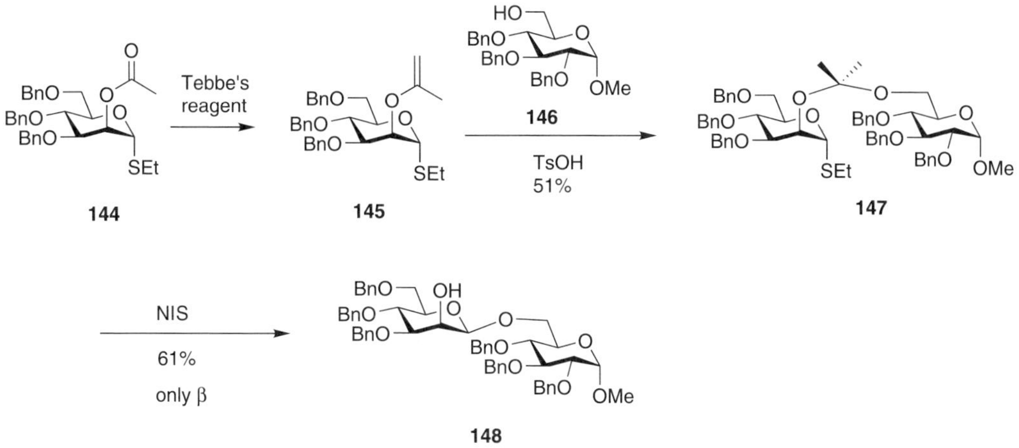

SCHEME 5.26 Intramolecular aglycone delivery using isopropylidene ketal tethering.

activated with MeOTf as a promoter to produce exclusively the β-mannosidically linked pentasaccharide (**152**). The unprotected *p*-methoxyphenyl glycoside of the core pentasaccharide (**153**) was obtained after deprotection and *N*-acetylation.

The use of the *p*-methoxybenzylidene acetal tethered intramolecular aglycone delivery was extended to polymer-supported oligosaccharide synthesis [77]. The method also proved to be successful in the synthesis of other difficult linkages, such as β-D-fructofuranosides [78], β-D-arabinofuranosides [79–81] and α-D-fucofuranosides [82].

Recently, an additional version of intramolecular aglycone delivery starting from 2-*O*-allyl derivatives and using iodonium ion-promoted tethering was described [83]. As illustrated in Scheme 5.28, the 2-*O*-allyl ether (**154**) was first isomerized to the propenyl ether (**155**). NIS-promoted coupling with the acceptor (**146**) then afforded the mixed acetal (**156**), which was transformed into the β-mannoside (**148**) by using NIS and AgOTf for activation.

This procedure was also employed using glycosyl fluorides as donors [84] and was applied for the synthesis of α-D-glucopyranosides [83,84]. However, only a modest yield was obtained when it was applied to the synthesis of a tetrasaccharide [85].

A different intramolecular glycosylation approach, which is related to the intramolecular aglycone delivery, has also been developed. In this approach, the donor and the acceptor are tethered at positions not directly involved in the glycoside bond formation. The acceptor has the hydroxyl to be glycosylated free, and the linker remains in the coupling product and has to be removed at a later stage. Numerous variations of the above intramolecular glycosylations were studied, including changes in the positions of linkers and variations in the length, the type and the rigidity of the linker. A detailed discussion of this "cyclo-glycosylation" approach is beyond the scope of this chapter and is covered in a recent review [86]. Some examples [87,88] are shown in Scheme 5.29 to illustrate the point that cyclo-glycosylation might be an effective tool in achieving stereocontrol.

As shown, the glycosylation reaction of the succinoyl tethered compounds (**157**, **159** and **161**) was performed by promotion with NIS and TMSOTf in acetonitrile. The (2′-3)-tethered compound (**157**) afforded the α-glucopyranoside (**158**) in excellent yield. Similarly, the (2′-3)-tethered (**159**) and also the (6′-3)-tethered (**161**) galactose derivatives afforded the *cis*-glycosides (**160** and **162**). The stereoselectivity of these reactions is noteworthy as both the acyl-type neighboring group in the donor and the nitrile type solvent favor *trans*-glycoside formation.

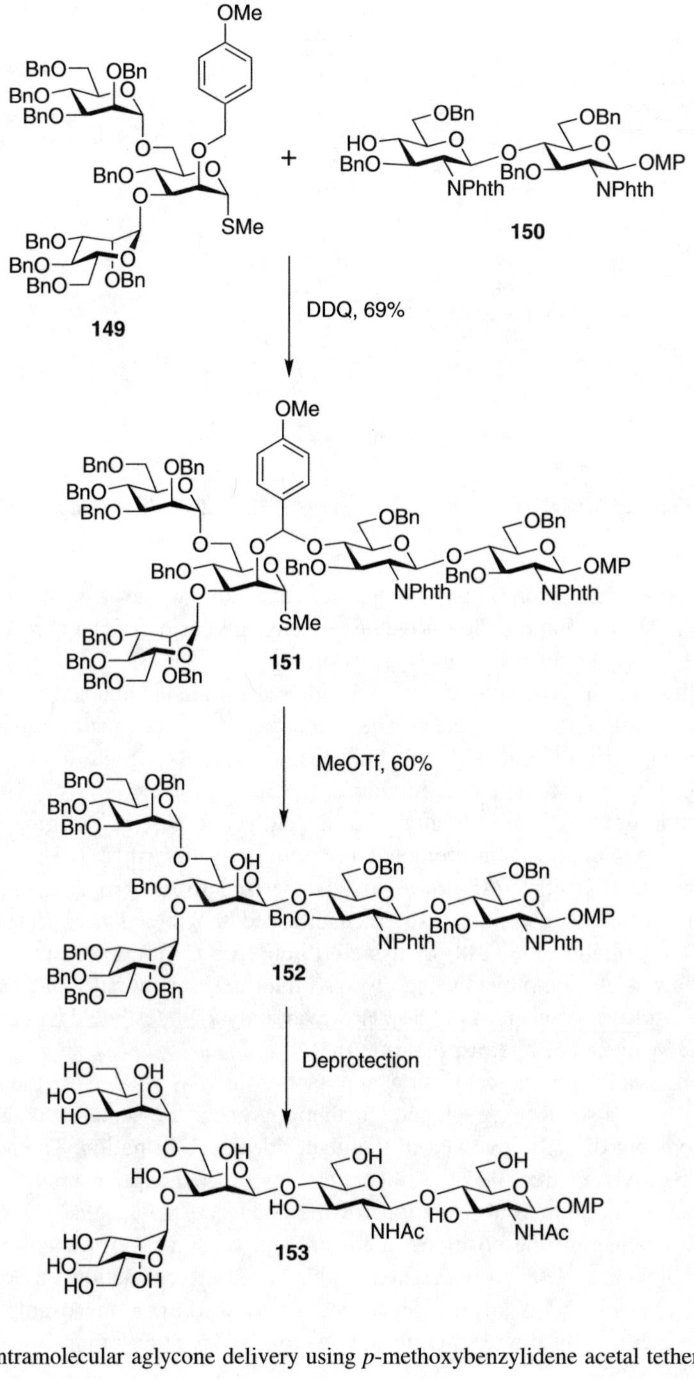

SCHEME 5.27 Intramolecular aglycone delivery using *p*-methoxybenzylidene acetal tethering.

5.5.2 One-Pot Multistep Glycosylations

In the traditional approaches of oligosaccharide synthesis, the product of a glycosylation reaction had to be isolated and it required some chemical transformation(s) to make it suitable for the next glycosylation reaction. In some of the synthetic strategies described in Section 5.4, such as in the sequential glycosylations and in the armed–disarmed glycosylations, the product of one

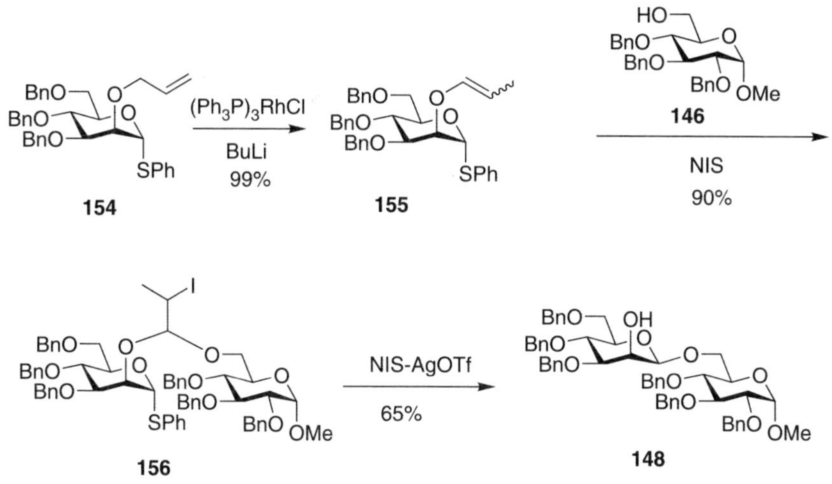

SCHEME 5.28 Intramolecular aglycone delivery using 2-*O*-allyl ethers.

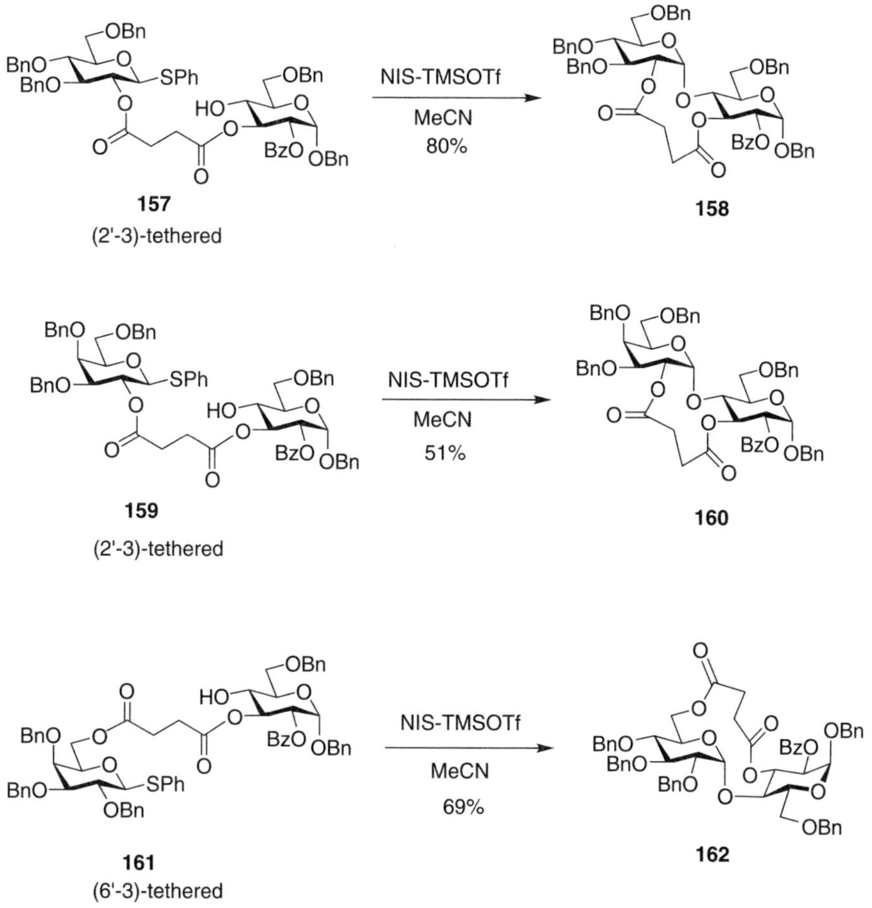

SCHEME 5.29 Intramolecular cyclo-glycosylations.

glycosylation reaction is used directly in the next coupling reaction. This opened the way for omitting the isolation step and performing multiple glycosylations in a one-pot manner.

The first one-pot chemical glycosylation was described by Raghavan et al. in 1993 [89]. In mechanistic studies they found that the rate-limiting step in glycosylations with glycosyl sulfoxides is the triflation of the sulfoxide; therefore, the reactivity of the sulfoxide can be manipulated by the substituent in the *para*-position of the phenyl aglycone. The reactivity difference is large enough that a *p*-methoxyphenyl sulfoxide can be selectively activated in the presence of the unsubstituted phenyl sulfoxide. In addition, silyl ethers are good accceptors when triflic acid is used as a promoter. However, they react more slowly than unprotected alcohols.

Based on these findings, a mixture of the three monosaccharides (**163**, **164** and **165**) was treated with triflic acid, and the reaction resulted in the formation of a 25% yield of the trisaccharide (**167**) (Scheme 5.30). It was also shown that the more reactive sulfoxide (**164**) initially reacts with the more reactive acceptor (**165**) to form the disaccharide (**166**). In the second stage of the reaction the less reactive silylated disaccharide (**166**) is glycosylated with the less reactive sulfoxide (**163**), giving the trisaccharide (**167**).

One-pot sequential glycosylations using different types of glycosyl donors have been introduced by Takahashi et al. [90]. Thus, the thioglycoside (**169**) was glycosylated with the glycosyl bromide (**168**) using AgOTf as a promoter to give the disaccharide (**170**) (Scheme 5.31). The thioglycoside function, which is stable in this step, is readily activated by adding a second activator to the reaction mixture. Thus, addition of NIS and the acceptor (**171**) resulted in the formation of the trisaccharide (**172**) in 84% yield for the two steps. It should be noted that the traditional synthesis using the same steps was not only more time consuming but gave the product in lower isolated yield.

Besides the above glycosyl bromide → thioglycoside → *O*-glycoside sequence, similar one-pot glycosylations were performed using glycosyl trichloroacetimidate → thioglycoside → *O*-glycoside and glycosyl fluoride → thioglycoside → *O*-glycoside sequences [90].

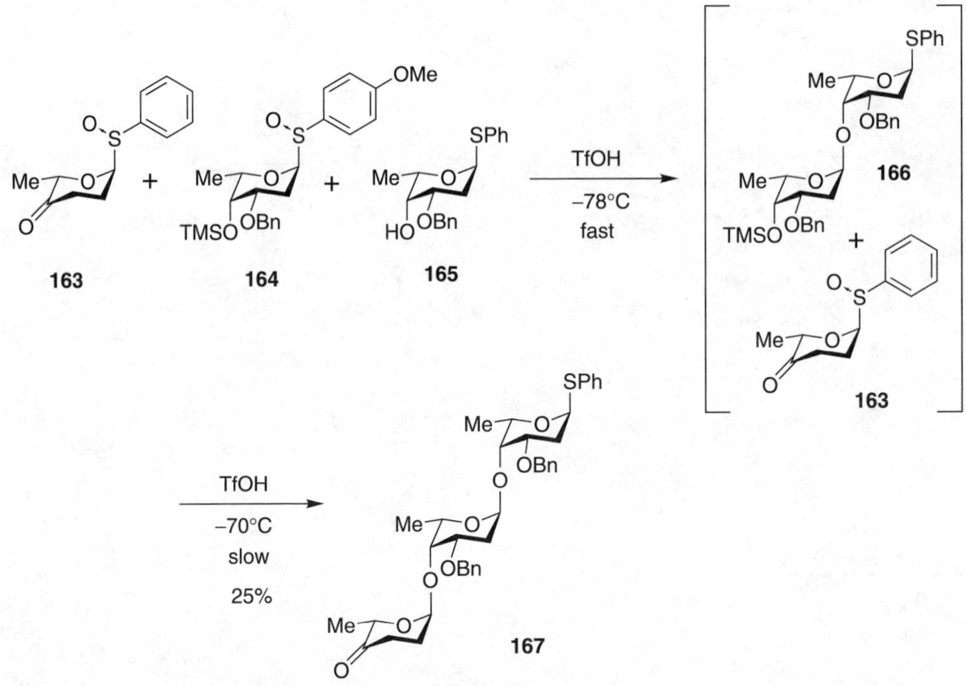

SCHEME 5.30 One-pot synthesis of the ciclamycin trisaccharide.

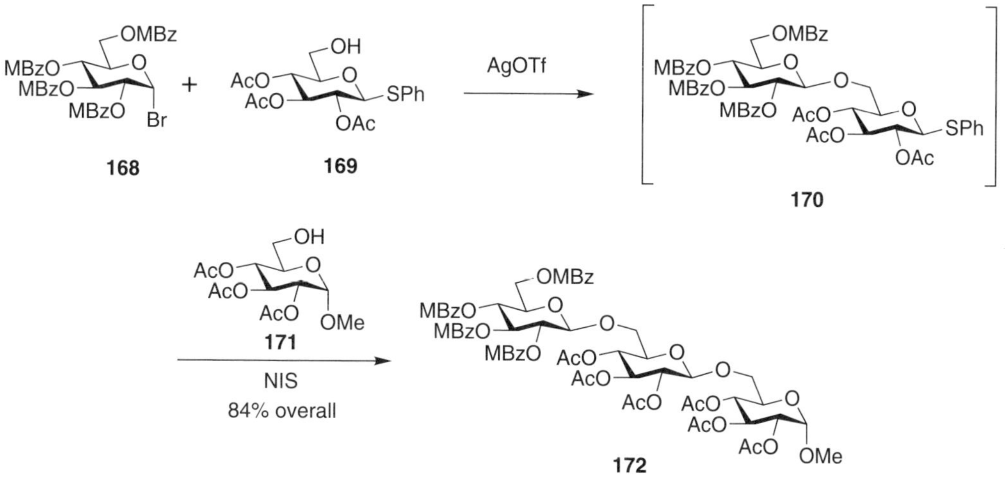

SCHEME 5.31 One-pot sequential glycosylation using different types of glycosyl donors.

Armed–disarmed type glycosylations can also be readily performed in a one-pot manner. As the glycosyl donors are of identical type in these reactions, often the same promoter can be used for all the glycosylation steps. An important development was the quantitation of relative donor reactivities and the introduction of programmable one-pot oligosaccharide synthesis [50,51], which allowed the generation of an oligosaccharide library. The preparation of the Lewis^y antigenic determinant (**173**) [91] is shown in Scheme 5.32 and Scheme 5.33 as an example of the programmable one-pot oligosaccharide synthesis.

As illustrated, the hexasaccharide (**173**) was synthesized from the monomers (**174–178**), having the relative reactivity values (RRVs) indicated in Scheme 5.32. The disaccharide building block (**180**) was prepared by coupling **175** with **177** followed by removal of the two levulinoyl groups from the resulting disaccharide (**179**). Similarly, the reducing end block (**182**) was obtained by glycosylation of **178** with the galactose donor (**176**) and cleavage of the *p*-methoxybenzyl ether from **181**. In the one-pot assembly of the oligosaccharide, fucosylation of the diol (**180**) was performed using NIS–TfOH as a promoter at − 70°C then, after the addition of the acceptor (**182**), raising the temperature to − 25°C. The protected hexasaccharide (**184**) was obtained in 44% yield, which was equivalent to 81% per glycosylation. Deprotection of **184** afforded the Lewis^y antigenic determinant (**173**).

An additional factor which was used to tune reactivity in armed–disarmed glycosylations is the dependency of glycosylation rates on the solvent [92]. Thus, one-pot glycosylations were accomplished by performing the first glycosylation in ether, which slows down the reaction, and performing the second glycosylation in dichloromethane [92].

One-pot sequential glycosylations were extended for the synthesis of branched oligosaccharides by exploiting the reactivity differences of the different hydroxyl groups in acceptors having two unprotected hydroxyls [93]. Thus, when a mixture of the glycosyl bromide (**185**) and the thioglycoside (**186**) donors and the 3,6-diol acceptor (**187**) was treated first with AgOTf, then with NIS–TfOH, the branched trisaccharide (**189**) was obtained in 76% yield (Scheme 5.34). In the first stage of the reaction only the glycosyl bromide (**185**) is activated and it reacts with the more reactive primary hydroxyl of the acceptor, giving the intermediate **188**. Subsequently, **188** is glycosylated in the second step by the thioglycoside at O-3.

In a similar manner, a combinatorial library was synthesized by one-pot glycosylations [94].

The high performance of one-pot multiple glycosylations is shown by the example in Scheme 5.35, in which six glycosidic linkages were synthesized [95].

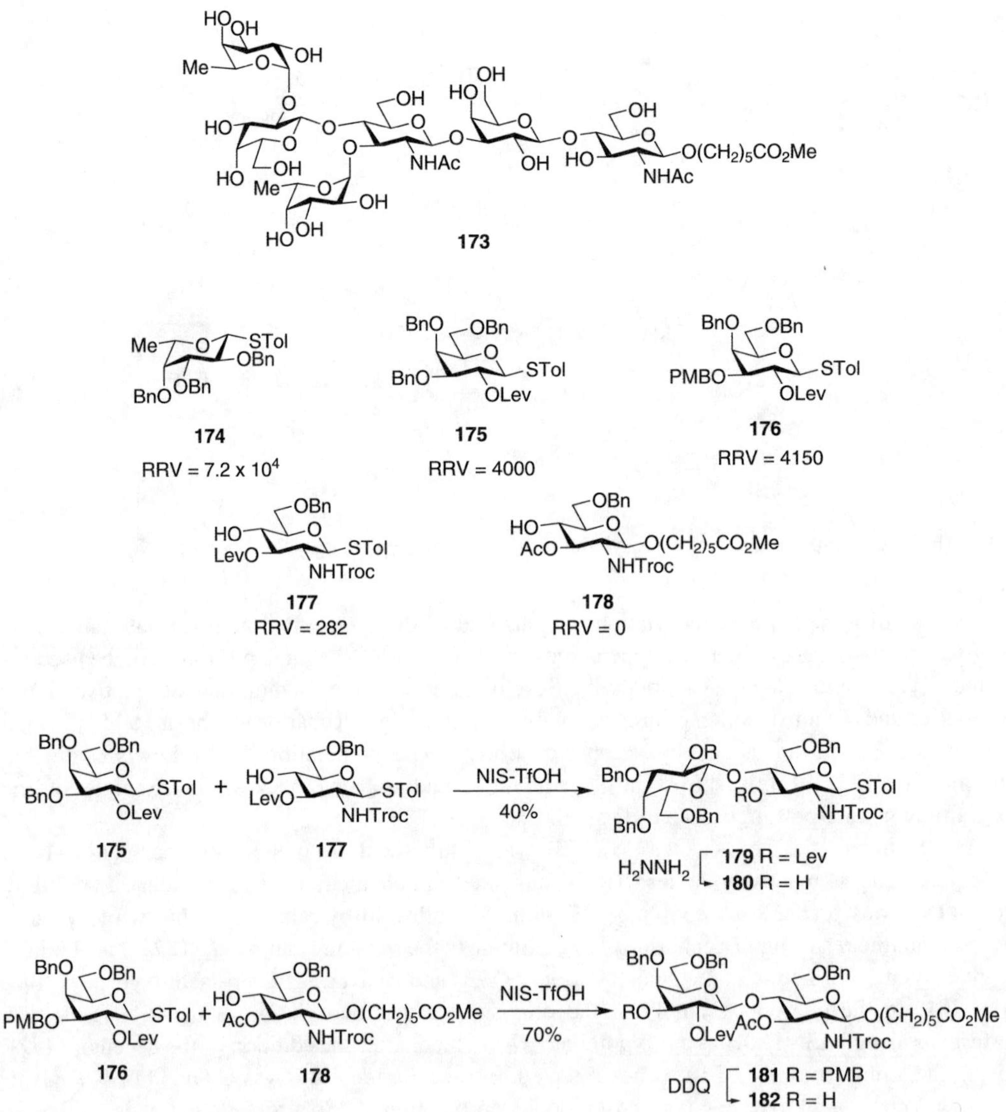

SCHEME 5.32 Building blocks for the one-pot synthesis of the Lewis^y antigenic determinant.

As shown, when **190** was sequentially treated with the monomers **185** and **191–195** in combination with the promoters given in the scheme, the branched heptasaccharide (**201**) was obtained in 24% overall yield. Formation of the heptasaccharide is understandable by considering the intermediates (**196–200**) formed after each step, which are shown in parentheses. Deprotection of the heptasaccharide (**201**) gave the methyl glycoside of the phytoelicitor heptasaccharide (**52**) (Schemes 5.10 and 5.11).

5.5.3 Polymer-Supported and Solid-Phase Oligosaccharide Synthesis

Inspired by the success of Merrifield on solid-phase peptide synthesis, studies on solid-phase oligosaccharide syntheses were initiated in the early 1970s [96–98]. Using various approaches, some small oligosaccharides were synthesized by different groups. However, the lack of reliable

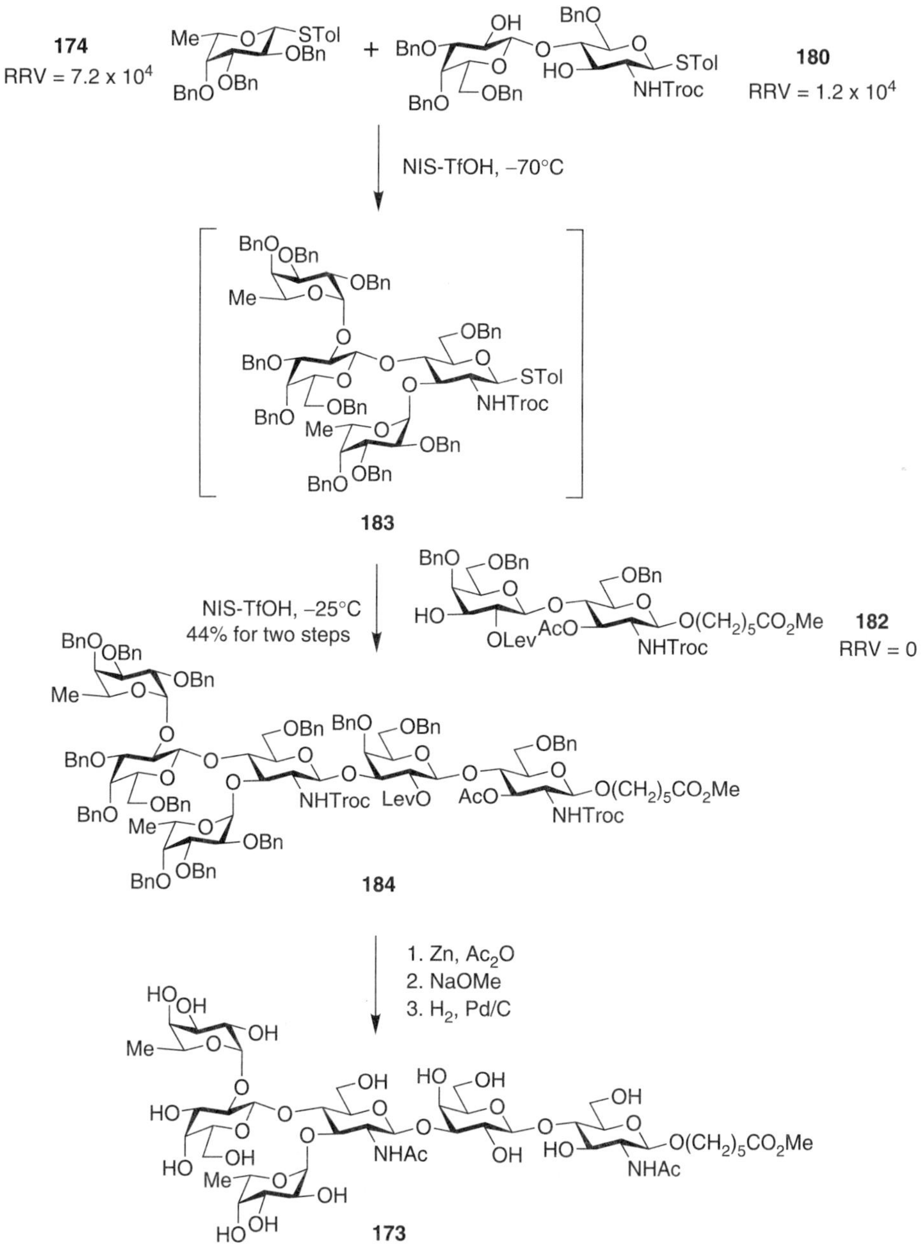

SCHEME 5.33 One-pot synthesis of the Lewis^y antigenic determinant.

and efficient glycosylation methods, and the unavailability of analytical methods for monitoring the reactions in the solid phase, prevented the development of a practically useful procedure. With the advent of new glycosylation methods in the 1980s, the largely dormant area became revitalized in the 1990s.

Polymer-supported liquid-phase synthesis was introduced for the preparation of oligosaccharides by Douglas, Whitfield and Krepinsky [99]. In their procedure, the carbohydrate is attached to the soluble polymer, poly(ethylene glycol)monomethyl ether (PEG), and all the chemical

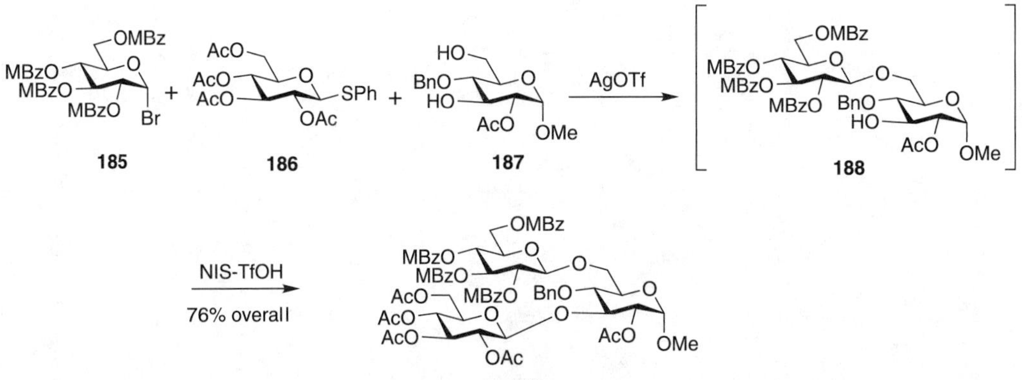

SCHEME 5.34 One-pot synthesis of a branched trisaccharide.

transformations are performed in solution. The polymer is precipitated by ethereal solvents after each step, and removal of any excess reagents is simply ensured by filtration and washing. A great advantage of the use of soluble polymers is that it eliminates the analytical problems associated with solid-phase synthesis, and reactions in liquid-phase synthesis can easily be monitored by NMR spectroscopy.

The polymer-supported liquid-phase synthesis is illustrated by the preparation of heparin-related oligosaccharides [100] (Scheme 5.36). The α-L-Ido*p*A-(1→4)-D-Glc*p* disaccharide derivative (**203**) is attached to the polymer via a succinoyl linker and chain elongation is performed with the imidate donor (**202**). One elongation cycle consists of the following consecutive steps: (1) removal of the temporary levulinoyl group with hydrazine, (2) TMSOTf-promoted coupling with the donor (**202**) and (3) capping of the unreacted 4-OH groups by acetylation. By repeating the cycle, oligomers up to dodecasaccharide (**208**) were prepared. The protected oligosaccharides (**205**–**208**) were converted to the sulfated oligosaccharides (**209**–**212**). First, the oligosaccharides were cleaved from the polymer by lithium peroxide with parallel removal of all the other ester groups. After catalytic hydrogenolysis of the benzyl groups, the free hydroxyls were persulfated to afford the heparin oligosaccharide analogs (**209**–**212**).

Solid-phase synthesis on insoluble supports has also been extensively studied [101–103]. As support matrices, Merrifield's resin (polystyrene cross-linked with 1% divinylbenzene), controlled pore glass or PEG grafted on polystyrene (TentaGel, ArgoGel) are commonly employed — the latter have more desirable swelling properties. Various linkers have been applied to attach the carbohydrate to the solid support, including silyl ethers and acid- or base-labile linkers such as the Wang or the succinoyl linkers, respectively. Substituted benzyl ether type linkers can be removed under mild conditions by oxidative or reductive processes. The recently introduced octenediol linker [104] is very versatile as it provides *n*-pentenyl glycosides after cleavage by olefin metathesis.

Synthetic strategies differ by binding either the glycosyl donor [105] or the glycosyl acceptor to the solid phase. A bidirectional approach [106], which allows for chain elongation in both directions, has also been reported.

A great variety of glycosyl donors and glycosylation methods have been tried in solid-phase synthesis, including glycals [105], glycosyl sulfoxides [107], thioglycosides [108,109], glycosyl trichloroacetimidates [110], *n*-pentenyl glycosides [111], glycosyl iodides [112] and glycosyl phosphates [104,113]. Oligosaccharides up to the size of a dodecamer [109,114] have been synthesized.

Scheme 5.37 illustrates solid-phase oligosaccharide synthesis using glycal-derived 1,2-anhydro sugars as glycosyl donors [105].

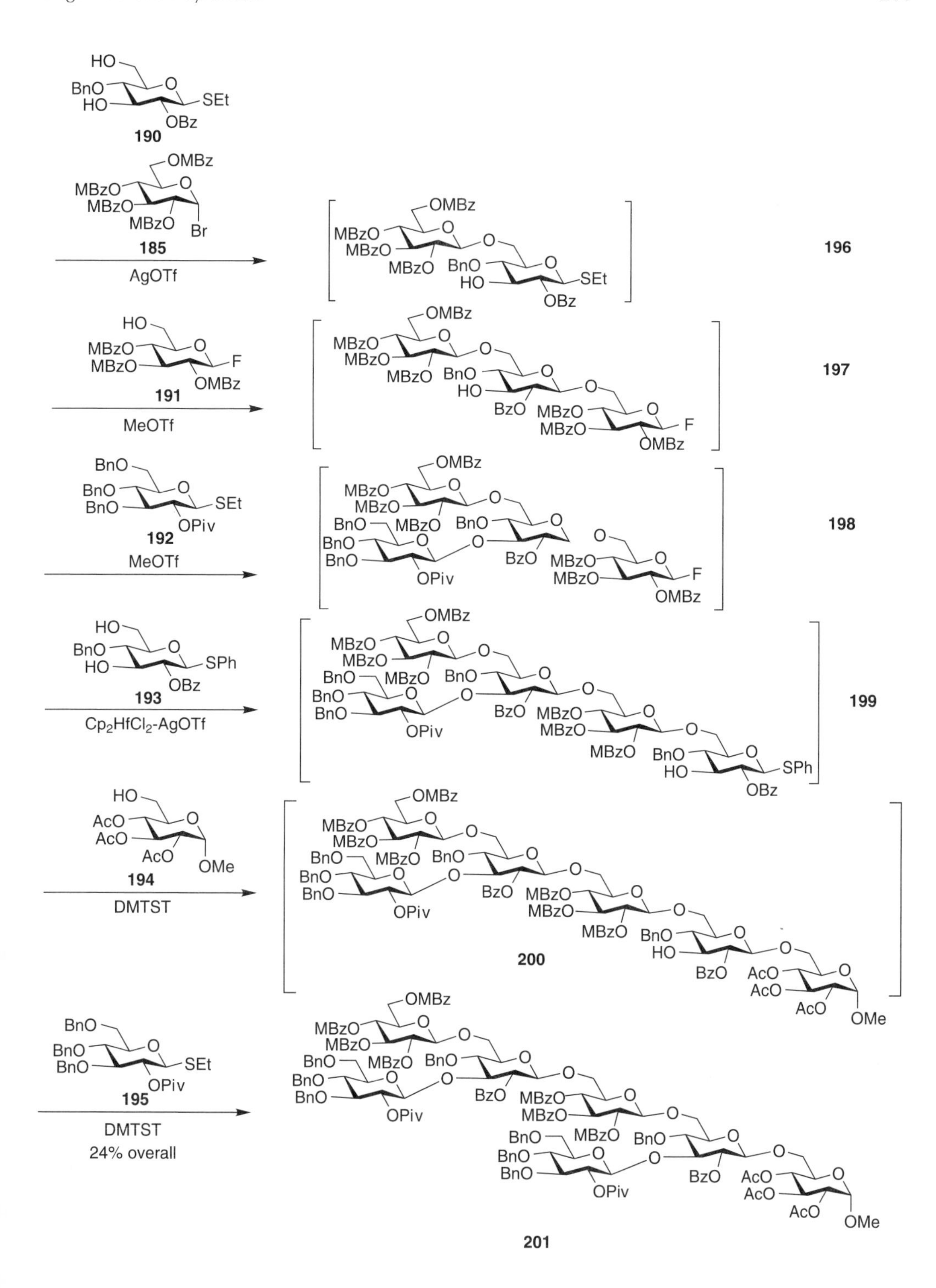

SCHEME 5.35 One-pot synthesis of branched heptasaccharide.

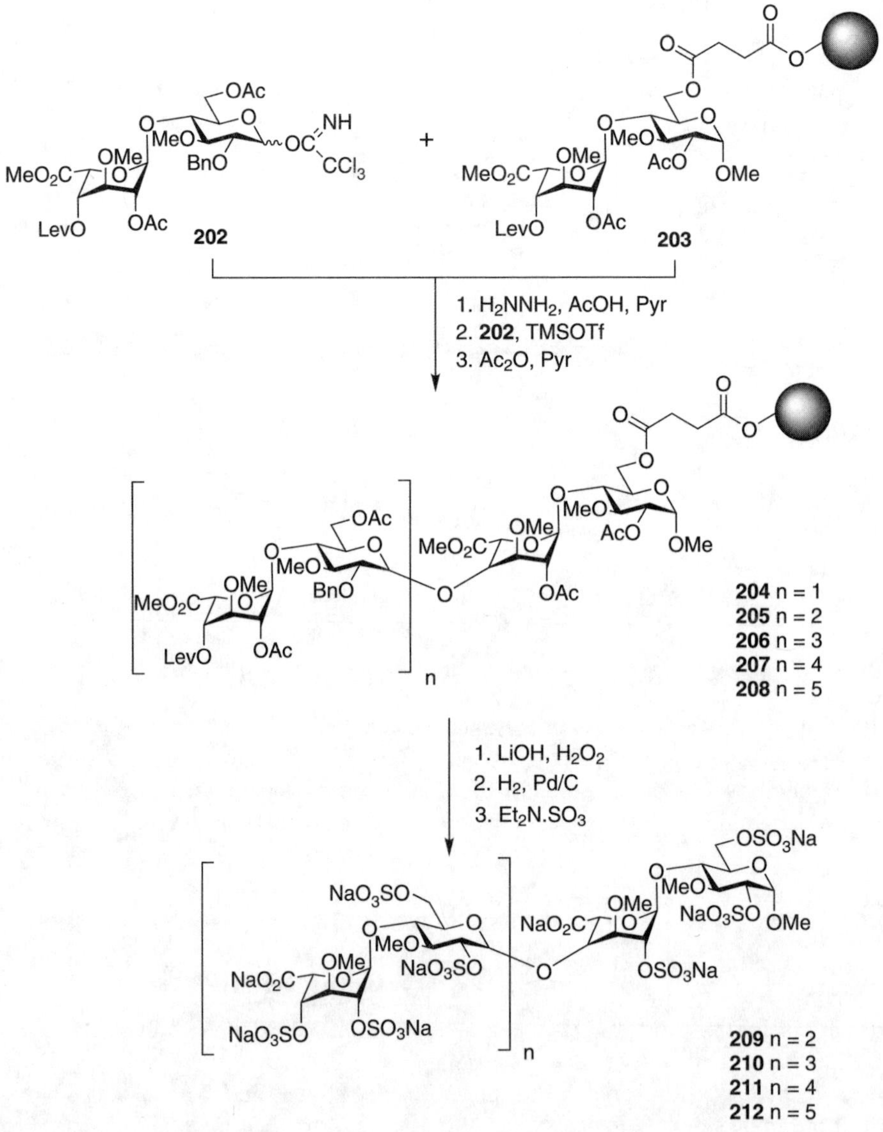

SCHEME 5.36 Polymer-supported liquid-phase synthesis of heparin-mimic oligosaccharides.

As illustrated, a D-galactal derivative was attached to Merrifield's resin via a diphenylsilyl linker. The glycal (**213**) was converted to the 1,2-anhydrosugar (**214**) with DMDO. The resin-bound glycosyl donor (**214**) reacted with the glycal acceptor (**215**) to give the β-(1→6)-linked disaccharide (**216**). Repetition of this procedure afforded the trisaccharide (**218**), which was activated again and coupled with the D-glucal derivative (**220**). After cleaving the resulting product (**221**) from the resin, the tetrasaccharide (**222**) was obtained in 32% overall yield.

In the above synthetic strategy, the glycosyl donor is bound to the solid support. As most side reactions during glycosylations (such as hydrolysis) involve the donor, this reduces the yield and might result in termination of chain elongation. This problem can be avoided in acceptor-bound strategies where the glycosyl donor can be used in excess to push glycosylations to completion.

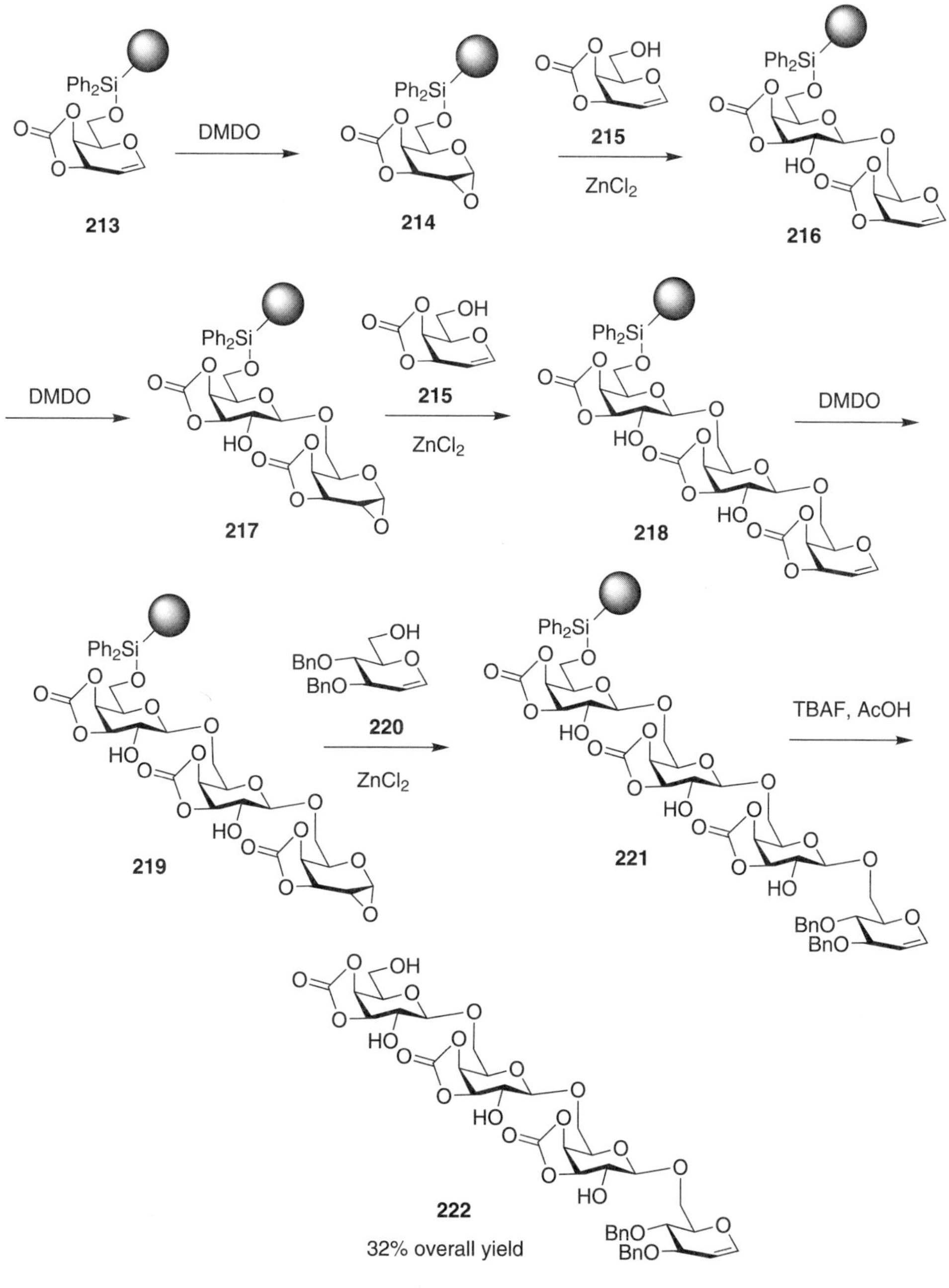

213 DMDO 214 215 ZnCl2 216

DMDO 217 215 ZnCl2 218 DMDO

219 220 ZnCl2 221 TBAF, AcOH

222 32% overall yield

SCHEME 5.37 Solid-phase synthesis using glycal-derived 1,2-anhydrosugars.

Solid-phase oligosaccharide synthesis offers good opportunities for automation, and using a modified peptide synthesizer automated oligosaccharide synthesis has been described [114]. As an example of automated oligosaccharide synthesis, the preparation of the core pentasaccharide of *N*-linked glycoproteins (**223**) is shown in Scheme 5.38 and Scheme 5.39 [115].

As illustrated, the pentasaccharide is synthesized from the trichloroacetimidate building blocks (**224–226**). To avoid problems in the synthesis of the critical β-D-mannosidic linkage in the solid phase, this bond was incorporated into the disaccharide building block **225**, which was prepared

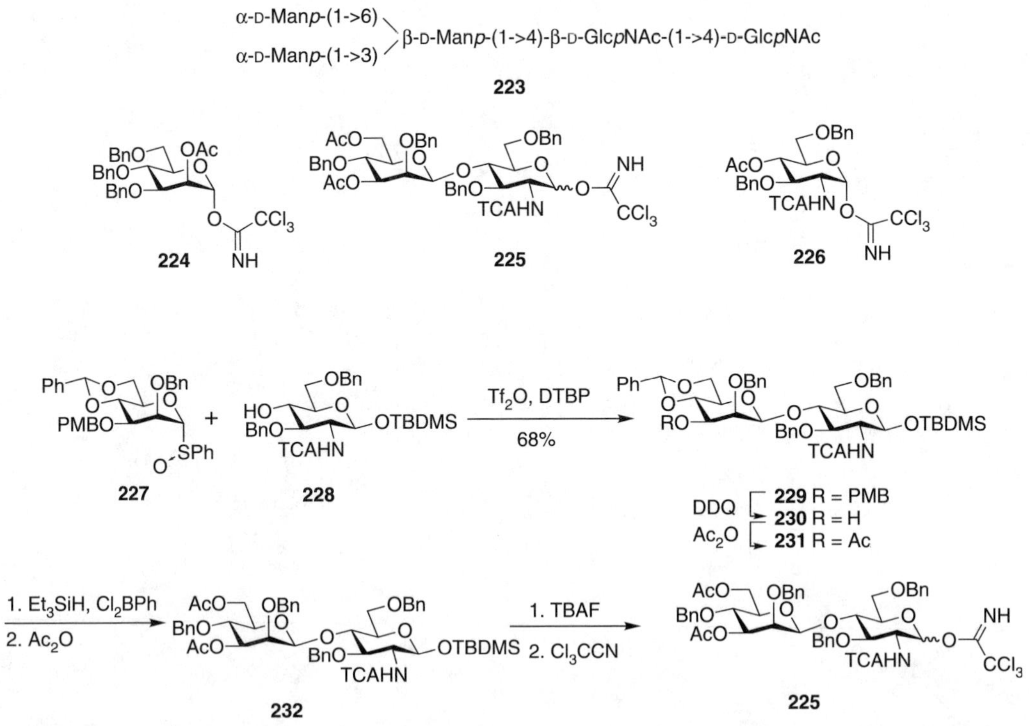

α-D-Man*p*-(1->6)
α-D-Man*p*-(1->3) β-D-Man*p*-(1->4)-β-D-Glc*p*NAc-(1->4)-D-Glc*p*NAc

223

SCHEME 5.38 Building blocks for the solid-phase synthesis of the core pentasaccharide of *N*-linked glycoproteins.

in solution. Glycosylation of the D-glucosamine acceptor (**228**) with the D-mannosyl sulfoxide (**227**) gave the β-linked disaccharide (**229**). Exchange of the *p*-methoxybenzyl group to acetyl (**229** → **231**) was followed by reductive opening of the benzylidene acetal and acetylation of the primary hydroxyl of the 4-*O*-benzyl ether to give the 3′,6′-di-acetate (**232**). Desilylation followed by trichloroacetimidate formation gave the disaccharide donor (**225**).

The solid-phase synthesis started with the glycosylation of the octenediol-functionalized Merrifield resin (**233**) with the glucosamine imidate (**226**) (Scheme 5.39). The coupling product (**234**) was deacetylated, and the resulting **235** was coupled with the disaccharide donor (**225**). The trisaccharide (**236**) was deacetylated and the diol (**237**) was bis-glycosylated with the mannosyl donor (**224**). The pentasaccharide (**238**) was cleaved from the resin using olefin metathesis with Grubb's catalyst to give the protected pentasaccharide as the pentenyl glycoside (**239**). HPLC analysis of the crude product showed that it contains 27% of the desired **239**, the remainder being $(n - 1)$ and $(n - 2)$ deletion sequences.

Automated solid-phase oligosaccharide synthesis shows great promise as a future technology for the generation of oligosaccharides. In order to be generally applicable, automated solid-phase oligosaccharide synthesis needs to be developed to reach consistently high coupling yields and elaborate a relatively small set of building blocks from which the highly diverse oligosaccharidic structures are all available.

5.6 SUMMARY AND OUTLOOK

A great number of different strategies and techniques have been developed for oligosaccharide synthesis. For easier understanding, the various methods are separated in this chapter; however,

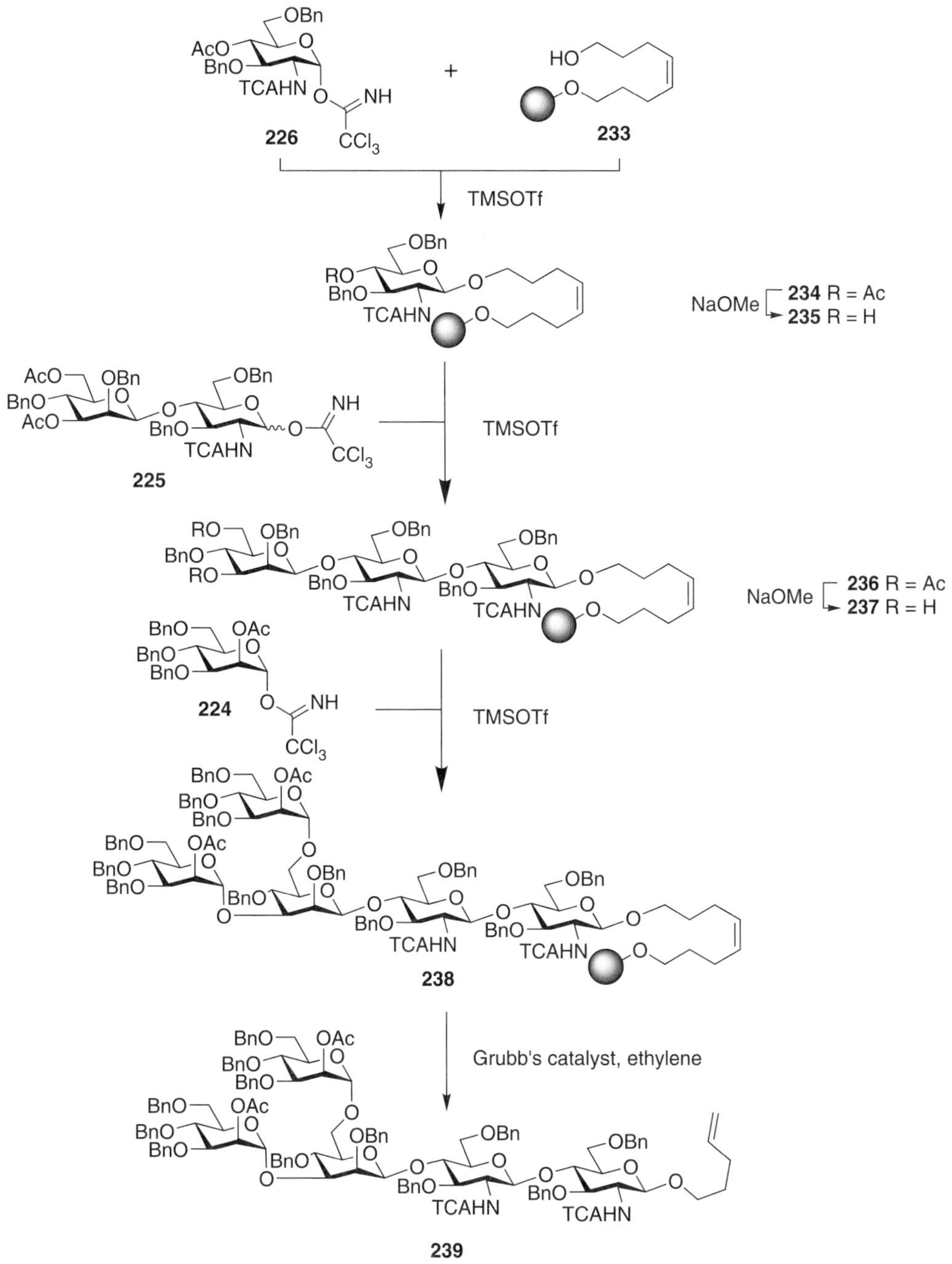

SCHEME 5.39 Automated solid-phase synthesis of the core pentasaccharide of *N*-linked glycoproteins.

in real-life situations a combination of the above strategies and techniques is often more advantageous. Numerous examples of this, and a more detailed discussion of the topics described in this chapter, can be found in a number of excellent reviews [61,86,101–103,116–121].

In recent years, tremendous progress has been made in the chemical synthesis of oligosaccharides. New glycosylation and protection methods resulted in new strategies and techniques for the preparation of oligosaccharides. As a result, large and complex oligosaccharides are synthetically accessible today with relative ease. With the additional driving force resulting from the increased interest in glycobiology, the formerly largely chemistry-driven area of oligosaccharide synthesis is anticipated to develop at an even faster pace.

REFERENCES

1. Dwek, R A, Glycobiology: toward understanding the function of sugars, *Chem. Rev.*, 96, 683–720, 1996.
2. Varki, A, Biological roles of oligosaccharides: all of the theories are correct, *Glycobiology*, 3, 97–130, 1993.
3. Sharon, N, *Complex Carbohydrates. Their Chemistry, Biosynthesis and Functions*, Addison-Wesley, Reading, MA, USA, 1975, pp. 6–9.
4. Laine, R A, A calculation of all possible oligosaccharide isomers both branched and linear yields 1.05×10^{12} structures for a reducing hexasaccharide: the *Isomer Barrier* to development of single-method saccharide sequencing or synthesis systems, *Glycobiology*, 4, 759–767, 1994.
5. Matsuzaki, Y, Ito, Y, Nakahara, Y, Ogawa, T, Synthesis of branched poly-*N*-acetyl-lactosamine type pentaantennary pentacosasaccharide: glycan part of a glycosyl ceramide from rabbit erythrocyte membrane, *Tetrahedron Lett.*, 34, 1061–1064, 1993.
6. Helferich, B, Klein, W, Zur Synthese von Disacchariden IV. Zwei Tetra-acetyl-β-D-Glucosen, *Justus Liebigs Ann. Chem.*, 450, 219–229, 1926.
7. Nicolaou, K C, Hummel, C W, Bockovich, N J, Wong, C H, Stereocontrolled synthesis of sialyl Le^x, the oligosaccharide binding ligand to ELAM-1 (sialyl = *N*-acetylneuramin), *J. Chem. Soc. Chem. Commun.*, 870–872, 1991.
8. Toepfer, A, Schmidt, R R, An efficient synthesis of the Lewis X (Le^x) antigen family, *Tetrahedron Lett.*, 33, 5161–5164, 1992.
9. Toepfer, A, Kinzy, W, Schmidt, R R, Glycosyl imidates. 65. Efficient synthesis of the Lewis antigen X (Le^x) family, *Liebigs Ann. Chem.*, 449–464, 1994.
10. Schmidt, R R, Toepfer, A, Glycosylimidates. 50. Glycosylation with highly reactive glycosyl donors: efficiency of the inverse procedure, *Tetrahedron Lett.*, 32, 3353–3356, 1991.
11. Bommer, R, Kinzy, W, Schmidt, R R, Glycosyl imidates. 49. Synthesis of the octasaccharide moiety of the dimeric Le^x antigen, *Liebigs Ann. Chem.*, 425–433, 1991.
12. Windmüller, R, Schmidt, R R, Glycosylimidates. 66. Efficient synthesis of *lactoneo* series antigens, H, Lewis X (Le^x) and Lewis Y (Le^y), *Tetrahedron Lett.*, 35, 7927–7930, 1994.
13. Hummel, G, Schmidt, R R, Glycosylimidates. 79. A versatile synthesis of the *lactoneo*-series antigens — synthesis of sialyl dimer Lewis X and of dimer Lewis Y, *Tetrahedron Lett.*, 38, 1173–1176, 1997.
14. Fügedi, P, Birberg, W, Garegg, P J, Pilotti, Å, Syntheses of a branched heptasaccharide having phyto-alexin-elicitor activity, *Carbohydr. Res.*, 164, 297–312, 1987.
15. Fügedi, P, Garegg, P J, Kvarnström, I, Svansson, L, Synthesis of a heptasaccharide, structurally related to the phytoelicitor active glucan of *Phytophthora megasperma* f.sp. *glycinea*, *J. Carbohydr. Chem.*, 7, 389–397, 1988.
16. Birberg, W, Fügedi, P, Garegg, P J, Pilotti, Å, Syntheses of a heptasaccharide β-linked to an 8-methoxycarbonyl-oct-1-yl linking arm and of a decasaccharide with structures corresponding to the phytoelicitor active glucan of *Phytophthora megasperma* f.sp. *glycinea*, *J. Carbohydr. Chem.*, 8, 47–57, 1989.
17. Nicolaou, K C, Dolle, R E, Papahatjis, D P, Randall, J L, Practical synthesis of oligosaccharides. Partial synthesis of avermectin B_{1a}, *J. Am. Chem. Soc.*, 106, 4189–4192, 1984.
18. Nicolaou, K C, Randall, J L, Furst, G T, Stereospecific synthesis of rhynchosporosides: a family of fungal metabolites causing scald disease in barley and other grasses, *J. Am. Chem. Soc.*, 107, 5556–5558, 1985.
19. Nicolaou, K C, Caulfield, T J, Kataoka, H, Stylianides, N A, Total synthesis of the tumor-associated Le^x family of glycoshingolipids, *J. Am. Chem. Soc.*, 112, 3693–3695, 1990.
20. Nicolaou, K C, Hummel, C W, Iwabuchi, Y, Total synthesis of sialyl dimeric Le^x, *J. Am. Chem. Soc.*, 114, 3126–3128, 1992.

21. Nicolaou, K C, Bockovich, N J, Carcanague, D R, Total synthesis of sulfated Le^x and Le^a-type oligosaccharide selectin ligands, *J. Am. Chem. Soc.*, 115, 8843–8844, 1993.
22. Nicolaou, K C, Caulfield, T, Kataoka, H, Kumazawa, T, A practical and enantioselective synthesis of glycosphingolipids and related-compounds. Total synthesis of globotriaosylceramide (Gb_3), *J. Am. Chem. Soc.*, 110, 7910–7912, 1988.
23. Nicolaou, K C, Caulfield, T J, Katoaka, H, Total synthesis of globotriaosylceramide (Gb_3) and lysoglobotriaosylceramide (lysoGb_3), *Carbohydr. Res.*, 202, 177–191, 1990.
24. Halcomb, R L, Danishefsky, S J, On the direct epoxidation of glycals: application of a reiterative strategy for the synthesis of β-linked oligosaccharides, *J. Am. Chem. Soc.*, 111, 6661–6666, 1989.
25. Sliedregt, L A J M, van der Marel, G A, van Boom, J H, Trimethylsilyl triflate mediated chemoselective condensation of arylsulfenyl glycosides, *Tetrahedron Lett.*, 35, 4015–4018, 1994.
26. Gildersleeve, J, Smith, A, Sakurai, K, Raghavan, S, Kahne, D, Scavenging byproducts in the sulfoxide glycosylation reaction: application to the synthesis of ciclamycin 0, *J. Am. Chem. Soc.*, 121, 6176–6182, 1999.
27. Xuereb, H, Maletic, M, Gildersleeve, J, Pelczer, I, Kahne, D, Design of an oligosaccharide scaffold that binds in the minor groove of DNA, *J. Am. Chem. Soc.*, 122, 1883–1890, 2000.
28. Martin-Lomas, M, Khiar, N, Garcia, S, Koessler, J-L, Nieto, P M, Rademacher, T W, Inositolphosphoglycan mediators structurally related to glycosyl phosphatidylinositol anchors: synthesis, structure and biological activity, *Chem. Eur. J.*, 6, 3608–3621, 2000.
29. Kanie, O, Ito, Y, Ogawa, T, Orthogonal glycosylation strategy in oligosaccharide synthesis, *J. Am. Chem. Soc.*, 116, 12073–12074, 1994.
30. Kanie, O, Ito, Y, Ogawa, T, Orthogonal glycosylation strategy in synthesis of extended blood group B determinant, *Tetrahedron Lett.*, 37, 4551–4554, 1996.
31. Ito, Y, Kanie, O, Ogawa, T, Orthogonal glycosylation strategy for rapid assembly of oligosaccharides on a polymer support, *Angew. Chem. Int. Ed. Engl.*, 35, 2510–2512, 1996.
32. Mootoo, D R, Konradsson, P, Udodong, U, Fraser-Reid, B, "Armed"and "disarmed" *n*-pentenyl glycosides in saccharide couplings leading to oligosaccharides, *J. Am. Chem. Soc.*, 110, 5583–5584, 1988.
33. Konradsson, P, Mootoo, D R, McDevitt, R E, Fraser-Reid, B, Iodonium ion generated *in situ* from *N*-iodosuccinimide and trifluoromethanesulphonic acid promotes direct linkage of "disarmed" pent-4-enyl glycosides, *J. Chem. Soc. Chem. Commun.*, 270–272, 1990.
34. Konradsson, P, Udodong, U E, Fraser-Reid, B, Iodonium promoted reactions of disarmed thioglycosides, *Tetrahedron Lett.*, 31, 4313–4316, 1990.
35. Veeneman, G H, van Boom, J H, An efficient thioglycoside-mediated formation of α-glycosidic linkages promoted by iodonium dicollidine perchlorate, *Tetrahedron Lett.*, 31, 275–278, 1990.
36. Veeneman, G H, van Leeuwen, S H, van Boom, J H, Iodonium ion promoted reactions at the anomeric centre. 2. An efficient thioglycoside mediated approach toward the formation of 1,2-*trans* linked glycosides and glycosidic esters, *Tetrahedron Lett.*, 31, 1331–1334, 1990.
37. Friesen, R W, Danishefsky, S J, On the controlled oxidative coupling of glycals: a new strategy for the rapid assembly of oligosaccharides, *J. Am. Chem. Soc.*, 111, 6656–6660, 1989.
38. Barrena, M I, Echarri, R, Castillon, S, Synthesis of disaccharides by selective metallocene promoted activation of glycosyl fluorides, *Synlett*, 675–676, 1996.
39. Cheung, M K, Douglas, N L, Hinzen, B, Ley, S V, Pannecoucke, X, One-pot synthesis of tetra- and pentasaccharides from monomeric building blocks using the principles of orthogonality and reactivity tuning, *Synlett*, 257–260, 1997.
40. Nguyen, H M, Poole, J L, Gin, D Y, Chemoselective iterative dehydrative glycosylation, *Angew. Chem. Int. Ed.*, 40, 414–417, 2001.
41. Hashimoto, S, Sakamoto, H, Honda, T, Ikegami, S, Oligosaccharide synthesis based on glycosyl donors and acceptors carrying phosphorus-containing leaving groups, *Tetrahedron Lett.*, 38, 5181–5184, 1997.
42. Hashimoto, S, Sakamoto, H, Honda, T, Abe, H, Nakamura, S, Ikegami, S, "Armed–disarmed" glycosidation strategy based on glycosyl donors and acceptors carrying phosphoroamidate as a leaving group: a convergent synthesis of globotriaosylceramide, *Tetrahedron Lett.*, 38, 8969–8972, 1997.

43. Chiba, H, Funasaka, S, Kiyota, K, Mukaiyama, T, Catalytic and chemoselective glycosylation between "armed" and "disarmed" glycosyl *p*-trifluoromethylbenzylthio-*p*-trifluoromethylphenyl formimidates, *Chem. Lett.*, 746–747, 2002.
44. Chiba, H, Mukaiyama, T, Highly α-stereoselective one-pot sequential glycosylation using glucosyl thioformimidate derivatives, *Chem. Lett.*, 32, 172–173, 2003.
45. Chiba, H, Funasaka, S, Mukaiyama, T, Catalytic and stereoselective glycosylation with glucosyl thioformimidates, *Bull. Chem. Soc. Jpn.*, 76, 1629–1644, 2003.
46. Fraser-Reid, B, Wu, Z, Andrews, C W, Skowronski, E, Bowen, J P, Torsional effects in glycoside reactivity: saccharide couplings mediated by acetal protecting groups, *J. Am. Chem. Soc.*, 113, 1434–1435, 1991.
47. Grice, P, Ley, S V, Pietruszka, J, Osborn, H M I, Priepke, H W M, Warriner, S L, A new strategy for oligosaccharide assembly exploiting cyclohexane-1,2-diacetal methodology: an efficient synthesis of a high mannose type nonasaccharide, *Chem. Eur. J.*, 3, 431–440, 1997.
48. Boons, G J, Geurtsen, R, Holmes, D, Chemoselective glycosylations. 1. Differences in size of anomeric leaving groups can be exploited in chemoselective glycosylations, *Tetrahedron Lett.*, 36, 6325–6328, 1995.
49. Geurtsen, R, Holmes, D S, Boons, G J, Chemoselective glycosylations. 2. Differences in size of anomeric leaving groups can be exploited in chemoselective glycosylations, *J. Org. Chem.*, 62, 8145–8154, 1997.
50. Zhang, Z, Ollmann, I R, Ye, X-S, Wischnat, R, Baasov, T, Wong, C-H, Programmable one-pot oligosaccharide synthesis, *J. Am. Chem. Soc.*, 121, 734–753, 1999.
51. Ye, X-S, Wong, C-H, Anomeric reactivity-based one-pot oligosaccharides synthesis: a rapid route to oligosaccharide libraries, *J. Org. Chem.*, 65, 2410–2431, 2000.
52. Roy, R, Andersson, F O, Letellier, M, "Active" and "latent" thioglycosyl donors in oligosaccharide synthesis. Application to the synthesis of α-sialosides, *Tetrahedron Lett.*, 33, 6053–6056, 1992.
53. Cao, S, Hernandez-Mateo, F, Roy, R, Scope and applications of "active and latent" thioglycosyl donors, part 4. *J. Carbohydr. Chem.*, 17, 609–632, 1998.
54. Cao, S, Gan, Z, Roy, R, Active–latent glycosylation strategy toward Lewis X pentasaccharide in a form suitable for neoglycoconjugate syntheses, *Carbohydr. Res.*, 318, 75–81, 1999.
55. Fraser-Reid, B, Wu, Z, Udodong, U E, Ottosson, H, Armed/disarmed effects in glycosyl donors: rationalization and sidetracking, *J. Org. Chem.*, 55, 6068–6070, 1990.
56. Allen, J G, Fraser-Reid, B, *n*-Pentenyl glycosyl orthoesters as versatile intermediates in oligosaccharide synthesis. The proteoglycan linkage region, *J. Am. Chem. Soc.*, 121, 468–469, 1999.
57. Boons, G J, Isles, S, Vinyl glycosides in oligosaccharide synthesis. 1. A new latent–active glycosylation strategy, *Tetrahedron Lett.*, 35, 3593–3596, 1994.
58. Boons, G J, Isles, S, Vinyl glycosides in oligosaccharide synthesis. 2. The use of allyl and vinyl glycosides in oligosaccharide synthesis, *J. Org. Chem.*, 61, 4262–4271, 1996.
59. Boons, G J, Heskamp, B, Hout, F, Vinyl glycosides in oligosaccharide synthesis: a strategy for the preparation of trisaccharide libraries based on latent–active glycosylations, *Angew. Chem. Int. Ed. Engl.*, 35, 2845–2847, 1996.
60. Bai, Y, Boons, G J, Burton, A, Johnson, M, Haller, M, Vinyl glycosides in oligosaccharide synthesis. 6. 3-Buten-2-yl 2-azido-2-deoxy glycosides and 3-buten-2-yl 2-phthalimido-2-deoxy glycosides as novel glycosyl donors, *J. Carbohydr. Chem.*, 19, 939–958, 2000.
61. Boons, G J, Strategies in oligosaccharide synthesis, *Tetrahedron*, 52, 1095–1121, 1996.
62. Kim, K S, Kim, J H, Lee, Y J, Lee, Y J, Park, J, 2-(Hydroxycarbonyl)benzyl glycosides: a novel type of glycosyl donors for highly efficient β-mannopyranosylation and oligosaccharide synthesis by latent–active glycosylation, *J. Am. Chem. Soc.*, 123, 8477–8481, 2001.
63. Kim, K S, Park, J, Lee, Y J, Seo, Y S, Dual stereoselectivity of 1-(2′-carboxy)benzyl 2-deoxyglycosides as glycosyl donors in the direct construction of 2-deoxyglycosyl linkages, *Angew. Chem. Int. Ed.*, 42, 459–462, 2003.
64. Kim, K S, Kang, S S, Seo, Y S, Kim, H J, Lee, Y J, Jeong, K-S, Glycosylation with 2′-carboxybenzyl glycosides as glycosyl donors: scope and application to the synthesis of a tetrasaccharide, *Synlett*, 1311–1314, 2003.
65. Barresi, F, Hindsgaul, O, Synthesis of β-mannopyranosides by intramolecular aglycon delivery, *J. Am. Chem. Soc.*, 113, 9376–9377, 1991.

66. Barresi, F, Hindsgaul, O, The synthesis of β-mannopyranosides by intramolecular aglycon delivery: scope and limitations of the existing methodology, *Can. J. Chem.*, 72, 1447–1465, 1994.
67. Stork, G, Kim, G, Stereocontrolled synthesis of disaccharides via the temporary silicon connection, *J. Am. Chem. Soc.*, 114, 1087–1088, 1992.
68. Stork, G, Laclair, J J, Stereoselective synthesis of β-mannopyranosides via the temporary silicon connection method, *J. Am. Chem. Soc.*, 118, 247–248, 1996.
69. Bols, M, Stereocontrolled synthesis of α-glucosides by intramolecular glycosidation, *J. Chem. Soc. Chem. Commun.*, 913–914, 1992.
70. Bols, M, Application of intramolecular glycosidation to the stereocontrolled synthesis of disaccharides containing α-gluco and α-galacto linkages, *J. Chem. Soc. Chem. Commun.*, 791–792, 1993.
71. Bols, M, Intramolecular glycosidation: stereocontrolled synthesis of α-glucosides from a 2-O-alkoxysilyl thioglucoside, *Acta Chem. Scand.*, 47, 829–834, 1993.
72. Bols, M, Efficient stereocontrolled glycosidation of secondary sugar hydroxyls by silicon tethered intramolecular glycosidation, *Tetrahedron*, 49, 10049–10060, 1993.
73. Ito, Y, Ogawa, T, A novel approach to the stereoselective synthesis of β-mannosides, *Angew. Chem. Int. Ed. Engl.*, 33, 1765–1767, 1994.
74. Ito, Y, Ohnishi, Y, Ogawa, T, Nakahara, Y, Highly optimized β-mannosylation via *p*-methoxybenzyl assisted intramolecular aglycon delivery, *Synlett*, 1102–1104, 1998.
75. Dan, A, Ito, Y, Ogawa, T, Stereocontrolled synthesis of the pentasaccharide core structure of asparagine-linked glycoprotein oligosaccharide based on a highly convergent strategy, *Tetrahedron Lett.*, 36, 7487–7490, 1995.
76. Dan, A, Ito, Y, Ogawa, T, A convergent and stereocontrolled synthetic route to the core pentasaccharide structure of asparagine-linked glycoproteins, *J. Org. Chem.*, 60, 4680–4681, 1995.
77. Ito, Y, Ogawa, T, Intramolecular aglycon delivery on polymer support: gatekeeper monitored glycosylation, *J. Am. Chem. Soc.*, 119, 5562–5566, 1997.
78. Krog-Jensen, C, Oscarson, S, Stereospecific synthesis of β-D-fructofuranosides using the internal aglycon delivery approach, *J. Org. Chem.*, 61, 4512–4513, 1996.
79. Bamhaoud, T, Sanchez, S, Prandi, J, 1,2,5-Ortho esters of D-arabinose as versatile arabinofuranosidic building blocks. Concise synthesis of the tetrasaccharidic cap of the lipoarabinomannan of *Mycobacterium tuberculosis*, *Chem. Commun.*, 659–660, 2000.
80. Sanchez, S, Bamhaoud, T, Prandi, J, A comprehensive glycosylation system for the elaboration of oligoarabinofuranosides, *Tetrahedron Lett.*, 41, 7447–77452, 2000.
81. Marotte, K, Sanchez, S, Bamhaoud, T, Prandi, J, Synthesis of oligoarabinofuranosides from the mycobacterial cell wall, *Eur. J. Org. Chem.*, 3587–3598, 2003.
82. Gelin, M, Ferrières, V, Lefeuvre, M, Plusquellec, D, First intramolecular aglycon delivery onto a D-fucofuranosyl entity for the synthesis of α-D-fucofuranose-containing disaccharides, *Eur. J. Org. Chem.*, 1285–1293, 2003.
83. Seward, C M P, Cumpstey, I, Aloui, M, Ennis, S C, Redgrave, A J, Fairbanks, A J, Stereoselective *cis* glycosylation of 2-*O*-allyl protected glycosyl donors by intramolecular aglycon delivery (IAD), *Chem. Commun.*, 1409–1410, 2000.
84. Cumpstey, I, Fairbanks, A J, Redgrave, A J, Stereospecific synthesis of 1,2-*cis* glycosides by allyl-mediated intramolecular aglycon delivery. 2. The use of glycosyl fluorides, *Org. Lett.*, 3, 2371–2374, 2001.
85. Fairbanks, A J, Intramolecular aglycon delivery (IAD): the solution to 1,2-*cis* stereocontrol for oligosaccharide synthesis? *Synlett*, 1945–1958, 2003.
86. Jung, K-H, Müller, M, Schmidt, R R, Intramolecular *O*-glycoside bond formation, *Chem. Rev.*, 100, 4423–4442, 2000.
87. Ziegler, T, Ritter, A, Hurttlen, J, Intramolecular glycosylation of prearranged glycosides. 5. α-(1→4)-Selective glucosylation of glucosides and glucosamines, *Tetrahedron Lett.*, 38, 3715–3718, 1997.
88. Ziegler, T, Dettmann, R, Ariffadhillah, Zettl, U, Prearranged glycosides. 8. Intramolecular α-galactosylation via succinoyl tethered glycosides, *J. Carbohydr. Chem.*, 18, 1079–1095, 1999.
89. Raghavan, S, Kahne, D, A one-step synthesis of the ciclamycin trisaccharide, *J. Am. Chem. Soc.*, 115, 1580–1581, 1993.

90. Yamada, H, Harada, T, Miyazaki, H, Takahashi, T, One-pot sequential glycosylation: a new method for the synthesis of oligosaccharides, *Tetrahedron Lett.*, 35, 3979–3982, 1994.
91. Mong, K-K T, Wong, C-H, Reactivity-based one-pot synthesis of a Lewis Y carbohydrate hapten: a colon-rectal cancer antigen determinant, *Angew. Chem. Int. Ed.*, 41, 4087–4090, 2002.
92. Lahmann, M, Oscarson, S, One-pot oligosaccharide synthesis exploiting solvent reactivity effects, *Org. Lett.*, 2, 3881–3882, 2000.
93. Yamada, H, Kato, T, Takahashi, T, One-pot sequential glycosylation: a new method for the synthesis of branched oligosaccharides, *Tetrahedron Lett.*, 40, 4581–4584, 1999.
94. Takahashi, T, Adachi, M, Matsuda, A, Doi, T, Combinatorial synthesis of trisaccharides via solution-phase one-pot glycosylation, *Tetrahedron Lett.*, 41, 2599–2603, 2000.
95. Tanaka, H, Adachi, M, Tsukamoto, H, Ikeda, T, Yamada, H, Takahashi, T, Synthesis of di-branched heptasaccharide by one-pot glycosylation using seven independent building blocks, *Org. Lett.*, 4, 4213–4216, 2002.
96. Frechet, J M, Schuerch, C, Solid-phase synthesis of oligosaccharides. 1. Preparation of the solid support. Poly[*p*-(1-propen-3-ol-1-yl)styrene], *J. Am. Chem. Soc.*, 93, 492–496, 1971.
97. Fréchet, J M, Schuerch, C, Solid-phase synthesis of oligosaccharides. 2. Steric control by C-6 substituents in glucoside synthesis, *J. Am. Chem. Soc.*, 94, 604–609, 1972.
98. Fréchet, J M, Schuerch, C, Solid-phase synthesis of oligosaccharides. 3. Preparation of some derivatives of di- and tri-saccharides via a simple alcoholysis reaction, *Carbohydr. Res.*, 22, 399–412, 1972.
99. Douglas, S P, Whitfield, D M, Krepinsky, J J, Polymer-supported solution synthesis of oligosaccharides, *J. Am. Chem. Soc.*, 113, 5095–5097, 1991.
100. Dreef-Tromp, C M, Willems, H A M, Westerduin, P, van Veelen, P, van Boeckel, C A A, Polymer-supported solution synthesis of heparan sulphate-like oligomers, *Bioorg. Med. Chem. Lett.*, 7, 1175–1180, 1997.
101. Osborn, H M I, Khan, T H, Recent developments in polymer supported syntheses of oligosaccharides and glycopeptides, *Tetrahedron*, 55, 1807–1850, 1999.
102. Seeberger, P H, Haase, W C, Solid-phase oligosaccharide synthesis and combinatorial carbohydrate libraries, *Chem. Rev.*, 100, 4349–4393, 2000.
103. Seeberger, P H, Solid phase oligosaccharide synthesis, *J. Carbohydr. Chem.*, 21, 613–643, 2002.
104. Andrade, R B, Plante, O J, Melean, L G, Seeberger, P H, Solid-phase oligosaccharide synthesis: preparation of complex structures using a novel linker and different glycosylating agents, *Org. Lett.*, 1, 1811–1814, 1999.
105. Danishefsky, S J, McClure, K F, Randolph, J T, Ruggeri, R B, A strategy for the solid-phase synthesis of oligosaccharides, *Science*, 260, 1307–1309, 1993.
106. Zhu, T, Boons, G J, A two-directional approach for the solid-phase synthesis of trisaccharide libraries, *Angew. Chem. Int. Ed.*, 37, 1898–1900, 1998.
107. Liang, R, Yan, L, Loebach, J, Ge, M, Uozumi, Y, Sekanina, K, Horan, N, Gildersleeve, J, Thompson, C, Smith, A, Biswas, K, Still, W C, Kahne, D, Parallel synthesis and screening of a solid phase carbohydrate library, *Science*, 274, 1520–1522, 1996.
108. Nicolaou, K C, Winssinger, N, Pastor, J, DeRoose, F, A general and highly efficient solid phase synthesis of oligosaccharides. Total synthesis of a heptasaccharide phytoalexin elicitor (HPE), *J. Am. Chem. Soc.*, 119, 449–450, 1997.
109. Nicolaou, K C, Watanabe, N, Li, J, Pastor, J, Winssinger, N, Solid-phase synthesis of oligosaccharides: construction of a dodecasaccharide, *Angew. Chem. Int. Ed.*, 37, 1559–1561, 1998.
110. Wu, X, Grathwohl, M, Schmidt, R R, Efficient solid-phase synthesis of a complex, branched *N*-glycan hexasaccharide: use of a novel linker and temporary-protecting-group pattern, *Angew. Chem. Int. Ed.*, 41, 4489–4493, 2002.
111. Rodebaugh, R, Joshi, S, Fraser-Reid, B, Geysen, H M, Polymer-supported oligosaccharides via *n*-pentenyl glycosides: methodology for a carbohydrate library, *J. Org. Chem.*, 62, 5660–5661, 1997.
112. Lam, S N, Gervay-Hague, J, Solution- and solid-phase oligosaccharide synthesis using glucosyl iodides: a comparative study, *Carbohydr. Res.*, 337, 1953–1965, 2002.
113. Palmacci, E R, Plante, O J, Seeberger, P H, Oligosaccharide synthesis in solution and on solid support with glycosyl phosphates, *Eur. J. Org. Chem.*, 595–606, 2002.

114. Plante, O J, Palmacci, E R, Seeberger, P H, Automated solid-phase synthesis of oligosaccharides, *Science*, 291, 1523–1527, 2001.
115. Ratner, D M, Swanson, E R, Seeberger, P H, Automated synthesis of a protected *N*-linked glycoprotein core pentasaccharide, *Org. Lett.*, 5, 4717–4720, 2003.
116. Boons, G-J, Synthetic oligosaccharides: recent advances, *Drug Discovery Today*, 1, 331–342, 1996.
117. Davis, B G, Recent developments in oligosaccharide synthesis, *J. Chem. Soc. Perkin. Trans. 1*, 2137–2160, 2000.
118. Sears, P, Wong, C-H, Toward automated synthesis of oligosaccharides and glycoproteins, *Science*, 291, 2344–2350, 2001.
119. Nicolaou, K C, Mitchell, H J, Adventures in carbohydrate chemistry: new synthetic technologies, chemical synthesis, molecular design, and chemical biology, *Angew. Chem. Int. Ed.*, 40, 1576–1624, 2001.
120. Fraser-Reid, B, Madsen, R, Campbell, A S, Roberts, C S, Merritt, J R, Chemical synthesis of oligosaccharides, In *Bioorganic Chemistry: Carbohydrates*, Hecht, S M, Ed., Oxford University Press, Oxford, pp. 89–133, 1999.
121. Seeberger, P H, Solid-phase oligosaccharide synthesis, In *Carbohydrate-Based Drug Discovery*, Wong, C H, Ed., Wiley-VCH Verlag, Weinheim, Germany, pp. 103–127, 2003.

Part II

From Sugars to Sugar-Like Structures to Non-Sugars

6 Functionalization of Sugars

Daniel E. Levy

CONTENTS

6.1 INTRODUCTION

The chemical modification of sugars to sugar derivatives or nonsugar products has been extensively studied as a strategy to prepare products ranging from rare sugars to novel chiral substances and to complex natural products. The methods utilized in the initial transformations are generally drawn from the principles of mainstream organic chemistry. In this chapter, many of these strategies are addressed. One should realize, however, that any one of the strategies presented herein could form the basis of a complete chapter in this book. In the interest of describing principles readily adaptable to the direct modification of sugars, this chapter makes extensive use of historical literature highlighting basic strategies. It is not the intent of this chapter to present an exhaustive analysis of this subject but rather to present ideas which, through the incorporation of basic organic chemistry principles, will lead the reader towards desired solutions to more complex problems.

6.1.1 Definition of Concept

Sugars, by nature, are highly polar and highly functionalized molecules bearing multiple adjacent stereogenic centers. These features render sugars extremely useful as chiral pool starting materials. Unfortunately, the utility and potential versatility of these compounds is frequently undermined by their high polarity and water solubility. These latter features contribute to the perception that sugars are inherently difficult to work with, and many chemists frequently decline to consider using these molecules, despite the fact that they may present solutions to difficult problems.

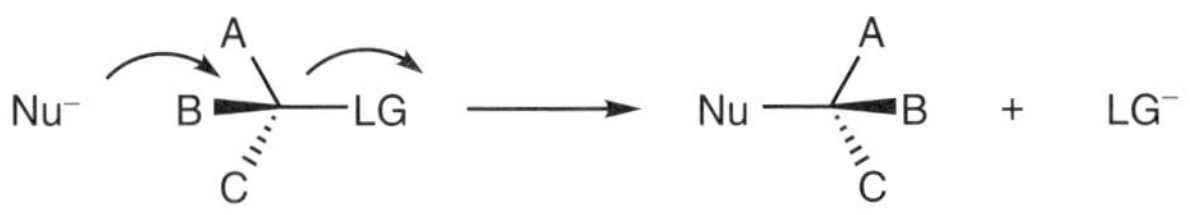

FIGURE 6.1 Mechanism of S_N2 reactions.

Chapter 3 of this book presents numerous strategies to the selective protection of sugar hydroxyl groups. In this chapter, both suitably protected and unprotected sugars are considered as substrates for a diverse range of chemical modifications. After reading this chapter, it is hoped that the reader will consider sugars under the category of conventional organic molecules useful as starting materials for synthetic applications ranging from drug design to the total synthesis of complex natural products.

6.1.2 S_N2 REACTIONS

By considering the structures of sugars, we recognize that the various hydroxyl groups react according to two general categories. The first — reactions at the anomeric center or C-1 — follow a set of rules described in Chapter 4 (Glycosylation Methods) and in Chapter 7 (Strategies Towards *C*-Glycosides). In the second category, all other hydroxyl groups are considered, and generally behave as standard primary and secondary alcohols. Moreover, the reactions utilized to modify sugars at these sites broadly relate back to S_N2 mechanisms (Figure 6.1). Figure 6.2 illustrates the different types of hydroxyl groups found in sugars. Although the illustrations in Figure 6.2 are hexose sugars, it is important to note that these types of hydroxyl groups and their relative relationships also apply to furanose (and all other) sugars. In this chapter, the functionalization of sugars will be addressed with respect to the modification or replacement of hydroxyl groups not present at the anomeric center.

6.2 SPECIAL CONSIDERATIONS WITH SUGARS

Although the chemical modifications to sugars and sugar derivatives described in this chapter generally follow S_N2 mechanisms, special consideration must be given to the cyclic nature of most

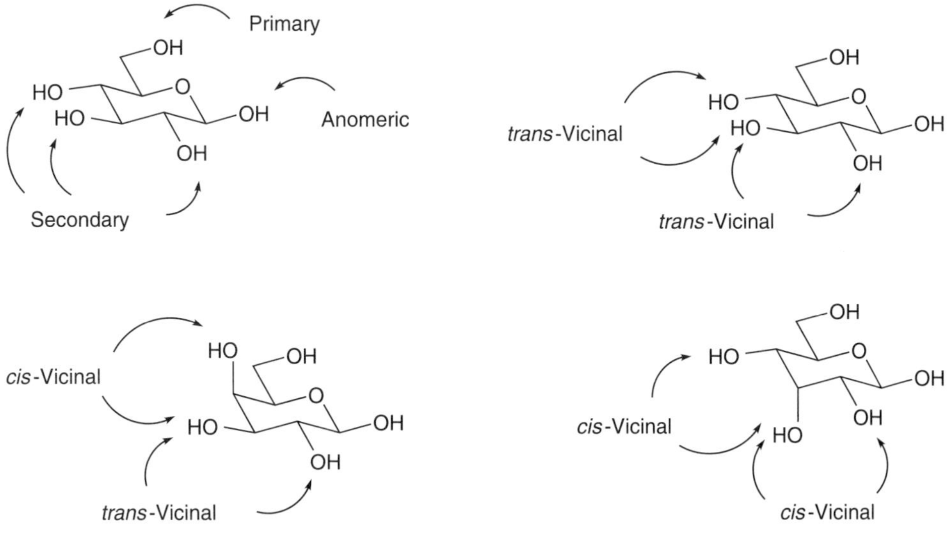

FIGURE 6.2 Types of hydroxyl groups found in sugars.

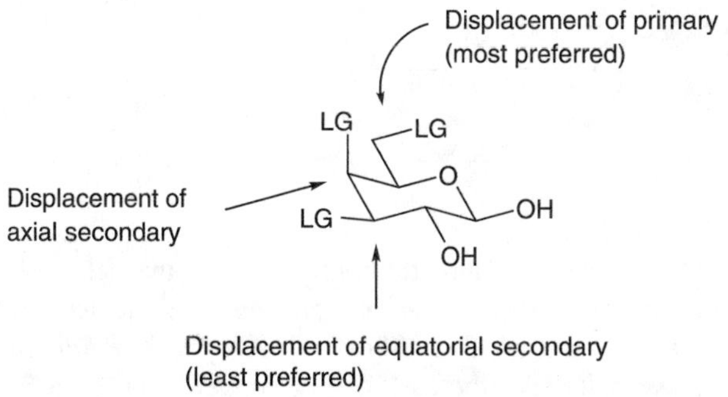

FIGURE 6.3 Preferred nucleophile trajectories.

sugars. In this respect, nucleophilic displacement of leaving groups must occur through the least sterically hindered trajectory of a given nucleophile. For example, as shown in Figure 6.3, nucleophilic displacement of primary leaving groups is preferred over displacement of axially oriented leaving groups. Moreover, displacement of axially oriented leaving groups is, in turn, preferred over displacement of equatorially oriented leaving groups.

The trends introduced in Figure 6.3 may seem direct. In fact, they are, when applied to six-membered rings bearing few substituents. However, when considering sugars, we cannot ignore the influence of the anomeric center.

6.2.1 Axial vs. Equatorial Approach

Because all sugars can exist in two anomeric configurations (Figure 6.4), the orientation of the anomeric substituent can have dramatic effects on the ability of nucleophiles to displace, leaving groups at specific sites. For example, as shown in Figure 6.5, the tosylate at C-3 of the β-glucopyranose is easily displaced by a variety of nucleophiles. However, the tosylate on the analogous α-glucopyranose cannot be displaced because of the 1,3-diaxial interaction between the approaching nucleophile and the anomeric substituent [1].

6.2.2 Substitution vs. Elimination

Because sugars are generally polyhydroxylated structures, conceptually, multiple hydroxyl groups may be converted to leaving groups and simultaneously displaced by given nucleophiles. However,

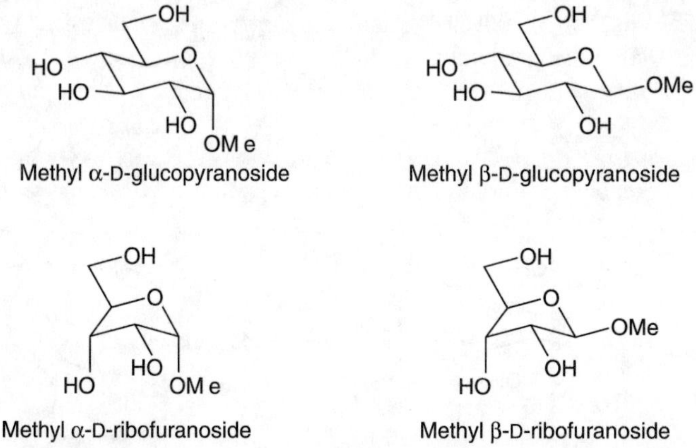

FIGURE 6.4 Anomeric configurations of glucose and ribose.

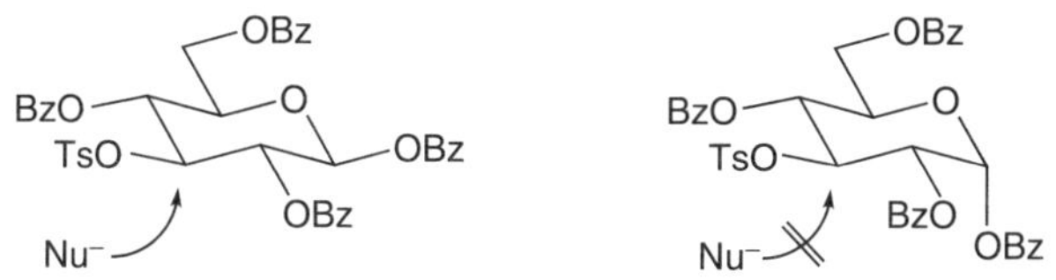

FIGURE 6.5 Effect of anomeric configuration on abilility of nucleophile to approach leaving group.

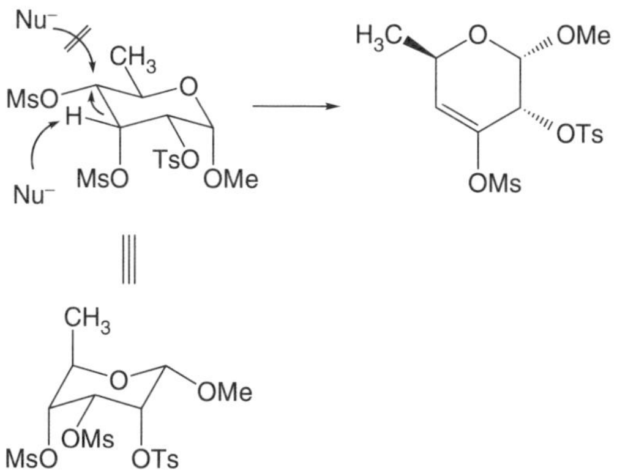

SCHEME 6.1 *cis*-Vicinal elimination mechanism illustrated through ring inversion generating *trans*-periplanar relationship between extracted proton and leaving group.

when considering the special case of *cis*-vicinal leaving groups, a frequent observation of attempted nucleophilic displacement is elimination with the generation of a double bond. The mechanism for this reaction, shown in Scheme 6.1, proceeds through extraction of a proton by the nucleophile followed by elimination of the leaving group oriented *trans*-periplanar to the developing negative charge [2].

6.2.3 Neighboring Group Participation

Again, referring to the polyfunctional nature of sugars, any discussion of reactions on the sugar ring would not be complete without addressing the effects of ring substituents present near the leaving group. We have already addressed steric effects generated by 1,3-diaxial interactions. Now, we will address the effects of nucleophilic substituents positioned at centers directly adjacent to designated leaving groups.

6.2.3.1 Esters

Esters are extremely common in sugar chemistry, partly because of the exploitation of acetate and benzoate protecting groups. If we consider a 1,2-diol that has been modified such that an acetate protects the hydroxyl group at C-1, and the hydroxyl group at C-2 is converted to a tosylate, the carbonyl of the acetate can displace the tosylate when the two groups are oriented *trans* to one another. As shown in Scheme 6.2, this results in the formation of an acyloxonium ion. Once formed, the acyloxonium ion can undergo reactions at three different sites [3,4]. For example, if a nucleophile reacts at C-2 and opens the acyloxonium ion, then the result is effectively a displacement of the tosylate with retention of the stereochemical configuration at C-2 (double inversion). However, if the nucleophile reacts at C-1 and similarly opens the acyloxonium ion, then

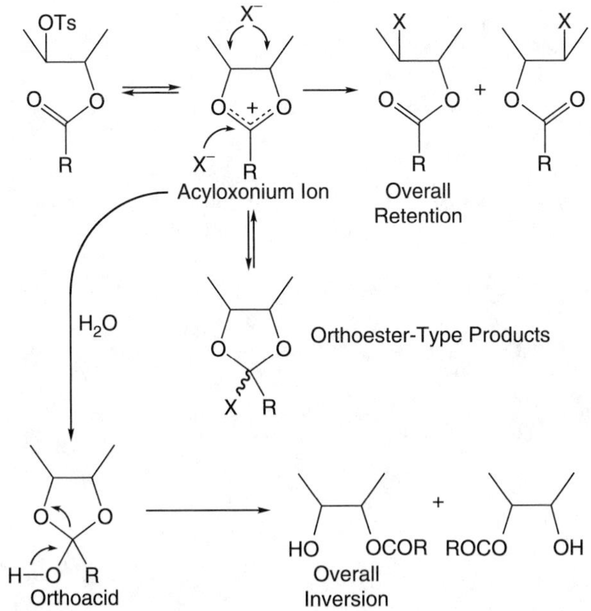

SCHEME 6.2 Neighboring group participation mechanisms with esters.

the result is the migration of the acetate from C-1 to C-2 with introduction of the nucleophile at C-1. Finally, if the nucleophile approaches the carbonyl center, then the result will be formation of an orthoester (in the case of alkoxy nucleophiles) or an orthoester-type product.

Further exploring the approach of nucleophiles to the carbonyl center, if the nucleophile is water, then orthoacids can be formed. However, orthoacids are very unstable and generally decompose to a mixture of two products. The first product is a derivative of the starting material with the tosylate displaced by hydroxide. In the second product, the acetate migrated from C-1 to C-2. With respect to mechanistic preferences, soft nucleophiles tend to favor ring opening, whereas hard nucleophiles tend to favor orthoester-type products.

6.2.3.2 Amides

Amides, being more nucleophilic than esters, can mirror esters and undergo similar types of reactions when present as neighboring groups. If we consider a 1,2-amino alcohol that has been modified such that the amino group is protected as an acetamide, and the hydroxyl group is converted to a tosylate, the carbonyl of the acetamide can displace the tosylate when the two groups are oriented *trans* to one another. As shown in Scheme 6.3, this results in the formation of an oxazolinium ion. Once formed, the oxazolinium ion can undergo reactions at three different sites. For example, if a nuclophile reacts at the former tosylate center and opens the oxazolinium ion, then the result is effectively a displacement of the tosylate with retention of the stereochemical configuration (double inversion). However, if the nucleophile approaches the carbonyl center, then the result will be formation of an orthoamide (in the case of alkoxy nucleophiles) or an orthoamide-type product.

Unlike acyloxonium ions, oxazolinium ions bear a proton capable of being removed under basic conditions. If these conditions are applied to the oxazolinium ion, then the result is formation of a stable oxazoline.

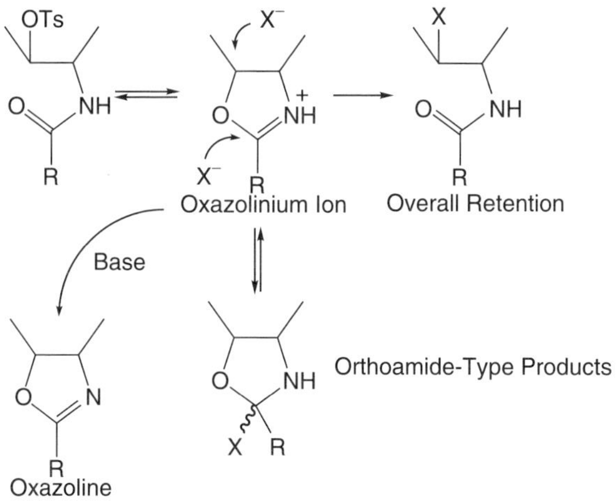

SCHEME 6.3 Neighboring group participation mechanisms with amides.

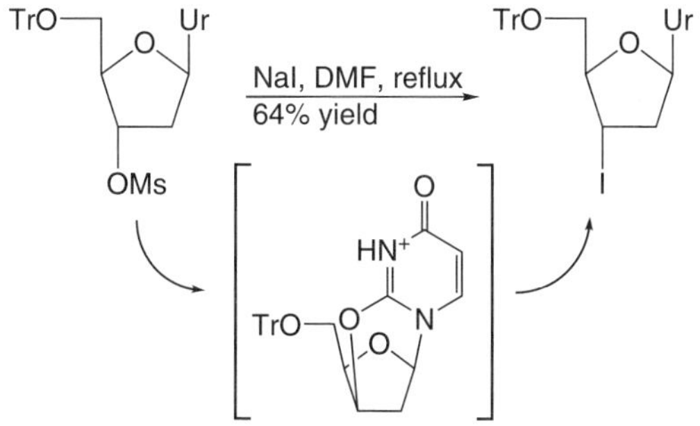

SCHEME 6.4 Neighboring group participation mechanism with nucleotides.

6.2.3.3 Nucleotides

Esters and amides are not the only functional groups capable of neighboring group participation. In fact, any functional group capable of delivering a carbonyl group to the intended site of reaction can become involved in this mechanism. For example, nucleotides are perfectly suited for such reactions. As shown in Scheme 6.4, uracil was shown to effectively displace the mesylate from the 3-position of the 2-deoxyribose drivative shown [5]. Upon introduction of sodium iodide, the cationic species was opened. Similar to other examples of neighboring group participation, the iodide was introduced with net retention of stereochemistry at the reaction center. As a final note, this example complements those shown in Schemes 6.2 and 6.3 by demonstrating that neighboring group participation need not be restricted to adjacent functionalities. In fact, with sugars, cross ring participation can be extremely useful.

6.3 FORMATION OF LEAVING GROUPS

Having addressed many of the issues that must be considered when designing strategies for the modification of sugars, attention is now directed towards the types of leaving groups compatible with S_N2 type reactions. Generally, the leaving groups can be considered in three categories: halides, sulfonates, and epoxides.

6.3.1 Halides as Leaving Groups

In most S_N2 reactions, halides routinely serve as convenient and readily available leaving groups. They are easily formed from alcohols under a variety of conditions. Because the application of these reactions to sugars is a major component of this chapter, a complete discussion of halogenation reactions will be presented in Section 6.4.

6.3.2 Sulfonates as Leaving Groups

Complementing the use of halide leaving groups is the exploitation of sulfonate leaving groups. This category includes mesylates, tosylates, triflates, and many other related groups. Representative examples of this category of leaving groups, easily formed from treatment of alcohols with sulfonyl chlorides or anhydrides [6], are illustrated in Figure 6.6.

6.3.3 Epoxysugars (Anhydro Sugars)

In discussing prospective sugar-derived electrophiles useful as substrates in nucleophilic substitution reactions, we cannot ignore epoxides or anhydro sugars. As described in Chapter 4, 1,2-anhydro sugars are useful for the generation of new glycosidic linkages. Since this chapter focuses on substitutions at sites other than the anomeric center, the anhydro sugars of interest are, in the case of hexoses, 2,3- and 3,4-anhydro sugars (Figure 6.7). For pentoses, only 2,3-anhydro sugars (Figure 6.7) will be discussed. Finally, the structures illustrated in Figure 6.7 are not intended to represent the only forms of anhydrosugars. In fact, 2,3- and 3,4-anhydrosugars may be prepared by almost any hexose or pentose (2,3- only). These variations thus lend themselves to broad possibilities for the use of these compounds as substrates for further transformations.

Anhydro sugars can be prepared in two general ways; by epoxidation of olefins, or nucleophilic displacement of a leaving group by an adjacent hydroxyl group. Regarding epoxidation of olefins, numerous efficient methods are available including, but certainly not limited to, use of *m*CPBA [7] and the Sharpless asymmetric epoxidation [8] (Scheme 6.5). The unsaturated sugars required in these reactions are readily available and their preparation will be addressed later in this chapter.

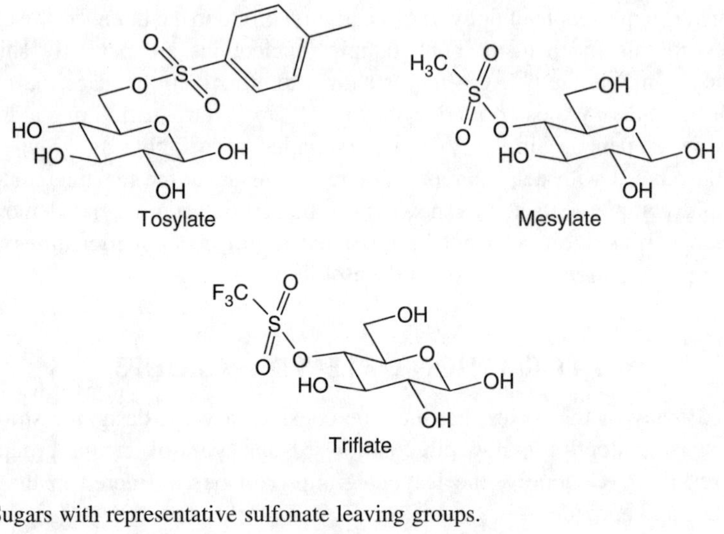

FIGURE 6.6 Sugars with representative sulfonate leaving groups.

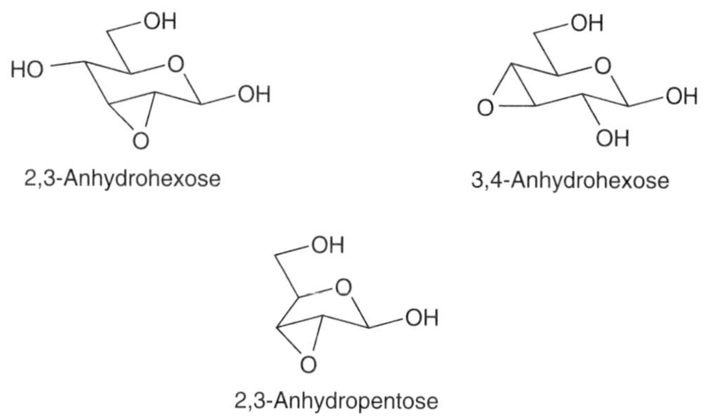

2,3-Anhydrohexose

3,4-Anhydrohexose

2,3-Anhydropentose

FIGURE 6.7 Examples of anhydrohexoses and anhydropentoses.

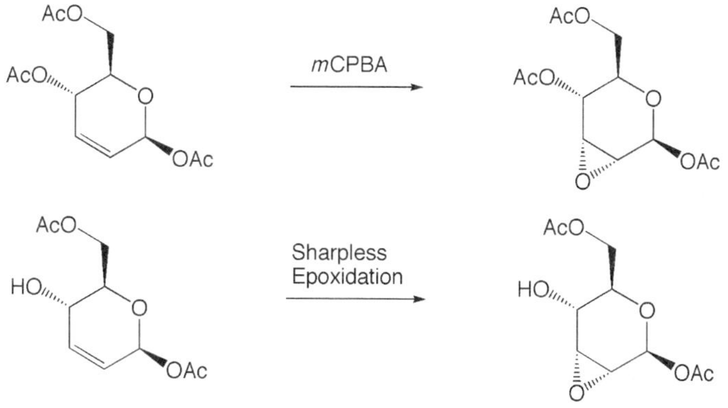

*m*CPBA

Sharpless Epoxidation

SCHEME 6.5 Epoxysugars via *m*CPBA or the Sharpless epoxidation.

Regarding epoxidation methods involving nucleophilic displacements, most sugars contain vicinal-diol units. When the hydroxyl groups of a given 1,2-diol are oriented *trans* to one another, they may be manipulated such that one hydroxyl group displaces the other when converted to a suitable leaving group [9–11] (Scheme 6.6). Perhaps the most readily available leaving groups are the sulfonates. However, if the hydroxyl groups of the 1,2-diol are oriented *cis* to one another, then conversion of a hydroxyl group to a sulfonate will not lead to formation of an epoxide. In this case, the stereochemistry of one of the hydroxyl groups must be inverted. In order to accomplish this, the halogenation methods described in the following section may be useful (Scheme 6.7).

6.3.4 Other Leaving Groups (Mitsunobu Reaction, Chlorosulfate Esters, Cyclic Sulfates)

In addition to the leaving groups described above, there are numerous reactions that possess mechanisms for creating leaving groups, and simultaneously induce nucleophilic substitutions. Among these is the Mitsunobu reaction [12]. This reaction, illustrated in Scheme 6.8, incorporates the use of triphenylphosphine and acidic nucleophiles, in conjunction with reagents such as diethylazodicarboxylate. As shown, the reaction proceeds through initial addition of triphenylphosphine to the azodicarboxylate. Subsequent combination of the alcohol with the phosphorus cation, followed by reaction with the nucleophile, introduces a new substituent at the site of the former hydroxyl group. This reaction is generally very efficient and driven by the generation of triphenylphosphine oxide and a substituted hydrazine.

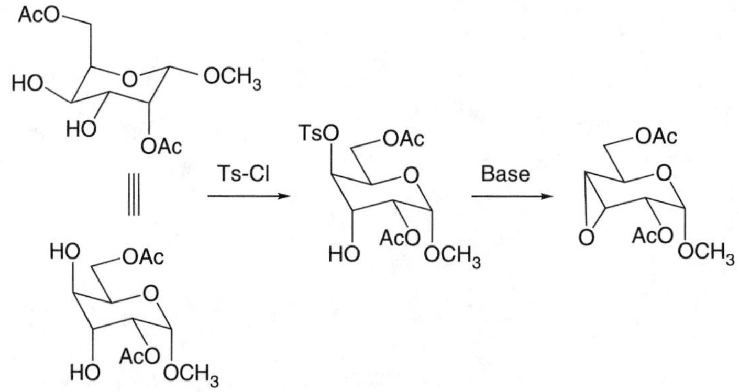

SCHEME 6.6 Epoxysugars from *trans*-1,2-diols.

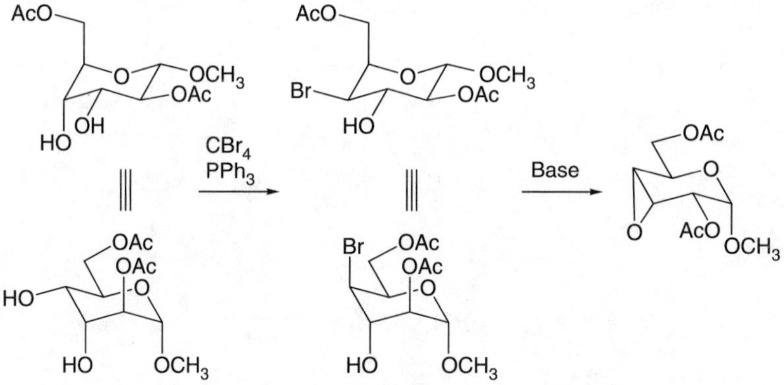

SCHEME 6.7 Epoxysugars from *cis*-1,2-diols.

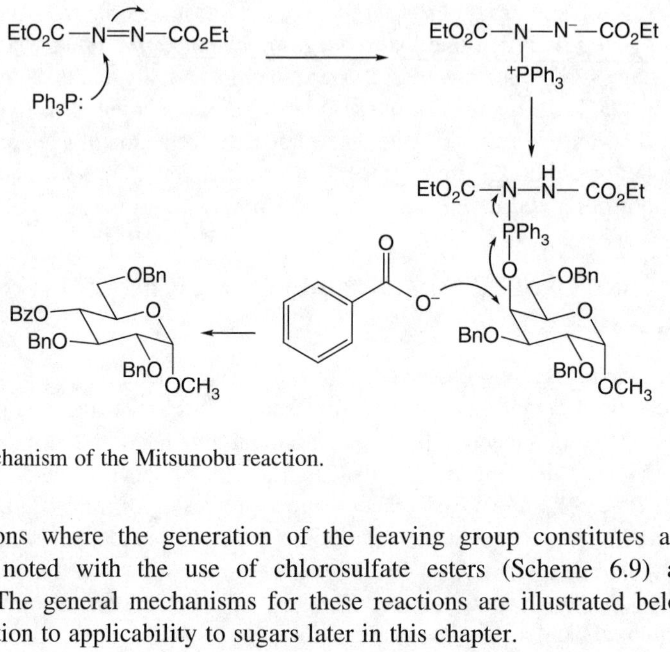

SCHEME 6.8 Mechanism of the Mitsunobu reaction.

Other reactions where the generation of the leaving group constitutes a portion of the mechanism are noted with the use of chlorosulfate esters (Scheme 6.9) and iminoesters (Scheme 6.10). The general mechanisms for these reactions are illustrated below, and will be discussed in relation to applicability to sugars later in this chapter.

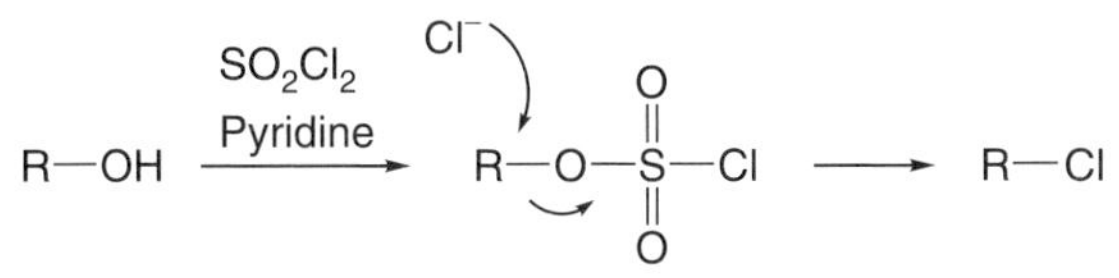

SCHEME 6.9 Chlorides from chlorosulfate esters.

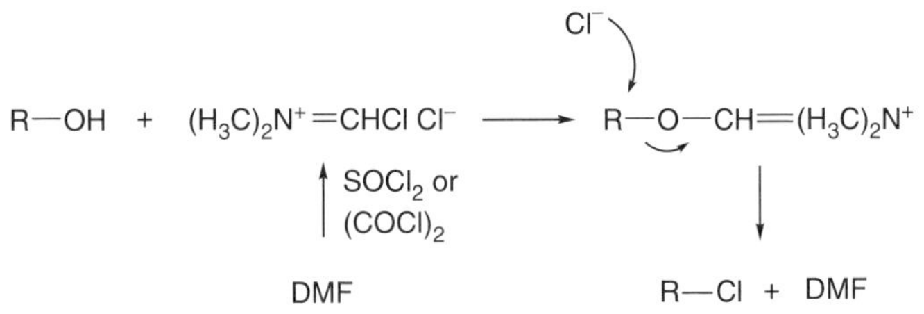

SCHEME 6.10 Chlorides from iminoesters.

6.4 HALOGENATION REACTIONS

Halogenation reactions are initiated by treating an electrophilic species with a nucleophilic halide. When applied to sugars, the electrophile is generally a hydroxyl group modified to a sulfonate leaving group or an epoxide. Additionally, cyclic sulfates are useful electrophilic species. Among the common nucleophiles are lithium-, sodium- or potassium-based halides, tetrabutylammonium halides and methylmagnesium halides. Although, with all permutations, this list is very broad, it is not implied to be inclusive. Finally, considering the types of electrophiles and nucleophiles listed above, these reactions proceed through standard S_N2 mechanisms, as well as mechanisms facilitated by various activating reagents. Examples of these reaction types are presented in the following paragraphs.

6.4.1 S_N2 Displacements of Sulfonates

In exploring halogenation reactions through the direct displacement of leaving groups, the products are generally defined by the stereochemistry and regiochemistry of the starting protected sugars. For example, as shown in Scheme 6.11, the tosyl group at C-6 of the illustrated protected glucoside is easily converted to various halides upon treatment with the appropriate lithium salts [13]. Furthermore, as shown in Scheme 6.12, similar reactions are available at secondary centers [14]. While the counterions in each of these reactions differ, they are similar because of the nucleophilic nature of the incoming halides. The carbohydrate literature is overflowing with additional examples highlighting various nucleophiles, nucleophile counterions and various sulfonates [15–17]. Consequently, this area will not be covered in additional detail here. Suffice it to say that, in general,

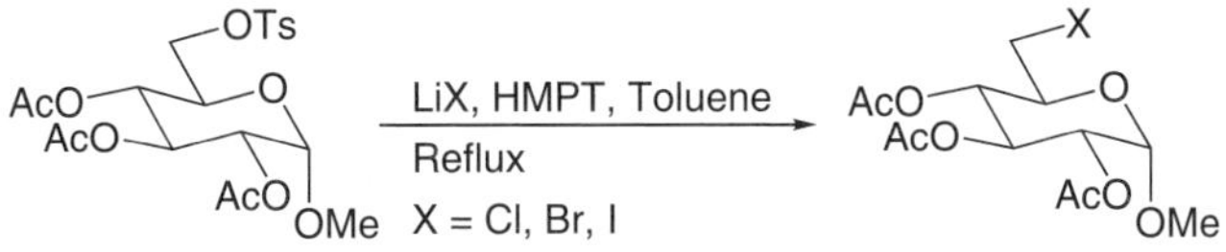

SCHEME 6.11 Halogenation reactions via displacement at primary centers.

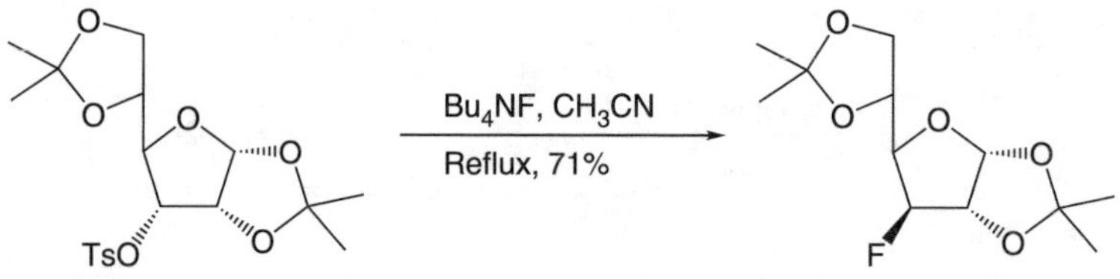

Bu$_4$NF, CH$_3$CN
Reflux, 71%

SCHEME 6.12 Halogenation reactions via displacement at secondary centers.

if an S_N2 reaction can be carried out on non-carbohydrate substrates, then the corresponding reaction with modified sugars will usually be successful.

6.4.2 S_N2 Opening of Epoxides

Unlike the displacement of sulfonate leaving groups, the ring opening of epoxides presents the issue of regiochemistry. In principle, epoxides possess two electrophilic sites (Figure 6.8) capable of attack by approaching nucleophiles. Generally, when dealing with epoxides fused to six-membered rings (pyranoses), the nucleophile will preferentially approach from the direction that results in the new substituent occupying a *trans* diaxial relationship with the liberated hydroxyl group (Figure 6.9). This generality is known as the Fürst–Plattner rule [18] and is discussed in greater detail in Chapter 11.

FIGURE 6.8 Epoxides possess multiple electrophilic sites.

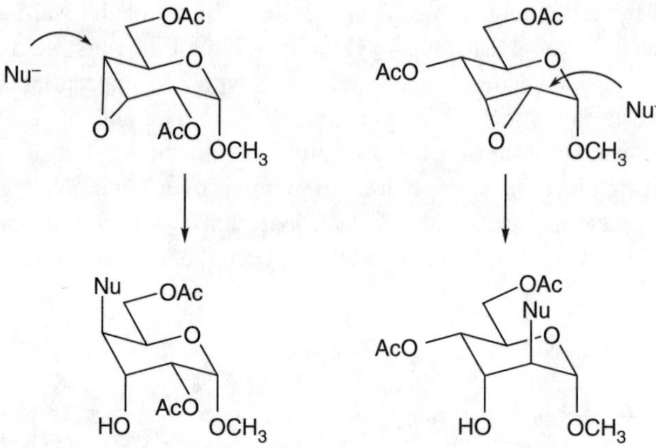

FIGURE 6.9 For pyranose sugars, the Fürst–Plattner rule predicts *trans*-diaxial relationships between nucleophiles and liberated hydroxyl groups.

Specific examples involving the ring opening of epoxides are illustrated in Schemes 6.13 and 6.14 [19–21]. As these examples suggest, while the Fürst–Plattner rule is a guide for predicting the regiochemical outcome of the nucleophilic opening of epoxides (Scheme 6.13) [17], this rule can be overridden by altering the specific reaction conditions. In fact, such variations can be as simple as replacing the counterion associated with a given nucleophile (Scheme 6.14) [19]. Finally, the utility of furanose anhydrosugars cannot be ignored. One example, illustrated in Scheme 6.15, shows the regiochemical introduction of an iodide to a nucleoside analog [19,20].

6.4.3 Use of Alkylphosphonium Salts

By virtue of their high affinity for oxygen, phosphines and related compounds have found utility in the direct conversion of alcohols to halides when used in conjunction with appropriate halide sources. The mechanism for these reactions, generally illustrated in Scheme 6.16 utilizing carbon tetrabromide as a halide source, proceeds through initial reaction of the halide source with the phosphine. Further reaction of the phosphine-halide adduct with an alcohol completes the oxidation of the phosphorus atom. Decomposition of this complex liberates an alkyl halide and a phosphine oxide. Because of the lower level of steric hindrance and the increased reactivity of primary hydroxyl groups, these reactions can be applied to sugars with no need for protecting groups. One example of

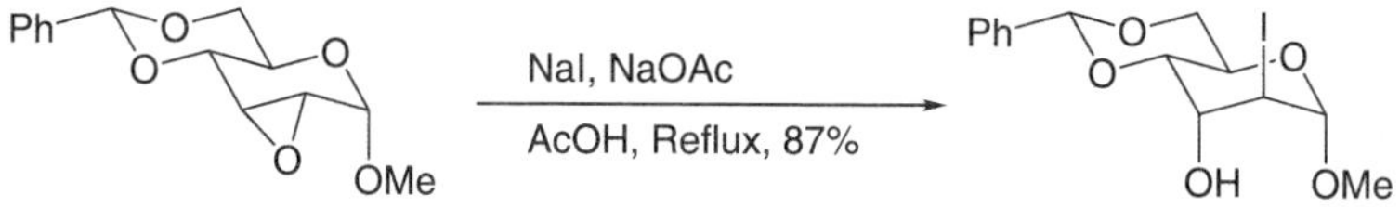

Ph O O O O OMe — NaI, NaOAc / AcOH, Reflux, 87% → Ph O O I O OH OMe

SCHEME 6.13 Example of the Fürst–Plattner product from a pyranose epoxide.

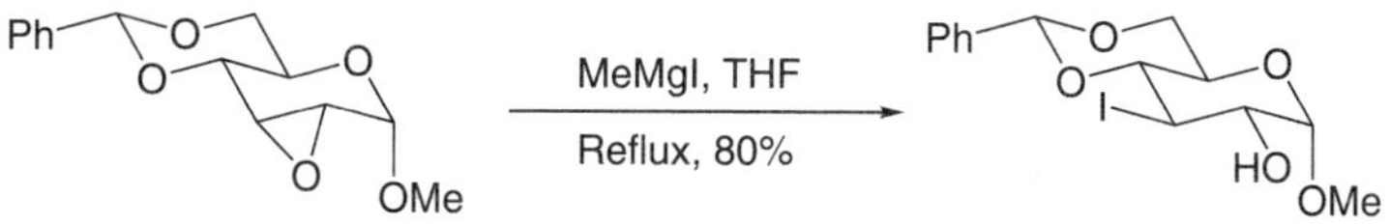

Ph O O O O OMe — MeMgI, THF / Reflux, 80% → Ph O O I O HO OMe

SCHEME 6.14 The Fürst–Plattner rule can be overridden.

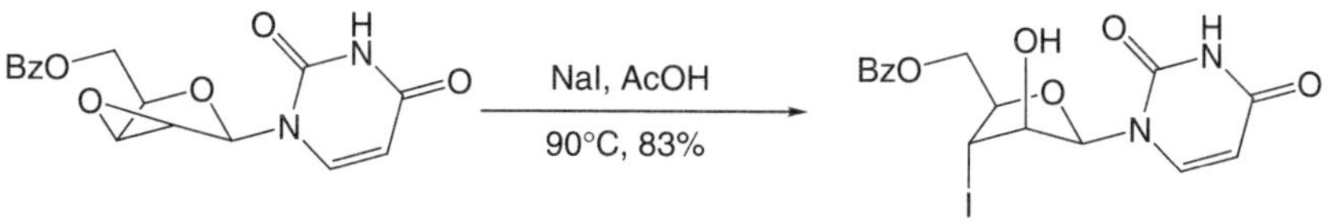

BzO O O N O H N O — NaI, AcOH / 90°C, 83% → BzO OH O I N O H N O

SCHEME 6.15 Furanose epoxides are substrates for nucleophilic ring opening.

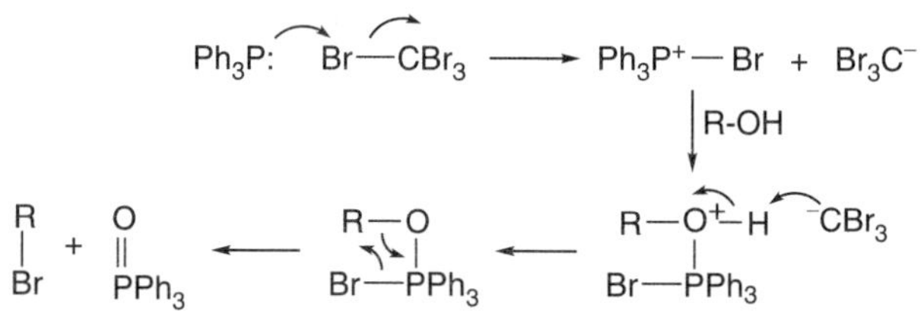

Ph_3P: Br—CBr_3 → Ph_3P^+—Br + Br_3C^-

R-OH

R—O^+—H $^-CBr_3$ / Br—PPh_3 → R—O / Br—PPh_3 → R—Br + O=PPh_3

SCHEME 6.16 Mechanism for phosphine mediated conversion of alcohols to halides.

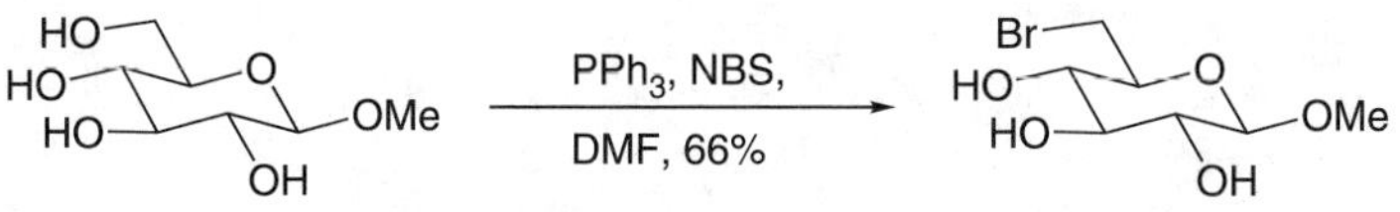

PPh_3, NBS, DMF, 66%

SCHEME 6.17 Phosphine mediated conversion of a primary hydroxyl group to a bromide.

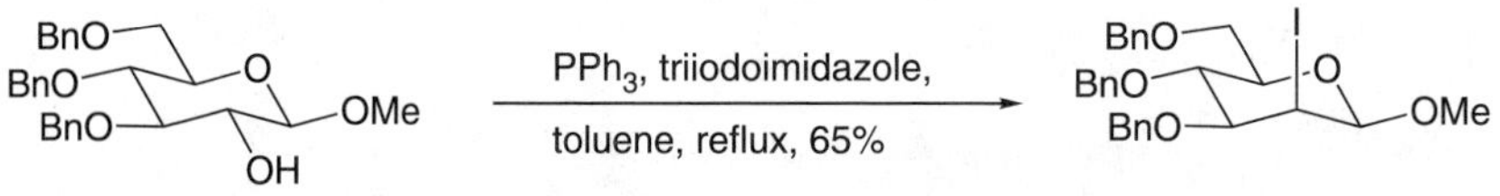

PPh_3, triiodoimidazole, toluene, reflux, 65%

SCHEME 6.18 Phosphine mediated conversion of a secondary alcohol to an iodide.

this reaction, applied to a glucose derivative, is shown in Scheme 6.17 [22,23]. Although this reaction utilizes *N*-bromosuccinimide as the bromide source, there are many others available including, but not limited to, carbon tetrabromide (Scheme 6.16), iodine and triiodoimidazole [24–26]. Finally, although these reactions demonstrate regioselectivity favoring primary hydroxyl groups, they can proceed at secondary centers when the sugar is suitably protected (Scheme 6.18) [27].

6.4.4 Use of Chlorosulfate Esters

Complementing triphenylphosphine mediated reactions is the use of chlorosulfate esters. Illustrated in Scheme 6.9, these reactions involve the initial addition of sulfonyl chloride to a hydroxyl group forming the intermediate chlorosulfate ester. The liberated chloride anion is then free to displace the chlorosulfate leaving group. Examples of this reaction, shown in Scheme 6.19 [28] and Scheme 6.20 [28,29], illustrate the application of this reaction to both unprotected and protected sugars. In these examples, it is important to note that the selectivity is dependant upon steric factors between the anomeric methoxy group and the approaching chloride nucleophile.

6.4.5 Use of Iminoesters and Sulfonylchlorides

Complementing the examples shown thus far is a wealth of chemistry capable of affecting the transformation of hydroxyl groups to halides. These include the use of iminoesters (Scheme 6.10) [30] and reagents such as methanesulfonyl chloride, which is commonly employed in the preparation of sulfonate leaving groups [31]. The latter example, highlighted in Scheme 6.21,

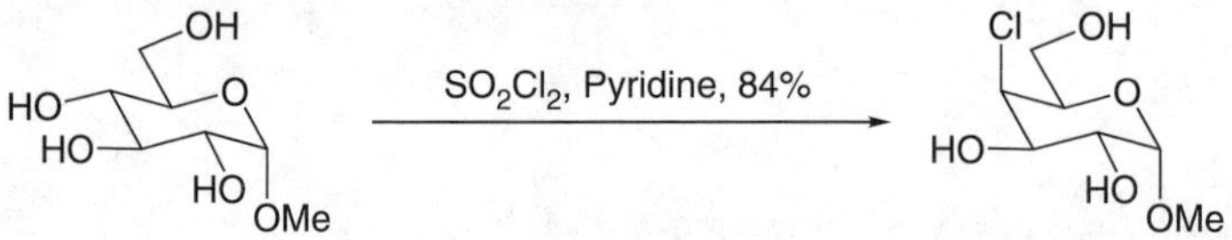

SO_2Cl_2, Pyridine, 84%

SCHEME 6.19 Chlorosulfate ester conversion of a secondary hydroxyl group to a chloride.

SO_2Cl_2, Pyridine, 47%

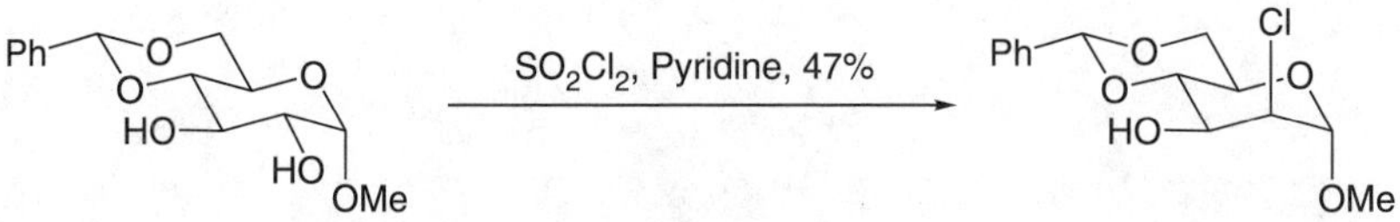

SCHEME 6.20 Chlorosulfate ester conversion of a secondary hydroxyl group to a chloride.

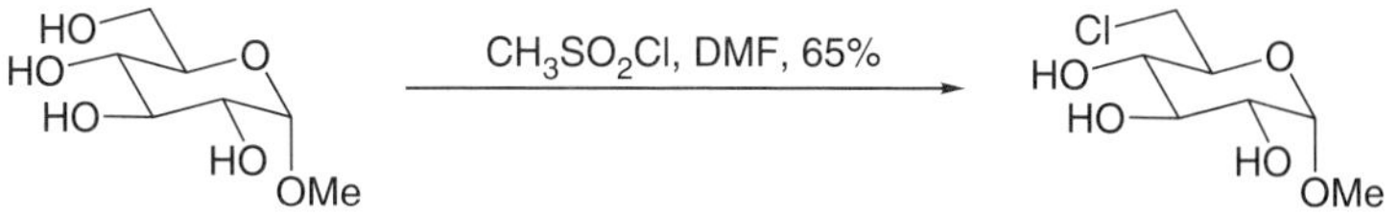

SCHEME 6.21 Direct conversion of a primary hydroxyl group to a chloride using methanesulfonyl chloride.

illustrates that halide anions liberated in the formation of sulfonates are capable of serving as tandem nucleophiles.

While further examples of these reactions are not presented here, these reactions (among others) were extensively cataloged by Larock [32] with respect to general organic chemistry. As such, it is important to note that while the literature may not frequently present these reactions in the context of carbohydrate chemistry, they are nonetheless extremely valuable tools for the introduction of halides to sugar ring systems.

6.4.6 Fluorination Reactions

Thus far, much has been discussed regarding the conversion of sugar hydroxyl groups to chlorides, bromides and iodides. However, the introduction of fluoride has yet to be addressed. Numerous reagents are available for the conversion of alcohols to fluorides with overall inversion of stereochemistry. Among the most useful is diethylaminosulfur trifluoride (DAST). This reagent has been extensively applied to sugar modifications with great success. Shown in Scheme 6.22, the mechanism involves initial displacement of fluoride, followed by the fluoride displacing the extremely active sulfoxy leaving group [33–37].

One published study applied DAST to a variety of glucose derivatives. As shown in Scheme 6.23, the regiochemistry of fluoride introduction was easily controlled through various protecting groups, anomeric configurations and cyclic forms [38]. Elaborating upon this observation, when the unprotected α-methyl glucopyranoside was subjected to DAST, fluorination occurred at C-4 and C-6. However, applying the same conditions to the β-methyl glucopyranoside resulted in fluorination at C-3 and C-6. Furthermore, tritylation at C-6 of the α-methyl glucopyranoside prevented fluorination at C-6, and the only observed reaction was at C-4. Finally, when a furanose form of glucose was used, fluorination occurred only at C-6.

6.4.7 Halogenation of *O*-Benzylidene Acetals

Benzylidene acetals are widely used protecting groups in the chemical modification of sugars because of their ease of formation, ease of cleavage and ability to simultaneously and selectively mask two hydroxyl groups. Additionally, given the stabilizing influences of the lone pairs of two oxygen atoms, benzylidene acetals are easily converted to 1-halo-2-benzoyloxy compounds via

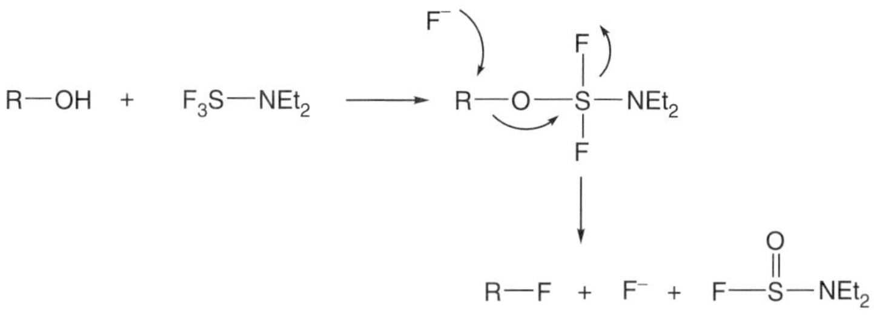

SCHEME 6.22 Mechanism for DAST mediated conversion of hydroxyl groups to fluorides.

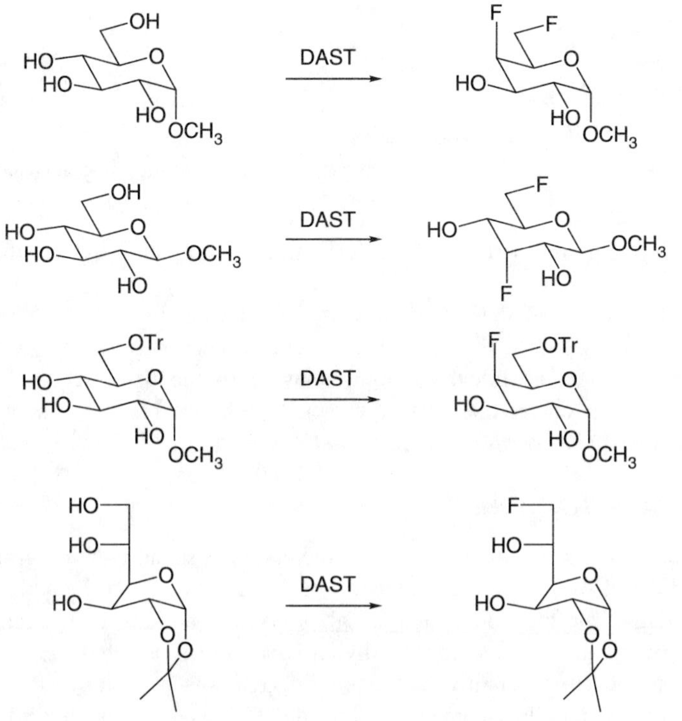

SCHEME 6.23 The regiochemistry of the DAST fluorination can be controlled through protecting groups, anomeric configuration and the cyclic form of the sugar.

cationic intermediates. The Hannessian–Hullar reaction [29,39–47], shown in Scheme 6.24, is a classic illustration of this chemistry. As shown, the acyloxonium ion, formed on treatment of the illustrated glucose benzylidene acetal with *N*-bromosuccinimide, is susceptible to nucleophilic ring opening by the liberated bromide ion. Delivery of the bromide to C-6 results in conversion of the benzoyloxonium ion to the illustrated benzoyl group at C-4. The net result is the selective conversion of a protecting group masking two hydroxyl groups to a combination of a leaving group and a new protecting group. This reaction was prominently used in the synthesis of daunosamine [48].

Because of the ready conversion of benzylidene acetals to acyloxonium ions, *N*-bromosuccinimide is not the only reagent used for effecting these types of reactions. In fact, Lewis acids may be used. In one example, illustrated in Scheme 6.25, triphenylmethyl tetrafluoroborate provides a convenient means of generating an acyloxonium ion in preparation for the introduction of a bromide [49]. As a final note, we should recognize that although the Hanessian–Hullar reaction depends on *N*-bromosuccinimide as both the ionizing and nucleophile

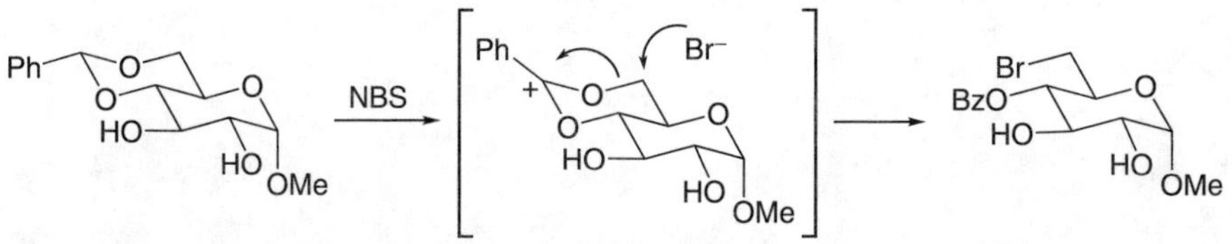

SCHEME 6.24 The Hanessian–Hullar reaction.

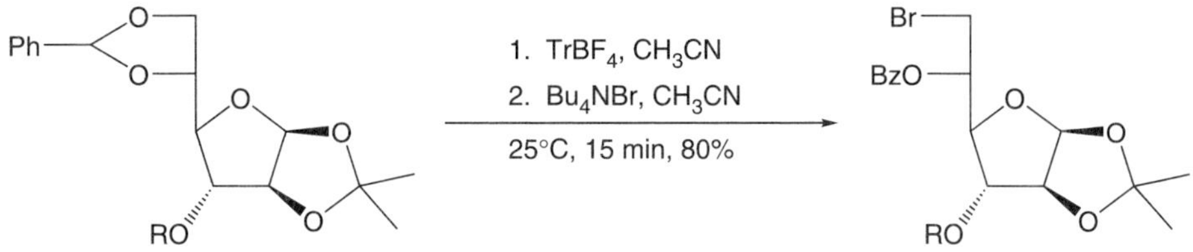

1. $TrBF_4$, CH_3CN
2. Bu_4NBr, CH_3CN
25°C, 15 min, 80%

SCHEME 6.25 Lewis acids can provide strategies towards Hanessian–Hullar type products.

source, this dependency is not present when Lewis acids are used. In fact, utilizing the strategy shown in Scheme 6.25, many different groups may be introduced by simply altering the nucleophile source.

6.4.8 Radical Processes

Before leaving the topic of halogenation reactions, we should acknowledge the fact that nucleophilic mechanisms are not the only reactions capable of the introduction of halides to sugars. In fact, the use of free radical mechanisms is a well-established alternative, and is entirely complementary to the use of nucleophilic strategies. Although no reactions of this type are presented here, the fact that an entire chapter can be devoted to this area warrants mention of this chemistry. In fact, this area has been recently reviewed with respect to all aspects and requirements of the reactions ranging from general reaction conditions and substrate types to regio/stereochemical considerations and reactions of the halogenated products [50].

6.5 Reactions Involving Nitrogen

The discussions presented thus far focus heavily on the conversion of hydroxyl groups to halo groups. However, it is important to note that many of these reactions can be adapted to the introduction of other types of functional groups involving oxygen, nitrogen, sulfur, and other elements. In this section, these variations, as well as other reaction types, will be discussed in the context of introducing nitrogen based functional groups to sugars.

6.5.1 S_N2 Reactions

The introduction of amine functionalities via S_N2 mechanisms follows the same rules as those discussed in the previous section regarding the introduction of halo groups. However, in addition to the common sulfonate (tosylate, mesylate, triflate, etc.) and epoxide electrophiles, halides serve as excellent leaving groups. As such, the strategies presented in Section 6.4 provide reasonable intermediates towards the introduction of nitrogen-based groups. Regarding appropriate nucleophilic species, commonly used groups include amines, azides and nitrites. Protected amines such as amides, imides and related functionalities are also useful.

The introduction of azido groups is particularly appealing as a method for the placement of nitrogen based groups on sugar ring systems because of its strong nucleophilicity and ease of reduction to amines under many conditions [32]. Schemes 6.26–6.28, illustrate the utility of this approach by combining sodium azide with sugar derived tosylates [51–53], mesylates [54], and epoxides [55]. One should note that reactivity is not limited to primary centers. Indeed, secondary centers are commonly used as hosts for leaving groups, and regiochemistry around the sugar ring can be controlled through steric factors and the trajectory of the nucleophile. In the case of the anhydrosugar (Scheme 6.28), the azide is delivered consistent with the Fürst–Plattner rule generating a *trans*-diaxial relationship between the azide and the new hydroxyl group [18].

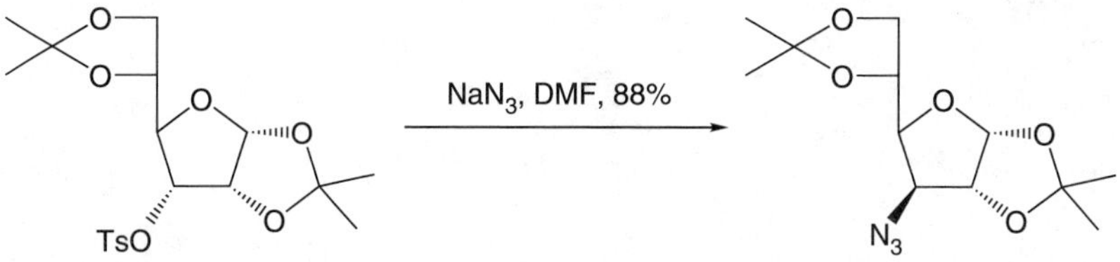
NaN$_3$, DMF, 88%
TsO
N$_3$

SCHEME 6.26 Sodium azide can displace secondary tosylates.

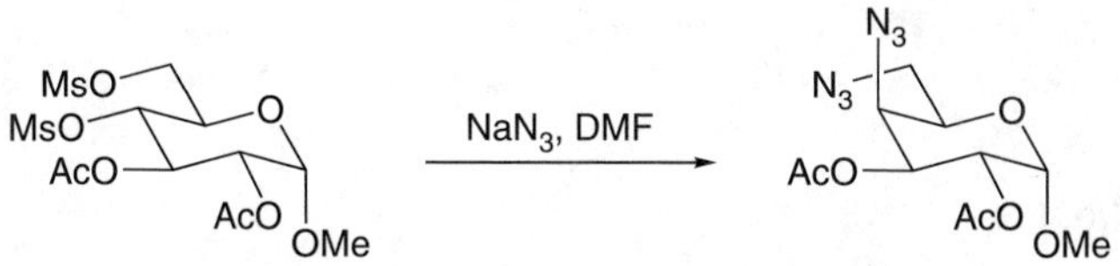
MsO
MsO
AcO
AcO
OMe
NaN$_3$, DMF
N$_3$
N$_3$
AcO
AcO
OMe

SCHEME 6.27 Sodium azide can displace primary and secondary mesylates.

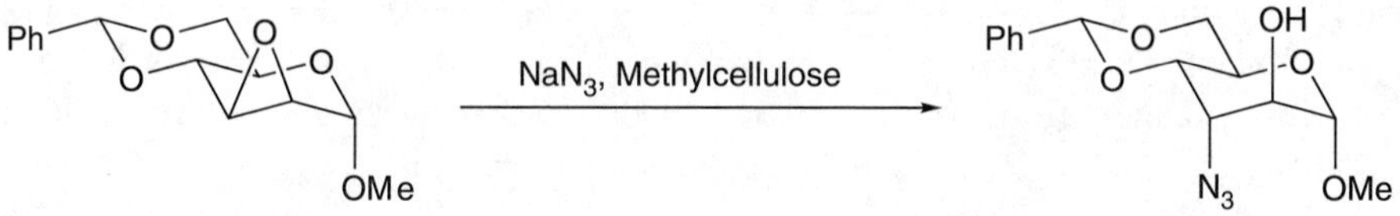
Ph
OMe
NaN$_3$, Methylcellulose
Ph
OH
N$_3$
OMe

SCHEME 6.28 Sodium azide opens epoxides according to the Fürst–Plattner rule.

Finally, while the examples shown in Schemes 6.26–6.28 may seem limited in scope, they illustrate the possibilities and should not be considered inclusive of all azide/leaving group combinations.

There are many instances in which the introduction of amines with specific substitution patterns is desired. With the abundance of structurally diverse commercially available amines from which to choose, the use of these compounds as nucleophiles is extremely valuable. While the example illustrated in Scheme 6.29 focuses on the hydrazine displacement of a secondary tosylate from a furanose sugar derivative [56], this type of reaction is easily achieved with most primary and secondary amines. Where the introduction of direct NH_2 groups is desired, the best methods rely on the reduction of azides or nitro groups, or the hydrolysis (hydrazinolysis) of imides [32].

As alluded to above, nitro-substituted sugars are useful sources of amines [32]. Utilizing sodium nitrite, nitro groups are easily introduced to sugar rings through nucleophilic mechanisms. One example, illustrated in Scheme 6.30, utilizes sodium nitrite to introduce a nitro group to the 6-position of a glucose derivative [57–59].

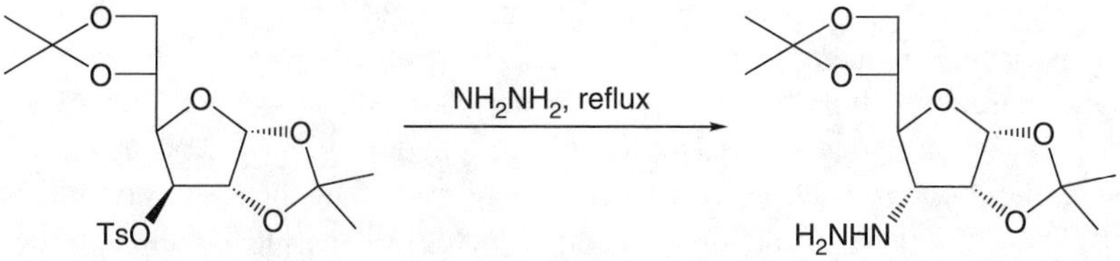
NH$_2$NH$_2$, reflux
TsO
H$_2$NHN

SCHEME 6.29 Hydrazine can displace secondary tosylates.

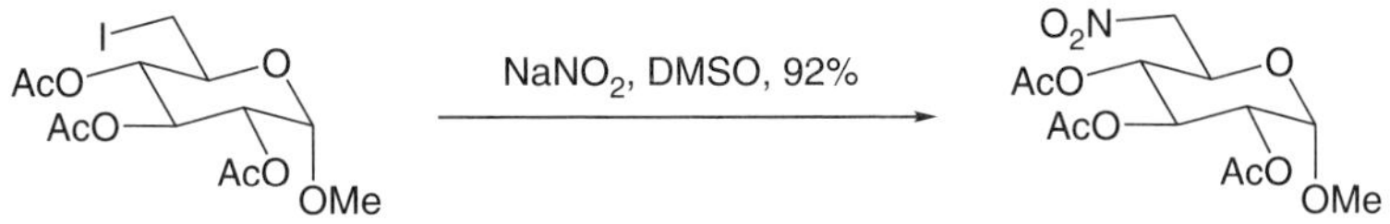

SCHEME 6.30 Sodium nitrite provides an easy entry to nitro-substituted sugars.

6.5.2 Formation of Nitrosugars

While the direct nucleophilic displacement of leaving groups provides valuable approaches towards the functionalization of sugars with nitrogen based substituents, numerous complementary approaches exist. Among these is the formation of nitrosugars through nitromethane condensation reactions [60,61]. Unlike nucleophilic displacements with sodium nitrite, these reactions require aldehyde functional groups as components in dual aldol condensations. One example, shown in Scheme 6.31, utilizes sodium periodate to oxidatively cleave a glucose derivative to a bis-aldehyde. The bis-aldehyde then undergoes condensation with nitromethane under basic conditions to regenerate the pyranose ring replacing the 3-hydroxy group with a nitro group. Although this condensation proceeds with little control over the stereochemical outcome, the illustrated 3-amino-3-deoxy mannoside was isolated following Raney nickel reduction of the nitro group.

6.5.3 The Mitsunobu Reaction

The Mitsunobu reaction has found wide use in the direct conversion of hydroxyl groups to other functionalities [12]. The mechanism for this reaction was previously discussed in Section 6.3.4 and illustrated in Scheme 6.8.

The nature of the nucleophile is critical when considering the Mitsunobu reaction as a means of introducing amine functionalities. Because of the proton transfer components illustrated in the reaction mechanism, only acidic nitrogens can be introduced. The most common nitrogen nucleophiles include phthalimide, hydrazoic acid and zinc azide [62]. Once placed, phthalimides are easily converted to amines utilizing hydrazine [6]. Moreover, azides are easily reduced to amines under numerous conditions [32]. Schemes 6.32 and 6.33 illustrate the application of this chemistry to nucleosides [63] and pyranosides [64], respectively.

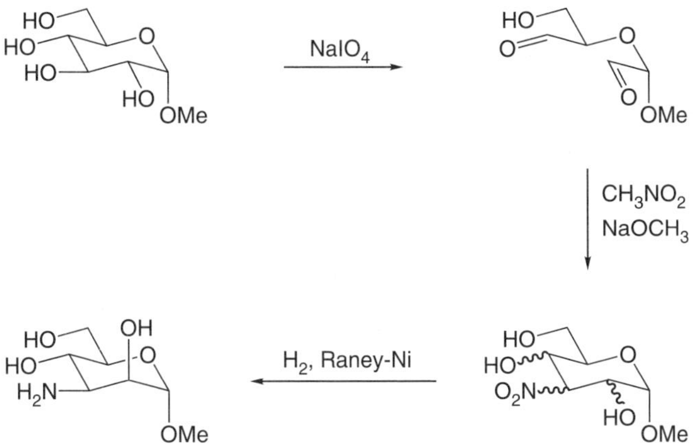

SCHEME 6.31 The nitromethane condensation can provide amine-substituted sugars and sugar precursors.

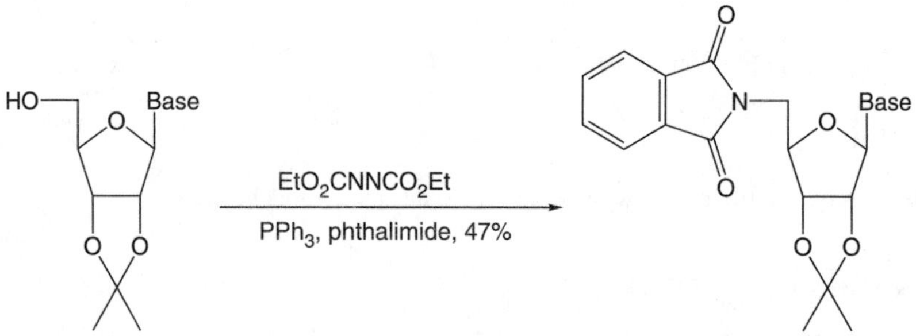

SCHEME 6.32 The Mitsunobu reaction can deliver phthalimides to sugars and sugar derivatives.

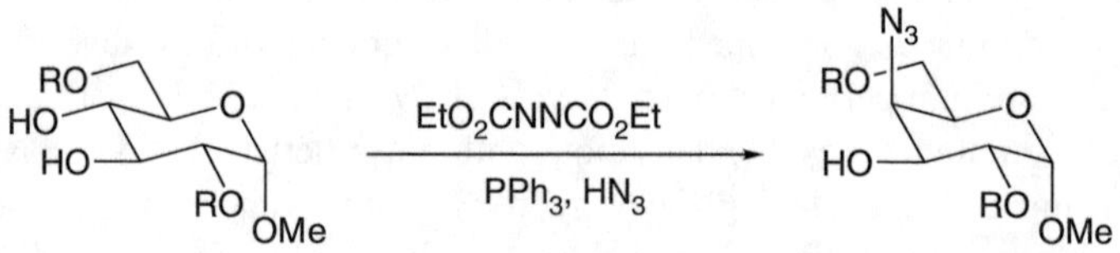

SCHEME 6.33 The Mitsunobu reaction can deliver azides to sugars and sugar derivatives.

6.6 REACTIONS INVOLVING OXYGEN AND SULFUR

As highly oxygenated molecules, sugars may be subjected to reactions designed to manipulate the stereochemistry of given centers. Additionally, the sugar hydroxyl groups are easily oxidized, dehydrated to epoxides or completely removed. In this section, the organic chemistry of sugars is discussed in the context of manipulating or removing sugar hydroxyl groups. Additionally, analogous chemistry will be applied to sulfur functional groups.

6.6.1 MANIPULATION OF SUGAR HYDROXYL GROUPS

6.6.1.1 Formation of Epoxysugars (Anhydrosugars)

As just mentioned, sugars are, by definition, highly oxygenated organic molecules. Consequently, the ability to convert the sugar hydroxyl groups into new and reactive functionalities is a useful strategy to the generation of non-sugar analogs. Unsaturated sugars, readily available via numerous methods included in (but not limited to) Section 6.6.2.2, are subject to most all of the epoxidation reactions applicable to olefins in general [32]. Moreover, as 1,2-diols are readily dehydrated to epoxides, the many variations of 1,2-diols present in sugar molecules are subject to similar reactions [9–11]. Examples of these reactions were discussed in Section 6.3.3 and will not be discussed in further detail here. Schemes 6.5 and 6.6 illustrate the formation of these epoxides, and the epoxidation of olefins is the only true oxygenation reaction that will be applied to sugars throughout this discussion.

Having illustrated the formation of sugar epoxides, the remaining discussion in this section will focus on the manipulation of oxygen atoms within sugar frameworks. These reactions will address both the ring opening of sugar epoxides and the direct inversion of the stereochemistry of sugar hydroxyl groups.

6.6.1.2 Ring Opening of Epoxides

Epoxides have already been discussed in the context of suitable electrophiles for the introduction of halides and amines to sugar scaffolds. Complementary to these discussions, epoxides are also useful

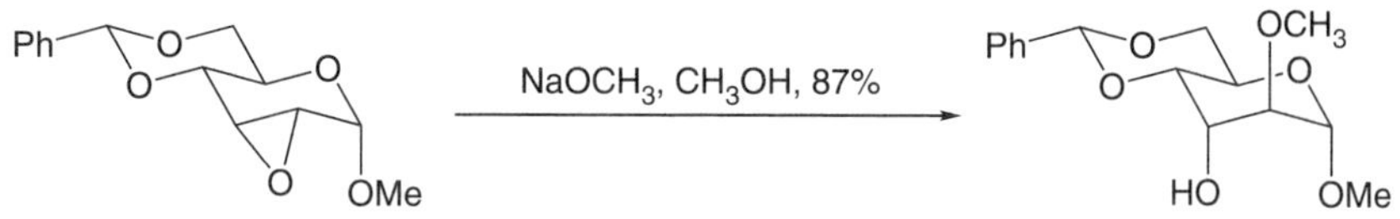

NaOCH$_3$, CH$_3$OH, 87%

SCHEME 6.34 Alkoxide ions can open anhydrosugars.

electrophiles for the introduction of oxygen-based functionalities. As shown in Scheme 6.34, this method was applied to the ring opening of a 2,3-anhydroallose derivative to give a derivative of 2-*O*-methyl-altrose [65]. The incoming nucleophile was a methoxide anion and was introduced via a trajectory consistent with the Fürst–Plattner rule [18].

6.6.1.3 Inversion of Stereocenters

In the previous example, sodium methoxide was used to open an epoxide and introduce an oxygen-based functionality to an anhydro sugar. As with all nucleophilic reactions, this one began with inversion of the stereochemistry at the reaction center. Moreover, as discussed above, epoxides are not the only electrophiles suitable for nucleophilic reactions on sugar rings. Other electrophiles include the complement of available sulfonates and halides. In Scheme 6.35, an example of the displacement of a tosylate group with sodium benzoate is illustrated [66]. Following hydrolysis of the benzoate group, it is easy to see how the stereochemistry of sugar hydroxyl groups can be conveniently inverted, providing access to alternate sugar configurations.

The Mitsunobu reaction is an important aspect of the stereochemical inversion of sugar hydroxyl groups. As discussed above, this reaction provides for the one-pot conversion of hydroxyl groups to benzoate (and related) esters [12]. The general reaction, illustrated in Scheme 6.8, is extremely versatile because of the number of acceptable commercially available carboxylic acids and should not be limited to the most commonly used *p*-nitrobenzoic acid.

6.6.2 Deoxygenation Reactions

Having addressed specific manipulations regarding the stereochemical inversion of sugar hydroxyl groups and the ring opening of anhydrosugars (sugar epoxides), discussion will now turn to methods of removing hydroxyl groups from sugar molecules. It will focus primarily on the direct removal of oxygens from the sugar ring, as well as the elimination and dehydration reactions resulting in olefin formation.

6.6.2.1 Removal of Hydroxyl Groups

The removal of hydroxyl groups from sugars is a reaction classified as a reduction. In general, these reactions are best accomplished via intermediate functionalization of the sugar ring. This type of

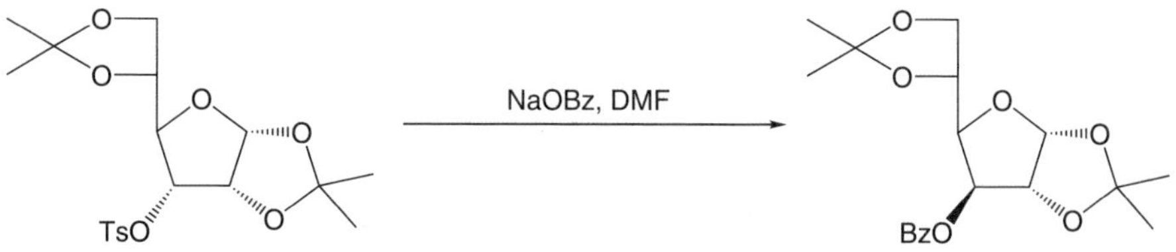

NaOBz, DMF

SCHEME 6.35 Oxygen-based nucleophiles can be used to invert the stereochemistry of sugar hydroxyl groups.

functionalization includes halogenation reactions, formation of sulfonates and the introduction of sulfur-based functional groups. As these reaction types are discussed in detail throughout this chapter, they will not be addressed further here. Additionally, since the step following primary functionalization requires reductive conditions, methods for the net removal of sugar hydroxyl groups will be discussed under "Reductions and Oxidations" (Section 6.8).

6.6.2.2 Elimination/Dehydration to Olefins

Numerous methods have been explored for the introduction of olefins to sugar rings. Many of these are generally applicable to any carbon–carbon bond within the sugar ring. However, there are some instances where the chemistry is specific to certain positions. In this section, the formation of unsaturated sugars is addressed progressing from 1,2-olefins (glycals) to 5,6-olefins.

6.6.2.2.1 1,2-Unsaturated Sugars (Glycals)

1,2-Unsaturated sugars (glycals and hydroxyglycals) are important to carbohydrate chemistry because of the ease by which they can be incorporated into more complex carbohydrate structures (Chapter 4 and Chapter 5). This class of olefinic sugars is readily prepared via elimination reactions directed from either C-1 or C-2. As shown in Scheme 6.36, subjecting glycosyl bromides to basic conditions accomplishes the desired transformation to a protected hydroxyglycal [67,68].

Anhydrosugars are easily converted to glycals. As shown in Scheme 6.37, this transformation is accomplished through initial ring opening of a 2,3-anhydrosugar with lithium iodide. Subsequent lithiation at C-1 followed by tandem elimination gives the glycal shown [17]. Moreover, as shown in Scheme 6.38, similar transformations can be accomplished via halogen–metal exchange at C-1 [67].

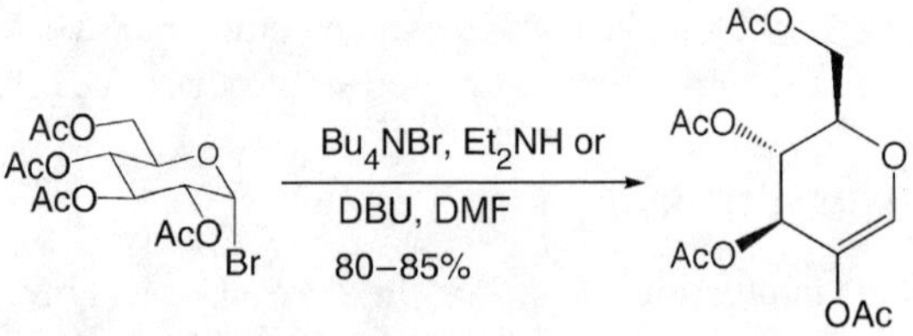
AcO
AcO
AcO
AcO
Br
O
Bu4NBr, Et2NH or
DBU, DMF
80–85%
AcO
AcO
AcO
O
OAc

SCHEME 6.36 Glycals are available from bromoglycosides.

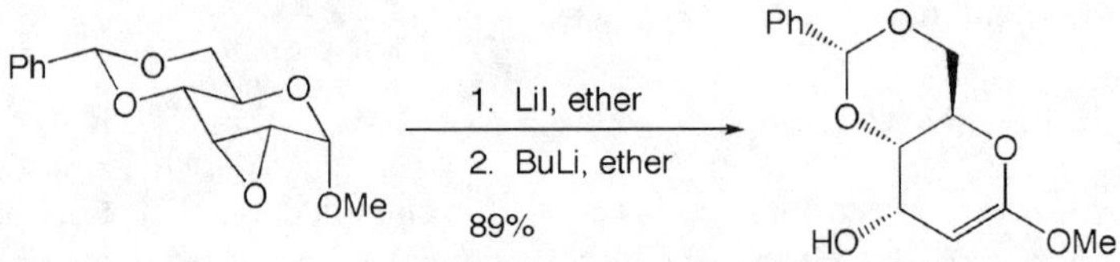
Ph
O
O
O
O
OMe
1. LiI, ether
2. BuLi, ether
89%
Ph
O
O
O
HO
OMe

SCHEME 6.37 Glycals are available from 2,3-anhydrosugars.

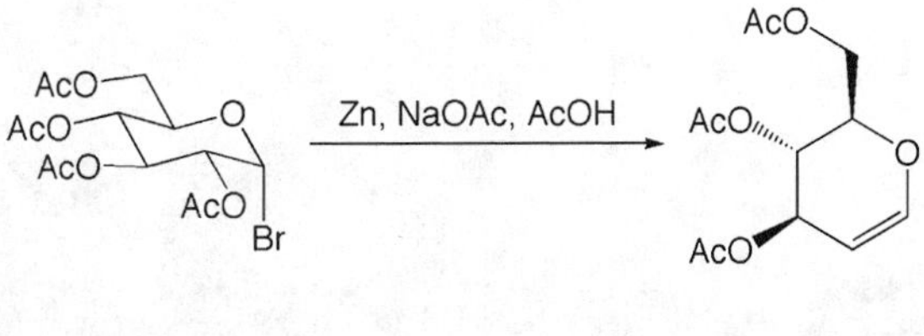
AcO
AcO
AcO
AcO
Br
O
Zn, NaOAc, AcOH
AcO
AcO
AcO
O

SCHEME 6.38 Glycals are available via reductive elimination of bromoglycosides.

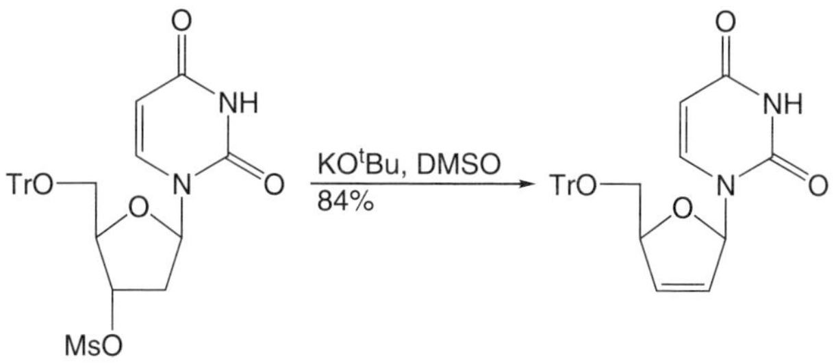

SCHEME 6.39 2,3-Unsaturated sugars are available via elimination reactions on 2-deoxysugar derivatives.

6.6.2.2.2 2,3-Unsaturated Sugars

When considering sugar ring unsaturation removed from the 1,2 position, many methods are available in addition to methods analogous to those used in glycal formation. As shown in Scheme 6.39, the direct elimination of sulfonates under basic conditions provides a useful strategy [20]. Moreover, in the case of *trans*-2,3-bis-sulfonates, initial displacement of one sulfonate with lithium iodide can be followed by metallation and tandem elimination, as shown in Scheme 6.40 [69].

Enamines are important functionalities in general organic chemistry; their structures being defined by a double bond with a vinyl amine. These structures are easily formed by the addition of amines to carbonyl compounds [70,71]. As shown in Scheme 6.41, the formation of enamines, when applied to sugars, can produce unsaturation at the 2,3 position [72]. In the example shown, a 2-deoxysugar derivative was condensed with pyrrolidine giving the illustrated enamine.

While there are many additional methods for the introduction of olefins at the 2,3 position of sugar rings, many of the methods available are generally applicable to other sites around the ring system. Therefore, the methods that will be presented in the remainder of this section should also be considered.

6.6.2.2.3 3,4-Unsaturated Sugars

The elimination of bis-sulfonates, described in the context of 2,3-unsaturated sugars (Scheme 6.40), was adapted to the formation of 3,4-unsaturated sugars [73]. In addition to this method, the direct conversion of vicinal hydroxyl groups was applied to a derivative of mannose. As shown in

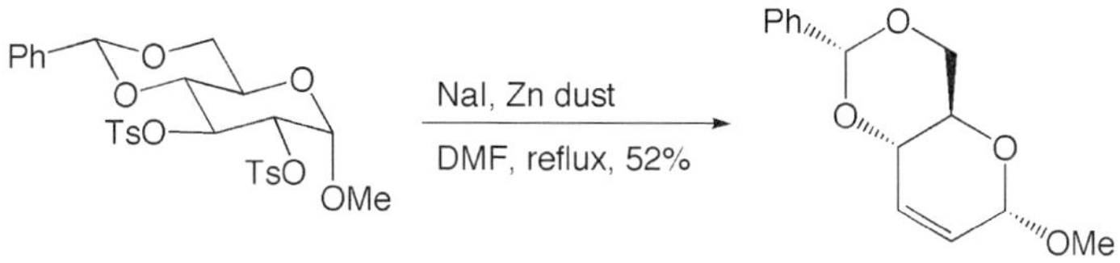

SCHEME 6.40 2,3-Unsaturated sugars are available from vicinal tosylates via initial displacement with iodide followed by metallation and elimination.

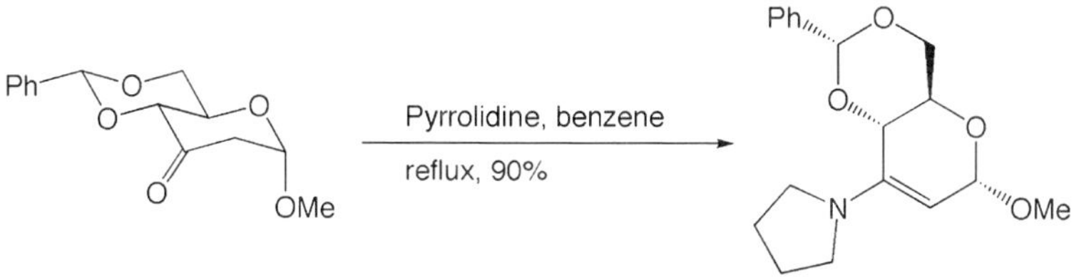

SCHEME 6.41 2,3-Unsaturated sugars are available via formation of enamines.

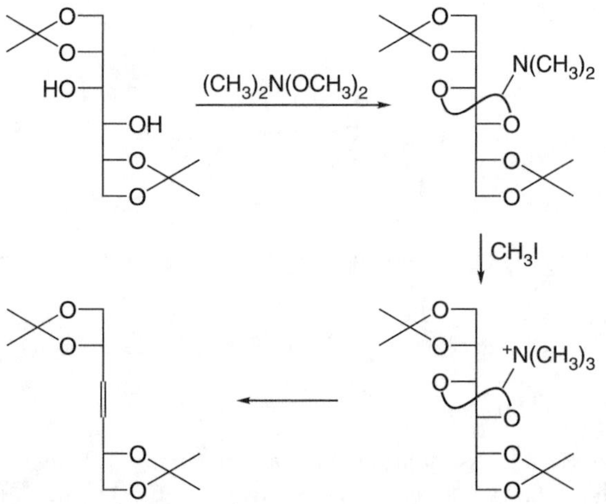

SCHEME 6.42 3,4-Unsaturated sugars are available via DMF acetals followed by elimination.

Scheme 6.42, diisopropylidene mannitol was treated with *N,N*-dimethylformamide dimethyl acetal followed by methyl iodide to effect this transformation [74,75]. The reaction proceeds through initial acetal transfer generating the illustrated mannitol acetal. Methyl iodide then alkylates the nitrogen, producing a quaternary ammonium salt. Loss of trimethylamine, followed by ring opening and decomposition, gives the 3,4-olefinic mannitol derivative shown.

6.6.2.2.4 4,5-Unsaturated Sugars

Aside from the general applicability of the methods for the formation of unsaturated sugars described thus far, the formation of 4,5-unsaturated sugars is unique when applied to uronic acids. By virtue of the acidity of protons adjacent to ester groups, uronic acid esters are easily converted to 4,5-unsaturated sugars when appropriate leaving groups are positioned at C-4. One example, shown in Scheme 6.43, illustrates this transformation applied to a glucuronic acid derivative [76,77].

6.6.2.2.5 5,6-Unsaturated Sugars

Hexose 5,6-unsaturated sugars or pentose 4,5-unsaturated sugars are the only examples discussed in this section where the unsaturation is endocyclic to the ring itself. Interestingly, this type of unsaturated sugar is also an enol ether, and can react as enol ethers generally do [78].

Previous sections presented approaches towards replacing the primary hydroxyl group at C-6 (hexoses) or C-5 (pentoses) with halides. These modified sugars are easily converted to unsaturated sugars utilizing reagents such as silver fluoride [79]. One early example, shown in Scheme 6.44, applies this chemistry to a glucose derivative.

It is important to remember that the examples presented in this section are only representative of the numerous methods available for the preparation of olefins from vicinal diols. Numerous other

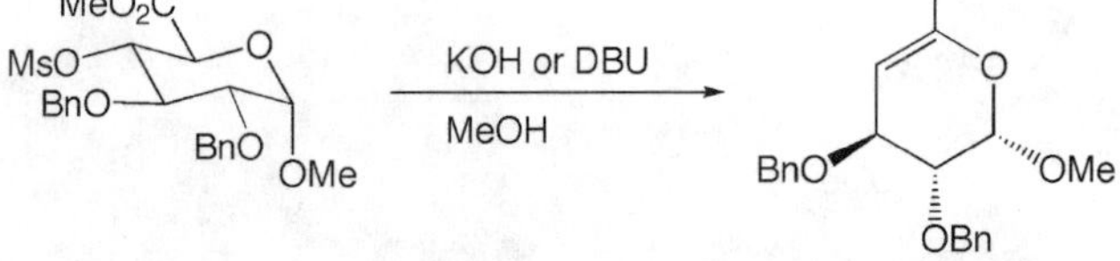

SCHEME 6.43 4,5-Unsaturated sugars are available via elimination of modified uronic acids.

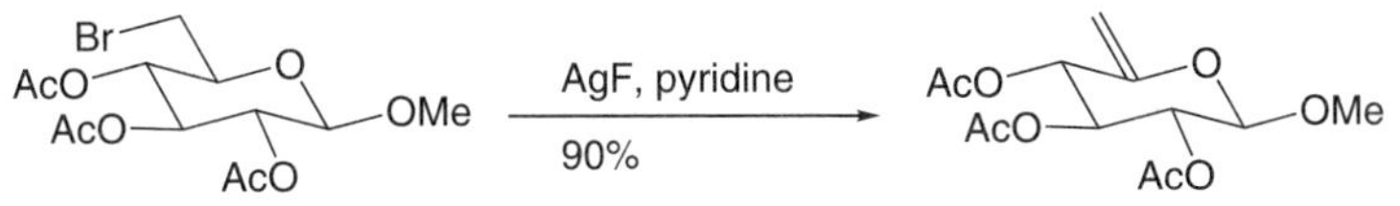

AcO, AcO, AcO, Br, OMe — AgF, pyridine, 90% → 5,6-unsaturated product (AcO, AcO, AcO, OMe)

SCHEME 6.44 5,6-Unsaturated sugars are available via elimination of primary halides.

methods exist, and the reader is encouraged to refer to the literature and general organic chemistry textbooks to supplement the strategies presented here.

6.6.3 Sulfuration Reactions

Any discussion concerning oxygen would not be complete without addressing sulfur. However, unlike the discussions of oxygenation reactions presented herein, the discussions of sulfuration reactions involve the introduction of sulfur atoms and sulfur based functional groups.

The introduction of sulfur-containing functional groups is similar to the nucleophilic-based processes discussed thus far. Like halogenation, amination and oxygenation reactions, the introduction of sulfur-based functionalities requires a nucleophilic sulfur species and an appropriate leaving group. As discussed above, appropriate electrophilic leaving groups include sulfonates, epoxides, and halides. Sulfur-based nucleophiles include, but are not limited to, thiocyanates, thioacetates, and thiophenols. Schemes 6.45–6.50 illustrate several combinations of nucleophiles and electrophiles utilized to introduce sulfur functionalities to sugar molecules [80–84]. While all these examples rely upon the conversion of one or more of the sugar hydroxyl groups to electrophilic leaving groups, one additional example, shown in Scheme 6.50, demonstrates that unprotected sugars with no activated hydroxyl groups can still be modified to incorporate thiols [85].

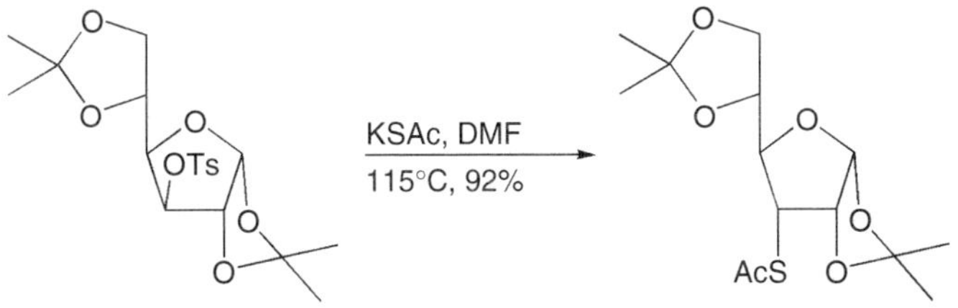

OTs — KSAc, DMF, 115°C, 92% → AcS

SCHEME 6.45 Potassium thioacetate is useful for the introduction of sulfur-based functionalities.

MsO, MsO, BzO, BzO, OMe — KSCN, DMF, 120°C, 39% → NCS, NCS, BzO, BzO, OMe

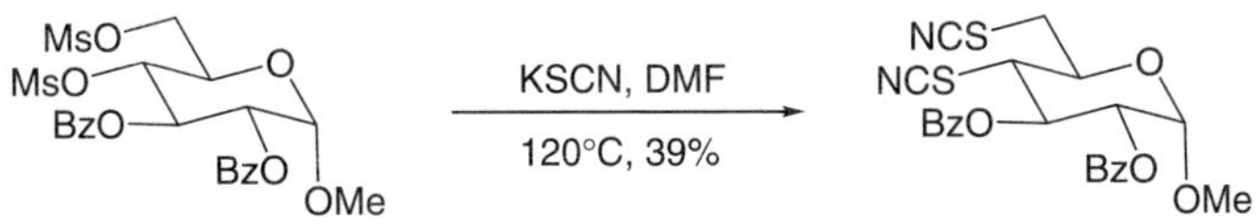

SCHEME 6.46 Potassium thiocyanate is a useful reagent for the introduction of sulfur-based functionalities.

Ph, TsO, HN, CBZ, OMe — PhSNa, DMF, 5°C, 90% → Ph, PhS, HN, CBZ, OMe

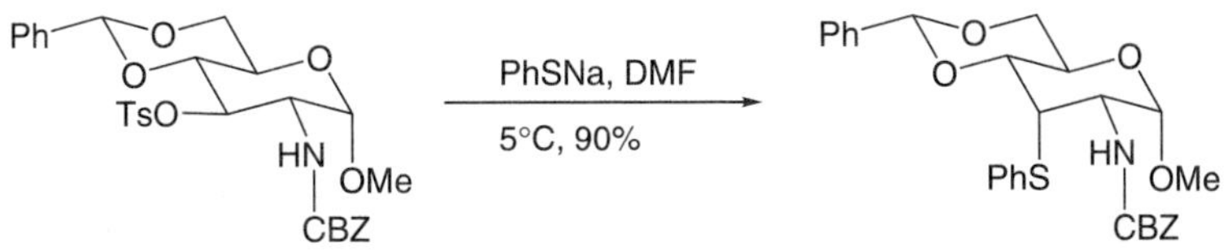

SCHEME 6.47 Sodium thiophenoxide is a useful reagent for the introduction of sulfur-based functionalities.

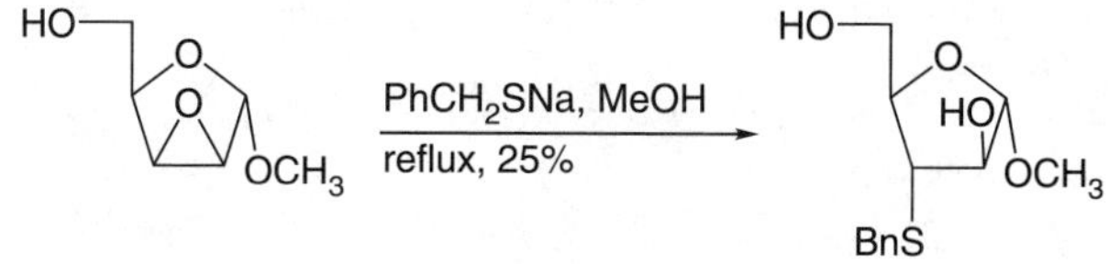

SCHEME 6.48 Sodium benzoxide is a useful reagent for the introduction of sulfur-based functionalities.

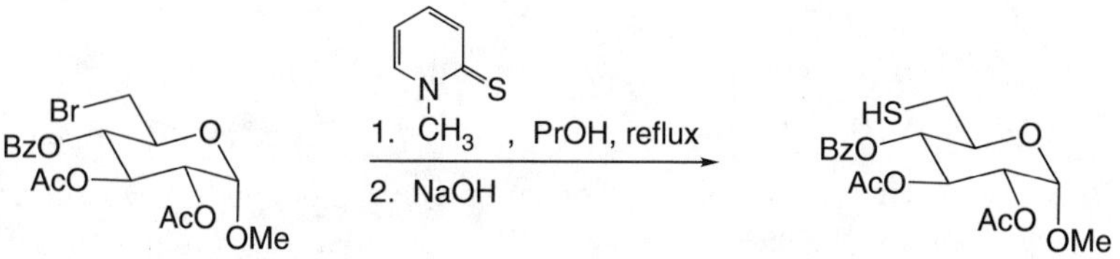

SCHEME 6.49 Thiols are available from reactions with thiopyridines.

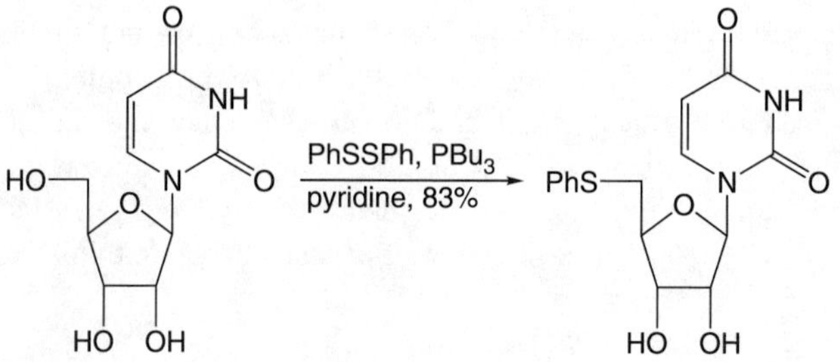

SCHEME 6.50 Diphenyldisulfide is a useful reagent for the introduction of sulfur-based functionalities.

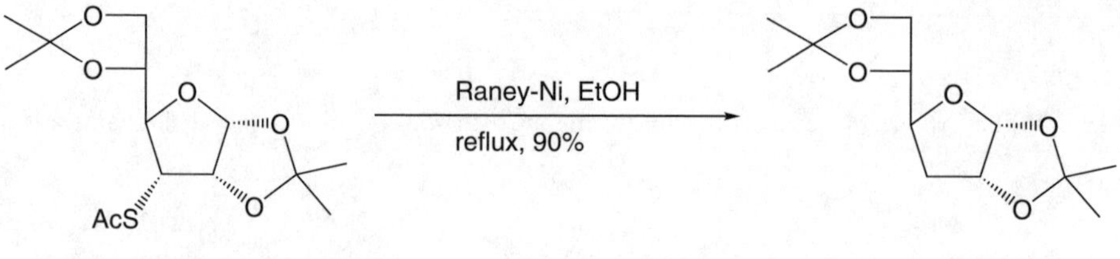

SCHEME 6.51 Raney nickel is a useful reagent for the removal of sulfur-based functionalities.

6.6.4 Desulfuration Reactions

Sulfur-based functionalities are useful intermediates susceptible to further modification. Moreover, they provide convenient strategies towards the complete removal of all functionality at given sugar ring positions. The ability to remove sulfur-based groups is at least as important as the ability to introduce them. In general, the method most preferred to remove sulfur groups is a reductive procedure utilizing Raney nickel as the reagent. Although Scheme 6.51 illustrates this procedure on a simple thioacetate [86], essentially identical conditions may be used on sulfides [87], thiocyanates [88], and any other sulfur-based group [32].

6.7 FORMATION OF CARBON–CARBON BONDS

The formation of carbon–carbon bonds is important to all aspects of organic chemistry from the design of small molecule drugs and pesticides to the total synthesis of complex natural products. When applied to any starting material, this reaction type may yield an infinite array of scaffold structures, capable of supporting every conceivable functional group in every conceivable combination. When applied to sugar molecules, the formation of carbon–carbon bonds fundamentally alters the sugar skeleton and yields new molecules easily transformed to more complex sugars, sugar-based natural products or, ultimately, noncarbohydrate structures. In this section, various methods used for the introduction of carbon–carbon bonds to sugars are introduced. Because the placement of carbon–carbon bonds at glycosidic centers constitutes a class of sugar derivatives known as *C*-glycosides, the reactions discussed in this section will focus on nonglycosidic sites. The chemistry of *C*-glycosides will be presented in Chapter 7.

6.7.1 ADDITION OF NUCLEOPHILES

Similar to the introduction of halides, amines, thiols and oxygen-based functionalities, the introduction of carbon–carbon bonds is commonly achieved through nucleophilic mechanisms. In these cases, the reactions may be perceived as introducing carbon-based functional groups. These nucleophilic groups are generally anionic in nature and vary in their respective nucleophilicity. Furthermore, their reactivity depends on the nature of the associated counterion. The first methods presented in this section involve the use of Grignard reagents, sodium/lithium reagents, cuprates and sulfur ylides.

6.7.1.1 Grignard Reagents

Grignard reagents are formed when alkyl halides are treated with magnesium metal. The resulting organometallic reagent possesses measurable stability and nucleophilicity sufficient to add into carbonyl systems in a 1,2 fashion. Although in their cyclic form, sugars do not possess carbonyl groups, their many hydroxyl groups can be isolated and easily oxidized to ketones. Because the oxidation of alcohols to ketones is accomplished through well-established methods, which are easily applied to sugars, examples of these reactions will not be discussed here. The examples presented in this section draw solely upon literature examples of sugar derivatives bearing carbonyl units. For more specific information regarding the preparation of these compounds, and oxidations in general, the reader is directed to the references provided in this section as well as standard organic texts [32,89].

Regarding the use of Grignard reagents to form carbon–carbon bonds on sugars, Scheme 6.52 illustrates the use of methylmagnesium iodide on a sugar derivative bearing a ketone at C-2 [90]. It is important to note that the resulting product bears the newly added methyl group in the equatorial position, while the newly formed hydroxyl group is axial. Similar results are observed in Scheme 6.53 where the Grignard reagent is derived from an acetylene unit [91]. The stereochemical outcome of this reaction is the result of the least hindered trajectory of the approaching nucleophile being from an equatorial direction. As will be noted with all additions to carbonyls in cyclic systems, this trend is a constant.

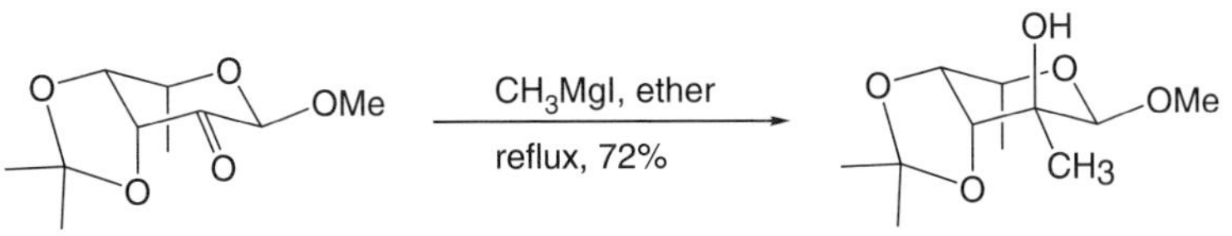

O O OMe O O → CH_3MgI, ether reflux, 72% → OH O O OMe O CH_3

SCHEME 6.52 Grignard reagents approach carbonyl groups from an equatorial trajectory.

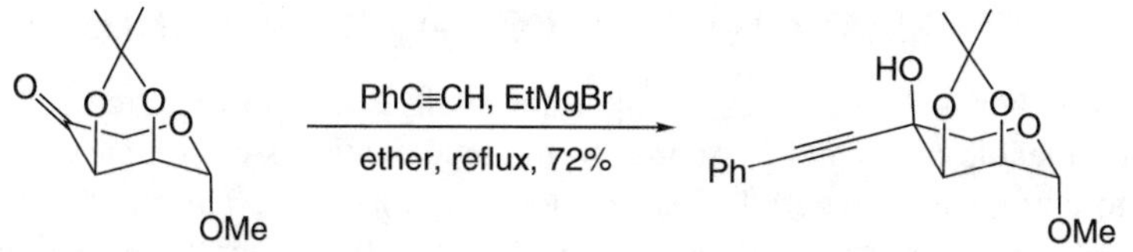

SCHEME 6.53 Grignard reagents approach carbonyl groups from an equatorial trajectory.

6.7.1.2 Organosodium, Organolithium and Organopotassium Reagents

Carbanionic nucleophiles are available in many forms other than Grignard reagents. In fact, the corresponding lithium, sodium, and potassium salts have found wide utility in the formation of carbon–carbon bonds throughout all facets of mainstream organic chemistry. As with Grignard reagents, these compounds easily add into carbonyl systems, producing both an alcohol and a new carbon–carbon bond, and approach is always from a pseudoequatorial trajectory. The example shown in Scheme 6.54 illustrates one application of the addition of an organolithium reagent to a sugar bearing a ketone unit [92]. This example further expands upon the utility of reagents capable of adding to carbonyl groups in a 1,2 fashion.

Departing from carbonyl-based systems, carbanionic nucleophiles have shown wide use in their ability to displace numerous leaving groups including halides, sulfonates and the ring opening of epoxides. Scheme 6.55 illustrates an example of the latter. In this example, a 2,3-anhydrosugar was treated with dimethylmalonate and sodium hydride. The resulting product showed successful incorporation of a malonate unit at C-2 of the sugar with liberation of a hydroxyl group at C-3 [93]. It should be noted that, like other epoxide ring-opening reactions, this example proceeded according to the Fürst–Plattner rule [18].

6.7.1.3 Cuprates

Cuprates have also been shown to be effective nucleophiles. Cuprates are formed when copper salts are added to organolithium reagents [89]. As shown in Scheme 6.56, these reagents, when added to anhydro sugars, readily form new carbon–carbon bonds with ring opening of the epoxide and

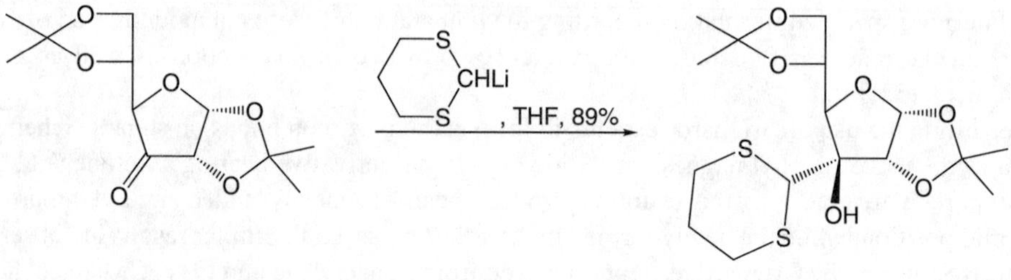

SCHEME 6.54 Other metallated carbon-based nucleophiles approach carbonyl groups from equatorial trajectories.

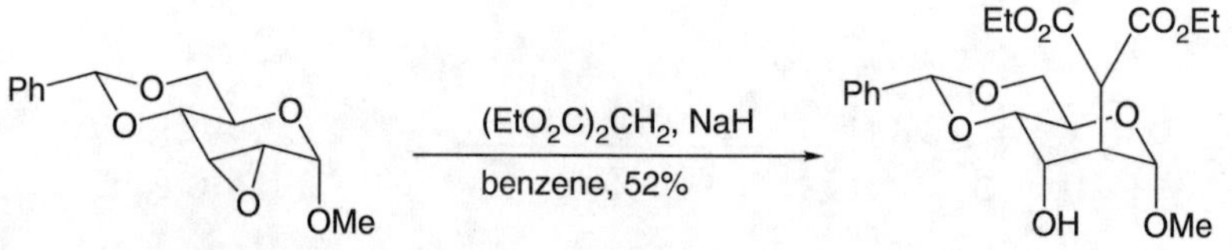

SCHEME 6.55 Carbon-based nucleophiles open epoxides according to the Fürst–Plattner rule.

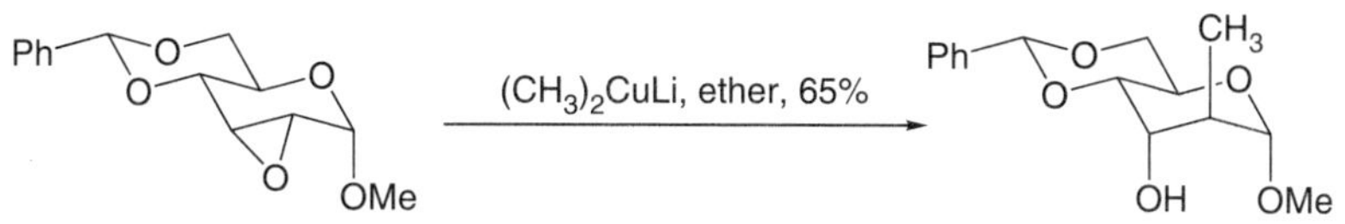

SCHEME 6.56 Cuprates open epoxides according to the Fürst–Plattner rule.

liberation of a new hydroxyl group. Again, the nucleophile adds to C-2 [94], consistent with the Fürst–Plattner rule [18].

Returning to carbonyl systems, all examples thus far have focused on 1,2 additions. However, cuprates are well known for their ability to add into conjugated carbonyl systems in a 1,4-fashion. As shown in Scheme 6.57, this chemistry was applied to a sugar-derived system with the net result of incorporating a new methyl group at C-2 and retaining the carbonyl unit for further reactions [95]. This illustrates the utility of various complementary chemistries to the preparation of complex sugar derivatives from readily accessible starting materials.

It should be noted that cuprates are not the only reagents capable of adding into conjugated systems. In fact, most electronically stabilized, readily formed nucleophiles undergo these reactions. As shown in Scheme 6.58, when treated with triethylamine, diethylmalonate is capable of undergoing 1,4-additions. The substrate, a 2,3-unsaturated-3-nitrosugar, received a new malonate unit at C-2 [96].

6.7.1.4 Sulfur Ylide Epoxidations/Wittig Reactions on Epoxides

To this point, much has been discussed regarding the 1,2 addition of nucleophiles to carbonyl systems and the nucleophilic ring opening of epoxides. Additionally, the preparation of sugar-based epoxides, or anhydrosugars, was discussed previously in this chapter. Considering the structure of anhydrosugars, the epoxide is located fused to the sugar ring. However, the introduction of epoxides joined to sugar rings in a spiro fashion provides unique opportunities for entry into otherwise inaccessible sugar derived structures. As shown in Scheme 6.59, spiro epoxysugars are readily available through the epoxidation of the corresponding sugar-derived ketones utilizing sulfur ylides [97]. As illustrated, the stereochemical progression of this reaction mirrors that of the approach of

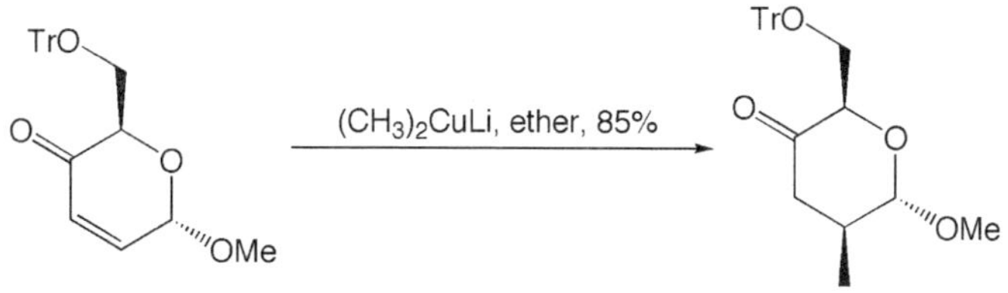

SCHEME 6.57 Cuprates add to conjugated carbonyl systems in a 1,4-fashion.

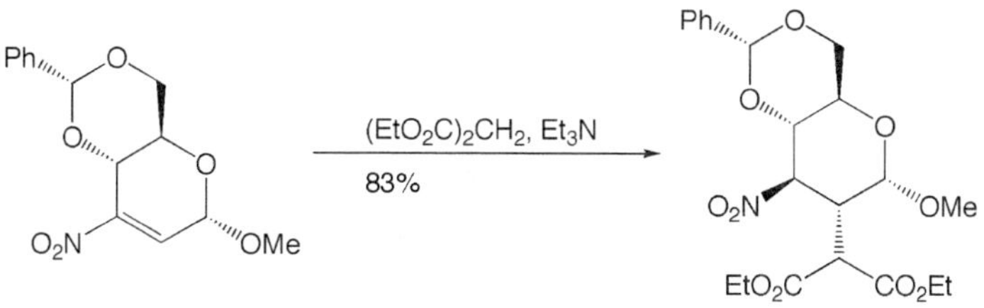

SCHEME 6.58 Malonate anions can add to conjugated systems in a 1,4-fashion.

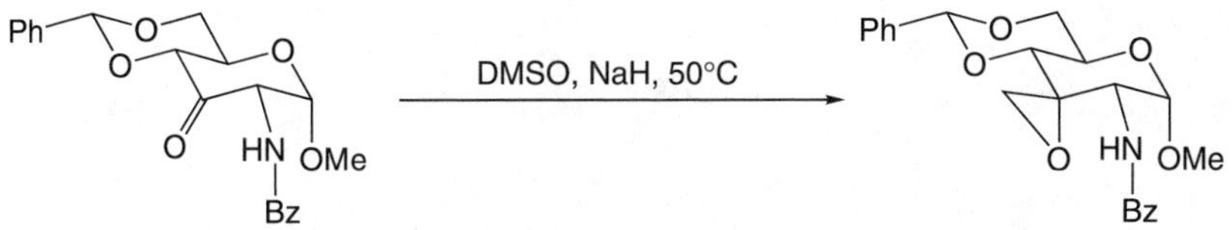

Ph
O
O
O
O
HN
OMe
Bz
DMSO, NaH, 50°C
Ph
O
O
O
O
HN
OMe
Bz

SCHEME 6.59 Sulfur ylides can form spiro epoxysugars.

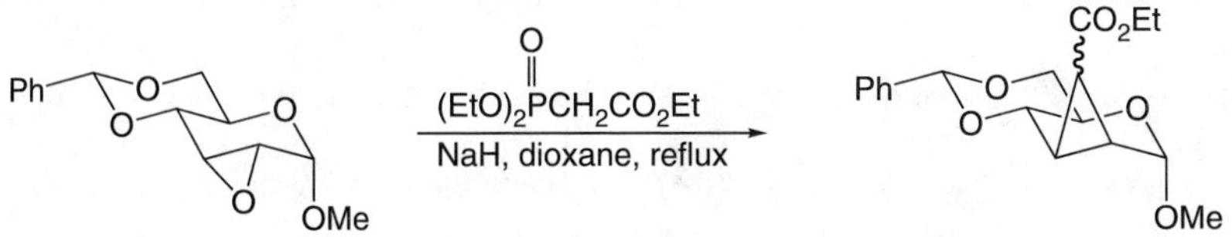

Ph
O
O
O
O
OMe
O
(EtO)2PCH2CO2Et
NaH, dioxane, reflux
CO2Et
Ph
O
O
O
OMe

SCHEME 6.60 Epoxysugars can be converted to cyclopropylsugars.

standard nucleophiles to ring systems in that the newly formed carbon–carbon bond is oriented in an equatorial position.

In one final example regarding the chemistry of anhydrosugars, the use of Horner–Emmons methodology is applied. It is well known that cyclopropane rings are available from epoxides utilizing this modification of the Wittig reaction. As shown in Scheme 6.60, this methodology is easily adapted to sugars [98]. Regarding the stereochemical outcome, it is important to note that the newly formed cyclopropane ring possesses the opposite stereochemical configuration to that of the starting epoxide.

6.7.2 Condensation Reactions

This section presents some basic condensation reactions that result in the chain elongation of sugars. The examples presented here are historically important to the general field of carbohydrate chemistry, and include the cyanohydrin chain extension and the nitromethane condensation.

6.7.2.1 The Cyanohydrin Chain Extension

The cyanohydrin chain extension results when a sugar is treated with sodium cyanide. The resulting cyanohydrin is then hydrolyzed to the corresponding carboxylic acid, resulting in a one-carbon

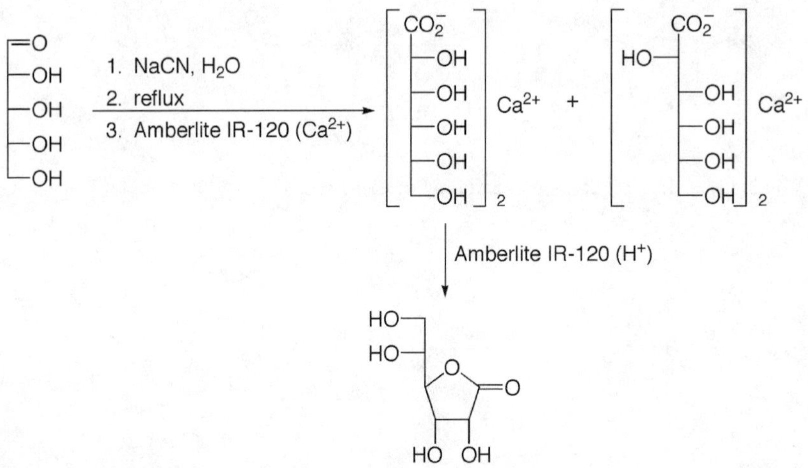

O
OH
OH
OH
OH
1. NaCN, H2O
2. reflux
3. Amberlite IR-120 (Ca2+)
CO2−
OH
OH
OH
OH
OH
2
Ca2+
+
CO2−
HO
OH
OH
OH
OH
2
Ca2+
Amberlite IR-120 (H+)
HO
HO
O
O
HO
OH

SCHEME 6.61 Cyanohydrins are useful for the chain elongation of sugars.

chain extension of the sugar. The example shown in Scheme 6.61 illustrates this reaction as applied to ribose and results in the formation of a new sugar-lactone [99].

In modifications to the cyanohydrin reaction, convenient entries into 2-amino-2-deoxysugars can be realized. One example, illustrated in Scheme 6.62, relies on the initial condensation of the aldehyde of ring-opened sugars with amines, thus forming imines. The resulting imines, as with the corresponding aldehydes, are susceptible to condensation with cyanide resulting in a one-carbon extension of the original sugar [100]. Additionally, one should remember that cyanohydrin condensations are not limited to specific types of aldehydes. As shown in Scheme 6.63, this reaction has been used to affect the chain clongation of sugars at the nonreducing end [101].

6.7.2.2 The Nitromethane Condensation

Another very useful method for the formation of carbon–carbon bonds that also results in the chain extension of sugars is the nitromethane condensation. This reaction involves the addition of nitromethane to the reducing end of sugars under basic conditions. Subsequent treatment with

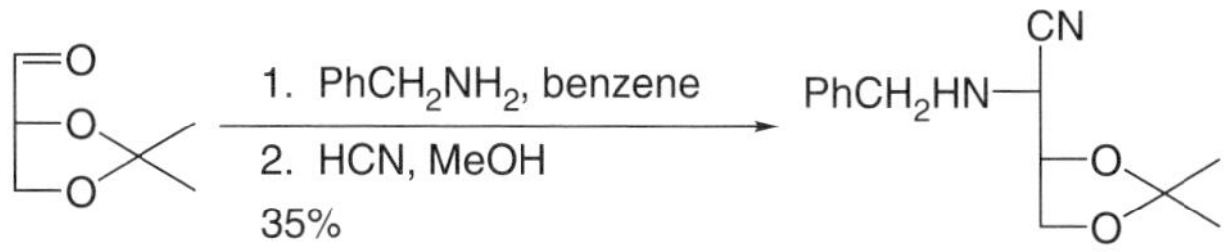

1. $PhCH_2NH_2$, benzene
2. HCN, MeOH
35%

CN
$PhCH_2HN$

SCHEME 6.62 Cyanohydrin reactions can generate 2-amino-2-deoxysugars.

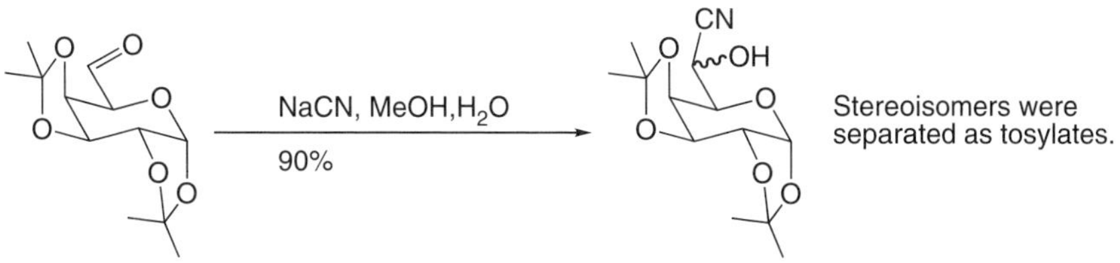

NaCN, MeOH,H_2O
90%

CN
OH

Stereoisomers were separated as tosylates.

SCHEME 6.63 Cyanohydrin reactions can elongate sugars at the nonreducing end.

1. $MeNO_2$, NaOMe, MeOH
2. Dowex - 50 (H^+)

CH_2NO_2 + CH_2NO_2

1. aq. NaOH
2. aq. H_2SO_4
3. Fractional crystallization

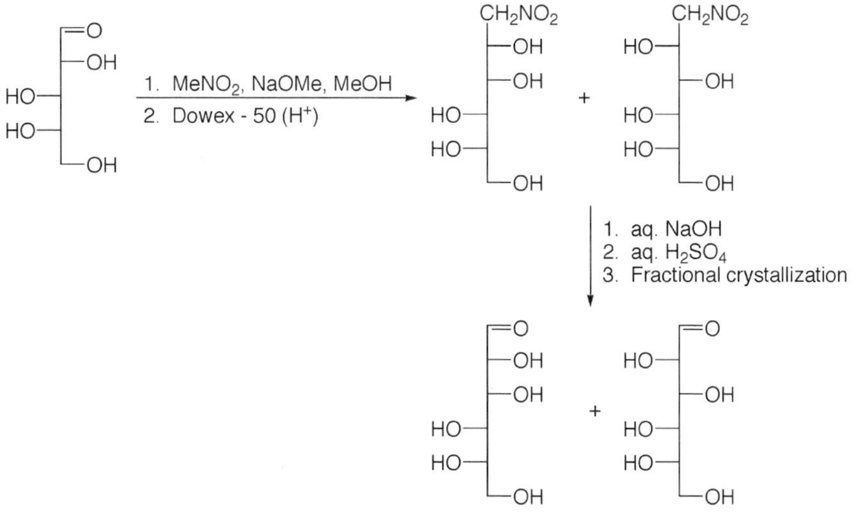

SCHEME 6.64 Nitromethane condensations are useful for the chain elongation of sugars.

sodium hydroxide, followed by sulfuric acid, results in removal of the nitro group with formation of a new aldehyde functionality defining the reducing end of the elongated sugar chain. Scheme 6.64 illustrates the application of this methodology to L-arabinose resulting in the formation of a mixture of L-glucose and L-mannose [102].

Although this chapter's discussion of condensation reactions is far from conclusive, it illustrates basic principles that generally apply to the chain extension of sugars at both the reducing and nonreducing ends. When considering the wealth of condensation and olefination reactions available within the realm of classical organic chemistry, many other methods for sugar chain extensions with formation of carbon–carbon bonds may be envisioned. Given that in their native cyclic forms, all sugars possess latent carbonyl groups and also that, through selective protection and oxidation sequences, new carbonyl groups may be introduced, excellent starting points may be found within aldol, Wittig and related methodologies.

6.7.3 Wittig/Horner–Emmons Reactions

As discussed in Sections 6.7.1 and 6.7.2, numerous methods exist for the introduction of carbon–carbon bonds to sugars. These methods both encompass the direct introduction of carbon-based nucleophiles as well as the extension of sugar chains at the reducing and nonreducing ends. An absolute complement to these methodologies is the use of the Wittig (and its related Horner–Emmons modification) reaction. Schemes 6.65 and 6.66 illustrate the utility of these methodologies as applied to the latter cases and as alluded to in the previous section [103]. What was not addressed is the utility of these reactions to introduce exocyclic olefinic functionalities to sugar rings. While fairly straightforward, the examples illustrated in Schemes 6.67 and 6.68 demonstrate that the formation of carbon–carbon bonds in sugar systems is not limited to the sole formation of single bonds [104,105].

6.7.4 Claisen Rearrangements

Complementary to the ionic methods for the introduction of carbon–carbon bonds discussed thus far is the use of electrocyclic rearrangements. These reactions include the Cope and Claisen

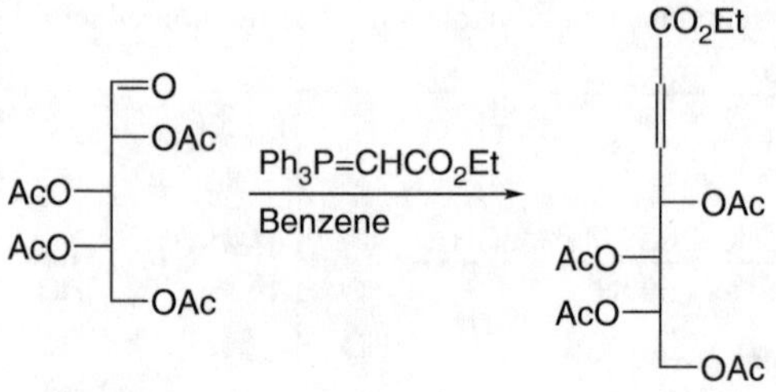

=O OAc AcO AcO OAc $Ph_3P{=}CHCO_2Et$ Benzene CO_2Et OAc AcO AcO OAc

SCHEME 6.65 The Wittig reaction is useful for the chain elongation of sugars at the reducing end.

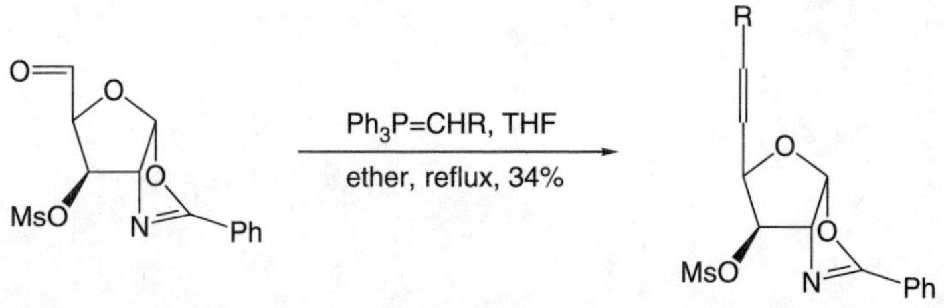

O= O O MsO N Ph $Ph_3P{=}CHR$, THF ether, reflux, 34% R O O MsO N Ph

SCHEME 6.66 The Wittig reaction is useful for the chain elongation of sugars at the nonreducing end.

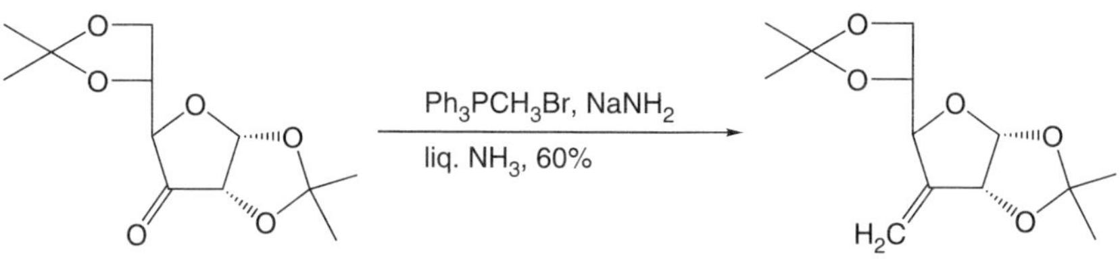

Ph_3PCH_3Br, $NaNH_2$
liq. NH_3, 60%

SCHEME 6.67 The Wittig reaction is generally useful as applied to sugar-derived ketones.

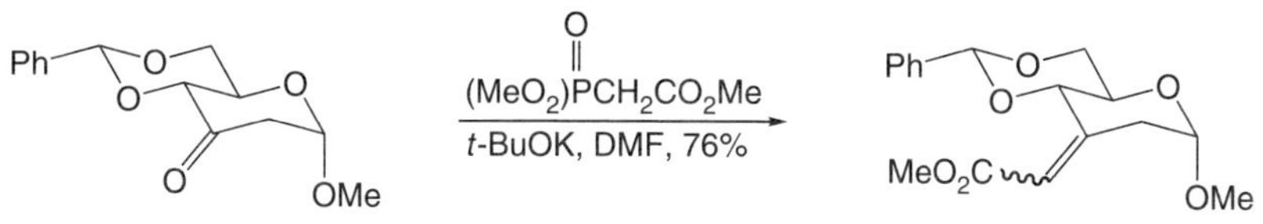

$(MeO_2)P(O)CH_2CO_2Me$
t-BuOK, DMF, 76%

SCHEME 6.68 The Horner–Emmons modification of the Wittig reaction is a generally useful alternative to the standard Wittig reaction.

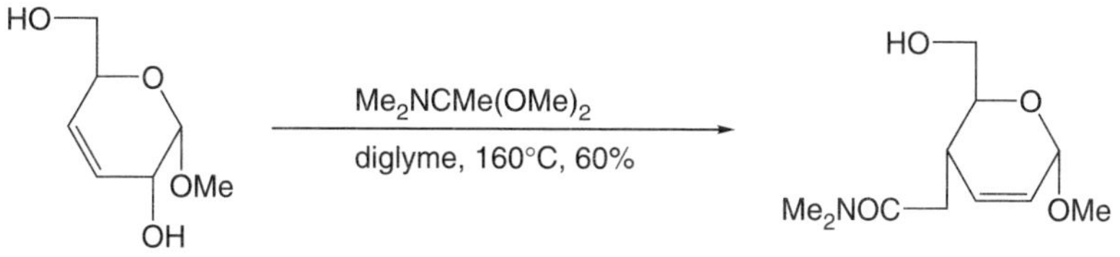

$Me_2NCMe(OMe)_2$
diglyme, 160°C, 60%

SCHEME 6.69 Electrocyclic rearrangements are useful for the introduction of new carbon–carbon bonds to sugars.

rearrangements and their corresponding orthoester and orthoamide modifications [106,107]. These reactions proceed through initial functionalization of an allylic alcohol to form a vinyl ether. Usually run at high temperatures, rearrangements occur with migration of the double bond and introduction of a new carbon–carbon bond. As discussed above, the incorporation of unsaturation in the sugar ring is readily achieved. Therefore, this methodology is a logical and useful next step in sugar ring modifications. One example, shown in Scheme 6.69, involves the treatment of the illustrated 3,4-unsaturated sugar with *N,N*-dimethylacetamide dimethyl acetal. Upon heating in diglyme, the resulting orthoamide Claisen rearrangement introduces a new amide group at C-4 with migration of the double bond to the 2,3 position [108].

6.8 REDUCTIONS AND OXIDATIONS

Having addressed numerous methods for the introduction of new functionalities to sugar rings, we now turn our attention to reduction and oxidation reactions.

6.8.1 Reduction Reactions

Beginning with reductions, we must remember that this class of reactions results in a decrease in the molecules oxidation state. In other words, there is a net loss of functionality on the molecule being reduced. This loss of functionality may be manifested in the conversion of a double bond to a single bond (olefins to alkanes and carbonyls to alcohols), or in the complete removal of heteroatom-based functionalities (deoxygenation, dehalogenation, deamination). In all cases, the net result of a reduction reaction is the formation of a simpler and less reactive structure from one that is more complex. In this section, reduction reactions are discussed in the context of halides, sulfonates, epoxides, and olefins.

6.8.1.1 Reduction of Halides and Sulfonates

As sugars are, by nature, highly oxygenated, a logical starting point is the direct deoxygenation of sugars. Throughout this chapter, methods have been presented for the conversion of alcohols to halides. Additionally, numerous references pointing to the conversion of alcohols to sulfonates were made. Taking advantage of these functionalizations is central to the most successful deoxygenation procedures available for the modification of sugar-based structures. In fact, where partial protection of sugars may leave multiple reactive hydroxyl groups, selectivity from various reductive techniques frequently allows one group to remain intact.

The ability of reducing agents, such as lithium aluminum hydride, to reduce halides and sulfonates to their corresponding alkyl derivatives is well known. Lithium aluminum hydride is also selective for the reduction of primary halides and sulfonates over secondary analogs. As shown in Scheme 6.70, this reaction was applied to a bis-tosylate with the major isolated product being the mono-tosylate with deoxygenation at C-6 [109].

While hydride-based reagents such as lithium aluminum hydride can selectively reduce at primary centers over secondary centers, reagents such as Raney nickel show little or no selectivity. As shown in Scheme 6.71, a bis-chloride was converted to the corresponding 4,6-dideoxy sugar using this reagent [110].

Finally, when the selective reduction of secondary centers is desired over primary centers, free radical chemistry provides the answer. Unlike ionic mechanisms, the formation of free radicals occurs more readily at tertiary centers, with secondary radicals being more stable than primary radicals. As shown in Scheme 6.72, tributyltin hydride sequentially removed the secondary chloride with complete reduction of both chlorides over extended reaction times [111].

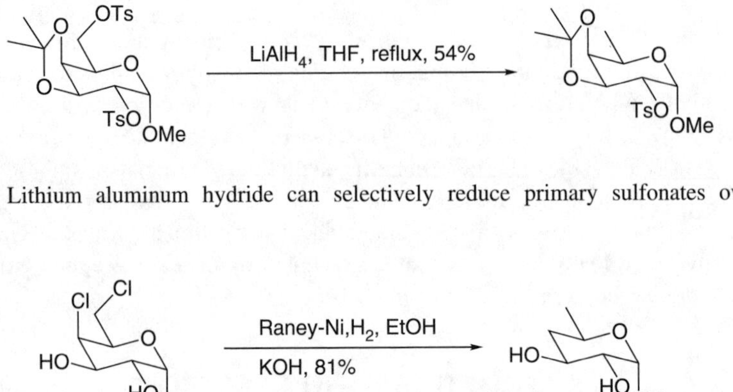

OTs O O O TsO OMe — $LiAlH_4$, THF, reflux, 54% → O O O TsO OMe

SCHEME 6.70 Lithium aluminum hydride can selectively reduce primary sulfonates over secondary sulfonates.

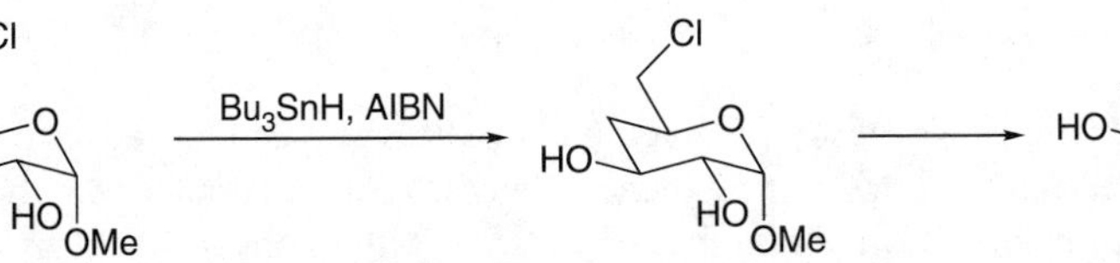

Cl Cl O HO HO OMe — Raney-Ni,H_2, EtOH / KOH, 81% → O HO HO OMe

SCHEME 6.71 Raney nickel is useful for the reduction of halides.

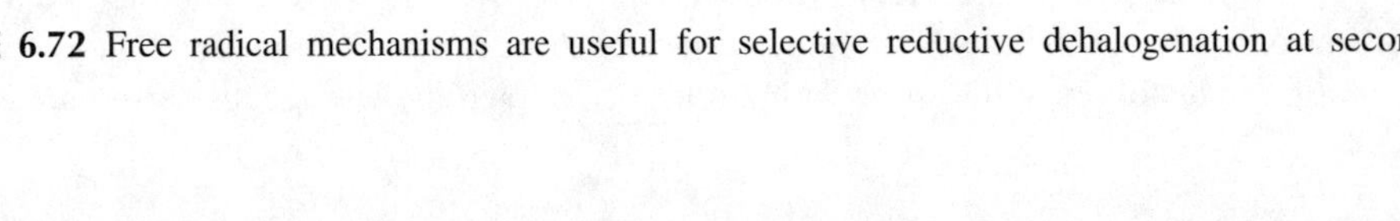

Cl Cl O HO HO OMe — Bu_3SnH, AIBN → Cl O HO HO OMe → O HO HO OMe

SCHEME 6.72 Free radical mechanisms are useful for selective reductive dehalogenation at secondary centers.

6.8.1.2 Reduction of Epoxides

Having discussed approaches towards the reduction of halides and sulfonates — two of the primary classes of electrophiles used to facilitate sugar modifications — it now seems fitting to address similar issues with respect to epoxides. Hydride-based reducing agents work through the direct delivery of hydrogen anions to electrophilic centers. As with all nucleophilic mechanisms discussed thus far, the hydride will approach through the least hindered trajectory. Because anhydrosugars can exist in various stereochemical configurations, the stereochemistry of the individual anhydrosugar will determine the regiochemical outcome of the hydride reduction. Schemes 6.73 and 6.74 illustrate this observation through the reduction of two isomeric epoxysugars. It is important to note that in both cases, the hydride is delivered through a trajectory consistent with the Fürst–Plattner rule [112–115].

6.8.1.3 Reduction of Olefins

Many examples have been presented for the generation of olefinic sugar derivatives. Given that olefins provide useful handles for further functionalization, discussion centered on the reduction of these compounds is warranted. In this section, both catalytic and boron-based reductions are addressed.

6.8.1.3.1 Catalytic Methods

It is well known that catalytic hydrogenations are useful for the conversion of olefins to their corresponding saturated derivatives. However, when the olefin is a part of a ring, there can be stereochemical consequences. As shown in Scheme 6.75, a tri-substituted endocyclic double bond

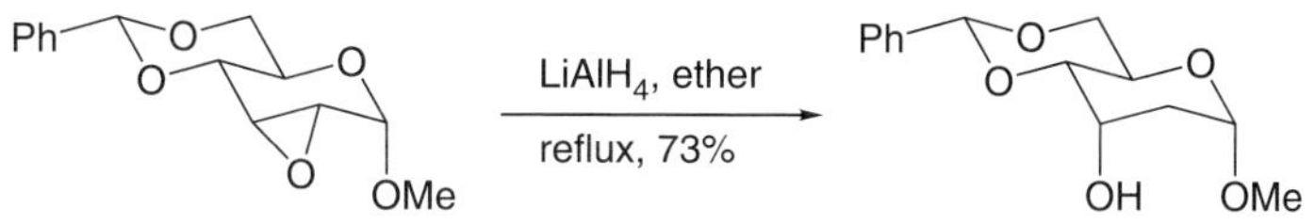

$LiAlH_4$, ether
reflux, 73%

SCHEME 6.73 Lithium aluminum hydride reduces epoxides producing products consistent with the Fürst–Plattner rule.

$LiAlH_4$, ether
reflux, 85%

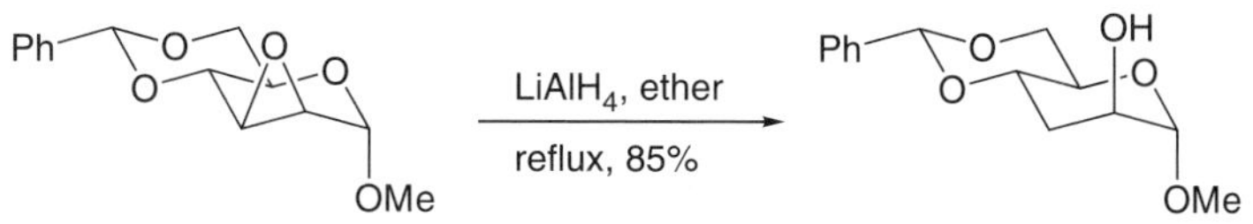

SCHEME 6.74 Lithium aluminum hydride reduces epoxides producing products consistent with the Fürst–Plattner rule.

20% $Pd(OH)_2/C$
H_2, EtOAc, 70%

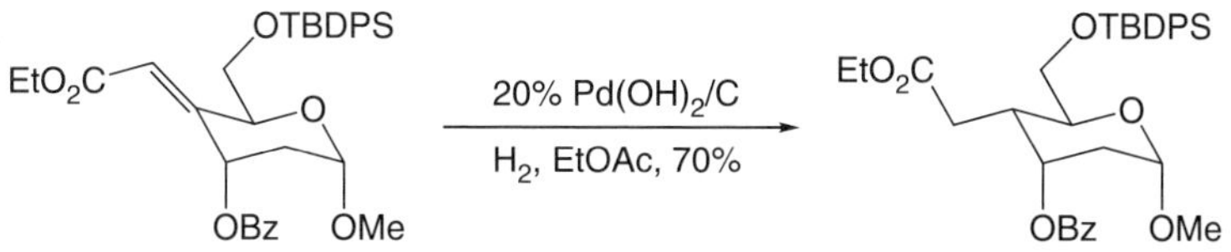

SCHEME 6.75 Catalytic hydrogenations reduce double and triple bonds through the least hindered face.

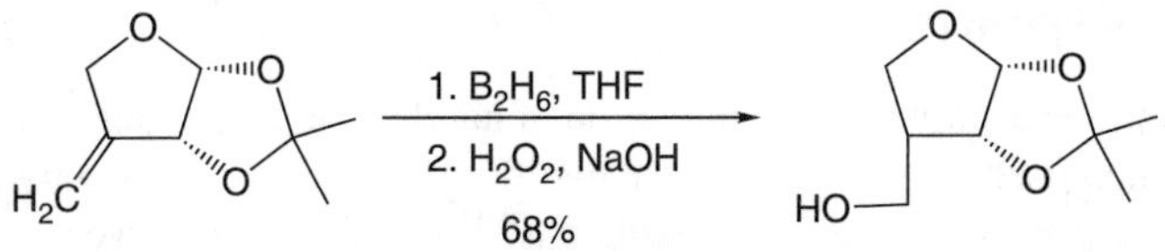

1. B_2H_6, THF
2. H_2O_2, NaOH
68%

SCHEME 6.76 Hydroboration reactions can reduce double bonds with the simultaneous introduction of new functionalities.

was reduced utilizing palladium hydroxide on activated carbon. The illustrated product was isolated as the major stereoisomer because of the hydrogenation occurring at the least hindered surface. Reduction from the lower face was blocked by the 3-benzoyloxy group [116].

6.8.1.3.2 Hydroborations

Complementary to catalytic hydrogenations, the use of diborane provides convenient routes towards the elimination of olefins. However, this reaction has the added advantage of introducing new functionality at the site of olefin reduction. Because of regiochemical considerations, this reaction is particularly useful when applied to exocyclic olefins. Scheme 6.76 shows that a sugar derivative was treated with diborane with an oxidative workup to yield the illustrated product bearing a new hydroxyl group exclusively at the primary center [117].

6.8.2 Oxidation Reactions

As discussed in Chapter 3, numerous methods are available for the selective protection of the highly oxygenated sugar rings. With the ability to leave select hydroxyl groups unprotected, any of the methods useful for the conversion of alcohols to aldehydes and ketones [32] may be applied to sugar-based systems. Because these reactions are well understood and generally applicable, they will not be discussed further here. In this section, the concept of oxidations will be addressed in the context of oxidative degradation reactions that are particularly important in classical carbohydrate chemistry.

Taking advantage of the oxidative lability of aldehydes, the reducing ends of ring-opened sugars are readily converted from aldehydes to carboxylic acids. These compounds, when treated with hydrogen peroxide and divalent iron salts, are readily degraded to lower sugars. This application of the Ruff degradation is illustrated in Scheme 6.77 in the conversion of a hexose to a pentose [118].

Since aldehydes are readily protected as thioacetals, it is not surprising that ring-opened sugars may be trapped as such species. Furthermore, utilizing peracids, thioacetals are readily oxidized to bis-sulfones. Under basic conditions, the sulfone degradation proceeds with a one-carbon reduction of the sugar length from the reducing end. An example of this reaction is illustrated in Scheme 6.78 [119–121].

As sugars generally possess 1,2-diols, we cannot ignore the ready conversion of these groups to aldehydes utilizing reagents such as lead tetraaectate and sodium periodate. Both of these reagents

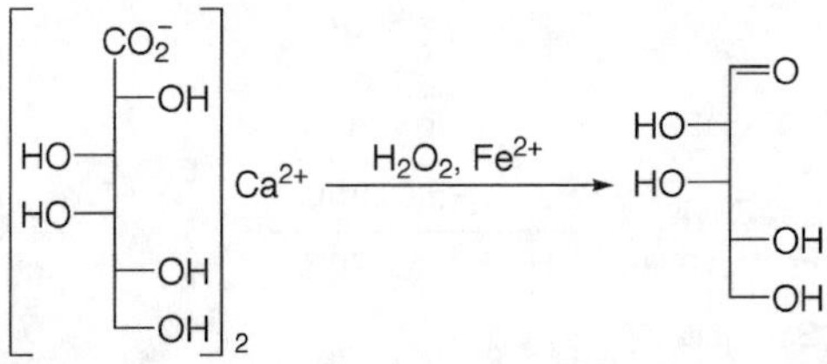

Ca^{2+} H_2O_2, Fe^{2+}

SCHEME 6.77 The Ruff degradation cleaves sugar-derived carboxylic acids to lower sugars.

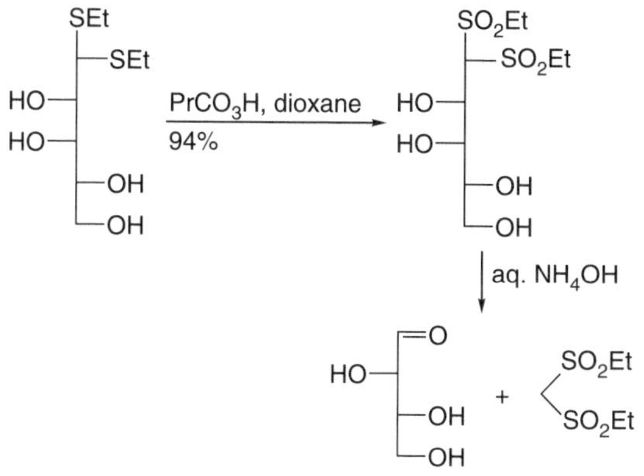

SCHEME 6.78 The sulfone degradation shortens sugar chains by exploiting the chemistry of aldehydes in the generation of active leaving groups.

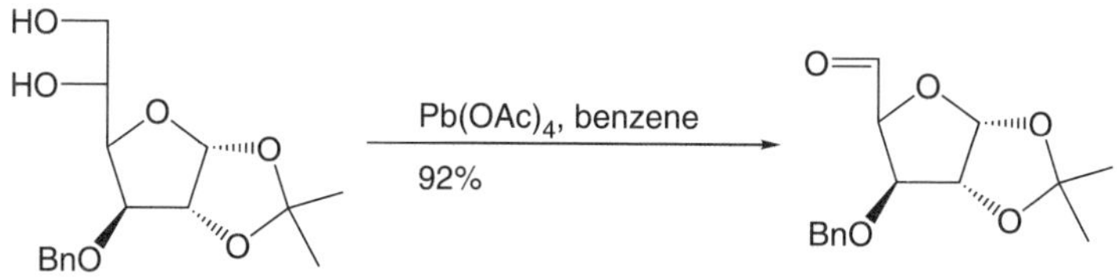

SCHEME 6.79 Lead tetraacetate cleaves 1,2-diols to aldehydes.

affect cleavage with similar mechanisms and an example of the lead-based reaction is shown in Scheme 6.79 [122]. Unlike the previous degradation examples, this approach is useful at any position in the sugar chain where 1,2-diols exist.

6.9 REARRANGEMENTS AND ISOMERIZATIONS

A discussion of classical rearrangements and isomerization reactions is a fitting end to this chapter. These types of reactions are extremely useful when considering their ability to convert readily available sugars to both scarce sugars and useful sugar derivatives. In this section, base-mediated isomerization reactions will be addressed, as will the Amadori rearrangement.

6.9.1 BASE CATALYZED ISOMERIZATIONS

By virtue of the aldehyde at the reducing end, sugars are susceptible to deprotonation and isomerization. The rearrangement from an aldose sugar to a ketose sugar, shown in Scheme 6.80, is a direct result of this property [123]. Based on the initial enolization step, this chemistry is easily applied to the direct C-2 epimerization of 2-deoxy-2-aminosugars. Scheme 6.81 illustrates this reaction in the conversion of *N*-acetyl-D-glucosamine to *N*-acetyl-D-mannosamine [124,125].

6.9.2 THE AMADORI REARRANGEMENT

Further expanding upon the effects bases have on sugars, the Amadori rearrangement incorporates new nitrogen functionalities with the generation of a ketose derivative. Scheme 6.82 illustrates this reaction applied to the combination of glucose with morpholine

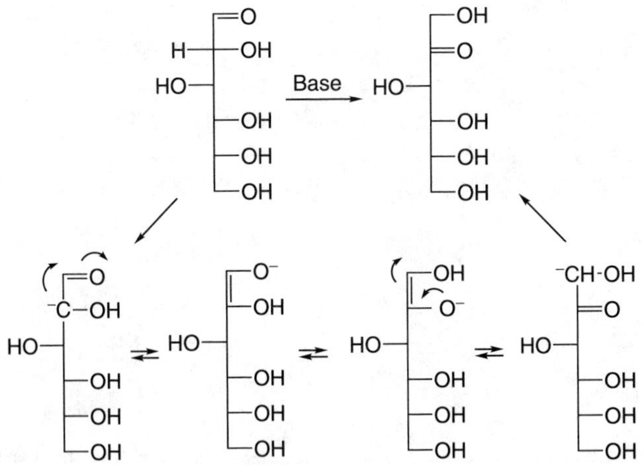

SCHEME 6.80 Aldoses can be converted to ketoses under basic conditions.

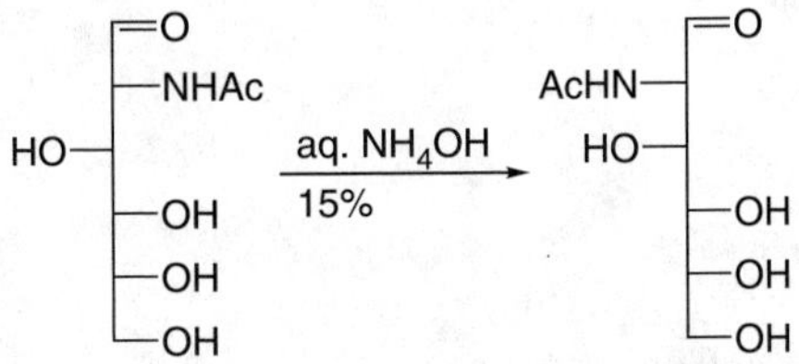

SCHEME 6.81 2-Amino-2-deoxysugars can be epimerized at C-2 under basic conditions.

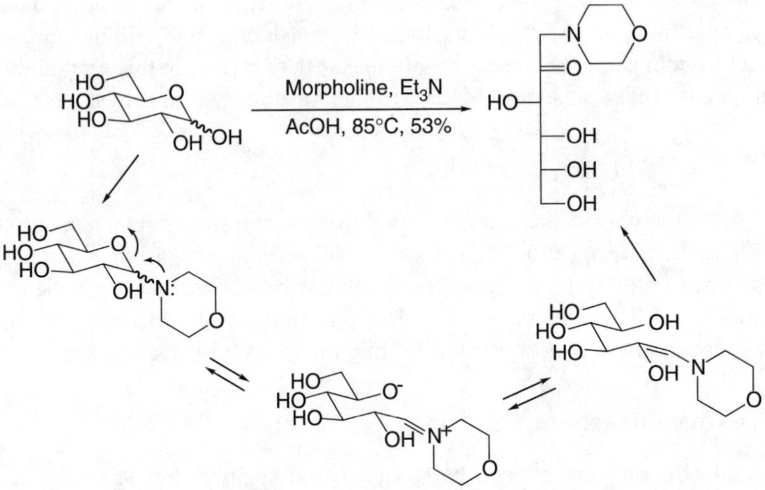

SCHEME 6.82 The Amadori rearrangement converts aldoses to 1-amino-1-deoxyketoses.

generating a 1-morpholino-ketose [126]. As illustrated, the reaction proceeds through initial formation of an enamine, followed by tautomerization of the enol to a C-2 carbonyl group.

6.10 CONCLUSION

Throughout this chapter, many methods for the direct modification of sugar rings were presented. Given the wealth of chemistry studied in these areas, the examples presented here are naturally far from conclusive. Nonetheless, it is hoped that these examples will be viewed as general concepts that can be further developed and tailored to individual needs. In fact, while each concept presented in this chapter is worthy of an individual chapter of its own, these examples all fall within the realm of mainstream organic chemistry. It was this realization, coupled with the fact that much of this chemistry is presented in greater detail throughout the remaining chapters of this book that led to the decision to present a more historical treatment of the functionalization of sugars. While the present chapter is intended as an introduction to the concept of sugars being useful organic molecules, it is hoped that the reader will study the remaining chapters and ultimately regard sugars as both starting materials and tools for the creative development of novel synthetic processes, and as useful synthetic scaffolds through multiple chemical disciplines.

REFERENCES

1. Hughes, N A, Speakman, P R H, Benzoate displacements on 3-*O*-toluene-*p*-sulphonyl-D-glucose derivatives: a new synthesis of D-allose, *J. Chem. Soc.*, 2236–2239, 1965.
2. Al-Radhi, A K, Brimacombe, J S, Tucker, L C N, Nucleophilic displacement reactions in carbohydrates. Part XIX. The reaction of methyl 6-deoxy-3-*O*-methyl-4-*O*-methylsulphonyl-2-*O*-*p*-tolylsulphonyl-α-D-allopyranoside with sodium benzoate in hexamethylphosphoric triamide: a vicinal axial substituent effect, *J. Chem. Soc. C.*, 2305–2310, 1971.
3. Pittman, C U Jr, McManus, S P, Larsen, J W, 1,3-Dioxolan-2-ylium and related heterocyclic cations, *Chem. Rev.*, 72, 357–438, 1972.
4. Goodman, L, Neighboring-group participation in sugars, *Adv. Carbohydr. Chem. Biochem.*, 22, 109–175, 1967.
5. Pfitzner, K E, Moffatt, J G, The synthesis and hydrolysis of 2,3-dideoxyuridine, *J. Org. Chem.*, 29, 1508–1511, 1964.
6. Greene, T W, Wutz, P G M, *Protective Groups in Organic Synthesis*, 3rd ed., Wiley-Interscience, New York, 1999.
7. Swern, D, *Organic Peroxides*, Vol. 2, Interscience, New York, 1971, pp. 355–533.
8. Gao, Y, Klunder, J M, Hanson, R M, Masamune, H, Ko, S Y, Sharpless, K B, Catalytic asymmetric epoxidation and kinetic resolution: modified procedures including in situ derivatization, *J. Am. Chem. Soc.*, 109, 5765–5780, 1987.
9. Neumann, H, Reaction of vicinal diols and dimethylformamide dimethyl acetal. New method for epoxide formation, *Chimia.*, 23, 267–269, 1969.
10. Guthrie, R D, Jenkins, I D, Yamasaki, R, Skelton, B W, White, A H, Epoxidations with triphenylphosphine and diethyl azodicarboxylate. Part 1. Synthesis of methyl 3,4-anhydro-D-tagatofuranosides, *J. Chem. Soc., Perkin. Trans 1*, 2328–2334, 1981.
11. Martin, J C, Franz, J A, Arhart, R J, Sulfuranes. XIV. Single-step syntheses of epoxides and other cyclic ethers by reaction of a diaryldialkoxysulfurane with diols, *J. Am. Chem. Soc.*, 96, 4604–4611, 1974.
12. Mitsunobu, O, The use of diethyl azodicarboxylate and triphenylphosphine in synthesis and transformation of natural products, *Synthesis*, 1–28, 1981.
13. Sinclair, H B, Displacement of carbohydrate sulfonates with halide ions of toluene-solubilized halides, *Carbohydr. Res.*, 15, 147–153, 1970.
14. Foster, A B, Hems, R, Deoxyfluoro sugars via displacement of sulfonyloxy groups with tetrabutylammonium fluoride, *Methods Carbohydr. Chem.*, 6, 197–200, 1972.

15. Goldsmith, D J, John, T K, Kwong, C D, Painter, G R III, Preparation and rearrangement of trichothecane-like compounds. Synthesis of aplysin and filiformin, *J. Org. Chem.*, 45, 3989–3993, 1980.
16. Hanessian, S, Vatèle, J M, Design and reactivity of organic functional groups: imidazolylsulfonate (imidazylate) — an efficient and versatile leaving group, *Tetrahedron Lett.*, 22, 3579–3582, 1981.
17. Lemieux, R U, Fraga, E, Watanabe, K A, Preparation of unsaturated carbohydrates. A facile synthesis of methyl 4,6-*O*-benzylidene-D-hex-2-enopyranosides, *Can. J. Chem.*, 46, 61–69, 1968.
18. Fürst, A, Plattner, P A, Abstracts of papers of *12th International Congress of Pure and Applied Chemistry*, 409, 1951.
19. Richards, G N, Wiggins, L F, The action of Grignard reagents on anhydro-sugars of ethylene oxide type. Part II. The behaviour of 4:6-benzylidine 2:3-anhydro-α-methyl-D-alloside towards ethyl- and phenyl-magnesium halides, *J. Chem. Soc.*, 2442–2446, 1953.
20. Horwitz, J P, Chua, J, Da Rooge, M A, Noel, M, Klundt, I L, Nucleosides. IX. The formation of 2,3-unsaturated pyrimidine nucleosides via a novel β-elimination reaction, *J. Org. Chem.*, 31, 205–211, 1966.
21. Hirata, M, Studies on nucleosides and nucleotides. IX. Nucleophilic substitution of secondary sulfonyloxy groups of pyrimidine nucleosides. II. Reaction of 2,2′-anhydro-1-(3′-*O*-tosyl-β-D-arabinofuranosyl)uracil with sodium bromide, sodium ethanethiol and sodium azide, *Chem. Pharm. Bull.*, 16, 291–295, 1968.
22. Hanessian, S, Ducharme, D, Masse, R, Capmau, M L, A one-flask preparation of methyl 6-azido-α-D-hexopyranosides, *Carbohydr. Res.*, 63, 265–269, 1978.
23. Hanessian, S, Ponpipom, M M, Lavallee, P, Procedures for the direct replacement of primary hydroxyl groups in carbohydrates by halogens, *Carbohydr. Res.*, 24, 45–56, 1972.
24. Garegg, P J, Some aspects of regio-, stereo-, and chemoselective reactions in carbohydrate chemistry, *Pure Appl. Chem.*, 56, 845–858, 1984.
25. Garegg, P J, Samuelsson, B, Novel reagent system for converting a hydroxy-group into an iodo-group in carbohydrates with inversion of configuration. Part 2, *J. Chem. Soc., Perkin. Trans 1*, 2866–2869, 1980.
26. Garegg, P J, Johansson, R, Samuelsson, B, Two reagent systems for converting hydroxy compounds into chlorides using the triphenylphosphine/imidazole system, *Synthesis*, 168–170, 1984.
27. Garegg, P J, Samuelsson, B, Novel reagent system for converting a hydroxy-group into an iodo-group in carbohydrates with inversion of configuration, *J. Chem. Soc., Chem. Commun.*, 978–980, 1979.
28. Jennings, H J, Jones, J K N, Reactions of sugar chlorosulfates. Part V. The synthesis of chlorodeoxy sugars, *Can. J. Chem.*, 43, 2372–2386, 1965.
29. Hanessian, S, Plessas, N R, Reaction of *O*-benzylidene sugars with *N*-bromosuccinimide. III. Applications to the synthesis of aminodeoxy and deoxy sugars of biological importance, *J. Org. Chem.*, 34, 1045–1053, 1969.
30. Hanessian, S, Plessas, N R, Reactions of carbohydrates with (halomethylene)dimethyliminium halides and related reagents. Synthesis of some chlorodeoxy sugars, *J. Org. Chem.*, 34, 2163–2170, 1969.
31. Evans, M E, Long, L, Parrish, F W, Reaction of carbohydrates with methylsulfonyl chloride in *N,N*-dimethylformamide. Preparation of some methyl 6-chloro-6-deoxyglycosides, *J. Org. Chem.*, 33, 1074–1076, 1968.
32. Larock, R C, *Comprehensive Organic Transformations: A Guide to Functional Group Preparations*, 2nd ed., Wiley–VCH, New York, 1999.
33. Nicolaou, K C, Dolle, R E, Papahatjis, D P, Practical synthesis of oligosaccharides. Partial synthesis of avermectin B1a, *J. Am. Chem. Soc.*, 106, 4189–4192, 1984.
34. Sharma, M, Korytnyk, W, A general and convenient method for synthesis of 6-fluoro-6-deoxyhexoses, *Tetrahedron Lett.*, 18, 573–576, 1977.
35. Sharma, R A, Kavai, I, Fu, Y L, Bobek, M, Synthesis of Image-difluorosaccharides, *Tetrahedron Lett.*, 18, 3433–3436, 1977.
36. Penglis, A A E, Fluorinated carbohydrates, *Adv. Carbohydr. Chem. Biochem.*, 38, 195–285, 1981.
37. Welch, J T, Tetrahedron report number 221: advances in the preparation of biologically active organofluorine compounds, *Tetrahedron*, 43, 3123–3197, 1987.

38. Card, P J, Reddy, G S, Fluorinated carbohydrates. 2. Selective fluorination of gluco- and mannopyranosides. Use of 2-D NMR for structural assignments, *J. Org. Chem.*, 48, 4734–4743, 1984.
39. Hanessian, S, The reaction of *O*-benzylidene sugars with *N*-bromosuccinimide, *Carbohydr. Res.*, 2, 86–88, 1966.
40. Hanessian, S, Deoxy sugars, *Adv. Carbohydr. Chem. Biochem.*, 23, 143–207, 1966.
41. Hanessian, S, Some approaches to the synthesis of halodeoxy sugars, *Adv. Chem. Ser.*, 74, 159–201, 1968.
42. Hanessian, S, Plessas, N R, Reaction of *O*-benzylidene sugars with *N*-bromosuccinimide. II. Scope and synthetic utility in the methyl 4,6-*O*-benzylidenehexopyranoside series, *J. Org. Chem.*, 34, 1035–1044, 1969.
43. Failla, D L, Hullar, T L, Siskin, S B, Selective transformation of *O*-benzylidene acetals into ω-bromo-substituted benzoate esters, *Chem. Commun.*, 716–717, 1966.
44. Hullar, T L, Siskin, S B, Facile bromination by *N*-bromosuccinimide of benzylidene acetals of carbohydrates. Application to the synthesis of 2,6-imino carbohydrates (substituted 2,5-oxazabicyclo[2.2.2] octanes), *J. Org. Chem.*, 35, 225–228, 1970.
45. Brimacombe, J S, Ching, O A, Stacey, M, A new synthesis of mycinose (6-deoxy-2,3-di-*O*-methyl-D-allose), *J. Chem. Soc. C.*, 197–198, 1969.
46. Howarth, G B, Szarek, W A, Jones, J K N, The synthesis of D-arcanose, *Carbohydr. Res.*, 7, 284–290, 1968.
47. Hanessian, S, Methyl 4-*O*-benzoyl-6-bromo-6-deoxy-hexopyranosides, *Methods Carbohydr. Chem.*, 6, 183–189, 1969.
48. Horton, D, Weckerle, W, A preparative synthesis of 3-amino-2,3,6-trideoxy-D-*lyxo*-hexose (daunosamine) hydrochloride from D-mannose, *Carbohydr. Res.*, 44, 227–240, 1975.
49. Hanessian, S, Staub, A P A, Applications diverses de reactifs d'arrachement d'hydrure aux glucides, *Tetrahedron Lett.*, 14, 3551–3554, 1973.
50. Somsák, L, Ferrier, R J, Radical-mediated brominations at ring positions of carbohydrates, *Adv. Carbohydr. Chem. Biochem.*, 49, 37–92, 1991.
51. Meyer, W, Synthese der 3-amino-3-desoxy- und 3,6-diamino-3,6-didesoxy-D-glucose, *Chem. Ber.*, 101, 3802–3807, 1968.
52. Brimacombe, J S, Bryan, J G H, Husain, A, Stacey, M, Tolley, M S, The oxidation of some carbohydrate derivatives, using acid anhydride-methyl sulphoxide mixtures and the Pfitzner–Moffatt reagent. Facile synthesis of 3-acetamido-3-deoxy-D-glucose and 3-amino-3-deoxy-D-xylose, *Carbohydr. Res.*, 3, 318–324, 1967.
53. Williams, D T, Jones, J K N, A new synthesis of 3-acetamido-3-deoxy-D-glucose, *Can. J. Chem.*, 45, 7–9, 1967.
54. Hill, S, Hough, L, Richardson, A C, Nucleophilic replacement reactions of sulphonates: Part I. The preparation of derivatives of 4,6-diamino-4,6-dideoxy-D-glucose and -D-galactose, *Carbohydr. Res.*, 8, 7–18, 1968.
55. Guthrie, R D, Liebmann, J A, Azidolysis of methyl 2,3-anhydro-4,6-*O*-benzylidene-D-glycosides, *Carbohydr. Res.*, 33, 355–358, 1974.
56. Wolfrom, M L, Bernsmann, J, Horton, D, Synthesis of amino sugars by reduction of hydrazine derivatives. 5-Amino-3,6-anhydro-5-deoxy-L-idose derivatives, *J. Org. Chem.*, 27, 4505–4509, 1967.
57. Sugihara, J M, Teerlink, W J, MacLeod, R, Dorrence, S M, Springer, C H, Direct syntheses of some cyano and nitro derivatives of carbohydrates by nucleophilic displacement, *J. Org. Chem.*, 28, 2079–2082, 1968.
58. Lindberg, B, Svensson, S, Methyl 6-deoxy-6-nitro-D-glucopyranosides, *Acta Chem. Scand.*, 21, 299–300, 1967.
59. Magerlein, B J, Lincomycin. X. The chemical synthesis of lincomycin, *Tetrahedron Lett.*, 11, 33–36, 1970.
60. Baer, H H, The nitro sugars, *Adv. Carbohydr. Chem. Biochem.*, 24, 67–138, 1969.
61. Richardson, A C, The nitro sugars, *Int. Rev. Sci.*, 7(Series 2), 131–147, 1976.
62. Viaud, M C, Rollin, P, Zinc azide mediated Mitsunobu substitution. An expedient method for the one-pot azidation of alcohols, *Synthesis*, 130–132, 1990.
63. Mitsunobu, O, Takizawa, S, Morimato, H, Benzoylamination of uridine derivatives, *J. Am. Chem. Soc.*, 98, 7858–7859, 1976.

64. Brandstetter, H H, Zbiral, E, Zur aktivierung partiell silylierter kohlenhydrate mittels triphenylphosphan/azodicarbonsaureester, *Helv. Chim. Acta*, 61, 1832–1841, 1978.
65. Robertson, G J, Griffith, C F, The conversion of derivatives of glucose into derivatives of altrose by simple optical inversions, *J. Chem. Soc.*, 1193–1201, 1935.
66. Meyer, W, Eine einfache synthese der 3-amino-3-desoxy-D-glucose (kanosamin), *Angew. Chem.*, 17, 1023 1966.
67. Roth, W, Pigman, W, D-Glucal and 6-deoxy-L-glucal, *Methods Carbohydr. Chem.*, 2, 405–408, 1963.
68. Rao, D R, Lerner, L M, Dehydrohalogenation of glycosyl halides with 1,8-diazabicyclo[5.4.0]undec-7-ene, *Carbohydr. Res.*, 22, 345–350, 1972.
69. Horton, D, Thomson, J K, Tindall, C G Jr., Unsaturated sugars via cyclic thionocarbonates, cyclic orthoformates, and disulfonic ester intermediates, *Methods Carbohydr. Chem.*, 6, 297–301, 1972.
70. Cook, A G, *Enamines: Synthesis, Structure and Reactions*, 2nd ed., Marcel Dekker, New York, 1987.
71. Dyke, S F, *The Chemistry of Enamines*, 1st ed., Cambridge University Press, New York, 1973.
72. Butterworth, R F, Overend, W G, Williams, N R, A new route to branched-chain sugars by *C*-alkylation of methyl glucopyranosiduloses, *Tetrahedron Lett.*, 28, 3239–3242, 1968.
73. Umezawa, S, Tsuchiya, T, Okazaki, Y, Synthesis of 3,4-dideoxy-3-enosides and the corresponding 3, 4-dideoxy sugars, *Bull. Chem. Soc. Jpn.*, 44, 3494 1977.
74. Hanessian, S, Bargiotti, A, LaRue, M, A mild and stereospecific conversion of vicinal diols into olefins, *Tetrahedron Lett*, 19, 737–740, 1978.
75. Eastwood, F W, Harrington, K J, Josan, J S, Pura, J L, The conversion of 2-dimethylamino-1,3-dioxolans into alkenes, *Tetrahedron Lett.*, 11, 5223–5224, 1970.
76. Kiss, J, Burkhardt, F, β-Eliminativer abbau bei 4-*0*-substituierten hexopyranosiduronat-derivaten, *Helv. Chim. Acta*, 53, 1000–1011, 1970.
77. Llewellyn, J W, Williams, J M, The synthesis of methyl 4-deoxy-β-D-*threo*-hex-4-enopyranoside, *Carbohydr. Res.*, 22, 221–224, 1972.
78. Fischer, P, Enol ethers — structure, synthesis and reactions, *Chem Ethers, Crown Ethers, Hydroxyl Groups Sulphur Analogues*, 2, 761–820, 1980.
79. Helferich, B, Himmen, E, Uber neue dicarbonyl-zucker, *Chem. Ber.*, 62, 2136–2141, 1929.
80. Nayak, U G, Whistler, R L, Nucleophilic displacement in 1,2:5,6-di-*O*-isopropylidene-3-*O*-(*p*-tolylsulfonyl)-α-D-glucofuranose, *J. Org. Chem.*, 34, 3819–3822, 1969.
81. Hill, J, Hough, L, Richardson, A C, Nucleophilic replacement reactions of sulphonates: Part II. The synthesis of derivatives of 4,6-dithio-D-galactose and D-glucose, and their conversion into 4,6-D-*xylo*-hexose, *Carbohydr. Res.*, 8, 19–28, 1968.
82. Haskell, T H, Woo, P W K, Watson, D R, Synthesis of deoxy sugar. Deoxygenation of an alcohol utilizing a facile nucleophilic displacement step, *J. Org. Chem.*, 42, 1302–1305, 1977.
83. Christensen, J E, Goodman, L, Preparation of some *trans*-aminomercaptofuranose sugars, *J. Org. Chem.*, 28, 2995–2999, 1963.
84. Yamada, M, Sotoya, K, Sakakibara, T, Takamoto, T, Sudoh, R, Studies on *N*-alkyl-2(1*H*)-pyridothione. 1. A new synthetic method for thiols, *J. Org. Chem.*, 42, 2180–2182, 1977.
85. Nakagawa, I, Hata, T, A convenient method for the synthesis of 5′-*S*-alkylthio-5′-deoxyribonucleosides, *Tetrahedron Lett.*, 16, 1409–1412, 1975.
86. Nayak, U G, Whistler, R L, Nucleophilic displacement in 1,2:5,6-di-*O*-isopropylidene-3-*O*-(*p*-tolylsulfonyl)-α-D-glucofuranose, *J. Org. Chem.*, 34, 3819–3822, 1969.
87. Anderson, C D, Goodman, L, Baker, B R, Potential anticancer agents XVI. Synthesis of 2-deoxy-β-D-ribofuranosides via the 2,3-episulfonium ion approach, *J. Am. Chem. Soc.*, 81, 898–902, 1959.
88. Gero, S D, Guthrie, R D, A simple synthesis of methyl 4-deoxy-α-D-*xylo*-hexopyranoside ("methyl 4-deoxy-α-D-glucoside"), *J. Chem. Soc. C.*, 1761–1762, 1967.
89. Smith, M B, March, J, *March's Advanced Organic Chemistry: Reactions, Mechanisms, and Structure*, 5th ed., Wiley-Interscience, New York, 2001.
90. King, R D, Overend, W G, Aspects of the p.m.r. spectra of branched-chain methyl glycopyranosides, *Carbohydr. Res.*, 9, 423–428, 1969.
91. Overend, W G, White, A C, Williams, N R, Branched-chain sugar: Part XII. Branched-chain sugars derived from methyl 2,3-*O*-isopropylidene-β-L-*erythro*-pentopyranosid-4-ulose, and a synthesis of L-apiose, *Carbohydr. Res.*, 15, 185–195, 1970.

92. Paulsen, H, Sinnwell, V, Stadler, P, Synthesis of branched carbohydrates with aldehyde side-chains. Simple synthesis of L-streptose and D-hamamelose, *Angew. Chem. Int. Ed. Engl.*, 11, 149–150, 1972.
93. Hanessian, S, Dextraze, P, Carbanions in carbohydrate chemistry: novel methods for chain extension and branching, *Can. J. Chem.*, 50, 226–232, 1972.
94. Hicks, D R, Fraser-Reid, B, The 2- and 3-*C*-methyl derivatives of methyl 2,3-dideoxy-α-D-*erythro*-hex-2-enopyranosid-4-ulose, *Can. J. Chem.*, 53, 2017–2023, 1974.
95. Yunker, M B, Plaumann, D E, Fraser-Reid, B, The stereochemistry of conjugate addition of lithium dialkyl cuprate reagents to some carbohydrate α-enones, *Can. J. Chem.*, 55, 4002–4009, 1977.
96. Baer, H H, Ong, K S, Raeactions of nitro sugars. IX. The synthesis of branched-chain dinitro sugars by Michael addition, *Can. J. Chem.*, 46, 2511–2517, 1968.
97. Jordaan, J H, Smedley, S, Synthesis of branched-chain amino sugars via a spiro-epoxide derivative of 2-amino-2-deoxy-D-glucose, *Carbohydr. Res.*, 16, 177–183, 1971.
98. Meyer, W, Kamprath-Scholtz, U, Synthese von 2,3-[carboxymethylen]- und 2,3-[2-hydroxy-athyliden]-2,3-didesoxy-D-mannose — monosacchariden mit einem cyclopropanring, *Chem. Ber.*, 105, 673–685, 1972.
99. Whistler, R L, BeMiller, J N, "α"-D-Glucosaccharino-1,4-lactone, *Methods Carbohydr. Chem.*, 2, 484–485, 1963.
100. Kuhn, R, Fischer, H, Uber D-threosamin und D-erythrosamin, *Liebigs Ann. Chem.*, 641, 152–160, 1961.
101. Saeki, H, Ohki, E, Synthesis of *N*-acetyllincosamine, *Chem. Pharm. Bull.*, 18, 789–802, 1970.
102. Sowden, J C, α-L-Glucose and L-mannose, *Methods Carbohydr. Chem.*, 1, 132–135, 1962.
103. Gigg, J, Gigg, R, Warren, C D, A synthesis of phytosphingosines from D-glucosamine, *J. Chem. Soc. C.*, 1872–1876, 1960.
104. Rosenthal, A, Sprinzl, M, Branched-chain sugar nucleosides. IV. 9-(3-deoxy-3-*C*-"hydroxymethyl"-β(and α)-D-allofuranosyl and ribofuranosyl)adenine, *Can. J. Chem.*, 47, 4477–4481, 1969.
105. Rosenthal, A, Catsoulacos, P, Synthesis of branched-chain sugars by the Wittig reaction, *Can. J. Chem.*, 46, 2868–2872, 1968.
106. Rhoads, S J, Raulins, N R, The Claisen and Cope rearrangements, *Org. React.*, 22, 1–252, 1975.
107. Bennett, G B, The Claisen rearrangement in organic synthesis; 1967 to January 1977, *Synthesis*, 589–606, 1977.
108. Corey, E J, Shibasaki, M, Simple, J K, Stereocontrolled synthesis of thromboxane B2 from D-glucose, *Tetrahedron Lett.*, 18, 1625–1626, 1977.
109. Brimacombe, J S, How, M J, Pneumococcus type V capsular polysaccharide: characterization of pneumosamine as 2-amino-2,6-dideoxy-L-talopyranose, *J. Chem. Soc.*, 5037–5040, 1962.
110. Lawton, B T, Szarek, W A, Jones, J K N, Synthesis of deoxy and aminodeoxy sugars by way of chlorodeoxy sugars, *Carbohydr. Res.*, 15, 397–402, 1970.
111. Arita, H, Ueda, N, Matsushima, Y, The reduction of chlorodeoxy sugars by tributyltin hydride, *Bull. Chem. Soc. Jpn.*, 45, 567–569, 1972.
112. Kovar, J, Dienstbierova, V, Jary, J, Uber aminozucker XII. Vergleich der substution sekundärer methansulfogruppen duch stickstoffhaltige gruppen, *Collect Czech Chem. Commun.*, 23, 2498–2503, 1976.
113. Prins, D A, Reduction of sugar epoxides to desoxysugars, *J. Am. Chem. Soc.*, 70, 3955–3957, 1948.
114. Rembarz, G, Kristallisierte 3-desoxy-D-mannose, *Chem. Ber.*, 93, 622–625, 1960.
115. Pratt, J W, Richtmyer, N K, Crystalline 3-deoxy-α-D-*ribo*-hexose. Preparation and properties of 1,6-anhydro-3-deoxy-α-D-*arabino*-hexopyranose, 1,6-anhydro-3-deoxy-β-D-*ribo*-hexopyranose and related compounds, *J. Am. Chem. Soc.*, 79, 2597–2600, 1957.
116. Hanessian, S, Lavellee, P, Total synthesis of (+)-thromboxane B_2 from D-glucose. A detailed account, *Can. J. Chem.*, 59, 870–877, 1981.
117. Ball, D H, Carey, F A, Klundt, I L, Long, L Jr., Apiose. II. 1,2:3,5-di-*O*-isopropylidene-α-D-apio-furanose and stereoselective syntheses of 3-deoxy-1,2-*O*-isopropylidene-α-D-apio-L-furanose and 3, 5-anhydro-1,2-*O*-isopropylidene-α-D-apio-L-furanose, *Carbohydr. Res.*, 10, 121–128, 1969.
118. Whistler, R L, BeMiller, J N, α-L-Lyxose, *Methods Carbohydr. Chem.*, 1, 79–80, 1962.
119. Hough, L, Taha, M I, The disulphones derived by oxidation of 2-amino-2-deoxy-D-glucose diethyl dithioacetal hydrochloride and its *N*-acetyl derivative with peroxypropionic acid, *J. Chem. Soc.*, 3564–3572, 1957.

120. Coxon, B, Hough, L, 3-Acetamido-3-deoxy-D-allose diethyl dithioacetal and its oxidation by peroxypropionic acid, *J. Chem. Soc.*, 1643–1649, 1961.
121. MacDonald, D L, β-D-Arabinose, *Methods Carbohydr. Chem.*, 1, 73–75, 1962.
122. Wolfrom, M L, Hanessian, S, The reaction of free carbonyl sugar derivatives with organometallic reagents. I. 6-Deoxy-L-idose and derivatives, *J. Org. Chem.*, 27, 1800–1804, 1962.
123. Whistler, R L, BeMiller, J N, "α"-D-Isosaccharino-1,4-lactone, *Methods Carbohydr. Chem.*, 2, 477–479, 1963.
124. Kuhn, R, Baschang, G, Uberfuhrung von 2-amino-2-desoxy-hexosen und -pentosen, *Liebigs Ann. Chem.*, 636, 164–173, 1960.
125. Wolfrom, M L, Chakravarty, P, Horton, D, Amino derivatives of starches. 2,6-Diamino-2,6-dideoxy-D-mannose dihydrochloride, *J. Org. Chem.*, 30, 2728–2731, 1965.
126. Hodge, J E, Fisher, B E, Amadori rearrangement products, *Methods Carbohydr. Chem.*, 2, 99–107, 1963.

7 Strategies towards *C*-Glycosides

Daniel E. Levy

CONTENTS

7.1 INTRODUCTION

Carbohydrates have long been a source of scientific interest because of their abundance in nature, and to the synthetic challenges posed by their polyhydroxylated structures. However, the commercial use of carbohydrates has been significantly limited by the hydrolytic lability of the glycosidic bond. With the advent of *C*-glycosides, this limitation promises to be overcome, thus paving the way for a new generation of carbohydrate-based products.

7.1.1 Definition and Nomenclature of *C*-Glycosides

C-glycosides occur when a carbon atom replaces the *exo*-oxygen atom of the glycosidic bond for any given *O*-glycoside. Additionally, for any furanose or pyranose, any exocyclic atoms linked to one or two ethereal carbons tetrahedrally will be only carbon or hydrogen, and there will be at least one carbon and one OH group on the ring. In accordance with these statements, structural definitions of *C* and *O*-glycosides are illustrated in Figure 7.1.

With the relatively minor structural modification involved in the transition from *O*-glycosides to *C*-glycosides comes the more involved question of how to name these compounds. Figure 7.2 shows a series of *C*-glycosides with their corresponding names. Compound **1** derives its name from the longest continuous chain of carbons. Since this is a seven carbon sugar, the name will be derived from heptitol. As the heptitol is cyclized between C-2 and C-6, it is designated as 2,6-anhydro. Finally, carbons 2–5 and carbons 6–7 bear the D-*gulo* and D-*glycero* configurations, respectively. By comparison, compound **4** is a 2,6-anhydro heptitol bearing the D-*glycero* configuration at carbons 6–7. However, carbons 2–5 bear the D-*ido* configuration and the molecule is now designated as such.

7.1.2 O-Glycosides vs. *C*-Glycosides: Comparisons of Physical Properties, Anomeric Effects, H-Bonding Abilities, Stabilities and Conformations

The structural and chemical similarities prevalent between *C* and *O*-glycosides are illustrated in Table 7.1 and are summarized as follows. Given that the bond lengths, van der Waals radii, electronegativities and bond rotational barriers are very similar between *O*- and *C*-glycosides,

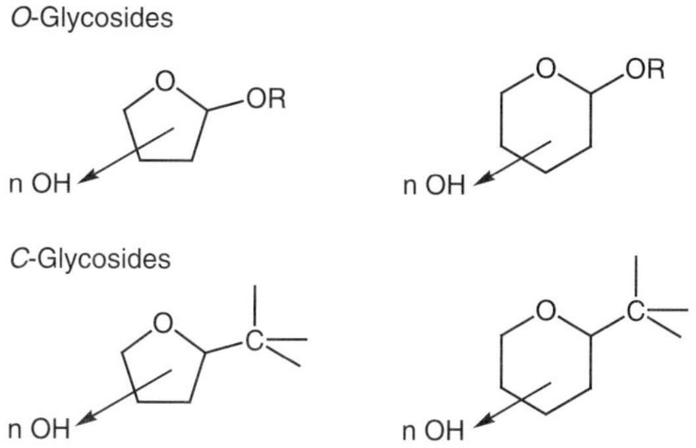

O-Glycosides
O
OR
n OH
O
OR
n OH
C-Glycosides
O
C
n OH
O
C
n OH

FIGURE 7.1 Definition of *C*-glycosides.

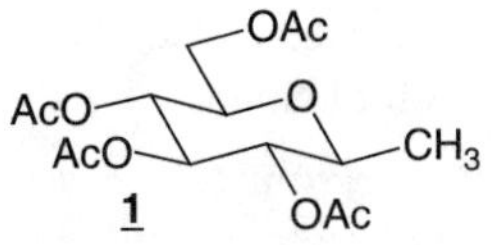

3,4,5,7-Tetra-*O*-acetyl-2,6-anhydro-1-deoxy-D-*glycero*-D-*gulo*-heptitol

- D-*glycero*: C_6 and C_7
- Anhydro: C_2 and C_6
- C_5,C_4,C_3,C_2: D-*gulo*stereochemistry
- Heptitol: seven carbons (1-deoxy)

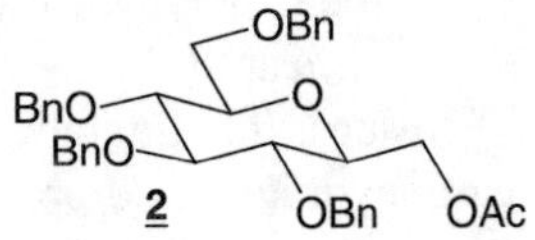

1-*O*-Acetyl-2,6-anhydro-3,4,5,7-tetra-*O*-benzyl-D-*glycero*-D-*gulo*-heptitol

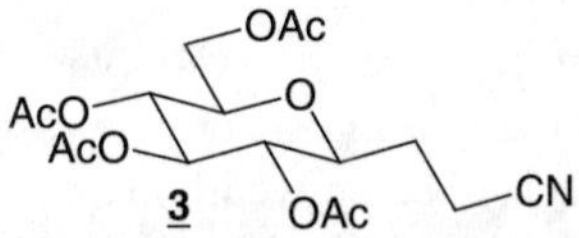

5,6,7,9-Tetra-*O*-acetyl-4,8-anhydro-2,3-dideoxy-D-*glycero*-D-*gulo*-nonanitrile

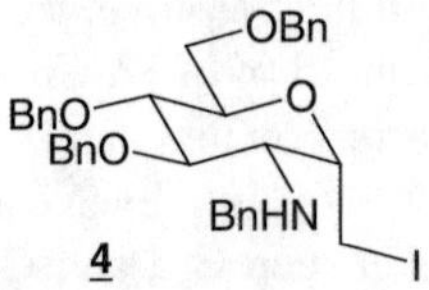

2,6-Anhydro-4,5,7-tri-*O*-benzyl-3-(N-benzylamino)-1,3-dideoxy-1-iodo-D-*glycero*-D-*ido*-heptitol

FIGURE 7.2 Representative *C*-glycosides and their nomenclature.

TABLE 7.1
Physical Properties of *O*- and *C*-Glycosides

	O-Glycosides	*C*-Glycosides
Bond length	O–C = 1.43 Å	O–C = 1.54 Å
Van der Waals radius	O = 1.52 Å	O = 2.0 Å
Electronegativity	O = 3.51	C = 2.35
Dipole moment	C–O = 0.74 D	C–C = 0.3 D
Bond rotational barrier	CH_3–O–CH_3 = 2.7 kcal/mol	CH_3–CH_3 = 2.88 kcal/mol
H-Bonding	Two	None
Anomeric effect	Yes	No
Exoanomeric effect	Yes	No
Stability	Cleaved by acids and enzymes	Stable to acids and enzymes
Conformation	$C_{1'}$–$C_{2'}$ antiperiplanar to O_1–C_1	$C_{1'}$–$C_{2'}$ antiperiplanar to C_1–C_2

we find the largest difference between physical constants in the dipole moments. However, although minor differences do exist, the conformations of both *O*- and *C*-glycosides are represented by similar antiperiplanar arrangements.

The major difference between *C*- and *O*-glycosides is found within chemical reactivities. Not only are *C*-glycosides absent of anomeric effects, they are also stable to acid hydrolysis and are incapable of forming hydrogen bonds.

Regarding conformations, it was determined that, as with *O*-glycosides, the $C_{1'}-C_{2'}$ bond is antiperiplanar to the C_1-C_2 bond in the corresponding *C*-glycosides [1,2]. Moreover, C_2 substituents were found to have no effect on *C*-glycosidic bond conformations and, considering 1,3-diaxial-like interactions, the *C*-aglyconic region will distort first. In summary, *C*-glycosides were found to be not conformationally rigid with conformations predictable based on steric effects. In fact, Kishi et al. [3] demonstrated that "it is possible to predict the conformational behavior around the glycosidic bond of given *C*-saccharides by placing the $C_{1'}-C_{2'}$ bond antiperiplanar to the $C_\alpha-C_n$ bond and then focusing principally on the steric interaction around the nonglycosidic bond." This postulate was demonstrated utilizing a diamond lattice to arrange the glycosidic conformations of the *C*-trisaccharide analogs of the blood group determinants [4,5].

7.1.3 Natural Occurring *C*-Glycosides

While *C*-glycoside chemistry has recently attracted much attention, *C*-glycosides were not invented by chemists. Many examples of *C*-glycosides (and *C*-nucleosides) have been isolated from nature. Some naturally occurring structures, shown in Figure 7.3, include the aloins [6–10], anthraquinone/flavone-based structures [11,12], showdomycin and B anformycin [13,14].

7.1.4 *C*-Glycosides as Stable Pharmacophores

With the rapid development of the chemistry of *C*-glycosides, the pharmaceutical and biotechnology industries have recognized that *C*-glycoside analogs of biologically active carbohydrates may be studied as stable pharmacophores. Currently, glycosides and saccharides are used in a variety of applications, ranging from foodstuffs to components of nucleic acids and cell surface glycoconjugates [15–18]. Realizing the extent to which these types of compounds are involved in everyday life, the advantages of using *C*-glycosides as drug candidates become

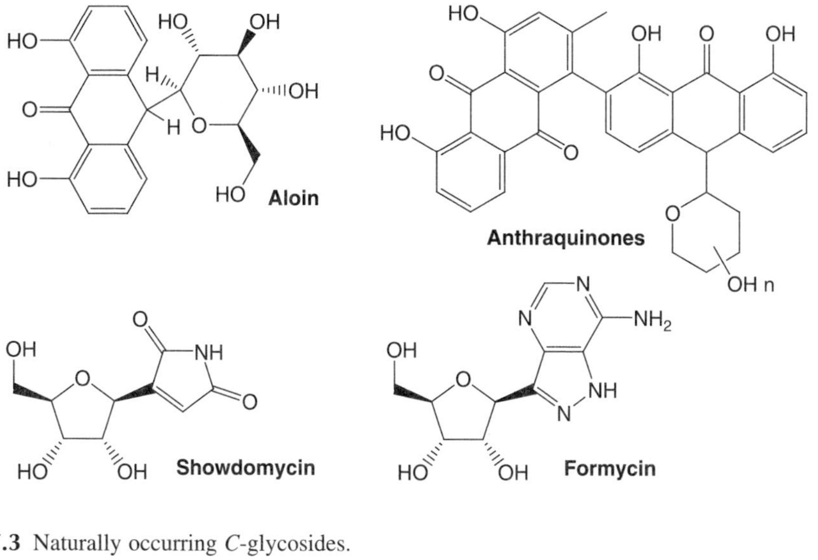

FIGURE 7.3 Naturally occurring *C*-glycosides.

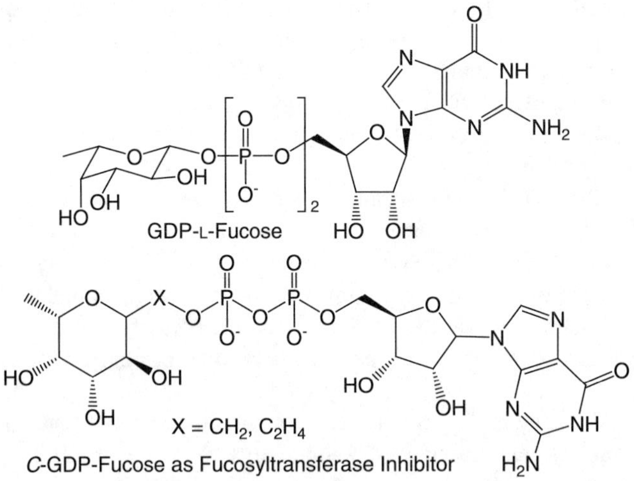

GDP-L-Fucose

$X = CH_2, C_2H_4$

C-GDP-Fucose as Fucosyltransferase Inhibitor

FIGURE 7.4 *C*-Glycosidic fucosyltransferase inhibitors.

apparent. These advantages include the stereochemically stable nature of *C*-glycosides and that chiral starting materials are readily available (sugar units). Moreover, *C*-glycosides as pharmacophores may yield novel enzyme inhibitors with structures that may be difficult or impossible to construct utilizing standard carbohydrate chemistry. This philosophy was extensively studied with respect to the inflammation pathway involving neutrophil recruitment.

Glycosyltransferases assemble polysaccharides from their monomeric units by adding activated nucleotide sugars to the growing carbohydrate chain [19]. The chemistry of glycosyltransferases is well known [20–22] and is illustrated in the biosynthesis of sialyl Lewis^x [23], an essential contributor to the cell adhesion process. In fact, without both the sialic acid and fucose components, the sialyl Lewis^x remnant is incapable of binding to selectins and effecting cell adhesion necessary for leukocyte extravasation [24–29].

Based on the above observations, inhibition of a key step in the synthesis of sialyl Lewis^x would provide a novel strategy towards the development of anti-inflammatory agents. This could be accomplished through the inhibition of α-1,3-fucosyltransferase, the enzyme responsible for transferring fucose to the sialyl Lewis^x epitope [30,31]. The natural substrate for this enzyme is GDP-L-fucose, the *C*-glycosidic analog of which was prepared by Luengo and Gleason [32] (Figure 7.4).

In addition to the inflammation pathway, the use of *C*-glycosides to inhibit biological processes was studied in areas ranging from other glucosidases including cellobiase [33], β-D-galactosidase [34], and others [35]. Additionally, inhibition of phosphorylases [36], modulation of immune pathways [37], lectin binding [38], and the blocking of pathogenesis [39–41] have also been addressed. Thus, through the exploration of novel and stable *C*-glycosidic structures, useful entities for application to and treatment of a wide range of immunological and pathological disorders may be identified.

7.2 SYNTHESIS OF *C*-GLYCOSIDES VIA ELECTROPHILIC SUBSTITUTIONS

One of the most widely used methods for the formation of *C*-glycosides involves electrophilic species derived from the sugar component. Some representative examples, shown in Figure 7.5, include oxonium compounds, lactones, enones, carbenes, and enol ethers. Once formed, nucleophiles can be induced to react with these reactive species, thus providing tremendous diversity in

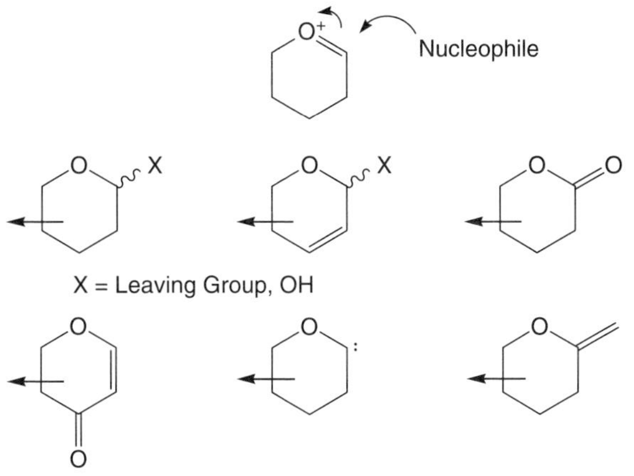

FIGURE 7.5 Examples of electrophilic substitutions.

the types of C-glycosides accessible. This section focuses on various approaches that are made use of in the formation of C-glycosides via the addition of nucleophiles to carbohydrate-derived electrophiles.

7.2.1 Anomeric Activating Groups and Stereoselectivity

When considering the formation of C-glycosides by electrophilic substitution, several direct comparisons can be made to the similar reactions utilized in the preparation of O-glycosides. Specifically, any activating group that is adequate for the formation of O-glycosides may be used in the preparation of C-glycosides. This rationale arises from the fact that C-glycosidations via electrophilic substitution occur through mechanisms similar to those observed in the corresponding O-glycosidations. Additionally, since C-glycoside technology generally utilizes slightly more stringent conditions, less reactive groups such as OAc and OTMS may also be used. Regarding stereochemical outcome, in the case of C-glycosides, because no anomeric effect exists, the stereochemistry is completely dictated by other stereoelectronic effects. Moreover, it should be noted that neighboring group participation is not a predominant factor. The major consequence of these observations is that axially (generally α) substituted C-glycosides are far more accessible than the corresponding equatorial (generally β) isomers.

7.2.2 Cyanation Reactions

Direct cyanations to sugars or sugar derivatives, represented in Figure 7.6, affect one-carbon extensions. The resulting nitriles are then available for a number of further reactions including cycloadditions, hydrolyses to acids, and reduction to aminomethyl groups. Unfortunately, there are presently no stereospecific methods available for these reactions. Requirements for this type of transformation include utilizing acetates, halides, and imidates as activating groups. Some Lewis acids used to effect these transformations include $SnCl_4$, TMSOTf, TMSCN, Et_2AlCN, and $Hg(CN)_2$. Additionally, both benzylated and acetylated sugars may be used.

FIGURE 7.6 Example of cyanation reactions.

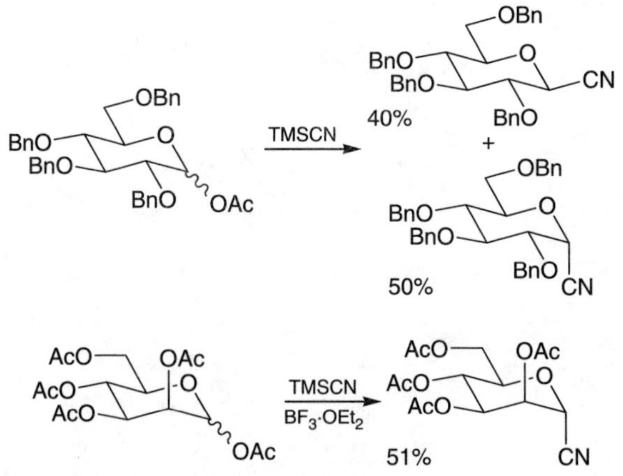

SCHEME 7.1 Cyanations of 1-*O*-acetoxyglycosides.

7.2.2.1 Cyanation Reactions on Activated Glycosyl Derivatives

Among the most basic methods for the formation of *C*-cyanoglycosides involves the reaction of anomeric *O*-acetyl derivatives with TMSCN. Examples of this reaction, shown in Scheme 7.1 [1,42], incorporate the use of both benzyl and acetyl protected sugars. Where benzylated sugars were used, the reaction provided a 1:1 ratio of anomers. However, when mannose pentaacetate was used, the only isolated compound possessed the α configuration. This observation can be explained by the involvement of the adjacent acetate group, as illustrated in Figure 7.7. The result of this intermediate is apparent in that the approach of nucleophiles is only possible from the α face. Similar chemistry was achieved using trichloroacetimidates [2] and glycosyl fluorides [43,44].

7.2.2.2 Cyanation Reactions on Glycals

Whereas the direct use of glycosides is widely useful, glycals have provided more complementary and, in many cases, more stereoselective results. For example, as shown in Scheme 7.2, Grynkiewicz and BeMiller [45] converted tri-*O*-acetyl-glucal to the illustrated cyanoglycoside.

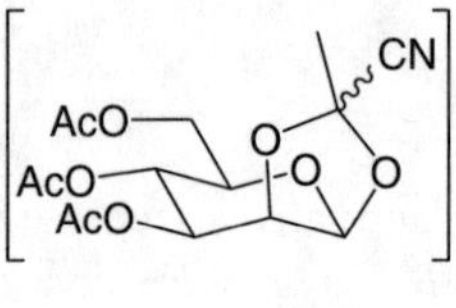

FIGURE 7.7 Cyanations with acetate participation.

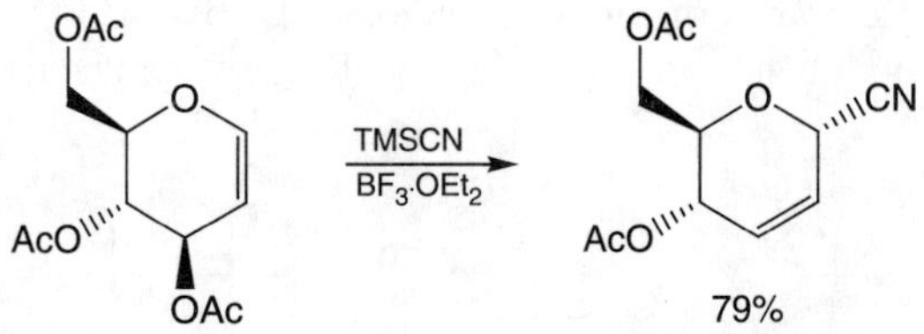

SCHEME 7.2 Cyanations of glycals.

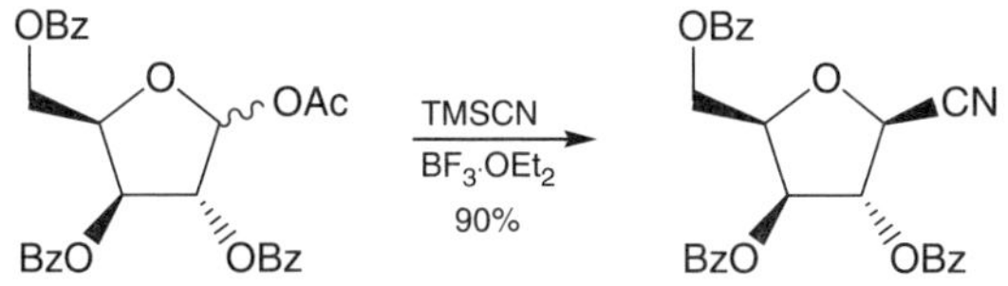

SCHEME 7.3 Cyanations of furanoses.

Similar results were noted with 1-methylglycals [46] and some arylsulfonium intermediate species [47].

7.2.2.3 Cyanation Reactions on Activated Furanosides

To date, only the chemistry regarding the addition of TMSCN to pyranose sugars has been discussed. However, such chemistry is easily applicable to the corresponding furanose forms. These reactions have been demonstrated with benzoyl protected furanoses catalyzed by boron-trifluoride etherate (Scheme 7.3) [1]. Additionally, benzyl protected furanose sugars [48] and furanosyl fluorides [44] are useful. In the latter case, although high yields are available, little or no stereochemical control is observed.

7.2.2.4 Cyanation Reactions with Metallocyanide Reagents

TMSCN is not the only reagent useful for the formation of cyanoglycosides. Complementary and perhaps equally important is the use of aluminum reagents. As applied to pyranose sugars, Grierson et al. [49] demonstrated that the stereochemical outcome of the cyanation reaction with diethylaluminum cyanide was temperature dependent. Thus, as shown in Scheme 7.4, at room temperature, a mixture of α and β isomers were formed while, at reflux, the predominant isomer possessed the α configuration. Additional examples of the use of aluminum reagents applied to the cyanation of glycosyl fluorides [43] are known, albeit having a common side reaction of the formation of the corresponding isocyanoglycosides [50].

7.2.2.5 Other Cyanation Reactions

Cyanoglycosides are accessible from a number of starting materials other than glycosides and glycals. In concluding this section, the formation of cyanoglycosides from vinylogous lactones is explored. As shown in Scheme 7.5, Tatsuta et al. [51] formed 2-deoxy cyanoglycosides utilizing

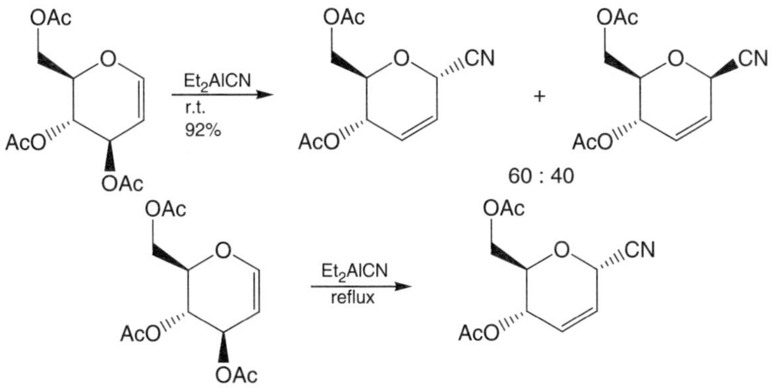

SCHEME 7.4 Glycal cyanations with Et_2AlCN.

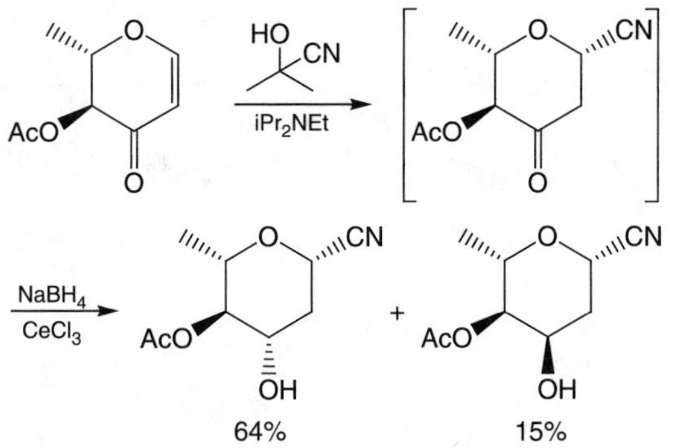

SCHEME 7.5 Vinyligous lactone cyanations with acetone cyanohydrin.

acetone cyanohydrin and Hünig's base. Following hydride reduction of the resulting ketone, a mixture of the 3-hydroxy isomers was isolated.

7.2.2.6 Cyanoglycoside Transformations

With the variety of methods available for the preparation of cyanoglycosides, a brief discussion of the versatility of these compounds is warranted. The relevant chemistry is directly derived from mainstream organic chemistry, and includes reductions of cyanoglycosides to aminomethylglycosides on treatment with lithium aluminum hydride [42]. Additionally, conversion of cyanoglycosides to amidoglycosides [52], carboxyglycosides [35], diazoketoglycosides [35], glycosyl amino acids [53], and various *C*-glycosidic heterocyclic structures [54,55] are known.

7.2.3 Alkylation, Allenylation, Allylation and Alkynation Reactions

Concerning *C*-glycoside chemistry in general, alkylation, allenylation, allylation, and alkynylation reactions, illustrated in Figure 7.8, are among the most useful methods for preparative scales. Moreover, they tend to yield the most valuable synthetic intermediates in that different functional groups of various lengths are easily obtained. As for the carbohydrate starting materials, many different glycosidic activating groups are useful in conjunction with a variety of Lewis acids. These reactions occur satisfactorily in many solvents including acetonitrile, nitromethane, and dichloromethane. Finally, both benzylated and acetylated sugars may be used and the predominant products bear the α configuration.

7.2.3.1 Use of Activated Glycosyl Derivatives and Anionic Nucleophiles

As early as 1974, Hanessian and Pernet [56] demonstrated the use of anion chemistry in the preparation of *C*-glycosides. As shown in Scheme 7.6, acetobromoglucose was treated with the anions of diethylmalonate and dibenzylmalonate, yielding two distinct *C*-glycosides. In the case of the dibenzyl analog, debenzylation, followed by treatment with the Meerwein reagent, gave a

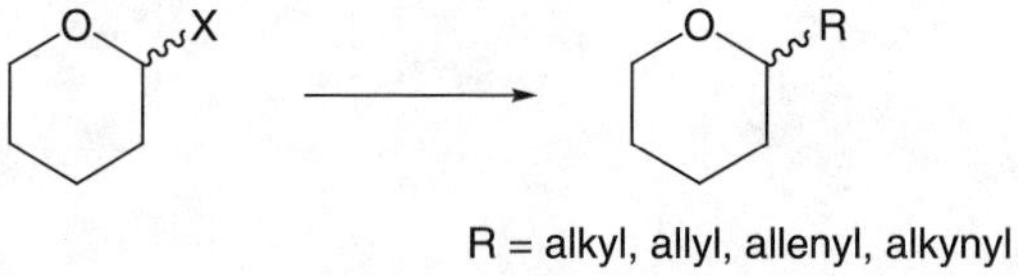

R = alkyl, allyl, allenyl, alkynyl

FIGURE 7.8 Examples of alkylations.

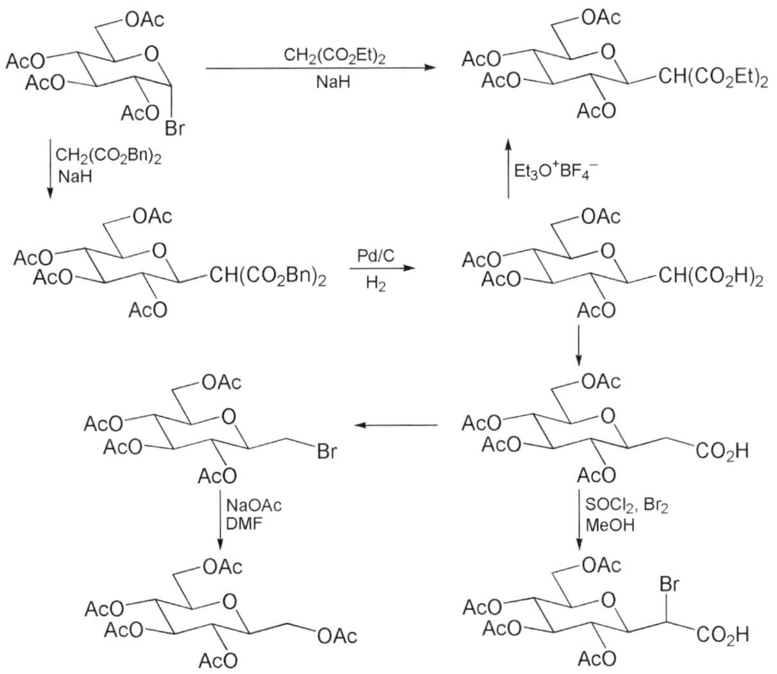

SCHEME 7.6 *C*-Glycosides via anionic nucleophiles.

product identical to that formed from diethylmalonate. Moreover, decarboxylation of the diacid gave the corresponding carboxylic acid which, when subjected to a modified Hundsdiecker reaction [57], yielded the illustrated β-bromopyranoside. Subsequent treatment of this bromide with sodium acetate in dimethylformamide gave the known β-acetoxymethyl-*C*-glycoside, thus demonstrating the accessibility of β anomers. One additional reaction, which demonstrated the feasibility of further functionalizing these *C*-glycosides, involved the α bromination of the previously described monocarboxylic acid.

7.2.3.2 Use of Activated Glycosyl Derivatives and Organometallic Reagents

Expanding upon the use of anionic nucleophiles, one cannot ignore the special properties and reactivities of organometallic reagents such as cuprates, Grignard reagents and organozinc compounds. In the following examples, the sugar units are activated in the form of glycosyl halides, epoxysugars, benzoates, sulfones, thioglycosides or glycals.

Beginning with copper-based reagents, Bihovsky et al. [58] (Scheme 7.7) utilized a variety of cuprates in the preparation of α-*C*-methylglycosides. The yields were relatively poor when acetyl protecting groups were used. However, substantial improvements in the yields were observed with methyl and benzyl protecting groups. In all cases, the α anomer was favored in ratios as high as 20:1. Additional examples applied to 1,2-anhydrosugars (epoxysugars) complement the use of glycosyl halides [59,60].

Complementary to the use of cuprates, Grignard reagents have found a useful place in the technology surrounding *C*-glycosidations. As shown in Scheme 7.8, Shulman et al. [61] utilized allyl and vinyl Grignard reagents to give β-*C*-glycosides in modest yields. Subsequent epoxidation of these products provided potential glucosidase inhibitors. In a later report, Hanessian et al. [62] confirmed the above results and postulated the intermediate epoxide, shown in Scheme 7.9, as an explanation of the stereoselectivity of this chemistry. Additional examples incorporating aryl [63]

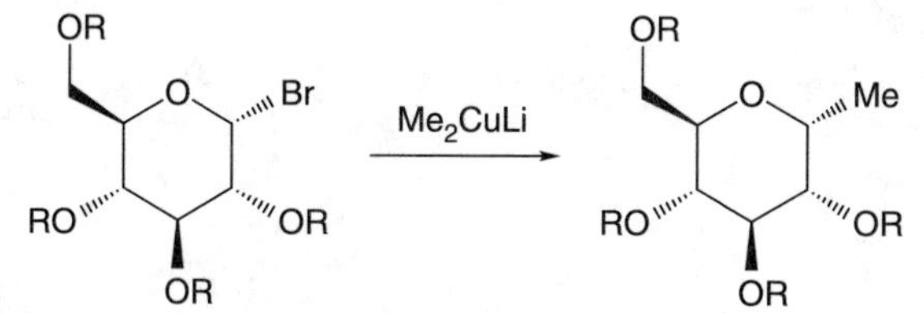

R	Yield (%)
Me	46
Bn	40
Ac	8

SCHEME 7.7 *C*-Glycosidations via cuprate nucleophiles.

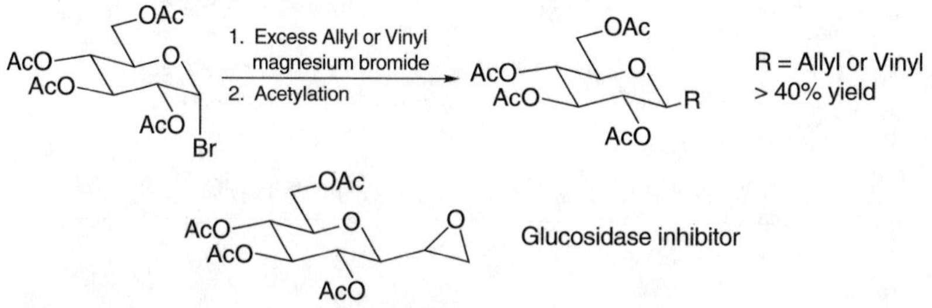

SCHEME 7.8 *C*-Glycosidations via Grignard reagents.

and acetylenic [64] units are known in conjunction with benzoate and sulfone activating groups, respectively.

Finally, the direct use of organozinc compounds is well known in the field of *C*-glycosidations. As shown in Scheme 7.10, Orsini and Pelizzoni [65] reacted acetylated glycals with *tert*-butoxycarbonylmethylzinc bromide. In the case of tri-*O*-acetyl-glucal, the reaction gave a 49% yield with a 2:1 ratio favoring the α anomer. Improved anomeric selectivities were noted when thioglycosides were treated with dialkyl zinc reagents [66].

7.2.3.3 Use of Modified Sugars and Anionic Nucleophiles

Although activating sugars is a common method to modify carbohydrate structures in preparation for further reactions, these changes generally retain the overall nature of the pyranose or furanose

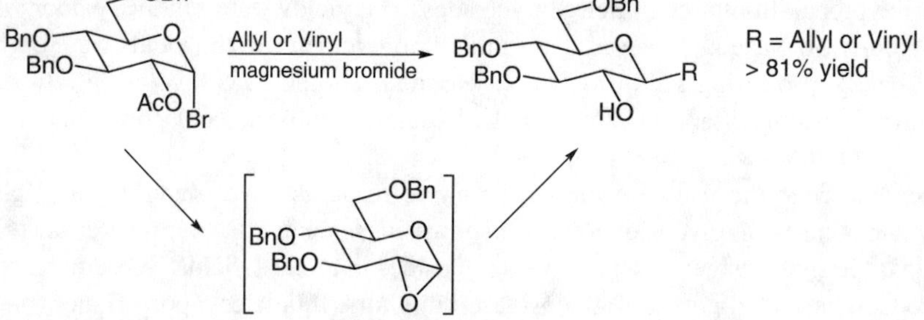

SCHEME 7.9 *C*-Glycosidations via Grignard reagents.

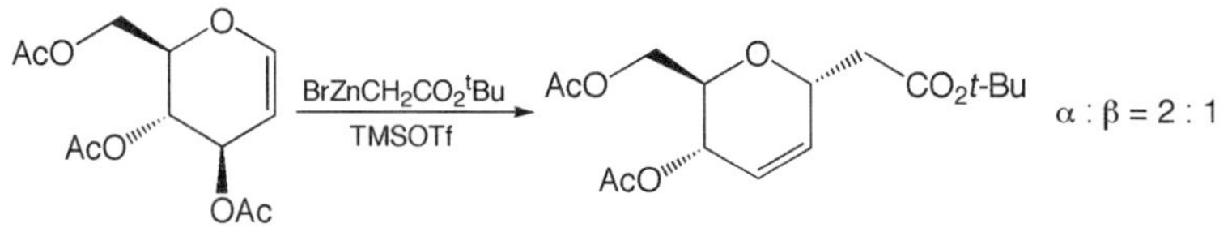

SCHEME 7.10 *C*-Glycosidations via organozinc reagents.

structure in that the inherent hemiacetal functionality is left intact in some masked form. However, there are alternate methods for modifying sugars prior to further reactions. Among the most common is oxidation to sugar lactones. By using sugar lactones as starting materials, organolithium compounds have found utility. As shown in Scheme 7.11, a ribose-derived lactone was treated with benzyloxymethyllithium to give the alkylated hemiacetal. Moreover, after removal of the MOM groups, the compound spontaneously rearranged to the illustrated pyranose form [67]. Although the isolated product is a ketose sugar, it bears some properties in common with *C*-glycosides and, as will be discussed later, can be converted to a true *C*-glycoside under reductive conditions. One should also note that this chemistry is compatible with other organometallic reagents, including Grignards [68].

7.2.3.4 Lewis Acid-Mediated Couplings with Olefins

Anionic nucleophiles provide valuable and versatile routes for the preparation of *C*-glycosides. However, the most extensively used technologies lie within the chemistry of Lewis acid-mediated reactions of carbohydrates with unsaturated hydrocarbons and derivatives thereof. In the next series of examples, reactions of sugars and sugar derivatives with olefins and their silyl-, stannyl-, and aluminum derivatives are discussed.

An early example of the reaction of protected furanoses with olefins is shown in Scheme 7.12. In this study, Cupps et al. [69] utilized tin tetrachloride to effect the formation of a 75% yield of the illustrated *C*-glycoside. This reaction showed little anomeric selectivity. Additionally, the chloride was identified as a byproduct arising from the intermediate carbocation.

Although the previous reaction involved the use of an unactivated olefin, this was apparently not the source of the lack of stereochemical induction. This is illustrated by work reported by Herscovici et al. [70,71], shown in Scheme 7.13, where several unactivated olefins were successfully used to produce α-*C*-glycosides in good to excellent yields. Additional work in this area resulted in a facile route to fused-ring *C*-glycosides [72] similar to the core structures of natural products including the halichrondrins [73], the herbicidins [74–76], and octosyl acid [77].

Although the use of unprotected olefins has proven fruitful in the preparation of *C*-glycosides, the most commonly utilized technology involves the use of silyl compounds. Reactions involving the coupling of sugar derivatives with allyl and acetylenic silanes have been initiated with a number of Lewis acids as well as a wide variety of glycosidic activating groups. The first of these to be addressed is the acetate activating group.

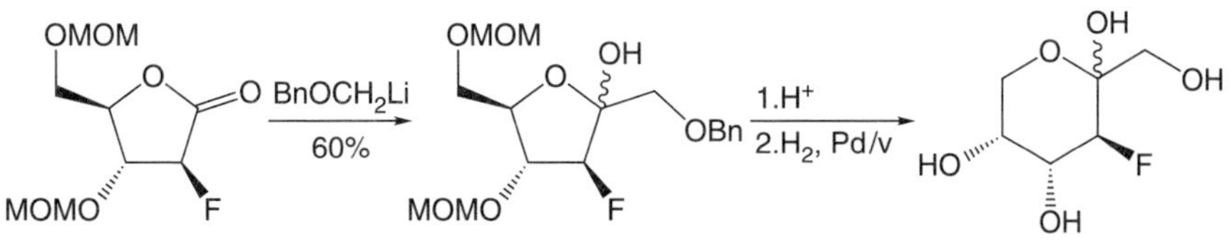

SCHEME 7.11 *C*-Glycosidations via organolithium reagents.

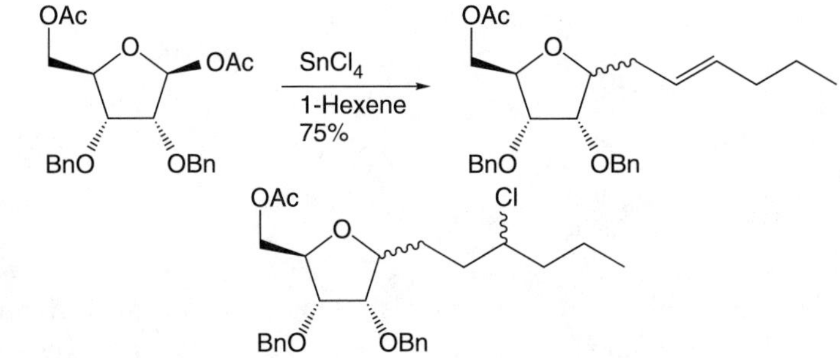

SCHEME 7.12 *C*-Glycosidations with 1-hexene.

An early report demonstrating the generality of Lewis acid-mediated *C*-glycosidations with allylsilanes was reported by Giannis and Sandhoff [78] and is shown in Scheme 7.14. In this study, allylations were catalyzed by borontrifluoride etherate utilizing either dichloroethane or acetonitrile as solvents. As observed, no stereoselectivity was noted in dichloroethane. However, the more polar acetonitrile, capable of coordinating to intermediate oxonium ions, induced fairly high selectivity. An additional reaction demonstrated the applicability of this technology to disaccharides. In a later report, similar observations were noted in the reaction of penta-*O*-acetyl-D-galactose with allyltrimethylsilane in nitromethane utilizing borontrifluoride etherate as the catalyst [52].

In addition to glycosyl acetates, 1-*O*-trifluoroacetyl [79] and 1-*O*-*p*-nitrobenzoyl esters [68], trichloroacetimidates [2], methyl glycosides [80–83], furanose glycosyl acetates [48] and ketose acetates [84] have been successfully modified in this manner. With respect to the use of 1-*O*-*p*-nitrobenzoyl esters, extended olefins [85,86], substituted olefins [87] and allenes [88] were incorporated onto sugar scaffolds.

Having addressed numerous activating groups compatible with the Lewis acid-mediated reaction with silyl-based reagents, we cannot ignore the utility of glycosyl fluorides, nitrates and glycals. This chemistry will be presented in the context of reactions with both allylsilanes and silyl enol ethers.

Beginning with glycosyl fluorides, as shown in Scheme 7.15, Nicolaou et al. [43] utilized the fluoride shown and effected the formation of a variety of *C*-glycosides. The example shown produced high yield with excellent stereoselectivity. Similar results were observed by Araki et al. [89].

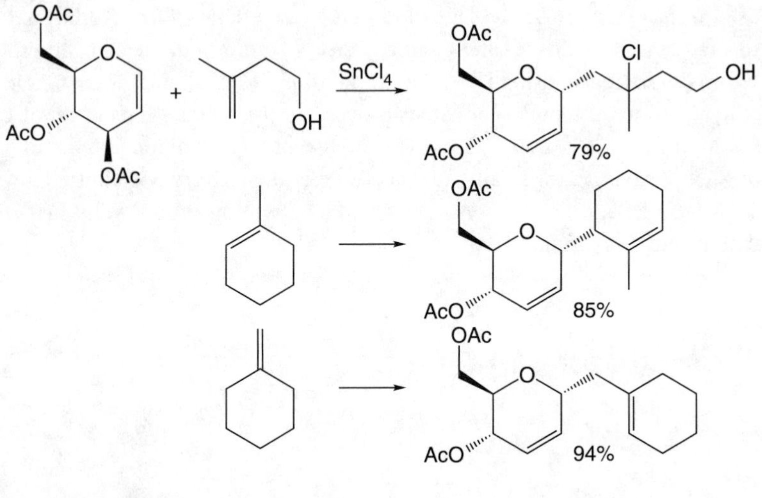

SCHEME 7.13 *C*-Glycosidations with olefins.

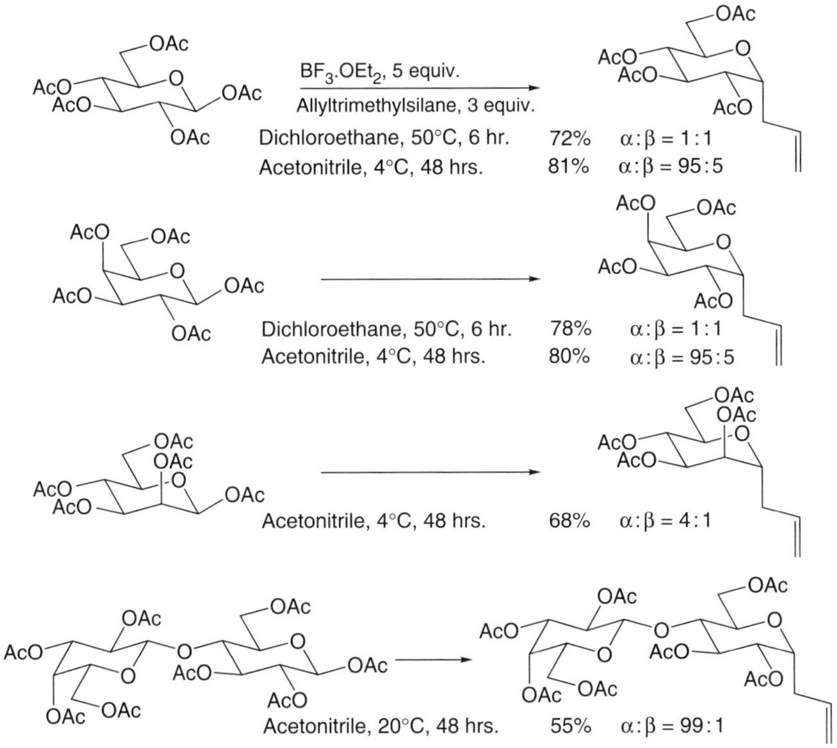

	Conditions	Yield	Selectivity
BF_3.OEt$_2$, 5 equiv. / Allyltrimethylsilane, 3 equiv.	Dichloroethane, 50°C, 6 hr.	72%	α:β = 1:1
	Acetonitrile, 4°C, 48 hrs.	81%	α:β = 95:5
	Dichloroethane, 50°C, 6 hr.	78%	α:β = 1:1
	Acetonitrile, 4°C, 48 hrs.	80%	α:β = 95:5
	Acetonitrile, 4°C, 48 hrs.	68%	α:β = 4:1
	Acetonitrile, 20°C, 48 hrs.	55%	α:β = 99:1

SCHEME 7.14 *C*-Glycosidations with allylsilanes.

Continuing with electron deficient activating groups, the utility of nitrosugars must be acknowledged. As shown in Scheme 7.16, Bertozzi and Bednarski [90] compared the reactivities and outcomes resulting from the variance of the activating group electronegativities. The data show that moving from methoxy to acetoxy to nitrate, both yields and reaction rates increase. The lesson of this study is that if *C*-glycosidations fail, changing to more electronegative anomeric activating groups may, in fact, allow the induction of otherwise difficult reactions.

Glycals serve as activated sugars because of the inherent reactivity of the endocyclic enol ether. Consequently, they have been extremely useful substrates in their complementarity to native activated sugars. Additionally, they have been used to demonstrate versatility not directly available from other sugar derivatives. In a good example of the utility of allylsilanes and silylacetylenes in *C*-glycoside chemistry, Ichikawa et al. [91] demonstrated a preference for the α anomer in all cases. The results, shown in Scheme 7.17, included a demonstration that a characteristic nOe can be used to confirm the stereochemical outcome of the reaction with *bis*-trimethylsilylacetylene.

With the advent of glycal chemistry being incorporated into *C*-glycoside technology came a host of new compounds previously difficult to prepare. As shown in Scheme 7.18, this statement is

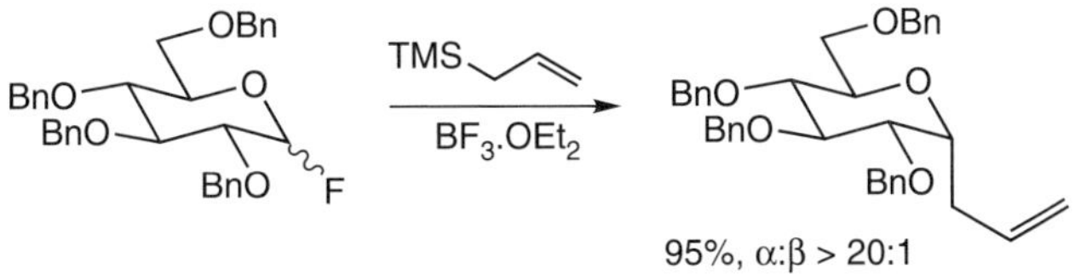

TMS, BF_3.OEt$_2$

95%, α:β > 20:1

SCHEME 7.15 *C*-Glycosidations with allylsilanes.

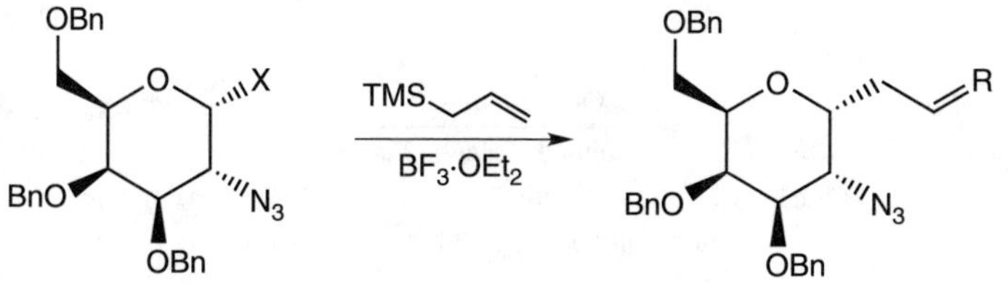

X	Resluts
OMe	No Reaction
OAc	Very Slow, R = CH_2
ONO_2	40% Yield, R = O after ozonolysis, α:β = 5:1

SCHEME 7.16 *C*-Glycosidations with allylsilanes.

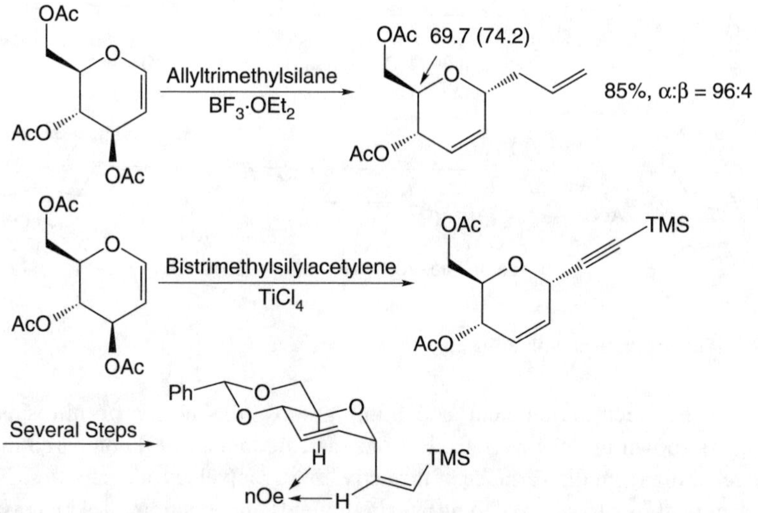

SCHEME 7.17 *C*-Glycosidations with allylsilanes and related compounds.

supported by a report by de Raadt and Stütz [92], in which a *C*-linked disaccharide was prepared by the direct coupling of an allylsilane modified sugar with a glycal. Although this example is not representative of the state of *C*-disaccharide chemistry, it does illustrate an extreme regarding the usefulness of glycals.

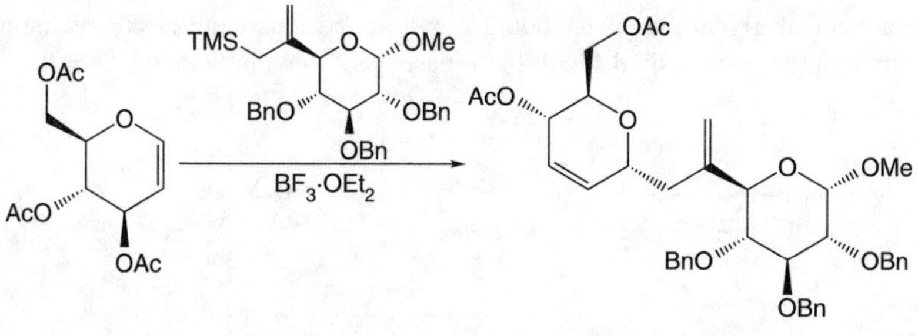

SCHEME 7.18 *C*-Glycosidations with allylsilanes.

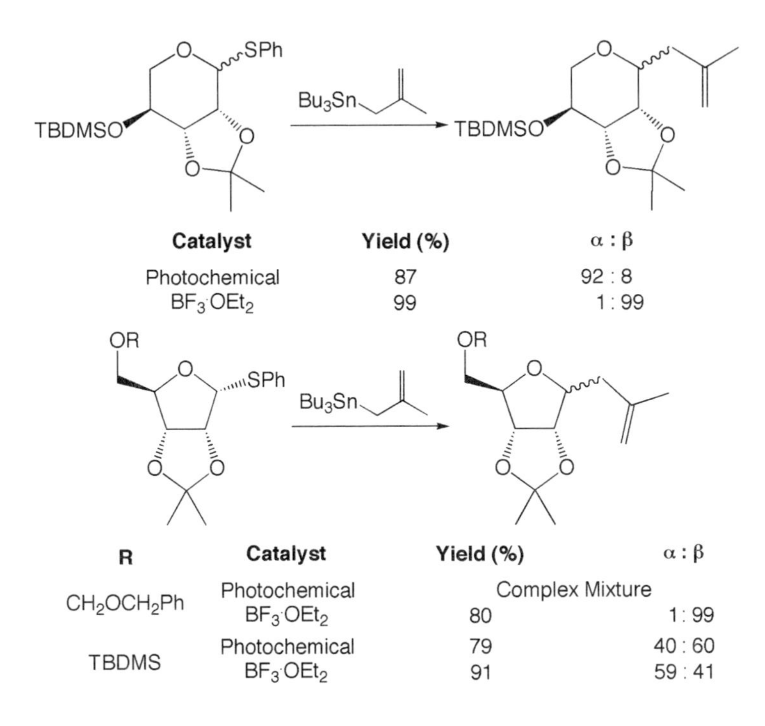

SCHEME 7.19 *C*-Glycosidation via allyltin reagents.

Considering glycals in general, glycals with distinct olefinic substitution patterns have found use. These substitutions are found in the form of 1-substituted glycals [46], 2-substituted glycals [93] and 2-acetoxyglycals [94].

Thus far, the nucleophiles discussed as viable reagents for *C*-glycosidations have centered upon silylated compounds. However, no discussion of *C*-glycosides would be complete without addressing the variety of nucleophiles available for the formation of such compounds. Therefore, the remainder of this section will focus on nonsilylated reagents and, specifically, tin and aluminum derived nucleophiles.

Allyltin derivatives are useful reagents for the preparation of *C*-glycosides. Specifically, their versatility is demonstrated by their ability to produce either α or β *C*-glycosides depending upon the reaction conditions used. As shown in Scheme 7.19, Keck et al. [95] initiated such reactions on phenylsulfide activated pyranose and furanose sugars. Aside from yields exceeding 80% in all cases, specific stereochemical consequences were observed. For example, utilizing pyranose sugars as *C*-glycosidation substrates, photochemical initiated *C*-glycosidations favored the formation of the α anomer, while Lewis acid-mediated reactions favored the β anomer. Similar results were observed for furanose sugars. However, an important deviance is the lack of stereospecificity observed under photolytic reaction conditions. Additional examples utilizing photochemical conditions are known [96,97] and generally yield ambiguous results at the anomeric center.

With the exception of photochemical reactions, the chemistry of allyltin compounds is analogous to that of allylsilanes and translates to the use of acetylenic tin reagents. As a final illustration of the utility of tin reagents to the formation of *C*-glycosides, the use of acetylenic tin reagents, shown in Scheme 7.20, is addressed. In the cited report, Zhai et al. [98] observed the formation of α-*C*-acetylenic glycosides in good yields. The reactions were catalyzed by zinc chloride and run in carbon tetrachloride.

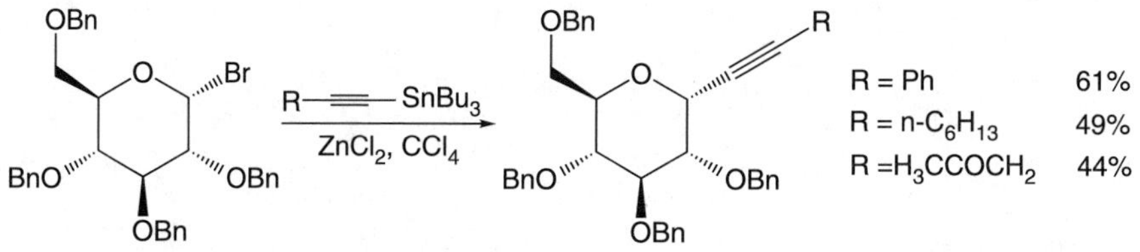

R	Yield
R = Ph	61%
R = n-C_6H_{13}	49%
R = H_3CCOCH_2	44%

SCHEME 7.20 *C*-Glycosidation via acetylenic tin reagents.

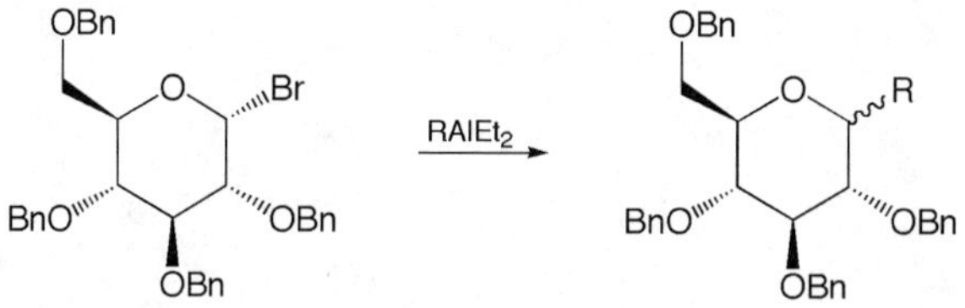

R	Yield (%)	α : β
TMS—C≡C	64	62 : 38
1-Naphthyl	48	100 : 0
CH_3CH_2	57	90 : 10
$(C_4H_9)_3Al$	63	8 : 92

SCHEME 7.21 *C*-Glycosidation via aluminum reagents.

Complementary to the chemistry of tin reagents as utilized in the formation of *C*-glycosides is the related chemistry surrounding organoaluminum compounds. The more common examples of this chemistry are applied to glycosyl halides as glycosidation substrates. Thus, as shown in Scheme 7.21, Tolstikov et al. [99] treated 2,3,4,6-tetra-*O*-benzyl-α-D-glucopyranosyl bromide with a variety of aluminum reagents effecting the formation of *C*-glycosides in good yields with demonstrated versatility in stereoselectivity. Specifically, no advantageous stereoselectivity was observed utilizing a trimethylsilylacetylene reagent, α selectivity was observed with ethyl and naphthyl reagents, and β selectivity was observed utilizing tri-*n*-butyl aluminum.

7.2.4 Arylation Reactions

In comparison with the reactions described in the previous section, arylation reactions, related to allylation reactions and illustrated in Figure 7.9, can be more convenient and versatile. For example, in most cases, Friedel–Crafts methodology applies. Additionally, not only is the conversion from *O*- to *C*-glycosides feasible, but both α and β anomers are accessible.

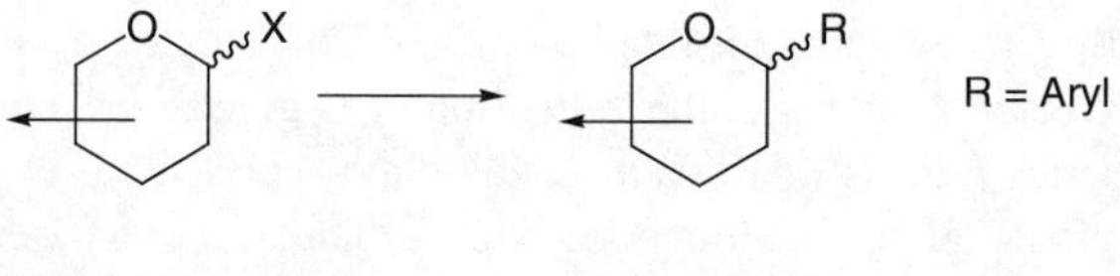

R = Aryl

FIGURE 7.9 Example of arylation reactions.

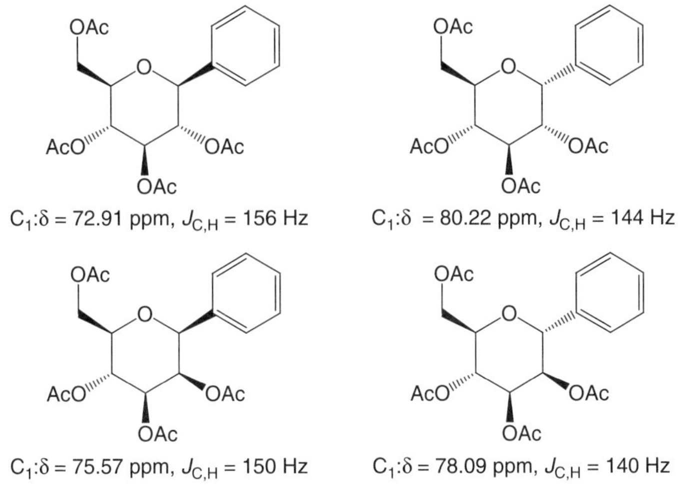

C_1:δ = 72.91 ppm, $J_{C,H}$ = 156 Hz

C_1:δ = 80.22 ppm, $J_{C,H}$ = 144 Hz

C_1:δ = 75.57 ppm, $J_{C,H}$ = 150 Hz

C_1:δ = 78.09 ppm, $J_{C,H}$ = 140 Hz

FIGURE 7.10 *C*-Arylglycoside chemical shifts and coupling constants.

Before exploring the various preparations of *C*-arylglycosides, it should be noted that an added feature of these compounds is the ease with which stereochemistry can be assigned. Specifically, a report by Bellosta et al. [100] showed that *C*-arylglucosides and *C*-arylmannosides exhibited characteristic differences in their chemical shifts and C–H coupling constants. As shown in Figure 7.10, the C_1 carbon of the α anomers consistently appeared downfield from the β anomers. Additionally, the ${}^1J_{C\text{-}1,H\text{-}1}$ coupling constants of the α anomers were approximately 10 Hz smaller than those for the β anomers.

7.2.4.1 Reactions with Metallated Aryl Compounds

When starting a discussion of the chemistry surrounding the preparation of *C*-arylglycosides, let us recall the work illustrated in Scheme 7.21 where Tolstikov et al. [99] treated 2,3,4,6-tetra-*O*-benzyl-α-D-glucopyranosyl bromide with a variety of aluminum reagents effecting the formation of *C*-glycosides in good yields with demonstrated versatility in stereoselectivity. One particular example from this report involved the use of 1-naphthyl-diethylaluminum. This particular reaction, in spite of its 48% yield, exhibited total α selectivity. Observations of the ability of arylaluminum species forming *C*-glycosides have also been noted with heterocyclic aromatic species [101]. Finally, aryltin reagents have been used similarly to arylaluminum compounds for the preparation of β *C*-glycosides from furanose-derived glycals catalyzed by palladium acetate [102,103].

7.2.4.2 Electrophilic Aromatic Substitutions

Although the use of aluminum and tin reagents have provided fruitful approaches to the preparation of *C*-arylglycosides, the majority of this chemistry centers upon reactions utilizing Lewis acid catalysis. Of particular importance are reactions occurring via electrophilic aromatic substitution. This type of chemistry has been widely applied to sugars bearing a variety of activating groups. These activating groups include, but are not limited to, acetoxy, trifluoroacetoxy, trichloroacetimido and halo substituents at the anomeric center. Additionally, glycals have found wide utility. In the following paragraphs, examples utilizing each of these groups are presented.

Beginning with glycosyl acetates, Hamamichi and Miyasaka [104] reported the coupling of an acetate activated ribofuranose with 1-methylnaphthalene (Scheme 7.22). This reaction produced a 79% yield with the formation of the α anomer comprising only 8.3% of the product mixture.

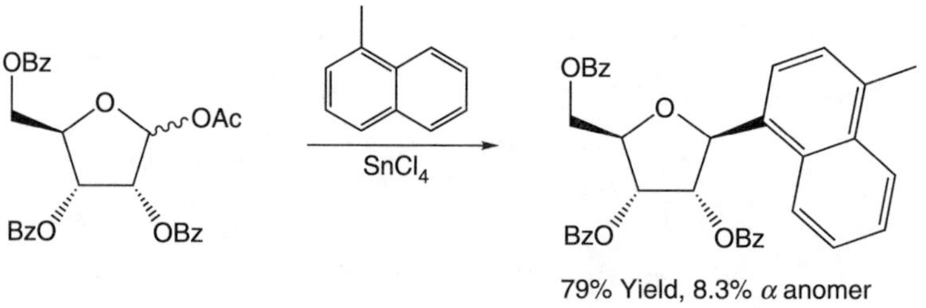

79% Yield, 8.3% α anomer

SCHEME 7.22 *C*-Glycosidation via electrophilic aromatic substitutions.

Additional examples resulting in poor anomeric selectivity are known. Notably, these anomeric mixtures could be equilibrated and separated following treatment with tin chloride [105,106].

Related to the acetate activating group is the more labile trifluoroacetate group. In a reaction between methylphenol and 2,3,4,6-tetra-*O*-benzyl-1-*O*-trifluoroacetyl-D-glucopyranose, shown in Scheme 7.23, Allevi et al. [79] observed the formation of a 50% yield of the β coupled product. Additional examples illustrating the dependence of the stereochemical outcome on the nature of the Lewis acid are known [107,108].

In one of the earliest demonstrations of *C*-arylation chemistry, Schmidt and Hoffman [109] utilized trichloroacetimidates to prepare the products shown in Scheme 7.24. In all cases, the β anomer was the preferred product. An interesting observation is the demonstrated ability to utilize both alkyl and silyl phenolic protecting groups. This chemistry was later expanded to include couplings with heterocyclic substrates [110,111].

Regarding glycosyl halides, Matsumoto et al. [112] published a report on a zirconocene dichloride–silver perchlorate complex as a promoter of Friedel–Crafts reactions in the preparation of *C*-glycosides. The reaction is shown in Scheme 7.25 and some of the results are summarized in Table 7.2. In this study, the first three runs resulted in glycosidation occurring at C_4, exclusively. Runs 4 to 6 resulted in glycosidation occurring at C_1, and runs 7 to 10 followed routes similar to the first three runs. What is important to note is the control over the anomeric configuration of the product. For example, where run 4 occurred with high β selectivity using five equivalents of the catalyst when only 0.2 equivalents were used, the selectivity was reversed giving high α selectivity. Additional examples diverging from glycosyl fluorides to other glycosyl halides are known [113].

To this point, all examples for the formation of *C*-aryl glycosides utilized Friedel–Crafts conditions with a propensity for selectively producing β anomers. These reactions were all directed at sugars bearing glycosidic activating groups. However, in 1988, Casiraghi et al. [114] demonstrated the selective accessibility of both α and β anomers directed by the nature of aromatic ring substitution. This study was applied to tri-*O*-acetyl-glucal and is illustrated in Scheme 7.26. Additionally, the NMR results defining coupling constants and nOe observations characteristic of each anomer are also shown. As stated, "in the D-series, for a given anomeric pair the more dextrorotatory member having $J_{4,5}$ of 8.7–10 Hz and relevant positive nOe between H_1 and H_5

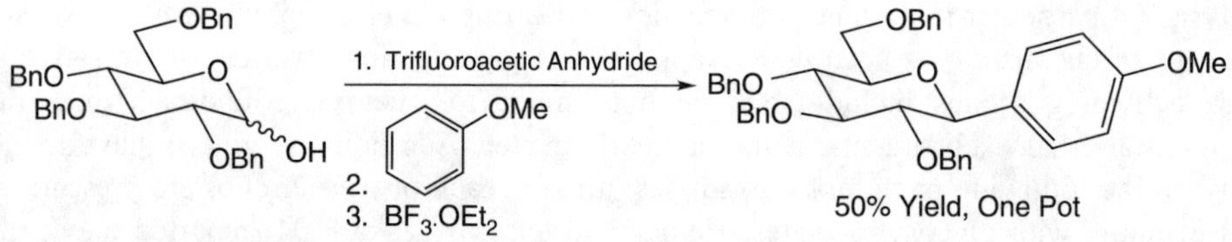

1. Trifluoroacetic Anhydride
2.
3. $BF_3.OEt_2$

50% Yield, One Pot

SCHEME 7.23 *C*-Glycosidation via electrophilic aromatic substitutions.

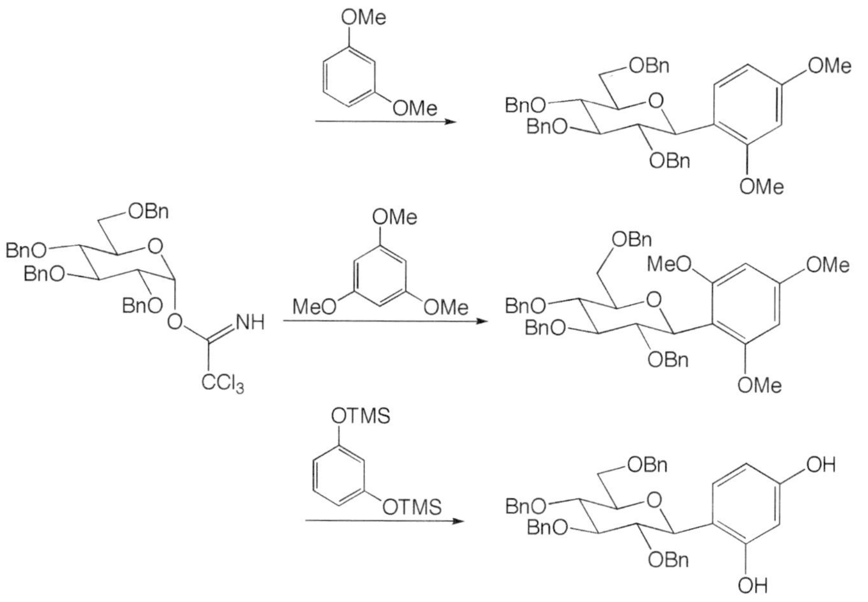

SCHEME 7.24 *C*-Glycosidation via electrophilic aromatic substitutions.

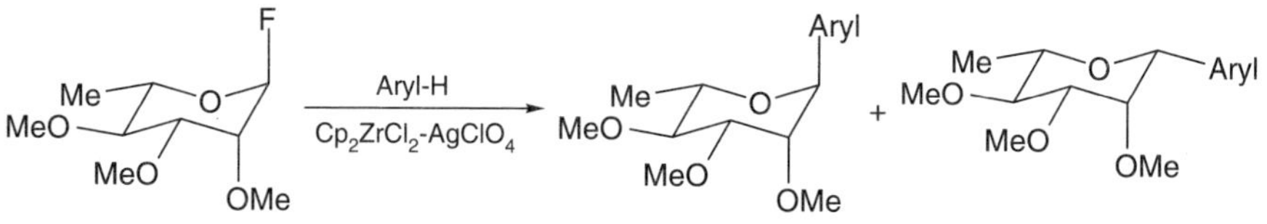

SCHEME 7.25 *C*-Glycosidation via electrophilic aromatic substitutions.

TABLE 7.2
***C*-Glycosidation via Electrophilic Aromatic Substitutions**

Run	Aryl-H	Lewis Acid Equivalents	Reaction Time	Yield (%)	α:β
1	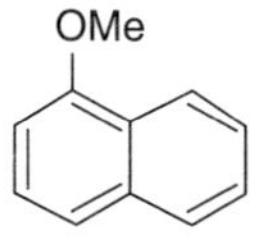	5.0	20 min	96	0:100
2		2.0	45 min	91	1:12
3		0.3	2 h	79	2:1
4	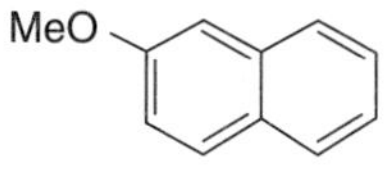	5.0	30 min	77	1:13
5		2.0	30 min	58	2:1
6		0.2	15 h	40	21:1
7	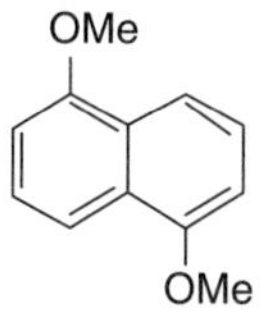	5.0	20 min	59	0:100
8		2.0	20 min	64	0:100
9		0.2	75 min	50	0:100

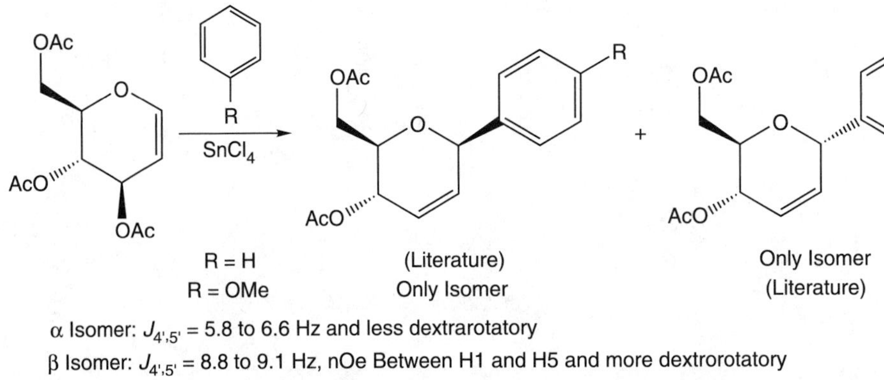
OAc
O
AcO
OAc
R
SnCl4
OAc
O
R
AcO
+
OAc
O
R
AcO
R = H
R = OMe
(Literature)
Only Isomer
Only Isomer
(Literature)
α Isomer: J4',5' = 5.8 to 6.6 Hz and less dextrarotatory
β Isomer: J4',5' = 8.8 to 9.1 Hz, nOe Between H1 and H5 and more dextrorotatory

SCHEME 7.26 *C*-Glycosidation via electrophilic aromatic substitutions.

should be assigned as β-D, whilst the other displaying a $J_{4,5}$ value in the range of 3–9 Hz and no nOe between H_1 and H_5 is to be named α-D." This chemistry was later applied to the coupling of glycals with heterocyclic substrates [91].

7.2.4.3 Intramolecular Electrophilic Aromatic Substitutions

Because electrophilic aromatic substitutions have provided access to *C*-arylglycosides via intermolecular reactions, similar technology is available for effecting the intramolecular delivery of the aromatic species. For example, utilizing the furanosyl fluoride shown in Scheme 7.27, Araki et al. [115] achieved formation of an 83% yield of the bicyclic product on treatment with borontrifluoride etherate. Similar chemistry is known for methyl glycosides [116,117] and glycosyl acetates [118].

7.2.4.4 *O*→*C* Migrations

Moving away from electrophilic aromatic substitutions as a means of forming *C*-arylglycosides, perhaps the most notable reaction is the *O*→*C* migration. This reaction is proposed to occur through initial formation of an *O*-glycoside. In the presence of Lewis acids, this *O*-glycoside then rearranges to a *C*-glycoside. Direct evidence for the *O*-glycoside intermediates was provided by Ramesh and Balasubramanian [119]. As shown in Scheme 7.28, *p*-methoxyphenol was coupled with the glycal to give the *O*-glycoside as a mixture of anomers. Regardless of the stereochemical outcome of the initial *O*-glycosidation, all stereochemistry at the anomeric center was lost on formation of the *C*-glycoside.

The study just cited was not the first example of *O*→*C* migrations. Rather, it was intended to introduce the reaction through a mechanistic insight. Early examples of this reaction were noted

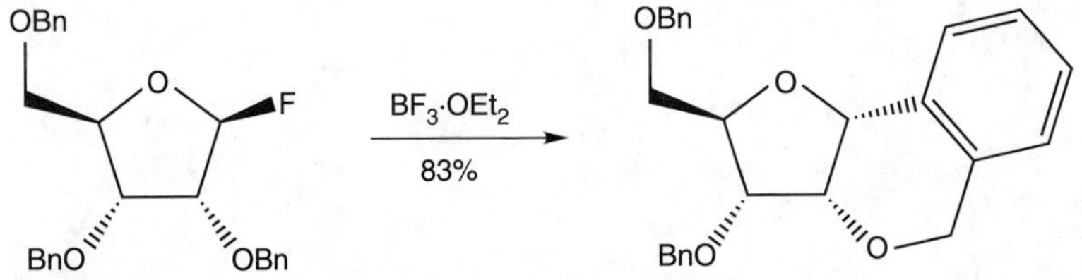
OBn
O
F
BnO
OBn
BF3·OEt2
83%
OBn
O
BnO
O

SCHEME 7.27 *C*-Glycosidation via intramolecular electrophilic aromatic substitutions.

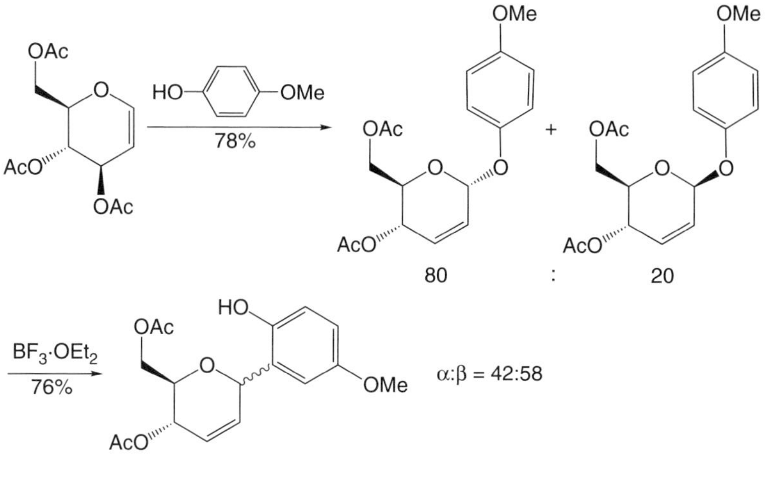

SCHEME 7.28 *C*-Glycosidation via *O*→*C* migrations.

using phenolic salts. An excellent example is contained in a report by Casiraghi et al. [120]. As shown in Scheme 7.29, a glycal derivative was treated with bromomagnesium phenoxide derivatives. In the two examples shown, the yields were relatively good, and the anomeric preference of this reaction favored the α configuration. Subsequent research further supported these observations [121]. This result provides a good comparison to the β selectivity observed utilizing electrophilic aromatic substitution reaction to effect the formation of *C*-arylglycosides. Finally, highlighting diversity of this chemistry, electron deficient 2-deoxy sugars [122–125] and furanose sugars [126] may be used.

7.2.5 Reactions with Enol Ethers, Silylenol Ethers and Enamines

As arylations are complementary to allylations, condensations of sugars with silylenol ethers and related compounds are analogous to condensations with allylsilanes. Consequently, the use of enol ethers, silylenol ethers and enamines in *C*-glycoside chemistry, illustrated in Figure 7.11, provide avenues to additional functionalities on the added groups. The incorporation of these compounds onto carbohydrates is accomplished through relatively general methods. The products are usually of the α configuration. However, the β isomers are accessible through epimerization. In all cases (except those involving glycols), benzylated sugars are preferred. Finally, the advantage of being

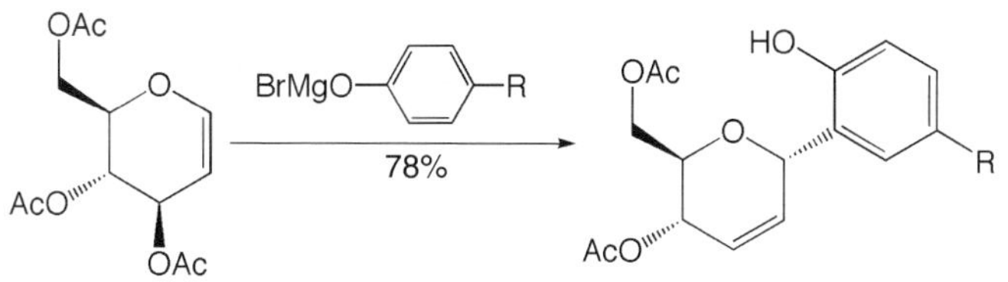

R	Yield (%)	α : β
H	63	29 : 1
OMe	82	100 : 1

SCHEME 7.29 *C*-Glycosidation via *O*→*C* migrations.

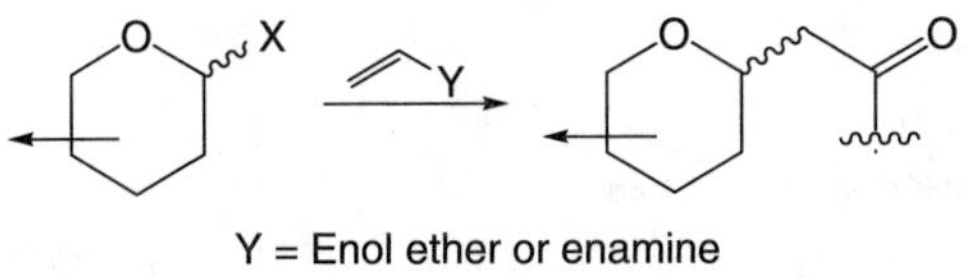
X
Y
O
Y = Enol ether or enamine

FIGURE 7.11 Examples of reactions with silylenol ethers and related compounds.

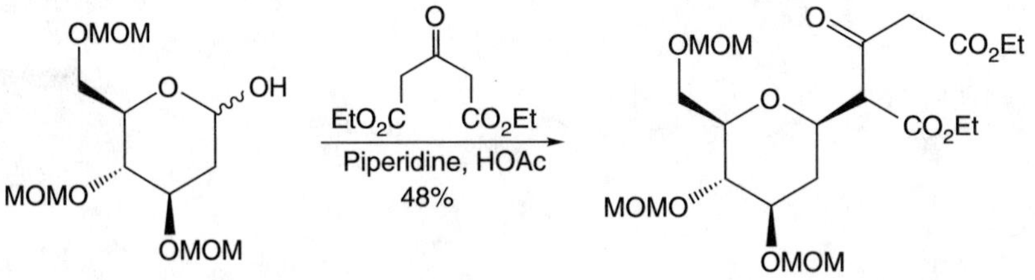
OMOM
OH
MOMO
OMOM
EtO2C
CO2Et
Piperidine, HOAc
48%
CO2Et
CO2Et

SCHEME 7.30 *C*-Glycosidations via enolates.

able to run these reactions at low temperature is offset by the need to activate the anomeric position in all cases.

7.2.5.1 Reactions with Enolates

The need for anomeric activation when utilizing enol ethers and related nucleophiles can be misleading. Examples have been reported illustrating the ability to couple anionic nucleophiles with unactivated sugars. For example, as shown in Scheme 7.30, Yamaguchi et al. [127] combined the illustrated unactivated 2-deoxy sugar with a keto-diester to give the isolated β-*C*-glycoside. However, this reaction does not occur through direct glycosidation pathways. Instead, as shown in Scheme 7.31, the nucleophile adds to the aldehyde of the open sugar. The resulting α,β-unsaturated carbonyl compound then recyclizes via 1,4-addition of a hydroxyl group allowing axial orientation of the extended unit. The resulting products are the illustrated β-*C*-glycosides. This mechanism will be revisited in the context of the Wittig reaction.

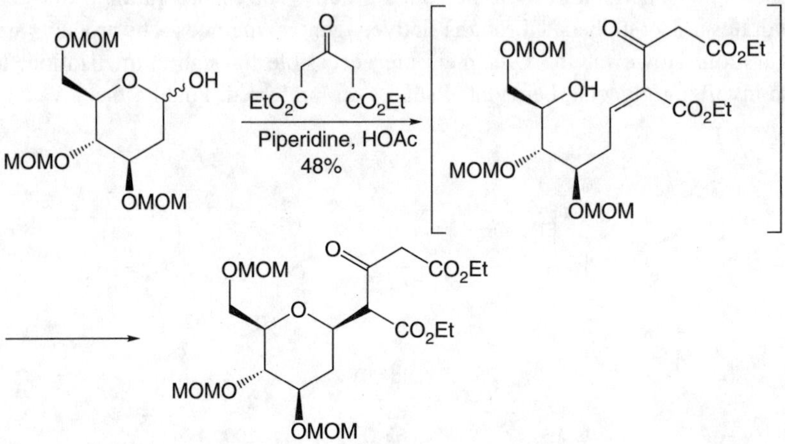
OMOM
OH
MOMO
OMOM
EtO2C
CO2Et
Piperidine, HOAc
48%
CO2Et
CO2Et

SCHEME 7.31 Mechanism of *C*-glycosidations via enolates.

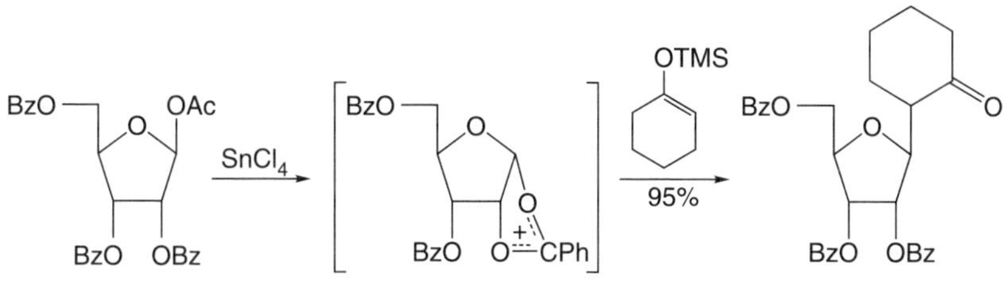

SCHEME 7.32 *C*-Glycosidations with silylenol ethers.

7.2.5.2 Reactions with Silylenol Ethers

Turing now to more conventional glycosidation methods, a variety of anomeric activating groups have been utilized in the direct coupling of silylenol ethers with sugars. The first example of such reactions, shown in Scheme 7.32, was reported by Hanessian et al. [128] in 1973. As illustrated, an acetate activated ribose derivative was treated with the TMS enolate of cyclohexanone in the presence of tin(IV) chloride yielding the *C*-riboside shown. One should note that this reaction resulted in a 95% yield with a strong preference for the β anomeric configuration. This specificity is caused by neighboring group assistance from the adjacent benzoyl group. In contrast, as shown in Scheme 7.33, Mukaiyama et al. [48] reported similar reactions with a variety of silylenol ethers. Utilizing the acetate activated ribose derivative, the reactions were catalyzed by trityl perchlorate and run in dimethoxyethane. In all cases, the yields exceeded 90% and, unlike the aforementioned couplings with enolates, the α anomer was overwhelmingly the predominant isomer. Although variations in the Lewis acids and solvents cannot be ignored, these related examples aptly illustrate the important role played by the functional groups adjacent to the anomeric center.

Further examples illustrating the utility of silylenol ethers in the preparation of *C*-glycosides are known. These include the ability of anomeric acetates to incorporate 2-deoxysugars [129,130]. Moreover, examples utilizing glycosyl trifluoroacetates [79], trichloroacetimidates [2], fluorides [89,131], chlorides [79,113,132], and glycals [133,134] are known. As the examples presented in the aforementioned studies rely on bimolecular interactions, one cannot ignore the advantages resulting from delivery of the silylenol ether from an intramolecular trajectory [135].

7.2.5.3 Reactions with Enamines

In a final example of the applicability of enol ether analogs to the synthesis of *C*-glycosides, Allevi et al. [136] explored a variety of enamines utilizing silver triflate as a catalyst. This study differed

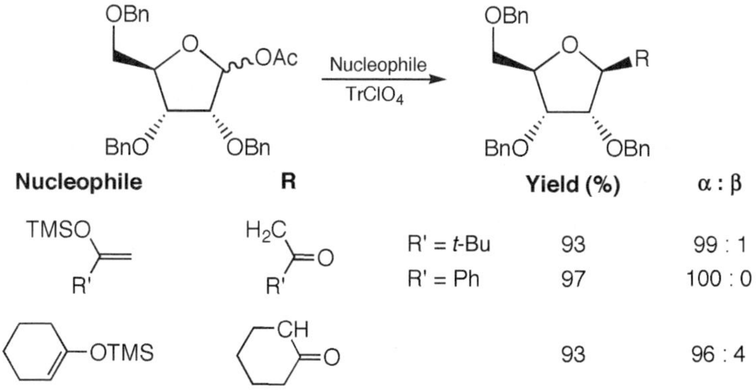

Nucleophile	R		Yield (%)	α : β
TMSO, R'	H_2C, =O, R'	R' = *t*-Bu	93	99 : 1
		R' = Ph	97	100 : 0
OTMS	CH, =O		93	96 : 4

SCHEME 7.33 *C*-Glycosidations with silylenol ethers.

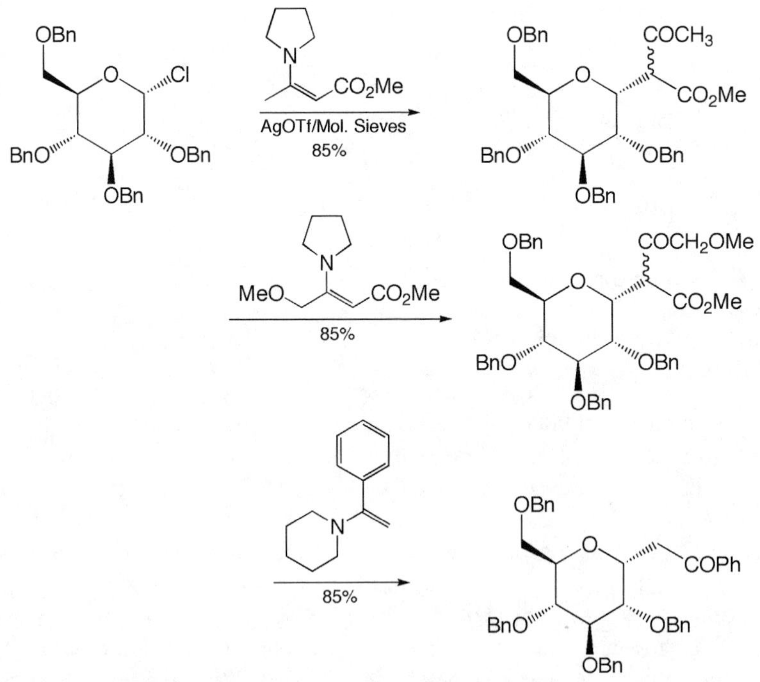

SCHEME 7.34 *C*-Glycosidations with enamines.

from that previously discussed in that the sugar is a glycosyl chloride and the enamines are conjugated to carbonyl functionalities. The results, shown in Scheme 7.34, are unlike those in the previous example, thus demonstrating a propensity for the formation of α anomers. Moreover, in all cases, the yields observed were approximately 85%. Thus, the conjugated enamines used in this report appear to have general applicability in condensations with glycosyl halides.

7.2.6 Nitroalkylation Reactions

The formation of nitroalkyl glycosides, illustrated in Figure 7.12, is limited in scope and few examples have been reported. The problems associated with this reaction involve the acidity of nitromethyl groups. Specifically, after formation of the *C*-glycoside, nitromethyl anions are easily formed. Moreover, when acetyl protecting groups are used, nitronates readily form. The few examples illustrated in this section are intended to convey the relatively limited accessibility of these compounds.

One specific example illustrating the utility of nitroalkylations was reported by Drew and Gross [137], and is shown in Scheme 7.35. In this report, a benzylidene protected glucose analog was converted in a 69% yield to a β-*C*-glycoside. Subsequent reduction of the nitro group gave the corresponding aminomethyl derivative suitable for further functionalization. It should be noted that a potential side reaction arises from nitromethane adding, via a 1,4-addition, to the intermediate olefin. Applied to furanose sugars, Koll et al. [138] reported the analogous reaction shown in Scheme 7.36.

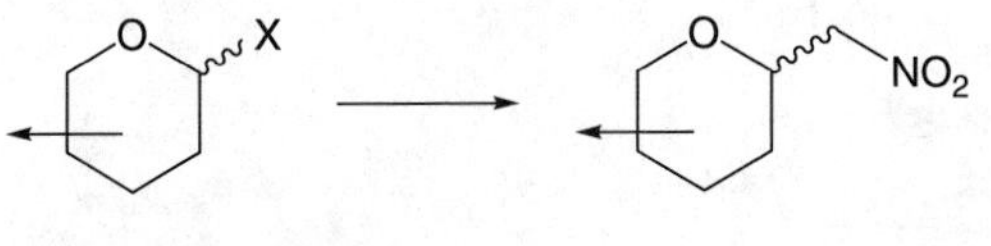

FIGURE 7.12 Example of nitroalkylation reactions.

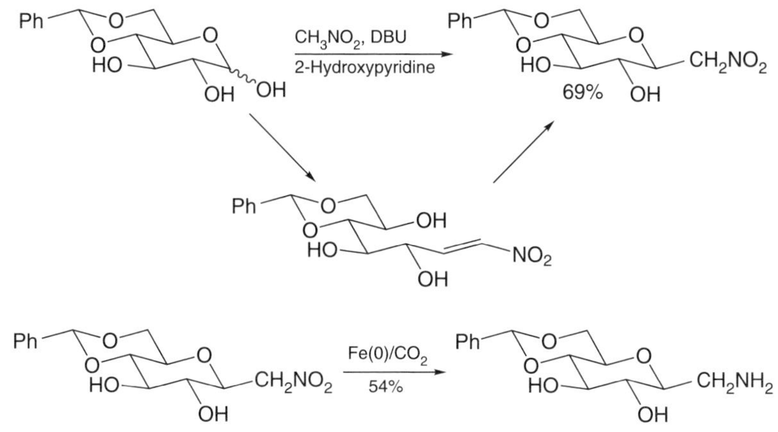

SCHEME 7.35 Nitroalkylation reactions.

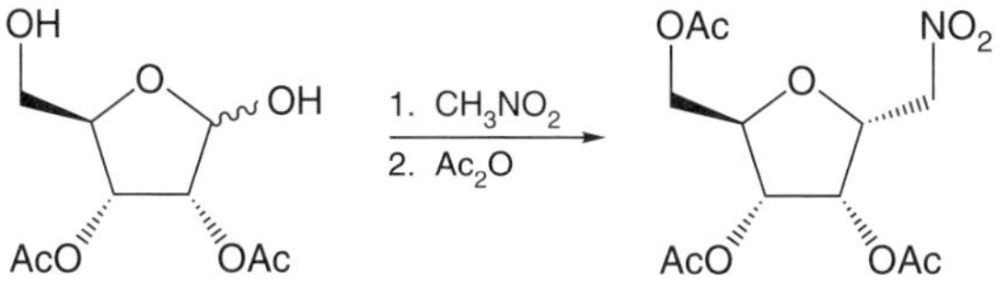

SCHEME 7.36 Nitroalkylation reactions.

Related to nitroalkylsugars are 1-deoxy-1-nitrosugars. These compounds are relatively easy to prepare via oximes. As shown in Scheme 7.37, Vasella et al. [139,140] effected this transformation via addition of an aldehyde to the oxime, shown. Subsequent treatment with ozone gave an anomeric mixture of the desired 1-deoxy-1-nitrosugar in a 73% overall yield.

The utility of 1-deoxy-1-nitrosugars was further demonstrated by Vasella et al. [141] in a study of nitroalkylation reactions. Summarized in Scheme 7.38, the study involved the addition of nitromethane and 2-nitropropane. The reactions gave an excess of 70% yield.

7.2.7 Reactions with Allylic Ethers

Unlike nitroalkylation reactions, employing allylic ethers as reagents in *C*-glycoside chemistry has proven to be substantially more useful. These reactions, illustrated in Figure 7.13, show a propensity for α selectivity and are able to provide additional asymmetric centers. However, these

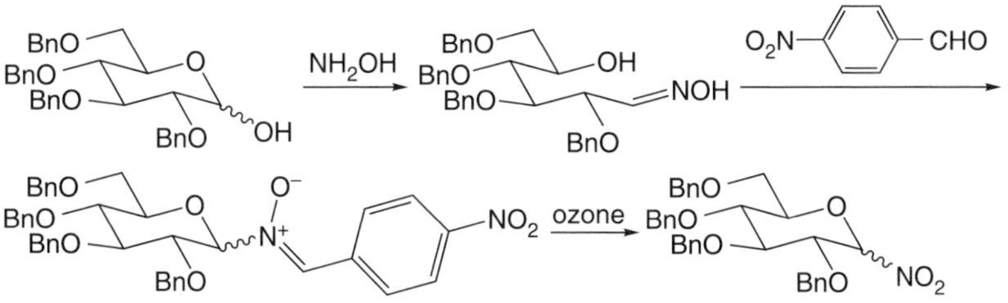

SCHEME 7.37 Synthesis of 1-deoxy-1-nitrosugars.

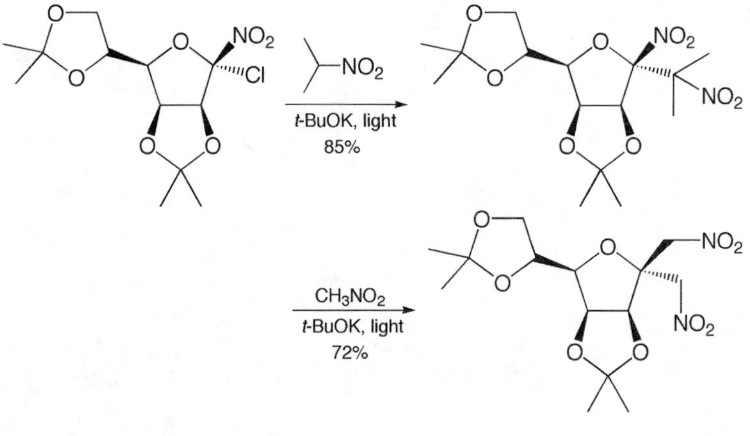

SCHEME 7.38 Nitroalkylation reactions.

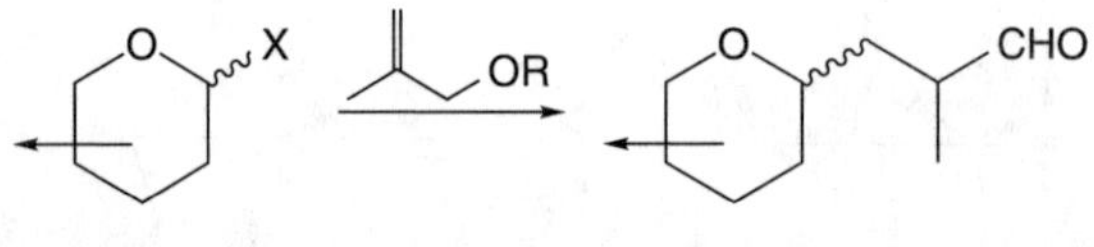

FIGURE 7.13 Example of reactions with allylic ethers.

reactions occur in an S_N2'-like fashion and are limited to glycals. Consequently, a wide variety of deoxy sugar derivatives are available.

An early example utilizing an allylic ether, shown in Scheme 7.39, was reported by Herscovici et al. [142], and is applicable to various glycals. As shown in Scheme 7.40, the mechanism of these transformations involves formation of the carbocation followed by the evolution of its equilibrium with the illustrated *O*-glycoside. As the carbocation reacts with the allylic ether, the reaction is driven to provide the desired product.

The fact that ketones, aldehydes and geminal diacetates are readily available from these reactions illustrates their complementarity to reactions with allylsilanes. Specifically, the equivalency of allylic ethers to homoenolates allows for the formation of compounds extended

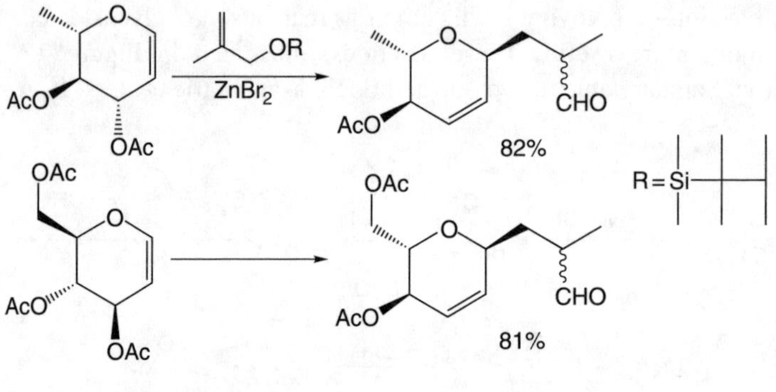

SCHEME 7.39 Reactions of glycals with allylic ethers.

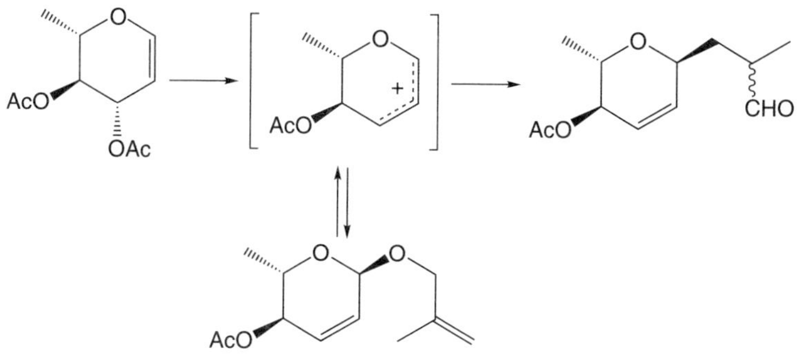

SCHEME 7.40 Reactions of glycals with allylic ethers.

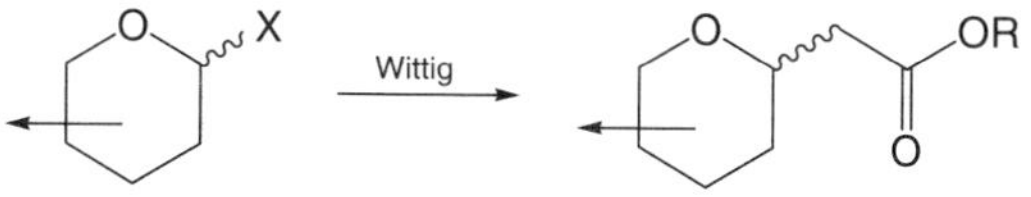

FIGURE 7.14 Example of Wittig reactions with sugars.

by one carbon unit when compared with the products of couplings with enolate equivalents already discussed.

7.2.8 Wittig Reactions with Lactols

While the use of allylic ethers provides convenient routes to the formation of ketones and aldehydes, the Wittig reaction, as illustrated in Figure 7.14, may be utilized in the formation of ester functionalized *C*-glycosides. This approach is either α or β selective and works with benzylated, acetylated and unprotected sugars. A particular advantage is that reaction intermediates can be intercepted and utilized. Moreover, glycosidic activation is not required for this type of reaction.

7.2.8.1 Wittig Reactions

In 1981, Acton et al. [143] performed the reaction shown in Scheme 7.41, to give a 3:1 ratio of anomeric esters. Shortly thereafter, Nicotra et al. [144] ran a similar reaction and studied the ^{13}C chemical shifts of the resulting α and β anomers. This study yielded an easy method for the assignment of the anomeric stereochemistry, with the anomeric carbon of the α anomer possessing a chemical shift of approximately 33 ppm and that of the β anomer possessing a chemical shift of approximately 37 ppm.

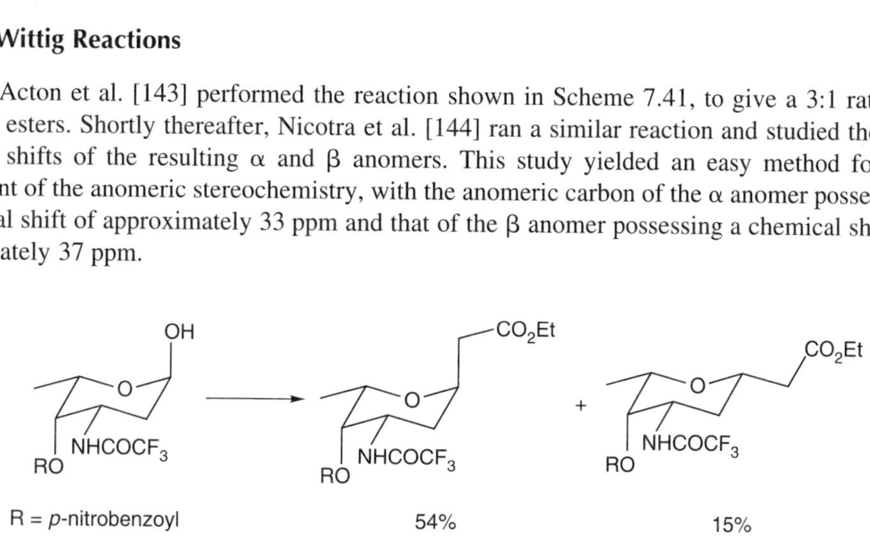

SCHEME 7.41 Wittig reactions with lactols.

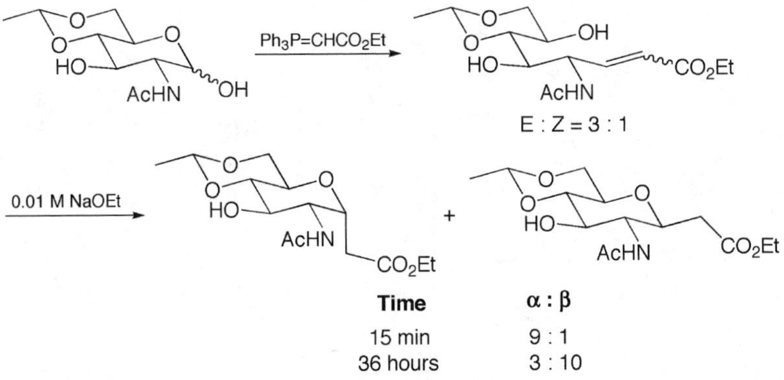

$Ph_3P{=}CHCO_2Et$

E : Z = 3 : 1

0.01 M NaOEt

Time	α : β
15 min	9 : 1
36 hours	3 : 10

SCHEME 7.42 Mechanism of Wittig reactions with lactols.

The mechanism of Wittig reactions applied to lactols is illustrated in Scheme 7.42 in a reaction reported by Giannis and Sandhoff [145]. As shown, the Wittig reagent reacts with the aldehyde of the straight-chain form of the sugar. The free 5-hydroxyl group then adds to the resulting α,β-unsaturated ester under basic conditions, giving the products shown. Moreover, the anomeric ratio of the product mixture is dependent upon the time of exposure to base, thus demonstrating the ability to convert the α anomer to the β anomer. In the cited report, the reaction shown was used in the synthesis of a potential β-D-hexosaminidase A inhibitor.

With the utility of Wittig reagents in the formation of *C*-glycosides, it is important to mention work reported by Dheilly et al. [146], in which benzyl protected glucose, galactose and mannose were treated with methyl bromoacetate, zinc dust and triphenylphosphine. The procedure described was reported by Shen et al. [147] as a one-pot formation and application of Wittig reagents. The original report utilized this methodology in transformations regarding aromatic and aliphatic aldehydes. Although all examples presented thus far utilize Wittig methodology in the context of pyranose sugars, one cannot ignore its applicability to the equally important furanoses [148].

7.2.8.2 Horner–Emmons Reactions

Complementary to the Wittig reaction is its Horner–Emmons modification. These reactions, unlike the phosphorus ylides utilized in Wittig reactions, make use of phosphonate anions as the nucleophilic species. As shown in Scheme 7.43, this methodology is applicable to the formation of *C*-glycosides from sugars. The reactions shown, reported by Allevi et al. [149], involve the coupling of triethylphosphonoacetate with 2,3,4,6-tetra-*O*-benzyl-D-mannopyranose, and produce high yields of the desired β-*C*-glycosides bearing ester functionalities. Although the reaction mechanism generally follows that noted in the previous section for Wittig reactions, one should be cautioned that some epimerization of the stereocenter at C_2 is possibly because of the relative acidity of the proton at C_2 considering the intermediate α,β-unsaturated ester. Related reactions involving sulfone phosphonates have also been reported [150,151].

7.2.8.3 Reactions with Sulfur Ylides

In addition to the utility of phosphonate anions and phosphorus ylides in the preparation of *C*-glycosides, a discussion of the related chemistry surrounding sulfur ylides is warranted. Unlike phosphorus ylides, sulfur ylides deliver methylene groups to carbonyls, thus forming epoxides. As shown in Scheme 7.44, Fréchou et al. [152] exploited this chemistry in the formation

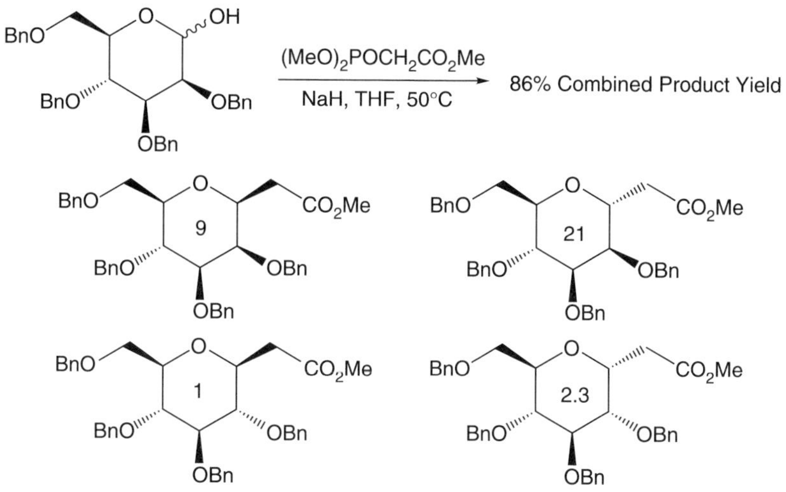
BnO
BnO
OBn
OBn
OH
(MeO)2POCH2CO2Me
NaH, THF, 50°C
86% Combined Product Yield
CO2Me
9
21
1
2.3

SCHEME 7.43 Horner–Emmons reactions with lactols.

of *C*-glycosides. As shown, the sulfur ylide was added to the benzyl-protected glucose to give the olefinic epoxide shown. Cyclization could then be triggered on treatment with base, thus forming the *C*-glycoside with opening of the epoxide. Similar results were observed utilizing the sulfoxide ylide and the acetonide protected furanose sugar shown. It should be noted that spontaneous cyclization occurred only when spontaneous elimination did not.

7.2.9 Nucleophilic Additions to Sugar Lactones Followed by Lactol Reductions

Sugar lactones are useful substrates for the formation of *C*-glycosides and are readily available from the oxidation of lactols. As shown in Figure 7.15, the use of this approach involves the addition of a nucleophile to the lactone followed by subsequent reduction of the resulting lactol. The two-step process provides *C*-glycosides in high yields and is useful with both benzyl and silyl protected sugar lactones. Moreover, as the hydride is generally delivered to the axial position, these reactions produce β-*C*-glycosides.

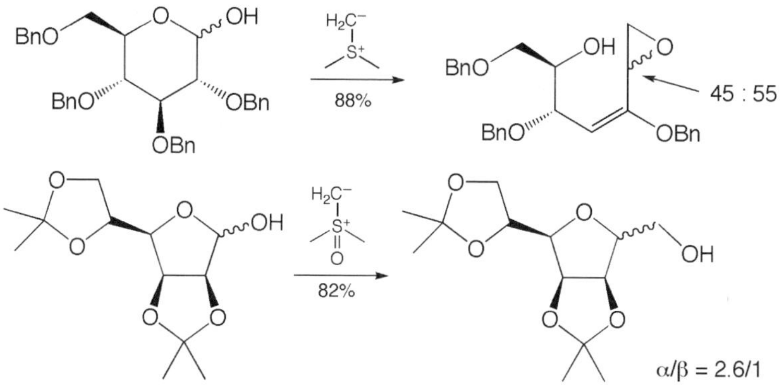
BnO
BnO
OBn
OBn
OH
88%
45 : 55
82%
α/β = 2.6/1

SCHEME 7.44 Reactions of sulfur ylides with lactols.

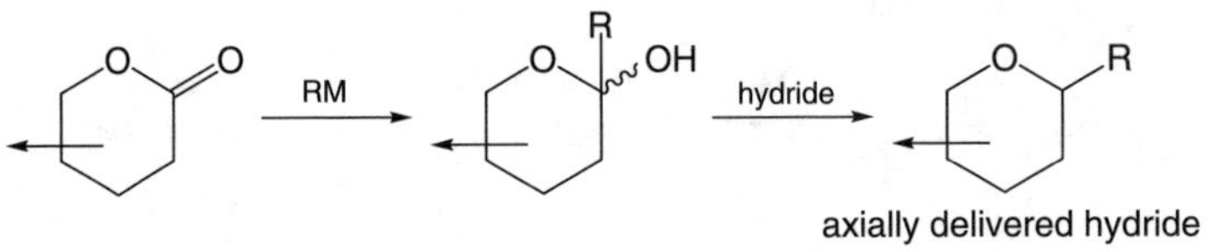

FIGURE 7.15 Example of *C*-glycosides from sugar lactones.

7.2.9.1 Lewis Acid-Trialkylsilane Reductions

As shown in Scheme 7.45, the versatility of products available from the addition of nucleophiles to sugar lactones was addressed by Horton and Priebe [153]. Following the illustrations in the scheme, gluconolactone was silylated and treated with a lithium reagent to give the observed lactol. This lactol was then acetylated and reduced, with Raney nickel, to the *C*-methylglycoside. Alternately, the lactol was partially acetylated. Subsequent Raney nickel reduction gave 1-methylglucose. In both cases, a byproduct of the acetylations was the open chain glucose derivative, **A**.

Moving away from alkyllithium species, Scheme 7.46 illustrates the use of lithium acetylide derivatives as useful nucleophiles for the formation of *C*-glycosides. As described by Lancelin et al. [154], lithium acetylides were added to sugar lactones giving lactols. The lactols were subsequently reduced to β-*C*-alkynylglycosides, with complete stereospecificity. This reaction was subsequently shown to be a general method for a variety of sugars and alkynyllithium reagents.

Utilizing aryl groups as nucleophiles, the related reaction shown in Scheme 7.47 was carried out by Kraus and Molina [155]. In this case, it is important to note that the predominant product, formed in greater than 80% yield, possesses the β configuration at the anomeric center.

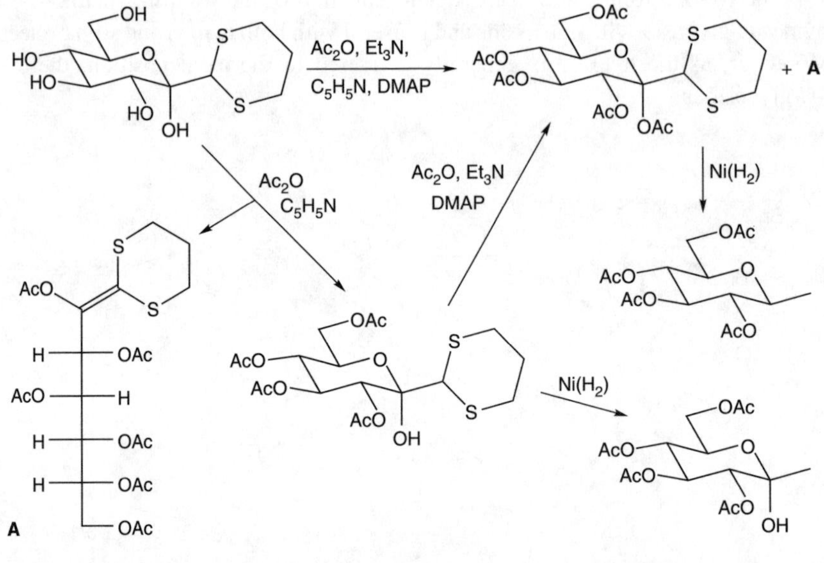

SCHEME 7.45 *C*-Glycosides from sugar lactones.

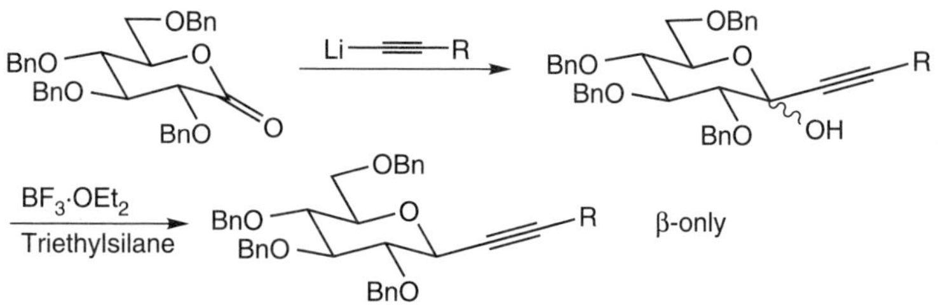

SCHEME 7.46 *C*-Glycosides from sugar lactones.

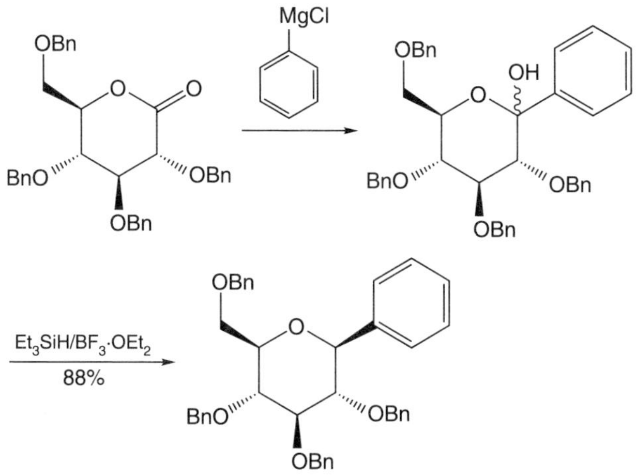

SCHEME 7.47 *C*-Glycosides from sugar lactones.

Related examples utilizing aryllithium reagents were reported by Czernecki and Ville [156] and Krohn et al. [157].

7.2.9.2 Other Reductions

Until now, all methods mentioned for the formation of *C*-glycosides from sugar lactones involve the incorporation of Lewis acid-trialkylsilane methodology for the deoxygenation. However, there are other methods for accomplishing this reaction. One particular example, shown in Scheme 7.48, was reported by Wilcox and Cowart [158] and involves the use of sodium cyanoborohydride in the presence of dichloroacetic acid. This reaction gave a 68% yield. Moreover, the use of *p*-toluenesulfonic acid instead of dichloroacetic acid totally blocked the desired transformation.

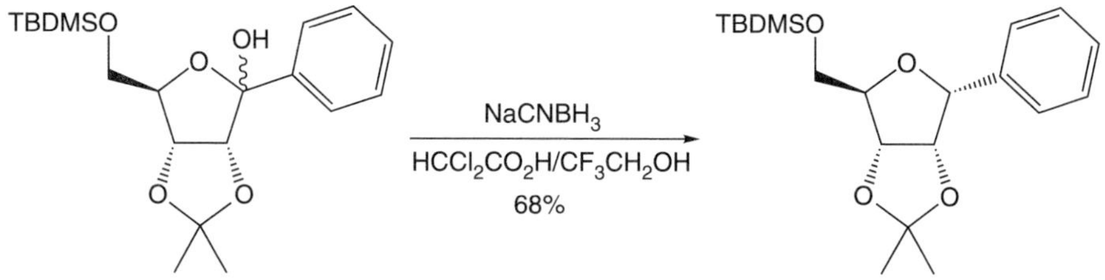

SCHEME 7.48 Lactol reduction with sodium cyanoborohydride.

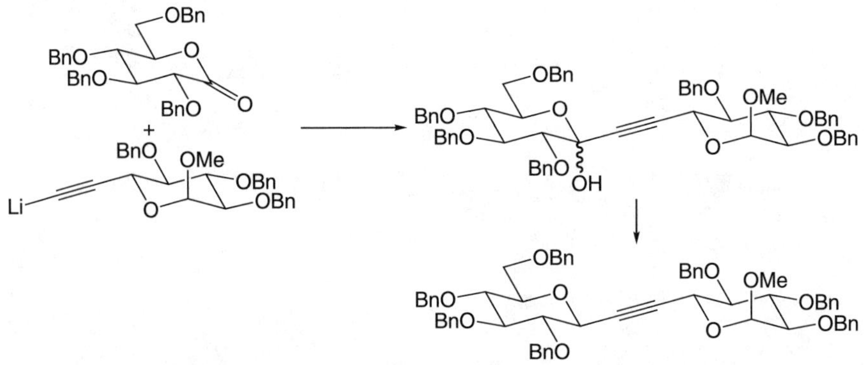

SCHEME 7.49 *C*-Glycosides from sugar lactones.

Perhaps the most interesting aspect of this reaction is the fact that the α configuration is made accessible through sugar lactone transformations.

7.2.9.3 Sugar–Sugar Couplings

Before closing this section, the application of nucleophilic additions to sugar lactones with respect to the joining of two sugar units deserves mention. As shown in Scheme 7.49, Rouzaud and Sinaÿ [159] utilized the lithium anion of a *C*-acetylenic glycoside to effect formation of an acetylene bridged *C*-disaccharide. Moreover, as shown in Scheme 7.50, Preuss and Schmidt [160] utilized a similar approach involving a carbohydrate-derived alkyllithium compound. These examples are presented to demonstrate that with the expansion of the chemistry surrounding *C*-glycosides, the preparation of increasingly more complicated structures becomes possible through substantially simpler processes.

7.2.10 Nucleophilic Additions to Sugars Containing Enones

Continuing from sugar lactones, a logical progression is the utilization of sugar-derived enones illustrated in Figure 7.16. These enones are readily available from the oxidation of glycals and the stereoselectivity of addition reactions is dictated entirely by stereoelectronic effects. These reactions are particularly useful in that 2-deoxy *C*-glycosides are the products.

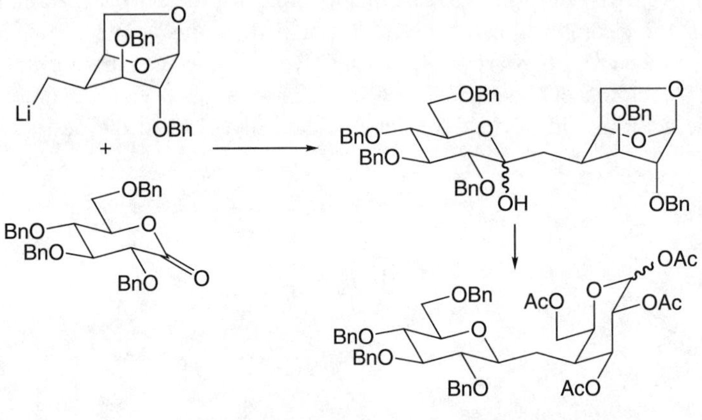

SCHEME 7.50 *C*-Glycosides from sugar lactones.

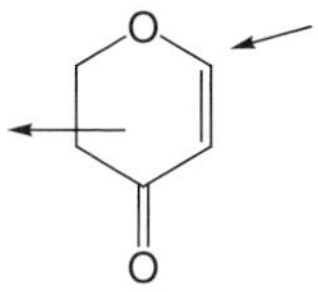

FIGURE 7.16 Sugar-derived enones or 3-ketoglycals.

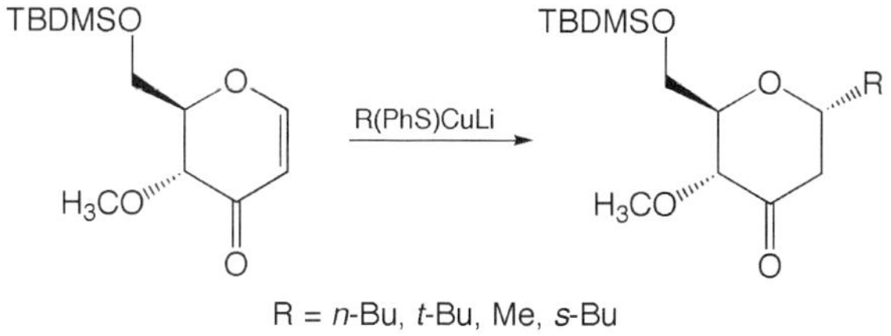
TBDMSO
H3CO
O
R(PhS)CuLi
TBDMSO
R
H3CO
O
R = n-Bu, t-Bu, Me, s-Bu

SCHEME 7.51 *C*-Glycosides from sugar-derived enones.

As early as 1983, Goodwin et al. [161] utilized various organocuprates to effect the formation of *C*-glycosides from sugar-derived enones. As shown in Scheme 7.51, a variety of alkyl groups, including the bulky *tert*-butyl group, were efficiently incorporated to the glycosidic center. In this report, the products exhibited the α configuration at the anomeric center. An additional example utilizing cuprates and presenting a rationalization of the stereochemical outcome was reported by Bellosta and Czernecki [162].

Complementary to the use of organocuprates, silylenol ethers have been useful as agents capable of adding to sugar-derived enones. As previously discussed, silylenol ethers have been exploited in *C*-glycoside technology with respect to their combination with anomerically-activated sugars. Because sugar-derived enones are highly activated sugars, the incorporation of silylenol ethers as reagents for effecting *C*-glycosidations is logical and complementary to methods previously addressed. As shown in Scheme 7.52, Kunz et al. [163] effected the use of silylenol ethers in the formation of a 90% yield of the *C*-glycoside shown. Compared with the use of organocuprates, the use of silylenol ethers allows the formation of β anomers further demonstrating advantages to having a variety of available *C*-glycosidation methods.

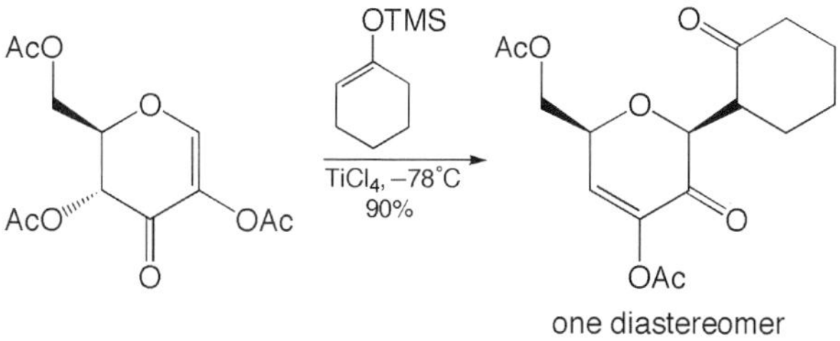
AcO
AcO
OAc
O
OTMS
TiCl4, −78°C
90%
AcO
O
O
OAc
one diastereomer

SCHEME 7.52 *C*-Glycosides from sugar-derived enones.

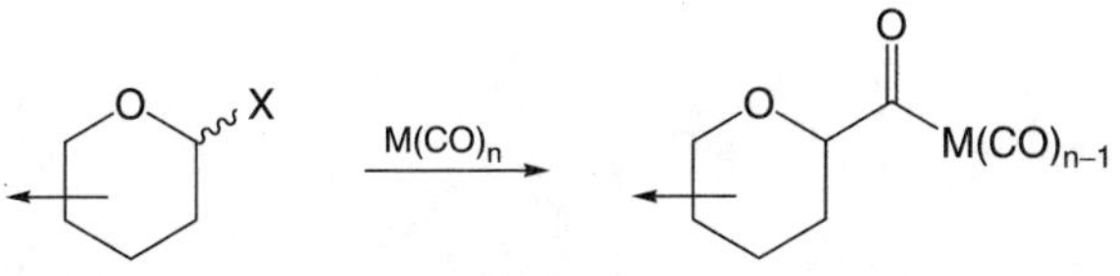

FIGURE 7.17 Example of transition metal-mediated carbon monoxide insertion reactions.

7.2.11 Transition Metal-Mediated Carbon Monoxide Insertions

Transition metals are capable of mediating reactions that produce *C*-glycosides. One such type of reaction is the transition metal-mediated insertion of carbon monoxide generalized in Figure 7.17. As a means of preparing *C*-glycosides this technology has, until recently, been relatively unexplored. The examples that have been reported demonstrate a propensity for β-selectivity. These reactions can be performed under catalytic conditions utilizing both benzylated and acylated sugars. Additionally, anomeric activation is not a prerequisite for reactions to take place.

7.2.11.1 Manganese Glycosides

In one of the earliest examples of carbon monoxide insertion reactions for the formation of *C*-glycosides, Deshong et al. [164] utilized the manganese glycoside shown in Scheme 7.53. This compound was reacted with carbon monoxide followed by the sodium salts of either methanol or thiophenol. The resulting products were the corresponding ester and thioester. Additional examples demonstrated that vinylic and acetylenic esters were also substrates for these reactions yielding the corresponding Michael products. Of particular convenience is the ready availability of manganese glycosides from the corresponding glycosylbromides [165].

7.2.11.2 Cobalt-Mediated Glycosidations

Manganese is not the only metal capable of effecting carbon monoxide insertion reactions. In 1988, Chatani et al. [166] demonstrated the utility of cobalt in the generation of *C*-glycosides. Specific examples, shown in Scheme 7.54, involve the conversion of glucose and galactose pentaacetates to

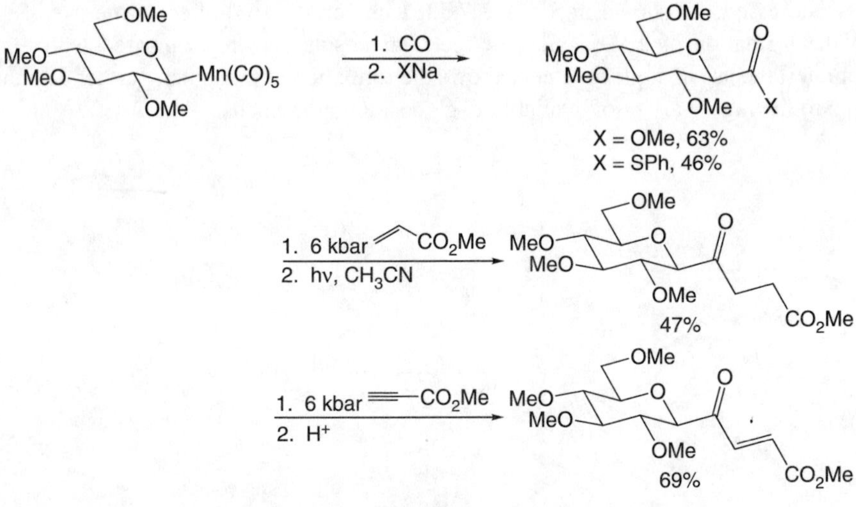

SCHEME 7.53 *C*-Glycosides from carbon monoxide insertions.

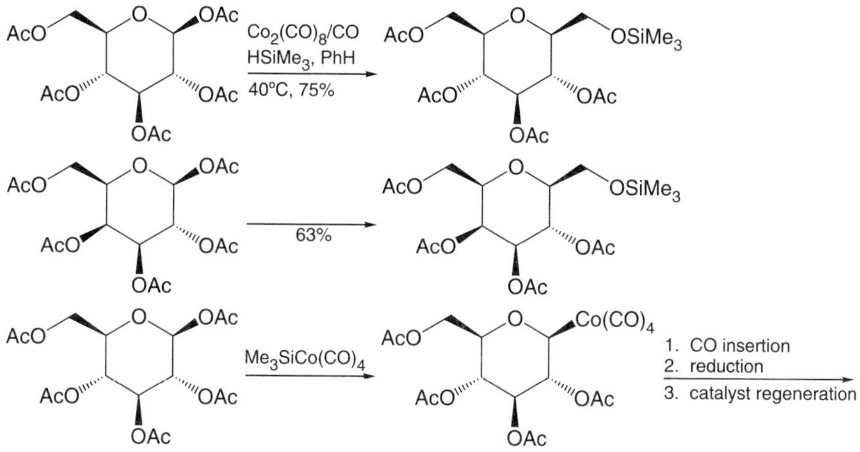

SCHEME 7.54 *C*-Glycosides from carbon monoxide insertions.

the trimethylsilyloxymethyl glycosides with β-selectivity. In the case of glucose pentaacetate, the use of trimethylsilylcobalt tetracarbonyl was illustrated. Additionally, this chemistry was applied to ketofuranoses with comparable results [32].

7.2.12 Reactions Involving Anomeric Carbenes

The incorporation of anomeric carbenes in *C*-glycoside technology has not received much attention. However, as shown in Figure 7.18, anomeric carbenes provide the only known method for the direct cyclopropanation of anomeric centers to date. Advantages to this method include general synthetic accessibility of the carbenes when benzylated sugars are utilized.

Two examples of the applicability of anomeric carbenes were reported by Vasella et al. [167,168]. The first demonstrated the utility of olefinic substrates such as *N*-phenylmaleimide, acrylonitrile and dimethylmaleate for the formation of glycosidic cyclopropanes. The carbene precursor in this example was the glucose-derived diazirine. As shown in Scheme 7.55, the use of dimethylmaleate produced a mixture of diastereomers with a combined yield of 72%. Although not presented as a schematic, the second and more dramatic example was used to functionalize fullerenes.

7.2.13 Reactions Involving Exoanomeric Methylenes

In concluding this section on electrophilic substitutions, the elaboration of exoanomeric methylenes (exoglycals) to more complex *C*-glycosides is addressed. These reactions, generalized in Figure 7.19, involve the addition of electrophiles to the anomeric olefin. Utilizing this technology makes double *C*-glycosidations possible. These reactions prefer benzyl protecting groups on

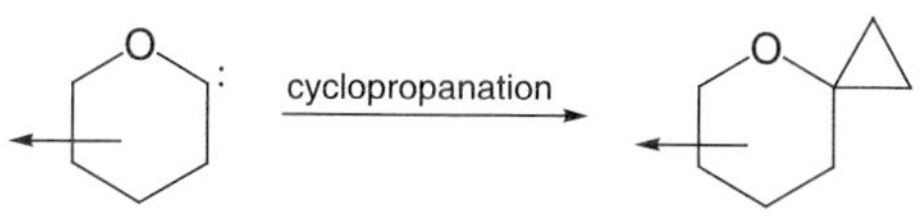

FIGURE 7.18 Example of glycosidic cyclopropanations.

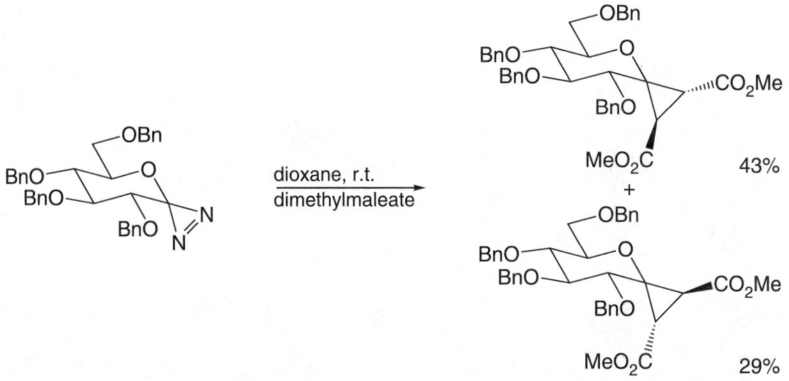

SCHEME 7.55 Glycosidic cyclopropanations.

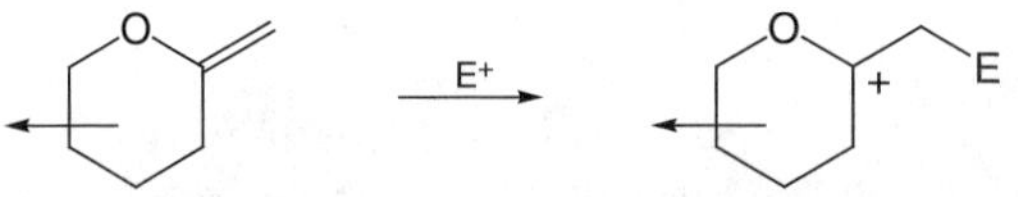

FIGURE 7.19 Example of electrophilic additions to exoanomeric methylene groups.

the sugars. However, the olefins are easily generated *in situ*, and reactions at the intermediate tertiary carbocation may be induced.

Technically, being *C*-glycosides themselves, sugars bearing exoanomeric methylene groups are easily prepared by reacting sugar lactones with Tebbe's reagent [169]. As shown in Scheme 7.56,

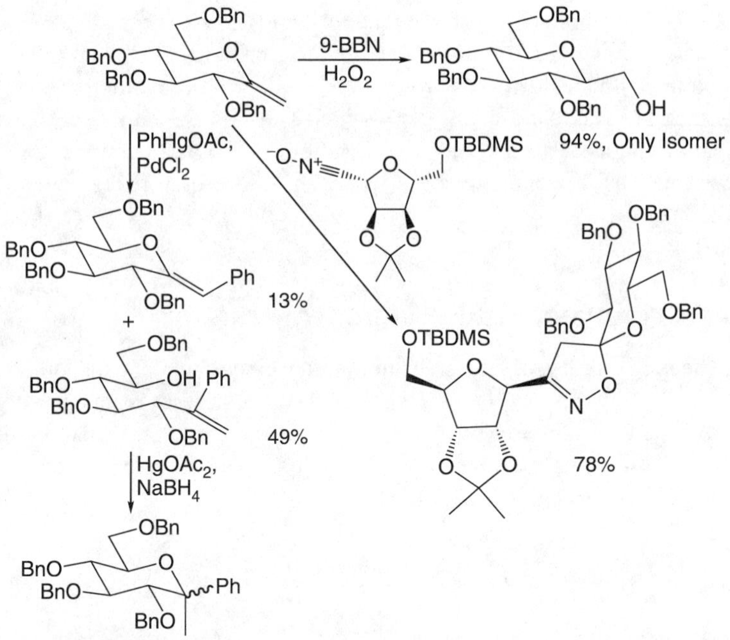

SCHEME 7.56 Reactions with exoanomeric methylene groups.

once formed, these sugar derivatives react under a variety of conditions yielding a broad spectrum of products. As demonstrated, alcohols may be introduced in high yields and high stereoselectivity. Interestingly, when less bulky agents such as a borane–THF complex were used, no stereoselectivity was observed. Moreover, 1,3-dipolar cycloaddition reactions and arylations utilizing heavy metal reagents were demonstrated. In the latter, nonspecific delivery of the phenyl group resulted in the formation of a mixture of products. However, the ring-opened compound will cyclize under reductive oxymercuration conditions. Additional examples of this chemistry were reported in the context of functionalization to hemiketal analogs [84] and difluoro olefins [170,171].

7.3 SYNTHESIS OF *C*-GLYCOSIDES VIA NUCLEOPHILIC SUGAR SUBSTITUTIONS

As thoroughly covered in Section 7.2, the use of electrophilic substitutions at the anomeric center has provided a substantially diverse variety of methods for the preparation of *C*-glycosides. However, a complementary approach to the preparation of these compounds, for which there is no counterpart in *O*-glycoside chemistry, is found through the use of sugar-derived nucleophiles. As shown in Scheme 7.57, C-1 lithiated carbohydrate analogs can act as nucleophiles in *C*-glycosidation reactions. This particular example, reported by Parker and Coburn [172], produced an intermediate compound capable of being converted to either a substituted glycal or a *C*-aryl glycoside. In this section, the formation of C-1 lithiated sugar derivatives and their incorporation into the preparation of *C*-glycosides are addressed. Additionally, the use of electron withdrawing stabilizing groups is discussed.

7.3.1 C-1 Lithiated Anomeric Carbanions by Direct Metal Exchange

The direct metallation of sugar derivatives, shown in Figure 7.20, is a useful method for the preparation of anomeric nucleophiles. Anomeric halides and glycals are excellent substrates for this reaction. Additionally, selective generation of α and β anomers is possible.

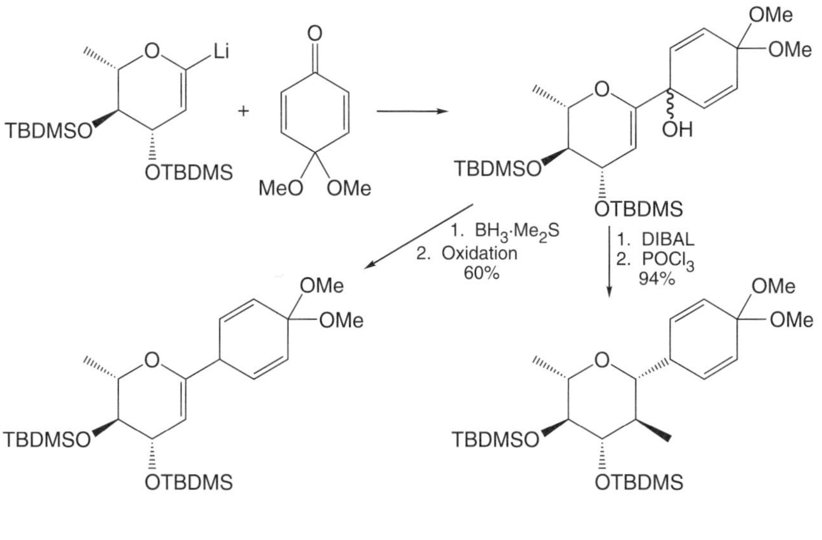

SCHEME 7.57 *C*-Glycosides from anomeric nucleophiles.

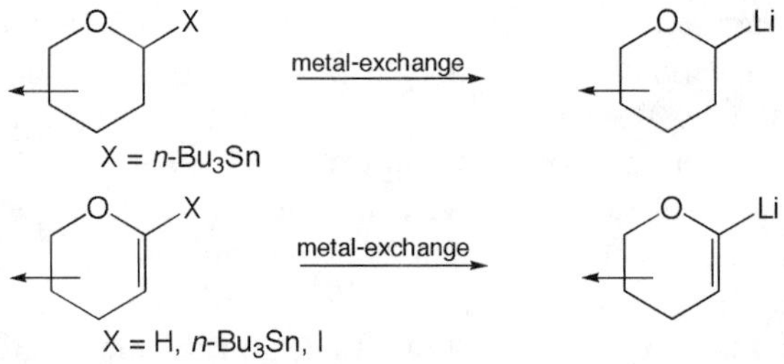

X
metal-exchange
Li
X = n-Bu3Sn
X
metal-exchange
Li
X = H, n-Bu3Sn, I

FIGURE 7.20 General metallation examples.

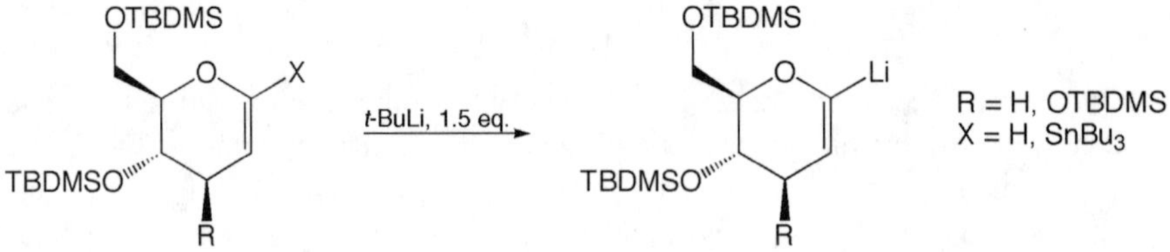

OTBDMS
X
t-BuLi, 1.5 eq.
OTBDMS
Li
TBDMSO
R
TBDMSO
R
R = H, OTBDMS
X = H, SnBu3

SCHEME 7.58 Glycal lithiations.

7.3.1.1 Hydrogen–Metal Exchanges

Direct applications of this methodology to the preparation of *C*-glycosides were demonstrated as early as 1986 when Sinaÿ et al. [173] converted 1-stannyl glycals to 1-lithioglycals on treatment with *tert*-butyllithium. These results, shown in Scheme 7.58, were complemented by Nicolaou's subsequent application to the generation of *C*-glycosides on treatment with allyl bromide [46]. Later work elaborating on the chemistry of glycals demonstrated the ease of formation of 1-stannyl glycals [174]. Finally, with respect to glycal metallation, 2-phenylsulfinyl derivatives have found utility on treatment with lithium diisopropylamide (LDA) [175]. This last example is advantageous in that stabilization of the anion is achieved by chelation of the sulfoxide to the metal.

Expanding upon the lithiation of 2-phenylsulfinyl glycals, Schmidt and Dietrich [33] effected coupling reactions with benzaldehyde (Scheme 7.59). As illustrated, this reaction resulted in an approximate yield of 86%. An important observation is the induction of stereochemistry at the newly formed center with a diastereomeric ratio of approximately 4:1.

Before moving to other means used in the preparation of C-1 lithioglycals, mention of a final example utilizing a hydrogen–metal exchange is warranted. The example of interest, shown in Scheme 7.60, was reported by Friesen and Loo [176] and involves the formation of C-1 iodinated

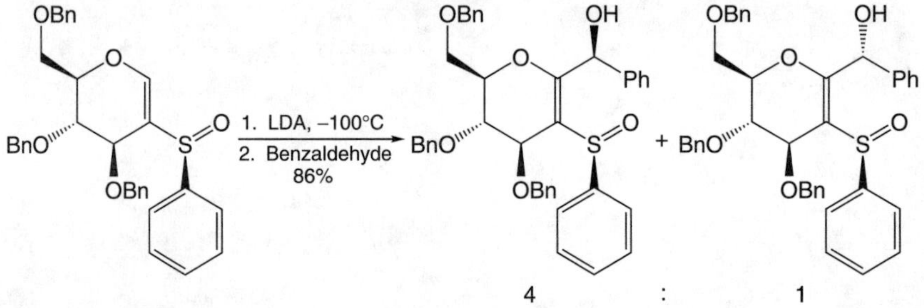

OBn
O
BnO
OBn
S
O
1. LDA, –100°C
2. Benzaldehyde
86%
OBn
OH
Ph
BnO
OBn
S
O
+
OBn
OH
Ph
BnO
OBn
S
O
4
:
1

SCHEME 7.59 Glycal lithiations.

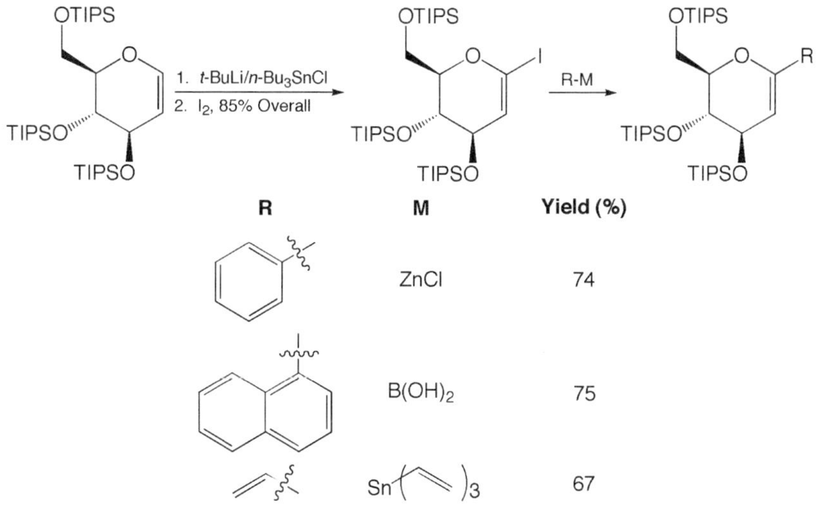

R	M	Yield (%)
[phenyl]	ZnCl	74
[1-naphthyl]	B(OH)2	75
[vinyl]	Sn(CH=CH2)3	67

SCHEME 7.60 *C*-Glycosides from C_1 iodo glycals.

glycals. As shown, direct lithiation was accompanied by treatment with tri-*n*-butylstannyl chloride. Subsequent treatment of the stannylated glycal with iodine gave the iodinated compound an 85% overall yield. These compounds have subsequently found use in coupling reactions with arylzinc, vinyltin, and arylboron compounds. Thus, a complement of unique *C*-glycosides is available through this methodology.

7.3.1.2 Metal–Metal Exchanges

Although the hydrogen–metal exchange has found use in the formation of C-1 lithiated glycals, sometimes the best method for the formation of these compounds rests in the ability to transmetallate from a different metallo glycal. These metal–metal exchanges have been extensively exploited in the derivatization of glycals and provide an efficient means of forming true anomeric anions from stannyl glycosides.

Continuing with previously described stannyl glycals, Hanessian et al. [174] demonstrated the ability to transform this glycal analog to a lithiated species in preparation for coupling with an aldehyde. The reaction, shown in Scheme 7.61, produced a 68% yield of the *C*-glycoside as a mixture of isomers at the newly formed stereogenic center. A similar use of this metal–metal exchange was applied to allylation reactions [177], conjugate additions [178] and the ring opening of epoxides [179]. The latter example involved transmetallation of a stannyl glycoside to a glycosyl cuprate.

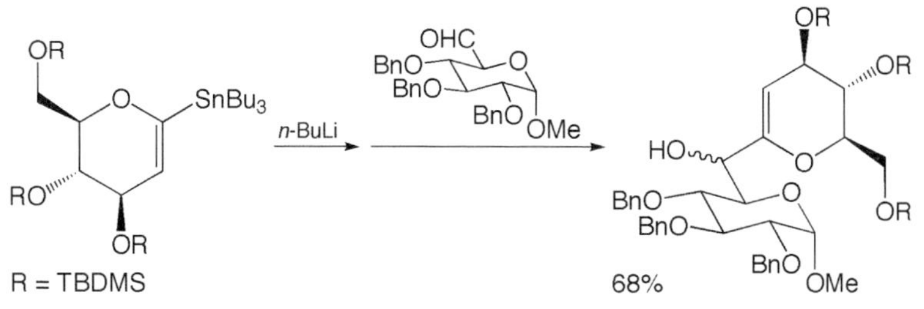

SCHEME 7.61 Glycal *trans*-metallations.

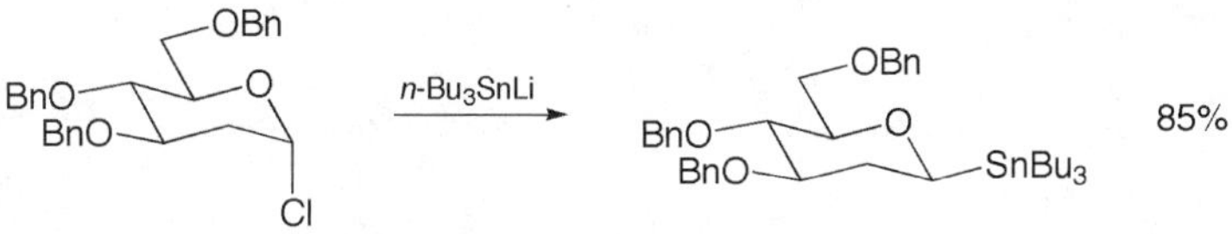

SCHEME 7.62 Glycosidic halogen–metal exchanges.

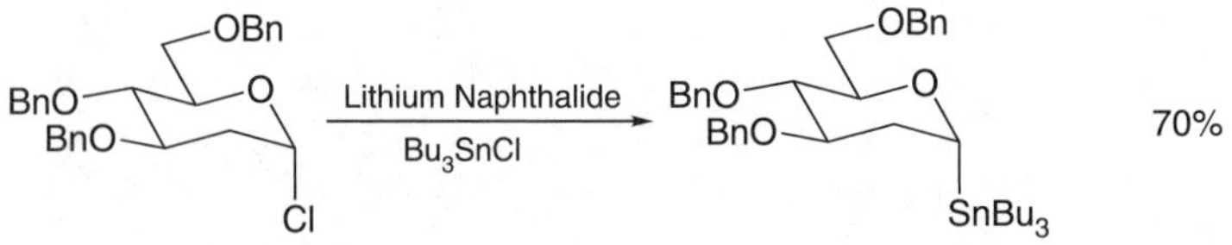

SCHEME 7.63 Glycosidic halogen–metal exchanges.

7.3.1.3 Halogen–Metal Exchanges

Complementary to hydrogen–metal exchanges and metal–metal exchanges are halogen–metal exchanges. This method has been particularly useful in conversions involving glycosyl chlorides. As shown in Scheme 7.62, Sinaÿ et al. [180] treated 2-deoxy-3,4,6-tri-*O*-benzyl-D-glucopyranosyl chloride with tri-*n*-butyl stannyllithium. The result was the formation of the desired stannyl glycoside in 85% yield with inversion of the stereochemistry at the anomeric center. Thus, the α-pyranosyl chloride yielded the β-stannyl glycoside.

7.3.2 C-1 Lithiated Anomeric Carbanions by Reduction

Complementary to the reaction shown in Scheme 7.62 is a method reported by the same group, demonstrating the formation of the corresponding α stannyl glycoside [180]. As shown in Scheme 7.63, this reaction provided a 70% yield of the desired product utilizing lithium naphthalide to form the intermediate C-1 lithiated sugar. Moreover, the intermediate lithioglycosides relevant to Scheme 7.62 and Scheme 7.63, shown in Figure 7.21, were demonstrated to be configurationally stable at −78°C. These observations are in complete agreement with the observed stereochemical stability of anomeric cuprates as reagents triggering the formation of *C*-glycosides on reaction with Michael acceptors [178].

The transient anomeric lithiation shown in Scheme 7.63 differs from standard halogen–metal exchanges in that a reductive mechanism is involved. The use of reductive methods illustrated in Figure 7.22 is complementary to the exploitation of metal exchange reactions for the formation of glycosidic nucleophiles given that strategies towards selectivity in the anomeric stereochemistry become available. Additional reductive anomeric metallations are known in the context of glycosidic sulfones as starting materials [181,182].

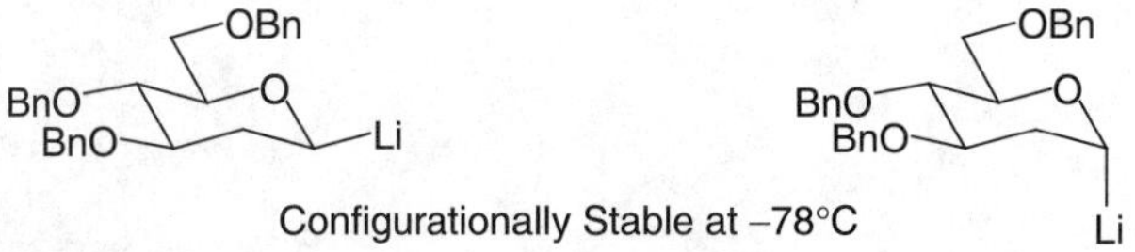

FIGURE 7.21 Stable lithioglycosides.

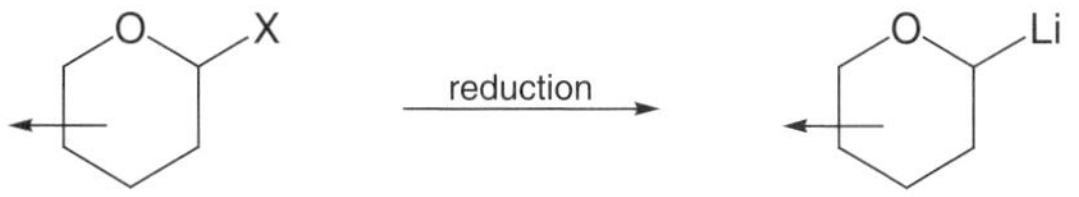

FIGURE 7.22 Reductive halogen–metal exchanges.

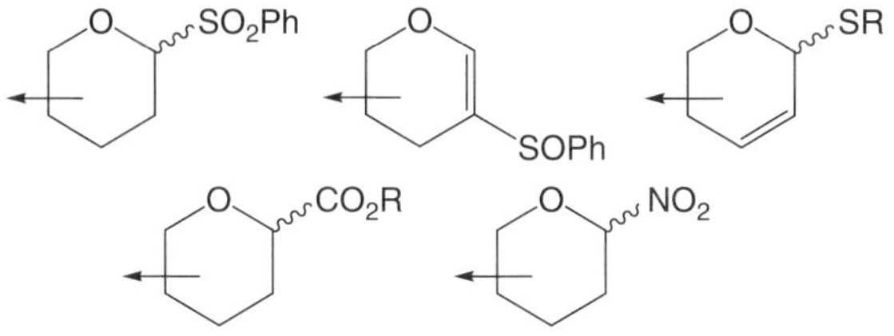

FIGURE 7.23 Electron deficient sugar derivatives.

7.3.3 C-1 Carbanions Stabilized by Sulfones, Sulfoxides, Carboxyl and Nitro Groups

The use of carbanions in the preparation of *C*-glycosides is often dependent on the ability to generate stable lithiated species. Consequently, the use of electron withdrawing groups as a means for anion stabilization is often employed. Figure 7.23 shows several representations of sugar derivatives containing electron-withdrawing groups in order to facilitate their C-1 deprotonation and subsequent utility in the formation of *C*-glycosides.

7.3.3.1 Sulfone Stabilized Anions

Of the examples shown in Figure 7.23, glycosyl phenylsulfones are extremely useful, particularly in conjunction with reductive methods discussed in Section 7.3.2. For example, Scheme 7.64 illustrates work reported by Beau and Sinaÿ [181], in which the illustrated phenylsulfone was utilized in the formation of β-lithiated sugars to be used in deuterium labeling experiments.

In the application of this methodology to the preparation of *C*-glycosides, Beau and Sinaÿ [183] utilized dimethyl carbonate as an electrophile. As shown in Scheme 7.65, the resulting product mixture favored the β oriented ester group in a ratio of 20:1. Subsequent reductive cleavage of the phenylsulfone was accomplished on treatment with lithium naphthalide giving the α-*C*-glycoside in 72% overall yield. Similar results were observed when phenyl benzoate was used as an electrophile.

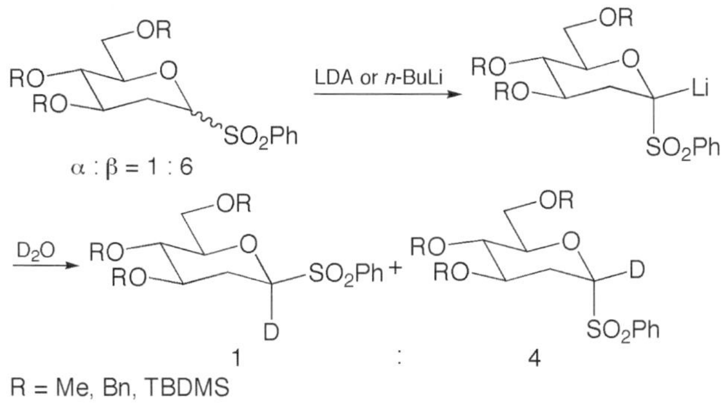

SCHEME 7.64 Lithiation of glycosyl phenylsulfones.

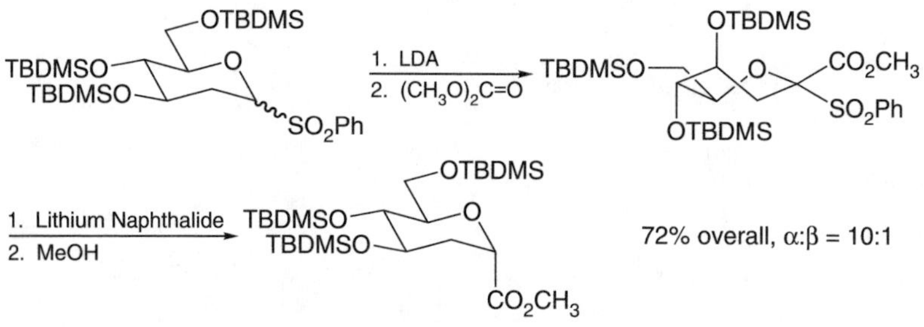

TBDMSO, TBDMSO, OTBDMS, O, SO$_2$Ph — 1. LDA 2. $(CH_3O)_2C{=}O$ → TBDMSO, OTBDMS, O, CO_2CH_3, SO_2Ph, OTBDMS — 1. Lithium Naphthalide 2. MeOH → TBDMSO, TBDMSO, OTBDMS, O, CO_2CH_3

72% overall, $\alpha:\beta$ = 10:1

SCHEME 7.65 Lithiation of glycosyl phenylsulfones.

At first glance, the results shown in Scheme 7.65 seem contrary to the stereochemical consequences of the reactions discussed thus far. This observation can be reconciled upon the realization that the resulting carbonyl-based substituent at the anomeric position can absorb the intermediate anion as an enolate. Subsequent quenching of the anion produces the observed α configuration. Additional work in this area was applied to the preparation of *C*-disaccharides [184].

7.3.3.2 Sulfide and Sulfoxide Stabilized Anions

To date, only glycosidic phenylsulfones have been discussed as substrates for the formation of glycosidic anions. However, as illustrated in Figure 7.23, other sugar derivatives are useful. Among these are glycal analogs. Early examples of the utilization of glycal analogs were reported by Schmidt et al. [185], and are shown in Scheme 7.66. In this report, the ability to directly lithiate protected glycals and introduce new glycosidic moieties was documented. As shown, metallations were accomplished utilizing butyllithium and potassium *tert*-butoxide at low temperatures.

The observations illustrated in Scheme 7.66 were complemented in the same report by the demonstration of the utility of 2-phenylsulfide activated glycals as *C*-glycosidation substrates. As shown in Scheme 7.67, 2-phenylsulfide-substituted glycals were easily prepared on treatment with 1,8-diazabicyclo[5.4.0]undec-7-ene (DBU) and chlorophenylsulfide. Subsequent metallations were then accomplished utilizing *tert*-butyllithium and LDA. Treatment of the anionic species with a variety of aldehydes triggered the formation of various *C*-glycosides. This work was later extended to the use of 2-phenylthio-substituted glycals [33] and phenyl thioglycosides [186].

7.3.3.3 Carboxy Stabilized Anions

As previously indicated, the carboxyl group is an excellent stabilizing group for anomeric anions. C-1 carboxyl glycosides are easily prepared from a number of methods discussed in Section 7.2. Additionally, many naturally occurring C-1 carboxyl glycosides are known. Among these is 3-deoxy-D-*manno*-2-octulosonic acid. This sugar has been identified as a component of the lipopolysaccharides of Gram-negative bacteria [187,188]. As shown in Scheme 7.68, Luthman et al. [189] demonstrated the ability to derivatize this sugar through C-1 metallation and subsequent

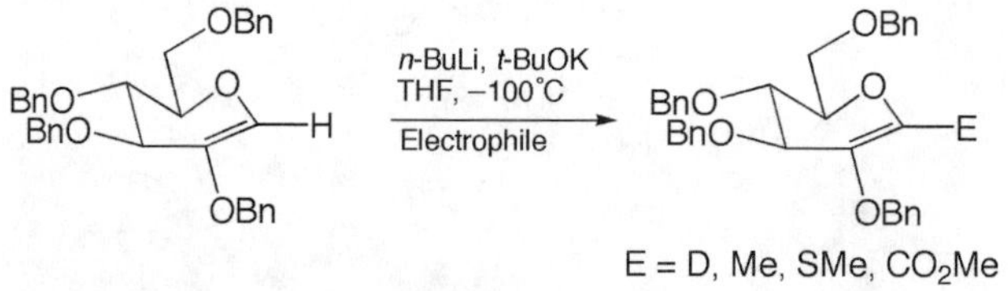

BnO, BnO, OBn, O, H, OBn — *n*-BuLi, *t*-BuOK THF, −100°C Electrophile → BnO, BnO, OBn, O, E, OBn

E = D, Me, SMe, CO_2Me

SCHEME 7.66 Glycal metallations.

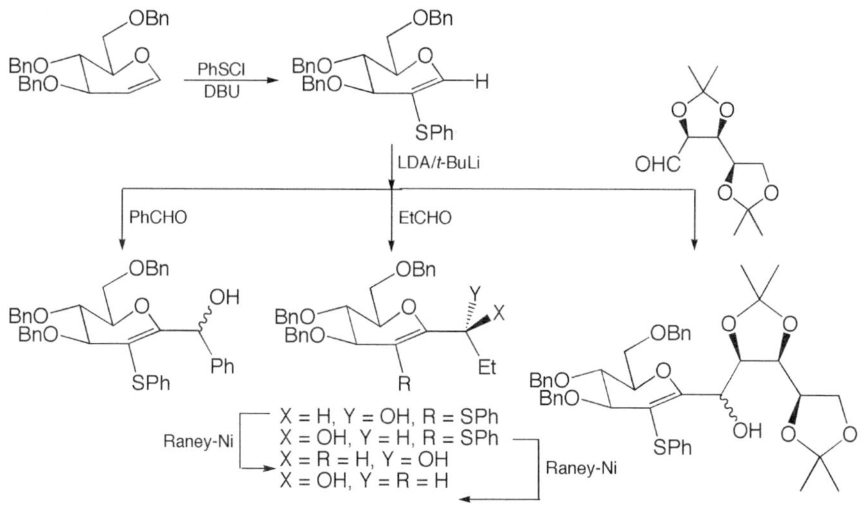

SCHEME 7.67 C_2 activated glycals in *C*-glycosidations.

alkylation. The results, obtained from both the methyl and ethyl ester derivatives of this sugar demonstrate moderate yields and high stereoselectivities. For example, utilizing *tert*-butyl bromoacetate, the yield of the desired alkylated sugar was only 30% with a ratio of 95:5 favoring the β anomer. However, utilizing formaldehyde afforded a 62% yield of the desired hydroxymethyl analog exhibiting a ratio of 90:10 favoring the β anomer. These results, accompanied by additional observations, are shown in Table 7.3. In all cases, metallation was accomplished utilizing LDA.

7.3.3.4 Nitro Stabilized Anions

Of the anionic activating groups shown in Figure 7.23, the nitro group has yet to be discussed. This group was among the first to be utilized in the preparation of *C*-glycosides with its usage appearing as early as 1983. At that time, Vasella et al. [139,140] demonstrated the preparation of these compounds, as shown in Scheme 7.69. Beginning with the oxime derived from 2,3,4,6-tetra-*O*-benzyl-D-glucopyranose, condensation with *p*-nitrobenzaldehyde afforded the imine shown. Subsequent ozonolysis afforded a 73% overall yield of the nitroglycoside as a mixture of anomers. Further exploration of the utility of these compounds as applied to the preparation of *C*-glycosides

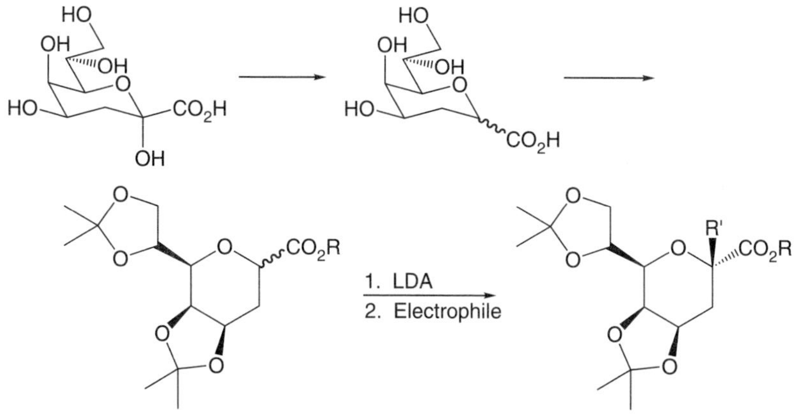

SCHEME 7.68 Carboxyglycosides in *C*-glycosidations.

TABLE 7.3
Carboxyglycosides in *C*-Glycosidations

R	R′	% α	% β	Total yield (%)
CN	CH_3	<5	>95	47
CN	CH_2CH_3	10	90	55
CH_2OH	CH_2CH_3	10	90	47
$COCH_3$	CH_2CH_3	5	95	62
CH_3	CH_2CH_3	<5	>95	50
$CH_2C{\equiv}CH$	CH_2CH_3	10	90	50
$CH_2CO_2^tBu$	CH_2CH_3	<5	>95	30
CH_2Ph	CH_3	5	95	67
$CH_2CH_2CO_2CH_3$	CH_2CH_3	25	75	27

showed their ease in deprotonation utilizing bases as mild as potassium carbonate. Furthermore, deprotonation in the presence of formaldehyde, followed by acetylation, gave the acetoxymethyl derivative shown. Final cleavage of the nitro substituent was easily accomplished in 87% yield utilizing radical conditions and providing the desired *C*-glycoside exclusively as the β anomer. As a final note, it is important to mention that utilizing nitro stabilized anomeric anions, β-elimination of the substituent at C-2 is rarely observed.

7.4 SYNTHESIS OF *C*-GLYCOSIDES VIA TRANSITION METAL-BASED METHODOLOGIES

Whereas the combination of electrophiles with sugar nucleophiles provides routes to *C*-glycosides, these conditions are not applicable when the electrophiles bear centers prone to epimerization. However, where acid/base chemistry is not applicable because of reactive stereogenic centers, the use of transition metal-mediated cross-coupling reactions may often be utilized. Substantial research in this area has been applied to the couplings between glycals with aryl and other π-conjugated aglycones. Some sugar-derived substrates for this type of reaction are shown in

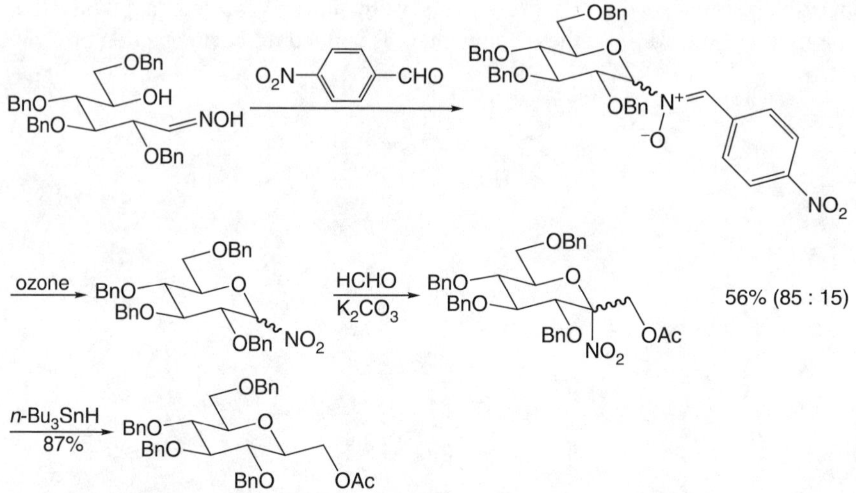

SCHEME 7.69 *C*-Glycosides from nitroglycosides.

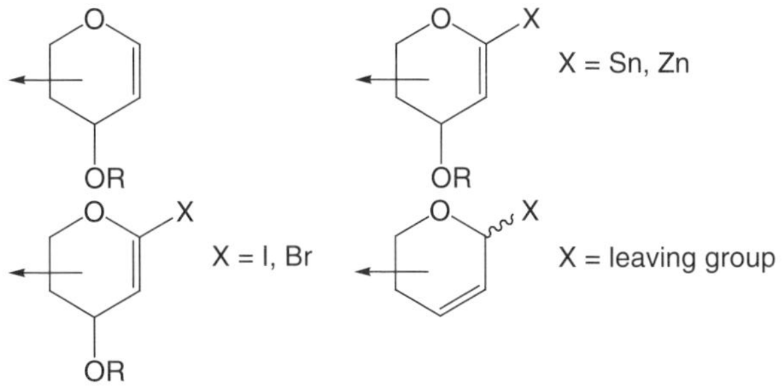

FIGURE 7.24 General substrates for transition metal-mediated *C*-glycosidations.

Figure 7.24 and include glycals, metallated glycals, halogenated glycals and sugar derivatives bearing allylic anomeric leaving groups.

7.4.1 Direct Coupling of Glycals with Aryl Groups

The utility of glycals as substrates for *C*-glycosidations is illustrated by their availability from commercial sources as well as the examples presented thus far. Consequently, the actual coupling of these compounds with both activated and unactivated aryl groups is complementary to the general chemistry surrounding *C*-glycosides. In this section, coupling reactions with unactivated, metallated and halogenated aromatic rings are discussed.

7.4.1.1 Arylation Reactions with Unsubstituted Aromatic Rings

Beginning with the simplest of cases, the direct coupling of glycals with benzene is discussed. As shown in Scheme 7.70, this reaction, reported by Czernecki and Gruy [190], was accomplished utilizing palladium acetate as a catalyst. The yields for this reaction were generally in the range of 40 to 90%, and resulted in the selective formation of the α configuration at the anomeric center. This chemistry was later adapted to substituted benzenes [191].

7.4.1.2 Arylation Reactions with Metallated Aromatic Rings

Given the relatively low yields for transition metal-mediated *C*-glycosidations with unactivated aromatic aglycones, the use of substituted aryl groups was explored. In Section 7.2, limited examples relating the coupling of glycals to aryltin compounds were explored. In this section, we expand upon this concept in the context of aryl mercury reagents, as illustrated in Scheme 7.71.

Salt additives have been used as a potential means of enhancing arylation reactions while decreasing the amount of palladium catalyst required. As shown in Scheme 7.72 [192], this concept was applied utilizing lithium chloride in the preparation of *C*-nucleoside analogs. Moreover, by varying the salt from lithium chloride to lithium acetate and sodium bicarbonate, product yield and

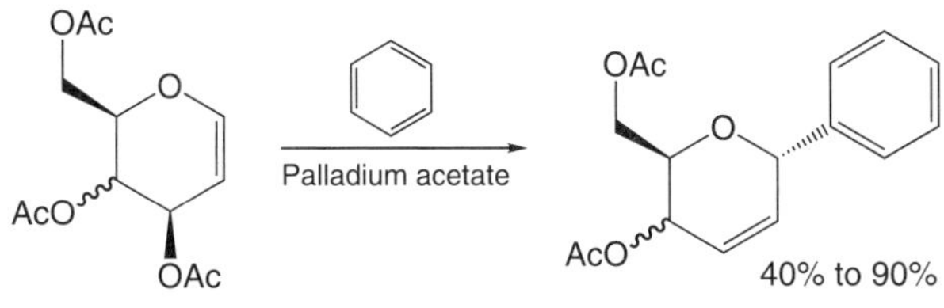

SCHEME 7.70 Palladium catalyzed coupling of glycals with benzene.

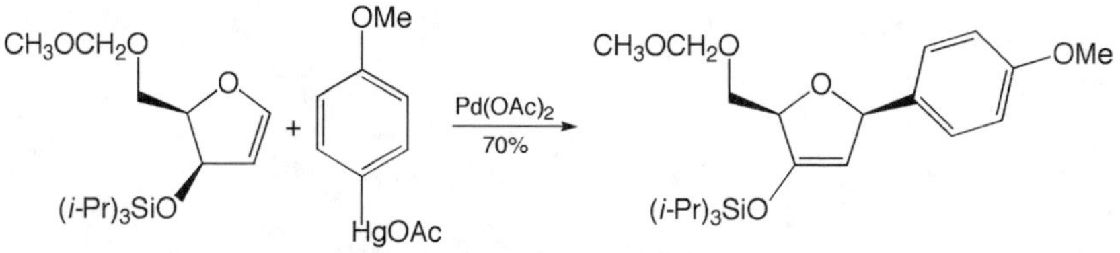

SCHEME 7.71 *C*-Glycosidation with aryl mercury reagents.

composition could be influenced [192,193]. Although this chemistry was broadly applied to acetate protected pyranose glycals, alternate protecting groups and furanose glycals were also useful [194].

7.4.1.3 Arylation Reactions with Halogenated Aromatic Rings

Where mercury-substituted aryl groups have proven useful in palladium-mediated *C*-glycosidations with glycals, aryl iodides have been shown to be complementary alternatives in these reactions. As shown in Scheme 7.73, Kwok et al. [195] produced couplings with both mercury and iodo compounds. In both cases, the yields were comparable and the reactions showed a consistent preference for α anomers. However, in the context of furanose glycals, this trend was noted to reverse, with products of these reactions exhibiting the β configuration [106]. Finally, with respect to the preparation of *C*-nucleosides, iodopyrimidines have provided valuable alternatives to the mercury reagents discussed in the previous section [196,197].

7.4.2 Coupling of Substituted Glycals with Aryl Groups

Although the direct use of glycals as *C*-glycosidation substrates has been fruitful, the results can often be improved upon utilizing C-1 activated glycals. Such activating groups include metals and halogens. Successful catalysts used to effect the coupling of these compounds with metal- or halo-substituted aryl compounds include nickel and palladium complexes.

Beginning with a discussion of the utility of tin-substituted glycals, Dubois and Beau [198] utilized 2,3,6-tri-*O*-benzyl-1-tri-*n*-butylstannyl-glucal in coupling reactions with various aromatic substrates. As shown in Scheme 7.74, tetrakis(triphenylphosphine)palladium catalyzed reaction with bromobenzene provided an 88% yield of the desired product. Additionally, when

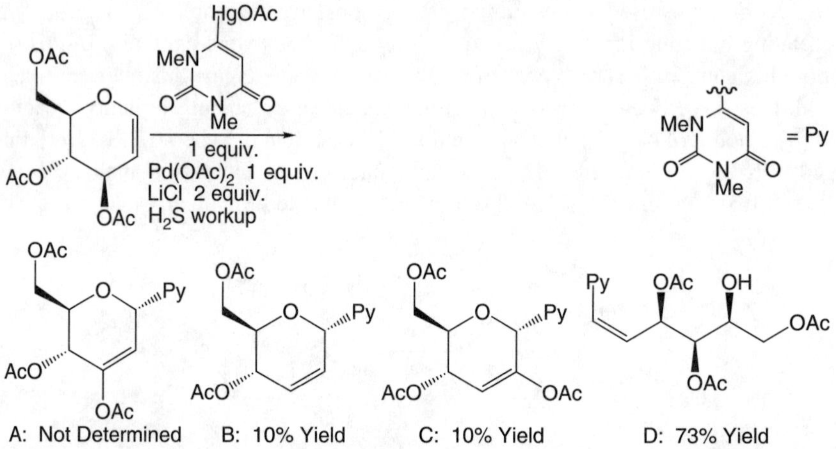

SCHEME 7.72 *C*-Glycosidation with aryl mercury reagents.

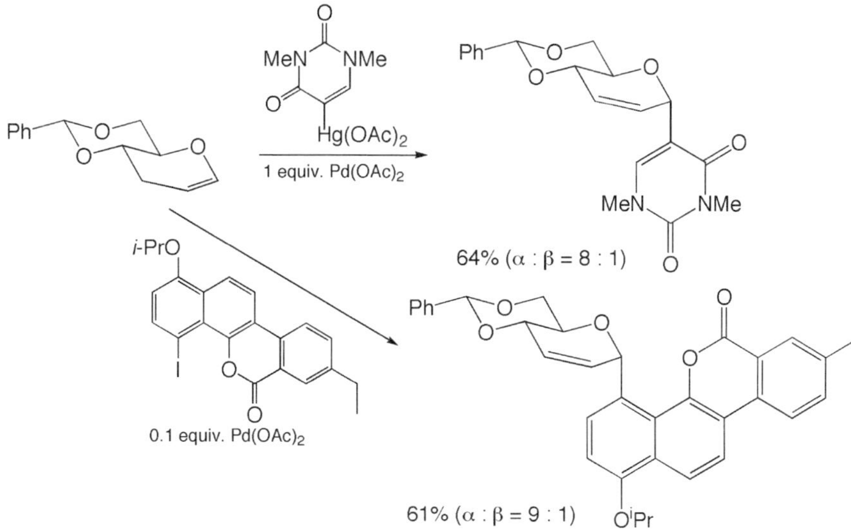

SCHEME 7.73 *C*-Glycosidation with aryl mercury/iodide reagents.

p-nitrobenzoyl chloride was used, dichloro dicyanopalladium produced the formation of the illustrated ketone in 71% yield. Further elaborations of this chemistry demonstrated its compatibility with both unprotected hydroxyl groups as well as very bulky aromatic bromides [199].

While C-1 stannylated glycals have been useful in the formation of *C*-glycosides, zinc- and iodo-substituted glycals have provided complementary avenues to these compounds. Utilizing either of these classes of glycals, the results of the coupling reactions were shown to be highly susceptible to the transition metal catalyst used [200]. Finally, centering specifically on C-1 iodo glycals, Friesen and Loo [176] initiated coupling reactions with both arylzinc and arylboron compounds.

7.4.3 Coupling of π-Allyl Complexes of Glycals

Thus far, all couplings have involved glycals/glycal derivatives and aryl compounds. However, the generality of this chemistry also applies to the use of 1,3-dicarbonyl compounds and other

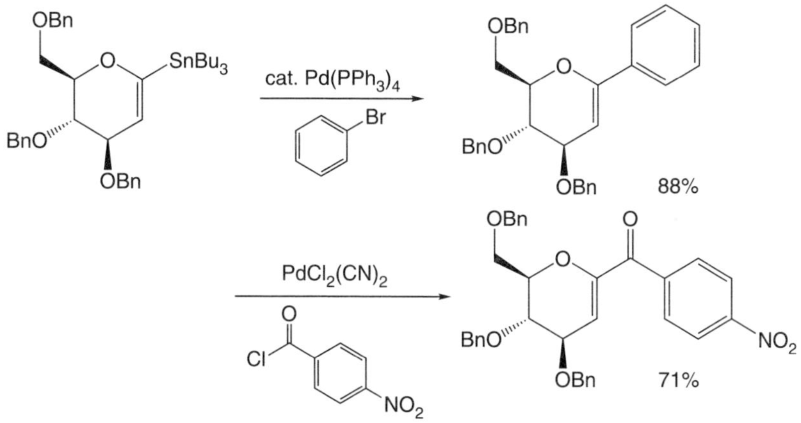

SCHEME 7.74 *C*-Glycosidation with stannyl glycals.

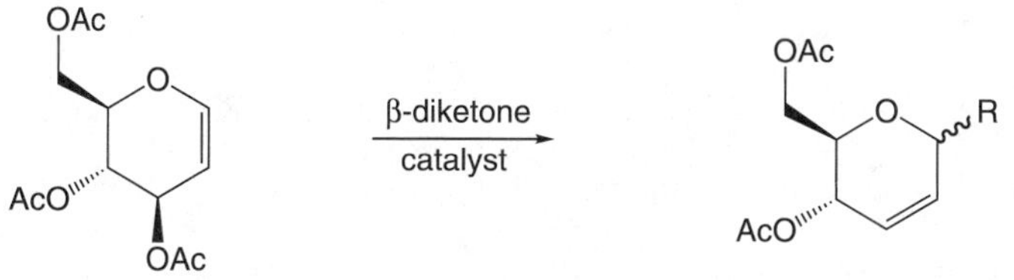

SCHEME 7.75 *C*-Glycosides from glycals and 1,3-diketones.

molecules with suitably electron deficient centers. These reactions take place with both the neutral and anionic forms of these compounds and generally exhibit moderate to high yields. Unlike the examples of the previous section, these reactions produce *C*-glycosides bearing unsaturation between C-2 and C-3. Moreover, the variety of compatible hydroxyl protecting groups includes acetyl, ketal and benzyl groups.

An early example of the utility of this approach to the formation of *C*-glycosides was presented by Yougai and Miwa [201], which compares the use of different Lewis acids and transition metal catalysts. The general reaction reported is shown in Scheme 7.75 and utilizes tri-*O*-acetyl-D-glucal as a substrate for coupling with a variety of dicarbonyl compounds. The results of the cited study are summarized in Table 7.4.

As the technology developed in this area, the advantages of utilizing glycal activating groups became apparent. Of interest is the use of intermediately formed 2,3-unsaturated *O*-phenylglycosides. Utilizing these intermediates, transition metal catalysis afforded a variety of 1,3-dicarbonyl compounds with anomeric ratios as high as 100% for either anomer [202,203]. An observation of particular interest is the retention of the anomeric configuration on conversion from the phenolic glycoside to the corresponding *C*-glycoside. Further work in this area demonstrated the utility of deprotonated 1,3-dicarbonyl compounds in comparable reactions [204], as well as 2,3-anhydrosugars [205] and glycosyl carbonates [206].

Mechanistically, the reactions described in this section occur via complexation with the catalyst forming a π-allylpalladium complex (Scheme 7.76). As the palladium approaches from the side of the sugar opposite to that of the glycosidic substituent, the incoming nucleophile approaches from the opposite side, giving the observed products.

TABLE 7.4
C-Glycosides from Glycals and 1,3-Diketones

β-Dicarbonyl Compound	*O*-Acetylated Glycal	Catalyst	% Yield	α:β
Acetylacetone	Glycal	$Pd(PhCN)_2Cl_2$	83	6:1
Acetylacetone	Glycal	BF_3	73	5:1
Acetylacetone	Galactal	$Pd(PhCN)_2Cl_2$	59	α only
Acetylacetone	Galactal	BF_3	72	α only
Acetylacetone	Allal	$Pd(PhCN)_2Cl_2$	62	5:1
Acetylacetone	Allal	BF_3	81	5:1
Methyl acetoacetate	Glucal	$Pd(PhCN)_2Cl_2$	85	4:1
Methyl acetoacetate	Xylal	$Pd(PhCN)_2Cl_2$	65	1:4
Ethyl benzoylacetate	Galactal	$Pd(PhCN)_2Cl_2$	65	α only
Ethyl benzoylacetate	Galactal	BF_3	81	α only
Ethyl 2-oxocyclohexane carboxylate	Glucal	$Pd(PhCN)_2Cl_2$	82	—
Ethyl 2-oxocyclohexane carboxylate	Glucal	BF_3	82	—

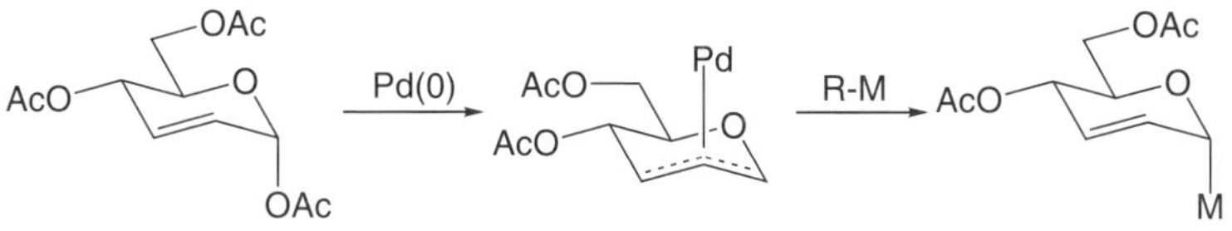

SCHEME 7.76 Mechanisms of palladium-mediated couplings.

7.5 SYNTHESIS OF C-GLYCOSIDES VIA ANOMERIC RADICALS

The *C*-glycosidations discussed thus far involve ionic mechanisms. However, neutral free radicals are also useful in the formation of *C*-glycosides. Specifically, the ability to form free radicals at anomeric centers, illustrated in Figure 7.25, provides an additional dimension to the developing *C*-glycosidation technologies. An interesting feature of free radical chemistry applied to *C*-glycosidations is the ability to couple these species with numerous readily available radical acceptors. The stereochemical course of these reactions generally provides α selectivity. This chemistry can be applied to acylated sugars and tolerates many substrate functional groups. Finally, intramolecular radical reactions provide avenues into compounds bearing difficult stereochemical relationships as well as fused ring systems.

7.5.1 Sources of Anomeric Radicals and Stereochemical Consequences

Anomeric free radicals can be generated from sugar substrates bearing a variety of activating groups. Additionally, the conditions required for the generation of these reactive species include both chemical and photolytic methods. Through combining specific reaction conditions and appropriate activating groups, both α and β oriented *C*-glycosides are available. In this section, the generation of anomeric radicals is discussed with respect to the utility of various activating groups. In addition, the stereochemical consequences of reaction conditions, as applied to different activating groups, are explored.

7.5.1.1 Nitroalkyl C-Glycosides as Radical Sources

One source of anomeric radicals and evidence for the preference of axial radical orientations has already been introduced in work reported by Vasella et al. [139,140]. The specific reaction, outlined in Scheme 7.77, involves the cleavage of a nitro group under radical forming conditions.

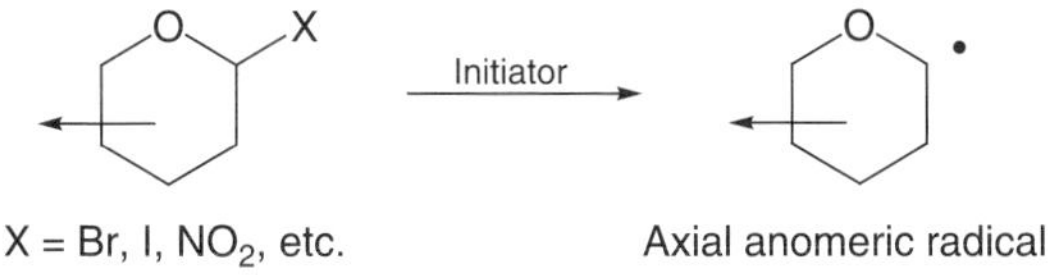

X = Br, I, NO_2, etc.   Axial anomeric radical

FIGURE 7.25 Anomeric radicals.

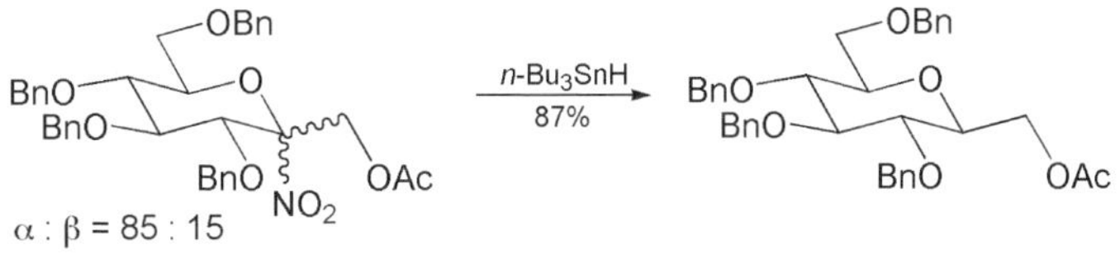

α : β = 85 : 15 n-Bu_3SnH, 87%

SCHEME 7.77 Axially oriented anomeric radicals.

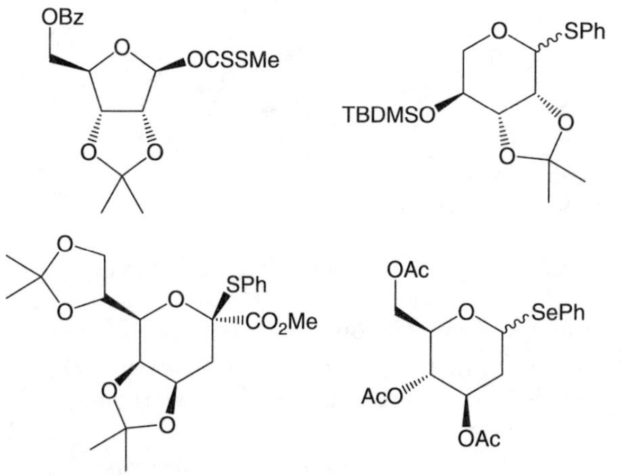

FIGURE 7.26 Substrates for anomeric radical formation.

The starting nitroglycoside existed as a mixture of anomers and the product was composed of the β anomer exclusively. This observation suggests the formation of an α oriented radical, which accepts a hydrogen radical from tributyltin hydride. Specific delivery of the hydrogen radical to the stable α anomeric radical thus explains the exclusive formation of the observed β-*C*-glycoside. Subsequent exploration of the actual stability of anomeric radicals was reported by Brakta et al. [207], and involves the migration of a free radical to the anomeric center of preformed *C*-glycosides.

7.5.1.2 Radicals from Activated Sugars

As indicated in the introduction to this chapter, glycosidic radicals prefer to adopt an α anomeric configuration. However, this preference can be manipulated through the use of a variety of anomeric activating groups and radical initiating conditions. Some glycosidic substrates used in the generation of anomeric free radicals are shown in Figure 7.26, which include oxygen-derived glycosides [208], thioglycosides [95,96] and selenoglycosides [209]. From these compounds, as well as utilizing either chemical or photolytic radical formations, both α and β *C*-glycosides have been shown to be accessible.

In addition to the use of substrates similar to those shown in Figure 7.26, glycosyl halides have found much use in the chemistry of glycosidic radicals. For example, glycosyl fluorides and bromides tend to produce α anomeric radicals, while chlorides and iodides show substantially less selectivity. A study involving the reduction of glycosyl bromides under various conditions was reported by Somsák et al. [210], and is outlined in Scheme 7.78. In both illustrated examples, the hydride was delivered with retention of the anomeric configurations, thus supporting the preferential formation of α anomeric radicals.

In addition to the above examples, anomeric radicals are subject to 1,2-shifts. This observation was extensively studied in the context of acetoxy and benzoyloxy groups migrating from C-2 to C-1 in a *cis* fashion providing access to products contrary to those expected via the anticipated α configuration of the native radical [211,212].

7.5.2 Anomeric Couplings with Radical Acceptors

The intermolecular reactions between glycosidic free radicals and radical acceptors have provided a variety of novel structures inaccessible from other glycosidation pathways. Although the types of

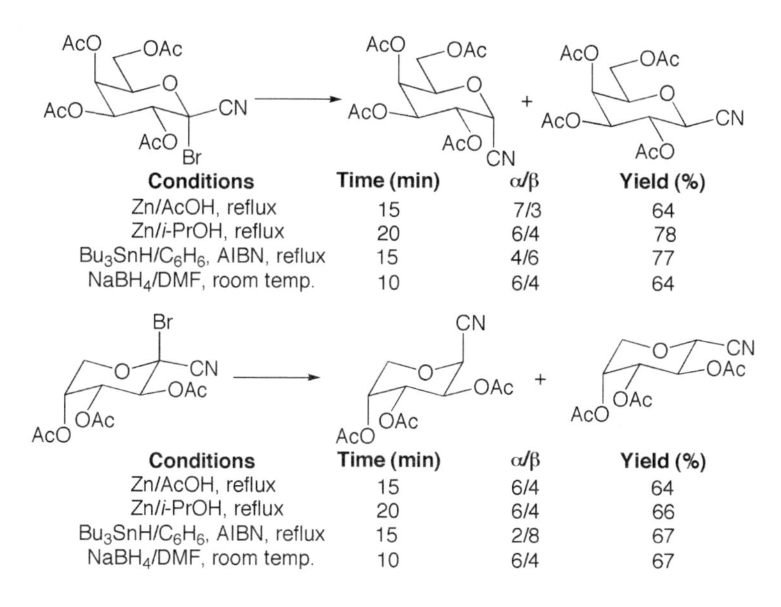

Conditions	Time (min)	α/β	Yield (%)
Zn/AcOH, reflux	15	7/3	64
Zn/*i*-PrOH, reflux	20	6/4	78
Bu_3SnH/C_6H_6, AIBN, reflux	15	4/6	77
$NaBH_4$/DMF, room temp.	10	6/4	64

Conditions	Time (min)	α/β	Yield (%)
Zn/AcOH, reflux	15	6/4	64
Zn/*i*-PrOH, reflux	20	6/4	66
Bu_3SnH/C_6H_6, AIBN, reflux	15	2/8	67
$NaBH_4$/DMF, room temp.	10	6/4	67

SCHEME 7.78 Reduction of glycosyl bromides.

radical acceptors useful are quite varied, the different anomeric activating groups used in the generation of free radicals substantially influence the yields and stereochemical outcome of the reactions. The two divisions within this section address the use of glycosyl halides and non-halogenated sugar derivatives in the exploitation of free radical *C*-glycosidations.

7.5.2.1 Non-Halogenated Radical Sources

Understanding the chemistry surrounding the formation and subsequent stereochemical preference of anomeric radicals, the various reactions utilizing these species in the preparation of *C*-glycosides can be discussed. Beginning with nitroglycosides, Dupuis et al. [213] produced reactions with acrylonitrile. As shown in Scheme 7.79, the radicals were formed utilizing tributyltin hydride. The observed products were isolated in approximately 50% yields and, in the case of the pyranose derivative, exhibited no stereochemical preference. Similar reactions beginning with glycosidic methylthiocarbonates [208] and phenylthioglycosides [95,96] are known.

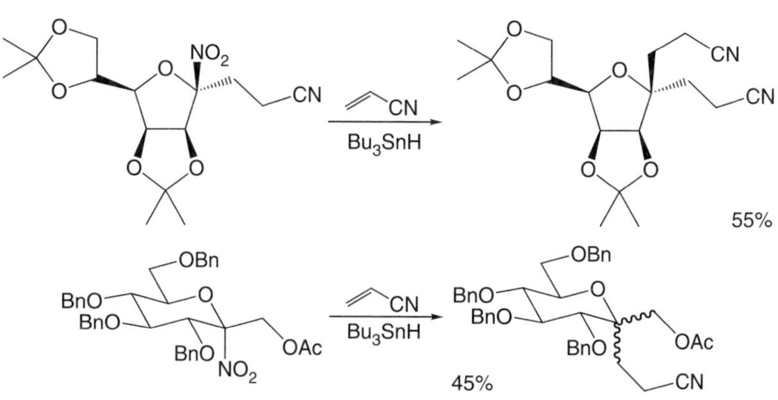

SCHEME 7.79 Radical *C*-glycosidations with nitroglycosides.

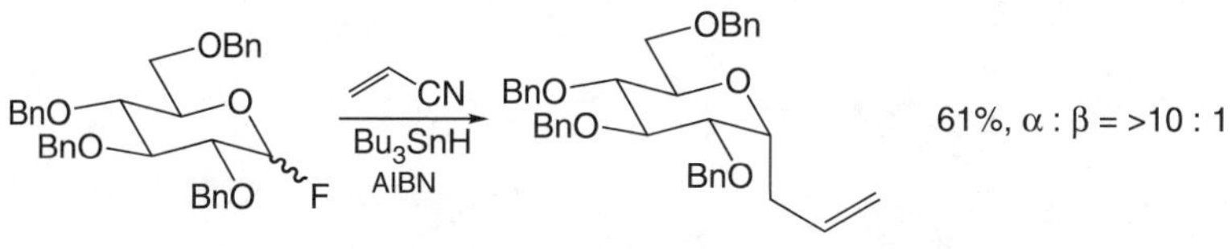

SCHEME 7.80 Radical *C*-glycosidations with glycosylfluorides.

7.5.2.2 Glycosyl Halides as Radical Sources

With all the anomeric activating groups useful in the generation of sugar-derived free radicals, glycosyl halides have been the most exploited. In the remainder of this section, various reactions involving glycosyl halides will be discussed. Additionally, the stereochemical consequences of these reactions will be analyzed.

Glycosyl fluorides, already addressed as substrates for cyanations and allylations, are excellent substrates for radical-mediated *C*-glycosidations. As shown in Scheme 7.80, Nicolaou et al. [43] demonstrated such reactions in the coupling of 2,3,4,6-tetra-*O*-benzyl-D-glucopyranosyl fluoride with acrylonitrile. The reaction gave a 61% yield with an anomeric ratio of greater than 10:1 favoring the α isomer.

Complementary to the use of glycosyl fluorides is the use of glycosyl chlorides under both ionic and radical conditions [97,214]. Similarly, reactions utilizing glycosyl bromides are known [215,216]. Interestingly, while low yields and poor anomeric selectivities are generally noted for pentose-derived glycosyl bromides, this observation does not translate to higher sugars [216]. Regarding glycosidic free radical generation, chemical methods are not limited to tin hydrides or irradiation. Catalysts such as vitamin B_{12} have also been used [217]. Finally, a wealth of information is available on the use of radical acceptors complementary to the already mentioned acrylonitrile. These acceptors range from simple molecules such as methyl vinyl ketone [217] to more complex conjugated carbonyl compounds capable of providing avenues to *C*-disaccharides [218,219] with one notable example reported by Giese et al. [220] (Figure 7.27).

7.5.3 Intramolecular Radical Reactions

Through the use of free radicals, otherwise inaccessible *C*-glycosidic structures are easily formed. However, the variability in yields and the frequently observed inconsistencies regarding the degree of stereochemical induction in intermolecular radical reactions have highlighted limitations in the general applicability of this technology. In order to alleviate some of these problems, the use of the intramolecular delivery of free radical acceptors was developed. An early example of intramolecular radical reaction is shown in Scheme 7.81. As illustrated, Giese et al. [221] stereospecifically formed bicyclic dideoxy sugars. The isolated yield was 53% and the stereochemical approach of the allyl group was entirely dictated by the stereochemistry at C-3. As shown in Figure 7.28, although several intermediate conformations of the radical species are

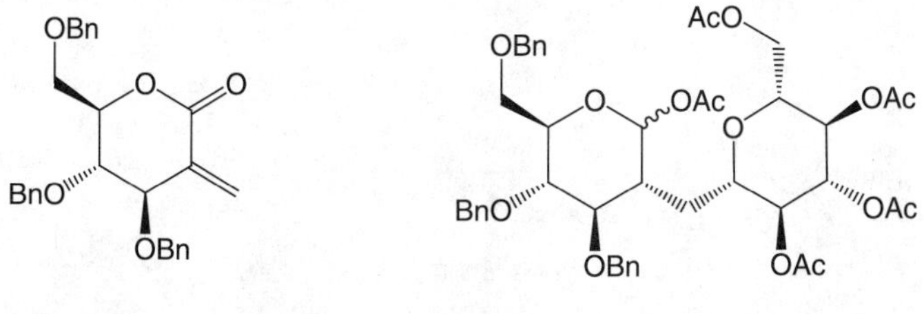

FIGURE 7.27 Radical acceptor and final *C*-disaccharide.

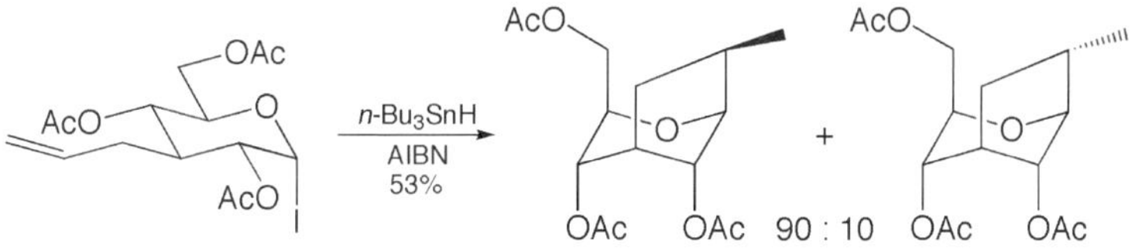

SCHEME 7.81 Intramolecular radical C-glycosidations.

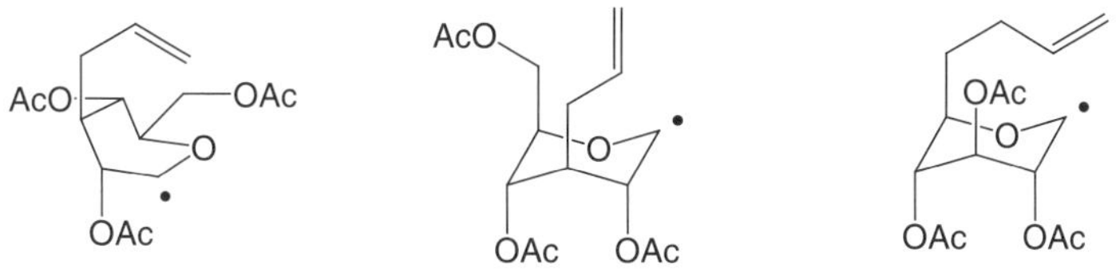

FIGURE 7.28 Proposed conformations of anomeric radicals.

possible, the actual intermediate radical species is postulated to be that bearing the boat conformation. The two other illustrated conformations are ruled out, as illustrated in Scheme 7.82, because of the inability of the 6-deoxy-6-allyl compound to cyclize under the same conditions used in Scheme 7.81.

Complementary to the direct modification of the sugar ring as a means of attaching radical receptors is the selective blocking of the sugar hydroxyl groups. As shown in Scheme 7.83, Mesmaeker et al. [222] substituted the 2-hydroxyl group of glucose derivatives with allyl and trimethylsilyl propargyl groups. Subsequent treatment with tributyltin hydride and a radical initiator produced yields in excess of 90% in both cases. The stereochemistry of the cyclizations

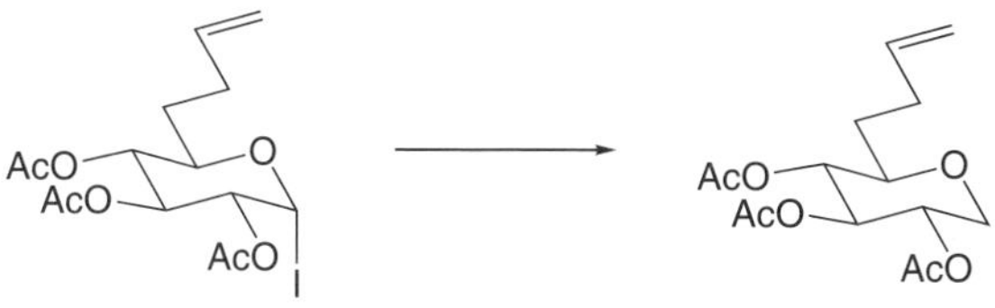

SCHEME 7.82 Attempted radical cyclization of 6-deoxy-6-allyl sugars.

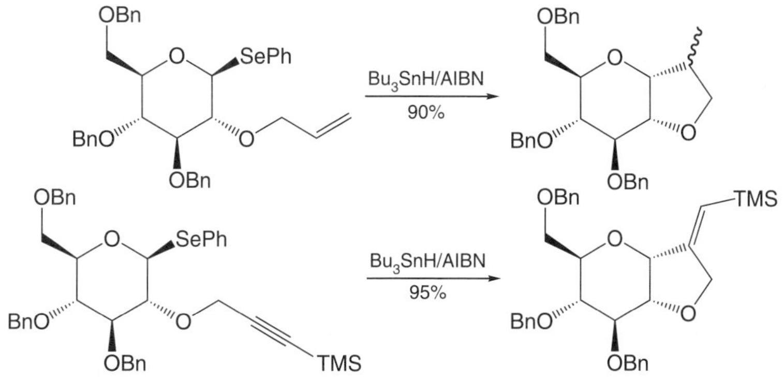

SCHEME 7.83 Intramolecular radical cyclizations of 2-allyl and 2-propargyl selenoglycosides.

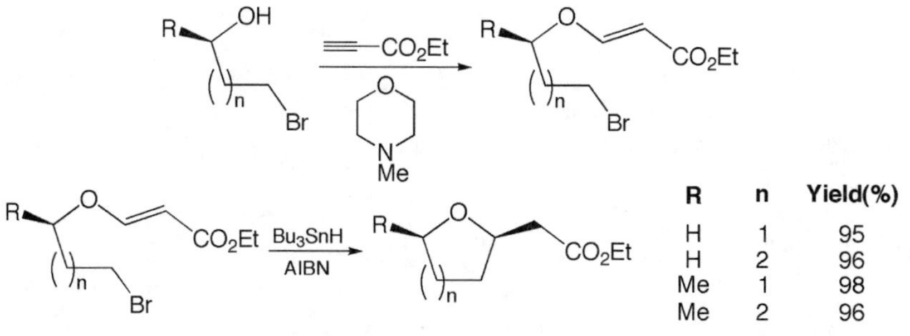

R	n	Yield(%)
H	1	95
H	2	96
Me	1	98
Me	2	96

SCHEME 7.84 Formation of *C*-glycosides via radical cyclizations of acyclic precursors.

took place as expected, with the approach of the radical acceptors forming *cis*-fused ring systems. Finally, unlike the previous examples presented in this section, the radicals in Scheme 7.83 were generated from phenylselenoglycosides. Related work by Stork et al. [223] successfully introduced styrene units to the anomeric position of various sugars via initial silicon tethering of the radical acceptors to the neighboring sugar hydroxyl groups.

In one final example of the utility of intramolecular delivery of free radical acceptors to the preparation of *C*-glycosides, Lee et al. [224] utilized radical chemistry to form the sugar ring from appropriate enol ethers. As shown in Scheme 7.84, the enol ethers were prepared following coupling of ethyl propargylacetate with various alcohols. In all cases, the radicals were generated from tin reagents and radical initiators. The yields observed exceeded 96% and the chemistry was adapted to reactions in which radicals were generated from acetylenes. Among others, this application is contained within the cited report.

7.6 SYNTHESIS OF *C*-GLYCOSIDES VIA REARRANGEMENTS AND CYCLOADDITIONS

As indicated above, the ability to direct reactions to the anomeric centers of modified sugars provides novel routes to *C*-glycosides. In Section 7.5, these reactions involved the intramolecular delivery of radical acceptors to anomeric free radicals. However, as the chemistry of *C*-glycosides developed, the use of a variety of cycloadditions and structural rearrangements became useful. Some substrates for these reactions — both intermolecular and intramolecular — are shown in Figure 7.29. By examining these structures, a variety of specific reaction types can be envisioned and recognized as applicable to the preparation of *C*-glycosides. Among these are Claisen rearrangements, Wittig rearrangements and carbenoid displacements. In this section, these and other reactions will be discussed through examples of their use in the formation of novel *C*-glycosidic structures.

7.6.1 REARRANGEMENTS BY SUBSTITUENT CLEAVAGE AND RECOMBINATION

Many rearrangements involve the cleavage of substituents from a given substrate followed by recombination of the cleaved group to yield a new product. A classical example of this type of reaction is the Wittig rearrangement. In the following paragraphs, this reaction will be presented in the context of its applicability to the preparation of *C*-glycosides. The Wittig rearrangement will also be compared with mechanistically similar rearrangements associated with carbenoids.

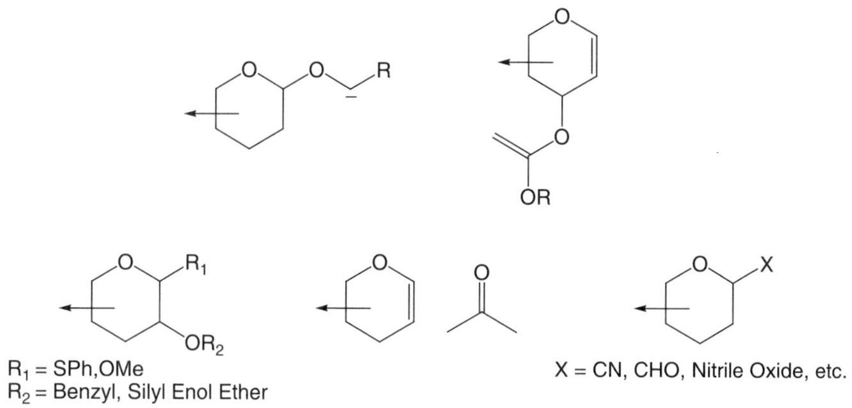

R_1 = SPh,OMe
R_2 = Benzyl, Silyl Enol Ether

X = CN, CHO, Nitrile Oxide, etc.

FIGURE 7.29 Rearrangement and cycloaddition substrates.

7.6.1.1 Wittig Rearrangements

The Wittig rearrangement [225], which is useful in the transformation of ethers to alcohols, has been used in the formation of a variety of *C*-hydroxyalkyl glycosides. The actual mechanism of this reaction is believed to involve a radical cleavage followed by recombination to the final product [226,227]. As shown in Scheme 7.85, Grindley and Wickramage [228] applied this methodology to the preparation of *C*-glycosides from their corresponding *O*-benzyl glycosides. Interestingly, the product composition was highly dependent upon the base used to initiate the reaction.

7.6.1.2 Carbenoid Rearrangements

Where the Wittig rearrangement provides products formed with retention of the configuration at the anomeric center, the use of carbenoids in the formation of *C*-glycosides provides a complementary

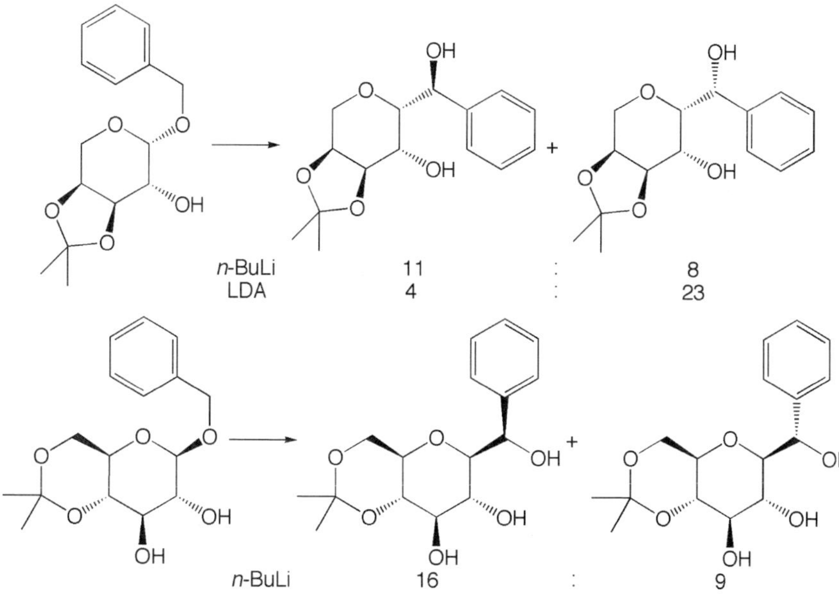

n-BuLi 11 : 8
LDA 4 : 23

n-BuLi 16 : 9

SCHEME 7.85 Wittig rearrangements.

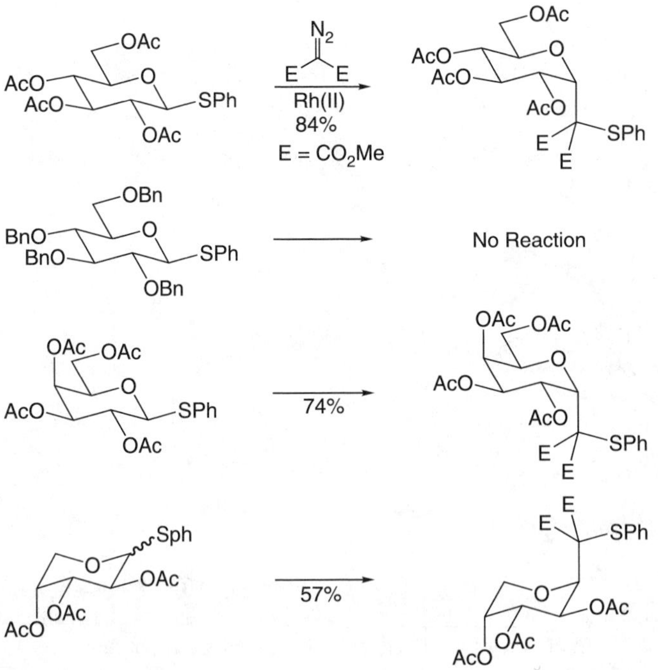

SCHEME 7.86 Carbenoid rearrangements.

approach, allowing the preparation of products with inverted anomeric configurations. As shown in Scheme 7.86, Kametani et al. [229] observed the formation of a variety of *C*-glycosides in yields as high as 84%. Interestingly, these reactions progressed well, utilizing peracetylated sugars, while no reaction was observed with the corresponding perbenzylated sugars.

7.6.2 Electrocyclic Rearrangements Involving Glycals

The rearrangement of modified glycals provides novel strategies for the preparation of *C*-glycosides. Such reactions involve the sigmatropic mechanisms associated with both Cope and Claisen reactions. In this section, these reactions will be overviewed in the context of 3,3-sigmatropic rearrangements.

Sigmatropic rearrangements of glycal derivatives are recognized as useful strategies for the preparation of *C*-glycosides. In fact, this methodology was used as early as 1984 when Tulshian and Fraser-Reid [230] applied the Cope and Claisen methodologies for the preparation of diverse *C*-glycosides from a single sugar analog (Scheme 7.87). Numerous additional examples directly pertaining to Claisen chemistry applied to glycals are known [231,232].

7.6.3 Rearrangements from the 2-Hydroxyl Group

Where Wittig rearrangements have provided *C*-glycosides via rearrangements from anomeric hydroxyl groups and Claisen rearrangements have provided *C*-glycosides via migrations from 3-hydroxyl groups, the delivery of substituents from 2-hydroxyl groups has yielded novel *C*-glycosidic structures. Two examples will be illustrated in this section.

The first example, reported by Craig and Munasinghem [135] and shown in Scheme 7.88, illustrates the types of strained fused-ring systems available from the delivery of nucleophilic species to adjacent centers.

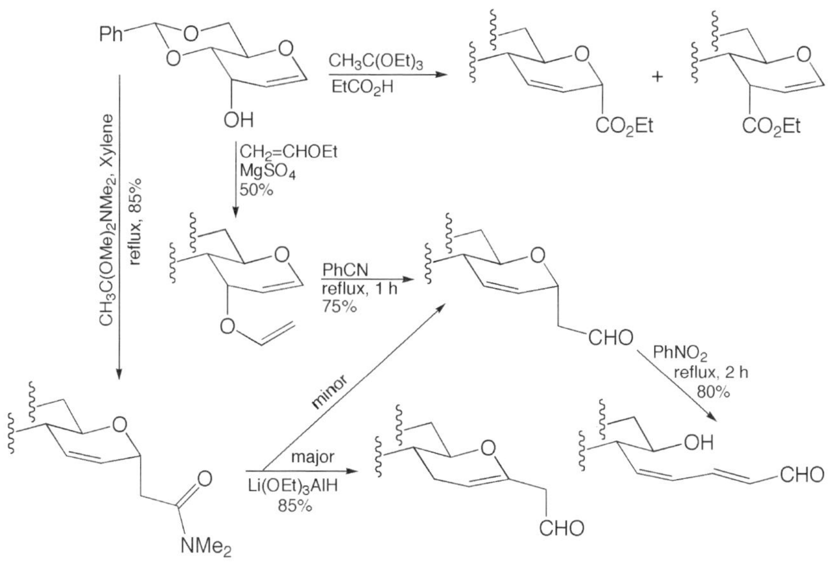

SCHEME 7.87 3,3-Sigmatropic rearrangements.

The second example, shown in Scheme 7.89, was reported by Martin [233] and illustrates how Lewis acid-mediated electrophilic aromatic substitutions can be exploited from an intramolecular strategy. Of particular importance is that following delivery of the aryl group, cleavage of the benzyl ether can be accomplished using hydrogenation techniques. The net result is a two-step rearrangement of a 2-benzylated sugar to a true *C*-glycoside. Although direct rearrangements from 2-hydroxyl groups to anomeric centers have not been well documented, the examples presented in this section provide strategies that are consistent with the rearrangements presented thus far.

7.7 SYNTHESIS OF *C*-GLYCOSIDES VIA FORMATION OF THE SUGAR RING

Each of the sections thus far has dealt with methods used in the conversion of natural sugars, and derivatives thereof, to *C*-glycosides. Deviating from this trend, this section addresses the formation of *C*-glycosides from non-carbohydrate substrates. Specifically, the formation of the sugar ring is the focus of the chemistry discussed. Several methods for the execution of this approach are

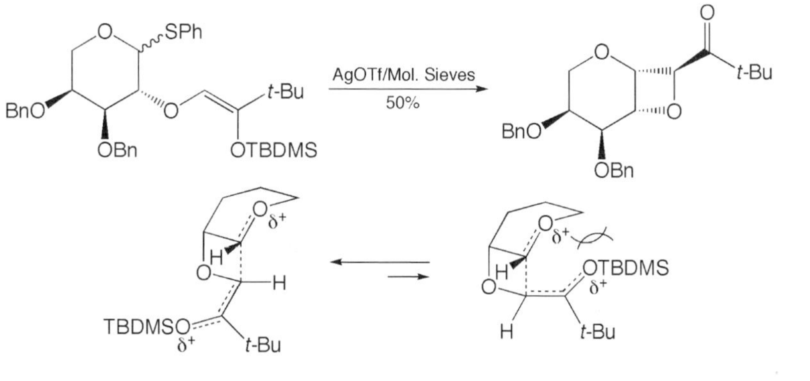

SCHEME 7.88 Rearrangements from 2-hydroxyl groups.

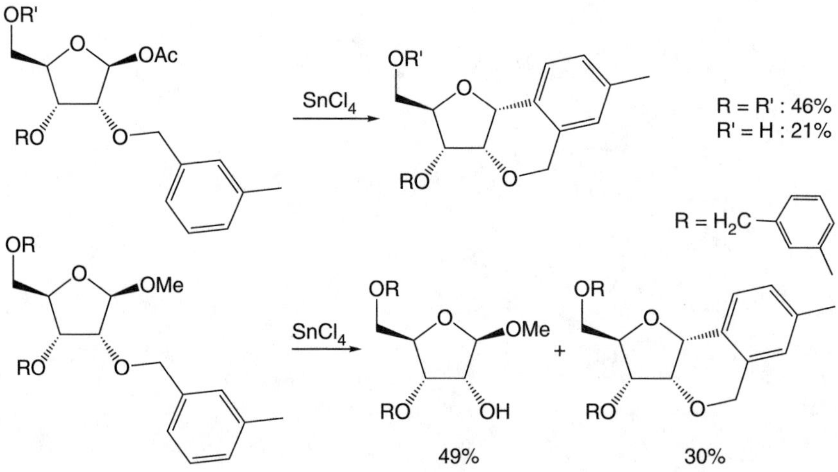

SCHEME 7.89 Rearrangements from 2-hydroxyl groups.

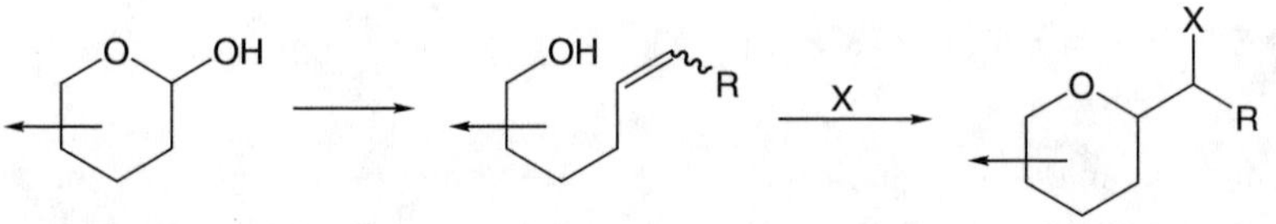

SCHEME 7.90 Olefination of ring-opened sugars.

presented in the following schemes, which include cyclizations of alcohols to olefins, alcohols participating in S_N2 displacements and cycloaddition reactions. The first of these examples, illustrated in Scheme 7.90, involves olefination of the ring-opened sugar followed by cyclization to the previously formed double bond. This addition is usually facilitated by the use of halogens or by converting the double bond to an epoxide, thus creating more electrophilic species. The second approach addressed, shown in Scheme 7.91, complements the first through the cyclization of hydroxy halides or hydroxy epoxides.

With respect to the final approach mentioned above, several types of cycloadditions are applicable to the preparation of *C*-glycosides. One particular example — the hetero Diels–Alder reaction — is illustrated in Scheme 7.92. Through this method, some versatility is added through the resulting site of unsaturation inadvertently formed.

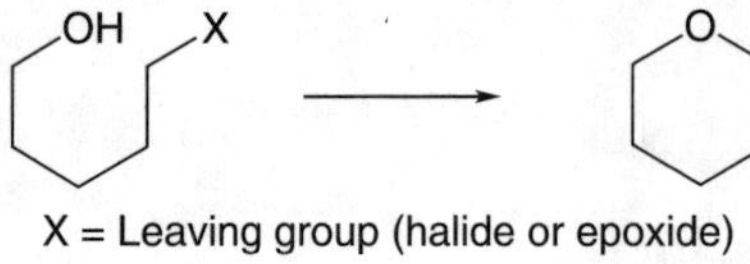

X = Leaving group (halide or epoxide)

SCHEME 7.91 Hydroxy halide/epoxide cyclizations.

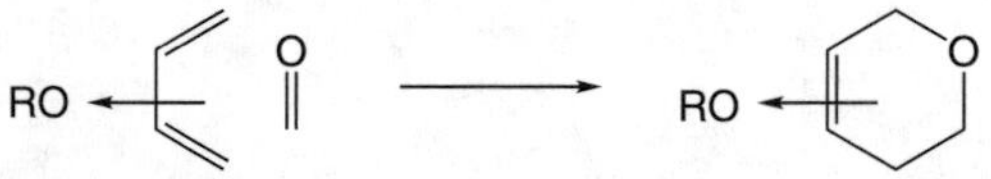

SCHEME 7.92 Hetero Diels–Alder reactions.

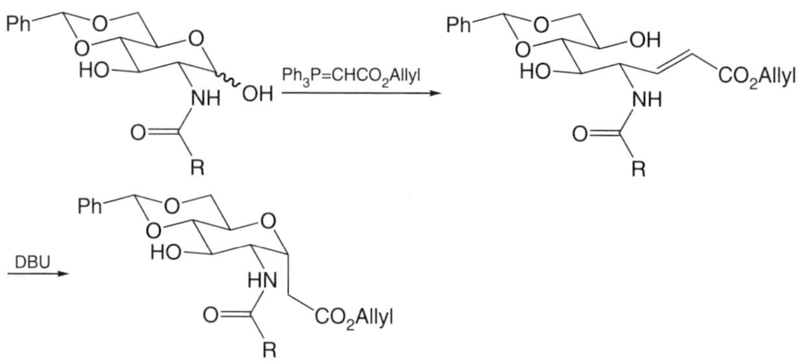

SCHEME 7.93 Wittig reactions followed by base cyclizations.

7.7.1 Wittig Reactions of Lactols Followed by Ring Closures

The Wittig reaction is a valuable method for the formation of olefins from aldehydes and ketones. In the case of carbohydrates, the aldehyde is masked in the form of a hemiacetal at the reducing end of a sugar or polysaccharide. However, because of the equilibrium between the hemiacetal and ring-opened form of sugars, the Wittig reaction can drive the equilibrium entirely to the ring-opened form, with the final product being a newly formed olefin. Once prepared, these hydroxyolefins can be cyclized under a variety of conditions allowing the formation of *C*-glycosides.

Because Michael-type reactions are among the most useful base-mediated cyclizations, the first examples presented in this chapter center around this chemistry. Thus, utilizing Wittig methodology, Vyplel et al. [37] prepared the hydroxyolefin (shown in Scheme 7.93) from the starting protected glucosamine. Treatment of this intermediate species with DBU effected cyclization of the 5-hydroxy group to the unsaturated ester in a Michael fashion. Notably, the cyclization provided the α anomer in 43% yield, with an additional 7% accounted for in the isolation of the β anomer.

Unlike base-mediated cyclizations, halocyclizations do not depend on the presence of Michael acceptors. As shown in Scheme 7.94, Armstrong and Teegarden [234] treated a hydroxyolefin (derived from a Wittig reaction applied to a protected arabinose) with *N*-bromosuccinimide and catalytic bromine to activate the illustrated cyclization. The reaction took place through an intermediate bromonium ion, providing a 52% yield of the β anomer. This observation is consistent with the preference for α anomers observed in base-mediated cyclizations. Moreover, the isolation of a bromo-substituted *C*-glycoside allows the preparation of new and novel structures. This chemistry was further adapted to the use of iodine instead of bromine [235].

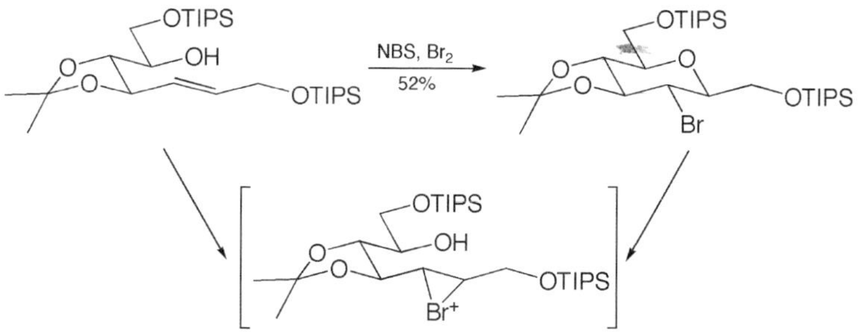

SCHEME 7.94 Wittig reactions followed by bromo cyclizations.

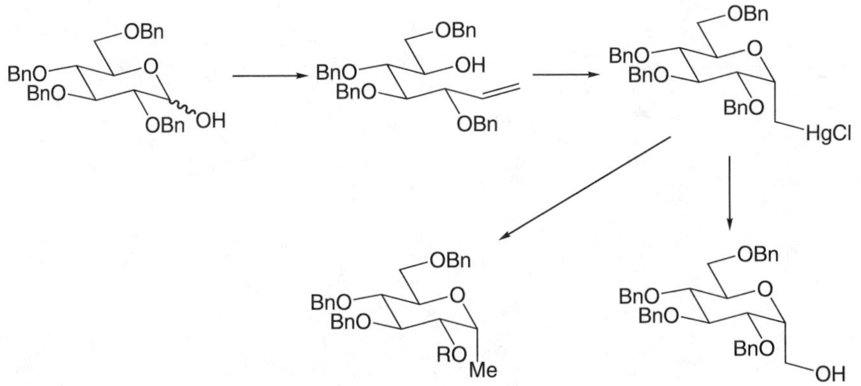

SCHEME 7.95 Wittig reactions followed by oxymercurations.

Methodologies that complement the halogen-mediated cyclizations are found within the chemistry surrounding mercury and selenium. Beginning with mercury, oxymercurations provide access to cyclic ethers via the co-ordination of a mercury reagent to olefins. The resulting products — as those isolated from halogen-mediated cyclizations — generally bear mercury substituents, which are easily modified or replaced under a variety of conditions. As shown in Scheme 7.95, Pougny et al. [36] utilized an oxymercuration in the cyclization of a hydroxyolefin, resulting from a Wittig reaction applied to 2,3,4,6-tetra-*O*-benzyl-D-glucopyranose. Subsequent cleavage of the mercury was accomplished under reductive conditions, yielding a *C*-methylglycoside. Moreover, oxidative conditions yielded the corresponding hydroxymethyl derivative.

Mercury-mediated cyclizations are not unique to the use of metals to initiate the formation of pyranose and furanose *C*-glycosides. As shown in Scheme 7.96, Lancelin et al. [236] performed similar reactions utilizing selenium reagents. As illustrated, the starting hydroxyolefin, prepared utilizing Wittig methodology, was treated with *N*-phenylselenophthalimide and camphor sulfonic acid (CSA). The resulting product mixture exhibited a 60% yield of the pyranosyl phenylselenide and a 30% yield of the furanosyl *C*-glycoside. Further reactions included the conversion of the

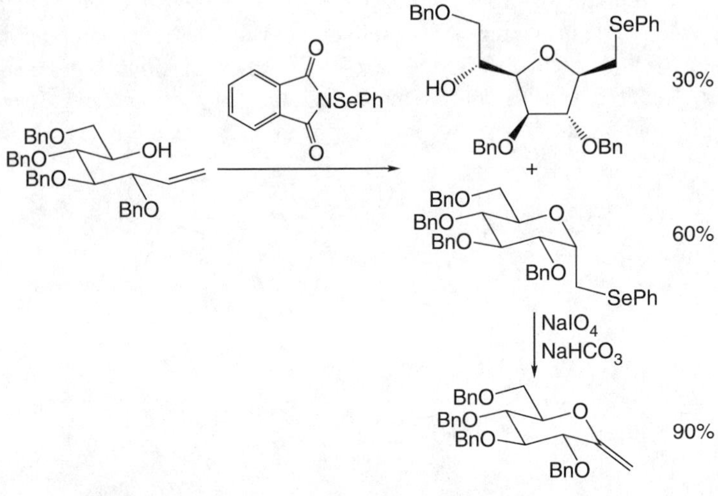

SCHEME 7.96 Selenium-mediated cyclization of Wittig products.

pyranosyl phenylselenide to the olefin following treatment with sodium periodate and sodium bicarbonate.

Given the variety of methods used for the formation of *C*-glycosides, it is helpful to compare them with one another. Several reports have accomplished this and the reader is referred to the cited references [237,238].

7.7.2 Addition of Grignard and Organozinc Reagents to Lactols

Wittig reactions are only one means of converting lactols to hydroxy olefins. Similar olefinations are available utilizing Grignard and organozinc reagents. Where Wittig reactions rely on the olefination of a reactive aldehyde, the use of vinylzinc and the corresponding magnesium reagents provide simple additions to carbonyl groups and effectively provide unprotected hydroxyl groups suitable for selective manipulations. Finally, the reagents ultimately utilized to effect ring closure of the resulting hydroxyolefins lie within those described in the previous section.

As shown in Scheme 7.97, Boschetti et al. [239] produced olefinations on a variety of furanoses utilizing divinyl zinc. In general, the initial additions gave high yields and stereoselectivity, with the vinyl group being delivered from the same face as the 2-benzyloxy groups. In this report, the final cyclizations were created utilizing mercury reagents and provided *C*-glycosidic hexoses in yields ranging from 51 to 71%. Similar work was reported by Lay et al. [240] and Carcano et al. [241].

7.7.3 Cyclization of Suitably Substituted Polyols

The chemistry presented in the previous two sections focussed on the formation of hydroxyolefins followed by cyclization to the desired *C*-glycosides. The latter steps in the examples shown rely largely upon the cyclization of protected polyols. In this section, cyclization methods including ether formations, ketal formations and halide displacements are presented.

7.7.3.1 Cyclizations via Ether Formations

Among the simplest of the cyclizations utilized in the formation of *C*-glycosides is the acid-mediated formation of ethers. Such reactions can be viewed as dehydration between two alcohol units, with the driving force for the reaction being the elimination of water. As shown in Scheme 7.98, Schmidt and Frick [242] formed the perbenzylated polyol by treating the lithiated

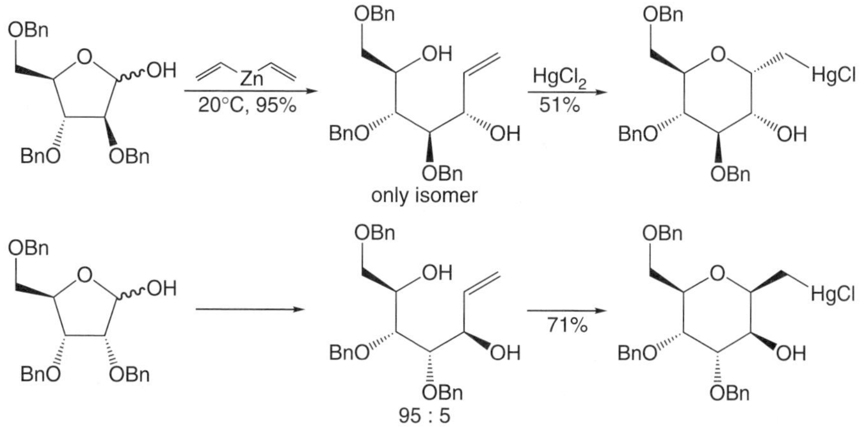

OBn OH OBn BnO OBn Zn 20°C, 95% BnO OH OBn only isomer $HgCl_2$ 51% HgCl BnO OH OBn

OBn OH BnO OBn BnO OH OBn 95 : 5 71% HgCl BnO OH OBn

SCHEME 7.97 Divinyl zinc additions and mercury cyclizations.

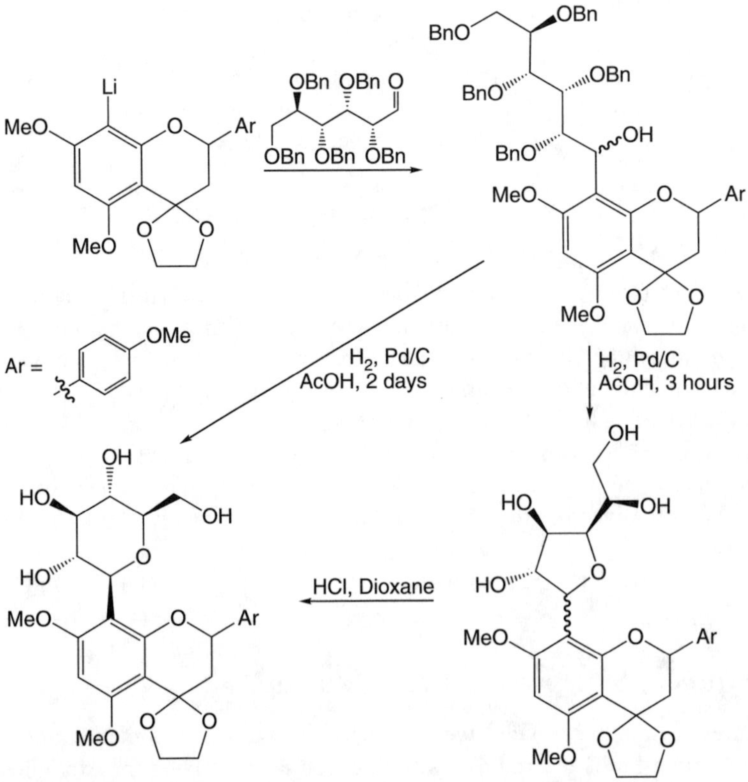

SCHEME 7.98 Acid-mediated cyclizations.

flavonoid with the ring-opened form of penta-*O*-benzylglucose. As illustrated, catalytic hydrogenation in acetic acid produced a *C*-furanoside after 3 h and a *C*-pyranoside after 2 days. Additionally, the furanoside was easily converted to the pyranoside on treatment with hydrochloric acid in dioxane.

7.7.3.2 Cyclizations via Ketal Formations

Ketals can be formed under acidic conditions. Consequently, this technology complements the acid-mediated cyclizations outlined in the previous example. As applied to the preparation of *C*-glycosides, the general approach is similar in that a nucleophile is added to a sugar derivative. As shown in Scheme 7.99, Schmidt and Frick [243] utilized a perbenzylated open-chain glucose. The resulting alcohol was then oxidized to the corresponding ketone, and the benzyl groups were removed under catalytic conditions. Following acetylation, the resulting polyol cyclized with formation of a single diastereomer of the papulacandin spiroketal. It should be noted that the ketals formed using this technology are anomerically alkylated *C*-glycosides, which resemble *O*- and *C*-glycosides in both their chemistry and appearance.

7.7.3.3 Cyclizations via Halide Displacements

Although technically ether formations, the displacement of halides by hydroxyl groups differs from the examples presented above in that these reactions are base mediated. As such, under these conditions, *C*-glycosidations complement the previously described cyclization strategies given that they allow reactions with acid sensitive substrates. As shown in Scheme 7.100, Schmid and Whitesides [244] used this approach in the preparation of *C*-furanosides. Specifically, the illustrated

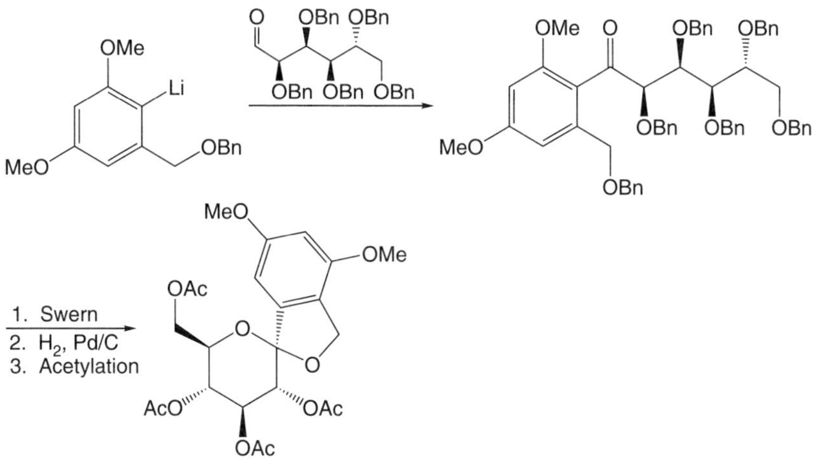

SCHEME 7.99 Cyclizations via ketal formations.

triol was protected with silyl groups. Subsequent treatment of the ketone with allylmagnesium chloride triggered the formation of a hydroxide, which readily displaced the terminal chloride. The final *C*-glycoside was isolated as a 2.7:1 mixture of diastereomers.

7.7.4 Rearrangements

Section 7.6 addressed the subject of the formation of *C*-glycosides via rearrangements. In the examples presented therein, the rearrangements were performed on substrates that already possessed suitable ring configurations for the desired *C*-glycosidic products. The rearrangements presented in this section differ from those in Section 7.6 in that the core ring required for the *C*-glycosides is either formed or substantially modified. The rearrangement types presented include electrocyclic reactions as well as ring contractions.

7.7.4.1 Electrocyclic Rearrangements

The preparation of *C*-glycosides via formation of the sugar ring can be achieved through the rearrangement of suitable substrates. For example, as shown in Scheme 7.101, Burke et al. [245] utilized the enolate Claisen rearrangement. Burke and colleagues converted a series of 6-alkenyl-1,4-dioxane-2-ones to dihydropyrans bearing *C*-methylcarboxyl glycosidic units. All reactions were stereospecific, and the yields and substitution patterns are shown in Table 7.5.

7.7.4.2 Ring Contractions

In addition to Claisen-type rearrangements, other methods are applicable to the formation of *C*-glycosides. As shown in Scheme 7.102, Fleet and Seymour [246] employed ring contraction

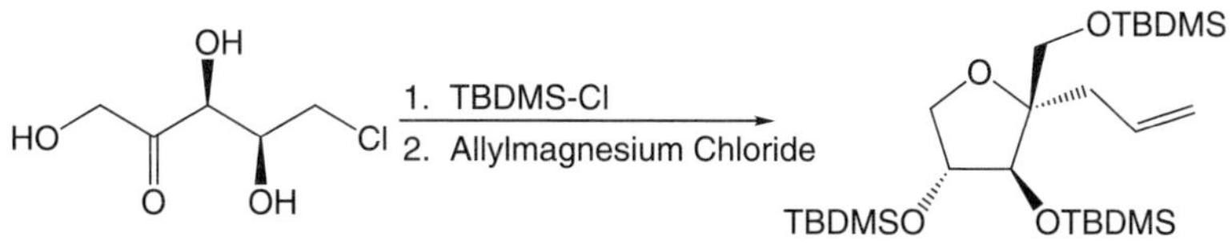

SCHEME 7.100 Cyclizations via halide displacements.

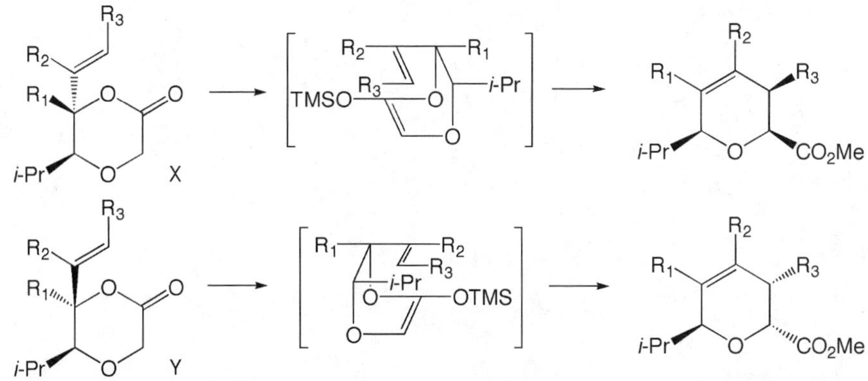

SCHEME 7.101 Enolate claisen rearrangement.

TABLE 7.5
Enolate Claisen Rearrangement Results

Substrate	Dihydropyran Product R_1	R_2	R_3	Yield (%)
X	H	H	H	67
X	H	Me	H	75
X	H	H	$SiMe_3$	52
X	H	H	Me	70
X	Me	H	H	90
X	Me	Me	H	80
Y	H	H	H	69
Y	H	Me	H	78
Y	H	H	$SiMe_3$	61
Y	H	H	Me	81
Y	Me	H	H	91

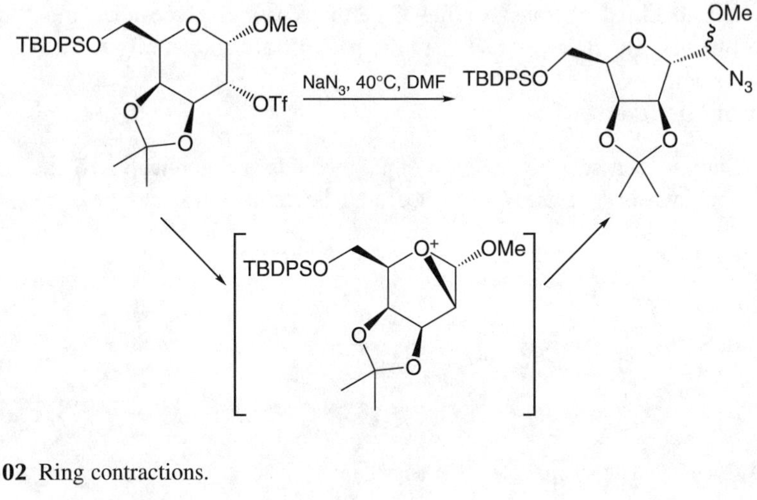

SCHEME 7.102 Ring contractions.

techniques in the transformation of methyl glycopyranosides to furanosyl *C*-glycosides. The reactions resulted in 51 to 91% yields, depending upon the anomeric configuration of the starting galactosides, and produced approximately 2:1 ratios of diastereomeric products. A potential mechanism involves the displacement of the triflate by the pyranose ring oxygen followed by opening of the three-membered intermediate by an azide group. In this respect, the mechanism resembles that belonging to the classical Favorskii rearrangement.

7.7.4.3 Other Rearrangements

In a final example of the utility of rearrangements in the formation of *C*-glycosides, Kuszmann et al. [247] reported reactions involving benzylidene acetals. As illustrated in Scheme 7.103, one reaction yielded two diastereomeric *C*-glycosides in a combined yield of approximately 30%. The mechanism associated with this reaction involves initial acetylation of one of the ring oxygens, followed by cleavage of the ring with formation of a conjugated carbocation. Migration of an acetate to the carbocation yields a species bearing an olefin on one side and an oxonium ion on the other. Ring closure followed by quenching of the resulting carbocation by an acetate group gives the observed products.

7.7.5 Cycloadditions

Perhaps the most useful method for the formation of substituted ring systems from acyclic substrates is the Diels–Alder reaction. As will be explored in this section, the Diels–Alder reaction fits well with the theme of this chapter through its use in the direct formation of ring systems capable of being converted to sugars. As shown in Scheme 7.104, Katagiri et al. [248] incorporated this methodology in transformations involving furan derivatives. As illustrated, the Diels–Alder reaction between 3,4-dibenzyloxyfuran and dimethyl (acetoxymethylene)malonate afforded a 57% yield of the bicyclic adduct. This product was then subjected to catalytic hydrogenation, followed by basic reductive conditions. The resulting cascade of reductions and rearrangements produced the dimethylmalonyl-*C*-lyxoside in an 85% isolated yield.

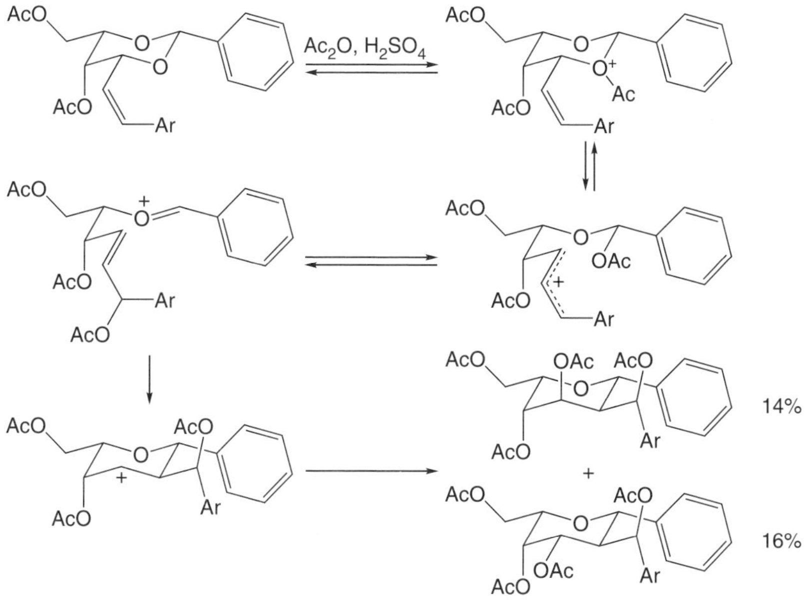

AcO AcO O O Ar Ac_2O, H_2SO_4 AcO AcO O O+ Ac Ar

AcO AcO AcO + O Ar AcO AcO O OAc + Ar

AcO AcO AcO O + Ar AcO OAc AcO O AcO Ar 14% + AcO AcO O OAc AcO Ar 16%

SCHEME 7.103 Benzylidene acetal rearrangements.

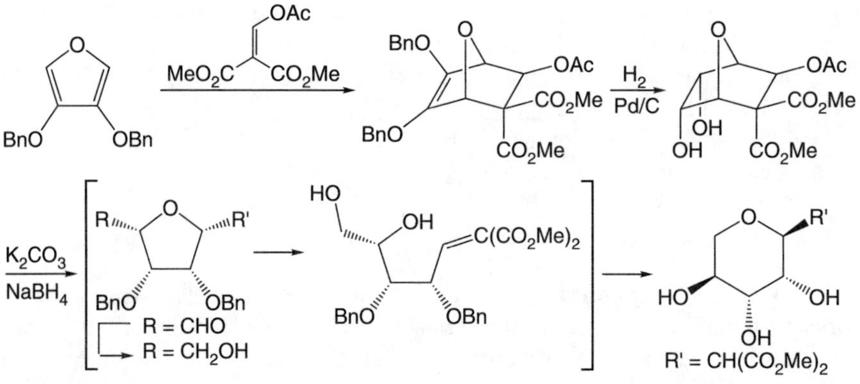

SCHEME 7.104 *C*-Glycosides via Diels–Alder reactions.

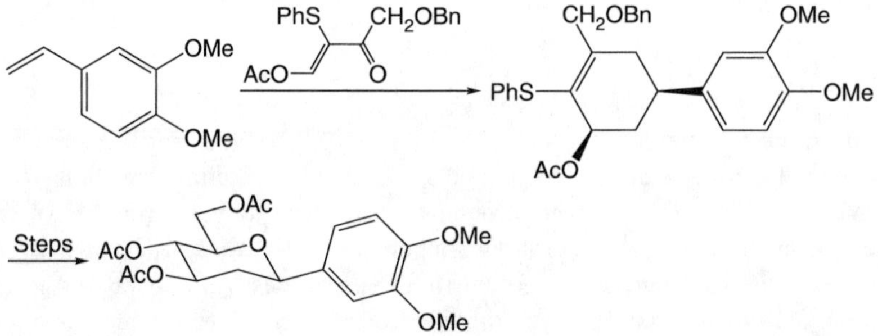

SCHEME 7.105 *C*-Glycosides via hetero Diels–Alder reactions.

As illustrated in the above-described example, the Diels–Alder reaction provides avenues into the formation of *C*-glycosides from furan-derived dienes. Similarly, the hetero Diels–Alder reaction allows for the direct formation of sugar rings from carbonyl groups. As shown in Scheme 7.105, Schmidt et al. [249] triggered reactions between conjugated carbonyl compounds and olefins. The reaction produced an 81% yield giving an adduct which, after further manipulations, was converted to a *C*-aryl glycoside. A complementary approach reported by Danishefsky et al. [250] incorporated carbonyl compounds as dienophiles.

7.7.6 Other Methods for the Formation of Sugar Rings

The methods described thus far represent only a small subset of the available techniques for the formation of *C*-furanosides and *C*-pyranosides. Other useful methods include cyclizations of halo olefins and ene-ynes [224] and the direct modification of pyran ring systems [251]. Although not discussed further here, these examples are mentioned to illustrate the breadth of work accomplished in this area.

7.8 FURTHER READING

This chapter introduced the nomenclature, physical properties, chemistry and potential uses of *C*-glycosides. Through specific examples and detailed analyses of synthetic strategies, the various chemical reactions associated with *C*-glycosides and methods for their preparation were explored.

In addition to the present compilation, three books [252–254] and numerous review articles [255–261] have been written covering a wide range of issues regarding *C*-glycosides. Because this chapter cannot be exhaustive, the reader is referred to the cited references herein for more detailed treatments.

ACKNOWLEDGMENTS

Material in this chapter, previously published in "The Chemistry of *C*-Glycosides" (DE Levy, C Tang, Tetrahedron Organic Chemistry Series, Volume 13, Oxford, Pergamon, 1995), was reproduced with permission from Elsevier Science.

REFERENCES

1. Kini, G D, Petrie, C R, Hennen, W J, Dalley, N K, Wilson, B E, Robbins, R K, Improved and large-scale synthesis of certain glycosyl cyanides. Synthesis of 2,5-anhydro-5-thio-D-allonitrile, *Carbohydr. Res.*, 159, 81–94, 1987.
2. Hoffman, M G, Schmidt, R R, *O*-Glycosyl imidates. 19. Reaction of glycosyl trichloroacetimidates with silylated C-nucleophiles, *Liebigs Ann. Chem.*, 2403–2419, 1985.
3. Wang, Y, Goekjian, P G, Ryckman, D M, Kishi, Y, Preferred conformation of *C*-glycosides. 4. Importance of 1,3-diaxial-like interactions around the nonglycosidic bond: prediction and experimental proof, *J. Org. Chem.*, 53, 4151–4153, 1988.
4. Haneda, T, Goekjian, P G, Kim, S H, Kishi, Y, Preferred conformation of *C*-glycosides. 10. Synthesis and conformational analysis of carbon trisaccharides, *J. Org. Chem.*, 57, 490–498, 1992.
5. Kishi, Y, Preferred solution conformation of marine natural product palytoxin and of *C*-glycosides and their parent glycosides, *Pure Appl. Chem.*, 65, 771–778, 1993.
6. Gruen, M, Franz, G, Isolation of two stereoisomeric aloins from aloe, *Pharmazie*, 34, 669–670, 1979.
7. Auterhoff, H, Graf, E, Eurisch, G, Alexa, M, Resolution of aloin into diastereomers and their characterization, *Arch. Pharm. (Weinheim, Ger)*, 313, 113–120, 1980.
8. Graf, E, Alexa, M, Stability of diastereomeric aloins A and B and their main decomposition product 4-hydroxyaloin, *Planta Med.*, 38, 121–127, 1980.
9. Rauwald, H W, Preparative separation of the diastereomeric aloins by droplet countercurrent chromatography (DCCC), *Arch. Pharm. (Weinheim, Ger.)*, 315, 769–772, 1982.
10. Rauwald, H W, Roth, K, New proton NMR analysis of the diastereomeric aloins, *Arch. Pharm. (Weinheim, Ger)*, 317, 362–367, 1984.
11. Adinolfi, M, Lanzetta, R, Marciano, C E, Parrilli, M, Giuliu, A D, A new class of anthraquinone-anthrone-*C*-glycosides from *Asphodelus ramosus* tubers, *Tetrahedron*, 47, 4435–4440, 1991.
12. Carte, B K, Carr, S, DeBrosse, C, Hemling, M E, Mackenzie, L, Offen, P, Berry, D, Aciculatin, a novel flavone-*C*-glycoside with DNA binding activity from *Chrysopogon aciculatis*, *Tetrahedron*, 47, 1815–1822, 1991.
13. Buchanan, J G, Wightman, R H, The *C*-nucleoside antibiotics, *Prog. Chem. Org. Nat. Prod.*, 44, 243–299, 1983.
14. Humber, D C, Mulholland, K R, Stoodley, R J, *C*-Nucleosides. Part 1. Preparation of tiazofurin and N-substituted tiazofurins from benzyl (2′,3′,5′-tri-*O*-benzoyl-β-D-ribofuranosyl)penicillinate, *J. Chem. Soc., Perkin Trans. 1*, 283–292, 1990.
15. Schmidt, R R, New methods of glycoside and oligosaccharide syntheses — are there alternatives to the Koenigs–Knorr method? *Angew. Chem.*, 98, 213–236, 1986.
16. Schmidt, R R, Recent developments in the synthesis of glycoconjugates, *Pure Appl. Chem.*, 61, 1257–1270, 1989.
17. Paulsen, H, Synthesis, conformation, and x-ray analysis of saccharide chains of glycoprotein core regions, *Angew. Chem.*, 102, 851–867, 1990.
18. Hakomori, S, Bifunctional role of glycosphingolipids. Modulators for transmembrane signaling and mediators for cellular interactions, *J. Biol. Chem.*, 265, 18713–18716, 1990.

19. Paulson, J C, Colley, K J, Glycosyltransferases. Structure, localization, and control of cell type-specific glycosylation, *J. Biol. Chem.*, 264, 17615–17618, 1989.
20. Kornfeld, R, Kornfeld, S, Assembly of asparagine-linked oligosaccharides, *Annu. Rev. Biochem.*, 54, 631–664, 1985.
21. Sadler, J E, Biosynthesis of glycoproteins: formation of O-linked oligosaccharides, *Biol. Carbohydr.*, 2, 199–288, 1984.
22. Basu, S, Basu, M, Expression of glycosphingolipid glycosyltransferases in development and transformation, *Glycoconjugates*, 3, 265–285, 1982.
23. Lowe, J B, Stoolman, L M, Nair, R P, Larsen, R D, Berhend, T L, Marks, R M, ELAM-1-dependent cell adhesion to vascular endothelium determined by a transfected human fucosyltransferase cDNA, *Cell*, 63, 475–484, 1990.
24. Berg, E L, Robinson, M K, Mansson, O, Butcher, E C, Magnani, J L, A carbohydrate domain common to both sialyl Lea and sialyl Lex is recognized by the endothelial cell leukocyte adhesion molecule ELAM-1, *J. Biol. Chem.*, 266, 14869–14872, 1991.
25. Tyrrell, D, James, P, Rao, N, Foxall, C, Abbas, S, Dasgupta, F, Nashed, M, Hasegawa, A, Kiso, M, Asa, D, Kidd, J, Brandley, B K, Structural requirements for the carbohydrate ligand of E-selectin, *Proc. Natl. Acad. Sci. USA*, 88, 10372–10376, 1991.
26. Polley, M J, Phillips, M L, Wayner, E, Nucelman, E, Kinghal, A K, Hakomori, S, Paulson, J C, CD62 and endothelial cell-leukocyte adhesion molecule 1 (ELAM-1) recognize the same carbohydrate ligand, sialyl-Lewis x, *Proc. Natl. Acad. Sci. USA*, 88, 6224–6228, 1991.
27. Imai, Y, Singer, M S, Fennie, C, Lasky, L A, Rosen, S D, Identification of a carbohydrate-based endothelial ligand for a lymphocyte homing receptor, *J. Cell. Biol.*, 113, 1213–1221, 1991.
28. Stoolman, L M, Rosen, S D, Possible role for cell-surface carbohydrate-binding molecules in lymphocyte recirculation, *J. Cell. Biol.*, 96, 722–729, 1983.
29. True, D D, Singer, M S, Lasky, L A, Rosen, S D, Requirement for sialic acid on the endothelial ligand of a lymphocyte homing receptor, *J. Cell. Biol.*, 111, 2757–2764, 1990.
30. Macher, B A, Holmes, E H, Swiedler, S J, Stults, C L M, Srnka, C A, Human α1-3-fucosyltransferases, *Glycobiology*, 1, 577–584, 1991.
31. Kuijpers, T W, Terminal glycosyltransferase activity: a selective role in cell adhesion, *Blood*, 81, 873–882, 1993.
32. Luengo, J I, Gleason, J G, Synthesis of *C*-fucopyranosyl analogs of GDP-L-fucose as inhibitors of fucosyltransferases, *Tetrahedron Lett.*, 33, 6911–6914, 1992.
33. Schmidt, R R, Dietrich, H, Amino substituted β-benzyl-C′-glycosides, novel β-glycosidase inhibitors, *Angew. Chem. Int. Ed. Engl.*, 30, 1328–1329, 1991.
34. Lehmann, J, Schlesselmann, P, Location of a proton-donating group at the re-face of a β-D-galactosidase-bound, diastereotopic substrate, *Carbohydr. Res.*, 113, 93–99, 1983.
35. Myers, R W, Lee, Y C, Synthesis of diazomethyl β-D-galactopyranosyl and β-D-glucopyranosyl ketones. Potential affinity-labeling reagents for carbohydrate-binding proteins, *Carbohydr. Res.*, 152, 143–158, 1986.
36. Pougny, J R, Nassr, M A M, Sinaÿ, P, Mercuricyclisation in carbohydrate chemistry: a highly stereo-selective route to α-D-*C*-glucopyranosyl derivatives, *J. Chem. Soc., Chem. Commun.*, 375–376, 1981.
37. Vyplel, H, Scholz, D, Macher, I, Schindlmaier, K, Schuetze, E, *C*-glycosidic analogs of lipid A and lipid X: synthesis and biological activities, *J. Med. Chem.*, 34, 2759–2767, 1991.
38. Nagy, J O, Wang, P, Gilbert, J H, Schaefer, M E, Hill, T G, Gallstrom, M R, Bednarski, M D, Carbohydrate materials bearing neuraminidase-resistant *C*-glycosides of sialic acid strongly inhibit the in vitro infectivity of influenza virus, *J. Med. Chem.*, 35, 4501–4502, 1992.
39. Bertozzi, C, Bednarski, M, *C*-glycosyl compounds bind to receptors on the surface of *Escherichia coli* and can target proteins to the organism, *Carbohydr. Res.*, 223, 243–253, 1992.
40. Eshdat, Y, Ofek, I, Yashouv-Gan, Y, Sharon, N, Mirelman, D, Isolation of a mannose-specific lectin from *Escherichia coli* and its role in the adherence of the bacteria to epithelial cells, *Biochem. Biophys. Res. Commun.*, 85, 1551–1559, 1978.
41. Firon, N, Ofek, I, Sharon, N, Interaction of mannose-containing oligosaccharides with the fimbrial lectin of *Escherichia coli*, *Biochem. Biophys. Res. Commun.*, 105, 1426–1432, 1982.
42. Lopez, M T G, Heras, F G D, Felix, A S, Cyanosugars. IV. Synthesis of α-D-glucopyranosyl and α-D-galactopyranosyl cyanides and related 1,2-*cis* *C*-glycosides, *J. Carbohydr. Chem.*, 6, 273–279, 1987.

43. Nicolaou, K C, Dolle, R E, Chucholowski, A, Randall, J L, Reactions of glycosyl fluorides. Synthesis of *C*-glycosides, *J. Chem. Soc., Chem. Commun.*, 1153–1154, 1984.
44. Araki, Y, Kobayashi, N, Watanabe, K, Ishido, Y, Synthetic studies by the use of fluorinated intermediates. Part 2. Synthesis of glycosyl cyanides and *C*-allyl glycosides by the use of glycosyl fluoride derivatives, *J. Carbohydr. Chem.*, 4, 565–585, 1985.
45. Grynkiewicz, G, BeMiller, J N, Synthesis of 2,3-dideoxy-D-hex-2-enopyranosyl cyanides, *Carbohydr. Res.*, 108, 229–235, 1982.
46. Nicolaou, K C, Hwang, C K, Duggan, M E, Stereospecific synthesis of 1,1-dialkylglycosides, *J. Chem. Soc., Chem. Commun.*, 925–927, 1986.
47. Smolyakova, I P, Smit, W A, Zal'chenko, E A, Chizhov, O S, Shashkov, A S, Arenesulfenyl chloride adducts of glucal derivatives in the preparation of *C*-glucosides, *Tetrahedron Lett.*, 34, 3047–3050, 1993.
48. Mukaiyama, T, Kobayashi, S, Shoda, S I, A facile synthesis of α-*C*-ribofuranosides from 1-*O*-acetylribose in the presence of trityl perchlorate, *Chem. Lett.*, 1529–1530, 1984.
49. Grierson, D S, Bonin, M, Husson, H P, Monneret, C, Florent, J C, An efficient synthesis of 1-cyano-2, 3-unsaturated sugars via the reaction of glycals with Et_2AlCN, *Tetrahedron Lett.*, 25, 4645–4646, 1984.
50. Drew, K N, Gross, P H, *C*-glycoside syntheses. 1. Glycosyl cyanides and isocyanides from glycosyl fluorides with full acetal protection, *J. Org. Chem.*, 56, 509–513, 1991.
51. Tatsuta, K, Hayakawa, J, Tatsuzawa, Y, Stereoselective syntheses of 2-deoxy-β-*C*-arabino- and ribopyranosides: 2-deoxy-β-arabino- and ribopyranosyl cyanides, *Bull. Chem. Soc. Jpn.*, 62, 490–492, 1989.
52. BeMiller, J N, Yadav, M P, Kalabokis, V N, Myers, R M, *N*-substituted (β-D-galactopyranosylmethyl)amines, *C*-β-D-galactopyranosylformamides, and related compounds, *Carbohydr. Res.*, 200, 111–126, 1990.
53. Pochet, S, Allard, P, Tam, H D, Igolen, J, Synthesis of *C*-glycosyl α-glycines, *J. Carbohydr. Chem.*, 1, 277–288, 1982–1983.
54. Poonian, M S, Nowoswiat, E F, A novel precursor for the synthesis of *C*-nucleoside analogs. Synthesis of the *C*-nucleoside analogs of ribavirin, bredinin, and related compounds, *J. Org. Chem.*, 45, 203–208, 1980.
55. El Khadem, H S, Kewai, J, 2-β-D-Ribofuranosylbenzoxazole from 2,5-anhydro-D-allonoimidate, and 1,3-dimethyl-8-β-D-ribofuranosylxanthine from 2,5-anhydro-D-allono-thioimidates and -dithioates, *Carbohydr. Res.*, 153, 271–283, 1986.
56. Hanessian, S, Pernet, A G, Carbanions in carbohydrate chemistry. Synthesis of *C*-glycosyl malonates, *Can. J. Chem.*, 52, 1266–1279, 1974.
57. Cristol, S J, Firth, W C, Convenient synthesis of alkyl halides from carboxylic acids, *J. Org. Chem.*, 26, 280 1961.
58. Bihovsky, R, Selick, C, Giusti, I, Synthesis of *C*-glucosides by reactions of glucosyl halides with organocuprates, *J. Org. Chem.*, 53, 4026–4031, 1988.
59. Bellosta, V, Czernecki, S, *C*-glycosyl compounds. Part X. Reaction of organocuprate reagents with protected 1,2-anhydro sugars. Stereocontrolled synthesis of 2-deoxy-*C*-glycosyl compounds, *Carbohydr. Res.*, 244, 275–284, 1993.
60. Bellosta, V, Czernecki, S, Stereocontrolled synthesis of *C*-glycosides by reaction of organocuprates with protected 1,2-anhydro sugars, and their transformation into 2-deoxy-*C*-glycosides, *J. Chem. Soc., Chem. Commun.*, 199–200, 1989.
61. Shulman, M L, Shiyan, S D, Khorlin, A Y, The synthesis of diasteromeric epoxy-(β-D-glucopyranosyl)ethanes and 1,2-epoxy-3-(β-D-glucopyranosyl)propanes, as irreversible inhibitors of β-D-glucosidase, *Carbohydr. Res.*, 33, 229–235, 1974.
62. Hanessian, S, Liak, T J, Dixit, D M, Synthesis of trans-fused perhydrofuropyrans and related α-methylene lactones: bicyclic ring-systems present in the ezomycins, the octosyl acids, and certain antitumor terpenoids, *Carbohydr. Res.*, 88, C14–C19, 1981.
63. Brown, D S, Ley, S V, Direct substitution of 2-benzenesulphonyl cyclic ethers using organozinc reagents, *Tetrahedron Lett.*, 29, 4869–4872, 1988.
64. Bolitt, V, Mioskowski, C, Falck, J R, Direct anomeric substitution of pyranyl esters using organocopper reagents, *Tetrahedron Lett.*, 30, 6027–6030, 1989.

65. Orsini, F, Pelizzoni, F, Synthesis of *C*-glycosyl compounds: the reaction of acetylated glycals with *tert*-butoxycarbonylmethylzinc bromide, *Carbohydr. Res.*, 243, 183–189, 1993.
66. Kozikowski, A P, Konoike, T, Ritter, A, Organometallics in organic synthesis. Applications of a new diorganozinc reaction to the synthesis of *C*-glycosyl compounds with evidence for an oxonium-ion mechanism, *Carbohydr. Res.*, 171, 109–124, 1987.
67. Bols, M, Szarek, W A, Synthesis of 3-deoxy-3-fluoro-D-fructose, *J. Chem. Soc., Chem. Commun.*, 445–446, 1992.
68. Lewis, M D, Cha, J K, Kishi, Y, Highly stereoselective approaches to α- and β-*C*-glycopyranosides, *J. Am. Chem. Soc.*, 104, 4976–4978, 1982.
69. Cupps, T L, Wise, D S, Townsend, L B, A further investigation of the stannic chloride-catalyzed condensation reaction of 1-hexene and 1,2,3,5-tetra-*O*-acyl-β-D-ribofuranoses, *Carbohydr. Res.*, 115, 59–73, 1983.
70. Herscovici, J, Muleka, K, Antonakis, K, Olefin addition to acetylated glycals. A new route to *C*-glycosides, *Tetrahedron Lett.*, 25, 5653–5656, 1984.
71. Herscovici, J, Muleka, K, Boumaiza, L, Antonakis, K, *C*-glycoside synthesis via glycal alkylation by olefinic derivatives, *J. Chem. Soc., Perkin Trans. 1*, 1995–2009, 1990.
72. Levy, D E, Dasgupta, F, Tang, P C, Synthesis of novel fused ring *C*-glycosides, *Tetrahedron Asym.*, 5, 2265–2268, 1994.
73. Cooper, A J, Salomon, R G, Total synthesis of halichondrins: enantioselective construction of a homochiral pentacyclic C1–C15 intermediate from D-ribose, *Tetrahedron Lett.*, 31, 3813–3816, 1990.
74. Newcombe, N J, Mahon, M F, Molloy, K C, Alker, D, Gallagher, T, Carbohydrate-based enolates as heterocyclic building blocks. Synthesis of the herbicidin glycoside, *J. Am. Chem. Soc.*, 115, 6430–6431, 1993.
75. Terahara, A, Haneishi, T, Arai, M, Hata, T, The revised structure of herbicidins, *J. Antibiot.*, 35, 1711–1714, 1982.
76. Yoshikawa, H, Takiguchi, Y, Terao, M, Terminal steps in the biosynthesis of herbicidins, nucleoside antibiotics, *J. Antibiot.*, 36, 30–35, 1983.
77. Kozaki, S, Sakanaka, O, Yasuda, T, Shimizu, T, Ogawa, S, Suami, T, Alternative total synthetic approach toward octosyl acid A, *J. Org. Chem.*, 53, 281–286, 1988.
78. Giannis, A, Sandhoff, K, Stereoselective synthesis of α-*C*-allylglycopyranosides, *Tetrahedron Lett.*, 26, 1479–1482, 1985.
79. Allevi, P, Anastasia, M, Ciuffreda, P, Fiecchi, A, Scala, A, The first direct method for *C*-glucopyranosyl derivatization of 2,3,4,6-tetra-*O*-benzyl-D-glucopyranose, *J. Chem. Soc., Chem. Commun.*, 1245–1246, 1987.
80. Nicotra, F, Panza, L, Russo, G, Stereoselective access to α- and β-D-fructofuranosyl *C*-glycosides, *J. Org. Chem.*, 52, 5627–5630, 1987.
81. Martin, O R, Rao, S P, Kurz, K G, El-Shenawy, H A, Intramolecular reactions of 2-*O*-organosilyl glycosides. Highly stereoselective synthesis of *C*-furanosides, *J. Am. Chem. Soc.*, 110, 8698–8700, 1988.
82. Bennek, J, Gray, G R, An efficient synthesis of anhydroalditols and allylic-glycosides, *J. Org. Chem.*, 52, 892–897, 1987.
83. Hosomi, A, Sakata, Y, Sakurai, H, Stereoselective synthesis of 3-(D-glycopyranosyl)propenes by use of allylsilanes, *Carbohydr. Res.*, 171, 223–232, 1987.
84. Wilcox, C S, Long, G W, Suh, H S, A new approach to *C*-glycoside congeners: metal carbene-mediated methylenation of aldonolactones, *Tetrahedron Lett.*, 25, 395–398, 1984.
85. Acton, E M, Ryan, K J, Tracy, M, Synthetic approach to anthracyclinone *C*-glycosides, *Tetrahedron Lett.*, 25, 5743–5746, 1984.
86. Martin, M G G, Horton, D, Preparative synthesis of *C*-(α-D-glucopyranosyl)-alkenes and -alkadienes: Diels–Alder reaction, *Carbohydr. Res.*, 191, 223–229, 1989.
87. Panek, J S, Sparks, M A, Oxygenated allylic silanes: useful homoenolate equivalents for the stereoselective *C*-glycosidation of pyranoside derivatives, *J. Org. Chem.*, 54, 2034–2038, 1989.
88. Babirad, S A, Wang, Y, Kishi, Y, Synthesis of *C*-disaccharides, *J. Org. Chem.*, 52, 1370–1372, 1987.

89. Araki, Y, Kobayashi, N, Ishido, Y, Synthetic studies with fluorinated intermediates. 3. Highly stereoselective *C*-α-D-ribofuranosylation. Reactions of D-ribofuranosyl fluoride derivatives with enol trimethylsilyl ethers and with allyltrimethylsilane, *Carbohydr. Res.*, 171, 125–139, 1987.
90. Bertozzi, C R, Bednarski, M D, The synthesis of 2-azido *C*-glycosyl sugars, *Tetrahedron Lett.*, 33, 3109–3112, 1992.
91. Ichikawa, Y, Isobe, M, Konobe, M, Goto, T, Synthesis of *C*-glycosyl compounds from 3,4,6-tri-*O*-acetyl-1,5-anhydro-D-arabino-hex-1-enitol and allyltrimethylsilane and bis(trimethylsilyl)acetylene, *Carbohydr. Res.*, 171, 193–199, 1987.
92. de Raadt, A, Stütz, A E, A one-step *C*-linked disaccharide synthesis from carbohydrate allylsilanes and tri-*O*-acetyl-D-glucal, *Carbohydr. Res.*, 220, 101–115, 1991.
93. Ferrier, R J, Petersen, P M, Unsaturated carbohydrates. Part 31. Trichothecene-related and other branched *C*-pyranoside compounds, *J. Chem. Soc., Perkin Trans. 1*, 2023–2028, 1992.
94. Tsukiyama, T, Isobe, M, *C*-glycosidation with silylacetylenes to D-glucals, *Tetrahedron Lett.*, 33, 7911–7914, 1992.
95. Keck, G E, Enholm, E J, Kachensky, D F, Two new methods for the synthesis of *C*-glycosides, *Tetrahedron Lett.*, 25, 1867–1870, 1984.
96. Waglund, T, Cleasson, A, Stereoselective synthesis of the α-allyl *C*-glycoside of 3-deoxy-D-manno-2-octulosonic acid (KDO) by use of radical chemistry, *Acta Chem. Scand.*, 46, 73 1992.
97. Nagy, J O, Bednarski, M D, The chemical-enzymatic synthesis of a carbon glycoside of *N*-acetylneuraminic acid, *Tetrahedron Lett.*, 32, 3953–3956, 1991.
98. Zhai, D, Zhai, W, Williams, R M, Alkynylation of mixed acetals with organotin acetylides, *J. Am. Chem. Soc.*, 110, 2501–2505, 1988.
99. Tolstikov, G A, Prokhorova, N A, Spivak, A Y, Khalilov, L M, Sultanmuratova, V R, Synthesis of *C*-glucopyranosides by reaction of organoaluminum compounds with 2,3,4,6-tetra-*O*-benzyl-α-D-glucopyranoside, *Zh. Org. Khim.*, 27, 2101–2106, 1991.
100. Bellosta, V, Chassagnard, C, Czernecki, S, *C*-glycosyl compounds. Part VIII. Structural studies by proton and carbon-13 NMR spectroscopy and circular dichroism of acetylated α- and β-D-gluco- and -manno-pyranosylarenes, *Carbohydr. Res.*, 219, 1–7, 1991.
101. Macdonald, S J F, Huizinga, W B, McKenzie, T C, Retention of configuration in the coupling of aluminated heterocycles with glycopyranosyl fluorides, *J. Org. Chem.*, 53, 3371–3373, 1988.
102. Outten, R A, Daves, D G Jr, Synthetic 1-methoxybenzo[*d*]naphtho[1,2-*b*]pyran-6-one *C*-glycosides, *J. Org. Chem.*, 52, 5064–5066, 1987.
103. Outten, R A, Daves, D G Jr, Benzo[*d*]naphtho[1,2-*b*]pyran-6-one *C*-glycosides. Aryltri-*n*-butylstannanes in palladium-mediated coupling with 2,3-dihydropyran and furanoid glycols, *J. Org. Chem.*, 54, 29–35, 1989.
104. Hamamichi, N, Miyasaka, T, Synthesis of methyl- and methoxy-substituted β-D-ribofuranosylnaphthalene derivatives by Lewis acid catalyzed ribofuranosylation, *J. Org. Chem.*, 56, 3731–3734, 1991.
105. Kwok, D I, Outten, R A, Huhn, R, Daves, G D Jr, 8-Ethyl-1-methoxybenzo[*d*]naphtho[1,2-*b*]pyran-6-one *C*-glycosides by acid-catalyzed glycosylation, *J. Org. Chem.*, 53, 5359–5361, 1988.
106. Farr, R N, Kwok, D I, Daves, G D Jr, 8-Ethenyl-1-hydroxy-4-β-D-ribofuranosylbenzo[*d*]naphtho[1,2-*b*]pyran-6-one and 8-ethenyl-1-hydroxy-4-(2′-deoxy-β-D-ribofuranosyl)benzo[*d*]naphtho[1,2-*b*]pyran-6-one. Synthetic *C*-glycosides related to the gilvocarcin, ravidomycin, and chrysomycin antibiotics, *J. Org. Chem.*, 57, 2093–2100, 1992.
107. Cai, M S, Qiu, D X, *C*-glycosyl compounds. XIV. Stereoselective and mild method for the synthesis of *C*-D-glucosylarenes in high yield, *Carbohydr. Res.*, 191, 125–129, 1989.
108. Cai, M S, Qiu, D X, Studies on *C*-glycosides. XIII. *C*-glucosides from *O*-glucosyl 3,5-dinitrobenzoate, *Synth. Commun.*, 19, 851–855, 1989.
109. Schmidt, R R, Hoffman, M, Glycosylimidates. Part 5. *C*-glycosides from *O*-glycosyl trichloroacetimidates, *Tetrahedron Lett.*, 23, 409–412, 1982.
110. Schmidt, R R, Effenberger, G, *O*-Glycosyl imidates. 29. Reaction of *O*-(glucopyranosyl) imidates with electron-rich heterocycles. Synthesis of *C*-glucosides, *Liebigs Ann. Chem.*, 29, 825–831, 1987.
111. El-Desoky, E I, Abdel-Rahman, H A R, Schmidt, R R, Glycosylation of 2,3-diphenylindole and derivatives synthesis of *O*-, *N*-, and *C*-glycopyranosides, *Liebigs Ann. Chem.*, 877–881, 1990.

112. Matsumoto, T, Katsuki, M, Suzuki, K, Bis(cyclopentadienyl)dichlorozirconium-silver perchlorate ($Cp2ZrCl_2-AgClO_4$): efficient promoter for the Friedel–Crafts approach to *C*-aryl glycosides, *Tetrahedron Lett.*, 30, 833–836, 1989.
113. Allevi, P, Anastasia, M, Ciuffreda, P, Fiecchi, A, Scala, A, *C*-glucopyranosyl derivatives from readily available 2,3,4,6-tetra-*O*-benzyl-α-D-glucopyranosyl chloride, *J. Chem. Soc., Chem. Commun.*, 101–102, 1987.
114. Casiraghi, G, Cornia, M, Colombo, L, Rassu, G, Fava, G G, Belicchi, M F, Zetta, L, 2,3-Unsaturated *C*-glucopyranosides: a guideline to the anomeric configurational assignment, *Tetrahedron Lett.*, 29, 5549–5552, 1988.
115. Araki, Y, Mokubo, E, Kobayashi, N, Nagasawa, J, Structural elucidation and a novel reductive cleavage of ribofuranosyl ring C-1–O bond of the intramolecular *C*-arylation product of tri-*O*-benzyl-β-D-ribofuranosyl fluoride, *Tetrahedron Lett.*, 30, 1115–1118, 1989.
116. Martin, O R, Hendricks, C A, Desphpande, P P, Cutler, A B, Kane, S A, Rao, S P, Synthesis of *C*-glycosylarenes by way of internal reactions of benzylated and benzoylated carbohydrate derivatives, *Carbohydr. Res.*, 196, 41–58, 1990.
117. Martin, O R, Mahnken, R E, Intramolecular *C*-arylation of benzylated sugars. The unexpected double *C*-glycosidation of *tris-O-m*-methoxybenzyl-β-D-ribofuranose derivatives, *J. Chem. Soc., Chem. Commun.*, 497–498, 1986.
118. Anastasia, M, Allevi, P, Ciuffreda, P, Fiecchi, A, Scala, A, A simple and ready internal *C*-glycosylation of 2,3,5-tri-*O*-benzylglycofuranoses promoted by the boron trifluoride–diethyl ether complex, *Carbohydr. Res.*, 208, 264–266, 1990.
119. Ramesh, N G, Balasubramanian, K K, Reaction of 2,3-unsaturated aryl glycosides with Lewis acids: a convenient entry to carbon-aryl glycosides, *Tetrahedron Lett.*, 33, 3061–3064, 1992.
120. Casiraghi, G, Cornia, M, Rassu, G, Zetta, L, Fava, G G, Belicchi, M F, A simple diastereoselective synthesis of 2′,3′-unsaturated aryl *C*-glucopyranosides, *Tetrahedron Lett.*, 29, 3323–3326, 1988.
121. Casiraghi, G, Cornia, M, Rassu, G, Zetta, L, Fava, G, Belicchi, F, Stereoselective arylation of pyranoid glycals, using bromomagnesium phenolates: an entry to 2,3-unsaturated *C*-α-glycopyranosylarenes, *Carbohydr. Res.*, 191, 243–251, 1989.
122. Matsumoto, T, Hosoya, T, Suzuki, K, Improvement in O → C-glycoside rearrangement approach to *C*-aryl glycosides. Use of 1-*O*-acetyl sugar as stable but efficient glycosyl donor, *Tetrahedron Lett.*, 32, 4629–4632, 1990.
123. Toshima, K, Matsuo, G, Ishizuka, T, Nakata, M, Kninoshita, M, *C*-Arylglycosidation of unprotected free sugar, *J. Chem. Soc., Chem. Commun.*, 1641–1642, 1992.
124. Toshima, K, Matsuo, G, Tatsuta, K, Efficient *C*-arylglycosylation of 1-*O*-methyl sugar by novel use of TMSOTf silver perchlorate catalyst system, *Tetrahedron Lett.*, 33, 2175–2178, 1992.
125. Matsumoto, T, Katsuki, M, Jona, H, Suzuki, K, Synthetic study toward vineomycins. Synthesis of *C*-aryl glycoside sector via hafnocene dichloride–silver perchlorate-promoted tactics, *Tetrahedron Lett.*, 30, 6185–6188, 1989.
126. Matsumoto, T, Hosoya, T, Suzuki, K, Total synthesis and absolute stereochemical assignment of gilvocarcin M, *J. Am. Chem. Soc.*, 114, 3568–3570, 1992.
127. Yamaguchi, M, Horiguchi, A, Ikegura, C, Minami, T, A synthesis of aryl *C*-glycosides via polyketides, *J. Chem. Soc., Chem. Commun.*, 434–436, 1992.
128. Ogawa, T, Pernet, A G, Hanessian, S, Nouvelles Methodes de *C*-Fonctionnalisation Anomerique: Acces aux Precurseurs Chimiques des *C*-Nucleosides, *Tetrahedron Lett.*, 37, 3543–3546, 1973.
129. Narasaka, K, Ichikawa, Y I, Kubota, H, Stereoselective preparation of β-*C*-glycosides from 2-deoxyribose utilizing neighboring participation by 3-*O*-methylsulfinylethyl group, *Chem. Lett.*, 2139–2142, 1987.
130. Ichikawa, Y, Kubota, H, Fujita, K, Okauchi, T, Narasaka, K, Stereoselective β-*C*- and β-*S*-glycosylation of 2-deoxyribofuranose derivatives controlled by the 3-hydroxy protective group, *Bull. Chem. Soc. Jpn.*, 62, 845–852, 1989.
131. Araki, Y, Watanabe, K, Kuan, F H, Itoh, K, Kobayashi, N, Ishido, Y, Synthetic studies by the use of fluorinated intermediates. Part I. A novel procedure for *C*-glycosylation involving Lewis acid-catalyzed coupling-reactions of glycosyl fluorides with enol trimethylsilyl ethers, *Carbohydr. Res.*, 127, C5–C9, 1984.

132. Allevi, P, Anastasia, M, Ciuffreda, P, Fiecchi, A, Scala, A, Epimerization of α- to β-*C*-glucopyranosides under mild basic conditions, *J. Chem. Soc., Perkin Trans. 1*, 1275–1280, 1989.
133. Dawe, R D, Fraser-Reid, B, α-*C*-glycopyranosides from Lewis acid catalysed condensations of acetylated glycals and enol silanes, *J. Chem. Soc., Chem. Commun.*, 1180–1181, 1981.
134. Kunz, H, Muller, B, Weissmuller, J, Stereoselective synthesis of *C*-glycosyl compounds via Michael addition of trimethylsilyl enol ethers and enamines to hex-1-enopyran-3-uloses, *Carbohydr. Res.*, 171, 25–34, 1987.
135. Craig, D, Munasinghem, V R N, Stereoselective template-directed *C*-glycosidation. Synthesis of bicyclic keto oxetanes via intramolecular cyclization reactions of (2-pyridylthio)glycosidic silyl enol ethers, *J. Chem. Soc., Chem. Commun.*, 901–903, 1993.
136. Allevi, P, Anastasia, M, Ciuffreda, P, Eicchim, A, Scala, A, Efficacious *C*-glucosidation of β-keto esters and ketones via enamines, *J. Chem. Soc., Chem. Commun.*, 57–58, 1988.
137. Drew, K N, Gross, P H, *C*-glycoside syntheses. II. Henry condensations of 4,6-*O*-alkylidene pyranoses with a 1,3-proton transfer catalyst — a route to blocked aminomethyl-*C*-glycosides, *Tetrahedron*, 47, 6113–6126, 1991.
138. Köll, P, Kopf, J, Wess, D, Brandenburg, H, Ein neuer Weg zur Darstellung von Glycofuranosylcyaniden (2,5-Anhydroaldononitrilen). Reduktion von acetylierten 2,5-Anhydro-1-desoxy-1-nitroalditolen mit Phosphortrichlorid, *Liebigs Ann. Chem.*, 685–693, 1988.
139. Aebischer, B, Vasella, A, Deoxynitrosugars. 4th Communication. Convenient synthesis of 1-deoxy-1-nitroaldoses, *Helv. Chim. Acta*, 66, 789–794, 1983.
140. Baumberger, F, Vasella, A, Deoxynitrosugars. 6th communication. Stereoelectronic control in the reductive denitration of tertiary nitro ethers. A synthesis of '*C*-glycosides', *Helv. Chim. Acta*, 66, 2210–2222, 1983.
141. Aebischer, B, Meuwly, R, Vasella, A, Chain elongation of 1-*C*-nitroglycosyl halides by substitution with some weakly basic carbanions, *Helv. Chim. Acta*, 67, 2236–2241, 1984.
142. Herscovici, J, Delatre, S, Antonakis, K, Lewis acid induced homoallylic *C*-alkylation. A new approach to *C*-glycosides, *J. Org. Chem.*, 52, 5691–5695, 1987.
143. Acton, E M, Ryan, K J, Smith, T H, *C*-Daunosaminyl derivatives by the Wittig reaction, *Carbohydr. Res.*, 97, 235–245, 1981.
144. Nicotra, F, Russo, G, Ronchetti, F, Toma, L, Stereospecific synthesis of ethyl (2-acetamido-2-deoxy-α-D-glucopyranosyl)acetate, *Carbohydr. Res.*, 124, C5–C7, 1983.
145. Giannis, A, Sandhoff, K, Diastereoselective synthesis of functionalized *C*-α- and -β-glycosyl derivatives of 2-acetamido-2-deoxy-D-glucose, *Carbohydr. Res.*, 171, 201–210, 1987.
146. Dheilly, L, Frechou, C, Beaupere, D, Uzan, R, Demailly, G, A new route to *C*-glycosyl compounds. Wittig-type reaction promoted by zinc, *Carbohydr. Res.*, 224, 301–306, 1992.
147. Shen, Y, Xin, Y, Zhao, J, A new method for carbon–carbon double bond formation promoted by tributylphosphine and zinc, *Tetrahedron Lett.*, 29, 6119–6120, 1988.
148. Sun, K M, Dawe, R D, Fraser-Reid, B, Synthesis and "anomerization" of *C*-glycosyl compounds related to some heteroxyxlic natural products, *Carbohydr. Res.*, 171, 35–47, 1987.
149. Allevi, P, Ciuffreda, P, Colombo, D, Monti, D, Speranza, G, Manitto, P, The Wittig–Horner reaction on 2,3,4,6-tetra-*O*-benzyl-D-mannopyranose and 2,3,4,6-tetra-*O*-benzyl-D-glucopyranose, *J. Chem. Soc., Perkin Trans. 1*, 1281–1283, 1989.
150. Barnes, J J, Davidson, A H, Hughes, L R, Procter, G, The synthesis of optically active tetrahydropyrans by the addition of a stabilized Wittig reagent to pyranose sugars, *J. Chem. Soc., Chem. Commun.*, 1292–1294, 1985.
151. Davidson, A H, Hughes, L R, Qureshi, S S, Wright, B, Wittig reactions on unprotected aldohexoses: formation of optically active tetrahydrofurans and tetrahydropyrans, *Tetrahedron Lett.*, 29, 693–696, 1988.
152. Fréchou, C, Dheilly, L, Beaupere, D, Demailly, R U, Reaction of sulfur ylides on reducing sugars: application to the synthesis of hydroxymethyl *C*-glycosides, *Tetrahedron Lett.*, 33, 5067–5070, 1992.
153. Horton, D, Priebe, W, Synthetic routes to higher-carbon sugars. Reaction of lactones with 2-lithio-1, 3-dithiane, *Carbohydr. Res.*, 94, 27–41, 1981.

154. Lancelin, J M, Zollo, P H A, Sinaÿ, P, Synthesis and conversions of *C*-(alkyn-1-yl)-β-D-glucopyranosides, *Tetrahedron Lett.*, 24, 4833–4836, 1983.
155. Kraus, G A, Molina, M T, A direct synthesis of *C*-glycosyl compounds, *J. Org. Chem.*, 53, 752–753, 1988.
156. Czernecki, S, Ville, G, *C*-glycosides. 7. Stereospecific *C*-glycosylation of aromatic and heterocyclic rings, *J. Org. Chem.*, 54, 610–612, 1989.
157. Krohn, K, Heins, H, Wielckens, K, Synthesis and cytotoxic activity of *C*-glycosidic nicotinamide riboside analogs, *J. Med. Chem.*, 35, 511–517, 1992.
158. Wilcox, C S, Cowart, M D, New approaches to synthetic receptors. Studies on the synthesis and properties of macrocyclic *C*-glycosyl compounds as chiral, water-soluble cyclophanes, *Carbohydr. Res.*, 171, 141–159, 1987.
159. Rouzaud, D, Sinaÿ, P, The first synthesis of a '*C*-disaccharide', *J. Chem. Soc., Chem. Commun.*, 1353–1354, 1983.
160. Preuss, R, Schmidt, R R, Synthesis of methylene bridged *C*-disaccharides, *J. Carbohydr. Chem.*, 10, 887–900, 1991.
161. Goodwin, T E, Crowder, C M, White, R B, Swanson, J S, Evans, F E, Meyer, W L, Stereoselective addition of organocopper reagents to a novel carbohydrate-derived 2,3-dihydro-4H-pyran-4-one, *J. Org. Chem.*, 48, 376–380, 1983.
162. Bellosta, V, Czernecki, S, *C*-glycosyl compounds. IV. Synthesis of (2-deoxy-α-D-glyc-2-enopyranosyl)arenes by stereospecific conjugate addition of organocopper reagents to peracetylated hex-1-enopyran-3-uloses, *Carbohydr. Res.*, 171, 279–288, 1987.
163. Kunz, H, Weismuller, J, Muller, B, Stereoselective *C*-glycoside synthesis through titanium(IV)-catalyzed addition of silyl enol ethers to 2-acyloxy-3-ketoglycal, *Tetrahedron Lett.*, 25, 3571–3574, 1984.
164. Deshong, P, Slough, G A, Elango, V, An organotransition metal based approach to the synthesis of *C*-glycosides, *J. Am. Chem. Soc.*, 107, 7788–7790, 1985.
165. Deshong, P, Slough, G A, Elango, V, Stereoselectivity in the formation of pentacarbonyl glycosylmanganese complexes, *Carbohydr. Res.*, 171, 342–345, 1987.
166. Chatani, N, Ikeda, T, Sano, T, Sonoda, N, Kurosawa, H, Kawasaki, Y, Murai, S, Catalytic siloxymethylation of glycosides by the $HSiR_3/CO/CO_2(CO)_8$ reaction. A new entry to *C*-glycosyl compounds, *J. Org. Chem.*, 53, 3387–3389, 1988.
167. Vasella, A, Waldraff, C A A, Glycosylidene carbenes. Part 3. Synthesis of spirocyclopropanes, *Helv. Chim. Acta*, 74, 585–593, 1991.
168. Vasella, A, Uhlmann, P, Waldraff, C A A, Diederich, F, Glycosylidenecarbenes. 9. Fullerene sugar: preparation of an enantiomerically-pure, spiro-bound *C*-glycoside of C_{60}, *Angew. Chem. Int. Ed. Engl.*, 31, 1388–1390, 1992.
169. RajanBabu, T V, Reddy, G S, 1-Methylene sugars as *C*-glycoside precursors, *J. Org. Chem.*, 51, 5458–5461, 1986.
170. Motherwell, W B, Tozer, M J, Ross, B C, A convenient method for replacement of the anomeric hydroxy group in carbohydrates by difluoromethyl functionality, *J. Chem. Soc., Chem. Commun.*, 1437–1439, 1989.
171. Motherwell, W B, Ross, B C, Tozer, M J, Some radical reactions of exocyclic carbohydrate difluoroenol ethers, *Synlett*, 68–70, 1989.
172. Parker, K A, Coburn, C A, Reductive aromatization of quinol ketals: a new synthesis of *C*-aryl glycosides, *J. Am. Chem. Soc.*, 113, 8516–8518, 1991.
173. Lesimple, P, Beau, J M, Jaurand, G, Sinaÿ, P, Preparation and use of lithiated glycals: vinylic deprotonation versus tin–lithium exchange from 1-tributylstannyl glycals, *Tetrahedron Lett.*, 27, 6201–6204, 1986.
174. Hanessian, S, Martin, M, Desai, R C, Formation of *C*-glycosides by polarity inversion at the anomeric center, *J. Chem. Soc., Chem. Commun.*, 926–927, 1986.
175. Schmidt, R R, Preuss, R, Synthesis of carbon bridged *C*-disaccharides, *Tetrahedron Lett.*, 30, 3409–3412, 1989.
176. Friesen, R W, Loo, R W, Preparation of *C*-aryl glucals via the palladium catalyzed coupling of metalated aromatics with 1-iodo-3,4,6-tri-*O*-(triisopropylsilyl)-D-glucal, *J. Org. Chem.*, 56, 4821–4823, 1991.

177. Grondin, R, Leblanc, Y, Hoogsteen, K, Synthesis of 2-amino *C*-glucosides and 2-amino *C*-glucoside spiroketals, *Tetrahedron Lett.*, 32, 5021–5024, 1991.
178. Hutchinson, D K, Fuchs, P L, Amelioration of the conjugate addition chemistry of α-alkoxycopper reagents: application to the stereospecific synthesis of *C*-glycosides, *J. Am. Chem. Soc.*, 109, 4930–4939, 1987.
179. Prandi, J, Audin, C, Beau, J M, Synthesis of *C*-glycosides: boron trifluoride etherate-promoted opening of epoxides by anomeric anions, *Tetrahedron Lett.*, 32, 769–772, 1991.
180. Lesimple, P, Beau, J M, Sinaÿ, P, Stereocontrolled preparation of 2-deoxy-*C*-α- and β-D-glucopyranosyl compounds from tributyl(2-deoxy-α- and -β-D-glucopyranosyl)stannanes, *Carbohydr. Res.*, 171, 289–300, 1987.
181. Beau, J M, Sinaÿ, P, Preparation and reductive lithiation of 2-deoxy-D-glucopyranosyl phenyl sulfones: a highly stereoselective route to *C*-glycosides, *Tetrahedron Lett.*, 26, 6185–6188, 1985.
182. Crich, D, Lim, L B L, Synthesis of 2-deoxy-β-*C*-pyranosides by diastereoselective hydrogen-atom transfer, *Tetrahedron Lett.*, 31, 1897–1900, 1990.
183. Beau, J M, Sinaÿ, P, D-Glycopyranosyl phenyl sulfones: acylation of their lithiated anions and reductive desulfonylation of the resulting acylated sulfones. A synthesis of α-D-*C*-glycosides, *Tetrahedron Lett.*, 26, 6193–6196, 1985.
184. Beau, J M, Sinaÿ, P, D-Glycopyranosyl phenyl sulfones: their use in a stereocontrolled synthesis of *cis*-2,6-disubstituted tetrahydropyrans (β-D-*C*-glycosides), *Tetrahedron Lett.*, 26, 6189–6192, 1985.
185. Schmidt, R R, Preuss, R, Betz, R, Vinyl carbanions. 33. C-1 lithiation of C-2 activated glucals, *Tetrahedron Lett.*, 28, 6591–6594, 1987.
186. Gomez, A M, Valverde, S, Fraser-Reid, B, A route to unsaturated spiroketals from phenylthiohex-2-enopyranosides via sequential alkylation, allylic rearrangement, and intramolecular glycosidation, *J. Chem. Soc., Chem. Commun.*, 1207–1208, 1991.
187. Allen, N E, Nonclassical targets for antibacterial agents, *Annu. Rep. Med. Chem.*, 20, 155–162, 1985.
188. Ray, P H, Kelsey, J E, Bigham, E C, Benedict, C D, Miller, T A, Synthesis and use of 3-deoxy-D-manno-2-octulosonate (KDO) in *Escherichia coli*. Potential sites of inhibition, *ACS Symp. Ser.*, 231, 141–169, 1983.
189. Luthman, K, Orbe, M, Waglund, T, Clesson, A, Synthesis of *C*-glycosides of 3-deoxy-D-manno-2-octulosonic acid (KDO). Stereoselectivity in an enolate reaction, *J. Org. Chem.*, 52, 3777–3784, 1987.
190. Czernecki, S, Gruy, E, New route to *C*-glycosides, *Tetrahedron Lett.*, 22, 437–440, 1981.
191. Czernecki, S, Dechavanne, V, Arylation of glycals catalyzed by palladium salts: novel synthesis of *C*-glycosides, *Can. J. Chem.*, 61, 533–540, 1983.
192. Cheng, J C Y, Daves, G D Jr, *C*-glycosides from palladium-mediated reactions of pyranoid glycals. Stereochemistry of formation of intermediate organopalladium adducts and factors affecting their stability and decomposition, *J. Org. Chem.*, 52, 3083–3090, 1987.
193. Arai, I, Lee, T D, Hanna, R, Daves, G D Jr, Palladium-catalyzed reactions of glycals with (1,3-dimethyl-2,4(1*H*,3*H*)-pyrimidinedion-5-yl)mercuric acetate. Facile regio- and stereospecific *C*-nucleoside syntheses, *Organometallics*, 1, 742–747, 1982.
194. Farr, R N, Daves, G D Jr, Efficient synthesis of 2′-deoxy-β-D-furanosyl *C*-glycosides. Palladium-mediated glycal-aglycone coupling and stereocontrolled β- and α-face reductions of 3-keto-furanosyl moieties, *J. Carbohydr. Chem.*, 9, 653–660, 1990.
195. Kwok, D I, Farr, R N, Daves, G D Jr, Palladium-mediated coupling of a 3-deoxy pyranoid glycal: stereochemistry of *C*-glycosyl bond formation, *J. Org. Chem.*, 56, 3711–3713, 1991.
196. Zhang, H C, Daves, G D Jr, Enantio- and diastereoisomers of 2,4-dimethoxy-5-(2,3-dideoxy-5-*O*-tritylribofuranosyl)pyrimidine. 2′,3′-Dideoxy pyrimidine *C*-nucleosides by palladium-mediated glycal–aglycon coupling, *J. Org. Chem.*, 58, 2557–2560, 1993.
197. Zhang, H C, Daves, G D Jr, Syntheses of 2′-deoxypseudouridine, 2′-deoxyformycin B, and 2′,3′-dideoxyformycin B by palladium-mediated glycal–aglycon coupling, *J. Org. Chem.*, 57, 4690–4696, 1992.
198. Dubois, E, Beau, J M, Formation of *C*-glycosides by a palladium-catalyzed coupling reaction of tributylstannyl glycals with organic halides, *J. Chem. Soc., Chem. Commun.*, 1191–1192, 1990.
199. Dubois, E, Beau, J M, Synthesis of *C*-glycopyranosyl compounds by a palladium-catalyzed coupling reaction of 1-tributylstannyl-D-glucals with organic halides, *Carbohydr. Res.*, 228, 103–120, 1992.

200. Tius, M A, Gomez, G J, Gu, X Q, Zaidi, J H, *C*-glycosylanthraquinone synthesis: total synthesis of vineomycinone B2 methyl ester, *J. Am. Chem. Soc.*, 113, 5775–5783, 1991.
201. Yougai, S, Miwa, T, *C*-glycopyranosides from the reaction of acetylated glycals with β-diketones, *J. Chem. Soc., Chem. Commun.*, 68–69, 1983.
202. Brakta, M, Le Borgne, F, Sinou, D, Stereospecific *C*-glycosidation catalyzed by a palladium (0) complex, *J. Carbohydr. Chem.*, 6, 307–311, 1987.
203. Brakta, M, Lhoste, P, Sinou, D, Palladium(0)-based approach to functionalized *C*-glycopyranosides, *J. Org. Chem.*, 54, 1890–1896, 1989.
204. RajanBabu, T V, Palladium(0)-catalyzed *C*-glycosylation: a facile alkylation of trifluoroacetylglucal, *J. Org. Chem.*, 50, 3642–3644, 1985.
205. Dunkerton, L V, Euske, J M, Serino, A J, Palladium-assisted reactions. III. Palladium(0)-assisted synthesis of *C*-glycopyranosyl compounds, *Carbohydr. Res.*, 171, 89–107, 1987.
206. Engelbrecht, G J, Holzapfel, C W, Palladium catalyzed reactions of unsaturated carbohydrates. A route to *C*-glycosides, *Heterocycles*, 32, 1267–1272, 1991.
207. Brakta, M, Lhoste, P, Sinou, D, Banoub, J, Functionalized *C*-glycosyl compounds. Part II. Synthesis of the anomeric ethyl 2-(4,6-di-*O*-benzyl-2,3-dideoxy-D-erythro-hex-2-enopyranosyl) acetates and their structural characterization, *Carbohydr. Res.*, 203, 148–155, 1990.
208. Araki, Y, Endo, T, Tanji, M, Nagasawa, J, Additions of a ribofuranosyl radical to olefins. A formal synthesis of showdomycin, *Tetrahedron Lett.*, 29, 351–354, 1988.
209. Giese, B, Ruckert, B, Groninger, K S, Muhn, R, Lindner, H J, Dimerization of carbohydrate radicals, *Liebigs Ann. Chem.*, 997–1000, 1988.
210. Somsák, L, Batta, G, Farkas, I, Preparation of 1,2-cis-glycosyl cyanides by the stereoselective reduction of acetylated 1-bromo-D-glycosyl cyanides, *Tetrahedron Lett.*, 27, 5877–5880, 1986.
211. Giese, B, Gilges, S, Groninger, K S, Lamberth, C, Witzel, T, Synthesis of 2-deoxy sugars, *Liebigs Ann. Chem.*, 615–617, 1988.
212. Giese, B, Groninger, K S, 1,3,4,6-Tetra-*O*-acetyl-2-deoxy-α-D-glycopyranose, *Org. Synth.*, 69, 66–71, 1990.
213. Dupuis, J, Giese, B, Hartung, J, Leising, M, Korth, H G, Sustmann, R, Electron transfer from trialkyltin radicals to nitrosugars: the synthesis of *C*-glycosides with tertiary anomeric carbon atoms, *J. Am. Chem. Soc.*, 107, 4332–4333, 1985.
214. Paulsen, H, Matschulat, P, Synthesis of *C*-glycosides of *N*-acetylneuraminic acid and other derivatives, *Liebigs Ann. Chem.*, 487–495, 1991.
215. Giese, B, Witzel, T, Synthesis of "*C*-disaccharides" by radical C–C bond formation, *Angew. Chem. Int. Ed. Engl.*, 25, 450–451, 1986.
216. Giese, B, Dupuis, J, Leising, M, Nix, M, Lindner, H J, Synthesis of *C*-pento, -hexo-, and -heptulopyranosyl compounds via radical carbon–carbon bond-formation reactions, *Carbohydr. Res.*, 171, 329–341, 1987.
217. Abrecht, S, Scheffold, R, Vitamin B12-catalyzed synthesis of *C*-glycosides, *Chimia*, 39, 211–212, 1985.
218. Bimwala, R M, Vogel, P, Synthesis of α-(1→3)- and α-(1→4)-linked *C*-disaccharides using (+)-7-oxabicyclo[2.2.1]hept-5-en-2-one (naked sugar), *Tetrahedron Lett.*, 32, 1429–1432, 1991.
219. Bimwala, R M, Vogel, P, Enantiomerically pure 7-oxabicyclo[2.2.1]hept-5-en-2-yl derivatives (naked sugars) as synthetic intermediates. 18. Synthesis of α-(1→2)-, α-(1→3)-, α-(1→4)-, and α-(1→5)-*C*-linked disaccharides through 2,3,4,6-tetra-*O*-acetylglucopyranosyl radical additions to 3-methylidene-7-oxabicyclo[2.2.1]heptan-2-one derivatives, *J. Org. Chem.*, 57, 2076–2083, 1992.
220. Giese, B, Hoch, M, Lamberth, C, Schmidt, R R, Synthesis of methylene bridged *C*-disaccharides, *Tetrahedron Lett.*, 29, 1375–1378, 1988.
221. Groninger, K S, Jager, K F, Giese, B, Cyclization reactions with allyl-substituted glucose derivatives, *Liebigs Ann. Chem.*, 731–732, 1987.
222. De Mesmaeker, A, Hoffmann, P, Ernst, B, Hug, P, Winkler, T, Stereoselective carbon–carbon bond formation in carbohydrates by radical cyclization reactions. III. Strategy for the preparation of C(1)-glycosides, *Tetrahedron Lett.*, 30, 6307–6310, 1989.
223. Stork, G, Suh, H S, Kim, G, The temporary silicon connection method in the control of regio- and stereochemistry. Applications to radical-mediated reactions. The stereospecific synthesis of *C*-glycosides, *J. Am. Chem. Soc.*, 113, 7054–7056, 1991.

224. Lee, E, Tae, J S, Lee, C, Park, C M, β-Alkoxyacrylates in radical cyclizations: remarkably efficient oxacycle synthesis, *Tetrahedron Lett.*, 34, 4831–4834, 1993.
225. Schoellkopf, U, Recent results in carbanion chemistry, *Angew. Chem. Int. Ed. Engl.*, 9, 763–773, 1970.
226. Lansbury, P, Pattison, V A, Sidler, J D, Bieber, J B, Mechanistic aspects of the rearrangement and elimination reactions of α-metalated benzyl alkyl ethers, *J. Am. Chem. Soc.*, 88, 78–84, 1966.
227. Hebert, E, Welvart, Z, Ghelfenstein, M, Swarc, H, Effect of high pressure on the Wittig rearrangement, *Tetrahedron Lett.*, 24, 1381–1384, 1983.
228. Grindley, T B, Wickramage, C, A novel approach to the synthesis of *C*-glycosyl compounds: the Wittig rearrangement, *J. Carbohydr. Chem.*, 7, 661–685, 1988.
229. Kametani, T, Kawamura, K, Honda, T, New entry to the *C*-glycosidation by means of carbenoid displacement reaction. Its application to the synthesis of showdomycin, *J. Am. Chem. Soc.*, 109, 3010–3017, 1987.
230. Tulshian, D B, Fraser-Reid, B, Routes to *C*-glycopyranosides via sigmatropic rearrangements, *J. Org. Chem.*, 49, 518–522, 1984.
231. Curran, D P, Suh, Y G, Selective mono-Claisen rearrangement of carbohydrate glycals. A chemical consequence of the vinylogous anomeric effect, *Carbohydr. Res.*, 171, 161–191, 1987.
232. Colombo, L, Casiraghi, G, Pittalis, A, Stereoselective synthesis of *C*-glycosyl α-amino acids, *J. Org. Chem.*, 56, 3897–3900, 1991.
233. Martin, O R, Intramolecular *C*-arylation of 2,3,5-tri-*O*-benzyl- and 2,3,5-tri-*O*-(3-methylbenzyl) pentofuranose derivatives, *Carbohydr. Res.*, 171, 211–222, 1987.
234. Armstrong, R W, Teegarden, B R, Synthesis of α-methyl 1′,2′-dideoxycellobioside: a novel *C*-disaccharide, *J. Org. Chem.*, 57, 915–922, 1992.
235. Nicotra, F, Panza, L, Ronchetti, F, Russo, G, Toma, L, Synthesis of *C*-glycosyl compounds by the Wittig iodocyclization procedure. Differences from mercuriocyclization, *Carbohydr. Res.*, 171, 49–57, 1987.
236. Lancelin, J M, Pougny, J R, Sinaÿ, P, Use of selenium in carbohydrate chemistry: preparation of *C*-glycoside congeners, *Carbohydr. Res.*, 136, 369–374, 1985.
237. Freeman, F, Robarge, K D, Electrophile-mediated cyclizations of 6-*O*-benzyl-1,2-dideoxy-3,4-*O*-isopropylidene-D-*ribo*-hex-1-enitol to derivatives of 2,5-anhydro-6-*O*-benzyl-3,4-*O*-isopropylidene-D-*altro*-hexitol, *Carbohydr. Res.*, 137, 89–97, 1985.
238. Freeman, F, Robarge, K D, Stereoselectivity in the electrophile-mediated cyclization of 2,3,5-tri-*O*-benzyl-1,2-dideoxy-D-*arabino*-hex-1-enitol. A stereocontrolled synthesis of 1-amino-2,5-anhydro-3, 4,6-tri-*O*-benzyl-1-deoxy-D-glucitol, *Carbohydr. Res.*, 171, 1–11, 1987.
239. Boschetti, A, Nicotra, F, Panza, L, Russo, G, Vinylation-electrophilic cyclization of aldopentoses: easy and stereoselective access to *C*-glycopyranosides of rare sugars, *J. Org. Chem.*, 53, 4181–4185, 1988.
240. Lay, L, Nicotra, F, Panza, L, Verani, A, Synthetic studies on α-*C*-glycosides of D-glucosamine, *Gazz. Chim. Ital.*, 122, 345–348, 1992.
241. Carcano, M, Nicotra, F, Panza, L, Russo, G, A new procedure for the synthesis of D-glucosamine α-*C*-glycosides, *J. Chem. Soc., Chem. Commun.*, 297–298, 1989.
242. Frick, W, Schmidt, R R, Convenient synthesis of *C*-β-D-glucopyranosyl arenes. Synthesis of 5,7,4′-tri-*O*-methylvitexin, *Liebigs Ann. Chem.*, 565–570, 1989.
243. Schmidt, R R, Frick, W, Short synthesis of *C*-aryl-glucopyranosides of the papulacandin type, *Tetrahedron*, 44, 7163–7169, 1988.
244. Schmid, W, Whitesides, G, A new approach to cyclitols based on rabbit-muscle aldolase (RAMA), *J. Am. Chem. Soc.*, 112, 9670–9671, 1990.
245. Burke, S D, Armistead, D M, Schoenen, F J, Fevig, J M, An enolate Claisen route to *C*-pyranosides. Development and application to an ionophore synthon, *Tetrahedron*, 42, 2787–2801, 1986.
246. Fleet, G W, Seymour, L C, Synthesis of intermediates for stereospecific synthesis of α- and β-*C*-nucleosides: ring contraction of protected 2-*O*-trifluoromethanesulphonates of galacto- and altro-pyranosides, *Tetrahedron Lett.*, 28, 3015–3018, 1987.
247. Kuszmann, J, Podányi, B, Jerkovich, G, Rearrangement of unsaturated 2,4-*O*-benzylidenehexitol derivatives into *C*-glycosylbenzene derivatives, *Carbohydr. Res.*, 232, 17–32, 1992.

248. Katagiri, N, Akatsuka, H, Haneda, T, Kaneko, C, Highly stereoselective total synthesis of β-ribofuranosylmalonate, *J. Org. Chem.*, 53, 5464–5470, 1988.
249. Schmidt, R R, Frick, W, Haag-Zeino, B, Apparao, S, De-novo synthesis of carbohydrates and related natural products. Part 28. *C*-arylglycosides and 3-deoxy-2-glyculosonates via inverse type hetero-Diels–Alder reaction, *Tetrahedron Lett.*, 28, 4045–4048, 1987.
250. Danishefsky, S, Philips, G, Ciufolini, M, A fully synthetic route to the papulacandins. Stereospecific spiroacetalization of a C-1-arylated methylglycoside, *Carbohydr. Res.*, 171, 317–327, 1987.
251. Hansson, S, Miller, J F, Liebskind, L S, Synthesis and reactions of enantiomerically pure molybdenum π-complexes of 2H-pyran. A general approach to the enantiospecific synthesis of *cis*-2,5-disubstituted-5,6-dihydro-2H-pyrans and *cis*-2,6-disubstituted tetrahydropyrans, *J. Am. Chem. Soc.*, 112, 9660–9661, 1990.
252. Levy, D E, Tang, C, *The Chemistry of* C*-Glycosides*, Elsevier, Oxford, 1995.
253. Postema, M H D, C*-Glycoside Synthesis*, CRC Press, London, 1995.
254. Hanessian, S, *Preparative Carbohydrate Chemistry*, Marcel Dekker, New York, 1997.
255. Herscovici, J, Antonakis, K, Recent developments in *C*-glycoside synthesis, *Stud. Nat. Prod. Chem.*, 10, 337–403, 1992.
256. Postema, M H D, Recent developments in the synthesis of *C*-glycosides, *Tetrahedron*, 48, 8545–8599, 1992.
257. Daves, G D Jr, *C*-glycoside synthesis by palladium-mediated glycal–aglycon coupling reactions, *Acc. Chem. Res.*, 23, 201–206, 1990.
258. Special Issue for *C*-Glycosides. *Carbohydr. Res.* 171, 1987.
259. Hacksell, U, Daves, G D Jr, The chemistry and biochemistry of *C*-nucleosides and *C*-arylglycosides, *Prog. Med. Chem.*, 22, 1–65, 1985.
260. Hanessian, S, Pernet, A G, Synthesis of naturally occurring *C*-nucleosides, their analogs, and functionalized *C*-glycosyl precursor, *Adv. Carbohydr. Chem. Biochem.*, 33, 111–188, 1976.
261. *Carbohydrate Chemistry*, Specialist Periodical Reports, Royal Chemical Society 1968–1990, Vol. 1–24, Sections on *C*-glycosides.

8 From Sugars to Carba-Sugars

Matthieu Sollogoub and Pierre Sinaÿ

CONTENTS

8.1 INTRODUCTION

In 1966, McCasland [1] first synthesized 2,3,4,5-tetrahydroxycyclohexanemethanol **1**, a talopyranose analog in which the ring oxygen is replaced by a methylene group (Figure 8.1). The term "pseudosugar" has been coined by him to designate any such analogs. As time passed, this terminology was besmirched and fraudulently employed for a large variety of carbocyclic sugar analogs, requiring a specification of the definition for different subclasses of mimetics.

S. Ogawa [2] proposed the use of the prefix "carba", preceded, where considered necessary, by the appropriate locant ("4a" for an aldofuranose, "5a" for an aldopyranose), followed by the name of the sugar. Unfortunately, even this unambiguous nomenclature was misconstrued by many authors to include carbocyclic analogs containing a nonsubstituted carbon, as in compounds **2** [3] and **3** [4] (Figure 8.2). In this chapter, we will restrict ourselves to compounds sticking to a strict definition: a carba-sugar is a carbocyclic analog of a true sugar in which the ring oxygen has been replaced by a methylene (CH_2) group. This implies that there are only 32 5a-carba-aldohexopyranoses and 16 4a-carba-aldopentofuranoses (Figure 8.2). If replacing the methylene group with an oxygen atom in a so-called carba-sugar does not restore a true sugar structure, as in **2**

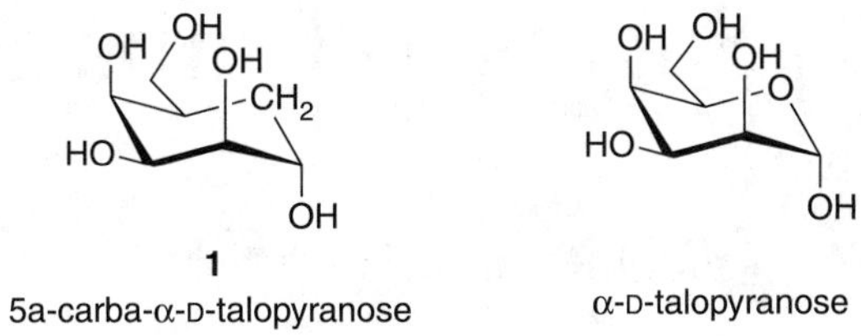

1
5a-carba-α-D-talopyranose

α-D-talopyranose

FIGURE 8.1 The first carba-sugar and its true parent sugar.

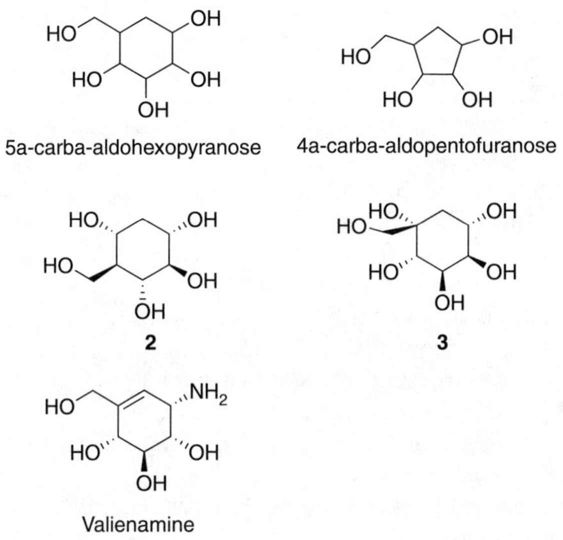

5a-carba-aldohexopyranose

4a-carba-aldopentofuranose

2

3

Valienamine

FIGURE 8.2 Real and false carba-sugars.

and **3**, or if the unsubstituted carbon is not tetrahedral anymore, as in valienamine (Figure 8.2), we propose to call it a pseudocarba-sugar.

The purpose of this chapter is to describe transformations of sugars into corresponding carba-sugars (endocyclic O → CH_2 replacement). Thus, it seems reasonable to first briefly answer the following question: why synthesize carba-sugars?

8.2 WHY SYNTHESIZE CARBA-SUGARS?

8.2.1 CARBA-MONOSACCHARIDES

The chemical total synthesis of natural products is a major and fascinating facet of organic chemistry. It has always been the driving force for the invention of new reactions and the development of elegant synthetic strategies. It is thus remarkable that 5a-carba-galactopyranose (Figure 8.3), a true carba-sugar, has been isolated from a fermentation broth of *Streptomyces* sp. MA-4145 [5] only a few years after its first chemical synthesis [6]. Unfortunately, this is the only natural carba-sugar known and its biological activity proved to be rather low. Thus, the hope that a new family of natural products with potentially interesting biological activities would be discovered was unfulfilled.

There are several logical reasons to build up carba-sugars. As depicted in Scheme 8.1, in a carba-sugar the acetal function has been transformed into a nonhydrolyzable ether, with the net result of making these compounds potential candidates for the competitive inhibition of glycosidases or glycosyltransferases.

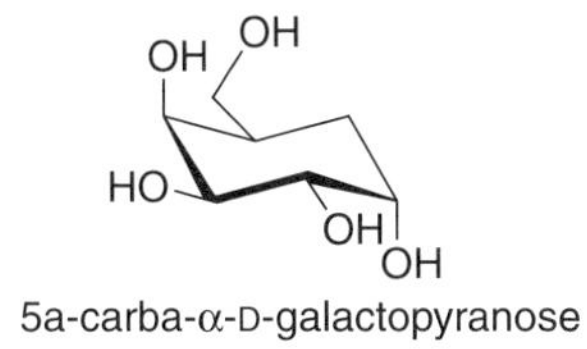

FIGURE 8.3 The only natural carba-sugar.

In the very first paper describing a carba-sugar, McCasland stated: "It is hoped that pseudosugars *(carba-sugars)* may be found acceptable in place of corresponding true sugars to some but not all enzymes or biological systems, and thus might serve to inhibit growth of malignant or pathogenic cells." Indeed, these analogs of the ground state conformation, as shown by most of the NMR data on 5a-carba-pyranoses, display rather weak inhibitory activities. For example, the 5a-carba-α-L-fucopyranose (Figure 8.4) binds to α-L-fucosidase (bovine kidney) but showed moderate inhibitory activity ($K_I = 84\ \mu M$) [7,8]. This result was attributed to hydrophobic interactions between the ring methylene group and the enzyme.

In order to improve the activity, some chemical modifications are necessary. Compound **4** (Figure 8.4), a carba-sugar analog of α-L-fucopyranosylamine, proved to be a much more potent inhibitor (more than 3000-fold increase in activity compared with the carba-sugar). In this case the justification of the presence of the methylene group is obvious: glycosylamines are sensitive molecules, whereas **4** is a stable derivative. An additional feature makes carbocyclic analogs even more potent: a double bond between C-5 and C-5a as in valienamine (Figure 8.2). However, together with **4**, these compounds are not *sensu stricto* carba-sugars anymore but enter the class of pseudocarba-sugars.

A nice property of carba-sugars that demonstrates their similarity with sugars is the fact that they taste sweet. 5a-Carba-α-glucopyranose (Figure 8.5) and 5a-carba-α-galactopyranose (Figure 8.3) are just about as sweet as D-glucose and sucrose, respectively [9]. The sweetness and structure relationship has been extensively studied using carba-sugars as model compounds [10]. Sweetness of all compounds was evaluated by five volunteers by tasting solutions of each sugar (approximately 10% aqueous) — a forgotten analytical method in current chemistry!

It is probably as stable probes of the specificity of enzymes that carba-sugars find their justification. A striking example of their utility is illustrated by the elucidation of the stereochemical selectivity of the reverse reaction of cellobiose phosphorylase, where only the 5a-carba-β-D-glucopyranose was a substrate of the reaction — not the α anomer (Scheme 8.2) [11].

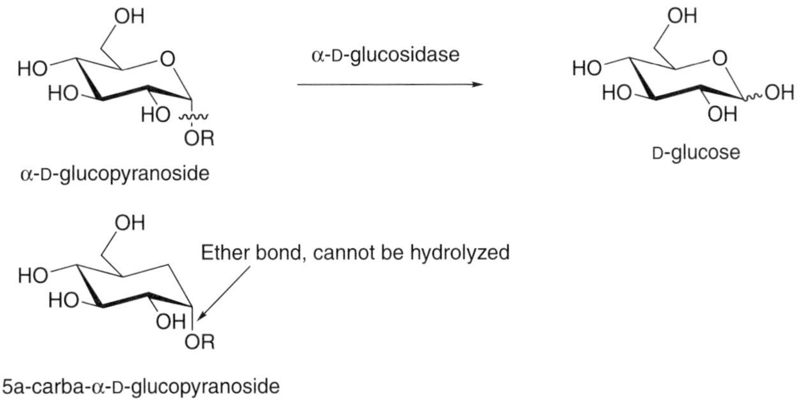

SCHEME 8.1 Carba-sugars as potential glycosidase inhibitors.

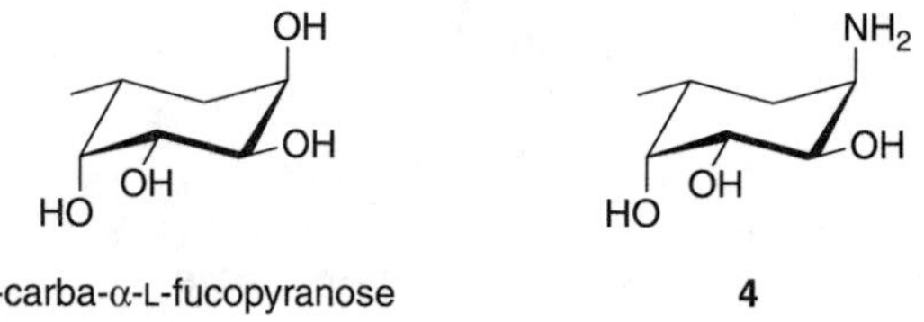

5a-carba-α-L-fucopyranose 4

FIGURE 8.4 Carba-fucose and an aminated analogue.

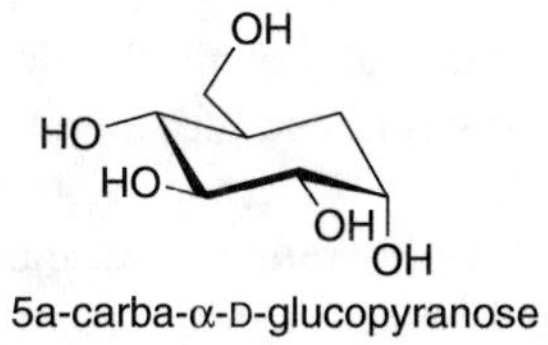

5a-carba-α-D-glucopyranose

FIGURE 8.5 5a-carba-α-D-glucopyranose.

This study was of course unrealizable with D-glucose itself because of mutarotation; a carba-sugar, not affected by this drawback, is therefore a very useful chemical probe.

In the case of 4a-carba-furanoses, very little is known regarding their potency as inhibitors, although the conformations of methyl 4a-carba-D-arabinofuranosides (Figure 8.6) have been thoroughly examined [12]. NMR studies show that the similarity of the distribution of conformers differs for each anomer (or pseudoanomer). For the α-isomers, there are significant differences in ring conformations between the glycoside and the carba-sugar. In contrast, the ring conformations of the β-isomer are similar to its furanoside counterpart. These observations qualify 4a-carba-furanoside derivatives as potential substrates for arabinosyltransferases.

8.2.2 Carba-Oligosaccharides

5′a-Carba-disaccharides are disaccharide mimics in which the ring oxygen of the nonreducing unit has been replaced by a methylene group (Figure 8.7). This feature transforms the acetalic interglycosidic linkage into an ether bond resistant to chemical and enzymatic hydrolysis.

NMR investigations and molecular mechanics calculations revealed that 5′a-carba-lactose (Figure 8.8) adopts the same conformation as its true parent sugar, lactose [13]. However,

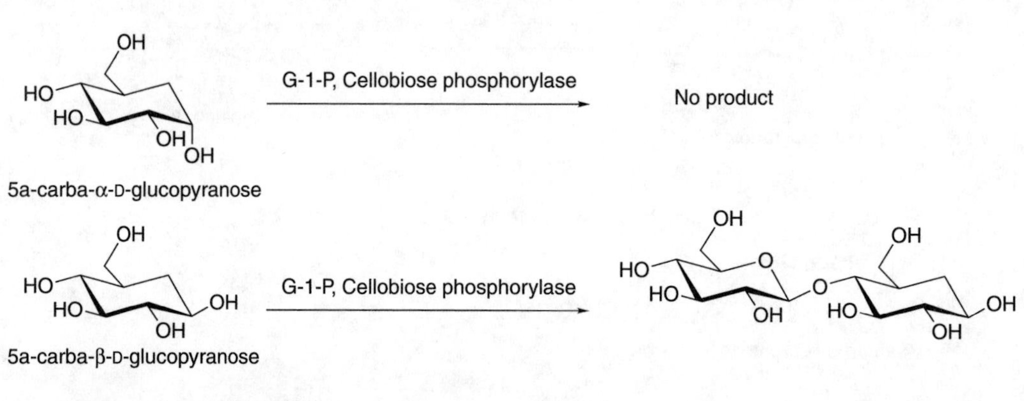

G-1-P, Cellobiose phosphorylase

No product

5a-carba-α-D-glucopyranose

G-1-P, Cellobiose phosphorylase

5a-carba-β-D-glucopyranose

SCHEME 8.2 Carba-sugars as stable chemical probes.

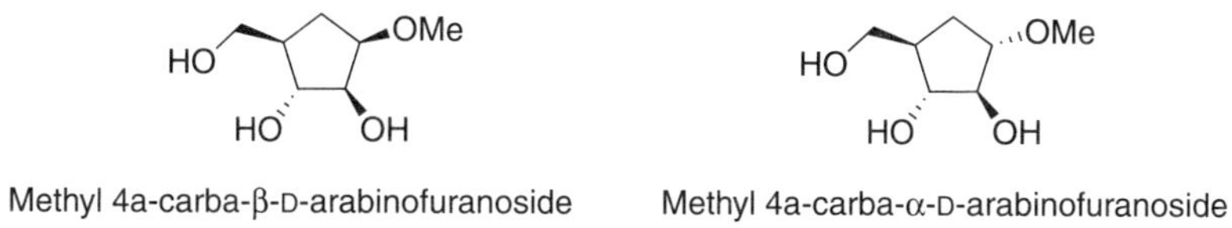

Methyl 4a-carba-β-D-arabinofuranoside Methyl 4a-carba-α-D-arabinofuranoside

FIGURE 8.6 4a-Carba-arabinofuranosides.

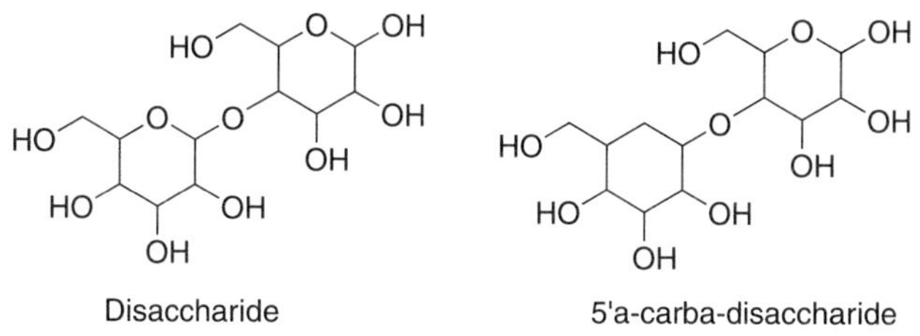

Disaccharide 5'a-carba-disaccharide

FIGURE 8.7 5′a-Carba-disaccharide and its parent disaccharide.

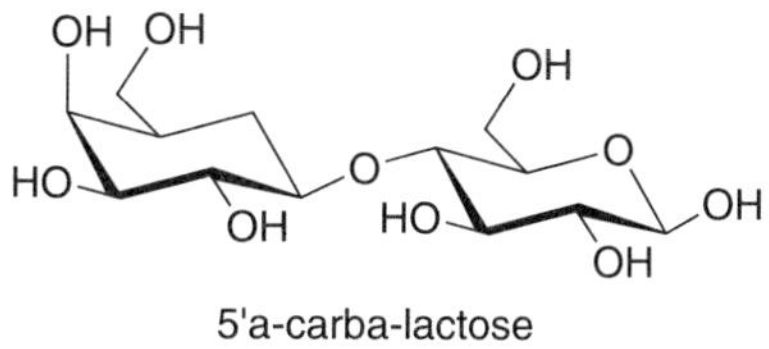

5'a-carba-lactose

FIGURE 8.8 5′a-Carba-lactose.

"glycosidic" linkages of carba-disaccharides are more flexible than those of the natural glycosides. The flexibility of these mimetics may pose limitations to their use as therapeutic agents; an entropy penalty has to be paid and conformers of local minima differing from that of the global minimum might be bound to biological receptors.

More complex carba-oligosaccharides such as **5** [14] and **6** [15] have also been synthesized (Figure 8.9). Inhibitory activities against jack-bean α-D-mannosidase and the kinetics of the reaction with GlcNAcT-V, respectively, were tested. Both results showed that the ether-linked carba-sugar analogs of oligosaccharides seem to act as substrates, rather than inhibitors, towards glycohydrolases and glycotransferases.

Here again, it is as probes that carba-oligosaccharides find their best applications. Compounds **7** and **8** were both submitted to the action of α-(1–3/4)-fucosyltransferase in the presence of GDP-fucose. Only **7** was accepted and fucosylated to give the Lewis^x analog **9** showing that this enzyme has a different mode of action regarding the synthesis of Lewis^a (Scheme 8.3) [16].

8.2.3 Carba-Glycosyl-Phosphates

Carba-fructofuranose 6-phosphate was found to be a good substrate for phosphofructokinase and 6-phosphofructo-2-kinase, with K_M values 5 to 20 times larger than that of fructose 6-phosphate (Scheme 8.4). The carba-sugar analog inhibits fructose-2,6-bisphosphatase with K_I values about 500 times higher than the K_M of the natural substrate [17]. It was found that this inhibition was probably caused by the stability of the resulting carba-fructofuranose 2,6-diphosphate compared with the corresponding sugar diphosphate [18]. However, this compound appeared to have a

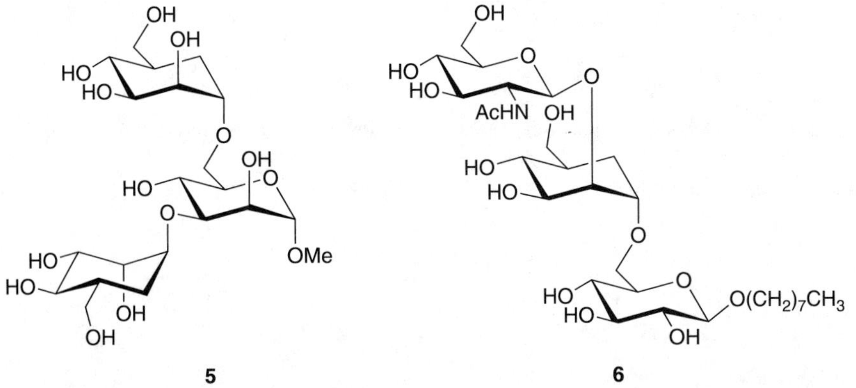

FIGURE 8.9 Complex carba-oligosaccharides.

lower affinity for the activation site of pyrophosphate-dependent 6-phosphofructo-1-kinase, compared with its parent sugar, indicating that the oxygen of the fructose ring is significant in the activation of the enzyme.

GDP-carba-fucose [19] and UDP-carba-galactose [20] have also been synthesized and tested as inhibitors of the corresponding glycosyltransferases (Figure 8.10). These compounds have been shown to be competitive inhibitors with K_I values similar to the K_M of the natural substrates.

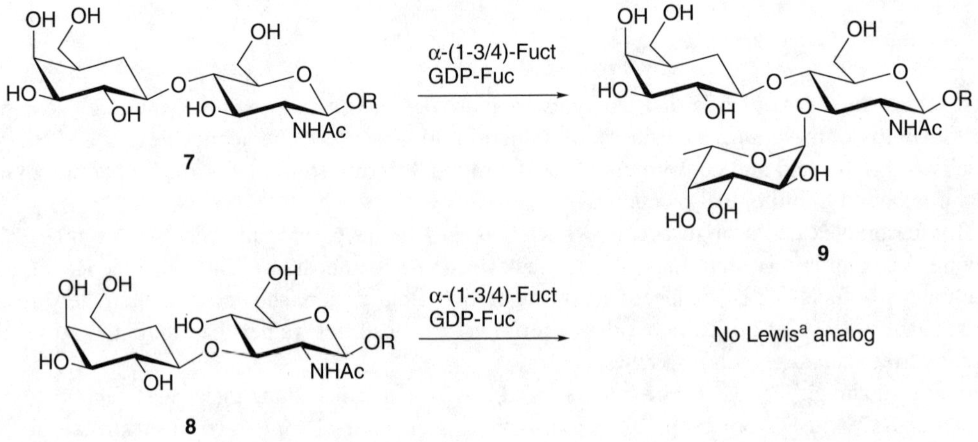

SCHEME 8.3 Carba-disaccharides as stable chemical probes.

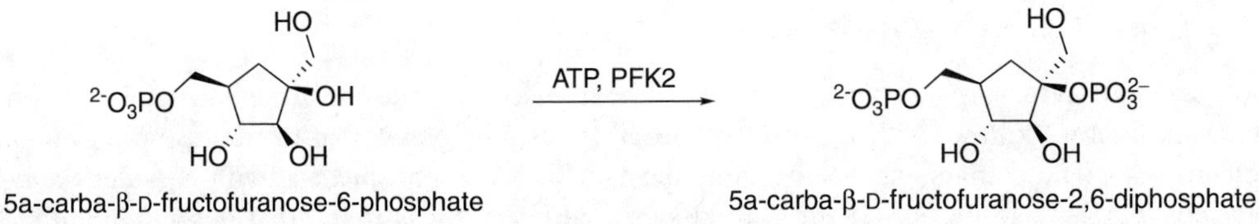

SCHEME 8.4 Action of PFK2 on a carba-fructose derivative.

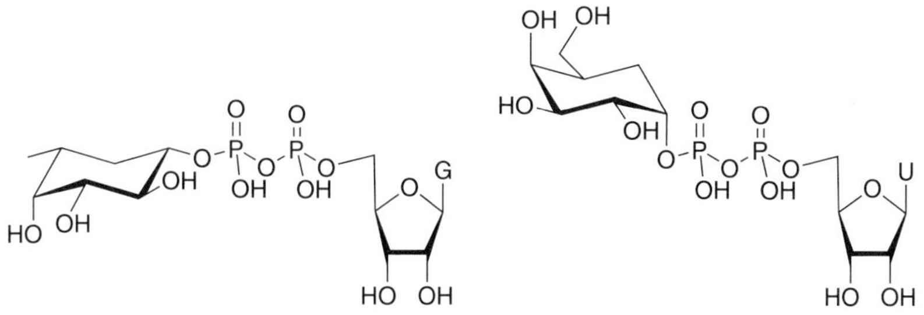

FIGURE 8.10 Carba-sugars for glycosyl-transferase inhibition.

8.3 SYNTHESIS OF CARBA-SUGARS FROM SUGARS

The chemistry of carba-sugars and closely related analogs or structures bearing a resemblance to sugars (carbohydrate mimics) collectively comprise a well documented field. Ogawa et al. synthesized and extensively studied a great number of carba-sugars, mainly taking advantage of the Diels–Alder reaction on nonsugar substrates followed by further transformations into appropriately functionalized carbocycles. As Ogawa does not use sugars as starting materials, his work will not be discussed here. However, the reader is referred to Ogawa's publications [2] as a supplement to this chapter.

With all the requisite stereochemistry in place and no need for extensive functionalization, sugars are the logical starting material for the synthesis of carba-sugars. Moreover, they are generally inexpensive and stereochemically pure compounds. To illustrate this point the starting sugar will be identified in each synthesis, even if the synthesis is described from a more advanced derivative.

The carbocyclization of sugars has been reviewed by Ferrier and Middleton [21]. In the context of these reactions, we describe herein only the pathways that were used for the synthesis of authentic carba-sugars. Moreover, we present only examples utilizing these transformations, and all carbocyclizations are presented along with appropriate yield and mechanistic information. Finally, we present this subject in the context of two families of synthetic strategies: cyclizations of open-chain sugars and rearrangements of cyclic sugars [22,23].

8.3.1 CYCLIZATION OF OPEN-CHAIN SUGARS

As reducing sugars are easily converted to their open-chain forms, various strategies have been developed to recyclize these compounds. If, for example, cyclization occurs through formation of a new O–C bond, the result is generation of a new reducing sugar. However, when cyclization results in the formation of a new C–C bond the product is a carba-sugar. In this section, various methods for the conversion of open-chain sugars to cyclic carba-sugars are presented.

8.3.1.1 Carbanionic Cyclizations

The generation of anionic species and their use as nucleophiles in the formation of C–C bonds is well established in the field of organic chemistry. As applied to the formation of carba-sugars, various stabilizing groups have gained particular attention. Among these are the use of malonates, nitro groups, phosphorus ylides and aldol precursors. The following paragraphs will address the utility of each of these groups.

8.3.1.1.1 Malonates

The use of malonate anions to recyclize open-chain sugars was first used by Suami et al. [24,25] beginning from L-arabinose (Scheme 8.5). This sugar, classically opened as a thioacetal, was protected by conversion of the primary alcohol to a trityl ether with subsequent benzylation of the secondary hydroxyl groups. Replacement of the trityl by a tosyl group afforded **10**. The aldehyde function was restored and the tosylate was substituted with iodide, giving the key iodo-aldehyde **11**. The cyclization of **11** with dimethyl malonate gave two pairs of diastereoisomers, **12** and **13**. Isomers **12** are the products of adding the malonate to the iodinated carbon followed by addition to the carbonyl. Alternatively, isomers **13** are the products of initial addition to the carbonyl followed by displacement of the iodo group by the resulting alkoxide. To finish the synthesis, **12** was acetylated and decarboxylated with elimination. The resulting unsaturated ester **14** was reduced and the double bond was separately hydroborated, affording 5a-carba-α-D-glucose and 5a-carba-β-L-altrose after deprotection.

The main drawback of this strategy is the low yield of the carbocyclization step, because of the competition between addition to the carbonyl and substitution of the iodide. To circumvent this problem, an alternate pathway incorporating the malonate moiety prior to the transformation of the primary hydroxyl group into a leaving group was explored. This alternative method was used for

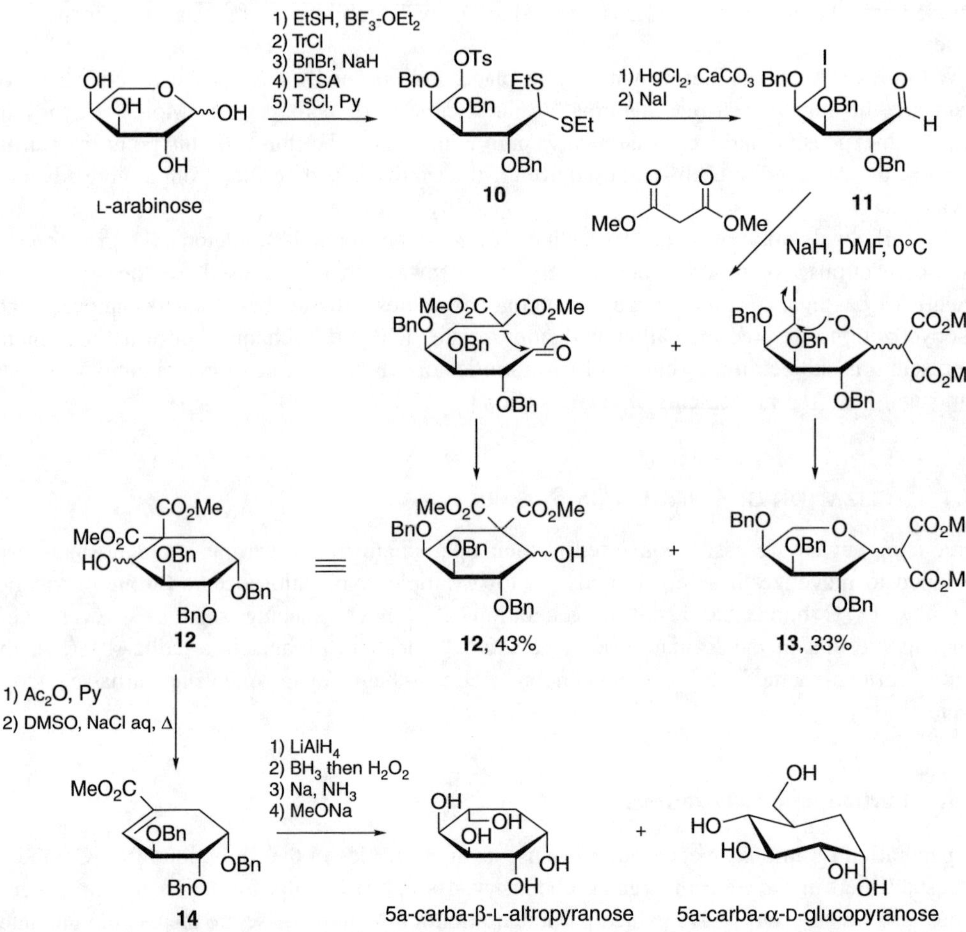

SCHEME 8.5 Synthesis of carba-sugars using malonates.

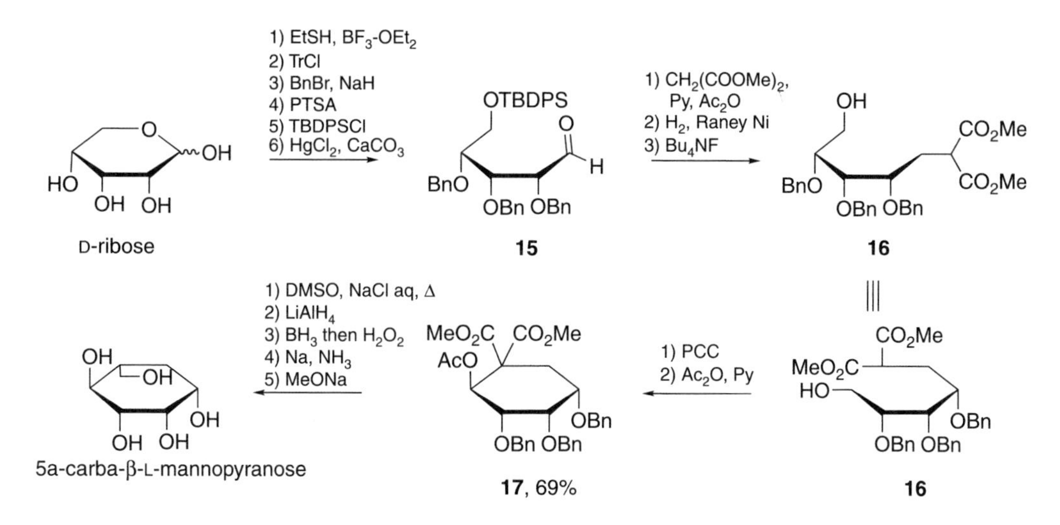

SCHEME 8.6 Improved synthesis of a carba-sugar using malonates.

the synthesis of carba-sugars from D-ribose [26] and D-xylose [27] (Scheme 8.6). As illustrated, the silylated derivative **15** was synthesized following the strategy reported for **10** (Scheme 8.5). Compound **15** then underwent a Knoevenagel condensation in the presence of an excess of dimethyl malonate in a mixture of acetic anhydride and pyridine to afford an α,β-unsaturated diester. This product was hydrogenated and desilylated to give the alcohol **16**. The key cyclization into a cyclohexane derivative was accomplished by the treatment of compound **16** with PCC followed by acetylation to afford a 69% yield of **17** as a single diastereoisomer. Only 5a-carba-β-L-mannopyranose was isolated utilizing the concluding sequence described in Scheme 8.5 (hydroboration being selective in this case).

Scheme 8.7 shows that the same methodology was used for the synthesis of carba-furanoses starting from D-erythrose. As illustrated, compound **18** was cyclized according to the methods described in Scheme 8.6, giving a 71% yield of the carbocycle **19**. The ultimate result of this sequence was isolation of 4a-carba-β-L-arabinofuranose.

8.3.1.1.2 The Use of Nitro Groups

Nitro groups are commonly used to stabilize carbanions on adjacent carbon atoms. This property was exploited in order to complementarily recyclize open-chain sugars to the chemistry in the previous section. As shown in Scheme 8.8, benzylated diacetone-D-glucose **20**, easily obtained from D-glucose, was regioselectively deprotected and the resulting diol reprotected at the primary position. Swern oxidation, followed by addition of nitromethane, afforded two alcohols **21**, which were separated at this stage. The application of identical reaction sequences to each diastereomer led to 5a-carba-D-glucopyranose and 5a-carba-L-idopyranose [28]. According to this sequence, the

CHO
H—OH
H—OH
CH2OH
D-erythrose
MeO2C
MeO2C
HO O O
18
1) PCC
2) Ac2O, Py
MeO2C
MeO2C
AcO O O
19, 71%
OH
HO OH
OH
4a-carba-β-L-arabinofuranose

SCHEME 8.7 Synthesis of a carba-furanose using malonates.

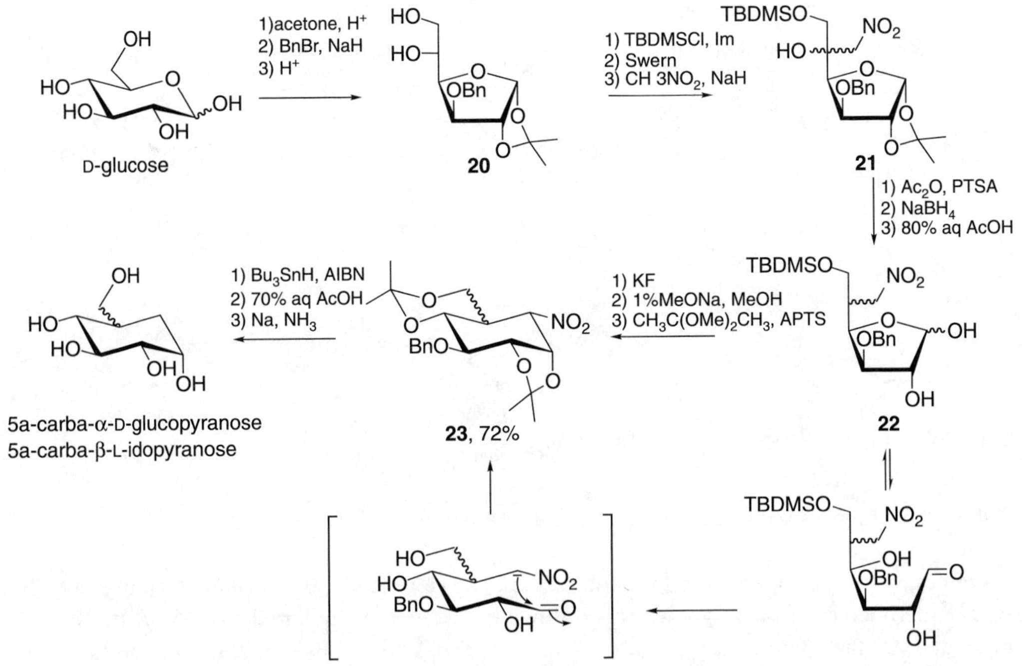

SCHEME 8.8 Synthesis of carba-pyranoses using a nitro group.

hydroxyl group was removed and the acetonide was cleaved. The resulting lactol **22** was then desilylated and cyclized under basic conditions. The isolated cyclohexane derivative was then protected as an acetonide with an overall yield of 72% over three steps. The conclusion of the synthesis consisted of the removal of the nitro group and deprotection of the hydroxyls of **23**.

Complementary to the preparation of carba-pyranosides, the use of nitro groups was also applied to the synthesis of carba-furanosides. As shown in Scheme 8.9, the previously prepared nitro derivative of D-glucose, **22**, was transformed into aldehyde **24** by oxidative cleavage. The cyclization

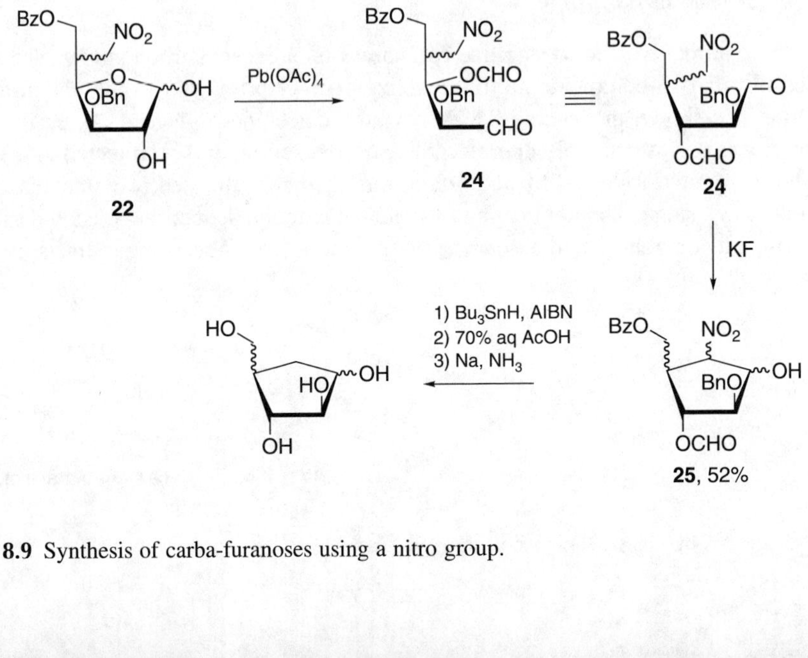

SCHEME 8.9 Synthesis of carba-furanoses using a nitro group.

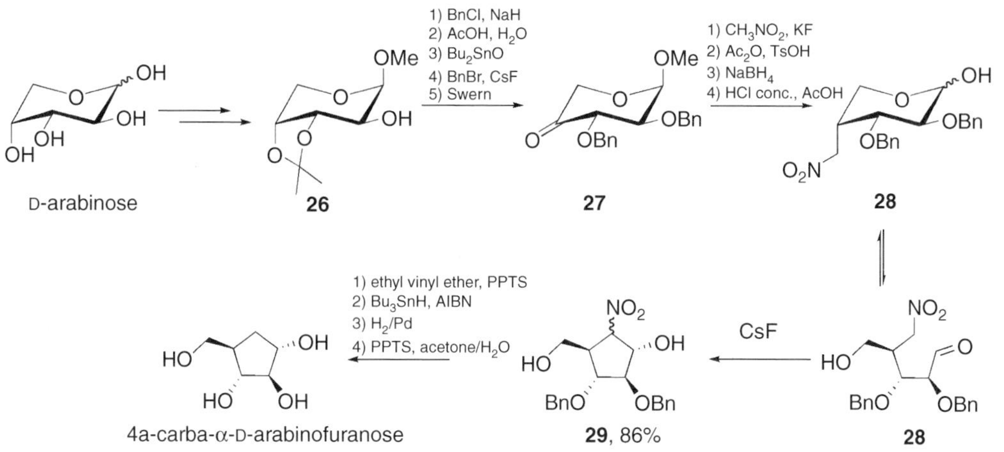

SCHEME 8.10 Synthesis of carba-arabinose using a nitro group.

reaction, performed using potassium fluoride (KF), afforded a mixture of isomeric cyclopentanes **25** in a yield of 52% [29]. The usual deprotection sequence afforded four 4a-carba-pentofuranoses.

A modified strategy allowed the stereoselective synthesis of 4a-carba-α-D-arabinofuranose from D-arabinose, as shown in Scheme 8.10 [30]. The D-arabinose derivative **26** was first benzylated and subsequent removal of the isopropylidene group was followed by regioselective benzylation via a stannylidene derivative. The free 4-hydroxyl group was then oxidized giving **27**. Addition of nitromethane to the carbonyl, followed by subsequent acetylation and stereoselective deacetoxyhydrogenation of the resulting alcohol with $NaBH_4$, gave, after acidic hydrolysis, the lactol **28** ready for carbocyclization. Indeed, in the presence of CsF, **28** cyclized in a yield of 86%. Subsequent deprotection of **29** afforded 4a-carba-α-D-arabinofuranose.

8.3.1.1.3 Phosphorus Ylides

In the context of preparing complex Wittig reagents, it is well established that phosphorus ylides can be alkylated. Applying this chemistry to the preparation of carba-sugars (Scheme 8.11), D-glucose was converted, in four steps, to aldehyde **30**. This aldehyde, on treatment with lithium dimethyl methylphosphonate, yielded **31**, following protecting group manipulations. Subsequent

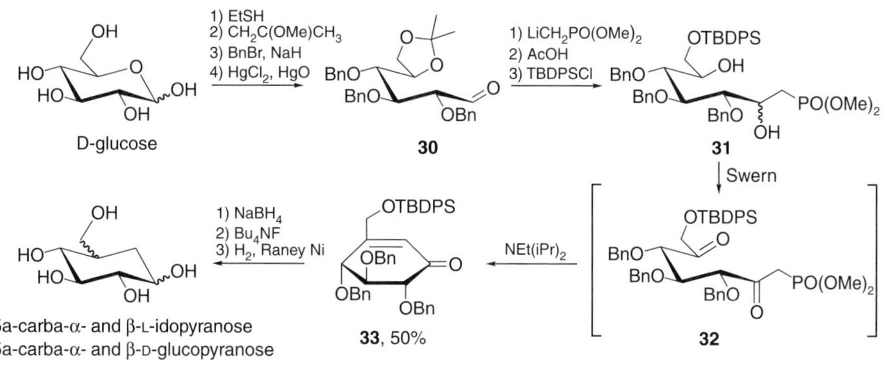

SCHEME 8.11 Synthesis of carba-pyranoses using a phosphorous ylide.

Swern oxidation provided the intermediate dicarbonyl derivative **32**, which spontaneously cyclized into cyclohexene **33** under the action of ethyl diisopropylamine in a 50% overall yield. The resulting unsaturated ketone was reduced, desilylated and hydrogenated to separately afford 5a-carba-α- and β-L-idopyranose, as well as 5a-carba-α- and β-L-glucopyranose [31].

8.3.1.1.4 Incorporation of Aldol Methodology

Thus far, all examples presented incorporate the use of stabilized anions in the critical ring-forming step of carba-sugar syntheses. To expand upon the illustrated examples, mention of the aldol reaction is certainly justified. As shown in Scheme 8.12, an aldol condensation was used in a cyclization that did not involve an opened sugar. Instead, the intact sugar ring was used to induce selectivity. As illustrated, D-glucose was transformed in its diacetonide derivative and the remaining hydroxyl group was oxidized yielding **34**. Olefination of the ketone and hydrogenation and reoxidation of the generated alcohol, afforded, after selective removal of an acetonide, the diol **35**. Subsequent oxidative cleavage with sodium periodate gave the di-carbonyl compound **36**. This molecule then underwent an aldolisation–crotonisation sequence to afford a 44% yield of the bicyclic derivative **37**. Epoxidation on the double bond, reduction of the ketone and opening of the epoxide afforded triol **38**. The final carba-sugar was obtained by benzylation of the triol, removal of the acetonide, reduction of the resulting lactol, oxidative cleavage of the obtained diol, reduction of the aldehyde and final hydrogenolysis. This strategy allowed the synthesis of 5a-carba-α-L-altropyranose [32] and, utilizing similar sequences, 5a-carba-glucopyranoses [33].

D-Glyceraldehyde, the smallest of the sugars, was also used to synthesize carba-sugars. As shown in Scheme 8.13, it was first protected as the acetonide and the carbonyl group in **39** was submitted to a vinylogous cross aldolisation with a furan silyloxy diene to selectively give

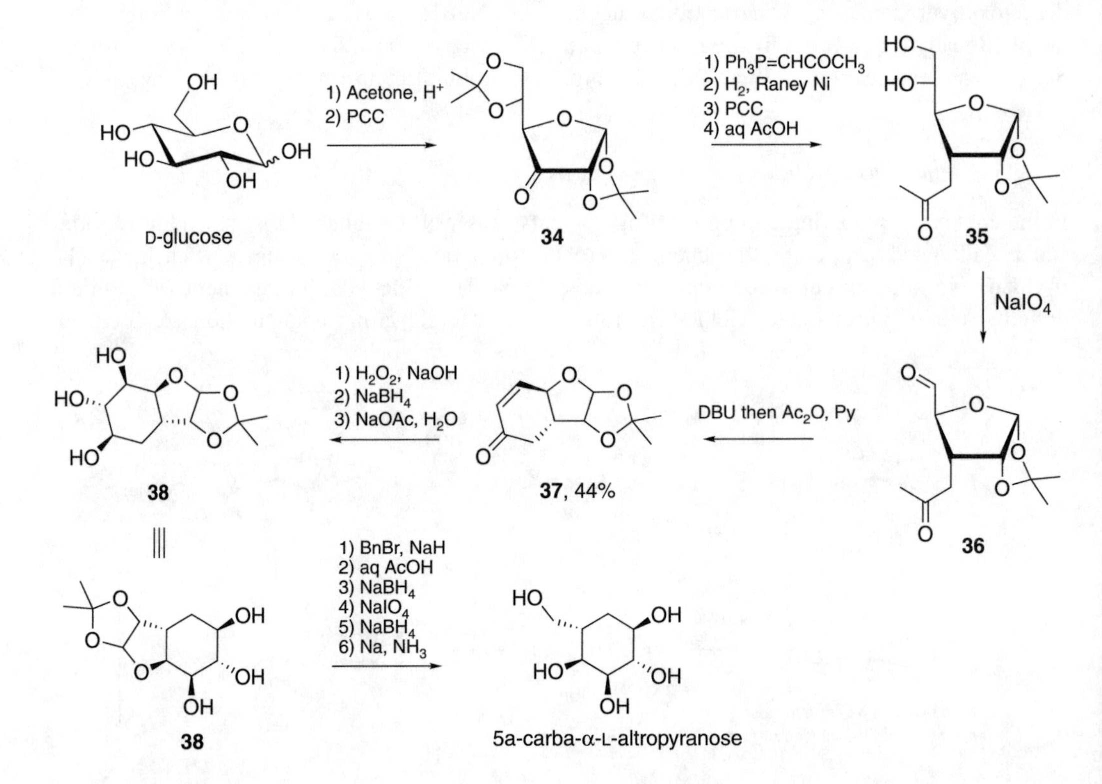

SCHEME 8.12 Synthesis of a carba-pyranose using an aldol reaction.

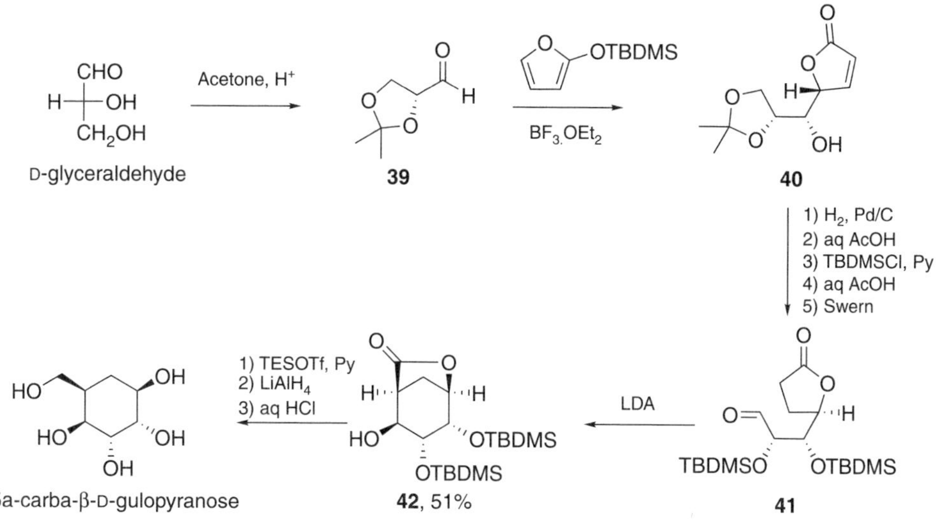

SCHEME 8.13 Synthesis of a carba-pyranose using an aldol reaction on D-glyceraldehyde.

the key alcohol **40**. The double bond in **40** was reduced, the acetonide cleaved and the resulting triol was protected with silyl groups. Mono-desilylation and Swern oxidation afforded the aldehyde **41**. Under basic conditions, **41** underwent a carbocyclization giving the bicyclic product **42** in a yield of 51% (this particular reaction was notably improved through the use of TBDMSOTf and DIPEA). Silylation of the generated alcohol **42**, reduction of the lactone and desilylation furnished 5a-carba-β-D-gulopyranose [34].

As illustrated in Scheme 8.14, it is possible to transform compound **40** into its isomer **43** under basic conditions. Thus, the same sequence described in Scheme 8.13 led to 5a-carba-β-L-mannopyranose [35].

A few 4a-carba-furanoses were synthesized using the same methodology. However, after saturation of the alcohol **40** the syntheses diverged when protection of the hydroxyl, cleavage of the acetonide and oxidative cleavage of the diol afforded the aldehyde **44** (Scheme 8.15). This compound underwent an aldolisation reaction whose stereoselectivity could be controlled by the reaction temperature. Specifically, at −90°C the bicyclic product **46** was obtained in a yield of 71%. Alternately, bicyclic derivative **45** was the principle product obtained when the reaction was run at room temperature (78%). Reduction of the lactone and desilylation afforded 4a-carba-β-D-xylofuranose from isomer **45**, whereas 4a-carba-β-D-ribofuranose was formed from **46**. Isomerisation of compound **40** led to the synthesis of two other isomeric carba-sugars: 4a-carba-β-L-lyxofuranose and 4a-carba-β-L-arabinofuranose [36].

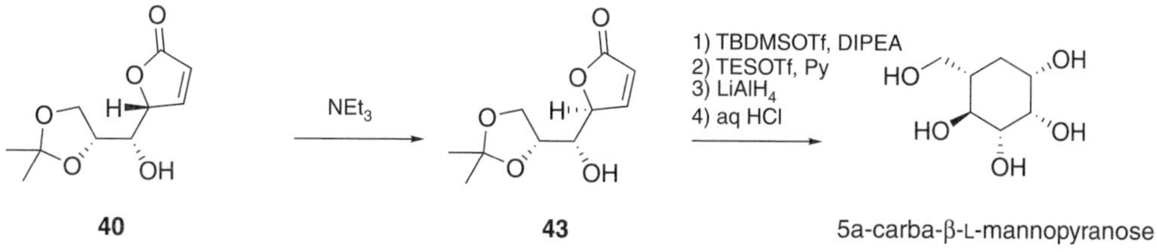

SCHEME 8.14 Isomerization of **40** and synthesis of an isomeric carba-sugar.

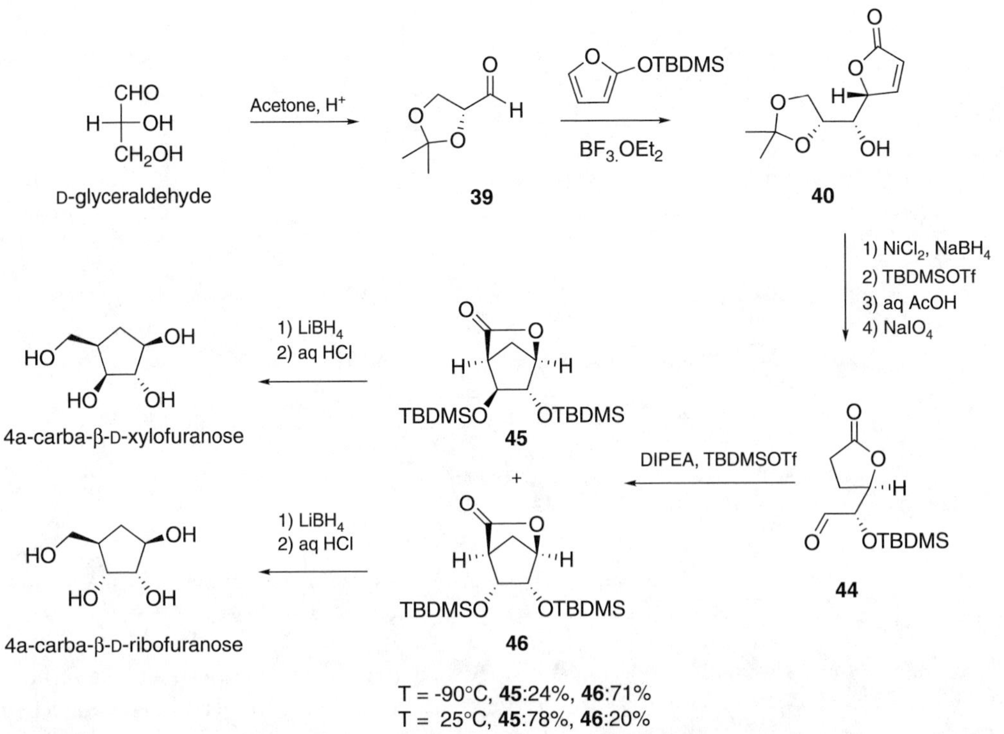

T = -90°C, **45**:24%, **46**:71%
T = 25°C, **45**:78%, **46**:20%

SCHEME 8.15 Synthesis of a carba-furanose using an aldol reaction on D-glyceraldehyde.

8.3.1.2 Carbocationic Cyclizations

An atypical synthesis of a carba-disaccharide using a carbocationic condensation is depicted in Scheme 8.16. It started from D-lyxose, which was glycosylated with benzyl alcohol, protected as the 2,3-acetonide and *C*-allylated into the branched chain sugar **47** via a free-radical C–C bond formation [37]. Sodium–ammonia reduction of the benzyl ether of compound **47** provided a hemiacetal, which was then treated with diiodobenzene diacetate (DIB) to afford **48**, as a distereoisomeric mixture, by fragmentation of anomeric alkoxy radicals. Acetal exchange with thiophenol, followed by basic hydrolysis and benzylation, provided **49**. Ozonolysis and subsequent oxidation of the double bond of **49** led to a mixture of acids, which were esterified by a protected derivative of glucose **50**. Tebbe olefination of esters **51** gave the key enol ethers **52**. Formation of the oxycarbenium ion **53** with methyl triflate induced the carbocyclization and elimination of the resulting carbocation **54**, and led to the carba-disaccharide precursor **55**. Stereoselective hydroboration and final deprotections afforded the 5′a-carba-lactose [38].

8.3.1.3 Radical Cyclizations

Free radical cyclizations have been used extensively to convert carbohydrates into carbocycles [39,40]. The general strategy consists of opening the sugar, transforming the carbonyl into a radical acceptor (alkene or alkyne) and converting one of the hydroxyls into a nucleophilic radical suitable to achieve cyclization.

One such example, depicted in Scheme 8.17, incorporates the classical dithioacetal opening of D-glucose. As illustrated, selective benzoylation of the primary hydroxyl, followed by di-acetalization and debenzoylation, afforded alcohol **56**. The primary alcohol in **56** was displaced

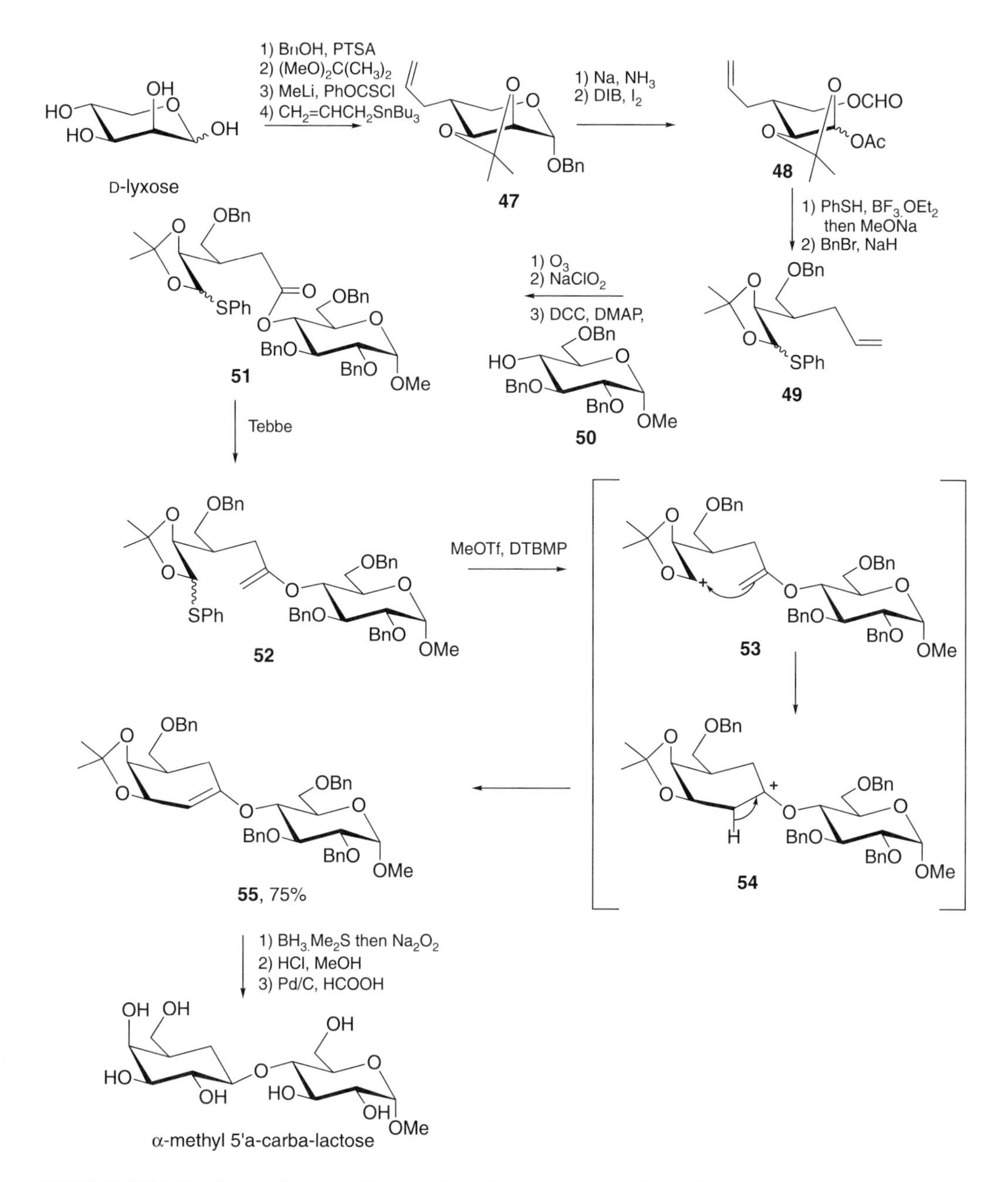

SCHEME 8.16 Synthesis of a carba-disaccharide using a carbocationic cyclization.

by bromine and the thioacetal cleaved. Olefination of the resulting aldehyde **57** gave the radical acceptor enol ether **58**, the nucleophilic radical being created by a tributyltin hydride mediated reaction with the bromide to afford a 60% yield of the major compound **59**. Cleavage of the acetonides yielded 6-*O*-methyl-5a-carba-α-L-glucopyranose [41].

Use of the Wittig reaction on a reducing sugar is a handy method to generate a double bond, which can then be used as a radical acceptor. As shown in Scheme 8.18, D-galactose was classically transformed into a selectively protected hemiacetal **60** by benzyl alcohol glycosylation, protection of the primary hydroxyl, 3,4-acetonide formation and debenzylation. The Wittig reaction followed by desilylation afforded the triol **61**, which was then selectively tosylated on the primary hydroxyl

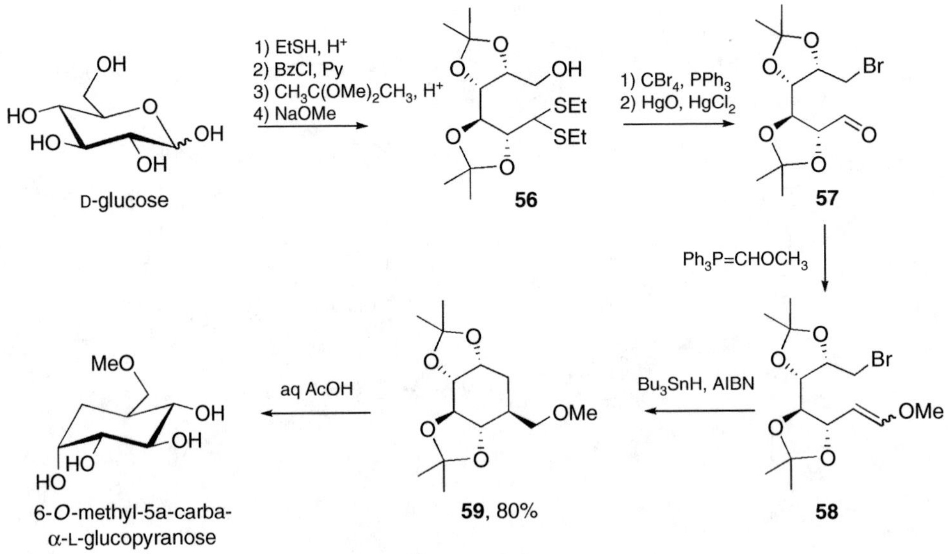

SCHEME 8.17 Synthesis of carba-pyranoses using a 6-*exo*-trig radical cyclization.

group, benzoylated on the other positions and transformed into iodide **62**. Tin hydride exposure then initiated a 6-*exo*-trig radical cyclization yielding, in a 90% yield, a 2:1 mixture of derivatives **63** and **64**. Following deprotection, L-rhamno and D-gulo carba-sugars were obtained [42].

A similar strategy, using an electron poor double bond, was used to synthesize carba-fructose [43]. As shown in Scheme 8.19, methyl glycosylation of D-arabinose, benzylation and hydrolysis of the glycoside afforded the furanosidic hemiacetal **65**. Wittig olefination of **65** followed by a Swern oxidation afforded the olefinic ketone **66** as a 3:2 mixture of *Z*:*E* isomers; the *Z* isomer was isolated by chromatography. To avoid 1,4 addition of $LiCHBr_2$, the ester was converted into the corresponding carboxylate, which then underwent a selective addition of $LiCHBr_2$ to the ketone, giving **67** as a single stereoisomer. Subsequent cyclization of the unsaturated geminal dibromide

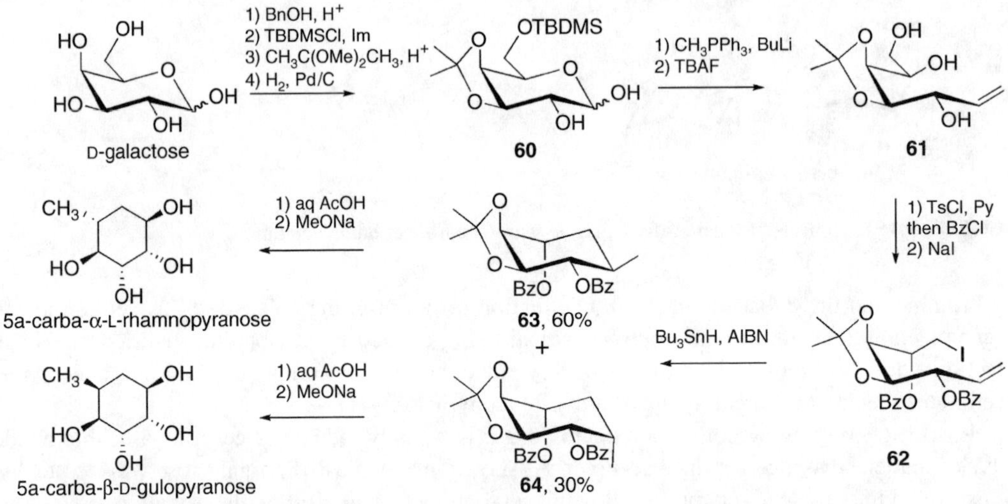

SCHEME 8.18 Synthesis of carba-pyranoses using a 6-*exo*-trig radical cyclization.

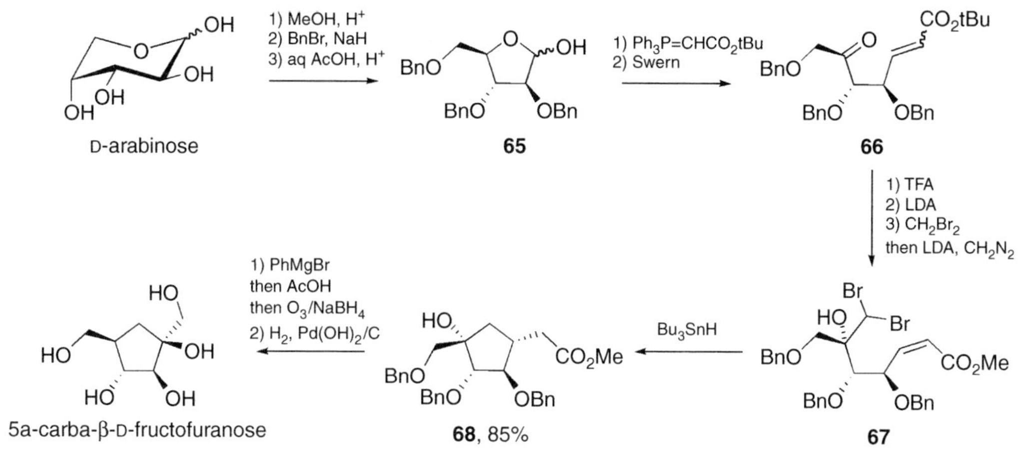

SCHEME 8.19 Synthesis of carba-fructofuranose using a 5-*exo*-trig radical cyclization.

provided the cyclopentane derivative **68** in a yield of 85%. Barbier–Wieland degradation of the ester and hydrogenolysis gave 5a-carba-β-D-fructofuranose.

Carba-sugars have also been synthesized through 6-*exo*-dig radical cyclizations. One example is shown in Scheme 8.20. As illustrated, the protected hemiacetal **69**, derived from D-ribose, was treated with trimethylsilylacetylide and the resulting diol was protected with MOM ethers to afford a 7:2 separable mixture of isomers. The silyl groups of the major isomer were removed and the resulting alcohol was converted into iodinated compound **70**. Radical cyclization of **70**, with the

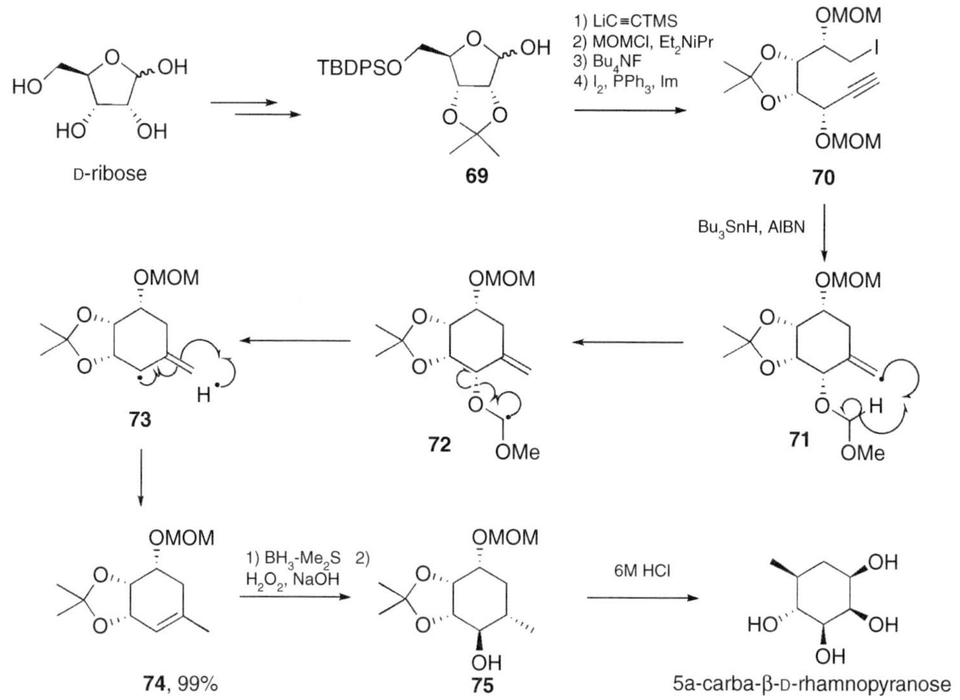

SCHEME 8.20 Synthesis of a carba-pyranose using a 6-*exo*-dig radical cyclization.

loss of an OMOM group, afforded the cyclohexene **74** in a yield of 99%. The formation of **74** was rationalized by cyclization of a primary radical in a 6-*exo* mode onto the alkyne, resulting in the formation of a vinyl radical **71**. Subsequent abstraction of hydrogen from the methylene carbon of the MOM group to give **72** was followed by β-scission, resulting in allylic radical **73**. Final migration of the double bond afforded the observed product **74**. Concluding the synthesis, hydroboration–oxidation stereoselectively provided **75** and deprotection yielded 5a-carba-β-D-rhamnopyranose [44].

A carba-mannose was synthesized using a variation on the theme illustrated in Scheme 8.20. As shown in Scheme 8.21, D-mannose was the starting material, a thiocarbonate was the radical precursor and a derivatized triple bond was the radical acceptor [45]. In the actual synthesis, D-mannose diacetonide **76**, prepared from mannose, was treated with lithium phenylacetylide to stereoselectively yield a diol, which was regioselectively silylated on the prop-2-ynilic hydroxyl group. The remaining alcohol was transformed into a phenyl thiocarbonate, yielding **77**. Application of a 6-*exo*-dig radical cyclization afforded a 2:3 mixture of cyclohexane derivatives **78** and **79** in a yield of 95%. Derivative **78** was separated from its isomer and the silyl group was exchanged with a benzyl group. Transformation of the double bond into a ketone was achieved by ozonolysis. Reduction of the ketone and deoxygenation of the resulting alcohol provided **80**. Final removal of the protecting groups gave 5a-carba-β-D-mannopyrannose. Similar strategies, alluded to in Scheme 8.22, allowed the synthesis of 5a-carba-α-D-allopyranose, -β-L-talopyranose and -α-L-gulopyranose [46].

Slight changes in protecting groups of the cyclization precursor (Scheme 8.22) allowed the syntheses of 5a-carba-α-D-glucopyranose and 5a-carba-α-D-galactopyranose. Thus, as shown in Scheme 8.23, compound **81** was obtained from D-mannose in a similar fashion as for derivative **77** and gave, under radical cyclization conditions, carbocyclic compounds **82** and **83** in 74% yield as a

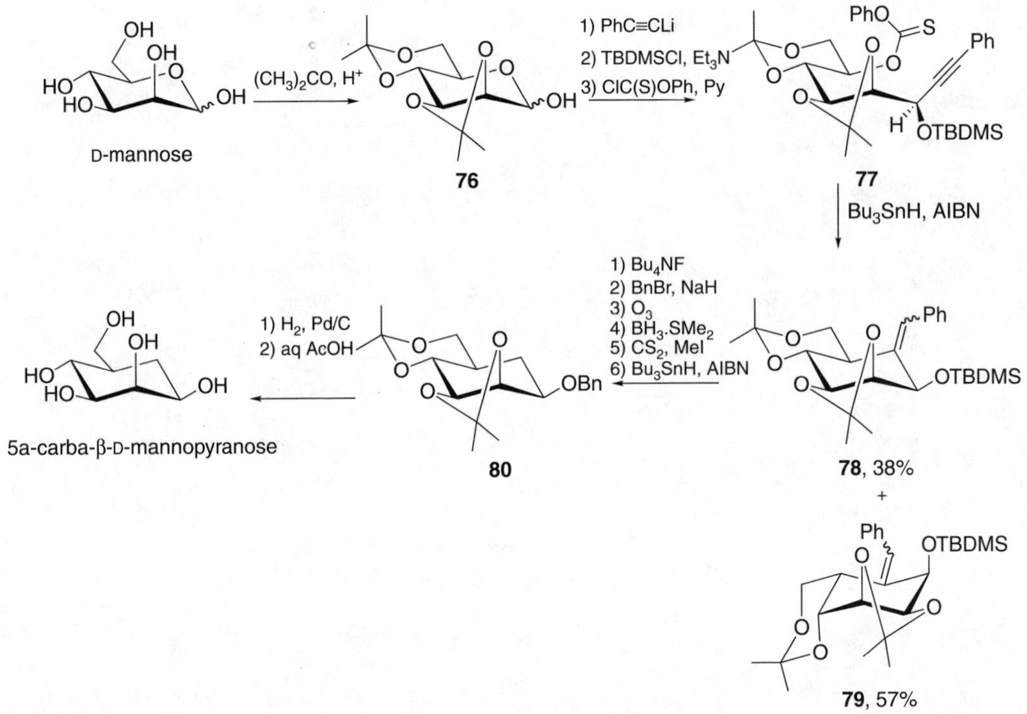

SCHEME 8.21 Synthesis of a carba-pyranose using a 6-*exo*-dig radical cyclization.

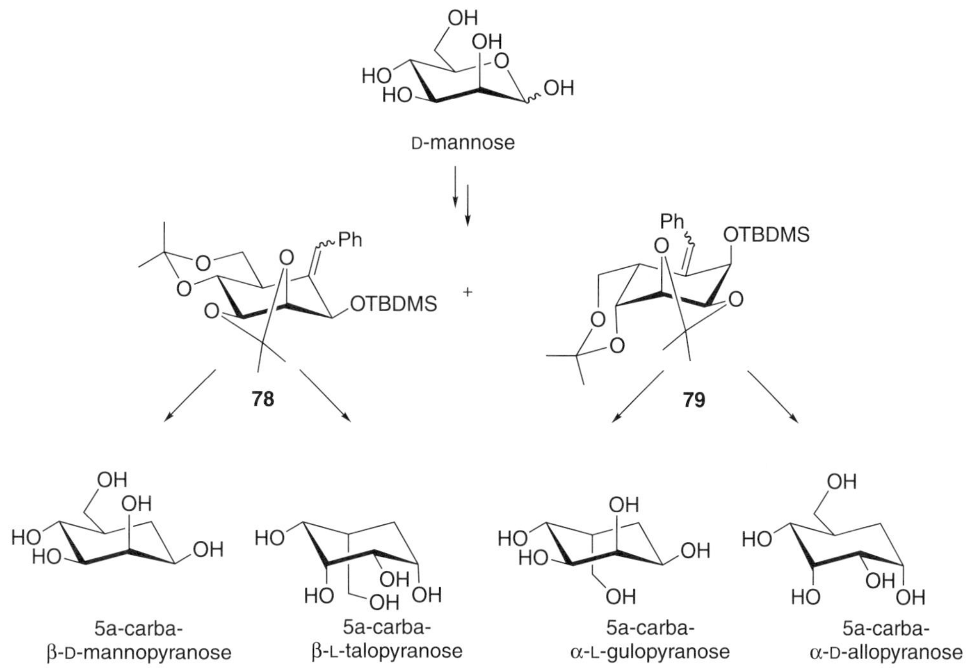

D-mannose

78 + 79

5a-carba-β-D-mannopyranose

5a-carba-β-L-talopyranose

5a-carba-α-L-gulopyranose

5a-carba-α-D-allopyranose

SCHEME 8.22 Synthesis of carba-pyranoses using a 6-*exo*-dig radical cyclization.

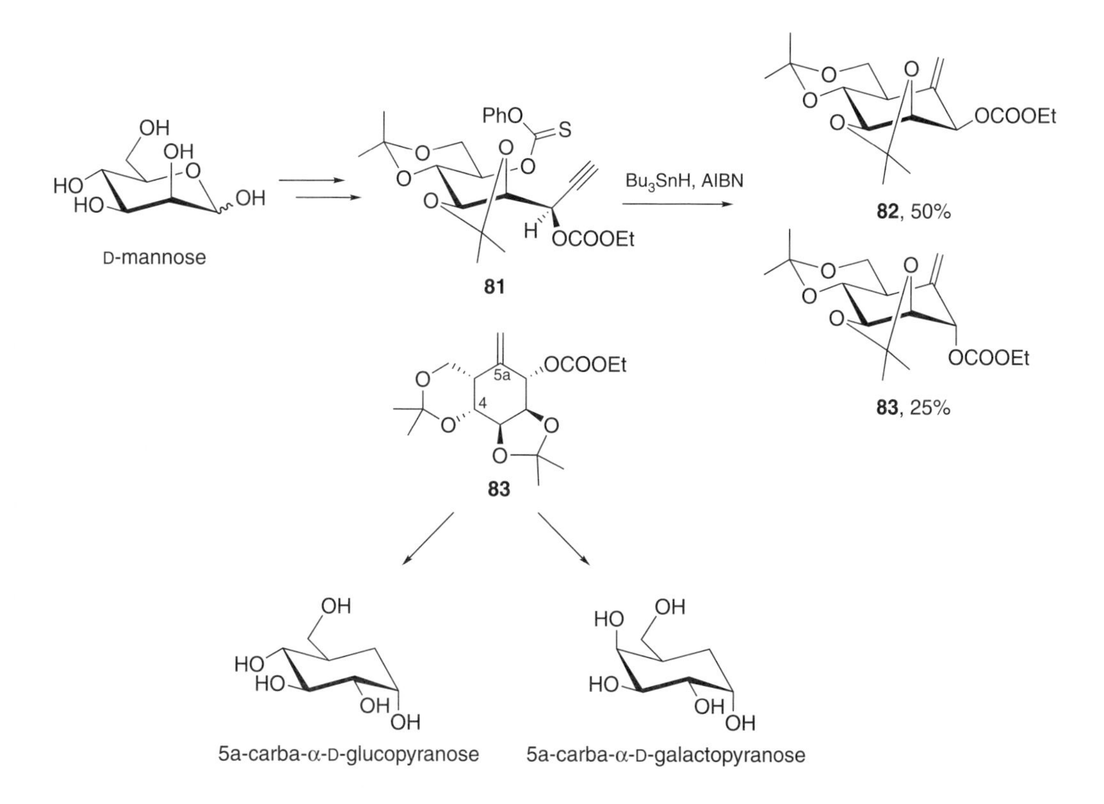

SCHEME 8.23 Synthesis of carba-pyranoses using a 6-*exo*-dig radical cyclization.

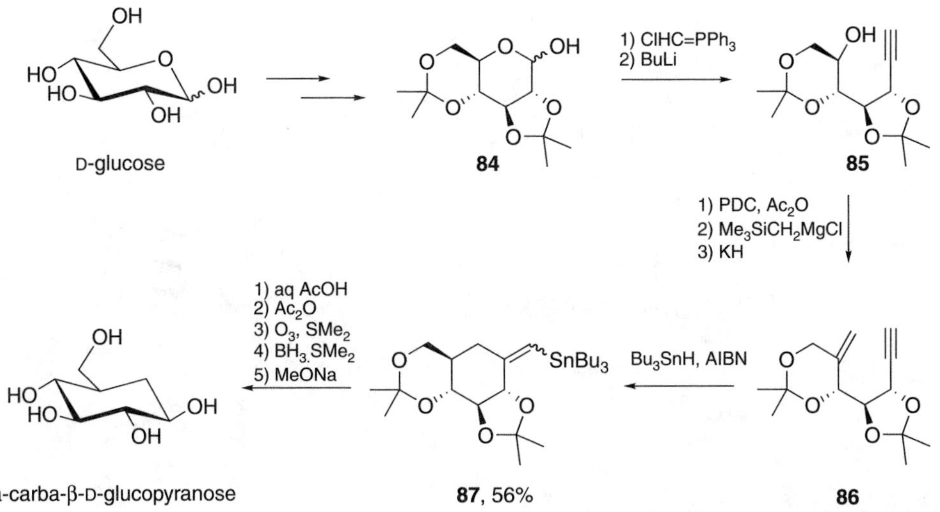

SCHEME 8.24 Synthesis of a carba-pyranose using a 6-*endo*-trig radical cyclization.

2:1 mixture. Dehydroxylation at position 4 or 5a of a compound obtained from **83** by ozonolysis and reduction, led to isolation of 5a-carba-α-D-glucopyranose, 5a-carba-β-L-gulopyranose and 5a-carba-α-D-galactopyranose [47].

6-*Endo*-trig radical cyclization of vinyl radicals was also investigated for the synthesis of carba-sugars [48]. As shown in Scheme 8.24, hemiacetal **84**, derived from D-glucose, was the starting point for the synthesis. It was first transformed into the alkyne **85** via a Wittig olefination with chloromethylene triphenylphosphorane followed by dehydrohalogenation [49]. The hydroxyl group in **85** was oxidized, and the resulting ketone was converted into the enyne **86**, utilizing the Peterson olefination. It then underwent radical cyclization to give alkenyl stannane **87** as the major compound in a yield of 56%. Acid hydrolysis and acetylation, followed by ozonolysis and reduction, afforded 5a-carba-β-D-glucopyranose. Similarly, when D-galactose was subjected to the same sequence of reactions, 5a-carba-β-D-galactopyranose was isolated in good yield.

Scheme 8.25 illustrates the elegant use of a cobalt-catalyzed radical cyclization to provide a 4a-carba-pentofuranose [50,51]. As illustrated, methyl α-D-glucopyranoside was classically 4,6-benzylidene-protected and benzylated on the remaining hydroxyls. The 6-OH was then deprotected by regioselective reductive cleavage of the acetal and converted into iodo compound **88**. Ring opening with zinc led to the unsaturated aldehyde **89**. Subsequent reduction of the aldehyde and conversion of the resulting alcohol to an iodide afforded **90**. The key reaction of this synthesis was the oxygenative 5-*exo*-trig cyclization, directly yielding the protected carba-sugar **91** (69%). Final hydrogenolysis gave the 4a-carba-α-D-arabinofuranose. The same sequence was also applied to D-galacto-, D-manno- and D-allo-hexopyranosides as starting materials. However, in light of the apparent generality of this process, the presence of an amine was found incompatible with this radical reaction.

8.3.1.4 Ring-Closing Metathesis

Ring-closing metathesis (RCM), a powerful method used to synthesize carbocycles, has been used for preparing 4a-carba-furanosides. An example is shown in Scheme 8.26, starting from D-mannose, which was transformed into the alkene **93** via the protected hemiacetal **92**. The hydroxyl of **93** was then oxidized, and the ketone was transformed into diene **94** through a second

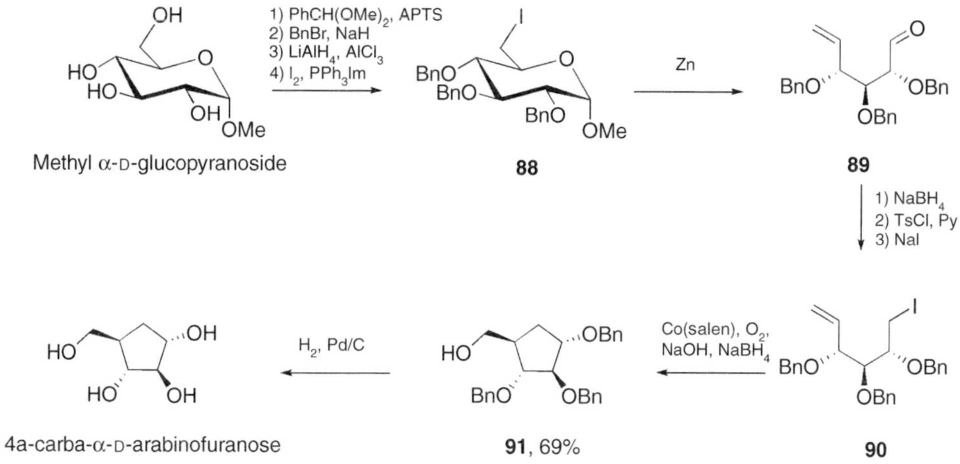

SCHEME 8.25 Synthesis of a carba-furanose using a 5-*exo*-trig radical cyclization.

Wittig olefination. An RCM reaction was then performed using Schrock's catalyst (as first generation Grubb's catalyst proved inefficient in this case) to afford the cyclized product **95** in a yield of 74%. Stereoselective hydrogenation, removal of the MOM group and methylation of the resulting hydroxyl provided methyl 4a-carba-β-D-arabinofuranoside after hydrogenolysis (Scheme 8.26) [52,53].

Starting now from D-arabinose, the usual sequence led to the hemiacetal **65**, which was oxidized into the corresponding lactone (Scheme 8.27). Subsequently, the ring was opened by the attack of two equivalents of methyl Grignard. The resulting diol was selectively acetylated on the secondary position and the tertiary alcohol was transformed into a chloride, which under basic conditions underwent elimination and simultaneous deacetylation; the resulting alcohol was oxidized to give **96**. Addition of vinyl magnesium bromide afforded a single tertiary alcohol, which was subsequently protected as benzyl ether. Oxidation of the diene in the presence of SeO_2 and

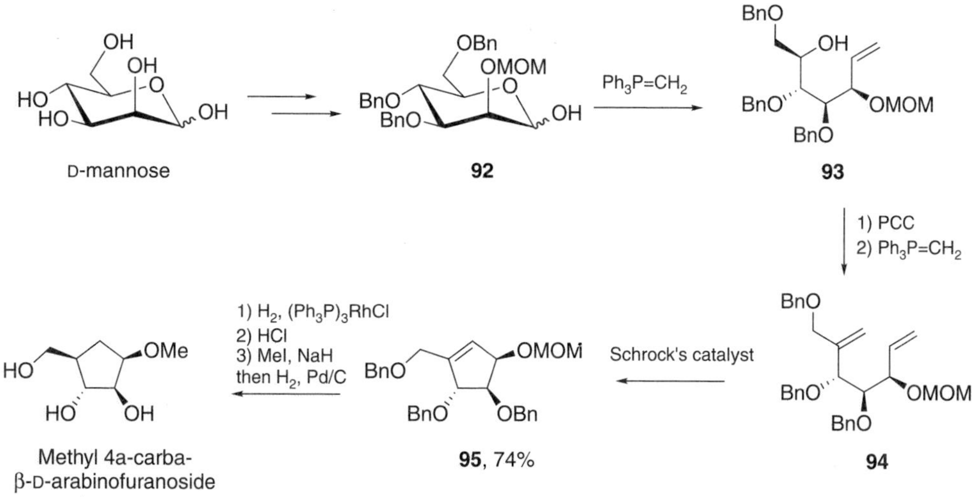

SCHEME 8.26 Synthesis of a carba-furanose using RCM.

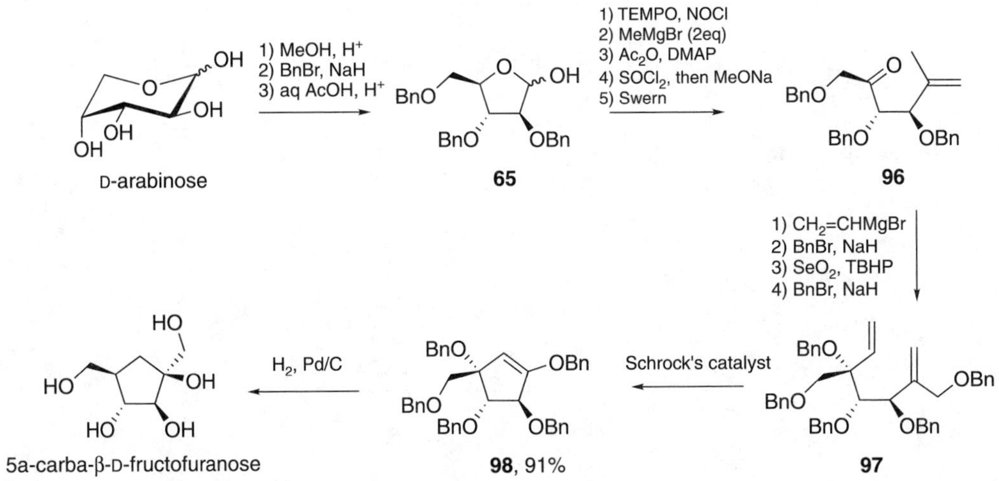

SCHEME 8.27 Synthesis of carba-fructofuranose using RCM.

TBHP yielded an allylic alcohol, which was benzylated to provide **97**. This highly functionalized diene underwent RCM with Schrock's catalyst affording cyclopentene **98** in a yield of 91%. Under the action of hydrogen and Pd/C, final reduction and debenzylation were performed simultaneously to give 5a-carba-β-D-fructofuranose [54].

The main drawback of the previous synthesis was the lengthy preparation of the di-alkene **96**. A nice way to circumvent this problem consisted of the use of a hydroformylation reaction. Indeed, only four steps (instead of nine) were necessary to obtain the key diene **99** from the same lactol **65** (Scheme 8.28). The classical sequence included subjecting **65** to the following reactions in sequence: Wittig reaction, oxidation of the generated alcohol, Grignard reaction on the ketone and benzyl protection of the tertiary alcohol. Schrock's catalyst was also used on isolated **99** to give the cyclopentene **100** in a yield of 98%. Wilkinson's catalyst and an atmosphere of carbon monoxide were used to perform the hydroformylation that yielded a separable mixture of the four possible products **101**, **102**, **103** and **104**. Hydride reduction of **101**, and subsequent hydrogenolysis, afforded the 5a-carba-β-D-fructofuranose. Even though the hydroformylation gave a mixture, the overall yield of the reaction was 32% from the commercial lactol **65** [55].

8.3.1.5 SmI_2-Mediated Pinacol Coupling

Another classical way to open a sugar is to reduce its carbonyl function. As shown in Scheme 8.29, D-mannitol, corresponding to reduced D-mannose, was easily transformed into **105** by tri-acetonation, partial hydrolysis, selective tritylation of the primary hydroxyl groups, benzylation of the remaining hydroxyls and detritylation. Swern oxidation and samarium diiodide-mediated pinacol coupling applied to **105** afforded **106** and **107** in 82% yield, as a 9:1 mixture. Diol **106** was transformed into cyclic sulfate **108**, the conformation of which was determined by the acetonide. This property allowed the regioselective formation of **109** by the action of a base on the proton that is *trans*-diaxially disposed to the sulfate leaving group. The transient vinyl sulfate was then hydrolyzed under acidic conditions. The resulting ketone, **109**, underwent Wittig olefination to afford the key alkene **110**. Removal of the acetonide and hydroboration-oxidation gave 5a-carba-α-L-galactopyranose. In demonstrating the diversity available from this sequence, 5a-carba-α-L-fucopyranose was isolated using the same sequence but with a silyl protection, followed by reduction of the double bond. It is worth noting that if the acetonide was not removed, the

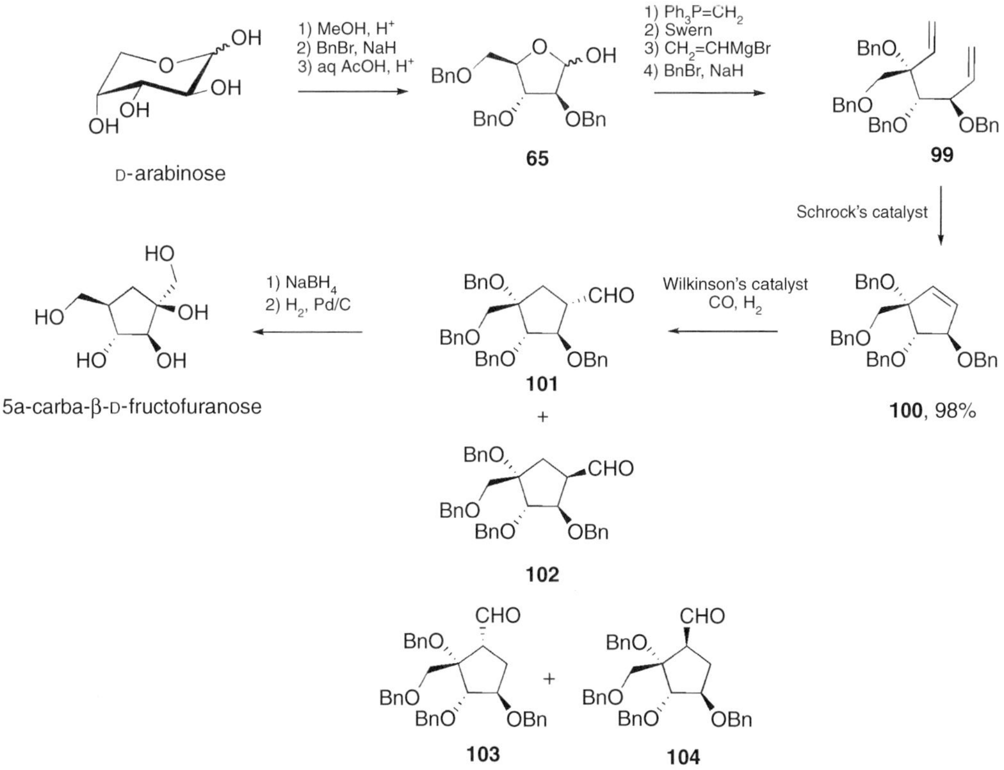

SCHEME 8.28 Synthesis of a carba-furanose using RCM and hydroformylation.

stereoselectivity of both hydroboration and hydrogenation were inverted to give 5a-carba-β-D-altropyranose and 6-deoxy-5a-carba-β-D-altropyranose, respectively, [56].

8.3.2 Rearrangements of Cyclic Sugars

A straightforward approach to carba-sugars is using direct rearrangements of cyclic sugars to carbocycles. These reactions generally rely on the fact that a cyclic sugar has a hidden electrophilic center — the anomeric carbon. The generation of a similarly masked nucleophilic function, when simultaneously liberated with the electrophilic center, allows execution of a direct carbocyclization [57].

8.3.2.1 The Ferrier-II Rearrangement

A well-established carbocyclization reaction of sugars to carba-sugars is the Ferrier-II reaction [58]. As shown in Scheme 8.30, Ferrier started with D-glucose, preparing the brominated analog **111** in two steps. This compound then underwent elimination to afford the key intermediate **112**, possessing a double bond between C-5 and C-6. **112** is an acetal enol ether. As such, it contains both a masked nucleophile and a masked electrophile. Both of these features were simultaneously liberated on treatment with mercury(II) acetate and water. Mechanistically, the transformation proceeded via hydromercuration of the double bond, resulting in the formation of an unstable hemiacetal **113**, followed by ring opening to dicarbonyl **114**. Cyclization then forms the hydroxy-ketone **115** in a yield of 83%. The basic feature of this reaction is the loss of the aglycon as a consequence of the mechanism that opens the sugar ring (Scheme 8.30) [59].

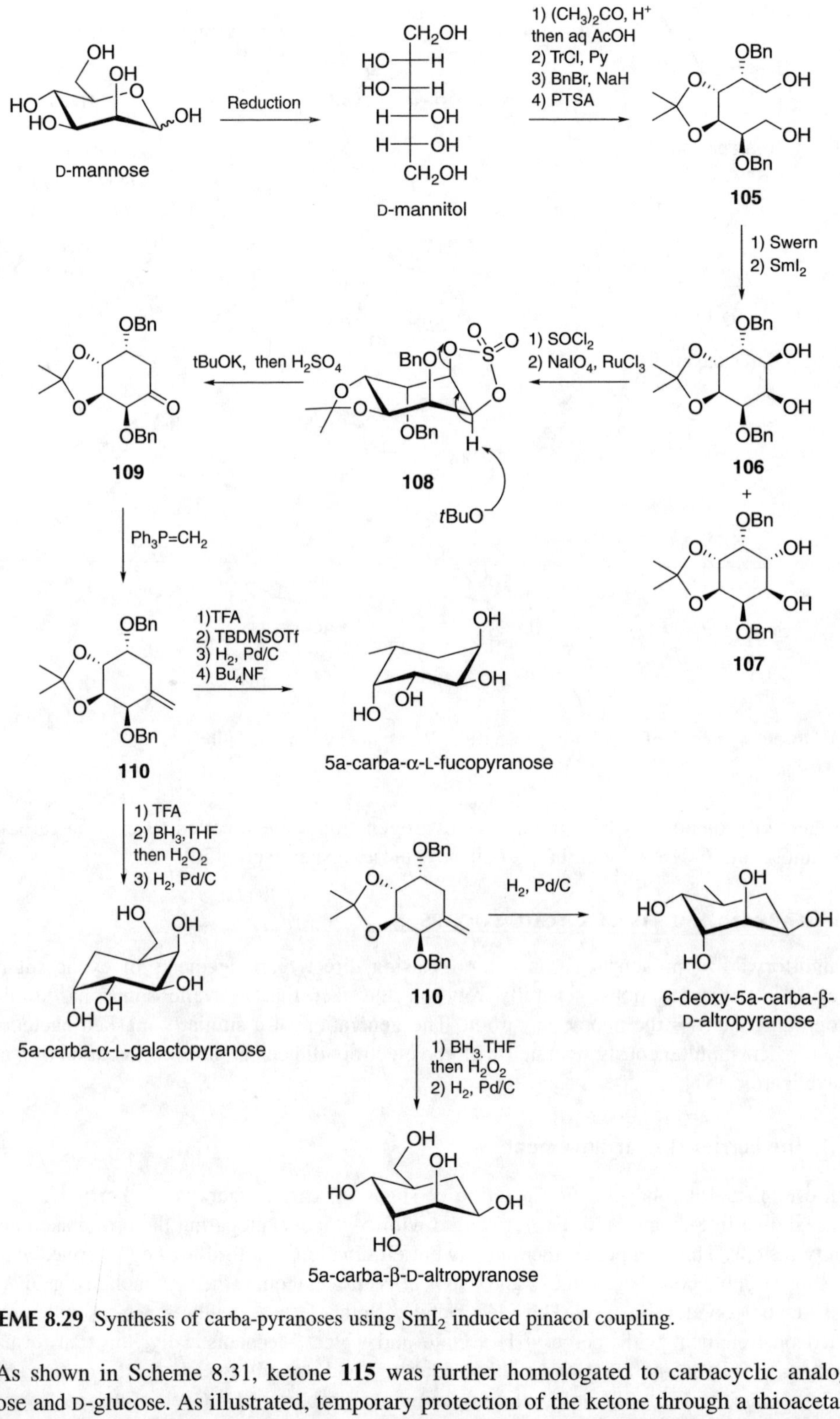

SCHEME 8.29 Synthesis of carba-pyranoses using SmI_2 induced pinacol coupling.

As shown in Scheme 8.31, ketone **115** was further homologated to carbacyclic analogs of L-idose and D-glucose. As illustrated, temporary protection of the ketone through a thioacetal was followed by the removal of the benzoyl groups. Protection with acetonides, and cleavage of the thioacetal, afforded the ketone **116**. This intermediate was transformed into carba-idopyranose by

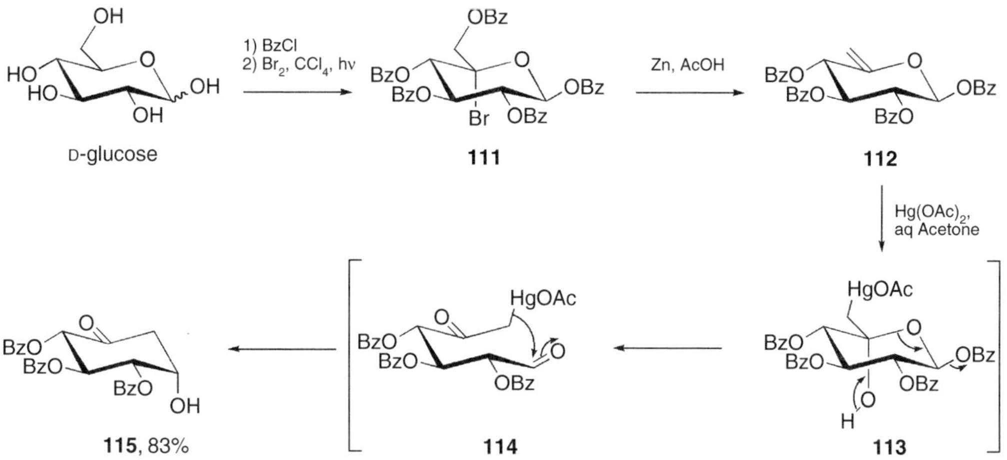

SCHEME 8.30 Ferrier-II rearrangement.

olefination, hydroboration and deprotection. The route to obtain the carba-glucopyranose is slightly longer, owing to the stereoselectivity of the hydroboration giving only the axial hydroxy methyl **117**. The hydroxyl was therefore oxidized to the aldehyde and epimerized under basic conditions. Final deprotection afforded the desired carba-glucopyranose [59].

The previous synthesis was slightly improved by the use of benzyl protecting groups and a Wittig olefination. As illustrated in Scheme 8.32, the 4,6-benzylidene analog of methyl α-D-glucopyranoside was benzylated and regioselectively opened to afford the alcohol **118**. Transformation to an iodide, followed by elimination, gave the key olefinic compound **119**. This unsaturated sugar was rearranged into cyclohexanones **120** and **121** as a 3:1 mixture in 80% yield. Wittig olefination of ketone **120** afforded the alkene **122**. Final hydrogenation provided 5a-carba-α-D-glucopyranose and -β-L-idopyranose as a 2:1 mixture [60].

In one additional example, illustrating the applicability of the Ferrier cyclization to amine-based substrates, the same strategy allowed the synthesis of carba-glucosamine (Scheme 8.33). Starting from Cbz protected methyl α-D-glucosaminide **123**, 4,6-benzylidene formation, di-benzylation, cleavage of the acetal, regioselective iodination and simultaneous elimination

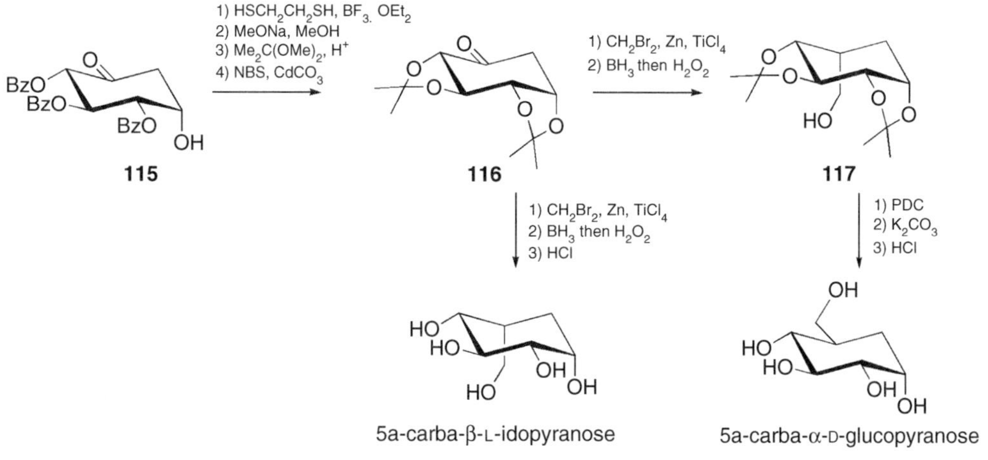

SCHEME 8.31 Synthesis of carba-pyranoses from the product of the Ferrier-II rearrangement.

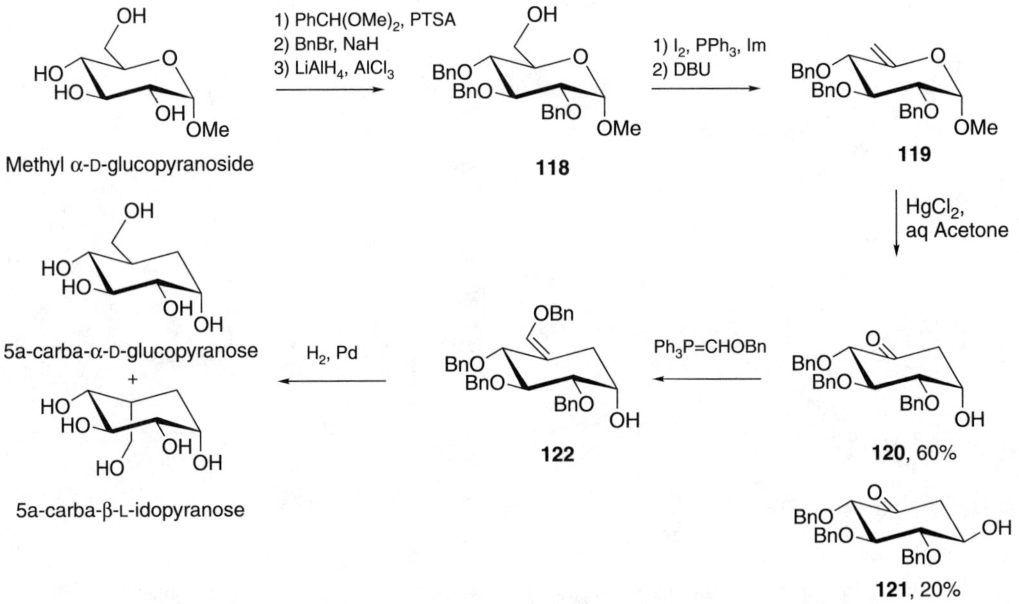

Methyl α-D-glucopyranoside

1) $PhCH(OMe)_2$, PTSA
2) BnBr, NaH
3) $LiAlH_4$, $AlCl_3$

118

1) I_2, PPh_3, Im
2) DBU

119

$HgCl_2$, aq Acetone

120, 60%

121, 20%

$Ph_3P{=}CHOBn$

122

H_2, Pd

5a-carba-α-D-glucopyranose
+
5a-carba-β-L-idopyranose

SCHEME 8.32 Synthesis of carba-pyranoses using the Ferrier-II rearrangement.

D-glucosamine

123

1) PhCHO, $ZnCl_2$
2) BnBr, NaH
3) aq AcOH
4) I_2, PPh_3, Im
5) BnBr, NaH

124

$HgCl_2$, aq Acetone

125, 64%

126, 11%

1) $Ph_3P{=}CHOMe$
2) $Hg(OAc)_2$ then $NaBH_4$

127

1) H_2, Pd
2) Ac_2O, Pyr

5a-carba-α-D-glucopyranosamine per-acetate

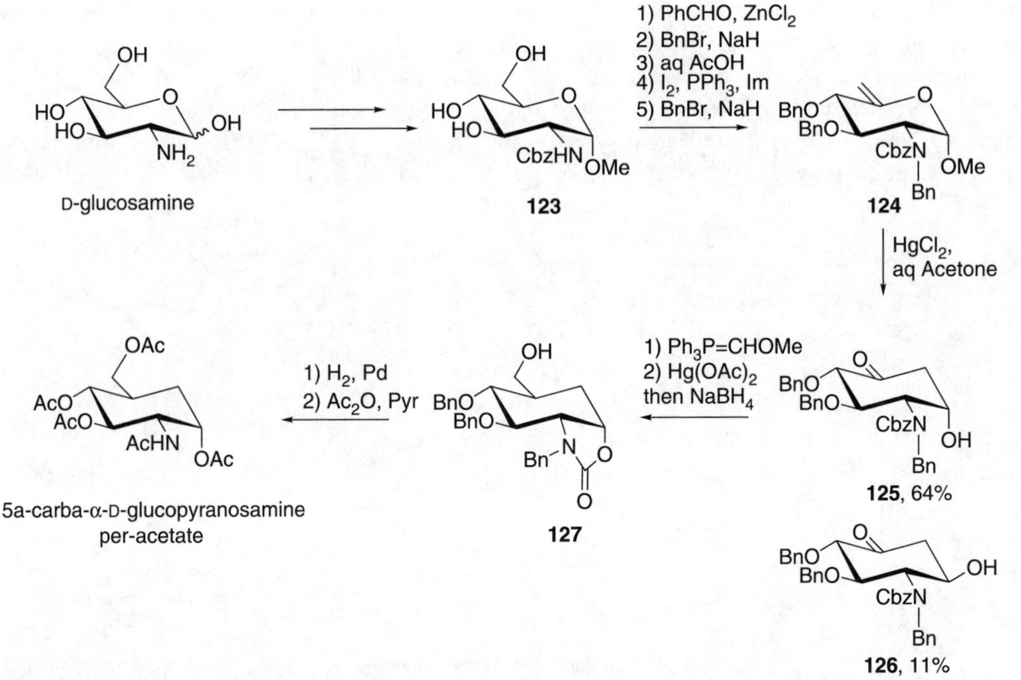

SCHEME 8.33 Synthesis of carba-glucasamine using the Ferrier-II rearrangement.

and benzylation gave the desired unsaturated compound **124**. Ferrier-II reaction on **124** yielded the cyclohexanones **125** and **126** in yields of 64% and 11%, respectively. The major compound **125** was olefinated with methoxymethylene-triphenylphosphorane. During this reaction, the basic conditions induced the unexpected formation of an oxazolidine ring. The vinyl ether was next treated with mercury acetate and the transient aldehyde was reduced with sodium borohydride. The major product of this reaction, **127**, was deprotected and acetylated to give the peracetylated carba-D-glucosamine [61].

8.3.2.2 Endocylic Cleavage Induced Rearrangements

Returning to substrate **119** (Scheme 8.32), it is possible to simultaneously unmask the nucleophilic and electrophilic centers in a different way. In fact, the action of a Lewis acid, such as triisobutylaluminum (TIBAL) [62] or $Cl_3TiO\textit{i}Pr$ [63], induces the endocyclic cleavage of the glycosidic acetal of **119**. Formation of complex **128** results in liberating an enolate and methoxycarbenium **129** and the attack of one onto the other directly affords the cyclohexane derivative **130** as a single isomer in 95% yield using $Cl_3TiO\textit{i}Pr$ (Scheme 8.34). An important feature of this process is the preservation of the aglycon. As in the previous syntheses, the carba-sugar may be obtained by homologation of the ketone **130**. In this case, Tebbe olefination, hydroboration and hydrogenolysis afforded methyl 5a-carba-β-L-idopyranoside [64].

As previously stated, a distinct feature of this rearrangement is the retention of the anomeric information. This advantage is nicely illustrated by the direct synthesis of a carba-disaccharide from a disaccharide [65]. As shown in Scheme 8.35, maltose was first transformed into the thioglycoside **131**. Manipulation to the free primary alcohol **132** was achieved using a classical sequence of reactions. At this point, glycosylation of methanol was realized in the presence of NCS to afford **133**, and the unsaturated derivative **134** was obtained after iodination and elimination. Triisobutylaluminum induced rearrangement nicely afforded the carba-disaccharide precursor **135** in a yield of 89% as a mixture of two isomers.

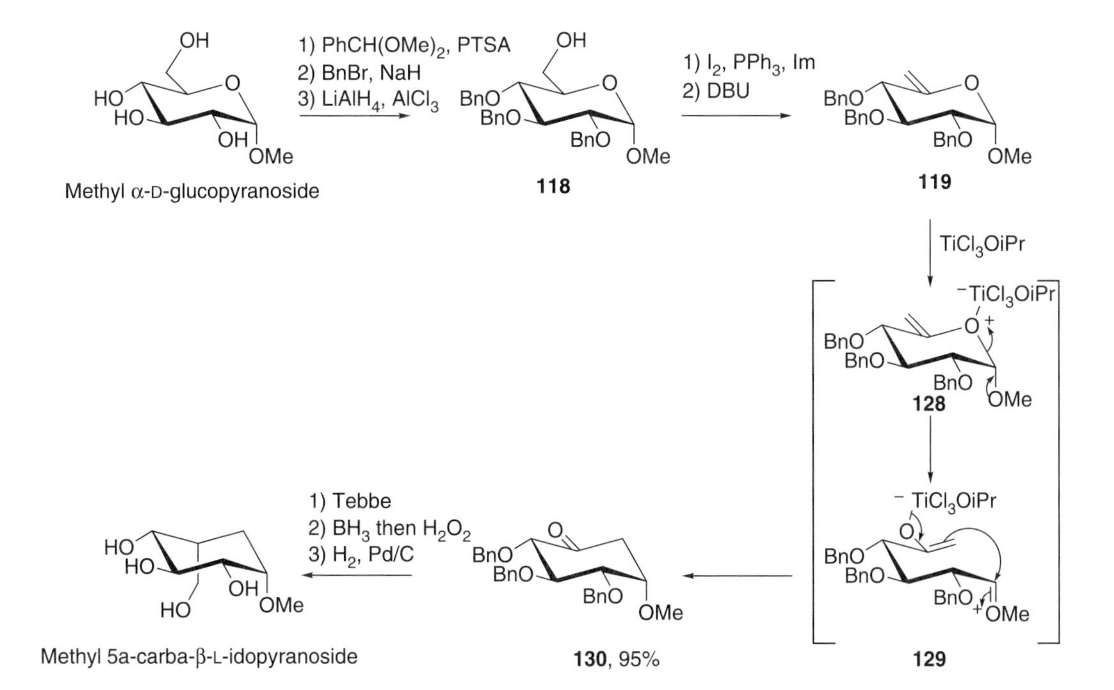

SCHEME 8.34 Synthesis of a carba-pyranoside using a Ti(IV) induced rearrangement.

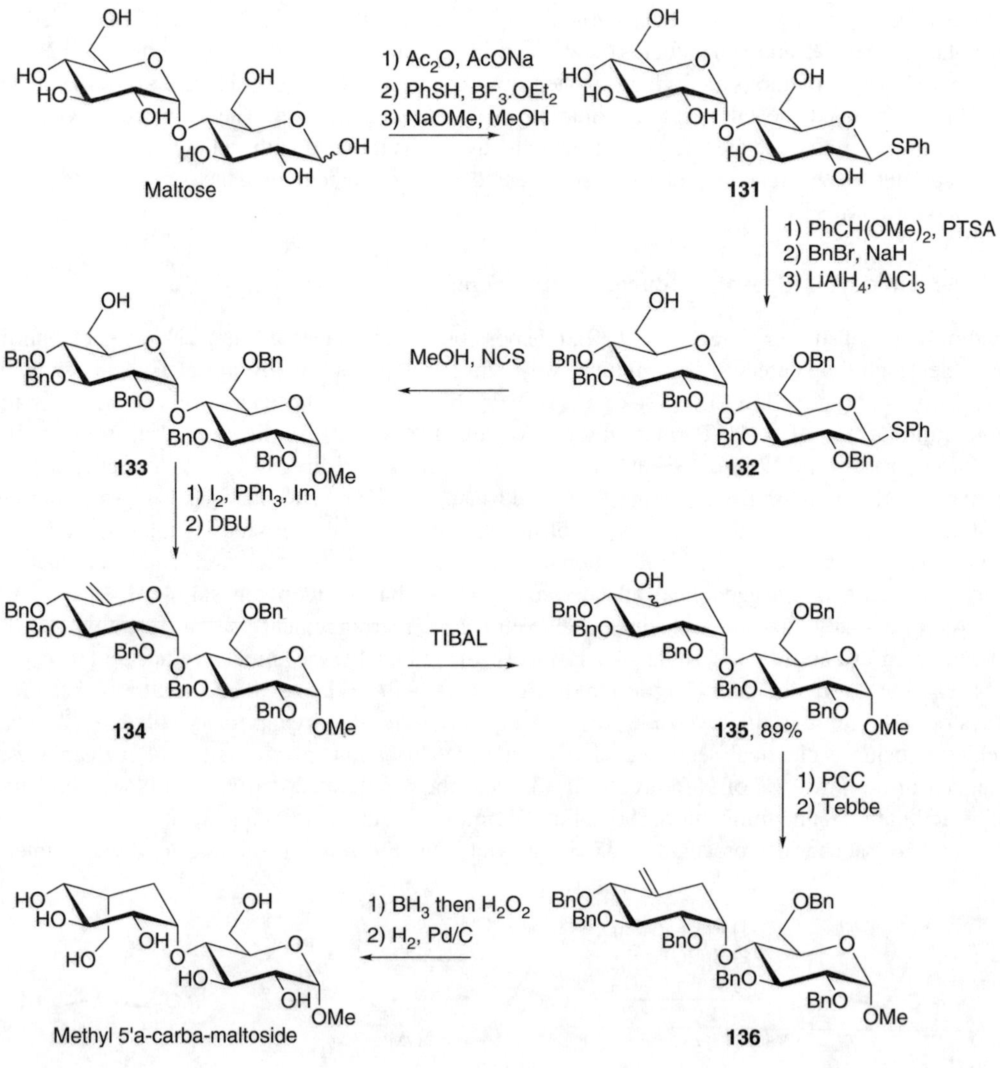

SCHEME 8.35 Synthesis of a carba-disaccharide using the TIBAL induced rearrangement.

This mixture was then oxidized and homologation of the resulting ketone provided **136**. After hydrogenolysis, methyl 5′α-carba-maltoside was isolated [64]. A characteristic of this method is that the rearrangement is done on the disaccharide itself. This method can thus also lead to carba-oligosaccharides through cascade rearrangements [66].

Another consequence of the endocyclic cleavage mechanism is the possibility of replacing the aglycon with any electron-donating group [67]. This idea has been applied to the synthesis of a carba-sugar. As shown in Scheme 8.36, methyl α-D-glucopyranoside was transformed into the protected *C*-glucoside **137** by benzylation, 1,6-acetolysis and glycosylation of 2-TMS-furane. Deprotection of the primary hydroxyl group, iodination and elimination afforded the unsaturated compound **138**. This *C*-glycoside nicely underwent triisobutylaluminum-promoted rearrangement to yield the carbocycle **139** (83%). This cyclohexane already presents the pattern of the carba-sugar with no need of homologation. Indeed, methylation of the hydroxyl, ozonolysis and esterification gave **140**, a direct precursor of carba-iduronic acid. Final reduction and hydrogenolysis furnished methyl 5a-carba-β-D-idopyranoside [64].

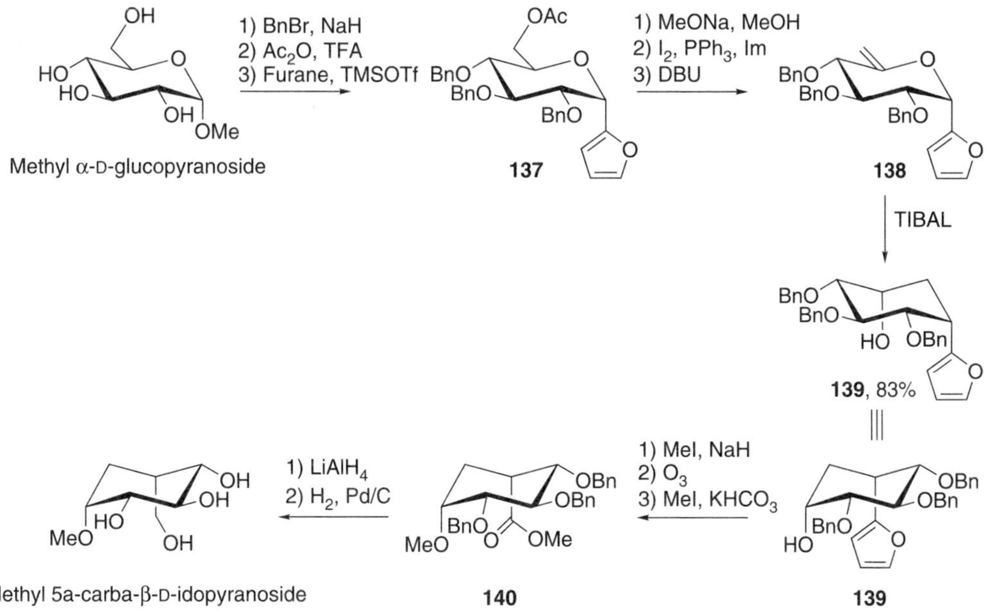

Methyl α-D-glucopyranoside

1) BnBr, NaH
2) Ac_2O, TFA
3) Furane, TMSOTf

137

1) MeONa, MeOH
2) I_2, PPh_3, Im
3) DBU

138

TIBAL

139, 83%

1) MeI, NaH
2) O_3
3) MeI, $KHCO_3$

140

1) $LiAlH_4$
2) H_2, Pd/C

Methyl 5a-carba-β-D-idopyranoside

SCHEME 8.36 Synthesis of a carba-pyranoside using the TIBAL induced rearrangement on a *C*-glycoside.

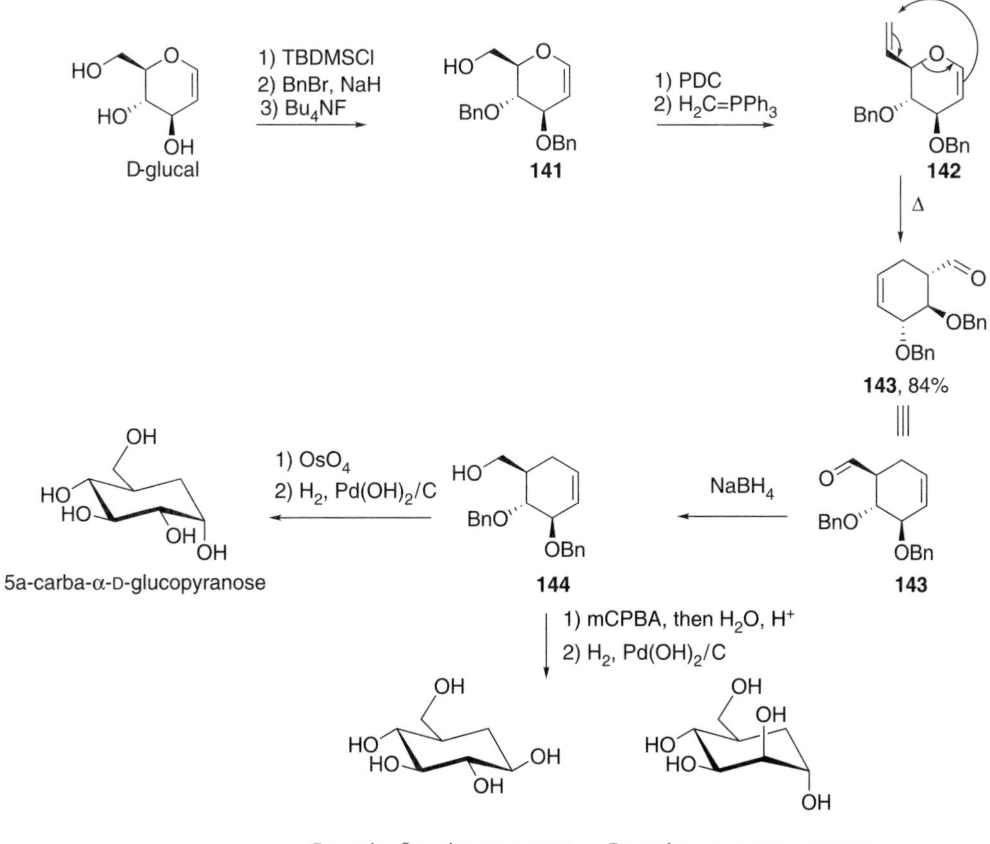

SCHEME 8.37 Synthesis of carba-pyranosides using the Claisen rearrangement.

8.3.2.3 The Claisen Rearrangement

An efficient synthesis of carba-sugars by carbocyclization utilized the Claisen rearrangement. As shown in Scheme 8.37, the key diene **142** was obtained from D-glucal by selective protection of the primary hydroxyl, 3,4-dibenzylation and desilylation affording **141**. Subsequent oxidation of the alcohol and olefination of the resulting aldehyde gave **142**. Sigmatropic rearrangement of diene **142** afforded the cyclohexene **143** in a yield of 84% and treatment with sodium borohydride gave alcohol **144**. The double bond in **144** was dihydroxylated and the resulting triol was deprotected to provide 5a-carba-α-D-glucopyranose. Alkene **144** was also converted into 5a-carba-β-D-glucopyranose and 5a-carba-α-D-mannopyranose by *trans*-hydroxylation and hydrogenolysis [68].

8.4 CONCLUSION

Carba-sugars are usually recognized by enzymes involving their parent sugars, but do not have sufficient affinity to act as efficient inhibitors. For this purpose, additional features such as amino groups or unsaturation are needed for the molecule to mimic a transition state, rather than a substrate. However, carba-sugars have been successfully used as probes to elucidate biosynthetic pathways and the topology of enzyme active sites.

Throughout this chapter, we have demonstrated that carba-sugars are generally relatively easily accessible sugar derivatives. Many methods have been proposed for the synthesis of carba-sugars, and some of them are now quite efficient. Even more complex molecules such as carba-disaccharides have been synthesized in a reasonable number of steps. While "real" carba-sugars do not seem to be very potent inhibitors, it is hoped that the chemistry used for their preparation will ultimately lead to the design and synthesis of novel structures possessing increasingly more interesting properties, ranging from the probes alluded to in this chapter to new classes of therapeutic agents.

REFERENCES

1. McCasland, G E, Furuta, S, Durham, L J, Alicyclic carbohydrates. XXIX. The synthesis of a pseudo-hexose (2,3,4,5-tetrahydrocyclohexanemethanol), *J. Org. Chem.*, 31, 1516–1521, 1966.
2. Suami, T, Ogawa, S, Chemistry of carba-sugars (pseudo-sugars) and their derivatives, *Adv. Carbohydr. Chem. Biochem.*, 48, 21–90, 1990.
3. Le Merrer, Y, Gravier-Pelletier, C, Maton, W, Numa, W M, Depezay, J-C, A concise route to carbasugars, *Synlett*, 8, 1322–1324, 1999.
4. Ohtake, H, Li, X L, Shiro, M, Ikegami, S, A highly efficient and shortcut synthesis of cyclitol derivatives via spiro sugar ortho esters, *Tetrahedron*, 56, 7109–7122, 2000.
5. Miller, T W, Arison, B H, Albers-Schonberg, G, Isolation of a cyclitol antibiotic: 2,3,4,5-tetrahydroxycyclohexanemethanol, *Biotech. Bioeng.*, 15, 1075–1080, 1973.
6. McCasland, G E, Furuta, S, Durham, L J, Alicyclic carbohydrates. XXIX. Epimerization of pseudo-alpha-DL-galactopyranose. Proton magnetic resonance studies, *J. Org. Chem.*, 33, 2841–2847, 1968.
7. Ogawa, S, Maruyama, A, Odagiri, T, Yuasa, H, Hashimoto, H, Synthesis and biological evaluation of alpha-L-fucosidase inhibitors: 5a-carba-alpha-L-fucopyranosylamine and related compounds, *Eur. J. Org. Chem.*, 5, 967–974, 2001.
8. Carpintero, M, Bastida, A, Garcia-Junceda, E, Jimenez-Barbero, J, Fernandez-Mayoralas, A, Synthesis of carba- and C-fucopyranosides and their evaluation as a-fucosidase inhibitors-analysis of an unusual conformation adopted by an amino-C-fucopyranoside, *Eur. J. Org. Chem.*, 4, 4127–4135, 2001.
9. Suami, T, Ogawa, S, Toyokuni, T, Sweet tasting pseudo-sugars, *Chem. Lett.*, 4, 611–612, 1983.

10. Ogawa, S, Uematsu, Y, Yoshida, S, Sasaki, N, Suami, T, Pseudo-sugars. 18. Synthesis and sweetness of pseudo-beta-D and L-fructopyranose, *J. Carbohydr. Chem.*, 6, 471–478, 1987.
11. Kitaoka, M, Ogawa, S, Taniguchi, H, A cellobiose phosphorylase from cellvibrio-gilvus recognizes only the beta-D-form of 5a-carba-glucopyranose, *Carbohydr. Res.*, 247, 355–359, 1993.
12. Callam, C S, Lowary, T L, Synthesis and conformational investigation of methyl 4a-carba-D-arabinofuranosides, *J. Org. Chem.*, 66, 8961–8972, 2001.
13. Carpintero, M, Fernandez-Mayoralas, A, Jimenez-Barbero, J, The conformational behaviour of fucosyl and carbafucosyl mimetics in the free and in the protein-bound states, *Eur. J. Org. Chem.*, 4, 681–689, 2001.
14. Ogawa, S, Sasaki, S, Tsunoda, H, Synthesis of carbocyclic analogs of the mannosyl trisaccharide — ether-linked and imino-linked methyl 3,6-bis(5a-carba-alpha-D-mannopyranosyl)-3,6-dideoxy-alpha-D-mannopyranosides, *Carbohydr. Res.*, 274, 183–196, 1995.
15. Ogawa, S, Furuya, T, Tsunoda, H, Hindsgaul, O, Stangier, K, Palcic, M M, Synthesis of beta-D-GlcpNAc-(1-2)-5a-carba-alpha-D-manp-(1-6)-beta-D-Glcp-$O(CH_2)_7CH_3$ — a reactive acceptor analog for *N*-acetylglucosaminyltransferase-V, *Carbohydr. Res.*, 271, 197–205, 1995.
16. Ogawa, S, Matsunaga, N, Li, H, Palcic, M M, Pseudosugars, 40 — synthesis of ether- and imino-linked octyl *N*-acetyl-5a′-carba-beta-lactosaminides and -isolactosaminides: acceptor substrates for alpha-(1→3/4)-fucosyltransferase, and enzymatic synthesis of 5a′-carbatrisaccharides, *Eur. J. Org. Chem.*, 631–642, 1999.
17. Wilcox, C S, Gaudino, J J, New approaches to enzyme regulators. Synthesis and enzymological activity of carbocyclic analogues of D-fructofuranose and D-fructofuranose 6-phosphate, *J. Am. Chem. Soc.*, 108, 3102–3104, 1986.
18. Fukusima, Y, Hayashi, M, Fujiwara, M, Miyagawa, T, Yoshikawa, G, Yano, T, Nakajima, H, Enzymatic synthesis of carbocyclic analogue of fructose 2,6-bisphosphate with 6-phosphofructo-2-kinase, *Chem. Lett.*, 575–576, 1998.
19. Cai, S, Stroud, M R, Hakomori, S, Toyokuni, T, Synthesis of carbocyclic analogs of guanosine 5′-(beta-L-fucopyranosyl diphosphate) (GDP-fucose) as potential inhibitors of fucosyltransferases, *J. Org. Chem.*, 57, 6693–6696, 1992.
20. Yuasa, H, Palcic, M M, Hindsgaul, O, Synthesis of the carbocyclic analog of uridine 5′-(alpha-D-galactopyranosyl diphosphate) (UDP-Gal) as an inhibitor of beta(1→4)-galactosyltransferase, *Can. J. Chem.*, 73, 2190–2195, 1995.
21. Ferrier, R J, Middleton, S, The conversion of carbohydrate-derivatives into functionalized cyclohexanes and cyclopentanes, *Chem. Rev.*, 93, 2779–2831, 1993.
22. Ferrier, R J, Direct conversion of 5,6-unsaturated hexopyranosyl compounds to functionalized cyclohexanones, *Top. Curr. Chem.*, 215, 277–291, 2001.
23. Dalko, P I, Sinaÿ, P, Recent advances in the conversion of carbohydrate furanosides and pyranosides into carbocycles, *Angew. Chem. Int. Ed.*, 38, 773–777, 1999.
24. Suami, T, Tadano, K, Kameda, Y, Iimura, Y, Synthesis of optically-active pseudo-alpha-D-glucose and pseudo-beta-L-altrose, *Chem. Lett.*, 1919–1922, 1984.
25. Tadano, K I, Kameda, Y, Iimura, Y, Suami, T, Syntheses of pseudo-alpha-D-glucopyranose and pseudo-beta-L-altropyranose from L-arabinose, *J. Carbohydr. Chem.*, 6, 231–244, 1987.
26. Tadano, K, Maeda, H, Hoshino, M, Iimura, Y, Suami, T, A new transformation of aldose derived synthons to pseudo-hexopyranose or pseudo-pentofuranose derivatives, *Chem. Lett.*, 1081–1084, 1986.
27. Tadano, K I, Maeda, H, Hoshino, M, Iimura, Y, Suami, T, A novel transformation of 4 aldoses to some optically pure pseudohexopyranoses and a pseudopentofuranose, carbocyclic analogs of hexopyranoses and pentofuranose — synthesis of derivatives of (1*S*,2*S*,3*R*,4*S*,5*S*)-2,3,4,5-tetrahydroxy-1-(hydroxymethyl)cyclohexanes, (1*S*,2*S*,3*R*,4*R*,5*S*)-2,3,4,5-tetrahydroxy-1-(hydroxymethyl)cyclohexanes, (1*R*,2*R*,3*R*,4*R*,5*S*)-2,3,4,5-tetrahydroxy-1-(hydroxymethyl)cyclohexanes, (1*S*,2*S*,3*R*,4*S*,5*R*)-2,3,4,5-tetrahydroxy-1-(hydroxymethyl)cyclohexanes and (1*S*,2*S*,3*S*,4*S*)-2,3,4-trihydroxy-1-(hydroxymethyl)cyclopentane, *J. Org. Chem.*, 52, 1946–1956, 1987.
28. Yoshikawa, M, Cha, B C, Nakae, T, Kitagawa, I, Syntheses of pseudo-alpha-D-glucopyranose and pseudo-beta-L-idopyranose, two optically-active pseudo-hexopyranoses, from D-glucose by using stereoselective reductive deacetoxylation with sodium-borohydride and cyclitol formation from nitrofuranose as key reactions, *Chem. Pharm. Bull.*, 36, 3714–3717, 1988.

29. Yoshikawa, M, Cha, B C, Okaichi, Y, Kitagawa, I, Synthesis of pseudo-alpha-D-arabinofuranose, pseudo-beta-D-arabinofuranose, and pseudo-beta-L-xylofuranose, three optically-active pseudo-pentofuranoses, from D-glucose, *Chem. Pharm. Bull.*, 36, 3718–3721, 1988.
30. Yoshikawa, M, Murakami, N, Inoue, Y, Hatakeyama, S, Kitagawa, I, A new approach to the synthesis of optically-active pseudo-sugar and pseudo-nucleoside — syntheses of pseudo-alpha-D-arabinofuranose, (+)-cyclaradine, and (+)-1-pseudo-beta-D-arabinofuranosyluracil from D-arabinose, *Chem. Pharm. Bull.*, 41, 636–638, 1993.
31. Paulsen, H, Vondeyn, W, Cyclitol reactions 13. Synthesis of pseudosugars from D-glucose by intramolecular Horner–Emmons olefination, *Liebigs Ann. Chem.*, 2, 125–131, 1987.
32. Suami, T, Tadano, K, Ueno, Y, Fukabori, C, A novel synthesis of pseudo-alpha-L-altropyranose from D-glucose, *Chem. Lett.*, 1557–1560, 1985.
33. Tadano, K, Ueno, Y, Fukabori, C, Hotta, Y, Suami, T, Construction of an optically-active 7-oxabicyclo(4.3.0)non-4-en-3-one skeleton from D-glucose, and its transformation to some pseudo-hexopyranoses, *Bull. Chem. Soc. Jpn.*, 60, 1727–1739, 1987.
34. Rassu, G, Auzzas, L, Pinna, L, Battistini, L, Zanardi, F, Marzocchi, L, Acquotti, D, Casiraghi, G, Variable strategy toward carbasugars and relatives. 1. Stereocontrolled synthesis of pseudo-beta-D-gulopyranose, pseudo-beta-D-xylofuranose, (pseudo-beta-D-gulopyranosyl)amine, and (pseudo-beta-D-xylofuranosyl)amine, *J. Org. Chem.*, 65(20), 6307–6318, 2000.
35. Zanardi, F, Battistini, L, Marzocchi, L, Acquotti, D, Rassu, G, Pinna, L, Auzzas, L, Zambrano, V, Casiraghi, G, Synthesis of a small repertoire of non-racemic 5a-carbahexopyranoses and 1-thio-5a-carbahexopyranoses, *Eur. J. Org. Chem.*, 1956–1965, 2002.
36. Rassu, G, Auzzas, L, Pinna, L, Zambrano, V, Battistini, L, Zanardi, F, Marzocchi, L, Acquotti, D, Casiraghi, G, Variable strategy toward carbasugars and relatives. 2. Diversity-based synthesis of beta-D-xylo, beta-D-ribo, beta-L-arabino, and beta-L-lyxo 4a-carbafuranoses and (4a-carbafuranosyl)thiols, *J. Org. Chem.*, 66, 8070–8075, 2001.
37. Keck, G E, Kachensky, D F, Enholm, E J, Pseudomonic acid C from L-lyxose, *J. Org. Chem.*, 50, 4317–4325, 1985.
38. Cheng, X H, Khan, N, Kumaran, G, Mootoo, D R, A convergent strategy for the synthesis of beta-carba-galacto-disaccharides, *Org. Lett.*, 3(9), 1323–1325, 2001.
39. Marco-Contelles, J, Martinez-Grau, A, Carbocycles from carbohydrates via free radical cyclizations: New synthetic approaches to glycomimetics, *Chem. Soc. Rev.*, 27, 155–162, 1998.
40. Marco-Contelles, J, Alhambra, C, Martinez-Grau, A, Carbocycles from carbohydrates via free radical cyclizations: synthesis and manipulation of annulated furanoses, *Synlett*, 693–699, 1998.
41. Marco-Contelles, J, Pozuelo, C, Jimeno, M I, Martinez, L, Martinez-Grau, A, 6-*Exo* free-radical cyclization of acyclic carbohydrate intermediates — a new synthetic route to enantiomerically pure polyhydroxylated cyclohexane derivatives, *J. Org. Chem.*, 57, 2625–2631, 1992.
42. Redlich, H, Sudau, W, Szardenings, A K, Vollerthun, R, 5-Carba analogs of sugars. 1. Radical cyclization of hept-1-enitols, *Carbohydr. Res.*, 226, 57–78, 1992.
43. Gaudino, J J, Wilcox, C S, A general approach to carbocyclic sugar analogs — preparation of a carbocyclic analog of beta-D-fructofuranose, *Carbohydr. Res.*, 206(2), 233–250, 1990.
44. Maudru, E, Singh, G, Wightman, R H, Radical cyclisation of carbohydrate alkynes: synthesis of highly functionalised cyclohexanes and carbasugars, *Chem. Commun.*, 1505–1506, 1998.
45. Gomez, A M, Danelon, G O, Moreno, E, Valverde, S, Lopez, J C, A novel entry to 5a-carba-hexopyranoses from carbohydrates based on a 6-exo-dig radical cyclization: synthesis of 5a-carba-beta-D-mannopyranose pentaacetate, *Chem. Commun.*, 175–176, 1999.
46. Gomez, A M, Moreno, E, Valverde, S, Lopez, J C, Stereodivergent synthesis of carbasugars from D-mannose. Syntheses of 5a-carba-alpha-D-allose, beta-L-talose, and alpha-L-gulose pentaacetates, *Synlett*, 891–894, 2002.
47. Gomez, A M, Moreno, E, Valverde, S, Lopez, J C, A general stereodivergent strategy for the preparation of carbasugars. Syntheses of 5a-carba-alpha-D-glucose, alpha-D-galactose, and beta-L-gulose pentaacetates from D-mannose, *Tetrahedron Lett.*, 43, 5559–5562, 2002.
48. Gomez, A M, Danelon, G O, Valverde, S, Lopez, J C, Regio- and stereocontrolled 6-endo-trig radical cyclization of vinyl radicals: a novel entry to carbasugars from carbohydrates, *J. Org. Chem.*, 63, 9626–9627, 1998.

49. Mella, M, Panza, L, Ronchetti, F, Toma, L, 1,2-Didcoxy-3,4-5,7-bis-*O*-(1-methylethylidene)-D-gluco and D-galacto-hept-1-ynitols — synthesis and conformational studies, *Tetrahedron*, 44, 1673–1678, 1988.
50. Désiré, J, Prandi, J, A new synthesis of carbapentofuranoses from carbohydrates, *Tetrahedron Lett.*, 38, 6189–6192, 1997.
51. Désiré, J, Prandi, J, The cobalt-catalyzed oxygenative radical route from hexopyranosides to carbapentofuranoses, *Eur. J. Org. Chem.*, 3075–3084, 2000.
52. Callam, C S, Lowary, T L, Total synthesis of both methyl 4a-carba-D-arabinofuranosides, *Org. Lett.*, 2, 167–169, 2000.
53. Callam, C S, Lowary, T L, Synthesis and conformational investigation of methyl 4a-carba-D-arabinofuranosides, *J. Org. Chem.*, 66, 8961–8972, 2001.
54. Seepersaud, M, Al-Abed, Y, Total synthesis of carba-D-fructofuranose via a novel metathesis reaction, *Org. Lett.*, 1, 1463–1465, 1999.
55. Seepersaud, M, Kettunen, M, Abu-Surrah, A S, Repo, T, Voelter, W, Al-Abed, Y, Hydroformylation of cyclopentenes, novel strategy for total synthesis of carba-D-fructofuranose, *Tetrahedron Lett.*, 43(10), 1793–1795, 2002.
56. Carpintero, M, Jaramillo, C, Fernandez-Mayoralas, A, Stereoselective synthesis of carba- and C-glycosyl analogs of fucopyranosides, *Eur. J. Org. Chem.*, 1285–1296, 2000.
57. Dalko, P I, Sinaÿ, P, Recent advances in the conversion of carbohydrate furanosides and pyranosides into carbocycles, *Angew. Chem. Int. Ed.*, 38, 773–777, 1999.
58. Ferrier, R J, Unsaturated carbohydrates 21. Carbocyclic ring-closure of a hex-5-enopyranoside derivative, *J. Chem. Soc. Perkin. Trans.*, 1, 1455–1458, 1979.
59. Blattner, R, Ferrier, R J, Crystalline pseudo-alpha-D-glucopyranose, *J. Chem. Soc. Chem. Comm.*, 1008–1009, 1987.
60. Barton, D H R, Gero, S D, Cleophax, J, Machado, A S, Quiclet-Sire, B, Synthetic methods for the preparation of D-pseudo-sugars and L-pseudo-sugars from D-glucose, *J. Chem. Soc. Chem. Comm.*, 1184–1186, 1988.
61. Barton, D H R, Augy-Dorey, S, Camara, J, Dalko, P, Delauméņy, J M, Géro, S D, Quiclet-Sire, B, Stütz, P, Synthetic methods for the preparation of basic D-pseudo-sugars and L-pseudo-sugars — synthesis of carbocyclic analogs of *N*-acetyl-muramyl-L-alanyl-D-isoglutamine (Mdp), *Tetrahedron*, 46, 215–230, 1990.
62. Das, S K, Mallet, J M, Sinaÿ, P, Novel carbocyclic ring closure of hex-5-enopyranosides, *Angew. Chem. Int. Ed.*, 36, 493–496, 1997.
63. Sollogoub, M, Mallet, J M, Sinaÿ, P, Titanium (IV) promoted rearrangement of 6-deoxy-hex-5-enopyranosides into cyclohexanones, *Tetrahedron Lett.*, 39, 3471–3472, 1998.
64. Sollogoub, M, Pearce, A J, Herault, A, Sinaÿ, P, Synthesis of carba-beta-D- and L-idopyranosides by rearrangement of unsaturated sugars, *Tetrahedron-Asymmetry*, 11, 283–294, 2000.
65. Pearce, A J, Sollogoub, M, Mallet, J M, Sinaÿ, P, Direct synthesis of pseudo-disaccharides by rearrangement of unsaturated disaccharides, *Eur. J. Org. Chem.*, 9, 2103–2117, 1999.
66. Pearce, A J, Mallet, J M, Sinaÿ, P, One-step synthesis of disaccharide mimetics via tandem rearrangement of unsaturated disaccharides, *Heterocycles*, 52, 819–826, 2000.
67. Sollogoub, M, Mallet, J M, Sinaÿ, P, Carbocyclic ring closure of unsaturated *S*-, *Se*-, and *C*-aryl glycosides, *Angew. Chem. Int. Ed.*, 39, 362–364, 2000.
68. Sudha, A V R L, Nagarajan, M, Carbohydrates to carbocycles: an expedient synthesis of pseudo-sugars, *Chem. Commun.*, 925–926, 1998.

9 Sugars with Endocyclic Heteroatoms Other than Oxygen

Peter Greimel, Josef Spreitz, Friedrich K. (Fitz) Sprenger, Arnold E. Stütz and Tanja M. Wrodnigg

CONTENTS

9.1 INTRODUCTION

For the past three decades, sugars and closely related mono- and bicyclic polyhydroxy compounds with heteroatoms other than oxygen in the ring, together with carbasugars in which the ring oxygen is replaced by a carbon atom, have accounted for a predominant portion of the carbohydrate literature not related to glycoside synthesis. This has occurred because of the diverse biological activities of these sugar mimetics and their eminent importance for carbohydrate chemistry and biochemistry as we understand it at the beginning of its second century.

Having been of purely academic interest in their very early days, it was soon recognized that both thiosugars (sugars with sulfur in the ring) and iminosugars (in which the ring oxygen is replaced by basic nitrogen) have extraordinarily interesting biological activities and tremendous diagnostic as well as therapeutic potential. This was deduced from the properties of natural sources containing such compounds as natural products. For example, 5-amino-5-deoxy-D-glucose **1** (Figure 9.1), coined nojirimycin, was found to be an antibiotic substance in 1966. Strong biological effects were also found in close relatives such as the corresponding 1-deoxy derivative **2**, a natural product in its own right, as well as in 2,5-dideoxy-2,5-imino-D-mannitol (DMDP, **3**) and bicyclic analogs bearing a "D-gluco" stereochemical motif such as the alkaloids castanospermine **4**, australine **5** and calystegine B_2 **6**, just to mention a few examples.

In the realm of thiosugars, the only pyranose known to be a natural product is 5-thio-D-mannose **7** (Figure 9.2), but more complex structures, salacinol **8** and cotalanol **9**, have recently been found in plants known for their antidiabetic properties in Indian folk medicine.

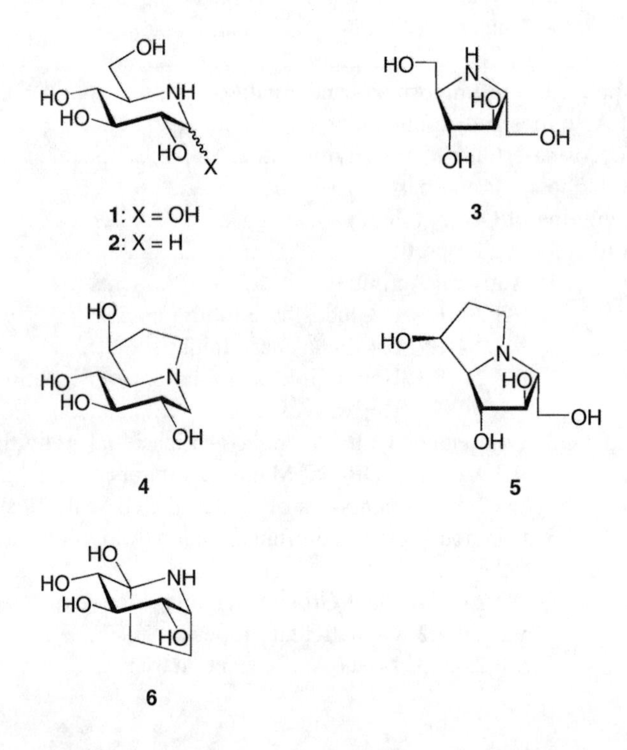

FIGURE 9.1 Some examples of natural sugars and related alkaloids with nitrogen in the ring.

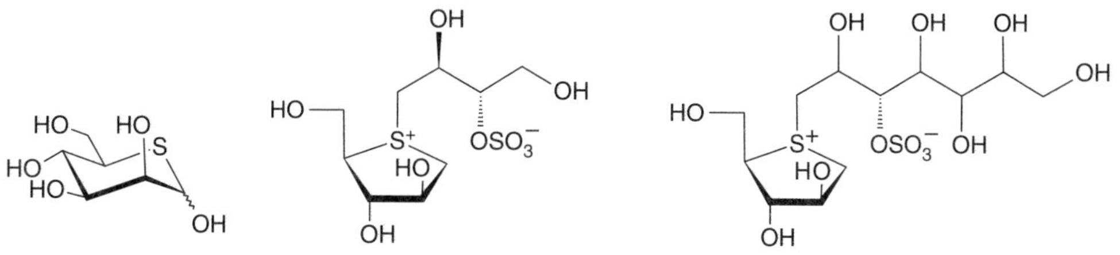

7 8 9

FIGURE 9.2 Natural sugar analogs with sulfur in the ring.

Displacement of the ring oxygen in a sugar by other heteroatoms such as nitrogen, sulfur, or phosphorus leads to pronounced changes of the physico-chemical properties and, consequently, to profound differences in the "information content" of the accordingly modified carbohydrate towards biochemical systems. The altered biological properties of such derivatives with sugar-utilizing or modifying enzymes, such as glycosidases, phosphorylases or glycosyl transferases, as well as other carbohydrate-recognizing biological entities can lead to useful applications as diagnostic or therapeutic agents. Sugar analogs with nitrogen instead of oxygen in the ring have attracted considerable and wide-spread attention because of their pronounced biological activities as inhibitors of glycosidases. Similar effects have been found with carbohydrates having sulfur in the ring. Imino- and thiosugars and their relatives thus have emerged as sought-after compounds for the synthetic challenges posed by their structures, their enzyme inhibitory properties and for their economic value as diagnostics and therapeutics.

Other elements introduced into this position, for example, phosphorus or selenium, change the physico-chemical properties more profoundly, leading to less stable and more sterically and electronically demanding derivatives and, thus, have had lesser impact on glycobiology and glycomimetics research. In spite of this vast and quickly growing field and the large number of relevant publications, this account is based on preselected pieces of relevant information that should allow the reader to get the basic picture. As there are excellent reviews available spanning practically all facets mentioned in this chapter, the interested reader may retrieve more in-depth information from more specialized sources.

9.2 THIOSUGARS WITH SULFUR IN THE RING

A large variety of organo-sulfur compounds exhibiting a wide range of different properties is abound in Nature. Sugars and sugar analogs with sulfur in the ring have attracted considerable interest, both in terms of synthetic challenges and as potential biological activities.

From the distinctly reduced mutarotation exhibited by 5-thioaldoses when compared with their oxygenated parent compounds it was concluded that a hemiacetal ring containing sulfur is more stable [1–3]. On the other hand, it was discovered that glycosides of such thiosugars are more prone to hydrolysis than their natural counterparts [4].

The ring size is generally dependent on the position of the thiol group, because of the higher nucleophilicity of sulfur compared with oxygen. Consequently, the pyranoid tautomers prevail over five-membered ring systems in 5-thiosugars, whereas in 4-thiosugars the furanoid form is found to be preferred over the pyranoid ring containing O-5. As with natural sugars, pyranoid 5-thiosugars have been found to be more stable than furanoid 4-thiosugars for thermodynamic reasons [5]. The smaller C–S–C bond angle in 5-thio-D-glucopyranose **10** (Figure 9.3), compared with the C–O–C bond angle in D-glucopyranose, makes the former more puckered [6]. Sulfoxides and sulfones have been prepared by oxidation of thiosugars but have found very limited use.

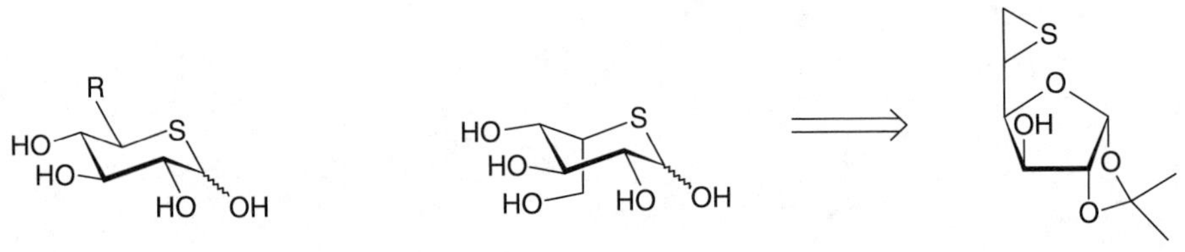

FIGURE 9.3 Early synthetic thiosugars.

As early as 1961, the first attempts to replace the ring oxygen in monosaccharides by a heteroatom were independently reported from the laboratories of Schwartz [2], Owen [3] and Whistler [7]. These researchers independently succeeded in preparing 5-thio-D-xylopyranose **11**, the first representative of the sulfur-in-the-ring family of sugars.

Adley and Owen [3] described the synthesis of 5-thio-L-idopyranose **12** by the ring opening of 5,6-dideoxy-5,6-epithio-1,2-*O*-isopropylidene-α-L-idofuranose **13**, with potassium acetate in a mixture of glacial acetic acid and acetic anhydride and subsequent deprotection. 5-Thio-D-glucopyranose **10** was first synthesized by Feather and Whistler [8].

Reist and coworkers [9,10] described the first sulfur-containing furanose rings. They prepared 4-thio-D-ribofuranose **14** (Figure 9.4) and 4-thio-L-ribofuranose via nucleophilic displacement of the tosylate in 2,3-*O*-isopropylidene-4-*O*-toluenesulfonyl-α-D- and -α-L-lyxopyranosides, respectively, with potassium thiobenzoate in DMF. Several nucleosides of 4-thiofuranoses were subsequently synthesized. For example, the 4′-thio analog **15** of natural thymidine and the corresponding derivative of cytidine were prepared by Whistler and coworkers [11].

9.2.1 Furanoid Systems

Replacement of the ring oxygen in naturally occurring, as well as synthetic, nucleosides for therapeutic reasons has been a major task of carbohydrate chemistry. Starting in the early 1960s, all diastereomeric 4-thio analogs of natural pentofuranoses, as well as some of their enantiomers, were prepared and, based on these approaches, a wide range of synthetic 4′-thionucleosides has been made available.

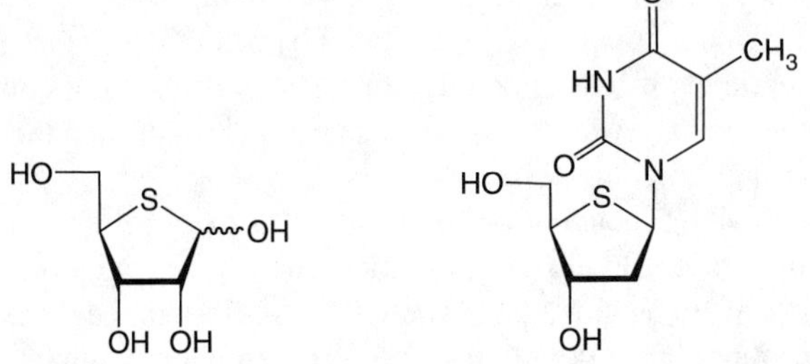

FIGURE 9.4 Thiofuranosides.

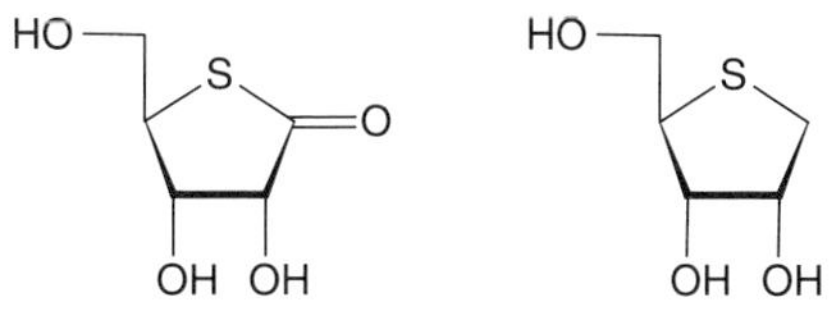

16 **17**

FIGURE 9.5 Compounds related to thiofuranoses.

9.2.1.1 4-Thioaldopentoses, 4-Thiopentonolactones and Derivatives

4-Thio-D- and -L-ribofuranosides, synthesized by Reist and coworkers [9,10], were among the first furanoid thiosugars with sulfur in the ring. Whistler and coworkers [12] reported the synthesis of methyl 4-thio-D-ribofuranoside from L-lyxose. The corresponding per-*O*-acetylated sugar served as an intermediate in the preparation of various purine and pyrimidine nucleosides [11]. Imbach and coworkers [13] provided a synthesis of free 4-thio-D-ribofuranose and prepared protected derivatives suitable for nucleoside syntheses [14]. Protected 4-thio-D-arabinofuranose, as well as the corresponding D-*xylo* configured sugar, was also synthesized by the Reist group [15]. Nayak and Whistler [16] achieved the synthesis of methyl 4-thio-β-D-arabinofuranoside and subsequently employed this compound for nucleoside syntheses [17].

Varela and Zunszain [18] synthesized 4-thio analogs **16** of D-ribonolactone as well as L-lyxonolactone (Figure 9.5). Approaches to 1,4-anhydro-4-thio-D- **17** and -L-ribitol were recently described by Altenbach and coworkers [19,20]. 1,4-Anhydro-4-thio-D-arabinitol was employed in the synthesis of a potential inhibitor of glycosyl transferases [21]. Access to 4-thiopentonolactones has also been provided, for example, by Dominguez and Owen [22].

9.2.1.2 4-Thioaldohexoses and Derivatives

In the series of the aldohexoses, 4-thio derivatives of D-glucose, **18** [23], D-mannose [24] and, more recently, D-galactose [25] have been synthesized (Figure 9.6). In addition, the 6-deoxy analogs of D-*gluco* [26], D-*gulo* [27] and D-*ido* [28] configured 4-thioaldohexoses have been prepared. For example, per-*O*-acetylated 6-deoxy-4-thio-D-galactofuranose **19** was obtained by acetolysis of methyl 6-deoxy-4-thio-D-galactopyranoside [29]. Garegg and coworkers [30] reported the

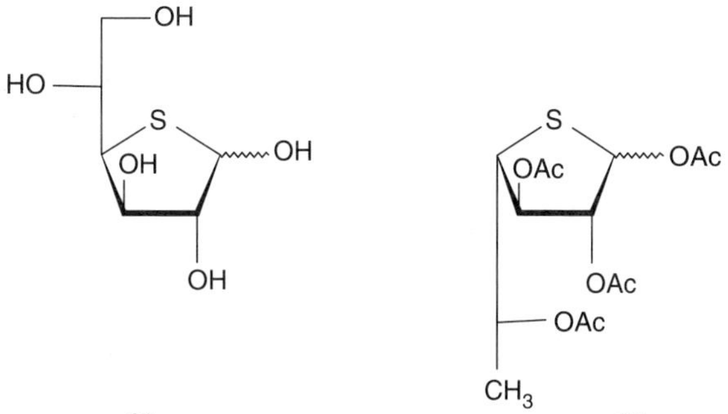

18 **19**

FIGURE 9.6 Examples of 4-Thiohexofuranoses.

synthesis of methyl 4-thio-α-D-talofuranoside. 4-Thiofuranoside derivatives of D-galactosamine are also known [31].

9.2.1.3 5-Thioketopentoses, 5-Thioketohexoses and Derivatives

Effenberger and coworkers reported a *de novo* approach to 5-thio-D-*threo*-2-pentulofuranose (5-thio-D-xylulose, **20**) (Figure 9.7) from 2-mercaptoacetaldehyde, employing rabbit muscle aldolase (RAMA; EC 4.1.2.13) and yeast transketolase (EC 2.2.1.1) as the catalysts [32].

5-Thio-D-fructofuranose **21** was synthesized by Chmielewski and Whistler [33]. The 2-deoxy derivative of 5-thio-β-D-fructofuranose, 2,5-anhydro-5-thio-D-mannitol **22** and the corresponding L-*ido* configured compound were prepared by ring opening reactions of D-mannitol-derived 1,2:5,6-diepoxides **23** followed by ring isomerization of the resulting thiepane **24**, and were found to be poor inhibitors of D-glucosidases [34].

Various protected bicyclic 2,5-anhydro derivatives of 5-thio-D-mannitol were obtained from methyl 2,3,4-tri-*O*-acetyl-6-*O*-mesyl-5-thio-α-D-glucopyranoside by transannular participation of the ring sulfur atom and ring opening of the thiiranium ion [35]. Similar reactions were earlier observed with, for example, a 4-*O*-methanesulfonyl-5-thio-D-xylopyranose [36].

9.2.2 Pyranoid Systems — 5-Thioaldohexoses, 6-Thioketohexoses and Derivatives

5-Thio-D-glucose **10** was one of the first examples of sugars with sulfur in the ring [8]. In a typical synthetic approach 1,2:5,6-di-*O*-isopropylidene-α-D-glucofuranose **25** was employed as the starting material [37] (Scheme 9.1). Through a sequence of standard protecting group manipulations, O-3 was protected as the benzyl ether **26**, the 5,6-acetal was removed by acid hydrolysis and O-6 was subsequently regioselectively protected with a benzoyl group giving **27**. Activation of C-5 was achieved by *O*-tosylation, leading to compound **28**. This intermediate was converted into the corresponding L-*ido* configured 5,6-oxirane **29** upon base treatment by intramolecular nucleophilic displacement of the tosylate. Reaction of **29** with thiourea gave the corresponding thiirane derivative **30** with inversion of configuration at C-5, which was regioselectively opened at C-6 with acetate. Removal of the protecting groups gave 5-thio-D-glucose **10** in ten steps and in generally fair yields.

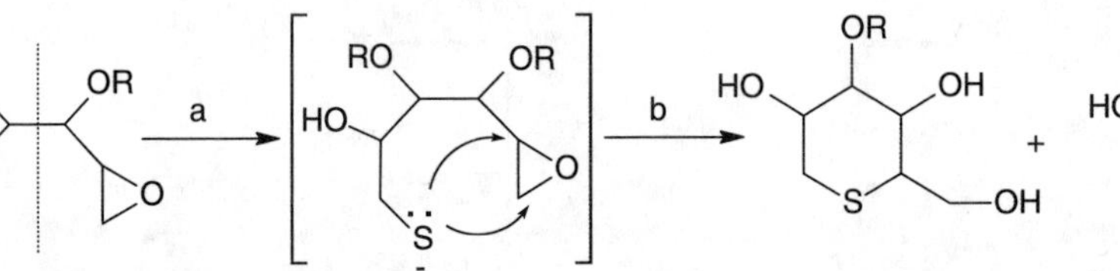

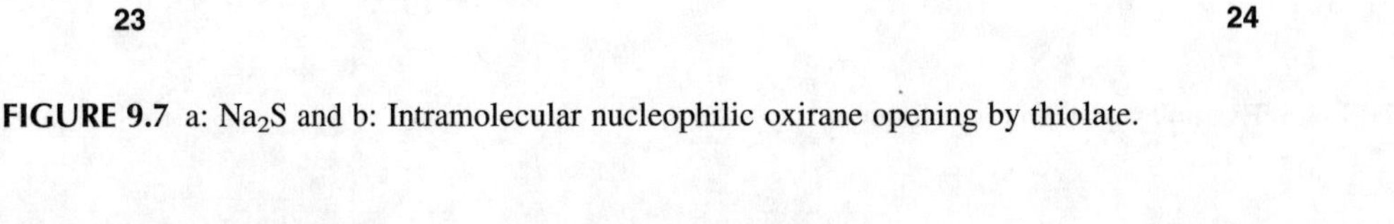

FIGURE 9.7 a: Na_2S and b: Intramolecular nucleophilic oxirane opening by thiolate.

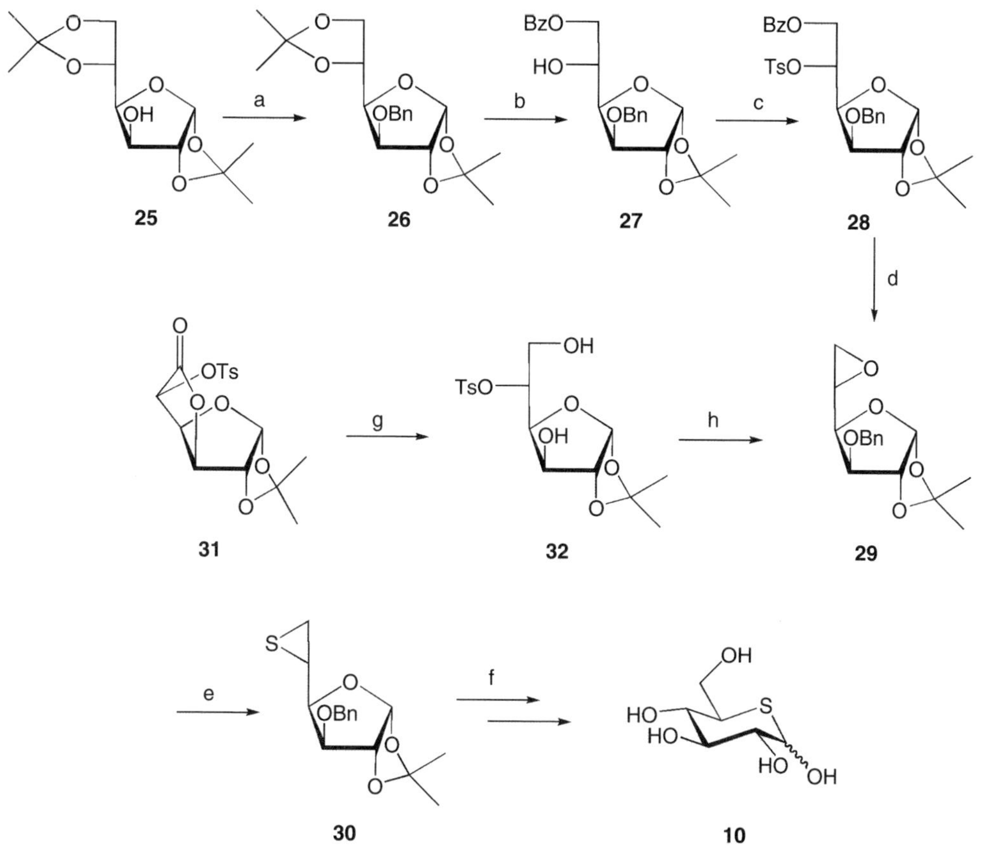

SCHEME 9.1 a: BnCl and NaH, b: H_3O^+ then BzCl, pyr., c: TsCl, pyr., d: base, e: thiourea, f: acetolysis and deprotection, g: BH_4^- and h: base.

Driguez and Henrissat devised an improved synthesis based on the same strategy [38]. These workers took advantage of the regiospecifically available C-5 position in 1,2-*O*-isopropylidene-α-D-glucofuranurono-6,3-lactone. This easily accessible starting material was *O*-sulfonylated, employing tosyl chloride, to furnish **31**; reduction of the lactone moiety afforded **32**, which was converted to the 5,6-oxirane **29**. The remaining steps to compound **10** were the same as in Whistler's approach. Application of the same general principle allows access to other 5-thioaldohexoses as was exemplified by Shin and Perlin in the D-*galacto* series [39].

A different chemical method for the preparation of 5-thioaldopyranosides was chosen by Hashimoto and coworkers [40] (Scheme 9.2). They converted per-*O*-benzylated methyl β-D-glucopyranoside **33** into the open-chain mixed acetal **34**, which, in turn, was treated with the Mitsunobu system triphenylphosphine and diethyl azodicarboxylate in the presence of benzoic acid to yield protected 5-thio-L-idopyranosides **35**.

An aldolase-based strategy conveniently provided access to thioketoses with sulfur in the ring. For example, RAMA-catalyzed carbon–carbon bond formation of a 3-thioglycerinaldehyde with dihydroxyacetone phosphate gave 6-thio-D-fructose **36** [32] (Figure 9.8), also available by enzymatic isomerization of 6-thio-D-glucose or 6-thio-L-sorbose with glucose isomerase (EC 5.3.1.5; *vide infra*) [32,41].

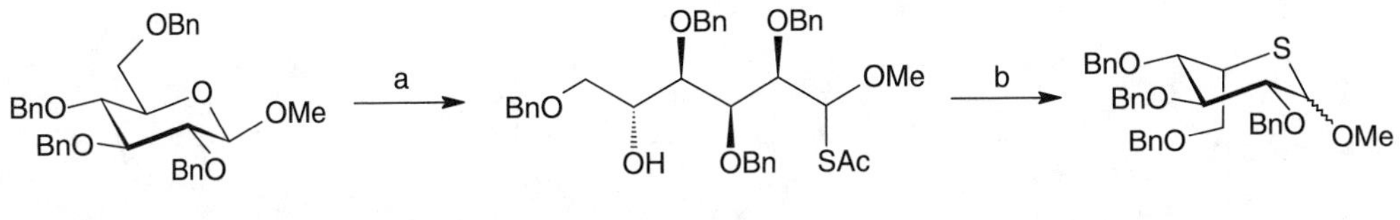

SCHEME 9.2 a: Me_2BBr then AcSH, $i Pr_2NEt$ and b: MsCl, pyr., then NaOMe.

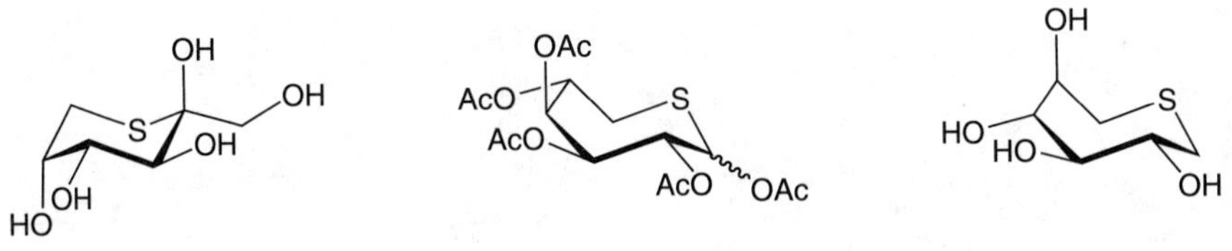

FIGURE 9.8 Thiofructose, thioseptanose and a 1,6-anhydro-6-thio-hexitol.

9.2.3 Septanoses and Derivatives

Despite the strong nucleophilicity of the thiol substituent, septanoses have not been found to spontaneously form from 6-thiosugars [5]. As with their natural counterparts, in cases where the formation of a pyranoid ring is not possible the 6-thiofuranose is favored over the 6-thioseptanose system [15,42].

Cox and Owen [42] and Whistler and Campbell [43] provided early synthetic approaches to the 6-thioseptanose system. The latter workers prepared a protected open-chain 6-thio dithioacetal of D-galactose that, upon deprotection at C-1 and S-6, could be cyclized to form the corresponding 1,2,3,4,5-penta-*O*-acetyl-D-galacto-6-thioseptanose **37**. From this intermediate, several derivatives, such as both anomers of the thioseptanosyl chloride as well as the methyl thioseptanosides, were synthesized. The methyl thioseptanoside was found to be more prone to hydrolysis than the corresponding sulfur containing pyranoid glycoside.

1,6-Anhydroalditol derivatives such as **38** were recently obtained by Le Merrer from D-mannitol [34]. Following earlier results featuring antithrombotic properties of 1,5-dithiopyranosides, Kuszmann and coworkers [44] prepared a range of 1,6-dithioseptanosides.

9.2.4 Examples of Glycomimetics with Sulfur in the Ring

9.2.4.1 Natural Products

Only a small number of sugars and related compounds with sulfur in the ring have been reported as natural products. In the following, some select and structurally quite dissimilar examples will be discussed.

5-Thio-D-mannose **7**, discovered as a metabolite of the marine sponge *Clathria pyramida* (Lendenfeld), is still unique as the only example of a 5-thioaldopyranose occurring in nature [45]. Shortly after its discovery, a synthesis from methyl 2,3:5,6-di-*O*-isopropylidene-α-D-mannofuranoside was reported [46], which essentially followed the principle of the approach depicted in Scheme 9.1. Earlier, 5-thio-D-mannose had been available by molybdate catalyzed epimerization [47,48] of 5-thio-D-glucose **10** [37].

The antibiotic thiolactomycin **39** (Figure 9.9), a fermentation product from a *Nocardia* species containing an unusual thiolactone moiety, was discovered in 1981 [49,50]. It resembles a

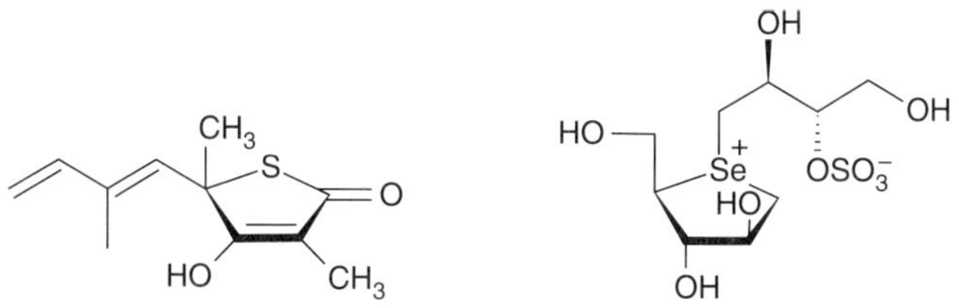

FIGURE 9.9 Thiolactomycin (**39**), the seleno-analog of salacinol (**40**).

sugar-derived α,β-unsaturated 4-thioglycono-1,4-lactone. Thiolactomycin was found to be a broad spectrum antibiotic [51] and was also shown to inhibit inducible β-lactamases [52]. A *de novo* synthesis of the racemate was reported in 1984 [53]. The antibiotic thiotetromycin is closely related structurally [54,55].

An interesting sugar relative, salacinol **8**, containing sulfur atoms in two different oxidation states was isolated and found to be the active component in the Indian plant *Salacia reticulate,* which is used in local traditional medicine for treatment of diabetes [56]. The compound was shown to be a potent inhibitor of intestinal α-glucosidases comparable in its activity with the powerful aminosugar derivative acarbose [57,58]. A closely related compound, cotalanol **9**, from the same source, exhibited similar inhibitory activity [59]. Consequently, two independent syntheses of these interesting thiosugars, as well as stereomers, were recently reported [60,61]. To get further insight into structure-activity relationships of the parent compounds, analogs in which the ring sulfur atom is replaced by nitrogen or selenium **40** have also been prepared and evaluated as glycosidase inhibitors, but were found to be weaker inhibitors than salacinol [62–64].

9.2.4.2 Synthetic Compounds

A wide variety of synthetic sugar and nucleoside mimicking compounds with sulfur in the ring have been prepared for basic research, with a view to producing biologically active compounds with improved biochemical properties. Various classes of compounds are discussed in the following paragraphs.

9.2.4.2.1 Glycosidase Inhibitors

As early as 1965 the inhibitory activity of 5-thio-D-xylopyranose **11** (Figure 9.10) against β-D-xylosidases was observed and a K_i value of 2 mM was reported for this reversible inhibition [65].

In-depth investigations [66–68] into the biological properties of 5-thio-D-glucose **10** rapidly led to the conclusion that the replacement of the ring oxygen with sulfur had a pronounced influence on sugar metabolism. Despite being nontoxic with an LD_{50} of 14 g/kg, diabetogenic effects of **10**

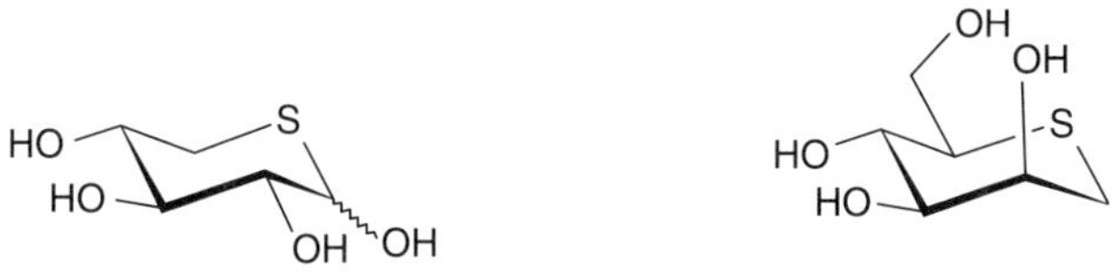

FIGURE 9.10 5-Thio-D-xylose (**11**), 1,5-anhydro-5-thio-D-mannitol (**41**).

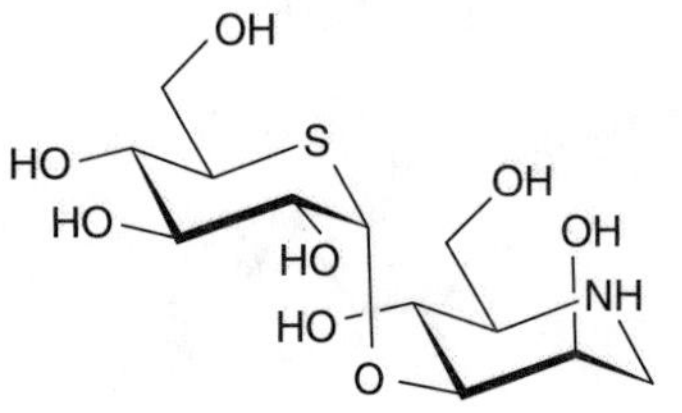

FIGURE 9.11 Thio analogs of glycosidase inhibitors.

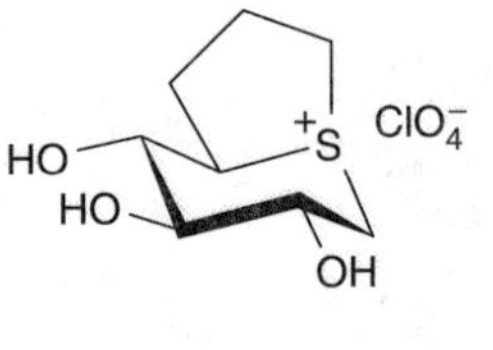

were reported by Hoffman and Whistler [66]. These workers found rapidly developing glucosuria and hyperglycemia in rats that had received this D-glucose mimic. Other effects observed were diminished glycogen contents in the liver and reduced membrane transport of D-glucose. Furthermore, this compound inhibited spermatogenesis in mice [69] and rats [70,71] and induced a reduction in the proliferation of trypanosoma [72]. Other effects included inhibition of cancer cell growth [73]. 5-Thio-D-glucose was also found to weakly inhibit yeast α-glucosidase with a K_i value of 750 μM, but was a poor inhibitor of the β-specific sweet almond enzyme [74].

The corresponding methyl α-pyranoside [74], as well as the 1-deoxy derivative **41** of thiomannose, also showed weak inhibitory effects on the former enzyme [34,75]. Significant effects on β-glucosidases could not be detected with any of these sugars or related derivatives [74,75]. The 3-*O*-(5-thio-α-D-glucopyranosyl) derivative of 1-deoxymannojirimycin **42** (Figure 9.11), a powerful D-mannosidase inhibitor, and related disaccharides were recently synthesized as potential inhibitors of *endo*-α-D-mannosidase of glycoprotein trimming [76,77], but did not show the activity hoped for (RG Spiro, private communication). The thio analog **43** of 1-deoxycastanospermine was reported by Pinto and coworkers [78].

Maltoside analogs with 5-thioglucose at the nonreducing end and the corresponding 1′-thioglycosides were prepared as potential glycosidase inhibitors by the same group [79,80].

Pinto and collaborators [81] synthesized several 5-thioglucopyranosylamines and found good inhibition of glucoamylase (K_i 4 μM), with disaccharide **44** (Figure 9.12) structurally resembling the cyclitol-4-aminoglucose subunit of acarbose. Related compounds were intended to be inhibitors of glycoprotein trimming mannosidases I and II [82]. An analog **45** of kojibioside was synthesized as a potential glucosidase I inhibitor by the same group [83]. Other thiosugar-containing oligosaccharides have been prepared as, for example, probes and potential inhibitors for *Trypanosoma cruzi* related research [84] and concanavalin A binders [85].

No pronounced biological effects were reported for 6-thio-D-fructopyranose **36**. Nonetheless, this compound was found to be unusually sweet and essentially nontoxic to mice [86]. In addition to its chemical synthesis [87], an enzymatic preparation from 6-thio-D-glucose employing glucose isomerase (EC 5.3.1.5) was reported [88]. In keeping with the general properties of pyranoid ring

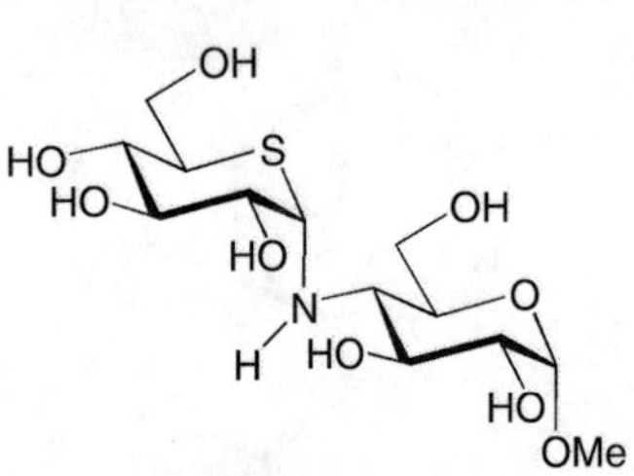

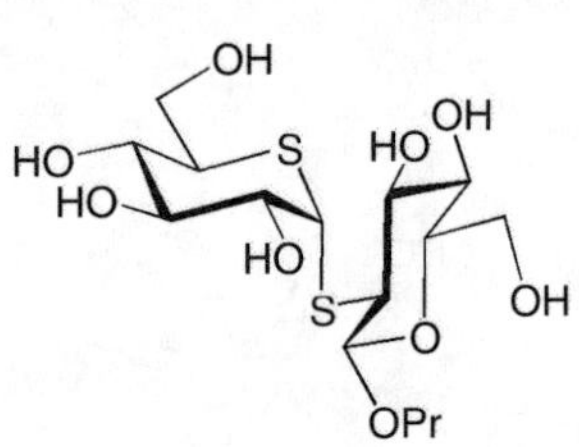

FIGURE 9.12 Glycosides of 5-thiosugars.

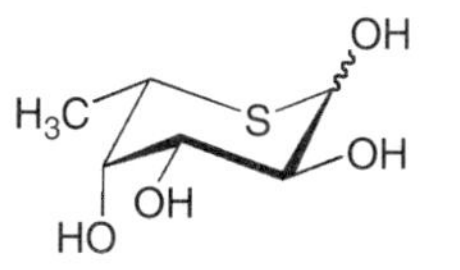

FIGURE 9.13 Examples of powerful glycosidase inhibitors with sulfur or nitrogen in the ring.

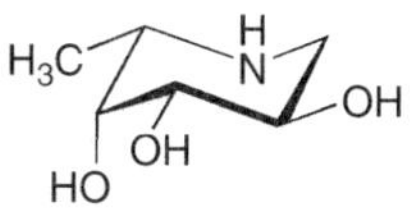

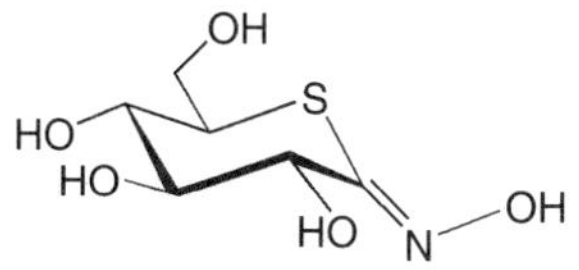

systems containing sulfur, the 6-thiofructopyranose is strongly favored in equilibrium allowing the isolation of this product in over 90% yield.

Hashimoto and coworkers [89] synthesized 5-thio-L-fucopyranose **46** (Figure 9.13), combining the synthetic approach depicted in Figure 9.8 with a reductive thiirane ring opening procedure previously reported in the synthesis of 6-deoxy-5-thio-D-glucose [35]. Compound **46** was a respectable α-L-fucosidase inhibitor with K_i values of 42 and 84 μM against the enzymes from bovine epididymis and kidneys, respectively. For comparison, it is about five orders of magnitude less powerful than 1-deoxy-L-fuconojirimycin **47** [90]. Recently, the same group reported the chemical synthesis of disaccharides containing the 5-thio-α-L-fucopyranosyl moiety [91]. These were found to exhibit inhibitory power comparable with that of **46**.

5-Thio-D-glucohydroximo-1,5-lactone **48** was prepared by Ermert and Vasella [92] as a potential inhibitor of β-glucosidases. However, it was found to be a weaker inhibitor than the oxygen-in-the-ring parent compound.

Non-natural mimics of neuraminic acids that inhibit viral sialidase have emerged as interesting anti-influenza drugs [93]. In this context, 6-thio analogs have been targeted by von Itzstein and coworkers [94] based on the chain extension of suitably protected 3-thio-*N*-acetyl-D-glucosamine derivatives as outlined by Mack and Brossmer [95,96]. Gratifyingly, one of the derivatives with sulfur in the ring, **49**, inhibited the influenza virus sialidase with an IC_{50} value in the low nanomolar range (Figure 9.14).

Finally, disaccharides with sulfur in the ring of the nonreducing sugar and basic nitrogen as interglycosidic linkage were prepared as potent glucoamylase inhibitors [97].

9.2.4.2.2 Other Compounds

The 5-thio analog of *N*-acetyl-D-glucosamine **50** was synthesized from suitably protected furanoid [98–100] or open-chain [101] D-glucosamine derivatives employing the general approach via the thiirane route [102], as well as from 5-thio-D-glucal **51** [103] via azidonitration [104].

Stachel and coworkers [105] prepared 4-thioascorbic acid **52** (Figure 9.15) and could demonstrate that this analog is a stronger reducing agent than the parent compound.

Orally administered 4-cyanophenyl 1,5-dithio-β-D-xylopyranoside **53** was reported to exhibit powerful antithrombotic activity [106]. Consequently, syntheses of several related compounds

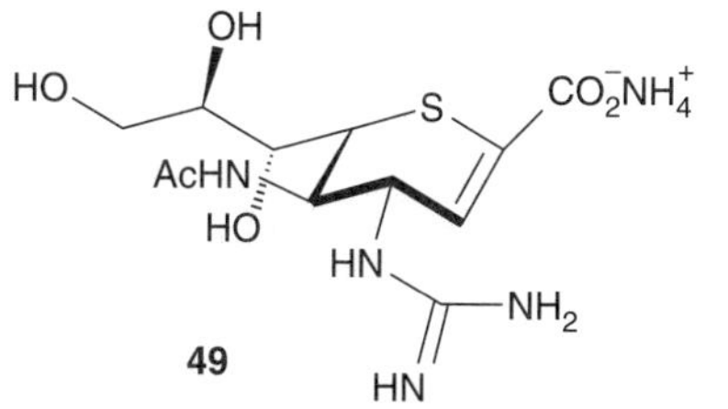

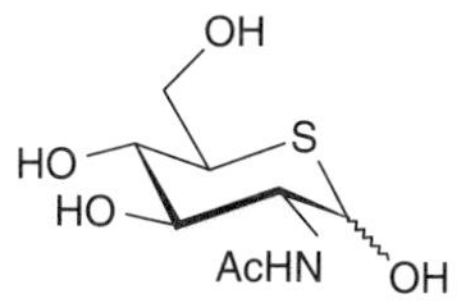

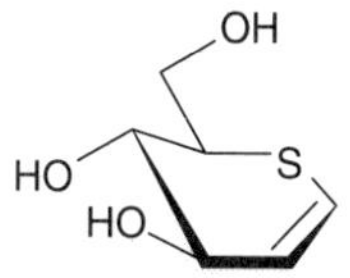

FIGURE 9.14 Thio analogs of biologically relevant sugars and a 5-thio-glycal.

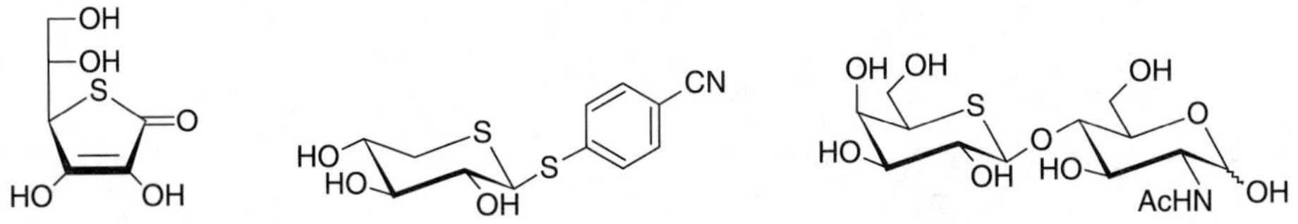

FIGURE 9.15 Examples of thio analogs of biologically relevant sugars.

exhibiting similar properties have been reported since [107–113]. The search for antithrombotic agents led also to the syntheses of *C*-aryl glycosides of 5-thio-D-xylopyranose [114,115].

9.2.4.2.3 Glycosyl Transferase Substrates

Wong and his group [116] investigated the enzymatic transfer of D-galactosyl moieties onto O-6 and O-4 of 5-thio-D-glucose. Successful and high yielding galactosyl transfer onto O-1 of free *N*-acetyl-5-thiogentosamine was reported by Thiem and coworkers [117]. A chemo-enzymatic approach to 5′-thio-*N*-acetyl-lactosamine **54** was performed by Hindsgaul and collaborators [118]. Recently, enzymatic transfer of activated *N*-acetyl-5-thio-D-glucosamine with lactose synthase was achieved [119].

9.2.4.2.4 Nucleosides

Early synthetic approaches to nucleosides with sulfur in the furanose ring were reported by Reist and coworkers [10,15], and Whistler and his group [11,17,120], mainly aiming at synthesizing antibacterial agents [121,122].

Walker and coworkers [123,124] synthesized, among others, 4′-thio derivatives of BVDU [*E*-5-(2-bromoethenyl)-2′-deoxyuridine] such as **55** (Figure 9.16) as antiviral compounds. 4′-Thio-AZT **56** and other thio analogs of antiviral nucleosides were prepared by Secrist [125] and other workers [126] but were subsequently found to be toxic [127]. This was also true for 2′-deoxy-4′-thio analogs of purine nucleosides [128]. Alternatively, the United States Food and Drug Administration has approved the 3′-thia analog **57** of L-cytidine as an antiHIV agent [129]. This was the first example of an L-enantiomer being more potent than the naturally configured compound. 4′-Thioarabino-nucleosides have also been reported to exhibit good activities against herpes and cytomegalo viruses [130].

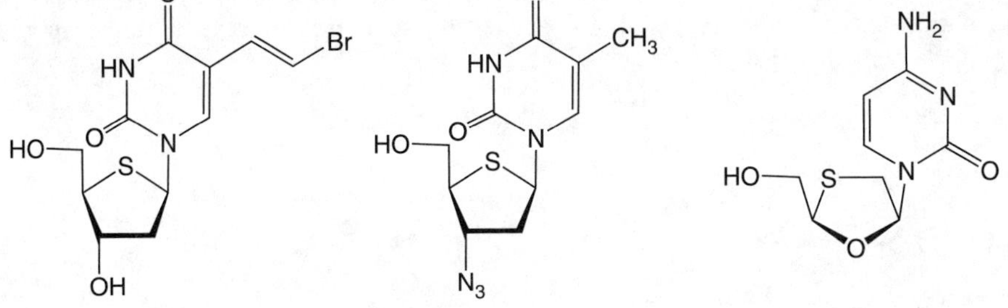

FIGURE 9.16 Thio analogs of biologically active nucleoside antibiotics.

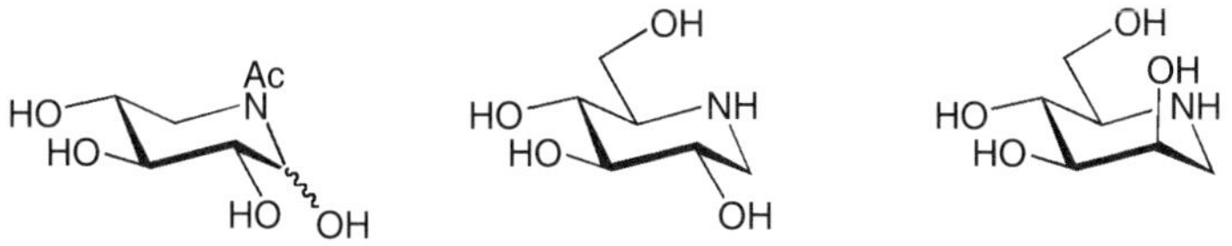

FIGURE 9.17 Early examples of synthetic iminosugars and iminoalditols.

9.3 IMINOSUGARS

Early investigations into carbohydrates with nitrogen instead of oxygen in the ring date back to the 1960s when Paulsen [131], Hanessian and Haskell [132] and JKN Jones and Szarek [133] independently prepared 5-acetamido-5-deoxy-D-xylopyranose **58** (Figure 9.17) as the first representative of this class of compounds. All pentose epimers with unprotected nitrogen in the ring were subsequently prepared by Paulsen and coworkers [134,135]. They studied their physicochemical behaviors and found that such compounds were stable only in neutral to weakly alkaline solutions, whereas other conditions led to stepwise elimination of water ultimately resulting in pyridines.

Subsequently, Paulsen and coworkers [136,137] cyclized 6-amino-6-deoxy-L-sorbose to obtain 1,5-dideoxy-1,5-imino-D-glucitol **2**. Concomitantly, Hanessian [138] synthesized 1,5-dideoxy-1,5-imino-D-mannitol **59** from D-fructose. Compound **2** was also obtained from the reduction [139] of the antibiotic nojirimycin (5-amino-5-deoxy-D-glucopyranose, **1**) [140], a compound isolated from fermentation broth of *Streptomyces* strains by Inouye and coworkers [141], and was coined 1-deoxynojirimycin. About a decade later, a BAYER group isolated compound **2** from fermentation broths of a range of *Bacillus* strains and discovered its powerful inhibitory activity against α-glucosidases [142,143]. This discovery spawned rapidly increasing and widespread interest in synthetic approaches to, and the biological activities of, sugar analogs with nitrogen instead of oxygen in the ring and led to the discovery of a wide range of related natural products between 1975 and the late 1990s. Amid quite a few others, this period saw the discoveries of such important glycosidase inhibitors as 1,4-dideoxy-1,4-imino-D-arabinitol **60** (Figure 9.18) [144–146], 2,5-dideoxy-2,5-imino-D-mannitol **3** [147], swainsonine **61** [148], castanospermine **4** [149] and epimers thereof [150,151]. In addition, australine **5** [152] and alexine **62** [153], the new class of calystegines (for example, **6** [154,155], featuring a *nor*-tropane skeleton) and the

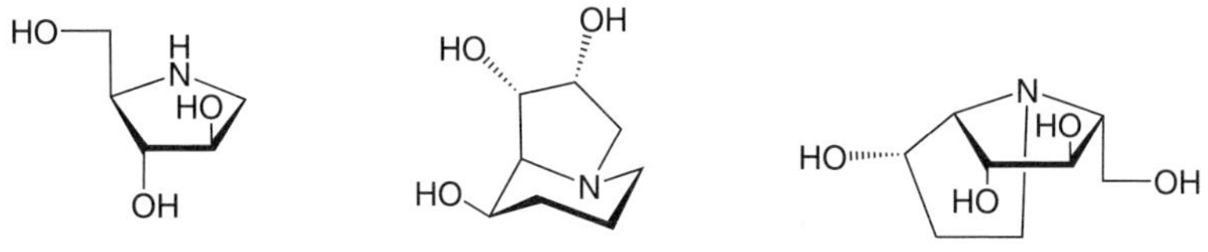

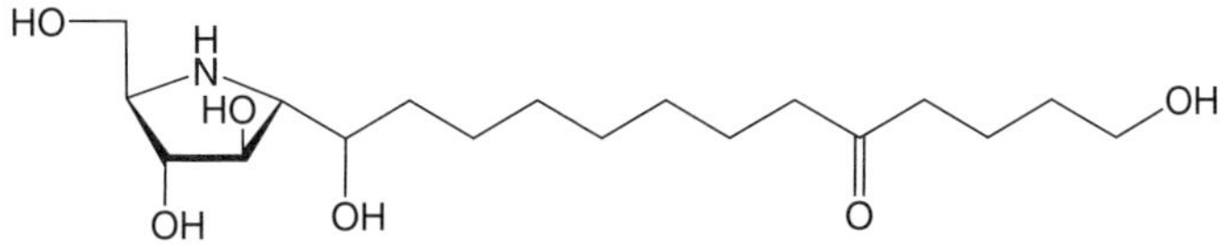

FIGURE 9.18 Examples of biologically active sugar-related natural products with nitrogen in the ring.

broussonetines [156] (a family of chain-extended analogs of compound **3**, with compound **63** as a typical representative) were found.

The large variety of biological activities found in these compounds and many of their derivatives was discovered to be based on their potent glycosidase [58,157], glycosyl transferase [158] and phosphorylase [159] inhibitory activities.

9.3.1 Typical Approaches to Iminosugars and Analogs

Because of the large diversity of iminosugars and relatives, a wide range of synthetic approaches to these structures has been devised. Most synthetically viable methods take advantage of the introduction of the "ring heteroatom" by conventional means followed by ring closure to the respective desired sugar analog.

The most frequently employed routes to iminosugars are based on simple principles, as exemplified in Scheme 9.3. Such approaches all rely on the introduction or liberation of an amino group or its equivalent at an appropriate distance to a carbonyl or a leaving group. Generally, the aldehyde of an aldose, the keto group of a ketose or the carboxyl group of a sugar lactone or ester have been employed, together with a suitably positioned free amine or azide or other amine equivalents, to cyclize in an intramolecular (reductive) amination or lactam formation. In the latter case, a final reductive step is necessary to access the desired iminoalditol. Instead of the carbonyl moiety, a leaving group such as a sulfonate, halide, cyclic

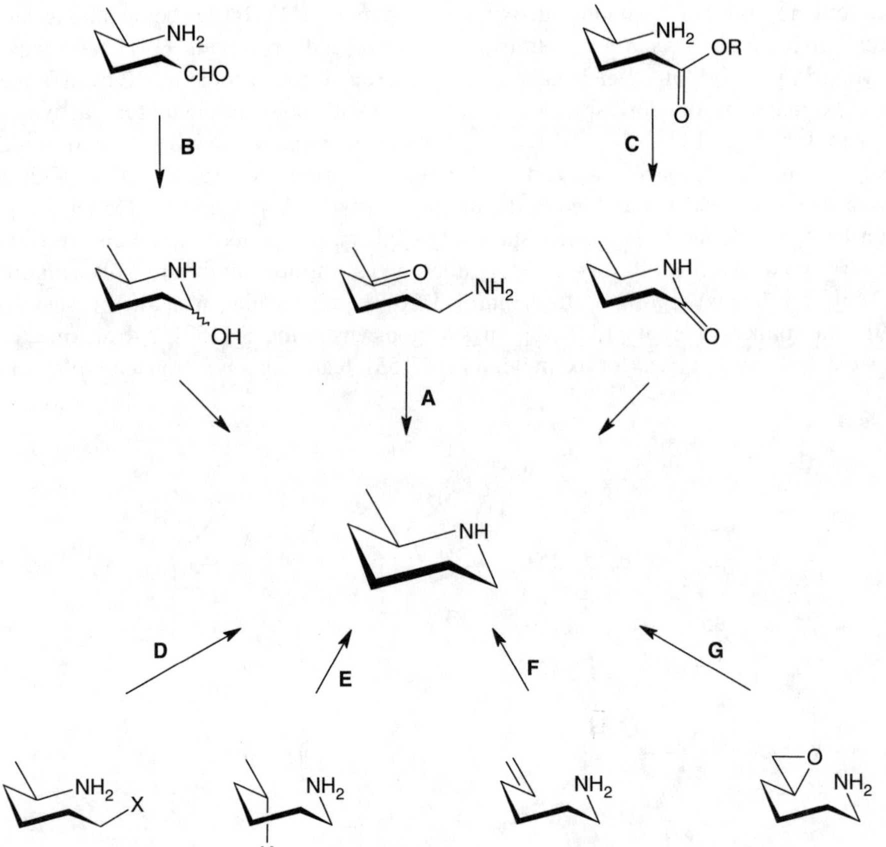

SCHEME 9.3 Typical synthetic approaches to sugar analogs with nitrogen in the ring.

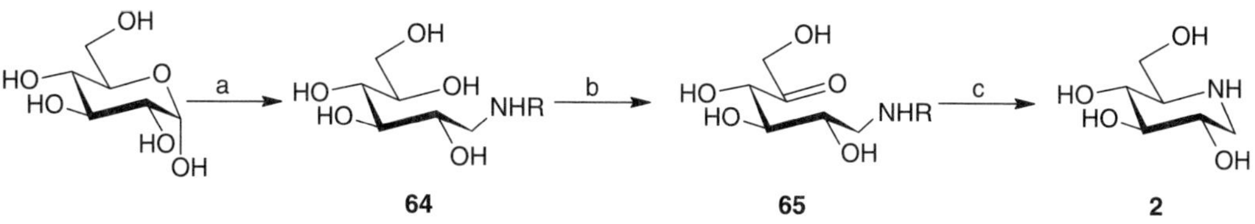

SCHEME 9.4 a: Reductive amination, *N*-protection, for example, formyl halide and b: *Gluconobacter oxidans*; c: H_2, Pd.

sulfate or oxirane, epimine and others may be attacked intramolecularly by the amino group to furnish the targeted heterocycle. Starting materials may stem from the chiral pool (carbohydrates, amino acids) or from *de novo* approaches, which are frequently based on biocatalytic functionalization steps (oxidation and stereoselective aldol addition). Examples are given in the following figures.

The paradigmatic route (Scheme 9.4) to, and basis for, the industrial production of 1-deoxynojirimycin **2** following Approach A was reported by a BAYER group. Reductive amination of D-glucose to furnish **64** was followed by regiospecific microbial oxidation of C-5 employing *Acetobacter oxidans* to give the corresponding 6-aminodeoxy-L-sorbose derivative **65**, which, in turn, was cyclized by intramolecular reductive amination to obtain the desired 1,5-dideoxy-1,5-imino-D-glucitol **2** in only four steps [160].

Employing slightly different protecting group strategies, the same approach based on the chemical modification of L-sorbose was taken, for example, by Demailly and his group (Scheme 9.5) [161] and Behling and coworkers (Scheme 9.6) [162].

De novo variations on the same theme, employing aldolase-based azidodeoxyketose syntheses, have been provided by Whitesides and coworkers [163], Effenberger and colleagues [164] and Wong and his group [165] to mention a few (Scheme 9.7).

An increasingly employed version of Approach A, the condensation of a dicarbonyl sugar with ammonia or a suitable amine, was, amongst others, reported by Reitz and Baxter [166] and is based on double reductive amination of glucose-derived 5-ulose **66** (Scheme 9.8).

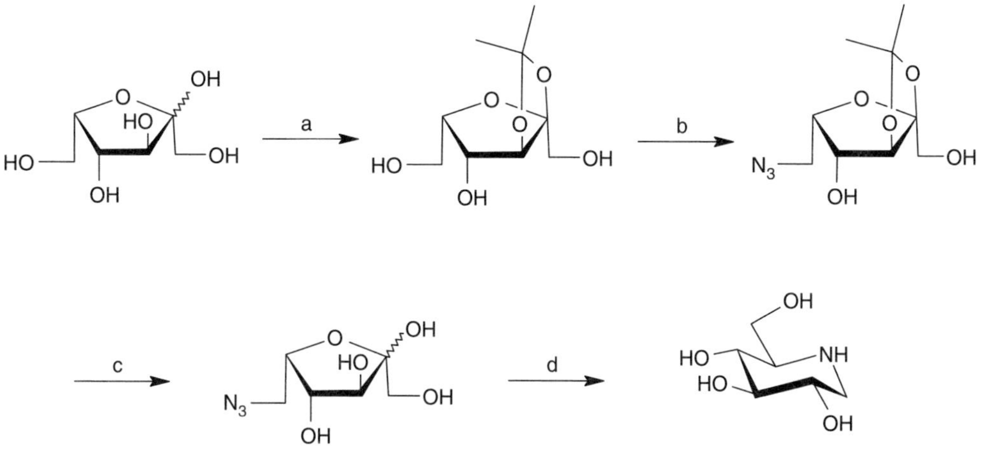

SCHEME 9.5 a: Acetone, H^+; b: Ph_3P, CBr_4, LiN_3, DMF, 120°C, 24 h, c: H_2O, Dowex 50X8-100, 60°C, 4 h and d: H_2, PtO_2, H_2O.

OH
O
HO
a
HO
OH
OH
O
O
HO
b
HO
OH
O
O
O
HO
N_3
OH
O

O
O
O
HO
c
H_2N
OH
O
d
OH
O
HO
H_2N
OH
OH
e
OH
HO
NH
HO
OH

SCHEME 9.6 a: Acetonedimethylacetal, H^+, b: Sulfonyl chloride, pyr., NaN_3, DMF, 100°C, 20 h, c: Ph_3P, THF, 20 h, then H_2O, d: Dowex H^+, H_2O and e: H_2, Pd/C.

O
HO
OPO_3^{2-}
+
O
N_3
H
OH
a
O
CH_2OR
HO
OH
N_3
OH
b
OH
HO
NH
HO
OH

SCHEME 9.7 a: Aldolase (EC 4.1.2.13), pH 6, 12 h, then phosphatase, pH 4.5, 48 h and b: H_2, Pd/C.

HO
HO
O
OH
O
O
a
HO
O
O
OH
O
O
b
O
OH
HO
OH
O
HO
66

OH
CHPh$_2$
c
HO
N
HO
OH
d
OH
HO
NH
HO
OH

SCHEME 9.8 a: Bu_2SnO, MeOH, Br_2, CH_2Cl_2, 0°C, b: Dowex 50, H_2O, c: Ph_2CHNH_2, HOAc, $NaBH_3CN$, MeOH and d: H_2, $Pd(OH)_2$, EtOH.

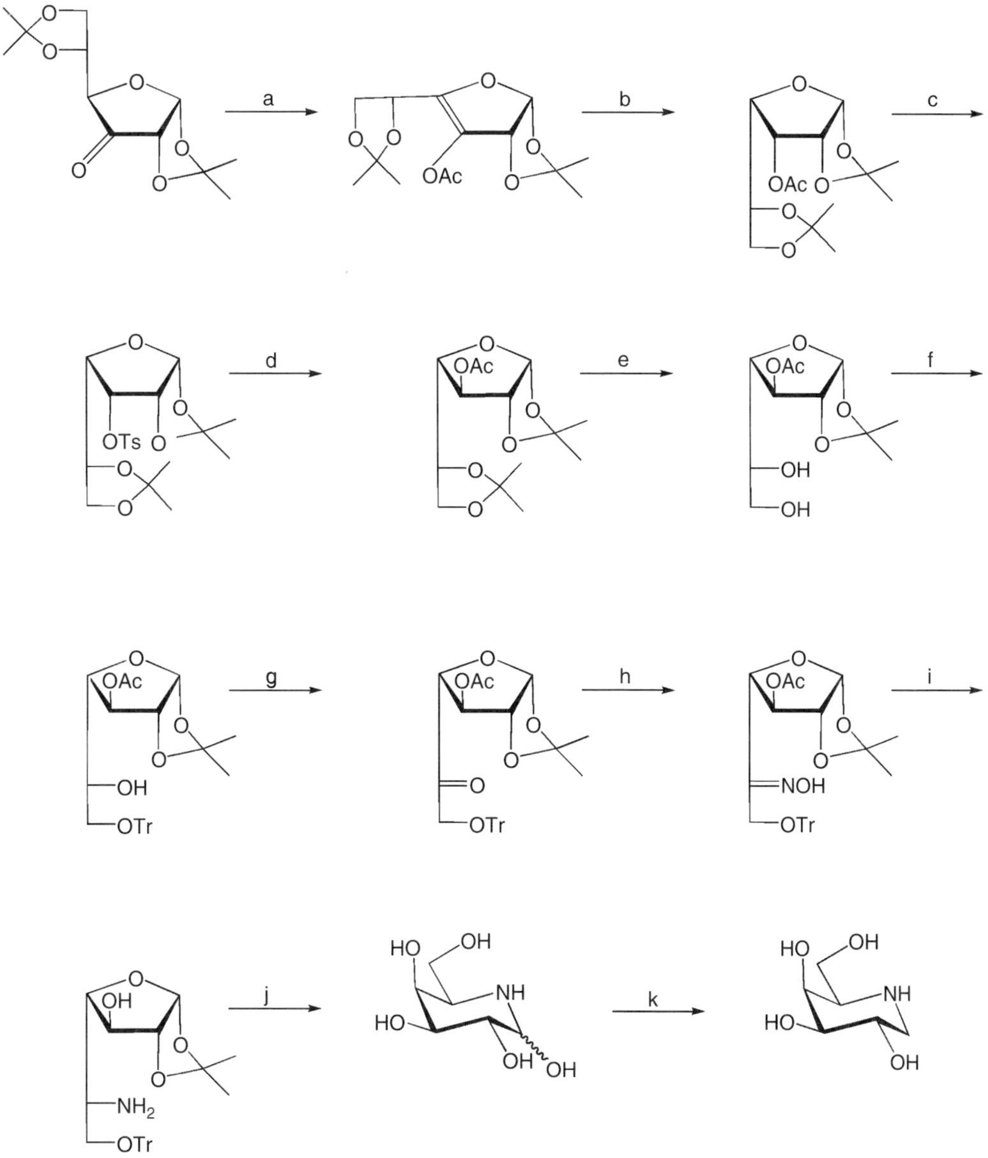

SCHEME 9.9 a: Ac_2O, pyr., b: H_2, $Pd(OH)_2/C$, MeOH, c: MeONa/MeOH, then TsCl, pyr., d: Bu_4NOAc, chlorobenzene, reflux, 5 h, e: HOAc, 5 h, f: TrCl, pyr., g: PDC, Ac_2O, h: NH_2OH, MeOH, i: NaOMe/MeOH, j: SO_2, MeOH, H_2O, 40 h, then $Ba(OH)_2$ and k: H_2, PtO_2, H_2O.

Examples for typical Approach B routes have been provided by Legler and Pohl (Scheme 9.9) in the 1,5-dideoxy-1,5-imino-D-galacto series [167], by Card and Hitz (Scheme 9.10) [168] in their synthesis of compound **3** and by Anzeveno and Creemer [169] who prepared compounds **1** and **2** (Scheme 9.11).

Excellent examples for chiral pool access to various iminoalditols via Approach C were provided, for example, by Lundt and coworkers (Scheme 9.12) [170]. Additionally, an interesting chlorobenzene-based *de novo* access to mannojirimycin was reported by Hudlicky and his group (Scheme 9.13) [171].

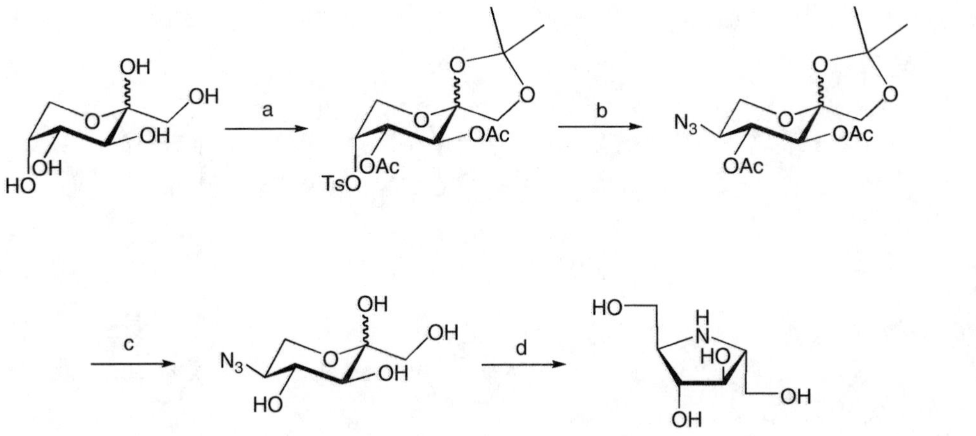

SCHEME 9.10 a: Acetonedimethylacetal, H^+, then TsCl, pyr., then Ac_2O, pyr., b: LiN_3, DMF, 70°C, 72 h, c: H_3O^+ and d: H_2, Pd/C, EtOH.

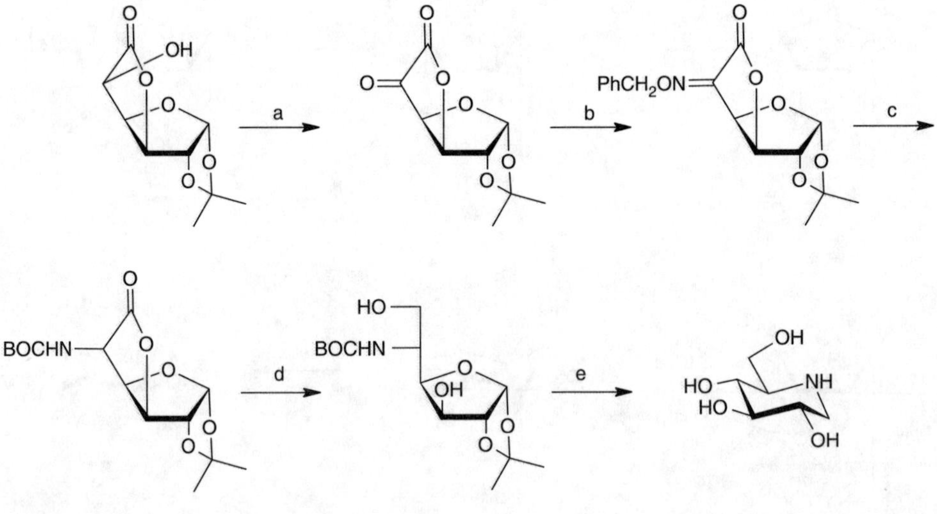

SCHEME 9.11 a: Swern, b: H_2NOBn, benzene, c: H_2, Pd/C, $(BOC)_2O$, EtOAc, d: LAH, THF, 0°C and e: H_2, Raney-Ni.

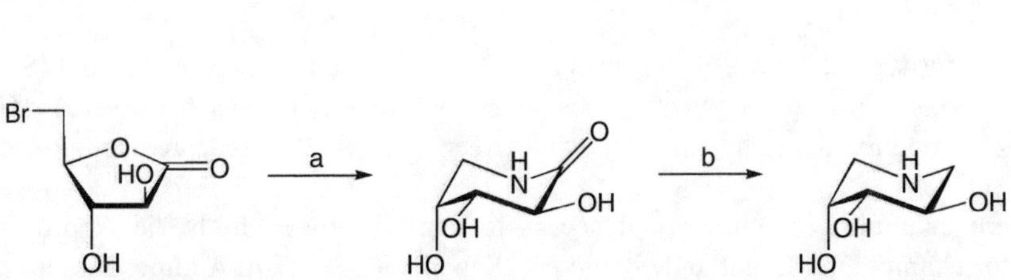

SCHEME 9.12 a: NH_4OH and b: $NaBH_4$, CF_3CO_2H.

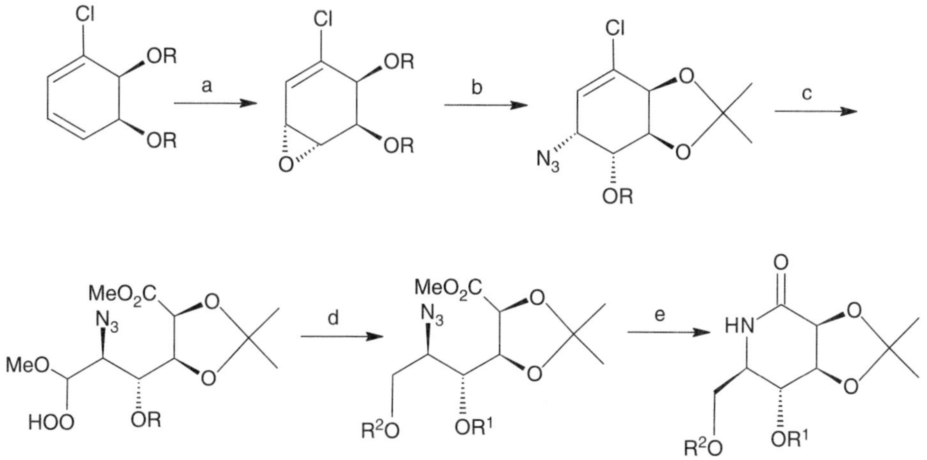

SCHEME 9.13 a: *m*CPBA, b: NaN_3, NH_4Cl, DME, EtOH, H_2O, c: ozonolysis, d: reduction and e: H_2, Pd/C.

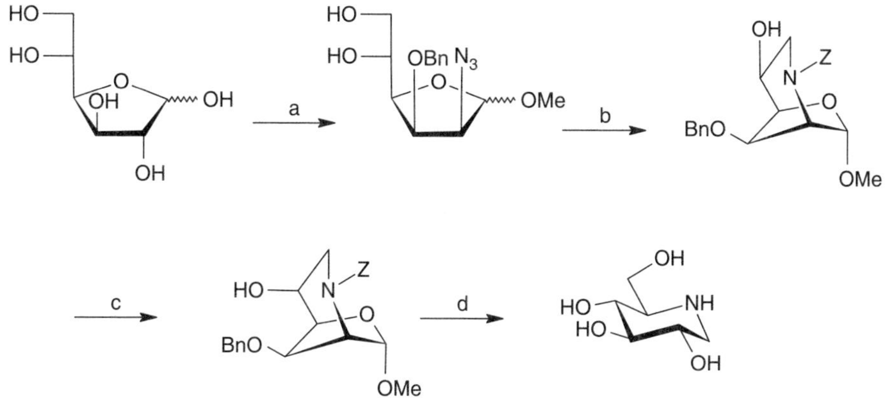

SCHEME 9.14 a: Steps, b: TsCl, pyr., then H_2, Pd, EtOH, 12 h, then BnOCOCl, c: PCC, then $NaBH_4$ and d: H_3O^+, $NaBH_4$, H_2, Pd/C.

Routes D and E have been frequently employed by Fleet and coworkers (Scheme 9.14) [172]. Symmetrical iminoalditols were made in a similar fashion by Duréault (Scheme 9.15) [173].

Examples of F and G Approaches were given by, amongst others, Bernotas and Ganem [174], Martin and coworkers [175] and Le Merrer and collaborators [176] (Schemes 9.16 and 9.17).

Among the approaches to iminosugar ring assembly, ring closing metathesis (RCM) is certainly one of the most interesting methods. By employing differently configured deoxyiodo sugars, a fairly general synthesis of calystegines (exemplified by the route to compound **6**) was conducted by Madsen and coworkers [177] (Scheme 9.18).

9.3.2 Biological Activities and Applications

9.3.2.1 Glycosidase Inhibitory Activities

Glycosidases are enzymes that are ubiquitous in organisms. They catalyze the hydrolysis of glycosidic bonds in saccharides and glycoconjugates to release reducing monosaccharides or small oligosaccharides. Depending on their mode of action, which is strongly influenced by the geometry of the active site, these enzymes are subdivided into retaining and inverting glycosidases.

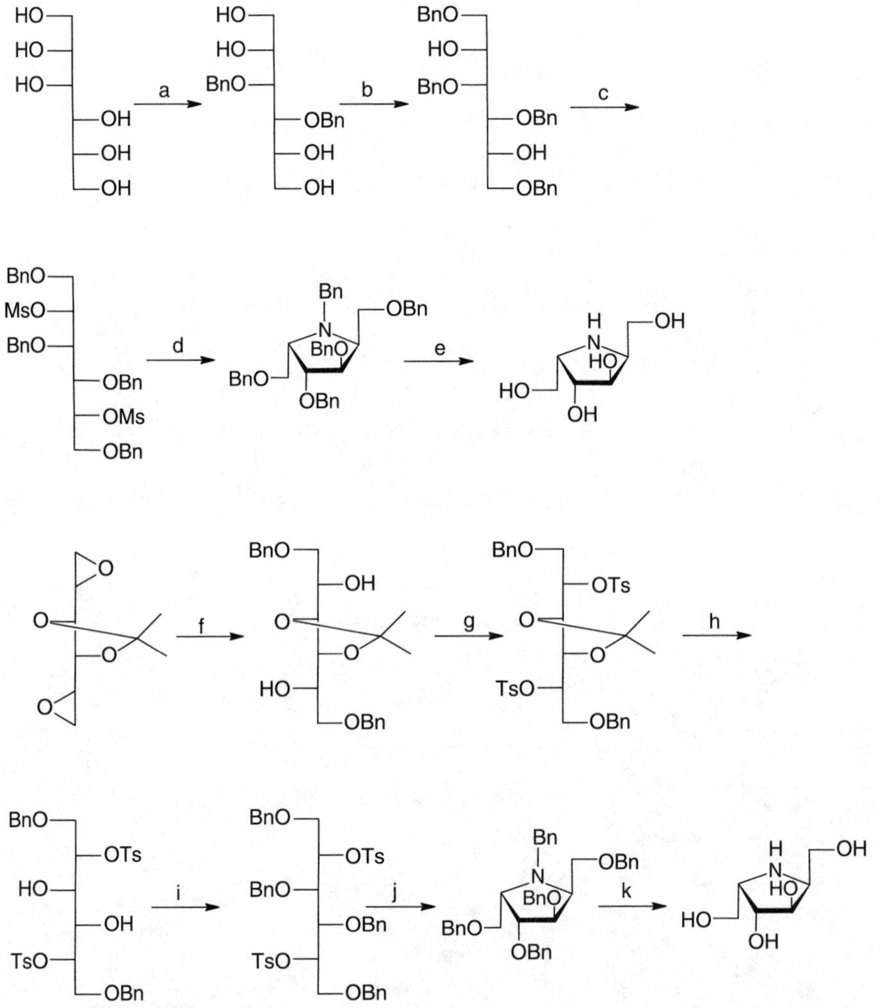

SCHEME 9.15 a: Acetone, H^+, then BnBr, NaH, then H_3O^+, b: Bu_2SnO, tol., then BnBr, Bu_4NI, c: MsCl, Et_3N, DMAP, CH_2Cl_2, d: $BnNH_2$, 120°C, 18 h, e: H_2, Pd, HOAc., f: NaH, BnOH, DMF, g: TsCl, NEt_3, DMAP, CH_2Cl_2, h: TFA, H_2O, i: benzyl trichloroacetimidate, H^+, CH_2Cl_2, j: $BnNH_2$, 120°C, 12 h and k: H_2, Pd, HOAc.

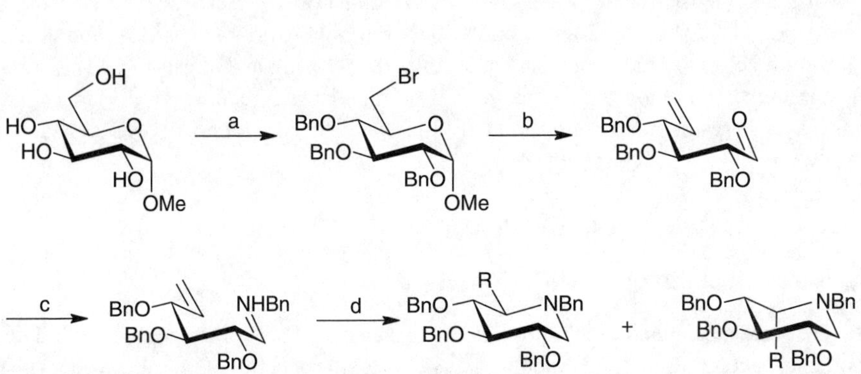

SCHEME 9.16 a: Steps, b: Zn, HOAc, c: $BnNH_2$, $NaBH_3CN$, PrOH, H_2O and d: Hg-trifluoroacetate, THF.

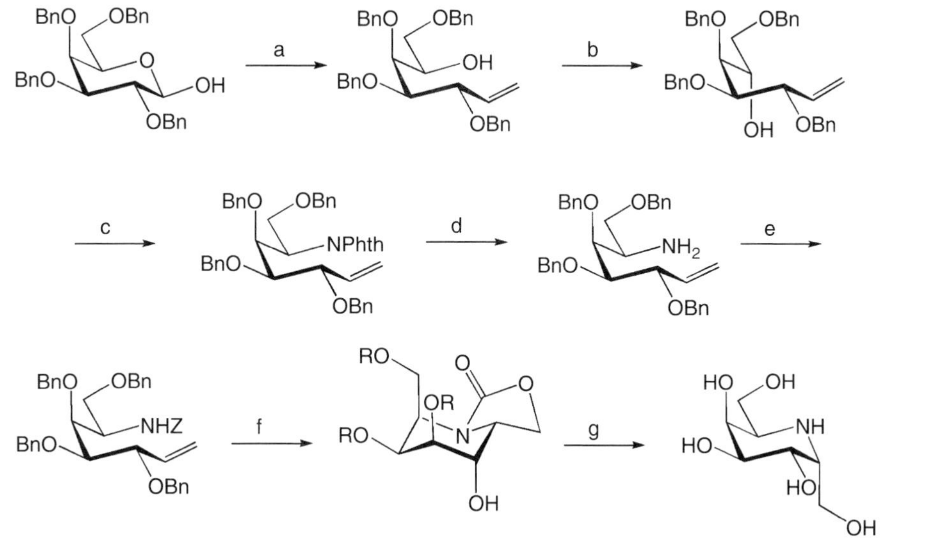

SCHEME 9.17 a: $Ph_3P = CH_2$, tol., b: Mitsunobu, then MeONa/MeOH, c: Phtalimide, Ph_3P, DEAD, d: hydrazine hydrate, MeOH, e: BnOCOCl, THF, f: Hg-trifluoroacetate, I_2, THF, then H_2, Pd/C and g: KOH, MeOH.

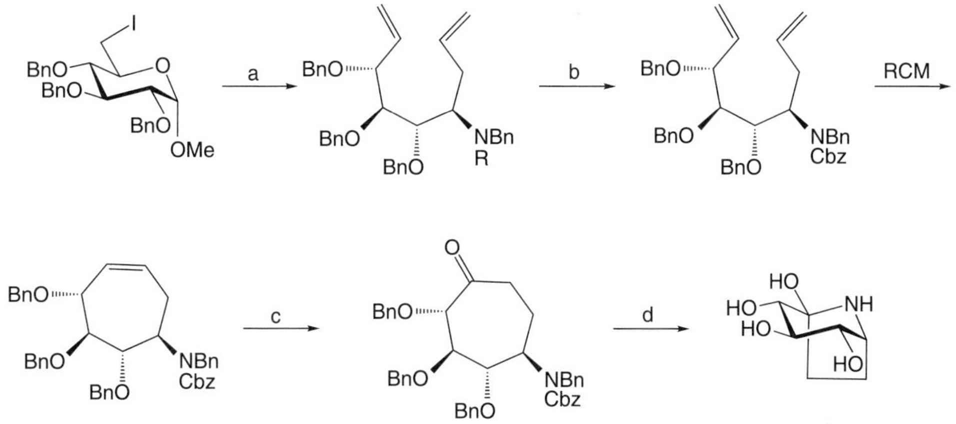

SCHEME 9.18 a: Zn, $BnNH_2$, $CH_2 = CHCH_2Br$, THF, b: CbzCl, $KHCO_3$, EtOAc, H_2O, c: hydroboration, Dess-Martin and d: H_2, Pd/C.

With retaining enzymes, the configurations at the anomeric center of the substrate (glycoside) and the product (free sugar) are the same after two nucleophilic attacks at this carbon, with overall double inversion of configuration (Scheme 9.19).

With inverting glycosidases, the anomeric configuration is changed from α in the substrate to β in the product, or vice versa during the hydrolysis reaction (Scheme 9.20).

Conveniently, glycosyl hydrolases have recently been classified into over 70 families, which are further grouped into "clans" based on their amino acid sequence similarities, reflecting structural as well as mechanistic relationships [178].

Inhibitors of glycosidases can serve as valuable tools and diagnostic agents for the investigation of enzyme mechanisms as well as enzyme deficiency effects and disorders. Sugar analogs with nitrogen instead of oxygen in the ring and structurally related alkaloidal substances have been recognized as powerful, reversible and usually competitive inhibitors of glycosidases [58,157,179].

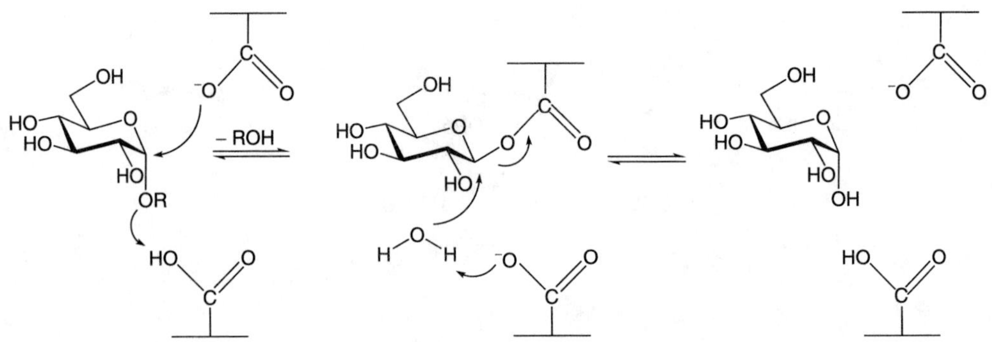

SCHEME 9.19 Mechanism of retaining glycosyl hydrolases.

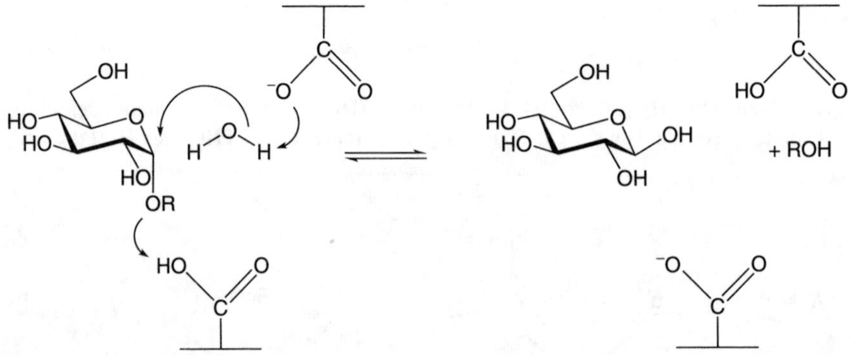

SCHEME 9.20 Mechanism of inverting glycosyl hydrolases.

Their inhibitory effects and potency of inhibition are generally related to their configurations as well as the properties of the ring nitrogen. For example, 1-deoxynojirimycin **2**, 2,5-dideoxy-2,5-imino-D-mannitol **3**, as well as their bicyclic analogs castanospermine **4** and australine **5**, and the alkaloid calystegine B2 **6** exhibit patterns of the hydroxyl groups around the respective carbon backbone, which are related to the D-*gluco* configuration. Incidentally, they have all been found to be powerful inhibitors of D-glucosidases, with K_i values in the micromolar range. The relationship between the arrangement of functional groups and biological activity is not always as apparent as in these cases. For example, swainsonine's **61** resemblance to a mannosyl moiety or mannosyl oxocarbenium ion was not obvious at first sight. Nonetheless, this compound has been found to be a strong inhibitor of a wide range of D-mannosidases [179].

9.3.2.2 Antidiabetic Properties

Following the discovery of 1-deoxynojirimycin **2** as a fermentation product with pronounced inhibitory effects on mammalian α-glucosidases, a novel therapeutic principle for the treatment of certain types of diabetes was postulated. Because of adverse side effects of the parent compound, a wide range of structurally altered analogs (Figure 9.19) have been synthesized and screened.

Of these, **67**, the *N*-hydroxyethyl derivative of compound **2**, exhibited suitable properties and has been marketed against noninsulin-dependent diabetes as Miglitol [180]. Similarly, Emiglitate **68** and the disaccharide mimetic MDL 73945 **69** show promising antidiabetic activities [181]. Other compounds such as 1,4-dideoxy-1,4-imino-D-arabinitol **60** have also been put forward as possible drug candidates in this context [159].

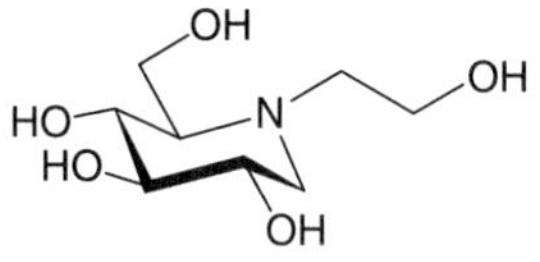

FIGURE 9.19 Antidiabetic iminosugars.

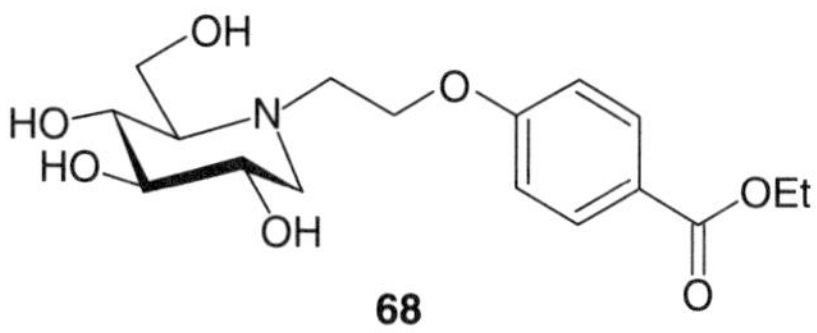

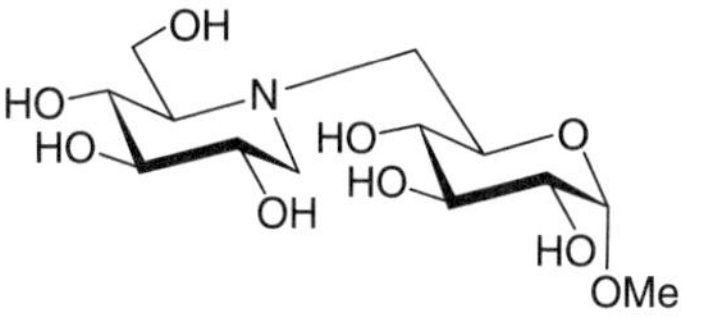

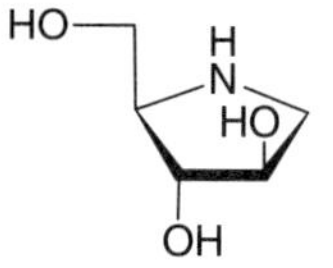

9.3.2.3 Inhibition of Glycoprotein Processing

A relatively small selection of iminosugars has been investigated concerning their effects on glycoprotein trimming glycosidases and, indeed, a few were found to inhibit these glycoprotein processing enzymes [182]. In the case of *N*-linked glycoproteins, the vital and highly ordered cascade of deglycosylation and reglycosylation steps occur on a common precursor oligosaccharide comprising fourteen monosaccharide units attached to an asparagine of the protein (Figure 9.20) [182]. Initial steps include consecutive removal of three glucose residues by glucosidases I and II followed by hydrolysis of mannosyl moieties by a set of mannosidases [182].

The biological activities of iminoalditols have been mainly attributed to interference with these early steps of the glycoprotein trimming process. For example, castanospermine **4**, 1-deoxynojirimycin **2** and some of its *N*-alkylated derivatives such as compounds **70**, **71** and **72** (Figure 9.21) are good inhibitors of both glucosidase I and II. Mannosidases ER Man I and II and Golgi Man IA and IB, all of which are responsible for the hydrolysis of the α1,2-bound mannosides, are inhibited by, for example, 1-deoxymannojirimycin, whereas Golgi Man II (α1,3- and 1,6-selectivity), Man II^{x}, as well as lysosomal mannosidase (α1,2-, α1,3- and also α1,6-) are rather selectively inhibited by swainsonine, **61** [182].

Incidentally, these particular iminosugars and alkaloids have been reported to exhibit a wide variety of anti-infective properties, anticancer and antimetastatic activities and immune modulatory functions (swainsonine), as will be mentioned in the following.

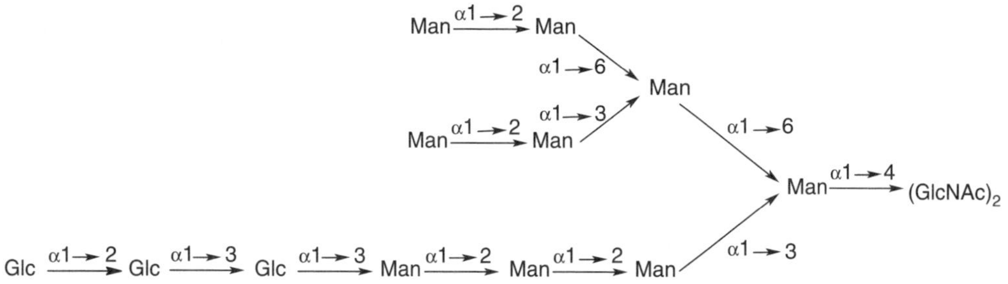

FIGURE 9.20 Substrate oligosaccharide of enzymatic glycoprotein trimming.

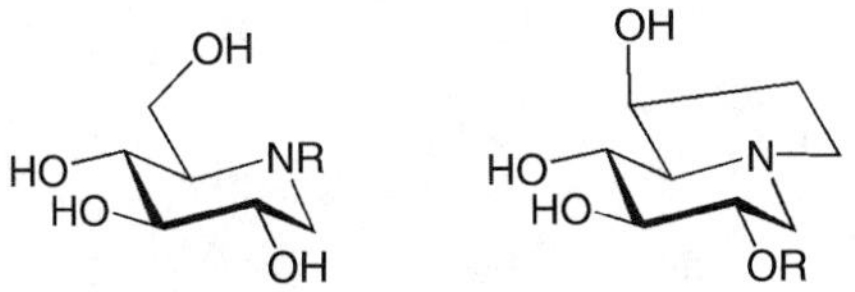

70: R = Me
71: R = Et
72: R = Bu

73: R = Butyryl

FIGURE 9.21 Iminosugars and analogs with antiviral properties

9.3.2.4 Anti-Infective Properties

9.3.2.4.1 Antiviral Activities

Because of their glycosidase inhibitory properties, quite a few iminosugars exhibit antiretroviral activities [183,184]. Amongst these compounds, the α-D-glucosidase and selected α-L-fucosidase inhibitors have gained most long-term attention. The glucosidase inhibitors under consideration have been found to interfere with the initial deglucosylation steps in the glycoprotein trimming process necessary for the maturation of *N*-linked glycoproteins [182] and, ultimately, for their correct folding and functioning.

9.3.2.4.1.1 α-Glucosidase Inhibitors. Glycoprotein gp120 of the HIV virus binds to the CD_4 antigen of T4-lymphocytes and the transmembrane glycoprotein gp41 anchors the envelope to the viral membrane. These glycoproteins play key roles in the early stages of viral infection (absorption, penetration, syncytium formation and spread of virus particles to neighboring cells). Consequently, compounds that can interfere with correct glycoprotein glycosylation can prevent binding and penetration. It has been shown that inhibitors of glucosidase I exhibit interesting antiretroviral activities caused by their interference with the correct glycosylation of viral envelope glycoproteins such as glycoproteins gp120 and gp41 in the case of HIV. Interestingly, inhibitors of α-L-fucosidases also exhibit antiHIV activities. Their mode of action has not been rigorously established to date.

In a Moloney murine leukemia virus assay that served as a model for HIV, it could be demonstrated that glucosidase inhibitors 1-deoxynojirimycin **2** and castanospermine **4** were active at concentrations of 1-2 μg/ml [185].

Of nearly 50 iminosugars probed for antiHIV activity [183], five were active at subcytotoxic levels. In particular, the α-glucosidase inhibitors 1-deoxynojirimycin **2** and the plant alkaloid castanospermine **4** were confirmed to be active at concentrations between 0.1 and 0.5 mg/ml. Even more pronounced were the activities of 1-deoxynojirimycin derivatives with enhanced lipophilic character such as *N*-methyl- **70**, *N*-ethyl- **71** and, in particular, *N*-butyl-1-deoxynojirimycin **72** in the same concentration range [183,186].

Compound **72** as well as the corresponding *N*-nonyl analog were also found to be active against bovine viral diarrhea virus (BVDV) by inducing misfolding of viral envelope proteins [187]. In addition, **72** was reported to inhibit human hepatitis B virus [188] as a novel inhibitor of glycolipid biosynthesis [189].

Interestingly, 1,4-dideoxy-1,4-imino-L-arabinitol **74** (Figure 9.22), an inhibitor [190] of α-glucosidase from yeast (IC_{50} 10 μM at pH 6.8), and other α-glucosidases [191] also exhibited considerable antiretroviral properties in this screen [183]. Yet another five-membered iminosugar exhibiting activity against retroviruses is nectrisine **75**, the aldimine form of 4-amino-4-deoxy-D-arabinose. This compound was initially discovered as a metabolite of the fungus *Nectria lucida* F-4490 and found to stimulate the capacity of immunosuppressed mice to produce antibodies against sheep red blood cells [192]. Subsequently, it was reported to be a potent

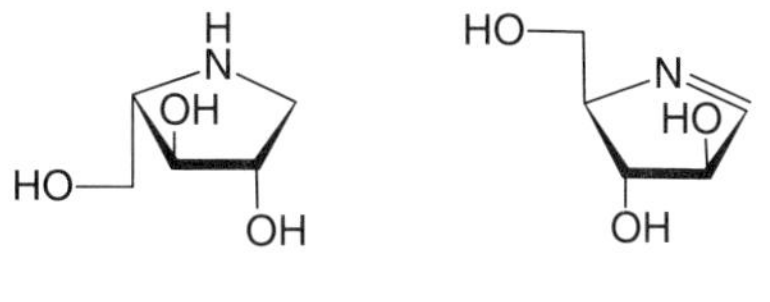

FIGURE 9.22 Iminosugars with antiviral properties.

inhibitor of α- and β-glucosidases and mannosidases with IC_{50} values in the low micromolar range [193]. In particular, α-glucosidase from yeast was inhibited with an IC_{50} value of 40 nM. Nectrisine was found to reduce the cytopathic effect of HIV by a factor of 30 at a concentration of 100 μg/ml.

Interestingly, the antiviral activity of the powerful D-glucosidase inhibitor 2,5-dideoxy-2,5-imino-D-mannitol (DMDP, **3**) was reported to be considerably lower than the values in the piperidine series [183].

The plant alkaloid [194] castanospermine **4** was discovered to inhibit HIV syncytium formation and virus replication [195]. Activity could be improved *in vitro* by a factor of 40, employing the corresponding 6-*O*-butanoyl derivative **73** [196]. Unfortunately, it turned out that the latter did not show any significant benefits *in vivo* [197], pointing to the fact that the beneficial butanoyl group is cleaved by nonspecific hydrolysis by esterases or lipases before the compound reaches the desired place of antiviral action. Castanospermine was also found to be active against the measles virus [198] by quantitative reduction of infectious measles particles in infected cells and by inhibition of syncytium formation. Both effects were attributed to aberrant folding of viral glycoproteins. The inhibitor also interferes with the expression of influenza virus, neuraminidase, on the cell surface [199] and with the correct folding of dengue virus envelope glycoprotein [200]. Furthermore, castanospermine reduced human cytomegalovirus infectivity in human embryo fibroblast cell cultures [201]. The 6-*O*-butanoyl derivative **73** blocked the growth of herpes simplex virus type-1 (strain SC16) [202] and type-2 at an IC_{50} level of 100 μM with de-acylated parent compound **4** found to be the predominant agent in the treated cells [203].

9.3.2.4.1.2 α-L-Fucosidase Inhibitors. Whereas the powerful α-L-fucosidase inhibitor 1,5-dideoxy-1,5-imino-L-fucitol **47** (1-deoxy-L-fuconojirimycin, K_i 0.04 nM) [90] did not show appreciable antiHIV activity, the corresponding *N*-methoxycarbonylpentyl derivative **76** (Figure 9.23) was found to be effective in the same range of activities as 1-deoxynojirimycin derivatives and also castanospermine **4** [183].

9.3.2.4.1.3 Neuraminidase Inhibitors — Siastatin B and Derivatives. Siastatin B, which can be regarded as a glycosyl acetamide of a branched chain 5-amino-5-deoxypentose, was isolated from a *Streptomyces* culture [204] and was subsequently shown, by total synthesis, to be **77** (Figure 9.24) [205].

The inhibition of bacterial *N*-acetylneuraminidases by this compound [206] made **77** an interesting lead compound and triggered the syntheses of a wide range of analogs, such as deoxy, dehydro and *N*-alkylated derivatives. These investigations were supported by molecular modeling calculations. Many derivatives exhibited similar or slightly superior activities when compared with

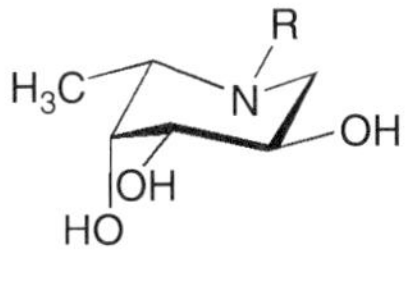

FIGURE 9.23 Antiviral L-fucosidase inhibitor.

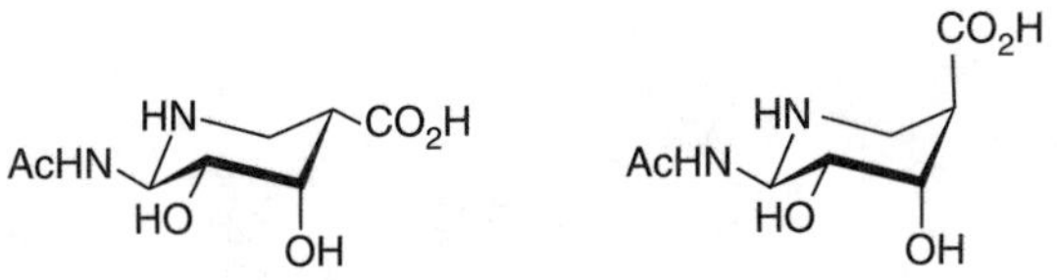

FIGURE 9.24 Antiviral neuraminidase inhibitors.

the parent compound [207]. Whereas quite a few of these substances inhibit only bacterial enzymes, 3-episiastatin B **78** was demonstrated to be a good inhibitor of influenza virus neuraminidases, thus reducing viral infectivity *in vitro* [208].

9.3.2.4.2 Compounds with (Potential) Antibacterial Activities

The very first iminosugar found as a natural product, 5-amino-5-deoxy-D-glucose (nojirimycin, **1**), exhibited antibacterial activities against *Xanthomonas oryzae*, *Shigella flexneri*, *Mycobacterium* 607 (the causative agent of bacterial diarrhea) and other bacteria [139]. Unfortunately, the chemical and biological stability of this compound was found to be too low to be taken advantage of as a therapeutic agent. The stable 1-deoxy analog, 1,5-dideoxy-1,5-imino-D-glucitol (1-deoxynojirimycin, **2**) does not exhibit appreciable antibacterial activity pointing to the fact that the reactive aldimine moiety is a key feature for the antibiotic properties of compound **1**.

Because of the fact that bacteria are quite demanding and variable targets of anti-infective chemotherapy, only a very limited number of iminosugars inhibiting bacterial enzyme systems *in vitro* or exhibiting established antibacterial properties have been discovered since. Nonetheless, several iminosugar-related structures featuring antibacterial properties have been discovered as natural products or have been recently synthesized.

9.3.2.4.2.1 UDP-Gal Mutase Inhibitors. Bacterial and parasitic cell walls, such as the ones of *Mycobacterium tuberculosis* or *Trypanosoma cruzi*, contain D-galactofuranosyl oligomers. These galactofuranose moieties result from the action of a specific enzyme, UDP-galactofuranose mutase, on UDP-galactopyranose and subsequent oligomerization by galactofuranosyl transferases (Figure 9.25). UDP-galactopyranose mutase is not found in humans but it is essential to the micro-organisms under consideration. Consequently, it has emerged as a viable target for novel

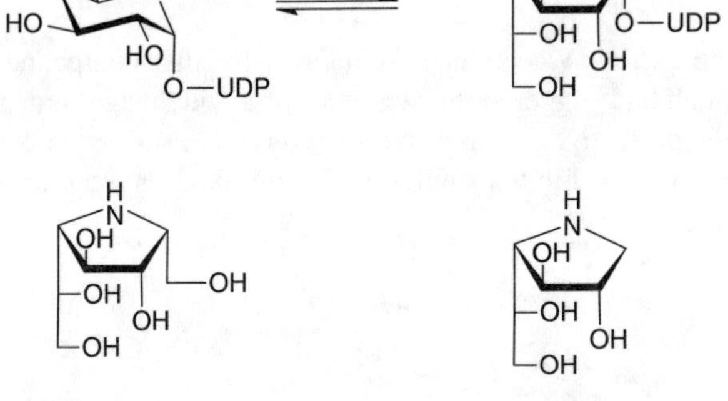

FIGURE 9.25 UDP-galactose mutase inhibiting iminoalditols with antibacterial properties.

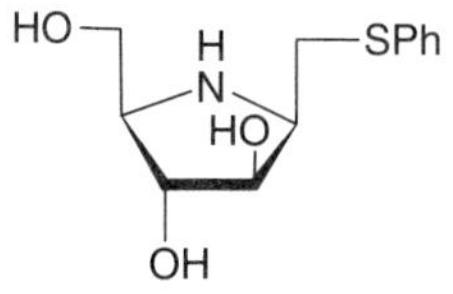

FIGURE 9.26 Inhibitors of bacterial cell wall biosynthesis.

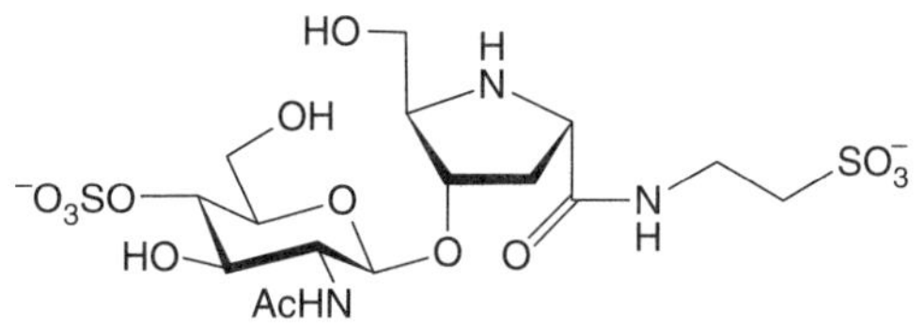

approaches to antibacterial and antiparasitic chemotherapy. A crystal structure of the corresponding *E. coli* enzyme was recently solved to a 2.4 Å resolution and was demonstrated to be a novel structural type [209].

Compounds **79** and **80**, the first inhibitors (inhibiting in the range of 60 to 80% at a concentration of 200 μg/ml) of this enzyme and, consequently, of mycobacterial galactan biosynthesis were reported by Fleet and coworkers. They suggested this approach as a potential new chemotherapeutic strategy against tuberculosis and leprosy [210]. Following this approach, combinatorial library generation of related and structurally extended galactofuranose mimics was subsequently reported by the same group [211]. Because the mechanism of the enzymatic pyranose-furanose interconversion will certainly be discovered in the near future, molecular modeling aided approaches to therapeutically useful inhibitors will be feasible in due course.

9.3.2.4.2.2 Inhibitors of Bacterial Cell Wall Biosynthesis. Interestingly, iminoalditol **81** (Figure 9.26) was found active against the *Mycobacterium avium* complex in an infected macrophage model at 4 μg/ml [212]. Furthermore, this compound increased tumor necrosis factor α (TNFα) production in the infected cells.

Bulgecins, for example, bulgecin A **82**, were reported in 1982 for the first time [213]. Their structures were established soon after [214] and they were found to interfere with the cell-wall synthesis of gram-negative bacteria due to a unique mechanism inhibiting soluble lytic transglycosylase (SLT) [215]. As this enzyme does not exist in mammals, potential specific inhibitors could serve as nontoxic antibiotics with no side effects against the gram-negative pathogenic organisms under consideration.

In the piperidine series, apart from the above-mentioned antibiotic nojirimycin **1**, siastatin **77** and derivatives thereof exhibit activity against bacterial neuraminidases, such as the enzymes from *Clostridium perfrigens*, with an IC_{50} of 18 μg/ml, and *Streptococcus* sp. with 6.3 μg/ml [206,207].

9.3.2.4.3 Compounds with Antifungal and/or Antiprotozoan Activities

A simple, benzyl substituted pyrrolidine, anisomycin **83** (Figure 9.27) from *Streptomyces griseolus* [216], was found active against pathogenic fungi as well as protozoa. *In vitro*, anisomycin inhibited *Trichomonas vaginalis*, *T. foetus*, *Endamoeba histolytica*, and *Candida albicans* at 3.12, 1.56 to 3.12, 1.56 and 1.56 to 12.5 μg/ml, respectively. Other pathogenic fungi and bacteria required 100 μg/ml or more. Anisomycin compared favorably in activity with 18 other antiprotozoan agents [217].

Iminosugar related nucleosides coined immucillins have recently been found to be powerful inhibitors of parasitic nucleoside hydrolases [218]. Compounds such as **84** (Figure 9.28) bind and inhibit protozoan nucleoside hydrolases with dissociation constants in the low nanomolar to picomolar range, thus promising to be antiprotozoan chemotherapeutics against parasite infections such as malaria, sleeping sickness and trypanosomiasis.

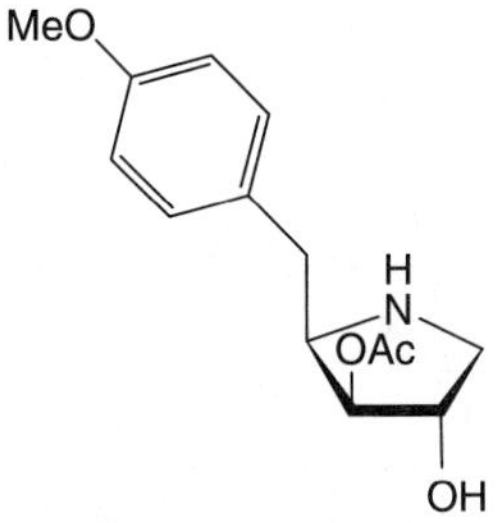

FIGURE 9.27 Antifungal natural product anisomycin.

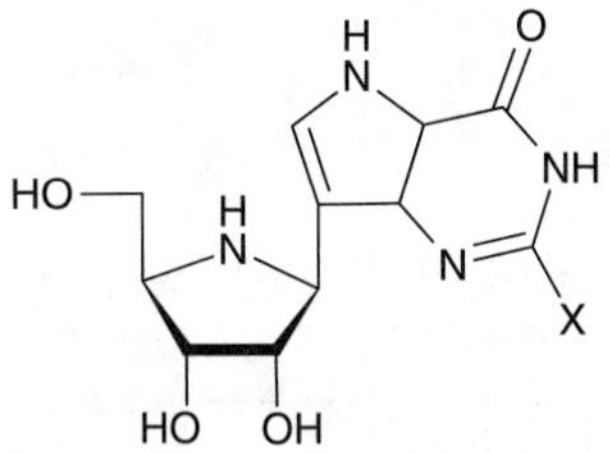

FIGURE 9.28 Potent nucleoside hydrolase inhibitors with antiprotozoan activities.

1-Deoxyfuconojirimycin **47** and close analogs thereof were recently found to interfere with fucosyl transferases of the parasite *Schistosoma mansoni* [219].

Plasmodium falciparum is the causative agent of cerebral malaria, which is caused by cytoadherence of infected erythrocytes to the venular endothelium in brain. Glycoprotein processing inhibitor castanospermine **4** reduced this "docking" in an *in vitro* model, employing human melanoma cells in an IC_{50} range of 0.6 to 0.7 mM. The corresponding 6-*O*-butanoyl derivative **73** was nearly two orders of magnitude more effective ($IC_{50} = 9$ μM) [220].

9.3.2.4.4 Others

9.3.2.4.4.1 Plant Growth Reducing Activity. Interestingly, iminosugars can regulate plant growth. For example, castanospermine has been found to inhibit root elongation in plants. Whereas with dicotyledons a growth reduction of 50% was observed at a concentration of 300 ppb, monocotyledons required a dose of 200 ppm for the same effect. Conversely, no such activity was found with swainsonine, **61** [221].

9.3.2.4.4.2 Nematicidal Properties. Crop damage by plant parasitic nematodes is estimated to amount to around 20% of the world's crop production causing serious economic damage. Experiments with 2,5-dideoxy-2,5-imino-D-mannitol (DMDP **3**) have revealed that potato cyst nematodes were considerably damaged in the presence of this compound. Its potential as a foliar spray, soil drench and seed coating has been suggested. Not clear, as yet, is DMDP's mode of action in this context [222].

9.3.2.4.4.3 Insect Antifeedant Activity. It has been recognized that iminosugars such as 1-deoxynojirimycin **2**, homonojirimycin **85** (Figure 9.29) and deoxymannojirimycin **59** markedly affect the maturation and feeding behavior of a number of economically significant pest insects,

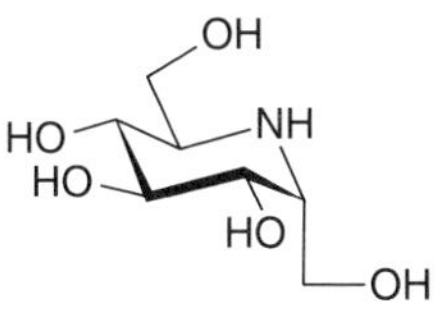

85

FIGURE 9.29 Insect antifeedant derivative of 1-deoxynojirimycin.

such as *Locusta migratoria* (migratory locust), *Schistocera gregaria* (desert locust), *Heliothis virescens* (tobacco budworm), *Tribolium confusum* (flour beetle), *Helicoverpa armigera* (cotton bollworm) and *Myzus persicae* (green peach aphid) [223]. In particular, the plant product 2,5-dideoxy-2,5-imino-D-mannitol (DMDP, **3**) showed a wide range of effects on insects.

9.4 OTHER HETEROATOMS IN THE RING

Reports on sugars analogs with heteroatoms other than sulfur and nitrogen are scattered in the literature but are relatively rare. Of the Group 16 (formerly Group VI) elements, sugar analogs with selenium in the ring have been reported by van Es and Whistler as early as 1967 [224]. Furanoid as well as pyranoid pentose derivatives were prepared by van Es and coworkers [225]. Other examples of selenosugars include the selenoderivative of salacinol **40** reported by Pinto and coworkers [64] and a range of selenopentoses [226].

A recent example for the preparation of tellurosugars was provided by Schiesser and coworkers [227].

Phosphosugars attracted interest for potential biochemical applications but have not met with expectations. Because of the lack of stability of the phosphane oxidation state (+3), the phosphane oxide is usually preferred in phosphosugar syntheses. The literature on phosphosugars until 1984 was reviewed by Yamamoto and Inokawa [228].

A typical synthetic approach based on the 1,4-addition of the phosphorus nucleophile to a sugar derived nitromethane condensation product is depicted in Scheme 9.21.

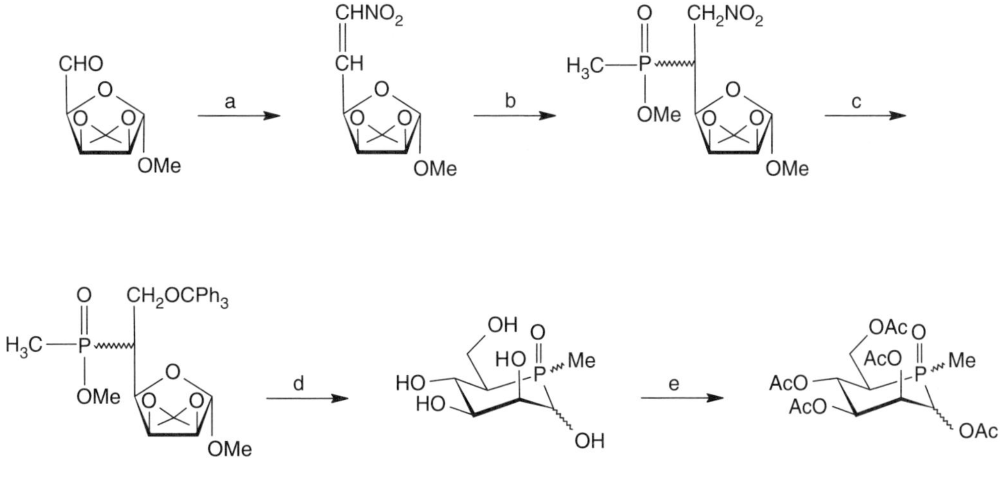

SCHEME 9.21 a: $MeNO_2$, then Ac_2O, NaOAc, b: MePh(=O)OMe, benzene, 80°C, c: H_2, PtO_2, then $NaNO_2$, HOAc, d: TrCl, pyr., e: sodium dihydrobis-(2-methoxyethoxy)aluminate, then HCl, EtOH and f: Ac_2O, pyr.

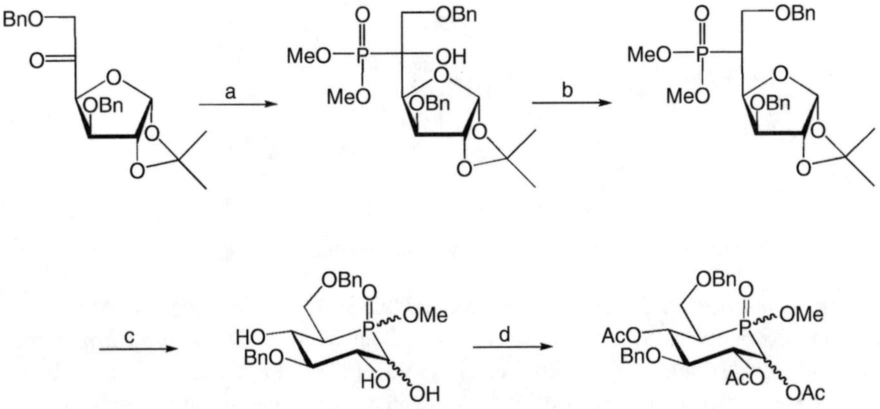

SCHEME 9.22 a: $O{=}PH(MeO)_2$, DBU, b: MeOCOCOCl, DMAP, MeCN, then Bu_3SnH, AIBN, tol., c: sodium dihydrobis-(2-methoxyethoxy)aluminate, HCl, H_2O_2 and d: Ac_2O, pyr., then CH_2N_2.

A variation is the thiophosphaneoxide reported by Yamamoto in 1987 [229]. The first D-mannopyranose with phosphorus in the ring was reported in 1989 [230] and an example of a fucose analog was provided in 1993 [231]. More recently, a new access to D-gluco and L-idopyranose analogs was reported starting from a suitable protected 5-ulose-derivative available from D-glucose as outlined in Scheme 9.22 [232]. Unfortunately, none of these sugar analogs were found to exhibit the interesting biological activities provided by iminosugars and carbohydrate analogs with sulfur in the ring.

9.5 FURTHER READING

Considering the size and number of diverse areas covered in this chapter, it was not possible to aim for comprehensiveness. Consequently, to fill the gaps between somewhat arbitrarily chosen examples, additional reading will guide the interested reader towards more specialized information for deeper insight.

An excellent review on thiosugars and analogs referring to well over 400 references has recently appeared [233]. A guiding selection of most significant contributions from leading laboratories and experts in the field of glycosidases and glycosidase inhibitors, including synthetic approaches, has been provided by Bols in 2002 [179a] and a collection of interesting aspects has also been described by Stütz [179b]. A review on 2,5-dideoxy-2,5-imino-D-mannitol (DMDP, **3**) and relatives has appeared very recently [234].

REFERENCES

1. Grimshaw, C E, Whistler, R L, Cleland, W W, Ring opening and closing rates for thiosugars, *J. Am. Chem. Soc.*, 101, 1521–1532, 1979.
2. Schwartz, J C P, Yule, K C, D-Xylothiapyranose: a sugar with sulfur in the ring, *Proc. Chem. Soc.*, 417, 1961.
3. Adley, T J, Owen, L N, Thio sugars with sulfur in the ring, *Proc. Chem. Soc.*, 418, 1961.
4. Whistler, R L, van Es, T, Solvolysis of methyl 5-thio-D-xylopyranosides and 2,3,4-tri-*O*-acetyl-5-thio-α-D-xylopyranosyl bromide, *J. Org. Chem.*, 28, 2303–2304, 1963.
5. Paulsen, H, Todt, K, Cyclic monosaccharides having nitrogen or sulfur in the ring, *Adv. Carbohydr. Chem. Biochem.*, 23, 115–232, 1968.

6. Lambert, J B, Wharry, S M, Conformational analysis of 5-thio-D-glucose, *J. Org. Chem.*, 46, 3193–3196, 1981.
7. Ingles, D L, Whistler, R L, Preparation of several methyl 5-thio-D-pentopyranosides, *J. Org. Chem.*, 27, 3896–3898, 1962.
8. Feather, M S, Whistler, R L, Derivatives of 5-deoxy-5-mercapto-D-glucose, *Tetrahedron Lett.*, 3, 667–668, 1962.
9. Reist, E J, Gueffroy, D E, Goodman, L, 4-Thio-L-ribose: a thiofuranose sugar, *J. Am. Chem. Soc.*, 85, 3715, 1963.
10. Reist, E J, Gueffroy, D E, Goodman, L, Synthesis of 4-Thio-D- and -L-ribofuranose and the corresponding adenine nucleosides, *J. Am. Chem. Soc.*, 86, 5658–5663, 1964.
11. Urbas, B, Whistler, R L, Synthesis of purine and pyrimidine nucleosides of thiopentoses, *J. Org. Chem.*, 31, 813–816, 1966.
12. Whistler, R L, Dick, W E, Ingle, T R, Rowell, R M, Urbas, B, Methyl 4-deoxy-4-mercapto-D-ribofuranoside, *J. Org. Chem.*, 29, 3723–3725, 1964.
13. Bellon, L, Barascut, J-L, Imbach, J-L, Efficient synthesis of 4-thio-D-ribofuranose and some 4′-thioribonucleosides, *Nucleos. Nucleot.*, 11, 1467–1479, 1992.
14. Leydier, C, Bellon, L, Barascut, J-L, Deydier, J, Maury, G, Pelicano, H, El Alaoui, M A, Imbach, J-L, A new synthesis of some 4′-thio-D-ribonucleosides and preliminary enzymatic evaluation, *Nucleos. Nucleot.*, 13, 2035–2050, 1994.
15. Reist, E J, Fisher, L V, Goodman, L, Thio Sugars. Synthesis of the adenine nucleosides of 4-thio-D-xylose and 4-thio-D-arabinose, *J. Org. Chem.*, 33, 189–192, 1968.
16. Whistler, R L, Nayak, U G, Perkins, A W Jr, Anomeric methyl 4-thio-D-arabinofuranosides, *J. Org. Chem.*, 35, 519–521, 1970.
17. Whistler, R L, Doner, L W, Nayak, U G, 4-Thio-D-arabinofuranosylpyrimidine nucleosides, *J. Org. Chem.*, 36, 108–110, 1971.
18. Varela, O, Zunszain, P A, First synthesis of aldopentono-1,4-thiolactones, *J. Org. Chem.*, 58, 7860–7864, 1993.
19. Altenbach, H-J, Merhof, G F, Synthesis of 1-Deoxy-4-thio-L-ribose starting from D-arabinitol, *Tetrahedron: Asymmetry*, 7, 3087–3090, 1996.
20. Altenbach, H-J, Brauer, D J, Merhof, G F, Synthesis of 1-deoxy-4-thio-D-ribose starting from thiophene-2-carboxylic acid, *Tetrahedron*, 53, 6019–6026, 1997.
21. Yuasa, H, Kajimoto, T, Wong, C-H, Synthesis of iminothiasugar as a potential transition-state analog inhibitor of glycosyltransfer reactions, *Tetrahedron Lett.*, 35, 8243–8246, 1994.
22. Dominguez, J N, Owen, L N, Approaches to the synthesis of sugar thiolactones, *Carbohydr. Res.*, 75, 101–107, 1979.
23. Vegh, L, Hardegger, E, 4-Thioglucose, *Helv. Chim. Acta*, 56, 2020–2025, 1973.
24. Shah, R H, Bose, J L, Bahl, O P, Synthesis of 4-thio-D-mannose, *Carbohydr. Res.*, 77, 107–115, 1979.
25. Varela, O, Cicero, D, de Lederkremer, R M, A convenient synthesis of 4-thio-D-galactofuranose, *J. Org. Chem.*, 54, 1884–1890, 1989.
26. Owen, L N, Ragg, P L, Thio-sugars. Part II. The thiofuranose ring, *J. Chem. Soc. C*, 1291–1296, 1966.
27. Boigegrain, R A, Gross, B, Syntheses in the 6-deoxy-4-thiohexose series. II. Synthesis of methyl 3,4,6-trideoxy-3,4-epithio-2-*O*-methylsulfonyl-α-D-allopyranoside and 1,2,3-tri-*O*-acetyl-4-*S*-acetyl-6-deoxy-4-thio-α- and β-D-gulopyranosides, *Carbohydr. Res.*, 41, 135–142, 1975.
28. Gross, B, Oriez, F X, Synthesis for the 6-deoxy-4-thio-D-altrose and -D-idose series, *Carbohydr. Res.*, 36, 385–391, 1974.
29. Cicero, D, Varela, O, de Lederkremer, R M, Synthesis of furanoid and pyranoid derivatives of 6-deoxy-4-thio-D-galactose, *Tetrahedron*, 46, 1131–1144, 1990.
30. Classon, B, Garegg, P J, Samuelsson, B, Liu, Z, Novel route to 4-thiofuranosides. Synthesis of methyl 4-thio-α-D-talofuranoside, *J. Carbohydr. Chem.*, 6, 593–597, 1987.
31. Fernández-Bolãnos, J G, Zafra, E, García, S, Fernández-Bolãnos, J, Fuentes, J, 4-Thiopyranoside and 4-thiofuranoside derivatives of D-galactosamine, *Carbohydr. Res.*, 305, 33–41, 1998.
32. Effenberger, F, Straub, A, Null, V, Stereoselektive Darstellung von Thiozuckern aus achiralen Vorstufen mittels Enzymen, *Liebigs Ann. Chem.*, 1297–1301, 1992.
33. Chmielewski, M, Whistler, R L, 5-Thio-D-fructofuranose, *J. Org. Chem.*, 40, 639–643, 1975.

34. Le Merrer, Y, Fuzier, M, Dosbaa, I, Foglietti, M-J, Depezay, J-C, Synthesis of thiosugars as weak inhibitors of glycosidases, *Tetrahedron*, 53, 16731–16746, 1997.
35. Bozó, E, Boros, S, Kuszmann, J, Gács-Baitz, E, Synthesis of 6-deoxy-5-thio-D-glucose, *Carbohydr. Res.*, 290, 159–173, 1996.
36. Clegg, W, Hughes, N A, Wood, C J, The formation of 4-thio-L-arabinofuranose derivatives by ring contraction of 1,2-*O*-isopropylidene-3,4-di-*O*-methanesulfonyl-5-thio-α-D-xylopyranose, *J. Chem. Soc., Chem. Commun.*, 300, 1975.
37. Whistler, R L, Lake, W C, 5-Thio-α-D-glucopyranose via conversion of a terminal oxirane to a terminal thiirane ring, *Methods Carbohydr. Chem.*, 6, 286–291, 1972.
38. Driguez, H, Henrissat, B, A novel synthesis of 5-thio-D-glucose, *Tetrahedron Lett.*, 22, 5061–5062, 1981.
39. Shin, J E N, Perlin, A S, Synthesis of 5-thio-D-galactose, *Carbohydr. Res.*, 76, 165–176, 1979.
40. Hashimoto, H, Kawanishi, M, Yuasa, H, New and facile synthetic routes to 5-thioaldohexopyranosides *via* aldose monothioacetal derivatives, *Tetrahedron Lett.*, 32, 7080–7087, 1991.
41. Chou, W-C, Chen, L, Fang, J-M, Wong, C-H, A new route to deoxythiosugars based on aldolases, *J. Am. Chem. Soc.*, 116, 6191–6194, 1994.
42. Cox, J M, Owen, L N, Thio-sugars. Part III. The thioseptanose ring, *J. Chem. Soc. C*, 1121–1130, 1967.
43. Whistler, R L, Campbell, C S, Synthesis of septanose derivatives of 6-deoxy-6-mercapto-D-galactose, *J. Org. Chem.*, 31, 816–818, 1966.
44. Bozó, E, Gáti, T, Demeter, A, Kuszmann, J, Synthesis of 4-cyano and 4-nitrophenyl 1,6-dithio-D-manno-L-ido- and D-glucoseptanosides possessing antithrombotic activity, *Carbohydr. Res.*, 337, 1351–1365, 2002.
45. Capon, R, MacLeod, J K, 5-Thio-D-mannose from the marine sponge *Clathria pyramida* (Lendenfeld). The first example of a naturally occurring 5-thiosugar, *J. Chem. Soc. Chem. Commun.*, 1200–1201, 1987.
46. Yuasa, H, Izukawa, Y, Hashimoto, H, Synthesis of 5-thio-D-mannose, *J. Carbohydr. Chem.*, 8, 753–763, 1989.
47. Bilik, V, Reactions of saccharides catalysed by molybdate ions. II. Epimerisation of D-glucose and D-mannose, *Chem. Zvesti.*, 26, 183–186, 1972.
48. Serianni, A S, Vuorinen, T, Bondo, P B, Stable isotopically-enriched D-glucose: strategies to introduce carbon, hydrogen and oxygen isotopes at various sites, *J. Carbohydr. Chem.*, 9, 513–541, 1990.
49. Oishi, H, Noto, T, Sasaki, H, Suzuki, K, Hayashi, T, Okazaki, H, Ando, K, Sawada, M, Thiolactomycin, a new antibiotic. I. Taxonomy of the producing organism, fermentation and biological properties, *J. Antibiot.*, 35, 391–395, 1982.
50. Sasaki, H, Oishi, H, Hayashi, T, Matsuura, I, Ando, K, Sawada, M, Thiolactomycin, a new antibiotic. II. Structure elucidation, *J. Antibiot.*, 35, 396–400, 1982.
51. Noto, T, Miyakawa, S, Oishi, H, Endo, H, Okazaki, H, Thiolactomycin, a new antibiotic. III. In vitro antibacterial activity, *J. Antibiot.*, 35, 401–410, 1982.
52. Miyakawa, S, Suzuki, K, Noto, T, Harada, Y, Okazaki, H, Thiolactomycin, a new antibiotic. IV. Biological properties and chemotherapeutic activity in mice, *J. Antibiot.*, 35, 411–419, 1982.
53. Wang, C-L J, Salvino, J M, Total synthesis of (+/−)-thiolactomycin, *Tetrahedron Lett.*, 25, 5243–5246, 1984.
54. Omura, S, Iwai, Y, Nakagawa, A, Iwata, R, Takahashi, Y, Shimizu, H, Tanaka, H, Thiotetromycin, a new antibiotic. Taxonomy, production, isolation, and physicochemical and biological properties, *J. Antibiot.*, 36, 109–114, 1983.
55. Omura, S, Nakagawa, A, Iwata, R, Hatano, A, Structure of a new antibiotic, thiotetromycin, *J. Antibiot.*, 36, 1781–1782, 1983.
56. Yoshikawa, M, Murakami, T, Shimada, H, Matsuda, H, Yamahara, J, Tanabe, G, Muraoka, O, Salacinol, potent antidiabetic principle with unique thiosugar sulfonium sulfate structure from the ayurvedic traditional medicine *Salacia reticulata* in Sri Lanka and India, *Tetrahedron Lett.*, 38, 8367–8370, 1997.

57. Yoshikawa, M, Morikawa, T, Matsuda, H, Tanabe, G, Muraoka, O, Absolute stereostructure of potent α-glucosidase inhibitor, salacinol, with unique thiosugar sulfonium sulfate inner salt structure from *Salacia reticulata*, *Bioorg. Med. Chem.*, 10, 1547–1554, 2002.
58. Legler, G, Glycoside hydrolases: mechanistic information from studies with reversible and irreversible inhibitors, *Adv. Carbohydr. Chem. Biochem.*, 48, 319–384, 1990.
59. Yoshikawa, M, Murakami, T, Yashiro, K, Matsuda, H, Kotalanol, a potent α-glucosidase inhibitor with thiosugar sulfonium sulfate structure, from antidiabetic Ayurvedic medicine *Salacia reticulata*, *Chem. Pharm. Bull.*, 46, 1339–1340, 1998.
60. Yuasa, H, Takata, J, Hashimoto, H, Synthesis of salacinol, *Tetrahedron Lett.*, 41, 6615–6618, 2000.
61. Ghavami, A, Johnston, B D, Pinto, B M, A new class of glycosidase inhibitor: synthesis of salacinol and its stereoisomers, *J. Org. Chem.*, 66, 2312–2317, 2001.
62. Ghavami, A, Johnston, B D, Jensen, M T, Svensson, B, Pinto, B M, Synthesis of nitrogen analogues of salacinol and their evaluation as glycosidase inhibitors, *J. Am. Chem. Soc.*, 123, 6268–6271, 2001.
63. Muraoka, O, Ying, S, Yoshikai, K, Matsuura, Y, Yamada, E, Minematsu, T, Tanabe, G, Matsuda, H, Yoshikawa, M, Synthesis of a nitrogen analogue of salacinol and its α-glucosidase inhibitory activity, *Chem. Pharm. Bull.*, 49, 1503–1505, 2001.
64. Johnston, B D, Ghavami, A S, Jensen, M T, Svensson, B, Pinto, B M, Synthesis of selenium analogues of the naturally occurring glycosidase inhibitor salacinol and their evaluation as glycosidase inhibitors, *J. Am. Chem. Soc.*, 124, 8245–8250, 2002.
65. Claeyssens, M, De Bruyne, C K, D-Xylose-derivatives with sulfur or nitrogen in the ring: powerful inhibitors of glycosidase-activities, *Naturwissenschaften*, 52, 515, 1965.
66. Hoffman, D J, Whistler, R L, Diabetogenic action of 5-thio-D-glucopyranose in rats, *Biochemistry*, 7, 4479–4483, 1968.
67. Whistler, R L, Lake, W C, Inhibition of cellular transport processes by 5-thio-D-glucopyranose, *Biochem. J.*, 130, 919–925, 1972.
68. Graham, L L, Whistler, R L, Uridine 5′-(5-Thio-α-D-glucopyranosyl pyrophosphate): chemical synthesis and activation of rat liver glycogen synthetase, *Biochemistry*, 15, 1189–1194, 1976.
69. Zysk, J R, Bushway, A A, Whistler, R L, Carlton, W W, Temporary sterility produced in male mice by 5-thio-D-glucose, *J. Reproduct. Fertil.*, 45, 69–72, 1975.
70. Nakamura, M, Hall, P F, Inhibition by 5-thio-D-glucopyranose of protein biosynthesis in vitro in spermatids from rat testis, *Biochim. Biophys. Acta*, 447, 474–483, 1976.
71. Nakamura, M, Hall, P F, Effect of 5-thio-D-glucose on protein synthesis in vitro by various types of cells from rat testes, *J. Reproduct. Fertil.*, 49, 395–397, 1977.
72. Bushway, A, Keenan, T W, 5-Thio-D-glucose is an acceptor for UDP-galactose: D-glucose 1-galactosyltransferase, *Biochem. Biophys. Res. Commun.*, 81, 305–309, 1978.
73. Bushway, A, Whistler, R L, Repression of cancer cell growth by 5-thio-D-glucose, *J. Carbohydr. Nucleos. Nucleot.*, 2, 399–405, 1975.
74. Kajimoto, T, Liu, K K-C, Pederson, R L, Zhong, Z, Ichikawa, Y, Porco, J A, Wong, C-H, Enzyme-catalyzed aldol condensation for asymmetric synthesis of azasugars: synthesis, evaluation, and modeling of glycosidase inhibitors, *J. Am. Chem. Soc.*, 113, 6187–6196, 1991.
75. Cubero, I I, López-Espinosa, M T P, Richardson, A C, Ortega, M D S, Enantiospecific synthesis of 1-deoxythiomannojirmycin from a derivative of D-glucose, *Carbohydr. Res.*, 242, 109–118, 1993.
76. Ding, Y, Hindsgaul, O, Syntheses of 1-deoxy-3-*S*-(1-thio-α-D-glucopyranosyl)-mannojirimycin and 1-deoxy-3-*O*-(5-thio-α-D-glucopyranosyl)-mannojirimycin as potential inhibitors of endo-α-D-mannosidase, *Bioorg. Med. Chem. Lett.*, 8, 1215–1220, 1998.
77. Izumi, M, Suhara, Y, Ichikawa, Y, Design and synthesis of potential inhibitors of Golgi endo-α-mannosidase: 5-Thio-D-glucopyranosyl-α(1-3)-1-deoxymannojirimycin and methyl 5-thio-D-glucopyranosyl-α(1-3)5-thio-α-D-mannopyranoside, *J. Org. Chem.*, 63, 4811–4816, 1998.
78. Svansson, L, Johnston, B D, Gu, J-H, Patrick, B, Pinto, B M, Synthesis and conformational analysis of a sulfonium-ion analogue of the glycosidase inhibitor castanospermine, *J. Am. Chem. Soc.*, 122, 10769–10775, 2000.
79. Mehta, S, Andrews, J S, Johnston, B D, Pinto, B M, Novel heteroanalogues of methyl maltoside containing sulfur and selenium as potential glycosidase inhibitors, *J. Am. Chem. Soc.*, 116, 1569–1570, 1994.

80. Mehta, S, Andrews, J S, Johnston, B D, Svensson, B, Pinto, B M, Synthesis and enzyme inhibitory activity of novel glycosidase inhibitors containing sulfur and selenium, *J. Am. Chem. Soc.*, 117, 9784–9790, 1995.
81. Randell, K D, Frandsen, T P, Stoffer, B, Johnson, M A, Svensson, B, Pinto, B M, Synthesis and glycosidase inhibitory activity of 5-thioglucopyranosylamines. Molecular modeling of complexes with glucoamylase, *Carbohydr. Res.*, 321, 143–156, 1999.
82. Johnston, B D, Pinto, B M, Synthesis of 1,2- and 1,3-*N*-linked disaccharides of 5-thio-α-D-mannopyranose as potential inhibitors of the processing mannosidase class I and mannosidase II enzymes, *J. Org. Chem.*, 63, 5797–5800, 1998.
83. Andrews, J S, Johnston, B D, Pinto, B M, Synthesis of a dithio analogue of *n*-propyl kojibioside as a potential glucosidase I inhibitor, *Carbohydr. Res.*, 310, 27–33, 1998.
84. Randell, K D, Johnston, B D, Lee, E E, Pinto, B M, Synthesis of oligosaccharide fragments of the glycosylinositolphospholipid of *Trypanosoma cruzi*: a new selenoglycoside glycosyl donor for the preparation of 4-thiogalactofuranosyl analogues, *Tetrahedron: Asymmetry*, 11, 207–222, 2000.
85. Yuasa, H, Matsuura, S, Hashimoto, H, Synthesis of 5-thiomannose-containing oligomannoside mimics: binding abilities to concanavalin A, *Bioorg. Med. Chem. Lett.*, 8, 1297–1300, 1998.
86. Pitts, M J, Chemielewski, M, Chen, M S, Abd El-Rahman, M M A, Whistler, R L, Metabolism of 5-thio-D-glucopyranose and 6-thio-D-fructopyranose in rats, *Arch. Biochem. Biophys.*, 169, 384–391, 1975.
87. Feather, M, Whistler, R L, Derivatives of 6-deoxy-6-mercapto-D-fructose, *J. Org. Chem.*, 28, 1567–1569, 1963.
88. Chmielewski, M, Chen, M-S, Whistler, R L, 6-Thio-β-D-fructopyranose, *Carbohydr. Res.*, 49, 479–481, 1976.
89. Hashimoto, H, Fujimori, T, Yuasa, H, Synthesis of 5-thio-L-fucose and its inhibitory effect on fucosidase, *J. Carbohydr. Chem.*, 9, 683–694, 1990.
90. Fleet, G W J, Shaw, A N, Evans, S V, Fellows, L E, Synthesis from D-glucose of 1,5-dideoxy-1,5-imino-L-fucitol, a potent α-L-fucosidase inhibitor, *J. Chem. Soc. Chem., Commun.*, 841–842, 1985.
91. Izumi, M, Tsuruta, O, Harayama, S, Hashimoto, H, Synthesis of 5-thio-L-fucose-containing disaccharides, as sequence-specific inhibitors, and 2′-fucosyllactose, as a substrate of α-L-fucosidases, *J. Org. Chem.*, 62, 992–998, 1997.
92. Ermert, P, Vasella, A, A new approach to 5-thiosugars: 5-Thio-D-glucohydroximo-1,5-lactone, synthesis and evaluation as β-glucosidase inhibitor, *Helv. Chim. Acta*, 76, 2687–2699, 1993.
93. von Itzstein, M, Wu, W-Y, Kok, G B, Pegg, M S, Dyason, J C, Jin, B, Phan, V T, Smythe, M L, White, H F, Oliver, S W, Colman, P M, Varghese, J N, Ryan, D M, Woods, J M, Bethell, R C, Hotham, V J, Cameron, J M, Penn, C R, Rational design of potent sialidase-based inhibitors of influenza virus replication, *Nature*, 363, 418–423, 1993.
94. Kok, G B, Campbell, M, Mackey, B, von Itzstein, M, Synthesis and biological evaluation of sulfur isosteres of the potent influenza virus sialidase inhibitors 4-amino-4-deoxy- and 4-deoxy-4-guanidino-Neu5Ac2en, *J. Chem. Soc., Perkin. Trans. 1*, 2811–2815, 1996.
95. Mack, H, Brossmer, R, Synthesis of 6-thiosialic acids and 6-thio-*N*-acetyl-D-neuraminic acid, *Tetrahedron Lett.*, 28, 191–194, 1987.
96. Mack, H, Brossmer, R, Synthese von 6-Thiosialinsäuren, *Tetrahedron*, 54, 4521–4538, 1998.
97. Andrews, J S, Weimar, T, Frandsen, T P, Svensson, B, Pinto, B M, Novel disaccharides containing sulfur in the ring and nitrogen in the interglycosidic linkage. Conformation of methyl 5′-thio-4-*N*-α-maltoside bound to glucoamylase and its activity as a competitive inhibitor, *J. Am. Chem. Soc.*, 117, 10799–10804, 1995.
98. Hasegawa, A, Kawai, Y, Kasugai, H, Kiso, M, Synthesis of 2-acetamido-2-deoxy-5-thio-D-glucopyranose, *Carbohydr. Res.*, 63, 131–137, 1978.
99. Guthrie, R D, O'Shea, K, The synthesis of 2-acetamido-2-deoxy-5-thio-D-glucopyranose and its derivatives, *Aust. J. Chem.*, 34, 2225–2230, 1981.
100. Bognar, R, Herczegh, P, Whistler, R L, Madumelu, E B, 2-Acetamido-2-deoxy-5-thio-D--glucopyranose (5-thio-*N*-acetyl-D-glucosamine), *Carbohydr. Res.*, 90, 138–143, 1981.

101. Tanahashi, E, Kiso, M, Hasegawa, A, A facile synthesis of 2-acetamido-2-deoxy-5-thio-D-glucopyranose, *Carbohydr. Res.*, 117, 304–308, 1983.
102. Csuk, R, Glänzer, B I, A short synthesis of 2-acetamido-2-deoxy-5-thio-D-glucose and -D-mannose from 5-thio-D-glucal, *J. Chem. Soc. Chem. Commun.*, 343–344, 1986.
103. Korytnyk, W, Angelino, N, Dodson-Simmons, O, Hanchak, M, Madson, M, Valentekovic-Horvath, S, Synthesis and conformation of 5-thio-D-glucal, an inhibitor of glycosidases, *Carbohydr. Res.*, 113, 166–171, 1983.
104. Lemieux, R U, Ratcliffe, R M, The azidonitration of tri-*O*-acetyl-D-galactal, *Can. J. Chem.*, 57, 1244–1251, 1979.
105. Stachel, H-D, Schachtner, J, Lotter, H, Synthesis of (+/−)-thioascorbic Acid, *Tetrahedron*, 49, 4871–4880, 1993.
106. Bellamy, F, Barberousse, V, Martin, N, Masson, P, Millet, J, Samreth, S, Sepulchre, C, Théveniaux, J, Horton, D, Thioxyloside derivatives as orally active venous antithrombotics, *J. Med. Chem.*, 38, 101s–115s, 1995.
107. Bozó, E, Boros, S, Kuszmann, J, Synthesis of 4-cyanophenyl 2-azido-2-deoxy- and 3-azido-3-deoxy-1,5-dithio-β-D-xylopyranosides, *Carbohydr. Res.*, 301, 23–32, 1997.
108. Bozó, E, Boros, S, Kuszmann, J, Synthesis of 4-cyanophenyl 4-azido-4-deoxy-1,5-dithio-β-D-xylopyranoside, *Carbohydr. Res.*, 302, 149–162, 1997.
109. Bozó, E, Boros, S, Kuszmann, J, Synthesis of 4-cyanophenyl 1,5-dithio-β-D-glucopyranoside and its 6-deoxy, as well as 6-deoxy-5-ene derivatives as oral antithrombotic agents, *Carbohydr. Res.*, 304, 271–280, 1997.
110. Bozó, E, Boros, S, Kuszmann, J, Gács-Baitz, E, Párkányi, L, An economic synthesis of 1,2,3,4-tetra-*O*-acetyl-5-thio-D-xylopyranose and its transformation into 4-substituted-phenyl 1,5-dithio-D-xylopyranosides possessing antithrombotic activity, *Carbohydr. Res.*, 308, 297–310, 1998.
111. Strumpel, M K, Buschmann, J, Szilágyi, L, Györgydeák, Z, Synthesis and structural studies of anomeric 2,3,4,6-tetra-*O*-acetyl-5-thio-D-glucopyranosyl azides, *Carbohydr. Res.*, 318, 91–97, 1999.
112. Collette, Y, Ou, K, Pires, J, Baudry, M, Descotes, G, Praly, J-P, Barberousse, V, An improved synthesis of an umbelliferyl 5-thioxylopyranoside, precursor of the antithrombotic drug Iliparcil, *Carbohydr. Res.*, 318, 162–166, 1999.
113. Bozó, E, Boros, S, Kuszmann, J, Synthesis of 4-cyanophenyl and 4-nitrophenyl 1,5-dithio-D-ribopyranosides as well as their 2-deoxy and 2,3-dideoxy derivatives possessing antithrombotic activity, *Carbohydr. Res.*, 321, 52–66, 1999.
114. Baudry, M, Barberousse, V, Descotes, G, Faure, R, Pires, J, Praly, J-P, Synthetic studies in the 5-thio-D-xylopyranose series part 1: a ready access to *C*-hetaryl 5-thio-D-xylopyranosides by electrophilic substitution, *Tetrahedron*, 54, 7431–7446, 1998.
115. Baudry, M, Barberousse, V, Descotes, G, Pires, J, Praly, J-P, Synthetic studies in the 5-thio-D-xylopyranose series part 2: coupling of 5-thio-D-xylopyranosyl donors with electron-rich aryl moieties: access to C-aryl 5-thio-D-xylopyranosides, *Tetrahedron*, 54, 7447–7456, 1998.
116. Wong, C-H, Krach, T, Gautheron-Le Narvor, C, Ichikawa, Y, Look, G C, Gaeta, F, Thompson, D, Nicolaou, K C, Synthesis of novel disaccharides based on glycosyltransferases: β1,4galactosyltransferase, *Tetrahedron Lett.*, 32, 4867–4870, 1991.
117. Nishida, Y, Wiemann, T, Thiem, J, Extension of the βgal1,1-transfer to *N*-acetyl-5-thio-gentosamine by galactosyltransferase, *Tetrahedron Lett.*, 34, 2905–2906, 1993.
118. Yuasa, H, Hindsgaul, O, Palcic, M M, Chemical-enzymatic synthesis of 5′-thio-*N*-acetyllactosamine: the first disaccharide with sulfur in the ring of the nonreducing sugar, *J. Am. Chem. Soc.*, 114, 5891–5892, 1992.
119. Tsuruta, O, Shinohara, G, Yuasa, H, Hashimoto, H, UDP-*N*-acetyl-5-thio-galactosamine is a substrate of lactose synthase, *Bioorg. Med. Chem. Lett.*, 7, 2523–2526, 1997.
120. Bobek, M, Whistler, R L, Bloch, A, Preparation and activity of the 4′-thio derivatives of some 6-substituted purine nucleosides, *J. Med. Chem.*, 13, 411–413, 1970.
121. Bobek, M, Whistler, R L, Bloch, A, Synthesis and biological activity of 4′-thio analogs of the antibiotic toyocamycin, *J. Med. Chem.*, 15, 168–171, 1972.
122. Bobek, M, Bloch, A, Parthasarathy, R, Whistler, R L, Synthesis and biological activity of 5-fluoro-4′-thiouridine and some related nucleosides, *J. Med. Chem.*, 18, 784–787, 1975.

123. Dyson, M R, Coe, P L, Walker, R T, The synthesis and antiviral properties of *E*-5-(2-bromovinyl)-4′-thio-2′-deoxyuridine, *J. Chem. Soc. Chem. Commun.*, 741–742, 1991.
124. Dyson, M R, Coe, P L, Walker, R T, The synthesis and antiviral activity of some 4′-thio-2′-deoxy nucleoside analogues, *J. Med. Chem.*, 34, 2782–2786, 1991.
125. Secrist, J A, Riggs, R M, Tiwari, K N, Montgomery, J A, Synthesis and anti-HIV activity of 4′-thio-2′,3′-dideoxynucleosides, *J. Med. Chem.*, 35, 533–538, 1992.
126. Tber, B, Fahmi, N-E, Ronco, G, Villa, P, Ewing, D F, Mackenzie, G, An alternative strategy for the synthesis of 3′-azido-2′,3′-dideoxy-4′-thionucleosides starting from D-xylose, *Carbohydr. Res.*, 267, 203–215, 1995.
127. Uenishi, J, Takahashi, K, Motoyama, M, Akashi, H, Sasaki, T, Syntheses and antitumor activities of D- and L-2′-deoxy-4′-thiopyrimidine nucleosides, *Nucleos. Nucleot.*, 13, 1347–1361, 1994.
128. van Draanen, N A, Freeman, G A, Short, Sa, Harvey, R, Jansen, R, Szczech, G, Koszalka, G W, Synthesis and antiviral activity of 2′-deoxy-4′-thio purine nucleosides, *J. Med. Chem.*, 39, 538–542, 1996.
129. Beach, J W, Jeong, L S, Alves, A J, Pohl, D, Kim, H O, Chang, C-N, Doong, S-L, Schinazi, R F, Cheng, Y-C, Chu, C K, Synthesis of enantiomerically pure (2′*R*,5′(S)-(-)-1-[2-(hydroxymethyl)oxathiolan-5-yl]cytosine as a potential antiviral agent against hepatitis B virus (HBV) and human immunodefficiency virus (HIV), *J. Org. Chem.*, 57, 2217–2219, 1992.
130. Yoshimura, Y, Watanabe, M, Satoh, H, Ashida, N, Ijichi, K, Sakata, S, Machida, H, Matsuda, A, A facile, alternative synthesis of 4′-thioarabinonucleosides and their biological activities, *J. Med. Chem.*, 40, 2177–2183, 1997.
131. Paulsen, H, Preparation of 5-acetamido-5-deoxy-D-xylopyranose, *Angew. Chem.*, 74, 901, 1962.
132. Hanessian, H, Haskell, T H, Synthesis of 5-acetamido-5-deoxypentoses. Sugar derivatives containing nitrogen in the ring, *J. Org. Chem.*, 28, 2604–2610, 1963.
133. Jones, J K N, Szarek, W A, Synthesis of a sugar derivative with nitrogen in the ring, *Can. J. Chem.*, 41, 636–639, 1963.
134. Paulsen, H, Leupold, F, Todt, K, Monosaccharides with N-containing ring. VIII. Synthesis and properties of 5-amino-5-deoxy-D-xylopyranose, *Ann. Chem.*, 692, 200–214, 1966.
135. Paulsen, H, Leupold, F, Monosaccharides with nitrogen-containing rings. XXIII. Preparation and reactions of the free 5-amino-5-deoxypentapyranoses, *Chem. Ber.*, 102, 2822–2834, 1969.
136. Paulsen, H, Carbohydrates containing nitrogen or sulfur in the hemiacetal ring, *Angew. Chem.*, 78, 501–506, 1966; *Angew. Chem. Int. Ed. Engl.*, 5, 495–511, 1966.
137. Paulsen, H, Sangster, I, Heyns, K, Synthese und Reaktionen von Keto-piperidinosen, *Chem. Ber.*, 100, 802–815, 1967.
138. Hanessian, S, Synthesis of hydropiperidines from carbohydrate precursors, *Chem. Ind. (Lond.)*, 51, 2126–2127, 1966.
139. Inouye, S, Tsuruoka, T, Ito, T, Niida, T, Structure and synthesis of nojirimycin, *Tetrahedron*, 24, 2125–2144, 1968.
140. Inouye, S, Tsuruoka, T, Niida, T, Structure of nojirimycin, sugar antibiotic with nitrogen in the ring, *J. Antibiot. Ser. A*, 19, 288–292, 1966.
141. Niwa, T, Inouye, S, Tsuruoka, T, Koaze, Y, Niida, T, Nojirimycin as a potent inhibitor of glucosidase, *Agr. Biol. Chem.*, 34, 966–968, 1970.
142. Schmidt, D D, Frommer, W, Müller, L, Truscheit, E, Glucosidase inhibitors from bacilli, *Naturwissenschaften*, 66, 584–585, 1979.
143. Truscheit, E, Frommer, W, Junge, B, Müller, L, Schmidt, D D, Wingender, W, Chemistry and biochemistry of microbial α-glucosidase inhibitors, *Angew. Chem. Int. Ed. Engl.*, 93, 738–755, 1981, *Angew. Chem. Int. Ed. Engl.*, 20, 744–761, 1981.
144. Furukawa, J, Okuda, S, Saito, K, Hatanaka, S I, 3,4-Dihydroxy-2-hydroxymethylpyrrolidine from Arachniodes standishii, *Phytochemistry*, 24, 593–594, 1985.
145. Nash, R J, Bell, E A, Williams, J M, 2-Hydroxymethyl-3,4-dihydroxypyrrolidine in fruits of *Angylocalyx boutiqueanus*, *Phytochemistry*, 24, 1620–1622, 1985.
146. Jones, D W C, Nash, R J, Bell, E A, Williams, J M, Identification of the 2-hydroxymethyl-3,4-dihydroxypyrrolidine (or 1,4-dideoxy-1,4-iminopentitol) from *Angylocalyx boutiqueanus* and from

Arachniodes standishii as the (2R,3R,4S)-isomer by the synthesis of its enantiomer, *Tetrahedron Lett.*, 26, 3125–3126, 1985.
147. Welter, A, Jadot, J, Dardenne, G, Marlier, M, Casimir, J, 2,5-Dihydroxymethyl-3,4-dihydroxypyrrolidine in the leaves of *Derris elliptica*, *Phytochemistry*, 15, 747–749, 1976.
148. Colgate, S M, Dorling, P R, Huxtable, C R, A spectroscopic investigation of swainsonine: an α-mannosidase inhibitor isolated from *Swainsona canescens*, *Aust. J. Chem.*, 32, 2257–2264, 1979.
149. Hohenschutz, L D, Bell, E A, Jewess, P J, Leworthy, D P, Pryce, R J, Arnold, E, Clardy, J, Castanospermine, a 1,6,7,8-tetrahydroxyoctahydroindolizine alkaloid from seeds of *Castanospermum australe*, *Phytochemistry*, 20, 811–814, 1981.
150. Molyneux, R J, Roitman, J N, Dunnheim, G, Szumilo, T, Elbein, A D, 6-Epicastanospermine, a novel indolizidine alkaloid that inhibits α-glucosidase, *Arch. Biochem. Biophys.*, 251, 450–457, 1986.
151. Molyneux, R J, Pan, Y T, Tropea, J E, Benson, M, Kaushal, G P, Elbein, A D, 6,7-Diepicastanospermine, a tetrahydroxyindolizidine alkaloid inhibitor of amyloglucosidase, *Biochemistry*, 30, 9981–9987, 1991.
152. Tropea, J E, Molyneux, R J, Kaushal, G P, Pan, Y T, Mitchell, M, Elbein, A D, Australine, a pyrolizidine alkaloid that inhibits amyloglucosidase and glycoprotein processing, *Biochemistry*, 28, 2027–2034, 1989.
153. Nash, R J, Fellows, L E, Dring, J V, Fleet, G W J, Derome, A E, Hamor, T A, Scofield, A M, Watkin, D J, Isolation from Alexa leiopetala and x-ray crystal structure of alexine, (1*R*,2*R*,3*R*,7*S*,8*S*)-3-hydroxymethyl-1,2,7-trihydroxypyrrolizidine [(2*R*,3*R*,4*R*,5*S*,6*S*)-2-hydroxymethyl-1-azabicyclo[3.3.0]octan-3,4,6-triol], a unique pyrrolizidine alkaloid, *Tetrahedron Lett.*, 29, 2487–2490, 1988.
154. Tepfer, D, Goldmann, A, Pamboukdjian, N, Maille, M, Lepingle, A, Chevalier, D, Denarie, J, Rosenberg, C, A plasmid of *Rhizobium meliloti* 41 encodes catabolism of two compounds from root exudates of *Calystegium sepium*, *J. Bacteriol.*, 170, 1153–1161, 1988.
155. Goldmann, A, Milat, M L, Ducrot, P H, Lallemand, J Y, Maille, M, Lepingle, A, Charpin, I, Tepfer, D, Tropane derivatives from *Calystegia sepium*, *Phytochemistry*, 29, 2125–2127, 1990.
156. Shibano, M, Kitagawa, S, Kusano, G, Studies on the constituents of *Broussonetia* species. I. Two new pyrrolidine alkaloids, broussonetines C and D, as β-galactosidase and β-mannosidase inhibitors from *Broussonetia kazinoki* Sieb, *Chem. Pharm. Bull.*, 45, 505–508, 1997.
157. Heightman, T D, Vasella, A T, Recent insights into inhibition, structure, and mechanism of configuration-retaining glycosidases, *Angew. Chem., Int. Ed.*, 38, 750–770, 1999.
158. Compain, P, Martin, O R, Carbohydrate mimetics-based glycosyltransferase inhibitors, *Bioorg. Med. Chem.*, 9, 3077–3092, 2001, and references cited therein.
159. Jakobsen, P, Lundbeck, J M, Kristiansen, M, Breinholt, J, Demuth, H, Pawlas, J, Candela, M P T, Andersen, B, Westergaard, N, Lundgren, K, Asano, N, Iminosugars: potential inhibitors of liver glycogen phosphorylase, *Bioorg. Med. Chem.*, 9, 733–744, 2001.
160. (a) Kinast, G, Schedel, M, Vierstufige 1-Desoxynojirimycin-Synthese mit einer Biotransformation als Zentralem Zwischenschritt, *Angew. Chem.*, 93, 799–800, 1981; (b) Schedel, M, *Regioselective Oxidation of Aminosorbitol with Gluconobacter oxydans, Key Reaction in the Industrial 1-Deoxynojirimycin Synthesis*, *Biotechnology*, Wiley-VCH, Weinheim, pp. 295–311, 2000.
161. Beaupere, D, Stasik, B, Uzan, R, Demailly, G, Azidation sélective du L-sorbose. Application à la synthèse rapide de la 1-dèsoxynojirimycine, *Carbohydr. Res.*, 191, 163–166, 1989.
162. Behling, J, Farid, P, Medich, J R, Scaros, M G, Prunier, M, A short and practical synthesis of 1-deoxynojirimycin, *Synth. Commun.*, 21, 1383–1386, 1991.
163. Hung, R R, Straub, J A, Whitesides, G M, α-Amino aldehyde equivalents as substrates for rabbit muscle aldolase: synthesis of 1,4-dideoxy-1,4-imino-D-arabinitol and 2(*R*),5(*R*)-bis(hydroxymethyl)-3(*R*),4(*R*)-dihydroxypyrrolidine, *J. Org. Chem.*, 56, 3849–3855, 1991.
164. Ziegler, T, Straub, A, Effenberger, F, Enzyme-catalyzed reactions. 3. Enzyme-catalyzed synthesis of 1-deoxymannojirimycin, 1-deoxynojirimycin and 1,4-dideoxy-1,4-imino-D-arabinitol, *Angew. Chem.*, 100, 737–738, 1988.
165. Von der Osten, C H, Sinskey, A J, Barbas, C F, Pederson, R L, Wang, Y F, Wong, C-H, Use of a recombinant bacterial fructose-1,6-diphophate aldolase in aldol reactions: preparative syntheses of

1-deoxynojirimycin, 1-deoxymannojirimycin, 1,4-dideoxy-1,4-imino-D-arabinitol, and fagomine, *J. Am. Chem. Soc.*, 111, 3924–3927, 1989.

166. Baxter, E W, Reitz, A B, Expeditious synthesis of azasugars by the double reductive amination of dicarbonyl sugars, *J. Org. Chem.*, 59, 3175–3185, 1994.
167. Legler, G, Pohl, S, Synthesis of 5-amino-5-deoxy-D-galactopyranose and 1,5-dideoxy-1,5-imino-D-galactitol, and their inhibition of α- and β-galactosidases, *Carbohydr. Res.*, 155, 19–129, 1986.
168. Card, P J, Hitz, W D, Synthesis of 2(*R*),5(*R*)-bis(hydroxymethyl)-3(*R*),4(*R*)-dihydroxypyrrolidine. A novel glycosidase inhibitor, *J. Org. Chem.*, 50, 891–893, 1985.
169. Anzeveno, P B, Creemer, L J, Efficient synthesis of (+)-nojirimycin and (+)-1-deoxynojirimycin, *Tetrahedron Lett.*, 31, 2085–2088, 1990.
170. Godskesen, M, Lundt, I, Madsen, R, Wichester, B, Deoxyiminoalditols from aldonolactones — V. Preparation of the four stereoisomers of 1,5-dideoxy-1,5-iminopentitols. Evaluation of these iminopentitols and three 1,5-dideoxy-1,5-iminoheptitols as glycosidase inhibitors, *Bioorg. Med. Chem.*, 4, 1857–1865, 1996.
171. Hudlicky, T, Rouden, J, Luna, H, Allen, S, Microbial oxidation of aromatics in enantiocontrolled synthesis. 2. Rational design of aza sugars (*endo*-nitrogenous). Total synthesis of (+)-kifunensine, mannojirimycin, and other glycosidase inhibitors, *J. Am. Chem. Soc.*, 116, 5099–5107, 1994.
172. Fleet, G W J, Fellows, L E, Smith, P W, Synthesis of deoxymannojirimycin, fagomine, deoxynojirimycin, 2-acetamido-1,5-imino-1,2,5-trideoxy-D-mannitol, 2-acetamido-1,5-imino-1,2,5-trideoxy-D-glucitol, 2*S*,3*R*,4*R*,5*R*-trihydroxypipecolic acid and 2*S*,3*R*,4*R*,5*S*-trihydroxypipecolic acid from methyl 3-*O*-benzyl-2,5-dideoxy-2,6-imino-α-D-mannofuranoside, *Tetrahedron*, 43, 979–990, 1987.
173. Duréault, A, Portal, M, Depezay, J C, Enantiospecific syntheses of 2,5-dideoxy-2,5-imino-D-mannitol and -L-iditol from D-mannitol, *Synlett*, 225–226, 1991.
174. Bernotas, R C, Ganem, B, Efficient preparation of enantiomerically pure cyclic aminoalditols, total synthesis of 1-deoxynojirimycin and 1-deoxymannojirimycin, *Tetrahedron Lett.*, 26, 1123–1126, 1985.
175. Martin, O R, Saavedra, O M, Xie, F, Liu, L, Picasso, S, Vogel, P, Kizu, H, Asano, N, α- and β-homogalactonojirimycins (α- and β-homogalactostatins): synthesis and further biological evaluation, *Bioorg. Med. Chem.*, 9, 1269–1278, 2001.
176. Le Merrer, Y, Poitout, L, Depezay, J-C, Dosbaa, I, Geoffroy, S, Foglietti, M-J, Synthesis of azasugars as potent inhibitors of glycosidases, *Bioorg. Med. Chem.*, 5, 519–533, 1997.
177. Skaanderup, P A, Madsen, R, Short syntheses of calystegine B2, B3, and B4, *Chem. Commun.*, 1106–1107, 2001.
178. (a) Henrissat, B, Glycosidase families, *Biochem. Soc. Trans.*, 26, 153–156, 1998; (b) Bourne, Y, Henrissat, B, Glycoside hydrolases and glycosyltransferases: families and functional modules, *Curr. Opin. Struct. Biol.*, 11, 593–600, 2001.
179. (a) Lillelund, V H, Jensen, H H, Liang, X, Bols, M, Recent developments of transition-state analogue glycosidase inhibitors of non-natural product origin, *Chem. Rev.*, 102, 515–553, 2002; (b) Stütz, A E, Ed., *Iminosugars as Glycosidase Inhibitors*, Wiley-VCH, Weinheim, 1999.
180. (a) Scott, L J, Spencer, C M, Miglitol: a review of its therapeutic potential in type 2 diabetes mellitus, *Drugs*, 59, 521–549, 2000; (b) Ahr, H J, Boberg, M, Brendel, E, Krause, H P, Steinke, W, Pharmacokinetics of miglitol. Absorption, distribution, metabolism, and excretion following administration to rats, dogs, and man, *Arzneimittel-Forschung*, 47, 734–745, 1997.
181. Robinson, K M, Begovic, M E, Rhinehart, B L, Heineke, E W, Ducep, J B, Kastner, P R, Franklin, F N, Darzin, C, New potent α-glucohydrolase inhibitor MDL 73945 with long duration of action on rats, *Diabetes*, 40, 825–830, 1991.
182. (a) Elbein, A D, Molyneux, R J, Inhibitors of glycoprotein processing, In *Iminosugars as Glycosidase Inhibitors*, Stütz, A E, Ed., Wiley-VCH, Weinheim, pp. 216–251, 1999; (b) Spiro, R G, Processing enzymes involved in the deglucosylation of *N*-linked oligosaccharides of glycoproteins: glucosidases I and II and endomannosidase, In *Carbohydrates in Chemistry and Biology. Part II*, Vol 3, Ernst, B, Hart, G W and Sinaÿ, P, Eds., Wiley-VCH, Weinheim, pp. 65–80, 2000; (c) Moremen, K W, α-Mannosidases in asparagine-linked oligosaccharide processing and catabolism, In *Carbohydrates in Chemistry and Biology. Part II*, Vol 3, Ernst, B, Hart, G W and Sinaÿ, P, Eds., Wiley-VCH, Weinheim, pp. 81–118, 2000.

183. (a) Fleet, G W J, Karpas, A, Dwek, R, Fellows, L E, Tyms, A S, Petursson, S, Namgoong, S K, Ramsden, N G, Smith, P W, Son, J C, Wilson, F, Witty, D R, Jacob, G S, Rademacher, T W, Inhibition of HIV replication by amino-sugar derivatives, *FEBS Lett.*, 237, 128–132, 1988; (b) Karpas, A, Fleet, G W J, Dwek, R A, Petursson, S, Namgoong, S K, Ramsden, N G, Jacob, G S, Rademacher, T W, Aminosugar derivatives as potential anti-human immunodeficiency virus agents, *Proc. Natl. Acad. Sci. USA*, 85, 9229–9233, 1988.
184. For recent reviews, see (a) van den Broek, L G A M, Azasugars: chemistry and their biological activity as potential anti-HIV drugs, In *Carbohydrates in Drug Design*, Witczak, Z J and Nieforth, K A, Eds., Marcel Dekker, New York, pp. 471–493, 1997; (b) van den Broek, L A G M, Vermaas, D J, Heskamp, B M, van Boeckel, C A A, Tan, M C A A, Bolscher, J G M, Ploegh, H L, van Kemenade, F J, de Goede, R E Y, Miedema, F, Chemical modification of azasugars, inhibitors of *N*-glycoprotein-processing glycosidases and of HIV-1 infection. Review and structure-activity relationships, *Recl. Trav. Chim. Pays-Bas*, 112, 82–94, 1993.
185. Sunkara, P S, Bowlin, T L, Liu, P S, Sjoerdsma, A, Antiretroviral activity of castanospermine and deoxynojirimycin, specific inhibitors of glycoprotein processing, *Biochem. Biophys. Res. Commun.*, 148, 206–210, 1987.
186. (a) Fischer, P B, Karlsson, G B, Dwek, R A, Platt, F M, *N*-butyldeoxynojirimycin-mediated inhibition of human immunodeficiency virus entry correlates with impaired gp120 shedding and gp41 exposure, *J. Virol.*, 70, 7153–7160, 1996; (b) Fischer, B P, Karlsson, G B, Butters, T D, Dwek, R A, Platt, F M, N-Butyldeoxynojirimycin-mediated inhibition of human immunodeficiency virus entry correlates with changes in antibody recognition of the V1/V2 region of gp120, *J. Virol.*, 70, 7143–7152, 1996; (c) Fischer, P B, Collin, M, Karlsson, G B, James, W, Butters, T D, Davis, S J, Gordon, S, Dwek, R A, Platt, F M, The α-glucosidase inhibitor *N*-butyldeoxynojirimycin inhibits human immunodeficiency virus entry at the level of post-CD_4 binding, *J. Virol.*, 69, 5791–5797, 1995.
187. (a) Branza-Nichita, N, Durantel, D, Carrouee-Durantel, S, Dwek, R A, Zitzmann, N, Antiviral effect of *N*-butyldeoxynojirimycin against bovine viral diarrhea virus correlates with misfolding of E2 envelope proteins and impairment of their association into E1–E2 heterodimers, *J. Virol.*, 75, 3527–3536, 2001; (b) Durantel, D, Branza-Nichita, N, Crrouee-Durantel, S, Butters, T D, Dwek, R A, Zitzmann, N, Study of the mechanism of antiviral action of iminosugar derivatives against bovine viral diarrhea virus, *J. Virol.*, 75, 8987–8998, 2001.
188. Block, T M, Lu, X, Platt, F M, Foster, G R, Gerlich, W H, Blumberg, B S, Dwek, R A, Secretion of human hepatitis B virus is inhibited by the imino sugar *N*-butyldeoxynojirimycin, *Proc. Natl. Acad. Sci. USA*, 91, 2235–2239, 1994.
189. Ganem, B, *N*-butyldeoxynojirimycin is a novel inhibitor of glycolipid biosynthesis. Secretion of human hepatitis B virus is inhibited by the imino sugar *N*-butyldeoxynojirimycin, *Chemtracts: Org. Chem.*, 7, 106–107, 1994.
190. Fleet, G W J, Nicholas, S J, Smith, P W, Evans, S V, Fellows, L E, Nash, R J, Potent competitive inhibition of α-galactosidase and α-glucosidase activity by 1,4-dideoxy-1,4-iminopentitols: syntheses of 1,4-dideoxy-1,4-imino-D-lyxitol and of both enantiomers of 1,4-dideoxy-1,4-iminoarabinitol, *Tetrahedron Lett.*, 26, 3127–3130, 1985.
191. Fleet, G W J, Smith, P W, The synthesis from D-xylose of the potent and specific enantiomeric glucosidase inhibitors, 1,4-dideoxy-1,4-imino-D-arabinitol and 1,4-dideoxy-1,4-imino-L-arabinitol, *Tetrahedron*, 42, 5685–5692, 1986.
192. (a) Shibata, T, Nakayama, O, Tsurumi, Y, Okohara, M, Terano, H, Kohsaka, M, A new immunomodulator, FR-900483, *J. Antibiot.*, 41, 296–301, 1988; (b) Kayakiri, H, Nakamura, K, Takase, S, Setoi, H, Uchida, I, Terano, H, Hashimoto, M, Tada, T, Koda, S, Structure and synthesis of nectrisine, a new immunomodulator isolated from a fungus, *Chem. Pharm. Bull.*, 39, 2807–2812, 1991.
193. Kim, Y J, Takatsuki, A, Kogoshi, N, Kitahara, T, Synthesis of nectrisine and related compounds, and their biological evaluation, *Tetrahedron*, 55, 8353–8364, 1999.
194. (a) Hohenschutz, L D, Bell, E A, Jewess, P J, Leworthy, D P, Pryce, R J, Arnold, E, Clardy, J, Castanospermine, a 1,6,7,8-tetrahydroxyoctahydroindolizine alkaloid, from seeds of *Castanospermum australe*, *Phytochemistry*, 20, 811–814, 1981; (b) Tyler, P C, Winchester, B G, Synthesis and

biological activity of castanospermine and close analogs, In *Iminosugars as Glycosidase Inhibitors*, Stütz, A E, Ed., Wiley-VCH, Weinheim, pp. 125–156, 1999.

195. (a) Gruters, R A, Neefjes, J J, Tersmette, M, de Goede, R E Y, Tulp, A, Huisman, H G, Miedema, F, Ploegh, H L, Interference with HIV-induced syncytium formation and viral infectivity by inhibitors of trimming glucosidase, *Nature*, 330, 74–77, 1987; (b) Montefiori, D C, Robinson, W E Jr, and Mitchell, W M, Role of protein *N*-glycosylation in pathogenesis of human immunodeficiency virus type 1, *Proc. Natl. Acad. Sci. USA*, 85, 9248–9252, 1988.
196. (a) Sunkara, P S, Taylor, D L, Kang, M S, Bowlin, T L, Liu, P S, Tyms, A S, Sjoerdsma, A, Anti-HIV activity of castanospermine analogues, *Lancet.*, 1206–1209, 1989; (b) Taylor, D L, Sunkara, P S, Liu, P S, Kang, M S, Bowlin, T L, Tyms, A S, 6-*O*-butanoylcastanospermine (MDL 28,574) inhibits glycoprotein processing and the growth of HIVs, *AIDS (London)*, 5, 669–693, 1991.
197. Ruprecht, R M, Bernard, L D, Bronson, R, Gama Sosa, M A, Mullaney, S, Castanospermine vs. its 6-*O*-butanoyl analog: a comparison of toxicity and antiviral activity in vitro and in vivo, *J. Acquired Immune Defic. Syndr.*, 4, 48–55, 1991.
198. (a) Bolt, G, Rode Pedersen, I, Blixenkrone-Moller, M, Processing of *N*-linked oligosaccharides on the measles virus glycoproteins: importance for antigenicity and for production of infectious virus particles, *Virus Res.*, 61, 43–51, 1999; (b) Bolt, G, The measles virus (MV) glycoproteins interact with cellular chaperones in the endoplasmatic reticulum and MV infection upregulates chaperone expression, *Arch. Virol.*, 146, 2055–2068, 2001.
199. Saito, T, Yanaguchi, I, Effect of glycosylation and glucose trimming inhibitors on the influenza A virus glycoproteins, *J. Vet. Med. Sci.*, 62, 575–581, 2000.
200. Courageot, M P, Frenkiel, M P, Dos Santos, C D, Deubel, V, Despres, P, Alpha-glucosidase inhibitors reduce dengue virus production by affecting the initial steps of virion morphogenesis in the endoplasmatic reticulum, *J. Virol.*, 74, 564–572, 2000.
201. Taylor, D L, Fellows, L E, Farrar, G H, Nash, R J, Taylor-Robinson, D, Mobberley, M A, Ryder, T A, Jeffries, D J, Tyms, A S, Loss of cytomegalovirus infectivity after treatment with castanospermine or related plant alkaloids correlates with aberrant glycoprotein synthesis, *Antiviral Res.*, 10, 11–26, 1988.
202. Bridges, C G, Ahmed, S P, Kang, M S, Nash, R J, Porter, E A, Tyms, A S, The effect of oral treatment with 6-*O*-butanoyl castanospermine (MDL 28,574) in the murine zosteriform model of HSV-1 infection, *Glycobiology*, 5, 249–253, 1995.
203. Ahmed, S P, Nash, R J, Bridges, C G, Taylor, D L, Kang, M S, Porter, E A, Tyms, A S, Antiviral activity and metabolism of the castanospermine derivative MDL 28,574, in cells infected with herpes simplex virus type 2, *Biochim. Biophys. Res. Commun.*, 208, 267–273, 1995.
204. Umezawa, H, Aoyagi, T, Komiyama, T, Morishima, H, Hamada, M, Takeuchi, T, Purification and characterisation of a sialidase inhibitor, siastatin, produced by *Streptomyces*, *J. Antibiot.*, 27, 963–969, 1974.
205. Nishimura, Y, Wang, W, Kondo, S, Aoyagi, T, Umezawa, H, Siastatin B, a potent neuraminidase inhibitor: the total synthesis and absolute configuration, *J. Am. Chem. Soc.*, 110, 7249–7250, 1988.
206. Kudo, T, Nishimura, Y, Kondo, S, Takeuchi, T, Syntheses and activities of *N*-substituted derivatives of siastatin B, *J. Antibiot.*, 45, 1662–1668, 1992.
207. (a) Kudo, T, Nishimura, Y, Kondo, S, Takeuchi, T, Syntheses of the potent inhibitors of neuraminidase, *N*-(1,2-dihydroxypropyl) derivatives of siastatin B and its 4-deoxy analogs, *J. Antibiot.*, 46, 300–309, 1993, and references cited there; (b) Shitara, E, Nishimura, Y, Nerome, K, Hiramoto, Y, Takeuchi, T, *Org. Lett.*, 2, 3837–3840, 2000, and references cited there; (c) For a recent review see: Lundt, I, Madsen, R, Isoiminosugars: glycosidase inhibitors with nitrogen at the anomeric position, In *Iminosugars as Glycosidase Inhibitors*, Stütz, A E, Ed., Wiley-VCH, Weinheim, pp. 112–124, 1999.
208. Nishimura, Y, Umezawa, Y, Kondo, S, Takeuchi, T, Mori, K, Kijima-Suda, I, Tomita, K, Sugawara, K, Nakamura, K, Synthesis of 3-episiastatin B analogues having anti-influenza virus activity, *J. Antibiot.*, 46, 1883–1889, 1993.
209. Sanders, D A R, Staines, A G, McMahon, S A, McNeil, M R, Whitfield, C, Naismith, J P, UDP-galactopyranose mutase has a novel structure and mechanism, *Nat. Struct. Biol.*, 8, 858–863, 2001, and references cited there.

210. Lee, R E, Smith, M D, Nash, R J, Griffiths, R C, McNeil, M, Grewal, R K, Yan, W, Besra, G S, Brennan, P J, Fleet, G W J, Inhibition of UDP-gal mutase and mycobacterial galactan biosynthesis by pyrrolidine analogs of galactofuranose, *Tetrahedron Lett.*, 38, 6733–6736, 1997.
211. Lee, R E, Smith, M D, Pickering, L, Fleet, G W J, An approach to combinatorial library generation of galactofuranose mimics as potential inhibitors of mycobacterial cell wall biosynthesis: Synthesis of a peptidomimetic of uridine 5′-diposphogalactofuranose (UDP-galf), *Tetrahedron Lett.*, 40, 8689–8692, 1999.
212. Maddry, J A, Bansal, N, Bermudez, L E, Comber, R N, Orme, I M, Suling, W J, Wilson, L N, Reynolds, R C, Homologated aza analogs of arabinose as antimycobacterial agents, *Bioorg. Med. Chem. Lett.*, 8, 237–242, 1998.
213. Imada, A, Kintaka, K, Nakao, M, Shinagawa, S, Bulgecin, a bacterial metabolite which in concert with beta-lactam antibiotics causes bulge formation, *J. Antibiot.*, 35, 1400–1403, 1982.
214. (a) Shinagawa, S, Maki, M, Kintaka, K, Imada, A, Asai, M, Isolation and characterization of bulgecins, new bacterial metabolites with bulge-inducing activity, *J. Antibiot.*, 38, 17–23, 1985; (b) Shinagawa, S, Kasahara, F, Wada, Y, Harada, S, Asai, M, Structures of bulgecins, bacterial metabolites with bulge-inducing activity, *Tetrahedron*, 40, 3465–3470, 1984.
215. Templin, M F, Edwards, D H, Höltje, J-V, A murein hydrolase is the specific target of bulgecin in *Escherichia coli*, *J. Biol. Chem.*, 267, 20039–20043, 1992.
216. Sobin, B A, Tanner, F W Jr, Anisomycin, a new antiprotozoan antibiotic, *J. Am. Chem. Soc.*, 76, 4053, 1954.
217. Lynch, J E, English, A R, Bauck, H, Deligianis, H, Studies on the in vitro activity of anisomycin, *Antibiot. Chemother.*, 4, 844–848, 1954.
218. (a) Kicska, G A, Tyler, P C, Evans, G B, Furneaux, R H, Schramm, V L, Kim, K, Purine-less death in *Plasmodium falciparum* induced by immucillin-H, a transition state analogue of purine nucleoside phosphorylase, *J. Biol. Chem.*, 277, 3226–3231, 2002; (b) Evans, G B, Furneaux, R H, Gainsford, G J, Schramm, V L, Tyler, P C, Synthesis of transition state analogue inhibitors for purine nucleoside phosphorylase and *N*-riboside hydrolases, *Tetrahedron*, 56, 3053–3062, 2000.
219. Marques, E T A, Ichikawa, Y, Strand, M, August, J T, Hart, G W, Schnaar, R L, Fucosyltransferases in *Schistosoma mansoni* development, *Glycobiology*, 11, 249–259, 2001.
220. Wright, P S, Cross-Doersen, D E, Schroeder, K K, Bowlin, T L, McCann, P P, Bitonti, A J, Disruption of *Plasmodium falciparum*-infected erythrocyte cytoadherence to human melanoma cells with inhibitors of glycoprotein processing, *Biochem. Pharmacol.*, 41, 1855–1861, 1991.
221. Stevens, K L, Molyneux, R J, Castanospermine — a plant growth regulator, *J. Chem. Ecol.*, 14, 1467–1473, 1988.
222. Birch, A N E, Robertson, W M, Geoghegan, I E, McGavin, W J, Alphey, T J W, Phillips, M S, Fellows, L E, Watson, A A, Simmonds, M S J, Porter, L E, DMDP — a plant-derived sugar analogue with systemic activity against plant parasitic nematodes, *Nematologica*, 39, 521–535, 1993.
223. (a) Evans, S V, Gatehouse, A M R, Fellows, L E, Detrimental effects of 2,5-dihydroxymethyl-3,4-dihydroxypyrrolidine in some tropical legume seeds on larvae of the bruchid *Callosobruchus maculatus*, *Entomol. Exp. App.*, 37, 257–261, 1985; (b) Simmonds, M S J, Blaney, W M, Fellows, L E, Behavioral and electrophysiological study of antifeedant mechanisms associated with polyhydroxy alkaloids, *J. Chem. Ecol.*, 16, 3167–3177, 1990; (c) Blaney, W M, Simmonds, M S J, Evans, S V, Fellows, L E, The role of the secondary plant compound 2,5-dihydroxymethyl-3,4-dihydroxypyrrolidine as a feeding inhibitor for insects, *Entomol. Exp. Appl.*, 36, 209–216, 1984; (d) Dreyer, D L, Jones, K C, Molyneux, R J, Feeding deterrency of some pyrrolizidine, indolizidine and quinolizidine alkaloids towards pea aphid (*Acyrthosiphon pisum*) and evidence for phloem transport of indolizidine alkaloid swainsonine, *J. Chem. Ecol.*, 11, 1045–1050, 1985.
224. van Es, T, Whistler, R L, Derivatives of 5-deoxy-5-seleno-D-xylose, *Tetrahedron*, 23, 2849–2853, 1967.
225. Blumberg, K, Fuccello, A, van Es, T, Selenium derivatives of L-arabinose, D-ribose and D-xylose, *Carbohydr. Res.*, 59, 351–362, 1977.
226. Lucas, M A, Nguyen, O T K, Schiesser, C H, Zheng, S-L, Prepartion of 5-selenopentopyranose sugars from pentose starting materials by samarium(II) iodide or (phenylseleno)formate mediated ring closures, *Tetrahedron*, 56, 3995–4000, 2000.

227. Nguyen, O T K, Schiesser, C H, Preparation of 5-telluropentopyranose sugars from common pentose starting materials, *Tetrahedron Lett.*, 43, 3799–3800, 2002.
228. Yamamoto, H, Inokawa, S, Sugar analogs having phosphorus in the hemiacetal ring, *Adv. Carbohydr. Chem. Biochem.*, 42, 135–191, 1984.
229. Yamamoto, H, Hanaya, T, Shigetoh, N, Kawamoto, H, Inokawa, S, Synthesis and characterisation of 5-deoxy-3-*O*-methyl-5-*C*-(*R*)- and (*S*)-phenylphosphinothioyl-α- and -β-D-xylopyranoses. The first sugar analogue having a phosphinothioyl group in the hemiacetal ring, *Chem. Lett.*, 2081–2084, 1987.
230. Yamamoto, H, Hanaya, T, Ohmori, K, Kawamoto, H, Synthesis of 5-deoxy-[(*R* and *S*) methylphosphinyl]-α,β-D-mannopyranoses. The first P-in-ring sugar analogues of D-mannose type, *Chem. Lett.*, 1471–1472, 1989.
231. Hanaya, T, Yasuda, K, Yamamoto, H, Yamamoto, H, Stereoselectivity in the preparation of 5,6-dideoxy-5-dimethoxyphosphinyl-D- and -L-hexofuranoses, and an efficient synthesis of 5,6-dideoxy-5-hydroxyphosphinyl-L-galactopyranose (a P-in-the-ring L-fucose analogue), *Bull. Chem. Soc. Jpn.*, 66, 2315–2322, 1993.
232. Hanaya, T, Fujii, Y, Ikejiri, S, Yamamoto, H, A new route for preparation of 5-deoxy-5-hydroxyphosphinyl-D-gluco- and L-idopyranose derivatives, *Heterocycles*, 50, 323–332, 1999.
233. Fernandez-Bolanos, J G, Al-Masoudi, N A L, Maya, I, Sugar derivatives having sulfur in the ring. *Adv. Carbohydr. Chem. Biochem.*, 57, 21–98, 2001.
234. Wrodnigg, T M, From lianas to glycobiology tools: twenty-five years of 2,5-dideoxy-2,5-imino-D-mannitol, In *Timely Research Perspectives in Carbohydrate Chemistry*, Schmid, W and Stütz, A E, Eds., Springer, Vienna, pp. 41–76, 2002.

Part III

Sugars as Tools, Chiral Pool Starting Materials and Formidable Synthetic Targets

10 Sugars as Chiral Auxiliaries

Norbert Pleuss, Gernot Zech, Bartlomiej Furman and Horst Kunz

CONTENTS

10.1 INTRODUCTION

Carbohydrates contain many functional groups and stereogenic centers in one molecular unit. They were the first of the major classes of natural products to become the subject of preparative chemistry through the fundamental work of Emil Fischer [1]. In the form of blood group substances, carbohydrate conjugates were identified as carriers of selective biological functions by Karl Landsteiner [2] 100 years ago. In spite of these early landmarks, the role of carbohydrates in terms of biological and chemical selectivity was largely ignored over a long period. Only during the past few decades has the involvement of carbohydrates in processes of biological selectivity been recognized, for example, as the ligands of selectins in inflammatory diseases [3] or as tumor-associated glycoprotein antigens [4]. Owing to the intensifying interest in the functions of glycan structures of glycolipids and glycoproteins in biological recognition phenomena, the synthesis of model compounds, oligosaccharides [5], glycolipids [6], and glycopeptides [7] received increasing attention. The development of chemical methods for the synthesis of glycoconjugates also induced concepts to utilize carbohydrates as tools in stereochemical differentiations, not only as the starting materials in ex-chiral pool syntheses of interesting enantiopure compounds [8], but also as chiral auxiliaries in asymmetric synthesis. Based on earlier reviews [9,10], recent developments in the use of sugars as chiral auxiliaries are outlined in this chapter.

10.2 ASYMMETRIC CYCLOADDITION REACTIONS

10.2.1 [2+1] CYCLOADDITIONS

The diastereoselective and enantioselective preparation of cyclopropanes has attracted attention since chiral cyclopropanes were found to occur in many natural products [11]. Moreover, cyclopropanes are useful intermediates in organic synthesis. There are many methods of cyclopropane ring opening that transfer stereochemical information from the substrate to acyclic products in a stereocontrolled manner [12]. Among the methods used for the preparation of cyclopropanes from olefins, the Simmons–Smith and related reactions as well as reactions of diazoalkanes catalyzed by rhodium, copper and cobalt salts have frequently been applied [13]. The preparatively simple Makosza reaction [14] has scarcely been used.

An efficient asymmetric synthesis of chiral cyclopropanes has been described by Charette et al. [15], who treated allyl β-D-glucopyranosides **1** with an excess of diethyl zinc/diiodomethane (Scheme 10.1). Cyclopropanes **2** were obtained in high yield and high diastereoselectivity (Table 10.1).

The unprotected 2-OH group plays a decisive role in controlling the diastereofacial differentiation of **1**. With diethyl zinc, it forms a coordinative anchor for the Simmons–Smith intermediate. As a consequence, the cyclopropanation proceeds almost exclusively *syn* to the free 2-OH group. A methodology for the cleavage from the auxiliary under mild basic conditions has been described also [15]. The other enantiomer of the cyclopropane is available by using the corresponding α-glycoside or by using the pseudoenantiomeric auxiliary, 6-deoxy-3,4-di-*O*-benzyl-L-glucose derived from L-rhamnose [16].

This method was utilized for the synthesis of the four stereoisomers of coronamic acid [17]. The (*E*)-allylic β-D-glucopyranoside **3** was cyclopropanated under Simmons–Smith conditions at −30°C with high diastereoselectivity (>100:1). The reaction of the corresponding (*Z*)-allylic

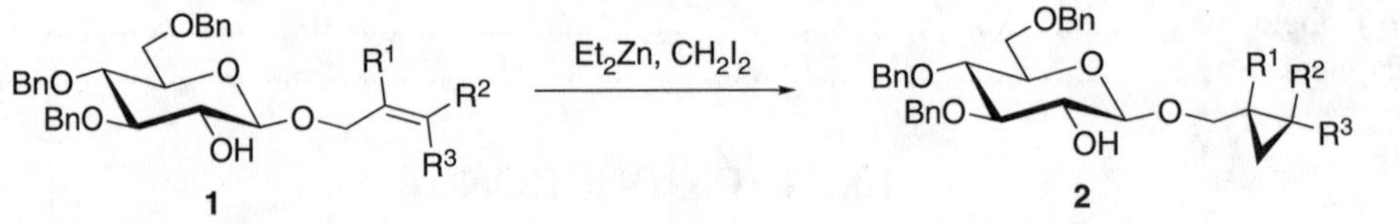

OBn BnO O BnO OH O R^1 R^2 R^3 **1** Et_2Zn, CH_2I_2 OBn BnO O BnO OH O R^1 R^2 R^3 **2**

SCHEME 10.1 Cyclopropanations using glucose as a chiral auxiliary.

TABLE 10.1
Diastereoselectivity in Syntheses of Cyclopropanes According to Scheme 10.1 (Ref. [15])

R^1	R^2	R^3	*T* (°C)	Diastereomeric Ratio
H	Pr	H	−35 to 0	124:1
H	Me	H	−35 to 0	>50:1
H	Ph	H	−35 to 0	130:1
H	H	Ph	−35 to 0	114:1
H	Me	Me	−35 to 0	111:1

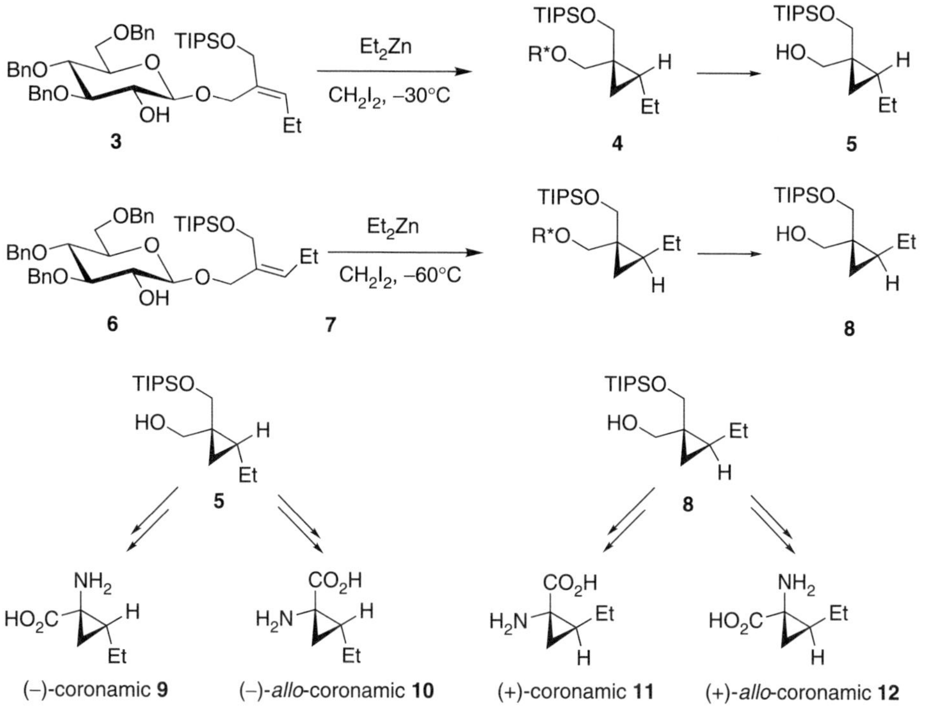

SCHEME 10.2 Application of glucose-directed cyclopropanations to the preparation of coronamic acid and *allo*-coronamic acid.

ether **6** proceeded with lower stereoselectivity. At $-60°C$, however, **7** was obtained in 98% yield and with a 66:1 diastereomeric ratio (Scheme 10.2). After cleavage from the chiral auxiliary, the cyclopropyl methanol **5** was transformed into either (−)-coronamic acid **9** or (−)-*allo*-coronamic acid **10**, while cyclopropane **8** gave the (+)-coronamic acid **11** or the (+)-*allo*-enantiomer **12** [17].

An asymmetric Simmons–Smith reaction was reported by Kang et al. [18]. The reaction of β-D-fructopyranoside **13** with α,β-unsaturated aldehydes gave *endo*-acetals **14** along with *exo*-isomers **15** in a ratio of about 1.5:1. The *endo*-acetals afforded the best selectivity, typically giving (2*R*,3*R*)-hydroxymethyl cyclopropanes **17** with up to 85% *ee*. It should be noted that the corresponding *exo*-acetals **15** underwent the cyclopropanation reaction with lower stereoselectivity. In these cases, the group R^1 cannot effectively block either side of the alkene in contrast to the *endo*-isomer [18] (Scheme 10.3).

Copper and rhodium complexes catalyze the reaction of alkenes with diazoacetate to give alkyl cyclopropanecarboxylates [13]. In the presence of $Cu(acac)_2$, the reaction of carbohydrate enol ether **20** with methyl diazoacetate afforded a 1:4 mixture of *cis*- and *trans*-cyclopropanes **21** and **22** (*cis*-product **21** was obtained with 95% *de*). When the reaction was catalyzed by CuOTf in the presence of ligand **23**, the *trans*-product **22** was obtained with 60% *de* (Scheme 10.4). The absolute configuration of the major diastereomer was not given [19].

10.2.2 [2 + 2] Cycloadditions

[2+2] Cycloadditions may proceed via concerted or nonconcerted mechanisms. Photochemical [2+2] cycloadditions take place via triplet intermediates [20]. Photochemical cycloaddition reactions of olefins to carbonyl compounds known as Paterno–Büchi reactions [21]

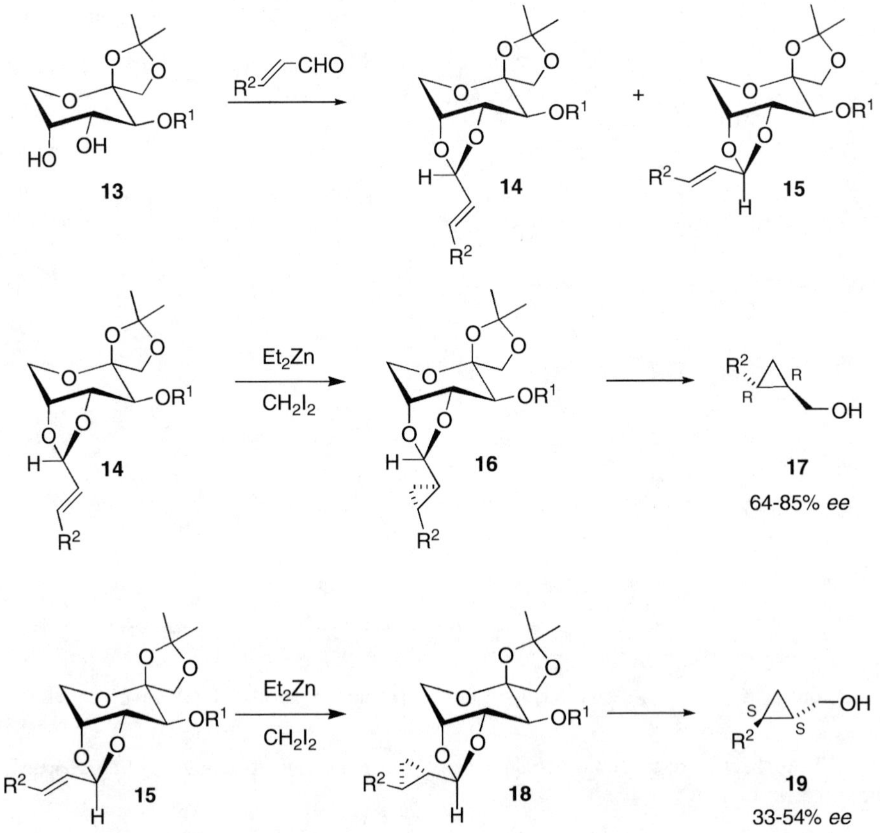

SCHEME 10.3 Simmons-Smith cyclopropanations directed by sugar-derived chiral auxiliaries.

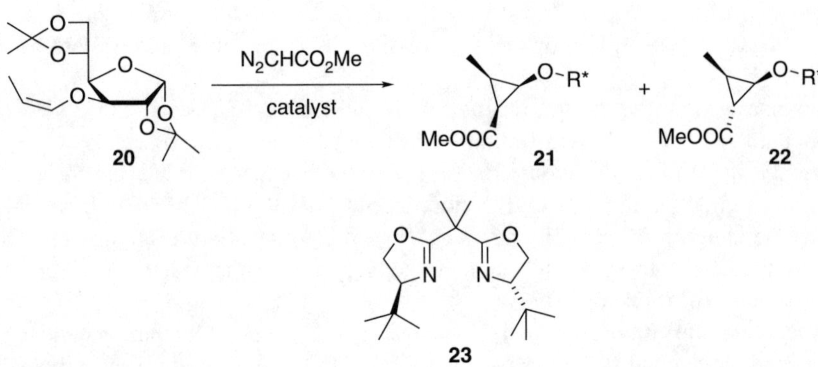

SCHEME 10.4 Copper and rhodium catalyzed cyclopropanation reactions directed by sugar-derived chiral auxiliaries.

furnish oxetanes. Carbohydrate derivatives have been applied successfully as chiral auxiliaries in Paterno–Büchi reactions [22]. In one example, irradiation of the carbohydrate phenyl glyoxylic ester **24** and furan gave a mixture of dioxabicycloheptanes **25** and **26** in a ratio of 1:9 (Scheme 10.5) [23].

The influence of temperature on the stereoselectivity of this reaction has been studied. It was found that elevation of the temperature led to greater diastereoselectivity [23].

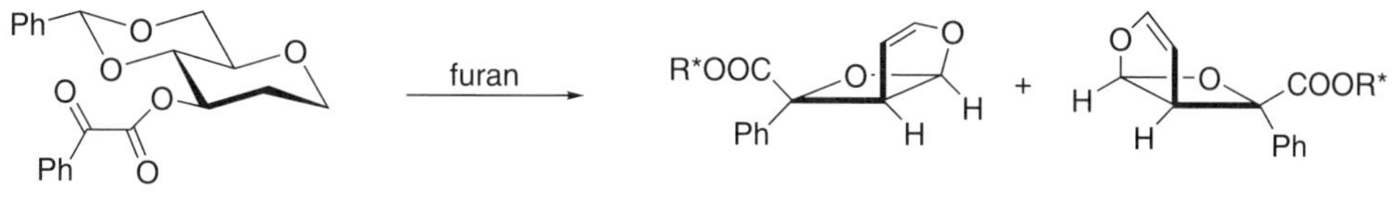

SCHEME 10.5 Use of sugar-derived chiral auxiliaries in Paterno-Büchi reactions.

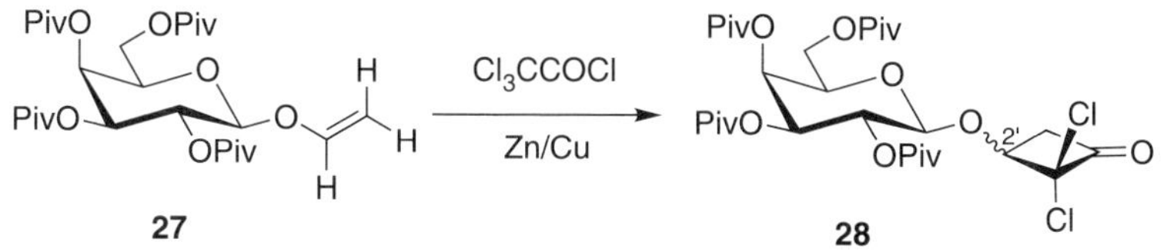

SCHEME 10.6 [2+2] Cycloadditions with dichloroketene directed by a galactose-derived auxiliary.

The thermal [2+2] cycloaddition of cumulenes with alkenes, imines or carbonyl compounds is one of the most useful methods of four-membered ring formation. The cycloaddition of ketenes with alkenes to give cyclobutanones represents a reaction of general importance. According to Woodward and Hoffmann, these reactions proceed via a $[_{\pi}2_s + _{\pi}2_a]$ pathway [24]. Dihaloketenes are more reactive than simple ketenes and readily react with electron-rich olefins [25].

O-Pivaloyl-D-galactopyranosides were shown to be efficient stereodifferentiating tools [9]. Thus, the [2+2] cycloaddition of dichloroketene to chiral vinyl galactoside **27** afforded the cyclobutanone **28** with reasonable stereoselectivity (dr 4:1) [26] (Scheme 10.6). The resulting 2,2-dichlorocyclobutanones are reactive and often cannot be isolated in pure form. More stable cyclobutanols were isolated after reduction of the keto group [26].

Reactions of ketenes with imines open a useful route to β-lactams [27]. The stereospecificity of these reactions has been extensively investigated given the practical importance of β-lactam antibiotics [28]. The generally accepted mechanism of this cycloaddition is outlined below (Scheme 10.7).

The addition of an imine to an alkene provides a zwitterion, which subsequently undergoes a conrotatory cyclization to give the four-membered ring. The configuration of the imine determines

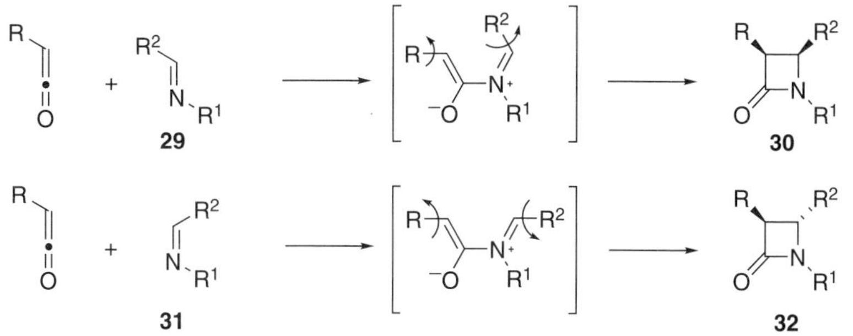

SCHEME 10.7 Mechanism of the formation of β-lactams via the [2+2] cycloaddition of imines with ketenes.

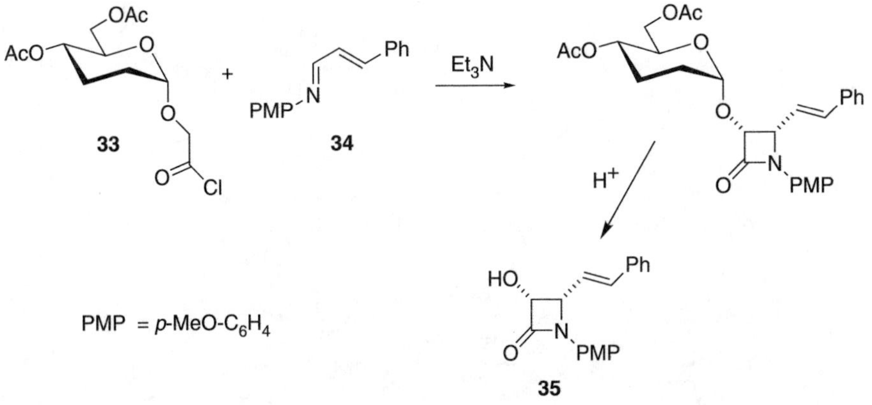

PMP = p-MeO-C_6H_4

SCHEME 10.8 Formation of β-lactams directed by sugar-derived chiral auxiliaries.

the *cis*- or *trans*-stereochemistry of the β-lactam [29]. In general, *cis*-β-lactams **30** are formed from acyclic (*E*)-imines **29**, whereas *trans*-β-lactams **32** are formed only from rigid cyclic (*Z*)-imines **31**. Chiral residues have been introduced on either the ketene or the imine.

Borer and Balogh [30] used the chiral ketene precursor **33** in the asymmetric [2+2] cycloaddition with imine **34** derived from cinnamaldehyde. After detachment from the chiral auxiliary under acidic conditions, the *cis*-β-lactam **35** was obtained in 52% overall yield and 70% ee (Scheme 10.8).

Efficient asymmetric induction was achieved in [2+2] cycloadditions when carbohydrate-derived oxazolidinones were used [31]. Treatment of the acid **36** with the Mukaiyama reagent (2-chloro-1-methylpyridinium iodide) generated the ketene, which was added to imines **37**, giving the *cis*-β-lactams **38** with excellent diastereomeric excess (Scheme 10.9, Table 10.2).

Under the same conditions, the [2+2] cycloaddition of **36** to cyclic imine **39** gave *trans*-β-lactam **40** as the only detectable isomer [31] (Scheme 10.10). Unfortunately, the authors did not assign the absolute configuration of the β-lactam **40**.

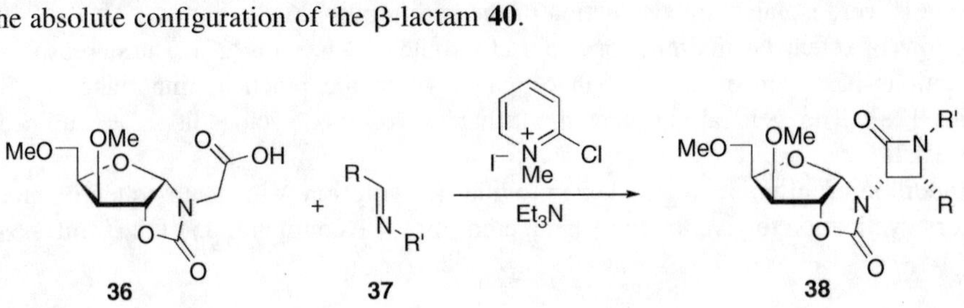

SCHEME 10.9 Formation of β-lactams directed by sugar-derived chiral auxiliaries.

TABLE 10.2
Results of Stereoselective β-Lactam Syntheses According to Scheme 10.9 (Ref. [31])

R	R′	Yield (%)	Diastereomeric Ratio
Phenyl	Benzyl	67	>99:1
(*E*)–CH=CH–Ph	Benzyl	71	>99:1
Phenyl	*p*-Methoxyphenyl	69	>99:1
(*E*)–CH=CH–Ph	*p*-Methoxyphenyl	58	>99:1

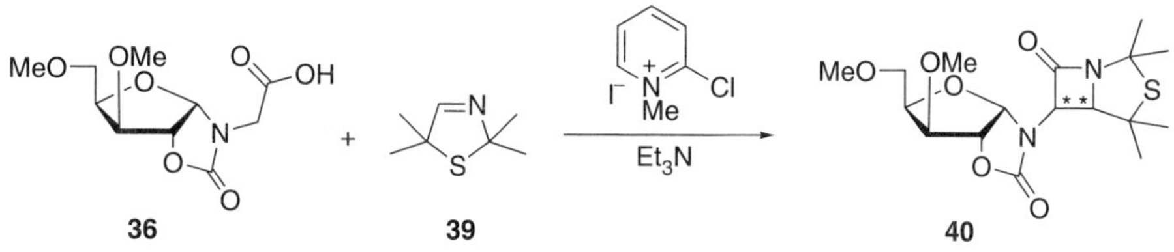

SCHEME 10.10 Formation of β-lactams directed by sugar-derived chiral auxiliaries.

Imines derived from optically active amines have also been used in asymmetric Staudinger syntheses of β-lactams. Georg et al. [32] employed *O*-acetylated galactosylamine as the chiral auxiliary for the synthesis of *cis*-β-lactams. The reaction between imines **41** and aryloxyacetyl chlorides **42** in the presence of Et_3N afforded *cis*-configurated *N*-galactosyl-β-lactams **43** in diastereomeric ratios not exceeding 66:34 (Scheme 10.11).

Chiral Schiff base **44** derived from glucosamine and cinnamaldehyde reacted with *N*-phthalimidoketene, generated from phthalimidoacetyl chloride **45a**, to give the (3*S*,4*S*)-β-lactam **46a** in high yield as the sole product [33] (Scheme 10.12). It should be noted that a similar reaction using methoxyketene generated from **45b** proceeded with only moderate selectivity (dr 6:3). The sugar auxiliary was removed by a β-elimination of the β-lactam.

An alternative route to chiral β-lactams was provided by reactions of electron-deficient isocyanates with chiral nucleophilic alkenes such as vinyl ethers or vinyl acetates. Chlorosulfonyl isocyanate (CSI), a commonly used reactive isocyanate [34], undergoes stereospecific *syn*-addition to alkenes. The chlorosulfonyl group can subsequently be reductively removed from the nitrogen atom. It has been shown that reactions between CSI and (*Z*)- and (*E*)-alkenyl ethers stereoselectively give *cis*-3,4-disubstituted azetidinones from (*Z*)-olefins and *trans*-3,4-disubstituted azetidinones from (*E*)-olefins. To explain the stereoselectivity of this

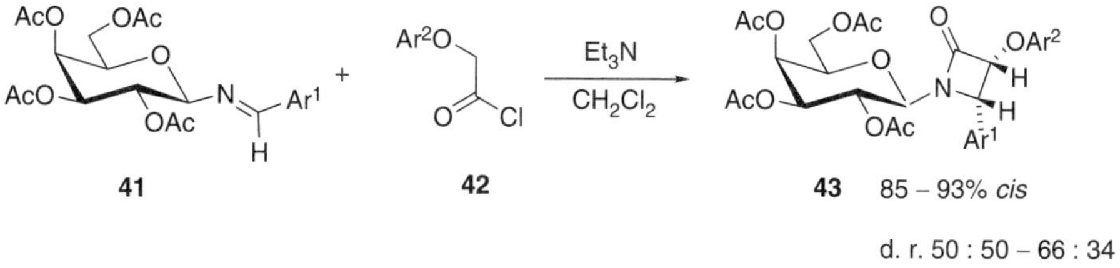

SCHEME 10.11 Asymmetric Staudinger preparation of β-lactams directed by sugar-derived chiral auxiliaries.

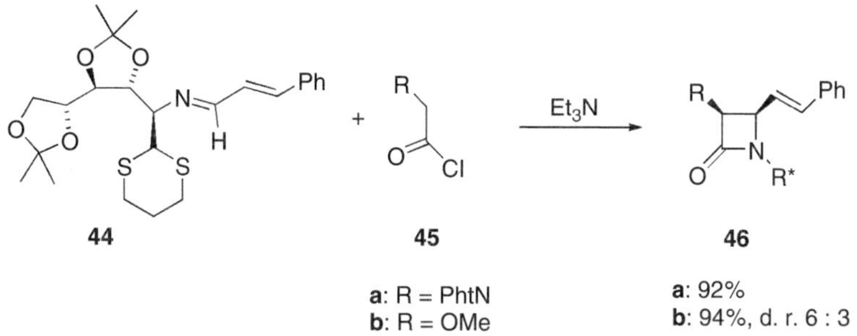

SCHEME 10.12 Formation of β-lactams from sugar-derived chiral Schiff bases.

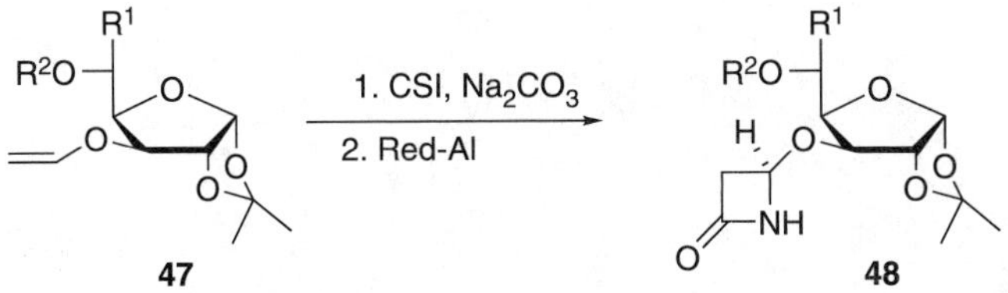

SCHEME 10.13 Formation of β-lactams on reaction with chlorosulfonyl isocyanate (CSI) with sugar vinyl ethers.

reaction, a concerted mechanism has been proposed [35]. This was later supported by molecular calculations [36].

In the presence of sodium carbonate as a base, CSI reacts with sugar vinyl ethers having a chiral center next to the oxygen atom [37]. In this sense, CSI reacts with vinyl ethers **47** derived from xylofuranose with efficient stereocontrol [38] (Scheme 10.13). Large substituents at C-4 of the furanose ring effectively shield the *re*-face of the enol ether resulting in the diastereoselective formation of β-lactams **48** (Figure 10.1, Table 10.3).

The same group [39] also reported highly stereoselective [2+2] cycloaddition reactions of CSI with 5-*O*-vinyl derivatives of glycofuranoses **49** (Scheme 10.14). The size of the groups at C-3 and C-5 obviously plays a decisive role in the diastereofacial differentiation. The presence of a small substituent R^1 or of a large substituent R^2 at C-3 in **49** resulted in excellent asymmetric induction (Table 10.4). This indicates that the attack of the isocyanate occurs from the side occupied by R^1. The (*S*)-configuration of the major diastereomer indicates that the *si*-face of the olefin is effectively blocked by the TIBS group of the auxiliary [39] (Figure 10.2).

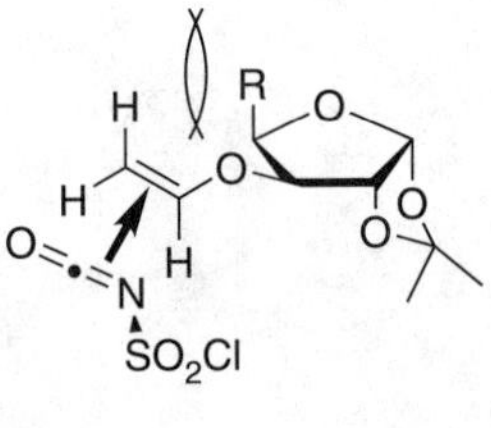

FIGURE 10.1 Diasterofacial differentiation in [2+2]-cycloaddition of chlorosulfonyl isocyanate with carbohydrate vinyl ethers **47**.

TABLE 10.3
Stereoselective Formation of β-Lactams from Carbohydrate Vinyl Ethers (Scheme 10.13, Ref. [38])

R^1	R^2	*de* (%)
H	Ts	33
H	TIBS	54
CH_2OTs	Ts	>95
H	Ph_3Si	>95

Ts=4-Me–C_6H_4–SO_2; TIBS=2,4,6-(^{i}Pr)$_3$–C_6H_2–SO_2.

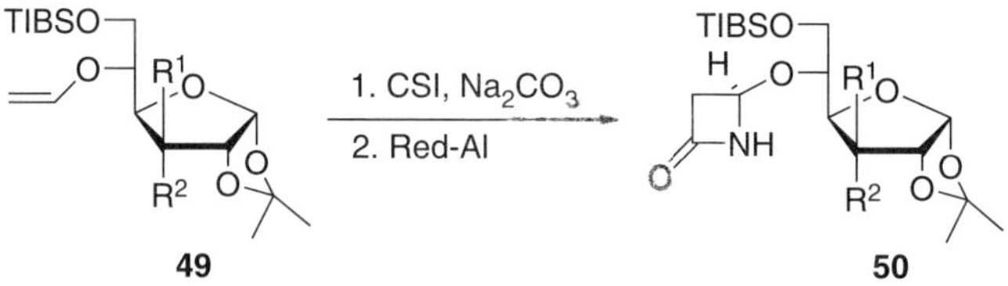

SCHEME 10.14 Formation of β-lactams on reaction with chlorosulfonyl isocyanate (CSI) with sugar vinyl ethers.

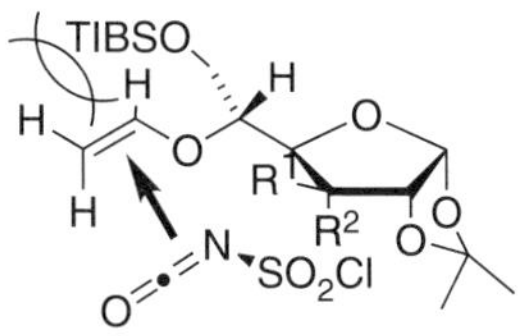

FIGURE 10.2 Supposed stereodifferentiation in the reactions of vinyl ethers **49** with CSI (Scheme 10.14, Ref. [39]).

TABLE 10.4
Diastereoselectivity of β-Lactam Formation According to Scheme 10.14 (Ref. [39])

R^1	R^2	*de* (%)
OBz	H	4
OMTM	H	38
OBn	H	72
OMe	H	92
H	H	>95
H	OMTM	>95

Bz=Ph–CO; Bn=Ph–CH_2; MTM=CH_3–S–CH_2.

10.2.3 [3 + 2] Cycloadditions

The 1,3-dipolar cycloaddition reactions proceed via a concerted suprafacial pathway, which ensures the complete transfer of stereochemical information from the substrates to the cycloadducts. Thus, the stereochemistry of the alkene is retained in the product [40] (Scheme 10.15). With these reactions, up to four stereogenic centers can be formed in one step. Stereodifferentiating groups can be introduced either in the dipole or in the dipolarophile.

10.2.3.1 Carbohydrate-Linked Nitrones

The 1,3-dipolar cycloaddition of nitrones to olefins gives 1,2-oxazolidines. Because nitrones can undergo (*Z*)/(*E*)-isomerization, diastereomers are formed, especially if the reaction is performed at elevated temperature (Scheme 10.16). In addition, 1,3-dipolar cycloadditions of nitrones to olefins may proceed via *endo*- or *exo*-transition states [41].

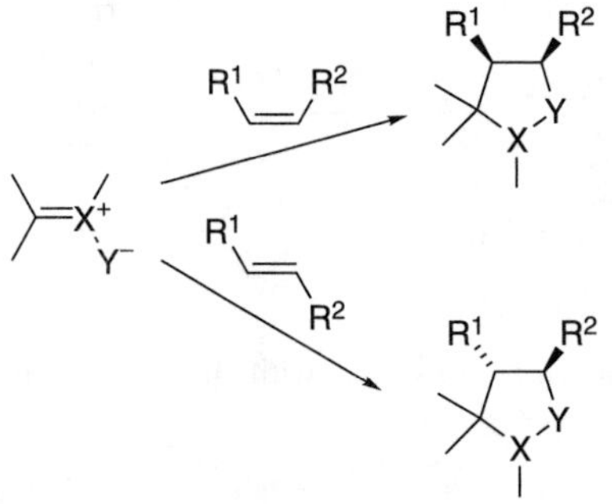

SCHEME 10.15 Stereochemical outcome of 1,3-dipolar cycloadditions with *cis* and *trans* olefins.

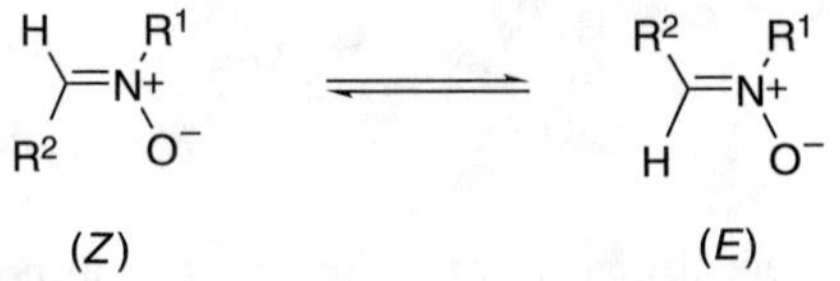

SCHEME 10.16 (*Z*)/(*E*)-Isomerization of nitrones.

A stereodifferentiating group can be introduced to the substituent at the nitrogen (*N*-chiral nitrones). Most of the nitrones obtained from sugars belong to this class of compounds. Vasella et al. [42,43] found that the 1,3-dipolar cycloaddition of *N*-glycosyl nitrones to methyl methacrylate leads to *N*-glycosylisoxazolidines **52** with high diastereoselectivity. This is exemplified below by the cycloaddition of D-mannofuranosylnitrone **51** with methyl methacrylate (Scheme 10.17).

The observed diastereoselectivity was rationalized on the basis of a kinetic anomeric effect increasing the reactivity of certain conformers, which control the direction of the attack [44]. Vasella also applied *N*-ribofuranosylnitrones in the 1,3-dipolar cycloaddition to alkenes [43]. However, the selectivities obtained from these nitrones proved to be lower than those obtained with **51**.

Recently, analogous nitrones have been used by others in the synthesis of (2*S*)-4-oxopipecolic acid [45] and (+)-negamycin [46]. Chiacchio et al. [47] applied the Vasella-type nitrone **53** in an enantioselective synthesis of isoxazolidinyl thymine **55** (Scheme 10.18). As illustrated, the *N*-furanosyl nitrone **53** reacted with vinyl acetate to give a 1:1 mixture of two isoxazolidines **54** epimeric at C-5. In contrast to the poor *cis*/*trans*-diastereoselectivity, the diastereofacial selectivity

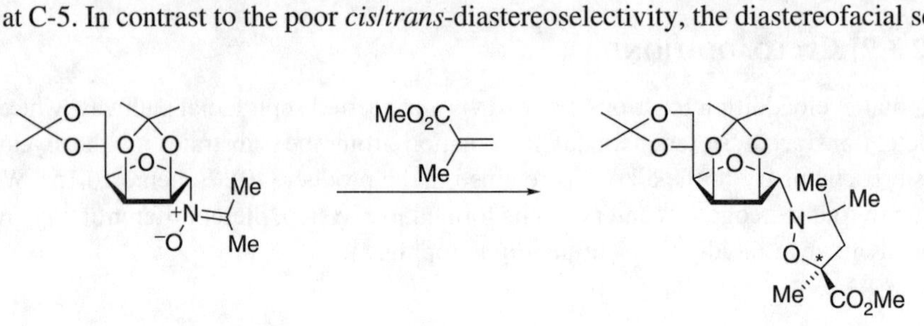

(5*S*) : (5*R*) 95 : 5

SCHEME 10.17 1,3-Dipolar cycloaddition of *N*-glycosylnitrones with olefins giving chiral *N*-glycosylisoxazolidines.

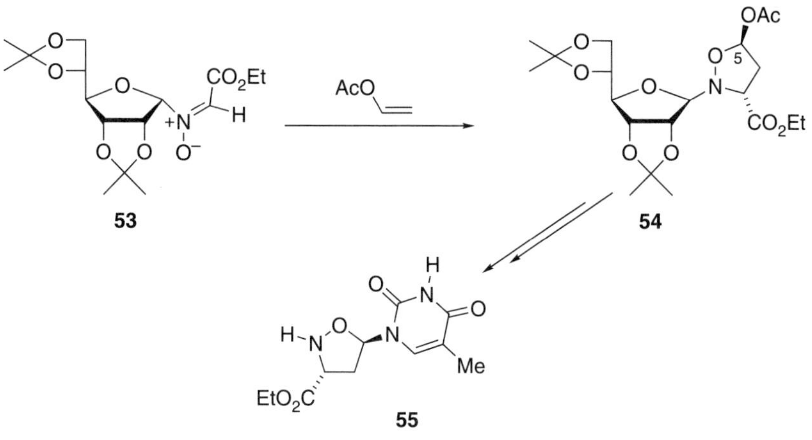

SCHEME 10.18 Use of *N*-glycosylnitrones in the preparation of chiral nucleoside analogs.

was high. The epimeric isoxazolidines were subsequently transformed into the *N*-unsubstituted nucleoside **55**.

Whitney and coworkers [48] reported diastereoselective reactions between chiral nitrones and chiral dipolarophiles. The reaction of Vasella's nitrone **56** with the protected vinylglycine derivative **57** afforded isoxazolidines **58** with high diastereoselectivity (19:1) (Scheme 10.19). After removal of the glycosyl residue, the major isoxazolidine was converted into the antibiotic acivicin [48].

Chiral nitrones **59** prepared from unprotected D-glucopyranosyl oxime underwent 1,3-dipolar cycloadditions with *N*-arylmaleimides to give the *anti*-isoxazolidinones **60** as the major isomers [49]. The *syn*-diastereomers **61** were formed with a diastereomeric excess of more than 90% only when the Ar^2 group was a 2,6-disubstituted arene (Scheme 10.20). The hydrogen bond between the

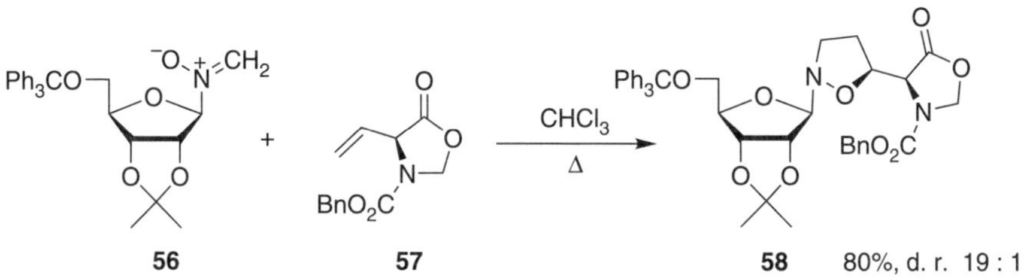

SCHEME 10.19 Application of *N*-glycosylnitrone cyclizations in the preparation of the antibiotic acivicin.

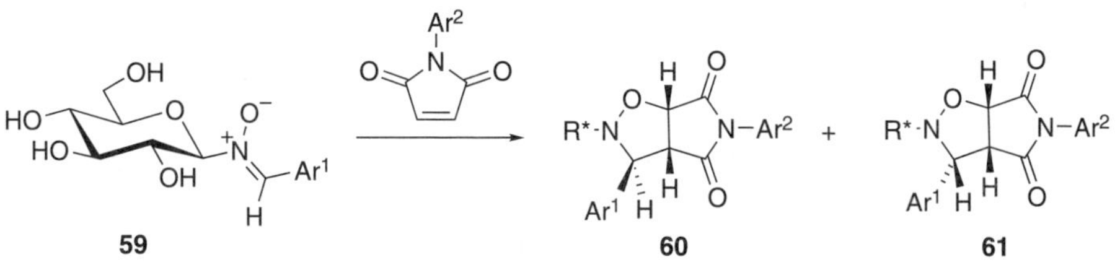

SCHEME 10.20 1,3-Dipolar cycloadditions of *N*-arylmaleimides with chiral nitrones derived fiom unprotected glucose.

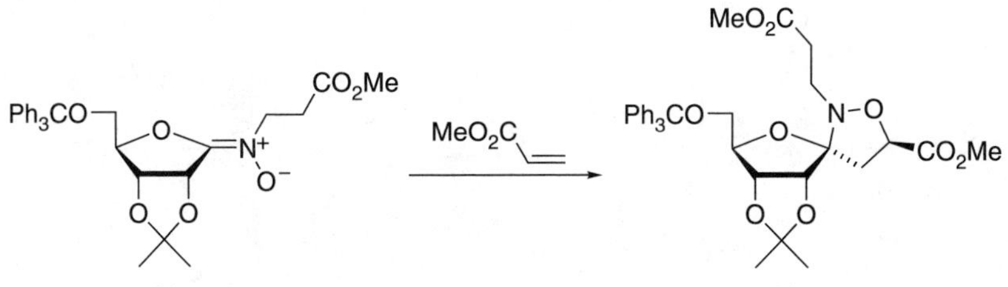

SCHEME 10.21 Generation of chiral spiro isoxazolidines from sugar-derived nitrones.

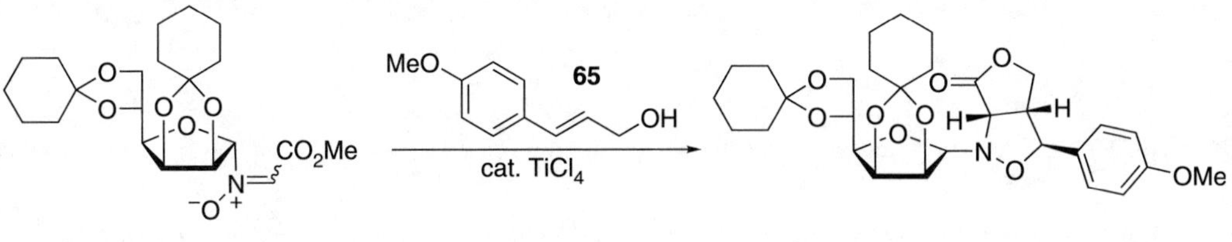

SCHEME 10.22 Intramolecular 1,3-dipolar cycloadditions with *N*-glycosylnitrones.

nitrone oxygen and the hydroxyl group at C-2 was assumed to control the dipole conformation and stereoselectivity of these reactions.

The chiral moiety of the nitrone can also be located at the carbon atom. Yokoyama et al. [50] used this approach in the asymmetric synthesis of spiro isoxazolidines (Scheme 10.21). The ribose-derived nitrone **62** is obtained from the corresponding oxime by Michael addition to methyl acrylate. With a second equivalent of methyl acrylate, the sugar nitrone **62** gave a single cycloadduct **63**, which was converted into the corresponding pyrrolidine by reduction.

Intramolecular 1,3-dipolar cycloadditions of alkenylnitrones have been widely applied in organic synthesis [41]. Tamura and Sakamoto [51] reported a short synthesis of the aminoacyl side chain of nikkomycin Bz through diastereoselective intramolecular nitrone 1,3-dipolar cycloaddition. Starting from L-gulonic-γ-lactone, they prepared the *N*-mannofuranosyl nitrone **64,** which underwent transesterification with (*E*)-*p*-methoxycinnamyl alcohol **65** and a catalytic amount of $TiCl_4$, followed by *in situ* 1,3-dipolar cycloaddition to give cycloadduct **66** as a single isomer in 75% yield (Scheme 10.22). However, analogous reactions of other allylic alcohols proceeded with poor diastereofacial selectivity.

10.2.3.2 Carbohydrate-Linked Dipolarophiles

A range of auxiliaries attached to dipolarophiles has been investigated in 1,3-dipolar cycloadditions. Carbohydrate auxiliaries have also been linked to the dipolarophile. Highly selective 1,3-dipolar cycloadditions were achieved using acryloyl esters of *chiro*-inositol derivatives as the auxiliaries [52]. The acryloyl ester **67** underwent 1,3-dipolar cycloaddition with nitrile oxides to give (*S*)-isoxazolines **68** with a diastereomeric excess of 90% (Scheme 10.23). The sugar auxiliary was removed by reduction of the ester group. In this study, the authors suggested that the major cycloadducts are formed from the *s-cis*-conformer **67** of the acrylate, whereas the minor ones stem from *s-trans*-conformer **69** (Figure 10.3). In any case, the bulky silyloxy group effectively hindered the attack from the *re*-face of the olefinic double bond [52].

Recently, Biao et al. investigated an asymmetric induction in 1,3-dipolar cycloaddition reactions of nitrile oxides to chiral alkenes [53]. They found that the reaction between aromatic (s-cis) R = tBuPh$_2$Si C_6H_6, 20°C 67 68 54-95%

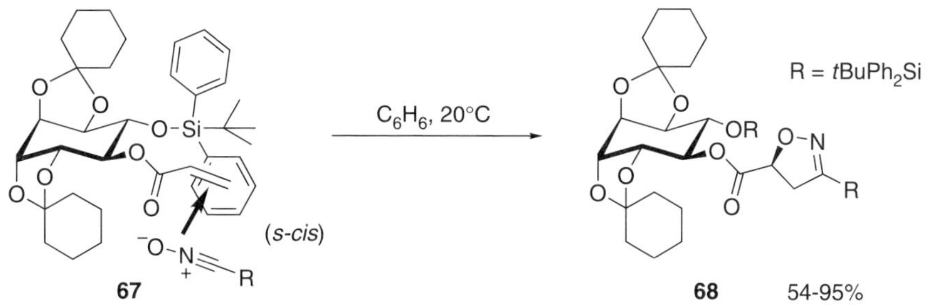

SCHEME 10.23 Generation of chiral isoxazolines by reaction of nitrile oxides with dipolarophiles bearing sugar-based auxiliaries.

(s-trans) 69

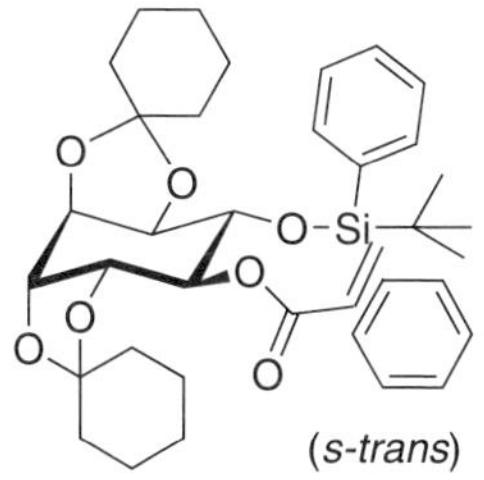

FIGURE 10.3 Minor (*s-trans*) conformer **69** of dipolarophiles **67** (Scheme 10.23, Ref. [25]).

nitrile oxides and acryloyl ester **70** derived from diacetone glucose provided adduct **71** with a diastereoselectivity up to 70% *de* (Scheme 10.24). In these reactions, the *re*-face of the double bond of **70** is more readily accessible than the *si*-face, which is blocked by the bulky 5,6-isopropylidene group [53].

10.2.4 [4 + 2] Cycloadditions (Diels–Alder Reactions)

The unique feature of the Diels–Alder reaction — generating up to four new stereogenic centers in one step — combined with the variety of useful reactants, has made this reaction a favorite methodology for constructing molecular diversity [54]. Therefore, asymmetric versions of this reaction represent an attractive tool for the diastereoselective synthesis of optically active compounds [55,56]. Various carbohydrate auxiliaries, either attached to the diene or to the dienophile component, have been successfully used to produce stereochemically pure cycloadducts [57].

Initially planned as a possible pathway to disaccharides, the thermal reaction of the benzyl protected glucofuranosyl butadienyl ether **72** with butyl glyoxylate affording 1-*O*-linked

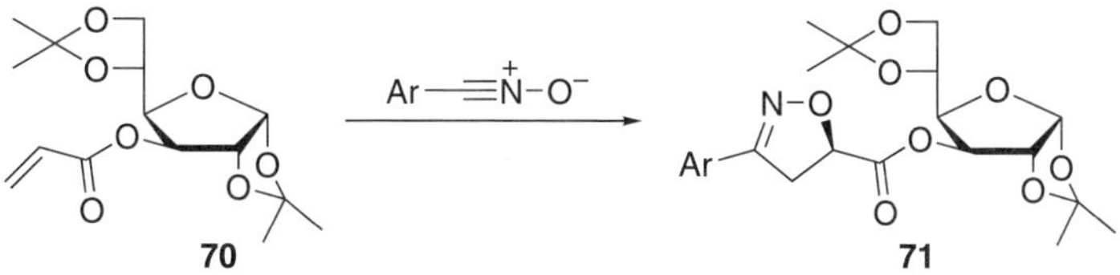

70 71

SCHEME 10.24 Generation of chiral isoxazolines by reaction of nitrile oxides with dipolarophiles bearing sugar-based auxiliaries.

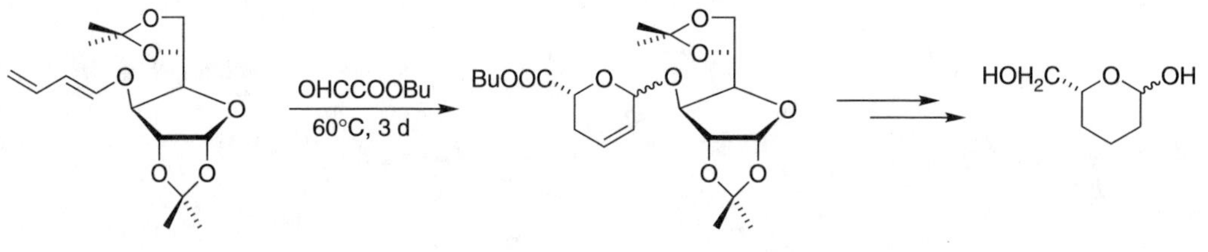

72 73 93% 74

endo : *exo* 78 : 22

endo : d. r. 77 : 23

SCHEME 10.25 The first example of a carbohydrate auxiliary-controlled Diels-Alder reaction.

2,3,4-trideoxyhexopyranose **73** can be considered as the first carbohydrate auxiliary-controlled Diels–Alder reaction described in the literature [58,59] (Scheme 10.25). Further manipulations and acidic hydrolytic cleavage of the glucose moiety gave the trideoxyhexopyranose **74**. The *endo*-selectivity and the diastereofacial differentiation were modest, possibly because of the high temperature applied and the fact that no activating and chelating Lewis acid was used.

Dienophiles linked to carbohydrates rapidly undergo Lewis acid-promoted asymmetric cycloadditions at low temperature. In these cases, the diastereofacial differentiation arises from the chiral framework of the carbohydrate coupled with the co-ordination of the dienophile to the Lewis acid. In this manner, the 3-*O*-acryloylglucofuranose derivative **75** stereoselectively reacted with cyclopentadiene in the presence of $TiCl_4$ furnishing the *endo*-adduct **77** as the (*R*)-diastereomer [60] (Scheme 10.26). The stereochemical outcome of this reaction could be rationalized by formation of reactive complex **76** in which the C_α-*re*-face of the dienophile is effectively shielded by the titanium moiety. Owing to the reduced Lewis acid potency of the catalyst bound as a titanate, the cycloaddition at low temperature only occurred with the reactive cyclopentadiene. This limitation was overcome by introduction of the 4,6-dipivaloylated 3-*O*-acryloyl-dihydro-L-glucal **78**, which underwent facile reaction with several dienes at 0°C under promotion with Lewis acids of varying strengths. The carbohydrate-bound cycloadducts (e.g. **79**) were formed with good to excellent stereoselectivity and complete *endo*-selectivity [61] (Scheme 10.27). In these cycloadditions, the titanium catalyst plays an important role — it coordinates to the carbonyl groups of both acryloyl and pivaloyl functionalities, thereby activating the dienophile and renders

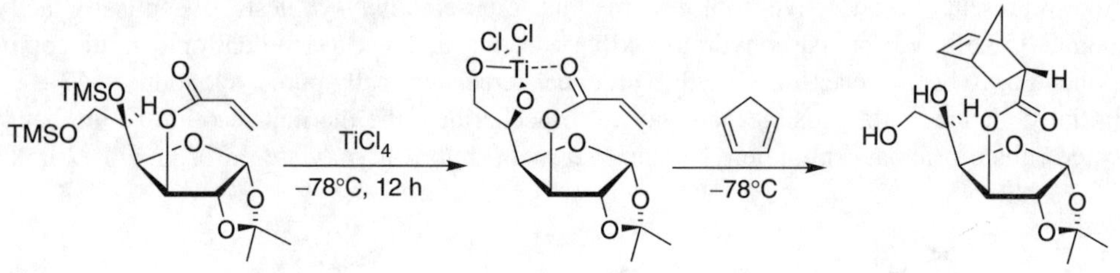

75 76 77 75%

endo : *exo* 100 : 0

endo: d. r. 93 : 7

SCHEME 10.26 Lewis acid coordination-assisted Diels-Alder reaction of a 3-*O*-acryloylglucofuranose derivative with cyclopentadiene.

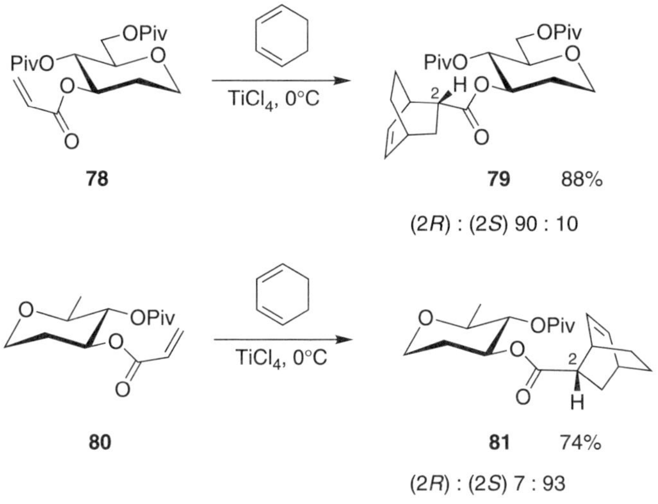

TiCl$_4$, 0°C

78

79 88%

(2*R*) : (2*S*) 90 : 10

TiCl$_4$, 0°C

80

81 74%

(2*R*) : (2*S*) 7 : 93

SCHEME 10.27 Lewis acid coordination-assisted Diels-Alder reaction of acryloylglycopyranose derivatives with cyclohexadiene.

the 4-*O*-pivaloyloxy group a bulky ligand, which effectively shields the *re*-face of the acrylate. By using the pseudoenantiomeric dihydro-L-rhamnal derivative **80**, complete reversal of stereoselectivity was observed, and the (*S*)-configurated *endo*-adduct **81** was formed [61] (Scheme 10.27).

A similar approach to enantiomerically pure norbornene derivatives was developed by Nouguier et al. who employed 1,3:2,4-di-*O*-methylene acetals of pentitols as chiral templates [62]. Hence, the 5-*O*-acryloyl-D-arabinatol derivative **82** underwent highly stereoselective Lewis acid catalyzed cycloaddition with cyclopentadiene, giving **83** (Scheme 10.28). The stereochemical outcome of the reaction was explained in terms of the chelate complex **84**, in which the chair-like dioxane ring and the acrylic moiety are fixed in two parallel planes, forcing the diene to approach the cisoid acrylate from the *si*-face. The synthesis and utility of various methylene protected glycosides have also been reported by this group [63–66].

Enhanced diastereofacial discrimination of an acrylic double bond was observed when *p*-methylbenzyl β-L-arabinopyranoside **85** bearing an η^6-chromium(0) tricarbonyl moiety reacted with dienes (e.g. isoprene) [67]. After decomplexation with pyridine, cycloadduct **86** was isolated as a mixture of diastereomers (90% *de*), whereas the uncomplexed counterpart showed inferior

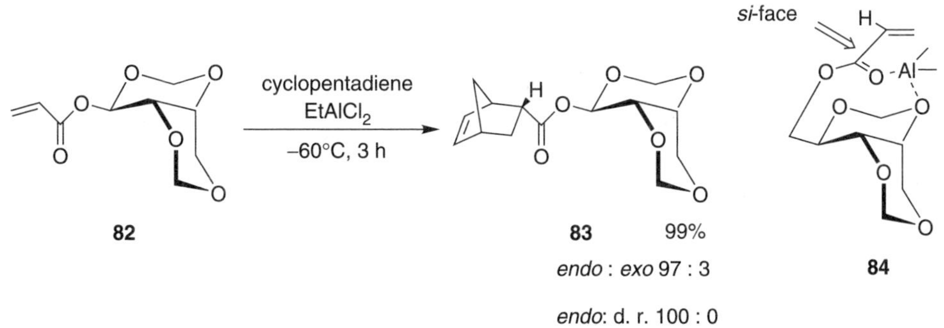

cyclopentadiene
EtAlCl$_2$
−60°C, 3 h

82

83 99%

endo : *exo* 97 : 3

endo: d. r. 100 : 0

si-face

84

SCHEME 10.28 Lewis acid coordination-assisted Diels-Alder reaction of a 5-*O*-acryloylarabinatol derivative with cyclopentadiene.

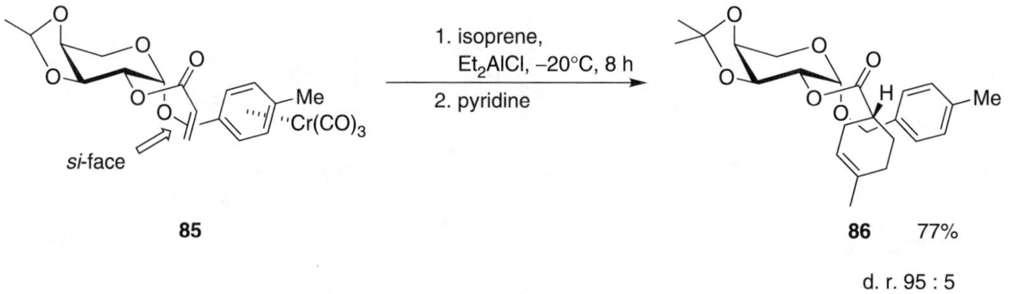

SCHEME 10.29 The stereochemical outcome of the Lewis acid assisted Diels-Alder reaction can be influenced by chromium complexation.

diastereoselectivity (56% *de*) in the Diels–Alder reaction owing to the diminished bulk and increased flexibility of the phenyl ligand (Scheme 10.29).

A striking solvent-dependent reversal of facial selectivity was found when the L-quebrachitol based acrylate **87** reacted with cyclopentadiene in the presence of Lewis acid catalysts such as $TiCl_4$ or $SnCl_4$ [68] (Scheme 10.30). When the reaction was performed in Et_2O, the (*S*)-*endo*-cycloadduct **88** was furnished exclusively, possibly because of the formation of complex **90** favoring *re*-face attack of the diene. In contrast, use of noncoordinating solvents such as toluene led predominantly to the (*R*)-configured adduct **89**, indicative of formation of chelate complex **91** and thereby preferred *si*-face attack of the diene (Figure 10.4). Related solvent effects were observed by Loupy [69] and Ferreira [70], who studied the stereodifferentiating potential of isosorbide- and galactose-derived acrylates, respectively, in the cycloaddition with cyclopentadiene.

Carbohydrate-based *N*-acryloyl, -cinnamoyl and -crotonoyl 1,3-oxazolidin-2-ones have also been shown to undergo highly stereoselective cycloadditions with cyclopentadiene. Hence, D-galactosyl-spirooxazolidinones **92** were converted to the corresponding *endo*-adducts **93** upon addition of Et_2AlCl [71] (Scheme 10.31). The origin of the high diastereofacial selectivity was rationalized by an intermediate aluminum chelate complex in which cyclopentadiene preferentially attacks from the C_α-*re*-face (front face as drawn) of the *all-cis*-configured imide. The adducts were cleaved from the chiral matrix using lithium benzyloxide also leading to partial epimerization about the spiro-carbon atom of the recovered galactosyl auxiliary. This drawback could be circumnavigated by investigation of 1,3-oxazin-2-ones **94** and **95** derived from 2,3:4,6-di-*O*-isopropylidene-2-keto-L-gulonic acid [72] and 2,3:4,5-di-*O*-isopropylidene-β-D-fructopyranose

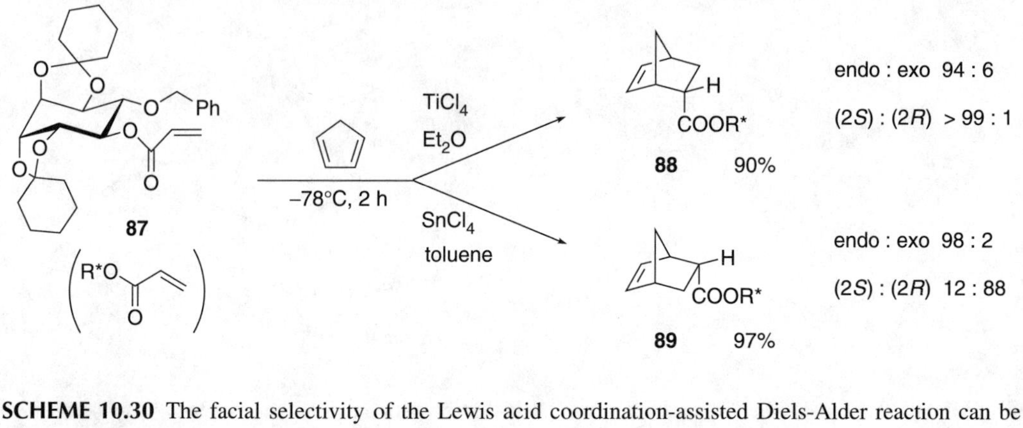

SCHEME 10.30 The facial selectivity of the Lewis acid coordination-assisted Diels-Alder reaction can be solvent-dependent.

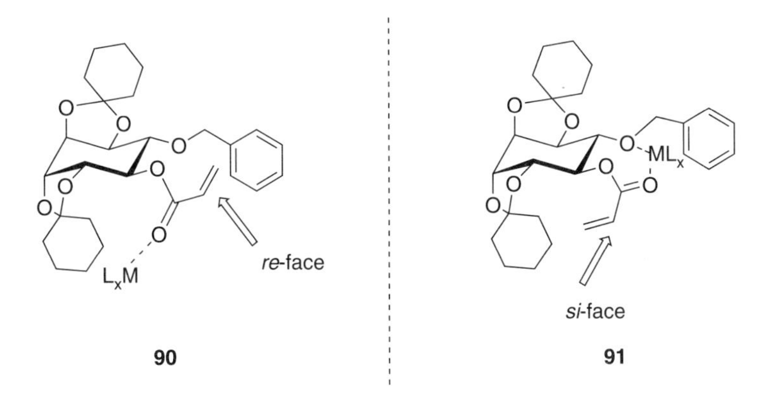

FIGURE 10.4 Solvent depending diastereodifferentiation on L-quebrachitol acrylates **87** (Scheme 10.30, Ref. [68]).

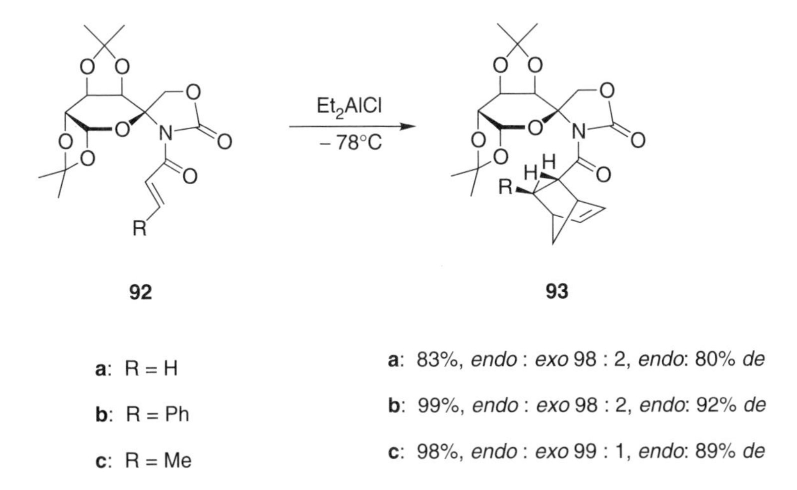

SCHEME 10.31 Carbohydrate-based 1,3-oxazolidin-2-ones can undergo highly stereoselective cycloadditions with cyclopentadiene.

[73], respectively (Figure 10.5). These auxiliaries exerted equally high levels of asymmetric induction (comparable to **92**), and could be recovered in enantiomerically pure form after cleavage of the resulting diastereomers.

A similar approach towards norbornane carboxylic acid derivatives was developed using pseudoenantiomeric *N*-acyl-D-galactosyl and D-arabinosyl-oxazolidinones **96** and **97**. Reaction with cyclopentadiene under promotion of Me_2AlCl, and subsequent hydrogenolysis of the double bond and basic hydrolysis of the imide moiety, furnished the 3-methyl-norbornane-2-carboxylic acids **98** and **99**, respectively, in high enantiomeric purity [74,75] (Scheme 10.32).

Horton et al. were interested in the synthesis of tetra-*C*-substituted carbocycles, and extensively studied asymmetric Diels–Alder reactions employing acyclic unsaturated carbohydrate derivatives. Thus, the thermal cycloaddition of the D-arabinose-derived *cis*-dienophile (*Z*)-**100** with cyclopentadiene gave *endo*-adduct **101** in excellent optical purity. The high diastereofacial differentiation in this reaction arises from a highly favored conformation at the allylic center [76,77] (Scheme 10.33). Therefore, conformer **102** seems to be more favored than **103**, where the latter suffers from severe allylic strain between the methoxycarbonyl and the C_4-acetoxy group

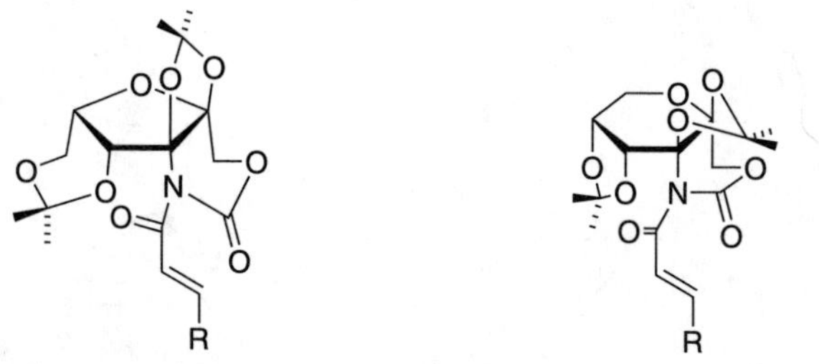

94 95

FIGURE 10.5 *N*-Acryloyl oxazinones derived from L-gulose and D-fructose Ref. [72,73].

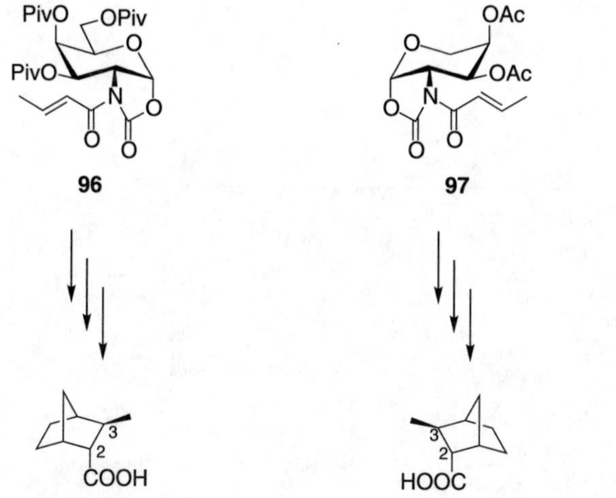

98 99

(2*S*,3*R*) : (2*R*,3*S*) 98 : 2 (2*S*,3*R*) : (2*R*,3*S*) 6 : 94

SCHEME 10.32 *N*-Acyl-D-glycosyl-oxazolidinones can undergo highly stereoselective cycloadditions with cyclopentadiene.

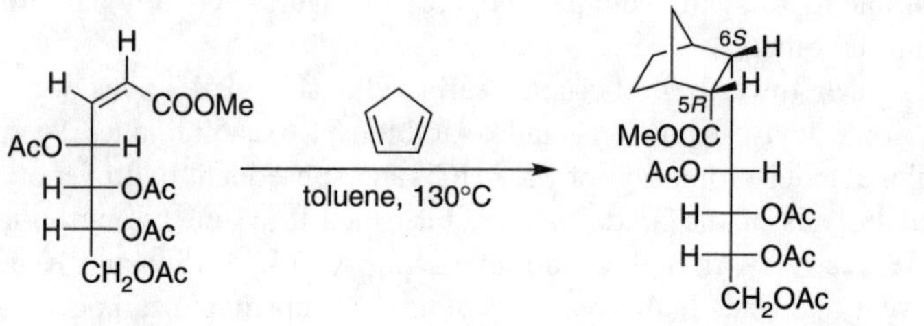

(*Z*)-**100** **101** quant.

d. r. 95 : 5

SCHEME 10.33 Acyclic sugar-derived dienophiles can undergo highly stereoselective cycloadditions with cyclopentadiene.

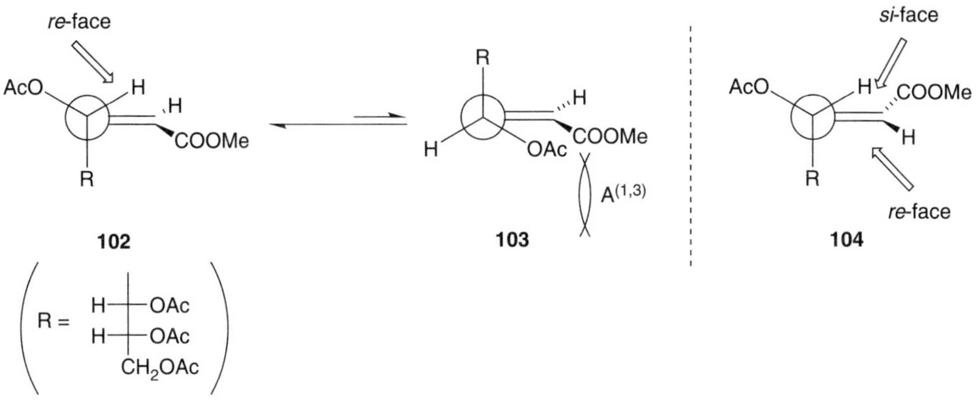

FIGURE 10.6 Diastereodifferentiation in Diels-Alder reactions of dienophile (*Z*)-**100** (Scheme 10.33, Ref. [76,77]).

(Figure 10.6). Hence, C_α-*re*-face attack of the diene preferentially occurs, resulting in formation of the (5*R*,6*S*)-configured adduct **101**. The conformational hypothesis aforementioned was reinforced by the Diels–Alder reaction of butenolide **105** providing a fixed conformation around the allylic oxygen atom (Scheme 10.34). In this case, the *endo*-adduct **106** was isolated as the pure (5*S*,6*R*)-diastereomer, along with some *exo*-isomer, indicating exclusive attack of the diene at the C_α-*si*-face of the enoate. The C_α-*re*-face is efficiently shielded by the steric bulk of the R moiety (Figure 10.7). In contrast, the cyloaddition of the *trans*-isomer (*E*)-**100** with cyclopentadiene furnished a mixture of the four possible stereoisomers (*endo*:*exo* 31:69, *si*:*re* 64:36) [78]. In this case, the diastereofacial differentiation was insufficient because of the conformational mobility of the sugar chain lacking any allylic strain (see **104** in Figure 10.6).

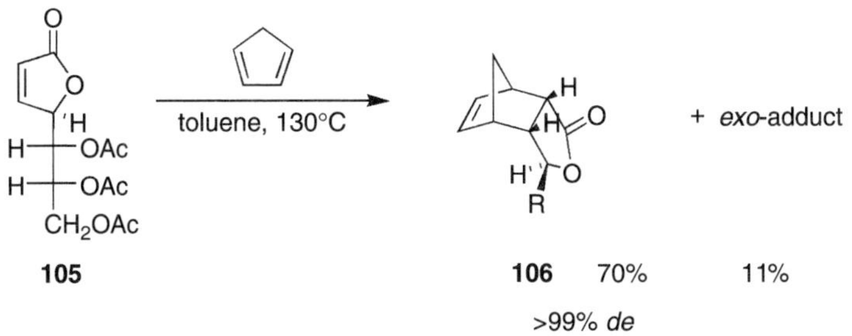

SCHEME 10.34 Acyclic sugar-derived dienophiles can undergo highly stereoselective cycloadditions with cyclopentadiene.

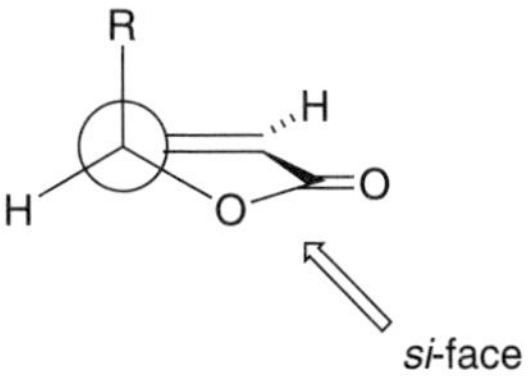

FIGURE 10.7 Diastereofacial differentiation on butenolide **105** (Scheme 10.34, Ref. [78]).

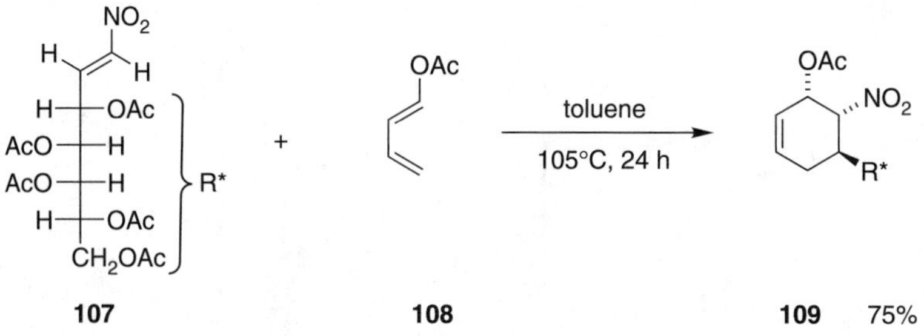

SCHEME 10.35 Reaction of chiral sugar-nitroalkanes in [4+2] cycloadditions.

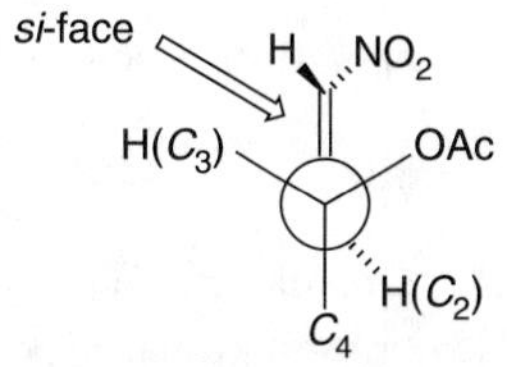

FIGURE 10.8 Diastereofacial differentiation on nitro alkene **107** (Scheme 10.35, Ref. [92]).

Chiral sugar-nitroalkenes have been shown to undergo thermal [4+2] cycloadditions with 2,3-dimethylbuta-1,3-diene [79] and 1-*O*-substituted buta-1,3-dienes [80]. In the latter case, reaction of D-*galacto* nitroalkene **107** with 1-acetoxy-buta-1,3-diene **108** furnished adduct **109** with complete regio-, *endo*- and diastereofacial selectivity [81] (Scheme 10.35). Thus, attack of the diene must occur at the less hindered face of the nitroolefin — that is, at the C_1-*si*-face in its presumably most stable conformer (Figure 10.8). This result parallels the empirical finding of Franck et al. [82] that the face selectivity in intermolecular Diels–Alder reactions is a predictable function of the configuration of the chiral center adjacent to the dienophilic double bond. The result is also in accordance with the stereochemical rule based on electrostatic considerations given by Kahn and Hehre [83].

The reaction of glucopyranosyl 1-oxy-1,3-butadienes with a series of dienophilic components was comprehensively studied by Stoodley et al. [84,85]. Thus, at ambient temperature, cycloaddition of the chiral diene **110a**, for example, with *p*-benzoquinone, resulted in the predominant formation of adduct **111a** (Scheme 10.36), arising from *endo*-addition of the dienophile to the less hindered top-face of the diene in conformer **112a** (Figure 10.9). The formation of the minor component is believed to occur via top-face attack to conformer **113a** (both conformers are in agreement with the *exo*-anomeric effect). When methylated diene **110b** was reacted with the quinone, enhanced facial discrimination was observed, showing that the 2-substituted counterpart offers less conformational flexibility. Conformer **113b** is more constrained than **113a** owing to severe interaction between the C_2-methyl group and the anomeric hydrogen [85].

Full regio- and stereoselectivity were also achieved in the cycloaddition of 5-glucopyranosyloxy-1,4-naphthoquinone **114** with Danishefsky's diene. The adduct **115** was isolated in high yield as a single isomer [86] (Scheme 10.37). The authors provided an explanation for the stereochemical outcome (1,6-stereoinduction) of this interesting reaction involving a boat-like geometry of the quinone ring, which effectively shields the *syn*-face of the C_2-C_3 double bond and therefore guarantees preferential *endo*-attack of the diene to the *anti*-face.

The methodology described above was recently exploited for an effective asymmetric total synthesis of (−)-4-*epi*-shikimic acid **118** [87]. The cycloaddition of maleic anhydride with glucopyranosyloxy diene **116** served as the key step in this approach, which generated bicyclic

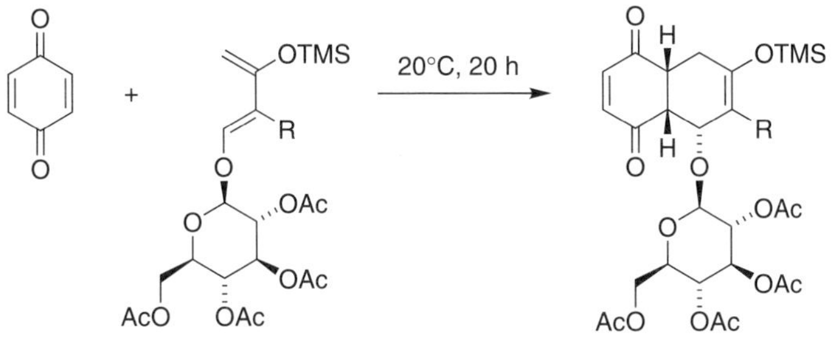

20°C, 20 h

110

a: R = H

b: R = Me

111

a: 46% d. r. 89 : 11

b: 69% d. r. >95 : 5

SCHEME 10.36 Reaction of glucopyranosyl 1 -oxy-1,4-butadienes with *p*-benzoquinone.

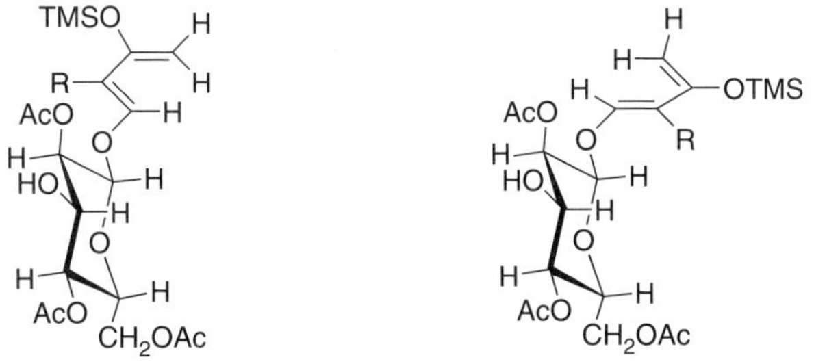

112

a: R = H

b: R = Me

113

a: R = H

b: R = Me

FIGURE 10.9 Diastereofacial differentiation on *O*-glucosyl butadienes (Scheme 10.36 [84,85]).

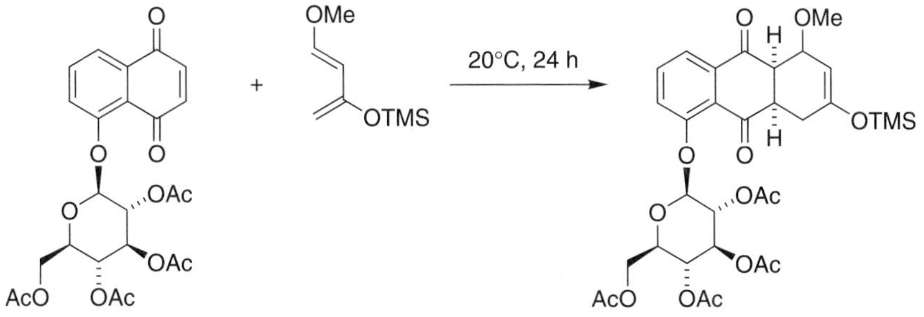

20°C, 24 h

114

115 92%

SCHEME 10.37 Reaction of 5-glucopyranosyloxy-1,4-naphthoquinone with Danishefsky's diene.

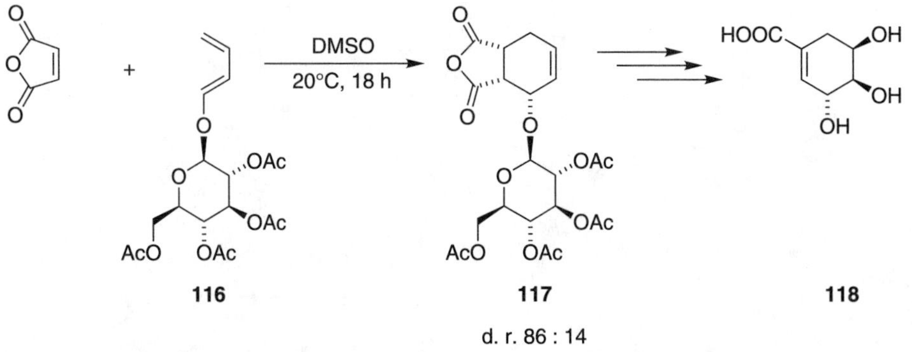

SCHEME 10.38 Application of sugar-directed Diels-Alder reactions in the preparation of (-)-4-*epi*-shikimic acid, **118**.

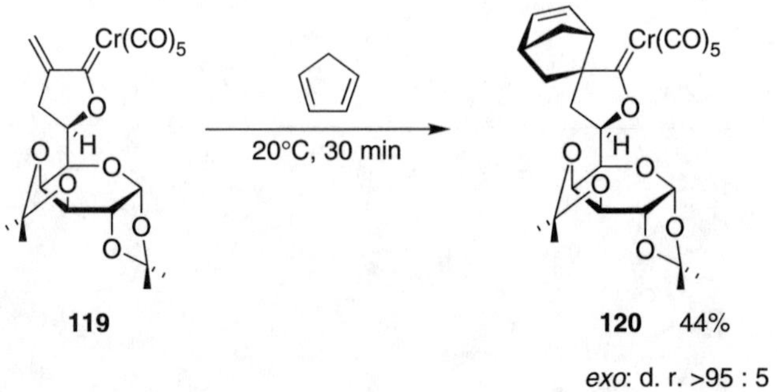

SCHEME 10.39 An example of an *exo*-selective Diels-Alder reaction directed by a sugar derivative and chromium complexation.

anhydride **117** with good stereoselectivity. This compound was further transformed to D-glucose and **118** (Scheme 10.38).

Carbohydrate-functionalized α-*exo*-methylene-2-oxacyclopentylidene chromium complexes [88,89] based on D-glyceraldehyde [90] and D-galactose [91], and their ability to undergo highly diastereoselective Diels–Alder reactions, were investigated by Doetz and coworkers. In the latter case, oxacyclopentylidene chromium complex **119**, readily obtained from 6-aldehydo-1,2:3,4-di-*O*-isopropylidene-α-D-galactopyranose by subsequent γ-addition of allenylmagnesiumbromide, cycloisomerization of the resulting ω-alkynol, separation of the diastereomers and α-*exo*-methylene functionalization was reacted with cyclopentadiene [91] (Scheme 10.39). The cycloadduct **120**, formed at room temperature as a single isomer, was determined to be the result of an *exo*-selective addition of the diene to the more readily accessible *si*-face of the C=C bond. This reflects the steric influence of both the bulky pentacarbonylchromium fragment and the arabinopyranosyl moiety (Figure 10.10).

10.2.5 HETERO DIELS–ALDER REACTIONS

Efficient hetero Diels–Alder reactions have been carried out using α-chloronitroso saccharides as dienophiles and various symmetrical and unsymmetrical dienes. Thus, 2,3:5,6-di-*O*-isopropylidene-1-*C*-nitroso-α-D-mannofuranosyl chloride **122**, easily prepared from hydroximolactone **121** and *tert*-butylhypochlorite, was proven to be a stable and highly reactive dienophilic compound [92].

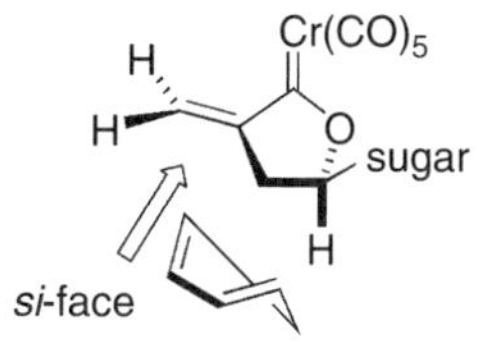

FIGURE 10.10 Face differentiation on methylene-2-oxacyclopentylidene chromium complexes **119** (Scheme 10.39, Ref. [91]).

Cycloaddition of **122** with *cis*-1,2-bisacetoxy-3,5-cyclohexadiene furnished bisacetoxy 3,6-dihydro-2*H*-1,2-oxazine **126**, which was isolated as its hydrochloride in high enantiomeric purity [93] (Scheme 10.40). The reaction is presumed to involve an *exo*-attack of the diene at the sterically less hindered *re*-face of the nitroso double bond oriented syn-periplanar to the C_1–O bond, favored

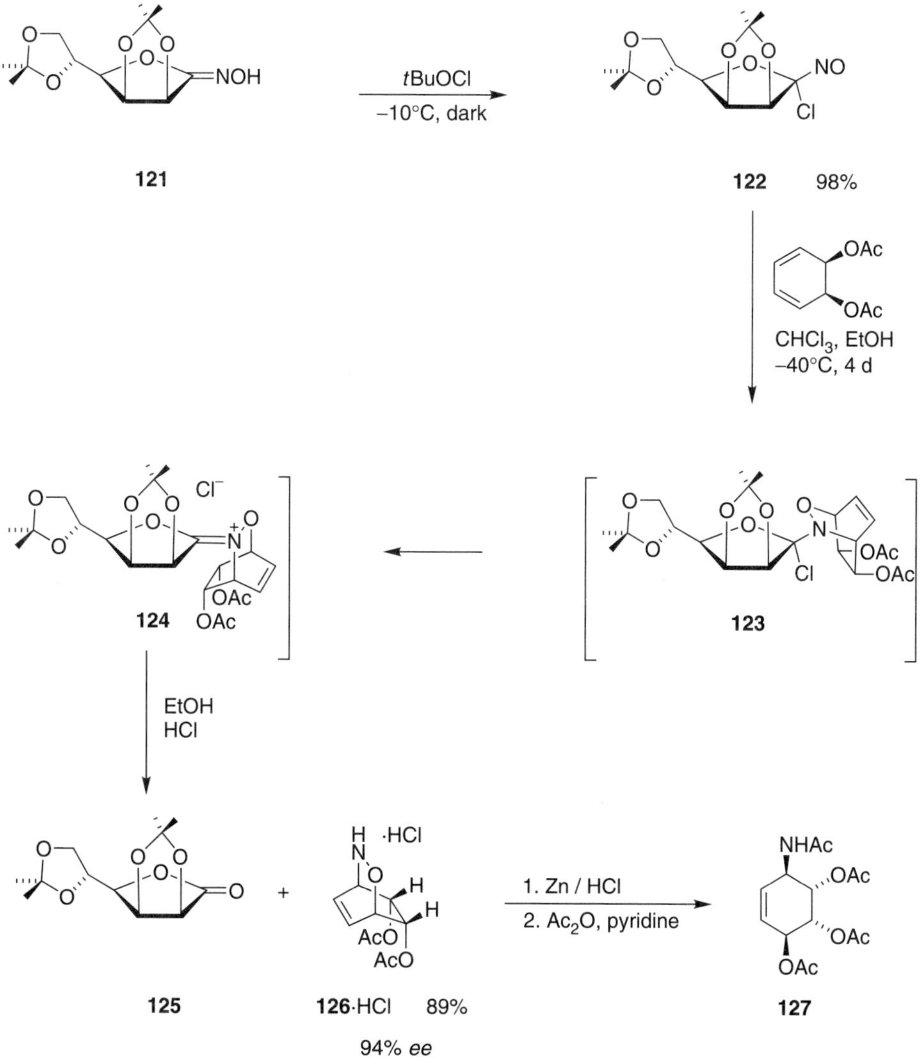

SCHEME 10.40 Application of a sugar-directed hetero Diels-Alder reaction to the preparation of conduramine A1 tetraacetate, **127**.

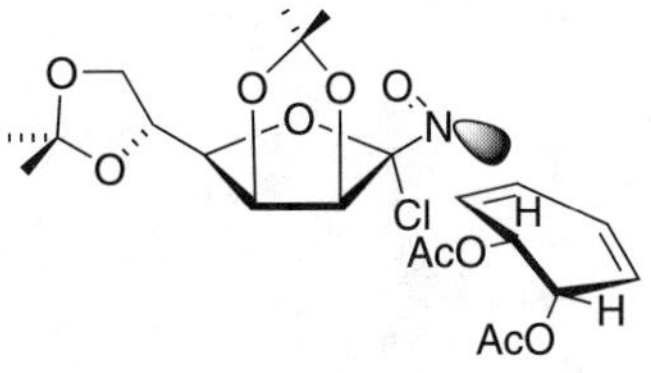

FIGURE 10.11 Diastereodifferentiation on 1-nitro mannofuranosyl chloride **122** (Ref. [92]).

by an overlap of the nitrogen n-orbital with the σ^*-orbital of the adjacent C_1–O bond (Figure 10.11). The 1-chloro-1-hydroxamino derivative **123** thus formed undergoes subsequent elimination of chloride. The intermediate iminium salt **124** is solvolyzed by ethanol yielding bicyclic dihydrooxazine **126** and mannono lactone **125**. Finally, reductive cleavage of the N–O bond using Zn/HCl and acetylation led to aminocyclitol conduramine A1 tetraacetate **127**. Similarly, employment of the D-*ribo* configured chloro nitroso ether **128**, a pseudoenantiomer of **122** (Figure 10.12), enabled the synthesis of a series of enantiomers of those obtained using the D-*manno* derivative **122** [94].

Compared with the reaction with cyclohexadiene, the diastereoselectivity of the Diels–Alder reaction of **122** with cyclopentadiene was found to be lower under a variety of reaction conditions. In the latter case, the cycloaddition afforded the [2.2.1] bicyclic salt **129**, which was further converted to **130** — the precursor of a carbocyclic L-nucleoside [95] (Scheme 10.41).

In situ generated, highly reactive nitroso olefins have been shown to undergo facile cycloadditions with enol ethers, for example, **131** derived from 1,2:5,6-di-*O*-isopropylidene-α-D-glucofuranose giving chiral 4,5-dihydro-6*H*-1,2-oxazine derivative **132** [96] (Scheme 10.42). It is important to note that (*E*)-enol ethers reacted much faster and more stereoselectively than the (*Z*)-configurated ones, allowing efficient stereocontrol even if (*E*)/(*Z*)-mixtures containing the (*Z*)-isomer as the minor component were used.

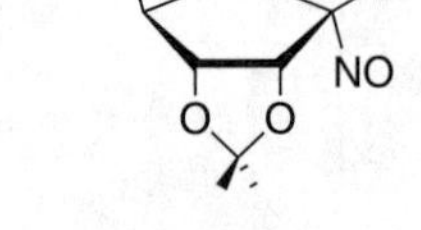

FIGURE 10.12 Protected 1-nitro ribofuranosyl chloride.

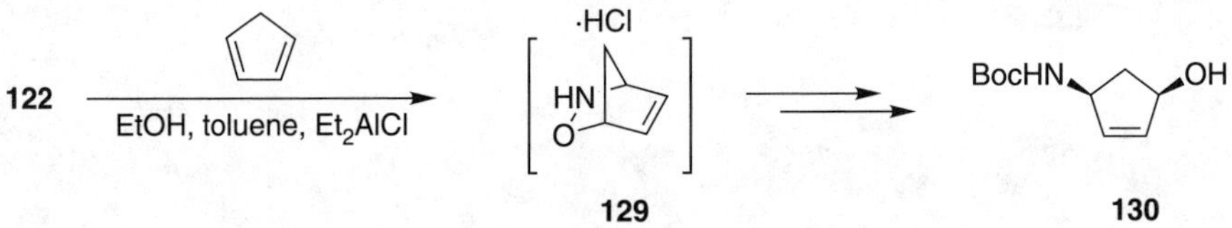

SCHEME 10.41 Preparation of a carbocyclic nucleoside precursor via a sugar-directed hetero Diels-Alder reaction.

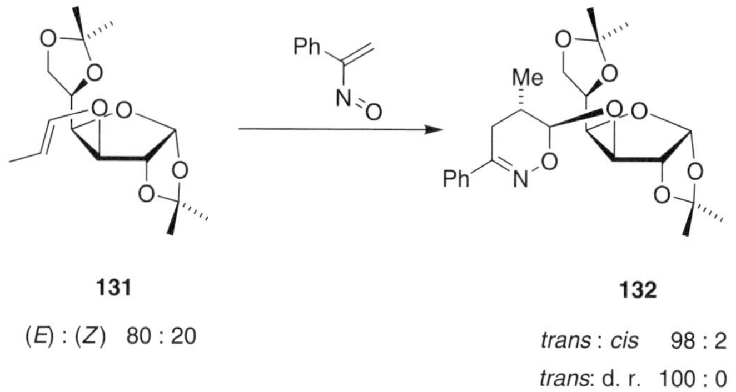
Ph
N=O
Me
Ph
N
O
131
(E) : (Z) 80 : 20
132
trans : cis 98 : 2
trans: d. r. 100 : 0

SCHEME 10.42 *In situ* generated nitroso olefins undergo facile cycloadditions with enol ethers.

Further progress towards the stereoselective synthesis of dihydrooxazines has also been reported by Reissig and coworkers who exploited the formal [3+3] reaction of lithiated methoxyallene with nitrones derived from (*R*)-glyceraldehyde [97,98]. Subsequent transformations enabled the synthesis of enantiomerically pure pyrrolidine and 2,5-dihydropyrrole derivatives incorporating one asymmetric unit of the chiral template.

A highly stereoselective variant of the tandem [4+2]/[3+2]-fused mode Denmark-type reaction [99] using a nitroalkenyl sugar as electron-poor heterodiene was described as a valuable methodology for the synthesis of bicyclic nitrogen heterocycles [100–103]. Thus, the three-component reaction of D-galactose-derived nitroalkene **133a** with ethyl vinyl ether **134** and methyl acrylate **135** furnished the bicyclic nitroso acetal **136a** with high facial selectivity at ambient temperature [102] (Scheme 10.43). The domino reaction involves preliminary regioselective hetero Diels–Alder reaction between the nitroalkene and the electron-rich enol ether and subsequent trapping of the resulting nitronate with the acrylate in a 1,3-dipolar cycloaddition. Structural assignments of the cycloadduct confirmed that the [4+2] process occurs via an *endo*-approach of the enol ether to the less hindered C_1-*si*-face of the heterodiene followed by an *exo*-attack of the acrylate at the C_1-*re*-face of the nitronate (Scheme 10.44). The *endo*-preference can be rationalized in terms of a stabilizing electrostatic interaction between the

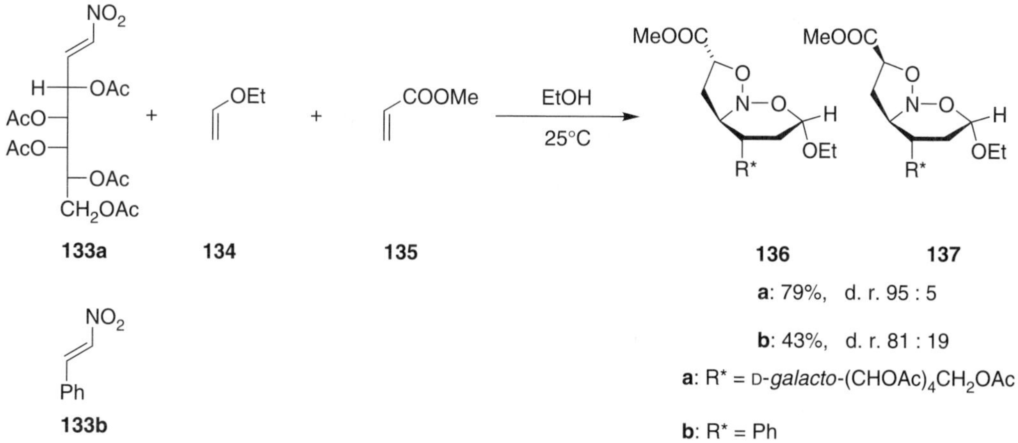
NO2
H—OAc
AcO—
AcO—
—OAc
CH2OAc
133a
+
OEt
134
+
COOMe
135
EtOH
25°C
MeOOC
N—O
H
OEt
R*
MeOOC
N—O
H
OEt
R*
136
137
a: 79%, d. r. 95 : 5
b: 43%, d. r. 81 : 19
a: R* = D-galacto-(CHOAc)4CH2OAc
b: R* = Ph
NO2
Ph
133b

SCHEME 10.43 A sugar directed tandem [4+2]/[3+2] Denmark-type reaction produces fused bicyclic structures with high stereoselectivity.

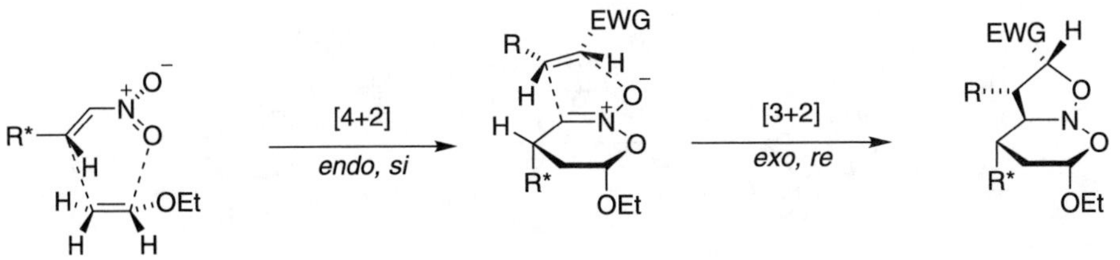

SCHEME 10.44 Mechanism of the tandem [4+2]/[3+2] Denmark-type cycloaddition.

oxygen on the enol ether and the charged nitrogen atom on the heterodiene. In contrast, a consideration of the competing transition structures strongly favored the *exo*-attack observed in the [3+2] reaction [103]. However, the chiral auxiliary exhibits a high level of diastereofacial differentiation in both cycloaddition steps. The analogous reaction of achiral β-nitrostyrene **133b** gave racemic bicycles *rac*-**136b** and *rac*-**137b**, respectively, in a diastereomeric ratio of only 81:19 (see Scheme 10.43).

Asymmetric aza-Diels–Alder reactions of *N*-galactosyl imines **138** with isoprene and similar dienes under promotion with zinc chloride etherate gave pure regioisomers of *N*-galactosyl piperidine derivatives **139** [104] (Scheme 10.45). Although the diastereoselectivity of these reactions was moderate reaching up to 90:10, pure diastereomers were usually obtained after flash chromatography. The synthesis and further applications of glycosyl imines will be described in Section 10.3.1.

Carbohydrate-directed asymmetric cycloaddition additions with aza-dienophiles were also explored by Stoodley and coworkers [105,106] and others [107]. Likewise, the reaction of β-D-glucopyranosyloxy-pentadienoate **141** with 4-phenyl-4*H*-1,2,4-triazoline-3,5-dione **140a** afforded the single diastereomer **142a** in good yield (Scheme 10.46). The absolute stereochemistry observed was rationalized by assuming an *endo*-attack of the triazoline to the less-hindered face of the diene (see Figure 10.9) [106]. Interestingly, when *N*-phenylmaleimide **140b** was reacted with **141**, an excess of dienophile and rather harsh conditions were required to furnish cycloadduct **142b** in a diastereomeric ratio of 75:25. This indicates that olefinic dienophiles show a diminished ability to discriminate the diastereotopic faces of a diene compared with their aza-dienophilic counterparts. This effect is caused by the shorter C=N bond enabling a closer approach of the aza-dienophile in the transition state of the reaction [108].

Oxa-Diels–Alder reactions between 1-oxy-3-silyloxybuta-1,3-dienes and aromatic aldehydes (either component being attached to a carbohydrate auxiliary) have been investigated by Stoodley et al. [109,110], drawing on the pioneering work of the Danishefsky group [111,112]. For example, the reaction of the carbohydrate-linked diene **143** with *p*-nitrobenzaldehyde in the presence of Eu(III) catalysts gave dihydropyrans **144–147** [109] (Scheme 10.47). When the chiral Eu complexes (+)-Eu(hfc)$_3$ and (−)-Eu(hfc)$_3$ were used, double stereodifferentiation was

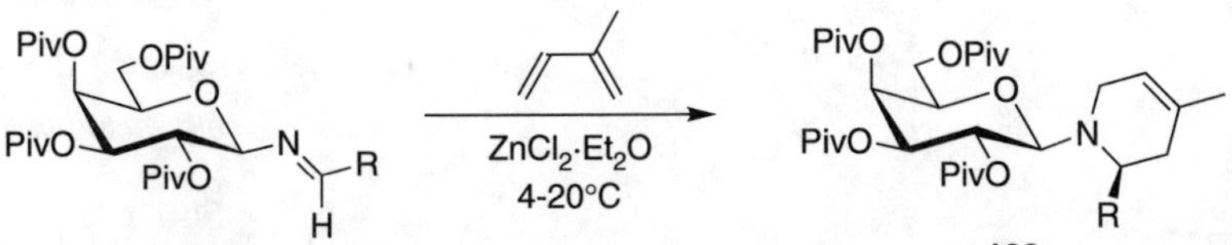

SCHEME 10.45 Asymmetric aza-Diels-Alder reactions can be achieved using sugar auxiliaries.

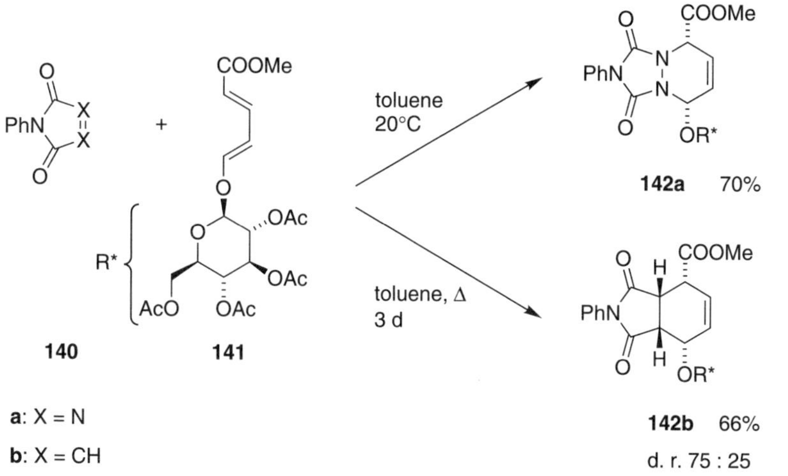

SCHEME 10.46 Carbohydrate-directed hetero Diels-Alder reactions with aza-dienophiles.

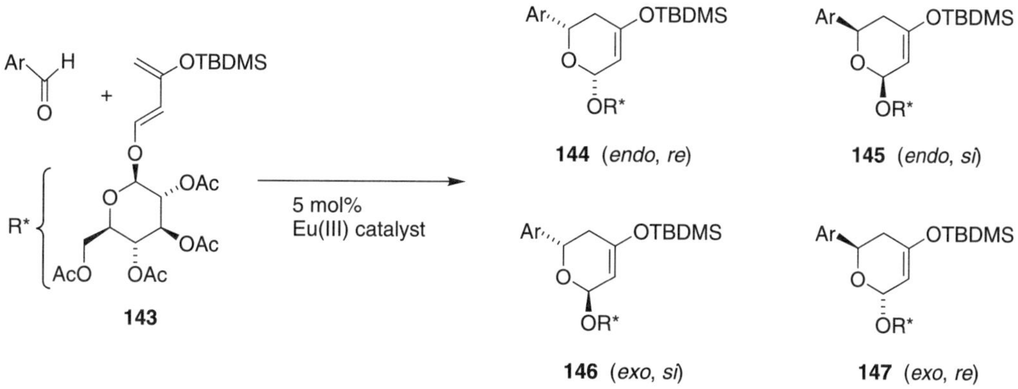

SCHEME 10.47 Sugar auxiliaries can influence the stereochemical outcome of oxa-Diels-Alder reactions.

observed yielding a 55:45 mixture of the *cis*-cycloadducts **144** and **145** for the mismatched case (using the dextrorotatory catalyst), and an 80:10:10 mixture of **144**, **146** and **147** for the matched case (using the levorotatory catalyst). The *cis*-isomers **144** and **145** arise from *endo*-attack of the aldehyde at the diene's C_1-*re*- and C_1-*si*-face, respectively. The finding that with increasing reaction time, *exo*-compound **146** was formed as the major product when the achiral $Eu(fod)_3$ was employed has been attributed to a catalyst-dependent epimerization of the initially formed *endo*-adduct **144** to give the thermodynamically more stable *trans*-dihydropyran **146**. In any case, preferential formation of cycloadduct **144** could be explained by the *exo*-anomeric model already outlined in Figure 10.9.

The same group also studied the $BF_3{\cdot}OEt_2$ catalyzed inverse reaction between chiral 2-oxybenzaldehyde **148** and Danishefsky's diene, showing a notable example of 1,5-asymmetric induction [110] (Scheme 10.48). After acidic workup, a 9:1 mixture of the 2-aryl-2,3-dihydropyran-4-ones **149*S*** and **149*R*** was isolated obviously formed via an aldol-cyclocondensation sequence.

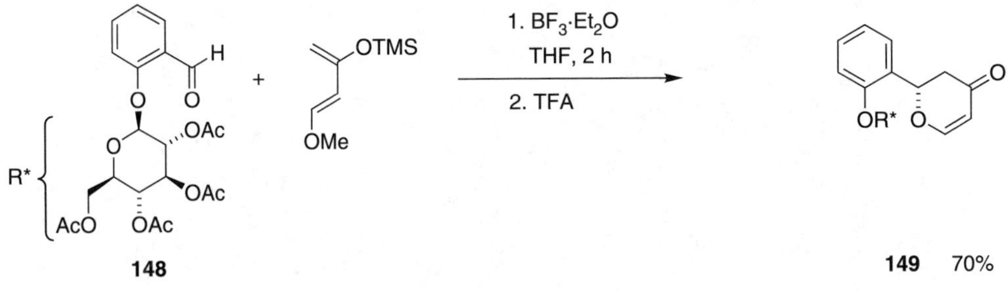

SCHEME 10.48 Sugar auxiliaries can influence the stereochemical outcome of oxa-Diels-Alder reactions.

10.3 STEREOSELECTIVE ADDITION AND SUBSTITUTION REACTIONS

10.3.1 ADDITIONS TO GLYCOSYL IMINES AND OTHER NUCLEOPHILIC ADDITIONS

Nucleophilic additions to *N*-glycosyl aldimines enable access to a wide variety of products, and have been extensively investigated [9,113,114]. In particular, 2,3,4,6-tetra-*O*-pivaloyl-β-D-galactopyranosylamine **150**, readily accessible in large quantities via a five-step protocol starting from D-galactose [115], has been proven to be a valuable chiral auxiliary for a variety of processes. The reaction with aromatic or aliphatic aldehydes either under acid-catalysis or using molecular sieves, respectively, delivered the pure β-anomeric (*E*)-configured aldimines **138**. The stereodifferentiating ability of imines **138** was first shown in Strecker syntheses using trimethylsilylcyanide and zinc chloride as a stoichiometrically added promotor, giving the α-amino nitriles **151** in high yields and diastereoselectivities [115,116] (Scheme 10.49). The diastereofacial differentiation in this reaction is ascribed to the steric, stereoelectronic and complexing effects of the carbohydrate. Therefore, preferential formation of the (*R*)-configured amino nitriles can be rationalized by formation of the zinc chelate complex **154**, in which the plane of the imine double bond is located perpendicular to the sugar plane (Figure 10.13). According to the *exo*-anomeric effect, this conformation is stabilized by $\pi \rightarrow \sigma^*$ delocalization of the C=N π electrons into the σ^*-orbital of the ring C–O bond. The coordination of the Lewis acid to both imine nitrogen and carbonyl oxygen of the equatorial C_2-pivaloyl group enables an effective

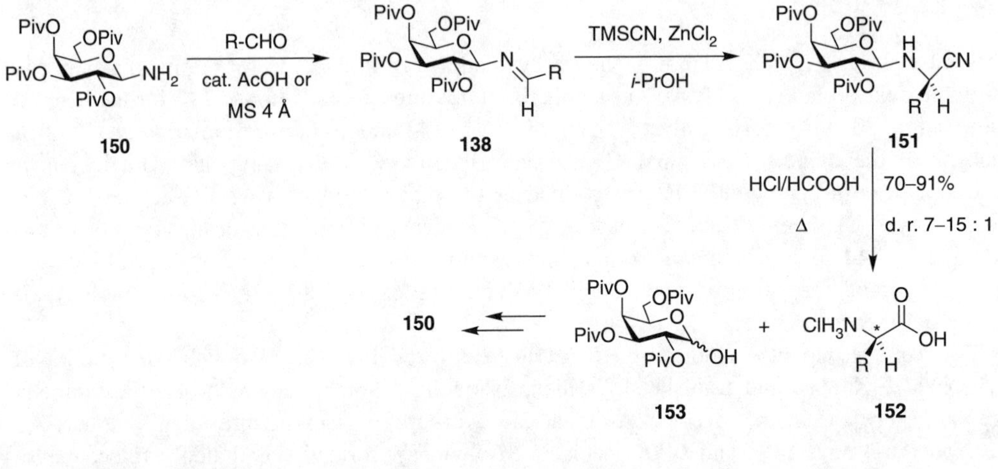

SCHEME 10.49 Stereoselective Strecker synthesis using glycosylamines.

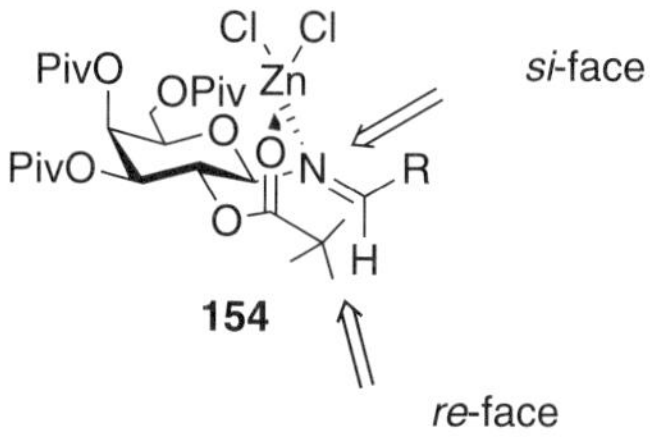

FIGURE 10.13 Diastereodifferentiation on *N*-galactosyl imines **138** (Scheme 10.49, Ref. [115–118).

shielding of the imine *re*-face, thus resulting in preferred nucleophilic attack of the released cyanide at the *si*-face (Figure 10.13). Subsequent cleavage from the auxiliary and hydrolysis of the nitrile were accomplished by treatment with HCl/formic acid to give pure D-amino acids **152** (also elucidating the absolute stereochemistry) and galactopyranose **153**, which could easily be regenerated to **150**.

The efficiency of this methodology was further demonstrated in asymmetric Ugi four-component condensations. Thus, galactosylamine **150** reacted with a variety of aldehydes, *tert*-butyl isocyanide and formic acid in the presence of zinc chloride under formation of the corresponding *N*-galactosyl-*N*-formyl-α-D-amino acid amides **155** [117,118] (Scheme 10.50). The amino acid derivatives **155** were obtained in high yield and excellent diastereoselectivity. They could be detached from the auxiliary by methanolysis of the *N*-formyl bond, followed by acid-catalyzed solvolysis of the *N*-glycoside and final hydrolysis of the amide bond, yielding the free D-amino acid hydrochlorides **152**. In contrast, analogous reactions carried out with pseudoenantiomeric 2,3,4-α-D-arabinopyranosylamine **156** (Figure 10.14) furnished L-amino acids *ent*-**152**, enabling access to both stereochemical series [119]. Similar results were obtained by Ugi and coworkers using tetra-*O*-alkyl-β-D-glucopyranosylamines [120] and 2-acetamido-3,4,6-tri-*O*-acetyl-1-amino-2-desoxy-β-D-glucopyranose [121], respectively.

An alternative cleavage methodology of the α-amino acids being generated via Ugi reactions with **150** and **156** was reported by Linderman et al. [122], who made use of isocyanide **157** (Scheme 10.51). In this approach, the *N*-glycosidic bond and the silyl protecting group of the

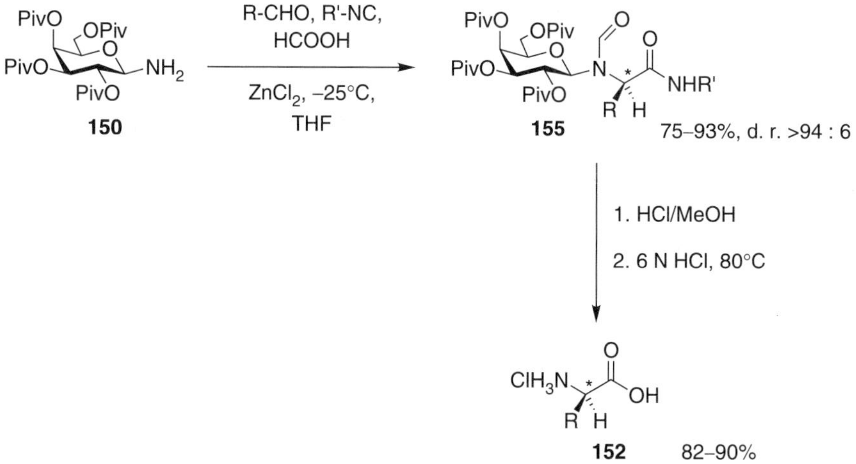

SCHEME 10.50 Stereoselective Ugi four-component condensation using glycosylamines.

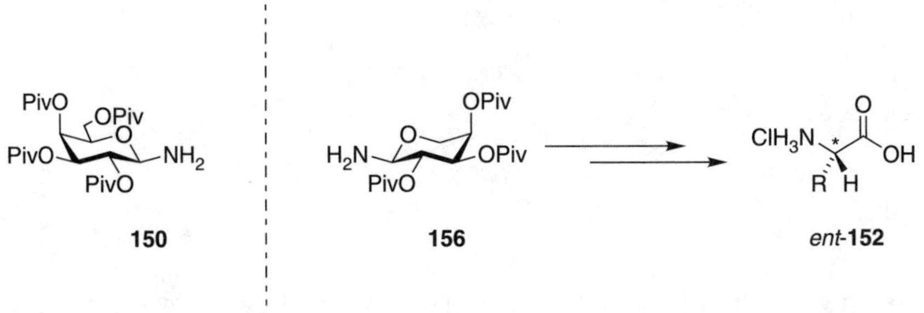

FIGURE 10.14 Pseudoenantiomers: D-galactopyranosylamine and D-arabinopyranosylamine (Ref. [119]).

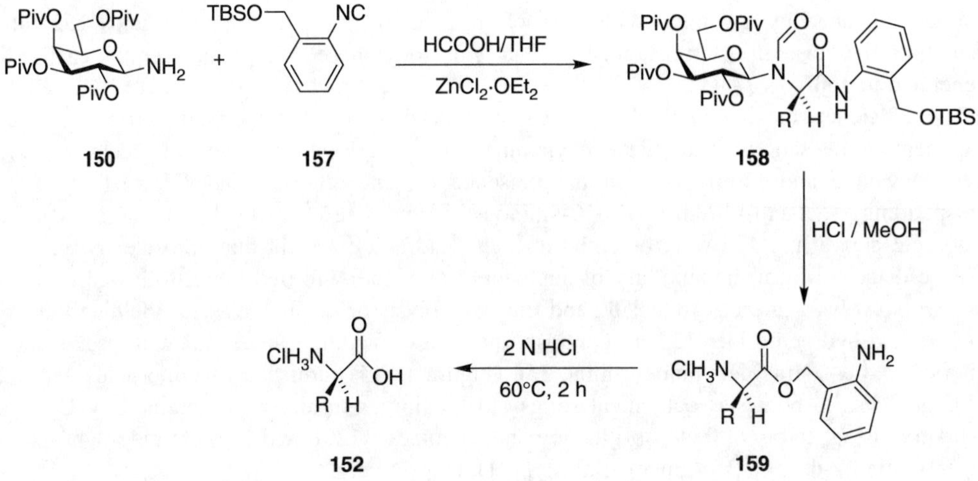

SCHEME 10.51 Stereoselective Ugi four-component condensation using glycosylamines.

primarily formed, Ugi condensation product **158** were cleaved simultaneously by methanolic HCl. The released hydroxyl function intramolecularly reacted with the amide to form benzyl ester **159**. This ester could be hydrolyzed with 2*N* HCl at elevated temperature to give pure amino acid hydrochlorides **152**.

The methodology outlined in Scheme 10.50 has also been adopted to a corresponding stereoselective solid-phase approach to α-amino acids using the galactosylamine **160** immobilized on a polystyrene-based resin, enabling the combinatorial synthesis of stereoisomerically pure compounds (Figure 10.15) [123].

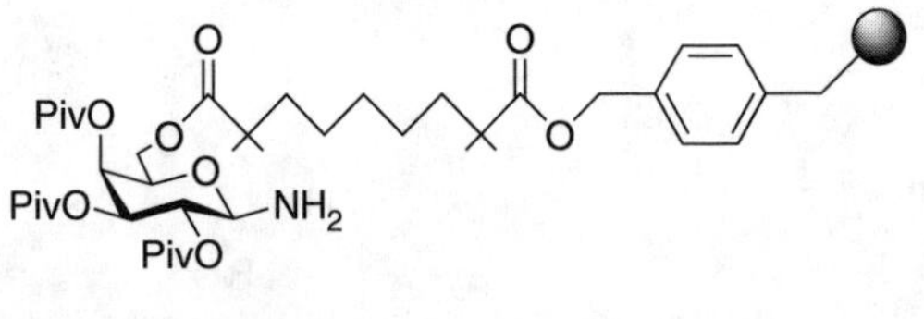

FIGURE 10.15 Polymer-linked galactosylamine for stereoselective combinatorial synthesis (Ref. [123]).

Carbohydrate-linked aldehydes or isocyanides have also been employed to perform Ugi [124–126], Strecker [127] and Asinger [128] reactions, but the diastereoselectivities observed had been unsatisfying.

Other nucleophiles that have been added to glycosyl imines include tri-*n*-butyl-allylstannane [129–131], diethylphosphite [132], *O*-silyl ketene acetals [133] and bis-*O*-silyl ketene acetals [134], giving chiral homoallylamines, α-amino phosphonic acids and β-amino acid derivatives, respectively. In the latter case, the Mannich reaction of prochiral monosubstituted, bis-*O*-trimethylsilyl ketene acetals with galactosyl imines **138**, promoted by $ZnCl_2$, furnished α,β-disubstituted β-amino acid derivatives **161** in good yields and excellent diastereoselectivities (Scheme 10.52). Conversion of the 2,3-diphenyl-β-alanine derivative **161a** to the corresponding 3-aminopropanol **162** proved the *syn*-relationship of the two phenyl substituents, indicating that a like-approach between the *si*-faces of the prochiral units is preferred (Scheme 10.53).

Further extension of the reaction pool of Schiff bases **138** was achieved by their reaction with *trans*-1-methoxy-3-(trimethylsilyloxy)-1,3-butadiene (Danishefsky's diene) to give 2-substituted 5,6-didehydro-piperidin-4-ones **164** [135,136] (Scheme 10.54). The reaction is considered to be a sequence of an initial Mannich reaction between the imine and the silyl enol ether, followed by an intramolecular Michael addition and subsequent elimination of methanol. If the reaction was terminated by dilute ammonium chloride solution, then the Mannich bases **163** could be isolated and further transformed to the dehydropiperidinones **164** by treatment with dilute hydrochloric acid. This result proved that the reaction pathway is not a concerted hetero Diels–Alder type process between the electron-rich diene and the activated imine. The use of hydrogen chloride as a terminating agent resulted in exclusive isolation of the piperidine derivatives **164** formed with

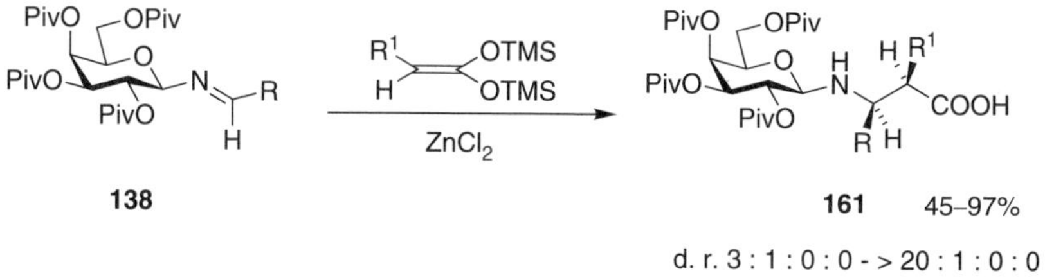

SCHEME 10.52 Mannich reaction with *N*-glycosylamines proceed with high diastereoselectivity.

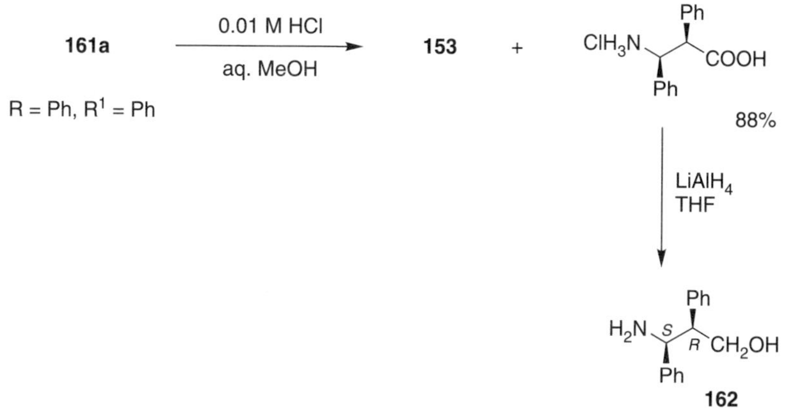

SCHEME 10.53 A *syn* approach between the *si* faces of the prochiral components of the sugar-directed Mannich reaction is preferred.

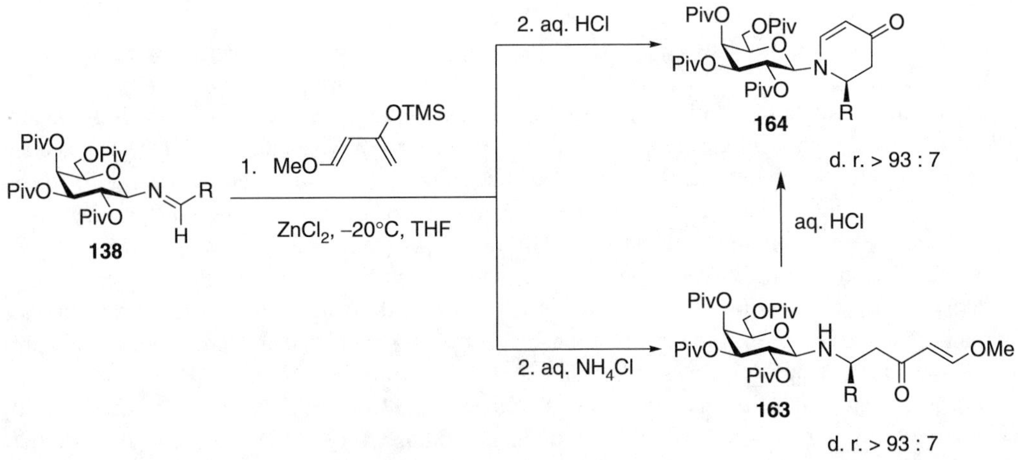

SCHEME 10.54 A tandem Mannich/Michael sequence, directed by a sugar auxiliary, produced Diels-Alder-type products with excellent diastereoselectivities.

excellent diastereoselectivity. These compounds are useful precursors for the synthesis of chiral nitrogen heterocycles and alkaloids (see Section 10.3.2.2).

Diastereoselective additions of nucleophiles to *N*-glycosylnitrones have been performed by Vasella et al. [44]. These authors extensively studied the reaction of lithium dimethyl phosphite to *N*-mannofuranosyl nitrones **165**, giving predominantly (*S*)-configured *N*-hydroxy-*N*-glycosylamino-phosphonates **167** (Scheme 10.55). The stereoselectivity observed (dr 94:6 for **165a** → **167a**) was explained in terms of the kinetic anomeric effect [44]. This implies that the stabilizing *exo*-anomeric effect present in those conformers of the product possessing a coplanar arrangement of the n_N-orbital and the σ^*-orbital of the C_1–O bond already affects the corresponding transition states, thereby lowering their energy. According to this principle, the interaction is possible in both conformers **169** and **170**. The latter conformer points the nitrone-oxygen *endo* towards the furanose and thus offers less steric interaction between the nitrone *C*-substituents and the glycosyl moiety than *O*-*exo*-conformer **169** (Figure 10.16). Steric reasons were also considered responsible for the selective antiattack at the (*Z*)-configurated nitrone double bond furnishing the (*S*)-configured product. Comparative studies with analogous *N*-pseudoglycosylnitrones (*N*-alkylnitrones) **166** showed that the ring oxygen is of major importance with regard to both the diastereoselectivity and the rate of reaction. Therefore, in this case, aminophosphonate **168a** was formed with much lower

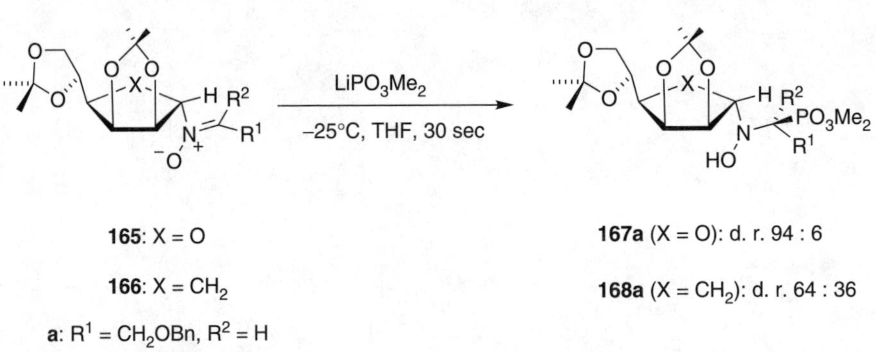

SCHEME 10.55 Diastereoselective addition of nucleophiles to *N*-glycosylnitrones can be achieved using sugar-derived auxiliaries.

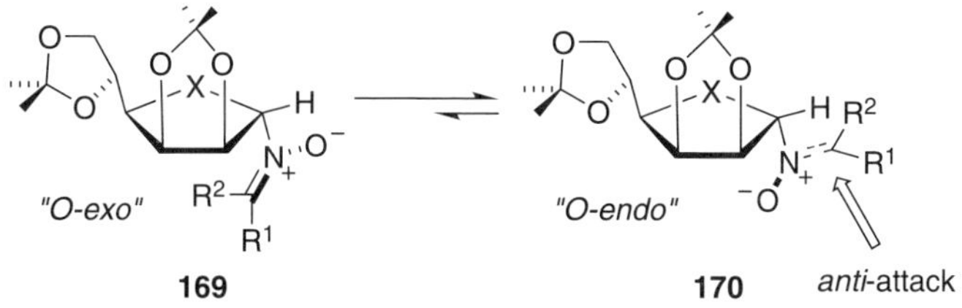

FIGURE 10.16 Stereodifferentiation on *N*-mannofuranosyl nitrones (Ref. [44]).

diastereoselectivity, indicating that no distinct conformational bias between *O-exo* and *O-endo* conformer is given.

The methodologies outlined in Scheme 10.55 and similar approaches were applied to gain access to enantiomerically enriched *N*-hydroxylamines [137,138]. In the Abbott approach, treatment of acetaldehyde-derived mannofuranosyl nitrone **165b** with metalated benzo[*b*]thiophene selectively furnished addition product **171**, which was further transformed to (+)-(*R*)-zileuton **172** [137] (Scheme 10.56), a potent selective 5-lipoxygenase inhibitor [139,140].

Highly diastereoselective α-hydroxyallylations of aldehydes using the mannopyranosyloxyallyl-tributylstannane **173** were efficiently carried out by Roush and coworkers [141]. These authors showed that the $BF_3{\cdot}OEt_2$-promoted double asymmetric reaction of **173** with α-chiral aldehydes (e.g., (*S*)-**174**) led to *all-syn*-configured **175** as the major product of this matched pairing (Scheme 10.57). The stereochemical outcome of the reaction was explained by a preferred Felkin–Anh approach of the Lewis acid–aldehyde complex to the *si*-face of the enol ether, which was essentially perpendicular to the pyran C–O bond as a consequence of the *exo*-anomeric effect. When the enantiomeric aldehyde (*R*)-**174** was chosen, a similar diastereomeric ratio in favor of the *anti,syn*-product **176** was observed, indicating that the enantioselectivity of the chiral stannane is sufficient to completely overcome the intrinsic diastereofacial bias of (*R*)-**174** in this mismatched situation. A real matched/mismatched pair relationship between the chiral auxiliary and the chiral aldehyde was implicated in the case where α-alkoxy aldehydes were used, which, in turn, upon addition of $MgBr_2$, enabled reversal of the stereochemical preference, and was interpreted as the result of a chelate-controlled reaction. Achiral aldehydes were found to afford very poor stereoselectivities. In contrast, modest to high diastereoselectivities for the α-hydroxyallylation of achiral aldehydes were observed employing D-glucal-derived auxiliary **177** bearing the allylstannane at the nonanomeric position [142]. Likewise, $AlCl_3$-promoted reaction of benzaldehyde with **177** furnished the *syn*-isomer **178** with high asymmetric induction (Scheme 10.58).

An alternative methodology for the synthesis of *syn*-1,2-diols, exploiting the diastereoselective addition of aromatic aldehydes to the oxyallylanion of a simple allyl arabinopyranoside, was reported by Takei [143]. Other asymmetric addition reactions to be mentioned in this context

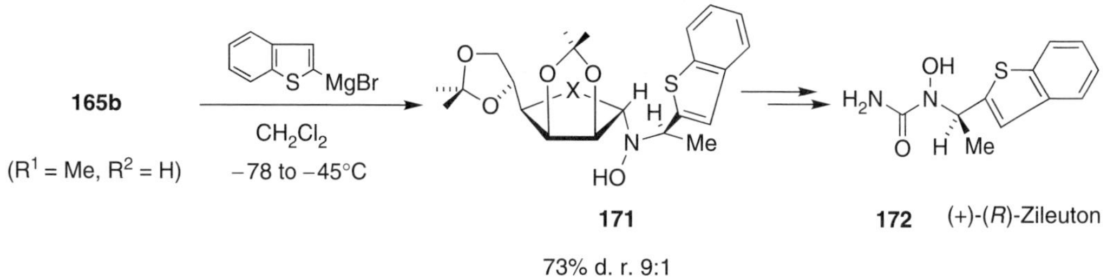

SCHEME 10.56 Use of sugar auxiliaries to direct the stereochemical outcomes of nucleophilic additions to *N*-glycosylnitrones was incorporated into the preparation of zileuton, **172**.

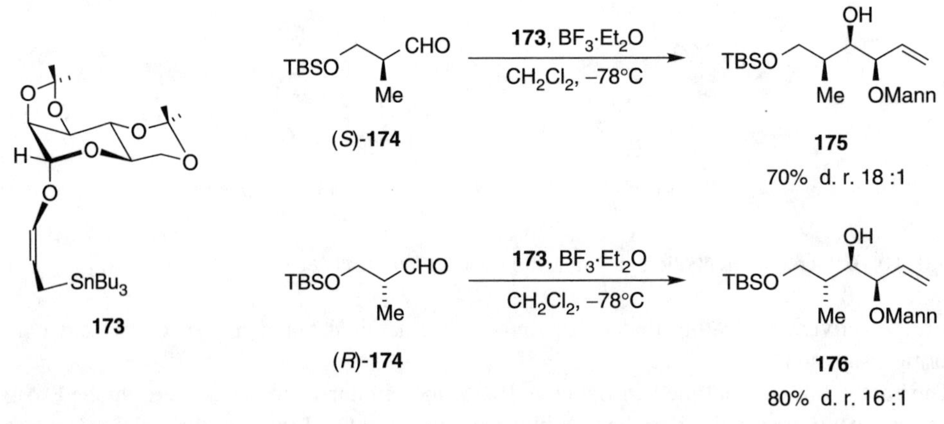

SCHEME 10.57 The stereochemical outcomes of α-hydroxyallylations of aldehydes can be influenced by sugar-derived auxiliaries.

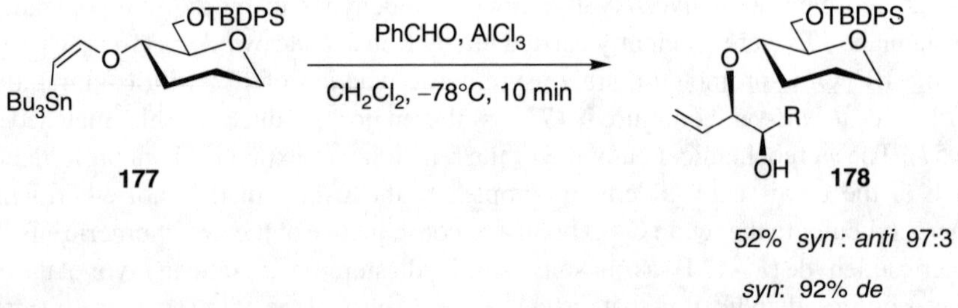

SCHEME 10.58 The stereochemical outcomes of α-hydroxyallylations of aldehydes can be influenced by sugar-derived auxiliaries.

include the α-methylallylation of aldehydes [144], the addition of Grignard reagents to β- and γ-alkoxyaldehydes [145] and the Hosomi–Sakurai reaction of allylsilanes containing arabinose-derived alcohols [146].

10.3.2 Conjugate Additions

10.3.2.1 Conjugate Additions to Bicyclic Carbohydrate Oxazolidinones

Bicyclic oxazolidinones derived from carbohydrates have been used as chiral auxiliaries in conjugate addition reactions [147]. After deprotonation with MeMgBr, the D-*galacto*-oxazolidin-2-one **178** and the D-*gluco*-oxazolidin-2-one **179** (Figure 10.17) were *N*-acylated with

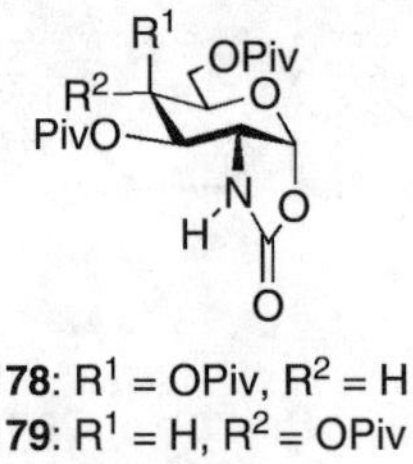

178: R^1 = OPiv, R^2 = H
179: R^1 = H, R^2 = OPiv

FIGURE 10.17 Bicyclic oxazolidinones, auxiliaries derived from galactosamine and glucosamine (Ref. [147, 148]).

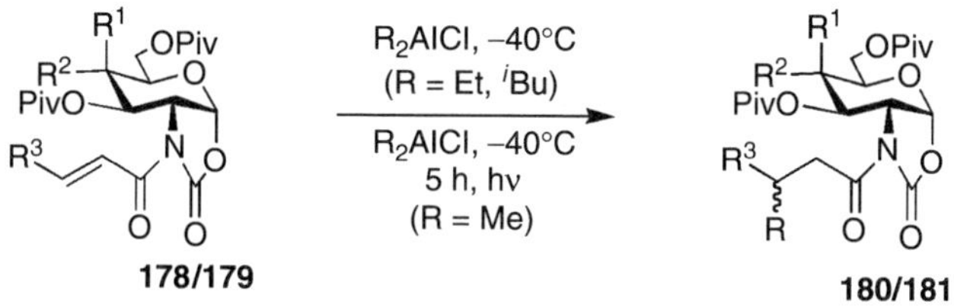

SCHEME 10.59 Bicyclic oxazolidinones derived fiom carbohydrates are efficient chiral auxiliaries in conjugate addition reactions of organoaluminum compounds.

TABLE 10.5
Stereoselective Conjugate Addition of Organoaluminium Compounds to Carbohydrate Derived N-Acyl Oxazolidinones According to Scheme 10.59 (Ref. [147,148])

R^1	R^2	R^3	R_2AlCl	Yield (%)	(*R*):(*S*)
H	OPiv	Ph	Et_2AlCl	90	90:10
H	OPiv	Ph	*i*-Bu_2AlCl	73	92:8
OPiv	H	Ph	Et_2AlCl	84	96:4
OPiv	H	Ph	Me_2AlCl	82	>98:2

α,β-unsaturated acid halides. The resulting imides reacted as Michael acceptors with several dialkyl aluminum chlorides at low temperatures (Scheme 10.59, Table 10.5).

Irradiation with UV light or the presence of oxygen were required for the transfer of a methyl group from Me_2AlCl to the Michael acceptors **178** and **179**, whereas the transfer of higher alkyl groups readily proceeded at low temperatures without irradiation. Aluminum complexes were proposed to explain the high stereoselectivity of the conjugate additions (higher than 19:1 in some cases). Owing to steric hindrance of the bulky pivaloyl protecting groups, the addition of the alkyl groups occurred from the *exo*-face of the acceptors **178** and **179**. The oxazolidinones derived from D-galactosamine gave higher selectivities than those prepared from D-glucosamine.

The resulting diastereomers of β-branched carboxylic acid derivatives **180** and **181** could be separated by chromatography or crystallization. The pure diastereomer **182** was cleaved from the carbohydrate auxiliary by treatment with $LiOH/H_2O_2$ furnishing the (*R*)-configured carboxylic acid **183** (Scheme 10.60). The oxazolidinone **178** was recovered almost quantitatively in this process.

The formal addition of an alkyl halide (RX) to the C=C double bond of the Michael acceptors proceeded in a one-pot process by initial 1,4-addition of the dialkyl aluminum chloride and subsequent trapping of the enolate intermediate with *N*-halosuccinimides [148]

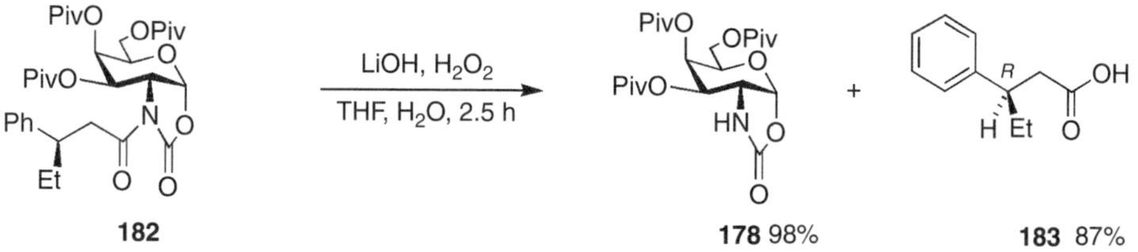

SCHEME 10.60 Cleavage of sugar-derived bicyclic oxazolidinone auxiliaries can be achieved on treatment with LiOH realeasing chiral carboxylic acids.

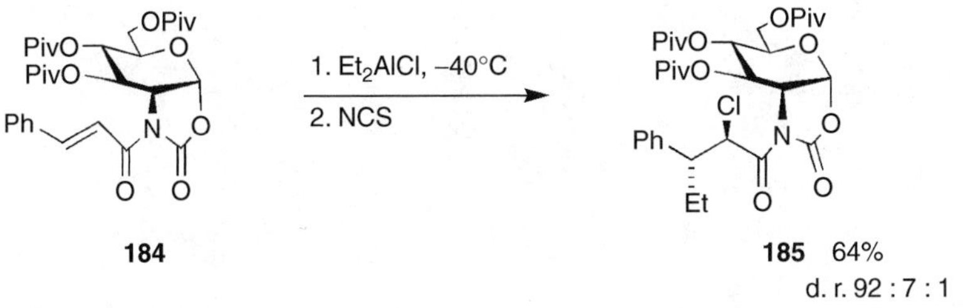

SCHEME 10.61 Stereochemical control over the formal addition of alkyl halides across double bonds can be achieved through the use of sugar-based auxiliaries.

(Scheme 10.61). The diastereoselectivity of the halogenation step was higher for the glucosyl oxazolidinone **184** than for the corresponding galactosyl oxazolidinone. The reaction of **184** with Et_2AlCl and *N*-chlorosuccinimide showed a high *anti*-selectivity, and gave a product mixture of diastereomers in a ratio of 92:7:1. The diastereomers were separated by chromatography or crystallization, and the α-halogenated, β-branched carboxylic acid derivative **185** was formed in an acceptable yield.

Conjugate additions of organocuprates to various carbohydrate linked α,β-unsaturated esters were described by Tadano et al. [149]. For example, the benzyl protected crotonyl ester **186**, synthesized from methyl-α-D-glucopyranoside, underwent conjugate addition of vinyl cuprate at −78°C in a mixture of THF and dimethyl sulfide giving diastereomeric adducts **187/188** (Scheme 10.62). Cleavage from the auxiliary was accomplished by saponification followed by conversion of the carboxylic acids into their benzyl esters. The β-vinyl butanoic esters **189** and **190** were thus obtained with good yields and high enantioselectivity. The diastereomeric excess increased with a combination of a larger 3-*O*-substituent and a smaller 6-*O*-substituent. Best results were obtained with an *O*-benzyl group at the 3-position of the glucopyranoside. Addition reactions of other organocuprates (e.g., $Et_2CuMgBr$, $EtCu{\cdot}BF_3$ and Et_2CuLi) to enoate **186** gave β-alkylated butanoic esters in similar yields and diastereomeric ratios.

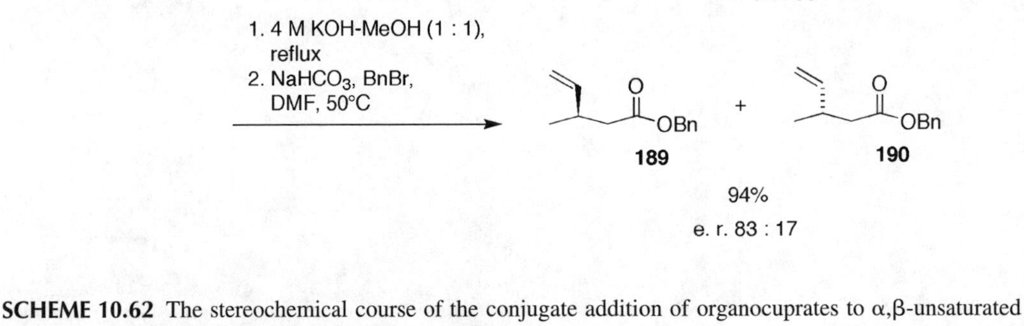

SCHEME 10.62 The stereochemical course of the conjugate addition of organocuprates to α,β-unsaturated esters can be influenced using sugar-based auxiliaries.

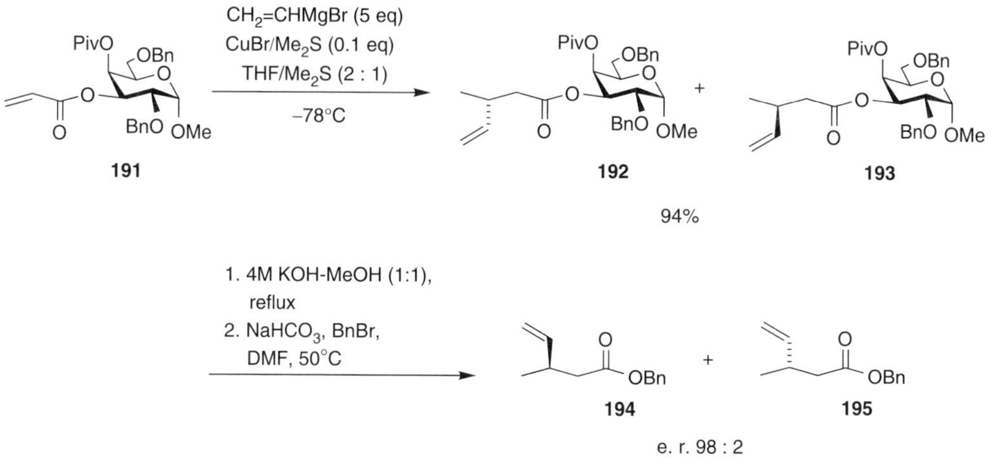

SCHEME 10.63 The stereochemical course of the conjugate addition of organocuprates to α,β-unsaturated esters can be influenced using sugar-based auxiliaries.

Methyl-α-D-mannopyranoside and methyl-α-D-galactopyranoside have also been used as chiral auxiliaries for conjugate additions of organocuprates to crotonyl esters. For the galactosyl derivative, the crotonyl ester was attached to the 3-hydroxyl group of the galactopyranoside. The addition of vinyl cuprate to ester **191** proceeded with high yield and stereoselectivity (Scheme 10.63). In this case, the diastereomeric esters **192** and **193** delivered the enantiomeric benzyl esters **194** and **195** with 96% *ee*.

The stereochemical outcome of the addition was explained in terms of steric interactions, as illustrated in Scheme 10.64. The carbonyl groups of both the crotonyl and pivaloyl esters are co-ordinated by the magnesium organocuprate or the magnesium halides formed during the reaction. The crotonyl ester **191** probably prefers an *s-trans*-conformation. As the *si*-face of the double bond is shielded by the bulky pivaloyl group, the cuprate reagent approaches from the less

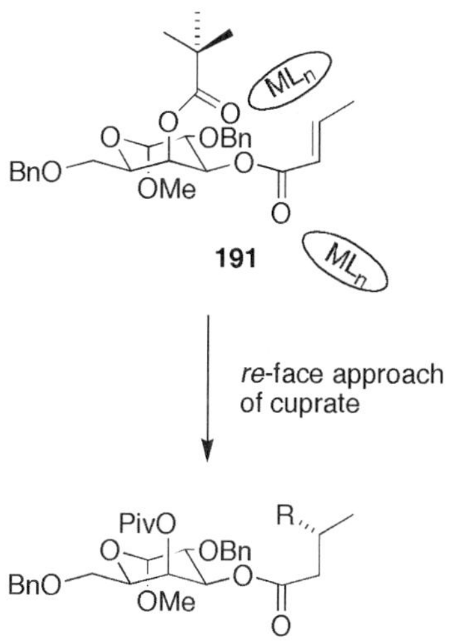

SCHEME 10.64 Rationalization of the stereochemical course of the conjugate addition of organocuprates to α,β-unsaturated esters as influenced by sugar-based auxiliaries.

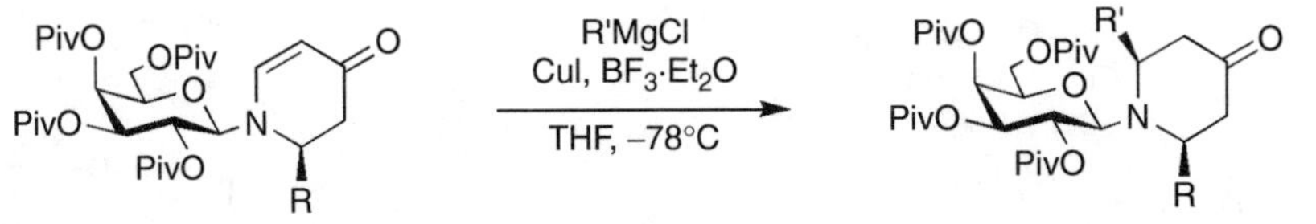

SCHEME 10.65 Lewis acids can facilitate the addition of organocuprates to chiral 2-substituted piperidinone derivatives prepared as previously described.

TABLE 10.6
Stereoselective Synthesis of 2,6-Disubstituted Piperidinones According to Scheme 10.65 (Ref. [136])

R	R′	Yield (%)	Diastereomeric Ratio
3-Furyl	*i*-Propyl	61	>12:1
5-Hexenyl	*n*-Propyl	81	>12:1
3-Furyl	*i*-Amyl	61	>8:1

hindered *re*-face. Similar transition state models were proposed for chiral auxiliaries derived from glucose and mannose.

The chiral 2-substituted piperidinone derivatives generated in the tandem-Mannich–Michael reaction described above are versatile precursors for the synthesis of a variety of heterocycles [136]. Although the conjugate addition of Grignard reagents, organocuprates and organolithium compounds to the dehydropiperidinones **164** did not occur, these vinylogous amides only reacted with a combination of soft nucleophiles (organocuprates) and hard electrophiles, for example, boron trifluoride and trimethylchlorosilane, furnishing 2,6-*cis*-disubstituted piperidinone derivatives **192** in good yields and high diastereomeric ratios (Scheme 10.65, Table 10.6).

10.3.2.2 Synthesis of Enantiomerically Pure Alkaloids Using Carbohydrate Auxiliaries

A sequence of tandem-Mannich–Michael reactions and conjugate additions of organocuprates has been used for the synthesis of enantiomerically pure alkaloids such as (−)-dihydro-pinidine **193**, the indolizidine alkaloid (5*R*,8a*R*)-gephyrotoxin 167B **194** and the decahydroquinoline alkaloid 4a-*epi*-pumiliotoxin C **195** (Figure 10.18) [150].

For the synthesis of the *trans*-annulated 4a-*epi*-pumiliotoxin C **195** (Scheme 10.66), the imine prepared from galactosyl amine **150** and 5-hexenal was reacted with Danishefsky's diene to give the 2-substituted dehydropiperidinone derivative **196** with high diastereoselectivity. The subsequent addition of propyl cuprate promoted by borontrifluoride etherate furnished the 2,6-*cis*-disubstituted

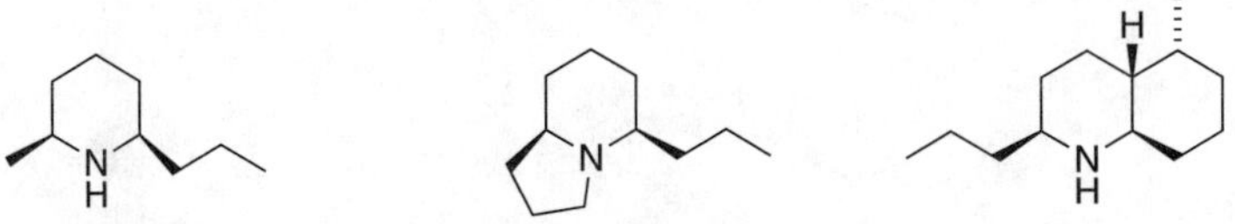

FIGURE 10.18 Piperidine alkaloids.

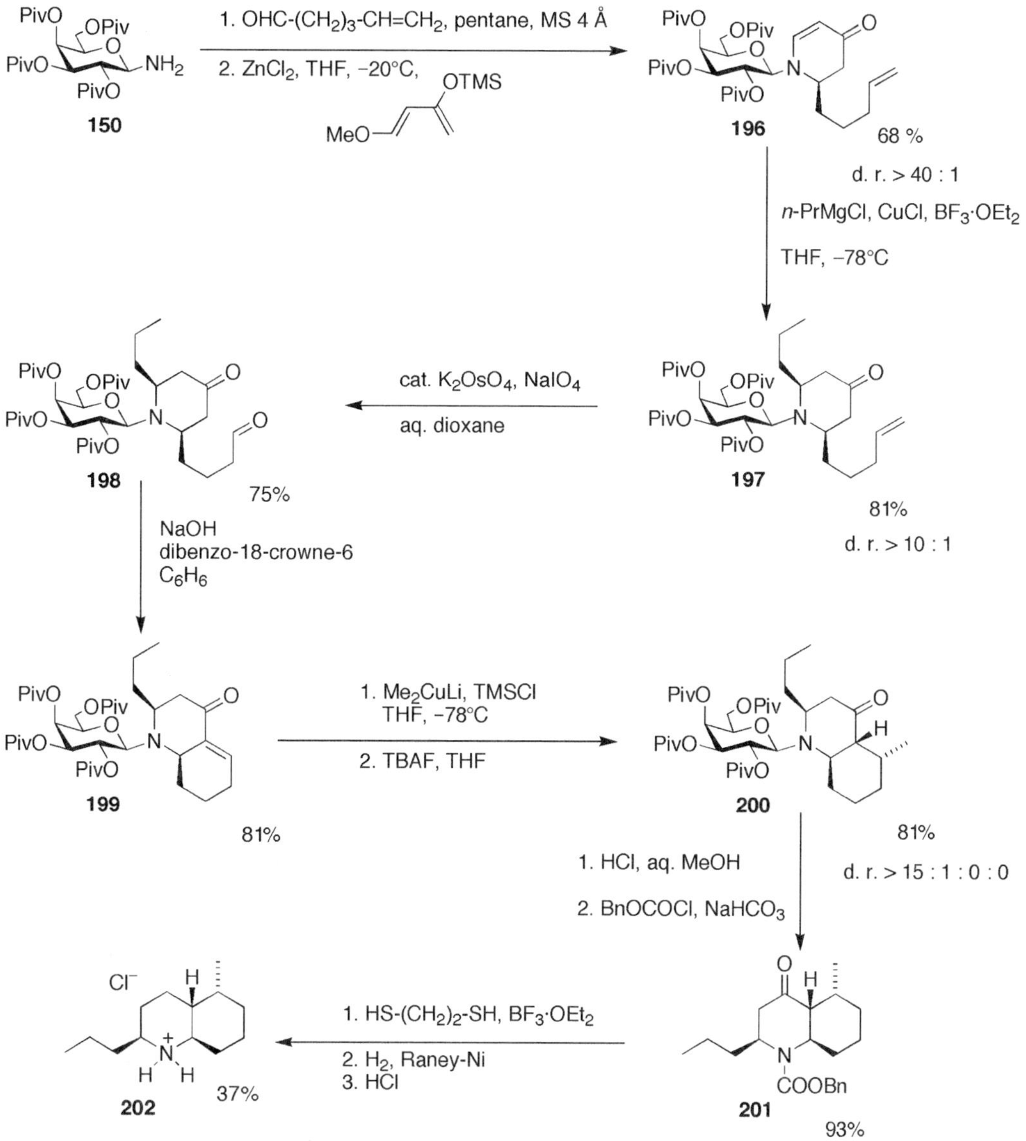

SCHEME 10.66 Preparation of the alkaloid 4a-*epi*-pumiliotoxin C, **202**, incorporating a sugar-directed tandem Mannich-Michael sequence.

piperidinone **197** with a diastereomeric excess higher than 10:1. The C=C double bond in the side chain was cleaved oxidatively with $NaIO_4$, and an intramolecular aldol condensation of the resulting aldehyde **198** furnished the pumiliotoxin C skeleton **199**. Addition of Me_2CuLi in the presence of trimethylchlorosilane and subsequent cleavage of the silyl enol ether with TBAF surprisingly gave the *trans*-annulated decahydroquinoline **200**. This result was unexpected because analogous octahydroquinolines bearing a phenoxycarbonyl protecting group at the nitrogen (instead of a carbohydrate skeleton) form *cis*-annulated decahydroquinolines in conjugate cuprate additions under identical conditions [151]. Obviously, the carbohydrate moiety directs the protonation of the enolate formed by the cuprate addition to **199** and fluoridolysis of the intermediate silyl enol ether towards the thermodynamically less favored *trans*-annulated decahydroquinoline. The heterocycle **200** was cleaved from the carbohydrate auxiliary by acidic hydrolysis, and the nitrogen was subsequently protected with the benzyloxycarbonyl group. The hydrochloride of 4a-*epi*-pumiliotoxin C **202** was obtained by transforming **201** into the dithiolane, followed by hydrogenolytic removal of both the dithioketal and the benzyloxycarbonyl protecting group and an acidic workup.

10.3.3 Reactions Involving Enolates

10.3.3.1 Alkylations

The alkylation of enolates constitutes a very powerful method for the formation of C–C bonds. Several methods have been developed in order to control the stereoselectivity of this process. First applications of carbohydrate auxiliaries in stereoselective alkylations of ester enolates were described by Heathcock et al. [152] in 1981.

To obtain highly stereoselective alkylation reactions of auxiliary linked ester or amide enolates some conditions have to be kept:

1. The enolate must be formed with high (*E*)- or (*Z*)-selectivity.
2. The stereotopic faces of the enolate double bond must be efficiently differentiated by the auxiliary.

The capability of the highly oxygenated carbohydrate auxiliaries to coordinate the counter-ion of the enolate allows the formation of chiral chelate complexes with a restricted flexibility of the enolate moiety. The cation complexation also increases the tendency of the carbohydrate to react as a leaving group. It has been found [153] that the enolate **204** generated by deprotonation of the carbohydrate linked ester **203** with LDA underwent an elimination of the carbohydrate moiety generating the alcohol chelate complex **205** and the ketene **206** (Scheme 10.67).

Alkylation reactions of ester derivatives of 1,2:5,6-di-*O*-isopropylidene-D-gulofuranose were investigated by Mulzer et al. [154]. The deprotonation of the ester **207** with LDA or LTMP and trapping of the enolate with trimethylchlorosilane furnished the silyl enol ether **208** with a very high (*E*)-selectivity (95:5), whereas the enol ether obtained by deprotonation with LHMDS was formed without any (*E*)/(*Z*)-selectivity (Scheme 10.68).

The alkylation of the enolates with methyl iodide furnished the product **209** with high diastereoselectivities and good yields when LDA and LTMP were used (Scheme 10.69). With LHMDS, the yield of the alkylation was only moderate and no stereoselectivity was obtained. This was surprising because a complete enolization of ester **207** was observed with LHMDS. The authors also found that the yields and stereoselectivity were not influenced by the addition of complexing reagents ($MgBr_2$ and $ZnCl_2$) or reagents that reduce complexation (TMEDA and

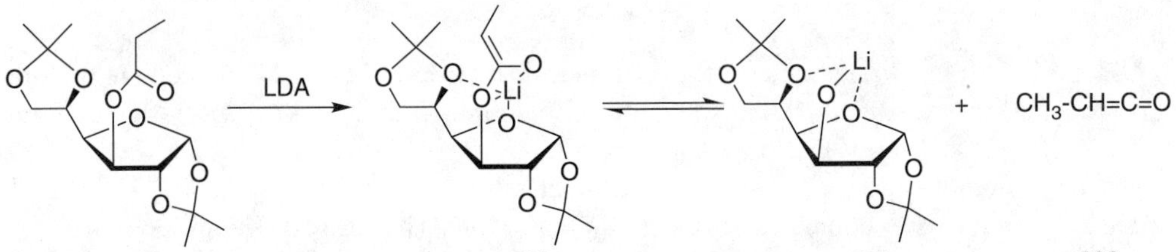

SCHEME 10.67 Generation of enolates bearing sugar-derived auxiliaries results in the formation of rigid and chiral chelate complexes capable of directing the stereochemical course of subsequent reactions. A common side reaction is the activation of the sugar auxiliary to act as a leaving group.

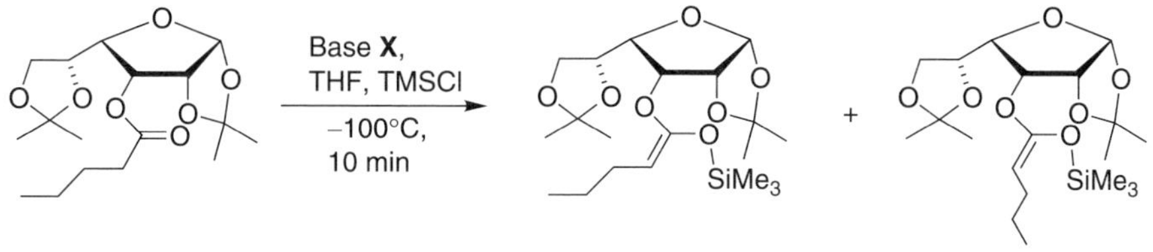

Base **X**, THF, TMSCl, −100°C, 10 min

207 (*E*)-**208** (*Z*)-**208**

X = LTMP: (*E*) : (*Z*) = 95 : 5
X = LHMDS: (*E*) : (*Z*) = 50 : 50

SCHEME 10.68 Silyl enol-ethers bearing sugar-derived auxiliaries can be generated with high *E* selectivity using LTMP as a base where no *E*/*Z* selectivity is realized using LHMDS.

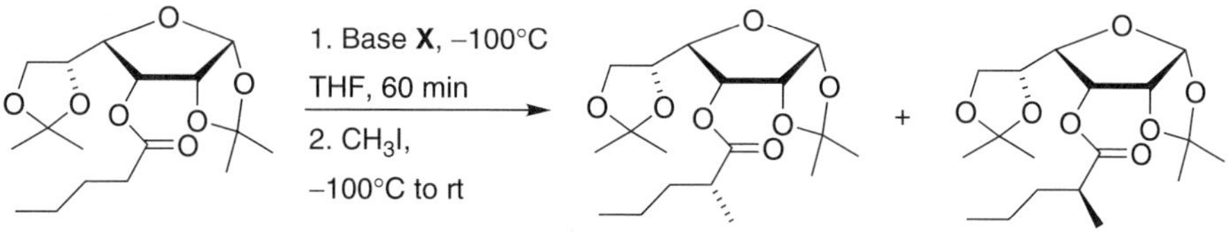

1. Base **X**, −100°C THF, 60 min
2. CH_3I, −100°C to rt

207 (*R*)-**209** (*S*)-**209**

X = LTMP: (*R*) : (*S*) = 90 : 10, 89%
X = LHMDS: (*R*) : (*S*) = 50 : 50, 47%

SCHEME 10.69 As the *E*/*Z* selectivity of enolate formation is base-dependent, so is the influence on the stereochemical outcomes of alkylation reactions.

HMPT). These results were explained by a "postenolization complex" **210** (Figure 10.19), in which the protonated amide base is still in close contact with the enolate. The attack of the methyl iodide preferentially takes place from the less hindered *si*-face of the double bond. With LHMDS, the approach of the electrophile is hindered by the bulky TMS-groups, and an internal proton return from the amide base to the enolate becomes a competing process. The consequences of the reprotonation are low yields of methylated products and the recovery of the corresponding amount of starting material.

D-Glucose-derived oxazinones, which can be considered as "chiral glycine" derivatives, were also employed in asymmetric alkylation reactions [155]. The oxazinone represents a rigid

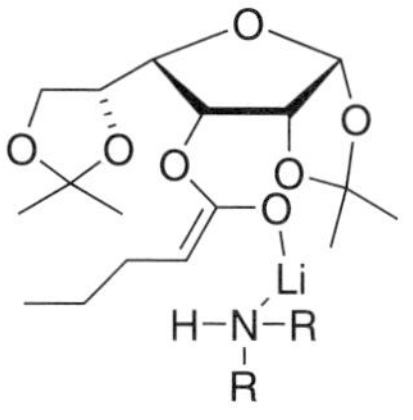

210

FIGURE 10.19 Proposed "postenolization complex" of gulofuranose ester enolates (Ref. [154])

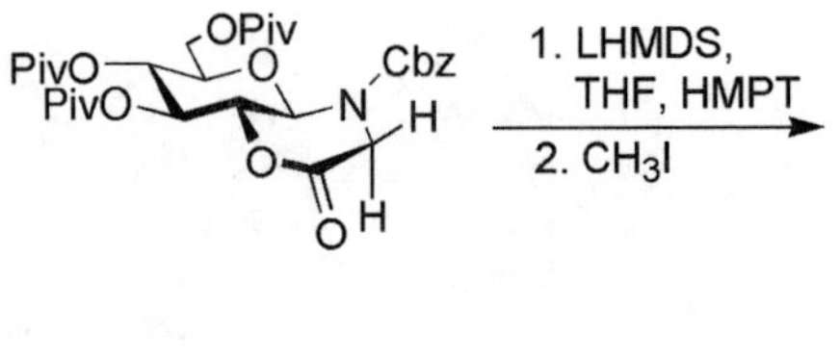

SCHEME 10.70 Chiral glycine derivatives such as D-glucose-derived oxazinones are useful in asymmetric alkylation reactions.

system for which no complexation is required to obtain a stereochemically defined enolate (Scheme 10.70). Complete deprotonation of the oxazinone **211** with LHMDS was obtained in THF but the enolate was not reactive towards electrophiles unless HMPT was added. In the presence of this additive, methyl iodide reacted with the enolate to give the methylated product **212** in a yield of 57% and with excellent stereoselectivity (98:2). With other electrophiles such as allyl bromide or benzyl bromide, dialkylated products and unreacted starting material were isolated in considerable amounts. The low yields of the alkylations were explained by an internal proton return process.

Chiral oxazolidinone auxiliaries derived from D-xylose were applied by Koell et al. [156]. The oxazolidinones were acylated with various acid halides furnishing imides, which are substrates for α-alkylation reactions. For example, the butyric acid derivative **213** was deprotonated with LDA to give the (*Z*)-configured enolate **214**, which was reacted with methyl iodide (Scheme 10.71). The methylated product **215** was formed in a moderate yield of 45% and a diastereomeric ratio of 7:1. The approach of the electrophile occurred from the less hindered *si*-face of the enolate

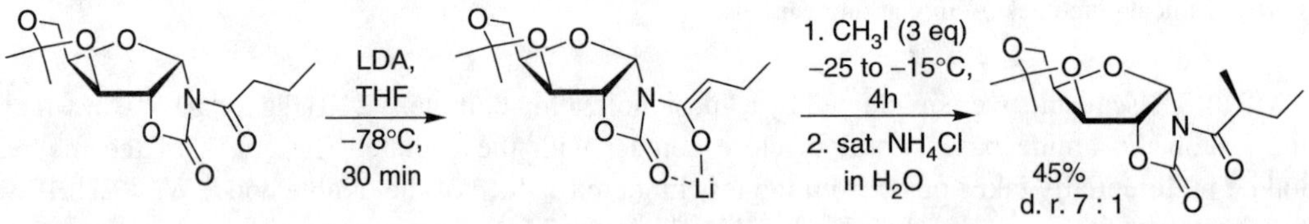

SCHEME 10.71 *N*-Acylated sugar-derived chiral oxazolidinones are substrates for asymmetric α-alkylation reactions.

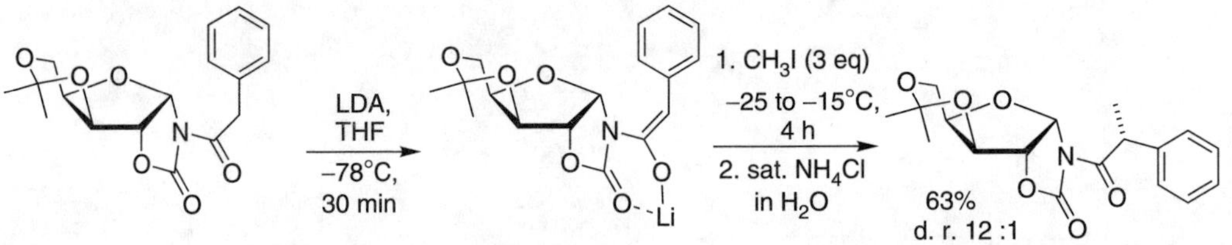

SCHEME 10.72 The stereochemical outcomes observed for α-alkylation of *N*-acylated sugar-derived chiral oxazolidinones are dependent upon the nature of the acyl substituents.

double bond. Surprisingly, an inverse stereoselectivity was observed for phenylacetimides such as **216** (Scheme 10.72). The (*E*)-enolate **217** was obviously formed, and the electrophile approached from the *re*-face of the enolate double bond. The addition of methyl iodide furnished the (*R*)-configured imide **218** with high diastereoselectivity. The authors explain the formation of the (*E*)-enolate by assuming a stereoelectronic interaction between the aromatic ring and the ring oxygen atom of the carbohydrate moiety. A short distance between the aromatic ring and the carbohydrate was detected by x-ray diffraction analysis.

10.3.3.2 Halogenations and Acylations

Diastereoselective halogenation reactions using a chiral glucose auxiliary were described by Duhamel et al. [157]. The ketene acetal **220** was prepared by trapping the lithium enolate of the ester **219** with trimethylchlorosilane (Scheme 10.73). Treatment of the ketene acetal **220** with *N*-chlorosuccinimide furnished the α-chloroester **221** with high diastereomeric excess. Cleavage from the auxiliary was accomplished by treatment with LiOOH, and the corresponding α-chloro acid **222** was isolated in a good yield and in an enantiomeric excess of 95%. The authors showed that the diastereofacial differentiation within the reaction was independent from the (*E*)- or (*Z*)-configuration of the ketene acetal. In a similar procedure using *N*-bromosuccinimide instead of NCS α-bromo carboxylic acids could be synthesized with lower stereoselectivities.

The chiral oxazolidinone auxiliaries introduced by Koell et al. [158] were used for halogenation reactions of various imides. The imide **223** was transformed into a boron enolate and reacted with NCS as the electrophile (Scheme 10.74). The (*S*)-configured α-chloro imide **224** was obtained in moderate yield with a diastereomeric ratio of 4:1. As in the alkylation reactions described above, an inversion of the stereoselectivity was observed for phenylacetimides.

Oxazolidinone auxiliaries were also employed for stereoselective acylation reactions of imide enolates with acid halides [158]. For example, the lithium enolate of the imide **223** reacted

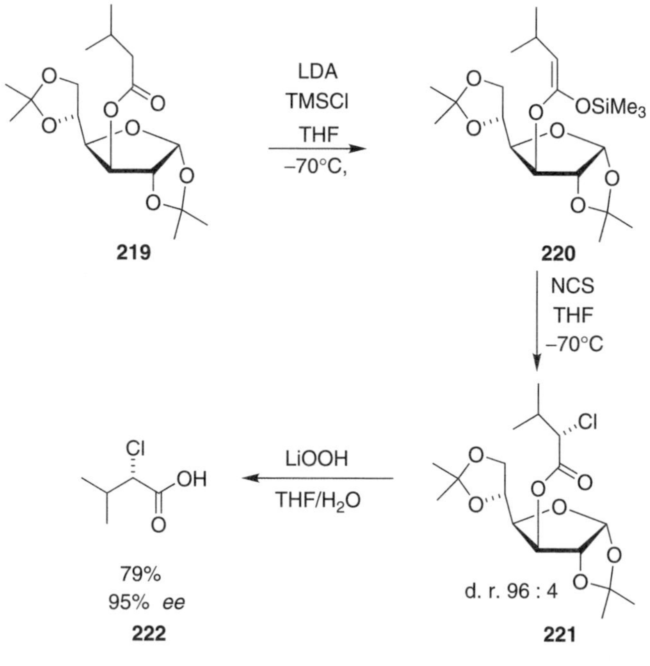

SCHEME 10.73 Chiral α-chlorocarboxylic acids can be generated using sugar-based auxiliaries.

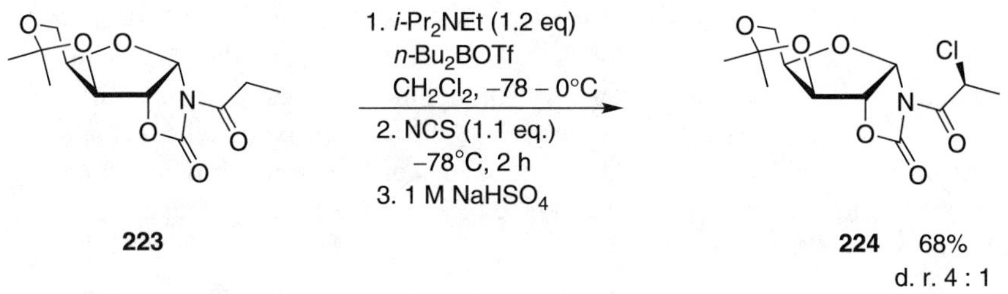

223

1. i-Pr_2NEt (1.2 eq)
n-Bu_2BOTf
CH_2Cl_2, −78 – 0°C
2. NCS (1.1 eq.)
−78°C, 2 h
3. 1 M $NaHSO_4$

224 68%
d. r. 4 : 1

SCHEME 10.74 Sugar-derived imides can undergo asymmetric halogenation reactions.

223

1. LHMDS, THF
−78°C, 30 min
2. PhCOCl (1.6 eq),
−78°C, 2 min
3. sat. NH_4Cl in H_2O

225 75%
d. r. 15 : 1

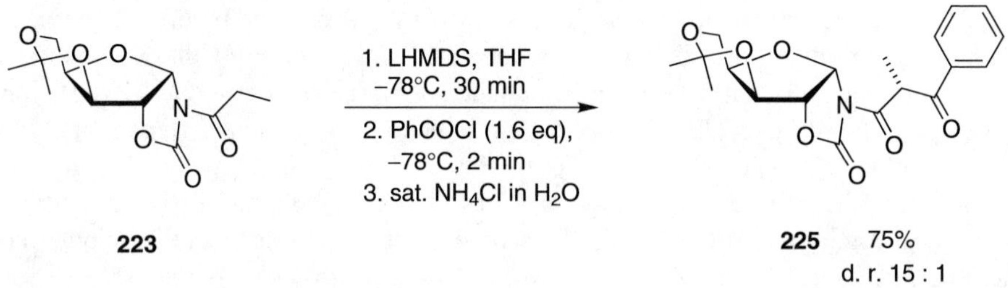

SCHEME 10.75 Asymmetric acylation reactions can be accomplished using sugar-derived oxazolidinone auxiliaries.

with benzoyl chloride furnishing the β-keto imide **225** with a high diastereomeric excess (Scheme 10.75). As in the alkylation and halogenation reactions with this chiral auxiliary, an inversion of the stereoselectivity was observed for imides of phenyl acetic acids. The β-keto imides, which are stable to epimerization even under the applied basic reaction conditions, were cleaved from the carbohydrate auxiliary by treatment with LiOOH.

10.3.3.3 Aldol Reactions

Early investigations of asymmetric aldol reactions with chiral carbohydrate auxliliaries were carried out by Heathcock [152] and Bandraege [159], but often only low stereoselectivities were observed. In additional studies, Banks et al. [73] used oxazinone auxiliaries for aldol reactions, which had been employed for other asymmetric reactions. The lithium enolate of the *N*-acylated oxazinone **226** reacted with benzaldehyde, furnishing exclusively the *syn*-aldols **227A** and **227B** in a ratio of 10:1 (Scheme 10.76).

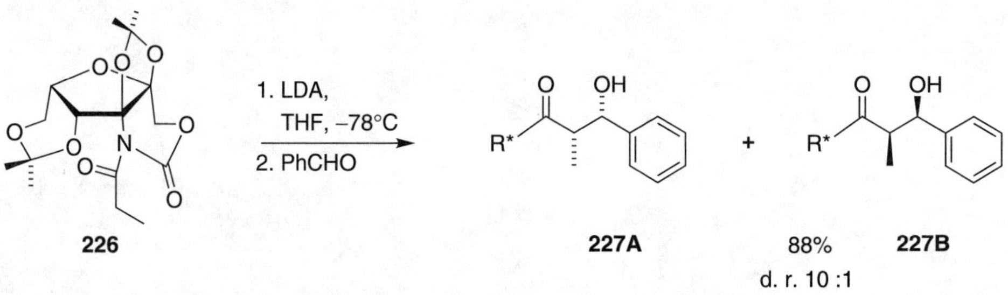

226

1. LDA,
THF, −78°C
2. PhCHO

227A + 227B
88%
d. r. 10 :1

SCHEME 10.76 Sugar-derived auxiliaries can influence the stereochemical outcomes of aldol reactions.

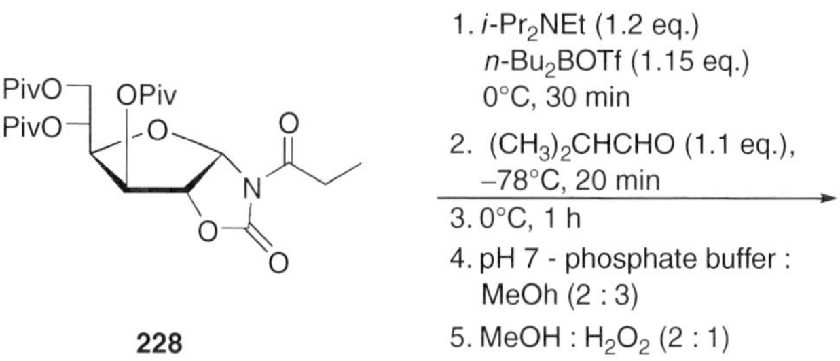

1. i-Pr_2NEt (1.2 eq.) n-Bu_2BOTf (1.15 eq.) 0°C, 30 min
2. $(CH_3)_2CHCHO$ (1.1 eq.), −78°C, 20 min
3. 0°C, 1 h
4. pH 7 - phosphate buffer : MeOh (2 : 3)
5. MeOH : H_2O_2 (2 : 1)

228 → 229 (59%, d. r. 16 : 1)

SCHEME 10.77 Sugar-derived auxiliaries can influence the stereochemical outcomes of aldol reactions.

Chiral oxazolidinone auxiliaries based on D-glucose were used for aldol reactions by Koell et al. [160]. The highest selectivities were observed with auxiliaries equipped with the pivaloyl protecting group. The pivaloylated oxazolidinone **228** was transformed into the boron enolate according to the procedure of Evans [161] and subsequently reacted with aliphatic and aromatic aldehydes. The best results were obtained with isobutyric aldehyde (Scheme 10.77). The *syn*-aldol **229** was formed in 16-fold excess over the *anti*-diastereomer and with an acceptable yield of 59%. The authors explain the stereoselectivity by a chair-like transition state according to Zimmermann–Traxler. The electrophile approaches at the less hindered *re*-face of the (*Z*)-configured enolate double bond. For *N*-phenacetyl substituents, an inversed stereoselectivity was observed as described above for these oxazolidinone auxiliaries.

In the examples presented thus far, the carbohydrate auxiliary was connected to the C–H-acidic component of the aldol reaction so that both diastereotopic faces of the enolate double bond were differentiated by the auxiliary. In a different approach, Ozaki et al. [162] linked the electrophile to a derivative of L-quebrachitol **230**, an optically active cyclitol. Therefore, L-quebrachitol was transformed into the α-keto ester **231**. The $SnCl_4$-promoted addition of various silyl enol ethers furnished α-hydroxy esters with good to excellent stereoselectivities. Best results were obtained when the ester **231** was reacted with the silyl enol ether **232** (Scheme 10.78). The product **233** was formed in a yield of 98% with a diastereomeric excess higher than 98:2. The authors propose the formation of a five-membered chelate ring **234** between the two carbonyl groups and the Lewis acid $SnCl_4$ (Figure 10.20). The attack of the nucleophile occurs from the sterically less hindered *re*-face of the chelate. The *si*-face of the complex is shielded by the bulky trialkylsilyl group. The generated α-hydroxy esters were cleaved from the auxiliary by saponification and subsequent re-esterification with diazomethane.

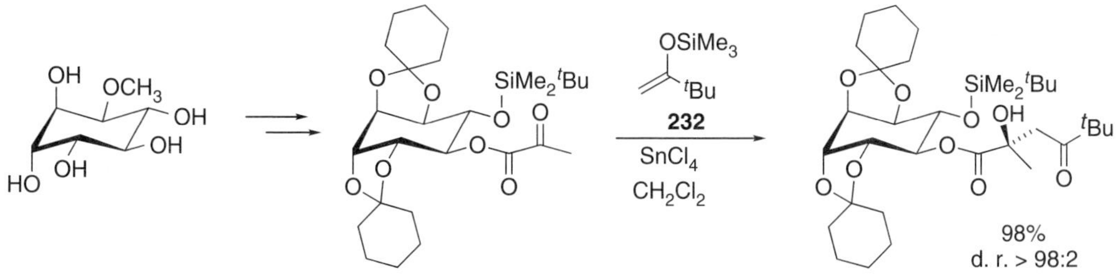

230 → 231 → (232, $SnCl_4$, CH_2Cl_2) → 233 (98%, d. r. > 98:2)

SCHEME 10.78 The stereochemical course of aldol reactions can be influenced through attachment of the chiral auxiliary to the electrophilic component.

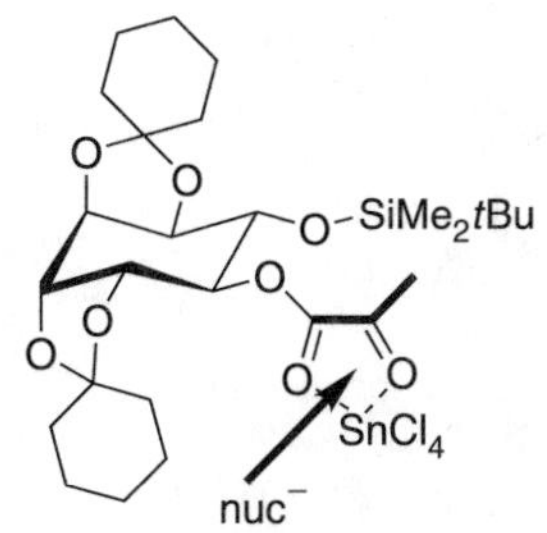

FIGURE 10.20 Chelate control in aldol reaction of L-quebrachitol pyruvate **231** (Scheme 10.78, Ref. [162]).

10.4 REARRANGEMENT REACTIONS

Rearrangement reactions are very important in organic synthesis [163] because they allow the formation of new bonds and stereogenic centers in a single step. Advances in the chemistry of rearrangement reactions have seen the use of carbohydrate auxiliaries applied for the incorporation of asymmetric control. In one example, a diacetone-D-glucose template was used for the asymmetric Overman rearrangement of trichloroacetimidates [164]. With this highly enantioselective methodology, both enantiomers of α-amino acids could be synthesized (Scheme 10.79). The precursor **235** was stereoselectively reduced with $LiAlH_4$ to give the (*E*)-olefin (*E*)-**236**, whereas the hydrogenation of **235** with a palladium catalyst furnished the (*Z*)-olefin (*Z*)-**236**. Both allylic alcohols were transformed into the corresponding trichloroacetimidates (*E*)-**237** and (*Z*)-**237**. The thermal rearrangement of the (*E*)-olefin (*E*)-**237** gave the (*R*)-configured trichloroacetamide (*R*)-**238** with a diastereomeric ratio higher than 94:6. The (*R*)-configuration indicates that the attack of the nitrogen atom took place from the *re*-face of the double bond of (*E*)-**237**. The (*S*)-configured trichloroacetamide (*S*)-**238** was formed from the (*Z*)-olefin as the only detectable product via an approach of nitrogen from the *si*-face of the double bond of (*Z*)-**237**. The enantiomeric amino acids D-alanine (*R*)-**239** and L-alanine (*S*)-**239** were obtained by the oxidation of (*R*)-**238** and (*S*)-**238** with RuO_4 followed by acid hydrolysis.

A D-glucose-derived auxiliary was successfully used for an asymmetric [2,3]-Wittig rearrangement [165]. The ether **240** was prepared from tetra-*O*-methyl-D-glucose in four steps in an overall yield of 60%. Deprotonation with *n*-BuLi generated the anion **241**, which rearranged to give the diastereomeric alcohols *syn*-**242** and *anti*-**242** with a diastereoselectivity of 9:1 and a yield of 94% (Scheme 10.80). The preferred formation of the *syn*-alcohol *syn*-**242** can be explained by the transition state models shown in Figure 10.21. Transition state **243** is favored because it involves less steric repulsion between the methyl group and the trimethylsilyl group than transition state **244**.

Cross-conjugated cyclopentenones can be synthezised enantioselectively via an asymmetric variant of the allene ether Nazarov cyclization using a D-glucose-derived auxiliary [166]. The auxiliary was connected with the allene moiety (Scheme 10.81). Deprotonation of the α-anomeric allene **245** with *n*-BuLi, the addition of the morpholino amide **246** and an acidic workup with HCl in ethanol furnished the cyclopentenone (*S*)-**247** with a moderate yield (67%) and enantioselectivity (67% *ee*). The proposed mechanism of the cyclization is shown in Scheme 10.82. Under acidic conditions, the tetrahedral intermediate **248**, generated by the addition of the anion of allene **245** to the morpholino amide **246**, irreversibly eliminates morpholine. The resulting ketone **249** is protonated and cation **250** cyclizes to give cation **252** with the formation of a new stereogenic center. The reaction is completed by the elimination of the glucosyl cation **254**. Because the

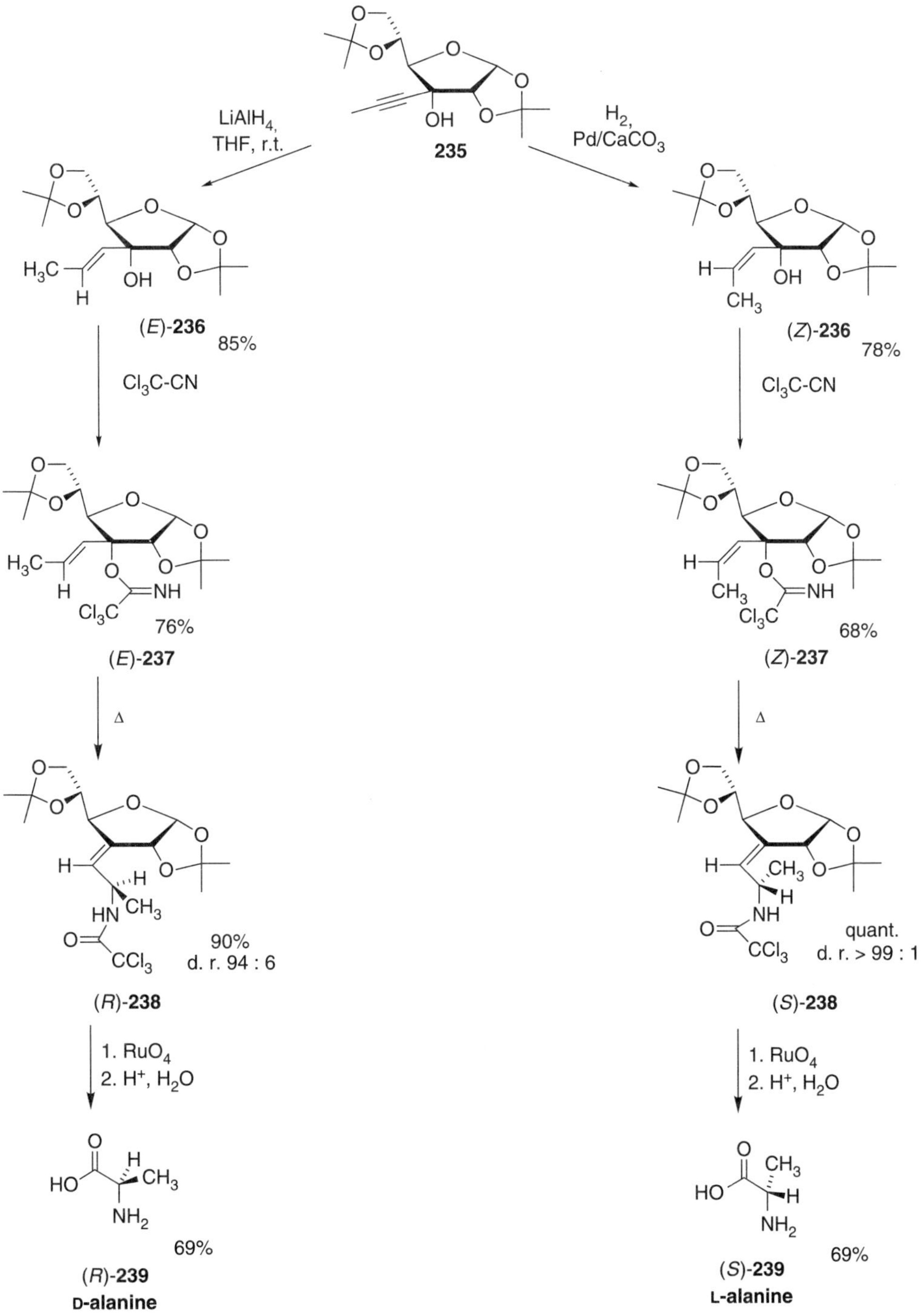

SCHEME 10.79 The stereochemical course of the Overman rearrangement of trichloroacetimidates can be influenced using sugar-derived auxiliaries.

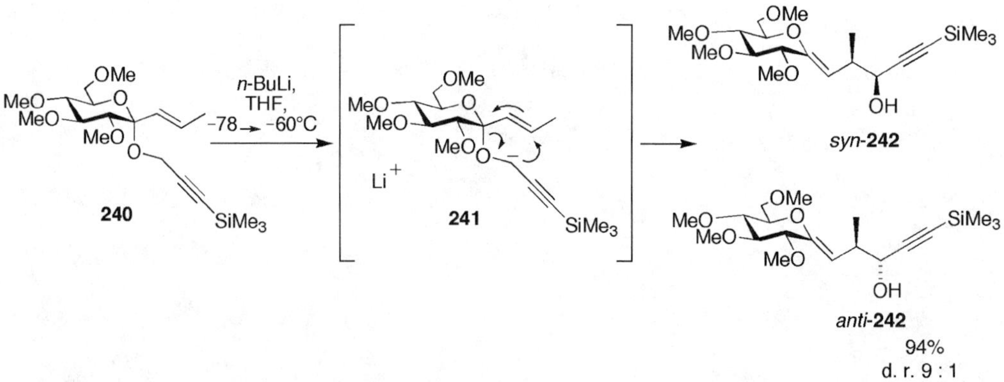

SCHEME 10.80 The stereochemical course of the [2,3]-Wittig rearrangement can be influenced using sugar-derived auxiliaries.

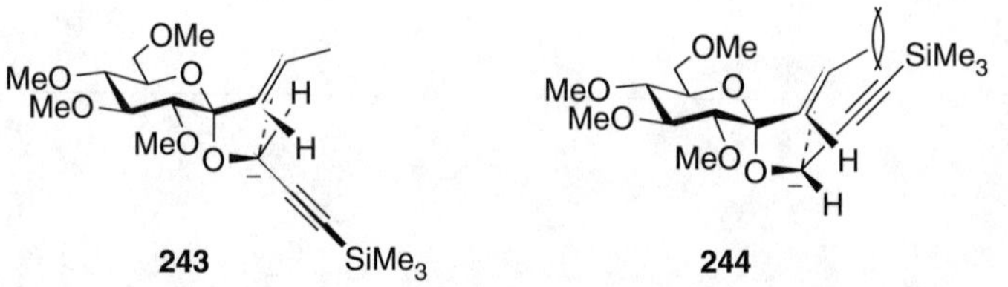

FIGURE 10.21 Diastereodifferentiation in [2,3] Wittig Rearrangement on a glucose-derived auxiliary (Scheme 10.80, Ref. [165]).

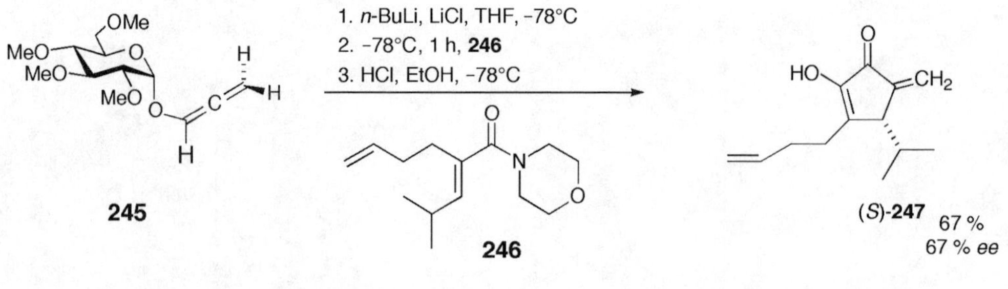

SCHEME 10.81 The stereochemical course of Nazarov cyclization can be influenced using sugar-derived auxiliaries.

cyclization of cation **251** is probably reversible, the stereoselectivity depends on the stability of the cations **250**, **251**, and **252**.

The corresponding reaction of the β-anomeric allene **255** furnished the enantiomeric cyclopentenone (*R*)-**247** with a similar yield and enantiomeric excess (Scheme 10.83). The authors found an increased enantioselectivity (up to 82% *ee* for the best entry) for almost all cyclization reactions when hexafluoroisopropanol was used as the solvent during the acidic workup.

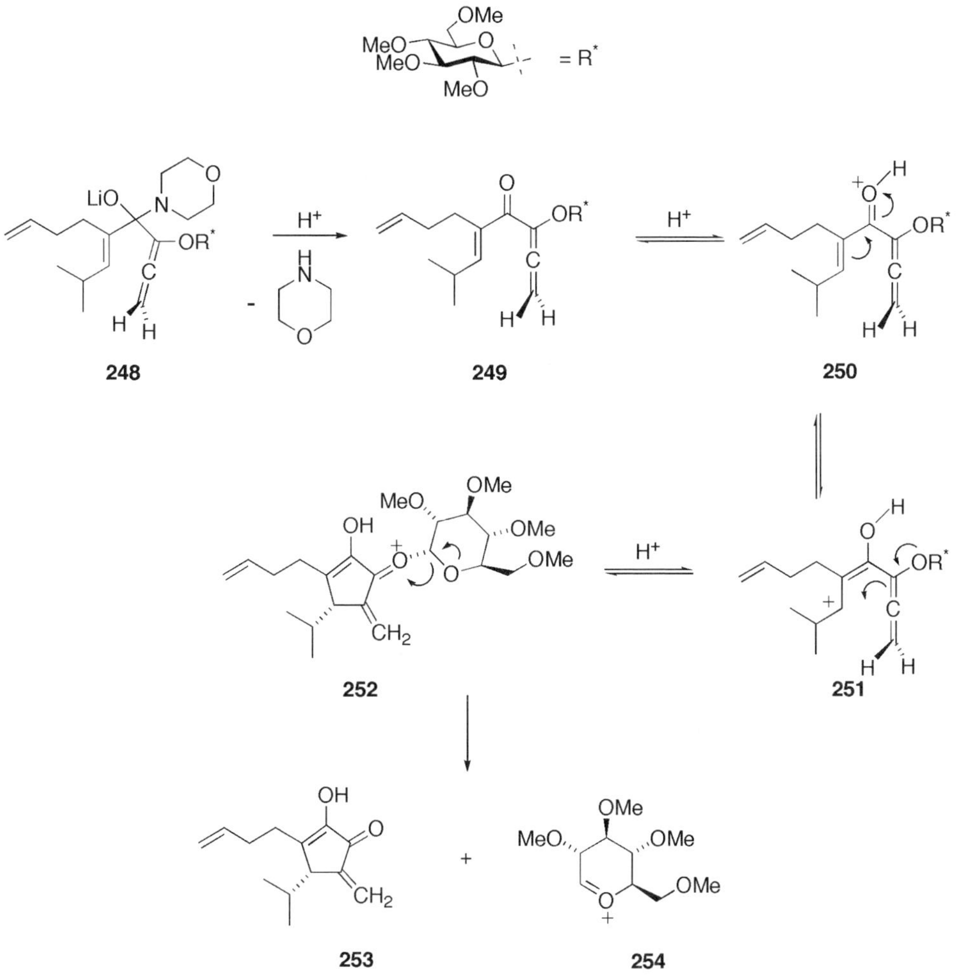

SCHEME 10.82 The proposed mechanism of the sugar-directed Nazarov cyclization.

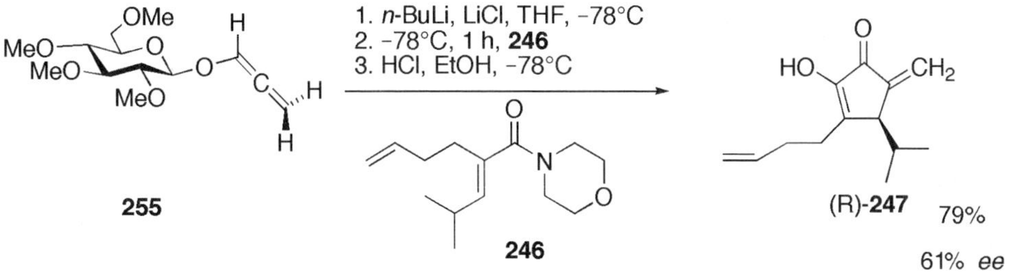

SCHEME 10.83 Cyclization reactions of β-anomeric allenes are subject to influence by chiral auxiliaries.

10.5 RADICAL REACTIONS

Radical reactions generally offer versatile alternatives to ionic reactions [167]. Many methods have been developed that allow chemoselective and regioselective radical transformations. More recently, stereoselective radical reactions have been reported [168].

In some examples, the stereochemistry of radical reactions was controlled by chiral carbohydrate auxiliaries. As a radical counterpart to the ionic conjugate additions discussed above, Garner et al. [169] prepared carbohydrate linked radicals that were reacted with α,β-unsaturated esters. The radical precursor, the carboxylic acid **256**, generated by the addition of (*S*)-methyl lactate to tri-*O*-benzyl-D-glucal and subsequent ester hydrolysis, was decarboxylated by Barton's procedure (Scheme 10.84) [170]. Trapping of the chiral radical **258** with methyl acrylate furnished the saturated ester **259** in 61% yield and with high diastereoselectivity (11:1). The auxiliary caused a preferential addition to the *si*-face of radical **258**, probably due to entropic effects. The ester **259** was transformed in acceptable yield to the γ-butyrolactone **261** by reductive removal of the thiopyridyl group followed by acid hydrolysis.

A diastereoselective cyclization of a 5-hexenyl radical linked to a carbohydrate scaffold was reported by Enholm et al. [171]. The authors used (+)-isosorbit and (−)-D-xylose as the chiral auxiliaries. The α,β-unsaturated bromo ester **262** derived from (+)-isosorbit was reacted with tributyltinhydride and Lewis acids. The influence of the reaction temperature and of the solvent on the yield and stereoselectivity of the cyclization were also examined. Best results were obtained when $ZnCl_2$ was used as Lewis acid at −78°C (Scheme 10.85). The cyclization furnished the ester

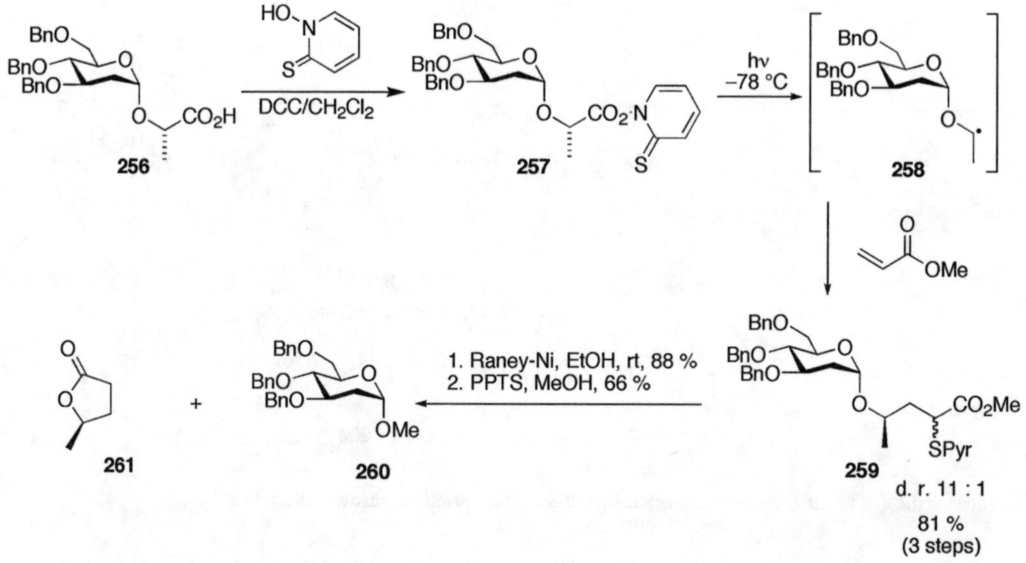

BnO BnO BnO O O CO2H 256 — HO N S, DCC/CH2Cl2 → BnO BnO BnO O O CO2N S 257 — hν, −78 °C → [BnO BnO BnO O O 258] — methyl acrylate (O, OMe) → BnO BnO BnO O O CO2Me SPyr 259 d. r. 11 : 1 81 % (3 steps) — 1. Raney-Ni, EtOH, rt, 88 % 2. PPTS, MeOH, 66 % → O O 261 + BnO BnO BnO O OMe 260

SCHEME 10.84 The stereochemical course of radical reactions can be influenced through incorporation of sugar-derived auxiliaries.

O O H O O H OBn Br 262 — n-Bu3SnH, ZnCl2, Et3B, O2, CH2Cl2/THF, -78°C → H O O H O O H OBn 263 85 % d. r. 100 : 1 — LiOH → HO H O 264

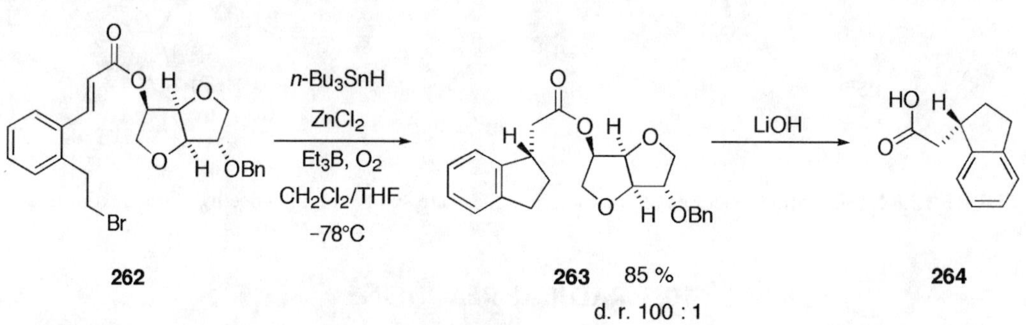

SCHEME 10.85 The stereochemical course of radical reactions can be influenced through incorporation of sugar-derived auxiliaries.

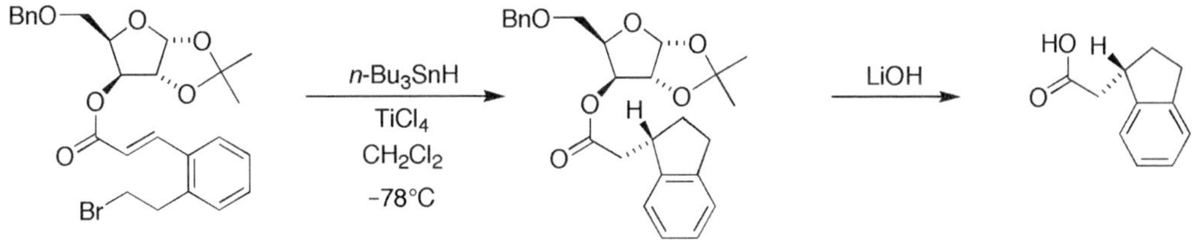

SCHEME 10.86 The preparation of (*S*)-(+)-indan acetic acid, **264**, using sugar-derived auxiliaries.

263 with good yield and excellent diastereoselectivity. The product was cleaved from the carbohydrate auxiliary with LiOH generating (*S*)-(+)-indan acid **264** with a high enantiomeric excess.

The precursor **265,** derived from (−)-D-xylose, was transformed into the saturated ester **266** with a similar diastereoselectivity when $TiCl_4$ was added (Scheme 10.86). After saponification of the carbohydrate ester **266,** (*S*)-(+)-indan acid **264** was isolated.

Enholm et al. [172] also examined free radical allylations of carbohydrate linked bromo esters. The D-xylose-derived ester **267** was reacted with allytributyltin and $ZnCl_2$ as a Lewis acid to give the ester **268** (Scheme 10.87). The diastereoselectivity and yield of the allylation reaction were high.

Although a large variety of reactions were successfully transferred to the solid phase there are only a few examples of free radical reactions on solid phase [173,174]. The free allyl radical transfer was achieved when the carbohydrate auxiliary was linked to a noncross-linked polystyrene polymer (NCPS) [175]. The use of NCPS instead of cross-linked supports has several advantages, including complete solubility in many organic solvents [176]. The polymer **269** was obtained in a radical polymerization of two equivalents of styrene and one equivalent *p*-chloromethylstyrene. The protected D-xylose **270** was covalently linked to the support by a Williamson synthesis generating compound **271** (Scheme 10.88).

The remaining hydroxyl group at C-3 of the xylose was esterified with 2-bromo propionic acid in good yield. An allyl transfer to bromo ester **272** was achieved under thermal conditions (refluxing benzene). All attempts at adding Lewis acids to the reaction mixture led to cleavage of the carbohydrate from the polymer backbone. The allylated product **273** was isolated in a yield of 93%. Saponification of **273** with LiOH gave (*R*)-(−)-2-methyl-4-pentenoic acid **274** with high yield and excellent enantiomeric excess (97% *ee*). The carbohydrate linked polymer **271** could be recovered with high yield. This is the first example of a diastereoselective radical reaction that is sterically controlled by a polymer linked carbohydrate auxiliary.

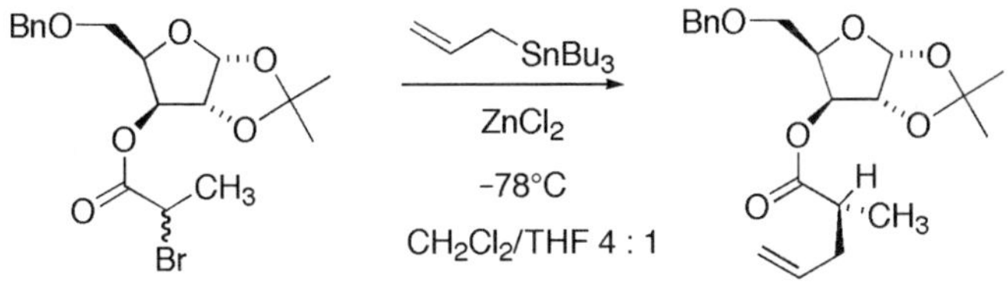

SCHEME 10.87 The stereochemical course of free radical allylations can be influenced through incorporation of sugar-derived auxiliaries.

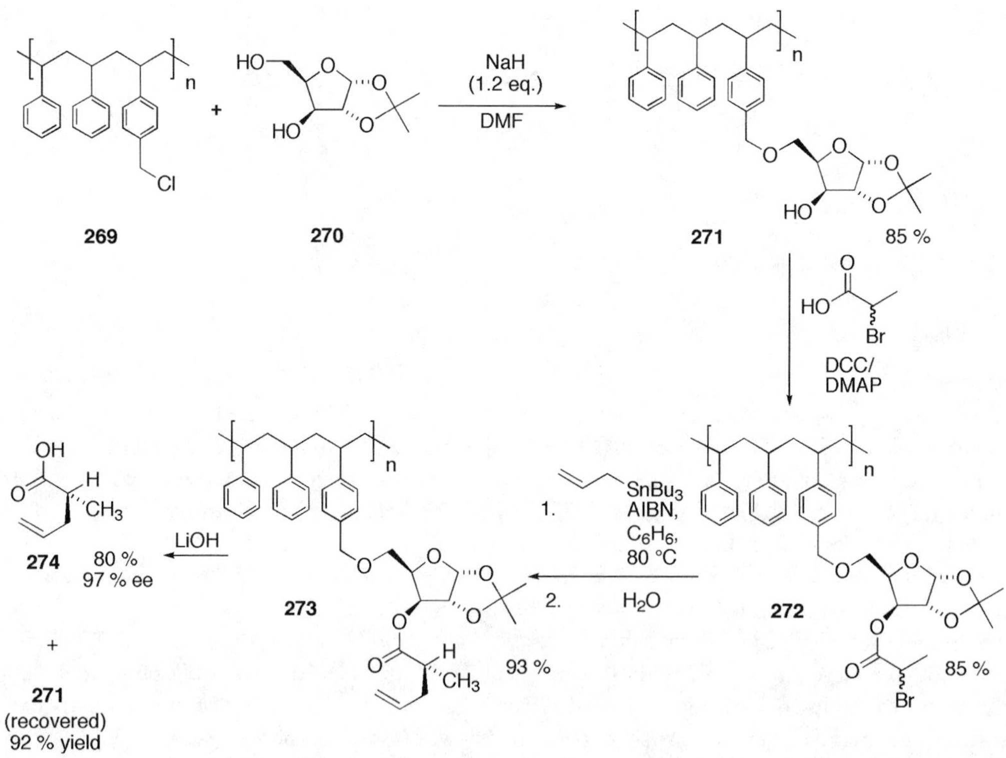

SCHEME 10.88 Sugar-directed asymmetric free radical allylations can proceed on a solid support.

10.6 MISCELLANEOUS APPLICATIONS OF CARBOHYDRATE AUXILIARIES

In the remainder of this chapter, some other applications of carbohydrate auxiliaries are outlined.

Hollingsworth et al. [177] reported on a stereoselective conversion of allyl alcohols to chiral alkane-1,2-diols using D-glucose as a chiral auxiliary. Fischer glycosidation of glucose (**275**) with allyl alcohol gave a mixture of the α- and β-anomers α-**276** and β-**276** (Scheme 10.89). After chromatographic separation of the anomers, the double bond of the α-anomer α-**276** was activated by the soft electrophile Hg^{2+}. The nucleophilic attack of the hydroxyl group at C-2 generated a mercurated alkoxy intermediate that was reduced by $NaBH_4$. The reaction furnished a mixture of diastereomers (91:9). The diastereomers were separated and the major diastereomer **277** was transformed into (*R*)-1,2-propanediol (*R*)-**278** by acetolysis and oxidative removal of the anomeric carbon with hydrogen peroxide. The diol was isolated in an almost enantiopure form (>99% *ee*).

The aforementioned procedure was also applied to β-anomer β-**276** (Scheme 10.90). In this case, the cyclization after mercuration was less diastereoselective (3:1 dr). After separation of the diastereomers, the major diastereomer was converted into (*S*)-1,2-propanediol (*S*)-**278** with more than 99% *ee*. The advantage of this method is that both configurations of the product can be synthesized from one chiral auxiliary by separation of the anomeric glycosides and transformation of each anomer into the corresponding diol with excellent enantioselectivity.

Chiral lactones were synthesized by stereoselective reduction reactions with a carbohydrate auxiliary [178]. The benzyl protected 3,6-anhydro-α-D-glucopyranoside **279** was acylated with acid halide **280**, and the product **281** was reduced with $Zn(BH_4)_2$ (Scheme 10.91). The alcohol **282** was obtained with a high diastereomeric excess (96:4). Cleavage from the auxiliary by

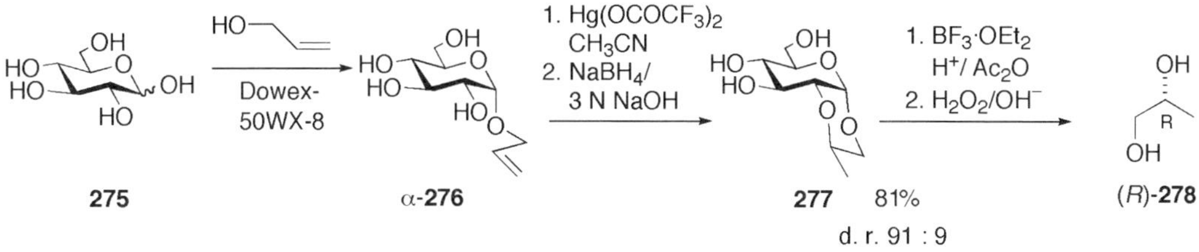

SCHEME 10.89 Allyl alcohols can be converted to chiral 1,2-diols using sugar-derived auxiliaries.

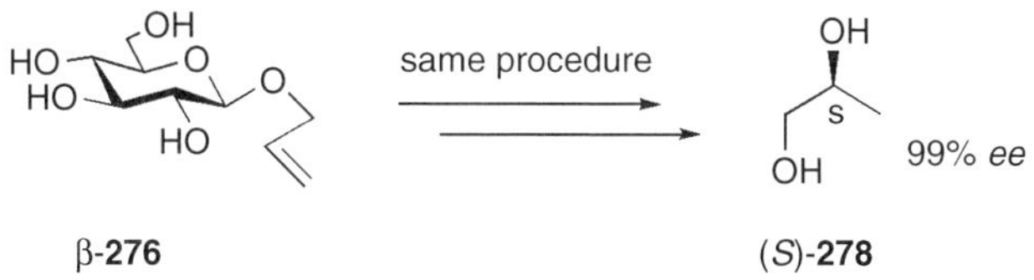

SCHEME 10.90 Allyl alcohols can be converted to chiral 1,2-diols using sugar-derived auxiliaries.

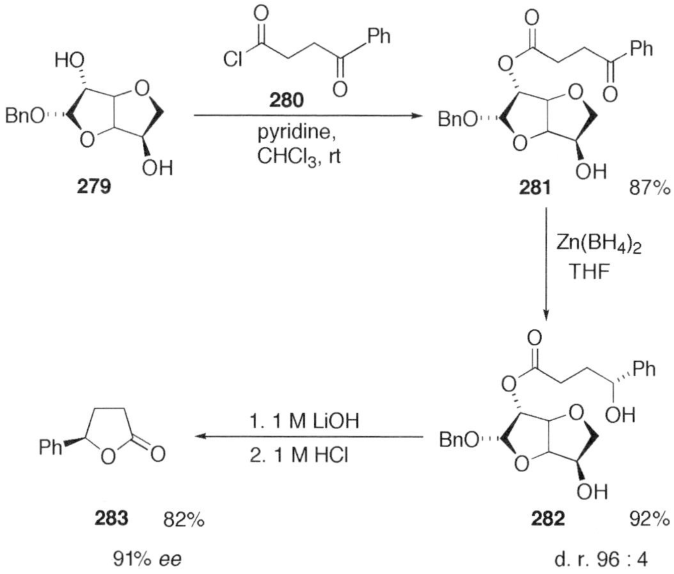

SCHEME 10.91 Sugar-derived auxiliaries can influence the stereochemical course of the reduction of carbonyls.

saponification and treatment with hydrochloric acid gave the chiral lactone **283** with 91% *ee*. The use of other reducing agents like $NaBH_4$, $LiBH_4$ and $Ce(BH_4)_3$ resulted in decreased enantiomeric excess.

Higher substituted five- and six-membered lactones were also prepared by this procedure, but the enantiomeric excesses of the products were lower. The authors explain the diastereoselectivity of the reduction by a chelate complex in which the metal ion coordinates the hydroxyl group at C-5, one ring oxygen atom and the carbonyl group, which is then attacked by the complex hydride.

Chung et al. [179] demonstrated that carbohydrate auxiliaries can also be employed for the preparation of chiral (arene)tricarbonyl complexes, which are useful intermediates in organic synthesis. The D-glucose-derived benzylidene acetal **284** was reacted with hexacarbonyl chromium

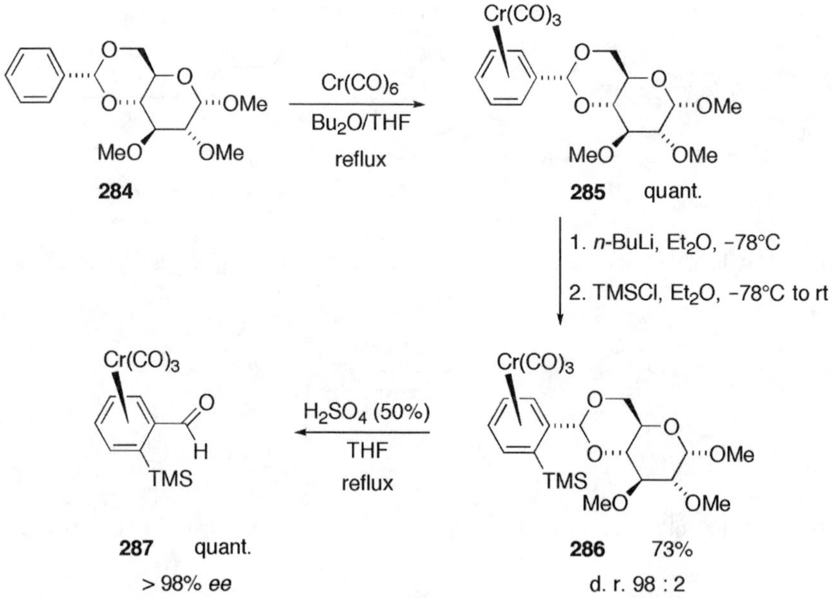

SCHEME 10.92 Chiral (arene)tricarbonyl complexes can be prepared through the influence of sugar-derived auxiliaries.

to furnish the complex **285** quantitatively (Scheme 10.92). Deprotonation of **285** with *n*-BuLi and trapping the anion with trimethylchlorosilane as the electrophile generated the complex **286** with excellent diastereoselectivity. The carbohydrate was cleaved by acid hydrolysis quantitatively liberating the *ortho*-substituted (benzaldehyde)tricarbonylchromium complex **287** in almost enantiopure form (>98% *ee*).

10.7 CONCLUSION

The examples outlined in this chapter show that carbohydrates are efficient stereodifferentiating auxiliaries, which offer possibilities for stereochemical discrimination in a wide variety of chemical reactions. Interesting chiral products are accessible, including chiral carbo- and heterocycles, α- and β-amino acid derivatives, β-lactams, branched carbonyl compounds and amines. Owing to the immense material published since the time of the earlier review articles on carbohydrates in asymmetric synthesis [9,10], the examples discussed in this chapter necessarily focused on the use of carbohydrates as auxiliaries covalently linked to and cleavable from the substrate. Given the scope of this chapter, a discussion of other interesting asymmetric reactions has not been permitted — for example, reactions in which carbohydrate-derived Lewis acids, such as cyclopentadienyl titanium carbohydrate complexes, exhibit stereocontrol in aldol reactions [180]. Similarly, processes in which *in situ* glycosylation induces reactivity and stereodifferentiation — for example, in Mannich reactions of imines [181] — have also been excluded from this discussion.

Finally, carbohydrate ligands of enantioselective catalysts have been described for a limited number of reactions. Bis-phosphites of carbohydrates have been reported as ligands of efficient catalysts in enantioselective hydrogenations [182] and hydrocyanations [183], and a bifunctional dihydroglucal-based catalyst was recently found to effect asymmetric cyanosilylations of ketones [184]. Carbohydrate-derived titanocenes have been used in the enantioselective catalysis of reactions of diethyl zinc with carbonyl compounds [113]. Oxazolinones of amino sugars have been shown to be efficient catalysts in enantioselective palladium(0)-catalyzed allylation reactions of *C*-nucleophiles [185].

It is anticipated that carbohydrates will be used more frequently for the construction of enantioselective catalysts in the future and the results of these efforts will justify a separate review.

REFERENCES

1. Fischer, E, Verbindungen des Phenylhydrazins mit den Zuckerarten, *Ber. Dtsch. Chem. Ges.*, 17, 579–588, 1884.
2. Landsteiner, K, *Wiener. Klin. Wochenschrift*, 14, 1132, 1901.
3. Somers, W T, Tang, J, Shaw, G D, Camphausen, R T, Insights into the molecular basis of leukocyte tethering and rolling revealed by structures of P- and E-selectin bound S Le(x) and PSGL-1, *Cell.*, 103, 467–479, 2000, and references cited therein.
4. Fukuda, M, Tsuboi, S, Mucin-type *O*-glycans and leukosialin, *Biochim. Biophys. Acta*, 1455, 205–217, 1999, and references cited therein.
5. Ernst, B, Hart, G W, Sinaÿ, P, *Carbohydrates in Chemistry and Biology*, Vol. 1, Wiley-VCH, Weinheim, 2000.
6. Vankar, Y D, Schmidt, R R, Chemistry of glycosphingolipids–carbohydrate molecules of biological significance, *Chem. Soc. Rev.*, 29, 201–216, 2000.
7. Herzner, H, Reipen, T, Schultz, M, Kunz, H, Synthesis of glycopeptides containing carbohydrate and peptide recognition motifs, *Chem. Rev.*, 100, 4495–4537, 2000.
8. Hollingsworth, R I, Wang, G, Towards a carbohydrate-based chemistry: progress in the development of general-purpose chiral synthons from carbohydrates, *Chem. Rev.*, 100, 4267–4282, 2000.
9. Rueck, K, Kunz, H, Carbohydrates as chiral auxiliaries in stereoselective synthesis, *Angew. Chem. Int. Ed.*, 32, 336–358, 1993.
10. Hultin, P G, Earle, M A, Sudharshan, M, Synthetic studies with carbohydrate-derived chiral auxiliaries, *Tetrahedron*, 53, 14823–14870, 1997.
11. Naumann, K, *Synthetic Pyrethroid Insecticides: Structure and Properties*, Springer, Berlin, 1990.
12. Reissig, H U, *The Chemistry of the Cyclopropyl Group.* Part 1, Wiley, Chichester, 1987.
13. Reissig, H U, Formation of C–C bonds by [2+1] cycloadditions, In *Methods of Organic Chemistry (Houben–Weyl)*, Vol. 21c, Helmchen, G, Hoffmann, R W, Mulzer, J, Schaumann, E, Eds., Thieme, Stuttgart, pp. 3179–3270, 1995.
14. Murali, R, Ramana, C V, Nagarajan, N, Synthesis of 1,2-cyclopropanated sugars from glycals, *J. Chem. Soc. Chem. Commun.*, 217–218, 1995.
15. Charette, A B, Coté, B, Marcoux, J F, Carbohydrates as chiral auxiliaries: asymmetric cyclopropanation reaction of acyclic olefins, *J. Am. Chem. Soc.*, 113, 8166–8167, 1991.
16. Charette, A B, Turcotte, N, Marcoux, J F, The use of α-D-glucopyranosides as surrogates for the β-L-glucopyranosides in the stereoselective cyclopropanation reaction, *Tetrahedron Lett.*, 35, 513–516, 1994.
17. Charette, A B, Coté, B, Stereoselective synthesis of all four isomers of coronamic acid: a general approach to 3-methanoamino acids, *J. Am. Chem. Soc.*, 117, 12721–12732, 1995.
18. Kang, J, Lim, G J, Yoon, S K, Kim, M Y, Asymmetric cyclopropanation using new chiral auxiliaries derived from D-fructose, *J. Org. Chem.*, 60, 564–577, 1995.
19. Schumacher, R, Reissig, H U, Stereoselective cyclopropanation of chiral carbohydrate-derived enol ethers, *Synlett*, 1121–1122, 1996.
20. Mattay, J, Conrads, R, Hoffmann, R, Formation of C–C bonds by light-induced [2+2] cycloadditions, In *Methods of Organic Chemistry (Houben–Weyl)*, Vol. 21c, Helmchen, G, Hoffmann, R W, Mulzer, J, Schaumann, E, Eds., Thieme, Stuttgart, pp. 3085–3178, 1995.
21. Carless, H A J, *Photochemistry in Organic Synthesis*, Spec. Publ. No. 57, Royal Society of Chemistry, London, 1986.
22. Jarosz, S, Zamojski, A, Asymmetric photocycloaddition between furan and optically active ketones, *Tetrahedron*, 38, 1453–1456, 1982.
23. Pelzer, R, Juetten, P, Scharf, H D, Isoselektivität bei der asymmetrischen Paterno–Buechi-Reaktion unter Verwendung von Kohlenhydraten als chirale Auxiliare, *Chem. Ber.*, 122, 487–491, 1989.
24. Woodward, R B, Hoffmann, R, *The Conservation of Orbital Symmetry*, Academic, New York, 1971.
25. Tidwell, T T, *Ketenes*, Wiley, New York, 1995, pp. 459–527.

26. Ganz, I, Kunz, H, Carbohydrates as chiral auxiliaries. [2+2] Cycloadditions to enol ethers, *Synthesis*, 1353–1357, 1994.
27. Georg, G I, Ravikumar, V T, In *The Organic Chemistry of β-Lactams*, Georg, G I, Ed., VCH Publishers, Inc., New York, pp. 295–381, 1993.
28. Thomas, R C, In *Recent Progress in the Chemical Synthesis of Antibiotics*, Lukacs, G, and Ohno, M, Eds., Springer, Berlin, pp. 533–562, 1990.
29. Cossio, F P, Arrieta, A, Lecea, B, Ugalde, J M, Chiral control in the Staudinger reaction between ketenes and imines. A theoretical SCF-MO study on asymmetric torquoselectivity, *J. Am. Chem. Soc.*, 116, 2085–2093, 1994.
30. Borer, B C, Balogh, D W, An asymmetric synthesis of a 3-hydroxy-β-lactam by ketene-imine cycloaddition: utilization of chiral ketenes from carbohydrates, *Tetrahedron Lett.*, 32, 1039–1040, 1991.
31. Saul, R, Kopf, J, Koell, P, Synthesis of a new chiral oxazolidinone auxiliary based on D-xylose and its application to the Staudinger reaction, *Tetrahedron: Asymmetry*, 11, 423–433, 2000.
32. Georg, G I, Akguen, E, Masheva, P M, Milstead, M, Ping, H, Wu, Z J, Velde, D V, Galactose-imines in the Staudinger reaction, *Tetrahedron Lett.*, 33, 2111–2114, 1992.
33. Barton, D H R, Gateau-Oleska, A, Anaya-Mateos, J, Cléophax, J, Géro, S D, Chiaroni, A, Riche, C, Asymmetric synthesis of 1,3,4-trisubstituted and 3,4-disubstituted 2-azetidinones: strategy based on use of D-glucosamines as a chiral auxiliary in the Staudinger reaction, *J. Chem. Soc. Perkin. Trans. 1*, 3211–3212, 1990.
34. Dhar, D N, Murthy, K S M, Recent advances in the chemistry of chlorosulfonyl isocyanate, *Synthesis*, 437–449, 1986.
35. Effenberger, F, Kiefer, G, Stereochemistry of the cycloaddition of sulfonyl isocyanates and *N*-sulfinylsulfonamides to enol ethers, *Angew. Chem. Int. Ed.*, 6, 951–952, 1967.
36. Cossio, F P, Roa, G, Lecea, B, Ugalde, J M, Substituent and solvent effect in the [2+2] cycloaddition reaction between olefins and isocyanates, *J. Am. Chem. Soc.*, 117, 12306–12313, 1995.
37. Chmielewski, M, Kaluza, Z, Furman, B, Stereocontrolled synthesis of 1-oxabicyclic β-lactam antibiotics via [2+2] cycloaddition of isocyanates to sugar vinyl ethers, *Chem. Commun.*, 2689–2696, 1996.
38. Kaluza, Z, Furman, B, Patel, M, Chmielewski, M, Asymmetric induction in [2+2] cycloaddition of chlorosulfonyl isocyanate to 1,2-*O*-isopropylidene-3-*O*-vinyl-glycofuranoses, *Tetrahedron: Asymmetry*, 5, 2179–2186, 1994.
39. Kaluza, Z, Furman, B, Chmielewski, M, Asymmetric induction in [2+2] cycloaddition of chlorosulfonyl isocyanate to 1,2-*O*-isopropylidene-5-*O*-vinyl-D-glycofuranoses, *Tetrahedron: Asymmetry*, 6, 1719–1730, 1995.
40. Cinquini, M, Cozzi, F, 1,3-Dipolar cycloadditions, In *Methods of Organic Chemistry (Houben–Weyl)*, Vol. 21c, Helmchen, G, Hoffmann, R W, Mulzer, J, Schaumann, E, Eds., Thieme, Stuttgart, pp. 2953–2987, 1995.
41. Gothelf, K V, Jorgensen, K A, Asymmetric 1,3-dipolar cycloaddition reactions, *Chem. Rev.*, 98, 863–909, 1998.
42. Vasella, A, Synthese und Umwandlungen von Isoxazolidin-Nucleosiden, *Helv. Chim. Acta*, 60, 426–446, 1977.
43. Vasella, A, Stereoselektivität und Reaktivität bei der 1,3-dipolaren Cycloaddition chiraler *N*-(Alkoxyalkyl)nitrone, *Helv. Chim. Acta*, 60, 1273–1294, 1977.
44. Huber, R, Vasella, A, The kinetic anomeric effect. Additions of nucleophiles and of dipolarophiles to *N*-glycosylnitrones and to *N*-pseudoglycosylnitrones, *Tetrahedron*, 46, 33–58, 1990.
45. Machetti, F, Cordero, F M, De Sarlo, F, Guarna, A, Brandi, A, A new synthesis of (2*S*)-4-oxopipecolic acid by thermal rearrangement of enantiopure spirocyclopropaneisoxazolidine, *Tetrahedron Lett.*, 37, 4205–4208, 1996.
46. Kasahara, K, Iida, H, Kibayashi, Ch., Asymmetric total synthesis of (+)-negamycin and (−)-3-epinegamycin via enantioselective 1,3-dipolar cycloaddition, *J. Org. Chem.*, 54, 2225–2233, 1989.
47. Chiacchio, U, Corsaro, A, Gumina, G, Rescifina, A, Iannazzo, D, Piperno, A, Romeo, G, Romeo, R, Homochiral α-D- and β-D-isoxazolidinylthymidines via 1,3-dipolar cycloaddition, *J. Org. Chem.*, 64, 9321–9327, 1999.
48. Mzengeza, S, Yang, C M, Whitney, R A, A total synthesis of acivicin, *J. Am. Chem. Soc.*, 109, 276–277, 1987.

49. Fisera, L, Al-Timara, U A R, Ertl, P, Prónayová, N, Preparation and stereoselectivity of 1,3-dipolar cycloaddition of D-glucose-derived nitrones to *N*-arylmaleimides, *Monatsh. Chem.*, 124, 1019–1029, 1993.
50. Yokoyama, M, Yamada, N, Togo, H, Synthesis of spiro sugar isoxazolidines via tandem Michael addition-1,3-dipolar cycloaddition, *Chem. Lett.*, 753–756, 1990.
51. Tamura, O, Mita, N, Kusaka, N, Suzuki, H, Sakamoto, M, Intramolecular cycloaddition of α-allyloxycarbonylnitrone bearing a chiral sugar auxiliary: a short-step synthesis of the *N*-terminal amino acid component of nikkomycin Bz, *Tetrahedron Lett.*, 38, 428–432, 1997.
52. Akiyama, T, Okada, K, Ozaki, S, The preparation of optically active Δ^2-isoxazolines via addition of nitrile oxides to chiral acryloyl esters bearing cyclitols as auxiliaries, *Tetrahedron Lett.*, 33, 5763–5766, 1992.
53. Ao, Z, Ying, K, Biao, J, Asymmetric 1,3-dipolar cycloaddition of nitrile oxides to chiral acryloyl esters bearing glucofuranose as auxiliary, *Chin. J. Chem.*, 18, 220–224, 2000.
54. Oppolzer, W, Intermolecular Diels–Alder reactions, In *Comprehensive Organic Synthesis*, Trost, B M, Fleming, I, Paquette, L A, Eds., Pergamon Press, Oxford, pp. 315–399, 1991.
55. Tietze, L F, Kettschau, G, Hetero-Diels–Alder reactions in organic chemistry, In *Topics in Current Chemistry: Stereoselective Heterocyclic Synthesis I*, Metz, P, Ed., Springer, Berlin, pp. 1–120, 1997.
56. Helmchen, G, Hoffmann, R W, Mulzer, J, Schaumann, E, Eds.,[4+2] cycloadditions (Chapter 1.6), In *Methods of Organic Chemistry (Houben–Weyl)*, Vol. 21c, Thieme, Stuttgart, 1995.
57. Rueck-Braun, K, Kunz, H, *Chiral Auxiliaries in Cycloadditions*, Wiley-VCH, Weinheim, 1999.
58. David, S, Eustache, J, Lubineau, A, Cycloaddition as a possible path to disaccharide synthesis: stereochemical course of the reaction of butyl glyoxylate with a chiral, protected dienyl ether of glucose, *J. Chem. Soc. Perkin. Trans. 1*, 2274–2278, 1974.
59. David, S, Lubineau, A, Thieffry, A, Stéréochimie de la cycloaddition sur les éthers butadiényliques d'alcools chiraux. Dérivés en 4 et 6 de glucosides perbenzylés, *Tetrahedron*, 34, 299–304, 1978.
60. Kunz, H, Mueller, B, Schanzenbach, D, Diastereoselective Diels–Alder reactions using carbohydrate templates, *Angew. Chem. Int. Ed.*, 26, 267–269, 1987.
61. Staehle, W, Kunz, H, Carbohydrates as chiral templates: stereoselective Diels–Alder synthesis with dienes of differing reactivity, *Synlett*, 260–262, 1991.
62. Gras, J L, Poncet, A, Nouguier, R, Ligand assisted asymmetric synthesis. II. Diastereoselective Diels–Alder additions with Lewis acid attracting auxiliaries derived from pentitols, *Tetrahedron Lett.*, 33, 3323–3326, 1992.
63. Nouguier, R, Gras, J L, Giraud, B, Virgili, A, Methyl 3,4-β-D-arabinoside as a new chiral template for the asymmetric Diels–Alder reaction, *Tetrahedron Lett.*, 32, 5529–5530, 1991.
64. Nouguier, R, Gras, J L, Giraud, B, Virgili, A, Induction asymétrique dans la réaction de Diels–Alder du cyclopentadiène sur des dérivés acryliques de *O*-méthylèneacétals de l'arabinose et du ribose, *Tetrahedron*, 48, 6245–6252, 1992.
65. Nouguier, R, Mignon, V, Gras, J L, Synthesis of methylene acetals in the D-glucose, D-galactose, D-mannose, and D-fructose series by an improved transacetalation reaction from dimethoxymethane, *Carbohydr. Res.*, 277, 339–345, 1995.
66. Nouguier, R, Mignon, V, Gras, J L, Synthesis of enantiomerically pure (2*S*)-2-endo-hydroxymethyl-5-norbornene by Diels–Alder reaction based on a new fructose-derived auxiliary, *J. Org. Chem.*, 64, 1412–1414, 1999.
67. Shing, T K M, Chow, H F, Chung, I H F, Sugarometallic chemistry: aglycone–chromium complex as chiral auxiliary in asymmetric Diels–Alder reaction, *Tetrahedron Lett.*, 37, 3713–3716, 1996.
68. Akiyama, T, Horiguchi, N, Ida, T, Ozaki, S, Stereodivergent Diels–Alder reactions employing cyclitols as chiral auxiliaries, *Chem. Lett.*, 975–976, 1995.
69. Loupy, A, Monteux, D, Asymmetric Diels–Alder: monobenzylated isosorbide and isomannide as highly effective chiral auxiliaries, *Tetrahedron Lett.*, 37, 7023–7026, 1996.
70. Ferreira, M L G, Pinheiro, S, Perrone, C C, Costa, P R R, Ferreira, V F, New carbohydrate-based auxiliaries in Diels–Alder reaction, *Tetrahedron: Asymmetry*, 9, 2671–2680, 1998.
71. Banks, M R, Blake, A J, Cadogan, J I G, Dawson, I M, Gaur, S, Gosney, I, Gould, R O, Grant, K J, Hodgson, P K G, (5R)-7,8,9,10-Di-*O*-isopropylidene-2,6-dioxa-4-azaspiro[4,5]-decan-3-one: a new chiral spirooxazolidin-2-one derived from D-(+)-galactose for use in asymmetric tranformations, *J. Chem. Soc. Chem. Commun.*, 1146–1148, 1993.

72. Banks, M R, Cadogan, J I G, Gosney, I, Gaur, S, Hodgson, P K G, Highly regio- and stereospecific preparation of a new carbohydrate-based 1,3-oxazin-2-one by the INIR method and its application in some asymmetric transformations, *Tetrahedron: Asymmetry*, 5, 2447–2458, 1994.
73. Banks, M R, Cadogan, J I G, Gosney, I, Gould, R O, Hodgson, P K G, McDougall, D, Preparation of enantiomerically pure fructose-derived 1,3-oxazin-2-one by INIR methodology and its application as a chiral auxiliary in some model asymmetric reactions, *Tetrahedron*, 54, 9765–9784, 1998.
74. Kunz, H and Engel, S, Unpublished results.
75. Engel, S, PhD dissertation, University of Mainz, 1994.
76. Horton, D, Koh, D, Generalized approach from sugars to enantiomerically pure tetra-C-substituted carbocycles, *Tetrahedron Lett.*, 34, 2283–2286, 1993.
77. Horton, D, Koh, D, Takagi, Y, Stereocontrol in Diels–Alder cycloaddition to unsaturated sugars: reactivities of *cis*-dienophiles with cyclopentadiene, *Carbohydr. Res.*, 250, 261–274, 1993.
78. Horton, D, Koh, D, Stereocontrol in Diels–Alder cycloaddition to unsaturated sugars: reactivities of acyclic seven-carbon trans-dienophiles derived from aldopentoses, *Carbohydr. Res.*, 250, 249–260, 1993.
79. Serrano, J A, Moreno, M C, Román, E, Arjona, O, Plumet, J, Jimenéz, J, Enantioselective synthesis of cyclohexene nitro aldehydes via Diels–Alder reactions with sugar nitroolefins, *J. Chem. Soc. Perkin. Trans 1*, 3207–3212, 1991.
80. Serrano, J A, Cáceres, L E, Román, E, Asymmetric Diels–Alder reactions of chiral sugar nitroalkene: diastereofacial selectivity and regioselectivity, *J. Chem. Soc. Perkin. Trans*, 1, 941–942, 1992.
81. Serrano, J A, Cáceres, L E, Román, E, Asymmetric Diels–Alder reactions between chiral sugar nitroalkenes and 1-*O*-substituted buta-1,3-dienes. Synthesis and reactivity of new cyclohexenyl derivatives, *J. Chem. Soc. Perkin. Trans 1*, 1863–1871, 1995.
82. Franck, R W, Argade, S, Subramanian, C S, Frechet, D M, The face selectivity directed by an allylic group of a diene in the Diels–Alder reaction is reversed from that of a dienophile, *Tetrahedron Lett.*, 26, 3187–3190, 1985.
83. Kahn, S D, Hehre, W J, Modeling chemical reactivity. 5. Facial selectivity in Diels–Alder cycloadditions, *J. Am. Chem. Soc.*, 109, 663–666, 1987.
84. Gupta, R C, Raynor, C M, Stoodley, R J, Slawin, A M Z, Williams, D J, Asymmetric Diels–Alder reactions. Part 1. Diastereofacial reactivity of (*E*)-3-trimethylsilyloxybuta-1,3-dienyl 2,3,4,6-tetra-*O*-acetyl-β-D-glucopyranoside towards cyclic dienophiles, *J. Chem. Soc. Perkin. Trans 1*, 1773–1785, 1988.
85. Gupta, R C, Larsen, D S, Stoodley, R J, Slawin, A M Z, Williams, D J, Asymmetric Diels–Alder reactions. Part 2. A model to account for the diastereofacial reactivity of (*E*)-1-(2′,3′,4′,6′-tetra-*O*-acetyl-β-D-glucopyranosyloxy)-3-trimethylsiloxybuta-1,3-diene and its 2-methyl derivative, *J. Chem. Soc. Perkin. Trans 1*, 739–749, 1989.
86. Beagley, B, Curtis, A D M, Pritchard, R G, Stoodley, R J, Asymmetric Diels–Alder reactions. Part 6. Regio- and stereo-selective cycloadditions of 5-′2′,3′,4′,6′-tetra-*O*-acetyl-β-D-glucopyranosyloxy)-1, 4-naphthoquinone, *J. Chem. Soc. Perkin. Trans 1*, 1981–1991, 1992.
87. Pornpakakul, S, Pritchard, R G, Stoodley, R J, Asymmetric synthesis of (−)-4-epi-shikimic acid, *Tetrahedron Lett.*, 41, 2691–2694, 2000.
88. Ehlenz, R, Neuss, O, Teckenbrock, M, Doetz, K H, Carbohydrate functionalized oxacyclopentylidene complexes, *Tetrahedron*, 53, 5143–5158, 1997.
89. Doetz, K H, Ehlenz, R, Carbohydrate-modified metal carbenes: synthesis and first applications, *Chem. Eur. J.*, 3, 1751–1756, 1997.
90. Weyershausen, B, Nieger, M, Doetz, K H, Chiral 2-oxacyclopentylidene complexes via metal-assisted cycloisomerization of carbohydrate-derived butynols, *Organometallics*, 17, 1602–1707, 1998.
91. Weyershausen, B, Nieger, M, Doetz, K H, Stereospecific exo-selective Diels–Alder reactions with carbohydrate-functionalized α-exo-methylene-2-oxacyclopentylidene chromium complexes, *J. Org. Chem.*, 64, 4206–4210, 1999.
92. Felber, H, Kresze, G, Prewo, R, Vasella, A, Diastereoselectivity and reactivity in the Diels–Alder reactions of α-chloronitroso ethers, *Helv. Chim. Acta*, 69, 1137–1146, 1986.
93. Werbitzky, O, Klier, K, Felber, H, Asymmetric induction of four chiral centres by hetero Diels–Alder reaction of a chiral nitroso dienophile, *Liebigs Ann. Chem.*, 267–270, 1990.

94. Braun, H, Felber, H, Kresze, G, Schmidtchen, F P, Prewo, R, Vasella, A, Diastereoselektive Diels–Alder-Reaktionen mit α-Chloronitrososaccharіden, *Liebigs Ann. Chem.*, 261–268, 1993.
95. Zhang, D, Sueling, C, Miller, M J, The hetero Diels–Alder reactions between D-mannose-derived halonitroso compounds and cyclopentadiene: scope and limitations, *J. Org. Chem.*, 63, 885–888, 1998.
96. Arnold, T, Orschel, B, Reissig, H U, Diacetoneglucose as auxiliary group for the asymmetric hetero-Diels–Alder reaction with nitrosoalkenes, *Angew. Chem. Int. Ed.*, 31, 1033–1035, 1992.
97. Schade, W, Reissig, H U, A new diastereoselective synthesis of enantiomerically pure 1,2-oxazines derivatives by addition of lithiated methoxyallene to chiral nitrones, *Synlett*, 632–634, 1999.
98. Pulz, R, Watanabe, T, Schade, W, Reissig, H U, A stereoconvergent synthesis of enantiopure 3-methoxypyrrolidines and 3-methoxy-2,5-dihydropyrroles from 3,6-dihydro-2*H*-1,2-oxazines, *Synlett*, 983–986, 2000.
99. Denmark, S E, Thorarensen, A, Tandem [4+2]/[3+2] cycloadditions of nitroalkenes, *Chem. Rev.*, 96, 137–165, 1996.
100. Avalos, M, Babiano, R, Cintas, P, Higes, F J, Jiménez, J L, Palacios, J C, Silva, M A, Substrate-controlled stereodifferentiation of tandem [4+2]/[3+2] cycloadditions by a vicinal carbohydrate-based template, *J. Org. Chem.*, 61, 1880–1882, 1996.
101. Avalos, M, Babiano, R, Cintas, P, Jiménez, J L, Palacios, J C, Silva, M A, Asymmetric tandem reactions based on nitroalkenes: a one-pot construction of functionalized chiral bicycles by a three-component reaction, *Chem. Commun.*, 459–460, 1998.
102. Avalos, M, Babiano, R, Cintas, P, Higes, F J, Jiménez, J L, Palacios, J C, Silva, M A, Diastereoselective cycloadditions of nitroalkenes as an approach to the assembly of bicyclic nitrogen heterocycles, *J. Org. Chem.*, 64, 1494–1502, 1999.
103. Avalos, M, Babiano, R, Bravo, J L, Cintas, P, Jiménez, J L, Palacios, J C, Silva, M A, Understanding diastereofacial selection in carbohydrate-based domino cycloadditions: Semiempirical and DFT calculations, *Chem. Eur. J.*, 6, 267–277, 2000.
104. Pfrengle, W, Kunz, H, Hetero Diels–Alder reactions on a carbohydrate template: stereoselective synthesis of (*S*)-anabasine, *J. Org. Chem.*, 54, 4261–4263, 1989.
105. Aspinall, I H, Cowley, P M, Mitchell, G, Stoodley, R J, Asymmetric synthesis of (3*S*)-2,3,4,5-tetrahydropyridazine-3-carboxylic acid, *J. Chem. Soc. Chem. Commun.*, 1179–1180, 1993.
106. Aspinall, I H, Cowley, P M, Stoodley, R J, Enhanced discrimination by aza dienophiles over their olefinic counterparts for the diastereotopic faces of methyl (*E,E*)-5-(2′,3′.4′,6′-tetra-*O*-acetyl-β-D-glucopyranosyloxy)penta-2,4-dienoate, *Tetrahedron Lett.*, 35, 3397–3400, 1994.
107. Avalos, M, Babiano, R, Cintas, P, Clemente, F R, Jiménez, J L, Palacios, J C, Sánchez, J B, Hetero Diels–Alder reactions of homochiral 1,2-diaza-1,3-butadienes with diethyl azodicarboxylate under microwave irradiation. Theoretical rationale of the stereochemical outcome, *J. Org. Chem.*, 64, 6297–6305, 1999.
108. McCarrick, M A, Wu, Y D, Houk, K N, Hetero Diels–Alder reaction transition structures: reactivity, stereoselectivity, catalysis, solvent effects and the exo-lone-pair effect, *J. Org. Chem.*, 58, 3330–3343, 1993.
109. Lowe, R F, Stoodley, R J, Hetero Diels–Alder reaction involving (*E*)-3-(*tert*-butyldimethylsiloxy)-1-(2′,3′,4′,6′-tetra-*O*-acetyl-β-D-glucopyranosyloxy)buta-1,3-diene and *p*-nitrobenzaldehyde, *Tetrahedron Lett.*, 35, 6351–6354, 1994.
110. Cousins, R P C, Curtis, A D M, Ding, W C, Stoodley, R J, 1,5-Asymmetric inductions in the reactions of 2-(2′,3′,4′,6′-tetra-*O*-acetyl-β-D-glucopyranosyloxy)benzaldehyde with Danishefsky's diene, *Tetrahedron Lett.*, 36, 8689–8692, 1995.
111. Bednarski, M, Danishefsky, S, On the interactivity of chiral auxiliaries with chiral catalysts in the hetero Diels–Alder reaction: a new route to L-glycolipids, *J. Am. Chem. Soc.*, 105, 6968–6969, 1983.
112. Bednarski, M, Danishefsky, S, Interactivity of chiral catalysts and chiral auxiliaries in the cycloaddition of activated dienes with aldehydes: a synthesis of L-glucose, *J. Am. Chem. Soc.*, 108, 7060–7067, 1986.
113. Kunz, H, Stereoselective syntheses using carbohydrates as chiral auxiliaries, *Pure Appl. Chem.*, 67, 1627–1635, 1995.

114. Kunz, H, Weymann, M, Follmann, M, Allef, P, Oertel, K, Schultz-Kukula, M, Hofmeister, A, Stereoselective syntheses of chiral heterocycles and alkaloids using carbohydrate auxiliaries, *Polish J. Chem.*, 73, 15–27, 1999.
115. Kunz, H, Sager, W, Schanzenbach, D, Decker, M, Carbohydrates as chiral templates: stereoselective Strecker synthesis of D-α-amino nitriles and acids using *O*-pivaloylated D-galactosylamine as the auxiliary, *Liebigs Ann. Chem.*, 649–654, 1991.
116. Kunz, H, Sager, W, Diastereoselective Strecker synthesis of α-amino nitriles using carbohydrate templates, *Angew. Chem. Int. Ed.*, 26, 557–559, 1987.
117. Kunz, H, Pfrengle, W, Asymmetric synthesis on carbohydrate templates: stereoselective Ugi synthesis of α-amino acid derivatives, *J. Am. Chem. Soc.*, 110, 651–652, 1988.
118. Kunz, H, Pfrengle, W, Carbohydrates as chiral templates: asymmetric Ugi-synthesis of α-amino acids using galactosylamines as the chiral matrices, *Tetrahedron*, 44, 5487–5494, 1988.
119. Kunz, H, Pfrengle, W, Sager, W, Carbohydrates as chiral templates: diastereoselective Ugi synthesis of (*S*)-amino acids using *O*-acylated D-arabinopyranosylamine as the auxiliary, *Tetrahedron Lett.*, 30, 4109–4110, 1989.
120. Goebel, M, Ugi, I, *O*-Alkyl-D-glucopyranosylamines and their derivatives, *Synthesis*, 1095–1098, 1991.
121. Lehnhoff, S, Goebel, M, Karl, R M, Kloesel, R, Ugi, I, Stereoselective syntheses of peptide derivatives with 2-acetamido-3,4,6-tri-*O*-acetyl-1-amino-2-deoxy-β-D-glucopyranose by 4-component condensation, *Angew. Chem. Int. Ed.*, 34, 1104–1107, 1995.
122. Linderman, R J, Binet, S, Petrich, S R, Enhanced diastereoselectivity in the asymmetric Ugi reaction using a new "convertible" isonitrile, *J. Org. Chem.*, 64, 336–337, 1999.
123. Oertel, K, Zech, G, Kunz, H, Stereoselective combinatorial Ugi-multicomponent synthesis on solid phase, *Angew. Chem. Int. Ed.*, 39, 1431–1433, 2000.
124. Ziegler, T, Schloemer, R, Koch, C, Passerini and Ugi reactions of anomeric glucosyl isonitriles, *Tetrahedron Lett.*, 39, 5957–5960, 1998.
125. Ziegler, T, Kaisers, H J, Schloemer, R, Koch, C, Passerini and Ugi reactions of benzyl and acetyl protected isocyanoglucoses, *Tetrahedron*, 55, 8397–8408, 1999.
126. Lockhoff, O, An access to glycoconjugate libraries through multicomponent reactions, *Angew. Chem. Int. Ed.*, 37, 3436–3439, 1998.
127. Vincent, S P, Schleyer, A, Wong, C H, Asymmetric Strecker synthesis of *C*-glycopeptide, *J. Org. Chem.*, 65, 4440–4443, 2000.
128. Schlemminger, I, Janknecht, H H, Maison, W, Saak, W, Martens, J, Synthesis of the first enantiomerically pure 3-thiazolines via Asinger reaction, *Tetrahedron Lett.*, 41, 7289–7292, 2000.
129. Laschat, S, Kunz, H, Carbohydrates as chiral templates: Stereoselective synthesis of chiral homoallyl amines and β-amino acids, *Synlett*, 51–52, 1990.
130. Laschat, S, Kunz, H, Carbohydrates as chiral templates: diastereoselective synthesis of *N*-glycosyl-*N*-homoallylamines and β-amino acids from imines, *J. Org. Chem.*, 56, 5883–5889, 1991.
131. Deloisy, S, Kunz, H, A novel synthesis of chain-extended amino sugar derivatives through Aza-Cope rearrangement of *N*-galactosyl-*N*-homoallylamines, *Tetrahedron Lett.*, 39, 791–794, 1998.
132. Laschat, S, Kunz, H, Carbohydrates as chiral templates: stereoselective synthesis of (*R*)- and (*S*)-α-aminophosphonic acid derivatives, *Synthesis*, 90–95, 1992.
133. Kunz, H, Schanzenbach, D, Carbohydrates as chiral templates: stereoselective synthesis of β-amino acids, *Angew. Chem. Int. Ed.*, 28, 1068–1069, 1989.
134. Kunz, H, Burgard, A, Schanzenbach, D, Asymmetric syntheses of β-amino acids with two new stereogenic centres at the α and β-positions, *Angew. Chem. Int. Ed.*, 36, 386–387, 1997.
135. Kunz, H, Pfrengle, W, Carbohydrates as chiral templates: stereoselective tandem Mannich–Michael reaction for the synthesis of piperidine alkaloids, *Angew. Chem. Int. Ed.*, 28, 1067–1068, 1989.
136. Weymann, M, Pfrengle, W, Schollmeyer, D, Kunz, H, Enantioselective syntheses of 2-alkyl, 2,6-dialkylpiperidines and indolizidine alkaloids through diastereoselective Mannich–Michael reactions, *Synthesis*, 1151–1160, 1997.
137. Basha, A, Henry, R, McLaughlin, M A, Ratajczyk, J D, Wittenberger, S J, Addition of organometallic reagents to *N*-glycosyl nitrones. Enantioselective syntheses of (+)-(*R*)- and (−)-(*S*)-zileuton, *J. Org. Chem.*, 59, 6103–6106, 1994.

138. Rohloff, J C, Alfredson, T V, Schwartz, M A, Enantioselective synthesis of 5-LO inhibitors using a gulofuranose auxiliary, *Tetrahedron Lett.*, 35, 1011–1014, 1994.
139. Stewart, A O, Bhatia, P A, Martin, J G, Summers, J B, Rodriques, K E, Martin, M B, Holms, J H, Moore, J L, Craig, R A, Kolasa, T, Ratajczyk, J D, Mazdiyasni, H, Kerdesky, F A J, DeNinno, S L, Maki, R G, Bouska, J B, Young, P R, Lanni, C, Bell, R L, Carter, G W, Brooks, C D W, Structure-activity relationships of *N*-hydroxyurea 5-lipoxygenase inhibitors, *J. Med. Chem.*, 40, 1955–1968, 1997.
140. Young, R N, Inhibitors of 5-lipoxygenase: a therapeutic potential yet to be fully realized? *Eur. J. Med. Chem.*, 34, 671–685, 1999.
141. Roush, W R, VanNieuwenhze, M S, [(Z)-γ-[(Diisopropylidene-α-D-mannopyranosyl)oxy)allyl)-tributylstannane: a new chiral reagent for the asymmetric α-hydroxyallylation of aldehydes, *J. Am. Chem. Soc.*, 116, 8536–8543, 1994.
142. Yamamoto, Y, Kobayashi, K, Okano, H, Kadota, I, Asymmetric synthesis of syn-1,2-diols via the reaction of aldehydes with chiral γ-(tetrahydropyranyloxy)allylstannanes, *J. Org. Chem.*, 57, 7003–7005, 1992.
143. Chika, J I, Takei, H, Arabinose as a chiral auxiliary for the asymmetric α-hydroxyallylation of aldehydes to syn-1,2-diols, *Tetrahedron Lett.*, 39, 605–608, 1998.
144. Felix, D, Szymoniak, J, Moïse, C, Asymmetric allylation of aldehydes using tiglyltitanocenes attached to a carbohydrate, *Tetrahedron*, 53, 16097–16106, 1997.
145. Yoshida, T, Chika, J I, Takei, H, Asymmetric nucleophilic addition to β- and γ-alkoxy aldehydes using carbohydrate as a chiral auxiliary, *Tetrahedron Lett.*, 39, 4305–4308, 1998.
146. Shing, T K M, Li, L H, Asymmetric Hosomi–Sakurai reaction of allylsilanes containing arabinose-derived alcohols as chiral auxiliaries, *J. Org. Chem.*, 62, 1230–1233, 1997.
147. Rueck, K, Kunz, H, Stereoselective conjugate addition of organoaluminum chlorides to α,β-unsaturated carboxylic acid derivatives, *Synthesis*, 1018–1028, 1993.
148. Rueck-Braun, K, Stamm, A, Engel, S, Kunz, H, β-branched α-halo carboxylic acid derivatives via stereoselective 1,4-addition of dialkylaluminum chlorides to α,β-unsaturated *N*-acyloxazolidinones, *J. Org. Chem.*, 62, 967–975, 1997.
149. Totani, K, Nagatsuka, T, Yamaguchi, S, Takao, K, Ohba, S, Tadano, K, Highly diastereoselective 1, 4-addition of an organocuprate to methyl α-D-gluco, α-D-manno, or α-D-galactopyranosides tethering an α,β-unsaturated ester, *J. Org. Chem.*, 66, 5965–5975, 2001.
150. Weymann, M, Schultz-Kukula, M, and Kunz, H, Auxiliary-controlled stereoselective enolate protonation: enantioselective synthesis of cis and trans annulated decahydroquinoline alkaloids, *Tetrahedron Lett.*, 39, 7835–7838.
151. Comins, D L, Dehghani, A, A short, asymmetric synthesis of (−)-pumiliotoxin C, *J. Chem. Soc. Chem. Commun.*, 1838–1839, 1993.
152. Heathcock, C H, White, C T, Morrison, J J, van Derveer, D, Double stereodifferentiation as a method for achieving superior Cram's rule selectivity in aldol condensations with chiral aldehydes, *J. Org. Chem.*, 46, 1296–1309, 1981.
153. Kunz, H, Mohr, J, Carbohydrates as chiral templates: reactivity and stereoselectivity of carboxylic ester enolates, *J. Chem. Soc. Chem. Commun.*, 1315–1317, 1988.
154. Mulzer, J, Hiersemann, M, Buschmann, J, Luger, P, 1,2:5,6-Di-*O*-isopropylidene-α-D-gulofuranose as a chiral auxiliary in the diastereoselective *C*-methylation of ester enolates, *Liebigs Ann. Chem.*, 649–654, 1995.
155. Keynes, M N, Earle, M A, Sudharshan, M, Hultin, P G, Synthesis and asymmetric alkylation of glucose-derived bicyclic oxazinones, *Tetrahedron*, 52, 8685–8702, 1996.
156. Koell, P, Luetzen, A, D-Xylose derived oxazolidin-2-ones as chiral auxiliaries in stereoselective alkylations, *Tetrahedron: Asymmetry*, 7, 637–640, 1996.
157. Angibaud, P, Chaumette, J L, Desmurs, J R, Duhamel, L, Ple, G, Valnot, J Y, Duhamel, P, Asymmetric synthesis of 2-chloro- and 2-bromo-alkanoic acids by halogenation of α-D-glucofuranose-derived silyl ketene acetals, *Tetrahedron: Asymmetry*, 6, 1919–1932, 1995.
158. Luetzen, A, Koell, P, D-Xylose derived oxazolidin-2-ones as chiral auxiliaries in stereoselective acylations and halogenations, *Tetrahedron: Asymmetry*, 8, 29–32, 1997.
159. Bandraege, S, Josephson, S, Moerch, L, Vallen, S, Asymmetric synthesis of β-hydroxyesters by Reformatsky reactions and amide base mediated condensations, *Acta Chem. Scand. Ser. B*, 35, 273–277, 1981.

160. Stoever, M, Luetzen, A, Koell, P, New glyco-oxazolidin-2-ones as chiral auxiliaries in boron-mediated asymmetric aldol reactions, *Tetrahedron: Asymmetry*, 11, 371–374, 2000.
161. Evans, D A, Bartroli, J, Shih, T L, Enantioselective aldol condensations: erythro-selective chiral aldol condensations via boron-enolates, *J. Am. Chem. Soc.*, 103, 2127–2129, 1981.
162. Akiyama, Y, Ishikawa, K, Ozaki, S, Asymmetric synthesis of functionalized tertiary alcohols by diastereoselective aldol reaction of silyl enol ether and ketene silyl acetals with α-keto esters bearing an optically active cyclitol as a chiral auxiliary, *Synlett*, 275–276, 1994.
163. Enders, D, Knopp, M, Schiffers, R, Asymmetric [3,3]-sigmatropic rearrangements in organic synthesis, *Tetrahedron: Asymmetry*, 7, 1847–1882, 1996.
164. Eguchi, T, Koudate, T, Kakinuma, K, The Overman-rearrangement on a diacetone-D-glucose template, *Tetrahedron*, 49, 4527–4540, 1993.
165. Tomooka, Y, Nakamura, Y, Nakai, T, [2,3]-Wittig-rearrangement using glucose as chiral auxiliary, *Synlett*, 321–322, 1995.
166. Harrington, P E, Tius, M A, Asymmetric cyclopentannelation — chiral auxiliary on the allene, *Org. Lett.*, 2, 2447–2450, 2000.
167. Leonard, J, Diez-Barra, E, Merino, S, Control of asymmetry through conjugate addition reactions, *Eur. J. Org. Chem.*, 2051–2061, 1998.
168. Murakata, M, Jono, T, Mizuno, Y, Hoshino, O, Construction of chiral quarternary carbon centers by catalytic enantioselective radical mediated allylation, *J. Am. Chem. Soc.*, 119, 11713–11714, 1997.
169. Garner, P P, Cox, P B, Klippenstein, S J, Auxiliary induced ρ-stereocontrol in acetaloxyalkyl radical addition reactions, *J. Am. Chem. Soc.*, 117, 4183–4184, 1995.
170. Barton, D H R, Crich, D, Kretzschmar, G, The invention of new radical chain reactions, part 9, *J. Chem. Soc. Perkin. Trans. 1*, 39–54, 1986.
171. Enholm, E J, Cottone, J S, Allais, F, Highly diastereoselective 5-hexenyl radical cyclizations with Lewis acids and carbohydrate scaffolds, *Org. Lett.*, 3, 145–147, 2001.
172. Enholm, E J, Gallagher, M E, Jiang, S, Batson, W A, Free radical allyl transfers utilizing soluble non-cross-linked polystyrene and carbohydrate scaffold supports, *Org. Lett.*, 2, 3355–3357, 2000.
173. Mendonca, A J, Xiao, X Y, Optimization of solid supports for combinatorial chemical synthesis, *Med. Res. Rev.*, 19, 451–462, 1999.
174. Sibi, M P, Chandramouli, S V, Intermolecular free radical reactions on solid support, allylation of esters, *Tetrahedron Lett.*, 38, 8929–8932, 1997.
175. Enholm, E J, Gallagher, M E, Lombardi, J S, Moran, K M, Schulte, J P, An allylstannane reagent on non-cross-linked polystyrene support, *Org. Lett.*, 1, 689–691, 1999.
176. Wentworth, P, Janda, K D, Liquid phase chemistry: recent advances in soluble polymer-supported catalysts, reagents and synthesis, *Chem. Commun.*, 1917–1924, 1999.
177. Huang, G, Hollingsworth, R I, The stereoselective conversion of 2-alkenyl alcohols to (*R*)- or (*S*)-alkane-1,2-diols using D-glucose as chiral auxiliary, *Tetrahedron Lett.*, 40, 581–584, 1999.
178. Nair, V, Prabhakaran, J, George, T G, A facile synthesis of optically active lactones using benzyl-3,6-anhydro glucofuranoside as chiral auxiliary, *Tetrahedron*, 44, 15061–15068, 1997.
179. Haan, J W, Son, S U, Chung, Y K, Chiral auxiliary directed asymmetric ortho-lithiation of (arene)tricarbonylchromium complexes, *J. Org. Chem.*, 62, 8264–8267, 1997.
180. Duthaler, R O, Hafner, A, Riediker, M, Asymmetric C–C bond formation with titanium carbohydrate complexes, *Pure Appl. Chem.*, 62, 631–642, 1990.
181. Allef, P, Kunz, H, Glycosylation-induced asymmetric synthesis: α-amino acid esters via Mannich reaction, *Tetrahedron: Asymmetry*, 11, 375–378, 2000.
182. Kumar, A, Oehme, G, Roque, J P, Schwarze, M, Selke, R, Increase in the enantioselectivity of asymmetric hydrogenation in water influenced by surfactants or polymerized micelles, *Angew. Chem. Int. Ed.*, 33, 2197–2199, 1994.
183. RajanBabu, T V, Casalunovo, A L, Tailored ligands for asymmetric catalysis: the hydrocyanation of vinylarenes, *J. Am. Chem. Soc.*, 114, 6265–6266, 1992.
184. Hamashima, Y, Kanai, M, Shibasaki, M, Catalytic enantioselective cyanosilylation of ketones, *J. Am. Chem. Soc.*, 122, 7412–7413, 2000.
185. Glaeser, B, Kunz, H, Enantioselective allylic substitution using a novel (Phosphino-α-D-glucopyrano-oxazoline)-palladium catalyst, *Synlett*, 53–54, 1998.

11 Sugars as Chiral Starting Materials in Enantiospecific Synthesis

Yves Chapleur and Françoise Chrétien

CONTENTS

11.1 INTRODUCTION

Since the synthesis of urea in 1828 by Friedrich Wöhler, total synthesis has always been a challenging area. Sugars were synthesized by Emil Fischer as early as 1890. Major breakthroughs in the synthesis of highly complex molecules were reported in the 1960s, including the total synthesis of vitamin B_{12} [1] and complex alkaloids like reserpine [2]. A more challenging task was then to achieve total syntheses capable of producing optically pure compounds identical to natural targets. In the context of enantiomerically pure compound synthesis (EPC synthesis), the use of chiral starting materials amenable to suitable chiral building blocks was attractive.

Carbohydrates are prominent among the chiral raw materials available from nature. This is because of several factors, such as the high enantiomeric purity of sugars and the number of available chiral centers. This last feature is an advantage in terms of optical purity because the probability of finding L-glucose in D-glucose is extremely low. Obviously this is not true for compounds having only one chiral center. Along the same line, D-glucose could be also contaminated with its diastereomers D-mannose or D-galactose, but their chemical reactivity is rather different and they could be easily detected in such a sample of D-glucose. Thus, at first glance, sugars can be regarded as almost enantiomerically pure (up to 99%) compounds. Provided that subsequent chemical manipulations do not involve any unwanted epimerization, one can consider the final product as enantiomerically, if not diastereomerically, pure.

The number of available chiral centers or functional groups can be regarded as a major inconvenience because of the protection requirements that increase the number of steps. The number of carbon atoms present in common sugars, usually four to seven carbon atoms, seems ideal for the elaboration of building blocks according to modern synthetic strategies. However, many natural product syntheses starting from sugars involve carbon–carbon bond cleavage to remove unnecessary carbon atoms (and chiral centers). Although this could be considered a loss of time and money, this loss is generally compensated for by the low cost of naturally occurring carbohydrates.

Last but not least, the existence of both enantiomeric forms of the same carbohydrate at affordable prices is a major advantage. This is true for most pentoses such as arabinose, xylose, lyxose and even ribose. Given the high "stereodiversity" of carbohydrates it is quite easy to find the most appropriate starting sugar to ensure a good fit between this raw material and the planned target. Several excellent reviews [3–11], book chapters [12–14] and books [15,16] have been published on these topics.

The purpose of this chapter is to summarize the organic chemistry of carbohydrates involved in total syntheses of enantiomerically pure compounds. In this context, an overview of the synthesis of chiral targets from sugars could be arranged according to target structure (heterocyclic compounds, macrocyclic compounds, acyclic molecules and so on) or starting sugars. Instead of describing several syntheses of different classes of compounds, we propose in this chapter a reaction-based arrangement, which will present several key points of representative syntheses. Each key point will rely on a particular transformation of the sugar template or starting material. Thus, many of the synthetic strategies described below are related to the synthesis of advanced intermediates *en route* to more complex structures. Subsequent steps of these syntheses, not relevant to carbohydrate chemistry, will not be reported in full. Some synthetic approaches to complex targets will also be reported, provided they include such key reactions. When available, the descriptions of different synthetic strategies toward the same target have been privileged. Whenever possible, for a given transformation, some references to seminal related papers not necessarily oriented toward a target total synthesis, will also be given. This chapter will concentrate on syntheses published after 1985, although some former and creative synthetic strategies will also be reviewed.

Through this chapter, the different key reactions are arranged by a wide type of transformations of the sugar; some examples are chain extension at the nonreducing end, formation of tetrahydropyran or furan and carbocycle formation. Nitrogen containing heterocycles directly evolved from sugars (the so-called azasugars) or carbocyclic analogs of sugars (the so-called carbasugars) are not detailed in this chapter.

Most of the key transformations described here involve carbon–carbon bond formation. Carbohydrate functional group manipulations are a long-standing research area in carbohydrate chemistry often reviewed in specialized books and series. We hope to provide the reader with a set of different chemical modifications of carbohydrate ring systems or acyclic carbohydrates available to devise new synthetic routes to biologically relevant targets.

11.2 CARBOHYDRATES AS SOURCES OF CARBON ATOMS IN TOTAL SYNTHESES

One of the earliest syntheses of a chiral compound from a carbohydrate was reported by Wolfrom, Lemieux and Olin, who described the preparation of optically active L-alanine from D-glucosamine [17]. This transformation was devised to establish the structure of alanine from the known D-glucosamine. Only a few functional group manipulations were needed in this synthesis, such as elaboration of the aldehyde group into the methyl group of alanine through the corresponding dithioacetal. Excision of three carbon atoms was also required (Scheme 11.1).

It should be noted that few chemical syntheses relying on the principle of using part of or the whole sugar backbone to transfer it into a chiral target have been reported since this seminal paper. Most of the papers dealing with the transformation of a sugar to a noncarbohydrate compound were focused on structural determinations [3]. However, the disclosure of the arachidonic acid cascade renewed interest in carbohydrates as starting materials. This cascade of biological events produced a number of hydroxylated lipid-type compounds, including prostaglandins and thromboxane B2. The latter is structurally close to a sugar and was an obvious target starting from glucose. Several syntheses of these compounds were disclosed and can be considered as evidence of synthetic chemists' passion for carbohydrates.

This approach has been rapidly put into new concepts and reviewed. The *Chiron approach* concept was introduced by Hanessian in a review [4] and, in more detail, in his book [15]. The carbohydrate is considered as a chiral synthon, which has to be found retrosynthetically in a

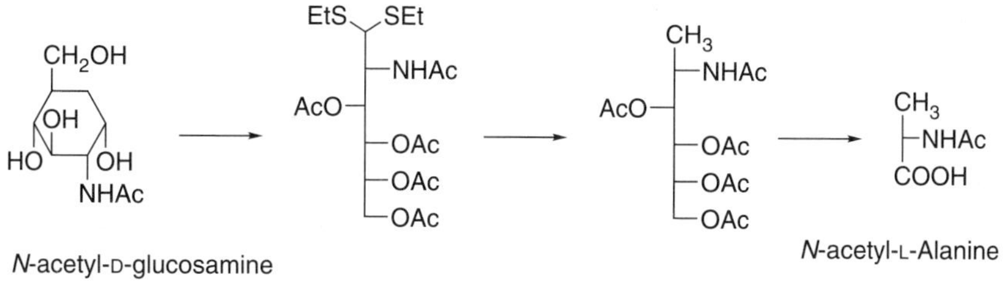

CH2OH OH HO OH NHAc
N-acetyl-D-glucosamine
EtS SEt NHAc AcO OAc OAc OAc
CH3 NHAc AcO OAc OAc OAc
CH3 NHAc COOH
N-acetyl-L-Alanine

SCHEME 11.1 Configurational correlation of D-glucosamine with L-alanine.

given target. This obviously involves two main analyses, depending on the target. The latter could include an apparent portion of sugar (mainly those with furanose or pyranose rings), or have a hidden similarity with a carbohydrate (for example, polyketide-type macrocycles or linear molecules). In this context computer-aided retrosynthetic analysis has been developed by Hanessian et al. [18]. Other terms have been coined to name this use of sugar backbones in syntheses of different classes of compounds. Chirality transfer is one of the most popular terms used [5].

Considering a carbohydrate template, it is clear that carbon–carbon bonds could be formed on different points of the carbon chain with quite different results in terms of target structures. Carbon–carbon bond formation on one of the secondary carbon atoms of the carbohydrate chain, hopefully with some stereocontrol, allows the construction of branched-chain structures. Extension at primary carbon atoms (C6 or C5), with or without creation of a new chiral center, opens the way to linear targets or macrocycles. Creation of carbon–carbon bonds at C1 is ideally suited for construction of *C*-glycosyl compounds, which can be regarded as oxygenated heterocycles such as tetrahydropyran or tetrahydrofuran ring systems. Finally, formation of carbocycles by ring closure of the carbohydrate chain or by annelation on the sugar template provides an efficient way to construct polycyclic targets of the terpenoid or alkaloid fields. These highly versatile sugar backbone manipulations allowed many successful syntheses of a large number of structures derived from multiple different biosynthetic pathways.

11.3 BRANCHING A CARBON CHAIN ON THE CARBOHYDRATE RING

The introduction of carbon chains on one or several of the secondary carbon atoms of a carbohydrate template is a key reaction in the use of carbohydrates as starting materials. The development of a few carbon–carbon bond-forming reactions has triggered interest in using carbohydrates in total syntheses during the last two decades. As a result, many research groups were focused on synthesizing more and more complex targets and contributed to the invention of new and creative methods for the formation of carbon–carbon bonds [19].

Obviously, a carbon–carbon bond-forming reaction is expected to take place with high stereocontrol in an efficient total synthesis. This requirement could be reached rather easily in carbohydrate chemistry by taking advantage of the template effect of furanose or pyranose ring systems. The major concern in such reactions is related to the reaction conditions, which need to be compatible with the intrinsic sensitivity of sugars (especially that of protected sugars). Elegant solutions have been found for this long standing problem in the last 20 years. In this section we found it more convenient to arrange the presentation according to the type of starting material used. However, for each type of starting material a few reactions are given and in each section a reaction type arrangement is then used.

11.3.1 USING EPOXIDES

Epoxides are highly reactive species capable of stereoselective ring opening reactions on exposure to nucleophiles. Sugar-derived epoxides are easily obtained from diols by suitable sequences of protection, activation and ring-closure. Moreover, judicious arrangement of the reaction sequences can provide two different epoxides from the same diol. Epoxides are also available from the corresponding olefins, often with high diastereocontrol. This chemistry has been well developed in connection with aminosugars and aminoglycosides [20]. However, carbon–carbon bond formation using epoxides is a much more difficult task. Fortunately, smooth carbon nucleophiles have become available, among which cuprates play a prominent role. Their reactions with epoxides occur at relatively low temperatures with high stereocontrol. As a rule, the nucleophilic attack occurs so that the two substituents resulting from the opening are in a *trans*-relationship, and the Fürst–Plattner rule predicts that the two substituents of the reaction product should be axially oriented on a pyranose ring [21]. This allows one to easily predict which carbon atom of the epoxide would undergo nucleophilic attack in a given ring conformation.

As shown in Scheme 11.2, the epoxide **1** can exist as two different conformers, designated $^4C_1$ and $^1C_4$. According to the preceding rules, the $^4C_1$ conformer will undergo nucleophilic attack at C2 to yield **2** whereas the $^1C_4$ one would give the C3 branched isomer **3**. Thus, providing the activation energies for reactions with the individual epoxide conformers are lower than conformational interconversion, the net result of the reaction depends on the conformer population. Provided that the ring conformation is locked by ring fusion, that is, via formation of a 4,6-*O*-benzylidene ring, opening of a 2,3-anhydro derivative could be achieved with very high and predictable, regio- and stereocontrol. As shown below, a simple change in the epoxide configuration (e.g., from the *allo* epoxide **1** to the *manno* epoxide **4**, both available from D-glucose), would completely reverse the situation and would give **5** (to compare with **2**).

Since its introduction in carbohydrate chemistry by Hicks and Fraser-Reid in 1975 [22], applications of the opening of epoxide with carbon nucleophiles abound in the literature. Just a few seminal examples will be mentioned here, other examples will be found in combination with other branching-chain methods.

As shown below, the attack of epoxide **6** with lithium dimethylcuprate is a key step of Hanessian et al.'s erythronolide synthesis [23]. This methodology was also applied to the preparation of other polyketide-derived macrolides. Specific to erythronolide, introduction of the methyl group at C2 was achieved according to Scheme 11.3.

Given the conformational bias of epoxide **6**, opening occurred with total regio- and stereocontrol to afford **7** in good yield. The absolute configuration at C2 of **7** was opposite to that required in erythronolide. A clever sequence of oxidation, followed by equilibration of the resulting ketone to the thermodynamically favored equatorial isomer **8**, allowed net inversion of configuration, the equilibrium being displaced by crystallization of **8** in the reaction medium. Related strategies have been used in alternate synthetic approaches to erythronolide [24–27].

Another example from the macrolide field is the synthesis of antimycin A3, which starts with opening of epoxide **9** [28]. Opening of this *manno* epoxide with butylmagnesium chloride afforded the C3 branched-derivative **10** with total regio- and stereo-selectivity. Again the Fürst–Plattner rule applies and the introduction of the substituent at C3 was expected. The remainder of the synthesis

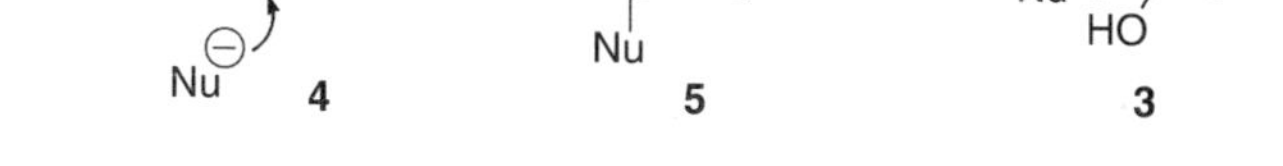

SCHEME 11.2 Epoxide ring opening.

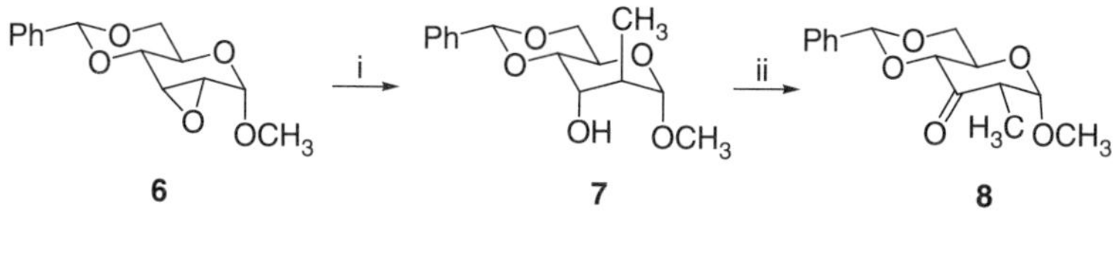

SCHEME 11.3 Reagents: i: Me_2CuLi; ii: (1) DMSO, Ac_2O and (2) MeONa, MeOH.

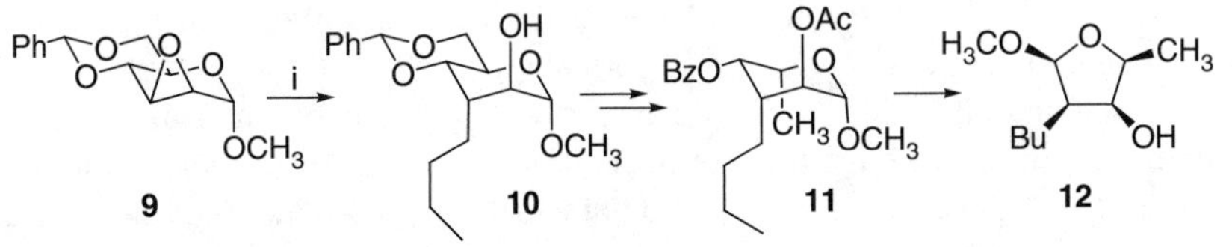

SCHEME 11.4 Reagent: i: BuMgCl.

consisted of oxidative cleavage of the C1–C2 bond to provide the key intermediate **12** (Scheme 11.4).

Application of this strategy allowed the stereocontrolled construction of the C10 to C14 part of the left wing of pseudomonic acids [29] (Scheme 11.5). Thus, opening of the *manno* epoxide **9** gave the branched-chain derivative **13**, an intermediate in the synthesis of a pheromone of *Scolytus multistriatus* [30]. As shown in Scheme 11.5, this alcohol was elaborated to the alcohol **14**, by an oxidation–reduction sequence. After alcohol protection, benzylidene opening under Hanessian–Hullar conditions gave the bromide derivative **15**. Subsequent Vasella ring opening reaction [31] gave olefin **16**. Further functional group manipulation afforded the vinylic chloride **17,** representing the left part of the pseudomonic acids.

Nucleophilic opening of furanose-fused epoxides is not under stereoelectronic control but is rather controlled by steric hindrance. However, high stereocontrol in the ring opening can also be achieved on furanose-derived epoxides but a careful choice of the starting sugar is needed. An illustrative example, found in the total synthesis of pseudomonic acids, is given below (Scheme 11.6). The chiral C11 to C14 chain was elaborated from methyl α-D-arabinofuranoside by Fleet et al. [32,33]. Epoxide **18** was prepared in four steps from the latter compound and underwent complete stereo- and regio-selective ring opening with a high order cyanocuprate establishing the two required configurations at C12 and C13 of the western part of pseudomonic acid C.

Upon considering this scheme, the ring opening of epoxide **18** must occur from the β-face either at C2 or C3. The use of the β anomer forced the attack to occur on the less hindered C3 carbon atom, giving alcohol **19**. The latter was then elaborated to the phosphonium salt **20** suitable for Wittig coupling with the tetrahydropyran part of pseudomonic acid. It would be of interest to compare these two synthetic strategies toward the left wing of pseudomonic acids in terms of overall length of synthesis and efficiency.

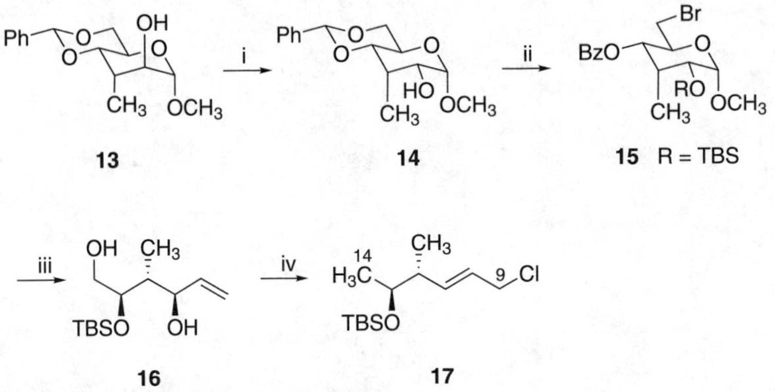

SCHEME 11.5 Reagents: i: (1) DMSO, TFA, DCCI and (2) $LiAlH_4$, Et_2O, 95%; ii: (1) TBSCl, imidazole, DMF, 83% and (2) NBS, $BaCO_3$, 75%; iii: (1) Zn, PrOH, H_2O and (2) $NaBH_4$, EtOH, 50%; iv: (1) TsCl, pyridine, (2) NaI, butanone, (3) DMSO, $NaBH_4$, 60%, (4) MeONa and (5) $SOCl_2$, pyridine, 88%.

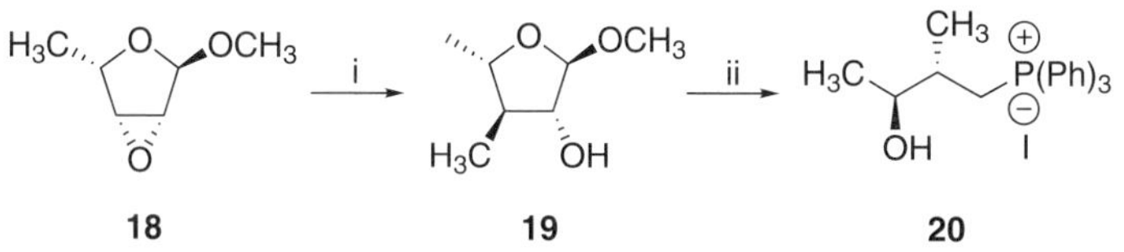

SCHEME 11.6 Reagents: i: $Me_2CuCNLi_2$, 71%; ii: (1) H^+, (2) $NaIO_4$ then $NaBH_4$, 60%, (3) trimsyl chloride, pyridine, (4) NaI, butanone 90% and (5) PPh_3, toluene, 89%.

The method described above is not limited to readily available organometallics such as dimethylcuprates or standard alkyl Grignard reagents under copper (I) catalysis. More complex organometallics, such as the one prepared from the chloride derivative **17**, have been successfully used for opening of the epoxide **21** giving access to **22**; an advanced intermediate in Sinaÿ's synthesis of pseudomonic acid C derivatives [29] (Scheme 11.7). Here again, the conformational bias of the D-xylose derived epoxide **21** allows complete regio- and stereochemical control of the reaction, thus establishing the correct configuration of all chiral centers of the core structure of pseudomonic acids A and C.

The clear-cut behavior in the opening of epoxides can suffer some exceptions. Such abnormal reactions have been reported in the opening of epoxide **21** with vinylmagnesium bromide. In this case, the vinyl residue was introduced at C2 as expected but with retention of configuration; this is because of preferential opening with the bromide ion followed by the displacement of the bromide atom with the carbon nucleophile [34].

As mentioned above, ring opening of conformationally unbiased epoxides is less regioselective and the two expected compounds can be obtained in a ratio depending on the conformational equilibrium between the $^1C_4$ and $^4C_1$ chairs. Obviously the use of the 4,6-acetal ring allows opening at C2 or C3 only. Nevertheless, branching of a carbon chain at C4 is possible by using 1,6-anhydro derivatives. In this case, epoxide openings obey the Fürst–Plattner rule giving the *trans*-diequatorially substituted compounds. As the 1,6-anhydro derivative is locked in the unusual $^1C_4$ conformation, the substituents introduced this way become equatorially oriented upon hydrolysis of the 1,6-acetal bridge. This subterfuge also allows substitution at C2 and C3 with opposite configuration and regio-chemical outcomes, as compared with the "normal" version (Scheme 11.8). Thus, epoxide **23**, obtained in two steps from readily available 1,6-anhydro-D-glucose (levoglucosan), was treated with allylmagnesium bromide in the presence of copper iodide to give the axially substituted alcohol **24**. Deoxygenation at C2 and further manipulation of the allyl group allowed the preparation of **25**, a key intermediate in the total synthesis of thromboxane B_2 [35]. It is worth noting that ring opening of the corresponding *galacto* epoxide **26** would mainly give the C3-substituted *gulo* derivative.

Alcohols resulting from a first epoxide opening can indeed be submitted to a second sequence, as shown in Procter's approach to the macrolide antibiotic rosaramycin [36]. Here again the 1,6-anhydro bridge was chosen to lock the epoxides in the proper conformation to ensure completely regioselective introduction of an allyl chain at C2 in **28** (Scheme 11.9). Going back to the usual $^4C_1$

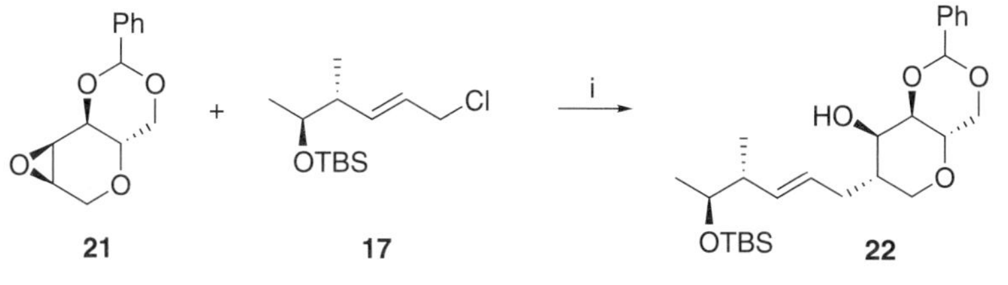

SCHEME 11.7 Reagent: i: CuI, THF, 43%.

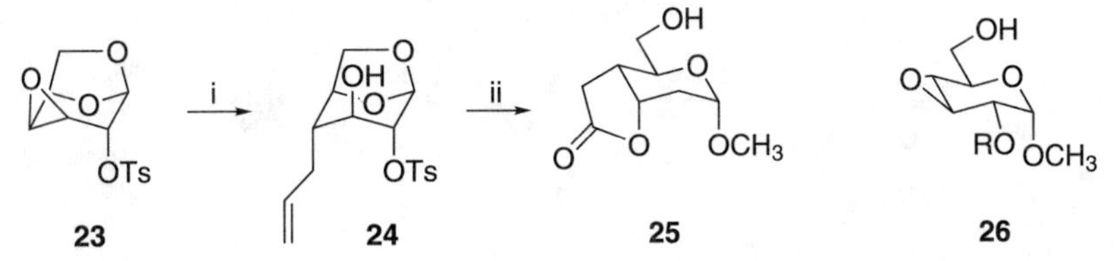

SCHEME 11.8 Reagents: i: $CH_2{=}CHCH_2MgCl$, CuI, THF, 88%; ii: (1) $LiBHEt_3$, (2) TsCl, 51% two steps, (3) RuO_2, $NaIO_4$, 80%.

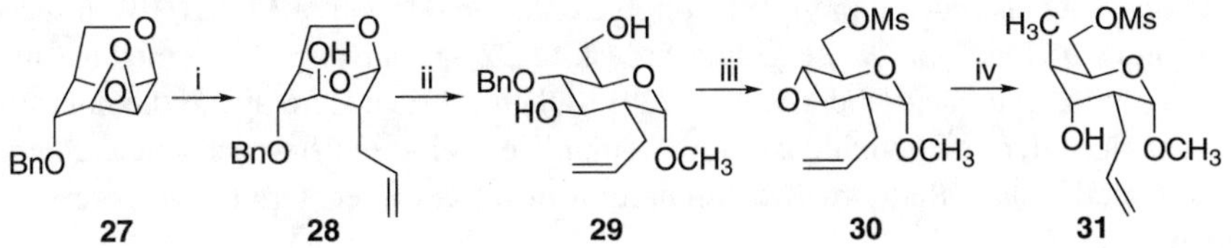

SCHEME 11.9 Reagents: i: $CH_2{=}CHCH_2MgCl$, Et_2O, 90%; ii: MeOH, HCl, 96%; iii: (1) MsCl, NEt_3, 92%, (2) Na, liq. NH_3 and (3) MeONa, 83%; iv: MeMgCl, CuBr, 92%.

conformation led to the preparation of **29,** which was further transformed into epoxide **30**. Although this compound is not conformationally biased, introduction of the methyl group occurred exclusively at C4 giving **31** the correct stereochemical configurations and requisite functionalities at the C2–C7 fragment of the target [37].

Epoxide ring opening reactions utilizing other carbon nucleophiles such as cyanide [38] have been reported in the synthesis of epithienamycin [39] and miharamycin [40]. Because of the expanse of literature surrounding such ring-opening reactions and the multitude of useful nucleophiles, it is almost impossible to give an exhaustive account of all relevant synthetic sequences. This strategy has been widely used in the synthesis of macrolide antibiotics and substructures thereof. Indeed, opening of an epoxide with dimethylcuprate and iterations of this sequence should lead, after some manipulations, to the basic triade of polypropionates. To mention just a few important targets, epoxide openings with methyl groups have found applications in the recent syntheses of the C22 to C27 fragment of FK506 [41], the F ring of spongistatin, [42,43], epothilone [44] and the synthesis of a subunit of rapamycin [45].

Aside from alkylcuprates, alternate methods for the introduction of a second methyl group via olefination reactions have been proposed, as shown below. An illustrative example of this strategy will be given in conjunction with synthetic approaches to the Prelog–Djerassi lactone [46,47], a degradation product of the macrolide antibiotic methymycin and a key intermediate on the way to polypropionate products.

11.3.2 Using Unsaturated Carbohydrates

Direct formation of carbon–carbon bonds by reacting carbon nucleophiles with carbohydrate electrophiles is not limited to epoxide opening reactions. Unsaturated sugars can be used with success to create a carbon–carbon bond on one of the olefinic carbons. Unsaturated sugar derivatives are readily available directly from diols, or starting from glycals and using the Ferrier rearrangement to 2,3-unsaturated products. Among the most useful unsaturated carbohydrates, enones and allylic esters occupy a prominent place as starting compounds.

11.3.2.1 Using Enones

1,4-Additions of alkylcuprates on sugar derived enones has been studied and employed in total syntheses. These reactions compare well with epoxide openings in terms of efficiency. Moreover, Michael additions on cyclic enones follow precise rules allowing good prediction of the stereochemical outcome with axial attack to the double bond being highly favored for stereoelectronic reasons. A good example of this strategy is found in Fraser-Reid's synthesis of α-multistriatin **108** (Scheme 11.26) [48]. Weiler's synthesis of the same compound will be described later in this chapter [49].

As illustrated in Scheme 11.10, enone **32**, prepared from D-glucal in a multistep sequence, was reacted with lithium dimethylcuprate. Clean axial attack of the reagent at C4 gave the substituted ketone **33**. Subsequent Wittig methylation and reduction was used to introduce the second methyl group of **34**, a key intermediate in the synthesis of α-multistriatin. The same sequence was used in the preparation of calcimycin (A23187), the 4-*C*-methyl synthon **33** being used to construct the two required chirons [50].

Although this synthetic sequence is efficient in terms of stereocontrol, the elaboration of ketone **32** remains lengthy. A shorter route has been proposed, which takes advantage of a smooth reaction of lithium dimethylcuprate with an enol ester [51]. As shown in Scheme 11.11, a clean attack on ester **35** was observed on exposure to this cuprate, triggering the formation of a keto group at C2 and fast elimination of an acetyloxy group at C4. The subsequently formed intermediate enone **36** reacted readily with an excess of reagent, to provide the β-methyl ketone **37**. As in the α-multistriatin syntheses, the carbonyl group of **37** was elaborated into a methyl group with high stereocontrol, opening the way to the preparation of propionate pathway compounds [51].

In addition to dimethylcuprates, various alternate cuprate reagents can be used. As shown in Scheme 11.12, a divinylcuprate was used in a 1,4-addition employed in the total synthesis of meroquinene **42** (Scheme 11.12), a degradation product of cinchonine and also an intermediate *en route* to cinchona alkaloids such as quinine [52,53]. As illustrated, enone **36**, available via acetoxyglucal **35** (Scheme 11.11), was treated with divinylcuprate to exclusively afford the axially substituted 4-vinyl derivative **38**. Trapping of this intermediate with methyl bromoacetate gave a mixture of C3 epimers readily equilibrated to **39** in the presence of triethylamine. Further manipulations of **39** gave the 2-deoxy derivative **40** and, in turn, the dialdehyde **41**. Cyclization of the latter to enantiomerically pure meroquinene **42** proceeded uneventfully.

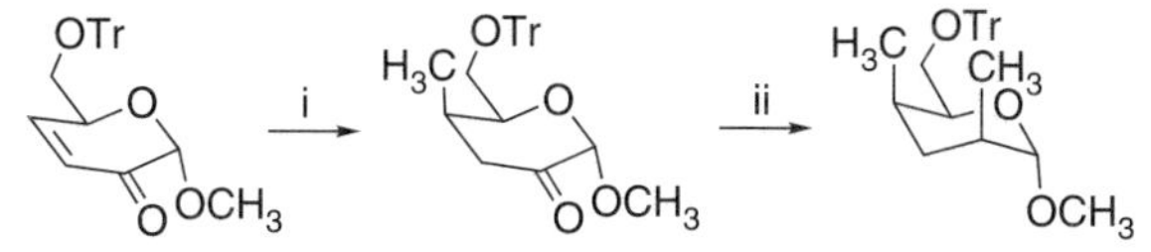

SCHEME 11.10 Reagents: i: MeLi, CuBr, Me_2S; ii: (1) Ph_3PCH_2 and (2) Raney–Ni, EtOAc.

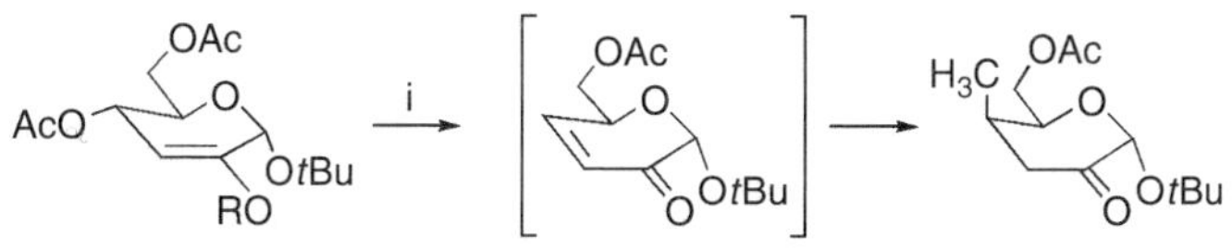

SCHEME 11.11 Reagent: i: Me_2CuLi.

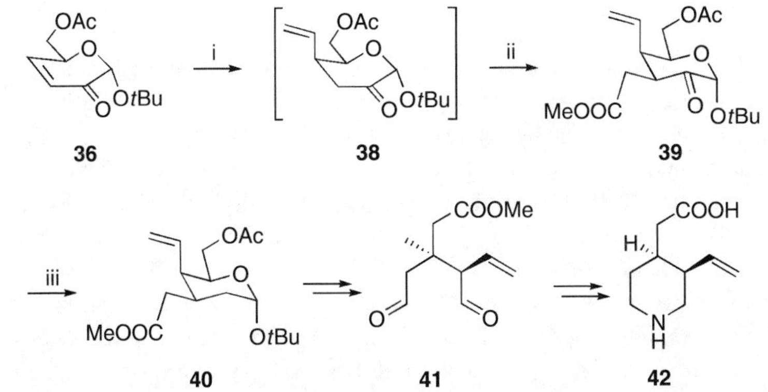

SCHEME 11.12 Reagents: i: $(CH_2{=}CH)_2$, CuCN $(MgBr)_2$; ii: (1) $BrCH_2COOMe$ and (2) Et_3N, DMF; iii: (1) $TsNHNH_2$, (2) $NaBH_3CN$ and (3) $A_cON_a{\cdot}3H_2O$.

A more economical use of **40**, in terms of carbon atoms and chiral centers, was devised to produce the heteroyohimbine alkaloid ajmalicine and its 19-epimer [54]. As shown in Scheme 11.13, deoxygenation of **40** at C6 followed by oxidative cleavage of the vinyl group gave the corresponding aldehyde **43**, which was equilibrated to the equatorial isomer **44**. Subsequent reaction with tryptamine gave the piperidinone **45**. Bisher–Napieralsky cyclization provided the alkaloid skeleton, which was elaborated to ajmalicine **46**.

11.3.2.2 Using Allylic Esters

As already mentioned, unsaturated sugars, particularly glycals, are readily available compounds. Ferrier rearrangement of glycals in the presence of Lewis acids and alcohols easily takes place to provide 2,3-unsaturated carbohydrates. The allylic esters obtained this way can be engaged in different reactions with organometallic reagents.

A pioneering study from Chapleur's group established the use of cyanocuprates to introduce a carbon chain at different branching-points of the sugar ring. The reaction takes advantage of the facile displacement of allylic acetates by cuprates. It occurs via an S_N2' mechanism. Thus, the desired stereochemistry can be adjusted by selecting the appropriate configuration at the

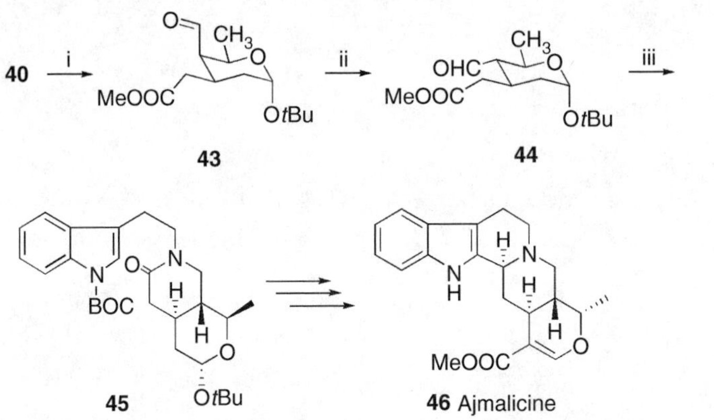

SCHEME 11.13 Reagents: i: (1) MeONa, (2) Ph_3P, CCl_4, 90%, (3) Bu_3SnH, 94% and (4) O_3, CH_2Cl_2, 86%; ii: DBU, DMF, 69%; iii: (1) tryptamine then MeOH, $NaBH_4$, 93% and (2) $(BOC)_2O$, DMAP, 99%.

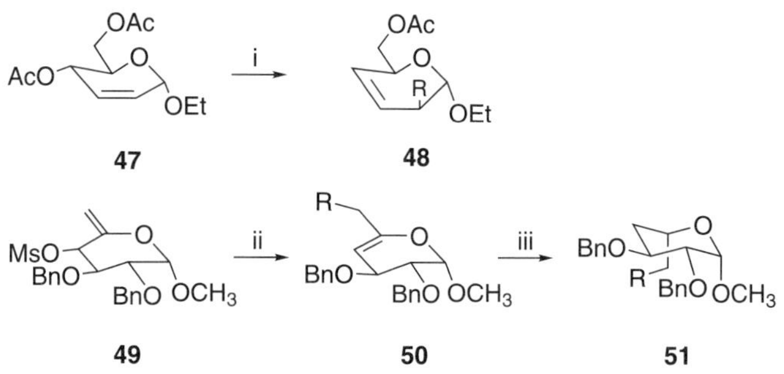

SCHEME 11.14 Reagents: i: MeCuCNLi; ii: RCuCNLi; iii: Wilkinson catalyst, H_2.

ester-bearing center [55]. Additional work by Valverde using allylic benzothiazoles has been published [56]. As shown in Scheme 11.14, allylic acetate **47** reacted cleanly with MeCuCNLi in diethyl ether to afford the 2-*C*-methyl derivative **48** as a single isomer. Chain extension at C6 has also been achieved by using 5,6-unsaturated sulfonates. As illustrated, the 5,6-hexenopyranoside **49** reacted with a cuprate to afford the 4,5-unsaturated compound **50**. Further reduction of the double bond gave only the L-sugar **51**. This useful reaction has been used in our group to assemble the lactonic part of mevinic acid with an analog of the hexahydronapthalene ring system [57]. A very close methodology was used in the synthesis of isoavenaciolide [58].

An additional application of this reaction, reported by Danishefsky, was featured in his synthesis of avermectin A_{1a} [59]. Scheme 11.15 shows that retrosynthetic analysis of the avermectin spiroketal structure **52** gave the *C*-glycoside **53** as a possible intermediate. The synthesis of this intermediate started with tri-*O*-pivaloyl-D-glucal, which was treated with a crotylsilane in the presence of a Lewis acid to afford the rearranged product **55** together with the C6 epimer. The crucial installation of the C2 methyl group was achieved using our methodology to produce **56** as a single isomer. Thus, a very short sequence allowed the construction of the right-hand side of the spiroketal unit of the target compound, including two chiral centers and the crucial C22–C23 double bond.

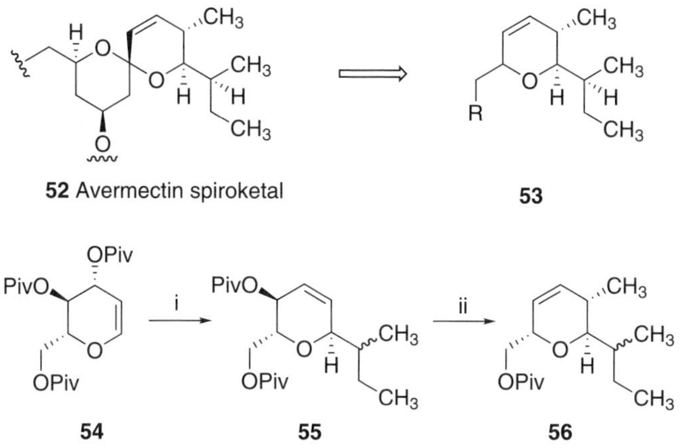

SCHEME 11.15 Reagents: i: (1) (*Z*)-triphenylcrotylsilane, BF_3–Et_2O, 90% and (2) H_2, Pd/C, MeOH + 10% pyridine, 90%; ii: Me_2CuLi, Et_2O, 70%.

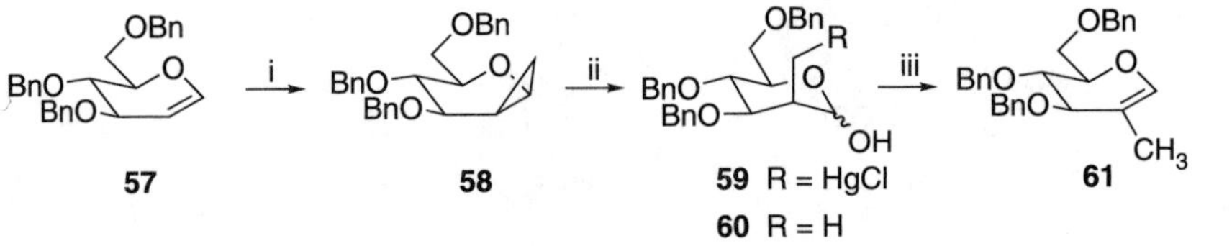

57 58 59 R = HgCl 60 R = H 61

SCHEME 11.16 Reagents: i: Et_2Zn, CH_2Cl_2, 96%; ii: (1) $Hg(CF_3COO)_2$, THF, then NaCl and (2) Bu_3SnH, 81%; iii: Ms_2O, Et_3N, 84%.

Another use of unsaturated sugars is found in their utility in cycloaddition reactions. In the following paragraphs, we will detail the cyclopropanation of glycals as a means of introducing C2 methyl groups. Other cycloaddition reactions leading to larger rings will be reviewed later.

Two types of glycal cyclopropanations, with different stereochemical courses, have been reported. In a modification of the Simmons–Smith procedure, zinc mediated cyclopropanations take place preferentially on the same face of the glycal as the C3 substituent, whereas with metal carbenes the reaction takes place from the opposite face. Ring opening of these strained rings gives C2 branched-derivatives and allows formation of an *O*-glycosidic bond in the presence of an alcohol [60]. This strategy, developed by many groups, was recently applied to 2-*C*-methyl branched-disaccharides using platinum-catalyzed ring opening procedures [61]. Applied to natural products, cyclopropanation of a D-glucal derivative constitutes the early steps of the total synthesis of the C29 to C44 portion of spongistatin 1, also known as altohyrtin [62].

The synthesis of the required compound **61** commenced with the D-glucal derivative **57** (Scheme 11.16), the cyclopropanation of which gave the β-derivative **58** as the major compound [63]. Opening the latter gave the intermediate mercurio derivative **59**, which was reduced to **60**. Final mesylation at the anomeric center followed by spontaneous elimination yielded the branched-glycal **61**. This compound was in turn debenzylated [62].

Continuing from the product of Scheme 11.16, the envisioned strategy was to use the allylic alcohol resulting from debenzylation to construct an allylic ester capable of undergoing an Ireland–Claisen rearrangement. This plan was not as efficient as expected, leading Heathcock's group to devise a new starting compound; again obtained by cyclopropanation of D-glucal. The new strategy, illustrated in Scheme 11.17, aimed to initially secure the proper configuration at C2 utilizing cyclopropanation from the α face. This was achieved by utilizing the C4 hydroxyl group of the D-glucal derivative **62** in a complex with a zinc reagent. Cyclopropanation occurred from the α face producing a yield of up to 94%. Subsequent manipulation of protecting groups gave **63**, which was opened in the presence of pyridinium tribromide to afford the 2-*C*-bromomethyl derivative. Reduction to the methyl derivative gave **64,** which was suitable for *C*-glycoside formation.

Comparing the strategies from Schemes 11.16 and 11.17 illustrates that fine tuning the starting compound allows one to choose the stereochemistry at C2 of the branched chain. It is clear that this method compares well with other 2-*C*-methylation processes in terms of length and stereoselectivity, and should find many applications in the field.

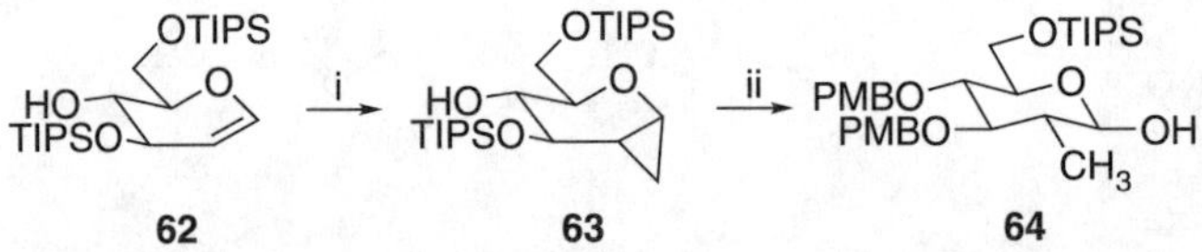

62 63 64

SCHEME 11.17 Reagents: i: (1) TESCl, imidazole, Et_3N, CH_2Cl_2 and (2) Et_2Zn, CH_2I_2, toluene; ii: (1) CSA, H_2O–THF, (2) NaH, PMBCl, DMF, (3) $PyrHBr_3$, pyridine, H_2O, THF and (4) Bu_3SnH, AIBN, toluene.

Complementing cyclopropanation reactions, the dichlorocyclopropanation of glycals has also been used as a means to introduce methyl groups at sugar C2 positions. An excellent illustration of this strategy was reported in connection with a synthesis of the C29 to C51 fragment of altohyrtin A [64].

Other complementary methodologies include the preparation of substituted cyclopropanes from glycals using rhodium acetate carbenoid additions [65,66]. Additionally, acid catalyzed cyclopropane opening reactions in alcoholic solutions afford the 2-*C*-branched-glycosides. These combined reactions were used to prepare a key intermediate in marine diterpene norrisolide synthesis from D-mannose [67].

Cyclopropanation reactions are not limited to glycals. In fact, cyclopropanations at other sites of hexoses are known [68]. However, ring opening reactions of these compounds generally yield ring enlarged carbohydrates [69,70]. Applying this chemistry, an elegant synthesis of chrysanthemic acid derivatives was reported by Fraser-Reid using epoxides as starting materials [71]. The C3 to C9 carbon chain of epothilone A has been elaborated using such a reaction of 2-*C*-methyl glycal, constructed via hetero-Diels–Alder cycloaddition [72].

Regarding cycloaddition reactions leading to larger rings, the formation of cyclobutanes is known. This strategy will be reviewed together with annulation reactions in Section 11.6.2. Additionally, radical cyclizations have found a wide range of applications in carbohydrate chemistry. In particular, radical additions to sugar double bonds have been explored. An example of the use of this reaction in the synthesis of isoavenaciolide will be discussed later.

11.3.3 Using Keto-Sugars

As shown above, many of the methods that are available to form carbon–carbon bonds rely on the reaction of a carbanion with a suitable electrophile. Keto-sugars are readily available by simple oxidation of hydroxyl groups. Thus, the reaction of carbohydrate-derived ketones or aldehydes with carbanions has been extensively explored. However, keto groups are suitable substrates in olefinations, such as Wittig reactions, leading to versatile intermediates for the construction of complex structures. This section will detail some application of keto-sugars in total syntheses along these two main lines.

11.3.3.1 Nucleophilic Additions to Keto-Sugars

The stereochemical outcome of the addition of carbanions to ketones yielding tertiary alcohols (or secondary alcohols in the case of aldehydes) is variable and depends on the substrate, the counterion and the solvent. Numerous applications of this strategy to natural product synthesis from carbohydrates can be found in the literature and this approach was fruitful in pioneering syntheses of polyketide-type products. Here again, the template effect of the sugar plays a tremendous role in the stereochemical outcome of the reaction. Chelation controlled nucleophilic addition can also be used to form chiral centers in a highly predictable way.

An example of the use of this reaction in total synthesis can be found in the construction of erythronolide from two glucose units [73]. Erythronolide encompasses two quaternary chiral centers. Preparation of the first, illustrated in Scheme 11.18, utilized ketone **65** obtained by a multistep process from epoxide **2**. Reaction of **65** with methyllithium was not stereospecific and gave a mixture of alcohols. This mixture was, in turn, methylated to provide **66** and **68,** which could be easily separated by fractional crystallization. These two chirons were used to construct the two subunits of the target compounds. Interestingly, the methoxy group of **66** was retained through the synthesis while that of **68** served as a leaving group in a β-elimination, giving **69**. The double bond of **69** was in turn reduced to secure the correct stereochemistries at C4–C5 of the pyranose ring of **70**.

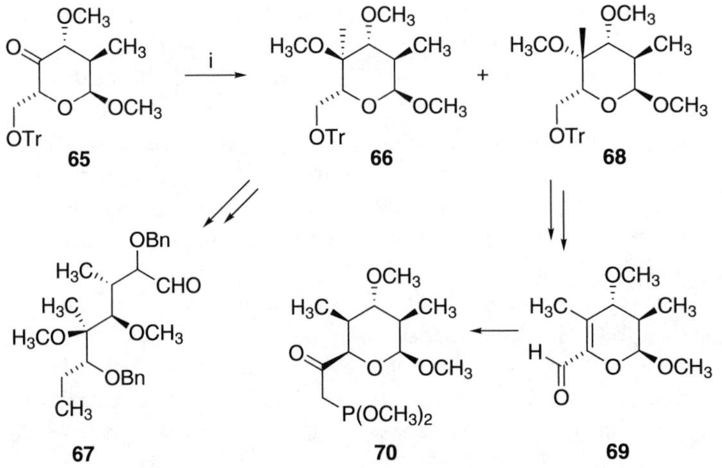

SCHEME 11.18 Reagents: i: (1) MeLi, Et_2O and (2) MeI, NaH.

The nucleophilic addition to keto groups of sugars is useful for the production of tertiary alcohols but can also be used to yield tertiary amines. One such example, in relation to the synthesis of the immunosuppressant compound myriocin, is given below. A comparison of several synthetic approaches to this deceptively simple molecule follows because of its interesting biological properties.

Previous preparations by Scolastico were based on the Strecker synthesis of aminonitrile and lacked steroselectivity [74,75]. More recently, two formal syntheses were reported from the same ketone **71**. In Rama Rao's synthesis (Scheme 11.19) [76], **71** was condensed with vinyl magnesium bromide to give the tertiary alcohol **72** as a single isomer. This compound was then transformed into the vinyl epoxide **73** that, under palladium catalysis, reacted with 4-methoxyphenyl isocyanate to produce the oxazolidinone **74** with retention of its configuration. The remainder of the synthesis consisted of heterocycle opening and adjustment of the oxidation level to provide the lactone **75**. Excision of two carbons was necessary to form the known aldehyde **76**, previously transformed into myriocin [74].

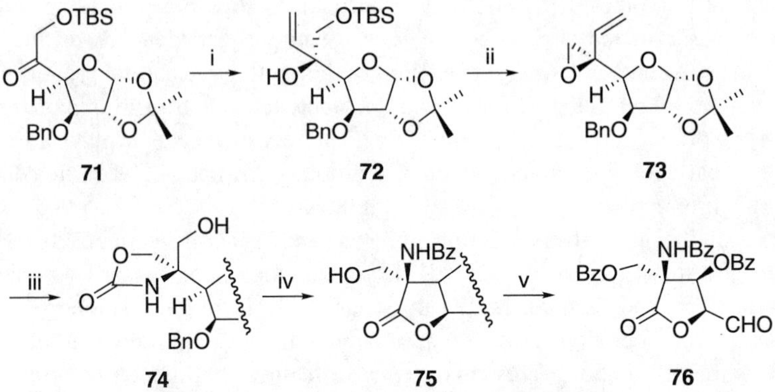

SCHEME 11.19 Reagents: i: $CH_2{=}CHMgBr$, THF; ii: (1) TBAF, THF and (2) Ph_3P, DEAD, toluene; iii: (1) 4-OMe–C_6H_4NCO, $Pd(Ph_3P)_4$, $(iPrO)_3P$, (2) O_3, CH_2Cl_2 then $NaBH_4$, (3) NaH, BnBr, THF and (4) CAN, CH_3CN, H_2O; iv: (1) 5*N* NaOH, THF, (2) Bz_2O, MeOH, (3) $RuCl_3$, MeOH, (4) CH_2N_2, (5) H_2, Pd/C, MeOH and (6) NH_3, MeOH; v: (1) AcOH, H_2O, (2) $NaBH_3CN$, THF and (3) $NaIO_4$, MeOH, H_2O.

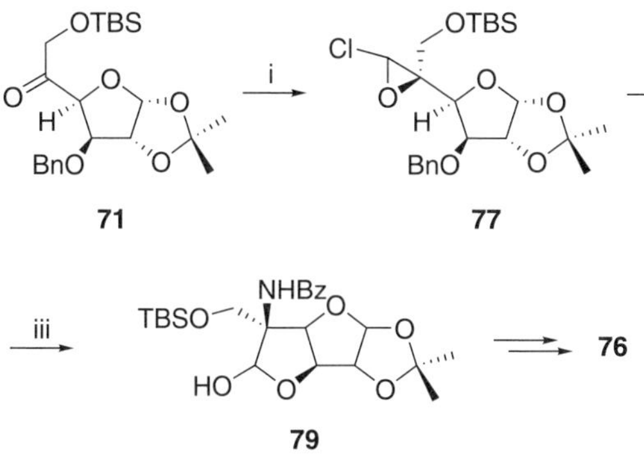

SCHEME 11.20 Reagents: i: LDA, THF, CH_2Cl_2; ii: NaN_3, DMPU, 15-crown-5 ether; iii: (1) H_2, Pd/C and (2) Bz_2O, MeOH.

Olesker's synthesis (Scheme 11.20), also utilizing **71**, incorporated reaction with the anion of dichloromethane, giving the chloroepoxide **77**. *In situ* ring opening on treatment with sodium azide gave the azido aldehyde **78** as a single isomer in 55% overall yield [77]. The nature of the protecting group at C6 proved important in the diastereoselectivity. All chiral centers and most of the functional groups of myriocin were present in **78**. Simultaneous removal of the benzyl group at O3 and reduction of the azide gave the desired amine **79**. Benzoylation and oxidation of the lactol and removal of the silyl groups gave intermediate **76**.

In comparison with the above strategies, Yoshikawa's route to myriocin [78,79] relied on the use of linear carbohydrate derived ketone **80** (Scheme 11.21). Previous experiments showed that the combination of protecting groups in **80** was beneficial for the stereochemical outcome of the addition of lithiated dichloromethane anion. This was because of the greater strain on the carbonyl group in **80,** in comparison with the analog **84**. Thus, the expected tertiary alcohol **81** was obtained as a single stereoisomer. Azido aldehyde **82** was then obtained by treatment with NaN_3 in the presence of 15-crown-5. The remainder of the synthesis consisted of functional group manipulations, that is, reductive opening of the benzylidene ring and elaboration of the amino-acid group to provide the key intermediate **83**. It is noteworthy that all the six carbon atoms of the starting 2-deoxy-D-glucose are retained in the target structure in this synthesis.

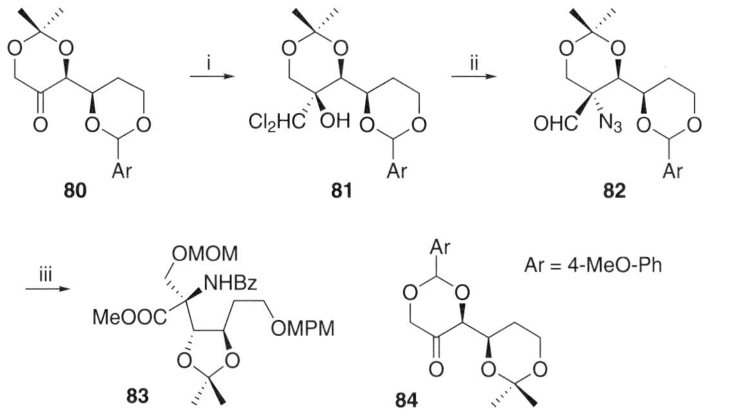

SCHEME 11.21 Reagents: i: LDA, CH_2Cl_2, THF; ii: NaN_3, HMPA, 15-crown-5 ether; iii: (1) $NaBH_4$, EtOH, (2) MOMCl, iPrEt_2N, (3) H_2, Pd/C, (4) BzCl, pyridine, (5) $NaBH_3CN$, TMSCl, (6) $(COCl)_2$, DMSO, Et_3N, (7) $NaClO_2$, NH_2SO_3H and (8) CH_2N_2.

Many uses for the addition of lithiated dichloromethane to keto-sugars have been illustrated thus far. With respect to diversity, this noteworthy addition to a 2-keto derivative of D-glucose was used as a key step in the synthesis of sphingofungin E [80]. Additionally, a recent paper by Sato [81] provides a new process for the synthesis of chloroepoxides.

Nucleophilic additions to keto-sugars have been used in the synthesis of the highly oxygenated 2,8-dioxabicyclo-[3.2.1] octane core of the zaragozic acids, also known as squalestatins [82]. These compounds are powerful inhibitors of squalene synthase and have been studied as modulators of blood cholesterol levels. This densely functionalized ring system, incorporating two tertiary alcohol groups, is closely related to a sugar, and keto-sugars were obvious starting compounds. One of the first reports of the synthesis of squalestatin 1 incorporated 1,6-anhydro galactose as the starting material [83] (Scheme 11.22). In a retrosynthetic analysis of squalestatin 1 (zaragozic acid A) **85**, chiron **86**, derived from acetal cleavage, clearly shows a double branched-chain carbohydrate structure. Its synthesis was envisioned by carbon–carbon bond formation between C4 and C5 (squalestatin numbering). This disconnection provided a new chiron and a 2-keto-idose (or -gulose) structure can be recognized in **87**. The synthesis actually started from the readily available 1,6-anhydro-D-galactose. Indeed, compound **88** exhibits important steric bulk on the β face and this ensured attack of the carbonyl group from the α face. The lithio derivative of furan was chosen as a nonchiral four-carbon building block. Condensation with the latter gave an excellent yield of the tertiary alcohol **89**, establishing the correct stereochemistry at C5 of squalestatin. It should be noted that the stereochemistry at the neighboring center C6 required inversion in order to match that of the target.

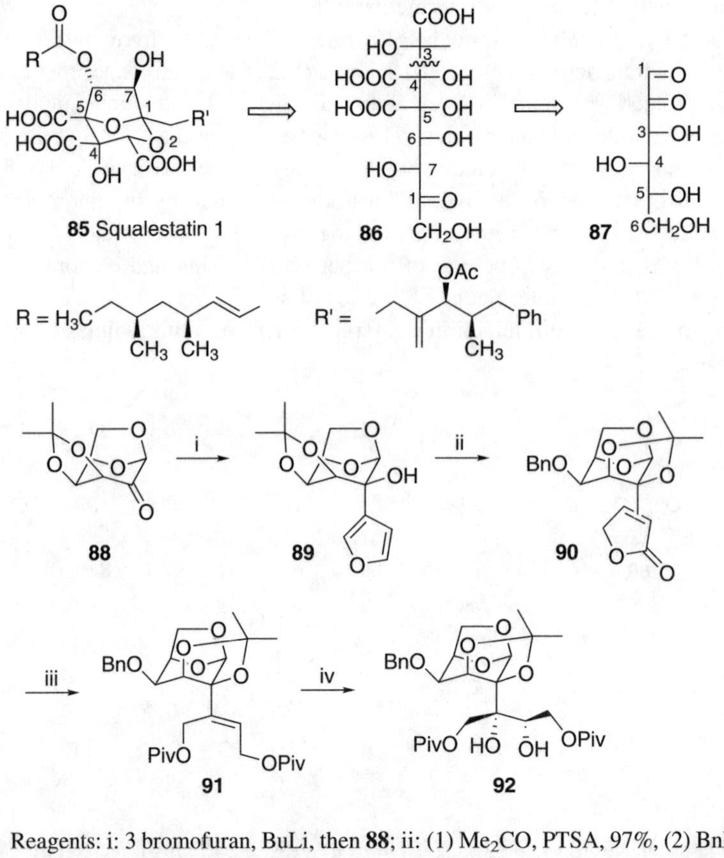

SCHEME 11.22 Reagents: i: 3 bromofuran, BuLi, then **88**; ii: (1) Me_2CO, PTSA, 97%, (2) BnBr, NaH, DMF, 93%, (3) Me_2CO, pyridine, H_2O, Br_2 and (4) HCl 10% Me_2CO; iii: (1) DIBAH, 82% and (2) PivCl, Et_3N, 99%; iv: OsO_4, $K_3Fe(CN)_6$, hydroquinine 4-chlorobenzoate.

In the concluding steps, manipulation of the furan ring of **89** gave **90** as a mixture of positional isomers. These were collectively converted to the unsaturated diol **91**. The last crucial step, installation of two hydroxyl groups on the double bond, was achieved using a standard osmylation reaction [84]. In a second approach for the same step, the Sharpless asymmetric dihydroxylation of **91** was used and yielded one diastereoisomer **92** almost exclusively [85]. This second approach concluded with the synthesis of a lactone containing all correct stereocenters of the squalestatin core with the exception of that at C6.

In his total synthesis of zaragozic acid A (squalestatin S1) Heathcock [86], also made use of a 1,6-anhydro derivative, that of 4-keto-D-glucose **93** [87]. As shown in Scheme 11.23, **93** underwent nucleophilic addition by a trialkylsilylmethyl magnesium chloride from the less hindered β face. Tamao oxidation of the intermediate silyl derivative gave the hydroxymethyl branched-chain sugar **94**. Opening of the 1,6-anhydro bridge and formation of an acetal, followed by lactol oxidation, gave lactone **95**. Nucleophilic addition of the cerium derivative of butyl-3-ene gave the expected lactol. This was, in turn, dehydrated to yield the required bicyclic system **96**.

Continuing, one carbon homologation of **96** was easily achieved by oxidation of the alcohol to the corresponding ketone and a subsequent enolisation–formylation sequence. The last lacking carbon was introduced by nucleophilic addition on the complex *keto-sugar* **97** using vinyl magnesium bromide in the presence of cerium salts. The diol **98** was further elaborated into a relay compound already prepared from squalestatin [88].

Grignard additions to carbohydrate-derived ketones are well known processes providing stereochemically defined quaternary alcohols. However, they are limited to accessible Grignard reagents. Alternate routes allowing the coupling of complex entities appeared recently with the discovery of samarium derivatives. Thus, the synthesis of methyl-α-caryophilloside **102**, a complex sugar found in the cell wall of *Pseudomonas caryphylli*, relies on the diiodosamarium-mediated coupling of acylchloride **99** with a ketone **100** (Scheme 11.24) [89]. Both partners were sugars derived from glucose by stepwise procedures. The key step afforded a good yield of the major equatorial adduct **101**. Subsequent functional group manipulations gave the target compound **102**. This high yielding coupling reaction, proceeding under mild conditions, is well suited for sensitive compounds and should find many applications. Earlier works described the anomeric coupling of two sugars units [90,91].

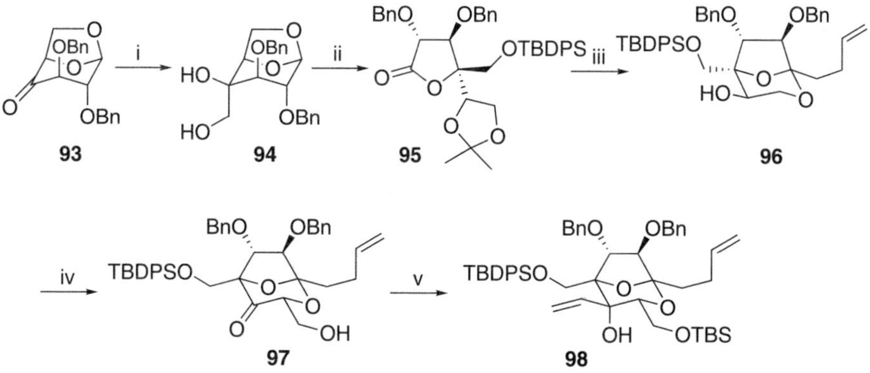

SCHEME 11.23 Reagents: i: (1) $Me_2(iPrO)SiCH_2MgCl$, THF, (2) H_2O_2, MeOH, $NaHCO_3$; ii: (1) TFA, Ac_2O, (2) MeONa, (3) Me_2CO, H^+ and (4) PDC, MS; iii: (1) $CH_2CHCH_2CH_2CeCl_2$, THF and (2) HCl, H_2O; iv: (1) DMSO, TFAA, Et_3N and (2) *t*BuLi, HCHO, THF; v: (1) TBSCl, imidazole, DMF and (2) $CH_2{=}CHMgBr$, $CeCl_3$.

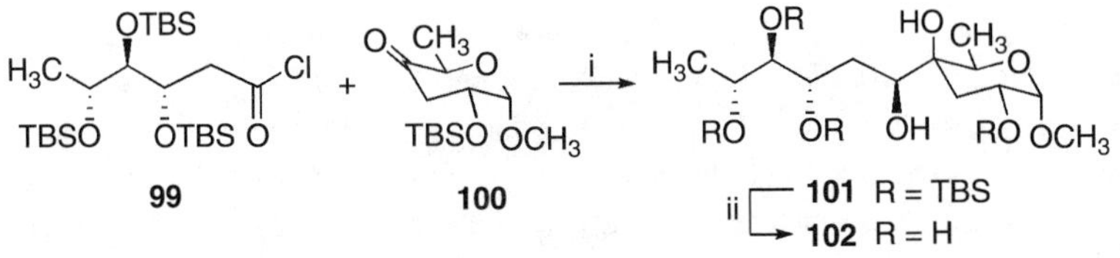

SCHEME 11.24 Reagents: i: SmI_2, THF, 63%; ii: HCl, MeOH, H_2O, 77%.

11.3.3.2 Wittig Type Olefinations

The Wittig olefination is probably one of the most widely used reactions in organic chemistry. Carbohydrates are well-suited substrates in this reaction because of the presence of the aldehyde group at the anomeric position. Additionally, aldehydes and ketones are easily obtained from the many hydroxyl groups [92]. Some seminal examples are given below showing the versatility of this reaction in the carbohydrate field. One of the most useful reactions in connection with total synthesis is methylenation. Thus, a two-step sequence involving Wittig olefination and subsequent reduction of the double bond allows stereoselective formation of carbon–carbon bonds by taking advantage of the template effect of the sugar ring. In certain cases, this is a good alternative or complement to epoxide opening.

Combined uses of epoxide opening and the above-mentioned sequence involving ketone formation, Wittig methylenation and reduction was exploited by Lukacs to produce all possible stereoisomers of methyl 2,4-dideoxy-2,4-dimethyl hexopyranosides (key compounds of macrolide antibiotics syntheses). The sequence, illustrated in Scheme 11.25, utilized the *altro* derivative **7**, which was transformed into a 3:1 mixture of **104** and **105** via olefin **103**.

Application of the preceding strategy to multistriatin synthesis (Scheme 11.26) started with compound **7**, which was deoxygenated at C3 and further elaborated to ketone **106** [49]. Wittig methylenation provided **107**, and subsequent double bond reduction using Wilkinson's catalyst afforded the dimethylated compound **34** [93]. Further manipulation yielded α-multistriatin **108**, the pheromone of *Scolytus multistriatus*. Other syntheses of this pheromon from sugars have been reported [48,94,95].

An inverse strategy was followed in two syntheses of the Prelog–Djerassi lactone, independently disclosed by Fraser-Reid [96] and Isobe [97], starting from tri-*O*-acetyl D-glucal. As shown in Scheme 11.27, glycoside **109** was submitted to a methyl Grignard addition to give the corresponding tertiary alcohol. Allylic oxidation gave olefin **110**, which was reduced to **111** as a single isomer. Wittig methylenation gave the expected olefin in excellent yield. After removal of the benzoyl group at C6, reduction of the double bond produced a mixture of the methylated compound (4:1) in favor of the equatorial epimer **112**. The remainder of Isobe's synthesis is relevant from chain extension at C6 and was successfully accomplished by transformation of the C6 aldehyde into a vinyl sulfone. Stereocontrolled methyllithium

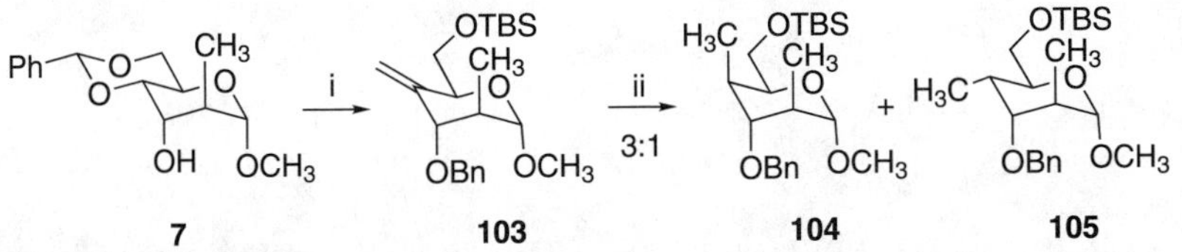

SCHEME 11.25 Reagents: i: (1) BnBr, NaH, (2) EtOH, HCl, (3) TBSCl, 92%, (4) PCC and (5) $Ph_3P{=}CH_2$; ii: H_2, Pd/C, EtOAc.

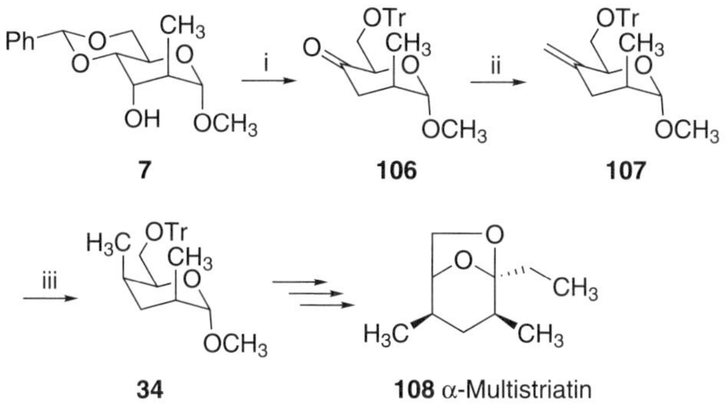

SCHEME 11.26 Reagents: i: (1) NaH, CS_2, MeI, Et_2O, (2) Bu_3SnH, toluene, (3) TsOH, MeOH, (4) Ph_3CCl, pyridine and (5) $CrO_3{\cdot}2py$, CH_2Cl_2; ii: $Ph_3P^+{-}CH_3$, Br^-, BuLi, Et_2O; iii: H_2, Ph_3PRhO_2, benzene.

addition, in the presence of phenylselenyl chloride, gave the adduct **113**, which was in turn elaborated to Prelog–Djerassi lactone **114**.

An alternate route, proposed a few years later by Jones and Wood [98], is illustrated in Scheme 11.28. The authors chose to introduce the methyl group at C2 first, by the now conventional epoxide route. Unfortunately, this did not give the correct configuration needed in the target lactone. The subsequent manipulation of **7** gave the 4-keto derivative, which was submitted to Wittig methylenation to yield **115**. Reduction of the double bond gave a 2:1 mixture of the expected isomers. This mixture was transformed into ketone **116**, which was treated with sodium methoxide to provide the equatorial derivative. Reduction of the carbonyl group and Barton–McCombie deoxygenation gave the dimethyl derivative **117**.

Compound **117** is also a key intermediate in the synthesis of diplodiatoxin, a toxin produced by *Diplodia maydis* on maize and responsible for cattle disease [99]. The two chiral centers present in **117** are transferred in the target compound and induce all other chiralities of this bicyclic structure constructed via an intramolecular Diels–Alder reaction (Scheme 11.29).

Analogous sequences have been used in the synthesis of the ansa chain of rifamycin [100,101] and in the synthesis of the avermectin B_{1a} spiroketal unit **52** [102]. The retrosynthetic plan used is described in Scheme 11.30 and leads to the same intermediate, **118**.

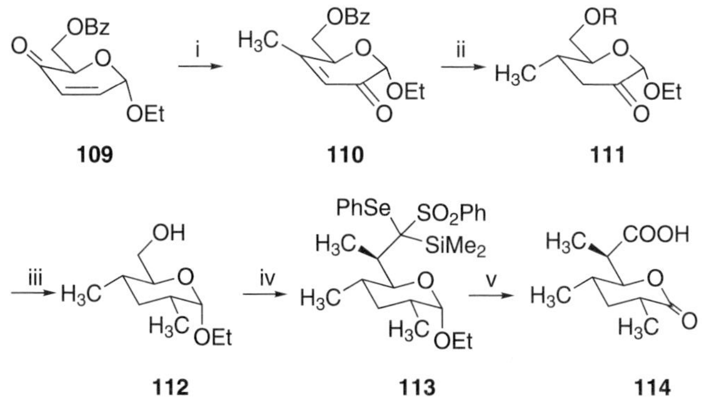

SCHEME 11.27 Reagents: i: (1) MeLi, iPr_2O and (2) CrO_3; ii: H_2, Pd/C; iii: (1) $Ph_3P{=}CH_2$, 90% and (2) H_2, Pd/C; iv: (1) $(COCl)_2$, DMSO, Et_3N, (2) $PhS(Me_3Si)_2CLi$, then *m*CPBA and (3) MeLi, then PhSeCl; v: (1) H_2O_2, H^+ and (2) Br_2, AcONa.

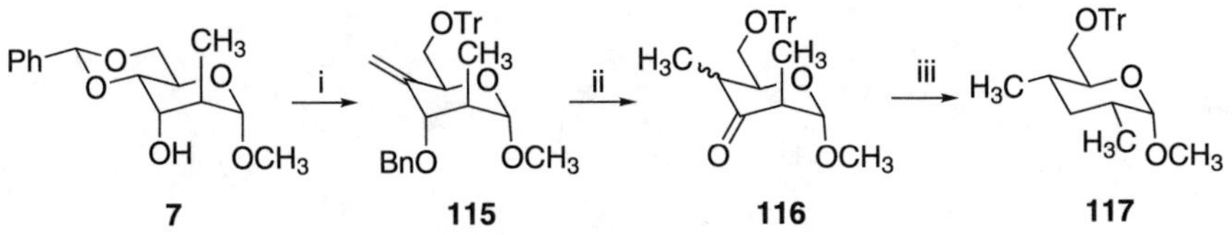

SCHEME 11.28 Reagents: i: (1) BnBr, NaH, 94%, (2) MeOH, PTSA, 80%, (3) TrCl, (4) $(COCl)_2$, DMSO, Et_3N and (5) $Ph_3P{=}CH_2$, 84%; ii: (1) H_2, Pd/C and (2) $(COCl)_2$, DMSO, Et_3N; iii: (1) MeONa, MeOH, (2) $NaBH_4$, MeOH and (3) Barton–McCombie.

A useful sequence using methylenation of a ketone was devised in Hanessian's synthesis of a fragment of the antibiotic boromycin. As shown in Scheme 11.31, 2-acetyloxy-D-glucal **119** was used as a chiral starting material. Ferrier rearrangement of this compound gave the expected enolester **35**. This reacted smoothly with methylene-triphenyl-phosphorane at the carbonyl of the enol ester to provide the intermediate enone **36.** Subsequent *in situ* reaction with excess phosphorane yielded the diene **120**, which was reduced to the corresponding 2-*C* methyl sugar. This sequence had the major advantage of presenting a very short route for extensive deoxygenation of sugars. This reinforces the potential of sugars as starting materials by making deoxygenation reactions simple processes, although it is "chirally expensive."

As seen from the examples thus far, almost all possible synthetic sequences to install methyl groups at C2 and C4 of pyranose rings, with or without C3 hydroxyl groups, have been reported. If high stereocontrol is achieved in the introduction of the first methyl groups, variable stereocontrol in the reduction of the methylene compound is observed. For example, various elements play preferential roles in the attack of double bonds at C4, orientation of neighboring groups at C3, the nature of protecting groups at O6 and even anomeric configurations.

Another popular olefination–reduction sequence starts with commercially available ketone **122**, transformed into olefin **123** using a Wittig or Wittig–Horner reaction [103] (illustrated in Scheme 11.32). Reduction of the double bond proceeds with high stereocontrol to yield the *allo* derivative **124**. The *cis* arrangement at C2–C3 allowed formation of lactone **125** and the synthesis of avenaciolide **126** [104,105].

The stereochemical course of the reduction of sugar olefins is primarily governed by the neighboring groups. This step proved to be highly stereoselective in the pyranose series, as shown in the synthesis of the thromboxane B_2 by Hanessian [106]. As shown in Scheme 11.33, ketone **127**,

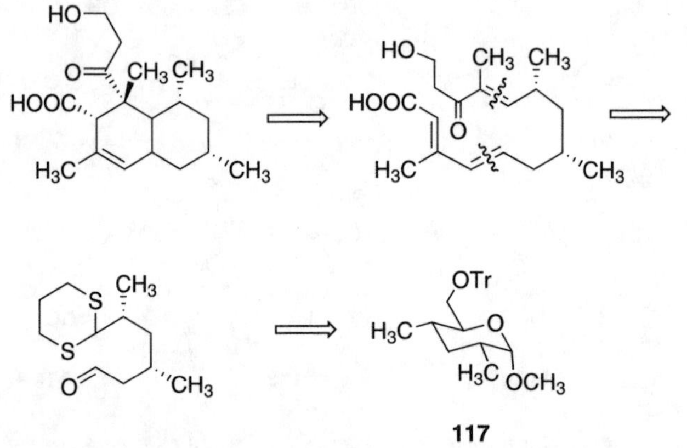

SCHEME 11.29 Retrosynthetic analysis of diplodiatoxin.

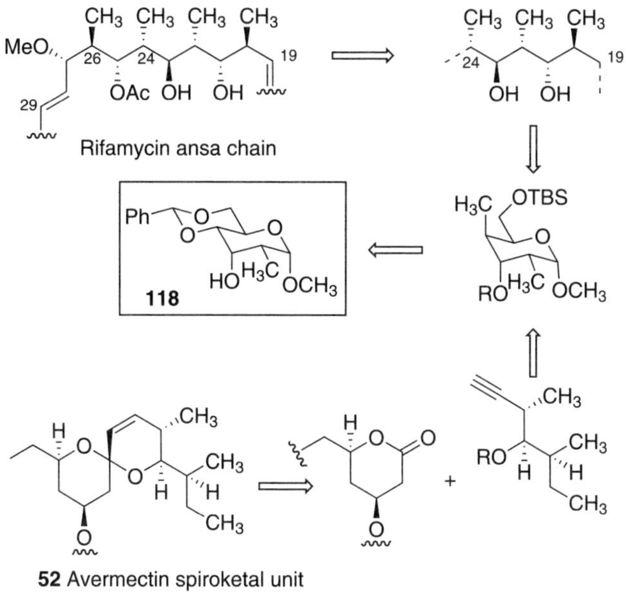

Rifamycin ansa chain

52 Avermectin spiroketal unit

SCHEME 11.30 Retrosynthetic analysis of avermectin spiroketal unit.

available from epoxide **2**, underwent Horner olefination and reduction into ester **128**. Here again, lactonization gave the advanced intermediate **129** *en route* to TB_2.

An alternative to the above sequence has been proposed in a formal synthesis of canadensolide [107]. In this case, the simple displacement of a secondary triflate by lithium *t*-butyl acetate can be carried out on a ribopyranose derivative in the presence of an epoxide.

Numerous examples of this synthetic sequence, not limited to stabilized phosphoranes, can be found in the literature. The method is supposed to be applicable to all secondary hydroxyl groups of sugar rings. Chain extension at the primary hydroxyl group or at the hemiacetalic carbon using Wittig olefination will be examined later on in this chapter (Section 11.4).

A number of more complex structures, useful in total syntheses, are accessible from olefins formed by Wittig olefinations of keto-sugars. A fruitful reaction is the dihydroxylation of double bonds. Obviously, methylene derivatives will give the hydroxymethyl branched-chain derivatives.

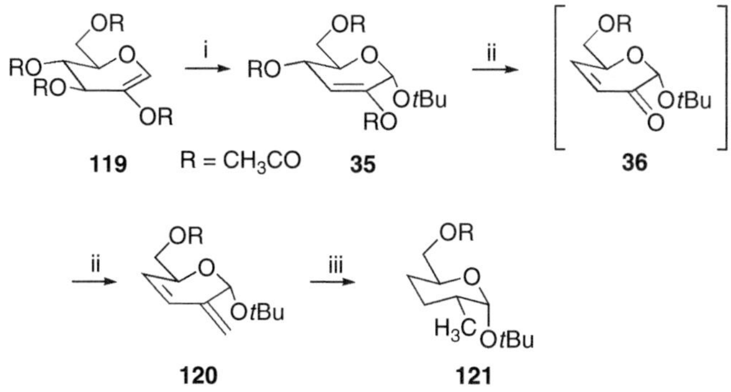

119 R = CH_3CO 35 36

120 121

SCHEME 11.31 Reagents: i: *t*BuOH, BF_3–Et_2O; ii: $Ph_3P{=}CH_2$; iii: H_2, Pd/C.

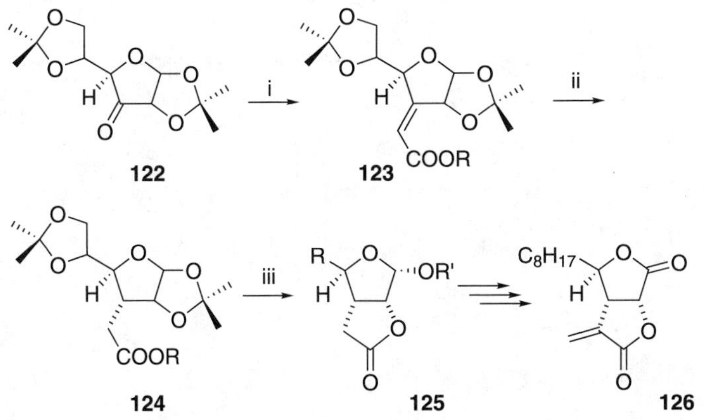

SCHEME 11.32 Reagents: i: $(MeO)_2P(O)CH_2CO_2Me$, *t*BuOK, DMF; ii: H_2, 10% Pd/C, EtOH; iii: (1) 8% aq. H_2SO_4, MeOH, (2) $NaIO_4$, H_2O, (3) $C_7H_{15}P^+Ph_3Br^-$, BuLi, THF, (4) H_2, 10% Pd/C and (5) 0.5% H_2SO_4, dioxane, H_2O.

It should be noted that this sequence could provide good alternatives to trialkylsilylmethyl Grignard reagent additions followed by Tamao oxidations or spiro epoxide openings [19]. Dihydroxylations of carboxymethylene branched-chain sugars are even more interesting and have been used in several syntheses of naturally occurring compounds. Several will be discussed in the following paragraphs.

The first example is relevant to the zaragozic acids field. As shown in Scheme 11.34, ketone **71** was treated with methyl triphenylphosphoranylidene acetate to give the *Z* olefin **130**. Osmium tetroxide dihydroxylation of the latter gave the two possible diols with **131** as the major product. Problems arose in the attempted formation of the required 2,8-dioxabicyclo[3.2.1]-octane, possibly because of the presence of the carboxy residue. Only 1,5-anhydro derivatives were formed upon transacetalation of **131** and **132**. Thus, allylic derivative **133**, synthesized by the same route, was submitted to dihydroxylation and gave a mixture of diols. In this case, upon treatment with *p*-toluene sulfonic acid, the diol **134** gave the expected ring system **135**. These synthetic studies illustrate the inherent problematic nature associated with synthetic approaches to zaragozic acids [108].

Another example of the use of Wittig products is shown in Scheme 11.35. Two total syntheses by Chida's group made use of the Overman rearrangement to introduce nitrogen functionality on a tertiary carbon atom [109,110]. This was illustrated by the construction of lactacystin **143**, a potent neurotrophic factor found in *Streptomyces* culture broth.

The synthesis commenced with the known 3-*C*-methyl derivative **136**, available from diacetone glucose via a well described sequence of oxidation, Wittig olefination and reduction. Protection at C6 followed by oxidation gave the 5-keto derivative, which was immediately submitted to Wittig

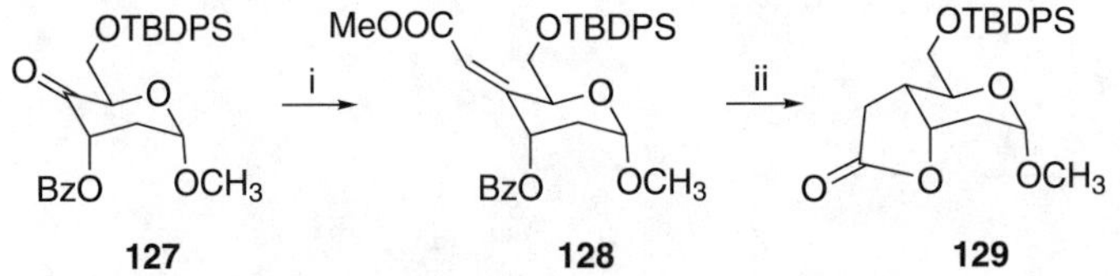

SCHEME 11.33 Reagents: i: $(MeO)_2P(O)CH_2CO_2Me$, *t*BuOK, 90% *E/Z* 1:1; ii: (1) H_2, Pd/C or $Pd(OH)_2/C$ and (2) MeONa, 76%.

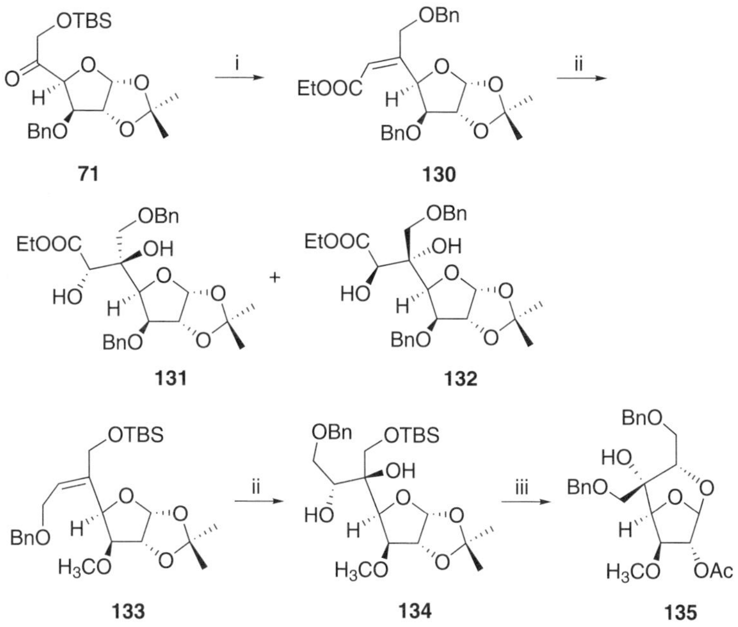

SCHEME 11.34 Reagents: i: (1) $Ph_3P{=}CHCO_2Et$, (2) Bu_4NF and (3) NaH, BnBr; ii: OsO_4, NMO; iii: (1) PTSA, $CHCl_3$ and (2) Ac_2O, pyridine.

olefination giving **137** as a 1:1 Z/E mixture of isomers. This inseparable mixture was reduced with DIBAH to the allylic alcohols and subsequently treated with trichloroacetonitrile to give the corresponding trichloroacetimidates. Upon heating at 140°C in toluene (sealed tube), an Overman rearrangement [111] occurred to give trichloroacetamide **139**. Acid hydrolysis gave a separable mixture of lactol **140** (72%) and its isomer (19%). It is interesting to note that **140** is always the major isomer irrespective of the *Z* or *E* configuration of the starting trichloroacetimidate **139**. Periodate oxidation of **140** gave the corresponding lactol, in equilibrium with the corresponding hemiaminal, which was then oxidized to the lactam **141**. Protecting group manipulations allowed the grafting of the isobutyl side chain from the aldehyde to give **142**. The next steps were devoted to oxidation of the allyl side chain to an acid and coupling with a suitable derivative of cystein. Final deprotection gave lactacystin **143**.

Recently, a similar strategy, shown in Scheme 11.36, was applied to ketone **145**, derived from D-mannose via the known glycal **144** [112]. Wittig olefination of **145**, followed by reduction of the ester group, gave the *E* allylic alcohol **146** as the major compound (>15:1). Trichloroacetimidate formation and Overman rearrangement gave the α,α-disubstituted amino acid precursor **147**. Here again, a 7:1 mixture of epimers was formed, even with the pure *E* isomer as the starting alcohol. This is explained by the fact that, although both faces of the olefin are accessible, there is substantial preference for reaction on only one face.

There are many other rearrangements utilizing Wittig products as substrates. These, and others, will continue to be addressed in later sections of this chapter because of their diversity and general applicability to carbohydrate chemistry.

11.3.4 Using Carbohydrates as Nucleophiles

As already mentioned, most carbon–carbon bond forming reactions in carbohydrate chemistry use the sugar as an electrophile reacting with a carbon nucleophile. Recently, carbohydrate chemists

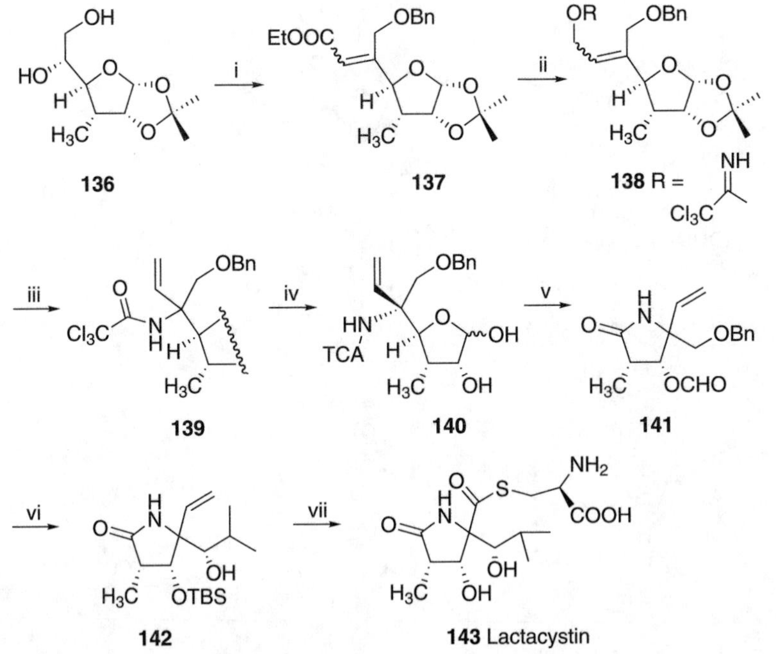

SCHEME 11.35 Reagents: i: (1) Bu_3SnO, toluene, reflux, then BnBr, CsF, DMF, (2) CrO_3, H_2SO_4, acetone and (3) $Ph_3P{=}CHCO_2Et$, toluene; ii: (1) DIBAL, CH_2Cl_2 and (2) Cl_3CCN, NaH, Et_2O; iii: heat, toluene, sealed tube; iv: TFA, H_2O; v: (1) $NaIO_4$, MeOH, H_2O, (2) CrO_3, H_2SO_4, acetone and (3) $NaBH_4$, MeOH; vi: (1) TBSOTf, 2,6-lutidine, CH_2Cl_2, (2) Na, liq. NH_3, THF, (3) DMSO, DCC, TFA, pyridine and (4) *i*PrMgBr, THF; vii: (1) TFA, H_2O, (2) O_3, CH_2Cl_2, then Me_2S, (3) $NaClO_2$, NaH_2PO_4, $HOSO_2NH_2$, *t*BuOH, H_2O, (4) bis(2-oxo-3-oxazolidinyl)phosphonic chloride, *N*-acetyl-L-cysteine allyl ester, Et_3N, CH_2Cl_2 and (5) $Pd(Ph_3P)_4$, HCO_2H, Et_3N, THF.

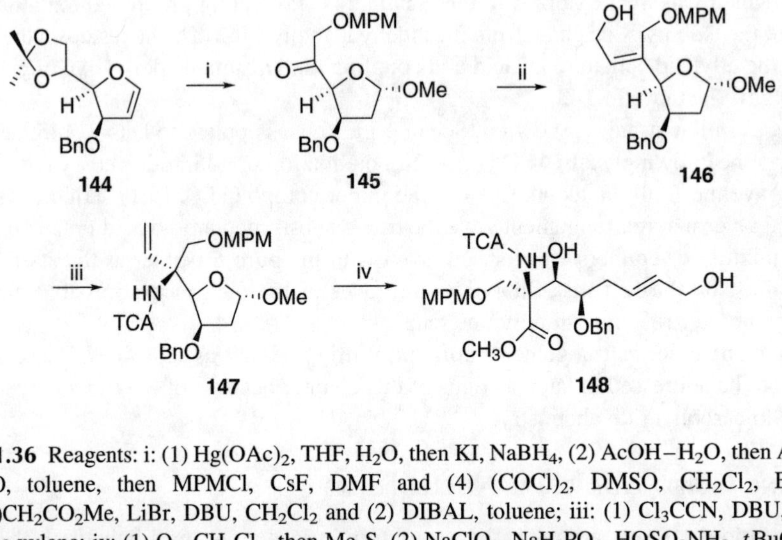

SCHEME 11.36 Reagents: i: (1) $Hg(OAc)_2$, THF, H_2O, then KI, $NaBH_4$, (2) AcOH–H_2O, then AcCl, MeOH, (3) Bu_2SnO, toluene, then MPMCl, CsF, DMF and (4) $(COCl)_2$, DMSO, CH_2Cl_2, Et_3N; ii: (1) $(MeO)_2P(O)CH_2CO_2Me$, LiBr, DBU, CH_2Cl_2 and (2) DIBAL, toluene; iii: (1) Cl_3CCN, DBU, CH_2Cl_2 and (2) K_2CO_3, *o*-xylene; iv: (1) O_3, CH_2Cl_2, then Me_2S, (2) $NaClO_2$, NaH_2PO_4, $HOSO_2NH_2$, *t*BuOH, H_2O, (3) 4*N* HCl, THF, then $Ph_3P{=}CHCO_2Et$, toluene and (4) DIBAL, THF, toluene.

realized that carbanions derived from sugars are quite stable and utilizable to build elaborated structures. Seminal work along this line came from our group. We showed that enolates derived from methyl di-*O*-benzylidene-D-mannopyranoside using butyllithium-mediated opening of the 2,3-*O*-benzylidene ring gave an enolate sufficiently stable to be alkylated *in situ* with different electrophiles [113,114]. Aldol reactions also took place with interesting diastereoselectivity [115]. This strategy allowed short constructions of complex assemblies [116].

Giese reported an interesting total synthesis of soraphen $A_{1\alpha}$ **149**, partly based on this strategy [117]. This macrolide is an inhibitor of lipid synthesis in fungi and thus a highly potent fungicide. A retrosynthetic analysis promptly led to two subunits, which would be linked by lactone formation and C9–C10 double bond formation. These disconnections provided the chiron **150**, which is structurally close to an L-ketose (Scheme 11.37). However, looking at this sugar differently led to the D-sugar analog **151**. A good precursor of this compound is aldehyde **156**.

Treatment of the known acetal **152** according to our procedure [113] gave the 2-*C*-methyl branched-sugar **8** as the only isomer. Subsequent reduction with DIBAH gave an excellent yield of the axial alcohol **153**. The stage was set for the introduction of the second methyl group at C4. A now classical epoxide opening was chosen. Replacement of the benzylidene acetal by a silyl protecting group at C6 left the 3,4-diol, which was treated with Viehe's salt to give the axial chloride **154**. Epoxide formation occurred upon treatment with base and opening of the epoxide under the action of methylmagnesium chloride in the presence of copper(I) bromide occurred in nearly quantitative yield. Protection of the resulting alcohol gave **155**. Removal of the silyl group and Swern oxidation gave the required aldehyde **156**. As shown here, the direct methylation of a carbohydrate enolate offers an excellent alternative, in terms of synthetic length and yields, to previously reported sequences.

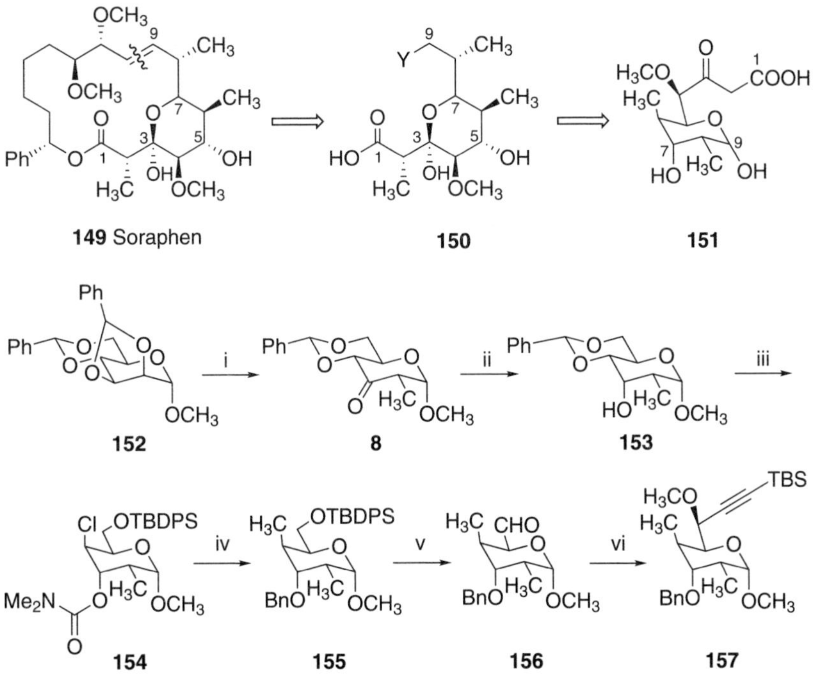

149 Soraphen **150** **151**

152 **8** **153**

154 **155** **156** **157**

SCHEME 11.37 Reagents: i: BuLi, THF, then MeI, HMPA, 50%; ii: DIBAH, 89%; iii: (1) H_2, Pd/C, EtOAc, (2) TBDPSCl, DMAP, CH_2Cl_2, 89% and (3) $Me_2N^+{=}CCl_2$, Cl^-, CH_2Cl_2, Et_3N; iv: (1) MeLi, TMEDA, THF, (2) MeMgCl, CuBr, THF, 94% and (3) NaH, BnBr, THF, 89%; v: (1) TBAF, THF, 87% and (2) $(COCl)_2$, DMSO, Et_3N, 91%; vi: (1) TBS–C≡C–MgBr, 87% and (2) Ag_2O, MeI, 89%.

The last steps of the synthesis, *en route* to synthon **157**, consisted of chain elongation at C6 using a suitable carbanion. An acetylenic carbanion, representing C2–C3 of **151**, was chosen because of a known propensity to add stereoselectively but also because they are good precursors of ketones. A brief study showed that the chelation controlled addition, that is, the use of magnesium derivatives, led to the required stereochemistry at C6. Other examples of such chelation controlled addition will be given in the following sections.

Applications of the same enolate alkylation procedure to furnish *gem*-dialkylated compounds have been used in the synthesis of onnamide A [118]. This type of *gem*-2-*C*-dimethyl group is also found in pederin and mycalamides A and B. As shown from the retrosynthetic analysis of some members of this class of compounds (Scheme 11.38), *C*-glycosidic structure **158** could be obtained from the α-glycoside **159**. The latter could be prepared from the ketone **160** via gem-dialkylation of the corresponding enolate, as described for the synthesis of **8**.

An alternate route to prepare 2-*C*-gem-dimethyl branched-chain sugars involves cyclopropanation of a 4-*C*-methylene derivative followed by ring opening as used in a mycalamide synthesis [119] or in the synthesis of the C18 to C23 subunit of lasonolide A [120].

Alkylations of sugar-derived carbanions have found applications in the zaragozic acid core structure syntheses. The key point of these syntheses is the formation of quaternary carbon centers. A possible approach would be to alkylate an enolate generated at C4 of a furanose ring (C5 in zaragozic acid numbering). This was considered by several groups. The first approach came from Gurjar's group, which started from 2-ketogulonic acid [121]. The latter was transformed into aldehyde **161** using standard reactions. As illustrated in scheme 11.39, upon treatment with formaldehyde in basic medium, clean hydroxymethylation occurred to give **162** in excellent yield. Successive acetal formation and elaboration of the primary alcohol to a vinyl group gave **163**. Dihydroxylation of the vinyl group, followed by protection of the primary hydroxyl group and oxidation, gave ketone **164**. Wittig homologation of this ketone was impossible, so vinyl Grignard addition was carried out to provide the desired allylic alcohol **165** as a single isomer. Dihydroxylation of the vinyl appendage also proceeded with excellent diastereoselectivity, establishing the correct configuration of all chiral centers of zaragozic acid as shown in **166**. The last steps were devoted to the construction of the acetal ring of **167**, which was formed uneventfully due to the correct protections installed all along the synthesis.

A related strategy starting from a uronic acid was reported to construct the core structure of zaragozic acid [122,123]. The key point was the formation of the C4–C5 bond using aldol condensation. As shown in Scheme 11.40, methyl uronate **168** was first treated with KHMDS at −100°C, and then treated with a premixed solution of cerium chloride and (*R*)-glyceraldehyde acetonide to give the expected aldol as a mixture of diastereoisomers in a yield of 65 to 70%.

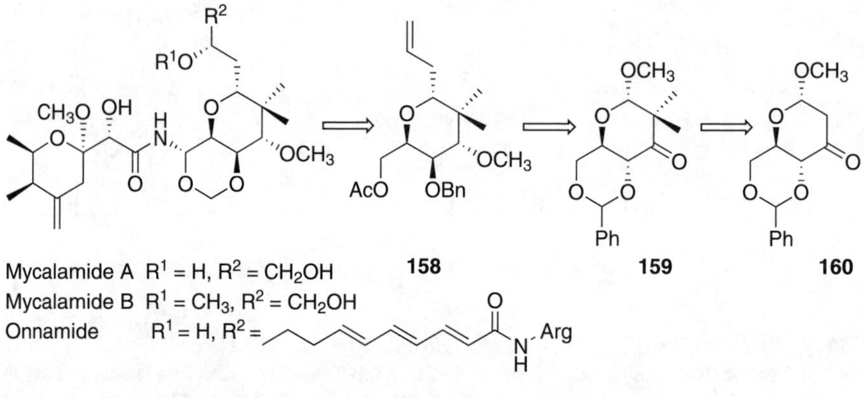

Mycalamide A $R^1 = H, R^2 = CH_2OH$
Mycalamide B $R^1 = CH_3, R^2 = CH_2OH$
Onnamide $R^1 = H, R^2 =$ …NHArg

158 **159** **160**

SCHEME 11.38 Retrosynthetic analysis of onnamide and mycalamide.

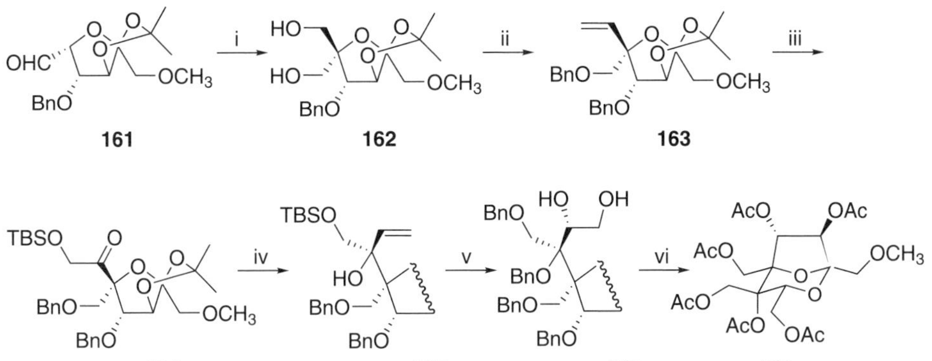

SCHEME 11.39 Reagents: i: HCHO, 1*N* NaOH, 90%; ii: (1) Li liq. NH_3, (2) $Me_2C(OMe)_2$, PTSA, (3) $(COCl)_2$, DMSO, Et_3N, (4) $Ph_3P{=}CH_2$, (5) H_2SO_4, MeOH and (6) NaH, BnBr, DMF; iii: (1) OsO_4, NMO, *t*BuOH/H_2O, (2) TBSCl, imidazole and (3) $(COCl)_2$, DMSO, Et_3N; iv: $CH_2CHMgBr$, THF, 90%; v: (1) TBAF, THF, (2) NaH, BnBr, THF and (3) OsO_4, NMO, *t*BuOH/H_2O; vi: (1) PTSA, $CHCl_3$, (2) Ac_2O, pyridine and (3) H_2O, $Pd(OH)_2$, MeOH, then Ac_2O, pyridine.

Oxidation of the hydroxyl group gave two epimeric ketones **169** and **170** in a 4:1 ratio. The major compound resulted from attack of the enolate from the α face. Reduction of the carbonyl functions of **169**, followed by protection as benzyl ethers, gave a mixture of epimers **171**. Removal of the acetal groups and simultaneous glycoside formation gave mainly **172** (57%) together with the seven-membered acetal (19%). The need for protection at O4 (zaragozic acid numbering) was once again pointed out. Another use of an uronate enolate to form gem-dialkylated compounds has been reported by Florent in the synthesis of punaglandin precursor [124].

Although no total synthesis is yet based on this reaction, a recent report by Lubineau [125] on the use of allylic halides in a Barbier reaction is worthy of note. This clever method is a good alternative to generate sugar nucleophiles in water. As shown in Scheme 11.41, standard Ferrier chemistry is used to produce the starting halide **173** from **47**. Treatment of **173** with indium powder in water at room temperature in the presence of benzaldehyde gave the C2-branched-derivative **174** in high yield as a single isomer. Coupling of two sugar units has been achieved this way and this opens the door to the *C*-analogs of disacccharides like **175**. The observed diastereoselectivity is

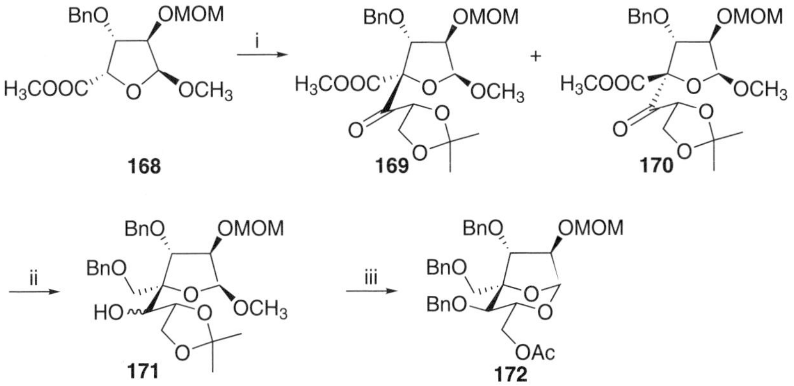

SCHEME 11.40 Reagents: i: (1) KHMDS, THF, $CeCl_3$, then D-glyceraldehyde acetonide 70% and (2) Dess–Martin periodinane 88%; ii: $LiAlH_4$, Et_2O, 70%; iii: (1) NaH, BnBr, DMF, 89%, (2) Dowex 50W H^+ and (3) Ac_2O, DMAP, CH_2Cl_2.

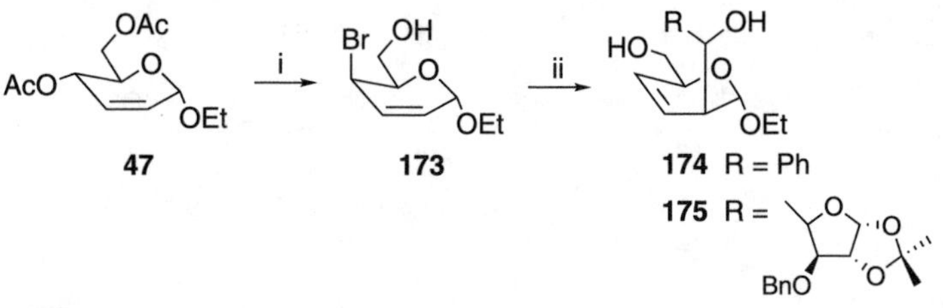

SCHEME 11.41 Reagents: i: (1) Et_3N, MeOH, H_2O, 96%, (2) TBSCl, pyridine, 90%, (3) Ph_3P-CBr_4, CH_2Cl_2, 75% and (4) TBAF, THF; ii: RCOH, In, H_2O.

explained in terms of chelation of the aldehyde with the allylindium, which directs the entering aldehyde on the same face. These remarkably mild and environmentally friendly conditions should allow the construction of complex structures. Here again, the transformation of the double bond into a diol via epoxidation is possible.

11.3.5 Using Rearrangements

Rearrangement reactions offer numerous advantages to construct carbon–carbon bonds. They are often triggered by heating under close to neutral conditions. Usually a rearrangement proceeds with high stereoselectivity, especially in ring systems. One of the most popular reactions of this type in carbohydrate chemistry is the Claisen rearrangement and its variants, key reactions of many total syntheses. The principle of this transformation is illustrated below.

In its simplest version (Scheme 11.42), an allylic alcohol like **177** is treated under Eschenmoser conditions to give, with retention of configuration, the acetamide **176**. Similarly, esterification to **178** and subsequent enolization to **179** sets up the Claisen conversion to **180**.

Application of Claisen methodology led to a number of successful total syntheses of natural compounds. Thromboxane B2 has been synthesized from a 4,5-unsaturated sugar derived from D-glucose [126]. The Ireland–Claisen rearrangement [127] is also the key reaction used to control the introduction of the right wing of pseudomonic acids [128].

A good example of the use of Claisen type rearrangements can be found in Fleet's synthesis of pseudomonic acid from arabinose, in which this strategy is incorporated twice [33]. The synthesis (Scheme 11.43) was envisioned from the very simple chiral allylic alcohol **181**, obtained from D-arabinose. The first Eschenmoser rearrangement gave the acetamido derivative **182** in 85% yield. Iodolactonisation followed by iodohydric acid elimination gave the glycal **183**, which was in turn transformed into a new allylic alcohol **184**. Treatment of **184** with *N,N*-dimethylformamide dimethylacetal in refluxing xylene gave the rearranged product **185** in a yield of 94%, thus controlling the introduction of the *C*-glycosidic appendage. The last key step was

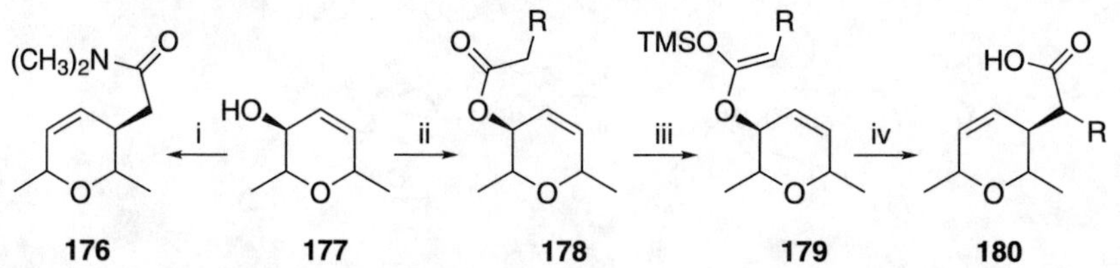

SCHEME 11.42 Reagents: i: $MeC(OMe)_2NMe_2$, heat; ii: RCH_2COCl, Et_3N; iii: LDA, then TMSCl low temperature; iv: back to RT or heat.

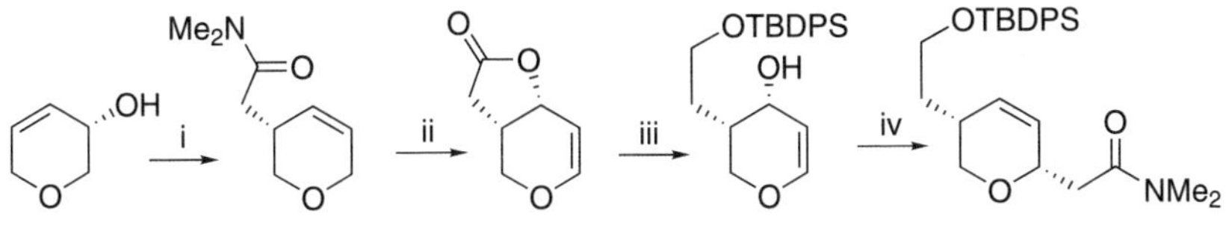

SCHEME 11.43 Reagents: i: $MeC(OMe)_2NMe_2$, xylene; ii: (1) I_2, THF, H_2O and (2) DBU, benzene; iii: (1) $NaBH_4$, EtOH and (2) TBDPSCl, imidazole, DMF; iv: $MeC(OMe)_2NMe_2$, xylene; v: (1) MeLi, THF, (2) OsO_4, NMO, (3) cyclohexanone, $CuSO_4$, PTSA and (4) NaH, $(EtO)_2P(O)CH_2CO_2Et$, THF; vi: (1) PCC, CH_2Cl_2, (2) $MeCH(OH)CH(Me)CH_2PPh_3$, BuLi and (3) 50% aq. AcOH.

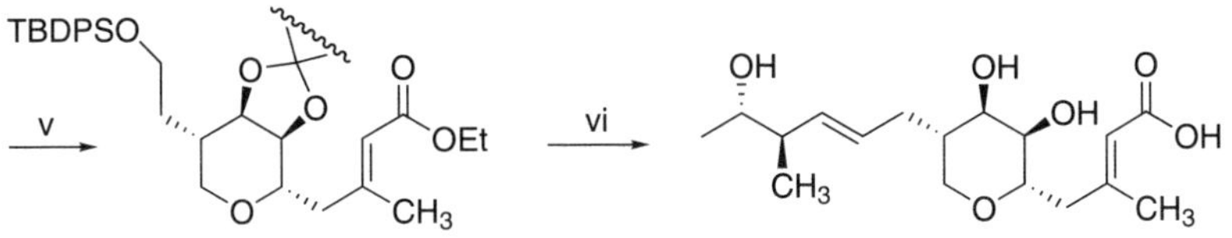

the introduction of the two hydroxyl groups via osmylation of the double bond. Further manipulations of the *C*-glycosyl chain gave **186,** featuring all functions of the right wing. Formation of an aldehyde, followed by Wittig olefination, allowed the grafting of the left wing of pseudomonic acid **187**.

Allylic alcohols derived from *exo*-olefins are attractive substrates for Claisen type rearrangements producing quaternary carbons, again with high stereocontrol. Pioneering studies by Fraser-Reid on the out-of-ring Claisen rearrangement paved the way to a number of densely functionalized carbohydrate structures and then to complex targets [129–131]. As shown in Scheme 11.44, an ethylvinyl ether was prepared from the Wittig olefination of ketosugar **188,** followed by reduction to the allyl alcohol **190**. Vinylation of the latter gave **191**, which rearranged into the aldehyde **192** upon heating.

Different situations were studied showing that the folding of the system controls attacks on the double bond from the β face of the sugar, thus placing the final acetaldehyde residue on this face and the pendant vinyl group on the α face. This strategy has been used to approach annelated pyranosides, the diquinane systems in particular. Indeed, derivatives of **192** are well suited for further reactions such as radical cyclizations involving intramolecular additions of radicals to the

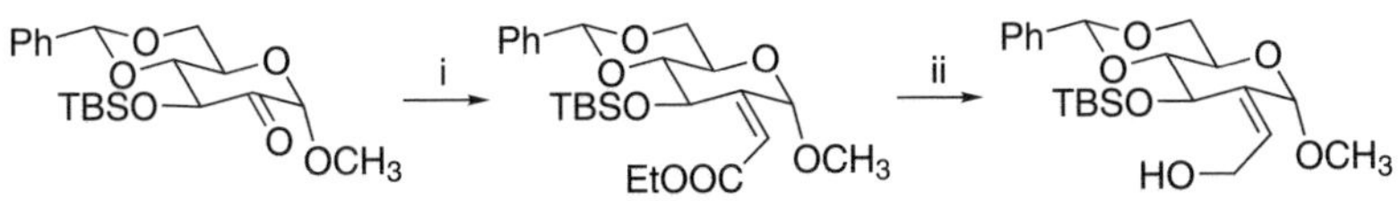

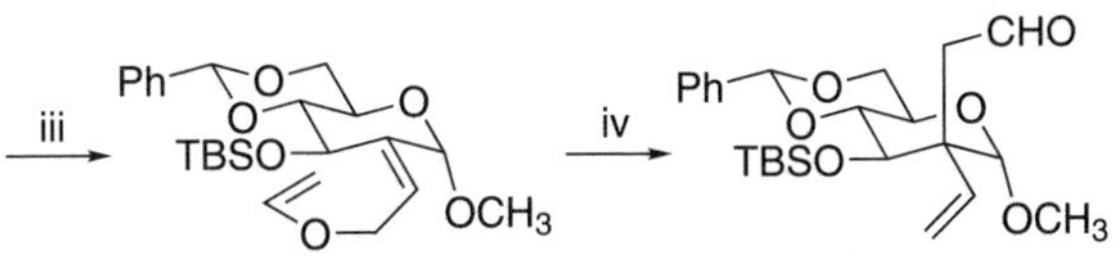

SCHEME 11.44 Reagents: i: $Ph_3P{=}CHCO_2Et$, CH_3CN; ii: $LiAlH_4$, Et_2O; iii: $CH_2{=}CHOEt$, $Hg(OAc)_2$; iv: PhCN, reflux.

vinyl group followed by radical additions on aldehyde carbonyl groups [131,132]. The synthesis of pipitzol described below (Scheme 11.45) highlights this strategy among others, including the alkylation of sugar enolates [133].

The synthesis started from the known ketone **160**, the enolate of which was alkylated with propargyl bromide to give **193**. The sequence described above involving the orthoester variant of the Claisen rearrangement was performed on this compound to give ester **194**. Transformation of this compound into the nitrile **195** was carried out using routine reactions. Addition of a tributyltin radical to the triple bond was then achieved giving a vinyl radical, which was added to the neighboring vinyl group producing a methylene radical. Final addition of the latter on the nitrile group gave the expected cyclopentanone **196**. The resulting *exo* methylene group was transformed into a cyclopropane and the carbonyl group was reduced to a methylene. Manipulation of the benzylidene ring gave the 6-iodo-4-*O*-benzyl derivative which underwent Vasella ring opening to provide aldehyde **197**. Further manipulations of this carbonyl group allowed one carbon excision. Vinyl group transformation allowed the formation of an aldehyde suitable for aldol ring closure, giving **198** and thus establishing the A ring of pipitzol. The last steps were the installation of a methyl group at C6, opening of the cyclopropane, installation of a C5–C6 double bond and generation of the hydroxyl group at C5 to furnish **199** and the target compound **200**.

The same strategy was exploited in the furanose series by Tadano's group [134]. In his seminal paper on this topic, Tadano explored this reaction on different isomeric alcohols (see compound **201** below) showing that the *E* isomer, on treatment with triethylorthoacetate (Johnson variant), smoothly rearranged from the β face, giving only one isomer. The reaction

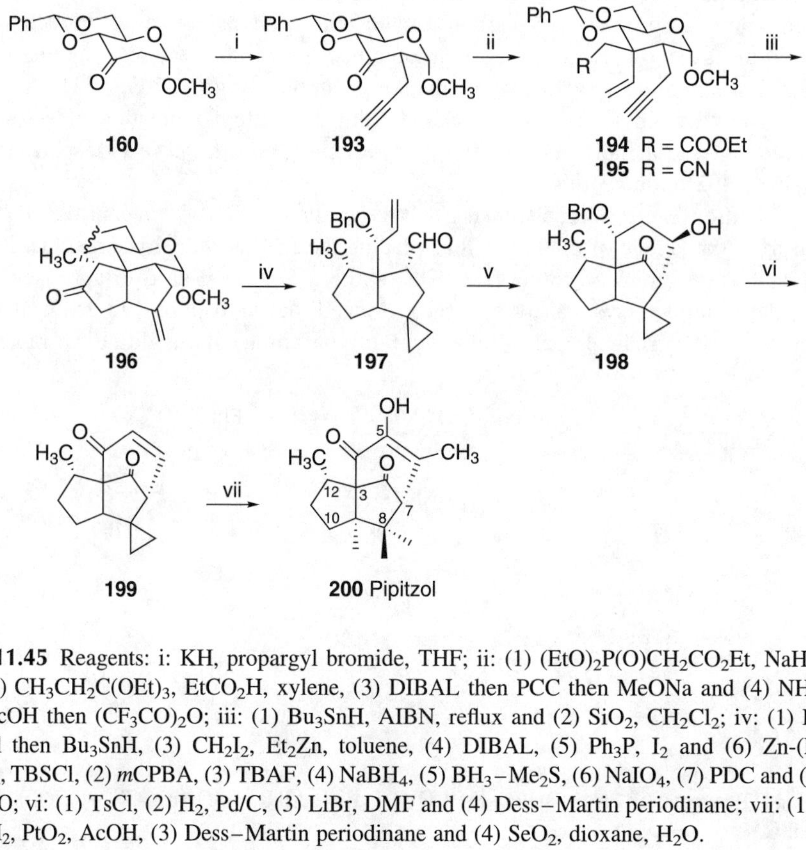

SCHEME 11.45 Reagents: i: KH, propargyl bromide, THF; ii: (1) $(EtO)_2P(O)CH_2CO_2Et$, NaH, THF then DIBAL, (2) $CH_3CH_2C(OEt)_3$, $EtCO_2H$, xylene, (3) DIBAL then PCC then MeONa and (4) NH_2OH–HCl, NaOAc, AcOH then $(CF_3CO)_2O$; iii: (1) Bu_3SnH, AIBN, reflux and (2) SiO_2, CH_2Cl_2; iv: (1) DIBAL, (2) PhOC(S)Cl then Bu_3SnH, (3) CH_2I_2, Et_2Zn, toluene, (4) DIBAL, (5) Ph_3P, I_2 and (6) Zn-(Hg); v: (1) $KN(TMS)_2$, TBSCl, (2) *m*CPBA, (3) TBAF, (4) $NaBH_4$, (5) BH_3–Me_2S, (6) $NaIO_4$, (7) PDC and (8) Na_2CO_3, MeOH, H_2O; vi: (1) TsCl, (2) H_2, Pd/C, (3) LiBr, DMF and (4) Dess–Martin periodinane; vii: (1) Me_2CuLi, THF, (2) H_2, PtO_2, AcOH, (3) Dess–Martin periodinane and (4) SeO_2, dioxane, H_2O.

with the *Z* isomer of **201** was sluggish and gave a mixture of epimers at C3. The influence of isopropylidene acetals was also examined. In the next paragraph, we will describe the synthesis of an insect sex attractant, (−)-anastrephin **207** [135]. The illustrated strategy led to a number of heterocyclic compounds such as (−)-acetomycin [136,137], asteltoxin [138] and eremantholide A [139,140]. Other uses of this strategy are found in the construction of annulated terpenoid structures described in Section 11.6.2.

As shown in Scheme 11.46, compound **202**, resulting from the *ortho* ester Claisen rearrangement on olefin **201** (available from diacetone-D-glucose), was employed as a key starting material [141]. This operation secured the stereochemistry of the quaternary center of the target molecule. A two carbon homologation of **202** and further manipulations gave olefin **203**. Protection of the primary hydroxyl group was followed by ozonolysis of the vinyl group. Wittig olefination of the resulting aldehyde function gave the α,β-unsaturated system. Deprotection of the primary alcohol and oxidation to aldehyde **204** set the stage for carbocycle formation, the other key step of the synthesis. A ketyl radical, generated by treating an aldehyde with samarium iodide, was then added to the unsaturated ester. Lactone **205** was obtained directly as the main product together with epimers at C7a and C3a (target numbering). The last steps of the synthesis consisted of transformation of the furanose ring into methyl and vinyl substituents at C4 to give **206**, which was in turn transformed into the target compound **207**.

The Ireland variant of the Claisen rearrangement relies on the generation of the vinyl moiety of the double unsaturated system by enolization of an ester. The principle of this technique is given in Scheme 11.42 and it allows the transformation of allylic alcohol **177** into **180** via ester **178** and the ketene acetal **179**. Application of this chemistry to the construction of the chiral quaternary carbon atom of the zaragozic acid core is shown in Scheme 11.47 [142]. Additional examples of this rearrangement are found in Section 11.5.2.2.

As illustrated in Scheme 11.47, the uronate **208** was treated with base at low temperature giving the corresponding enolate trapped *in situ* with trimethylsilyl chloride. The resulting silyl ketene

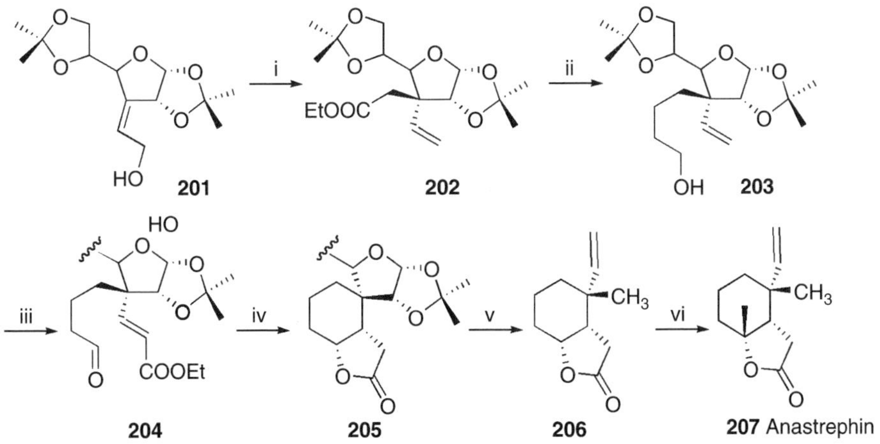

SCHEME 11.46 Reagents: i: $CH_3C(OEt)_3$, C_2H_5COOH; ii: (1) $LiAlH_4$, THF, (2) Ph_3P, DEAD, MeI, (3) $CH_2(CO_2Me)_2$, NaH, THF, (4) NaCl, DMSO, H_2O and (5) $LiAlH_4$, THF; iii: (1) TBSCl, imidazole, DMF, (2) O_3 then Ph_3P, CH_2Cl_2, (3) $(EtO)_2P(O)CH_2CO_2Et$, NaH, DMF, (4) TBAF, H_2O, THF and (5) PCC, CH_2Cl_2; iv: SmI_2, *i*PrOH, HMPA, THF, 35%; v: (1) 1*N* HCl, THF, (2) $NaIO_4$, MeOH, H_2O, (3) $NaBH_4$, MeOH, (4) BnBr, NaH, DMF, (5) 60% aq. TFA, (6) BnBr, NaH, DMF, (7) EtSH, conc. HCl, (8) Raney–Ni, EtOH, (9) $NaIO_4$, MeOH, H_2O and (10) $Ph_3P{=}CH_2$, THF; vi: (1) $LiAlH_4$, THF, (2) TBDPSCl, imidazole, DMF, (3) PCC, CH_2Cl_2, (4) Zn/CH_2Br_2, $TiCl_4$, CH_2Cl_2, (5) *m*CPBA, $NaHCO_3$, CH_2Cl_2, (6) $LiAlH_4$, THF and (7) PCC, CH_2Cl_2.

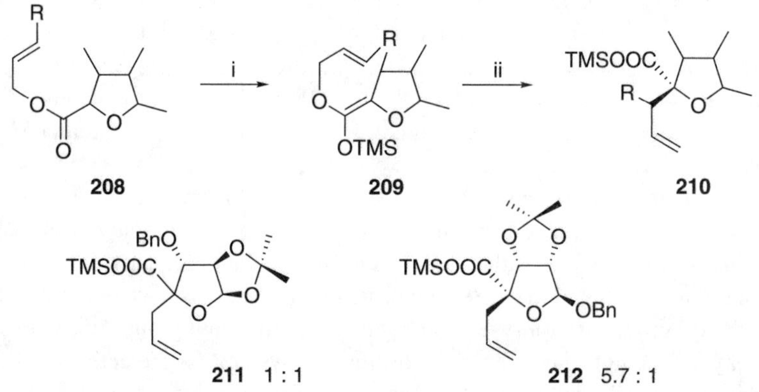

SCHEME 11.47 Reagents: i: LDA, TMSCl, HMPA, − 100°C; ii: room temperature (if TMSOOC).

acetal **209** rearranges upon warming to give **210** together with some elimination product. This reaction was performed on a glucose derivative giving the rearranged compound **211** as a 1.5:1.0 mixture of epimers at C4. In the mannose series the reaction gave **212** as the major product (5.7:1.0 mixture). The latter was further extended to an advanced intermediary *en route* to zaragozic acid. An analogous strategy was recently applied to the synthesis of (−)-cinatrin B, an inhibitor of phospholipase found in *Circinotrichum falcatisporum* [143].

It is worth noting that the net result of this [3,3]-sigmatropic rearrangement is an intramolecular enolate allylation. The process described above may be compared with the intermolecular aldol reaction developed by Fraisse et al. [122]. Wittig rearrangements have been successfully used in the construction of the quaternary carbon of zaragozic acid [144,145]. Thus, condensation of an acetylenic carbanion with a lactone, followed by *O*-glycosylation with a suitable allylic alcohol, gave the required unsaturated system which, on treatment with butyllithium, rearranged to the *C*-glycoside.

As illustrated in this section, a number of methods are available allowing access to branched-chain sugar derivatives with new substituents at different positions of the sugar skeleton. The enormous advantage of these methods is the high stereocontrol provided by the template effect, although it appears limited to a few carbons. This drawback can be circumvented by applying the concept of pyranosidic homologation introduced by Fraser-Reid in the 1980s [146]. This strategy consists of constructing a second pyranose ring fused to the first one, suitable for making new branching. Final opening of the system would afford a highly functionalized long chain carbohydrate. A synthesis of the chain of streptovaricin containing nine contiguous chiral centers was reported according to this strategy [147], with various methods for such pyranosidic homologations available [148–151].

11.4 CHAIN EXTENSIONS OF SUGARS

It is clear that chain extensions of carbohydrate skeletons are crucial in the construction of many important targets. Convergent syntheses rely on the assembly of small building blocks. Examination of a few reported total syntheses clearly shows that five to eight carbon building blocks are frequently used. In this sense, sugars are suitable starting compounds. Nevertheless extensions of the carbon chain, either at C1 or at C6 (C5), may contribute to the enlargement of the building block by a few supplementary atoms. However, the assembly of building blocks requires them to be suitably arranged to create new carbon–carbon bonds. This prompted many studies aimed at the creation of carbon–carbon bonds at the primary carbon atom of sugars. The atoms of these newly generated bonds may contain chiral centers.

Some methods dealing with chain extensions are reviewed in this section. Additional methods dealing with extensions at C1 are then detailed. Selected methods designed for the formation of *C*-glycosidic bonds, that is, for the formation of tetrahydro-pyrans or furan rings, are described in Section 11.5.

11.4.1 Chain Extensions at the Primary Carbon Atom

In principle, chain extensions at C6 of hexoses or at C5 of pentoses seem straightforward because the C6 and C5 hydroxyl groups are easily isolated for reactions. The main problems associated with this chemistry remain in the concomitant creation of chiral centers. Three groups of reactions can be proposed, based on the reaction type involved for the formation of the carbon–carbon bond. The first reaction type is based on the addition of suitable nucleophiles to an aldehyde. From the same aldehyde it is also possible to form olefins using Wittig type reactions. Finally, one can take advantage of the reactivity at the primary carbon atoms by substitution of suitable leaving groups with carbon nucleophiles.

11.4.1.1 Using Nucleophilic Additions to Aldehydes

The nucleophilic addition of a carbanion to a carbohydrate aldehyde group is a well-established method for the creation of carbon–carbon bonds with the formation of new chiral centers at C6. Good to excellent stereocontrol is always observed, possibly because of chelation control with the ring oxygen [152]. This observation was exploited in the construction of higher carbohydrates from hexoses. For example, useful syntheses of LD-Hep*p*, a key component of the core oligosaccharide of LPS gram negative bacteria, from D-mannose were reported [153].

Let us focus on an example of chain extension at C6 of a dideoxy D-glucose derivative reported by two groups in connection with their FK506 total synthesis programs. The first, report by Rama Rao [154], described the elaboration of aldehyde **213** obtained from glucose according to a known procedure [101]. As illustrated in Scheme 11.48, condensation of this aldehyde with the chiral Grignard reagent prepared from **214** in ether at room temperature gave **215** as a single isomer. The configuration at the newly created chiral center is explained by the formation of a five-membered chelate between magnesium and the oxygen atoms of the aldehyde and the sugar ring. The *Re* face of the aldehyde is thus shielded by the sugar ring.

A single diastereoisomer was also obtained by Ireland in the same context but using a chiral lithium derivative obtained from **216** by metal–halogen exchange [155]. However, the presence of magnesium bromide is needed to achieve up to 20:1 diastereoselectivity.

A variant of this process was used to achieve a chain extension at C5 of a pentose derivative, allowing the construction of the oxahydrinden part of avermectins [59,156]. As shown in Scheme 11.49, the D-ribose derived aldehyde **217** was treated with (*E*)-trimethylcrotylsilane in the presence of BF_3–Et_2O [157]. This produced a mixture of three isomers in an 8.9:1.1:1.0 ratio, from which compound **218** was isolated in a yield of 78%. The subsequent steps allowed deoxygenation of the sugar ring to give the diol **219**. This compound was, in turn, transformed into the

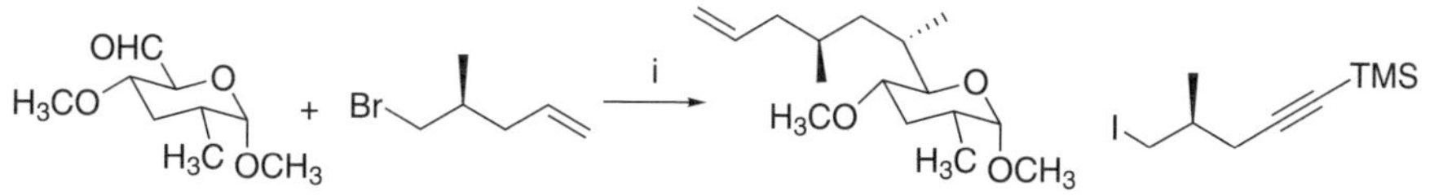

OHC H3CO O H3C OCH3 + Br i H3CO O H3C OCH3 I TMS

213 214 215 216

SCHEME 11.48 Reagent: i: Mg, Et_2O.

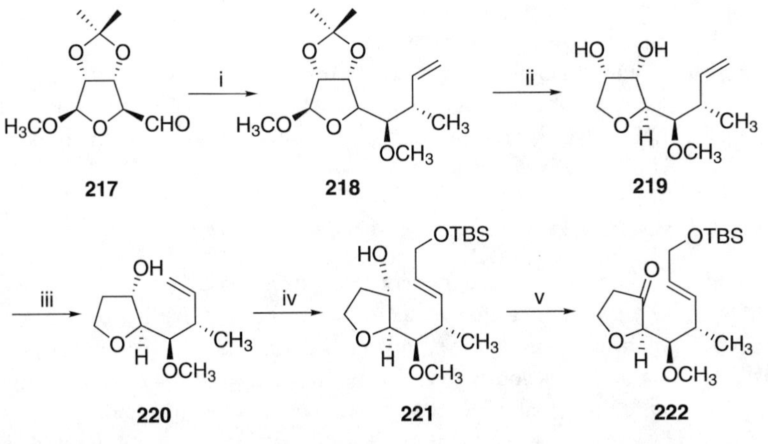

SCHEME 11.49 Reagents: i: trimethylcrotylsilane then NaH, MeI; ii: (1) HCl, MeOH and (2) BF_3–Et_2O, Et_3SiH, CH_2Cl_2; iii: (1) $(CH_3)_2C(OAc)COBr$, CH_2Cl_2, (2) Amberlite IRA 400(OH^-), MeOH and (3) $LiBHEt_3$; iv: (1) O_3 then Zn–AcOH, (2) Ph_3P=$CHCO_2Me$ then DIBAL and iii. TBSCl; v: PCC.

corresponding epoxide. Opening of this epoxide allowed deoxygenation at C2 and afforded **220**. The latter was homologated using ozonolysis, Wittig olefination, reduction with DIBAH and protection of the resulting allylic alcohol, giving **221**. Final oxidation gave the desired ketone **222** ready for aldol reaction with the northern part of avermectin.

It is of interest to compare this route to that described by Williams in the synthesis of the same type of compound from 1,4-anhydro-sorbitol [158]. As shown in Scheme 11.50, an eight step protection sequence from this starting compound gave the required aldehyde **223**, already encompassing the C5 chiral carbon atom of avermectin. Elaboration of this aldehyde to the ketone **224** was achieved by a methyl Grignard addition and oxidation sequence. Not unexpectedly, Grignard addition of 3-butenylmagnesium bromide to the carbonyl proceeded with a modest 3:1 diastereoselectivity. In contrast to the preceding example, the presence of a methoxy group α to the carbonyl group is not sufficient to completely control the stereochemical outcome. The major isomer was transformed into lactone **225** and then into ketone **226**. The next crucial step was the six-membered ring formation by intramolecular aldol reaction to give **227**. The chirality present at C4

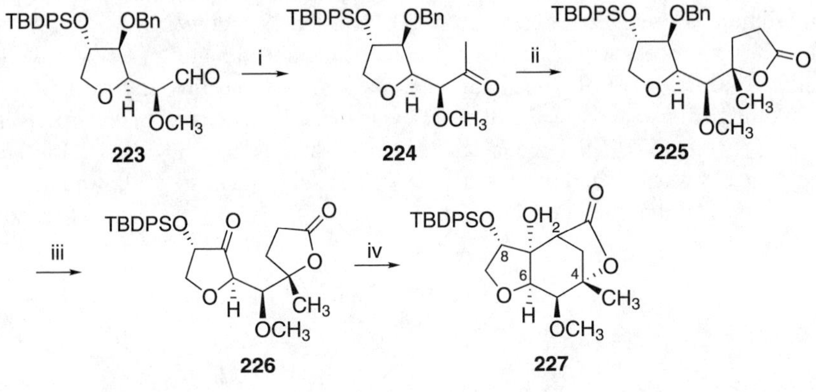

SCHEME 11.50 Reagents: i: (1) MeMgCl, THF and (2) PCC, Al_2O_3; ii: (1) CH_2=$CHCH_2CH_2MgBr$, THF, (2) O_3, CH_2Cl_2, pyridine then Ph_3P and (3) PCC, Al_2O_3; iii: (1) H_2, Pd black, EtOH and (2) PCC, Al_2O_3; iv: LDA, THF.

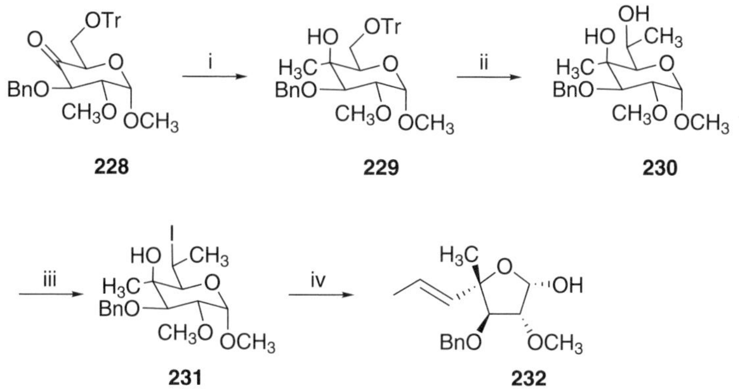

SCHEME 11.51 Reagents: i: Me_2CuLi, MeLi, Et_2O; ii: (1) CSA, MeOH, (2) SO_3, pyridine, DMSO and (3) MeMgBr, THF; iii: (1) $MeSO_2Cl$, pyridine and (2) Bu_4NI, benzene; iv: Zn, EtOH.

controls that transferred to C2, the chirality at C7 being dictated by the presence of the *O*-silyl group at C8. This was clear from the reaction of the other epimer at C4 obtained in the Grignard addition.

Another example of Grignard additions to C6 aldehyde groups is found in Kallmerten's approach to the C11 to C18 part of herbimycin, a benzoquinoid macrocyclic lactam (Scheme 11.51) [159]. The first interesting step of this synthesis is the equatorial addition of lithium dimethylcopper–methylithium combination on ulose **228,** giving **229** in excellent yield and diastereoselectivity. The other Grignard addition was performed with equally high stereocontrol on the C6 aldehyde derived from **229** to give **230** in good yield. This compound was transformed into the corresponding iodo derivative **231**. The latter, treated under Vasella conditions [31], was opened to furnish the expected aldehyde, which was cyclized to furanose **232**. Subsequent steps of this synthesis are also of interest and provide solutions to the manipulation of chiral linear systems using Wittig rearrangements. A synthesis of herbimycin was also reported by Tatsuta [160].

An alternate example using the techniques described above was reported by the same group in the synthesis of an advanced intermediate of rapamycin [45]. As shown in Scheme 11.52, aldehyde **233** was prepared, via a multistep sequence, from the well-known 2-*C*-methyl glucose derivative **7**. It was treated with propynyl Grignard reagent, and the resulting product was oxidized to the

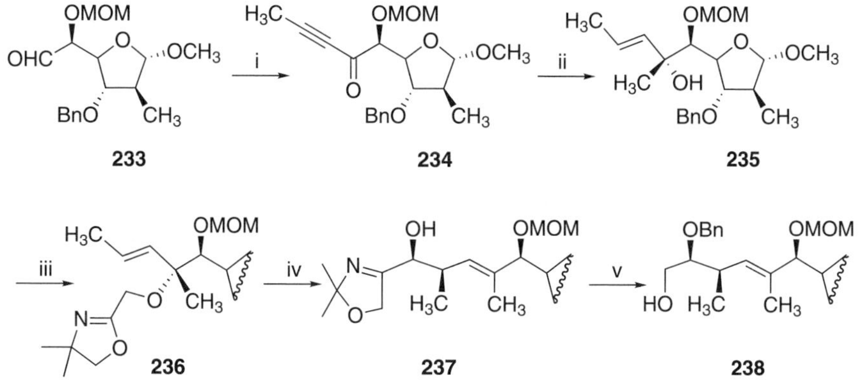

SCHEME 11.52 Reagents: i: (1) MeC≡CMgBr, THF and (2) $(COCl)_2$, DMSO, Et_3N, CH_2Cl_2; ii: (1) MeMgBr, THF and (2) $LiAlH_4$, THF; iii: ICH, chloromethyloxazoline, DME; iv: BuLi, THF; v: (1) KF, BnBr, DME and (2) TFA, H_2O then $LiAlH_4$.

ketone **234**. A *chelation controlled* methyl Grignard addition on the carbonyl group gave the tertiary alcohol as a single diastereoisomer. Final reduction of the triple bond gave **235**. Because of the close structural similarities, it is interesting to compare this Grignard addition to **234** with that carried out on **224**, which was considerably less selective and gave the opposite attack. The substrate clearly differed by the orientation of the *O*-benzyl group at C3 and the ketone substituent. Chelation between this oxygen and that of the carbonyl group could play a role as large ring chelates have been suggested to explain the stereochemical outcome of Grignard additions [161,162].

Alkylation of the tertiary alcohol of **235** with a chloromethyloxazolidine gave the corresponding ether **236**, which was treated with *n*BuLi at low temperature to promote the [2,3]-sigmatropic rearrangement to **237**, isolated as a 4.5:1.0 mixture of epimers. Protection of the hydroxyl group and oxazolidine reduction gave the primary alcohol **238**, which was separated from its epimer. This compound, representing the C23 to C32 subunit of rapamycin, was again submitted to chain extension using a similar rearrangement.

Dithiane-derived carbanions are also of interest for the chain elongation of sugars. These nucleophiles are useful in epoxide opening reactions but also react with aldehydes. This approach was well illustrated by Paulsen's and by Redlich's groups. A typical example can be found in the synthesis of talaromycin, which proceeds by elongation at C6 of glucofuranose using a lithiodithiane condensation on the C6 aldehyde. Oxidation of the resulting alcohol and Wittig olefination gave the methylene derivative, hydroboration of which gave the primary alcohol [163]. The reactivity of aldehydes in hetero-Diels–Alder reaction has been exploited to extend the carbon skeleton of sugars. A good example could be found in the total synthesis of a complex nucleoside, octosyl acid [164]. The same compound has also been synthesized using a reverse approach, which involved the formation of a 5-nitro derivative at C5′ of a nucleoside. In this example, carbanion formation and condensation with D-glyceraldehyde gave the hydroxy nitro compound. This compound underwent a Nef reaction to form the keto derivative [165]. Another strategy was developed by Knapp to reach this type of compound by the stepwise modification of anomeric cyanide [166].

11.4.1.2 Using Wittig Olefinations

As alternatives to nucleophilic additions to C6 or C5 aldehydes, olefinations of the same compounds offer a repertoire of solutions to chain extensions or to the coupling of two carbohydrate building blocks. As shown in the above section, chain extensions at C6 are often followed by the introduction of new functional groups and thus new chiral centers. In this regard, olefins obtained via Wittig-type reactions are excellent starting compounds.

A wonderful example of such olefin functionalizations is given in the total synthesis of palytoxin. This accomplishment by Kishi's group is certainly a monument in organic synthesis [167,168]. The enormous accumulation of chiral centers (64 + 7 double bonds) together with the number of heterocycles constitutes probably one of the most challenging synthetic problems resolved so far. Palytoxin consists of four highly oxygenated tetrahydropyrans tethered to each other by four to ten carbon links containing hydroxyl groups [169]. These heterocycles look like *C*-glycosyl derivatives and chain extensions at C6, and *C*-glycosylation methods have been developed for the circumstance. In addition, formation of carbon–carbon bonds between vinyl iodides and aldehydes, the Kishi–Nozaki method, proved fruitful. In this section, we will focus only on the dihydroxylation of sugar-derived olefins, which was thoroughly studied by Kishi's group in the frame of this total synthesis [170]. *C*-Glycosylation methods will be described in Section 11.5.

An example of chain elongation developed in palytoxin synthesis is given below. As shown in Scheme 11.53, compound **239** was reacted with the chiral phosphorane **240** to furnish the *Z* isomer

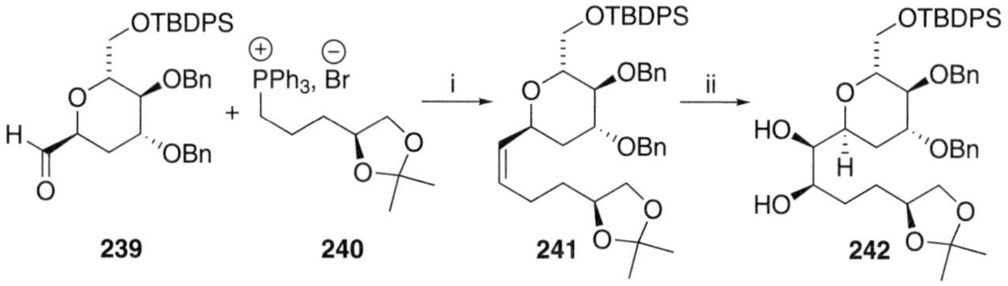

SCHEME 11.53 Reagents: i: BuLi; ii: OsO_4, NMO, *t*BuOH.

241 as the major compound. Dihydroxylation of the double bond of **241** proceeded with a good selectivity (7:1) in favor of diol **242**.

Double bond dihydroxylation was also developed by Brimacombe's group for the synthesis of long chain sugars [171]. All dihydroxylations were carried out with osmium tetraoxide as a catalytic or stoichiometric reagent. The stereochemical outcome of these reactions is controlled by an already existing chiral center close to the double bond. Kishi's empirical rules have been proposed, which stipulate that the relative stereochemistry between the pre-existing hydroxyl or alkoxy group at the adjacent stereocenter and the newly introduced hydroxyl group of the major product will be *erythro* [172].

The existence of natural complex nucleoside antibiotics has prompted interest in their syntheses [173]. These compounds are composed of a C5′ extended ribonucleoside. This extra chain of two or more carbons carries chiral centers. Some of these nucleosides, like tunicamycins or octosyl acids, can be regarded as *C*-linked disaccharides and their syntheses are not straightforward. Liposidomycin constitutes another class of complex nucleosides, which has been partially synthesized using a Wittig chain extension at C5 of the furanose derivative **243** [174].

As shown in Scheme 11.54, the synthesis began with di-*O*-isopropylidene-D-allose **243**. Protection at C3 and removal of the 5,6-*O*-isopropylidene group gave a diol subsequently cleaved with sodium periodate. The resulting aldehyde was olefinated to give **244** in an overall yield of 66%. The key of the synthesis is the construction of the amino acid moiety. This was realized by reduction of the ester function of **244** to give the corresponding allylic alcohol. Sharpless

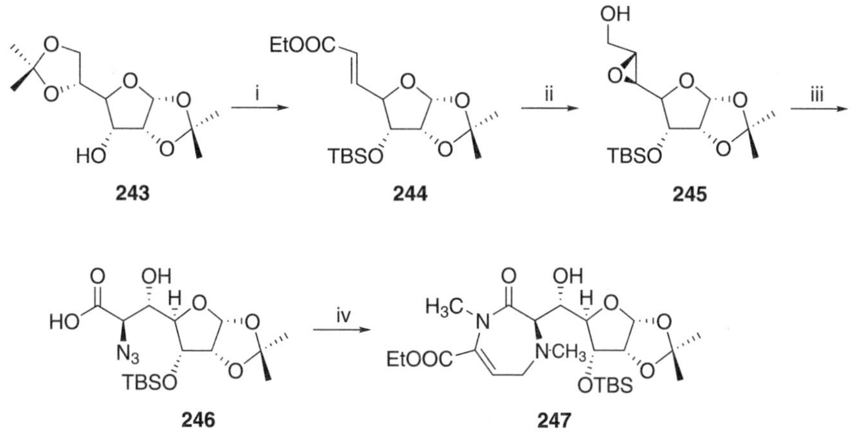

SCHEME 11.54 Reagents: i: (1) TBSCl, (2) AcOH, H_2O, (3) $NaIO_4$, H_2O, THF and (4) $Ph_3P{=}CHCO_2Me$, CH_2Cl_2; ii: (1) DIBAL, THF and (2) D-DIET, $Ti(OiPr)_4$, *t*BuOOH; iii: (1) $RuCl_3$, $NaIO_4$ and (2) NaN_3, MeOH; iv: (1) EEDQ, $CH_2CHCH(O_2CPh)CH(CO_2Et)NHCH_3$ and (2) O_3 then Ph_3P then $NaHB(OAc)_3$.

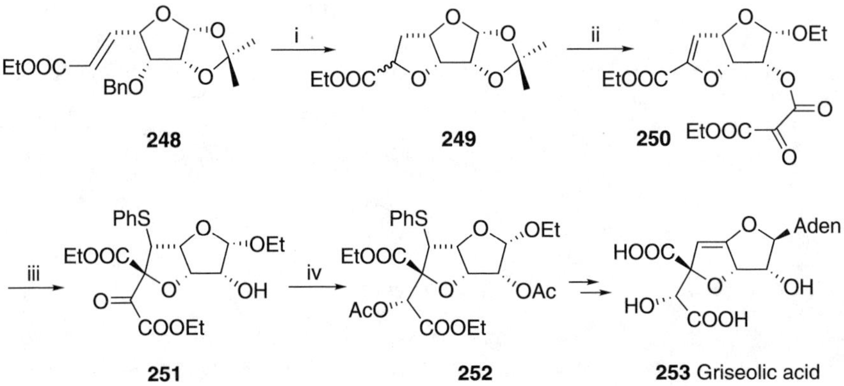

SCHEME 11.55 Reagents: i: (1) H_2, 20% Pd $(OH)_2$ and (2) LDA, Br_2; ii: (1) LDA, $PhSO_2SPh$, 85%, (2) H_2SO_4, EtOH, 90%, (3) *m*CPBA, 77% and (4) EtO_2CCOCl, pyridine 95%; iii: PhS^-Na^+, 65%; iv: (1) Ac_2O, pyridine, 95%, (2) $NaBH_4$, 72% and (3) Ac_2O, pyridine, 97%.

epoxidation mediated by (−)-diethyl tartarate gave the 5*S*,6*R* epoxy alcohol **245** as a single isomer. Oxidation of the primary alcohol and treatment of the epoxy acid with sodium azide secured the introduction of the azido function with the correct regio and stereochemistry. The remainder of the synthesis was the construction of the diazepanone from the amino acid moiety. The absolute configuration at C5 and C6 of the compound obtained this way did not match with the natural product. The same synthesis was carried out using the *Z* isomer of **244**, giving the 5*S*,6*S* isomer of **246**. Elaboration of the heterocycle of **247** allowed comparison with the natural product and thus established the correct stereostructure of the liposidomycins.

The total synthesis of griseolic acid **253** in the field of complex nucleosides will now be described [175]. This compound has been derived from D-glucose by successive chain extensions at C6 to construct the eight carbon atoms chain of griseolic acid. The synthesis, shown in Scheme 11.55, commenced with compound **248** derived from diacetone-D-glucose in three steps [176]. This compound was then transformed into the anhydro derivative **249**. Introduction of a C6–C7 double bond was then carried out by sulfenylation of the ester enolate. Replacement of the isopropylidene group by an ethyl glycoside, followed by sulfur oxidation and elimination, gave the corresponding alcohol. Esterification at O2 with ethyl oxalyl chloride led to **250**. Thiolate addition on the double bond generated the ester enolate, which reacted with the oxalate to produce **251**. Reduction of the α-keto ester gave mainly the protected alcohol as its acetate **252**. From this compound, griseolic acid was obtained by sequential introduction of the adenine moiety via a Vorbrüggen reaction, and sulfoxide elimination to create the enol ether bond. Final deprotection gave griseolic acid **253**.

Although the introduction of hydroxyl groups is well developed, the formation of a carbon–carbon bond along the chain extension is not straightforward. An efficient way to construct such a chiral center at C6 of a carbohydrate, using unsaturated sugars, was proposed by Weiler in his synthesis of the C2 to C15 fragment **257** of ionomycin [177]. The principle of this reaction is summarized below and illustrated in Scheme 11.56. As shown, a cyclic lactone is constructed by an intramolecular Wadsworth–Emmons reaction using a phosphonate grafted at O4, as in compound **254**. Obviously, the reaction gave only the *Z* isomer **255**. Given the stereoelectronic control of the cuprate 1,4-addition, it was very easy to introduce a methyl group at C6 with high stereocontrol, in favor of the axial isomer, to produce **256**. This procedure may also be of interest for introducing heteroatoms.

A number of syntheses make use of olefination reactions to connect two building blocks. In this regard, the use of C6 aldehyde groups has been widely developed. A good example of this strategy

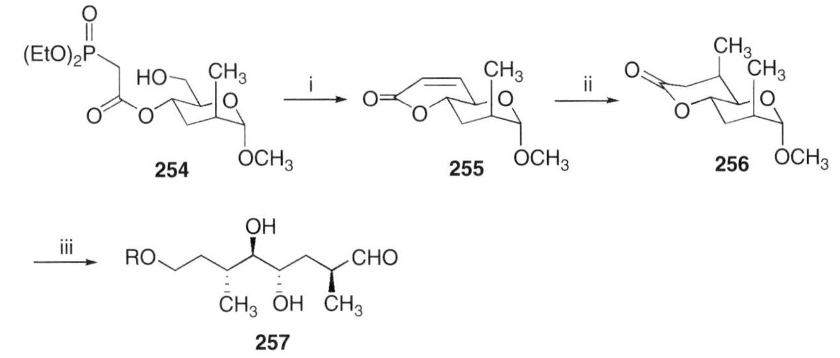

SCHEME 11.56 Reagents: i: (1) DMSO, DCC, $Cl_2CHCOOH$ and (2) NaH, THF; ii: $Me_2Cu(CN)Li$; iii: (1) DIBAL, Et_2O, (2) PivCl, pyridine, (3) $(COCl)_2$, DMSO, Et_3N, CH_2Cl_2 and (4) Ph_3PRhCl_3, H_2, benzene.

is given below with the synthesis of nafuredin, a polyene lactone isolated from the fermentation broth of *Aspergillus niger* [178]. This potential antihelmintic compound was prepared from glucose by two sequential double bond formations via Wittig and Julia reactions for the grafting of the polyunsaturated chain.

As shown in Scheme 11.57, a multistep sequence gave 4-*C*-methyl branched-chain derivative **258**. Debenzylation followed by sequential protection and deprotection liberated the primary hydroxyl group of **259**. Dess–Martin oxidation and Horner olefination gave the unsaturated ester **260**. Reduction of the ester function gave the corresponding alcohol, which reacted under Mitsunobu conditions with a thiol to give, after sulfur oxidation, the sulfone **261**. Deprotonation of this compound with KHMDS and condensation with the aldehyde representing the C9 to C18 part of the target gave **262** as the single *E* isomer in a yield of 79%. The last steps were dedicated to the formation of the epoxide and the lactone function of the target compound narefudin.

The reverse strategy, which consists of using a carbohydrate derived phosphorane, is seldom used because some elimination problems arise during the generation of sugar-derived phosphoranes. One of the first syntheses of such compounds was reported by Secrist [179].

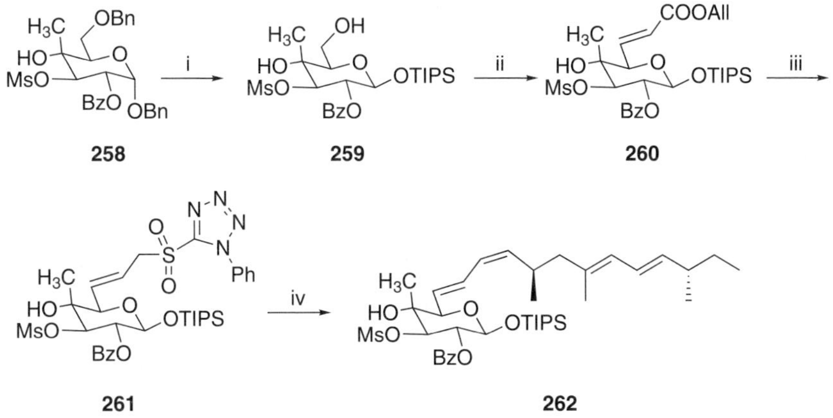

SCHEME 11.57 Reagents: i: (1) H_2, $Pd(OH)_2$, EtOH, (2) TBSCl, *i*Pr_2NEt, DMF and (3) TIPSOTf, 2,6-lutidine, CH_2Cl_2; ii: (1) TBAF, BF_3–Et_2O, CH_3CN, (2) Dess–Martin periodinane, CH_2Cl_2 and (3) $(EtO)_2P(O)CH_2CO_2$allyl, *i*Pr_2NEt, LiCl, CH_3CN; iii: (1) $(Ph_3P)_4Pd$, $NaBH_4$, Et_2O, (2) MeO_2CCl, Et_3N, THF then LiAlH(*t*BuO)$_3$, (3) Bu_3P, DEAD, 1-phenyl-1*H*-tetrazole-5-thiol, THF and (4) H_2O_2, cat. $Mo_7O_{24}(NH_4)_6$–4 H_2O, EtOH; iv: KHMDS, THF then **262**.

This methodology is well suited for the synthesis of long-chain sugars and has been applied for the creation of the C98–C99 double bond of palytoxin [170,180].

A synthetic approach toward tunicamycin **267**, based on this principle, has been reported. Tunicamycin shows a direct carbon link between C6 of a galactosamine residue and C5 of a uridine moiety. The formation of this link has been carried out by Wittig reactions on model compounds using Secrist's phosphorane [181]. As shown in Scheme 11.58, the phosphonium salt **263** was treated with lithium hexamethyldisilazane to generate the phosphorane, which was reacted with aldehyde **264**. Reduction of the double bond and benzyl hydrogenolysis of **265** was followed by acetylation to provide the model compound **266**.

A total synthesis of anamarine involves the coupling of a phosphorane located at C6 of an unsaturated δ-lactone obtained from D-glucose with a chiral aldehyde [182]. The coupling of sugar aldehydes with C6 sugar phosphoranes or phosphonates was reported by Jarosz [183–185]. The construction of oligosaccharide mimics using an iterative Wittig olefination between a glycosyl aldehyde, prepared using the *thiazole methodology*, and a glycosyl phosphorane, derived from glucose, has been thoroughly studied by Marra and Dondoni [186,187].

11.4.1.3 Direct Substitutions at the Primary Carbon Atom

Direct formation of carbon–carbon bonds at primary carbon atoms has been seldom exploited in carbohydrate chemistry. The basic principle is the use of powerful leaving groups at C6 that could be displaced by a suitable carbanion. The major drawback is the possible elimination resulting from H5 abstraction by the carbanion. Nevertheless, this approach proved successful using cuprates or Grignard reagents in the presence of copper salts, and tosylates were good enough to undergo such displacements [188]. Recently, the synthesis of the simple bioactive lactone 2,3-dihydroxy-tetradecan-5-olide from D-glucose was reported, based on the displacement of the tosylate group of methyl 2,3-di-*O*-benzyl-6-*O*-tosyl-α-D-glucopyranoside with octylmagnesium bromide in the presence of copper (I) iodide in a yield of 77% [189]. A review on this type of reaction has been published [190].

Two key steps are involved in a recent synthesis of dysiherbaine, a neuroexcitotoxin from the Micronesian marine sponge *Dysidea herbacea*: the formation of a carbon–carbon bond at C6 of a sugar and further manipulation of the chain introduced this way [191]. As shown in Scheme 11.59, the known epoxide **268** was used to introduce an azide group at C3 to give **269**. Opening of the 1,6-anhydro bridge and reduction of the hemiacetal group gave, after protection, the diacetate **270**.

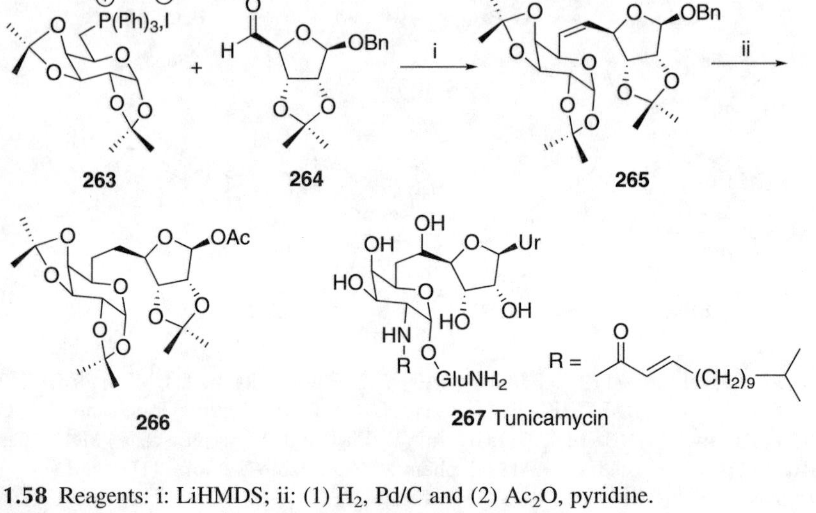

SCHEME 11.58 Reagents: i: LiHMDS; ii: (1) H_2, Pd/C and (2) Ac_2O, pyridine.

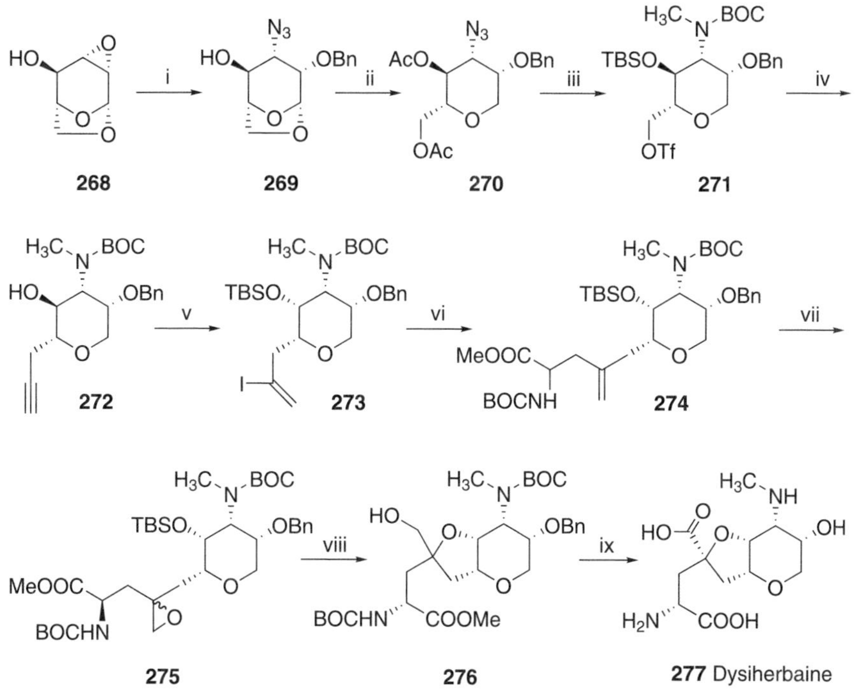

SCHEME 11.59 Reagents: i: (1) NaH, BnBr, DMF and (2) NaN_3, NH_4Cl, $MeOCH_2CH_2OH$, H_2O; ii: (1) BF_3-Et_2O, then Ac_2O, 84% and (2) Et_3SiH, TMSOTf, BF_3-Et_2O, CH_2Cl_2, CH_3CN; iii: (1) MeONa, MeOH, (2) TBSOTf, 2,6-lutidine, CH_2Cl_2, (3) Ph_3P, THF, then H_2O, (4) BOC_2O, Et_3N, CH_2Cl_2, (5) NaH, MeI, DMF, (6) CSA, CH_2Cl_2, MeOH and (7) Tf_2O, 2,6-lutidine, CH_2Cl_2; iv: (1) HC≡C–$SiMe_3$, BuLi, THF–HMPA and (2) TBAF, THF; v: (1) $(COCl)_2$, DMSO, $i Pr_2NEt$, CH_2Cl_2, (2) $NaBH_4$, THF, MeOH, (3) TBSOTf, 2,6-lutidine, CH_2Cl_2 and (4) β-iodo-9-BBN, pentane, then AcOH; vi: $PdCl_2(PPh_3)_2$, $MeO_2CCH(NHBOC)CH_2ZnI$, THF, DMA; vii: (1) TBAF, THF, (2) BOC_2O, Et_3N, CH_2Cl_2 and (3) *m*CPBA, CH_2Cl_2; viii: (1) CSA, CH_2Cl_2, (2) 1*N* NaOH, THF, (3) TPAP, NMO, CH_3CN and (4) $TMSCHN_2$, MeOH; ix: (1) H_2, Pd/C, MeOH and (2) 6*N* HCl.

Protecting group manipulations allowed the introduction of a triflyl group at O6 to yield **271**. The latter was treated with the trimethylsilylacetylene anion to give **272** in a yield of 76% over the two steps.

Inversion of configuration was then achieved via an oxidation and reduction sequence. The vinylic iodide **273** was prepared from the triple bond and subsequent chain extension was carried out in fair yield using an organozinc derivative obtained from iodoalanine under palladium catalysis. The last chemical manipulation of the chain was the double bond epoxidation, which proceeded without any selectivity to provide **275**. Subsequent ring closure gave **276** and adjustment of oxidation levels gave dysiherbaine **277**. Incidentally, the C4 epimer of the natural product was devoid of biological activity. A shorter synthesis was reported by Snider using a *C*-allyl glycosylation strategy [192].

The use of a direct carbon–carbon bond formation allowing further coupling reactions has been reported by Salomon [193] in the total synthesis of halicondrin B. As shown in Scheme 11.60, the iodo derivative **278** treated with α-lithioacetonylidene-triphenylphosphorane gave the stabilized phosphorane **279** in good yield. The latter was then ready to be coupled with aldehyde **280**, thus achieving the coupling of two building blocks, a key reaction in total synthesis of large molecules, and giving **281**. Cleavage of the glycosidic bond and hydrolysis of the acetals, followed by Triton B-catalyzed intramolecular 1,4-addition secured heterocycle formation and subsequent acetylation, gave the protected **282**, representing a large part of the A–E ring system of the target.

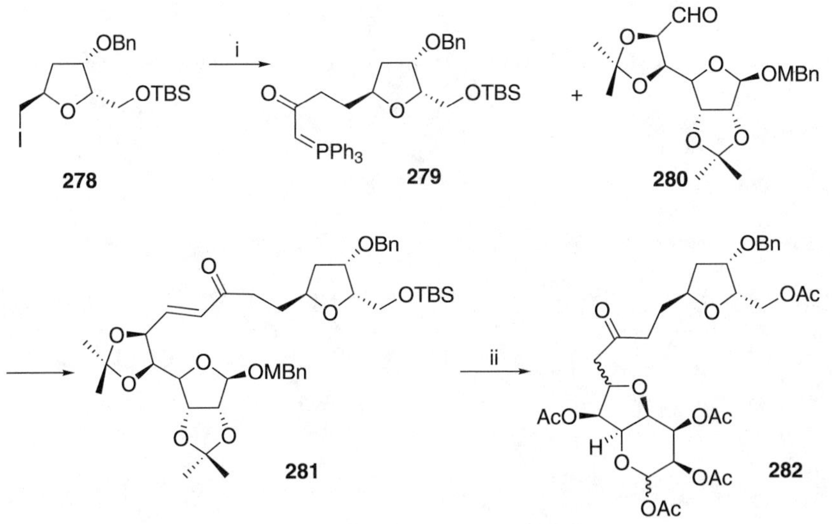

SCHEME 11.60 Reagents: i: lithioacetonylidene triphenyl phosphorane; ii: (1) CAN, H_2O, 87%, (2) AcOH, H_2O, H:1, (3) Triton B and (4) Ac_2O, pyridine.

It should be noted that direct substitution of sulfonate or iodide derivatives at C6 is sensitive to steric hindrance. Kishi reported a failure in the substitution of a primary tosylate by vinyl cuprate in the synthesis of mycalamides where C4 had two geminal methyl substituents. An alternate five step route was developed to introduce the vinyl substituent [119]. Direct substitution of a primary triflate by an aryl Grignard reagent is a key step in the formal synthesis of (+)-apicularen A from D-glucal [194]. Displacement of a triflate with an alkynyllithium has also been used in the total synthesis of (+)-panaxacol [195].

Radical reactions are well known for their mildness and the compatibility of the reaction conditions with a large number of functional and protecting groups present on carbohydrate building blocks. Since the recognition of the practical use of radical reactions for the formation of carbon–carbon bonds in the late eighties, many methodologies and some total syntheses make use of them [196]. A beautiful example of an efficient use of a radical reaction to perform chain extensions at C6 of a carbohydrate, with concomitant formation of a chiral center, is given here.

The total synthesis of tunicamycin V and its 5′-epimer, reported by Myers [197,198], started with the 5,6-unsaturated disaccharide **283** prepared by C6-selenoxide elimination (Scheme 11.61). The other partner was obtained by treatment of aldehyde **284** with benzeneselenol in pyridine and the resulting adduct was treated with an excess of dimethyldichlorosilane. Coupling with **283** achieved the construction of the key intermediate **285**. Radical cyclization was performed in the presence of tributyltin hydride using triethylborane as an initiator to give **286**. The stereochemistry at C5′ of the nucleoside was found to strongly depend on the nature of the 3′ protecting group. Actually, the absence of protecting groups efficiently favored the correct stereochemistry (*R*/*S* 7.5:1.0). A second chiral center was formed on the galactose residue with high stereocontrol because the intermediate radical is preferably trapped in an axial orientation. As already mentioned, only radical reactions would accommodate the presence of free hydroxyl groups (and many other sensitive groups) during carbon–carbon bond formation. This type of temporary silyl ketal tether has found many applications, especially in the field of *C*-disaccharides [199–202]. It has also been used in a synthetic approach to herbicidin [203].

The possible use of organocuprate reactions with allylic esters as a way to create carbon–carbon bonds at C6 positions of pyranoses has already been mentioned [55]. This has been applied

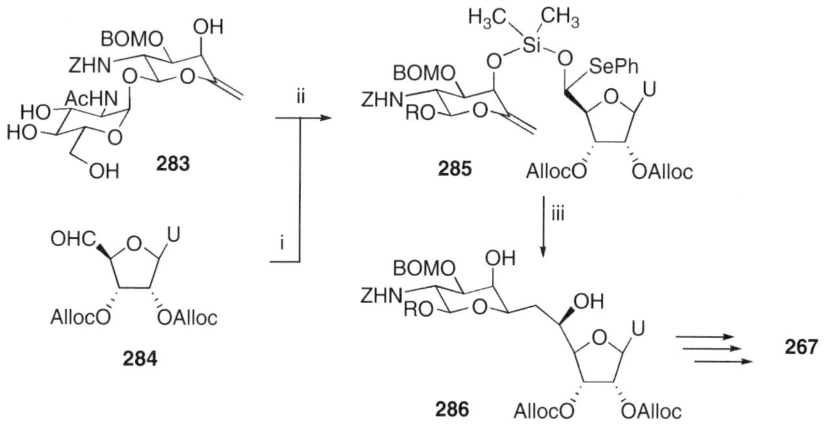

SCHEME 11.61 Reagents: i: (1) PhSeH, pyridine; ii: $(CH_3)_2SiCl_2$; iii: (1) $Cl_2Pd(PPh_3)_2$, Bu_3SnH, 85%, (2) Et_3B, Bu_3SnH and (3) KF, H_2O, MeOH.

in our group for the synthesis of mevinic acid analogs [57]. Additionally, the application of this methodology to a formal synthesis of the antifungal metabolite isoavenaciolide was reported by Dugger [58].

As shown in Scheme 11.62, the protected ribose derivative **287** was treated with sodium iodide and the reaction of the resulting 5-iodo derivative with DBU promoted elimination of hydroiodic acid, giving **288**. This allylic carbonate underwent S_N2' reaction, on treatment with heptyllithium cyanocuprate, to provide **289** in a yield of 85%. As already demonstrated by our group, only monoalkylcyanocuprate was efficient [55]. The last step of the synthesis relies on branching at C3 of the furanose skeleton. Bromoacetal **290**, prepared from **289** using standard procedures, was treated with tributyltin hydride in benzene to give the expected *cis*-fused five-membered ring **291**. Obviously, radical attack of the double bond occurred from the α face to form an intermediate radical located at C4, which reacted with tributyltin hydride from the β face, opposite to the C3 substituent. Acetal deprotection and final oxidation of the two hemiacetal functions gave **292**, a known precursor of isoavenaciolide **293**.

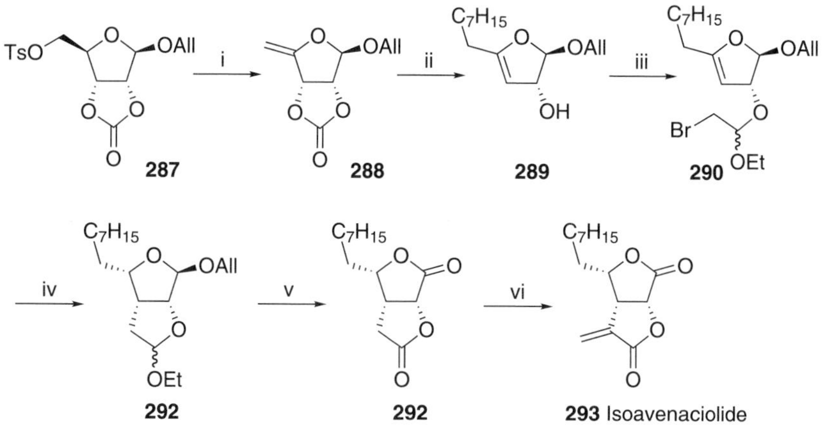

SCHEME 11.62 Reagents: i: (1) NaI, DMF, (2) DBU, DMF and (3) $CO(Im)_2$, DMF; ii: $C_7H_{15}CuCNLi$, Et_2O; iii: $BrCH_2CH(OEt)Br$, Et_3N, CH_2Cl_2; iv: Bu_3SnH, AIBN, benzene; v: (1) $[COD]Ir(Ph_2MeP)_2PF_6$ and (2) aq. H_2SO_4; vi: CrO_3 pyridine.

11.4.2 CHAIN EXTENSIONS AT THE ANOMERIC CENTER

There are two ways to extend the sugar skeleton at C1. The first way, which will be summarized in this section, uses the aldehyde group, or an equivalent, for the formation of a carbon–carbon bond and results in the formation of a linear sugar. In the second way, the new bond is formed at the anomeric center, exploiting its peculiar reactivity although the heterocyclic nature of the sugar is retained. This second method, by far more investigated, will be the subject of Section 11.5.

Although they are quite similar in their results, chain extensions at the sugar reducing end are less used than C6 extensions, for obvious reasons. First, the aldehyde group is mostly involved in a furanose or pyranose ring, thus masking its reactivity as a carbonyl group. Second, reactions performed on linear sugars are more difficult to control in terms of diastereoselectivity. Third, because extensions at C6 are preferred for the preceding reasons, syntheses are often designed to favor this strategy versus C1 extensions. Nevertheless, some examples can be found in total syntheses using sugars as starting materials.

11.4.2.1 Using Dithioacetals

An example of the use of dithioacetals to perform C1 chain extensions can be found in Rama Rao's approach to the synthesis of FK506 [154,204]. One of the most challenging problems of the FK506 synthesis is the existence of a 1,2,3-tricarbonyl function at C8–C10 of the target. Dithioacetals are easily formed from reducing sugars under the action of thiols or, more advantageously, 1,3-dithiols. The easy deprotonation of dithianes by strong bases is well known and thus opens the way for the alkylation of sugars at C1.

Thus, as shown in Scheme 11.63, compound **215** was treated with propanedithiol in the presence of boron trifluoride etherate at 0°C to give dithiane **294** in 80% yield. Treatment of the latter with Schlosser's base (butyllithium/*t*BuOK) in pentane at −78°C followed by addition of oxalate **295** gave a 60% yield of the coupling product **296**. Subsequent removal of the dithiane ring afforded the hemiacetal **297** in good yield. This method is the *umpolung* version of the method used by Ireland to solve the same synthetic problem (see Section 11.4.2.2) [155]. Another example of

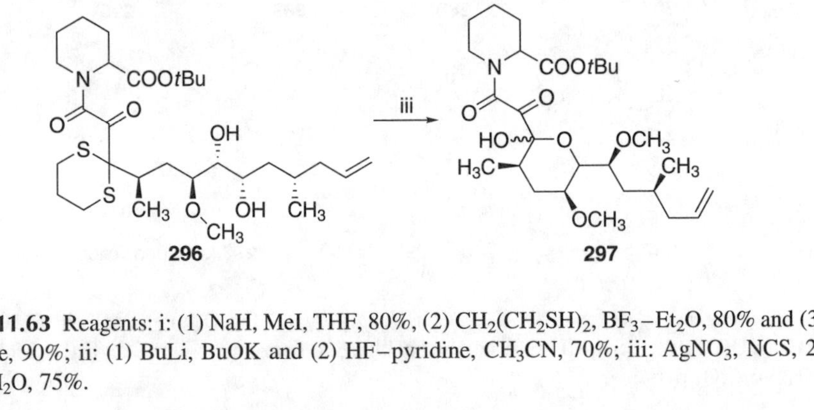

SCHEME 11.63 Reagents: i: (1) NaH, MeI, THF, 80%, (2) $CH_2(CH_2SH)_2$, BF_3–Et_2O, 80% and (3) TBSOTf, 2,6-lutidine, 90%; ii: (1) BuLi, BuOK and (2) HF–pyridine, CH_3CN, 70%; iii: $AgNO_3$, NCS, 2,6-lutidine, CH_3CN, H_2O, 75%.

a total synthesis using dithiane formation as one of the key steps is the preparation of sesbanimide [205,206].

Methods of chain extensions generally use linear sugars with free aldehydes mostly prepared via dithioacetal formation, protection of the remaining hydroxyl groups and aldehyde recovery. This allows one to use all methods applicable to aldehydes. Among others, the condensation of ethyl diazoacetate has been studied in detail by López–Herrera [207] in a new synthesis of KDO by a two carbon homologation of an open-chain mannose derivative.

11.4.2.2 Using Wittig Olefinations

The reducing end of sugars is, in principle, prone to Wittig type olefinations. The most widely used reaction is that of stabilized phosphoranes leading to α,β-unsaturated esters. The drawback of this approach is the facile ring closure by 1,4-addition of the ring oxygen on the activated double bond. This reaction is nevertheless advantageously used for the synthesis of *C*-glycosides as shown in the next section. The direct methylenation of the aldehyde group of hexoses has been introduced by Sinaÿ [208]. Deprotonation of the hemiacetal is required before reacting with methylenetriphenylphosphorane. The final product is an open-chain sugar but most of its use is centered on its ring closure to form *C*-glycosides. Application of this methodology by Nicotra's group led to *C*-glycosyl phosphonate analogs of the biologically relevant glycosyl phosphates [209].

As mentioned above, Wittig extension at the reducing end is one of the key steps of Ireland's synthesis of the bottom half of FK506 [155], which is to be compared with the strategy described in Section 11.4.2.1 [154]. As shown in Scheme 11.64, the anomeric free sugar **298** was transformed into an ester in a yield of 87% (E/Z 20:1) using the appropriate Wittig reagent in refluxing benzene. Reduction of the ester function gave the allylic alcohol **299**. The next step was the functionalization of the double bond to introduce an oxygen atom at C2. Sharpless epoxidation of the double bond was preferred for the sake of simplifying the spectral analysis of the products. Opening of the epoxide gave the β-*C*-glycoside **300**. The subsequent transformations of this part of the building block involved oxidation of the primary

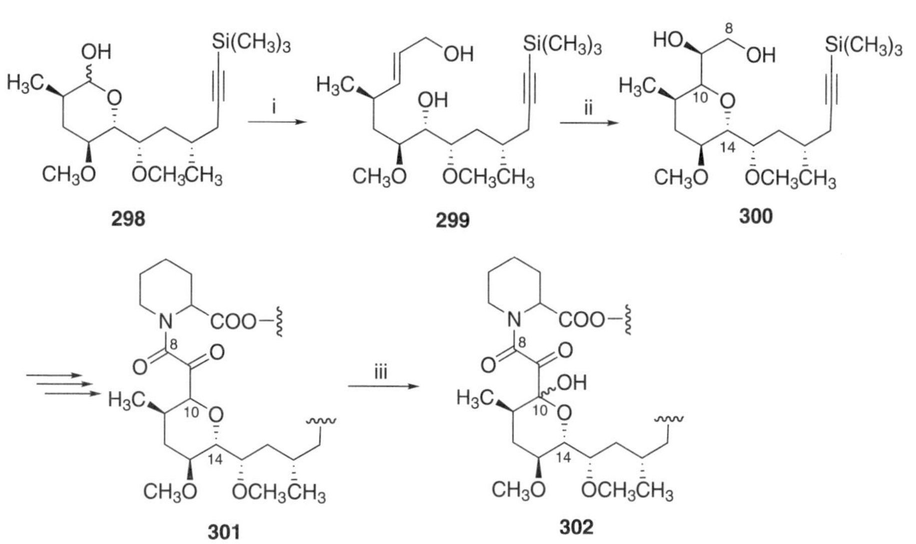

SCHEME 11.64 Reagents: i: (1) $Ph_3P{=}CHCO_2Et$, benzene and (2) DIBAL; ii: $Ti(OiPr)_4$, (+)-DET, TBHP; iii: (1) $Li(sBu)_3BH$, $NaBO_3$, (2) Ph_3P, I_2, imidazole, 70%, (3) KHMDS, TESCl, (4) C_8K, $ZnCl_2$, AgOAc and (5) dimethyldioxirane.

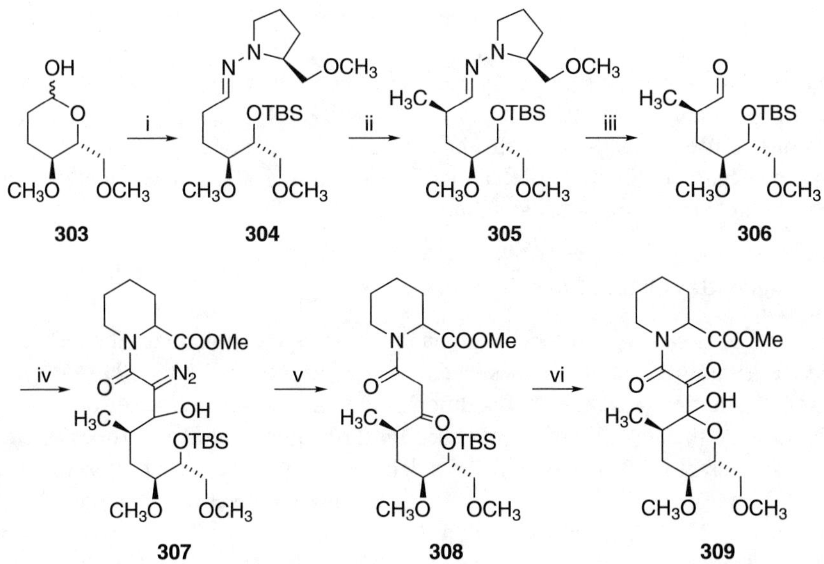

SCHEME 11.65 Reagents: i: (1) MeI, NaH, THF, 92%, (2) AcOH, H_2O, 97%, (3) SAMP, CH_2Cl_2, 86% and (4) TBSCl, imidazole, DMF, 75%; ii: LDA, MeI, THF, 75%; iii: O_3, CH_2Cl_2, 73%; iv: lithio derivative of *N*-diazoacetyl (*S*)-pipecolinate methyl ester, THF; v: $Rh_2(OAc)_4$, DME; vi: (1) $(MeO)_2CHNMe_2$, 93%, (2) O_3, CH_2Cl_2, 70% and (3) HF, CH_3CN, 95%.

alcohol to an acid, and, after coupling with the appropriate amine, oxidation of the secondary alcohol to produce **301**. This allowed the introduction of the required hemiacetal group to secure the 1,2,3-tricarbonyl sequence of FK506, as in **302**. This would appear to be a tedious process but should be considered in light of the very sensitive tricarbonyl group chemistry. The synthesis of (+)-colletodiol also took advantage of a Wittig chain elongation [210].

11.4.2.3 Miscellaneous Methods

The well-known formation of hydrazones from reducing sugars has been exploited in Kocienski's synthesis of the C1 to C5 fragment of FK506 [211]. As shown in Scheme 11.65, compound **303**, derived from tri-*O*-acetyl-D-glucal, was reacted with Enders hydrazine to give **304** after alcohol protection. Diastereoselective alkylation of **304** gave **305**, which was treated with ozone to give the aldehyde **306**. Aldol condensation of **306** with a lithio derivative of *N*-diazoacetyl (*S*)-pipecolinate ester gave **307**, which was in turn elaborated to **308** by oxidation of the alcohol and decomposition of the diazoketone. Introduction of the third carbonyl group, achieved according to reported procedures, yielded **309**, which was isolated as a mixture of isomers because of a restricted rotation around the amide bond.

11.5 CREATION OF *C*-GLYCOSIDIC BONDS

Chain extension at C1 of reducing sugars with retention of the sugar ring system has been the subject of intense research since the early discovery of naturally occurring *C*-glycosyl compounds. These analogs of *O*-glycosides are of tremendous importance as building blocks for the synthesis of carbohydrate mimics as well as to the field of natural product synthesis. Indeed, the high number of tetrahydropyran or tetrahydrofuran ring systems in natural products explains the quantity of results obtained in this area.

To organize the review of this vast research field, we will draw a distinction between *C*-glycosyl compounds with and without an anomeric hydroxyl group. As will be discussed, the latter can be prepared from the former by reduction at the anomeric center. Many excellent reviews have appeared on the methods of *C*-glycoside formation [212–219] and their properties [220].

11.5.1 Creation of C-Glycosidic Bonds with Retention of the Anomeric Hydroxyl Group

Many natural compounds include heterocyclic systems constructed by hemiacetal formation between a hydroxyl group and a keto group of the chain in a 1,4- or 1,5-relative disposition. It is advantageous in a retrosynthetic analysis to take this point into account and to devise a disconnection next to the carbonyl group. Moreover, the synthetic connection does not involve chiral center formation. An obvious translation of this principle in carbohydrate chemistry is the formation of a carbon–carbon bond at the anomeric center by nucleophilic addition to lactones. However, other methods have also been devised to reach this goal.

A plethora of examples can be found in the carbohydrate literature in which a sugar lactone is opposed to a suitable carbanion to create a carbon–carbon bond. One advantage of this approach is the strong tendency of the hemiacetal formed in the first condensation to retain its cyclic form, thus preventing overcondensation, which would result in the formation of a tertiary alcohol.

We will illustrate this strategy with an example found in the total synthesis of lysocellin by Yonemitsu [221,222]. As shown in Scheme 11.66, the C1 to C9 fragment was prepared by condensation of lactone **313** with the lithium salt of benzyl acetate. This part of the synthesis also serves as an illustration of sugar branching to reach a Prelog–Djerassi lactone type intermediate already mentioned in Section 11.3. A detailed discussion of this scheme follows.

Compound **310**, prepared from D-glucose using known chemistry, was treated with borane–THF to carry out the hydroboration of the double bond and also to achieve unusual opening of the furanose ring in the same step. The resulting linear alcohol in **311** was reduced to a methyl group via tosylation and lithium aluminum hydride reduction. After acetal removal, the diol function was cleaved to the aldehyde. The latter was submitted to Wadsworth–Emmons homologation giving, in one step, the unsaturated lactone **312**. Nine further steps were required to achieve stereoselective

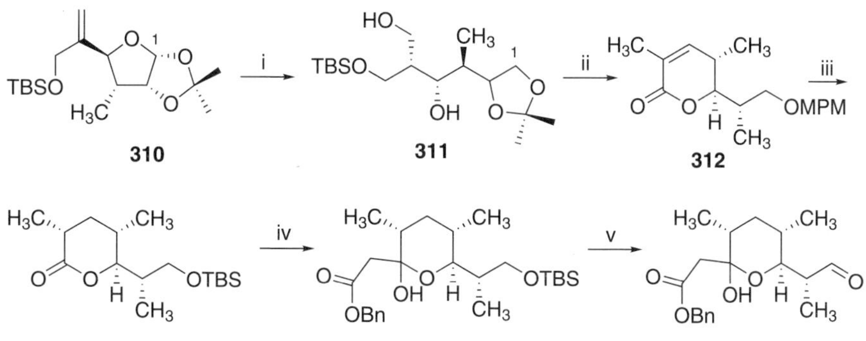

TBSO H3C O 1 O O 310 —i→ HO CH3 1 TBSO OH O O 311 —ii→ H3C CH3 O O H CH3 OMPM 312 —iii→

H3C CH3 O O H CH3 OTBS 313 —iv→ H3C CH3 O OBn OH O H CH3 OTBS 314 —v→ H3C CH3 O OBn OH O H CH3 O 315

SCHEME 11.66 Reagents: i: (1) BH_3, THF and (2) *t*BuOOH, NaOH; ii: (1) TsCl, CH_2Cl_2, pyridine, (2) $LiAlH_4$, Et_2O, (3) NaH, MPMCl, THF, (4) 2*N*, H_2SO_4, MeOH, (5) $NaIO_4$ and (6) BuLi, $(MeO)_2P(O)CH(Me)CO_2Me$; iii: (1) NaH, BnCl, DMSO, THF, (2) 1*N* H_2SO_4, dioxane, (3) PCC, 3Å M.S., (4) H_2, $Pd(OH)_2$, AcOEt and (5) TBSCl, imidazole; iv: LDA, CH_3CO_2Bn, THF, HMPA; v: (1) 1*N* HCl, THF and (2) SO_3–pyridine, DMSO, Et_3N.

reduction of the double bond to secure the stereochemistry at C4 of the target. This sequence included reduction to the lactol, isopropyl glycoside formation, stereocontrolled double bond reduction and finally reoxidation to the lactone. The last steps involved condensation of the lithio acetate derivative on **313** to introduce the C1–C2 branch of **314**. Standard steps allowed formation of the aldehyde **315** prepared for subsequent coupling with the eastern part of lysocellin.

The formation of a new chiral center during formation of a carbon–carbon bond with good stereocontrol is possible using this strategy. An interesting example was reported by Burke in the synthesis of the C1 to C14 subunit of halicondrin B [223,224]. This synthesis also involves two strategies to construct the second type of *C*-glycosyl compounds, the preparation of which is reviewed in the next section.

As shown in Scheme 11.67, the halicondrin B synthesis commenced with the condensation of the chiral α-alkoxy carbanion, derived from stannane **321**, on the protected derivative **316**, prepared from the well-known D-*glycero*-D-*gulo*-heptonolactone. The resulting diol was mesylated on the secondary alcohol to give **317**. The formation of a single diastereoisomer in this condensation is noteworthy. This compound was treated with ethylmagnesium bromide in refluxing benzene to promote a pinacol rearrangement. Subsequent reduction of the intermediate ketone at C7 (halicondrin numbering) gave **318** with the required configuration. This clever transformation created a *C*-glycosyl derivative with high stereocontrol. The next steps of this synthesis included ozonolysis of the vinyl terminus to an aldehyde and Wittig homologation to ester **319**. Upon treatment with a soluble methoxide ion in benzene, 1,4-addition of the alkoxide to the activated double bond occurred to give the thermodynamically more stable isomer **320**. This last method is widely used to create *C*-glycosydic bonds and additional examples are reviewed below.

Many other syntheses incorporate such nucleophilic additions to lactones with subsequent reduction to *C*-glycosyl compounds because of the facile reduction of the tertiary anomeric hydroxyl. Condensation of acetylenic carbanions with lactones, followed by reduction, has been used as a starting point to construct the anthraquinone ring system of vineomycinone B_2 [225].

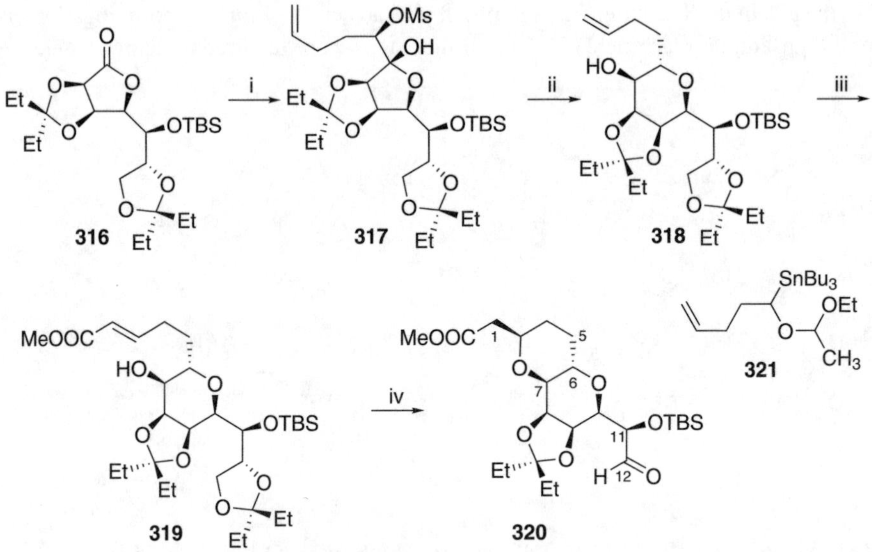

SCHEME 11.67 Reagents: i: (1) **321**, BuLi, THF, (2) 1.8*N* HCl and (3) MsCl, Et_3N, DMAP, CH_2Cl_2, 58%; ii: (1) EtMgBr, benzene, 47% and (2) DIBAH, THF, 99%; iii: (1) O_3, CH_2Cl_2, then PPh_3 and (2) $Ph_3P{=}CHCO_2Me$; iv: $BnMe_3NOMe$, benzene, 90%.

11.5.2 Creation of *C*-Glycosidic Bonds with Replacement of the Anomeric Hydroxyl Group

The direct formation of carbon–carbon bonds at anomeric centers is by far more difficult than the preceding strategy. Good stereocontrol is desirable but forming such a bond was still a challenging issue until recently. Having a look at naturally occurring complex structures quickly convinces us that creation of *C*-glycosyl structures is of paramount importance. For example, extraordinarily complex polyether macrocycles such as palytoxin, brevetoxin, maitotoxin, halicondrin, spongistatin and congeners can be regarded as an accumulation of *C*-glycosidic structures. Some of these compounds succumb to total synthesis, relying on new carbon–carbon bond forming reactions as well as existing ones adapted to relevant synthetic problems.

Among the solutions to the *C*-glycosylation problem, the well-known Wittig olefination – Michael addition sequence occupies a prominent place. This "old" reaction was widely exploited in connection with exploration of *C*-nucleosides in the 1970s. The growing demand for *C*-glycosyl structures prompted the invention of new methods, some of which are reviewed below.

11.5.2.1 Cyclization Reactions

As already seen, the Michael addition of an oxygen nucleophile to an activated double bond is an excellent way to construct a *C*-glycoside. One useful approach to this strategy is to perform a Wittig reaction on a reducing sugar. In general, the open chain derivative is not isolated but undergoes *in situ* cyclization. Equilibration to the thermodynamically stable isomer occurs. An example related to the synthesis of halicondrin is reviewed in the foregoing section. The result of this procedure is the formation of a methylene group next to the anomeric center. Here we will discuss a variant of this strategy, which allows the introduction of a hydroxyl group.

Nicolaou's total synthesis of brevetoxin is also a milestone in the total synthesis of naturally occurring compounds from sugars [226,227]. This polyether marine natural product is a toxin, associated with the "red tide" phenomenon, which interferes with sodium channel systems and causes cell damage. This compound embodies 11 *trans*-fused oxygenated rings, including tetrahydropyran, oxepane and oxocene, and 23 stereogenic centers. The most challenging problem is heterocyclization which, from a carbohydrate chemist's point of view, corresponds to the formation of a *C*-glycosidic bond. Five of these bonds have been established by hydroxy-epoxide cyclization as illustrated in Scheme 11.68 in the synthesis of the ABC ring system of brevetoxin [228].

A multistep route from tri-*O*-acetyl-D-glucal led to a *C*-glycosyl compound **322**, which represents the C ring of the target compound. One carbon homologation at C6, via displacement of the O6 tosylate by cyanide ion, followed by reduction to the corresponding aldehyde gave **323**. Subsequent Wittig olefination yielded **324**. Oxidation of the secondary hydroxyl group of **324** gave the corresponding ketone. Treatment of this compound with excess trimethylaluminum afforded the tertiary alcohol **325** accompanied by its epimer (5:1). Subsequent treatment with methyllithium gave a 1:6 mixture of epimers. The crucial step was the opening of the 6-*endo* hydroxy epoxide, which was prepared as follows. Protection of the tertiary alcohol was followed by reduction of the ester group to an allylic alcohol. Epoxidation of this compound according to the Sharpless procedure gave **326**, which was further elaborated to the dibromoolefin **327** using standard procedures. This 6-*endo* activated epoxide was then ready for opening by the tertiary alcohol upon treatment with camphorsulfonic acid at 0°C to give **328** [229]. The last steps involved the transformation of the dibromoolefin into the acetylenic derivative and then to **329**, ready for the construction of the A ring of brevetoxin. This apparently long sequence had the merit of securing two chiral centers at the same time, which are defined by the sense of the Sharpless epoxidation.

A recent example of a Wittig-intramolecular Michael addition with formation of two chiral centers is given below [230]. In the synthesis of luminacins C1 and C2 from D-glucal, the sugar-like

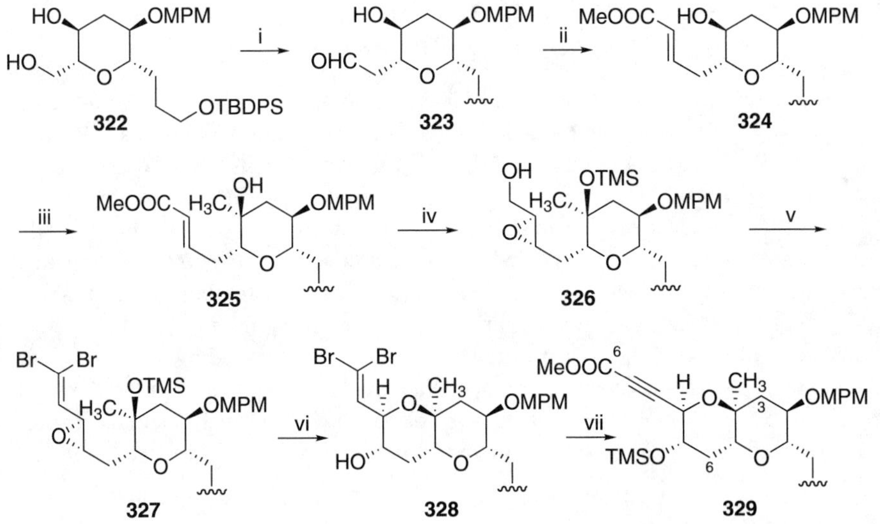

SCHEME 11.68 Reagents: i: (1) TsCl, pyridine, 85%, (2) NaCN, DMSO, 91% and (3) DIBAL, CH_2Cl_2, then 1*N* HCl, 95%; ii: $Ph_3P{=}CHCO_2Me$, benzene, 89%; iii: (1) $(COCl)_2$, DMSO, CH_2Cl_2, Et_3N and (2) $AlMe_3$, CH_2Cl_2, 94%; iv: (1) TMS imidazole, CH_2Cl_2, (2) DIBAL, CH_2Cl_2 and (3) (−)DET, $Ti(O\textit{i}Pr)_4$, *t*BuOOH, CH_2Cl_2; v: (1) SO_3·pyridine, Et_3N, DMSO, CH_2Cl_2, 95% and (2) Ph_3P, CCl_4, CH_2Cl_2; vi: TBAF, THF; vii: (1) TMS imidazole, CH_2Cl_2 and (2) MeLi, THF, then $ClCO_2Me$, 75%.

structure was actually constructed via *C*-glycosylation to introduce the C2′–C3′ part of the target. As illustrated in Scheme 11.69, functional group manipulation of tri-*O*-acetyl-D-glucal led to **330** in six steps. Wittig olefination at C4 followed by acetal hydrolysis gave **331**. Another Wittig reaction at the anomeric center, performed in toluene at 100°C, gave **332** as a single isomer. This compound cyclized upon treatment with catalytic sodium methoxide in hot toluene to produce **333** as a single isomer. The anomeric configuration was found to be α, in contrast to the "normal" behavior in which the equatorial anomer is favored. Compound **333** should exist in a $^1C_4$ conformation with the glycosidic appendage in an equatorial orientation. This was explained by strong repulsions between the olefin at C6′ and the C5′ and C7′ substituents. However, the stereochemistry at C2′ was opposite to that of the target. Equilibration in basic medium (MeONa, MeOH at 60°C) allowed the formation

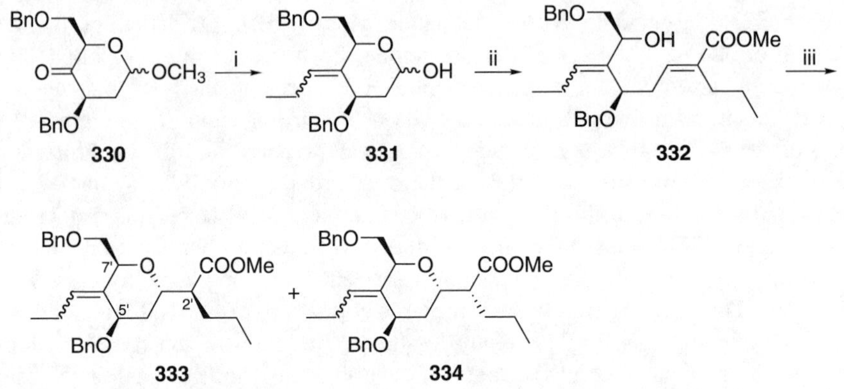

SCHEME 11.69 Reagents: i: (1) $PrPPh_3Br$, BuLi, THF, 84% and (2) $AcOH{-}H_2O$, 80%; ii: $CH_3CH_2CH_2C(CO_2Me){=}PPh_3$, 76%; iii: MeONa, toluene, 80%.

of a 1:1 mixture of **333** and the requisite **334**. The remainder of the synthesis consisted of dihydroxylation of the olefin and excision of carbon C7′.

This strategy was adopted in the synthesis of halicondrin [231]. A related strategy was also developed in Kishi's synthesis of halicondrin [232,233]. Finally, it should be noted that this same strategy can be performed on ketoses, thus forming *gem*-di-substituted *C*-glycosyl compounds. The ketose variation was used in an approach to the synthesis of verrucosidin [234].

11.5.2.2 Direct *C*-Glycosylations

11.5.2.2.1 Allylations

O-Glycoside formation is the result of the reaction of an alcohol with an appropriate activated glycosyl donor. The *carbon* version of the same reaction was a more difficult challenge, solved by the reaction of the sugar hemiacetals with allylsilanes [235]. As with *O*-glycosylations, this *C*-glycosylation strategy is promoted by Lewis acids and the involvement of oxonium ions is likely. This reaction was well studied and successfully used in total syntheses. As shown below, the high versatility of the allyl group led to the development of very efficient procedures for functionalization of the chain.

This reaction was originally developed in connection with the palytoxin synthetic challenge [236]. In addition to reducing sugars, *O*-glycosides or glycosyl esters can be used [237]. Several variants have been proposed over the years. Simple olefins react with 2-*O*-benzyl-glycals under Lewis acid catalysis to provide fused *C*-glycosides [238]. Substituted allylsilanes, [239] silylketene acetals [240] and other Lewis acid catalysts [241] have also been used. This strategy allows the elaboration of α-*C*-glycosyl compounds **336** from **335** (Scheme 11.70). Such allylations have also been used, *inter alia*, in the synthesis of the ABC [228] and the IJK [242] ring systems of brevetoxin B, the C1 to C12 fragment of halichondrin [233], in the construction of the C18 to C32 fragment of swinholide [243] and in the construction of the LM and NO subunits of maitotoxin [244]

Another method to achieve allylation at C1 can be realized by nucleophilic addition of allylmagnesium bromide to a lactone followed by reduction (**337** to **338**). This method, also illustrated in Scheme 11.70, was shown to produce the β-derivatives, in sharp contrast to the preceding α-selective method. An application of this *C*-glycosylation strategy to the synthesis of the AB ring system of ciguatoxin CTX3C is given below [245]. This polyether with 12 heterocyles is a toxin isolated from ciguatera food poisoning. As in the case of the brevetoxin synthesis, one of the most challenging problems is the construction of *C*-glycosidic six-membered ring structures, as well as the construction of seven-, eight- or nine-membered rings present in the toxin.

An example of the use of allyl groups in ring-closing metathesis (RCM) is given below [246]. As illustrated in Scheme 11.71, the AB ring system was constructed from allylglycoside **342** prepared from **341** according to Kishi's procedure. This compound was treated with iodine to selectively remove the 2-*O*-benzyl group, as shown by Nicotra [247]. The resulting free hydroxyl group was then protected as allyl ether to give **343**. This compound was treated with Grubbs

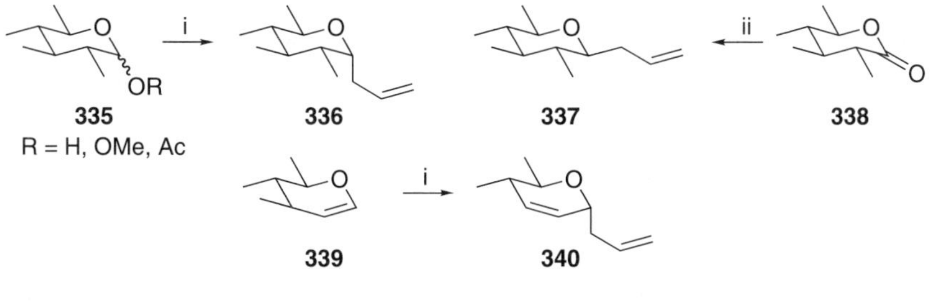

335 R = H, OMe, Ac; 336; 337; 338; 339; 340

SCHEME 11.70 Reagents: i: Allyl-Si(Me)$_3$, BF_3–Et_2O; ii: (1) AllylMgBr and (2) Et_3SiH, BF_3–Et_2O.

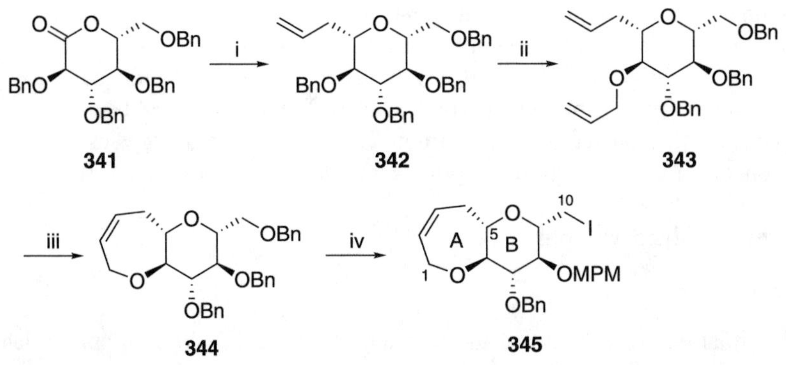

SCHEME 11.71 Reagents: i: (1) AllylMgBr, Et_2O and (2) Et_3SiH, BF_3–Et_2O, CH_3CN; ii: (1) I_2, CH_2Cl_2, (2) Zn, AcOH, Et_2O, MeOH and (3) AllylBr, KH, THF, 83%; iii: $(PCy_3)Cl_2Ru$=CHPh, CH_2Cl_2, 97%; iv: (1) Na, NH_3, (2) 4-OMe–$C_6H_4CH(OMe)_2$, PTSA, (3) BnBr, NaH, 84%, (4) DIBAL, CH_2Cl_2, 82% and (5) I_2, PPh_3, imidazole, toluene, 82%.

catalyst to promote RCM to the required seven-membered ring **344** in a yield of 97%. The last steps allowed the transformation of **344** to the iodo derivative **345**, which represents the C1 to C10 part of the target. Other RCM reactions were required to couple the AB ring system with the DE part, thus building the C ring.

C-Allylation has also been used in the construction of the BCD ring system of ciguatoxin [248]. Finally, the anomeric allylation reaction works equally well with ketose glycosides to yield bis substituted *C*-glycosyl derivatives, which were used in the total synthesis of (+)-citreoviral [249].

Ferrier rearrangement-based *C*-glycosylation has also been proposed as an efficient way to introduce allyl groups at glycal anomeric positions with concomitant migration of the double bond at C2–C3 (**339** to **340**; Scheme 11.70) [250]. Unprotected glycals have also been successfully used [251]. Dihydroxylation of the external double bond of **340** type compounds was possible, as shown in model studies toward the avermectin spiroketal subunit [252]. Introduction of more complex appendages by this route was also employed in the total synthesis of avermectin previously shown [59]. Carbohydrate-derived allylsilanes were also effective [253]. Additionally, silylated allylic alcohols [254] and olefins [255] can be used as alkylating agents of glycals in the presence of Lewis acids. Several syntheses of natural products include this anomeric *C*-allylation of glycals. In fact, an approach to the ring system of forskolin by Hanna started from a *C*-allyl D-glucal derivative [256]. In another example, azadirachtin synthetic studies involved *C*-allylation of 2-bromo-glycal and subsequent elaboration of the allylic side chain [257]. Finally, a short synthesis of dysiherbaine based on this strategy was reported by Snider [192].

11.5.2.2.2 Anomeric Anions

As already mentioned in this chapter, nucleophilic carbohydrates have, until recently, been poorly investigated [218]. However, several groups have reported, at almost the same time, on the properties and availability of anomeric carbanions. Glycals were the first substrates to be investigated in this manner because no elimination can occur.

As shown in Scheme 11.72, direct lithiation of protected glucal **346** using *t*BuLi to form **347** has been proposed [258]. From the lithium anion **347** it was also possible to prepare the vinyl stannane **348**, which can be isolated and retreated with *n*BuLi to give the lithio derivative **347** [259,260]. This anion has been condensed with aldehydes and esters. Lithiation of 2-phenylthioglycal was also effective [261]. A good example of the use of such a dihydropyran anion could be found in a pederin synthesis [262].

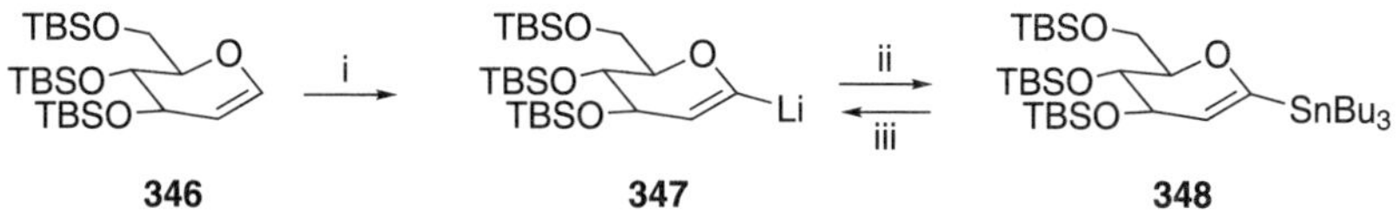

SCHEME 11.72 Reagents: i: *t*BuLi; ii: Bu_3SnCl; iii: BuLi.

To illustrate the possible use of sugar-derived vinylic anions, we will focus on the synthesis of herbicidin, a complex nucleoside isolated from strains of *Streptomyces* showing interesting herbicidal properties (see Scheme 11.74). The most challenging part of the synthesis is obviously the construction of a carbon–carbon bond between two sugar units.

In an approach using a vinylic anion (Scheme 11.73), compound **350**, obtained from the stannane **349**, was condensed with aldehyde **351** to give alcohols **354** as a mixture of epimers in a yield of 31% [263,264]. These compounds were further transformed into ketone **355** and the deoxy compound **356**. In terms of yield, the reaction of **350** with the acyl chloride **352** was not better, but the triflate **353** gave interesting results (41%). The major drawback of this approach was the need for further functionalization of the double bond to install the keto group at C7 and to secure the anomeric configuration at C6.

In another synthesis, Gallagher described a fruitful approach based on the use of a carbohydrate enolate, which takes advantage of the presence of the C7 keto group in the target compound [265–267]. In this approach, an enolate had to be formed at the anomeric position of the *gluco* chiron, stable toward β-elimination. Moreover, the use of a strained system, as in **357**, should force enolization to occur on the correct side.

As shown in Scheme 11.74, condensation of the potassium enolate of **357** with aldehyde **358** gave, after dehydration, the enone **359**. Reduction of the double bond secured the correct stereochemistry at the anomeric center of the *gluco* moiety and allowed formation of the hemiacetal. The last steps of the synthesis were devoted to opening the 3,6-anhydro bridge via radical bromination to give **360**. Silver assisted hydrolysis of the bromoether gave the corresponding aldehyde, which was further oxidized to the ester. Final hydrolysis of the isopropylidene group gave **361**.

Anomeric organosamarium derivatives are stable toward β-elimination. They can be prepared from 2-pyridyl sulfones, are good nucleophiles and react with aldehydes and ketones to form *C*-glycosides [90,91,268]. A recent synthesis of herbicidin B was reported by Matsuda's group using a related strategy, namely, regioselective formation of an anomeric enolate by diiodosamarium reduction of 2-keto-thioglycoside [269]. As shown in Scheme 11.75, selective protection of a β-thiophenyl glycoside gave ketone **362**, which was treated with SmI_2 in THF and then with aldehyde **363** to give **364** as a mixture of diastereoisomers (79/21). Dehydration gave enone **365**, which was reduced to the β-anomer because of the half-chair conformation of **365**. To circumvent this drawback the O8′–O9′ protecting groups were changed to bulkier silyl ether

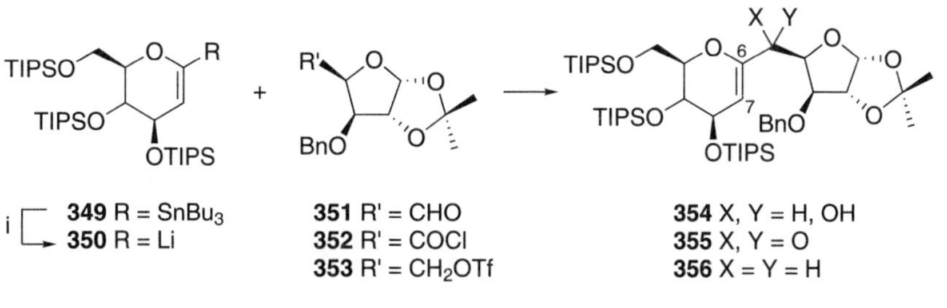

SCHEME 11.73 Reagent: i: BuLi.

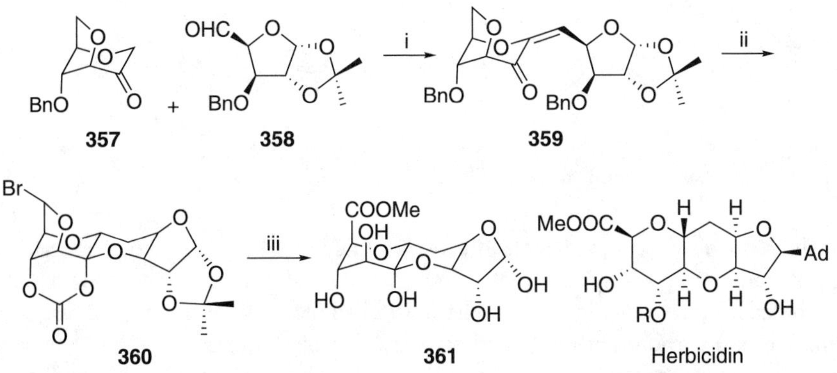

SCHEME 11.74 Reagents: i: *t*BuOK, THF, 60%; ii: (1) H_2, 10% Pd/C, EtOAc, 60%, (2) Im_2CO, CH_2Cl_2 and (3) Br_2, hγ, CCl_4, 65%; iii: (1) MeONa, MeOH, (2) Ag_2CO_3, (3) I_2, KOH, MeOH and (4) TFA/H_2O.

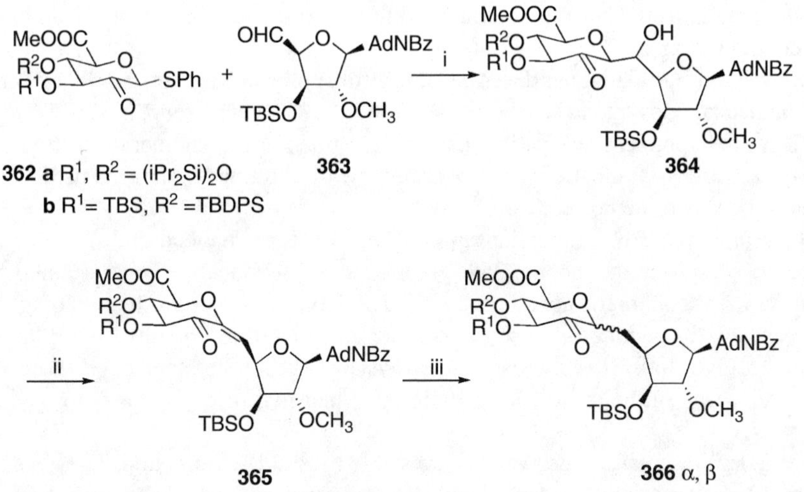

SCHEME 11.75 Reagents: i: SmI_2, THF; ii: MeO_2CNSO_2, Et_3N, toluene; iii: Pd/C, HCO_2NH_4, MeOH.

groups as in **362b**. This was believed to induce steric repulsions between the two groups and thus induce a conformational change to the $^1C_4$ conformation (sugar numbering). This proved to be the case and a 3:1 α/β mixture of **366** was obtained upon hydrogen transfer reduction of the enone **365**. It is interesting to note that this enolization procedure is compatible with the presence of an *N*-benzoyl protected adenine derivative. Further functional group manipulation gave herbicidine.

11.5.2.2.3 Rearrangements

Here again, rearrangement reactions proved useful for the formation of carbon–carbon bonds at the anomeric center. This reaction was originally devised for the construction of *C*-glycosides [270–272], but was used with some success in several syntheses like those of nonactic acid [273], tirandamycic acid [274] and streptolic acid [275].

To illustrate this process, we will give some details of a synthesis of the C27 to C38 segment of halicondrin [232]. As shown in Scheme 11.76, selective protection of D-galactal gave diester **367**. Treatment of the latter in basic medium in the presence of HMPA led to the rearranged product **368** as the major compound. It is interesting to note that in the absence of HMPA the other isomer at

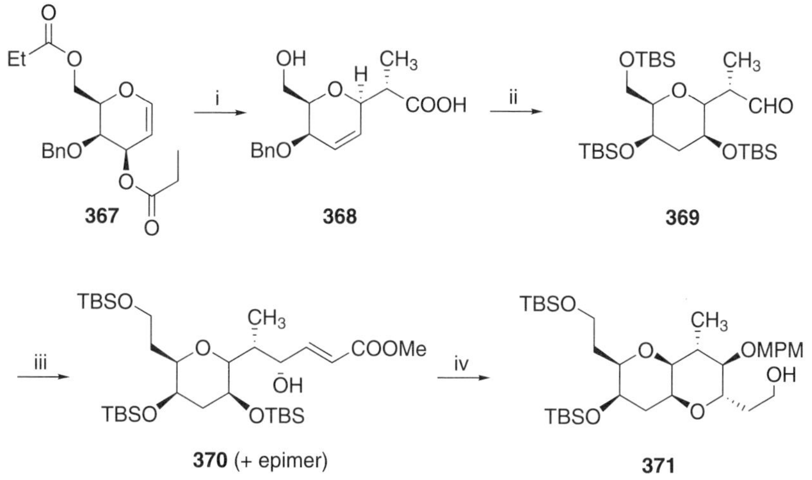

SCHEME 11.76 Reagents: i: (1) LiHMDS, TBSCl, HMPA, THF, then benzene reflux and (2) 1*N* NaOH, H_2O, THF; ii: (1) I_2, KI, $NaHCO_3$, (2) Bu_3SnH, AIBN, benzene, (3) DIBAL, THF, (4) PTSA, MeOH, (5) Tf_2O, pyridine, CH_2Cl_2, then NaCN, DMF, (6) DIBAL then $NaBH_4$, MeOH, (7) H_2, $Pd(OH)_2$, (8) EtSH, $BF_3{\cdot}Et_2O$ and (9) TBSOTf, Et_3N, I_2, $NaHCO_3$; iii: $ICH{=}CHCO_2Me$, $NiCl_2{-}CrCl_2$, THF; iv: (1) PPh_3, $4\text{-}NO_2C_6H_4CO_2H$ then $EtO_2CN{=}NCO_2Et$, (2) K_2CO_3, MeOH, (3) $4\text{-}OMe{-}C_6H_4CH_2OC({=}NH)CCl_3$, $BF_3{-}Et_2O$, (4) HF, pyridine, CH_3CN, (5) $MeC(OMe)_2Me$, PPTS, (6) PPTS, MeOH, (7) TBSOTf, Et_3N, CH_2Cl_2 and (8) $LiAlH_4$, Et_2O.

C3 was obtained. Indeed, the configuration at the anomeric center was dictated by the *galacto* configuration of the starting compound. Iodolactonisation of **368** gave the corresponding lactone, which was reduced to the aldehyde and the free hydroxyl groups were protected as silyl ethers to give **369**. Kishi–Nozaki coupling of this aldehyde with a suitable vinyliodide gave the expected alcohols **370** as a 2:3 mixture. One of the epimers was transformed into the desired one using Mitsunobu inversion. The last steps of this synthetic study involved Michael addition to form the second fused tetrahydropyran yielding **371**.

11.5.2.2.4 Radical Reactions

Several strategies involving the addition of anomeric radicals to activated olefins have been proposed [276]. Keck developed an allylation reaction to construct pseudomonic acids by reacting an anomeric radical, derived from L-lyxose, with allyltributyltin [277]. Radical additions to olefins generally require large excesses of the latter to observe significant rates. If it is easy with allyltin or acrylonitrile, it is considerably more difficult with complex olefins. The problem could be circumvented using an intramolecular radical addition. Such radical reactions have been previously described in this chapter in connection with the chain elongation at C6 of sugars for the synthesis of tunicamycin (Section 11.4.1.3). We can now consider the reaction from the opposite side and we give here some details on the cyclization of an anomeric radical to an enol ether as used in an approach to the synthesis of herbicidin [203].

Scheme 11.77 shows that the two subunits **372** and **373** were linked by a silyl ketal tether to give **374**. An anomeric radical was then produced by treatment with tributyltin hydride. This radical reacted with the enol ether acceptor to give the cyclic derivative **375** as the major product in a yield of 43% together with two of the three possible isomers in yields of 6 and 13%. The combined yield shows that more than 56% of the radical attack occurred from the α face of the *gluco* residue. However, the intermediate radical, located at C4, is mainly trapped by the α face of the furanose moiety. Although this approach is attractive, further elaboration to the

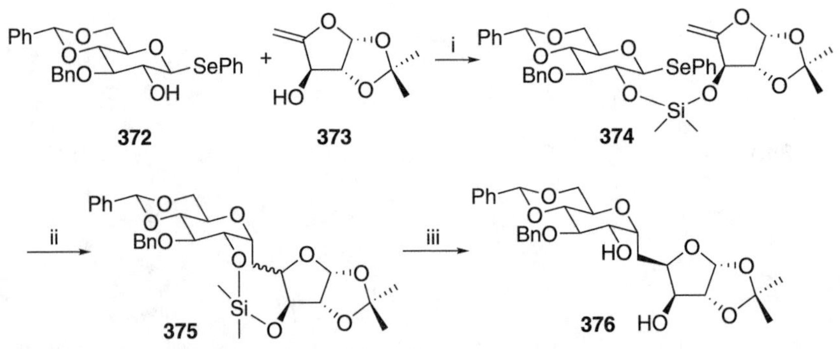

SCHEME 11.77 Reagents: i: BuLi, Me_2SiCl_2 then Et_3N; ii: Bu_3SnH, AIBN, toluene; iii: TBAF, THF.

target raises the problem of the distinction between the two secondary hydroxyl groups of **376** after removal of the tether.

11.5.2.2.5 Direct Olefination of Lactones

Olefination of sugar lactones is a recently discovered process, which seems promising for the creation of anomeric double bonds. As seen above in the synthesis of the herbicidin skeleton, stereoselective reduction of these *exo-glycals* is possible. Although a few examples of the intramolecular Wittig reaction were known [278], the direct olefination of esters remained unexplored. Our group contributed to the development of the Wittig olefination of sugar lactones. Dichloro- and dibromo-olefins are easily synthesized using triphenylphosphine and carbon tetrachloride [279–281] or bromomethylenetriphenylphosphorane [282]. Stabilized phosphoranes also react at elevated temperature with lactones to provide *exo*-glycals [283–285]. Reduction of these *exo*-glycals opens the way to *C*-glycosides [285–287] and allows a straightforward synthesis of muscarines by stereoselective introduction of a methyl group at the anomeric position [288].

We describe below a short synthesis of goniofufurone based on an intramolecular Wittig olefination [289]. As shown in Scheme 11.78, this cytotoxic lactone was prepared from the known aldehyde **351**, which was condensed with phenylmagnesium bromide to give the alcohol **377** as the major isomer (16:1). Inversion of the newly formed chiral center was needed, and was carried out

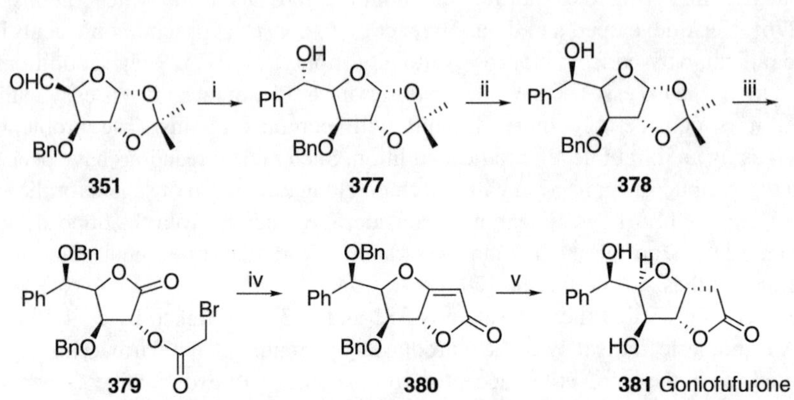

SCHEME 11.78 Reagents: i: PhMgBr, Et_2O, 78%; ii: (1) PCC, CH_2Cl_2 and (2) $NaBH_4$, $CeCl_3 \cdot 7H_2O$, MeOH, 67%; iii: (1) BnBr, NaH, THF, 87%, (2) TFA, H_2O, 85%, (3) Br_2, $BaCO_3$, dioxane–H_2O, 52% and (4) $BrCO_2CH_2Br$, pyridine, Et_2O, 90%; iv: PPh_3, CH_3CN then DBU; v: H_2, 10% Pd/C.

by oxidation of the ketone and stereoselective reduction with sodium borohydride and cerium trichloride (**377**, **378** 1:8). Protection at O5 and removal of the acetal gave the corresponding diol, which was selectively oxidized to the lactone. Esterification at O2 by bromoacetic acid gave **379**. Reaction with triphenylphosphine in acetonitrile gave the phosphonium salt, which upon treatment with DBU cyclized to **380**. Excess DBU promoted epimerization at C2. The last steps to reach the target **381** were reduction of the double bond and removal of the benzyl groups (58%). Some C5 deoxy compound (22%) resulting from hydrogenolysis of the wrong side of the bis-benzyl ether was observed.

Complementing the above strategy, we found that *C*-glycosylidene compounds undergo facile 1,4-addition of alcohols to the double-bond forming anhydro sugars. This methodology was used to construct the dioxabicyclic system of the zaragozic acids [290].

11.5.2.2.6 Miscellaneous Methods

Some variants of the above methods have been successfully utilized in total syntheses. For example, anomeric vinyl stannanes were found useful for coupling with suitable vinyl or aromatic iodides. This allowed the total synthesis of the anthracycline derivative vineomycinone B2 using palladium catalysis [291].

Sugar-derived enol triflates were also effective in β-alkyl Suzuki coupling reactions with sugar boronates and provide a route to fused polycyclic ether like ciguatoxin [292].

Some reactions described for the branching of sugars apply to the synthesis of *C*-glycosides. In the same context the reaction of anomeric epoxides with Grignard reagents or cuprates occurs at low temperature to give good yields of the β-*C*-glycosyl derivatives as the major product [293]. Additionally, the 1,4-addition of cuprates [280] or acetylenic carbanions [294] to 1-ene-3-ones has also been described.

The direct linkage of phenolic aromatic rings in the presence of hafnocene dichloride-silver perchlorate as a promoter has been described in the synthesis of gilvocarcin M and V [168,295,296].

Anomeric *C*-arylation by nucleophilic addition of suitable aryl anions to a lactone, followed by Lewis acid catalyzed reduction of the resulting hemiacetal, has been used in the synthesis of (−)-altholactone from D-gulonolactone [297].

The reaction of anomeric sulfones with vinyl magnesium bromide in the presence of zinc bromide allows direct vinylation at the anomeric center. This reaction was the starting point of an approach for the synthesis of the C29 to C51 fragment of altohyrtin A [64]. The C29 to C44 (EF) portion of this compound has been prepared by a stepwise sequence involving anomeric cyanation, using the reaction of anomeric acetate in the presence of BF_3–Et_2O. The anomeric cyanide was then elaborated to the Weinreb amide, which was condensed with an appropriate vinylic carbanion representing the C29 to C37 chain [298].

11.6 FORMATION OF CARBOCYCLES

Many naturally occurring compounds such as terpenes and alkaloids are carbocycles. They often exhibit novel biological properties, which stimulate interest in their synthesis. The chiral pool does not offer a large variety of cyclic precursors, and most of these precursors are poorly functionalized. It is not surprising that carbocyclization of carbohydrates has been a research area of interest. Until recently, no general solution existed to this problem. However, the 1979 report by Ferrier [299] of the facile ring closure of 5,6-unsaturated carbohydrates into cyclohexanones paved the way for many syntheses of carbocyclic structures. Alternatively, the development of radical reactions furnished a good and quite general solution to the synthesis of five-membered ring systems. A few syntheses of natural compounds using these two main approaches and some related methods are described in the first part of this section.

Other methods that rely on the use of the carbohydrate template to construct fused carbocycles have been successfully investigated as strategies for the formation of carbocycles from sugars. Chirality transfer from the sugar to the new carbocycle is observed for these reactions. These annulation reactions are the subject of the second part of this section.

11.6.1 Carbocyclization of the Sugar Backbone

11.6.1.1 Ferrier Carbocyclization

Treatment of 5,6-unsaturated sugar **382** (Scheme 11.79) with mercury(II) salts was shown to induce what is called the second Ferrier rearrangement. The reaction proceeds via mercuration of the double bond and, in the presence of water, the alkyl glycoside **383** is then hydrolyzed to the aldehyde **384**, which undergoes intramolecular aldol reaction to produce a β-hydroxy cyclohexanone **385**. This reaction produces only six-membered rings. An excellent review by Ferrier and Middleton was published some years ago [300].

Variants of this reaction used to construct inositols were introduced by Prestwich [301] and Sinaÿ [302]. These new methods have been reviewed elsewhere [303]. Additionally, a chapter of this book deals with methods of carbocyclization to form carbasugars from sugars.

In a synthetic approach to rapamycin (Scheme 11.80), Danishefsky reported the synthesis of the C28 to C49 part of the target [304]. The substituted cyclohexane ring was constructed from methyl α-D-glucopyranoside derived compound **386**. Hanessian–Hullar opening of the benzylidene ring was carried out to give the 6-bromo derivative, which was further modified to protect O4 as a benzyl ether. Elimination gave the 5,6-unsaturated key intermediate **387**. Carbocyclization occurred upon treatment with mercury salts giving cyclohexanone **388**.

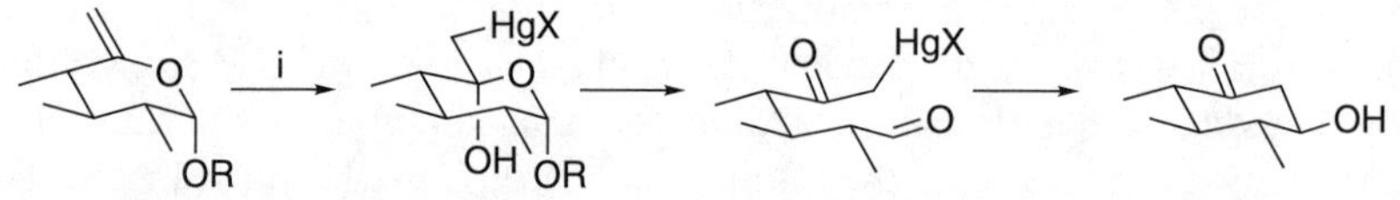

SCHEME 11.79 Reagents: i: $HgCl_2$, H_2O, acetone.

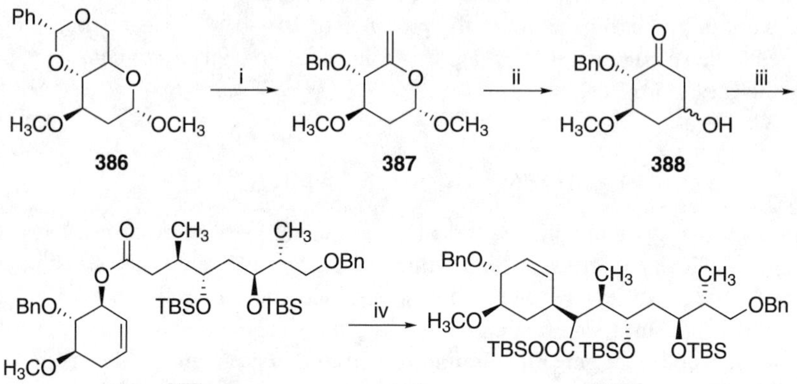

SCHEME 11.80 Reagents: i: (1) NBS, $BaCO_3$, (2) MeONa, MeOH and (3) NaH, BnBr; ii: $HgCl_2$; iii: (1) MsCl, pyridine, (2) $LiBH_4$, $CeCl_3{\cdot}7H_2O$ and (3) $HO_2CCH_2CH(CH_3)CH(OTBS)CH_2CH(OTBS)CH(CH_3)CH_2OBr$, EDCl; iv: LDA, HMPA, TBSCl, and then heat.

Dehydration and reduction of the resulting cyclohexenone under Luche's conditions gave the corresponding allylic alcohol. Esterification with an acid representing the C39 to C26 part of rapamycin gave compound **389**. Finally, Ireland–Claisen rearrangement gave the branched-chain cyclohexane **390**.

Ferrier carbocyclization can form only cyclohexanones. Nevertheless, chemical manipulations of the latter can provide us with seven-membered rings. This is illustrated by the synthesis of (−)-calystegine B_2 by Boyer and Lallemand [305]. These natural polyhydroxylated nortropane compounds are growth stimulators of nitrogen fixing bacteria.

As shown in Scheme 11.81, the known unsaturated sugar **391** was transformed into cyclohexanone **392** according to Ferrier. Protection of the hydroxyl group, followed by silyl enol ether formation, gave **393**. Cyclopropanation of the double bond gave **394**. Ring opening by $FeCl_3$ gave the ring-enlarged ketone, which underwent hydrochloric acid elimination to the enone **395**. Complete reduction of the enone group, followed by displacement of the hydroxyl group with azide, gave **396**. The last steps were devoted to oxidation and ring formation upon hydrogenolysis of the azide group to furnish the target compound calystegine **397**.

This carbocyclization found excellent applications in the field of alkaloids. The highly oxygenated *Amaryllidaceae* alkaloid family presents interesting biological activities, and these alkaloids have been studied as antitumor agents. Three members of this family attracted attention. Their C rings are closely related to an inositol and carbocyclization of a sugar was an obvious strategy. A total synthesis of lycoricidine was published by Ogawa's group based on such a carbocycle construction [306,307].

As shown in Scheme 11.82, the basic cyclohexane ring was obtained from D-glucose by epoxide opening to introduce an azide group at C2. Protection of the hydroxyl groups at C3 and C4, followed by introduction of the double bond at C5 to C6 using standard procedures, gave **398**. Ferrier carbocyclization was achieved using mercury trifluoroacetate according to the method published by our group for the carbocyclization of the same type of compounds [308]. Elimination of water gave the cyclohexenone **399**. Reduction of the carbonyl group of **399** gave the corresponding alcohol protected as a 4-methoxybenzyl ether. After reduction of the azide group, amide bond formation with a suitable bromobenzoic acid gave **400**. The next crucial step was the Heck reaction to achieve ring closure, which proceeded in a 68% yield to give **401**. It is interesting to note that the same reaction performed with benzyl protecting groups and on the

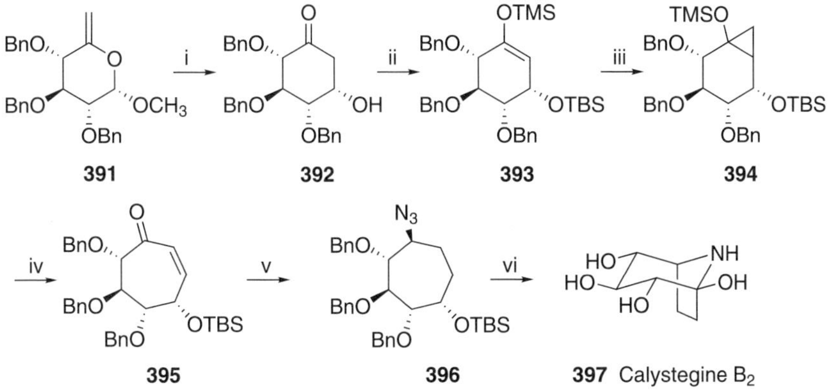

SCHEME 11.81 Reagents: i: $Hg(OAc)_2$, AcOH (1%), acetone, H_2O, 64%; ii: (1) TBSOTf, 2,6-lutidine, 90% and (2) LDA, TMSCl, 70%; iii: Et_2Zn, CH_2I_2, toluene; iv: (1) $FeCl_3$, DMF and (2) AcONa, MeOH. v: (1) H_2, 10% Pd/C, EtOH, 90%, (2) $NaBH_4$, dioxane, (3) MsCl, DMAP, pyridine and (4) NaN_3, DMF; vi: (1) TBAF, THF, (2) PCC, CH_2Cl_2, (3) H_2, 10% Pd/C, aq. AcOH and (4) NaOH, H_2O.

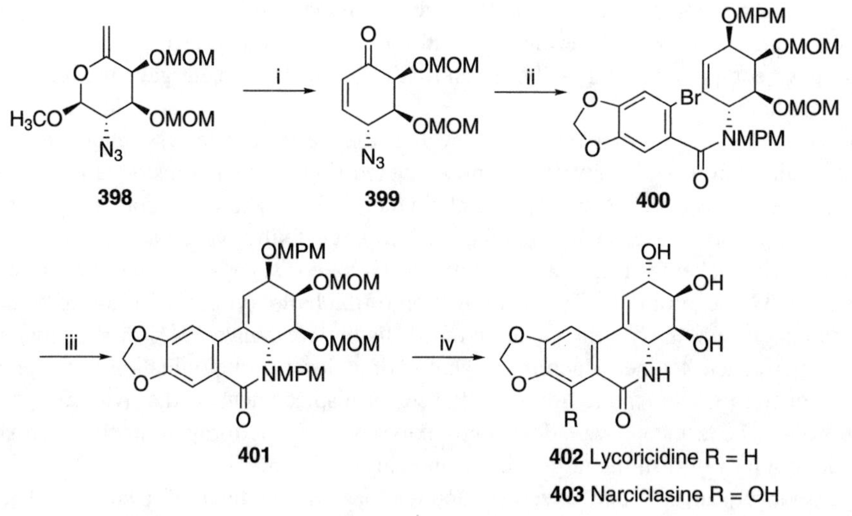

398 399 400 401

402 Lycoricidine R = H

403 Narciclasine R = OH

SCHEME 11.82 Reagents: i: (1) $(CF_3CO_2)_2Hg$, acetone, H_2O and (2) MsCl, Et_3N, CH_2Cl_2; ii: (1) $NaBH_4$, $CeCl_3{\cdot}7H_2O$, MeOH, (2) NaH, MPMCl, DMF, (3) $LiAlH_4$, Et_2O, (4) 6-bromopiperonylic acid, $(EtO)_2P(O)CN$, Et_3N and (5) NaH, MPMCl, DMF; iii: $Pd(OAc)_2$, DIPHOS, TlOAc, DMF; iv: (1) DDQ, CH_2Cl_2, H_2O, (2) Ph_3P, DEAD, PhCOOH, (3) MeONa, MeOH, (4) 1*N* HCl, THF then Ac_2O pyridine and (5) TFA $CHCl_3$.

C2 epimer failed [309]. Inversion of configuration at C2 was thus performed and final deprotection gave enantiomerically pure lycoricidine **402**.

Pancratistatin **408** is a close analog of lycoricidine, which was also the subject of intense investigations. In a synthetic approach (Scheme 11.83), Danishefsky studied the Ferrier carbocyclization as a means to construct the C ring, but prior introduction of the aromatic ring at C6 of the sugar was achieved [310]. Thus, aldehyde **404** was condensed with the lithium anion of **405** to give a mixture of the expected alcohols. Dehydration occurred on heating in pyridine to give **406**. Though carbocyclization of the latter proceeded well to give **407**, problems arose from

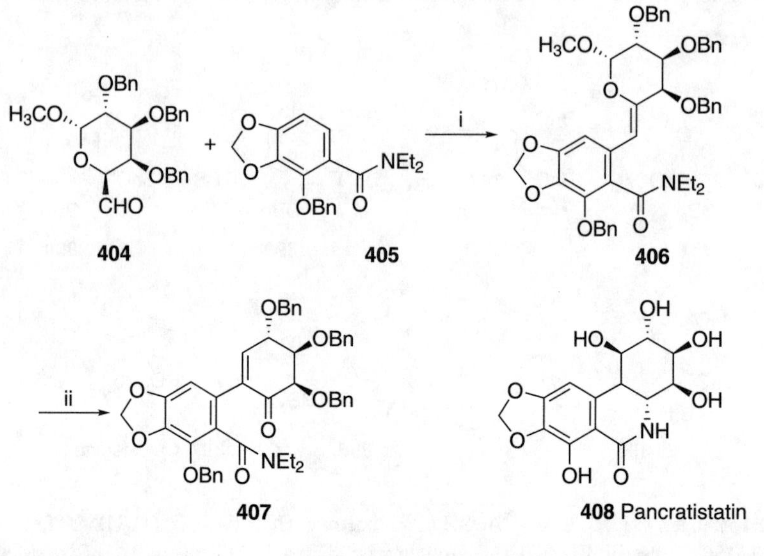

404 405 406 407

408 Pancratistatin

SCHEME 11.83 Reagents: i: (1) *s*BuLi, TMEDA, THF, (2) Ph_3PBr_2, K_2CO_3, CH_2Cl_2, 75% and (3) pyridine, reflux; ii: (1) $HgCl_2$, H_2O, CH_3CN and (2) Ac_2O, DMAP, pyridine.

the unexpectedly complex manipulation of the tertiary amide used on the aromatic ring. This problem was solved in our own synthetic approach to narciclasine **403** [311].

Branchaud [312,313] has reported other synthetic approaches to pancratistatin, which are also based on a Ferrier carbocyclization to produce the C ring. In these approaches, arylation of the C ring was studied.

Two synthetic approaches to both C and A and A and D rings of taxol have been reported by Ermolenko [314,315] and Chida [316]. In each case, the carbocyclization occurred early in the synthesis to produce a mostly deoxygenated cyclohexanone, which was elaborated using standard transformations. An enantiospecific synthesis of a taxol A building unit has be reported, which used a single chiral center of L-arabinose [317]. Based on the same approach, other natural products that have been synthesized include hygromycin [318], valienamine [319], cyclophellitol [320], mesembranol [321], FR65814 [322] and paniculide [323]. In this last case, Ferrier carbocyclization was carried out on an annulated sugar (with a five-membered lactone ring branched at C4 and O3). It is worth noting that the reaction was incomplete, probably because of the presence of the fused ring system [324,325].

11.6.1.2 Radical Cyclizations

Radical reactions can be performed under mild conditions compatible with many functional groups, and protecting groups are generally not required. The intramolecular addition of radicals to double bonds is a favored process, often used in carbohydrate chemistry. With this methodology, the formation of carbocycles from sugars can be easily accomplished. Although the 5-*exo* mode of cyclization is favored, it is also possible to construct six-membered rings by a 6-*exo* pathway. Examples of six-membered ring formations are reviewed below.

In connection with a program on developing free radical methodology, Fraser-Reid described several approaches to densely functionalized carbocycles. This led to the construction of the carbocyclic core of tetrodotoxin [326]. The synthesis of cyclophellitol **414** and its *epi* derivative, described here, illustrates the principles of this strategy [327].

As shown in Scheme 11.84, glucal derivative **409** was oxidized to the corresponding aldehyde, which was treated with the lithium derivative of phenylacetylene to give a mixture of alcohols protected as their acetates **410**. Iodoglycosylation of the glycal with 4-methoxybenzyl alcohol gave

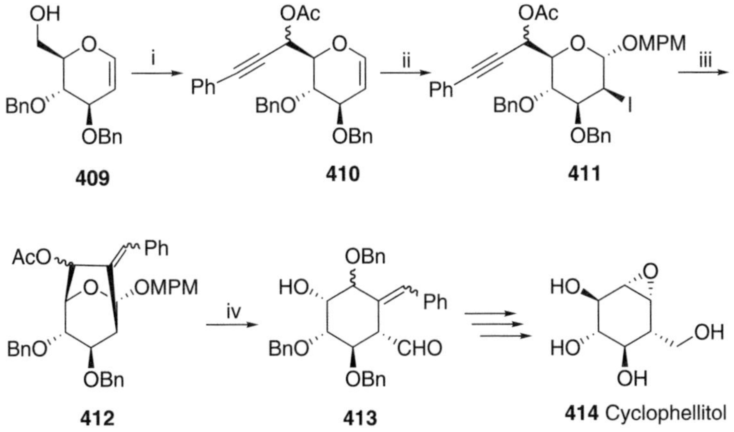

OH, BnO, OBn, **409**; i; OAc, Ph, BnO, OBn, **410**; ii; OAc, OMPM, Ph, BnO, I, OBn, **411**; iii

AcO, Ph, OMPM, BnO, OBn, **412**; iv; OBn, HO, Ph, BnO, CHO, OBn, **413**; HO, O, OH, HO, OH, **414** Cyclophellitol

SCHEME 11.84 Reagents: i: (1) $(COCl)_2$, DMSO, (2) BuLi, $HCCC_6H_5$ and (3) Ac_2O, pyridine; ii: NIS, 4-OMe–$C_6H_4CH_2OH$, CH_3CN, 92%; iii: (1) AIBN, Bu_3SnH, benzene, (2) MeONa, MeOH and (3) NaH, BnBr; iv: DDQ, CH_2Cl_2, H_2O.

the key compound **411**. Treatment of this iodo compound with tributyltin hydride formed a radical at C2, which was added to the triple bond to form **412**, together with the 2-deoxy derivative of **411** (4:1). Opening of the pyran ring gave the carbocycle **413**, subsequently elaborated to the Tatsuta intermediate of the cyclophellitol synthesis [328]. A review on cyclophellitol syntheses appeared recently [329].

A second example is drawn from the *Amaryllidaceae* alkaloid field. The key step of the synthesis of 7-deoxy pancratistatin is a radical addition to a nitrone derived from a sugar [330,331]. As shown in Scheme 11.85, D-gulonolactone **415** was transformed into oxime **416** in a seven-step sequence. The connection between the sugar skeleton and the aromatic part was realized by esterification of 4-bromo piperonylic alcohol with acid **416** to give **417**. Treatment of the latter with butyllithium promoted rearrangement to a benzylic alcohol, which was immediately oxidized to the aldehyde **418**. Hemiacetal formation between this aldehyde and the C1 hydroxyl group gave **419**. Ketone reduction gave a mixture of alcohols, which were transformed into the imidazolyl thionocarbonate **420**. This radical precursor was then reacted with tributyltin hydride to promote cyclization, giving **421** with the correct stereochemistry. Lactone to lactam exchange and protecting groups' removal gave the target 7-deoxy-pancratistatin **422**. Other examples of radical cyclization leading to six-membered carbocycles relevant to cyclitols chemistry have been reported and were reviewed by Marco Contelles [332].

Intramolecular 5-*exo* radical additions to double bonds forming five-membered rings have been widely explored over the last few decades, and are probably the methods of choice to form such rings. The radicals can be produced in different ways. The classical tributyltin-mediated formation of radicals from halides, thionocarbonates, and so on, has been widely used. More recently, diiodosamarium-mediated radical formations have been explored. In these cases, aldehydes are reacted with SmI_2 to generate ketyl radicals, which can be added to activated double bonds.

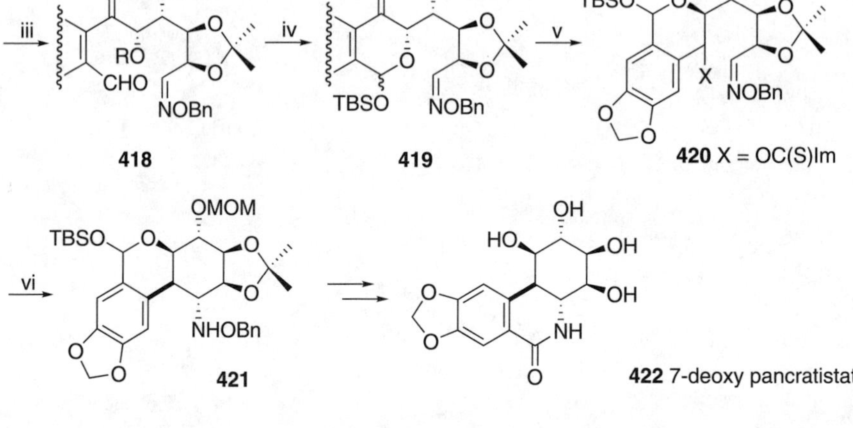

SCHEME 11.85 Reagents: i: (1) TBSCl, imidazole, (2) DIBAL, (3) $BnONH_2$·HCl, (4) MOMCl, DIEA, (5) HF·pyridine, (6) TPAP, NMO and (7) $NaClO_2$, KH_2PO_4; ii: Ph_3P, DEAD, 4-bromo-5-(hydroxymethyl)-1,2-(methylenedioxy)benzene; iii: (1) BuLi and (2) TPAP, NMO; iv: (1) HF·pyridine and (2) TBSCl; v: (1) $NaBH_4$, MeOH and (2) $CS(Im)_2$, DMAP, pyridine; vi: Bu_3SnH, AIBN.

Although a lot of methodological investigations have been reported [332], total syntheses of natural compounds relying on this strategy are scarce.

Naturally occurring cyclopentylamines are found in allosamidin, mannostatin and trehazolin, all of which can be regarded as sugar mimics, exhibiting inhibitory activities toward glycosidases [333].

Radical additions to *O*-alkyl aldoximes, easily accessible from reducing sugars, have been exploited for the synthesis of allosamizoline, the aglycon of allosamidin [334]. The C ring of anguidine has also been prepared by reductive carbocyclization of an aldehyde on an unsaturated ester [335].

Two interesting syntheses of trehazolamine [336,337] appeared at the same time, based on a diiodosamarium-promoted reductive carbocyclization. A rapid description of both methods will shed light on this strategy.

Giese's synthesis (Scheme 11.86) started with oxime **423**, easily obtained from D-glucose [336]. The key cyclization was carried out by formation of a ketyl radical at C5 on treatment with diiodosamarium. Addition of this radical on the oxime double bond gave **424** as a single diastereoisomer. Inversion of configuration at C4 was achieved to give **425**. Protecting group removal gave trehazolamine **426**.

Chiara's synthesis (Scheme 11.86) is based on a pinacol reaction that is also promoted by diiodosamarium [338]. Keto aldehyde **427**, obtained in a few steps from glucose, was treated with SmI_2 to give **428** as a 1:1 mixture of isomers [337]. Attempts to introduce the nitrogen group by direct substitution of the secondary alcohol failed; thus, the unsaturated derivative **429** was prepared from the above mixture. Sharpless epoxidation with (−)-diisopropyl tartarate gave epoxide **430** with the correct configuration, which was opened with sodium azide. Concomitant azide reduction and debenzylation gave trehazolamine **426**. Additional examples of model reactions of carbocyclizations can be found in two recent reviews [303,332].

11.6.1.3 Aldolisation and Related Methods

Intramolecular aldol reactions and other olefinations are valuable methods for carbocycle formation. However, possible limitations are the often strongly basic reaction conditions needed for such

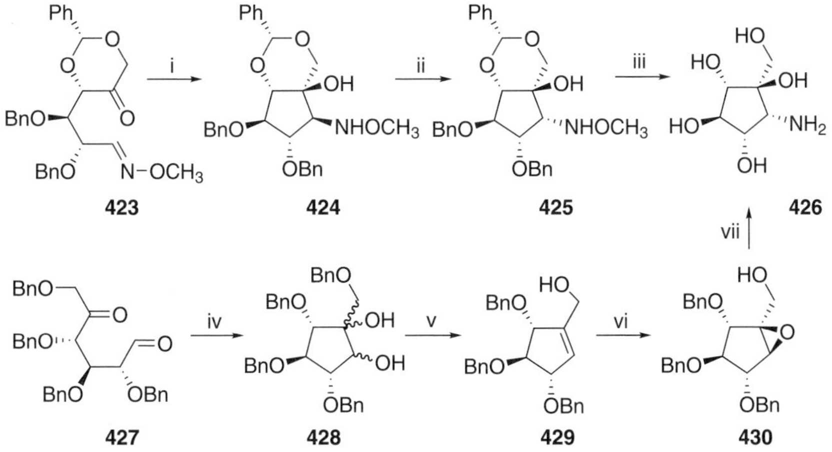

SCHEME 11.86 Reagents: i: SmI_2, *t*BuOH, THF; ii: (1) Ac_2O, pyridine, (2) $Pb(OAc)_4$, benzene, (3) K_2CO_3, MeOH and (4) $LiAlH_4$, MeONa, THF; iii: Na, NH_3; iv: SmI_2, THF, *t*BuOH; v: (1) $CS(Im)_2$, toluene, (2) Ac_2O, TMSOTf, (3) $(EtO)_3P$, heat and (4) MeONa, CH_2Cl_2, MeOH; vi: L-DIPT, $Ti(OiPr)_4$, *t*BuOOH, CH_2Cl_2; vii: (1) LiN_3, NH_4Cl, DMF and (2) H_2, $Pd(OH)_2$.

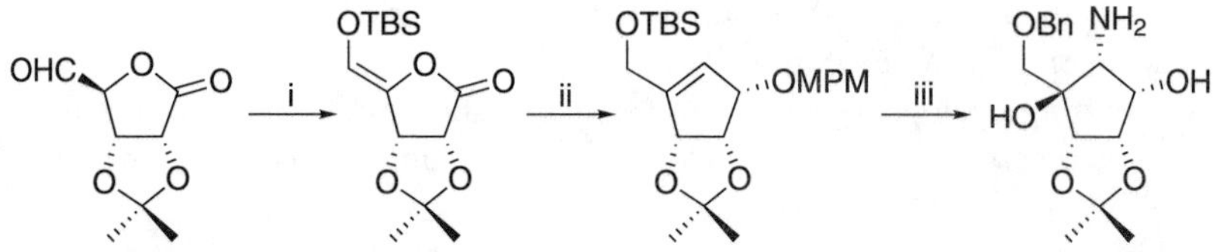

SCHEME 11.87 Reagents: i: DABCO, TBSCl, DMF, 43%; ii: (1) LiCH=PPh_3, THF, (2) $NaBH_4$, $CeCl_3 \cdot 7H_2O$, EtOH, 89% and (3) 4-OMe–$C_6H_4CH_2Cl$, NaH, DMF; iii: (1) TBAF, THF, (2) BnBr, NaH, DMF, (3) *m*CPBA, CH_2Cl_2, (4) DDQ, CH_2Cl_2, H_2O, (5) NaN_3, NH_4Cl, DMF and (6) Ph_3P, THF then H_2O.

transformations. Some examples have been reported, including the synthesis of the unsaturated carbanucleoside neplanocin A using an intramolecular Wadsworth–Emmons reaction [339]. Phosphorus-based carbanions have also been used by Vasella in his synthesis of a glyoxylase I inhibitor 2-crotonyl-(4*R*, 5*R*, 6*R*)-4,5,6-trihydroxylcyclohex-2-enone (COTC) by reacting an enol lactone with a suitable phosphonate [340]. Cyclohexenones are available by this route.

Enol lactones have been used to construct cyclopentenones [341], which have been used to prepare mannostatin A [342]. More recently, a related strategy was applied to the synthesis of trehazolin analogs [343].

As shown in Scheme 11.87, aldehyde **431** was transformed into silylenol ether **432**, which was treated with methylene triphenylphosphorane to give the expected cyclopentenone. Reduction of the keto group and protection of the resulting alcohol gave **433**. As in Chiara's synthesis, epoxidation and opening of the latter with sodium azide completed the synthesis of 6-*epi*-trehazoline **434**.

An aldol type reaction was used in a recent synthesis of COTC [344]. In this case (Scheme 11.88), a vinyl sulfone was prepared from ribonolactone derivative **435** by nucleophilic addition of a methylphenyl sulfone anion. Tin chloride-mediated ring closure of **436** produced the α-sulfonyl cyclohexenone **437**. Conjugate addition of tributylstannyllithium and quenching with formaldehyde gave, after elimination of the tributylstannyl group, the hydroxymethylated derivative **438**. Deprotection and esterification gave the target compound COTC.

Gabosines have been isolated from *Streptomyces* and are structurally related to COTC. Gabosine I has been recently synthesized from glucose by Lubineau, taking advantage of the Kishi–Nozaki coupling between the aldehyde at C1 and a vinyl bromide branched at C5. The vinyl bromide resulted from a Wittig reaction of the 5-keto derivative [345]. Other methods have been specially devised to construct the cyclohexane ring of (−)-fumagillol by internal nucleophilic displacement of a primary tosylate by a suitable enolate [346].

To conclude this nonexhaustive review of carbocyclization methods, we should mention intramolecular cycloaddition as a useful tool to construct carbocycles. Nitrile oxide–olefin cycloaddition seems to be a useful strategy to construct carbocycles from carbohydrates because

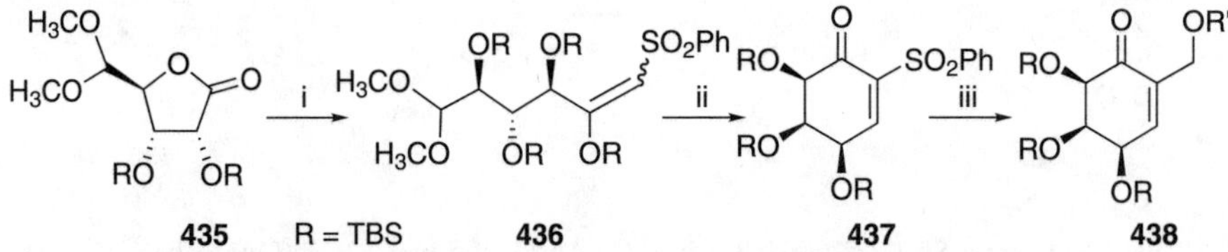

SCHEME 11.88 Reagents: i: (1) $MeSO_2Ph$, BuLi, THF and (2) TBSOTf, 2,6-lutidine; ii: $SnCl_4$, CH_2Cl_2; iii: (1) Bu_3SnLi, THF then HCHO gas and (2) SiO_2, benzene.

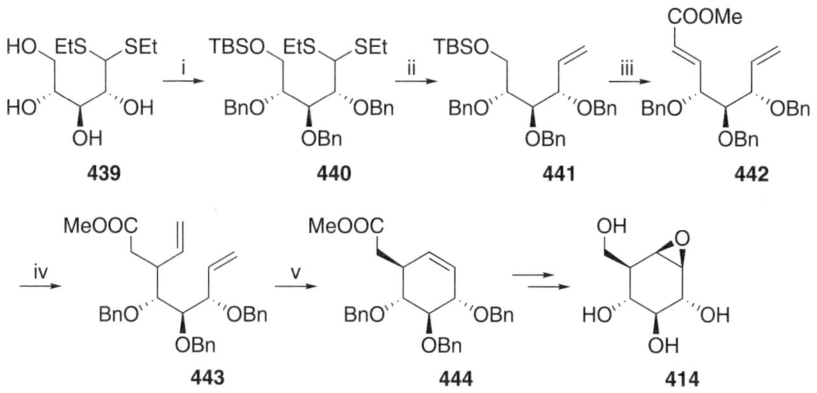

SCHEME 11.89 Reagents: i: (1) TBSCl, imidazole, CH_2Cl_2, 90% and (2) NaH, BnBr, Bu_4NI, 80%; ii: (1) HgO, $HgCl_2$, acetone, H_2O, 80% and (2) $Cp_2TiClAlMe_3$, pyridine, toluene, THF, 80%; iii: (1) TBAF, THF, 88%, (2) DMSO, $(COCl)_2$, Et_3N, CH_2Cl_2 and (3) Ph_3PCHCO_2Me, CH_2Cl_2; iv: $(CH_2{=}CH)_2CuMgBr$, TMSCl, THF, 90%; v: 15 mol% $(Cy_3P)_2RuCl_2(CHPh)$, CH_2Cl_2, 92%.

nitrile oxides are available from oximes easily derived from reducing sugars. Two applications of this strategy have been reported. The first one is the synthesis of cyclophellitol in about 15 steps from L-glucose [328], and the second one is the synthesis of allosamizoline from D-glucosamine [347]. In both cases, good diastereoselectivity was observed, and the method is suitable to form both five- and six-membered rings. Gabosines C and E have been prepared from D-ribose in this way [348]. Chain extension at C1 of the reducing sugar by condensation of vinylmagnesium bromide yielded the olefin. The nitrile oxide was then constructed by oxidation of the primary hydroxyl group to an aldehyde followed by oxime formation. Classical elaboration of the nitrile oxide and cycloaddition gave the cycloadduct with high stereocontrol.

A variant of the above method is the nitrone–olefin cycloaddition, exploited for the synthesis of shikimic acid [349] and for the synthesis of the carbocyclic nucleoside neplanocin [350]. RCM has been recently exploited in Ziegler's synthesis of cyclophellitol [351]. This synthesis encompasses a number of steps described throughout this chapter, including chain extensions at C1, at C6 and branching by Michael addition followed by carbocyclization.

The synthesis shown in Scheme 11.89 began with xylose dithioacetal **439**. Protection of the primary alcohol as a silyl ether and benzylation of the secondary hydroxyl groups gave **440**. Aldehyde formation and Tebbe methylenation yielded **441**. Chain extension at C5 was performed by classical Wittig homologation to afford ester **442**. The vinyl group, required to carry out RCM, was introduced via the copper-assisted addition of vinylmagnesium bromide, giving **443**. Treatment of **443** with Grubbs catalyst led to olefin **444**. Five steps were then needed to remove one extra carbon and establish the epoxide function of **414**.

11.6.2 Annulation Reactions on the Sugar Template

This final section deals with the construction of carbocycles, fused or spiro, to the sugar ring. Such constructions are often called annelated sugars. As soon as a second cycle is fused to the sugar, the template effect is operating, and highly stereocontrolled reactions can be carried out on this second cycle. One should note that most of the carbocyclization methods described above can be employed to construct an annelated sugar. The use of this approach in total synthesis is of interest for the construction of polycyclic molecules but could be also used to construct a single carbocycle in which parts of the sugar backbone become substituents.

11.6.2.1 Cycloadditions

Cycloadditions are widely used methods for carbocycle construction [352]. In this context, the Diels–Alder reaction has been studied in carbohydrate chemistry. Using this approach, the sugar can be used as the dienophile or as the diene with a large preference for the former.

Easily prepared from glycols, enones have been investigated as dienophiles. They react with butadiene under Lewis acid catalysis to form chiral cyclohexenes used in the synthesis of compactin analogs [353]. Levoglucosenone has been used in a Diels–Alder reaction with acetoxybutadiene to construct a part of the indole alkaloid reserpine [354], and in synthetic studies toward tetrodotoxin [355]. Analogs of the anthracycline rhodomycinone have been similarly prepared [356]. [4 + 2]-Cycloaddition of the same enone with silyloxydiene allowed the creation of the fused ring system present in actinobolin [357].

The intramolecular Diels–Alder reaction (IMDA) is often more rewarding in terms of stereocontrol. Many disfavored cycloadditions can be achieved intramolecularly. Enone derived from glycals have been elaborated to naphtopyrans by IMDA [358]. A key intermediate *en route* to forskolin was prepared by an IMDA of an enone [359]. Our group applied a related strategy to construct the decalinic structures present in the mevinic acids by combining IMDA reactions and Ferrier carbocyclizations [324]. The diene was tethered to the sugar 2,3-ene-4-one via a glycosidic bond. This dictated the stereochemistry at C2 and C3, which represented the *cis* decalin ring junction needed in the natural compound [325].

An approach to the construction of the forskolin tricyclic system has been proposed by Hanna, taking advantage of the IMDA reaction [256,360]. This creative approach cumulates a number of efficient reactions mentioned in the preceding sections. As shown in Scheme 11.90, the synthesis started with allyl-*C*-glycosyl derivative **445**, which underwent an Ireland–Claisen rearrangement to furnish **446** in 92% yield. Chemical manipulation of the allyl group allowed the formation of a protected aldehyde group, while the ester chain was converted into the other aldehyde group of **447**. Wittig homologation introduced the diene part of **448**. Deprotection of the dimethyl acetal of **448** gave the corresponding aldehyde, which was treated with allyl Grignard. Subsequent Dess–Martin oxidation gave the ketone **449**. Finally, IMDA reaction occurred at 0°C in the presence of boron trifluoride etherate, and was highly stereoselective giving **450** as a single isomer. Heating at 160°C

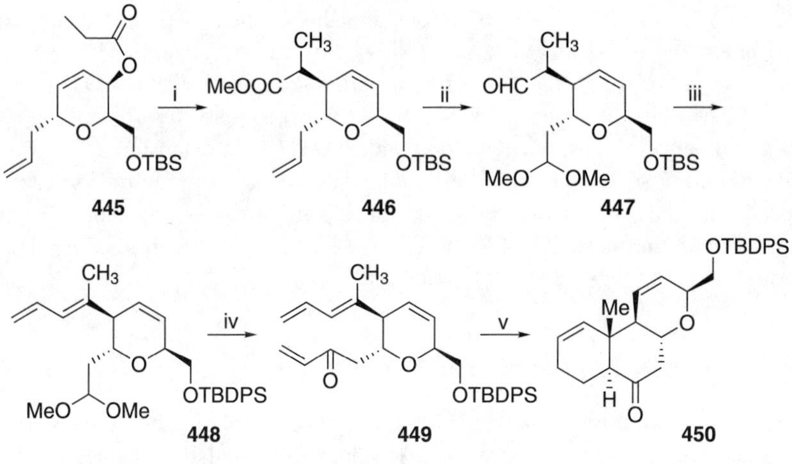

SCHEME 11.90 Reagents: i: (1) HMDS, BuLi, THF, TBSCl, DMPU, (2) toluene, reflux and (3) KF, $KHCO_3$, MeI; ii: (1) OsO_4, NMO, (2) $NaIO_4$, EtOH, H_2O, (3) $HC(OMe)_3$, PPTS, CH_2Cl_2, (4) DIBAL, CH_2Cl_2 and (5) Dess–Martin periodinane; iii: (1) $CH_2{=}CHMgBr$ and (2) $POCl_3$, pyridine; iv: (1) TFA, H_2O, (2) TBDPSCl, DMAP, pyridine, CH_2Cl_2, (3) $H_2C{=}CHMgBr$, THF and (4) Dess–Martin periodinane; v: $BF_3{-}Et_2O$, CH_2Cl_2, 50%.

in toluene was found to be far less selective. Attempts to introduce the *gem* dimethyl substituent by using the appropriate dienophile chain were unsuccessful. This highlights one of the limitations of such cycloaddition reactions. An excellent application of the IMDA reaction of a complex *C*-glycoside with an enone derived from L-rhamnal can be found in a synthetic approach to azadirachtin by Fraser-Reid's group [257].

The nitrile oxide–olefin cycloaddition has been used to construct the oxahydrindene part of avermectins. The olefin was introduced at C3 of a glucofuranose unit, whereas the nitrile oxide was created by chain elongation at C5 [361].

RCM has also been investigated as a tool to form annulated sugars. Two successive *C*-branchings at C3 and C4 of allyl chains are needed. RCM then allows the formation of a cyclohexene ring [362]. This technique has not yet been applied to natural product synthesis.

11.6.2.2 Anionic Cyclizations

The Robinson annulation has been widely used in organic synthesis to construct carbocycles. Carbohydrate chemistry has also been concerned by this strategy and several cases have been studied. In the first case, detailed here, the two partners of the reaction — that is, the nucleophile and the electrophile — are located on two chains grafted on the sugar skeleton. The complex ring structure of trichothecene was constructed using this strategy [363,364]. The total synthesis of verrucarol, a member of this class of compounds, was reported by the same group [365].

As shown in Scheme 11.91, the synthesis started from compound **202**, prepared via a Claisen rearrangement. This compound was homologated to the malonate **451**. Cleavage of the 5,6-*O*-isopropylidene group and Malaprade oxidation of the resulting diol gave the expected aldehyde. With both partners in place, cyclization was performed by brief treatment with sodium methoxide to yield **452**. The vinyl appendage was ozonized with reductive workup to introduce the primary hydroxyl function protected as a methoxymethyl (MOM) ether. Further elimination and removal of one of the ester groups then gave the bicyclic compound **453**. Subsequent manipulations transformed the ester function into the methyl group required in the target. Furanose ring modification was then carried out — deoxygenation at C2 and oxidation at C1 — to obtain the lactone **454**.

The total synthesis of verrucarol needed the elaboration of the C ring and an enlargement of the B ring. This problem was solved by Dieckman condensation between the lactone and an appropriate ester group was introduced by sequential alkylation of **454**. Methylation of the lactone enolate, followed by condensation with 4-*O*-(*tert*-butyldiphenylsilyloxy)-butanal, gave **455** as a mixture of diastereoisomers. The synthesis was continued with the (*R*) isomer. Six steps were needed to modify the side chain and to introduce the ester function. Potassium hexamethyl disilazane-mediated Dieckman condensation gave the mixture of diasteroisomers **456** (5:4 ratio). Protection of the hemiacetal function and Barton–Crich decarboxylation gave **457** as a mixture of isomers at the alcohol. Mesylation of this alcohol and removal of the acetal protecting group allowed ring expansion to **458**. Transformation into verrucarol required an additional seven steps.

Another approach to the tricothecene skeleton proposed by Fraser-Reid takes advantage of carbohydrate enolate chemistry to construct the A ring [366]. The C ring was formed by intramolecular sulfonate displacement by an appropriate amide enolate. A route to the same skeleton involving [2 + 2] cycloaddition on the double bond of an enone obtained from a glycal has been proposed by Fetizon [367]. Ring expansion of the cyclobutene obtained this way can be effected to construct the C ring of these compounds.

A bis-annulated carbohydrate has been reported by Pipelier and Ermolenko [368] in their synthetic route to quadrone. This complex structure was elaborated from levoglucosan by two successive branchings of allyl groups at C2 and C4 using epoxide opening methodology. A keto group, formed from the alcohol at C3, allowed initial cyclopentenone construction. Further manipulation of this ketone allowed a second Robinson annulation to form the second fused six-membered carbocycle.

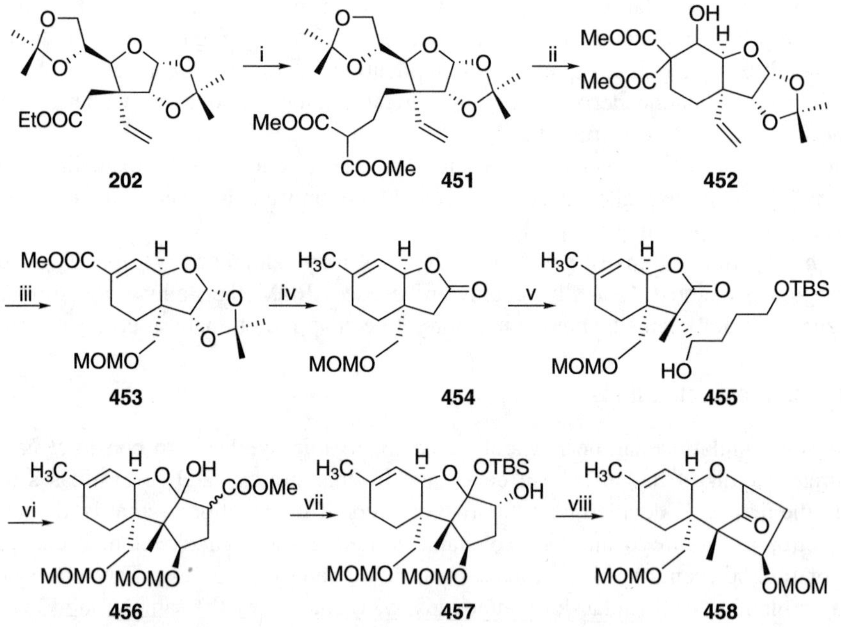

SCHEME 11.91 Reagents: i: (1) $LiAlH_4$, THF, (2) Ph_3P, DEAD, MeI, THF and (3) NaH, $CH_2(CO_2Me)_2$, THF; ii: (1) 60% aq. AcOH, 99%, (2) $NaIO_4$, H_2O, MeOH and (3) MeONa, MeOH; iii: (1) Ac_2O, pyridine, 92%, (2) O_3, MeOH, CH_2Cl_2 then $NaBH_4$, 98%, (3) NaCl, DMSO, H_2O, heat, 47% and (4) MOMCl, *i*Pr_2NEt, 83%; iv: (1) DIBAL, CH_2Cl_2, (2) Ph_3P, CCl_4, benzene, 88%, (3) Bu_3SnH, AIBN, toluene, (4) 60% aq. TFA, (5) PTSA, MeOH, (6) NaH, imidazole, then CS_2, then MeI, (7) Bu_3SnH, AIBN, benzene and (8) Jones reagent, acetone; v: (1) LDA, THF then MeI, (2) LDA, THF then 4-*O*-(*t*butyldiphenylsilyloxy)butanol; vi: (1) TBAF, THF, (2) PivCl, pyridine, (3) MOMCl, *i*Pr_2NEt, CH_2Cl_2, (4) MeONa, MeOH, (5) Jones reagent, acetone, (6) CH_2Cl_2, Et_2O, $CHCl_3$ and (7) KHMDS, THF; vii: (1) TBSOTf, 2,6-lutidine, CH_2Cl_2, (2) 4M KOH, MeOH, (3) WSC, DMAP, *N*-hydroxypyridine-2-thione, *t*BuSH, O_2, CH_2Cl_2, (4) MsCl, pyridine and (5) TBAF, THF.

Robinson annulation of carbohydrates has been reported based on the chemistry of sugar enolates. This annulation process has been used for the synthesis of the C ring of taxanes. As shown in Scheme 11.92, ketone **459** was enolized with LiTMP and treated with (trimethylsilyl)but-3-en-2-one to yield adduct **460**, which was then cyclized to **461** in basic medium [369,370]. This enone was reduced to the corresponding allylic alcohol. Bromoacetal formation and radical mediated ring closure gave **462** [371,372]. The hydroxyl groups, resulting from a Tamao oxidation, were silylated to give **463**. Benzylidene ring opening gave the 6-bromo derivative, which was treated under Vasella's conditions to yield **464** [373]. The latter was elaborated to a chiral branched-cyclohexane **465**, a precursor of the taxane C ring.

Some constructions of aromatic rings fused to carbohydrates have been reported in connection with the synthesis of quinone type antibiotics. A strategy of this type was involved in the synthesis of olivin. 1,4-Addition of a toluyl anion to a sugar enone was followed by an intramolecular aldol to close the ring [374]. This strategy was also applied to the synthesis of cryptosporin using a nitro sugar as Michael acceptor [375]. The construction of spiro-cyclopentane on carbohydrates using intramolecular aldol reactions have been reported *en route* to punaglandin IV [124].

11.6.2.3 Radical Cyclizations

It is probably superfluous to say that radical reactions are suitable for the construction of carbocycles. What is working to achieve carbohydrate ring closure is obviously suitable to form

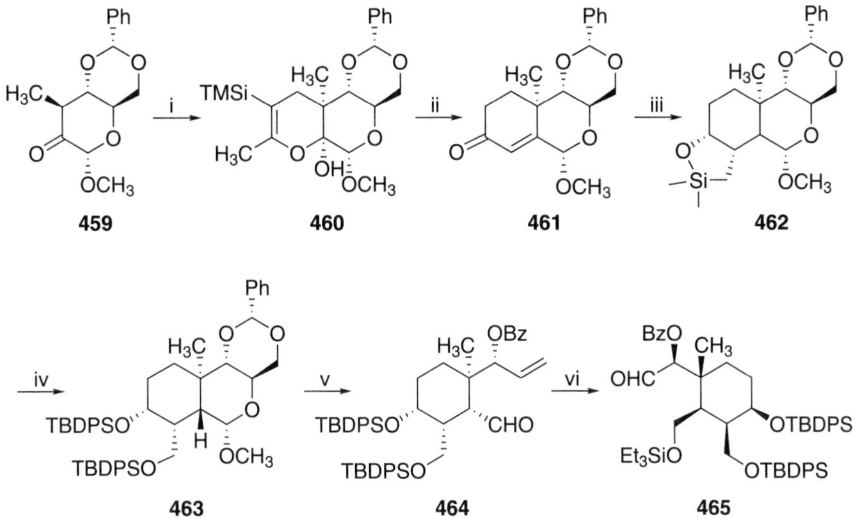

SCHEME 11.92 Reagents: i: LiTMP, Et_2O, then β-(trimethylsilyl)but-3-en-2-one; ii: KOH, MeOH; iii: (1) L-selectride, toluene, (2) $ClSiMe_2CH_2Br$, Et_3N and (3) Bu_3SnH, AIBN; iv: (1) H_2O_2, KF and (2) TBDPSCl, imidazole, CH_2Cl_2; v: (1) NBS, $BaCO_3$, CCl_4 and (2) Zn, *i*PrOH, 61%; vi: (1) $NaBH_4$, *i*PrOH, (2) Et_3SiCl, imidazole, CH_2Cl_2 and (3) O_3, CH_2Cl_2 then Me_2S.

carbocyclic structures. In particular, cascade radical reactions have been used to construct diquinane skeletons [131–133]. Details of such approaches have been given in an earlier section. A synthesis of Woodward's densely functionalized carbocyclic precursor to reserpine has been reported based upon serial radical 5-*exo*/6-*exo* cyclizations of dienic hexopyranoses [376].

11.7 CONCLUSIONS

It is clear that carbohydrate chemistry has reached a high level of sophistication, mainly because of the recognition that a sugar backbone can be incorporated into a large number of compounds close to, but often far away from, sugar structures. It is interesting to note that extremely complex molecules such as polyether toxins succumbed to total synthesis using carbohydrates as starting materials. The successful syntheses of enantiomerically pure compounds from sugars are now well established and, since the beginning of the development of this concept, numerous methods have been proposed. The organization of this review by a variety of reactions or methods clearly leads us to the conclusion that several synthetic strategies are used. It is also clear that emerging synthetic reactions can be successfully applied to carbohydrates. Radical reactions proved to be outstanding methods for many transformations. This also seems true for olefin metathesis. Specific methods for carbohydrate carbocyclization, such as the Ferrier reaction, have largely extended the scope of carbohydrate-using synthesis. Undoubtedly, other recently discovered methods for the synthesis of chiral multifunctional cyclopentanes (or cycloctanes) will find application in total syntheses in the near future. All types of naturally occurring structures could probably be synthesized from sugars. Nevertheless, the number of steps could be prohibitive depending on the nature of the target. It is clear that the development of new synthetic transformations of sugars with only a limited number of protecting groups is still needed. As in other fields of chemistry, there is still room for further methodological investigation to improve stereoselectivity, efficacy and to find new environmentally friendly reagents.

REFERENCES

1. Woodward, R B, The total synthesis of vitamin B 12, *Pure Appl. Chem.*, 33, 145–177, 1973.
2. Woodward, R B, Bader, F E, Bickel, H, Frey, A J, Kierstead, R W, The total synthesis of reserpine, *J. Am. Chem. Soc.*, 78, 2023–2035, 1956.
3. Inch, T D, The use of carbohydrates in the synthesis and configurational assignment of optically active non-carbohydrate compounds, *Adv. Carbohydr. Chem. Biochem.*, 27, 191–225, 1972.
4. Hanessian, S, Approaches to the total synthesis of natural products using chiral templates derived from carbohydrates, *Acc. Chem. Res.*, 159–165, 1979.
5. Fraser-Reid, B, Anderson, R C, Carbohydrate derivatives in the asymmetric synthesis of natural products, *Fortschr. Chem. Org. Naturst.*, 39, 1–61, 1980.
6. Fraser-Reid, B, Sun, K M, Tam, T F, Carbohydrate derivatives in the asymmetric synthesis of natural products: some applications of furanose sugars, *Bull. Soc. Chim. Fr.*, 238–246, 1981.
7. Inch, T D, Formation of convenient chiral intermediates from carbohydrates and their use in synthesis, *Tetrahedron*, 40, 3161–3213, 1984.
8. Williams, N R, Davison, B E, Ferrier, R J, Furneaux, R H, Synthesis of enantiomerically pure noncarbohydrate compounds, *Carbohydr. Chem.*, 17, 244–255, 1985.
9. Krohn, K, Chiral building blocks from carbohydrates, *Nachr. Chem. Tech. Lab.*, 35, 1155–1160, 1987.
10. Hollingsworth, R I, Wang, G, Toward a carbohydrate-based chemistry: progress in the development of general-purpose chiral synthons from carbohydrates, *Chem. Rev.*, 100, 4267–4282, 2000.
11. Nicolaou, K C, Mitchell, H J, Adventures in carbohydrate chemistry: new synthetic technologies, chemical synthesis, molecular design, and chemical biology, *Angew. Chem. Int. Ed.*, 40, 1576–1624, 2001.
12. Vasella, A, Chiral building blocks in enantiomer synthesis from sugars, In *Modern Synthetic Methods*, Vol. 2, Scheffold, R, Ed., Otto Salle Verlag, Frankfurt/Main, pp. 173–267, 1980.
13. Scott, J W, Readily available chiral carbon fragments and their use in synthesis, In *Asymmetric Synthesis*, Vol. 4, Morrison, J D and Scott, J W, Eds., New York, Academic Press, p. 1, 1980.
14. Fraser-Reid, B, Tsang, R, Carbocycles from carbohydrates: the annulated sugar approach, In *Strategies and Tactics in Organic Synthesis*, Vol. 2, Lindberg, T, Ed., Academic Press, New York, 1988.
15. Hanessian, S, *Total Synthesis of Natural Products: The Chiron Approach*, Pergamon Press, New York, 1983.
16. Boons, G-J, Hale, K J, *Organic Synthesis with Carbohydrates*, Sheffield Academic Press, Sheffield, 2000.
17. Wolfrom, M L, Lemieux, R U, Olin, S M, Configurational correlation of L-(levo)-glyceraldehyde with natural (dextro)-alanine by a direct chemical method, *J. Am. Chem. Soc.*, 71, 2870–2873, 1949.
18. Hanessian, S, Franco, J, Larouche, B, The psychobiological basis of heuristic synthesis planning-man, machine and the chiron approach, *Pure Appl. Chem.*, 62, 1887–1910, 1990.
19. Chapleur, Y, Chrétien, F, Selected methods for branched-chain carbohydrates synthesis, In *Modern Synthetic Methods in Carbohydrate Chemistry*, Hanessian, S, Ed., Marcel Dekker, New York, pp. 207–252, 1996.
20. For a review, see Williams, N R, Oxirane derivatives of aldoses, *Adv. Carbohydr. Chem. Biochem.*, 25, 109–180, 1970.
21. Fürst, A, Plattner, P A, Abstracts of papers, *12th Int. Congr. Pure & Appl. Chem.*, 409, 1951.
22. Hicks, D R, Fraser-Reid, B, The 2- and 3-*C*-methyl derivatives of methyl 2,3-dideoxy-α-D-*erythro*-hex-2-enopyranosid-4-ulose, *Can. J. Chem.*, 53, 2107–2123, 1975.
23. Hanessian, S, Rancourt, G, Guindon, Y, Assembly of the carbon skeletal framework of erythronolide A, *Can. J. Chem.*, 56, 1843–1846, 1978.
24. Oikawa, Y, Nishi, T, Yonemitsu, O, Chiral synthesis of polyketide-derived natural product. Chemical correlation of chiral synthons, derived from D-glucose for the synthesis of erythromycin A, with chemical cleavage products of the natural antibiotic, *J. Chem. Soc., Perkin. Trans. 1*, 27–33, 1985.
25. Tone, H, Nishi, T, Oikawa, Y, Hikota, M, Yonemitsu, O, A stereoselective total synthesis of (9*S*)-9-dihydroerythronolide A from D-glucose, *Tetrahedron Lett.*, 28, 4569–4572, 1987.

26. Sviridov, A F, Borodkin, V S, Ermolenko, M S, Yashunsky, D V, Kochetkov, N K, Stereocontrolled synthesis of erythronolides A and B in a (C5-C9) + (C3-C4) + (C11-C13) sequence from 1,6-anhydro-β-D-glycopyranose I, *Tetrahedron*, 47, 2291–2316, 1991.
27. Sviridov, A F, Borodkin, V S, Ermolenko, M S, Yashunsky, D V, Kochetkov, N K, Stereocontrolled synthesis of erythronolides A and B in a (C5-C9) + (C3-C4) + (C1-C2) + (C11-C13) sequence from 1,6 anhydro-β-D-glucopyranose (Levoglucosan), *Tetrahedron*, 47, 2317–2336, 1991.
28. Aburaki, S, Kinoshita, M, Improved synthesis of antimycin A3, *Bull. Chem. Soc. Jpn.*, 52, 198–203, 1979.
29. Beau, J.-M, Aburaki, S, Pougny, J R, Sinaÿ, P, Total synthesis of methyl (+)-pseudomonate C from carbohydrates, *J. Am. Chem. Soc.*, 621–622, 1983.
30. Pougny, J R, Sinaÿ, P, (3*S*,4*S*)-4-methylheptan-3-ol, a pheromone component of the smaller european alm bark beetle: synthesis from D-glucose, *J. Chem. Res.*, 1, 1982.
31. Bernet, B, Vasella, A, Carbocyclische verbindungen aus monosacchariden. I Umsetzung in der glucosereihe, *Helv. Chim. Acta*, 62, 1990–2016, 1979.
32. Fleet, G W J, Shing, T K M, Enantiospecific synthesis of (3*S*-hydroxy-2*S*-methyl) butyl triphenylphosphonium iodide: a precursor for the chiral side chain of pseudomonic acid, *Tetrahedron Lett.*, 24, 3657–3660, 1983.
33. Fleet, G W J, Gough, M J, Shing, T K M, Enantiospecific total synthesis of pseudomonic acids from arabinose, *Tetrahedron Lett.*, 24, 3661–3664, 1983.
34. Brockway, C, Kocienski, P, Pant, C, Unusual stereochemistry in the copper-catalyzed ring opening of a carbohydrate oxirane with vinylmagnesium, *J. Chem. Soc., Perkin. Trans. 1*, 875–878, 1984.
35. Kelly, A, Roberts, J S, A simple, stereocontrolled synthesis of thromboxane B2 synthon, *J. Chem. Soc., Chem. Commun.*, 228–229, 1980.
36. Challenger, S, Procter, G, Anhydro-D-glucopyranose in organic synthesis: preparation of a fragment for a synthesis of rosaramycin, *Tetrahedron Lett.*, 27, 391–394, 1986.
37. Procter, G, Genin, D, Challenger, S, 1,6-Anhydro-β-glucopyranose in organic synthesis: preparation of a fragment for the synthesis of rosaramycin, *Carbohydr. Res.*, 202, 81–92, 1990.
38. Mubarak, A, Fraser-Reid, B, Synthetic routes to 3-*C*-cyano-3-deoxy-D-galactopyranose derivatives, *J. Org. Chem.*, 47, 4265–4268, 1982.
39. Knierzinger, A, Vasella, A, Synthesis of 6-epithienamycin, *J. Chem. Soc., Chem. Commun.*, 9–11, 1984.
40. Rauter, A, Ferreira, M, Borges, C, Duarte, T, Piedade, F, Silva, M, Santos, H, Construction of a branched chain at C-3 of a hexopyranoside. Synthesis of miharamycin sugar moiety analogs, *Carbohydr. Res.*, 325, 1–15, 2000.
41. Linde, R G II, Egbertson, M, Coleman, R S, Jones, A B, Danishefsky, S J, Efficient preparation of intermediates corresponding to C22–C27 and C28–C34 of FK506, *J. Org. Chem.*, 55, 2771–2776, 1990.
42. Kary, P D, Roberts, S M, Watson, D J, Studies towards the total synthesis of altohyrtin A: a convergent approach to the C38–C51 carbon framework, *Tetrahedron: Asymmetry*, 10, 213–216, 1999.
43. Kary, P D, Roberts, S M, A second generation pyranose-based approach to the F ring (C38-C45) of altohyrtin A, *Tetrahedron: Asymmetry*, 10, 217–219, 1999.
44. Ermolenko, M S, Potier, P, Synthesis of epothilones B and D from D-glucose, *Tetrahedron Lett.*, 43, 2895–2898, 2002.
45. Sin, N and Kallmerten, J, Diastereoselective [2,3] Wittig rearrangement of carbohydrate-derived tertiary allylic ethers. 2. Synthesis of an advanced rapamycin intermediate from D-glucose, *Tetrahedron Lett.*, 34, 753–756, 1993.
46. Anliker, R, Dvornik, D, Gubler, K, Heusser, H, Prelog, V, Metabolic products of actinomycetes V. The lactone of β-hydroxy-α,α,γ-trimethylpimelic acid, a degradation product of narbomycin, picromycin, and methymycin, *Helv. Chim. Acta*, 39, 1785–1790, 1956.
47. Djerassi, C, Zderic, J A, The structure of the antibiotic methymycin, *J. Am. Chem. Soc.*, 78, 2907–2908, 1956.
48. Plaumann, D E, Fitzsimmons, B J, Ritchie, B M, Fraser-Reid, B, Synthetic routes to 6,8-dioxabicyclo[3.2.1]octyl pheromones from D-glucose derivatives. 4. Synthesis of (−)-α-multistriatin, *J. Org. Chem.*, 47, 941–946, 1982.

49. Sum, P-E, Weiler, L, Stereoselective synthesis of (−)-α-multistriatin from D-glucose, *Can. J. Chem.*, 60, 327–334, 1982.
50. Nakahara, Y, Fujita, A, Beppu, K, Ogawa, T, Total synthesis of antibiotic A23187 (calcimycin) from D-glucose, *Tetrahedron*, 42, 6465–6476, 1986.
51. Hanessian, S, Tyler, P C, Chapleur, Y, Reaction of lithium dimethylcuprate with conformationnaly biased acyloxy enol esters. Regio and stereocontrolled access to functionalized six-carbon chiral synthons, *Tetrahedron Lett.*, 22, 4583–4586, 1981.
52. Udodong, U E, Fraser-Reid, B, Formal total synthesis of 1β-methylcarbapenem via a novel route to deoxyamino sugars, *J. Org. Chem.*, 54, 2103–2112, 1989.
53. Hanessian, S, Faucher, A-M, Leger, S, Total synthesis of (+)-meroquinene, *Tetrahedron*, 46, 231–243, 1990.
54. Hanessian, S, Faucher, A, A general stereocontrolled strategy for the heteroyohimbine alkaloids: The total synthesis of (6)-ajmalicine and (+)-19-epiajmalicine, *J. Org. Chem.*, 56, 2947–2949, 1991.
55. Chapleur, Y, Grapsas, Y, Stereospecific formation of C–C bonds on ethyl 4,6-di-*O*-acetyl-2,3 dideoxy-D-*ribo*-hex-3-enopyranoside, *Carbohydr. Res.*, 141, 153–158, 1985.
56. Valverde, S, Bernabe, M, Gomez, A M, Puebla, P, Cross coupling reactions of 2-(allyloxy(thio))-benzothiazoles with organocopper reagents in dihydropyranoid systems — mechanistic implications of the substrate and the reagent — regiocontrolled and stereocontrolled access to branched-chain sugars, *J. Org. Chem.*, 57, 4546–4550, 1992.
57. Boquel, P, Chapleur, Y, A new strategy for the synthesis of mevinic acids analogues, *Tetrahedron Lett.*, 31, 1369–1372, 1990.
58. McDonald, C E, Dugger, R W, A formal total synthesis of (−)-isoavenaciolide, *Tetrahedron Lett.*, 29, 2413–2416, 1988.
59. Danishefsky, S J, Armistead, D M, Wincott, F E, Selnick, H G, Hungate, R, The total synthesis of the aglycon of avermectin A1a, *J. Am. Chem. Soc.*, 109, 8117–8119, 1987.
60. Cousins, G S, Hoberg, J, Synthesis and chemistry of cyclopropanated carbohydrates, *Chem. Soc. Rev.*, 29, 165–174, 2000.
61. Beyer, J, Skaanderup, P R, Madsen, R, Platinum-catalyzed ring opening of 1,2-cyclopropanated sugars with *O*-nucleophiles. Convenient synthesis of 2-*C*-branched carbohydrates, *J. Am. Chem. Soc.*, 122, 9575–9583, 2000.
62. Wallace, G A, Scott, R W, Heathcock, C H, Synthesis of the C29-C44 portion of spongistatin 1 (altohyrtin A), *J. Org. Chem.*, 65, 4145–4152, 2000.
63. Scott, R W, Heathcock, C H, An efficient synthesis of 3,4,6-tri-*O*-benzyl-2-*C*-methyl-D-glucal, *Carbohydr. Res.*, 291, 205–208, 1996.
64. Fernandez Megia, E, Gourlaouen, N, Ley, S V, Rowlands, G J, Studies towards the synthesis of the C29–C51 fragment of altohyrtin A, *Synlett*, 991–993, 1998.
65. Timmers, C M, Leeuwenburgh, M A, Verheijen, J C, van der Marel, G A, van Boom, J H, Rhodium(II) catalyzed asymmetric cyclopropanation of glycals with ethyl diazoacetate, *Tetrahedron: Asymmetry*, 7, 49–52, 1996.
66. Hoberg, J O, Claffey, D J, Cyclopropanation of unsaturated sugars with ethyl diazoacetate, *Tetrahedron Lett.*, 37, 2533–2536, 1996.
67. Kim, C, Hoang, R, Theodorakis, E A, Synthetic studies on norrisolide: enantioselective synthesis of the norrisane side chain, *Org. Lett.*, 1, 1295–1297, 1999.
68. Duchaussoy, P, Di Cesare, P, Gross, B, Synthesis of sugars containing the dihalocyclopropane moiety: a new access to the heptose series, *Synthesis*, 198–200, 1979.
69. Hoberg, J O, Formation of seven-membered oxacycles through ring expansion of cyclopropanated carbohydrates, *J. Org. Chem.*, 62, 6615–6618, 1997.
70. Ramana, C V, Murali, R, Nagarajan, M, Synthesis and reactions of 1,2-cyclopropanated sugars, *J. Org. Chem.*, 62, 7694–7703, 1997.
71. Fitzsimmons, B J, Fraser-Reid, B, Annulated pyranosides as chiral synthons for carbocyclic systems. Enantiospecific routes to both (+)- and (−)-chrysanthemum dicarboxylic acids from a single progenitor, *J. Am. Chem. Soc.*, 101, 6123–6125, 1979.
72. Meng, D F, Sorensen, E J, Bertinato, P, Danishefsky, S J, Studies toward a synthesis of epothilone A: Use of hydropyran templates for the management of acyclic stereochemical relationships, *J. Org. Chem.*, 61, 7998–7999, 1996.

73. Hanessian, S, Rancourt, G, Approaches to the total synthesis of natural products from carbohydrates, *Pure Appl. Chem.*, 1201–1214, 1977.
74. Banfi, L, Beretta, M G, Colombo, L, Gennari, C, Scolastico, C, Total synthesis of (+)-thermozymocidin (myriocin) from D-fructose, *J. Chem. Soc., Chem. Commun.*, 488–490, 1982.
75. Banfi, L, Beretta, M G, Colombo, L, Gennari, C, Scolastico, C, 2-Benzoylamino-2-deoxy-2-hydroxymethyl-D-hexono-1,4-lactones: synthesis from D-fructose and utilization in the total synthesis of thermozymocidin (myriocin), *J. Chem. Soc., Perkin. Trans. 1*, 1613–1619, 1983.
76. Rama Rao, A V, Gurjar, M K, Devi, T R, Kumar, K R, A formal synthesis of a novel immunosuppressant ISP-1 stereocontrolled Pd(0) catalyzed cis hydroxyamination of carbohydrate derived vinyl epoxide, *Tetrahedron Lett.*, 34, 1653–1656, 1993.
77. Deloisy, S, Tha, T T, Olesker, A, Lukacs, G, Synthesis of α-azido aldehydes. Stereoselective formal access to the immunosuppressant myriocin, *Tetrahedron Lett.*, 35, 4783–4786, 1994.
78. Yoshikawa, M, Yokokawa, Y, Okuno, Y, Murakami, N, Total synthesis of a novel immunosuppressant, myriocin (thermozymocidin, ISP-I), and Z-myriocin, *Chem. Pharm. Bull.*, 42, 994–996, 1994.
79. Yoshikawa, M, Yokokawa, Y, Okuno, Y, Murakami, N, Total syntheses of myriocin and Z-myriocin, two potent immunosuppressants, from 2-deoxy-D-glucose, *Tetrahedron*, 51, 6209–6228, 1995.
80. Nakamura, T, Shiozaki, M, Total synthesis of sphingofungin E, *Tetrahedron Lett.*, 42, 2701–2704, 2001.
81. Sato, K-i, Sekiguchi, T, Hozumi, T, Yamazaki, T, Akai, S, Improved synthetic method for preparing spiro α-chloroepoxides, *Tetrahedron Lett.*, 43, 3087–3090, 2002.
82. Nadin, A, Nicolaou, K C, Chemistry and biology of the zaragozic acids (Squalestatins), *Angew. Chem. Int. Ed. Engl.*, 35, 1623–1656, 1996.
83. Abdel-Rahman, H, Adams, J P, Boyes, A L, Kelly, M J, Mansfield, D J, Procopiou, P A, Roberts, S M, Slee, D H, Watson, N S, A synthetic approach to squalestatin 1, *J. Chem. Soc., Chem. Commun.*, 1839–1841, 1993.
84. Abdel-Rahman, H, Adams, J P, Boyes, A L, Kelly, M J, Mansfield, D J, Procopiou, P A, Roberts, S M, Slee, D H, Sidebottom, P D, Sik, V, Watson, N S, Synthetis of the bicyclic core structure of squalestatin 1, *J. Chem. Soc., Chem. Commun.*, 1841–1843, 1993.
85. Abdel-Rahman, H, Adams, J P, Boyes, A L, Kelly, M J, Lamont, R B, Mansfield, D J, Procopiou, P A, Roberts, S M, Slee, D H, Watson, N S, Synthesis of a novel spirocyclic lactone in a potential route to squalestatin 1, *J. Chem. Soc., Perkin. Trans. 1*, 1259–1261, 1994.
86. Caron, S, Stoermer, D, Mapp, A K, Heathcock, C H, Total synthesis of zaragozic acid A (squalestatin S1). Synthesis of the relay compound, *J. Org. Chem.*, 61, 9126–9134, 1996.
87. Caron, S, McDonald, A I, Heathcock, C H, An improved synthesis of 1,6-anhydro-2,3-di-*O*-benzyl-β-D-*xylo*-hexopyranos-4-ulose, *Carbohydr. Res.*, 281, 179–182, 1996.
88. Stoermer, D, Caron, S, Heathcock, C H, Total synthesis of zaragozic acid A (squalestatin S1). Degradation to a relay compound and reassembly of the natural product, *J. Org. Chem.*, 61, 9115–9125, 1996.
89. Prandi, J, Couturier, G, Synthesis of methyl α-caryophylloside, *Tetrahedron Lett.*, 41, 49–52, 2000.
90. Mazeas, D, Skrydstrup, T, Beau, J M, A highly stereoselective synthesis of 1,2-*trans*-*C*-glycosides via glycosyl samarium(III) compounds, *Angew. Chem. Int. Ed. Engl.*, 34, 909–912, 1995.
91. Jarreton, O, Skrydstrup, T, Beau, J M, The stereospecific synthesis of methyl α-*C*-mannobioside: a potential inhibitor of M-tuberculosis binding to human macrophages, *Chem. Commun.*, 1661–1662, 1996.
92. Zhdanov, Y A, Alekseev, Y E, Alekseeva, V G, Wittig reaction in carbohydrate chemistry, *Adv. Carbohydr. Chem.*, 27, 227–299, 1972.
93. For an alternate route to this compound see: Chapleur, Y, Germain, F, Aubry, A, Bayeul, D, Elimination reactions of 2-*C*-methyl-2-deoxy sugars: application to the synthesis of methyl 2,3, 4-trideoxy-2,4-di-*C*-methyl-6-*O*-triphenylmethyl-D-*lyxo-hexo-pyranoside* and its X-ray crystal structure, *J. Carbohydr. Chem.*, 3, 443–459, 1984.
94. Mori, M, Chuman, T, Kato, K, Mori, K, A stereoselective synthesis of natural (4*S*,6*S*,7*S*)-serricornin, the sex pheromone of cigaret beetle, from levoglucosenone, *Tetrahedron Lett.*, 23, 4593–4596, 1982.

95. Lagrange, A, Olesker, A, Costa, S S, Lukacs, G, Tong, T T, A route from D-galactose to the aggregation pheromone component (−)-α-multistriatin, *Carbohydr. Res.*, 110, 159–164, 1982.
96. Jarosz, S, Fraser-Reid, B, A route to Prelog-Djerassi lactone from methyl α-D-glucopyranoside, *Tetrahedron Lett.*, 22, 2533–2534, 1981.
97. Isobe, M, Ichikawa, Y, Goto, T, Total synthesis of (+)-Prelog-Djerassi lactonic acid, *Tetrahedron Lett.*, 22, 4287–4290, 1981.
98. Jones, K, Wood, W W, An approach to the total synthesis of the Prelog-Djerassi lactone, *J. Chem. Soc., Perkin. Trans. 1*, 537–545, 1987.
99. Ichibara, A, Kawagishi, H, Tokugawa, N, Sakamura, S, Stereoselective total synthesis and stereochemistry of diplodiatoxin, a mycotoxin from *Diplodia maydis*, *Tetrahedron Lett.*, 27, 1347–1350, 1986.
100. Hanessian, S, Pougny, J R, Boessenkool, I K, Total synthesis of the C19–C29 aliphatic segment of (+)-rifamycin S — A formal synthesis of the antibiotic, *J. Am. Chem. Soc.*, 104, 6164–6166, 1982.
101. Hanessian, S, Pougny, J R, Boessenkool, I K, Assembly of the C19-C29 aliphatic segment of rifamycin S from D-glucose by the chiron approach, *Tetrahedron*, 40, 1289–1301, 1984.
102. Hanessian, S, Ugolini, A, Therien, M, Stereocontrolled synthesis of the spiro ketal unit of avermectin B1a aglycon, *J. Org. Chem.*, 48, 4427–4430, 1983.
103. Rosenthal, A, Nguyen, L, Branched-chain sugar nucleosides I. 9-(3-deoxy-3-*C*-(2′-hydroxyethyl)-β-D-allofuranosyl)-adenine and 9-(3-deoxy-3-*C*-(2′-hydroxyethyl)-β-D-ribofuranosyl)-adenine, *J. Org. Chem.*, 34, 1029–1034, 1969.
104. Anderson, R C, Fraser-Reid, B, Synthesis of optically active avenaciolide from D-glucose. Correct stereochemistry of the natural product, *J. Am. Chem. Soc.*, 97, 3780–3781, 1975.
105. Ohrui, H, Emoto, S, Stereoselective syntheses of (+)- and (−)-avenaciolide from D-glucose the correct absolute configuration of natural avenaciolide, *Tetrahedron Lett.*, 16, 3657–3660, 1975.
106. Hanessian, S, Lavallee, P, Total synthesis of (+)-thromboxane B2 from D-glucose. A detailed account, *Can. J. Chem.*, 59, 870–877, 1981.
107. Al-Abed, Y, Naz, N, Mootoo, D, Voelter, W, 4-*O*-TfO-2,3-anhydro-β-L-ribopyranosides as chiron: a formal synthesis of canadensolide, *Tetrahedron Lett.*, 37, 8641–8642, 1996.
108. Gurjar, M K, Das, S K, Sadalapure, K S, Zaragozic acid: unusual stereochemical dependence in anhydride formation of carbohydrate templates, *Tetrahedron Lett.*, 36, 1933–1936, 1995.
109. Chida, N, Takeoka, J, Tsutsumi, N, Ogawa, S, Total synthesis of (+)-lactacystin from D-glucose, *J. Chem. Soc., Chem. Commun.*, 793–794, 1995.
110. Chida, N, Takeoka, J, Ando, K, Tsutsumi, N, Ogawa, S, Stereoselective total synthesis of (+)-lactacystin from glucose, *Tetrahedron*, 53, 16287–16298, 1997.
111. Overman, L E, Thermal and mercuric ion catalyzed [3,3]-sigmatropic rearrangement of allylic trichloroacetimidates. 1,3 Transposition of alcohol and amine functions, *J. Am. Chem. Soc.*, 96, 597–599, 1974.
112. Oishi, T, Ando, K, Chida, N, Stereoselective total synthesis of (+)-myriocin from D-mannose, *Chem. Commun.*, 1932–1933, 2001.
113. Chapleur, Y, A short synthesis of 2-*C*-alkyl-2-deoxy sugars from D-mannose, *J. Chem. Soc., Chem. Commun.*, 141–142, 1983.
114. Tsang, R, Fraser-Reid, B, *O*- and *C*-Acylation of some carbohydrate enolates, *J. Chem. Soc., Chem. Commun.*, 60–62, 1984.
115. Chapleur, Y, Longchambon, F, Gillier, H, Aldolisation of a carbohydrate enolate: stereochemical outcome and X-ray crystal structure determination of an aldol product, *J. Chem. Soc., Chem. Commun.*, 564–566, 1988.
116. Wei, A, Haudrechy, A, Audin, C, Jun, H S, Haudrechy-Bretel, N, Kishi, Y, Preferred conformation of *C*-glycosides 14. Synthesis and conformational analysis of carbon analogs of the blood group determinant H-type II, *J. Org. Chem.*, 60, 2160–2169, 1995.
117. Abel, S, Faber, D, Huter, O, Giese, B, Total synthesis of soraphen A(1α), *Synthesis*, 188–197, 1999.
118. Hong, C Y, Kishi, Y, Total synthesis of onnamide A, *J. Am. Chem. Soc.*, 113, 9693–9694, 1991.
119. Hong, C Y, Kishi, Y, Total synthesis of mycalamides A and B, *J. Org. Chem.*, 55, 4242–4245, 1990.

120. Gurjar, M K, Kumar, P, Rao, B V, Stereocontrolled synthesis of spirocyclopropane sugars and their application to asymmetric formation of tertiary chiral centres: a route to 2,2′-dialkylated pyranose subunit (C-18–C-23) of lasonolide A, *Tetrahedron Lett.*, 37, 8617–8620, 1996.
121. Gurjar, M K, Das, S K, Kunwar, A C, Studies towards the synthesis of the highly oxygenated bicyclic core of zaragozic acid: incorporation of three quaternary chiral carbon centres, *Tetrahedron Lett.*, 36, 1937–1940, 1995.
122. Fraisse, P, Hanna, I, Lallemand, J Y, Short synthesis of the bicyclic core of the zaragozic acids, *Tetrahedron Lett.*, 39, 7853–7856, 1998.
123. Fraisse, P, Hanna, I, Lallemand, J Y, Prange, T, Ricard, L, An approach to the bicyclic core of the zaragozic acids via the aldol reaction between methyl (α-D-xylofuranoside)uronate and D-(*R*)-glyceraldehyde acetonide, *Tetrahedron*, 55, 11819–11832, 1999.
124. Kuhn, C, Florent, J-C, A carbohydrate approach to 4-hydroxy-2-cyclopentenone moiety of antitumor prostanoid punaglandin IV via alkylation of ester uronate, *Tetrahedron Lett.*, 39, 4247–4250, 1998.
125. Canac, Y, Levoirier, E, Lubineau, A, New access to *C*-branched sugars and *C*-disaccharides under indium promoted Barbier-type allylations in aqueous media, *J. Org. Chem.*, 66, 3206–3210, 2001.
126. Corey, E J, Shibasaki, M, Knolle, J, Simple stereocontrolled synthesis of thromboxane B2 from D-glucose, *Tetrahedron Lett.*, 1625–1626, 1977.
127. Ireland, R E, Mueller, R H, The Claisen rearrangement of allyl esters, *J. Am. Chem. Soc.*, 94, 5897–5899, 1972.
128. Curran, D P, An approach to the enantiocontrolled synthesis of pseudomonic acids via a novel mono Claisen rearrangement, *Tetrahedron Lett.*, 23, 4309–4310, 1982.
129. Fraser-Reid, B, Tsang, R, Tulshian, D B, Sun, K M, Highly stereoselective routes to functionalized geminal alkyl derivatives of carbohydrates, *J. Org. Chem.*, 46, 3764–3767, 1981.
130. Tulshian, D B, Tsang, R, Fraser-Reid, B, Out-of-ring Claisen rearrangements are highly stereoselective in pyranoses: routes to *gem*-dialkylated sugars, *J. Org. Chem.*, 49, 2347–2355, 1984.
131. For an interesting discussion on the diastereoselectivity of Claisen rearrangement on pyranose ring, see Pak, H, Dickson, J K, Fraser-Reid, B, Serial radical cyclisation of branched carbohydrates, *J. Org. Chem.*, 54, 5357–5364, 1989.
132. Dickson, J K, Tsang, R, Llera, J M, Fraser-Reid, B, Serial radical cyclisation of branched carbohydrates 1. Simple pyranosides diquinanes, *J. Org. Chem.*, 54, 5350–5356, 1989.
133. Pak, H, Canalda, I I, Fraser-Reid, B, Carbohydrates to carbocyles: a synthesis of (−)-α-pipitzol, *J. Org. Chem.*, 55, 3009–3011, 1990.
134. For a review, see Tadano, K, Natural product synthesis starting with carbohydrates based on the Claisen rearrangement protocol, In *Studies in Natural Products Chemistry, Vol. 10: Stereoselective Synthesis, Part F*, Atta-ur-Rahman, Ed., Elsevier, Amsterdam, The Netherlands, pp. 405–455, 1992.
135. Tadano, K, Isshiki, Y, Minami, M, Ogawa, S, Samarium(II) iodide-mediated reductive cyclization approach to total synthesis of the insect sex attractant (−)-anastrephin, *Tetrahedron Lett.*, 33, 7899–7902, 1992.
136. Tadano, K, Ishihara, J, Ogawa, S, Total synthesis of (−)-acetomycin, *Tetrahedron Lett.*, 31, 2609–2612, 1990.
137. Ishihara, J, Tomita, K, Tadano, K, Ogawa, S, Total syntheses of (−)-acetomycin and its 3-stereoisomers at C-4 and C-5, *J. Org. Chem.*, 57, 3789–3798, 1992.
138. Tadano, K, Yamada, H, Idogaki, Y, Ogawa, S, Suami, T, Total synthesis of asteltoxin, *Tetrahedron*, 46, 2353–2366, 1990.
139. Takao, K, Ochiai, H, Hashizuka, T, Koshimura, H, Tadano, K, Ogawa, S, A total synthesis of (+)-eremantholide A, *Tetrahedron Lett.*, 36, 1487–1490, 1995.
140. Takao, K, Ochiai, H, Yoshida, K, Hashizuka, T, Koshimura, H, Tadano, K, Ogawa, S, Novel total synthesis of (+)-eremantholide A, *J. Org. Chem.*, 60, 8179–8193, 1995.
141. Tadano, K, Idokagi, Y, Yamada, H S T, Orthester Claisen rearrangements of three 3-*C*-(hydroxymethyl)methylene derivatives of hexofuranoses: stereoselective introduction of a quaternary center on C-3 of D-*ribo*-, L-*lyxo*-, and D-*arabino*-hexofuranoses, *J. Org. Chem.*, 52, 1201–1210, 1987.

142. Mann, R K, Parsons, J G, Rizzacasa, M A, Towards the synthesis of the squalestatins/zaragozic acids: synthesis of an advanced intermediate and introduction of the C-1 sidechain, *J. Chem. Soc., Perkin. Trans. 1*, 1283–1293, 1998.
143. Cuzzupe, A N, Di Florio, R, Rizzacasa, M A, Enantiospecific synthesis of the phospholipase A2 inhibitor (−)-cinatrin B, *J. Org. Chem.*, 67, 4392–4398, 2002.
144. Tomooka, K, Kikuchi, M, Igawa, K, Suzuki, M, Keong, P H, Nakai, T, Stereoselective total synthesis of zaragozic acid A based on an acetal [1,2] Wittig rearrangement, *Angew. Chem. Int. Ed.*, 39, 4502–4505, 2000.
145. Tomooka, K, Kikuchi, M, Igawa, K, Keong, P H, Nakai, T, Stereochemical features of the [1, 2]-Wittig rearrangement of *O*-glycosides derived from D-galactono- and D-xylono-gamma-lactones: a new approach to the core part of zaragozic acids, *Tetrahedron Lett.*, 40, 1917–1920, 1999.
146. Molino, B F, Magdzinski, L, Fraser-Reid, B, Pyranosidic homologation: Part I: extending the carbohydrate template via C-6 and C-4 + Part II, *Tetrahedron Lett.*, 24, 5819–5822, 1983.
147. Mootoo, D R, Fraser-Reid, B, The nine contiguous chiral centers in streptovaricin A via pyranosidic homologation, *J. Org. Chem.*, 54, 5548–5550, 1989.
148. Fraser-Reid, B, Magdzinski, L, Molino, B, New strategy for carbohydrate-based syntheses of multichiral arrays: pyranosidic homologation. 2, *J. Am. Chem. Soc.*, 106, 731–734, 1984.
149. Chapleur, Y, Euvrard, M N, Hetero Diels–Alder reactions in carbohydrate chemistry: a new strategy for multichiral arrays synthesis, *J. Chem. Soc., Chem. Commun.*, 884–885, 1987.
150. Ghini, A A, Burnouf, C, Lopez, J C, Olesker, A, Lukacs, G, Intramolecular Diels–Alder reactions on pyranose trienes. Stereoselective access to bis-annulated pyranosides, *Tetrahedron Lett.*, 31, 2301–2304, 1990.
151. Mayon, P, Euvrard, M.-N, Moufid, N, Chapleur, Y, Diastereoselective anionic and radical conjugate addition on a chiral enone: a route to multichiral arrays, *J. Chem. Soc., Chem. Commun.*, 399–401, 1994.
152. Wolfrom, M L, Hanessian, S, The reaction of free carbonyl sugar derivatives with organometallic reagents. I. 6-Deoxy-L-idose and derivatives, *J. Org. Chem.*, 27, 1800–1804, 1962.
153. Dasser, M, Chrétien, F, Chapleur, Y, A facile and stereospecific synthesis of L-*glycero*-D-*manno*-heptose and some derivatives, *J. Chem. Soc., Perkin. Trans. 1*, 3091–3094, 1990, and references cited.
154. Rama Rao, A V, Chakraborty, T K, Laxma Reddy, K, Studies directed towards the synthesis of immunosuppressive agent FK-506: synthesis of the entire bottom-half, *Tetrahedron Lett.*, 32, 1251–1254, 1991.
155. Ireland, R E, Gleason, J L, Gegnas, L D, Highsmith, T K, A total synthesis of FK-506, *J. Org. Chem.*, 61, 6856–6872, 1996.
156. Armistead, D M, Danishefsky, S J, Model studies directed toward the avermectins: a route to the oxahydrindene subunit, *Tetrahedron Lett.*, 28, 4959–4962, 1987.
157. For a detailed study of this reaction, see Danishefsky, S J, Deninno, M P, Plillips, G B, Zelle, R E, Lartey, P A, On the communication of chirality from furanose and pyranose rings to monosaccharide side chains: anomalous results in the glucose series, *Tetrahedron*, 42, 2809–2819, 1986.
158. Williams, D R, Klinger, F D, Dabral, V, Synthesis of the optically active hexahydrobenzofuran nucleus of the avermectins, *Tetrahedron Lett.*, 29, 3415–3418, 1988.
159. Eshelman, J E, Epps, J L, Kallmerten, J, Diastereoselective [2,3] Wittig rearrangement of carbohydrate-derived tertiary allylic ethers. 1. Synthesis of the C11-C18 subunit of herbimycin a from D-glucose, *Tetrahedron Lett.*, 34, 749–752, 1993.
160. Nakata, M, Osumi, T, Ueno, A, Kimura, T, Tamai, T, Tatsuta, K, Total synthesis of herbimycin A, *Tetrahedron Lett.*, 32, 6015–6018, 1991.
161. Carchon, G, Chrétien, F, Chapleur, Y, Stereoselective nucleophilic addition on sugar hemiacetal controlled by large-ring chelates, *Carbohydr. Lett.*, 2, 17–22, 1996.
162. Marco Contelles, J, de Opazo, E, Arroyo, N, Synthesis of higher-carbon sugars by addition of organometallic reagents to aldehydes or lactols derived from carbohydrates, *Tetrahedron*, 57, 4729–4739, 2001.
163. Redlich, H, Lenfers, J B, Thormälhen, S, Synthesis of (−)-talaromycin A. The use of the Δ2 effect for stereocontrol at spiroacetalic centres, *Synthesis*, 1112–1117, 1992.

164. Danishefsky, S J, Hungate, R, Schulte, G, Total synthesis of octosyl acid A. Intramolecular Williamson reaction via a cyclic stannylene derivative, *J. Am. Chem. Soc.*, 110, 7434–7440, 1988.
165. Kozaki, S, Sakana, O, Yasuda, T, Shimizu, T, Ogawa, S, Suami, T, An alternative total synthetic approach toward octosyl acid A, *J. Org. Chem.*, 53, 281–286, 1988.
166. Knapp, S, Shieh, W C, Jaramillo, C, Trilles, R V, Nandan, S R, Synthesis of the ezomycin octosyl nucleoside, *J. Org. Chem.*, 59, 946–948, 1994.
167. Kishi, Y, Natural products synthesis: palytoxin, *Pure Appl. Chem.*, 61, 313–324, 1989.
168. For a review of the strategy of this total synthesis, see Nicolaou, K C, Sorensen, E J, *Classics in Total Synthesis*, VCH, Weinheim, 1996, p. 798.
169. Ko, S S, Finan, J M, Yonaga, M, Kishi, Y, Uemura, D, Hirata, Y, Stereochemistry of palytoxin. Part 2. C1–C6, C47–C74, and C77–C83 segments, *J. Am. Chem. Soc.*, 104, 7364–7367, 1982.
170. Armstrong, R W, Beau, J-M, Cheon, S H, Christ, W J, Fujioka, H, Ham, W H, Hawkins, L D, Jin, H, Kang, S H, Kishi, Y, Martinelli, M J, McWhorter, W W Jr., Mizuno, M, Nakata, M, Stutz, A, Talamas, F X, Taniguchi, M, Tino, J A, Ueda, K, Uenishi, J-I, White, J B, Yonaga, M, Total synthesis of a fully protected palytoxin carboxylic acid, *J. Am. Chem. Soc.*, 111, 7525–7530, 1989.
171. Brimacombe, J S, Rahman, K M M, Branched-chain sugars. Part 18. Syntheses of D-rubranitrose (2,3,6-trideoxy-3-*C*-methyl-4-*O*-methyl-3-nitro-D-*xylo*-hexopyranose) and a derivative of D-kijanose (2,3,4,6-tetradeoxy-4-methoxycarbonylamino-3-methyl-3-nitro-α-D-*xylo*-hexopyraose), *J. Chem. Soc., Perkin. Trans. 1*, 1073–1079, 1985, and previous references of this series.
172. Cha, J K, Christ, J, Kishi, Y, On stereochemistry of osmium tetraoxide oxidation of allylic alcohol systems, *Tetrahedron*, 40, 2247–2255, 1984.
173. For a review, see Knapp, S, Synthesis of complex nucleoside antibiotics, *Chem. Rev.*, 95, 1859–1876, 1995.
174. Knapp, S, Morriello, G J, Doss, G A, Synthesis of the liposidomycin diazepanone nucleoside, *Org. Lett.*, 4, 603–606, 2002.
175. Tulshian, D B, Czarniecki, M, Total synthesis of griseolic acid, *J. Am. Chem. Soc.*, 117, 7009–7010, 1995.
176. Tulshian, D, Doll, R J, Stansberry, M F, McPhail, A T, Total synthesis of griseolic acid derivatives from D-glucose, *J. Org. Chem.*, 56, 6819–6822, 1991.
177. Nicoll-Griffith, D A, Weiler, L, Introduction of a chiral centre on C-6 of a carbohydrate unit: application to the synthesis of the C-2 to C-15 fragment of lonomycin, *Tetrahedron*, 47, 2733–2750, 1991.
178. Takano, D, Nagamitsu, T, Ui, H, Shiomi, K, Yamaguchi, Y, Masuma, R, Kuwajima, I, Omura, S, Total synthesis of nafuredin, a selective NADH-fumarate reductase inhibitor, *Org. Lett.*, 3, 2289–2291, 2001.
179. Secrist, J A III, and Wu, S-R, Construction of long-chain carbohydrates. Synthesis and chemistry of galactose 6-phosphorane, *J. Org. Chem.*, 44, 1434–1438, 1979.
180. Armstrong, R W, Beau, J-M, Cheon, S H, Christ, W J, Fujioka, H, Ham, W H, Hawkins, L D, Jin, H, Kang, S H, Kishi, Y, Martinelli, M J, McWhorter, W W Jr., Mizuno, M, Nakata, M, Stutz, A, Talamas, F X, Taniguchi, M, Tino, J A, Ueda, K, Uenishi, J-i, White, J B, Yonaga, M, Total synthesis of palytoxin carboxylic acid and palytoxin amide, *J. Am. Chem. Soc.*, 111, 7530–7533, 1989.
181. Karpiesiuk, W, Banaszek, A, Simple approach to *O*-protected deaminotunicaminyluracil, *Tetrahedron*, 50, 2965–2974, 1994.
182. Lichtenthaler, F W, Lorenz, K, Ma, W, A convergent total synthesis of (−)-anamarine from D-glucose, *Tetrahedron Lett.*, 28, 47–50, 1987.
183. For a review, see Jarosz, S, Synthesis of higher carbon sugars *via* coupling of simple monosaccharides — Wittig, Horner–Emmons, and related methods, *J. Carbohydr. Chem.*, 20, 93–107, 2001.
184. Jarosz, S, Mach, M, Phosphonate versus phosphorane method in the synthesis of higher carbon sugars. Preparation of D-*erythro*-L-*manno*-D-*gluco*-dodecitol, *J. Chem. Soc., Perkin. Trans. 1*, 3943–3948, 1998.
185. Jarosz, S, Salanski, P, Mach, M, Application of stabilized sugar-derived phosphoranes in the synthesis of higher carbon monosaccharides. First synthesis of a C-21-dialdose, *Tetrahedron*, 54, 2583–2594, 1998.

186. Dondoni, A, Marra, A, Mizuno, M, Giovannini, P P, Linear total synthetic routes to D-*C*-(1,6)-linked oligoglucoses and oligogalactoses up to pentaoses by iterative Wittig olefination assembly, *J. Org. Chem.*, 67, 4186–4199, 2002.
187. Dondoni, A, Kleban, M, Zuurmond, H, Marra, A, Synthesis of (1 → 6)-*C*-oligogalactosides by iterative Wittig olefination, *Tetrahedron Lett.*, 39, 7991–7994, 1998.
188. Pougny, J.-R, Carbon–carbon bond formation using primary tosylates derived from carbohydrates, *Tetrahedron Lett.*, 25, 2363–2366, 1984.
189. Toshima, H, Sato, H, Ichihara, A, Total synthesis of (2*S*,3*R*,5*S*)-(−)-2,3-dihydroxytetradecan-5-olide, a new biologically active δ-lactone produced by seiridium unicorne, *Tetrahedron*, 55, 2581–2590, 1999.
190. Erdik, E, Copper(I) catalyzed reactions of organolithiums and Grignard reagents, *Tetrahedron*, 40, 641–657, 1984.
191. Sasaki, M, Koike, T, Sakai, R, Tachibana, K, Total synthesis of (−)-dysiherbaine, a novel neuroexcitotoxic amino acid, *Tetrahedron Lett.*, 41, 3923–3926, 2000.
192. Snider, B B, Hawryluk, N A, Synthesis of (−)-dysiherbaine, *Org. Lett.*, 2, 635–638, 2000.
193. Cooper, A J, Pan, W, Salomon, R G, Total synthesis of halichondrin b from common sugars: an F-ring intermediate from D-glucose and efficient construction of the C1 to C21 segment, *Tetrahedron Lett.*, 34, 8193–8196, 1993.
194. Lewis, A, Stefanuti, I, Swain, S A, Smith, S A, Taylor, R J K, Carbohydrate-based routes to salicylate natural products: formal total synthesis of (+)-apicularen A from D-glucal, *Tetrahedron Lett.*, 42, 5549–5552, 2001.
195. Kotsuki, H, Kadota, I, Ochi, M, An efficient method for the alkylation of chiral triflates with alkynyllithium reagents. A highly concise total synthesis of (+)-panaxacol, *Tetrahedron Lett.*, 31, 4609–4612, 1990.
196. Giese, B, *Radical in Organic Synthesis: Formation of Carbon–Carbon Bonds*, Pergamon Press, Oxford, 1986.
197. Myers, A G, Gin, D Y, Rogers, D H, A convergent synthetic route to the tunicamycin antibiotics — synthesis of (+)-tunicamycin-V, *J. Am. Chem. Soc.*, 115, 2036–2038, 1993.
198. Myers, A G, Gin, D Y, Rogers, D H, Synthetic studies of the tunicamycin antibiotics. Preparation of (+)-tunicaminyluracil, (+)-tunicamycin-V, and 5′-*epi*-tunicamycin-V, *J. Am. Chem. Soc.*, 116, 4697–4718, 1994.
199. Vauzeilles, B, Cravo, D, Mallet, J M, Sinaÿ, P, An expeditious synthesis of a *C*-disaccharide using a temporary ketal connection, *Synlett*, 522–524, 1993.
200. Xin, Y C, Mallet, J M, Sinaÿ, P, An expeditious synthesis of a *C*-disaccharide using a temporary silaketal connection, *J. Chem. Soc., Chem. Commun.*, 864–865, 1993.
201. Mallet, A, Mallet, J M, Sinaÿ, P, The use of selenophenyl galactopyranosides for the synthesis of alpha and beta-(1 → 4)-*C*-disaccharides, *Tetrahedron: Asymmetry*, 5, 2593–2608, 1994.
202. Sinaÿ, P, Synthesis of oligosaccharide mimetics, *Pure Appl. Chem.*, 69, 459–463, 1997.
203. Fairbanks, A J, Perrin, E, Sinaÿ, P, Synthesis of the carbon skeleton of the herbicidins *via* a temporary silaketal tether, *Synlett*, 679–681, 1996.
204. Rama Rao, A V, Chakraborty, T K, Laxma Reddy, K, Studies directed towards the synthesis of immunosuppressive agent FK-506: construction of the tricarbonyl moiety, *Tetrahedron Lett.*, 31, 1439–1442, 1990.
205. Fleet, G W J, Shing, T, An approach to the enantiospecific synthesis of sesbanimide and its enantiomer from D-glucose, *J. Chem. Soc., Chem. Commun.*, 835–837, 1984.
206. Tomioka, K, Hagiwara, A, Koga, K, Total synthesis of (+)-sesbanimide a from D-glucose, *Tetrahedron Lett.*, 29, 3095–3096, 1988.
207. Lopez-Herrera, F J, Sarabia-Garcia, F, Condensation of D-mannosaldehyde derivatives with ethyl diazoacetate. An easy and stereoselective chain elongation methodology for carbohydrates: application to new syntheses for KDO and 2-deoxy-β-KDO, *Tetrahedron*, 53, 3325–3346, 1997. For other syntheses of KDO using analogous homologation, see Refs. [10,11] of this paper.
208. Pougny, J R, Nassr, M A M, Sinaÿ, P, Mercuricyclization in carbohydrate chemistry. A highly stereoselective route to α-D-*C*-glucopyranosyl derivatives, *J. Chem. Soc., Chem. Commun.*, 375–376, 1981.

209. Nicotra, F, Synthesis of glycosyl phosphate mimics, In *Carbohydrate Mimics*, Ch. 4, Chapleur, Y, Ed., Wiley-VCH, Weinheim, pp. 67–85, 1999.
210. Keck, G E, Boden, E P, Wiley, M R, Total synthesis of (+)-colletodiol: New methodology for the synthesis of macrolactones, *J. Org. Chem.*, 54, 896–906, 1989.
211. Kocienski, P, Stocks, M, Donald, D, Cooper, M, Manners, A, A synthesis of the C(1)-C(15) segment of tsukubaenolide (FK506), *Tetrahedron Lett.*, 29, 4481–4484, 1988.
212. Postema, M H D, Recent developments in the synthesis of *C*-glycosides, *Tetrahedron*, 48, 8545 1992.
213. Herscovici, J and Antonakis, K, Recent developments in *C*-glycoside synthesis, In *Studies in Natural Products Chemistry, Vol. 10, Stereoselective Synthesis, Part F*, Atta-ur-Rahman, Ed., Elsevier, Amsterdam, The Netherlands, pp. 337–403, 1992.
214. Jaramillo, C, Knapp, S, Synthesis of *C*-aryl glycosides, *Synthesis*, 1–20, 1994.
215. Witczak, Z J, Synthesis of *C*-glycosyl compounds and other natural-products from levoglucosenone, *Pure Appl. Chem.*, 66, 2189–2192, 1994.
216. Levy, D E, Tang, C, *The Chemistry of* C-*Glycosides*, Pergamon–Elsevier, Kidlington, UK, 1995.
217. Postema, M H D, C-*Glycoside Synthesis*, CRC, Boca Raton, FL, 1995.
218. For a review, see Beau, J-M, Gallagher, T, Nucleophilic *C*-glycosyl donors for *C*-glycoside synthesis, *Top Curr. Chem.*, 1–54, 1997.
219. Du, Y G, Linhardt, R J, Vlahov, I R, Recent advances in stereoselective *C*-glycoside synthesis, *Tetrahedron*, 54, 9913–9959, 1998.
220. Jimenez-Barbero, J, Espinosa, J F, Asensio, J L, Canada, F J, Poveda, A, The conformation of *C*-glycosyl compounds, *Adv. Carbohydr. Chem. Biochem.*, 56, 235–284, 2000.
221. Horita, K, Inoue, T, Tanaka, K, Yonemitsu, O, Total synthesis of the polyether antibiotic lysocellin. 1. Stereocontrolled synthesis of the C1–C9 and C16–C23 fragments, *Tetrahedron Lett.*, 33, 5537–5540, 1992.
222. Horita, K, Inoue, T, Tanaka, K, Yonemitsu, O, Stereoselective total synthesis of lysocellin, the representative polyether antibiotic of the lysocellin family. Part 1. Synthesis of C1–C9 and C16–C23 subunits, *Tetrahedron*, 52, 531–550, 1996.
223. Burke, S D, Jung, K W, Phillips, J R, Perri, R E, An expeditious synthesis of the C(1)–C(14) subunit of halichondrin b, *Tetrahedron Lett.*, 35, 703–706, 1994.
224. Burke, S D, Jung, K W, Lambert, W T, Phillips, J R, Klovning, J J, Halichondrin B: synthesis of the C(1)–C(15) subunit, *J. Org. Chem.*, 65, 4070–4087, 2000.
225. Bollit, V, Mioskowski, C, Kollah, R O, Manna, S, Rajapaska, D, Falck, J R, Total synthesis of vineomycinone B2 methyl ester *via* double Bradsher cyclisation, *J. Am. Chem. Soc.*, 113, 6320-6321, 1991.
226. Nicolaou, K C, The total synthesis of brevetoxin B: a twelve-year odyssey in organic synthesis, *Angew. Chem. Int. Ed. Engl.*, 35, 589–607, 1996.
227. Nicolaou, K C, Postema, M H D, Miller, N D, Recent advances in the synthesis of cyclic polyether marine natural products, In *Carbohydrate Mimics*, Vol. 1, Chapleur, Y, Ed., Wiley-VCH, Weinheim, pp. 1–18, 1998.
228. Nicolaou, K C, Duggan, M E, Hwang, C K, Synthesis of the ABC ring system of brevetoxin B, *J. Am. Chem. Soc.*, 111, 6666–6689, 1989.
229. Nicolaou, K C, Prasad, C V C, Somers, P K, Hwang, C K, Activation of 7-*endo* over 6-*exo* epoxide openings. Synthesis of oxepane and tetrahydropyran systems, *J. Am. Chem. Soc.*, 111, 5335–5340, 1989.
230. Tatsuta, K, Nakano, S, Narazaki, F, Nakamura, Y, The first total synthesis and establishment of absolute structure of luminacins C1 and C2, *Tetrahedron Lett.*, 42, 7625–7628, 2001.
231. Kim, S, Salomon, R G, Total synthesis of halicondrins: highly stereoselective construction of a homochiral pentasubstituted H-ring pyran intermediate from α-D-glucose, *Tetrahedron Lett.*, 30, 6279–6282, 1989.
232. Aicher, T D, Buszek, K R, Fang, F G, Forsyth, C J, Jung, S H, Kishi, Y, Matelich, M C, Scola, P M, Spero, D M, Yoon, S K, Total synthesis of halicondrin B and norhalicondrin B, *J. Am. Chem. Soc.*, 114, 3162–3164, 1992.
233. Duan, J J, Kishi, Y, Synthetic studies on halichondrins: a new practical synthesis of the C-1–C-12 segment, *Tetrahedron Lett.*, 34, 7541–7544, 1993.

234. Nishiyama, S, Shizuri, Y, Shigemori, H, Yamamura, S, Synthetic studies on verucosidin and its absolute configuration, *Tetrahedron Lett.*, 27, 723–726, 1986.
235. Lewis, M D, Cha, J K, Kishi, Y, Highly stereoselective approaches to α- and β-*C*-glycopyranosides, *J. Am. Chem. Soc.*, 104, 4976–4978, 1982.
236. McWorther, W W Jr, Kang, S M, Kishi, Y, Synthetic studies on palitoxin: stereocontrolled practical synthesis of the C85–C98 fragment, *Tetrahedron Lett.*, 24, 2243–2246, 1983.
237. Hung, S C, Lin, C C, Wong, C H, One-pot synthesis of 1-allyl- and 1-allenyl-6-*O*-acetyl-2,3, 4-tri-*O*-benzyl-α-D-glycosides from methyl tetra-*O*-benzyl-α-D-glycosides, *Tetrahedron Lett.*, 38, 5419–5422, 1997.
238. Levy, D E, Dasgupta, F, Tang, P C, Synthesis of novel fused ring *C*-glycosides, *Tetrahedron: Asymmetry*, 5, 2265–2268, 1994.
239. Jegou, A, Pacheco, C, Veyrieres, A, Stereoselective synthesis of α-*C*-glycopyranosyl isoprenoid compounds, *Synlett*, 81–83, 1998.
240. Minehan, T G, Kishi, Y, β-selective *C*-glycosidations: Lewis-acid mediated reactions of carbohydrates with silyl ketene acetals, *Tetrahedron Lett.*, 38, 6815–6818, 1997.
241. Kozikowski, A P, Sorgi, K L, A mild method for the synthesis of anomerically allylated *C*-glycopyranosides and *C*-glycofuranosides, *Tetrahedron Lett.*, 23, 2281–2284, 1982.
242. Nicolaou, K C, Duggan, M E, Hwang, C K, Synthesis of the IJK ring system of brevetoxin B, *J. Am. Chem. Soc.*, 111, 6682–6690, 1989.
243. Richter, P K, Tomaszewski, M J, Miller, R A, Patron, A P, Nicolaou, K C, Stereoselective construction of the C18–C32 fragment of swinholide, *J. Chem. Soc., Chem. Commun.*, 1151–1152, 1994.
244. Sasaki, M, Nonomura, T, Murata, M, Tachibana, K, Synthesis and stereochemical confirmation of the *cis*-fused L/M and N/O ring systems of maitotoxin, *Tetrahedron Lett.*, 35, 5023–5026, 1994.
245. Maruyama, M, Inoue, M, Oishi, T, Oguri, H, Ogasawara, Y, Shindo, Y, Hirama, M, Convergent synthesis of the ABCDE ring system of ciguatoxin CTX3C, *Tetrahedron*, 58, 1835–1851, 2002.
246. For a review, see Jorgensen, M, Hadwiger, P, Madsen, R, Stutz, A E, Wrodnigg, T M, Olefin metathesis in carbohydrate chemistry, *Curr. Org. Chem.*, 4, 565–588, 2000.
247. Cipolla, L, Lay, L, Nicotra, F, New and easy access to *C*-glycosides of glucosamine and mannosamine, *J. Org. Chem.*, 62, 6678–6681, 1997.
248. Kira, K, Hamajima, A, Isobe, M, Synthesis of the BCD-ring of ciguatoxin 1B using an acetylene cobalt complex and vinylsilane strategy, *Tetrahedron*, 58, 1875–1888, 2002.
249. Suh, H, Wilcox, C S, Chemistry of F1,F0-ATPase inhibitors. Stereoselective total syntheses of (+)-citreoviral and (−)-citreovirin, *J. Am. Chem. Soc.*, 110, 470–481, 1988.
250. Danishefsky, S, Kervin, J F Jr, On the addition of allyltrimethylsilane to glycal acetates, *J. Org. Chem.*, 47, 3803–3805, 1982.
251. Toshima, K, Ishizuka, T, Matsuo, G, Nakata, M, Allyl *C*-glycosidations of totally unprotected glycals and allyltrimethylsilane with trimethylsilyl trifluoromethanesulfonate (TMSOTf), *Tetrahedron Lett.*, 35, 5673–5676, 1994.
252. Wincott, F E, Danishefsky, S J, Schulte, G, Model studies directed toward the avermectins: a route to the spiroketal subunit, *Tetrahedron Lett.*, 28, 4951–4954, 1987.
253. De Raadt, A, Stuetz, A E, A one-step *C*-linked disaccharide synthesis from carbohydrate allylsilanes and tri-*O*-acetyl-D-glucal, *Carbohydr. Res.*, 220, 101–115, 1991.
254. Herscovici, J, Delatre, S, Antonakis, K, Lewis acid induced homoallylic *C*-alkylation. A new approach to *C*-glycosides, *J. Org. Chem.*, 52, 5691–5695, 1987.
255. Herscovici, J, Muleka, K, Boutmaîza, L, Antonakis, K, *C*-glycoside synthesis *via* glycal alkylation by olefinic derivatives, *J. Chem. Soc., Perkin. Trans. 1*, 1995–2009, 1990.
256. Hanna, I, Wlodyka, P, A new synthetic approach to forskolin: construction of the ABC ring system from D-galactose, *J. Org. Chem.*, 62, 6985–6990, 1997.
257. Haag, D, Chen, X T, Fraser-Reid, B, Carbohydrate based IMDA/aldol strategy towards the densely functionalized trans-decalin subunit of azadirachtin, *Chem. Commun.*, 2577–2578, 1998.
258. Nicolaou, K C, Hwang, C-K, Duggan, M E, Stereospecific synthesis of 1,1-dialkylglycosides, *J. Chem. Soc., Chem. Commun.*, 925–926, 1986.

259. Hanessian, S, Martin, M, Desai, R C, Formation of *C*-glycosides by polarity inversion at the anomeric centre, *J. Chem. Soc., Chem. Commun.*, 926–927, 1986.
260. Lesimple, P, Beau, J -M, Jaurand, G, Sinaÿ, P, Preparation and use of lithiated glycals: vinylic deprotonation *versus* tin–lithium exchange from 1-tributylstannyl glycals, *Tetrahedron Lett.*, 27, 6201–6204, 1986.
261. Schmidt, R R, Preuss, R, Betz, R, *C*-1 lithiation of *C*-2 activated glucals, *Tetrahedron Lett.*, 28, 6591–6594, 1987.
262. Jarowicki, K, Kocienski, P, Marczak, S, Willson, T, A synthesis of (+) pederin, the metallated dihydropyran approach, *Tetrahedron Lett.*, 31, 3433–3436, 1990.
263. Bearder, J R, Dewis, M L, Whiting, D A, Methods for the formation of the C1 (hexose)–C5 (pentose) bond: an approach to the synthesis of undecoses related to herbicidin, *Synlett*, 805–806, 1993.
264. Bearder, J R, Dewis, M L, Whiting, D A, Short synthetic route to congeners of the undecose antibiotic herbicidin, *J. Chem. Soc., Perkin. Trans. 1*, 227–233, 1995.
265. Cox, P, Mahon, M F, Molloy, K C, Lister, S, Gallagher, T, Lithiated dihydropyrans as ketone enolate equivalents: a model study for the herbicidins, *Tetrahedron Lett.*, 29, 1993–1996, 1988.
266. Newcombe, N J, Mahon, M F, Molloy, K C, Alker, D, Gallagher, T, Carbohydrate-based enolates as heterocyclic building blocks — synthesis of the herbicidin glycoside, *J. Am. Chem. Soc.*, 115, 6430–6431, 1993.
267. Binch, H M, Gallagher, T, Studies directed towards the synthesis of herbicidins. Model study based on late stage *N*-glycosylation, *J. Chem. Soc., Perkin. Trans. 1*, 401–402, 1996.
268. Skrydstrup, T, Mazeas, D, Elmouchir, M, Doisneau, G, Riche, C, Chiaroni, A, Beau, J M, 1,2-*cis*-*C*-glycoside synthesis by samarium diiodide-promoted radical cyclizations, *Chem. Eur. J.*, 3, 1342–1356, 1997, and references therein.
269. Ichikawa, S, Shuto, S, Matsuda, A, The first synthesis of herbicidin B. Stereoselective construction of the tricyclic undecose moiety by a conformational restriction strategy using steric repulsion between adjacent bulky silyl protecting groups on a pyranose ring, *J. Am. Chem. Soc.*, 121, 10270–10280, 1999.
270. Ireland, R E, Mueller, R H, Willard, A K, Ester enolate Claisen rearrangement — stereochemical control through stereoselective enolate formation, *J. Am. Chem. Soc.*, 98, 2868–2877, 1976.
271. Burke, S D, Armistead, D M, Schoenen, F J, Fevig, J M, An enolate Claisen route to *C*-pyranosides. Development and application to an ionophore synthon, *Tetrahedron*, 42, 2787–2801, 1986.
272. For a recent application, see Vidal, T, Haudrechy, A, Langlois, Y, Claisen–Ireland rearrangement: a new route to *C*-glycosides, *Tetrahedron Lett.*, 40, 5677–5680, 1999.
273. Ireland, R E, Vevert, J P, A chiral total synthesis of (−)- and (+)-nonactic acids from carbohydrate precursors and the definition of the transition for the enolate claisen rearrangement in heterocyclic systems, *J. Org. Chem.*, 4259–4260, 1980.
274. Ireland, R E, Wuts, P G M, Ernst, B, 3-Acyltetramic acid antibiotics. 1. Synthesis of tirandamycic acid, *J. Am. Chem. Soc.*, 103, 3205–3207, 1981.
275. Ireland, R E, Smith, M G, 3-Acytetramic acid antibiotics. 2. Synthesis of (+)-streptolic acid, *J. Am. Chem. Soc.*, 110, 854–860, 1988.
276. Praly, J P, Structure of anomeric glycosyl radicals and their transformations under reductive conditions, *Adv. Carbohydr. Chem. Biochem.*, 56, 65–152, 2000.
277. Keck, G E, Kachensky, D F, Enholm, E J, Pseudomonic acid C from L-lyxose, *J. Org. Chem.*, 50, 4317–4325, 1985.
278. Murphy, P J, Brennan, J, The Wittig olefination reaction with carbonyl compounds other than aldehydes and ketones, *Chem. Soc. Rev.*, 17, 1–30, 1988.
279. Chapleur, Y, A convenient synthesis of substituted tetrahydrofurans from sugar lactones, *J. Chem. Soc., Chem. Commun.*, 449–450, 1984.
280. Bellosta, V, Czernecki, S, *C*-glycosyl compounds IV. Synthesis of (2-deoxy-α-D-glyc-2-enopyranosyl)arenes by stereospecific conjugate addition of organocopper reagents to peracetylated hex-1-enopyran-3-uloses, *Carbohydr. Res.*, 171, 279–288, 1987.
281. Lakhrissi, M, Chapleur, Y, Dichloromethylenation of lactones 6: efficient synthesis of dichloroolefins from lactones and acetates using triphenylphosphine and tetrachloromethane, *J. Org. Chem.*, 59, 5752–5755, 1994.

282. Lakhrissi, Y, Taillefumier, C, Chrétien, F, Chapleur, Y, Facile dibromoolefination of lactones using bromomethylene triphenylphosphorane, *Tetrahedron Lett.*, 42, 7265–7268, 2001.
283. Lakhrissi, M, Chapleur, Y, Wittig olefination of lactone carbonyl groups, *Angew. Chem. Int. Ed. Engl.*, 35, 750–752, 1996.
284. Lakhrissi, M, Taillefumier, C, Chaouch, A, Didierjean, C, Aubry, A, Chapleur, Y, Wittig olefination of partially protected sugar lactones: a direct entry to dioxabicyclic and trioxatricyclic systems, *Tetrahedron Lett.*, 39, 6457–6460, 1998.
285. Lakhrissi, Y, Taillefumier, C, Lakhrissi, M, Chapleur, Y, Efficient conditions for the synthesis of *C*-glycosylidenes derivatives: a direct and stereoselective route to *C*-glycosyl compounds, *Tetrahedron: Asymmetry*, 9, 417–421, 2000.
286. Molina, A, Czernecki, S, Xie, J, Stereocontrolled synthesis of β-*C*-glycosides and amino β-*C*-glycosides by Wittig olefination of perbenzylated glyconolactones derivatives, *Tetrahedron Lett.*, 39, 7507–7510, 1998.
287. Bandzouzi, A, Chapleur, Y, Dichloromethylenation of sugar lactones: a new access to 1-deoxy-1-*C*-methyl-*C*-glycosyl compounds, *Carbohydr. Res.*, 171, 13–24, 1987.
288. Bandzouzi, A, Chapleur, Y, Dichloromethylenation of sugar lactones: a stereospecific synthesis of L-(+)-muscarine and L-(+)-epimuscarine toluene-p-sulphonates, *J. Chem. Soc., Perkin. Trans. 1*, 661–664, 1987.
289. Murphy, P J, Dennison, S T, The total synthesis of goniofufurone, *Tetrahedron*, 49, 6695–6700, 1993.
290. Taillefumier, C, Lakhrissi, M, Chapleur, Y, Two novel strategies for the construction of the core acetal of the zaragozic acids, *Synlett*, 697–700, 1999.
291. Tius, M A, Gomez-Galeno, J, Gu, X, Zaidi, J H, *C*-glycosylanthraquinone synthesis: total synthesis of vineomycinone B2 methyl ester, *J. Am. Chem. Soc.*, 113, 5775–5783, 1991.
292. Sasaki, M, Ishikawa, M, Fuwa, H, Tachibana, K, A general strategy for the convergent synthesis of fused polycyclic ethers via β-alkyl Suzuki coupling: synthesis of the ABCD ring fragment of ciguatoxins, *Tetrahedron*, 58, 1889–1911, 2002.
293. Allwein, S P, Cox, J M, Howard, B E, Johnson, H W, Rainier, J D, *C*-Glycosides to fused polycyclic ethers, *Tetrahedron*, 58, 1997–2009, 2002.
294. Kirschning, A, Harders, J, Addition of *C*-nucleophiles to carbohydrate-derived 2,3-dihydro-4*H*-pyran-4-ones: a new entry to thromboxane analogues, *Tetrahedron*, 53, 7867–7876, 1997.
295. Matsumoto, T, Hosoya, T, Suzuki, K, Total synthesis and absolute stereochemical assignment of gilvocarcin M, *J. Am. Chem. Soc.*, 114, 3568–3570, 1992.
296. Hosoya, T, Takashiro, E, Matsumoto, T, Suzuki, K, Total synthesis of the gilvocarcins, *J. Am. Chem. Soc.*, 116, 1004–1015, 1994.
297. Gillhouley, J G, Shing, T M K, Enantiospecific syntheses of (+) and (−) altholactone (goniothalenol), *J. Chem. Soc., Chem. Commun.*, 976–979, 1988.
298. Terauchi, T, Morita, M, Kimijima, K, Nakamura, Y, Hayashi, G, Tanaka, T, Kanoh, N, Nakata, M, Synthetic studies on altohyrtins (spongistatins): synthesis of the C29–C44 (EF) portion, *Tetrahedron Lett.*, 42, 5505–5508, 2001.
299. Ferrier, R J, Unsaturated carbohydrates. Part 21. A carbocyclic ring closure of a hex-5-enopyranoside derivative, *J. Chem. Soc., Perkin. Trans. 1*, 1455–1458, 1979.
300. Ferrier, R J, Middleton, S, The conversion of carbohydrate derivatives into functionalized cyclohexanes and cyclopentanes, *Chem. Rev.*, 93, 2779–2831, 1993.
301. Estevez, V A, Prestwich, G D, Synthesis of enantiomerically pure, P-1-tethered inositol tetrakis(phosphate) affinity labels via a Ferrier rearrangement, *J. Am. Chem. Soc.*, 113, 9885–9887, 1991.
302. Sollogoub, M, Mallet, J M, Sinaÿ, P, Titanium (IV) promoted rearrangement of 6-deoxy-hex-5-enopyranosides into cyclohexanones, *Tetrahedron Lett.*, 39, 3471–3472, 1998.
303. Dalko, P I, Sinaÿ, P, Recent advances in the conversion of carbohydrate furanosides and pyranosides into carbocycles, *Angew. Chem. Int. Ed.*, 38, 773–777, 1999.
304. Fisher, M J, Myres, C D, Joglar, J, Chen, S H, Danishefsky, S J, Synthetic studies toward rapamycin: a solution to a problem in chirality merger through use of the Ireland reaction, *J. Org. Chem.*, 56, 5826–5834, 1991.

305. Boyer, F D, Lallemand, J Y, Synthetic studies on the 1-hydroxy nortropane system, Part 3. Total enantioselective synthesis of (−)-calystegine-b2, *Synlett*, 969–971, 1992.
306. Chida, N, Ohtsuka, M, Ogawa, S, Stereoselective total synthesis of (+)-lycoricidine, *Tetrahedron Lett.*, 32, 4525–4528, 1991.
307. Chida, N, Ohtsuka, M, Ogawa, S, Total synthesis of (+)-lycoricidine and its 2-epimer from D-glucose, *J. Org. Chem.*, 58, 4441–4447, 1993.
308. Chrétien, F, Chapleur, Y, Chiral cyclohexanes from carbohydrates: successful carbocyclisation of a D-*arabino-hex*-5 enopyranoside derivative, *J. Chem. Soc., Chem. Commun.*, 1268–1269, 1984.
309. Chapleur, Y, Chrétien, F, Khaldi, M, Recent developments in the enantiospecific synthesis of Amaryllidacae alkaloids, In *Antibiotics and Antiviral Compounds*, Krohn, K, Kirst, E and Maas, H, Eds., Verlag Chemie, Weinheim, pp. 380–388, 1993.
310. Park, T K, Danishefsky, S J, A concise route to enantiomerically pure 2-arylcyclohexenones of relevance to the pancratistatin problem, *Tetrahedron Lett.*, 36, 195–196, 1995.
311. Khaldi, M, Chrétien, F, Chapleur, Y, A short route to enantiomerically pure narciclasine derivatives, *Tetrahedron Lett.*, 36, 3003–3006, 1995.
312. Friestad, G K, Branchaud, B P, A new approach to the pancratistatin C-ring from D-glucose: Ferrier rearrangement, pseudoinversion and Pd-catalyzed cyclizations, *Tetrahedron Lett.*, 38, 5933–5936, 1997.
313. Grubb, L M, Dowdy, A L, Blanchette, H S, Friestad, G K, Branchaud, B P, An approach to (+)-pancratistatin from D-glucose: a conformational lock solves a stereochemical problem, *Tetrahedron Lett.*, 40, 2691–2694, 1999.
314. Ermolenko, M S, Shekharam, T, Lukacs, G, Potier, P, A convergent carbohydrate approach to the synthesis of taxol 1. Ring A subunit, *Tetrahedron Lett.*, 36, 2461–2464, 1995.
315. Ermolenko, M S, Lukacs, G, Potier, P, A convergent carbohydrate approach to the synthesis of taxol 2. Ring C subunit, *Tetrahedron Lett.*, 36, 2465–2468, 1995.
316. Momose, T, Setoguchi, M, Fujita, T, Tamura, H, Chida, N, Chiral synthesis of the CD ring unit of paclitaxel from D-glucal, *Chem. Commun.*, 2237–2238, 2000.
317. Pettersson, L, Frejd, T, Magnusson, G, An enantiospecific synthesis of a taxol A-ring building unit, *Tetrahedron Lett.*, 28, 2753–2756, 1987.
318. Chida, N, Ohtsuka, M, Nakazawa, K, Ogawa, S, Total synthesis of hygromycin A, *J. Chem. Soc., Chem. Commun.*, 436–438, 1989.
319. Knapp, S, Naughton, A B J, Dhar, T G M, Intramolecular amino delivery reactions for the synthesis of valienamine and analogues, *Tetrahedron Lett.*, 33, 1025–1028, 1992.
320. Sato, K, Bokura, M, Moriyama, H, Igarashi, T, Total synthesis of a novel β-glucosidase inhibitor, cyclophellitol starting from D-glucose, *Chem. Lett.*, 37–40, 1994.
321. Chida, N, Sugihara, K, Ogawa, S, Chiral synthesis of (−)-mesembranol starting from D-glucose, *J. Chem. Soc., Chem. Commun.*, 901–902, 1994.
322. Amano, S, Ogawa, N, Ohtsuka, M, Chida, N, Chiral and stereoselective total synthesis of novel immunosuppressant FR65814 from D-glucose, *Tetrahedron*, 55, 2205–2224, 1999.
323. Amano, S, Takemura, N, Ohtsuka, M, Ogawa, S, Chida, N, Total synthesis of paniculide A from glucose, *Tetrahedron*, 55, 3855–3870, 1999.
324. Taillefumier, C, Chapleur, Y, Bayeul, D, Aubry, A, An entry to enantiomerically pure *cis* decalinic structures from carbohydrates, *J. Chem. Soc., Chem. Commun.*, 937–938, 1995.
325. For other examples of sluggish Ferrier carbocyclization, see Taillefumier, C, Chapleur, Y, Enantiomerically pure decalinic structures from carbohydrates using intramolecular Diels–Alder and Ferrier carbocyclization, *Can. J. Chem.*, 78, 708–722, 2000.
326. Alonso, R A, Burgey, C S, Rao, B V, Vite, G D, Vollerthun, R, Zottola, M A, Fraser-Reid, B, Carbohydrates to carbocycles — synthesis of the densely functionalized carbocyclic core of tetrodotoxin by radical cyclization of an anhydro sugar precursor, *J. Am. Chem. Soc.*, 115, 6666–6672, 1993.
327. McDevitt, R E, Fraser-Reid, B, A divergent route for a total synthesis of cyclophellitol and epicyclophellitol from a [2.2.2]oxabicyclic glycoside prepared from D-glucal, *J. Org. Chem.*, 59, 3250–3252, 1994.
328. Tatsuta, K, Niwata, Y, Umezawa, K, Toshima, K, Nakata, M, Enantiospecific total synthesis of a β-D-glucosidase inhibitor, cyclophellitol, *Tetrahedron Lett.*, 31, 1171–1172, 1990.

329. Marco Contelles, J, Cyclohexane epoxides — chemistry and biochemistry of (+)-cyclophellitol, *Eur. J. Org. Chem.*, 1607–1618, 2001.
330. Keck, G E, McHardy, S F, Murry, J A, Total synthesis of (+)-7-deoxypancratistatin: a radical cyclization approach, *J. Am. Chem. Soc.*, 117, 7289–7290, 1995.
331. Keck, G E, McHardy, S F, Murry, J A, Diastereoselective 6-*exo* radical cyclizations of oxime ethers: total synthesis of 7-deoxypancratistatin, *J. Org. Chem.*, 64, 4465–4476, 1999.
332. Martinez-Grau, A, Marco-Contelles, J, Carbocycles from carbohydrates via free radical cyclizations: new synthetic approaches to glycomimetics, *Chem. Soc. Rev.*, 27, 155–162, 1998.
333. Ganem, B, Glycomimetics that inhibit carbohydrate metabolism, In *Carbohydrate Mimics*, Ch. 13, Chapleur, Y, Ed., Wiley-VCH, Weinheim, pp. 239–258, 1999.
334. Simpkins, N S, Stokes, S, Whittle, A J, An enantiospecific synthesis of allosamizoline, *J. Chem. Soc., Perkin. Trans. 1*, 2471–2477, 1992.
335. Enholm, E J, Satici, H, Trivellas, A, Samarium(II) iodide mediated carbocycles from carbohydrates: application to the synthesis of the C ring of anguidine, *J. Org. Chem.*, 54, 5841–5843, 1989.
336. Boiron, A, Zillig, P, Faber, D, Giese, B, Synthesis of trehazolin from D-glucose, *J. Org. Chem.*, 63, 5877–5882, 1998.
337. Storch de Garcia, I, Dietrich, H, Bobo, S, Chiara, J L, A highly efficient pinacol coupling approach to trehazolamine starting from D-glucose, *J. Org. Chem.*, 63, 5883–5889, 1998.
338. Chiara, JL, New reductive carbocyclisations of carbohydrate derivatives promoted by samarium diiodide, In *Carbohydrate Mimics*, Ch. 7, Chapleur, Y, Ed., Wiley-VCH, Weinheim, pp. 123–141, 1999.
339. Marquez, V E, Lim, M, Tseng, C K H, Markovac, A, Priest, M A, Khan, M S, Kaskar, B, Total synthesis of neplanocin A, *J. Org. Chem.*, 53, 5709–5714, 1987.
340. Mirza, S, Molleyres, L P, Vasella, A, Synthesis of a glyoxylase I inhibitor from *Streptomyces griseosporeus* Niida et Ogasawara, *Helv. Chim. Acta*, 68, 988–996, 1985.
341. Bélanger, P, Prasit, P, Carbocycles from carbohydrates: a simple route to an enantiomerically pure prostaglandin intermediate, *Tetrahedron Lett.*, 29, 5521–5524, 1988.
342. Li, C, Fuchs, P L, Methanesulfenyl triflate promoted iminosulfenylation of an allylic trichloroacetimidate. An efficient and stereospecific total synthesis of (+) mannostatin A, *Tetrahedron Lett.*, 35, 5121–5124, 1994.
343. Shiozaki, M, Kobayashi, Y, Arai, M, Haruyama, H, Synthesis of 6-*epi*-trehazolin from D-ribonolactone — evidence for the non-existence of a 5,6-ring fused structural isomer of 6-*epi*-trehazolin, *Tetrahedron Lett.*, 35, 887–890, 1994.
344. Tatsuta, K, Yasuda, S, Araki, N, Takahashi, M, Kamiya, Y, Total synthesis of a glyoxalase I inhibitor and its precursor, (−)-KD16-U1, *Tetrahedron Lett.*, 39, 401–402, 1998.
345. Lubineau, A, Billault, I, New access to unsaturated keto carba sugars (gabosines) using an intramolecular Nozaki–Kishi reaction as the key step, *J. Org. Chem.*, 63, 5668–5671, 1998.
346. Kim, D, Ahn, S K, Bae, H, Choi, W J, Kim, H S, An asymmetric total synthesis of (−)-fumagillol, *Tetrahedron Lett.*, 38, 4437–4440, 1997.
347. Nakata, M, Akazawa, S, Kitamura, S, Tatsuta, K, Enantiospecific total synthesis of (−)allosazoline, an aminocyclitol moiety of the insect chitinase inhibitor allosamidin, *Tetrahedron Lett.*, 32, 5363–5366, 1991.
348. Lygo, B, Swiatyj, M, Trabsa, H, Voyle, M, Synthesis of (+)-gabosines C and E from D-ribose, *Tetrahedron Lett.*, 35, 4197–4200, 1994.
349. Jiang, S D, Mekki, B, Singh, G, Wightman, R H, Enantiospecific synthesis of (−)-5-*epi*-shikimic acid and a new route to (−)-shikimic acid, *Tetrahedron Lett.*, 35, 5505–5508, 1994.
350. Vanhessche, K, Gonzales Bello, C, Vandewalle, M, Total synthesis of (−)-neplanocin A from L-ribulose, *Synlett*, 921–922, 1991.
351. Ziegler, F E, Wang, Y Z, A synthesis of (+)-cyclophellitol from D-xylose, *J. Org. Chem.*, 63, 426–427, 1998.
352. For a review, see Giuliano, RM, Cycloaddition reactions in carbohydrate chemistry — an overview, In *Cycloaddition Reactions in Carbohydrate Chemistry*, Giuliano, RM, Ed., American Chemical Society, Washington, DC, pp. 1–23, 1992.

353. Prasad, J S, Clive, D L J, Da Silva, G V J, Synthetic studies to compactin: use of tri-*O*-acetyl-D-glucal for the preparation of chiral cyclohexenes, *J. Org. Chem.*, 51, 2717–2721, 1986.
354. Isobe, M, Fukami, N, Nishikawa, T, Goto, T, Synthesis of chiral cyclohexanes from levoglucosenone and its application to an indole alkaloid reserpine, *Heterocycles*, 25, 521–532, 1987.
355. Isobe, M, Nishikawa, T, Pikul, S, Goto, T, Synthetic studies on tetrodotoxin (1) stereocontrolled synthesis of the cyclohexane moiety, *Tetrahedron Lett.*, 28, 6485–6488, 1987.
356. Dyong, I, Hagedorn, H-W, Thiem, J, Diastereoselective aufbau eines rhodomycinone-modells aus D-glucose, *Liebigs Ann. Chem.*, 551–563, 1986.
357. Abdur Rahman, M, Fraser-Reid, B, Actinobolin via the anomeric effect, *J. Am. Chem. Soc.*, 107, 5576–5578, 1985.
358. Herscovici, J, Delatre, S, Antonakis, K, Enantiospecific naphthopyran synthesis by intramolecular Diels–Alder cyclisation of 4-keto 2,3-unsaturated *C*-glycosides, *Tetrahedron Lett.*, 32, 1183–1186, 1991.
359. Tsang, R, Fraser-Reid, B, Pyranose α-enones provide ready access to functionalized *trans* decalins via bis annulated pyranosides obtained by intramolecular Diels–Alder reactions; a key intermediate for forskolin, *J. Org. Chem.*, 57, 1065–1067, 1992.
360. Hanna, I, Lallemand, J Y, Wlodyka, P, A synthetic approach to the tricyclic system of forskolin from D-galactose, *Tetrahedron Lett.*, 35, 6685–6688, 1994.
361. Prashad, M, Fraser-Reid, B, The oxahydrindene component of the avermectins, *J. Org. Chem.*, 50, 1564–1566, 1985.
362. Holt, D J, Barker, W D, Jenkins, P R, Panda, J, Ghosh, S, Stereoselective preparation of enantiomerically pure annulated carbohydrates using ring-closing metathesis, *J. Org. Chem.*, 65, 482–493, 2000.
363. Tadano, K, Kanazawa, S, Takao, K, Ogawa, S, Intramolecular aldol condensation applied to D-glucose-derived δ-ketoaldehydes — access to enantiomerically pure 6-membered carbocycles, *Tetrahedron*, 48, 4283–4300, 1992.
364. Ishihara, J, Nonaka, R, Terasawa, Y, Tadano, K, Ogawa, S, Synthetic studies on the trichothecene family from D-glucose, *Tetrahedron: Asymmetry*, 5, 2217–2232, 1994.
365. Ishihara, J, Nonaka, R, Terasawa, Y, Shiraki, R, Yabu, K, Kataoka, H, Ochiai, Y, Tadano, K.-i, Total synthesis of (−)-verrucarol, a component of naturally occurring verrucarin A, *Tetrahedron Lett.*, 38, 8311–8314, 1997.
366. Tsang, R, Fraser-Reid, B, A route to optically active tricothecane skeleton by bisannulation of a pyranose derivative, *J. Org. Chem.*, 50, 4661–4663, 1985.
367. Fetizon, M, Duc, D K, Nguyen, D T, An approach to the synthesis of optically active trichothecenes from tri-*O*-acetyl-D-glucal, *Tetrahedron Lett.*, 27, 1777–1780, 1986.
368. Ermolenko, M S, Pipelier, M, A carbohydrate-based synthetic approach to quadrone, *Tetrahedron Lett.*, 38, 5975–5976, 1997.
369. Bonnert, R V, Jenkins, P R, The first example of a Robinson annulation on a carbohydrate derivative, *J. Chem. Soc., Chem. Commun.*, 6–7, 1987.
370. Bonnert, R V, Howarth, J, Jenkins, P R, Lawrence, N J, Robinson annulation on a carbohydrate derivative, *J Chem Soc, Perkin Trans. 1*, 1225–1229, 1991.
371. Bonnert, R V, Davies, M J, Horwath, J, Jenkins, P R, The stereochemistry of the Stork silyl methylene radical cyclisation in an annulated sugar derivative, *J. Chem. Soc., Chem. Commun.*, 148–149, 1990.
372. Bonnert, R V, Davies, M J, Howarth, J, Jenkins, P R, Lawrence, N J, Stereoselective reductive hydroxymethylation of an annulated sugar derivative via radical cyclisation, *J Chem Soc, Perkin Trans. 1*, 27–29, 1992.
373. Boa, A N, Clark, J, Jenkins, P R, Lawrence, N J, Construction of a key intermediate in the asymmetric synthesis of chiral taxanes, *J. Chem. Soc., Chem. Commun.*, 151–152, 1993.
374. Franck, R W, Bhat, V, Subramanian, C S, Stereoselective total synthesis of the natural enantiomer of olivin trimethyl ether, *J. Am. Chem. Soc.*, 108, 2455–2457, 1986.
375. Brade, W, Vasella, A, Synthesis of naphto[2,3,b] pyrandiones: (−)cryptosporin, *Helv. Chim. Acta*, 72, 1649–1657, 1989.
376. Gomez, A M, Lopez, J C, Fraser-Reid, B, Serial radical cyclization of pyranose-derived dienes in the stereocontrolled synthesis of Woodward's reserpine precursor, *J. Org. Chem.*, 60, 3859–3870, 1995.

12 Synthesis of Carbohydrate Containing Complex Natural Compounds

Kazunobu Toshima

CONTENTS

Abbreviations: Ac, acetyl; AIBN, 2,2′-azobisisobutyronitrile; All, allyl; Alloc, allyloxycarbonyl; Bn, benzyl; BOM, benzyloxymethyl; BPC, *p*-phenylbenzoyl; Bu, butyl; *t*-Bu, *tert*-butyl; Bz, benzoyl; CAN, ammonium hexanitratocerate (IV); Cbz, bezyloxycarbonyl; CDI, carbonyldiimizazole; ClAc, chloroacetyl; Cp, cyclopentadienyl; CSA, DL-10-camphorsulfonic acid; DABCO, 1,4-diazabicyclo[2.2.2]octane; DAST, diethylaminosulfur trifluoride; dba, dibenzylideneacetone; DBU, 1,8-diazabicyclo[5.4.0]undec-7-ene; DCC, *N,N*-dicyclohexylcarbodiimide; DDQ, 2,3-dichloro-5,6-dicyano-1,4-benzoquinone; DEAD, diethyl azodicarboxylate; DEIPS, diethylisopropylsilyl; DIBAL, diisobutylaluminum hydride; DMAP,

4-dimethylaminopyridine; DMA, *N,N*-dimethylacetamide; DMF, dimethylformamide; DNP, dinitrophenyl; DTBMP, 2,6-di-*tert*-butyl-4-methylpyridine; Et, ethyl; $Eu(fod)_3$, tris(6,6,7,7,8,8,8-heptafluoro-2,2-dimethyl-3,5-octanedionato)europium; FMOC, 9-fluorenylmethoxycarbonyl; HMPA, hexamethyl phosphoric triamide; Im, imidazolyl; Ipc, isopinocamphenyl; IPDMS, isopropyldimethylsilyl; KHMDS, potassium bis(trimethylsilyl)amide; LAH, lithium aluminum hydride; LDA, lithium diisopropylamide; LiHMDS, lithium bis(trimethylsilyl)amide; *m*CPBA, *m*-chloroperoxybenzoic acid; Me, methyl; MOM, methoxymethyl; Ms, methanesulfonyl; NB, *o*-nitrobenzyl; NBS, *N*-bromosuccinimide; NCS, *N*-chlorosuccinimide; NIS, *N*-iodosuccinimide; NMO, *N*-methylmorpholine *N*-oxide; NMP, *N*-methyl-2-pyrrolidinone; PCC, pyridinium chlorochromate; Ph, phenyl; Phth, phthaloyl; PMB, *p*-methoxybenzyl; Pr, propyl; *i*-Pr, isopropyl; PPTS, pyridinium *p*-toluenesulfonate; Pv, pivaloyl; Py, pyridine; rt, room temperature; TBAF, tetrabutylammonium fluoride; TBAI, tetrabutylammonium iodide; TBS, *tert*-butyldimethylsilyl; TEOC, 2-(trimethylsilyl)ethoxycarbonyl; TES, triethylsilyl; Tf, trifluoromethanesulfonyl; THF, tetrahydrofuran; TIPS, triisopropylsilyl; TMEDA, *N,N,N′,N′*-tetramethylethylenediamine; TMS, trimethylsilyl; TPAP, tetra-*n*-propylammonium perruthenate; Tr, triphenylmethyl; TPS, *tert*-butyldiphenylsilyl; Ts, *p*-toluenesulfonyl.

12.1 INTRODUCTION

Many carbohydrate-containing complex natural compounds are found in nature as important biological substances [1]. A large number of recent biological studies on these glycosubstances, which possess mono- and oligosaccharides, such as proteoglycans, glycoproteins, glycolipid and antibiotics at the molecular level have shed light on the biological significance of their carbohydrate parts (glycons) in molecular recognition for the transmission of biological information. Therefore, it is now recognized that carbohydrates are at the heart of a multitude of biological events. With the stimulant biological background, the efficient synthesis of not only the carbohydrate itself but also carbohydrate-containing complex natural compounds is becoming more and more important in the field of organic chemistry and chemical biology [2]. For the efficient total or partial synthesis of mono- and oligosaccharide-containing complex natural products, there are several issues to overcome. The first one is the moment of glycosidation [3], the carbohydrate attachment, within the planned sequence so as to optimize efficiency. The second issue relates to the choice of the glycosidation method to be used with regard to the yield and the stereoselectivity. The third issue is associated with the choice of appropriate protecting groups [4]. This chapter focuses on highlights of the carbohydrate chemistry leading to the successful total synthesis of some representative natural products possessing both mono- and oligosaccharide side chains and complex noncarbohydrate components, except for glycoproteins and glycolipids. In this survey of the current advances in this area, the carbohydrate-containing complex natural compounds are arbitrarily classified into three groups: (1) *O*-glycoside antibiotics, (2) *C*-glycoside antibiotics and (3) others.

12.2 *O*-GLYCOSIDE ANTIBIOTICS

12.2.1 METHYMYCIN

The total synthesis of the 12-membered macrolide antibiotic, methymycin, by Masamune [5] (Scheme 12.1) involves the glycosidation of the macrocyclic lactone **1** with 1-α-bromo-2-acetyldesosamine hydrobromide **2**, using lutidine in chloroform to give the desired β-glycoside **3** as the major anomer. This is converted into methymycin **4** by deacetylation, employing Et_3N in MeOH. The regioselectivity of the glycosidation results from the difference in the reactivity between the C-3 secondary alcohol and the C-10 tertiary alcohol in **1**. The β-stereoselectivity of the glycosidation comes from the participation of the C-2′ acetoxy neighboring group in **2**. It was noted that retaining the *N,N*-dimethylamino group as its hydrobromide salt during the reaction was essential for successful glycoside formation, and that the use of Ag and Hg salts and other bases stronger than lutidine led to the complete destruction of the sugar moiety.

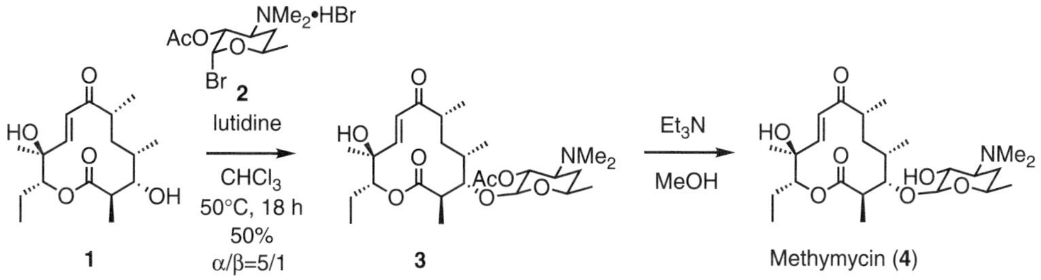

2, lutidine, $CHCl_3$, 50°C, 18 h, 50%, α/β=5/1 — 1 → 3; Et_3N, MeOH — 3 → Methymycin (4)

SCHEME 12.1 Synthesis of methymycin by Masamune [5].

12.2.2 Erythromycin A

The first highlight of carbohydrate chemistry in Woodward's total synthesis of erythromycin A [6] (Scheme 12.2) is the glycosidation of the 14-membered macrolide lactone **5** with *S*-pyrimidyl D-desosaminide **6**, using AgOTf as the activator. The glycosidation proceeded regio- and stereoselectively with C-2′ neighboring group participation to give the β-glycoside **7**. The second glycosidation, reaction of **7** with the *S*-pyridyl L-cladinoside **8** using $Pb(ClO_4)_2$ as the promoter, followed by the deprotection of the C-2′ methoxycarbonyl group by methanolysis, and the C-4″ acetyl and the C-9 *p*-phenylbenzoyl groups by using Na–Hg in MeOH, afforded (9*S*)-erythromycylamine **10**. The subsequent conversion of the C-9 amine in **10** into the ketone, via the imine **12**, furnished erythromycin A (**13**). In this historical synthesis, it was also clarified that the stereoselective introduction of the acid-sensitive 2,6-dideoxy sugar, L-cladinose, to the unreactive C-3 hydroxy group in the aglycon was one of the most difficult tasks after the aglycon synthesis.

After 10 years, Toshima and Tatsuta overcame this problem [7] and demonstrated the effective synthesis of erythromycin A (**13**) from its aglycon, (9*S*)-9-dihydroerythronolide A (**14**), which was synthesized by Kinoshita and Nakata using the 2,6-anhydro-2-thio sugar strategy developed by Toshima (Scheme 12.3). Thus, the glycosidation of **17** (obtained from **14** in three steps) with the *S*-pyrimidyl D-desosaminide **6** using a modified Woodward procedure gave the β-glycoside **18**. After oxidation of the *N,N*-dimethylamino group of **18**, the resulting **19** was effectively glycosylated with the 2,6-anhydro-2-thio sugar **20** (corresponding to L-cladinose) using NIS and TfOH to furnish the α-glycoside **21** in excellent yield. The extremely high α-stereoselectivity must arise from the repulsive interaction between the sulfur atom in the rigid conformation of 2,6-anhydro-2-thio sugar **20** and the approaching hydroxyl group in **19**. The conversion of the 2,6-anhydro-2-thio system in **21** into the desired 2,6-dideoxy structure by hydrogenolysis with Raney–Ni, followed by the regioselective oxidation of the C-9 hydroxyl group of **23**, afforded erythromycin A (**13**).

12.2.3 Tylosin

Tatsuta's total synthesis of tylosin [8] (Scheme 12.4) involves the regio- and stereoselective introduction of the amino disaccharide, 4-*O*-(α-L-mycarosyl)-D-mycaminose and D-mycinose onto the C-5 and C-23 hydroxy groups, respectively, of the 16-membered macrolide aglycon. The first glycosidation of the suitably protected aglycon derivative **25**, which was derived from D-glucose, with D-mycaminosyl bromide **26** using HgO and $HgBr_2$ as the activators, followed by methanolysis, yielded the β-glycoside **27**. After selective acetylation of the C-2′ hydroxy group in **27** by its own basicity, the glycosidation of the resultant **28** was accomplished by Tatsuta's method using the glycal of mycarose **29** and 1,3-dibromo-5,5-dimethylhydantoin to selectively give the 2-bromo-α-glycoside **30**. It was confirmed that the presence of a bromine atom at the C-2″ position of the mycaroside moiety induced the glycoside bond to resist even drastic acid hydrolysis. Thus, glycosidation of **31**, which was obtained after deprotection by acid hydrolysis followed by reductive debromination, with the mycinosyl bromide **32** under Helferich conditions using

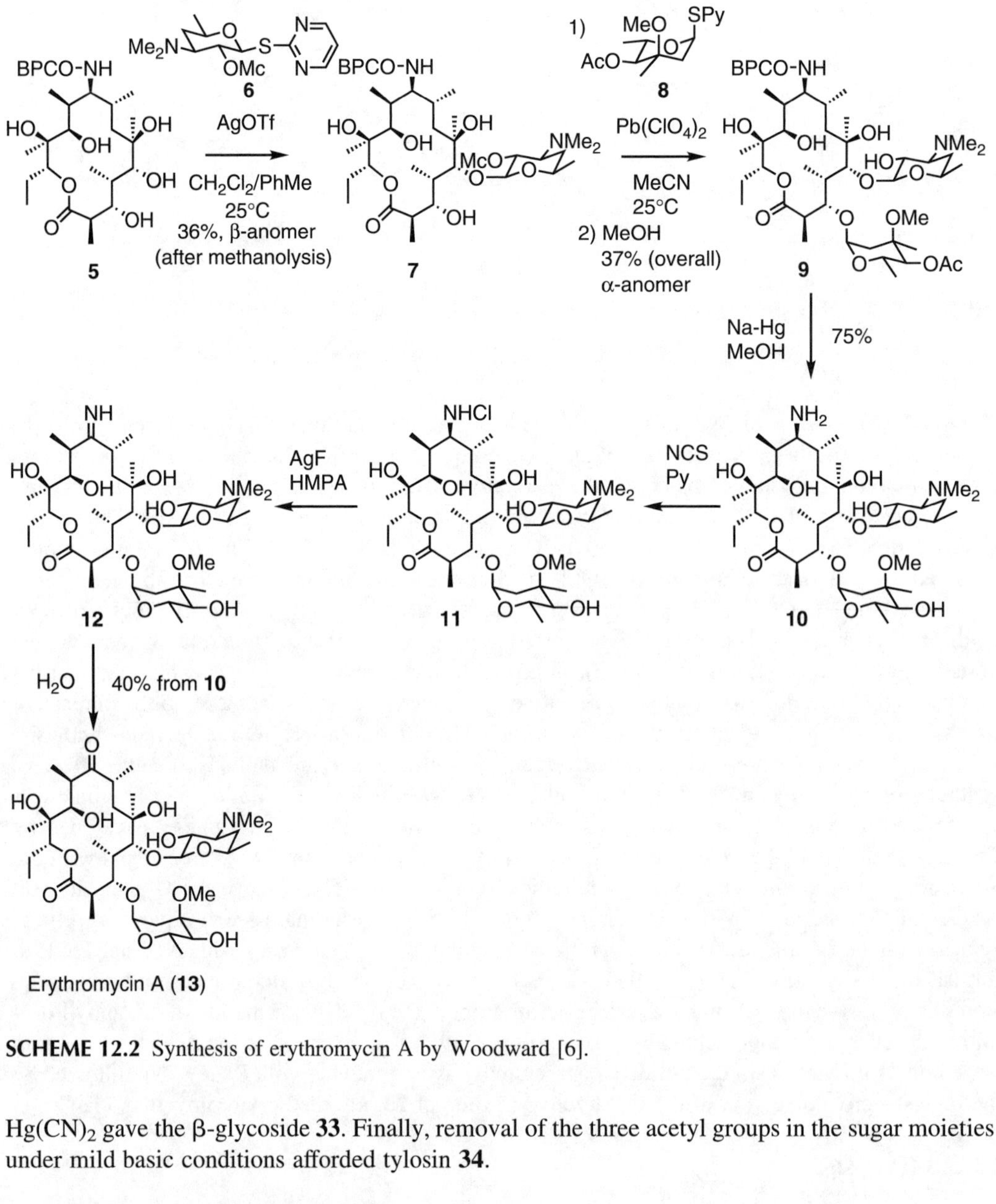

SCHEME 12.2 Synthesis of erythromycin A by Woodward [6].

$Hg(CN)_2$ gave the β-glycoside **33**. Finally, removal of the three acetyl groups in the sugar moieties under mild basic conditions afforded tylosin **34**.

12.2.4 Mycinamicins IV and VII

The total syntheses of mycinamicins IV and VII (Scheme 12.5) were achieved by Suzuki [9] via glycosidations, using glycosyl fluorides as glycosyl donors and Cp_2MCl_2–$AgClO_4$ (M = Zr or Hf) as new activators for the stereoselective introduction of the β-D-mycinosyl and β-D-desosaminyl moieties. Stereoselective glycosylation of the 16-membered macrolactonic alcohol **35** with the glycosyl fluoride **36** in the presence of Cp_2HfCl_2–$AgClO_4$, followed by deprotection gave mycinamicin VII (**38**). The glycosylation of **39**, which was obtained by the selective protection of the C-2′ hydroxyl group with methyl chloroformate without additional base, was accomplished using the fluoride **40** and Cp_2ZrCl_2–$AgClO_4$ in benzene without neighboring group assistance. Mycinamicin IV (**42**) was obtained after deprotection of the desired β-glycoside **41** under mild basic hydrolytic conditions.

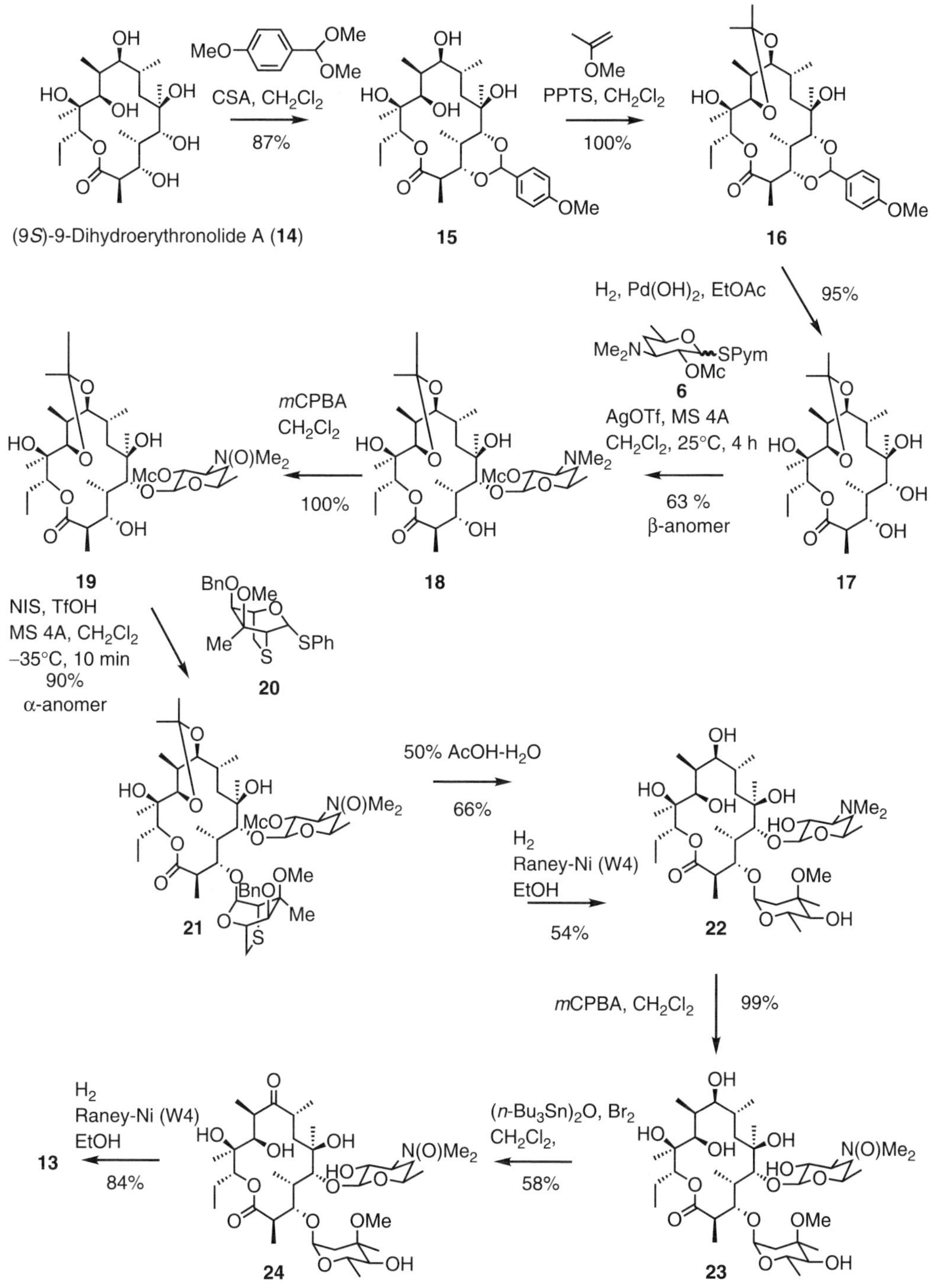

SCHEME 12.3 Synthesis of erythromycin A by Toshima-Tatsuta [7].

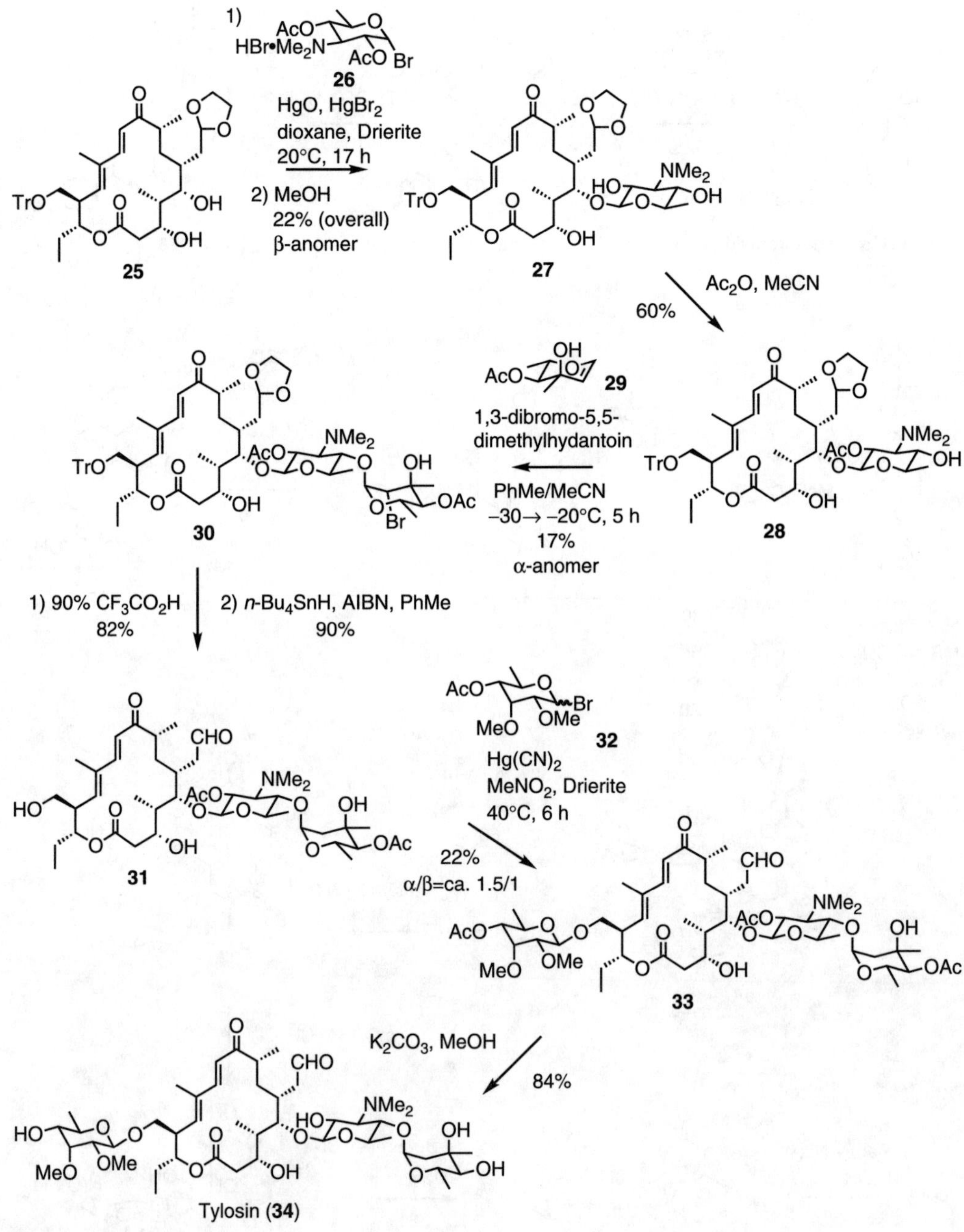

SCHEME 12.4 Synthesis of tylosin by Tatsuta [8].

12.2.5 AVERMECTINS

The synthesis of avermectin B_{1a} from its aglycon derivative by Nicolaou [10] (Scheme 12.6) involves the synthesis of the oligosaccharide from a phenylthio sugar via a glycosyl fluoride. He found that glycosyl fluorides were effectively prepared from the corresponding phenyl thioglycosides using DAST–NBS or HF·Py–NBS. The glycosyl fluoride **44** was chemoselectively activated by the Mukaiyama method, using $SnCl_2$–$AgClO_4$, to be stereoselectively attached to the

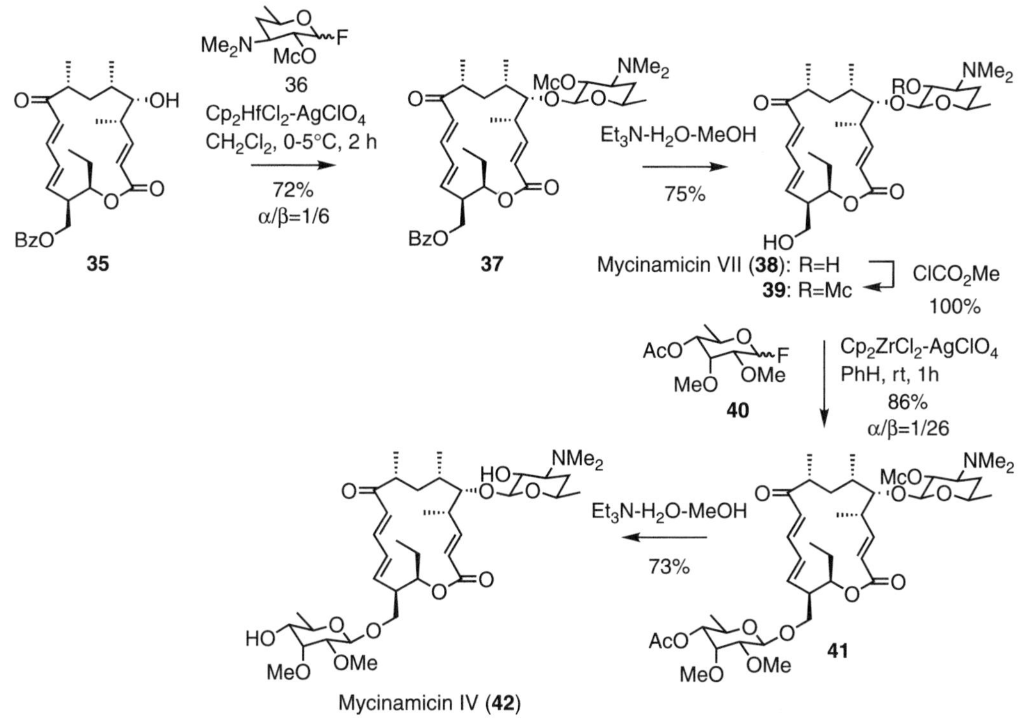

SCHEME 12.5 Synthesis of mycinamicins IV and VII by Suzuki [9].

thioglycoside **43**. The resultant α-disaccharide **45** was converted into the corresponding glycosyl fluoride **46** using DAST–NBS. The glycosidation of **46** and the avermectin B_{1a} aglycon derivative **47**, which was prepared from avermectin B_{1a} by acid hydrolysis, followed by selective monosilylation using $SnCl_2$–$AgClO_4$ gave avermectin B_{1a} (**49**) after deprotection of the silyl groups in the α-glycoside **48**.

Hanessian's synthesis of avermectin B_{1a} [11] is accomplished via the glycosidation of the pyrimidyl thioglycoside **50**, which is obtained from avermectin B_{1a}, and the avermectin B_{1a} aglycon derivative **51** (Scheme 12.7). Thus, **50** was coupled with **51** in the presence of AgOTf to selectively afford the desired α-anomer **52**. The crucial deconjugation of the C-2,3 double bond was accomplished by trimethylsilylation of the C-7 hydroxy group, followed by formation and quenching the TMS enolate of the ester to give the desired C-3,4 double bond isomer. The subsequent desilylation completed the synthesis of avermectin B_{1a} (**49**).

The highlight on carbohydrate chemistry in Danishefsky's total synthesis of avermectin A_{1a} [12] is the use of glycals as glycosyl donors in the α-stereoselective glycosidations (Scheme 12.8). The glycosyl acceptor **54**, and the glycosyl donor, glycal **55**, were prepared from the common intermediate **53**, which was derived from the Danishefsky diene and acetaldehyde followed by oxidation. The coupling of **54** and **55** using NIS gave the desired α-glycoside **56**. The configuration of the formed glycosidic bond and the C-2′ iodine arises from a *trans*-diaxial addition. The disaccharide **56** was converted into the corresponding phenyl thioglycoside **57** by Hanessian's method. Following oxidation, thermolysis and reductive deiodination, the disaccharide glycal **60** was isolated. The glycosidation of **60** and the avermectin A_{1a} aglycon derivative **61** using NIS afforded the α-glycoside **62**, which, after reductive deiodination and deacetylation, led to avermectin A_{1a} (**63**).

The glycosylation chemistry in White's total synthesis of avermectin B_{1a} [13] is similar to Hanessian's synthesis of avermectin B_{1a} (Scheme 12.9). Thus, the glycosidation of the

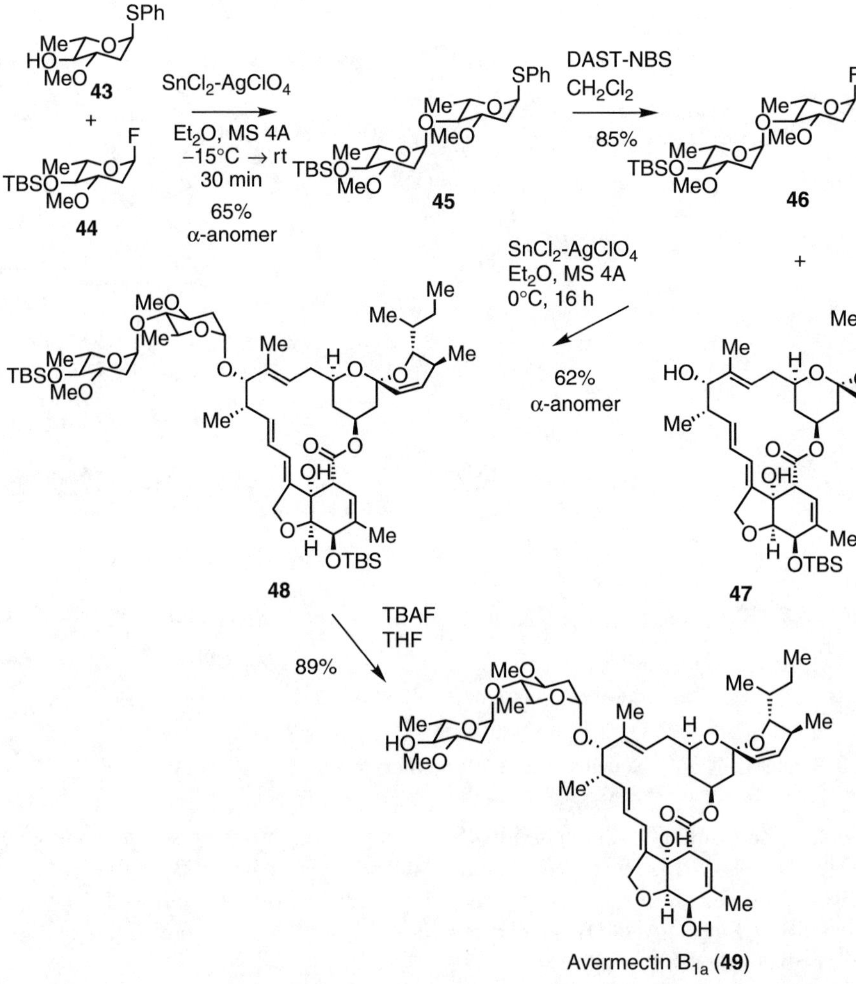

SCHEME 12.6 Synthesis of avermectin B_{1a} by Nicolaou [10].

thioglycoside **50** and the avermectin B_{1a} derivative **47** provided avermectin B_{1a} (**49**) after desilylation of **48**.

Imidazolylcarbonyl and imdazolylthiocarbonyl derivatives were effectively employed in Ley's total synthesis of avermectin B_{1a} [14] (Scheme 12.10). Oleandrose **64** reacted with AcOH activated by CDI to give a 1:1 mixture of the diacetate **65** and the C-1 monoacetate **66**, the former of which was selectively reduced using super-hydride to afford the C-4 monoacetate **67**. The glycosidation of the C-1 monoacetate **66** and the C-4 monoacetate **67**, via the imidazolylcarbonyl derivative **68** using $AgClO_4$, furnished the disaccharide **69**, which was subjected to selective super-hydride deacetylation to give **70**. A second glycosylation between the imidazolylthiocarbonyl glycoside **71**, obtained from **70**, and the avermectin B_{1a} aglycon derivative **72** using $AgClO_4$ and $CaCO_3$ gave avermectin B_{1a} (**49**) after deacetylation using super-hydride.

12.2.6 Efrotomycin

The two-stage activation procedure for oligosaccharide synthesis, which was developed by Nicolaou and applied to his synthesis of avermectin B_{1a} [10], was also employed in his total synthesis of efrotomycin [15] (Scheme 12.11). The chemo- and stereoselective glycosidation of the

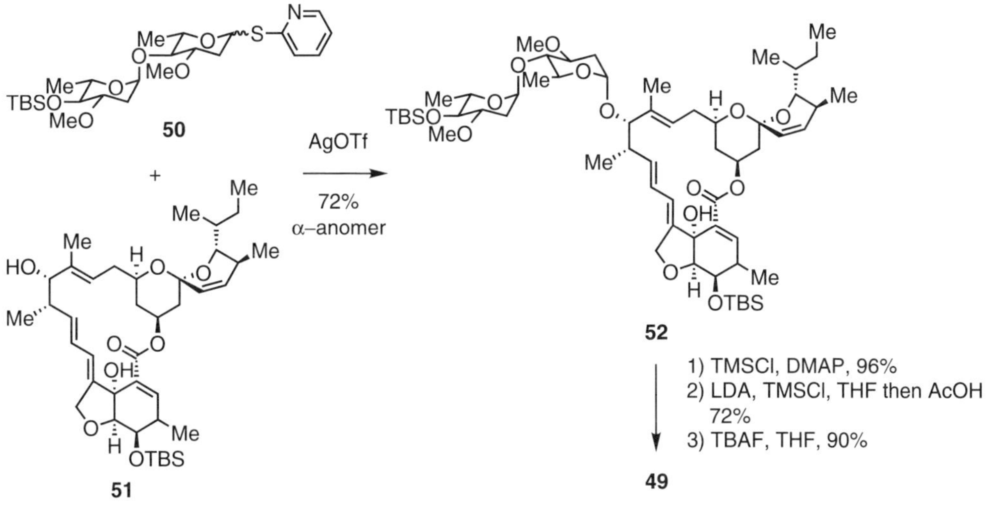

SCHEME 12.7 Synthesis of avermectin B_{1a} by Hanessian [11].

phenyl thioglycoside **74** and the glycosyl fluoride **75** using $SnCl_2$–$AgClO_4$ gave the phenylthio disaccharide **76**, which was converted into the corresponding fluoride **77** employing DAST–NBS. Because the β-stereoselectivity of the next glycosidation of **77** and **80** was low, the glycosyl donor **77** was converted into another glycosyl donor **79** possessing two acetyl groups as the protecting groups. The reaction of **79** and **80** using $SnCl_2$–$AgClO_4$ proceeded stereoselectively to afford the desired β-glycoside **82**, after deacetylation. The subsequent coupling of the lactone **82** and the amine **83**, followed by desilylation and de-*p*-methoxybenzylation, furnished efrotomycin (**84**).

12.2.7 Amphotericin B

One of the highlights of carbohydrate chemistry in Nicolaou's total synthesis of the polyene macrolide antibiotic, amphotericin B [16], is the stereoselective introduction of D-mycosamine into the aglycon derivative (Scheme 12.12). The stereoselective construction of a 1,2-*cis*-β-D-glycosidic bond, a typical situation in β-D-mannosides, is a task long recognized as difficult in carbohydrate chemistry. In order to prepare the desired β-glycosidic linkage, an indirect approach was planned using a well-established 1,2-*trans* glycoside construction, followed by an appropriate inversion at C-2′ of the sugar moiety. The glycosidation of the amphotericin B aglycon derivative **85** and the glycosyl trichloroacetimidate **86** using PPTS in hexane under high dilution conditions give a 1:1 mixture of the desired β-glycoside **87** and the orthoester of **86**. The deprotection and inversion at the C-2′ position of the sugar moiety of **87** by an oxidation–reduction sequence produced the desired β-D-mycosamine structure **90** via the ketone **89**. The subsequent desilylation and azide reduction in the sugar part, and the removal of the isopropylidene and methyl groups in the aglycon moiety, led to amphotericin B methyl ester **93**, which was saponified to amphotericin B (**94**).

12.2.8 Elaiophylin

Kinoshita et al.'s total synthesis of elaiophylin (azalomycin B) [17] (Scheme 12.13) involves the glycosidation of the ethyl ketone **96**, which was synthesized from D-glucose, with the suitably protected fucal **95**, using CSA to give the 2-deoxy-α-glycoside **97** directly. The aldol reaction of the 16-membered lactonic aldehyde **98**, which was also derived from D-glucose, and the *Z*-dibutylboron enolate of the glycosylated ethyl ketone **97** furnished the desired aldol **99** as one of

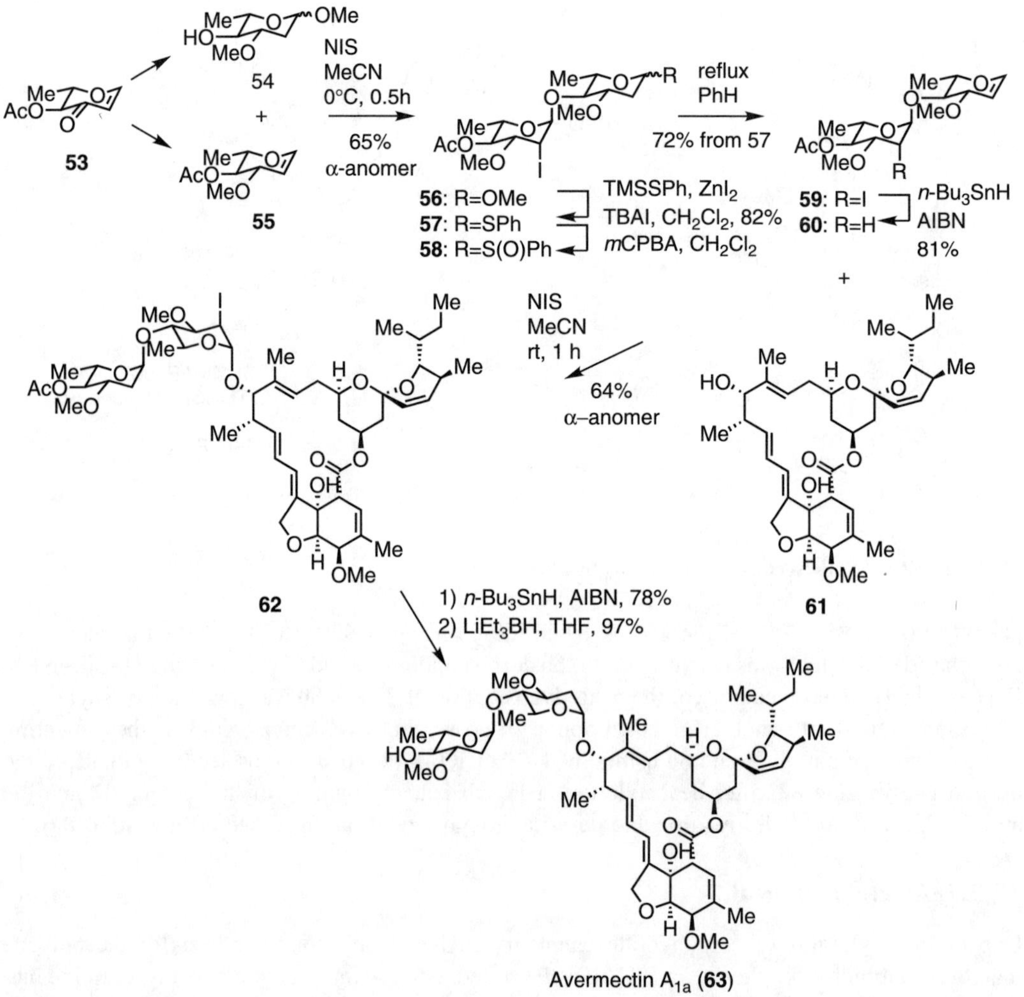

SCHEME 12.8 Synthesis of avermectin A_{1a} by Danishefsky [12].

three diastereomers. The effective and careful deprotection of the silyl groups in **99** under mild acidic conditions gave the labile antibiotic, elaiophylin (azalomycin B) **100**. In this total synthesis, it was found that the newly developed DEIPS group was very useful for synthesis of complex and labile natural products.

12.2.9 Cytovaricin

The highlight of carbohydrate chemistry in Evans' total synthesis of cytovaricin [18] (Scheme 12.14) is the introduction of its sugar part, D-cymarose, into its polyol subunit at the middle stage of this total synthesis. Treatment of the alcohol **101** and the glycosyl acetate **102** in toluene at −20°C with trityl perchlorate, followed by warming of the heterogeneous mixture to −3°C, produced a 1:4 equilibrium mixture in favor of the desired β-glycoside **103**. When the temperature was maintained at −20°C during the reaction, a 3:1 mixture of glycosides favoring the undesired α-anomer of **103** was provided. The undesired α-anomer was recovered and resubmitted to the reaction conditions to give a 4:1 mixture of glycosides favoring the β-anomer **103**. The addition of MeLi to the *N*-methoxy-*N*-methylamide of **103** afforded the methyl ketone which was

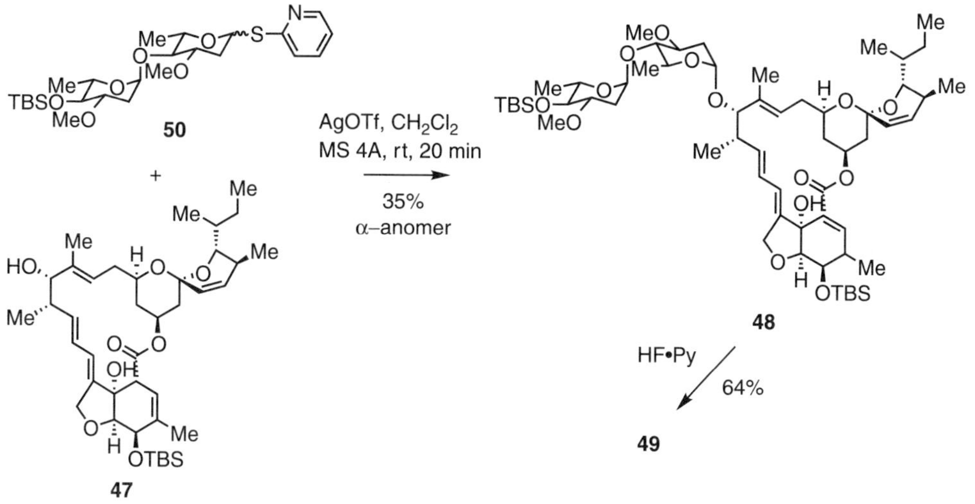

50 + 47

AgOTf, CH_2Cl_2
MS 4A, rt, 20 min
35%
α–anomer

48

HF•Py
64%

49

SCHEME 12.9 Synthesis of avermectin B_{1a} by White [13].

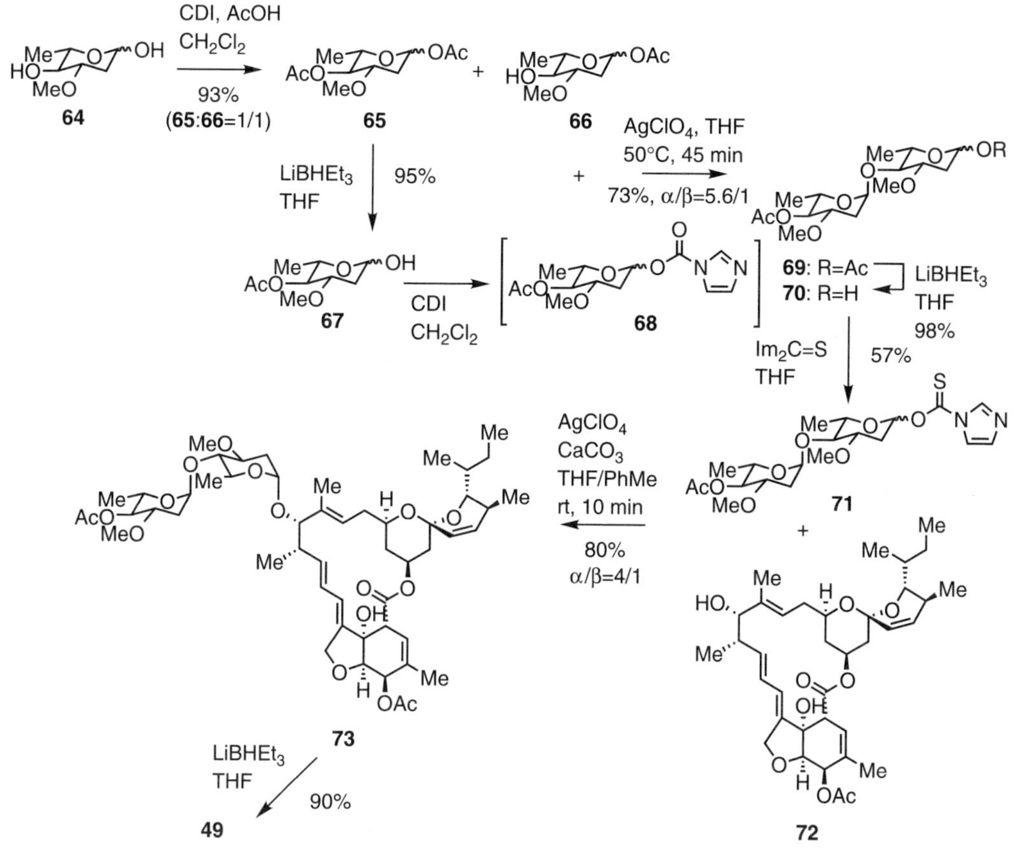

64

CDI, AcOH
CH_2Cl_2
93%
(**65**:**66**=1/1)

65 + 66

LiBHEt$_3$
THF
95%

67

CDI
CH_2Cl_2

68

AgClO$_4$, THF
50°C, 45 min
73%, α/β=5.6/1

69: R=Ac
70: R=H

LiBHEt$_3$
THF
98%

Im$_2$C=S
THF
57%

71 + 72

AgClO$_4$
CaCO$_3$
THF/PhMe
rt, 10 min
80%
α/β=4/1

73

LiBHEt$_3$
THF
90%

49

SCHEME 12.10 Synthesis of avermectin B_{1a} by Ley [14].

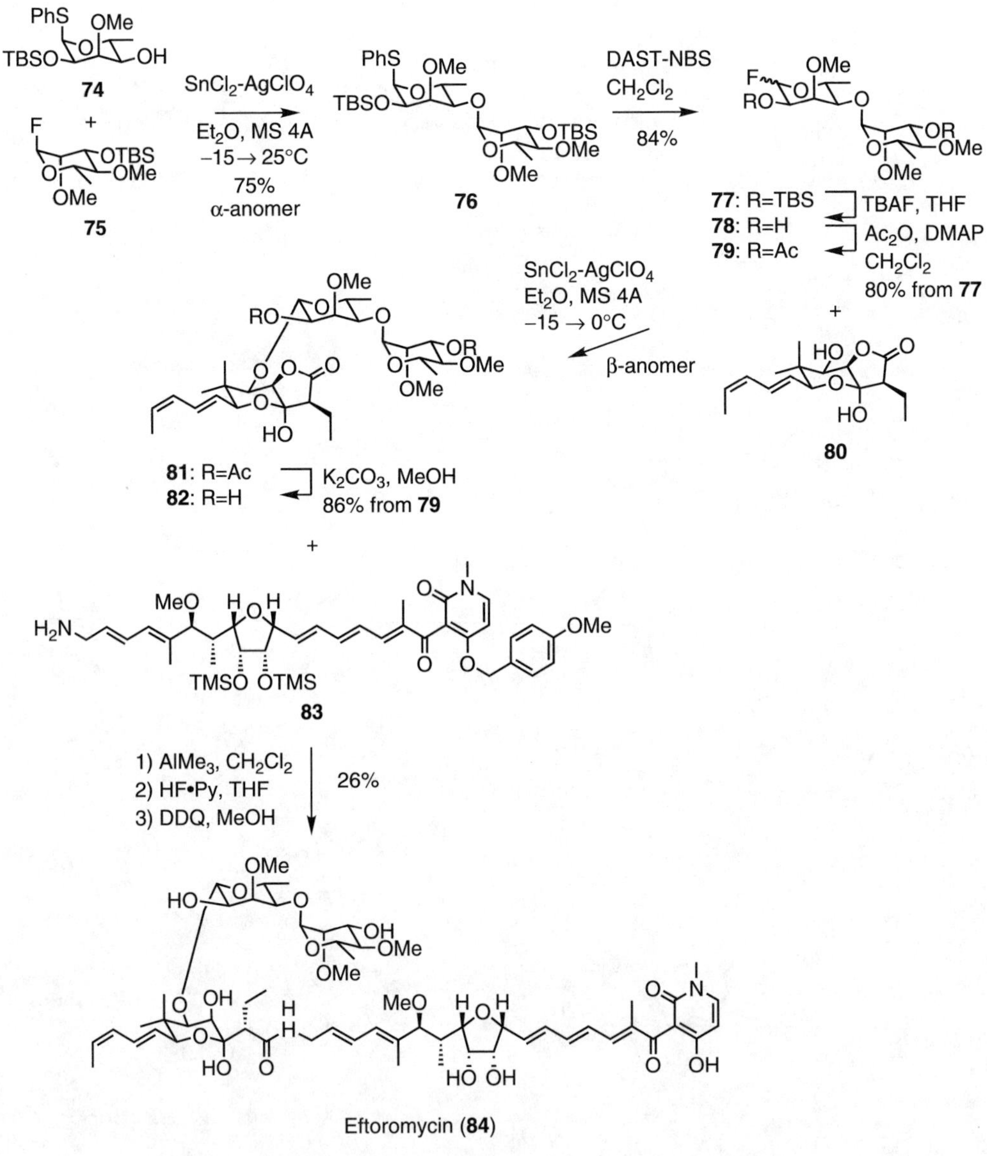

SCHEME 12.11 Synthesis of efrotomycin by Nicolaou [15].

subjected to cleavage of the *p*-methoxybenzyl group with DDQ, an α-chelation-controlled Grignard reaction employing $PhS(CH_2)_4MgBr$, and silylation of the resultant alcohol with TES to furnish **104**. The subsequent diastereoselective dihydroxylation of the terminal olefin of **104** with concomitant formation of a sulfone function, followed by Swern oxidation of the resultant primary hydroxyl group, silylation of the remaining tertiary alcohol and Wittig homologation using $(EtO)_2POCH_2CO_2TMS$ gave the carboxylic acid **105**. The Julia coupling of the sulfone **105** and the aldehyde **106** using $LiNEt_2$ gave a mixture of β-hydroxy sulfones, which was acetylated and treated with $NaHCO_3$ to yield the sodium carboxylate. The latest reaction was performed to prevent deconjugative cleavage of the C-4 oxygen bond under the strong reducing conditions necessary for the elimination of the β-acetoxy sulfone. Exposure of the acetoxy sulfone to 6% Na–Hg, accompanied by cleavage of the C-21 (2,2,2-trichloroethoxy) methoxy group, gave the desired

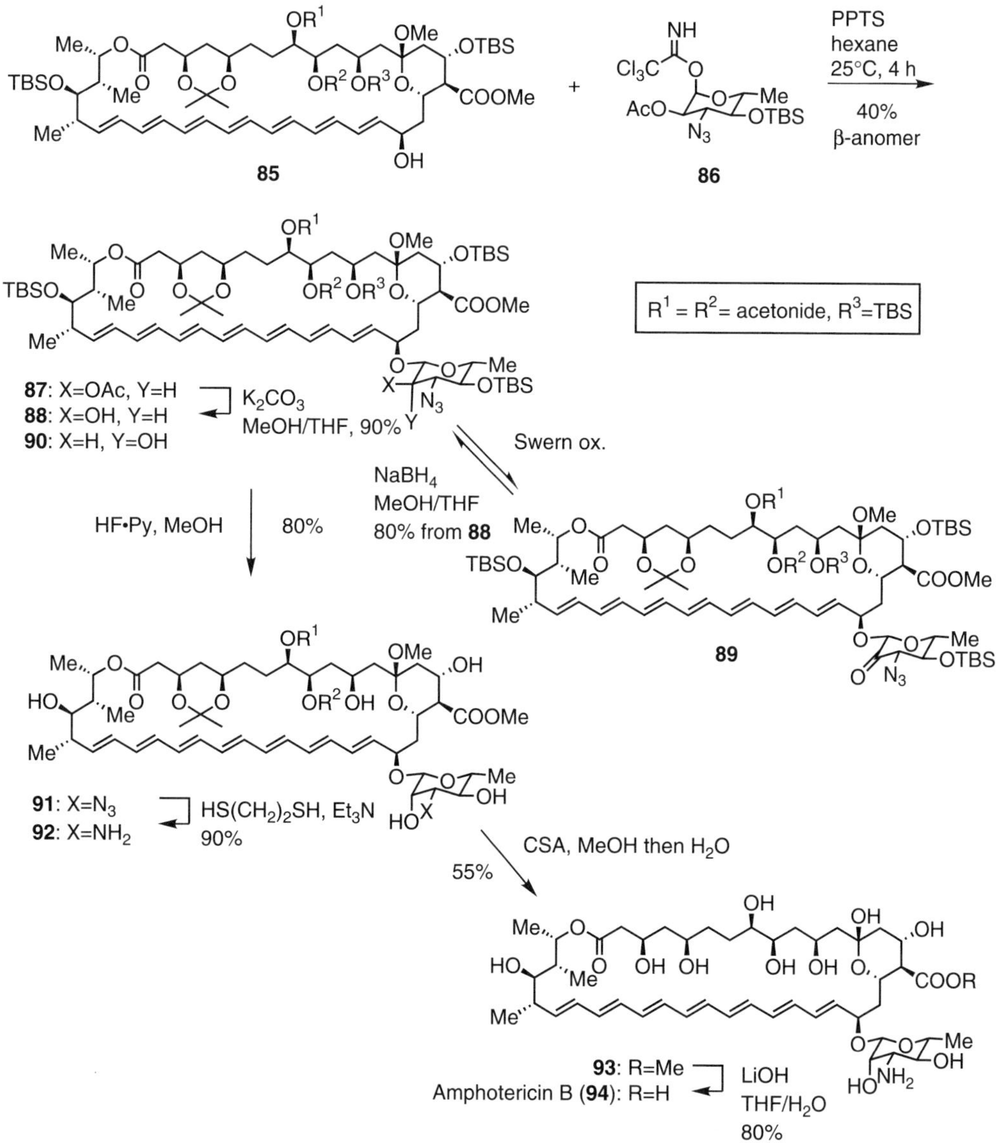

SCHEME 12.12 Synthesis of amphotericin B by Nicolaou [16].

trans-olefin **107**. The macrolactonization of the seco acid **107** by Keck's method using DMAP·HCl, DMAP and DCC furnished cytovaricin **111**, through cleavage of the *p*-methoxybenzyl group of **108**, Dess-Martin oxidation of **109** and desilylation of **110**.

12.2.10 Calicheamicin γ_1^I

Nicolaou's total synthesis of the enediyne antibiotic, calicheamicin γ_1^I [19], (Scheme 12.15 to Scheme 12.19) involves several new aspects in carbohydrate chemistry. The calicheamicin γ_1^I oligosaccharide synthesis began with the construction of the D-ring intermediate from L-rhamnose. Thus, L-rhamnose **112** was converted into the glycosyl donor, trichloroacetimidate **118**, through the

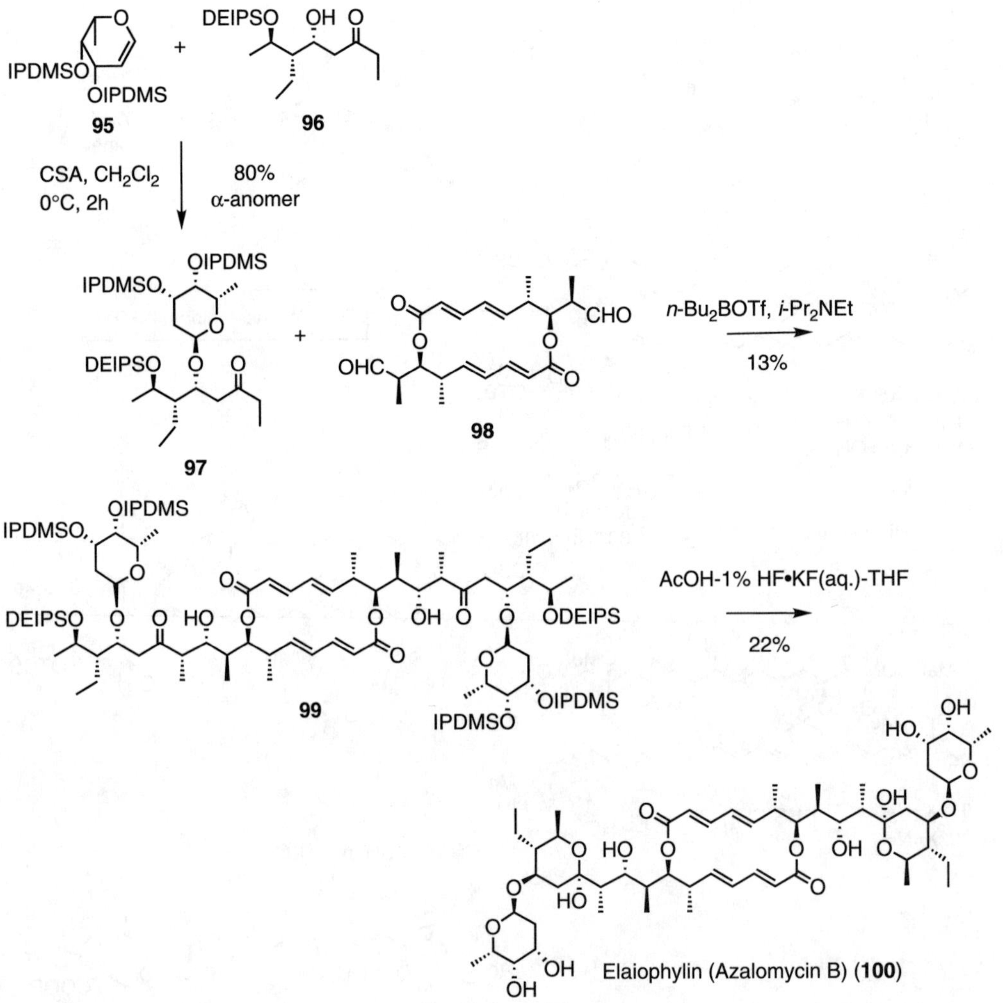

SCHEME 12.13 Synthesis of elaiophylin by Kinoshita-Toshima-Tatsuta [17].

tentative formation of the phenylthio glycoside **114** and the regioselective methylation of the C-3 hydroxyl group using *n*-Bu_2SnO, MeI and CsF, followed by deprotection of the anomeric phenylthio group employing NBS and H_2O. The aryl glycosidation of the trichloroacetimidate **118** and the fully substituted phenol **119** using $BF_3 \cdot Et_2O$ in CH_2Cl_2 occurred stereoselectively with neighboring-group participation to give the aryl α-glycoside **120**, which was converted into the suitably protected acid chloride **122** in six steps. The E-ring intermediate, glycosyl fluoride **129**, was prepared from the methyl ester of L-serine hydrochloride **123** by Brown's stereoselective allylation using optically active diisopinocamphenylborane and anomeric fluorination by means of DAST as the key steps. On the other hand, the A-ring subunit, *o*-nitrobenzyl β-glycoside, **133**, was synthesized from D-fucose by standard procedures in five steps. The glycal **134** provided the B-ring intermediate **142** through the formation of the double bond between the C-3 and C-4 positions, the introduction of the 3-chlorobenzoyl group at the C-2 position by ester migration from the C-1 position and the construction of the free hydroxylamine at the C-1 position via the phthalimide **141** (Scheme 12.16). Thus, the assembly of each subunit for the calicheamicin γ_1^I oligosaccharide synthesis was effectively performed.

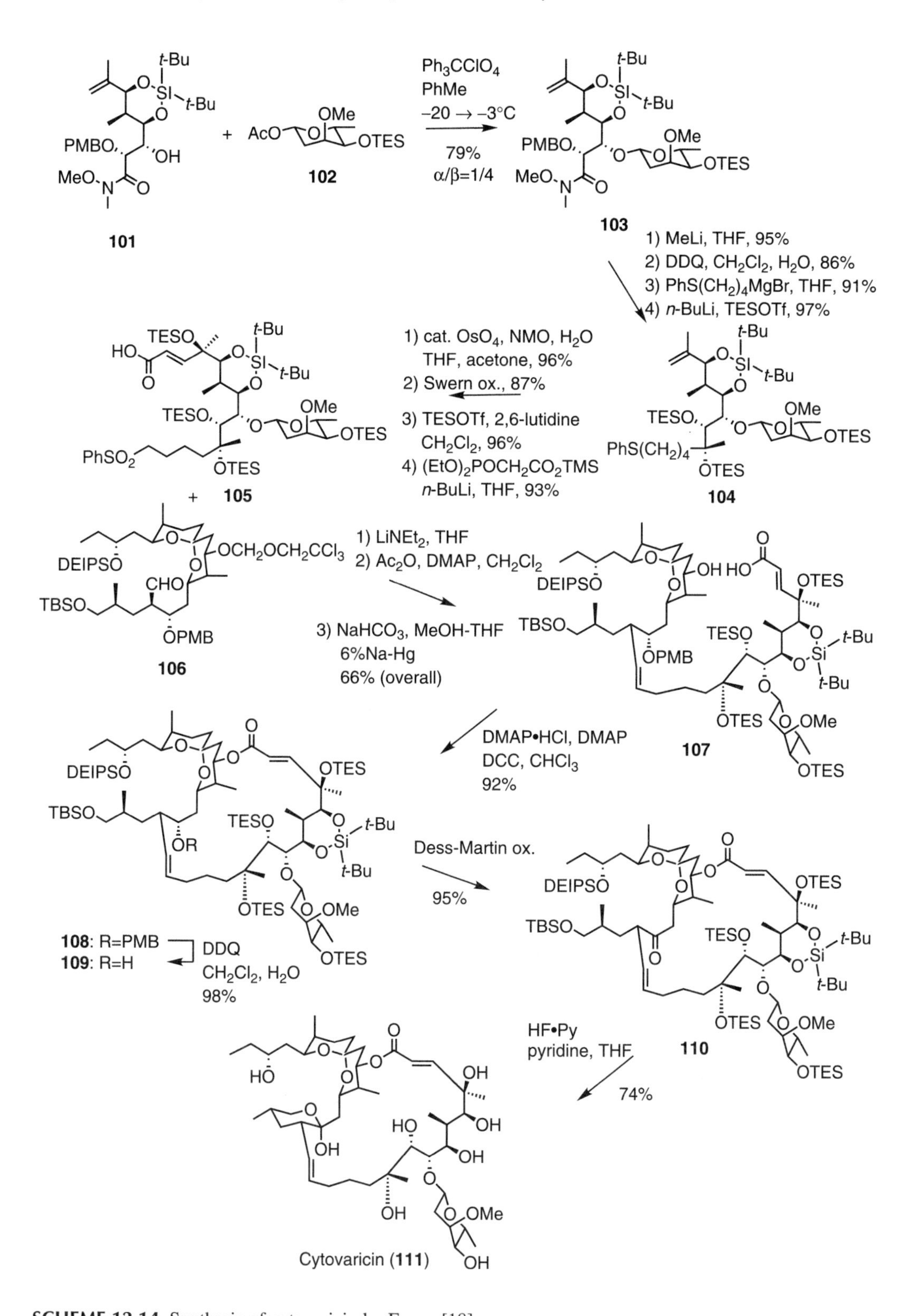

SCHEME 12.14 Synthesis of cytovaricin by Evans [18].

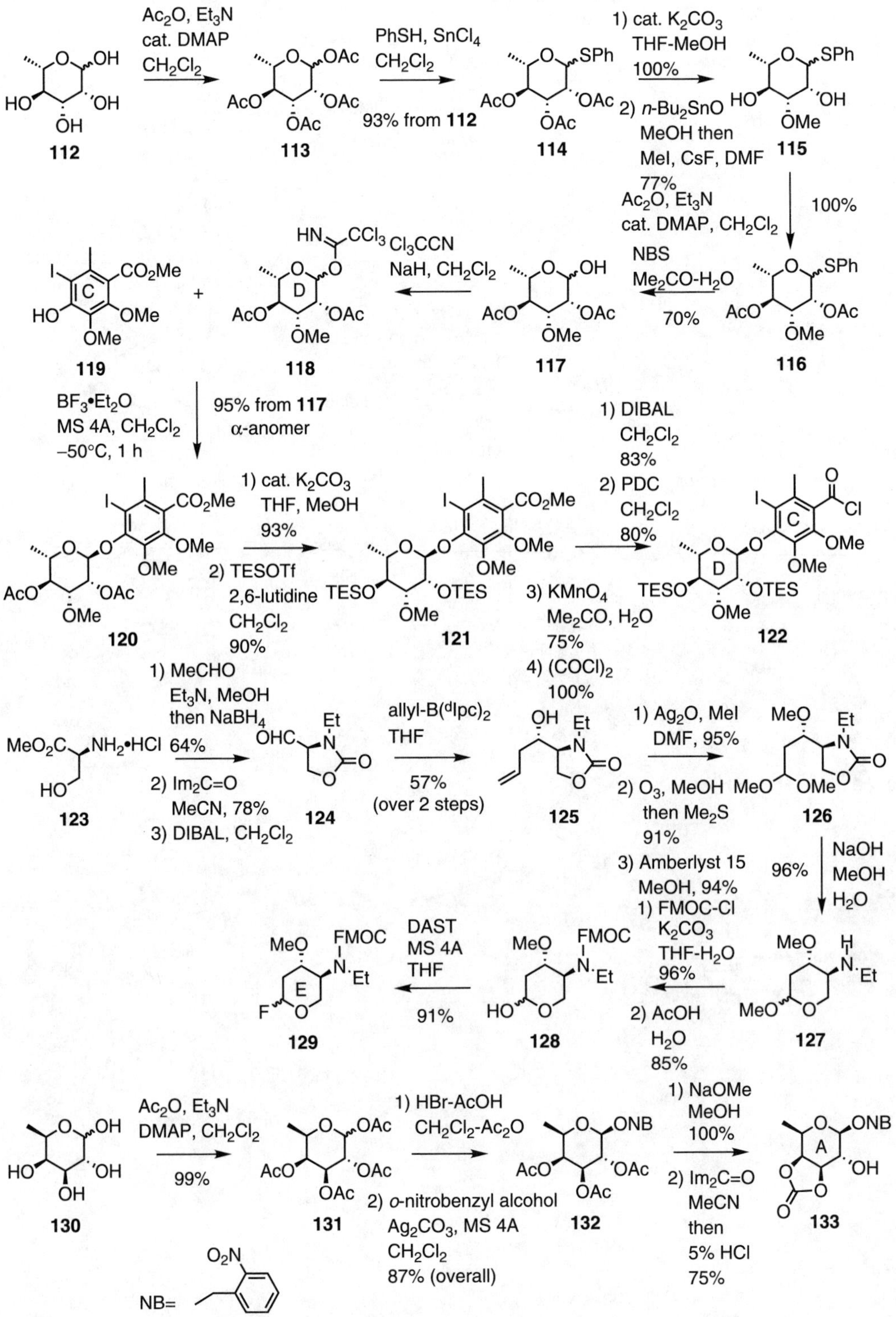

SCHEME 12.15 Synthesis of calicheamicin γ_1^{I} by Nicolaou (1) [19].

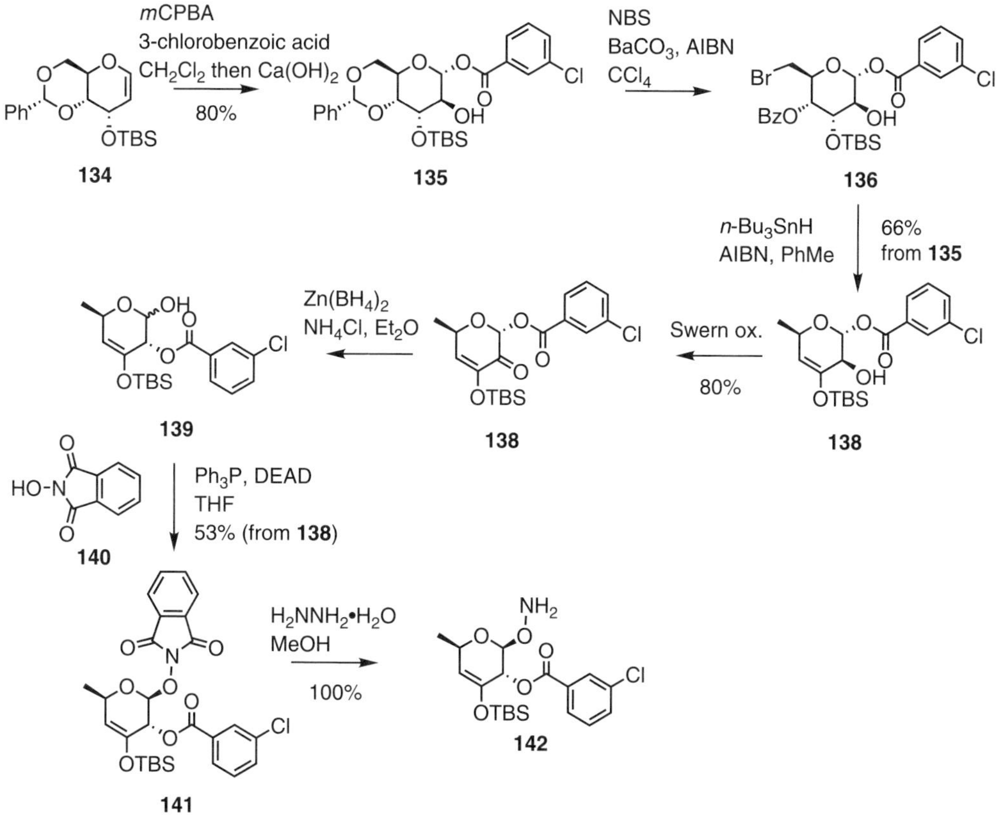

SCHEME 12.16 Synthesis of calicheamicin $\gamma_1{}^I$ by Nicolaou (2) [19].

The glycosidation of the glycosyl fluoride **129** and the alcohol **133** by the Mukaiyama method using $AgClO_4$ and $SnCl_2$ in THF furnished a 4.5:1.0 mixture of axial and equatorial isomers in favor of the desired axial glycoside **143**, which was converted into the ketone **145** after deprotection of the diol carbonate and regioselective oxidation of the resultant C-3 hydroxy group in the A-ring (Scheme 12.17). The subsequent coupling of the ketone **145** and the amine **142** under acidic conditions provided the oxime ether **146**. After silylation and thiocarbonylimidazole formation, the resultant **147** was subjected to thermolysis which induced the [3,3]-sigmatropic rearrangement providing **148**. The condensation of the thiol **149**, derived from **148**, and the acid chloride **122** gave the pentasaccharide **151** after desilylation, reduction and silylation. The final glycosidation of the trichloroacetimidate **152**, which was obtained from **151** via the photolytic cleavage of the *o*-nitrobenzyl group, and the calicheamicin γ_1^I aglycon derivative **153** using $BF_3{\cdot}Et_2O$ furnished the desired β-glycoside **154** along with the corresponding desilylated product which could be converted into **154** by silylation (Scheme 12.18). The trisulfide formation using AcSH and *N*-(methyldithio)phthalimide, the reduction of the oxime bond employing $NaCNBH_4{-}BF_3{\cdot}Et_2O$ and the deprotection of the silyl, FMOC and the ethylene acetal groups led to the total synthesis of calicheamicin γ_1^I **160** (Scheme 12.19).

Danishefsky's total synthesis of calicheamicin γ_1^I [20] (Scheme 12.20 to Scheme 12.22) involves the construction of the oligosaccharide moiety using glycal strategy. Di-*O*-acetyl-L-rhamnal **161** was converted into the trichloroacetimidate **170** through the Lewis acid-catalyzed Ferrier rearrangement of **161** with benzyl alcohol, the dihydroxylation of the unsaturated sugar **164** and the regioselective methylation of the C-3 hydroxy group of **165**. The subsequent aryl glycosidation of **170** and **171** using $BF_3{\cdot}Et_2O$ gave the aryl α-glycoside **172**, whose nitrile group

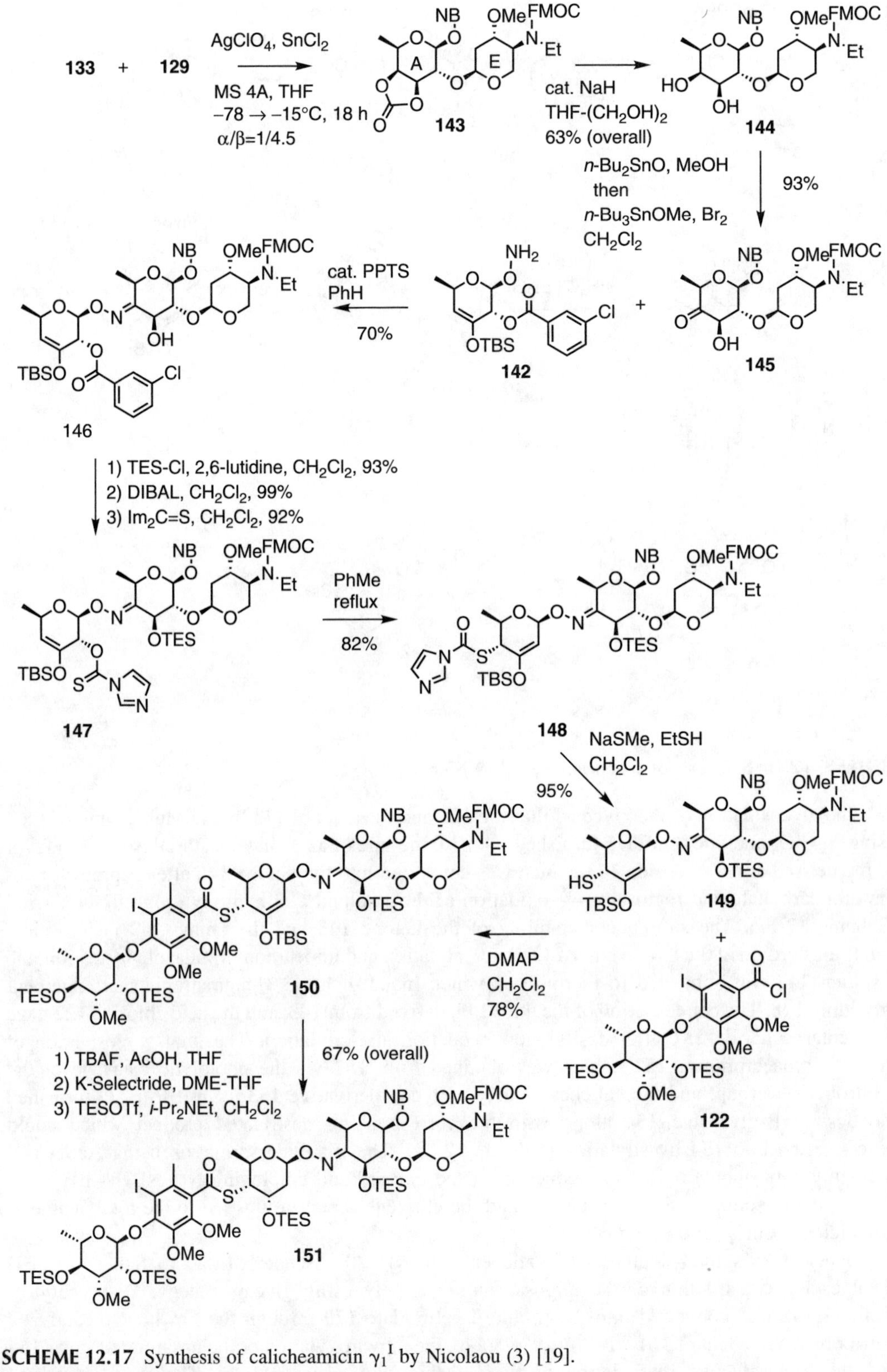

SCHEME 12.17 Synthesis of calicheamicin γ_1^I by Nicolaou (3) [19].

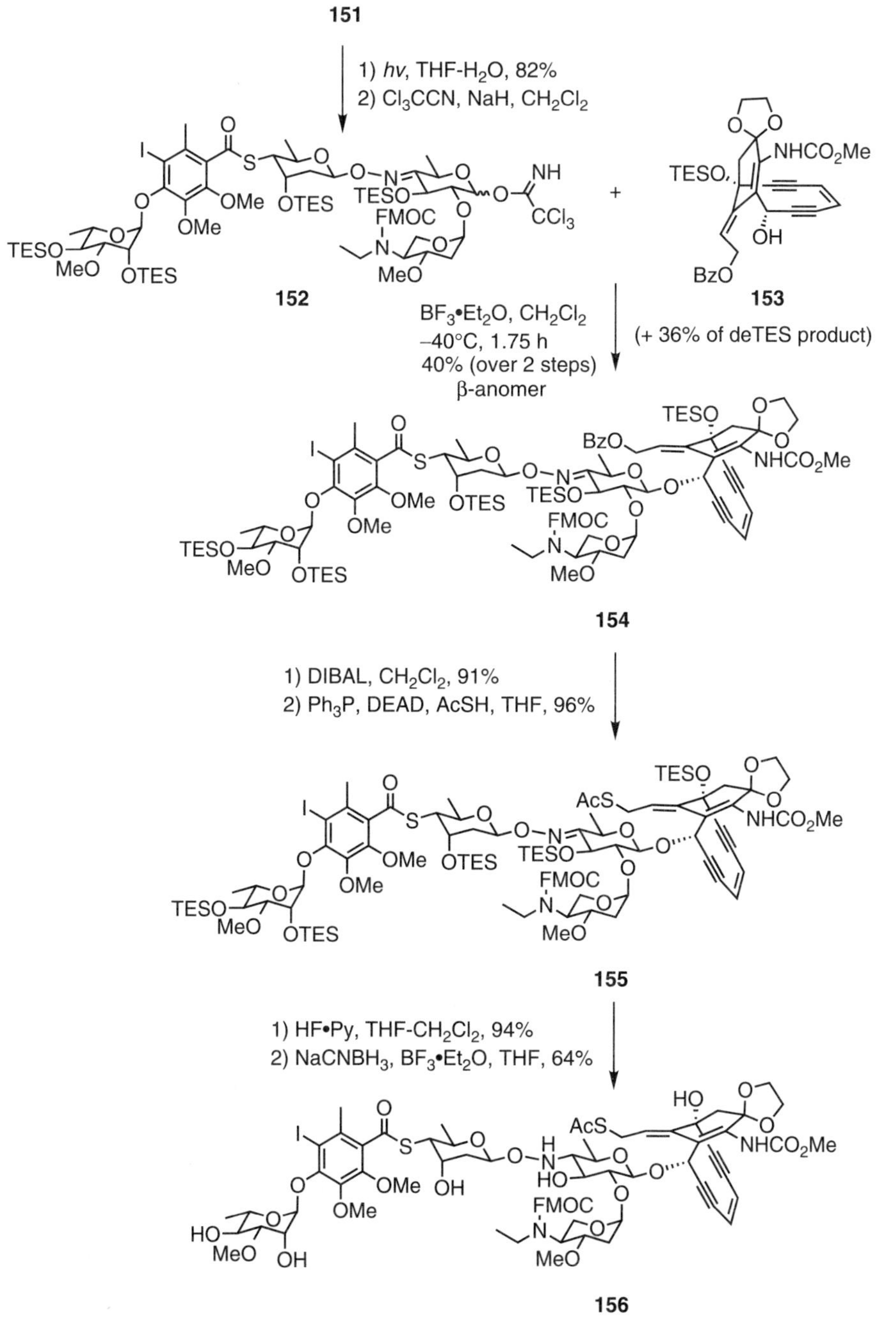

151

1) $h\nu$, THF-H_2O, 82%
2) Cl_3CCN, NaH, CH_2Cl_2

152 + 153

BF_3•Et_2O, CH_2Cl_2
–40°C, 1.75 h
40% (over 2 steps)
β-anomer
(+ 36% of deTES product)

154

1) DIBAL, CH_2Cl_2, 91%
2) Ph_3P, DEAD, AcSH, THF, 96%

155

1) HF•Py, THF-CH_2Cl_2, 94%
2) $NaCNBH_3$, BF_3•Et_2O, THF, 64%

156

SCHEME 12.18 Synthesis of calicheamicin γ_1^{I} by Nicolaou (4) [19].

was transformed into acid chloride to afford the C–D-fragment **176**. The glycal **182** corresponding to the E-fragment was prepared from the mesylate **177**. The introduction of an azide group, followed by its conversion into the FMOC-protected *N*-ethylamine, and the formation of glycal via the phenylthio glycoside **181** furnished the desired glycal **182**. On the other hand, the A-fragment **187** was obtained from the glycal **183** by regioselective *p*-methoxybenzylation of the C-3 hydroxy

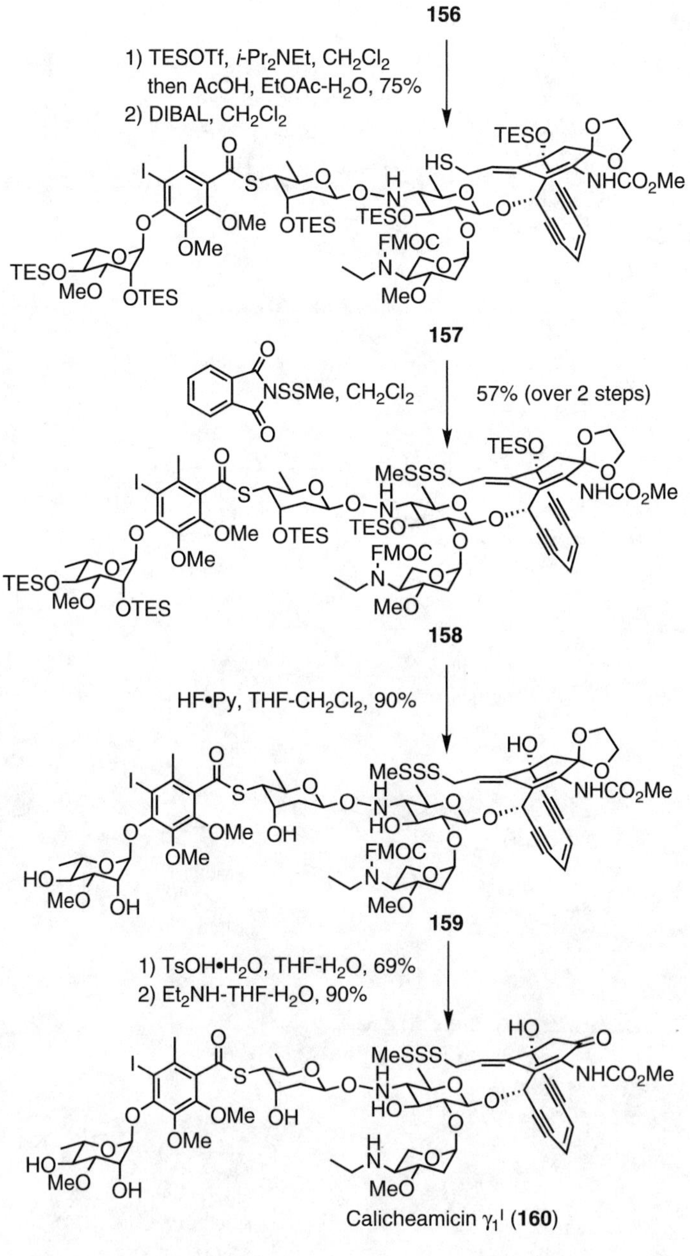

156

1) TESOTf, *i*-Pr_2NEt, CH_2Cl_2
then AcOH, EtOAc-H_2O, 75%
2) DIBAL, CH_2Cl_2

157

NSSMe, CH_2Cl_2
57% (over 2 steps)

158

HF•Py, THF-CH_2Cl_2, 90%

159

1) TsOH•H_2O, THF-H_2O, 69%
2) Et_2NH-THF-H_2O, 90%

Calicheamicin γ_1^{I} (**160**)

SCHEME 12.19 Synthesis of calicheamicin γ_1^{I} by Nicolaou (5) [19].

group of **183**, and the coupling reaction of glycal **184** and *p*-methoxybenzyl alcohol via the 1,2-epoxy intermediate **185**. The B-fragment, thiol **196**, was synthesized from the glycal **188** (Scheme 12.21). The 3,4-unsaturated phenylthio glycoside formation by Ferrier rearrangement of **188** with thiophenol, regioselective deoxygenation of the C-6 primary hydroxyl group and the introduction of thiol and hydroxylamine functions at the C-4 and C-1 positions, respectively, via the glycosidation of glycal **193** and TEOC–HNOH afforded **196**. Coupling of **196** with the acid chloride **176** afforded the aryldisaccharide **197**. On the other hand, the alcohol **187**, corresponding

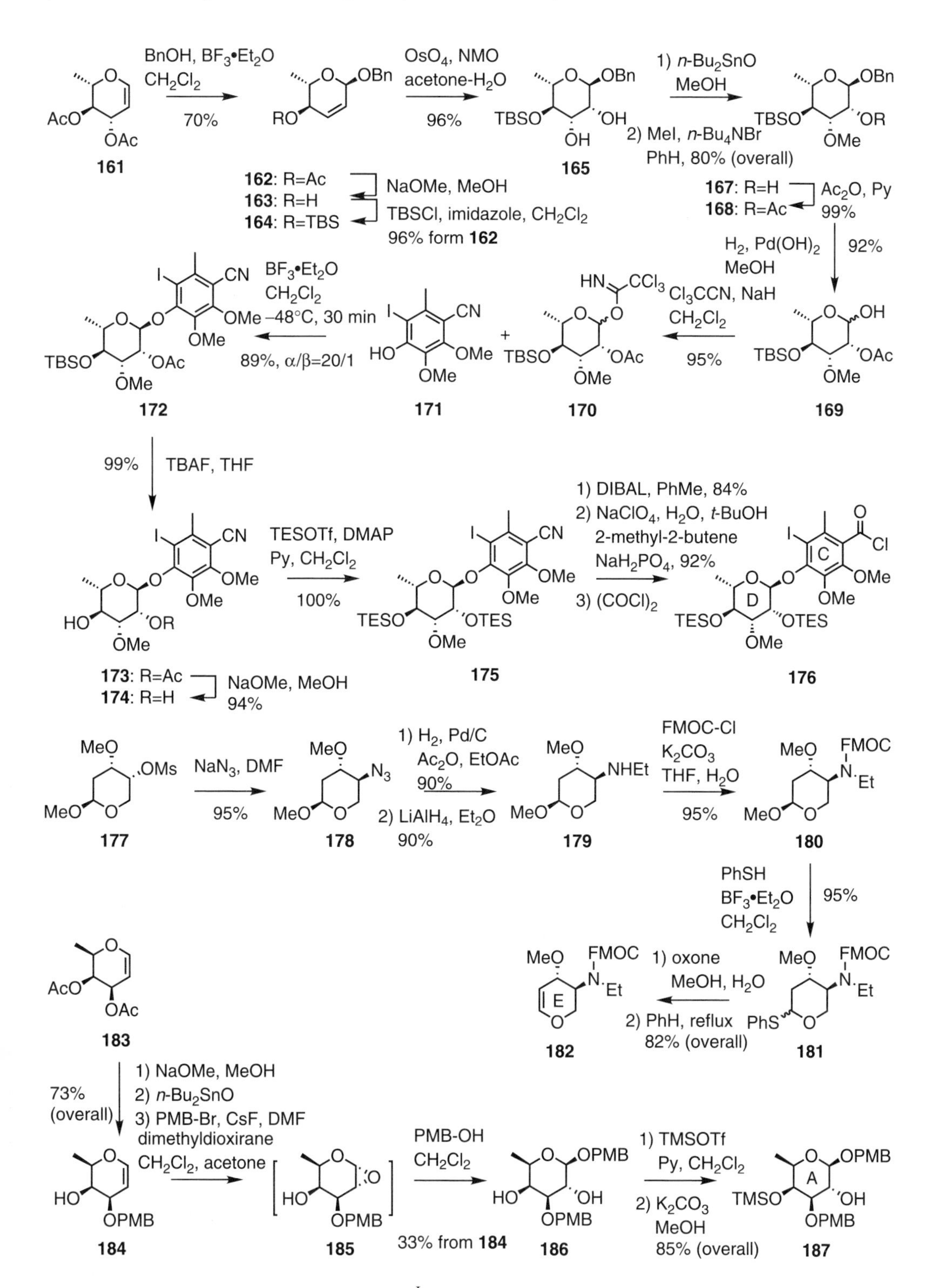

SCHEME 12.20 Synthesis of calicheamicin $\gamma_1{}^{I}$ Danishefsky (1) [20].

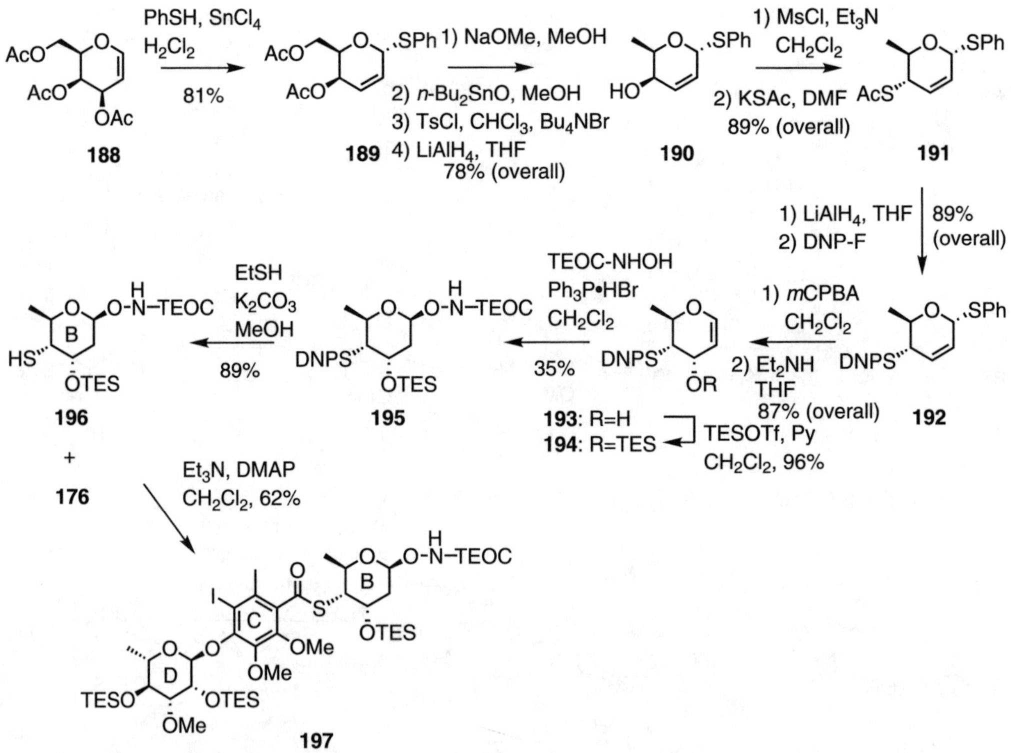

SCHEME 12.21 Synthesis of calicheamicin $\gamma_1{}^I$ Danishefsky (2) [20].

to the A-fragment, was glycosylated with the glycal **182**, corresponding to the E-fragment, to yield the disaccharide **198** after acid hydrolysis (Scheme 12.22). Radical deiodination and triflate formation gave **200**. Treatment of **197** and **200** under Kahne's coupling conditions gave **201**, which was converted into the glycosyl donor, trichloroacetimidate **203**, via deprotection of the *p*-methoxybenzyl protecting group using DDQ in the presence of a pH 7 buffer. The final β-stereoselective glycosidation of **203** and the calicheamicin γ_1^I aglycon derivative **204** using AgOTf furnished calicheamicin γ_1^I (**160**), after deprotection of the ethylene acetal and silyl groups.

12.2.11 Neocarzinostatin Chromophore

The highlight of carbohydrate chemistry in Myers' total synthesis of the enediyne neocarzinostatin chromophore [21] (Scheme 12.23) is the α-stereoselective glycosidation of the very labile aglycon **206** with the glycosyl donor, trichloroacetimidate **207**. Using $BF_3 \cdot Et_2O$ in the presence of MS 3A in PhMe at −30°C the glycosidation provided the protected α-glycoside **208**, which was converted into neocarzinostatin chromophore **209** by immediate desilylation. The α-stereoselectivity of the glycosylation may come from the direction of the glycosyl acceptor to the α-face of the oxonium intermediate by interaction with the amino group at the C-2 position (possible hydrogen-bonding or internal basic group).

12.2.12 Eleutherobin

Nicolaou's total synthesis of a marine antitumor agent, eleutherobin [22] (Scheme 12.24), involves the attachment of D-arabinose to the aglycon intermediate. Thus, the glycosidation of the (+)-carvone-derived hydroxy aldehyde **210** and the trichloroacetimidate **211**, derived from

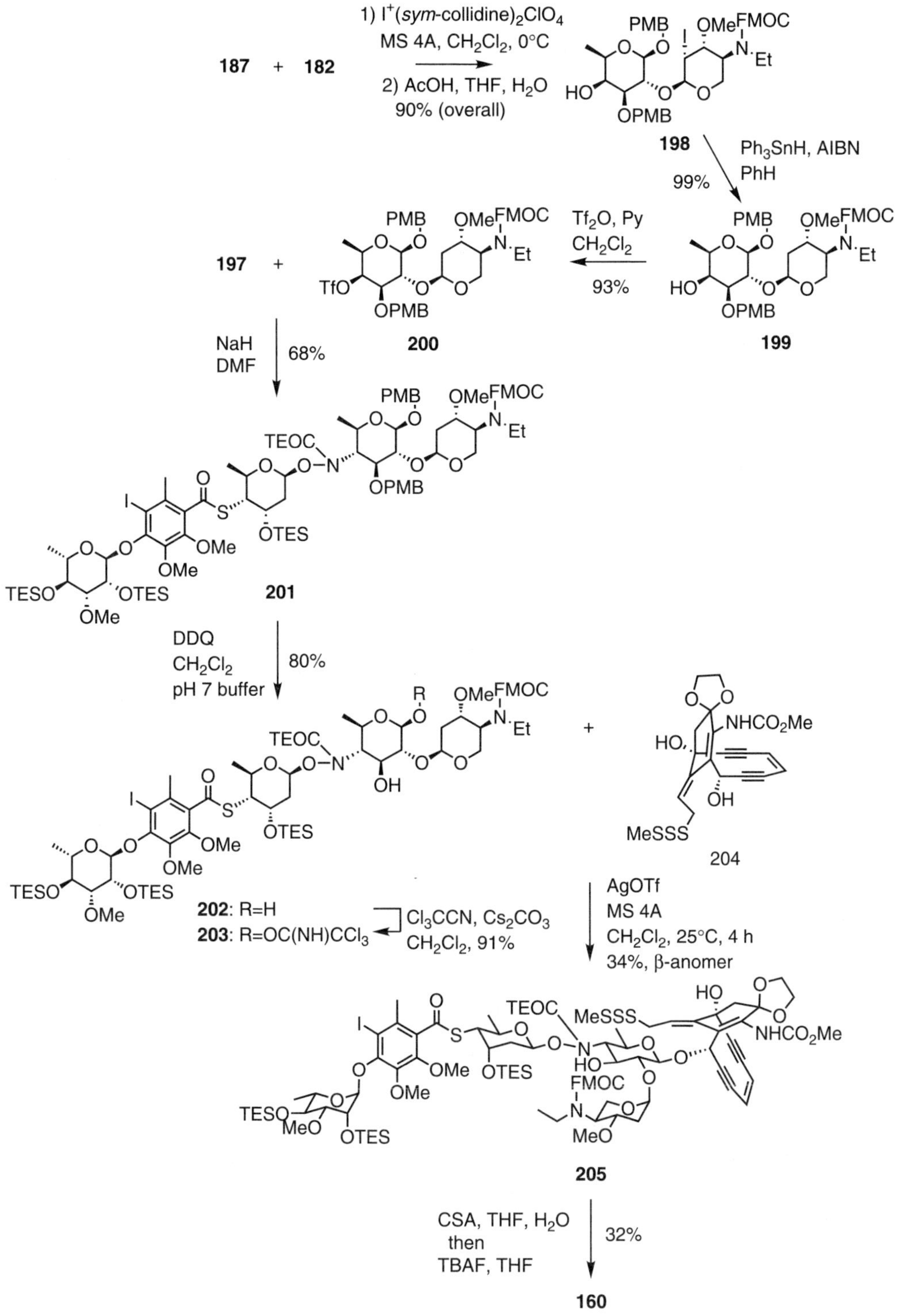

SCHEME 12.22 Synthesis of calicheamicin γ_1^{I} Danishefsky (3) [20].

SCHEME 12.23 Synthesis of neocarzinostatin chromophore by Myers [21].

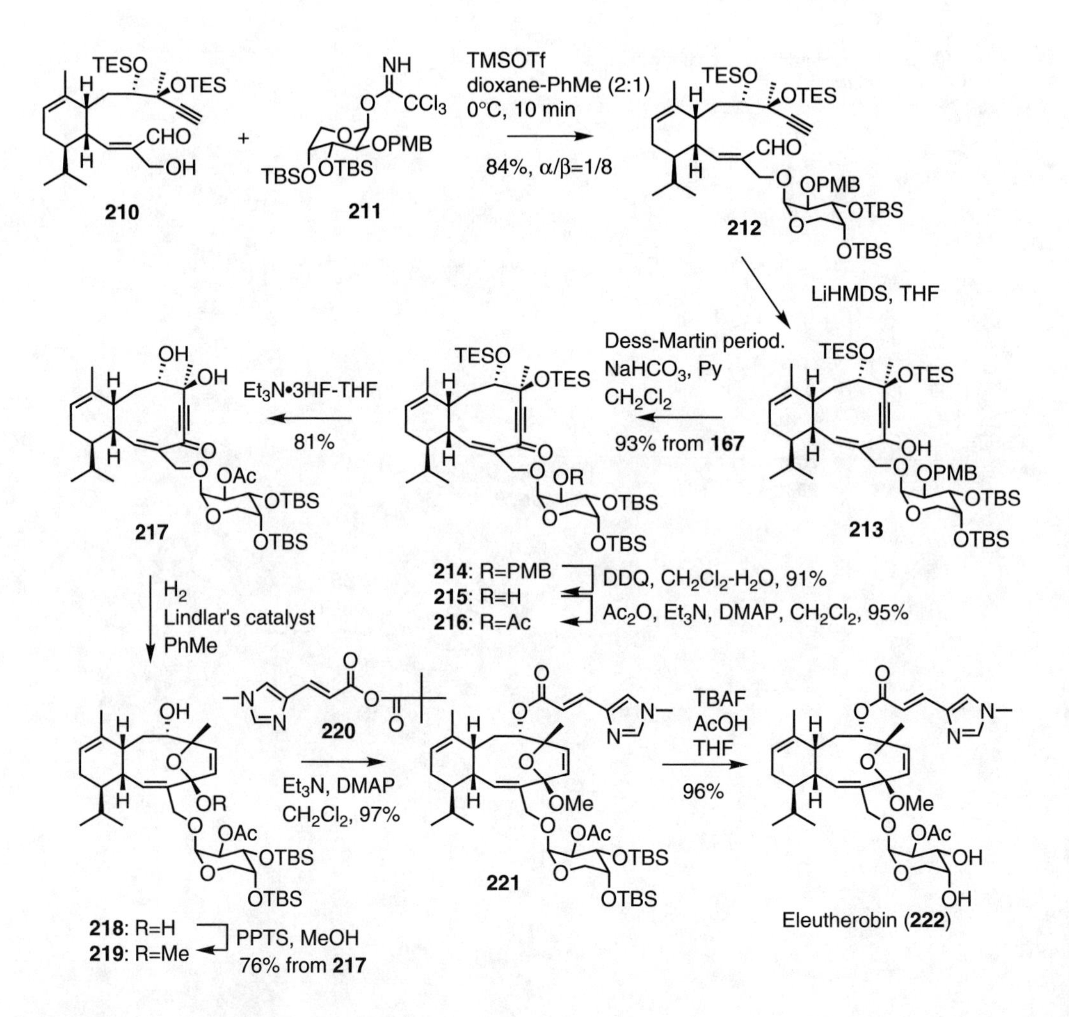

SCHEME 12.24 Synthesis of eleutherobin by Nicolaou [22].

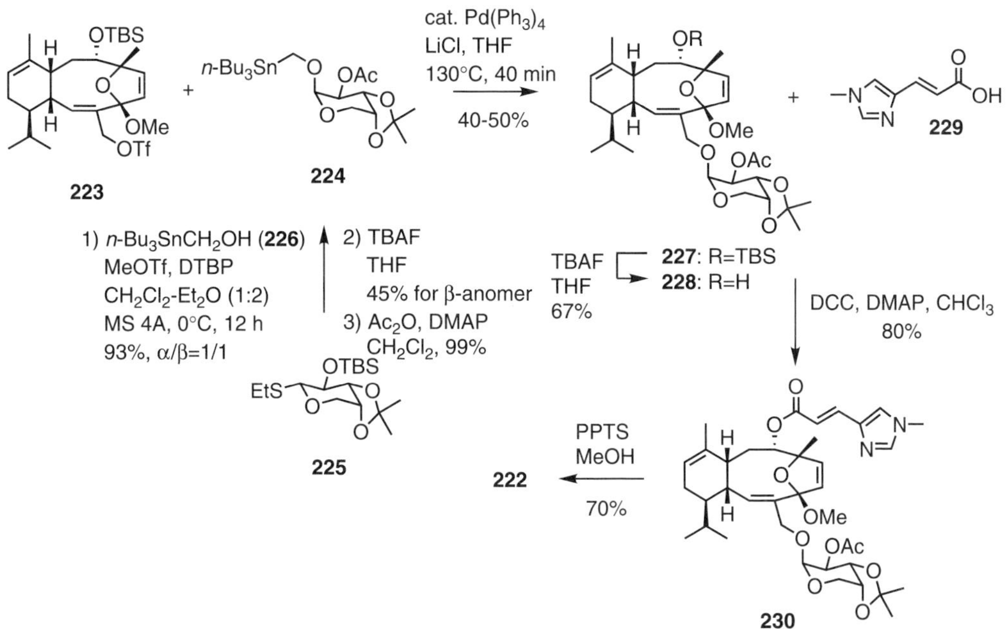

SCHEME 12.25 Synthesis of eleutherobin by Danishefsky [23].

D-arabinose tetraacetate using TMSOTf in dioxane–toluene (2:1) at 0°C, gave an 8:1 mixture of products in favor of the desired β-glycoside **212**, which was treated with LiHMDS to yield the cyclized product **213**. The subsequent oxidation of the resultant alcohol **213**, followed by protecting group manipulations, afforded the diol **217**. The selective hydrogenation of the acetylene moiety of **217** using Lindlar's catalyst produced the desired lactol **219** after methoxy ketalization. Finally, the attachment of the mixed anhydride **220** onto the alcohol **219**, employing Et_3N and DMAP, furnished eleutherobin **222** after desilylation.

Danishefsky's total synthesis of eleutherobin [23] (Scheme 12.25) is characterized by a novel oxycarbaglycosidation for coupling the aglycon and the carbohydrate moieties. Thus, the Pd-mediated reaction of the vinyl triflate **223** and the tributylstannylated glycosyl donor **224**, which was prepared by the glycosidation of the ethylthio glycoside **225** and tributylstannylmethanol **226** using MeOTf, gave the Stille coupling product **227**. After removal of the silyl group of **227**, condensation with the carboxylic acid **229** and removal of the isopropylidene group of the resultant **230**, eleutherobin **222** was obtained.

12.2.13 Olivomycin A

Roush's total synthesis of olivomycin A [24] (Scheme 12.26 and Scheme 12.27) involves the construction and assembly of the five carbohydrate subunits of this antitumor antibiotic. Reaction of the glycal **231** and PhSCl, followed by hydrolysis of the intermediate glycosyl chloride, gave the 2-thiophenyl pyranose, which was converted into the glycosyl trichloroacetimidate **232** possessing a thiophenyl group at the C-2 position as a neighboring group. On the other hand, the glycal **233** derived from **231** was coupled with the glycosyl acetate **234** by TMSOTf to yield the α-disaccharide **235**, which was transformed into the glycosyl trichloroacetimidate **236** in a way similar to that for **232** from **231**. The glycosidation of **232** and the olivomycin A aglycon derivative **237** using TMSOTf stereoselectively gave the desired β-glycoside **238**, with the assistance of neighboring group participation. After protecting group manipulations, the resultant **239** was

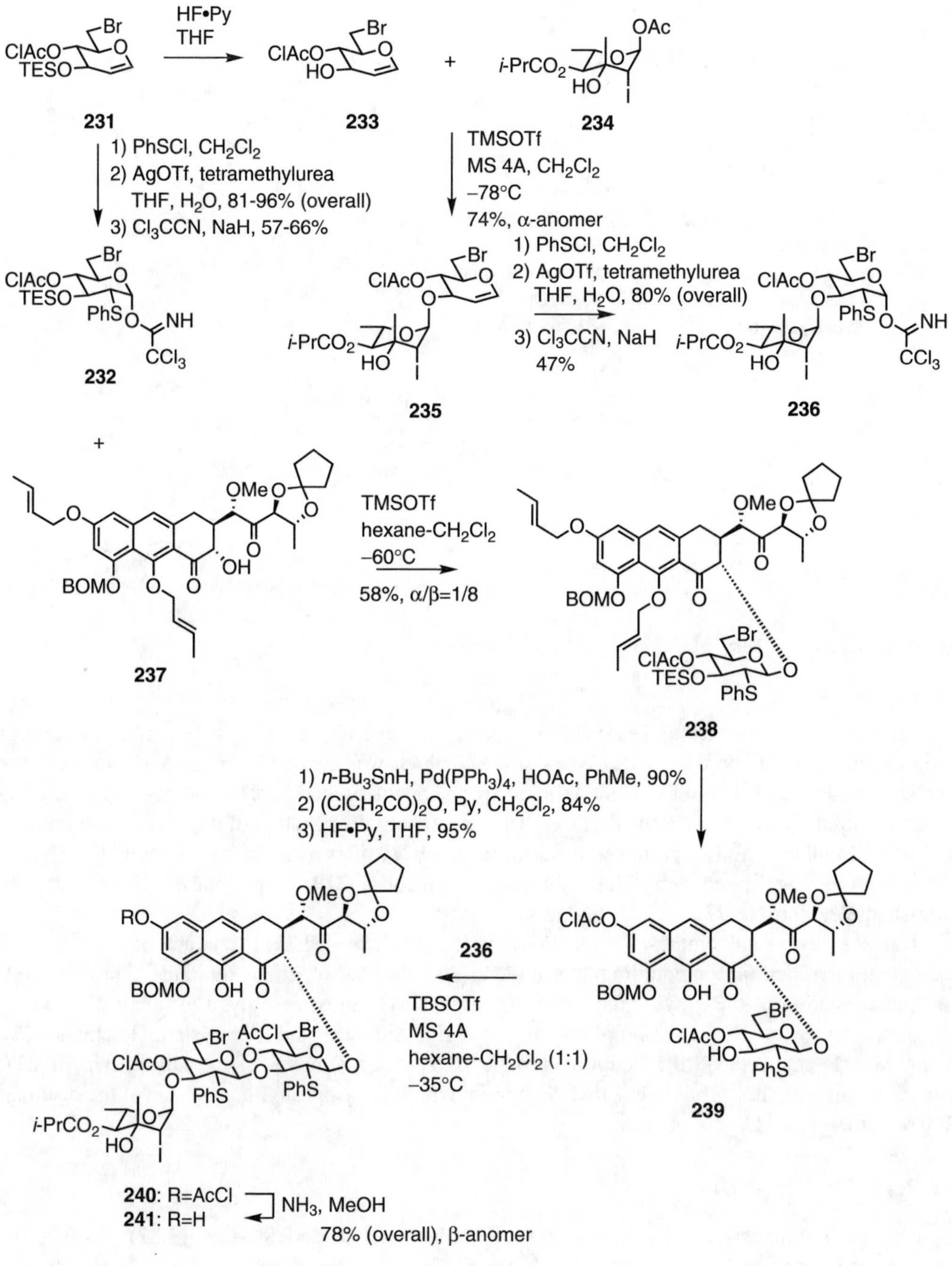

SCHEME 12.26 Synthesis of olivomycin A by Roush (1) [24].

stereoselectively β-glycosylated with the trichloroacetimidate **236** by TBSOTf to give the trisaccharide **241** after selective deprotection of the chloroacetyl group of the aromatic moiety in **240**.

The A–B disaccharide moiety **247** was constructed by the glycosidation of the glycosyl acetate **244**, which was prepared from the glycal **242** and the glycal **245** using I^+(*sym*-collidine)$_2ClO_4$ as

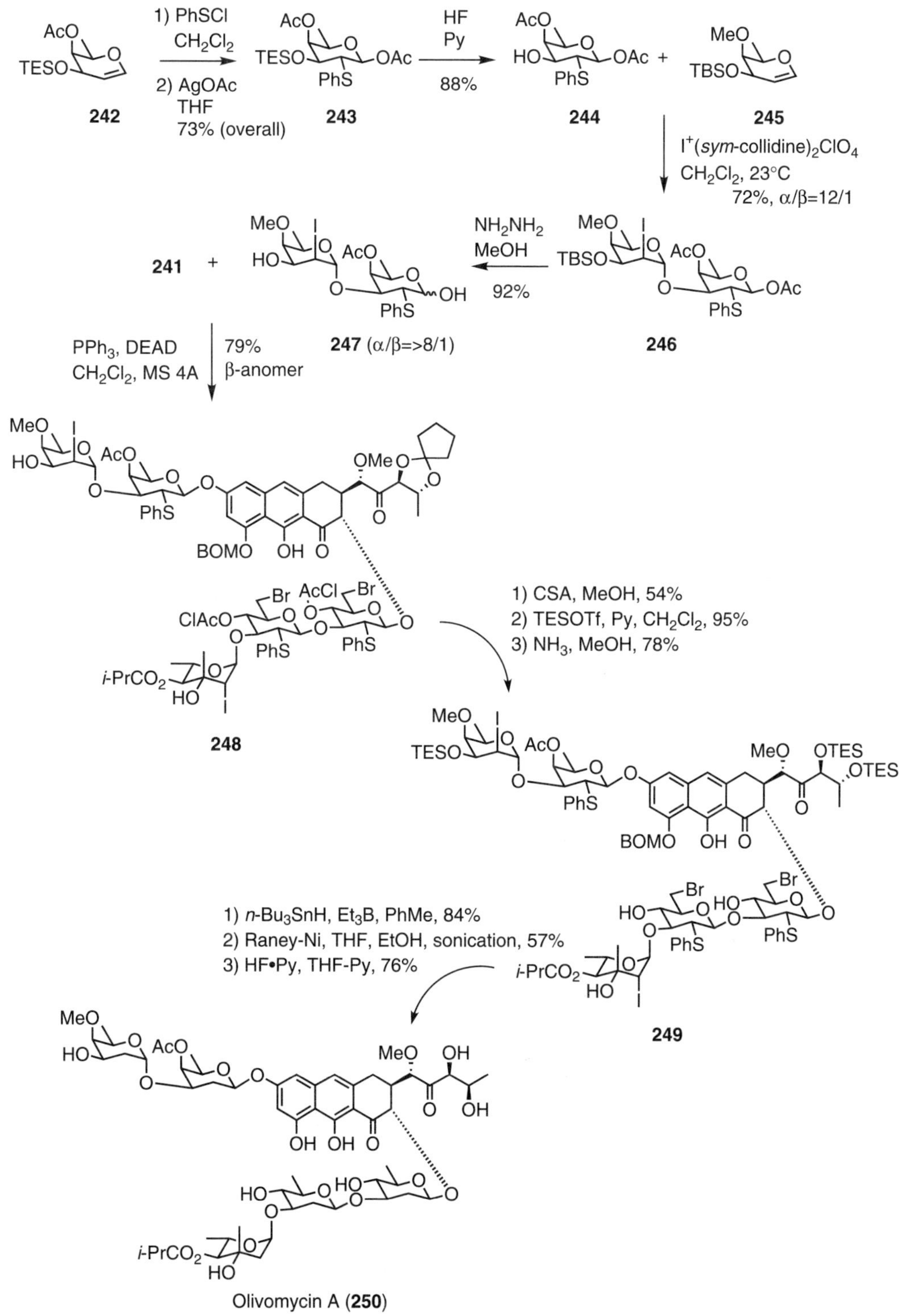

SCHEME 12.27 Synthesis of olivomycin A by Roush (2) [24].

the key step. The final coupling reaction of **241** and **247** by Mitsunobu methodology using PPh_3 and DEAD afforded the desired β-glycoside **248**, which was subsequently subjected to deacetalization, silylation and dechloroacetylation to give **249**. Finally, the reductive dehalogenation of **249** followed by the removal of thiophenyl and BOM groups, by hydrogenolysis with sonication, and desilylation provided olivomycin A (**250**).

12.2.14 Everninomicin 13,284-1

Nicolaou's total synthesis of everninomicin 13,284-1 (Scheme 12.28 to Scheme 12.33) [25], a member of the orthosomicin class of antibiotics, was accomplished by the construction of a complex oligosaccharide containing two sensitive orthoester moieties and terminating with two aromatic esters using several novel synthetic strategies and methods. The $SnCl_2$-mediated glycosidation of the glycosyl fluoride **251**, corresponding to the B-ring, and the alcohol **252**, corresponding to the C-ring, gave the desired β-glycoside **254** after reductive desulfurization. The subsequent bis-desilylation and selective monoallylation of the resultant diol afforded the alcohol **255**, which was activated with *n*-BuLi and coupled with the acyl fluoride **256** corresponding to the A_1-ring to give the ester **257**. The deprotection of the allyl group followed by silylation gave **259**, whose *p*-methoxybenzyl groups were removed using PhSH and $BF_3 \cdot Et_2O$. Subsequently, acetyl groups were installed in their positions to afford the diacetate **260**. The selective cleavage of the C-1 acetate of **260** and the formation of a trichloroacetimidate at this position gave **262**, which was treated with PhSeH in the presence of $BF_3 \cdot Et_2O$ to give the β-phenylseleno glycoside with participation of the C-2 acetate. After deprotection of the silyl group, the resultant alcohol was glycosylated with the glycosyl fluoride **264** corresponding to the A-ring using $SnCl_2$ to furnish the desired glycoside **265**; the formation of the α-glycosidic bond was controlled by the anomeric effect. Hydrolysis of **265** gave the C-2 hydroxy compound **266**, which was treated with DAST to produce the 2-phenylselenoglycosyl fluoride **267** with an inversion of the stereochemistry at the C-2 position.

On the other hand, the glycosidation of the glycosyl sulfoxide **268** corresponding to the D-ring and the alcohol **269** corresponding to the E-ring, using Tf_2O and DTBMP, gave the desired β-glycoside **270** via the S_N2 attack of **269** on the α-glycosyl triflate intermediate. After the *p*-methoxybenzyl group was removed, the resultant alcohol **271** was oxidized to the corresponding ketone, which was treated with MeLi to produce the desired tertiary alcohol **272** as a single isomer. The deoxygenation of the C-6 hydroxy group in the D-ring was carried out by standard procedures in three steps. The subsequent regioselective mono *p*-methoxybenzylation of **275**, followed by desilylation and acetylation, gave the triacetate **277**. Finally, the trichloroacetimidate **279** was obtained by selective removal of the anomeric acetate and trichloroacetimidate formation at this position.

Continuing with the synthesis (Scheme 12.30), the assembly of the F-, G- and H-rings began with the regio- and stereoselective glycosidation of the tin acetal **280** corresponding to the F-ring and the trichloroacetimidate **281** corresponding to the G-ring. The coupling of **280** and **281** using TMSOTf, followed by PPTS-mediated cleavage of the intermediate TMS ether, gave the desired disaccharide **282**. The protecting group manipulations of **282** involving methylation, debenzoylation, benzylation, de-*p*-methoxybenzylation, silylation, deallylation and monochloroacetylation gave the suitably protected alcohol **286**, which was coupled with the 2-phenylseleno glycosyl fluoride **287** corresponding to the H-ring, employing $SnCl_2$ to afford the trisaccharide **288**. After dechloroacetylation, the resultant **289** was subjected to Sinaÿ's orthoester formation. Thus, the oxidation of the phenylseleno group of **289**, followed by heating the resultant selenoxide in a mixture of vinyl acetate–toluene–diisopropylamine in a sealed tube, caused the sequential *syn*-elimination and ring closure at the anomeric position to give the ortho ester **291** via **290**. The selective desilylation of TIPS of **291** and the benzoylation of the resultant alcohol

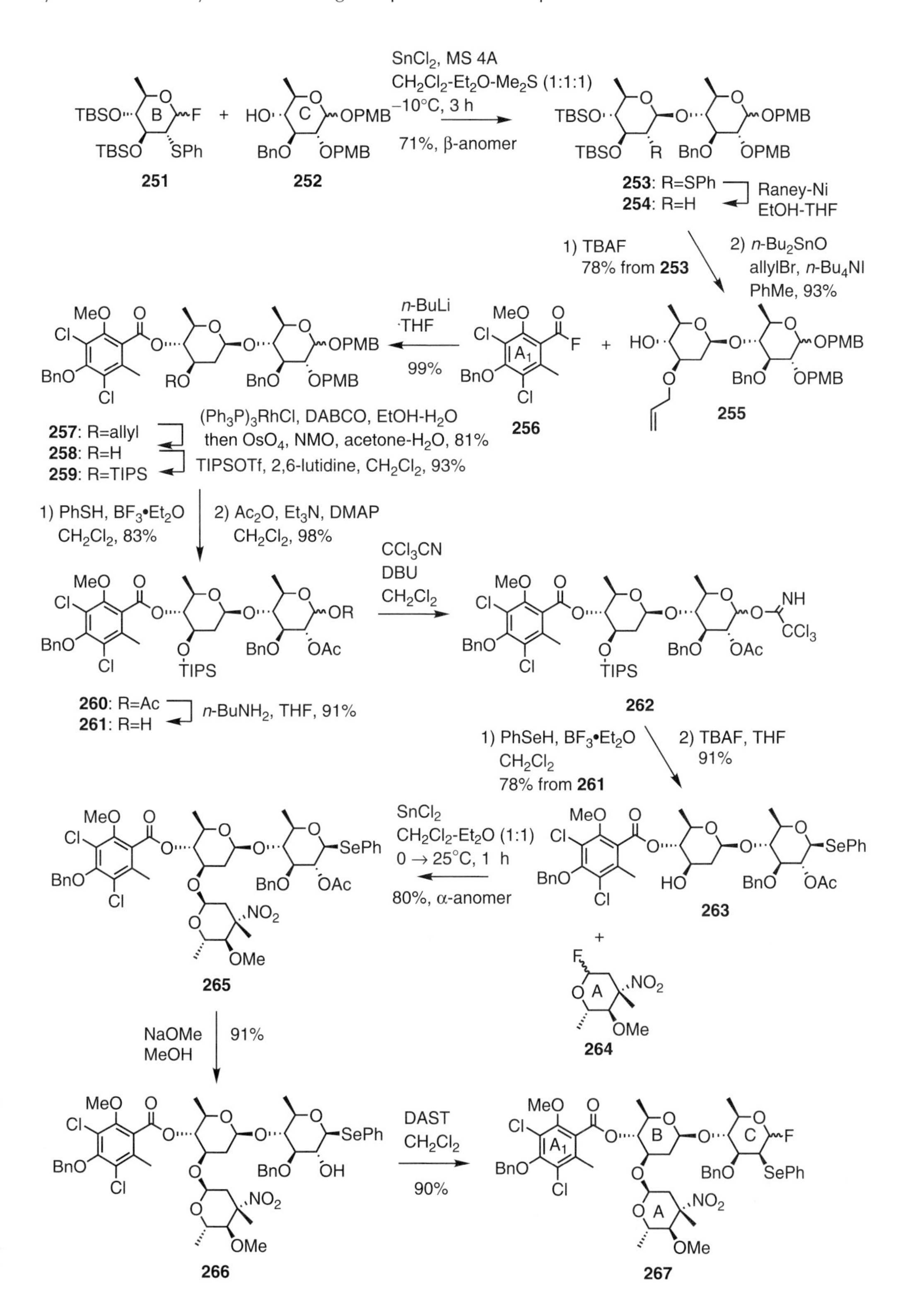

SCHEME 12.28 Synthesis of everninomicin 13,284-1 by Nicolaou (1) [25].

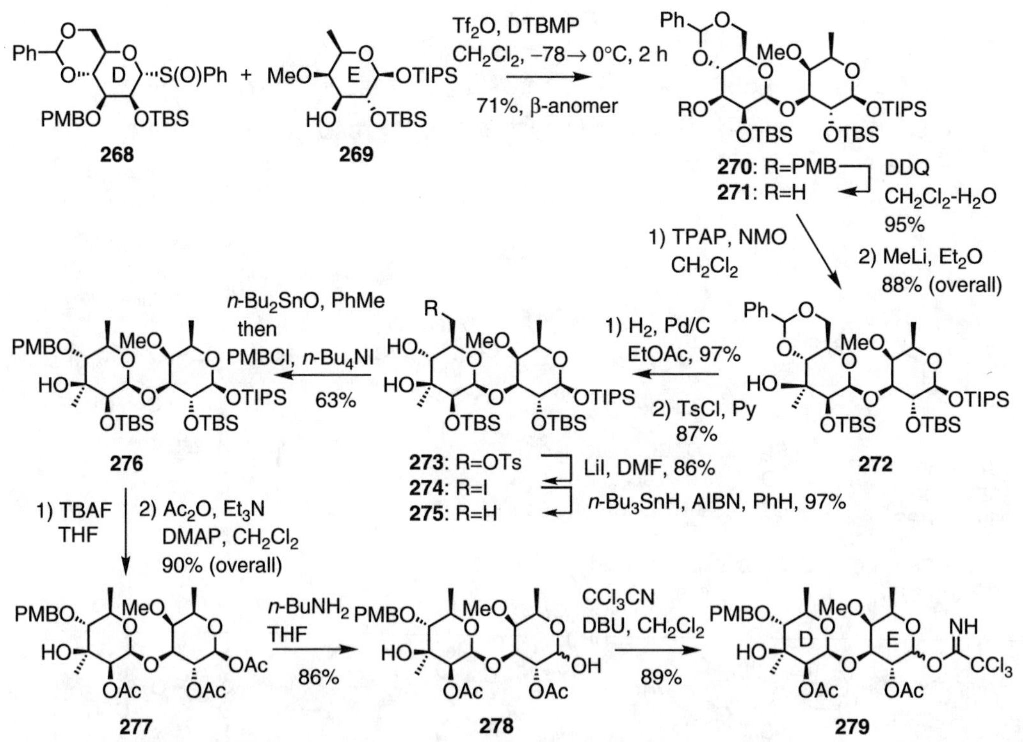

SCHEME 12.29 Synthesis of everninomicin 13,284-1 by Nicolaou (2) [25].

furnished **292**, which was subjected to desilylation and dehydration with the Martin sulfurane to afford the olefin **293**. After the exchange of the benzoyl group with the silyl group, the resultant olefin **295** was dihydroxylated using NMO and OsO_4 in the presence of quinuclidine to give the desired diol **296** as a major product.

The inversion of the stereochemistry of the C-2 hydroxy group in the H-ring was accomplished via the regioselective monobenzoylation and the oxidation–reduction steps to furnish the alcohol **298**. The benzoyl group was removed and the resultant diol **299** was converted into the methylene acetal **300**. The DDQ-mediated removal of the *p*-methoxybenzyl group of **300**, the esterification of **301** with the acyl fluoride **302** corresponding to the A_2-ring and the deprotection of the silyl group of **303** led to the alcohol **304**.

The glycosidation of the trichloroacetimidate **279** and the alcohol **304** using $BF_3{\cdot}Et_2O$ gave the oligosaccharide **305** possessing the desired β-glycoside bond. The protecting group exchanges involving deacetylation, benzylation, de-*p*-methoxybenzylation, chloroacetylation, debenzylation, silylation and dechloroacetylation furnished the suitably protected alcohol **311**.

The final $SnCl_2$-mediated glycosidation of the 2-phenylseleno glycosyl fluoride **267** and the alcohol **311** gave the oligosaccharide **312** with complete stereocontrol, which was subjected to Sinaÿ's orthoester formation to afford everninomicin 13,384-1 (**314**) after successful deprotections, including reductive debenzylation and TBAF-induced desilylation.

12.2.15 Polycavernoside A

The total synthesis of polycavernoside A by Murai [26] (Scheme 12.34) involves the construction of L-fucosyl-D-xylose and the attachment of this disaccharide to the aglycon. The glycosyl

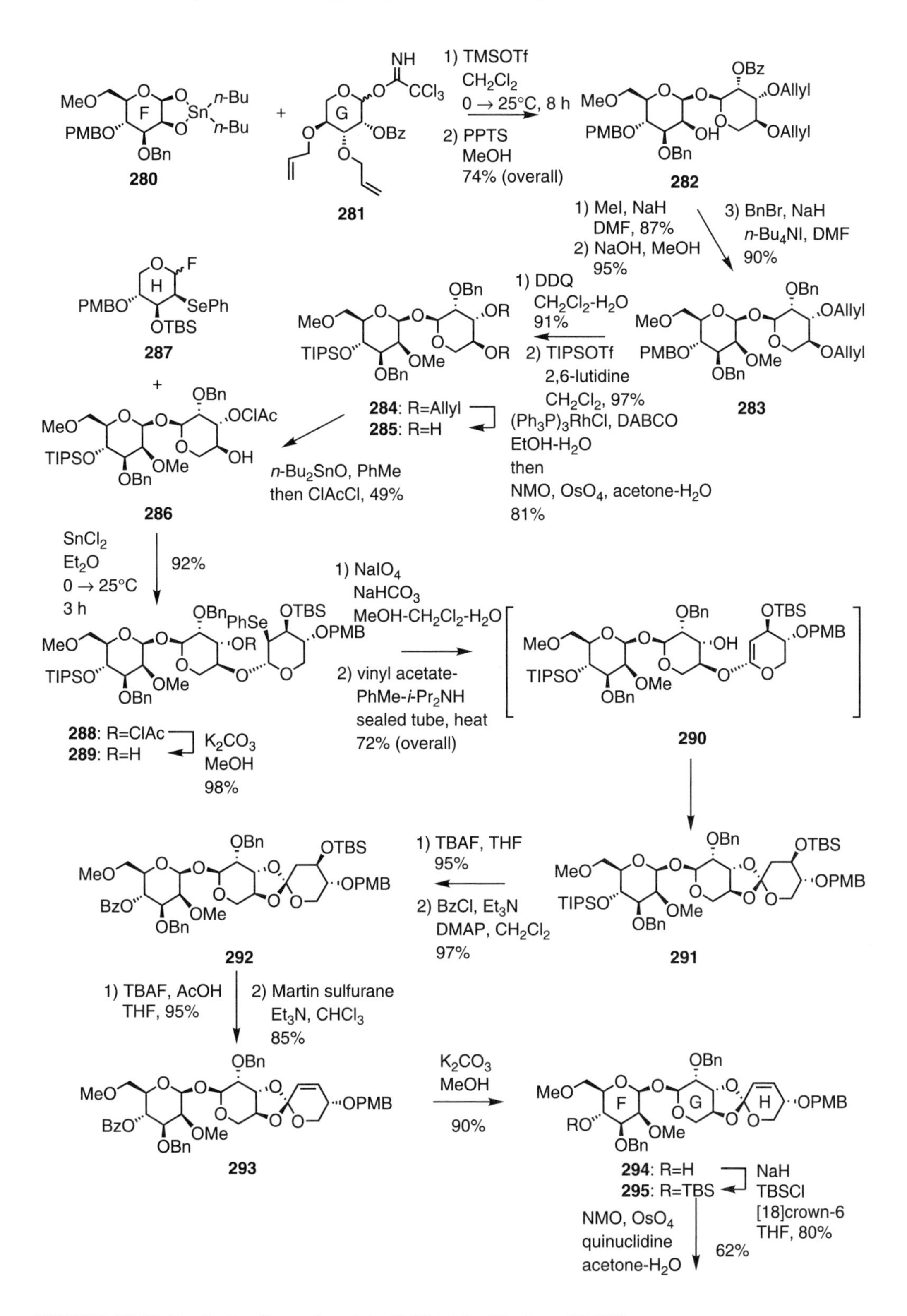

SCHEME 12.30 Synthesis of everninomicin 13,284-1 by Nicolaou (3) [25].

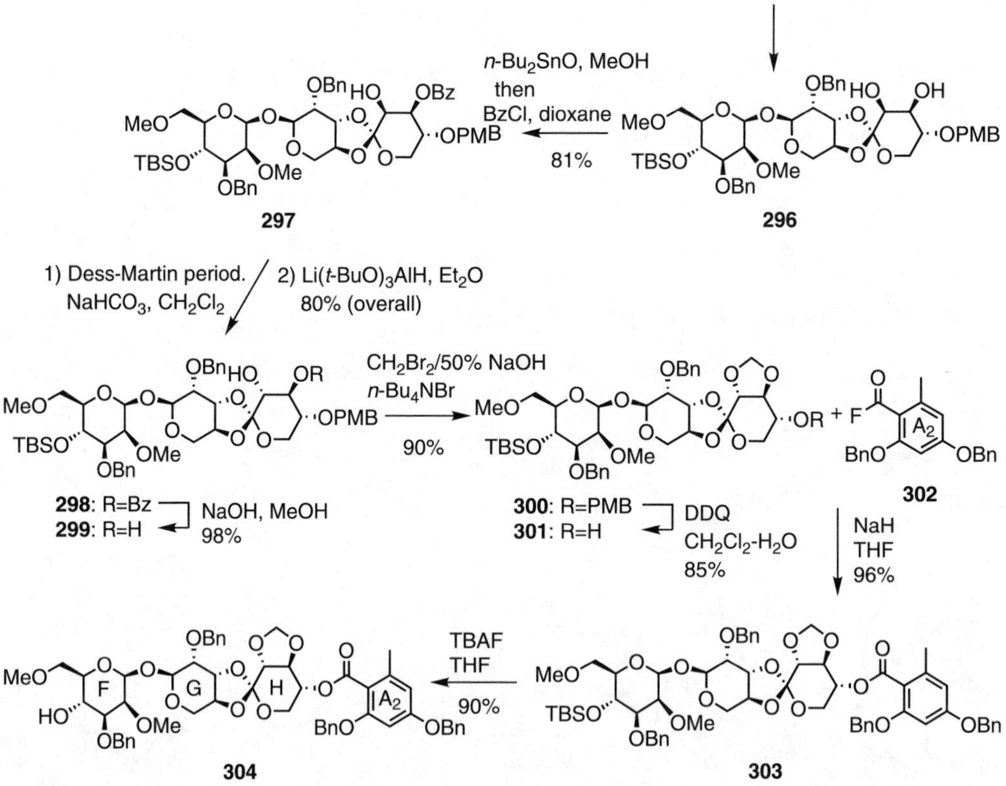

SCHEME 12.31 Synthesis of everninomicin 13,284-1 by Nicolaou (4) [25].

fluoride **316** was chemoselectively activated by $BF_3{\cdot}Et_2O$ and coupled with the phenylthio glycoside **315** to give the desired α-glycoside **317** with moderate stereoselectivity. The second glycosidation, reaction of **317** with the aglycon intermediate **318** using NBS, selectively proceeded to furnish the β-glycoside **319**. After deprotection of the benzyl group with DDQ, the resultant vinyl iodide **320** was coupled with dienyl mercury **321** in the presence of $Pd(PPh_3)_4$ to afford polycavernoside A (**322**).

Paquette's total synthesis of polycavernoside A [27] (Scheme 12.35) was accomplished via the glycosidation of the phenylthio disaccharide **323** and the aglycon derivative **318** using NBS. This reaction provided the glycoside **324** with the desired β-glycoside linkage, which was deprotected to give the vinyl iodide **325**. The cross-coupling of **325** and the dienylstannane **326** by the Stille method using $PdCl_2(MeCN)_2$ furnished polycavernoside A (**322**).

12.2.16 VANCOMYCIN

Nicolaou's total synthesis of vancomycin [28] (Scheme 12.36) was achieved via the subsequent introduction of glucose and vancosamine to the suitably protected aglycon. The reaction of the vancomycin aglycon derivative **327** and the glucose-derived trichloroacetimidate **328** was accomplished using $BF_3{\cdot}Et_2O$, giving the desired β-glycoside **329** as a major anomer whose Alloc protecting group was removed to furnish the alcohol **330**. The second glycosidation, reaction of the alcohol **330** and the glycosyl fluoride **331** corresponding to vancosamine was achieved employing $BF_3{\cdot}Et_2O$ to give the desired glycoside **332** with α-glycosidic linkage.

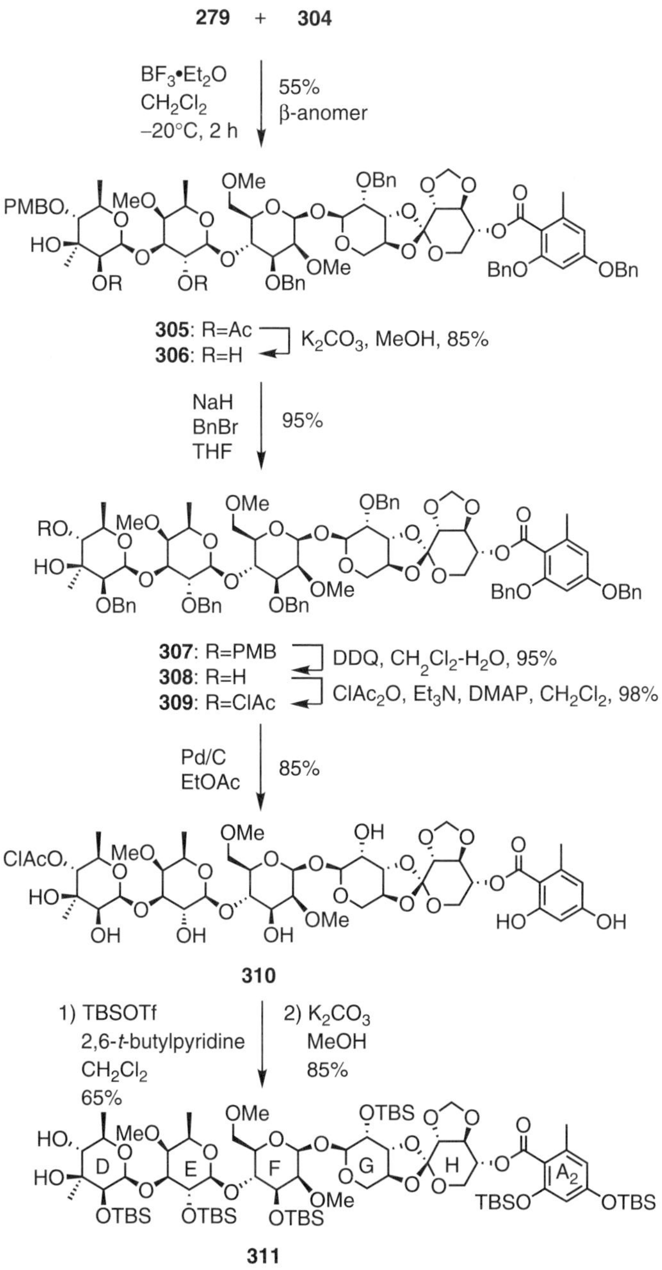

SCHEME 12.32 Synthesis of everninomicin 13,284-1 by Nicolaou (5) [25].

The deprotection of the silyl, acetyl and Cbz groups, followed by hydrolysis of the methyl ester, furnished vancomycin **336**.

Kahne's synthesis of vancomycin from the aglycon [29] (Scheme 12.37) utilizes glycosidations employing glycosyl sulfoxides. Treatment of the vancomycin aglycon derivative **337**, prepared from natural vancomycin and the glycosyl sulfoxide **338** with Tf_2O, DTBMP and $BF_3{\cdot}Et_2O$, gave the β-glycoside **340** after deprotection. The glycosidation of **340** and the glycosyl sulfoxide **341** using Tf_2O and $BF_3{\cdot}Et_2O$ afforded the desired α-glycoside **342**, leading to vancomycin **336** after deprotection.

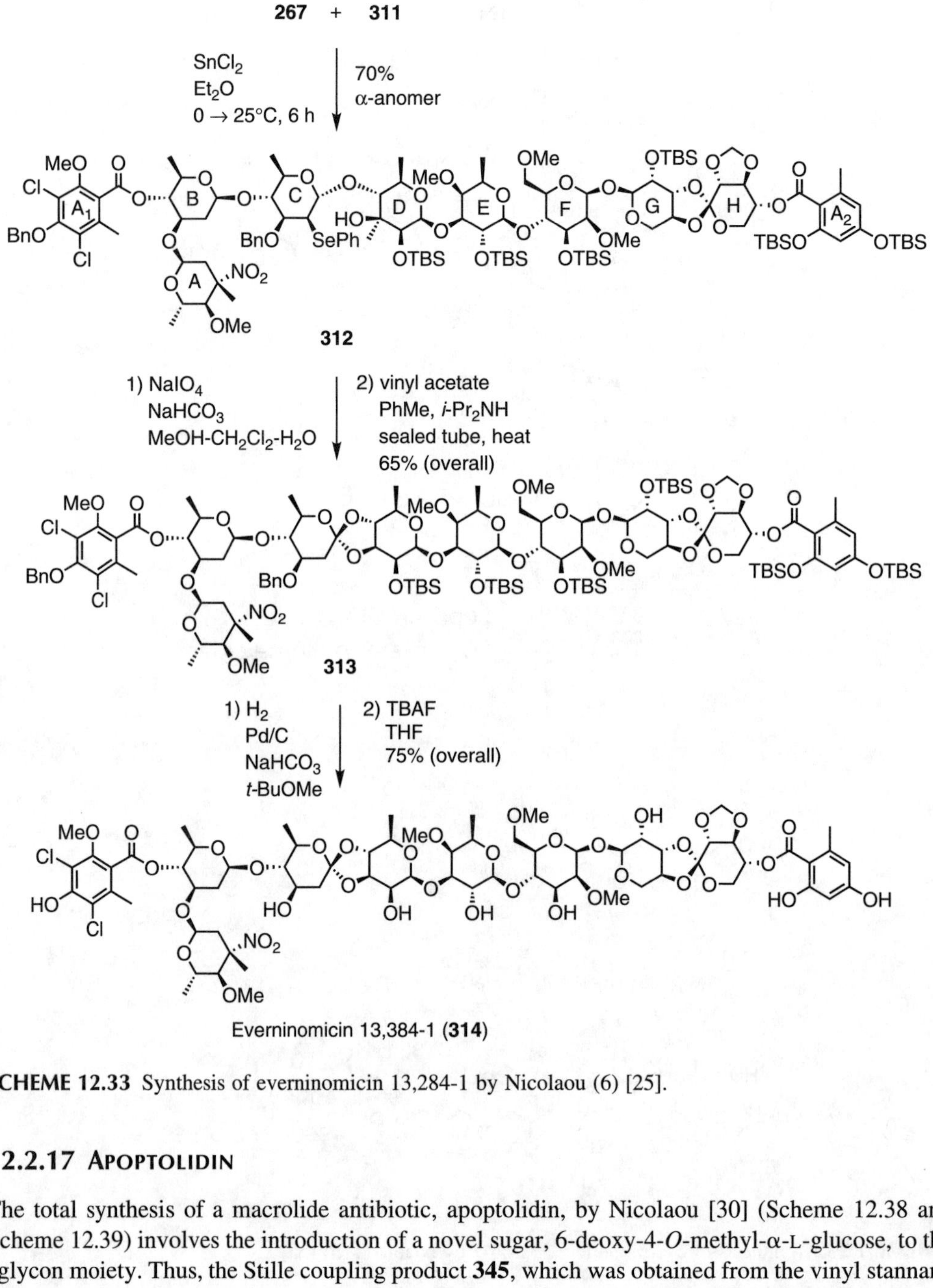

267 + **311**

$SnCl_2$, Et_2O, 0 → 25°C, 6 h; 70%, α-anomer

312

1) $NaIO_4$, $NaHCO_3$, $MeOH$-CH_2Cl_2-H_2O; 2) vinyl acetate, PhMe, *i*-Pr_2NH, sealed tube, heat; 65% (overall)

313

1) H_2, Pd/C, $NaHCO_3$, *t*-BuOMe; 2) TBAF, THF; 75% (overall)

Everninomicin 13,384-1 (**314**)

SCHEME 12.33 Synthesis of everninomicin 13,284-1 by Nicolaou (6) [25].

12.2.17 Apoptolidin

The total synthesis of a macrolide antibiotic, apoptolidin, by Nicolaou [30] (Scheme 12.38 and Scheme 12.39) involves the introduction of a novel sugar, 6-deoxy-4-*O*-methyl-α-L-glucose, to the aglycon moiety. Thus, the Stille coupling product **345**, which was obtained from the vinyl stannane **343** and the vinyl iodide **344**, was glycosylated with the glycosyl sulfoxide **346** by Tf_2O and DTBMP to give the α-glycoside **347**. After alkaline hydrolysis using KOH, the resultant dihydroxy carboxylic acid **348** was subjected to Yamaguchi macrolactonization to furnish the desired 20-membered macrolide **349**. The remaining free hydroxyl group was protected with a chloroacetyl group and the triethylsilyl group was selectively removed to afford the alcohol **350**. On the other hand, the β-D-oleandrosyl-α-L-olivomycose disaccharide moiety was constructed from **351** and **352**. The glycosidation of the glycosyl fluoride **351** and the diol **352** using $SnCl_2$ gave the desired β-disaccharide **353** together with the 3-*O*-glycoside. The remaining tertiary alcohol was silylated

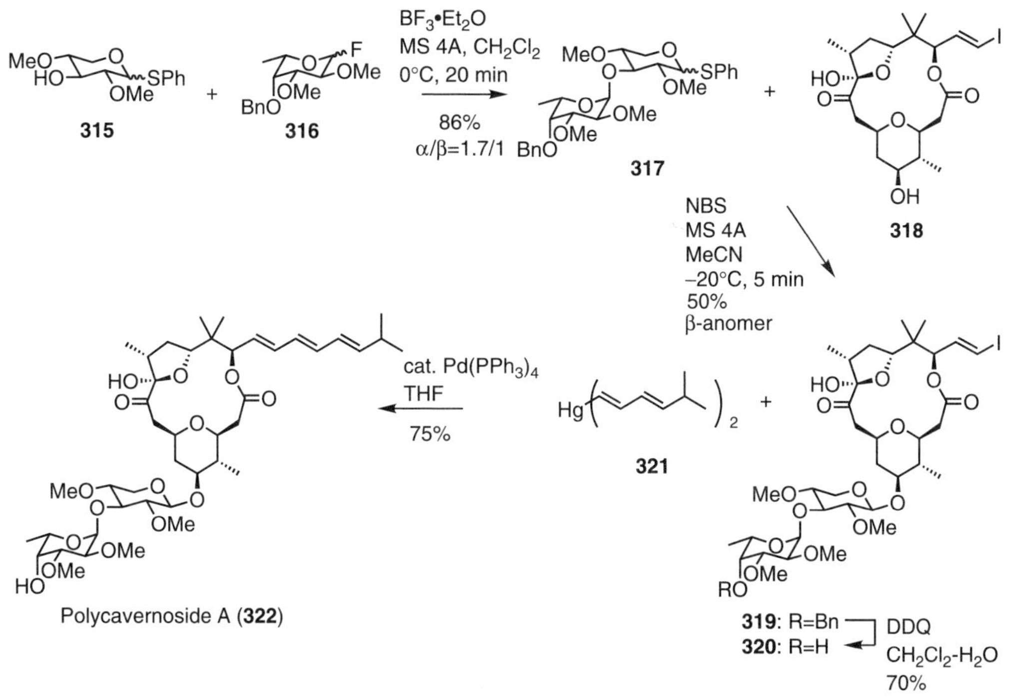

SCHEME 12.34 Synthesis of polycavernoside A by Murai [26].

and the resultant **354** was treated with Raney–Ni in the presence of H_2 to furnish the 2-deoxy disaccharide lactol **355**. The lactol **355** was converted into the glycosyl fluoride **356** by DAST.

The final glycosidation of the alcohol **350** with the glycosyl fluoride **356**, employing $SnCl_2$, gave the desired α-glycoside **357**, which was subjected to careful deprotection of the silyl and chloroacetyl groups and cleavage of the methyl glycoside to provide the labile antibiotic, apoptolidin **358**.

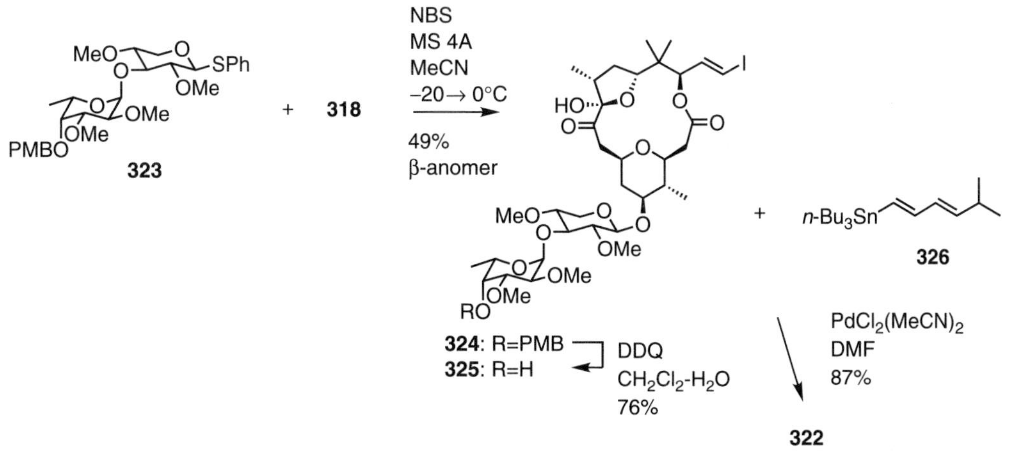

SCHEME 12.35 Synthesis of polycavernoside A by Paquette [27].

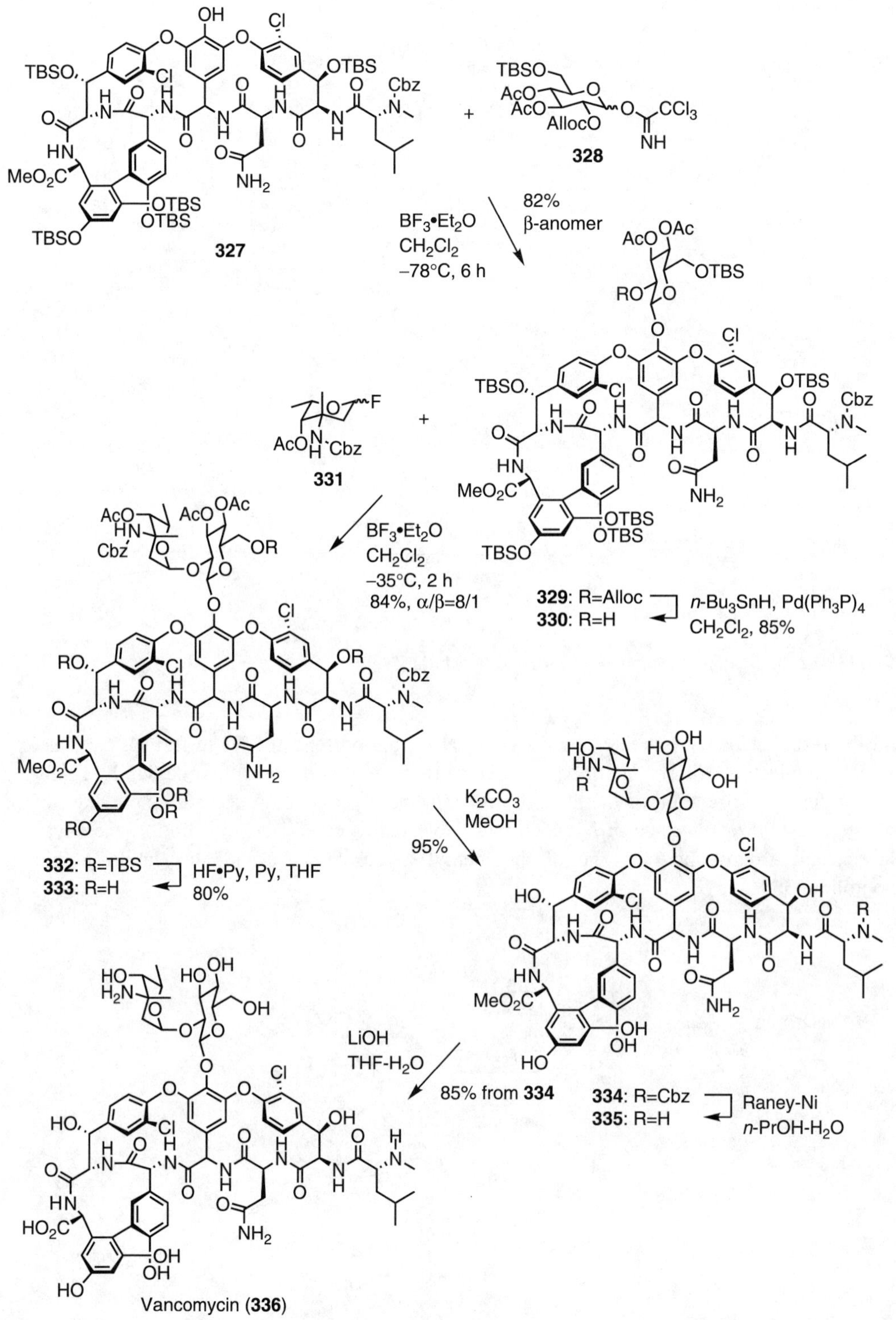

SCHEME 12.36 Synthesis of vancomycin by Nicolaou [28].

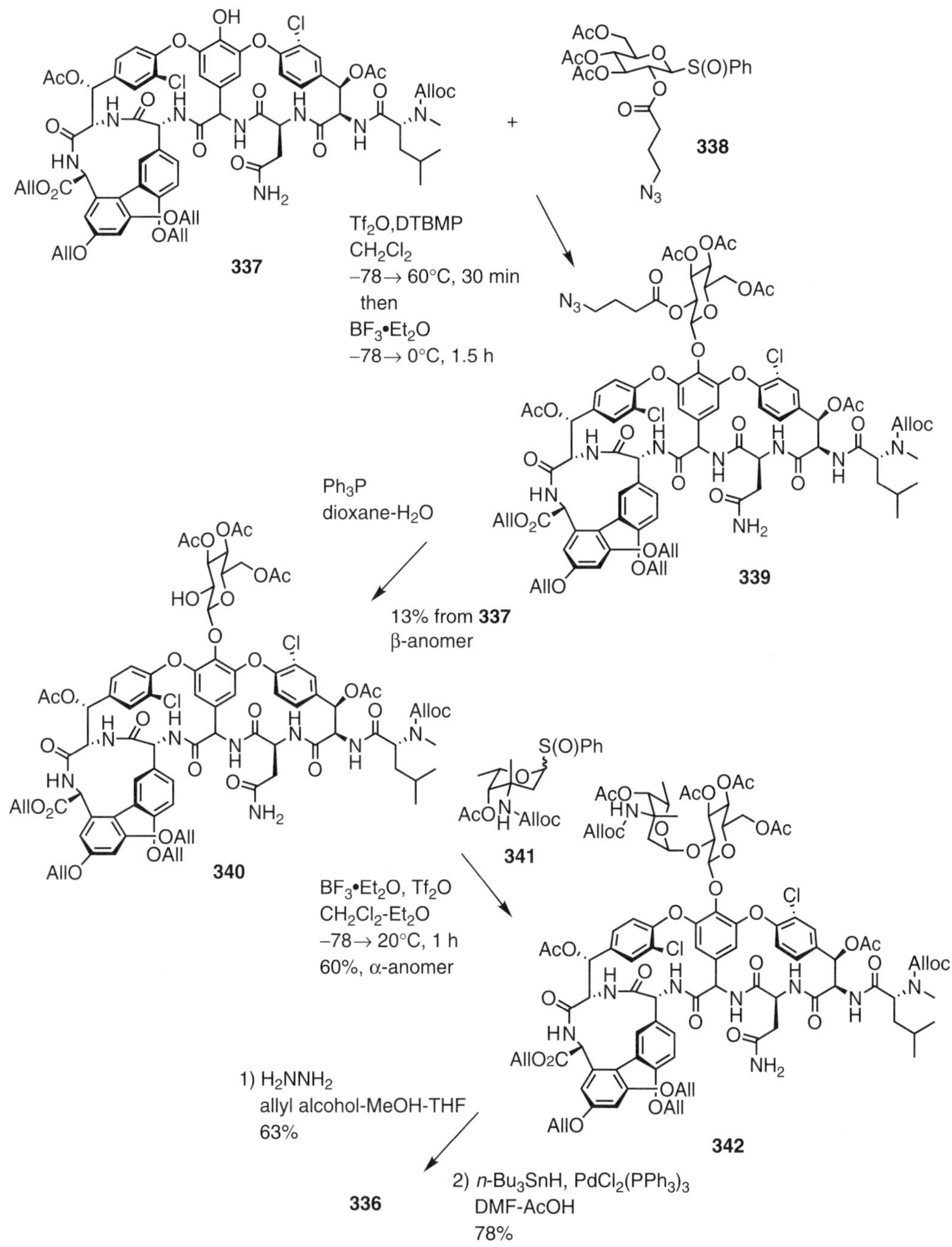

SCHEME 12.37 Synthesis of vancomycin by Kahne [29].

12.3 *C*-GLYCOSIDE ANTIBIOTICS

12.3.1 VINEOMYCINONE B_2 METHYL ESTER

Danishefsky's total synthesis of vineomycinone B_2 methyl ester [31] (Scheme 12.40) involves the construction of the sugar moiety possessing a *C*-glycosidic bond by a hetero Diels–Alder

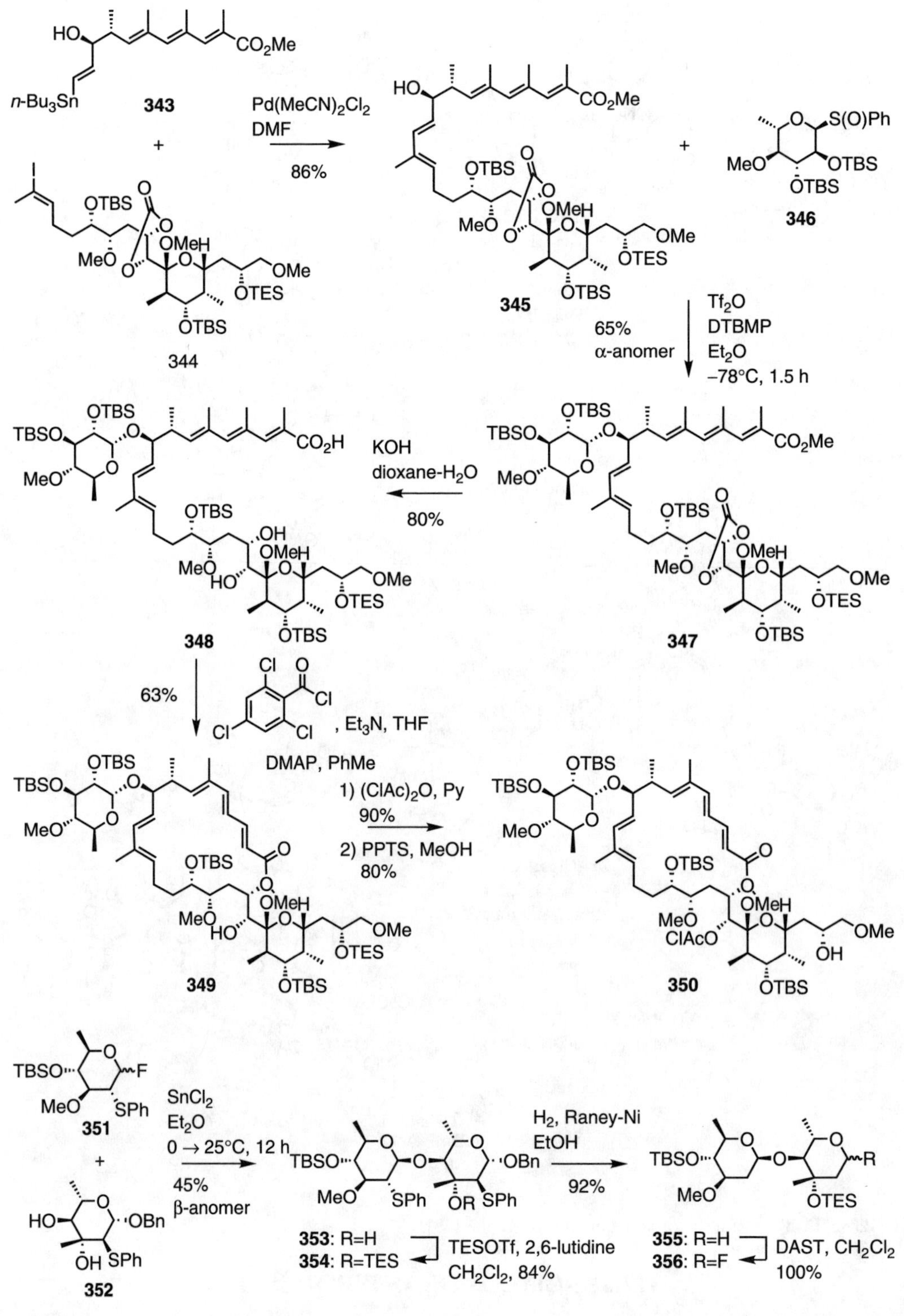

SCHEME 12.38 Synthesis apoptolidin by Nicolaou (1) [30].

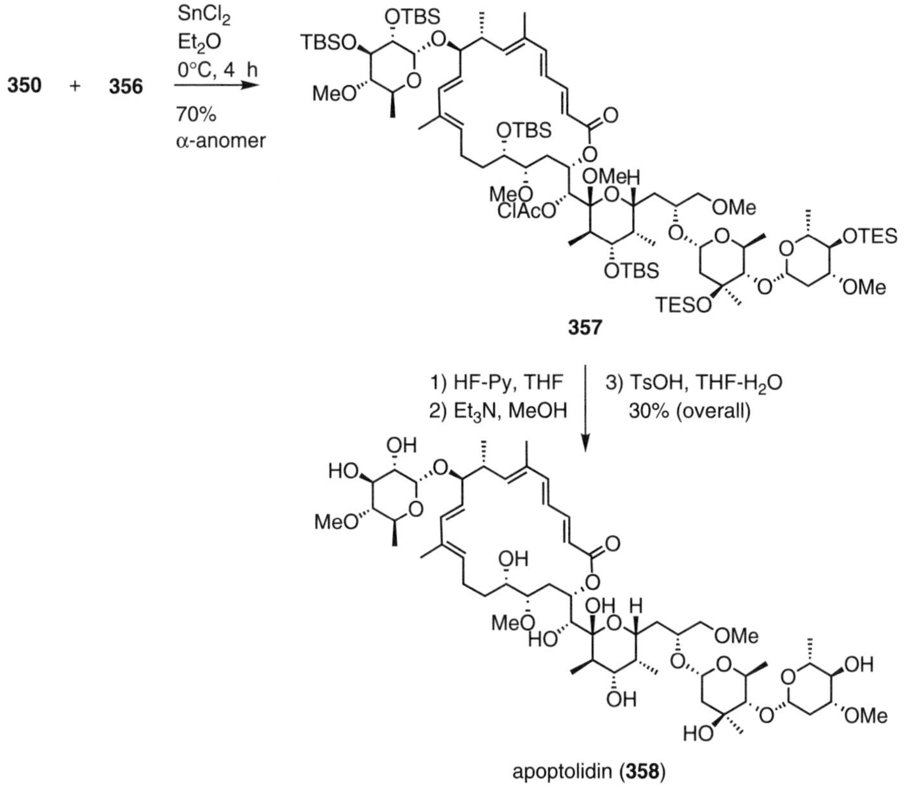

SCHEME 12.39 Synthesis apoptolidin by Nicolaou (2) [30].

reaction using a silyloxydiene. Thus, the reaction of the diene **359** with the aldehyde **360** in the presence of $Eu(fod)_3$ gave racemic **361**. Hydroboration of the silyl enol ether **361**, employing $BH_3 \cdot SMe_2$, followed by oxidative workup yielded **362**, which was subjected to demethylation to furnish **363**. After the coupling of **363** with l-menthyl ester bromomagnesium salt, the enatiomerically pure and desired **364** was obtained after separation of the resultant four isomers. Finally, transesterification of **364** with MeOH under basic conditions afforded vineomycinone B_2 methyl ester (**365**).

Suzuki's total synthesis of vineomycinone B_2 methyl ester [32] (Scheme 12.41) is characterized by an $O \rightarrow C$ glycoside rearrangement. Thus, the aryl *C*-glycosidation of the glycosyl fluoride **366** and the anthrol **367** using $Cp_2HfCl_2{-}AgClO_4$ proceeded regio- and stereoselectively to give the desired aryl β-*C*-glycoside **368** via $O \rightarrow C$ glycoside rearrangement. The phenolic hydroxy group was protected as a methyl ether and the benzoyl protecting groups in the sugar moiety were replaced by silyl groups to give the anthracene **369**. After **369** was lithiated, the resultant arylmetal intermediate was reacted with Me_2SnCl to yield the stannane **370**. The coupling of the stannane **370** and the aldehyde **371** gave the adduct **372**, which was subjected to benzoylation, oxidation and debenzylation to furnish **374**. Deoxygenation at the benzylic position of **374** produced the anthraquinone **375**. The final conversion of the terminal olefin of **375** into the methyl ester was carried out via ozonolysis in MeOH and deprotection of the two methyl and two silyl groups of the resultant **377** using BBr_3 furnished vineomycinone B_2 methyl ester (**365**).

In one additional example, the transition-metal-catalyzed coupling reaction of a glycal is used for creating a *C*-glycosidic bond in the total synthesis of vineomycinone B_2 methyl ester by Tius [33]

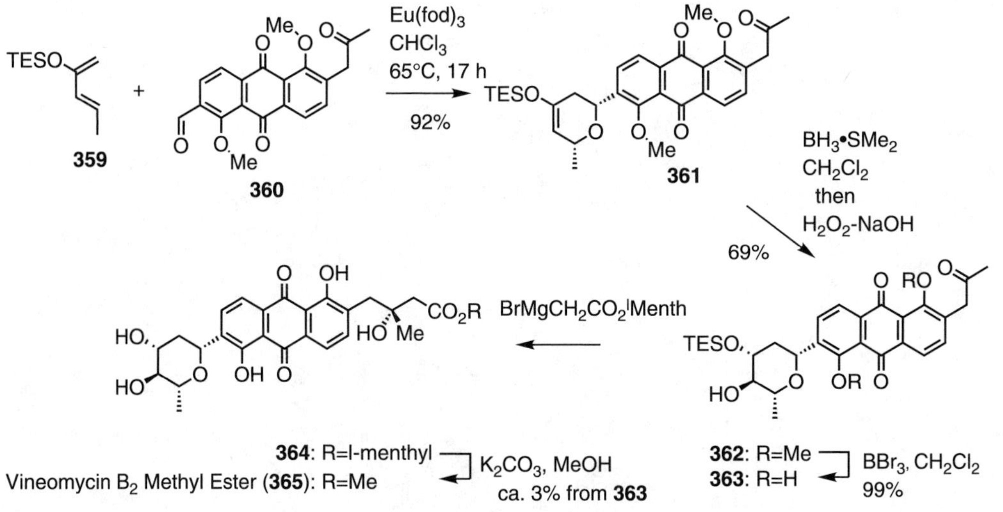

SCHEME 12.40 Synthesis of vineomycinone B_2 methyl ester by Danishefsky [31].

(Scheme 12.42). Thus, the chlorozinc derivative **379** generated from the glycal **378** was coupled with the iodoanthracene **380** using $Pd(PPh_3)_2Cl_2$ and DIBAL to give the adduct **381**. The stereoselective reduction of **381**, employing methanolic HCl and $NaBH_3CN$, provided the desired aryl β-*C*-glycoside **382** as a single isomer, which was reacted with Bu_3SnCl in the presence of *n*-BuLi and TMEDA to give the stannane **383**. The Pd-mediated reaction of the stannane **383** and allyl bromide **384** afforded the adduct **385**. The addition of Me_2CuLi to **385** proceeded stereoselectively to yield **386**, which furnished vineomycinone B_2 methyl ester (**365**) via oxidation of **386** using bis(pyridine)silver permanganate and deprotection of **387** employing HCl in MeOH.

12.3.2 Medermycin

Tatsuta's total synthesis of medermycin [34] (Scheme 12.43) involves the use of a lactone sugar derivative to form a *C*-glycosidic bond. Thus, the coupling of the lactol **388** and the aryl bromide **389** in the presence of *n*-BuLi gave the aryl β-*C*-glycoside **391** after oxidation of the acetal of **390**. The reduction of **391** afforded the 1′-deoxy compound, which was converted into the *N,N*-diethylamide and subjected to reductive debromination to yield the 2-deoxy aryl β-*C*-glycoside **392**. After protecting group manipulations, the resultant **394** was regioselectively formylated and hydrolyzed to produce the hemiacetal **395**. Treatment of **395** with thiophenol followed by oxidation gave the sulfone compound, whose diol was protected as a cyclohexylidene acetal to yield **397** as a mixture. The subsequent coupling of **397** and the enone **398** in the presence of *n*-BuLi gave the hydroquinone **399**, which was *O*-methylated and reduced to furnish the desired alcohol **400**. After acid hydrolysis, the resultant hemiacetal was subjected to Wittig reaction using $Ph_3P{=}CHCO_2Et$ to give a mixture of the lactone **401** and the ester **402** through Michael cyclization of the Wittig product, with or without lactonization. The less hindered 3′-hydroxy groups were protected with silyl groups to give **403** and **404**, the latter of which was recycled to a mixture of **401** and **402** under acidic conditions through retro-Michael and Michael reactions. After the desired **403** was converted into the ketone **405**, the introduction of the amino group and its dimethylation yielded the *N,N*-dimethyl compound,

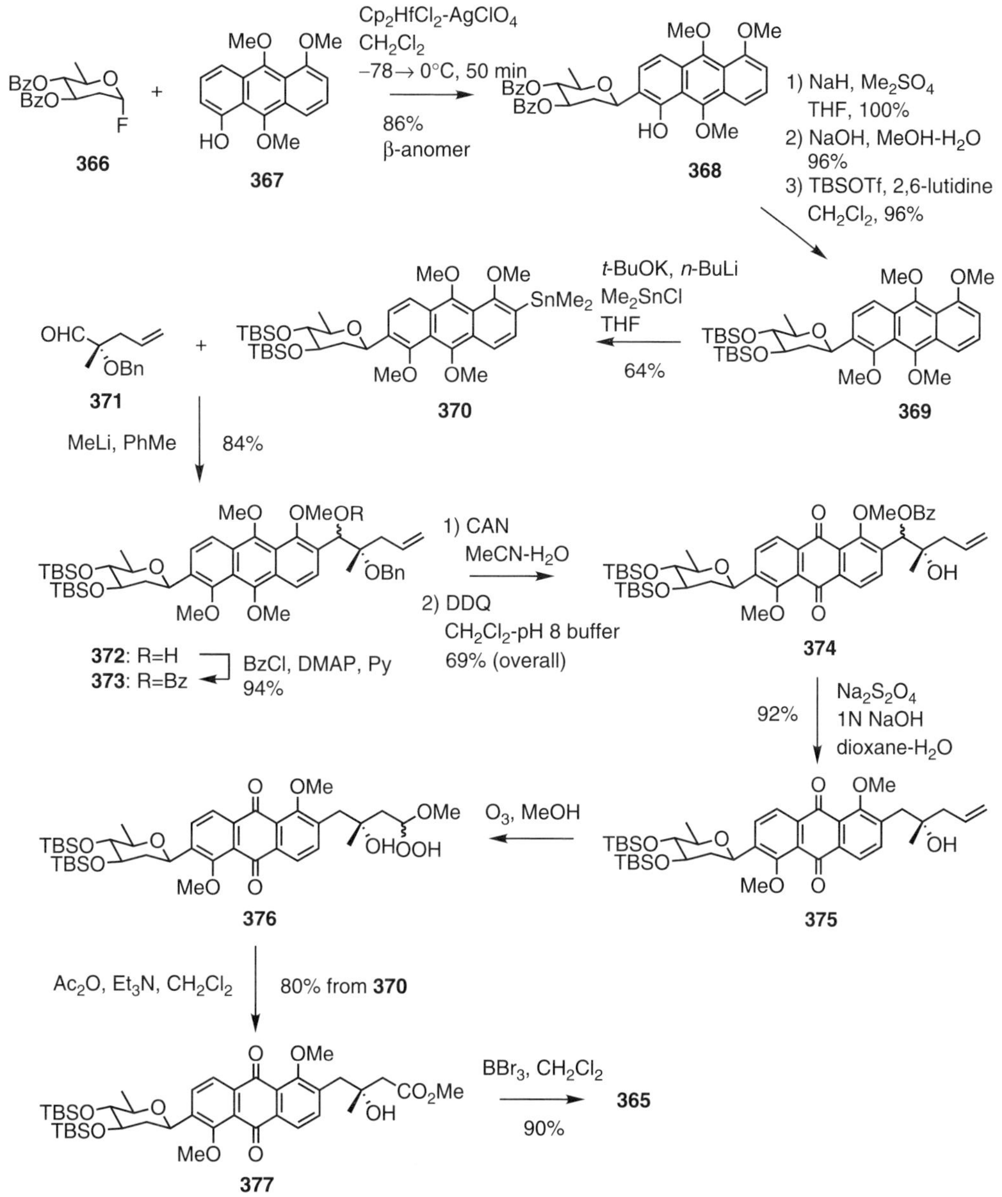

SCHEME 12.41 Synthesis of vineomycinone B_2 methyl ester by Suzuki [32].

which was oxidized employing CAN to yield the quinone **407**. Finally, the deprotection of the methyl and methoxymethyl groups in **407** using $AlCl_3$ furnished medermycin (**408**), which was shown to be identical with lactoquinomycin.

12.3.3 Urdamycinone B

The total synthesis of (−)-urdamycinone B, the enantiomer of natural urdamycinone B, by Yamaguchi [35] (Scheme 12.44) involves the utilization of a *C*-glycoside polyketide as a

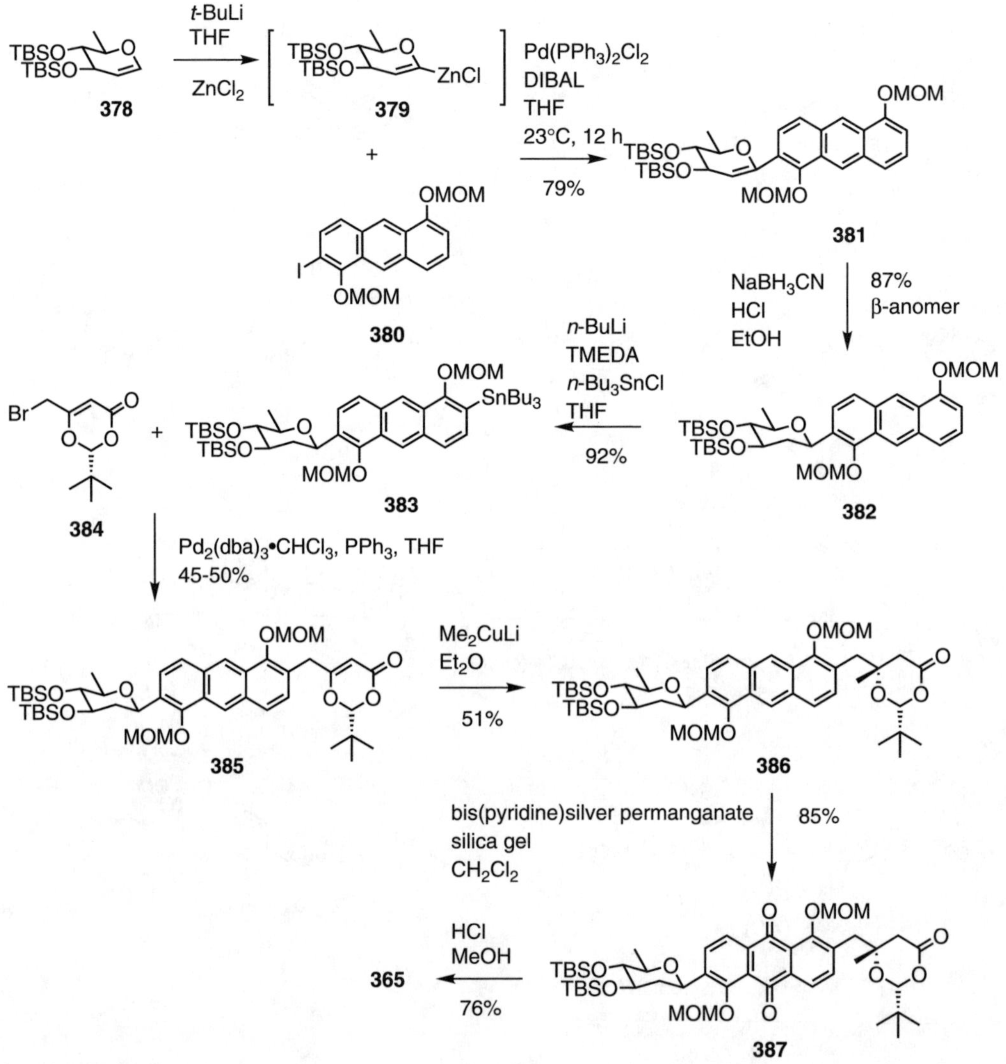

SCHEME 12.42 Synthesis of vineomycinone B_2 methyl ester by Tius [33].

promising approach. Condensation of the lactol **409** and the β-oxoglutarate under Knoevenagel conditions, followed by reduction, gave the β-hydroxyglutarate **412**. The polyketide **413**, generated by the reaction of **412** and the acetoacetate dianion in THF–HMPA, was aromatized with $Ca(OAc)_2$ to yield the naphthalene-1,8-diol **414** with a β-*C*-glycosidic linkage as a single isomer. The glutarate **414** was coupled with lithiated 2-methyl-3-buten-2-yl acetate to afford **415**, which was dealkoxycarbonylated with a Pd-catalyst and protected with MOM groups to furnish the enol lactone **416**. The keto aldehyde **417** prepared from **416** was treated with acetylacetone monothioketal and aromatized using K_2CO_3 to provide the anthracene **418**. Deprotection, oxidation and dethioacetalization gave the diketone **419**, which was cyclized by aldol reaction under basic conditions to furnish (−)-urdamycinone B (**420**) along with its *C*-3 epimer **421**.

Sulikowski's total synthesis of urdamycinone B [36] (Scheme 12.45) is characterized by *C*-glycosidation of a carbohydrate lactone and Diels–Alder reaction of a *C*-glycosyl

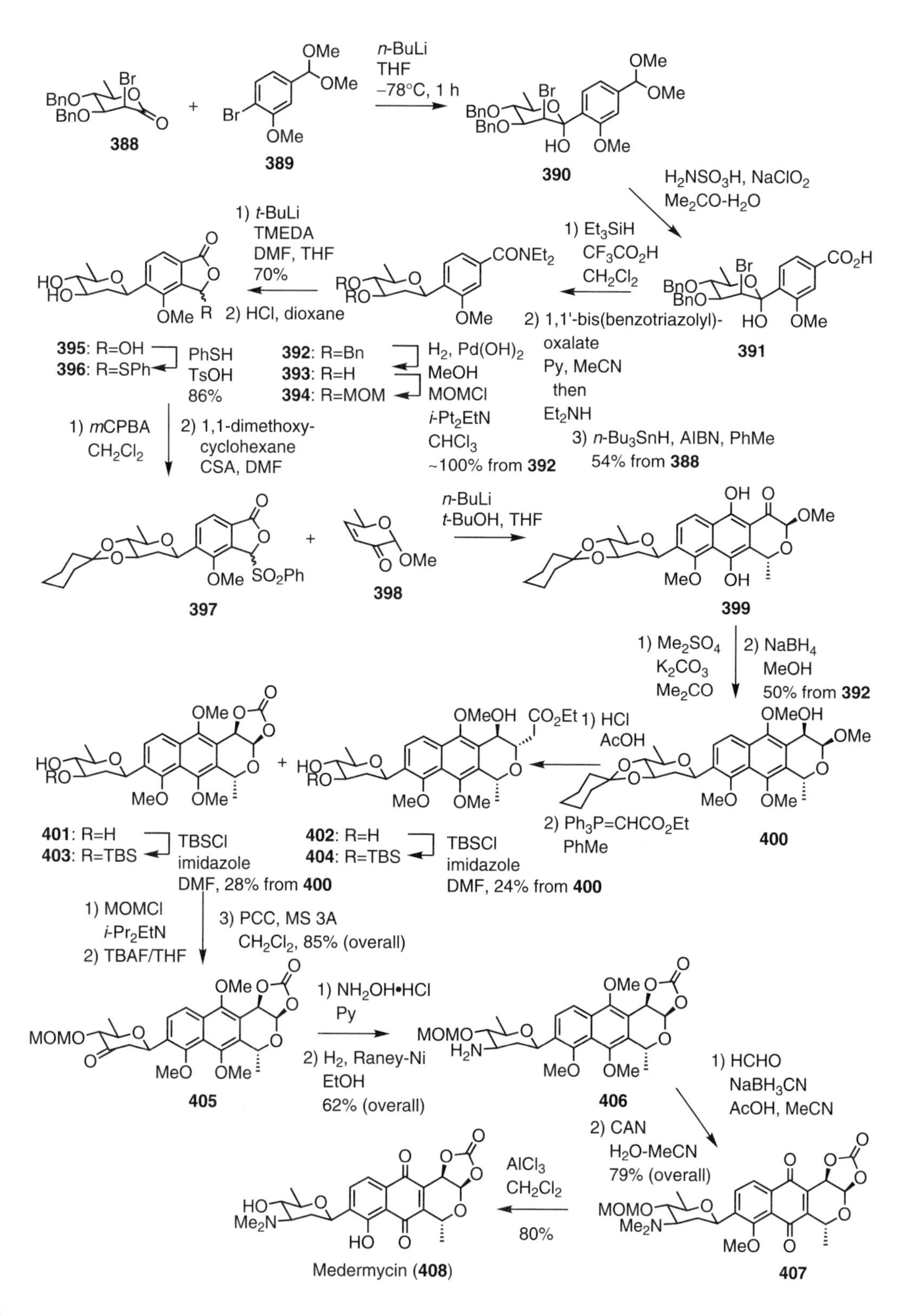

SCHEME 12.43 Synthesis of medermycin by Tatsuta [34].

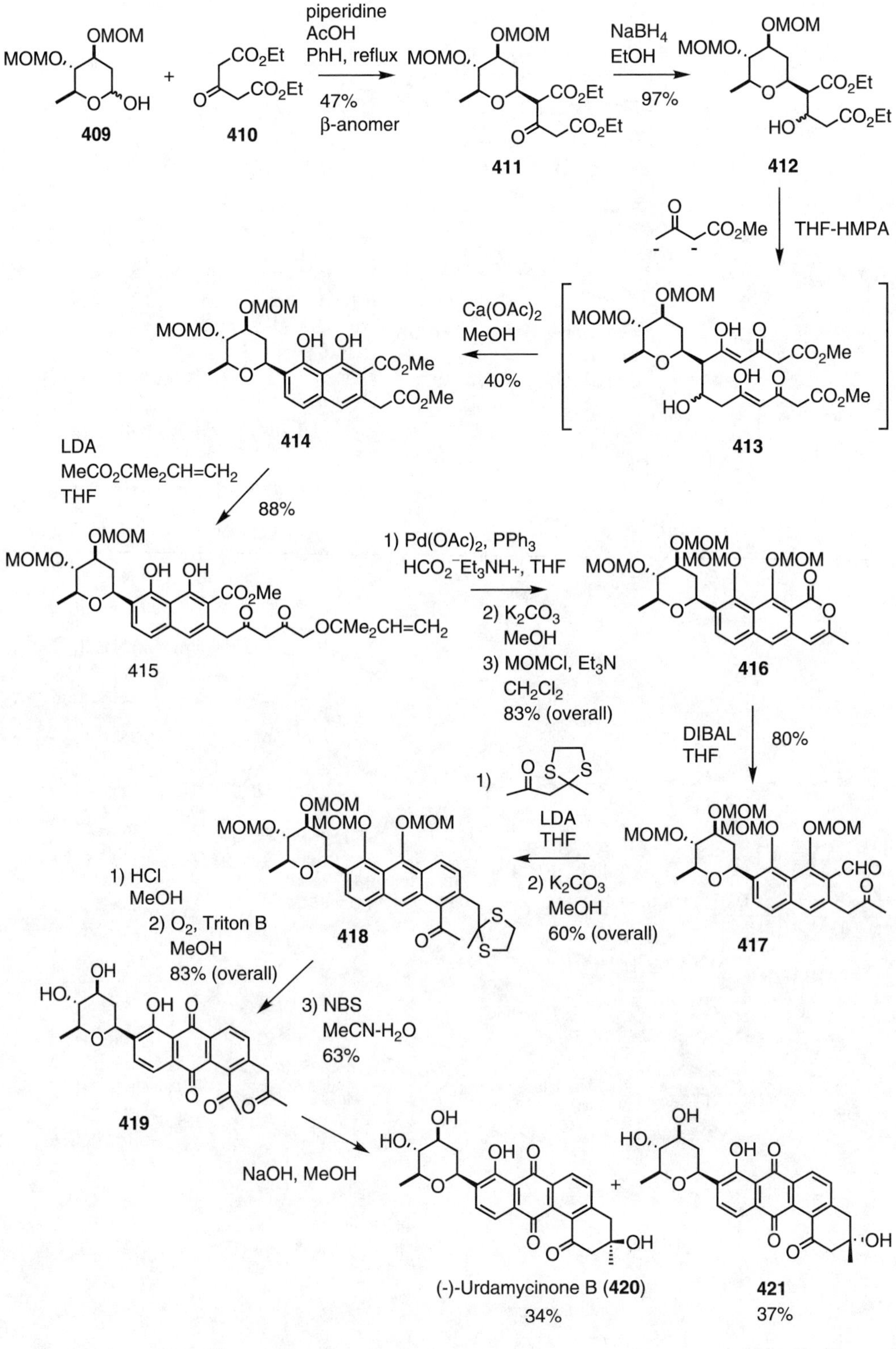

SCHEME 12.44 Synthesis of urdamycinone B enantiomer by Yamaguchi [35].

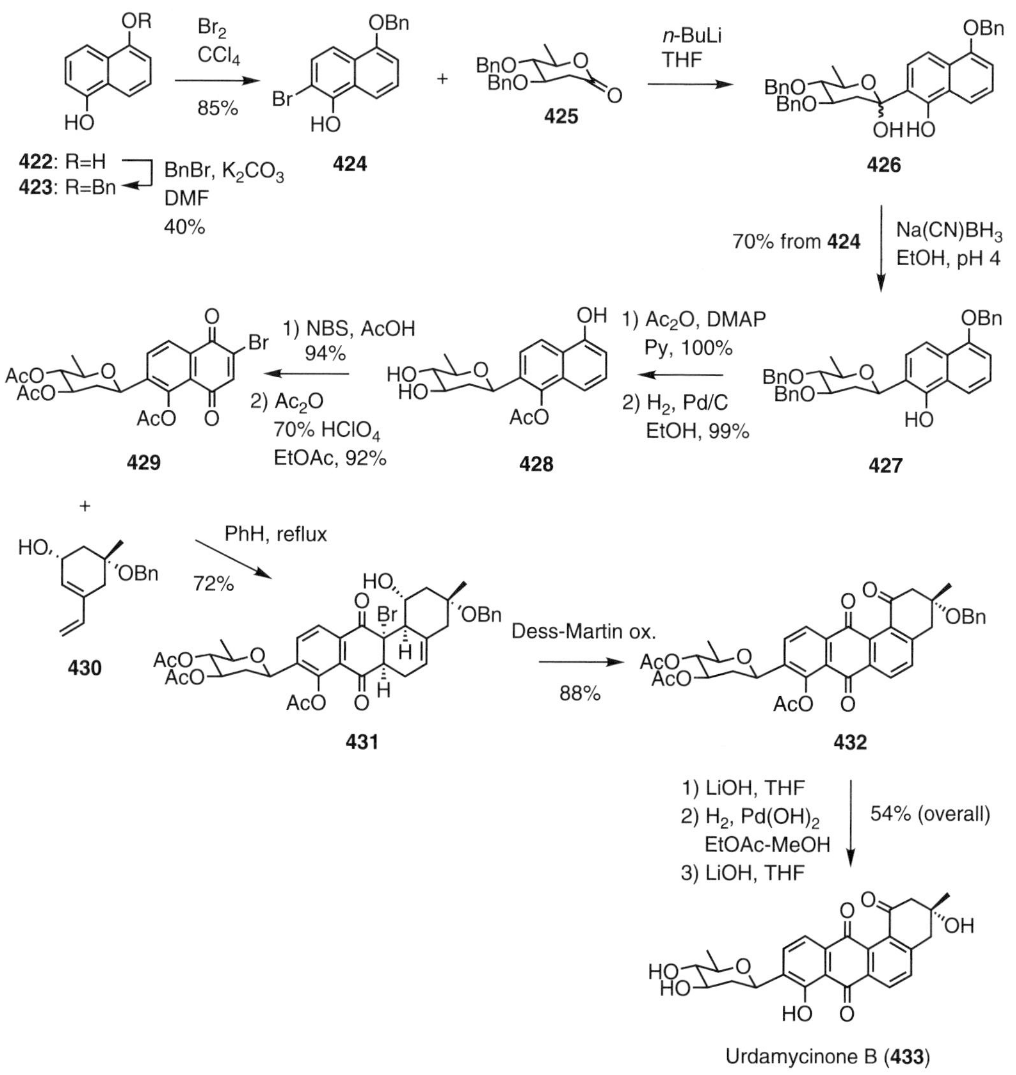

SCHEME 12.45 Synthesis of urdamycinone B by Sulikowski [36].

juglone derivative. Thus, the coupling of the bromonaphthalene **424** and the lactone **425** in the presence of *n*-BuLi gave the hemiketal **426**, which was reduced using $Na(CN)BH_3$ to afford the aryl β-*C*-glycoside **427**. After acetylation and debenzylation, the resultant **428** was subjected to oxidation employing NBS, followed by peracetylation, to yield the bromojuglone **429**. The subsequent Diels–Alder reaction of the bromojuglone **429** and the diene **430** gave the adduct **431**, which was oxidized by the Dess-Martin method to furnish the anthraquinone **432**. The removal of the protecting groups of **432** in three steps, avoiding the β-elimination of the C-3 tertiary alcohol, afforded natural urdamycinone B (**433**).

Toshima's total synthesis of urdamycinone B [37] (Scheme 12.46) was accomplished using an unprotected sugar. Thus, the unprotected *C*-glycosyljuglone **436** was synthesized in two steps involving the aryl *C*-glycosidation of 1,5-naphthalenediol **422** and the totally unprotected D-olivose **434**, using TMSOTf, and the subsequent regioselective photo-oxygenation of the resultant *C*-glycosylnaphthalenediol **435**. The regioselective Diels–Alder reaction of **436** and the diene **437**

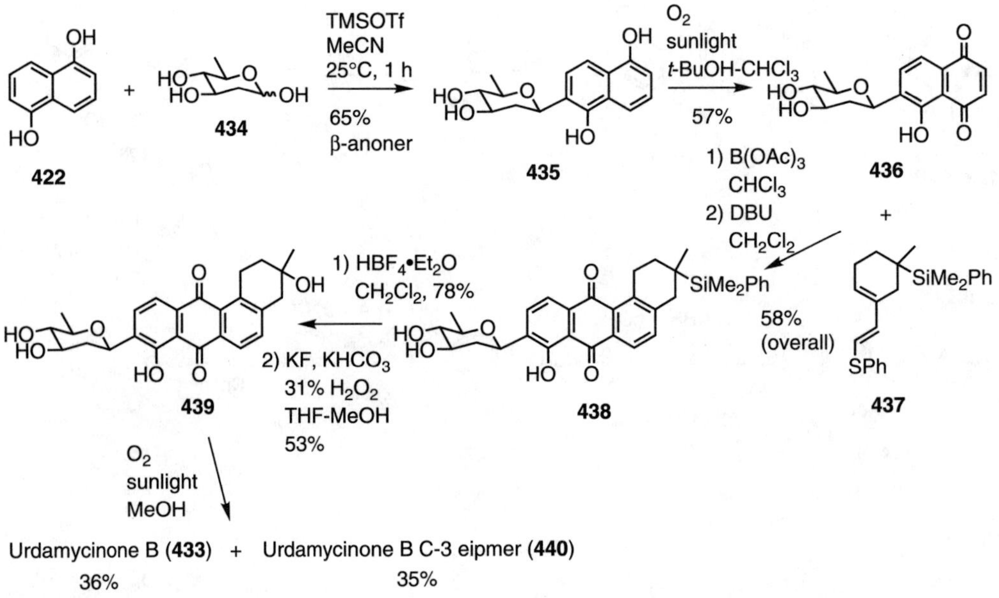

Urdamycinone B (**433**) + Urdamycinone B C-3 eipmer (**440**)
36% 35%

SCHEME 12.46 Synthesis of urdamycinone B by Toshima [37].

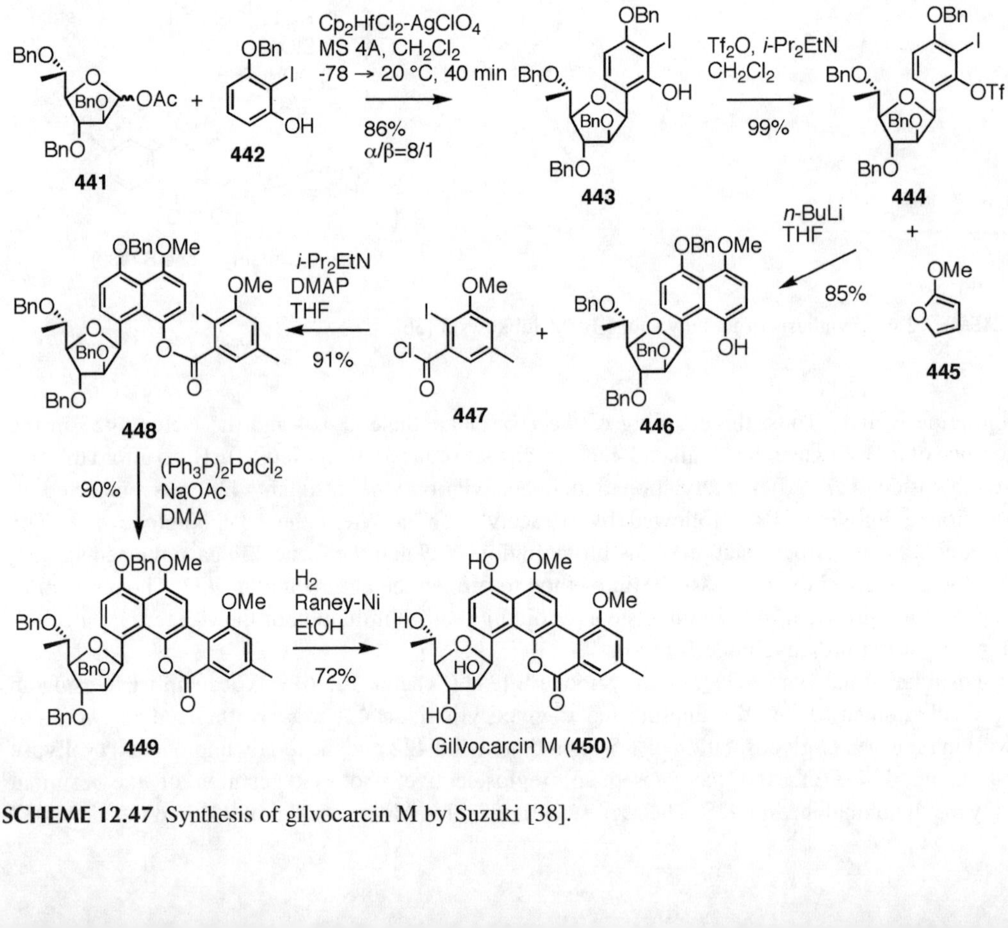

SCHEME 12.47 Synthesis of gilvocarcin M by Suzuki [38].

using $B(OAc)_3$, followed by the conversion of the silyl group to the tertiary alcohol and the regioselective introduction of the ketone function at the C-1 position, led to the total synthesis of urdamycinone B (**433**).

12.3.4 Gilvocarcin M

Suzuki's total synthesis of gilvocarcin M [38] (Scheme 12.47) involves a Cp_2HfCl_2–$AgClO_4$ mediated *C*-glycosidation. Thus, the coupling of the glycosyl acetate **441** and the iodophenol **442** using Cp_2HfCl_2–$AgClO_4$ gave the desired aryl α-*C*-glycoside **443**. After triflation, the resultant **444** was treated with *n*-BuLi and 2-methoxyfuran **445** to yield the adduct **446** via [4 + 2] cycloaddition and aromatization. Acylation of **446** with the acid chloride **447** afforded the ester **448**, which was subjected to the intramolecular biaryl coupling using a Pd-catalyst to provide the tetracycle **449**. Finally, hydrogenolysis of **449** in the presence of Raney–Ni to remove the benzyl groups furnished gilvocarcin M (**450**).

12.4 OTHERS

12.4.1 Bidesmosidic Triterpene Saponin

The synthesis of a bidesmosidic triterpene saponin by Yu and Hui [39] (Scheme 12.48) represents a conceptual advance in its use of two flasks and the combination of one-pot glycosidation and two-directional glycosidation methods. Thus, the glycosidation of the oleanolic ester **451** and the trichloroacetimidate **452** using TMSOTf provided the desired glycoside **453**, which was converted to the acid **454** upon warming to room temperature. Addition of the trichloroacetimidate **455** into the above mixture gave the bis-glycoside **456**. In another flask, the disaccharide **459** was synthesized by the glycosidation of the phenyl thioglycoside **457** and the trichloroacetimidate **458** using TMSOTf. Addition of the mixture into the first flask, followed by the addition of TMSOTf and NIS, afforded bidesmosidic triterpene saponin **461** after deprotection.

12.4.2 Digitoxin

McDonald's synthesis of digitoxin [40] (Scheme 12.49) employs a tungsten-catalyzed *endo*-selective alkynol cyclization method. The glycosidation of the glycal **462** and the alkynol alcohol **463**, using Ph_3P–HBr in PhMe, proceeded stereoselectively to yield the 2-deoxy-β-glycoside **464**, which was subjected to debenzoylation and tungsten carbonyl catalyzed *endo*-selective cycloisomerization to give the disaccharide **466**. After protecting group manipulation, the resultant **468** was coupled with **463** by the same glycosidation protocol to furnish the desired β-linked disaccharide **470** after deacylation. The subsequent cyclization of **470** using tungsten carbonyl, followed by acetylation, provided the trisaccharide **472**. The final glycosidation of the glycal **472** and the digitoxigenin aglycon **473**, using Ph_3P–HBr in $CHCl_3$, gave the protected digitoxin derivative **474**, which was deprotected to furnish digitoxin (**476**).

12.5 CONCLUDING REMARKS

Many carbohydrate-containing complex natural compounds have effectively been synthesized using a good command of cutting-edge carbohydrate chemistry. In spite of great progress in the chemical synthesis of carbohydrate-containing complex natural compounds, the general

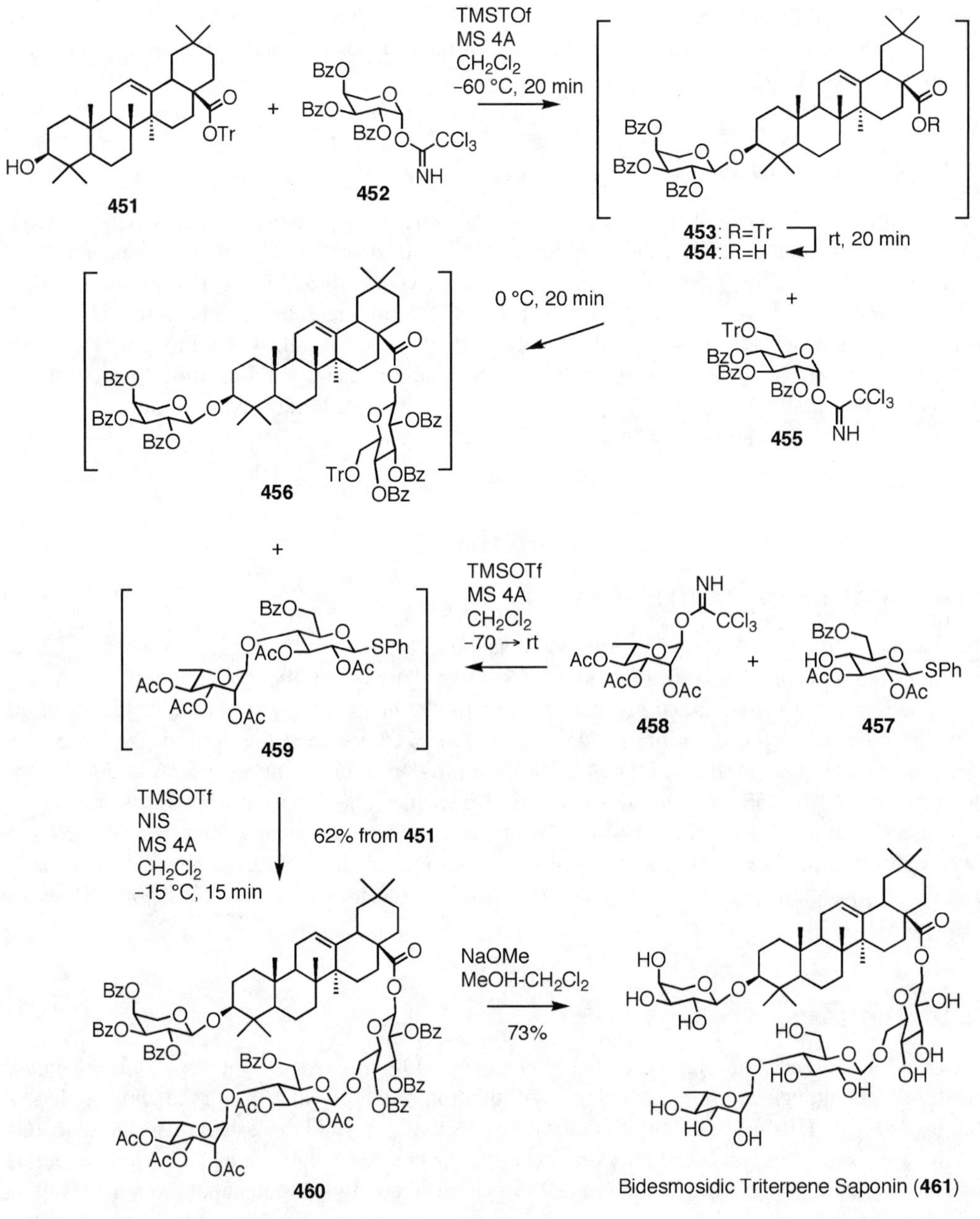

SCHEME 12.48 Synthesis of bidesmosidic triterpene saponin by Yu-Hui [39].

aspects of chemical yield and stereoselectivity have not yet been ascertained. Furthermore, a general chemical method for glycosidation has still not been realized. A major breakthrough may be needed for synthesizing any given carbohydrate-containing natural compound through fully controlled chemistry. In addition, environmentally benign chemistry on carbohydrates, *Green Carbohydrate Chemistry*, must be established in this century. Since carbohydrates are indispensable biosubstances in our life activity, the study of carbohydrate chemistry will continue for a long time.

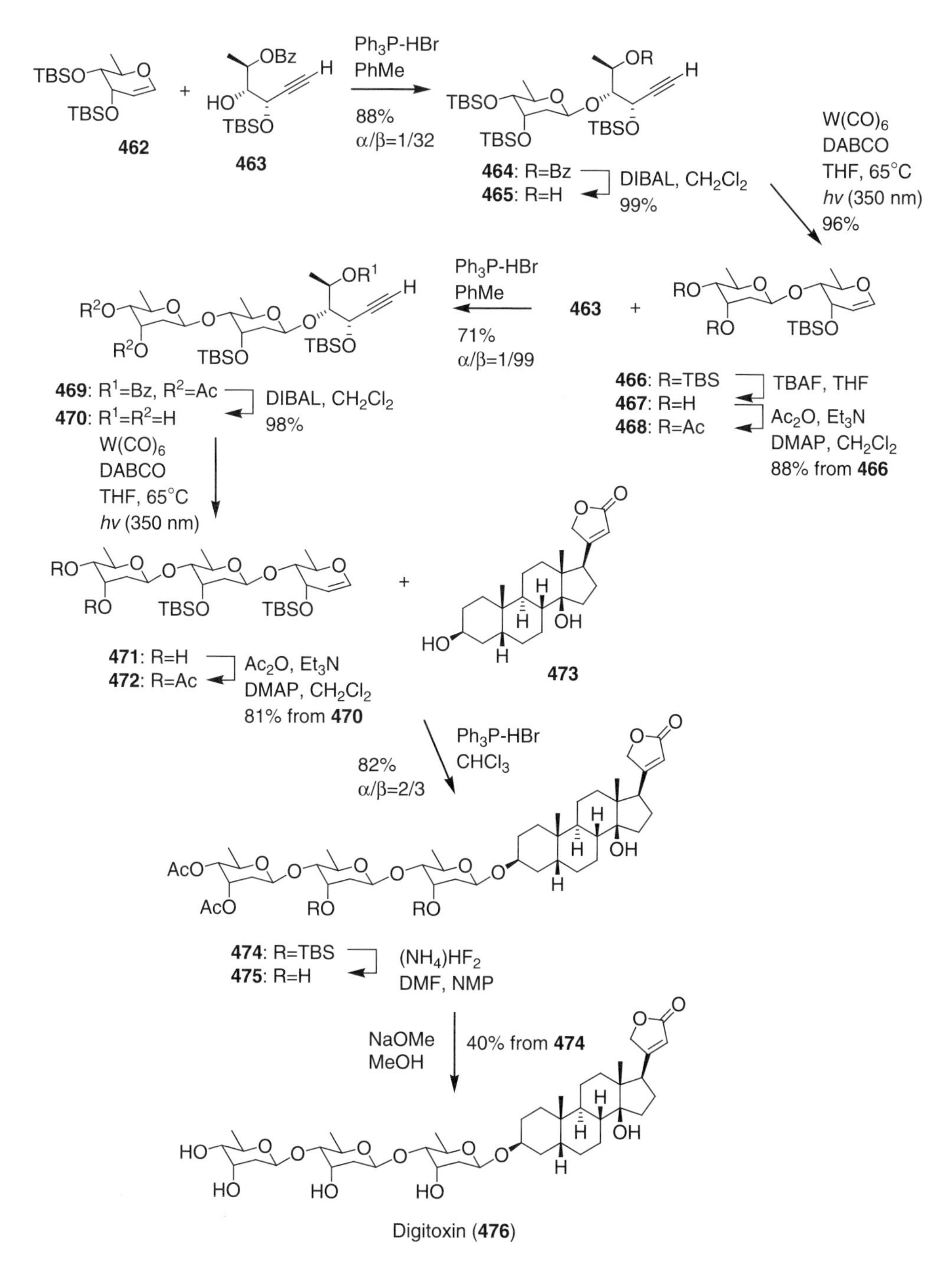

SCHEME 12.49 Synthesis of digitoxin by McDonald [40].

REFERENCES

1. (a) Ernst, B, Hart, G, Sinaÿ, P, Eds., *Carbohydrates in Chemistry and Biology*, Wiley-VCH, New York, 2000; (b) Fraser-Reid, B O, Tatsuta, K, Thiem, J, Eds., *Glycoscience: Chemistry and Chemical Biology I–III*, Springer, Berlin, 2001; (c) Dumitriu, S, Ed., *Polysaccharides in Medicinal Applications*, Marcel Dekker, New York, 1996.
2. (a) Hanessian, S, Ed., *Preparative Carbohydrate Chemistry*, Marcel Dekker, New York, 1997; (b) Nicolaou, K C, Mitchell, H J, Adventures in carbohydrate chemistry: new synthetic technologies, chemical synthesis, molecular design, and chemical biology, *Angew. Chem. Int. Ed.*, 40, 1576–1624, 2001.
3. (a) Schmidt, R R, New methods for the synthesis of glycosides and oligosaccharide-are there alternatives to the Koenigs–Knorr method? *Angew. Chem. Int. Ed.*, 25, 212–235, 1986; (b) Sinaÿ, P, Recent advances in glycosylation reactions, *Pure Appl. Chem.*, 63, 519–528, 1991; (c) Toshima, K, Tatsuta, K, Recent progress in *O*-glycosidation methods and its application to natural products synthesis, *Chem. Rev.*, 93, 1503–1531, 1993; (d) Boons, G J, Strategies in oligosaccharide synthesis, *Tetrahedron*, 52, 1095–1121, 1996; (e) Davis, B G, Recent developments in oligosaccharide synthesis, *J. Chem. Soc. Perkin. Trans. 1*, 2137–2160, 2000.
4. (a) Greene, T W, Wuts, P G M, *Protective Groups in Organic Synthesis*, 3rd ed., Wiley, New York, 1999; (b) Kocienski, P J, *Protecting Groups*, Georg Thieme, Stuttgart, 1994.
5. (a) Masamune, S, Kim, C U, Wilson, K E, Spessard, G O, Georghiou, P E, Bates, G S, Syntheses of macrolide antibiotics. I. Methymycin, *J. Am. Chem. Soc.*, 97, 3513–3515, 1975; (b) Masamune, S, Yamamoto, H, Kamata, S, Fukuzawa, A, Syntheses of macrolide antibiotics. II. Methymycin, *J. Am. Chem. Soc.*, 97, 3513–3515, 1975.
6. (a) Woodward, R B, Logusch, E, Namibiar, K P, Sakan, K, Ward, D E, Au-Yeung, B W, Balaram, P, Browne, L J, Card, P J, Chen, C H, Chênevert, R B, Fliri, A, Frobel, K, Gais, H J, Garratt, D G, Hayakawa, K, Heggie, W, Hesson, D P, Hoppe, D, Hoppe, I, Hyatt, J A, Ikesa, D, Jacobi, P A, Kim, K S, Kobuke, Y, Kojima, K, Krowicki, K, Lee, V J, Leutert, T, Malchenko, S, Martens, J, Matthews, R S, Ong, B S, Press, J B, Rajan Babu, T V, Rousseau, G, Sauter, H M, Suzuki, M, Tatsuta, K, Tolbert, L M, Truesdale, E A, Uchida, I, Ueda, Y, Uyehara, T, Vasella, A T, Vladuchick, W C, Wade, P A, Williams, R M, Wong, H N C, Asymmetric total synthesis of erythromycin. 1. Synthesis of an erythronolide A seco acid derivative via asymmetric induction, *J. Am. Chem. Soc.*, 103, 3210–3213, 1981; (b) Woodward, R B, Logusch, E, Namibiar, K P, Sakan, K, Ward, D E, Au-Yeung, B W, Balaram, P, Browne, L J, Card, P J, Chen, C H, Chênevert, R B, Fliri, A, Frobel, K, Gais, H J, Garratt, D G, Hayakawa, K, Heggie, W, Hesson, D P, Hoppe, D, Hoppe, I, Hyatt, J A, Ikesa, D, Jacobi, P A, Kim, K S, Kobuke, Y, Kojima, K, Krowicki, K, Lee, V J, Leutert, T, Malchenko, S, Martens, J, Matthews, R S, Ong, B S, Press, J B, Rajan Babu, T V, Rousseau, G, Sauter, H M, Suzuki, M, Tatsuta, K, Tolbert, L M, Truesdale, E A, Uchida, I, Ueda, Y, Uyehara, T, Vasella, A T, Vladuchick, W C, Wade, P A, Williams, R M, Wong, H N C, Asymmetric total synthesis of erythromycin. 2. Synthesis of an erythronolide A lactone system, *J. Am. Chem. Soc.*, 103, 3213–3215, 1981; (c) Woodward, R B, Logusch, E, Namibiar, K P, Sakan, K, Ward, D E, Au-Yeung, B W, Balaram, P, Browne, L J, Card, P J, Chen, C H, Chênevert, R B, Fliri, A, Frobel, K, Gais, H J, Garratt, D G, Hayakawa, K, Heggie, W, Hesson, D P, Hoppe, D, Hoppe, I, Hyatt, J A, Ikesa, D, Jacobi, P A, Kim, K S, Kobuke, Y, Kojima, K, Krowicki, K, Lee, V J, Leutert, T, Malchenko, S, Martens, J, Matthews, R S, Ong, B S, Press, J B, Rajan Babu, T V, Rousseau, G, Sauter, H M, Suzuki, M, Tatsuta, K, Tolbert, L M, Truesdale, E A, Uchida, I, Ueda, Y, Uyehara, T, Vasella, A T, Vladuchick, W C, Wade, P A, Williams, R M, Wong, H N C, Asymmetric total synthesis of erythromycin. 3. Total synthesis of erythromycin, *J. Am. Chem. Soc.*, 103, 3215–3217, 1981.
7. (a) Toshima, K, Mukaiyama, S, Yoshida, T, Tamai, T, Tatsuta, K, Application of efficient glycosylation of 2,6-anhydro-2-thio sugar to the total synthesis of erythromycin A, *Tetrahedron Lett.*, 32, 6155–6158, 1991; (b) Toshima, K, Nozaki, Y, Mukaiyama, S, Tamai, T, Nakata, M, Tatsuta, K, Kinoshita, M, Application of highly stereocontrolled glycosidations employing 2,6-anhydro-2-thio sugars to the syntheses of erythromycin A and olivomycin A trisaccharide, *J. Am. Chem. Soc.*, 117, 3717–3727, 1995.

8. (a) Tatsuta, K, Amemiya, Y, Kanemura, Y, Kinoshita, M, Total synthesis of tylonolide, an aglycone of tylosin, *Tetrahedron Lett.*, 22, 3997–4000, 1981; (b) Tatsuta, K, Amemiya, Y, Kanemura, Y, Takahashi, H, Kinoshita, M, Total synthesis of tylosin, *Tetrahedron Lett.*, 23, 3375–3378, 1982.
9. (a) Suzuki, K, Matsumoto, T, Tomooka, K, Matsumoto, K, Tsuchihashi, G, Stereocontrolled first total synthesis of mycinolide IV, *Chem. Lett.*, 113–116, 1987; (b) Matsumoto, T, Maeta, H, Suzuki, K, Tsuchihashi, G, First total synthesis of mycinamicin IV and VII. Successful application of new glycosidation reaction, *Tetrahedron Lett.*, 29, 3575–3578, 1988.
10. Nicolaou, K C, Dolle, R E, Papahatjis, D P, Randall, J L, Practical synthesis of oligosaccharides, Partial synthesis of avermectin B_{1a}. *J. Am. Chem. Soc.*, 106, 4189–4192, 1984.
11. (a) Hanessian, S, Ugolini, A, Dubê, D, Hodges, P J, André, C, Synthesis of (+)-avermectin B_{1a}, *J. Am. Chem. Soc.*, 108, 2776–2778, 1986; (b) Hanessian, S, Ugolini, A, Hodges, P J, Beaulieu, P, Dubê, D, André, C, Progress in natural product chemistry by the chiron and related approaches-synthesis of avermectin B_{1a}, *Pure Appl. Chem.*, 59, 299–316, 1987.
12. (a) Danishefsky, S J, Armistead, D M, Wincott, F E, Selnick, H G, Hungate, R, The total synthesis of the aglycon of avermectin A_{1a}, *J. Am. Chem. Soc.*, 109, 8117–8119, 1987; (b) Danishefsky, S J, Selnick, H G, Armistead, D M, Wincott, F E, The total synthesis of avermectin A_{1a}. New protocols for the synthesis of novel 2-deoxypyranose systems and their axial glycosides, *J. Am. Chem. Soc.*, 109, 8119–8120, 1987; (c) Danishefsky, S J, Armistead, D M, Wincott, F E, Selnick, H G, Hungate, R, The total synthesis of avermectin A_{1a}, *J. Am. Chem. Soc.*, 111, 2967–2980, 1989.
13. White, J D, Bolton, G L, Dantanarayana, A P, Fox, C M J, Hinter, R N, Jackson, R W, Sakuma, K, Warrier, U S, Total synthesis of the antiparastic agent avermectin B_{1a}, *J. Am. Chem. Soc.*, 117, 1908–1939, 1995.
14. (a) Ford, M J, Kinght, J G, Ley, S V, Vile, S, Total synthesis of avermectin B_{1a}: synthesis of the carbohydrate bis-oleandrose fragment and coupling to the avermectin B_{1a} aglycone, *Synlett*, 331–332, 1990; (b) Ley, S V, Armstrong, A, Díez-Martin, D, Ford, M J, Grice, P, Knight, J G, Kolb, H C, Madin, A, Marby, C A, Mukherjee, S, Shaw, A N, Slawin, A M Z, Vile, S, White, A D, Williams, D J, Woods, M, Total synthesis of the anthelmintic macrolide avermectin B_{1a}, *J. Chem. Soc. Perkin. Trans. 1*, 667–692, 1991.
15. (a) Dolle, R E, Nicolaou, K C, Carbohydrate-based synthesis of the goldinonolactone and the tetrahydrofuran fragment of aurodox and efrotomycin, *J. Chem. Soc. Chem. Commun.*, 1016–1985, 1985; (b) Dolle, R E, Nicolaou, K C, Total synthesis of elfamycins: aurodox and efrotomycin. 1. Strategy and construction of key intermediates, *J. Am. Chem. Soc.*, 107, 1691–1694, 1985; (c) Dolle, R E, Nicolaou, K C, Total synthesis of Elfamycins: aurodox and efrotomycin. 2. Coupling of key intermediates and completion of the synthesis, *J. Am. Chem. Soc.*, 107, 1695–1698, 1985.
16. (a) Nicolaou, K C, Daines, R A, Uenishi, J, Li, W S, Papahatjis, D P, Chakraborty, T K, Stereocontrolled construction of key building blocks for the total synthesis of amphoteronolide B and amphotericin B, *J. Am. Chem. Soc.*, 109, 2205–2208, 1987; (b) Nicolaou, K C, Daines, R A, and Chakraborty, T K, Total synthesis of anphoteronolide B, *J. Am. Chem. Soc.*, 109, 2208–2210, 1987; (c) Nicolaou, K C, Chakraborty, T K, Ogawa, Y, Daines, R A, Simpkins, N S, Furst, G T, Chemistry of amphotericin B. Degradation studies and preparation of amphoteronoloide B, *J. Am. Chem. Soc.*, 110, 4660–4672, 1988; (d) Nicolaou, C, Daines, R A, Uenishi, J, Li, W S, Papahatjis, D P, Chakraborty, T K, Total synthesis of amphoteronolide B and amphotericin B. 1. Strategy and stereocontrolled construction of key building blocks, *J. Am. Chem. Soc.*, 110, 4672–4685, 1988; (e) Nicolaou, K C, Daines, R A, Chakraborty, T K, Ogawa, Y, Total synthesis of amphoteronolide B and amphotericin B. 2. Total synthesis of amphoteronolide B, *J. Am. Chem. Soc.*, 110, 4685–4696, 1988; (f) Nicolaou, K C, Daines, R A, Ogawa, Y, Chakraborty, T K, Total synthesis of amphotericin B. 3. The final stages, *J. Am. Chem. Soc.*, 110, 4696–4705, 1988.
17. (a) Toshima, K, Tatsuta, K, Kinoshita, M, Total synthesis of elaiophylin (azalomycin B), *Tetrahedron Lett.*, 27, 4741–4744, 1986; (b) Toshima, K, Tatsuta, K, Kinoshita, M, Total synthesis of elaiophylin (azalomycin B), *Bull. Chem. Soc. Jpn.*, 61, 2369–2381, 1988.
18. Evans, D A, Kaldor, S W, Jones, T K, Clardy, J, Stout, T J, Total synthesis of the macrolide antibiotic cytovaricin, *J. Am. Chem. Soc.*, 112, 7001–7031, 1990.
19. (a) Nicolaou, K C, Groneberg, R D, Miyazaki, T, Stylianides, N A, Schulze, T J, Stahl, W, Total synthesis of the oligosaccharide fragment of calicheamicin λ_1^I, *J. Am. Chem. Soc.*, 112, 8193–8195, 1990; (b) Smith, A L, Hwang, C K, Pitsinos, E, Scarlato, G R, Nicolaou, K C, Enantioselective total

synthesis of (−)-calicheamicinone, *J. Am. Chem. Soc.*, 114, 3134–3136, 1992; (c) Nicolaou, K C, Hummel, C W, Pitsinos, E N, Nakada, M, Smith, A L, Shibayama, K, Saimoto, H, Total synthesis of calicheamicin λ_1^I, *J. Am. Chem. Soc.*, 114, 10082–10084, 1992; (d) Groneberg, R D, Miyazaki, T, Stylianides, N A, Schulze, T J, Stahl, W, Schrener, E P, Suzuki, T, Iwabuchi, Y, Smith, A L, Nicolaou, K C, Total synthesis of calicheamicin g_I^I. 1. Synthesis of the oligosaccharide fragment, *J. Am. Chem. Soc.*, 115, 7593–7611, 1993; (e) Smith, A L, Pitsinos, E N, Hwang, C K, Mizumo, Y, Saimoto, H, Scarlato, G R, Suzuki, T, Nicolaou, K C, Total synthesis of calicheamicin g_I^I. 2. Development of an enantioselective route to (−)-calicheamicinone, *J. Am. Chem. Soc.*, 115, 7612–7624, 1993; (f) Nicolaou, K C, Hummel, C W, Nakada, M, Shibayama, K, Pitsinos, E N, Saimoto, H, Mizumo, Y, Baldenius, K U, Smith, A L, Total synthesis of calicheamicin λ_1^I. 3. The final stages, *J. Am. Chem. Soc.*, 115, 7625–7635, 1993.

20. (a) Hitchcock, S A, Boyer, S H, Chu-Moyer, M Y, Olson, S H, Danishefsky, S J, A convergent total synthesis of calicheamicin λ_1^I, *Angew. Chem. Int. Ed.*, 33, 858–862, 1994; (b) Halcomb, R L, Boyer, S H, Wittman, M D, Olson, S H, Denhart, D J, Liu, K K C, Danishefsky, S J, Studies related to the carbohydrate sectors of esperamicin and calicheamicin: definition of the stability limits of the esperamicin domain and fashioning of a glycosyl donor from the calicheamicin domain, *J. Am. Chem. Soc.*, 117, 5720–5749, 1995; (c) Hitchcock, S A, Chu-Moyer, M Y, Boyer, S H, Olson, S H, Danishefsky, S J, A remarkable glycosidation reaction: the total synthesis of calicheamicin λ_1^I, *J. Am. Chem. Soc.*, 117, 5750–5756, 1995.
21. (a) Myers, A G, Hammond, M, Wu, Y, Xiang, J N, Harrington, P M, Elaine, E Y, Enantioselective synthesis of neocarzinostatin chromophore aglycon, *J. Am. Chem. Soc.*, 118, 10006–10007, 1996; (b) Myers, A G, Liang, J, Hannomd, M, Harrington, P M, Wu, Y, Kuo, E Y, Total synthesis of (+)-neocarzinostatin chromohore, *J. Am. Chem. Soc.*, 120, 5319–5320, 1998.
22. (a) Nicolaou, K C, van Delft, F, Ohshima, T, Vourloumis, D, Xu, J, Hosokawa, S, Pfefferkorn, J, Kim, S, Li, T, Total synthesis of eleutherobin, *Angew. Chem. Int. Ed.*, 36, 2520–2524, 1997; (b) Nicolaou, K C, Oshima, T, Hosokawa, S, van Delft, F L, Vourloumis, D, Xu, J Y, Pfefferkorn, J, Kim, S, Total synthesis of eleutherobin and eleuthosides A and B, *J. Am. Chem. Soc.*, 120, 8674–8680, 1998.
23. Chen, X T, Bhattacharya, S K, Zhou, B, Gutteridge, C E, Pettus, T R R, Danishefsky, S J, The total synthesis of eleutherobin, *J. Am. Chem. Soc.*, 121, 6563–6579, 1999.
24. (a) Roush, W R, Murphy, M, An improved synthesis of naphthoate precursors to olivine, *J. Org. Chem.*, 57, 6622–6629, 1992; (b) Roush, W R, Lin, X F, Studies on the synthesis of aureolic acid antibiotics: Highly stereoselective synthesis of aryl 2-deoxy-β-glycosides via the Mitsunobu reaction and synthesis of the olivomycin A-B disaccharide, *J. Am. Chem. Soc.*, 117, 2236–2250, 1995; (c) Roush, W R, Hartz, R A, Gustin, D J, Total synthesis of olivomycin A, *J. Am. Chem. Soc.*, 121, 1990–1991, 1999.
25. (a) Nicolaou, K C, Mitchell, H J, Suzuki, H, Rodrígues, R M, Baudoin, O, Fylaktakidou, K C, Total synthesis of everninomicin 13,384-1—part 1: synthesis of the $A_1B(A)C$ fragment, *Angew. Chem. Int. Ed.*, 38, 3334–3339, 1999; (b) Nicolaou, K C, Rodríguez, R M, Fylatakidou, K C, Suzuki, H, Mitchell, H J, Total synthesis of everninomicin 13,384-1—part 2: synthesis of the $FGHA_2$ fragment, *Angew. Chem. Int. Ed.*, 38, 3340–3345, 1999; (c) Nicolaou, K C, Mitchell, H J, Rodríguez, R M, Fylaktakidou, K C, Suzuki, H, Total synthesis of everninomicin 13,384-1—part 3: synthesis of the DE fragment and completion of the total synthesis, *Angew. Chem. Int. Ed.*, 38, 3345–3350, 1999.
26. (a) Fujiwara, K, Amano, S, Murai, A, Relative configuration of a marine toxin polycavernoside A, *Chem Lett*, 855–856, 1995; (b) Fujiwara, K, Murai, A, Yotsu-Yamashita, M, Yasumoto, T, Total synthesis of and absolute configuration of polycavernoside A, *J. Am. Chem. Soc.*, 120, 10770–10771, 1988.
27. (a) Johnston, J N, Paquette, L A, Studies directed toward the total synthesis of polycavernoside A, Enantioselective synthesis of the disaccharide component. *Tetrahedron Lett.*, 36, 4341–4344, 1995; (b) Paquette, L A, Barriault, L, Pissarnitski, D, A convergent total synthesis of the macrolactone disaccharide toxin (−)-polycavernoside A, *J. Am. Chem. Soc.*, 121, 4542–4543, 1999.
28. (a) Nicolaou, K C, Natarajan, S, Li, H, Jain, N F, Hughes, R, Michael, E, Ramanjulu, J M, Boddy, C N C, Christopher, N C, Takayanagi, M, Total synthesis of vancomycin aglycon—part 1: synthesis of amino acids 4–7 and construction of the AB-COD ring skeleton, *Angew. Chem. Int. Ed.*, 37, 2708–2714, 1998; (b) Nicolaou, K C, Jain, N F, Natarajan, S, Hughes, R, Solomon, M E, Li, H, Joshi, M, Takayanagi, M,

Koumbis, A E, Bando, T, Total synthesis of vancomycin aglycon—part 2: synthesis of amino acids 1-3 and construction of the AB-COD-DOE ring skeleton, *Angew. Chem. Int. Ed.*, 37, 2714–2716, 1998; (c) Nicolaou, K C, Takayanagi, M, Jain, N, Natarajan, S, Koumbis, A E, Bando, T, Ramanjulu, J M, Total synthesis of vancomycin aglycon—part 3: final stages, *Angew. Chem. Int. Ed.*, 37, 2717–2719, 1998; (d) Nicolaou, K C, Mitchell, H J, Jain, N F, Winssinger, N, Hughes, R, Bando, T, Total synthesis of vancomycin, *Angew. Chem. Int. Ed.*, 38, 240–244, 1999; (e) Nicolaou, K C, Li, H, Christopher, N C, Ramanjulu, J M, Yue, T Y, Natarajan, S, Chu, X J, Brase, S, and Rubsam, F, Total synthesis of vancomycin—part 1: design and development of methodology, *Chem. Eur. J.*, 5, 2584–2601, 1999; (f) Nicolaou, K C, Boddy, C N C, Li, H, Alexandros, E, Hughes, R, Natarajan, S, Jain, N F, Ramanjulu, J M, Brase, S, Solomon, M E, Total synthesis of vancomycin—part 2: retrosynthetic analysis, synthesis of amino acid building blocks and strategy evaluations, *Chem. Eur. J.*, 5, 2602–2621, 1999; (g) Nicolaou, K C, Koumbis, A E, Takayanagi, M, Natarajan, S, Jain, N F, Bando, T, Li, H, Hughes, R, Total synthesis of vancomycin—part 3: synthesis of the aglycon, *Chem. Eur. J.*, 5, 2622–2647, 1999; (h) Nicolaou, K C, Mitchell, H J, Nareshkumar, F, Bando, T, Hughes, R, Winssinger, N, Natarajan, S, Koumbis, A E, Total synthesis of vancomycin—part 4: attachment of the sugar moieties and completion of the synthesis, *Chem. Eur. J.*, 5, 2648–2667, 1999.

29. Thompson, C, Ge, M, Kahne, D, Synthesis of vancomycin from the aglycon, *J. Am. Chem. Soc.*, 121, 1237–1244, 1999.
30. (a) Nicolaou, K C, Li, Y, Fylaktakidou, K C, Mitchell, H J, Wei, H X, Weyershausen, B, Total synthesis of apoptolidin: part 1. Retrosynthetic analysis and construction of building blocks, *Angew. Chem. Int. Ed.*, 40, 3849–3854, 2001; (b) Nicolaou, K C, Li, Y, Fylaktakidou, K C, Mitchell, H J, Sugita, K, Total synthesis of apoptolidin: part 2. Coupling of key building blocks and completion of the synthesis, *Angew. Chem. Int. Ed.*, 40, 3854–3857, 2001.
31. Danishefsky, S J, Uang, B J, Quallich, G, Total synthesis of vineomycinone B_2 methyl ester, *J. Am. Chem. Soc.*, 107, 1285–1293, 1995.
32. Matsumoto, T, Katsuki, M, Jona, H, Suzuki, K, Convergent total synthesis of vineomycinone B_2 methyl ester and its C-12-epimer, *J. Am. Chem. Soc.*, 113, 6982–6992, 1991.
33. Tius, M A, Gomez-Galeno, J, Gu, X, Zaidi, J H, *C*-glycosylanthraquinone synthesis: total synthesis of vineomycinone B_2 methyl ester, *J. Am. Chem. Soc.*, 113, 5775–5783, 1991.
34. Tatsuta, K, Ozaki, H, Yamaguchi, M, Tanaka, M, Okui, T, Enantioselective total synthesis of medermycin (lactoquinomycin), *Tetrahedron Lett.*, 31, 5495–5498, 1990.
35. (a) Yamaguchi, M, Horiguchi, A, Ikeura, C, Minami, T, A synthesis of aryl *C*-glycosides via polyketides, *J. Chem. Soc. Chem. Commun.*, 434–436, 1992; (b) Yamaguchi, M, Okumura, T, Horiguchi, A, Ikeura, C, Minami, T, Total synthesis of (−)-urdamycinone B through polyketide condensation, *J. Org. Chem.*, 57, 1647–1649, 1992.
36. (a) Boyd, V A, Drake, B E, Sulikowski, G A, Preparation of 2-deoxy-β-*C*-arylglycosides and *C*-arylglycosides from carbohydrate lactones, *J. Org. Chem.*, 58, 3191–3193, 1993; (b) Boyd, V A, Sulikowski, G A, Total synthesis of the angucycline antibiotics urdamycinone B and 104-2 via a common synthetic intermediate, *J. Am. Chem. Soc.*, 117, 8472–8473, 1995.
37. (a) Matsuo, G, Miki, Y, Nakata, M, Matsumura, S, Toshima, K, Total synthesis of urdamycinone B via *C*-glycosidation of an unprotected sugar and Diels–Alder reaction of *C*-glycosyl juglone, *Chem. Commun.*, 225–226, 1996; (b) Matsuo, G, Matsumura, S, Toshima, K, Two-step synthesis of *C*-glycosyl juglones from unprotected sugars: a novel approach to angucycline antibiotics, *Chem. Commun.*, 2173–2174, 1996; (c) Matsuo, G, Miki, Y, Nakata, M, Matsumura, S, Toshima, K, Total synthesis of *C*-glycosylangucycline, urdamycinone B, using an unprotected sugar, *J. Org. Chem.*, 64, 7101–7106, 1999.
38. (a) Matsumoto, T, Hosoya, T, Suzuki, K, Total synthesis and absolute stereochemical assignment of gilvocarcin M, *J. Am. Chem. Soc.*, 114, 3568–3570, 1992; (b) Hosoya, T, Takashiro, E, Matsumoto, T, Suzuki, K, *J. Am. Chem. Soc.*, 116, 1004–1015, 1994.
39. Yu, B, Xie, J, Deng, S, Hui, Y, First synthesis of a bidesmosidic triterpene saponin by a highly efficient procedure, *J. Am. Chem. Soc.*, 121, 12196–12197, 1999.
40. McDonald, F E, Reddy, K S, Convergent synthesis of digitoxin: stereoselective synthesis and glycosylation of the digoxin trisaccharide glycal, *Angew. Chem. Int. Ed.*, 40, 3653–3655, 2001.

13 Total Asymmetric Synthesis of Monosaccharides and Analogs

Pierre Vogel

CONTENTS

13.1 INTRODUCTION

The total synthesis of carbohydrates and analogs has kept chemists occupied since 1861 when Butlerow [1] discovered the *formose reaction*, which generates mixtures of racemic aldoses and ketoses by oligomerization of formaldehyde in the presence of $Ca(OH)_2$. Currently, with the advent of highly stereoselective and enantioselective methods, almost any natural or non-natural carbohydrate can be readily obtained from inexpensive starting materials in an enantiomerically pure form. D-Glucose, D-mannose, D-glucosamine, D- and L-arabinose from natural sources, are certainly cheaper than those from total synthesis. However, when dealing with unnatural enantiomers of common carbohydrates, or with unusual derivatives in which hydroxy groups are replaced by other functionalities (amino-, alkoxy-, thio-, halogeno-, carbon-substituents, etc.), total synthesis from non-carbohydrate precursors may be easier and advantageous. By total synthesis, carbohydrates can be directly prepared in suitably protected forms. In contrast, syntheses starting from natural sugars sometimes require several delicate chemical operations.

This review describes the most important synthetic approaches for the preparation of monosaccharides and their derivatives developed during the last 20 years. It concentrates on techniques for generating enantiomerically enriched or pure carbohydrates and analogs. For earlier work, the reader is referred to available reviews [2], including a recent one by this author [3]. In the latter, the material was presented by classifying the targeted sugars, starting from three-carbon systems and moving toward larger systems, rather than classifying according to the synthetic principles [3]. Here, the material is organised according to the synthetic concepts applied.

Aldoses, alditols and their derivatives are considered, including aza- and thiosugars (with nitrogen and sulfur in the pyranose or furanose rings).

13.2 THE FORMOSE REACTION

The formose reaction was developed by Loew [4] and Fischer [5] who isolated *rac*-fructose osazone from the reaction mixture. The reaction exhibits an induction period during which small amounts of glycolaldehyde, glyceraldehyde, and dihydroxyacetone are formed. These are believed to act as catalytic species in the subsequent steps via complexation with calcium ions. The yield of formose sugars reaches a maximum at the so-called yellowing point [6]. On further reaction, branched sugars are formed involving aldol condensations, followed by cross-Cannizarro reactions [7]. Depending on the nature of the base and additives used to induce the formaldehyde oligomerization, various proportions of trioses, tetroses, pentoses, hexoses, and long-chain aldoses and ketoses are obtained [8]. The addition of glycolaldehyde or a higher aldose to the reaction mixture considerably reduces the induction period for the oligomerization. Umpolung catalysts of the thiamin-type also reduce the induction period [9]. When carried out in *N,N*-dimethylformamide (DMF), considerable control in the product distribution of the formose reaction is possible by adjustment of the water content (Scheme 13.1). When, for instance, formaldehyde is heated to 75°C for 1 h with Et_3N and thiamin hydrochloride in 8:1 DMF/H_2O, D,L-2-*C*-hydroxymethyl-3-pentulose, characterized as tetraacetate **1**, is produced in 28% yield [10].

Eschenmoser and coworkers [11] studied the aldomerization of glycoaldehyde phosphate leading to mixtures containing mostly racemates of the two diastereomeric tetrose-2,4-diphosphates and eight hexose-2,4,6-triphosphates (Scheme 13.2, Route A). At 20°C, in the absence of air, a 0.08 M solution of glycolaldehyde phosphate **2** in 2 M NaOH gave 80% yield of a 1:10 mixture of tetrose **3** and hexose **4** derivatives with DL-allose-2,4,6-triphosphate comprising up to 50% of the mixture of sugar phosphates.

In the presence of formaldehyde (0.5 mol equiv.), sugar phosphates were formed in up to 45% yield, with pentose-2,4-diphosphates dominating over hexose triphosphates by a ratio of 3:1 (Scheme 13.2, Route B). The major component was found to be D,L-ribose-2,4-diphosphate with the ratios of ribose-, arabinose-, lyxose-, and xylose-2,4-diphosphates being 52:14:23:11, respectively. The aldomerization of **2** in the presence of H_2CO is a variant of the formose reaction. It avoids the formation of complex product mixtures as a consequence of the fact that aldoses, which are phosphorylated at the C(2) position, cannot undergo aldose–ketose tautomerization. The preference for ribose-2,4-diphosphate **5** and allose-2,4,6-triphosphate formation might be relevant to a discussion of the origin of ribonucleic acids.

The "classical" formose reaction gives a very large number of carbohydrates including both straight- and branched-chain isomers [8]. Straight-chain carbohydrates (Scheme 13.3) such as trioses, tetroses, pentoses, and hexoses are readily obtained in good yield by a reaction of formaldehyde with syngas in the presence of $RhCl(CO)(PPh_3)_2$ and tertiary amines [12].

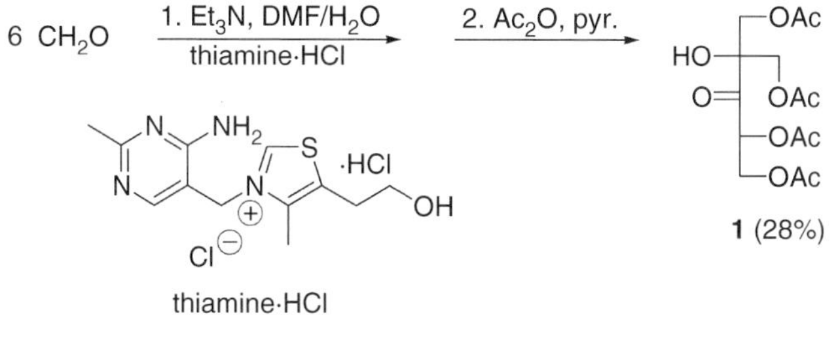

SCHEME 13.1 Examples of selective formose reactions.

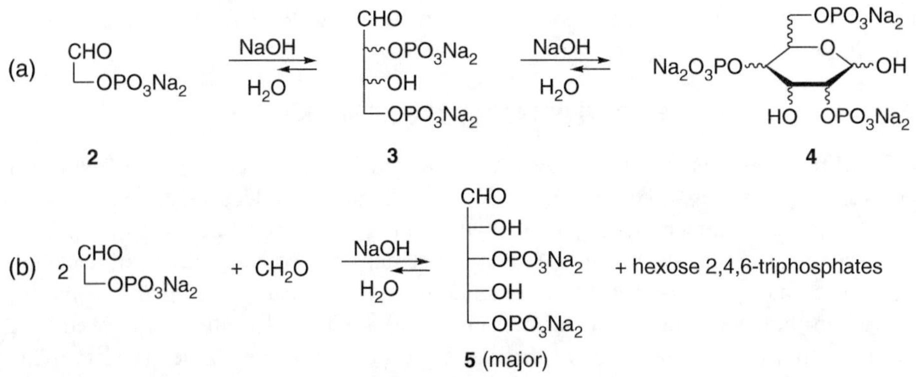

(a) **2** CHO–OPO_3Na_2 $\underset{H_2O}{\overset{NaOH}{\rightleftharpoons}}$ **3** CHO, OPO_3Na_2, OH, OPO_3Na_2 $\underset{H_2O}{\overset{NaOH}{\rightleftharpoons}}$ **4** Na_2O_3PO, OPO_3Na_2, O, OH, HO, OPO_3Na_2

(b) 2 CHO–OPO_3Na_2 + CH_2O $\underset{H_2O}{\overset{NaOH}{\rightleftharpoons}}$ CHO, OH, OPO_3Na_2, OH, OPO_3Na_2 **5** (major) + hexose 2,4,6-triphosphates

SCHEME 13.2 Selective condensations of glycolaldehyde phosphate alone (a) or in the presence of formaldehyde (b).

$$CH_2O + CO + H_2 \xrightarrow{cat.} HOCH_2CHO$$

$$2\ CH_2O + CO + H_2 \xrightarrow{''} C_3H_6O_3$$

$$2\ CH_2O + 2\ CO + H_2 \xrightarrow{''} C_4H_8O_4$$

$$3\ CH_2O + 2\ CO + 2\ H_2 \xrightarrow{''} C_5H_{10}O_5$$

$$3\ CH_2O + 3\ CO + 3\ H_2 \xrightarrow{''} C_6H_{12}O_6$$

cat: $Rh(CO)(Ph_3P)_2Cl$ and tertiary amines

SCHEME 13.3 Rh(I)-catalyzed condensations of formaldehyde with syngas giving linear carbohydrates.

13.3 PREBIOTIC SYNTHESIS OF CARBOHYDRATES

The formation of Earth from a diffuse cloud of cosmic gas and dust occurred some 4.6×10^9 years ago. It is proposed that ca. 4.0×10^9 years ago, bodies of water were formed and organic chemistry became established. The oldest known fossils date back ca. 3.6×10^9 years and resemble modern blue green algae. Biogenesis from organic chemistry to a primitive cell must therefore have occurred within a period of ca. 0.4×10^9 years. It is accepted that there was no free oxygen until the advent of photosynthetic bacteria ca. 2.7×10^9 years ago. Under these (reductive) conditions, energy required for chemical synthesis would have been available from the sun in the form of ultraviolet radiation, now blocked by the ozone layer. Water, ammonia, HCN, acetonitrile, acrylonitrile, cyanogen, cyanoacetylene, and formaldehyde are believed to have been the building blocks for nature. Laboratory experiments have shown that HCN is formed in good yield from gaseous mixtures of N_2, H_2 and NH_3 in spark discharge experiments by the action of ultraviolet radiation on mixtures of CH_4 and NH_3, gases abundant in outer space. A spark discharge passed through CH_4 and N_2, or through HCN-produced cyanoacetylene and cyanogen, respectively. Similar experiments have demonstrated the formation of formaldehyde [13]. Shevlin and coworkers [14] have reported that cocondensation of carbon with H_2O and NH_3 at 77 K generates amino acids. They also demonstrated that atomic carbon generated by vaporizing in an arc under high pressure reacts with water at 77 K to form low yields of straight-chain aldoses with up to five carbon centers. A mechanism (Scheme 13.4) involving hydroxymethylene species was supported by deuterium labeling studies [15].

There is a debate over whether the "classical" formose reaction [3–5] might have played a role in the prebiotic synthesis of carbohydrates. When a slurry of carbonate-apatite is boiled with 0.5 M formaldehyde at pH 8.5 a yield lower then 40% in sugars is reached after a few hours.

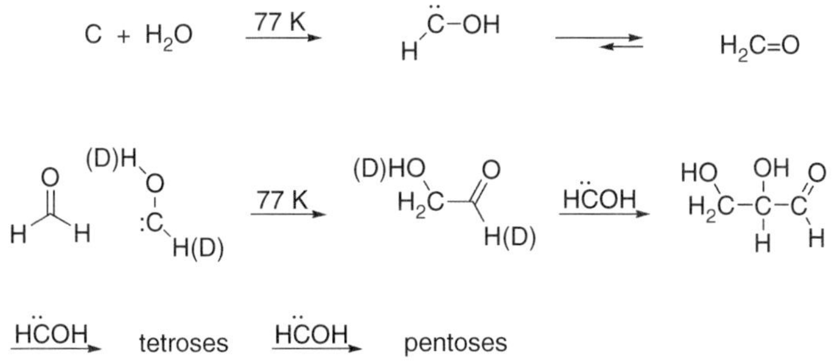

SCHEME 13.4 Reaction of carbon atoms with water. Formation of aldoses.

Prolonged heating decomposes the carbohydrates. Sugars have been detected from 0.01 M formaldehyde but not from a 0.001 M solution. Thus, it appears that the "classical" formose model for prebiotic accumulation of sugars is not plausible because it requires concentrated solutions of formaldehyde, and the sugars formed are rapidly decomposed [16]. Iron(III)hydroxide oxide [Fe(OH)O] has been shown to catalyze the condensation of 25 mM D,L-glyceraldehyde to ketohexoses at 15°C (pH 5 to 6). After 16 days, a mixture comprised of 15.2% sorbose, 13.9% fructose, 6.1% psicose, 5.6% tagatose, and 2.5% dendroketose is obtained. After 96 days at 15°C, this mixture was not decomposed. [Fe(OH)O] also catalyses the isomerization of glyceraldehyde into dihydroxyacetone and of dihydroxyacetone into lactic acid [17] (Scheme 13.5).

The "classical formose" conditions are not capable of producing large amounts of ribose (for RNA synthesis) or any other individual sugar. In contrast, the reduced sugar pentaerythritol is formed with great selectivity by ultraviolet irradiation of 0.1 M formaldehyde. This compound may have played an important role in prebiotic chemistry [18]. The seminal work of Eschenmoser and coworkers [11] (Scheme 13.2) suggests that the "initial RNA world" might have involved glycolaldehyde phosphate [19]. In order to explain the concentration process required, one can envisage that double-layer hydroxide minerals might have played a decisive role; in particular, those incorporating sodium sulfite, which can absorb formaldehyde, glycoaldehyde, and glyceraldehyde by adduct formation with the immobilized sulfite anions. This translates into

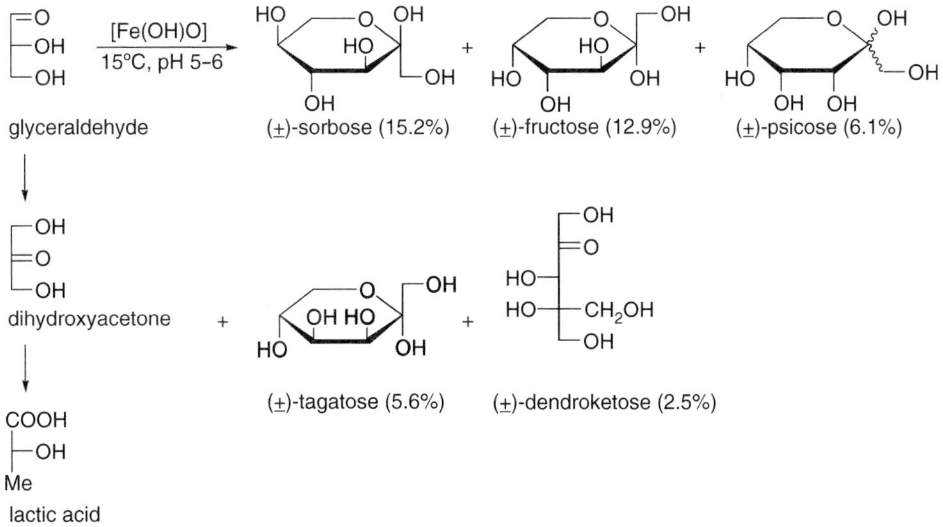

SCHEME 13.5 [Fe(OH)O]-catalyzed reactions of D,L-glyceraldehyde.

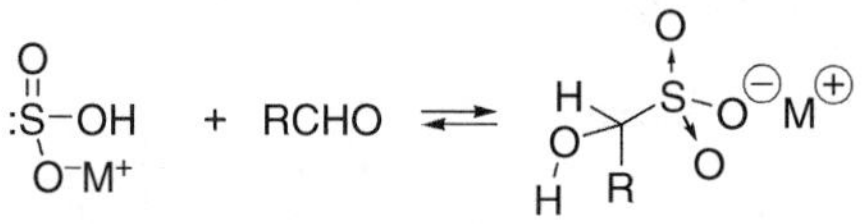

SCHEME 13.6 Sulfite anion and aldehyde adduct formation. A possible concentration process in double-layer hydroxide minerals such as $Mg_2Al(OH)_6^+[SO_3H^-(H_2O)_2]$.

observable uptake at concentrations ≥ 50 mM [20] (Scheme 13.6). Thus, sugars have been proposed as the optimal biosynthetic carbon substrates of aqueous life throughout the universe [21].

13.4 ALDOLASE-CATALYZED ASYMMETRIC ALDOL CONDENSATIONS

The enzymatic aldol reaction represents a useful method for the synthesis of various sugars and sugar-like structures. More than 20 different aldolases have been isolated (see Table 13.1 for examples) and several of these have been cloned and overexpressed. They catalyze the stereospecific aldol condensation of an aldehyde with a ketone donor. Two types of aldolases are known. Type I aldolases, found primarily in animals and higher plants, do not require any cofactor. The x-ray structure of rabbit muscle aldolase (RAMA) indicates that Lys-229 is responsible for Schiff-base formation with dihydroxyacetone phosphate (DHAP) (Scheme 13.7a). Type II aldolases, found primarily in micro-organisms, use Zn^{++} as a cofactor, which acts as a Lewis acid enhancing the electrophilicity of the ketone (Scheme 13.7b). In both cases, the aldolases accept a variety of natural (Table 13.1) and non-natural acceptor substrates (Scheme 13.8).

13.4.1 Resolution of Racemic Aldehydes

Fructose-1,6-diphosphate (FDP) aldolase catalyzes the reversible aldol addition of DHAP and D-glyceraldehyde-3-phosphate (G3P) to form D-fructose-1,6-diphosphate (FDP), for which $K_{eq} \approx 10^4\ M^{-1}$ in favor of FDP formation (Scheme 13.9). RAMA accepts a wide range of aldehyde acceptor substrates with DHAP as the donor to stereospecifically generate 3*S*,4*S* vicinal diols (Scheme 13.8). The diastereoselectivity exhibited by FDP aldolase depends on the reaction conditions. Racemic mixtures of non-natural aldehyde acceptors can be partially resolved only under conditions of kinetic control. When six-membered hemiacetals can be formed, racemic mixtures of aldehydes can be resolved under conditions of thermodynamic control (Scheme 13.10).

D,L-Glyceraldehyde and 1,3-dihydroxyacetone are obtained from glycerol via mild oxidation with, for example, hydrogen peroxide in the presence of catalytic ferrous salts [23]. The selective formation of trioses has been observed in the formose reaction when α-ketols bearing electron-withdrawing substituents were added to the reaction mixture [24]. In the presence of thiazolium salts, selective conversion of formaldehyde into 1,3-dihydroxyacetone has been reported [25]. Hydration of halopropargyl alcohol, followed by hydrolysis, gives 1,3-dihydroxyacetone [26]. DHAP can be generated by three different procedures: (1) *in situ* from fructose 1,6-diphosphate with the enzyme triosephosphate isomerase; (2) from the dimer of dihydroxyacetone by chemical phosphorylation with $POCl_3$ (Scheme 13.3); or (3) from dihydroxyacetone by enzymatic phosphorylation using ATP and glycerol kinase, with *in situ* generation of ATP using phosphoenol pyruvate (PEP) or acetyl phosphate as the phosphate donor [23] (Scheme 13.11).

13.4.2 One-Pot Total Syntheses of Carbohydrates

A one-pot procedure has been proposed for the conversion of dihydroxyacetone and PEP into D-tagatose-1,6-diphosphate **6** (Scheme 13.12). The reaction mixture contains glycerolkinase, pyruvate kinase, triose phosphate isomerase, and a D-tagatose 1,6-diphosphate aldolase [27].

TABLE 13.1
Examples of Enzymes Catalyzing the Equilibria of Natural Products with Various Aldol Donors and Various Aldehydes (the Wavy Line Indicating the C–C Bond Involved in the Reversible Aldol Reaction)

Aldol donor (nucleophiles)

DHAP	$^{\ominus}$OOC–C(OP)=CH$_2$	
FDP aldolase	DAHP synthetase	3-deoxy-2-oxo-6-P-gluconate aldolase
fuculose-1-P-aldolase	KDO synthetase	KDO aldolase
	sialic acid synthetase	sialic acid aldolase
tagatose-1,6-P$_2$ aldolase	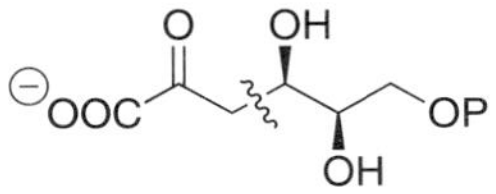	3-deoxy-2-oxo-6-P-galactonate aldolase

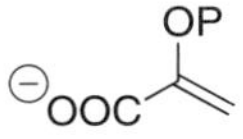

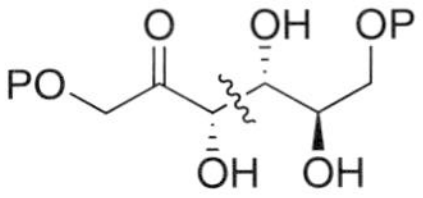

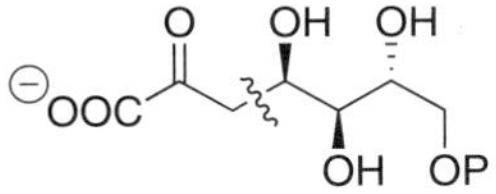

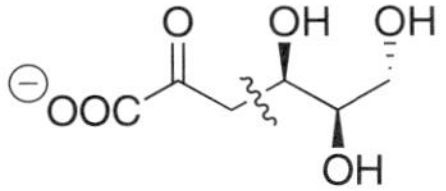

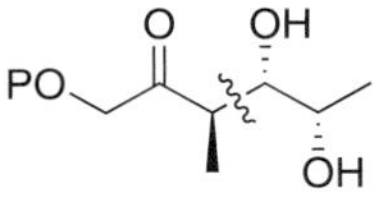

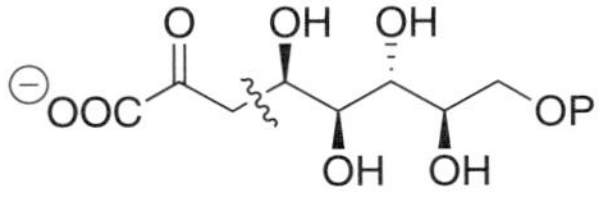

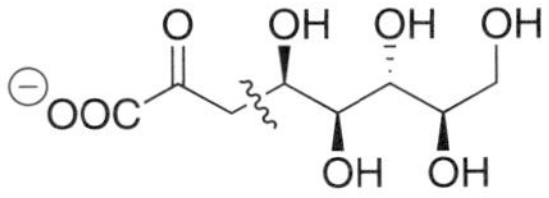

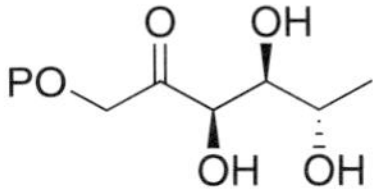

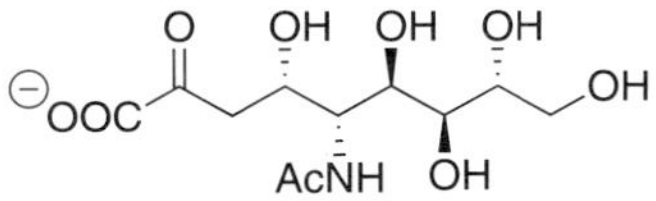

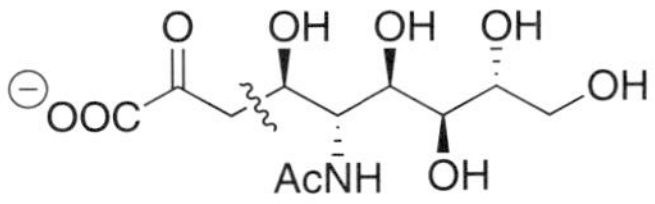

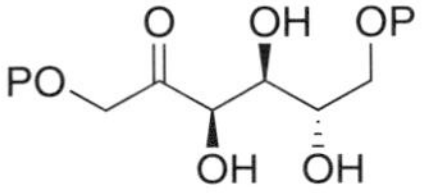

Continued

TABLE 13.1
Continued

Aldol donor:

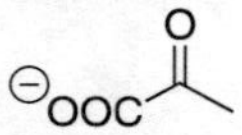

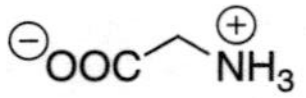

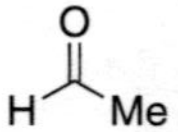

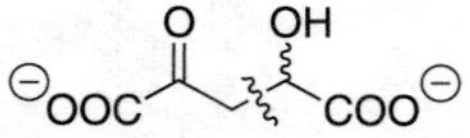

Pyruvate		Acetaldehyde	Glycine
4-hydroxy-2-oxo-glutarate aldolase	3-deoxy-2-oxo-L-arabinoate aldolase	2-deoxyribose-5-P-aldolase	D-Thr aldolase
4-hydroxy-4-methyl-2-oxoglutarate aldolase	3-deoxy-2-oxo-D-pentanoate aldolase	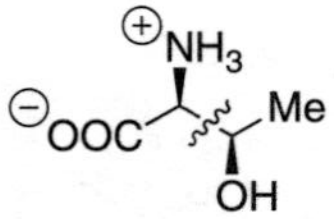	L-Thr aldolase
3-deoxy-2-oxo-D-glucarate aldolase	hydroxybutyrate aldolase	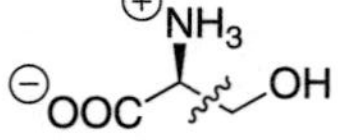	Ser-hydroxymethyl transferase

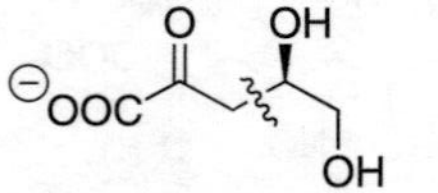

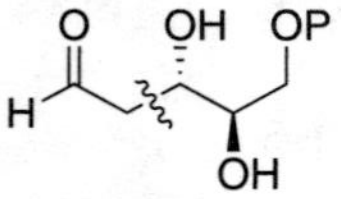

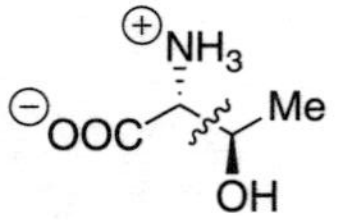

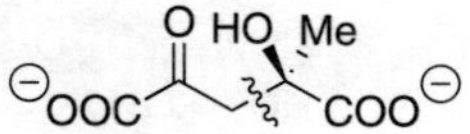

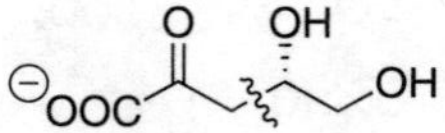

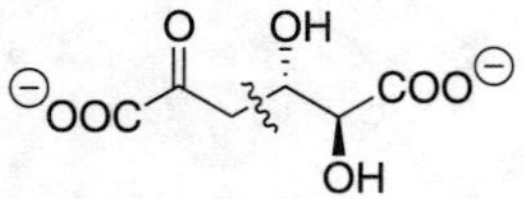

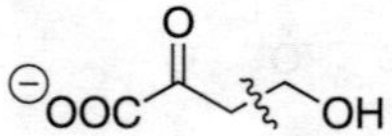

FDP = fructose-1,6-diphosphate; DHAP = dihydroxyacetone phosphate; KDO = 3-deoxy-D-*manno*-octulosonate; P = $^{-}O_3P$.

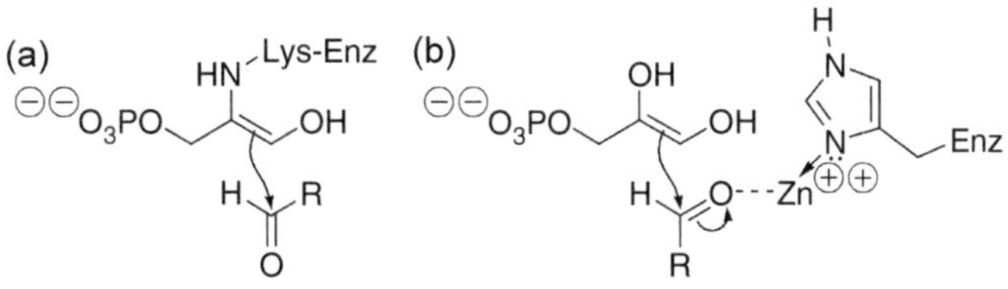

SCHEME 13.7 (a) Type I aldolases form enamine nucleophiles (donor); (b) type II aldolases use Zn^{2+} as a cofactor activating the aldehyde (acceptor).

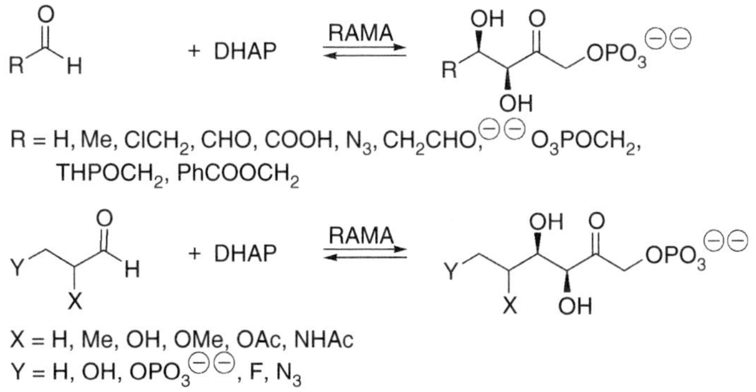

R = H, Me, $ClCH_2$, CHO, COOH, N_3, CH_2CHO, $^{\ominus\ominus}O_3POCH_2$, $THPOCH_2$, $PhCOOCH_2$

X = H, Me, OH, OMe, OAc, NHAc
Y = H, OH, $OPO_3^{\ominus\ominus}$, F, N_3

SCHEME 13.8 Examples of RAMA-catalyzed aldol condensations.

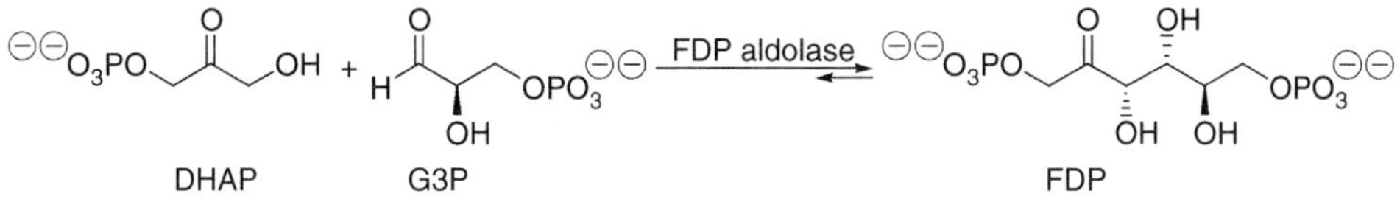

SCHEME 13.9 Stereospecific FDPaldolase-catalyzed aldol reaction of DHAP + G3P $\rightleftarrows$ FDP.

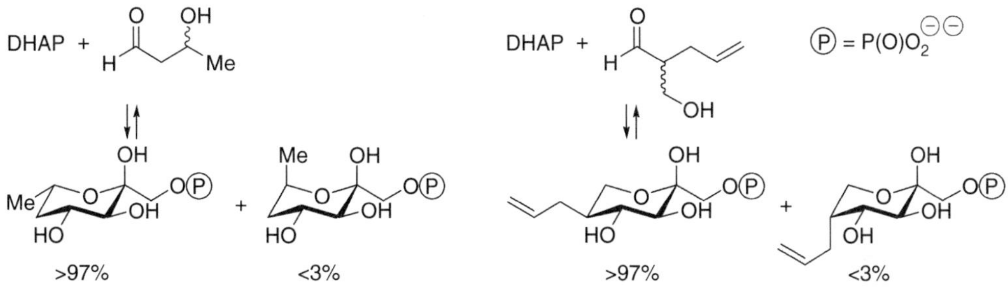

SCHEME 13.10 Thermodynamically controlled resolution of racemic aldehydes with FDP aldolase from RAMA.

An efficient asymmetric total synthesis of L-fructose combines the Sharpless asymmetric dihydroxylation with an enzyme-catalyzed aldol reaction. L-Glyceraldehyde, prepared from acrolein, is condensed to DHAP in a buffered water suspension of lysed cells of K12 *Escherichia coli* containing an excess of L-rhamnulose-1-phosphate (Rha) aldolase (*E. coli* raised on L-rhamnose as sole carbon source). The L-fructose phosphate obtained is hydrolyzed to L-fructose with acid phosphatase. Similarly, the RAMA-catalyzed condensation of D-glyceraldehyde with DHAP,

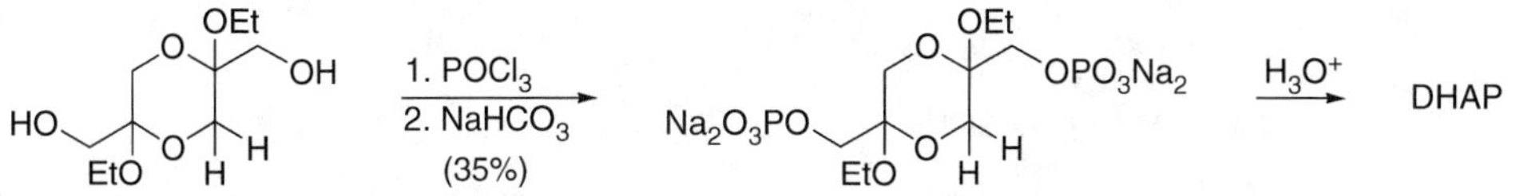

SCHEME 13.11 Chemical synthesis of dihydroxyacetone phosphate (DHAP).

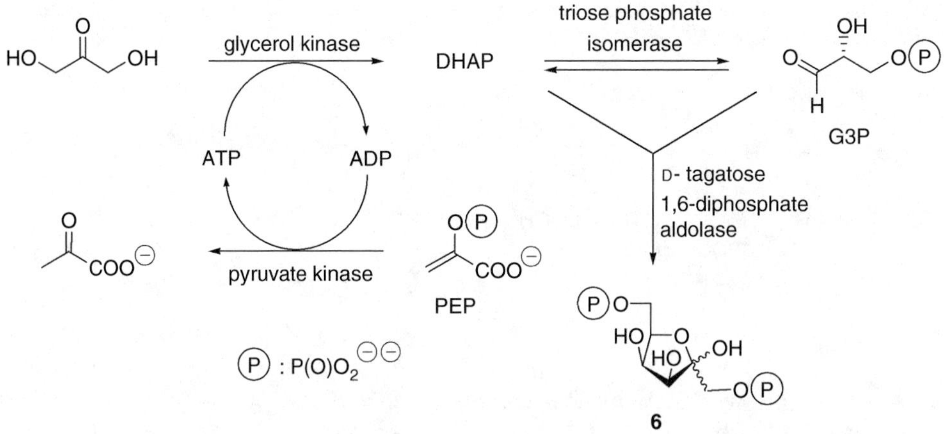

SCHEME 13.12 One-pot synthesis of D-tagatose 1,6-diphosphate.

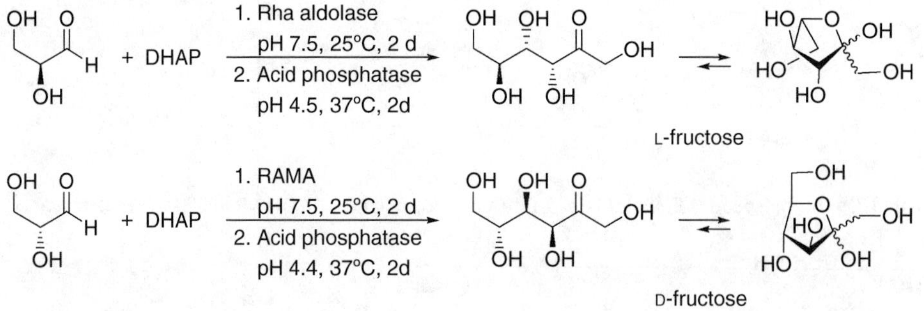

SCHEME 13.13 Syntheses of L- and D-fructose.

followed by acid phosphatase-catalyzed hydrolysis, furnishes D-fructose (Scheme 13.13). The method may also be used to generate 6-deoxy-D- and L-galacto-2-heptulose from (*E*)-crotonaldehyde, and 6-phenyl-D- and L-galacto-2-hexulose from (*E*)-cinnamaldehyde [8].

The isomerization of L-fructose catalyzed by fucose isomerase (available from commercial recombinant *E. coli* strains) furnishes L-glucose [29] (Scheme 13.14).

13.4.3 Synthesis of 1,5-Dideoxy-1,5-Iminoalditols

Two potent glycosidase inhibitors, (−)-1-deoxymannonojirimycin (−)-**7** and (+)-1-deoxynojirimycin (+)-**8**, are readily obtained in three steps utilizing RAMA as a catalyst in the key C–C bond forming step [22,30]. From racemic 3-azido-2-hydroxypropanal and dihydroxyacetone monophosphate (DHAP), diastereomeric 6-azidoketones are formed. Following the acid phosphatase-catalyzed removal of phosphate and subsequent reductive amination (Scheme 13.15), the products are isolated in a 4:1 ratio favoring the *manno* derivative. A similar result is obtained with

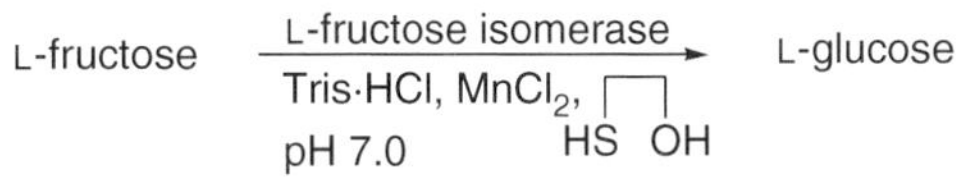

L-fructose —(L-fructose isomerase; Tris·HCl, $MnCl_2$, HS OH; pH 7.0)→ L-glucose

SCHEME 13.14 Isomerization of L-fructose into L-glucose.

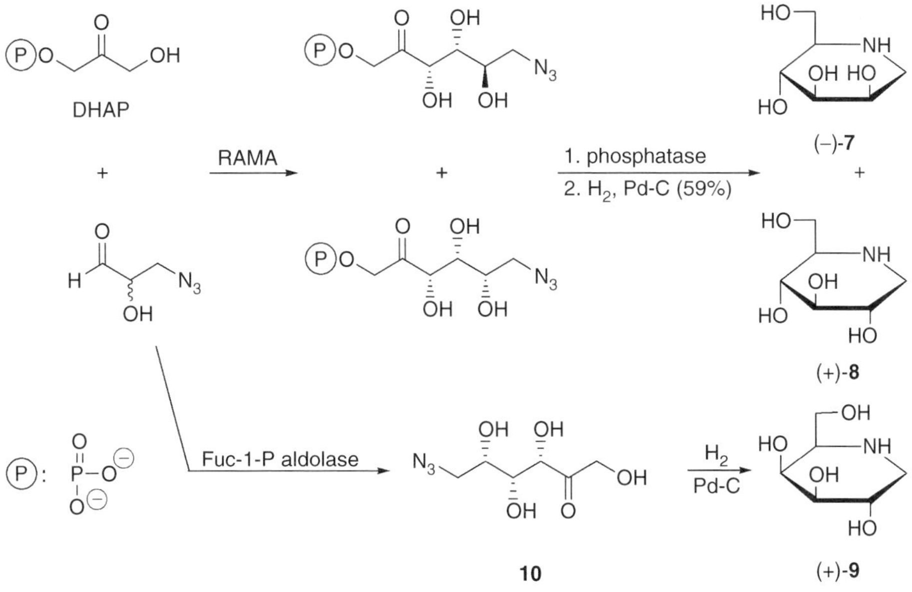

SCHEME 13.15 Chemoenzymatic synthesis of 1,5-dideoxy-1,5-imino-alditols.

fructose-1,6-diphosphate (FDP) aldolase from *E. coli* [31]. Exclusive formation of (−)-**7** and (+)-**8** is observed if the respective enantiomerically pure azidoaldehydes are used as starting materials. An analogous RAMA-catalyzed aldol reaction/reductive amination procedure has been used in the total synthesis of 2-acetylamino-1,2,5-trideoxy-1,5-imino-D-glucitol and 2-acetylamino-1,2,5-trideoxy-1,5-imino-D-mannitol from (*S*)- and (*R*)-3-azido-2-acetamidopropanal, respectively [32]. The 6-deoxy analogs of the 1,5-dideoxy-1,5-iminohexitols were obtained by direct reductive amination of the aldol products prior to removal of the phosphate group [22]. Fuculose-1-phosphate (Fuc-1-P) aldolase catalyzed the aldol reaction between DHAP and (±)-3-azido-2-hydroxy-propanal leading to the ketose-1-phosphate **10**, which contained the L-enantiomer of the 2-hydroxypropanal derivative (Scheme 13.15). Reduction of the azide generated an amine, which cyclized to an imine that was hydrogenated with high diastereoselectivity, providing (+)-1-deoxygalactostatine (+)-**9** [22].

13.4.4 Synthesis of 2,5-Dideoxy-2,5-Iminoalditols

When 2-azidoaldehydes are used as substrates in the RAMA-catalyzed aldol reaction with dihydroxyacetone phosphate (DHAP), the azidoketones thus obtained can be reduced into the corresponding primary amines. Subsequent equilibration to imine intermediates, followed by reduction, generates the corresponding pyrrolidines (Scheme 13.16) [22,33]. 1,4-Dideoxy-1,4-imino-D-arabinitol **11** was prepared from azidoacetaldehyde. Both (2*R*,5*R*)- and (2*S*,5*R*)-bis(hydroxymethyl)-(3*R*,4*R*)-dihydroxypyrrolidine (**12** and **13**) were derived from racemic 2-azido-3-hydroxypropanol. The aldol product resulting from kinetic control was converted into the (2*R*,2*R*) derivative **12**, whereas the product resulting from thermodynamic control gave the

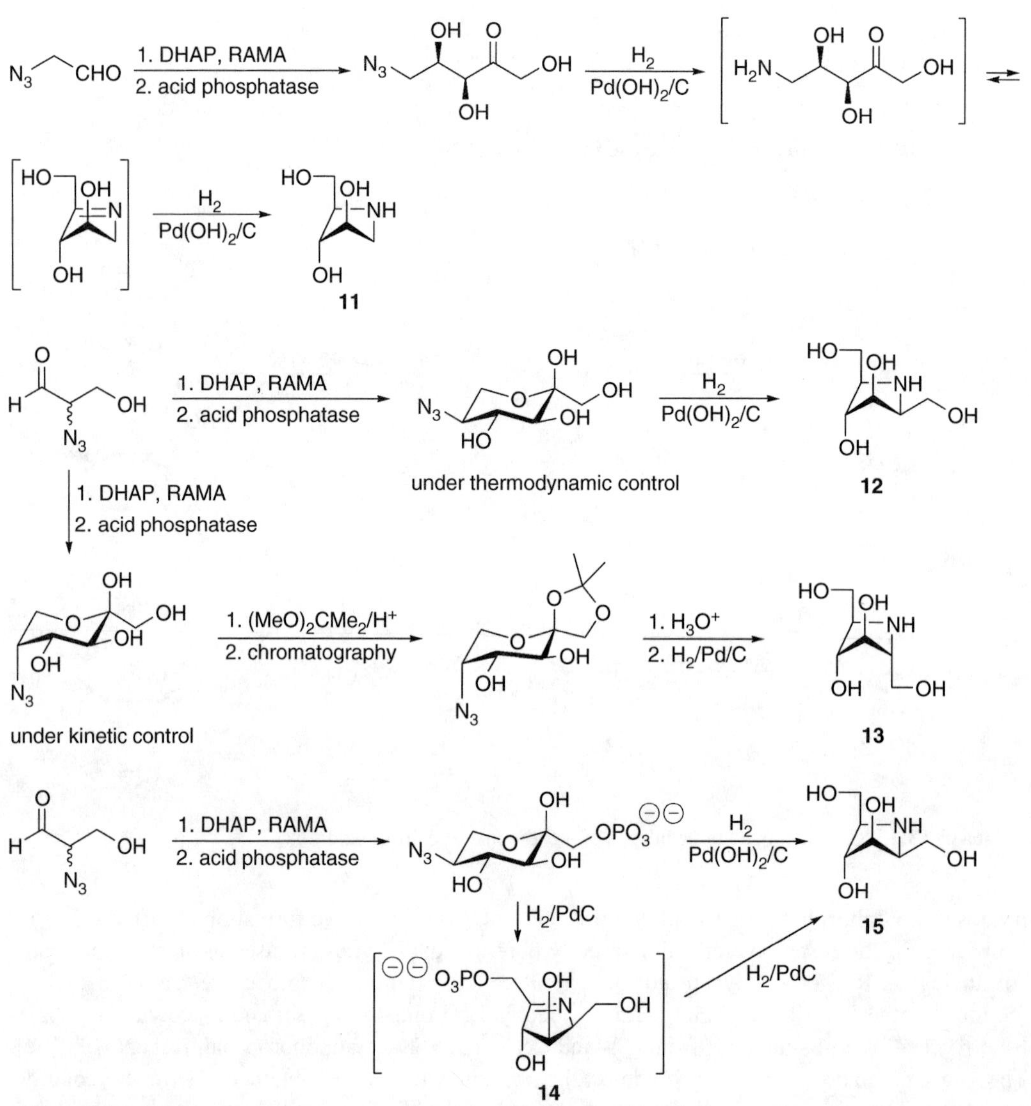

SCHEME 13.16 Examples of chemoenzymatic syntheses of 2,5-dideoxy-2,5-iminoalditols based on RAMA-catalyzed aldol reactions.

(2*S*,5*R*)-stereoisomer **13** [33]. Similar transformations with 3-acetamido-2-azidopropanal gave azasugars that were structurally related to *N*-acetylglucosamine [34]. The Pd-catalyzed reductive aminations of the azidoketones are stereoselective. 6-Deoxy-azasugars and their analogs can also be prepared by direct reductive amination of the aldol products prior to removal of the phosphate group. The reaction is thought to involve an imine 6-phosphate intermediate **14**, as exemplified by the synthesis of **15** (Scheme 13.16).

One of the most efficient methods for the generation of 2,5-dideoxy-2,5-iminogalactitol **16** relies on the fuculose-1-phosphate aldolase-catalyzed aldol condensation of 2-azido-3-hydroxypropanal with dihydroxyacetone monophosphate (Scheme 13.17). The same method, applied to (2*R*)-2-azidopropanal (*R*)-**17** and to (2*S*)-2-azido-propanal (*S*)-**17**, allows for the preparation of 2,5,6-trideoxy-2,5-imino-D-allitol **18** and 2,5,6-trideoxy-2,5-imino-L-talitol **19**, respectively [22].

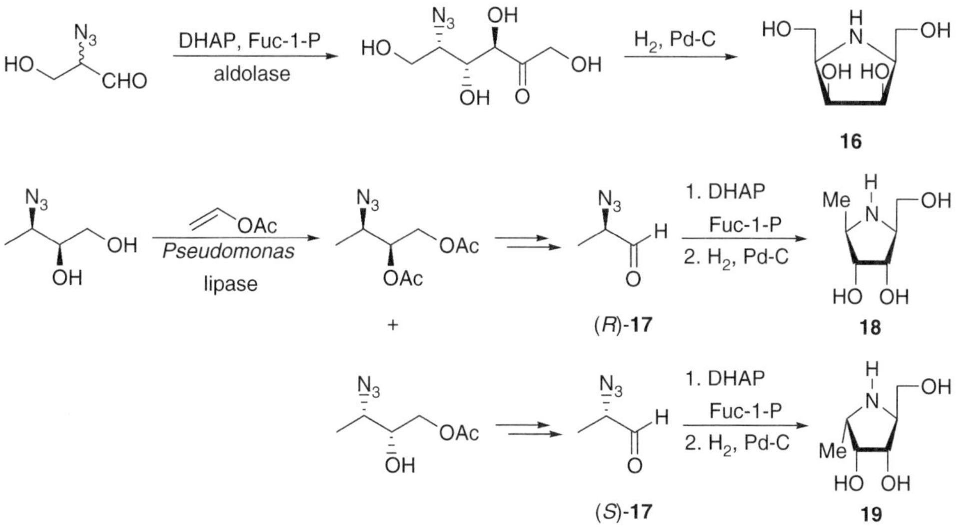

SCHEME 13.17 Examples of chemoenzymatic syntheses of 2,5-dideoxy-2,5-iminoalditols based on fuculose-1-phosphate aldolase-catalyzed aldol reactions.

13.4.5 Synthesis of Deoxy-Thiohexoses

The aldolase-catalyzed aldol reaction illustrated in Scheme 13.18 has been very successful [35]. The required (*R*)-3-thioglyceraldehyde (**20**) is obtained from the regioselective epoxide ring opening of (*S*)-glycidaldehyde diethyl acetal with thioacetic acid and its potassium salt. Condensation of the thioaldehyde **20** with DHAP, catalyzed by rabbit muscle fructose-1,6-diphosphate aldolase, followed by removal of the phosphate group using acid phosphatase, yields thio-L-sorbose **21**. Acetylation of **21** generates the tetraacetate **22**, which is subsequently reduced under ionic conditions to the peracetate of 1-deoxy-5-thio-D-glucopyranose **23**. Applying similar techniques, 1-deoxy-5-thio-D-galactose, 1-deoxy-5-thio-L-altrose, 1-deoxy-5-thio-D-mannose, 1-deoxy-5-thio-L-mannose and 2-deoxy-5-thio-D-ribose were prepared [35].

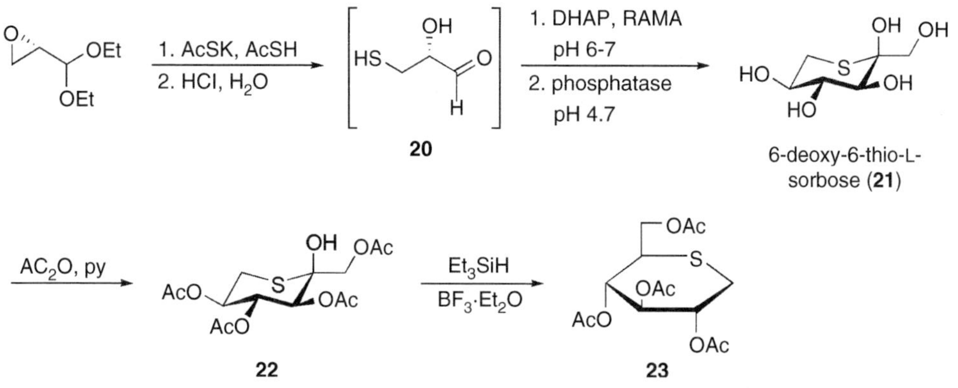

SCHEME 13.18 Synthesis of deoxythiosugars based on a RAMA-catalyzed aldol reaction.

13.5 CHAIN ELONGATION OF ALDEHYDES THROUGH NUCLEOPHILIC ADDITIONS

Chemical asymmetric cross-aldol condensations using enantiomerically pure Lewis acids (instead of an aldolase) as promoters have been used to prepare monosaccharides and analogs [36,37]. If enantiomerically pure aldehydes (such as diol-protected D- or L-glyceraldehyde) are available, then they can be chain elongated by one, two or more carbon centers with high diastereoselectivities. The classical Kiliani–Fischer cyanohydrin synthesis [38] is a milestone in carbohydrate chemistry and has been used in numerous applications [39]. Nevertheless, the diastereoselectivity of the nucleophile addition is often low, and the harsh reaction conditions that are required to reveal the chain-elongated aldose from either their aldonic acid or directly from the cyanohydrin are serious drawbacks. Currently, there are many more flexible methods to carry out one-carbon homologations of aldehydes including reactions at the reductive end of aldoses, presented below. Aldehyde allylation with allyl boronates [40,41] or allylstannanes [42–44] are quite useful because of their high diastereoselectivity and the diversity of modifications that can be applied to the allylic alcohols. With achiral aldehydes, enantiomerically pure allylic and allenyl stannanes can be used in the asymmetric synthesis of monosaccharides and analogs [43,44].

13.5.1 Total Synthesis of D- and L-Glyceraldehyde and Other C-3 Aldose Derivatives

D- and L-glyceraldehyde derivatives are chirons that have been extensively exploited in the total synthesis of monosaccharides and analogs. The acetonide of D-glyceraldehyde ((*R*)-**24**, 2,3-*O*-isopropylidene-D-glyceraldehyde) is easily obtained from D-mannitol [45]. D-Glyceraldehyde has also been derived from D-fructose, while L-glyceraldehyde has been obtained from L-sorbose [46]. The acetonide of L-glyceraldehyde ((*S*)-**24**, (*S*)-1:2,3-*O*-isopropylidene-L-glyceraldehyde) is usually derived from ascorbic acid [47].

The aldehydes (*R*)-**24** and (*S*)-**24** are not stable as monomers and undergo racemization on storage. Derivative (*R*)-**25** (2-*O*-benzylglyceraldehyde) has been proposed as an alternative to (*R*)-**24**. It is obtained from (*S*,*S*)-tartaric acid, as shown in Scheme 13.19 [48]. Enantiomer (*S*)-**25** can be derived from (*R*,*R*)-tartaric acid in the same way. (*R*,*R*)-Tartaric acid is obtained in large quantities from potassium hydrogen tartrate, a waste product of wineries. Racemic tartaric acid is synthesized

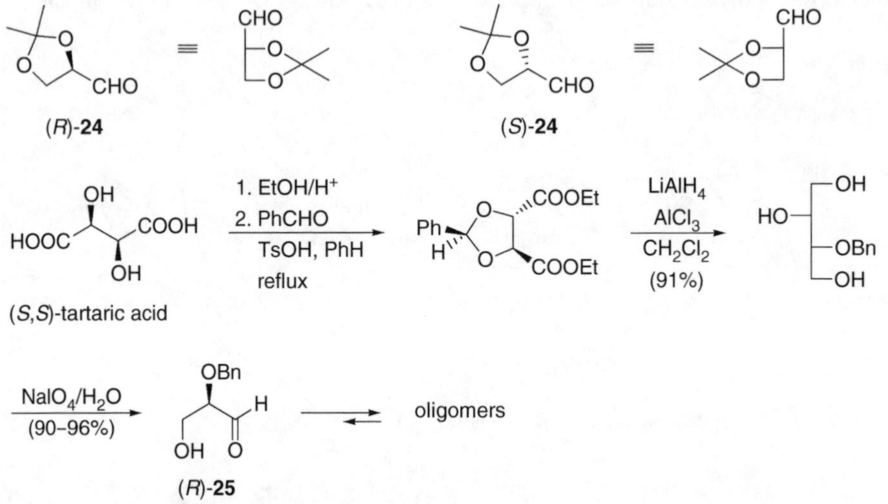

(*R*)-**24** (*S*)-**24**

(*S*,*S*)-tartaric acid → 1. EtOH/H+ 2. PhCHO, TsOH, PhH, reflux → LiAlH4, AlCl3, CH2Cl2 (91%) → NaIO4/H2O (90–96%) → (*R*)-**25** ⇌ oligomers

SCHEME 13.19 Synthesis of 2-*O*-benzyl-D-glyceraldehyde.

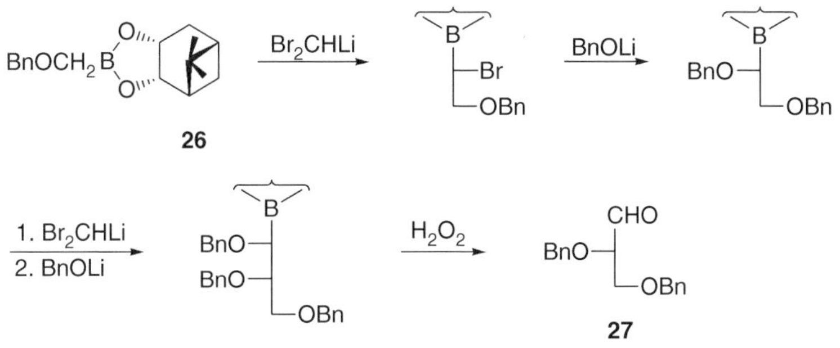

SCHEME 13.20 Asymmetric chain elongation of dibromomethyllithium.

[49] in large scale from maleic anhydride and H_2O_2. Its resolution is carried out either by crystallization or by enzymatic or microbiological enantiodifferentiating conversions. Thus, both (*S*,*S*)- and (*R*,*R*)-tartaric acid are inexpensive and readily available [50].

Chain extensions using an insertion reaction of dichloromethyllithium or dibromomethyllithium with (*S*)-pinanediol [(benzyloxy)methyl]boronate **26** has been used to generate L-C_3-, L-C_4- and L-C_5-aldoses [51]. In order to obtain 2,3-di-*O*-benzyl-L-glyceraldehyde **27**, the insertion reaction has to be applied twice (Scheme 13.20). By repeating the process two more times, L-ribose has been prepared with high enantiomeric purity [51].

The synthesis of 3-*O*-methyl-D-glyceraldehyde starts with D-fructose [52]. The preparation of 2-*O*-methyl-D-glyceraldehyde and 2-*O*-benzyl-D-glyceraldehyde (*R*)-**25** starts from D-mannitol [53].

Enantiomerically pure derivatives of glycerol can be prepared in large quantities through the lipase (pig pancreas; EC 3.1.1.3)-catalyzed hydrolysis of prochiral diacetate **28**. The procedure gives (*R*)-**29** (45% yield, 88% ee), which can be converted into the crystalline derivatives (*R*)-**30** or (*S*)-**30** (>99% ee) as shown in Scheme 13.21 [54].

The (*R*)- and (*S*)-benzyl epoxypropyl ether (*R*)-**31** and (*S*)-**31** were derived from *O*-benzyl-L-serine (Scheme 13.22) [55].

Stable and easily handled, protected forms of L- and D-glyceraldehyde were obtained by the Sharpless asymmetric dihydroxylation [56] of the benzene-1,2-dimethanol acetal of acrolein (Scheme 13.23). The method produces either diol (*R*)-**32** or (*S*)-**32** with 97% ee after recrystallization from benzene. These diols can be converted into useful C-3 chiral building blocks such as epoxides (*R*)-**33** and (*S*)-**33**, respectively [28,57].

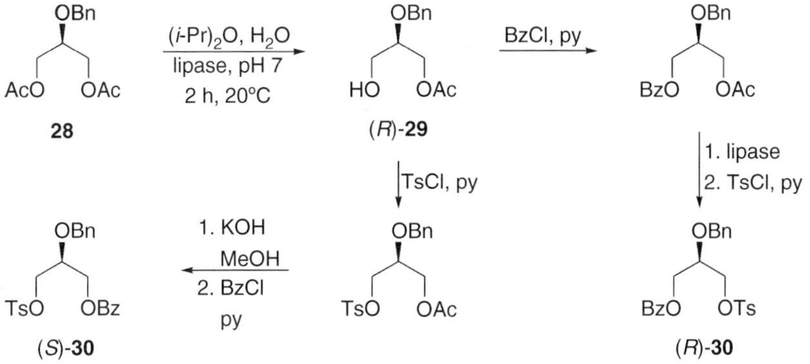

SCHEME 13.21 Desymmetrization of meso-diacetate by lipase-catalyzed hydrolysis: synthesis of C_3-alditol derivatives.

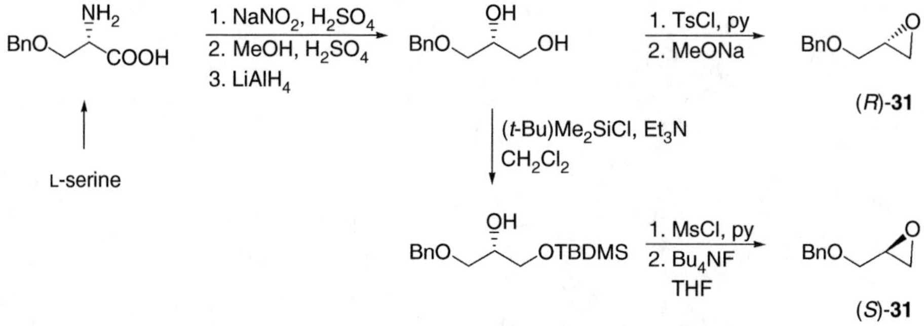

SCHEME 13.22 Syntheses of enantiomerically pure epoxides.

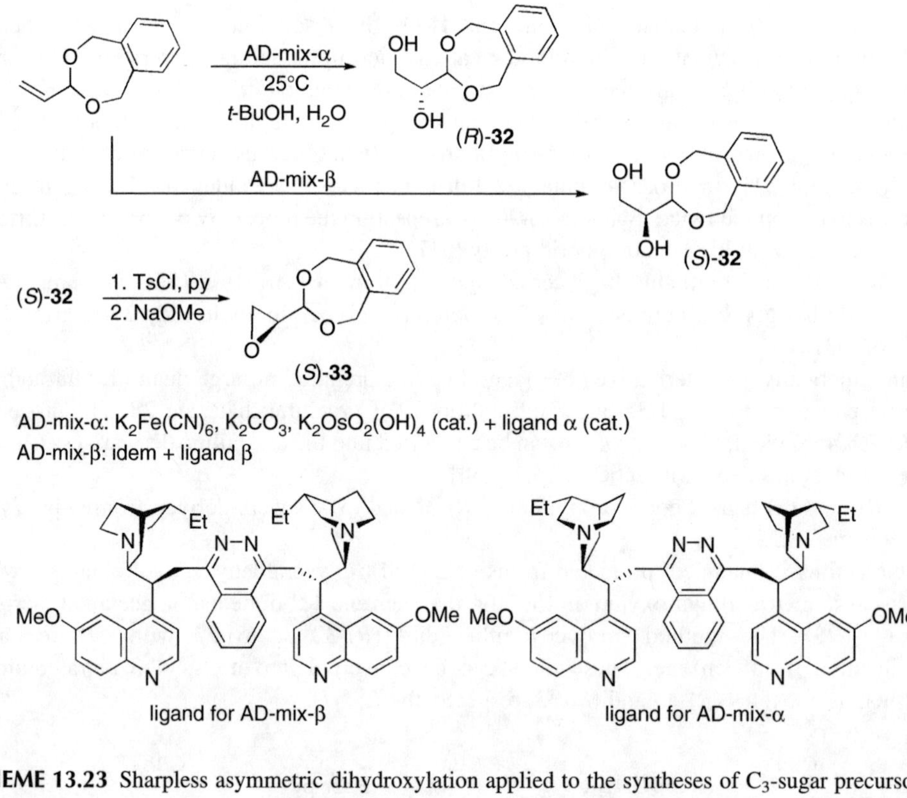

AD-mix-α: $K_2Fe(CN)_6$, K_2CO_3, $K_2OsO_2(OH)_4$ (cat.) + ligand α (cat.)
AD-mix-β: idem + ligand β

ligand for AD-mix-β

ligand for AD-mix-α

SCHEME 13.23 Sharpless asymmetric dihydroxylation applied to the syntheses of C_3-sugar precursors.

Derivatives of D- and L-glyceraldehydes, such as 2-amino-2-deoxyglyceraldehyde (serinal), 3-deoxyglyceraldehyde (2-hydroxypropanal) and 2,3-dideoxy-2-aminoglyceraldehyde (2-aminopropanal), have been used extensively to construct rare monosaccharides and analogs through chain elongations, applying nucleophilic additions to their carbonyl moieties. Semiprotected (*R*)- and (*S*)-2-hydroxypropanols are most simply derived from the readily available D-(−)-lactic and L-(+)-lactic acids, respectively. (*S*)-2-Benzyloxypropanal can be obtained via benzylation of ethyl L-lactate, followed by $LiAlH_4$ reduction and employment of the Swern oxidation [58]. *N*-(*t*-Butoxycarbonyl)-L-alaninal can be obtained with high enantiomeric purity by $LiAlH_4$ reduction of the *N*-methoxy-*N*-methyl-α-(*t*-butoxycarbonylamino)carboxamide of alanine [59]. Alternatively, *N*-9-(9-phenylfluorenyl)-L-alaninal (L-**34**, Figure 13.1) was derived from L-alanine [60].

The *N*-ethyloxazolidinone **38** (Scheme 13.24) is obtained from L-serine by treating (*S*)-serine methyl ester hydrochloride with Et_3N, acetaldehyde and $NaBH_4$ to give *N*-ethylamine **36**.

FIGURE 13.1 *N*-9-(9-Phenylfluorenyl)-L-alanine (L-**34**).

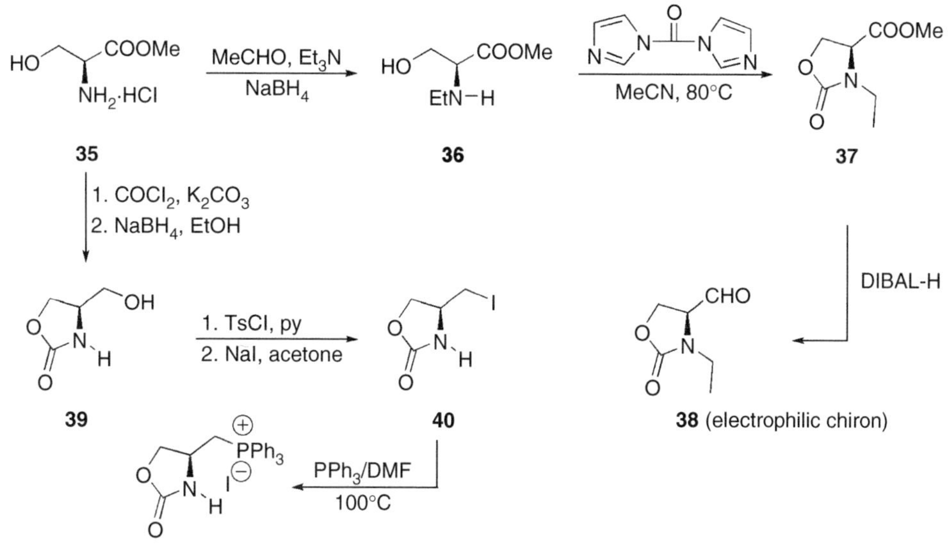

SCHEME 13.24 Syntheses of electrophilic and nucleophilic C_3-chiron containing masked 2-amino moieties.

Oxazolidinone formation with carbonyldiimidazole leads to **37**, the reduction of which generates aldehyde **38** [61]. A nucleophilic alaninol synthon **41** was derived from **35** by protection of the alcohol and amine moieties as a carbamate on treatment with phosgene. Reduction of the ester gave the corresponding alaninol **39**, which was tosylated. The tosylate was then successively displaced with iodide and triphenylphosphine yielding **41** [62] (Scheme 13.24)

13.5.2 One-Carbon Homologation of Aldoses: The Thiazole-Based Method

Dondoni and coworkers [63] have shown that homologation of α-hydroxycarbaldehydes can be achieved with high antiselectivity by addition of 2-(trimethylsilyl)thiazole (**42**) (Scheme 13.25). For instance, D-glyceraldehyde acetonide (*R*)-**24** reacts with **42** giving **43** in 96% yields with the *anti* vs. *syn* diastereoselectivity better than 95:5. Release of the aldehyde requires protection of the alcohol as a benzyl ether, methylation of the thiazole generates intermediate **43 Me** that is not isolated but reduced *in situ* with $NaBH_4$ to give thiazoline **43 H**. Mercury(II)-catalyzed hydrolysis liberate the semiprotected D-erythrose derivative D-**45** in 62% overall yield [64]. Methylation of the thiazole moiety can also be achieved with methyl triflate instead of MeI, and copper(II)chloride can be used instead of mercury(II)chloride [65].

The iterative addition and unmasking protocols were repeated over several consecutive cycles so that the chain elongation of the triose (*R*)-**24** was brought up to the nonose derivative **46** (all-*anti* configuration of the polyol) (Scheme 13.26).

For the preparation of *syn* isomers, alcohol **43**, resulting from the thiazole addition to the starting aldehyde, has to be oxidized into the corresponding ketone **43 K**, which is reduced by K-Selectride to the *syn* isomer **47** (Scheme 13.27). The α-amino aldehyde L-**50** derived from

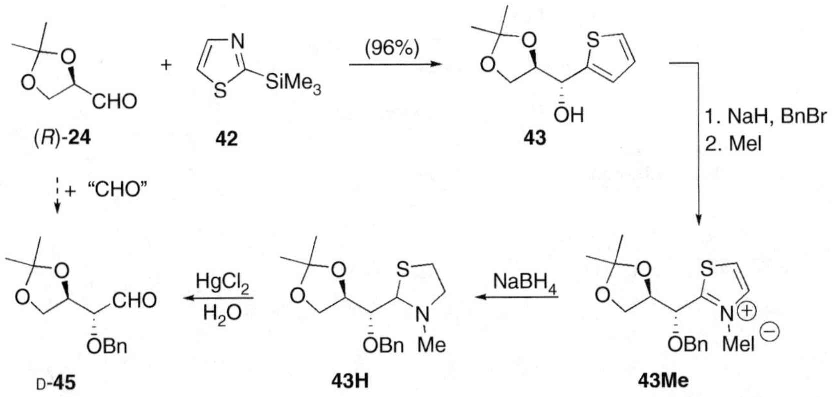

SCHEME 13.25 Dondoni's one-carbon chain elongation.

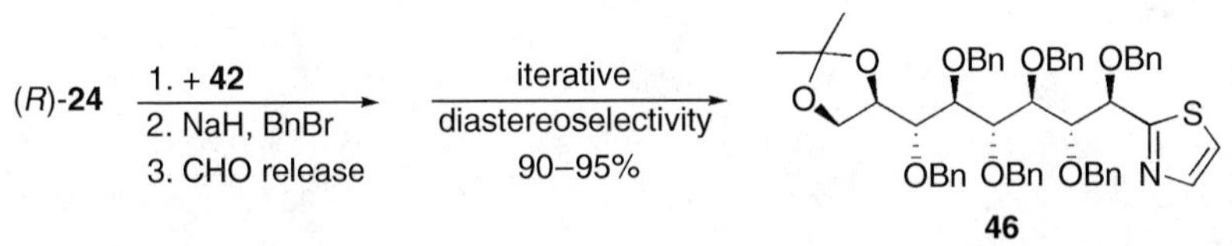

SCHEME 13.26 Dondoni's iterative aldose chain elongation [63].

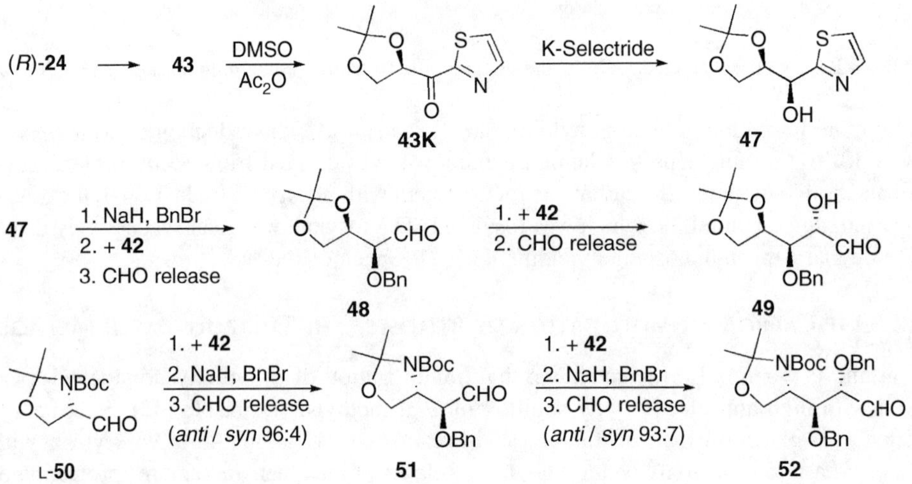

SCHEME 13.27 Examples of syntheses of aldoses by Dondoni's one-carbon chain homologation.

L-serine was converted into aminotetrose and pentose derivatives **51** and **52**, respectively [66]. The *anti* diastereoselectivity observed for addition of **42** to the *N*,*N*-diprotected α-amino aldehyde L-**50** can be reversed to *syn* selectivity by using a *N*-monoprotected derivative [66b].

An aminohomologation of carbaldehydes was developed by Dondoni and coworkers, thus remarkably extending the scope of their one-carbon chain-elongation method (Scheme 13.28). For example, the *N*-benzylnitrone **53**, derived from D-glyceraldehyde acetonide (*R*)-**24**, was added to 2-lithiothiazole giving the *syn* adduct **54** with 92% diastereoselectivity. Interestingly, the same reaction applied to **53** precomplexed with Et_2AlCl or $TiCl_4$ gave the *anti* diastereomer **55**

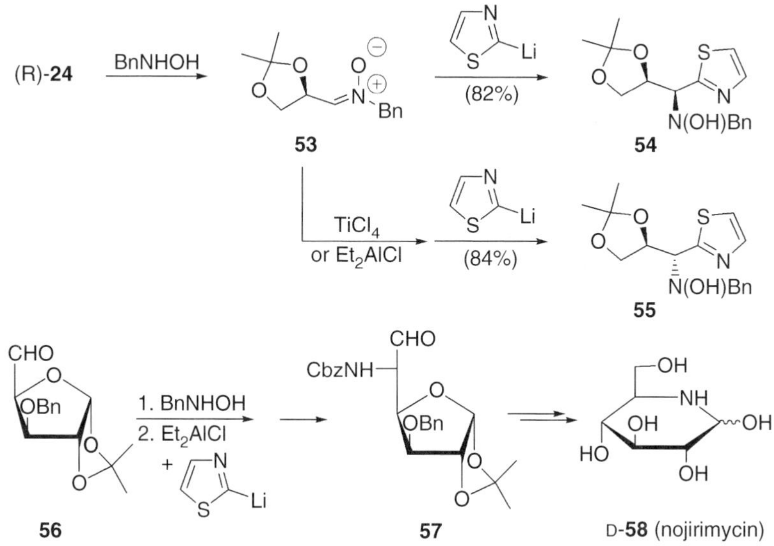

SCHEME 13.28 Dondoni's synthesis of amino sugars.

preferentially in high yield. The method has been applied to the synthesis of all kinds of aminosugars including D-nojirimycin D-**58**, derived from the dialdehyde sugar derivative **56** (Scheme 13.28) [67a].

The aminohomologation of (*R*)-**24** via nitronate **53** can use the addition of 2-lithiofuran instead of 2-lithiothiazole. The furyl moiety can then be oxidized to the corresponding 2-aminoaldonic acids [67b]. Alternatively, the nucleophilic addition of alkoxy-methyllithium derivatives to nitrones of type **53** are either *syn* or *anti* selective in the absence or the presence of Et_2AlCl, respectively. The adducts thus obtained were converted into C-4 building blocks and β-hydroxy-α-aminoacids [67c].

13.5.3 Other Methods of One-Carbon Chain Elongation of Aldoses

An alternative method (Scheme 13.29) for the homologation of the D-glyceraldehyde derivative (*R*)-**24** to derivatives of D-erythrose **60** and D-threose **61** was proposed by Kusakabe and Sato [68]. Reaction of (*R*)-**24** with appropriate 1-(trimethylsilyl)vinyl-copper reagents lead to either *anti* or

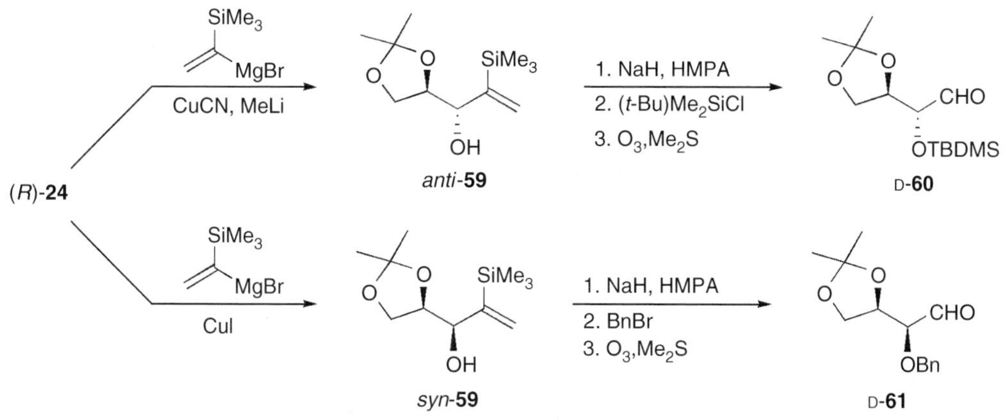

SCHEME 13.29 *Anti* and *syn* diastereoselective vinylations of D-glyceraldehyde acetonide.

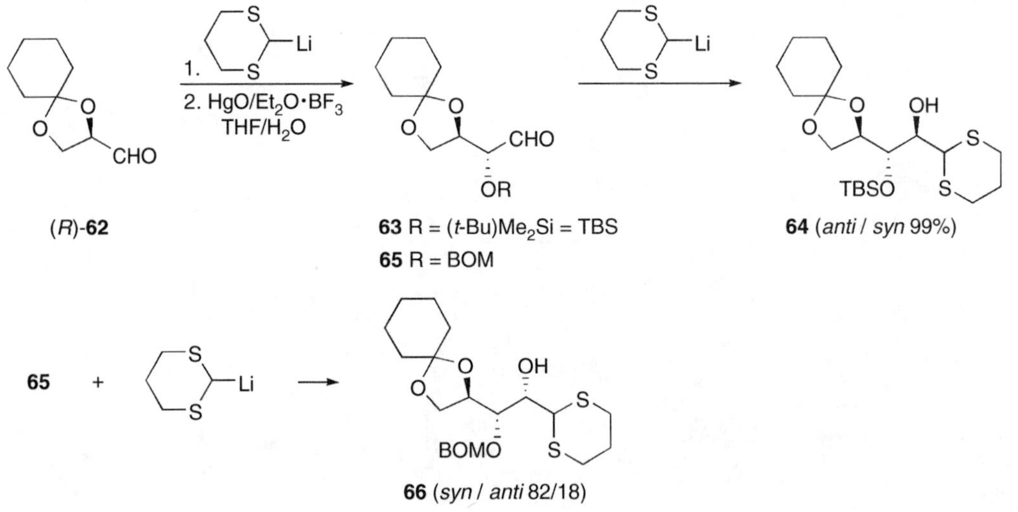

1. 2-lithio-1,3-dithiane
2. $HgO/Et_2O\cdot BF_3$
THF/H_2O

(*R*)-**62**

63 R = (*t*-Bu)Me_2Si = TBS
65 R = BOM

64 (*anti* / *syn* 99%)

65 + 2-lithio-1,3-dithiane →

66 (*syn* / *anti* 82/18)

SCHEME 13.30 Diastereoselective 2-lithio-1,3-dithiane additions.

syn stereoselective adducts **59** *anti* (*anti*/*syn* 20:1) or **59** *syn* (*syn*/*anti* 98:2) in 87% yield. Alcohol protection, followed by ozonolysis, furnished D-**60** and D-**61**, respectively.

The nitroaldol condensation with nitromethane (Henry's reaction), followed by Nef decomposition of the resultant nitronate under strongly acidic conditions, has been used to elongate aldehydes. For instance, *N*-acetyl-D-mannosamine has been converted into *N*-acetylneuraminic acid applying this method iteratively [69]. Chikashita and coworkers [70] have reported good levels of *anti* diastereoselectivity better than 99% in an iterative homologation sequence using 2-lithio-1,3-dithiane [71] with 2,3-*O*-cyclohexylidene-D-glyceraldehyde (*R*)-**62**. In the case of the BOM-protected tetrose derivative, the addition of 2-lithio-1,3-dithiane was *syn* selective (*syn*/*anti* 82:18) (Scheme 13.30).

13.5.4 Additions of Enantiomerically Pure One-Carbon Synthons

The addition of (+)-(*R*)-methyl *p*-tolylsulfoxide to carboxylic esters gives the corresponding β-keto sulfoxides. When applied to α,β-unsaturated esters, the ketosulfoxides thus obtained are reduced with high diastereoselectivity (better than 95:5) with $LiAlH_4$ or $(i\text{-Bu})_2AlH$ giving optically active β-hydroxy sulfoxides such as (*R*,*S*)-**67** and (*R*,*R*)-**67** (Scheme 13.31) [72]. When applied to ester **68**, this method generated allylic alcohol **69** that underwent highly diastereoselective osmium-catalyzed dihydroxylation giving **70**. A Pummerer rearrangement and subsequent reduction with $(i\text{-Bu})_2AlH$ and acetylation furnished the L-arabinitol derivative **71** (Scheme 13.31) [73].

13.5.5 Two-Carbon Chain Elongation of Aldehydes

13.5.5.1 Asymmetric Aldol Reactions

Enantiomerically pure glycolaldehyde derivatives **72** undergo aldol condensations in the presence of Et_3N giving mixtures of erythrose and threose derivatives **73** (Scheme 13.32) for which the erythrose/threose ratio reaches 58:42 and the L/D ratio reaches 62:38 [74].

A very elegant asymmetric synthesis of D-ribose from achiral starting materials has been presented by Mukaiyama and coworkers [36]. It is based on the cross-aldolization of crotonaldehyde and enoxysilane **74** in the presence of an enantiomerically pure diamine **75**, the chiral inducer (Scheme 13.33). High diastereoselectivity (*anti*/*syn* $>$ 98:2) and high

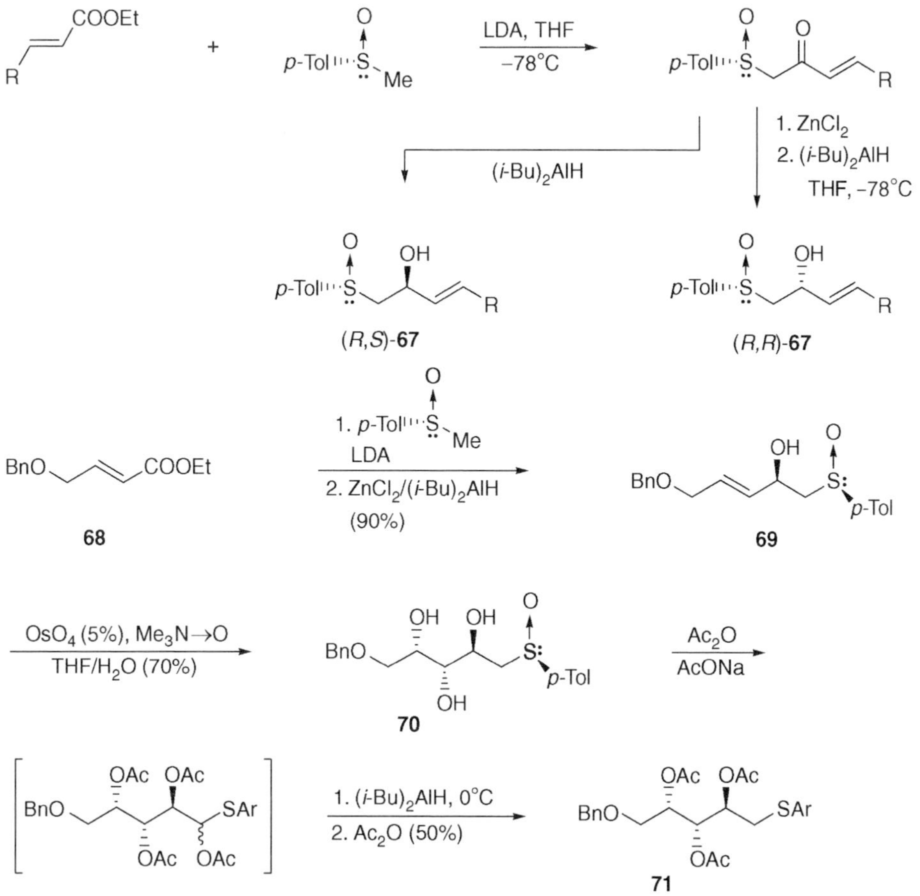

SCHEME 13.31 Asymmetric synthesis of L-arabinitol derivatives via stereoselective dihydroxylation of an enantiomerically pure allylic alcohol.

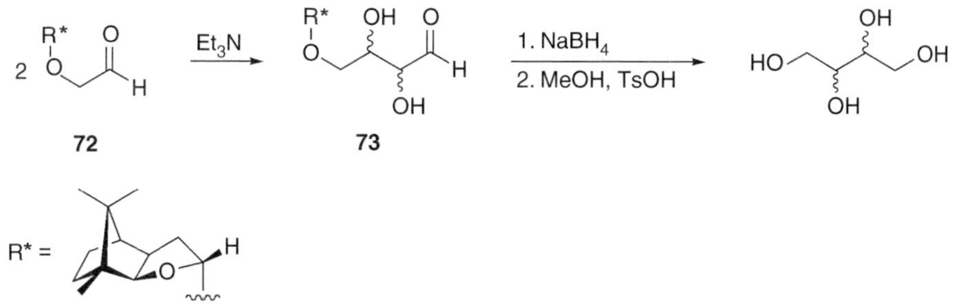

SCHEME 13.32 Asymmetry induced by a chiral auxiliary in the aldol condensation of glycoaldehyde.

enantioselectivity (>97% ee for *anti* aldol) are observed. The antialdol **76** is then doubly hydroxylated with moderate facial selectivity to give a 72:28 mixture of aldonolactones **77** and **78**. Reduction of the major lactone **77** provides **79**, the debenzylation of which furnishes D-ribose. The same method has been applied to the preparation of 4-*C*-methyl-D-ribose and 6-deoxy-L-talose starting with methacrolein and (*E*)-but-2-enal ((*E*)-crotonaldehyde), respectively (Scheme 13.33).

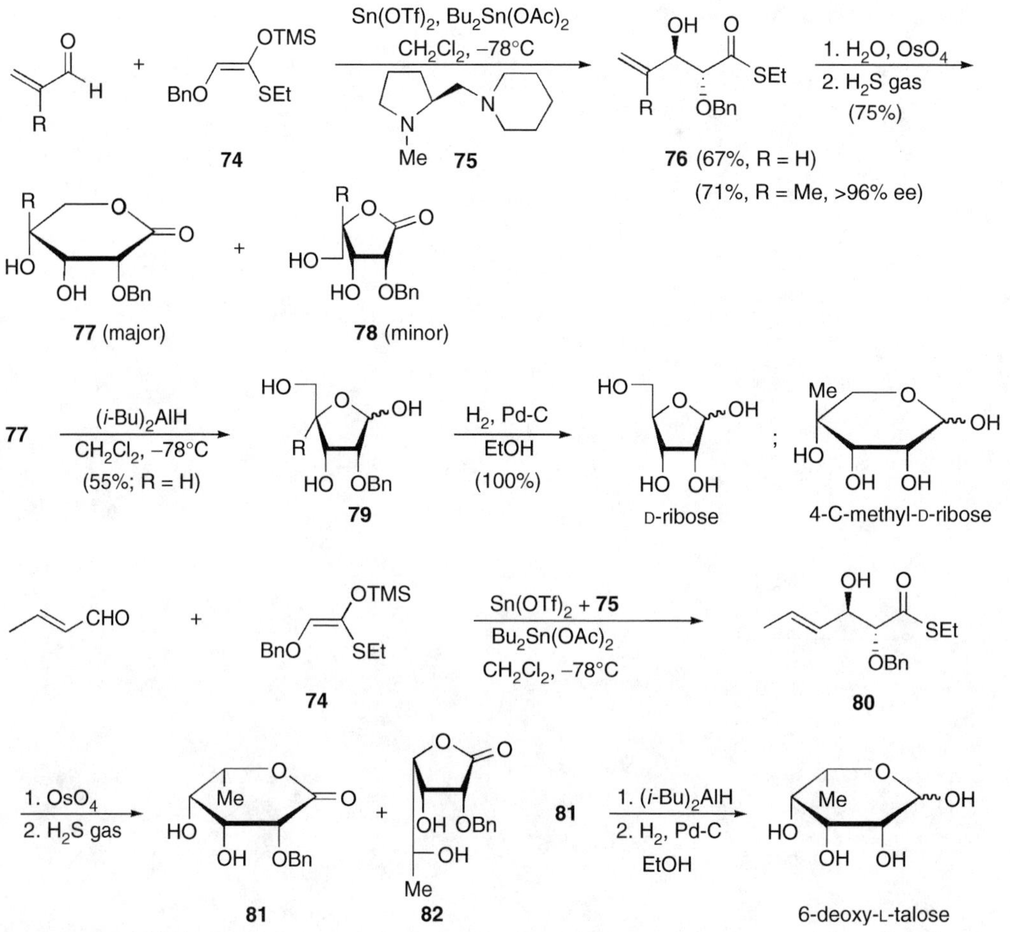

O H R + OTMS BnO SEt **74** → $Sn(OTf)_2$, $Bu_2Sn(OAc)_2$, CH_2Cl_2, −78°C, N Me **75** N → OH O SEt R OBn **76** (67%, R = H) (71%, R = Me, >96% ee) → 1. H_2O, OsO_4 2. H_2S gas (75%)

R O =O HO OH OBn **77** (major) + R O O HO HO OBn **78** (minor)

77 → $(i\text{-}Bu)_2AlH$, CH_2Cl_2, −78°C (55%; R = H) → HO O OH R HO OBn **79** → H_2, Pd-C, EtOH (100%) → HO O OH HO OH D-ribose ; Me O OH HO OH OH 4-C-methyl-D-ribose

CHO + OTMS BnO SEt **74** → $Sn(OTf)_2$ + **75**, $Bu_2Sn(OAc)_2$, CH_2Cl_2, −78°C → OH O SEt OBn **80**

→ 1. OsO_4 2. H_2S gas → O Me =O HO OH OBn **81** + O O OH OBn OH Me **82**

81 → 1. $(i\text{-}Bu)_2AlH$ 2. H_2, Pd-C EtOH → O Me OH HO OH OH 6-deoxy-L-talose

SCHEME 13.33 Mukaiyama's asymmetric aldol reactions: total synthesis of D-ribose, 4-*C*-methyl-D-ribose and 6-deoxy-L-talose.

Applying an analogous method, Kobayashi and Kawasuji [37] have prepared L-fucose from (*E*)-crotonaldehyde and the ketene acetal **83** in four steps and 49% overall yield (Scheme 13.34). The asymmetric aldol condensation is catalyzed by a complex made of $Sn(OTf)_2$ and chiral diamine **84**.

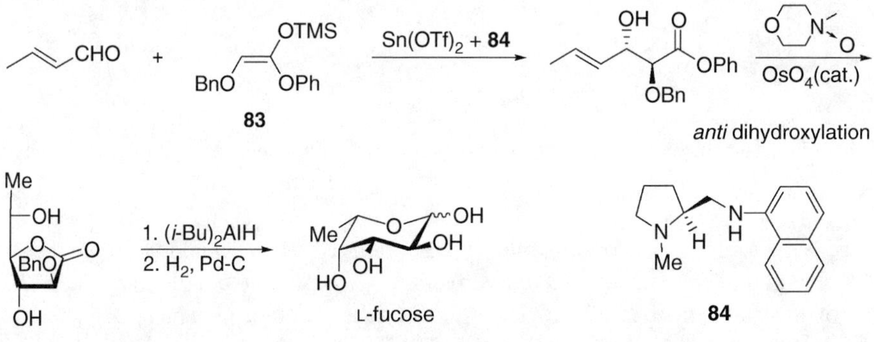

CHO + OTMS BnO OPh **83** → $Sn(OTf)_2$ + **84** → OH O OPh OBn → O N O (OsO_4(cat.)) *anti* dihydroxylation

Me OH O BnO O OH → 1. $(i\text{-}Bu)_2AlH$ 2. H_2, Pd-C → Me O OH OH OH HO L-fucose

N H N H Me **84**

SCHEME 13.34 Kobayashi's total synthesis of L-fucose.

Finally, recent work published by the MacMillan Group [74] demonstrates the feasibility of utilizing selective aldol reactions to prepare carbohydrates in as little as two steps.

13.5.5.2 Nucleophilic Additions to Enantiomerically Pure Aldehydes

Mukaiyama and coworkers have pioneered many routes to the total syntheses of rare carbohydrates such as the 2-amino-2-deoxypentoses [75]. In 1982, they reported that the potassium enolate derived from the magnesium salt of the (*R*)-atrolactic acid derivative **85** adds to 2,3-*O*-isopropylidene-D-glyceraldehyde in a highly stereoselective manner giving, after alcohol protection, imine hydrolysis and amine protection, the D-arabinopentoate derivative **86** (Scheme 13.35). Further elaboration leads to 2-acetamido-2-deoxy-D-arabinose **87**. In a similar fashion, starting from (*S*)-atrolactic acid, 2-acetamido-2-deoxy-D-ribose **88** was prepared [76].

Several syntheses of aminodeoxypentoses have employed a similar approach in which a three-carbon starting material is condensed with a two-carbon entity. For instance, the nucleophilic addition of methyl isocyanate to (*R*)-**24** is highly diastereoselective giving a mixture of *erythro* and *threo* adducts (Scheme 13.36). The major adduct is then converted into methyl 2-amino-2-deoxy-D-arabonate [77].

A synthetic equivalent of the glycolaldehyde anion, the dioxaborolene **89** (Scheme 13.37), has been used for the carbon chain elongation of aldehydes. Thus, L-ribose is prepared from the addition of 2,3-*O*-cyclohexylidene-L-glyceraldehyde (*S*)-**62** to **80**. Double addition and higher addition reactions yielding polymers is alleviated by using a polymer-supported reagent [78].

Ethyl (*S*)-lactate has been the primary source of chirality in several syntheses of aminodeoxycarbohydrates. The derivative **91** of 2-amino-2-deoxy-L-lyxonic acid is the major product of condensation of 2-*O*-benzyl L-lactaldehyde with the silyl ketene acetal **90**. The derived ester **91** can be converted into lactone **92** (Scheme 13.38) [79], an intermediate for the synthesis of L-daunosamine and L-vancosamine.

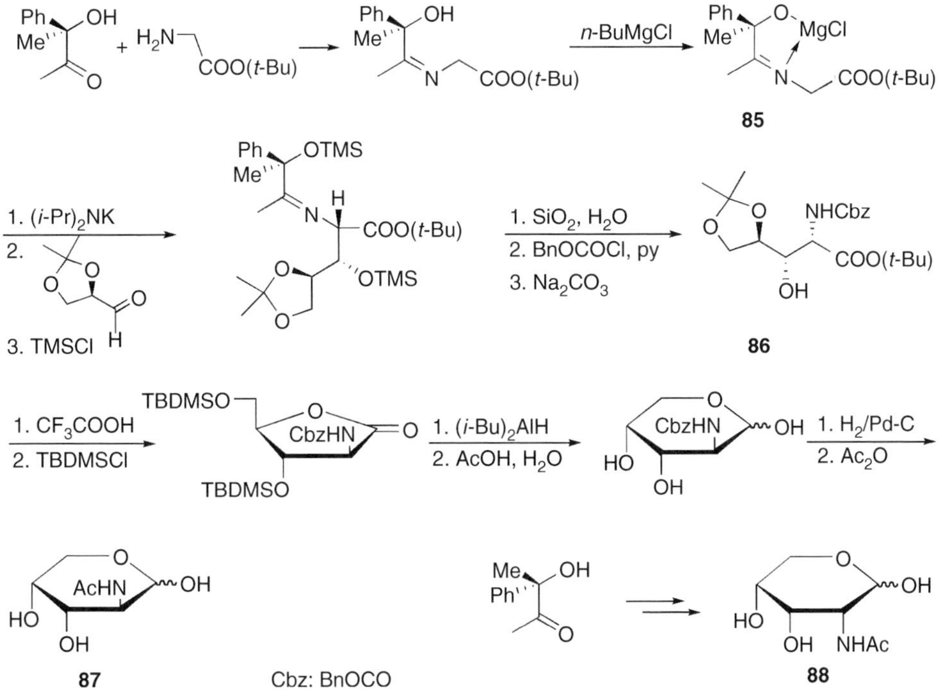

SCHEME 13.35 Syntheses of 2-amino-2-deoxypentoses.

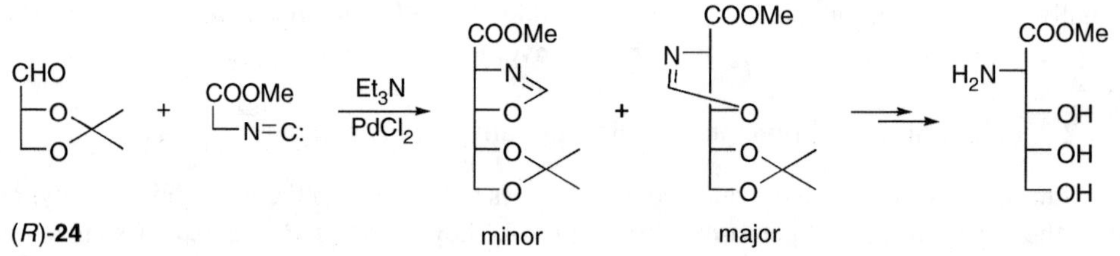

SCHEME 13.36 Synthesis of methyl 2-amino-2-deoxy-D-arabonate.

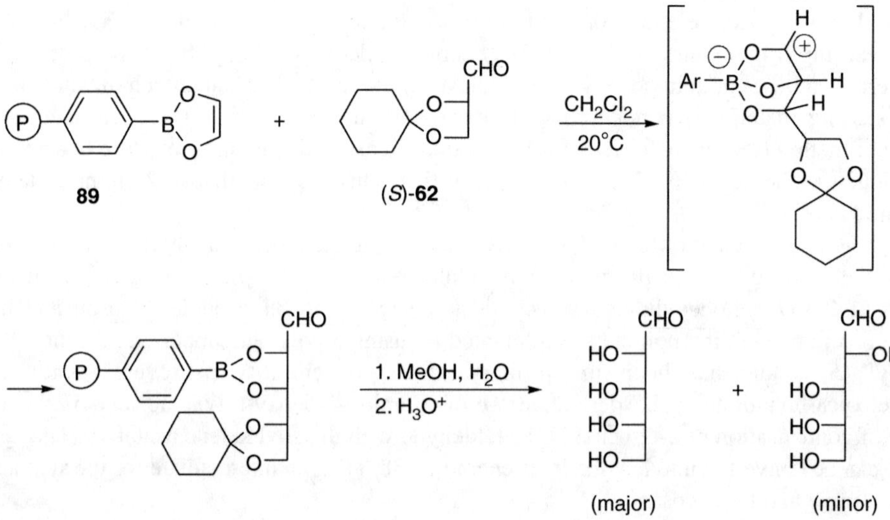

SCHEME 13.37 Synthesis of L-ribose with a polymer-supported glycoaldehyde anion equivalent.

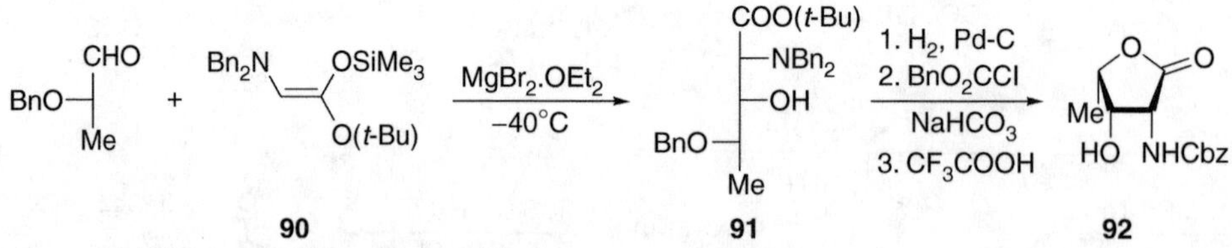

SCHEME 13.38 Synthesis of a 2,5-dideoxy-2-aminoaldonolactone.

13.5.5.3 Nitro-Aldol Condensations

The syntheses of D- and L-2-amino-2-deoxy-arabinose and of 1,4-dideoxy-1,4-imino-D-lyxitol **94** have been achieved via the nitro-aldol condensation (Henry's reaction) of 2-*O*-benzyl-D-glyceraldehyde (*R*)-**25** and the diethyl acetal of nitroacetaldehyde (Scheme 13.39), which gives a 88:12 mixture of the *arabino* and *ribo*-adducts. Their reduction and subsequent protection of the amines thus obtained, and then selective tosylation of the primary alcohol and hydrogenolysis, gives **93**, which is subsequently converted into **94** [80].

13.5.5.4 Nucleophilic Additions of Enantiomerically Pure Enolates

Braun's enantiomerically pure acetate **95** ((*R*)-"HYTRA") [81] can be converted to the lithium enolate **96**. Subsequent addition to acrolein predominantly gives (1′*R*,3*R*)-**97**

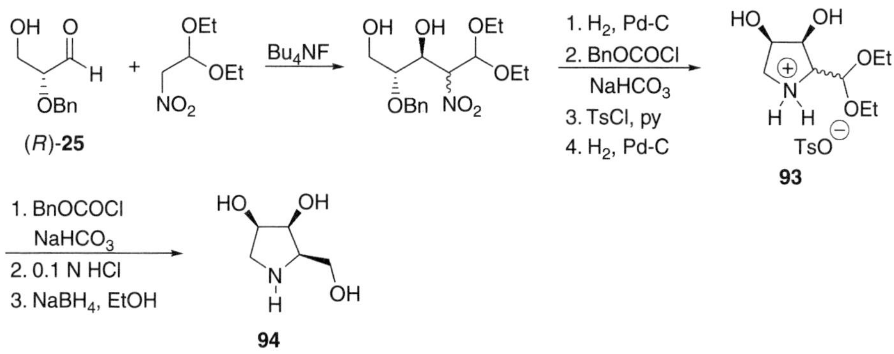

SCHEME 13.39 Synthesis of 1,4-dideoxy-1,4-imino-D-lyxitol.

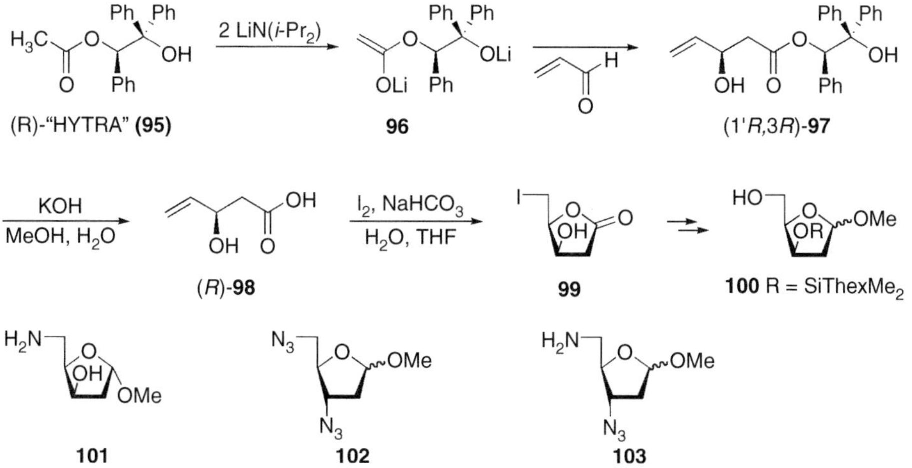

SCHEME 13.40 Diastereoselective aldol reaction with (*R*)-"HYTRA".

(diastereoselectivity: 92:8). Alkaline hydrolysis of (1′*R*,3*R*)-**97** provides (*R*)-**98** with 83% ee (Scheme 13.40). On treatment of (*R*)-**98** with (*S*)-1-phenylethylamine and recrystallization, (*R*)-**98** is isolated in 42% overall yield and >99% ee. This compound was converted into **99** via iodolactonization. Compound **99** is a precursor of the 2-deoxyfuranosides **101** to **103** [82].

D-Digitoxose, a component of cardiac glycosides found in *Digitalis purpurea* and other higher plants, can be prepared following a similar method starting from ethyl (*R*)-lactate and (*R*)-HYTRA [83] (Scheme 13.41).

The chiral titanated bislactim ether **104** undergoes 1,2-addition to α,β-unsaturated aldehydes. With enal **105**, it gives adduct **106**, the epoxidation and subsequent hydrolysis of which forms the branched-chain amino-acid derivative **107** [84] (Scheme 13.42).

The phenylalanine-derived oxazolidinone **108** undergoes a diastereoselective aldol reaction with crotonaldehyde to give the *syn* product **109**. Formation of the Weinreb amide, followed by silylation of the crude secondary alcohol, provides **110**. The chiral auxiliary is recovered at this stage. Product **110** is a potential precursor of all kinds of monosaccharides and analogs, including 1-deoxynojirimycin (Scheme 13.43) [85].

The C_{33}–C_{37}-unit of (+)-calyculin A (a marine natural product) is an amide derived from 5-*O*-methyl-4-deoxy-4-dimethylamino-D-ribonic acid, which has been prepared by Evans and

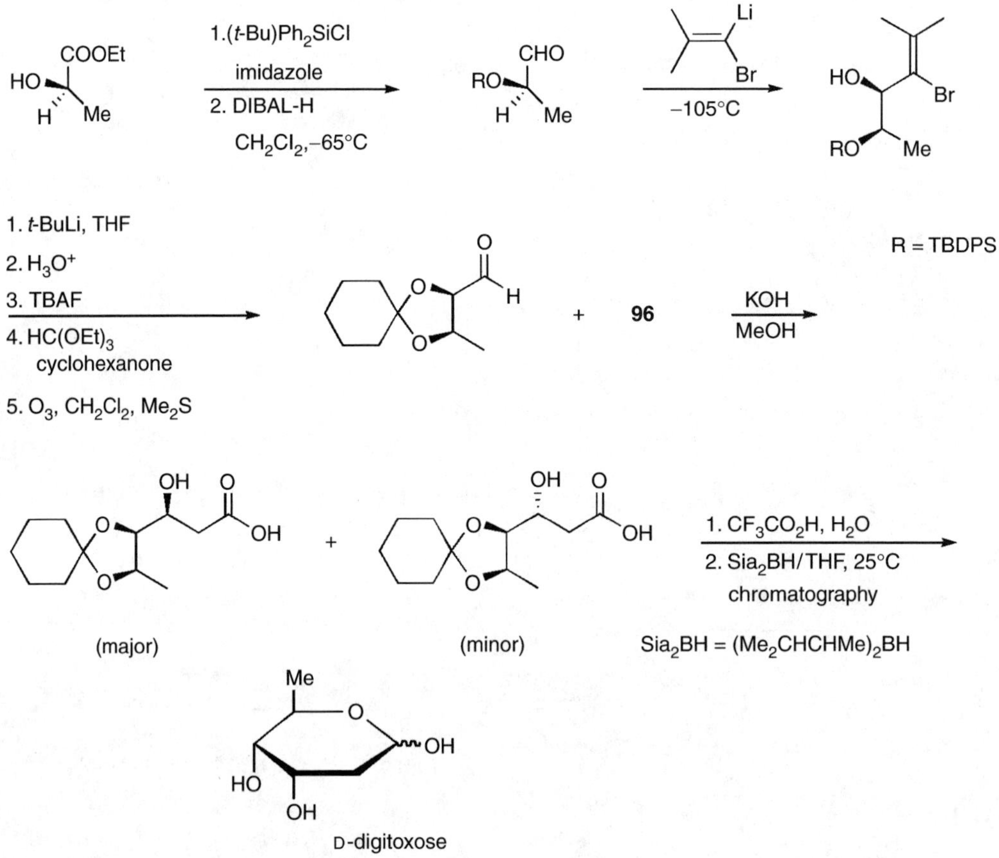

SCHEME 13.41 Braun's synthesis of D-digitoxose.

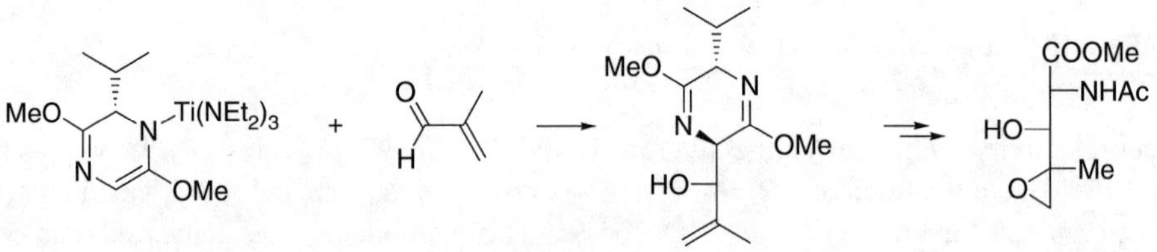

SCHEME 13.42 Use of an enantiomerically pure bislactim ether.

coworkers [86]. *N*-Protection of sarcosine as benzyl carbamate affords acid **111**, which is activated and used to *N*-acylate the (*S*)-phenylalanine-derived oxazolidinone. This gives **112**, which is methoxymethylated diastereoselectively (98:2) to give **113**. Reductive removal of the chiral auxiliary, followed by Swern oxidation, forms aldehyde **114** with little racemization if Hünig's base (i-Pr_2NEt) is used instead of Et_3N. Enolization of imide **115**, followed by addition of tetramethylethylenediamine and aldehyde **114**, produces the antialdol **116** (60%) accompanied by 24% of other diastereomers (Scheme 13.44). Compound **116** has been used for amide formation with primary amines.

An enantioselective synthesis of 3-deoxypentoses from (−)-myrtenal has been proposed by Franck-Neumann and coworkers [42] (Scheme 13.45). It features the Mukaiyama

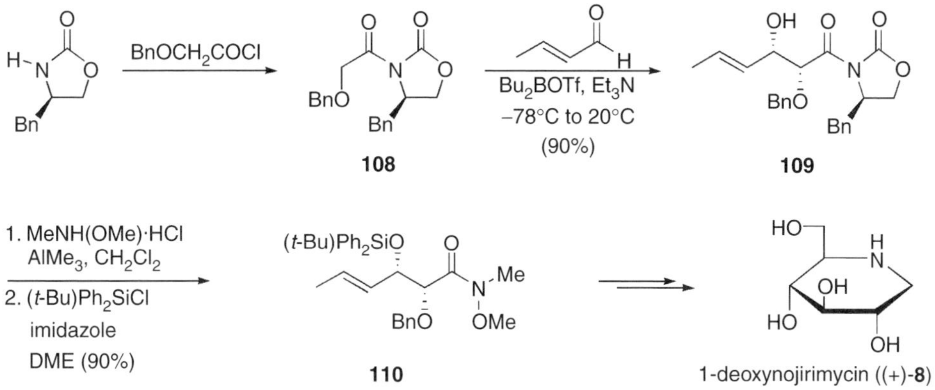

SCHEME 13.43 Use of an Evans' homochiral enolate.

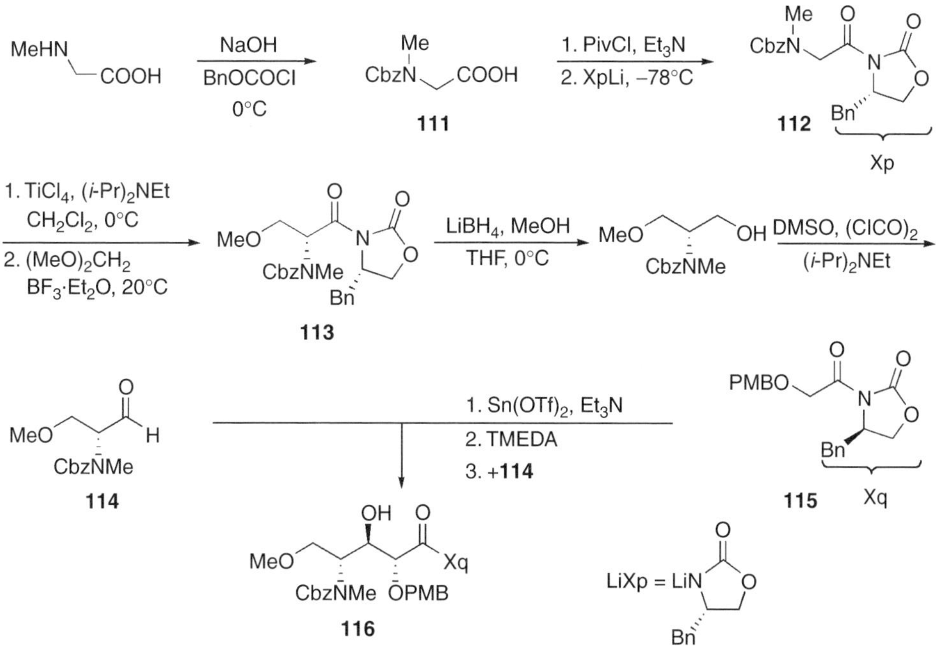

SCHEME 13.44 Synthesis of a 4-deoxy-4-dimethylamino-D-ribonic acid derivative.

cross-aldolization of benzyloxyacetaldehyde and the tricarbonyliron complex **117** derived from the condensation of (−)-myrtenal with acetone. The diastereomeric aldols **118** and **119** are separated and converted (Scheme 13.45) into protected derivatives of the 3-deoxypentoses **120** and **121** [87].

Enders and Jegelka [88] have used 1,3-dioxan-5-one **122**, a protected dihydroxyacetone derivative, to construct enantiomerically pure C_5- to C_9-deoxycarbohydrates. For example, reaction of **122** with SAMP gives the hydrazone **123**, which is deprotonated and alkylated with methyl iodide to yield **124**. The monoalkylated hydrazone is then alkylated in the same manner with chloromethyl benzyl ether to form **125**. Cleavage of the hydrazone with ozone furnishes the protected ulose **126** (>98% de, >98% ee), which is deprotected to (−)-5-deoxy-L-*threo*-3-pentulose **127**. Reduction of **126** with L-Selectride, followed by deprotection, provides 5-deoxy-D-arabinitol **128** (>95% de, >95% ee) (Scheme 13.46).

(–)-myrtenal

1. (acetone) 2. $Fe_2(CO)_9$

117 $Fe(CO)_3$

1. TMSOTf, Et_3N, CH_2Cl_2, 0°C
2. $BnOCH_2CHO$, $BF_3{\cdot}Et_2O$

118 (39%) + **119** (23%)

1. $BH_3{\cdot}Me_2S$, THF, 20°C
2. Ac_2O, Et_3N, DMAP, CH_2Cl_2, 20°C
3. $Ce(NH_4)_2(NO_3)_6$, MeCN, H_2O
4. O_3, MeOH, –78°C
5. Me_2S

idem

120 **121**

SCHEME 13.45 Franck-Neumann's synthesis of 3-deoxypentoses.

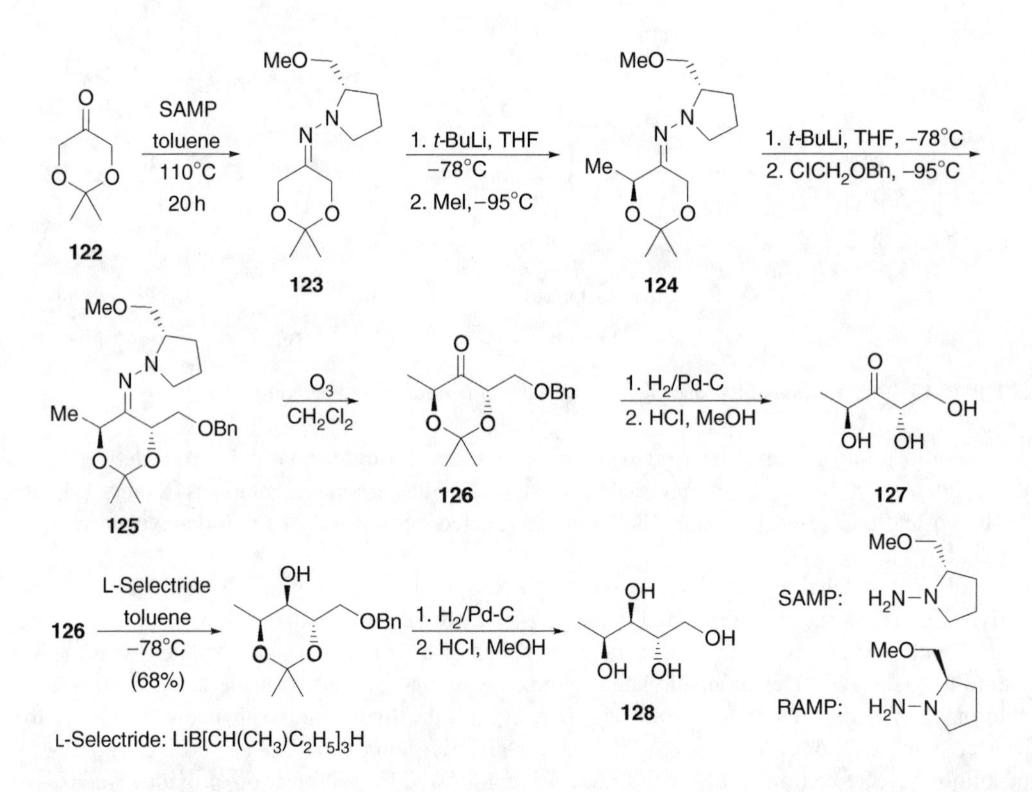

SCHEME 13.46 Enders' synthesis of 5-deoxy-D-arabinitol.

13.5.5.5 Aldehyde Olefinations and Asymmetric Epoxidations

After reduction of the enal with diisobutylaluminium hydride, the Wittig olefination of D-glyceraldehyde acetonide (*R*)-**24** with $Ph_3P{=}CHCHO$ gives the (*E*)-allylic alcohol **129**. The Katsuki–Sharpless enantioselective epoxidation [89] applied to **129** allows the preparation of D-arabinitol (= D-lyxitol) and ribitol, a *meso* alditol (Scheme 13.47). Similarly, Wittig olefination of (*R*)-**24** with $Ph_3P{=}CHCH(OEt)_2$, followed by acidic hydrolysis of the diethyl acetal and subsequent reduction of the enal with diisobutylaluminium hydride, provides the (*Z*)-allylic alcohol **130**. Diastereoselective epoxidation and hydrolysis leads to D-arabinitol or xylitol, another *meso* alditol [90a].

The Katsuki–Sharpless asymmetric epoxidation of (*E*)-allylic alcohols is the key step in the total synthesis of all tetroses and hexoses developed by Sharpless and Masamune [90c], which are summarized in Scheme 13.48 for the L-series. The epoxides obtained from the allylic alcohol precursors undergo a Payne rearrangement in the presence of NaOH, giving terminal epoxides that regioselectively open to phenylsulfides on treatment with PhSNa. Following protection of the diols as acetonides, the sulfides are oxidized with *m*-chloroperbenzoic acid into the corresponding sulfoxides. These sulfoxides then undergo Pummerer rearrangement on treatment with Ac_2O and AcONa liberating, after hydrolysis, the corresponding aldose derivatives. Thus, (*Z*)-but-2-ene-1,4-diol can be converted into 8 tetroses and 16 hexoses if one considers the

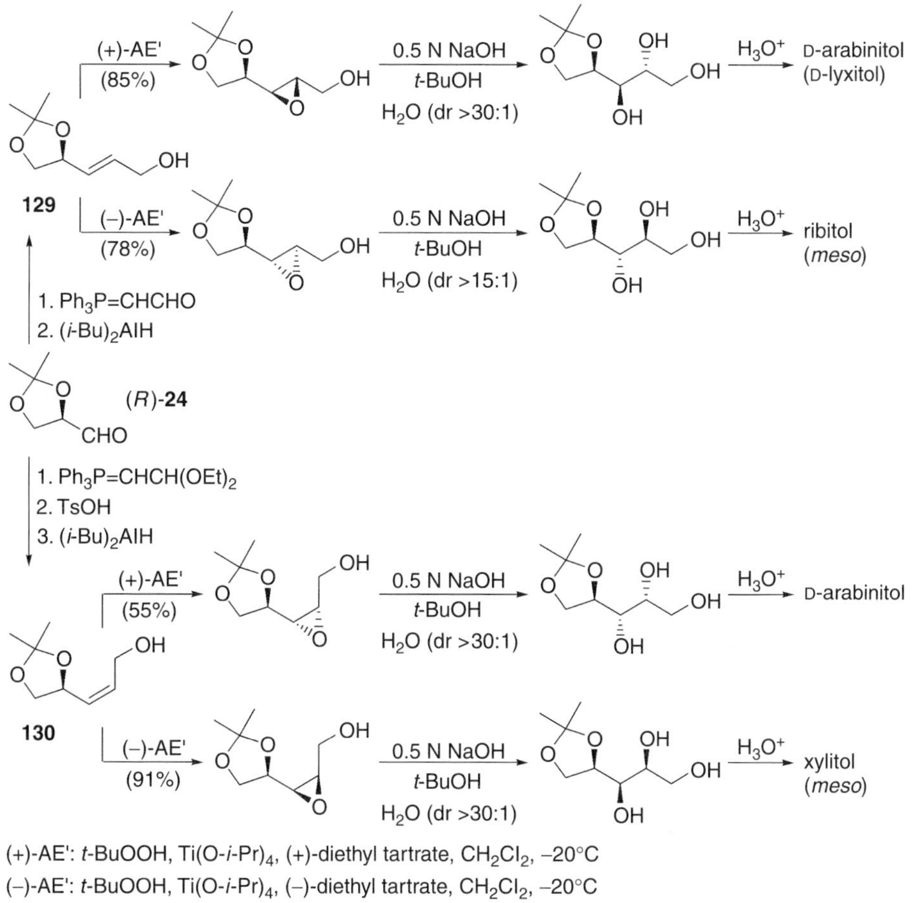

(+)-AE': *t*-BuOOH, Ti(O-*i*-Pr)$_4$, (+)-diethyl tartrate, CH_2Cl_2, −20°C
(−)-AE': *t*-BuOOH, Ti(O-*i*-Pr)$_4$, (−)-diethyl tartrate, CH_2Cl_2, −20°C

SCHEME 13.47 Wittig olefination and Katsuki–Sharpless asymmetric epoxidation applied in the conversion of D-glyceraldehyde into pentitols.

base-catalyzed isomerization of *cis*-disubstituted dioxolane into the more stable *trans* isomers (Scheme 13.48).

The methodology of Wittig–Horner–Emmons olefinations to convert aldehydes into their corresponding two-carbon chain-elongated allylic alcohols and their subsequent asymmetric epoxidation has been used to prepare (+)-galactonojirimycin, (+)-nojirimycin, 1-deoxygalactonojirimycin and 1-deoxynojirimycin. The method involves the regioselective opening of the epoxides with azide ion [90d].

13.5.5.6 Aldehyde Olefinations and Dihydroxylations

Allylic alcohol **129** can be protected as a silyl ether and then subjected to the Sharpless asymmetric dihydroxylation giving other alditol stereomers. The latter can be converted into all kinds of C_5-monosaccharide derivatives. An example is given in Scheme 13.49 [91].

Koskinen and Otsomaa [92] have converted the L-threonine-derived aldehyde **131** (Scheme 13.50) into methyl 4-amino-4,6-dideoxy-*gulo*-pyranosides **134**. A modified Horner–Wadsworth–Emmons olefination leads to the (*Z*)-enoate **132**. Acidic hydrolysis of the aminal protection affects lactonization giving **133**. Double hydroxylation to the sterically less hindered face of the alkene moiety, followed by lactone reduction and methyl glycosidation furnishes **134** in 42% overall yield based on **131**.

Ikemoto and Schreiber [93] have prepared (−)-hikizimycin starting from L-(+)-tartaric acid for the hikosamine portion and from D-glucose for the kanosamine part. The synthesis of a suitably protected form of hikosamine follows a two-directional chain strategy with terminus differentiation [94] (Scheme 13.51). L-(+)-Diisopropyl tartrate, which will provide the C(6) and C(7) stereocenters of the undecose, is benzylated. In the same pot, reduction with DIBAL-H and Wittig–Horner–Emmons double-chain elongation provides **135**. Double dihydroxylation of **135** follows Kishi's rule giving a tetrol with high diastereoselectivity that is protected as its silyl tetraether **136** [95]. Desymmetrization of the diethyl octadioate **136** is possible with DIBAL-H in CH_2Cl_2 and generates alcohol **137** as the main product. This can be attributed to an entropy effect; once DIBAL-H, a dimeric reagent, has reacted with one of the two carboxylic moieties, a highly polar intermediate (aluminium alcoholate) is formed, which blocks a large number of solvent molecules

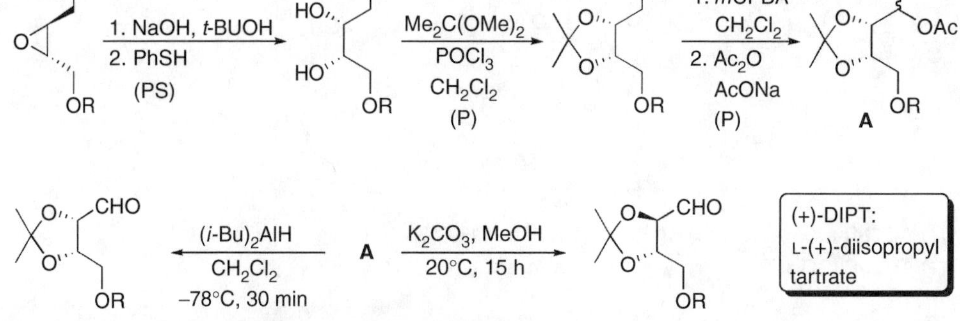

SCHEME 13.48 Sharpless and Masamune's syntheses of tetroses and hexoses.

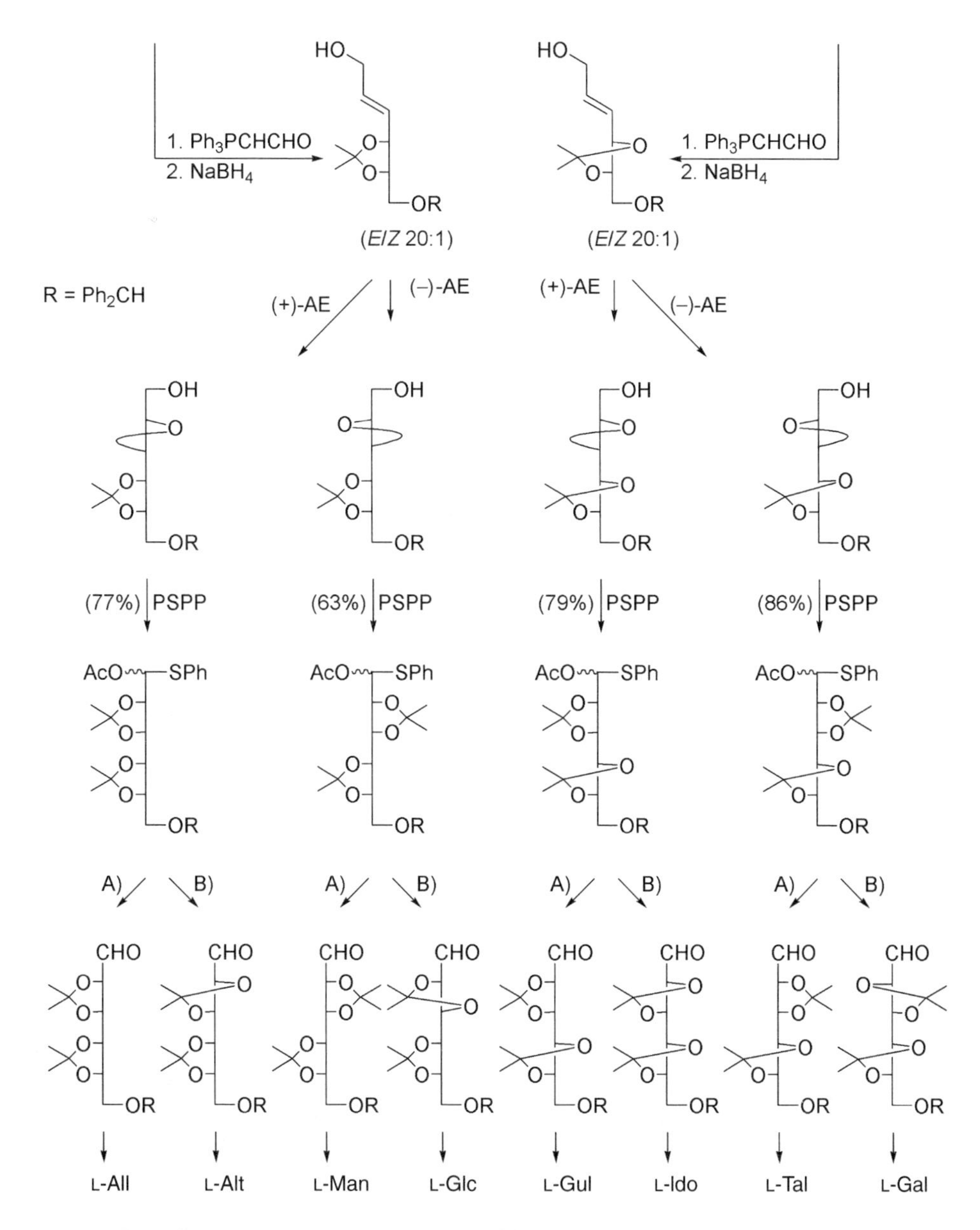

(+)-AE: $Ti(i\text{-}PrO)_4$, L-(+)-diisopropyl tartrate: *t*-BuOOH (see Scheme 13.47)

(–)-AE: $Ti(i\text{-}PrO)_4$, D-(–)-diisopropyl tartrate: *t*-BuOOH

PSPP: 1) NaOH, *t*-BuOH (Payne rearr.); 2) PhSH (substitution); 3) $Me_2C(OMe)_2$, TsOH (protection); 4) mCPBA, CH_2Cl_2, –78°C; 5) Ac_2O, AcONa (Pummerer reaction)

A) $(i\text{-}Bu)_2AlH$

B) K_2CO_3/MeOH

R = Ph_2CH (benzhydryl); PCC: pyridinium chlorochromate

SCHEME 13.48 (Continued)

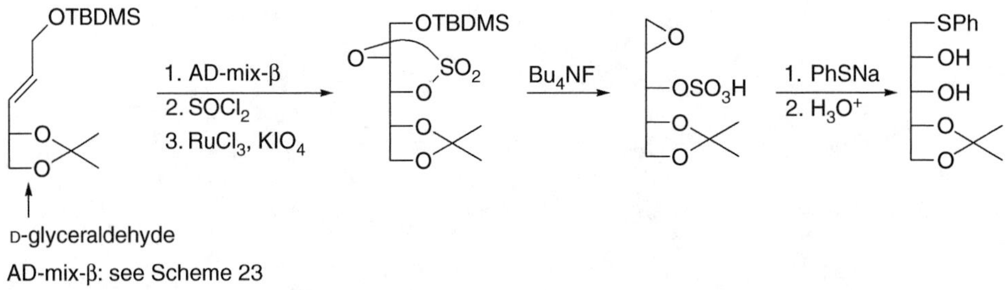

SCHEME 13.49 Sharpless asymmetric dihydroxylation of an allylic silyl ether; configuration inversion via regioselective intramolecular displacement of cyclic sulfates.

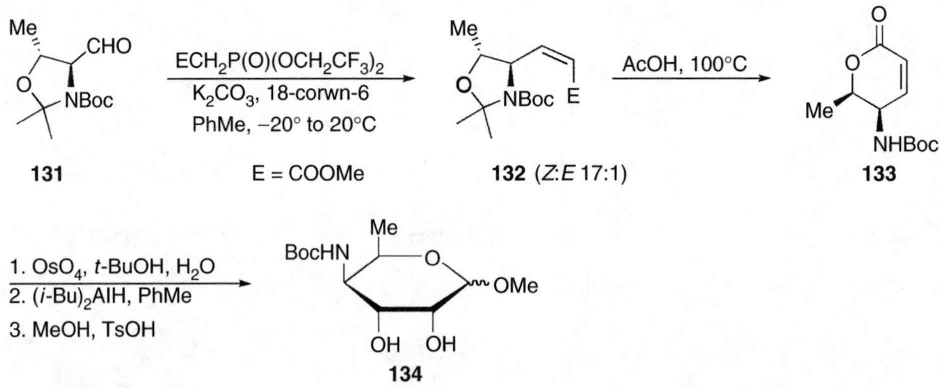

SCHEME 13.50 Conversion of L-threonine into a 4-amino-4,6-dideoxy-D-*gulo*-pyranoside.

(CH_2Cl_2, "Napalm effect") and thus dramatically increases the mass of the system compared with the starting material. The reaction of the second carboxylic moiety is thus retarded because of a more negative entropy of condensation. (The larger the masses of two reactants, the more negative the entropy change of their condensation. If the rate constant ratio of the two successive reductions were $k_1/k_2 = 2$, then a maximum yield of 49% would have been obtained for **137**.) Swern oxidation of **137**, followed by Tebbe vinylation, ester group reduction into a primary alcohol, Swern oxidation into the corresponding aldehyde and Wittig–Horner–Emmons olefination generates **138**. The four silyl ethers are exchanged for two acetonides giving **139**. Double hydroxylation in the presence of dihydroquinine *p*-chlorobenzoate leads to tetrol **140** with a good diastereoselectivity. Acidic treatment and diol protection affords the γ-lactone **141**, which is reduced into the corresponding furanose. Treatment with benzoyl chloride generates pyranoside **142** with an unprotected 4-hydroxy group (steric hindrance). The latter is converted to a triflate and then displaced with azide anion to give **143**. This operation introduces the nitrogen moiety with the required configuration. Acetonide methanolysis, followed by methanolysis of the benzoates and acetylation, generates **144**, which undergoes Vorbrüggen's glycosidation with bis(trimethylsilyl)cytosine, giving an intermediate that is acetylated and oxidized with dichlorodicyanoquinone (DDQ). This leads to a site-selective debenzylation of the 6-benzyloxy moiety giving **145**. Glycosidation of **145** with **146** under Kahne's conditions [96] gives **147**. Deprotection and hydrogenation of **147** furnishes (−)-hikizimycin.

13.5.5.7 Aldehyde Olefinations and Conjugate Additions

The first asymmetric total synthesis of acosamine and daunosamine starting from nonsugar precursor was reported by Fuganti and coworkers [97]. They found that baker's yeast catalyzes the asymmetric pinacolic cross-coupling of cinnamaldehyde and ethenal giving the *anti* diol **148** [98]. This diol was protected as an acetonide and submitted to ozonolysis giving L-**149**. Olefination of L-**149** with MeOOCCH=PPh_3, followed by treatment with ammonia, provided **150**, which was then converted into *N*-trifluoroacetylacosamine **151** (Scheme 13.52).

Methyl 3-*epi*-D-daunosaminide **154** has been derived from D-**149**′ via a Wittig-type olefination using (2-thiazolylmethylene)triphenylphosphorane (Scheme 13.53). A 1:1 mixture of (*E*)- and (*Z*)-alkenes is obtained, which is isomerized in the presence of iodine into a 9:1 mixture of (*E*)-**152** and (*Z*)-**152**. Methylation of the thiazole moiety increases the electrophilicity of the alkene, which then accepts nucleophiles such as benzylamine. The adduct is treated with $NaBH_4$ to give a thiazolidine. Acetylation and mercury-mediated hydrolysis of the thiazolidine ring generates **153**, which, on acidic treatment in methanol, yields the *N*-benzyl 3-*epi*-D-daunosaminide **154** [99].

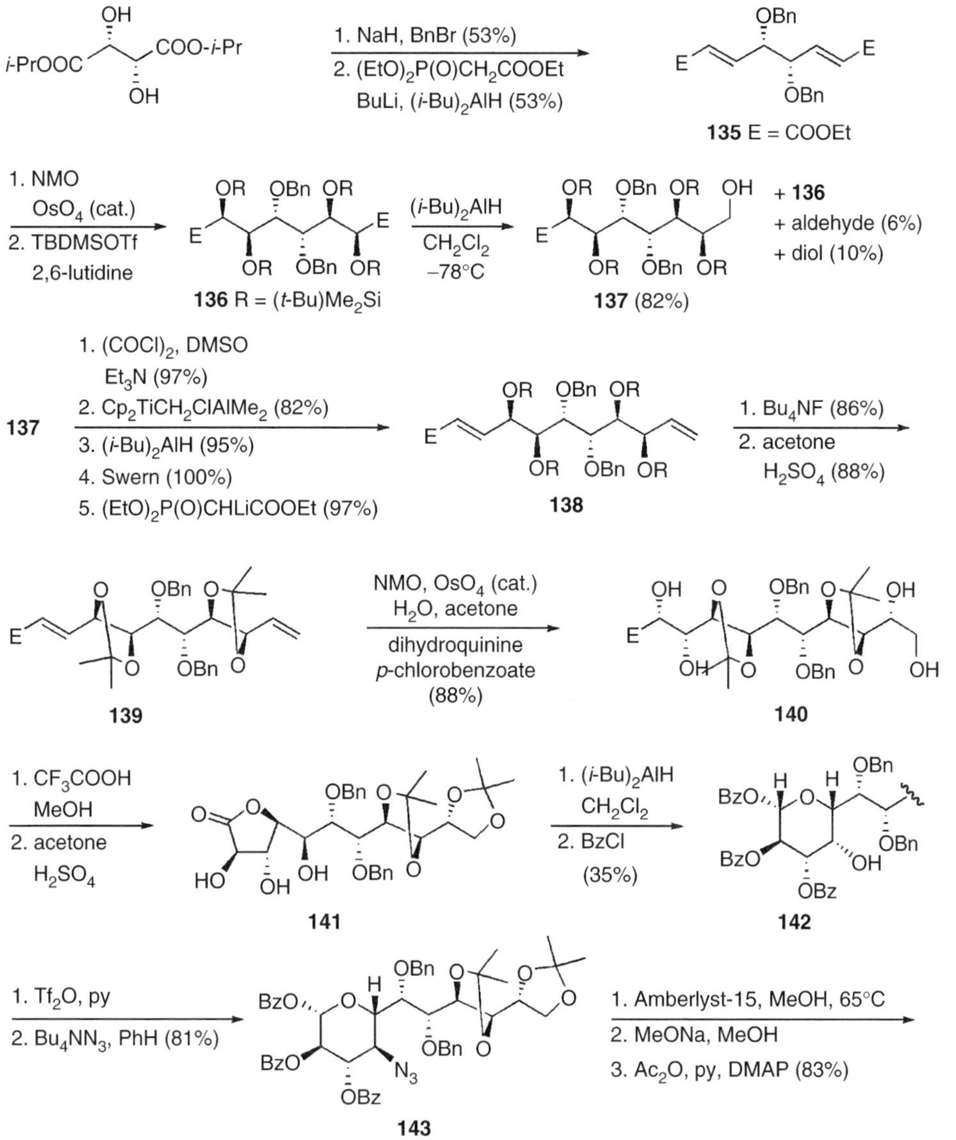

SCHEME 13.51 Schreiber's synthesis of (−)-hikizimycin featuring a two-directional chain-elongation strategy involving an aldehyde olefination and face-selective dihydroxylations.

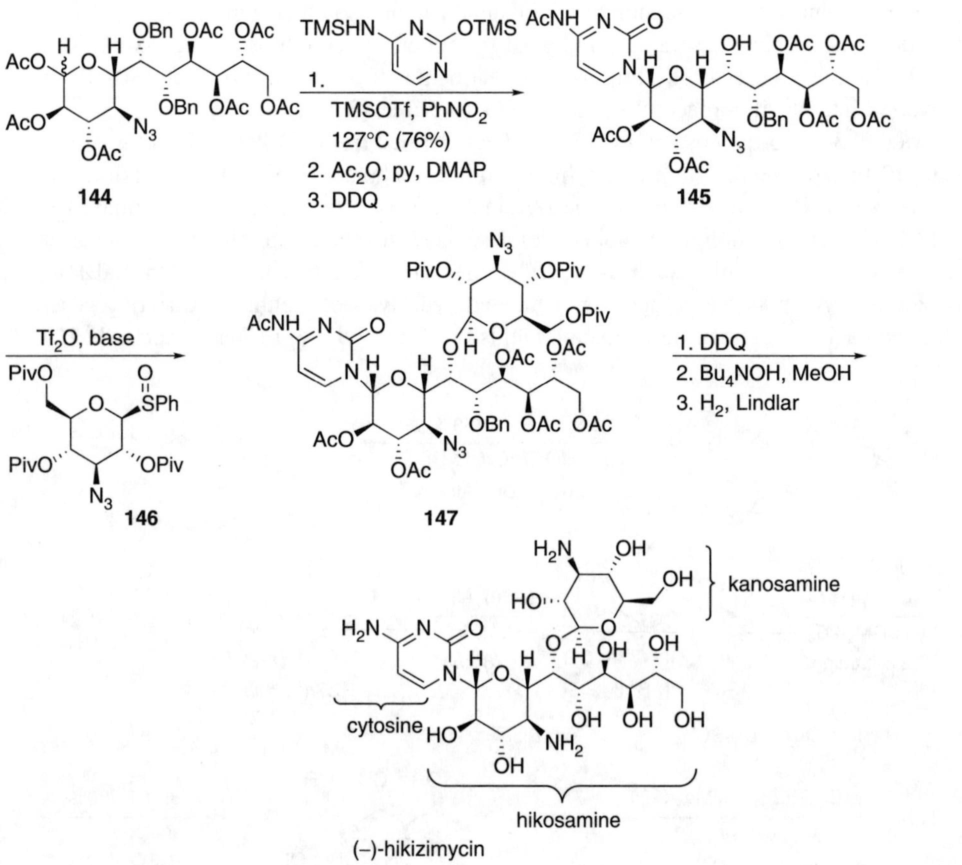

SCHEME 13.51 (Continued)

13.5.5.8 Allylation and Subsequent Ozonolysis

The biologically important 2-deoxypentoses can be prepared readily by the two-carbon chain elongation of 2,3-*O*-isopropylidene-D-glyceraldehyde following Roush's allylation method (Scheme 13.54), relying on the highly diastereoselective additions of enantiomerically pure allylboronates derived from (*R*,*R*)- and (*S*,*S*)-tartaric acid [100]. Similarly, the 2,6-dideoxyhexose derivative **155** was obtained by Roush and Straub (Scheme 13.54) [101].

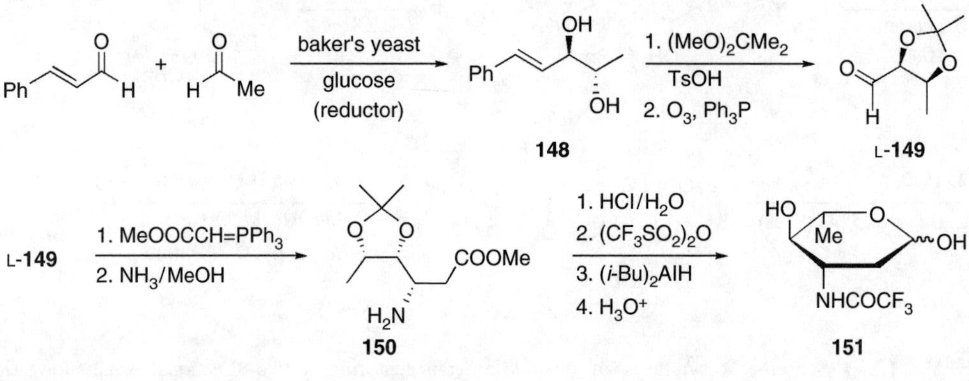

SCHEME 13.52 Fuganti's synthesis of *N*-trifluoroacetyl-L-acosamine.

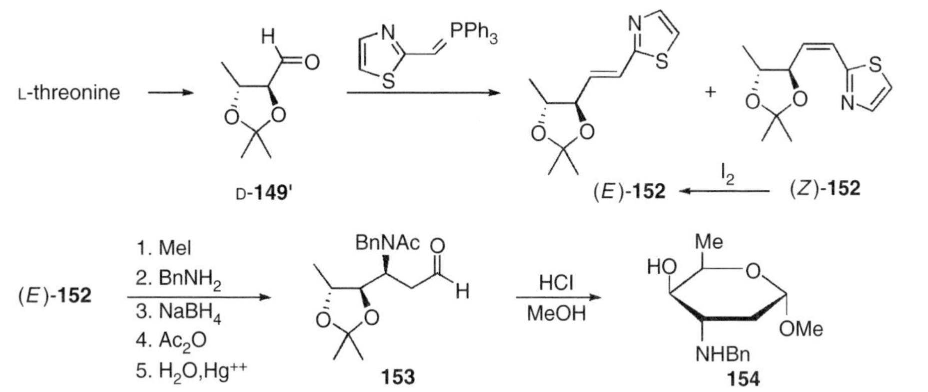

SCHEME 13.53 Dondoni's synthesis of methyl *N*-benzyl-3-*epi*-D-daunosaminide.

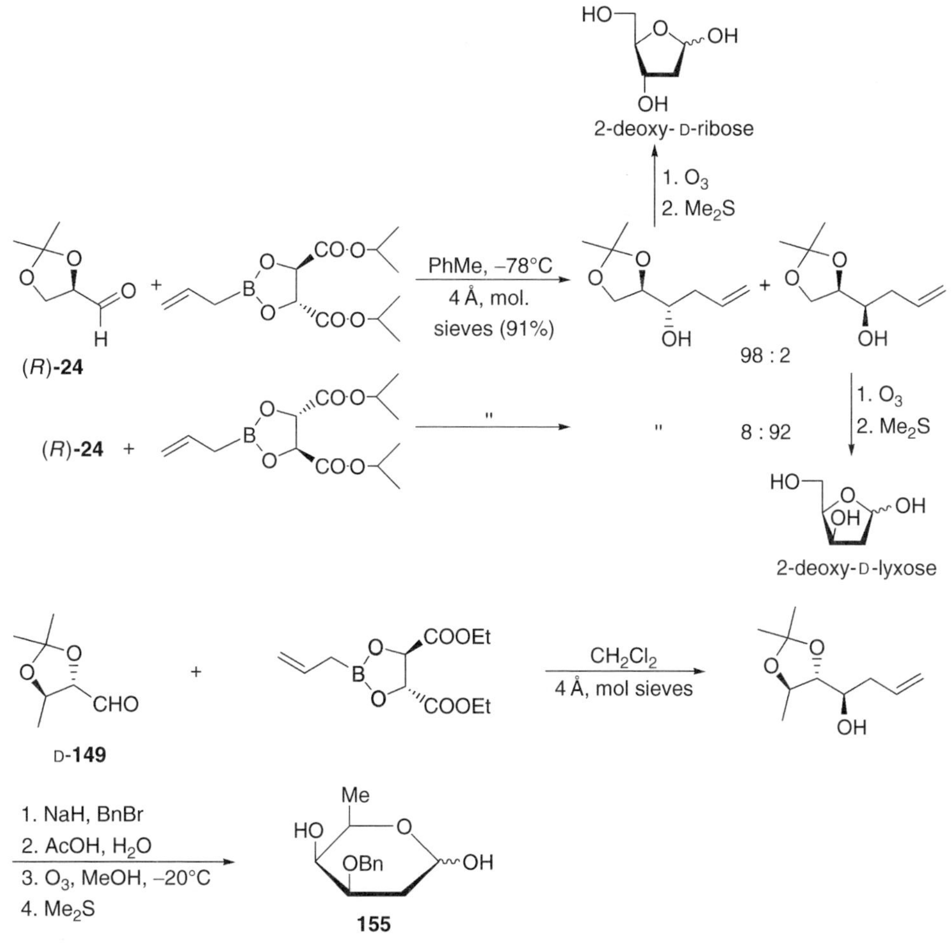

SCHEME 13.54 Roush's synthesis of 2-deoxyaldoses.

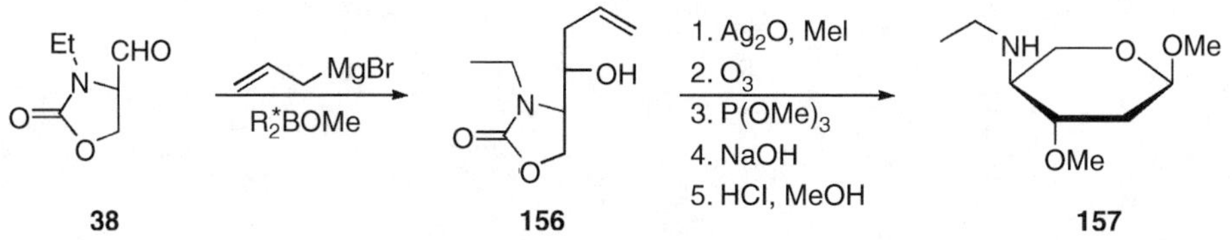

SCHEME 13.55 Asymmetric allylation catalyzed by (−)-β-methoxydiisopinocamphenylborane.

The addition of allylmagnesium bromide to the serinal derivative **38** is *syn* selective in the presence of (−)-β-methoxydiisopinocampheylborane, giving the *erythro* derivative **156** (Scheme 13.55). Methylation of the alcoholic moiety of **156**, followed by ozonolysis with reductive workup, hydrolysis of the carbamate and acidic treatment forms the methyl glycoside **157** of the E-ring moiety of calicheamicin [61]. A similar approach was proposed by Roush starting from the L-serinal derivative **50** [102].

The aldehyde L-**149′** reacts with diallylzinc to give **158** as the major product, the ozonolysis of which, when followed by acidic hydrolysis, furnishes 2-deoxy-L-fucose, which is a component of several antibiotics [103]. Similarly, **159** can be converted into L-mycarose (2,6-dideoxy-3-*C*-methyl-L-*ribo*-hexopyranose). The same *anti* preference is observed for all these diallylzinc additions (Scheme 13.56). Using allylmagnesium bromide leads to mixtures of *anti* and *syn* adducts. The *syn* adducts derived from L-**149′** and **159** were converted as above into 2,6-dideoxy-L-*arabino*-hexopyranose **160** and into L-boivinose **161**, respectively. For other examples of tetrose derivative allylations, see Ref. [104].

The two-carbon chain elongation of protected D-glyceraldehyde (*R*)-**24** and (*R*)-**62** can be realized via (*E*)-γ-alkoxyallylboronate additions, followed by alkene ozonolysis. The allylboronate **162** generated *in situ* adds to (*R*)-**62** giving a single stereoisomer **163**, which can be converted to D-arabinose derivatives following ozonolysis (Scheme 13.57). Similarly, **164** added to (*R*)-**24** gave a major adduct **165** that was then converted into 2-*O*-methyl-D-arabinose derivatives [105].

The first noncarbohydrate-based asymmetric synthesis of kedarosamine uses the *N*,*O*-protected D-threonine **166**. It is first converted into the corresponding Weinreb amide via the acyl chloride. Subsequent coupling with the allyl Grignard reagent provides **167**. The nonchelation controlled reduction of ketone **167** with $NaBH_4$ is *syn* selective, whereas 1,2-chelation controlled reduction

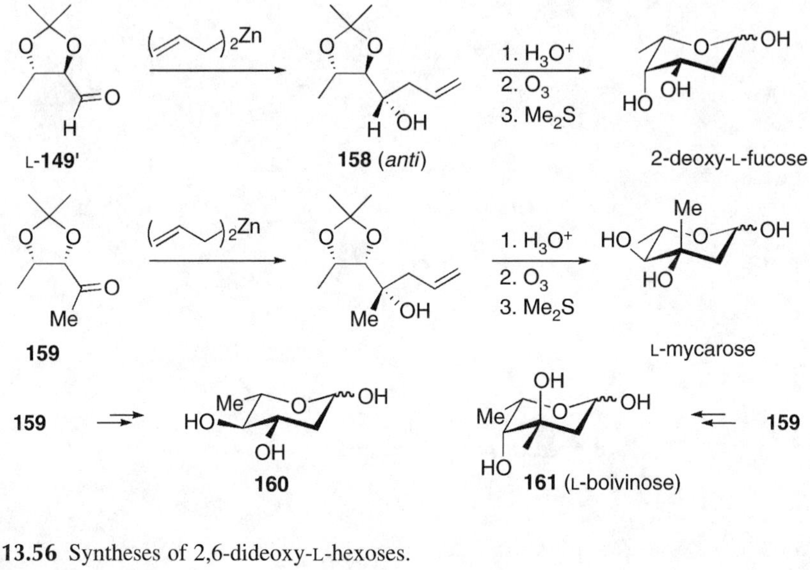

SCHEME 13.56 Syntheses of 2,6-dideoxy-L-hexoses.

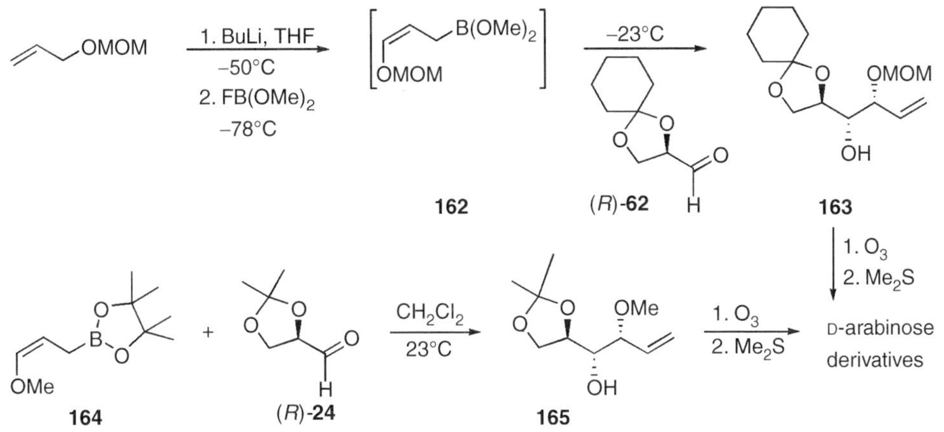

SCHEME 13.57 Roush's syntheses of D-arabinose derivatives.

with $Zn(BH_4)_2$ is *anti* selective. Deketalization of **167**, followed by reduction with the Evans' reagent, gives the *anti* alcohol **168**. The olefin is then cleaved by ozonolysis. Subsequent ring closure to the corresponding pyranoses **169** occurs spontaneously. Fischer glycosidation of **169** with methanol, followed by amine deprotection and methylation, yields methyl α-L-kedarosaminide [106] (Scheme 13.58).

13.5.6 Three-Carbon Chain Elongations

13.5.6.1 Allylmetal Additions

Roush's three-carbon chain-elongation method applied to 2,3-*O*-cyclohexylidene-D-glyceraldehyde (*R*)-**62** is an efficient approach to D-glucitol derivatives, which relies on the *syn*-selective epoxidation of homoallylic alcohol **170** (Scheme 13.59) [105,107].

Addition of allyl bromide to *N*-benzyl-*N*-carbobenzoxy-*O*-*tert*-butyldimethyl-D-serinal **171** in the presence of $SnCl_2$ and NaI preferentially gives the *anti* adduct **172** (Scheme 13.60). After protection of the alcohol, dihydroxylation affords a 1:1 diastereomeric mixture of diols.

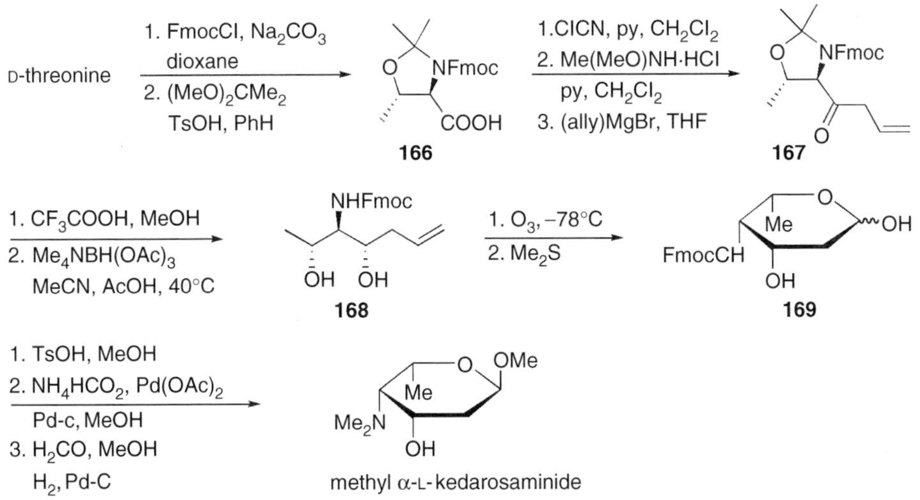

SCHEME 13.58 Kihlberg's synthesis of methyl α-L-kedarosaminide.

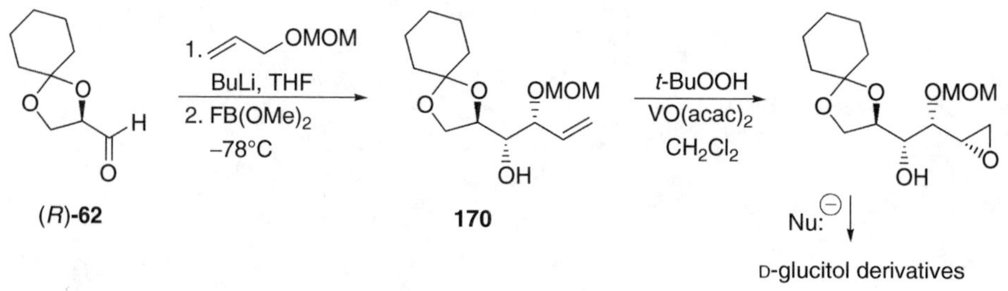

SCHEME 13.59 Roush's conversion of D-glyceraldehyde into D-glucitol derivatives.

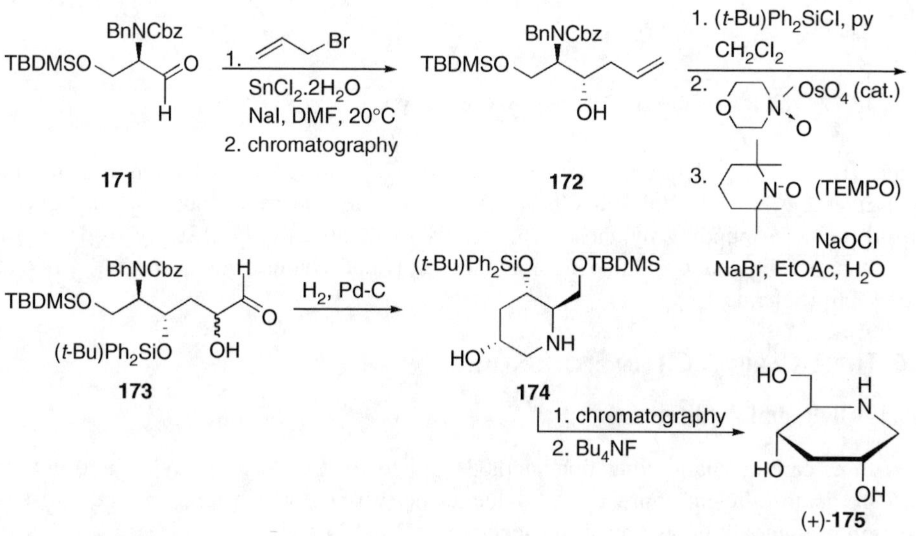

SCHEME 13.60 Jurczak's synthesis of 1,3-dideoxynojirimycin.

Oxidation with TEMPO gives α-hydroxy-aldehydes **173** that are hydrogenated to **174**. Chromatographic separation, followed by desilylation, provides enantiomerically pure 1,3-dideoxynojirimycin (+)-**175** [108].

13.5.6.2 Wittig–Horner–Emmons Olefinations

Wittig–Horner–Emmons olefination of serinal derivative L-**50** (see Scheme 13.27) with carbonylphosphorane **176** generates enone **177**. Subsequent conversion to (−)-nojirimycin L-**58** proceeds via *anti* dihydroxylation and reduction with $NaBH_4$ (Scheme 13.61), whereas reduction with Red-Al leads to (−)-mannonojirimycin [109]. A similar approach has been applied to generate galactostatine [110].

13.5.6.3 Aldol Reactions

The total syntheses of 3-deoxy-D-*ribo*- and 3-deoxy-D-*arabino*-hexose were realized via the cross-aldolization of 2,3-*O*-isopropylene-D-glyceraldehyde (*R*)-**24** and 1,1-dimethoxyacetone (Scheme 13.62). The key step of the synthesis is the diastereoselective reduction of one of the aldols **179** via boron chelates. Treatment of **179** with triisobutylborane, and then with $NaBH_4$ gives *syn*-1,3 **181** and *anti*-1,3-diol **182** in a ratio 95:5. Acidic hydrolysis provides

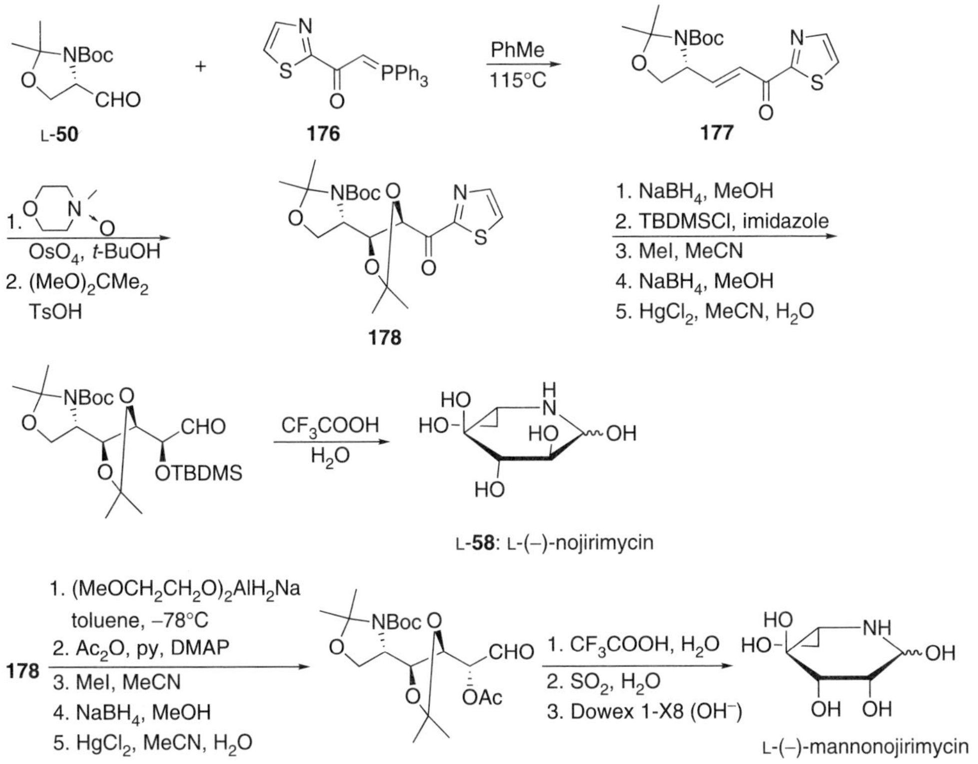

SCHEME 13.61 Dondoni's synthesis of azasugars.

3-deoxy-D-*ribo*-hexose. If aldol **179** is treated first with an equimolar amount of aluminium triisopropoxide, diol **182** is obtained in 62% yield (together with 15% of **181**). Compound **182** is then converted into 3-deoxy-D-*arabino*-hexose [111].

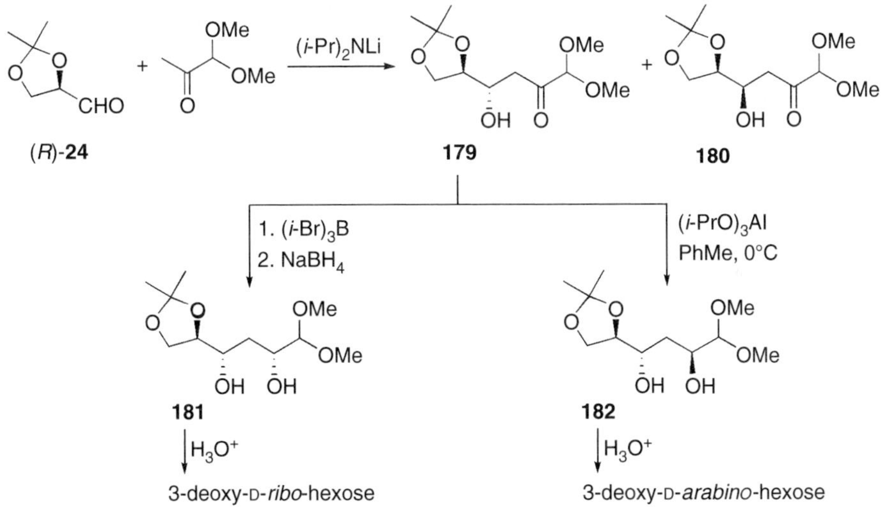

SCHEME 13.62 Conversion of D-glyceraldehyde into 3-deoxy-D-hexoses.

13.5.6.4 Propenyllithium Additions to Carboxylic Esters

Polt and Sames [112] have converted L-serine into *N*-methylfucosamine (Scheme 13.63). The method relies on the diastereoselective addition of propenyllithium to the aldehyde derived from protected L-serine derivative **183**, giving allylic alcohol **184**. Catalytic osmylation of **184** gives a 6:1 mixture of *anti,syn* **185** and *syn,syn*-aminotriols. Protection of the triol **185** as its triacetate, reductive methylation and desilylation provides **186**. Successive Swern oxidation, methanolysis and benzhydryl group hydrogenation leads to *N*-methylfucosamine **187**.

13.5.7 Four-Carbon Chain Elongations

13.5.7.1 But-2-en-1-yl Metal Additions

The addition of (*E*)-crotylboronate **188** to aldehyde (*R*)-**24** is highly *anti* diastereoselective [107]. The addition of (*R*)-**189**, a crotylstannane derivative, is also highly diastereoselective in the presence of $MgBr_2{\cdot}OEt_2$ as promoter, giving the *syn:syn* product **190** as single adduct [113] (Scheme 13.64). The protected threose derivative **190** adds to the γ-alkoxyallylstannane (*S*)-**192**, giving the corresponding *syn,syn* product of allylation using $MgBr_2{\cdot}OEt_2$ as promoter, and the *syn,anti* adduct **193** in the presence of $BF_3{\cdot}Et_2O$ promoter (S_N2' reaction, open transition state, Felkin–Anh rule) [114].

Treatment of α-alkoxyallylstannane with $InCl_3$ generates the corresponding γ-alkoxyallylindium dichloride (*R*)-**195**, which adds to aldehyde **191** (six-membered transition state) with high *anti,anti* diastereoselectivity, producing **196** [115]. The octitol derivative **201** has been prepared via BF_3 etherate-promoted addition of (*R*)-γ-OTBDMS allylic stannane (*R*)-**189** to enal **198** derived from L-serine (Scheme 13.65) [116]. This leads to a single *syn* adduct **199**, which is converted to the silyl ether **200**. This compound adopts probably the Saito conformation, which makes the double *cis*-dihydroxylation doubly diastereoselective with the formation of tetrol **201** as major product. Selective oxidative cleavage of tetrol **201** affords a lactol. Desilylation followed by acetonide hydrolysis generates **202**, a precursor of destomic acid [117].

13.5.7.2 Nucleophilic Additions of α-Furyl Derivatives

A synthesis of (+)-KDO has been reported by Martin and Zinke [118], which relies on the highly stereoselective addition of 2-furyllithium to 2,3-*O*-isopropylidene-D-glyceraldehyde (*R*)-**24**.

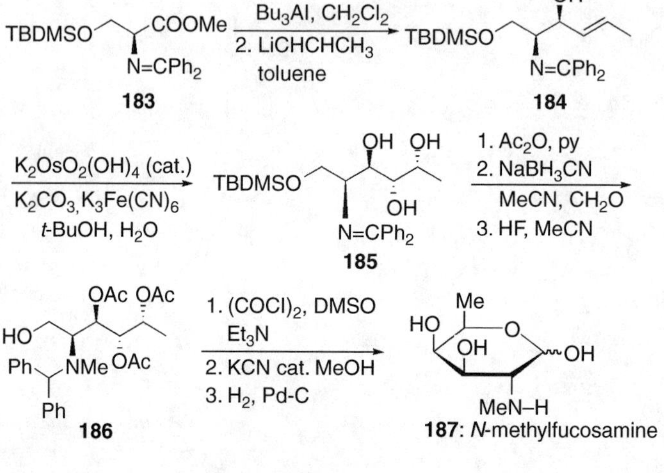

SCHEME 13.63 Polt and Sames' synthesis of *N*-methylfucosamine.

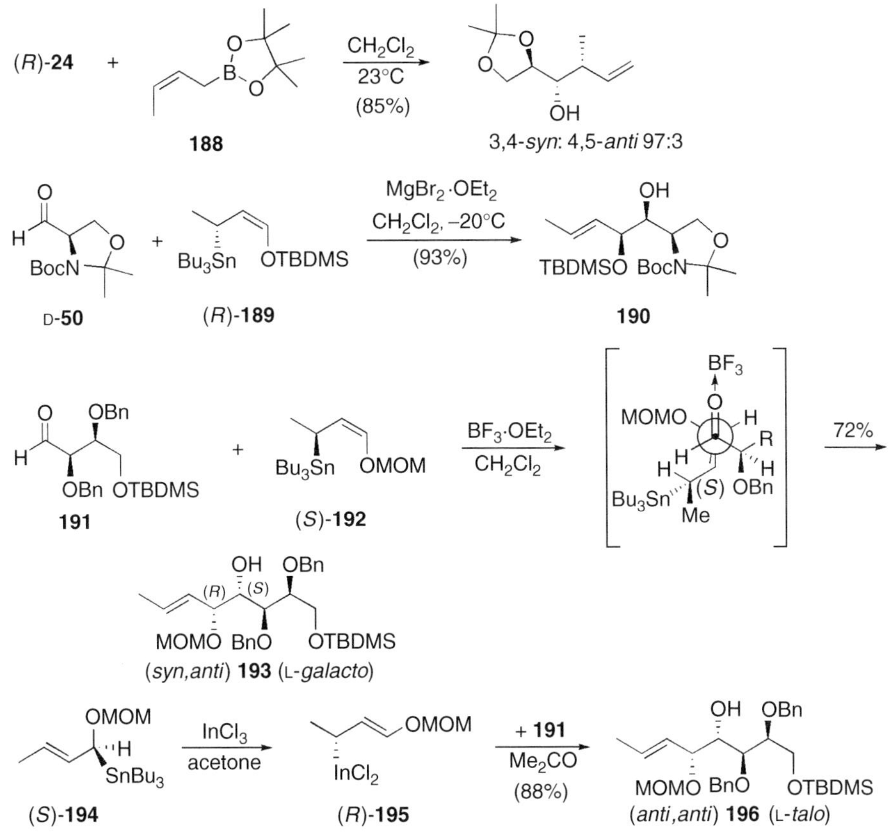

SCHEME 13.64 Diastereoselective crotyl group additions.

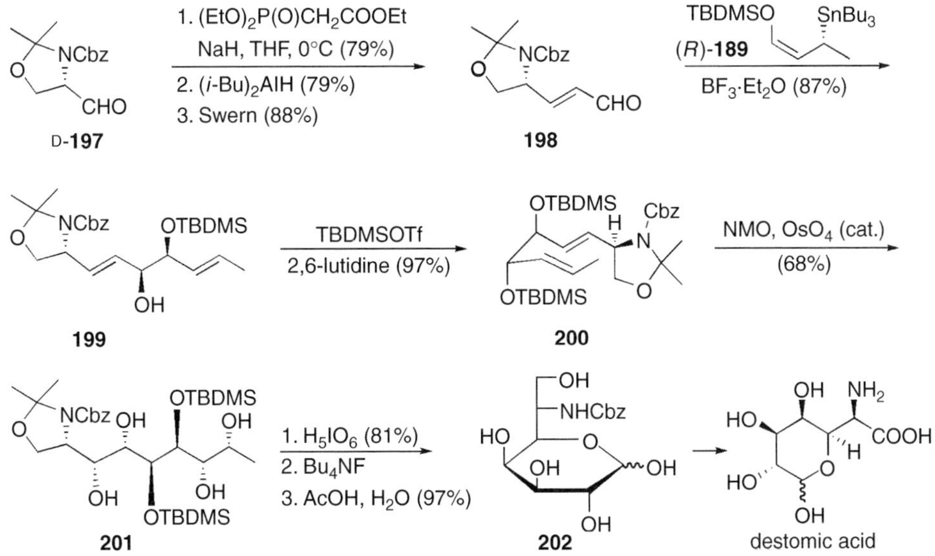

SCHEME 13.65 Marshall's allylation applied to the synthesis of destomic acid.

Trapping of the alcohol intermediate with (*t*-Bu)Me_2SiCl gives **203** in 53% overall yield. Metalation of the furan ring of **203** and sequential addition of benzyl chloromethyl ether and desilylation provides **204**. Treatment of **204** with *t*-BuOOH in the presence of $VO(acac)_2$ and subsequent *O*-methylation of the resulting lactols (α/β 4.5:1) delivers methyl glycoside **205** together with its β-anomer. Stereoselective ketone reduction with K-Selectride gives **206** (9.8:1 diastereoselectivity), which is converted into carbamate **207**. Iodocyclization of **207** furnishes carbonate **208**. Deiodination followed by debenzylation provides alcohol **209** (78%). Swern oxidation of **209**, followed by oxidation of the intermediate aldehyde under conditions that induce hydrolysis of the carbonate moiety, give **210** and its deprotection concludes the synthesis of (+)-KDO (Scheme 13.66).

Condensation of 2,3-*O*-isopropylidene-D-glyceraldehyde-*N*-benzylimine **211** with 2-(trimethylsilyloxy)furan in the presence of a Lewis acid generates a mixture of butenolides **212** that is converted into D-*ribo*-*N,N*-diprotected derivative **213** in 68% yield (Scheme 13.67) [119]. Dihydroxylation of **213**, followed by diol protection and lactone reduction, furnishes the 5-amino-5-deoxyheptose derivative **214**. Hydrogenolysis in MeOH generates **215** and its deprotection furnishes enantiomerically pure 1,5-dideoxy-1,5-imino-D-*glycero*-D-*allo*-heptitol **216** [120].

13.5.7.3 Hydroxyalkylation of Pyrrole Derivatives

The Lewis acid-promoted condensations of *N*-(*t*-butylcarbonyl)-2-(*t*-butyldimethylsiloxy)pyrrole with 2,3-*O*-isopropylidene-D-glyceraldehyde (*R*)-**24** generate pyrrolidones with high diastereoselectivity. For instance, with 1.5 equivalents of $SnCl_4$ in ether the reaction gives crystalline D-*arabino*-configurated α,β-unsaturated γ-lactam **217** as the sole product in 80% yield, whereas, in the presence of one equivalent of BF_3 etherate, reversal of the stereochemical preference is observed, resulting in predominant formation of crystalline D-*ribo*-epimer **218** (70% yield), along with less than 20% of **217**. Almost quantitative epimerization **217** → **218** occurs on treatment with

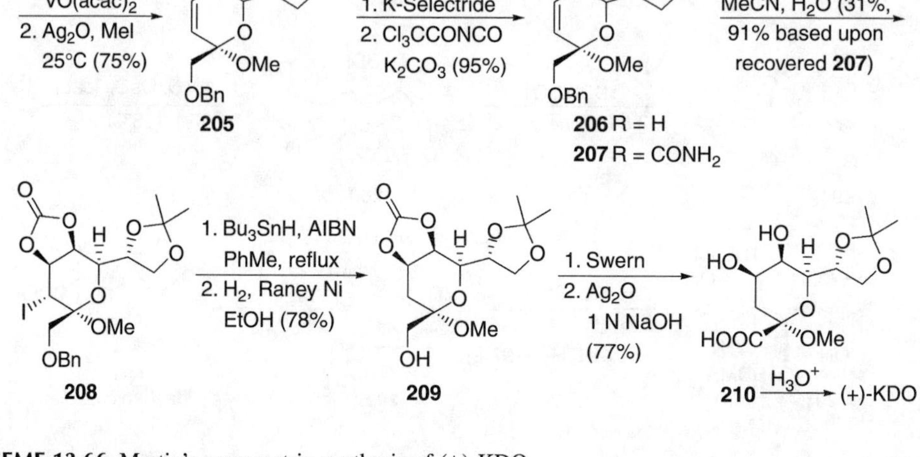

SCHEME 13.66 Martin's asymmetric synthesis of (+)-KDO.

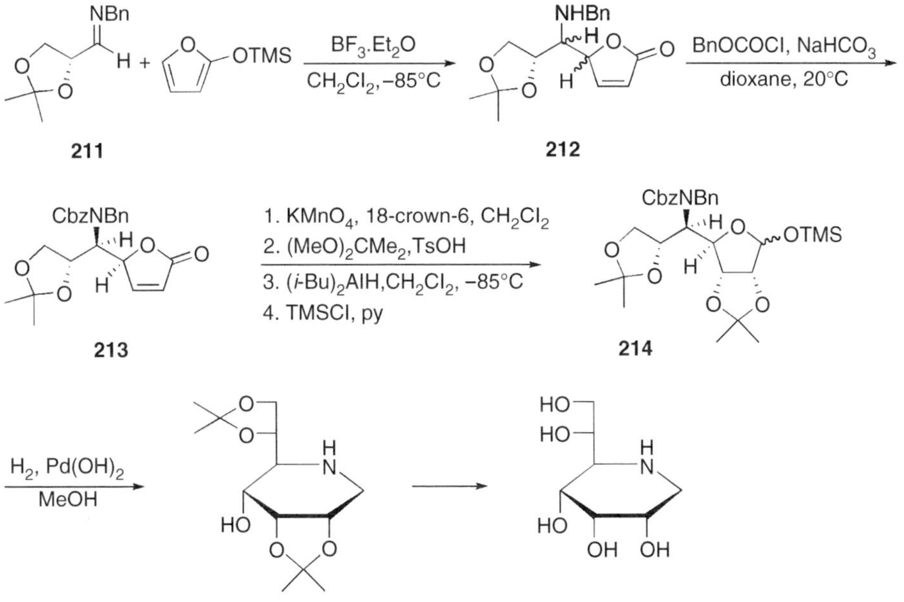

SCHEME 13.67 Casiraghi's synthesis of a 1,5-dideoxy-1,5-iminoheptitol.

Et_3N in CH_2Cl_2 in the presence of 4-dimethylaminopyridine (Scheme 13.68). Dihydroxylations of the trimethylsilyl ethers of **217** and **218** generate the 4-amino-4-deoxy-heptono-1,4-lactam derivatives **219** and **220**, respectively [121]. Lactam hydrolysis of **219** with LiOH, followed by the Malaprade diol cleavage with $NaIO_4$ and further oxidation and deprotection, allows the preparation of 4-*epi*-polyoxamic acid [122]. Lactam **217** and its enantiomer derived from (*S*)-**24** have been converted into all four stereomers of *cis*-1,2-dihydroxypyrrolizidine [123]. Compounds **217** and **218** have been used also to prepare the *trans*-2,3-*cis*-3,4-dihydroxyprolines [124,125].

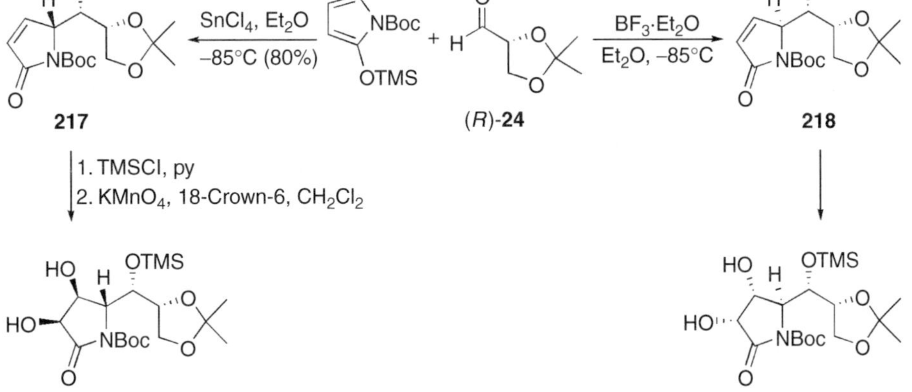

SCHEME 13.68 Aminodeoxyaldonolactams through hydroxyalkylation of a pyrrole derivative.

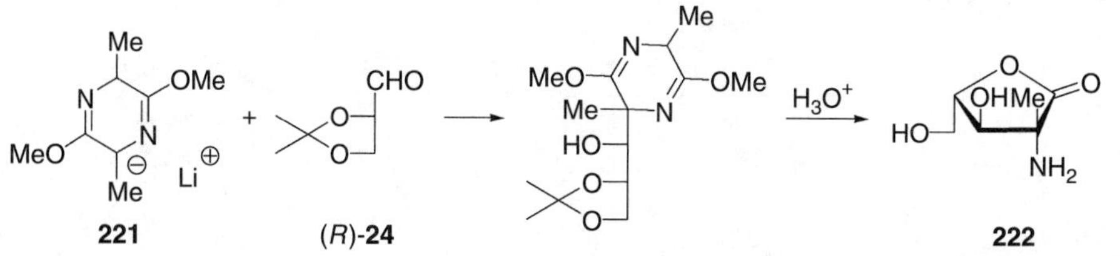

SCHEME 13.69 Synthesis of a branched-chain amino aldonolactone.

13.5.8 Synthesis of Branched-Chain Monosaccharides from C_3-Aldoses

Most of the methods presented for the *de novo* synthesis of "linear" monosaccharides can be used to prepare branched-chain sugars and analogs and some examples are given above and below. The branched-chain aminolactone **222** was prepared by two-carbon chain elongation via addition of **221** to 2,3-*O*-isopropylidene-L-glyceraldehyde (Scheme 13.69) [126].

Aldol reaction of (*S*)-2-benzyloxypropanal with the lithium enolate of methyl 2-methoxy-propanoate gives a 7:2:1 mixture of β-hydroxyesters (Scheme 13.70). After protection of the alcohol moiety, the isomeric mixture is reduced with $LiAlH_4$ and the resulting primary alcohols separated by chromatography on silica gel. Oxidation of the major alcohol **223** (isolated in 40% yield) into an aldehyde is followed by Wittig methylenation. This provides **224**. Hydroboration of **224** gives a primary alcohol that is oxidized (Swern) into aldehyde **225**. Hydrogenation yields L-cladinose, a saccharide moiety of erythromycin A [127].

The isomeric methyl 3,6-dideoxy-3-*C*-methylhexofuranosides **230** and **232** (Scheme 13.71) were derived from (*S*)-2-benzyloxypropanal via homoaldol reactions in three steps. Addition of α-titanated (*E*)-2-butenylcarbamate (±)-**226** to (*S*)-2-benzyloxypropanal gives in 80% yield a 53:47 mixture of enantiomerically pure (*Z*)-3,4-antiadducts **227** and **228**. These are separated by chromatography on silica gel. Face-selective epoxidation of **227** yields **229**, the acidic methanolysis of which forms **230**. Similarly, adduct **228** is converted into epoxide **231** and then into methyl furanoside **232** in high yield [128].

L-Arcanose and L-olimycose were prepared in enantiomerically pure forms and with high stereoselectivity by the Lewis-acid promoted addition of (*S*)-2-benzyloxypropanal to

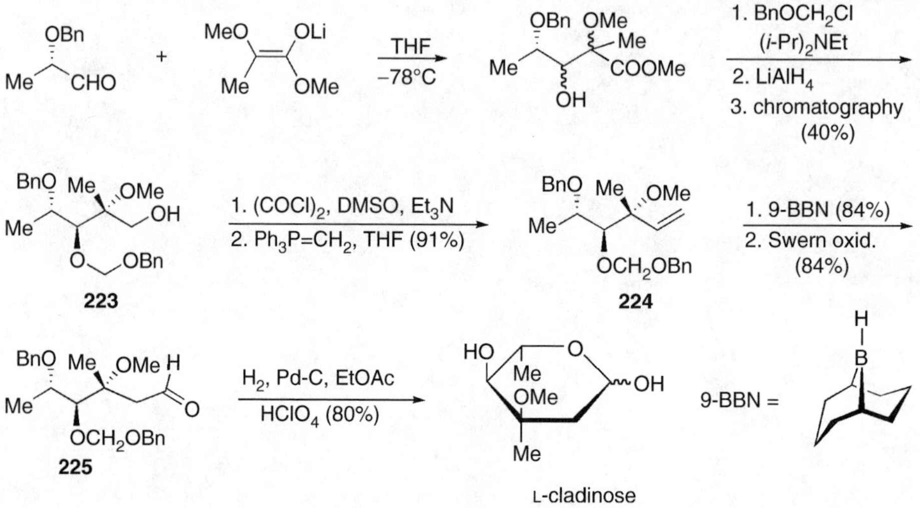

SCHEME 13.70 Heathcock's synthesis of L-cladinose.

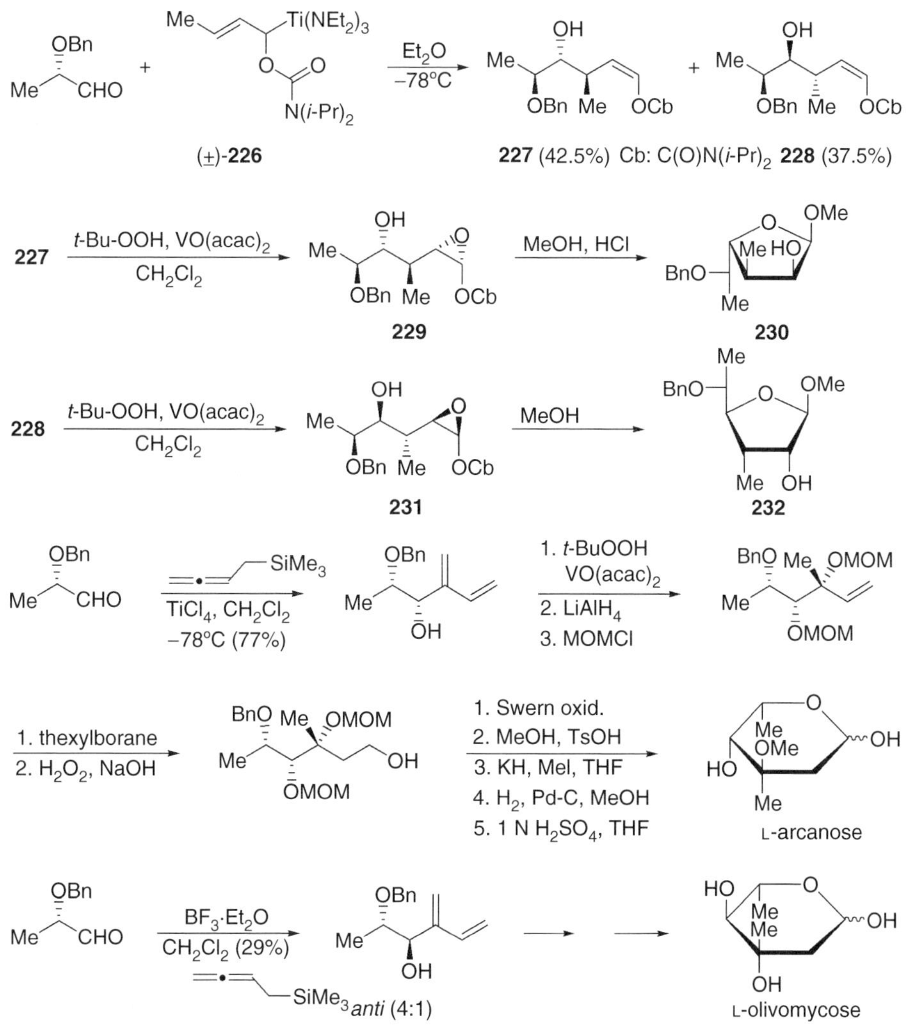

SCHEME 13.71 Syntheses of branched-chain monosaccharides starting from (*S*)-2-benzyloxypropanal.

1-trimethylsilyl-2,3-butadiene (Scheme 13.71). Depending on the nature of the Lewis acid either the *syn* (with $TiCl_4$) or the *anti* adduct (with $BF_3{\cdot}Et_2O$) can be obtained. Epoxidation with lateral control by the allylic alcohol moieties and standard reactions lead to the unprotected monosaccharides [129].

Mukaiyama and coworkers [37] (Scheme 13.72) have prepared 4-*C*-methyl-D-ribose via their asymmetric cross-aldolization between α-methylacrolein and enoxysilane **74** (Scheme 13.33).

Total syntheses of 2,3-dideoxy-3-*C*-methyl-D-*manno*-heptose **235** and of 2,3-dideoxy-2,3-di-*C*-methyl-D-*glycero*-D-*galacto*-heptose **236** have been realized by addition of 2-(trimethylsiloxy)furan to 2,3-*O*-isopropylidene-D-glyceraldehyde (*R*)-**24**. In the presence of $BF_3{\cdot}Et_2O$, the α,β-unsaturated lactone **233** is obtained, which undergoes Michael addition of Me_2CuLi with high facial selectivity to give lactone **234**. Reduction followed by acidic workup leads to **275** (Scheme 13.73). α-Methylation of **234** and subsequent reduction furnishes **236** [230].

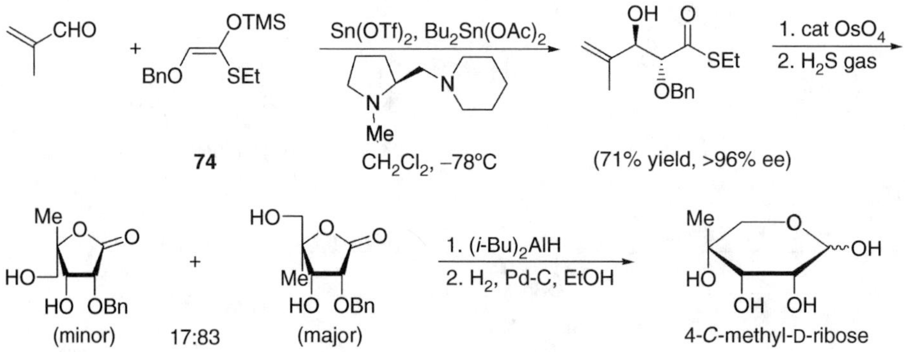

SCHEME 13.72 Mukaiyama's synthesis of a branched-chain aldose.

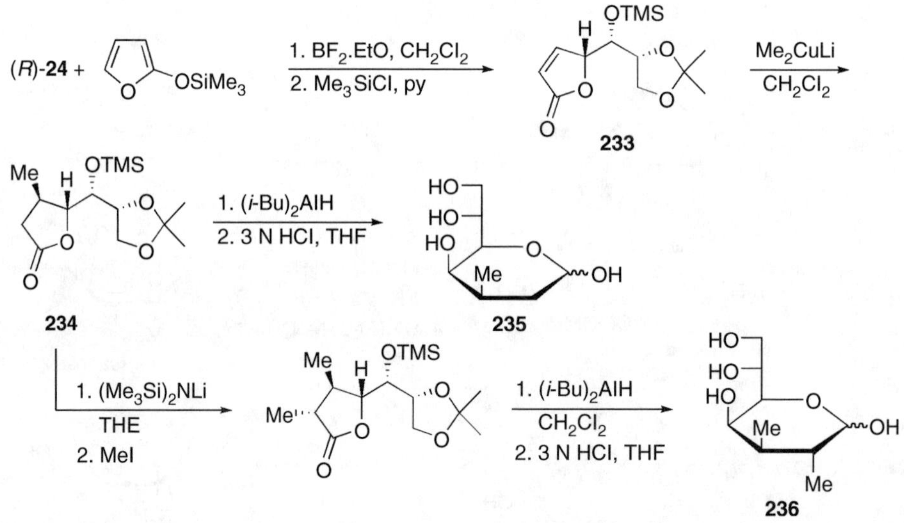

SCHEME 13.73 Casiraghi's synthesis of branched-chain heptoses.

13.6 HETERO DIELS–ALDER ADDITIONS

13.6.1 Achiral Aldehydes as Dienophiles

The cycloadditions of achiral oxy-substituted dienes with aldehydes in the presence of enantiomerically pure (+)-Eu(hfc)$_3$ shows only modest enantioselectivities. Similarly, modest diastereoselectivities are observed for the reactions of chiral oxy-substituted dienes (e.g., **237**) with aldehydes in the presence of achiral Eu(fod)$_3$. However, the combination of chiral dienes with chiral (+)-Eu(hfc)$_3$ catalyst shows interesting interactivities, resulting in some instances in diastereoselectivities of 97: 3. This is exemplified with Bednarski and Danishefsky's total synthesis of L-glucose (Scheme 13.74) [131]. In this synthesis, diene **237** having the 1-phenylmenthyloxy substituent [132] adds to benzaldehyde in the presence of catalytic Eu(hfc)$_3$ giving a 25:1 mixture of adducts **238** and **239**. Treatment of pure **238** (obtained by crystallization) with CF_3COOH gives L-dihydropyrone **240**. Oxidation of **240** with Mn(OAc)$_4$ [133] introduces the 4-acetoxy group from the less sterically hindered face of the enone. Reduction of enone **241** under Luche's conditions [134] gives rise to the L-glucal analog **242**. Osmylation and subsequent acetylation generates **243**. Ozonolysis of **243**, followed by oxidative treatment with H_2O_2, gives the protected L-glucuronic

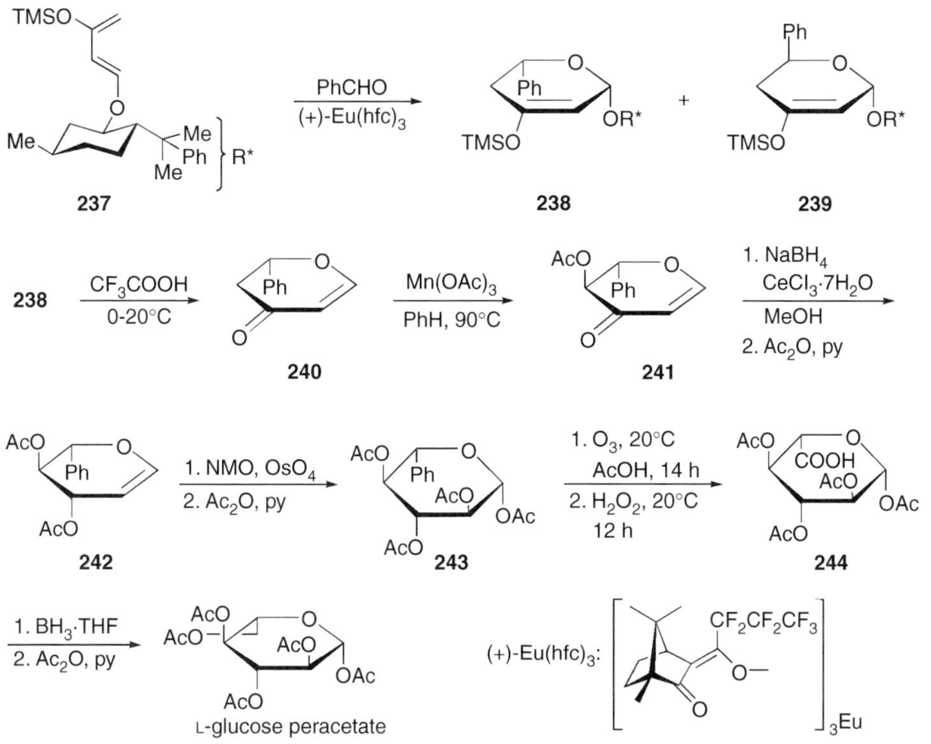

SCHEME 13.74 Danishefsky's synthesis of L-glucose.

acid **244**, the borane reduction of which furnishes the peracetate of L-glucose [131] following acetylation.

By applying a similar approach, Wu and coworkers [135] observed a highly double-stereoselective hetero Diels–Alder addition between diene **245** and ethyl glyoxylate. This reaction was catalyzed by the (salen)Co^{II} complex **246** (Scheme 13.75). The major adduct **247** is hydroborated to give alcohol **248**. Stereochemical inversion is achieved via oxidation and subsequent reduction to **249**. Acid methanolysis followed by diol protection furnishes **250**. Quenching of the lithium enolate of **250** with phenyl disulfide and subsequent oxidation with *N*-bromosuccinimide forms, after deprotection, (+)-KDO (3-deoxy-D-*manno*-oct-2-ulosonic acid) [136a].

The reaction of 5-nitrofuran-2-carboxaldehyde and 2,3,4,6-tetra-*O*-acetyl-β-D-glucopyranosides of (*E*,*E*)-4-ethoxy-2-[(*tert*-butyl)dimethylsilyloxy]butadiene can be highly diastereoselective depending on the nature of the lanthanide Lewis acid used to promote the cycloaddition ($Yb(fod)_3$, $La(fod)_3$). The adducts thus obtained are readily converted into β-D-glucopyranosyl (1→4)-linked glycals [136b].

13.6.2 Chiral Aldehydes as Dienophiles: Synthesis of Long-Chain Sugars

The first total synthesis of (+)-KDO was presented by Danishefsky and coworkers [137]. The hetero Diels–Alder addition of α-selenoaldehyde **251** to the α-furyl-substituted diene **252** gives an adduct mixture that, on treatment with CF_3COOH, delivers a 5:1 mixture of *cis*/*trans* dihydropyrones **253** and **254**. Reduction of pure **253** generates the 1-substituted glycal **255**, which reacts with methanol as expected. After benzoylation **256** is obtained. Oxidative elimination of the phenylseleno group gives alkene **257**, which is dihydroxylated to a diol and subsequently benzoylated, thus affording

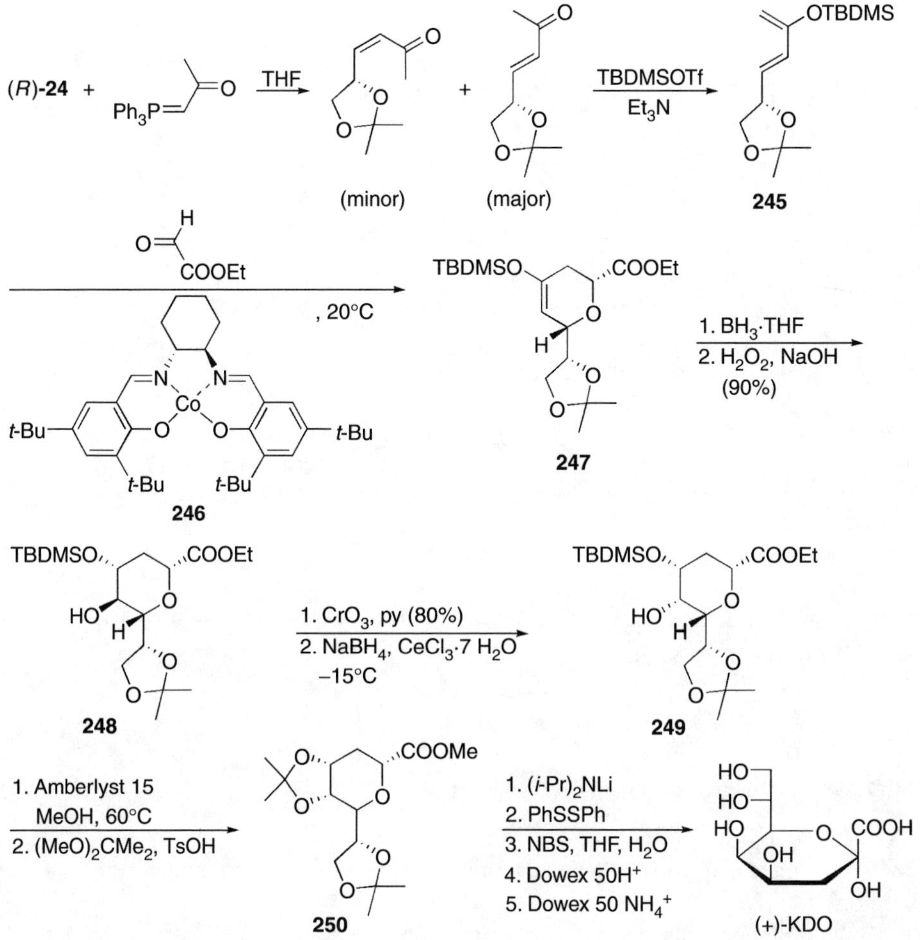

SCHEME 13.75 Wu's synthesis of (+)-KDO.

the tetrabenzoate **258**. Oxidation with RuO_4, followed by esterification with diazomethane, generates methyl KDO α-methyl glycoside from which (+)-KDO was liberated on deprotection (Scheme 13.76).

The octosyl acids isolated from *Streptomyces cacaoi* are part of a broader group of polyoxin antifungal nucleosides [138]. Danishefsky and coworkers [139] reported a total synthesis of octosyl acid A featuring the hetero Diels–Alder addition of Danishefsky's diene **258** to D-ribose-derived aldehyde **259** (which can be obtained enantiomerically pure via the "naked sugar" methodology). The pyrone **260** (Scheme 13.77) is reduced under Luche's conditions to give the corresponding allylic alcohol protected as its *p*-methoxybenzyl ether **261**. Oxidative double-bond cleavage of dihydropyran **261** generates a lactol that is oxidized to γ-lactone **262**. Hydrolysis of **262**, followed by benzylation of the liberated alcohol and esterification of the free carboxylic acid, gives **263**. Oxidative debenzylation liberates the 7-hydroxy group, which is mesylated and sequentially subjected to acid methanolysis and treatment with acetic acid to furnish the triacetate **264**. Vorbrüggen's glycosidation of 2,4-bis(trimethylsilyloxy)-5-carbomethoxypyrimidine with **264** generates **265**. Alkaline methanolysis liberates the 2,3-diol that reacts with Bu_2SnO to give the cyclic stannylene derivative **266**, which, on treatment with CsF, undergoes selective S_Ni substitution with formation of the anhydro-octonic acid derivative **267**. Deprotection of **267** furnishes octosyl acid A.

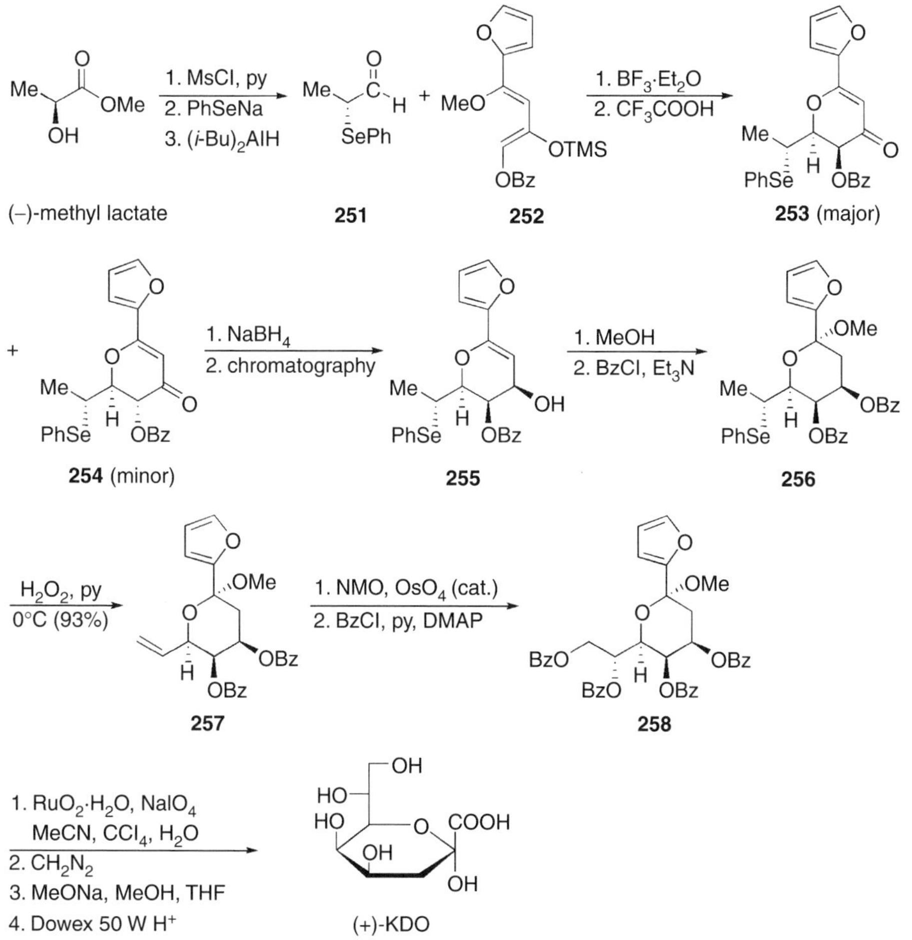

SCHEME 13.76 Danishefsky's total synthesis of (+)-KDO.

As its methyl thioglycoside, lincosamine (6-amino-6,8-dideoxy-D-*erythro*-D-*galacto*-octopyranose) is the saccharide portion of the clinically important antibiotic lincomycin. Racemic β-methyl lincosaminide was prepared by Danishefsky and coworkers [140] who utilize the hetero Diels–Alder addition of crotonaldehyde to the Danishefsky's diene **268** to afford (*E*)-*cis*-2-(1-propenyl)-3-(benzyloxy)-2,3-dihydro-4-pyrone. The $ZnBr_2$-catalyzed cycloaddition of **268** to *N*-carbobenzoxy-*O*-protected-D-*allo*-threoninal gives a 2:1 mixture of adducts **270** and **271** that has been converted into **272**, a protected form of 6-amino-6,8-dideoxy-D-*erythro*-L-*galacto*-octopyranose (Scheme 13.78). Analogously, the $ZnBr_2$-promoted cycloaddition of the D-threoninal derivative **273** to **268** leads to a 3:1 mixture of adducts. The major adduct **274** was transformed into the 6-amino-6,8-dideoxy-L-*threo*-L-*galacto*-octopyranose derivative **275** [141].

An elegant total synthesis of the semiprotected form of lincosamine was realized by Marshall and Beaudoin [116]. An aldehyde derived from destomic acid (6-amino-6-deoxy-L-*glycero*-D-*galacto*-heptonic acid) was derived in a similar way from a L-serinal derivative via hetero Diels–Alder addition to 1-ethoxy-3-[(trimethylsilyl)oxy]-4-benzyloxy-1,3-butadiene [142]. A similar method was applied to the preparation of a semiprotected form of anhydrogalantinic acid, a component of the antibiotic galantin I [142].

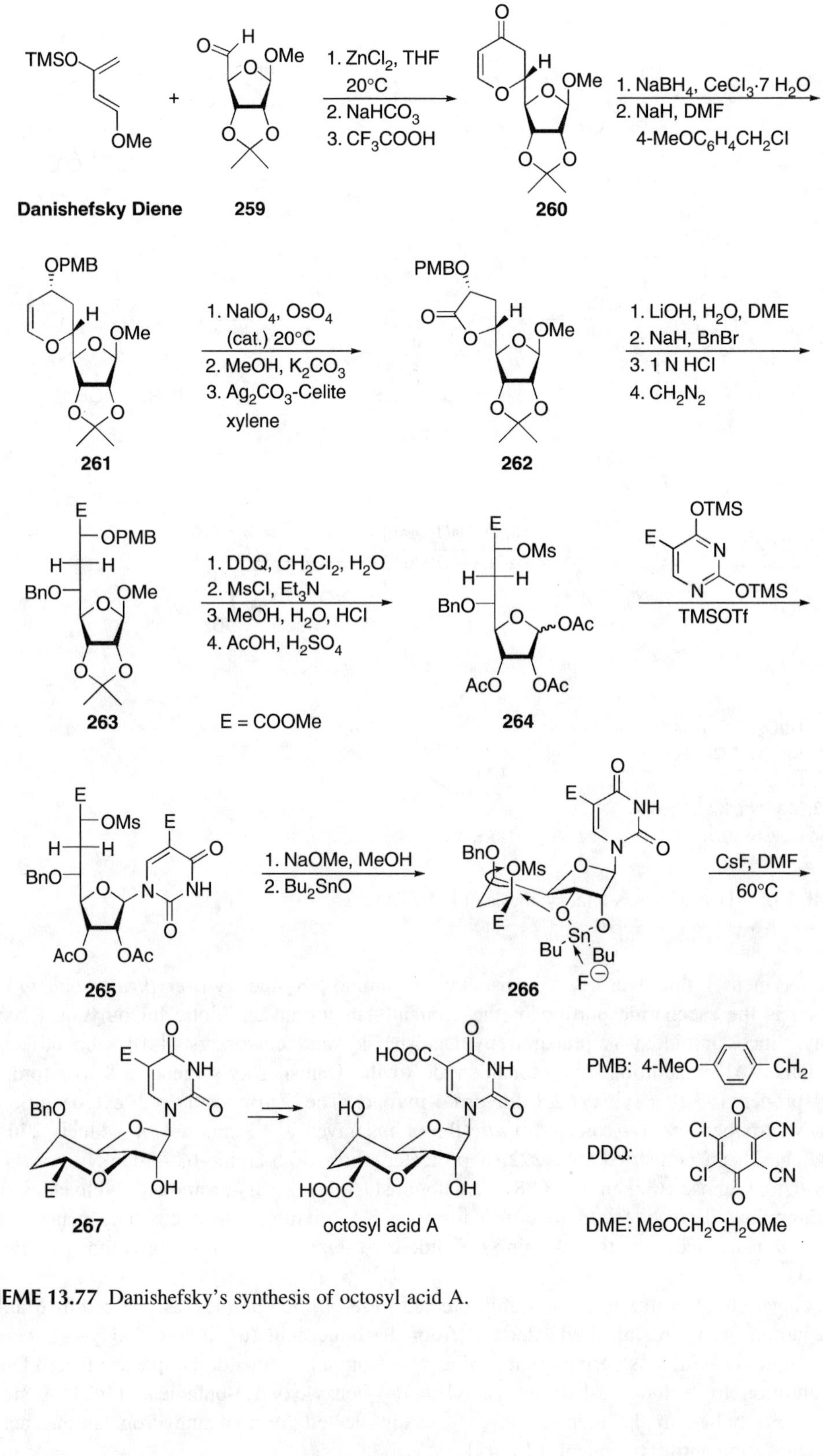

Danishefsky Diene + 259 → (1. $ZnCl_2$, THF, 20°C; 2. $NaHCO_3$; 3. CF_3COOH) → 260

260 → (1. $NaBH_4$, $CeCl_3 \cdot 7H_2O$; 2. NaH, DMF, $4\text{-}MeOC_6H_4CH_2Cl$) → 261

261 → (1. $NaIO_4$, OsO_4 (cat.) 20°C; 2. MeOH, K_2CO_3; 3. Ag_2CO_3-Celite, xylene) → 262

262 → (1. LiOH, H_2O, DME; 2. NaH, BnBr; 3. 1 N HCl; 4. CH_2N_2) → 263

263 → (1. DDQ, CH_2Cl_2, H_2O; 2. MsCl, Et_3N; 3. MeOH, H_2O, HCl; 4. AcOH, H_2SO_4) → 264

E = COOMe

264 → (silylated pyrimidine (OTMS, OTMS, E), TMSOTf) → 265

265 → (1. NaOMe, MeOH; 2. Bu_2SnO) → 266

266 → (CsF, DMF, 60°C) → 267 → octosyl acid A

PMB: 4-MeO–C_6H_4–CH_2

DDQ: 2,3-dichloro-5,6-dicyano-1,4-benzoquinone

DME: $MeOCH_2CH_2OMe$

SCHEME 13.77 Danishefsky's synthesis of octosyl acid A.

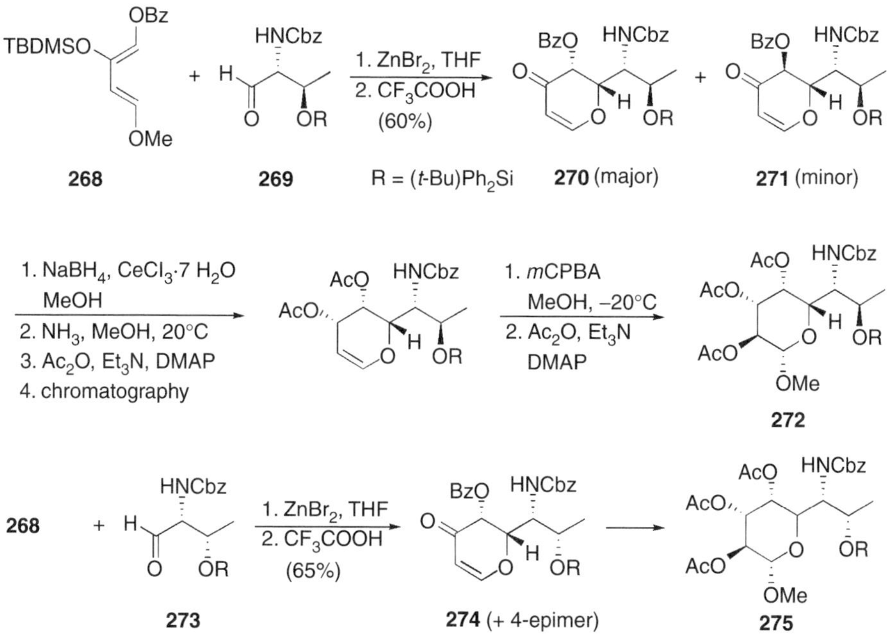

SCHEME 13.78 Danishefsky's synthesis of lincosamine and Jurczak's synthesis of a lincosamine isomer.

13.6.3 Hetero Diels–Alder Additions of 1-Oxa-1,3-Dienes

13.6.3.1 With Chiral 1-Oxa-1,3-dienes

The [4 + 2] cycloaddition of α,β-unsaturated aldehydes and ketones (1-oxa-1,3-dienes) to enol ethers (Diels–Alder addition with inverse electron demand) has been an attractive route for the synthesis of 3,4-dihydro-2*H*-pyrans [143–146], which can be converted into deoxy- and dideoxypyranosides [147,148].

The enantiopure 1-oxa-1,3-diene **278** is prepared by acylation of benzyl vinyl ether with oxalyl chloride. This generates acyl chloride **276**, which acylates the lithium salt of 2-oxazolidinone **277**. In the presence of Me_2AlCl, diene **278** adds to (*Z*)-1-acetoxy-2-ethoxyethene giving mostly adduct **279**, whereas, when using Me_3SiOTf as promoter of the hetero Diels–Alder addition, diastereomer **280** is the major adduct (Scheme 13.79). Adducts **279** and **280** have been converted into ethyl β-D-mannopyranoside and ethyl-β-L-mannopyranoside, respectively [149].

The chiral oxadiene **281** adds to ethyl vinyl ether giving a 2:1 mixture of *endo* adducts (+)-**282** and (−)-**283**. Deacylation of pure (+)-**282** delivers **284**, which is then protected as a silyl ether to give **285**. Raney nickel reduction and deprotection provides ethyl glycoside **286**. Protection of **284** as a benzyl ether, followed by desulfurization and subsequent hydroboration and deprotection, provides (−)-L-olivose [150] (Scheme 13.80).

The Diels–Alder addition of enantiomerically pure oxadienes to (*Z*)-2-ethoxyvinyl acetate are highly diastereoselective and have allowed Tietze and coworkers to prepare derivatives of 4-deoxy-D-*lyxo*-hexose [146].

13.6.3.2 With Chiral Enol Ethers as Dienophiles

Alternatively, Dujardin et al. [151] prepared derivatives of 2-deoxy-L-*arabino*-hexose via the Diels–Alder addition of achiral oxadiene **287** to *O*-vinyl mandelic ester (−)-**288** (Scheme 13.81).

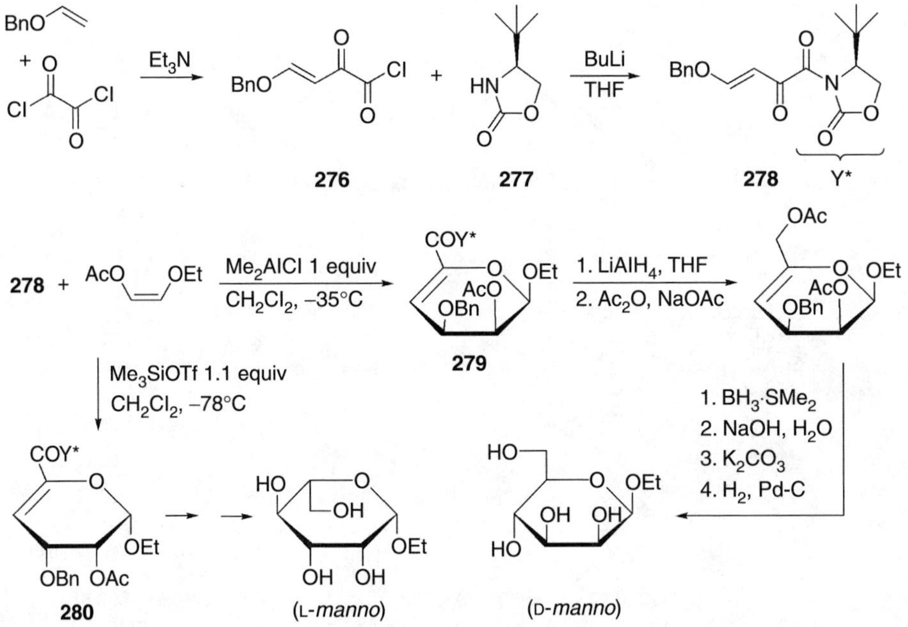

SCHEME 13.79 Tietze's syntheses of L-hexoses.

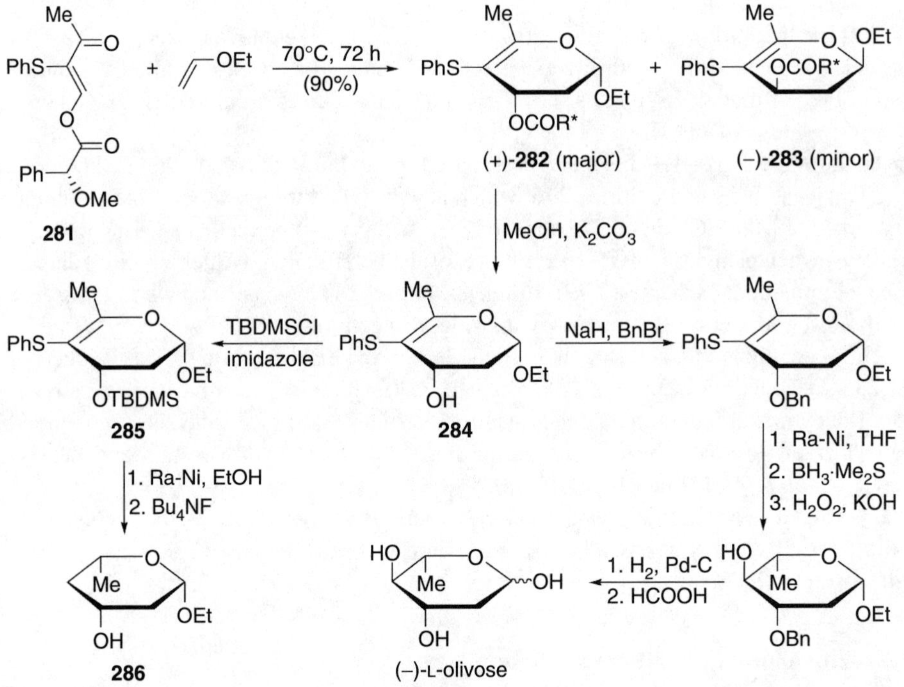

SCHEME 13.80 Schmidt's synthesis of (−)-L-olivose.

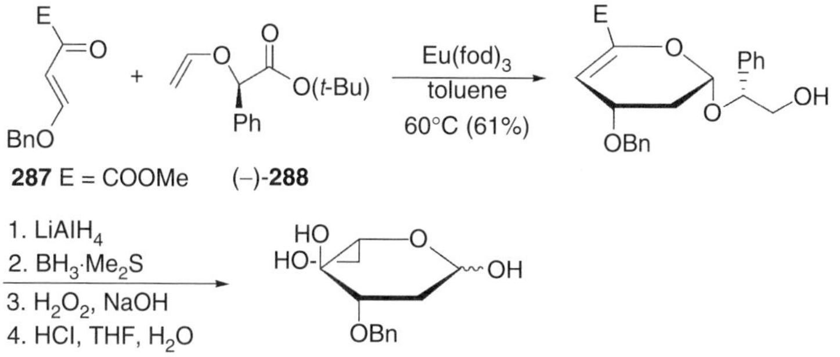

SCHEME 13.81 Dujardin's synthesis of 2-deoxy-L-*arabino*-hexose.

13.6.3.3 Induced Asymmetry by Lewis-Acid Catalysts

Independently, the groups of Evans [152] and Jørgensen [153] have shown that β,γ-unsaturated α-keto esters react with ethyl vinyl ether in the presence of enantiomerically pure bisoxazoline copper(II) complexes as catalysts leading to enantiomerically enriched dihydropyrans. For instance **289** and **290** in ether at 20°C and in the presence of complex **291** add to give the *endo* adduct **292** in 60% yield and with an enantiomeric excess of 96.5%. The latter was then converted into ethyl β-D-mannopyranoside tetraacetate **293** as shown in Scheme 13.82 [154a]. Related cyclization reactions, catalyzed by $SnCl_4$, were reported by the Hanessian Group [154b].

The first catalytic, highly enantioselective hetero Diels–Alder reactions of thiabutadienes with an acyloxazolidinone dienophile using homochiral copper and nickel triflate and perchlorate bis(oxazoline) and bis(imine)complex catalysts to generate dihydrothiopyrans were reported by Saito et al. [155].

13.6.4 Nitroso Dienophiles: Synthesis of Azasugars

Homochiral nitroso dienophiles can add to conjugated dienes in the hetero Diels-Alder mode, sometimes with good diastereoselectivities [156]. For instance, the chloronitroso dienophile **294** derived from D-mannose [157] adds to diene **295** in MeOH/HC(OMe)$_3$ to give adduct **296**, which has been converted into the 1,5,6-trideoxy-1,5-iminoalditols and the aldonolactames as shown in Scheme 13.83 [156c,158].

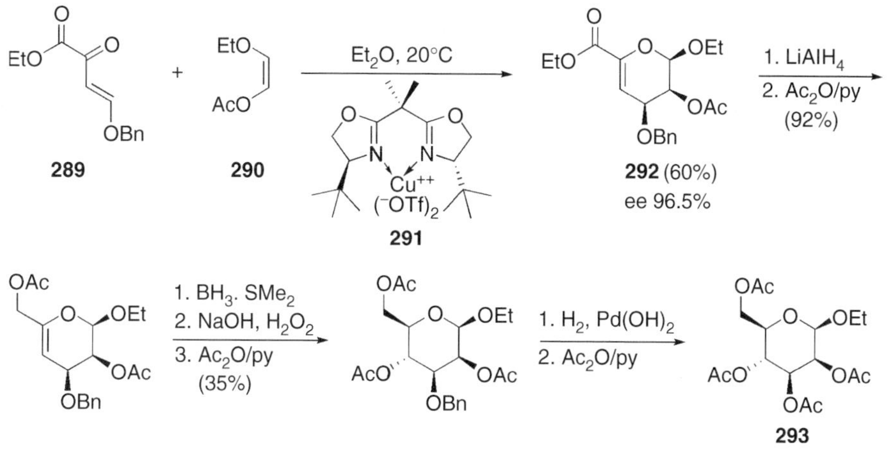

SCHEME 13.82 Jørgensen's asymmetric synthesis of an ethyl β-D-mannopyranoside derivative.

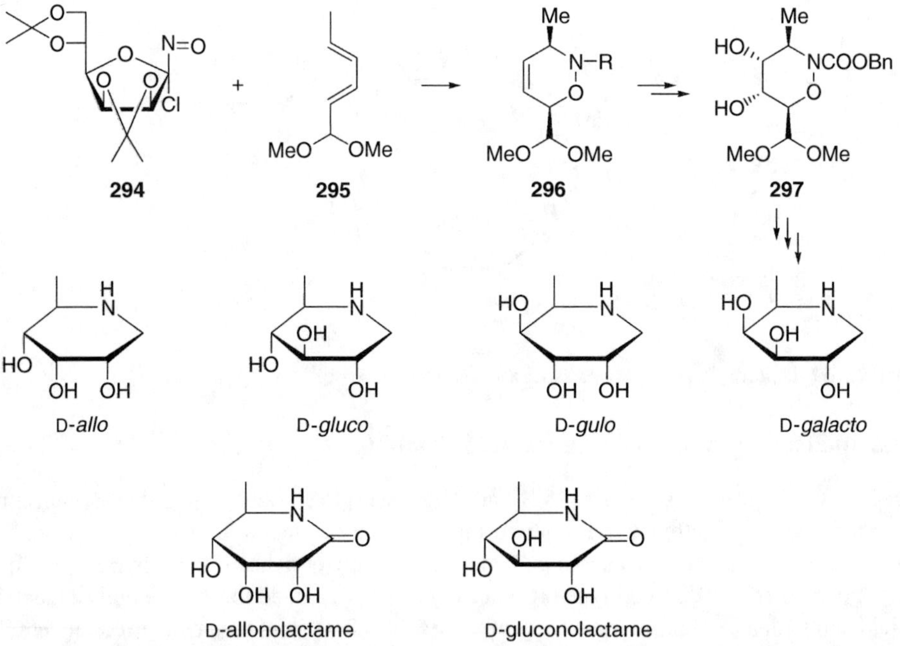

SCHEME 13.83 Streith's synthesis of 1,5,6-trideoxy-1,5-iminoalditols and aldonolactames.

Intermediate **297** was also converted into the potent α-L-fucosidase and α-galactosidase inhibitor **298** (Scheme 13.84) [159].

Good diastereoselectivities were observed for the hetero Diels-Alder addition of homochiral 1,3-dienes with achiral acyl-nitroso dienophiles. An example is shown in Scheme 13.85 for the total synthesis of 4-amino-4,5-dideoxy-L-lyxose derivatives **300** and **301**, potent inhibitors of α-L-fucosidase [160].

13.6.5 *N*-Methyltriazoline-3,5-Dione as a Dienophile: Synthesis of 1-Azafagomine

Methyl urazol **302** was oxidized into *N*-methyl triazoline-3,5-dione with *tert*-butyl hypochlorite. Without isolation, it was then reacted with (*E*)-penta-2,4-diene-1-ol giving adduct (±)-**303** as a racemic mixture. On treatment of (±)-**303** with a lipase (R/Novozym 435) in vinyl acetate, 38% of ester (*S*)-**304** (86% ee) and 29% of alcohol (*R*)-**303** (59% ee) were isolated. Epoxidation of (*S*)-**304**

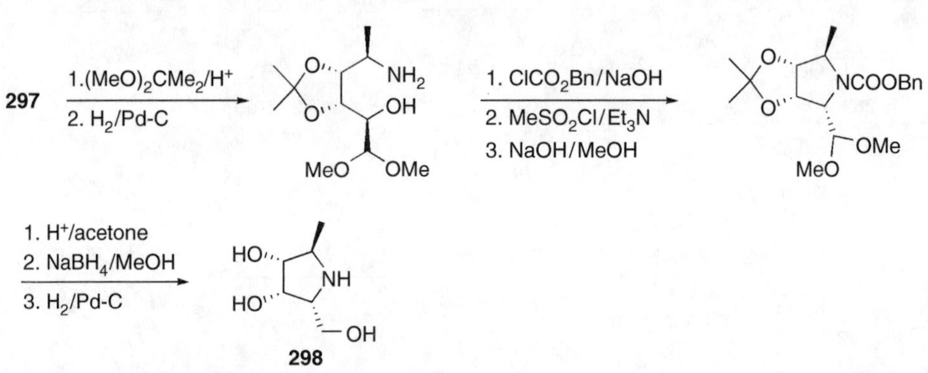

SCHEME 13.84 Defoin's synthesis of 2,5-imino-2,5,6-trideoxy-D-altritol.

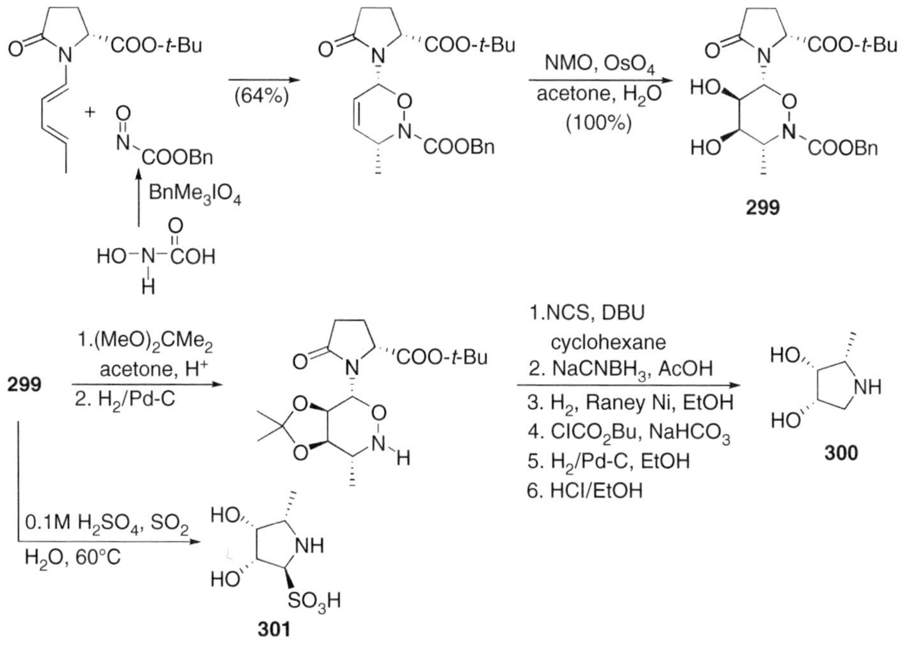

SCHEME 13.85 Defoin's syntheses of 4-amino-4,5-dideoxy-L-lyxose derivatives.

provided 13% of (−)-**305** and 68% of (−)-**306**. The latter epoxide was hydrolyzed under acidic conditions to provide **307** that reacted with hydrazine to give (−)-1-azafagomine enantiomerically pure (Scheme 13.86) [161]. This compound is a slow binding inhibitor of almond β-glucosidase ($K_i = 0.33\ \mu M$). A similar sequence of reactions converted (*R*)-**303** into (+)-1-azafagomine. The latter is not an inhibitor of β-glucosidases [162].

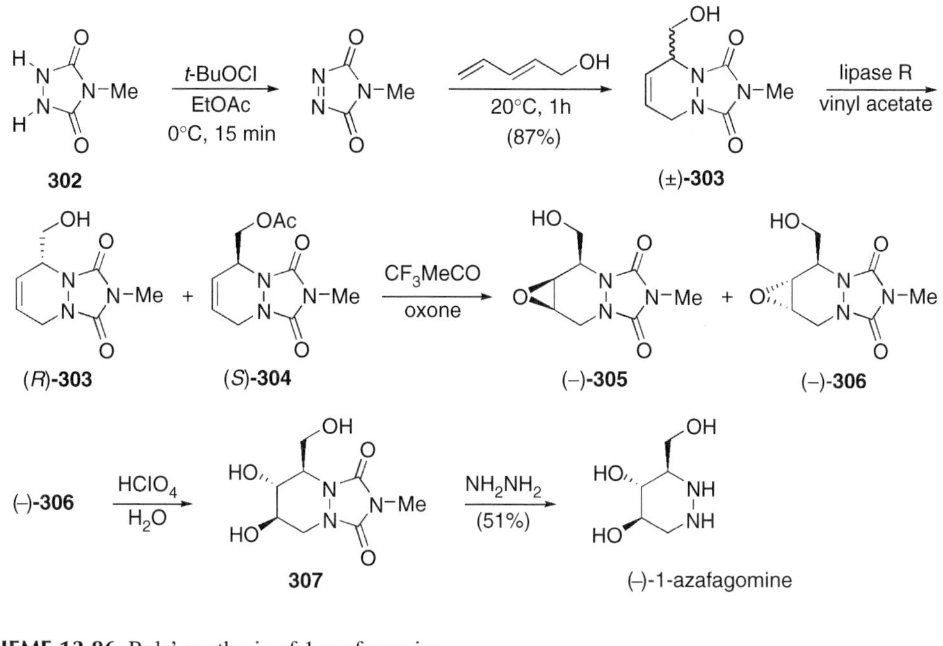

SCHEME 13.86 Bols' synthesis of 1-azafagomine.

13.7 CYCLOADDITIONS OF FURANS

13.7.1 DIELS–ALDER ADDITIONS

Just and his group pioneered the use of 7-oxabicyclo[2.2.1]hept-2-enes as starting materials in the synthesis of carbohydrates and analogs [163]. These bicyclic systems have the advantage of being able to undergo highly face-selective reactions. They are easily obtained by simple Diels–Alder additions of furan to alkene dienophiles. For example (Scheme 13.87), the mixture of racemic adducts **308** and **308′**, obtained by addition of furan to methyl (*E*)-2-nitroacrylate, is dihydroxylated into diols **309** and **309′**, separated by crystallization. Treatment of the acetonide of **309** with a base affords alkene **310**. Ozonolysis of **310** and reduction leads to a mixture of epimeric triols **311**, the cleavage of which with sodium periodate forms 2,5-anhydro-3,4-*O*-isopropylidene-D,L-allose **312** in 15% yield based on the starting methyl (*E*)-2-nitroacrylate. The same allose derivative was obtained from the Diels–Alder adduct **313** of furan to dimethyl acetylenedicarboxylate [163].

The adducts **308** and **308′** were also transformed into epoxide **314**. Ring opening of the epoxide, followed by ozonolysis and reduction, furnishes keto-ester **315**. Compounds **312** and **315** were converted into a variety of racemic *C*-nucleosides [164].

Furan adds to vinylene carbonate to produce an *exo*/*endo* mixture of Diels–Alder adducts [165]. Double hydroxylation of the 7-oxanorbornene double bond is highly *exo* face selective (Scheme 13.88). The diol thus obtained is protected as an acetonide. Saponification of the carbonate liberates a mixture of diols that is oxidized into *meso*-1,5-anhydroallaric acid derivative **316**. Treatment of **316** with Ac_2O generates the anhydride **317**. Subsequent reaction of **317** with methanol gives racemic **318** that can be resolved by fractional crystallization with brucine or by chromatographic separation of the (*R*)-1-(β-naphthyl)ethylamides. The individual isomers of **318** each react with ClCOOEt and Me_3SiN_3 *in situ* to provide enantiomerically pure D- and L-riboside derivatives [166].

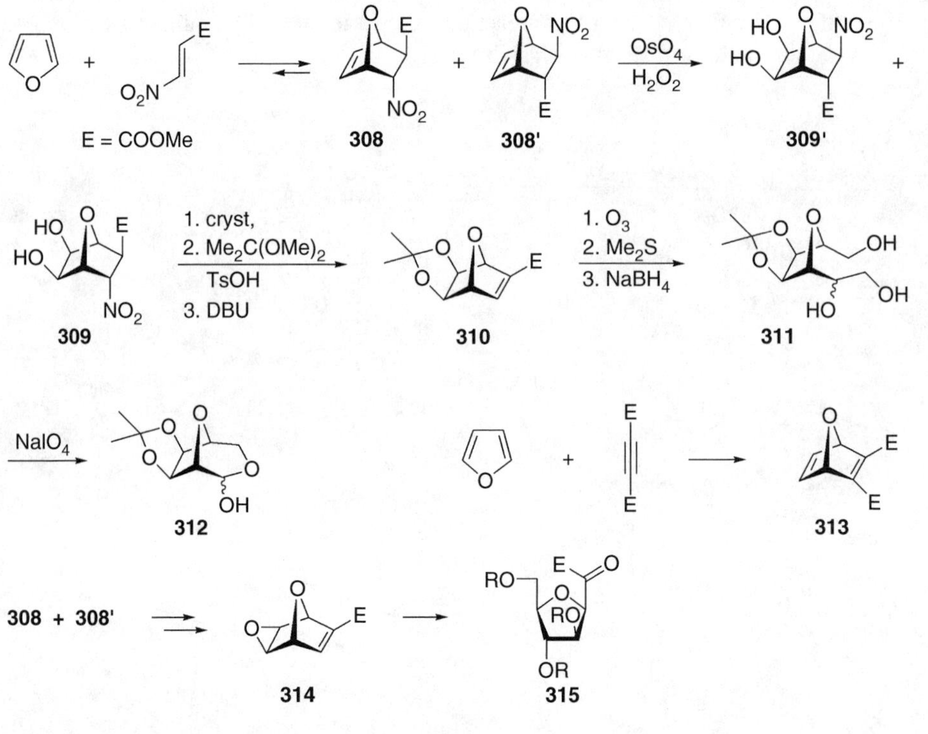

SCHEME 13.87 Just's synthesis of *C*-nucleoside precursors.

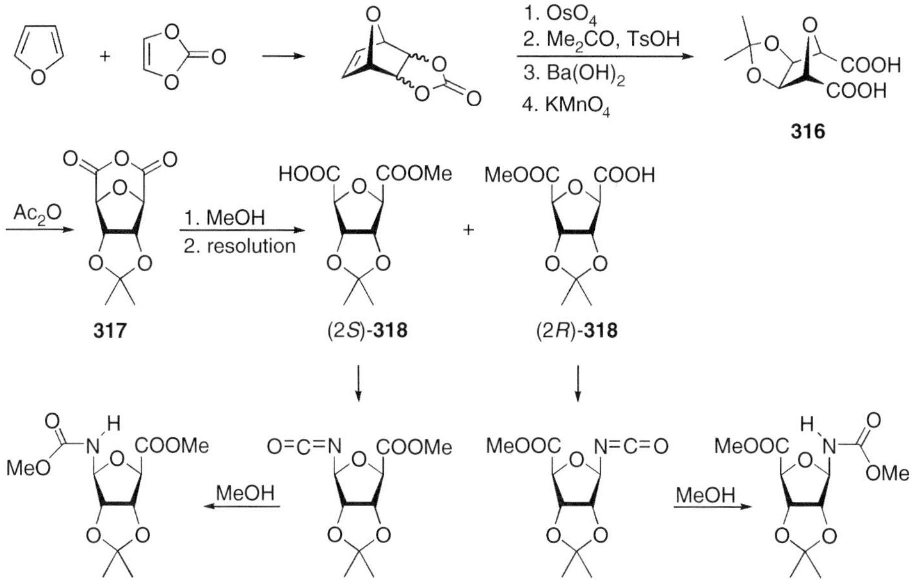

SCHEME 13.88 Schmidt's total synthesis of D- and L-riboside derivatives.

13.7.2 The "Naked Sugars of the First Generation"

13.7.2.1 Total Synthesis of Pentoses and Hexoses

The 1-cyanovinyl (1′*S*)-camphanate (derived from (1*S*)-camphanic acid and pyruvonitrile) adds to furan in the presence of ZnI_2 as catalyst. After 7 days at room temperature a mixture of four possible diastereomeric Diels–Alder adducts is formed (95%) from which pure adduct **319** can be isolated by crystallization. Unreacted furan is recovered and the diastereomeric mixture left from the crystallization is heated to give furan and 1-cyanovinyl (1′*S*)-camphanate. These components are easily recycled allowing the preparation of additional diastereomerically pure **319** (the reversibility of the furan Diels–Alder addition is exploited here). Starting from commercially available (1*R*)-camphanic acid, pure adduct **320** can readily be prepared in large quantities [167]. Camphanic acid auxiliaries can be replaced by the chiral auxiliaries RADO(Et)OH or SADO(Et)OH derived from (*R,R*)-tartaric acid and (*S,S*)-tartaric acid [168]. Using 1-cyanovinyl acetate as the dienophile, a racemic mixture of adducts is obtained. Subsequent hydrolysis provides a mixture of cyanohydrines easily resolved via brucine complexation. Reaction of the diastereomerically pure complex with Ac_2O provides the enantiomerically pure 7-oxanorbornene derivatives **321** (Scheme 13.89) [169,170].

Enantiomerically pure 7-oxanorbornenyl derivatives **319**, **320**, **321**, their saponification products (recovery of the chiral auxiliary in the aqueous phase) and ketones (+)-**322** and (−)-**322** are coined "naked sugars of the first generation" because they are chirons (= enantiomerically pure synthetic intermediates) like those derived from natural hexoses. Like natural sugars, they are enantiomerically pure. However, unlike natural sugars, they possess three unsubstituted (naked) carbon centers, the substitution of which follows highly stereoselective methods giving polysubstituted 7-oxabicyclo[2.2.1]heptane-2-ones that can be oxidized into the corresponding uronolactones as illustrated below.

Double hydroxylation of **319** gives a diol that is directly converted into acetonide (−)-**323**. Saponification followed by treatment with formalin liberates ketone (+)-**324** and allows recovery of the chiral auxiliary ((−)-camphanic acid) by simple extraction. The Baeyer–Villiger oxidation of

(+)-**324** is highly regioselective giving uronolactone (−)-**325** that is converted into uronic acid (−)-**326** in a one-pot operation. Oxidative decarboxylation of (−)-**326** gives bromide (−)-**327**, a precursor of anticancer agent 5′-deoxy-5-fluorouridine [171]. Displacement of the bromide by water produces the D-ribose derivative (−)-**328** (Scheme 13.89).

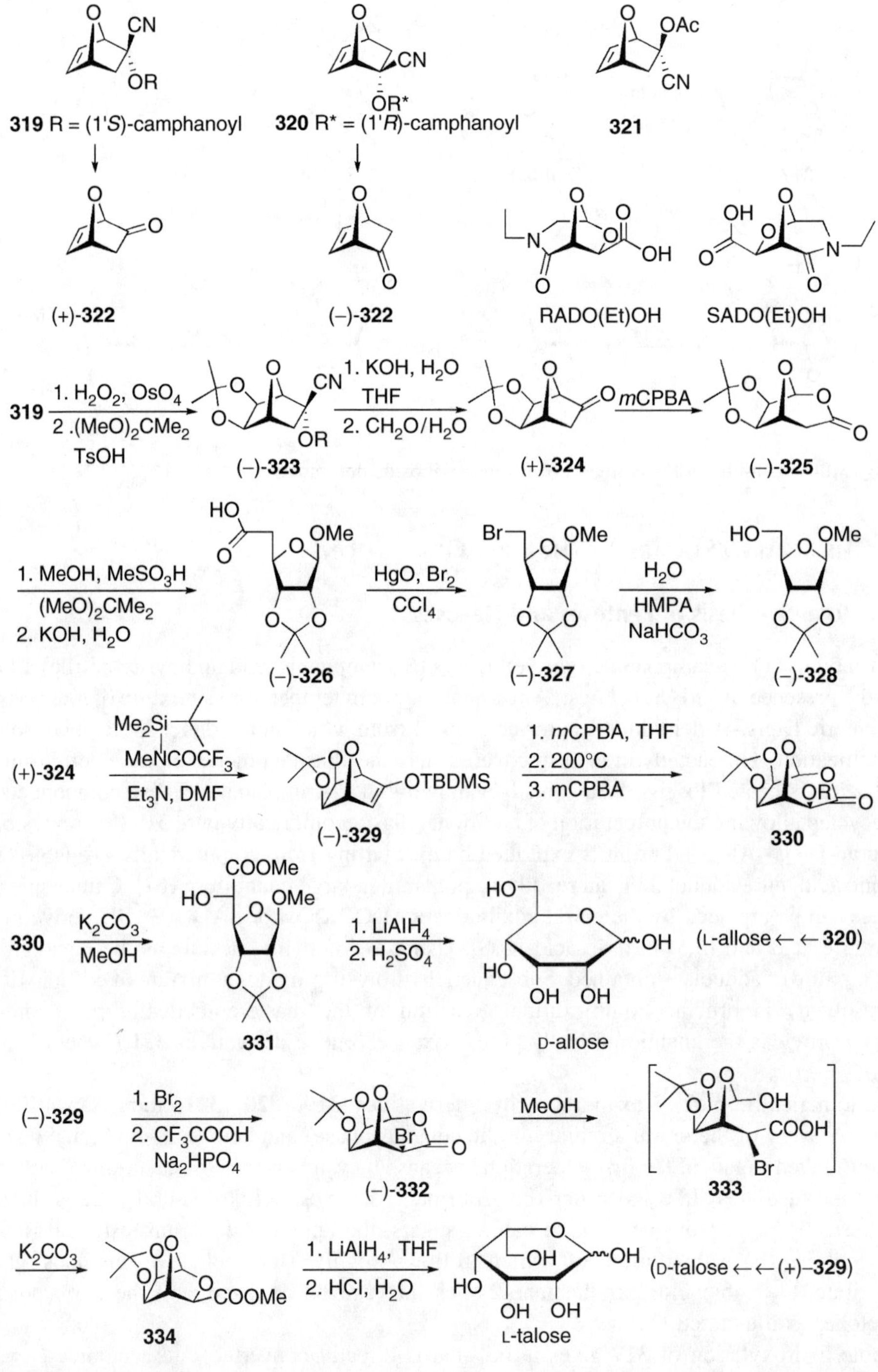

SCHEME 13.89 Naked sugars of the first generation. Asymmetric total syntheses of carbohydrates with early recovery of the chiral auxiliary.

The silyl enol ether (−)-**329** derived from ketone (+)-**324** is epoxidized stereospecifically on its *exo* face with *m*-chloroperbenzoic acid. The intermediate hemiacetal is heated to give an α-(metachlorobenzoyloxy) ketone that undergoes a highly regioselective Baeyer–Villiger oxidation to the uronolactone **330**. It is the electron-releasing ability of the 7-oxa ethereal moiety which is responsible for this regioselectivity [172]. The uronolactone **330** reacts with methanol in the presence of anhydrous potassium carbonate producing the uronic ester **331**, the reduction of which, followed by acid hydrolysis of the acetal moieties, delivers unprotected D-allose. The same sequence of reactions transforms **320** into L-allose [173]. If, instead of epoxidizing enol ether (−)-**329**, then one reacts it with bromine, an α-bromoketone is obtained that undergoes a highly regioselective Baeyer–Villiger oxidation into uronolactone (−)-**332**. Methanolysis generates the unstable furanose **333**, which undergoes an intramolecular S_N2 displacement with formation of the anhydrosugar **334**. After reduction of the ester moiety and acidic hydrolysis, L-talose is obtained. Starting with (+)-**329**, D-talose is prepared with the same ease [173] (Scheme 13.89).

Acetals of (+)-**322** are epoxidized into **335** (Scheme 13.90). Acid-promoted ring opening of the epoxides of **335** with participation of the *endo* OR group leads to the *trans*-5,6-dioxy-substituted 7-oxabicyclo[2.2.1]heptan-2-ones **337** [174]. The reaction sequences of Scheme 13.89 applied to **337**, and to its enantiomer obtained from (−)-**332**, generate D- and L-altrose or L- and D-galactose, respectively.

Electrophilic addition of PhSeCl to alkene **319** is highly stereo- and regioselective (steric effects) and provides adduct **338** if carried out in the presence of an acetate nucleophile. Methanolysis followed by treatment with formalin and protection of the *endo* alcohol gives ketone **339**. Low-temperature oxidation of the selenide **339** generates a selenoxide, which does not

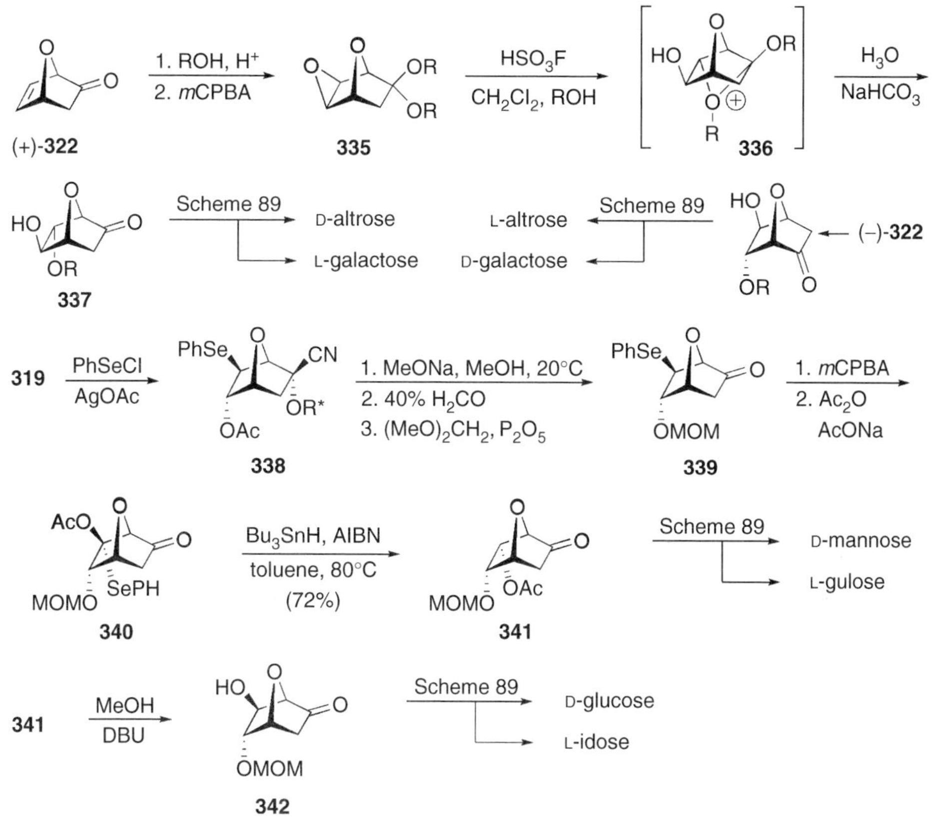

SCHEME 13.90 Asymmetric syntheses of hexoses starting from furan.

eliminate readily but has the time to undergo a seleno-Pummerer rearrangement in the presence of acetic anhydride and sodium acetate, leading to **340**. Reductive elimination of the seleno moiety with tin hydride forms **341** bearing two *endo* hydroxy groups protected in an orthogonal fashion (Scheme 13.90) [175]. Applying the reactions of Scheme 13.89 to **341** and to its enantiomer generates, in principle, hexoses with a *manno* or *gulo* configuration. Methanolysis of **341** occurs with complete epimerization at C(6) (retro-aldol aldolization) giving **342**, a potential precursor of D-glucose and L-idose derivatives (Scheme 13.90).

13.7.2.2 Total Synthesis of Deoxyhexoses

Under conditions of kinetic control, the cyanoesters **319** add soft electrophiles E^+X^- to generate the corresponding adducts **344** with high *exo* face selectivity and high regioselectivity. In these cases, the bridged ion intermediates **343** are attacked preferentially by the nucleophile X^- at C(5), the least sterically hindered center and also the one that can support the highest partial positive charge (field effect of the electron-withdrawing CN and ester substituents at C(2)). With the 7-oxabicyclo[2.2.1]hept-5-en-2-one ((+)-**322**, derived from **319**), electrophiles E^+X^- add with opposite regioselectivity and give the corresponding adducts **345**. The nucleophile's (X^-) preference for carbon center C(6) is attributed to the electron-releasing effect of the carbonyl function in positively charged intermediates **346** ↔ **346′** because of favorable n(CO) ↔ σC(1,2) ↔ pC(6)$^+$ hyperconjugative interaction (frangomeric effect of the carbonyl n electron pairs) [176]. The principle described here (Scheme 13.91) has been applied to convert the "naked sugars" into 2-, 3- and 4-deoxyhexoses and derivatives in highly stereoselective manners.

Addition of benzeneselenyl chloride to **319**, followed by oxidative elimination of the selenium and saponification of the cyanoester moiety, generates chloroenone **348** in 94% yield. Reduction of the ketone is highly *exo* face selective, leading to exclusive formation of *endo* alcohol **349**, which can be protected as a benzyl ether. Ozonolysis of the chloralkene, followed by reductive workup, yields the fully protected methyl 2,5-anhydro-4-deoxy-L-*xylo*-hexuronate **350** in 90% yield. This sugar can be converted into the partially protected 2,5-anhydro-4-deoxy-D-*xylo*-hexonic acid **351** (Scheme 13.92). The triflate derived from *endo* alcohol **349** can be displaced by BnOLi to give the *exo* benzylic ether **352**. The same sequence of reactions as for the conversion of **349** → **351** provides the 2,5-anhydro-3-deoxy-D-*ribo*-hexonic acid derivative **353**. Enantiomers of **351** and **353** can be readily prepared starting from furan adduct **320**.

The addition of PhSeCl to enone (−)-**322** (obtained by saponification of **320**) exclusively gives adduct **354**. After oxidative elimination of the phenylseleno group and subsequent reduction, *endo* alcohol **355** is obtained in high yield. Applying the reaction sequence described in Scheme 13.92, **355** is converted into derivatives of 2,5-anhydro-3-deoxy-L-*xylo*-hexuronate **356** and 2,5-anhydro-4-deoxy-D-*xylo*-hexonic acid **357** (Scheme 13.93). Inversion of configuration of alcohol **355** via

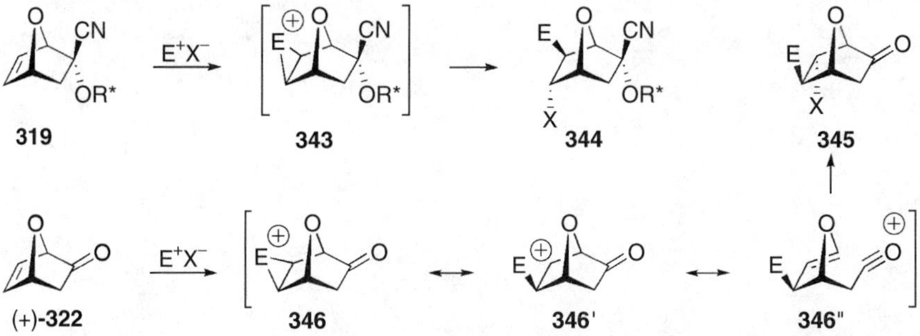

SCHEME 13.91 The highly stereo- and regioselective electrophilic additions of the "naked sugars."

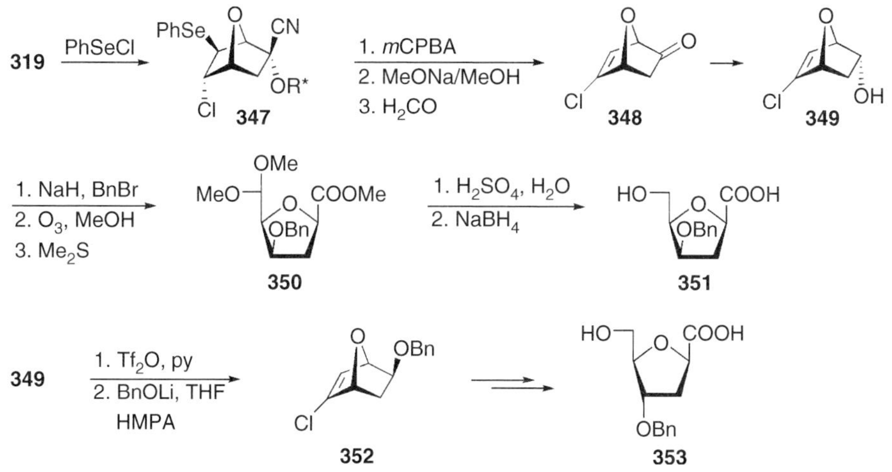

SCHEME 13.92 Total synthesis of 2-deoxy-2,5-anhydrohexonic acids.

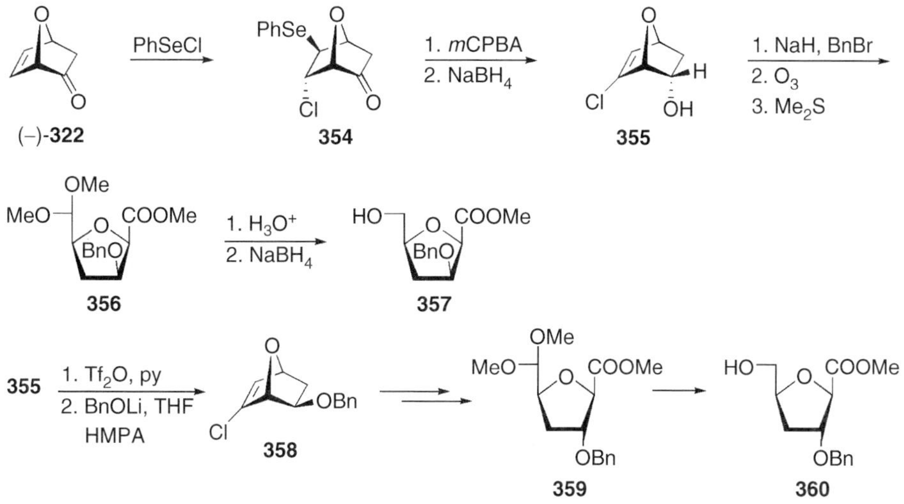

SCHEME 13.93 Total syntheses of 3-deoxy-2,5-anhydrohexonic acids.

displacement of its triflate and application of the reaction sequence of Scheme 13.90 furnishes **359** and **360** (Scheme 13.93) [177].

Addition of PhSeBr to enone (−)-**322** gives adduct **361**, which undergoes oxidative elimination of the phenylseleno group to give bromoenone **362**. Ketone reduction and protection of the *endo* alcohol furnishes **363**, and the double hydroxylation of its chloralkene unit generates **364** after acetylation. Baeyer–Villiger oxidation yields uronolactone **365**, which was converted into 3-deoxy-α-D-*arabino*-hexopyranoside [178] and into methyl 4-deoxyhexonate **368**. Finally, it was converted into 4-deoxy-D-*lyxo*-hexose derivative **369** (Scheme 13.94) [179].

The "naked sugars of the first generation" (**319**, **320**, **321**, (+)-**322**, (−)-**322**) are useful chirons for preparing all kinds of rare sugars derivatives [180] and analogs such as *C*-linked disaccharides [181]. They have been converted into conduritols, cyclitols and carbahexose derivatives [174,182]. The bicyclic ketones can also be converted into their enoxysilanes and then cleaved by ozonolysis. Depending on the workup conditions, 2,5-anhydrohexaldaric acid or 2,5-anhydrohexonic acid derivatives can be readily prepared in both enantiomerically pure forms [183].

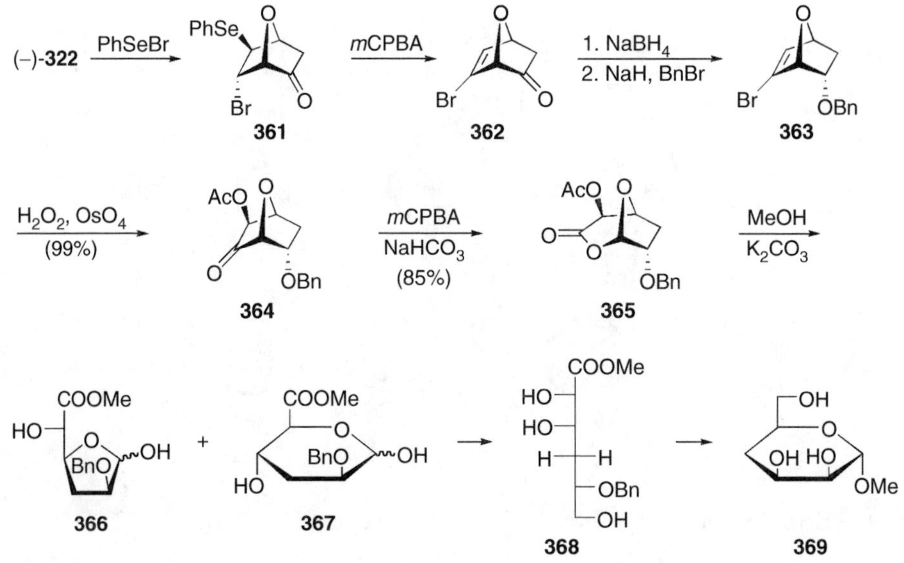

SCHEME 13.94 Total synthesis of a 4-deoxypyranoside and derivatives.

13.7.2.3 Total Synthesis of Aminodeoxyhexoses and Derivatives

Aziridination of the benzyl acetal (+)-**370** furnishes **371**, which undergoes ring opening under acidic conditions with the participation and the migration of the *endo*-benzyloxy group (possible intermediate **372**). This allows the stereo- and regiospecific 5-*exo*-amino and 6-*endo*-oxy substitution of the 7-oxanorbornanone system as in (+)-**373**. Formation of the corresponding enoxysilane and oxidation with *m*CPBA provides ketone **374**. Ketone **374** undergoes Baeyer–Villiger oxidation to give an uronolactone, which is methanolysed and silylated to **375**. Uronic ester reduction and deprotection furnishes aminosugar **376** (Scheme 13.95) [184].

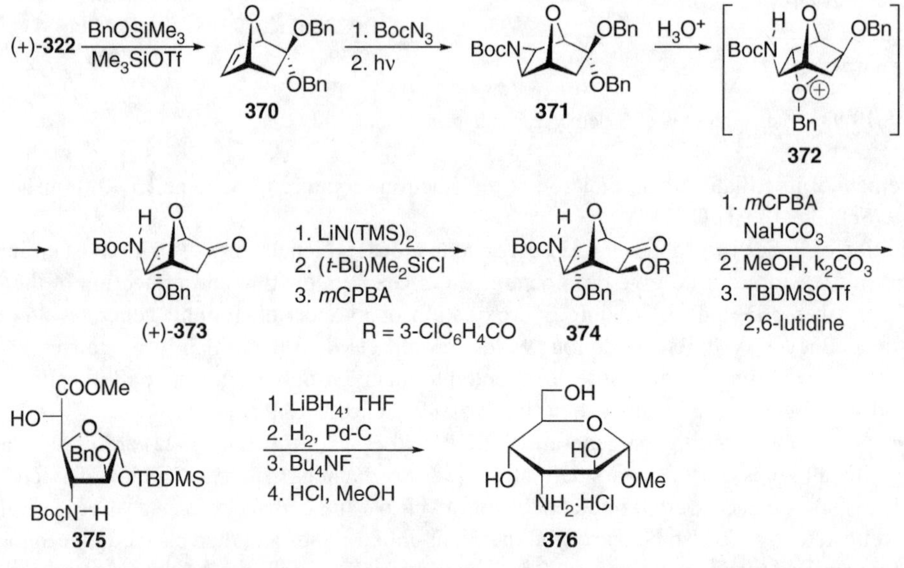

SCHEME 13.95 Total synthesis of methyl 3-amino-3-deoxy-D-*altro*-pyranoside.

Aziridination of **319** to an *N*-benzoylarizidine gives (−)-**377**, which can be rearranged under acidic conditions to provide (+)-**379** exclusively. The rearrangement implies the intermediacy of **378**, which is displaced intramolecularly into an oxazolidine. Saponification and treatment with formaline liberates ketone (+)-**379**. Bromination of ketone (+)-**379** gives **380**, which is then oxidized into an uronolactone, adding MeOH to yield the furanose derivative **381**. The latter displaces the α-bromouronic ester to generate the anhydro sugar **382**. Reduction and deprotection provides **383** (Scheme 13.96) [185].

The "naked sugar" methodology has also provided deoxypolyoxin C (Scheme 13.97). The α-bromouronolactone (−)-**332** (see Scheme 13.89) gives allyl uronic ester **384** on treatment with allyl alcohol and triflic acid. Selective hydrolysis of the allyl ester, followed by azide displacement of the bromide with retention of configuration, generates an intermediate cesium carboxylate that reacts with benzyl bromide to give **385**. Conversion of **385** into triacetate **386**, followed by Vorbrüggen's glycosidation, reduction and deprotection, yields deoxypolyoxin C [186]. Intermediate **385** has been converted in two steps into (+)-D-idonojirimycin (Scheme 13.97) [187]. L-Daunosamine [180a] and 2,3-dideoxy-D-*arabino*-hexose [188] have also been derived from the "naked sugars" **319** and (+)-**322**, respectively.

13.7.2.4 Long-Chain Carbohydrates and Analogs

Mukaiyama cross-aldolizations of 2,3-*O*-isopropylidene-D-glyceraldehyde (*R*)-**24** with bicyclic ketones (+)-**322** and (−)-**322** are highly diastereoselective and lead to aldols (+)-**387** and (−)-**388**, respectively. For steric reasons, only the *exo* face of the enoxysilanes (−)-**329** and (+)-**329** (see Scheme 13.89) can be attacked by the complexes [(*R*)-**24**]·$TiCl_4$. Depending on the configuration of the enoxysilanes, the (*R*)-**24** complex involving the 2-*O* or that involving the 3-*O* co-ordination with $TiCl_4$ reacts (transition state **389** vs. **390**; Scheme 13.98) to minimize steric repulsions. Baeyer–Villiger oxidation of aldol (+)-**387** generates the corresponding uronolactone that reacts with benzyl alcohol in DMSO and in the presence of cesium fluoride (base catalyst), giving the benzyl uronate (+)-**391**. On standing at room temperature in the

SCHEME 13.96 Total synthesis of 3-amino-3-deoxy-L-talose.

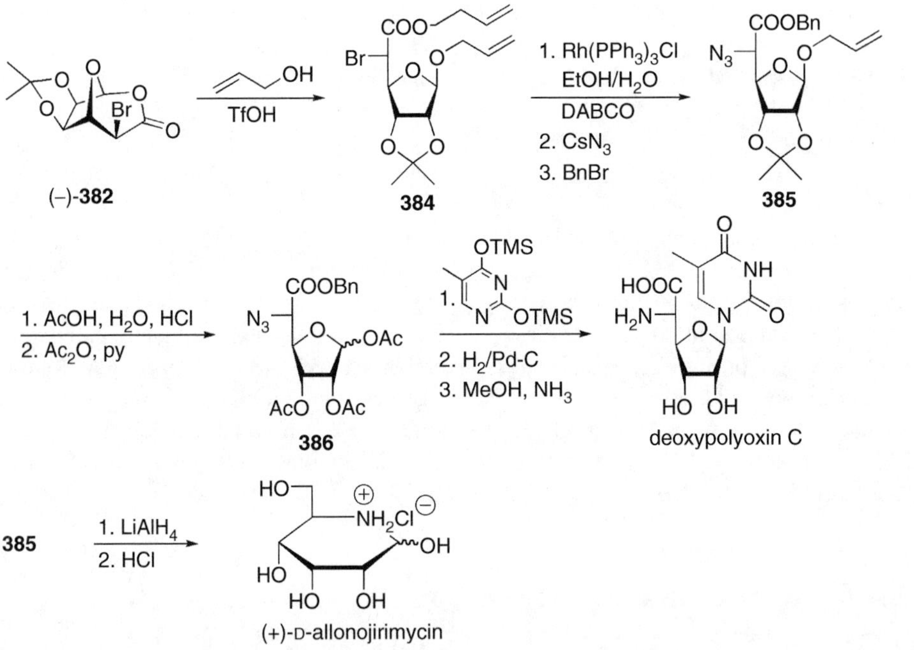

SCHEME 13.97 Total synthesis of deoxypolyoxin C and of (+)-D-allonojirimycin.

presence of CsF (or alternate bases such as K_2CO_3), (+)-**391** is epimerized into the more stable uronic ester (−)-**392**. At equilibrium in DMSO, a 1:13.5 ratio is found for (+)-**391**/(−)-**392**. In benzyl alcohol, the product ratio is nearly 1:1, suggesting that chelation between the hydroxy group at C(6) and the carboxylate group is responsible for the higher stability of (−)-**392** compared with that of (+)-**391** in DMSO. Silylation of (+)-**391**, followed by hydrogenation of the benzyl ester and subsequent Curtius rearrangement (one-pot operation), provides the partially protected 5-amino-5-deoxy-D-*erythro*-D-*talo*-octofuranosides (+)-**393**. Desilylation of (+)-**393** and subsequent hydrogenation liberates the amine that creates an imine with the aldose, which is hydrogenated, giving (−)-1,5-dideoxy-1,5-imino-2,3:7,8-di-*O*-isopropylidene-D-*erythro*-D-*talo*-octitol (−)-**394**. Acidic hydrolysis gives the unprotected 1,5-dideoxy-1,5-imino-octitol (−)-**395**, which can be converted into the unprotected pentahydroxyindolizidine (−)-**396** in one step under Mitsunobu conditions. Applying the same methods to the benzyl uronate leads to the 5-deoxy-5-amino-D-*erythro*-L-*allo*-octofuranoside derivatives (+)-**397**, which can be converted into (−)-1,5-dideoxy-1,5-imino-D-*erythro*-L-*allo*-octitol (−)-**398** and pentahydroxyindolizidine (−)-**399**. Following similar routes, aldol (+)-**398** was converted into 5-amino-5-deoxy-D-*threo*-L-*talo*-octofuranosides **400** and 5-amino-5-deoxy-D-*threo*-D-*allo*-octofuranosides **403**, which, in turn, can be readily transformed into the corresponding 1,5-dideoxy-1,5-imino-alditols (+)-**401** and (+)-**404**, and finally to the pentahydroxyindolizidines (+)-**402** and (+)-**405**, respectively. With the possibility of substituting the 7-oxabicyclo[2.2.1]heptan-2-one at C(5) and C(6) with two protected hydroxy groups in all possible configurations (see Scheme 13.90) or with other groups (see Scheme 13.92 to Scheme 13.96), the reaction sequence of Scheme 13.98 makes possible the preparation of a large number of octose derivatives. This demonstrates the high efficiency and versatility of the "naked sugar" methodology for the *de novo* synthesis of unusual and complicated monosaccharides [189].

13.7.2.5 "Naked Sugars of the Second Generation": Synthesis of Doubly Branched-Chain Sugars

Doubly branched heptono-1,4-lactones and polypropionate fragments were obtained from 2,4-dimethylfuran **406** via Diels–Alder addition to 1-cyanovinyl-(1′*R*)-camphanate **407** [190]. Without solvent, the ZnI_2-catalyzed and reversible cycloaddition leads to a major crystalline diastereomeric adduct (+)-**408** (99.5% de). Double hydroxylation of the alkene moiety of (+)-**408**, followed by diol protection as an acetonide, provides (−)-**409**. Methanolysis followed by treatment with formaline liberates ketone (+)-**410**. Baeyer–Villiger oxidation and subsequent α-methylation generates the *exo*-α-methyluronolactone (−)-**411**. Quenching of the lithium enolate of (−)-**411** with MeOH at −50°C gives the *endo*-α-methyluronolactone (−)-**412**. Acid hydrolysis of (−)-**412** and subsequent silylation and reduction forms (+)-**413** as major heptono-1,4-lactone (Scheme 13.99). Similarly, enantiomers of this doubly branched sugar can be

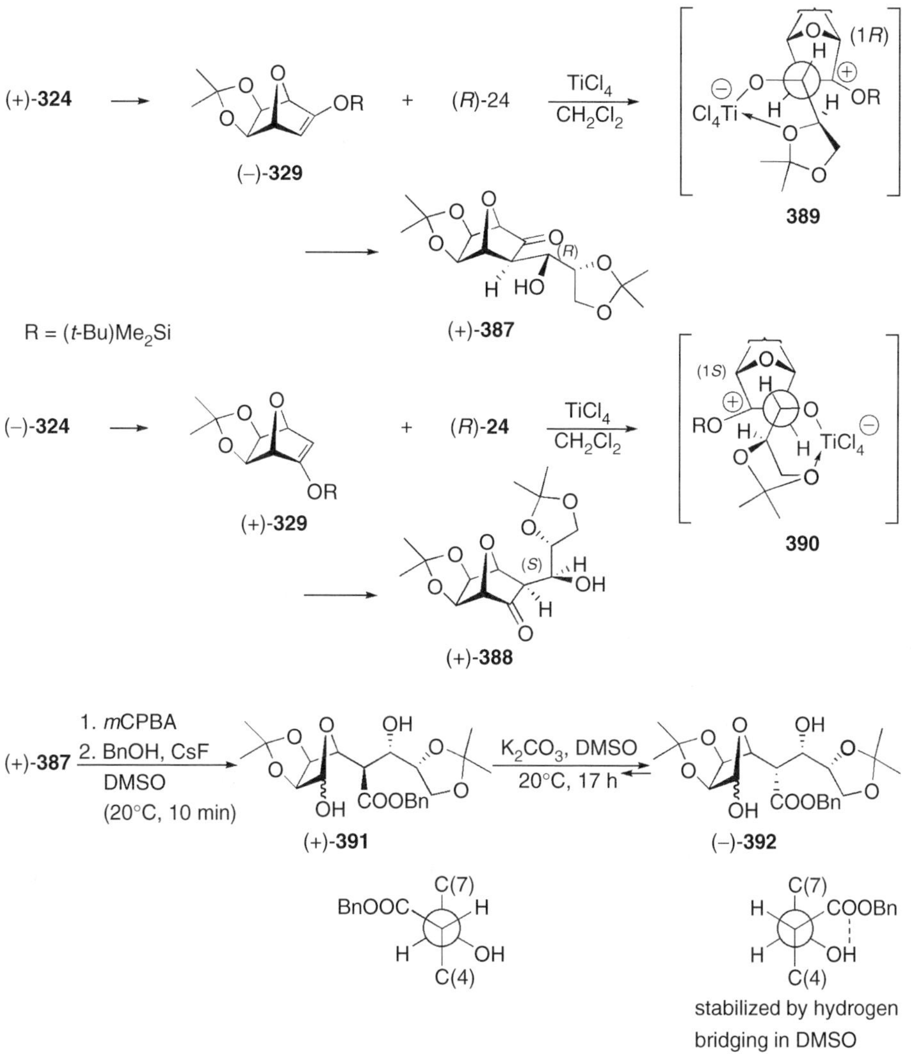

SCHEME 13.98 Cross-aldol reactions with "naked sugars." Syntheses of long-chain carbohydrates, indolizidines and analogs.

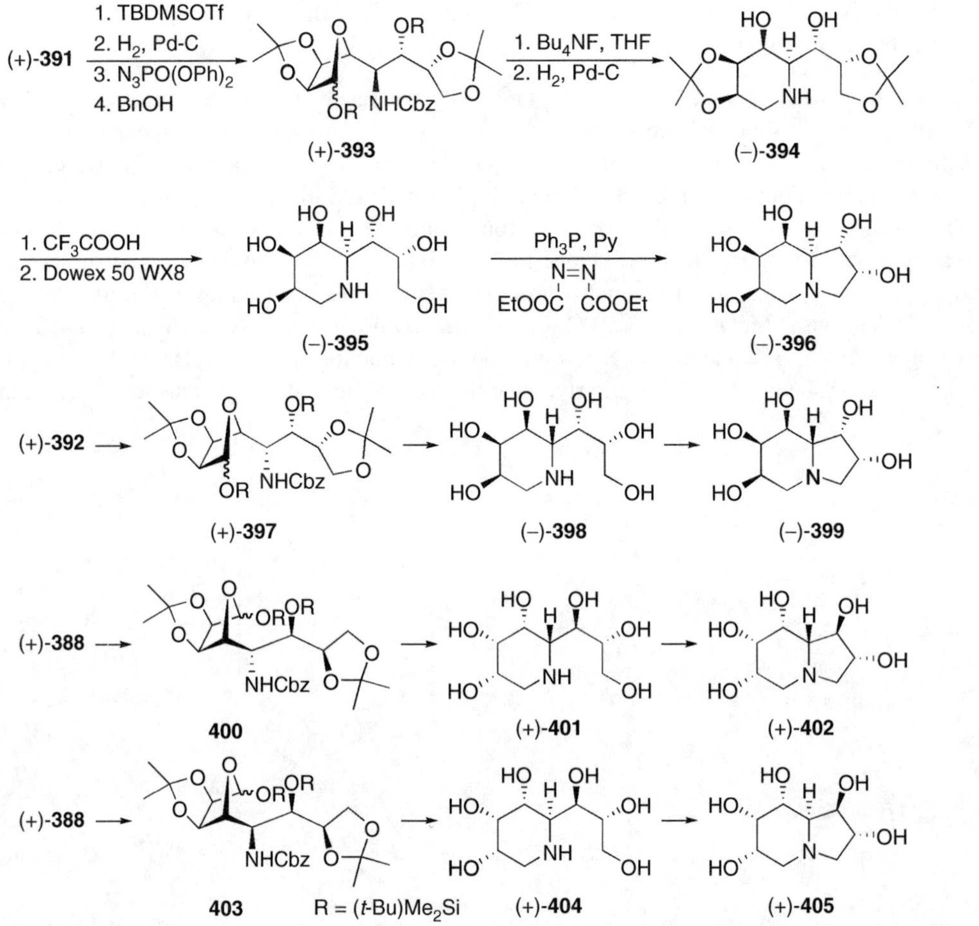

SCHEME 13.98 (Continued).

prepared starting from adduct (−)-**414** obtained by addition of 2,4-dimethylfuran to 1-cyanovinyl (1′*S*)-camphanate **407′** (Scheme 13.99). After conversion of (−)-**414** into dimethyl acetal **415**, regio- and *exo* face-selective hydroboration and further transformations generate the doubly branched uronic acid (−)-**418** [191].

The method of thermodynamic diastereoselection (through diastereoselective crystallization of equilibrating adducts; see Scheme 13.99) was applied to furan derivatives bearing readily recoverable chiral auxiliaries. For instance, the acetal of (2*S*,3*S*)-butane-2,3-diol and furfural is equilibrated in molten maleic anhydride with one major crystalline product [192]. In a similar way (Scheme 13.100), (*S*)-camphanate of furfuryl alcohol undergoes Diels–Alder addition in molten maleic anhydride, giving one major crystalline adduct (+)-**419** [193] that has been converted into doubly branched carba-hexopyranoses and derivatives [194], and into the new 2,6-dideoxy-2,6-iminoheptitol **420** (Scheme 13.100) [195].

13.7.3 Dipolar Cycloadditions of Furans

In 1982, Vasella and Voeffray [196] presented a synthesis of (+)-nojirimycin (5-amino-5-deoxy-D-glucose) using the 1,3-dipolar cycloaddition of a D-mannose-derived nitrone to furan. The chiral source in that synthesis (D-mannose) is not recovered. Jäger and Müller used

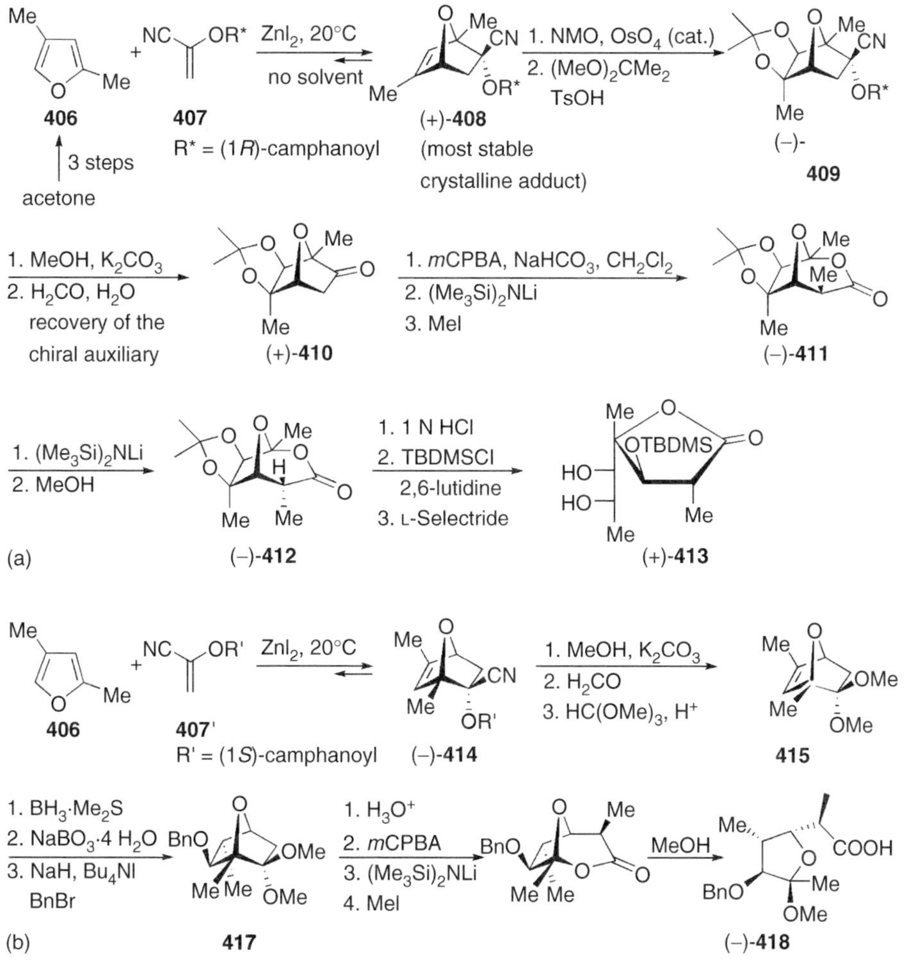

SCHEME 13.99 Applications of the "naked sugars of the second generation" to the synthesis of branched-chain sugar derivatives.

cycloadditions of nitrile-oxides to furan to produce racemic isoxazolines, which were then transformed into racemic 5-amino-5-deoxyfuranosides and aminopolyols [197]. Isoxazoline-3-carboxaldehyde (±)-**422** was resolved by the technique of Alexakis and Mangeney using (−)-(1*S*,2*S*)-diphenylethanediamine [198] (Scheme 13.101). Aldehyde (±)-**422** is obtained from the reaction of 2-nitroethanal diethyl acetal and furan, giving adduct (±)-**421**. Face-selective dihydroxylation of the enol ether, followed by isopropylidination, provides (±)-**422**. Following resolution, reduction of (+)-**422** with $NaBH_4$, followed by acetonide hydrolysis, generates furanose (+)-**423**, which is hydrogenated to provide (+)-1-deoxynojirimycin. The same reaction sequence applied to (−)-**422** furnished (−)-1-deoxynojirimycin in the shortest approach reported to date. In contrast, the most efficient synthesis converts D-glucose into (+)-1-deoxynojirimycin in four steps with 60% yield [199].

With $LiAlH_4$, both the aldehyde and isoxazoline moieties of (+)-**422** are reduced with formation of aminotriol (−)-**424**. Treatment of (−)-**424** with acid generates the 1,6-anhydro-*ido*-piperidinose (+)-**425**, the hydrogenation of which provides (−)-L-1-deoxyidonojirimycin. The same sequence of reactions applied to (−)-**422** gives (+)-D-1-deoxyidonojirimycin [200].

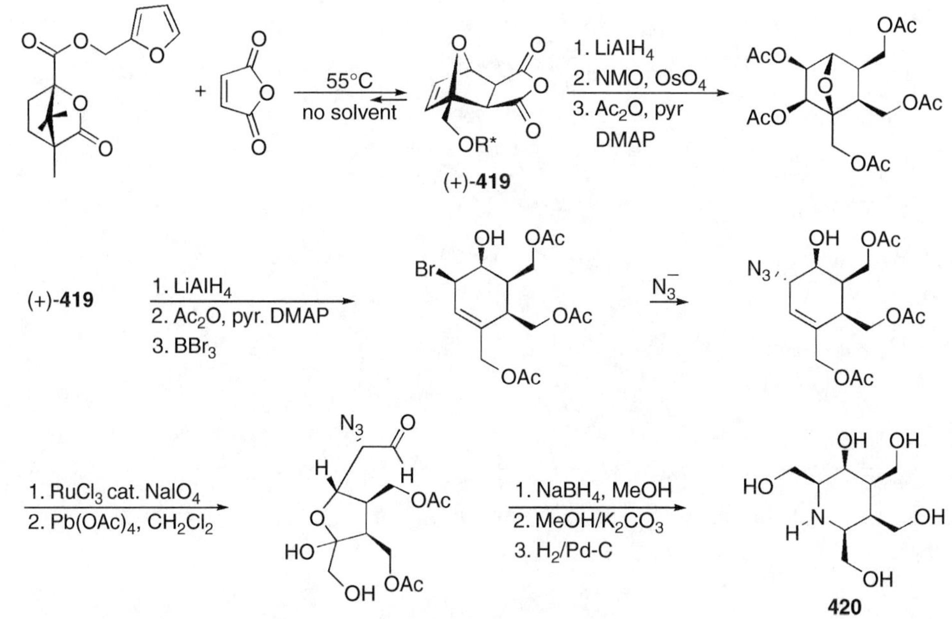

SCHEME 13.100 Synthesis of a doubly branched iminoalditol.

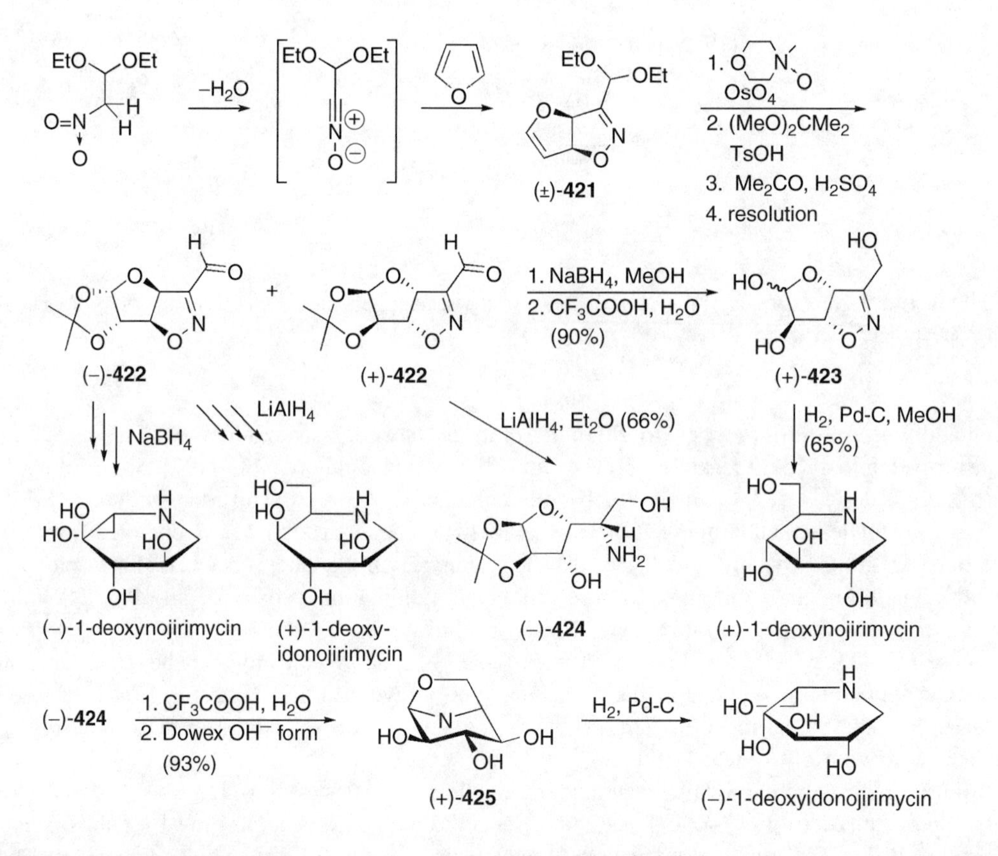

SCHEME 13.101 Total syntheses of (−)- and (+)-1-deoxynojirimycin and of (+)- and (−)-1-deoxyidonojirimycin starting from furan.

13.7.4 [4 + 3]-CYCLOADDITIONS OF FURAN

After dehalogenation with the zinc/copper couple, the [4 + 3] cycloaddition of furan to the 1,3-dibromo-2-oxyallyl cation generates 8-oxabicyclo[3.2.1]oct-6-en-3-one **426**. Double hydroxylation of the alkene moiety of **426**, followed by diol protection, generates the meso ketone **427**. Enantioselective deprotonation with homochiral lithium (*R*,*R*)-bis(1-phenylethyl)amide in the presence of Me_3SiCl gives the enoxysilane **428** with high yield (up to 98%) and 85% ee. Oxidation of **428** with PhIO generates the corresponding *exo*-α-hydroxyketone **429**. Treatment of **429** with $Pb(OAc)_4$ in MeOH, followed by reduction, provides the semiprotected methyl 3,6-anhydro-2-deoxy-L-*allo*-heptonate **430** (Scheme 13.102) [201].

Hoffmann and coworkers [202] developed general procedures that convert **426** into a number of 2,6-anhydroheptitols as enantiomerically pure isomers (Scheme 13.103). The treatment of **426** with (+)-bis[(*R*)-1-phenylethyl]amine, *n*BuLi and Et_3SiCl/Et_3N (*in situ* quench) in THF gives the corresponding silyl enol ether, which is oxidized with *m*CPBA in THF/H_2O into (−)-**431**. The resulting β-hydroxyketone is silylated and enolized on treatment with lithium diisopropylamide in the presence of Et_3SiCl/Et_3N. The intermediate enol ether is then oxidized with *m*CPBA. After treatment with $CF_3COOH/THF/H_2O$, (+)-**432** is isolated in 62% overall yield. The reduction of (+)-**432** with $NaBH(OAc)_3$ gives *endo* alcohol (−)-**433**, whereas reduction with $NaBH_4$ and $MgBr_2$ in MeOH affords the *exo* alcohol **435**. Oxidative cleavage of the endocyclic alkene moieties of **433** and **435**, with subsequent reduction with $NaBH_4$ and acetalization with benzaldehyde dimethyl acetal provides alditols **434** and **436**, respectively. Selective monobenzoylation of **435**, followed by inversion of alcohol at C-2, provides **437**, which is then converted into the 2,6-anhydroheptitol derivative **438** (Scheme 13.103).

Enantiomerically pure 3-oxo-8-oxabicyclo[3.2.1]octyl-2-yl derivatives were obtained by [4 + 3] cycloaddition of furan with chiral 1,2-dioxyallyl cation engendered *in situ* by acid-catalyzed heterolysis of enantiomerically pure, mixed acetals derived from 1,1-dimethoxy-acetone and enantiomerically pure, secondary benzyl alcohols [203]. For instance, mixed acetal **439** is converted into the silyl enol ether **440**. In the presence of a catalytic amount of trimethylsilyl triflate, **440** generates a cationic intermediate that adds to furan at −95°C, giving

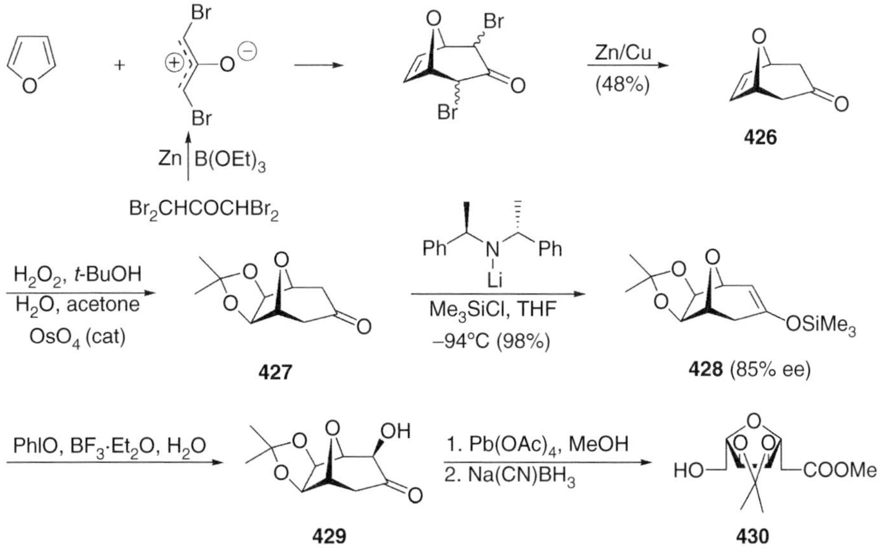

Br
Zn | B(OEt)$_3$
$Br_2CHCOCHBr_2$
Zn/Cu (48%)
426
H_2O_2, t-BuOH
H_2O, acetone
OsO_4 (cat)
427
Ph N Ph
Li
Me_3SiCl, THF
−94°C (98%)
OSiMe$_3$
428 (85% ee)
PhIO, $BF_3 \cdot Et_2O$, H_2O
OH
429
1. $Pb(OAc)_4$, MeOH
2. $Na(CN)BH_3$
HO COOMe
430

SCHEME 13.102 Simpkins's synthesis of methyl 3,6-anhydro-2-deoxy-L-*allo*-heptonate.

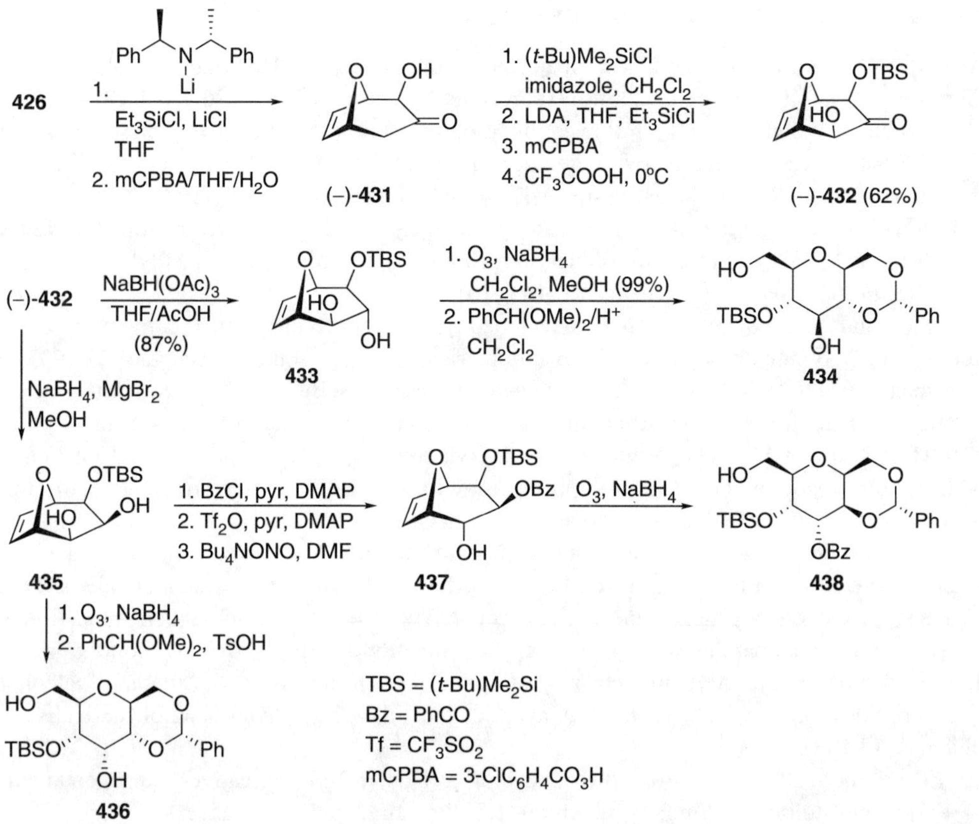

SCHEME 13.103 Hoffmann's syntheses of 2,6-anhydroheptitols starting from 8-oxabicyclo[3.2.1]octan-3-one.

adduct (−)-**441** (pure single diastereomer after chromatography). The latter can be regioselectively enolized (Scheme 13.104) and oxidized with *m*CPBA giving, after esterification, pivalate **442**. Reduction with $NaBH_4$ and subsequent alkene cleavage generates **443**. Treatment of pivalate **442** with DBU and ultrasound produces the 2-epimer **444** (85% yield). Its reduction with $NaBH_4$ gives **445**. Subsequent alkene cleavage generates **446**. Inversion of the alcohol moiety C-3 of **446**, followed by alkene cleavage, furnishes the 2,8-anhydroheptitol **447** (Scheme 13.104) [202].

13.8 CARBOHYDRATES AND ANALOGS FROM ACHIRAL HYDROCARBONS

13.8.1 From Cyclopentadiene

Xylitol has been derived from the product of photo-oxidation of cyclopentadiene [204], (*Z*)-(4*RS*)-4,5-epoxypent-2-enal (Scheme 13.105). Chemoselective reduction of the formyl group gives *cis*-hydroxyepoxypentene **448**, which is directly acetylated into **449**. Treatment of **449** with tetrabutylammonium acetate in Ac_2O opens the epoxide with formation of **450**. De-*O*-acetylation gives **451**, the epoxidation of which with *p*-nitroperbenzoic acid generates a 3:7 mixture of epoxides **452** and **453**, isolated as peracetates. The major epoxide **453** is hydrolyzed into xylitol via the orthoester **454**.

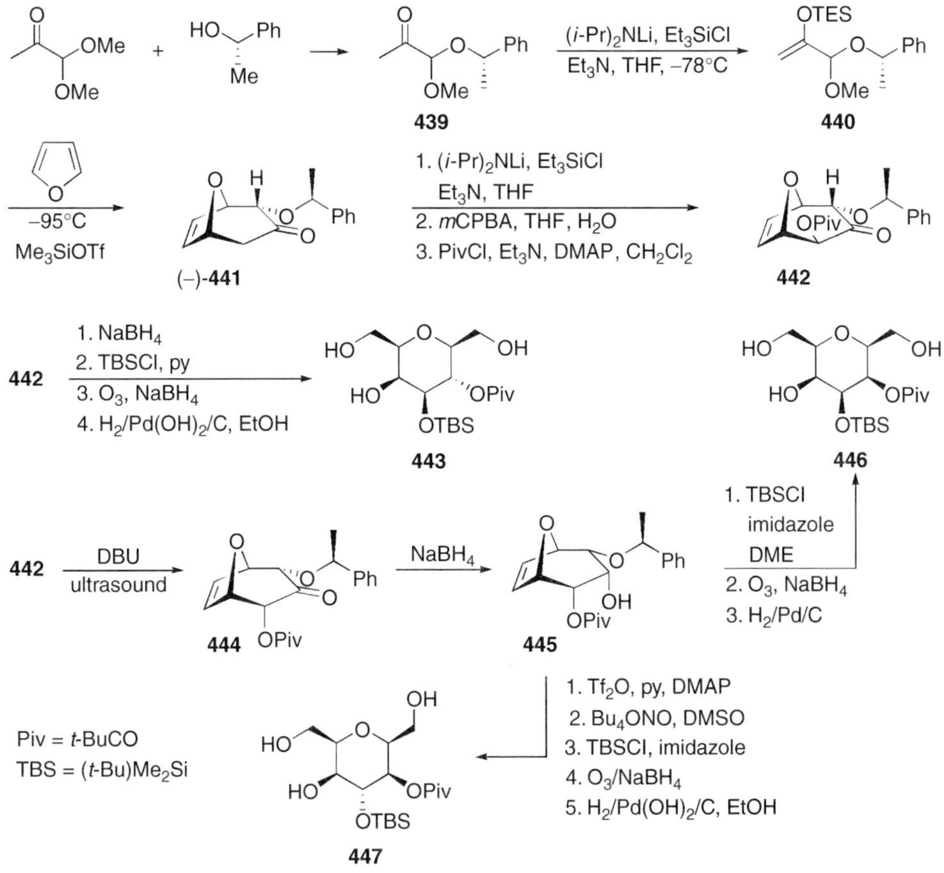

SCHEME 13.104 Hoffmann's asymmetric [4 + 3] cycloaddition of furan. Total syntheses of 2,6-anhydroheptitols.

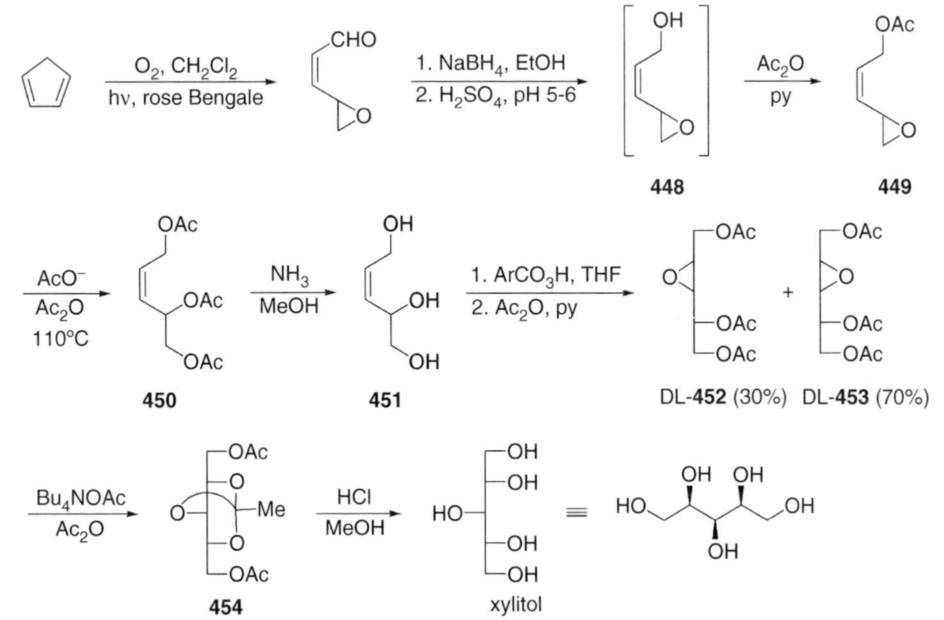

SCHEME 13.105 Synthesis of xylitol.

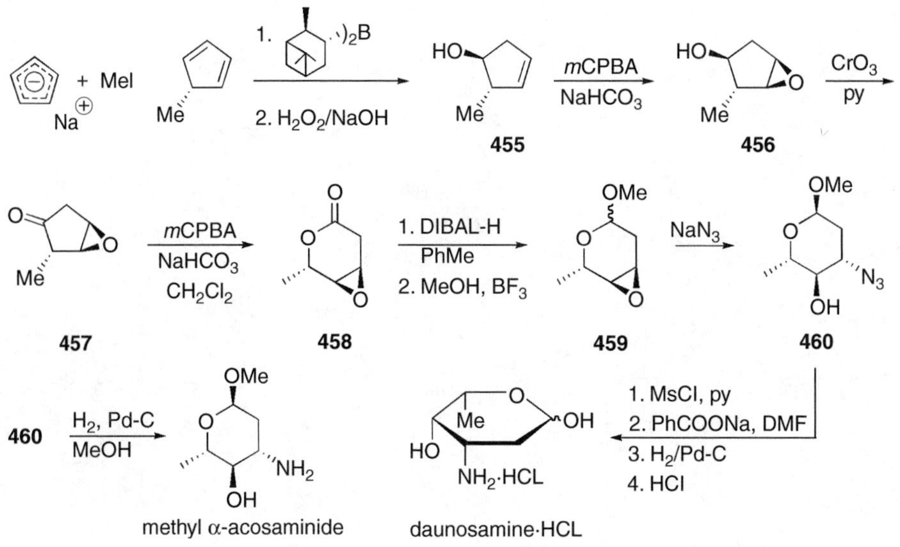

SCHEME 13.106 Roche's syntheses of α-acosamine and daunosamine.

An elegant total synthesis of acosamine and daunosamine (13 steps, 15% overall yield) was developed at Roche, starting from cyclopentadiene [205]. Monomethylation of cyclopentadiene gives 5-methylcyclopentadiene, which is asymmetrically hydroborated with (−)-di-3-pinanylborane giving (*S*)-alcohol **455** (Scheme 13.106). Epoxidation *syn* with respect to the homoallylic alcohol (lateral control by the hydroxy group) provides **456**, which is oxidized into ketone **457**. Baeyer–Villiger oxidation of **457** is regioselective, with the oxygen insertion involving the migration of the secondary (rather than the primary) α-carbon center as expected. Lactone **458** is reduced to the pyranose and Fischer glycosidation with methanol furnishes **459**. The epoxide unit of **459** is opened by azide anion, giving azido-alcohol **460**, the reduction of which leads to methyl α-acosaminide. Inversion of the alcohol at C(4) of **460** via mesylation and treatment with sodium benzoate, followed by hydrogenation and treatment with acid, provides daunosamine·HCl.

A total synthesis of 1,3-dideoxynojirimycin starting from cyclopentadiene was proposed by Johnson et al. [206]. Photooxidation of cyclopentadiene and reductive workup with thiourea generates *cis*-cyclopent-2-ene-1,4-diol, which is monoacylated with high enantioselectivity (>99% ee) with isoprenyl acetate and *Candida antarctica* lipase B (Novo Nordisk SP 435) to give **461**. After silylation of **461** and subsequent treatment with KOH and oxidation, enantiomerically pure enone **462** is obtained [207].

Treatment of **462** with iodine and pyridine leads to α-iodination [208]. Successive reduction under Luche conditions, alcohol silylation, carbonylation of the iodoalkene, reduction of the obtained enal and alcohol silylation leads to **463**. Ozonolysis of **463** gives the corresponding keto-aldehyde, which is then transformed into **464** via reductive amination with high diastereoselectivity (*syn*:*anti* > 20:1). Deprotection delivers (+)-**465** (Scheme 13.107). A similar approach converted enantiomerically pure (2*R*,3*R*)-2,3-isopropylidenedioxycyclopent-4-en-1-one derived from cyclopentadiene [209] into (−)-1-deoxymannonojirimycin and (−)-1-deoxytalonojirimycin (1,5-dideoxy-1,5-imino-D-talitol) [210]. Enone **462** was converted to (2*R*,3*S*)-2,3-bis (*tert*-butyl)dimethylsilyloxy]cyclopent-4-en-1-one and then to (+)-1-deoxynojirimycin [211].

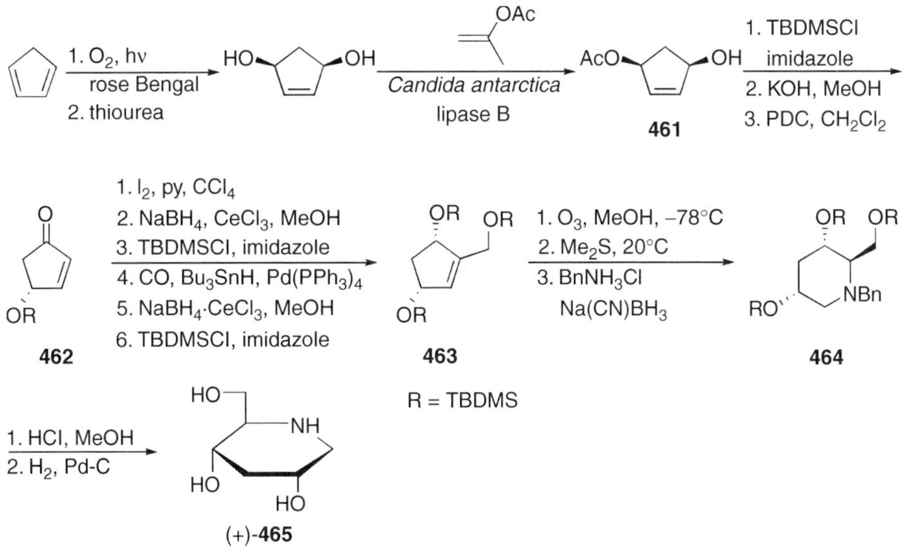

1. O_2, hν, rose Bengal; 2. thiourea
Candida antarctica lipase B
461
1. TBDMSCl, imidazole; 2. KOH, MeOH; 3. PDC, CH_2Cl_2
462
1. I_2, py, CCl_4; 2. $NaBH_4$, $CeCl_3$, MeOH; 3. TBDMSCl, imidazole; 4. CO, Bu_3SnH, $Pd(PPh_3)_4$; 5. $NaBH_4{\cdot}CeCl_3$, MeOH; 6. TBDMSCl, imidazole
463
R = TBDMS
1. O_3, MeOH, –78°C; 2. Me_2S, 20°C; 3. $BnNH_3Cl$, $Na(CN)BH_3$
464
1. HCl, MeOH; 2. H_2, Pd-C
(+)-**465**

SCHEME 13.107 Johnson's synthesis of 1,3-dideoxynojirimycin.

Mehta and coworkers recently presented a synthesis of racemic deoxynojirimycin analogs and isofagomin analogs [211b,211c], starting from the Diels–Alder adduct of vinyl acetate to 5,5-dimethoxy-1,2,3,4-tetrachlorocyclopentadiene.

13.8.2 From Benzene and Derivatives

Dioxygenases present in the blocked mutants of *Pseudomonas putida*, a soil bacterium, degrades benzene and its derivatives into cyclohexa-3,5-diene-1,2-diols (Scheme 13.108). With chlorobenzene, diol **466** is obtained with an enantiomeric excess greater than 99%. This compound is converted, in a few chemical steps, into tetrose, pentose and hexose derivatives (Scheme 13.108a) [212]. Further applications of this methodology are shown in Scheme 13.108b [212b]. A recombinant strain of *Escherichia coli* expressing naphthalene dioxygenase from *Pseudomonas* sp NCIB 9816-4 was found to specifically oxidize *N*-methyl-2-pyridone into the *cis*-5,6-dihydro-5,6-dihydro derivative [212c].

13.8.3 From Cycloheptatriene

Cycloheptatriene was converted into L-glucose via *Pseudomonas cepacia* lipase-mediated desymmetrization of a *meso*-3-*O*-protected cyclohept-6-ene-1,3,5-triol using isopropenyl acetate as solvent (Scheme 13.109). Cycloheptatriene is oxidized to tropone by hydride transfer to the trityl cation. Reduction of tropone with $NaBH_4$ generates cyclohepta-3,5-dienol (Scheme 13.109). After protection of the alcohol moiety, singlet oxygen adds to the diene forming endoperoxide **468**, the reduction of which furnishes **469**. Enzyme-catalyzed monoacetylation leads to **470** in 40% yield and high enantiomeric purity. Protection of the alcohol moiety of **470**, methanolysis of its acetate and oxidation provides enone **471**. After enolization, the corresponding silyl ether **472** is oxidized to give **473**. Reduction of the ketone and diol protection furnishes **474**. Double hydroxylation and protection of the diol generates **475**. Conversion of the benzyloxymethoxy group into a mesyloxy group, followed by desilylation and

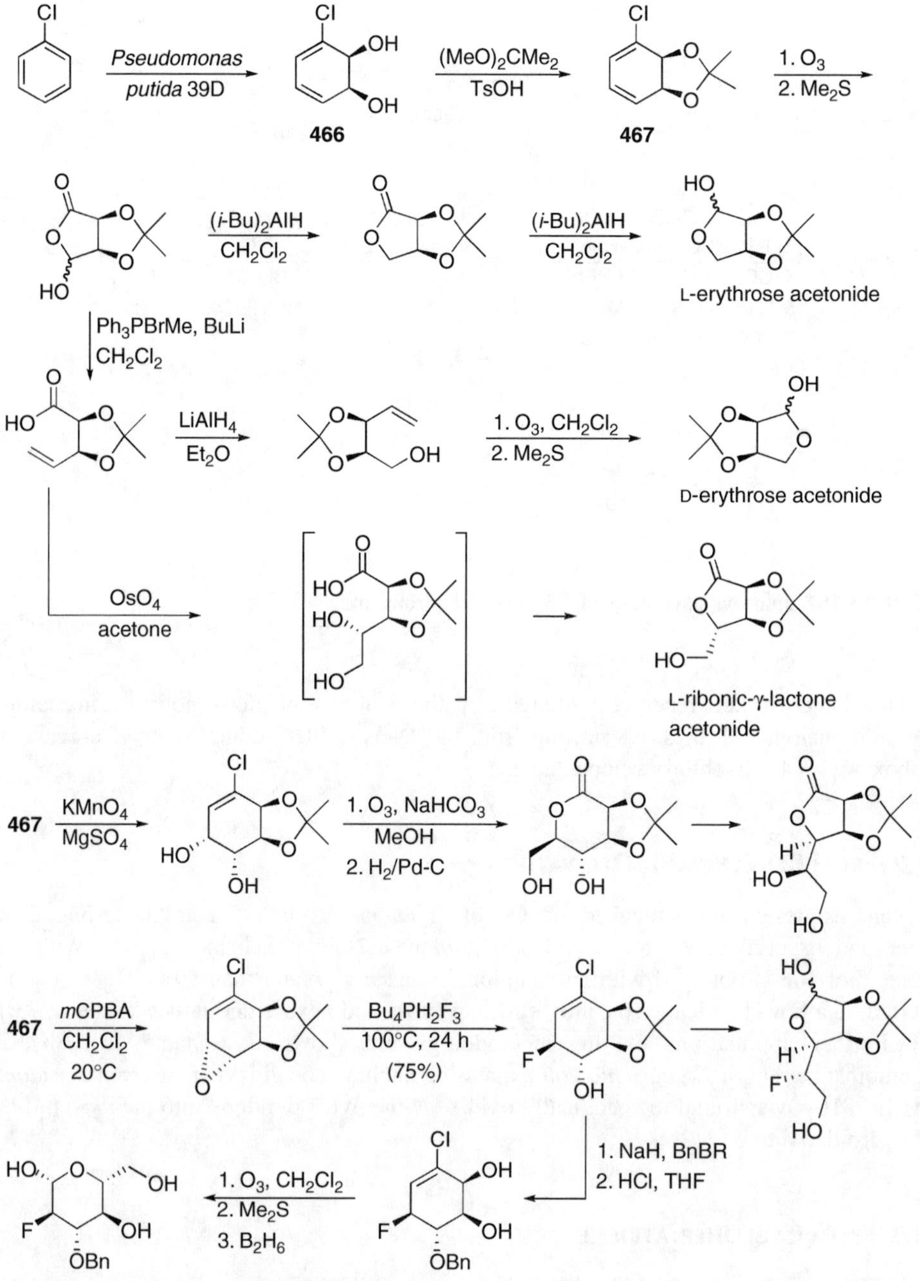

SCHEME 13.108a Hudlicky's syntheses of tetrose, pentose and hexose derivatives.

Swern oxidation, provides enone **476**. After reduction of ketone **476** to the allylic alcohol, ozonolysis and reduction give triol **477**, which is oxidatively cleaved to 2,3:4,5-di-*O*-isopropylidene-L-glucose. On treatment with acid, L-glucose is released [213a].

Alcohol **470** was also converted to 3-deoxy-D-*arabino*-heptulosonic acid [213b] and to enantio pure 2-deoxyhexoses [213c]. Cycloheptatrienone (tropone) has been converted into heptitol derivatives via the optical resolution of the (tropone)Fe(CO)$_3$ complex [213d].

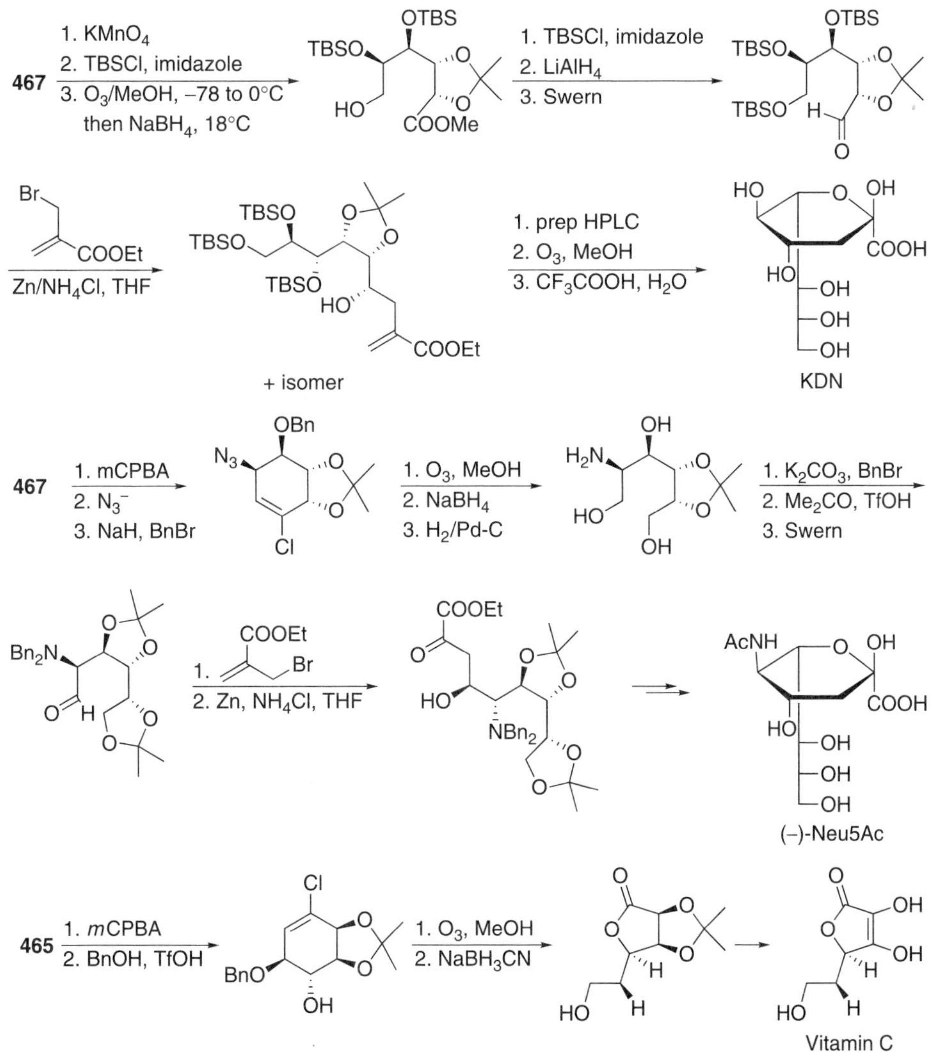

SCHEME 13.108b Banwell's total syntheses of KDN, Neu5Ac and Vitamin C.

13.8.4 From Penta-1,4-Diene

When applied to penta-1,4-diene, the Sharpless asymmetric dihydroxylation forms a 1:1 mixture of (2*S*,4*S*)- and (2*S*,4*R*)-penta-1,2,4,5-tetrols **478** and **479**, which can be converted to diepoxides **480** and **481**, respectively [214] (Scheme 13.110). A stereo- and enantioselective synthesis of **480** is possible starting from 1,5-dichloropenta-2,4-dione applying Noyori's asymmetric hydrogenation [215].

13.9 ENANTIOSELECTIVE EPOXIDATION OF ALLYLIC ALCOHOLS

Although this powerful method has already been presented in Scheme 13.47 and Scheme 13.48, additional examples are presented in this section.

AZT (3′-azido-3′-deoxythymidine) and other modified nucleosides were obtained by Jung and coworkers [216a] starting from crotonaldehyde (Scheme 13.111). The chirality is introduced via Katsuki–Sharpless epoxidation. Enolization of crotonaldehyde with TMSCl and Et_3N gives a

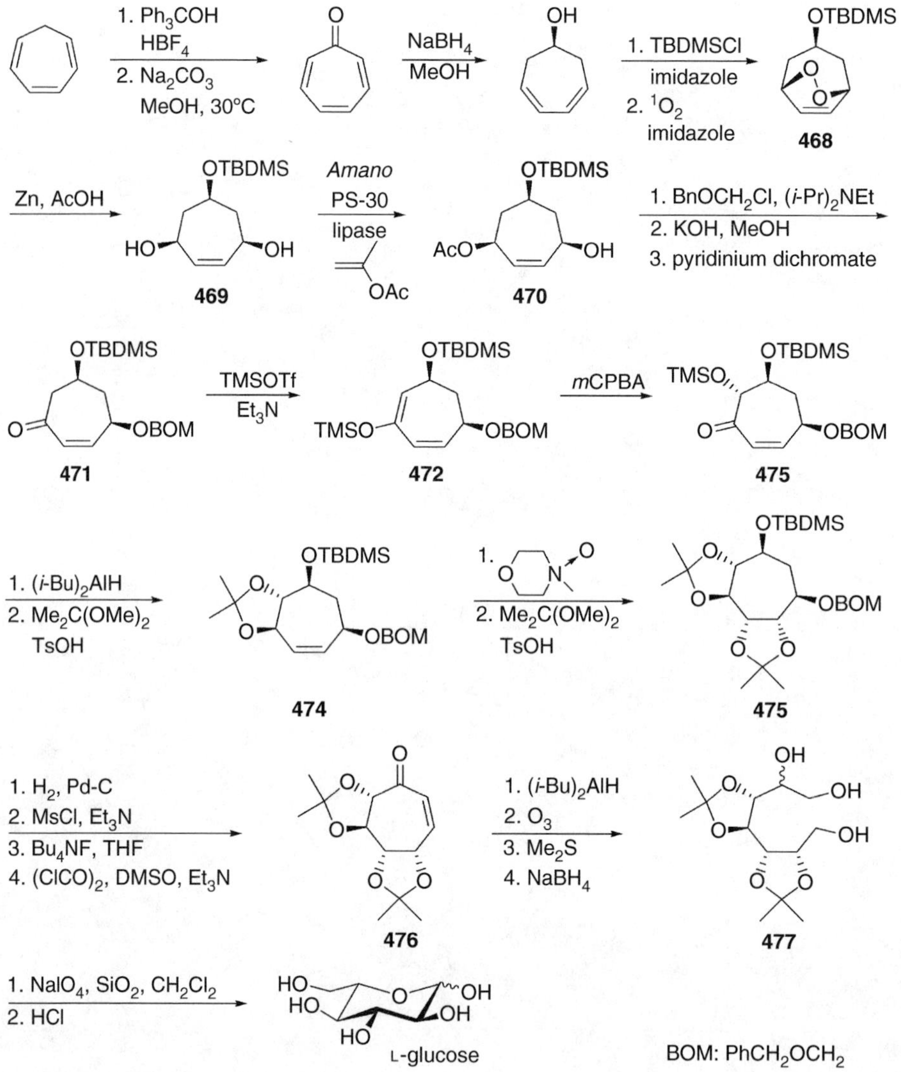

SCHEME 13.109 Johnson's total synthesis of L-glucose.

mixture of (*E*)- and (*Z*)-1-(trimethylsilyloxy)butadienes. Condensation with methyl orthoformate using catalytic $ZnCl_2$ gives enal acetal **485**. Reduction of **485** leads to the allylic alcohol **486**, the epoxidation of which affords epoxy alcohol **487** (>95% ee, 74% yield). Opening of the epoxide with $TMSN_3/Et_2AlF$ in CH_2Cl_2 provides **488**. Acid hydrolysis of **488** gives the corresponding 2,3-dideoxypentose derivatives **489** [216b].

13.9.1 DESYMMETRIZATION OF *MESO* DIENOLS

Katsuki–Sharpless desymmetrization of penta-1,4-dien-3-ol gives the monoepoxide **490**, which can be converted into 1,4-dideoxy-1,4-imino-D-lyxitol **491** (Scheme 13.112) [217].

By applying a Katsuki–Sharpless asymmetric epoxidation, Schreiber and coworkers [218] obtained (+)-KDO from the diallyl alcohol **492** (Scheme 13.113). The method generates

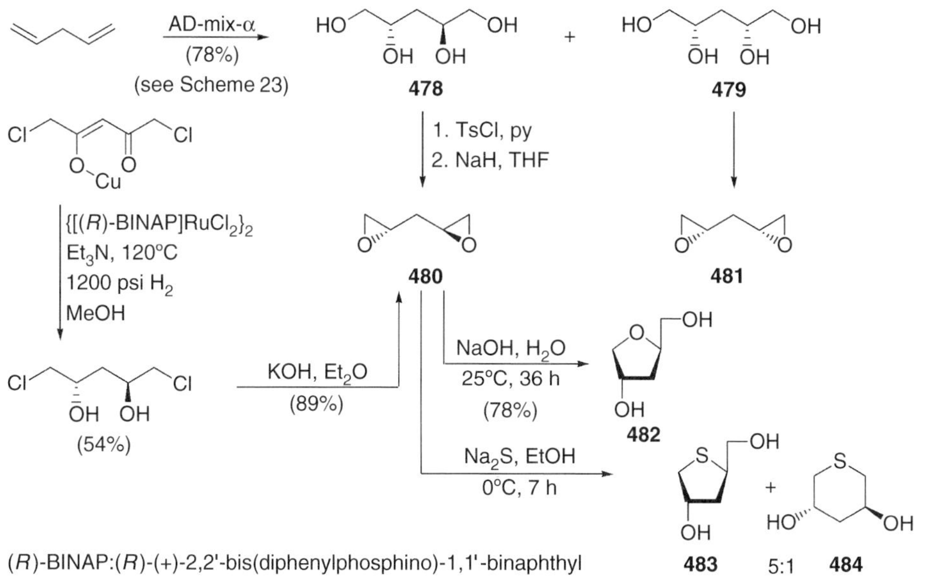

SCHEME 13.110 Conversion of penta-1,4-diene into pentitol derivatives.

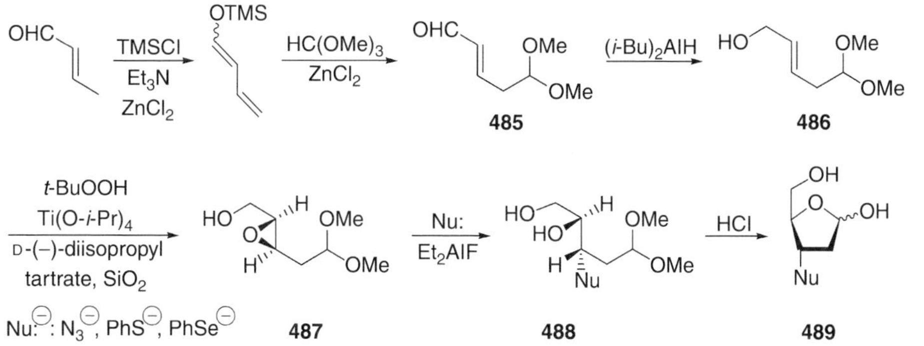

SCHEME 13.111 Jung's total syntheses of 2,3-deoxy-3-substituted D-ribose derivatives.

2,3,4,5-tetra-*O*-benzyl-D-arabinose **493**, which is coupled with $CrCl_2$ to benzyl α-(bromomethyl)-acrylate giving a 1:1 mixture of alcohols **494** and **495**. Chromatographic separation and silylation followed by ozonolysis, hydrogenation and treatment with acid provides (+)-KDO.

13.9.2 Kinetic Resolution of Racemic Allylic Alcohols

The Katsuki–Sharpless asymmetric epoxidation of racemic diol (±)-**496** (obtained by allylation of (*E*)-crotonaldehyde) gives, after chromatographic separation, the *erythro*-epoxide (+)-**497** (33% yield, >95% ee). Its urethane undergoes assisted epoxide ring opening under acidic conditions, providing a 10:1 mixture of *arabino*-carbonate (+)-**498** and the *ribo*-stereomer. Carbonate hydrolysis and subsequent ozonolysis generates D-olivose (Scheme 13.114). Asymmetric epoxidation of the kinetically resolved dienol (−)-**496** (72%) leads to (−)-**497**

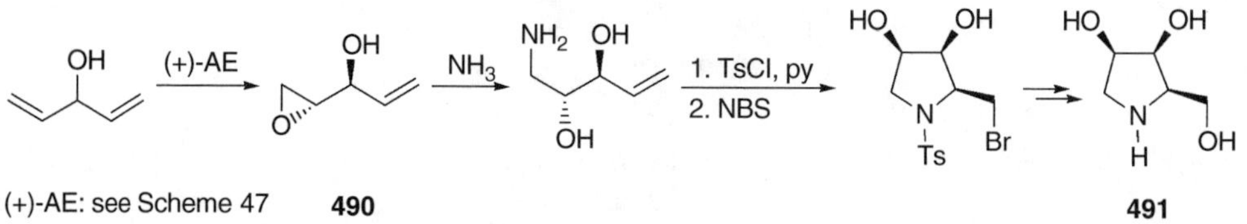

SCHEME 13.112 Jäger's syntheses of 1,4-dideoxy-1,4-iminopentitols.

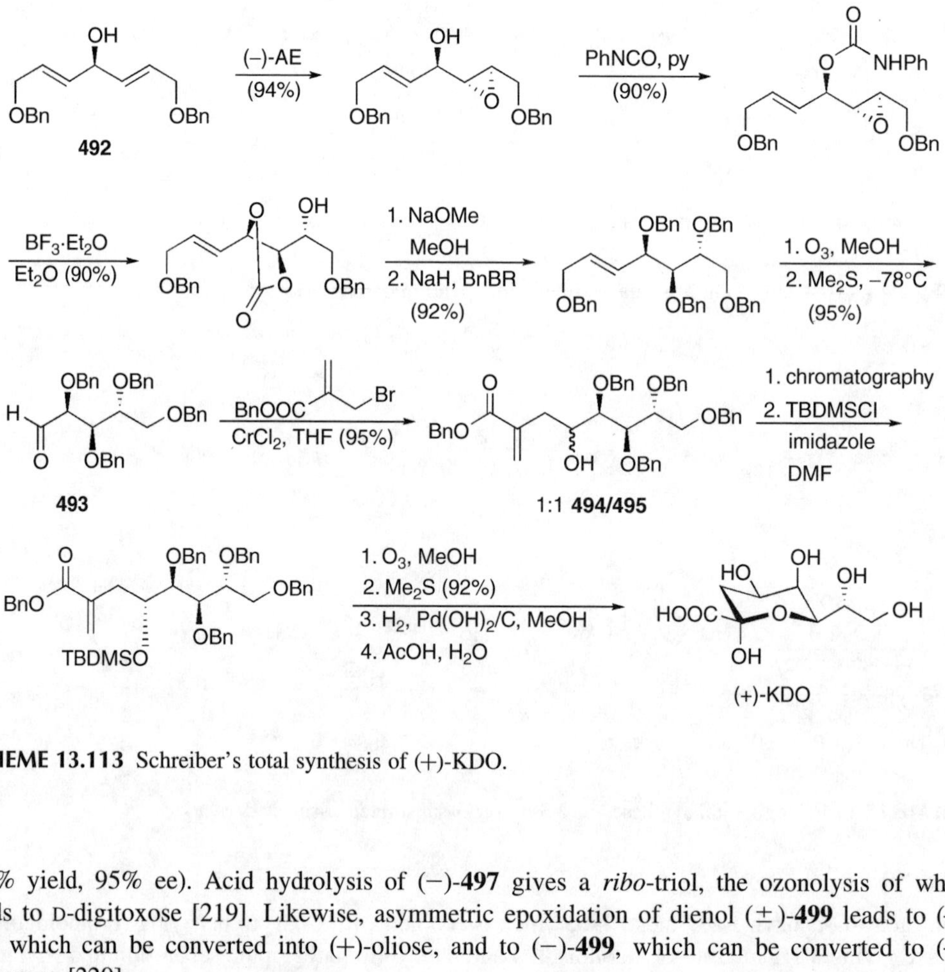

SCHEME 13.113 Schreiber's total synthesis of (+)-KDO.

(75% yield, 95% ee). Acid hydrolysis of (−)-**497** gives a *ribo*-triol, the ozonolysis of which leads to D-digitoxose [219]. Likewise, asymmetric epoxidation of dienol (±)-**499** leads to (+)-**500** which can be converted into (+)-oliose, and to (−)-**499**, which can be converted to (+)-cymarose [220].

The preparations of D- and L-chalcose were achieved in a similar way to the racemic mixtures of *threo*-dipropenylglycol (±)-**502**, obtained by reductive dimerization of crotonaldehyde [221].

A simple, divergent, asymmetric synthesis of the four stereoisomers of the 3-amino-2,3,6-trideoxy-L-hexose family was proposed by Dai and coworkers [222], which is based on the Katsuki–Sharpless asymmetric epoxidation of allylic alcohols (Scheme 13.115). Recently, *N*-trifluoroacetyl-L-daunosamine, *N*-trifluoroacetyl-L-acosamine, *N*-benzoyl-D-acosamine and *N*-benzoyl-D-nistosamine were derived from methyl sorbate via the methyl 4,5-epoxy-(*E*)-hex-2-enoates obtained via a chemoenzymatic method [223].

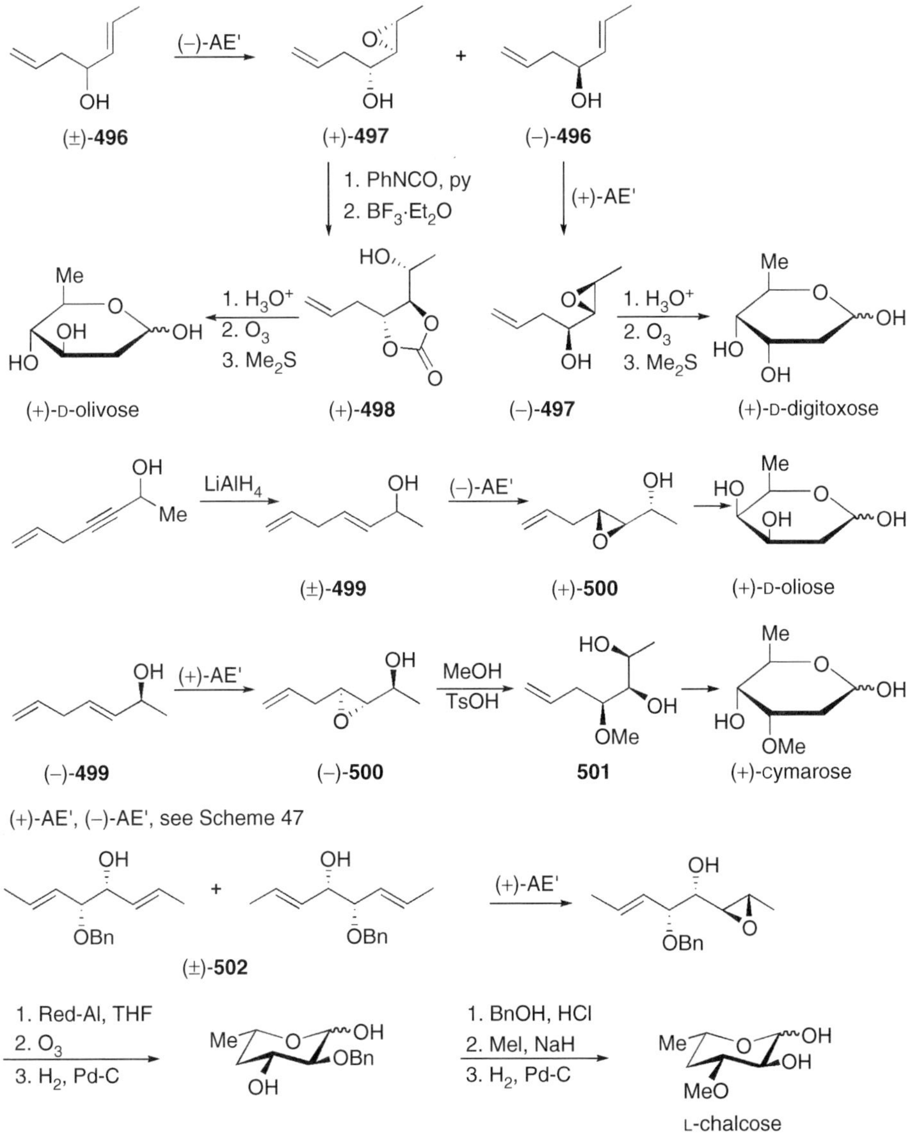

SCHEME 13.114 Total syntheses of 2,6-dideoxyhexoses.

Application of the Katsuki–Sharpless enantioselective epoxidation to racemic mono-*O*-benzylated divinylglycol allowed the preparation of enantiomerically pure L-*lyxo* and D-*lyxo*-pentoses and analogs [224,225].

13.10 ENANTIOSELECTIVE SHARPLESS DIHYDROXYLATIONS AND AMINOHYDROXYLATIONS

This extremely powerful method, already presented in Scheme 13.23, warrants the presentation of select additional examples in this section. For instance (see Scheme 13.116), tetritol and tetrose

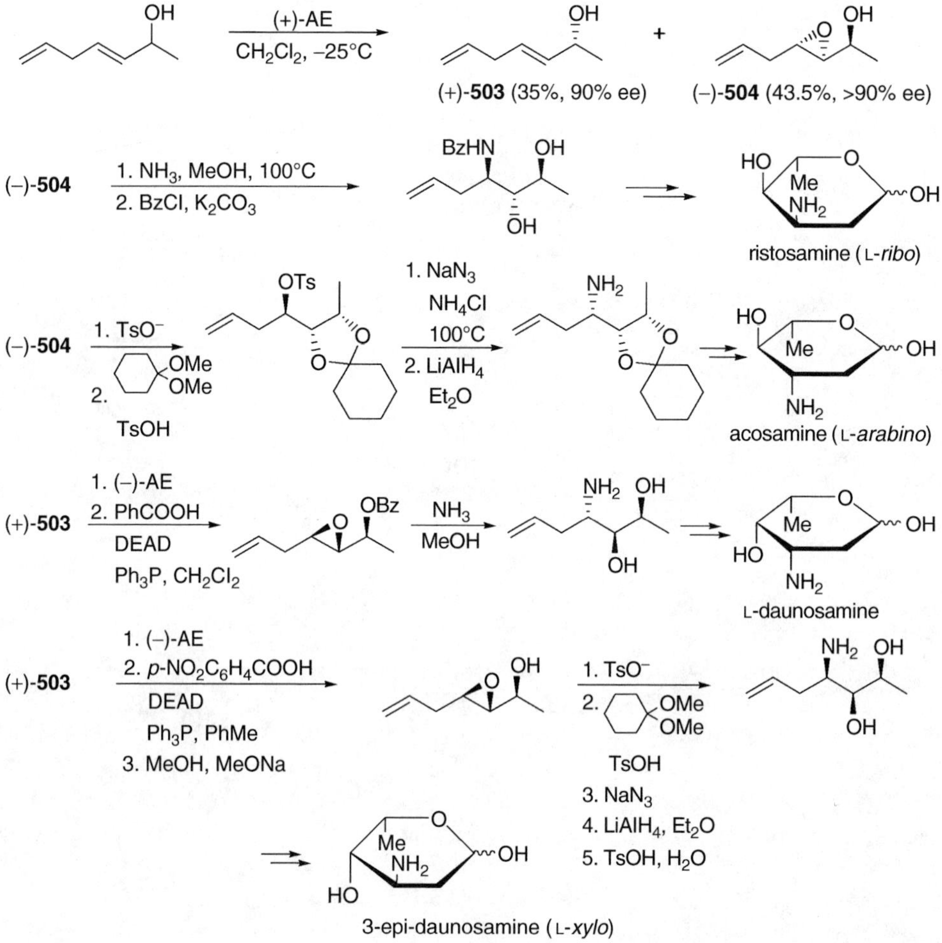

SCHEME 13.115 Total syntheses of 3-amino-2,3,6-trideoxyhexoses.

derivatives are readily obtained from asymmetric dihydroxylation of (*E*)-but-2-ene-1,4-diol [226], and 4-deoxy D- and L-threose are derived from benzene-1,2-dimethyl acetal of (*E*)-crotonaldehyde [28].

Asymmetric dihydroxylation of the dimethyl acetal of 5-[(*tert*-butyldiphenylsilyl)oxy]-(*E*)-pent-3-enal generates a diol that is converted in three steps into 2-deoxy-xylofuranosides with high enantiomeric excess [227]. Asymmetric dihydroxylation of 2-vinylfuran gives diol (+)-**504**. Its oxidation with *m*CPBA generates an enone that eliminates an equivalent of water with the formation of (+)-isolevoglucosenone. This compound can be isomerized into (−)-levoglucosenone, as shown in Scheme 13.117. The L-hexose derivatives (−)-isolevoglucosenone and (+-)-levoglucosenone are obtained with the same ease.

Applying the same route, D- and L-mannose were obtained in a five-step synthesis (39% overall yield) from furfural. Similarly, the same methodology was extended to the preparation of D- and L-gulose and D- and L-talose (19% yield) [229].

The Sharpless asymmetric aminohydroxylation [230] of the electron-deficient 2-vinylfuran **505** gives a 7:1 mixture of semiprotected amino alcohols **506** and **507** (41%). The major product (ee $>$ 86%) is reduced with diisobutylaluminum hydride, giving diol **508** [231], which can be converted into the β-hydroxyfurylamine derivative **509**. This compound is an

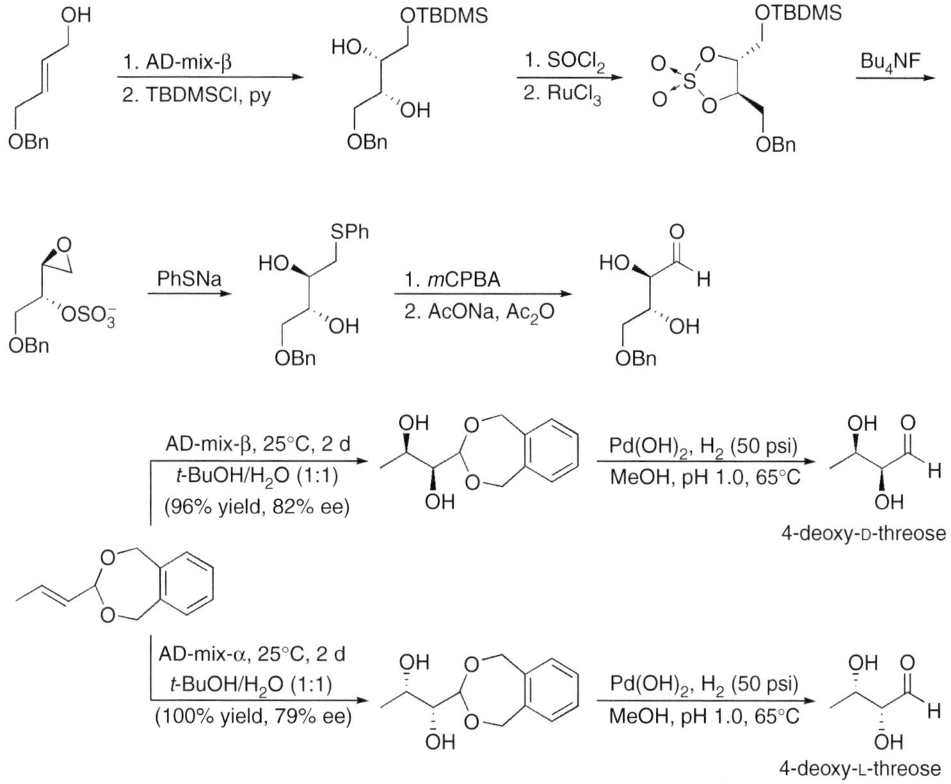

OH; 1. AD-mix-β 2. TBDMSCl, py; OTBDMS HO OH OBn; 1. $SOCl_2$ 2. $RuCl_3$; OTBDMS; Bu_4NF; OBn

PhSNa; SPh HO OH OBn; 1. *m*CPBA 2. AcONa, Ac_2O; HO O H OH OBn

AD-mix-β, 25°C, 2 d *t*-BuOH/H_2O (1:1) (96% yield, 82% ee); $Pd(OH)_2$, H_2 (50 psi) MeOH, pH 1.0, 65°C; 4-deoxy-D-threose

AD-mix-α, 25°C, 2 d *t*-BuOH/H_2O (1:1) (100% yield, 79% ee); $Pd(OH)_2$, H_2 (50 psi) MeOH, pH 1.0, 65°C; 4-deoxy-L-threose

SCHEME 13.116 Total asymmetric syntheses of tetritol and tetrose derivatives.

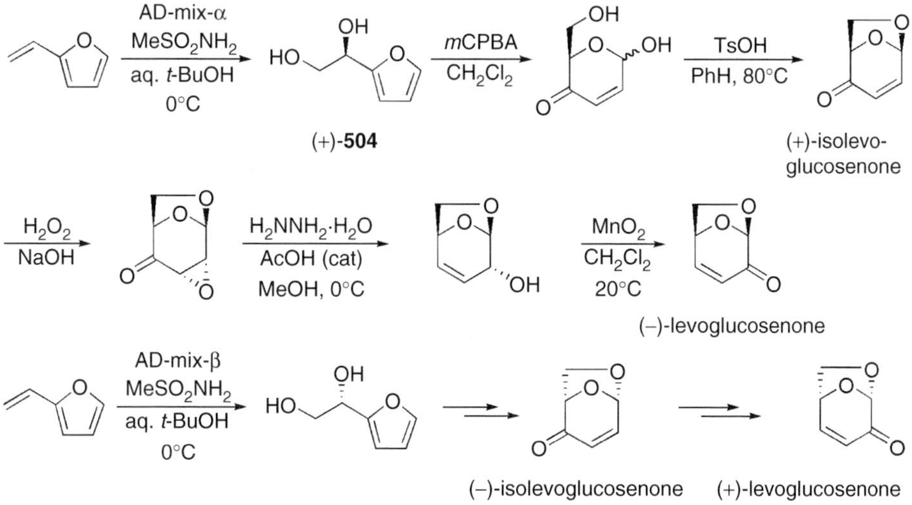

AD-mix-α $MeSO_2NH_2$ aq. *t*-BuOH 0°C; (+)-**504**; *m*CPBA CH_2Cl_2; TsOH PhH, 80°C; (+)-isolevo-glucosenone

H_2O_2 NaOH; $H_2NNH_2 \cdot H_2O$ AcOH (cat) MeOH, 0°C; MnO_2 CH_2Cl_2 20°C; (−)-levoglucosenone

AD-mix-β $MeSO_2NH_2$ aq. *t*-BuOH 0°C; (−)-isolevoglucosenone; (+)-levoglucosenone

SCHEME 13.117 Total syntheses of (+)- and (−)-isolevoglucosenone and of (+)- and (−)-levoglucosenone [228].

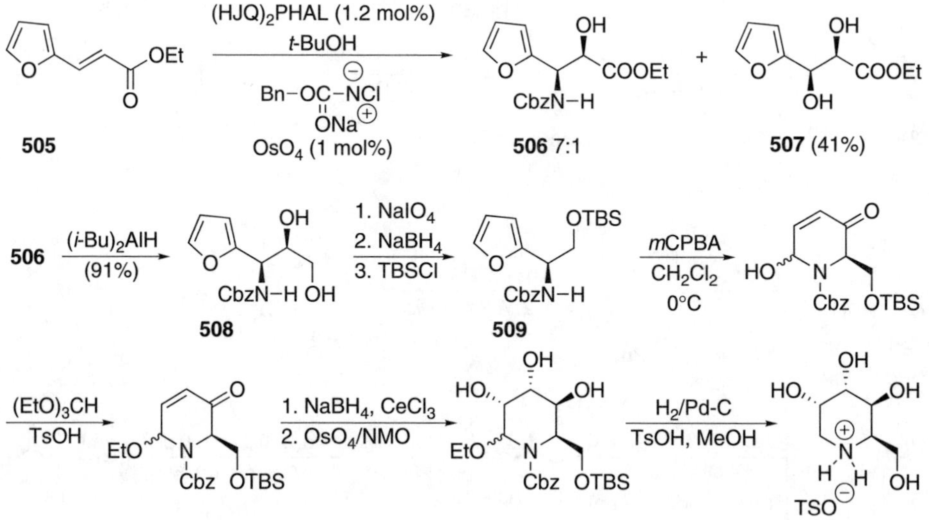

SCHEME 13.118 Application of the Sharpless asymmetric aminohydroxylation and of the aza-Achmatowicz reaction to the synthesis of a 1,5-dideoxy-1,5-iminoalditol.

important synthetic building block for various biologically important molecules [232] including 1,5-dideoxy-1,5-imino-alditols [232] (Scheme 13.118). A less regioselective, but shorter way to intermediate **509** is the direct asymmetric aminohydroxylation of vinylfuran [232–234].

13.11 CONCLUSION

For a long time, enantiomerically pure monosaccharides and analogs were difficult targets for synthetic chemists because of the complexity arising from their high stereochemical information and their multifunctionality. During the last 20 years, the synthetic methodologies have become so powerful that almost any targets can now be reached with satisfactory overall yield and high enantiomeric purity, including sugars and their analogs of biological interest. These include the aminodeoxy and aminodideoxyhexoses, long-chain sugars, azasugars, and thiosugars. Total asymmetric synthesis allows one to prepare both enantiomeric forms of almost any kind of monosaccharide and, frequently, in a more economical fashion than following classical approaches using natural carbohydrates. Depending on the target, pure chemical procedures, chemoenzymatic methods or a combination of both methodologies can be successfully applied.

REFERENCES

1. (a) Butlerow, M A, Formation synthétique d'une substance sucrée, *CR Séances Acad. Sci.*, 53, 145–147, 1861; (b) Butlerow, M A, Bildung einer zuckerartigen substanz durch synthese, *Ann. Chem.*, 120, 295–298, 1861; (c) Jones, J K N, Szarek, W A, *Total Synthesis of Natural Products*, Vol. 1, ApSimon, J, ed., Wiley-Interscience, New York, pp. 1–80, 1973; (d) Zamojski, A, Grynkiewicz, G, *Total Synthesis of Natural Products*, Vol. 6, ApSimon, J, ed., Wiley-Interscience, New York, pp. 141–235, 1984; (e) McGarvey, G J, Kimura, M, Oh, T, Williams, J M, Acyclic stereoselective synthesis of carbohydrates, *J. Carbohydr. Chem.*, 3, 125–188, 1984.
2. (a) Schmidt, R R, De novo synthesis of carbohydrates and related natural products, *Pure Appl. Chem.*, 59, 415–424, 1987; (b) Zamojski, A, Total synthesis of sugars, In *Preparative Carbohydrate*

Chemistry, Hanessian, S, ed., Dekker, New York, pp. 615–636, 1997; (c) Kirschning, A, Jesberger, M, Schoning, K-U, Concepts for the total synthesis of deoxy sugars, *Synthesis*, 507–540, 2001.
3. Vogel, P, De novo syntheses of monosaccharides, In *Encyclopedia of Glycosciences*, Vol. 2, Fraser-Reid, B, Tatsuta, K, Thiem, J, eds., Springer, Berlin, pp. 1023–1174, 2001, chap. 4.4.
4. (a) Loew, O, Über Bildung von Zuckerarten aus Formaldehyd, *Ber. Deutsch. Chem. Ges.*, 22, 470–478, 1889; (b) Loew, O, Nachträgliche Bemerkungen über Formose, *Ber. Deutsch. Chem. Ges.*, 22, 478–483, 1889.
5. (a) Fischer, E, Passmore, F, Bildung von Acrose aus Formaldehyd, *Ber. Deutsch. Chem. Ges.*, 22, 359–361, 1889; (b) Fischer, E, Synthese der Mannose und Lävulose, *Ber. Deutsch. Chem. Ges.*, 23, 370–394, 1890.
6. Mizuno, T, Mori, T, Shiomi, N, Nakatsuji, N, Studies on synthesis and utilization of formose. Part 1. Sugar formation by the formaldehyde condensation in the presence of inorganic or organic bases, *Nippon Nogli Kagaku Kaishi. J. Agr. Chem. Soc. Jpn.*, 44, 324–331, 1970.
7. Shigemasa, Y, Nagae, O, Sakazawa, C, Nakashima, R, Matsuura, T, A selective formose reaction, *J. Am. Chem. Soc.*, 100, 1309–1310, 1978.
8. (a) Weiss, A H, Socha, R F, Likholobov, V A, Sakharov, M M, Polyols from formaldehyde, *CHEMTECH*, 10, 643–647, 1980; (b) Socha, R F, Weiss, A H, Sakharov, M M, Homogeneously catalyzed condensation of formaldehyde to carbohydrates. VII. An overall formose reaction model, *J. Catal.*, 67, 207–217, 1981; (c) Decker, P, Did evolution begin with formaldehyde? *Umschau.*, 73, 733–734, 1973.
9. (a) Castells, J, Geijo, F, López-Calahorra, F, The "formion reaction": a promising entry to carbohydrates from formaldehyde, *Tetrahedron Lett.*, 21, 4517–4520, 1980; (b) Castells, J, López-Calahorra, F, Geijo, F, The formion reaction, *Carbohydr. Res.*, 116, 197–207, 1983; (c) Matsumoto, T, Inove, S, Selective formation of triose from formaldehyde catalysed by ethylbenzothiazolium bromide, *J. Chem. Soc. Chem. Commun.*, 171–172, 1983; (d) Matsumoto, T, Yamamoto, H, Inove, S, Selective formation of triose from formaldehyde catalyzed by thiazolium salt, *J. Am. Chem. Soc.*, 106, 4829–4832, 1984; (e) Shigemasa, Y, Sasaki, Y, Ueda, N, Nakashima, R, Formose reaction. XXI. A selective formation of dihydroxyacetone in the formose reaction in *N*, *N*-dimethylformamide, *Bull. Chem. Soc. Jpn.*, 57, 2761–2767, 1984.
10. Shigemasa, Y, Ueda, T, Saimoto, H, First synthesis of DL-2-*C*-hydroxymethyl-3-pentulose in the formose reaction, *J. Carbohydr. Chem.*, 8, 669–673, 1989.
11. Müller, D, Pitsch, S, Kittaka, A, Wagner, E, Wintner, C E, Eschenmoser, A, Chemie von α-Aminonitrilen. Aldomerisierung von Glycolaldehyd-phosphat zu racemischen Hexose-2,4,6-triphosphaten und (in Gegenwart von Formaldehyd) racemischen Pentose-2,4-diphosphaten: *rac*-Allose-2,4,6-triphosphat und *rac*-Ribose-2,4-diphosphat sind die Reaktionhauptprodukte, *Helv. Chim. Acta*, 73, 1410–1468, 1990.
12. Okano, T, Ito, H, Konishi, H, Kiji, J, One-step synthesis of straight-chain carbohydrates from formaldehyde and syngas, *Chem. Lett.*, 1731–1734, 1986.
13. Sutherland, J D, Whitfield, J N, Prebiotic chemistry: a bioorganic perspective, *Tetrahedron*, 53, 11493–11527, 1997.
14. Shevlin, P B, McPherson, D W, Melius, P, Reaction of atomic carbon with ammonia. The mechanism of formation of amino acid precursors, *J. Am. Chem. Soc.*, 105, 488–491, 1983.
15. Flanagam, G, Ahmed, S N, Shevlin, P B, Formation of carbohydrates in the reaction of atomic carbon with water, *J. Am. Chem. Soc.*, 114, 3892–3896, 1992.
16. Reid, C, Orgel, L E, Synthesis of sugars in potentially prebiotic conditions, *Nature (London)*, 216, 455, 1967.
17. Weber, A L, Prebiotic sugar synthesis: hexose and hydroxy acid synthesis from glyceraldehyde catalyzed by iron(III)hydroxide oxide, *J. Mol. Evol.*, 35, 1–6, 1992.
18. Schwartz, A W, De Graaf, R M, The prebiotic synthesis of carbohydrates: a reassessment, *J. Mol. Evol.*, 36, 101–106, 1993.
19. Pitsch, S, Pombo-Villar, E, Eschenmoser, A, Chemie von α-Aminonitrilen. Über die Bildung von 2-Oxoethyl-phosphaten ("Glycoaldehyd-phosphaten") aus *rac*-Oxirancarbonitril and anorganischem Phosphat und über (formate) Konstitutionelle Zusammenhänge zwischen 2-Oxoethyl-Phosphaten und Oligo(hexo- and pentopyranosyl)nucleotid-Rückgraten, *Helv. Chim. Acta*, 77, 2251–2285, 1994.

20. Pitch, S, Krishnamurthy, R, Arrhenius, G, Concentration of simple aldehydes by sulfit-containing double-layer hydroxide minerals: implication for biopoesis, *Helv. Chim. Acta*, 83, 2398–2424, 2000.
21. Weber, A L, Sugars as the optimal biosynthetic carbon substrate of aqueous life throughout the universe, *Origins. Life Evol. Biosphere*, 30, 33–43, 2000.
22. Wong, C H, Halcomb, R L, Ichikawa, Y, Kajimoto, T, Enzymes in organic synthesis: application to the problems of carbohydrate recognition (part 1), *Angew. Chem. Int. Ed. Engl.*, 34, 412–432, 1995.
23. Witzemann, E J, The isolation of crystalline DL-glyceric aldehyde from a syrup obtained by the oxidation of glycerol, *J. Am. Chem. Soc.*, 36, 2223–2234, 1914.
24. Morozov, A A, Selective production of trioses from formadehyde, *React. Kinet. Catal. Lett.*, 46, 71–77, 1992; Chem Abstr 116, 174545, 1992.
25. (a) Saimoto, H, Kotani, K, Shigemasa, Y, Suzuki, M, Harada, K-I, Isolation of an intermediate of formose reaction catalyzed by thiamin · HCl, *Tetrahedron Lett.*, 30, 2553–2554, 1989; (b) Yamashita, K, Wakao, N, Nango, M, Tsuda, K, Formose reaction by polymer-supported thiazolium salts, *J. Polym. Sci. Part A: Polym. Chem*, 30, 2247–2250, 1992; Chem Abstr 117, 1992 151580z.
26. (a) Ando, T, Shioi, S, Nakagawa, M, Total synthesis of carbohydrates. I. Dihydroxyacetone and DL-erythrulose, *Bull. Chem. Soc. Jpn.*, 45, 2611–2615, 1972; (b) Sonogashira, K, Nakagawa, M, Total synthesis of carbohydrates. II. DL-Erythrose and DL-threose, *Bull. Chem. Soc. Jpn.*, 45, 2616–2620, 1972.
27. Eyrisch, O, Sinerius, G, Fessner, W-D, Facile enzymic *de novo* synthesis and NMR spectroscopic characterization of D-tagatose 1,6-bisphosphate, *Carbohydr. Res.*, 238, 287–306, 1993.
28. Henderson, I, Sharpless, K B, Wong, C-H, Synthesis of carbohydrates via tandem use of the osmium-catalyzed asymmetric dihydroxylation and enzyme-catalyzed aldol addition reactions, *J. Am. Chem. Soc.*, 116, 558–561, 1994.
29. Alajarín, R, García-Junceda, E, Wong, C-H, A short enzymatic synthesis of L-glucose from dihydroxyacetone phosphate and L-glyceraldehyde, *J. Org. Chem.*, 60, 4294–4295, 1995.
30. (a) Ziegler, T, Straub, A, Effenberger, F, Enzyme-catalyzed synthesis of 1-deoxymannojirimycin, 1-deoxynojirimycin, and 1,4-dideoxy-1,4-imino-D-arabinitol, *Angew. Chem. Int. Ed. Engl.*, 27, 716–717, 1988; (b) Straub, A, Effenberger, F, Fisher, P, Aldolase-catalyzed C–C bond formation for stereoselective synthesis of nitrogen-containing carbohydrates, *J. Org. Chem.*, 55, 3926–3932, 1990; (c) Effenberger, F, Null, V, Eine neue effiziente synthese von fagomin, *Liebigs Ann. Chem.*, 1211–1212, 1992; (d) Lemaire, M, Valentin, M L, Hecquet, L, Demuynck, C, Bolte, J, Chemo-enzymatic synthesis of new protected aldoketoses: intermediates in the biosynthesis and chemical syntheis of nojirimycin and mannojirimycin, *Tetrahedron Asymmetry*, 6, 67–70, 1995; (e) Gijsen, H J M, Qiao, L, Fitz, W, Wong, C-H, Recent advances in the chemoenzymatic synthesis of carbohydrates and carbohydrate mimetics, *Chem. Rev.*, 96, 443–473, 1996.
31. von der Osten, C H, Sinskey, A J, Barbas, C F III, Pederson, R L, Wang, Y-F, Wong, C-H, Use of recombinant bacterial fructose-1,6-diphosphate aldolase in aldol reactions: preparative syntheses of 1-deoxynojirimycin, 1-deoxymannojirimycin, 1,4-dideoxy-1,4-imino-D-arabinitol, and fagomine, *J. Am. Chem. Soc.*, 111, 3924–3927, 1989.
32. Kajimoto, T, Liu, K K-C, Pederson, R L, Zhong, Z, Ichikawa, Y, Porco, J A Jr, Wong, C-H, Enzyme-catalyzed aldol condensation for asymmetric synthesis of azasugars: synthesis, evaluation, and modeling of glycosidase inhibitors, *J. Am. Chem. Soc.*, 113, 6187–6196, 1991.
33. (a) Hung, R R, Straub, J A, Whitesides, G M, α-Amino aldehyde equivalents as substrates for rabbit muscle aldolase: synthesis of 1,4-dideoxy-D-arabinitol and 2(*R*),5(*R*)-bis(hydroxymethyl)-3(*R*),4(*R*)-dihydroxypyrrolidine, *J. Org. Chem.*, 56, 3849–3855, 1991; (b) Liu, K K-C, Kajimoto, T, Chen, L, Zhong, Z, Ichikawa, Y, Wong, C-H, Use of dihydroxyacetone phosphate dependent aldolases in the synthesis of deoxyazasugars, *J. Org. Chem.*, 56, 6280–6289, 1991; (c) Kajimoto, T, Chen, L, Liu, K K-C, Wong, C-H, Palladium-mediated stereocontrolled reductive amination of azido sugars prepared from enzymatic aldol condensation: a general approach to the synthesis of deoxy aza sugars, *J. Am. Chem. Soc.*, 113, 6678–6680, 1991.
34. Takaoka, Y, Kajimoto, T, Wong, C-H, Inhibition of *N*-acetyl-glucosaminyltransfer enzymes: chemical-enzymatic synthesis of new five-membered acetamido azasugars, *J. Org. Chem.*, 58, 4809–4812, 1993.

35. Chou, W-C, Chen, L, Fang, J-M, Wong, C-H, A new route to deoxythiosugars based on aldolases, *J. Am. Chem. Soc.*, 116, 6191–6194, 1994.
36. Mukaiyama, T, Shiina, I, Kobayashi, S, A convenient and versatile route for the stereoselective synthesis of monosaccharides via key chiral synthons prepared from achiral sources, *Chem. Lett.*, 2201–2204, 1990.
37. Kobayashi, S, Kawasuji, T, A new synthetic route to monosaccharides from simple achiral compounds by using a catalytic asymmetric aldol reaction as a key step, *Synlett*, 911–913, 1993.
38. (a) Kiliani, H, Über das Cyanhydrin der Lävulose, *Ber. Deutsch. Chem. Ges.*, 18, 3066–3070, 1885; (b) Fischer, E, Reduction von Säuren der Zuckergruppe, *Ber. Deutsch. Chem. Ges.*, 22, 2204–2209, 1889; (c) Lichtenthaler, F W, Emil Fischer's proof of the configuration of sugars: a centemial tribute, *Angew. Chen. Int. Ed. Engl.*, 31, 1541–1556, 1992.
39. Stanek, J, Cerny, M, Kocourek, J, Pacák, J, *The Monosaccharides*, Academic Press, New York, 1965, p. 144.
40. (a) Roush, W R, Hoong, L K, Palmer, M A J, Straub, J A, Palkowitz, A D, Asymmetric synthesis using tartrate ester modified allylboronates. 2. Single and double asymmetric reactions with alkoxy-substituted aldehydes, *J. Org. Chem.*, 55, 4117–4126, 1990; (b) Roush, W R, Ando, K, Powers, D B, Palkowitz, A D, Halterman, R L, Asymmetric synthesis using diisopropyl tartrate modified (*E*)- and (*Z*)-crotylboronates: preparation of the chiral crotylboronates and reactions with achiral aldehydes, *J. Am. Chem. Soc.*, 112, 6339–6348, 1990.
41. (a) Roush, W R, Straub, J A, VanNieuwenhze, M S, A stereochemically general synthesis of 2-deoxyhexoses via the asymmetric allylboration of 2,3-epoxy aldehydes, *J. Org. Chem.*, 56, 1636–1648, 1991; (b) Roush, W R, Lin, X, Straub, J A, A highly stereoselective synthesis of the AB disaccharide unit of olivomycin A, *J. Org. Chem.*, 56, 1649–1655, 1991.
42. Maruyama, K, Ishihara, Y, Yamamoto, Y, A short and stereoselective synthesis of (±)-Prelog-Djerassi lactonic acid, *Tetrahedron Lett.*, 22, 4235–4238, 1981.
43. Marshall, J A, Luke, G P, Diastereoselective addition of enantioenriched (γ-alkoxyallyl)stannanes to α-alkoxy aldehydes. A synthetic route to carbohydrates, *J. Org. Chem.*, 56, 483–485, 1991.
44. Marshall, J A, Chiral allylic and allenic stannanes as reagents for asymmetric synthesis, *Chem. Rev.*, 96, 31–47, 1996.
45. Jurczak, J, Pikul, S, Bauer, T, (*R*)- and (*S*)-2,3-isopropylideneglyceraldehyde in stereoselective organic synthesis, *Tetrahedron*, 42, 447–448, 1986.
46. Perlin, A S, *D-, L- and DL-Glyceraldehyde*, *Methods in Carbohydrate Chemistry*, Vol. 1, Academic Press, New York, pp. 1962, 61–63.
47. Hubschwerlen, C, A convenient synthesis of L-(*S*)-glyceraldehyde acetonide from L-ascorbic acid, *Synthesis*, 962–964, 1986.
48. Jäger, V, Wehner, V, 2-O-benzylglyceraldehyde: a synthetic building block available in both enantiomeric forms and configurationally stable owing to rapid oligomerization, *Angew. Chem. Int. Ed. Engl.*, 28, 469–470, 1989.
49. Bewsey, J A, Synthetic tartaric acid and the economics of food acidulants, *Chem. Ind.*, 119–120, 1977.
50. Seebach, D, Hungerbühler, E, Syntheses of enantiomerically pure compounds (EPC-syntheses). Tartaric acid, an ideal source of chiral building blocks for syntheses? In *Modern Synthetic Methods*, Vol. 2, Scheffold, R, ed., Otto Salle Verlag, Verlag Sauerländer, Frankfurt am Main, pp. 91–171, 1980.
51. Matteson, D S, Peterson, M L, Synthesis of L-(+)-ribose via (*S*)-pinanediol (α*S*)-α-bromo boronic esters, *J. Org. Chem.*, 52, 5116–5121, 1987.
52. Bischofberger, N, Waldmann, H, Saito, T, Simon, E S, Lees, W, Bednarski, M D, Whitesides, G M, Synthesis of analogues of 1,3-dihydroxyacetone phosphate and glyceraldehyde 3-phosphate for use in studies of fructose-1,6-diphosphate aldolase, *J. Org. Chem.*, 53, 3457–3465, 1988.
53. Ballou, C E, Fischer, H O L, The synthesis of D-glyceraldehyde-3-phosphate, *J. Am. Chem. Soc.*, 77, 3329–3331, 1995.
54. Kerscher, V, Kreiser, W, Enantiomeren reine glycerin-derivate durch enzymatische hydrolyse prochiraler ester, *Tetrahedron Lett.*, 28, 531–534, 1987.
55. De Witt, P D, Misiti, D, Zappia, G, Synthesis of (2*R*)- and (2*S*)-benzyl-2,3-epoxypropyl ether from a common precursor: O-benzyl-L-serine, *Tetrahedron Lett.*, 30, 5505–5506, 1989.

56. Kolb, H C, VanNieuwenhze, M S, Sharpless, K B, Catalytic asymmetric dihydroxylation, *Chem. Rev.*, 94, 2483–2547, 1994.
57. Oi, R, Sharpless, K B, Asymmetric dihydroxylation of acrolein acetals: synthesis of stable equivalents of enantiopure glyceraldehyde and glycidaldehyde, *Tetrahedron Lett.*, 33, 2095–2098, 1992.
58. Wuts, P G M, Bigelow, S S, Total synthesis of oleandrose and the avermectin disaccharide, benzyl α-L-oleandrosyl-α-L-4-acetoxyoleandroside, *J. Org. Chem.*, 48, 3489–3493, 1983.
59. Lubell, W D, Rapoport, H, Configurational stability of *N*-protected α-amino aldehydes, *J. Am. Chem. Soc.*, 109, 236–239, 1987.
60. Fehrentz, J A, Castro, B, An efficient synthesis of optically active α-(*t*-butoxycarbonylamino)-aldehydes from α-aminoacids, *Synthesis*, 676–678, 1983.
61. Nicolaou, K C, Gronenberg, R D, Stylianides, N A, Miyazaki, T, Synthesis of the *CD* and *E* ring systems of the calicheamicin γ_1^2 oligosaccharide, *J. Chem. Soc. Chem. Commun.*, 1275–1277, 1990.
62. Sibi, M P, Renhowe, P A, A new nucleophilic alaninol synthon from serine, *Tetrahedron Lett.*, 31, 7407–7410, 1990.
63. Dondoni, A, Marra, A, Thiazole-based one-carbon extension of carbohydrate derivatives, In *Preparative Carbohydrate Chemistry*, Hanessian, S, ed., Marcel Dekker, New York, pp. 173–205, 1997.
64. Dondoni, A, Merino, P, Diastereoselective homologation of D-(*R*)-glyceraldehyde acetonide using 2-(trimethylsilyl)thiazole: 2-*O*-benzyl-3,4-O-isopropylidene-D-erythrose, *Org. Synth.*, 72, 21–28, 1993.
65. Dondoni, A, Marra, A, Perrone, D, Efficacious modification of the procedure for the aldehyde release from 2-substituted thiazoles, *J. Org. Chem.*, 58, 275–277, 1993.
66. (a) Dondoni, A, Fantin, G, Fogagnolo, M, Pedrini, P, Stereochemistry associated with the addition of 2-(trimethylsilyl)thiazole to differentially protected α-amino aldehydes. Applications toward the synthesis of amino sugars and sphingosines, *J. Org. Chem.*, 55, 1439–1446, 1990; (b) Dondoni, A, Perrone, D, Merino, P, Chelation- and non-chelation-controlled addition of 2-(trimethylsilyl)thiazole to α-aminoaldehydes. Stereoselective synthesis of β-amino-α-hydroxy aldehyde intermediate for the preparation of the human immunodeficiency virus proteinase inhibitor Ro31-8959, *J. Org. Chem.*, 60, 8074–8080, 1995.
67. (a) Dondoni, A, Franco, S, Junquera, F, Merchán, F L, Merino, P, Tejero, T, Bertolasi, V, Stereoselective homologation-amination of aldehydes by addition of their nitrones to C-2 metalated thiazoles. A general entry to α-amino aldehydes and aminosugars, *Chem. Eur. J.*, 1, 505–520, 1995; (b) Dondoni, A, Franco, S, Merchán, F L, Merino, P, Tejero, T, Stereocontrol by diethylaluminum chloride in the addition of 2-lithiofuran and N-methyl-2-lithioimidazole to α-alkoxy nitrones. Total synthesis of 5-*O*-carbamoylpolyoxamic acid, *Tetrahedron Lett.*, 34, 5479–5482, 1993; (c) Merino, P, Franco, S, Merchán, F L, Revuella, J, Tejero, T, Efficient synthesis of (2*R*,3*S*)- and (2*S*,3*S*)-2-amino-1,3-4-butanetriols through stereodivergent hydroxymethylation of D-glyceraldehyde nitrones, *Tetrahedron Lett.*, 43, 459–462, 2002.
68. Kusakabe, M, Sato, F, Stereoselective synthesis of D-erythrose and D-threose derivatives from D-glyceraldehyde acetonide and their reactions with 1-(trimethylsiyl)vinyl cuprate reagent. Synthesis of allitol acetate, *Chem. Lett.*, 1473–1476, 1986.
69. Benzing-Nguyen, L, Perry, M B, Stepwise synthesis of *N*-acetylneuraminic acid and *N*-acetyl [1-^{13}C]neuraminic acid, *J. Org. Chem.*, 43, 551–554, 1978.
70. Chikashita, H, Nikaya, T, Itoh, K, Iterative and stereoselective one-carbon homologation of 1,2-O-cyclohexylidene-D-glyceraldehyde to aldose derivatives by employing 2-lithio-1,3-dithiane as a formyl anion equivalent, *Nat. Prod. Lett.*, 2, 183–190, 1993.
71. (a) Seebach, D, Nucleophile acylierung mit 2-lithium-1,3-dithianen bzw. -1,3,5-trithianen, *Synthesis*, 5, 17–36, 1969; (b) Bulman Page, P C, van Niel, M B, Prodger, J C, Synthetic uses of the 1,3-dithiane grouping from 1977 to 1988, *Tetrahedron*, 45, 7643–7677, 1989.
72. Solladié, G, Demailly, G, Greck, C, Reduction of β-keto sulfoxides: a highly efficient asymmetric synthesis of both enantiomers of allylic alcohols, *J. Org. Chem.*, 50, 1552–1554, 1985.
73. Solladié, G, Frechou, C, Hutt, J, Demailly, G, Stereospecific hydroxylation of chiral allylic β-hydroxysulfoxides: asymmetric synthesis of L-arabinitol, *Bull. Soc. Chim. Fr.*, 827–836, 1987.

74. (a) Noe, C R, Knollmüller, M, Ettmayer, P, Ein Weg zur asymmetrischen Induktion bei der Bildung von Zuckern, *Liebigs Ann. Chem.*, 637–643, 1989; (b) Wulff, G, Hansen, A, 2-Substituted 1,3,2-dioxaboroles as synthetic equivalents for the glycoaldehyde anion, *Angew. Chem. Int. Ed.*, 25, 560–562, 1986; (c) Northrup, A B, Mangion, I K, Hettche, F, MacMillan, D W C, Enantioselective organocatalytic direct aldol reactions of alpha-oxyaldehydes: step one in a two-step synthesis of carbohydrates, *Angew. Chem. Int. Ed.*, 43, 2152–2154, 2004; (d) Northrup, A B, MacMillan, D W C, Two-step synthesis of carbohydrates by selective aldol reactions, *Science*, 305, 1752–1755, 2004; (e) Mangion, I K, MacMillan, D W C, Total synthesis of brasoside and littoralisone, *J. Am. Chem. Soc.*, 127, 3696–3697, 2005.
75. Mukaiyama, T, A fresh approach to the synthesis of carbohydrates, In *Trends in Synthetic Carbohydrate Chemistry, ACS Symposium Series 386*, Horton, D, Hawkins, L D, McGarvey, G D, eds., American Chemical Society, Washington, pp. 278–289, 1989, chap. 15.
76. Mukaiyama, T, Miwa, T, Nakatsuka, T, A stereoselective synthesis of 2-amino-2-deoxy-D-arabinose and D-ribose, *Chem. Lett.*, 145–148, 1982.
77. Yamamoto, Y, Kirihata, M, Ichimoto, I, Ueda, H, Stereoselective synthesis of 2-amino-2-deoxy-D-arabinose and 2-deoxy-D-ribose, *Agric. Biol. Chem.*, 49, 1435–1439, 1985.
78. Wulff, G, Hansen, A, Synthesis of monosaccharides with the aid of a new synthetic equivalent for the glycolaldehyde anion, *Carbohydr. Res.*, 164, 123–140, 1987.
79. Banfi, L, Cardani, S, Potenza, D, Scholastico, C, Stereoselective synthesis of *t*-butyl 2-amino-2,5-dideoxy-L-*lyxo*-pentanoate formal synthesis of L-daunosamine, *Tetrahedron*, 43, 2317–2322, 1987.
80. (a) Wehner, V, Jäger, V, Synthesis of D- and L-2-amino-2-deoxyarabinose and 1,4-dideoxy-1,4-iminolyxitol by ($C_2 + C_3$)-nitroaldol addition with 2-O-benzylglyceraldehyde, *Angew. Chem. Int. Ed. Engl.*, 29, 1169–1171, 1990; (b) Hanessian, S, Kloss, J, Total synthesis of biologically important amino sugars via thenitroaldol reaction, *Tetrahedron Lett.*, 26, 1261–1264, 1985.
81. Devant, R, Mahler, U, Braun, H, Stereoselektive aldolreaktionen mit (*R*)- und (*S*)-(2-hydroxy-1,2,2-triphenylethyl)acetat und verwandten glycolmonoacetaten, *Chem. Ber.*, 121, 397–406, 1988.
82. Gräf, S, Braun, M, De novo synthesis of enantiomerically pure deoxy- and aminodeoxyfuranosides, *Liebigs Ann. Chem.*, 1091–1098, 1993.
83. Braun, M, Mortiz, J, A convenient route to both enantiomers of digitoxose, *Synlett*, 750–752, 1991.
84. Schöllkopf, U, Tiller, T, Bardenhagen, J, Asymmetric synthesis of (enantiomerically and diastereomerically virtually pure) methyl 2-amino-4,5-epoxy-3-hydroxy-alkanoates and methyl 2-amino-3-hydroxy-4,5-methylene-alkanoates by the bislactam ether method, *Tetrahedron*, 44, 5293–5305, 1988.
85. Rudge, A J, Collins, I, Holmes, A B, Baker, R, An enantioselective synthesis of deoxynojirimycin, *Angew. Chem. Int. Ed. Engl.*, 33, 2320–2322, 1994.
86. Evans, D A, Gage, J R, Leighton, J L, Total synthesis of (+)-calyculin A, *J. Am. Chem. Soc.*, 114, 9434–9453, 1992.
87. Franck-Neumann, M, Bissinger, P, Geoffroy, P, Aldol condensation reactions of (dienone)tricarbonyliron complexes. 4. Enantioselective total synthesis of 3-deoxypentoses from (−)-myrtenal, *Tetrahedron Lett.*, 38, 4477–4478, 1997.
88. Enders, D, Jegelka, U, 1,3-Dioxan-5-one as C_3-building block for the diastereo- and enantioselective synthesis of C_5- to C_9-deoxy sugars using the SAMP-/RAMP-hydrazone method, *Tetrahedron Lett.*, 34, 2453–2456, 1993.
89. (a) Katsuki, T, Sharpless, K B, The first practical method for asymmetric epoxidation, *J. Am. Chem. Soc.*, 102, 5974–5976, 1980; (b) Pfenninger, A, Asymmetric epoxidation of allylic alcohols: the Sharpless epoxidation, *Synthesis*, 89–116, 1986; (c) Jorgensen, K A, Transition-metal-catalyzed epoxidations, *Chem. Rev.*, 89, 431–458, 1989; (d) Sharpless, K B, The discovery of the asymmetric epoxidation, *Chem. Br.*, 38–44, 1986.
90. (a) Katsuki, T, Lee, A W M, Ma, P, Martin, V S, Masamune, S, Sharpless, K B, Tuddenham, D, Walker, F J, Synthesis of saccharides and related polyhydroxylated natural products. 1. Simple alditols, *J. Org. Chem.*, 47, 1373–1378, 1982; (b) Ma, P, Martin, V S, Masamune, S, Sharpless, K B, Viti, S M, Synthesis of saccharides and related polyhydroxylated natural products. 2. Simple deoxyalditols, *J. Org. Chem.*, 47, 1378–1380, 1982; (c) Ko, S Y, Lee, A W M, Masamune, S, Reed, L A III, Sharpless, K B, Walker, F J, Total synthesis of the L-hexoses, *Tetrahedron*, 46, 245–264, 1990; (d) Aoyagi, S, Fujimaki, S, Yamazaki, N, Kibayashi, C, Total syntheses of galactosidase

inhibitors (+)-galactostatin and (+)-1-deoxygalactostatin, *J. Org. Chem.*, 56, 815–819, 1991; (e) Iida, H, Yamazaki, N, Kibayashi, C, Total synthesis of (+)-nojirimycin and (+)-1-deoxynojirimycin, *J. Org. Chem.*, 52, 3337–3342, 1987.

91. (a) Sharpless, K B, Amberg, W, Bennani, Y L, Crispino, G A, Hartiung, J, Jung, K-S, Kwong, H-L, Morikawa, K, Wang, Z-M, Xu, D, Zhang, X-L, The osmium-catalyzed asymmetric dihydroxylation. A new ligand class and a process improvement, *J. Org. Chem.*, 57, 2768–2771, 1992; (b) Becker, H, Sharpless, K B, A new ligand class for asymmetric dihydroxylation of olefins, *Angew. Chem. Int. Ed. Engl.*, 35, 448–451, 1996; (c) see also Li, G, Chang, H-T, Sharpless, K B, Catalytic asymmetric aminohydroxylation (AA), *Angew. Chem. Int. Ed. Engl.*, 35, 451–454, 1996.
92. Koskinen, A M P, Otsomaa, L A, A new access to enantiomerically pure deoxy aminohexoses: methyl 4-amino-4,6-dideoxygulopyranoside and *epi*-tolyposamine, *Tetrahedron*, 53, 6473–6484, 1997.
93. Ikemoto, N, Schreiber, S L, Total synthesis of (−)-hikizimycin employing the strategy of two-directional chain synthesis, *J. Am. Chem. Soc.*, 114, 2524–2536, 1992.
94. Poss, C S, Schreiber, S L, Two-directional chain synthesis and terminus differentiation, *Acc. Chem. Res.*, 27, 9–17, 1994.
95. Cha, J K, Christ, W J, Kishi, Y, On stereochemistry of osmium tetraoxide oxidation of allylic alcohol systems, *Tetrahedron*, 40, 2247–2255, 1984.
96. Kahne, D, Walker, S, Cheng, Y, Van Enger, D, Glycosylation of unreactive substrates, *J. Am. Chem. Soc.*, 111, 6881–6882, 1989.
97. (a) Fronza, G, Fuganti, C, Grasselli, P, Synthesis of *N*-trifluoroacetyl-L-acosamine and -L-daunosamine, *J. Chem. Soc. Chem. Commun.*, 442–444, 1980; (b) see also Fuganti, C, Grasselli, P, Pedrocchi-Fantoni, G, Synthesis of *N*-benzoyl-L-daunosamine from D-threonine, *Tetrahedron Lett.*, 22, 4017–4020, 1981.
98. (a) Fuganti, C, Grasselli, P, Stereospecific synthesis from non-carbohydrate precursors of the deoxy- and methyl-branched deoxy-sugars L-amicetose, L-mycarose, and L-olivomycose, *J. Chem. Soc. Chem. Commun.*, 299–300, 1978; (b) Fronza, G, Fuganti, C, Grasselli, P, Marinoni, G, Synthesis of L- and D-2,3,6-trideoxy-3-amino-*xylo*-hexose from non-carbohydrate precursors, *Tetrahedron Lett.*, 20, 3883–3886, 1979.
99. Dondoni, A, Fantin, G, Fogagnolo, M, Merino, P, Regio- and stereoselective conjugate addition of nitrogen nucleophiles to 2-alkenyl *N*-methylthiazolium iodides. Synthesis of 3-epi-daunosamine and some lincosamine analogues, *Tetrahedron*, 46, 6167–6184, 1990; see also Ref. [97b].
100. (a) Roush, W R, Walts, A E, Hoong, L K, Diastereo- and enantioselective aldehyde addition reactions of 2-allyl-1,3,2-dioxaborolane-4,5-dicarboxylic esters, a useful class of tartrate ester modified allylboronates, *J. Am. Chem. Soc.*, 107, 8186–8190, 1985; (b) see also Roush, W R, Halterman, R L, Diisopropyl tartrate modified (*E*)-enolate equivalents, *J. Am. Chem. Soc.*, 108, 294–296, 1986.
101. Roush, W R, Straub, J A, Total synthesis of the AB disaccharide unit of olivomycin A, *Tetrahedron Lett.*, 27, 3349–3352, 1986.
102. Roush, R W, Hunt, J A, Asymmetric allylboration of 2-*N*,3-*O*-isopropylidene-*N*-Boc-L-serinal: diastereoselective synthesis of the calicheamicin γ_1^1 aminosugar, *J. Org. Chem.*, 60, 798–806, 1995.
103. (a) Fronza, G, Fuganti, C, Grasselli, P, Pedrocchi-Fantoni, G, Zirotti, C, On the steric course of the addition of diallylzinc onto α,β-dialkoxy chiral carbonyl compounds: stereospecific synthesis of 2,6-dideoxysugars of the L-series, *Tetrahedron Lett.*, 23, 4143–4146, 1982; (b) Fronza, G, Fuganti, C, Grasselli, P, Pedrocchi-Fantoni, G, Zirotti, C, On the steric course of the addition of crotyl metals onto (2*S*,3*S*)-2,3-isopropylidenedioxybutyraldehyde and (3*S*,4*S*)-3,4-isopropylidenedioxypentanone. Synthesis of 2,6-dideoxy-2-C-methyl branched sugars of the L-series, *Chem. Lett.*, 335–338, 1984.
104. Williams, D R, Klingler, F D, Stereoselective allylation for preparation of L-hexose derivatives, *Tetrahedron Lett.*, 28, 869–872, 1987.
105. Roush, W R, Michaelides, M R, Studies on the total synthesis of sesbanimide: a highly diastereoselective synthesis of the AB ring system, *Tetrahedron Lett.*, 27, 3353–3356, 1986.
106. Vuljanic, T, Kihlberg, J, Somfai, P, Diastereoselective synthesis of methyl α-kedarosaminide, a carbohydrate moiety of the enediyne antitumor antibiotic kedaricidin chromophore, *Tetrahedron Lett.*, 35, 6937–6940, 1994.

107. Roush, W R, Adam, M A, Walts, A E, Harris, D J, Stereochemistry of the reactions of substituted allylboronates with chiral aldehydes. Factors influencing aldehyde diastereofacial selectivity, *J. Am. Chem. Soc.*, 108, 3422–3434, 1986.
108. Gryko, D, Jurczak, J, Total synthesis of 1,3-dideoxynojirimycin, *Tetrahedron Lett.*, 38, 8275–8278, 1997.
109. Dondoni, A, Merino, P, Perrone, D, Totally chemical synthesis of azasugars via thiazole intermediates. Stereodivergent routes to (−)-nojirimycin, (−)-mannojirimycin and their 3-deoxy derivatives from serine, *Tetrahedron*, 49, 2939–2956, 1993.
110. Dondoni, A, Perrone, D, Total synthesis of (+)-galactostatin. An illustration of the utility of the thiazole-aldehyde synthesis, *J. Org. Chem.*, 60, 4749–4754, 1995.
111. Nasaraka, K, Pai, F-C, Stereoselective reduction of β-hydroxyketones to 1,3-diols. Highly selective 1,3-asymmetric induction via boron chelates, *Tetrahedron*, 40, 2233–2238, 1984.
112. Sames, D, Polt, R, An enantioselective synthesis of *N*-methylfucosamine via tandem C–C/C–O bond formation, *J. Org. Chem.*, 59, 4596–4601, 1994.
113. Marshall, J A, Seletsky, B M, Coan, P S, Highly *syn*-selective additions of allylic stannanes to protected α-amino aldehydes, *J. Org. Chem.*, 59, 5139–5140, 1994.
114. Marshall, J A, Seletsky, B M, Luke, G P, Synthesis of protected carbohydrate derivatives through homologation of threose and erythrose derivatives with chiral γ-alkoxy allylic stannanes, *J. Org. Chem.*, 59, 3413–3420, 1994.
115. Marshall, J A, Chiral allylic and allenic stannanes as reagents for asymmetric synthesis, *Chem. Rev.*, 96, 31–47, 1996.
116. Marshall, J A, Beaudoin, S, Stereoselective synthesis of differentially protected derivatives of the higher amino sugars destomic acid and lincosamine from serine and threonine, *J. Org. Chem.*, 61, 581–586, 1996.
117. (a) See also Hashimoto, H, Asano, K, Fujii, F, Yoshimura, J, Synthesis of destomic acid and epi-destomic acid, and their C-6 epimers, *Carbohydr. Res.*, 104, 87–104, 1982; (b) Dondoni, A, Franco, S, Merchán, F, Merino, P, Tejero, T, Stereocontrolled addition of 2-thiazolyl organometallic reagents to *C*-galactopyranosylnitrone. A formal synthesis of destomic acid and lincosamine, *Synlett*, 78–80, 1993.
118. Martin, S F, Zinke, P W, The furan approach to higher monosaccharides. A concise total synthesis of (+)-KDO, *J. Am. Chem. Soc.*, 111, 2311–2313, 1989.
119. Casiraghi, G, Zanardi, F, Appendino, G, Rassu, G, The vinylogous aldol reaction: a valuable, yet understated carbon–carbon bond forming maneuver, *Chem. Rev.*, 100, 1929–1972, 2000.
120. Rassu, G, Pinna, L, Spanu, P, Culeddu, N, Casiraghi, G, Gaspari Fava, G, Belicchi, M, Pelosi, G, Total synthesis of 1,5-dideoxy-1,5-iminoalditols, *Tetrahedron*, 48, 727–742, 1992.
121. (a) Casiraghi, G, Rassu, G, Spanu, P, Pinna, L, *N*-(*tert*-Butoxycarbonyl)-2-(*tert*-butyldimethylsilyl)pyrrole: a promising compound for synthesis of chiral nonracemic hydroxylated pyrrolidine derivatives, *J. Org. Chem.*, 57, 3760–3763, 1992; (b) Rassu, G, Casiraghi, G, Spannu, P, Pinna, L, Homochiral α,β-unsaturated γ-lactams: versatile templates, *Tetrahedron Asymmetry*, 3, 1035–1048, 1992.
122. Casiraghi, G, Rassu, G, Spanu, P, Pinna, L, *N-tert*-Butoxycarbonyl-2-(*tert*-butyldimethylsilyloxy)pyrrole as a glycine anion equivalent: a flexible enantioselective access to polyhydroxy-α-amino acids, *Tetrahedron Lett.*, 35, 2423–2426, 1994.
123. Casiraghi, G, Spanu, P, Rassu, G, Pinna, L, Ulgheri, F, Total syntheses of all four isomers of *cis*-1,2-dihydroxypyrrolizidine, *J. Org. Chem.*, 59, 2906–2909, 1994.
124. Zanardi, F, Battistini, L, Nespi, M, Rassu, G, Spanu, P, Cornia, M, Casiraghi, G, Total synthesis of both enantiomers of *trans*-2,3-*cis*-3,4-dihydroxyproline, *Tetrahedron Asymmetry*, 7, 1167–1180, 1996.
125. Rassu, G, Zanardi, F, Battistini, L, Casiraghi, G, The synthetic utility of furan-, pyrrole- and thiophene-based 2-silyloxy dienes, *Chem. Soc. Rev.*, 29, 109–118, 2000.
126. Depezay, J-C., Duréault, A, Double induction asymétrique: hydroalkylation diastéréosélective de la L-alanine par le L-glycéraldéhyde, *Tetrahedron Lett.*, 25, 1459–1462, 1984; see also Ref. [84].
127. Heathcock, C H, Montgomery, S H, A cyclic stereoselection. 28. Use of stereoselective aldol methodology in the total synthesis of cladinose, *Tetrahedron Lett.*, 26, 1001–1004, 1985.

128. Hoppe, D, Tarara, G, Wikens, M, Jones, P G, Schmidt, D, Stezowski, J J, Enantioselective synthesis of methyl furanosides of unnatural 3,6-dideoxy-3-methylaldohexoses form lactates by homoaldol reactions, *Angew. Chem. Int. Ed. Engl.*, 26, 1034–1035, 1987.
129. Hatakeyama, S, Sugawara, K, Takano, S, Enantio- and stereocontrolled syntheses of branched-chain sugars, L-arcanose and L-olivomycose based on the chemistry of 1-trimethylsilyl-2,3-butadiene, *Tetrahedron Lett.*, 32, 4513–4516, 1991.
130. Casiraghi, G, Pinna, L, Rassu, G, Spanu, P, Ulgheri, F, Total synthesis of 2,3-dideoxy-C-methylheptose derivatives, *Tetrahedron Asymmetry*, 4, 681–686, 1993.
131. Bednarski, M, Danishefsky, S, Interactivity of chiral catalysts and chiral auxiliaries in the cycloaddition of activated dienes with aldehydes: a synthesis of L-glucose, *J. Am. Chem. Soc.*, 108, 7060–7067, 1986.
132. Danishefsky, S, Bednarski, M, Izawa, T, Maring, C, A general route to 1-alkoxy-3-[(trimethylsilyl)oxy]-1,3-butadienes: vinylogous transesterification, *J. Org. Chem.*, 49, 2290–2292, 1984.
133. Danishefsky, S, Bednarski, M, On the acetoxylation of 2,3-dihydro-4-pyrones: a concise, fully synthetic route to the glucal stereochemical series, *Tetrahedron Lett.*, 26, 3411–3412, 1985.
134. Luche, J-L, Lanthanides in organic chemistry. 1. Selective 1,2-reductions of conjugated ketones, *J. Am. Chem. Soc.*, 100, 2226–2227, 1978.
135. Hu, Y-J, Huang, X-D, Yao, Z-J, Wu, Y-L, Formal synthesis of 3-deoxy-D-*manno*-2-octulosonic acid (KDO) via a highly double stereoselective hetero-Diels–Alder reaction directed by a (salen)Co^{II} catalyst and chiral diene, *J. Org. Chem.*, 63, 2456–2461, 1998.
136. (a) Lubineau, A, Augé, J, Lubin, N, New strategy in the synthesis of 3-deoxy-D-*manno*-2-octulosonic acid (KDO), 2-deoxy-KDO and thioglycoside of KDO, *Tetrahedron*, 49, 4639–4650, 1993; (b) Cousins, R P C, Pritchard, R G, Raynor, C M, Smith, M, Stoodley, R J, Stereocontrolled cycloaddition route to β-D-glucopyranosyl (1→4)-linked glycals, *Tetrahedron Lett.*, 43, 489–492, 2002.
137. Danishefsky, S J, DeNinno, M P, Chen, S-H, Stereoselective total syntheses of the naturally occurring enantiomers of *N*-acetylneuraminic acid and 3-deoxy-D-*manno*-2-octulosonic acid. A new and stereospecific approach to sialo and 3-deoxy-D-*manno*-2-octulosonic acid conjugates, *J. Am. Chem. Soc.*, 110, 3929–3940, 1988.
138. Isono, K, Crain, P F, McCloskey, J A, Isolation and structure of octosyl acids. Anhydrooctose uronic acid nucleosides, *J. Am. Chem. Soc.*, 97, 943–945, 1975.
139. Danishefsky, S J, Hungate, R, Schulte, G, Total synthesis of octosyl acid. A. Intramolecular Williamson reaction via a cyclic stannylene derivative, *J. Am. Chem. Soc.*, 110, 7434–7440, 1988.
140. Danishefsky, S J, Larson, E, Springer, J P, A totally synthetic route to lincosamine: some observations on the diastereofacial selectivity of electrophilic reactions on the double bonds of various 5-(1-alkenyl)arabinopyranosides, *J. Am. Chem. Soc.*, 107, 1274–1280, 1985.
141. Golebiowsky, A, Jurczak, J, The cyclocondensation reaction of 1-benzoyloxy-2-(*tert*-butyl)dimethylsilyloxy-4-methoxy-1,3-butadiene with *N,O*-protected D-threoninals and D-*allo*-threoninals, *Tetrahedron*, 47, 1045–1052, 1991.
142. Golebiowski, A, Kosak, J, Jurczak, J, Syntheses of destomic acid and anhydrogalantinic acid from L-serinal, *J. Org. Chem.*, 56, 7344–7347, 1991.
143. (a) Desimoni, G, Tacconi, G, Heterodiene syntheses with α,β-unsaturated carbonyl compounds, *Chem. Rev.*, 75, 651–692, 1975; (b) Boger, D L, Weinreb, S M, *Hetero-Diels–Alder Methodology in Organic Synthesis*, Academic Press, New York, 1987.
144. Snider, B B, Lewis-acid-catalyzed ene reactions, *Acc. Chem. Res.*, 13, 426–432, 1980.
145. (a) Tietze, L F, Beifuss, U, Sequential transformations in organic chemistry: a synthetic strategy with a future, *Angew. Chem. Int. Ed. Engl.*, 32, 131–163, 1993; (b) Tietze, L F, Kettschau, G, Gewert, J A, Schuffenhauer, A, Hetero-Diels–Alder reactions of 1-oxa-1,3-dienes, *Curr. Org. Chem.*, 2, 19–62, 1998.
146. Tietze, L F, Schneider, C, Montenbruck, A, Asymmetric 1,6-induction in hetero-Diels–Alder reactions of chiral oxabutadienes for a de novo synthesis of enantiomerically pure carbohydrates: Lewis acid dependent reversal of facial selectivity, *Angew. Chem. Int. Ed. Engl.*, 33, 980–982, 1994.
147. (a) Ismail, Z M, Hoffmann, H M R, New dihydropyrans: Lewis acid catalyzed cycloadditions of α,β-unsaturated acyl cyanides to simple unactivated olefins and dienes: a readily accessible route to derivatives of rose oxide, *Angew. Chem. Int. Ed. Engl.*, 21, 859–860, 1982; (b) El-Abed, D, Jellal, A,

Santelli, M, Cyanures d'acyles éthyléniques V: additions conjuguées des ethers d'énols triméthysilylés, *Tetrahedron Lett.*, 25, 4503–4504, 1984; (c) Tietze, L F, Voss, E, Synthesis of 3-amino sugars of the daunosamine type through hetero-Diels–Alder reaction of enaminones, *Tetrahedron Lett.*, 27, 6181–6184, 1986; (d) Apparao, S, Maier, M E, Schmidt, R R, Reactivity and diastereoselectivity in inverse-type hetero-Diels–Alder dihydropyran syntheses, *Synthesis*, 900–904, 1987; (e) Schmidt, R R, De novo synthesis of carbohydrates and related natural products, *Pure Appl. Chem.*, 59, 415–424, 1987; (f) Tietze, L F, Voss, E, Harms, K, Sheldrick, G M, Hetero-Diels–Alder reaction of enaminecarbaldehydes. An entry to branched aminosugars, *Tetrahedron Lett.*, 26, 5273–5276, 1985; (g) Tietze, L F, Hartfiel, U, Hetero-Diels–Alder reaction of substituted 1-oxabutadienes and 2-ethoxyvinyl acetate. An entry to various natural occurring carbohydrates, *Tetrahedron Lett.*, 31, 1697–1700, 1990; (h) Hayman, C M, Larsen, D S, Brooker, S, facile, A, A facile syntheses of 2,6-dideoxy-6,6,6-trifluorinated carbohydrate analogues, *Aust. J. Chem.*, 51, 545–553, 1998.

148. Boger, D L, Robarge, K D, A divergent de novo synthesis of carbohydrates based on an accelerated inverse electron demand Diels–Alder reaction of 1-oxa-1,3-butadienes, *J. Org. Chem.*, 53, 5793–5796, 1988.
149. (a) Tietze, L F, Montenbruck, A, Schneider, C, De novo synthesis of enantiopure carbohydrates: preparation of ethyl β-D- and β-L-mannopyranosides by an asymmetrically induced hetero-Diels–Alder reaction, *Synlett*, 509–510, 1994; (b) see also Tietze, L F, Schneider, C, Gote, A, Stereodivergent hetero-Diels–Alder reactions of chiral 1-oxa-1,3-butadienes through a conformational switch induced by Lewis acids, *Chem. Eur. J.*, 2, 139–148, 1996.
150. (a) Schmidt, R R, Maier, M, Diastereospecific synthesis of 2,6-dideoxy- and 2,4,6-trideoxy-sugars via hetero-Diels–Alder reaction, *Tetrahedron Lett.*, 26, 2065–2068, 1985; (b) see also De Gaudenzi, L, Apparao, S, Schmidt, R R, De novo synthesis of carbohydrates and related natural products. Part 33. Synthesis of 2-deoxy-L- and D-galacto-heptose via inverse type hetero-Diels–Alder reaction, *Tetrahedron*, 46, 277–290, 1990.
151. (a) Dujardin, G, Rossignol, S, Brown, E, An improved dienophile-induced access to enantiopure 2,4-dideoxysugar lactones via hetero-Diels–Alder reaction: synthesis of the (±)-lactone moiety of compactin, *Synthesis*, 4, 763–770, 1998; (b) Dujardin, G, Rossignol, S, Brown, E, Chiral enol ethers in carbohydrate chemistry: de novo synthesis of protected L-2-deoxy hexoses, *Tetrahedron Lett.*, 37, 4007–4010, 1996.
152. (a) Evans, D A, Johnson, J S, Catalytic enantioselective hetero-Diels–Alder reactions of α,β-unsaturated acyl phosphonates with enol ethers, *J. Am. Chem. Soc.*, 120, 4895–4896, 1998; (b) Evans, D A, Olhava, E J, Johnson, J S, Janey, J M, Chiral C_2-symmetric Cu^{II} complexes as catalysts for enantioselective hetero-Diels–Alder reactions, *Angew. Chem. Int. Ed.*, 37, 3372–3375, 1998; (c) Evans, D A, Johnson, J S, Olhava, E J, Enantioselective synthesis of dihydropyrans. Catalysis of hetero-Diels–Alder reactions by bis(oxazoline)copper(II)complexes, *J. Am. Chem. Soc.*, 122, 1635–1649, 2000.
153. Thorhauge, J, Johannsen, M, Jørgensen, K A, Highly enantioselective catalytic hetero-Diels–Alder reaction with inverse electron demand, *Angew. Chem. Int. Ed.*, 37, 2404–2406, 1998.
154. (a) Audrain, H, Thorhauge, J, Hazell, R G, Jørgensen, K A, A novel catalytic and highly enantioselective approach for the synthesis of optically active carbohydrate derivatives, *J. Org. Chem.*, 65, 4487–4497, 2000; (b) Hanessian, S, Compain, P, Lewis acid promoted cyclocondensations of α-ketophosphonoenoates with dienes-from Diels-Alder to hetero Diels-Alder reactions, *Tetrahedron*, 58, 6521–6529, 2002.
155. Saito, T, Takekawa, K, Takahashi, T, The first catalytic, highly enantioselective hetero-Diels–Alder reaction of thiabutadienes, *Chem. Commun.*, 1001–1002, 1999.
156. (a) Werbitzky, O, Klier, K, Fleber, H, Asymmetric induction of four chiral centers by hetero-Diels–Alder reaction of a chiral nitroso dienophile, *Liebigs Ann. Chem.*, 267–270, 1990; (b) Braun, H, Fleber, H, Kresze, G, Schmidtchen, F P, Prewo, R, Vasella, A, Diastereoselektive Diels–Alder-reactionen mit α-chlornitrososacchariden, *Liebigs Ann. Chem.*, 261–268, 1993; (c) Streith, J, Defoin, A, Azasugar syntheses and multistep cascade rearrangements via hetero-Diels–Alder cycloadditions with nitroso dienophiles, *Synlett*, 189–200, 1996; (d) Defoin, A, Joubert, M, Heuchel, J-M, Strehler, C, Streith, J, Enantioselective Diels–Alder reaction with an α-chloronitroso dienophile derived from 5-*O*-acetyl-2,3-isopropylidenedioxy-D-ribose, *Synthesis*, 1719–1726, 2000.

157. Fleber, H, Kresze, G, Prewo, R, Vasella, A, Diastereoselectivity and reactivity in the Diels–Alder reactions of α-chloronitro ethers, *Helv. Chim. Acta*, 69, 1137–1146, 1986.
158. (a) Defoin, A, Sarazin, H, Streith, J, Asymmetric synthesis of 6-deoxy-D-*allo*-nojirimycin, D-*fuco*nojirimycin and their 1-deoxy derivatives, *Synlett*, 1, 1187–1188, 1995; (b) Defoin, A, Sarazin, H, Streith, J, Synthesis of 1,6-dideoxynojirimycin, 1,6-dideoxy-D-*allo*-nojirimycin, and 1,6-dideoxy-D-*gulo*-nojirimycin via asymmetric hetero-Diels–Alder reactions, *Helv. Chim. Acta*, 79, 560–567, 1996; (c) Defoin, A, Sarazin, H, Streith, J, 6-Deoxy-*allo*-nojirimycin in the racemic and D-series, 6-deoxy-D,L-*talo*-nojirimycin, their 1-deoxy-derivatives and 6-deoxy-2-D, L-allosamine via hetero-Diels–Alder cycloadditions, *Tetrahedron*, 53, 13769–13782, 1997.
159. Sifferlen, T, Defoin, A, Streith, J, Le Novën, D, Tarnus, C, Dosbaâ, I, Foglietti, M-J, Chiral 5-methyl-trihydroxypyrrolidines. Preparation from 1,2-oxazines and glycosidase inhibitory properties, *Tetrahedron*, 56, 971–978, 2000.
160. (a) Joubert, M, Defoin, A, Tarnus, C, Streith, J, Synthesis of potent α-D-mannosidase and α-L-fucosidase inhibitors: 4-amino-4-deoxy-D-erythrose and 4-amino-4,5-dideoxy-L-lyxose, *Synlett*, 1366–1368, 2000; (b) see also Defoin, A, Sifferelen, Th, Streith, J, Dosbaâ, I, Foglietti, M-J, Potent glycosidase inhibitors via hetero-Diels–Alder reactions: asymmetric synthesis of 5-methyl-trihydroxypyrrolidines, *Tetrahedron Asymmetry*, 8, 363–366, 1997; (c) Bach, P, Bols, M, Synthesis of an 1-azaglucose analogue with ring-oxygen retained, *Tetrahedron Lett.*, 40, 3461–3464, 1999.
161. Liang, X, Bols, M, Chemoenzymatic synthesis of enantiopure 1-azafagomine, *J. Org. Chem.*, 64, 8485–8488, 1999.
162. Ernholt, B V, Thomsen, I. b B, Lohse, A, Plesner, I W, Jensen, K B, Hazell, R G, Liang, X, Jakobsen, A, Bols, M, Enantiospecific synthesis of 1-azafagomine, *Chem. Eur. J.*, 6, 278–287, 2000.
163. (a) Just, G, Martel, A, *C*-Nucleosides and related compounds. Synthesis of D,L-3,4-isopropylidene-2, 5-anhydroallose: a novel periodate cleavage, *Tetrahedron Lett.*, 5, 1517–1520, 1973; (b) Just, G, Grozinger, K, A correction to "a novel periodate cleavage", *Tetrahedron Lett.*, 4165–4168, 1974; (c) Just, G, Grozinger, K, A total synthesis of dimethyl 2,3-*O*-isopropylidene-1-*O*-oxalyl-β-DL-*ribo*-hexofuran-5-ulosuronate, *Can. J. Chem.*, 53, 2701–2706, 1975; (d) Just, G, Martel, A, Grozinger, K, Ramjeesingh, M, *C*-Nucleosides and related compounds. IV. The synthesis and chemistry of D,L-2,5-anhydroallose derivatives, *Can. J. Chem.*, 53, 131–137, 1975; (e) Just, G, Ramjeesingh, M, Liak, T J, Syntheses of *C*-nucleosides. IX. Reactions of D,L-3,4-di-*O*-isopropylidene-2,5-anhydroallose with Wittig reagents. Syntheses of bis-homo anhydro-*C*-nucleosides, *Can. J. Chem.*, 54, 2940–2947, 1976.
164. (a) Just, G, Lim, M-I, *C*-Nucleosides and related compounds. XII. Synthesis of D,L-2′-deoxy-showdomycin, *Can. J. Chem.*, 55, 2993–2997, 1977; (b) Just, G, Liak, T J, Lim, M-I, Potvin, P, Tsantrizos, Y S, *C*-Nucleosides and related compounds. XV. The synthesis of D,L-2′-*epi*-showdomycin and D,L-showdomycin, *Can. J. Chem.*, 58, 2024–2033, 1980; (c) see also Kozikowski, A P, Floyd, W C, Borohydride induced cleavage of azo derivatives of β-ketoesters. A useful variant of the Japp–Klingemann reaction for *C*-nucleoside synthesis, *Tetrahedron Lett.*, 19–22, 1978.
165. Kowarski, C R, Sarel, S, A total stereoselective synthesis of *myo*-, *allo*-, *neo*-, and *epi*-inositols, *J. Org. Chem.*, 38, 117–119, 1973.
166. Schmidt, R R, Lieberknecht, A, Funktionelle D- and L-ribose-derivate über eine racematspaltung mit rückführung, *Angew. Chem.*, 90, 821–822, 1978.
167. Vieira, E, Vogel, P, The preparation of optically pure 7-oxabicyclo[2.2.1]hept-2-ene derivatives. The CD spectrum of (+)-(1*R*)-7-oxabicyclo[2.2.1]hept-5-en-2-one, *Helv. Chim. Acta*, 66, 1865–1871, 1983.
168. (a) Reymond, J-L, Vogel, P, New chiral auxiliaries and new optically pure ketene equivalents derived form tartaric acids. Improved synthesis of (−)-7-oxabicyclo[2.2.1]hept-5-en-2-one, *Tetrahedron Asymmetry*, 1, 729–736, 1990; (b) see also Foster, A, Kovac, T, Mosimann, H, Renaud, P, Vogel, P, Resolution of 7-oxabicyclo[2.2.1]hept-5-en-2-one via cyclic aminals, *Tetrahedron Asymmetry*, 10, 567–571, 1999.
169. Warm, A, Vogel, P, Syntheses of (+)- and (−)-methyl 8-epinonactate and (+)- and (−)-methyl nonactate, *Helv. Chim. Acta*, 70, 690–700, 1987.

170. For other enantiomerically enriched 7-oxabicyclo[2.2.1]heptane derivatives, see Vogel, P, Cossy, J, Plumet, J, Arjona, O, Derivatives of 7-oxabicyclo[2.2.1]heptane in nature and as useful synthetic intermediates, *Tetrahedron*, 55, 13521–13642, 1999.
171. Kiss, J, D'Souza, R, van Koeveringe, J A, Arnold, W, Stereosepezifische synthese des cancerostatikums 5′-deoxy-5-fluor-uridin (5-DFUR) und seiner 5′-deuterierten derivate, *Helv. Chim. Acta*, 65, 1522–1537, 1982.
172. Arvai, G, Fattori, D, Vogel, P, Substituent effects on the regioselectivity of the Bayer–Villiger oxidation of 7-oxabicyclo[2.2.1]heptan-2-ones, *Tetrahedron*, 48, 10621–10636, 1992.
173. Auberson, Y, Vogel, P, Total synthesis of L-allose, L-talose, and derivatives, *Helv. Chim. Acta*, 72, 278–286, 1989.
174. (a) Le Drian, C, Vogel, P, Acid-catalyzed rearrangements of 5,6-*exo*-epoxy-7-oxabicyclo[2.2.1]hept-2-yl derivatives. Migratory aptitudes of acyl *vs*. alkyl groups in Wagner–Meerwein transpositions, *Helv. Chim. Acta*, 70, 1703–1720, 1987; (b) Le Drian, C, Vieira, E, Vogel, P, Synthesis of (−)-conduritol C (1L-cyclohex-5-ene-1,2,3,4-tetrol), *Helv. Chim. Acta*, 72, 338–347, 1989; (c) Le Drian, C, Vogel, P, Total syntheses of (−)-conduritol B ((−)-1L-cyclohex-5-ene-1,3/2,4-tetrol) and of (+)-conduritol F ((+)-1D-cyclohex-5-ene-1,2,3/3-tetrol). Determination of the absolute configuration of (+)-leucanthemitol, *Helv. Chim. Acta*, 73, 161–168, 1990.
175. Emery, F, Vogel, P, Regioselective synthesis of 5-*endo*,6-*endo*-dihydroxy-7-oxabicyclo[2.2.1]heptan-2-one with differential protection of the diol group, *Synlett*, 420–422, 1995.
176. (a) Carrupt, P-A, Vogel, P, Regioselective additions of electrophiles to olefins remotely perturbed. The carbonyl group as a homoconjugated electron-donating substituent, *Tetrahedron Lett.*, 23, 2563–2566, 1982; (b) Carrupt, P-A, Vogel, P, The carbonyl group as homoconjugated electron-donating substituent. Ab initio STO 3G-MO calculations, *Tetrahedron Lett.*, 25, 2879–2882, 1984; (c) Carrupt, P-A, Vogel, P, *Ab initio* MO calculations on the rearrangements of 7-oxa-2-bicyclo[2.2.1]heptyl cations. The facile migration of acyl group in Wagner–Meerwein rearrangements, *J. Phys. Org. Chem.*, 1, 287–298, 1988; (d) Carrupt, P-A, Vogel, P, The carbonyl group as homocojugated electron-releasing substituent. Regioselective electrophilic additions at bicyclo[2.2.1]hept-5-en-2-one, bicyclo[2.2.2]oct-5-en-2-one, and derivatives, *Helv. Chim. Acta*, 72, 1008–1028, 1989; (e) Carrupt, P-A, Vogel, P, Through-bond interactions of β-carbonyl and β-imine lone pairs with cationic 2p orbital. Quantum calculations on bicyclo[2.2.2]oct-1-yl cation and derivatives, *J. Org. Chem.*, 55, 5696–5700, 1990; (f) Gerber, P, Vogel, P, Stereoselective synthesis of new 8-oxabicyclo[3.2.1]oct-6-en-2-one and 8-oxabicyclo[3.2.1]octa-3,6-dien-2-one derivatives. The electron-releasing and electron-withdrawing effects of homoconjugated ketones, *Indian J. Chem.*, 40B, 898–904, 2001.
177. Gasparini, F, Vogel, P, Highly stereoselective total syntheses of 2,5-anhydro-3-deoxy- and 4-deoxy-D-hexonic acids and of the related deoxyadenosines-C, *J. Org. Chem.*, 55, 2451–2457, 1990.
178. Fattori, D, de Guchteneere, E, Vogel, P, The electron-releasing homoconjugated carbonyl group. Applications to the total synthesis of 3-deoxy-, 4-deoxy-hexose, lividosamine and derivatives, *Tetrahedron Lett.*, 30, 7415–7418, 1989.
179. Fattori, D, Vogel, P, Total asymmetric syntheses of 3- and 4-deoxy-hexoses and derivatives, *Tetrahedron*, 48, 10587–10602, 1992.
180. (a) Warm, A, Vogel, P, "Naked sugars" as synthetic intermediates. Total synthesis of L-daunosamine, *J. Org. Chem.*, 51, 5348–5353, 1986; (b) Vogel, P, Fattori, D, Gasparini, F, Le Drian, C, Optically pure 7-oxabicyclo[2.2.1]hept-5-en-2-yl derivatives ("naked sugars") as new chirons, *Synlett*, 173–185, 1990; (c) Vogel, P, Synthesis of rare carbohydrates and biomolecules from furan, *Bull. Soc. Chim. Belg.*, 99, 395–439, 1990; (d) Vogel, P, Synthesis of rare carbohydrates and analogues starting from enantiomerically pure 7-oxabicyclo[2.2.1]heptyl derivatives ("naked sugars"), *Curr. Org. Chem.*, 4, 455–480, 2000.
181. (a) Bimwala, R M, Vogel, P, Synthesis of (α1→2); (α1→3); (α1→4); and (α1→5)-*C*-linked disaccharides through 2,3,4,6-tetra-O-acetylglucopyranosyl radical additions to 3-methylidene-7-oxabicyclo[2.2.1]heptan-2-one derivatives, *J. Org. Chem.*, 57, 2076–2083, 1992; (b) Ferritto, R, Vogel, P, Synthesis of α-D-(1→3) and α-D-(1→4)-*C*-linked galactosides of D-mannose derivatives. Conformation of α-*C*-galactosides, *Tetrahedron Asymmetry*, 5, 2077–2092, 1994; (c) Emery, F, Vogel, P, Synthesis of undeculofuranoside derivatives of the herbicidins and of analogues, *J. Org. Chem.*, 60, 5843–5854, 1995; (d) Cossy, J, Ranaivosata, J-L, Bellosta, V, Ancerewicz, J, Ferrito, R,

Vogel, P, Reductive oxa ring opening of 7-oxabicyclo[2.2.1]heptan-2-ones. Synthesis of *C*-α-galactosides of carbapentopyranoses, *J. Org. Chem.*, 60, 8351–8359, 1995; (e) Ferritto, R, Vogel, P, Stereoselective synthesis of α-*C*-galactopyranosides of conduritols and aminoconduritols, *Synlett*, 281–282, 1996; (f) Jeanneret, V, Meerpoel, L, Vogel, P, *C*-Glycosides and *C*-disaccharides precursors through carbonylative Stille coupling reactions, *Tetrahedron Lett.*, 38, 543–546, 1997; (g) Baudat, A, Vogel, P, Aza-*C*-disaccharides: synthesis of 6-deoxygalactonojirimycin β-*C*(1→3) linked with D-altrofuranosides and D-galactose, *J. Org. Chem.*, 62, 6252–6260, 1997; (h) Kraehenbuehl, K, Picasso, S, Vogel, P, Synthesis of *C*-linked imino disaccharides (= aza-*C*-disaccharides) with a pyrrolidine-3,4-diol moiety attached at C(3) of galactose via a hydroxymethylene linker and of a 7-(1,2,3-trihydroxypropyl)-oxtahydroxyindolizine-1,2,6,8-tetrol, *Helv. Chim. Acta*, 81, 1439–1479, 1998; (i) Marquis, C, Picasso, S, Vogel, P, Total asymmetric synthesis of methyl 3-deoxy-3-(1′,2′,6′-trideoxy-2′,6′-imino-D-galactitol-1-yl)-α-*manno*-pyranoside (a *C*-linked iminodisaccharide): conformational analysis and glycosidase inhibition, *Synthesis*, 1441–1452, 1999; (j) Pasquarello, C, Picasso, S, Demange, R, Malissard, M, Berger, E G, Vogel, P, The *C*-disaccharide α-*C*(1→3)mannopyranoside of *N*-acetylgalactosamine is an inhibitor of glycohydrolases and of human α-1,3-fucosyltransferase VI. Its epimer α-(1→3)-mannopyranoside of *N*-acetyltalosamine is not, *J. Org. Chem.*, 65, 4251–4260, 2000; (k) Gerber, P, Vogel, P, Stereoselective synthesis of 8-oxabicyclo[3.2.1]octane-2,3,4,6,7-pentols and total asymmetric synthesis of 2,6-anhydrohepturonic acid derivatives and of β-*C*-manno-pyranosides suitable for the construction of (1→3)-*C*,*C*-linked trisaccharides, *Helv. Chim. Acta*, 84, 1363–1393, 2001.

182. Allemann, S, Vogel, P, Enantioselective synthesis of (−)-conduramine C_1 and aminobromocyclitol derivatives, *Helv. Chim. Acta*, 77, 1–9, 1994.
183. (a) Gasparini, F, Vogel, P, Highly stereoselective total syntheses of 2,5-anhydro-4-deoxy-D-ribo-hexonic acid and of (1*S*)-1-*C*-(6-amino-7*H*-purin-8-yl)-1,4-anhydro-3-deoxy-D-*erythro*-pentitol (= cordycepin C), *Helv. Chim. Acta*, 72, 271–277, 1989; (b) Bimwala, R M, Vogel, P, Total synthesis of 2-(β-D-ribofuranosyl)-thiazole-4-carboxamide (tiazofurin) and of precursors of *ribo*-C-nucleosides, *Helv. Chim. Acta*, 72, 1825–1832, 1989; (c) Jeanneret, V, Gasparini, F, Péchy, P, Vogel, P, Total, asymmetric synthesis of (1*R*)-1-C-(6′-amino-7′*H*-purin-8′-yl)-1,4-anhydro-3-azido-2,3-dideoxy-D-*erythro*-pentitol, *Tetrahedron*, 48, 10637–10644, 1992; (d) Péchy, P, Gasparini, F, Vogel, P, Stereoselective synthesis of (1*R*)-1-*C*-(6-amino-7*H*-purin-8-yl)-1,4-anhydro-2,3-dideoxy-3-fluoro-D-*erythro*-pentitol, *Synlett*, 676–678, 1992.
184. Nativi, C, Reymond, J-L, Vogel, P, Acid-catalyzed rearrangement of 3-aza-8-oxatricyclo[3.2.1.$0^{2,4}$]octan-6-one acetals. Highly stereoselective total synthesis of 3-amino-3-deoxy-D-altrose and derivatives, *Helv. Chim. Acta*, 72, 882–891, 1989.
185. Hünenberger, P, Allemann, S, Vogel, P, Total asymmetric synthesis of 3-amino-3-deoxy-L-talose and derivatives, *Carbohydr. Res.*, 257, 175–187, 1994.
186. Auberson, Y, Vogel, P, Total, asymmetric synthesis of deoxypolyoxin C, *Tetrahedron*, 46, 7019–7032, 1990.
187. Auberson, Y, Vogel, P, Total synthesis of allonojirimycin (5-amino-5-deoxy-D-allopyranose), *Angew. Chem. Int. Ed. Engl.*, 28, 1498–1499, 1989.
188. (a) de Guchteneere, E, Fattori, D, Vogel, P, Total asymmetric synthesis of D-lividosamine and 2-acetamido-2,3-dideoxy-D-*arabino*-hexose derivatives, *Tetrahedron*, 48, 10603–10620, 1992; (b) see also Durgnat, J-M, Vogel, P, Total synthesis of the gastroprotective substance AI-77-B and of analogues, *Helv. Chim. Acta*, 76, 222–240, 1993.
189. (a) Jeganathan, S, Vogel, P, Highly stereoselective total syntheses of octose and derivatives, *J. Org. Chem.*, 56, 1133–1142, 1991; (b) Chen, Y, Vogel, P, Total asymmetric syntheses of 1,5-dideoxy-1,5-iminooctitols and 1,2,6,7,8-pentahydroxyindolizidines, *J. Org. Chem.*, 59, 2487–2496, 1994.
190. Kernen, P, Vogel, P, Total asymmetric synthesis of polypropionate fragments and doubly-branched heptono-1,4-lactones, *Tetrahedron Lett.*, 34, 2473–2476, 1993.
191. Sevin, A-F, Vogel, P, A new stereoselective and convergent approach to the synthesis of long-chain polypropionate fragments, *J. Org. Chem.*, 59, 5820–5926, 1994.
192. Guidi, A, Theurillat-Moritz, V, Vogel, P, Pinkerton, A A, Enantiomerically pure Diels–Alder adducts of maleic anhydride to furfural acetals through thermodynamic control. Single crystal and

molecular structure of (1S,4R,4′S,5′S)-1-(4′,5′-dimethyldioxolan-2′-yl)-5,6-dimethylidene-7-oxabicyclo[2.2.1]hept-2-ene, *Tetrahedron Asymmetry*, 7, 3153–3162, 1996.
193. Theurillat-Moritz, V, Vogel, P, Synthesis of enantiomerically pure 7-oxabicyclo[2.2.1]hept-2-enes precursors in the preparation of taxol analogues, *Tetrahedron Asymmetry*, 7, 3163–3168, 1996.
194. Jotterand, N, Vogel, P, Schenk, K, Total asymmetric synthesis of doubly branched carbahexopyranoses and amino derivatives starting from the Diels–Alder adducts of maleic anhydride to furfuryl esters, *Helv. Chim. Acta*, 82, 821–847, 1999.
195. Jotterand, N, Vogel, P, Synthesis of a 5-hydroxypiperidine-2,3,4,6-tetramethanol, a new 2,6-dideoxy-2,6-iminoheptitol derivative, *J. Org. Chem.*, 64, 8973–8975, 1999.
196. Vasella, A, Voeffray, R, Total Syntheses von Nojirimycin, *Helv. Chim. Acta*, 65, 1134–1144, 1982.
197. (a) Müller, I, Jäger, V, Synthesis of amino sugars via isoxazolines. The concept and one application: nitrile oxide/furan adducts, *Tetrahedron Lett.*, 23, 4777–4780, 1982; (b) Jäger, V, Müller, I, Synthesis of amino sugars via isoxazolines. Nitrile oxide-furan adducts as key intermediates in the isoxazoline route towards novel amino sugar derivatives, *Tetrahedron*, 41, 3519–3528, 1985; (c) Müller, R, Leibold, T, Pätzel, M, Jäger, V, A new synthesis of 1,3,4-trideoxy-1,4-iminoglycitols of varying chain length by ($C_3 + C_n$)-coupling of allyl halides with glycononitrile oxides, *Angew. Chem. Int. Ed. Engl.*, 33, 1295–1298, 1994.
198. Mangeney, P, Alexakis, A, Normant, J-F, Resolution and determination of enantiomeric excesses of chiral aldehydes via chiral imidazolidines, *Tetrahedron Lett.*, 29, 2677–2680, 1988.
199. Kinast, G, Schedel, M, A four-step synthesis of 1-deoxynojirimycin with a biotransformation as cardinal reaction step, *Angew. Chem. Int. Ed. Engl.*, 20, 805–806, 1981.
200. Schaller, C, Vogel, P, Jäger, V, Total synthesis of (+)- and (−)-1-deoxynojirimycin (1,5-dideoxy-1,5-imino-D- and L-glucitol) and of (+)- and (−)-1-deoxyidonojirimycin (1,5-dideoxy-1,5-imino-D- and L-iditol) via furoisoxazoline-3-aldehydes, *Carbohydr. Res.*, 314, 25–35, 1998.
201. Cox, P J, Simpkins, N S, An enantioselective deprotonation route to a versatile intermediate for *C*-nucleoside synthesis, *Synlett*, 321–323, 1991.
202. Hoffmann, H M R, Dunkel, R, Mentzel, M, Reuter, H, Stark, C B W, The total synthesis of *C*-glycosides with completely resolved seven-carbon backbone polyol stereochemistry: stereochemical correlations and access to L-configured and other rare carbohydrates, *Chem. Eur. J.*, 7, 4771–4789, 2001.
203. Stark, C B W, Pierau, S, Wartchow, R, Hoffmann, H M R, Chiral allyl cations are captured by furan with 100%. Stereoselectivity: synthesis of enantiopure 2-alkoxy-8-oxabicyclo[3.2.1]oct-6-en-3-ones by low temperature [4 + 3] cycloaddition, *Chem. Eur. J.*, 6, 684–691, 2000.
204. Holland, D, Stoddart, J F, Regioselective and stereoselective methods for the synthesis of the pentitols, *J. Chem. Soc. Perkin Trans. 1*, 1553–1571, 1983.
205. Grethe, G, Sereno, J, Williams, T H, Uskokovic, M R, Asymmetric synthesis of daunosamine, *J. Org. Chem.*, 48, 5315–5317, 1983.
206. Johnson, C R, Golebiowski, A, Braun, M P, Sundram, H, Chemoenzymatic synthesis of 1,3-dideoxynojirimycin, *Tetrahedron Lett.*, 35, 1833–1834, 1994.
207. Johnson, C R, Braun, M P, A two-step, three component synthesis of PGE_1: utilization of α-iodoenones in Pd(0)-catalyzed cross-coupling of organoboranes, *J. Am. Chem. Soc.*, 115, 11014–11015, 1993.
208. Johnson, C R, Adams, J P, Braun, M P, Senanayake, C B W, Workulich, P W, Uskoković, M R, Direct α-iodination of cycloalkenones, *Tetrahedron Lett.*, 33, 917–918, 1992.
209. (a) Johnson, C R, Penning, T D, Triply convergent synthesis of (−)-prostaglandin E2 methyl ester, *J. Am. Chem. Soc.*, 110, 4726–4735, 1988; (b) Parry, R J, Haridas, K, De Jong, R, Johnson, C R, Biosynthesis of aristeromycin: evidence for the intermediacy of a 4β-hydroxymethyl-1α,2α,3α-trihydroxycyclopentanetriol, *Tetrahedron Lett.*, 31, 7549–7552, 1990.
210. Johnson, C R, Golebiowski, A, Schoffers, E, Sundram, H, Braun, M P, Chemoenzymatic synthesis of azasugars: D-*talo*- and D-*manno*-1-deoxynojirimycin, *Synlett*, 313–314, 1995.
211. (a) Johnson, C R, Nerurkar, B M, Golebiowski, A, Sundram, H, Esker, J L, Chemoenzymatic synthesis of *trans*-4,5-dihydroxycyclopent-2-enones: conversion to D-1-deoxynojirimycin, *J. Chem. Soc., Chem. Commun.*, 1139–1140, 1995; (b) Mehta, G, Mohl, N, A norbornyl route to azasugars: a new synthesis of deoxynojirimycin analogues, *Tetrahedron Lett.*, 41, 5741–5745, 2000;

(c) Mehta, G, Mohal, N, A norbornyl route to azasugars: stereoselective synthesis of isofagomine analogues, *Tetrahedron Lett.*, 41, 5747–5751, 2000.

212. (a) Hudlicky, T, Entwistle, D A, Pitzer, K K, Thorpe, A J, Modern method of monosaccharide synthesis from non-carbohydrate sources, *Chem. Rev.*, 96, 1195–1220, 1996; (b) Banwell, M, De Savi, C, Watson, K, First total synthesis of (+)-3-deoxy-D-*glycero*-D-*galacto*-2-nonulosonic acid (KDN) from a non-carbohydrate source, *Chem. Commun.*, 1189–1190, 1998; (c) Banwell, M, De Savi, C, Watson, K, Diastereoselective synthesis of (−)-*N*-acetylneuraminic acid (Neu5Ac) from a non-carbohydrate source, *J. Chem. Soc. Perkin Trans. 1*, 2251–2252, 1998; (d) Banwell, M C, Blakey, S, Harfoot, G, Longmore, R W, *Cis*-1,2-dihydrocatechols in chemical synthesis: first synthesis of L-ascorbic acid (vitamin C) from a non-carbohydrate source, *Aust. J. Chem.*, 52, 137–142, 1999; (e) Modyanova, L, Azerad, R, Dioxygenase-catalyzed formation of dihydrodiol metabolites of N-methyl-2-pyridone, *Tetrahedron Lett.*, 41, 3865–3869, 2000.
213. (a) Johnson, J R, Golebiowski, A, Steensma, D H, Enzymatic asymmetrization in organic media: synthesis of unnatural glucose from cycloheptatriene, *J. Am. Chem. Soc.*, 114, 9414–9418, 1992; (b) Johnson, C R, Kozak, J, Chemoenzymatic synthesis of 3-deoxy-D-*arabino*-heptulosonic acid from cycloheptatriene, *J. Org. Chem.*, 59, 2910–2912, 1994; (c) Johnson, C R, Golebiowski, A, Kozak, J, From cycloheptatriene to enantiopure sugars: synthesis of 2-deoxyhexoses, *Carbohydr. Res.*, 309, 331–335, 1998; (d) Pearson, A J, Katiyar, S, Approaches to the synthesis of heptitol derivatives via stereocontrolled functionalization of cycloheptatrienone using organo-iron chemistry, *Tetrahedron*, 56, 2297–2304, 2000.
214. Jung, M E, Kretschik, O, Enantiospecific total synthesis of L-2′,3′-dideoxyisonucleosides via regioselective opening of optically active C_2-symmetric 1,4-pentadiene bis-epoxide, *J. Org. Chem.*, 63, 2975–2981, 1998.
215. Rychnovsky, S D, Griesgaber, G, Zeller, S, Skalitzky, D J, Optically pure 1,3-diols from (2*R*,4*R*)- and (2*S*,4*S*)-1,2: 4,5-diepoxypentane, *J. Org. Chem.*, 56, 5161–5169, 1991.
216. (a) Jung, M E, Gardiner, J M, Synthetic approaches to 3′-azido-3′-deoxythymidine and other modified nucleosides, *J. Org. Chem.*, 56, 2614–2615, 1991; (b) Jung, M E, Gardiner, J M, Synthesis of antiviral nucleosides from crotonaldehyde. Part 3. Total synthesis of didehydrodideoxythymidine (d4T), *Tetrahedron Lett.*, 33, 3841–3844, 1992; (c) for further applications, see Matsushima, Y, Nakayama, T, Tohyama, S, Eguchi, T, Kakinuma, K, Versatile route to 2,6-dideoxyamino sugars from non-sugar materials: syntheses of vicenisamine and kedarosamine, *J. Chem. Soc. Perkin Trans. 1*, 569–577, 2001; (d) Martín, R, Moyano, A, Pericàs, M A, Riera, A, A concise enantioselective entry to the syntheses of deoxy-azasugars, *Org. Lett.*, 2, 93–95, 2000; (e) Díaz, Y, Bravo, F, Castillón, S, Synthesis of purine and pyrimidine isodideoxynucleosides from (*S*)-glycydol using iodoetherification as key step. Synthesis of (*S*,*S*)-iso-ddA, *J. Org. Chem.*, 64, 6508–6511, 1999.
217. Jäger, V, Hümmer, W, Cyclization of *N*-protected 1-amino-4-pentene-2,3-diols to *lyxo*-configurated deoxyimino sugars (*cis*-dihydroxypyrrolidines); synthesis of potential glycosidase inhibitors, *Angew. Chem. Int. Ed. Engl.*, 29, 1171–1173, 1990.
218. Smith, D S, Wang, Z, Schreiber, S L, The asymmetric epoxydation of divinyl carbinols: theory and applications, *Tetrahedron*, 46, 4793–4808, 1990.
219. (a) Roush, W R, Brown, R J, Total synthesis of carbohydrates: stereoselective syntheses of 2,6-dideoxy-D-*arabino*-hexose and 2,6-dideoxy-D-*ribo*-hexose, *J. Org. Chem.*, 47, 1371–1373, 1982; (b) Roush, W R, Brown, R J, DiMare, M, Total synthesis of carbohydrates. 2. Regiochemical control of nucleophilic ring opening of acylated 2,3-epoxy alcohols, *J. Org. Chem.*, 48, 5083–5093, 1983; (c) Roush, W R, Straub, J A, Van Nieuwenhze, M S, A stereochemically general synthesis of 2-deoxyhexoses via asymmetric allylboration of 2,3-epoxy aldehydes, *J. Org. Chem.*, 56, 1636–1648, 1991.
220. (a) Roush, W R, Brown, R J, Total synthesis of carbohydrates. 3. Efficient, enantioselective syntheses of 2,6-dideoxyhexoses, *J. Org. Chem.*, 48, 5093–5101, 1983; (b) Nicolaou, K C, Rodríguez, R M, Mitchell, H J, van Delft, F L, *Angew. Chem. Int. Ed.*, 37, 1874–1876, 1998.
221. Küfner, U, Schmidt, R R, Synthesis of deoxyhexoses form divinylglycols—the synthesis of D- and L-chalcose, *Angew. Chem. Int. Ed.*, 25, 89, 1986.
222. Dai, L-x, Lou, B-I, Zhang, Y-z, A simple, divergent, asymmetric syntheis of all members of the 2,3, 6-trideoxy-3-aminohexose family, *J. Am. Chem. Soc.*, 110, 5195–5196, 1988.

223. Ono, M, Saotome, C, Akita, H, Total syntheses of *N*-fluoroacetyl-L-daunosamine, *N*-trifluoroacetyl-L-acosamine, *N*-benzoyl-D-acosamine, and *N*-benzoyl-D-ristosamine from an achiral precursor, methyl sorbate, *Heterocycles*, 45, 1257–1261, 1997.
224. Schmidt, R R, Frische, K, Synthese von partiell geschützten Zuckern aus *meso*-Divinylglycol, *Liebigs Ann. Chem.*, 209–214, 1988.
225. (a) For other applications, see Xu, Y-M, Zhou, W-S, A new approach to 1-deoxy-azasugars: asymmetric synthesis of deoxymannojirimycin, *Tetrahedron Lett.*, 37, 1461–1462, 1996; (b) Marshall, J A, Tang, Y, Enantioselective synthesis of carbohydrate precursors via 1,2:2,3-bis-epoxide intermediates, *J. Org. Chem.*, 59, 1457–1464, 1994; (c) Jäger, V, Hümmer, W, Cyclization of *N*-protected 1-amino-4-pentene-2,3-diols to *lyxo*-configurated deoxyimino sugars (*cis*-dihydroxypyrrolidines); synthesis of potential glycosidase inhibitors, *Angew. Chem. Int. Ed.*, 29, 1171–1173, 1990.
226. Ko, S Y, Malik, M, Synthesis of carbohydrates and related polyhydroxylated compounds employing asymmetric dihydroxylation. 1. An access to *erythro*-diols, *Tetrahedron Lett.*, 34, 4675–4678, 1993.
227. Jung, M E, Gardiner, J M, Asymmetric synthesis of carbohydrates: synthesis of 2-deoxy-D- and 2-deoxy-L-xylofuranosides from a simple achiral precursor, *Tetrahedron Lett.*, 35, 6755–6758, 1994.
228. Taniguchi, T, Ohnishi, H, Ogasawara, K, An expedient preparation of chiral building blocks having levoglucosenone chromophore: a new enantiocontrolled route to (−)-β-multistriatin and (+)-exo-brevicomin, *Chem. Commun.*, 1477–1478, 1996.
229. (a) Harries, J M, Keranen, M D, O'Doherty, G A, Syntheses of D- and L-mannose, gulose, and talose via diastereoselective and enantioselective dihydroxylation reactions, *J. Org. Chem.*, 64, 2982–2983, 1999; (b) see also Harris, J M, O'Doherty, G A, Enantioselective synthesis of 5-substituted α,β-unsaturated δ-lactones: application to the synthesis of styryllactones, *Tetrahedron Lett.*, 41, 183–187, 2000.
230. Li, G, Chang, H-T, Sharpless, K B, Catalytic asymmetric aminohydroxylation (AA) of olefins, *Angew. Chem. Int. Ed. Engl.*, 35, 451–454, 1996.
231. Bushey, M L, Haukass, M H, O'Doherty, G A, Asymmetric aminohydroxylation of vinylfuran, *J. Org. Chem.*, 64, 2985, 1999.
232. (a) Yang, C-F, Xu, Y-M, Liao, L-X, Zhou, W-S, Asymmetric total syntheis of (+)-desoxoprosophylline, *Tetrahedron Lett.*, 39, 9227–9228, 1998; (b) Ciufolini, M A, Hermann, C Y W, Dong, Q, Shimizu, T, Swaminathan, S, Xi, N, Nitrogen heterocycles from furans: the aza-Achmatowiz reaction, *Synlett*, 105–114, 1998.
233. Haukaas, M H, O'Doherty, G A, Synthesis of D- and L-deoxymannojirimycin via an asymmetric aminohydroxylation of vinylfuran, *Org. Lett.*, 3, 401–404, 2001.
234. For further applications of asymmetric dihydroxylation, see Lemaire-Audoire, S, Vogel, P, Total, asymmetric synthesis of D-*ido*-4-heptulosuronate derivatives starting from diethyl 4-oxopimelate, *Tetrahedron Asymmetry*, 10, 1283–1293, 1999.

Part IV

Additional Topics

14 Combinatorial Carbohydrate Chemistry

Prabhat Arya and Bugga VNBS Sarma

CONTENTS

14.1 INTRODUCTION

Early breakthroughs in peptide and oligonucleotide synthesis played an important role in providing pure samples that were used as chemical probes to understand their roles in biological processes. Solid-phase synthetic methods developed for these two classes of biopolymers could be successfully transferred to automation, providing easy access to a wide variety of derivatives. This turned out to be crucial in understanding various cellular responses related to proteins and DNAs/RNAs. During this period, a third class of natural biopolymers — carbohydrates and carbohydrate conjugates — were found to play important roles in intra- and intercellular events [1]. Owing to the high degree of difficulty involved in obtaining their pure derivatives, it was not easy to understand the exact nature of their involvement in biological events. In general, the field of "glycobiology" severely suffers from the lack of pure samples for use as biological probes.

In recent years, interest in chemical glycobiology has grown significantly because it is now well established that carbohydrates and carbohydrate conjugates are involved in the modulation of protein functions, fertilization, chronic inflammation, immune responses, and cancer metastases. Carbohydrate derivatives (e.g., glycolipids and glycoproteins) that are present on host cell surfaces provide specific binding sites for the attachment of bacterial and viral pathogens, leading to infectious diseases. Several groups have succeeded in identifying specific complex carbohydrate conjugates as tumor markers with unique structural features. These moieties have become highly attractive, challenging synthetic targets in developing chemically well-defined synthetic vaccines for cancer, and in the design of agents that could specifically deliver anticancer drugs to tumor cell surfaces [2].

Unlike polypeptides and nucleic acids, which are linear biopolymers, carbohydrates are nonlinear molecules and these derivatives present tremendous challenges in developing their total syntheses. In addition to their branched nature, the linkages between two glycoside moieties could present significant synthetic challenges as they can exist in both α- or β-anomeric configurations. Another daunting task with carbohydrates is the use of efficient orthogonal protection–deprotection strategies, as well as difficult stereoselective glycosyl coupling reactions. All of these factors have attributed to slow, tedious and laborious syntheses of carbohydrate derivatives [3]. These challenges need to be addressed before efficient automated processes for the synthesis of complex carbohydrates and their conjugates can be developed. During the past two decades, tremendous progress has been made in developing solid-phase syntheses of complex carbohydrates and carbohydrate conjugates [4]. In most cases, the solid-phase methods are not general in nature, and several carbohydrate-derived coupling reactions do not produce the required products in stereoselective manners.

The lack of chemical stability and bioavailability associated with peptides and nucleic acids means that the field of peptide and nucleic acid mimics has grown significantly. There are several solid-phase synthesis methods that allow rapid access to these derivatives with relative ease. The design and synthesis of agents that are relatively simple and mimic the structural and functional aspects of peptides and nucleic acids, has been an attractive strategy in the search for drug-like candidates. Similarly, the development of small molecule mimics of complex carbohydrates and glycoconjugates has triggered parallel developments in the design and synthesis of inhibitors of oligosaccharide functions [5]. In many cases, it has been established that, despite the complexity of the oligosaccharide moieties of carbohydrate conjugates, the terminal sugars (two to four residues) and their conformations are critical for biological activities. In cases like this, where the chemical complexity of the synthetic target(s) is relatively simple, the use of revolutionary synthetic strategies such as combinatorial chemistry allows rapid access to potential carbohydrate mimics.

Organic synthesis has always been a limiting factor when it comes to searching for bioactive compounds with drug-like properties. To meet the growing demand for the economical synthesis of large numbers of diverse chemical compounds in a relatively short time, solid-phase synthesis and combinatorial synthesis are emerging technologies in the arena of medicinal chemistry [6]. Over the years, solid-phase synthesis has continued to undergo refinement, and has been successfully extended to the synthesis of small organic molecules [7]. Although this field has existed for nearly 10 years, there are still major challenges to overcome before the benefits of combinatorial chemistry are broadly realized [8]. Nevertheless, this area of research is an important part of most drug discovery-based research programs. The branched nature of complex carbohydrates and the lack of highly stereoselective solid-phase reactions to obtain carbohydrate derivatives have precluded the rapid and efficient generation of oligosaccharide libraries either by solution or solid-phase synthesis [9].

In recent years, several groups have reported novel combinatorial approaches to the synthesis of oligosaccharides and carbohydrate mimics. In most cases, the development of efficient solid-phase reactions has been crucial to the library synthesis, as is discussed in this chapter.

14.2 SOLUTION-PHASE LIBRARY SYNTHESIS OF CARBOHYDRATES

The polyvalent nature, the requirement for several orthogonal protection and deprotection steps, and the lack of a general glycosylation protocol make the rapid assembly of carbohydrates a challenging process. During the last few years, these challenges have continued to be addressed, resulting in novel approaches for the rapid assembly of oligosaccharides leading to combinatorial oligosaccharide libraries.

14.2.1 HINDSGAUL'S RANDOM GLYCOSYLATION

One of the daunting tasks in carbohydrate chemistry is the appropriate use of protecting groups and their deprotection. This is certainly not simple when it comes to complex carbohydrate synthesis in solution, let alone the solid-phase attempts. With this in mind, Hindsgaul et al. [10] undertook the challenge of working with fully deprotected glycosyl acceptors. If successful, their approach could provide several coupled derivatives in one step. The next hurdle to overcome was the development of an efficient purification method. In early attempts, an example of which is shown in Scheme 14.1, the perbenzylated trichloroacetimidate donor **1** was reacted with the unprotected disaccharide acceptor **2** with a hydrophobic group at the anomeric position. The use of the hydrophobic group facilitated purification. To their surprise, unlike the conventional thinking that hydroxyl groups on carbohydrates possess different reactivity profiles, six trisaccharides were formed in nearly equal yields as major products, while β-linked fucosylated trisaccharides were also formed in minor quantities. Interestingly, only 30% of the disaccharide acceptor was fucosylated. The trisaccharides were separated by high pressure liquid chromatography (HPLC) and then characterized by nuclear magnetic resonance (NMR). The outcome of this study demonstrated that this method faced problems of uncontrolled glycosylation reactions, low yields and a requirement for extensive purifications at the end.

14.2.2 BOONS'S LATENT-ACTIVE GLYCOSYLATION

Using solution synthesis, Boons et al. [11] developed a novel glycosylation strategy that utilized latent-active reactivity principles. In this approach, the major building block, 3-buten-2-yl glycoside **3** (Scheme 14.2) was converted into a glycosyl donor as well as an acceptor. For example, in one case, the terminal olefinic moiety in glycosyl derivative **3** was subjected to isomerization using BuLi/$(Ph_3P)_3RhCl$, giving the corresponding glycosyl donor **4**. Alternatively, the acetate group at C-6 in **3** was subjected to deacetylation (NaOMe), producing the glycosyl acceptor **5**. The glycosylation coupling reaction between **4** and **5** provided the disaccharide **6** in ca. 90% yield with no control over the stereochemistry at the anomeric position. This method was efficiently utilized

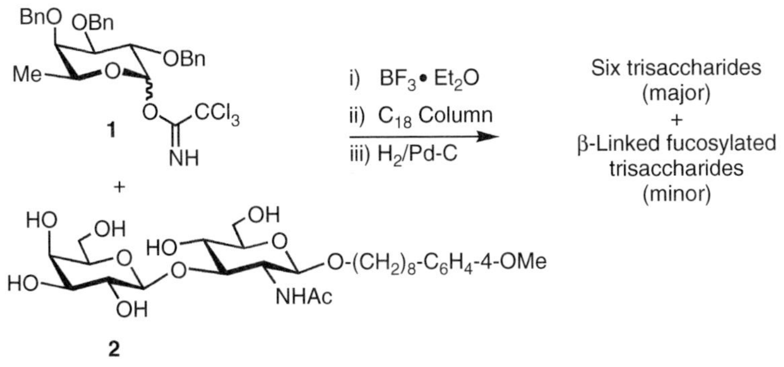

i) $BF_3 \cdot Et_2O$
ii) C_{18} Column
iii) H_2/Pd-C

Six trisaccharides (major) + β-Linked fucosylated trisaccharides (minor)

$O-(CH_2)_8-C_6H_4$-4-OMe

SCHEME 14.1 Hindsgaul's random glycosylation.

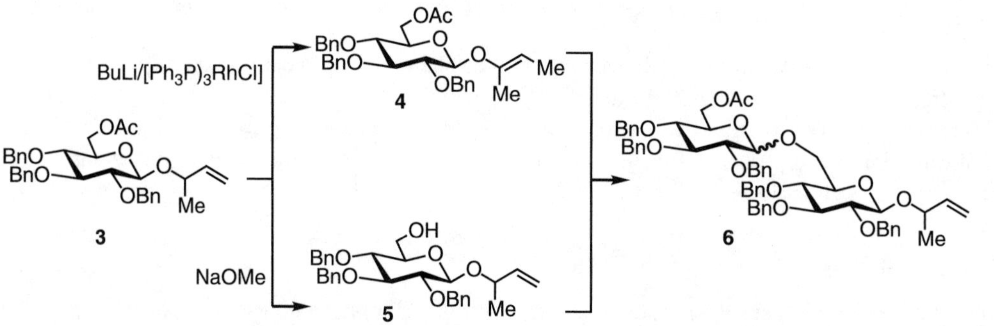

SCHEME 14.2 Boons's latent-active glycosylation.

in the solution-phase synthesis of linear and branched trisaccharide libraries. The purification of saccharide mixtures was achieved by chromatography.

14.2.3 Ichikawa's Stereoselective (and Nonregioselective) Glycosylation

In a different approach, Ichikawa et al. [12] used a method that involves the "divergent" synthesis of an oligosaccharide library, employing stereoselective glycosylation combined with random glycosylation strategies. In their studies, the glycosyl acceptor had hydroxyl groups at C-3 and C-4 available for coupling reactions. In one example, the monosaccharide, 6-deoxy-3,4-di-*O*-trimethylsilyl-L-glucal **7** (Scheme 14.3) was converted into the corresponding iodo derivative **8** following desilylation of both hydroxyl groups. This glycosylation acceptor (with two hydroxyl groups) was reacted with glucal **7** as a donor. Surprisingly, the glycosylation reactions were stereoselective and α-(1→4)- and β-(1→3)-linked disaccharides were formed in near equal yields. The stereochemistry of the α-anomeric glycosidic linkage was controlled by a glycosylation reaction under iodonium ion-catalyzed conditions. By repeating the above steps, these two disaccharides were converted into the corresponding acceptors and then subjected to repeated glycosylation reactions. As before, a mixture of eight trisaccharides was formed in ca. 70% yield. All of them have an iodo group at the C-2 position that could be further reduced or subjected to substitution to enhance the library diversity.

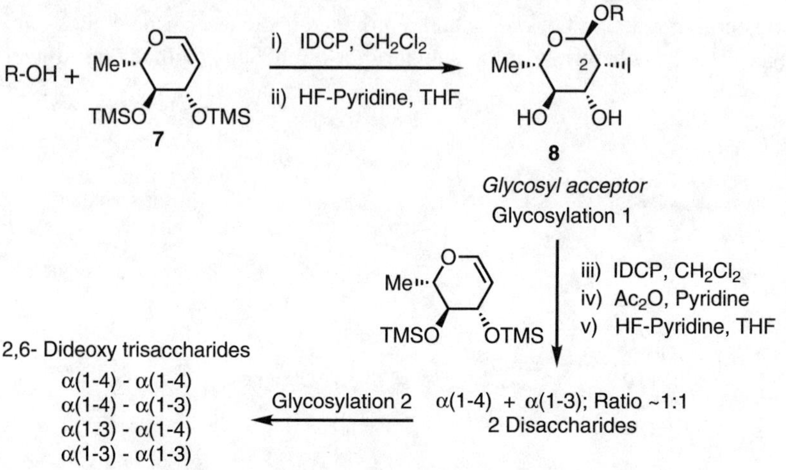

SCHEME 14.3 Ichikawa's stereoselective (nonregioselective) glycosylation.

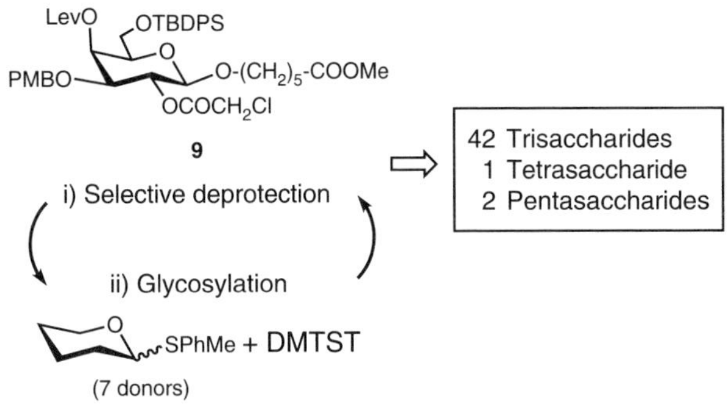

LevO OTBDPS
PMBO O-$(CH_2)_5$-COOMe
$OCOCH_2Cl$
9
i) Selective deprotection
ii) Glycosylation
SPhMe + DMTST
(7 donors)

42 Trisaccharides
1 Tetrasaccharide
2 Pentasaccharides

SCHEME 14.4 Wong's orthogonally protected building-block approach.

14.2.4 Orthogonal Protection in Library Synthesis

With the goals of developing methods that could provide stereochemically well-defined oligosaccharide derivatives, Wong et al. [13] relied upon orthogonal protecting groups on glycoside acceptors and thioglycosides as donors. In their approach, they utilized the monosaccharide building block **9** with four selectively removable hydroxyl protecting groups (Scheme 14.4). These groups were chloroacetyl (ClAc) at C-2, *p*-methoxybenzyl (PMB) at C-3, levulinoyl (Lev) at C-4 and *tert*-butyldiphenylsilyl (TBDPS) at C-6. The anomeric hydroxyl at C-1 was protected by the hydrocarbon chain with a terminal carboxymethylester group. At a given time, following the selective deprotection of the hydroxyl group, this was subjected to coupling with seven thioglycoside donors in the presence of dimethyl(methylthio)sulfonium triflate (DMTST). This allowed the solution synthesis of 45 derivatives in total. Out of these, 42 derivatives were found to be trisaccharides. Of the remainder, one was a tetrasaccharide and two were pentasaccharides.

Moreover, the researchers developed a computerized database, *OptiMer*, having the anomeric reactivity values for orthogonally protected thioglycosides [14]. This database turned out to be crucial in predicting the reactivity profiles of different thioglycosides and led to the development of one-pot glycosylation assembly strategies. The solution-phase chemoselective one-pot glycosylation approach may prove to be a valuable strategy for combinatorial solid-phase oligosaccharide library synthesis. The use of Wong's OptiMer database for selection of glycosyl donors and acceptors would certainly rival solid-phase approaches for the rapid synthesis of oligosaccharide libraries.

In a different approach to one-pot oligosaccharide synthesis, the reactivity difference between hydroxyls of the glycosyl acceptor was taken advantage of (Scheme 14.5) [15], in combination with the one-pot activation of thioglycosides and glycosyl bromides. Regioselective glycosylation of 3,6-diol acceptors, first with glycosyl bromides at O-6, then a second glycosylation of the resulting disaccharides with free O-3, afforded a 54 compound library of (1→3,6)-branched trisaccharides. Alternatively, sequential glycosylation of O-6 unprotected thioglycosides with glycosyl bromides, followed by one-pot activation of the thioglycoside, gave a library of 18 (1→6)-linked trisaccharides [15].

14.3 SOLID-PHASE LIBRARY SYNTHESIS OF CARBOHYDRATES

Solid-phase approaches may ease the product isolation and purification in the generation of larger oligosaccharide libraries. However, solid-support oligosaccharide synthesis requires thorough

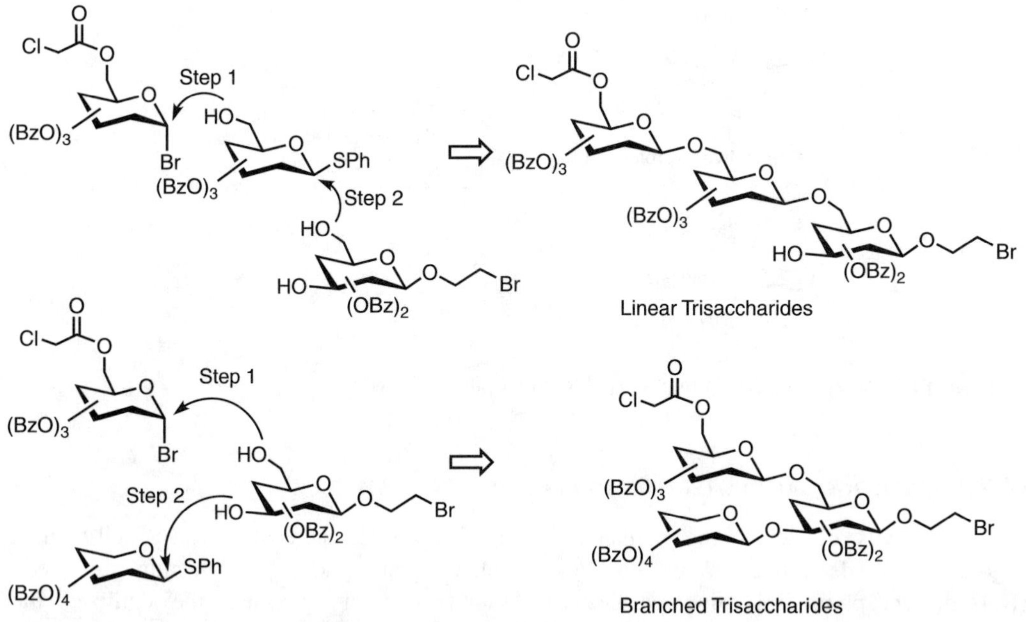

SCHEME 14.5 Takahashi's one-pot glycosylation approach.

optimization of steps for adapting solution synthesis to the solid support. This can be attributed to the additional challenges of solid support-derived approaches together with the inherent nature of oligosaccharides. Because of these challenges, very few oligosaccharide libraries have been synthesized successfully on solid support.

14.3.1 Kahne's Split-Mix Approach to Glycosylation

Unlike the solution-phase approaches discussed above, Kahne et al. [16] decided to explore solid-phase split-mix strategies for the synthesis of carbohydrate libraries. In order to obtain efficient glycoside coupling reactions that are stereoselective in nature, the researchers utilized the well-established sulfoxide derivatives as glycosyl donors. In the past, these types of glycosyl donors have been found to be readily activated at low temperatures. This appears to be independent of the nature of the protecting groups on the glycosyl donor. This method results in high yields of glycoside coupling reactions with a high degree of stereoselection. Scheme 14.6 shows this novel coupling method in combination with the split-mix approach and a library of approximately 1300 di- and tri-saccharides was synthesized in three steps. For the library synthesis by a split-mix strategy, the Still encoded method was utilized for deconvolution. The six glycosyl acceptors were anchored onto TentaGel amine resin and then coupled with 12 glycosyl donors by the sulfoxide method. After three steps, which included amine acylation and deprotection of the glycoside hydroxyl groups, a coded library anchored onto a solid support was obtained. The one bead–one compound library was tested to study the binding of glycoside derivatives to the lectin *Bahuinia purpurea* using a colorimetric assay. The anchoring of the ligand on the bead may mimic the presentation of the ligand on cell surfaces. Thus, this approach may prove to be valuable when searching for new cell surface antiadhesive agents.

14.3.2 Boons's Two-Directional Approach

Following solid-phase efforts from Kahne's group, Boons et al. [17] reported a small glycoside library that was synthesized using a two-directional split-mix method. As shown in Scheme 14.7,

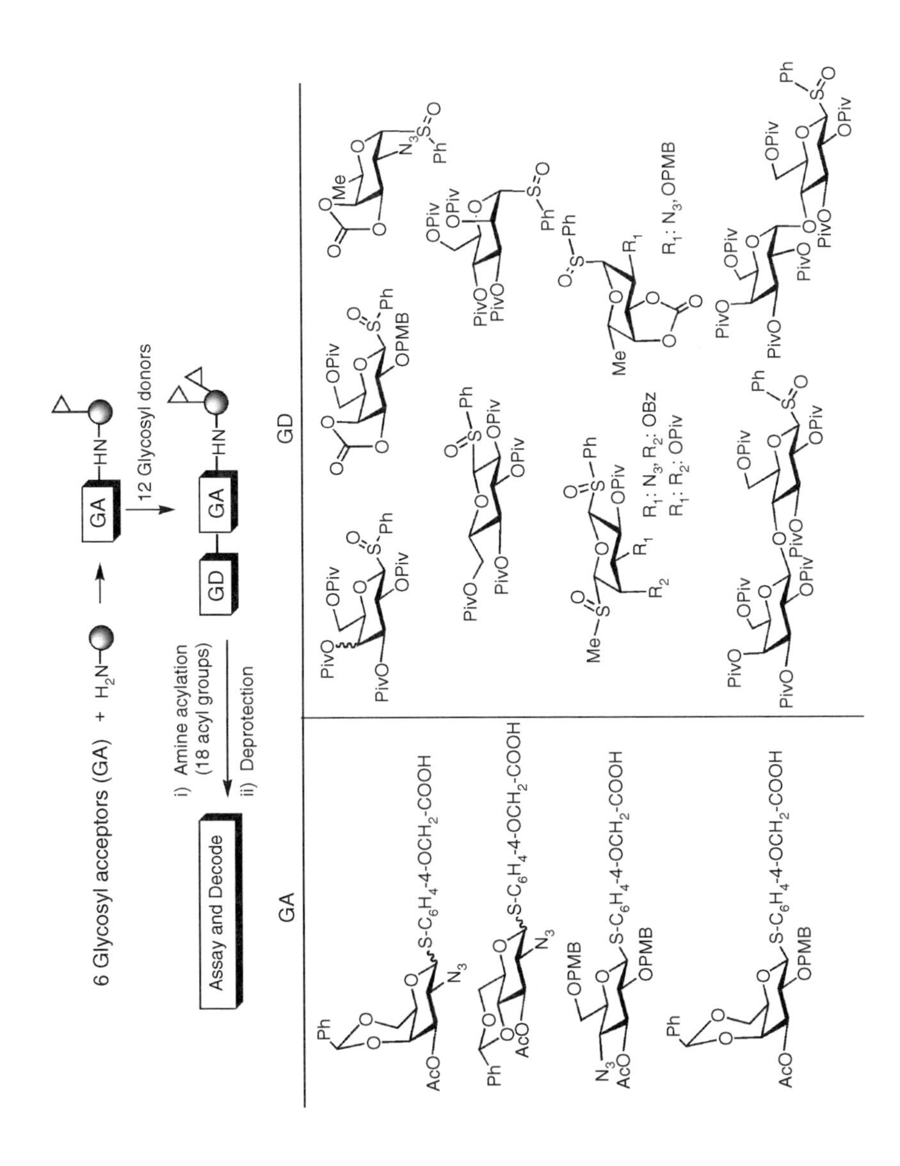

SCHEME 14.6 Kahne's split-mix glycosylation.

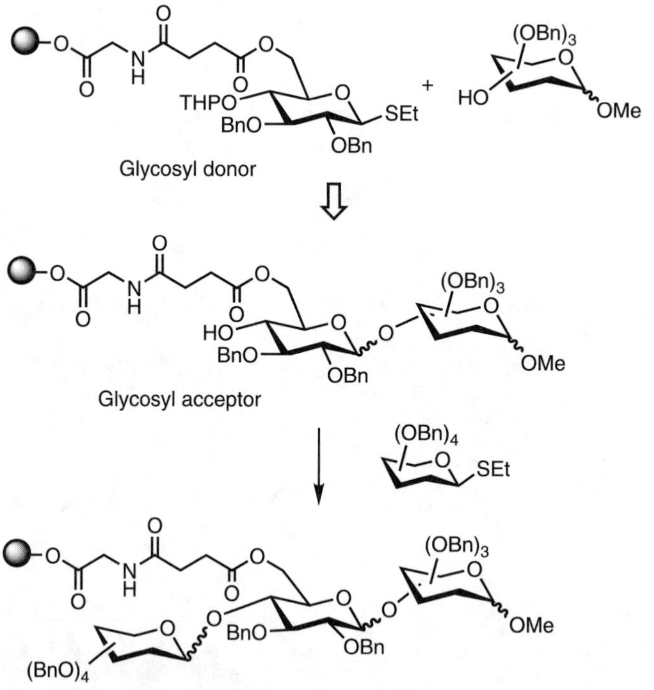

SCHEME 14.7 Boons's two-directional glycosylation.

the thioethylglycoside derivative, useful as both a donor and an acceptor, was immobilized onto TentaGel resin with glycine as the terminal amino group. The thioglycoside donor was anchored onto the resin via the succinamide linker. In order to eliminate the formation of oligomeric side products during *N*-iodosuccinimide (NIS)/trimethylsilyl trifluoromethanesulfonate (TMSOTF)-based glycosylation reactions, the hydroxyl group at C-4 on the thioglycoside donor was protected as the tetrahydropyranyl (THP) ether. Following the coupling, the group was easily removed on solid phase. Subsequent glycosylations with additional donors were then carried out. The glycoside coupling reactions on solid phase were not stereoselective and resulted in mixtures. A small, 12-compound library was synthesized using this approach. The trisaccharides were purified by chromatography following cleavage from the support.

14.3.3 Ito's Capture and Release Strategy

The development of efficient and reliable glycosidic coupling reactions on solid phase is an ongoing challenge and results are still far from optimal. In most cases, it has been observed that these reactions on heterogeneous supports are sluggish and not very stereoselective. To overcome these hurdles, the use of soluble polyethylene glycol (PEG) polymers is becoming an attractive alternative [18]. The short-chain PEG polymers dissolve in certain organic solvents. This allows the study of glycosidic reactions in solutions. Following the coupling reaction, the product coupled with the soluble support could be separated by precipitation using an appropriate solvent. To explore the use of two types of supports, Ito et al. [19] recently developed a solid-phase capture and release strategy for the synthesis of oligosaccharides on a soluble polymer support. Scheme 14.8 shows the glycosyl acceptor **10**, bound to a low molecular weight PEG support that was glycosylated with thioglycoside donor **11** with a chloroacetyl group at C-3. Following the coupling reaction, the PEG-bound component was recovered by filtration through a pad of silica gel to remove excess donor **11** and the side products. The coupled product **12**, combined with the

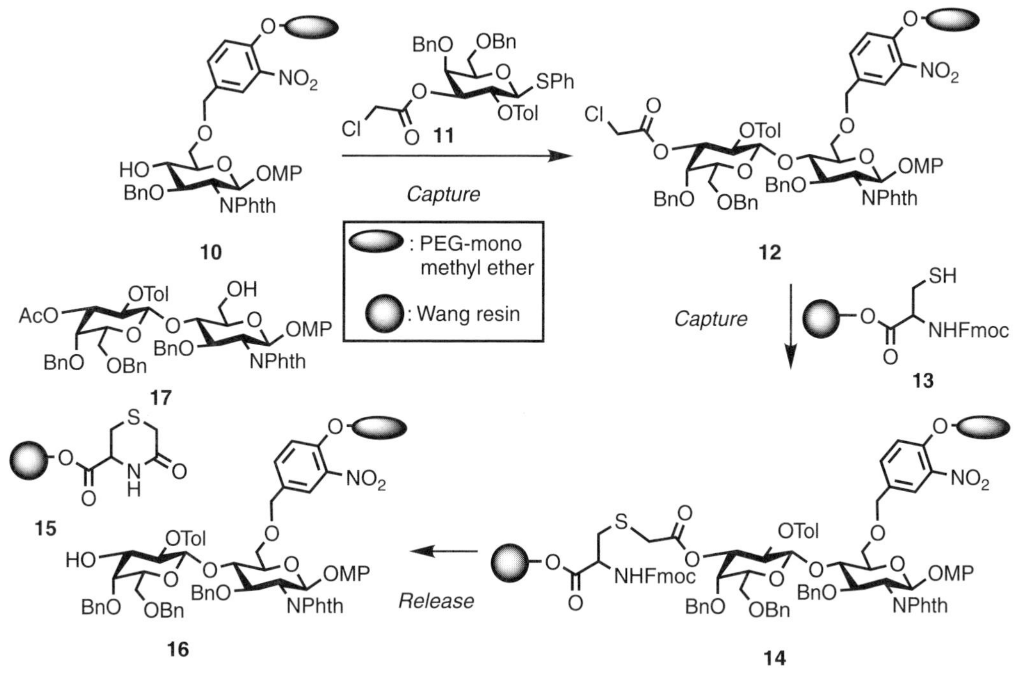

SCHEME 14.8 Ito's capture–release approach to oligosaccharides.

unreacted acceptor **10**, was subjected to a chemoselective reaction on a polystyrene-based solid support. Thus, the reaction of the chloroacetyl group from the PEG-based support with resin bound thiol **13** resulted in derivative **14** with mixed supports. After fluorenylmethyoxycarbonyl (Fmoc) removal, the free amine undergoes spontaneous intracyclization releasing the PEG-bound disaccharide **16** and resin-supported product, **15**. These two were easily separated and the released PEG-bound product **16** was further cleaved, giving the disaccharide derivative **17**. Disaccharide **16** was subjected to two additional cycles of the capture and release strategy for the synthesis of tetrasaccharides. By combining the use of two types of supports, the authors were able to develop an elegant approach to the synthesis of oligosaccharides.

Using the trichloroacetamidate glycosylation coupling and alkyl thiol-based resin, Schmidt et al. [20] developed a method that enables repetitive glycosylation and deprotection reactions in high yields. In this repetitive solid-phase glycosylation approach (Scheme 14.9), the fully protected 2-*O*-acetyl-D-mannosyl trichloroacetamidate donor **19** was coupled to the thiol polymer **18** using TMSOTf, giving the α-glycosylated product **20** exclusively. The *O*-2-acetyl group in mannose led to α-product formation owing to neighboring group participation. In addition, the *O*-acetyl group provides the required temporary protection for chain extension. Deacetylation of **20** was carried out under standard conditions giving **21**, which was then glycosylated with **19**. The repetitive glycosylation of **21** with **19** and deacetylation was continued for up to five cycles. Final cleavage from solid support using NBS provided methyl α-(1→2)-D-mannopentaoside **22** in high yields. The product was analyzed by MALDI-TOF.

14.3.4 Linkers in Solid-Phase Synthesis

Fraser–Reid et al. [21,22] reported the synthesis of polymer supported oligosaccharides using *n*-pentenyl glycosides, while Seeberger et al. [23] accomplished the synthesis of β-(1→4)- and β-(1→6)-linked oligosaccharides using glycosyl phosphates combined with an octenediol linker.

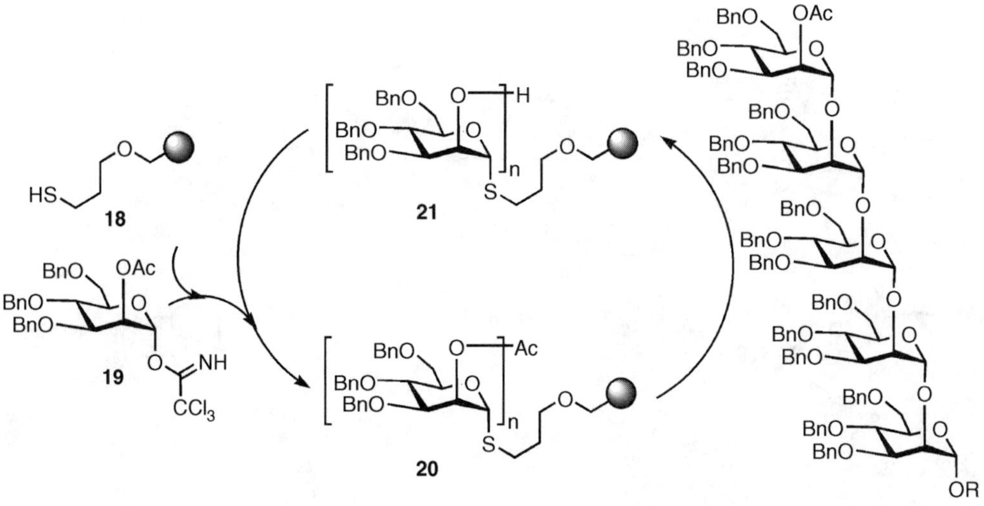

SCHEME 14.9 Schmidt's repetitive solid-phase glycosylation.

Hindsgaul et al. [24] achieved the synthesis of thio-oligosaccharides by nucleophilic substitution of triflate-activated glycosides from resin-bound glycoside-1-thiolates containing unprotected hydroxyl groups. Using the photo-cleavable aglycone linker, Nicolaou et al. [25] reported the synthesis of heptasaccharides and dodecasaccharides on solid phase.

14.4 DYNAMIC COMBINATORIAL CHEMISTRY

Lehn's [26] dynamic combinatorial approach toward carbohydrate library generation differs fundamentally from the solution and solid-phase library approaches discussed above. This dynamic combinatorial library (DCL) strategy involves the transient formation of compounds *in situ* using reversible reactions. These compounds are generated in the presence of a receptor (adaptive combinatorial library), which then selects for ligand(s) possessing the highest affinity. Alternately, the receptor can be added after equilibration is achieved (pre-equilibrated dynamic library). Lehn et al. developed a dynamic combinatorial library against the lectin concanavalin A (conA). The natural ligand for this lectin is the branched trisaccharide **23** (Scheme 14.10). To generate an

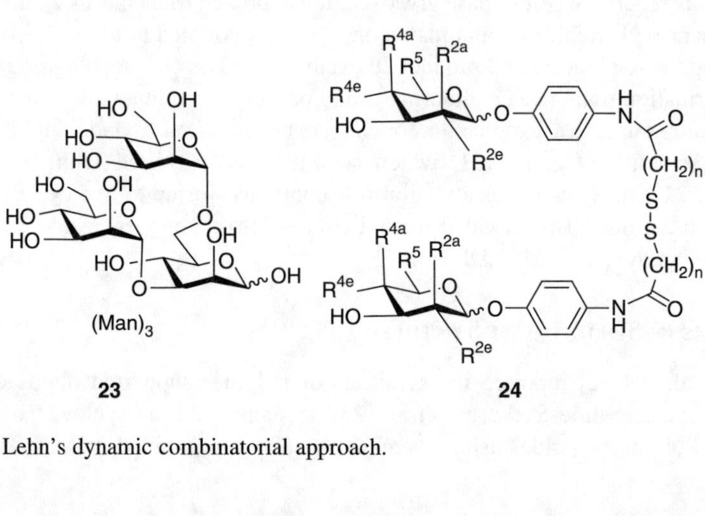

SCHEME 14.10 Lehn's dynamic combinatorial approach.

appropriate spacer between two monosaccharide units, a chemoselective thiol disulfide reaction was selected. The monosaccharides of D-mannose, D-galactose, L-arabinose, and D-xylose were derivatized with the phenylamido group and **24** was attached to two linkers differing only by a CH_2 group. Both homodimers (pH > 7) and heterodimers (in the presence of dithiothreitol) were generated. The libraries were produced either in the presence of conA, or conA was added after pre-equilibration where conA was immobilized onto sepharose beads for isolation of the bound compound. A *bis*-mannose derivative was preferentially selected by the receptor and demonstrated the potential for such simultaneous library generation and screening capabilities of this approach.

14.5 CARBOHYDRATE SCAFFOLDS IN COMBINATORIAL CHEMISTRY

The complex and polyfunctional nature of carbohydrates has made the development of automated methods for oligosaccharide libraries extremely slow. However, carbohydrates can serve as excellent library scaffolds [27]. Several chiral hydroxyl groups, upon appropriate modifications and utilization, could be utilized further in the display of chemical diversities. This approach could provide a novel entry to different classes of compounds to be explored for drug-like properties.

With these goals, the first solid-phase library synthesis that utilized carbohydrate scaffolds was reported by Sofia et al. [28] in 1998. An important feature of their approach was that the scaffold **25** (Scheme 14.11) had three distinct functionalities: (i) the carboxylic acid, (ii) the free hydroxyl group, and (iii) the protected amino group. This derivative was loaded onto amino acid functionalized trityl-based TentaGel resin. From **26**, several libraries were synthesized using the IRORI radiofrequency tagged split-mix method [29]. The library was generated from the sugar

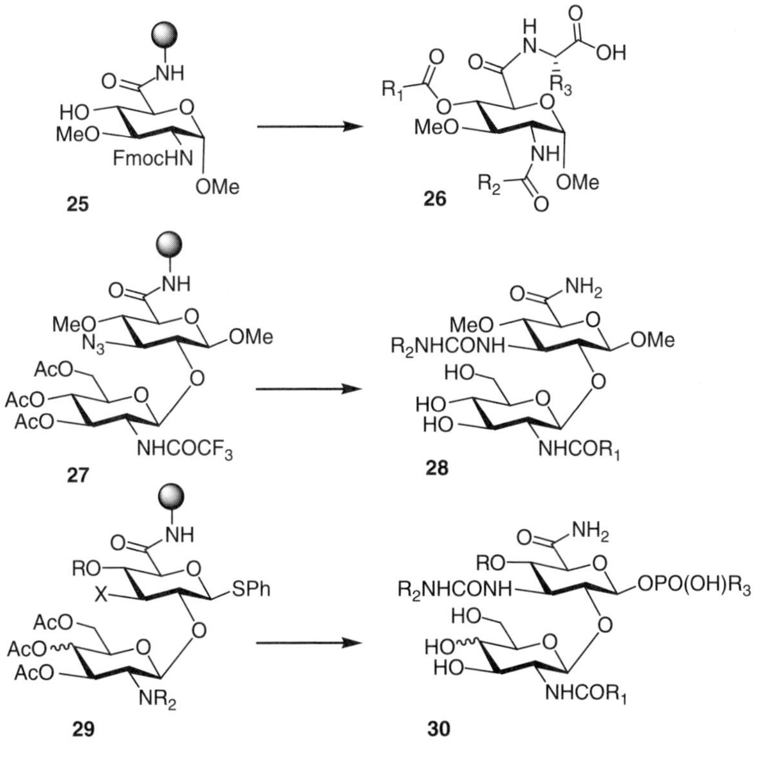

SCHEME 14.11 Carbohydrate-based small molecule scaffolds.

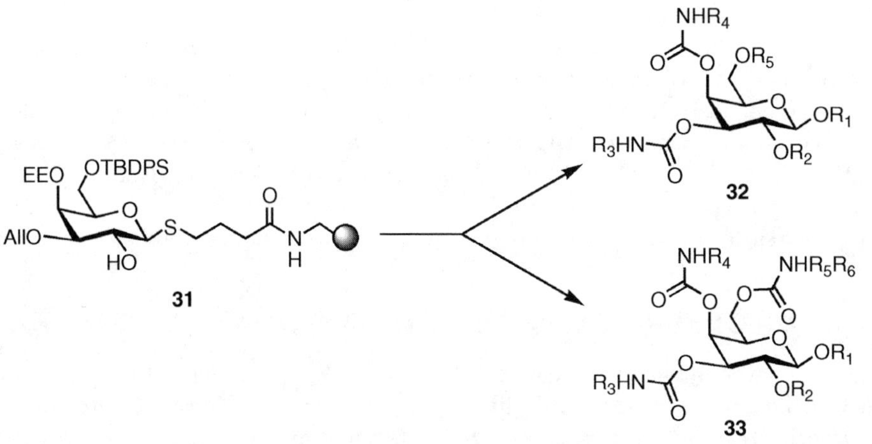

SCHEME 14.12 Kunz's orthogonally protected scaffold.

scaffold via modification of the above described functionalities as diversity points, including (i) functionalization of amino group at C-2, (ii) esterification of the hydroxyl group at C-4, and (iii) amino acid diversity from the –COOH group at C-6. The libraries were analyzed for purities by high pressure liquid chromatography-mass spectrometry (HPLC-MS). In another example, they achieved the solid-phase synthesis of a library of β-disaccharides on Rink amide resin using the disaccharide scaffold **27**. The key feature of this approach was the use of phenylsulfenyl 2-deoxy-2-trifluoroacetamido-glycopyranosides as glycosyl donors in the synthesis of β-linked disaccharides. Disaccharide **27** was derivatized using six isocyanates and eight carboxylic acids to generate a 48-membered library **28**. The glycoside donor discussed above and its galactosamine counterpart were used in the synthesis of meonomycin A disaccharide templated libraries **29**. A combinatorial library of 1300 disaccharides **30** was made by introducing diversities in **29** at C-1, C-3, and C-2′. The library was screened for inhibition of bacterial cell-wall biosynthesis and bacterial growth, and novel classes of potent inhibitors for both processes were identified.

Kunz et al. [30] used an orthogonally protected thioglycoside **31** (Scheme 14.12) as a scaffold. The protecting groups utilized were *tert*-butyldiphenylsilyl (TBDPS), 1-ethoxyethyl (EE) and allyl (All) groups. An important feature of the scaffold was the use of functionalized thioglycosides, which not only served as glycosyl donors but also as linkers for immobilization onto aminomethyl polystyrene resins. This combinatorial method was extended to the galactopyranose scaffold **31** containing five diversity sites. An array of structurally diverse compounds **32** and **33** were successfully synthesized using sequential deprotection and derivatization protocols.

The solid-phase synthesis of 1,6-anhydro sugar derivatives, starting with the 2-*O*-*p*-toluenesulfonyl-β-D-galactopyranoside derivative **34** (Scheme 14.13), was reported by a

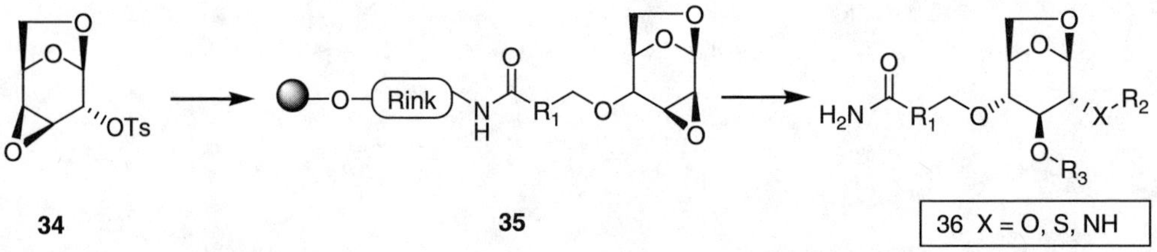

SCHEME 14.13 Solid-phase synthesis of levoglucosan derivatives.

Novartis group [31]. The scaffold **35** was derivatized at the C-2, C-3, and C-4 positions to several diversity sites (e.g., ether, thioether, and amino groups). These groups were further derivatized in the library synthesis of **36**. An important feature was the utilization of stereo- and regioselective epoxide ring-opening reactions. In addition, no protection/deprotection reactions were required. The opening of epoxides by oxygen, sulfur and nitrogen nucleophiles was studied in this approach.

14.6 CARBOHYDRATE/GLYCOCONJUGATE-LIKE COMPOUNDS (GLYCOMIMETICS) BY COMBINATORIAL CHEMISTRY

14.6.1 MULTIPLE COMPONENT CONDENSATIONS (MCC)

In combinatorial chemistry, the development of multicomponent reactions leading to product formation is an attractive strategy because relatively complex molecules can be assembled with fewer steps and in shorter periods. For example, the Ugi multicomponent reaction involving the combination of an isocyanide, an aldehyde, an amine, and a carboxylic acid results in the synthesis of α-acyl amino amide derivatives [32]. The scope of this reaction has been explored in solid-phase synthesis and it allows the generation of a large number of compounds with relative ease. This reaction has been employed in the synthesis of a library of *C*-glycoside conjugated amino amides [33]. Scheme 14.14 shows that, on reaction with carboxylic acids **38**, isocyanides **39**, and Rink amide resin derivatized with different amino acids **40**, the *C*-fucose aldehyde **37** results in the library synthesis of *C*-linked fucosyl amino acids **41** as potential mimics of sialyl Lewis^x.

A similar Ugi approach was utilized by Wong et al. [34] in the library synthesis of artificial glycopeptide-based small-molecule mimics of the aminoglycoside antibiotic neomycin. They utilized a soluble PEG polymer in the synthesis. According to the researchers, the use of a neamine moiety **42** (Scheme 14.15) that has the terminal saccharide unit is critical because it is the major glycoside component involved in the binding of human immunodeficiency virus (HIV)–ribonucleic acid (RNA) transactivator protein. Thus, in their plan, the crucial glycoside component was kept constant and the diversity was introduced as amino acid residues **43**. The four-component condensation reaction with **44** and **45** provided the neomycin mimic library **46**. The use of the PEG linker facilitated the Ugi reaction on the soluble support and the products could be easily separated by precipitation. One of the limiting factors in this Ugi approach is the lack of stereocontrol at the newly formed stereogenic center.

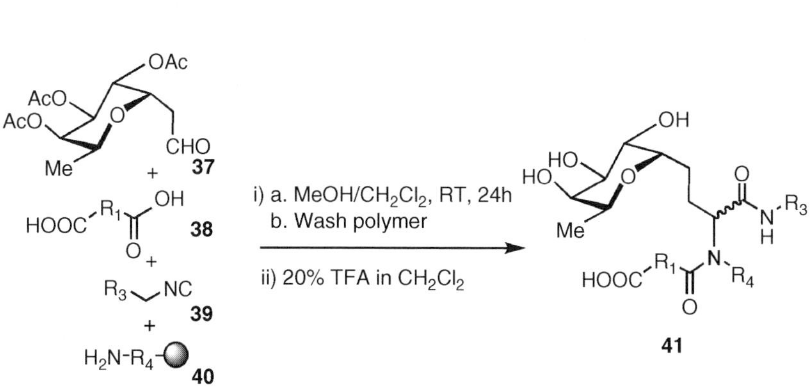

i) a. MeOH/CH_2Cl_2, RT, 24h
b. Wash polymer
ii) 20% TFA in CH_2Cl_2

SCHEME 14.14 Synthesis of glycomimetics by Ugi four-component condensation.

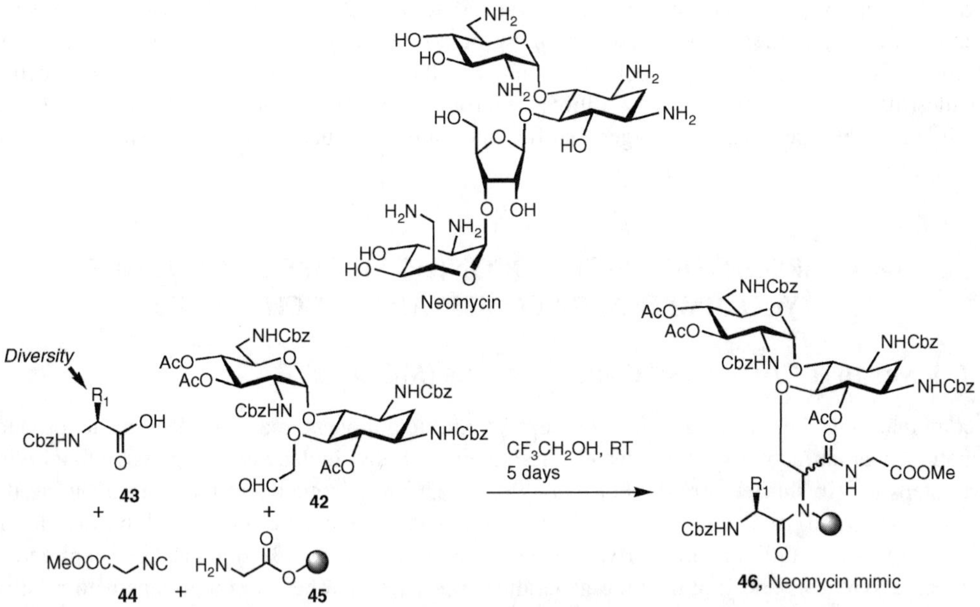

SCHEME 14.15 Wong's Ugi-multicomponent combinatorial synthesis of neomycin mimics.

A recent report by Kunz et al. [35] describes the first stereoselective, combinatorial Ugi multi-component synthesis on solid support. As shown in Scheme 14.16, *O*-pivaloylated-galactosamine **47**, linked to Wang resin via the $\alpha,\alpha,\alpha',\alpha'$-tetramethyl azelaic acid spacer, was utilized in the library synthesis giving glycosylated amino acids **48** with a high degree of diastereoselection. The products were cleaved from the support upon acid treatment and were compared with the authentic samples prepared by solution synthesis.

14.6.2 Glycohybrids

To explore the concept of using terminal carbohydrate moieties as crucial elements, Hindsgaul et al. [36] developed a new class of compounds — glycohybrids. These compounds represent a unique combination of glycoside moieties combined with amino acids and do not occur in natural systems. Cognizant with the typical weak carbohydrate-protein bindings, the idea was to assist carbohydrate-based weak interactions by additional amino acids that may provide the extra contact points for specific bindings.

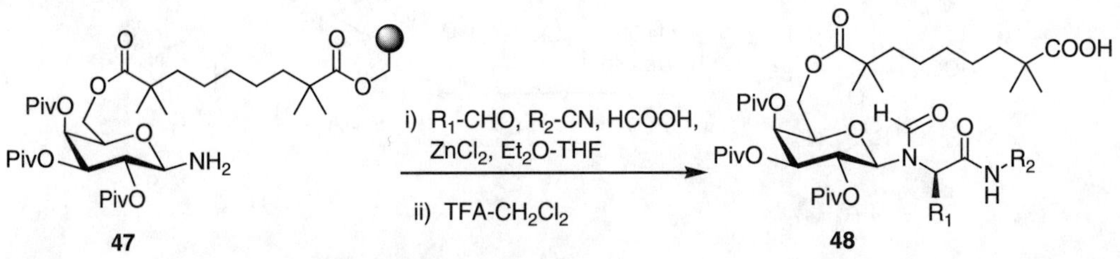

SCHEME 14.16 Kunz's stereoselective Ugi combinatorial synthesis on solid phase.

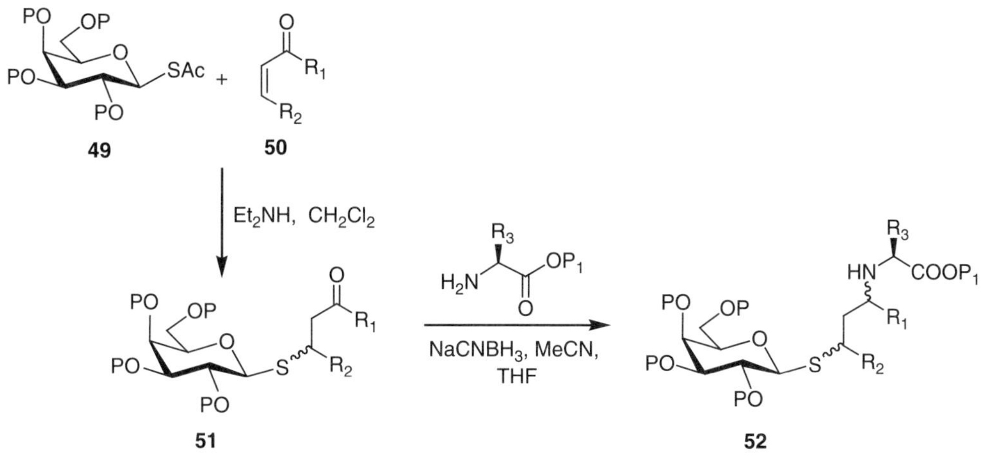

SCHEME 14.17 Hindsgaul's approach to glycohybrids.

A library of glycohybrids was synthesized by the reaction of 1-thio-β-D-galactopyranose derivatives with electron deficient olefins in a Michael-type manner. Following the addition, the plan was to use the carbonyl functionality in a reductive amination. This would allow the coupling of amino acids onto glycoside moieties. The researchers also used hydrophobic protecting groups on glycoside hydroxyls to facilitate reverse-phase purification after the library synthesis. In one example (Scheme 14.17), the 1-thiosugar with the hydroxyl groups protected as *O*-laurates, **49**, was reacted in a Michael fashion with various acceptors, **50**, giving the corresponding keto-derivatives, **51**, as diastereomeric mixtures in high yields. These were then reacted with several amino acid ester derivatives by reductive amination. Thus, a library of thirty compounds, **52**, in which each product contains a mixture of four diastereomers was produced. This library was screened for inhibitors of β-galactosidases.

Thioglycosides were also utilized in the development of solid-phase syntheses of thio-oligosaccharides with free hydroxyl groups on glycosides. A key feature of this approach relied on the high reactivity of nucleophilic glycoside-1-thiolates that could preferentially react with triflate-activated glycosides with protected disulfides at C-1 in the presence of free hydroxyl groups. The required free thiolates were then generated from the protected unsymmetrical ethyl disulfide at C-1. As shown in Scheme 14.18, the immobilized thiolate **53**, via trityl-resin, was then glycosylated with triflate derived 1-dithioethylgalactoside **54** giving the immobilized disaccharide **55** [37]. Following debenzoylation (NaOMe in THF), the anomeric disulfide was then reduced with dithio-threitol (DTT) to obtain the fully deprotected disaccharide **56**, which was then subjected to glycosylation with the triflate derivative **54** in the presence of a crown ether. Thus compound **57** was obtained following the complete deprotection of hydroxyl groups as before and after cleavage from the resin.

14.7 GLYCOPEPTIDE-LIKE DERIVATIVES BY COMBINATORIAL CHEMISTRY

The endocyclic nitrogen-containing carbohydrate derivatives (also known as azasugars) are important because many of them are found to be selective inhibitors of carbohydrate processing enzymes, particularly glycosidases. With the goals of developing a combinatorial chemistry program that allows rapid access to azasugar based artificial glycopeptides, Bols et al. [38] synthesized a library of 125 compounds **60** containing a 1-azafagomine moiety **58** (Scheme 14.19). In their design, the tripeptide library **59** was coupled with the 1-azafagomine derivative possessing

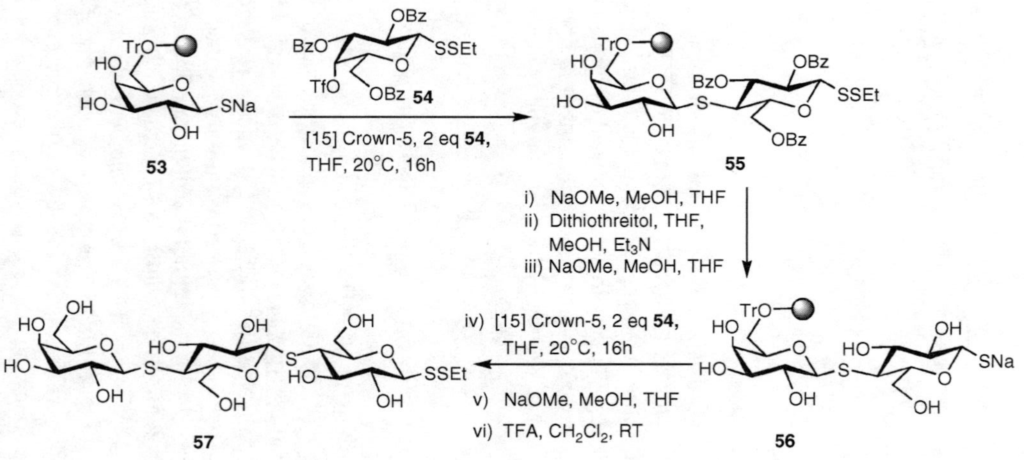

53

54

[15] Crown-5, 2 eq **54**, THF, 20°C, 16h

55

i) NaOMe, MeOH, THF
ii) Dithiothreitol, THF, MeOH, Et_3N
iii) NaOMe, MeOH, THF

56

iv) [15] Crown-5, 2 eq **54**, THF, 20°C, 16h
v) NaOMe, MeOH, THF
vi) TFA, CH_2Cl_2, RT

57

SCHEME 14.18 Hindsgaul's solid-phase synthesis of thio-oligosaccharides.

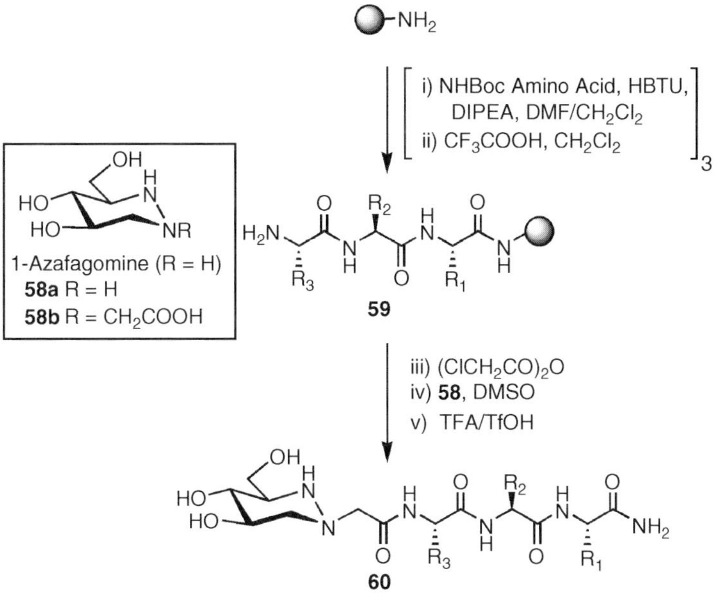

SCHEME 14.19 Bols's approach to azasugar-based artificial glycopeptides.

a carboxyl group via an amide bond acetic acid linker. The tripeptide library was synthesized by a standard split/mix peptide library-synthesis approach prior to coupling with the azasugar moiety. Following cleavage from the support, the library was screened for the inhibition of β-glycosidase, α-glucosidase, and the glycogen phosphorylases. Several compounds were found to display β-glycosidase inhibition. Following the deconvolution of the library, it was found that several compounds caused inhibition, but the compound with three hydroxyproline residues as the peptide fragment showed the best activity.

14.7.1 Glycosylated Amino Acids as Building Blocks

The rapid assembly of glycopeptides by solid-phase synthesis has been achieved by use of many glycosylated Fmoc-protected amino acid pentafluorophenyl esters (OPfp) or free amino acids (Figure 14.1) [39]. A number of solid-support methods have been explored to generate parallel arrays of glycopeptides with native and isosteric substituted glycosidic linkages. In an elegant approach by St. Hilaire and colleagues [39], the synthesis of combinatorial glycopeptide libraries with unambiguous characterization of active compounds (Scheme 14.20) was achieved. The 30,000

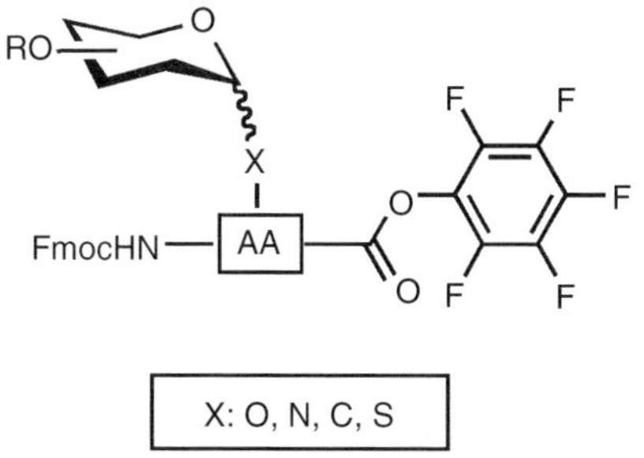

FIGURE 14.1 Glycosylated amino acids for glycopeptides.

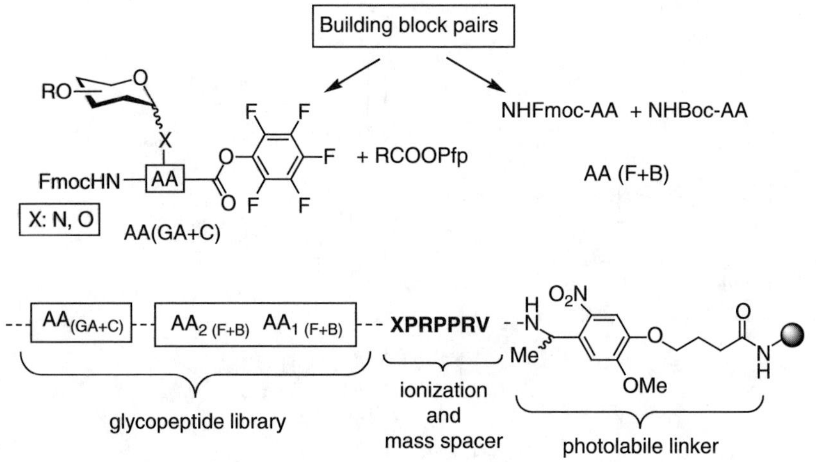

SCHEME 14.20 Solid-phase synthesis of glycopeptide-based mimics.

membered heptaglycopeptide encoded library was synthesized by a split-mix approach on PEGA resin containing a photolabile linker. The glycopeptide-based resins were then screened against a fluorescent-labelled lectin from *Lathyrus odoratus*. The active compounds, on fluorescent beads, were identified by irradiation using a MALDI/TOF-MS laser with concurrent analysis of the ladder of terminal fragments, and directly gave the sequence as well as providing the structure of the glycopeptides. Significantly, all active glycopeptides contained the terminal mannose moiety, an important feature for the lectin recognition.

Following the building-block approach, artificial glycopeptides were synthesized by Wong et al. [40], as small-molecule mimics of sialyl Lewis^x. They utilized the fucosylated amino acid *O*-allyl carboxylic ester building block (Scheme 14.21). The properly protected, threonine

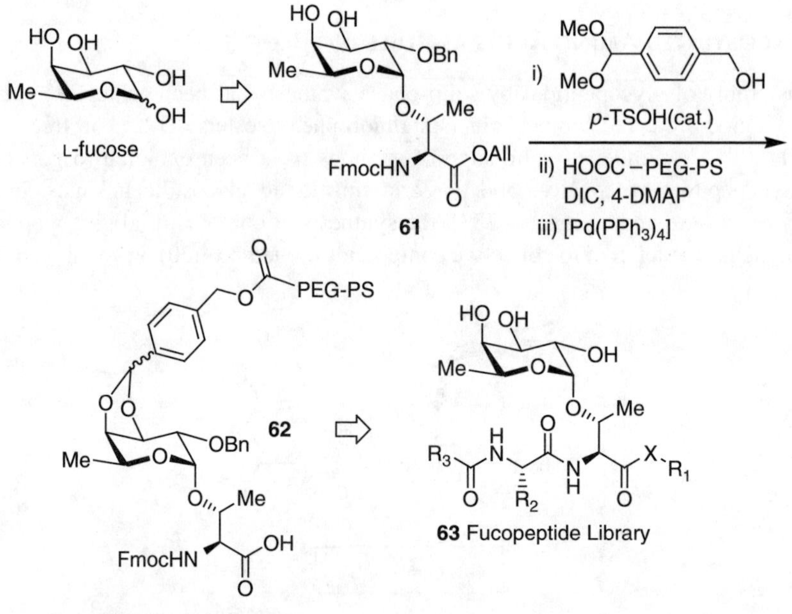

SCHEME 14.21 Wong's fucopeptide-based sialyl Lewis^x mimics.

glycosylated fucose derivative **61** was immobilized onto the carboxy functionalized PEG-PS resin via an acid sensitive *p*-(acyloxymethyl)benzylidene acetal (*p*-AMBA). After removal of the allyl protecting group, the fucopeptide library **63** was obtained from the building block **62**. The library synthesis by parallel approach had the diversity at the amino acid side chain and both at the *N*-and *C*-termini of the polypeptide backbone. In this library, the fucose moiety was kept constant because it is the crucial glycoside section required for the recognition of sialyl Lewis^x by E-selectin. Following cleavage and purification, the derivatives showed moderate binding affinities with E- and P-selectins.

14.7.2 Cyclic Artificial Glycopeptides

Parallel robot-assisted solid-phase syntheses of cyclic sugar, amino acid-based, artificial glycopeptides were accomplished by van Boom et al. [41]. These cyclic artificial glycopeptides resemble cyclodextrins in their structures. Scheme 14.22 shows that the linear derivative **66** was first synthesized using building block **64** on solid phase. The linear derivative **66** was then subjected to acid-catalyzed cyclization, giving the cyclic artificial glycopeptides **67a** and **67b**. Following purification, the products were characterized by NMR.

14.7.3 Automated Synthesis of Artificial Glycopeptides

With the goal of developing automated methods to the synthesis of artificial glycopeptides, Arya et al. [42] were interested in stable, artificial glycopeptides that possessed pertinent features of

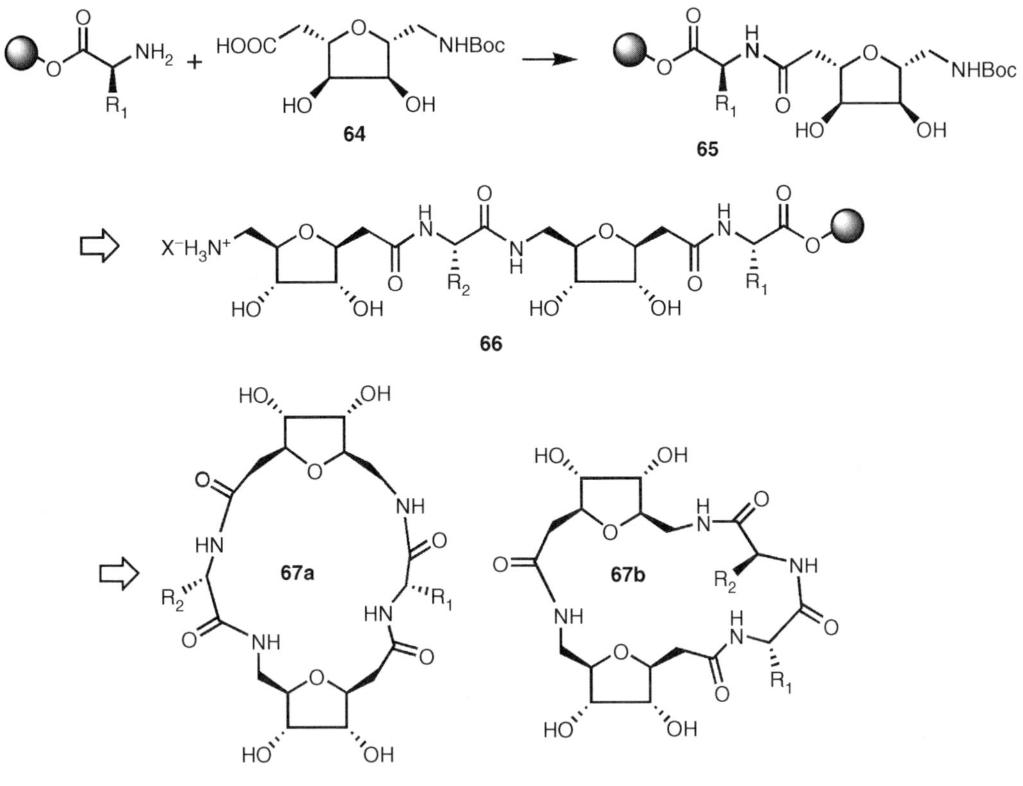

SCHEME 14.22 Sugar amino acid-derived cyclic derivatives.

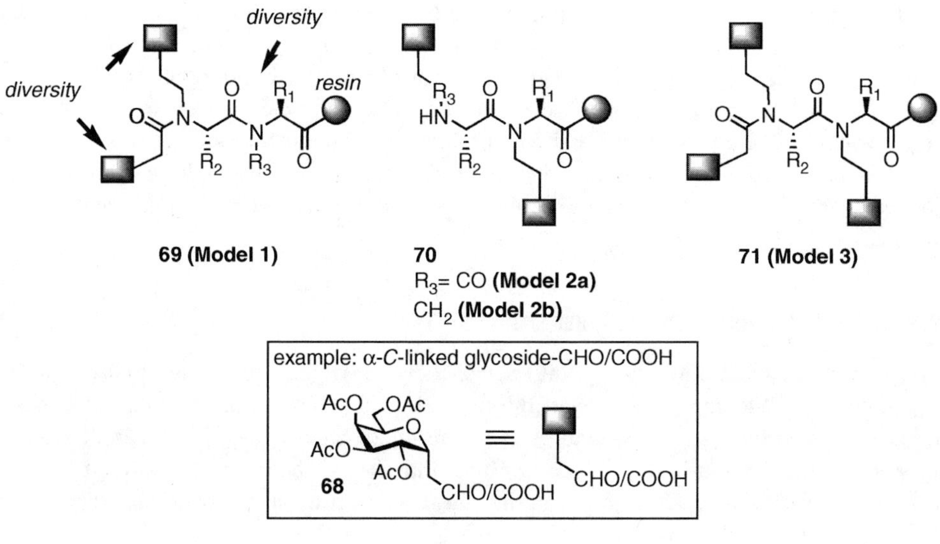

SCHEME 14.23 Arya's automated approach to artificial glycopeptides.

the underlying protein as well as the carbohydrate moiety or glycoform of the glycoconjugates. This idea led to the development of carbohydrate diversity on a dipeptide/pseudopeptide scaffold, which demonstrates the flexibility of this approach (see Model 1, Model 2 and Model 3, **69** through **71**, Scheme 14.23). The carbohydrates were incorporated as α- and β-linked *C*-glycosides, which are stable isosteres of native terminal sugars. The versatility of this approach is reflected by the fact that a variety of sugars can be independently incorporated as peracetylated ethanal- or ethanoic acid derivatives **68**. The building blocks (as carbon-linked sugar aldehydes and carbon-linked sugar acids) can be incorporated either at the *N* terminal moiety or at the internal amide nitrogen of a short peptide/pseudopeptide scaffold. This can be done in a highly flexible and controlled manner. Using this approach, libraries of artificial glycopeptides could be easily synthesized for probing carbohydrate–protein interactions. The libraries display two (i.e., homogeneous or heterogeneous) copies of carbohydrates, whereas the dipeptide scaffold may contribute to secondary interactions with the biological target.

Using TentaGel S RAM resin as solid support, the artificial glycopeptide libraries, based upon Model 2a and Model 2b, were successfully synthesized on a multiple organic synthesizer (MOS). The success of the methodology was dependent on the optimization of the reductive amination reaction of the acetylated *C*-glycoside ethanol derivatives with the amino group of the anchored amino acid. After several attempts at the synthesis of glycosyl amino derivative on solid phase, the reductive amination product could be obtained in high yields (70 to 95%) by using a relatively low excess of the glycoside derivatives. The next task — the coupling of the resulting secondary amine to the next amino acid — was accomplished after a series of optimization with HATU. Deacetylation of the sugar derivatives was achieved with hydrazine hydrate/*N*,*N*-dimethylformamide mixture at pH 9–10 for 4.5 h.

Using the method discussed above, four 96-compound artificial glycopeptide libraries were synthesized in a fully automated manner. The artificial glycopeptide libraries show a combination of two glycosides: (i) α-galactoside and β-galactoside, and (ii) α-glucoside and α-mannoside (Figure 14.2). The compounds in the artificial glycopeptide libraries were used to test the ability of these derivatives to inhibit the reglucosylation of *N*-linked glycoproteins by a glucose processing enzyme. Reglucosylation of *N*-linked glycoproteins appears to be a critical step in *N*-glycoprotein biosynthesis, protein folding and trafficking pathways [43]. In a different study, the artificial glycopeptide libraries were tested in enzyme systems that convert a glucose

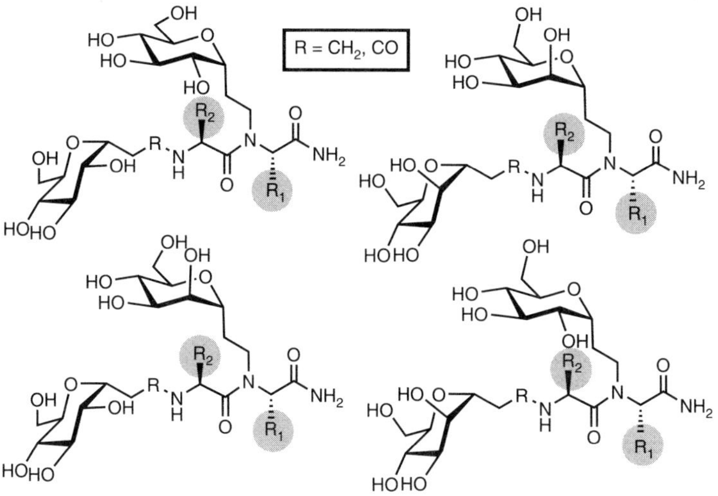

R = CH_2, CO

R_1	R_2
CH_3	CH_2Ph
CH_3	$CH_2CH(CH_3)_2$
CH_3	$CH(CH_3)_2$
CH_3	$CH(CH_3)CH_2CH_3$
CH_3	Ph
CH_3	CH_2OH
CH_3	$CH(CH_3)OH$
CH_3	BnOH
CH_3	CH_2COOH
CH_3	$(CH_2)_2COOH$
CH_2COOH	Bn
CH_2COOH	$CH_2CH(CH_3)_2$
CH_2COOH	Ph
CH_2COOH	CH_3
CH_2OH	CH_2COOH
CH_2OH	Bn
CH_2OH	CH_3
CH_2OH	$CH_2CH(CH_3)_2$
$(CH_2)_2COOH$	BnOH
$(CH_2)_2COOH$	$(CH_2)_2COOH$
CH_2Ph	CH_3
$CH(CH_3)OH$	CH_3CH_2OH
$CH(CH_3)OH$	BnOH

FIGURE 14.2 Examples of artificial glycopeptide libraries prepared by automation.

moiety into rhamnose prior to incorporation of the rhamnose unit during the biosynthesis of the mycobacterium cell wall. The inhibition of this step may play an important role in the development of novel carbohydrate-derived therapies to combat *Mycobacterium tuberculosis* cell-wall biosynthesis [44]. Further, inhibition of this biosynthetic pathway may lead to the development of compounds with a specific action, because this particular biotransformation does not occur in mammalian systems. To date, few artificial glycopeptide derivatives as potential glycoside-based inhibitors containing at least one negatively charged, amino acid residue have been identified [45].

14.8 SUMMARY AND OUTLOOK

Our ability to rapidly access pure peptide and nucleic acid derivatives has played an important role in understanding the biological functions of these two classes of natural biopolymers. The field of glycobiology has not advanced to that level, mainly because of the difficulty of accessing carbohydrate-based derivatives. Over the years, significant progress has been made in carbohydrate-oriented solution and solid-phase synthesis, including combinatorial chemistry. Several combinatorial approaches covered in this chapter have the potential to make carbohydrate-based compounds accessible. However, some methods may require further improvement before they are accepted by the community. In the future, automated synthetic methods will probably grow and greatly impact the field of chemical glycobiology by adding to our understanding of the biological functions of carbohydrates and the use of carbohydrate-based chemical probes.

ACKNOWLEDGMENTS

The authors thank Dr Nallareddy Babu for providing technical assistance.

REFERENCES

1. (a) Bertozzi, C R, Kiessling, L L, Chemical glycobiology, *Science*, 291, 2357–2365, 2001; (b) Dwek, R A, Glycobiology: Toward understanding the function of sugars, *Chem. Rev.*, 96, 683–720, 1996; (c) Varki, A, Biological roles of oligosaccharides: All of the theories are correct, *Glycobiology*, 3, 97–130, 1993; (d) Branza-Nichita, N, Petrescu, A J, Negroiu, G, Dwek, R A, Petrescu, S-M, *N*-Glycosylation processing and glycoprotein folding- lessons from tyrosine-related proteins, *Chem. Rev.*, 100, 4697–4711, 2000; (e) Butters, T D, Dwek, R A, Platt, F M, Inhibition of glycosphingolipid biosynthesis: Application to lysosomal storage disorders, *Chem. Rev.*, 100, 4683–4696, 2000; (f) Duus, J Ø, Hilaire, P M,. St, Meldal, M, Bock, K, Carbohydrate chemistry: Synthetic and structural challenges towards the end of the 20th century, *Pure Appl. Chem.*, 71, 755–765, 1999.
2. (a) Danishefsky, S J, Roberge, J Y, Advances in the development of convergent schemes for the synthesis of biologically important glycoconjugates, *Pure Appl. Chem.*, 67, 1647–1662, 1995; (b) Gege, C, Vogel, J, Bendas, G, Rothe, U, Schmidt, R R, Synthesis of the sialyl Lewis X epitope attached to glycolipids with different core structures and their selectin-binding characteristics in a dynamic test system, *Chem. Eur. J.*, 6, 111–122, 2000; (c) Seeberger, P H, Danishefsky, S J, Solid phase synthesis of oligosaccharides and glycoconjugates by the glycal assembly method: A five year retrospective, *Acc. Chem. Res.*, 31, 685–695, 1998; (d) Meinjohanns, E, Meldal, M, Paulsen, H, Dwek, R A, Bock, K, Novel sequential solid-phase synthesis of *N*-linked glycopeptides from natural sources, *J. Chem. Soc. Perkin. Trans.* 1, 549–560, 1998; (e) Sames, D, Chen, X-T, Danishefsky, S J, Convergent total synthesis of a tumour-associated mucin motif, *Nature*, 389, 587–591, 1997; (f) Seitz, O, Wong, C-H, Chemoenzymatic solution- and solid-phase synthesis of *O*-glycopeptides of the mucin domain of MAdCAM-1. A general route to *O*-LacNAc, *O*-sialyl-LacNAc, and *O*-sialyl-Lewis-X peptides, *J. Am. Chem. Soc.*, 119, 8766–8776, 1997; (g) Danishefsky, S J, Bilodeau, M T, Glycals in organic synthesis: The evolution of comprehensive strategies for the assembly of oligosaccharides and glycoconjugates of biological consequence, *Angew. Chem. Int. Ed. Engl.*, 35, 1380–1417, 1996; (h) Wang, Z-G, Zhang, X-F, Ito, Y, Nakahara, Y, Ogawa, T, Stereocontrolled syntheses of *O*-glycans of core class 2 with a linear tetrameric lactosamine chain and with three lactosamine branches, *Carbohydr. Res.*, 295, 25–39, 1996; (i) Danishefsky, S J, Allen, J A, From the laboratory to the clinic: A retrospective on fully synthetic carbohydrate-based anticancer vaccines, *Angew. Chem. Int. Ed.*, 39, 836–863, 2000, and references therein.
3. (a) See Hanessian, S, Ed., In *Preparative Carbohydrate Chemistry*, Marcel Dekker, New York, 1997; (b) Benjamin, G, Davis, J, Recent developments in oligosaccharide synthesis, *J. Chem. Soc. Perkin. Trans.* 1, 2137–2160, 2000; (c) Toshima, K, Tatsuta, K, Recent progress in *O*-glycosylation methods and its application to natural products synthesis, *Chem. Rev.*, 93, 1503–1531, 1999; (d) Hanessian, S, Lou, B, Stereocontrolled glycosyl transfer reactions with unprotected glycosyl donors, *Chem. Rev.*, 100, 4443–4463, 2000; (e) Boons, G-J, Demchenko, A V, Recent advances in *O*-sialylation, *Chem. Rev.*, 100, 4539–4565, 2000; (f) Jung, K-H, Muller, M, Schmidt, R R, Intramolecular *O*-glycoside bond formation, *Chem. Rev.*, 100, 4423–4442, 2000.
4. (a) For recent reviews on solid-phase oligosaccharide synthesis and programmed synthetic approaches to complex carbohydrates and carbohydrate conjugates, see Seeberger, P H, Haase, W-C, Solid-phase oligosaccharide synthesis and combinatorial carbohydrate libraries, *Chem. Rev.*, 100, 4349–4393, 2000; (b) Koeller, K M, Wong, C-H, Synthesis of complex carbohydrates and glycoconjugates: enzyme-based and programmable one-pot strategies, *Chem. Rev.*, 100, 4465–4493, 2000; (c) Herzner, H, Reipen, T, Schultz, M, Kunz, H, Synthesis of glycopeptides containing carbohydrate and peptide recognition motifs, *Chem. Rev.*, 100, 4495–4537, 2000.
5. (a) Marcaurelle, L A, Bertozzi, C R, New directions in the synthesis of glycopeptide mimetics, *Chem. Eur. J.*, 5, 1384–1390, 1999; (b) Simanek, E E, McGarvey, G J, Jablonowski, J A, Wong, C-H, Selectin-carbohydrate interactions: From natural ligands to designed mimics, *Chem. Rev.*, 98, 833–862, 1998; (c) Sears, P, Wong, C-H, Carbohydrate mimetics: A new strategy for tackling the problem of carbohydrate-mediated biological recognition, *Angew. Chem. Int. Ed.*, 38, 2300–2324, 1999; (d) Hilaire, P M, St, Meldal, M, Glycopeptide and oligosaccharide libraries, *Angew. Chem. Int. Ed.*, 39, 1162–1179, 2000; (e) Dondoni, A, Marra, A, Methods for anomeric carbon-linked and

fused sugar amino acid synthesis: The gateway to artificial glycopeptides, *Chem. Rev.*, 100, 4395–4421, 2000.

6. (a) Thompson, L A, Ellman, J A, Synthesis and applications of small molecule libraries, *Chem. Rev.*, 95, 555–600, 1996; (b) Gordon, E M, Gallop, M A, Patel, D V, Strategy and tactics in combinatorial organic synthesis. Applications to drug discovery, *Acc. Chem. Res.*, 29, 144–154, 1996; (c) Lam, K S, Lebel, M, Krchnak, V, The one-bead-one-compound combinatorial library method, *Chem. Rev.*, 97, 411–448, 1997; (d) Dolle, R E, Nelson, K H, Jr., Comprehensive survey of combinatorial library synthesis: 1998, *J. Comb. Chem.*, 1, 235–282, 1999; (e) Balkenhohl, F, von dem Bussche-Hunnefeld, C, Lansky, A, Zechel, C, Combinatorial synthesis of small organic molecules, *Angew. Chem. Int. Ed.*, 35, 2288–2337, 1996; (f) Guillier, F, Orain, D, Bradley, M, Linkers and cleavage strategies in solid-phase organic synthesis and combinatorial chemistry, *Chem. Rev.*, 100, 2091–2157, 2000.
7. (a) Merrifield, R B, Solid phase peptide synthesis. I. The synthesis of a tetrapeptide, *J. Am. Chem. Soc.*, 85, 2149–2154, 1963; (b) Furka, A, Sebestyen, F, Asgedom, M, Dibo, G, General method for rapid synthesis of multicomponent peptide mixtures, *Int. J. Peptide Protein Res.*, 37, 487–493, 1991; (c) Houghten, R A, Pinilla, C, Blondelle, S E, Appel, J R, Dooley, C T, Cuervo, J H, Generation and use of synthetic peptide combinatorial libraries for basic research and drug discovery, *Nature*, 354, 84–86, 1991.
8. (a) Gordon, E M, Kerwin, J F, Jr., Eds., *Combinatorial Chemistry and Molecular Diversity in Drug Discovery*, Wiley, New York, 1998; (b) Chaiken, I M, Janda, K D, Eds.,*Molecular Diversity and Combinatorial Chemistry — Libraries and Drug Discovery*, *ACS Series*, 1996; (c) Obrecht, D Villalgordo, J M, Eds., *Solid-Supported Combinatorial and Parallel Synthesis of Small-Molecular-Weight Compound Libraries*, Pergamon Press, Oxford, 1998; (d) Burgess, K, Ed., *Solid Phase Organic Synthesis*, Wiley-Interscience, New York, 2000; (e) See, a special issue on combinatorial chemistry, *Chem. Rev.*, 2, 1997; (f) Arya, P, Joseph, R, Chou, D T H, Toward high-throughput synthesis of complex natural product-like compounds in the genomics and proteomics age, *Chem. Biol.*, 9, 145–156, 2002; (g) Arya, P, Baek, M-G, Natural-product-like chiral derivatives by solid-phase synthesis, *Curr. Opin. Chem. Biol.*, 5, 292–301, 2001; (h) Arya, P, Chou, D T H, and Baek, M-G, Diversity-based organic synthesis in the era of genomics and proteomics, *Angew. Chem. Int. Ed. Engl.*, 40, 339–442, 2001.
9. (a) Sofia, M J, Wang, Z-G, Carbohydrate-based combinatorial libraries, *Mol. Diversity*, 3, 75–94, 1998, For early reviews on combinatorial efforts in complex carbohydrates, carbohydrate conjugates and carbohydrate mimics, see; (b) Hindsgaul, O, Combinatorial carbohydrate chemistry, *Glycoimmunology*, 2, 219–236, 1998; (c) Schweizer, F, Hindsgaul, O, Combinatorial synthesis of carbohydrates, *Curr. Opin. Chem. Biol.*, 3, 291–298, 1999; (d) Arya, P, Ben, R N, Combinatorial chemistry for the synthesis of carbohydrate libraries, *Angew. Chem. Int. Ed.*, 36, 1280–1282, 1997; (e) Arya, P, Ben, R N, Kutterer, K M K, *Combinatorial Chemistry for the Synthesis of Carbohydrates/ Carbohydrate Mimics Libraries*, *Organic Synthesis Highlights*, Vol. 4, 2000, pp. 337–342; (f) Barkley, A, Arya, P, Combinatorial chemistry toward understanding the function(s) of carbohydrates and carbohydrate mimics, *Chem. Eur. J.*, 7, 555–563, 2001; (g) Marcaurelle, L A, Seeberger, P H, Combinatorial carbohydrate chemistry, *Curr. Opin. Chem. Biol.*, 6, 289–296, 2002; (h) Schweizer, F, Glycosamino acids: Building blocks for combinatorial synthesis — Implications for drug discovery, *Angew. Chem. Int. Ed.*, 41, 230–253, 2002.
10. (a) Kanie, O, Barresi, F, Ding, Y, Labbe, J, Otter, A, Forsberg, L S, Ernst, B, Hindsgaul, O, A strategy of random glycosylation for the production of oligosaccharide libraries, *Angew. Chem. Int. Ed. Engl.*, 34, 2720–2722, 1995; (b) Ding, Y, Labbe, J, Kanie, O, Hindsgaul, O, Towards oligosaccharide libraries: A study of the random galactosylation of unprotected *N*-acetylglucosamine, *Bioorg. Med. Chem.*, 4, 683–692, 1996.
11. (a) Boons, G-J, Isles, S, Vinyl glycosides in oligosaccharide synthesis. 2. The use of allyl and vinyl glycosides in oligosaccharide synthesis, *J. Org. Chem.*, 61, 4262–4271, 1996; (b) Johnson, M, Arles, C, Boons, G-J, Vinyl glycosides in oligosaccharide synthesis (Part 5): A latent-active glycosylation strategy for the preparation of branched trisaccharide libraries, *Tetrahedron Lett.*, 39, 9801–9804, 1998.
12. Izumi, M, and Ichikawa, Y, Combinatorial synthesis of oligosaccharide library of 2,6-dideoxysugars, *Tetrahedron Lett.*, 39, 2079–2082, 1998.

13. Wong, C-H, Ye, X-S, Zhang, Z, Assembly of oligosaccharide libraries with a designed building block and an efficient orthogonal protection-deprotection strategy, *J. Am. Chem. Soc.*, 120, 7137–7138, 1998.
14. (a) Zhang, Z, Ollmann, I R, Ye, X-S, Wischnat, R, Bassov, T, Wong, C-H, Programmable one-pot oligosaccharide synthesis, *J. Am. Chem. Soc.*, 121, 734–753, 1999; (b) Ye, X-S, Wong, C-H, Anomeric reactivity-based one-pot oligosaccharide synthesis: A rapid route to oligosaccharide libraries, *J. Org. Chem.*, 65, 2410–2431, 2000.
15. Takahashi, T, Adachi, M, Matsuda, A, Doi, T, Combinatorial synthesis of trisaccharides via solution-phase one-pot glycosylation, *Tetrahedron Lett.*, 41, 2599–2603, 2000.
16. (a) Liang, R, Yan, L, Loebach, J, Ge, M, Uozumi, Y, Sekanina, K, Horan, N, Glidersleeve, J, Thompson, C, Smith, A, Biswas, K, Still, W C, Kahne, D, Parallel synthesis and screening of a solid phase carbohydrate library, *Science*, 274, 1520–1522, 1996; (b) Horan, N, Yan, L, Isobe, H, Whitesides, G W, Kahne, D, Nonstatisical binding of a protein to clustered carbohydrates, *Proc. Natl. Acad. Sci. USA*, 96, 11782–11786, 1999.
17. Zhu, T, and Boons, G-J, A two-directional approach for the solid-phase synthesis of trisaccharide libraries, *Angew. Chem. Int. Ed.*, 37, 1898–1900, 1998.
18. Douglas, S P, Whitfield, D M, Krepinsky, J J, Polymer-supported solution synthesis of oligosaccharides using a novel versatile linker for the synthesis of D-mannopentaose, a structural unit of D-mannans of pathogenic yeasts, *J. Am. Chem. Soc.*, 117, 2116–2117, 1995.
19. Ito, Y, Kanie, O, Ogawa, T, Orthogonal glycosylation strategy for rapid assembly of oligosaccharides on a polymer support, *Angew. Chem. Int. Ed. Engl.*, 35, 2510–2512, 1996.
20. Rademann, J, Schmidt, R R, Repetitive solid phase glycosylation on an alkyl thiol polymer leading to sugar oligomers containing 1,2-*trans*- and 1,2-*cis*-glycosidic linkages, *J. Org. Chem.*, 62, 3650–3653, 1997.
21. Rodebaugh, R, Joshi, S, Fraser-Reid, B, Geysen, H M, Polymer-supported oligosaccharides via *n*-pentenyl glycosides: Methodology for a carbohydrate library, *J. Org. Chem.*, 62, 5660–5661, 1997.
22. Anilkumar, G, Nair, L G, Fraser-Reid, B, Targeted glycosyl donor delivery for site-selective glycosylation, *Org. Lett.*, 2, 2587–2589, 2000.
23. (a) Andrade, R B, Plante, O J, Melean, L G, Seeberger, P H, Solid-phase oligosaccharide synthesis: Preparation of complex structures using a novel linker and different glycosylating agents, *Org. Lett.*, 1, 1811–1814, 1999; (b) Zheng, C, Seeberger, P H, Danishefsky, S J, Solid support oligosaccharide synthesis: construction of β-linked oligosaccharides by coupling of glycal derived thioethyl glycosyl donors, *J. Org. Chem.*, 63, 1126–1130, 1998.
24. Hummel, G, Hindsgaul, O, Solid-phase synthesis of thio-oligosaccharides, *Angew. Chem. Int. Ed.*, 38, 1782–1784, 1999.
25. (a) Nicolaou, K C, Winssinger, N, Pastor, J, DeRoose, F, A general and highly efficient solid phase synthesis of oligosaccharides. Total synthesis of a heptasaccharide phytoalexin elicitor (HPE), *J. Am. Chem. Soc.*, 119, 449–450, 1997; (b) Nicolaou, K C, Watanabe, N, Li, J, Pastor, J, Winssinger, N, Solid-phase synthesis of oligosaccharides: Construction of a dodecasaccharide, *Angew. Chem. Int. Ed.*, 37, 1559–1561, 1998.
26. (a) Ramstrom, O, Lehn, J-M, In Situ generation and screening of a dynamic combinatorial carbohydrate library against concanavalin A, *Chem. BioChem.*, 1, 41–48, 2000; (b) Lehn, J-M, Dynamic combinatorial chemistry and virtual combinatorial libraries, *Chem. Eur. J.*, 5, 2455–2463, 1999; (c) Cousins, G R L, Poulsen, S-A, Sanders, J K M, Molecular evolution: Dynamic combinatorial libraries, autocatalytic networks and the quest for molecular function, *Curr. Opin. Chem. Biol.*, 4, 270–279, 2000.
27. (a) Hirschmann, R, Yao, W, Cascieri, M A, Strader, C D, Maechler, L, Cichy-Knight, M A, Hynes, J, Jr., van Rijn, R D, Sprengeler, P A, Smith, A B III, Synthesis of potent cyclic hexapeptide NK-1 antagonists. Use of a minilibrary in transforming a peptidal somatostatin receptor ligand into an NK-1 receptor ligand via a polyvalent peptidomimetic, *J. Med. Chem.*, 39, 2441–2448, 1996; (b) Hirschmann, R, Hynes, J, Jr., Cichy-Knight, M A, van Rijn, R D, Sprengeler, P A, Spoors, P G, Shakespeare, W C, Pietranico-Cole, S, Barbosa, J, Liu, J, Yao, W, Rohrer, S, Smith, A B, III, Modulation of receptor and receptor subtype affinities using diastereomeric and enantiomeric monosaccharide scaffolds as a means to structural and biological diversity. A new route to ether synthesis, *J. Med. Chem.*, 41, 1382–1391, 1998.

28. (a) Sofia, M J, The generation of carbohydrate-based combinatorial libraries for drug discovery, *Med. Chem. Res.*, 8, 362–378, 1998; (b) Sofia, M J, Hunter, R, Chan, T Y, Vaughan, A, Dulina, R, Wang, H, Gange, D, Carbohydrate-based small-molecule scaffolds for the construction of universal pharmacophore mapping libraries, *J. Org. Chem.*, 63, 2802–2803, 1998; (c) Silva, D J, Wang, H, Allanson, N M, Jain, R K, Sofia, M J, Stereospecific solution- and solid-phase glycosylations. Synthesis of β-linked saccharides and construction of disaccharide libraries using phenylsulfenyl 2-deoxy-2-trifluoroacetamido glycopyranosides as glycosyl donors, *J. Org. Chem.*, 64, 5926–5929, 1999; (d) Sofia, M J, Allanson, N, Hatzenbuhler, N T, Jain, R, Kakarla, R, Kogan, N, Liang, R, Liu, D, Silva, D J, Wang, H, Gange, D, Anderson, J, Chen, A, Chi, F, Dulina, R, Huang, B, Kamau, M, Wang, C, Baizman, E, Branstrom, A, Bristol, N, Goldman, R, Han, K, Longley, C, Midha, S, Axelrod, H R, Discovery of novel disaccharide antibacterial agents using a combinatorial library approach, *J. Med. Chem.*, 42, 3193–3198, 1999.
29. Nicolaou, K C, Xiao, X Y, Parandoosh, Z, Senyei, A, Nova, M P, Radiofrequency encoded combinatorial chemistry, *Angew. Chem. Int. Ed. Engl.*, 34, 2289–2291, 1995.
30. (a) Wunberg, T, Kallus, C, Optaz, T, Henke, S, Schmidt, W, Kunz, H, Carbohydrates as multifunctional chiral scaffolds in combinatorial synthesis, *Angew. Chem. Int. Ed.*, 37, 2503–2505, 1998; (b) Kallus, C, Opatz, T, Wunberg, T, Schmidt, W, Henke, S, Kunz, H, Combinatorial solid-phase synthesis using D-galactose as a chiral five-dimension-diversity scaffold, *Tetrahedron Lett.*, 40, 7783–7786, 1999.
31. (a) Brill, W K-D, De Mesmaeker, A, Wendeborn, S, Solid-phase synthesis of levoglucosan derivatives, *Synlett*, 1085–1089, 1998; (b) Brill, W K-D, Tirefort, D, Opening of levoglucosane derived epoxides with oxygen, nitrogen and sulfur nucleophiles, *Tetrahedron Lett.*, 39, 787–790, 1998; (c) Wendeborn, S, De Mesmaeker, A, Brill, W K-D, Berteina, S, Synthesis of diverse and complex molecules on the solid phase, *Acc. Chem. Res.*, 33, 215–224, 2000.
32. (a) Ugi, I, From isocyanides via four — component condensation to antibiotic synthesis, *Angew. Chem. Int. Ed. Engl.*, 21, 810–819, 1982; (b) Domling, A, The discovery of new isocyanide-based multi-component reactions, *Curr. Opin. Chem. Biol.*, 4, 318–323, 2000; (c) Domling, A, Ugi, I, Multicomponent reactions with isocyanides, *Angew. Chem. Int. Ed.*, 39, 3168–3210, 2000.
33. Sutherlin, D P, Stark, T M, Hughes, R, Armstrong, R W, Generation of *C*-glycoside peptide ligands for cell surface carbohydrate receptors using a four-component condensation on solid support, *J. Org. Chem.*, 61, 8350–8354, 1996.
34. Park, W K C, Auer, M, Jaksche, H, Wong, C-H, Rapid combinatorial synthesis of aminoglycoside antibiotic mimetics: Use of a polyethylene glycol-linked amine and a neamine-derived aldehyde in multiple component condensation as a strategy for the discovery of new inhibitors of the HIV RNA Rev responsive element, *J. Am. Chem. Soc.*, 118, 10150–10155, 1996.
35. Oertel, K, Zech, G, Kunz, H, Stereoselective combinatorial Ugi-multicomponent synthesis on solid phase, *Angew. Chem. Int. Ed.*, 39, 1431–1433, 2000.
36. Nilsson, U J, Fournier, E J-L, Hindsgaul, O, Solid-phase extraction on C18 silica as a purification strategy in the solution synthesis of a 1-thio-α-galactopyranoside library, *Bioorg. Med. Chem.*, 6, 1563–1575, 1998.
37. Hummel, G, Hindsgaul, O, Solid-phase synthesis of thio-oligosaccharides, *Angew. Chem. Int. Ed.*, 38, 1782–1784, 1999.
38. (a) Bols, M, 1-Aza sugars, apparent transition state analogues of equatorial glycoside formation/cleavage, *Acc. Chem. Res.*, 31, 1–8, 1998, and references cited therein; (b) Byrgesen, E, Nielsen, J, Willert, M, Bols, M, Combinatorial chemistry of piperidine based carbohydrate mimics, *Tetrahedron Lett.*, 38, 5697–5700, 1997; (c) Lohse, A, Jensen, K B, Lundgren, K, Bols, M, Synthesis and deconvolution of the first combinatorial library of glycosidase inhibitors, *Bioorg. Med. Chem.*, 7, 1965–1971,
1999.
39. (a) Meldal, M, St. Hilaire, P M, Synthetic methods of glycopeptide assembly, and biological analysis of glycopeptide products, *Curr. Opin. Chem. Biol.*, 1, 552–563, 1997; (b) St. Hilaire, P M, Lowary, T L, Meldal, M, Bock, K, Oligosaccharide mimetics obtained by novel, rapid screening of carboxylic acid encoded glycopeptide libraries, *J. Am. Chem. Soc.*, 120, 13312–13320, 1998.

40. Lampe, T F J, Weitz-Schmidt, G, Wong, C-H, Parallel synthesis of sialyl Lewis X mimetics on a solid phase: Access to a library of fucopeptides, *Angew. Chem. Int. Ed.*, 37, 1707–1710, 1998.
41. van Well, R M, Overkleeft, H S, Overhand, M, Vang Carstenen, E, van der Marel, G A, van Boom, J H, Parallel synthesis of cyclic sugar amino acid/amino acid hybrid molecules, *Tetrahedron Lett.*, 41, 9331–9335, 2000.
42. (a) Kutterer, K M K, Barnes, M L, Arya, P, Automated, solid-phase synthesis of *C*-neoglycopeptides: Coupling of glycosyl derivatives to resin-bound peptides, *J. Comb. Chem.*, 1, 28–31, 1999; (b) Arya, P, Kutterer, K M K, Barkley, A, Glycomimetics: A programmed approach toward neoglycopeptide libraries, *J. Comb. Chem.*, 2, 120–126, 2000; (c) Arya, P, Barkley, A, Randell, K D, Automated high-throughput synthesis of artificial glycopeptides. Small-molecule probes for chemical glycobiology, *J. Comb. Chem.*, 4, 193–198, 2002.
43. (a) Ellgarrd, L, Molinari, M, Helenius, A, Setting the standards: Quality control in the secretory pathway, *Science*, 286, 1882–1888, 1999; (b) Zapun, A, Jakob, C A, Thomas, D Y, Bergeron, J J M, Protein folding in a specialized compartment: The endoplasmic reticulum, *Structure*, R173–R182, 1999.
44. McNeil, M, In *Genetics of Bacterial Polysaccharides*, Goldberg, J B, Ed., CRC Press, Boca Raton, FL, 1999, pp. 207–223.
45. Arya, P, Barkley A, Randell, K D, Unpublished results.

15 Glycopeptides

Mickael Mogemark and Jan Kihlberg

CONTENTS

15.1 STRUCTURES AND BIOLOGICAL FUNCTIONS OF PROTEIN-LINKED CARBOHYDRATES

The majority of all proteins are post-translationally modified, glycosylation being the most common modification. Therefore, almost all eukaryotic, some bacterial and many viral proteins carry carbohydrates that can be divided into two principal groups. In *N*-glycosides the saccharides are covalently attached to the side-chain of asparagine, whereas the hydroxyl group in amino acids such as serine, threonine, hydroxylysine and tyrosine are glycosylated in the structurally more diverse *O*-linked glycoproteins [1–3].

In nature, the majority of *N*-linked glycoproteins are assembled by posttranslational enzymatic transfer of a common oligosaccharide, having the composition $Glc_3Man_9GlcNAc_2$, from dolichol phosphate to asparagine side chains in the protein [4]. Only asparagine residues in the consensus sequence Asn-Xaa-Ser/Thr, where Xaa is any amino acid except proline, can be glycosylated [5]. After being transferred, the oligosaccharide is trimmed to a pentasaccharide by various glycosyl hydrolases (glycosidases) and then additional glycosylations give rise to larger *N*-linked carbohydrate chains of a few different types (cf. **1–3**, Figure 15.1) [6]. As a consequence, *N*-linked glycoproteins obtained through this biosynthetic pathway have a common pentasaccharide core [Manα6(Manα3)Manβ4GlcNAcβ4GlcNAc] in which the chitobiose moiety (GlcNAcβ4Glc-NAc) is bound to asparagine. During recent years, linkages between asparagine and some other monosaccharides such as D-glucose [7], D-*N*-acetylgalactosamine [8] and L-rhamnose [9] have been found in bacterial glycoproteins.

In contrast to the large *N*-linked oligosaccharides, *O*-linked glycoproteins are built up by sequential attachment of monosaccharides by different enzymes to hydroxylated amino acids in the protein [4]. Therefore no single, common core exists for *O*-glycoproteins. In mucins, glycosylation of serine and threonine with an α-linked D-*N*-acetylgalactosamine residue, followed by subsequent glycosylation gives several different core structures (cf. **4–9**, Figure 15.2), which in turn often carry additional mono- or oligosaccharides [10,11]. Mucins are an important and abundant class of extensively glycosylated proteins that are expressed on the surface of epithelial cells, or secreted

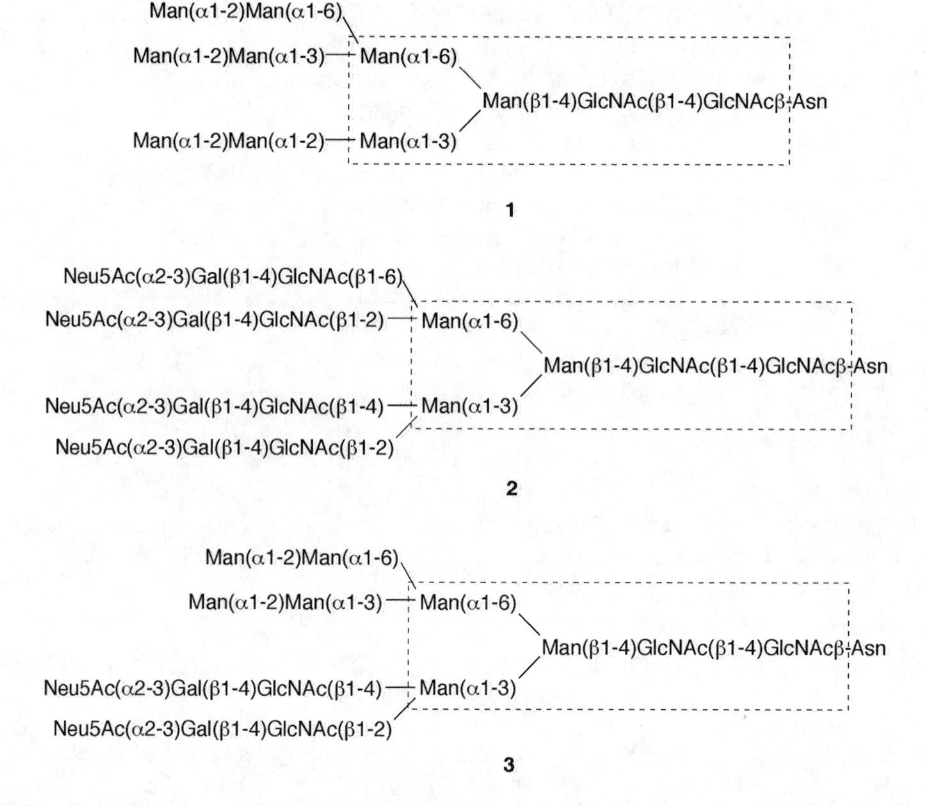

FIGURE 15.1 Three examples of *N*-linked carbohydrate chains found on glycoproteins: oligomannose type (**1**), *N*-acetyllactosamine type (or complex type, **2**) and hybrid type (**3**). The common pentasaccharide core is indicated.

FIGURE 15.2 Examples of *O*-linked carbohydrates found on glycoproteins. Mucin type *O*-linked carbohydrates have several different cores (cf. **4–9**) which often carry further oligosaccharides. Proteoglycans also have a core structure (**10**), on which long carbohydrate chains are assembled. Other types of *O*-glycosides are found, for example, in nuclear and cytoplasmic proteins (**11**), the collagens (**12**), proteins involved in blood coagulation (**13**), glycogenin (**14**) and proteins produced in yeast (**15**).

from these cells. In epithelial cancers, such as those affecting the breast, ovary, lung and colon, mucins display altered structures constituting potential targets for development of cancer vaccines [12–14]. A second important group of *O*-glycoproteins, the proteoglycans, contain a β-D-xylosyl residue linked to serine. In proteoglycans such as heparin, chondroitin sulfate and dermatan sulfate long carbohydrate chains are linked to the protein backbone via a trisaccharide core (**10**; Figure 15.2) [15]. Recently, it has been found that serine and threonine residues in nuclear pore proteins, transcription factors and cytoskeletal proteins may carry β-linked D-*N*-acetylglucosamine residues (**11**, Figure 15.2) [16]. This glycosylation occurs in a dynamic manner and seems to be involved in transcriptional regulation, analogous to regulation by phosphorylation. The collagens, which are structural proteins, contain hydroxylysine (Hyl) which is frequently glycosylated with β-D-galactosyl or α-D-glucosyl-(1 → 2)-β-D-galactosyl moieties (**12**, Figure 15.2) [17]. α-L-Fucosyl residues linked to serine and threonine (**13**, Figure 15.2) are rarely found in glycoproteins but occur on some essential proteins such as human factor IX, in various fibrinolytic and coagulation proteins [18] and in an insect neuropeptide [19]. Glucose linked to tyrosine (**14**, Figure 15.2) occurs in glycogenin (α-linked) and also in crystalline surface layers of *Clostridium thermohydrosulfuricum* (β-linked) [20,21]. Many glycoproteins produced by yeasts have α-D-mannosyl residues linked to serine and threonine (**15**, Figure 15.2) [22]. This type of glycosylation may also occur on recombinant human proteins produced in yeasts, even though it is not found in the native human proteins [23].

As revealed in the preceding paragraphs, glycoproteins display a multitude of different glycosylation patterns. Owing to the biosynthetic pathways, the carbohydrate moieties of glycoproteins are microheterogeneous and the resulting glycoforms thereby provide additional structural diversity. During the last decades the spectrum of protein-bound carbohydrates' biological roles has slowly began to emerge (reviewed in Refs. [1,3,24,25]). In general, the biological functions of protein glycosylation can be divided into two classes. The first class consists of the structural and modulatory properties provided by glycans, whereas the second involves specific interactions between the carbohydrate moieties and proteins. The latter class may be further divided into events of recognition by protein receptors within the same organism, and those resulting from recognition by exogenous agents, such as toxins, viruses and bacteria.

Protein-linked glycans are now known to have many protective, stabilizing and organizational functions. Importantly, glycans are involved in providing the correct folding for newly synthesized proteins in the endoplasmic reticulum (ER), and also in subsequent maintenance of protein solubility and conformation [24]. Consequently, proteins with incorrect glycosylation may not fold correctly and may fail to exit the ER. Because the glycans are located on the external parts of glycoproteins they may serve to protect the adjacent parts of the protein from enzymatic proteolysis or recognition by antibodies. It is also known that glycosylation of matrix proteins, such as collagens and proteoglycans, is important for maintenance of tissue structure, porosity and integrity.

The above functions are obviously critical for the function of the organism, but they do not explain the large structural variety found among protein-linked glycans. This may instead reflect more specific functions, a number of which have been elucidated to date. Some of the first examples of events relying on specific recognition of glycans concern intracellular trafficking, clearance and turnover of glycoproteins. Phosphorylation of mannose residues in oligomannose-type *N*-linked glycans is now known to be required for selective transport of catabolic enzymes from the ER to the lysosomes, where the final degradation of most cellular macromolecules takes place [26]. Most serum glycoproteins contain terminal sialic acid residues, which may be lost during aging resulting in exposure of underlying, non-reducing galactose or galactosamine. This results in a rapid Gal/GalNAc specific uptake from serum by receptors found in the liver, thereby allowing clearance of aged glycoproteins from the circulation [27]. Another well established example of glycan function is their role as blood group determinants, which are recognized by antibodies in different blood groups. Specific antibody interactions may also be of importance in immune surveillance of malignant cells, for instance, epithelial cells which display mucins with altered, tumor-specific carbohydrate moieties [12]. Critical functions of leukocytes, for instance in inflammatory responses, are mediated by interactions between members of the selectins found on leukocytes and carbohydrate ligands, for example, the tetrasaccharide sialyl Lewis^x found in *O*-linked glycoproteins on epithelial cells [28]. Recently, T cell recognition of glycopeptide fragments from type II collagen, which contained β-galactosylated hydroxylysine, was found to be of importance in mouse models for rheumatoid arthritis as well as in a group of severely ill patients [29–31].

It is well known that a variety of viruses, bacteria and parasites use glycoproteins as points of attachment and that this step is critical in pathogenesis [32]. One of the most studied examples is the hemagglutinin of influenza viruses, which bind to terminal, nonreducing sialic acid residues in glycoproteins. Influenza viruses that infect humans are known to bind to α2,6-linked sialic acid moieties whereas those that infect animals recognize α2,3-linked sialic acid. Another example concerns development of urinary tract infections by *Escherichia coli* that produce type I fimbriae. The fimbriae contain protein adhesins, which mediate binding to mannose residues of glycoproteins found on the surface of eptithelial cells in the urinary bladder [33].

Isolation of well-defined glycoproteins or glycopeptides from natural sources is difficult or even impossible. Efforts to understand the role of glycosylation of proteins, or to utilize glycopeptides as tools in drug discovery and drug design, therefore rely on synthesis to a significant extent. Fortunately, considerable progress in development of methodology for synthesis of glycopeptides has been made during recent decades (reviewed in Refs. [34–38]). This chapter attempts to summarize the overall principles, as well as current knowledge on the synthesis of different types of glycopeptides. First some general strategic considerations, and an overview of suitable protecting groups, will be given. Then methods for stereoselective attachment of carbohydrates to amino acids are discussed, with particular emphasis being put on the preparation of building blocks for use in solid-phase glycopeptide synthesis based on the 9-fluorenylmethoxycarbonyl (Fmoc) protective group strategy. Finally, chemoenzymatic approaches to glycopeptides and also synthesis of glycoproteins will be discussed briefly.

15.2 GENERAL ASPECTS OF GLYCOPEPTIDE SYNTHESIS

15.2.1 Strategic Considerations

Glycopeptides can, in principle, be prepared by two different approaches. The first one consists of attachment of the carbohydrate moiety to the target peptide after the peptide has been assembled, while the second relies on use of glycosylated amino acids as building blocks together with amino acids in stepwise peptide synthesis. Condensations between glycosyl amines and peptides containing aspartic or glutamic acid residues, according to the first approach, have been accomplished in syntheses of *N*-linked glycopeptides [39–41]. However, glycosylamines often undergo anomerization [42] resulting in formation of the undesired α-glycosides, which may be most difficult to remove when preparing glycopeptides. Attempts towards direct *O*-glycosylation of serine and threonine residues in peptides have, in general, resulted in only limited success [43–45]. This has been suggested to be due to the low solubility of peptides under conditions used for glycoside synthesis. Recently, encouraging progress was made for glycosylation of serine residues in simple model peptides attached to a solid phase, but it remains to be seen if the approach is compatible with more complex peptides [46]. At present, the use of glycosylated amino acids as building blocks in stepwise assembly therefore constitutes a more reliable and efficient approach for synthesis of both *N*- and *O*-linked glycopeptides than attempts to attach carbohydrates directly to a peptide.

As for peptides, yields obtained in solution synthesis of glycopeptides are often only modest and isolation of intermediates usually makes the approach cumbersome when performed on a small scale. A way to circumvent this problem was suggested in a recent report which describes the use of unprotected oligosaccharide moieties to facilitate workup by mediating precipitation in solution-phase glycopeptide synthesis [47]. However, solid-phase synthesis does currently constitute the most reliable and efficient method for preparation of glycopeptides (reviewed in Refs. [34–38]). Consequently, efficient synthetic routes to glycosylated amino acids for use as building blocks in solid-phase synthesis are of central importance in research dealing with preparation of glycopeptides.

15.2.2 Selection of Protecting-Groups

When choosing protecting groups for glycosylated amino acids the additional complexity of glycopeptides, as compared to ordinary peptides, must be taken into consideration. Importantly, glycosidic bonds are labile towards strong acids and glycopeptides may undergo side-reactions, such as β-elimination of glycosylated serine or threonine residues and epimerization of peptide stereocenters upon treatment with a strong base (Figure 15.3).

15.2.2.1 Protection of the α-Amino Group

In the *tert*-butoxycarbonyl (Boc) strategy for solid-phase peptide synthesis, cleavage from the solid-phase and side-chain deprotection is performed with strong acids such as hydrogen fluoride. Treatment of a glycopeptide with strong acid results in partial or, in most cases, complete cleavage

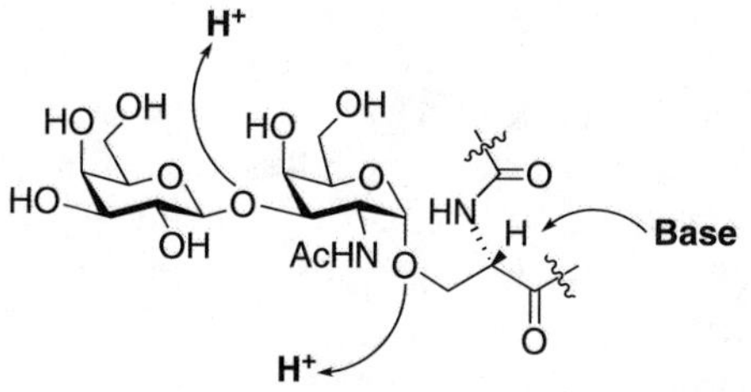

FIGURE 15.3 *O*-Glycosides are cleaved by strong acids and *O*-linked derivatives of serine and threonine undergo base-catalyzed β-elimination.

of *O*-glycosidic bonds and the Boc approach is therefore less suitable for solid-phase glycopeptide synthesis. Instead, the use of the 9-fluorenylmethoxycarbonyl (Fmoc) group [48], which is removed by weak bases such as piperidine, morpholine or 1,7-diaza-[5,4,0]-bicycloundec-1,8-ene (DBU), provides a more versatile alternative (Figure 15.4). In the Fmoc strategy, cleavage from the solid-phase and simultaneous deprotection of amino acid side-chains, after assembly of the glycopeptide, is usually carried out with the moderately strong trifluoroacetic acid (TFA). As discussed in Section 15.2.2.3, the glycosidic bonds of most saccharides are stable towards TFA, but it may be necessary to use acyl protecting groups to obtain sufficient stability for some saccharides. Both morpholine [34] and piperidine have been used extensively for Fmoc removal in glycopeptide synthesis and they do not cause β-elimination of saccharide moieties linked to serine or threonine. Piperidine, which is more basic than morpholine, gives a faster and more efficient Fmoc cleavage. Consequently, use of piperidine has been found to give improved purity in synthesis of glycopeptides with difficult sequences [49]. Numerous synthetic examples have now established that the Fmoc protecting group protocol is sufficiently mild to allow efficient and reliable solid-phase assembly of glycopeptides.

15.2.2.2 Protection of the α-Carboxyl Group

Glycosylated amino acids for use in glycopeptide synthesis have predominantly carried allyl, benzyl, phenacyl or *tert*-butyl protecting groups for the α-carboxyl group [35,50]. These esters can be removed selectively in presence of the Fmoc group by palladium-catalyzed allyl transfer, hydrogenolysis, zinc reduction or TFA treatment, respectively. After deprotection, the liberated carboxyl group is activated before the building block is attached to a peptide. An attractive alternative takes advantage of the fact that Fmoc amino acid pentafluorophenyl esters, which are in general used as acylating agents in peptide synthesis, are stable enough to survive glycosylation and subsequent purification [35]. Protecting group manipulations are thus minimized since the

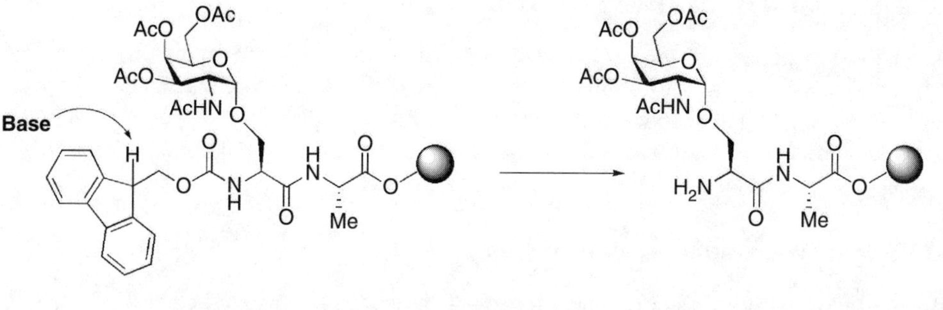

FIGURE 15.4 The Fmoc strategy, which relies on mild Fmoc cleavage with morpholine or piperidine, is commonly employed for synthesis of glycopeptides.

pentafluorophenyl ester plays a dual role, first as a protecting group and then as an acylating agent. Protecting group manipulations can be further decreased by direct *O*-glycosylation of Fmoc amino acids having unprotected carboxyl groups [51,52]. This method has been used to prepare numerous 1,2-trans-*O*-glycosylated Fmoc amino acids as well as 1,2-*cis*-linked glycosides of fucose.

15.2.2.3 Protection of the Carbohydrate Hydroxyl Groups

Acetyl and benzoyl groups are often used for protection of the carbohydrate hydroxyls in glycosylated amino acids. Because of their electron-withdrawing nature they have been advocated in order to stabilize *O*-glycosidic bonds during the acid catalyzed cleavage from the solid phase. It has, however, been shown that *O*-glycosidic linkages of common saccharides, such as glucose, galactose, mannose, *N*-acetylglucosamine and *N*-acetylgalactosamine, survive treatment with TFA for a limited period of time (≤ 2 h) in the absence of acyl protection from the hydroxyl groups [41,53,54]. Acyl protection of these saccharides is therefore not necessary, but it should be kept in mind that prolonged treatment with acid, especially in the presence of nucleophiles such as water, may result in decomposition of unprotected oligosaccharide moieties. In contrast, glycopeptides containing the 6-deoxysugar L-fucose, which undergoes acid catalyzed hydrolysis five to six times faster than the corresponding nondeoxygenated monosaccharide galactose, have been shown to require stabilizing acyl protecting groups during cleavage with TFA [55,56]. Thus, even though acyl protection of carbohydrate hydroxyl groups is not an absolute necessity, it can be considered as a suitable precaution in order to avoid degradation during cleavage from the solid phase. Furthermore, in many cases the synthetic route to the glycosylated amino acid automatically provides the acyl protected derivatives.

It has been emphasized that side-reactions, such as β-elimination and epimerization of peptide stereocenters, could be encountered during base-mediated removal of *O*-acyl protecting groups from glycopeptides [34]. When investigated in a model study, conditions which are in common use for deacetylation of glycopeptides caused neither β-elimination nor epimerization of the *O*-linked model glycopeptide **16** (Scheme 15.1) [57]. Therefore, with very few exceptions [56,58], these concerns appear to be unwarranted for removal of acetyl groups, unless the α-amino group of the glycosylated serine or threonine is *N*-methylated which increases the rate of β-elimination [59]. In contrast, benzoyl groups require more drastic conditions for their removal, which when applied to **16** resulted in both β-elimination (→**18**) and some epimerization (→**19** and **20**; Scheme 15.1) [57], suggesting that it may be desirable to avoid benzoyl protection of glycopeptides [58,60,61].

Recently, the use of fluorobenzoyl protecting groups instead of ordinary benzoates was found to suppress β-elimination in synthesis of *O*-linked glycopeptides such as **24**, thereby providing an alternative to benzoates (Scheme 15.2) [62]. According to this study, fluorobenzoyl groups appear to combine the advantages of the benzoyl group in formation of glycosidic bonds (i.e., high stereoselectivity and low levels of orthoester formation) with the ease of removal characteristic of the acetyl group.

Acid-labile protecting groups constitute an alternative to acyl protection for glycosylated amino acids. Several such groups including trimethylsilyl (TMS), *tert*-butyldimethylsilyl (TBDMS), *tert*-butyldiphenylsilyl (TBDPS), *p*-methoxybenzyl and isopropylidene have also been used in syntheses of both *N*- and *O*-linked building blocks [29,63–67]. When used in glycopeptide synthesis, these protecting groups were rapidly and completely removed simultaneously with TFA-catalyzed cleavage from the solid phase, thereby eliminating the need for separate deprotection of the carbohydrate moiety. It should be pointed out that acid-labile protecting groups cannot be used for fucosides due to extensive hydrolysis of the latter on cleavage with TFA [55,56].

Benzyl ethers are often employed as protecting groups in oligosaccharide synthesis and they have also been used in solid-phase glycopeptide synthesis [68,69]. Benzyl ethers are usually removed by hydrogenolysis, which is incompatible with glycopeptides containing cysteine, methionine and to some extent tryptophan [70]. Moreover, TFA catalyzed cleavage of

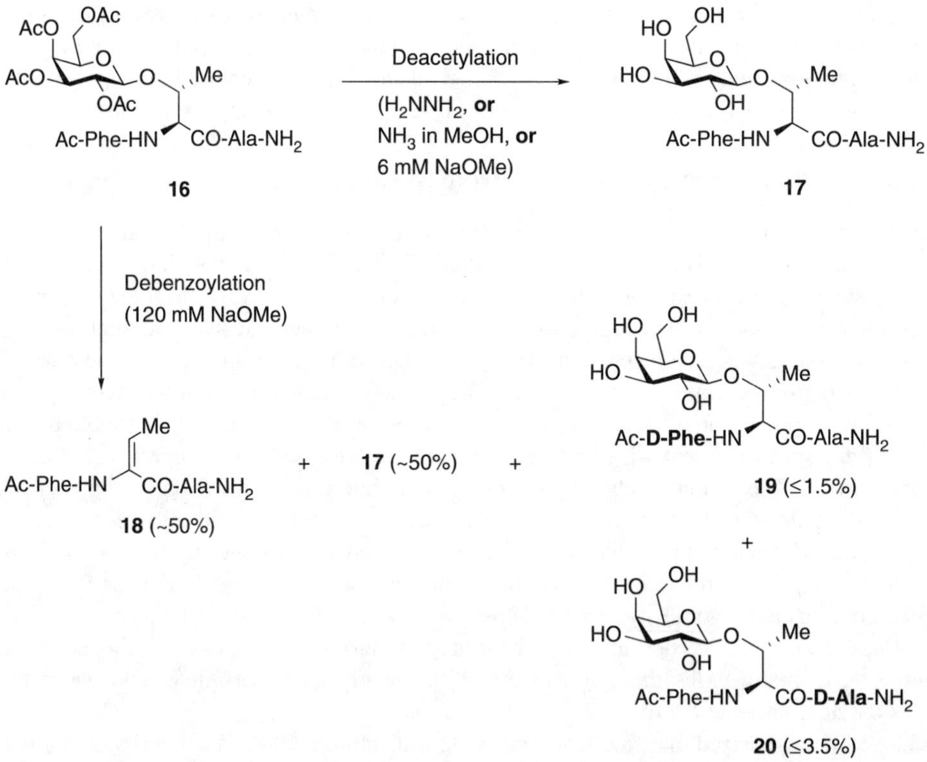

SCHEME 15.1 Deacetylation of model glycopeptide **16** caused neither β-elimination nor epimerization of peptide stereocenters. Both types of side-reactions were, however, encountered under more drastic conditions often employed for removal of benzoyl groups.

O-benzylated glycopeptides from the solid phase was found to result in formation of several partially de-*O*-benzylated products, thereby complicating the final debenzylation and purification. However, one report suggests that this problem may be circumvented by performing the deprotection in two acid-catalyzed steps [71].

15.2.3 Practical Aspects of Solid-Phase Synthesis

As indicated in the preceding sections, glycopeptides can be synthesized on solid phase by the Fmoc strategy under conditions identical to those employed for synthesis of peptides [72]. Consequently, linkers and resins used in Fmoc peptide synthesis can usually be employed without problems in the synthesis of glycopeptides. Glycopeptides having a C-terminal carboxylic acid are often prepared directly on a 4-alkoxybenzyl alcohol resin [73] or via a 4-hydroxymethylphenoxyacetic acid linker, both of which are cleaved with TFA [74]. The TFA labile Rink linker, *p*-[α-(fluoren-9-ylmethoxyformamido)-2,4-dimethoxybenzyl]phenoxyacetic acid, is suitable for synthesis of *C*-terminal amides [75]. An allylic linker termed HYCRON, which allows Pd(0) catalyzed cleavage under almost neutral conditions, has also found use in synthesis of glycopeptides [76]. This linker allows the synthesis of glycopeptides which do not withstand the basic and acidic conditions normally employed in the Fmoc-strategy.

The carbohydrate moiety of a glycosylated amino acid could, potentially, impose steric hindrance in stepwise glycopeptide synthesis, thereby reducing coupling efficiency and hampering

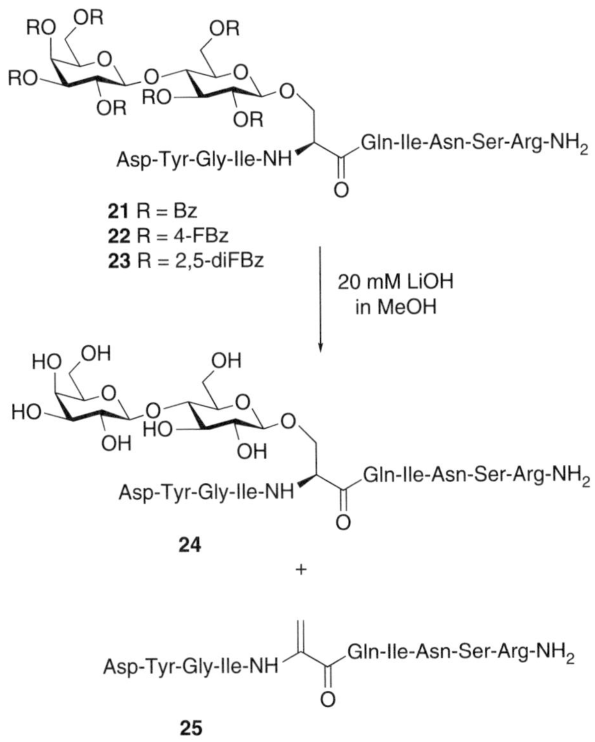

21 R = Bz
22 R = 4-FBz
23 R = 2,5-diFBz

20 mM LiOH in MeOH

24

+

25

R	reaction time	amount of 25
Bz	8 h	~50%
4-F-Bz	8 h	~40%
2,5-diF-Bz	0.5 h	0%

SCHEME 15.2 Influence of the type of benzoyl group on the extent of β-elimination during base-catalyzed deacylation.

further elongation of the peptide chain. However, an overview of glycopeptides prepared to date shows that no such problems exist. This conclusion is independent of both the amino acid and the carbohydrate moiety, as revealed by the fact that amino acids which carry oligosaccharides as large as undecasaccharides have been found to couple as well as ordinary nonglycosylated amino acids [77].

Large efforts are usually invested in the preparation of glycosylated amino acids and the smallest possible excess should therefore be used in each coupling. Fortunately, as little as 1 to 1.5 equivalents of glycosylated amino acids, relative to the capacity of the peptide resin, can be used provided that the minimum amount of solvent is employed [64,78]. Even though any method for activation and coupling used in the synthesis of peptides should also be suitable for a glycosylated amino acid, two methods have predominantly been used: the *N,N'*-diisopropylcarbodiimide/1-hydroxybenzotriazole method and the pentafluorophenyl (Pfp) ester method [35,79]. More recently, 1-hydroxybenzotriazole has begun to be replaced with the more potent additive, 1-hydroxy-7-azabenzotriazole [80]. When a minimal amount of glycosylated amino acid is being coupled, monitoring of the coupling becomes important. This can be performed visually, or in a spectrophotometer using either 3,4-dihydro-3-hydroxy-4-oxo-1,2,3-benzotriazine (Dhbt-OH) [81] or Bromophenol Blue [82] as indicators of unreacted amino groups on the solid phase.

15.3 SYNTHESIS OF *O*-LINKED GLYCOPEPTIDES

15.3.1 1,2-*TRANS*-*O*-LINKED GLYCOPEPTIDES

Stereoselectivity is usually ensured through neighboring-group participation in the synthesis of 1,2-trans-*O*-linked glycosylated amino acids (Scheme 15.3). This means that acetyl or benzoyl protecting groups are employed at O-2 of glycosyl donors, such as the glucose and mannose derivatives **26** and **31**. Derivatives of *N*-acetylglucosamine (e.g., **27**) appear to have such a participating group in place at C-2. However, low yields are usually obtained due to the low reactivity of oxazoline derivative **30**, which is formed during the glycosylation. Therefore, temporary amino protecting-groups are often employed when building blocks corresponding to β-D-GlcNAc-(1→O)-Ser/Thr are prepared (cf. Section 15.3.1.5). Synthesis of α-mannosylated serine and threonine derivatives such as **32** is often straightforward since α-selectivity is favored both by the use of a participating group at O-2 and by the anomeric effect.

15.3.1.1 The β-D-Xyl-(1→O)-Ser Linkage

Proteoglycans such as heparin, heparan sulfate, chondroitin and dermatan sulfate all have large glycosaminoglycan chains attached to a protein core [15]. A tetrasaccharide, β-D-GlcA-(1→3)-β-D-Gal-(1→3)-β-D-Gal-(1→4)-β-D-Xyl, common to several proteoglycans joins the polysaccharide to serine residues in the protein. Sulfation of the two galactose residues, as well as phosphorylation of the xylose moiety, in the tetrasaccharide increases the complexity of proteoglycans further. In an early study, galactosyl donor **33** was coupled to xyloside **34**, which gave disaccharide **35** after some further synthetic transformations (Scheme 15.4) [83].

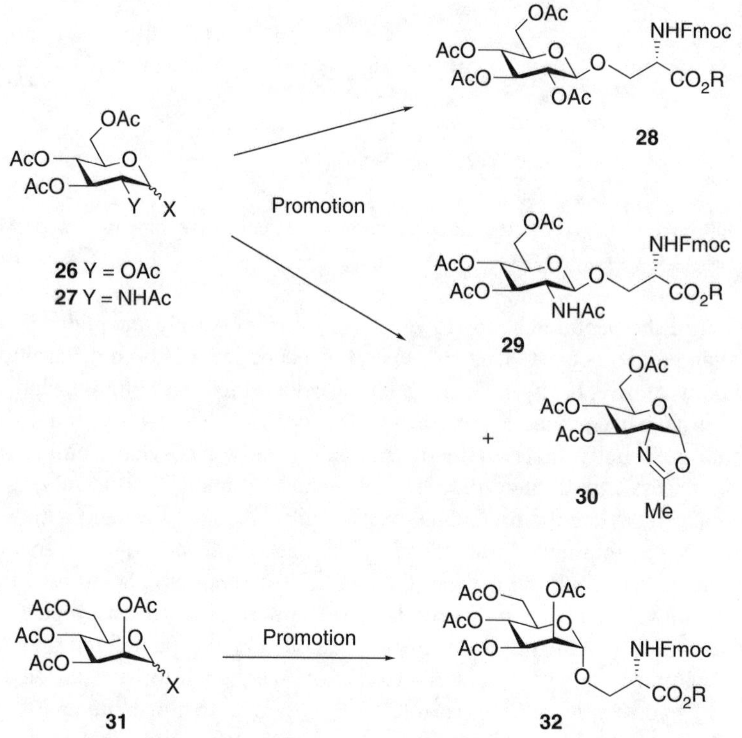

OAc AcO AcO O Y X

26 Y = OAc
27 Y = NHAc

Promotion

OAc AcO AcO O O NHFmoc CO_2R OAc
28

OAc AcO AcO O O NHFmoc CO_2R NHAc
29

\+ OAc AcO AcO O N O Me
30

AcO OAc AcO AcO O X
31

Promotion

AcO OAc AcO AcO O O NHFmoc CO_2R
32

SCHEME 15.3 1,2-*trans*-*O*-Linked glycosylated amino acids are often prepared using a participating protecting group at C-2 of the glycosyl donor. However, oxazoline formation (cf. **30**) usually leads to low yields for *N*-acetylglucosamine.

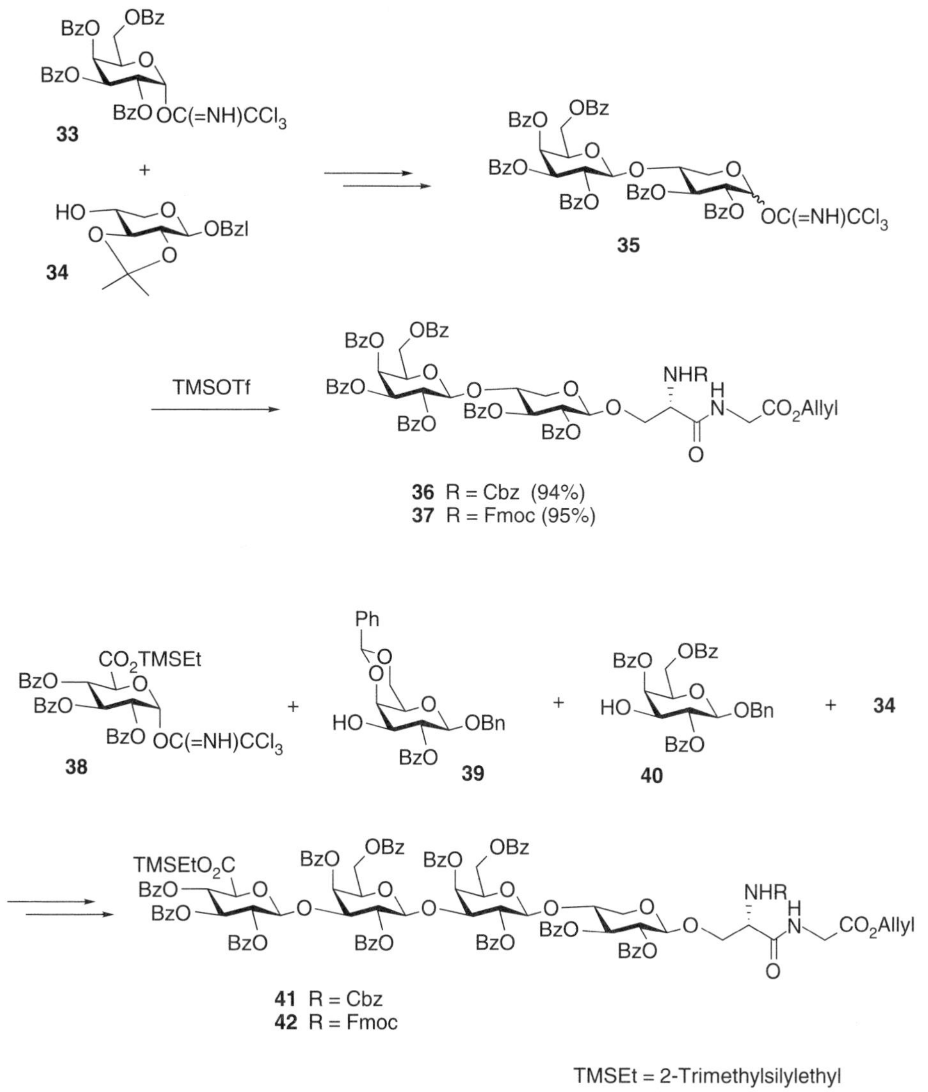

SCHEME 15.4 Glycosylation of a Ser-Gly dipeptide using trichloroacetimidate donors; installation of the 1,2-trans linkages between xylose and serine.

Using trimethylsilyl triflate as a promoter trichloroacetimidate **35** was then coupled to protected Ser–Gly dipeptides to give **36** and **37** in excellent yields. Deprotection of **36** and **37** at the *N*- or *C*-terminus then allowed construction of partially or fully glycosylated proteoglycan oligopeptides in solution. This synthetic approach was later extended so that monosaccharide building blocks **38**, **39**, **40** and **34** were assembled into a tetrasaccharide, analogous to **35** [84]. Use of this tetrasaccharide donor allowed synthesis of glycosylated dipeptides **41** and **42**, which were employed in the synthesis of even larger glycopeptides. A further extension included sulfation at O-4 in one of the galactose residues and phosphorylation at O-2 of the xylose unit [85].

15.3.1.2 The β-D-Glc-(1→O)-Ser Linkage

Proteins involved in coagulation and fibrinolysis carry oligosaccharides consisting of xylose residues linked to serine via a β-D-glucose moiety [86]. Building block **44**, in which glucose is β-linked to serine, was obtained by boron trifluoride etherate promoted glycosylation of Fmoc serine with β-D-glucose pentaacetate (**43**, Scheme 15.5) [87]. Since both starting materials are commercially available and inexpensive, this constitutes a very simple one-step approach to **44**, even though the yield is modest. The analogous building block **46** was obtained in a higher yield by silver triflate promoted glycosylation of Fmoc serine pentafluorophenyl ester with the benzoylated glucosyl bromide **45** [61]. When compound **46** was used in the synthesis of a hexapeptide segment from blood-clotting factor IX, it was found that base-catalyzed removal of the *O*-benzoyl groups

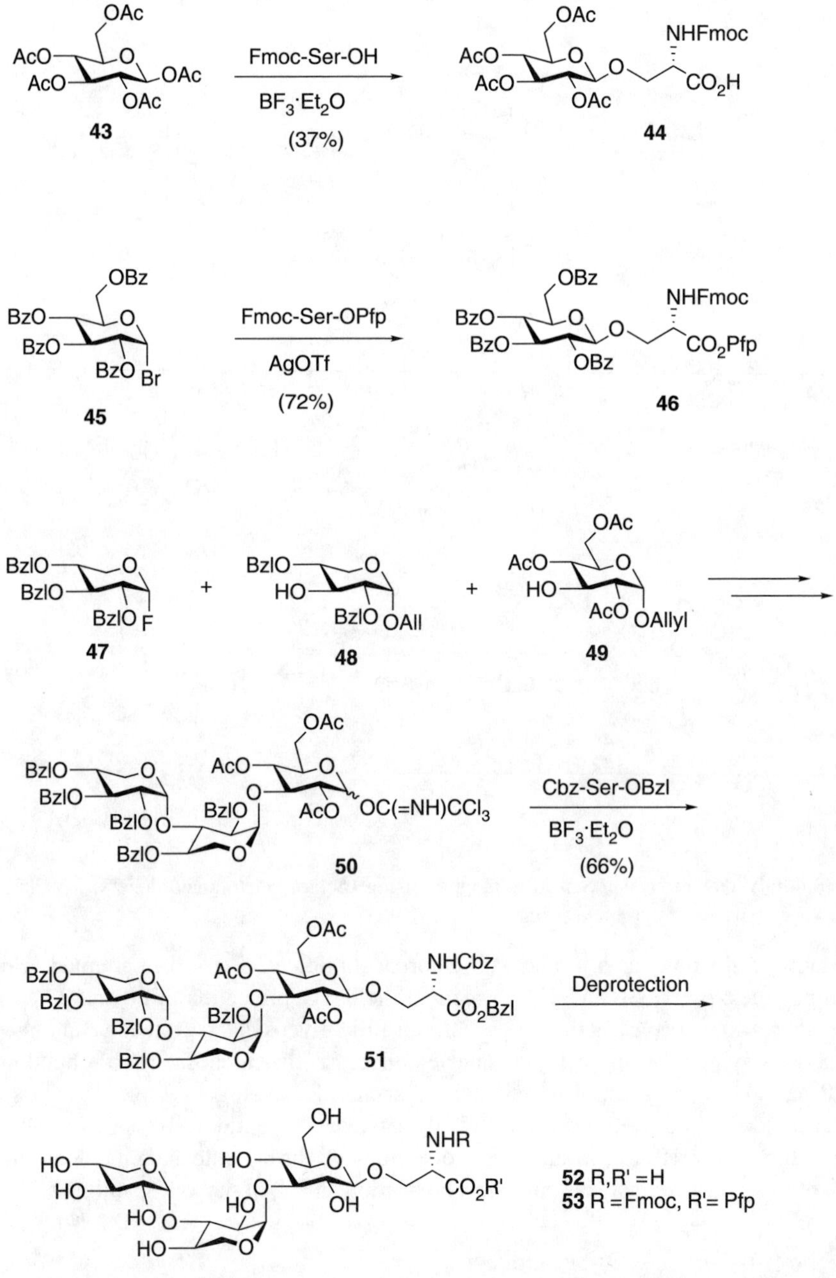

43 → Fmoc-Ser-OH, $BF_3{\cdot}Et_2O$ (37%) → **44**

45 → Fmoc-Ser-OPfp, AgOTf (72%) → **46**

47 + **48** + **49** → → **50**

50 → Cbz-Ser-OBzl, $BF_3{\cdot}Et_2O$ (66%) → **51**

51 → Deprotection → **52** R,R' = H; **53** R = Fmoc, R' = Pfp

SCHEME 15.5 β-Glucosylation of serine building blocks having the carboxyl group unprotected or protected with a pentafluorophenyl (Pfp) or a benzyl (Bzl) group.

had to be carried out under carefully controlled conditions in order to avoid β-elimination of the glucose moiety. This example illustrates that use of benzoyl groups may cause problems since rather harsh basic conditions are required for their removal (cf. discussion in Section 15.2.2.3).

Preparation of building block **53**, in which the serine carries an α-D-Xyl-(1→3)-α-D-Xyl-(1→3)-β-D-Glc trisaccharide moiety, was achieved by a more complex synthetic route [61,88]. Trisaccharide trichloroacetimidate **50** was first assembled from building blocks **47**–**49** and then linked to benzyloxycarbonyl serine benzyl ester to give **51** [88]. Since the protecting groups of the serine moiety are unsuitable for solid-phase glycopeptide synthesis, compound **51** was completely deprotected to give **52**. The serine residue was then reprotected with an Fmoc group, after which the carboxyl group was activated as a pentafluorophenyl ester. Glycosylated amino acid **52** was only available in a small quantity and the hydroxyl groups of the carbohydrate moiety were left unprotected in order to minimize the number of synthetic steps. Building block **53** was then successfully employed in glycopeptide synthesis, including TFA catalyzed cleavage from the solid phase [61]. This constitutes an early example that acyl protecting groups do not have to be used for the carbohydrate moiety of glycopeptides during synthesis and acid-catalyzed cleavage (cf. Section 15.2.2.3).

15.3.1.3 The β-D-Gal-(1→O)-Hyl Linkage

In collagen, lysine residues located in the sequence Gly-Xaa-Lys can become posttranslationally hydroxylated and then glycosylated, either with a β-D-galactosyl or an α-D-glucosyl-(1→2)-β-D-galactosyl moiety [17,89]. Galactosylated derivatives of (*5R*)-5-hydroxy-L-lysine suitable for use in solid-phase synthesis have been prepared by two efficient routes (Scheme 15.6). In the first route, protected hydroxylysine derivative **55**, prepared from hydroxylysine in four steps, was glycosylated with galactosyl bromide **54** using silver silicate as promoter [90]. Deallylation of the ester (86%) then gave a building block that was used in synthesis of a glycopeptide from type II collagen. The second route relied on sodium hydride induced condensation of **54** with copper complex **58**, which was obtained from hydroxylysine in two steps [91]. Dissociation of the copper complex and Fmoc protection of the α-amino group (50%) provided glycosylated hydroxylysine **60**, which was used in the synthesis of a peptide sequence derived from type IV collagen.

The synthesis of a hydroxylysine derivative which carries an α-D-glucosyl-(1→2)-β-D-galactosyl moiety and its use in glycopeptide synthesis has also been reported (Scheme 15.6) [66]. Treatment of hydroxylysine derivative **55** with the α-1,2-anhydrogalactose **61** in the presence of zinc chloride gave derivative **62**. In this case, predominant formation of a 1,2-*trans*-glycosidic bond was ensured by opening the epoxide in an S_N2-like manner, instead of relying on a participating acyl group at O-2 of the galactose residue. This left HO-2 of galactose unprotected, that is, ready for glycosylation which was achieved with thioglucoside **63** using *N*-iodosuccinimide and silver triflate as promoters. Recently, extensive immunological studies have been carried out with synthetic glycopeptides prepared from building blocks **57** and **65**, as well as mono-deoxygenated analogs of **57**. These studies revealed that galactosylated peptides from type II collagen were recognized by T cells obtained in mouse models for autoimmune rheumatoid arthritis [29–31] and also by T cells from a group of severely ill patients suffering from rheumatoid arthritis [31].

15.3.1.4 The α-D-Man-(1→O)-Ser/Thr Linkage

Glycoproteins produced naturally in yeasts, or recombinant proteins expressed in yeasts, may carry saccharides linked to serine and threonine by an α-D-mannose residue. Oligosaccharides linked to serine or threonine by a mannose residue have also been found in glycoproteins of the nervous system [92]. Synthesis of α-mannosyl serine and threonine derivatives is usually straightforward. For example, a participating acetyl group ensures that the desired α-linkage is obtained in high yield when mannosyl bromide **66** is coupled with benzyl esters of Fmoc serine and threonine to give

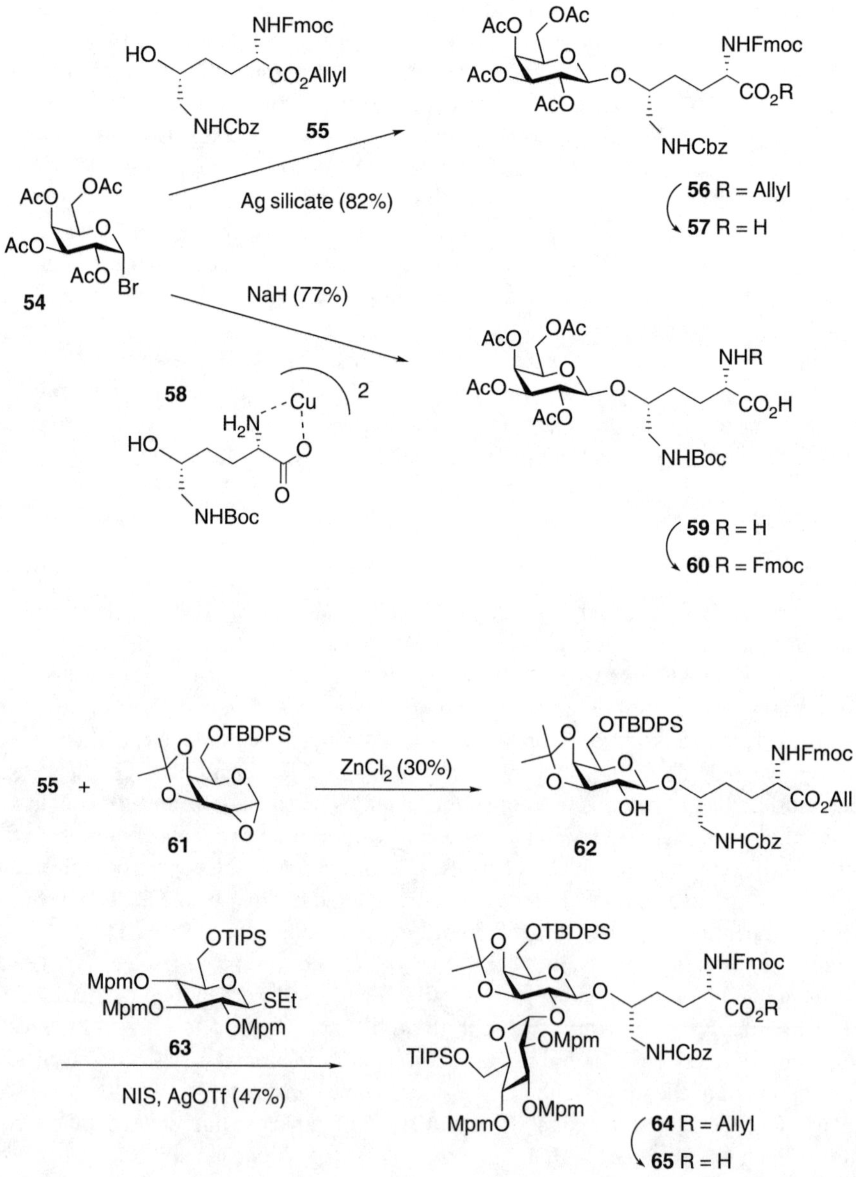

SCHEME 15.6 Construction of the 1,2-trans linkage between galactose and hydroxylysine by glycosylation of **55** with glycosyl bromide **54** using an insoluable silver silicate promoter, by nucleophilic S_N2-like displacement of **54** with copper complex **58**, or by opening of α-1,2-anhydrogalactoside **61** with **55**.

67 and **68** (Scheme 15.7) [93]. In an alternative synthesis, peracetylated mannose was used as glycosyl donor in a boron trifluoride etherate promoted glycosylation of *C*-terminally unprotected Fmoc-threonine [87]. Use of a participating group at O-2 of the mannosyl donor is not an absolute requirement since formation of α-mannosides is also favored because of the anomeric effect. The synthesis of dimannosylated threonine and serine derivatives **71** and **72** through silver triflate promoted glycosylation with dimannosyl bromide **70**, obtained from **66** and **69**, relied on the anomeric effect to provide the correct configuration at C-1 [94]. Building blocks **71** and **72** were used to prepare two glycosylated heptadecapeptide fragments from human insulin-like growth factor 1 (IGF-1). In this study the corresponding nonglycosylated fragments were also prepared,

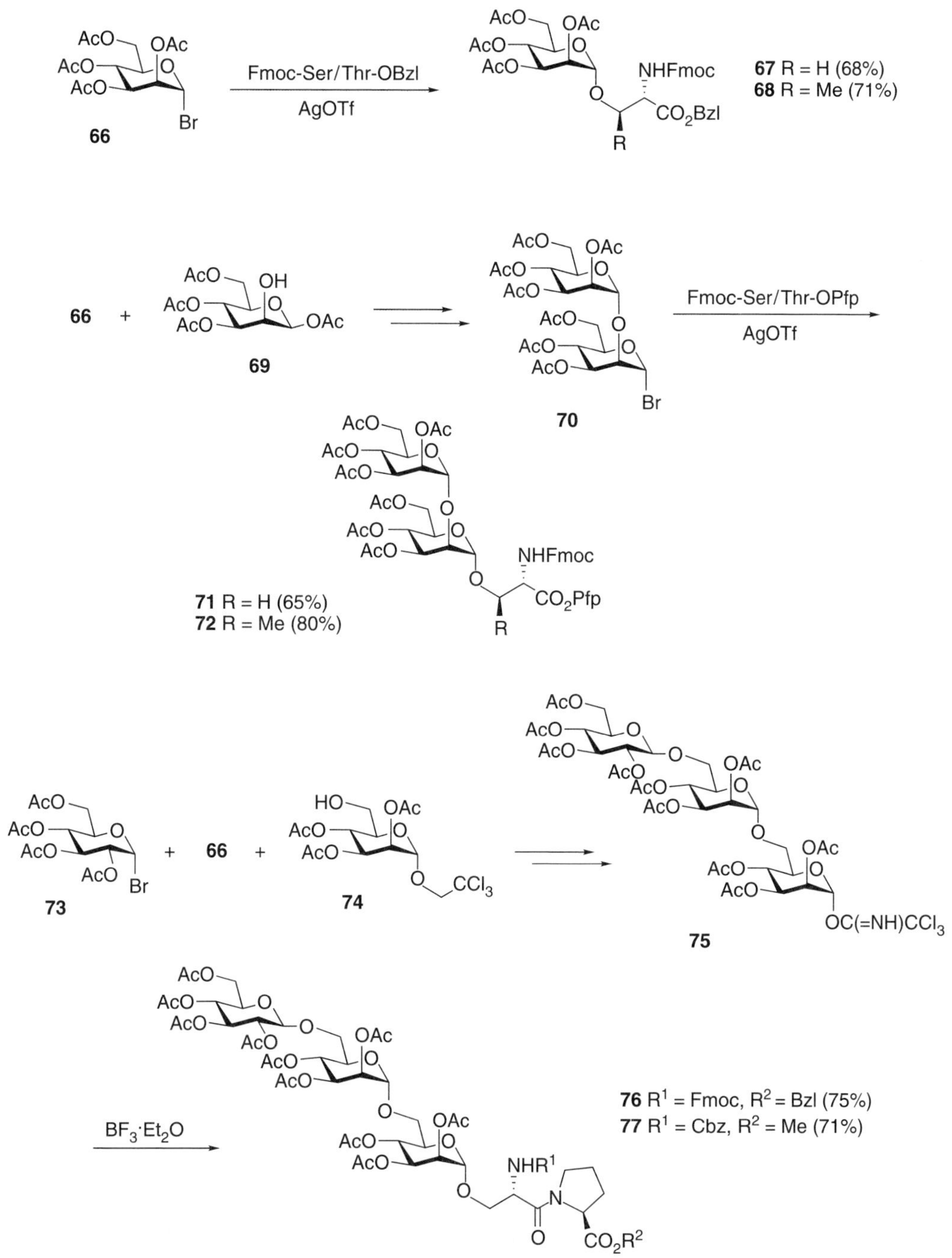

SCHEME 15.7 α-Glycosylation of serine and threonine building blocks with mannosyl donors carrying *O*-acetyl participating groups favors formation of an α-mannoside.

and comparative NMR studies showed that the sugar moiety had only a limited effect on the conformation of the peptide backbone.

Some more complex oligosaccharides with α-mannosyl serine or threonine linkages have also been prepared. For instance, triglycosyl dipeptides **76** and **77** were prepared by assembly of building blocks **73**, **66** and **74** to give trisaccharide donor **75**, followed by boron trifluoride etherate promoted coupling to appropriately protected derivatives of the dipeptide Ser-Pro [95,96]. The two dipeptides were then deprotected at their *C*- or *N*-termini, respectively, and then coupled to each other to give a hexaglycosyl tetrapeptide. This glycopeptide was intended to be used as a model compound for studies of a glycoprotein which elicits the synthesis of phytoalexins, that is, antimicrobial substances, in plants. More recently, the syntheses of derivatives of serine and threonine, which carried the tetrasaccharide α-D-Neu5Ac-(2→3)-β-D-Gal-(1→4)-β-D-GlcNAc-(1→2)-α-D-Man, were reported [97]. This *O*-linked carbohydrate is found in the membrane-associated glycoprotein α-dystroglycan, which binds to extracellular matrix components such as laminin.

15.3.1.5 The β-D-GlcNAc-(1→O)-Ser/Thr Linkage

N-Acetylglucosamine is β-linked to serine and threonine residues in many proteins, but predominantly in those found in the nucleoplasm and the cytoplasm [16]. It is believed that a dynamic equilibrium between *O*-GlcNAc glycosylation and phosphorylation of serine or threonine influences the nucleocytoplasmic transport of glycoproteins. *O*-GlcNAc-containing glycopeptides were first prepared 40 years ago, and those early attempts relied on the use of 2-acetamido-2-deoxy-3,4,6-tri-*O*-protected glucosyl halogenides as donors [98,99]. However, the poor reactivity of

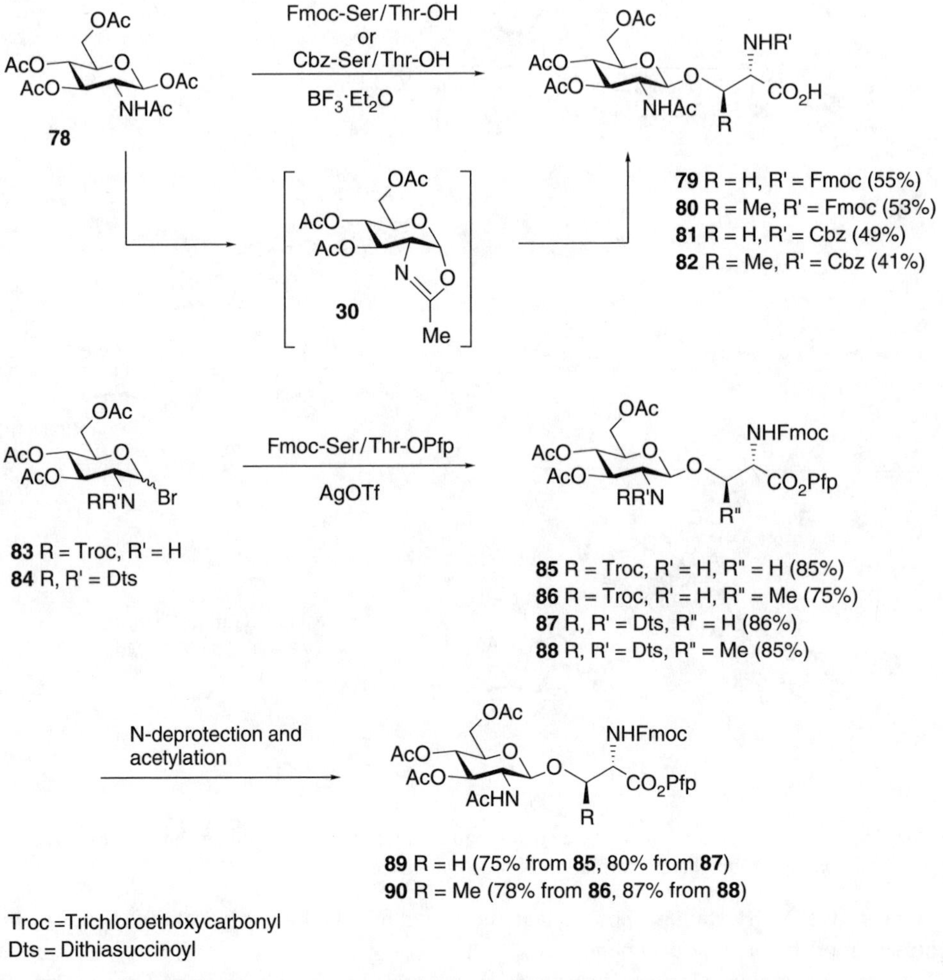

Fmoc-Ser/Thr-OH or Cbz-Ser/Thr-OH, $BF_3 \cdot Et_2O$

78 → [30] → 79–82

79 R = H, R' = Fmoc (55%)
80 R = Me, R' = Fmoc (53%)
81 R = H, R' = Cbz (49%)
82 R = Me, R' = Cbz (41%)

Fmoc-Ser/Thr-OPfp, AgOTf

83 R = Troc, R' = H
84 R, R' = Dts

85 R = Troc, R' = H, R" = H (85%)
86 R = Troc, R' = H, R" = Me (75%)
87 R, R' = Dts, R" = H (86%)
88 R, R' = Dts, R" = Me (85%)

N-deprotection and acetylation

89 R = H (75% from 85, 80% from 87)
90 R = Me (78% from 86, 87% from 88)

Troc = Trichloroethoxycarbonyl
Dts = Dithiasuccinoyl

SCHEME 15.8 β-Glycosylation of serine and threonine building blocks with peracetylated β-*N*-acetylglucosamine **78**, or with *N*-Troc/*N*-Dts protected glycosyl bromides **83** and **84**.

oxazolines such as **30** (Scheme 15.8), which are formed as intermediates during the glycosylation, caused the target glycosylated amino acids to form in low yields. Later it was found that oxazoline reactivity may be increased by the use of Lewis acid catalysts, and reaction of *N*-acetyl-β-D-glucosamine tetraacetate **78** with protected derivatives of serine and threonine using boron trifluoride etherate as a promoter allowed the preparation of building blocks **79**–**82** (Scheme 15.8) [100,101]. The yields were still modest (ca. 50%) but this is compensated by the fact that Fmoc-protected derivatives such as **79** and **80** may be used directly in solid-phase glycopeptide synthesis. Tetraacetate **78** has also been employed, under boron trifluoride etherate promotion, to glycosylate a serine derivative which was protected with the *p*-(phenylacetoxy)benzyloxycarbonyl (PhAcOZ) group [102]. The resulting building block was then employed in the synthesis of a phosphorylated and glycosylated peptide fragment from the large subunit of mammalian RNA polymerase II, and cleavage of the PhAcOZ group was performed enzymatically with penicillin G acylase.

Replacement of the *N*-acetyl group of GlcNAc with an electron-withdrawing *N*-allyloxy-carbonyl (Aloc) [103], *N*-trichloroethoxycarbonyl (Troc) [87,104–106] or *N*-dithiasuccinoyl (Dts) [107,108] protecting group reduced the extent of oxazoline formation, thus permitting highly efficient glycosylations of protected derivatives of serine and threonine. This, for instance, allowed preparation of the glycosylated amino acid pentafluorophenyl esters **85**–**88** in high yields (Scheme 15.8). Reductive cleavage of the Troc or Dts group can then be performed in the presence of acetic anhydride to give **89** and **90**, thus minimizing the number of synthetic steps and potential side-reactions. This is of particular importance with serine and threonine pentafluorophenyl esters, since these active esters react readily with an unprotected amino group on the GlcNAc moiety in an intramolecular acylation reaction. Alternatively, removal of the Dts group and subsequent acetylation can be carried out after incorporation of the protected building block in a glycopeptide [107,108]. Glycosylated amino acids **89** and **90** have been used in solid-phase synthesis of peptide fragments derived from serum response factor and RNA polymerase II [107,108].

Recently, a new method for preparation of GlcNAc containing serine and threonine building blocks was reported [109]. A protected D-glucal was, in a one-pot procedure, reacted with Cbz protected threonine methyl ester and *N*-(TMS)acetamide to give an *O*-GlcNAc-threonine building block, which was not used in further synthesis of glycopeptides.

15.3.2 1,2-*cis*-*O*-Linked Glycopeptides

1,2-*cis*-*O*-Linked glycosylated amino acids are, with some exceptions, prepared using a nonparticipating group at C-2 of the glycosyl donor (Scheme 15.9). For these glycosylated amino acids stereoselectivity is ensured by different parameters, such as the anomeric effect, which

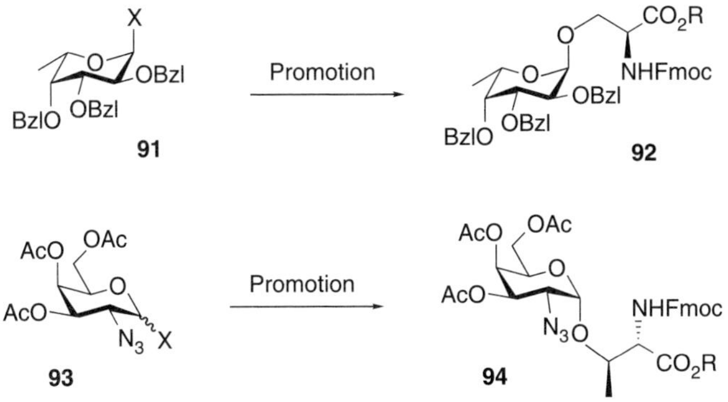

SCHEME 15.9 1,2-*cis*-*O*-Linked glycosylated amino acids are predominantly prepared using nonparticipating *O*-benzyl or azido groups at C-2 of the glycosyl donor.

influence the transition state and usually lead to the formation of a more stable α-glycosidic product. For α-fucosylated derivatives of serine and threonine (e.g., **92**) this means that benzyl ether-type protecting groups may be employed for the hydroxyl groups of fucosyl donor **91**. When 2-acetamido-2-deoxy-D-galactose is α-*O*-glycosidically bound to serine or threonine most syntheses have utilized the nonparticipating azido group at C-2 of the glycosyl donor (cf. transformation of **93** into **94**; Scheme 15.9).

15.3.2.1 The α-L-Fuc-(1→O)-Ser/Thr Linkage

Proteins involved in coagulation and fibrinolysis may carry L-fucosyl residues α-linked to serine and threonine [18]. One approach to fucosylated amino acids involved glycosylation of Fmoc-protected threonine *tert*-butyl ester with the *O*-methoxybenzylated thiofucoside **95** under DMTST promotion to give **96** (Scheme 15.10) [110]. Because of the acid-lability of the α-fucosidic linkage (e.g., to TFA) the methoxybenzyl groups then had to be replaced by acetyl groups in order to avoid decomposition during cleavage of the target glycopeptide from the solid phase. This was accomplished in three steps, by deprotection of the hydroxyl groups with ceric ammonium nitrate, followed by *O*-acetylation and cleavage of the *tert*-butyl ester with formic acid. The resulting fucosylated threonine **97** was then used in solid-phase synthesis of a 36-amino-acid peptide which is found in the insect *Locusta migratoria* and reported to be a serine protease inhibitor [110].

An analogous route, involving use of a benzylated fucosyl donor and reprotection of the hydroxyl groups with acetyl groups, has been used to prepare a fucosylated serine building block [56]. This was employed to prepare a glycopeptide fragment from human factor IX. More direct and rapid access to fucosylated building blocks has been achieved by treatment of Fmoc-protected serine and threonine with peracetylated fucose **98** under boron trifluoride etherate promotion [111]. During the reaction β-fucosides are formed under the influence of the participating acetyl group at

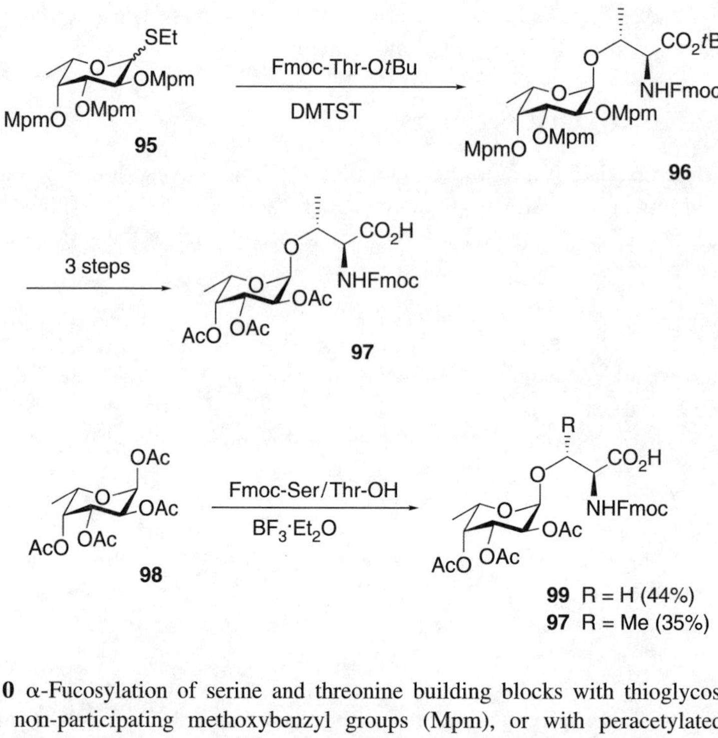

SEt
OMpm
MpmO OMpm
95
Fmoc-Thr-O*t*Bu
DMTST
CO_2tBu
NHFmoc
OMpm
MpmO OMpm
96
3 steps
CO_2H
NHFmoc
OAc
AcO OAc
97
OAc
OAc
AcO OAc
98
Fmoc-Ser/Thr-OH
$BF_3 \cdot Et_2O$
R
CO_2H
NHFmoc
OAc
AcO OAc
99 R = H (44%)
97 R = Me (35%)

SCHEME 15.10 α-Fucosylation of serine and threonine building blocks with thioglycoside **95**, which is protected with non-participating methoxybenzyl groups (Mpm), or with peracetylated **98** which, via anomerisation, gives the thermodynamically more stable α-fucosides **97** and **99**.

O-2 of **98**. Because of the lability of the glycosidic bond of *O*-fucosides these intermediates then rearranged to the thermodynamically more stable α-fucosides **99** and **97**, which were obtained in yields of 44 and 35%, respectively.

15.3.2.2 The α-D-GalNAc-(1→O)-Ser/Thr Linkage

2-Acetamido-2-deoxy-D-galactose (GalNAc) bound to serine or threonine through an α-glycosidic bond is an abundant glycosylation motif found in various *O*-linked glycoproteins, such as different types of mucins. Biosynthesis of mucins is initiated by the addition of *N*-acetylgalactosamine to serine and threonine residues in the mucin polypeptide backbone by a family of *N*-acetylgalactosaminyltransferases, thus giving the Tn antigen (α-D-GalNAc-Ser/Thr) [112]. Galactosyl residues are then added to the *N*-acetylgalactosamine moiety to form the type 1 core β-D-Gal-(1→3)-α-D-GalNAc-Ser/Thr (also called the T antigen). In normal cells the type 1 core acts as a substrate for core 2 β1,6GlcNAc transferases so that the type 2 core trisaccharide {β-D-Gal-(1→3)-[β-D-GlcNAc-(1→6)]-GalNAc} is formed [11,12]. Finally, oligo *N*-acetyllactosamine moieties are added, after which addition of sialic acid or fucose residues terminates the growth of the mucin-linked oligosaccharide structure. In contrast, in several different epithelial cancers, such as those affecting the breast, ovary, lung and colon, low expression of core 2 β1,6GlcNAc transferases together with increased levels of sialyltransferases result in mucins that display simpler *O*-linked carbohydrates [11,12]. These include the Tn and T antigens, as well as their mono- and disialylated derivatives. The observations of these tumor-associated carbohydrate antigens [12,113–115] on the surface of cancer cells has stimulated significant synthetic efforts. Overall, these efforts may be viewed as directed towards development of synthetic carbohydrate-based anticancer vaccines (reviewed in Refs. [13,14]). This section therefore aims to give an overview of the synthesis of glycopeptides which carry the tumor-associated Tn and T antigens and their mono- and disialylated derivatives. Recently two excellent reviews with a wider perspective of glycopeptides with an α-GalNAc-Ser/Thr linkage have appeared [37,116].

15.3.2.2.1 Synthesis of the Tn Antigen (α-D-GalNAc-Ser/Thr)

At present three different approaches to establish the α-glycosidic bond in the Tn antigen can be discerned. The vast majority of the syntheses of Tn antigen building blocks have utilized the nonparticipating azido group at C-2 of the GalNAc donor (reviewed in Refs. [79,117]), thereby relying on pioneering studies in the laboratories of Paulsen [118,119] and Lemieux [120]. Often, 1-bromo- [121] or 1-chlorosugars [49,122,123] such as **100** have been employed in Koenigs–Knorr-type glycosylations of Fmoc-protected serine and threonine esters (Scheme 15.11). Protected derivatives of threonine often give somewhat better α-selectivities than those of serine, and use of a promoter such as silver perchlorate usually results in high α-selectivities. In order to obtain the desired Tn building blocks the azido group in compounds such as **101** or **102** has to be reduced and subsequently acetylated, after which cleavage of the ester gives **103** or **104**. Alternatively, reductive acetylation can be performed in one step with thioacetic acid [49,121]. A drawback of the nonparticipating azido group is that azidonitration of tri-*O*-acetyl-D-galactal according to the original procedure [120] gives only moderate yields and requires lengthy purification of intermediates by chromatography. However, an improved procedure which avoids chromatography of the intermediates and provides a high overall yield has been reported [124]. The efficiency in using the azido group at C-2 of GalNAc donors was recently highlighted in the synthesis of glycopeptide **107** (Scheme 15.11) [46]. Synthesis of **107** involves glycosylation of a serine in a resin-bound peptide, which constitutes a significant challenge. Glycosylation of **106**, in which the hydroxyl group of serine is unprotected, was performed with 2-azido galactosyl trichloroacetimidate **105**. Use of an excess of **105** allowed glycopeptide **107** to be isolated in a yield of 78% after TFA catalyzed cleavage from the solid phase.

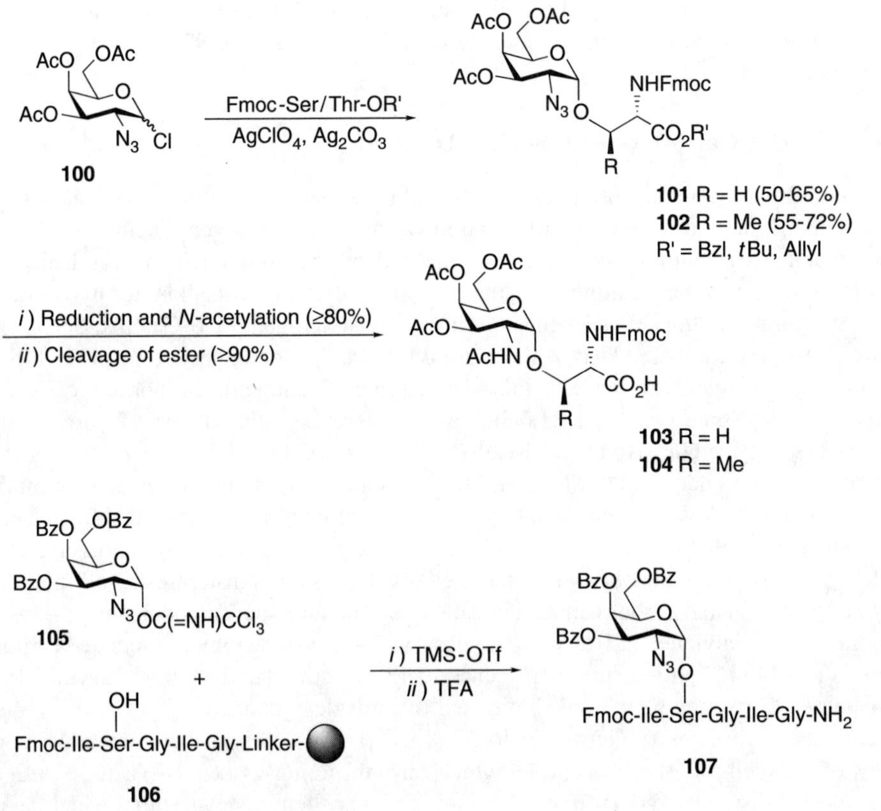

AcO OAc
AcO
O
N_3 Cl
100

Fmoc-Ser/Thr-OR'
$AgClO_4$, Ag_2CO_3

AcO OAc
AcO
O
N_3 O
NHFmoc
CO_2R'
R
101 R = H (50-65%)
102 R = Me (55-72%)
R' = Bzl, *t*Bu, Allyl

i) Reduction and *N*-acetylation (≥80%)
ii) Cleavage of ester (≥90%)

AcO OAc
AcO
O
AcHN O
NHFmoc
CO_2H
R
103 R = H
104 R = Me

BzO OBz
BzO
O
N_3 OC(=NH)CCl_3
105
+
OH
Fmoc-Ile-Ser-Gly-Ile-Gly-Linker-
106

i) TMS-OTf
ii) TFA

BzO OBz
BzO
O
N_3 O
Fmoc-Ile-Ser-Gly-Ile-Gly-NH_2
107

SCHEME 15.11 Preparation of the Tn antigen building blocks by α-glycosylation of serine and threonine building blocks, or by direct α-glycosylation of a solid-phase supported peptide. In both cases galactosyl donors, i.e. **100** and **105**, which have the 2-acetamido functionality masked as an azide were used.

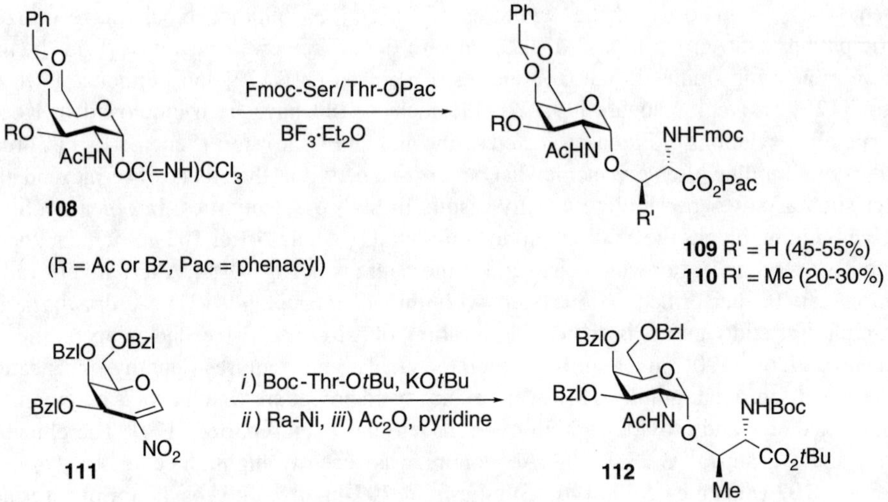

Ph
O
O
O
RO
AcHN OC(=NH)CCl_3
108
(R = Ac or Bz, Pac = phenacyl)

Fmoc-Ser/Thr-OPac
BF_3·Et_2O

Ph
O
O
O
RO
AcHN O
NHFmoc
CO_2Pac
R'
109 R' = H (45-55%)
110 R' = Me (20-30%)

BzlO OBzl
BzlO
O
NO_2
111

i) Boc-Thr-O*t*Bu, KO*t*Bu
ii) Ra-Ni, *iii*) Ac_2O, pyridine

BzlO OBzl
BzlO
O
AcHN O
NHBoc
CO_2tBu
Me
112

SCHEME 15.12 Preparation of the Tn antigen building blocks by α-glycosylation of serine and threonine building blocks with GalNAc donor **108**, which retains the 2-acetamido functionality at C-2, or with 2-nitrogalactal **111**.

An alternative route to Tn building blocks depends on the three-step conversion of N-acetylgalactosamine to trichloroacetimidate **108**, which is then used in boron trifluoride etherate promoted glycosylations of Fmoc protected derivatives of serine and threonine (Scheme 15.12) [125]. Using this approach serine building block, **109** is obtained in yields of 45–55%, whereas threonine derivative **110** is unfortunately obtained in a substantially lower yield (20 to 30%). This constitutes one of the most interesting routes to serine-based Tn antigen building blocks since the somewhat time-consuming synthesis of chlorosugar **100** from tri-*O*-acetyl-D-galactal is avoided. Remarkably, only traces of β-glycoside and oxazoline are formed in this reaction, in spite of the presence of the *N*-acetyl group in the GalNAc donor. The origin of the α-selectivity was attributed to the presence of a 1,3-dioxane ring that positions O-4 and O-6 so that a transient oxocarbonium ion is stabilized, thereby promoting formation of the thermodynamically favored α-glycoside [125].

Some years ago, yet another conceptually different route to Tn antigen building blocks was reported [126]. Instead of performing a glycosylation, the α-glycosidic bond was formed in a Michael-type reaction between 2-nitrogalactal **111** and derivatives of serine and threonine protected with Boc and tert-butyl groups (Scheme 15.12). Unfortunately, reduction of the nitro group is not compatible with use of an N^{α}-Fmoc group, thus requiring use of a Boc-group which subsequently has to be replaced by a Fmoc group before **112** can be used in glycopeptide synthesis.

15.3.2.2.2 Synthesis of the T Antigen [β-D-Gal-(1→3)-α-D-GalNAc-Ser/Thr]

Syntheses of T antigen building blocks were reported 20 years ago [122,127,128], and a substantial amount of publications concerning these building blocks is now available (reviewed in Refs. [11,79]). Just as for the Tn antigen, three different approaches have been used for coupling of β-D-Gal-(1→3)-D-GalNAc building blocks to serine and threonine: routes relying on a nonparticipating azido group at C-2 of the GalNAc residue, restriction of the conformational freedom of GalNAc with a 4,6-benzylidene acetal and use of a Michael-type addition to a 2-nitrogalactal (Scheme 15.13). In a recent variation of one of the original approaches, 2-azido galactose **114** was glycosylated in high yield with peracetylated trichloroacetimidate **113**, using trimethylsilyl triflate as promoter [129]. The resulting disaccharide **115** was then transformed into glycosyl bromide **116** which was coupled directly, or after conversion into the corresponding chloride, with Fmoc serine and threonine pentafluorophenyl esters to give building blocks **117** and **118**, respectively. Anomeric mixtures were obtained in the glycosylations, which accounts for the fact that the yields were only in the 50–60% range.

In an alternative route, disaccharide trichloroacetimidate **119**, in which the GalNAc moiety carries a 4,6-benzylidene acetal, has been used in boron trifluoride etherate-promoted glycosylations of Fmoc protected derivatives of serine and threonine [130]. Advantages of this approach are that donor **119** is obtained in a short, five-step synthetic route from *N*-acetylgalactosamine and that glycosylation of serine proceeds in good yield. However, as in the analogous synthesis of the Tn antigen, a significantly lower yield is obtained on glycosylation of threonine. The T antigen has also been prepared by addition of Boc-Ser/Thr-O*t*Bu to Michael acceptor **122** [131]. α-Glycosides **123** and **124** were obtained stereoselectively in close to 80% yields after reduction of the nitro group and acetylation. Removal of the protecting groups from the amino acid part, and reprotection of the α-amino group with an Fmoc group, was then performed in 85% yield for threonine derivative **124**.

It is well known that the stereoselectivity of *O*-glycosylations becomes less predictable as the complexity and size of the glycosyl donor increases [132]. Simultaneously, problems in separation of anomeric mixtures usually become greater. Therefore, syntheses of the T antigen have also been performed by addition of a galactose residue to a protected Tn antigen building block (Scheme 15.14), instead of attaching a β-D-Gal-(1→3)-D-GalNAc equivalent to serine or threonine. The former approach was recently termed the "cassette approach" by Danishefsky [121] but it had been in practice several years before this term was introduced. Reliable syntheses have, for instance, utilized galactosyl trichloroacetimidate **125** or bromide **126** as donors in glycosylation

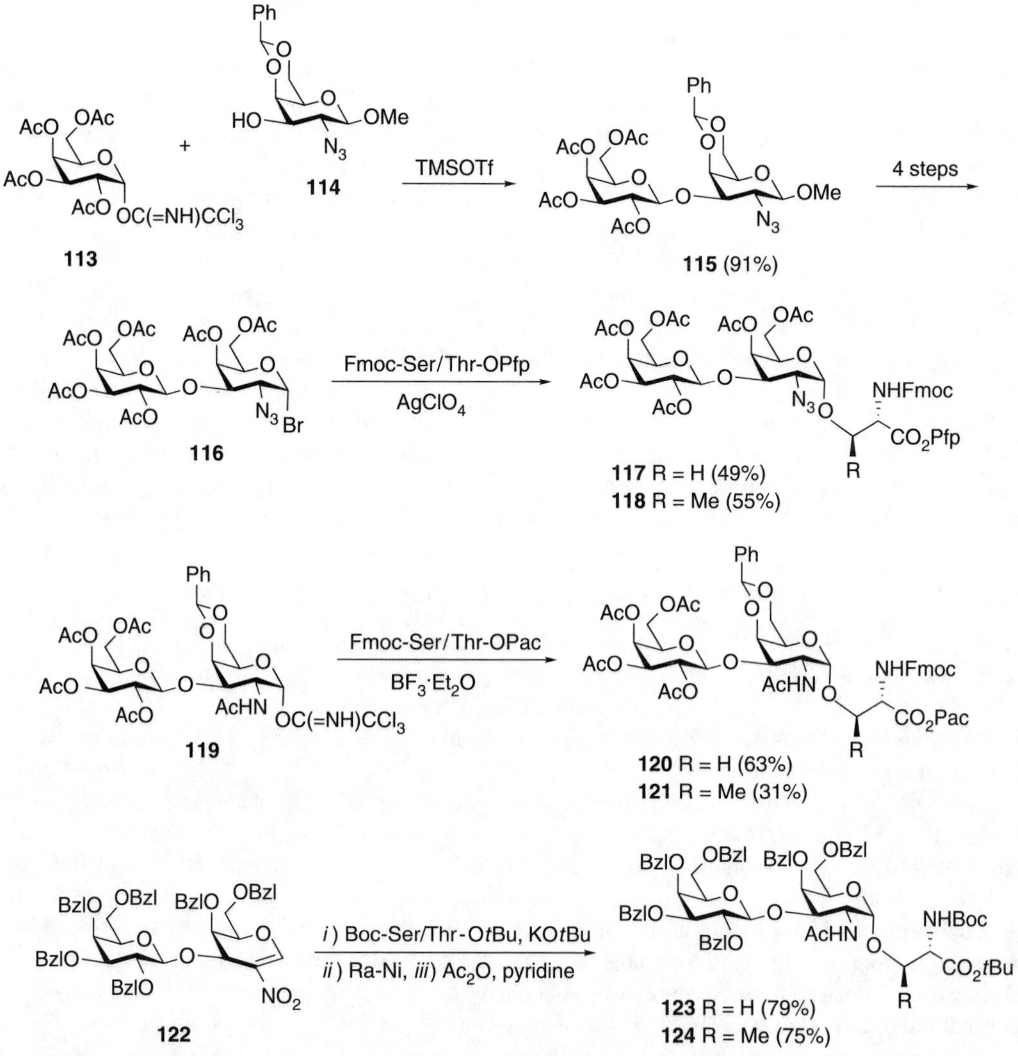

SCHEME 15.13 The cassette approach; synthesis of T antigen building blocks by α-glycosylation of serine and threonine building blocks with three different disaccharide glycosyl donors.

of the Tn antigen building blocks **127** [133,134]. Use of galactal **130** constitutes an even better alternative in view of the high yield obtained in coupling to building block **131** [121]. The cassette approach has also been used to extend the saccharide chain of glycopeptide **133**, while still being attached to a solid support [45]. Evaluation of different donors and reaction conditions revealed that the best results were obtained when donor **134** was activated using freshly distilled trimethylsilyl triflate. However, it should be noted that as much as eight equivalents of **134** were used to provide **135** in a yield of 67% after cleavage from the solid phase.

15.3.2.2.3 Synthesis of the Sialyl-Tn Antigen [α-NeuNAc-(2→6)-α-D-GalNAc-Ser/Thr]

The first synthesis of the sialyl Tn antigen was reported by Ogawa et al. and involved the synthesis of a disaccharide donor consisting of an anomeric trichloroacetimidate of α-NeuNAc-(2→6)-α-D-GalNAc [135]. The donor was then used in glycosylation of *N*-(benzyloxycarbonyl)-L-serine

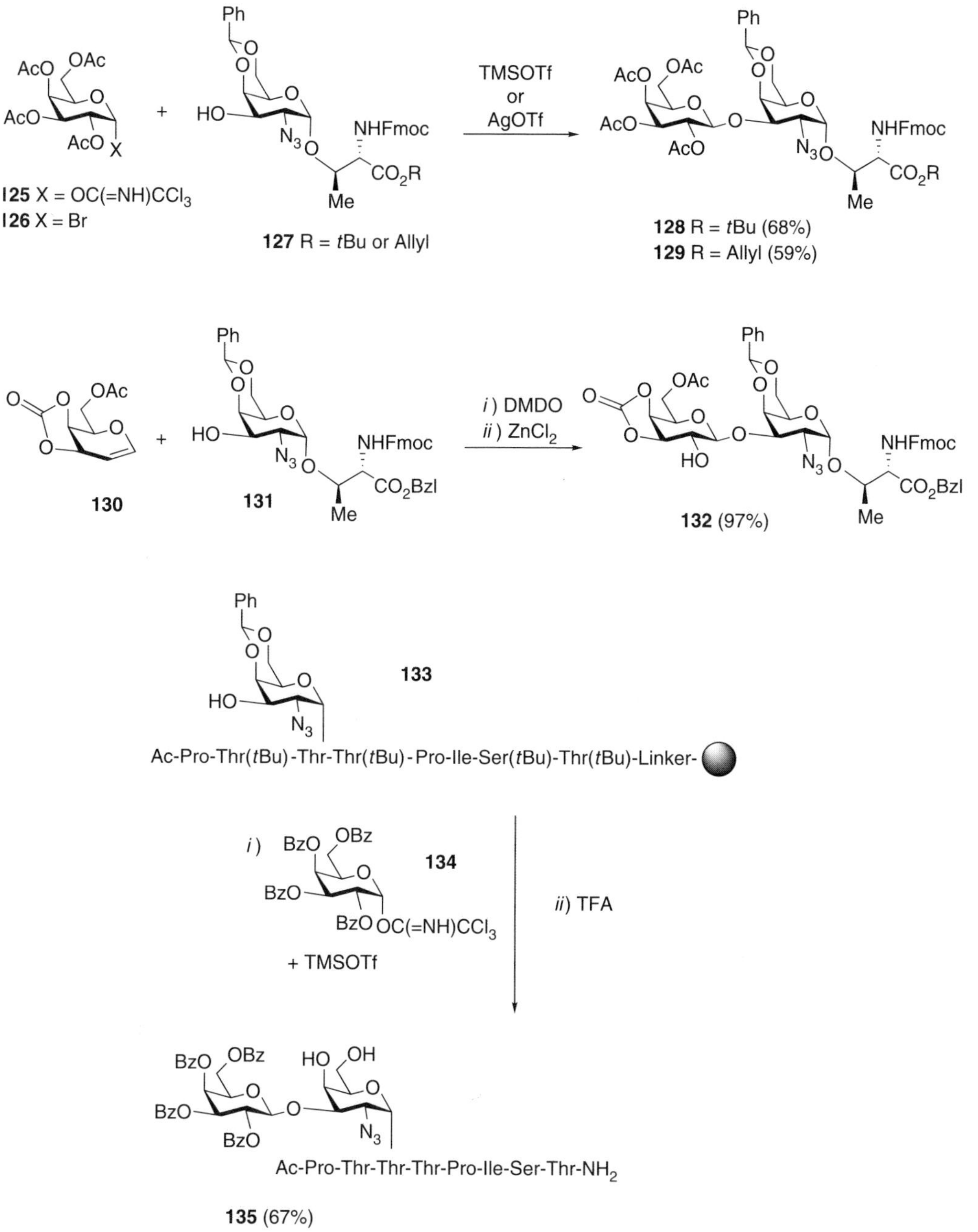

125 X = OC(=NH)CCl_3
126 X = Br
127 R = *t*Bu or Allyl
TMSOTf or AgOTf
128 R = *t*Bu (68%)
129 R = Allyl (59%)

130 **131**
i) DMDO
ii) $ZnCl_2$
132 (97%)

133
Ac-Pro-Thr(*t*Bu)-Thr-Thr(*t*Bu)-Pro-Ile-Ser(*t*Bu)-Thr(*t*Bu)-Linker-

i) **134** BzO OC(=NH)CCl_3 + TMSOTf
ii) TFA

Ac-Pro-Thr-Thr-Thr-Pro-Ile-Ser-Thr-NH_2
135 (67%)

SCHEME 15.14 Synthesis of T antigen building blocks by β-glycosylations of Tn antigen building blocks **127** and **131**, and preparation of T antigen peptide **135** by solid-phase β-glycosylation of Tn antigen peptide **133**.

benzyl ester to give an anomeric mixture in poor stereoselectivity (α/β *ca.* 2:1). Unfortunately, the mixture could not be separated until after reduction of the azido group at C-2 of the GalNAc residue, which provided the target sialyl-Tn antigen building block in a yield of 34% based on the donor used in the glycosylation. This synthesis thus illustrates the problems with stereoselectivity and separation of stereoisomers discussed above for the T antigen.

Later syntheses have attempted to avoid such problems by using a cassette approach in which Tn antigen building blocks are sialylated (Scheme 15.15). In one route the acetyl groups were first removed from Tn building block **136** by using sodium methoxide under carefully controlled conditions in order to avoid simultaneous Fmoc cleavage [136,137]. In the key step, triol **137** was then glycosylated regio- and stereoselectively with sialyl xanthate **138** using methylsulfenyl bromide in combination with silver triflate as promoters. Use of low temperature and acetonitrile as solvent allowed the α-2,6-linked sialyl-Tn derivative **139** to be isolated in a yield of 32%. After acetylation of the two unprotected hydroxyl groups, and cleavage of the *tert*-butyl group, this building block was used in synthesis of a glycopeptide from the MUC-1 mucin [136,137].

An alternative strategy for the synthesis of the sialyl-Tn antigen was reported in the same year as the synthesis of **139** [64]. In this case, the hydroxyl groups of Tn building block **140** were first deprotected and an isoproylidene acetal was then used to protect O-3 and O-4. Sialylation of **141** was again performed with xanthate **138** under conditions that were almost identical to those used in the synthesis of **139**. This provided sialyl-Tn antigen **142** in a higher yield (49%) than that obtained for **139**, but it should be noted that the Tn-antigen building block **140** is somewhat more cumbersome to prepare than **136**. Removal of the benzyl ester from **142** then gave a building block, which was used in the solid-phase synthesis of a glycopeptide from HIV gp120 [64]. Recently, two other routes to the sialyl-Tn antigen, based on Tn antigen

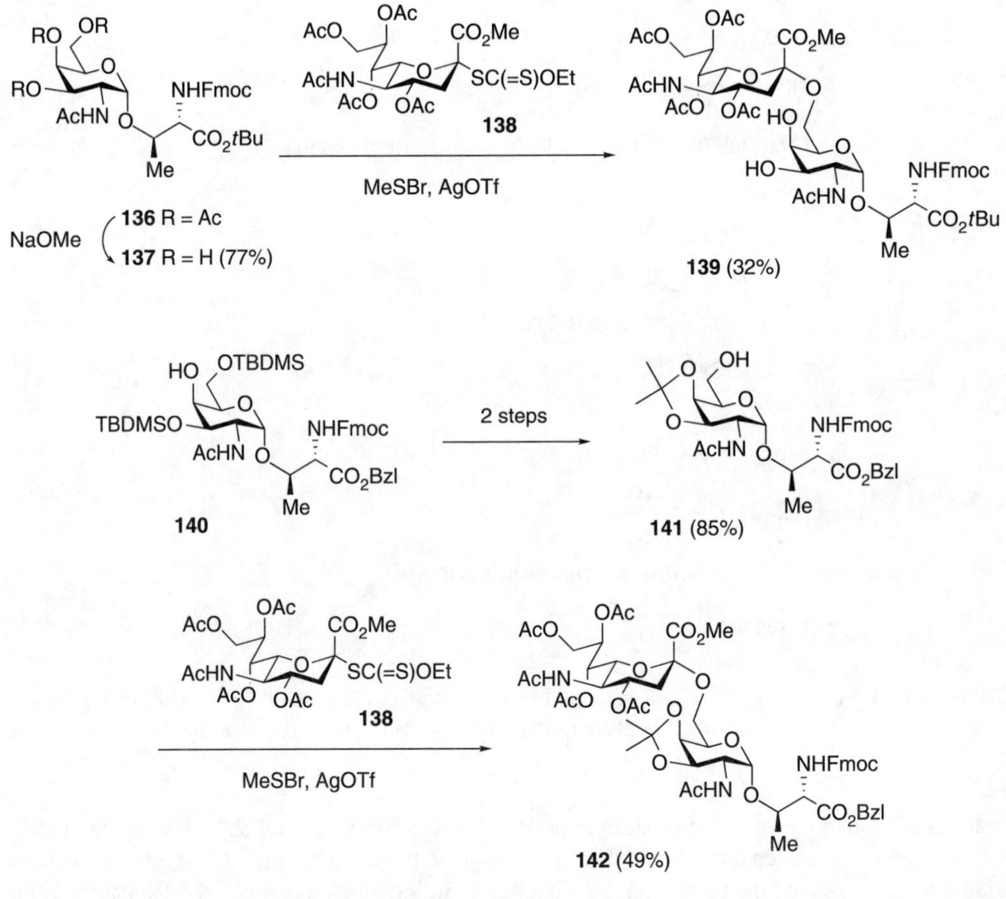

SCHEME 15.15 Synthesis of sialyl-Tn antigen building blocks by sialylation of the Tn antigen building blocks **137** and **141** with sialyl xanthate **138**.

building blocks having O-3 and O-4 of the GalNAc moiety protected, have been reported [131, 138]. In these syntheses the yields in the key sialylation step ranged from 37 to 65%. A chemoenzymatic synthesis of a glycopeptide from the MUC-1 mucin, in which the sialyl moiety was added enzymatically to a peptide carrying the Tn antigen, has also been recently described (cf. Section 15.5).

15.3.2.2.4 Synthesis of the 2,3-Sialyl-T Antigen [α-NeuNAc-(2→3)-β-D-Gal-(1→3)-α-D-GalNAc-Ser/Thr]

The first synthesis of a 2,3-sialyl-T antigen building block [71] and its subsequent use in solid-phase glycopeptide synthesis [139] was reported in 1997 and 1998, respectively. The crucial step in the preparation of the building block was glycosylation of 2-azidogalactosyl serine **144** with the sialylated galactose unit **143** (Scheme 15.16). By using boron trifluoride etherate as promoter, β-glycoside **145** was obtained in a yield of 54% together with a minor amount (7%) of the corresponding α-glycoside. The stereoselectivity of this step is noteworthy since nonparticipating benzyl ether is found at O-2 of the glycosyl donor. A further four steps were then required to transform **145** into 2,3-sialyl-T antigen building block **146**, which was used in solid-phase synthesis of a 27-mer glycopeptide from human α2HS-glycoprotein [139]. After completion of the synthesis, and cleavage from the solid support with TFA, the benzyl ethers of the carbohydrate moiety were completely removed by treatment with trimethylsilyl triflate in TFA containing thioanisole. A related cassette approach to the 2,3-sialyl-T antigen was described somewhat later [138].

A different route to the 2,3-sialyl-T antigen has also been reported [140]. In this case, trisaccharide building block **149** was first assembled by glycosylation of 2-azidogalactoside **148** with disaccharide **147**. After manipulation of the anomeric position, donor **150** was used in silver triflate-promoted glycosylations of Fmoc serine and threonine as their pentafluorophenyl esters. This gave the desired α-linked, 2,3-sialyl-T antigen building blocks **151** and **152** in yields of 33 and 41% respectively, but also considerable amounts of the corresponding β-glycosides (24 and 10%). This example again illustrates that low stereoselectivities may be obtained in *O*-glycosylations as the complexity and size of the glycosyl donor increases. When building blocks **151** and **152** were used in solid-phase synthesis of two neoglycopeptides for further use in investigations of T cell responses, it was found that the removal of the methyl ester and the acetyl groups required unexpectedly harsh conditions [141]. This involved treatment with methanolic sodium methoxide at pH 14, which had to be conducted at −30°C to avoid β-elimination, followed by additional treatment with hydrazine hydrate.

Chemoenzymatic syntheses of glycopeptides, which carry the 2,3-sialyl-T antigen, have been reported and are discussed in Section 15.5.

15.3.2.2.5 Synthesis of the 2,6-Sialyl-T Antigen {β-D-Gal-(1→3)-[α-NeuNAc-(2→6)]-α-D-GalNAc-Ser/Thr}

A synthesis of the 2,6-sialyl-T antigen was reported more than a decade ago, but the yield and stereoselectivity were low for steps introducing the sialic acid and for linking the 2,6-sialyl-T trisaccharide to serine [142]. A more practical approach to the 2,6-sialyl-T antigen based on the glycal assembly strategy has been reported for the purpose of developing anticancer vaccines (Scheme 15.17) [138,143]. In this case, building blocks **130**, **153** and **154** were assembled into glycal **155**, which was then transformed into glycosyl donors **156**, **157** and **158** in two steps each (*ca.* 45% yields from **155**). Glycosylation of Fmoc threonine benzyl ester with bromide **156**, activated with silver perchlorate, or with trichloroacetimidate **157** and using boron trifluoride as a promoter, gave exclusively α-linked **159** in 74% and approximately

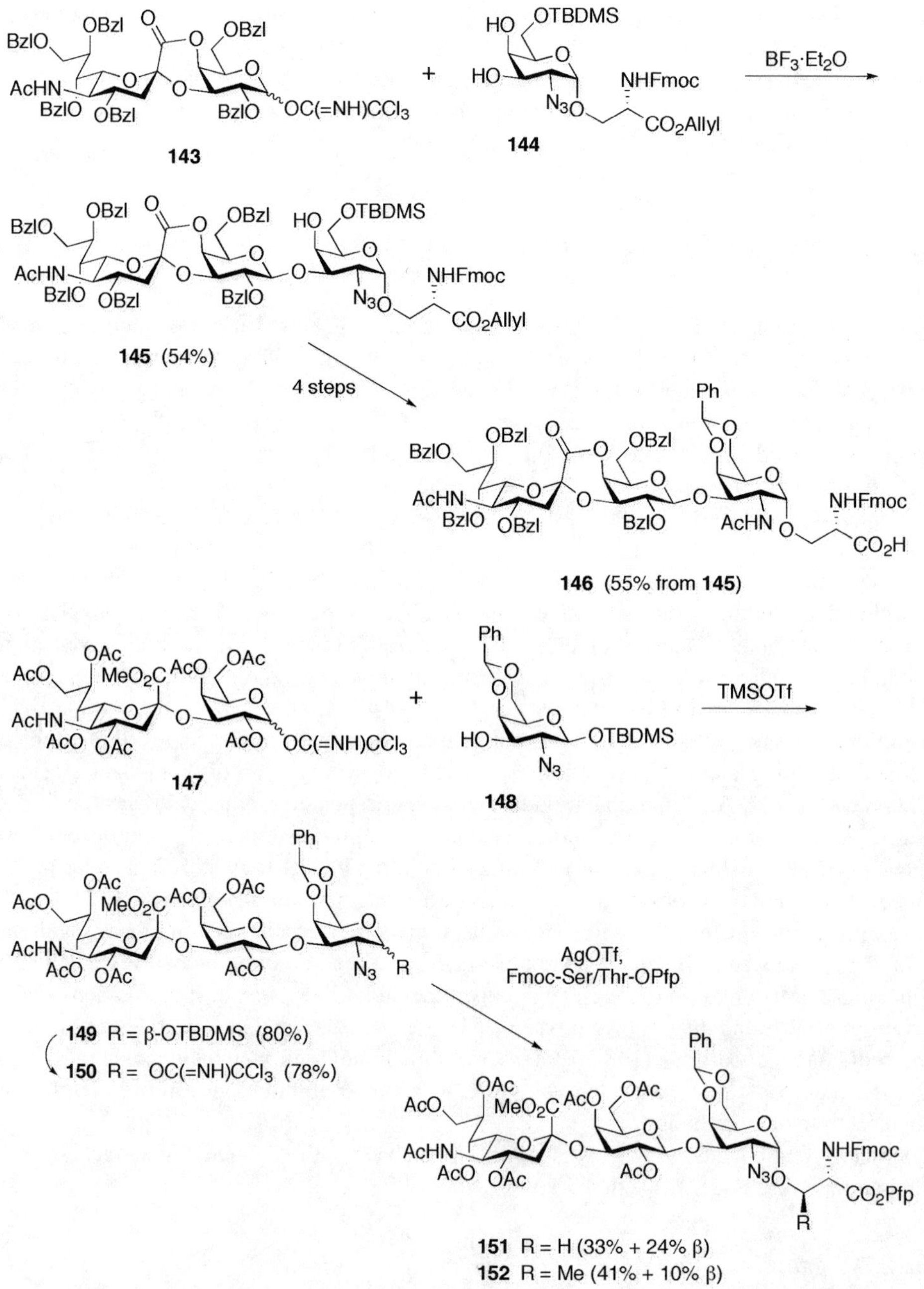

SCHEME 15.16 Synthesis of 2,3-sialyl-T antigen building blocks by β-glycosylation of Tn antigen **144**, or by α-glycosylation of serine and threonine building blocks with trisaccharide donor **150**.

60% yields, respectively. In contrast, when Fmoc serine benzyl ester was used as acceptor the stereoselectivity varied from α:β 2.6:1 to 30:1, whereas the yields of the α/β-mixtures ranged from 30 to 70%. The stereoselectivities are thus significantly higher than in the analogous synthesis of the 2,3-sialyl T antigens **151** and **152** (Scheme 15.16), which may be due to the presence of the bulky sialic acid moiety at O-6 of the 2-azido galactose moiety in donors **156–158**.

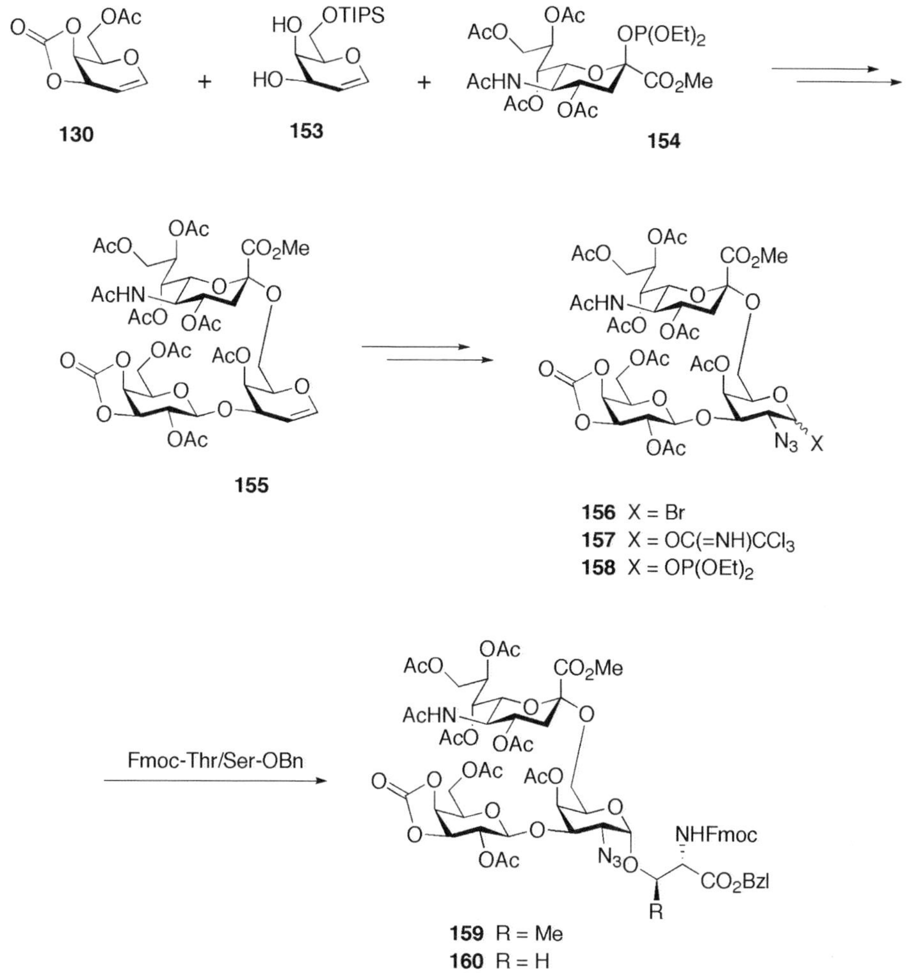

130 + 153 + 154

155

156 X = Br
157 X = OC(=NH)CCl_3
158 X = $OP(OEt)_2$

Fmoc-Thr/Ser-OBn

159 R = Me
160 R = H

SCHEME 15.17 Synthesis of 2,6-sialyl-T antigen building blocks by α-glycosylation of serine and threonine building blocks with trisaccharide donors **156-158**.

15.4 SYNTHESIS OF *N*-LINKED GLYCOPEPTIDES

Great effort has been focused on the synthesis of *N*-linked glycopeptides because of the important role of *N*-linked saccharides in glycoproteins, with impressive results being reported during the last 10 to 20 years (reviewed in Refs. [37,38,144,145]). As mentioned in Section 15.2.1, two basic strategies exist for preparation of *N*-linked glycopeptides:

1. The building-block approach, where a glycosyl asparagine derivative is used in stepwise assembly of the glycopeptide [34,146–151].
2. The convergent approach, in which the peptide and glycosyl amine are prepared independently and condensed by formation of an amide linkage late in the synthesis [39–41].

In both strategies an amide bond is formed between a glycosyl amine and an activated aspartic acid derivative to obtain the *N*-glycosidic linkage, instead of glycosylation of asparagine (Figure 15.5). However, the coupling of glycosyl amines and aspartic acid residues in peptides

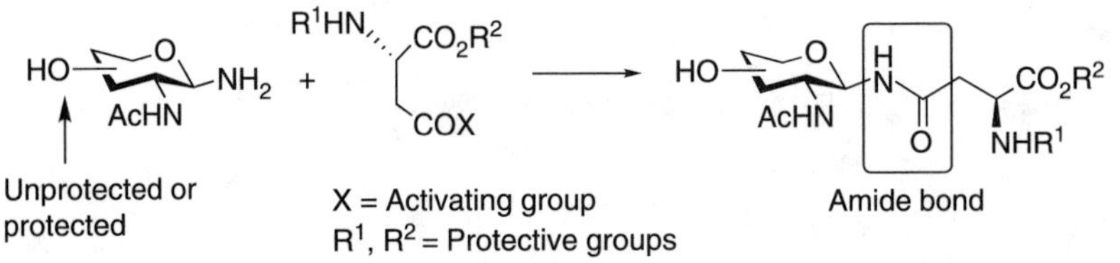

FIGURE 15.5 A schematic description of how *N*-linked glycosylated amino acids are often prepared by the formation of an amide bond between a glycosyl amine and an activated aspartic acid derivative.

suffer from a side reaction where aspartic acid can undergo aspartimide formation during activation (Figure 15.6) [152]. This side reaction is particularly facile for Asp-Gly, Asp-Ala, and Asp-Ser sequences. Aspartimides can then give rise to peptides that are linked either through the α- or the β-carboxy group, which may make purification of the final glycopeptide difficult and reduces the overall yield [39,41,153]. Owing to the problems associated with the convergent strategy, the building-block approach is currently the most frequently employed method for synthesis of *N*-linked glycopeptides.

Facile routes for preparation of glycosyl amines are a prerequisite to gain access to *N*-glycosylated amino acid building blocks and glycopeptides. Reduction of glycosyl azides and treatment of reducing sugars with saturated ammonium hydrogen carbonate solution are two of the most commonly employed procedures for the synthesis of glycosyl amines. Glycosyl azides may preferably be obtained from the corresponding glycosyl halides [154–156], and several procedures exist for their reduction to glycosyl amines [155,157–162]. An alternative route to glycosyl azides has also been reported, which involves the use of glycals as starting materials [159,163,164]. A standard procedure for preparation of *N*-linked glycosylated amino acids via a glycosyl azide is outlined in Scheme 15.18. This involves synthesis of glycosyl azide **162** from the corresponding glycosyl halide **161** in a high yield [156] and subsequent hydrogenation to afford glycosyl amine **163** [165]. Coupling of **163** to an aspartic acid derivative under activation by dicyclohexylcarbodiimide (DCC) and 1-hydroxybenzotriazole (HOBt) gave building-block **164** [166].

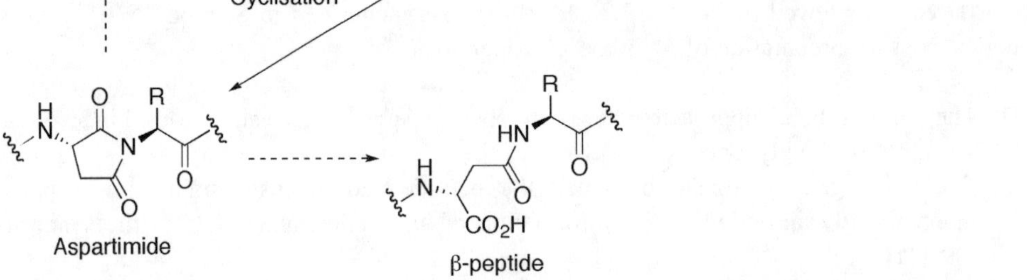

FIGURE 15.6 Aspartic acid residues in peptides often undergo aspartimide formation during activation, prior to attempts to couple to glycosyl amines. This side-reaction gives rise to formation of peptides that are linked either through the α- or the β-carboxy group as by-products.

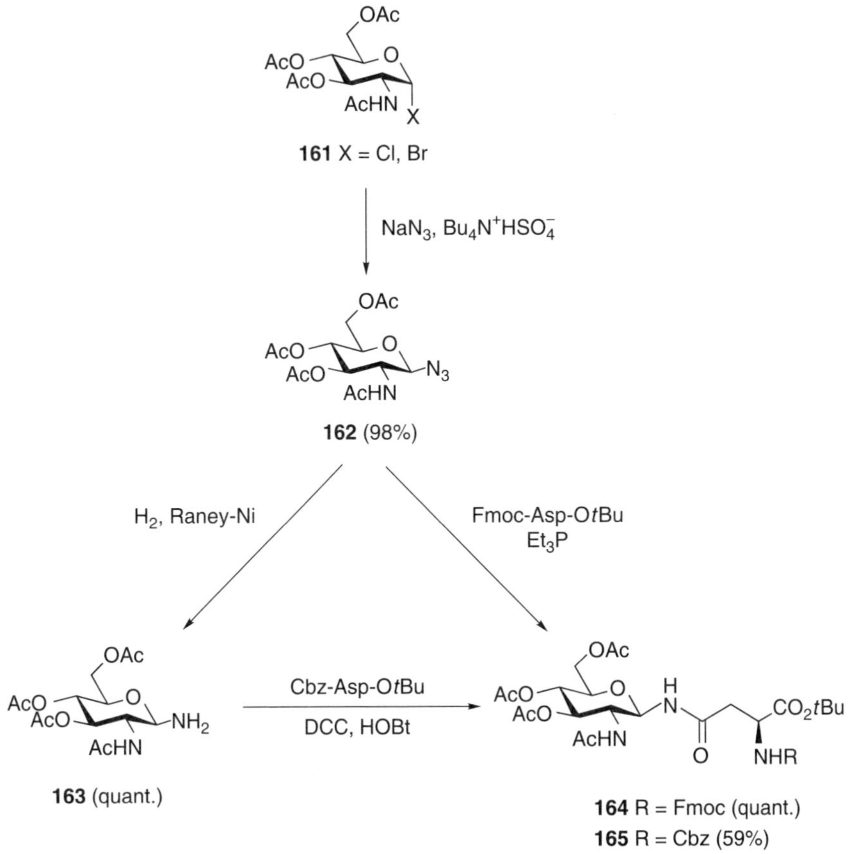

OAc
AcO
AcO
O
AcHN
X
161 X = Cl, Br

NaN_3, $Bu_4N^+HSO_4^-$

OAc
AcO
AcO
O
N_3
AcHN
162 (98%)

H_2, Raney-Ni

Fmoc-Asp-O*t*Bu
Et_3P

OAc
AcO
AcO
O
NH_2
AcHN
163 (quant.)

Cbz-Asp-O*t*Bu
DCC, HOBt

OAc
AcO
AcO
O
H
N
CO_2tBu
AcHN
O
NHR

164 R = Fmoc (quant.)
165 R = Cbz (59%)

SCHEME 15.18 Preparation of β-linked aspargine glycosides **164** and **165** by reduction of glucosaminyl azide **162** and coupling to aspartic acid derivatives.

Glycosyl azides often undergo anomerization during reduction resulting in the formation of undesired α-glycosides [42]; therefore, the conditions used for hydrogenation have to be carefully optimized to suppress this side-reaction. Since the presence of acids and noble metal catalysts accelerates the anomerization process, use of basic conditions and procedures that avoid noble metals may minimize the α-glycoside formation [162]. Anomerization may be avoided by direct coupling of glycosyl azides, for example, **162**, to an aspartic acid derivative in the presence of a trialkylphosphine to give the aspargine glycoside **165** via a concerted intramolecular reaction [167–170].

In the other main route to glycosyl amines, the amino group is introduced at the reducing end of an unprotected carbohydrate using saturated ammonium bicarbonate [171]. Thus, chitobiose **166** was treated with saturated ammonium bicarbonate to give **167** (Scheme 15.19). The glycosylamine was subsequently reacted with the pentafluorophenyl (Pfp) ester of Fmoc-Asp(OPfp)-O*t* Bu, without competing acylation of the unprotected hydroxyl groups of **167**. After cleavage of the *tert*-butyl ester with TFA, the chitobiosyl-asparagine derivative **168** was used directly in peptide synthesis [172].

An elegant application of this methodology has been reported for synthesis of complex *N*-linked glycosyl asparagine building blocks using saccharides obtained from natural glycoproteins (Scheme 15.20) [77]. A key step consisted of a mild hydrazinolysis of the glycoprotein bovine fetuin to release the triantennary, complex-type β-glycosyl hydrazine derivative **169**. *N*-Acetylation, followed by treatment with copper acetate, yielded the reducing oligosaccharide **170**, which gave glycosyl amine **171** in excellent yield (95%) upon treatment with

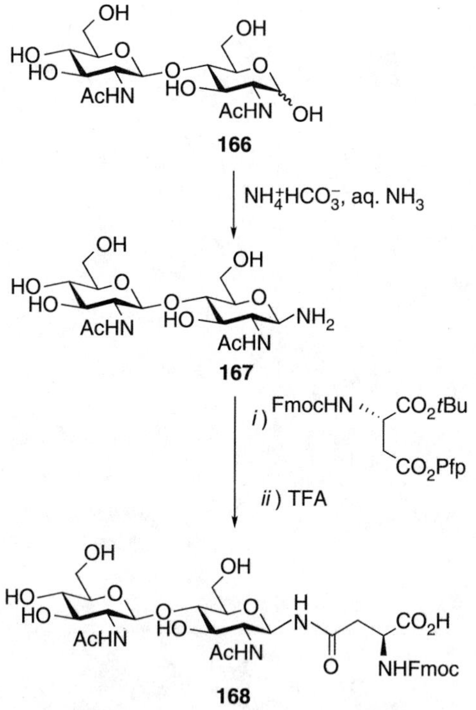

SCHEME 15.19 Synthesis of glycosyl amine **167** by amination of chitobiose and subsequent transformation into chitobiosyl-aspargine derivative **168**.

saturated ammonium bicarbonate in DMSO. Coupling of unprotected oligosaccharide **171** to the activated ester Fmoc-Asp(ODhbt)-O*t*Bu, followed by acetylation and acidolysis of the *tert*-butyl ester, furnished the triantennary building block **172**, which was successfully employed in solid-phase glycopeptide synthesis.

Finally, it should be pointed out that anomeric glycosyl isothiocyanates, which can be prepared from the corresponding glycosyl bromides, have also been employed in direct formation of the amide linkage in glycosylated asparagines (Scheme 15.21) [173–175]. Following this procedure, the glycosyl isothiocyanate **173** was reacted with aspartic acid **174** to produce thiocarbamic acid derivative **176** via **175**, which resembles a mixed anhydride. Thiocarbamic acid **176** then reacted with amino acid anhydride **177**, which is formed in the reaction leading to **176**, to provide aspartyl amide **178** in a one-pot reaction [174].

Many impressive studies have been carried out towards the synthesis of complex and high mannose *N*-linked oligosaccharide structures [176–178]. The core pentasaccharide of *N*-glycans offers a synthetic challenge on its own, since it contains a β-mannosidic linkage which is most difficult to obtain through conventional glycosylation protocols. Initial studies directed towards the Asn-core pentasaccharide utilized mannosyl halides in combination with silver-silicate as an insoluble promoter to enhance formation of the β-mannosidic linkage [69,179,180]. However, the reaction was still accompanied by formation of the corresponding nonnatural α-mannoside to give a 2:3 α/β anomeric mixture, which could be separated by column chromatography. A more reliable approach to the β-mannosidic linkage relies on glycosylation of HO-4′ of chitobiose with a gluco-configured halide with an acyl protecting group on O-2. This protecting group then directs, the formation of the β-linkage through neighboring group participation. After removal of the acyl group, the hydroxyl group can be oxidized to the corresponding ketone, which is then reduced to give the *manno* configuration [181].

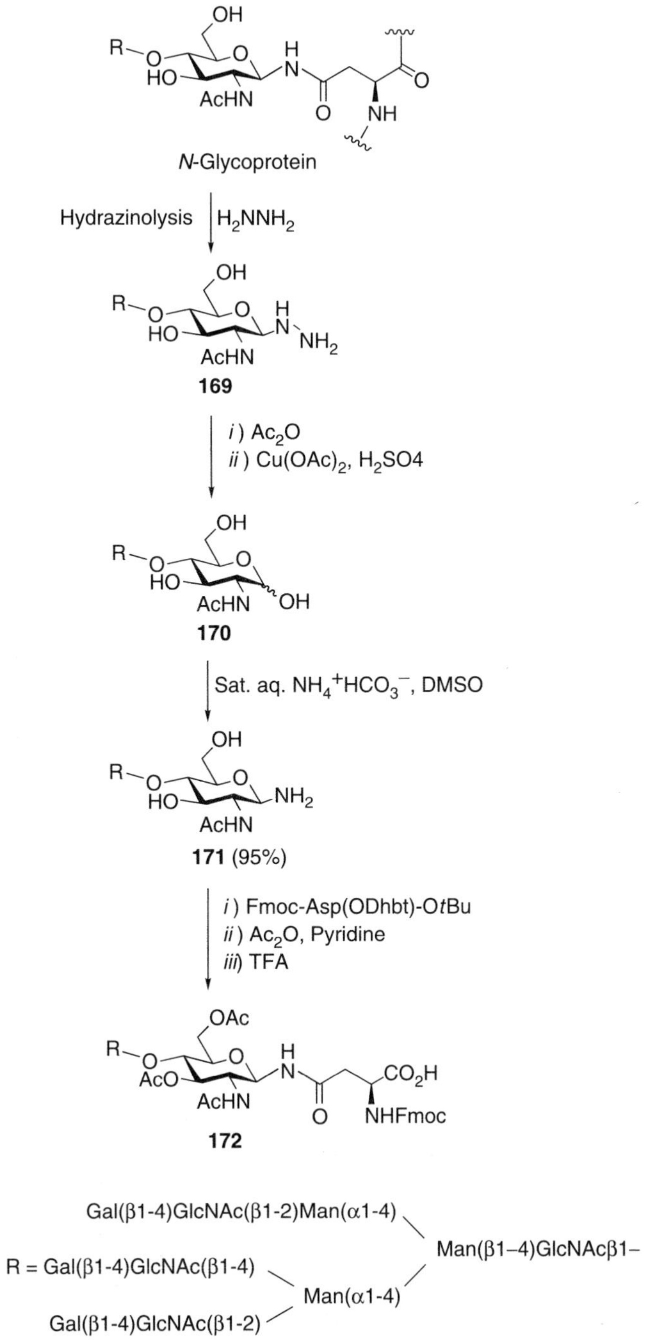

N-Glycoprotein

Hydrazinolysis H_2NNH_2

169

i) Ac_2O
ii) $Cu(OAc)_2$, H_2SO4

170

Sat. aq. $NH_4^+HCO_3^-$, DMSO

171 (95%)

i) Fmoc-Asp(ODhbt)-O*t*Bu
ii) Ac_2O, Pyridine
iii) TFA

172

Gal(β1-4)GlcNAc(β1-2)Man(α1-4)
Man(β1–4)GlcNAcβ1–
R = Gal(β1-4)GlcNAc(β1-4)
Man(α1-4)
Gal(β1-4)GlcNAc(β1-2)

SCHEME 15.20 Release of a tri-antennary complex type oligosaccharide from bovine fetuin allowed preparation of asparagine building block **172**.

In an impressive chemoenzymatic synthesis of an asparagine-linked undecasaccaride [162], the β-mannosidic linkage was obtained through neighboring group participation of the acetyl protecting group at O-2 in donor **180** (Scheme 15.22) [182,183]. After deacetylation and reprotection, HO-2 of the glucose moiety in **181** was transformed into a triflate, which was

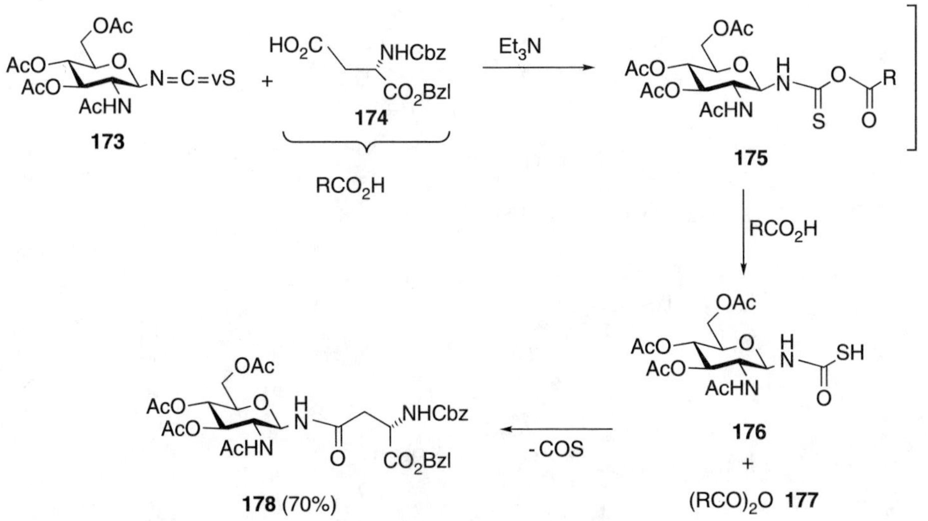

SCHEME 15.21 Preparation of β-linked aspargine glycoside **178** from glycosyl isothiocyanate **173**.

substituted by the adjacent carbamoyl group upon heating in DMF. Subsequent hydrolysis of the β-*manno*-configurated imidocarbonate furnished the corresponding carbonate, which was subjected to methanolysis to give trisaccharide **182** with the desired *manno* configuration. After glycosylation with disaccharide **183** to give **184**, followed by a series of protecting group manipulations and further elongation with disaccharide **183**, heptasaccharide **185** was obtained. The azide in **185** was then reduced and the resulting amine **186** was coupled to the activated pentafluorophenyl ester of Cbz-Asp(OPfp)-OBzl to give the β-linked *N*-glycoside **187** (Scheme 15.23). Hydrogenation resulted in a completely deblocked heptasaccharide–asparagine conjugate **188**, which was subjected to enzymatic galactosidation to incorporate β-linked galactose residues in both of the two arms. Subsequent (α2-6)-sialyltransferase-catalyzed sialylation of the two galactose moieties was accomplished, by incubation of the crude nonasaccharide–aspargine conjugate with the enzyme and CMP-Neu5Ac as substrate, to provide the undecasaccharide–asparagine conjugate **189** in a one-pot reaction [162,184,185].

In spite of the potential problems with aspartimide formation in the convergent syntheses of *N*-linked glycopeptides, this strategy is still appealing since it offers higher diversity at a later stage in synthesis than the building block approach. The scope and limitations of the convergent approach were investigated in a comprehensive model study involving coupling of mono- to heptasaccharide glycosyl amines to aspartic acid residues in penta- and hexapeptides [39]. After optimization of the conditions for coupling to the glycosyl amines, aspartimide formation could be suppressed and a set of *N*-linked glycopeptides was synthesized in high yields. However, when coupling involved large glycosyl amines the peptide always had a large leucine residue attached to the α-carboxyl group of the activated aspartic acid moiety, which suppresses aspartimide formation. Peptides that are prone to undergo aspartimide formation, that is, peptides with a small alanine residue *C*-terminal to the activated aspartic acid, were only condensed with small, reactive glycosyl amines. Even though these results are encouraging, it should be pointed out that it could be expected that the convergent strategy will still be limited by problems with aspartimide formation.

Use of the 2-hydroxy-4-methoxybenzyl (Hmb) group as a backbone amide protecting group constitutes an appealing way to prevent aspartimide formation [41]. The versatility of this strategy was examined using the hexapeptide Ac-Glu-Asp-Ala-Ser-Lys-Ala-NH_2, in which the alanine was Hmb protected as a model compound (cf. resin **190**, Scheme 15.24). To prevent potential side reactions the Hmb group was first *O*-acetylated, after which the allyl group was removed from the

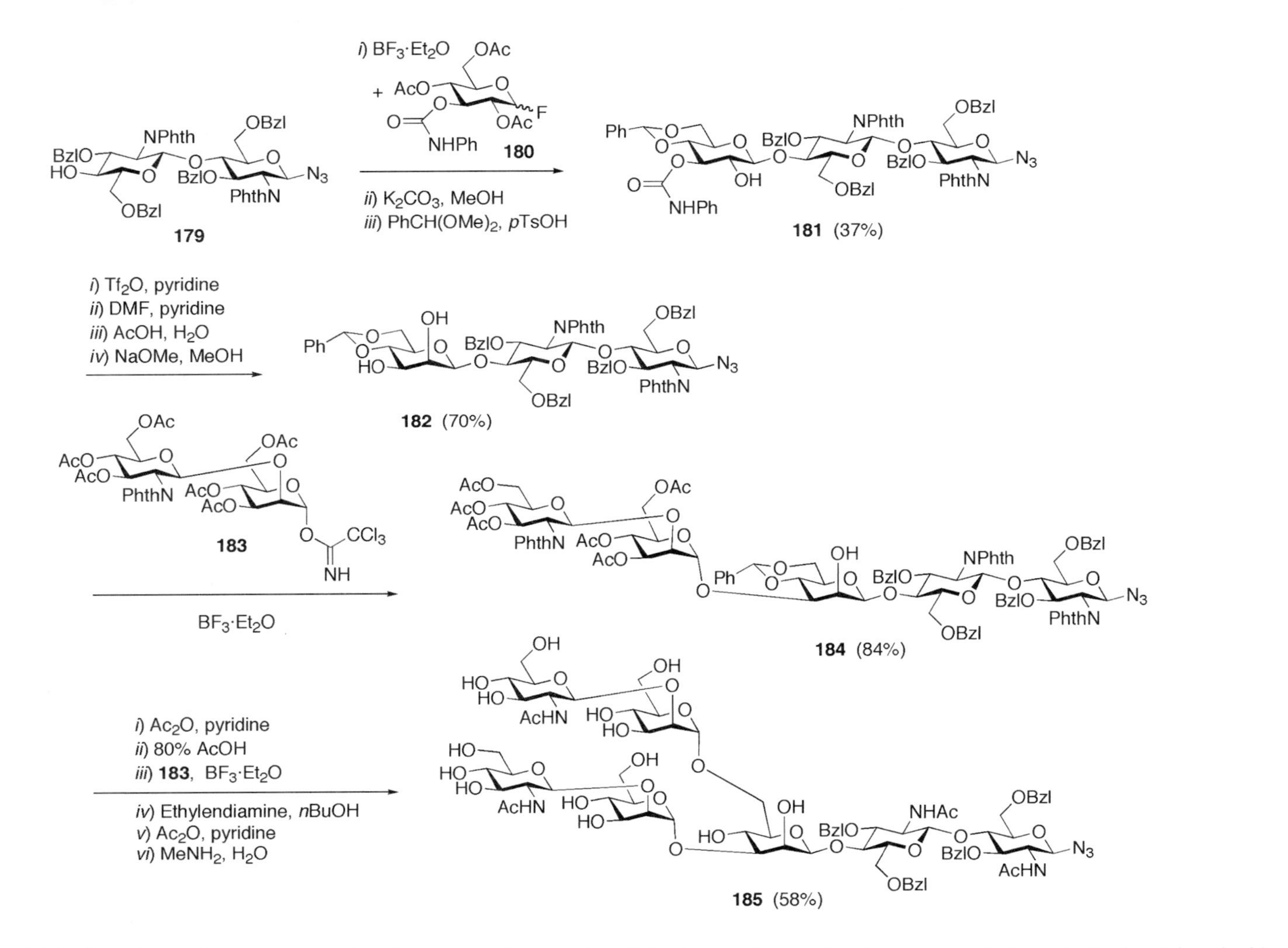

SCHEME 15.22 Total synthesis of complex type 2 undecasaccharide-aspargine **189**. Construction of the β-mannosidic linkage in the key trisaccharide **182** and preparation of core heptasaccharide **185**.

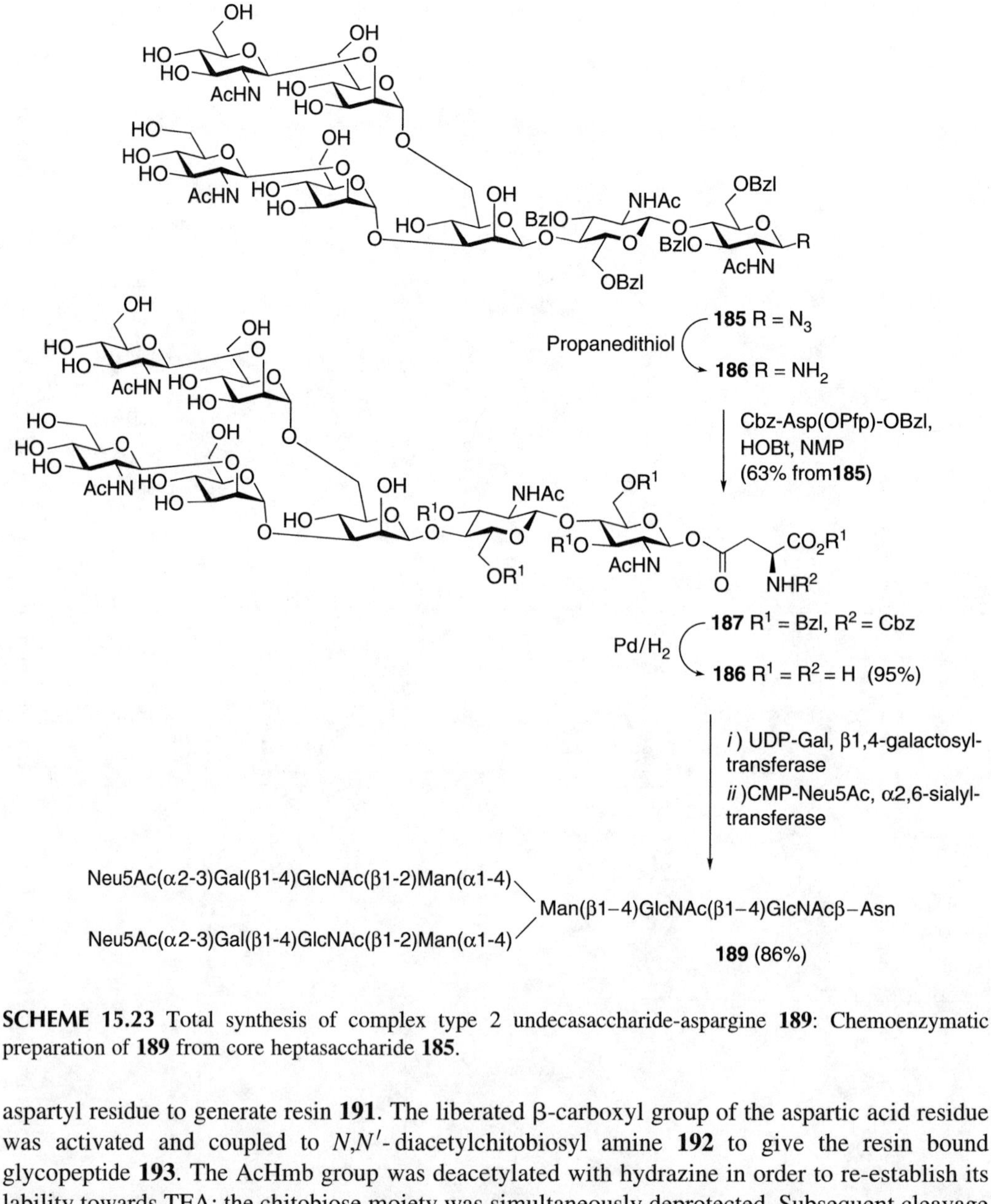

SCHEME 15.23 Total synthesis of complex type 2 undecasaccharide-aspargine **189**: Chemoenzymatic preparation of **189** from core heptasaccharide **185**.

aspartyl residue to generate resin **191**. The liberated β-carboxyl group of the aspartic acid residue was activated and coupled to *N,N'*-diacetylchitobiosyl amine **192** to give the resin bound glycopeptide **193**. The AcHmb group was deacetylated with hydrazine in order to re-establish its lability towards TFA; the chitobiose moiety was simultaneously deprotected. Subsequent cleavage from the resin and simultaneous deprotection of the peptide protecting groups was performed with TFA, and glycopeptide **194** was isolated in an overall yield of 30%. However, as for all convergent approaches, formation of the undesired α-*N*-glycoside as a byproduct may render final purification difficult. In addition, since the Hmb group and the carbohydrate part are simultaneously deacetylated prior to the TFA treatment, acid labile glycosidic linkages such as the α-fucosidic will not survive the final TFA-catalyzed cleavage and deprotection.

15.5 CHEMOENZYMATIC SYNTHESIS OF GLYCOPEPTIDES

Chemical synthesis of glycopeptides which carry large oligosaccharide residues is often demanding and time-consuming. In such cases, use of enzymatic techniques for the elaboration of existing,

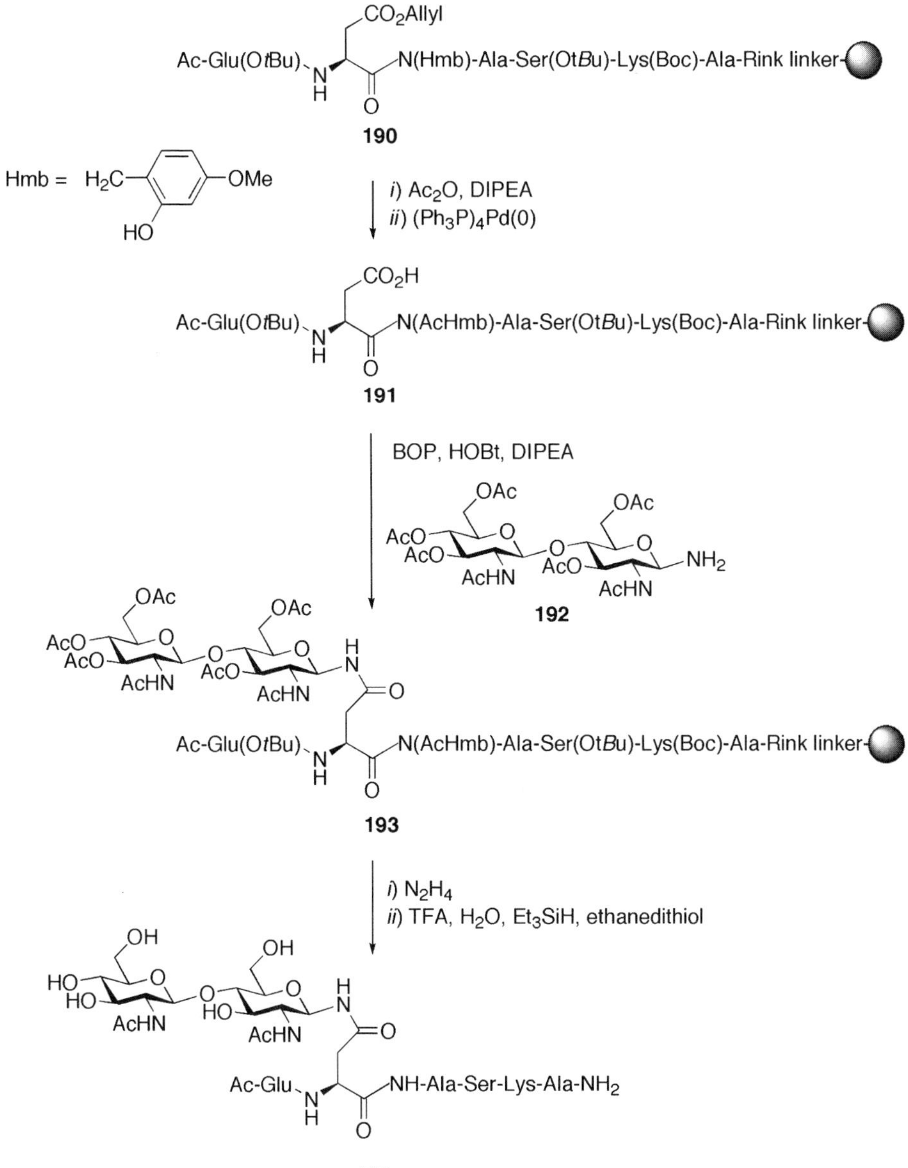

SCHEME 15.24 Convergent synthesis of *N*-linked glycopeptides by coupling of solid-supported peptide **191** to chitobiosyl amine **192** followed by cleavage and deprotection.

smaller glycans on glycopeptides has proved to be a successful alternative. To date, chemoenzymatic approaches to glycopeptides have mainly dealt with synthesis of *N*-linked glycopeptides [162,185–187], glycopeptides which carry the sialyl Lewis^{x} tetrasaccharide or fragments thereof [101,188–191] and sialylated glycopeptides of the mucin type [134].

The synthesis of an undecasaccharide *N*-linked to asparagine by a combination of chemical and glycosyl-transferase mediated glycosylations (cf. Scheme 15.22 and Scheme 15.23 in Section 15.4) is an impressive illustration of the power of the chemoenzymatic approach [162,185]. In another route to *N*-linked glycopeptides, the enzymes EndoA and EndoM were used for transglycosylations of oligosaccharides, such as $Man_9GlcNAc$ obtained from naturally derived $Man_9GlcNAc_2$-Asn, to chemically synthesized glycopeptides containing a GlcNAc-Asn moiety [186,187,192,193]. Despite its advantages, this route unfortunately suffers from the drawback that the yields in the transglycosylation step are only 10 to 30%.

In a display of the power of glycosyltransferases for elaboration of glycopeptides, two glycoforms of the sulfated *N*-terminal domain of P-selectin glycoprotein ligand-1 (PSGL-1) were prepared on a small scale (Scheme 15.25) [188]. This was achieved by chemical synthesis of glycopeptide **195** which was converted to **196** using core 1 β1,3-galactosyltransferase, which had been purified from rat liver. Glycopeptide **196** was then converted to **197** and **198** by application of five glycosyltransferases and one sulfotransferase, which were either expressed from recombinant sources or obtained from commercial sources. It was found that **197**, but not the isomeric **198**,

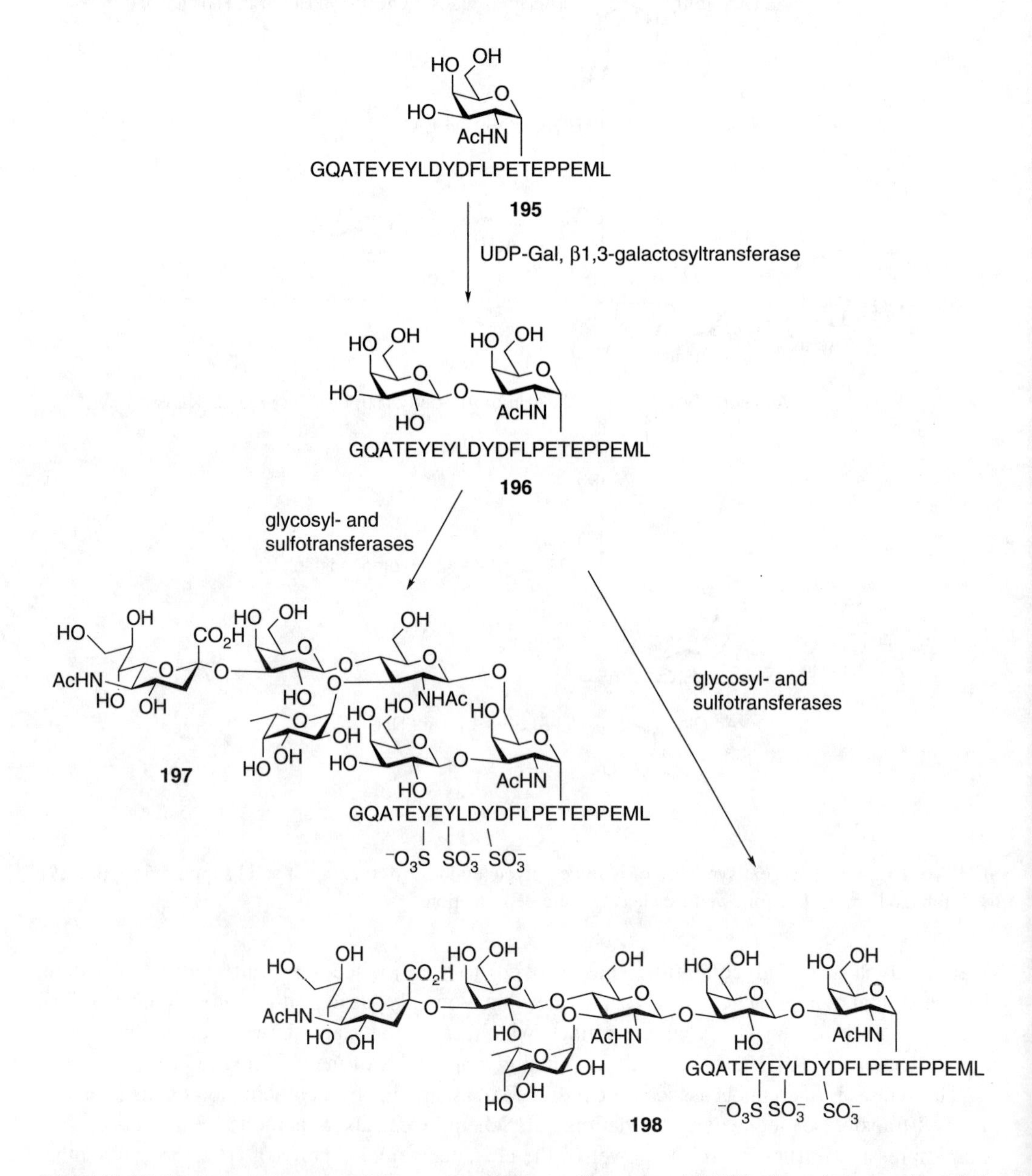

SCHEME 15.25 Enzymatic glycosylations and sulfations of glycopeptide **195** allowed preparation of the isomeric glycopeptides **197** and **198**.

bound well to P-selectin; a recognition event that is crucial for adhesion of leukocytes to the vascular cell wall during inflammation.

In a recent example dealing with the synthesis of mucin-type glycopeptides, building blocks corresponding to the simple Tn and T antigens were assembled into glycopeptides **199** and **201** by solid-phase synthesis (Scheme 15.26) [134]. Recombinant sialyltransferases and core 2 β1,6-*N*-acetylglucosaminyltransferase were then used to convert these readily available glycopeptides into the more complex **200**, **202** and **203**, which contain the sialyl-Tn and 2,3-sialyl-T antigens as well as the core 2 trisaccharide.

Chemoenzymatic synthesis can thus be an attractive approach for preparation of glycopeptides which contain larger oligosaccharide moieties. The approach is usually based on the fact that cassette-like building blocks corresponding to amino acids glycosylated with mono- or disaccharides can be prepared readily via chemical synthesis. Such building blocks allow site-specific incorporation of the carbohydrate moieties at selected positions in a polypeptide, in contrast to most enzymes which attach carbohydrates to peptides [112]. Both natural and nonnatural glycopeptides can be prepared since the glycosyl transferases which modify glycans on polypeptides often [134], but not always [194], show little specificity for the sequence of

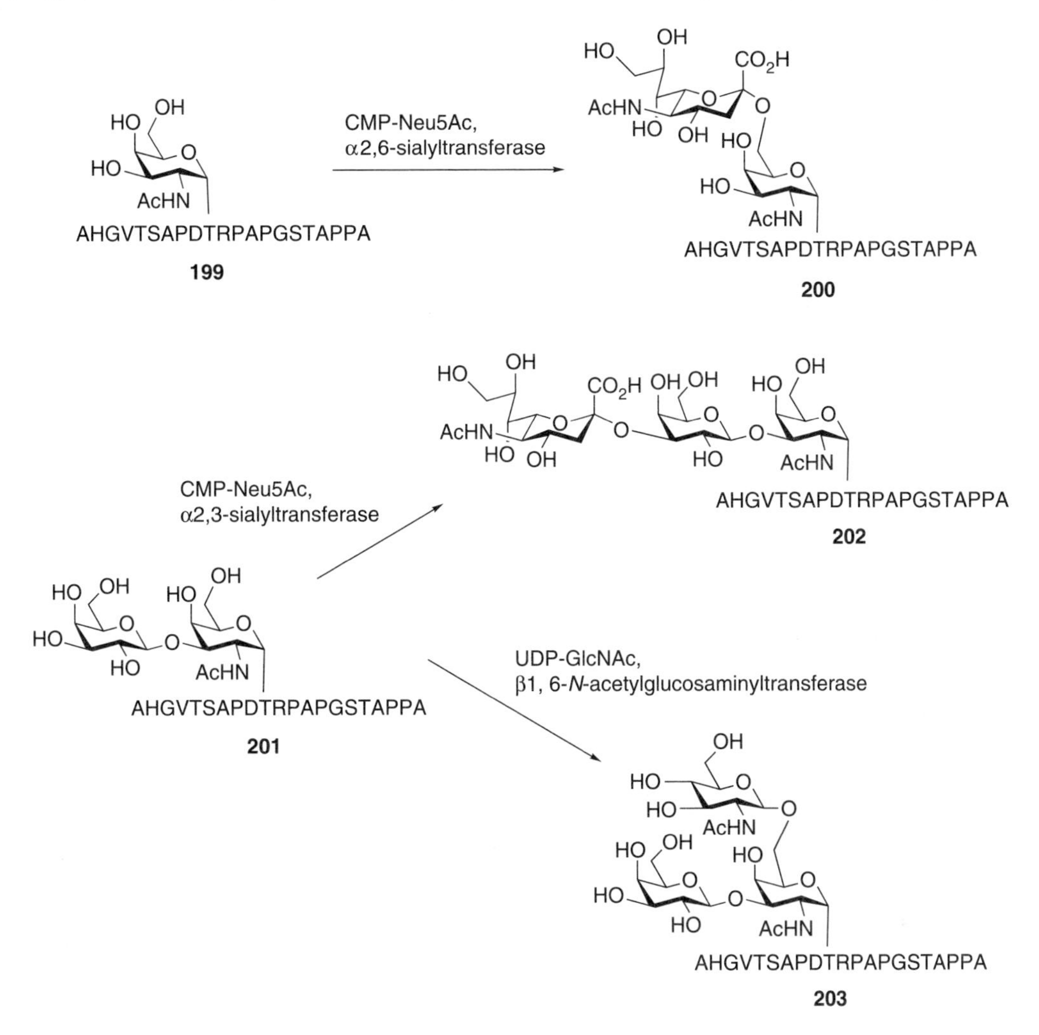

SCHEME 15.26 Enzymatic sialylations of the mucin-type Tn and T antigen glycopeptides **199** and **201** gave glycopeptides **200**, **202** and **203** which contained the sialyl-Tn and the sialyl-T antigens.

the peptide. If the enzymatic glycosylations are carried out in solution after cleavage of the glycopeptide from the solid phase, problems associated with the enzyme penetrating the cross-linked solid support can be avoided [195], as can incompatibilities between the enzyme and the solid support or the linker used as attachment for the glycopeptide [101]. However, it should be noted that the availability of glycosyl transferases is sometimes a limiting factor in a chemoenzymatic strategy.

15.6 SYNTHESIS OF GLYCOPROTEINS

The development of methodology for the synthesis of glycopeptides outlined in this chapter, as well as advances in fields such as enzymology and molecular and cell biology, has now brought synthesis of well-defined glycoproteins with natural glycosylation patterns within reach for the glycobiologist (reviewed in Ref. [196]). Direct chemical synthesis using glycosylated amino acids as building blocks in the solid-phase has been used for synthesis of "small" glycoproteins such as PMP-C, a serine protease inhibitor isolated from the insect *Locusta migratoria* [110,197]. Larger glycoproteins may be obtained by preparation of shorter glycopeptides, which are then condensed with other (glyco)polypeptides using ligation techniques. Native chemical ligation of a *C*-terminal thioester with a *N*-terminal cysteinyl peptide was, for instance, applied to synthesise the antimicrobial *O*-linked glycoprotein diptericin [198], whereas protease-catalyzed ligation was used for the synthesis of a single, unnatural glycoform of ribonuclease B (RNase-B) [199]. Enzyme-catalyzed techniques for elaboration of existing glycans on glycoproteins have also proved to be successful. This may involve use of glycosyltransferases to extend existing glycans, or trimming of these by glycosidases followed by glycosyltransferase-catalyzed assembly to give alternative structures in a process termed "glycoprotein remodeling" [199]. Bacterial systems used for expression of proteins do not usually provide glycosylation. Use of eukaryotic systems can circumvent this problem but heterogeneous products are often obtained. In the future, "glycosylation engineering" with regulated levels of glycosyltransferases, for instance, by introduction of extra glycosyltransferase-expressing genes, may contribute to solve this problem. Glycosyltransferase inhibitors, or inhibitors of glycosidases involved in trimming of *N*-linked oligosaccharides, have also been used to simplify the glycosylation pattern of glycoproteins produced in cell systems.

REFERENCES

1. Lis, H, Sharon, N, Protein glycosylation. Structural and functional aspects, *Eur. J. Biochem.*, 218, 1–27, 1993.
2. Sharon, N, Lis, H, *Glycoproteins: Structure and Function*, Gabius, H-J and Gabius, S, Eds., Chapman & Hall, Weinheim, pp. 133–162, 1997.
3. Dwek, R A, Glycobiology: toward understanding the function of sugars, *Chem. Rev.*, 96, 683–720, 1996.
4. Brockhausen, I, Schachter, H, Glycosyltransferases involved in *N*- and *O*-Glycan biosynthesis, In *Glycosciences: Status and Perspectives*, Gabius, H-J, and Gabius, S, Eds., Chapman & Hall, Weinheim, pp. 79–113, 1997.
5. Hart, G W, Brew, K, Grant, G A, Bradshaw, R A, Lennarz, W J, Primary structural requirements for the enzymatic formation of the *N*-glycosidic bond in glycoproteins, *J. Biol. Chem.*, 254, 9747–9753, 1979.
6. Kornfeld, R, Kornfeld, S, Assembly of asparagine-linked oligosaccharides, *Ann. Rev. Biochem.*, 54, 631–664, 1985.
7. Wieland, F, Heitzer, R, Schaefer, W, Asparaginylglucose: novel type of carbohydrate linkage, *Proc. Natl. Acad. Sci. USA*, 80, 5470–5474, 1983.
8. Paul, G, Lottspeich, F, Wieland, F, Asparaginyl-*N*-acetylgalactosamine: linkage unit of halobacterial glycosaminoglycan, *J. Biol. Chem.*, (1986), 1020–1024, 1986.

9. Messner, P, Sleytr, U B, Asparaginyl-rhamnose: a novel type of protein-carbohydrate linkage in a eubacterial surface-layer glycoprotein, *FEBS Lett.*, 228, 317–320, 1988.
10. Carraway, K L, Hull, S R, Cell surface mucin-type glycoproteins and mucin-like domains, *Glycobiology*, 1, 131–138, 1991.
11. Hanisch, F-G, Müller, S, Muc1: the polymorphic appearance of a human mucin, *Glycobiology*, 10, 439–449, 2000.
12. Taylor-Papadimitriou, J, Burchell, J, Miles, D W, Dalziel, M, MUC1 and cancer, *Biochem. Biophys. Acta*, 1455, 301–313, 1999.
13. Toyokuni, T, Singhal, A K, Synthetic carbohydrate vaccines based on tumour-associated antigens, *Chem. Soc. Rev.*, 231–242, 1995.
14. Danishefsky, S J, Allen, J R, From the laboratory to the clinic: a retrospective on fully synthetic carbohydrate-based anticancer vaccines, *Angew. Chem. Int. Ed.*, 39, 836–863, 2000.
15. Kjellén, L, Lindahl, U, Proteoglycans: structure and interactions, *Ann. Rev. Biochem.*, 60, 443–475, 1991.
16. Zachara, N E, Hart, G W, The emerging significance of *O*-GlcNAc in cellular regulation, *Chem. Rev.*, 102, 431–438, 2002.
17. Kivirikko, K I, Myllylä, R, Post-translational modifications, In *Collagen in Health and Disease*, Weiss, J B, and Jayson, M I V, Eds., Churchill Livingstone, Edinburgh, pp. 101–120, 1982.
18. Gohlke, M, Baude, G, Nuck, R, Grunow, D, Kannicht, C, Bringmann, P, Donner, P, Reutter, W, *O*-Linked L-fucose is present in *Desmodus rotundus* salivary plasminogen activator, *J. Biol. Chem.*, 271, 7381–7386, 1996.
19. Nakakura, N, Hietter, H, van Dorsselaer, A, Luu, B, Isolation and structural determination of three peptides from the insect *Locusta migratoria*, *Eur. J. Biochem.*, 204, 147–153, 1992.
20. Smythe, C, Caudwell, F B, Ferguson, M, Cohen, P, Isolation and structural analysis of a peptide containing the novel tyrosyl-glucose linkage in glycogenin, *EMBO J.*, 7, 2681–2686, 1988.
21. Bock, K, Schuster-Kolbe, J, Altman, E, Allmaier, G, Stahl, B, Christian, R, Sleytr, U B, Messner, P, Primary structure of the *O*-glycosidically linked glycan chain of the crystalline surface layer glycoprotein of *Thermoanaerobacter thermohydrosulfuricus* L111-69, *J. Biol. Chem.*, 269, 7137–7144, 1994.
22. Gunnarsson, A, Svensson, B, Nilsson, B, Svensson, S, Structural studies on the *O*-glyosidically linked carbohydrate chains of glucoamylase G1 from *Aspergillus niger*, *Eur. J. Biochem.*, 145, 463–467, 1984.
23. Gellerfors, P, Axelsson, K, Helander, A, Johansson, S, Kenne, L, Lindqvist, S, Pavlu, B, Skottner, A, Fryklund, L, Isolation and characterization of a glycosylated form of human insulin-like growth factor I produced in *Saccharomyces cerevisiae*, *J. Biol. Chem.*, 264, 11444–11449, 1989.
24. Varki, A, Biological roles of oligosaccharides: all of the theories are correct, *Glycobiology*, 3, 97–130, 1993.
25. Varki, A, Cummings, R, Esko, J, Freeze, H, Hart, G, Marth, J, *Essentials of Glycobiology*, Cold Spring Harbor, New York, 1999.
26. Kornfeld, S, Mellman, I, The biogenesis of lysosomes, *Ann. Rev. Cell. Biol.*, 5, 483–525, 1989.
27. Ashwell, G, Harford, J, Carbohydrate-specific receptors of the liver, *Ann. Rev. Biochem.*, 51, 531–554, 1982.
28. Phillips, M L, Nudelman, E, Gaeta, F C A, Perez, M, Singhal, A K, Hakomori, S-i, Paulson, J C, ELAM-1 mediates cell adhesion by recognition of a carbohydrate ligand, sialyl-Le^x, *Science*, 250, 1130–1132, 1990.
29. Broddefalk, J, Bäcklund, J, Almqvist, F, Johansson, M, Holmdahl, R, Kihlberg, J, T cells recognize a glycopeptide derived from type II collagen in a model for rheumatoid arthritis, *J. Am. Chem. Soc.*, 120, 7676–7683, 1998.
30. Holm, B, Bäcklund, J, Recio, M A F, Holmdahl, R, Kihlberg, J, Glycopeptide specificity of helper T cells obtained in mouse models for rheumatoid arthritis, *Chem. BioChem.*, 3, 1209–1222, 2002.
31. Bäcklund, J, Carlsen, S, Höger, T, Holm, B, Fugger, L, Kihlberg, J, Burkhardt, H, Holmdahl, R, Predominant selection of T cells specific for the glycosylated collagen type II epitope (263–270) in humanized transgenic mice and in rheumatoid arthritis, *Proc. Natl. Acad. Sci. USA*, 99, 9960–9965, 2002.

32. Karlsson, K-A, Microbial recognition of target-cell glycoconjugates, *Curr. Opin. Struct. Biol.*, 5, 622–635, 1995.
33. Soto, G E, Hultgren, S J, Bacterial adhesins: common themes and variations in architecture and assembly, *J. Bacteriol.*, 181, 1059–1071, 1999.
34. Kunz, H, Synthesis of glycopeptides, partial structures of biological recognition components, *Angew. Chem. Int. Ed.*, 26, 294–308, 1987.
35. Meldal, M, Glycopeptide synthesis, In *Neoglycoconjugates: Preparation and Applications*, Lee, Y C, and Lee, R T, Eds., Academic Press, Inc., San Diego, pp. 145–198, 1994.
36. Kihlberg, J, Elofsson, M, Solid-phase synthesis of glycopeptides: immunological studies with T cell stimulating glycopeptides, *Curr. Med. Chem.*, 4, 79–110, 1997.
37. Herzner, H, Reipen, T, Schultz, M, Kunz, H, Synthesis of glycopeptides containing carbohydrate and peptide recognition motifs, *Chem. Rev.*, 100, 4495–4537, 2000.
38. Seitz, O, Glycopeptide synthesis and the effects of glycosylation on protein structure and activity, *Chem. BioChem.*, 1, 214–246, 2000.
39. Cohen-Anisfeld, S T, Lansbury, P T Jr, A practical, convergent method for glycopeptide synthesis, *J. Am. Chem. Soc.*, 115, 10531–10537, 1993.
40. Vetter, D, Tumelty, D, Singh, S K, Gallop, M A, A versatile solid-phase synthesis of *N*-linked glycopeptides, *Angew. Chem. Int. Ed.*, 34, 60–62, 1995.
41. Offer, J, Quibell, M, Johnson, T, On-resin solid-phase synthesis of asparagine *N*-linked glycopeptides: use of *N*-(2-acetoxy-4-methoxybenzyl) (AcHmb) aspartyl amide-bond protection to prevent unwanted aspartimide formation, *J. Chem. Soc. Perkin Trans. 1*, 175–182, 1996.
42. Paul, B, Korytnyk, W, Synthesis of 2-acetamido-3,4,6-tri-*O*-acetyl-2-deoxy-β-D-glucopyranosylamine and dimer formation, *Carbohydr. Res.*, 67, 457–468, 1978.
43. Hollósi, M, Kollát, E, Laczkó, I, Medzihradszky, K F, Thurin, J, Otvos, L Jr, Solid-phase synthesis of glycopeptides: glycosylation of resin-bound serine-peptides by 3,4,6,-tri-*O*-acetyl-D-glucose-oxazoline, *Tetrahedron Lett.*, 32, 1531–1534, 1991.
44. Andrews, D M, Seale, P W, Solid-phase synthesis of *O*-mannosylated peptides: two strategies compared, *Int. J. Pept. Protein Res.*, 42, 165–170, 1993.
45. Paulsen, H, Schleyer, A, Mathieux, N, Meldal, M, Bock, K, New solid-phase oligosaccharide synthesis on glycopeptides bound to a solid phase, *J. Chem. Soc. Perkin Trans. 1*, 281–293, 1997.
46. Schleyer, A, Meldal, M, Manat, R, Paulsen, H, Bock, K, Direct solid-phase glycosylations of peptide templates on a novel PEG-based resin, *Angew. Chem. Int. Ed.*, 36, 1976–1978, 1997.
47. Wen, S, Guo, Z, Unprotected oligosaccharides as phase tags: solution-phase synthesis of glycopeptides with solid-phase workups, *Org. Lett.*, 3, 3773–3776, 2001.
48. Carpino, L A, Han, G Y, The 9-fluorenylmethoxycarbonyl amino-protecting group, *J. Org. Chem.*, 37, 3404–3409, 1972.
49. Vuljanic, T, Bergquist, K-E, Clausen, H, Roy, S, Kihlberg, J, Piperidine is preferred to morpholine for Fmoc cleavage in solid phase glycopeptide synthesis as exemplified by preparation of glycopeptides related to HIV gp120 and mucins, *Tetrahedron*, 52, 7983–8000, 1996.
50. Garg, H G, von dem Bruch, K, Kunz, H, Development in the synthesis of glycopeptides containing glycosyl L-asparagine, L-serine, and L-threonine, *Adv. Carbohydr. Chem. Biochem.*, 50, 277–310, 1994.
51. Elofsson, M, Walse, B, Kihlberg, J, Building blocks for glycopeptide synthesis: glycosylation of 3-mercaptopropionic acid and Fmoc amino acids with unprotected carboxyl groups, *Tetrahedron Lett.*, 32, 7613–7616, 1991.
52. Kihlberg, J, Elofsson, M, Salvador, L A, Direct synthesis of glycosylated amino acids from carbohydrate peracetates and Fmoc amino acids: solid-phase synthesis of biomedicinally interesting glycopeptides, *Methods Enzymol.*, 289, 221–245, 1997.
53. Urge, L, Otvos, L Jr, Lang, E, Wroblewski, K, Laczko, I, Hollosi, M, Fmoc-protected, glycosylated asparagines potentially useful as reagents in the solid phase synthesis of *N*-glycopeptides, *Carbohydr. Res.*, 235, 83–93, 1992.
54. Urge, L, Jackson, D C, Gorbics, L, Wroblewski, K, Graczyk, G, Otvos, L Jr, Synthesis and conformational analysis of *N*-glycopeptides that contain extended sugar chains, *Tetrahedron*, 50, 2373–2390, 1994.

55. Unverzagt, C, Kunz, H, Synthesis of glycopeptides and neoglycoproteins containing the fucosylated linkage region of *N*-glycoproteins, *Bioorg. Med. Chem.*, 2, 1189–1201, 1994.
56. Peters, S, Lowary, T L, Hindsgaul, O, Meldal, M, Bock, K, Solid-phase synthesis of a fucosylated glycopeptide of human factor IX with a fucose-α-(1-*O*)-serine linkage, *J. Chem. Soc. Perkin Trans. 1*, 3017–3022, 1995.
57. Sjölin, P, Elofsson, M, Kihlberg, J, Removal of acyl protective groups from glycopeptides: base does not epimerize peptide stereocenters and β-elimination is slow, *J. Org. Chem.*, 61, 560–565, 1996.
58. Sjölin, P, George, S K, Bergquist, K-E, Roy, S, Svensson, A, Kihlberg, J, Synthesis of deoxy and alanine-substituted derivatives of a T cell stimulating glycopeptide — an investigation of conditions for cleavage from the solid phase and deprotection, *J. Chem. Soc. Perkin Trans. 1*, 1731–1742, 1999.
59. Sjölin, P, Kihlberg, J, Deacetylation of N^{α}-methylated glycopeptides reveals that aza-enolates provide protection against β-elimination of carbohydrates *O*-linked to serine, *Tetrahedron Lett.*, 41, 4435–4439, 2000.
60. Paulsen, H, Schultz, M, Klamann, J-D, Waller, B, Paal, M, Building units of oligosaccharides, LXVI. Synthesis of *O*-glycopeptide blocks of glycophorine, *Liebigs Ann. Chem.*, 2028–2048, 1985.
61. Reimer, K B, Meldal, M, Kusumoto, S, Fukase, K, Bock, K, Small-scale solid-phase *O*-glycopeptide synthesis of linear and cyclized hexapeptides from blood-clotting factor IX containing *O*-(α-D-Xyl-(1→3)-α-D-Xyl-(1→3)-β-D-Glc)-L-Ser, *J. Chem. Soc. Perkin Trans. 1*, 925–932, 1993.
62. Sjölin, P, Kihlberg, J, Use of fluorobenzoyl protective groups in synthesis of glycopeptides: β-elimination of *O*-linked carbohydrates is suppressed, *J. Org. Chem.*, 66, 2957–2965, 2001.
63. Christiansen-Brams, I, Jansson, A M, Meldal, M, Breddam, K, Bock, K, Silyl protection in the solid-phase synthesis of *N*-linked glycopeptides. Preparation of glycosylated fluorogenic substrates for subtilisin, *Bioorg. Med. Chem.*, 2, 1153–1167, 1994.
64. Elofsson, M, Salvador, L A, Kihlberg, J, Preparation of Tn and sialyl Tn building blocks used in Fmoc solid-phase synthesis of glycopeptide fragments from HIV gp120, *Tetrahedron*, 53, 369–390, 1997.
65. Broddefalk, J, Bergquist, K-E, Kihlberg, J, Use of acid-labile protective groups for carbohydrate moieties in synthesis of glycopeptides related to type II collagen, *Tetrahedron*, 54, 12047–12070, 1998.
66. Broddefalk, J, Forsgren, M, Sethson, I, Kihlberg, J, Preparation of a diglycosylated hydroxylysine building block used in solid-phase synthesis of a glycopeptide from type II collagen, *J. Org. Chem.*, 64, 8948–8953, 1999.
67. Holm, B, Linse, S, Kihlberg, J, Synthesis of an *N*-linked glycopeptide from vitamin K-dependent protein S, *Tetrahedron*, 54, 11995–12006, 1998.
68. Nakahara, Y, Nakahara, Y, Ogawa, T, Solid-phase synthesis of an *O*-linked glycopeptide based on a benzyl-protected glycan approach, *Carbohydr. Res.*, 292, 71–81, 1996.
69. Guo, Z-W, Nakahara, Y, Nakahara, Y, Ogawa, T, Solid-phase synthesis of the CD52 glycopeptide carrying an *N*-linked core pentasaccharide structure, *Angew. Chem. Int. Ed.*, 36, 1464–1466, 1997.
70. Bodanszky, M, Martinez, J, Side reactions in peptide synthesis, *Synthesis*, 333–356, 1981.
71. Nakahara, Y, Nakahara, Y, Ito, Y, Ogawa, T, Total synthesis of B-chain of human α2HS glycoprotein, *Tetrahedron Lett.*, 38, 7211–7214, 1997.
72. Chan, W C, White, P D, *Fmoc Solid Phase Peptide Synthesis*, Oxford University, Oxford, 2000.
73. Wang, S-S, p-Alkoxybenzyl alcohol resin and p-alkoxy benzyloxycarbonyl hydrazine resin for solid phase synthesis of protected peptide fragments, *J. Am. Chem. Soc.*, 94, 1328–1333, 1973.
74. Sheppard, R C, Williams, B J, Acid-labile resin linkage agents for use in solid phase peptide synthesis, *Int. J. Pept. Protein Res.*, 20, 451–454, 1982.
75. Rink, H, Solid-phase synthesis of protected peptide fragments using a trialkoxy-diphenyl-methylester resin, *Tetrahedron Lett.*, 28, 3787–3790, 1987.
76. Seitz, O, Kunz, H, HYCRON, an allylic anchor for high-efficiency solid phase synthesis of protected peptides and glycopeptides, *J. Org. Chem.*, 62, 813–826, 1997.
77. Meinjohanns, E, Meldal, M, Paulsen, H, Dwek, R A, Bock, K, Novel sequential solid-phase synthesis of *N*-linked glycopeptides from natural sources, *J. Chem. Soc. Perkin Trans. 1*, 549–560, 1998.
78. Jansson, A M, Jensen, K J, Meldal, M, Lomako, J, Lomako, W M, Olsen, C E, Bock, K, Solid-phase glycopeptide synthesis of tyrosine-glycosylated glycogenin fragments as substrates for glucosylation by glycogenin, *J. Chem. Soc. Perkin Trans. 1*, 1001–1006, 1996.

79. Norberg, T, Lüning, B, Tejbrant, J, Solid-phase synthesis of *O*-glycopeptides, *Methods Enzymol.*, 247, 87–106, 1994.
80. Carpino, L A, 1-Hydroxy-7-azabenzotriazole. An efficient peptide coupling additive, *J. Am. Chem. Soc.*, 115, 4397–4398, 1993.
81. Cameron, L R, Holder, J L, Meldal, M, Sheppard, R C, Peptide synthesis. Part 13. Feedback control in solid phase synthesis. Use of fluorenylmethoxycarbonyl amino acid 3,4-dihydro-4-oxo-1,2,3-benzotriazin-3-yl esters in a fully automated system, *J. Chem. Soc. Perkin Trans. 1*, 2895–2901, 1988.
82. Flegel, M, Sheppard, R C, A sensitive general method for quantitative monitoring of continuous flow solid phase peptide synthesis, *J. Chem. Soc. Chem. Commun.*, 536–538, 1990.
83. Rio, S, Beau, J-M, Jacquinet, J-C, Synthesis of glycopeptides from the carbohydrate-protein linkage region of proteoglycans, *Carbohydr. Res.*, 219, 71–90, 1991.
84. Rio, S, Beau, J-M, Jacquinet, J-C, Total synthesis of the carbohydrate-protein linkage region common to several mammalian proteoglycans, *Carbohydr. Res.*, 244, 295–313, 1993.
85. Rio, S, Beau, J-M, Jacquinet, J-C, Synthesis of sulfated and phosphorylated glycopeptides from the carbohydrate-protein linkage region of proteoglycans, *Carbohydr. Res.*, 255, 103–124, 1994.
86. Nishimura, H, Takao, T, Hase, S, Shimonishi, Y, Iwanaga, S, Human factor IX has a tetrasaccharide *O*-glycosidically linked to serine 61 through the fucose residue, *J. Biol. Chem.*, 267, 17520–17525, 1992.
87. Salvador, L A, Elofsson, M, Kihlberg, J, Preparation of building blocks for glycopeptide synthesis by glycosylation of Fmoc amino acids having unprotected carboxyl groups, *Tetrahedron*, 51, 5643–5656, 1995.
88. Fukase, K, Hase, S, Ikenaka, T, Kusumoto, S, Synthesis of new serine-linked oligosaccharides in blood-clotting factors VII and IX and protein Z. The syntheses of *O*-α-D-xylopyranosyl-(1→3)-D-glucopyranose, *O*-α-D-xylopyranosyl-(1→3)-O-α-D-xylopyranosyl-(1→3)-D-glucopyranose, and their conjugates with serine, *Bull. Chem. Soc. Jpn.*, 65, 436–445, 1992.
89. Spiro, R G, The structure of the disaccharide unit of the renal glomerular basement membrane, *J. Biol. Chem.*, 242, 4813–4823, 1967.
90. Holm, B, Broddefalk, J, Flodell, S, Wellner, E, Kihlberg, J, An improved synthesis of a galactosylated hydroxylysine building block and its use in solid-phase glycopeptide synthesis, *Tetrahedron*, 56, 1579–1586, 2000.
91. Malkar, N B, Lauer-Fields, J L, Fields, G B, Convenient synthesis of glycosylated hydroxylysine derivatives for use in solid-phase peptide synthesis, *Tetrahedron Lett.*, 41, 1137–1140, 2000.
92. Yuen, C-T, Chai, W, Loveless, R W, Lawson, A M, Margolis, R U, Feizi, T, Brain contains HNK-1 immunoreactive *O*-glycans of the sulfoglucuronyl lactosamine series that terminate in 2-linked or 2,6-linked hexose (mannose), *J. Biol. Chem.*, 272, 8924–8931, 1997.
93. Vegad, H, Gray, C J, Somers, P J, Dutta, A S, Glycosylation of Fmoc amino acids: preparation of mono- and di-glycosylated derivatives and their incorporation into Arg-Gly-Asp (RGD)-containing glycopeptides, *J. Chem. Soc. Perkin Trans. 1*, 1429–1441, 1997.
94. Jansson, A M, Meldal, M, Bock, K, Solid-phase synthesis and characterization of *O*-dimannosylated heptadecapeptide analogues of human insulin-like growth factor 1 (IGF-1), *J. Chem. Soc. Perkin Trans. 1*, 1699–1707, 1992.
95. Takeda, T, Kanemitsu, T, Ishiguro, M, Ogihara, Y, Matsubara, M, Synthesis of a glycopeptide with phytoalexin elicitor activity I. Syntheses of a triglycosyl L-serine and a triglycosyl L-seryl-L-proline dipeptide, *Carbohydr. Res.*, 256, 59–69, 1994.
96. Takeda, T, Kanemitsu, T, Shimizu, N, Ogihara, Y, Matsubara, M, Synthesis of triglycosyl tetrapeptides and hexaglycosyl tetrapeptide, *Carbohydr. Res.*, 283, 81–93, 1996.
97. Seifert, J, Ogawa, T, Ito, Y, Syntheses of α-dystroglycan derived sialylated glycosyl amino acids carrying a novel mannosyl serine/threonine linkage, *Tetrahedron Lett.*, 40, 6803–6807, 1999.
98. Micheel, F, Köchling, H, Darstellung von glykosiden des D-glucosamins mit aliphatischen und aromatischen alkoholen und mit serin nach der oxazolin-methode, *Chem. Ber.*, 91, 673–676, 1958.
99. Jones, J K N, Perry, M B, Shelton, B, Walton, D J, The carbohydrate–protein linkage in glycoproteins. Part I. The synthesis of some model substituted amides and an L-seryl-D-glucosaminide, *Can. J. Chem.*, 39, 1005–1016, 1961.

100. Arsequell, G, Krippner, L, Dwek, R A, Wong, S Y C, Building blocks for solid-phase glycopeptide synthesis: 2-acetamido-2-deoxy-β-D-glycosides of FmocSerOH and FmocThrOH, *J. Chem. Soc. Chem. Commun.*, 2383–2384, 1994.
101. Seitz, O, Wong, C-H, Chemoenzymatic solution- and solid-phase synthesis of *O*-glycopeptides of the mucin domain of MAdCAM-1. A general route to *O*-LacNAc, *O*-sialyl-LacNAc, and *O*-sialyl-Lewis-X peptides, *J. Am. Chem. Soc.*, 119, 8766–8776, 1997.
102. Pohl, T, Waldmann, H, Chemoenzymatic synthesis of a characteristic phosphorylated and glycosylated peptide fragment of the large subunit of mammalian RNA polymerase II, *J. Am. Chem. Soc.*, 119, 6702–6710, 1997.
103. Vargas-Berenguel, A, Meldal, M, Paulsen, H, Bock, K, Convenient synthesis of *O*-(2-acetamido-2-deoxy-β-D-glucopyranosyl)-serine and -threonine building blocks for solid-phase glycopeptide assembly, *J. Chem. Soc. Perkin Trans. 1*, 2615–2619, 1994.
104. Schultz, M, Kunz, H, Enzymatic glycosylation of *O*-glycopeptides, *Tetrahedron Lett.*, 33, 5319–5322, 1992.
105. Meinjohanns, E, Meldal, M, Bock, K, Efficient synthesis of *O*-(2-acetamido-2-deoxy-β-D-glucopyranosyl)-Ser/Thr building blocks for SPPS of *O*-GlcNAc glycopeptides, *Tetrahedron Lett.*, 36, 9205–9208, 1995.
106. Saha, U K, Schmidt, R R, Efficient synthesis of *O*-(2-acetamido-2-deoxy-β-D-glucopyranosyl)-serine and -threonine building blocks for glycopeptide formation, *J. Chem. Soc. Perkin Trans.*, 1, 1855–1860, 1997.
107. Meinjohanns, E, Vargas-Berenguel, A, Meldal, M, Paulsen, H, Bock, K, Comparison of *N*-Dts and *N*-Aloc in the solid-phase synthesis of *O*-GlcNAc glycopeptide fragments of RNA-polymerase II and mammalian neurofilaments, *J. Chem. Soc. Perkin Trans. 1*, 2165–2175, 1995.
108. Jensen, K J, Hansen, P R, Venugopal, D, Barany, G, Synthesis of 2-acetamido-2-deoxy-β-D-glucopyranose *O*-glycopeptides from *N*-dithiasuccinoyl-protected derivatives, *J. Am. Chem. Soc.*, 118, 3148–3155, 1996.
109. Liu, J, Gin, D Y, C2-Amidoglycosylation. Scope and mechanism of nitrogen transfer, *J. Am. Chem. Soc.*, 124, 9789–9797, 2002.
110. Hietter, H, Schultz, M, Kunz, H, Solid-phase synthesis of a 36 amino acid *O*-fucosyl-peptide with the uncommon Thr-Fuc linkage, *Synlett*, 1219–1220, 1995.
111. Elofsson, M, Roy, S, Salvador, L A, Kihlberg, J, Building blocks for glycopeptide synthesis: preparation of α-*O*-fucosylated Fmoc serine and threonine in one step from L-fucose tetraacetate, *Tetrahedron Lett.*, 37, 7645–7648, 1996.
112. Wandall, H H, Hassan, H, Mirgorodskaya, E, Kristensen, A K, Roepstorff, P, Bennett, E P, Nielsen, P A, Hollingsworth, M A, Burchell, J, Taylor-Papadimitriou, J, Clausen, H, Substrate specificities of three members of the human UDP-*N*-acetyl-α-D-galactosamine:polypeptide *N*-acetylgalactosaminyltransferase family, GalNAc-T1, -T2, and -T3, *J. Biol. Chem.*, 272, 23503–23514, 1997.
113. Springer, G F, T and Tn, general carcinoma autoantigens, *Science*, 224, 1198–1206, 1984.
114. Hakomori, S-I, Aberrant glycosylation in tumors and tumor-associated carbohydrate antigens, *Adv. Cancer Res.*, 52, 257–331, 1989.
115. Irimura, T, Denda, K, Iida, S-I, Takeuchi, H, Kato, K, Diverse glycosylation of MUC1 and MUC2: potential significance in tumor immunity, *J. Biochem. (Tokyo)*, 126, 975–985, 1999.
116. Brocke, C, Kunz, H, Synthesis of tumor-associated glycopeptide antigens, *Bioorg. Med. Chem.*, 10, 3085–3112, 2002.
117. Nakahara, Y, Iijima, H, Ogawa, T, Stereocontrolled approaches to *O*-glycopeptide synthesis, In *Synthetic Oligosaccharides*, Kovác, P, Ed., American Chemical Society, Washington DC, *ACS Symp. Ser.*, Vol. 560, pp. 249–266, 1994.
118. Paulsen, H, Stenzel, W, Bausteine von oligosacchariden. Synthese α-glykosidish verknüpfter 2-aminozucker-oligosaccharide, *Angew. Chem.*, 87, 547–548, 1975.
119. Paulsen, H, Stenzel, W, Stereoselektive synthese α-glycosidisch verknüpfter di- und oligosaccharide der 2-amino-2-desoxy-D-glucopyranose, *Chem. Ber.*, 111, 2334–2347, 1978.
120. Lemieux, R U, Ratcliffe, R M, The azidonitration of tri-*O*-acetyl-D-galactal, *Can. J. Chem.*, 57, 1244–1251, 1979.
121. Kuduk, S D, Schwarz, J B, Chen, X-T, Glunz, P W, Sames, D, Ragupathi, G, Livingston, P O, Danishefsky, S J, Synthetic and immunological studies on clustered modes of mucin-related Tn and

TF *O*-linked antigens: the preparation of a glycopeptide-based vaccine for clinical trials against prostate cancer, *J. Am. Chem. Soc.*, 120, 12474–12485, 1998.

122. Kunz, H, Birnbach, S, Synthesis of *O*-glycopeptides of the tumor-associated Tn- and T-antigen type and their binding to bovine serum albumin, *Angew. Chem. Int. Ed.*, 25, 360–362, 1986.
123. Paulsen, H, Adermann, K, Synthesis of *O*-glycopeptides of the *N*-terminus of interleukin-2, *Liebigs Ann. Chem.*, 751–769, 1989.
124. Broddefalk, J, Nilsson, U, Kihlberg, J, An improved synthesis of 3,4,6-tri-*O*-acetyl-2-azido-2-deoxy-α-D-galactopyranosyl bromide: a key component for synthesis of glycopeptides and glycolipids, *J. Carbohydr. Chem.*, 13, 129–132, 1994.
125. Yule, J E, Wong, T C, Gandhi, S S, Qiu, D, Riopel, M A, Koganty, R R, Steric control of *N*-acetylgalactosamine in glycosidic bond formation, *Tetrahedron Lett.*, 36, 6839–6842, 1995.
126. Winterfeld, G A, Ito, Y, Ogawa, T, Schmidt, R R, A novel and efficient route towards α-GalNAc-Ser and α-GalNAc-Thr building blocks for glycopeptide synthesis, *Eur. J. Org. Chem.*, 1167–1171, 1999.
127. Paulsen, H, Hölck, J-P, Synthese der glycopeptide *O*-β-D-galactopyranosyl-(1→3)-*O*-(2-acetamido-2-desoxy-α-D-galactopyranosyl)-(1→3)-L-serin und L-threonin, *Carbohydr. Res.*, 109, 89–107, 1982.
128. Verez Bencomo, V, Jacquinet, J-C, Sinay, P, The synthesis of derivatives of O-β-D-galactopyranosyl-(1→3)-*O*-(2-acetamido-2-deoxy-α-D-galactopyranosyl-L-serine and -L-threonine, *Carbohydr. Res.*, 110, C9–C11, 1982.
129. Paulsen, H, Peters, S, Bielfeldt, T, Meldal, M, Bock, K, Synthesis of the glycosyl amino acids N^{α}-Fmoc-Ser[Ac_4-β-D-Galp-(1→3)-Ac_2-α-D-$GalN_3p$]-OPfp and N^{α}-Fmoc-Thr[Ac_4-β-D-Galp-(1→3)-Ac_2-α-D-$GalN_3p$]-OPfp and the application in the solid-phase peptide synthesis of multiply glycosylated mucin peptides with T^n and T antigenic structures, *Carbohydr. Res.*, 268, 17–34, 1995.
130. Qiu, D, Gandhi, S S, Koganty, R R, βGal(1-3)GalNAc block donor for the synthesis of TF and αsialyl(2-6)TF as glycopeptide building blocks, *Tetrahedron Lett.*, 37, 595–598, 1996.
131. Winterfeld, G A, Schmidt, R R, Nitroglycal concatenation: a broadly applicable and efficient approach to the synthesis of complex *O*-glycans, *Angew. Chem. Int. Ed.*, 40, 2654–2657, 2001.
132. Paulsen, H, Advances in selective chemical syntheses of complex oligosaccharides, *Angew. Chem. Int. Ed.*, 21, 155–173, 1982.
133. Mathieux, N, Paulsen, H, Meldal, M, Bock, K, Synthesis of glycopeptide sequences of repeating units of the mucins MUC 2 and MUC 3 containing oligosaccharide side-chains with core 1, core 2, core 3, core 4 and core 6 structure, *J. Chem. Soc. Perkin Trans. 1*, 2359–2368, 1997.
134. George, S K, Schwientek, T, Holm, B, Reis, C A, Clausen, H, Kihlberg, J, Chemoenzymatic synthesis of sialylated glycopeptides derived from mucins and T cell stimulating peptides, *J. Am. Chem. Soc.*, 123, 11117–11125, 2001.
135. Iijima, H, Ogawa, T, Total synthesis of 3-*O*-[2-acetamido-6-*O*-(*N*-acetyl-α-D-neuraminyl)-2-deoxy-α-D-galactosyl]-L-serine and a stereoisomer, *Carbohydr. Res.*, 172, 183–193, 1988.
136. Liebe, B, Kunz, H, Solid-phase synthesis of a tumor-associated sialyl-T_N antigen glycopeptide with a partial sequence of the "tandem repeat" of the MUC-1 mucin, *Angew. Chem. Int. Ed.*, 36, 618–621, 1997.
137. Liebe, B, Kunz, H, Solid-phase synthesis of a sialyl-Tn-glycoundecapeptide of the MUC1 Repeating Unit, *Helv. Chim. Acta*, 80, 1473–1482, 1997.
138. Schwarz, J B, Kuduk, S D, Chen, X-T, Sames, D, Glunz, P W, Danishefsky, S J, A broadly applicable method for the efficient synthesis of α-*O*-linked glycopeptides and clustered sialic acid residues, *J. Am. Chem. Soc.*, 121, 2662–2673, 1999.
139. Nakahara, Y, Nakahara, Y, Ito, Y, Ogawa, T, Solid-phase synthesis of the B-chain of human α2HS glycoprotein, *Carbohydr. Res.*, 309, 287–296, 1998.
140. Komba, S, Meldal, M, Werdelin, O, Jensen, T, Bock, K, Convenient synthesis of Thr and Ser carrying the tumor associated sialyl-(2→3)-T antigen as building blocks for solid-phase glycopeptide synthesis, *J. Chem. Soc. Perkin Trans. 1*, 415–419, 1999.
141. Komba, S, Werdelin, O, Jensen, T, Meldal, M, Synthesis of tumor associated sialyl-T-glycopeptides and their immunogenicity, *J. Pept. Sci.*, 6, 585–593, 2000.
142. Iijima, H, Ogawa, T, Synthesis of a mucin-type *O*-glycosylated amino acid, β-Gal-(1→3)-[α-Neu5Ac-(2→6)]-α-GalNAc-(1→3)-Ser, *Carbohydr. Res.*, 186, 95–106, 1989.

143. Sames, D, Chen, X-T, Danishefsky, S J, Convergent total synthesis of a tumour-associated mucin motif, *Nature*, 389, 587–591, 1997.
144. Paulsen, H, Syntheses, conformations and X-ray structure analyses of the saccharide chains from the core regions of glycoproteins, *Angew. Chem. Int. Ed.*, 29, 823–839, 1990.
145. Arsequell, G, Valencia, G, Recent advances in the synthesis of complex *N*-glycopeptides, *Tetrahedron Asymmetry*, 10, 3045–3094, 1999.
146. Kunz, H, Dombo, B, Solid phase synthesis of peptides and glycopeptides on polymeric supports with allylic anchor groups, *Angew. Chem. Int. Ed.*, 27, 711–713, 1988.
147. Otvos, L Jr, Wroblewski, K, Kollat, E, Perczel, A, Hollosi, M, Fasman, G D, Ertl, H C J, Thurin, J, Coupling strategies in solid-phase synthesis of glycopeptides, *Pept. Res.*, 2, 362–366, 1989.
148. Otvos, L Jr, Urge, L, Hollosi, M, Wroblewski, K, Graczyk, G, Fasman, G D, Thurin, J, Automated solid-phase synthesis of glycopeptides. Incorporation of unprotected mono- and disaccharide units of *N*-glycoprotein antennae into T cell epitopic peptides, *Tetrahedron Lett.*, 31, 5889–5892, 1990.
149. Meldal, M, Bock, K, Pentafluorophenyl esters for temporary carboxyl group protection in solid phase synthesis of *N*-linked glycopeptides, *Tetrahedron Lett.*, 31, 6987–6990, 1990.
150. Christiansen-Brams, I, Meldal, M, Bock, K, Protected-mode synthesis of *N*-linked glycopeptides: Single-step preparation of building blocks as peracetyl glycosylated Nα Fmoc asparagine OPfp esters, *J. Chem. Soc. Perkin Trans. 1*, 1461–1471, 1993.
151. Meldal, M, Recent developments in glycopeptide and oligosaccharide synthesis, *Curr. Opin. Struct. Biol.*, 4, 710–718, 1994.
152. Bodanszky, M, Natarajan, S, Side reactions in peptide synthesis. II. Formation of succinimide derivatives from aspartyl residues, *J. Org. Chem.*, 40, 2495–2499, 1975.
153. Lloyd-Williams, P, Albericio, F, Giralt, E, *Chemical Approaches to the Synthesis of Peptides and Proteins*, CRC Press LLC, New York, pp. 61–62, 1997.
154. Kunz, H, Waldman, H, März, J, Synthesis of partial structures of *N*-glycopeptides representing the linkage regions of the transmembrane neuraminidase of an influenza virus and of factor B of the human complement system, *Liebigs Ann. Chem.*, 45–49, 1989.
155. Thiem, J, Wiemann, T, Combined chemoenzymic synthesis of *N*-glycoprotein synthons, *Angew. Chem.*, 102, 78–80, 1990.
156. Tropper, F D, Andersson, F O, Braun, S, Roy, R, Phase transfer catalysis as a general and stereoselective entry into glycosyl azides from glycosyl halides, *Synthesis*, 618–620, 1992.
157. Marks, G S, Neuberger, A, Synthetic studies relating to the carbohydrate-protein linkage in egg albumin, *J. Chem. Soc.*, 4872–4879, 1961.
158. Nakabayashi, S, Warren, C D, Jeanloz, R W, The preparation of a partially protected heptasaccharide–asparagine intermediate for glycopeptide synthesis, *Carbohydr. Res.*, 174, 279–289, 1988.
159. McDonald, F E, Danishefsky, S J, A stereoselective route from glycals to asparagine-linked *N*-protected glycopeptides, *J. Org. Chem.*, 57, 7001–7002, 1992.
160. von dem Bruch, K, Kunz, H, Synthesis of *N*-glycopeptide clusters with Lewis^x antigen side chains and their binding of carrier proteins, *Angew. Chem. Int. Ed.*, 33, 101–103, 1994.
161. Saha, U K, Roy, R, First synthesis of *N*-linked-glycopeptoid as new glycopeptidomimetics, *Tetrahedron Lett.*, 36, 3635–3638, 1995.
162. Unverzagt, C, Chemoenzymatic synthesis of a sialylated undecasaccharide–asparagine conjugate, *Angew. Chem. Int. Ed.*, 35, 2350–2353, 1996.
163. Wang, Z-G, Zhang, X, Live, D, Danishefsky, S J, From glycals to glycopeptides: a convergent and stereoselective total synthesis of a high mannose *N*-linked glycopeptide, *Angew. Chem. Int. Ed.*, 39, 3652–3656, 2000.
164. Iserloh, U, Dudkin, V, Wang, Z-G, Danishefsky, S J, Reducing oligosaccharides via glycal assembly: on the remarkable stability of anomeric hydroxyl groups to global deprotection with sodium in liquid ammonia, *Tetrahedron Lett.*, 43, 7027–7030, 2002.
165. Kunz, H, Unverzagt, C, Protecting-group-dependent stability of intersaccharide bonds — Synthesis of a fucosyl-chitobiose glycopeptide, *Angew. Chem. Int. Ed.*, 27, 1697–1699, 1988.
166. Clark, R S, Banerjee, S, Coward, J K, Yeast oligosaccharyltransferase: glycosylation of peptide substrates and chemical characterization of the glycopeptide product, *J. Org. Chem.*, 55, 6275–6285, 1990.

167. Garcia, J, Urpi, F, Vilarrasa, J, New synthetic tricks. Triphenylphosphine-mediated amide formation from carboxylic acids and azides, *Tetrahedron Lett.*, 25, 4841–4844, 1984.
168. Urpi, F, Vilarrasa, J, New synthetic tricks. Advantages of using triethylphosphine in some phosphorus-based reactions, *Tetrahedron Lett.*, 27, 4623–4624, 1986.
169. Maunier, V, Boullanger, P, Lafont, D, A one-pot synthesis of glycosyl amides from glycosyl azides using a modified Staudinger reaction, *J. Carbohydr. Chem.*, 16, 231–235, 1997.
170. Mizuno, M, Muramoto, I, Kobayashi, K, Yaginuma, H, Inazu, T, A simple method for the synthesis of N^{β}-glycosylated asparagine and -glutamine derivatives, *Synthesis*, 162–165, 1999.
171. Likhosherstov, L M, Novikova, O S, Derevitskaja, V A, Kochetkov, N K, A new simple synthesis of amino sugar β-D-glycosylamines, *Carbohydr. Res.*, 146, c1–c5, 1986.
172. Urge, L, Kollat, E, Hollosi, M, Laczko, I, Wroblewski, K, Thurin, J, Otvos, L Jr, Solid-phase synthesis of glycopeptides: synthesis of N^{α}-fluorenylmethoxycarbonyl L-asparagine N^{β}-glycosides, *Tetrahedron Lett.*, 32, 3445–3448, 1991.
173. Zurabyan, S E, Macharadze, R G, Khorlin, A Y, Synthesis of 2-acetamido-2-deoxy-β-D-glucopyranosyl-L-asparagine derivative, *Bioorg. Khim.*, 4, 1135–1136, 1978.
174. Khorlin, A Y, Zurabyan, S E, Macharadze, R G, Synthesis of glycosylamides and 4-*N*-glycosyl-L-asparagine derivatives, *Carbohydr. Res.*, 85, 201–208, 1980.
175. Günther, W, Kunz, H, Synthesis of a β-mannosyl–chitobiosyl–asparagine conjugate: a central core region element of *N*-glycoproteins, *Angew. Chem.*, 102, 1050–1051, 1990.
176. Ogawa, T, Yamamoto, H, Nukada, T, Kitajima, T, Sugimoto, M, Synthetic approach to glycan chains of a glycoprotein and a proteoglycan, *Pure Appl. Chem.*, 56, 779–795, 1984.
177. Paulsen, H, Wilkens, R, Brockhausen, I, Building units of oligosaccharides. CIV. Synthesis of branched tetrasaccharide and pentasaccharide structures of *N*-glycoproteins methylated at 4′-OH of the branching unit, *Liebigs Ann. Chem.*, 1303–1313, 1992.
178. Guo, Z-W, Ito, Y, Nakahara, Y, Ogawa, T, Synthetic study on a novel Asn-linked core structure: synthesis of a pentasaccharide α-D-Man-(1-3)-[α-D-Man-(1-6)]-β-D-Man-(1-4)-[β-D-GlcNAc-(1-6)]-β-D-GlcNAc-OMp, *Carbohydr. Res.*, 306, 539–544, 1998.
179. Ogawa, T, Sugimoto, M, Kitajima, T, Sadozai, K K, Nukada, T, Synthetic studies on cell surface glycans. 51. Total synthesis of an undecasaccharide. A typical carbohydrate sequence for the complex type of glycan chains of a glycoprotein, *Tetrahedron Lett.*, 27, 5739–5742, 1986.
180. Guo, Z-W, Nakahara, Y, Nakahara, Y, Ogawa, T, Solid-phase synthesis of CD52 glycopeptide and an efficient route to Asn-core pentasaccharide conjugate, *Bioorg. Med. Chem.*, 5, 1917–1924, 1997.
181. Ekborg, G, Lindberg, B, Lönngren, J, Synthesis of β-D-mannopyranosides, *Acta Chem. Scand.*, 26, 3287–3292, 1972.
182. Kunz, H, Günther, W, β-Mannoside synthesis by configuration inversion of β-glucosides by intramolecular nucleophilic substitution, *Angew. Chem.*, 100, 1118–1119, 1988.
183. Günther, W, Kunz, H, Synthesis of β-D-mannosides from β-D-glucosides via an intramolecular SN2 reaction at C-2, *Carbohydr. Res.*, 228, 217–241, 1992.
184. Unverzagt, C, Synthesis of a branched heptasaccharide by regioselective glycosylation, *Angew. Chem. Int. Ed.*, 33, 1102–1104, 1994.
185. Unverzagt, C, Chemoenzymic synthesis of a sialylated diantennary *N*-glycan linked to asparagine, *Carbohydr. Res.*, 305, 423–431, 1998.
186. Wang, L-X, Tang, M, Suzuki, T, Kitajima, K, Inoue, Y, Inoue, S, Fan, J-Q, Lee, Y C, Combined chemical and enzymatic synthesis of a *C*-glycopeptide and its inhibitory activity toward glycoamidases, *J. Am. Chem. Soc.*, 119, 11137–11146, 1997.
187. Mizuno, M, Haneda, K, Iguchi, R, Muramoto, I, Kawakami, T, Aimoto, S, Yamamoto, K, Inazu, T, Synthesis of a glycopeptide containing oligosaccharides: chemoenzymatic synthesis of eel calcitonin analogues having natural *N*-linked oligosaccharides, *J. Am. Chem. Soc.*, 121, 284–290, 1999.
188. Leppänen, A, Mehta, P, Ouyang, Y-B, Ju, T, Helin, J, Moore, K L, van Die, I, Canfield, W M, McEver, R P, Cummings, R D, A novel glycosulfopeptide binds to P-selectin and inhibits leukocyte adhesion to P-selectin, *J. Biol. Chem.*, 274, 24838–24848, 1999.
189. Sallas, F, Nishimura, S-I, Chemo-enzymatic synthesis of glycopolymers and sequential glycopeptides bearing lactosamine and sialyl Lewis^{x} unit pendant chains, *J. Chem. Soc. Perkin Trans. 1*, 2091–2103, 2000.

190. Koeller, K M, Smith, M E B, Huang, R-F, Wong, C-H, Chemoenzymatic synthesis of a PSGL-1 *N*-terminal glycopeptide containing tyrosine sulfate and α-*O*-linked sialyl Lewis^x, *J. Am. Chem. Soc.*, 122, 4241–4242, 2000.
191. Matsuda, M, Nishimura, S-I, Nakajima, F, Nishimura, T, Heterobifunctional ligands: practical synthesis of a cell adhesive glycopeptide that interacts with both selectins and integrins, *J. Med. Chem.*, 44, 715–724, 2001.
192. Haneda, K, Inazu, T, Tamamoto, K, Kumagai, H, Nakahara, Y, Kobata, A, Transglycosylation of intact sialo complex-type oligosaccharides to the *N*-acetylglucosamine moieties of glycopeptides by *Mucor hiemalis* endo-β-*N*-acetylglucosaminidase, *Carbohydr. Res.*, 292, 61–70, 1996.
193. Yamamoto, K, Fujimori, K, Haneda, K, Mizuno, M, Inazu, T, Kumugai, H, Chemoenzymatic synthesis of a novel glycopeptide using a microbial endoglycosidase, *Carbohydr. Res.*, 305, 415–422, 1998.
194. Granovsky, M, Bielfeldt, T, Peters, S, Paulsen, H, Meldal, M, Brockhausen, J, Brockhausen, I, UDP galactose:glycoprotein—*N*-acetyl-D-galactosamine 3-β-D-galactosyltransferase activity synthesizing *O*-glycan core 1 is controlled by the amino acid sequence and glycosylation of glycopeptide substrates, *Eur. J. Biochem.*, 221, 1039–1046, 1994.
195. Meldal, M, Auzanneau, F-I, Hindsgaul, O, Palcic, M M, A PEGA resin for use in the solid-phase chemical-enzymatic synthesis of glycopeptides, *J. Chem. Soc. Chem. Commun.*, 1849–1850, 1994.
196. Davis, B J, Synthesis of glycoproteins, *Chem Rev*, 102, 579–601, 2002.
197. Mer, G, Hietter, H, Lefèvre, J-F, Stabilization of proteins by glycosylation examined by NMR analysis of a fucosylated proteinase inhibitor, *Nature Struct. Biol.*, 3, 45–53, 1996.
198. Shin, Y, Winans, K A, Backes, B J, Kent, S B. H, Ellman, J A, Bertozzi, C R, Fmoc-based synthesis of peptide-$^{\alpha}$thioesters: application to the total chemical synthesis of a glycoprotein by native chemical ligation, *J. Am. Chem. Soc.*, 121, 11684–11689, 1999.
199. Witte, K, Sears, P, Martin, R, Wong, C-H, Enzymatic glycoprotein synthesis: Preparation of ribonuclease glycoforms via enzymatic glycopeptide condensation and glycosylation, *J. Am. Chem. Soc.*, 119, 2114–2118, 1997.

16 Carbohydrate Mimetics in Drug Discovery

Beat Ernst, Hartmuth C. Kolb and Oliver Schwardt

CONTENTS

16.1 INTRODUCTION

Over the past two decades, research has uncovered numerous biological roles for carbohydrates, ranging from functions as simple as energy storage to complex processes that regulate transport, protein function, intercellular adhesion, signal transduction, malignant transformation, and viral and bacterial cell-surface recognition [1]. As carbohydrate-related drug-discovery targets are increasingly discovered and validated, the tremendous pharmaceutical potential of carbohydrates is beginning to be exploited by the pharmaceutical industry. In this review, we can only cover a fraction of the potential drug-discovery applications; we focus on type 2 diabetes, viral infections and inflammatory diseases. In addition, we review carbohydrate-based antagonists of the myelin-associated glycoprotein (MAG), which show promising effects in axon regeneration. Numerous other applications in the areas of thrombosis [2], bacterial infections [3] and anticancer vaccines [4] are not covered here because they have recently been reviewed. Because of their stereochemical diversity, carbohydrates are valuable tools for drug discovery, and they have been used as chiral scaffolds in combinatorial chemistry. Applications of this kind have recently been reviewed [5]; it is thus unnecessary to discuss these aspects in this chapter.

16.2 MAG ANTAGONISTS

16.2.1 Biological Rationale

Axons of the adult mammalian central nervous system (CNS) possess a very limited ability to regenerate after injury, largely because of environmental factors that prevent their growth [6]. Much of this inhibition is caused by myelin, the insulating lipid and protein material that is wrapped around axons to ensure rapid transmission of electrical signals along central nerve fibers. In the CNS, myelin is produced by supporting glial cells called oligodendrocytes. These cells also produce growth-inhibitory proteoglycans that become embedded within the myelin sheath [7]. Previous work has shown that axon regeneration can be accomplished in mice by immunizing the animals with myelin before inflicting a spinal cord injury [8]. These observations suggest that neutralization of growth inhibitors in myelin immediately after injury may promote regeneration, with the glial scar forming around these regenerated axons.

To date, at least three inhibitors of regeneration have been identified: MAG [9]; Nogo-A, named for its inhibitory action on axonal growth [10]; and oligodendrocyte-myelin glycoprotein (OMgp) [11]. All three of these inhibitors are located in a myelin membrane immediately adjacent to the axon, which is an optimal location for mediating axon–glia interactions [12]. Recently, it has been reported that all three growth inhibitors appear to bind to the same inhibitor, the Nogo receptor (NgR; Figure 16.1) [11,13,14].

NgR is a leucine-rich repeat (LRR) protein that is attached to the extracellular surface of the neuronal membrane by glycophosphatidylinositol (GPI) [13b]. Several observations suggest that NgR is indeed a key target for stimulating axon growth. Removal of NgR by cleaving its GPI membrane anchor results in loss of the growth-inhibitory action of all three proteins. In contrast, introduction of exogenous NgR into neurons that are unresponsive to the growth-inhibitory proteins renders them responsive [11,13,14]. Thus, the receptor NgR seems to act as the major convergence point on the surface of growth cones for detecting many of the inhibitory influences of CNS myelin. Because NgR has no transmembrane or intracytoplasmic domains, it must

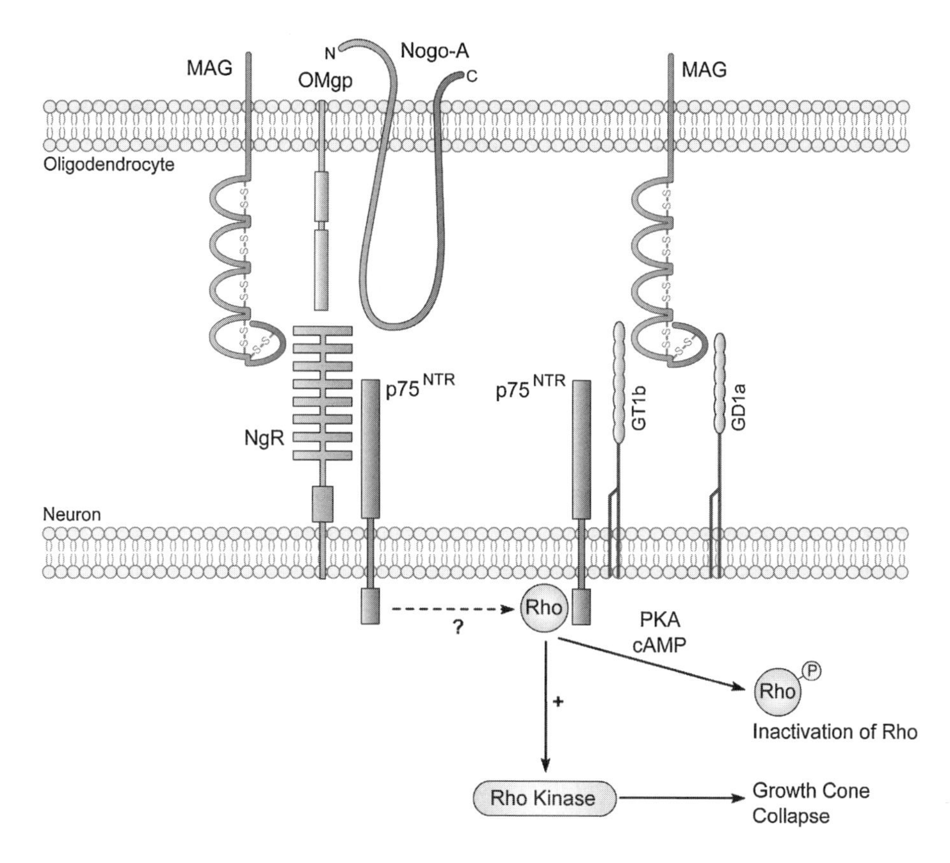

FIGURE 16.1 Blocking axonal growth. The three growth inhibitory proteins MAG, OMgp and Nogo-A are produced by oligodendrocytes in the CNS. They all bind to the same neuronal receptor NgR, resulting in the activation of signaling pathways that block axonal growth and induce growth cone collapse [15] (drawing by Alexander Vögtli).

produce inhibition by binding to a currently unidentified membrane-bound coreceptor that transduces the extracellular signal and activates intracellular signaling cascades, leading to the collapse of the growth cone [15].

Nogo-A has two inhibitory domains: a long *N*-terminal region [10b] and an oligodendrocyte surface domain called Nogo-66 [10c], which binds to NgR [13b]. OMgp is a GPI-anchored protein that was recently found to potently inhibit neurite outgrowth *in vitro* [11]. MAG, a member of the siglec family of sialic acid (Neu5Ac) binding I-type lectins [16] (see Section 16.2.2), is expressed by myelinating glial cells in the central and peripheral nervous systems [9a]. Its mechanism of function is not yet fully understood. Although it inhibits axonal growth in various *in vitro* assays, deleting the MAG gene, surprisingly, does not promote neuronal regeneration in mice [17]. MAG was found to bind to NgR with high affinity [13a,14]. Interestingly, the NgR binding sites for OMpg and Nogo-66 appear to overlap, whereas MAG and Nogo-66 bind to different sites on NgR [13a].

In vitro binding studies reveal that MAG not only binds to NgR but also to certain gangliosides that contain the Neu5Ac-α(2-3)-Gal-β(1-3)-Gal*N*Ac glycan structure, for example, GD1a and GT1b [18]. Gangliosides are sialic acid-bearing glycosphingolipids that comprise major determinants on mammalian nerve cells [19]. Ganglioside action requires the low-affinity nonselective neurotropin receptor p75, which, upon complexation with GT1b, binds to MAG [20]. Thus, MAG seems to have two independent receptors (Figure 16.1).

The signaling molecules downstream of NgR or the GT1b/p75 complex, which transduce the intracellular activation signals of Nogo-A, MAG and OMpg, are largely unknown. Experimental evidence suggests that signal transduction proceeds via Rho (Figure 16.1), a small membrane-bound guanosine triphosphatase (GTPase). Neurites were shown to grow on MAG substrates *in vitro* when Rho-GTPase activity was blocked with the enzyme C3 from *Clostridium botulinum* [21]. In addition, it is possible to overcome the GT1b-mediated blockade of axonal elongation by inhibiting Rho-kinase, a target of Rho that interacts with the cytoskeleton. It has also been shown that MAG-induced inhibition can be reversed by deactivating Rho through phosphorylation by protein kinase A (PKA), activated by cyclic adenosine monophosphate (cAMP) [22]. In line with these observations, it has recently been found that cAMP promotes neurite growth after spinal cord injury *in vivo* [23].

In summary, NgR is a key receptor involved in blocking neurite growth. By interfering with its action, either directly by inhibiting the receptor/ligand interactions or indirectly by interfering with the signal transduction cascade, it may be possible to develop new therapeutic treatments for neuronal injuries.

16.2.2 The Siglec Family

MAG (Siglec-4a) is a member of the siglec family of sialic acid (Neu5Ac) binding I-type lectins (carbohydrate binding proteins) within the immunoglobulin superfamily IgSF (Table 16.1) [16].

All siglecs share similar structural features not commonly found in other members of the IgSF. These features include an unusual V-set domain at the *N*-terminus and 1 (CD33) to 16 (sialoadhesin; Sn) C-2-set domains, followed by the membrane anchor and the cytoplasmic *C*-terminus. The binding sites of Sn and CD22 possess a high degree of sequence similarity [24]. In the center of this site is an arginine residue, which is present in all members of the siglec family. An X-ray crystallographic analysis of the *N*-terminal domain of Sn complexed with α(2-3)-sialyllactose (Figure 16.2) shows a network of close interactions between the protein and the sialic acid and galactose residues, while the glucose residue is bonded via water molecules [25].

The binding model shown in Figure 16.2 is probably also representative for MAG, SMP and CD33, because the seven amino acid residues of the binding site involved in interactions with the glycan ligand correspond to highly conserved residues within the siglec family.

TABLE 16.1
Established Members of the Siglec Family [16e]

Siglec	Alternate Name	Tissue/Cell Type Distribution	Minimal Carbohydrate Structure(s) Recognized
Siglec-1	Sn (Sialoadhesin)	Macrophages in spleen, lymph nodes, and bone marrow	Neu5Ac-α(2-3)-Gal-β(1-3/4)-Glc*N*Ac- Neu5Ac-α(2-3)-Gal-β(1-3)-Gal*N*Ac-
Siglec-2	CD22	B cells	Neu5Ac-α(2-6)-Gal-β(1-4)-Glc*N*Ac-
Siglec-3	CD33	Myeloid cell lineage	Neu5Ac-α(2-3)-Gal-β(1-3/4)-Glc*N*Ac- Neu5Ac-α(2-3)-Gal-β(1-3)-Gal*N*Ac-
Siglec-4a	MAG	Peripheral and central nerve system	Neu5Ac-α(2-3)-Gal-β(1-3)-Gal*N*Ac-
Siglec-4b	SMP (Schwann cell myelin protein)	Schwann cells in quail	Neu5Ac-α(2-3)-Gal-
Siglec-5	SMP (Schwann cell myelin protein)	Granulocytes and monocytes	Neu5Ac-α(2-3/6)-Gal-

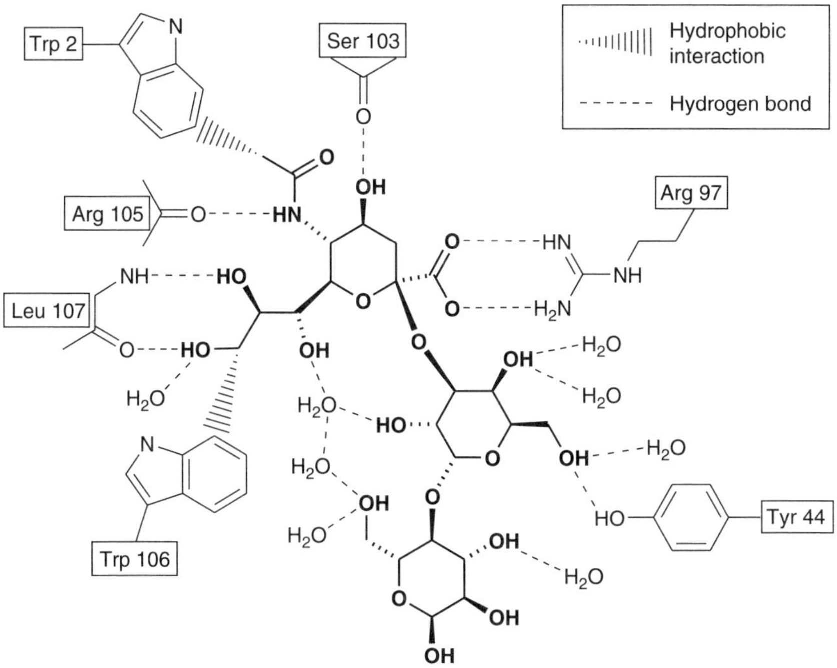

FIGURE 16.2 Defined interactions of specific amino acid residues in Sn with α(2-3)-sialyllactose [25].

16.2.3 MAG Antagonists

Based on the observation that gangliosides carry 75 to 80% of the sialic acid in the brain, and the major brain gangliosides GD1a and GT1b (Figure 16.3) express the preferred Neu5Ac-α(2-3)-Gal-β(1-3)-Gal*N*Ac terminal target determinant for MAG [18], it was hypothesized that brain gangliosides are functional nerve-cell surface ligands responsible for MAG-mediated inhibition of nerve regeneration [26].

This hypothesis is supported by a considerable amount of experimental evidence. First, MAG binds with high affinity and specificity to GD1a, GT1b and similar gangliosides [27]. Second, the binding of MAG with gangliosides is blocked by mAb 513, a conformationally specific anti-MAG antibody that also blocks MAG-neuron binding [27b]. Finally, mice that genetically lack the Neu5Ac-α(2-3)-Gal-β(1-3)-Gal*N*Ac terminus on gangliosides (but not on glycoproteins) suffer from axon degeneration and demyelination similar to that in MAG knockout mice [28].

In order to elucidate the extended glycoconjugate binding specifities of MAG and other siglecs (Schwann cell myelin protein [SMP] and sialoadhesin [Sn]), Schnaar et al. [26a,27b] measured the siglec-mediated half-maximal cell adhesion to several immobilized natural and synthetically modified gangliosides (Table 16.2) [29].

Apart from the two major brain gangliosides GD1a and GT1b (entries 7 and 8), which adhere to MAG with moderate potency, other compounds like GM3 and GM4 (entries 1 and 2) also support significant adhesion of MAG, albeit at more than tenfold higher concentrations. In contrast, gangliosides missing the terminal α(2-3)-linked sialic acid such as GD3, GM1 and GM1α (entries 3, 4 and 5) do not support adhesion of MAG. In an effort to identify the substructural determinants of sialic acid required for MAG binding, chemically modified derivatives of GM3 and GD1a were tested (not shown in Table 16.2). However, none of the modifications, including 4-, 7-, 8-, and 9-deoxy derivatives, 1-alcohols as well as several esters or amides, supported binding to MAG.

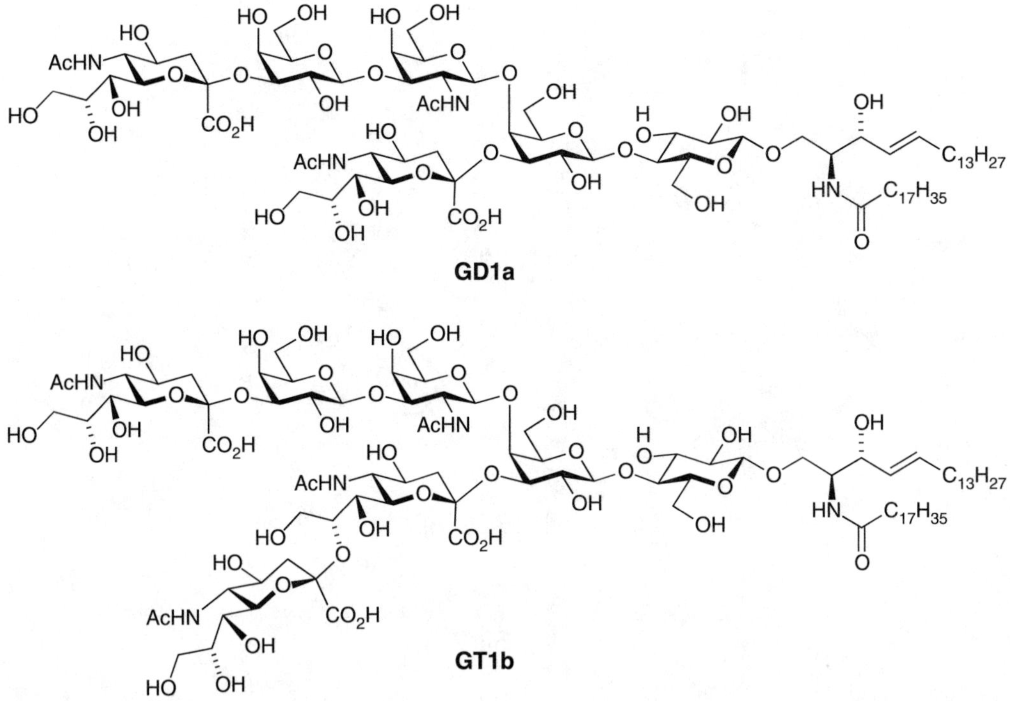

FIGURE 16.3 The major brain gangliosides GD1a and GT1b.

GM1α, GD1α, GT1aα, and GQ1bα are members of a less abundant family of brain gangliosides called *Chol-1*, based on their reactivity towards a polyclonal antiserum raised against cholinergic neurons [30]. They bear an α(2-6)-linked sialic acid on the Gal*N*Ac residue of the gangliotetraose core, making them part of the *α-series* ganglioside family [31]. In contrast to GM1α, the other α-series gangliosides displayed enhanced avidity for MAG. A comparison of binding with the Chol-1 gangliosides GD1α, GT1aα and GQ1bα (entries 9–11) and the closely related gangliosides, GM1b, GD1a and GT1b (entries 6–8), which lack the α(2-6)-Neu5Ac residue, reveals that an α(2-6)-bound sialic acid leads to a three- to tenfold enhanced adhesion of MAG. GQ1bα (Figure 16.4) represents the most potent, natural MAG antagonist discovered to date.

The low abundance of GQ1bα and the other α-series gangliosides (0.5 mg kg^{-1} of brain) compared with the major brain ganglioside GD1a (1200 mg kg^{-1}) prompted several synthetic efforts. The first total synthesis on a milligram scale was achieved in 1995 by Kiso et al. [29c].

The increased ability of Chol-1 gangliosides, which all carry an α(2-6)-Neu5Ac residue, to support MAG-mediated adhesion suggests this residue is in direct contact with the siglec. A detailed structure-activity analysis (not shown in Table 16.2) based on GD1α revealed that the hydroxyl groups on the exocyclic glycerol chain of the α(2-6)-linked sialic acid are not strictly required for binding. In fact, the entire α(2-6)-Neu5Ac moiety of GD1α can be replaced by a sulfate group (entry 12) without deleterious effect on MAG binding, suggesting that the negative charge at that position is important for binding. Moreoever, the internal α(2-3)-linked sialic acid can also be replaced in this manner, giving rise to the most potent MAG antagonist tested to date [26a]; a GT1aα mimic with two sulfate moieties (entry 13), which displays a ten times higher binding affinity for MAG than GT1aα.

Further GM1b-based structure-activity studies, which addressed alterations in the gangliotetraose core [Glc*N*Ac instead of Gal*N*Ac (entry 14) or Gal-β(1-6)- instead of Gal-β(1-3)-linkage (entry 15)], did not lead to improved binding affinities compared with GM1b.

TABLE 16.2
MAG-Mediated Cell Adhesion to Natural and Modified Gangliosides [26a, 27b]

Entry	Ganglioside	Concentrations of Ganglioside Supporting Half-Maximal Cell Adhesion [pmol well^{-1}]	[μmol l^{-1}]
1	Neu5Ac-α(2-3)-Gal-β-Cer GM4	>100[a]	>0.5
2	Neu5Ac-α(2-3)-Gal-β(1-4)-Glc-β-Cer GM3	>100	>0.5
3	Neu5Ac-α(2-8)-Neu5Ac-α(2-3)-Gal-β(1-4)-Glc-β-Cer GD3	n.d.[b]	–
4	Gal-β(1-3)-Gal*N*Ac-β(1-4)-Gal-β(1-4)-Glc-β-Cer Neu5Ac-α(2-3) (on Gal) GM1	n.d.	–
5	Neu5Ac-α(2-6) (on Gal*N*Ac) Gal-β(1-3)-Gal*N*Ac-β(1-4)-Gal-β(1-4)-Glc-β-Cer GM1α	n.d.	–
6	Neu5Ac-α(2-3)-Gal-β(1-3)-Gal*N*Ac-β(1-4)-Gal-β(1-4)-Glc-β-Cer GM1b	80	0.4
7	Neu5Ac-α(2-3)-Gal-β(1-3)-Gal*N*Ac-β(1-4)-Gal-β(1-4)-Glc-β-Cer Neu5Ac-α(2-3) (on Gal) GD1a	50	0.25
8	Neu5Ac-α(2-3)-Gal-β(1-3)-Gal*N*Ac-β(1-4)-Gal-β(1-4)-Glc-β-Cer Neu5Ac-α(2-8)-Neu5Ac-α(2-3) (on Gal) GT1b	50	0.25
9	Neu5Ac-α(2-6) (on Gal*N*Ac) Neu5Ac-α(2-3)-Gal-β(1-3)-Gal*N*Ac-β(1-4)-Gal-β(1-4)-Glc-β-Cer GD1α	19	0.095
10	Neu5Ac-α(2-6) (on Gal*N*Ac) Neu5Ac-α(2-3)-Gal-β(1-3)-Gal*N*Ac-β(1-4)-Gal-β(1-4)-Glc-β-Cer Neu5Ac-α(2-3) (on Gal) GT1aα	17	0.085
11	Neu5Ac-α(2-6) (on Gal*N*Ac) Neu5Ac-α(2-3)-Gal-β(1-3)-Gal*N*Ac-β(1-4)-Gal-β(1-4)-Glc-β-Cer Neu5Ac-α(2-8)-Neu5Ac-α(2-3) (on Gal) GQ1bα	6.0	0.03
12	**HO_3S-(6)** (on Gal*N*Ac) Neu5Ac-α(2-3)-Gal-β(1-3)-Gal*N*Ac-β(1-4)-Gal-β(1-4)-Glc-β-Cer	22	0.11
13	**HO_3S-(6)** (on Gal*N*Ac) Neu5Ac-α(2-3)-Gal-β(1-**4**)-Gal*N*Ac-β(1-4)-Gal-β(1-4)-Glc-β-Cer **HO_3S-(3)** (on Gal)	1.5	0.0075
14	Neu5Ac-α(2-3)-Gal-β(1-3)-**Glc*N*Ac**-β(1-4)-Gal-β(1-4)-Glc-β-Cer	240	1.2
15	Neu5Ac-α(2-3)-Gal-β(1-**6**)-Gal*N*Ac-β(1-4)-Gal-β(1-4)-Glc-β-Cer	87	0.435

[a] Low but statistically significant adhesion over background.
[b] No detectable adhesion at >100 pmol/well^{-1}.

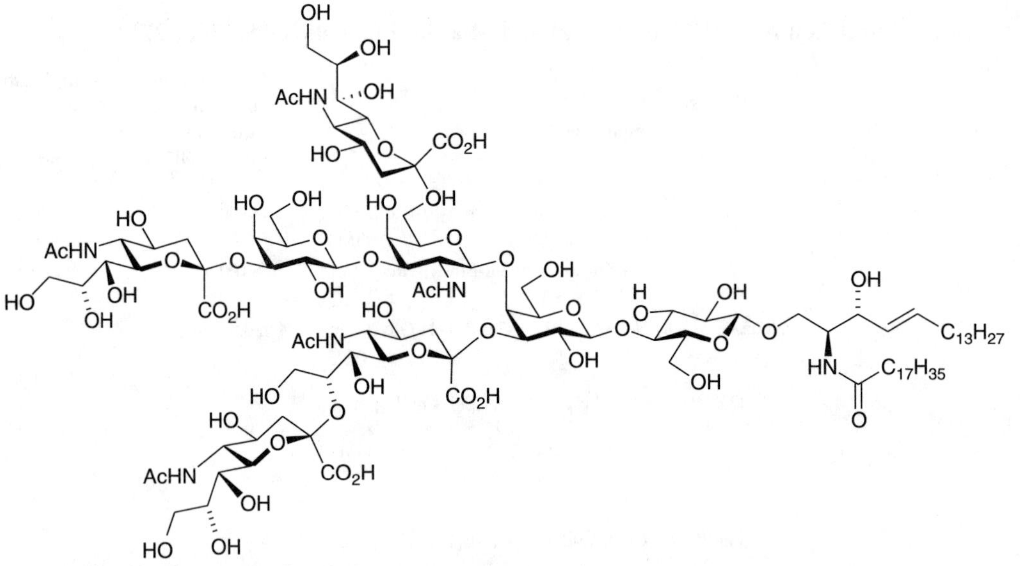

GQ1bα

FIGURE 16.4 The Chol-1 ganglioside GQ1bα.

In an effort to elucidate the glycan specificity of MAG and sialoadhesin, Kelm et al. [32] studied their interactions with a series of synthetic monovalent oligosaccharides using a hapten inhibition assay (Table 16.3). Native human erythrocytes were incubated with antibody-complexed Fc-MAG_{d1-3} in the presence of the monovalent oligosaccharides at different concentrations. The results are quoted as relative inhibitory potencies (rIPs), calculated by dividing the IC_{50} of the indicated reference compound by the IC_{50} of the compound of interest. Consequently, antagonists inhibiting better than the reference compound have rIP values above 1.0.

In contrast to the results obtained by Schnaar and coworkers [26a], where gangliosides missing the terminal α(2-3)-linked sialic acid appeared to be nonbinders to MAG, the study of Kelm et al. revealed oligosaccharides with terminal α(2-6)-bound Neu5Ac (entries 2 and 4 in Table 16.3) to exhibit modest inhibition of the siglec. These (among other) discrepancies are probably caused by the different assays used. Schnaar et al. measured binding of full-length cell surface MAG to saccharides oriented on an apposing membrane monolayer, whereas the Kelm group measured the site affinity of soluble saccharides for a soluble MAG-Fc chimera engineered to have only the *N*-terminal three (of five) Ig-like domains.

In general, MAG-bound oligosaccharides with α(2-3)-linked sialic acid are about eight times more active than their isomers with α(2-6)-bound Neu5Ac (entries 1–4). This ratio is independent of the subterminal glycan structure, demonstrating the importance of the α(2-3)-linked sialic acid for MAG binding. Furthermore, the underlying glycan appears to interact with the protein because the rIP value of the pentasaccharide (entry 3) is six times higher than that of α(2-3)-sialyl lactose (entry 1) and about twice as large as that of the trisaccharide Neu5Ac-α(2-3)-Gal-β(1-4)-Glc*N*Ac-β-SE (entry 12) with a similar terminal sequence, suggesting that the contact site of MAG extends beyond the three terminal saccharide units.

Modifications at positions 4 (not shown in Table 16.3), 8 (entry 6) and 9 (entry 7) of the terminal sialic acid are detrimental to activity, whereas deoxygenation at C-7 (entry 5) slightly improves the activity compared with α(2-3)-sialyl lactose (entry 1). Interestingly, the replacement of the entire terminal Neu5Ac moiety by Kdn (2-keto-3-deoxy-D-*glycero*-D-*galacto*-nononic acid; entry 8) resulted in a 6.5-fold improved inhibition compared with the Neu5Ac-containing analog (entry 13).

TABLE 16.3
Relative inhibitory potencies (rIPs) of Oligosaccharide Inhibitors of MAG [32]

Entry	Compound	Reference Compound	rIP
1	Neu5Ac-α(**2-3**)-Gal-β(1-4)-Glc	1	1
2	Neu5Ac-α(**2-6**)-Gal-β(1-4)-Glc	1	0.12
3	Neu5Ac-α(**2-3**)-Gal-β(1-4)-All*N*Ac-β(1-3)-Gal-β(1-4)-Glc-β-SE[a]	1	6.3
4	Neu5Ac-α(**2-6**)-Gal-β(1-3)-Glc*N*Ac-β(1-3)-Gal-β(1-4)-Glc-β-SE	1	0.82
5	**7-deoxy**-Neu5Ac-α(2-3)-Gal-β(1-4)-Glc-β-EtN$_3$	1	1.56
6	**8-deoxy**-Neu5Ac-α(2-3)-Gal-β(1-4)-Glc-β-EtN$_3$	1	n.a.[b]
7	**9-*O*-methyl**-Neu5Ac-α(2-3)-Gal-β(1-4)-Glc-β-EtN$_3$	1	0.18
8	**Kdn**-α(2-3)-Gal-β(1-4)-Glc*N*Ac-β(1-3)-Gal-β(1-4)-Glc-β-SE	13	6.47
9	Neu5Ac-α(2-3)-**(4-deoxy)**Gal-β(1-4)-Glc*N*Ac-β(1-3)-Gal-β(1-4)-Glc-β-SE	3	2.25
10	Neu5Ac-α(2-3)-**(6-deoxy)**Gal-β(1-4)-Glc*N*Ac-β(1-3)-Gal-β(1-4)-Glc-β-SE	3	1.5
11	Neu5Ac-α(2-3)-Gal-β**(1-3)-Gal*N*Ac**-β-SE	1	1.73
12	Neu5Ac-α(2-3)-Gal-β**(1-4)-Glc*N*Ac**-β-SE	1	2.71
13	Neu5Ac-α(2-3)-Gal-β(1-4)-**Glc*N*Ac**-β(1-3)-Gal-β(1-4)-Glc-β-SE	13	1
14	Neu5Ac-α(2-3)-Gal-β(1-4)-**Glc*N*octanoyl**-β(1-3)-Gal-β(1-4)-Glc-β-SE	3	1.69
15	**Neu5Ac-α(2-6)** (on Gal*N*Ac) Neu5Ac-α(2-3)-Gal-β(1-3)-Gal*N*Ac-β(1-4)-Gal-β(1-4)-Glc-β-SE	13	3.06
16	**Neu5Ac-α(2-6)** (on Gal*N*Ac) Neu5Ac-α(2-3)-Gal-β(1-3)-Gal*N*Ac-β(1-4)-Gal-β(1-4)-Glc-β-SE **Neu5Ac-α(2-3)** (on Gal)	13	0.95

[a] SE = 2-(trimethylsilyl)ethyl.
[b] n.a.: not applicable since less than 50% inhibition at the highest concentration was tested.

In order to investigate the role of the subterminal saccharides, oligosaccharides with modifications in the Gal moiety were tested. Replacement of the hydroxyl groups at C-4 or C-6 by a hydrogen (entries 9 and 10) or an amino group slightly improved the inhibition of MAG, whereas additional substituents or the replacement of the ring oxygen by an *N*-alkyl group (data not shown) had deleterious effects. In contrast to the observations by Schnaar et al. (entries 14 and 15 in Table 16.2), this study suggests the linkage to (1-3 or 1-4) and the nature of the second subterminal monosaccharide (Gal*N*Ac or Glc*N*Ac) to be of minor importance (entries 11 and 12). Nevertheless, the *N*-acetyl moiety does appear to be beneficial for the inhibitory potency because these trisaccharides are better inhibitors than α(2-3)-sialyl lactose (entry 1). Compounds bearing hydrophobic substituents such as octanoyl (entry 14), benzoyl or phthaloyl on the nitrogen of glucosamine showed moderately improved inhibition of MAG. In contrast, the 3-position of Glc*N*Ac does not appear to be important because the rIPs of the 3-deoxy, 3-*O*-methyl/3-Fuc derivatives and the 3-epimer (All*N*Ac; entry 3) are quite similar, suggesting that the binding pocket in this area is relatively open, accepting even a fucose residue.

Inhibition experiments with synthetically generated oligosaccharide fragments of gangliosides GD1α and GT1α (entries 15 and 16) revealed that an additional α(2-6)-linked sialic acid enhances the inhibition to MAG significantly (entry 15), whereas the introduction of a sialic acid residue at position 3 of the underlying Gal moiety (entry 16) reverses this enhancement, giving an rIP value similar to that of α(2-3)-sialyl lactose (entry 1). This result conflicts with the data obtained by Schnaar et al. [26a], where GT1a was reported to be a much better inhibitor than GM3 [α(2-3)-sialyl lactose] (Table 16.2).

TABLE 16.4
Relative inhibitory potencies (rIPs) of Sialoside Inhibitors of MAG [33]

Entry	Compound	rIP
1	Neu5Ac-α-Me	1.00
2	Neu5Ac-α-Bn	9.80
3	Neu5Propyl-α-Me	1.56
4	Neu5ThioAc-α-Me	3.85
5	Neu5FAc-α-Me	16.94
6	Neu5ClAc-α-Me	7.00
7	Neu5F_3Ac-α-Me	4.04
8	9-Deoxy-Neu5Ac-α-Me	n.a.
9	9-Cl-Neu5Ac-α-Me	n.a.
10	9-Thio-Neu5Ac-α-Me	n.a.
11	9-NH_2-Neu5Ac-α-Me	2.98

n.a.: not applicable since less than 50% inhibition at the highest concentration tested.

In order to examine the substructural determinants of sialic acid required for MAG binding, a second study by Kelm and coworkers [33] evaluated the inhibitory potential of simple sialic acid glycosides towards MAG and sialoadhesin using their hapten inhibition assay (Table 16.4).

Neu5Ac-α-Bn (entry 2) was found to be ten times more potent than the corresponding methyl sialoside (entry 1), which was used as the reference compound in all assays. This finding suggests that the benzyl group participates in additional hydrophobic or steric interactions with the protein.

Because the substituents at position 5 and position 9 of sialic acid are critical for binding to siglecs [16b,27b,34], the rIPs of a series of methyl sialosides with modifications at C-5 were compared. Replacement of acetyl by propionyl (entry 3) enhances the rIP only slightly, whereas an *N*-thioacetyl residue (entry 4) leads to a marked increase in the affinity for MAG. Halogen substitution also led to improved binding affinities, as shown by the chloro (entry 6) and trifluoro (entry 7) derivatives, which possess seven- and fourfold higher affinities for MAG, respectively. By far the most potent inhibitor for MAG in this study was NeuFAc-α-Me (entry 5) with an rIP value of 17. It is possible that the electronegative halogens mediate an additional interaction with the binding site of MAG. The substituent in position 9 appears to be essential for binding to the siglec via hydrogen-bonding because the substitution of the 9-hydroxy function with H (entry 8), Cl (entry 9) or SH (entry 10) leads to a complete loss of inhibition, whereas the 9-amino derivative (entry 11) was a more potent inhibitor than the reference sialoside (entry 1).

16.2.4 Summary of the Structure Affinity Relationship

Although the data obtained in the previously described MAG binding studies (Section 16.2.3) are somewhat contradictory, they do shed light on the structure–activity relationship of glycan recognition by MAG, especially when interpreted in view of the recent X-ray crystal structure of the complex between the closely related siglec Sn and α(2-3)-sialyllactose [25] (Figure 16.2, Section 16.2.2). The structure–activity data are compiled for the tetrasaccharide Neu5Ac-α(2-3)-Gal-β(1-3)-[Neu5Ac-α(2-6)]-Gal*N*Ac (Figure 16.5).

16.2.4.1 Neuraminic Acid Residues

MAG has a high preference for the Neu5Ac-α(2-3)-Gal epitope. In contrast, a terminal α(2-6)-linked sialic acid only provides weak interactions, which are obviously not sufficient to mediate stable binding of MAG to cells [32]. Thus, the spatial orientation of the carboxyl group of sialic acid

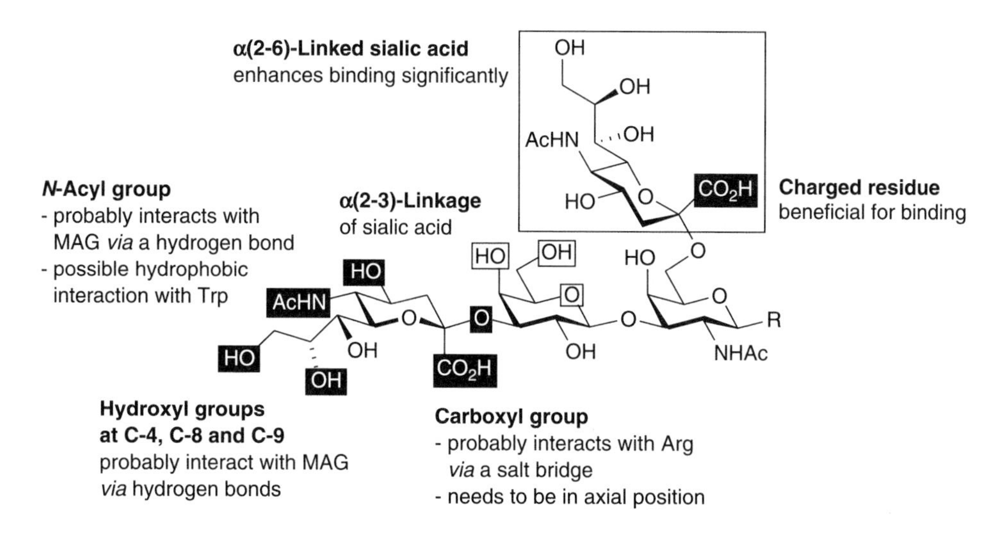

FIGURE 16.5 Functional groups of the tetrasaccharide Neu5Ac-α(2-3)-Gal-β(1–3)-[Neu5Ac-α(2-6)]-Gal*N*Ac involved in the binding to MAG. Residues contributing to the binding strength are printed white on a black background. Groups in areas with possible steric constraints are marked with boxes.

is essential for binding. This is further corroborated by the crystal structure of sialoadhesin [25] (Figure 16.2), which shows the carboxyl group interacts with an arginine residue in the binding site of siglecs.

16.2.4.2 The Hydroxyl Groups on Sialic Acid

The experiments have demonstrated that the hydroxyl groups at C-4, C-8 and C-9 of sialic acid are important for binding, whereas the 7-OH is not necessary [33]. Most likely, these beneficial OH groups function as hydrogen-bond donors to the siglec, while the hydroxyl group at C-7 appears not to be in contact with the protein, according to the crystal structure [25].

16.2.4.3 The *N*-Acetyl Residue at C-5 of Sialic Acid

The crystal structure shows that a tryptophan residue in the siglec's binding site interacts specifically with the methyl group of the acetyl moiety [25]. It is not surprising, therefore, that acyl substituents have a profound effect on the binding affinity. Generally, sialosides with halogenated acetyl residues at C-5 showed strongly enhanced binding to MAG (Table 16.4) [33]. The enhanced affinity correlates with the electronegativity of the halogen atom and can be rationalized by halogen-mediated additional contacts with the protein or by electronic effects on the amide, which result in a weaker hydrogen bond-acceptor quality of the carbonyl oxygen and a significantly stronger hydrogen bond-donating ability of the amide NH [33].

16.2.4.4 The Subterminal Saccharide

The results of the MAG binding assays imply that not only the terminal sialic acid residue but also the subterminal oligosaccharides contribute to binding. Additional substituents at position 6 of the subterminal galactose caused a significant drop in inhibition, whereas the 4- and 6-deoxy compounds showed increased activity (entries 9 and 10 in Table 16.3) [32], although the crystal structure [25] suggests a hydrogen bond between the hydroxyl group at C-6 of Gal and Tyr44 (Figure 16.2). However, these effects may simply be a result of conformational changes of the whole glycan caused by these modifications, resulting in a less suitable presentation of sialic acid.

Glycoproteins and -lipids carry the terminal disaccharide unit, Neu5Ac-α(2-3)-Gal, either β(1-3)-linked to Gal*N*Ac or Glc*N*Ac or β(1-4)-linked to Glc*N*Ac. The role of the underlying monosaccharide residue in binding to MAG remains unclear because available experimental data is currently contradictory. Schnaar and coworkers (Table 16.2) found a strong decrease of inhibition if the terminal Neu5Ac-α(2-3)-Gal unit was linked in β(1-6) fashion to Gal*N*Ac or in β(1-4) fashion to Glc*N*Ac [26a]. Alternatively, the Kelm group (Table 16.3) observed only minor changes in activity for structures with the terminal disaccharide linked in β(1-3) or β(1-4) fashion to Gal*N*Ac or Glc*N*Ac [32]. An additional sialic acid at position 6 of the Gal*N*Ac residue strongly enhances inhibition of MAG, probably caused by additional binding interactions with the protein. Indeed, the crystal structure (Figure 16.2) shows a water-filled cavity, which may accommodate this residue. According to Schnaar et al., the α(2-6)-Neu5Ac moiety may be replaced by sulfate (entry 12 in Table 16.2) without loss in binding affinity to MAG, stressing the importance of negative charge at that position. In contrast, the glycerol side chain is of minor importance [26a].

16.2.5 Summary and Outlook

The inhibition of the interaction of nerve cells with growth inhibitors MAG, Nogo-A and OMgp promises to be of great therapeutic value for the treatment of injuries to the nervous system. Several compounds, mostly gangliosides and mimics thereof, have been found to be suitable as ligands and inhibitors of MAG. However, these compounds are still at a very early stage, and further studies are thus necessary to test their effect on neurite outgrowth *in vivo* for concept validation purposes. In addition, it is necessary to develop *drug-like* noncarbohydrate mimics of the carbohydrate-based lead structures, which are unsuitable for therapeutic application, because of their structural complexity and high physiological lability.

16.3 GLYCOSIDASE INHIBITORS

16.3.1 Biological Rationale

The biosynthesis of membrane-bound glycoproteins and glycolipids occurs in the endoplasmic reticulum (ER) and the Golgi apparatus. This process is mediated by glycosyltransferases that catalyze the glycosylation of a specific acceptor at a specific position, using activated monosaccharides (Leloir sugar nucleotides) as glycosyl donors. The opposite transformation (i.e., the hydrolytic cleavage of glycosidic linkages) is catalyzed by glycosidases. These biologically important enzymes have been thoroughly investigated in the past decade [35–38] and are believed to be well understood today.

Glycosidase inhibitors — compounds that inhibit carbohydrate-processing enzymes — are of considerable therapeutic value. Inhibitors of intestinal α-glycosidases are used to treat type 2 diabetes mellitus by blocking oligosaccharide hydrolysis and glucose uptake [39] (see Section 16.3.2). Glycosidases that are involved in the trimming of cell- and viral-surface oligosaccharides are also useful as therapeutic targets. Inhibitors of these glycosidases disrupt the biosynthesis of cell-surface oligosaccharides and, as a consequence, may interfere with cell–cell or cell–virus recognition processes. Proof of concept stems from neuraminidase inhibitors that have recently been developed for the treatment of influenza [40] (see Section 16.3.3). α-Glycosidase inhibitors may also prove useful for the treatment of HIV [41], Gauchers disease, hepatitis and cancer [42].

Since the beginning of the 1960s, a number of potent glycosidase inhibitors have been found in plants or micro-organisms. These discoveries have prompted work on the synthesis of these natural products and their analogs, leading to the development of new types of very potent synthetic inhibitors. Given that numerous reviews have been published on this topic [43], we will focus our discussion on the use of α-glucosidase inhibitors in the treatment of type 2 diabetes (Section 16.3.2) and of neuraminidase inhibitors for the treatment and prevention of influenza (Section 16.3.3).

16.3.2 α-Glucosidase Inhibitors

16.3.2.1 Background: Diabetes Mellitus

Diabetes mellitus is a chronic disease characterized by inherited or acquired deficiency in insulin production by the pancreas or by resistance of tissues to insulin. Traditionally, the disorder has been divided into type 1 (insulin-dependent) and type 2 (noninsulin dependent) forms, although this distinction is not always clear [44–46].

Type 1 diabetes mellitus is a heterogeneous and polygenic disorder, and patients suffering from this disease require treatment with insulin for their survival. To date, no specific environmental agents responsible for triggering the disease have been identified, although risk factors may include viral infections, early infant diet and toxins. Other potential disease modifying factors include vaccines, psychological stress and climatic influences [44].

Patients with type 2 diabetes display not only a relative deficiency of this hormone, but also a pronounced resistance to insulin. The disease has a strong genetic basis, with environmental factors and age playing prominent roles [46]. Glucose tolerance usually deteriorates with advancing age, and there is a close relationship between obesity and type 2 diabetes. The primary goal in the treatment of individuals with type 2 diabetes is the correction of hyperglycemia because tight glycemic control has been shown to significantly reduce the risk of long-term complications, including both microvascular (notably retinopathy and nephropathy) and macrovascular (cardiovascular, cerebrovascular, and peripheral vascular disease) effects [45]. When dietary and lifestyle modifications alone fail to achieve glycemic control, antihyperglycemic therapy is usually initiated with a single oral agent [45,47]. However, because type 2 diabetes is a progressive disease, it is usually necessary to resort to a multidrug regimen after single-drug treatment ceases to be effective [45,48]. In the later stages of the disease, many patients suffering from type 2 diabetes require insulin to ensure adequate glycemic control [49].

Four major classes of oral antidiabetic agents have been developed for the treatment of type 2 diabetes (Figure 16.6) [45]. These include:

1. *Sulfonylureas* (e.g., glipizide **1**) and *meglitidines* (e.g., repaglinide **2**) are insulin secretagogues, which act directly on pancreatic islet β-cells to close ATP-sensitive potassium ion channels in the cell membrane. This increases intracellular calcium and indirectly facilitates secretion of insulin in response to the presence of glucose [50].
2. The antidiabetic action of the *biguanide* metformin **3** is thought to result primarily from a reduction in insulin resistance, but it also has been shown to inhibit hepatic gluconeogenesis [50].
3. *Thiazolidinediones* (e.g., rosiglitazone **4**) reduce insulin resistance (insulin sensitizers). These compounds stimulate peroxisome-proliferator-activated receptors (PPARs) on the nuclear surface, leading to increased glucose uptake and reduced hepatic gluconeogenesis [50].
4. *α-Glucosidase inhibitors* (e.g., acarbose **5**) exhibit mild antihyperglycemic activity through competitive inhibition of the α-glucosidases in the small intestine, leading to a reduction in the rate of absorption of carbohydrates [39,50].

16.3.2.2 α-Glucosidase Inhibitors for Treatment of Type 2 Diabetes

Among the di- and oligosaccharidases that are produced by the mammalian gastrointestinal tract are pancreatic α-amylase and α-glucosidase enzymes in the brush border of the small intestine, such as glucoamylase, maltase, sucrase, and dextrinase. These enzymes digest the complex dietary carbohydrates to monosaccharides, which are subsequently absorbed through the intestinal wall. Their complete or partial inhibition causes the absorption of glucose to be

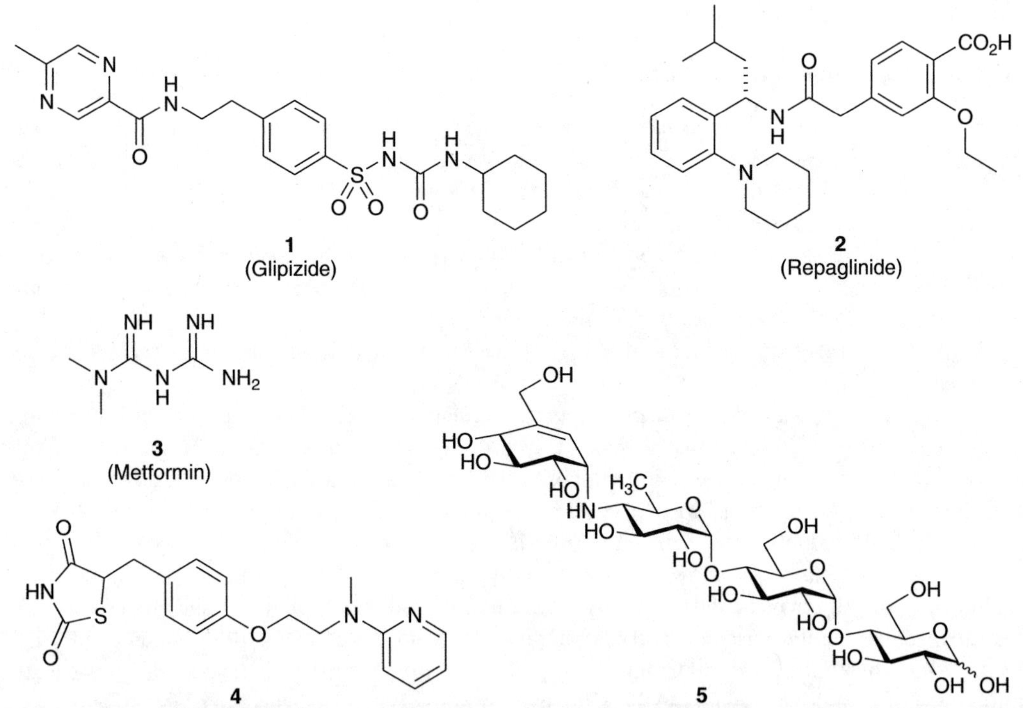

1 (Glipizide)

2 (Repaglinide)

3 (Metformin)

4 (Rosiglitazone)

5 (Acarbose)

FIGURE 16.6 Examples of oral antidiabetic agents.

reduced, because of incomplete carbohydrate digestion. As a consequence, this leads to a reduction of the postprandial rise in blood glucose levels as well as a decrease in glucose-induced insulin secretion [39,51].

The rational design of potent glycosidase inhibitors may be based on the transition state structures of the enzymatic reaction because this is the point of greatest stabilization by the enzyme — in the case of glycosidases with K_Ds up to 10^{-20} M. However, the true catalytic mechanism of glycosidase-mediated hydrolysis has not been fully understood until recently [37,38].

Acidic glycoside hydrolysis is believed to be an A1 process, which proceeds via an intermediate oxocarbenium ion **6**, generated by protonation and elimination of the exocyclic oxygen at the anomeric center (Scheme 16.1). The transition state is believed to be late and similar in structure to **6** [43d].

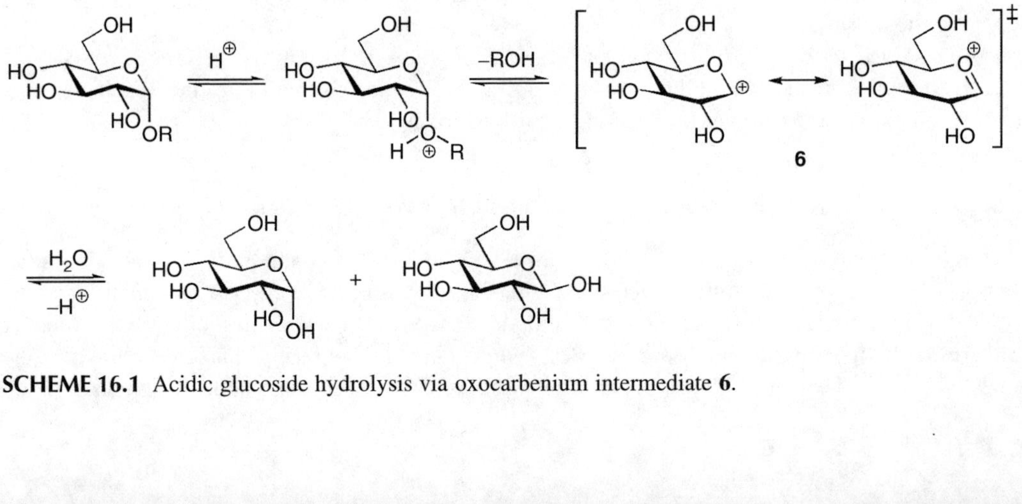

SCHEME 16.1 Acidic glucoside hydrolysis via oxocarbenium intermediate **6**.

In general, there are two mechanistic pathways, marked “a” and “b” in Scheme 16.2, with different stereochemical outcomes: inversion or retention of the anomeric configuration. Both mechanisms involve oxocarbenium ion-like transition states and a pair of carboxylic acids at the active site of the glycosidase. In stereochemistry-retaining enzymes (path “a”), these two residues are located on average approximately 5.5 Å apart [37], and the reaction proceeds via a double displacement mechanism wherein one carboxylic acid acts as a general base and the other as a general acid (Scheme 16.2a). In inverting glycosidases (path “b”), the two carboxylic acid residues have a distance of approximately 10 Å [37], and the reaction occurs via a single displacement mechanism (Scheme 16.2b).

In the late 1960s, scientists at Bayer initiated the search for inhibitors of intestinal sucrase with the goal of finding a treatment for diabetes. These efforts led to the discovery of the pseudotetrasaccharide acarbose **5** (Figure 16.7) from the fermentation broth of the *Actinoplanes*

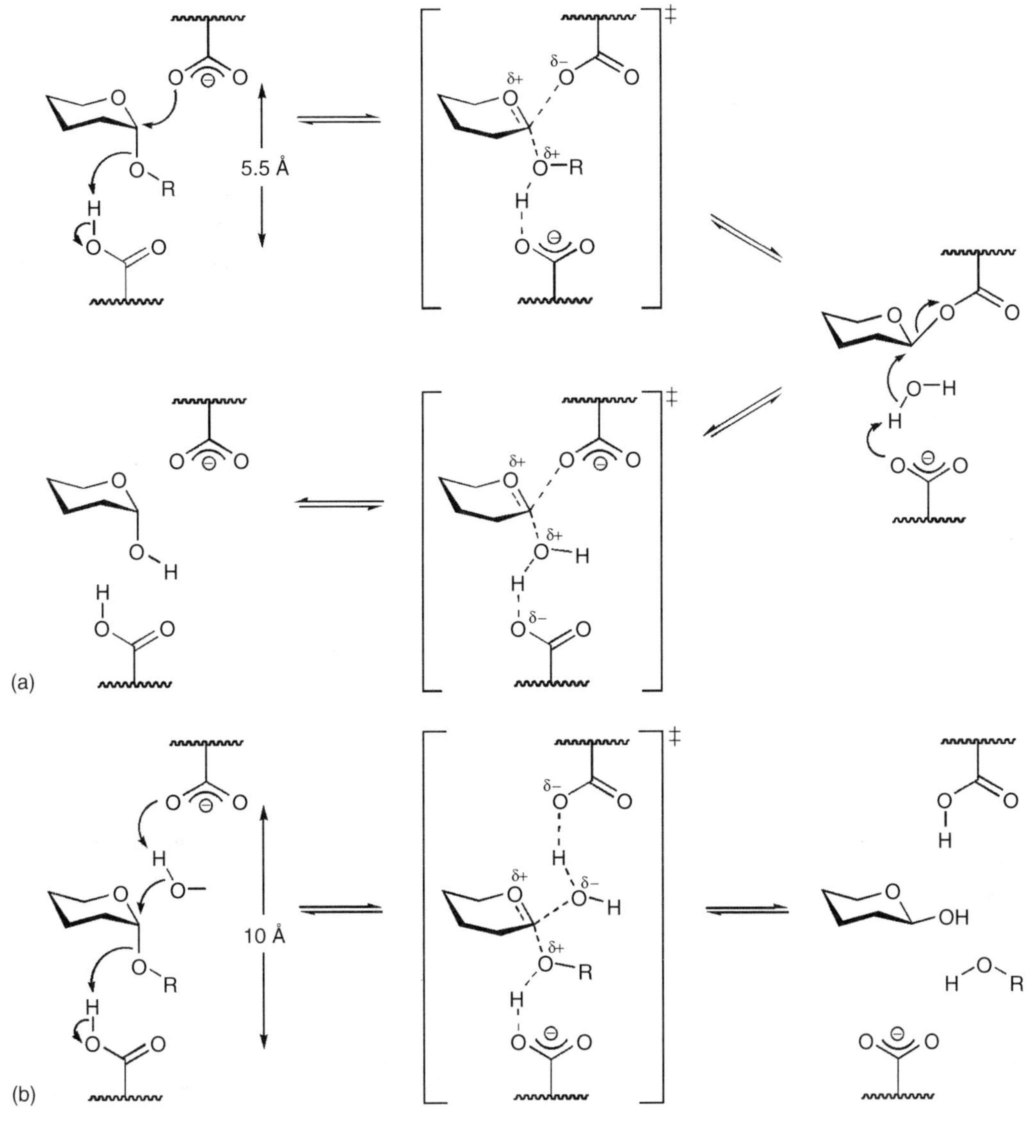

SCHEME 16.2 Glycosidase mechanisms for (a) a retaining α-glycosidase and (b) an inverting α-glycosidase proceeding through an oxocarbenium ion-like intermediate.

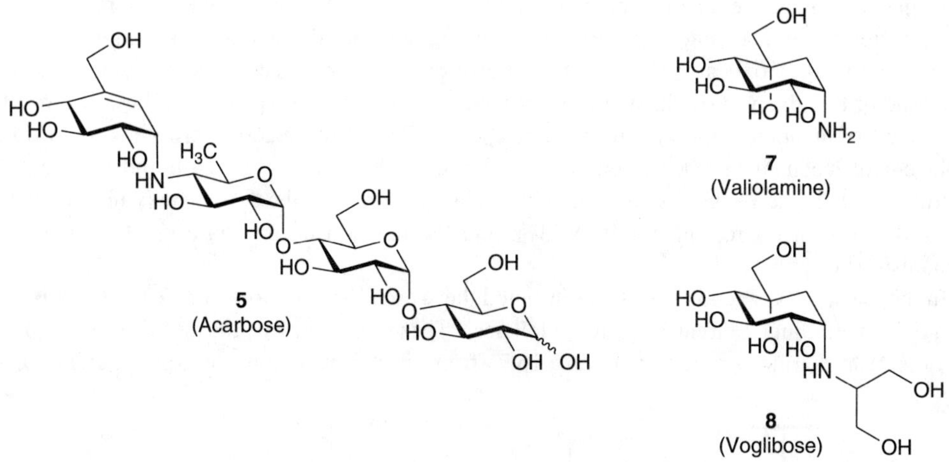

FIGURE 16.7 The commercially available α-glucosidase inhibitors acarbose **5** and voglibose **8**, a derivative of valiolamine **7**.

strain SE 50 [52]. Acarbose is a potent inhibitor of pig intestinal sucrase, with an IC_{50} value of 0.5 μM, and it is also effective in carbohydrate loading tests on rats and healthy volunteers. It acts as a competitive inhibitor of pancreatic α-amylase and intestinal membrane-bound α-glucoside hydrolases [53]. The characteristic terminal trihydroxy(hydroxymethyl)cyclohexene moiety, bonded by way of an imino linkage at the allylic position to a 4-amino-4,6-dideoxy-D-glucopyranose, can be regarded as a transition-state analog of oxocarbenium ion **6** (Scheme 16.1). In clinical studies, which lasted up to 1 year, significant reductions of blood glucose and glycated hemoglobin (HbA_{1c}) levels in patients with type 2 diabetes were observed for both acarbose monotherapy and combination therapy with other oral antidiabetic drugs or insulin [54]. Acarbose was introduced into the market in Germany (in 1990 as Glucobay™) and the U.S.A. (in 1998 as Precose™) as the first drug for the treatment of type 2 diabetes. It acts locally in the gastrointestinal (GI) tract and less than 2% is absorbed and excreted in the urine. This is highly desirable because it causes most of the drug to remain in the GI, the site of action. A portion of the drug is inactivated by intestinal bacteria within 14 to 24 h after dosing and the rest is eliminated unchanged in the feces [54].

Valiolamine **7**, a substituted cyclohexane similar to the active subunit of acarbose **5**, was isolated from the culture broth of *Streptomyces hygroscopicus* variant *limoneus* [55a] (Figure 16.7). The compound is a potent inhibitor of pig intestinal maltase and sucrase, with IC_{50} values of 2.2 and 0.049 μM [55b]. In an effort to enhance the inhibitory activity *in vitro*, Horii et al. [55b] synthesized numerous *N*-substituted valiolamine analogs, and obtained the simple derivative voglibose (AO-128) **8** by reductive amination of valiolamine with dihydroxyacetone. Moreover, **8** is an extremely potent inhibitor of both pig intestinal α-glucosidases, with IC_{50} values of 15 nM for maltase and 5 nM for sucrase. Voglibose is now on the market in Japan (marketed as Basen™).

In traditional Chinese medicine, mulberry leaves have been used to treat diabetes. Mulberry extracts are able to suppress the rise of blood glucose levels that follows eating, and the active compound has been identified as 1-deoxynojirimycin (DNJ, moranoline) **9** (Figure 16.8), isolated from the roots of mulberry trees [56]. The first chemical syntheses of DNJ started from L-sorbofuranose [57] and nojirimycin (NJ) **10** [58]. The latter compound was discovered in 1966 as the first natural glucose mimic with a nitrogen atom in place of the ring oxygen [59]. NJ has antibiotic properties and was shown to be a potent inhibitor of various α- and β-glucosidases [43a].

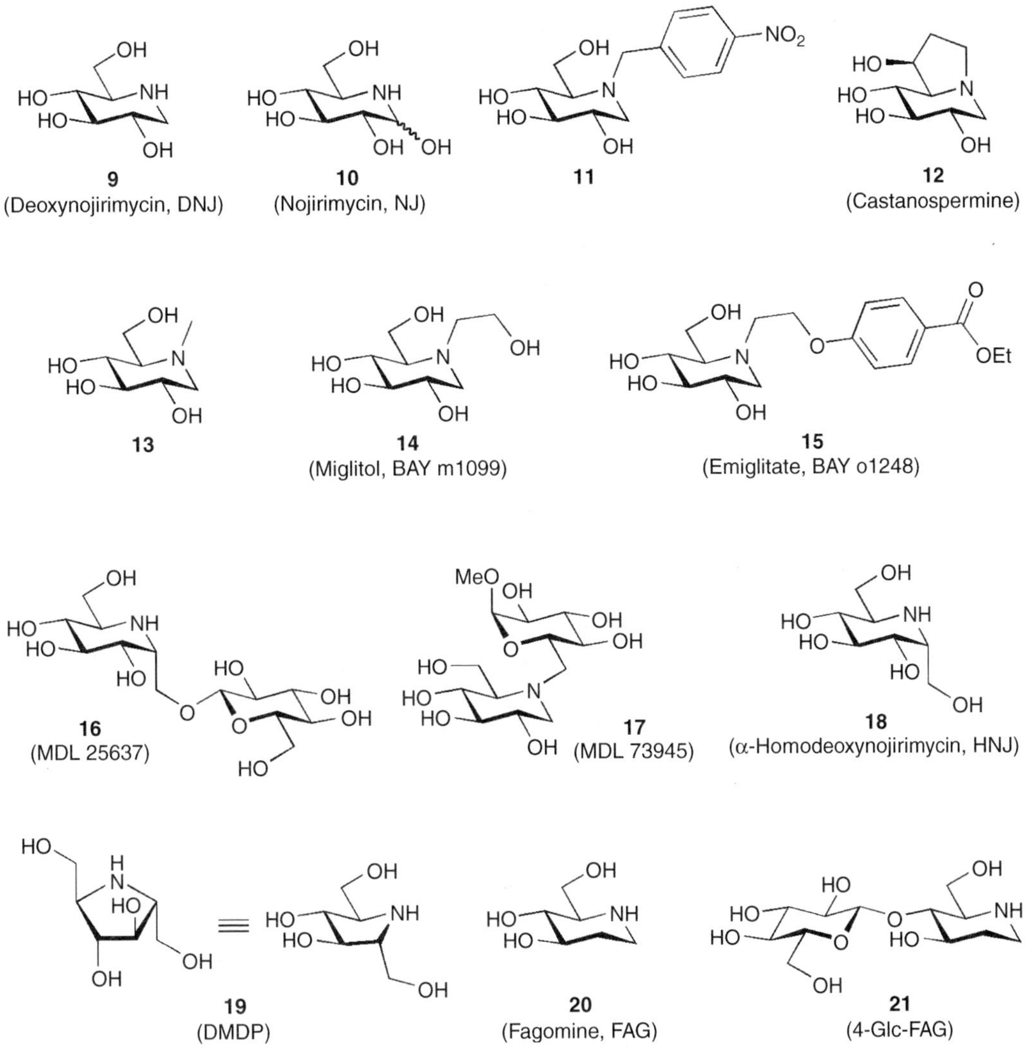

9 (Deoxynojirimycin, DNJ) **10** (Nojirimycin, NJ) **11** **12** (Castanospermine)

13 **14** (Miglitol, BAY m1099) **15** (Emiglitate, BAY o1248)

16 (MDL 25637) **17** (MDL 73945) **18** (α-Homodeoxynojirimycin, HNJ)

19 (DMDP) **20** (Fagomine, FAG) **21** (4-Glc-FAG)

FIGURE 16.8 Deoxynojirimycin **9** and derivatives acting as α-glucosidase inhibitors.

It is controversial whether DNJ **9** can be regarded as a true transition-state analog. Although protonated DNJ does mimic the charge development of the oxocarbenium intermediate **6** (Scheme 16.1), it adopts a chair conformation instead of the expected half-chair conformation of **6**. A linear free energy relationship analysis (k_{cat}/K_m vs. K_i), which tested the resemblance of DNJ with the transition-state of β-glucosidase hydrolysis, suggested that the compound was not a transition-state analog [37c]. Nevertheless, the fact that *N*-substituted derivative **11** can be used as a transition-state analog to obtain catalytic antibodies indicates that the compound does have some transition-state characteristics [60].

The discovery of the inhibitory effects of DNJ and similar compounds on mammalian α-glucosidases prompted the search for therapeutic applications. However, these efforts are hampered by target-specificity issues and the fact that there are marked differences between species [61], rendering *in vivo* testing unreliable. A large number of naturally occurring or synthetic DNJ analogs were thoroughly tested, leading to the realization that DNJ **9** and its derivatives, such as castanospermine **12**, *N*-methyl DNJ **13**, miglitol (BAY m1099) **14**, emiglitate (BAY o1248)

15, MDL 25637 **16**, and MDL 73945 **17** (Figure 16.8), are effective in delaying the hyperglycemic response to oral starch and sucrose in normal and diabetic (streptozotocin-induced) rats and mice [62].

In a recent study, the antihyperglycemic effects of the naturally occurring compounds DNJ **9**, α-homodeoxynojirimycin (HNJ) **18**, castanospermine **12**, DMDP (2,5-dideoxy-2,5-imino-D-mannitol) **19**, 1,2-dideoxynojirimycin (fagomine [FAG]) **20**, and several derivatives of FAG, for example, 4-*O*-β-D-glucopyranosylfagomine (4-Glc-FAG) **21** (Figure 16.8), were evaluated in streptozotocin-diabetic mice [63]. FAG and DMDP, applied in doses of 150 μmol kg^{-1}, significantly reduced the blood glucose level 2 h after intraperitoneal injection and the antihyperglycemic activity was sustained over a period of 2 to 6 h after administration. The onset of action with 4-Glc-FAG **21** was delayed compared with FAG, and the antihyperglycemic effects proved dose-dependent (75 to 300 μmol kg^{-1}). It is feasible that 4-Glc-FAG is not the pharmacologically active species, and the time lag may reflect its hydrolysis to active FAG by β(1-4)-glucosidases. Interestingly, despite the excellent α-glucosidase inhibitory potency of the parent compounds DNJ and HNJ *in vitro*, their *in vivo* efficacy was only moderate [64]. These findings suggest that the mechanism of action of these sugar mimics involves more than simply α-glucosidase inhibition. Additional mechanistic work studied the effects of FAG and 4-Glc-FAG on immunoreactive insulin (IRI) release from the perfused pancreas of normal rats. In a concentration-dependent manner (0.1 to 1 mM), FAG significantly enhanced the total amount of IRI that was released for 7 min by 8.3 mM glucose stimulation, whereas this potentiating effect was not observed with 4-Glc-FAG at a concentration of 1 mM. This finding lends further support to the hypothesis that 4-Glc-FAG needs to be hydrolyzed *in vivo* to provide pharmacologically active FAG [63].

The DNJ derivatives miglitol **14**, emiglitate **15**, MDL 25637 **16**, and MDL 73945 **17** were selected for further evaluation. MDL 25637 and MDL 73945 appeared to be more effective on postprandial glucose and insulin response when administered 30–60 min *before* a sucrose load rather than when given simultaneously with sucrose [62e]. The long-lasting effect of MDL 73945 **17** has been suggested to be caused by quasi-irreversible binding to α-glucosidases [62e]. Sucrose loading tests on rats were employed to determine the inhibitor doses at which the postprandial increase in blood glucose is reduced by 50% (ED_{50}). Miglitol **14** and emiglitate **15** showed similar *in vivo* potencies, with ED_{50} values of 0.24 and 0.16 mg kg^{-1} body weight, respectively [53]. These studies also revealed the *in vivo* effect for both compounds to be a long-lasting effect. Miglitol **14** proved to significantly decrease the postprandial blood glucose as well as glycated hemoglobin (HbA_{1c}) levels in several clinical studies with approximately 4000 type 2 diabetes patients [65]. In 1999, miglitol was licensed in Europe and the U.S.A. (as Diastabol™ and Glyset™, respectively) for the treatment of type 2 diabetes mellitus. As with acarbose, miglitol acts locally in the GI tract, but, in contrast to acarbose, the majority of the orally taken drug is absorbed in the GI tract and excreted unchanged in the urine. Only a small amount is eliminated with the feces. The proportion of the drug that is absorbed decreases with increasing doses. Thus, single doses of up to 25 mg are absorbed to more than 95%, whereas only 60% are absorbed at a dose of 100 mg [66].

16.3.2.3 Summary and Outlook

Despite considerable research in the field of glycosidase inhibitors over the last 30 years, which has provided a variety of potential inhibitors [43], only three α-glucosidase inhibitors — acarbose, miglitol, and voglibose — are currently available on the market for the treatment of type 2 diabetes mellitus. The main challenges that research in this area must contend with are target-specificity and the unreliable nature of *in vivo* models because of marked differences between species [61].

Several clinical studies have proven the efficacy of α-glucosidase inhibitors for the treatment of noninsulin dependent diabetes mellitus as adjuncts to diet or in combination with other oral agents [45]. However, their main disadvantages are the high costs of the treatment and the high incidence of gastrointestinal side effects, such as flatulence, diarrhea and abdominal cramps, caused by nonresorbed carbohydrates in the large intestine. These carbohydrates have an osmotic action and are degraded to gases by bacteria [54].

16.3.3 Neuraminidase Inhibitors

16.3.3.1 Background: Influenza

Influenza is an acute respiratory tract infection caused by either influenza A or B viruses, members of the *Orthomyxoviridae* family [67]. The disease affects millions of people each year, most often occurring in the winter and early spring. According to the Center for Disease Control and Prevention (CDC), influenza causes 114,000 hospitalizations and 20,000 deaths annually, especially amongst elderly people or patients with other chronic diseases, and the combination of influenza and pneumonia is the sixth leading cause of death in the U.S.A. [40c,d]. Despite considerable mechanistic knowledge of viral infectivity, there are currently no therapeutic measures that completely control the disease. Although vaccination is the most effective approach to preventing influenza illness, it provides only limited protection because of the ease with which the virus mutates to escape the immune system. Consequently, vaccines must be reformulated each year because of this high antigenic drift [40]. Until recently, options for the therapeutic treatment of influenza have been limited to the antiviral drugs amantadine **22** and rimantadine **23** (Figure 16.9), which interfere with the influenza A virus M2 protein, the matrix protein of the viral envelope. By blocking the ion-channel function of the viral M2 protein, the adamantane amines inhibit the removal of the viral protein capsid (uncoating), thus preventing penetration of the host cell. Because influenza B strains do not contain M2 protein in their envelope, the compounds are inactive against these viruses [40c].

Clinical use of these drugs is limited not just because of their ineffectiveness against influenza B viruses, but also because of the rapid emergence of resistance [40]. A new class of agents, the neuraminidase inhibitors, has recently been developed that possesses significant activity against both influenza A and B strains.

16.3.3.2 The Role of Neuraminidase in Influenza Virus Replication

The influenza virus reaches the respiratory epithelium through aerosolized droplet transmission generated by coughing, sneezing or talking [40,68]. Hemagglutinin and neuraminidase are viral surface glycoproteins that affect the ability of the virus to cause disease [40]. Hemagglutinin is responsible for the attachment of the influenza virus to the host cell by binding to terminal sialic acid residues on the surface of respiratory epithelial cells, thus allowing the virus to penetrate the plasma membrane of the cell, uncoat and enter the cytoplasm. Following cell entry, viral ribonucleic acid (RNA) strands are replicated in the nucleus of the infected host cell and then packaged into new virions. In the final step of the process, a new generation of virions buds off from the host cell's surface membrane, ready to propagate the infection by invading adjacent cells. Because the viral envelope is formed from the host cell's membrane, the newly generated virions are covered with sialic acid residues. Neuraminidase is required to destroy these ligands for viral hemagglutinin by cleaving the terminal sialic acid residues from the host cell receptors and from the newly formed virions, allowing the release of progeny viruses. In the absence of neuraminidase, the

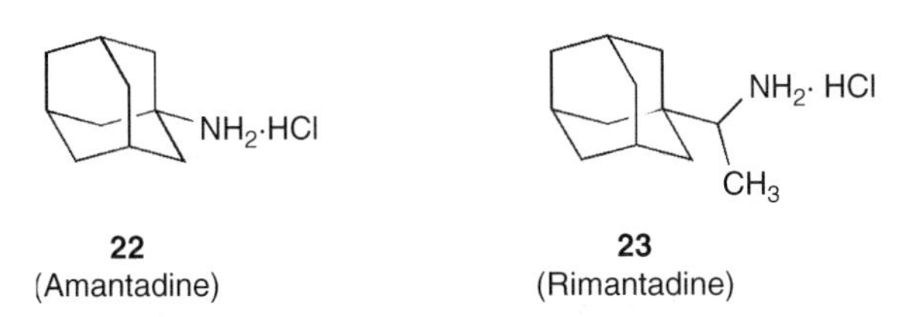

FIGURE 16.9 The antiviral adamantane amines amantadine **22** and rimantadine **23**.

virus particles would be unable to escape from binding to the host cell surface, which would cause their aggregation on the cell surface, resulting in the inhibition of viral release [69,70]. Thus, neuraminidase appears to protect the virus from its own hemagglutinin [40b]. An additional function of this enzyme may be to prevent viral aggregation and inactivation by sialic acid residues in respiratory mucus [70]. It has also been postulated that neuraminidase enhances viral pathogenicity by mechanisms not directly associated with its effect on sialic acid cleavage, including stimulation of the production of proinflammatory cytokines and increased cellular apoptosis [70b,71]. Consequently, neuramidase is a well-validated target for the therapeutic intervention against influenza.

16.3.3.3 Structure Affinity Relationship

Type A influenza viruses carry at least nine neuraminidase subtypes, while type B viruses seem to express only a single form [40c]. Although the sequence homologies of influenza A neuraminidase subtypes can be quite low, their active sites are not only highly conserved, but are also largely independent of antigenic change [72], which is highly beneficial for drug development. The influenza neuraminidases have molecular weights of approximately 240 kDa. Their mushroom-shaped heads are made up of four identical subunits in a square-planar arrangement, which are anchored to the viral coat at the *N*-terminus by a long thin stalk [40b,73]. Sialic acid (*N*-acetylneuraminic acid [Neu5Ac]), the product of catalysis, binds in a large pocket on the upper surface of the subunit, which is lined with an unusually large number of charged amino acid residues. These amino acids are believed to be involved in the hydrolysis process and in the stabilization of the transition-state, the sialosyl cation **24** (Scheme 16.3) [74].

The exact mechanism of enzyme catalysis as well as the structural and energetic aspects of substrate and inhibitor binding have been studied with the aid of molecular modeling, based on the complex of Neu5Ac with influenza virus A/Tokyo/3/67 neuraminidase (Scheme 16.4) [74c].

Enzyme catalysis proceeds in four steps. The first step is the binding of the sialoside to the neuraminidase, which causes a considerable distortion of the Neu5Ac pyranose ring from a $^2C_5$ chair conformation to pseudoboat conformation. The second step leads to the formation of the endocyclic sialosyl cation transition-state via proton transfer from the solvent to the substrate. The final two steps are the formation and release of Neu5Ac [74c].

Neu5Ac2en **25** (Scheme 16.3), which can be regarded as an analog of the transition-state leading to cation **24**, is a potent inhibitor of all neuraminidases, with activities typically in the micromolar range [75]. The selectivity for a particular neuraminidase with potentially higher inhibition rates may be potentially increased by preparing structurally modified derivatives of Neu5Ac2en.

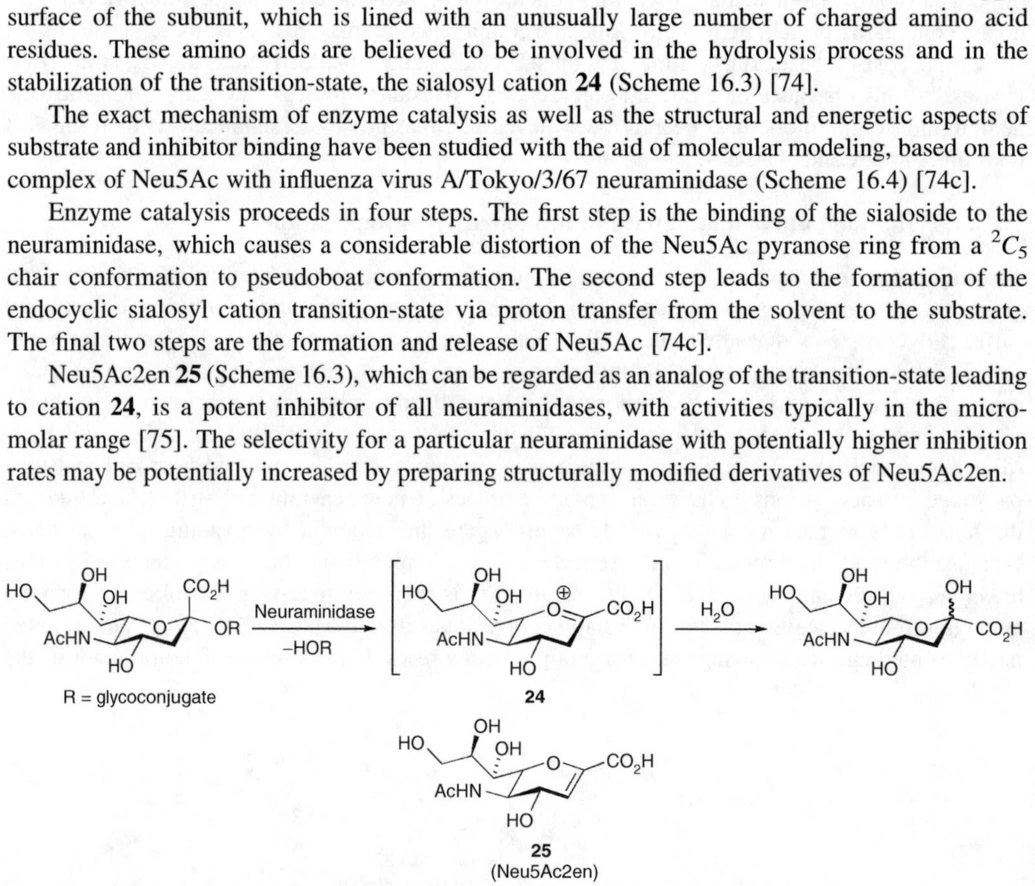

SCHEME 16.3 Sialoside hydrolysis via sialosyl cation **24**.

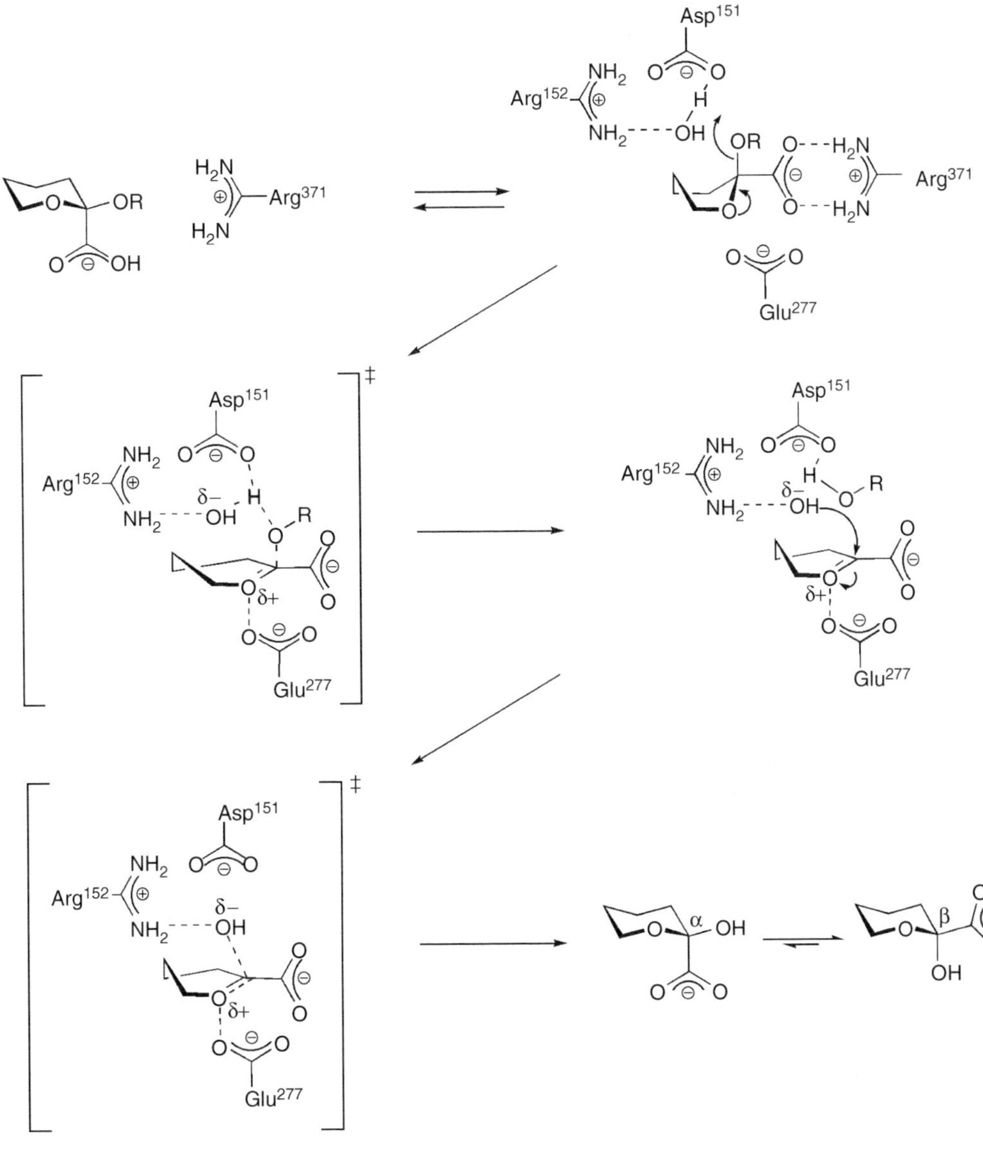

SCHEME 16.4 Proposed mechanism of catalysis of influenza virus neuraminidase [74c].

16.3.3.4 Mimetics of Neu5Ac2en as Neuraminidase Inhibitors

In an early study, deoxyneuraminic derivative **26** (Figure 16.10) was found to be a micromolar inhibitor of influenza neuraminidase [76]. X-Ray crystallographic studies with **26** were the basis for the design of compounds **27** and **28** by von Itzstein and coworkers [74]. These molecules take advantage of proposed salt bridges between the positively charged amino or guanidino group at C-4 and two negatively charged glutamic acid residues in the enzyme's active site [74,77]. These additional binding interactions significantly enhance inhibition, resulting in subnanomolar K_i values (Figure 16.10).

Compound **28**, named zanamivir, was selected for clinical studies for the prophylaxis and treatment of influenza virus. Zanamivir was found to be suitable as a drug because it is highly selective for influenza type A and B viruses, which it inhibits with high potency

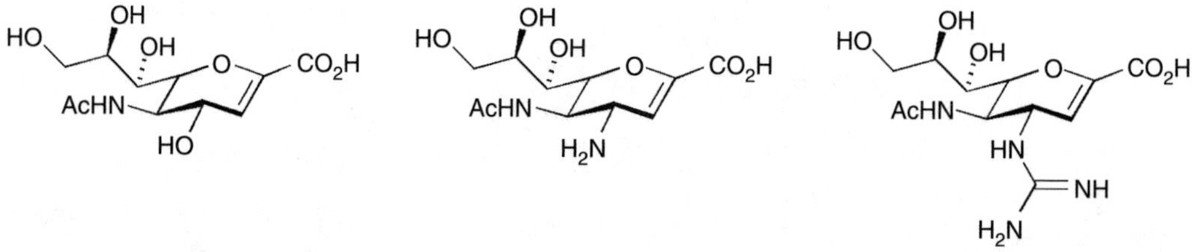

26 $K_i = 4\ \mu M$

27 $K_i = 10^{-8}$ M

28 (Zanamivir, Relenza) $K_i = 10^{-10}$ M

FIGURE 16.10 Structural improvements of Neu5Ac2en **26** leading to zanamivir **28**.

($IC_{50} \approx 10^{-11}$ M) [74]. In 2000, zanamivir (Relenza™) was approved by the Food and Drug Administration (FDA) for treatment of influenza virus A and B infections.

Ironically, the guanidino group, which is responsible for zanamivir's high potency, is also the cause of the drug's poor gastrointestinal absorption, because it is a highly polar zwitterionic compound [74a,78]. As a consequence, zanamivir needs to be administered as a dry powder by oral inhalation. Between 70 and 87% of the inhaled dose is deposited in the oropharynx, between 7 and 21% reaches the lungs and between 4 and 17% is absorbed systemically [40c]. Systemically absorbed drug has a plasma half-life of 2.5 to 5.1 h with limited plasma protein binding (<10%) and is excreted unchanged in the urine within 24 h, whereas unabsorbed, swallowed drug is excreted in the feces [40c].

Prompted by the promising results obtained with zanamivir, intense research efforts were directed at finding derivatives with improved pharmacokinetic properties. However, these efforts have mostly met with limited success, and derivatives of zanamivir with variations of the guanidino group [79] (Figure 16.11, compounds **29**) or of the substituent at C-5 (compound **30**) [80] as well as

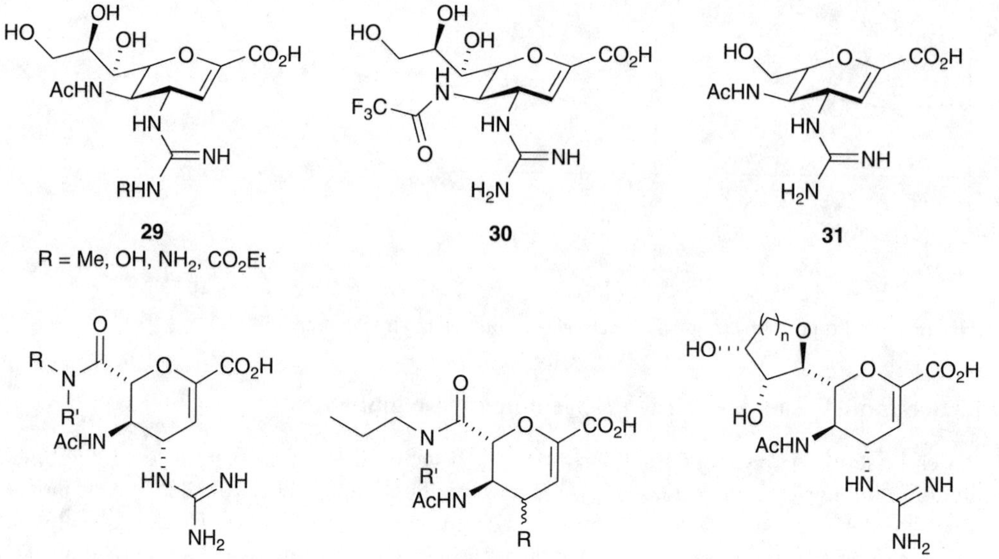

29 R = Me, OH, NH_2, CO_2Et

30

31

32 R, R' = alkyl, aryl

33 R = H, OH, NH_2; R' = *i*-Pr, $Ph(CH_2)_2$

34 n = 1, 2, 3

FIGURE 16.11 Derivatives of zanamivir with variations of the guanidino group (compounds **29**, **33**) of the substituent at C-5 (**30**), or of the glycerol side chain (**31**–**34**).

modifications of the glycerol side chain (compounds **31** [81] and compound **32** [82]) were less potent than the drug.

Interestingly, Neu5Ac2en mimetics of the general structure **33**, bearing a carboxamide residue instead of the glycerol side chain and lacking the guanidino group at C-4 were found to be quite potent neuraminidase inhibitors ($IC_{50} \approx 10^{-8}$ M) [83]. This suggests that the key binding interactions of these compounds also involve, apart from the crucial carboxylate group, the hydrophobic carboxamide group at C-6 [83]. Recently, the bicyclic ether derivatives **34** of zanamivir were reported to exhibit inhibitory activities against influenza virus A neuraminidase in the nanomolar range, comparable with those of zanamivir. Furthermore, compound **34** ($n = 2$) showed high oral efficacy in a mouse infection model [84].

The search for anti-influenza drugs with improved pharmacokinetic properties has led to the development of noncarbohydrate-based mimics of zanamivir **28**, the most significant one being the sialylmimetic oseltamivir **35**, discovered in the late 1990s (Figure 16.12) [40a,85]. Oseltamivir **35** was found to bind as potently as zanamivir to influenza viruses A and B [85], probably via a cyclohexene ring conformation that resembles the transition-state leading to intermediate **24** (Scheme 16.3). Oseltamivir does retain the negatively charged carboxylate group of sialic acid, which provides for key binding interactions to neuraminidase, but it also possesses a hydrophobic hydrocarbon chain, which interacts with a hydrophobic patch in the active site of the enzyme.

Oseltamivir came on the market in 2000 under the name of Tamiflu™ **36**, which is the phosphate salt of the ethyl ester prodrug [86]. Oseltamivir phosphate is well absorbed in the GI tract and can therefore be administered orally. Approximately 80% of an oral dose reaches

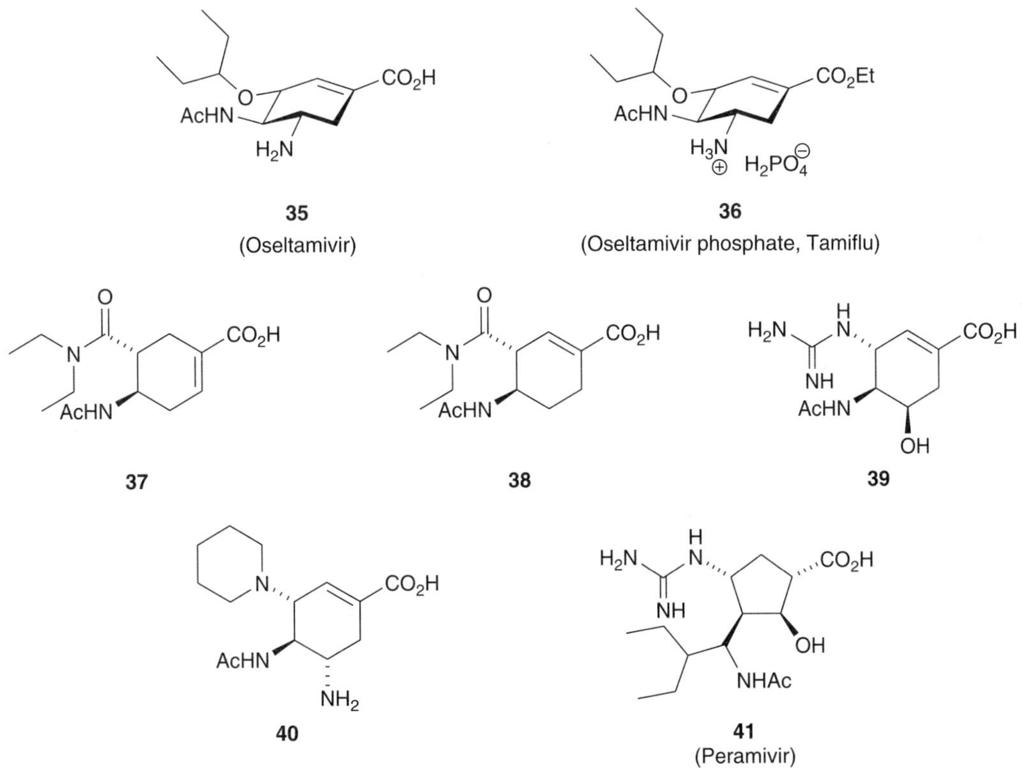

CO_2H AcHN O H_2N **35** (Oseltamivir)

CO_2Et AcHN O $H_3N^{\oplus}$ $H_2PO_4^{\ominus}$ **36** (Oseltamivir phosphate, Tamiflu)

O N CO_2H AcHN **37**

O N CO_2H AcHN **38**

H_2N H N NH CO_2H AcHN OH **39**

N CO_2H AcHN NH_2 **40**

H_2N H N NH CO_2H OH NHAc **41** (Peramivir)

FIGURE 16.12 Oseltamivir **35** and other carbocyclic neuraminidase inhibitors.

the systemic circulation after conversion to the active compound, predominantly by hepatic esterases. The active compound has a plasma half-life of 6 to 10 h, with low plasma protein binding (3%). Oseltamivir carboxylate is eliminated unchanged in the urine, and less than 20% of an oral dose of oseltamivir phosphate is eliminated in the feces [40c].

Prompted by the success achieved with simple carbocyclic compounds (e.g., **35**), which are potent inhibitors of influenza neuraminidase, several other Neu5Ac2en mimetics of this kind were studied. The isomeric carbocyclic derivatives **37** and **38** are selective for influenza A sialidase, but only **38** showed significant inhibition ($IC_{50} \approx 2 \times 10^{-8}$ M) [87]. Other carbocyclic neuraminidase inhibitors include the guanidino substituted cyclohexene **39** [88] and amine derivatives [89] such as **40**, which was found to be an inhibitor of influenza B neuraminidase comparable with oseltamivir **35**, but was not as potent against influenza A virus [89a].

These efforts have culminated in the five-membered carbocyclic compound peramivir **41**, which is an orally available influenza neuraminidase inhibitor, obtained by rational design [90]. In a comprehensive study, **41** has been shown to possess neuraminidase activity comparable with both zanamivir **28** and oseltamivir **35** against a number of influenza strains. Furthermore, **41** was found to be very selective for influenza neuraminidase over mammalian, bacterial or other viral neuraminidases. Peramivir **41** is currently undergoing clinical evaluation as an oral treatment for influenza in humans [90b].

Another study identified a class of hydrophobic benzoic acid derivatives as potent inhibitors of influenza neuraminidase (Figure 16.13) [91].

Based on the lead compound BANA 113 **42**, which inhibits influenza A neuraminidase with a K_i of 2.5 μM [91a,b], several novel 2-pyrrolidinone derivatives were designed. Whereas compound **43** (BANA 205) is a relatively weak inhibitor of influenza A and B neuraminidases, with IC_{50} values in the low micromolar range, compound **44** (BANA 206) is much more potent, with an IC_{50} value of 48 nM against influenza A neuraminidase. However, this compound is not as potent against influenza B virus sialidase [91d,e].

Interestingly, the comparison of the X-ray crystal structures of the complexes of influenza A neuraminidase with zanamivir **28** and BANA 113 **42** revealed that, while the carboxylate and *N*-acetyl groups occupy the same subsites, the guanidino functionalities interact with the enzyme in different ways (Figure 16.14) [91d].

In the complex of zanamivir, the C-8 and C-9 hydroxyls of the glycerol moiety form hydrogen bonds to Glu 278 within subsite 6, whereas the guanidino functionality occupies subsite 4, where it interacts with Asp 152 and Glu 229 through salt bridges [92]. In contrast, BANA 113 **42**, which lacks the glycerol moiety, places the guanidine group into subsite 6, where it forms a salt bridge with Glu 278 [91d].

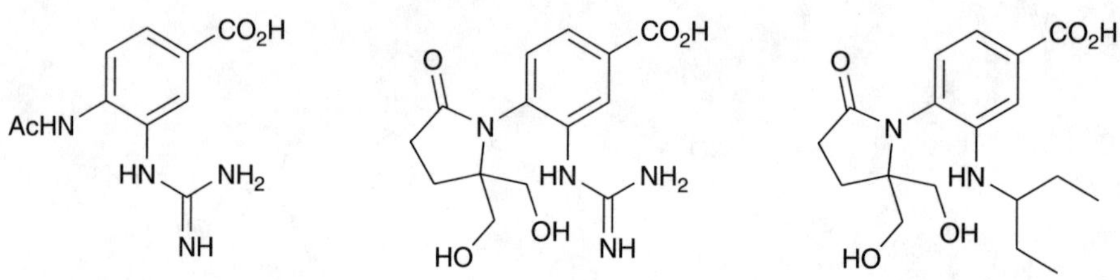

FIGURE 16.13 The BANAs, a class of hydrophobic benzoic acid derivatives, acting as potent inhibitors of influenza neuraminidase.

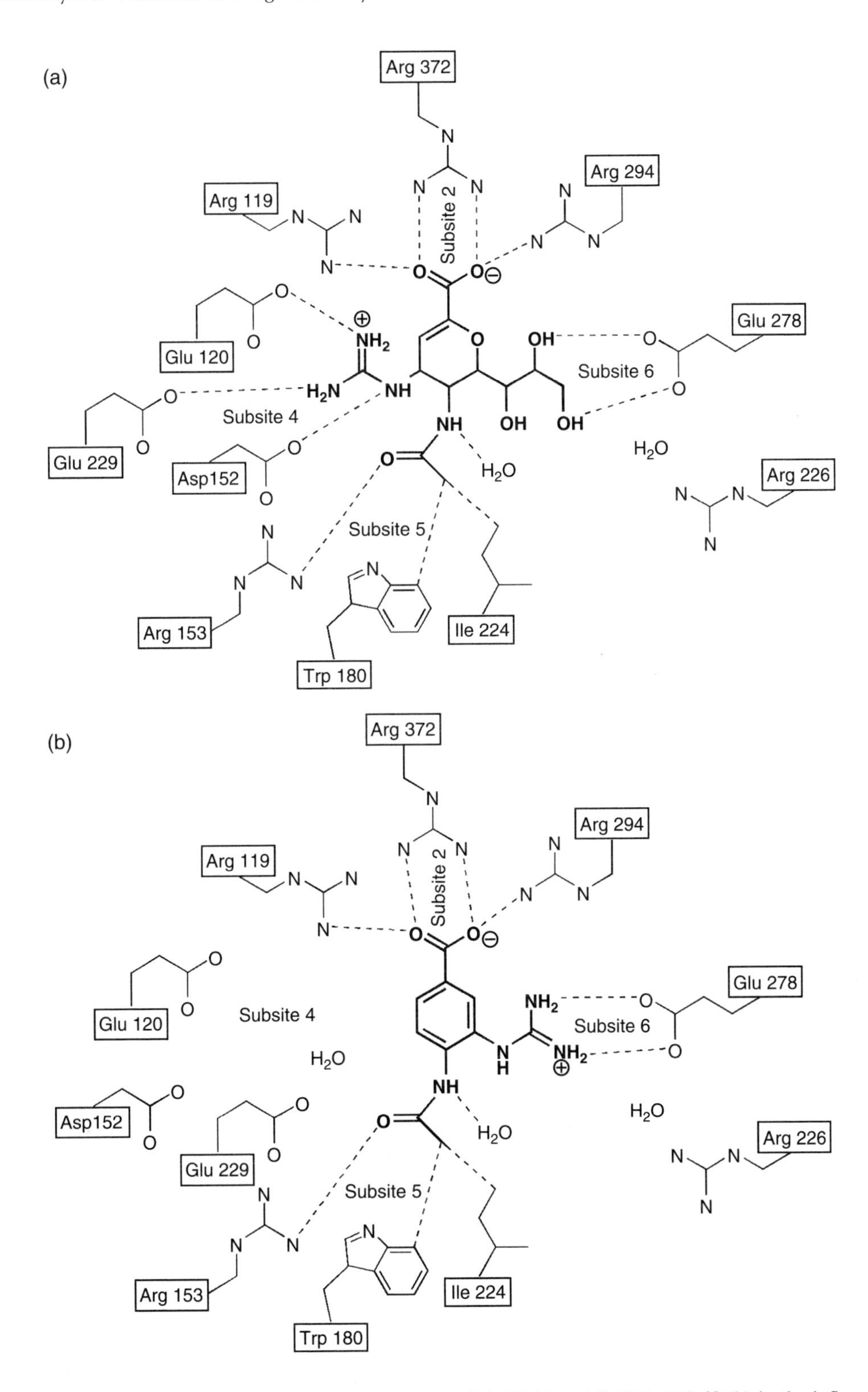

FIGURE 16.14 Depiction of the interactions of zanamivir **28** (a) and BANA 113 **42** (b) in the influenza A neuraminidase binding site (N9 numbering) [91d,92].

16.3.3.5 Summary and Outlook

Several potent influenza neuraminidase inhibitors have been synthesized, and some of them have proven to be effective in treating influenza virus infections in humans. Table 16.5 summarizes the chemical structures and names of these compounds as well as their IC_{50} values.

Two of these antiviral agents, zanamivir **28** and oseltamivir **35**, have recently become available on the market for the treatment of influenza. These compounds are superior to the older antiviral drugs amantadine and rimantadine (Figure 16.9) because they are active also against influenza virus B, have fewer side effects (especially in older patients) and are less sensitive to viral mutations [40c]. Zanamivir, which is poorly absorbed in the GI tract, is formulated as a powder for oral inhalation, whereas oseltamivir **35** is administered orally as a capsule. Both drugs have to be taken within 2 days of symptom onset in order to reduce the duration of uncomplicated influenza illness by approximately 1 to 1.5 days [40c]. The two compounds appear to be equally effective, although no direct comparative studies have been reported to date. Both drugs have shown efficacy in the prevention of influenza, although they are not currently approved for this use [40].

Several problems still remain, providing scope for future research in this field. First, the successful treatment of influenza with neuraminidase inhibitors requires the drugs to be taken at a very early stage of the infection because they are ineffective once an influenza infection has developed the full symptoms. This is not always practical, however, since it is often difficult to distinguish the early symptoms of influenza from those of a common cold. Second, it is questionable whether the shortening of the disease by 1 day justifies the high cost of treatment with these neuraminidase inhibitors. The cost factor would be even more dramatic if these agents were to be used for extended influenza prophylaxis [40d].

Because of these drawbacks, vaccination is still considered to be the most effective approach to prevent influenza illness. Inoculation with a vaccine based on the most common strains of influenza A and B in 1999/2000 has been shown to prevent influenza in 70 to 90% of treated patients. When patients do develop disease after vaccination, the illness is generally less severe, with a marked decrease in secondary complications or death [40d].

16.4 SELECTIN ANTAGONISTS

16.4.1 BIOLOGICAL RATIONALE

Leukocyte recruitment to sites of injury or infection is an important defense mechanism of the immune system. However, the excessive infiltration of inflamed tissues by leukocytes can cause acute or chronic conditions such as reperfusion injury, stroke, psoriasis, rheumatoid arthritis and respiratory diseases [93]. Consequently, there is a great medical need for preventing excessive leukocyte recruitment, which takes place in a regulated chain of events: rolling, firm adhesion, diapedesis and transendothelial migration (Figure 16.15). The first step in the inflammatory cascade — leukocyte rolling — is mediated largely by the selectin family of adhesion molecules and their carbohydrate ligands [94].

The selectins are a family of three calcium-dependent cell-adhesion molecules (P-, E- and L-selectin). They are transmembrane glycoproteins containing an amino-terminal lectin domain, an epidermal growth factor domain and a variable number of complement regulatory-like repeats. There is greater than 50% homology between the lectin domains of the three human selectins. P-selectin is stored in secretory granules of endothelial cells and platelets, ready to be transported to the cell surface within minutes after stimulation with cytokines [95]. In contrast, E-selectin expression on endothelial cells takes several hours upon inflammatory stimulation [96]. Finally, L-selectin is constitutively expressed on the surface of most leukocytes [97]. It has been proposed that the ability of the selectins to mediate the rapid and transient interactions typical of leukocyte tethering and rolling under shear conditions requires them to bind to their physiological ligands

TABLE 16.5
Chemical Structures, Names and IC_{50} Values of Potent Influenza Neuraminidase Inhibitors

Chemical structure	Chemical Name	Abbreviation	IC_{50} (n*M*)	
			NA	NB
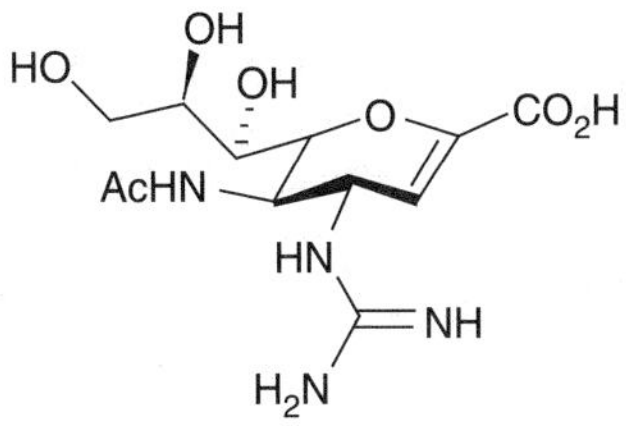 **28** [40]	2,3-Didehydro-2,4-dideoxy-4-guanidino-*N*-acetylneuraminic acid	Zanamivir GG167	0.3–2.3[a]	1.6–17.0[a]
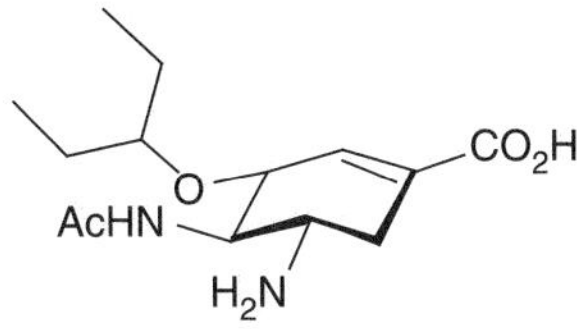 **35** [51]	(3*R*,4*R*,5*S*)-4-Acetamido-5-amino-3-(1′-ethylpropoxy)-cyclohex-1-ene-1-carboxylic acid	Oseltamivir GS4071	0.01–2.2[a]	6.4–24.3[a]
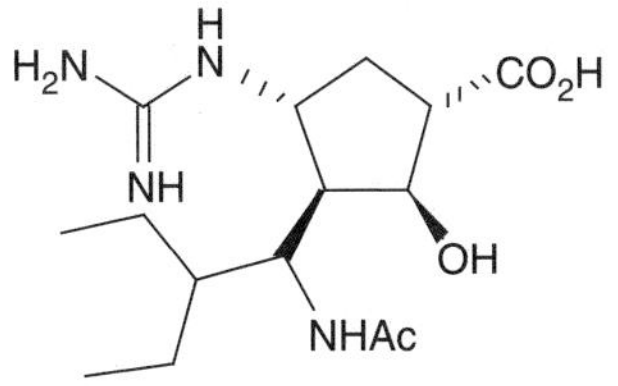 **41** [56]	(1*S*,2*S*,3*R*,4*R*,1′*S*)-3-(1′-Acetamido-2′-ethyl)butyl-4-guanidino-2-hydroxycyclopentane-1-carboxylic acid	Peramivir BCX-1812	0.09–1.4[a]	0.60–11[a]
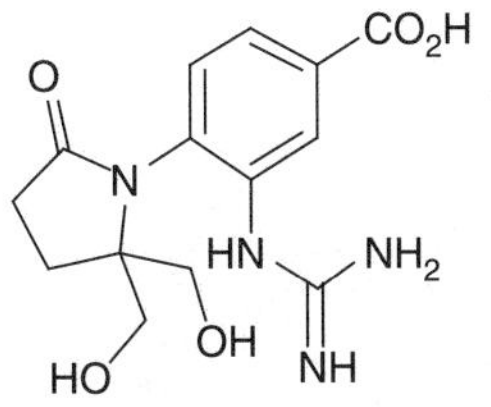 **43** [57]	1-(4-Carboxy-2-guanidinophenyl)-5,5-bis(hydroxymethyl)pyrrolidin-2-one	BANA205	5000[b]	8000[b]
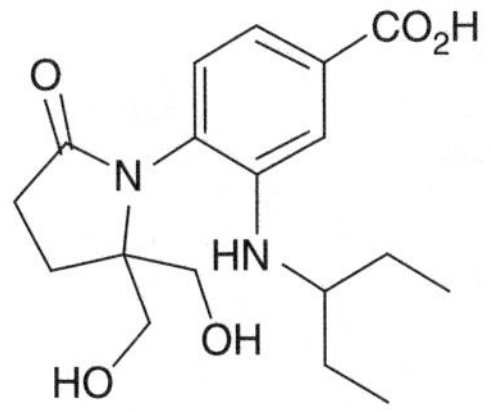 **44** [57]	1-[4-Carboxy-2-(3-pentylamino)phenyl]-5,5-bis(hydroxymethyl)pyrrolidin-2-one	BANA206	48[b]	224000[b]

[a] Range of inhibitory potency against 15 different influenza A neuraminidases and 8 different influenza B neuraminidases [63Wb].

[b] Determined on strains A/tern/Australia/G70C/75(N9) and B/Lee/40 [63Xd].

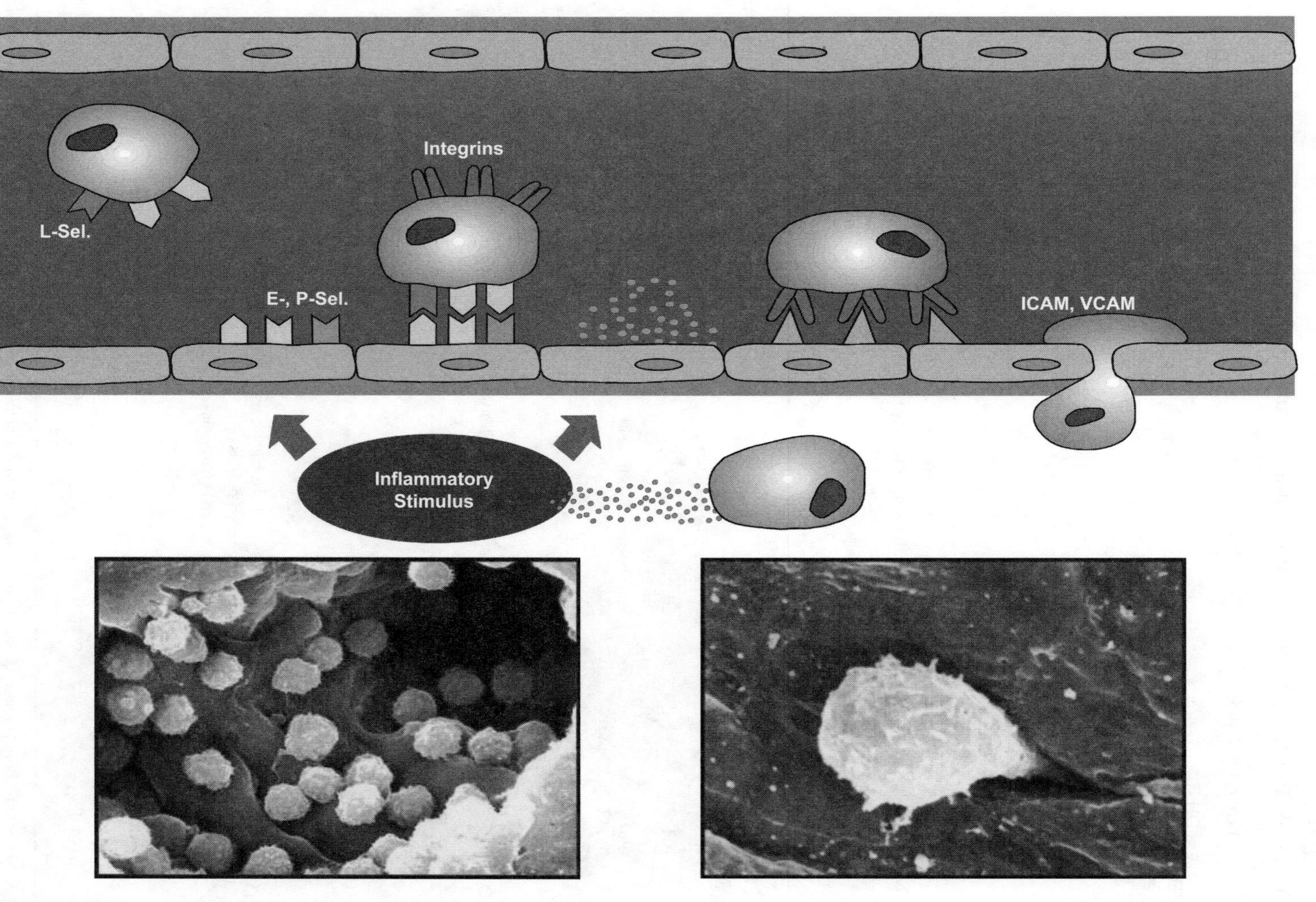

FIGURE 16.15 The inflammatory cascade.

with fast kinetics or a high tensile strength [98]. To test this hypothesis, the affinities and kinetics of three selectin-ligand interactions that can mediate rolling were recently measured using surface plasmon resonance (SPR) technology: L-selectin binding to glycosylation-dependent cell-adhesion molecule-1 (GlyCam-1) [99], P-selectin binding to P-selectin glycoprotein ligand-1 (PSGL-1) [100] and E-selectin binding to E-selectin ligand-1 (ESL-1) [101] (Figure 16.16). The exceptionally high association rate constant of the P-selectin-PSGL-1 interaction is in good agreement with the well-documented function of PSGL-1 as a selectin ligand capable of mediating leukocyte capturing under flow conditions *in vivo* as well as *in vitro* [102].

The observations that selectin knockout mice [103a–c] and LAD-type-2 patients (*leukocyte adhesion deficiency*) [103d], which lack the endogenous ligands for E-selectin, show impaired inflammatory responses, serve as genetic validation of the selectins as drug-discovery targets. Although the exact nature of the carbohydrate ligands and their interaction with the selectins is not fully understood, it is accepted that the expression of fucosylated and sialylated glycans such as sialyl Lewis^x (sLe^x) is required for function [104]. Soluble sLe^x not only inhibits leukocyte binding to selectins *in vitro*, but also has beneficial effects in inflammatory disease models [105].

16.4.2 Structure Affinity Relationship

Because the tetrasaccharide sLe^x **45** (Figure 16.17) is the carbohydrate epitope recognized by all three selectins [104], it serves as a lead structure for the design of selectin antagonists [106]. The pharmacophores required for selectin recognition are the carboxylic acid function, the 3- and 4-hydroxyl groups of the fucose and the 4- and 6-hydroxyl groups of the galactose [107]. Additional information was obtained by saturation transfer difference nuclear magnetic resonance (STD-NMR) [108] (Figure 16.17).

Projects for the design of functionally active sLe^x mimics were aided by molecular modeling right from the beginning. These modeling studies were aimed at understanding and predicting the three-dimensional (3-D) structure of the selectins and their ligands in solution, as well as in the bioactive conformation. They were supported by X-ray crystallography and nuclear magnetic resonance (NMR) methods.

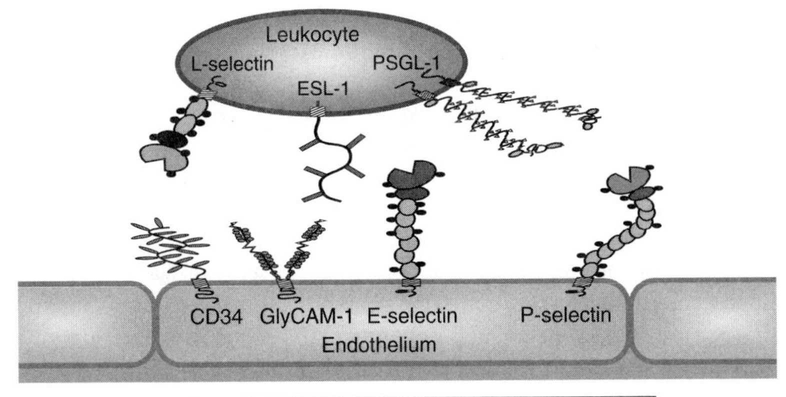

	K_D[μm]	k_{on}[$m^{-1}s^{-1}$]	k_{off}[s^{-1}]	$t_{1/2}$[s]
L-selectin	108	> 10^5	>10	0.06
E-selectin	62	$4 \cdot 10^4$	3.0	0.2
P-selectin	0.3	$4 \cdot 10^6$	1.4	0.5

FIGURE 16.16 Selectins and their physiological ligands: affinity and kinetics.

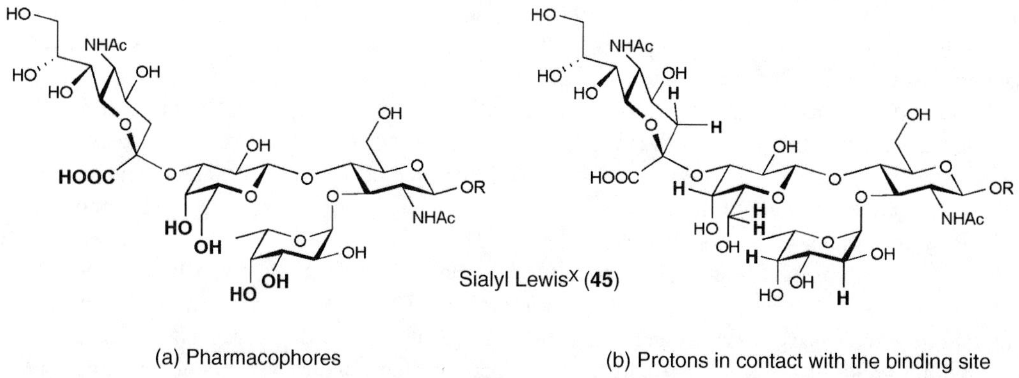

Sialyl Lewis^{X} (**45**)

(a) Pharmacophores (b) Protons in contact with the binding site

FIGURE 16.17 Pharmacophores of sialyl Lewis^{x} **45** (sLe^{x}). (a) Functional groups required for binding as determined by chemical SAR study are shown in boldface [107]; (b) Identification of the binding epitope by STD-NMR [108]; protons in close contact to the binding site are highlighted in large boldface; protons in the 3-position of neuraminic acid are only in loose contact to the binding site and therefore highlighted only in small boldface.

All drug-discovery approaches have so far been focused on blocking the lectin domain. Given that the role of the EGF and CR domains in ligand binding is still unclear [109], our structural considerations will concentrate exclusively on the lectin domain and the selectin ligands.

16.4.2.1 Selectin Structure

The crystal structure of the complex between mannose and the C-type lectin domain of the rat mannose-binding protein A (MBP-A) [110] has had a great impact on the development of selectin antagonists because of the high degree of homology between MBP-A and the selectins. In contrast to other crystal structures of oligosaccharide/lectin complexes [111] where the role of calcium is to maintain the structural integrity of the binding site, the mannose/MBP-A complex is special in that there is a direct interaction between the carbohydrate and the calcium ion. This observation and the structural similarity of mannose and fucose led to the hypothesis that the 2- and 3-hydroxyls of fucose bind to the calcium of E-selectin. This structural assumption was further supported by Weis et al. [112], who solved the crystal structure of sLe^{x} bound to a selectin-like mutant of the MBP-A. Based on this information, various models of the sLe^{x}–E-selectin interaction have been developed [113,114c,115]. Each of the models has predicted some interactions correctly, but others incorrectly. In all models, however, the interaction of fucose predicted in the analogy of the binding mode of mannose [112] turned out to be incorrect.

When the crystal structure of the lec-EGF domain of E-selectin was solved by Graves et al. [116], some important differences between MBP-A and E-selectin became apparent. In E-selectin, the co-ordination sphere of calcium contains only four amino acid residues (Glu80, Asn82, Asn105 and Asp106) instead of five, and the remaining co-ordination site is filled by a water molecule. Recently, Camphausen et al. [117] significantly advanced our understanding of selectin/carbohydrate binding by determining the X-ray crystal structures of complexes of sLe^{x} with the lectin/EGF domains of human P- and E-selectin. Their work not only provided interesting new insights, but also a few surprises. The majority of the contacts appear to be electrostatic in nature. The contact area between sLe^{x} and the selectins is relatively small (approximately 550 Å^{2}) and the calcium interacts with the 3- and 4-hydroxyls of fucose (Figure 16.18).

Based on semiempirical molecular orbital calculations with an implicit treatment of the water environment, Pichierri and Matsuo [118] recently suggested that the sLe^{x} tetrasaccharide binds to

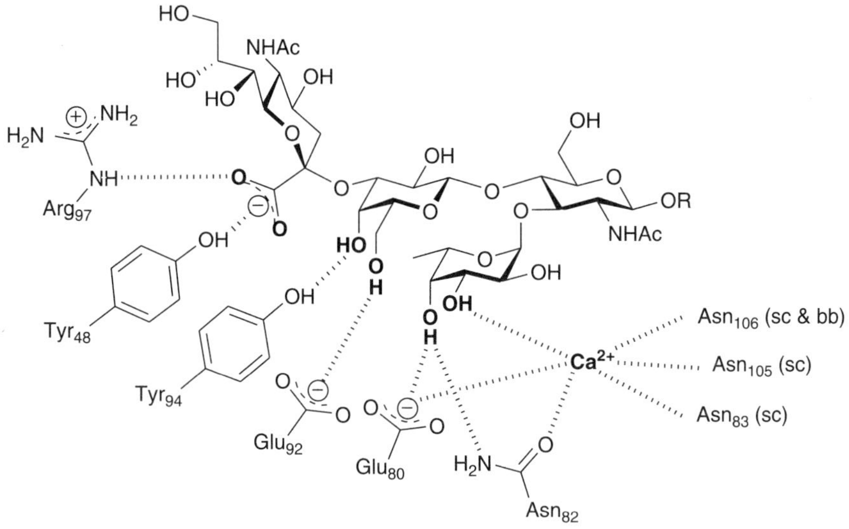

FIGURE 16.18 Interactions in the E-selectin binding side with the pharmacophores of sLe^x [117] (sc: side chain, bb: backbone).

E-selectin in the neutral (not anionic) state, an observation that could open up new perspectives in the design of sLe^x mimics.

16.4.2.2 Solution and Bioactive Conformation of sLe^x

The conformational preferences of sLe^x in solution and bound to the selectins were elucidated primarily with the aid of NMR spectroscopy. Early work [119] suggested that sLe^x adopts a single conformation in solution. However, subsequent NMR and molecular dynamics studies revealed a certain degree of flexibility of the glycosidic linkages [114,120; see also Section 16.4.3.2].

The bioactive conformation of sLe^x was determined by Cooke et al. [114a], Hensley et al. [121], Peters et al. [122], Poppe et al. [114c] and Homans et al. [114d] based on a careful analysis of its nOe (nuclear Overhauser effect) signals in solution and its transfer-NOE signals when bound to E-selectin. The dihedral angles Φ and Ψ about each glycosidic linkage, as determined by NMR experiments, are given in Table 16.6. For comparison, the values obtained by Camphausen et al. [117] from the crystal structure are also included in Table 16.6.

16.4.3 Families of Antagonists Identified So Far

Early drug-discovery approaches utilized carbohydrate fragments of the physiological ligands as potential anti-inflammatory therapeutics. This strategy led to the development of Cylexin, a sLe^x pentasaccharide, which initially showed promising biological results [105]. However, Cylexin turned out to be unsuccessful as a drug candidate [124], probably because of its low biological affinity [125] and limited bioavailability [123c]. Because of its complex structure, the availability of Cylexin was initially very limited, chemical synthesis providing only gram quantities [126]. This problem, however, was elegantly solved by Wong et al. [119b], who developed a multi-enzyme system for the large-scale production of sLe^x.

Despite their ultimate failure, these early studies did provide the *proof of principle*, which prompted intense efforts towards finding more *drug-like* and synthetically accessible selectin

TABLE 16.6
Dihedral Angles Φ and Ψ for the Glycosidic Bonds of sLe^x 1 [114a,c,d,121,122] Determined by NMR and Protein Crystallography [117]

References	Selectin	Φ/Ψ Neu5Ac(2-3)Gal	Φ/Ψ Gal(1-4)GlcNAc	Φ/Ψ Fuc(1-3)GlcNAc
NMR [122a]	E	−76°/+6°	+39°/+12°	+38°/+26°
NMR [114c]	E	−58°/−22°	+24°/+34°	+71°/+14°
	P	−85°/−4°	+45°/+18°	+61°/+26°
NMR [114d]	E	−43°/−12°	+45°/+19°	+29°/+41°
X-ray [117]	E	−65°/−12°	+34°/+16°	+41°/+22°
	P	−65°/−8°	+40°/+8°	+40°/+16°
	P[a]	−55°/−11°	+40°/+16°	+70°/+20°

[a] Dihedral angles measured on the sLe^x moiety of PSGL-1 bound to P-selectin.

antagonists with improved affinity and bioavailability. The approaches can be divided into two categories:

1. *De novo* design of antagonists based on the spatial orientation of the pharmacophores of sLe^x in the bioactive conformation (see Section 16.4.3.1).
2. Structural modification of sLe^x tetra- or pentasaccharide, with the goal of identifying suitable L-fucose, *N*-acetyl-D-glucosamine, D-galactose or *N*-acetyl-neuraminic acid replacements (see Section 16.4.3.2).

The IC_{50} values (concentration to achieve 50% displacement of competitive ligand) that are reported in the literature are difficult to compare directly because they were obtained with conceptually different *in vitro* assays [127]. Therefore, this chapter cites only the *relative* affinities of the test compounds compared with sLe^x, which has an $IC_{50}^{E\text{-selectin}}$ value of approximately 1 mM. For example, a relative IC_{50} value (rIC_{50}) of 0.1 corresponds to an IC_{50} value of approximately 100 μM (for more details with respect to *in vitro* and *in vivo* evaluation of selectin antagonists, see Section 16.4.4).

16.4.3.1 Mimetics of sLe^x

The conformational analysis of sLe^x using Monte Carlo and molecular dynamics methods in conjunction with an implicit water model (GB/SA) [120] suggested that the pharmacophores in solution were preorganized in an orientation similar to the bound conformation, which had been determined by transfer-NOE NMR [122] (Figure 16.19). In this conformation, the fucose and galactose subunits (which both contain groups that interact with the protein) are stacked above each other, and the GlcNAc-moiety plays the role of a linker that holds them in the correct spatial orientation (core conformation angle 30°; Figure 16.20). In addition, the carboxylic acid function of Neu5Ac points out of the projection plane leading to an acid orientation angle of approximately −120° (Figure 16.20).

SLe^x mimics **46** to **48**, which had the GlcNAc unit replaced by (*R*,*R*)-cyclohexan-1,2-diol and sialic acid substituted by glycolic acid or (*S*)-cyclohexyl lactic acid, were active as antagonists (Figure 16.21). These observations are in agreement with predictions based on computational studies that suggested these compounds are able to adopt the bioactive conformation. These computational simulations further predicted that the (*R*)-lactic acid isomers were inactive because of their inability to properly orient the pharmacophores. Indeed, replacement of sialic acid by

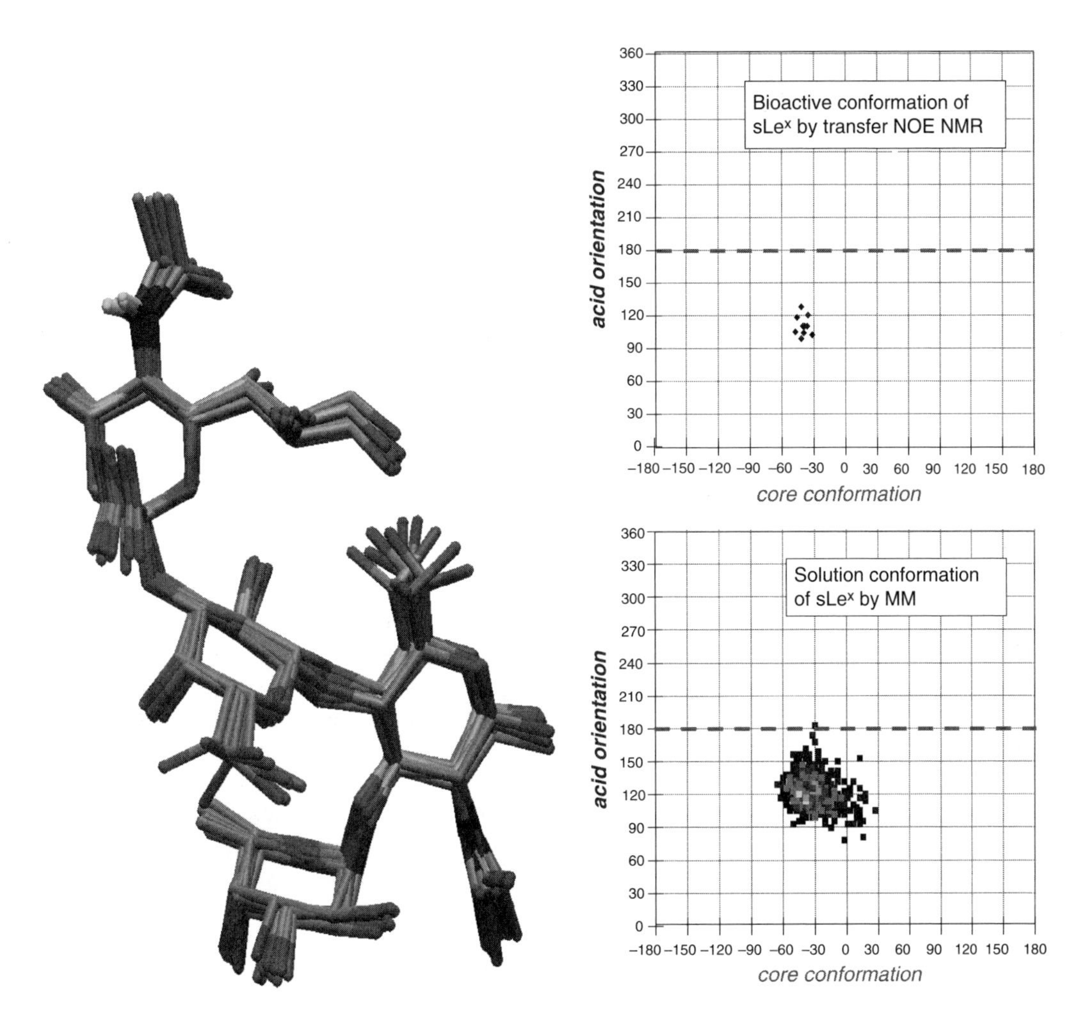

FIGURE 16.19 Comparison of the solution conformation of sLe^x determined by molecular modeling (MM) [120] and the bioactive conformation determined by transfer nOe NMR [122].

(*R*)-lactic acid and derivatives thereof gave rise to inactive compounds [120]. Likewise, compound **51** (Figure 16.21) was inactive, probably because its carboxylic acid group is unable to adopt the correct bioactive orientation [128]. Compounds with improved affinity were obtained by rigidifying the core structure sterically with an adjacent methyl group. Thus, compound **49** [129] is 100 times more active than sLe^x in a competitive binding assay, corresponding to an $rIC_{50}^{E\text{-}sel.}$ of 0.01 and an $IC_{50}^{E\text{-}sel.}$ of approximately 10 μM. Surprisingly, the rigidification of the Lewis^x core by incorporating the D-galactose and L-fucose moieties in a macrocyclic ring system **50** did not enhance the biological affinity ($rIC_{50} = 1$) [130].

An alternative approach utilized a pharmacophore model based on the bound conformation of sLe^x. Three-dimensional pharmacophore searches by Kondo et al. [127] and Brandley et al. [131] led to the antagonists **52** ($rIC_{50}^{E\text{-}sel} \approx 0.15$) and **53** ($rIC_{50}^{P\text{-}sel} \approx 0.5$) (Figure 16.22) [127].

Kogan et al. [132] designed a series of sLe^x mimetics by replacing L-fucose with D-mannose (Figure 16.23). The carboxylic acid function was linked to the mannose through a biphenyl spacer, which replaced the Gal-β(1-4)-Glc*N*Ac disaccharide (→**54**) of the lead structure. Although these compounds lack the important 4- and 6-hydroxyl groups of the D-galactose moiety, they turned out to be slightly more potent than sLe^x. The activity could be further improved by modifying the linker between the carboxylic acid function and the biphenyl core (→**55** and **56**) [133]. A breakthrough was

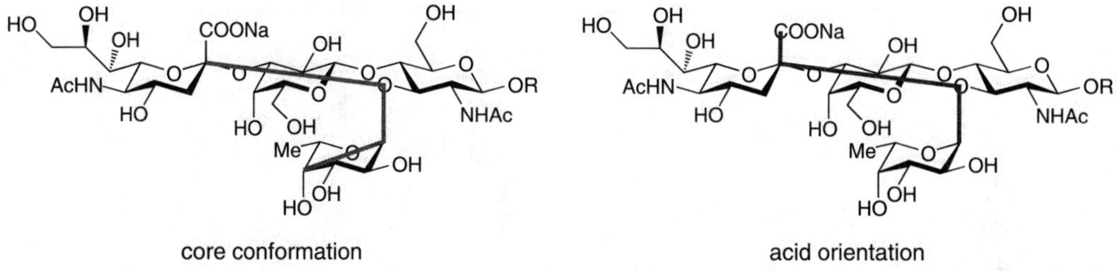

core conformation

acid orientation

FIGURE 16.20 Core conformation and acid orientation are used as parameters to describe the spatial orientation of the saccharide moieties of sLe^x.

achieved with the dimers of **54**, which were found to be from 6- to 50-fold more active than sLe^x in their ability to inhibit the binding of HL-60 cells to E- and P-selectin/IgG, respectively. These compounds were inspired by previous reports on the improved E-selectin binding of extended dimeric sialyl $Lewis^x$ derivatives [134]. One candidate, **57** (TBC1269) [135], is currently undergoing clinical development for the treatment of asthma [136].

Following a similar approach, Taylor et al. [137] predicted one of the low-energy conformers of the tetralin derivative **58** to position its pharmacophores in an orientation similar to the bioactive conformer of sLe^x [122]. Supporting this hypothesis, compound **58** proved to be twice as active as sLe^x against E-selectin. Finally, Gravel et al. [138] adopted an analogous strategy, which led to the *cis*-decalinic mimic **59** showing a biological activity comparable with sLe^x for both E- and P-selectin.

To further improve the activity of selectin antagonists, various groups have attempted to access additional (preferentially hydrophobic) interactions with the receptor. DeFrees et al. [139] showed

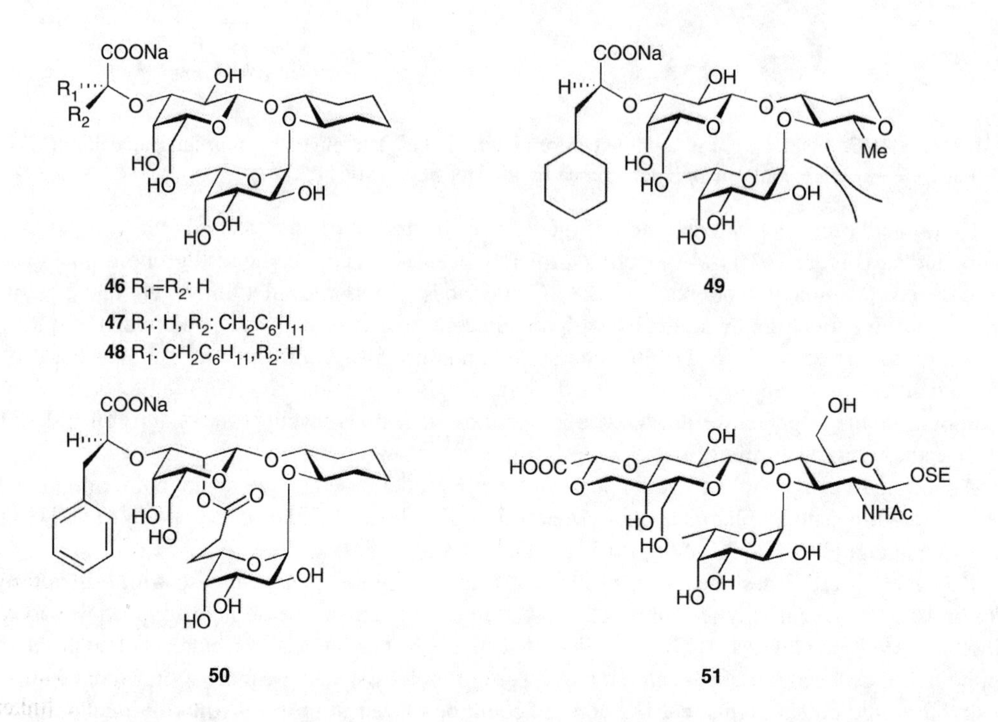

FIGURE 16.21 Selectin antagonists designed to be preorganized in the bioactive conformation.

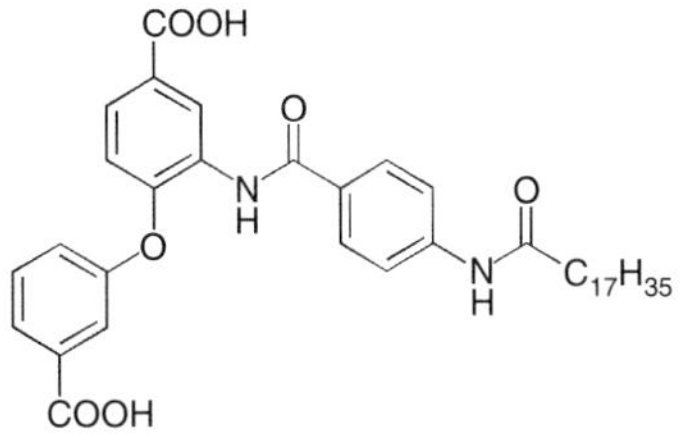

FIGURE 16.22 Selectin antagonists identified using 3D-pharmacophores to search 3D-database.

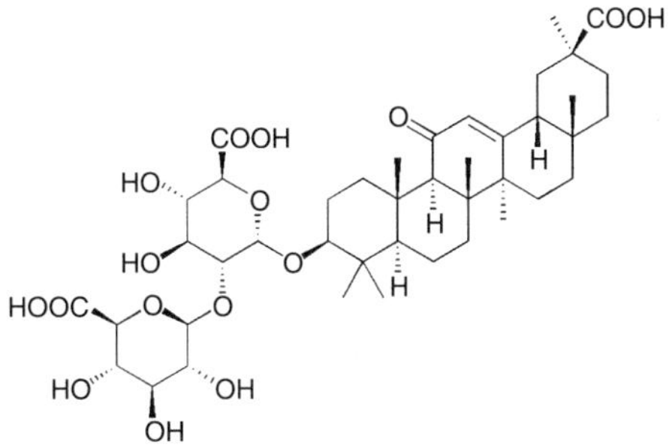

that higher affinities are obtained when the acetyl group on the glucosamine moiety of sLe^x is replaced by benzoate (**60**, rIC$_{50}^{\text{E-sel}}$: 0.3) or naphthoate (**61**, rIC$_{50}^{\text{E-sel}}$: 0.08) (Figure 16.24). However, the structural basis for this improvement in activity is not fully understood, despite the availability of the crystal structure of Lec-EGF domain of E-selectin [140].

Tsujishita et al. [141] found **62** to be superior to sLe^x against E- and P-selectin in a competitive binding assay. They identified three critical interactions with E-selectin using molecular dynamic simulations: (1) co-ordination of the 2- and 3-hydroxy groups of L-fucose to calcium, (2) sulfate binding to a basic protein residue, and (3) an extended hydrophobic interaction between the long-branched alkyl chains at the reducing end of **62** and the protein surface. The molecular dynamic simulations further suggested that the interaction between the sugar and calcium was weak, and that the contact with either of the two hydrophobic areas on the E-selectin surface (Tyr44, Pro46 and Tyr48 or Ala9, Leu114, and the alkyl portions of Lys111,112,113) was strong.

We have further explored hydrophobic interactions during our search for high-affinity selectin ligands. Selected examples are shown in Figure 16.25 [115]. Whereas the sulfonamide derivative

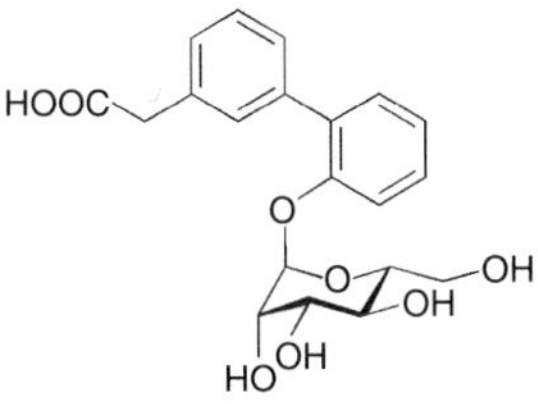

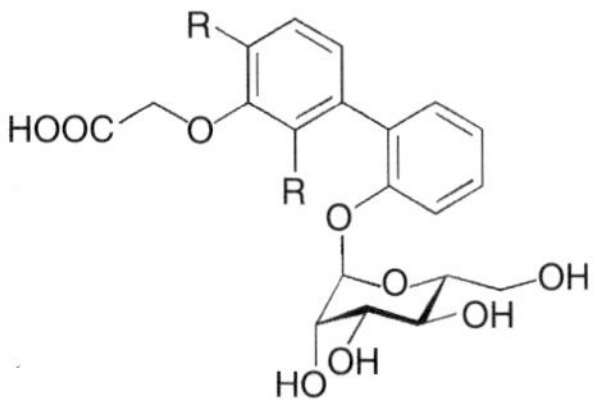

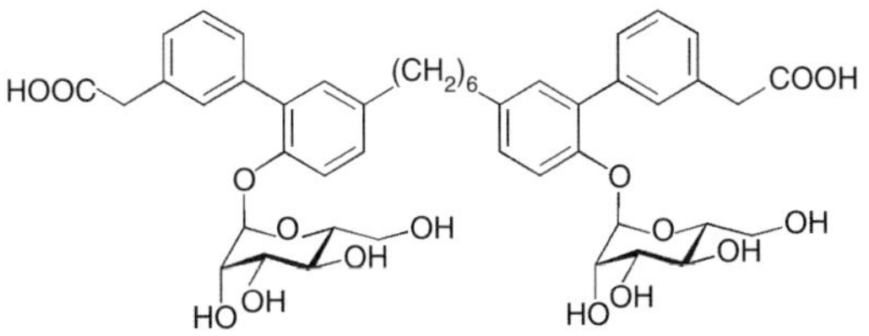

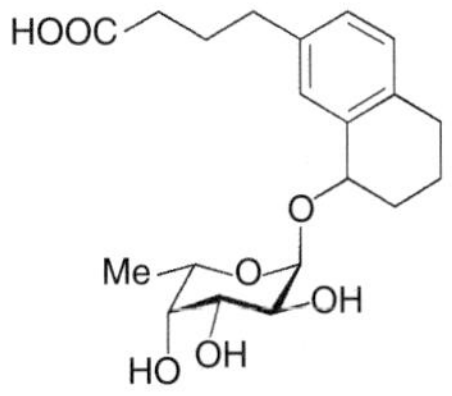

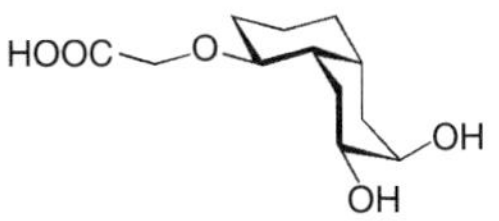

FIGURE 16.23 Replacement of Gal-β(1-4)Glc*N*Ac disaccharide mimics.

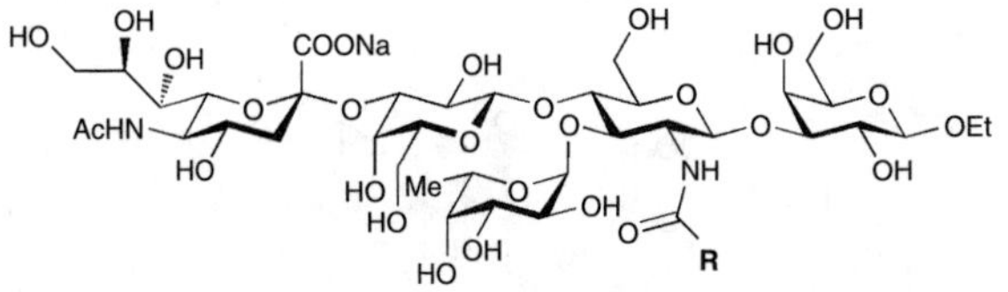

60 (R= benzoyl)
61 (R=naphthoyl)

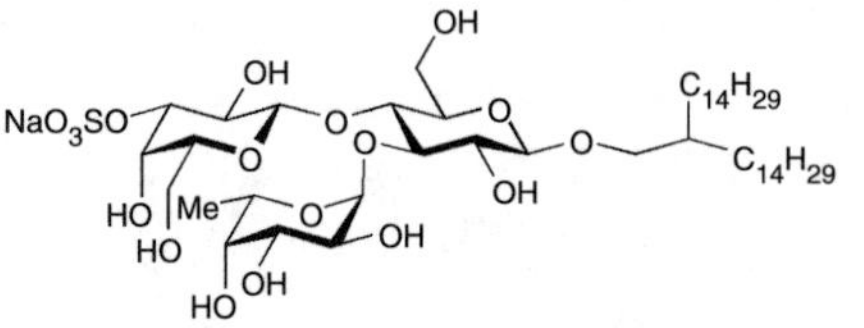

62

FIGURE 16.24 Selectin antagonists containing additional hydrophobic substituents.

63b, the thiourea derivative **63c**, the glycolate **63d** or the sulfonic acid derivative **63e** gave similarly or less active antagonists, activity was substantially increased by aromatic (**63f,g**) or heteroaromatic (**63h–j**) acyl groups.

16.4.3.2 Modifications of sLe^x

Modifications of the different carbohydrate components of sLe^x have been extensively investigated and a comprehensive summary has recently been published [106a,142]. Therefore, we will focus only on selected examples that illustrate the efforts of numerous research groups in academia and the pharmaceutical industry to develop *drug-like* selectin antagonists with acceptable pharmacodynamic and pharmacokinetic properties.

16.4.3.2.1 GlcNAc-Moiety

According to the X-ray analysis by Camphausen [117], the GlcNAc-moiety of sLe^x is a structural element that keeps the adjacent L-fucose and D-galactose in the correct spatial orientation.

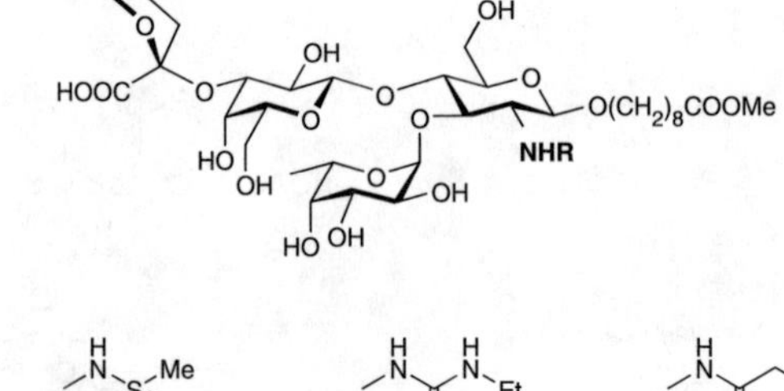

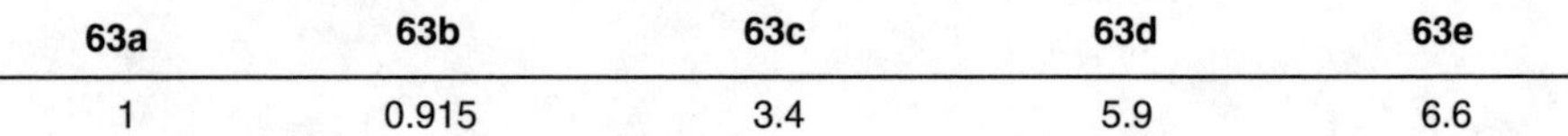

NHR =	63a	63b	63c	63d	63e
rIC$_{50}$	1	0.915	3.4	5.9	6.6

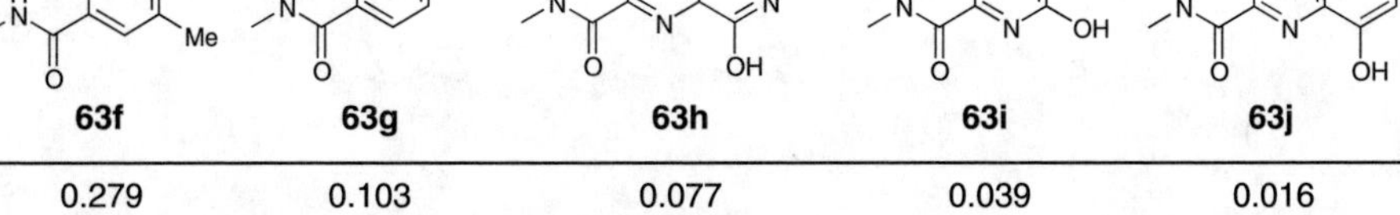

NHR =	63f	63g	63h	63i	63j
rIC$_{50}$	0.279	0.103	0.077	0.039	0.016

FIGURE 16.25 Modification of the acyl group of the glucosamine moiety of sLe^x.

Therefore, it is not surprising that its replacement by geometrically correct building blocks leads to active mimetics. Some examples are summarized in Figure 16.26. Conversely, highly flexible or geometrically inaccurate linkers result in compounds with considerably lower affinity for E-selectin (Figure 16.26; compounds **67–70**). Interestingly, P-selectin affinity does not follow the same pattern, and **67**, which is inactive in the cell-free E-selectin binding assay, proved to be more active against P-selectin than sLe^x [143].

16.4.3.2.2 Gal-β(1-4)-GlcNAc Moiety

Various noncarbohydrate replacements for this disaccharide unit have been investigated. These include *ortho*- and *para*-benzenedimethanol (→**71** and **72**) [148], a lactam diol (→**73**) [146], several flexible linkers (→**74–76**) [149–151] and aryl-cyclohexyl ethers (→**77** and **78**) [152,153]. In all cases, the application of noncarbohydrate building blocks led to inactive derivatives (Figure 16.27), presumably because of entropy costs associated with the conformational flexibility of these spacers and the absence of pharmacophoric groups that are equivalent to those found on the galactose moiety.

16.4.3.2.3 Glycopeptides

Wong et al. designed the *O*-fucopeptides **79** and **80** (Figure 16.28), which contain α-*O*-fucosyl-aminocyclohexanol and α-*O*-fucosyl-L-threonine as core structures. In these inhibitors,

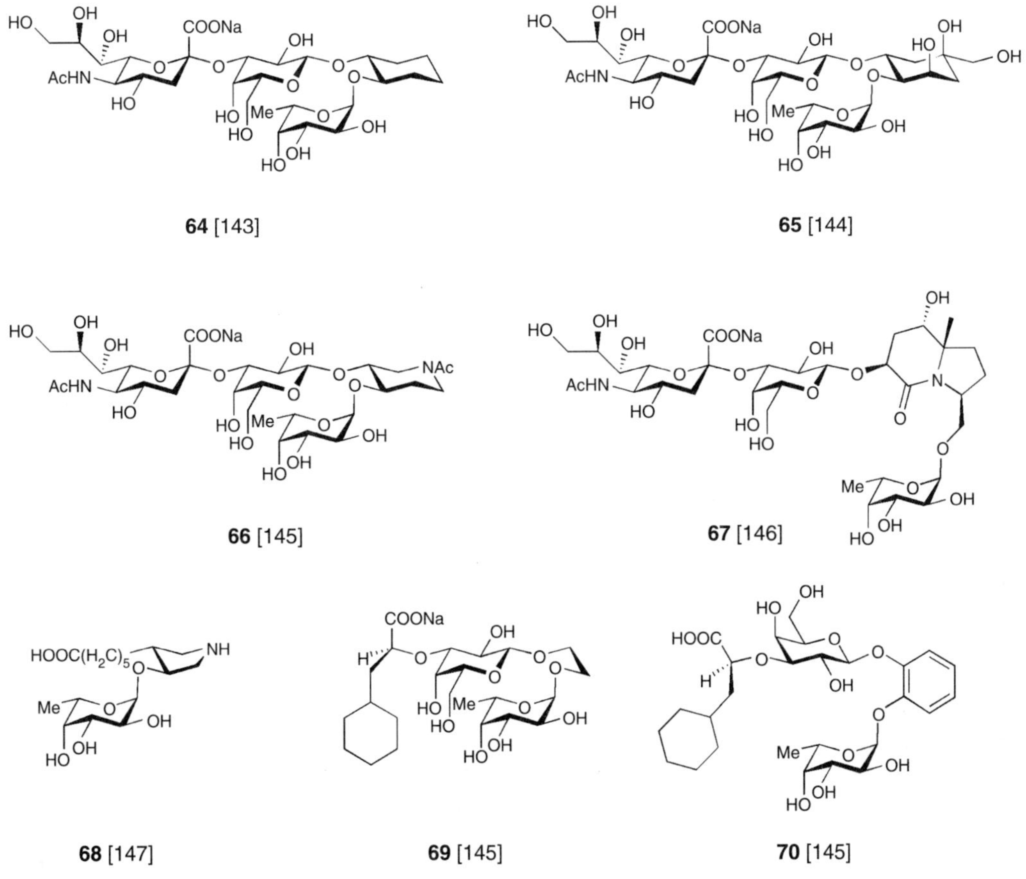

64 [143]

65 [144]

66 [145]

67 [146]

68 [147]

69 [145]

70 [145]

FIGURE 16.26 Replacement of the Glc*N*Ac moiety in sLe^x.

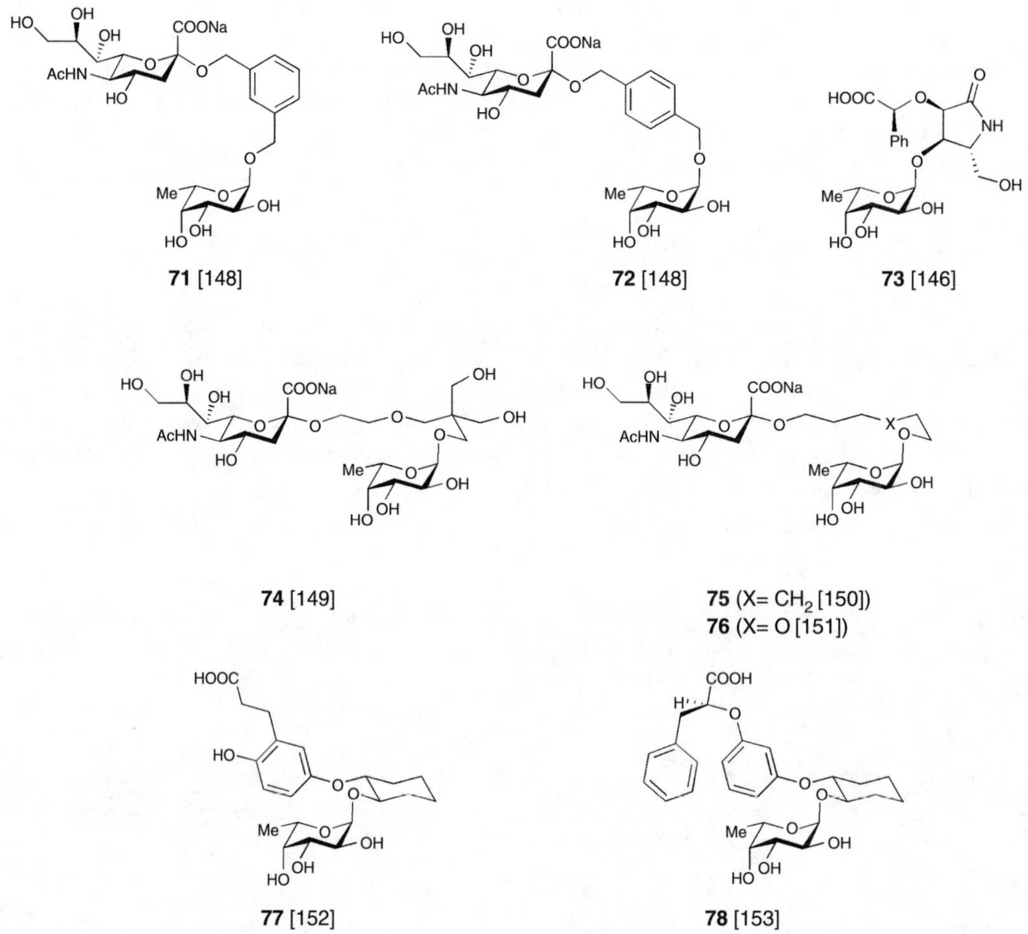

71 [148] **72** [148] **73** [146]

74 [149] **75** (X= CH_2 [150]) **76** (X= O [151])

77 [152] **78** [153]

FIGURE 16.27 Replacement of the Gal-β(1-4)-Glc*N*Ac moiety.

(*R*,*R*)-aminocyclohexanol and L-threonine replace Glc*N*Ac, and the amino acids contain groups that mimic the 4- and 6-hydroxyls of Gal [154]. Wong et al. also took advantage of the increased stability of *C*-glycosides (compared with *O*-glycosides) against endogenous glycosidases, leading to fucopeptides, in which the galactose residue was replaced by 4-hydroxy L-threonine (**81**; inhibitory potency comparable with sLe^x) or L-threonine [155]. In a next step, the α-L-fucoside moiety was substituted by the much cheaper α-D-mannoside, giving rise to compounds with strong affinity for P-selectin. One candidate, **82**, is 15 times more potent than sLe^x [156]. Recently, the cyclic analog **83** was found to be approximately 2000-fold more active than sLe^x [157].

16.4.3.3 Peptidic Antagonists

Two strategies for the development of peptidic selectin antagonists have been pursued based on the following concepts: (1) peptides that are functional mimics of the carbohydrate ligands may be competitive selectin inhibitors, and (2) peptides that are homologous to sequences of the carbohydrate recognition domain should block selectin binding by *masking* the carbohydrate ligands.

An interesting contribution to the former strategy was made by Martens et al. [158], who screened recombinant peptide libraries to identify and optimize novel E-selectin ligands.

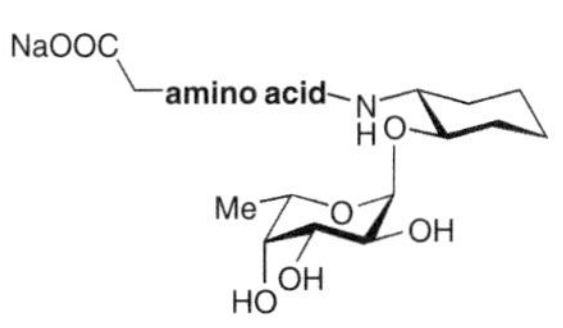

79 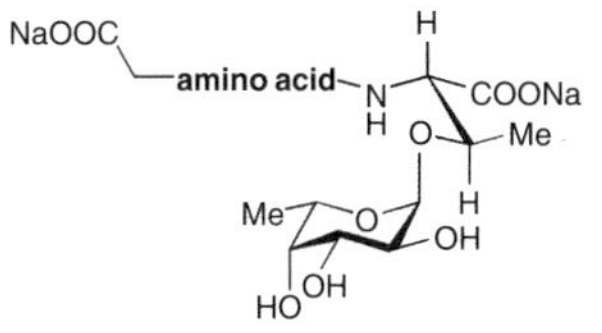 80 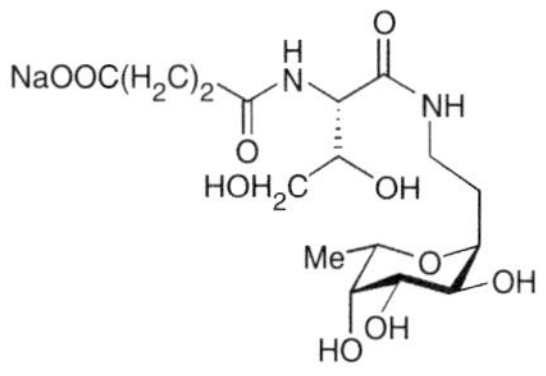 81

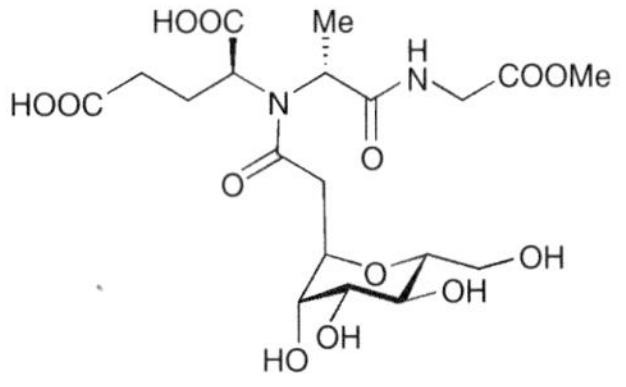

82 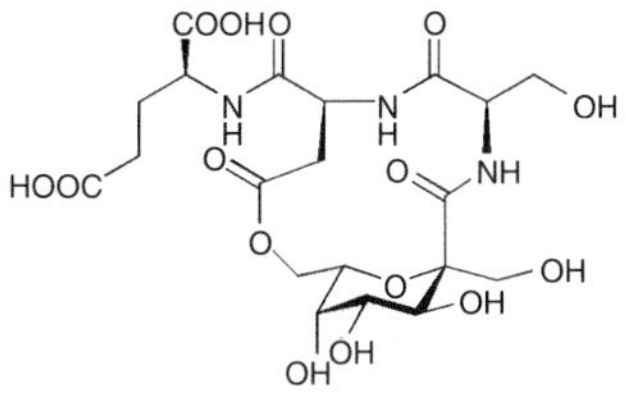 83

FIGURE 16.28 *O*-Fucosides, *C*-fucosides and *C*-mannosides.

Several peptides (e.g., DITWDQLWDLMK) were found to inhibit E-selectin with K_D values in the low nanomolar range and to block E-selectin-mediated adhesion to neutrophils in static and flow-cell assays. *In vivo* experiments with an acute inflammatory response model in mice demonstrated that the best candidate was able to noticeably lower the extent of neutrophil transmigration to the site of inflammation. However, it was shown that the mechanism of action does not involve inhibition of E-selectin binding to sLe^x.

Blaszczyk-Thurin et al. [159] generated an array of 163 unique peptides by combinatorial library panning, starting from the weak E-selectin antagonist DLWDWVVGKPAG ($rIC_{50} \approx 7$ mM; competitive binding assay) [160]. The replacement of a single amino acid residue at position 5 (Phe5Tyr) led to the peptide DLWD**F**VVGKPAG with much improved binding affinity (rIC_{50}: 200 μM).

Fukuda et al. [161] employed phage peptide libraries in a similar approach to the isolation of anti-idiotype antibodies to identify peptides that bind to anticarbohydrate antibodies recognizing E-selectin ligands. The peptidic leads were found to adhere to E-, P- and L-selectin in a calcium-dependent manner, and a phage harboring the heptapeptide IELLQAR showed the best activity. For lead confirmation, both the heptapeptide and the circular analog CIELLQARC were synthesized, and their affinity (in the range of monomeric sLe^x) was measured in a cell-free assay. However, although IELLQAR reduces inflammation in a dose-dependent manner, it has no *in vivo* effect on rat heart allograft survival [162].

Finally, Molenaar et al. [163] used phage display libraries to identify P-selectin-specific peptides. Peptide truncation and alanine-scanning studies revealed the EWVDV consensus motif to be responsible for effective inhibition. The best activity (IC_{50}: 2 μM) was obtained with the 17-mer CDV**EWVDV**SSLEWDLPC. Because EWVDV-containing peptides inhibit the binding of glycoside ligands to P-selectin, it is conceivable that EWVDV interacts with, or in close proximity to, the actual carbohydrate recognition domain of P-selectin. Intriguingly, the apparent avidity of the peptides was increased 200-fold when they were presented in a tetrameric form.

A second strategy for finding inhibitors of selectin-dependent cell adhesion (*cf.* introduction to Section 16.4.3.3) is based on the hypothesis that peptides derived from the lectin domains may *mask*

	1	11	21	31	41	51
E-selectin	WSYNTSTEAM	TYDEASAYCQ	QRYTHLVAIQ	NKEEIEYLNS	ILSYSPSYYW	IGIRKVNNVW
P-selectin	WTYHYSEKPM	NWQRARRFCR	DNYTDLVAIQ	NKAEIEYLEK	TLPFSRSYYW	IGIRKIGGGI
L-selectin	WTYHYSTKAY	SWNISRKYCQ	NRYTDLVAIQ	NKNEIDYLNK	VLPYYSSYYW	IGIRKNNKTW

23-30 [74] **54-63 [74]** **48-55 [73]** **37-50 [75]**

	61	71	81	91	101	111
E-selectin	VWVGTQKPLT	EEAKNWAPGE	PNNRQKDEDC	VEIYIKREKD	VGMWNDERCS	KKKLALCYTA
P-selectin	TWVGTNKSLT	EEAENWGDGE	PNNKKSKEDC	VEIYIKRERD	SGKWNDDACH	KRKAALCYTA
L-selectin	TWVGTKKALT	NEAENWADNE	PNNKRNNEDC	VEIYIKSPSA	PGKWNDEHCL	KKKHALCYTA

70-79 [74] **109-118 [76]**

FIGURE 16.29 Amino acid sequence of the lectin domain of E-, L- and P-selectin.

selectin ligands by binding to them. To investigate the role of various regions of the three lectin domains (Figure 16.29), a series of peptides were synthesized and tested for their ability to inhibit the adhesion of neutrophiles or myeloid HL-60 cells to immobilized selectins [164,165], selectin-transfected COS-7 cells [164] or IL-1β-treated HUVECs [165].

McEver et al. [164] found that peptides corresponding to residues 23 to 30, 54 to 63 and 70 to 79 of the *N*-terminal lectin domain of P-selectin inhibit leukocyte adhesion to P-selectin. Peptides corresponding to the homologous 23 to 30 and 54 to 63 regions of E- and L-selectin also prevent cell binding to P-selectin in a Ca^{2+}-dependent manner. The IC_{50} values ranged from 50 to 700 μM. Control experiments showed that peptides with scrambled sequences failed to block cell adhesion, indicating that the inhibitory effects of the native sequences were not caused by nonspecific effects.

Macher et al. [165] have synthesized a series of peptides based on the conserved regions 48 to 55 of E-, L- and P-selectin and successfully tested their ability to block selectin-mediated cell adhesion *in vitro* (YYWIGIR: $IC_{50} = 10$ μM) and neutrophil recruitment to sites of inflammation *in vivo* (thioglycolate- and IL-1α-induced neutrophil infiltration).

Finally, Heavner et al. reported that the P-selectin regions 37 to 50 [167] and the regions E-, L- and P-selectin 109 to 118 [168] inhibit the leukocyte adhesion to P- and E-selectin chimeras in the low μM range.

Based on the functional similarities between the pertussis toxin and the selectins (including their ability to bind to similar sialic acid-containing oligosaccharides structures), Rozdzinski et al. [166] and Heerze et al. [169] synthesized peptides from the regions of sequence similarity in an effort to find molecules that adhere to sialic acid-containing glycoconjugates. The recombinant subunits S2 and S3 of pertussis toxin and the synthetic peptides competitively inhibited adherence of neutrophils to selectin-coated surfaces and to endothelial cells *in vitro*.

In summary, peptide fragments of the three selectins display very interesting biological activities *in vitro* as well as *in vivo*. However, their translation into peptidomimetics, a necessity due to the inadequate plasma stability of peptidic antagonists, has not yet been reported.

16.4.4 Biological Evaluation

Selectins regulate the initial attachment and rolling phase of leukocyte recruitment. Because this process takes place under flow conditions, fast-binding kinetics is required to overcome the shear forces. Interestingly, selectins show only weak overall affinity to their physiological ligands, resulting from only a few, mostly electrostatic, interactions [101]. Consequently, cell-free assays must rely on a multimeric presentation of the receptor or the ligand to ensure that sufficient interactions survive the washing steps in order to produce adequate signals for measuring inhibition of binding.

A number of different competitive cell-free and cell-based binding assays under static and hydrodynamic flow conditions have been used to obtain affinity data for selectin ligands. In addition to the fact that different positive controls have been used, this makes a direct comparison of reported binding affinities difficult. Therefore, in this chapter, we quote *relative* affinities wherever possible (Section 16.4.3). Another problem that has a negative effect on assay reliability has been encountered in cases where acidic ion exchange resins are used in the final step of the antagonist synthesis [170]. Small amounts of polyanions released from the resin were found to be potent selectin inhibitors, especially for P-selectin. These polyanions are difficult to remove and are not detectable by routine analysis. As a result, published assay data for P-selectin antagonists should be considered with caution.

Cell-free competitive binding assays under static conditions are available in various different formats. In one approach, a solution of the inhibitor and a preformed complex of selectin-Ab chimera, a biotinylated anti-FC Ab and a streptavidin alkaline phosphatase conjugate are added to a glycolipid (e.g., sLe^x-ceramide, sulfatide) immobilized on microtiter plates. After washing, a colorimetric assay using *p*-nitrophenyl phosphate is performed to quantify the extent of binding [131,171]. In another cell-free assay, recombinant soluble E-, P- or L-selectin/IgG is immobilized on microtiter plates and incubated with the test compound and a polymer (polyacrylate [172], polylysin [173]) bearing sLe^a and biotin. After unbound ligand and polymer are washed from the plate, streptavidin/horseradish peroxidase conjugate is added to enable colorimetric determination of binding.

A frequently used competitive cell/molecule-based assay relies on the interaction of sLe^x-bearing HL-60 cells containing immobilized recombinant selectin/IgG. Solutions containing HL-60 cells and test compound are added to wells coated with E-, P- or L-selectin. After washing the plates to remove unbound cells, the HL-60 cells are lysed and assayed for myeloperoxidase activity using *o*-phenylenediamine and H_2O_2. The percentage of inhibition is determined by comparing the absorbance at 492 nm with wells containing no test molecule [174].

Unfortunately, results from different cell-based assays are difficult to compare, as illustrated in the following paragraphs for P-selectin. Table 16.7 provides an overview of the many assay conditions that have been developed. They vary in the nature of the P-selectin ligand PSGL-1 and P-selectin, as well as the blocking agent, and the duration and temperature of incubation.

Given that the selectin-mediated rolling of leukocytes on the activated endothelium is a nonequilibrium process, cell-based flow assays are required to properly characterize selectin antagonists [195]. Dynamic *in vitro* assays monitor the rolling of polymorphonuclear neutrophils (PMN) on stimulated human umbilical vein endothelial cells (HUVEC) under hydrodynamic shear-stress conditions, analogous to those observed *in vivo*. Although this assay is very useful for studying the biophysics of rolling [196], its throughput is too low for it to be useful as a routine assay for evaluating antagonists. Recently, a rolling assay that utilizes polystyrene beads coated with sLe^x (instead of whole cells) has been reported [197].

Leukocyte rolling is a prerequisite for subsequent leukocyte extravasation and tissue injury. Intravital microscopy (IVM) has been successfully employed to observe how leukocytes travel from the microcirculation to inflamed tissue, according to the regulated chain of events that includes rolling, firm adhesion, diapedesis, and migration through the interstitium [198]. An advantage of this method is that IVM allows the *in vivo* characterization of selectin antagonists for

TABLE 16.7
P-selectin Static Cell Adhesion Assay Formats

Entry	P-Selectin Ligand	P-Selectin	Blocking Agent	Incubation Time	Incubation Temperature	References
1	HL-60	P-selectin[a]	1% BSA	1 h	RT	[175]
2	HL-60	P-selectin[a]	1% BSA	15 min	RT	[176]
3	HL-60	P-selectin[a]	0.1% Casein	20 min	22°C	[177]
4	HL-60	P-selectin[a]	–	10 min	37°C	[178]
5	HL-60	P-selectin beads	3% BSA	10 min	RT	[179]
6	HL-60	P-selectin COS[b]	–	20 min	4°C	[180]
7	HL-60	P-selectin CHO[c]	1% FBS	20 min	25°C	[181]
8	HL-60	P-selectin/hIgG	0.1% Casein	20 min	RT	[182]
9	HL-60	HUVEC	2% BSA	45 min	4°C	[183]
10	Neutrophils	P-selectin CHO[c]	0.5% HAS	15 min	37°C	[184]
11	PMN	HUVEC	1% BSA	30 min	4°C	[185]
12	PMN	P-selectin/hIgG	5% BLOTTO	30 min	RT	[186]
13	PBL[d]	P-selectin CHO[c]	–	10 min	4 °C	[187]
14	Monocytes	Member P-selectin[e]	0.5% HAS	15 min	25°C	[188]
15	Leukocyte	P-selectin[a]	1% BSA	30 min	4°C	[189]
16	Lymphocytes	P-selectin/hIgG	2% BSA	30 min	RT	[190]
17	Myeloid cells	P-selectin/hIgG	0.1% BSA	20 min	RT	[191]
18	PSGL-1 CHO[f]	P-selectin[a]	1% HSA	20 min	RT	[192]
19	PSGL-1 CHO[f]	P-selectin/hIgG	0.3% BSA	12 min	RT	[193]
20	PSGL-1/hIgG	P-selectin CHO[c]	1% BSA	8 min	RT	[194]

[a] Purified by affinity chromatography from human platelets.
[b] P-selectin transfected COS-cells.
[c] P-selectin transfected CHO-cells.
[d] Periferal blood T-cells.
[e] Membranous P-selectin from P-selectin transfected CHO-cells.
[f] PSGL-1, FucTIII and core 2 transfected CHO-cells.

each selectin separately by using the corresponding knockout mice or the appropriate antiselectin antibodies [199].

Finally, routine animal models for inflammatory diseases can be used for testing the *in vivo* efficacy of selectin antagonists; the murine peritonitis model and the ear edema model are the most common. In the murine peritonitis model [200], the migration of leukocytes in response to an acute inflammatory stimulus is assessed by intraperitoneal injection of thioglycolate. In the arachidonic acid- or croton oil-induced ear edema model [131,201], inflammation is measured as neutrophil infiltration, represented by myeloperoxidase activity in ear biopsy samples.

16.4.5 Summary and Outlook

A promising strategy for the development of novel anti-inflammatory therapeutics is based on blocking the interaction between leukocytes and the endothelium. Because the selectin-mediated initial rolling step is a prerequisite for the inflammatory cascade to take place, the selectins are a viable target. sLe^x serves as a good lead structure for the design of selectin antagonists because it is the common carbohydrate epitope present in all physiological selectin ligands as PSGL-1, ESL-1, GlyCAM-1, and so on.

Based on this lead, several academic and industrial groups have designed selectin inhibitors with improved pharmacodynamic and pharmacokinetic properties. The major goals (besides improvement of affinity) were to reduce molecular weight, decrease the synthetic complexity and reduce the carbohydrate nature. However, to date, none of these efforts have led to an approved drug. The main reason for this is the low binding affinity (because of the flat nature of the selectin binding sites), which offers interactions that are predominantly electrostatic in nature so that the contribution of hydrophilic interactions to binding are compensated by the penalty for desolvation.

Recent X-ray crystal structures of human E-, P- and L-selectin co-crystallized with sLe^x have provided valuable insights into the nature of the binding site and the areas surrounding it. This new information, along with the knowledge that a second, potentially drugable, binding site exists in close proximity to the sLe^x binding site opens new possibilities for the design and development of a new generation of selectin antagonists.

ACKNOWLEDGMENTS

We acknowledge the fruitful discussions with Dr. John L. Magnani/GlycoMimetics, Rockville, on cell-based P-selectin assays.

REFERENCES

1. (a) Alper, J, Searching for medicine's sweet spot, *Science*, 291, 2338–2343, 2001; (b) McAuliffe, J C, Hindsgaul, O, Carbohydrate drugs — an ongoing challenge, *Chem. Ind.*, 170–174, 1997; (c) Hounsell, E F, Glycosylation targets for drug design, *Carbohydr. Res.*, 300, 47–48, 1997; (d) Nagai, Y, A challenge to the riddle of the carbohydrate chain, *Pure Appl. Chem.*, 69, 1893–1896, 1997; (e) Dwek, R A, Glycobiology: toward understanding the function of sugars, *Chem. Rev.*, 96, 683–720, 1996; (f) Simon, P M, Pharmaceutical oligosaccharides, *Drug Discov. Today*, 1, 522–528, 1996.
2. (a) Petitou, M, Herbert, J-M, A new generation of antithrombotics based on synthetic oligosaccharides, In *Carbohydrate-Based Drug Discovery*, Wong, C-H, Ed., Wiley-VCH, Weinheim, pp. 441–460, 2003. (b) Petitou, M, Casu, B, Lindahl, U, 1976–1983, a critical period in the history of heparin: the discovery of the antithrombin binding site, *Biochemie.*, 85, 83–89, 2003; (c) Petitou, M, van Boeckel, C A A, A synthetic antithrombin III binding pentasaccharide is now a drug! What comes next? *Angew. Chem. Int. Ed.*, 43, 3118–3133, 2004; for clinical aspects on Arixtra®; (d) Petitou, M, Duchaussoy, P, Herbert, J-M, Duc, G, El Hajji, M, Branellec, J-F, Donat, F, Necciari, J, Cariou, R, Bouthier, J, Garrigou, E, The synthetic pentasaccharide fondaparinux: First in the class of antithrombotic agents that selectively inhibit coagulation factor Xa, *Semin. Thromb. Hemost.*, 28, 393–402, 2002.
3. (a) Reviews on Glycopeptide Antibiotics: Williams, D H, Bardsley, B, The vancomycin group of antibiotics and the fight against resistant bacteria, *Angew. Chem. Int. Ed.*, 38, 1172–1193, 1999; (b) Kaplan, J, Korty, B D, Axelsen, P H, Loll, P J, The role of sugar residues in molecular recognition by vancomycin, *J. Med. Chem.*, 44, 1837–1840, 2001; (c) Nicolaou, K C, Boddy, C N C, Bräse, S, Winssinger, N, Chemistry, biology, and medicine of the glycopeptide antibiotics, *Angew. Chem. Int. Ed.*, 38, 2096–2152, 1999; (d) Reviews on Aminoglycosides: Ferrier, R J, Blattner, R, Field, R A, Furneaux, R H, Gardiner, J M, Hoberg, J, Kartha, K P R, Tilbrook, D M G, Tyler, P C, Wightmen, R H, Antibiotics, *Carbohydr. Chem.*, 33, 257–274, 2002; (e) Dworkin, R J, Aminoglycosides for the treatment of Gram-negative infections: therapeutic use, resistance and future outlook, *Drug Resist Update*, 2, 173–179, 1999; (f) Zembower, T R, Noskin, G A, Postelnick, M J, Nguyen, C, Peterson, L R, The utility of aminoglycosides in an era of emerging drug resistance, *Int. J. Antimicrob. Agents*, 10, 95–105, 1998.
4. (a) Danishefsky, S J, Allen, J R, From the laboratory to the clinic: a retrospective on fully synthetic carbohydrate-based anticancer vaccines, *Angew. Chem. Int. Ed.*, 39, 836–863, 2000; (b) Koganty, R R, Reddish, M A, Longenecker, B M, Glycopeptide- and carbohydrate-based synthetic vaccines for the immunotherapy of cancer, *Drug Discov. Today*, 1, 190–198, 1996.

5. (a) Gruner, S A W, Locardi, E, Lohof, E, Kessler, H, Carbohydrate-based mimetics in drug design: sugar amino acids and carbohydrate scaffolds, *Chem. Rev.*, 102, 491–514, 2002; (b) Schweizer, F, Glycosamino acids: building blocks for combinatorial synthesis — implications for drug discovery, *Angew. Chem. Int. Ed.*, 41, 230–253, 2002.
6. (a) Schwab, M E, Bartholdi, D, Degeneration and regeneration of axons in the lesioned spinal cord, *Physiol. Rev.*, 76, 319–370, 1996; (b) Qiu, J, Cai, D, Filbin, M T, Glial inhibition of nerve regeneration in the mature mammalian CNS, *Glia.*, 29, 166–174, 2000; (c) Fournier, A E, Strittmatter, S M, Repulsive factors and axon regeneration in the CNS, *Curr. Opin. Neurobiol.*, 11, 89–94, 2001.
7. (a) Huber, A B, Schwab, M E, Review: Nogo-A, a potent inhibitor of neurite outgrowth and regeneration, *Biol. Chem.*, 381, 407–419, 2000; (b) Brittis, P A, Flanagan, J G, Nogo domains and a Nogo receptor: implications for axon regeneration, *Neuron.*, 30, 11–14, 2001.
8. Huang, D W, McKerracher, L, Braun, P E, David, S, A therapeutic vaccine approach to stimulate axon regeneration in the adult mammalian spinal cord, *Neuron.*, 24, 639–647, 1999.
9. (a) Trapp, B D, Myelin-associated glycoprotein: location and potential functions, *Ann. NY Acad. Sci.*, 605, 29–43, 1990; (b) Mukhopadhyay, G, Doherty, P, Walsh, F S, Crocker, P R, Filbin, M T, A novel role for myelin-associated glycoprotein as an inhibitor of axonal regeneration, *Neuron.*, 13, 757–767, 1994; (c) McKerracher, L, David, S, Jackson, D L, Kottis, V, Dunn, R J, Braun, P E, Identification of myelin-associated glycoprotein as a major myelin-derived inhibitor of neurite growth, *Neuron.*, 13, 805–811, 1994.
10. (a) Prinjha, R, Moore, S E, Vinson, M, Blake, S, Morrow, R, Christie, G, Michalovich, D, Simmons, D L, Walsh, F S, Neurobiology: inhibitor of neurite outgrowth in humans, *Nature*, 403, 383–384, 2000; (b) Chen, M S, Huber, A B, van der Haar, M E, Frank, M, Schnell, L, Spillmann, A A, Christ, F, Schwab, M E, Nogo-A is a myelin-associated neurite outgrowth inhibitor and an antigen for monoclonal antibody IN-1, *Nature*, 403, 434–439, 2000; (c) GrandPré, T, Nakamura, F, Vartanian, T, Strittmatter, S M, Identification of the Nogo inhibitor of axon regeneration as a Reticulon protein, *Nature*, 403, 439–444, 2000.
11. Wang, K C, Koprivica, V, Kim, J A, Sivasankaran, R, Guo, Y, Neve, R L, He, Z, Oligodendrocyte-myelin glycoprotein is a Nogo receptor ligand that inhibits neurite outgrowth, *Nature*, 417, 941–944, 2002.
12. (a) Baumann, N, Pham-Dinh, D, Biology of oligodendrocyte and myelin in the mammalian central nervous system, *Physiol. Rev.*, 81, 871–927, 2001; (b) Huber, A B, Weinmann, O, Brosamle, C, Oertle, T, Schwab, M E, Patterns of Nogo mRNA and protein expression in the developing and adult rat and after CNS lesions, *J. Neurosci.*, 22, 3553–3567, 2002.
13. (a) Liu, B P, Fournier, A, GrandPré, T, Strittmatter, S M, Myelin-associated glycoprotein as a functional ligand for the Nogo-66 receptor, *Science*, 297, 1190–1193, 2002; (b) Fournier, A E, GrandPré, T, Strittmatter, S M, Identification of a receptor mediating Nogo-66 inhibition of axonal regeneration, *Nature*, 409, 341–346, 2001.
14. Domeniconi, M, Cao, Z, Spencer, T, Sivasankaran, R, Wang, K C, Nikulina, E, Kimura, N, Cai, H, Deng, K, Gao, Y, He, Z, Filbin, M T, Myelin-associated glycoprotein interacts with the Nogo66 receptor to inhibit neurite outgrowth, *Neuron.*, 35, 283–290, 2002.
15. Woolf, C J, Bloechlinger, S, It takes more than two to Nogo, *Science*, 297, 1132–1134, 2002.
16. (a) Kelm, S, Pelz, A, Schauer, R, Filbin, M T, Song, T, de Bellard, M E, Schnaar, R L, Mahoney, J A, Hartnell, A, Bradfield, P, Sialoadhesin, myelin-associated glycoprotein and CD22 define a new family of sialic acid-dependent adhesion molecules of the immunoglobulin superfamily, *Curr. Biol.*, 4, 965–972, 1994; (b) Powell, L D, Varki, A, I-type lectins, *J. Biol. Chem.*, 270, 14243–14246, 1995; (c) Crocker, P R, Varki, A, Siglecs in the immune system, *Immunology*, 103, 137–145, 2001; (d) Schachner, M, Bartsch, U, Multiple functions of the myelin-associated glycoprotein MAG (siglec-4a) in formation and maintenance of myelin, *Glia.*, 29, 154–165, 2000; (e) Powell, L, I-type lectins, In *Essentials of Glycobiology*, Varki, A, Cummings, R, Esko, J, Freeze, H, Hart, G, Marth, J, Eds., Cold Spring Harbor Laboratory Press, Cold Spring Harbor, pp. 363–378, 1999.
17. Bartsch, U, Bandtlow, C E, Schnell, L, Bartsch, S, Spillmann, A A, Rubin, B P, Hillenbrand, R, Montag, D, Schwab, M E, Schachner, M, Lack of evidence that myelin-associated glycoprotein is a major inhibitor of axonal regeneration in the CNS, *Neuron.*, 15, 1375–1381, 1995.

18. Crocker, P R, Kelm, S, Hartnell, A, Freeman, S, Nath, D, Vinson, M, Mucklow, S, Sialoadhesin and related cellular recognition molecules of the immunoglobulin superfamily, *Biochem. Soc. Trans.*, 24, 150–156, 1996.
19. Schnaar, R L, Glycobiology of the nervous system. In *Carbohydrates in Chemistry and Biology, Part II: Biology of Saccharides*, Wiley-VCH, Weinheim, pp. 1013–1027, 2000.
20. Yamashita, T, Higuchi, H, Tohyama, M, The p75 receptor transduces the signal from myelin-associated glycoprotein to Rho, *J. Cell. Biol.*, 157, 565–570, 2002.
21. Lehmann, M, Fournier, A, Selles-Navarro, I, Dergham, P, Sebok, A, Leclerc, N, Tigyi, G, McKerracher, L, Inactivation of Rho signaling pathway promotes CNS axon regeneration, *J. Neurosci.*, 19, 7537–7547, 1999.
22. Lang, P, Gesbert, F, Delespine-Carmagnat, M, Stancou, R, Pouchelet, M, Bertoglio, J, Protein kinase A phosphorylation of RhoA mediates the morphological and functional effects of cyclic AMP in cytotoxic lymphocytes, *EMBO J.*, 15, 510–519, 1996.
23. (a) Qiu, J, Cai, D, Dai, H, McAtee, M, Hoffman, P N, Bregman, B S, Filbin, M T, Spinal axon regeneration induced by elevation of cyclic AMP, *Neuron.*, 34, 895–903, 2002; (b) Neumann, S, Bradke, F, Tessier-Lavigne, M, Basbaum, A I, Regeneration of sensory axons within the injured spinal cord induced by intraganglionic cAMP elevation, *Neuron.*, 34, 885–893, 2002.
24. (a) Vinson, M, van der Merwe, P A, Kelm, S, May, A, Jones, E Y, Crocker, P R, Characterization of the sialic acid-binding site in sialoadhesin by site-directed mutagenesis, *J. Biol. Chem.*, 271, 9267–9272, 1996; (b) van der Merwe, P A, Crocker, P R, Vinson, M, Barclay, A N, Schauer, R, Kelm, S, Localization of the putative sialic acid-binding site on the immunoglobulin superfamily cell-surface molecule CD22, *J. Biol. Chem.*, 271, 9273–9280, 1996.
25. May, A P, Robinson, R C, Vinson, M, Crocker, P R, Jones, E Y, Crystal structure of the *N*-terminal domain of sialoadhesin in complex with 3′-sialyllactose at 1.85 Å resolution, *Mol. Cell.*, 1, 719–728, 1998.
26. (a) Collins, B E, Ito, H, Sawada, N, Ishida, H, Kiso, M, Schnaar, R L, Enhanced binding of the neural siglecs, myelin-associated glycoprotein and Schwann cell myelin protein, to Chol-1 (α-series) gangliosides and novel sulfated Chol-1 analogs, *J. Biol. Chem.*, 274, 37637–37643, 1999; (b) Vyas, A A, Schnaar, R L, Brain gangliosides: functional ligands for myelin stability and the control of nerve regeneration, *Biochimie*, 83, 677–682, 2001; (c) Vyas, A A, Patel, H V, Fromholt, S E, Heffer-Lauc, M, Vyas, K A, Dang, J, Schachner, M, Schnaar, R L, From the cover: gangliosides are functional nerve cell ligands for myelin-associated glycoprotein (MAG), an inhibitor of nerve regeneration, *Proc. Natl. Acad. Sci. USA*, 99, 8412–8417, 2002.
27. (a) Yang, L J S, Zeller, C B, Shaper, N L, Kiso, M, Hasegawa, A, Shapiro, R E, Schnaar, R L, Gangliosides are neuronal ligands for myelin-associated glycoprotein, *Proc. Natl. Acad. Sci. USA*, 93, 814–818, 1996; (b) Collins, B E, Kiso, M, Hasegawa, A, Tropak, M B, Roder, J C, Crocker, P R, Schnaar, R L, Binding specificities of the sialoadhesin family of I-type lectins. Sialic acid linkage and substructure requirements for binding of myelin-associated glycoprotein, Schwann cell myelin protein, and sialoadhesin, *J. Biol. Chem.*, 272, 16889–16895, 1997.
28. Sheikh, K A, Sun, J, Liu, Y, Kawai, H, Crawford, T O, Proia, R L, Griffin, J W, Schnaar, R L, Mice lacking complex gangliosides develop Wallerian degeneration and myelination defects, *Proc. Natl. Acad. Sci. USA*, 96, 7532–7537, 1999.
29. (a) Prabhanjan, H, Kameyama, A, Ishida, H, Kiso, M, Hasegawa, A, Regio- and stereo-selective synthesis of ganglioside GM1b and some positional analogs, *Carbohydr. Res.*, 220, 127–143, 1991; (b) Prabhanjan, H, Aoyama, K, Kiso, M, Hasegawa, A, Synthesis of disialoganglioside GD1 and its positional isomer, *Carbohydr. Res.*, 233, 87–99, 1992; (c) Hotta, K, Ishida, H, Kiso, M, Hasegawa, A, Synthetic studies on sialoglycoconjugates 66: first total synthesis of a cholinergic neuron-specific ganglioside GQ1bα, *J. Carbohydr. Chem.*, 14, 491–506, 1995; (d) Ito, H, Ishida, H, Kiso, M, Hasegawa, A, Synthetic studies on sialoglycoconjugates, Part 106. Synthesis of sialyl Le^x ganglioside analogs modified at C-6 of the galactose residue to elucidate the mechanism of selection recognition, *Carbohydr. Res.*, 306, 581–585, 1998.
30. Richardson, P J, Walker, J H, Jones, R T, Whittaker, V P, Identification of a cholinergic-specific antigen Chol-1 as a ganglioside, *J. Neurochem.*, 38, 1605–1614, 1982.

31. Hirabayashi, Y, Nakao, T, Irie, F, Whittaker, V P, Kon, K, Ando, S, Structural characterization of a novel cholinergic neuron-specific ganglioside in bovine brain, *J. Biol. Chem.*, 267, 12973–12978, 1992.
32. Strenge, K, Schauer, R, Bovin, N, Hasegawa, A, Ishida, H, Kiso, M, Kelm, S, Glycan specificity of myelin-associated glycoprotein and sialoadhesin deduced from interactions with synthetic oligosaccharides, *Eur. J. Biochem.*, 258, 677–685, 1998.
33. Kelm, S, Brossmer, R, Isecke, R, Gross, H J, Strenge, K, Schauer, R, Functional groups of sialic acids involved in binding to siglecs (sialoadhesins) deduced from interactions with synthetic analogues, *Eur. J. Biochem.*, 255, 663–672, 1998.
34. Sjoberg, E R, Powell, L D, Klein, A, Varki, A, Natural ligands of the B cell adhesion molecule CD22 beta can be masked by 9-*O*-acetylation of sialic acids, *J. Cell. Biol.*, 126, 549–562, 1994.
35. (a) Legler, G, Glycoside hydrolases: mechanistic information from studies with reversible and irreversible inhibitors, *Adv Carbohydr Chem Biochem*, 48, 319–385, 1990; (b) Legler, G, Glycosidase inhibition by basic sugar analogs and the transition state of enzymatic glycoside hydrolysis, In *Iminosugars as Glycosidase Inhibitors*, Stütz, A E, Ed., Wiley-VCH, Weinheim, pp. 31–36, 1999.
36. (a) Sinnott, M L, Catalytic mechanisms of enzymic glycosyl transfer, *Chem. Rev.*, 90, 1171–1202, 1990; (b) Davies, G, Sinnott, M L, Withers, S G, Glycosyl transfer, In *Comprehensive Biological Catalysis*, Sinnott, M L, Ed., Academic Press, New York, pp. 119–208, 1998.
37. (a) Namchuk, M N, Withers, S G, Mechanism of agrobacterium β-glucosidase: kinetic analysis of the role of noncovalent enzyme/substrate interactions, *Biochemistry*, 34, 16194–16202, 1995; (b) Mosi, R, Sham, H, Uitdehaag, J C M, Ruiterkamp, R, Dijkstra, B W, Withers, S G, Reassessment of acarbose as a transition state analogue inhibitor of cyclodextrin glycosyltransferase, *Biochemistry*, 37, 17192–17198, 1998; (c) Withers, S G, Namchuk, M, Mosi, R, Potent glycosidase inhibitors: transition state mimics or simply fortuitous binders?, In *Iminosugars as Glycosidase Inhibitors*, Stütz, A E, Ed., Wiley-VCH, Weinheim, pp. 188–206, 1999; (d) Ly, H, Withers, S G, Mutagenesis of glycosidases, *Annu. Rev. Biochem.*, 68, 487–522, 1999; (e) Zechel, D L, Withers, S G, Glycosidase mechanisms: anatomy of a finely tuned catalyst, *Acc. Chem. Res.*, 33, 11–18, 2000; (f) Rye, C S, Withers, S G, Glycosidase mechanisms, *Curr. Opin. Chem. Biol.*, 4, 573–580, 2000.
38. (a) Davies, G, Henrissat, B, Structures and mechanisms of glycosyl hydrolases, *Structure*, 3, 853–859, 1995; (b) White, A, Rose, D R, Mechanism of catalysis by retaining beta-glycosyl hydrolases, *Curr. Opin. Struct. Biol.*, 7, 645–651, 1997; (c) Wolfenden, R, Lu, X, Young, G, Spontaneous hydrolysis of glycosides, *J. Am. Chem. Soc.*, 120, 6814–6815, 1998; (d) Heightmann, T D, Vasella, A T, Recent insights into inhibition, structure, and mechanism of configuration-retaining glycosidases, *Angew. Chem. Int. Ed.*, 38, 750–770, 1999; (e) Vasella, A, Davies, G J, Böhm, M, Glycosidase mechanisms, *Curr. Opin. Chem. Biol.*, 6, 619–629, 2002.
39. Truscheit, E, Frommer, W, Junge, B, Müller, L, Schmidt, D D, Wingender, W, Chemistry and biochemistry of microbial α-glucosidase inhibitors, *Angew. Chem.*, 93, 738–755, 1981.
40. (a) For recent reviews: see Laver, W G, Bischofberger, N, Webster, R G, Disarming flu viruses, *Sci. Am.*, 280, 78–87, 1999; (b) Varghese, N, Development of neuraminidase inhibitors as anti-influenza virus drugs, *Drug Dev. Res.*, 46, 176–196, 1999; (c) Gonzalez, L S, New neuraminidase inhibitors for influenza: current and potential therapeutic roles, *Formulary*, 35, 812–831, 2000; (d) Dreitlein, W B, Maratos, J, Brocavich, J, Zanamivir and oseltamivir: two new options for the treatment and prevention of influenza, *Clin. Ther.*, 23, 327–355, 2001, and references cited.
41. (a) Groopman, J E, Current advances in the diagnosis and treatment of AIDS: an introduction, *Rev. Infect. Dis.*, 12, 908–911, 1990; (b) de Clercq, E, Antiviral drugs: current state of the art, *J. Clin. Virol.*, 22, 73–89, 2001.
42. Zitzmann, N, Mehta, A S, Carrouée, S, Butters, T D, Platt, F M, McCauley, J, Blumberg, B S, Dwek, R A, Block, T M, Imino sugars inhibit the formation and secretion of bovine viral diarrhea virus, a pestivirus model of hepatitis C virus: implications for the development of broad spectrum anti-hepatitis virus agents, *Proc. Natl. Acad. Sci. USA*, 96, 11878–11882, 1999.
43. For the most recent reviews: see (a) Asano, N, Nash, R J, Molyneux, R J, Fleet, G W J, Sugar-mimic glycosidase inhibitors: natural occurrence, biological activity and prospects for therapeutic application, *Tetrahedron: Asymmetry*, 11, 1645–1680, 2000; (b) Asano, N, Alkaloidal

sugar mimetics: biological activities and therapeutic applications, *J. Enzyme Inhib.*, 15, 215–234, 2000; (c) Wrodnigg, T M, From lianas to glycobiology tools: twenty-five years of 2,5-dideoxy-2,5-imino-D-mannitol, *Monatshefte für Chemie*, 133, 393–426, 2002; (d) Lillelund, V H, Jensen, H H, Liang, X, Bols, M, Recent developments of transition-state analogue inhibitors of non-natural product origin, *Chem. Rev.*, 102, 515–553, 2002, and references cited.

44. Atkinson, M A, Eisenbarth, G S, Type 1 diabetes: new perspectives on disease pathogenesis and treatment, *Lancet*, 358, 221–229, 2001.
45. Charpentier, G, Oral combination therapy for type 2 diabetes, *Diabetes Metab. Res. Rev.*, 18, S70–S76, 2002.
46. International Diabetes Center, http://www.parknicollet.com/diabetes (Accessed May 8, 2003).
47. Edelman, S V, Type II diabetes mellitus, *Adv. Intern. Med.*, 43, 449–500, 1998.
48. Riddle, M, Combining sulfonylureas and other oral agents, *Am. J. Med.*, 108, 15S–22S, 2000.
49. Yki-Järvinen, H, Combination therapy with insulin and oral agents: optimizing glycemic control in patients with type 2 diabetes mellitus, *Diabetes Metab. Res. Rev.*, 18, S77–S81, 2002.
50. DeFronzo, R A, Pharmacologic therapy for type 2 diabetes mellitus, *Ann. Intern. Med.*, 131, 281–303, 1999.
51. Creutzfeld, W, Effects of the α-glucosidase inhibitor acarbose on the development of long-term complications in diabetic animals: patho-physiological and therapeutic implications, *Diabetes Metab. Res. Rev.*, 15, 289–296, 1999.
52. Schmidt, D D, Frommer, W, Müller, L, Junge, B, Wingender, W, Truscheit, E, Schäfer, D, α-Glucosidase inhibitors. New complex oligosaccharides of microbial origin, *Naturwissenschaften*, 64, 535–536, 1977.
53. Puls, W, Keup, U, Krause, H P, Thomas, G, Hoffmeister, F, Glucosidase inhibition. A new approach to the treatment of diabetes, obesity, and hyperlipoproteinemia, *Naturwissenschaften*, 64, 536–537, 1977.
54. (a) Balfour, J A, McTavish, D, Acarbose — an update of its pharmacology and therapeutic use in diabetes mellitus, *Drugs*, 46, 1025–1054, 1993; (b) Hoffmann, J, Spengler, M, Efficacy of 24-week monotherapy with acarbose, glibenclamide or placebo in NIDDM patients, *Diabetes Care*, 6, 561–566, 1994; (c) Chiasson, J L, Josse, R G, Hunt, J A, Palmason, C, Rodger, N W, Ross, S A, Ryan, E A, Tan, M H, Wolever, T M, The efficacy of acarbose in the treatment of patients with non-insulin dependent diabetes mellitus, *Ann. Intern. Med.*, 121, 928–935, 1994.
55. (a) Kameda, Y, Asano, N, Yoshikawa, M, Takeuchi, M, Yamaguchi, T, Matsui, K, Horii, S, Fukase, H, Valiolamine, a new α-glucosidase inhibiting aminocyclitol produced by *Streptomyces hygroscopicus*, *J. Antibiot.*, 37, 1301–1307, 1984; (b) Horii, S, Fukase, H, Matsuo, T, Kameda, Y, Asano, N, Matsui, K, Synthesis and α-D-glucosidase inhibitory activity of *N*-substituted valiolamine derivatives as potential oral antidiabetic agents, *J. Med. Chem.*, 29, 1038–1046, 1986.
56. Yagi, M, Kuono, T, Aoyagi, Y, Murai, H, The structure of moranoline, a piperidine alkaloid from Morus species, *Nippon Nog Kag Kaish*, 50, 571–572, 1976.
57. Paulsen, H, Todt, K, Monosaccharides with a nitrogen-containing ring. XV. Nuclear magnetic resonance spectroscopic studies of hindered rotation of monosaccharides with a nitrogen-containing ring, *Chem. Ber.*, 100, 3397–3404, 1967.
58. Inouye, S, Tsuruoka, T, Ito, T, Niida, T, Structure and synthesis of nojirimycin, *Tetrahedron*, 24, 2125–2144, 1968.
59. Inouye, S, Tsuruoka, T, Niida, T, Structure of nojirimycin, sugar antibiotic with nitrogen in the ring, *J. Antibiot.*, 19, 288–292, 1966.
60. Yu, J, Choi, S Y, Moon, K D, Chung, H H, Youn, H J, Jeong, S, Park, H, Schultz, P G, A glycosidase antibody elicited against a chair-like transition state analog by *in vitro* immunization, *Proc. Natl. Acad. Sci. USA*, 95, 2880–2884, 1998.
61. Winchester, B, Fleet, G W J, Amino-sugar glycosidase inhibitors: versatile tools for glycobiologists, *Glycobiology*, 2, 199–210, 1992.
62. (a) Lembcke, B, Folsch, U R, Creutzfeldt, W, Effect of 1-desoxynojirimycin derivatives on small intestinal disaccharide activities and on active transport in vitro, *Digestion*, 31, 120–127, 1985; (b) Taylor, R H, Barker, H M, Bowey, E A, Canfield, J E, Regulation of the absorption of dietary carbohydrate in man by two new glucosidase inhibitors, *Gut*, 27, 1471–1478, 1986; (c) Rhinehart, G L, Robinson, K M, Payne, A J, Wheately, M E, Fisher, J L, Liu, P S, Cheng, W, Castanospermine

blocks the hyperglycemic response to carbohydrates *in vivo*: a result of intestinal disaccharidase inhibition, *Life Sci.*, 41, 2325–2331, 1987; (d) Samulitis, B K, Goda, T, Lee, S M, Koldovski, O, Inhibitory mechanism of acarbose and 1-deoxynojirimycin derivatives on carbohydrases in rat small intestine, *Drugs Exp. Clin. Res.*, 13, 517–524, 1987; (e) Robinson, K M, Begovic, M E, Rhinehart, G L, Heineke, E W, Ducep, J B, Kastner, P R, Marshall, F N, Danzin, C, New potent α-glucohydrolase inhibitor MDL 73945 with long duration of action in rats, *Diabetes*, 40, 825–830, 1991.

63. Nojima, H, Kimura, I, Chen, F, Sugihara, Y, Haruno, M, Kato, A, Asano, N, Antihyperglycemic effects of *N*-containing sugars from *Xanthocercis zambesiaca Morus bombycis*, *Aglaonema treubii*, and *Castanospermum australe* in streptozotocin-diabetic mice, *J. Nat. Prod.*, 61, 397–400, 1998.
64. (a) Asano, N, Oseki, K, Kizu, H, Matsui, K, Nitrogen-in-the-ring pyranoses and furanoses: structural basis of inhibition of mammalian glycosidases, *J. Med. Chem.*, 37, 3701–3706, 1994; (b) Asano, N, Nishida, M, Kato, A, Kizu, H, Matsui, K, Shimada, Y, Itoh, T, Baba, M, Watson, A A, Nash, R J, de Q Lilley, P M, Watkin, D J, Fleet, G W J, Homonojirimycin isomers and *N*-alkylated homonojirimycins: structural and conformational basis of inhibition of glycosidases, *J. Med. Chem.*, 41, 2565–2571, 1998.
65. (a) Johnston, P S, Feig, P U, Coniff, R F, Krol, A, Davidson, J A, Haffner, S M, Long-term titrated-dose alpha-glucosidase inhibition in non-insulin-requiring Hispanic NIDDM patients, *Diabetes Care*, 21, 409–415, 1998; (b) Johnston, P S, Feig, P U, Coniff, R F, Krol, A, Kelley, D E, Mooradian, A D, Chronic treatment of African-American type 2 diabetic patients with alpha-glucosidase inhibition, *Diabetes Care*, 21, 416–422, 1998; (c) Mitrakou, A, Tountas, N, Raptis, A E, Bauer, R J, Schulz, H, Raptis, S A, Long-term effectiveness of a new alpha-glucosidase inhibitor (BAY m1099-miglitol) in insulin-treated type 2 diabetes mellitus, *Diabet. Med.*, 15, 657–660, 1998.
66. Ahr, H J, Boberg, M, Brendel, E, Krause, H P, Steinke, W, Pharmacokinetics of miglitol. Absorption, distribution, metabolism, and excretion following administration to rats, dogs, and man, *Arzneimittel-Forschung*, 47, 734–745, 1997.
67. Betts, R F, Influenza virus, In *Mandell, Douglas, and Bennett's Principles and Practice of Infectious Diseases*, Mandell, G L, Bennett, J E, Dolin, R, Eds., Churchill Livingstone, New York, pp. 1546–1567, 1995.
68. Hwang, M Y, Do you have the flu?, *JAMA*, 281, 962, 1999.
69. (a) Palese, P, Schulman, J L, Bodo, G, Meindl, P, Inhibition of influenza and parainfluenza virus replication in tissue culture by 2-deoxy-2,3-dehydro-*N*-trifluoroacetylneuraminic acid (FANA), *Virology*, 59, 490–498, 1974; (b) Palese, P, Tobita, K, Ueda, M, Compans, R W, Characterization of temperature-sensitive influenza virus mutants defective in neuraminidase, *Virology*, 61, 397–410, 1974; (c) Palese, P, Compans, R W, Inhibition of influenza virus replication in tissue culture by 2-deoxy-2,3-dehydro-*N*-trifluoroacetylneuraminic acid (FANA): mechanism of action, *J. Gen. Virol.*, 33, 159–163, 1976.
70. (a) Colman, P M, Influenza virus neuraminidase: structure, antibodies, and inhibitors, *Protein Sci.*, 3, 1687–1696, 1994; (b) Calfee, D P, Hayden, F G, New approaches to influenza chemotherapy: neuraminidase inhibitors, *Drugs*, 56, 537–553, 1998; (c) Air, G M, Ghate, A A, Stray, S J, Influenza neuraminidase as target for antivirals, *Adv. Virus Res.*, 54, 375–402, 1999; (d) Finley, J B, Atigadda, V R, Duarte, F, Zhao, J J, Brouillette, W J, Air, G M, Luo, M, Novel aromatic inhibitors of influenza virus neuraminidase make selective interactions with conserved residues and water molecules in the active site, *J. Mol. Biol.*, 293, 1107–1119, 1999.
71. Houde, M, Arora, D J, Stimulation of tumor necrosis factor secretion by purified influenza virus neuraminidase, *Cell Immunol.*, 129, 104–111, 1990.
72. (a) Woods, J M, Bethell, R C, Coates, J A V, Healy, N, Hiscox, S A, Pearson, B A, Ryan, D M, Ticehurst, J, Tilling, J, 4-Guanidino-2,4-dideoxy-2,3-dehydro-*N*-acetylneuraminic acid is a highly effective inhibitor both of the sialidase (neuraminidase) and of growth of a wide range of influenza A and B viruses in vitro, *Antimicrob. Agents Chemother.*, 37, 1473–1479, 1993; (b) von Itzstein, M, Dyason, J C, Oliver, S W, White, H F, Wu, W Y, Kok, G B, Pegg, M S, A study of the active site of influenza virus sialidase: an approach to the rational design of novel anti-influenza drugs, *J. Med. Chem.*, 39, 388–391, 1996.

73. (a) Coleman, P M, Varghese, J N, Laver, W G, Structure of the catalytic and antigenic sites in influenza virus neuraminidase, *Nature*, 303, 41–44, 1983; (b) Varghese, N, Colman, P M, Three-dimensional structure of the neuraminidase of influenza virus A/Tokyo/3/67 at 2.2 Å resolution, *J. Mol. Biol.*, 221, 473–486, 1991; (c) Burmeister, W P, Ruigrok, R W H, Cusack, S, The 2.2 Å resolution crystal structure of influenza B neuraminidase and its complex with sialic acid, *EMBO J.*, 11, 49–56, 1992.
74. (a) von Itzstein, M, Wu, W Y, Kok, G B, Pegg, M S, Dyason, J C, Jin, B, Phan, T V, Smythe, M L, White, H F, Oliver, S W, Colman, P M, Varghese, J N, Ryan, D M, Wood, J M, Bethell, R C, Hotham, V J, Cameron, J M, Penn, C R, Rational design of potent sialidase-based inhibitors of influenza virus replication, *Nature*, 363, 418–423, 1993; (b) Holzer, C T, von Itzstein, M, Jin, B, Pegg, M S, Stewart, W P, Wu, W Y, Inhibition of sialidases from viral, bacterial and mammalian sources by analogues of 2-deoxy-2,3-didehydro-*N*-acetylneuraminic acid modified at the C-4 position, *Glycoconjugate J.*, 10, 40–44, 1993; (c) Taylor, N R, von Itzstein, M, Molecular modeling studies on ligand binding to sialidase from influenza virus and the mechanism of catalysis, *J. Med. Chem.*, 37, 616–624, 1994; (d) von Itzstein, M, Wu, W Y, Jin, B, The synthesis of 2,3-didehydro-2,4-dideoxy-4-guanidinyl-*N*-acetylneuraminic acid: a potent influenza virus sialidase inhibitor, *Carbohydr. Res.*, 259, 301–305, 1994; (e) Kiefel, M J, von Itzstein, M, Influenza virus sialidase: a target for drug discovery, *Prog. Med. Chem.*, 36, 1–28, 1999; (f) Kiefel, M J, von Itzstein, M, Recent advances in the synthesis of sialic acid derivatives and sialylmimetics as biological probes, *Chem. Rev.*, 102, 471–490, 2002.
75. (a) Saito, M, Yu, R K, Biochemistry and function of sialidases, In *Biology of the Sialic Acids*, Rosenberg, A, Ed., Plenum Press, New York, pp. 261–318, 1995; (b) Dowle, M D, Howes, P D, Recent advances in sialidase inhibitors, *Expert Opin. Ther. Pat.*, 8, 1461–1478, 1998.
76. Meindl, P, Bodo, G, Palese, P, Schulman, J, Tuppy, H, Inhibition of neuraminidase activity by derivatives of 2-deoxy-2,3-dehydro-*N*-acetylneuraminic acid, *Virology*, 58, 457–463, 1974.
77. (a) Kamerling, J P, Schauer, R, Shukla, A K, Stoll, S, van Beek, H, Vliegenthart, J F G, Migration of *O*-acetyl groups in *N,O*-acetylneuraminic acids, *Eur. J. Biochem.*, 162, 601–607, 1987; (b) Hiramatsu, Y, Tsukida, T, Nakai, Y, Inoue, Y, Kondo, H, Study on selectin blocker. 8. Lead discovery of a non-sugar antagonist using a 3D-pharmacophore model, *J. Med. Chem.*, 43, 1476–1483, 2000.
78. Ryan, D M, Ticehurst, J, Dempsey, M H, Penn, C R, Inhibition of influenza virus replication in mice by GG 167 is consistent with extracellular activity of viral neuraminidase, *Antimicrob. Agents Chemother.*, 38, 2270–2275, 1994.
79. Chandler, M, Bamford, M J, Conroy, R, Lamont, B, Patel, B, Patel, V K, Steeples, I P, Storer, R, Weir, N G, Wright, M, Williamson, C, Synthesis of the potent influenza neuraminidase inhibitor 4-guanidino-Neu5Ac2en. X-ray molecular structure of 5-acetamido-4-amino-2,6-anhydro-3,4,5-trideoxy-D-*erythro*-L-*gluco*-nonionic acid, *J. Chem. Soc., Perkin. Trans. 1*, 1173–1180, 1995.
80. Smith, P W, Starkey, I D, Howes, P D, Sollis, S L, Keeling, S P, Cherry, P C, von Itzstein, M, Wu, W Y, Jin, B, Synthesis and influenza virus sialidase inhibitory activity of analogs of 4-guanidino-Neu5Ac2en (GG167) with modified 5-substituents, *Eur. J. Med. Chem.*, 31, 143–150, 1996.
81. Bamford, M J, Pichel, J C, Husman, W, Patel, B, Storer, R, Weir, N G, Synthesis of 6-, 7- and 8-carbon sugar analogs of potent anti-influenza 2,3-didehydro-2,3-dideoxy-*N*-acetylneuraminic acid derivatives, *J. Chem. Soc., Perkin. Trans. 1*, 1181–1187, 1995.
82. Smith, P W, Sollis, S L, Howes, P D, Cherry, P C, Starkey, I D, Cobley, K N, Weston, H, Scicinski, J, Merritt, A, Whittington, A, Wyatt, P, Taylor, N, Green, D, Bethell, R, Madar, S, Fenton, R J, Morley, P J, Pateman, T, Beresford, A, Dihydropyrancarboxamides related to zanamivir: a new series of inhibitors of influenza virus sialidases. 1. Discovery, synthesis, biological activity, and structure-activity relationships of 4-guanidino- and 4-amino-4*H*-pyran-6-carboxamides, *J. Med. Chem.*, 41, 787–797, 1998.
83. Wyatt, P G, Coomber, B A, Evans, D N, Jack, T I, Fulton, H E, Wonacott, A J, Colman, P, Varghese, J, Sialidase inhibitors related to zanamivir. Further SAR studies of 4-amino-4*H*-pyran-2-carboxylic acid-6-propylamides, *Bioorg. Med. Chem. Lett.*, 11, 669–673, 2001.
84. Masuda, T, Shibuya, S, Arai, M, Yoshida, S, Tomozawa, T, Ohno, A, Yamashita, M, Honda, T, Synthesis and anti-influenza evaluation of orally active bicyclic ether derivatives related to zanamivir, *Bioorg. Med. Chem. Lett.*, 13, 669–673, 2003.

85. (a) Kim, C U, Lew, W, Williams, M A, Liu, H, Zhang, L, Swaminathan, S, Bischofberger, N, Chen, M S, Mendel, D B, Tai, C Y, Laver, W G, Stevens, R C, Influenza neuraminidase inhibitors possessing a novel hydrophobic interaction in the enzyme active site: design, synthesis, and structural analysis of carbocyclic sialic acid analogues with potent anti-influenza activity, *J. Am. Chem. Soc.*, 119, 681–690, 1997; (b) Kim, C U, Lew, W, Williams, M A, Wu, H, Zhang, L, Chen, X, Mendel, D B, Laver, W G, Stevens, R C, Structure-activity relationship studies of novel carbocyclic influenza neuraminidase inhibitors, *J. Med. Chem.*, 41, 2451–2460, 1998; (c) Mendel, D B, Tai, C Y, Escarpe, P A, Li, W, Sidwell, R W, Huffman, J H, Sweet, C, Jakeman, K J, Merson, J, Lacy, S A, Lew, W, Williams, M A, Zhang, L, Chen, M S, Bischofberger, N, Kim, C U, Oral administration of a prodrug of the influenza virus neuraminidase inhibitor GS 4071 protects mice and ferrets against influenza infection, *Antimicrob. Agents Chemother.*, 42, 640–646, 1998; (d) Lew, W, Chen, X W, Kim, C U, Discovery and development of GS 4104 (oseltamivir): an orally active influenza neuraminidase inhibitor, *Curr. Med. Chem.*, 7, 663–672, 2000.
86. McClellan, K, Perry, C M, Oseltamivir: a review of its use in influenza, *Drugs*, 61, 261–283, 2001.
87. Kerrigan, S A, Smith, P W, Stoodley, R J, (4*R*,5*R*)-4-acetylamino-5-diethylcarbamoylcyclohex-1-ene-1-carboxylic acid and (3*R*,4*R*)-4-acetylamino-3-diethylcarbamoylcyclohex-1-ene-1-carboxylic acid: new inhibitors of influenza virus sialidases, *Tetrahedron Lett.*, 42, 4709–4712, 2001.
88. Bianco, A, Brufani, M, Manna, F, Melchioni, C, Synthesis of a carbocyclic sialic acid analogue for the inhibition of influenza virus neuraminidase, *Carbohydr. Res.*, 332, 23–31, 2001.
89. (a) Lew, W, Wu, H, Chen, X, Graves, B J, Escarpe, P A, MacArthur, H L, Mendel, D B, Kim, C U, Carbocyclic influenza neuraminidase inhibitors possessing a C3-cyclic amine side chain: synthesis and inhibitory activity, *Bioorg. Med. Chem. Lett.*, 10, 1257–1260, 2000; (b) Hochgürtel, M, Biesinger, R, Kroth, H, Piecha, D, Hofmann, M W, Krause, S, Schaaf, O, Nicolau, C, Eliseev, A E, Ketones as building blocks for dynamic combinatorial libraries: highly active neuraminidase inhibitors generated via selection pressure of the biological target, *J. Med. Chem.*, 46, 356–358, 2003.
90. (a) Babu, Y S, Chand, P, Bantia, S, Kotian, P, Dehghani, A, El-Kattan, Y, Lin, T H, Hutchison, T L, Elliot, A J, Parker, C D, Ananth, S L, Horn, L L, Laver, G W, Montgomery, J A, BCX-1812 (RWJ-270201): discovery of a novel, highly potent, orally active, and selective influenza neuraminidase inhibitor through structure-based drug design, *J. Med. Chem.*, 43, 3482–3486, 2000; (b) Bantia, S, Parker, C D, Ananth, S L, Horn, L L, Andries, K, Chand, P, Kotian, P L, Dehghani, A, El-Kattan, Y, Lin, T, Hutchison, T L, Montgomery, J A, Kellog, D L, Babu, Y S, Comparison of the anti-influenza virus activity of RWJ-270201 with those of oseltamivir and zanamivir, *Antimicrob. Agents Chemother.*, 45, 1162–1167, 2001.
91. (a) Jedrzejas, M J, Singh, S, Brouillette, W J, Laver, W G, Air, G M, Luo, M, Structures of aromatic inhibitors of influenza virus neuraminidase, *Biochemistry*, 34, 3144–3151, 1995; (b) Singh, S, Jedrzejas, M J, Air, G M, Luo, M, Laver, W G, Brouillette, W J, Structure-based inhibitors of influenza virus sialidase. A benzoic acid lead with novel interaction, *J. Med. Chem.*, 38, 3217–3225, 1995; (c) Atigadda, V R, Brouillette, W J, Duarte, F, Babu, Y S, Bantia, S, Chand, P, Chu, N, Montgomery, J A, Walsh, D A, Sudbeck, E, Finley, J, Air, G M, Luo, M, Laver, G W, Hydrophobic benzoic acids as inhibitors of influenza neuraminidases, *Bioorg. Med. Chem.*, 7, 2487–2497, 1999; (d) Finley, J B, Atigadda, V R, Duarte, F, Zhao, J J, Brouillette, W J, Air, G M, Luo, M, Novel aromatic inhibitors of influenza virus neuraminidase make selective interactions with conserved residues and water molecules in the active site, *J. Mol. Biol.*, 293, 1107–1119, 1999; (e) Atigadda, V R, Brouillette, W J, Duarte, F, Ali, S M, Babu, Y S, Bantia, S, Chand, P, Chu, N, Montgomery, J A, Walsh, D A, Sudbeck, E A, Finley, J, Luo, M, Air, G M, Laver, G W, Potent inhibition of influenza sialidase by a benzoic acid containing a 2-pyrrolidinone substituent, *J. Med. Chem.*, 42, 2332–2343, 1999.
92. Varghese, J N, Epa, V C, Colman, P M, Three-dimensional structure of the complex of 4-guanidino-Neu5ac2en and influenza virus neuraminidase, *Protein Sci.*, 4, 1081–1087, 1995.
93. (a) Mousa, S A, Recent advances in cell adhesion molecule (CAM) research and development: targeting CAMs for therapeutic and diagnostic applications, *Drugs Fut.*, 21, 283–289, 1996; (b) Mousa, S A, Cheresh, D A, Recent advances in cell adhesion molecules and extracellular matrix proteins: potential clinical implications, *Drug Discov. Today*, 2, 187–199, 1997; (c) Cines, D B, Pollak, E S, Buck, C A, Loscalzo, J, Zimmermann, G A, McEver, R P, Pober, J S, Wick, T M,

Konkle, B A, Schwartz, B S, Barnathan, E S, McCrae, K R, Hug, B A, Schmidt, A.-M., Stern, D M, Endothelial cells in physiology and in the pathophysiology of vascular disorders, *Blood*, 91, 3527–3561, 1998.

94. (a) Lobb, R R, *Adhesion. Its Role in Inflammatory Disease*, Harlan, J M, Liu, D Y, Eds., W.H. Freeman, New York, p. 1, 1992, chap. 1; (b) Paulson, J C, *Adhesion. Its Role in Inflammatory Disease*, Harlan, J M, Liu, D Y, Eds., W.H. Freeman, New York, p. 19, 1992, chap. 2; (c) Kansas, G S, Selectins and their ligands: current concepts and controversies, *Blood*, 88, 3259–3287, 1996; (d) Varki, A, Selectin ligands, *Proc. Natl. Acad. Sci. USA*, 91, 7390–7397, 1994.
95. (a) Geng, J G, Bevilacqua, M P, Moore, K L, McIntyre, T M, Prescott, S M, Kim, J M, Bliss, G A, Zimmermann, G A, McEver, R E, Rapid neutrophil adhesion to activated endothelium mediated by GMP-140, *Nature*, 343, 757–760, 1990; (b) Hattori, R, Hamilton, K K, Fugate, R D, McEver, R P, Sims, P J, Stimulated secretion of endothelial von Willebrand factor is accompanied by rapid redistribution to the cell surface of the intracellular granule membrane protein GMP-140, *J. Biol. Chem.*, 264, 7768–7771, 1989.
96. Bevilacqua, M P, Pober, J S, Mendrick, D L, Cotran, R S, Gimbrone, M A Jr, Identification of an inducible endothelial-leukocyte adhesion molecule, *Proc. Natl. Acad. Sci. USA*, 84, 9238–9242, 1987.
97. (a) Gallatin, W M, Weissman, I L, Butcher, E C, A cell-surface molecule involved in organ-specific homing of lymphocytes, *Nature*, 304, 30–34, 1983; (b) Lewinsohn, D M, Bargatze, R F, Butcher, E C, Leukocyte-endothelial cell recognition: evidence of a common molecular mechanism shared by neutrophils, lymphocytes, and other leukocytes, *J. Immunol.*, 138, 4313–4321, 1987.
98. (a) Lawrence, M B, Springer, T A, Leukocytes roll on a selectin at physiologic flow rates: distinction from and prerequisite for adhesion through integrins, *Cell*, 65, 859–873, 1991; (b) Alon, R, Hammer, D A, Springer, T A, Lifetime of the P-selectin-carbohydrate bond and its response to tensile force in hydrodynamic flow, *Nature*, 374, 539–542, 1995.
99. Nicholson, M W, Barclay, A N, Singer, M S, Rosen, S D, van der Merwe, P A, Affinity and kinetic analysis of L-selectin (CD62L) binding to glycosylation-dependent cell-adhesion molecule-1, *J. Biol. Chem.*, 273, 763–770, 1998.
100. Mehta, P, Cummings, R D, McEver, R P, Affinity and kinetic analysis of P-selectin binding to P-selectin glycoprotein ligand-1, *J. Biol. Chem.*, 273, 32506–32513, 1998.
101. Wild, M K, Huang, M.-C., Schulze-Horsel, U, van der Merwe, P A, Vestweber, D, Affinity, kinetics and thermodynamics of E-selectin binding to E-selectin ligand-1, *J. Biol. Chem.*, 276, 31602–31612, 2001.
102. (a) Moore, K L, Patel, K D, Bruehl, R E, Li, F, Johnson, D A, Lichtenstein, H S, Cummings, R D, Bainton, D F, McEver, R P, P-selectin glycoprotein ligand-1 mediates rolling of human neutrophils on P-selectin, *J. Cell. Biol.*, 128, 661–671, 1995; (b) Patel, K D, McEver, R P, Comparison of tethering and rolling of eosinophils and neutrophils through selectins and P-selectin glycoprotein ligand-1, *J. Immunol.*, 159, 4555–4565, 1997; (c) Norman, K E, Moore, K L, McEver, R P, Ley, K, Leukocyte rolling *in vivo* is mediated by P-selectin glycoprotein ligand-1, *Blood*, 86, 4417–4421, 1995.
103. (a) Tedder, T F, Steeber, D A, Chen, A, Engel, P, The selectins: vascular adhesion molecules, *FASEB J*, 9, 866–873, 1995; (b) McEver, R P, Moore, K L, Cummings, R D, Leukocyte trafficking mediated by selectin-carbohydrate interactions, *J. Biol. Chem.*, 270, 11025–11028, 1995; (c) Lasky, L, Selectin-carbohydrate interactions and the initiation of the inflammatory response, *Annu. Rev. Biochem.*, 64, 113–139, 1995; (d) von Andrian, U H, Berger, E M, Ramezani, L, Chambers, J D, Ochs, H D, Harlan, J M, Paulson, J C, Etzioni, A, Arfors, K-E, In vivo behavior of neutrophils from two patients with distinct inherited leukocyte adhesion deficiency syndromes, *J. Clin. Invest.*, 91, 2893–2897, 1993.
104. (a) Phillips, L, Nudelman, E, Gaeta, F C A, Perez, M, Singhal, A K, Hakomori, S-I, Paulson, J C, ELAM-1 mediates cell adhesion by recognition of a carbohydrate ligand, sialyl-Le^X, *Science*, 250, 1130–1132, 1990; (b) Walz, G, Aruffo, A, Kolanus, W, Bevilacqua, M, Seed, B, Recognition by ELAM-1 of the sialyl-Le^X determinant on myeloid and tumor cells, *Science*, 250, 1132–1135, 1990; (c) Berg, E L, Robinson, M K, Mansson, O, Butcher, E C, Magnani, J L, A carbohydrate domain common to both sialyl Le^a is recognized by the endothelial cell leukocyte adhesion molecule ELAM-1, *J. Biol. Chem.*, 266, 14869–14872, 1991.

105. (a) Mihelcic, D, Schleiffenbaum, B, Tedder, T F, Sharar, S R, Harlan, J M, Winn, R K, Inhibition of leukocyte L-selectin function with a monoclonal antibody attenuates reperfusion injury to the rabbit ear, *Blood*, 84, 2322–2328, 1994; (b) Flynn, D M, Buda, A J, Jeffords, P R, Lefer, D J, A sialyl Lewis^x-containing carbohydrate reduces infarct size: role of selectins in myocardial reperfusion injury, *Am. J. Physiol.*, 271, H2086–H2096, 1996; (c) Sagara, H, Ra, C, Okada, T, Shinohara, S, Fukuda, T, Okumura, K, Makino, S, Sialyl Lewis^x analog inhibits eosinophil accumulation and late asthmatic response in a guinea-pig model of asthma, *Int. Arch. Allergy Immunol.*, 111, 32–36, 1996; (d) Park, I Y, Lee, D S, Song, M H, Kim, W, Won, J M, Cylexin: a P-selectin inhibitor prolongs heart allograft survival in hypersensitized rat recipients, *Transplant. Proc.*, 30, 2927–2928, 1998.
106. For reviews, see. (a) Simanek, E E, McGarvey, G J, Jablonowski, J A, Wong, C-H, Selectin-carbohydrate interactions: from natural ligands to designed mimics, *Chem. Rev.*, 98, 833–862, 1998; (b) Bertozzi, C R, Cracking the carbohydrate code for selectin recognition, *Chem. Biol.*, 2, 703–708, 1995; (c) Musser, J H, Anderson, M B, Levy, D E, Glycomimetics as selectin inhibitors, *Curr. Pharm. Des.*, 1, 221–232, 1995; (d) Giannis, A, The sialyl Lewis^x group and its analogues as ligands for selectins: chemoenzymatic syntheses and biological functions, *Angew. Chem. Int. Ed.*, 33, 178–190, 1994.
107. Role of hydroxyls of L-Fuc moiety. (a) Ramphal, J Y, Zheng, Z.-L., Perez, C, Walker, L E, DeFrees, S A, Gaeta, F C A, Structure–activity relationships of sialyl Lewis X-containing oligosaccharides.1. Effect of modifications of the fucose moiety, *J. Med. Chem.*, 37, 3459–3463, 1994; (b) Brandley, B K, Kiso, M, Abbas, S, Nikrad, P, Srivasatava, O, Foxall, C, Oda, Y, Hasegawa, A, Structure–function studies on selectin carbohydrate ligands. Modifications to fucose, sialyl acid and sulphate as sialyl acid replacement, *Glycobiology*, 3, 633–639, 1993; (c) Stahl, W, Sprengard, U, Kretschmar, G, Kunz, H, Synthesis of deoxy sialyl Lewis^x analogs. Potential selectin antagonists, *Angew. Chem. Int. Ed.*, 33, 2096–2098, 1994. Role of the hydroxyls of D-Gal moiety; (d) Baenteli, R, Ernst, B, Synthesis of sialyl Lewis^x mimetics. 2. Modification of the 6-position of galactose, *Bioorg. Med. Chem. Lett.*, 11, 459–462, 2001. Role of the substituent of the Neu5Ac moiety; (e) Tyrrell, D, James, P, Rao, N, Foxall, C, Abbas, S, Dasgupta, F, Nashed, M, Hasegawa, A, Kiso, M, Asa, D, Kidd, J, Brandley, B K, Structural requirements for the carbohydrate ligand of E-selectin, *Proc. Natl. Acad. Sci. USA*, 88, 10372–10376, 1991; (f) Ohmoto, H, Nakamura, K, Inoue, T, Kondo, N, Inoue, Y, Yoshino, K, Kondo, H, Ishida, H, Kiso, M, Hasegawa, A, Studies on selectin blocker. 1. Structure–activity relationships of sialyl Lewis X analogs, *J. Med. Chem.*, 39, 1339–1343, 1996. Role of the substituents of the GlcNAc moiety; (g) DeFrees, S A, Gaeta, F C A, Lin, Y C, Ichikawa, Y, Wong, C.-H., Ligand recognition by E-selectin: analysis of conformation and activity of synthetic monomeric and bivalent sialyl Lewis X analogs, *J. Am. Chem. Soc.*, 115, 7549–7550, 1993; (h) Hiramatsu, Y, Tsujishita, H, Kondo, H, Studies on selectin blocker. 3. Investigation of the carbohydrate ligand sialyl Lewis^x recognition site of P-selectin, *J. Med. Chem.*, 39, 4547–4553, 1996; (i) Wada, Y, Saito, T, Matsuda, N, Ihmoto, H, Yoshino, K, Ohashi, M, Kondo, H, Ishida, H, Kiso, M, Hasegawa, A, Studies on selectin blockers. 2. Novel selectin blocker as potential therapeutics for inflammatory disorders, *J. Med. Chem.*, 39, 2055–2059, 1996.
108. Rinnbauer, M, Ernst, B, Wagner, B, Magnani, J, Benie, A J, Peters, T, Epitope mapping of sialyl Lewis^x bound to E-selectin using saturation transfer difference NMR experiments, *Glycobiology*, 13, 435–443, 2003.
109. (a) Siegelman, M H, Cheng, I C, Weissman, I L, Wakeland, E K, The mouse lymph node homing receptor is identical with the lymphocyte cell surface marker Ly-22: role of the EGF domain in endothelial binding, *Cell.*, 61, 611–622, 1990; (b) Watson, S R, Imai, Y, Fennie, C, Geoffrey, J, Singer, M, Rosen, S D, Lasky, L A, The complement binding-like domains of the murine homing receptor facilitate lectin activity, *J. Cell. Biol.*, 115, 235–243, 1991; (c) Kansas, G S, Saunders, K B, Ley, K, Zakrzewics, A, Gibson, R M, Furie, B C, Tedder, T F, A role for the epidermal growth factor-like domain of P-selectin in ligand recognition and cell adhesion, *J. Cell. Biol.*, 124, 609–618, 1994; (d) Li, S H, Burns, D K, Rumberger, J M, Presky, D H, Wilkinson, V L, Anostario, M Jr., Wolitzky, B A, Norton, C R, Familleti, P C, Kim, K J, Goldstein, A L, Cox, D C, Huang, K.-S., Consensus repeat domains of E-selectin enhance ligand binding, *J. Biol. Chem.*, 269, 4431–4437, 1994; (e) Bargatze, R F, Kurk, S, Watts, G, Kishimoto, T K, Speer, C A, Jutila, M A, *In vivo* and *in vitro* functional examination of a conserved epitope of L- and E-selectin crucial for leukocyte-endothelial cell interactions,

J. Immunol., 152, 5814–5825, 1994; (f) Gibson, R M, Kansas, G S, Tedder, T F, Furie, B, Furie, B C, Lectin and epidermal growth factor domains of P-selectin at physiologic density are the recognition unit for leukocyte binding, *Blood*, 85, 151–158, 1995; (g) Kolbinger, F, Patton, J T, Geisenhoff, G, Aenis, A, Li, X, Katopodis, A G, The carbohydrate-recognition domain of E-selectin is sufficient for ligand binding under both static and flow conditions, *Biochemistry*, 35, 6385–6392, 1996.

110. Weis, W I, Drickamer, K, Hendrickson, W A, Structure of a C-type mannose-binding protein complexed with an oligosaccharide, *Nature*, 360, 127–134, 1992.
111. (a) Wright, C S, Refinement of the crystal structure of wheat germ agglutinin isolectin 2 at 1.8 Å resolution, *J. Mol. Biol.*, 194, 501–529, 1987; (b) Bourne, Y, Rouge, P, Cambillau, C, X-Ray Structure of a (α-Man(1-3)β-Man(1-4)GlcNAc)-lectin complex at 2.1-Å resolution. The role of water in sugar–lectin interaction, *J. Biol. Chem.*, 265, 18161–18165, 1990; (c) Shaanan, B, Lis, H, Sharon, N, Structure of a legume lectin with an ordered *N*-linked carbohydrate in complex with lactose, *Science*, 254, 862–866, 1991; (d) Naismith, J H, Emmerich, C, Habash, J, Harop, S J, Hellewell, J R, Hunter, W N, Rafetery, J, Kalb, A J, Yariv, J, Refined structure of concanavalin a complexed with methyl α-D-mannopyranoside at 2.0 Å resolution and comparison with the saccharide-free structure, *Acta Crystallogr.*, D50, 847–858, 1994.
112. Ng, K. K.-S., Weis, W I, Structure of a selectin-like mutant of mannose-binding protein complexed with sialylated and sulfated Lewis oligosaccharides, *Biochemistry*, 36, 979–988, 1997.
113. (a) Hiramatsu, Y, Tsujishita, H, Kondo, H, Studies on selectin blocker. 3. Investigation of the carbohydrate ligand sialyl Lewis X recognition site of P-selectin, *J. Med. Chem.*, 39, 4547–4553, 1996; (b) Kogan, T P, Revelle, B M, Tapp, S, Scott, D, Beck, P J, A single amino acid residue can determine the ligand specificity of E-selectin, *J. Biol. Chem.*, 270, 14047–14055, 1995.
114. (a) Cooke, R M, Hale, R S, Lister, S G, Shah, G, Weir, M P, The conformation of the sialyl Lewis X ligand changes upon binding to E-selectin, *Biochemistry*, 33, 10591–10596, 1994; (b) Rutherford, T J, Spackman, D G, Simpson, P J, Homans, S W, 5 Nanosecond molecular dynamics and NMR study of conformational transitions in the sialyl-Lewis X antigen, *Glycobiology*, 4, 59–68, 1994; (c) Poppe, L, Brown, G S, Philo, J S, Nikrad, P V, Shah, B H, Conformation of sLe^x tetrasaccharide, free in solution and bound to E-, P-, and L-selectin, *J. Am. Chem. Soc.*, 119, 1727–1736, 1997; (d) Harris, R, Kiddle, G R, Field, R A, Milton, M J, Ernst, B, Magnani, J L, Homans, S W, Stable-isotope-assisted NMR studies on ^{13}C-enriched sialyl Lewis^x in solution and bound to E-selectin, *J. Am. Chem. Soc.*, 121, 2546–2551, 1999.
115. (a) Ernst, B, Dragic, Z, Marti, S, Müller, C, Wagner, B, Jahnke, W, Magnani, J L, Norman, K E, Oehrlein, R, Peters, T, Kolb, H C, Design and synthesis of E-selectin antagonists, *Chimia*, 55, 268–274, 2001; (b) Thoma, G, Magnani, J L, Patton, J T, Synthesis and biological evaluation of a sialyl Lewis X mimic with significantly improved E-selectin inhibition, *Bioorg. Med. Chem. Lett.*, 11, 923–925, 2001.
116. Graves, B J, Crowther, R L, Chandran, C, Rumberger, J M, Li, S, Huang, K-S, Presky, D H, Familletti, P C, Wolitzky, B A, Burns, B K, Insight into E-Selectin/ligand interaction from the crystal structure and mutagenesis of the lec/EGF domains, *Nature*, 367, 532–538, 1994.
117. Somers, W S, Tang, J, Shaw, G D, Camphausen, R T, Insights into the molecular basis of leukocyte tethering and rolling revealed by structures of P- and E-selectin bound to sLe^X and PSGL-1, *Cell*, 103, 467–479, 2000.
118. Pichierri, F, Matsuo, Y, Effect of protonation of the *N*-acetyl neuraminic acid residue of sialyl Lewis^x: a molecular orbital study with insights into its binding properties toward the carbohydrate recognition domain of E-selectin, *Bioorg. Med. Chem. Lett.*, 10, 2751–2757, 2002.
119. (a) Ball, G E, O'Neill, R A, Schultz, J E, Lowe, J B, Weston, B W, Nagy, J O, Brown, E G, Hobbs, C J, Bednarski, M D, Synthesis and structural analysis using 2-D NMR of sialyl Lewis X (SLe^x) and Lewis X (Le^x) oligosaccharides: ligands related to E-selectin [ELAM-1] binding, *J. Am. Chem. Soc.*, 114, 5449–5545, 1992; (b) Ishikawa, Y, Lin, C-Y, Dumas, D P, Shen, G-J, Garcia-Junceda, E, Williams, M A, Bayer, R, Ketcham, C, Walker, L E, Paulson, J C, Wong, C-H, Chemical-enzymic synthesis and conformational analysis of sialyl Lewis X and derivatives, *J. Am. Chem. Soc.*, 114, 9283–9298, 1992; (c) Lin, C-Y, Hummel, C W, Huang, D-H, Ishikawa, Y, Nicolaou, K C, Wong, C-H, Conformational studies of sialyl Lewis X in aqueous solution, *J. Am. Chem. Soc.*, 114, 5452–5454, 1992.

120. (a) Kolb, H C, Ernst, B, Development of tools for the design of selectin antagonists, *Chem Eur J*, 3, 1571–1578, 1997; (b) Kolb, H C, Ernst, B, Recent progresses in the glycodrug area, *Pure Appl. Chem.*, 69, 1879–1884, 1997.
121. Hensley, P, McDervitt, P J, Brooks, I, Trill, J J, Field, J A, McNulty, D E, Connor, J R, Griswold, D E, Kumar, N V, Kopple, K D, Carr, S A, Dalton, B J, Johanson, K J, The soluble form of E-selectin is an asymmetric monomer. Expression, purification, and characterization of the recombinant protein, *J. Biol. Chem.*, 269, 23949–23958, 1994.
122. (a) Scheffler, K, Ernst, B, Katopodis, A, Magnani, J L, Wang, W T, Weisemann, R, Peters, T, Determiation of the bioactive conformation of the carbohydrate ligand in the E-selectin/sialyl Lewis^x complex, *Angew. Chem. Int. Ed.*, 34, 1841–1844, 1995; (b) Scheffler, K, Brisson, J-R, Weisemann, R, Magnani, J L, Wang, W T, Ernst, B, Peters, T, Application of homonuclear 3D and pseudo-3D NMR experiments to elucidate fine details in 2D-trNOESY spectra of sialyl Lewis^x bound to E-selectin, *Biomol. NMR*, 9, 423–436, 1997.
123. (a) Birnbaum, Y, Patterson, M, Kloner, R A, The effect of CY-1503, a sialyl Lewis^x analog blocker of the selectin adhesion molecules, on infarct size and no-reflow in the rabbit model of acute myocardial infarction/reperfusion, *J. Mol. Cell. Cardiol.*, 29, 2013–2025, 1997; (b) Gill, E A, Kong, Y, Horwitz, L D, An oligosaccharide sialyl-Lewis^x analog does not reduce myocardial infarct size after ischemia and reperfusion in dogs, *Circulation.*, 94, 542–546, 1996; (c) Lefer, D J, Flynn, D M, Phillips, M L, Ratcliffe, M, Buda, A J, A novel sialyl Lewis^x analog attenuates neutrophil accumulation and myocardial necrosis after ischemia and reperfusion, *Circulation*, 90, 2390–2401, 1994.
124. In the first quarter of 1999, Cytel Corporation, La Jolla, CA announced the results from the Cylexin (CY1503, sialyl Lewis^x analogue) Phase II/III clinical trials (treatment of reperfusion injury in infants undergoing corrective surgery for congenital heart disease), which indicated that Cylexin showed no benefit over the placebo.
125. Koenig, A, Jain, R, Vig, R, Norgard-Sumnicht, K E, Matta, K L, Varki, A, Selectin inhibition: synthesis and evaluation of novel sialylated, sulfated and fucosylated oligosaccharides, including the major capping group of GlyCAM-1, *Glycobiology*, 7, 79–93, 1997.
126. (a) Kameyama, A, Ishida, H, Kiso, M, Hasegawa, A, Synthetic studies on sialoglycoconjugates. Part 21. Total synthesis of sialyl Lewis X, *Carbohydr. Res.*, 209, C1–C4, 1991; (b) Nicolaou, K C, Hummel, C W, Iwabuchi, Y, Total synthesis of sialyl dimeric Le^X, *J. Am. Chem. Soc.*, 114, 3126–3128, 1992; (c) Danishefsky, S J, Gervay, J, Peterson, J M, McDonald, F E, Koseki, K, Oriyama, T, Griffith, D A, Wong, C H, Dumas, D P, Remarkable regioselectivity in the chemical glycosylation of glycal acceptors: a concise solution to the synthesis of sialyl-Lewis X glycal, *J. Am. Chem. Soc.*, 114, 8329–8331, 1992.
127. Hiramatsu, Y, Tsukida, T, Nakai, Y, Inoue, Y, Kondo, H, Study on selectin blocker. 8. Lead discovery of a non-sugar antagonist using a 3D-pharmacophore model, *J. Med. Chem.*, 43, 1476–1483, 2000.
128. Thoma, G, Schwarzenbach, F, Duthaler, R O, Synthesis of sialyl Lewis^x mimic with carboxylic acid group: chemical approach toward the elucidation of the bioactive conformation of sialyl Lewis^x, *J. Org. Chem.*, 61, 514–524, 1996.
129. Thoma, G, Magnani, J L, Patton, J T, Ernst, B, Jahnke, W, Preorganization of the bioactive conformation of sialyl Lewis^x analogues correlates with their affinity to E-selectin, *Angew. Chem. Int. Ed.*, 40, 1941–1945, 2001.
130. Kolb, H C, Design and synthesis of a macrocyclic E-selectin antagonist, *Bioorg. Med. Chem. Lett.*, 7, 2629–2634, 1997.
131. Rao, B N N, Anderson, M B, Musser, J H, Gilbert, J H, Schaefer, M E, Foxall, C, Brandley, B K, Sialyl Lewis^x mimics derived from a pharmacophore search are selectin inhibitors with anti-inflammatory activity, *J. Biol. Chem.*, 269, 19663–19666, 1994.
132. (a) Kogan, T P, Dupre, B, Keller, K M, Scott, I L, Bui, H, Market, R V, Beck, P J, Voytus, J A, Revelle, B M, Scott, D, Rational design and synthesis of small molecule, non-oligosaccharide selectin inhibitors: (α-D-mannopyranosyloxy) biphenyl-substituted carboxylic acids, *J. Med. Chem.*, 38, 4976–4984, 1995; (b) Kogan, T P, Dupre, B, Bui, H, McAbee, K L, Kassir, J A, Scott, I L, Hu, X, Vanderslice, P, Beck, P J, Dixon, R A F, Novel synthetic inhibitors of selectin-mediated cell adhesion: synthesis of 1,6-bis[3-(3-carboxymethylphenyl)-4-(2-α-D-mannopyranosyloxy) phenyl]-hexane (TBC1269), *J. Med. Chem.*, 41, 1099–1111, 1998.

133. Dupre, B, Bui, H, Scott, I L, Market, R V, Keller, K M, Beck, P J, Kogan, T P, Glycomimetic selectin inhibitors: (α-D-mannopyranosyloxy) methylbiphenyls, *Bioorg. Med. Chem. Lett.*, 6, 569–572, 1996.
134. (a) Patel, T P, Goelz, S E, Lobb, R R, Parekh, R B, Isolation and characterization of natural protein-associated carbohydrates for E-selectin, *Biochemistry*, 33, 14815–14824, 1994; (b) Miyauchi, H, Yuri, M, Tanaka, M, Kawamura, N, Hayashi, M, Synthesis and inhibitory effects of bivalent sialyl Lewis X analogs at preventing cell adhesion, *Bioorg. Med. Chem. Lett.*, 7, 989–992, 1997.
135. Palma-Vargas, J M, Toledo-Pereyra, L, Dean, R E, Harkema, J M, Dixon, R A F, Kogan, T P, Small-molecule selectin inhibitor protects against liver inflammatory response after ischemia and reperfusion, *J. Am. Coll. Surg.*, 185, 365–372, 1997.
136. Press releases by Revotar Biopharmaceuticals AG and Texas Biotechnology Corporation, February 6 and August 5, 2003.
137. Murphy, P V, Hubbard, R E, Manallack, D T, Wills, R E, Montana, J G, Taylor, R J K, The design, synthesis, and evaluation of novel conformationally rigid analogs of sialyl Lewis^x, *Bioorg. Med. Chem. Lett.*, 41, 2421–2439, 1998.
138. De Vleeschauwer, M, Vaillancourt, M, Goudreau, N, Guindon, Y, Gravel, D, Design and synthesis of a new sialyl Lewis X mimetic: how selective are the selectin receptors? *Bioorg. Med. Chem. Lett.*, 11, 1109–1112, 2001.
139. Ramphal, J Y, Hiroshige, M, Lou, B, Gaudino, J J, Hayashi, M, Chen, S M, Chiang, L C, Gaeta, F C, DeFrees, S A, Ligand interactions with E-selectin. Identification of a new binding site for recognition of *N*-acyl aromatic glucosamine substituents of sialyl Lewis X, *J. Med. Chem.*, 39, 1357–1360, 1996.
140. Graves, B J, Crowther, R L, Chandran, C, Rumberger, J M, Li, S, Huang, K-S., Presky, D H, Familletti, P C, Wolitzky, B A, Burns, D K, Insight into E-selectin/ligand interaction from the crystal structure and mutagenesis of the lec/EGF domains, *Nature*, 367, 532–538, 1994.
141. Tsujishita, H, Hiramatsu, Y, Kondo, N, Ohmoto, H, Kondo, H, Kiso, M, Hasegawa, A, Selectin–ligand interactions revealed by molecular dynamics simulation in solution, *J. Med. Chem.*, 40, 362–369, 1997.
142. (a) Kaila, N, Thomas, B E, Design and synthesis of sialyl Lewis^x mimics as E- and P-selectin inhibitors, *Med Res Rev*, 22, 566–601, 2002; (b) Simanek, E E, McGarvey, G J, Jablonowski, J A, Wong, C.-H., Selectin-carbohydrate interactions: from natural ligands to designed mimics, *Chem. Rev.*, 98, 833–862, 1998.
143. Toepfer, A, Kretzschmar, G, Bartnik, E, Synthesis of novel mimetics of the sialyl Lewis^x determinant, *Tetrahedron Lett.*, 36, 9161–9164, 1995.
144. Hanessian, S, Reddy, G V, Huynh, H K, Pan, J, Pedatella, S, Ernst, B, Kolb, H C, Design and synthesis of sialyl Le^X mimetics based on carbocyclic scaffolds derived from (-) quinic acid, *Bioorg. Med. Chem. Lett.*, 7, 2729–2734, 1997.
145. Kolb, HC, Ernst, B, unpublished results.
146. Hanessian, S, Reddy, G V, Huynh, H K, McNaughton-Smith, G, Ernst, B, Kolb, H C, Magnani, J L, Sweeley, C, Exploration of β-turn scaffolding motifs as components of sialyl Le^x mimetics and their relevance to P-selectin, *Bioorg. Med. Chem. Lett.*, 8, 2803–2808, 1998.
147. Huang, H, Wong, C-H, Synthesis of biologically active sialyl Lewis^x mimetics, *J. Org. Chem.*, 60, 3100–3106, 1995.
148. Kaila, N, Yu, H-A, Xiang, Y, Design and synthesis of novel sialyl Lewis^x mimics, *Tetrahedron Lett.*, 36, 5503–5506, 1995.
149. Hanessian, S, Prabhanjan, H, Design and synthesis of glycomimetic prototypes — a model sialyl Lewis^x ligand for E-selectin, *Synlett*, 868–870, 1994.
150. Dekany, G, Wright, K, Ward, P, Toth, I, Synthesis of sialyl Lewis^x analogues, *J. Carbohydr. Chem.*, 15, 383–398, 1996.
151. Dasgupta, F, Musser, J H, Levy, D E, and Tang, P C, Preparation of Fucose-Containing Sialic Acids as Selectin Inhibitors, U.S. Patent 5658880, 1997.
152. Liu, A, Dillon, K, Campell, R M, Cox, D C, Huryn, D M, Synthesis of E-selectin inhibitors: use of an aryl-cyclohexyl ether as a disaccharide scaffold, *Tetrahedron Lett.*, 37, 3785–3788, 1996.
153. Baenteli, R, Ernst, B, Synthesis of sialyl Lewis^x mimics: replacement of galactose by aromatic spacers, *Tetrahedron Lett.*, 38, 4059–4062, 1997.

154. (a) Wang, R, Wong, C-H, Synthesis of sialyl Lewis X mimetics: use of *O*-α-fucosyl-(1*R*,2*R*)-2-aminocyclohexanol as core structure, *Tetrahedron Lett.*, 37, 5427–5430, 1996; (b) Lin, C-C, Shimazaki, M, Heck, M-P, Aoki, S, Wang, R, Kimura, T, Ritzen, H, Takayama, S, Wu, S-H, Weitz-Schmidt, G, Wong, C-H, Synthesis of sialyl Lewis X mimetics and related structures using the glycosyl phosphite methodology and evaluation of E-selectin inhibition, *J. Am. Chem. Soc.*, 118, 6826–6840, 1996; (c) Lin, C-C, Kimura, T, Wu, S-H, Weitz-Schmidt, G, Wong, C-H, Liposome-like fucopeptides as sialyl Lewis^x mimetics, *Bioorg. Med. Chem. Lett.*, 6, 2755–2760, 1996; (d) Wu, S-H, Shimazaki, M, Lin, C-C, Qiao, L, Moree, W J, Weitz-Schmidt, G, Wong, C-H, Synthesis of fucopeptides as sialyl Lewis^x mimetics, *Angew. Chem. Int. Ed.*, 35, 88–90, 1996; (e) Lampe, T F J, Weitz-Schmidt, G, Wong, C-H, Parallel synthesis of sialyl Lewis X mimetics on a solid phase: access to a library of fucopeptides, *Angew. Chem. Int. Ed.*, 37, 1707–1711, 1998.
155. (a) Uchiyama, T, Vassilev, V P, Kajimoto, T, Wong, W, Huang, H, Lin, C-C, Wong, C-H, Design and synthesis of sialyl Lewis^x mimetics, *J. Am. Chem. Soc.*, 117, 5395–5396, 1995; (b) Uchiyama, T, Woltering, T J, Wong, W, Lin, C-C, Kajimoto, T, Takebayashi, M, Weitz-Schmidt, G, Asakura, T, Noda, M, Wong, C-H, Design and synthesis of *C*-linked fucosides as inhibitors of E-selectin, *Bioorg. Med. Chem.*, 4, 1149–1165, 1996.
156. Tsai, C-Y, Park, W K C, Weitz-Schmidt, G, Ernst, B, Wong, C-H, Synthesis of sialyl Lewis^x mimetics using the Ugi four-component reaction, *Bioorg. Med. Chem. Lett.*, 8, 2333–2338, 1998.
157. Tsai, C-Y, Huang, X, Wong, C-H, Design and synthesis of cyclic sialyl Lewis X mimetics: a remarkable enhancement of inhibition by pre-organizing all essential functional groups, *Tetrahedron Lett.*, 41, 9499–9503, 2000.
158. Martens, C L, Cwirla, S E, Lee, R Y-W, Whitehorn, E, Chen, E Y-F, Bakker, A, Martin, E L, Wagstrom, C, Gopalan, P, Smith, C W, Tate, E, Koller, K J, Schatz, P J, Dower, W J, Barrett, R W, Peptides which bind to E-selectin and block neutrophil adhesion, *J. Biol. Chem.*, 270, 21129–21136, 1995.
159. Insug, O, Otvos, L Jr, Kieber-Emmons, T, Blaszczyk-Thurin, M, Role of SA-Le^x and E-selectin in metastasis assessed with peptide antagonists, *Peptides*, 23, 999–1010, 2002.
160. Insug, O, Kieber-Emmons, T, Otvos, L Jr., and Blaszczyk-Thurin, M, Peptide mimicking sialyl-Lewis^a with anti-inflammatory activity, *Biochem. Biophys. Res. Commun.*, 268, 106–111, 2000.
161. Fukuda, M N, Ohyama, C, Lowitz, K, Matsuo, O, Pasqualini, R, Ruoslahti, E, Fukuda, M, A peptide mimic of E-selectin ligand inhibits sialyl Lewis^x-dependent lung colonization of tumor cells, *Cancer Res.*, 60, 450–456, 2000.
162. Renkonen, R, Fukuda, M N, Petrov, L, Paavonen, T, Renkonen, J, Häyry, P, Fukuda, M, A peptide mimic of selectin ligands abolishes *in vivo* inflammation but has no effect on the rat heart allograft survival, *Transplantation*, 74, 2–6, 2002.
163. Molenaar, T J M, Appeldoorn, C C M, de Haas, S A M, Michon, I N M, Bonnefoy, A, Hoylaerts, M F, Pannekoek, H, van Berkel, T J C, Kuiper, J, Biessen, E A L, Specific inhibition of P-selectin-mediated cell adhesion by phage display-derived peptide antagonists, *Blood*, 100, 3570–3577, 2002.
164. (a) Geng, J-G, Heavner, G A, McEver, R P, Lectin domain peptides from selectins interact with both cell surface ligands and Ca^{2+} ions, *J. Biol. Chem.*, 267, 19846–19853, 1992; (b) Geng, J-G, Moore, K L, Johnson, A E, McEver, R P, Neutrophil recognition requires a $Ca^{(2+)}$-induced conformational change in the lectin domain of GMP-140, *J. Biol. Chem.*, 266, 22313–22318, 1991.
165. (a) Briggs, J B, Oda, Y, Gilbert, J H, Schaefer, M E, Macher, B A, Peptides inhibit selectin-mediated cell adhesion *in vitro*, and neutrophil influx into inflammatory sites *in vivo*, *Glycobiology*, 5, 583–588, 1995; (b) Briggs, J B, Larsen, R A, Harris, R B, Sekar, K V S, Macher, B A, Structure/activity studies of anti-inflammatory peptides based on a conserved peptide region of the lectin domain of E-, L- and P-selectin, *Glycobiology*, 6, 831–836, 1996.
166. Rozdzinski, E, Burnette, W N, Jones, T, Mar, V, Tuomanen, E, Prokaryotic peptides that block leukocyte adherence to selectins, *J. Exp. Med.*, 178, 917–924, 1993.
167. Kruszynski, M, Nakada, M T, Tam, S H, Taylor, A H, Fieles, W E, Heavner, G A, Determination of the core sequence of an antagonist of selectin-dependent leukocyte adhesion and correlation of its structure with molecular modeling studies, *Arch. Biochem. Biophys.*, 331, 23–30, 1996.

168. Tam, S H, Nakada, M T, Kruszynski, M, Fieles, W E, Taylor, A H, Mervic, M, Heavner, G A, Structure-function studies on synthetic peptides derived from the 109–118 lectin domain of selectins, *Biochem. Biophys. Res. Commun.*, 227, 712–717, 1996.
169. Heerze, L D, Smith, R H, Wang, N, Armstrong, G D, Utilization of sialic acid-binding synthetic peptide sequences derived from pertussis toxin as novel anti-inflammatory agents, *Glycobiology*, 5, 427–433, 1995.
170. Kretzschmar, G, Toepfer, A, Hüls, C, Krause, M, Pitfalls in the synthesis and biological evaluation of sialyl Lewis^x mimetics as potential selectin antagonists, *Tetrahedron*, 53, 2485–2494, 1997.
171. (a) Foxall, C, Watson, S R, Dowbenko, D, Fennie, C, Lasky, L A, Kiso, M, Hasegawa, A, Asa, D, Brandley, B K, The three members of the selectin receptor family recognize a common carbohydrate epitope, the sialyl Lewis^x oligosaccharide, *J. Cell. Biol.*, 117, 895–902, 1992; (b) Jacob, G S, Kirmaier, C, Abbas, S Z, Howard, S C, Steininger, C N, Welply, J K, Scudder, P, Binding of sialyl Lewis X to E-selectin as measured by fluorescence polarization, *Biochemistry*, 34, 1210–1217, 1995; (c) Galustian, C, Childs, R A, Yuen, C-T, Hasegawa, A, Kiso, M, Lubineau, A, Shaw, G, Feizi, T, Valency dependent patterns of binding of human L-selectin toward sialyl and sulfated oligosaccharides of Le^a and Le^x types: relevance to anti-adhesion therapeutics, *Biochemistry*, 36, 5260–5266, 1997; (d) Marinier, A, Martel, A, Banville, J, Bachand, C, Remillard, R, Lapointe, P, Turmel, B, Menard, M, Harte, W E Jr., Wright, J J K, Todderud, G, Tramposch, K M, Bajorath, J, Hollenbaugh, D, Aruffo, A, Sulfated galactocerebrosides as potential antiinflammatory agents, *J. Med. Chem.*, 40, 3234–3247, 1997.
172. (a) Weitz-Schmidt, G, Stokmaier, D, Scheel, G, Nifant'ev, N E, Tuzikov, A B, Bovin, N V, An E-selectin binding assay based on a polyacrylamide-type glycoconjugate, *Anal. Biochem.*, 238, 184–190, 1996; (b) Weitz-Schmidt, G, Gong, K W, Wong, C-H, Selectin/glycoconjugate binding assays for the identification and optimization of selectin antagonists, *Anal. Biochem.*, 273, 81–88, 1999.
173. Thoma, G, Magnani, J L, Öhrlein, R, Ernst, B, Schwarzenbach, F, Duthaler, R O, Synthesis of oligosaccharide-polylysine conjugates: a well characterized sialyl Lewis^a polymer for ELISA, *J. Am. Chem. Soc.*, 119, 7414–7415, 1997.
174. (a) DeFrees, S A, Kosch, W, Way, W, Paulson, J C, Sabesan, S, Halcomb, R L, Huang, D-H, Ichikawa, Y, Wong, C-H, Ligand recognition by E-selectin: synthesis, inhibitory activity, and conformational analysis of bivalent sialyl Lewis^x analogs, *J. Am. Chem. Soc.*, 117, 66–79, 1995; (b) Kretschmar, G, Sprengard, U, Kunz, H, Bartnik, E, Schmidt, W D, Toepfer, A, Hörsch, B, Krause, M, Seiffge, D, Oligosaccharide recognition by selectins: synthesis and biological activity of multivalent sialyl Lewis-X ligands, *Tetrahedron*, 51, 13015–13030, 1995.
175. Zhou, Q, Moore, K L, Smith, D F, Varki, A, McEver, R P, Cummings, R D, The selectin GMP-140 binds to sialylated, fucosylated lactosaminoglycans on both myeloid and nonmyeloid cells, *J. Cell. Biol.*, 115, 557–564, 1991.
176. Ohnishi, M, Koike, H, Kawamura, N, Tojo, S J, Hayashi, M, Morooka, S, Role of P-selectin in the early stage of the arthus reaction, *Immunopharmacology*, 34, 161–170, 1996.
177. Norgard, K E, Moore, K L, Diaz, S, Stults, N L, Ushiyama, S, McEver, R P, Cummings, R D, Varki, A, Characterization of a specific ligand for P-selectin on myeloid cells. A minor glycoprotein with sialylated *O*-linked oligosaccharides, *J. Biol. Chem.*, 268, 12764–12774, 1993.
178. Yokota, S, Nunn, M F, Morooka, S, Cross-linking of the ninth consensus repeat domain of P-selectin (GMP-140, CD62P) with a monoclonal antibody enhanced leukocyte adhesive activity, *Biochem. Biophys. Res. Commun.*, 218, 709–713, 1996.
179. Revelle, B M, Scott, D, Beck, P J, Single amino acid residues in the E- and P-selectin epidermal growth factor domains can determine carbohydrate binding specificity, *J. Biol. Chem.*, 271, 16160–16170, 1996.
180. Kansas, G S, Saunders, K B, Ley, K, Zakrzewicz, A, Gibson, R M, Furie, B C, Furie, B, Tedder, T F, A role for the epidermal growth factor-like domain of P-selectin in ligand recognition and cell adhesion, *J. Cell. Biol.*, 124, 609–618, 1994.
181. Freedman, S J, Sanford, D G, Bachovchin, W W, Furie, B C, Baleja, J D, Furie, B, Structure and function of the epidermal growth factor domain of P-selectin, *Biochemistry*, 35, 13733–13744, 1996.

182. Ushiyama, S, Laue, T M, Moore, K L, Erickson, H P, McEver, R P, Structural and functional characterization of monomeric soluble P-selectin and comparison with membrane P-selectin, *J. Biol. Chem.*, 268, 15229–15237, 1993.
183. Tyrell, D, James, P, Rao, N, Foxall, C, Abbas, S, Dasgupta, F, Nashed, M, Hasegawa, A, Kiso, M, Asa, D, Kidd, J, Brandley, B K, Structural requirements for the carbohydrate ligand of E-selectin, *PNAS*, 88, 10372–10376, 1991.
184. Patel, K D, Nollert, M U, McEver, R P, P-selectin must extend a sufficient length from the plasma membrane to mediate rolling of neutrophils, *J. Cell. Biol.*, 131, 1893–1902, 1995.
185. Nelson, R M, Dolich, S, Aruffo, A, Cecconi, O, Bevilacqua, M P, Higher-affinity oligosaccharide ligands for E-selectin, *J. Clin. Invest.*, 91, 1157–1166, 1993.
186. Elstad, M R, La Pine, T R, Cowley, F S, McEver, R P, McIntyre, T M, Prescott, S M, Zimmermann, G A, P-selectin regulates platelet-activating factor synthesis and phagocytosis by monocytes, *J. Immunol.*, 155, 2109–2122, 1995.
187. Wein, M, Sterbinsky, S A, Bickel, C A, Schleiner, R P, Bochner, B S, Comparison of human eosinophil and neutrophil ligands for P-selectin: ligands for P-selectin differ from those for E-selectin, *Am. J. Respir. Cell. Mol.*, 12, 315–319, 1995.
188. Todderud, G, Alford, J, Millsap, K A, Aruffo, A, Tramposch, T K, PMN binding to P-selectin is inhibited by sulfatide, *J. Leukocyte Biol.*, 52, 85–88, 1992.
189. Alon, R, Rossiter, H, Wang, X, Springer, T A, Kupper, T S, Distinct cell surface ligands mediate T lymphocyte attachment and rolling on P- and E-selectin under physiological flow, *J. Cell. Biol.*, 127, 1485–1495, 1994.
190. Vachino, G, Chang, X-J, Veldman, G M, Kumar, R, Sako, D, Fouser, L A, Berndt, M C, Cumming, D A, P-selectin glycoprotein ligand-1 is the major counter-receptor for P-selectin on stimulated T cells and is widely distributed in non-functional form on many lymphocytic cells, *J. Biol. Chem.*, 270, 21966–21974, 1995.
191. Aruffo, A, Kolanus, W, Walz, G, Fredman, P, Seed, B, CD62/P-selectin recognition of myeloid and tumor cell sulfatides, *Cell.*, 67, 35–44, 1991.
192. Li, F, Wilkins, P P, Crawley, S, Weinstein, J, Cummings, R D, McEver, R P, Post-translational modifications of recombinant P-selectin glycoprotein ligand-1 required for binding to P- and E-selectin, *J. Biol. Chem.*, 271, 3255–3264, 1996.
193. Kumar, R, Camphausen, R T, Sullivan, F X, Cumming, D A, Core2 β-1,6-*N*-acetylglucosaminyltransferase enzyme activity is critical for P-selectin glycoprotein ligand-1 binding to P-selectin, *Blood*, 88, 3872–3879, 1996.
194. Sako, D, Comess, K M, Barone, K M, Camphausen, R T, Cumming, D A, Shaw, G D, A sulfated peptide segment at the amino terminus of PSGL-1 is critical for P-selectin binding, *Cell.*, 83, 323–331, 1995.
195. (a) Jutila, M A, Bargatze, R F, Kurk, S, Warnock, R A, Ehsani, N, Watson, S R, Walcheck, B, Cell surface P- and E-selectin support shear-dependent rolling of bovine γ/δ T cells, *J. Immunol.*, 153, 3917–3928, 1994; (b) Puri, K D, Finger, E B, Springer, T A, The faster kinetics of L-selectin than of E-selectin and P-selectin rolling at comparable binding strength, *J. Immunol.*, 158, 405–413, 1997; (c) Thoma, G, Patton, J T, Magnani, J L, Ernst, B, Öhrlein, R, Duthaler, R O, Versatile functionalization of polylysine: synthesis, characterization, and use of neoglycoconjugates, *J. Am. Chem. Soc.*, 121, 5919–5929, 1999; (d) Bänteli, R, Herold, P, Bruns, C, Patton, J T, Magnani, J L, Thoma, G, Potent E-selectin antagonists, *Helv. Chim. Acta*, 83, 2893–2907, 2000.
196. (a) Tempelman, L A, Hammer, D A, Receptor-mediated binding of IgE-sensitized rat basophilic leukemia cells to antigen-coated substrates under hydrodynamic flow, *Biophys. J.*, 66, 1231–1243, 1994; (b) Alon, R, Hammer, D A, Springer, T A, Lifetime of the P-selectin-carbohydrate bond and its response to tensile force in hydrodynamic flow, *Nature*, 374, 539–542, 1995.
197. Brunk, D K, Goetz, D J, Hammer, D A, Sialyl Lewis(X)/E-selectin-mediated rolling in a cell-free system, *Biophys. J.*, 71, 2902–2907, 1996.
198. (a) Olofsson, M, Arfors, K-E, Ramezani, L, Wolizky, B A, Butcher, E C, van Andrian, U H, E-selectin mediates leukocyte rolling in interleukin-1-treated rabbit mesentery venules, *Blood*, 84, 2749–2758, 1994; (b) Kubes, P, Jutila, M, Payne, D, Therapeutic potential of inhibiting leukocyte rolling in ischemia/reperfusion, *J. Clin. Invest.*, 95, 2510–2519, 1995; (c) Norman, K E, Moore, K L, McEver, R P, Ley, K, Leukocyte rolling *in vivo* is mediated by P-selectin

glycoprotein ligand-1, *Blood*, 86, 4417–4421, 1995; (d) Damiano, E R, Westheider, J, Tözeren, A, Ley, K, Variation in the velocity, deformation, and adhesion energy density of leukocytes rolling within venules, *Circ. Res.*, 79, 1122–1130, 1996; (e) Kanwar, S, Bullard, D C, Hickey, M J, Smith, C W, Beaudet, A L, Wolitzky, B A, Kubes, P, The association between α_4-integrin, P-selectin, and E-selectin in an allergic model of inflammation, *J. Exp. Med.*, 185, 1077–1087, 1997.

199. (a) Norman, K E, Anderson, G P, Kolb, H C, Ley, K, Ernst, B, Sialyl Lewis^x (sLe^x) and an sLe^x mimetic, CGP69669A, disrupt E-selectin-dependent leukocyte rolling *in vivo*, *Blood*, 91, 475–483, 1998; (b) Piccio, L, Rossi, B, Scarpini, E, Laudanna, C, Giagulli, C, Issekutz, A C, Vestweber, D, Butcher, E C, Constantin, G, Molecular mechanisms involved in lymphocyte recruitment in inflamed brain microvessels: critical roles for P-selectin glycoprotein ligand-1 and heterotrimeric G_i-linked receptors, *J. Immunol.*, 168, 1940–1949, 2002.
200. Bosse, R, Vestweber, D, Only simultaneous blocking of the L- and P-selectin completely inhibits neutrophil migration into mouse peritoneum, *Eur. J. Immunol.*, 24, 3019–3024, 1994.
201. (a) Harris, R R, Mackin, W M, Batt, D G, Rakich, S M, Collins, R J, Bruin, E M, Ackerman, N R, Cellular and biochemical characterization of the anti-inflammatory effects of DuP 654 in the arachidonic acid murine skin inflammation model, *Skin Pharmacol.*, 3, 29–40, 1990; (b) Yawalkar, S, Wiesenberg-Boettcher, I, Gibson, J R, Siskin, S B, Pignat, W, Dermatopharmacologic investigation of halobetasol propionate in comparison with clobetasol 17-propionate, *J. Am. Acad. Dermatol.*, 25, 1137–1144, 1991.

Index

B

C

D

E

J

K

L

M

N

O

P

Q

R

T

U

V

W

X

Y

Z